The Universal Phylogenetic Tree of Life

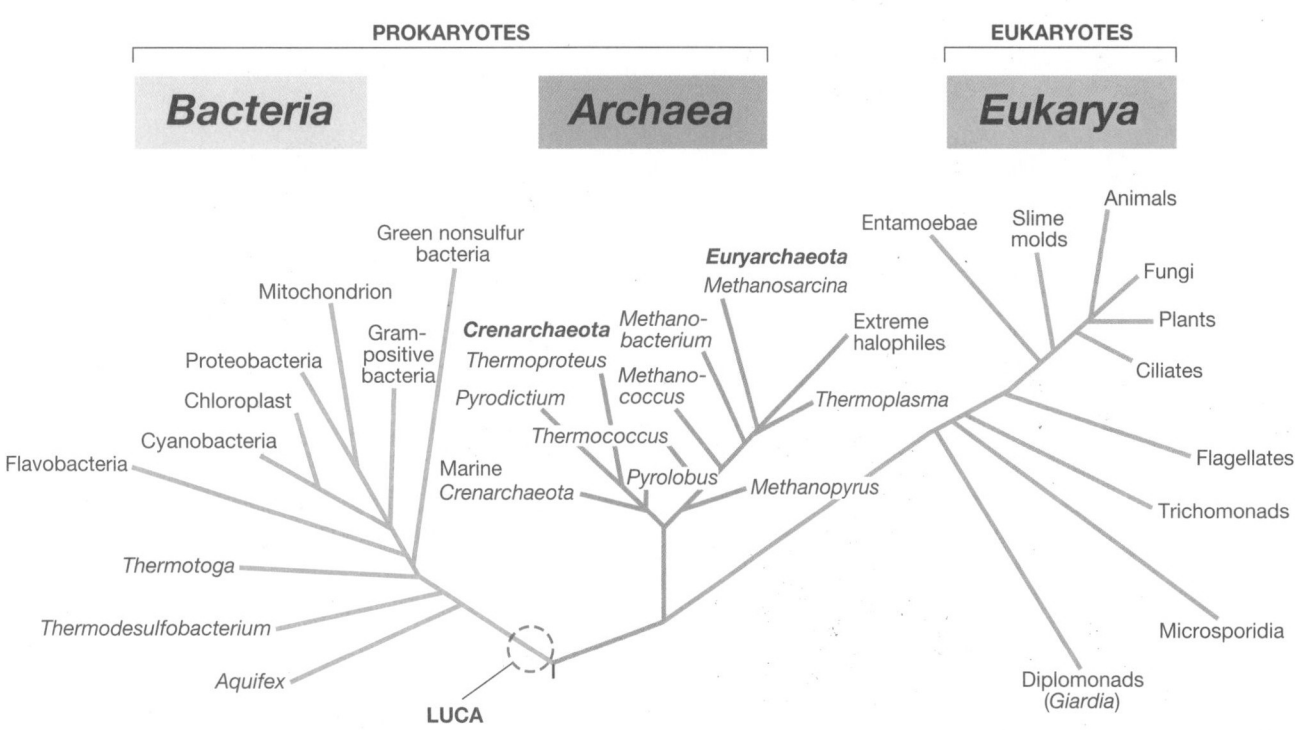

The Phylogeny of *Bacteria*

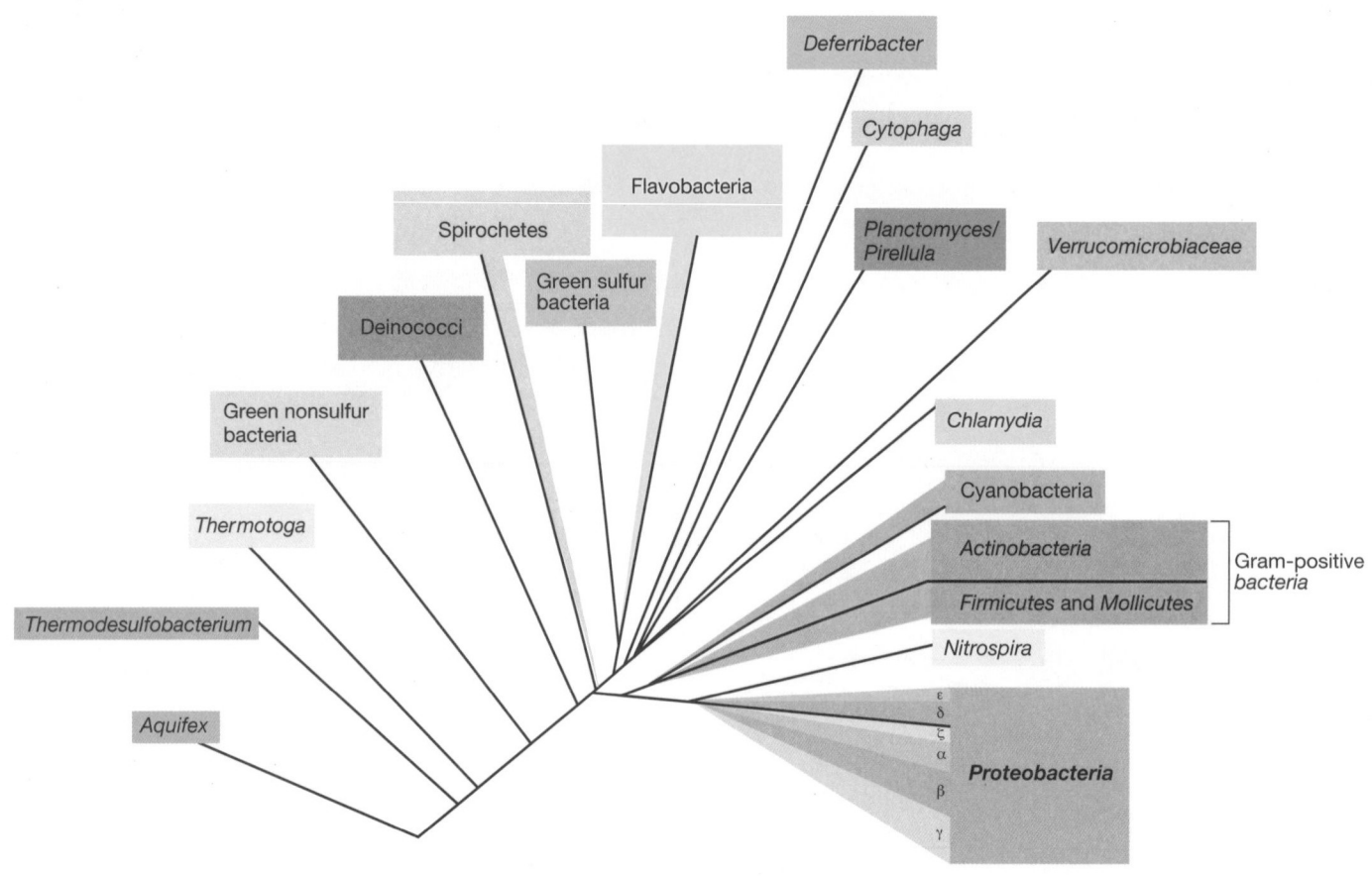

Your steps to success.

Access included with any new book.

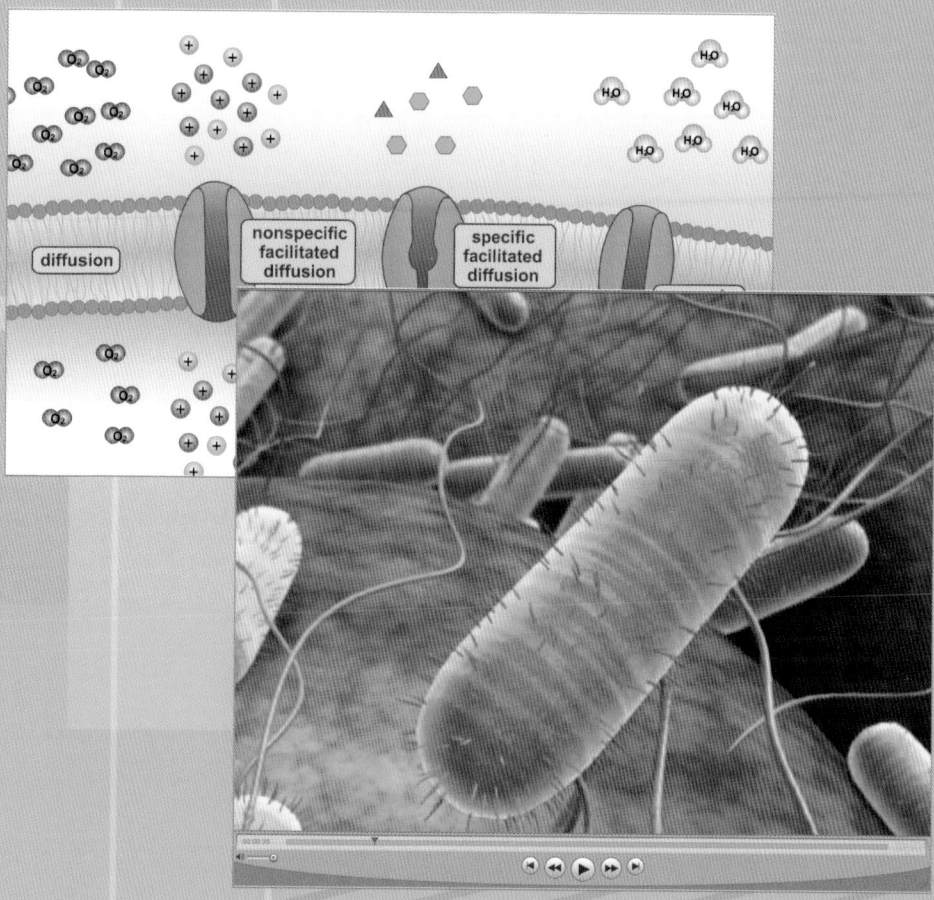

www.microbiologyplace.com

TO REGISTER

1. Go to www.microbiologyplace.com
2. Click "Register."
3. Follow the on-screen instructions to create your login name and password.

Your Access Code is:

*

*Note: If there is no silver foil covering the access code, it may already have been redeemed, and therefore may no longer be valid. In that case, you can purchase access online using a major credit card or PayPal account. To do so, go to www.microbiologyplace.com, **click on** "Buy Access," **and follow the on-screen instructions.***

TO LOG IN

1. Go to www.microbiologyplace.com
2. Click "Log In."
3. Pick your book cover.
4. Enter your login name and password.
5. Click "Log In."

Hint:
Remember to bookmark the site after you log in.

Technical Support:
http://247pearsoned.custhelp.com

Brock

Biology of Microorganisms

Global Edition
Thirteenth Edition

Michael T. Madigan
Southern Illinois University Carbondale

John M. Martinko
Southern Illinois University Carbondale

David A. Stahl
University of Washington Seattle

David P. Clark
Southern Illinois University Carbondale

PEARSON

Boston Columbus Indianapolis New York San Francisco Upper Saddle River
Amsterdam Cape Town Dubai London Madrid Milan Munich Paris Montréal Toronto
Delhi Mexico City São Paulo Sydney Hong Kong Seoul Singapore Taipei Tokyo

If you purchased this book within the United States or Canada you should be aware that it has been imported without the approval of the Publisher or the Author.

Executive Editor: Deirdre Espinoza
International Senior Acquisitions Editor: Laura Dent
Project Editor: Katie Cook
Associate Project Editor: Shannon Cutt
Development Editor: Elmarie Hutchinson
Art Development Manager: Laura Southworth
Art Editor: Elisheva Marcus
Managing Editor: Deborah Cogan
Production Manager: Michele Mangelli
Production Supervisor: Karen Gulliver and David Novak
Copyeditor: Anita Wagner

Art Coordinator: Jean Lake
Photo Researcher: Maureen Spuhler
Director, Media Development: Lauren Fogel
Media Producers: Sarah Young-Dualan, Lucinda Bingham, and Ziki Dekel
Art: Imagineering Media Services, Inc.
Text Design: Riezebos Holzbaur Design Group
Senior Manufacturing Buyer: Stacey Weinberger
Senior Marketing Manager: Neena Bali
International Marketing Director: Ann Oravetz
Compositor: Progressive Information Technologies
Cover Design: Jodi Notowitz

Cover Image: Sinclair Stammers/Science Photo Library

Credits for selected images can be found on page 1105.

ISBN 10: 0-321-73551-X
ISBN 13: 978-0-321-73551-5
2 3 4 5 6 7 8 9 10—CKV—15 14 13 12

About the Authors

Michael T. Madigan received his B.S. in Biology and Education from Wisconsin State University–Stevens Point (1971) and his M.S. (1974) and Ph.D. (1976) in Bacteriology from the University of Wisconsin–Madison. His graduate research was on the hot spring bacterium *Chloroflexus* in the laboratory of Thomas Brock. Following a three-year postdoctoral in the Department of Microbiology, Indiana University, Mike moved to Southern Illinois University–Carbondale, where he has been a professor of microbiology for 32 years. He has coauthored *Biology of Microorganisms* since the fourth edition (1984) and teaches courses in introductory microbiology, bacterial diversity, and diagnostic and applied microbiology. In 1988 Mike was selected as the Outstanding Teacher in the College of Science and in 1993, the Outstanding Researcher. In 2001 he received the SIUC Outstanding Scholar Award. In 2003 he received the Carski Award for Distinguished Undergraduate Teaching from the American Society for Microbiology (ASM), and he is an elected Fellow of the American Academy of Microbiology. Mike's research is focused on bacteria that inhabit extreme environments, and for the past 12 years he has studied the microbiology of permanently ice-covered lakes in the McMurdo Dry Valleys, Antarctica. In addition to his research papers, he has edited a major treatise on phototrophic bacteria and served for over a decade as chief editor of the journal *Archives of Microbiology*. He currently serves on the editorial board of the journals *Environmental Microbiology* and *Antonie van Leeuwenhoek*. Mike's nonscientific interests include forestry, reading, and caring for his dogs and horses. He lives beside a peaceful and quiet lake with his wife, Nancy, five shelter dogs (Gaino, Snuffy, Pepto, Peanut, and Merry), and four horses (Springer, Feivel, Gwen, and Festus).

John M. Martinko received his B.S. in Biology from Cleveland State University. He then worked at Case Western Reserve University, conducting research on the serology and epidemiology of *Streptococcus pyogenes*. His doctoral work at the State University of New York–Buffalo investigated antibody specificity and antibody idiotypes. As a postdoctoral fellow, he worked at Albert Einstein College of Medicine in New York on the structure of major histocompatibility complex proteins. Since 1981, he has been in the Department of Microbiology at Southern Illinois University–Carbondale where he was Associate Professor and Chair, and Director of the Molecular Biology, Microbiology, and Biochemistry Graduate Program. He retired in 2009, but remains active in the department as a researcher and teacher. His research investigates structural changes in major histocompatibility proteins. He teaches an advanced course in immunology and presents immunology and host defense lectures to medical students. He also chairs the Institutional Animal Care and Use Committee at SIUC. He has been active in educational outreach programs for pre-university students and teachers. For his educational efforts, he won the 2007 SIUC Outstanding Teaching Award. He is also an avid golfer and cyclist. John lives in Carbondale with his wife Judy, a high school science teacher.

David A. Stahl received his B.S. degree in Microbiology from the University of Washington–Seattle, later completing graduate studies in microbial phylogeny and evolution with Carl Woese in the Department of Microbiology at the University of Illinois–Champaign-Urbana. Subsequent work as a postdoctoral fellow with Norman Pace, then at the National Jewish Hospital in Colorado, focused on early applications of 16S rRNA-based sequence analysis to the study of natural microbial communities. In 1984 Dave joined the faculty at the University of Illinois–Champaign-Urbana, holding appointments in Veterinary Medicine, Microbiology, and Civil Engineering. In 1994 he moved to the Department of Civil Engineering at Northwestern University, and in 2000 returned to his alma mater, the University of Washington–Seattle, as a professor in the Departments of Civil and Environmental Engineering and Microbiology. Dave is known for his work in microbial evolution, ecology, and systematics, and received the 1999 Bergey Award and the 2006 Procter & Gamble Award in Applied and Environmental Microbiology from the ASM; he is also an elected Fellow of the American Academy of Microbiology. His main research interests are the biogeochemistry of nitrogen and sulfur compounds and the microbial communities that sustain these nutrient cycles. His laboratory was first to culture ammonia-oxidizing *Archaea*, a group now believed to be the main mediators of this key process in the nitrogen cycle. He has taught several courses in environmental microbiology, is one of the co-founding editors of the journal *Environmental Microbiology*, and has served on many advisory committees. Outside teaching and the lab, Dave enjoys hiking, bicycling, spending time with family, reading a good science fiction book, and, with his wife Lin, renovating an old farmhouse on Bainbridge Island, Washington.

David P. Clark grew up in Croydon, a London suburb. He won a scholarship to Christ's College, Cambridge, where he received his B.A. degree in Natural Sciences in 1973. In 1977 he received his Ph.D. from Bristol University, Department of Bacteriology, for work on the effect of cell envelope composition on the entry of antibiotics into *Escherichia coli*. He then left England on a postdoctoral studying the genetics of lipid metabolism in the laboratory of John Cronan at Yale University. A year later he moved with the same laboratory to the University of Illinois at Urbana-Champaign. David joined the Department of Microbiology at Southern Illinois University–Carbondale in 1981. His research has focused on the growth of bacteria by fermentation under anaerobic conditions. He has published numerous research papers and graduated over 20 Masters and Doctoral students. In 1989 he won the SIUC College of Science Outstanding Researcher Award. In 1991 he was the Royal Society Guest Research Fellow at the Department of Molecular Biology and Biotechnology, Sheffield University, England. In addition to *Brock Biology of Microorganisms*, David is the author of four other science books: *Molecular Biology Made Simple and Fun*, now in its fourth edition; *Molecular Biology: Understanding the Genetic Revolution*; *Biotechnology: Applying the Genetic Revolution*; and *Germs, Genes, & Civilization: How Epidemics Shaped Who We Are Today*. David is unmarried and lives with two cats, Little George, who is orange and very nosey, and Mr. Ralph, who is mostly black and eats cardboard.

Dedications

Michael T. Madigan dedicates this book to the memory of his children who rest on Boot Hill: Andy, Marcy, Willie, Plum, Teal, and Sugar. Whether in good times or bad, they always greeted him with tails a waggin'.

John M. Martinko dedicates this book to his daughters Sarah, Helen, and Martha, and to his wife Judy. Thanks for all of your support!

David A. Stahl dedicates this book to his wife, Lin. My love, and one that helps me keep the important things in perspective.

David P. Clark dedicates this book to his father, Leslie, who set him the example of reading as many books as possible.

Preface

The authors and Benjamin Cummings Publishers proudly present the 13th edition of *Brock Biology of Microorganisms* (*BBOM* 13/e). This book is truly a milestone in the annals of microbiology textbooks. *Brock Biology of Microorganisms*, and its predecessor, *Biology of Microorganisms*, has introduced the field of microbiology to students for 41 years, more than any other textbook of microbiology. Nevertheless, although this book goes back over four decades, its two main objectives have remained firm since the first edition was published in 1970: (1) to present the principles of microbiology in a clear and engaging fashion, and (2) to provide the classroom tools necessary for delivering outstanding microbiology courses. The 13th edition of *BBOM* fulfills these objectives in new and exciting ways.

Veteran textbook authors Madigan, Martinko, and Clark welcome our new coauthor, Dave Stahl, to this edition of *BBOM*. Dave is one of the world's foremost experts in microbial ecology and has masterfully crafted an exciting new view of the ecology material in *BBOM*, including a new chapter devoted entirely to microbial symbioses, a first for any textbook of microbiology. Users will find that the themes of ecology and evolution that have permeated this book since its inception reach new heights in the 13th edition. These fundamental themes also underlie the remaining content of the book—the basic principles of microbiology, the molecular biology and genetics that support microbiology today, the huge diversity of metabolisms and organisms, and the medical and immunological facets of microbiology. It is our belief that outstanding content coupled with outstanding presentation have come together to make *BBOM* 13/e the most comprehensive and effective textbook of microbiology available today.

What's New in the 13th Edition?

In terms of content and pedagogy, instructors who have used *BBOM* previously will find the 13th edition to be the same old friend they remember; that is, a book loaded with accurate, up-to-the-minute content that is impeccably organized and visually enticing. The 36 chapters in *BBOM* 13/e are organized into modules by numbered head, which allows instructors to fine-tune course content to the needs of their students. In addition, study aids and review tools are an integral part of the text. Our new MiniQuiz feature, which debuts in the 13th edition, is designed to quiz students' comprehension as they work their way through each chapter. Also new to this edition is the end-of-chapter review tool called "Big Ideas." These capsule summaries pull together the key concepts from each numbered section in a wrap-up style that is certain to be a big hit with students, especially the night before examinations! Our end-of-chapter key terms list, two detailed appendices, a comprehensive glossary, and a thorough index complete the hard copy learning package. Many additional learning resources are available online (see below).

In terms of presentation, *BBOM* 13/e will easily draw in and engage the reader. The book has been designed in a beautiful yet simple fashion that gives the art and pedagogical elements the breathing room they need to be effective and the authors the freedom to present concepts in a more visually appealing way. Supporting the narrative are spectacular illustrations, with every piece of art rendered in a refreshing new style. Moreover, the art complements, and in many cases integrates, the hundreds of photos in *BBOM*, many of which are new to the 13th edition. And, as users of *BBOM* have come to expect, our distinctive illustrations remain the most accurate and consistent of those in any microbiology textbook today.

The authors are keenly aware that it is easy to keep piling on new material and fattening up a textbook. In response to this trend, *BBOM* 13/e went on a diet. With careful attention to content and presentation, *BBOM* 13/e is actually a shorter book than *BBOM* 12/e. The authors have carefully considered every topic to ensure that content at any point in the book is a reflection of both what the student already knows and what the student needs to know in a world where microbiology has become the most exciting and relevant of the biological sciences. The result is a more streamlined and exciting treatment of microbiology that both students and instructors will appreciate.

Revision Highlights:

Chapter 1

- Find new coverage on the evolution and major habitats of microorganisms—Earth's most pervasive and extensive biomass.
- A more visually compelling presentation of the impacts of microorganisms on humans better emphasizes the importance of microorganisms for the maintenance of all life on Earth.

Chapter 2

- New coverage of cell biology and the nature of the chromosome in prokaryotic and eukaryotic cells is complemented by a visually engaging overview of the microbial world.

Chapter 3

- The cell chemistry chapter that previously held this position is now available online (www.microbiologyplace.com). The new Chapter 3 explores cell structure and function with strong new visuals to carry the text and new coverage of the lipids and cell walls of *Bacteria* and *Archaea*.

Chapter 4

- Find updated coverage of catabolic principles along with an overview of essential anabolic reactions.
- Newly rendered and more instructive art makes mastering key metabolic pathways and bioenergetic principles a more visual experience.

Chapter 5

- Updated coverage of the events in cell division and their relation to medical microbiology connects basic science to applications.
- Newly rendered art throughout makes the important concepts of cell division and population growth more vivid, engaging, and interactive.

Chapter 6

- The concise primer on molecular biology that every student needs to know is updated and now includes an overview of the structures of nucleic acids and proteins and the nature of chromosomes and plasmids.

Chapter 7

- Find new coverage of the latest discoveries in the molecular biology of *Archaea* and comparisons with related molecular processes in *Bacteria*.
- A new section highlights the emerging area of regulation by microRNA in eukaryotes.

Chapter 8

- Review major updates on the regulation of gene expression—one of the hottest areas in microbiology today—including expanded coverage of cell sensing capacities and signal transduction.
- Enjoy the new Microbial Sidebar featuring CRISPR, the newly discovered form of RNA-based regulation used by *Bacteria* and *Archaea* to ward off viral attack.

Chapter 9

- Major updates of the principles of virology are complemented with an overview of viral diversity.
- New art reinforces the relevance and importance of viruses as agents of genetic exchange.

Chapter 10

- The fundamental principles of microbial genetics are updated and supplemented with new coverage that compares and contrasts bacterial and archaeal genetics.

Chapter 11

- Find "one-stop shopping" for coverage of molecular biological methods, including cloning and genetic manipulations, as a prelude to the genomics discussion in the next chapter.
- Enjoy the colorful new Microbial Sidebar on new fluorescent labeling methods that can differentiate even very closely related bacteria.

Chapter 12

- Extensive updates on microbial genomics and transcriptomics will be found along with new coverage of the emerging related areas of metabolomics and interactomics.

Chapter 13

- The two chapters covering metabolic diversity have been revised and moved up to Chapters 13 and 14 to precede rather than follow coverage of microbial diversity, better linking these two important and often related areas.
- This chapter is loaded with reworked art and text that highlight the unity and diversity of the bioenergetics underlying phototrophic and chemolithotrophic metabolisms.

Chapter 14

- Restyled and impeccably consistent art showcases the comparative biochemistry of the aerobic and anaerobic catabolism of carbon compounds.

Chapter 15

- This retooled chapter combines the essentials of industrial microbiology and biotechnology, including the production of biofuels and emerging green microbial technologies.

Chapter 16

- Find new coverage of the origin of life and how the evolutionary process works in microorganisms.
- Microbial phylogenies from small subunit ribosomal RNA gene analyses are compared with those from multiple-gene and full genomic analyses.

Chapters 17–19

- Coverage of the diversity of *Bacteria* and *Archaea* better emphasizes phylogeny with increased focus on phyla of particular importance to plants and animals and to the health of our planet.
- Spectacular photomicrographs and electron micrographs carry the reader through prokaryotic diversity.

Chapter 20

- A heavily revised treatment of the diversity of microbial eukaryotes is supported by many stunning new color photos and photomicrographs.
- Find an increased emphasis on the phylogenetic relationships of eukaryotes and the "bacterial nature" of eukaryotic organelles.

Chapter 21

- Viruses, the most genetically diverse of all microorganisms, come into sharper focus with major updates on their diversity.
- A new section describes viruses in nature and their abundance in aquatic habitats.

Chapter 22

- This chapter features a major new treatment of the latest molecular techniques used in microbial ecology, including CARD-FISH, ARISA, biosensors, NanoSIMS, flow cytometry, and multiple displacement DNA amplification.
- Find exciting new coverage of methods for functional analyses of single cells, including single-cell genomics and single-cell stable isotope analysis, and expanded coverage of methods for analyses of microbial communities, including metagenomics, metatranscriptomics, and metaproteomics.

Chapter 23

- A comparison of the major habitats of *Bacteria* and *Archaea* is supported by spectacular new photos and by art that summarizes the phylogenetic diversity and functional significance of prokaryotes in each habitat.
- Find broad new coverage of the microbial ecology of microbial mat communities and prokaryotes that inhabit the deep subsurface.

Chapter 24

- Revised coverage of the classical nutrient cycles is bolstered by new art, while new coverage highlights the calcium and silica cycles and how these affect CO_2 sequestration and global climate.
- Improved integration of biodegradation and bioremediation shows how natural microbial processes can be exploited for the benefit of humankind.

Chapter 25

- This new chapter focuses entirely on microbial symbioses, including bacterial–bacterial symbioses and symbioses between bacteria and their plant, mammal, or invertebrate hosts. Find coverage here of all of the established as well as more recently discovered symbioses, including the human gut and how its microbiome may control obesity, the rumen of animals important to agriculture, the hindgut of termites, the light organ of the squid, the symbioses between hydrothermal vent animals and chemolithotrophic bacteria, the essential bacterial symbioses of insects, medicinal leeches, reef-building corals, and more, all supported by spectacular new color photos and art.
- Learn how insects have shaped the genomes of their bacterial endosymbionts.
- Marvel at the new Microbial Sidebar that tells the intriguing story of the attine ants and their fungal gardens.

Chapter 26

- Key updates will be found on microbial drug resistance and are supported by new art that reveals the frightening reality that several human pathogens are resistant to all known antimicrobial drugs.

Chapter 27

- Extensively reworked sections on the normal microbial flora of humans include new coverage of the human microbiome and a molecular snapshot of the skin microflora.
- Find revised coverage of the principles of virulence and pathogenicity that connect infection and disease.

Chapter 28

- Here we present the perfect overview of immunology for instructors who wish to cover only the fundamental concepts and how the immune system resists the onslaught of infectious disease.
- Find late-breaking practical information on the immune response, including vaccines and immune allergies.

Chapter 29

- Built on the shoulders of the previous chapter, here is a more detailed probe of the mechanisms of immunity with emphasis on the molecular and cellular interactions that control innate and adaptive immunity.

Chapter 30

- This short chapter presents an exclusively molecular picture of immunology, including receptor–ligand interactions (the "triggers" of the immune response), along with genetics of the key proteins that drive adaptive immunity.

Chapter 31

- Find revised and expanded coverage of molecular analyses in clinical microbiology, including new enzyme immunoassays, reverse transcriptase PCR, and real-time PCR.

Chapter 32

- Review major updates of the principles of disease tracking, using 2009 pandemic H1N1 influenza as a model for how newly emerging infectious diseases are tracked.
- Find updated coverage throughout, especially of the HIV/AIDS pandemic.

Chapter 33

- Read all about the origins and history of pandemic H1N1 influenza and how the H1N1 virus is related to strains of influenza that already existed in animal populations.
- Hot new coverage of immunization strategies for HIV/AIDS.

Chapter 34

- Follow the emergence, rapid dispersal, and eventual entrenchment of West Nile virus as an endemic disease in North America.
- Expanded coverage of malaria—the deadliest human disease of all time—includes the promise of new antiparasitic drugs and disease prevention methods.

Chapter 35

- Find updates of water microbiology, including new rapid methods for detecting specific indicator organisms.

Chapter 36

- Explore new methods of food processing, including aseptic and high-pressure methods that can dramatically extend the shelf-life and safety of perishable foods and drinks.

Cutting Edge Coverage Includes the Most Current Presentation of Microbial Ecology

The 13th edition enhances the themes of ecology and evolution throughout, and is the only book on the market to include specialized coverage of archael and eukaryotic molecular biology. The book represents the most current research in the field, with special attention paid to the microbial ecology chapters:

Chapter 22, Methods in Microbial Ecology, is heavily updated to present the latest molecular techniques used in microbial ecology, including CARD-FISH, ARISA, biosensors, NanoSIMS, flow cytometry, and multiple displacement DNA amplification. It also includes exciting new coverage of methods for functional analyses of single cells, including single-cell genomics and single-cell stable isotope analysis, and expanded coverage of methods for analyses of microbial communities, including metagenomics, metatranscriptomics, and metaproteomics.

Chapter 23, Major Microbial Habitats and Diversity, compares the major habitats of *Bacteria* and *Archaea* and is supported by spectacular new photos and art that summarize the phylogenetic diversity and functional significance of prokaryotes in each habitat.

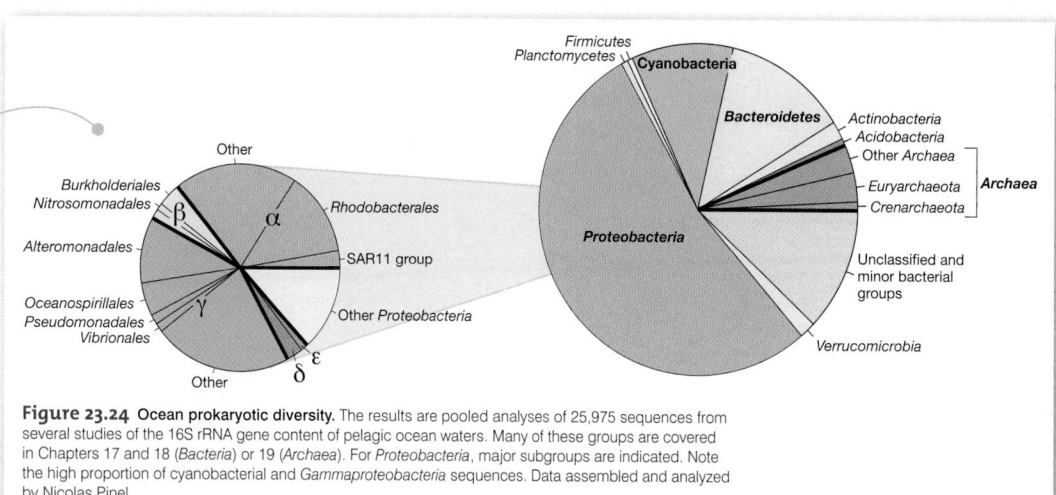

Figure 23.24 Ocean prokaryotic diversity. The results are pooled analyses of 25,975 sequences from several studies of the 16S rRNA gene content of pelagic ocean waters. Many of these groups are covered in Chapters 17 and 18 (*Bacteria*) or 19 (*Archaea*). For *Proteobacteria*, major subgroups are indicated. Note the high proportion of cyanobacterial and *Gammaproteobacteria* sequences. Data assembled and analyzed by Nicolas Pinel.

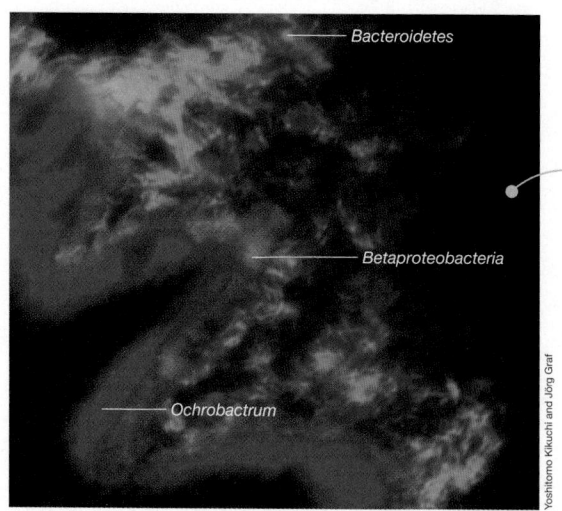

Figure 25.40 Micrograph of a FISH-stained microbial community in the bladder of *Hirudo verbana*. A probe (red) targeted at the 16S rRNA of *Betaproteobacteria* and a probe (green) targeted at the 16S rRNA of *Bacteroidetes* reveal distinct layers of different bacteria in the lumen of the bladder. Staining with DAPI (blue), which binds to DNA, reveals the intracellular alphaproteobacterium *Ochrobactrum* and host nuclei.

Chapter 24, Nutrient Cycles, Biodegradation, and Bioremediation. Exciting updates of all the nutrient cycles that form the heart of environmental microbiology and microbial ecology.

Chapter 25, Microbial Symbioses, is a completely new chapter focused entirely on microbial symbioses, including bacterial–bacterial symbioses and symbioses between bacteria and their plant, mammal, or invertebrate hosts. Find coverage here of all the established as well as more recently discovered symbioses—including the human gut and how its microbiome may control obesity, the rumen of animals important to agriculture, the hindgut of termites, the light organ of the squid, the symbioses between hydrothermal vent animals and chemolithotrophic bacteria, and the essential bacterial symbioses of insects, medicinal leeches, reef-building corals, and more.

For a detailed list of chapter-by-chapter updates, see page 5 of the Preface.

Thoroughly Updated and Revised Art

The art has been revised and updated throughout the book to give students a clear view into the microbial world. Color and style conventions are used consistently to make the art accessible and easy to understand.

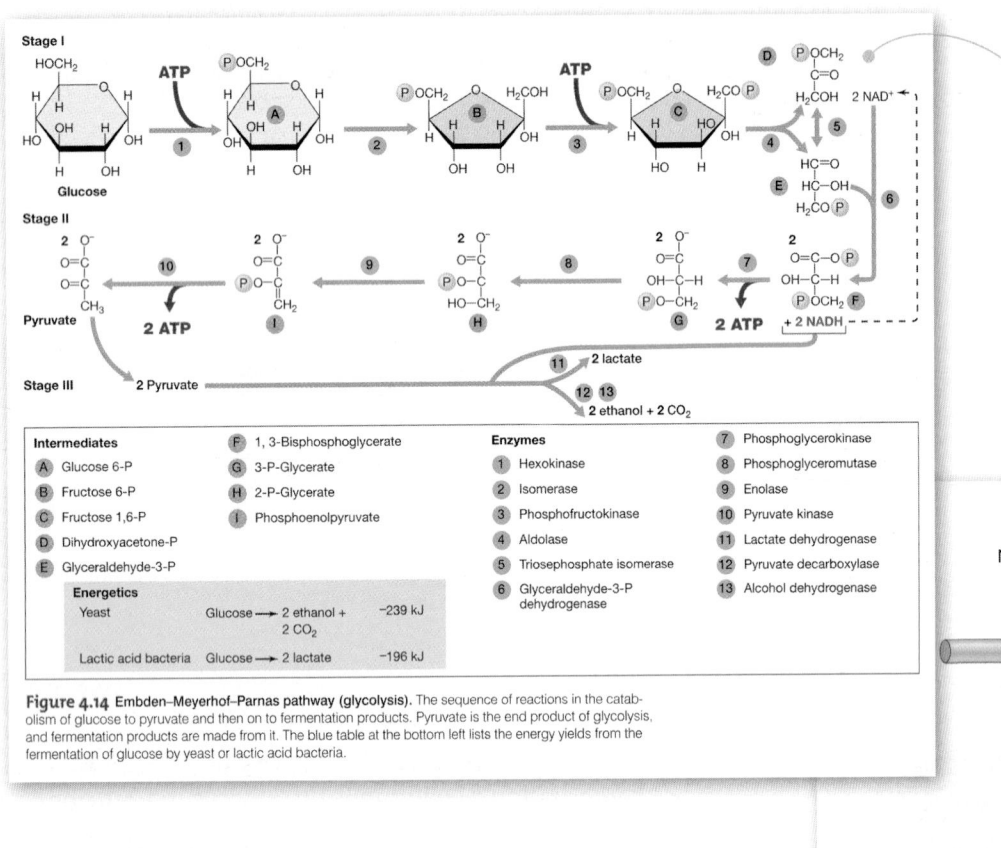

Carefully redesigned new art clearly guides students through challenging concepts. The style for metabolic figures and other pathway processes has been simplified, and color-coded steps and chemical structures increase student comprehension.

Figure 4.14 Embden–Meyerhof–Parnas pathway (glycolysis). The sequence of reactions in the catabolism of glucose to pyruvate and then on to fermentation products. Pyruvate is the end product of glycolysis, and fermentation products are made from it. The blue table at the bottom left lists the energy yields from the fermentation of glucose by yeast or lactic acid bacteria.

Dimensionality has been added to some figures, lending more realism and vivacity to the presentation. Figures in which nucleic acids or cells are depicted are now more dimensional to clearly identify key genes and cell structures.

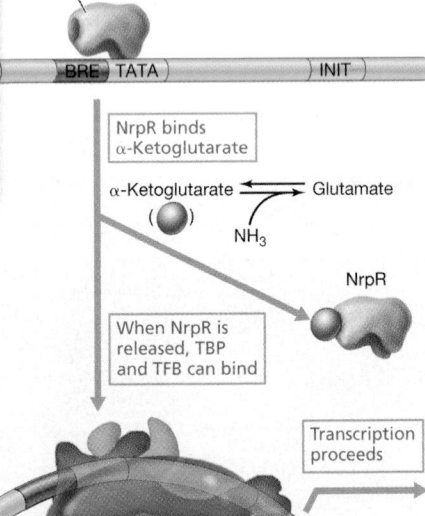

Figure 8.15 Repression of genes for nitrogen metabolism in Archaea. The NrpR protein of *Methanococcus maripaludis* acts as a repressor. It blocks the binding of the TFB and TBP proteins, which are required for promoter recognition, to the BRE site and TATA box, respectively. If there is a shortage of ammonia, α-ketoglutarate is not converted to glutamate. The α-ketoglutarate accumulates and binds to NrpR, releasing it from the DNA. Now TBP and TFB can bind. This in turn allows RNA polymerase to bind and transcribe the operon.

Illustrations and photos are often paired to give an idealized view next to a realistic view and to reinforce the connection between theory and practice.

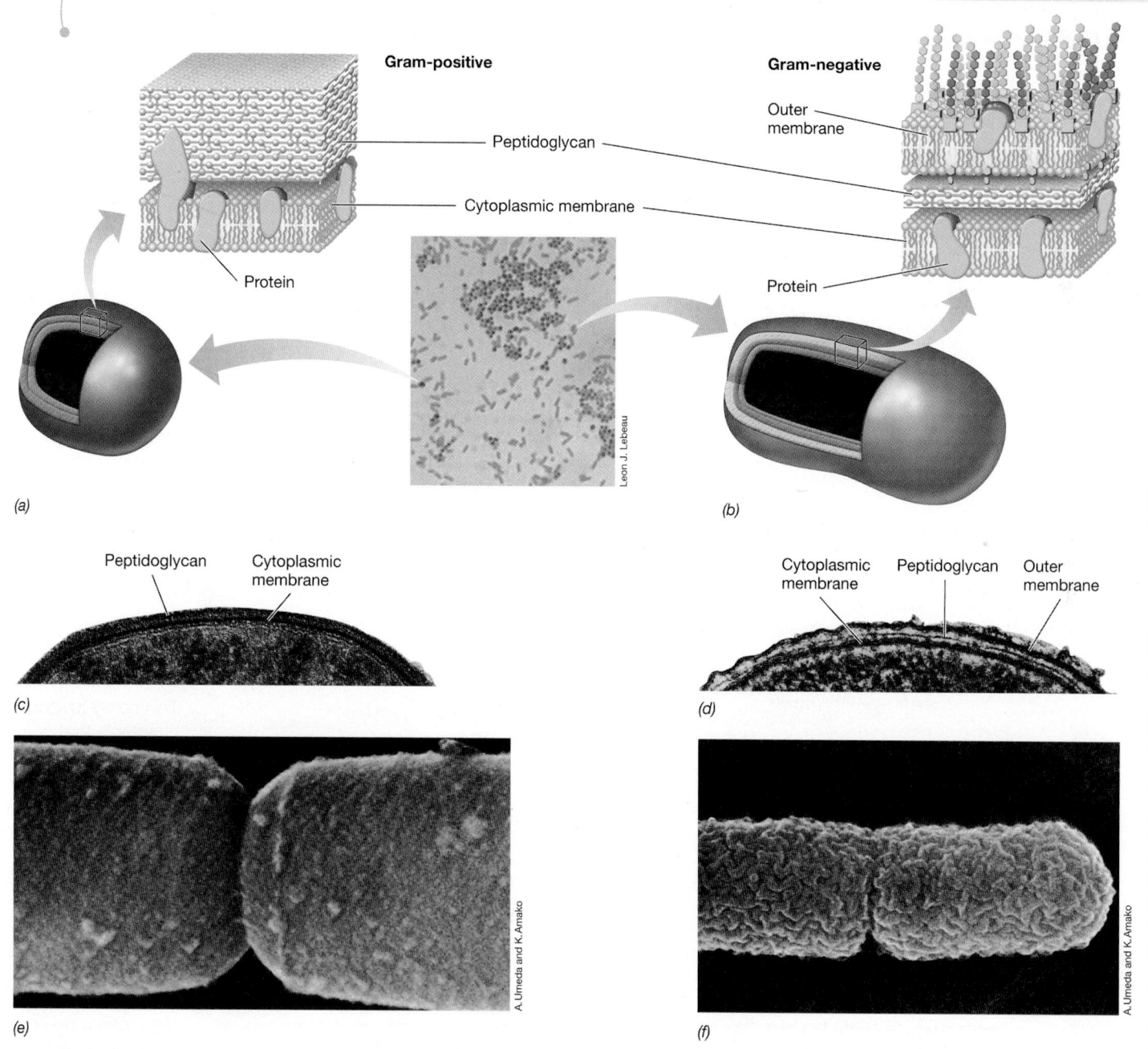

Figure 3.15 Cell walls of *Bacteria*. *(a, b)* Schematic diagrams of gram-positive and gram-negative cell walls. The Gram stain photo in the center shows cells of *Staphylococcus aureus* (purple, gram-positive) and *Escherichia coli* (pink, gram-negative). *(c, d)* Transmission electron micrographs (TEMs) showing the cell wall of a gram-positive bacterium and a gram-negative bacterium. *(e, f)* Scanning electron micrographs of gram-positive and gram-negative bacteria, respectively. Note differences in surface texture. Each cell in the TEMs is about 1 μm wide.

Conceptual Framework Helps Students Focus on the Key Concepts

The first twelve chapters cover the principles of microbiology. Basic principles are presented early on and then used as the foundation to tackle the material in greater detail later.

Brief Contents

New chapter on symbiosis ties together the core concepts of the book—health, diversity, and the human ecosystem.

This newly revised chapter is the perfect overview for instructors who wish to cover immunology at a generalized level including the fundamental concepts of how the immune system resists the onslaught of infectious disease. Instructors who like to go into more detail can build on the core principles taught in Chapter 28 by covering Immune Mechanisms (Ch. 29) and Molecular Immunology (Ch. 30).

Information on metabolic diversity precedes the coverage of microbial diversity, better linking these important and often related areas.

Big Ideas

2.1

Microscopes are essential for studying microorganisms. Bright-field microscopy, the most common form of microscopy, employs a microscope with a series of lenses to magnify and resolve the image.

2.2

An inherent limitation of bright-field microscopy is the lack of contrast between cells and their surroundings. This problem can be overcome by the use of stains or by alternative forms of light microscopy, such as phase contrast or dark field.

2.3

Differential interference contrast microscopy and confocal scanning laser microscopy allow enhanced three-dimensional imaging or imaging through thick specimens. The atomic force microscope gives a very detailed three-dimensional image of live preparations.

2.4

Electron microscopes have far greater resolving power than do light microscopes, the limits of resolution being about 0.2 nm. The two major forms of electron microscopy are transmission, used primarily to observe internal cell structure, and scanning, used to examine the surface of specimens.

2.7

Comparative rRNA gene sequencing has defined three domains of life: *Bacteria*, *Archaea*, and *Eukarya*. Molecular sequence comparisons have shown that the organelles of *Eukarya* were originally *Bacteria* and have spawned new tools for microbial ecology and clinical microbiology.

2.8

All cells need sources of carbon and energy for growth. Chemoorganotrophs, chemolithotrophs, and phototrophs use organic chemicals, inorganic chemicals, or light, respectively, as their source of energy. Autotrophs use CO_2 as their carbon source, while heterotrophs use organic compounds. Extremophiles thrive under environmental conditions of high pressure or salt, or extremes of temperature or pH.

2.9

Several phyla of *Bacteria* are known, and an enormous diversity of cell morphologies and physiologies are represented. *Proteobacteria* are the largest group of *Bacteria* and contain many well-known bacteria, including *Escherichia coli*. Other major phyla include gram-positive bacteria, cyanobacteria, spirochetes, and green bacteria.

MiniQuiz

- What are the primary response regulator and the primary sensor kinase for regulating chemotaxis?
- Why is adaptation during chemotaxis important?
- How does the response of the chemotaxis system to an attractant differ from its response to a repellent?

Additional Resources

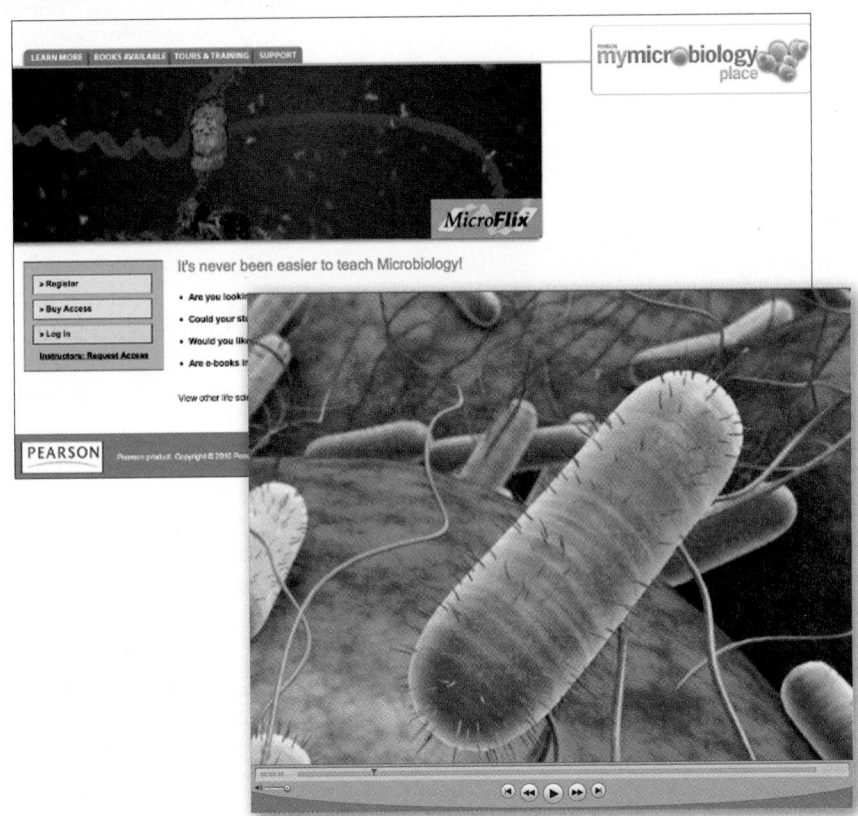

The MyMicrobiologyPlace website is rich with media assets to give students extra practice. It includes chapter quizzes, new quantitative questions, animations, and additional tutorials.
www.microbiologyplace.com

Quantitative Questions

1. **Number of genes in plasmid R100.** The *Escherichia coli* plasmid R100 is a circular molecule of DNA containing 93.4 kbp. The average *E. coli* protein contains 300 amino acids; assume that the same is true for R100 proteins. With this assumption, calculate how many genes are in this plasmid.

2. **Compare DNA polymerases.** *Escherichia coli* contains at least five different DNA polymerases. The three most characterized are DNA Pol I, Pol II, and Pol III. Polymerase I and II replicate DNA at about 20–40 nucleotides/sec whereas Pol III replicates at 250 to 1000 nucleotides/sec. The genome of *E. coli* strain K-12 is 4,639,221 bp. At the higher rates, how long does it take to reproduce the chromosome? How do these numbers agree with the roles of these DNA polymerases?

Instructor Resource DVD (IR-DVD)
0-321-72086-5 / 978-0-321-72086-3

The IR-DVD offers a wealth of media resources including all the art from the book in both JPEG and PPT formats, PowerPoint lecture outlines, computerized test bank, and answer keys all in one convenient location. The animations help bring lectures to life, while the select step-edit figures help break down complicated processes.

Instructor Manual and Test Bank
0-321-72834-3 / 978-0-321-72834-0

by W. Matthew Sattley and Christopher A. Gulvik

The electronic Instructor Manual/Test Bank provides chapter summaries that help with class preparation as well as the answers to the end-of-chapter review and application questions. The test bank contains 3,000 questions for use in quizzes, tests, and exams.

Acknowledgments

A book of this stature is not the product of its authors alone but instead is the collective effort of the many people who comprise the book team. These include folks both inside and outside of Benjamin Cummings. Executive editor Deirdre Espinoza and project editor Katie Cook, both of Benjamin Cummings, were the workhorses in editorial. Deirdre paved the way for the 13th edition and skillfully maneuvered the book around the occasional roadblocks that accompany any major textbook project. Katie ran the day-to-day operations of the *BBOM* team in a highly professional manner, expertly managing reviews and many other details and keeping all facets of the project on track.

The *BBOM* 13/e production and design team was headed up by Michele Mangelli (Mangelli Productions) who oversaw Yvo Riezebos (Riezebos Holzbaur Design Group), and Laura Southworth (Benjamin Cummings). Michele managed the production team and did a great job of keeping everyone on mission and on budget. The artistic magic of Yvo is clearly visible in the beautiful text and cover designs of *BBOM* 13/e. Laura created the new art look for *BBOM* 13/e, one that readers should immediately appreciate for its clarity, consistency, and modern style. The authors are extremely grateful to Michele, Yvo, and Laura, as well as to the artist team at the studio of Imagineering (Toronto), for helping the authors produce such a beautiful book. Others in production included Karen Gulliver, Jean Lake, and Maureen Spuhler. Karen was our excellent production editor who ensured that a polished book emerged from a raw manuscript, while Jean was our art coordinator, tracking and routing art and handling interactions with the art studio. Maureen was our photo researcher who helped the authors locate photos that met the exacting standards of *BBOM*. The authors are extremely grateful to Karen, Jean, and Maureen for transforming literally thousands of pages of text and art manuscript into a superb learning tool.

The authors wish to give special thanks to four other members of the production team: Elmarie Hutchinson, Anita Wagner, Elisheva (Ellie) Marcus, and Elizabeth McPherson. Our developmental editor Elmarie was a key contributor early in the project, helping the authors better link text and art and massaging the text to improve readability. Anita was our absolutely spectacular copyeditor; the authors could not have asked for a brighter or more effective person in this key position on the book team. Anita improved the accuracy, clarity, and consistency of the text and rendered her editorial services in a style that the authors found both helpful and time saving. Ellie (Benjamin Cummings) was our art liaison on this project, translating for the art house the intentions of the authors. Ellie has the unique gift of viewing art from both an artistic and a scientific perspective. Therefore, the consistency, clarity, and accuracy of the art in *BBOM* 13/e are in large part due to her superb efforts. Elizabeth (University of Tennessee) was our manuscript accuracy checker; her eagle eye, extensive knowledge of all areas of microbiology, prompt service, and knack for editorial troubleshooting greatly improved the accuracy and authority of the final product.

The authors also wish to acknowledge the excellent contributions of Dr. Matt Sattley, Indiana Wesleyan University. Matt, a former doctoral student of MTM, composed the Instructor's Manual that accompanies *BBOM* 13/e. The manual should greatly assist instructors of any vintage to better organize their microbiology courses and select review questions for student assignments. We also thank Christopher Gulvik, University of Tennessee, for revising the test bank questions for this edition.

No textbook in microbiology could be published without thorough reviewing of the manuscript and the gift of new photos from experts in the field. We are therefore extremely grateful for the kind help of the many individuals who provided general or technical reviews of the manuscript or who supplied new photos. They are listed below. And last but not least, the authors thank the women in their lives—Nancy (MTM), Judy (JMM), Linda (DAS), and Donna (DPC)—for the sacrifices they have made the past two years while this book was in preparation and for simply putting up with them during the ordeal that is "a *BBOM* revision."

F.C. Thomas Allnutt
Daniel Arp, *Oregon State University*
Marie Asao, *Ohio State University*
Tracey Baas, *University of Rochester*
Zsuzsanna Balogh-Brunstad, *Hartwick College*
Teri Balser, *University of Wisconsin–Madison*
Tamar Barkay, *Rutgers University*
John Baross, *University of Washington*
Douglas Bartlett, *Scripps Institute of Oceanography*
Carl Bauer, *Indiana University*
David Bechhofer, *Mount Sinai School of Medicine*
Mercedes Berlanga, *University of Barcelona (Spain)*
Werner Bischoff, *Wake Forest University School of Medicine*
Luz Blanco, *University of Michigan*
Robert Blankenship, *Washington University–St. Louis*
Antje Boetius, *Max Planck Institute for Marine Microbiology (Germany)*
Jörg Bollmann, *University of Toronto (Canada)*
Andreas Brune, *Universität Marburg (Germany)*
Don Bryant, *Penn State University*
Richard Calendar, *University of California–Berkeley*
Donald Canfield, *University of Southern Denmark*
Centers for Disease Control and Prevention Public Health Image Library, *Atlanta, Georgia*

Kee Chan, *Boston University*

Jiguo Chen, *Mississippi State University*

Randy Cohrs, *University of Colorado Health Sciences Center*

Morris Cooper, *Southern Illinois University School of Medicine*

Amaya Garcia Costas, *Penn State University*

Lluïsa Cros Miguel, *Institut de Ciències del Mar (Spain)*

Laszlo Csonka, *Purdue University*

Diana Cundell, *Philadelphia University*

Philip Cunningham, *Wayne State University*

Cameron Currie, *University of Wisconsin*

Holger Daims, *University of Vienna (Austria)*

Dayle Daines, *Mercer University School of Medicine*

Richard Daniel, *Newcastle University Medical School*

Edward F. DeLong, *Massachusetts Institute of Technology*

James Dickson, *Iowa State University*

Kevin Diebel, *Metropolitan State College of Denver*

Nancy DiIulio, *Case Western Reserve University*

Nicole Dubilier, *Max Planck Institute for Marine Microbiology (Germany)*

Paul Dunlap, *University of Michigan*

Tassos Economou, *Institute of Molecular Biology and Biotechnology, Iraklio-Crete (Greece)*

Siegfried Engelbrecht-Vandré, *Universität Osnabrück (Germany)*

Jean Euzéby, *École Nationale Vétérinaire de Toulouse (France)*

Tom Fenchel, *University of Copenhagen (Denmark)*

Matthew Fields, *Montana State University*

Jed Fuhrman, *University of Southern California*

Daniel Gage, *University of Connecticut*

Howard Gest, *Indiana University*

Steve Giovannoni, *Oregon State University*

Veronica Godoy-Carter, *Northeastern University*

Gerhard Gottschalk, *University of Göttingen, Germany*

Jörg Graf, *University of Connecticut*

Dennis Grogan, *University of Cincinnati*

Ricardo Guerrero, *University of Barcelona (Spain)*

Hermie Harmsen, *University of Groningen (The Netherlands)*

Terry Hazen, *Lawrence Berkeley National Laboratory*

Heather Hoffman, *George Washington University*

James Holden, *University of Massachusetts–Amherst*

Julie Huber, *Marine Biological Laboratories, Woods Hole*

Michael Ibba, *Ohio State University*

Johannes Imhoff, *University of Kiel (Germany)*

Kazuhito Inoue, *Kanagawa University (Japan)*

Rohit Kumar Jangra, *University of Texas Medical Branch*

Ken Jarrell, *Queen's University (Canada)*

Glenn Johnson, *Air Force Research Laboratory*

Deborah O. Jung, *Southern Illinois University*

Marina Kalyuzhnaya, *University of Washington*

Deborah Kelley, *University of Washington*

David Kehoe, *Indiana University*

Stan Kikkert, *Mesa Community College*

Christine Kirvan, *California State University–Sacramento*

Kazuhiko Koike, *Hiroshima University (Japan)*

Martin Konneke, *Universität Oldenburg (Germany)*

Allan Konopka, *Pacific Northwest Laboratories*

Susan F. Koval, *University of Western Ontario*

Lee Krumholz, *University of Oklahoma*

Martin Langer, *Universität Bonn (Germany)*

Amparo Latorre, *Universidad de València (Spain)*

Mary Lidstrom, *University of Washington*

Steven Lindow, *University of California–Berkeley*

Wen-Tso Liu, *University of Illinois*

Zijuan Liu, *Oakland University*

Jeppe Lund Nielsen, *Aalborg University (Denmark)*

John Makemson, *Florida International University*

George Maldonado, *University of Minnesota*

Linda Mandelco, *Bainbridge Island, Washington*

William Margolin, *University of Texas Health Sciences Center*

Willm Matens-Habbena, *University of Washington*

Margaret McFall-Ngai, *University of Wisconsin*

Michael McInerney, *University of Oklahoma*

Elizabeth McPherson, *University of Tennessee*

Aubrey Mendonca, *Iowa State University*

William Metcalf, *University of Illinois*

Duboise Monroe, *University of Southern Maine*

Katsu Murakami, *Penn State University*

Eugene Nester, *University of Washington*

Tullis Onstott, *Princeton University*

Aharon Oren, *Hebrew University, Jerusalem*

Victoria Orphan, *California Institute of Technology*

Jörg Overmann, *Universität Munich (Germany)*

Hans Paerl, *University of North Carolina*

Vijay Pancholi, *Ohio State University College of Medicine*

Matthew Parsek, *University of Washington*

Nicolas Pinel, *University of Washington*

Jörg Piper, *Bad Bertrich (Germany)*

Thomas Pistole, *University of New Hampshire*

Edith Porter, *California State University–Los Angeles*

Michael Poulsen, *University of Wisconsin*

James Prosser, *University of Aberdeen (Scotland)*

Niels Peter Revsbech, *University of Aarhus (Denmark)*

Jackie Reynolds, *Richland College*

Kelly Reynolds, *University of Arizona*

Anna-Louise Reysenbach, *Portland State University*

Gary Roberts, *University of Wisconsin*

Melanie Romero-Guss, *Northeastern University*

Vladimir Samarkin, *University of Georgia*

Kathleen Sandman, *Ohio State University*

W. Matthew Sattley, *Indiana Wesleyan University*

Gene Scalarone, *Idaho State University*

Bernhard Schink, *Universität Konstanz (Germany)*

Tom Schmidt, *Michigan State University*

Timothy Sellati, *Albany Medical College*

Sara Silverstone, *Nazareth College*

Christopher Smith, *College of San Mateo*

Joyce Solheim, *University of Nebraska Medical Center*

Evan Solomon, *University of Washington*

John Spear, *Colorado School of Mines*

Nancy Spear, *Murphysboro, Illinois*

John Steiert, *Missouri State University*

Selvakumar Subbian, *University of Medicine and Dentistry of New Jersey*
Karen Sullivan, *Louisiana State University*
Jianming Tang, *University of Alabama–Birmingham*
Yi-Wei Tang, *Vanderbilt University*
Ralph Tanner, *University of Oklahoma*
J.H. Theis, *School of Medicine University of California–Davis*
Abbas Vafai, *Center for Disease Control and Prevention*
Alex Valm, *Woods Hole Oceanographic Institution*
Esta van Heerden, *University of the Free State (South Africa)*
Michael Wagner, *University of Vienna (Austria)*
David Ward, *Montana State University*
Gerhard Wanner, *Universität Munich (Germany)*
Ernesto Weil, *University of Puerto Rico*
Dave Westenberg, *Missouri University of Science and Technology*
William Whitman, *University of Georgia*
Fritz Widdel, *Max Planck Institute for Marine Microbiology (Germany)*
Arlene Wise, *University of Pennsylvania*
Carl Woese, *University of Illinois*
Howard Young
Vladimir Yurkov, *University of Manitoba (Canada)*
John Zamora, *Middle Tennessee State University*
Davide Zannoni, *University of Bologna (Italy)*
Stephen Zinder, *Cornell University*

International Contributors of select Microbial Sidebar features:
Luke Alderwick, *University of Birmingham*
Beatrix Fahnert, *University of Cardiff*
Yue Jiang, *Hong Kong Baptist University*
Mike Osta, *American University of Beirut*
Eng Lee Tan, *Singapore Polytechnic*

International Reviewers:
Jianzhong He, *National University of Singapore*
Stanley Lau, *Hong Kong University of Science and Technology*
Queck Choon Lau, *Ngee Ann Polytechnic*
Robin May, *University of Birmingham*
Stefan Schmidt, *University of KwaZulu Natal*

As hard as a publishing team may try, no textbook can ever be completely error free. Although we are confident the reader will be hard pressed to find errors in *BBOM* 13/e, any errors that do exist, either of commission or omission, are solely the responsibility of the authors. In past editions, users have been kind enough to contact us when they found an error. Users should feel free to continue to do so and to contact the authors directly about any errors, concerns, or questions they may have about the book. We will do our best to address them.

Michael T. Madigan (madigan@micro.siu.edu)
John M. Martinko (martinko@micro.siu.edu)
David A. Stahl (dastahl@u.washington.edu)
David P. Clark (clark@micro.siu.edu)

Brief Contents

Contents

22   Contents

UNIT 7 Microbial Ecology

UNIT 8   Antimicrobial Agents and Pathogenicity

UNIT 9   Immunology

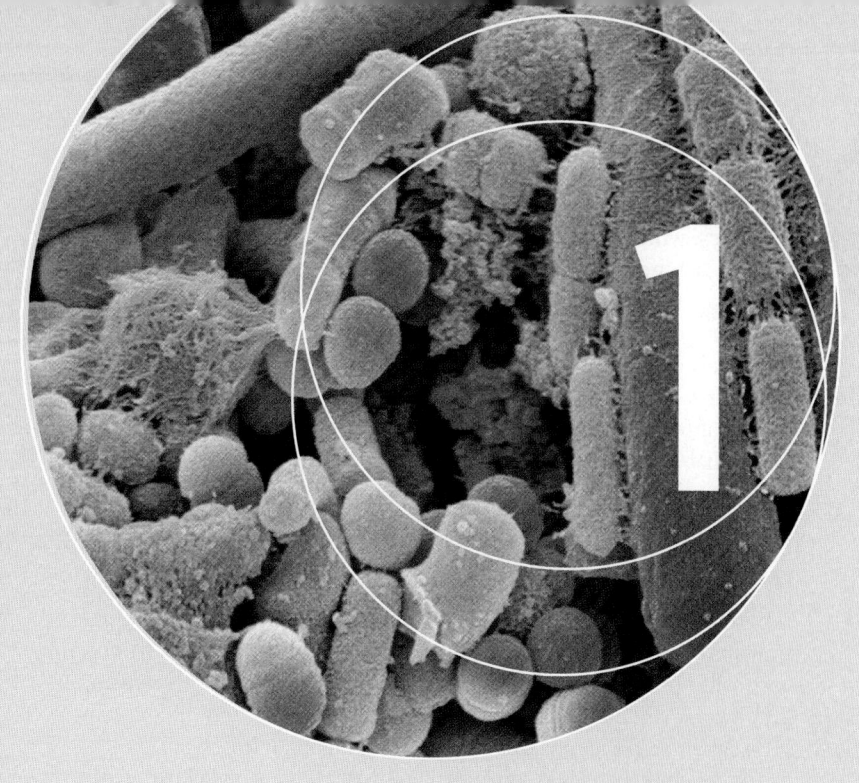

Microorganisms and Microbiology

Bacteria, such as these scraped from the surface of a human tongue, are independent microorganisms that live and interact with other microorganisms in microbial communities.

Microbiology is the study of microorganisms. **Microorganisms** are all single-celled microscopic organisms and include the viruses, which are microscopic but not cellular. Microbial cells differ in a fundamental way from the cells of plants and animals in that microorganisms are independent entities that carry out their life processes independently of other cells. By contrast, plant and animal cells are unable to live alone in nature and instead exist only as parts of multicellular structures, such as the organ systems of animals or the leaves of plants.

What is the science of microbiology all about? Microbiology is about microbial cells and how they work, especially the bacteria, a very large group of very small cells (**Figure 1.1**) that, collectively, have enormous basic and practical importance. Microbiology is about diversity and evolution of microbial cells, about how different kinds of microorganisms arose and why. It is also about what microorganisms do in the world at large, in soils and waters, in the human body, and in animals and plants. One way or another, microorganisms affect and support all other forms of life, and thus microbiology can be considered the most fundamental of the biological sciences.

This chapter begins our journey into the microbial world. Here we discover what microorganisms are and their impact on planet Earth. We set the stage for consideration of the structure and evolution of microorganisms that will unfold in the next chapter. We also place microbiology in historical perspective, as a process of scientific discovery. From the landmark contributions of both early microbiologists and scientists practicing today, we can see the effects that microorganisms have in medicine, agriculture, the environment, and other aspects of our daily lives.

Ⓘ Introduction to Microbiology

In the first five sections of this chapter we introduce the field of microbiology, look at microorganisms as cells, examine where and how microorganisms live in nature, survey the evolutionary history of microbial life, and examine the impact that microorganisms have had and continue to have on human affairs.

1.1 The Science of Microbiology

The science of microbiology revolves around two interconnected themes: (1) understanding the living world of microscopic organisms, and (2) applying our understanding of microbial life processes for the benefit of humankind and planet Earth.

As a *basic* biological science, microbiology uses and develops tools for probing the fundamental processes of life. Scientists have obtained a rather sophisticated understanding of the chemical and physical basis of life from studies of microorganisms because microbial cells share many characteristics with cells of multicellular organisms; indeed, *all* cells have much in common. But unlike plants and animals, microbial cells can be grown to extremely high densities in small-scale laboratory cultures (Figure 1.1), making them readily amenable to rapid biochemical and genetic study. Collectively, these features make microorganisms excellent experimental systems for illuminating life processes common to multicellular organisms, including humans.

As an *applied* biological science, microbiology is at the center of many important aspects of human and veterinary medicine, agriculture, and industry. For example, although animal and plant infectious diseases are typically microbial, many microorganisms are absolutely essential to soil fertility and domestic animal welfare. Many large-scale industrial processes, such as the production of antibiotics and human proteins, rely heavily on microorganisms. Thus microorganisms affect the everyday lives of humans in both beneficial and detrimental ways.

Although microorganisms are the smallest forms of life, collectively they constitute the bulk of biomass on Earth and carry out many necessary chemical reactions for higher organisms. In the absence of microorganisms, higher life forms would never have evolved and could not now be sustained. Indeed, the very oxygen we breathe is the result of past microbial activity (as we will see in Figure 1.6). Moreover, humans, plants, and animals are intimately tied to microbial activities for the recycling of key nutrients and for degrading organic matter. It is safe to say that no

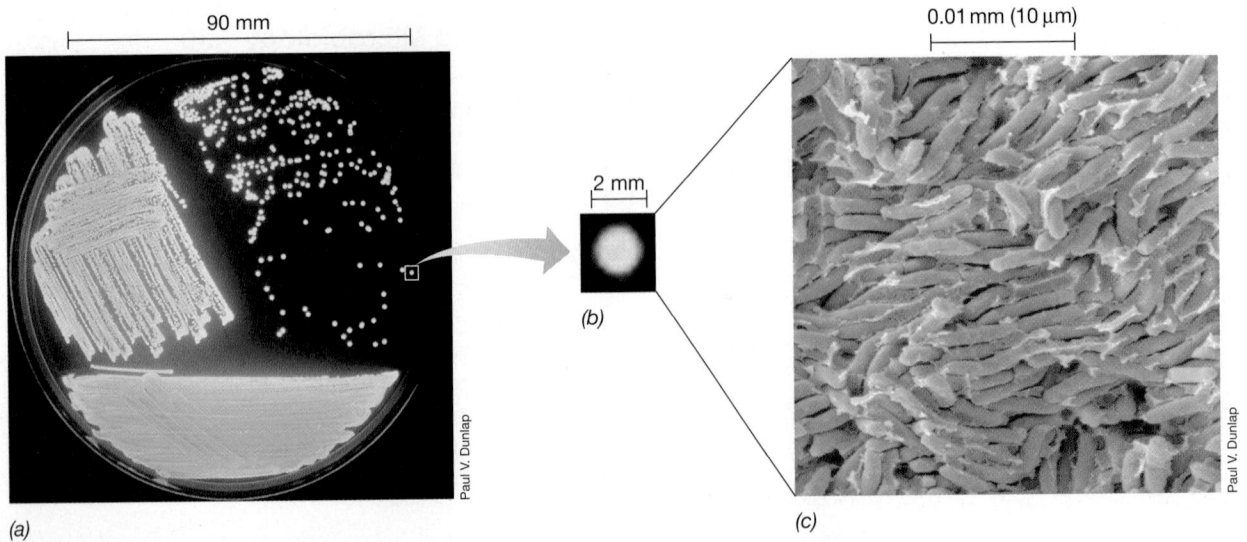

90 mm

0.01 mm (10 μm)

2 mm

(a) Paul V. Dunlap

(b)

(c) Paul V. Dunlap

Figure 1.1 Microbial cells. *(a)* Bioluminescent (light-emitting) colonies of the bacterium *Photobacterium* grown in laboratory culture on a Petri plate. *(b)* A single colony can contain more than 10 million (10^7) individual cells. *(c)* Scanning electron micrograph of cells of *Photobacterium*.

other life forms are as important as microorganisms for the support and maintenance of life on Earth.

Microorganisms existed on Earth for billions of years before plants and animals appeared, and we will see later that the genetic and physiological diversity of microbial life greatly exceeds that of the plants and animals. This huge diversity accounts for some of the spectacular properties of microorganisms. For example, we will see how microorganisms can live in places that would kill other organisms and how the diverse physiological capacities of microorganisms rank them as Earth's premier chemists. We will also trace the evolutionary history of microorganisms and see that three groups of cells can be distinguished by their evolutionary relationships. And finally, we will see how microorganisms have established important relationships with other organisms, some beneficial and some harmful.

We begin our study of microbiology with a consideration of the cellular structure of microorganisms.

MiniQuiz

- As they exist in nature, why can it be said that microbial cells differ fundamentally from the cells of higher organisms?
- Why are microbial cells useful tools for basic science?

1.2 Microbial Cells

A basic tenet of biology is that the **cell** is the fundamental unit of life. A single cell is an entity isolated from other such entities by a membrane; many cells also have a cell wall outside the membrane (**Figure 1.2**). The membrane defines the compartment that is the cell, maintains the correct proportions of internal constituents, and prevents leakage, while the wall lends structural strength to the cell. But the fact that a cell is a compartment does not mean that it is a *sealed* compartment. Instead, the membrane is semipermeable and thus the cell is an open, dynamic structure. Cells can communicate, move about, and exchange materials with their environments, and so they are constantly undergoing change.

Properties of Cellular Life

What essential properties characterize cells? **Figure 1.3** summarizes properties shared by all cellular microorganisms and additional properties that characterize only some of them. All cells show some form of **metabolism**. That is, they take up nutrients from the environment and transform them into new cell materials and waste products. During these transformations, energy is conserved in a form that can be drawn upon by the cell to support the synthesis of key structures. Production of the new structures culminates in the division of the cell to form two cells. The metabolic capabilities of cells can differ dramatically, but the final result of any cell's metabolic activities is to form two cells. In microbiology, we typically use the term **growth**, rather than "reproduction," to refer to the increase in cell number from cell division.

All cells undergo **evolution**, the process of descent with modification in which genetic variants are selected based on their reproductive fitness. Evolution is typically a slow process but can occur rapidly in microbial cells when selective pressure is strong. For example, we can witness today the selection for antibiotic resistance in pathogenic (disease-causing) bacteria by the indiscrimi-

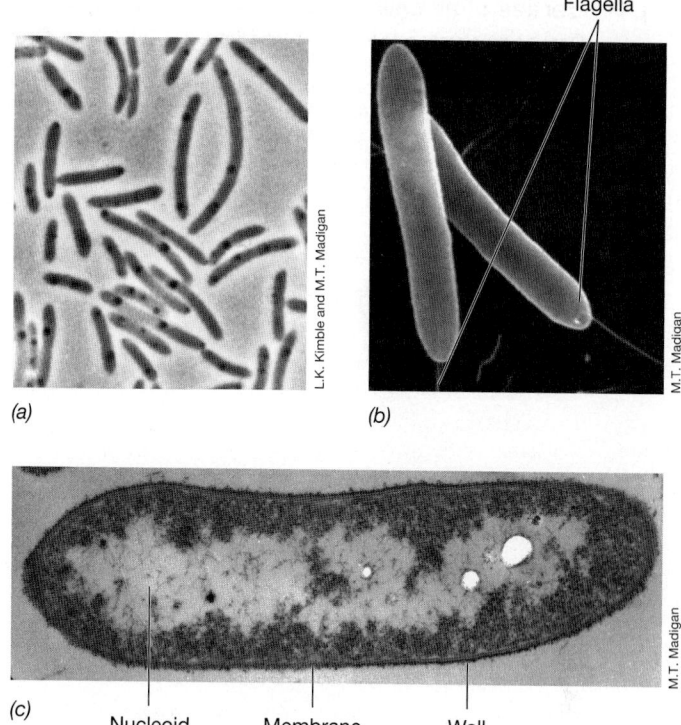

Flagella

(a)   *(b)*

L.K. Kimble and M.T. Madigan

M.T. Madigan

(c)   Nucleoid   Membrane   Wall

M.T. Madigan

Figure 1.2 Bacterial cells and some cell structures. *(a)* Rod-shaped cells of the bacterium *Heliobacterium modesticaldum* as seen in the light microscope; a single cell is about 1 μm in diameter. *(b)* Scanning electron micrograph of the same cells as in part a showing flagella, structures that rotate like a propeller and allow cells to swim. *(c)* Electron micrograph of a sectioned cell of *H. modesticaldum*. The light area is aggregated DNA, the nucleoid of the cell.

nate use of antibiotics in human and veterinary medicine. Evolution is *the* overarching theme of biology, and the tenets of evolution—variation and natural selection based on fitness—govern microbial life forms just as they do multicellular life forms.

Although all cells metabolize, grow, and evolve, the possession of other common properties varies from one species of cell to another. Many cells are capable of **motility**, typically by self-propulsion (Figure 1.2*b*). Motility allows cells to move away from danger or unfavorable conditions and to exploit new resources or opportunities. Some cells undergo **differentiation**, which may, for example, produce modified cells specialized for growth, dispersal, or survival. Some cells respond to chemical signals in their environment including those produced by other cells of either the same or different species. Responses to these signals may trigger new cellular activities. We can thus say that cells exhibit **communication**. As more is learned about this aspect of microbial life, it is quite possible that cell–cell communication will turn out to be a universal property of microbial cells.

Cells as Biochemical Catalysts and as Genetic Entities

The routine activities of cells can be viewed in two ways. On one hand, cells can be viewed as biochemical catalysts, carrying out the chemical reactions that constitute metabolism (**Figure 1.4**). On the other hand, cells can be viewed as genetic coding devices,

I. Properties of all cells

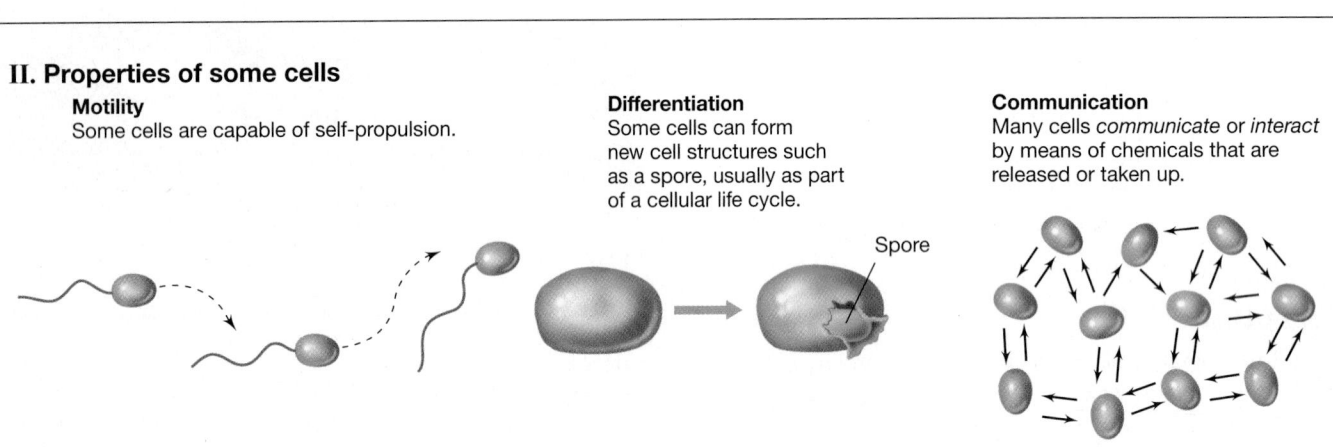

Compartmentalization and metabolism
A cell is a compartment that takes up nutrients from the environment, transforms them, and releases wastes into the environment. The cell is thus an *open* system.

Growth
Chemicals from the environment are turned into new cells under the genetic direction of preexisting cells.

Evolution
Cells contain genes and *evolve* to display new biological properties. Phylogenetic trees show the evolutionary relationships between cells.

Cell

Environment

Ancestral cell

Distinct species

Distinct species

II. Properties of some cells

Motility
Some cells are capable of self-propulsion.

Differentiation
Some cells can form new cell structures such as a spore, usually as part of a cellular life cycle.

Communication
Many cells *communicate* or *interact* by means of chemicals that are released or taken up.

Spore

Figure 1.3 The properties of cellular life.

replicating DNA and then processing it to form the RNAs and proteins needed for maintenance and growth under the prevailing conditions. DNA processing includes two main events, the production of RNAs (transcription) and the production of proteins (translation) (Figure 1.4).

Cells coordinate their catalytic and genetic functions to support cell growth. In the events that lead up to cell division, all constituents in the cell double. This requires that a cell's catalytic machinery, its **enzymes**, supply energy and precursors for the biosynthesis of all cell components, and that its entire complement of genes (its **genome**) replicates (Figure 1.4). The catalytic and genetic functions of the cell must therefore be highly coordinated. Also, as we will see later, these functions can be regulated to ensure that new cell materials are made in the proper order and concentrations and that the cell remains optimally tuned to its surroundings.

MiniQuiz

- What does the term "growth" mean in microbiology?
- List the six major properties of cells. Which of these are universal properties of all cells?
- Compare the catalytic and genetic functions of a microbial cell. Why is neither of value to a cell without the other?

1.3 Microorganisms and Their Environments

In nature, microbial cells live in populations in association with populations of cells of other species. A population is a group of cells derived from a single parental cell by successive cell divisions. The immediate environment in which a microbial population lives is called its **habitat**. Populations of cells interact with other populations in **microbial communities** (**Figure 1.5**). The diversity and abundance of microorganisms in microbial communities is controlled by the resources (foods) and conditions (temperature, pH, oxygen content, and so on) that prevail in their habitat.

Microbial populations interact with each other in beneficial, neutral, or harmful ways. For example, the metabolic waste products of one group of organisms can be nutrients or even poisons to other groups of organisms. Habitats differ markedly in their characteristics, and a habitat that is favorable for the growth of one organism may actually be harmful for another. Collectively, we call all the living organisms, together with the physical and chemical components of their environment, an **ecosystem**. Major microbial ecosystems are aquatic (oceans, ponds, lakes, streams, ice, hot springs), terrestrial (surface soils, deep subsurface), and other organisms, such as plants and animals.

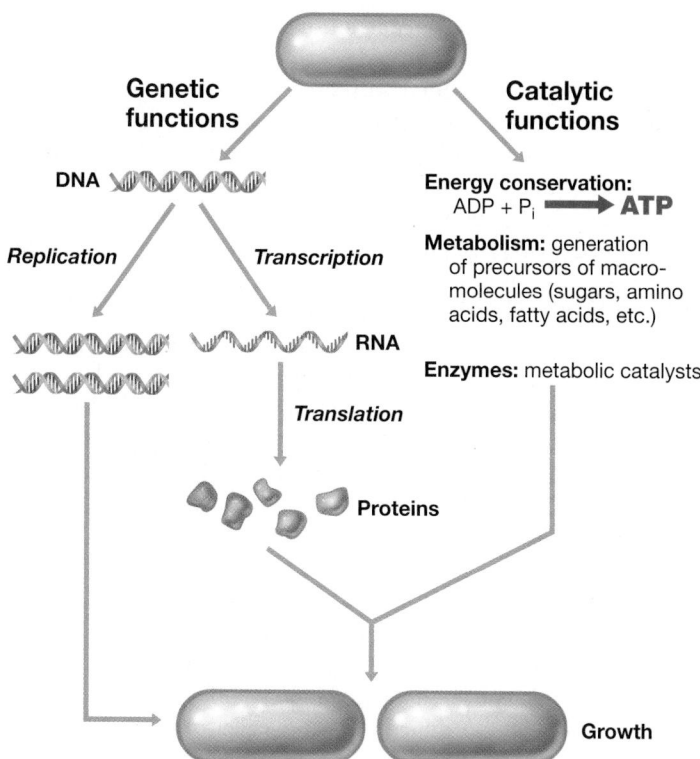

Figure 1.4 The catalytic and genetic functions of the cell. For a cell to reproduce itself there must be energy and precursors for the synthesis of new macromolecules, the genetic instructions must be replicated such that upon division each cell receives a copy, and genes must be expressed (transcribed and translated) to produce proteins and other macromolecules. Replication, transcription, and translation are the key molecular processes in cells.

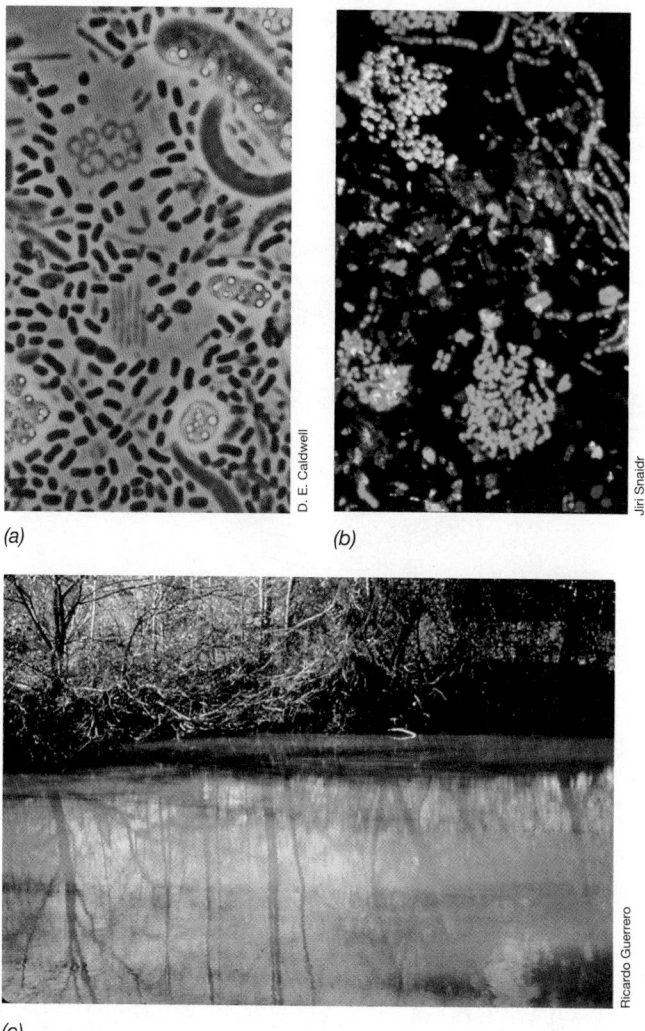

Figure 1.5 Microbial communities. *(a)* A bacterial community that developed in the depths of a small lake (Wintergreen Lake, Michigan), showing cells of various green and purple (large cells with sulfur granules) phototrophic bacteria. *(b)* A bacterial community in a sewage sludge sample. The sample was stained with a series of dyes, each of which stained a specific bacterial group. From *Journal of Bacteriology* 178: 3496–3500, Fig. 2b. © 1996 American Society for Microbiology. *(c)* Purple sulfur bacteria like that shown in part a (see also Figure 1.7a) that formed a dense bloom in a small Spanish lake.

An ecosystem is greatly influenced and in some cases even controlled by microbial activities. Microorganisms carrying out metabolic processes remove nutrients from the ecosystem and use them to build new cells. At the same time, they excrete waste products back into the environment. Thus, microbial ecosystems expand and contract, depending on the resources and conditions available. Over time, the metabolic activities of microorganisms gradually change their ecosystems, both chemically and physically. For example, molecular oxygen (O_2) is a vital nutrient for some microorganisms but a poison to others. If aerobic (oxygen-consuming) microorganisms remove O_2 from a habitat, rendering it anoxic (O_2 free), the changed conditions may favor the growth of anaerobic microorganisms that were formerly present in the habitat but unable to grow. In other words, as resources and conditions change in a microbial habitat, cell populations rise and fall, changing the habitat once again.

In later chapters, after we have learned about microbial structure and function, genetics, evolution, and diversity, we will return to a consideration of the ways in which microorganisms affect animals, plants, and the whole global ecosystem. This is the study of **microbial ecology**, perhaps the most exciting subdiscipline of microbiology today.

MiniQuiz

- How does a microbial community differ from a microbial population?

- What is a habitat? How can microorganisms change the characteristics of their habitats?

1.4 Evolution and the Extent of Microbial Life

Microorganisms were the first entities on Earth with the properties of living systems (Figure 1.3), and we will see that a particular group of microorganisms called the *cyanobacteria* were pivotal

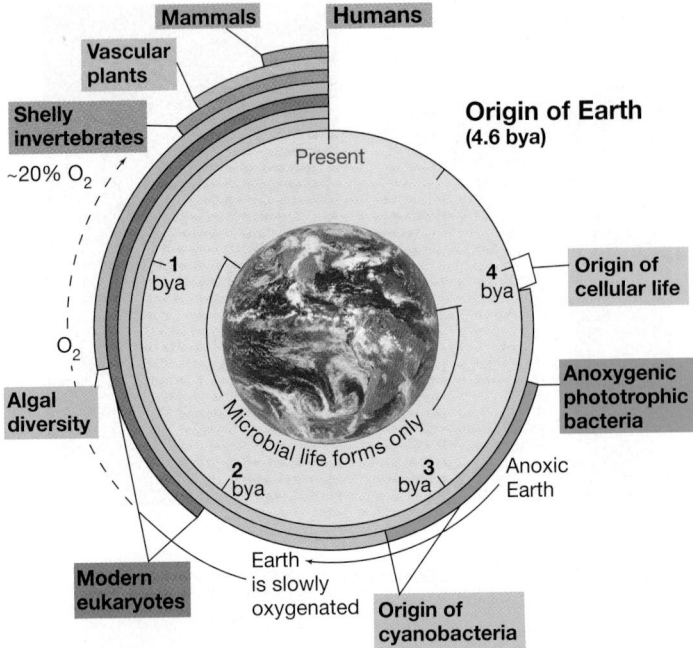

(a)

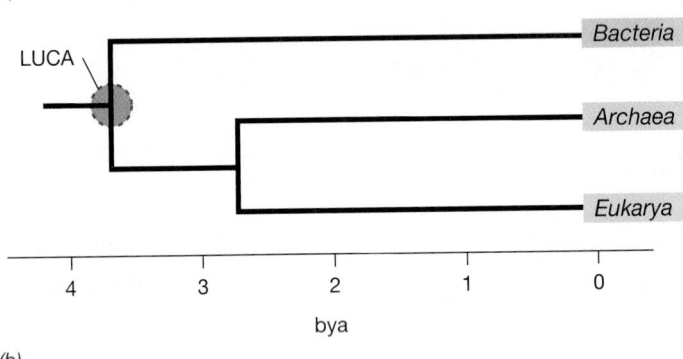

(b)

Figure 1.6 A summary of life on Earth through time and origin of the cellular domains. *(a)* Cellular life was present on Earth about 3.8 billion years ago (bya). Cyanobacteria began the slow oxygenation of Earth about 3 bya, but current levels of O_2 in the atmosphere were not achieved until 500–800 million years ago. Eukaryotes are nucleated cells and include both microbial and multicellular organisms. (Shelly invertebrates have shells or shell-like parts.) *(b)* The three domains of cellular organisms are *Bacteria*, *Archaea*, and *Eukarya*. The latter two lineages diverged long before nucleated cells with organelles (labeled as "modern eukaryotes" in part a) appear in the fossil record. LUCA, last universal common ancestor. Note that 80% of Earth's history was exclusively microbial.

in biological evolution because oxygen (O_2)—a waste product of their metabolism—prepared planet Earth for more complex life forms.

The First Cells and the Onset of Biological Evolution

How did cells originate? Were cells as we know them today the first self-replicating structures on Earth? Because all cells are constructed in similar ways, it is thought that all cells have descended from a common ancestral cell, the *last universal common ancestor* (LUCA). After the first cells arose from nonliving

materials, a process that occurred over hundreds of millions of years, their subsequent growth formed cell populations, and these then began to interact with other populations in microbial communities. Evolution selected for improvements and diversification of these early cells to eventually yield the highly complex and diverse cells we see today. We will consider this complexity and diversity in Chapters 2 and 17–21. We consider the topic of how life originated from nonliving materials in Chapter 16.

Life on Earth through the Ages

Earth is 4.6 billion years old. Scientists have evidence that cells first appeared on Earth between 3.8 and 3.9 billion years ago, and these organisms were exclusively microbial. In fact, microorganisms were the only life on Earth for most of its history (**Figure 1.6**). Gradually, and over enormous periods of time, more complex organisms appeared. What were some of the highlights along the way?

During the first 2 billion years or so of Earth's existence, its atmosphere was anoxic; O_2 was absent, and nitrogen (N_2), carbon dioxide (CO_2), and a few other gases were present. Only microorganisms capable of anaerobic metabolisms could survive under these conditions, but these included many different types of cells, including those that produce methane, called *methanogens*. The evolution of phototrophic microorganisms—organisms that harvest energy from sunlight—occurred within a billion years of the formation of Earth. The first phototrophs were relatively simple ones, such as purple bacteria and other anoxygenic (non-oxygen-evolving) phototrophs (**Figure 1.7a**; see also Figure 1.5), which are still widespread in anoxic habitats today. Cyanobacteria (oxygenic, or oxygen-evolving, phototrophs) (Figure 1.7b) evolved from anoxygenic phototrophs nearly a billion years later and began the slow process of oxygenating the atmosphere. Triggered by increases in O_2 in the atmosphere, multicellular life forms eventually evolved and continued to increase in complexity, culminating in the plants and animals we know today (Figure 1.6). We will

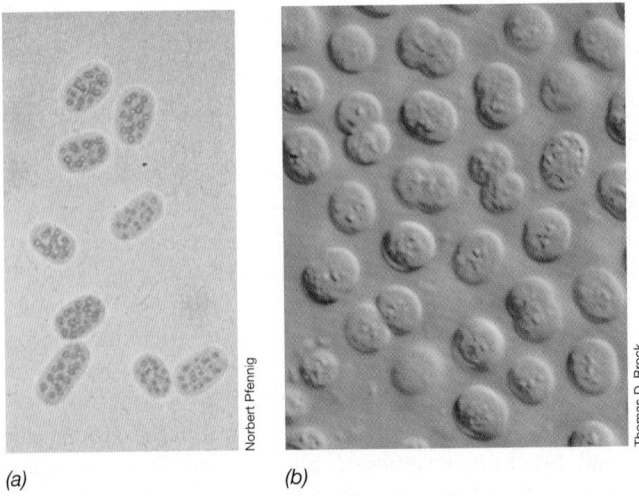

(a)   (b)

Figure 1.7 Phototrophic microorganisms. *(a)* Purple sulfur bacteria (anoxygenic phototrophs). *(b)* Cyanobacteria (oxygenic phototrophs). Purple bacteria appeared on Earth long before oxygenic phototrophs evolved (see Figure 1.6a).

explore the evolutionary history of life later, but note here that the events that unfolded beyond LUCA led to the evolution of three major lineages of microbial cells, the *Bacteria*, the *Archaea*, and the *Eukarya* (Figure 1.6*b*); microbial *Eukarya* were the ancestors of the plants and animals.

How do we know that evolutionary events unfolded as summarized in Figure 1.6? The answer is that we may never know that all details in our description are correct. However, scientists can reconstruct evolutionary transitions by using *biomarkers*, specific molecules that are unique to particular groups in present-day microorganisms. The presence or absence of a given biomarker in ancient rocks of a known age therefore reveals whether that particular group was present at that time.

One way or the other and over enormous periods of time (Figure 1.6), natural selection filled every suitable habitat on Earth with one or more populations of microorganisms. This brings us to the question of the current distribution of microbial life on Earth. What do we know about this important topic?

The Extent of Microbial Life

Microbial life is all around us. Examination of natural materials such as soil or water invariably reveals microbial cells. But unusual habitats such as boiling hot springs and glacial ice are also teeming with microorganisms. Although widespread on Earth, such tiny cells may seem inconsequential. But if we could count them all, what number would we reach?

Estimates of total microbial cell numbers on Earth are on the order of 2.5×10^{30} cells. The total amount of carbon present in this very large number of very small cells equals that of all plants on Earth (and plant carbon far exceeds animal carbon). But in addition, the collective contents of nitrogen and phosphorus in microbial cells is more than 10 times that in all plant biomass.

Thus, microbial cells, small as they are, constitute the major fraction of biomass on Earth and are key reservoirs of essential nutrients for life. Most microbial cells are found in just a few very large habitats. For example, most microbial cells do not reside on Earth's *surface* but instead lie underground in the oceanic and terrestrial *subsurface* (**Table 1.1**). Depths up to about 10 km under Earth's surface are clearly suitable for microbial life. We will see later that subsurface microbial habitats support diverse populations of microbial cells that make their livings in unusual ways and grow extremely slowly. By comparison to the subsurface, surface soils and waters contain a relatively small percentage of the total microbial cell numbers, and animals (including humans), which can be heavily colonized with microorganisms (see Figure 1.10), collectively contain only a tiny fraction of the total microbial cells on Earth (Table 1.1).

Because most of what we know about microbial life has come from the study of surface-dwelling organisms, there is obviously much left for future generations of microbiologists to discover and understand about the life forms that dominate Earth's biology. And when we consider the fact that surface-dwelling organisms already show enormous diversity, the hunt for new microorganisms in Earth's unexplored habitats should yield some exciting surprises.

Table 1.1 *Distribution of microorganisms in and on Earth[a]*

Habitat	Percent of total
Marine subsurface	66
Terrestrial subsurface	26
Surface soil	4.8
Oceans	2.2
All other habitats[b]	1.0

[a]Data compiled by William Whitman, University of Georgia, USA; refer to total numbers (estimated to be about 2.5×10^{30} cells) of *Bacteria* and *Archaea*. This enormous number of cells contain, collectively, about 5×10^{17} grams of carbon.
[b]Includes, in order of decreasing numbers: freshwater and salt lakes, domesticated animals, sea ice, termites, humans, and domesticated birds.

MiniQuiz

- What is LUCA and what major lineages of cells evolved from LUCA? Why were cyanobacteria so important in the evolution of life on Earth?

- How old is Earth, and when did cellular life forms first appear? How can we use science to reconstruct the sequence of organisms that appeared on Earth?

- Where are most microbial cells located on Earth?

1.5 The Impact of Microorganisms on Humans

Through the years microbiologists have had great success in discovering how microorganisms work, and application of this knowledge has greatly increased the beneficial effects of microorganisms and curtailed many of their harmful effects. Microbiology has thus greatly advanced human health and welfare. Besides understanding microorganisms as agents of disease, microbiology has made great advances in understanding the role of microorganisms in food and agriculture, and in exploiting microbial activities for producing valuable human products, generating energy, and cleaning up the environment.

Microorganisms as Agents of Disease

The statistics summarized in **Figure 1.8** show microbiologists' success in preventing infectious diseases since the beginning of the twentieth century. These data compare today's leading causes of death in the United States with those of 100 years ago. At the beginning of the twentieth century, the major causes of death in humans were infectious diseases caused by microorganisms called **pathogens**. Children and the aged in particular succumbed in large numbers to microbial diseases. Today, however, infectious diseases are much less deadly, at least in developed countries. Control of infectious disease has come from an increased understanding of disease processes, improved sanitary and public health practices, and the use of antimicrobial agents, such as antibiotics. As we will see from the next sections, the development of microbiology as a science can trace important aspects of its roots to studies of infectious disease.

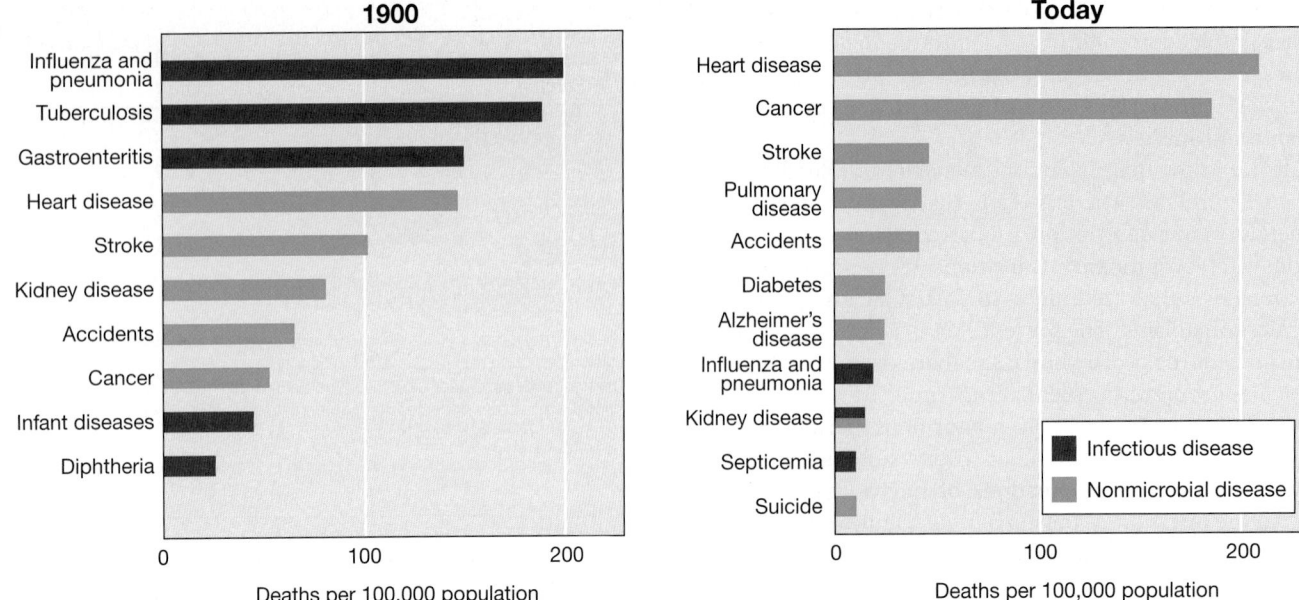

Figure 1.8 Death rates for the leading causes of death in the United States: 1900 and today. Infectious diseases were the leading causes of death in 1900, whereas today they account for relatively few deaths. Kidney diseases can be the result of microbial infections or systemic sources (diabetes, certain cancers, toxicities, metabolic diseases, etc.). Data are from the United States National Center for Health Statistics and the Centers for Disease Control and Prevention and are typical of recent years.

Although many infectious diseases can now be controlled, microorganisms can still be a major threat, particularly in developing countries. In the latter, microbial diseases are still the major causes of death, and millions still die yearly from other microbial diseases such as malaria, tuberculosis, cholera, African sleeping sickness, measles, pneumonia and other respiratory diseases, and diarrheal syndromes. In addition to these, humans worldwide are under threat from diseases that could emerge suddenly, such as bird or swine flu, or Ebola hemorrhagic fever, which are primarily animal diseases that under certain circumstances can be transmitted to humans and spread quickly through a population. And if this were not enough, consider the threat to humans worldwide from those who would deploy microbial bioterrorism agents! Clearly, microorganisms are still serious health threats to humans in all parts of the world.

Although we should obviously appreciate the powerful threat posed by pathogenic microorganisms, in reality, most microorganisms are not harmful to humans. In fact, most microorganisms cause no harm but instead are beneficial—and in many cases even essential—to human welfare and the functioning of the planet. We turn our attention to these microorganisms now.

Microorganisms, Digestive Processes, and Agriculture

Agriculture benefits from the cycling of nutrients by microorganisms. For example, a number of major crop plants are legumes. Legumes live in close association with bacteria that form structures called *nodules* on their roots. In the root nodules, these bacteria convert atmospheric nitrogen (N_2) into ammonia (NH_3) that the plants use as a nitrogen source for growth (**Figure 1.9**).

Thanks to the activities of these nitrogen-fixing bacteria, the legumes have no need for costly and polluting nitrogen fertilizers. Other bacteria cycle sulfur compounds, oxidizing toxic sulfur species such as hydrogen sulfide (H_2S) into sulfate (SO_4^{2-}), which is an essential plant nutrient (Figure 1.9c).

Also of major agricultural importance are the microorganisms that inhabit ruminant animals, such as cattle and sheep. These important domesticated animals have a characteristic digestive vessel called the *rumen* in which large populations of microorganisms digest and ferment cellulose, the major component of plant cell walls, at neutral pH (Figure 1.9d). Without these symbiotic microorganisms, cattle and sheep could not thrive on cellulose-rich (but otherwise nutrient-poor) food, such as grass and hay. Many domesticated and wild herbivorous mammals—including deer, bison, camels, giraffes, and goats—are also ruminants.

The ruminant digestive system contrasts sharply with that of humans and most other animals. In humans, food enters a highly acidic stomach where major digestive processes are chemical rather than microbial. In the human digestive tract, large microbial populations occur only in the colon (large intestine), a structure that comes after the stomach and small intestine and which lacks significant numbers of cellulose-degrading bacteria. However, other parts of the human body can be loaded with bacteria. In addition to the large intestine, the skin and oral cavity (**Figure 1.10**) contain a significant normal microbial flora, most of which benefits the host or at least does no harm.

In addition to benefiting plants and animals, microorganisms can also, of course, have negative effects on them. Microbial diseases of plants and animals used for human food cause major

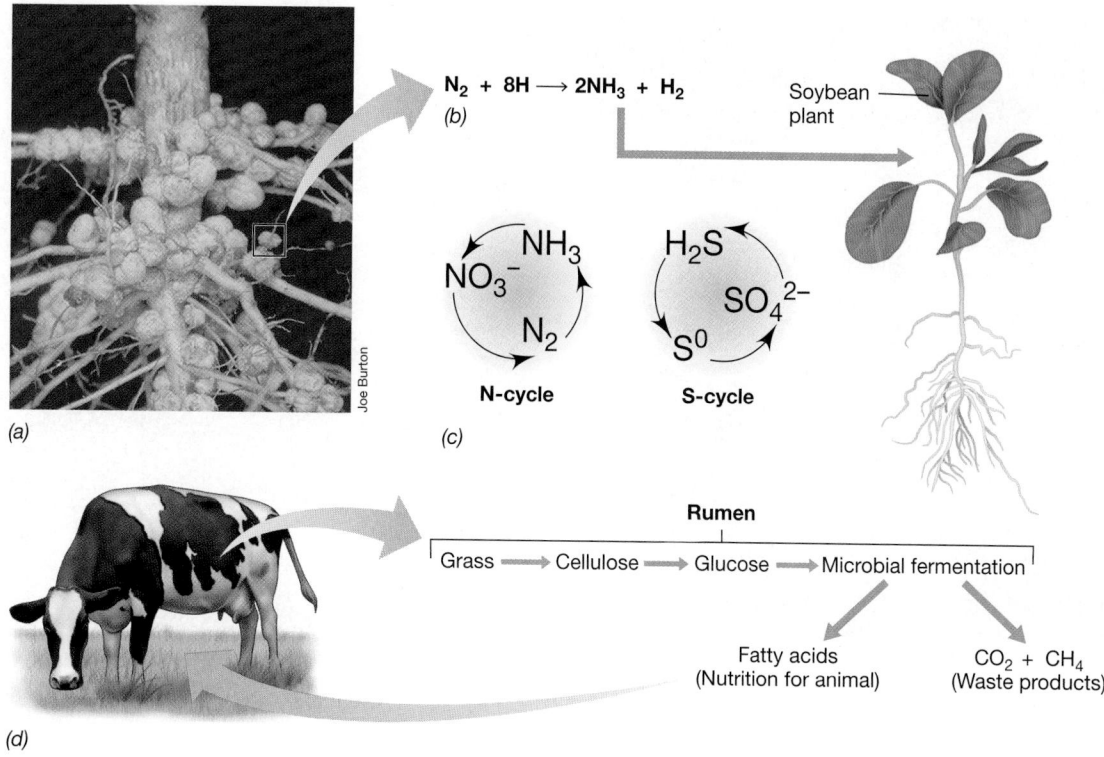

$$N_2 + 8H \longrightarrow 2NH_3 + H_2$$
(b)

Soybean plant

NH₃
NO₃⁻
N₂
N-cycle

H₂S
SO₄²⁻
S⁰
S-cycle
(c)

Joe Burton

(a)

Rumen

Grass → Cellulose → Glucose → Microbial fermentation

Fatty acids
(Nutrition for animal)

$CO_2 + CH_4$
(Waste products)

(d)

Figure 1.9 **Microorganisms in modern agriculture.** *(a, b)* Root nodules on this soybean plant contain bacteria that fix molecular nitrogen (N_2) for use by the plant. *(c)* The nitrogen and sulfur cycles, key nutrient cycles in nature. *(d)* Ruminant animals. Microorganisms in the rumen of the cow convert cellulose from grass into fatty acids that can be used by the animal.

economic losses in the agricultural industry every year. In some cases a food product can cause serious human disease, such as when pathogenic *Escherichia coli* or *Salmonella* is transmitted from infected meat, or when microbial pathogens are ingested with contaminated fresh fruits and vegetables. Thus microorganisms significantly impact the agriculture industry both positively and negatively.

Microorganisms and Food, Energy, and the Environment

Microorganisms play important roles in the food industry, including in the areas of spoilage, safety, and production. After plants and animals are produced for human consumption, the products must be delivered to consumers in a wholesome form. Food spoilage alone results in huge economic losses each year. Indeed, the canning, frozen food, and dried-food industries were founded as means to preserve foods that would otherwise easily undergo microbial spoilage. Food safety requires constant monitoring of food products to ensure they are free of pathogenic microorganisms and to track disease outbreaks to identify the source(s).

However, not all microorganisms in foods have harmful effects on food products or those who eat them. For example, many dairy products depend on the activities of microorganisms, including the fermentations that yield cheeses, yogurt, and buttermilk. Sauerkraut, pickles, and some sausages are also products of microbial fermentations. Moreover, baked goods and alcoholic

beverages rely on the fermentative activities of yeast, which generate carbon dioxide (CO_2) to raise the dough and alcohol as a key ingredient, respectively. Many of these fermentations are discussed in Chapter 14.

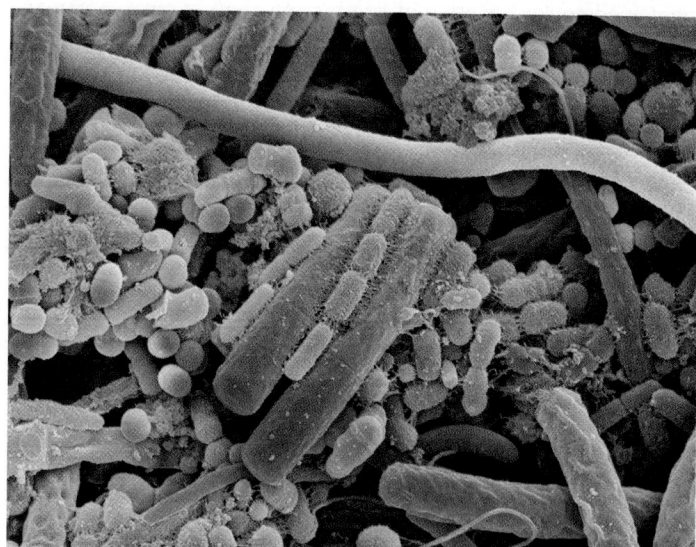

Figure 1.10 **Human oral bacterial community.** The oral cavity of warm-blooded animals contains high numbers of various bacteria, as shown in this electron micrograph (false color) of cells scraped from a human tongue.

(a)   *(b)*

Figure 1.11 Biofuels. *(a)* Natural gas (methane) is collected in a funnel from swamp sediments where it was produced by methanogens and then ignited as a demonstration experiment. *(b)* An ethanol plant in the United States. Sugars obtained from corn or other crops are fermented to ethanol for use as a motor fuel extender.

Some microorganisms produce *biofuels*. Natural gas (methane) is a product of the anaerobic degradation of organic matter by methanogenic microorganisms (**Figure 1.11**). Ethyl alcohol (ethanol), which is produced by the microbial fermentation of glucose from feedstocks such as sugarcane or cornstarch, is a major motor fuel in some countries (Figure 1.11*b*). Waste materials such as domestic refuse, animal wastes, and cellulose can also be converted to biofuels by microbial activities and are more efficient feedstocks for ethanol production than is corn. Soybeans are also used as biofuel feedstocks, as soybean oils can be converted into biodiesel to fuel diesel engines. As global oil production is waning, it is likely that various biofuels will take on a greater and greater part of the global energy picture.

Microorganisms are used to clean up human pollution, a process called *microbial bioremediation*, and to produce commercially valuable products by *industrial microbiology* and *biotechnology*. For example, microorganisms can be used to consume spilled oil, solvents, pesticides, and other environmentally toxic pollutants. Bioremediation accelerates cleanup in either of two ways: (1) by introducing specific microorganisms to a polluted environment, or (2) by adding nutrients that stimulate preexisting microorganisms to degrade the pollutants. In both cases the goal is to accelerate metabolism of the pollutant.

In industrial microbiology, microorganisms are grown on a large scale to make products of relatively low commercial value, such as antibiotics, enzymes, and various chemicals. By contrast, the related field of biotechnology employs *genetically engineered* microorganisms to synthesize products of high commercial value, such as human proteins. **Genomics** is the science of the identification and analysis of genomes and has greatly enhanced

biotechnology. Using genomic methods, biotechnologists can access the genome of virtually any organism and search in it for genes encoding proteins of commercial interest.

At this point the influence of microorganisms on humans should be apparent. Microorganisms are essential for life and their activities can cause significant benefit or harm to humans. As the eminent French scientist Louis Pasteur, one of the founders of microbiology, expressed it: "The role of the infinitely small in nature is infinitely large." We continue our introduction to the microbial world in the next section with an historical overview of the contributions of Pasteur and a few other key scientists.

MiniQuiz
- List two ways in which microorganisms are important in the food and agricultural industries.
- Which biofuel is widely used in many countries as a motor fuel?
- What is biotechnology and how might it improve the lives of humans?

ⓘ Pathways of Discovery in Microbiology

The future of any science is rooted in its past accomplishments. Although microbiology claims very early roots, the science did not really develop in a systematic way until the nineteenth century. Since that time, microbiology has expanded in a way

Table 1.2 *Giants of the early days of microbiology and their major contributions*

Investigator	Nationality	Dates[a]	Contributions
Robert Hooke	English	1664	Discovery of microorganisms (fungi)
Antoni van Leeuwenhoek	Dutch	1684	Discovery of bacteria
Edward Jenner	English	1798	Vaccination (smallpox)
Louis Pasteur	French	Mid- to late 1800s	Mechanism of fermentation, defeat of spontaneous generation, rabies and other vaccines, principles of immunization
Joseph Lister	English	1867	Methods for preventing infections during surgeries
Ferdinand Cohn	German	1876	Discovery of endospores
Robert Koch	German	Late 1800s	Koch's postulates, pure culture microbiology, discovery of agents of tuberculosis and cholera
Sergei Winogradsky	Russian	Late 1800s to mid-1900s	Chemolithotrophy and chemoautotrophy, nitrogen fixation, sulfur bacteria
Martinus Beijerinck	Dutch	Late 1800s to 1920	Enrichment culture technique, discovery of many metabolic groups of bacteria, concept of a virus

[a]The year in which the key paper describing the contribution was published, or the date range in which the investigator was most scientifically active.

unprecedented by any of the other biological sciences and has spawned several new but related fields. We retrace these pathways of discovery now and discuss a few of the major contributors (**Table 1.2**).

1.6 The Historical Roots of Microbiology: Hooke, van Leeuwenhoek, and Cohn

Although the existence of creatures too small to be seen with the naked eye had long been suspected, their discovery was linked to the invention of the microscope. Robert Hooke (1635–1703), an English mathematician and natural historian, was also an excellent microscopist. In his famous book *Micrographia* (1665), the first book devoted to microscopic observations, Hooke illustrated, among many other things, the fruiting structures of molds (**Figure 1.12**). This was the first known description of microorganisms. The first person to see bacteria was the Dutch draper and amateur microscope builder Antoni van Leeuwenhoek (1632–1723). In 1684, van Leeuwenhoek, who was well aware of the work of Hooke, used extremely simple microscopes of his own construction (**Figure 1.13**) to examine the microbial content of natural substances.

Van Leeuwenhoek's microscopes were crude by today's standards, but by careful manipulation and focusing he was able to see bacteria, microorganisms considerably smaller than molds (molds are fungi). He discovered bacteria in 1676 while studying pepper–water infusions. He reported his observations in a series of letters to the prestigious Royal Society of London, which published them in 1684 in English translation. Drawings of some of van Leeuwenhoek's "wee animalcules," as he referred to them, are shown in Figure 1.13*b*, and a photo taken through such a microscope is shown in Figure 1.13*c*.

As years went by, van Leeuwenhoek's observations were confirmed by many others. However, primarily because of the lack of experimental tools, little progress in understanding the nature and importance of the tiny creatures was made for almost 150 years. Only in the nineteenth century did improved microscopes and some simple tools for growing microoorganisms in the laboratory become available, and using these, the extent and nature of microbial life became more apparent.

In the mid- to late nineteenth century major advances in the science of microbiology were made because of the attention given to two major questions that pervaded biology and medicine at the time: (1) Does spontaneous generation occur? and (2) What is the nature of infectious disease? Answers to these seminal questions emerged from the work of two giants in the fledgling field of microbiology: the French chemist Louis Pasteur and the German physician Robert Koch. But before we explore their work, let us briefly consider the groundbreaking efforts of a German botanist, Ferdinand Cohn, a contemporary of Pasteur and Koch, and the founder of the field we now call *bacteriology*.

Ferdinand Cohn (1828–1898) was born in Breslau (now in Poland). He was trained as a botanist and became an excellent microscopist. His interests in microscopy led him to the study of unicellular algae and later to bacteria, including the large sulfur bacterium *Beggiatoa* (**Figure 1.14**). Cohn was particularly interested in heat resistance in bacteria, which led to his discovery that some bacteria form endospores. We now know that bacterial endospores are formed by differentiation from the mother (vegetative) cell (Figure 1.3) and that endospores are extremely heat-resistant. Cohn described the life cycle of the endospore-forming bacterium *Bacillus* (vegetative cell → endospore → vegetative cell) and showed that vegetative cells but not endospores were killed by boiling.

Cohn is credited with many other accomplishments. He laid the groundwork for a system of bacterial classification, including an early attempt to define a bacterial species, an issue still unresolved today, and founded a major scientific journal of plant and microbial biology. He strongly advocated use of the techniques and research of Robert Koch, the first medical microbiologist. Cohn devised simple but effective methods for preventing the contamination of culture media, such as the use of cotton for closing flasks and tubes. These methods were later used by Koch and allowed him to make rapid progress in the isolation and characterization of several disease-causing bacteria (Section 1.8).

1.7 Pasteur and the Defeat of Spontaneous Generation

The late nineteenth century saw the science of microbiology blossom. The theory of spontaneous generation was crushed by the brilliant work of the Frenchman Louis Pasteur (1822–1895).

Optical Isomers and Fermentations

Pasteur was a chemist by training and was one of the first to recognize the significance of *optical isomers*. A molecule is optically active if a pure solution or crystal diffracts light in only one direction. Pasteur studied crystals of tartaric acid that he separated by hand into those that bent a beam of polarized light to the left and those that bent the beam to the right (**Figure 1.15**). Pasteur found that the mold *Aspergillus* metabolized D-tartrate, which bent light to the right, but did not metabolize its optical isomer, L-tartrate. The fact that a living organism could discriminate between optical isomers was of profound significance to Pasteur, and he began to see living organisms as inherently asymmetric entities.

Pasteur's thinking on the asymmetry of life carried over into his work on fermentations and, eventually, spontaneous generation. At the invitation of a local industrialist who was having problems making alcohol from the fermentation of beets, Pasteur studied the mechanism of the alcoholic fermentation, at that time thought to be a strictly chemical process. The yeast cells in the fermenting broth were thought to be a complex chemical substance formed by the fermentation. Although ethyl alcohol does not form optical isomers, one of the side products of beet fermentation is amyl alcohol, which does, and Pasteur tested the fermenting juice and found the amyl alcohol to be of only one optical isomer. From his work on tartrate metabolism this suggested to Pasteur that the beet fermentation was a biological process. Microscopic observations and other simple but rigorous experiments convinced Pasteur that the alcoholic fermentation was catalyzed by living organisms, the yeast cells. Indeed, in Pasteur's own words: ". . . fermentation is associated with the life and structural integrity of the cells and not with their death and decay." From this foundation, Pasteur began a series of classic experiments on spontaneous generation, experiments that are forever linked to his name and to the science of microbiology.

Spontaneous Generation

The concept of **spontaneous generation** had existed since biblical times and its basic tenet can be easily grasped. For example, if

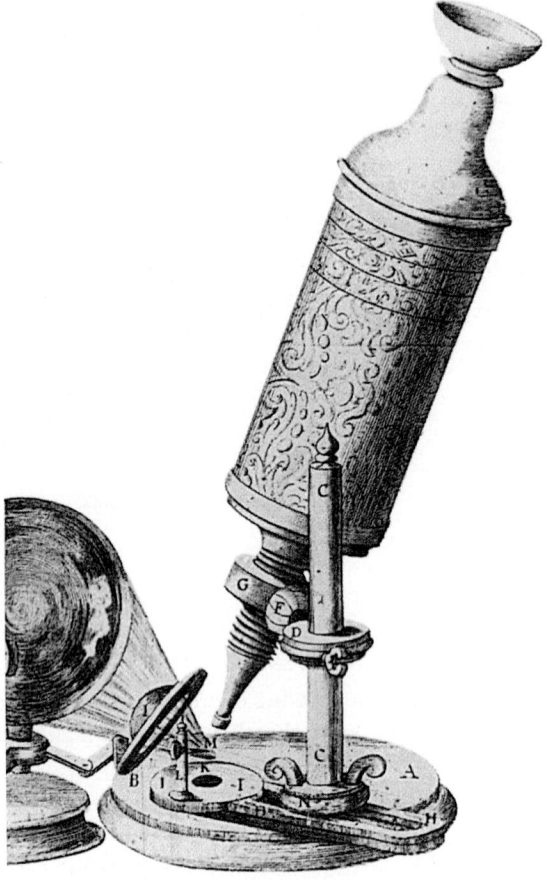

(a)

(b)

Figure 1.12 Robert Hooke and early microscopy. *(a)* A drawing of the microscope used by Robert Hooke in 1664. The lens was fitted at the end of an adjustable bellows (G) and light focused on the specimen by a separate lens (1). *(b)* This drawing of a mold that was growing on the surface of leather, together with other drawings and accompanying text published by Robert Hooke in *Micrographia* in 1665, were the first descriptions of microorganisms. The round structures contain spores of the mold. Compare Hooke's microscope with that of van Leeuwenhoek's shown in Figure 1.13.

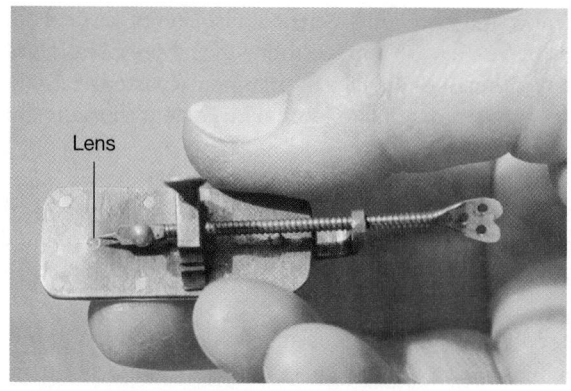

(a)

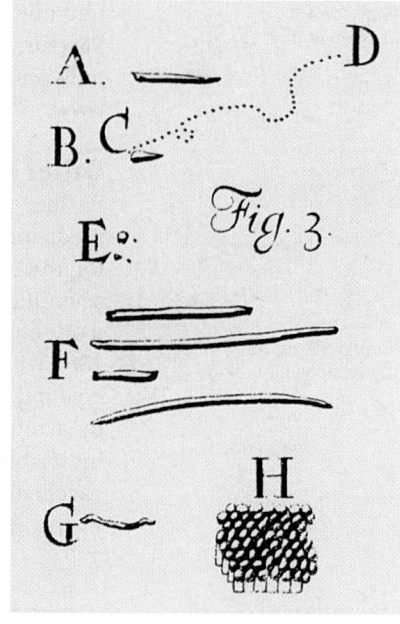

(b)

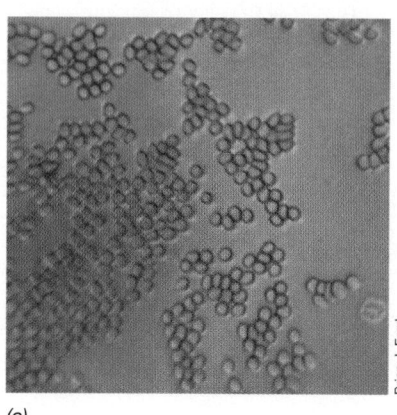

(c)

Figure 1.13 The van Leeuwenhoek microscope. *(a)* A replica of Antoni van Leeuwenhoek's microscope. *(b)* Van Leeuwenhoek's drawings of bacteria, published in 1684. Even from these simple drawings we can recognize several shapes of common bacteria: A, C, F, and G, rods; E, cocci; H, packets of cocci. *(c)* Photomicrograph of a human blood smear taken through a van Leeuwenhoek microscope. Red blood cells are clearly apparent.

food is allowed to stand for some time, it putrefies. When examined microscopically, the putrefied food is seen to be teeming with bacteria and perhaps even maggots and worms. From where do these organisms not apparent in the fresh food originate? Some people said they developed from seeds or germs that entered the food from air. Others said they arose spontaneously from nonliving materials, that is, by *spontaneous generation.* Who was right? Keen insight was necessary to solve this controversy, and this was exactly the kind of problem that appealed to Louis Pasteur.

Pasteur became a powerful opponent of spontaneous generation. Following his discoveries about fermentation, Pasteur predicted that microorganisms observed in putrefying materials are also present in air and that putrefaction resulted from the activities of microorganisms that entered from the air or that had been present on the surfaces of the containers holding the decaying materials. Pasteur further reasoned that if food were treated in such a way as to destroy all living organisms contaminating it, that is, if it were rendered **sterile** and then protected from further contamination, it should not putrefy.

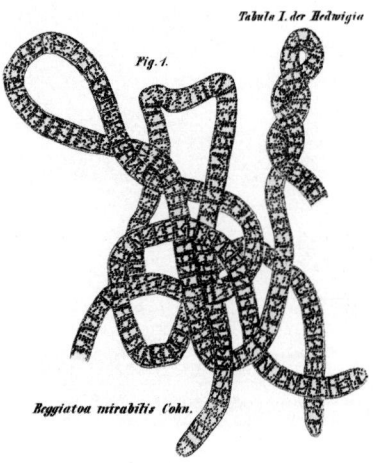

Figure 1.14 Drawing by Ferdinand Cohn of large filamentous sulfur-oxidizing bacteria *Beggiatoa mirabilis.* The small granules inside the cells consist of elemental sulfur, produced from the oxidation of hydrogen sulfide (H_2S). Cohn was the first to identify the granules as sulfur in 1866. A cell of *B. mirabilis* is about 15 µm in diameter. Compare with Figure 1.22b. *Beggiatoa* moves on solid surfaces by a gliding mechanism and in so doing, cells often twist about one another.

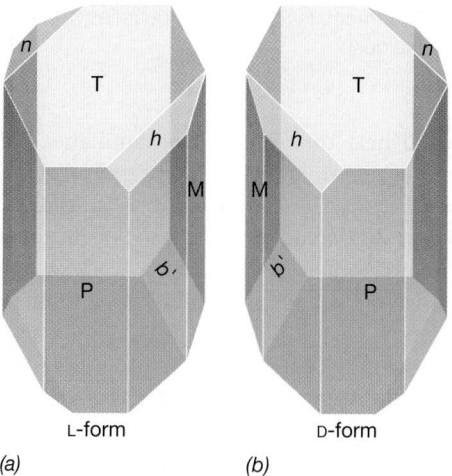

(a)   (b)

Figure 1.15 Louis Pasteur's drawings of tartaric acid crystals from his famous paper on optical activity. *(a)* Left-handed crystal (bends light to the left). *(b)* Right-handed crystal (bends light to the right). Note that the two crystals are mirror images of one another, a hallmark of optical isomers.

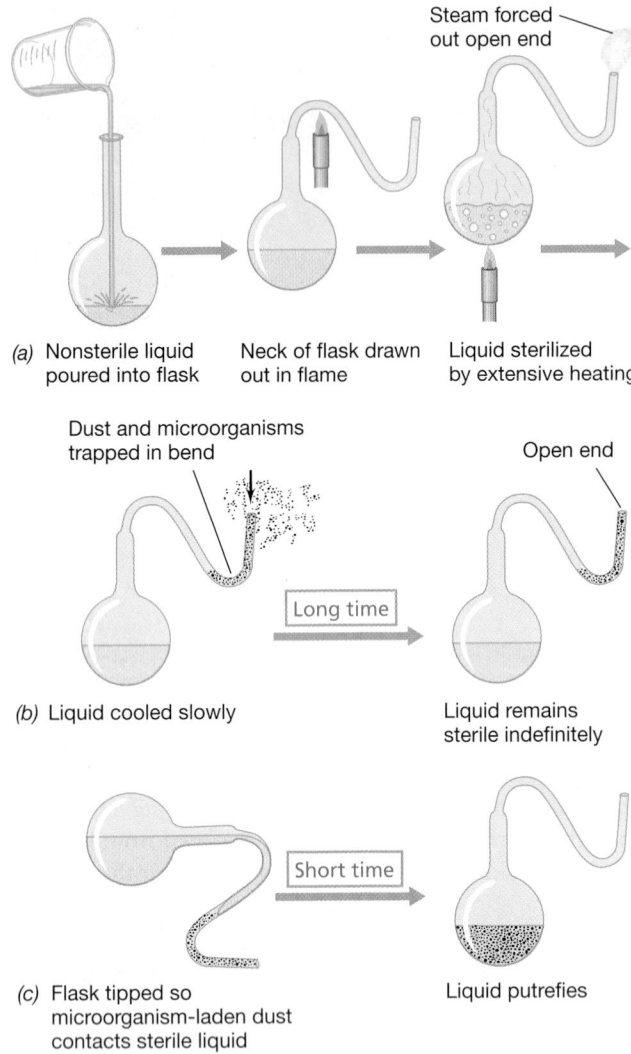

(a) Nonsterile liquid poured into flask

Neck of flask drawn out in flame

Liquid sterilized by extensive heating

Steam forced out open end

Dust and microorganisms trapped in bend

Open end

Long time

(b) Liquid cooled slowly

Liquid remains sterile indefinitely

Short time

(c) Flask tipped so microorganism-laden dust contacts sterile liquid

Liquid putrefies

Figure 1.16 The defeat of spontaneous generation: Pasteur's swan-necked flask experiment. In (c) the liquid putrefies because microorganisms enter with the dust.

Pasteur used heat to eliminate contaminants. Killing all the bacteria or other microorganisms in or on objects is a process we now call *sterilization*. Proponents of spontaneous generation criticized such experiments by declaring that "fresh air" was necessary for the phenomenon to occur. In 1864 Pasteur countered this objection simply and brilliantly by constructing a swan-necked flask, now called a *Pasteur flask* (**Figure 1.16**). In such a flask nutrient solutions could be heated to boiling and sterilized. However, after the flask was cooled, air was allowed to reenter, but the bend in the neck prevented particulate matter (including microorganisms) from entering the nutrient solution and causing putrefaction.

The teeming microorganisms observed after particulate matter was allowed to enter at the end of this simple experiment (Figure 1.16c) effectively settled the controversy, and microbiology was able to bury the idea of spontaneous generation for good and move ahead on firm footing. Incidentally, Pasteur's work also led to the development of effective sterilization procedures that were eventually refined and carried over into both basic and applied

microbiological research. Food science also owes a debt to Pasteur, as his principles are applied today in the preservation of milk and many other foods by heat treatment (pasteurization). www.microbiologyplace.com **Online Tutorial 1.1: Pasteur's Experiment**

Other Accomplishments of Louis Pasteur

Pasteur went on to many other triumphs in microbiology and medicine. Some highlights include his development of vaccines for the diseases anthrax, fowl cholera, and rabies during a very scientifically productive period from 1880 to 1890. Pasteur's work on rabies was his most famous success, culminating in July 1885 with the first administration of a rabies vaccine to a human, a young French boy named Joseph Meister who had been bitten by a rabid dog. In those days, a bite from a rabid animal was invariably fatal. News spread quickly of the success of Meister's vaccination, and of one administered shortly thereafter to a young shepherd boy, Jean Baptiste Jupille (**Figure 1.17**). Within a

(a)

(b)

Figure 1.17 Louis Pasteur and some symbols of his contributions to microbiology. (a) A French 5-franc note honoring Pasteur. The shepherd boy Jean Baptiste Jupille is shown killing a rabid dog that had attacked children. Pasteur's rabies vaccine saved Jupille's life. In France, the franc preceded the euro as a currency. (b) The Pasteur Institute, Paris, France. Today this structure, built for Pasteur by the French government, houses a museum that displays some of the original swan-necked flasks used in his experiments.

year several thousand people bitten by rabid animals had traveled to Paris to be treated with Pasteur's rabies vaccine.

Pasteur's fame from his rabies research was legendary and led the French government to establish the Pasteur Institute in Paris in 1888 (Figure 1.17*b*). Originally established as a clinical center for the treatment of rabies and other contagious diseases, the Pasteur Institute today is a major biomedical research center focused on antiserum and vaccine research and production. The medical and veterinary breakthroughs of Pasteur were not only highly significant in their own right but helped solidify the concept of the germ theory of disease, whose principles were being developed at about the same time by a second giant of this era, Robert Koch.

MiniQuiz

- Define the term sterile. How did Pasteur's experiments using swan-necked flasks defeat the theory of spontaneous generation?
- Besides ending the controversy over spontaneous generation, what other accomplishments do we credit to Pasteur?

1.8 Koch, Infectious Disease, and Pure Culture Microbiology

Proof that some microorganisms cause disease provided the greatest impetus for the development of microbiology as an independent biological science. Even as early as the sixteenth century it was thought that something that induced disease could be transmitted from a diseased person to a healthy person. After the discovery of microorganisms, it was widely believed that they were responsible, but definitive proof was lacking. Improvements in sanitation by Ignaz Semmelweis and Joseph Lister provided indirect evidence for the importance of microorganisms in causing human diseases, but it was not until the work of a German physician, Robert Koch (1843–1910) (**Figure 1.18**), that the concept of infectious disease was given experimental support.

The Germ Theory of Disease and Koch's Postulates

In his early work Koch studied anthrax, a disease of cattle and occasionally of humans. Anthrax is caused by an endospore-forming bacterium called *Bacillus anthracis*. By careful microscopy and by using special stains, Koch established that the bacteria were always present in the blood of an animal that was succumbing to the disease. However, Koch reasoned that the mere association of the bacterium with the disease was not proof of cause and effect. He sensed an opportunity to study cause and effect experimentally using anthrax. The results of this study formed the standard by which infectious diseases have been studied ever since.

Koch used mice as experimental animals. Using appropriate controls, Koch demonstrated that when a small amount of blood from a diseased mouse was injected into a healthy mouse, the latter quickly developed anthrax. He took blood from this second animal, injected it into another, and again observed the characteristic disease symptoms. However, Koch carried this experiment a critically important step further. He discovered that the anthrax bacteria could be grown in nutrient fluids *outside the host* and that even after many transfers in laboratory culture, the bacteria still caused the disease when inoculated into a healthy animal.

Figure 1.18 Robert Koch. The German physician and microbiologist is credited with founding medical microbiology and formulating his famous postulates.

On the basis of these experiments and others on the causative agent of tuberculosis, Koch formulated a set of rigorous criteria, now known as **Koch's postulates**, for definitively linking a specific microorganism to a specific disease. Koch's postulates state the following:

1. The disease-causing organism must always be present in animals suffering from the disease but not in healthy animals.

2. The organism must be cultivated in a pure culture away from the animal body.

3. The isolated organism must cause the disease when inoculated into healthy susceptible animals.

4. The organism must be isolated from the newly infected animals and cultured again in the laboratory, after which it should be seen to be the same as the original organism.

Koch's postulates, summarized in **Figure 1.19**, were a monumental step forward in the study of infectious diseases. The postulates not only offered a means for linking the cause and effect of an infectious disease, but also stressed the importance of *laboratory culture* of the putative infectious agent. With these postulates as a guide, Koch, his students, and those that followed them discovered the causative agents of most of the important

KOCH'S POSTULATES

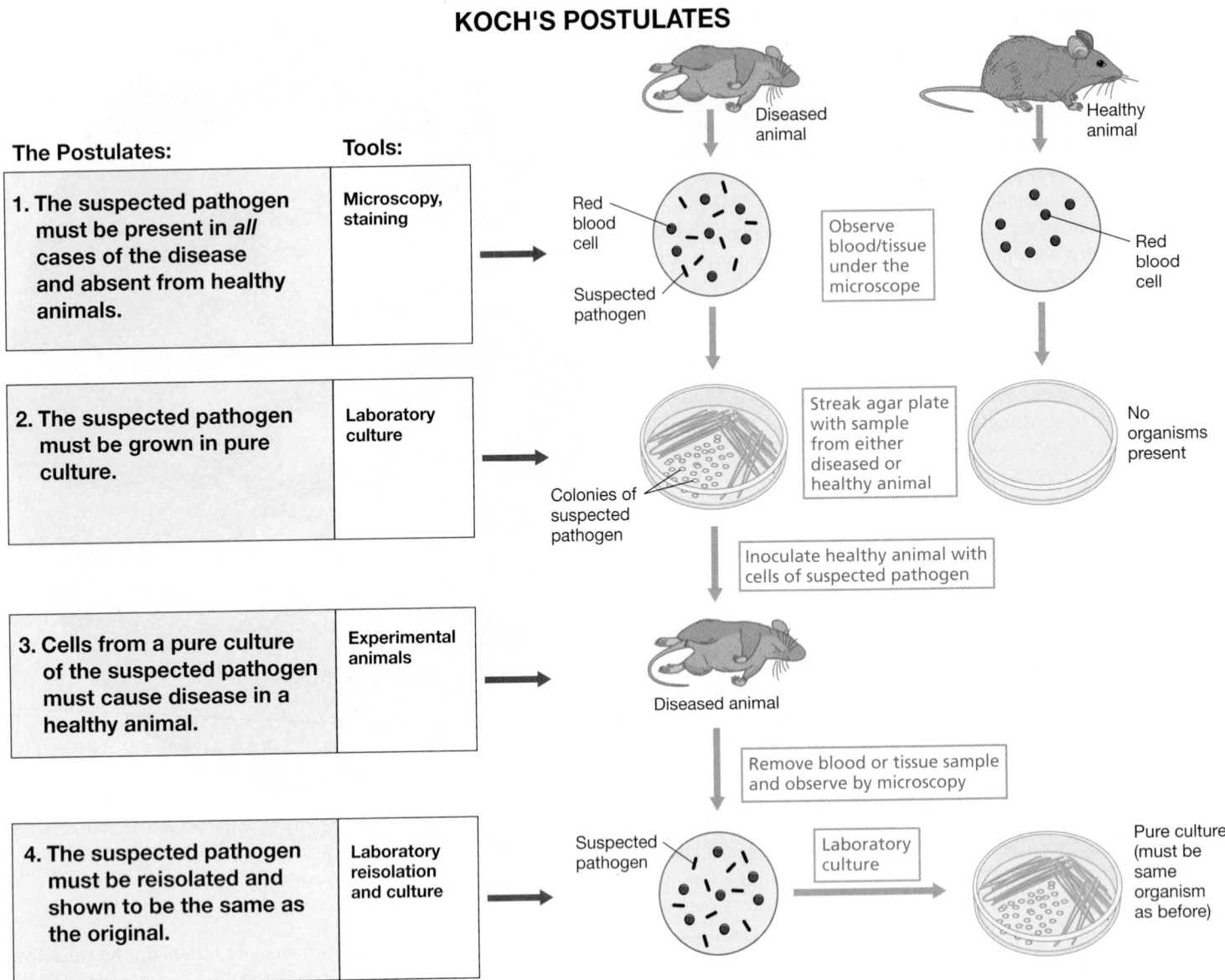

Figure 1.19 Koch's postulates for proving cause and effect in infectious diseases. Note that following isolation of a pure culture of the suspected pathogen, the cultured organism must both initiate the disease and be recovered from the diseased animal. Establishing the correct conditions for growing the pathogen is essential; otherwise it will be missed.

infectious diseases of humans and domestic animals. These discoveries led to the development of successful treatments for the prevention and cure of many of these diseases, thereby greatly improving the scientific basis of clinical medicine and human health and welfare (Figure 1.8).

Koch and Pure Cultures

To satisfy the second of Koch's postulates, the suspected pathogen must be isolated and grown away from other microorganisms in laboratory culture; in microbiology we say that such a culture is *pure*. The importance of this was not lost on Robert Koch in formulating his famous postulates, and to accomplish this goal, he and his associates developed several simple but ingenious methods of obtaining and growing bacteria in **pure culture**.

Koch started by using solid nutrients such as a potato slice to culture bacteria, but quickly developed more reliable methods, many of which are still in use today. Koch observed that when a solid surface was incubated in air, bacterial colonies developed, each having a characteristic shape and color. He inferred that each colony had arisen from a single bacterial cell that had fallen on the surface, found suitable nutrients, and multiplied. Each colony was a population of identical cells, or in other words, a pure culture, and Koch quickly realized that solid media provided an easy way to obtain pure cultures. However, because not all organisms grow on potato slices, Koch devised more exacting and reproducible nutrient solutions solidified with gelatin and, later, with agar—laboratory techniques that remain with us to this day (see the Microbial Sidebar, "Solid Media, Pure Cultures, and the Birth of Microbial Systematics").

Solid Media, Pure Cultures, and the Birth of Microbial Systematics

Robert Koch was the first to grow bacteria on solid culture media. Koch's early use of potato slices as solid media was fraught with problems. Besides the problem that not all bacteria can grow on potatoes, the slices were frequently overgrown with molds. Koch thus needed a more reliable and reproducible means of growing bacteria on solid media, and he found the answer for solidifying his nutrient solutions in agar.

Koch initially employed gelatin as a solidifying agent for the various nutrient fluids he used to culture bacteria, and he kept horizontal slabs of solid gelatin free of contamination under a bell jar or in a glass box (see Figure 1.20c). Nutrient-supplemented gelatin was a good culture medium for the isolation and study of various bacteria, but it had several drawbacks, the most important of which was that it did not remain solid at 37°C, the optimum temperature for growth of most human pathogens. Thus, a different solidifying agent was needed.

Agar is a polysaccharide derived from red algae. It was widely used in the nineteenth century as a gelling agent. Walter Hesse, an associate of Koch, first used agar as a solidifying agent for bacteriological culture media (Figure 1). The actual suggestion that agar be used instead of gelatin was made by Hesse's wife, Fannie. She had used agar to solidify fruit jellies. When it was tried as a solidifying agent for microbial media, its superior gelling qualities were immediately evident. Hesse wrote to Koch about this discovery, and Koch quickly adapted agar to his own studies, including his classic studies on the isolation of the bacterium *Mycobacterium tuberculosis,* the cause of the disease tuberculosis (see text and Figure 1.20).

Agar has many other properties that make it desirable as a gelling agent for microbial culture media. In particular, agar remains solid at 37°C and, after melting during the sterilization process, remains liquid to about 45°C, at which time it can be poured into sterile vessels. In addition, unlike gelatin,

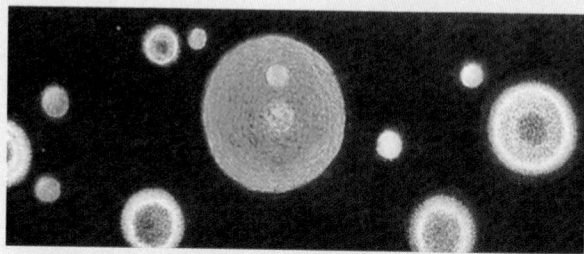

Figure 1 A hand-colored photograph taken by Walter Hesse of colonies formed on agar. The colonies include those of molds and bacteria obtained during Hesse's studies of the microbial content of air in Berlin, Germany, in 1882. From Hesse, W. 1884. "Ueber quantitative Bestimmung der in der Luft enthaltenen Mikroorganismen," in Struck, H. (ed.), *Mittheilungen aus dem Kaiserlichen Gesundheitsamte.* August Hirschwald.

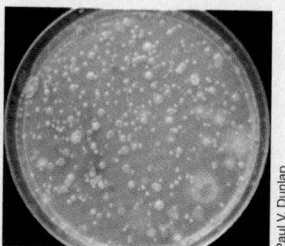

Figure 2 Photo of a Petri dish containing colonies of marine bacteria. Each colony contains millions of bacterial cells descended from a single cell.

agar is not degraded by most bacteria and typically yields a transparent medium, making it easier to differentiate bacterial colonies from inanimate particulate matter. For these reasons, agar found its place early in the annals of microbiology and is still used today for obtaining and maintaining pure cultures.

In 1887 Richard Petri, a German bacteriologist, published a brief paper describing a modification of Koch's flat plate technique (Figure 1.20c). Petri's enhancement, which turned out to be amazingly useful, was the development of the transparent double-sided dishes that bear his name (Figure 2). The advantages of Petri dishes were immediately apparent. They could easily be stacked and sterilized separately from the medium, and, following the addition of molten culture medium to the smaller of the two dishes, the larger dish could be used as a cover to prevent contamination. Colonies that formed on the surface of the agar in the Petri dish retained access to air without direct exposure to air and could easily be manipulated for further study. The original idea of Petri has not been improved on to this day, and the Petri dish, constructed of either glass or plastic, is a mainstay of the microbiology laboratory.

Koch quickly grasped the significance of pure cultures and was keenly aware of the implications his pure culture methods had for classifying microorganisms. He observed that colonies that differed in color, morphology, size, and the like (see Figure 2) bred true and could be distinguished from one another. Cells from different colonies typically differed in size and shape and often in their temperature or nutrient requirements as well. Koch realized that these differences among microorganisms met all the requirements that biological taxonomists had established for the classification of larger organisms, such as plant and animal species. In Koch's own words (translated from the German): "All bacteria which maintain the characteristics which differentiate one from another when they are cultured on the same medium and under the same conditions, should be designated as species, varieties, forms, or other suitable designation." Such insightful thinking was important for the rapid acceptance of microbiology as a new biological science, rooted as biology was in classification at the time of Koch. It has since had a profound impact on the diagnosis of infectious diseases and the field of microbial diversity.

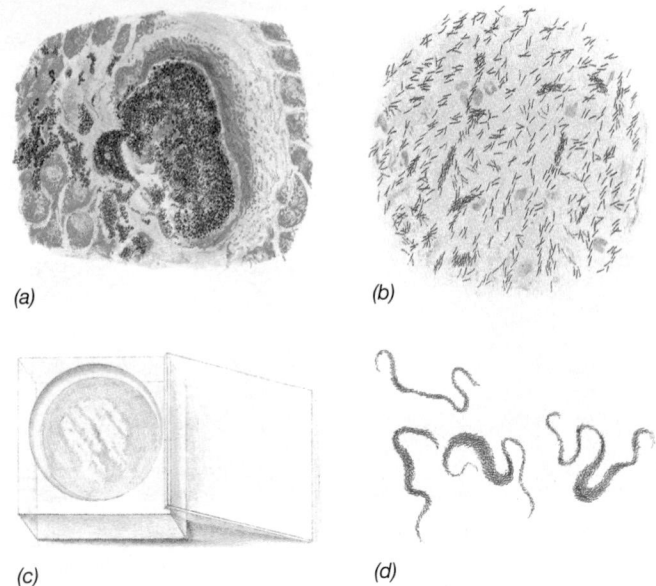

(a)   (b)

(c)   (d)

Figure 1.20 Robert Koch's drawings of *Mycobacterium tuberculosis.* *(a)* Section through infected lung tissue showing cells of *M. tuberculosis* (blue). *(b) M. tuberculosis* cells in a sputum sample from a tubercular patient. *(c)* Growth of *M. tuberculosis* on a glass plate of coagulated blood serum stored inside a glass box to prevent contamination. *(d) M. tuberculosis* cells taken from the plate in part c and observed microscopically; cells appear as long cordlike forms. Original drawings from Koch, R. 1884. "Die Aetiologie der Tuberkulose." *Mittheilungen aus dem Kaiserlichen Gesundheitsamte* 2:1–88.

Tuberculosis: The Ultimate Test of Koch's Postulates

Koch's crowning accomplishment in medical bacteriology was his discovery of the causative agent of tuberculosis. At the time Koch began this work (1881), one-seventh of all reported human deaths were caused by tuberculosis (Figure 1.8). There was a strong suspicion that tuberculosis was a contagious disease, but the suspected agent had never been seen, either in diseased tissues or in culture. Koch was determined to demonstrate the cause of tuberculosis, and to this end he brought together all of the methods he had so carefully developed in his previous studies with anthrax: microscopy, staining, pure culture isolation, and an animal model system.

As is now well known, the bacterium that causes tuberculosis, *Mycobacterium tuberculosis*, is very difficult to stain because of the large amounts of a waxy lipid present in its cell wall. But Koch devised a staining procedure for *M. tuberculosis* cells in tissue samples; using this method, he observed blue, rod-shaped cells of *M. tuberculosis* in tubercular tissues but not in healthy tissues (**Figure 1.20**). However, from his previous work on anthrax, Koch realized that he must *culture* this organism in order to prove that it was the cause of tuberculosis.

Obtaining cultures of *M. tuberculosis* was not easy, but eventually Koch was successful in growing colonies of this organism on a medium containing coagulated blood serum. Later he used agar, which had just been introduced as a solidifying agent (see

the Microbial Sidebar). Under the best of conditions, *M. tuberculosis* grows slowly in culture, but Koch's persistence and patience eventually led to pure cultures of this organism from human and animal sources.

From this point it was relatively easy for Koch to use his postulates (Figure 1.19) to obtain definitive proof that the organism he had isolated was the cause of the disease tuberculosis. Guinea pigs can be readily infected with *M. tuberculosis* and eventually succumb to systemic tuberculosis. Koch showed that diseased guinea pigs contained masses of *M. tuberculosis* cells in their lungs and that pure cultures obtained from such animals transmitted the disease to uninfected animals. Thus, Koch successfully satisfied all four of his postulates, and the cause of tuberculosis was understood. Koch announced his discovery of the cause of tuberculosis in 1882 and published a paper on the subject in 1884 in which his postulates are most clearly stated. For his contributions on tuberculosis, Robert Koch was awarded the 1905 Nobel Prize for Physiology or Medicine. Koch had many other triumphs in medicine, including discovering the organism responsible for the disease cholera and developing methods to diagnose exposure to *M. tuberculosis* (the tuberculin test).

Koch's Postulates Today

For human diseases in which an animal model is available, it is relatively easy to use Koch's postulates. In modern clinical medicine, however, this is not always so easy. For instance, the causative agents of several human diseases do not cause disease in any known experimental animals. These include many of the diseases associated with bacteria that live only *within* cells, such as the rickettsias and chlamydias, and diseases caused by some viruses and protozoan parasites. So for most of these diseases cause and effect cannot be unequivocally proven. However, the clinical and epidemiological (disease tracking) evidence for virtually every infectious disease of humans lends all but certain proof of the specific cause of the disease. Thus, although Koch's postulates remain the "gold standard" in medical microbiology, it has been impossible to satisfy all of his postulates for every human infectious disease.

MiniQuiz

- How do Koch's postulates ensure that cause and effect of a given disease are clearly differentiated?
- What advantages do solid media offer for the isolation of microorganisms?
- What is a pure culture?

1.9 The Rise of Microbial Diversity

As microbiology moved into the twentieth century, its initial focus on basic principles, methods, and medical aspects broadened to include studies of the microbial diversity of soil and water and the metabolic processes that organisms in these habitats carried out. Two giants of this era included the Dutchman Martinus Beijerinck and the Russian Sergei Winogradsky.

Martinus Beijerinck and the Enrichment Culture Technique

Martinus Beijerinck (1851–1931), a professor at the Delft Polytechnic School in Holland, was originally trained in botany, so he began his career in microbiology studying plants. Beijerinck's greatest contribution to the field of microbiology was his clear formulation of the **enrichment culture technique**. In enrichment cultures microorganisms are isolated from natural samples using highly selective techniques of adjusting nutrient and incubation conditions to favor a particular metabolic group of organisms. Beijerinck's skill with the enrichment method was readily apparent when, following Winogradsky's discovery of the process of nitrogen fixation, he isolated the aerobic nitrogen-fixing bacterium *Azotobacter* from soil (**Figure 1.21**).

Using the enrichment culture technique, Beijerinck isolated the first pure cultures of many soil and aquatic microorganisms,

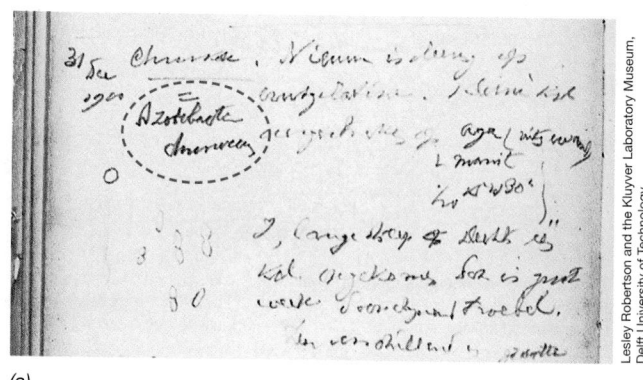

(a)

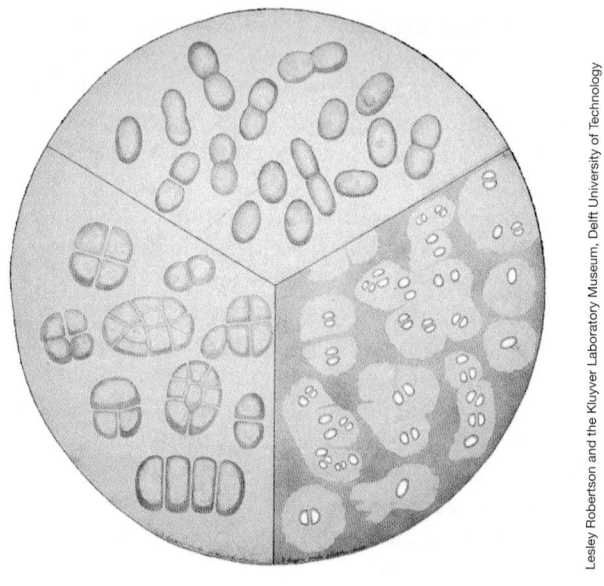

(b)

Figure 1.21 Martinus Beijerinck and *Azotobacter.* *(a)* A page from the laboratory notebook of M. Beijerinck dated 31 December 1900 describing the aerobic nitrogen-fixing bacterium *Azotobacter chroococcum* (name circled in red). Compare Beijerinck's drawings of pairs of *A. chroococcum* cells with the photomicrograph of cells of *Azotobacter* in Figure 17.18*a*. *(b)* A painting by M. Beijerinck's sister, Henriëtte Beijerinck, showing cells of *Azotobacter chroococcum*. Beijerinck used such paintings to illustrate his lectures.

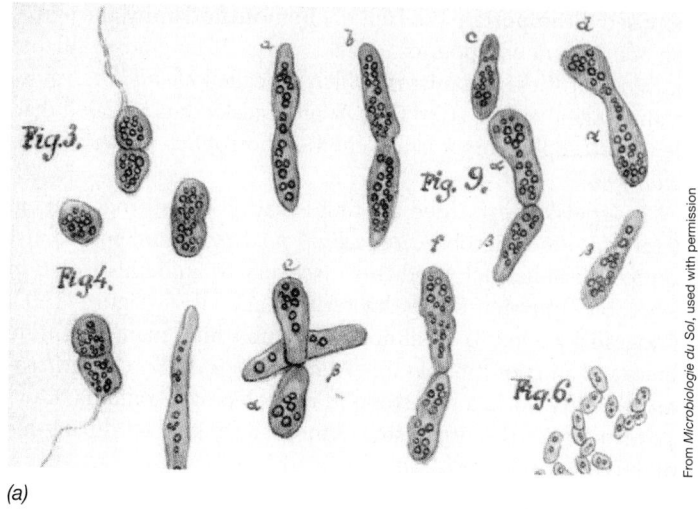

(a)

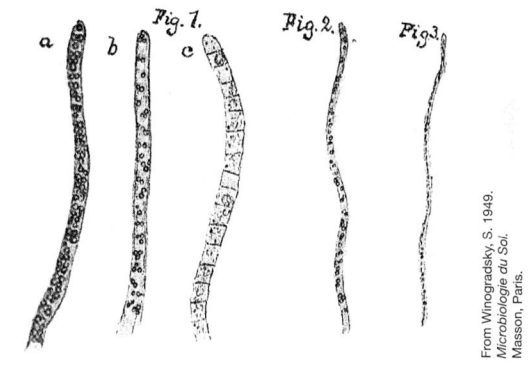

(b)

Figure 1.22 Sulfur bacteria. The original drawings were made by Sergei Winogradsky in the late 1880s and then copied and hand-colored by his wife Hélène. *(a)* Purple sulfur phototrophic bacteria. Figures 3 and 4 show cells of *Chromatium okenii* (compare with photomicrographs of *C. okenii* in Figures 1.5*a* and 1.7*a*). *(b) Beggiatoa*, a sulfur chemolithotroph (compare with Figure 1.14).

including sulfate-reducing and sulfur-oxidizing bacteria, nitrogen-fixing root nodule bacteria (Figure 1.9), *Lactobacillus* species, green algae, various anaerobic bacteria, and many others. In his studies of tobacco mosaic disease, Beijerinck used selective filtering techniques to show that the infectious agent (a virus) was smaller than a bacterium and that it somehow became incorporated into cells of the living host plant. In this insightful work, Beijerinck not only described the first virus, but also the basic principles of virology, which we present in Chapters 9 and 21.

Sergei Winogradsky, Chemolithotrophy, and Nitrogen Fixation

Sergei Winogradsky (1856–1953) had interests similar to Beijerinck's—the diversity of bacteria in soils and waters—and was highly successful in isolating several key bacteria from natural samples. Winogradsky was particularly interested in bacteria that cycle nitrogen and sulfur compounds, such as the nitrifying bacteria and the sulfur bacteria (**Figure 1.22**). He showed that these bacteria catalyze specific chemical transformations in nature and proposed the important concept of **chemolithotrophy**, the oxidation of *inorganic* compounds to yield energy. Winogradsky further showed that these organisms, which he called *chemolithotrophs,* obtained their carbon from CO_2. Winogradsky thus revealed that, like phototrophic organisms, chemolithotrophic bacteria were *autotrophs.*

Winogradsky performed the first isolation of a nitrogen-fixing bacterium, the anaerobe *Clostridium pasteurianum,* and as just mentioned, Beijerinck used this discovery to guide his isolation of aerobic nitrogen-fixing bacteria years later (Figure 1.21). Winogradsky lived to be almost 100, publishing many scientific papers and a major monograph, *Microbiologie du Sol* (*Soil Microbiology*). This work, a milestone in microbiology, contains drawings of many of the organisms Winogradsky studied during his lengthy career (Figure 1.22).

MiniQuiz

- What is meant by the term "enrichment culture"?
- What is meant by the term "chemolithotrophy"? In what way are chemolithotrophs like plants?

1.10 The Modern Era of Microbiology

In the twentieth century, the field of microbiology developed rapidly in two different yet complementary directions—*applied* and *basic.* During this period a host of new laboratory tools became available, and the science of microbiology began to mature and spawn new subdisciplines. Few of these subdisciplines were purely applied or purely basic. Instead, most had both discovery (basic) and problem-solving (applied) components. **Table 1.3** summarizes these major subdisciplines of microbiology that arose in the twentieth century.

Several microbiologists are remembered for their key contributions during this period. In the early twentieth century many remained focused on medical aspects of microbiology, and even today, many dedicated microbiologists grapple with the impacts of microorganisms on human, animal, and plant disease. But following World War II, an exciting new emphasis began to take hold with studies of the genetic properties of microorganisms. From roots in microbial genetics has emerged "modern biology," driven by molecular biology, genetic engineering, and genomics. This molecular approach has revolutionized scientific thinking in the life sciences and has driven experimental approaches to the most compelling problems in biology. Some key Nobel laureates and their contributions to the molecular era of microbiology are listed in **Table 1.4**.

Many of the advances in microbiology today are fueled by the genomics revolution; that is, we are clearly in the era of "molecular microbiology." Rapid progress in DNA sequencing technology and improved computational power have yielded huge amounts of genomic information that have supported major advances in medicine, agriculture, biotechnology, and microbial ecology. For example, to obtain the sequence of the entire genome of a bacterium takes only a few hours (although sequence analysis is a much more time-consuming process). The fast-paced field of

Table 1.3 *The major subdisciplines of microbiology*

Subdiscipline	Focus
I. Basic emphases[a]	
Microbial physiology	Nutrition, metabolism
Microbial genetics	Genes, heredity, and genetic variation
Microbial biochemistry	Enzymes and chemical reactions in cells
Microbial systematics	Classification and nomenclature
Virology	Viruses and subviral particles
Molecular biology	Nucleic acids and protein
Microbial ecology	Microbial diversity and activity in natural habitats; biogeochemistry
II. Applied emphases[a]	
Medical microbiology	Infectious disease
Immunology	Immune systems
Agricultural/soil microbiology	Microbial diversity and processes in soil
Industrial microbiology	Large-scale production of antibiotics, alcohol, and other chemicals
Biotechnology	Production of human proteins by genetically engineered microorganisms
Aquatic microbiology	Microbial processes in waters and wastewaters, drinking water safety

[a]None of these subdisciplines are devoted entirely to basic science or applied science. However, the subdisciplines listed in I tend to be more focused on discovery and those in II more focused on solving specific problems or synthesizing commercial products from microbial sources.

genomics has itself spawned highly focused new subdisciplines, such as *transcriptomics, proteomics,* and *metabolomics,* which explore, respectively, the patterns of RNA, protein, and metabolic expression in cells. The concepts of genomics, transcriptomics, proteomics, and metabolomics are all developed in Chapter 12.

All signs point to a continued maturation of molecular microbiology as we enter a period where technology is almost ahead of our ability to formulate exciting scientific questions. In fact, microbial research today is very close to defining the *minimalist genome*—the minimum complement of genes necessary for a living cell. When such a genetic blueprint is available, microbiologists should be able to define, at least in biochemical terms, the prerequisites for life. When that day arrives, can the laboratory creation of an actual living cell from nonliving components, that is, spontaneous generation under controlled laboratory conditions, be far off? Almost certainly not. Stay tuned, as much exciting science is on the way!

MiniQuiz

- For each of the following topics, name the subdiscipline of microbiology that focuses on it: metabolism, enzymology, nucleic acid and protein synthesis, microorganisms and their natural environments, microbial classification, inheritance of characteristics.

Table 1.4 *Some Nobel laureates in the era of molecular microbiology[a]*

Investigator(s)	Nationality	Discovery/Year[b]
George Beadle, Edward Tatum	American	One gene–one enzyme hypothesis/1941
Max Delbrück, Salvador Luria	German/Italian	Inheritance of characteristics in bacteria/1943
Joshua Lederberg	American	Conjugation and transduction in bacteria/1946/1952
James Watson, Francis Crick, Maurice Wilkins	American/British	Structure of DNA/1953
François Jacob, Jacques Monod, Andre Lwoff	French	Gene regulation by repressor proteins, operon concept/1959
Sydney Brenner	British	Messenger RNA, ribosomes as site of protein synthesis/1961
Marshall Nirenberg, Robert Holley, H. Gobind Khorana	American/Indian	Genetic code/1966
Howard Temin, David Baltimore, and Renato Dulbecco	American/Italian	Retroviruses and reverse transcriptase/1969
Hamilton Smith, Daniel Nathans, Werner Arber	American/Swiss	Restriction enzymes/1970
J. Michael Bishop, Harold Varmus	American	Cancer genes (oncogenes) in retroviruses/1972
Paul Berg	American	Recombinant DNA technology/1973
Roger Kornberg	American	Mechanism of transcription in eukaryotes/1974
Fred Sanger	British	Structure and sequencing of proteins, DNA sequencing 1958/1977
Carl Woese[c]	American	Discovery of *Archaea*/1977
Stanley Prusiner	American	Discovery and characterization of prions/1981
Sidney Altman, Thomas Cech	American	Catalytic properties of RNA/1981
Barry Marshall, Robin Warren	Australian	*Helicobacter pylori* as cause of peptic ulcers/1982
Luc Montagnier, Françoise Barré-Sinoussi, Harald zur Hausen	French/German	Discovery of human immunodeficiency virus as cause of AIDS/1983
Kary Mullis	American	Polymerase chain reaction/1985
Andrew Fire, Craig Mello	American	RNA interference/1998

[a]This select list covers major accomplishments since 1941. In virtually every case, the laureates listed had important coworkers that did not receive the Nobel Prize.
[b]Year indicates the year in which the discovery awarded with the Nobel Prize was published.
[c]Recipient of the 2003 Crafoord Prize in Biosciences, equivalent in scientific stature to the Nobel Prize.

Big Ideas

1.1

Microorganisms, which include all single-celled microscopic organisms and the viruses, are essential for the well-being of the planet and its plants and animals.

1.2

Metabolism, growth, and evolution are necessary properties of living systems. Cells must coordinate energy production and consumption with the flow of genetic information during cellular events leading up to cell division.

1.3

Microorganisms exist in nature in populations that interact with other populations in microbial communities. The activities of microorganisms in microbial communities can greatly affect and rapidly change the chemical and physical properties of their habitats.

1.4

Diverse microbial populations were widespread on Earth for billions of years before higher organisms appeared, and cyanobacteria in particular were important because they oxygenated the atmosphere. The cumulative microbial biomass on Earth exceeds that of higher organisms, and most microorganisms reside in the deep subsurface. *Bacteria, Archaea,* and *Eukarya* are the major phylogenetic lineages of cells.

1.5

Microorganisms can be both beneficial and harmful to humans, although many more microorganisms are beneficial or even essential than are harmful.

1.6

Robert Hooke was the first to describe microorganisms, and Antoni van Leeuwenhoek was the first to describe bacteria. Ferdinand Cohn founded the field of bacteriology and discovered bacterial endospores.

1.7

Louis Pasteur is best remembered for his ingenious experiments showing that living organisms do not arise spontaneously from nonliving matter. He developed many concepts and techniques central to the science of microbiology, including sterilization.

1.8

Robert Koch developed a set of criteria anchored in experimentation—Koch's postulates—for the study of infectious diseases and developed the first methods for growth of pure cultures of microorganisms.

1.9

Beijerinck and Winogradsky studied bacteria that inhabit soil and water. Out of their work came the enrichment culture technique and the concepts of chemolithotrophy and nitrogen fixation.

1.10

In the middle to latter part of the twentieth century, basic and applied subdisciplines of microbiology emerged; these have led to the current era of molecular microbiology.

Review of Key Terms

Cell the fundamental unit of living matter

Chemolithotrophy a form of metabolism in which energy is generated from inorganic compounds

Communication interactions between cells using chemical signals

Differentiation modification of cellular components to form a new structure, such as a spore

Ecosystem organisms plus their nonliving environment

Enrichment culture technique a method for isolating specific microorganisms from nature using specific culture media and incubation conditions

Enzyme a protein (or in some cases an RNA) catalyst that functions to speed up chemical reactions

Evolution descent with modification leading to new forms or species

Genome an organism's full complement of genes

Genomics the identification and analysis of genomes

Growth in microbiology, an increase in cell number with time

Habitat the environment in which a microbial population resides

Koch's postulates a set of criteria for proving that a given microorganism causes a given disease

Metabolism all biochemical reactions in a cell

Microbial community two or more populations of cells that coexist and interact in a habitat

Microbial ecology the study of microorganisms in their natural environments

Microorganism a microscopic organism consisting of a single cell or cell cluster or a virus

Motility the movement of cells by some form of self-propulsion

Pathogen a disease-causing microorganism

Pure culture a culture containing a single kind of microorganism

Spontaneous generation the hypothesis that living organisms can originate from nonliving matter

Sterile free of all living organisms (cells) and viruses

Review Questions

1. What is the difference between basic and applied microbiology (Section 1.1)?

2. Cells can be thought of as both catalysts and genetic entities. Explain how these two attributes of a cell differ (Section 1.2).

3. What is an ecosystem? What effects can microorganisms have on their ecosystems (Section 1.3)?

4. Why did the evolution of cyanobacteria change Earth forever (Section 1.4)?

5. How would you convince a friend that microorganisms are much more than just agents of disease (Section 1.5)?

6. For what contributions are Hooke, van Leeuwenhoek, and Ferdinand Cohn most remembered in microbiology (Section 1.6)?

7. Explain the principle behind the use of the Pasteur flask in studies on spontaneous generation (Section 1.7).

8. What is a pure culture and how can one be obtained? Why was knowledge of how to obtain a pure culture important for development of the science of microbiology (Section 1.8)?

9. What are Koch's postulates and how did they influence the development of microbiology? Why are Koch's postulates still relevant today (Section 1.8)?

10. In contrast to those of Robert Koch, what were the major microbiological interests of Martinus Beijerinck and Sergei Winogradsky (Section 1.9)?

11. How does the genomics revolution propel advances in microbiology (Section 1.10)?

Application Questions

1. Pasteur's experiments on spontaneous generation contributed to the methodology of microbiology, understanding of the origin of life, and techniques for the preservation of food. Explain briefly how Pasteur's experiments affected each of these topics.

2. Describe the lines of proof Robert Koch used to definitively associate the bacterium *Mycobacterium tuberculosis* with the disease tuberculosis. How would his proof have been flawed if any of the tools he developed for studying bacterial diseases had not been available for his study of tuberculosis?

3. Imagine that all microorganisms suddenly disappeared from Earth. From what you have learned in this chapter, why do you think that animals would eventually disappear from Earth? Why would plants disappear? If by contrast, all higher organisms suddenly disappeared, what in Figure 1.6 tells you that a similar fate would not befall microorganisms?

Need more practice? Test your understanding with Quantitative Questions; access additional study tools including tutorials, animations, and videos; and then test your knowledge with chapter quizzes and practice tests at **www.microbiologyplace.com**.

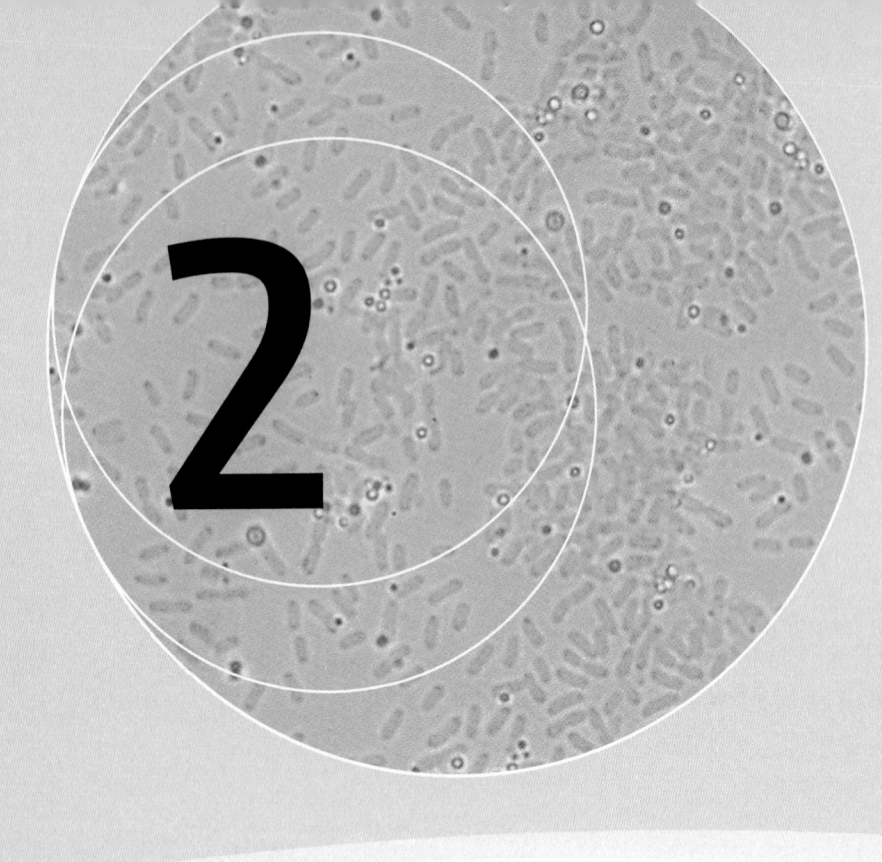

A Brief Journey to the Microbial World

Green sulfur bacteria are
phototrophic microorganisms
that form their own phylogenetic lineage and were some
of the first phototrophs to
evolve on Earth.

Seeing the Very Small

Historically, the science of microbiology blossomed as the ability to see microorganisms improved; thus, *microbiology* and *microscopy* advanced hand-in-hand. The microscope is the microbiologist's most basic tool, and every student of microbiology needs some background on how microscopes work and how microscopy is done. We therefore begin our brief journey to the microbial world by considering different types of microscopes and the applications of microscopy to imaging microorganisms.

2.1 Some Principles of Light Microscopy

Visualization of microorganisms requires a microscope, either a *light* microscope or an *electron* microscope. In general, light microscopes are used to examine cells at relatively low magnifications, and electron microscopes are used to look at cells and cell structures at very high magnification.

All microscopes employ lenses that magnify (enlarge) the image. Magnification, however, is not the limiting factor in our ability to see small objects. It is instead **resolution**—the ability to distinguish two adjacent objects as distinct and separate—that governs our ability to see the very small. Although magnification can be increased virtually without limit, resolution cannot, because resolution is a function of the physical properties of light.

We begin with the light microscope, for which the limits of resolution are about 0.2 μm (μm is the abbreviation for micrometer, 10^{-6} m). We then proceed to the electron microscope,

for which resolution is considerably greater than that of the light microscope.

The Compound Light Microscope

The light microscope uses visible light to illuminate cell structures. Several types of light microscopes are used in microbiology: *bright-field*, *phase-contrast*, *differential interference contrast*, *dark-field*, and *fluorescence*.

With the bright-field microscope, specimens are visualized because of the slight differences in contrast that exist between them and their surrounding medium. Contrast differences arise because cells absorb or scatter light to varying degrees. The compound bright-field microscope is commonly used in laboratory courses in biology and microbiology; the microscopes are called *compound* because they contain two lenses, *objective* and *ocular*, that function in combination to form the image. The light source is focused on the specimen by the condenser (**Figure 2.1**). Bacterial cells are typically difficult to see well with the bright-field microscope because the cells themselves lack significant contrast with their surrounding medium. Pigmented microorganisms are an exception because the color of the organism itself adds contrast, thus improving visualization (**Figure 2.2**). For cells lacking pigments there are ways to boost contrast, and we consider these methods in the next section.

Magnification and Resolution

The total magnification of a compound light microscope is the *product* of the magnification of its objective and ocular lenses

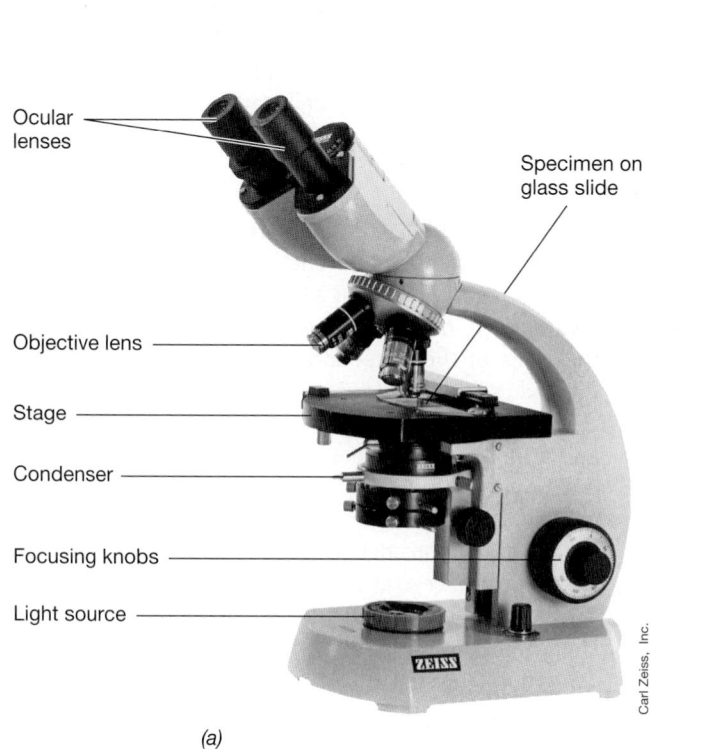

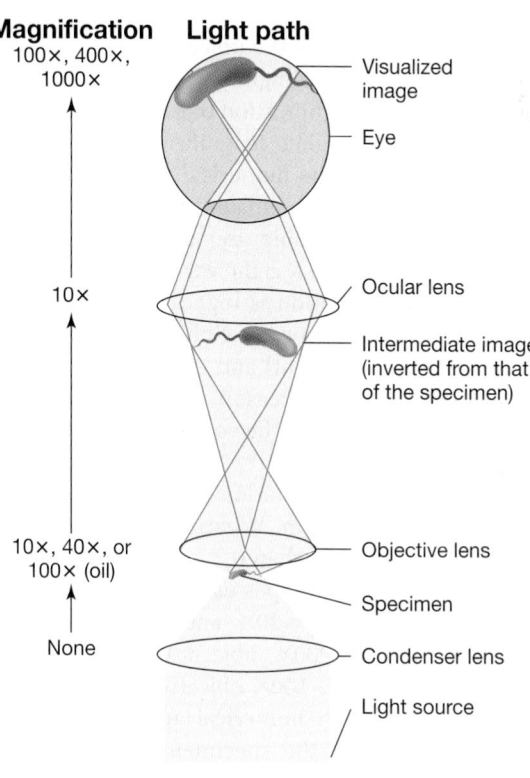

(a)   (b)

Figure 2.1 Microscopy. *(a)* A compound light microscope. *(b)* Path of light through a compound light microscope. Besides 10×, ocular lenses are available in 15–30× magnifications.

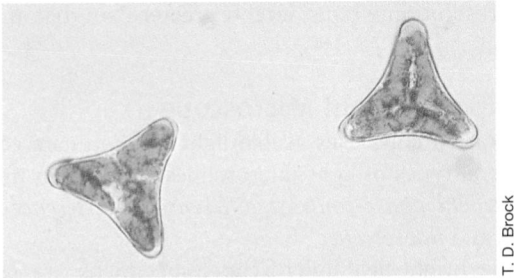

(a)

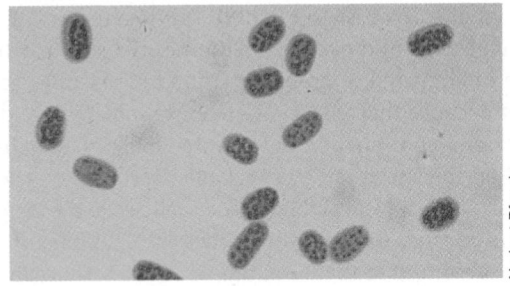

(b)

T. D. Brock

Norbert Pfennig

Figure 2.2 Bright-field photomicrographs of pigmented microor-ganisms. *(a)* A green alga (eukaryote). The green structures are chloro-plasts. *(b)* Purple phototrophic bacteria (prokaryote). The algal cell is about 15 μm wide, and the bacterial cells are about 5 μm wide. We contrast prokaryotic and eukaryotic cells in Section 2.5.

(Figure 2.1*b*). Magnifications of about 2000X are the upper limit for light microscopes. At magnifications above this, resolution does not improve. Resolution is a function of the wavelength of light used and a characteristic of the objective lens known as its *numerical aperture*, a measure of light-gathering ability. There is a correlation between the magnification of a lens and its numeri-cal aperture: Lenses with higher magnification typically have higher numerical apertures (the numerical aperture of a lens is stamped on the lens alongside the magnification). The diameter of the smallest object resolvable by any lens is equal to 0.5λ/numerical aperture, where λ is the wavelength of light used. Based on this formula, resolution is highest when blue light is used to illuminate a specimen (because blue light is of a shorter wavelength than white or red light) and the objective has a very high numerical aperture. For this reason, many light microscopes come fitted with a blue filter over the condenser lens to improve resolution.

As mentioned, the highest resolution possible in a compound light microscope is about 0.2 μm. What this means is that two objects that are closer together than 0.2 μm cannot be resolved as distinct and separate. Microscopes used in microbiology have ocular lenses that magnify 10–20X and objective lenses of 10–100X (Figure 2.1*b*). At 1000X, objects 0.2 μm in diameter can just be resolved. With the 100X objective, and with certain other objectives of very high numerical aperture, an optical-grade oil is placed between the specimen and the objective. Lenses on which oil is used are called *oil-immersion* lenses. Immersion oil increases the light-gathering ability of a lens by allowing some of the light rays emerging from the specimen at

angles (that would otherwise be lost to the objective lens) to be collected and viewed.

MiniQuiz
• Define and compare the terms magnification and resolution.
• What is the useful upper limit of magnification for a bright-field microscope? Why is this so?

2.2 Improving Contrast in Light Microscopy

In microscopy, improving contrast typically improves the final image. Staining is an easy way to improve contrast, but there are many other approaches.

Staining: Increasing Contrast for Bright-Field Microscopy

Dyes can be used to stain cells and increase their contrast so that they can be more easily seen in the bright-field microscope. Dyes are organic compounds, and each class of dye has an affinity for specific cellular materials. Many dyes used in microbiology are positively charged, and for this reason they are called *basic dyes*. Examples of basic dyes include methylene blue, crystal violet, and safranin. Basic dyes bind strongly to negatively charged cell com-ponents, such as nucleic acids and acidic polysaccharides. Because cell surfaces tend to be negatively charged, these dyes also combine with high affinity to the surfaces of cells, and hence are very useful general-purpose stains.

To perform a simple stain one begins with dried preparations of cells (**Figure 2.3**). A clean glass slide containing a dried sus-pension of cells is flooded for a minute or two with a dilute solution of a basic dye, rinsed several times in water, and blot-ted dry. Because their cells are so small, it is common to observe dried, stained preparations of bacteria with a high-power (oil-immersion) lens.

Differential Stains: The Gram Stain

Stains that render different kinds of cells different colors are called *differential* stains. An important differential-staining pro-cedure used in microbiology is the **Gram stain** (**Figure 2.4**). On the basis of their reaction to the Gram stain, bacteria can be divided into two major groups: *gram-positive* and *gram-negative*. After Gram staining, gram-positive bacteria appear purple-violet and gram-negative bacteria appear pink (Figure 2.4*b*). The color difference in the Gram stain arises because of differences in the cell wall structure of gram-positive and gram-negative cells, a topic we will consider in Chapter 3. After staining with a basic dye, typically crystal violet, treatment with ethanol decolorizes gram-negative but not gram-positive cells. Following counter-staining with a different-colored stain, typically safranin, the two cell types can be distinguished microscopically by their different colors (Figure 2.4*b*).

The Gram stain is one of the most useful staining procedures in microbiology. Typically, one begins the characterization of a new bacterium by determining whether it is gram-positive or

I. Preparing a smear

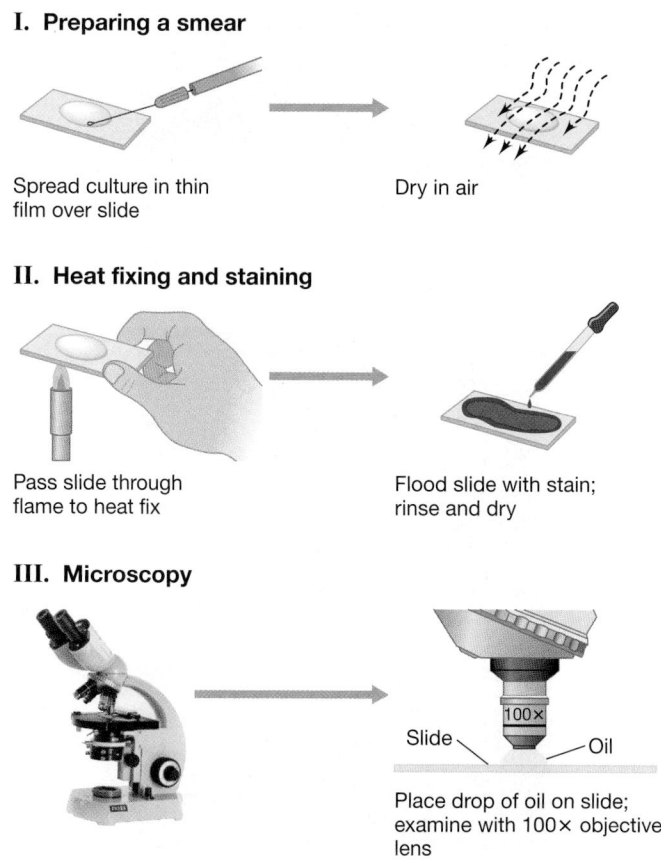

Spread culture in thin film over slide

Dry in air

II. Heat fixing and staining

Pass slide through flame to heat fix

Flood slide with stain; rinse and dry

III. Microscopy

Slide — 100× — Oil

Place drop of oil on slide; examine with 100× objective lens

Figure 2.3 Staining cells for microscopic observation. Stains improve the contrast between cells and their background.

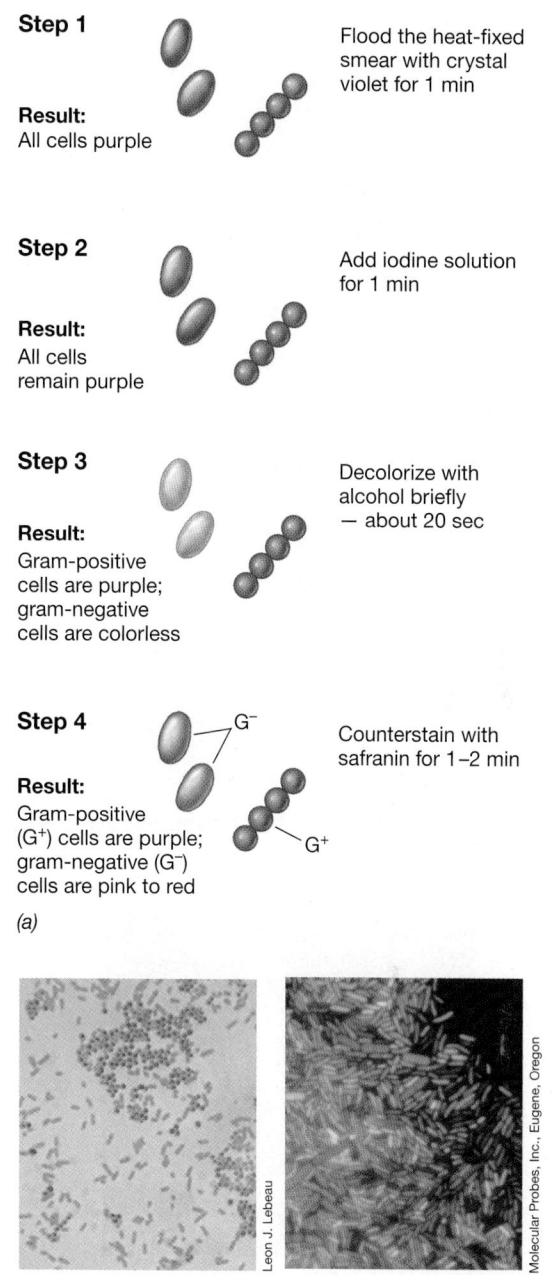

Step 1

Result: All cells purple

Flood the heat-fixed smear with crystal violet for 1 min

Step 2

Result: All cells remain purple

Add iodine solution for 1 min

Step 3

Result: Gram-positive cells are purple; gram-negative cells are colorless

Decolorize with alcohol briefly — about 20 sec

Step 4

G⁻

Result: Gram-positive (G⁺) cells are purple; gram-negative (G⁻) cells are pink to red

G⁺

Counterstain with safranin for 1–2 min

(a)

Leon J. Lebeau

Molecular Probes, Inc., Eugene, Oregon

(b)   *(c)*

Figure 2.4 The Gram stain. *(a)* Steps in the procedure. *(b)* Microscopic observation of gram-positive (purple) and gram-negative (pink) bacteria. The organisms are *Staphylococcus aureus* and *Escherichia coli*, respectively. *(c)* Cells of *Pseudomonas aeruginosa* (gram-negative, green) and *Bacillus cereus* (gram-positive, orange) stained with a one-step fluorescent staining method. This method allows for differentiating gram-positive from gram-negative cells in a single staining step.

gram-negative. If a fluorescent microscope, discussed below, is available, the Gram stain can be reduced to a one-step procedure in which gram-positive and gram-negative cells fluoresce different colors (Figure 2.4c).

Phase-Contrast and Dark-Field Microscopy

Staining, although a widely used procedure in light microscopy, kills cells and can distort their features. Two forms of light microscopy improve image contrast without the use of stain, and thus do not kill cells. These are phase-contrast microscopy and dark-field microscopy (**Figure 2.5**). The phase-contrast microscope in particular is widely used in teaching and research for the observation of wet-mount (living) preparations.

Phase-contrast microscopy is based on the principle that cells differ in refractive index (a factor by which light is slowed as it passes through a material) from their surroundings. Light passing through a cell thus differs in phase from light passing through the surrounding liquid. This subtle difference is amplified by a device in the objective lens of the phase-contrast microscope called the *phase ring*, resulting in a dark image on a light background (Figure 2.5b). The ring consists of a phase plate that amplifies the minute variation in phase. The development of phase-contrast microscopy stimulated other innovations in microscopy, such as fluorescence and confocal microscopy (discussed below), and greatly increased use of the light microscope in microbiology.

The dark-field microscope is a light microscope in which light reaches the specimen from the sides only. The only light that reaches the lens is that scattered by the specimen, and thus the specimen appears light on a dark background (Figure 2.5c). Resolution by dark-field microscopy is somewhat better than by light microscopy, and objects can often be resolved by dark-field that cannot be resolved by bright-field or even phase-contrast

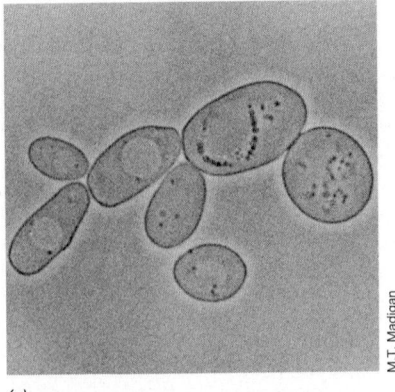

(a)

M.T. Madigan

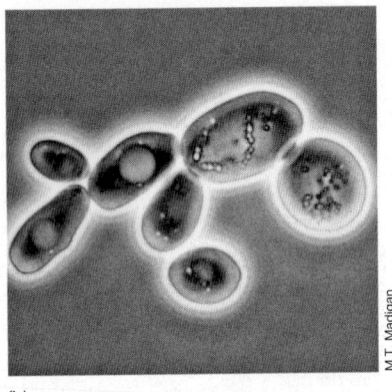

(b)

M.T. Madigan

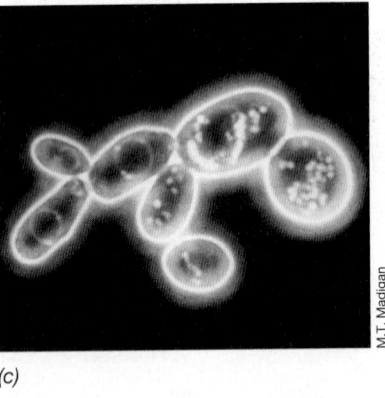

(c)

M.T. Madigan

Figure 2.5 Cells visualized by different types of light microscopy. The same field of cells of the baker's yeast *Saccharomyces cerevisiae* visualized by (a) bright-field microscopy, (b) phase-contrast microscopy, and (c) dark-field microscopy. Cells average 8–10 μm wide.

microscopes. Dark-field microscopy is also an excellent way to observe microbial motility, as bundles of flagella (the structures responsible for swimming motility) are often resolvable with this technique (Figure 3.40a).

Fluorescence Microscopy

The fluorescence microscope is used to visualize specimens that fluoresce—that is, emit light of one color following absorption of light of another color (**Figure 2.6**). Cells fluoresce either because they contain naturally fluorescent substances such as chlorophyll

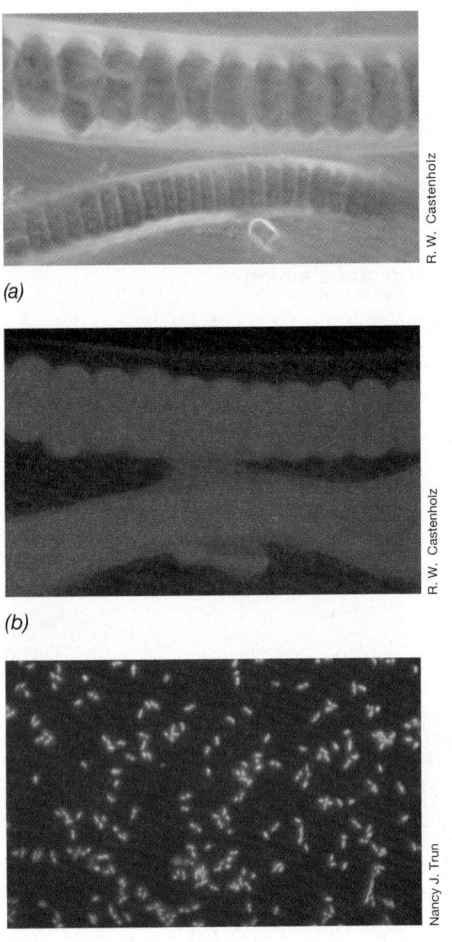

(a)

R. W. Castenholz

(b)

R. W. Castenholz

(c)

Nancy J. Trun

Figure 2.6 Fluorescence microscopy. (a, b) Cyanobacteria. The same cells are observed by bright-field microscopy in part a and by fluorescence microscopy in part b. The cells fluoresce red because they contain chlorophyll a and other pigments. (c) Fluorescence photomicrograph of cells of *Escherichia coli* made fluorescent by staining with the fluorescent dye DAPI.

or other fluorescing components, a phenomenon called *autofluorescence* (Figure 2.6a, b), or because the cells have been stained with a fluorescent dye (Figure 2.6c). DAPI (4′,6-diamidino-2-phenylindole) is a widely used fluorescent dye, staining cells bright blue because it complexes with the cell's DNA (Figure 2.6c). DAPI can be used to visualize cells in various habitats, such as soil, water, food, or a clinical specimen. Fluorescence microscopy using DAPI or related stains is therefore widely used in clinical diagnostic microbiology and also in microbial ecology for enumerating bacteria in a natural environment or, as in Figure 2.6c, in a cell suspension.

MiniQuiz

- What color will a gram-negative cell be after Gram staining by the conventional method?
- What major advantage does phase-contrast microscopy have over staining?
- How can cells be made to fluoresce?

2.3 Imaging Cells in Three Dimensions

Up to now we have considered forms of microscopy in which the images obtained are essentially two-dimensional. How can this limitation be overcome? We will see in the next section that the scanning electron microscope offers one solution to this problem, but certain forms of light microscopy can also improve the three-dimensional perspective of the image.

Differential Interference Contrast Microscopy

Differential interference contrast (DIC) microscopy is a form of light microscopy that employs a polarizer in the condenser to produce polarized light (light in a single plane). The polarized light then passes through a prism that generates two distinct beams. These beams traverse the specimen and enter the objective lens where they are recombined into one. Because the two beams pass through different substances with slightly different refractive indices, the combined beams are not totally in phase but instead create an interference effect. This effect visibly enhances subtle differences in cell structure. Thus, by DIC microscopy, cellular structures such as the nucleus of eukaryotic cells (**Figure 2.7**), or endospores, vacuoles, and granules of bacterial cells, appear more three-dimensional. DIC microscopy is typically used for observing unstained cells because it can reveal internal cell structures that are nearly invisible by the bright-field technique (compare Figure 2.5a with Figure 2.7a).

Atomic Force Microscopy

Another type of microscope useful for three-dimensional imaging of biological structures is the atomic force microscope (AFM). In atomic force microscopy, a tiny stylus is positioned extremely close to the specimen such that weak repulsive forces are established between the probe on the stylus and atoms on the surface of the specimen. During scanning, the stylus surveys the specimen surface, continually recording any deviations from a flat surface. The pattern that is generated is processed by a series of detectors that feed the digital information into a computer, which then outputs an image (Figure 2.7b).

Although the images obtained from an AFM appear similar to those from the scanning electron microscope (compare Figure 2.7b with Figure 2.10c), the AFM has the advantage that the specimen does not have to be treated with fixatives or coatings. The AFM thus allows living specimens to be viewed, something that is generally not possible with electron microscopes.

Confocal Scanning Laser Microscopy

A confocal scanning laser microscope (CSLM) is a computerized microscope that couples a laser source to a fluorescent microscope. This generates a three-dimensional image and allows the viewer to profile several planes of focus in the specimen (**Figure 2.8**). The laser beam is precisely adjusted such that only a particular layer within a specimen is in perfect focus at one time. By precisely illuminating only a single plane of focus, the CSLM eliminates stray light from other focal planes. Thus, when observing a relatively thick specimen such as a microbial biofilm (Figure 2.8a), not only are cells on the surface of the biofilm apparent, as would be the case with conventional light microscopy, but cells in

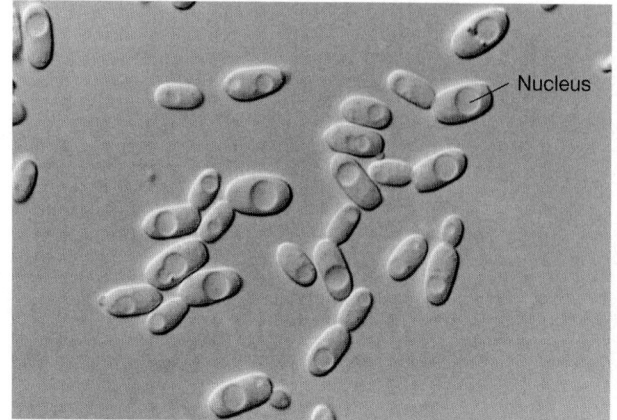

(a)

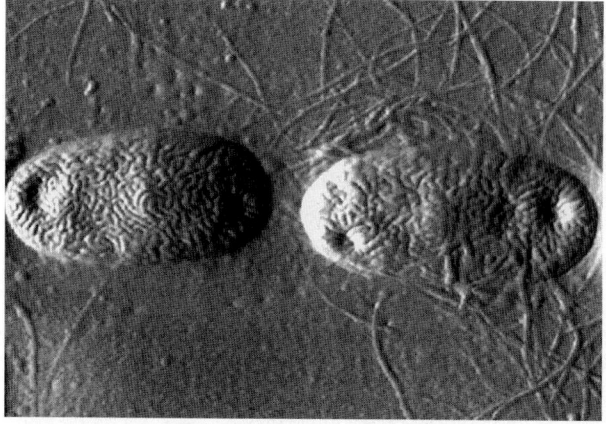

(b)

Figure 2.7 Three-dimensional imaging of cells. *(a)* Differential interference contrast and *(b)* atomic force microscopy. The yeast cells in part a are about 8 μm wide. Note the clearly visible nucleus and compare to Figure 2.5a. The bacterial cells in part b are 2.2 μm long and are from a biofilm that developed on the surface of a glass slide immersed for 24 h in a dog's water bowl.

the various layers can also be observed by adjusting the plane of focus of the laser beam. Using CSLM it has been possible to improve on the 0.2-μm resolution of the compound light microscope to a limit of about 0.1 μm.

Cells in CSLM preparations are typically stained with fluorescent dyes to make them more distinct (Figure 2.8). Alternatively, false-color images of unstained preparations can be generated such that different layers in the specimen are assigned different colors. The CLSM comes equipped with computer software that assembles digital images for subsequent image processing. Thus, images obtained from different layers can be digitally overlaid to reconstruct a three-dimensional image of the entire specimen (Figure 2.8).

CSLM has found widespread use in microbial ecology, especially for identifying populations of cells in a microbial habitat or for resolving the different components of a structured microbial habitat, such as a biofilm (Figure 2.8a). CSLM is particularly useful anywhere thick specimens are assessed for microbial content with depth.

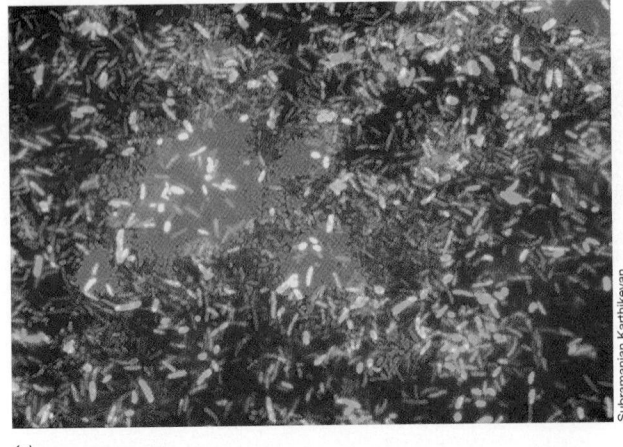

(a)

Subramanian Karthikeyan

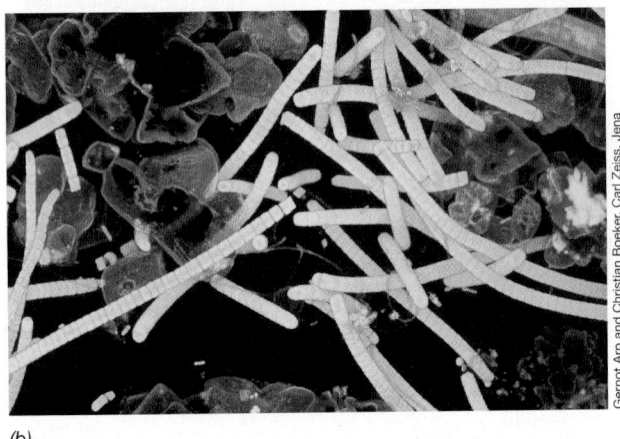

(b)

Gernot Arp and Christian Boeker, Carl Zeiss, Jena

Figure 2.8 Confocal scanning laser microscopy. *(a)* Confocal image of a microbial biofilm community cultivated in the laboratory. The green, rod-shaped cells are *Pseudomonas aeruginosa* experimentally introduced into the biofilm. Other cells of different colors are present at different depths in the biofilm. *(b)* Confocal image of a filamentous cyanobacterium growing in a soda lake. Cells are about 5 μm wide.

MiniQuiz

- What structure in eukaryotic cells is more easily seen in DIC than in bright-field microscopy? (*Hint:* Compare Figures 2.5a and 2.7a).
- How is CSLM able to view different layers in a thick preparation?

2.4 Electron Microscopy

Electron microscopes use electrons instead of visible light (photons) to image cells and cell structures. Electromagnets function as lenses in the electron microscope, and the whole system operates in a vacuum (**Figure 2.9**). Electron microscopes are fitted with cameras to allow a photograph, called an *electron micrograph*, to be taken.

Transmission Electron Microscopy

The transmission electron microscope (TEM) is used to examine cells and cell structure at very high magnification and resolution. The resolving power of a TEM is much greater than that of the

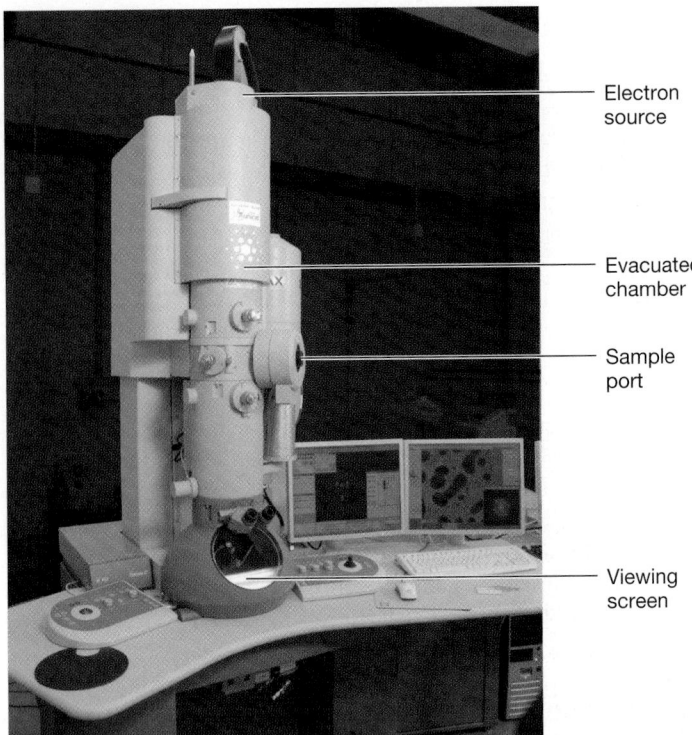

Electron source

Evacuated chamber

Sample port

Viewing screen

Figure 2.9 The electron microscope. This instrument encompasses both transmission and scanning electron microscope functions.

light microscope, even enabling one to view structures at the molecular level. This is because the wavelength of electrons is much shorter than the wavelength of visible light, and wavelength affects resolution (Section 2.1). For example, whereas the resolving power of a high-quality light microscope is about 0.2 *micrometer*, the resolving power of a high-quality TEM is about 0.2 *nanometer* (nm, 10^{-9} m). With such powerful resolution, even individual protein and nucleic acid molecules can be visualized in the transmission electron microscope (**Figure 2.10**, and see Figure 2.14*b*).

Unlike visible light, however, electron beams do not penetrate very well; even a single cell is too thick to reveal its internal contents directly by TEM. Consequently, special techniques of thin sectioning are needed to prepare specimens before observing them. A single bacterial cell, for instance, is cut into many, very thin (20–60 nm) slices, which are then examined individually by TEM (Figure 2.10*a*). To obtain sufficient contrast, the preparations are treated with stains such as osmic acid, or permanganate, uranium, lanthanum, or lead salts. Because these substances are composed of atoms of high atomic weight, they scatter electrons well and thus improve contrast.

Scanning Electron Microscopy

If only the external features of an organism are to be observed, thin sections are unnecessary. Intact cells or cell components can be observed directly by TEM with a technique called *negative staining* (Figure 2.10*b*). Alternatively, one can image the specimen using a *scanning electron microscope* (SEM) (Figure 2.9).

In scanning electron microscopy, the specimen is coated with a thin film of a heavy metal, such as gold. An electron beam then

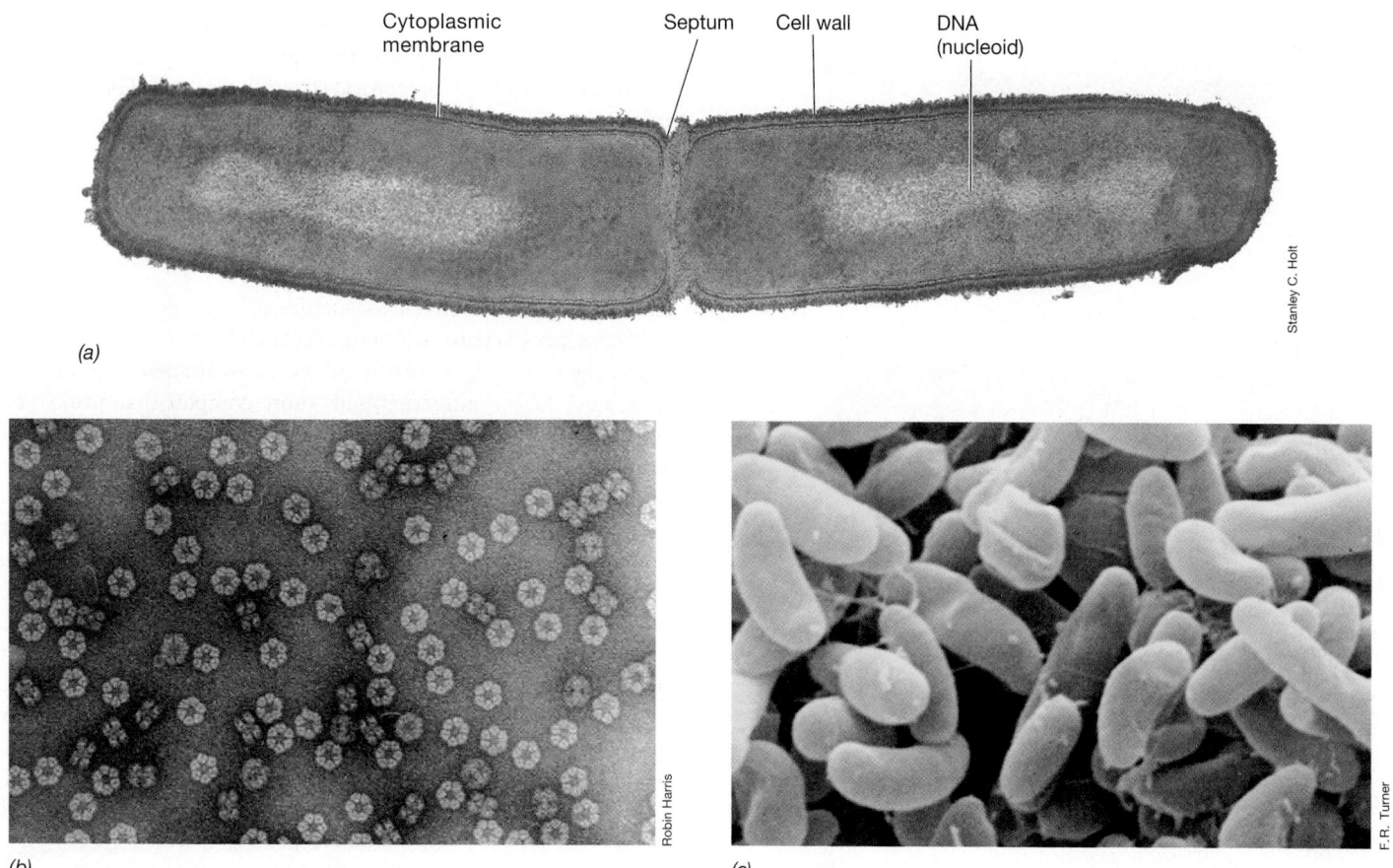

Cytoplasmic membrane   Septum   Cell wall   DNA (nucleoid)

(a)

Stanley C. Holt

(b)   Robin Harris

(c)   F.R. Turner

Figure 2.10 Electron micrographs. *(a)* Micrograph of a thin section of a dividing bacterial cell, taken by transmission electron microscopy (TEM). Note the DNA forming the nucleoid. The cell is about 0.8 μm wide. *(b)* TEM of negatively stained molecules of hemoglobin. Each hexagonal-shaped molecule is about 25 nanometers (nm) in diameter and consists of two doughnut-shaped rings, a total of 15 nm wide. *(c)* Scanning electron micrograph of bacterial cells. A single cell is about 0.75 μm wide.

scans back and forth across the specimen. Electrons scattered from the metal coating are collected and activate a viewing screen to produce an image (Figure 2.10*c*). In the SEM, even fairly large specimens can be observed, and the depth of field (the portion of the image that remains in sharp focus) is extremely good. A wide range of magnifications can be obtained with the SEM, from as low as 15× up to about 100,000×, but only the *surface* of an object is typically visualized.

Electron micrographs taken by either TEM or SEM are black-and-white images. Often times, false color is added to these images to boost their artistic appearance by manipulating the micrographs with a computer. But false color does not improve resolution of the micrograph or the scientific information it yields; resolution is set by the magnification used to take the original micrograph.

MiniQuiz

- What is an electron micrograph? Why do electron micrographs have so much greater resolution than light micrographs?

- What type of electron microscope would be used to view a cluster of cells? What type would be used to observe internal cell structure?

Ⅱ Cell Structure and Evolutionary History

We now consider some basic concepts of microbial cell structure that underlie many topics in this book. We first compare the internal architecture of microbial cells and differentiate eukaryotic from prokaryotic cells and cells from viruses. We then explore the evolutionary tree of life to see how the major groups of microorganisms that affect our lives and our planet are related.

2.5 Elements of Microbial Structure

All cells have much in common and contain many of the same components. For example, all cells have a permeability barrier called the **cytoplasmic membrane** that separates the inside of the cell, the **cytoplasm**, from the outside (**Figure 2.11**). The cytoplasm is an aqueous mixture of macromolecules—proteins, lipids, nucleic acids, and polysaccharides—small organic molecules (mainly precursors of macromolecules), various inorganic ions, and **ribosomes**, the cell's protein-synthesizing structures.

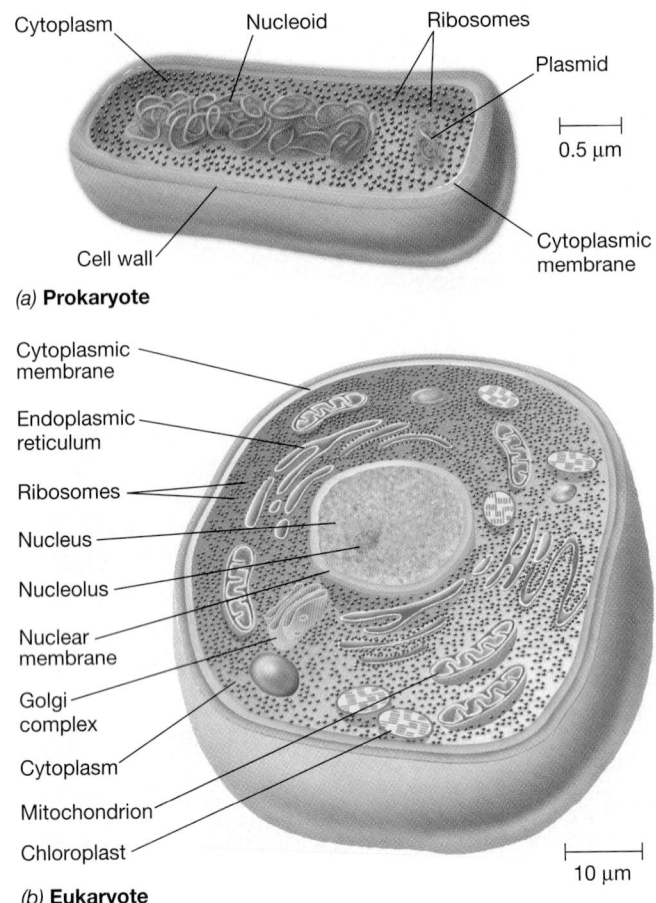

Cytoplasm   Nucleoid   Ribosomes

Plasmid

0.5 µm

Cell wall

Cytoplasmic membrane

(a) **Prokaryote**

Cytoplasmic membrane

Endoplasmic reticulum

Ribosomes

Nucleus

Nucleolus

Nuclear membrane

Golgi complex

Cytoplasm

Mitochondrion

Chloroplast

10 µm

(b) **Eukaryote**

Figure 2.11 **Internal structure of cells.** Note differences in scale and internal structure between the prokaryotic and eukaryotic cells.

Ribosomes interact with cytoplasmic proteins and messenger and transfer RNAs in the key process of protein synthesis (translation).

The **cell wall** lends structural strength to a cell. The cell wall is relatively permeable and located outside the membrane (Figure 2.11*a*); it is a much stronger layer than the membrane itself. Plant cells and most microorganisms have cell walls, whereas animal cells, with rare exceptions, do not.

Prokaryotic and Eukaryotic Cells

Examination of the internal structure of cells reveals two distinct patterns: **prokaryote** and **eukaryote** (**Figure 2.12**). Eukaryotes house their DNA in a membrane-enclosed **nucleus** and are typically much larger and structurally more complex than prokaryotic cells. In eukaryotic cells the key processes of DNA replication, transcription, and translation are partitioned; replication and transcription (RNA synthesis) occur in the nucleus while translation (protein synthesis) occurs in the cytoplasm. Eukaryotic microorganisms include algae and protozoa, collectively called *protists*, and the fungi and slime molds. The cells of plants and animals are also eukaryotic cells. We consider microbial eukaryotes in detail in Chapter 20.

A major property of eukaryotic cells is the presence of membrane-enclosed structures in the cytoplasm called **organelles**. These include, first and foremost, the nucleus, but also mitochondria and chloroplasts (the latter in photosynthetic cells only) (Figures 2.2*a* and 2.12*c*). As mentioned, the nucleus houses the cell's genome and is also the site of RNA synthesis in eukaryotic cells. Mitochondria and chloroplasts are dedicated to energy conservation and carry out respiration and photosynthesis, respectively.

In contrast to eukaryotic cells, prokaryotic cells have a simpler internal structure in which organelles are absent (Figures 2.11*a*

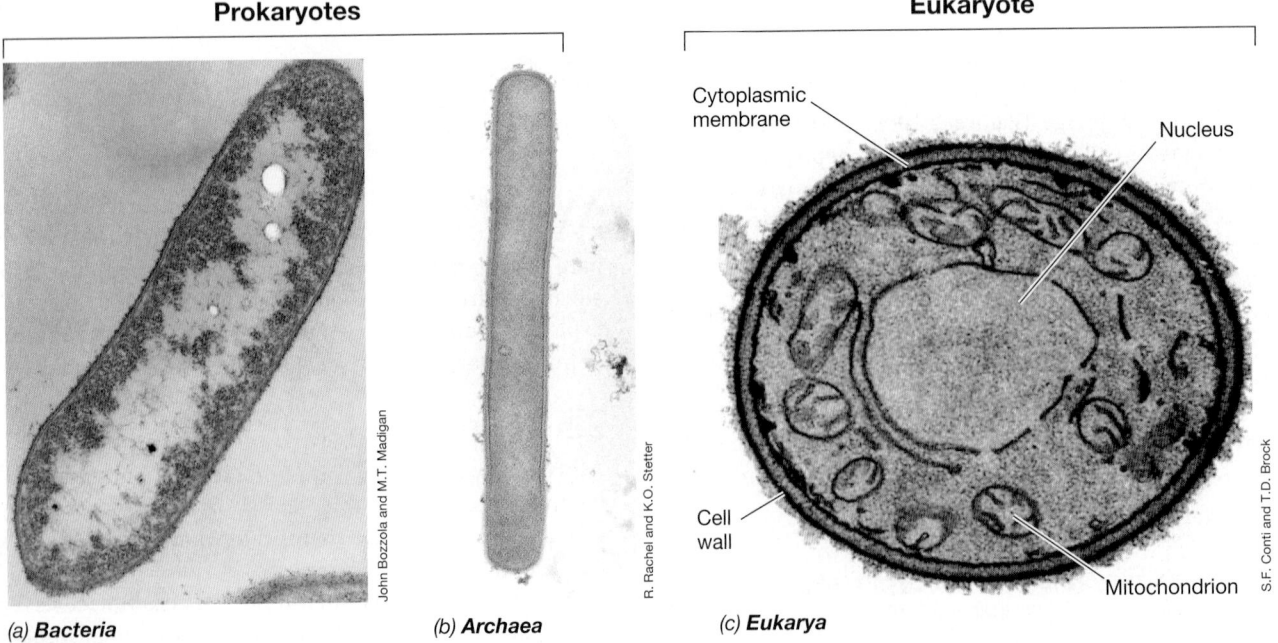

Prokaryotes

Eukaryote

Cytoplasmic membrane

Nucleus

Cell wall

Mitochondrion

(a) **Bacteria**   *(b)* **Archaea**   *(c)* **Eukarya**

John Bozzola and M.T. Madigan

R. Rachel and K.O. Stetter

S.F. Conti and T.D. Brock

Figure 2.12 **Electron micrographs of sectioned cells from each of the domains of living organisms.** *(a) Heliobacterium modesticaldum;* the cell measures 1 × 3 µm. *(b) Methanopyrus kandleri;* the cell measures 0.5 × 4 µm. Reinhard Rachel and Karl O. Stetter, 1981. *Archives of Microbiology* 128:288–293. © Springer-Verlag GmbH & Co. KG. *(c) Saccharomyces cerevisiae;* the cell measures 8 µm in diameter.

and 2.12*a*, *b*). However, prokaryotes differ from eukaryotes in many other ways as well. For example, prokaryotes can couple transcription directly to translation because their DNA resides in the cytoplasm and is not enclosed within a nucleus as in eukaryotes. Moreover, in contrast to eukaryotes, most prokaryotes employ their cytoplasmic membrane in energy-conservation reactions and have small, compact genomes consisting of circular DNA, as discussed in the next section. In terms of cell size, a typical rod-shaped prokaryote is 1–5 μm long and about 1 μm wide, but considerable variation is possible (Table 3.1). The range of sizes in eukaryotic cells is quite large. Eukaryotic cells are known with diameters as small as 0.8 μm or as large as several hundred micrometers. We revisit the subject of cell size in more detail in Section 3.2.

Despite the many clear-cut *structural* differences between prokaryotes and eukaryotes, it is very important that the word "prokaryote" not be given an *evolutionary* connotation. As was touched on in Chapter 1, the prokaryotic world consists of two evolutionarily distinct groups, the *Bacteria* and the *Archaea*. Moreover, the word "prokaryote" should not be considered synonymous with "primitive," as all cells living today—whether prokaryotes or eukaryotes—are highly evolved and closely adapted to their habitat. In Chapters 6 and 7 we compare and contrast the molecular biology of *Bacteria* and *Archaea*, highlighting their similarities and differences and relating them to molecular processes in eukaryotes.

Viruses

Viruses are a major class of microorganisms, but they are not cells (**Figure 2.13**). Viruses are much smaller than cells and lack many of the attributes of cells (Figure 1.3). Viruses vary in size, with the smallest known viruses being only about 10 nm in diameter.

Instead of being a dynamic open system, a virus particle is static and stable, unable to change or replace its parts by itself. Only when a virus infects a cell does it acquire the key attribute of a living system—replication. Unlike cells, viruses have no metabolic capabilities of their own. Although they contain their own genomes, viruses lack ribosomes. So to synthesize proteins, viruses depend on the biosynthetic machinery of the cells they have infected. Moreover, unlike cells, viral particles contain only a single form of nucleic acid, either DNA or RNA (this means, of course, that some viruses have RNA genomes).

Viruses are known to infect all types of cells, including microbial cells. Many viruses cause disease in the organisms they infect. However, viral infection can have many other effects on cells, including genetic alterations that can actually improve the capabilities of the cell. We discuss the field of virology and viral diversity in detail in Chapters 9 and 21, respectively.

MiniQuiz

- What important functions do the following play in a cell: cytoplasmic membrane, ribosomes, cell wall?
- By looking inside a cell how could you tell if it was a prokaryote or a eukaryote?
- How are viruses like cells, and in which major ways do they differ?

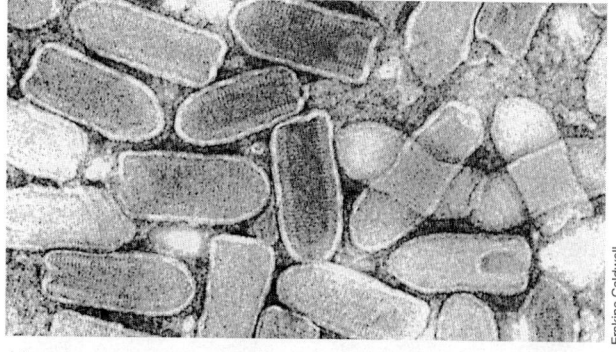

(a)

(b)

Figure 2.13 Viruses. *(a)* Particles of rhabdovirus (a virus that infects plants and animals). A single virus particle, called a *virion*, is about 65 nm (0.065 μm) wide. *(b)* Bacterial virus (bacteriophage) lambda. The head of each lambda virion is also about 65 nm wide. Viruses are composed of protein and nucleic acid and do not have structures such as walls or a cytoplasmic membrane.

2.6 Arrangement of DNA in Microbial Cells

The life processes of any cell are governed by its complement of genes, its *genome*. A gene is a segment of DNA (or RNA in RNA viruses) that encodes a protein or an RNA molecule. Here we consider how genomes are organized in prokaryotic and eukaryotic cells and consider the number of genes and proteins present in a model prokaryotic cell.

Nucleus versus Nucleoid

The genomes of prokaryotic and eukaryotic cells are organized differently. In most prokaryotic cells, DNA is present in a circular molecule called the *chromosome*; a few prokaryotes have a linear instead of a circular chromosome. The chromosome aggregates within the cell to form a mass called the **nucleoid**, visible in the electron microscope (**Figure 2.14**; see also Figure 2.10*a*).

Most prokaryotes have only a single chromosome. Because of this, they typically contain only a single copy of each gene and are therefore genetically *haploid*. Many prokaryotes also contain one or more small circles of DNA distinct from that of the chromosome, called **plasmids**. Plasmids typically contain genes that confer a special property (such as a unique metabolism) on a cell, rather than essential genes. This is in contrast to genes on the chromosome, most of which are needed for basic survival.

In eukaryotes, DNA is arranged in linear molecules within the membrane-enclosed nucleus; the DNA molecules are packaged

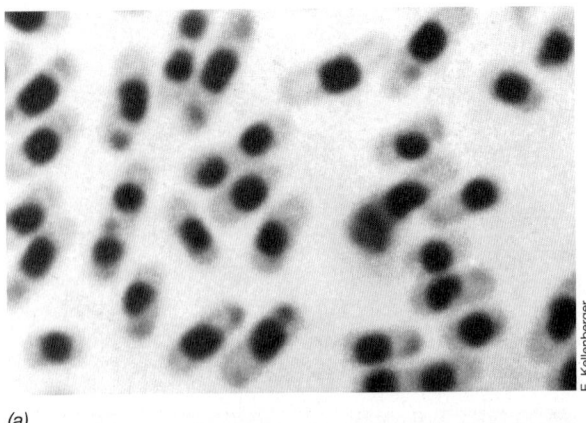

(a)

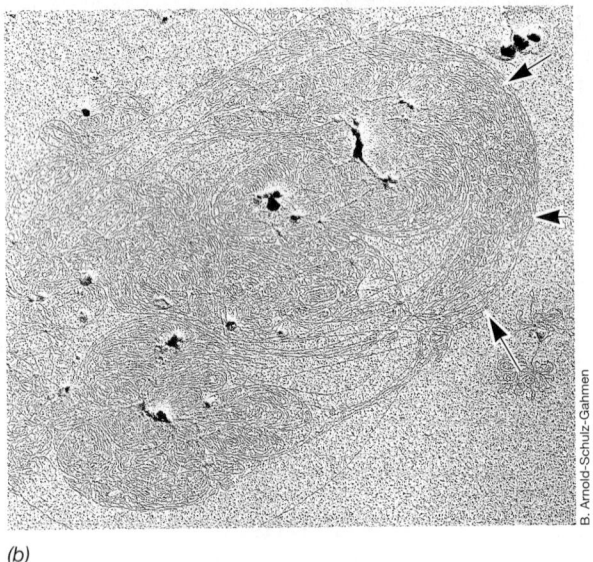

(b)

Figure 2.14 The nucleoid. *(a)* Photomicrograph of cells of *Escherichia coli* treated in such a way as to make the nucleoid visible. A single cell is about 3 μm and a nucleoid about 1 μm long. *(b)* Transmission electron micrograph of an isolated nucleoid released from a cell of *E. coli*. The cell was gently lysed to allow the highly compacted nucleoid to emerge intact. Arrows point to the edge of DNA strands.

with proteins and organized to form **chromosomes**. Chromosome number varies by organism. For example, a diploid cell of the baker's yeast *Saccharomyces cerevisiae* contains 32 chromosomes arranged in 16 pairs while human cells contain 46 chromosomes (23 pairs). Chromosomes in eukaryotes contain proteins that assist in folding and packing the DNA and other proteins that are required for transcription. A key genetic difference between prokaryotes and eukaryotes is that eukaryotes typically contain two copies of each gene and are thus genetically *diploid*. During cell division in eukaryotic cells the nucleus divides (following a doubling of chromosome number) in the process called *mitosis* (**Figure 2.15**). Two identical daughter cells result, with each daughter cell receiving a full complement of genes. The diploid genome of eukaryotic cells is halved in the process of *meiosis* to form haploid gametes for sexual reproduction. Fusion of two gametes during zygote formation restores the cell to the diploid state.

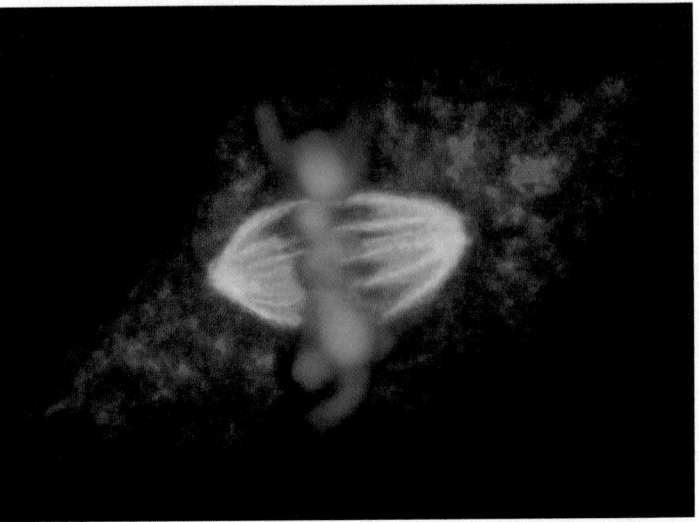

Figure 2.15 Mitosis in stained kangaroo rat cells. The cell was photographed while in the metaphase stage of mitotic division; only eukaryotic cells undergo mitosis. The green color stains a protein called tubulin, important in pulling chromosomes apart. The blue color is from a DNA-binding dye and shows the chromosomes.

Genes, Genomes, and Proteins

How many genes and proteins does a cell have? The genome of *Escherichia coli*, a model bacterium, is a single circular chromosome of 4,639,221 base pairs of DNA. Because the *E. coli* genome has been completely sequenced, we also know that it contains 4288 genes. The genomes of a few prokaryotes have three times this many genes, while the genomes of others contain fewer than one-twentieth as many. Eukaryotic cells typically have much larger genomes than prokaryotes. A human cell, for example, contains over 1000 times as much DNA as a cell of *E. coli* and about seven times as many genes.

Depending somewhat on growth conditions, a cell of *E. coli* contains about 1900 different kinds of proteins and about 2.4 million individual protein molecules. However, some proteins in *E. coli* are very abundant, others are only moderately abundant, and some are present in only one or a very few copies per cell. Thus, *E. coli* has mechanisms for regulating its genes so that not all genes are *expressed* (transcribed and translated) at the same time or to the same extent. Gene regulation is important to all cells, and we focus on the major mechanisms of gene regulation in Chapter 8.

MiniQuiz
- Differentiate between the nucleus and the nucleoid.
- What does it mean to say that a bacterial cell is haploid?
- Why does it make sense that a human cell would have more genes than a bacterial cell?

2.7 The Evolutionary Tree of Life

Evolution is the process of descent with modification that generates new varieties and eventually new species of organisms. Evolution occurs in any self-replicating system in which variation is

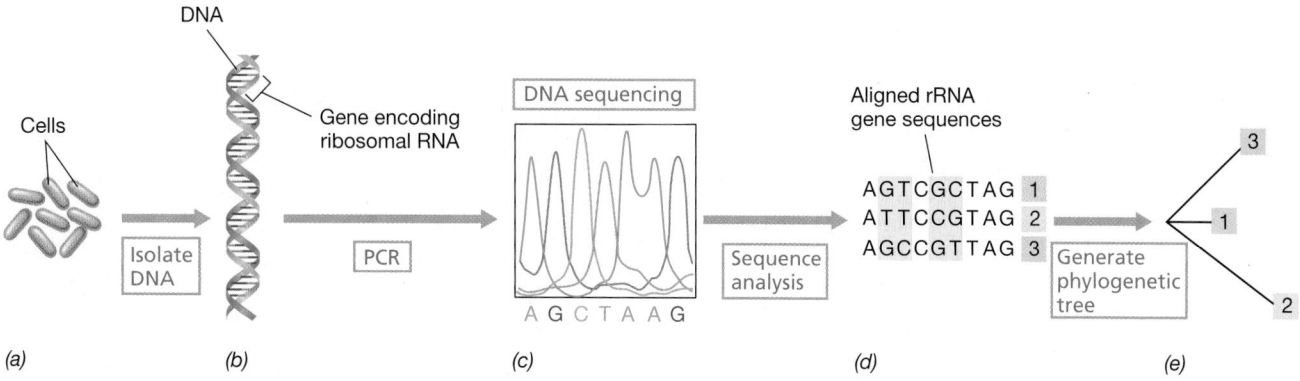

Figure 2.16 Ribosomal RNA (rRNA) gene sequencing and phylogeny. (a) DNA is extracted from cells. (b) Many identical copies of a gene encoding rRNA are made by the polymerase chain reaction (Section 6.11). (c, d) The gene is sequenced and the sequence aligned with rRNA sequences from other organisms. A computer algorithm makes pairwise comparisons at each base and generates a phylogenetic tree (e) that depicts evolutionary divergence. In the example shown, the sequence differences are highlighted in yellow and are as follows: organism 1 versus organism 2, three differences; 1 versus 3, two differences; 2 versus 3, four differences. Thus organisms 1 and 3 are closer relatives than are 2 and 3 or 1 and 2.

the result of mutation and selection is based on differential fitness. Thus, over time, both cells and viruses evolve.

Determining Evolutionary Relationships

The evolutionary relationships between organisms are the subject of **phylogeny**. Phylogenetic relationships between cells can be deduced by comparing the genetic information (nucleotide or amino acid sequences) that exists in their nucleic acids or proteins. For reasons that will be presented later, macromolecules that form the ribosome, in particular *ribosomal RNAs (rRNA)*, are excellent tools for discerning evolutionary relationships. Because all cells contain ribosomes (and thus rRNA), this molecule can and has been used to construct a phylogenetic tree of all cells, including microorganisms (see Figure 2.17). Carl Woese, an American microbiologist, pioneered the use of comparative rRNA sequence analysis as a measure of microbial phylogeny and, in so doing, revolutionized our understanding of cellular evolution. Viral phylogenies have also been determined, but because these microorganisms lack ribosomes, other molecules have been used for evolutionary metrics.

The steps in generating an RNA-based phylogenetic tree are outlined in **Figure 2.16**. In brief, genes encoding rRNA from two or more organisms are sequenced and the sequences aligned and scored, base-by-base, for sequence differences and identities using a computer; the greater the sequence variation between any two organisms, the greater their evolutionary divergence. Then, using a treeing algorithm, this divergence is depicted in the form of a phylogenetic tree.

The Three Domains of Life

From comparative rRNA sequencing, three phylogenetically distinct cellular lineages have been revealed. The lineages, called **domains**, are the **Bacteria** and the **Archaea** (both consisting of prokaryotic cells) and the **Eukarya** (eukaryotes) (**Figure 2.17**). The domains are thought to have diverged from a common ancestral organism (LUCA in Figure 2.17) early in the history of life on Earth.

The phylogenetic tree of life reveals two very important evolutionary facts: (1) As previously stated, all prokaryotes are *not* phylogenetically closely related, and (2) *Archaea* are actually more closely related to *Eukarya* than to *Bacteria* (Figure 2.17). Thus, from the last universal common ancestor (LUCA) of all life forms on Earth, evolutionary diversification diverged to yield the ancestors of the *Bacteria* and of a second main lineage (Figure 1.6). The latter once again diverged to yield the ancestors of the *Archaea*, a lineage that retained a prokaryotic cell structure, and the *Eukarya*, which did not. The universal tree of life shows that LUCA resides very early within the *Bacteria* domain (Figure 2.17).

Eukarya

Because the cells of animals and plants are all eukaryotic, it follows that eukaryotic microorganisms were the ancestors of multicellular organisms. The tree of life clearly bears this out. As expected, microbial eukaryotes branch off early on the eukaryotic lineage, while plants and animals branch near the crown of the tree (Figure 2.17). However, molecular sequencing and several other lines of evidence have shown that eukaryotic cells contain genes from cells of two domains. In addition to the genome in the chromosomes of the nucleus, mitochondria and chloroplasts of eukaryotes contain their own genomes (this DNA is arranged in a circular fashion, as in most prokaryotes), and ribosomes. Using molecular phylogenetic analyses (Figure 2.16), these organelles have been shown to be highly derived ancestors of specific lineages of *Bacteria* (Figure 2.17 and Section 2.9). Mitochondria and chloroplasts are therefore descendants of what are thought to have been free-living bacterial cells that developed an intimate intracellular association with cells of the *Eukarya* domain eons ago. The theory of how this stable arrangement of cells led to the modern eukaryotic cell with organelles has been called **endosymbiosis** (*endo* means "inside") and is discussed in Chapters 16 and 20.

Contributions of Molecular Sequencing to Microbiology

Molecular phylogeny has not only revealed the evolutionary connections between all cells—prokaryotes and eukaryotes—it has

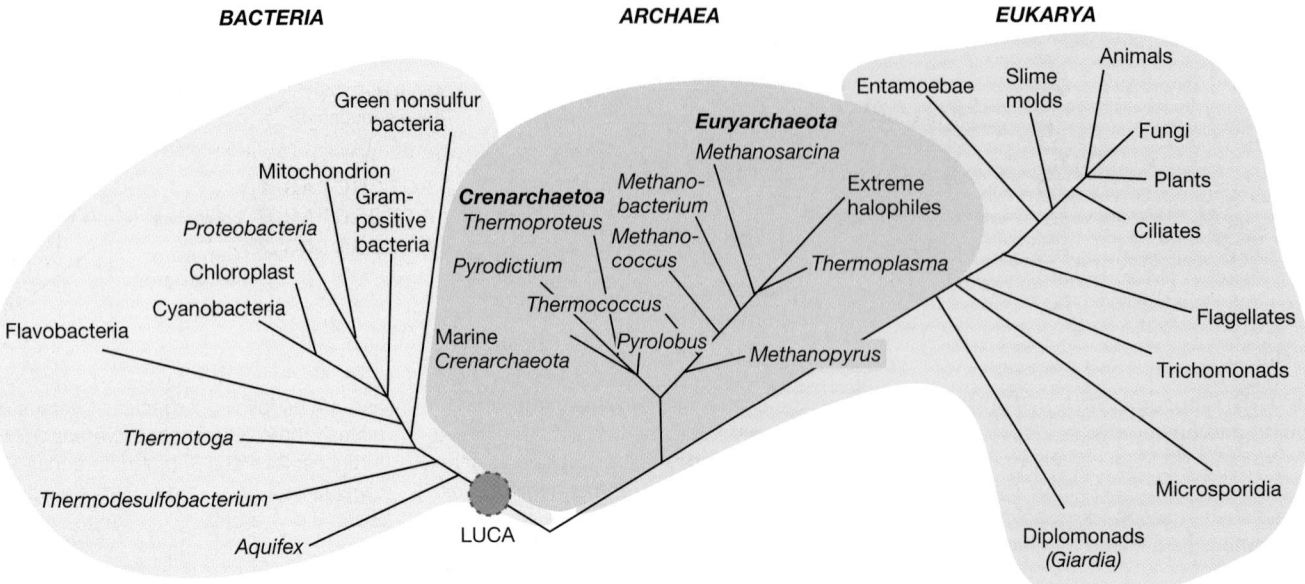

Figure 2.17 **The phylogenetic tree of life as defined by comparative rRNA gene sequencing.** The tree shows the three domains of organisms and a few representative groups in each domain. All *Bacteria* and *Archaea* and most *Eukarya* are microscopic organisms; only plants, animals, and fungi contain macro-organisms. Phylogenetic trees of each domain can be found in Figures 2.19, 2.28, and 2.32. LUCA, last universal common ancestor.

formed the first evolutionary framework for the prokaryotes, something that the science of microbiology had been without since its inception. In addition, molecular phylogeny has spawned exciting new research tools that have affected many subdisciplines of microbiology, in particular, microbial systematics and ecology, and clinical diagnostics. In these areas molecular phylogenetic methods have begun to shape our concept of a bacterial species and given microbial ecologists and clinical microbiologists the capacity to identify organisms without actually culturing them. This has greatly improved our picture of microbial diversity and has led to the staggering conclusion that most of the microbial diversity that exists on Earth has yet to be brought into laboratory culture.

MiniQuiz
- How can species of *Bacteria* and *Archaea* be distinguished by molecular criteria?
- What is endosymbiosis, and in what way did it benefit eukaryotic cells?

⬤Ⅲ Microbial Diversity

The diversity of microorganisms we see today is the result of nearly 4 billion years of evolution. Microbial diversity can be seen in many ways besides phylogeny, including cell size and morphology (shape), physiology, motility, mechanism of cell division, pathogenicity, developmental biology, adaptation to environmental extremes, and so on. In the following sections we paint a picture of microbial diversity with a broad brush. We then return to reconsider the topic in more detail in Chapters 16–21.

Our discussion of *microbial* diversity begins with a brief consideration of *metabolic* diversity. The two topics are closely linked. Through eons, microorganisms, especially the prokaryotes, have come to exploit every means of "making a living" consistent with the laws of chemistry and physics. This enormous metabolic versatility has allowed prokaryotes to thrive in every potential habitat on Earth suitable for life.

2.8 Metabolic Diversity

All cells require an energy source and a metabolic strategy for conserving energy from it to drive energy-consuming life processes. As far as is known, energy can be tapped from three sources in nature: organic chemicals, inorganic chemicals, and light (**Figure 2.18**).

Chemoorganotrophs

Organisms that conserve energy from chemicals are called *chemotrophs*, and those that use *organic* chemicals are called **chemoorganotrophs** (Figure 2.18). Thousands of different organic chemicals can be used by one or another microorganism. Indeed, all natural and even most synthetic organic compounds can be metabolized. Energy is conserved from the *oxidation* of the compound and is stored in the cell in the energy-rich bonds of the compound adenosine triphosphate (ATP).

Some microorganisms can obtain energy from an organic compound only in the presence of oxygen; these organisms are called *aerobes*. Others can obtain energy only in the absence of oxygen (*anaerobes*). Still others can break down organic compounds in either the presence or absence of oxygen. Most microorganisms that have been brought into laboratory culture are chemoorganotrophs.

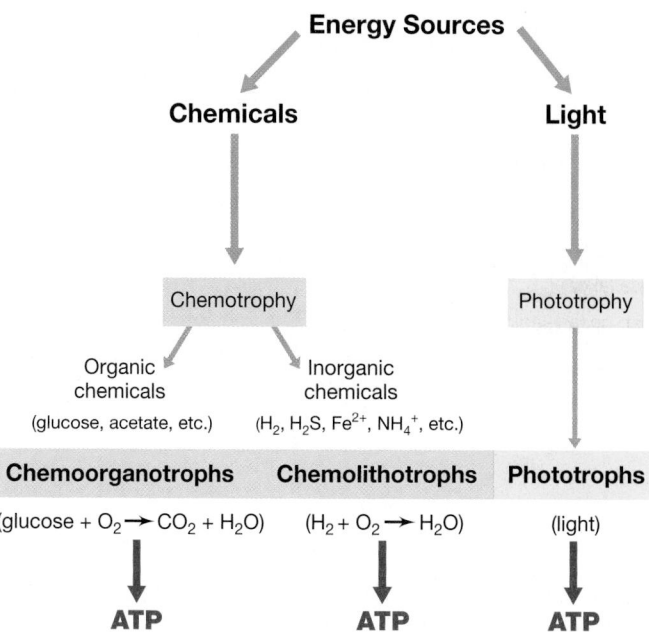

Figure 2.18 Metabolic options for conserving energy. The organic and inorganic chemicals listed here are just a few of the chemicals used by one organism or another. Chemotrophic organisms oxidize organic or inorganic chemicals, which yields ATP. Phototrophic organisms use solar energy to form ATP.

Chemolithotrophs

Many prokaryotes can tap the energy available from the oxidation of *inorganic* compounds. This form of metabolism is called *chemolithotrophy* and was discovered by the Russian microbiologist Winogradsky (⟳ Section 1.9). Organisms that carry out chemolithotrophic reactions are called **chemolithotrophs** (Figure 2.18). Chemolithotrophy occurs only in prokaryotes and is widely distributed among species of *Bacteria* and *Archaea*. Several inorganic compounds can be oxidized; for example, H_2, H_2S (hydrogen sulfide), NH_3 (ammonia), and Fe^{2+} (ferrous iron). Typically, a related group of chemolithotrophs specializes in the oxidation of a related group of inorganic compounds, and thus we have the "sulfur" bacteria, the "iron" bacteria, and so on.

The capacity to conserve energy from the oxidation of inorganic chemicals is a good metabolic strategy because competition from chemoorganotrophs, organisms that require organic energy sources, is not an issue. In addition, many of the inorganic compounds oxidized by chemolithotrophs, for example H_2 and H_2S, are actually the waste products of chemoorganotrophs. Thus, chemolithotrophs have evolved strategies for exploiting resources that chemoorganotrophs are unable to use, so it is common for species of these two physiological groups to live in close association with one another.

Phototrophs

Phototrophic microorganisms contain pigments that allow them to convert light energy into chemical energy, and thus their cells appear colored (Figure 2.2). Unlike chemotrophic organisms, then, **phototrophs** do not require chemicals as a source of energy. This is a significant metabolic advantage because competition with chemotrophic organisms for energy sources is not an issue and sunlight is available in many microbial habitats on Earth.

Two major forms of phototrophy are known in prokaryotes. In one form, called *oxygenic* photosynthesis, oxygen (O_2) is produced. Among microorganisms, oxygenic photosynthesis is characteristic of cyanobacteria and algae. The other form, *anoxygenic* photosynthesis, occurs in the purple and green bacteria and the heliobacteria, and does not yield O_2. However, both oxygenic and anoxygenic phototrophs have great similarities in their mechanism of ATP synthesis, a result of the fact that oxygenic photosynthesis evolved from the simpler anoxygenic form, and we return to this topic in Chapter 13.

Heterotrophs and Autotrophs

All cells require carbon in large amounts and can be considered either **heterotrophs**, which require organic compounds as their carbon source, or **autotrophs**, which use carbon dioxide (CO_2) as their carbon source. Chemoorganotrophs are by definition heterotrophs. By contrast, most chemolithotrophs and phototrophs are autotrophs. Autotrophs are sometimes called *primary producers* because they synthesize new organic matter from CO_2 for both their own benefit and that of chemoorganotrophs. The latter either feed directly on the cells of primary producers or live off products they excrete. Virtually all organic matter on Earth has been synthesized by primary producers, in particular, the phototrophs.

Habitats and Extreme Environments

Microorganisms are present everywhere on Earth that will support life. These include habitats we are all familiar with—soil, water, animals, and plants—as well as virtually any structures made by humans. Indeed, sterility (the absence of life forms) in a natural sample is extremely rare.

Some microbial habitats are ones in which humans could not survive, being too hot or too cold, too acidic or too caustic, or too salty. Although such environments would pose challenges to any life forms, they are often teeming with microorganisms. Organisms inhabiting such extreme environments are called **extremophiles**, a remarkable group of microorganisms that collectively define the physiochemical limits to life (**Table 2.1**).

Extremophiles abound in such harsh environments as volcanic hot springs; on or in the ice covering lakes, glaciers, or the polar seas; in extremely salty bodies of water; in soils and waters having a pH as low as 0 or as high as 12; and in the deep sea, where hydrostatic pressure can exceed 1000 times atmospheric. Interestingly, these prokaryotes do not just *tolerate* their particular environmental extreme, they actually *require* it in order to grow. That is why they are called extremophiles (the suffix *-phile* means "loving"). Table 2.1 summarizes the current "record holders" among extremophiles and lists the terms used to describe each class and the types of habitats in which they reside. We will revisit many of these organisms in later chapters and examine the special properties that allow for their growth in extreme environments.

Table 2.1 *Classes and examples of extremophiles*[a]

Extreme	Descriptive term	Genus/species	Domain	Habitat	Minimum	Optimum	Maximum
Temperature							
High	Hyperthermophile	*Methanopyrus kandleri*	*Archaea*	Undersea hydrothermal vents	90°C	**106°C**	122°C[b]
Low	Psychrophile	*Psychromonas ingrahamii*	*Bacteria*	Sea ice	−12°C	**5°C**	10°C
pH							
Low	Acidophile	*Picrophilus oshimae*	*Archaea*	Acidic hot springs	−0.06	**0.7**[c]	4
High	Alkaliphile	*Natronobacterium gregoryi*	*Archaea*	Soda lakes	8.5	**10**[d]	12
Pressure	Barophile (Piezophile)	*Moritella yayanosii*[e]	*Bacteria*	Deep ocean sediments	500 atm	**700 atm**	>1000 atm
Salt (NaCl)	Halophile	*Halobacterium salinarum*	*Archaea*	Salterns	15%	**25%**	32% (saturation)

[a]The organisms listed are the current "record holders" for growth at a particular extreme condition.
[b]Anaerobe showing growth at 122°C only under several atmospheres of pressure.
[c]*P. oshimae* is also a thermophile, growing optimally at 60°C.
[d]*N. gregoryi* is also an extreme halophile, growing optimally at 20% NaCl.
[e]*M. yayanosii* is also a psychrophile, growing optimally near 4°C.

MiniQuiz

- In terms of energy generation, how does a chemoorganotroph differ from a chemolithotroph?
- In terms of carbon acquisition, how does an autotroph differ from a heterotroph?
- What are extremophiles?

2.9 Bacteria

As we have seen, prokaryotes have diverged into two phylogenetically distinct domains, the *Archaea* and the *Bacteria* (Figure 2.17). We begin with the *Bacteria*, because most of the best-known prokaryotes reside in this domain.

Proteobacteria

The domain *Bacteria* contains an enormous variety of prokaryotes. All known disease-causing (pathogenic) prokaryotes are *Bacteria*, as are thousands of nonpathogenic species. A large variety of morphologies and physiologies are also observed in this domain. The **Proteobacteria** make up the largest phylum of *Bacteria* (**Figure 2.19**). Many chemoorganotrophic bacteria are *Proteobacteria*, including *Escherichia coli*, the model organism of microbial physiology, biochemistry, and molecular biology. Several phototrophic and chemolithotrophic species are also *Proteobacteria* (**Figure 2.20**). Many of these use H_2S in their metabolism, producing elemental sulfur (S^0) that is stored either inside or outside the cell (Figure 2.20). Sulfur is an oxidation product of H_2S and is further oxidized to sulfate (SO_4^{2-}). Sulfide and sulfur are oxidized to fuel important metabolic functions such as CO_2 fixation (autotrophy) or energy conservation (Figure 2.18).

Several other common prokaryotes of soil and water, and species that live in or on plants and animals in both harmless and disease-causing ways, are *Proteobacteria*. These include species of *Pseudomonas*, many of which can degrade complex or toxic natural and synthetic organic compounds, and *Azotobacter*, a bacterium that fixes nitrogen (utilizes gaseous nitrogen as a

nitrogen source, ↩ Figure 1.9). A number of key pathogens are *Proteobacteria*, including *Salmonella* (gastrointestinal diseases), *Rickettsia* (typhus and Rocky Mountain spotted fever), *Neisseria* (gonorrhea), and many others. And finally, the key respiratory organelle of eukaryotes, the mitochondrion, has evolutionary roots within the *Proteobacteria* (Figure 2.17).

Gram-Positive Bacteria

As we learned in Section 2.2, bacteria can be distinguished by the Gram-staining procedure, a technique that stains cells either gram-positive or gram-negative. The gram-positive phylum of *Bacteria* (Figure 2.19) contains many organisms that are united by their common phylogeny and cell wall structure. Here we find the endospore-forming *Bacillus* (discovered by Ferdinand Cohn,

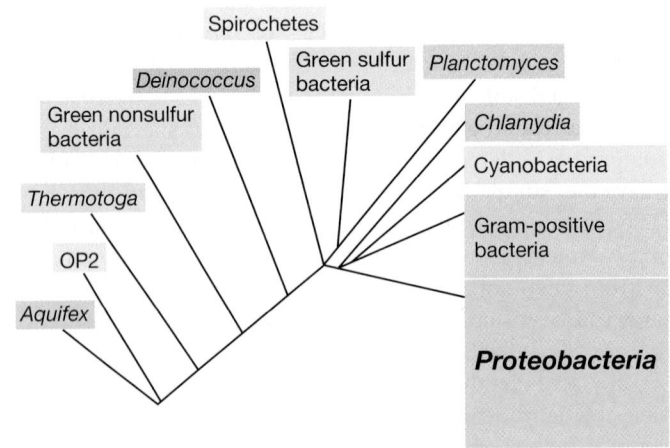

Figure 2.19 Phylogenetic tree of some representative *Bacteria*. The *Proteobacteria* are by far the largest phylum of *Bacteria* known. The lineage on the tree labeled OP2 does not represent a cultured organism but instead is an rRNA gene isolated from an organism in a natural sample. In this example, the closest known relative of OP2 would be *Aquifex*. Many thousands of other environmental sequences are known, and they branch all over the tree. Environmental sequences are also called *phylotypes*, and the technology for deriving them is considered in Section 22.4.

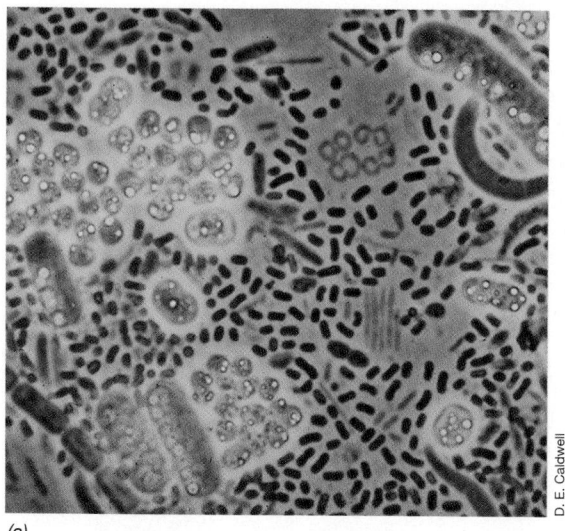

(a)

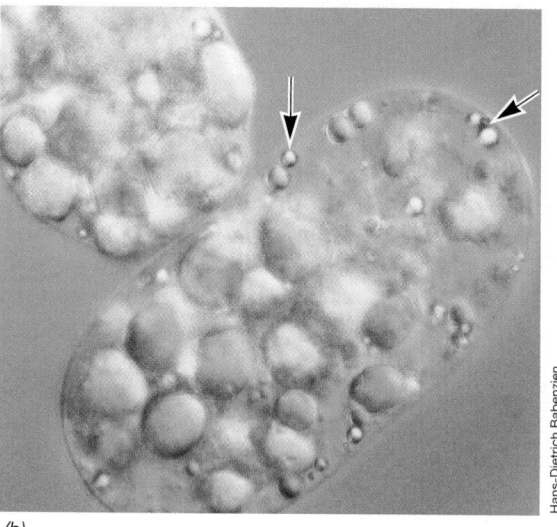

(b)

Figure 2.20 **Phototrophic and chemolithotrophic *Proteobacteria.*** *(a)* The phototrophic purple sulfur bacterium *Chromatium* (the large, red-orange, rod-shaped cells in this photomicrograph of a natural microbial community). A cell is about 10 μm wide. *(b)* The large chemolithotrophic sulfur-oxidizing bacterium *Achromatium*. A cell is about 20 μm wide. Globules of elemental sulfur can be seen in the cells (arrows). Both of these organisms oxidize hydrogen sulfide (H_2S).

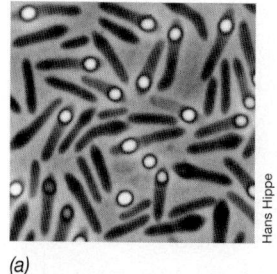

(a)

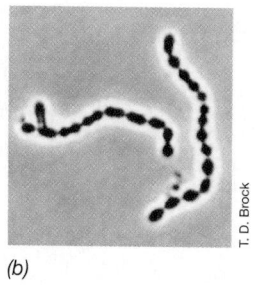

(b)

Figure 2.21 Gram-positive bacteria. *(a)* The rod-shaped endospore-forming bacterium *Bacillus*. Note the presence of endospores (bright refractile structures) inside the cells. Endospores are extremely resistant to heat, chemicals, and radiation. Cells are about 1.6 μm in diameter. *(b)* *Streptococcus*, a spherical cell that forms cell chains. Streptococci are widespread in dairy products, and some are potent pathogens. Cells are about 0.8 μm in diameter.

Cyanobacteria

The **cyanobacteria** are phylogenetic relatives of gram-positive bacteria (Figure 2.19) and are oxygenic phototrophs. The photosynthetic organelle of eukaryotic phototrophs, the chloroplast (Figure 2.2*a*), is related to the cyanobacteria (Figure 2.17). Cyanobacteria were key players in the evolution of life, as they were the first oxygenic phototrophs to evolve on Earth. The production of O_2 on an originally anoxic Earth paved the way for the evolution of cells that could respire using oxygen. The development of higher organisms, such as the plants and animals, followed billions of years later when Earth had a more oxygen-rich environment (↩ Figure 1.6). Cells of some cyanobacteria join to form filaments (**Figure 2.22**). Many other morphological forms of cyanobacteria are known, including unicellular, colonial, and heterocystous. Species in the latter group contain special structures called *heterocysts* that carry out nitrogen fixation.

Other Major Phyla of *Bacteria*

Several phyla of *Bacteria* contain species with unique morphologies and almost all of these stain gram-negatively. These lineages include the aquatic planctomycetes, characterized by cells with a distinct stalk that allows the organisms to attach to a solid substratum (**Figure 2.23**), and the helically shaped spirochetes (**Figure 2.24**). Several diseases, most notably syphilis and Lyme disease, are caused by spirochetes.

Two other major phyla of *Bacteria* are phototrophic: the green sulfur bacteria and the green nonsulfur bacteria (*Chloroflexus* group) (**Figure 2.25**). Species in both of these lineages contain similar photosynthetic pigments and are also autotrophs. *Chloroflexus* is a filamentous phototroph that inhabits hot springs and associates with cyanobacteria to form microbial mats, which are laminated microbial communities containing both phototrophs and chemotrophs. *Chloroflexus* is also noteworthy because its ancient relatives may have been the first phototrophic bacteria on Earth.

Other major phyla of *Bacteria* include the *Chlamydiae* and *Deinococcus-Thermus* groups (Figure 2.19). The phylum *Chlamydiae* harbors respiratory and sexually transmitted pathogens of humans. Chlamydia are intracellular parasites, cells

↩ Section 1.6) (**Figure 2.21**) and *Clostridium* and related spore-forming bacteria, such as the antibiotic-producing *Streptomyces*. Also included here are the lactic acid bacteria, common inhabitants of decaying plant material and dairy products that include organisms such as *Streptococcus* (Figure 2.21*b*) and *Lactobacillus*. Other interesting bacteria that fall within the gram-positive bacteria are the mycoplasmas. These bacteria lack a cell wall and have very small genomes, and many of them are pathogenic. *Mycoplasma* is a major genus of pathogenic bacteria in this medically important group. Cells of some *Archaea*, such as *Thermoplasma* (see Figure 2.31) and *Ferroplasma*, also lack cell walls.

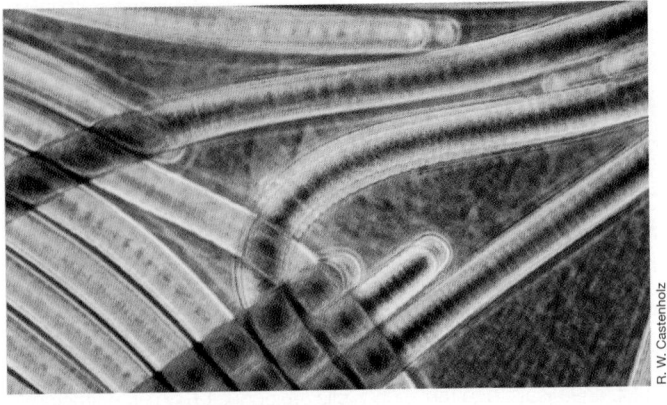

(a)

R. W. Castenholz

(b)

R. W. Castenholz

Figure 2.22 Filamentous cyanobacteria. *(a) Oscillatoria, (b) Spirulina.* Cells of both organisms are about 10 μm wide. Cyanobacteria are oxygenic phototrophs.

that live *inside* the cells of higher organisms, in this case, human cells. Several other pathogenic bacteria (for example, *Rickettsia,* described previously, and the gram-positive *Mycobacterium tuberculosis,* the cause of tuberculosis) are also intracellular pathogens. By living inside their host's cells, these pathogens avoid destruction by the host's immune response.

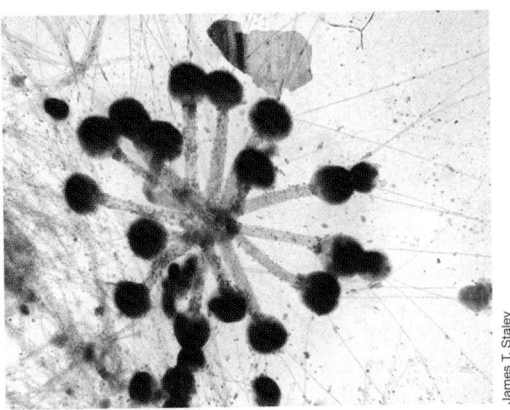

James T. Staley

Figure 2.23 The morphologically unusual stalked bacterium *Planctomyces.* Shown are several cells attached by their stalks to form a rosette. Cells are about 1.4 μm wide.

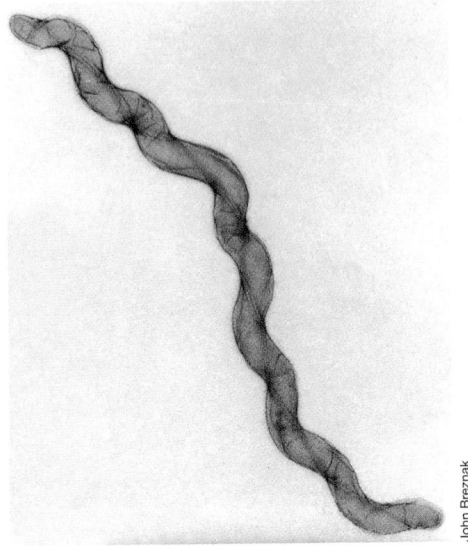

John Breznak

Figure 2.24 Spirochetes. Scanning electron micrograph of a cell of *Spirochaeta zuelzerae.* The cell is about 0.3 μm wide and tightly coiled.

The phylum *Deinococcus-Thermus* contains species with unusual cell walls and an innate resistance to high levels of radiation; *Deinococcus radiodurans* (**Figure 2.26**) is a major species in this group. This organism can survive doses of radiation many times greater than that sufficient to kill humans and can actually reassemble its chromosome after it has been shattered by intense radiation. We learn more about this amazing organism in Section 18.17.

Finally, several phyla branch off early in the phylogenetic tree of *Bacteria* (Figure 2.19). Although phylogenetically distinct, these groups are unified by their ability to grow at very high temperatures (*hyperthermophily,* Table 2.1). Organisms

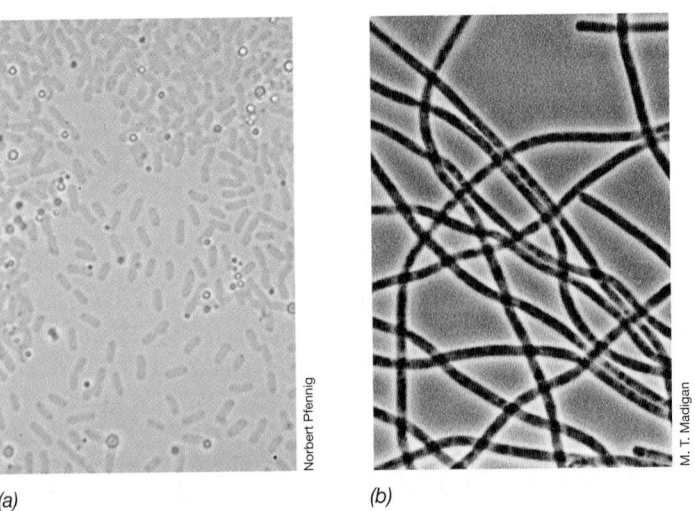

(a)   Norbert Pfennig

(b)   M. T. Madigan

Figure 2.25 Phototrophic green bacteria. *(a) Chlorobium* (green sulfur bacteria). A single cell is about 0.8 μm wide. *(b) Chloroflexus* (green nonsulfur bacteria). A filament is about 1.3 μm wide. Despite sharing many features such as pigments and photosynthetic membrane structures, these two genera are phylogenetically distinct (Figure 2.19).

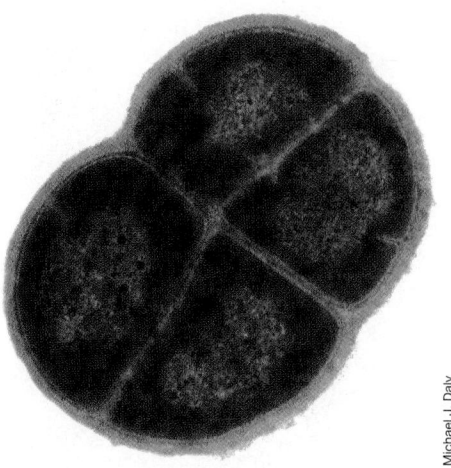

Figure 2.26 The highly radiation-resistant bacterium *Deinococcus radiodurans.* Cells of *D. radiodurans* divide in two planes to yield clusters of cells. A single cell is about 2.5 μm wide.

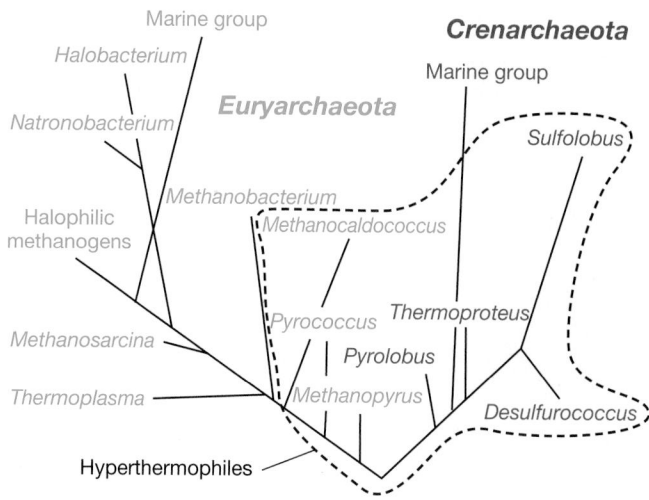

Figure 2.28 **Phylogenetic tree of some representative *Archaea.*** The organisms circled are hyperthermophiles, which grow at very high temperatures. The two major phyla are the *Crenarchaeota* and the *Euryarchaeota.* The "marine group" sequences are environmental rRNA sequences from marine *Archaea*, most of which have not been cultured.

such as *Aquifex* (**Figure 2.27**) and *Thermotoga* grow in hot springs that are near the boiling point. The early branching of these phyla on the phylogenetic tree (Figure 2.19) is consistent with the widely accepted hypothesis that the early Earth was much hotter than it is today. Assuming that early life forms were hyperthermophiles, it is not surprising that their closest living relatives today would also be hyperthermophiles. Interestingly, the phylogenetic trees of both *Bacteria* and *Archaea* are in agreement here; hyperthermophiles such as *Aquifex, Methanopyrus,* and *Pyrolobus* lie near the root of their respective phylogenetic trees.

MiniQuiz

- What is the largest phylum of *Bacteria*?
- In which phylum of *Bacteria* does the Gram stain reaction predict phylogeny?
- Why can it be said that the cyanobacteria prepared Earth for the evolution of higher life forms?
- What is physiologically unique about *Deinococcus*?

2.10 *Archaea*

Two phyla exist in the domain *Archaea,* the *Euryarchaeota* and the *Crenarchaeota* (**Figure 2.28**). Each of these forms a major branch on the archaeal tree. Most cultured *Archaea* are extremophiles, with species capable of growth at the highest temperatures, salinities, and extremes of pH known for any microorganism. The organism *Pyrolobus* (**Figure 2.29**), for example, is a hyperthermophile capable of growth at up to 113°C, and the methanogen *Methanopyrus* can grow up to 122°C (Table 2.1).

Although all *Archaea* are chemotrophic, *Halobacterium* can use light to make ATP but in a way quite distinct from that of phototrophic organisms (see later discussion). Some *Archaea* use

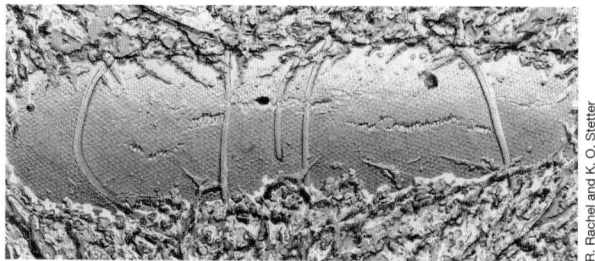

Figure 2.27 The hyperthermophile *Aquifex.* This hot spring organism uses H$_2$ as its energy source and can grow in temperatures up to 95°C. Transmission electron micrograph using a technique called freeze-etching, where a frozen replica of the cell is made and then visualized. The cell is about 0.5 μm wide.

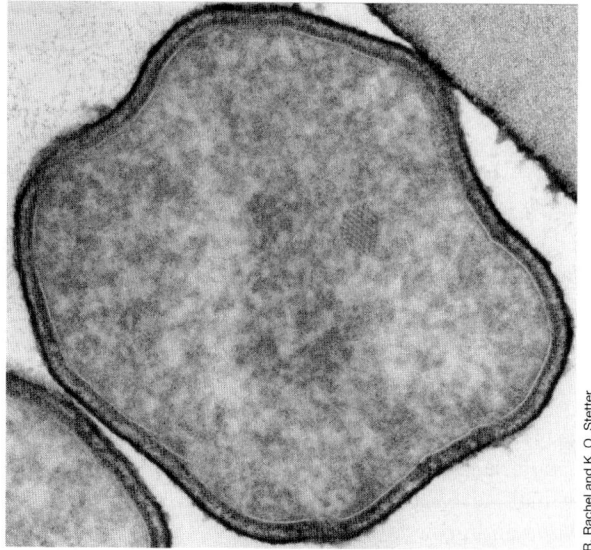

Figure 2.29 *Pyrolobus.* This hyperthermophile grows optimally above the boiling point of water. The cell is 1.4 μm wide.

Figure 2.30 Extremely halophilic Archaea. A vial of brine with precipitated salt crystals contains cells of the extreme halophile, *Halobacterium*. The organism contains red and purple pigments that absorb light and lead to ATP production. Cells of *Halobacterium* can also live within salt crystals themselves.

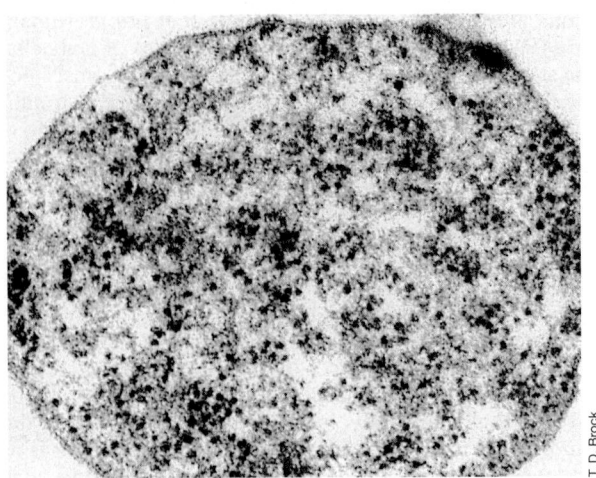

Figure 2.31 Extremely acidophilic Archaea. The organism *Thermoplasma* lacks a cell wall. The cell measures 1 μm wide.

organic compounds in their energy metabolism, while many others are chemolithotrophs, with hydrogen gas (H_2) being a widely used inorganic substance. Chemolithotrophic metabolisms are particularly widespread among hyperthermophilic *Archaea*.

Euryarchaeota

The *Euryarchaeota* branch on the tree of *Archaea* (Figure 2.28) contains four groups of organisms, the methanogens, the extreme halophiles, the thermoacidophiles, and some hyperthermophiles. Some of these require O_2 whereas others are actually killed by it, and some grow at the upper or lower extremes of pH (Table 2.1). For example, methanogens such as *Methanobacterium* are strict anaerobes and cannot tolerate even very low levels of O_2. The metabolism of methanogens is unique in that energy is conserved during the production of methane (natural gas). Methanogens are important organisms in the anaerobic degradation of organic matter in nature, and most of the natural gas found on Earth is a result of their metabolism.

The extreme halophiles are relatives of the methanogens (Figure 2.28), but are physiologically distinct from them. Unlike methanogens, which are killed by oxygen, most extreme halophiles require oxygen, and all are unified by their requirement for very large amounts of salt (NaCl) for metabolism and reproduction. It is for this reason that these organisms are called *halophiles* (salt lovers). In fact, organisms like *Halobacterium* are so salt loving that they can actually grow on and within salt crystals (**Figure 2.30**).

As we have seen, many prokaryotes are phototrophic and can generate adenosine triphosphate (ATP) using light energy (Section 2.8). Although *Halobacterium* species do not produce chlorophyll, they do synthesize a light-activated pigment that can trigger ATP synthesis (↩ Section 19.2). Extremely halophilic *Archaea* inhabit salt lakes, salterns (salt evaporation ponds), and other very salty environments. Some extreme halophiles, such as *Natronobacterium*, inhabit soda lakes, environments characterized by high levels of salt and high pH. Such organisms are *alkaliphilic* and grow at the highest pH of all known organisms (Table 2.1).

The third group of *Euryarchaeota* are the thermoacidophiles, organisms that grow best at high temperatures plus acidic pH.

These include *Thermoplasma* (**Figure 2.31**), an organism that like *Mycoplasma* (Section 2.9) lacks a cell wall. *Thermoplasma* grows best at 60–70°C and pH 2. The thermoacidophiles also include *Picrophilus*, the most acidophilic (acid-loving) of all known prokaryotes (Table 2.1).

The final group of *Euryarchaeota* consists of hyperthermophilic species, organisms whose growth temperature optimum lies above 80°C. These organisms show a variety of physiologies including methanogenesis (*Methanopyrus*), sulfate reduction (*Archaeoglobus*), iron oxidation (*Ferroglobus*) and sulfur reduction (*Pyrococcus*). Most of these organisms obtain their cell carbon from CO_2 and are thus autotrophs.

Crenarchaeota

The vast majority of cultured *Crenarchaeota* are hyperthermophiles (Figure 2.29). These organisms are either chemolithotrophs or chemoorganotrophs and grow in hot environments such as hot springs and hydrothermal vents (ocean floor hot springs). For the most part cultured *Crenarchaeota* are anaerobes (because of the high temperature, their habitats are typically anoxic), and many of them use H_2 present in their habitats as an energy source.

Some *Crenarchaeota* inhabit environments that contrast dramatically with thermal environments. For example, many of the prokaryotes suspended in the open oceans are *Crenarchaeota*, in an environment that is fully oxic and cold (~3°C). Some marine *Crenarchaeota* are chemolithotrophs that use ammonia (NH_3) as their energy source, but we know little about the metabolic activities of most marine *Archaea*. *Crenarchaeota* have also been detected in soil and freshwaters and are thus widely distributed in nature.

MiniQuiz

• What are the major phyla of *Archaea*?

• What is unusual about the genus *Halobacterium*? What group of *Archaea* is responsible for producing natural gas?

2.11 Phylogenetic Analyses of Natural Microbial Communities

Although thus far we have cultured only a small fraction of the *Archaea* and *Bacteria* that exist in nature, we still know a lot about their diversity, which is extensive. This is because it is possible to do phylogenetic analyses on rRNA genes obtained from cells in a natural sample without first having to culture the organisms that contained them. If a sample of soil or water contains rRNA, it is because organisms that made that rRNA are present in the sample. Thus, if we isolate all of the different rRNA genes from a natural sample, a relatively easy task, we can use the techniques described in Figure 2.16 to place them on the phylogenetic tree. Conceptually, this is equivalent to isolating pure cultures of every organism in the sample (a task that is currently not possible) and then extracting and analyzing their rRNA genes. These powerful techniques of molecular microbial community analysis bypass the culturing step—often the bottleneck in microbial diversity studies—and instead focus on the rRNA genes themselves.

From studies carried out using molecular community analysis it has become clear that microbial diversity far exceeds that which laboratory cultures have revealed. For example, a sampling of virtually any habitat will show that the vast majority of microorganisms present there have never been obtained in laboratory cultures. The phylogeny of these uncultured organisms, known as they are only from environmental rRNA gene sequences (phylotypes), is depicted in phylogenetic trees as lineages identified by letters or numbers (Figure 2.19, and lumped together in Figure 2.28 as "marine groups") instead of actual genus and species names.

In addition to sending the clear message that the breadth of microbial diversity is staggering, molecular microbial community analyses have stimulated innovative new culturing techniques to grow the "uncultured majority" of prokaryotes that we know exist. Moreover, full genomic analyses of uncultured *Archaea* and *Bacteria* (environmental genomics, ↻ Section 22.7) are also possible. Using environmental genomics to display the full complement of genes in uncultured organisms often reveals important secrets about their metabolic capacities that point to ways to bring them into laboratory culture.

MiniQuiz
- How can we know the microbial diversity of a natural habitat without first isolating and growing the organisms it contains?

2.12 Microbial *Eukarya*

Eukaryotic microorganisms are related by cell structure and phylogenetic history. The phylogeny of *Eukarya* based on ribosomal RNA sequencing (**Figure 2.32**) shows plants and animals to be farthest out on the branches of the tree; such late-branching groups are said to be the "most derived." By contrast, some of the earlier-branching *Eukarya* are structurally simple eukaryotes, lacking mitochondria and some other organelles. We will see in Chapter 20 that it has proven difficult to accurately track the phylogeny of eukaryotes using ribosomal RNA sequencing alone, so

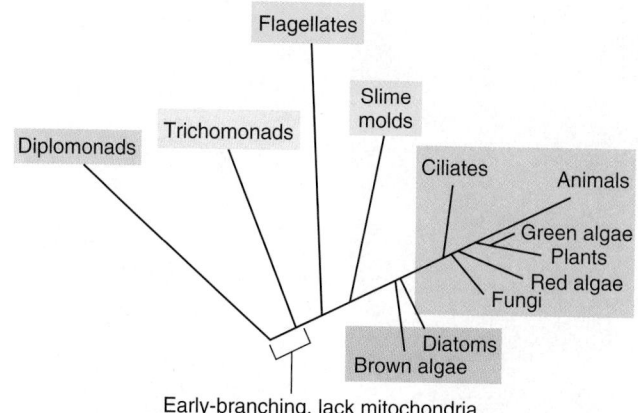

Figure 2.32 Phylogenetic tree of some representative *Eukarya*. This tree is based only on comparisons of genes encoding ribosomal RNA. Some early-branching species of *Eukarya* lack organelles other than the nucleus. Note that plants and animals branch near the apex of the tree. Not all known lineages of *Eukarya* are depicted.

other techniques have been used to supplement the general picture we present here.

Eukaryotic Microbial Diversity

The major groups are **protists** (algae and protozoa), fungi, and slime molds. Some protists, such as the algae (**Figure 2.33a**), are phototrophic. Algae contain chloroplasts and can live in environments containing only a few minerals (for example, K, P, Mg, N, S), water, CO_2, and light. Algae inhabit both soil and aquatic habitats and are major primary producers in nature. Fungi (Figure 2.33b) lack photosynthetic pigments and are either unicellular (yeasts) or filamentous (molds). Fungi are major agents of decomposition in nature and recycle much of the organic matter produced in soils and other ecosystems.

Cells of algae and fungi have cell walls, whereas the protozoa (Figure 2.33c) and slime molds do not. Protozoans are typically motile, and different species are widespread in nature in aquatic habitats or as pathogens of humans and other animals. Examples of protozoa are found throughout the phylogenetic tree of *Eukarya*. Some, like the flagellates, are fairly early-branching species, whereas others, like the ciliates such as *Paramecium* (Figure 2.33c), appear later on the phylogenetic tree (Figure 2.32).

The slime molds resemble protozoa in that they are motile and lack cell walls. However, slime molds differ from protozoa in both their phylogeny and by the fact that their cells undergo a complex life cycle. During the slime mold life cycle, motile cells aggregate to form a multicellular structure called a *fruiting body* from which spores are produced that yield new motile cells. Slime molds are the earliest branching organisms on the tree of *Eukarya* to show the cellular cooperation needed to form multicellular structures.

Lichens are leaflike structures often found growing on the surfaces of rocks and trees (**Figure 2.34**). Lichens are an example of a microbial mutualism, a partnership in which two organisms live together for mutual benefit. Lichens consist of a fungus and a phototrophic partner organism, either an alga (a eukaryote) or a

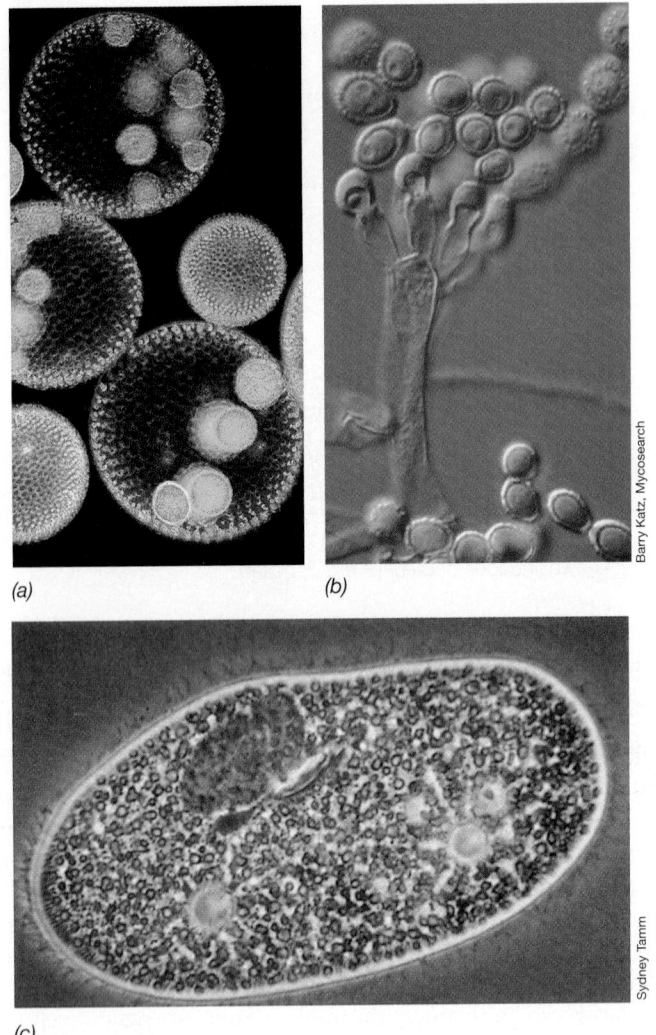

(a)   (b)

(c)

Figure 2.33 Microbial *Eukarya*. *(a)* Algae; dark-field photomicrograph of the colonial green alga *Volvox*. Each spherical cell contains several chloroplasts, the photosynthetic organelle of phototrophic eukaryotes. *(b)* Fungi; interference-contrast photomicrograph of spores of a typical mold. Each spore can give rise to a new filamentous fungus. *(c)* Protozoa; phase-contrast photomicrograph of the ciliated protozoan *Paramecium*. Cilia function like oars in a boat, conferring motility on the cell.

(a)

(b)

Figure 2.34 Lichens. *(a)* An orange-pigmented lichen growing on a rock, and *(b)* a yellow-pigmented lichen growing on a dead tree stump, Yellowstone National Park, USA. The color of the lichen comes from the pigmented (algal) component. Besides chlorophyll(s), lichen algae contain carotenoid pigments, which can be yellow, orange, brown, red, green, or purple.

cyanobacterium (a prokaryote). The phototrophic component is the primary producer while the fungus provides an anchor for the entire structure, protection from the elements, and a means of absorbing nutrients. Lichens have thus evolved a successful strategy of mutualistic interaction between two quite different microorganisms.

Postscript

Our tour of microbial diversity here is only an overview. The story expands in Chapters 16–21. In addition, the viruses were excluded because they are not cells. Nevertheless, viruses show enormous genetic diversity, and cells in all domains of life have viral parasites. So we devote some of Chapter 9 and all of Chapter 21 to this important topic.

We proceed now from our brief tour of microbial diversity to study some of the key remaining principles of microbiology: cell structure and function (Chapter 3), metabolism (Chapter 4), growth (Chapter 5), molecular biology (Chapters 6–8), and genetics and genomics (Chapters 9–12). Once we have mastered these important basics, we will be better prepared to revisit microbial diversity and many other aspects of microbiology in a more thorough way.

MiniQuiz
- List at least two ways algae differ from cyanobacteria.
- List at least two ways algae differ from protozoa.
- How do each of the components of a lichen benefit each other?

Big Ideas

2.1

Microscopes are essential for studying microorganisms. Bright-field microscopy, the most common form of microscopy, employs a microscope with a series of lenses to magnify and resolve the image.

2.2

An inherent limitation of bright-field microscopy is the lack of contrast between cells and their surroundings. This problem can be overcome by the use of stains or by alternative forms of light microscopy, such as phase contrast or dark field.

2.3

Differential interference contrast microscopy and confocal scanning laser microscopy allow enhanced three-dimensional imaging or imaging through thick specimens. The atomic force microscope gives a very detailed three-dimensional image of live preparations.

2.4

Electron microscopes have far greater resolving power than do light microscopes, the limits of resolution being about 0.2 nm. The two major forms of electron microscopy are transmission, used primarily to observe internal cell structure, and scanning, used to examine the surface of specimens.

2.5

All microbial cells share certain basic structures, such as their cytoplasmic membrane and ribosomes; most bacterial cells have a cell wall. Two structural patterns of cells are recognized: the prokaryote and the eukaryote. Viruses are not cells and depend on cells for their replication.

2.6

Genes govern the properties of cells, and a cell's complement of genes is called its genome. DNA is arranged in cells as chromosomes. Most prokaryotic species have a single circular chromosome; eukaryotic species have multiple chromosomes containing DNA arranged in linear fashion.

2.7

Comparative rRNA gene sequencing has defined three domains of life: *Bacteria*, *Archaea*, and *Eukarya*. Molecular sequence comparisons have shown that the organelles of *Eukarya* were originally *Bacteria* and have spawned new tools for microbial ecology and clinical microbiology.

2.8

All cells need sources of carbon and energy for growth. Chemoorganotrophs, chemolithotrophs, and phototrophs use organic chemicals, inorganic chemicals, or light, respectively, as their source of energy. Autotrophs use CO_2 as their carbon source, while heterotrophs use organic compounds. Extremophiles thrive under environmental conditions of high pressure or salt, or extremes of temperature or pH.

2.9

Several phyla of *Bacteria* are known, and an enormous diversity of cell morphologies and physiologies are represented. *Proteobacteria* are the largest group of *Bacteria* and contain many well-known bacteria, including *Escherichia coli*. Other major phyla include gram-positive bacteria, cyanobacteria, spirochetes, and green bacteria.

2.10

Two major phyla of *Archaea* are known, the *Euryarchaeota* and the *Crenarchaeota*, and most cultured representatives are extremophiles.

2.11

Retrieval and analysis of rRNA genes (phylotypes) from cells in natural samples have shown that many phylogenetically distinct *Bacteria* and *Archaea* exist in nature but remain to be cultured.

2.12

Microbial eukaryotes are a diverse group that includes algae and protozoa (protists), fungi, and slime molds. Some algae and fungi have developed mutualistic associations called lichens.

Review of Key Terms

Archaea one of two known domains of prokaryotes; compare with *Bacteria*

Autotroph an organism able to grow with carbon dioxide (CO_2) as its sole carbon source

Bacteria one of two known domains of prokaryotes; compare with *Archaea*

Cell wall a rigid layer present outside the cytoplasmic membrane; confers structural strength to the cell and protection from osmotic lysis

Chemolithotroph an organism that obtains its energy from the oxidation of inorganic compounds

Chemoorganotroph an organism that obtains its energy from the oxidation of organic compounds

Chromosome a genetic element containing genes essential to cell function

Cyanobacteria prokaryotic oxygenic phototrophs

Cytoplasm the aqueous internal portion of a cell, bounded by the cytoplasmic membrane

Cytoplasmic membrane the cell's permeability barrier to the environment; encloses the cytoplasm

Domain the highest level of biological classification

Endosymbiosis the theory that mitochondria and chloroplasts originated from *Bacteria*

Eukarya the domain of life that includes all eukaryotic cells

Eukaryote a cell having a membrane-enclosed nucleus and usually other membrane-enclosed organelles

Evolution change in a line of descent over time leading to new species or varieties within a species

Extremophile an organism that grows optimally under one or more environmental extremes

Gram stain a differential staining technique in which bacterial cells stain either pink (gram-negative) or purple (gram-positive) depending upon their structural makeup

Heterotroph an organism that requires organic carbon as its carbon source

Nucleoid the aggregated mass of DNA that constitutes the chromosome of cells of *Bacteria* and *Archaea*

Nucleus a membrane-enclosed structure that contains the chromosomes in eukaryotic cells

Organelle a membrane-enclosed structure, such as a mitochondrion or chloroplast, present in the cytoplasm of eukaryotic cells

Phototroph an organism that obtains its energy from light

Phylogeny the evolutionary relationships between organisms

Plasmid an extrachromosomal genetic element nonessential for growth

Prokaryote a cell that lacks a membrane-enclosed nucleus and other organelles

Proteobacteria a large phylum of *Bacteria* that includes many of the common gram-negative bacteria, such as *Escherichia coli*

Protists algae and protozoa

Resolution in microbiology, the ability to distinguish two objects as distinct and separate under the microscope

Ribosome a cytoplasmic particle that functions in protein synthesis

Virus a genetic element that contains either a DNA or an RNA genome, has an extracellular form (the virion), and depends on a host cell for replication

Review Questions

1. What are basic dyes? Why are they useful (Sections 2.1 and 2.2)?

2. What is the main difference between a phase-contrast microscope and a light microscope (Section 2.2)?

3. What is the major advantage of electron microscopes over light microscopes? What type of electron microscope would be used to view the three-dimensional features of a cell (Section 2.4)?

4. Which domains of life have a prokaryotic cell structure? Is prokaryotic cell structure a predictor of phylogenetic status (Section 2.5)?

5. What are the main characteristics that differentiate a eukaryote from a prokaryote (Section 2.5)?

6. How do viruses resemble cells? How do they differ from cells (Section 2.5)?

7. How do you determine the evolutionary relationships between organisms (Section 2.7)?

8. How many genes does an organism such as *Escherichia coli* have? How does this compare with the number of genes in one of your cells (Section 2.6)?

9. What is meant by the word endosymbiosis (Section 2.7)?

10. How would you explain the fact that many proteins of *Archaea* resemble their counterparts in eukaryotes more closely than those of *Bacteria* (Section 2.7)?

11. From the standpoint of energy metabolism, how do chemoorganotrophs differ from chemolithotrophs? What carbon sources do members of each group use? Are they heterotrophs or autotrophs (Section 2.8)?

12. What domain contains the phylum *Proteobacteria*? What is notable about the *Proteobacteria* (Section 2.9)?

13. What is unusual about the organism *Pyrolobus* (Sections 2.8 and 2.10)?

14. What similarities and differences exist between the following three organisms: *Pyrolobus, Halobacterium,* and *Thermoplasma* (Section 2.10)?

15. How have rRNA sequencing studies improved our understanding of microbial diversity (Section 2.11)?

16. What are the major similarities and differences between protists, fungi, and the slime molds (Section 2.12)?

Application Questions

1. Calculate the size of the smallest resolvable object if 600-nm light is used to observe a specimen with a 100× oil-immersion lens having a numerical aperture of 1.32. How could resolution be improved using this same lens?

2. Explain why a bacterium containing a plasmid can typically be "cured" of the plasmid (that is, the plasmid can be permanently removed) with no ill effects, whereas removal of the chromosome would be lethal.

3. It has been said that knowledge of the evolution of macroorganisms greatly preceded that of microorganisms. Why do you think that reconstruction of the evolutionary lineage of horses, for example, might have been an easier task than doing the same for any group of prokaryotes?

4. Examine the phylogenetic tree shown in Figure 2.16. Using the sequence data shown, describe why the tree would be incorrect if its branches remained the same but the positions of organisms 2 and 3 on the tree were switched.

5. Explain why even though microbiologists have cultured a great diversity of microorganisms, they know that an even greater diversity exists, despite having never seen or grown them in the laboratory.

6. What data from this chapter could you use to convince your friend that extremophiles are not just organisms that were "hanging on" in their respective habitats?

7. Defend this statement: If cyanobacteria had never evolved, life on Earth would have remained strictly microbial.

Cell Structure and Function in *Bacteria* and *Archaea*

Bacteria are keenly attuned to their environment and respond by directing their movements toward or away from chemical and physical stimuli.

I Cell Shape and Size

In this chapter we examine key structures of the prokaryotic cell: the cytoplasmic membrane, the cell wall, cell surface structures and inclusions, and mechanisms of motility. Our overarching theme will be structure and function. We begin this chapter by considering two key features of prokaryotic cells—their shape and small size. Prokaryotes typically have defined shapes and are extremely small cells. Shape is useful for differentiating cells of the *Bacteria* and the *Archaea* and size has profound effects on their biology.

3.1 Cell Morphology

In microbiology, the term **morphology** means cell shape. Several morphologies are known among prokaryotes, and the most common ones are described by terms that are part of the essential lexicon of the microbiologist.

Major Cell Morphologies

Examples of bacterial morphologies are shown in **Figure 3.1**. A bacterium that is spherical or ovoid in morphology is called a *coccus* (plural, cocci). A bacterium with a cylindrical shape is called a *rod* or a *bacillus*. Some rods twist into spiral shapes and are called *spirilla*. The cells of many prokaryotic species remain together in groups or clusters after cell division, and the arrangements are often characteristic of certain genera. For instance, some cocci form long chains (for example, the bacterium *Streptococcus*), others occur in three-dimensional cubes (*Sarcina*), and still others in grapelike clusters (*Staphylococcus*).

Several groups of bacteria are immediately recognizable by the unusual shapes of their individual cells. Examples include spirochetes, which are tightly coiled bacteria; appendaged bacteria, which possess extensions of their cells as long tubes or stalks; and filamentous bacteria, which form long, thin cells or chains of cells (Figure 3.1).

The cell morphologies shown here should be viewed with the understanding that they are *representative* shapes; many variations of these key morphologies are known. For example, there are fat rods, thin rods, short rods, and long rods, a rod simply being a cell that is longer in one dimension than in the other. As we will see, there are even square bacteria and star-shaped bacteria! Cell morphologies thus form a continuum, with some shapes, such as rods, being very common and others more unusual.

Morphology and Biology

Although cell morphology is easily recognized, it is in general a poor predictor of other properties of a cell. For example, under the microscope many rod-shaped *Archaea* look identical to rod-shaped *Bacteria*, yet we know they are of different phylogenetic

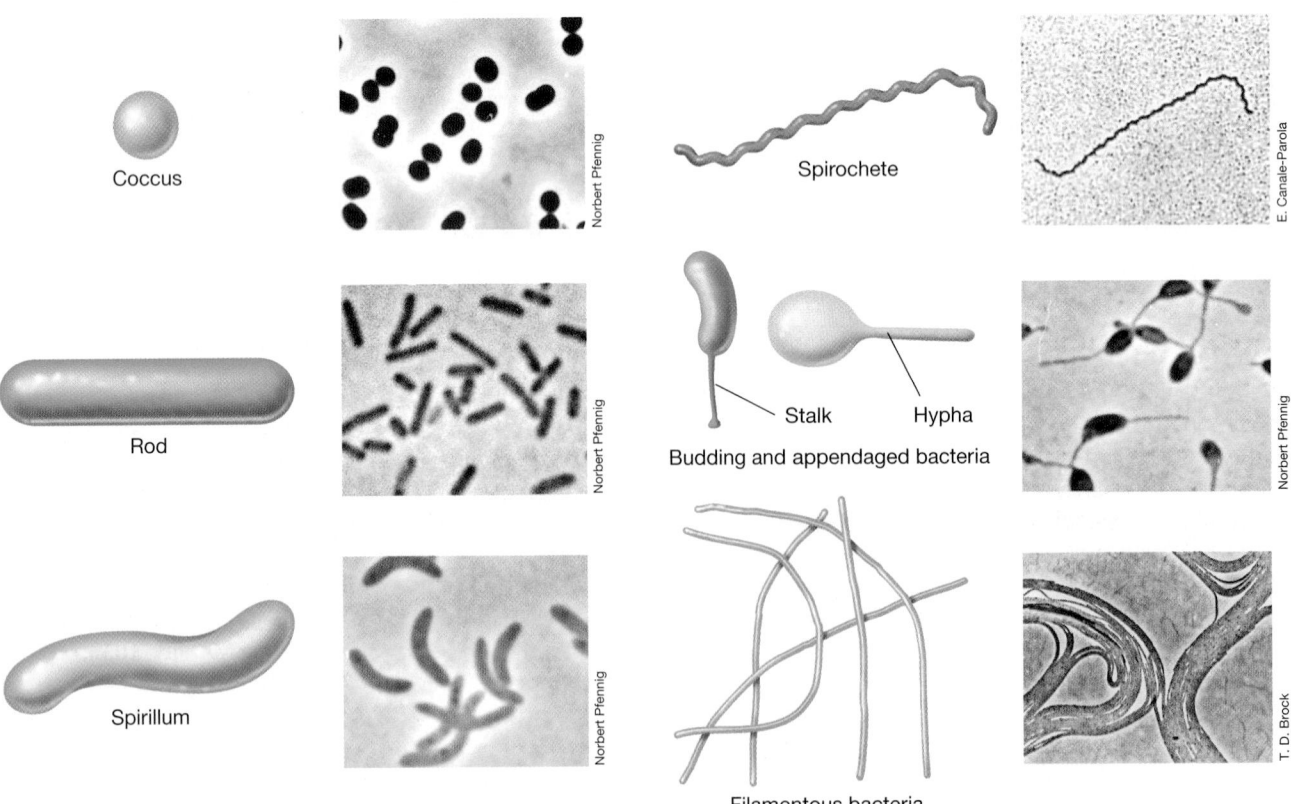

Figure 3.1 Representative cell morphologies of prokaryotes. Next to each drawing is a phase-contrast photomicrograph showing an example of that morphology. Organisms are coccus, *Thiocapsa roseopersicina* (diameter of a single cell = 1.5 μm); rod, *Desulfuromonas acetoxidans* (diameter = 1 μm); spirillum, *Rhodospirillum rubrum* (diameter = 1 μm); spirochete, *Spirochaeta stenostrepta* (diameter = 0.25 μm); budding and appendaged, *Rhodomicrobium vannielii* (diameter = 1.2 μm); filamentous, *Chloroflexus aurantiacus* (diameter = 0.8 μm).

domains (↩ Section 2.7). Thus, with very rare exceptions, it is impossible to predict the physiology, ecology, phylogeny, or virtually any other property of a prokaryotic cell, by simply knowing its morphology.

What sets the morphology of a particular species? Although we know something about *how* cell shape is controlled, we know little about *why* a particular cell evolved the morphology it has. Several selective forces are likely to be in play in setting the morphology of a given species. These include optimization for nutrient uptake (small cells and those with high surface-to-volume ratios), swimming motility in viscous environments or near surfaces (helical or spiral-shaped cells), gliding motility (filamentous bacteria), and so on. Thus morphology is not a trivial feature of a microbial cell. A cell's morphology is a genetically directed characteristic and has evolved to maximize fitness for the species in a particular habitat.

MiniQuiz

- How do cocci and rods differ in morphology?
- Is cell morphology a good predictor of other properties of the cell?

3.2 Cell Size and the Significance of Smallness

Prokaryotes vary in size from cells as small as about 0.2 μm in diameter to those more than 700 μm in diameter (**Table 3.1**). The vast majority of rod-shaped prokaryotes that have been cultured in the laboratory are between 0.5 and 4 μm wide and less than 15 μm long, but a few very large prokaryotes, such as *Epulopiscium fishelsoni*, are huge, with cells longer than 600 μm (0.6 millimeter) (**Figure 3.2**). This bacterium, phylogenetically related to the endospore-forming bacterium *Clostridium* and found in the gut of the surgeonfish, is interesting not only because it is so large, but also because it has an unusual form of cell division and contains multiple copies of its genome. Multiple offspring are formed and are then released from the *Epulopiscium* "mother cell." A mother cell of *Epulopiscium* contains several thousand genome copies, each of which is about the same size as the genome of *Escherichia coli* (4.6 million base pairs). The many copies are apparently necessary because the cell volume of *Epulopiscium* is so large (Table 3.1) that a single copy of its genome would not be sufficient to support the transcriptional and translational needs of the cell.

Cells of the largest known prokaryote, the sulfur chemolithotroph *Thiomargarita* (Figure 3.2*b*), can be 750 μm in diameter, nearly visible to the naked eye. Why these cells are so large is not well understood, although for sulfur bacteria a large cell size may be a mechanism for storing sulfur (an energy source). It is hypothesized that problems with nutrient uptake ultimately dictate the upper limits for the size of prokaryotic cells. Since the metabolic rate of a cell varies inversely with the square of its size, for very large cells nutrient uptake eventually limits metabolism to the point that the cell is no longer competitive with smaller cells.

Very large cells are not common in the prokaryotic world. In contrast to *Thiomargarita* or *Epulopiscium* (Figure 3.2), the

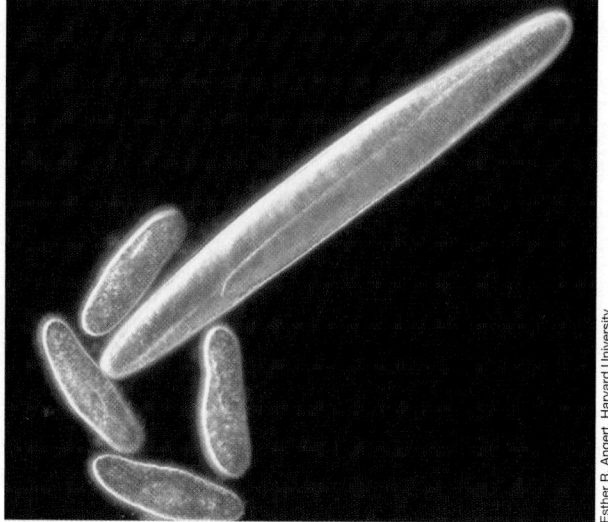

(a)

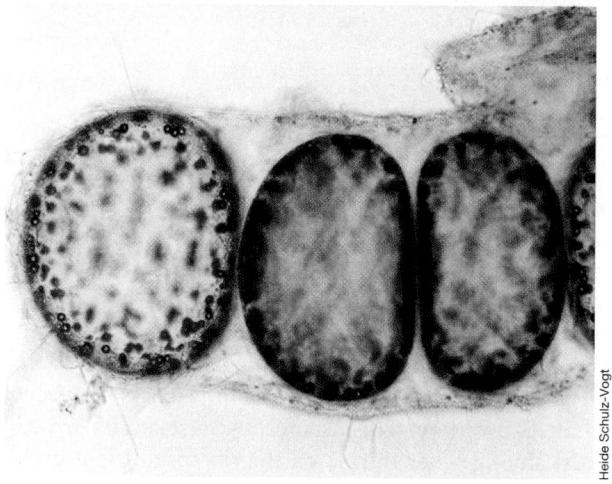

(b)

Figure 3.2 Some very large prokaryotes. *(a)* Dark-field photomicrograph of a giant prokaryote, *Epulopiscium fishelsoni*. The rod-shaped cell in this field is about 600 μm (0.6 mm) long and 75 μm wide and is shown with four cells of the protist (eukaryote) *Paramecium*, each of which is about 150 μm long. *E. fishelsoni* is a species of *Bacteria*, phylogenetically related to *Clostridium*. *(b) Thiomargarita namibiensis*, a large sulfur chemolithotroph (phylum *Proteobacteria* of the *Bacteria*) and currently the largest known prokaryote. Cell widths vary from 400 to 750 μm.

dimensions of an average rod-shaped prokaryote, the bacterium *E. coli*, for example, are about 1×2 μm; these dimensions are typical of most prokaryotes. For comparison, average eukaryotic cells can be 10 to more than 200 μm in diameter. In general, then, it can be said that prokaryotes are very small cells compared with eukaryotes.

Surface-to-Volume Ratios, Growth Rates, and Evolution

There are significant advantages to being small. Small cells have more surface area relative to cell volume than do large cells; that is, they have a higher *surface-to-volume ratio*. Consider a spherical coccus. The volume of such a cell is a function of the cube of

Table 3.1 *Cell size and volume of some prokaryotic cells, from the largest to the smallest*

Organism	Characteristics	Morphology	Size[a] (μm)	Cell volume (μm³)	E. coli volumes
Thiomargarita namibiensis	Sulfur chemolithotroph	Cocci in chains	750	200,000,000	100,000,000
Epulopiscium fishelsoni[a]	Chemoorganotroph	Rods with tapered ends	80 × 600	3,000,000	1,500,000
Beggiatoa species[a]	Sulfur chemolithotroph	Filaments	50 × 160	1,000,000	500,000
Achromatium oxaliferum	Sulfur chemolithotroph	Cocci	35 × 95	80,000	40,000
Lyngbya majuscula	Cyanobacterium	Filaments	8 × 80	40,000	20,000
Thiovulum majus	Sulfur chemolithotroph	Cocci	18	3,000	1500
Staphylothermus marinus[a]	Hyperthermophile	Cocci in irregular clusters	15	1,800	900
Magnetobacterium bavaricum	Magnetotactic bacterium	Rods	2 × 10	30	15
Escherichia coli	Chemoorganotroph	Rods	1 × 2	2	1
Pelagibacter ubique[a]	Marine chemoorganotroph	Rods	0.2 × 0.5	0.014	0.007
Mycoplasma pneumoniae	Pathogenic bacterium	Pleomorphic[b]	0.2	0.005	0.0025

[a]Where only one number is given, this is the diameter of spherical cells. The values given are for the largest cell size observed in each species. For example, for *T. namibiensis*, an average cell is only about 200 μm in diameter. But on occasion, giant cells of 750 μm are observed. Likewise, an average cell of *S. marinus* is about 1 μm in diameter. The species of *Beggiatoa* here is unclear and *E. fishelsoni* and *P. ubique* are not formally recognized names in taxonomy.
[b]*Mycoplasma* is a cell wall–less bacterium and can take on many shapes (*pleomorphic* means "many shapes").
Source: Data obtained from Schulz, H.N., and B.B. Jørgensen. 2001. *Ann. Rev. Microbiol.* 55: 105–137.

its radius ($V = \frac{4}{3}\pi r^3$), while its surface area is a function of the square of the radius ($S = 4\pi r^2$). Therefore, the S/V ratio of a spherical coccus is $3/r$ (**Figure 3.3**). As a cell increases in size, its S/V ratio decreases. To illustrate this, consider the S/V ratio for some of the cells of different sizes listed in Table 3.1: *Pelagibacter ubique*, 22; *E. coli*, 4.5; and *E. fishelsoni*, 0.05.

The S/V ratio of a cell affects several aspects of its biology, including its evolution. For instance, because a cell's growth rate depends, among other things, on the rate of nutrient exchange, the higher S/V ratio of smaller cells supports a faster rate of nutrient exchange per unit of cell volume compared with that of larger cells. Because of this, smaller cells, in general, grow faster

than larger cells, and a given amount of resources (the nutrients available to support growth) will support a larger population of small cells than of large cells. How can this affect evolution?

Each time a cell divides, its chromosome replicates. As DNA is replicated, occasional errors, called *mutations*, occur. Because mutation rates appear to be roughly the same in all cells, large or small, the more chromosome replications that occur, the greater the total number of mutations in the population. Mutations are the "raw material" of evolution; the larger the pool of mutations, the greater the evolutionary possibilities. Thus, because prokaryotic cells are quite small and are also genetically haploid (allowing mutations to be expressed immediately), they have, in general, the capacity for more rapid growth and evolution than larger, genetically diploid cells. In the latter, not only is the S/V ratio smaller but the effects of a mutation in one gene can be masked by a second, unmutated gene copy. These fundamental differences in size and genetics between prokaryotic and eukaryotic cells underlie the fact that prokaryotes can adapt quite rapidly to changing environmental conditions and can more easily exploit new habitats than can eukaryotic cells. We will see this concept in action in later chapters when we consider, for example, the enormous metabolic diversity of prokaryotes, or the spread of antibiotic resistance.

Lower Limits of Cell Size

From the foregoing discussion one might predict that smaller and smaller bacteria would have greater and greater selective advantages in nature. However, this is not true, as there are lower limits to cell size. If one considers the volume needed to house the essential components of a free-living cell—proteins, nucleic acids, ribosomes, and so on—a structure of 0.1 μm in diameter or less is simply insufficient to do the job, and structures 0.15 μm

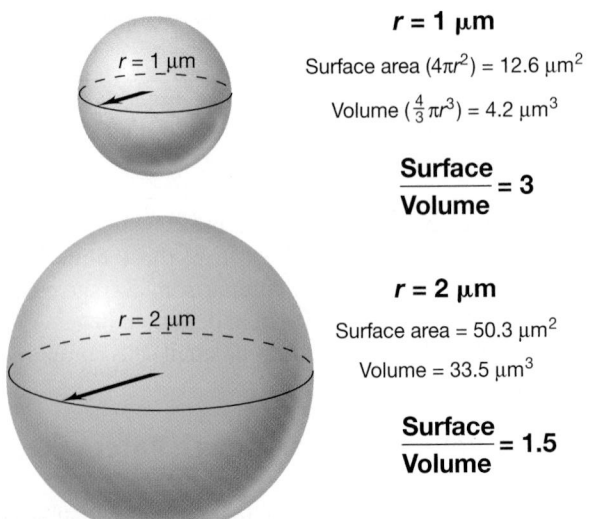

r = 1 μm

Surface area ($4\pi r^2$) = 12.6 μm²

Volume ($\frac{4}{3}\pi r^3$) = 4.2 μm³

$$\frac{\text{Surface}}{\text{Volume}} = 3$$

r = 2 μm

Surface area = 50.3 μm²

Volume = 33.5 μm³

$$\frac{\text{Surface}}{\text{Volume}} = 1.5$$

Figure 3.3 Surface area and volume relationships in cells. As a cell increases in size, its *S/V* ratio decreases.

in diameter are marginal. Thus, structures occasionally observed in nature of 0.1 μm or smaller that "look" like bacterial cells are almost certainly not so. Despite this, many very small prokaryotic cells are known and many have been grown in the laboratory. The open oceans, for example, contain 10^4–10^5 prokaryotic cells per milliliter, and these tend to be very small cells, 0.2–0.4 μm in diameter. We will see later that many pathogenic bacteria are also very small. When the genomes of these pathogens are examined, they are found to be highly streamlined and missing many genes whose functions are supplied to them by their hosts.

MiniQuiz

- What physical property of cells increases as cells become smaller?
- How can the small size and haploid genetics of prokaryotes accelerate their evolution?

The Cytoplasmic Membrane and Transport

We now consider the structure and function of a critical cell component, the cytoplasmic membrane. The cytoplasmic membrane plays many roles, chief among them as the "gatekeeper" for substances that enter and exit the cell.

3.3 The Cytoplasmic Membrane

The **cytoplasmic membrane** is a thin barrier that surrounds the cell and separates the cytoplasm from the cell's environment. If the membrane is broken, the integrity of the cell is destroyed, the cytoplasm leaks into the environment, and the cell dies. We will see that the cytoplasmic membrane confers little protection from osmotic lysis but is ideal as a selective permeability barrier.

Composition of Membranes

The general structure of the cytoplasmic membrane is a phospholipid bilayer. Phospholipids contain both hydrophobic (fatty acid) and hydrophilic (glycerol–phosphate) components and can be of many different chemical forms as a result of variation in the groups attached to the glycerol backbone (**Figure 3.4**) As phospholipids aggregate in an aqueous solution, they naturally form bilayer structures. In a phospholipid membrane, the fatty acids point inward toward each other to form a hydrophobic environment, and the hydrophilic portions remain exposed to the external environment or the cytoplasm (Figure 3.4*b*).

The cell's cytoplasmic membrane, which is 6–8 nanometers wide, can be seen with the electron microscope, where it appears as two dark-colored lines separated by a lighter area (Figure 3.4*c*). This *unit membrane*, as it is called (because each phospholipid leaf forms half of the "unit"), consists of a phospholipid bilayer with proteins embedded in it (**Figure 3.5**). Although in a diagram the cytoplasmic membrane may appear rather rigid, in reality it is somewhat fluid, having a consistency approximating that of a low-viscosity oil. Some freedom of movement of proteins within the membrane is possible, although it remains unclear exactly

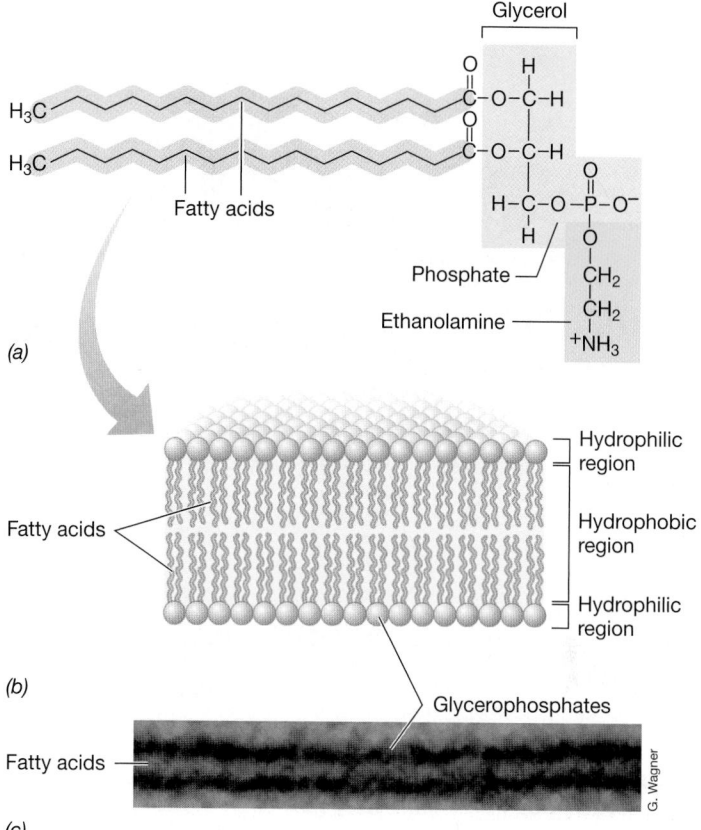

(a)

(b)

(c)

Figure 3.4 **Phospholipid bilayer membrane.** *(a)* Structure of the phospholipid phosphatidylethanolamine. *(b)* General architecture of a bilayer membrane; the blue balls depict glycerol with phosphate and (or) other hydrophilic groups. *(c)* Transmission electron micrograph of a membrane. The light inner area is the hydrophobic region of the model membrane shown in part b.

how extensive this is. The cytoplasmic membranes of some *Bacteria* are strengthened by molecules called *hopanoids*. These somewhat rigid planar molecules are structural analogs of sterols, compounds that strengthen the membranes of eukaryotic cells, many of which lack a cell wall.

Membrane Proteins

The major proteins of the cytoplasmic membrane have hydrophobic surfaces in their regions that span the membrane and hydrophilic surfaces in their regions that contact the environment and the cytoplasm (Figures 3.4 and 3.5). The *outer* surface of the cytoplasmic membrane faces the environment and in gram-negative bacteria interacts with a variety of proteins that bind substrates or process large molecules for transport into the cell (periplasmic proteins, see Section 3.7). The *inner* side of the cytoplasmic membrane faces the cytoplasm and interacts with proteins involved in energy-yielding reactions and other important cellular functions.

Many membrane proteins are firmly embedded in the membrane and are called *integral* membrane proteins. Other proteins have one portion anchored in the membrane and extramembrane regions that point into or out of the cell (Figure 3.5). Still

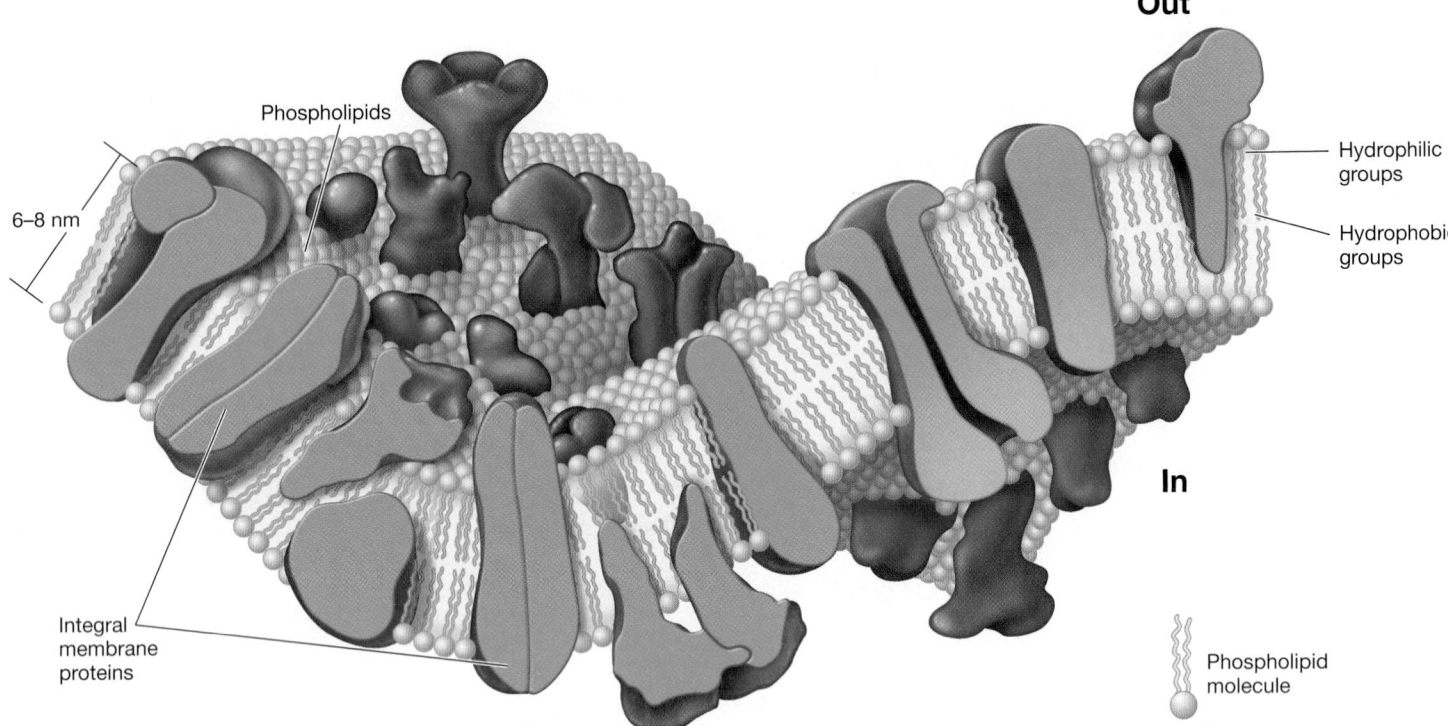

Figure 3.5 Structure of the cytoplasmic membrane. The inner surface (**In**) faces the cytoplasm and the outer surface (**Out**) faces the environment. Phospholipids compose the matrix of the cytoplasmic membrane with proteins embedded or surface associated. Although there are some chemical differences, the overall structure of the cytoplasmic membrane shown is similar in both prokaryotes and eukaryotes (but an exception to the bilayer design is shown in Figure 3.7e).

other proteins, called *peripheral* membrane proteins, are not membrane-embedded but nevertheless remain firmly associated with membrane surfaces. Some of these peripheral membrane proteins are lipoproteins, molecules that contain a lipid tail that anchors the protein into the membrane. Peripheral membrane proteins typically interact with integral membrane proteins in important cellular processes such as energy metabolism and transport.

Proteins in the cytoplasmic membrane are arranged in clusters (Figure 3.5), a strategy that allows proteins that need to interact to be adjacent to one another. The overall protein content of the membrane is quite high, and it is thought that the variation in lipid bilayer thickness (6–8 nm) is necessary to accommodate thicker and thinner patches of membrane proteins.

Archaeal Membranes

In contrast to the lipids of *Bacteria* and *Eukarya* in which *ester* linkages bond the fatty acids to glycerol, the lipids of *Archaea* contain *ether* bonds between glycerol and their hydrophobic side chains (**Figure 3.6**). Archaeal lipids lack true fatty acid side chains and instead, the side chains are composed of repeating units of the hydrophobic five-carbon hydrocarbon isoprene (Figure 3.6c).

The cytoplasmic membrane of *Archaea* can be constructed of either glycerol diethers (**Figure 3.7a**), which have 20-carbon side chains (the 20-C unit is called a *phytanyl* group), or diglycerol tetraethers (Figure 3.7b), which have 40-carbon side chains. In the tetraether lipid, the ends of the phytanyl side chains that

point inward from each glycerol molecule are covalently linked. This forms a lipid *monolayer* instead of a lipid *bilayer* membrane (Figure 3.7d, e). In contrast to lipid bilayers, lipid monolayer membranes are extremely resistant to heat denaturation and are therefore widely distributed in hyperthermophiles, prokaryotes that grow best at temperatures above 80°C. Membranes with a mixture of bilayer and monolayer character are also possible, with some of the inwardly opposing hydrophobic groups covalently bonded while others are not.

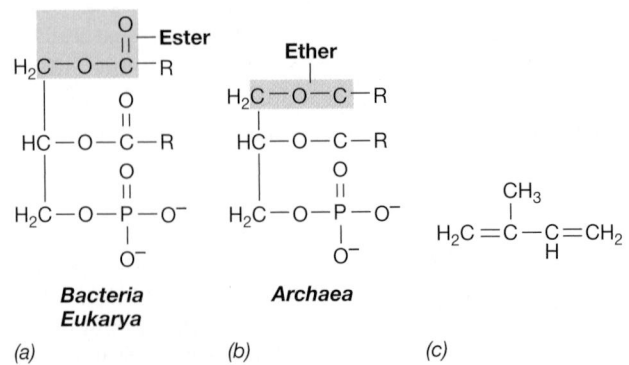

Figure 3.6 General structure of lipids. (a) The ester linkage and (b) the ether linkage. (c) Isoprene, the parent structure of the hydrophobic side chains of archaeal lipids. By contrast, in lipids of *Bacteria* and *Eukarya,* the side chains are composed of fatty acids (see Figure 3.4a).

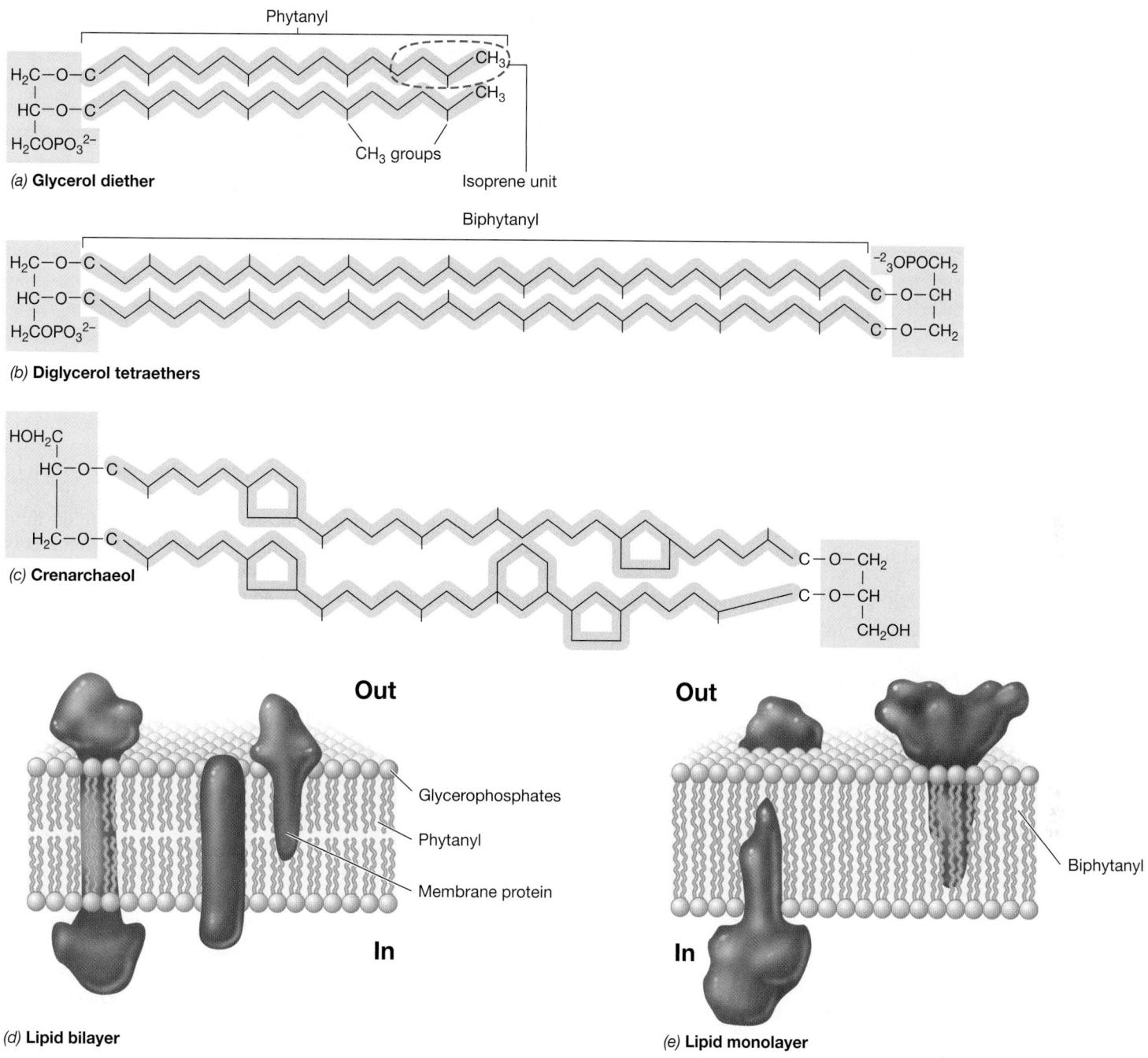

Figure 3.7 Major lipids of *Archaea* and the architecture of archaeal membranes. *(a, b)* Note that the hydrocarbon of the lipid is attached to the glycerol by an ether linkage in both cases. The hydrocarbon is phytanyl (C_{20}) in part a and biphytanyl (C_{40}) in part b. *(c)* A major lipid of *Crenarchaeota* is crenarchaeol, a lipid containing 5- and 6-carbon rings. *(d, e)* Membrane structure in *Archaea* may be bilayer or monolayer (or a mix of both).

Many archaeal lipids also contain rings within the hydrocarbon chains. For example, *crenarchaeol*, a lipid widespread among species of *Crenarchaeota* (Section 2.10), contains four cyclopentyl rings and one cyclohexyl ring (Figure 3.7c). The predominant membrane lipids of many *Euryarchaeota*, such as the methanogens and extreme halophiles, are glycolipids, lipids with a carbohydrate bonded to glycerol. Rings formed in the hydrocarbon side chains affect the properties of the lipids (and thus

overall membrane function), and considerable variation in the number and position of the rings has been discovered in the lipids of different species.

Despite the differences in chemistry between the cytoplasmic membranes of *Archaea* and organisms in the other domains, the fundamental construction of the archaeal cytoplasmic membrane—inner and outer hydrophilic surfaces and a hydrophobic interior—is the same as that of membranes in *Bacteria* and

Eukarya. Evolution has selected this design as the best solution to the main function of the cytoplasmic membrane—permeability—and we consider this problem now.

MiniQuiz

- Draw the basic structure of a lipid bilayer and label the hydrophilic and hydrophobic regions.
- How are the membrane lipids of *Bacteria* and *Archaea* similar, and how do they differ?

3.4 Functions of the Cytoplasmic Membrane

The cytoplasmic membrane is more than just a barrier separating the inside from the outside of the cell. The membrane plays critical roles in cell function. First and foremost, the membrane functions as a *permeability barrier*, preventing the passive leakage of solutes into or out of the cell (**Figure 3.8**). Secondly, the membrane is an *anchor* for many proteins. Some of these are enzymes that catalyze bioenergetic reactions and others transport solutes into and out of the cell. We will learn in the next chapter that the cytoplasmic membrane is also a major *site of energy conservation* in the cell. The membrane has an energetically charged form in which protons (H^+) are separated from hydroxyl ions (OH^-) across its surface (Figure 3.8). This charge separation is a form of energy, analogous to the potential energy present in a charged battery. This energy source, called the *proton motive force*, is responsible for driving many energy-requiring functions in the cell, including some forms of transport, motility, and biosynthesis of ATP.

The Cytoplasmic Membrane as a Permeability Barrier

The cytoplasm is a solution of salts, sugars, amino acids, nucleotides, and many other substances. The hydrophobic portion of the cytoplasmic membrane (Figure 3.5) is a tight barrier to diffusion of these substances. Although some small hydrophobic molecules pass the cytoplasmic membrane by diffusion, polar and charged molecules do not diffuse but instead must be transported. Even a substance as small as a proton (H^+) cannot diffuse across the membrane.

Table 3.2 *Comparative permeability of membranes to various molecules*

Substance	Rate of permeability[a]	Potential for diffusion into a cell
Water	100	Excellent
Glycerol	0.1	Good
Tryptophan	0.001	Fair/Poor
Glucose	0.001	Fair/Poor
Chloride ion (Cl^-)	0.000001	Very poor
Potassium ion (K^+)	0.0000001	Extremely poor
Sodium ion (Na^+)	0.00000001	Extremely poor

[a]Relative scale—permeability with respect to permeability to water given as 100. Permeability of the membrane to water may be affected by aquaporins (see text).

One substance that does freely pass the membrane in both directions is water, a molecule that is weakly polar but sufficiently small to pass between phospholipid molecules in the lipid bilayer (**Table 3.2**). But in addition, the movement of water across the membrane is accelerated by dedicated transport proteins called *aquaporins*. For example, aquaporin AqpZ of *Escherichia coli* imports or exports water depending on whether osmotic conditions in the cytoplasm are high or low, respectively. The relative permeability of the membrane to a few biologically relevant substances is shown in Table 3.2. As can be seen, most substances cannot diffuse into the cell and thus must be transported.

Transport Proteins

Transport proteins do more than just ferry substances across the membrane—they *accumulate* solutes against the concentration gradient. The necessity for carrier-mediated transport is easy to understand. If diffusion were the only mechanism by which solutes entered a cell, cells would never achieve the intracellular concentrations necessary to carry out biochemical reactions; that is, their rate of uptake and intracellular concentration would never exceed the external concentration, which in nature is often quite low (**Figure 3.9**). Hence, cells must have mechanisms for accumulating solutes—most of which are vital nutrients—to levels higher than those in their habitats, and this is the job of transport proteins.

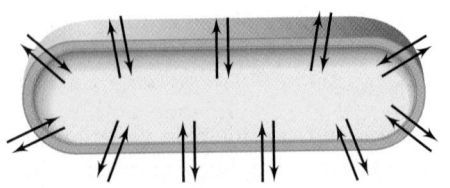

(a) **Permeability barrier:**
Prevents leakage and functions as a gateway for transport of nutrients into, and wastes out of, the cell

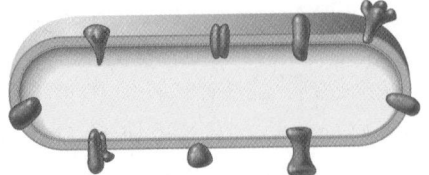

(b) **Protein anchor:**
Site of many proteins that participate in transport, bioenergetics, and chemotaxis

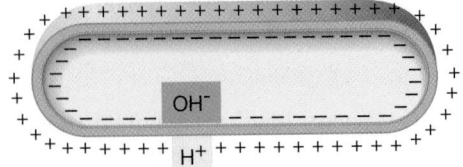

(c) **Energy conservation:**
Site of generation and use of the proton motive force

Figure 3.8 The major functions of the cytoplasmic membrane. Although structurally weak, the cytoplasmic membrane has many important cellular functions.

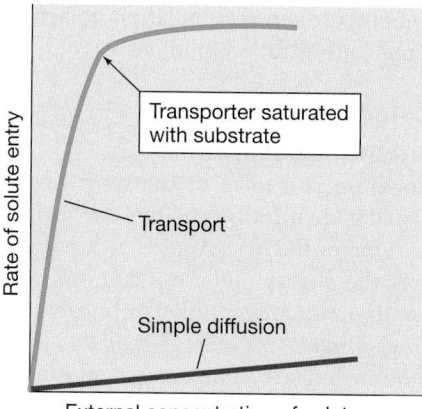

Figure 3.9 Transport versus diffusion. In transport, the uptake rate shows saturation at relatively low external concentrations.

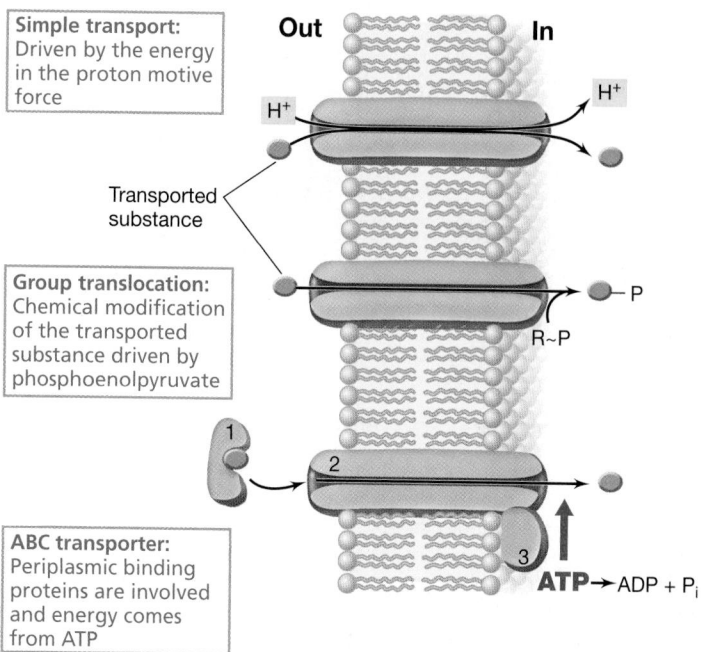

Figure 3.10 The three classes of transport systems. Note how simple transporters and the ABC system transport substances without chemical modification, whereas group translocation results in chemical modification (in this case phosphorylation) of the transported substance. The three proteins of the ABC system are labeled 1, 2, and 3.

Transport systems show several characteristic properties. First, in contrast with diffusion, transport systems show a *saturation effect*. If the concentration of substrate is high enough to saturate the transporter, which can occur at even the very low substrate concentrations found in nature, the rate of uptake becomes maximal and the addition of more substrate does not increase the rate (Figure 3.9). This characteristic feature of transport proteins is essential for a system that must concentrate nutrients from an often very dilute environment. A second characteristic of carrier-mediated transport is the *high specificity* of the transport event. Many carrier proteins react only with a single molecule, whereas a few show affinities for a closely related class of molecules, such as sugars or amino acids. This economy in uptake reduces the need for separate transport proteins for each different amino acid or sugar.

And finally, a third major characteristic of transport systems is that their biosynthesis is typically *highly regulated* by the cell. That is, the specific complement of transporters present in the cytoplasmic membrane of a cell at any one time is a function of both the resources available and their concentrations. Biosynthetic control of this type is important because a particular nutrient may need to be transported by one type of transporter when the nutrient is present at high concentration and by a different, higher-affinity transporter, when present at low concentration.

MiniQuiz

- List two reasons why a cell cannot depend on diffusion as a means of acquiring nutrients.
- Why is physical damage to the cytoplasmic membrane such a critical issue for the cell?

3.5 Transport and Transport Systems

Nutrient transport is a vital process. To fuel metabolism and support growth, cells need to import nutrients and export wastes on a continuous basis. To fulfill these requirements, several different mechanisms for transport exist in prokaryotes, each with its own unique features, and we explore this subject here.

Structure and Function of Membrane Transport Proteins

At least three transport systems exist in prokaryotes: simple transport, group translocation, and ABC transport. **Simple transport** consists only of a membrane-spanning transport protein, group translocation involves a series of proteins in the transport event, and the ABC system consists of three components: a substrate-binding protein, a membrane-integrated transporter, and an ATP-hydrolyzing protein (**Figure 3.10**). All transport systems require energy in some form, either from the proton motive force, or ATP, or some other energy-rich organic compound.

Figure 3.10 contrasts these transport systems. Regardless of the system, the membrane-spanning proteins typically show significant similarities in amino acid sequence, an indication of the common evolutionary roots of these structures. Membrane transporters are composed of 12 alpha helices that weave back and forth through the membrane to form a channel. It is through this channel that a solute is actually carried into the cell (**Figure 3.11**). The transport event requires that a conformational change occur in the membrane protein following binding of its solute. Like a gate swinging open, the conformational change then brings the solute into the cell.

Actual transport events can be of three types: uniport, symport, and antiport (Figure 3.11). *Uniporters* are proteins that transport a molecule unidirectionally across the membrane, either in or out. *Symporters* are cotransporters; they transport one molecule along with another substance, typically a proton. *Antiporters* are proteins that transport one molecule into the cell while simultaneously transporting a second molecule out of the cell.

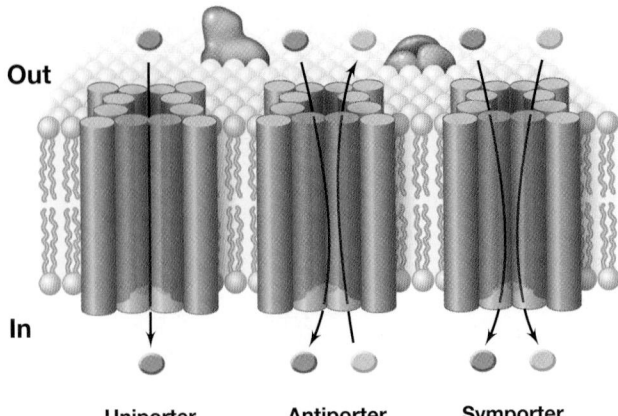

Figure 3.11 Structure of membrane-spanning transporters and types of transport events. Membrane-spanning transporters are made of 12 α-helices (each shown here as a cylinder) that aggregate to form a channel through the membrane. Shown here are three different transport events; for antiporters and symporters, the cotransported substance is shown in yellow.

Simple Transport: Lac Permease of *Escherichia coli*

The bacterium *Escherichia coli* metabolizes the disaccharide sugar lactose. Lactose is transported into cells of *E. coli* by the activity of a simple transporter, *lac permease*, a type of symporter. This is shown in **Figure 3.12**, where the activity of lac permease is compared with that of some other simple transporters, including uniporters and antiporters. We will see later that lac permease is one of three proteins required to metabolize lactose in *E. coli* and that the synthesis of these proteins is highly regulated by the cell (♂♀ Section 8.5).

As is true of all transport systems, the activity of lac permease is energy-driven. As each lactose molecule is transported into the cell, the energy in the proton motive force (Figure 3.8c) is diminished by the cotransport of protons into the cytoplasm. The membrane is reenergized through energy-yielding reactions that we will describe in Chapter 4. Thus the net result of lac permease

activity is the energy-driven accumulation of lactose in the cytoplasm against the concentration gradient.

Group Translocation: The Phosphotransferase System

Group translocation is a form of transport in which the substance transported is chemically modified during its uptake across the membrane. One of the best-studied group translocation systems transports the sugars glucose, mannose, and fructose in *E. coli*. These compounds are modified by phosphorylation during transport by the *phosphotransferase system*.

The phosphotransferase system consists of a family of proteins that work in concert; five proteins are necessary to transport any given sugar. Before the sugar is transported, the proteins in the phosphotransferase system are themselves alternately phosphorylated and dephosphorylated in a cascading fashion until the actual transporter, Enzyme II_c, phosphorylates the sugar during the transport event (**Figure 3.13**). A small protein called *HPr*, the enzyme that phosphorylates HPr (Enzyme I), and Enzyme II_a are all cytoplasmic proteins. By contrast, Enzyme II_b lies on the inner surface of the membrane and Enzyme II_c is an integral membrane protein. HPr and Enzyme I are nonspecific components of the phosphotransferase system and participate in the uptake of several different sugars. Several different versions of Enzyme II exist, one for each different sugar transported (Figure 3.13). Energy for the phosphotransferase system comes from the energy-rich compound phosphoenolpyruvate, which is a key intermediate in glycolysis, a major pathway for glucose metabolism present in most cells (♂♀ Section 4.8).

Periplasmic Binding Proteins and the ABC System

We will learn a bit later in this chapter that gram-negative bacteria contain a region called the *periplasm* that lies between the cytoplasmic membrane and a second membrane layer called the *outer membrane*, part of the gram-negative cell wall (Section 3.7). The periplasm contains many different proteins, several of which function in transport and are called *periplasmic binding proteins*.

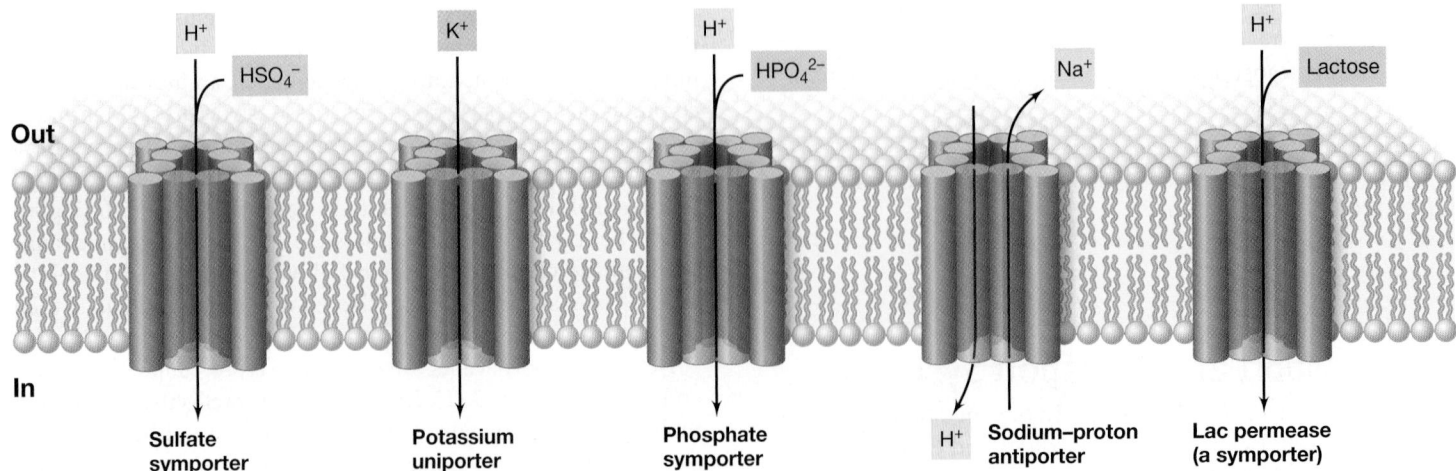

Figure 3.12 The lac permease of *Escherichia coli* and several other well-characterized simple transporters. Note the different classes of transport events depicted.

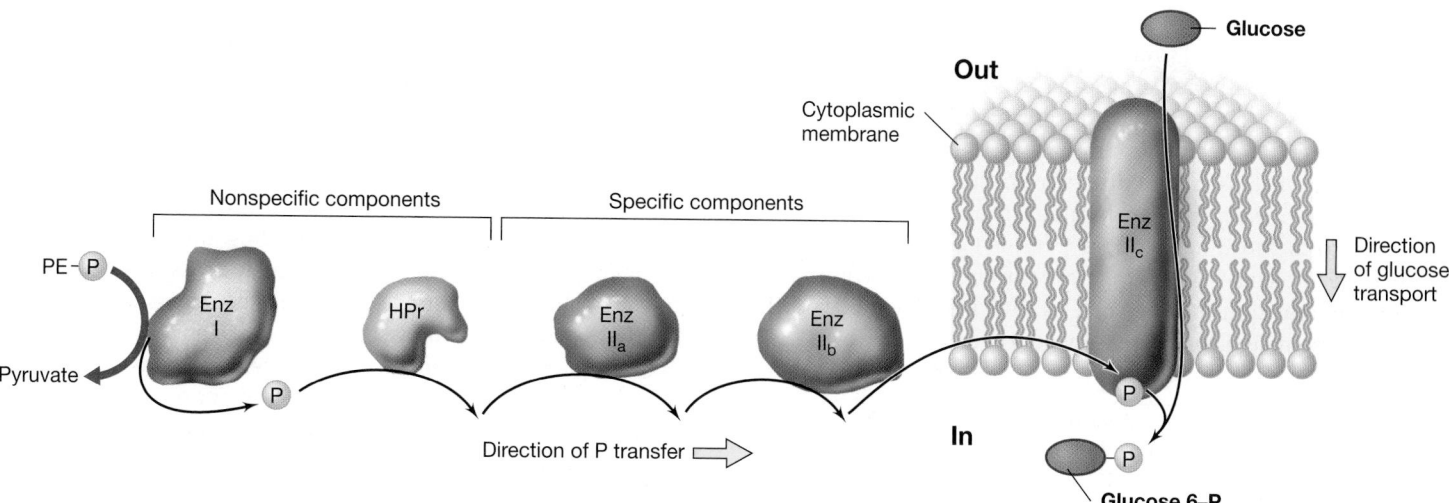

Figure 3.13 Mechanism of the phosphotransferase system of *Escherichia coli*. For glucose uptake, the system consists of five proteins: Enzyme (Enz) I, Enzymes II$_a$, II$_b$, and II$_c$, and HPr. A phosphate cascade occurs from phosphoenolpyruvate (PE-P) to Enzyme II$_c$ and the latter actually transports and phosphorylates the sugar. Proteins HPr and Enz I are nonspecific and transport any sugar. The Enz II components are specific for each particular sugar.

Transport systems that employ periplasmic binding proteins along with a membrane transporter and ATP-hydrolyzing proteins are called **ABC transport systems**, the "ABC" standing for *A*TP-*b*inding *c*assette, a structural feature of proteins that bind ATP (**Figure 3.14**). More than 200 different ABC transport systems have been identified in prokaryotes. ABC transporters exist for the uptake of organic compounds such as sugars and amino acids, inorganic nutrients such as sulfate and phosphate, and trace metals.

A characteristic property of periplasmic binding proteins is their high substrate affinity. These proteins can bind their substrate(s) even when they are at extremely low concentration; for example, less than 1 micromolar (10^{-6} M). Once its substrate is bound, the periplasmic binding protein interacts with its respective membrane transporter to transport the substrate into the cell driven by ATP hydrolysis (Figure 3.14).

Even though gram-positive bacteria lack a periplasm, they have ABC transport systems. In gram-positive bacteria, however, substrate-binding proteins are anchored to the external surface of the cytoplasmic membrane. Nevertheless, once these proteins bind substrate, they interact with a membrane transporter to catalyze uptake of the substrate at the expense of ATP hydrolysis, just as they do in gram-negative bacteria (Figure 3.14).

Protein Export

Thus far our discussion of transport has focused on small molecules. How do large molecules, such as proteins, get out of cells? Many proteins need to be either transported outside the cytoplasmic membrane or inserted in a specific way into the membrane in order to function properly. Proteins are exported through and inserted into prokaryotic membranes by the activities of other proteins called *translocases*, a key one being the Sec (*sec* for *sec*retory) system. The Sec system both exports proteins and inserts integral membrane proteins into the membrane. Proteins destined for transport are recognized by the Sec system

because they are tagged in a specific way. We discuss this process later (Section 6.21).

Protein export is important to bacteria because many bacterial enzymes are designed to function outside the cell (exoenzymes). For example, hydrolytic exoenzymes such as amylase or cellulase are excreted directly into the environment where they cleave starch or cellulose, respectively, into glucose; the glucose is then used by the cell as a carbon and energy source. In gram-negative

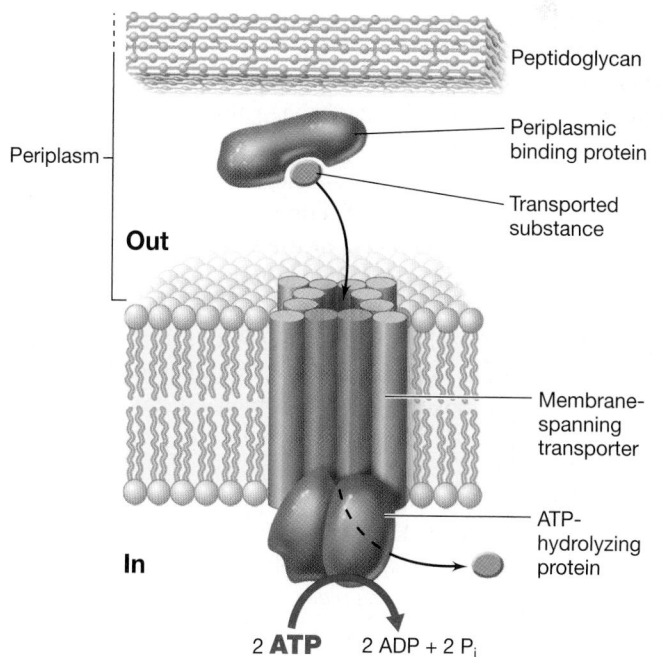

Figure 3.14 Mechanism of an ABC transporter. The periplasmic binding protein has high affinity for substrate, the membrane-spanning proteins form the transport channel, and the cytoplasmic ATP-hydrolyzing proteins supply the energy for the transport event.

bacteria, many enzymes are periplasmic enzymes, and these must traverse the cytoplasmic membrane in order to function. Moreover, many pathogenic bacteria excrete protein toxins or other harmful proteins into the host during infection. Many toxins are excreted by a second translocase system called the *type III secretion system*. This system differs from the Sec system in that the secreted protein is translocated from the bacterial cell directly into the host, for example, a human cell. However, all of these large molecules need to move through the cytoplasmic membrane, and translocases such as SecYEG and the type III secretion system assist in these transport events.

MiniQuiz

- Contrast simple transporters, the phosphotransferase system, and ABC transporters in terms of (1) energy source, (2) chemical alterations of the solute transported, and (3) number of proteins involved.

- Which transport system is best suited for the transport of nutrients present at extremely low levels, and why?

- Why is protein excretion important to cells?

Cell Walls of Prokaryotes

3.6 The Cell Wall of *Bacteria*: Peptidoglycan

Because of the activities of transport systems, the cytoplasm of bacterial cells maintains a high concentration of dissolved solutes. This causes a significant osmotic pressure—about 2 atmospheres in a typical bacterial cell. This is roughly the same as the pressure in an automobile tire. To withstand these pressures and prevent bursting (cell lysis), bacteria employ cell walls. Besides protecting against osmotic lysis, cell walls also confer shape and rigidity on the cell.

Species of *Bacteria* can be divided into two major groups, called **gram-positive** and **gram-negative**. The distinction between gram-positive and gram-negative bacteria is based on the **Gram stain** reaction (↩ Section 2.2). But differences in cell wall structure are at the heart of the Gram stain reaction. The surface of gram-positive and gram-negative cells as viewed in the electron microscope differs markedly, as shown in **Figure 3.15**. The gram-negative cell wall, or cell envelope as it is sometimes called, is chemically complex and consists of at least two layers, whereas the gram-positive cell wall is typically much thicker and consists primarily of a single type of molecule.

The focus of this section is on the polysaccharide component of the cell walls of *Bacteria*, both gram-positive and gram-negative. In the next section we describe the special wall components present in gram-negative *Bacteria*. And finally, in Section 3.8 we briefly describe the cell walls of *Archaea*.

Peptidoglycan

The walls of *Bacteria* have a rigid layer that is primarily responsible for the strength of the wall. In gram-negative bacteria, additional layers are present outside this rigid layer. The rigid layer, called **peptidoglycan**, is a polysaccharide composed of two sugar derivatives—*N-acetylglucosamine* and *N-acetylmuramic acid*—and a few amino acids, including L-alanine, D-alanine, D-glutamic acid, and either lysine or the structurally similar amino acid analog, diaminopimelic acid (DAP). These constituents are connected to form a repeating structure, the *glycan tetrapeptide* (**Figure 3.16**).

Long chains of peptidoglycan are biosynthesized adjacent to one another to form a sheet surrounding the cell (see Figure 3.18). The chains are connected through cross-links of amino acids. The glycosidic bonds connecting the sugars in the glycan strands are covalent bonds, but these provide rigidity to the structure in only one direction. Only after cross-linking is peptidoglycan strong in both the X and Y directions (**Figure 3.17**). Cross-linking occurs to different extents in different species of *Bacteria*; more extensive cross-linking results in greater rigidity.

In gram-negative bacteria, peptidoglycan cross-linkage occurs by peptide bond formation from the amino group of DAP of one glycan chain to the carboxyl group of the terminal D-alanine on the adjacent glycan chain (Figure 3.17). In gram-positive bacteria, cross-linkage may occur through a short peptide interbridge, the kinds and numbers of amino acids in the interbridge varying from species to species. For example, in the gram-positive *Staphylococcus aureus*, the interbridge peptide is composed of five glycine residues, a common interbridge amino acid (Figure 3.17*b*). The overall structure of peptidoglycan is shown in Figure 3.17*c*.

Peptidoglycan can be destroyed by certain agents. One such agent is the enzyme *lysozyme*, a protein that cleaves the β-1,4-glycosidic bonds between *N*-acetylglucosamine and *N*-acetylmuramic acid in peptidoglycan (Figure 3.16), thereby weakening the wall; water can then enter the cell and cause lysis. Lysozyme is found in animal secretions including tears, saliva, and other body fluids, and functions as a major line of defense against bacterial infection. When we consider peptidoglycan biosynthesis in Chapter 5 we will see that the important antibiotic penicillin also targets peptidoglycan, but in a different way from that of lysozyme. Whereas lysozyme destroys preexisting peptidoglycan, penicillin instead prevents its biosynthesis, leading eventually to osmotic lysis.

Diversity of Peptidoglycan

Peptidoglycan is present only in species of *Bacteria*—the sugar *N*-acetylmuramic acid and the amino acid analog DAP have never been found in the cell walls of *Archaea* or *Eukarya*. However, not all *Bacteria* examined have DAP in their peptidoglycan; some have lysine instead. An unusual feature of peptidoglycan is the presence of two amino acids of the D stereoisomer, D-alanine and D-glutamic acid. Proteins, by contrast, are always constructed of L-amino acids.

More than 100 different peptidoglycans are known, with diversity typically governed by the peptide cross-links and interbridge. In every form of peptidoglycan the glycan portion is constant; only the sugars *N*-acetylglucosamine and *N*-acetylmuramic acid are present and are connected in β-1,4 linkage (Figure 3.16). Moreover, the tetrapeptide shows major variation in only one amino acid, the lysine–DAP alternation. Thus, although the

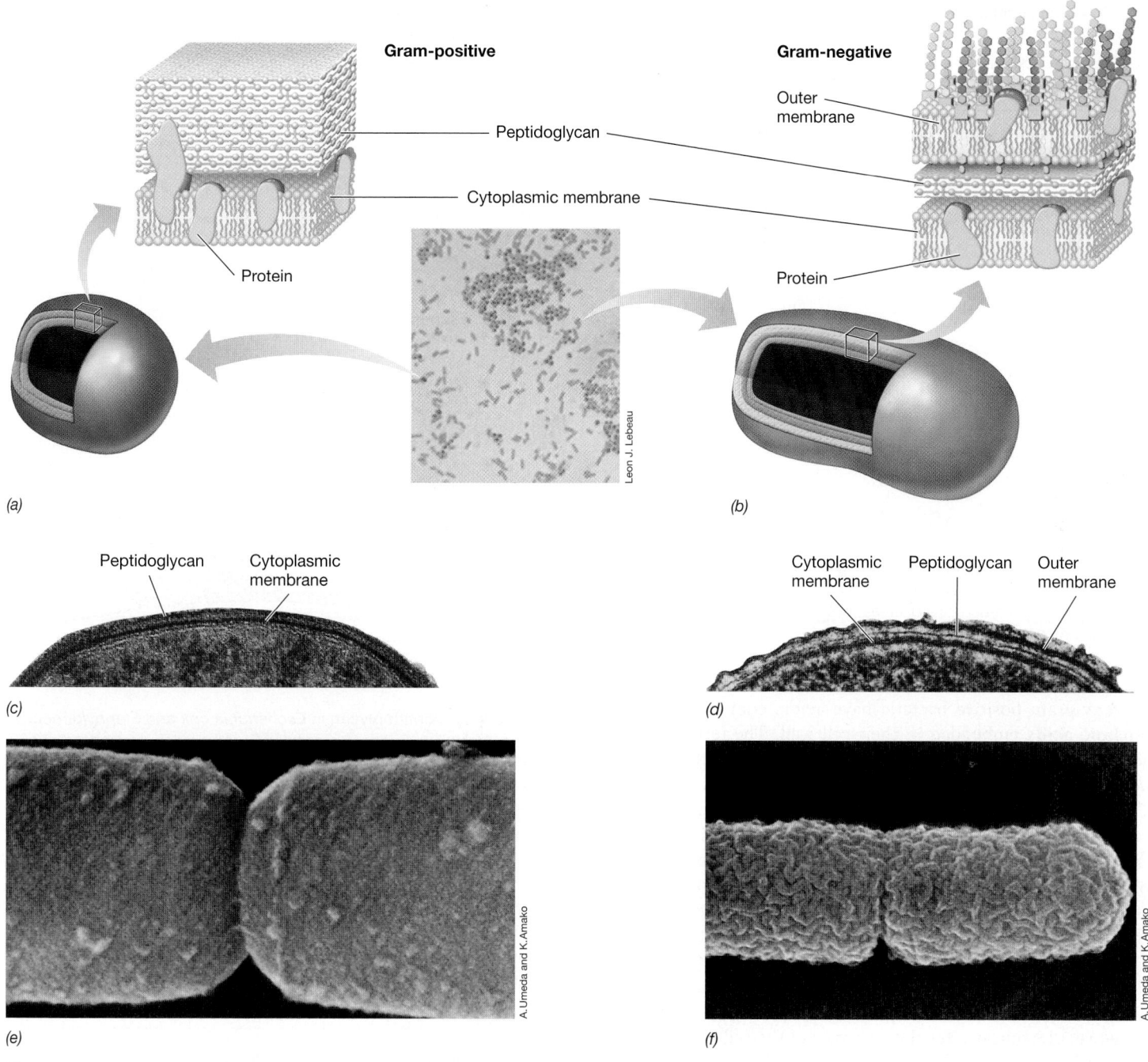

Figure 3.15 Cell walls of *Bacteria*. *(a, b)* Schematic diagrams of gram-positive and gram-negative cell walls. The Gram stain photo in the center shows cells of *Staphylococcus aureus* (purple, gram-positive) and *Escherichia coli* (pink, gram-negative). *(c, d)* Transmission electron micrographs (TEMs) showing the cell wall of a gram-positive bacterium and a gram-negative bacterium. *(e, f)* Scanning electron micrographs of gram-positive and gram-negative bacteria, respectively. Note differences in surface texture. Each cell in the TEMs is about 1 μm wide.

peptide composition of peptidoglycan can vary, the peptidoglycan backbone—alternating repeats of *N*-acetylglucosamine and *N*-acetylmuramic acid—is invariant.

The Gram-Positive Cell Wall

In gram-positive bacteria, as much as 90% of the wall is peptidoglycan. And, although some bacteria have only a single layer of

peptidoglycan surrounding the cell, many gram-positive bacteria have several sheets of peptidoglycan stacked one upon another (Figure 3.15*a*). It is thought that the peptidoglycan is laid down by the cell in "cables" about 50 nm wide, with each cable consisting of several cross-linked glycan strands (**Figure 3.18*a***). As the peptidoglycan "matures," the cables themselves become cross-linked to form an even stronger cell wall structure.

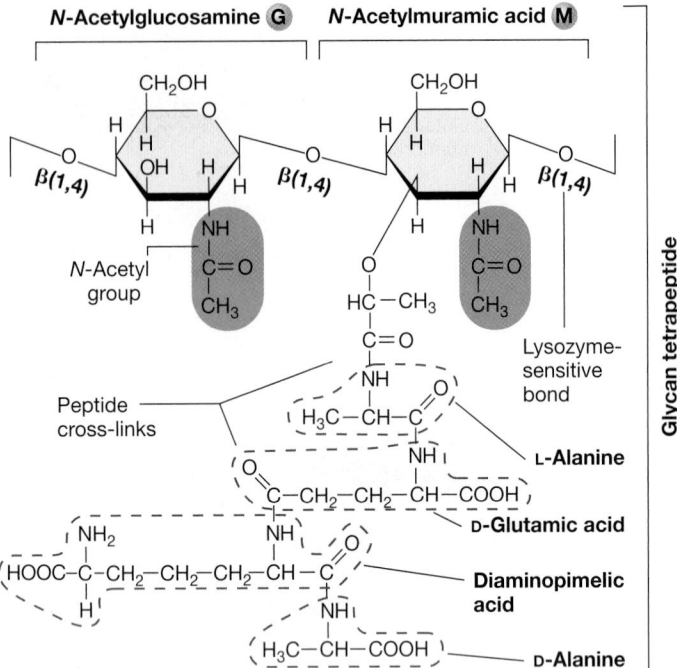

Figure 3.16 Structure of the repeating unit in peptidoglycan, the glycan tetrapeptide. The structure given is that found in *Escherichia coli* and most other gram-negative *Bacteria*. In some *Bacteria*, other amino acids are present as discussed in the text.

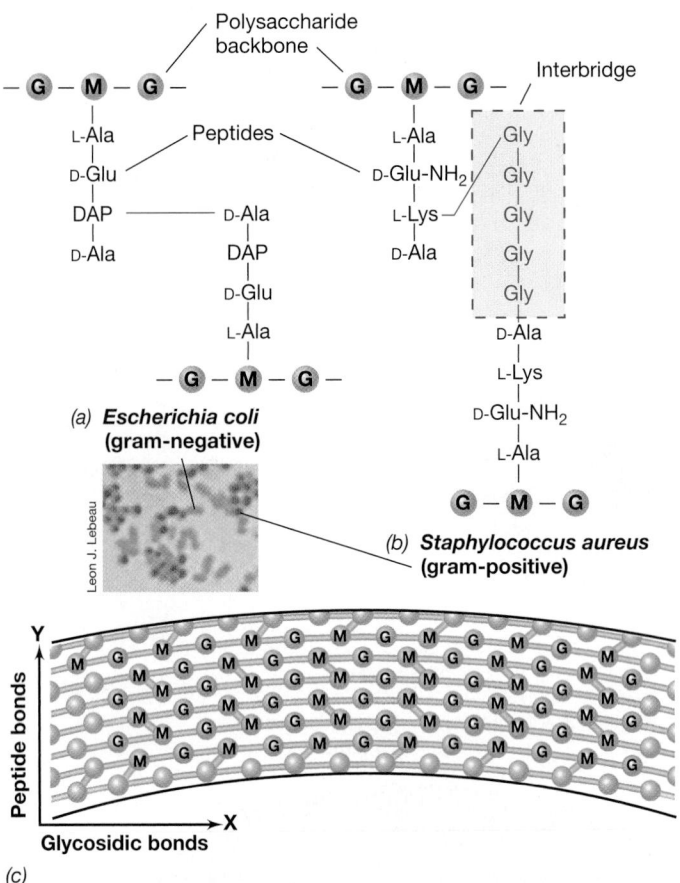

Figure 3.17 Peptidoglycan in *Escherichia coli* and *Staphylococcus aureus*. (a) No interbridge is present in *E. coli* peptidoglycan nor that of other gram-negative *Bacteria*. (b) The glycine interbridge in *S. aureus* (gram-positive). (c) Overall structure of peptidoglycan. G, N-acetylglucosamine; M, N-acetylmuramic acid. Note how glycosidic bonds confer strength on peptidoglycan in the X direction whereas peptide bonds confer strength in the Y direction.

Many gram-positive bacteria have acidic components called **teichoic acids** embedded in their cell wall. The term "teichoic acids" includes all cell wall, cytoplasmic membrane, and capsular polymers composed of glycerol phosphate or ribitol phosphate. These polyalcohols are connected by phosphate esters and typically contain sugars or D-alanine (Figure 3.18b). Teichoic acids are covalently bonded to muramic acid in the wall peptidoglycan. Because the phosphates are negatively charged, teichoic acids are at least in part responsible for the overall negative electrical charge of the cell surface. Teichoic acids also function to bind Ca^{2+} and Mg^{2+} for eventual transport into the cell. Certain teichoic acids are covalently bound to membrane lipids, and these are called *lipoteichoic* acids (Figure 3.18c).

Figure 3.18 summarizes the structure of the cell wall of gram-positive *Bacteria* and shows how teichoic acids and lipoteichoic acids are arranged in the overall wall structure. It also shows how the peptidoglycan cables run perpendicular to the long axis of a rod-shaped bacterium.

Cells That Lack Cell Walls

Although most prokaryotes cannot survive in nature without their cell walls, some do so naturally. These include the mycoplasmas, a group of pathogenic bacteria that causes several infectious diseases of humans and other animals, and the *Thermoplasma* group, species of *Archaea* that naturally lack cell walls. These bacteria are able to survive without cell walls because they either contain unusually tough cytoplasmic membranes or because they live in osmotically protected habitats such as the animal body. Most mycoplasmas have sterols in their

cytoplasmic membranes, and these probably function to add strength and rigidity to the membrane as they do in the cytoplasmic membranes of eukaryotic cells.

MiniQuiz

- Why do bacterial cells need cell walls? Do all bacteria have cell walls?
- Why is peptidoglycan such a strong molecule?
- What does the enzyme lysozyme do?

3.7 The Outer Membrane

In gram-negative bacteria only about 10% of the total cell wall consists of peptidoglycan (Figure 3.15b). Instead, most of the wall is composed of the **outer membrane**. This layer is effectively a second lipid bilayer, but it is not constructed solely of phospholipid and protein, as is the cytoplasmic membrane (Figure 3.5). The gram-negative cell outer membrane also contains polysaccharide. The lipid and polysaccharide are linked in the outer

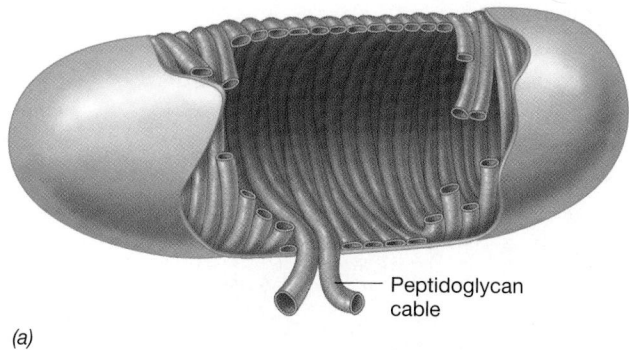

Peptidoglycan
cable

(a)

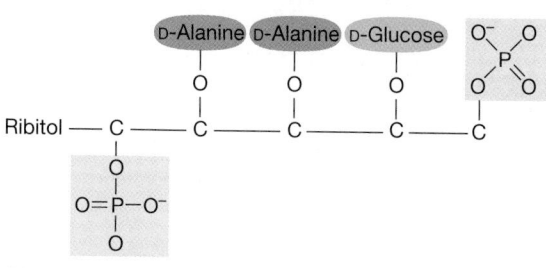

(b)

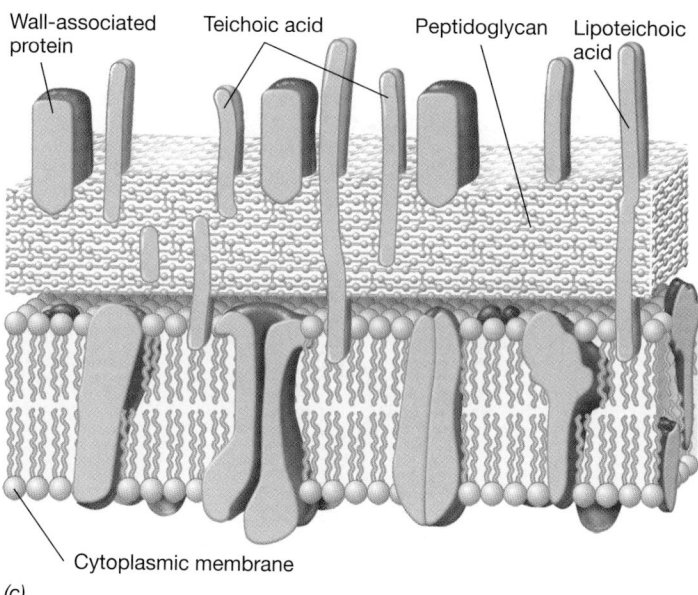

(c)

Figure 3.18 Structure of the gram-positive bacterial cell wall.
(a) Schematic of a gram-positive rod showing the internal architecture
of the peptidoglycan "cables." *(b)* Structure of a ribitol teichoic acid. The
teichoic acid is a polymer of the repeating ribitol unit shown here.
(c) Summary diagram of the gram-positive bacterial cell wall.

membrane to form a complex. Because of this, the outer membrane is also called the **lipopolysaccharide** layer, or simply **LPS**.

Chemistry and Activity of LPS

The chemistry of LPS from several bacteria is known. As seen in
Figure 3.19, the polysaccharide portion of LPS consists of two
components, the *core polysaccharide* and the *O-polysaccharide*.
In *Salmonella* species, where LPS has been best studied, the
core polysaccharide consists of ketodeoxyoctonate (KDO), various seven-carbon sugars (heptoses), glucose, galactose, and
N-acetylglucosamine. Connected to the core is the O-polysaccharide, which typically contains galactose, glucose, rhamnose, and
mannose, as well as one or more dideoxyhexoses, such as abequ-
ose, colitose, paratose, or tyvelose. These sugars are connected in
four- or five-membered sequences, which often are branched.
When the sequences repeat, the long O-polysaccharide is formed.

The relationship of the LPS layer to the overall gram-negative
cell wall is shown in **Figure 3.20**. The lipid portion of the LPS,
called *lipid A*, is not a typical glycerol lipid (see Figure 3.4*a*), but
instead the fatty acids are connected through the amine groups
from a disaccharide composed of glucosamine phosphate (Figure
3.19). The disaccharide is attached to the core polysaccharide
through KDO (Figure 3.19). Fatty acids commonly found in lipid
A include caproic (C_6), lauric (C_{12}), myristic (C_{14}), palmitic (C_{16}),
and stearic (C_{18}) acids.

LPS replaces much of the phospholipid in the outer half of the
outer membrane bilayer. By contrast, lipoprotein is present on
the inner half of the outer membrane, along with the usual phospholipids (Figure 3.20*a*). Lipoprotein functions as an anchor
tying the outer membrane to peptidoglycan. Thus, although the
overall structure of the outer membrane is considered a lipid
bilayer, its structure is distinct from that of the cytoplasmic
membrane (compare Figures 3.5 and 3.20*a*).

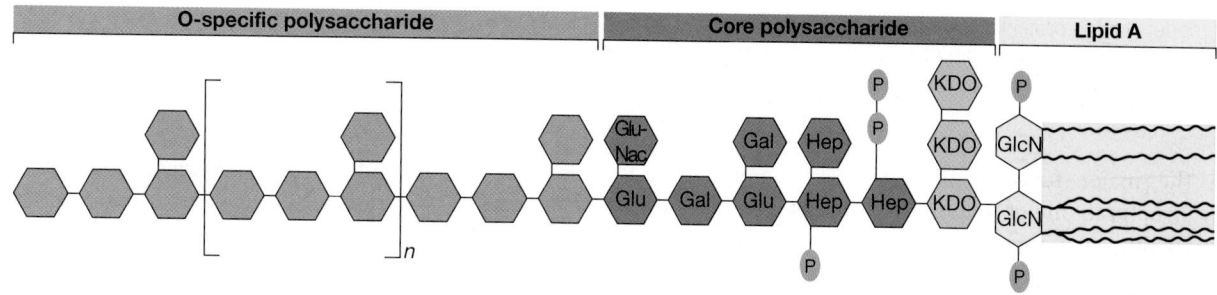

**Figure 3.19 Structure of the lipopolysaccha-
ride of gram-negative** *Bacteria***.** The chemistry of
lipid A and the polysaccharide components varies
among species of gram-negative *Bacteria*, but the
major components (lipid A–KDO–core–O-specific)
are typically the same. The O-specific polysaccharide varies greatly among species. KDO,
ketodeoxyoctonate; Hep, heptose; Glu, glucose;
Gal, galactose; GluNac, *N*-acetylglucosamine;
GlcN, glucosamine; P, phosphate. Glucosamine
and the lipid A fatty acids are linked through the
amine groups. The lipid A portion of LPS can be
toxic to animals and comprises the endotoxin
complex. Compare this figure with Figure 3.20 and
follow the LPS components by the color-coding.

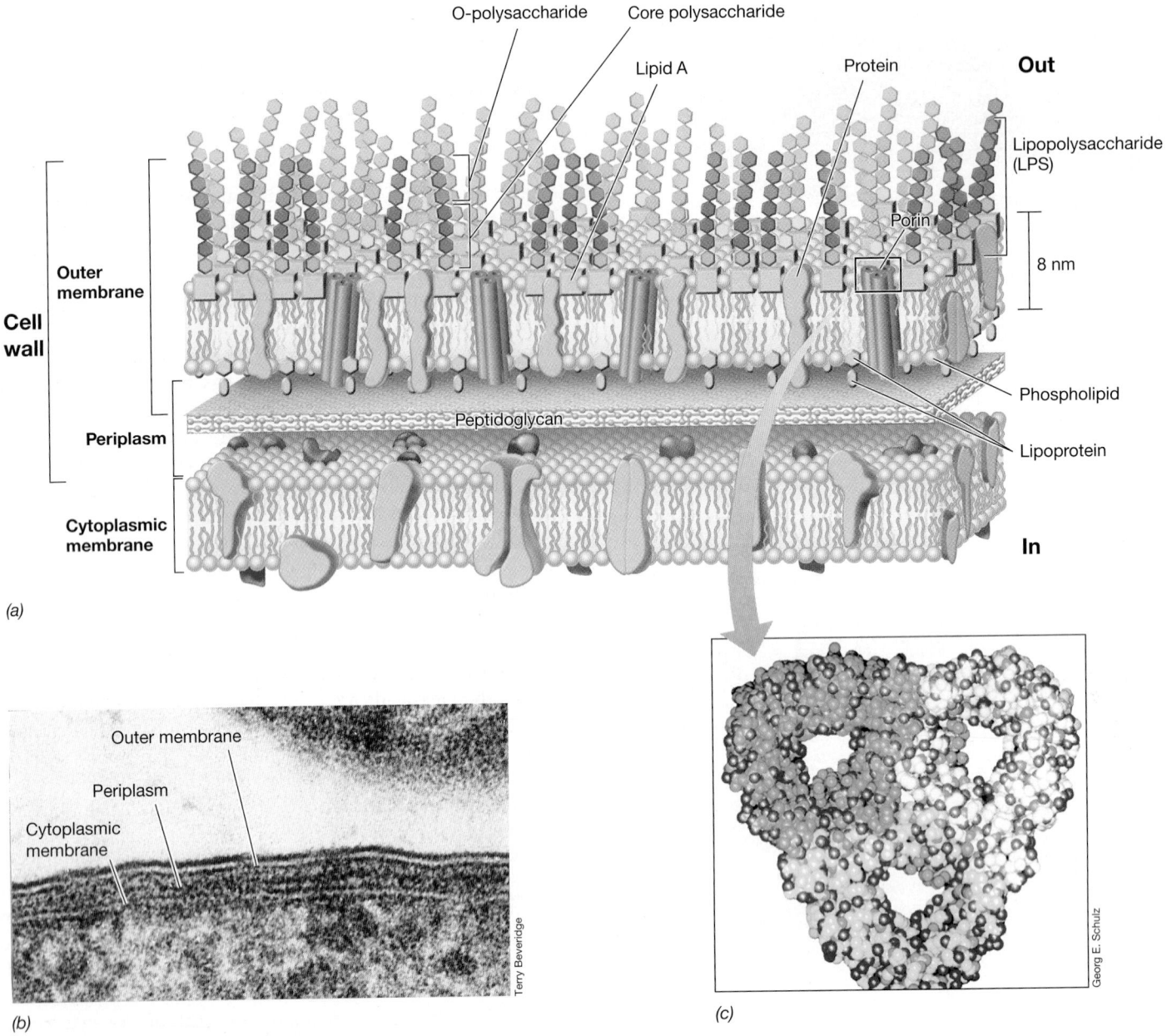

Figure 3.20 **The gram-negative cell wall.** *(a)* Arrangement of lipopolysaccharide, lipid A, phospholipid, porins, and lipoprotein in the outer membrane. See Figure 3.19 for details of the structure of LPS. *(b)* Transmission electron micrograph of a cell of *Escherichia coli* showing the cytoplasmic membrane and wall. *(c)* Molecular model of porin proteins. Note the four pores present, one within each of the proteins forming a porin molecule and a smaller central pore between the porin proteins. The view is perpendicular to the plane of the membrane.

Although the major function of the outer membrane is undoubtedly structural, one of its important biological activities is its toxicity to animals. Gram-negative bacteria that are pathogenic for humans and other mammals include species of *Salmonella*, *Shigella*, and *Escherichia*, among many others, and some of the intestinal symptoms these pathogens elicit are due to toxic outer membrane components. Toxicity is associated with the LPS layer, in particular, lipid A. The term *endotoxin* refers to this toxic component of LPS. Some endotoxins cause violent symptoms in humans, including gas, diarrhea, and vomiting, and the endotoxins produced by *Salmonella* and enteropathogenic strains of *E. coli* transmitted in contaminated foods are classic examples of this.

The Periplasm and Porins

Although permeable to small molecules, the outer membrane is not permeable to proteins or other large molecules. In fact, one of the major functions of the outer membrane is to keep proteins whose activities occur outside the cytoplasmic membrane from diffusing away from the cell. These proteins are present in a region called the **periplasm** (see Figure 3.20). This space, located between the outer surface of the cytoplasmic membrane and the inner surface of the outer membrane, is about 15 nm wide. The periplasm is gel-like in consistency because of the high concentration of proteins present there.

Depending on the organism, the periplasm can contain several different classes of proteins. These include hydrolytic enzymes, which function in the initial degradation of food molecules; binding proteins, which begin the process of transporting substrates (Section 3.5); and chemoreceptors, which are proteins involved in the chemotaxis response (Section 3.15). Most of these proteins reach the periplasm by way of the Sec protein-exporting system in the cytoplasmic membrane (Section 3.5).

The outer membrane of gram-negative bacteria is relatively permeable to small molecules even though it is a lipid bilayer. This is due to *porins* embedded in the outer membrane that function as channels for the entrance and exit of solutes (Figure 3.20). Several porins are known, including both specific and nonspecific classes.

Nonspecific porins form water-filled channels through which any small substance can pass. By contrast, specific porins contain a binding site for only one or a small group of structurally related substances. Porins are transmembrane proteins that consist of three identical subunits. Besides the channel present in each barrel of the porin, the barrels of the porin proteins associate in such a way that a hole about 1 nm in diameter is formed in the outer membrane through which very small solutes can travel (Figure 3.20c).

Relationship of Cell Wall Structure to the Gram Stain

The structural differences between the cell walls of gram-positive and gram-negative *Bacteria* are thought to be responsible for differences in the Gram stain reaction. In the Gram stain, an insoluble crystal violet–iodine complex forms inside the cell. This complex is extracted by alcohol from gram-negative but not from gram-positive bacteria (Section 2.2). As we have seen, gram-positive bacteria have very thick cell walls consisting primarily of peptidoglycan (Figure 3.18); these become dehydrated by the alcohol, causing the pores in the walls to close and preventing the insoluble crystal violet–iodine complex from escaping. By contrast, in gram-negative bacteria, alcohol readily penetrates the lipid-rich outer membrane and extracts the crystal violet–iodine complex from the cell. After alcohol treatment, gram-negative cells are nearly invisible unless they are counterstained with a second dye, a standard procedure in the Gram stain (Figure 2.4).

MiniQuiz
- What components constitute the outer membrane of gram-negative bacteria?
- What is the function of porins and where are they located in a gram-negative cell wall?
- What component of the cell has endotoxin properties?
- Why does alcohol readily decolorize gram-negative but not gram-positive bacteria?

3.8 Cell Walls of *Archaea*

Peptidoglycan, a key biomarker for *Bacteria*, is absent from the cell walls of *Archaea*. An outer membrane is typically lacking in *Archaea* as well. Instead, a variety of chemistries are found in the cell walls of *Archaea*, including polysaccharides, proteins, and glycoproteins.

Pseudomurein and Other Polysaccharide Walls

The cell walls of certain methanogenic *Archaea* contain a molecule that is remarkably similar to peptidoglycan, a polysaccharide called *pseudomurein* (the term "murein" is from the Latin word for "wall" and was an old term for peptidoglycan; **Figure 3.21**). The backbone of pseudomurein is composed of alternating repeats of *N*-acetylglucosamine (also found in peptidoglycan) and *N*-acetyltalosaminuronic acid; the latter replaces the *N*-acetylmuramic acid of peptidoglycan. Pseudomurein also differs from peptidoglycan in that the glycosidic bonds between the sugar derivatives are β-1,3 instead of β-1,4, and the amino acids are all of the L stereoisomer. It is thought that peptidoglycan and pseudomurein either arose by convergent evolution after *Bacteria* and *Archaea* had diverged or, more likely, by evolution from a common polysaccharide present in the cell walls of the common ancestor of the domains *Bacteria* and *Archaea*.

Cell walls of some other *Archaea* lack pseudomurein and instead contain other polysaccharides. For example, *Methanosarcina* species have thick polysaccharide walls composed of polymers of glucose, glucuronic acid, galactosamine uronic acid, and acetate. Extremely halophilic (salt-loving) *Archaea* such as *Halococcus*, which are related to *Methanosarcina*, have similar cell walls that

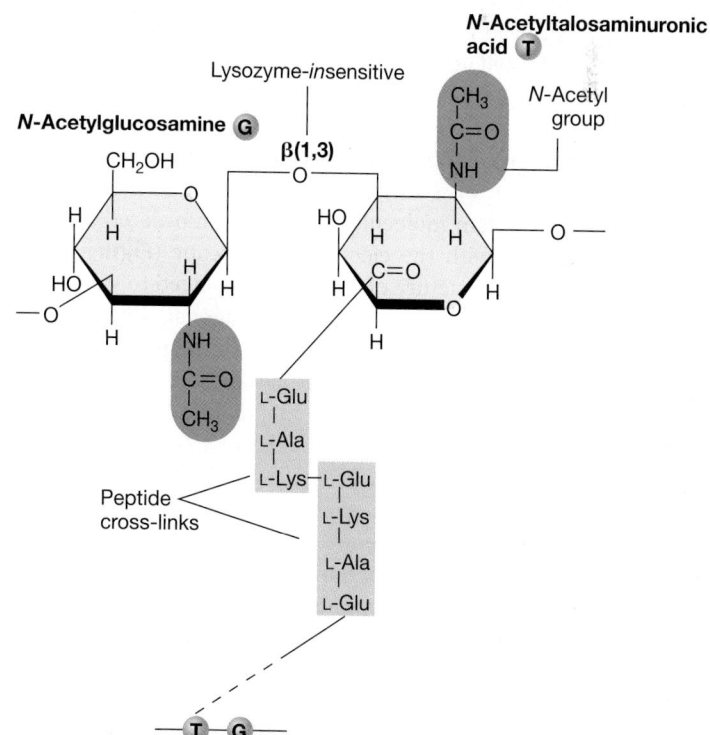

Figure 3.21 Pseudomurein. Structure of pseudomurein, the cell wall polymer of *Methanobacterium* species. Note the similarities and differences between pseudomurein and peptidoglycan (Figure 3.16).

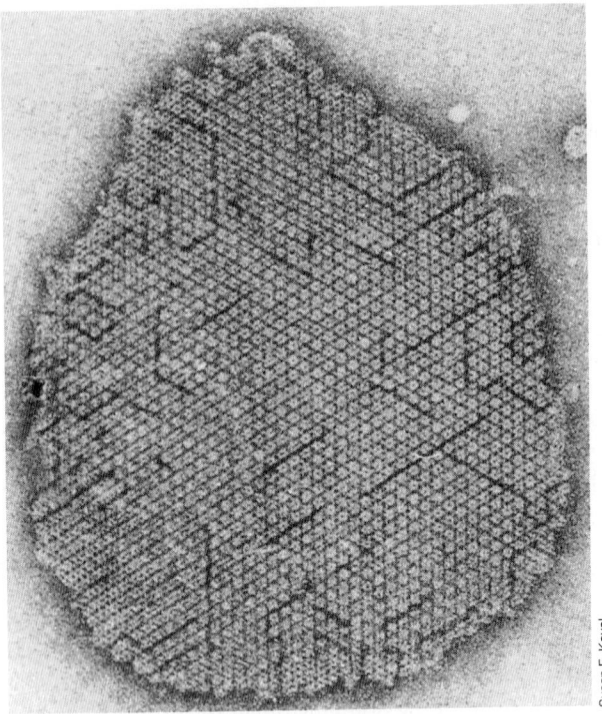

Figure 3.22 The S-layer. Transmission electron micrograph of an S-layer showing the paracrystalline structure. Shown is the S-layer from *Aquaspirillum serpens* (a species of *Bacteria*); this S-layer shows hexagonal symmetry as is common in S-layers of *Archaea* as well.

Susan F. Koval

also contain sulfate (SO_4^{2-}). The negative charge on the sulfates bind the high concentration of Na^+ present in the habitats of *Halococcus*, salt evaporation ponds and saline seas and lakes; this helps stabilize the cell wall in such strongly polar environments.

S-Layers

The most common cell wall in species of *Archaea* is the paracrystalline surface layer, or **S-layer**. S-layers consist of interlocking protein or glycoprotein molecules that show an ordered appearance when viewed with the electron microscope (**Figure 3.22**). The paracrystalline structure of S-layers is arranged to yield various symmetries, such as hexagonal, tetragonal, or trimeric, depending upon the number and structure of the protein or glycoprotein subunits of which they are composed. S-layers have been found in representatives of all major lineages of *Archaea* and also in several species of *Bacteria* (Figure 3.22).

The cell walls of some *Archaea*, for example the methanogen *Methanocaldococcus jannaschii*, consist only of an S-layer. Thus, S-layers are themselves sufficiently strong to withstand osmotic bursting. However, in many organisms S-layers are present in addition to other cell wall components, usually polysaccharides. For example, in *Bacillus brevis*, a species of *Bacteria*, an S-layer is present along with peptidoglycan. However, when an S-layer is present along with other wall components, the S-layer is always the *outermost* wall layer, the layer that is in direct contact with the environment.

Besides serving as protection from osmotic lysis, S-layers may have other functions. For example, as the interface between the cell and its environment, it is likely that the S-layer functions as a selective sieve, allowing the passage of low-molecular-weight solutes while excluding large molecules and structures (such as viruses). The S-layer may also function to retain proteins near the cell surface, much as the outer membrane (Section 3.7) does in gram-negative bacteria.

We thus see several cell wall chemistries in species of *Archaea*, varying from molecules that closely resemble peptidoglycan to those that totally lack a polysaccharide component. But with rare exception, all *Archaea* contain a cell wall of some sort, and as in *Bacteria*, the archaeal cell wall functions to prevent osmotic lysis and gives the cell its shape. In addition, because they lack peptidoglycan in their cell walls, *Archaea* are naturally resistant to the activity of lysozyme (Section 3.6) and the antibiotic penicillin, agents that either destroy peptidoglycan or prevent its proper synthesis.

MiniQuiz

- How does pseudomurein resemble peptidoglycan? How do the two molecules differ?
- What is the composition of an S-layer?
- Why are *Archaea* insensitive to penicillin?

Ⅳ Other Cell Surface Structures and Inclusions

In addition to cell walls, prokaryotic cells can have other layers or structures in contact with the environment. Moreover, cells often contain one or more types of cellular inclusions. We examine some of these here.

3.9 Cell Surface Structures

Many prokaryotes secrete slimy or sticky materials on their cell surface. These materials consist of either polysaccharide or protein. These are not considered part of the cell wall because they do not confer significant structural strength on the cell. The terms "capsule" and "slime layer" are used to describe these layers.

Capsules and Slime Layers

Capsules and slime layers may be thick or thin and rigid or flexible, depending on their chemistry and degree of hydration. Traditionally, if the layer is organized in a tight matrix that excludes small particles, such as India ink, it is called a **capsule** (**Figure 3.23**). By contrast, if the layer is more easily deformed, it will not exclude particles and is more difficult to see; this form is called a *slime layer*. In addition, capsules typically adhere firmly to the cell wall, and some are even covalently linked to peptidoglycan. Slime layers, by contrast, are loosely attached and can be lost from the cell surface.

Polysaccharide layers have several functions in bacteria. Surface polysaccharides assist in the attachment of microorganisms to solid surfaces. As we will see later, pathogenic microorganisms that enter the animal body by specific routes usually do so by first binding specifically to surface components of host tissues, and this binding is often mediated by bacterial cell surface polysaccharides. Many nonpathogenic bacteria also bind to solid surfaces in nature, sometimes forming a thick layer of cells called a *biofilm*. Extracellular polysaccharides play a key role in

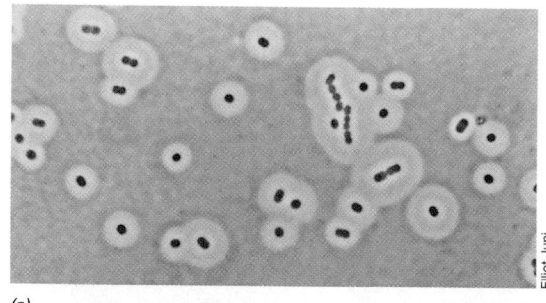

(a)

Elliot Juni

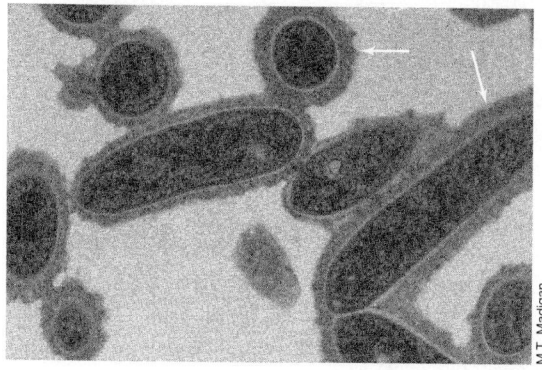

(b)

M.T. Madigan

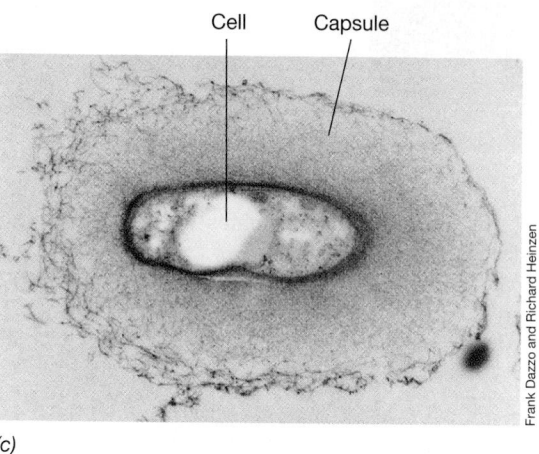

Cell Capsule

(c)

Frank Dazzo and Richard Heinzen

Figure 3.23 Bacterial capsules. *(a)* Capsules of *Acinetobacter* species observed by phase-contrast microscopy after negative staining of cells with India ink. India ink does not penetrate the capsule and so the capsule appears as a light area surrounding the cell, which appears black. *(b)* Transmission electron micrograph of a thin section of cells of *Rhodobacter capsulatus* with capsules (arrows) clearly evident; cells are about 0.9 μm wide. *(c)* Transmission electron micrograph of *Rhizobium trifolii* stained with ruthenium red to reveal the capsule. The cell is about 0.7 μm wide.

the development of biofilms.

Capsules can play other roles as well. For example, encapsulated pathogenic bacteria are typically more difficult for phagocytic cells of the immune system to recognize and subsequently destroy. In addition, because outer polysaccharide layers bind a significant amount of water, it is likely that these layers play some role in resistance of the cell to desiccation.

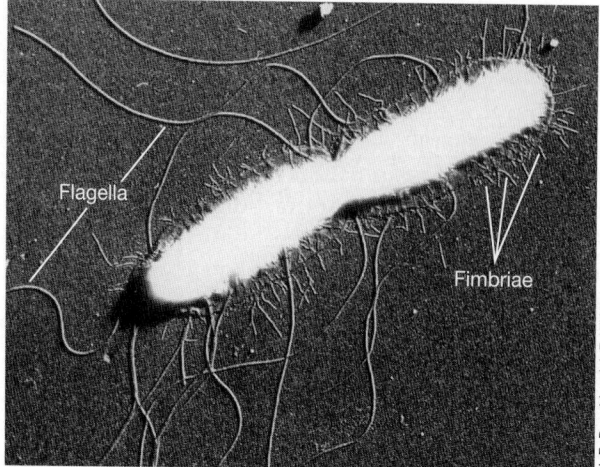

Flagella

Fimbriae

J. P. Duguid and J. F. Wilkinson

Figure 3.24 Fimbriae. Electron micrograph of a dividing cell of *Salmonella typhi*, showing flagella and fimbriae. A single cell is about 0.9 μm wide.

Fimbriae and Pili

Fimbriae and pili are filamentous structures composed of protein that extend from the surface of a cell and can have many functions. *Fimbriae* (**Figure 3.24**) enable cells to stick to surfaces, including animal tissues in the case of pathogenic bacteria, or to form pellicles (thin sheets of cells on a liquid surface) or biofilms on surfaces. Notorious human pathogens in which fimbriae assist in the disease process include *Salmonella* species (salmonellosis), *Neisseria gonorrhoeae* (gonorrhea), and *Bordetella pertussis* (whooping cough).

Pili are similar to fimbriae, but are typically longer and only one or a few pili are present on the surface of a cell. Because pili can be receptors for certain types of viruses, they can best be seen under the electron microscope when they become coated with virus particles (**Figure 3.25**). Many classes of pili are known, distinguished by their structure and function. Two very important functions of pili include facilitating genetic exchange between cells in a process called conjugation (Figure 3.25) and in the adhesion of pathogens to specific host tissues and subsequent invasion. The latter function has been best studied in gram-negative pathogens such as *Neisseria*, species of which cause gonorrhea and meningitis, but pili are also present on certain gram-positive pathogens such as *Streptococcus pyogenes*, the cause of strep throat and scarlet fever.

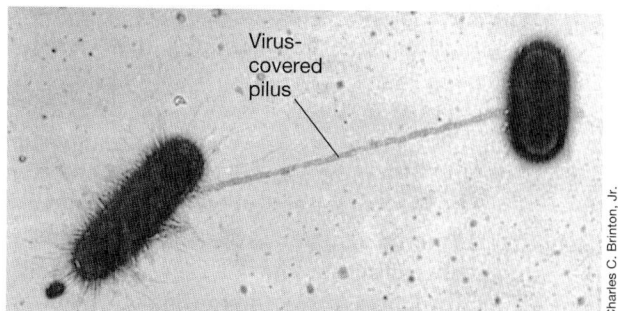

Virus-covered pilus

Charles C. Brinton, Jr.

Figure 3.25 Pili. The pilus on an *Escherichia coli* cell that is undergoing conjugation (a form of genetic transfer) with a second cell is better resolved because viruses have adhered to it. The cells are about 0.8 μm wide.

One important class of pili, called *type IV pili*, assist cells in adhesion but also allow for an unusual form of cell motility called *twitching motility*. Type IV pili are 6 nm in diameter and present only at the poles of those rod-shaped cells that contain them. Twitching motility is a type of gliding motility, movement along a solid surface (Section 3.14). In twitching motility, extension of pili followed by their retraction drags the cell along a solid surface, with energy supplied by ATP. Certain species of *Pseudomonas* and *Moraxella* are well known for their twitching motility.

Type IV pili have also been implicated as key colonization factors for certain human pathogens, including *Vibrio cholerae* (cholera) and *Neisseria gonorrhoeae* (gonorrhea). The twitching motility of these pathogens presumably assists the organism to locate specific sites for attachment to initiate the disease process. Type IV pili are also thought to mediate genetic transfer by the process of transformation in some bacteria, which, along with conjugation and transduction, are the three known means of horizontal gene transfer in prokaryotes (Chapter 10).

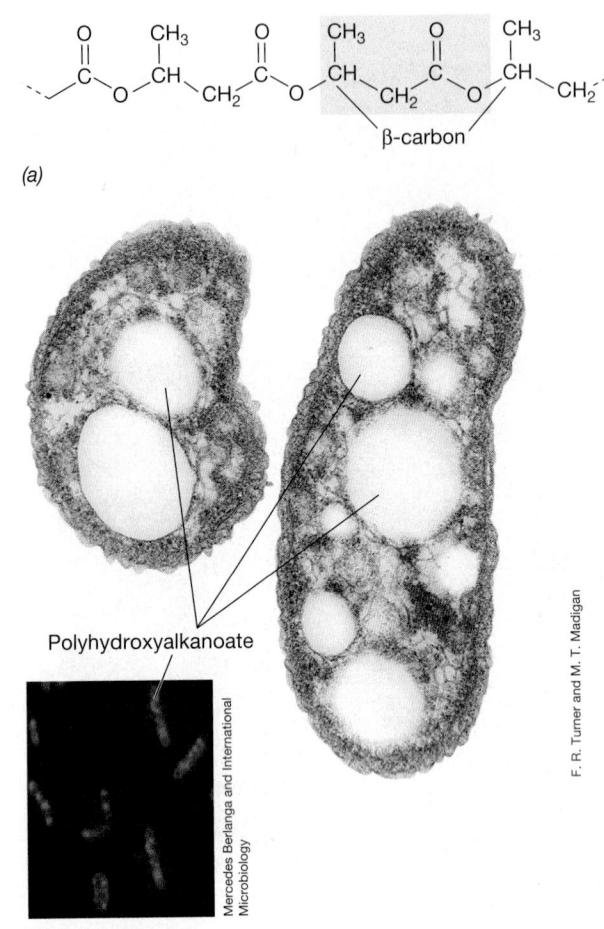

(a)

Polyhydroxyalkanoate

(b)

Figure 3.26 Poly-β-hydroxyalkanoates. *(a)* Chemical structure of poly-β-hydroxybutyrate, a common PHA. A monomeric unit is shown in color. Other PHAs are made by substituting longer-chain hydrocarbons for the –CH_3 group on the β carbon. *(b)* Electron micrograph of a thin section of cells of a bacterium containing granules of PHA. Color photo: Nile red–stained cells of a PHA-containing bacterium.

MiniQuiz
• Could a bacterial cell dispense with a cell wall if it had a capsule? Why or why not?
• How do fimbriae differ from pili, both structurally and functionally?

3.10 Cell Inclusions

Granules or other inclusions are often present in prokaryotic cells. Inclusions function as energy reserves and as reservoirs of structural building blocks. Inclusions can often be seen directly with the light microscope and are usually enclosed by single layer (nonunit) membranes that partition them off in the cell. Storing carbon or other substances in an insoluble inclusion confers an advantage on the cell because it reduces the osmotic stress that would be encountered if the same amount of the substance was dissolved in the cytoplasm.

Carbon Storage Polymers

One of the most common inclusion bodies in prokaryotic organisms is **poly-β-hydroxybutyric acid (PHB)**, a lipid that is formed from β-hydroxbutyric acid units. The monomers of PHB bond by ester linkage to form the PHB polymer, and then the polymer aggregates into granules; the latter can be observed by either light or electron microscope (**Figure 3.26**).

The monomer in the polymer is not only hydroxybutyrate (C_4) but can vary in length from as short as C_3 to as long as C_{18}. Thus, the more generic term *poly-β-hydroxyalkanoate* (PHA) is often used to describe this class of carbon- and energy-storage polymers. PHAs are synthesized by cells when there is an excess of carbon and are broken down for biosynthetic or energy purposes when conditions warrant. Many prokaryotes, including species of both *Bacteria* and *Archaea*, produce PHAs.

Another storage product is *glycogen*, which is a polymer of glucose. Like PHA, glycogen is a storehouse of both carbon and energy. Glycogen is produced when carbon is in excess in the environment and is consumed when carbon is limited. Glycogen

resembles starch, the major storage reserve of plants, but differs slightly from starch in the manner in which the glucose units are linked together.

Polyphosphate and Sulfur

Many microorganisms accumulate inorganic phosphate (PO_4^{3-}) in the form of granules of *polyphosphate* (**Figure 3.27a**). These granules can be degraded and used as sources of phosphate for nucleic acid and phospholipid biosyntheses and in some organisms can be used to make the energy-rich compound ATP. Phosphate is often a limiting nutrient in natural environments. Thus if a cell happens upon an excess of phosphate, it is advantageous to be able to store it as polyphosphate for future use.

Many gram-negative prokaryotes can oxidize reduced sulfur compounds, such as hydrogen sulfide (H_2S). The oxidation of sulfide is linked to either reactions of energy metabolism (chemolithotrophy) or CO_2 fixation (autotrophy). In either case, *elemental sulfur* (S^0) may accumulate in the cell in microscopically visible globules (Figure 3.27b). This sulfur remains as long as the source of reduced sulfur from which it was derived is still

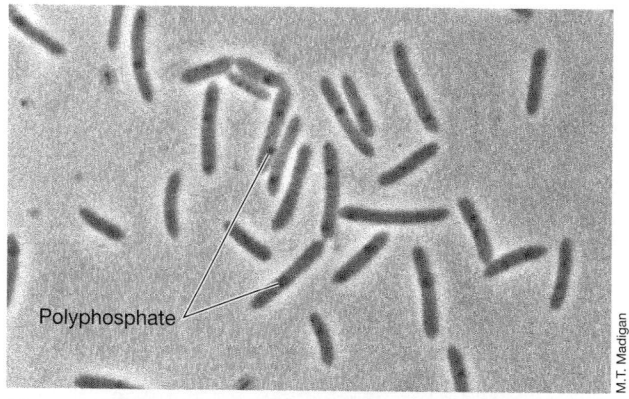

(a)

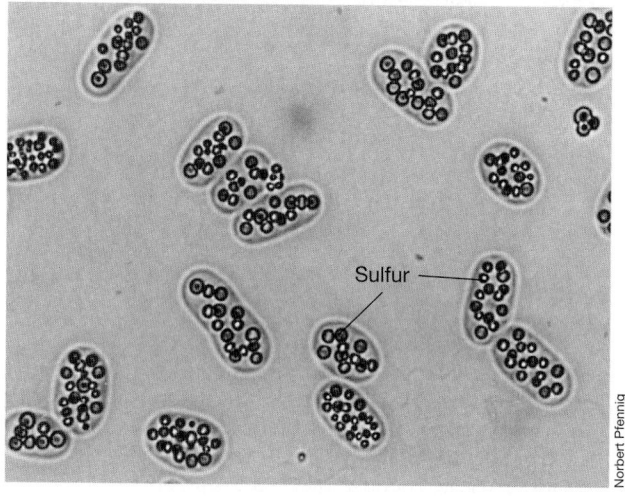

(b)

Figure 3.27 Polyphosphate and sulfur storage products. *(a)* Phase-contrast photomicrograph of cells of *Heliobacterium modesticaldum* showing polyphosphate as dark granules; a cell is about 1 μm wide. *(b)* Bright-field photomicrograph of cells of the purple sulfur bacterium *Isochromatium buderi*. The intracellular inclusions are sulfur globules formed from the oxidation of hydrogen sulfide (H_2S). A single cell is about 4 μm wide.

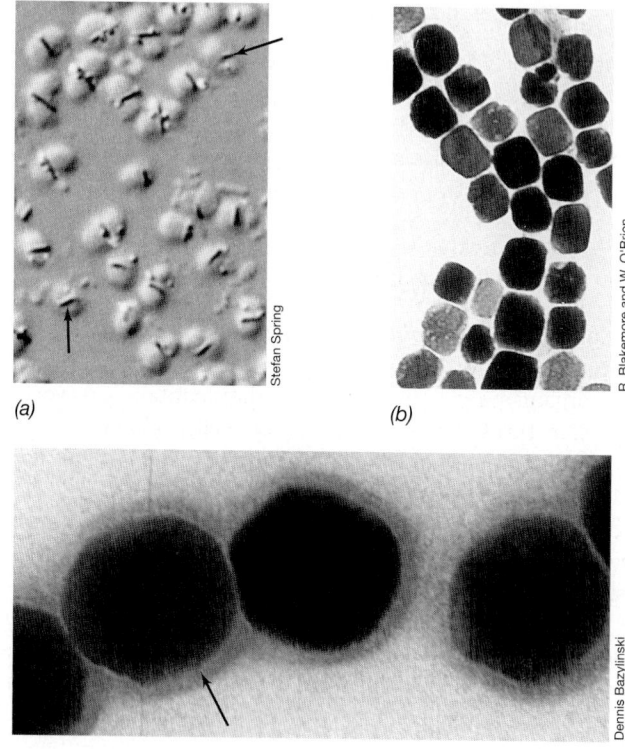

(a)   *(b)*

(c)

Figure 3.28 Magnetotactic bacteria and magnetosomes. *(a)* Differential interference contrast micrograph of coccoid magnetotactic bacteria; note chains of magnetosomes (arrows). A single cell is 2.2 μm wide. *(b)* Magnetosomes isolated from the magnetotactic bacterium *Magnetospirillum magnetotacticum*; each particle is about 50 nm wide. *(c)* Transmission electron micrograph of magnetosomes from a magnetic coccus. The arrow points to the membrane that surrounds each magnetosome. A single magnetosome is about 90 nm wide.

present. However, as the reduced sulfur source becomes limiting, the sulfur in the granules is oxidized to sulfate (SO_4^{2-}), and the granules slowly disappear as this reaction proceeds. Interestingly, although the sulfur globules appear to be in the cytoplasm they actually reside in the periplasm. The periplasm expands outward to accommodate the globules as H_2S is oxidized to S^0 and then contracts inward as S^0 is oxidized to SO_4^{2-}.

Magnetic Storage Inclusions: Magnetosomes

Some bacteria can orient themselves specifically within a magnetic field because they contain **magnetosomes**. These structures are intracellular particles of the iron mineral magnetite—Fe_3O_4 (**Figure 3.28**). Magnetosomes impart a magnetic dipole on a cell, allowing it to respond to a magnetic field. Bacteria that produce magnetosomes exhibit *magnetotaxis*, the process of orienting and migrating along Earth's magnetic field lines. Although the suffix "-taxis" is used in the word magnetotaxis, there is no evidence that magnetotactic bacteria employ the sensory systems of

chemotactic or phototactic bacteria (Section 3.15). Instead, the alignment of magnetosomes in the cell simply imparts a magnetic moment that orients the cell in a particular direction in its environment.

The major function of magnetosomes is unknown. However, magnetosomes have been found in several aquatic organisms that grow best in laboratory culture at low O_2 concentrations. It has thus been hypothesized that one function of magnetosomes may be to guide these primarily aquatic cells downward (the direction of Earth's magnetic field) toward the sediments where O_2 levels are lower.

Magnetosomes are surrounded by a thin membrane containing phospholipids, proteins, and glycoproteins (Figure 3.28b, c). This membrane is not a true unit (bilayer) membrane, as is the cytoplasmic membrane (Figure 3.5), and the proteins present play a role in precipitating Fe^{3+} (brought into the cell in soluble form by chelating agents) as Fe_3O_4 in the developing magnetosome. A similar nonunit membrane surrounds granules of PHA. The morphology of magnetosomes appears to be species-specific, varying in shape from square to rectangular to spike-shaped in different species, forming into chains inside the cell (Figure 3.28).

3.11 Gas Vesicles

Some prokaryotes are *planktonic*, meaning that they live a floating existence within the water column of lakes and the oceans. These organisms can float because they contain **gas vesicles**. These structures confer buoyancy on cells, allowing them to position themselves in a water column in response to environmental cues.

The most dramatic examples of gas-vesiculate bacteria are cyanobacteria that form massive accumulations called *blooms* in lakes or other bodies of water (**Figure 3.29**). Gas-vesiculate cells rise to the surface of the lake and are blown by winds into dense masses. Many primarily aquatic bacteria have gas vesicles and the property is found in both *Bacteria* and *Archaea*. By contrast, gas vesicles have never been found in eukaryotic microorganisms.

General Structure of Gas Vesicles

Gas vesicles are spindle-shaped structures made of protein; they are hollow yet rigid and of variable length and diameter (**Figure 3.30**). Gas vesicles in different organisms vary in length from about 300 to more than 1000 nm and in width from 45 to 120 nm, but the vesicles of a given organism are more or less of constant size. Gas vesicles may number from a few to hundreds per cell and are impermeable to water and solutes but permeable to gases. The presence of gas vesicles in cells can be determined either by light microscopy, where clusters of vesicles, called *gas vacuoles*, appear as irregular bright inclusions, or by transmission electron microscopy (Figure 3.30).

Figure 3.29 Buoyant cyanobacteria. Flotation of gas-vesiculate cyanobacteria that formed a bloom in a freshwater lake, Lake Mendota, Madison, Wisconsin (USA).

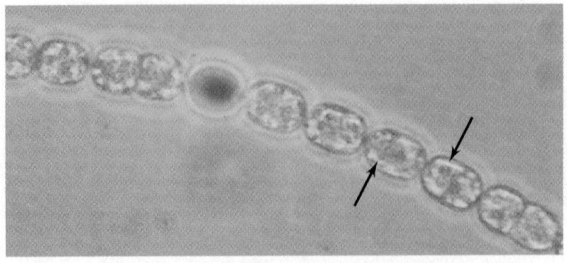

(a)

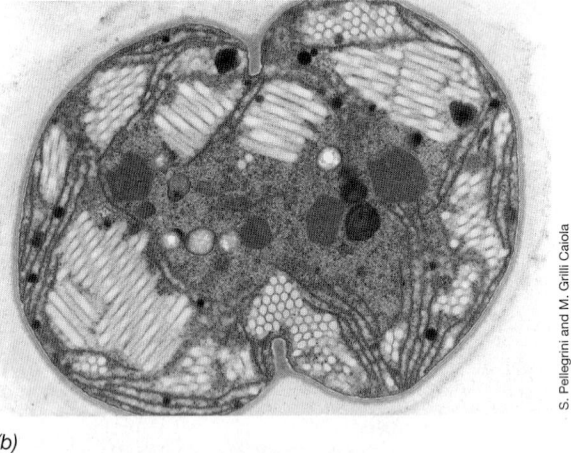

(b)

Figure 3.30 Gas vesicles of the cyanobacteria *Anabaena* and *Microcystis*. (a) Phase-contrast photomicrograph of *Anabaena*. Clusters of gas vesicles form phase-bright gas vacuoles (arrows). (b) Transmission electron micrograph of *Microcystis*. Gas vesicles are arranged in bundles, here seen in both longitudinal and cross section.

Molecular Structure of Gas Vesicles

The conical-shaped gas vesicle is composed of two different proteins. The major protein, called *GvpA*, forms the vesicle shell itself and is a small, hydrophobic, and very rigid protein. The rigidity is essential for the structure to resist the pressures exerted on it from outside. The minor protein, called *GvpC*, functions to strengthen the shell of the gas vesicle by cross-linking copies of GvpA (**Figure 3.31**).

Gas vesicles consist of copies of GvpA that align to yield parallel "ribs" that form the watertight shell. The ribs are then clamped by the GvpC protein, which binds the ribs at an angle to group several GvpA molecules together (Figure 3.31). Gas vesicles vary in shape in different organisms from long and thin to short and fat (compare Figures 3.30 and 3.31*a*), and shape is governed by how the GvpA and GvpC proteins interact to form the intact vesicle.

How do gas vesicles confer buoyancy, and what ecological benefit does buoyancy confer? The composition and pressure of the gas inside a gas vesicle is that of the gas in which the organism is suspended. However, because an inflated gas vesicle has a density of only about 10% of that of the cell proper, gas vesicles decrease cell density, thereby increasing its buoyancy. Phototrophic organisms in particular benefit from gas vesicles because they allow cells to adjust their vertical position in a water column to reach regions where the light intensity for photosynthesis is optimal.

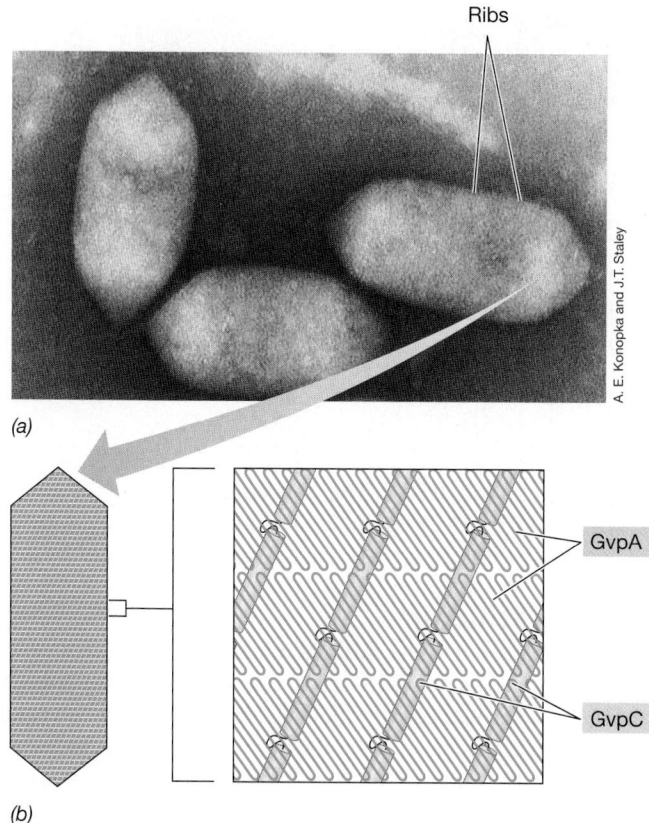

Ribs

GvpA

GvpC

(a)

(b)

A. E. Konopka and J.T. Staley

Figure 3.31 Gas vesicle architecture. Transmission electron micrographs of gas vesicles purified from the bacterium *Ancylobacter aquaticus* and examined in negatively stained preparations. A single vesicle is about 100 nm in diameter. *(b)* Model of how gas vesicle proteins GvpA and GvpC interact to form a watertight but gas-permeable structure. GvpA, a rigid β-sheet, makes up the rib, and GvpC, an α-helix structure, is the cross-linker.

MiniQuiz

- What gas is present in a gas vesicle? Why might a cell benefit from controlling its buoyancy?

- How are the two proteins that make up the gas vesicle, GvpA and GvpC, arranged to form such a water-impermeable structure?

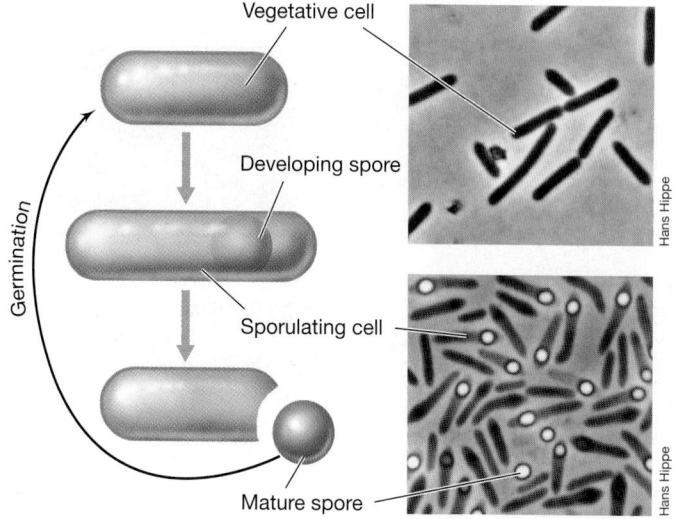

Vegetative cell

Developing spore

Sporulating cell

Mature spore

Germination

Hans Hippe

Hans Hippe

Figure 3.33 The life cycle of an endospore-forming bacterium. The phase-contrast photomicrographs are of cells of *Clostridium pascui*. A cell is about 0.8 μm wide.

3.12 Endospores

Certain species of *Bacteria* produce structures called **endospores** (**Figure 3.32**) during a process called *sporulation*. Endospores (the prefix *endo* means "within") are highly differentiated cells that are extremely resistant to heat, harsh chemicals, and radiation. Endospores function as survival structures and enable the organism to endure unfavorable growth conditions, including but not limited to extremes of temperature, drying, or nutrient depletion. Endospores can thus be thought of as the dormant stage of a bacterial life cycle: vegetative cell → endospore → vegetative cell. Endospores are also easily dispersed by wind, water, or through the animal gut. Endospore-forming bacteria are commonly found in soil, and species of *Bacillus* are the best-studied representatives.

Endospore Formation and Germination

During endospore formation, a vegetative cell is converted into a nongrowing, heat-resistant structure (**Figure 3.33**). Cells do not sporulate when they are actively growing but only when growth ceases owing to the exhaustion of an essential nutrient. Thus,

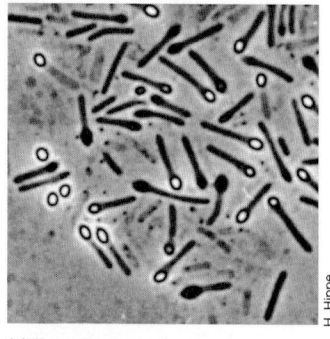

(a) **Terminal spores**

H. Hippe

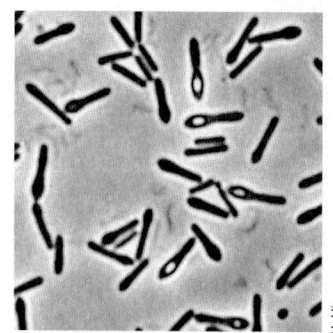

(b) **Subterminal spores**

H. Hippe

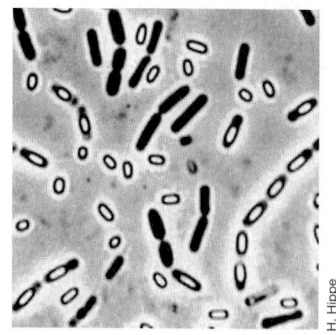

(c) **Central spores**

H. Hippe

Figure 3.32 The bacterial endospore. Phase-contrast photomicrographs illustrating endospore morphologies and intracellular locations in different species of endospore-forming bacteria. Endospores appear bright by phase-contrast microscopy.

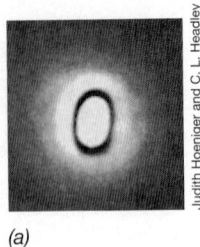

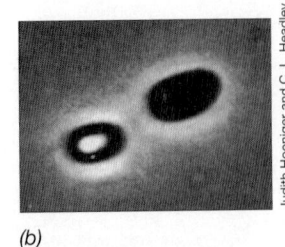

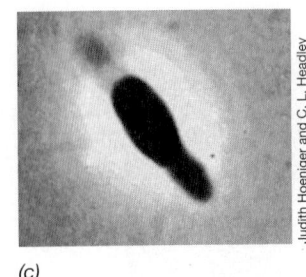

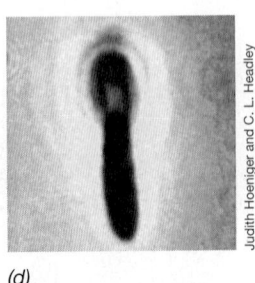

(a)   *(b)*   *(c)*   *(d)*

Judith Hoeniger and C. L. Headley

Figure 3.34 Endospore germination in *Bacillus*. Conversion of an endospore into a vegetative cell. The series of phase-contrast photomicrographs shows the sequence of events starting from *(a)* a highly refractile free endospore. *(b)* Activation: Refractility is being lost. *(c, d)* Outgrowth: The new vegetative cell is emerging.

cells of *Bacillus*, a typical endospore-forming bacterium, cease vegetative growth and begin sporulation when, for example, a key nutrient such as carbon or nitrogen becomes limiting.

An endospore can remain dormant for years, but it can convert back to a vegetative cell relatively rapidly. This process involves three steps: *activation*, *germination*, and *outgrowth* (**Figure 3.34**). Activation occurs when endospores are heated for several minutes at an elevated but sublethal temperature. Activated endospores are then conditioned to germinate when placed in the presence of specific nutrients, such as certain amino acids. Germination, typically a rapid process (on the order of several minutes), involves loss of microscopic refractility of the endospore, increased ability to be stained by dyes, and loss of resistance to heat and chemicals. The final stage, outgrowth, involves visible swelling due to water uptake and synthesis of

RNA, proteins, and DNA. The cell emerges from the broken endospore and begins to grow, remaining in vegetative growth until environmental signals once again trigger sporulation.

Endospore Structure

Endospores stand out under the light microscope as strongly refractile structures (see Figures 3.32–3.34). Endospores are impermeable to most dyes, so occasionally they are seen as unstained regions within cells that have been stained with basic dyes such as methylene blue. To stain endospores, special stains and procedures must be used. In the classical endospore-staining protocol, malachite green is used as a stain and is infused into the spore with steam.

The structure of the endospore as seen with the electron microscope differs distinctly from that of the vegetative cell (**Figure 3.35**). In particular, the endospore is structurally more complex in that it has many layers that are absent from the vegetative cell. The outermost layer is the *exosporium*, a thin protein covering. Within this are the *spore coats*, composed of layers of spore-specific proteins (Figure 3.35*b*). Below the spore coat is the *cortex*, which consists of loosely cross-linked peptidoglycan, and inside the cortex is the *core*, which contains the core wall, cytoplasmic membrane, cytoplasm, nucleoid, ribosomes, and other cellular essentials. Thus, the endospore differs structurally from the vegetative cell primarily in the kinds of structures found outside the core wall.

One substance that is characteristic of endospores but absent from vegetative cells is **dipicolinic acid** (**Figure 3.36**), which accumulates in the core. Endospores are also enriched in calcium (Ca^{2+}), most of which is complexed with dipicolinic acid (Figure 3.36*b*). The calcium–dipicolinic acid complex represents about

Exosporium
Spore coat
Core wall
Cortex
DNA

(a)   *(b)*

Figure 3.35 Structure of the bacterial endospore. *(a)* Transmission electron micrograph of a thin section through an endospore of *Bacillus megaterium*. *(b)* Fluorescent photomicrograph of a cell of *Bacillus subtilis* undergoing sporulation. The green color is a dye that specifically stains a sporulation protein in the spore coat.

(a)

(b)   Carboxylic acid groups

Figure 3.36 Dipicolinic acid (DPA). *(a)* Structure of DPA. *(b)* How Ca^{2+} cross-links DPA molecules to form a complex.

The *Mycobacterium tuberculosis* Cell Wall:
A Complex Architecture of Lipids and Carbohydrates

Every second, someone in the world becomes infected with tuberculosis (TB). The causative agent, *Mycobacterium tuberculosis,* is a bacterium responsible for approximately 3 million deaths each year worldwide.[1] *M. tuberculosis* is unique among prokaryotes in having evolved a plethora of defense and virulence strategies that enable it to reside within the lungs of its host and evade the onslaught of the ensuing immune response. The *M. tuberculosis* cell envelope provides a formidable protective barrier against drugs and host defense mechanisms, such as antibiotics and the bacteriocidal environment of the macrophage (⮁ Section 18.5).

First discovered in 1882 by Robert Koch (⮁ Section 1.8), *M. tuberculosis* is a rod-shaped filamentous bacterium with an unusual waxy cell wall that renders the cell impervious to conventional Gram staining. Instead, these bacilli turn red when subjected to a special mycobacterial stain, the Ziehl-Neelsen stain, and the key cell wall component necessary for this "acid-fastness" is a unique lipid called mycolic acid.

M. tuberculosis has a chemically complex cell wall dominated by three essential macromolecules: (1) peptidoglycan (PG), (2) arabinogalactan (AG), and (3) mycolic acids (MA) all of which are covalently attached, forming the mycolyl arabinogalactan-peptidoglycan (mAGP) cell wall core[2] (Figure 1). PG is composed of alternating *N*-acetylglucosamine (NAG) and *N*-acetylmuramic acid (NAM) residues forming long glycan chains which surround the cell (extensively cross-linked via stem peptides) to provide a rigid foundation for the attachment of arabinogalactan via a disaccharide linker unit (Figure 1). AG is a complex polysaccharide composed entirely of arabinofuranose (Ara*f*) and galactofuranose (Gal*f*) sugars, which are otherwise rarely found in nature (Figure 1). A single linear polysac-

charide of galactan is decorated with three domains of arabinan, a highly branched polysaccharide that serves as a molecular scaffold for the attachment site of mycolic acids.

Mycolic acids are long-chain fatty acids (~60 to 90 carbon atoms in length) and are extensively deployed over the surface of the TB bacilli. The cell wall–bound mycolates form the inner leaflet of an asymmetrical pseudo "myco-membrane," and it is thought that the hydrocarbon chains of these lipids interact in a tightly packed, parallel arrangement with other "free" non-covalently bound complex lipids, glycolipids, and lipoglycans which make up the outer leaflet of the "myco-membrane." Many of these components, such as trehalose dimycolate and lipoarabinomannan, are anti-

genic and contribute to the modulation of the host immune response during infection (Figure 1). The importance of the mAGP to the persistence and pathogenicity of the TB bacillus is underscored by the fact that several frontline anti-TB drugs, such as isoniazid and ethambutol, target the formation of mycolic acids and arabinogalactan, respectively.[3, 4]

With the global incidence of multi-drug-resistant strains of TB rapidly increasing, the need for new drugs is overwhelming. An international community of TB researchers are currently investigating fundamental aspects of *M. tuberculosis* cell wall biochemistry and how it interacts with the host immune system, in an effort to develop new therapies against this terrible burden to humankind.

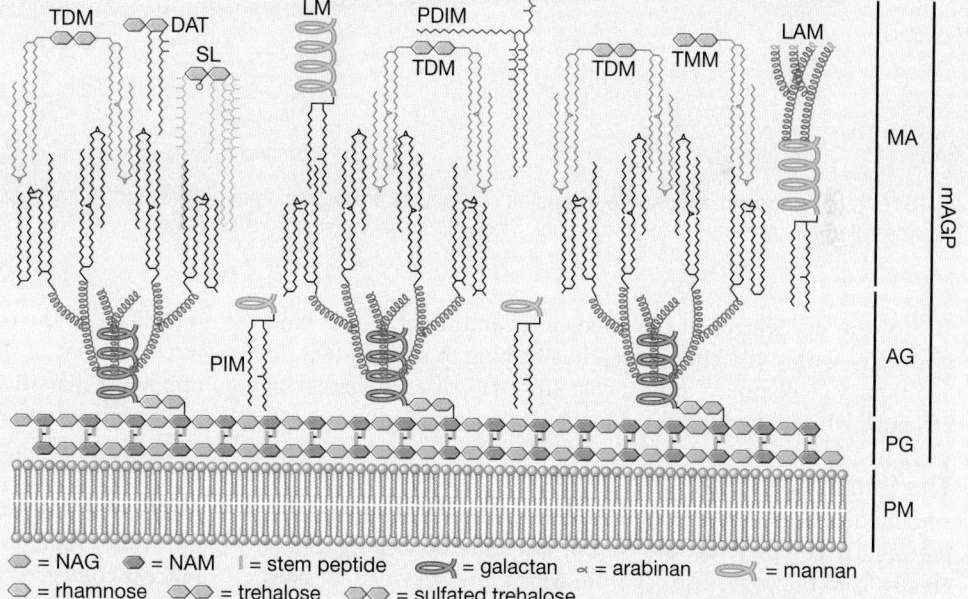

Figure 1 Structural representation of the *Mycobacterium tuberculosis* cell wall. The cell wall core (mAGP) is composed of peptidoglycan (PG) connected to mycolic acids (MA) via the connective polysaccharide arabinogalactan (AG). Other non-covalently associated cell wall components include complex lipids such as trehalose dimycolate (TDM), trehalose monomycolate (TMM), sulfolipid (SL), diacyl trehalose (DAT), and phthiocerol dimycocerosate (PDIM). Glycolipids and lipoglycans such as lipoarabinomannan (LAM), lipomannan (LM), and phosphatidylinositol mannosides (PIM) are interspersed throughout the outer "myco-membrane."

[1]Alderwick, L. J., *et al*. 2007. Structure, function and biosynthesis of the *Mycobacterium tuberculosis* cell wall: arabinogalactan and lipoarabinomannan assembly with a view to discovering new drug targets. *Biochem. Soc. Trans.* 35: 1325–1328.

[2]Bhatt, A., *et al*. 2007. The *Mycobacterium tuberculosis* FAS-II condensing enzymes: their role in mycolic acid biosynthesis, acid-fastness, pathogenesis and in future drug development. *Mol. Microbiol.* 64: 1442–1454.

[3]Banerjee, A., *et al*. 1994. inhA, a gene encoding a target for isoniazid and ethionamide in *Mycobacterium tuberculosis*. *Science* 263: 227–230.

[4]Belanger, A. E., *et al*. 1996. The embAB genes of *Mycobacterium avium* encode an arabinosyl transferase involved in cell wall arabinan biosynthesis that is the target for the antimycobacterial drug ethambutol. *Proc. Natl Acad. Sci. USA* 93: 11919–11924.

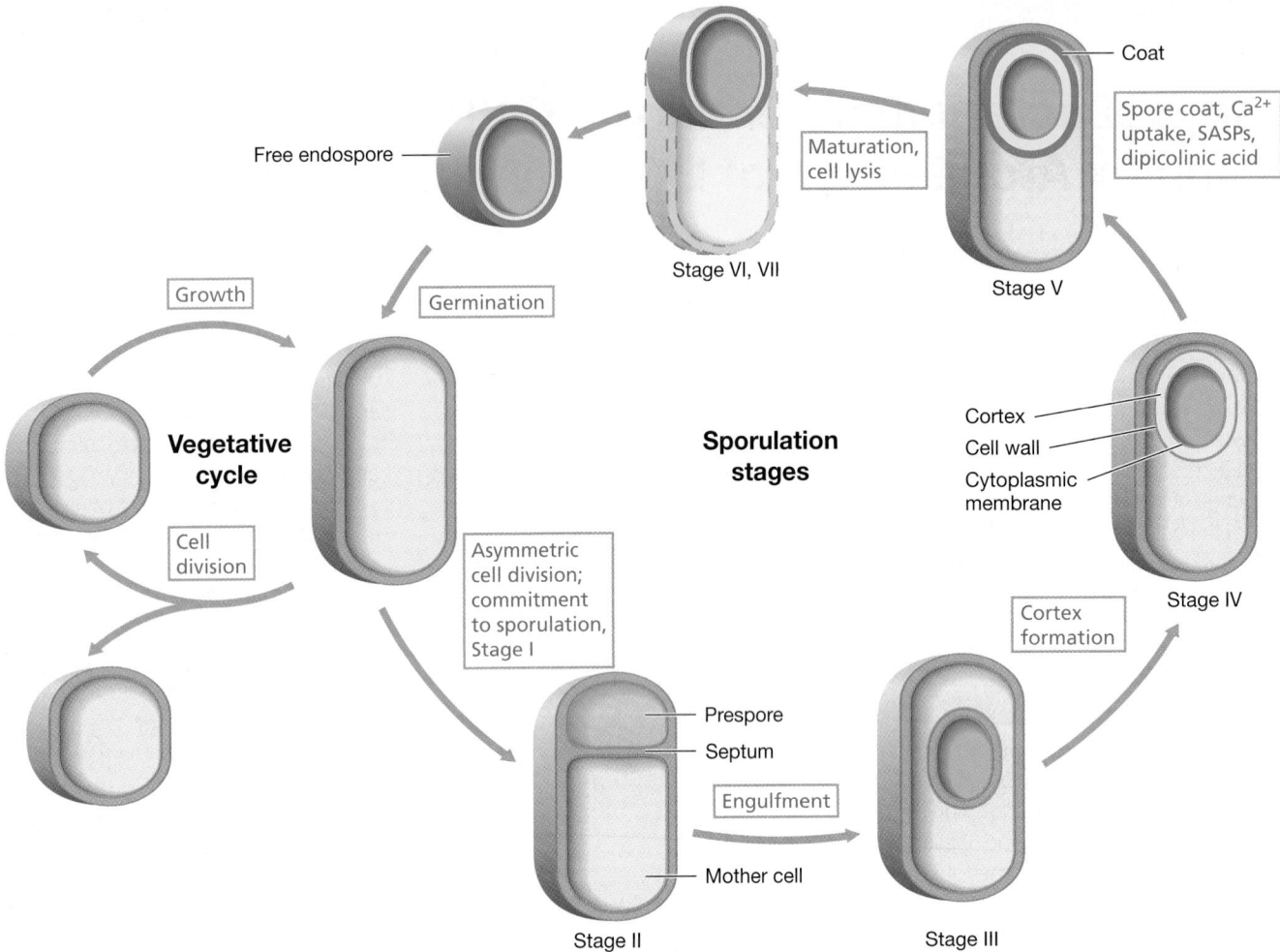

Figure 3.37 Stages in endospore formation. The stages are defined from genetic and microscopic analyses of sporulation in *Bacillus subtilis*, the model organism for studies of sporulation.

10% of the dry weight of the endospore, and functions to bind free water within the endospore, thus helping to dehydrate it. In addition, the complex intercalates (inserts between bases) in DNA, which stabilizes DNA against heat denaturation.

The Endospore Core and SASPs

Although both contain a copy of the chromosome and other essential cellular components, the core of a mature endospore differs greatly from the vegetative cell from which it was formed. Besides the high levels of calcium dipicolinate (Figure 3.36), which help reduce the water content of the core, the core becomes greatly dehydrated during the sporulation process. The core of a mature endospore has only 10–25% of the water content of the vegetative cell, and thus the consistency of the core cytoplasm is that of a gel. Dehydration of the core greatly increases the heat resistance of macromolecules within the spore. Some bacterial endospores survive heating to temperatures as high as 150°C, although 121°C, the standard for microbiological sterilization (121°C is autoclave temperature, ⇄ Section 26.1), kills the endospores of most species. Boiling has essentially no effect on endospore viability. Dehydration has also been shown to confer resistance in the endospore to chemicals, such as hydrogen peroxide (H_2O_2), and causes enzymes

remaining in the core to become inactive. In addition to the low water content of the endospore, the pH of the core is about one unit lower than that of the vegetative cell cytoplasm.

The endospore core contains high levels of *small acid-soluble proteins* (SASPs). These proteins are made during the sporulation process and have at least two functions. SASPs bind tightly to DNA in the core and protect it from potential damage from ultraviolet radiation, desiccation, and dry heat. Ultraviolet resistance is conferred when SASPs change the molecular structure of DNA from the normal "B" form to the more compact "A" form. A-form DNA better resists pyrimidine dimer formation by UV radiation, a means of mutation (⇄ Section 10.4), and resists the denaturing effects of dry heat. In addition, SASPs function as a carbon and energy source for the outgrowth of a new vegetative cell from the endospore during germination.

The Sporulation Process

Sporulation is a complex series of events in cellular differentiation; many genetically directed changes in the cell underlie the conversion from vegetative growth to sporulation. The structural changes occurring in sporulating cells of *Bacillus* are shown in **Figure 3.37**. Sporulation can be divided into several stages. In

Bacillus subtilis, where detailed studies have been done, the entire sporulation process takes about 8 hours and begins with asymmetric cell division (Figure 3.37). Genetic studies of mutants of *Bacillus*, each blocked at one of the stages of sporulation, indicate that more than 200 spore-specific genes exist. Sporulation requires a significant regulatory response in that the synthesis of many vegetative proteins must cease while endospore proteins are made. This is accomplished by the activation of several families of endospore-specific genes in response to an environmental trigger to sporulate. The proteins encoded by these genes catalyze the series of events leading from a moist, metabolizing, vegetative cell to a relatively dry, metabolically inert, but extremely resistant endospore (**Table 3.3**). In Section 8.12 we examine some of the molecular events that control the sporulation process.

Diversity and Phylogenetic Aspects of Endospore Formation

Nearly 20 genera of *Bacteria* form endospores, although the process has only been studied in detail in a few species of *Bacillus* and *Clostridium*. Nevertheless, many of the secrets to endospore survival, such as the formation of calcium–dipicolinate complexes (Figure 3.36) and the production of endospore-specific proteins, seem universal. Although some of the details of sporulation may vary from one organism to the next, the general principles seem to be the same in all endosporulating bacteria.

From a phylogenetic perspective, the capacity to produce endospores is found only in a particular sublineage of the gram-positive bacteria. Despite this, the physiologies of endospore-forming bacteria are highly diverse and include anaerobes, aerobes, phototrophs, and chemolithotrophs. In light of this physiological diversity, the actual triggers for endospore formation may vary with different species and could include signals other than simple nutrient starvation, the major trigger for endospore formation in *Bacillus*. No *Archaea* have been shown

to form endospores, suggesting that the capacity to produce endospores evolved sometime after the major prokaryotic lineages diverged billions of years ago (↶ Figure 1.6).

MiniQuiz

• What is dipicolinic acid and where is it found?

• What are SASPs and what is their function?

• What happens when an endospore germinates?

Microbial Locomotion

We finish our survey of microbial structure and function by considering cell locomotion. Most microbial cells can move under their own power, and motility allows cells to reach different parts of their environment. In nature, movement may present new opportunities and resources for a cell and be the difference between life and death.

We examine here the two major types of cell movement, *swimming* and *gliding*. We then consider how motile cells are able to move in a directed fashion toward or away from particular stimuli (phenomena called *taxes*) and present examples of these simple behavioral responses.

3.13 Flagella and Motility

Many prokaryotes are motile by swimming, and this function is due to a structure called the **flagellum** (plural, flagella) (**Figure 3.38**). The flagellum functions by rotation to push or pull the cell through a liquid medium.

Flagella of *Bacteria*

Bacterial flagella are long, thin appendages free at one end and attached to the cell at the other end. Bacterial flagella are so thin (15–20 nm, depending on the species) that a single flagellum can be seen with the light microscope only after being stained with special stains that increase their diameter (Figure 3.38). However, flagella are easily seen with the electron microscope (**Figure 3.39**).

Flagella can be attached to cells in different places. In **polar flagellation**, the flagella are attached at one or both ends of a cell. Occasionally a group of flagella (called a *tuft*) may arise at one end of the cell, a type of polar flagellation called *lophotrichous* (Figure 3.38c). Tufts of flagella can often be seen in unstained

Table 3.3 *Differences between endospores and vegetative cells*		
Characteristic	**Vegetative cell**	**Endospore**
Microscopic appearance	Nonrefractile	Refractile
Calcium content	Low	High
Dipicolinic acid	Absent	Present
Enzymatic activity	High	Low
Respiration rate	High	Low or absent
Macromolecular synthesis	Present	Absent
Heat resistance	Low	High
Radiation resistance	Low	High
Resistance to chemicals	Low	High
Lysozyme	Sensitive	Resistant
Water content	High, 80–90%	Low, 10–25% in core
Small acid-soluble proteins	Absent	Present

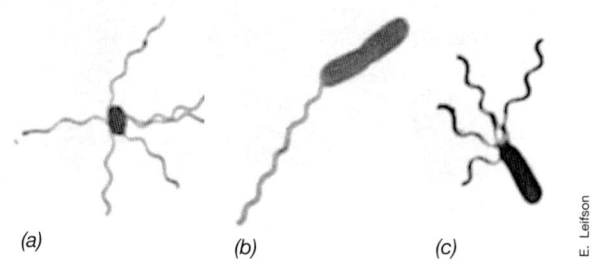

(a)   *(b)*   *(c)*

E. Leifson

Figure 3.38 Bacterial flagella. Light photomicrographs of prokaryotes containing different arrangements of flagella. Cells are stained with Leifson flagella stain. *(a)* Peritrichous. *(b)* Polar. *(c)* Lophotrichous.

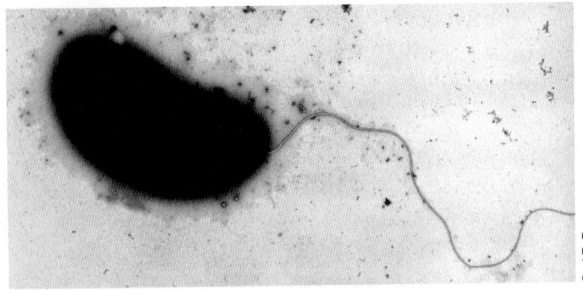

(a)

Carl E. Bauer

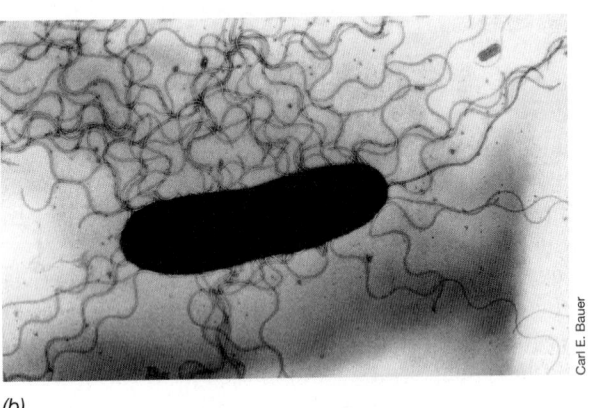

(b)

Carl E. Bauer

Figure 3.39 Bacterial flagella as observed by negative staining in the transmission electron microscope. *(a)* A single polar flagellum. *(b)* Peritrichous flagella. Both micrographs are of cells of the phototrophic bacterium *Rhodospirillum centenum*, which are about 1.5 μm wide. Cells of *R. centenum* are normally polarly flagellated but under certain growth conditions form peritrichous flagella. See Figure 3.49*b* for a photo of colonies of *R. centenum* cells that move toward an increasing gradient of light (phototaxis).

cells by dark-field or phase-contrast microscopy (**Figure 3.40**). When a tuft of flagella emerges from both poles of the cell, flagellation is called *amphitrichous*. In **peritrichous flagellation** (Figures 3.38*a* and 3.39*b*), flagella are inserted at many locations around the cell surface. The type of flagellation, polar or peritrichous, is a characteristic used in the classification of bacteria.

Flagellar Structure

Flagella are not straight but are actually helical. When flattened, flagella show a constant distance between adjacent curves, called the *wavelength*, and this wavelength is characteristic for the flagella of any given species (Figures 3.38–3.40). The filament of a bacterial flagellum is composed of many copies of a protein called *flagellin*. The shape and wavelength of the flagellum are in part determined by the structure of the flagellin protein and also to some extent by the direction of rotation of the filament. Flagellin is highly conserved in amino acid sequences in species of *Bacteria*, suggesting that flagellar motility evolved early and has deep roots within this domain.

A flagellum consists of several components and moves by rotation, much like a propeller of a boat motor. The base of the flagellum is structurally different from the filament. There is a wider region at the base of the filament called the *hook*. The hook consists of a single type of protein and connects the filament to the motor portion in the base (**Figure 3.41**).

The motor is anchored in the cytoplasmic membrane and cell wall. The motor consists of a central rod that passes through a series of rings. In gram-negative bacteria, an outer ring, called the *L ring*, is anchored in the lipopolysaccharide layer. A second ring, called the *P ring*, is anchored in the peptidoglycan layer of the cell wall. A third set of rings, called the *MS* and *C rings*, are located within the cytoplasmic membrane and the cytoplasm,

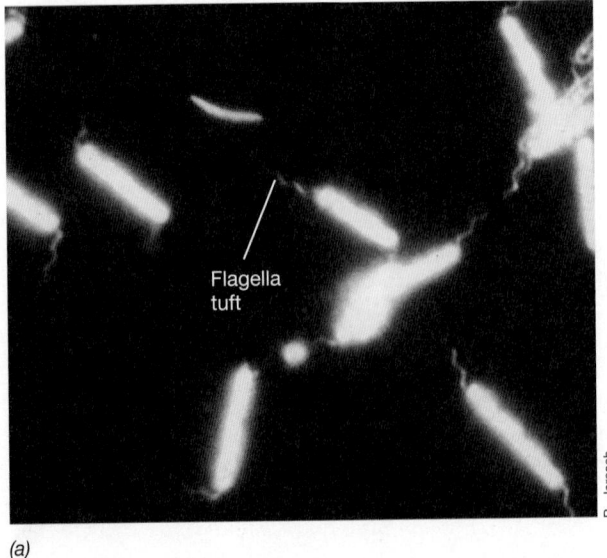

Flagella tuft

(a)

R. Jarosch

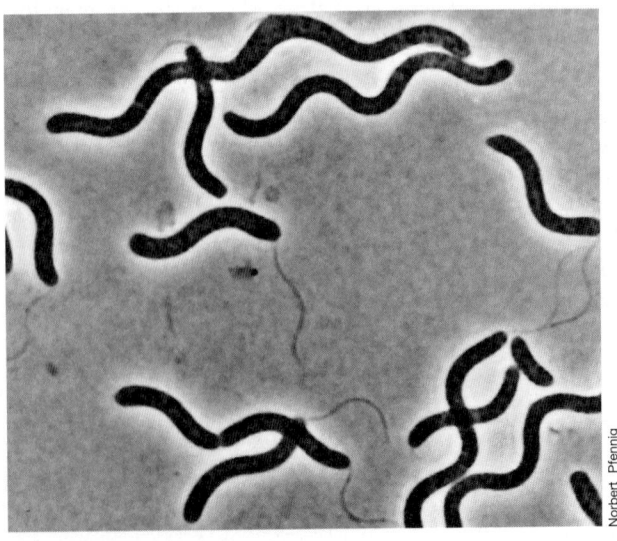

(b)

Norbert Pfennig

Figure 3.40 Bacterial flagella observed in living cells. *(a)* Dark-field photomicrograph of a group of large rod-shaped bacteria with flagellar tufts at each pole (amphitrichous flagellation). A single cell is about 2 μm wide. *(b)* Phase-contrast photomicrograph of cells of the large phototrophic purple bacterium *Rhodospirillum photometricum* with a tuft of lophotrichous flagella that emanate from one of the poles. A single cell measures about 3 × 30 μm.

UNIT 1

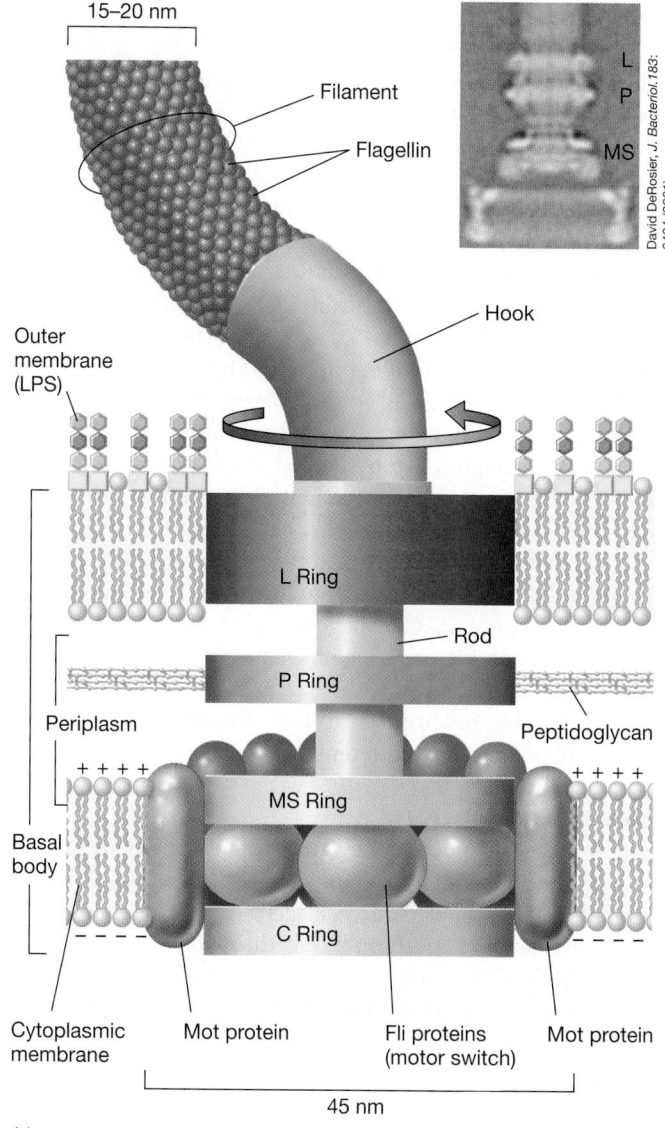

David DeRosier, J. Bacteriol.183: 6404 (2001)

(a)

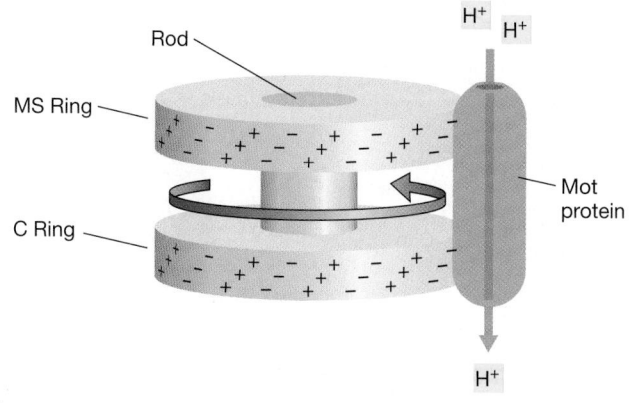

(b)

Figure 3.41 Structure and function of the flagellum in gram-negative *Bacteria*. *(a)* Structure. The L ring is embedded in the LPS and the P ring in peptidoglycan. The MS ring is embedded in the cytoplasmic membrane and the C ring in the cytoplasm. A narrow channel exists in the rod and filament through which flagellin molecules diffuse to reach the site of flagellar synthesis. The Mot proteins function as the flagellar motor, whereas the Fli proteins function as the motor switch. The flagellar motor rotates the filament to propel the cell through the medium. Inset: transmission electron micrograph of a flagellar basal body from *Salmonella enterica* with the various rings labeled. *(b)* Function. A "proton turbine" model has been proposed to explain rotation of the flagellum. Protons, flowing through the Mot proteins, may exert forces on charges present on the C and MS rings, thereby spinning the rotor.

respectively (Figure 3.41*a*). In gram-positive bacteria, which lack an outer membrane, only the inner pair of rings is present. Surrounding the inner ring and anchored in the cytoplasmic membrane are a series of proteins called *Mot proteins*. A final set of proteins, called the *Fli proteins* (Figure 3.41*a*), function as the motor switch, reversing the direction of rotation of the flagella in response to intracellular signals.

Flagellar Movement

The flagellum is a tiny rotary motor. How does this motor work? Rotary motors contain two main components: the *rotor* and the *stator*. In the flagellar motor, the rotor consists of the central rod and the L, P, C, and MS rings. Collectively, these structures make up the **basal body**. The stator consists of the Mot proteins that surround the basal body and function to generate torque.

Rotation of the flagellum is imparted by the basal body. The energy required for rotation of the flagellum comes from the proton motive force (⮌ Section 4.10). Proton movement across the cytoplasmic membrane through the Mot complex drives rotation of the flagellum (Figure 3.41). About 1000 protons are translocated per rotation of the flagellum, and a model for how this could work is shown in Figure 3.41*b*. In this model called the proton turbine model, protons flowing through channels in the Mot proteins exert electrostatic forces on helically arranged charges on the rotor proteins. Attractions between positive and negative charges would then cause the basal body to rotate as protons flow though the Mot proteins. www.microbiologyplace.com **Online Tutorial 3.1: The Prokaryotic Flagellum**

Archaeal Flagella

Besides *Bacteria*, flagellar motility is also widespread among species of *Archaea*; major genera of methanogens, extreme halophiles, thermoacidophiles, and hyperthermophiles are all capable of swimming motility. Archaeal flagella are roughly half the diameter of bacterial flagella, measuring only 10–13 nm in width (**Figure 3.42**), but impart movement to the cell by rotating, as do flagella in *Bacteria*. However, unlike *Bacteria*, in which a single type of protein makes up the flagellar filament, several different flagellin proteins are known from *Archaea*, and their amino acid sequences and genes that encode them bear no relationship to those of bacterial flagellin.

Studies of swimming cells of the extreme halophile *Halobacterium* show that they swim at speeds only about one-tenth that of cells of *Escherichia coli*. Whether this holds for all *Archaea* is

Labels for figure (a):
15–20 nm
Filament
Flagellin
Hook
Outer membrane (LPS)
L Ring
Rod
P Ring
Periplasm
Peptidoglycan
Basal body
MS Ring
Cytoplasmic membrane
C Ring
Mot protein
Fli proteins (motor switch)
Mot protein
45 nm
L
P
MS

Labels for figure (b):
Rod
MS Ring
C Ring
H⁺
H⁺
Mot protein
H⁺

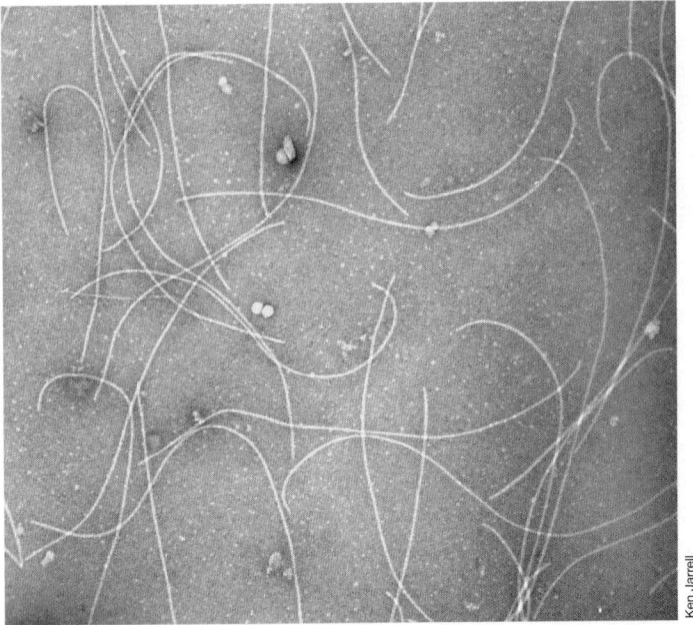

Figure 3.42 Archaeal flagella. Transmission electron micrograph of flagella isolated from cells of the methanogen *Methanococcus maripaludis*. A single flagellum is about 12 nm wide.

unknown, but the significantly smaller diameter of the archaeal flagellum compared with the bacterial flagellum would naturally reduce the torque and power of the flagellar motor such that slower swimming speeds would be expected. Moreover, from biochemical experiments with *Halobacterium* it appears that archaeal flagella are powered directly by ATP rather than by the proton motive force, the source of energy for the flagella of *Bacteria* (Figure 3.41). If this holds for the flagella of all motile *Archaea*, it would mean that the flagellar motors of *Archaea* and *Bacteria* employ fundamentally different mechanisms. Coupled with the clear differences in flagellar protein structure, this suggests that flagellar motility in *Bacteria* and *Archaea* evolved after the two prokaryotic domains had diverged over 3 billion years ago (↩ Figure 1.6*b*).

Flagellar Synthesis

Several gene products are required to support motility in *Bacteria*. In *Escherichia coli* and *Salmonella enterica* (*typhimurium*), where studies have been most extensive, over 50 genes are linked to motility. These genes have several functions, including encoding structural proteins of the flagellum and motor apparatus, export of flagellar proteins through the cytoplasmic membrane to the outside of the cell, and regulation of the many biochemical events surrounding the synthesis of new flagella.

A flagellar filament grows not from its base, as does an animal hair, but from its tip. The MS ring is synthesized first and inserted into the cytoplasmic membrane. Then other anchoring proteins are synthesized along with the hook before the filament forms (**Figure 3.43**). Flagellin molecules synthesized in the cytoplasm pass up through a 3-nm channel inside the filament and add on at the terminus to form the mature flagellum. At the end of the growing flagellum a protein "cap" exists. Cap proteins assist flagellin molecules that have diffused through the channel to organize at the flagellum termini to form new filament (Figure 3.43). Approximately 20,000 flagellin protein molecules are needed to make one filament. The flagellum grows more or less continuously until it reaches its final length. Broken flagella still rotate and can be repaired with new flagellin units passed through the filament channel to replace the lost ones.

Cell Speed and Motion

In *Bacteria*, flagella do not rotate at a constant speed but instead increase or decrease their rotational speed in relation to the strength of the proton motive force. Flagella can rotate at up to 300 revolutions per second and propel cells through a liquid at up to 60 cell lengths/sec. By contrast, the fastest known animal, the cheetah, moves at a maximum rate of about 25 body lengths/sec. Thus, when size is taken into account, a bacterial cell swimming at 60 lengths/sec is actually moving twice as fast as the fastest animal!

The swimming motions of polarly and lophotrichously flagellated organisms differ from those of peritrichously flagellated organisms, and these can be distinguished microscopically (**Figure 3.44**). Peritrichously flagellated organisms typically move in a

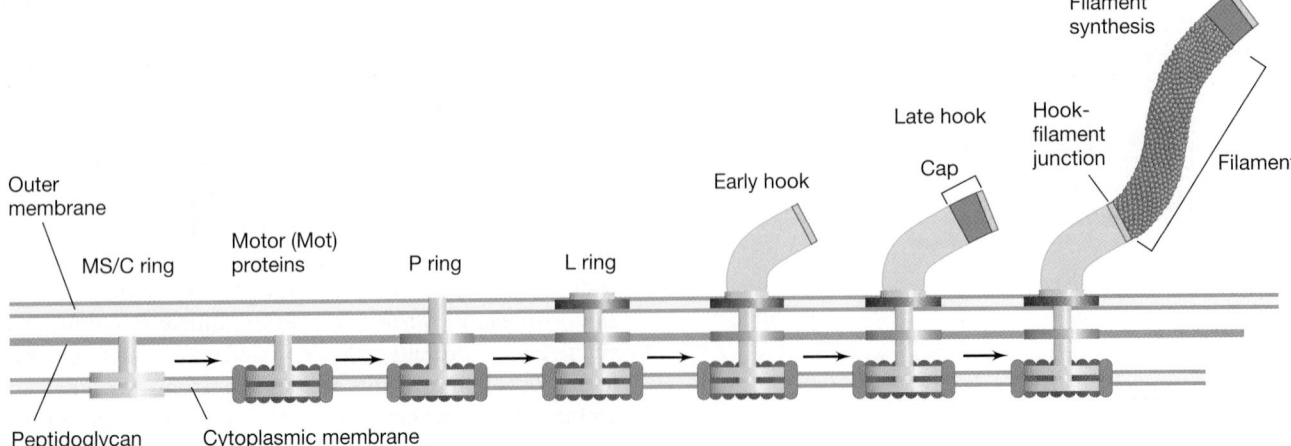

Figure 3.43 Flagella biosynthesis. Synthesis begins with assembly of MS and C rings in the cytoplasmic membrane, followed by the other rings, the hook, and the cap. Flagellin protein flows through the hook to form the filament and is guided into position by cap proteins.

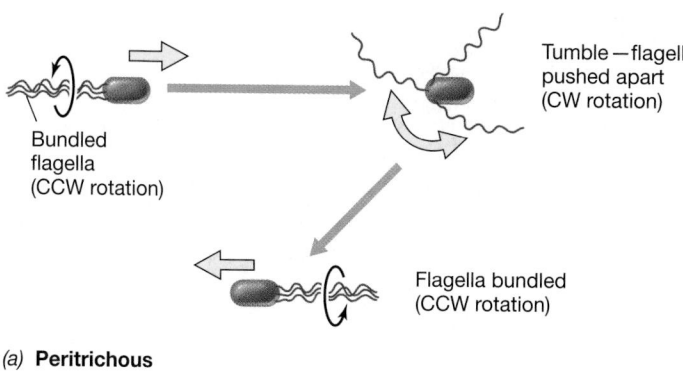

(a) **Peritrichous**

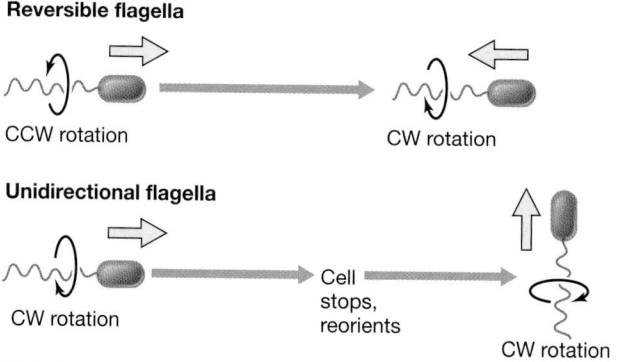

(b) **Polar**

Figure 3.44 Movement in peritrichously and polarly flagellated prokaryotes. *(a)* Peritrichous: Forward motion is imparted by all flagella rotating counterclockwise (CCW) in a bundle. Clockwise (CW) rotation causes the cell to tumble, and then a return to counterclockwise rotation leads the cell off in a new direction. *(b)* Polar: Cells change direction by reversing flagellar rotation (thus pulling instead of pushing the cell) or, with unidirectional flagella, by stopping periodically to reorient, and then moving forward by clockwise rotation of its flagella. The yellow arrows show the direction the cell is traveling.

straight line in a slow, deliberate fashion. Polarly flagellated organisms, on the other hand, move more rapidly, spinning around and seemingly dashing from place to place. The different behavior of flagella on polar and peritrichous organisms, including differences in reversibility of the flagellum, is illustrated in Figure 3.44.

Swimming speed is a genetically governed property because different motile species, even different species that are the same cell size, can swim at different maximum speeds. When assessing the capacity of a laboratory culture of a bacterium for swimming motility and swimming speed, observations should only be made on young cultures. In old cultures, otherwise motile cells often stop swimming and the culture may appear to be nonmotile.

MiniQuiz

- Cells of the rod-shaped *Salmonella* are peritrichously flagellated, those of the rod-shaped *Pseudomonas* polarly flagellated, and those of *Spirillum* lophotrichously flagellated. Sketch the three different cells here, showing how their flagella are arranged.

- Compare the flagella of *Bacteria* and *Archaea* in terms of their structure and function.

3.14 Gliding Motility

Some prokaryotes are motile but lack flagella. Most of these non-swimming yet motile bacteria move across solid surfaces in a process called *gliding*. Unlike flagellar motility, in which cells stop and then start off in a different direction, gliding motility is a slower and smoother form of movement and typically occurs along the long axis of the cell.

Diversity of Gliding Motility

Gliding motility is widely distributed among *Bacteria* but has been well studied in only a few groups. The gliding movement itself—up to 10 μm/sec in some gliding bacteria—is considerably slower than propulsion by flagella but still offers the cell a means of moving about its habitat.

Gliding prokaryotes are filamentous or rod-shaped cells (**Figure 3.45**), and the gliding process requires that the cells be in contact with a solid surface. The morphology of colonies of a typical gliding bacterium are distinctive, because cells glide out and move away from the center of the colony (Figure 3.45c). Perhaps the best-known gliding bacteria are the filamentous cyanobacteria (Figure 3.45a, b), certain gram-negative *Bacteria* such as *Myxococcus* and other myxobacteria, and species of *Cytophaga* and *Flavobacterium* (Figure 3.45c, d). No gliding *Archaea* are known, but once some of the *Archaea* that have been detected in soil using molecular techniques (⇌ Section 2.11) are isolated, gliding species would not be surprising.

Mechanisms of Gliding Motility

Although no gliding mechanism is thoroughly understood, it is clear that more than one mechanism is responsible for gliding motility. Cyanobacteria (phototrophic bacteria, Figure 3.45a, b) glide by secreting a polysaccharide slime on the outer surface of the cell. The slime contacts both the cell surface and the solid surface against which the cell moves. As the excreted slime adheres to the surface, the cell is pulled along. This mechanism is supported by the identification of slime-excreting pores on the cell surface of gliding filamentous cyanobacteria. The nonphototrophic gliding bacterium *Cytophaga* also moves at the expense of slime excretion, rotating along its long axis as it does.

Cells capable of "twitching motility" also display a form of gliding motility using a mechanism by which repeated extension and retraction of type IV pili propel the cell along a surface (Section 3.9). The gliding myxobacterium *Myxococcus xanthus* has two forms of gliding motility. One form is driven by type IV pili whereas the other is distinct from either the type IV pili or the slime extrusion methods. In this form of *M. xanthus* motility a protein adhesion complex is formed at one pole of the rod-shaped cell and remains at a fixed position on the surface as the cell glides forward. This means that the adhesion complex moves in the direction opposite that of the cell, presumably fueled by some sort of cytoplasmic motility engine perhaps linked to the cell cytoskeleton (⇌ Section 5.3). These different forms of motility can be expressed at the same time and are somehow coordinated by the cell, presumably in response to various signals from the environment (Section 3.15).

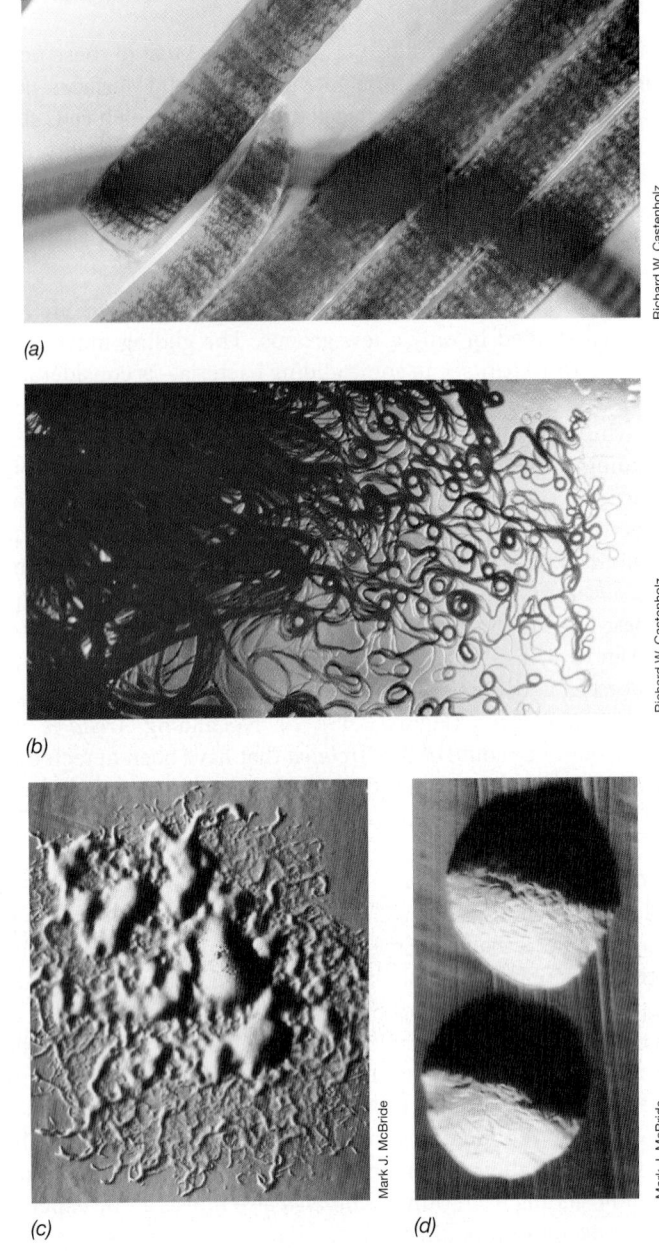

(a)

(b)

Richard W. Castenholz

(c)

Mark J. McBride

(d)

Mark J. McBride

Figure 3.45 Gliding bacteria. *(a, b)* The filamentous cyanobacterium *Oscillatoria* has cells about 35 μm wide. *(b) Oscillatoria* filaments gliding on an agar surface. *(c)* Masses of the bacterium *Flavobacterium johnsoniae* gliding away from the center of the colony (the colony is about 2.7 mm wide). *(d)* Nongliding mutant strain of *F. johnsoniae* showing typical colony morphology of nongliding bacteria (the colonies are 0.7–1 mm in diameter). See also Figure 3.46.

Neither slime extrusion nor twitching is the mechanism of gliding in other gliding bacteria. In *Flavobacterium johnsoniae* (Figure 3.45*c*), for example, no slime is excreted and the cells lack type IV pili. Instead, the movement of proteins on the cell surface may be the mechanism of gliding in this organism. Specific motility proteins anchored in the cytoplasmic and outer membranes are thought to propel cells of *F. johnsoniae* forward by a ratcheting mechanism (**Figure 3.46**). Movement of gliding-specific proteins in the cytoplasmic membrane is driven by energy from

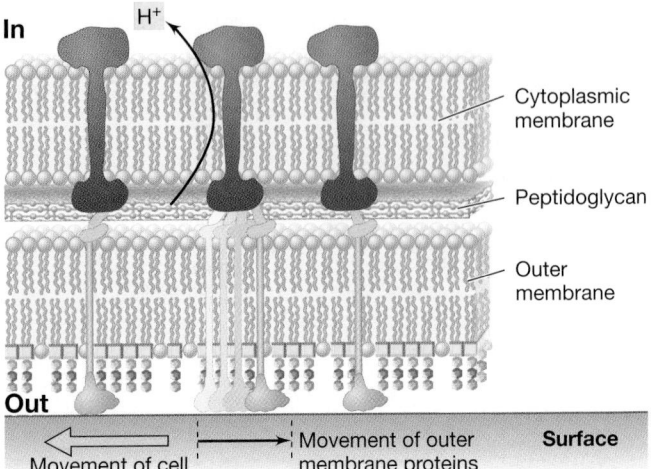

Figure 3.46 Gliding motility in *Flavobacterium johnsoniae.* Tracks (yellow) exist in the peptidoglycan that connect cytoplasmic proteins (brown) to outer membrane proteins (orange) and propel the outer membrane proteins along the solid surface. Note that the outer membrane proteins and the cell proper move in opposite directions.

the proton motive force that is somehow transmitted to gliding-specific proteins in the outer membrane. It is thought that movement of these proteins against the solid surface literally pulls the cell forward (Figure 3.46).

Like other forms of motility, gliding motility has significant ecological relevance. Gliding allows a cell to exploit new resources and to interact with other cells. In the latter regard, it is of interest that myxobacteria, such as *Myxococcus xanthus*, have a very social and cooperative lifestyle. In these bacteria gliding motility may play an important role in the cell-to-cell interactions that are necessary to complete their life cycle (Section 17.17).

MiniQuiz

- How does gliding motility differ from swimming motility in both mechanism and requirements?
- Contrast the mechanism of gliding motility in a filamentous cyanobacterium and in *Flavobacterium*.

3.15 Microbial Taxes

Prokaryotes often encounter gradients of physical or chemical agents in nature and have evolved means to respond to these gradients by moving either toward or away from the agent. Such a directed movement is called a *taxis* (plural, taxes). **Chemotaxis,** a response to chemicals, and **phototaxis,** a response to light, are two well-studied taxes. Here we discuss these taxes in a general way. In Section 8.8 we examine the mechanism of chemotaxis and its regulation in *Escherichia coli* as a model for all prokaryotic taxes.

Chemotaxis has been well studied in swimming bacteria, and much is known at the genetic level concerning how the chemical state of the environment is communicated to the flagellar assembly. Our discussion here will thus deal solely with swimming bacteria. However, some gliding bacteria (Section 3.14) are also

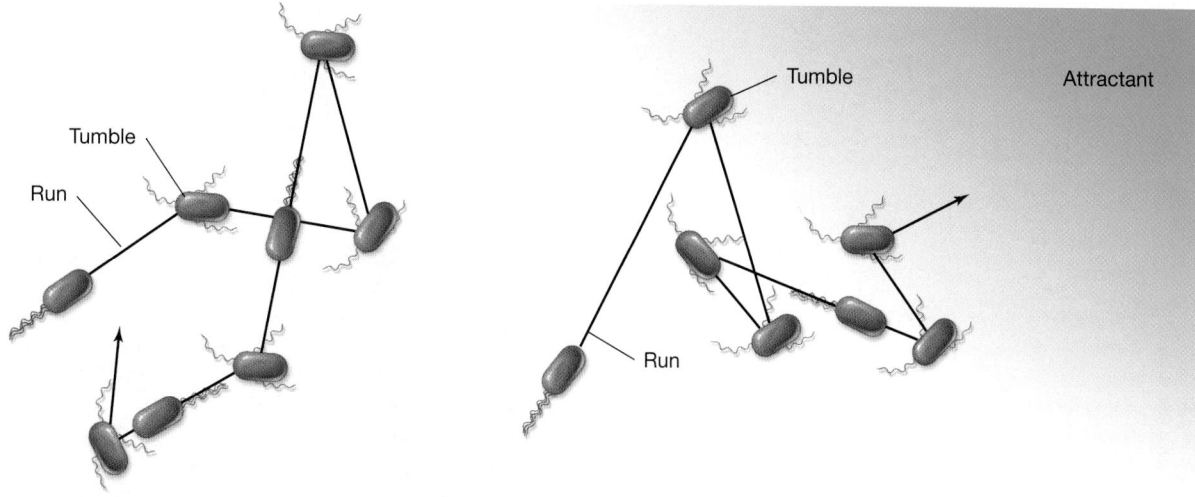

(a) **No attractant present: Random movement**   *(b)* **Attractant present: Directed movement**

Figure 3.47 Chemotaxis in a peritrichously flagellated bacterium such as *Escherichia coli.* *(a)* In the absence of a chemical attractant the cell swims randomly in runs, changing direction during tumbles. *(b)* In the presence of an attractant runs become biased, and the cell moves up the gradient of the attractant. The attractant gradient is depicted in green, with the highest concentration where the color is most intense.

chemotactic, and there are phototactic movements in filamentous cyanobacteria (Figure 3.45*b*). In addition, although they reside in a different evolutionary domain, many species of *Archaea* are also chemotactic and many of the same types of proteins that control chemotaxis in *Bacteria* are present in motile *Archaea* as well.

Chemotaxis

Much research on chemotaxis has been done with the peritrichously flagellated bacterium *E. coli.* To understand how chemotaxis affects the behavior of *E. coli*, consider the situation in which a cell experiences a gradient of some chemical in its environment (**Figure 3.47**). In the absence of a gradient, cells move in a random fashion that includes *runs*, in which the cell is swimming forward in a smooth fashion, and *tumbles*, when the cell stops and jiggles about. During forward movement in a run, the flagellar motor rotates counterclockwise. When flagella rotate clockwise, the bundle of flagella pushes apart, forward motion ceases, and the cells tumble (Figure 3.47).

Following a tumble, the direction of the next run is random. Thus, by means of runs and tumbles, the cell moves about its environment in a random fashion but does not really go anywhere. However, if a gradient of a chemical attractant is present, these random movements become biased. As the organism senses that it is moving toward higher concentrations of the attractant, runs become longer and tumbles are less frequent. The result of this behavioral response is that the organism moves up the concentration gradient of the attractant (Figure 3.47*b*). If the organism senses a repellent, the same general mechanism applies, although in this case it is the decrease in concentration of the repellent (rather than the increase in concentration of an attractant) that promotes runs.

How are chemical gradients sensed? Prokaryotic cells are too small to sense a gradient of a chemical along the length of a single cell. Instead, while moving, the cell monitors its environment, comparing its chemical or physical state with that sensed a few moments before. Bacterial cells are thus responding to *temporal* rather than *spatial* differences in the concentration of a chemical as they swim. Sensory information is fed through an elaborate cascade of proteins that eventually affect the direction of rotation of the flagellar motor. The attractants and repellents are sensed by a series of membrane proteins called *chemoreceptors*. These proteins bind the chemicals and begin the process of sensory transduction to the flagellum (Section 8.8). In a way, chemotaxis can be considered a type of sensory response system, analogous to sensory responses in the nervous system of animals.

Chemotaxis in Polarly Flagellated Bacteria

Chemotaxis in polarly flagellated cells shows similarities to and differences from that in peritrichously flagellated cells such as *E. coli.* Many polarly flagellated bacteria, such as *Pseudomonas* species, can reverse the direction of rotation of their flagella and in so doing reverse their direction of movement (Figure 3.44*b*). However, some polarly flagellated bacteria, such as the phototrophic bacterium *Rhodobacter sphaeroides*, have flagella that rotate only in a clockwise direction. How do such cells change direction, and are they chemotactic?

In cells of *R. sphaeroides*, which have only a single flagellum inserted subpolarly, rotation of the flagellum stops periodically. When it stops, the cell becomes reoriented in a random way by Brownian motion. As the flagellum begins to rotate again, the cell moves in a new direction. Nevertheless, cells of *R. sphaeroides* are strongly chemotactic to certain organic compounds and also show tactic responses to oxygen and light. *R. sphaeroides* cannot reverse its flagellar motor and tumble as *E. coli* can, but there is a

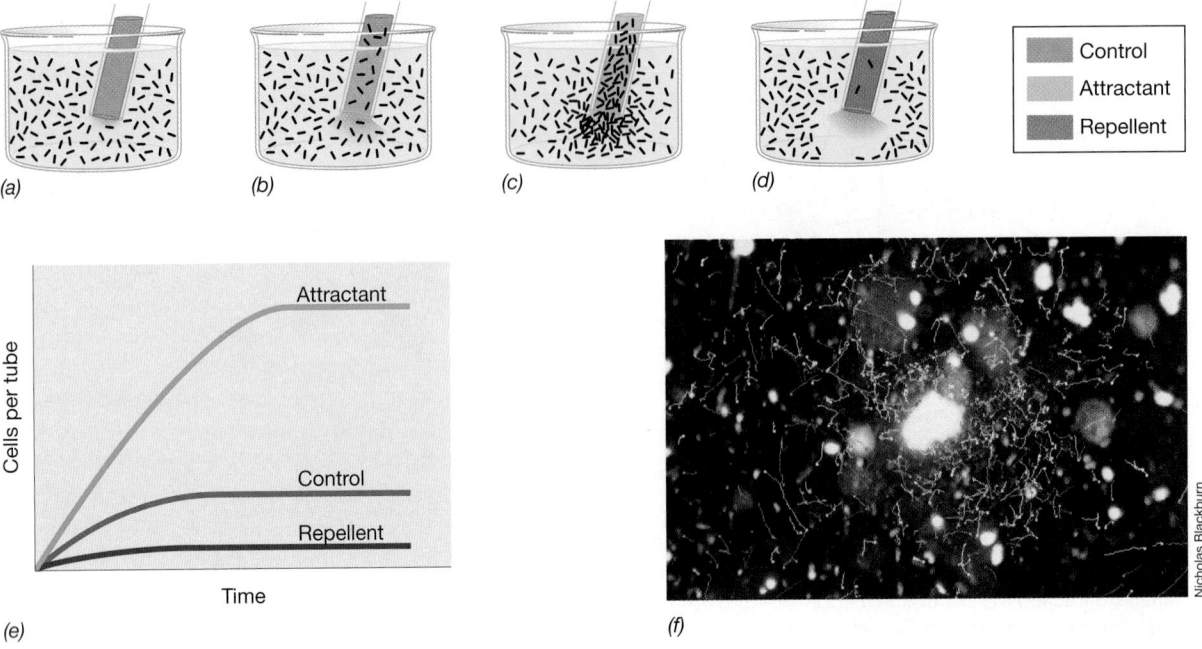

Figure 3.48 Measuring chemotaxis using a capillary tube assay. *(a)* Insertion of the capillary into a bacterial suspension. As the capillary is inserted, a gradient of the chemical begins to form. *(b)* Control capillary contains a salt solution that is neither an attractant nor a repellent. Cell concentration inside the capillary becomes the same as that outside. *(c)* Accumulation of bacteria in a capillary containing an attractant. *(d)* Repulsion of bacteria by a repellent. *(e)* Time course showing cell numbers in capillaries containing various chemicals. *(f)* Tracks of motile bacteria in seawater swarming around an algal cell (large white spot, center) photographed with a tracking video camera system attached to a microscope. The bacterial cells are showing positive aerotaxis by moving toward the oxygen-producing algal cell. The alga is about 60 μm in diameter.

similarity in that the cells maintain runs as long as they sense an increasing concentration of attractant; movement ceases if the cells sense a decreasing concentration of attractant. By random reorientation, a cell eventually finds a path of increasing attractant and maintains a run until either its chemoreceptors are saturated or it begins to sense a decrease in the level of attractant.

Measuring Chemotaxis

Bacterial chemotaxis can be demonstrated by immersing a small glass capillary tube containing an attractant in a suspension of motile bacteria that does not contain the attractant. From the tip of the capillary, a gradient forms into the surrounding medium, with the concentration of chemical gradually decreasing with distance from the tip (**Figure 3.48**). When an attractant is present, the bacteria will move toward it, forming a swarm around the open tip (Figure 3.48*c*) with many of the bacteria swimming into the capillary itself. Of course, because of random movements some bacteria will move into the capillary even if it contains a solution of the same composition as the medium (control solution, Figure 3.48*b*). However, when an attractant is present, movements become biased, and the number of bacteria within the capillary can be many times higher than external cell numbers. If the capillary is removed after a time period and the cells within the capillary are counted and compared with that of the control, attractants can easily be identified (Figure 3.48*e*).

If the inserted capillary contains a repellent, just the opposite occurs; the cells sense an increasing gradient of repellent and the appropriate chemoreceptors affect flagellar rotation to move the cells away from the repellent. In this case, the number of bacteria within the capillary will be fewer than in the control (Figure 3.48*d*). Using the capillary method, it is possible to screen chemicals to see if they are attractants or repellents for a given bacterium.

Chemotaxis can also be observed under a microscope. Using a video camera that captures the position of bacterial cells with time and shows the motility tracks of each cell, it is possible to see the chemotactic movements of cells (Figure 3.48*f*). This method has been adapted to studies of chemotaxis of bacteria in natural environments. In nature it is thought that the major chemotactic agents for bacteria are nutrients excreted from larger microbial cells or from live or dead macroorganisms. Algae, for example, produce both organic compounds and oxygen (O_2, from photosynthesis) that can trigger chemotactic movements of bacteria toward the algal cell (Figure 3.48*f*).

Phototaxis

Many phototrophic microorganisms can move toward light, a process called *phototaxis*. The advantage of phototaxis for a phototrophic organism is that it allows it to orient itself most efficiently to receive light for photosynthesis. This can be seen if a light spectrum is spread across a microscope slide on which there are motile phototrophic purple bacteria. On such a slide the bacteria accumulate at wavelengths at which their photosynthetic pigments absorb (**Figure 3.49**; ∂∂ Sections 13.1–13.5 cover photosynthesis). These pigments include, in particular, bacteriochlorophylls and carotenoids.

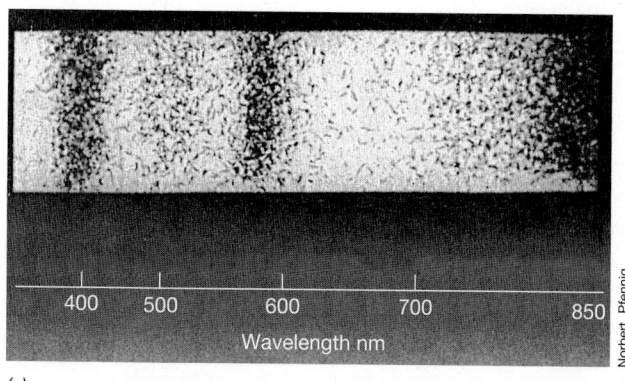

400 500   600   700   850
Wavelength nm

Norbert Pfennig

(a)

Light

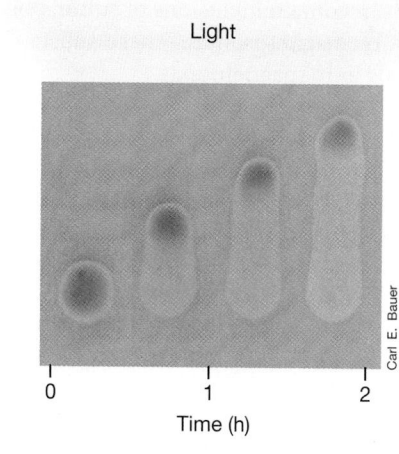

0   1   2
Time (h)

Carl E. Bauer

(b)

Figure 3.49 **Phototaxis of phototrophic bacteria.** *(a)* Scotophobic accumulation of the phototrophic purple bacterium *Thiospirillum jenense* at wavelengths of light at which its pigments absorb. A light spectrum was displayed on a microscope slide containing a dense suspension of the bacteria; after a period of time, the bacteria had accumulated selectively and the photomicrograph was taken. The wavelengths at which the bacteria accumulated are those at which the photosynthetic pigment bacteriochlorophyll *a* absorbs (compare with Figure 13.3*b*). *(b)* Phototaxis of an entire colony of the purple phototrophic bacterium *Rhodospirillum centenum*. These strongly phototactic cells move in unison toward the light source at the top. See Figure 3.39 for electron micrographs of flagellated *R. centenum* cells.

Two different light-mediated taxes are observed in phototrophic bacteria. One, called *scotophobotaxis*, can be observed only microscopically and occurs when a phototrophic bacterium happens to swim outside the illuminated field of view of the microscope into darkness. Entering darkness negatively affects the energy state of the cell and signals it to tumble, reverse direction, and once again swim in a run, thus reentering the light. Scotophobotaxis is presumably a mechanism by which phototrophic purple bacteria avoid entering darkened habitats when they are moving about in illuminated ones, and this likely improves their competitive success.

True phototaxis differs from scotophobotaxis; in phototaxis, cells move up a gradient of light from lower to higher intensities. Phototaxis is analogous to chemotaxis except the attractant in this case is light instead of a chemical. In some species, such as the highly motile phototrophic organism *Rhodospirillum centenum* (Figure 3.39), entire colonies of cells show phototaxis and move in unison toward the light (Figure 3.49*b*).

Several components of the regulatory system that govern chemotaxis also control phototaxis. This conclusion has emerged from the study of mutants of phototrophic bacteria defective in phototaxis; such mutants show defective chemotaxis systems as well. A *photoreceptor*, a protein that functions similar to a chemoreceptor but senses a gradient of light instead of chemicals, is the initial sensor in the phototaxis response. The photoreceptor then interacts with the same cytoplasmic proteins that control flagellar rotation in chemotaxis, maintaining the cell in a run if it is swimming toward an increasing intensity of light. Thus, although the stimulus in chemotaxis and phototaxis is different—chemicals versus light—the same molecular machinery processes both signals. We discuss this cytoplasmic machinery in detail in Section 8.8.

Other Taxes

Other bacterial taxes, such as movement toward or away from oxygen (*aerotaxis*, see Figure 3.48*f*) or toward or away from conditions of high ionic strength (*osmotaxis*), are known among various swimming prokaryotes. In some gliding cyanobacteria an unusual taxis, *hydrotaxis* (movement toward water), has also been observed. Hydrotaxis allows gliding cyanobacteria that inhabit dry environments, such as soils, to glide toward a gradient of increasing hydration.

It should be clear from our consideration of microbial taxes that motile prokaryotes do not just swim around at random, but instead remain keenly attuned to the chemical and physical state of their habitat. When gradients of virtually any nutrient form in nature, motile cells are "constantly on the move" exploiting them, and by so doing, improve their chances for survival. And from a mechanistic standpoint, prokaryotic cells monitor these gradients by periodically sampling their environment for chemicals, light, oxygen, salt, or other substances, and then processing the results through a common network of proteins that ultimately control the direction of flagellar rotation. By being able to move toward or away from various stimuli, prokaryotic cells have a better chance of competing successfully for resources and avoiding the harmful effects of substances that could damage or kill them.

MiniQuiz

- Define the word chemotaxis. How does chemotaxis differ from aerotaxis?
- What causes a run versus a tumble?
- How can chemotaxis be measured quantitatively?
- How does scotophobotaxis differ from phototaxis?

Big Ideas

3.1
Prokaryotic cells can have many different shapes; rods, cocci, and spirilla are common cell morphologies. Morphology is a poor predictor of other cell properties and is a genetically directed characteristic that has evolved to best serve the ecology of the cell.

3.2
Prokaryotes are typically smaller in size than eukaryotes, although some very large and some very small prokaryotes are known. The typical small size of prokaryotic cells affects their physiology, growth rate, ecology, and evolution. The lower limit for the diameter of a coccus-shaped cell is about 0.15 μm.

3.3
The cytoplasmic membrane is a highly selective permeability barrier constructed of lipids and proteins that form a bilayer, hydrophobic inside and hydrophilic outside. In contrast to *Bacteria* and *Eukarya*, *Archaea* contain ether-linked lipids, and hyperthermophilic species have membranes of monolayer construction.

3.4
The major functions of the cytoplasmic membrane are permeability, transport, and energy conservation. To accumulate nutrients against the concentration gradient, transport mechanisms are employed that are characterized by their specificity, saturation effect, and biosynthetic regulation.

3.5
At least three types of transporters are known: simple transporters, phosphotransferase systems, and ABC systems. Transport requires energy from either ATP directly or from the proton motive force to accumulate solutes in the cell against the concentration gradient.

3.6
The cell walls of *Bacteria* contain peptidoglycan. Peptidoglycan is a polysaccharide consisting of an alternating repeat of *N*-acetylglucosamine and *N*-acetylmuramic acid, the latter in adjacent strands cross-linked by tetrapeptides. One to several sheets of peptidoglycan can be present, depending on the organism. The enzyme lysozyme and the antibiotic penicillin target peptidoglycan, leading to cell lysis.

3.7
In addition to peptidoglycan, gram-negative bacteria have an outer membrane consisting of LPS, protein, and lipoprotein. Proteins called porins allow for permeability across the outer membrane. The gap between the outer and cytoplasmic membranes is called the periplasm and contains proteins involved in transport, sensing chemicals, and other important cell functions.

3.8
Cell walls of *Archaea* can be of several types, including pseudomurein, various polysaccharides, and S-layers, which are composed of protein or glycoprotein. As for *Bacteria*, the walls of *Archaea* protect the cell from osmotic lysis.

3.9
Many prokaryotic cells contain capsules, slime layers, pili, or fimbriae. These structures have several functions, including attachment, genetic exchange, and twitching motility.

3.10
Prokaryotic cells can contain inclusions of sulfur, polyphosphate, carbon polymers, or magnetosomes. These substances function as storage materials or in magnetotaxis.

3.11
Gas vesicles are cytoplasmic gas-filled structures that confer buoyancy on cells. Gas vesicles are composed of two different proteins arranged to form a gas-permeable but watertight structure.

3.12
The endospore is a highly resistant and differentiated bacterial cell produced by certain gram-positive *Bacteria*. Endospores are dehydrated and contain various protective agents such as calcium dipicolinate and small acid-soluble proteins, absent from vegetative cells. Endospores can remain dormant indefinitely but can germinate quickly when conditions warrant.

3.13
Swimming motility is due to flagella. The flagellum is a complex structure made of several proteins anchored in the cell wall and cytoplasmic membrane. The flagellum filament is made of a single kind of protein in *Bacteria* and rotates at the expense of the proton motive force. The flagella of *Archaea* and *Bacteria* differ in structure and probably also in their rotational mechanism.

3.14
Bacteria that move by gliding motility do not employ rotating flagella but instead creep along a solid surface by employing any of several different mechanisms.

3.15
Motile bacteria respond to chemical and physical gradients in their environment. In swimming bacteria, movement of a cell is biased either toward or away from a stimulus by controlling the lengths of runs and frequency of tumbles. Tumbles are controlled by the direction of rotation of the flagellum, which in turn is controlled by a network of sensory and response proteins.

Review of Key Terms

ABC (ATP-binding cassette) transport system a membrane transport system consisting of three proteins, one of which hydrolyzes ATP; the system transports specific nutrients into the cell

Basal body the "motor" portion of the bacterial flagellum, embedded in the cytoplasmic membrane and wall

Capsule a polysaccharide or protein outermost layer, usually rather slimy, present on some bacteria

Chemotaxis directed movement of an organism toward (positive chemotaxis) or away from (negative chemotaxis) a chemical gradient

Cytoplasmic membrane the permeability barrier of the cell, separating the cytoplasm from the environment

Dipicolinic acid a substance unique to endospores that confers heat resistance on these structures

Endospore a highly heat-resistant, thick-walled, differentiated structure produced by certain gram-positive *Bacteria*

Flagellum a long, thin cellular appendage capable of rotation and responsible for swimming motility in prokaryotic cells

Gas vesicles gas-filled cytoplasmic structures bounded by protein and conferring buoyancy on cells

Gram-negative a bacterial cell with a cell wall containing small amounts of peptidoglycan, and an outer membrane containing lipopolysaccharide, lipoprotein, and other complex macromolecules

Gram-positive a bacterial cell whose cell wall consists chiefly of peptidoglycan; it lacks the outer membrane of gram-negative cells

Gram stain a differential staining procedure that stains cells either purple (gram-positive cells) or pink (gram-negative cells)

Group translocation an energy-dependent transport system in which the substance transported is chemically modified during the process of being transported by a series of proteins

Lipopolysaccharide (LPS) a combination of lipid with polysaccharide and protein that forms the major portion of the outer membrane in gram-negative *Bacteria*

Magnetosome a particle of magnetite (Fe_3O_4) enclosed by a nonunit membrane in the cytoplasm of magnetotactic *Bacteria*

Morphology the *shape* of a cell—rod, coccus, spirillum, and so on

Outer membrane a phospholipid- and polysaccharide-containing unit membrane that lies external to the peptidoglycan layer in cells of gram-negative *Bacteria*

Peptidoglycan a polysaccharide composed of alternating repeats of *N*-acetylglucosamine and *N*-acetylmuramic acid arranged in adjacent layers and cross-linked by short peptides

Periplasm a gel-like region between the outer surface of the cytoplasmic membrane and the inner surface of the lipopolysaccharide layer of gram-negative *Bacteria*

Peritrichous flagellation having flagella located in many places around the surface of the cell

Phototaxis movement of an organism toward light

Pili thin, filamentous structures that extend from the surface of a cell and, depending on type, facilitate cell attachment, genetic exchange, or twitching motility

Polar flagellation having flagella emanating from one or both poles of the cell

Poly-β-hydroxybutyrate (PHB) a common storage material of prokaryotic cells consisting of a polymer of β-hydroxybutyrate or another β-alkanoic acid or mixtures of β-alkanoic acids

S-layer an outermost cell surface layer composed of protein or glycoprotein present on some *Bacteria* and *Archaea*

Simple transport system a transporter that consists of only a membrane-spanning protein and is typically driven by energy from the proton motive force

Teichoic acid a phosphorylated polyalcohol found in the cell wall of some gram-positive *Bacteria*

Review Questions

1. What are the major morphologies of prokaryotes? Draw cells for each morphology you list (Section 3.1).

2. How do mutations contribute to the evolution of prokaryotes (Section 3.2)?

3. What is a cytoplasmic membrane? Why is it important to a typical bacterial cell (Section 3.3)?

4. What are the two major functions of cytoplasmic membrane (Section 3.3)?

5. Explain in a single sentence why ionized molecules do not readily pass through the cytoplasmic membrane of a cell. How do such molecules get through the cytoplasmic membrane (Sections 3.4 and 3.5)?

6. Cells of *Escherichia coli* take up lactose via lac permease, glucose via the phosphotransferase system, and maltose via an ABC-type transporter. For each of these sugars describe: (1) the components of the transport system and (2) the source of energy that drives the transport event (Section 3.5).

7. What is the characteristic of the bacterial cell wall that differentiates gram-positive from gram-negative bacteria (Section 3.6)?

8. List several functions of the outer membrane in gram-negative *Bacteria*. What is the chemical composition of the outer membrane (Section 3.7)?

9. What cell wall polysaccharide common in *Bacteria* is absent from *Archaea*? What is unusual about S-layers compared to other cell walls of prokaryotes ? What types of cell walls are found in *Archaea* (Section 3.8)?

10. What function(s) do polysaccharide layers outside the cell wall have in prokaryotes (Section 3.9)?

11. What types of cytoplasmic inclusions are formed by prokaryotes? How does an inclusion of poly-β-hydroxybutyric acid differ from a magnetosome in composition and metabolic role (Section 3.10)?

12. What is the function of gas vesicles? How are these structures made such that they can remain gas tight (Section 3.11)?

13. In a few sentences, indicate how the bacterial endospore differs from the vegetative cell in structure, chemical composition, and ability to resist extreme environmental conditions (Section 3.12).

14. Define the following terms: mature endospore, vegetative cell, and germination (Section 3.12).

15. Describe the structure and function of a bacterial flagellum. What is the energy source for the flagellum? How do the flagella of *Bacteria* differ from those of *Archaea* in both size and composition (Section 3.13)?

16. How do the mechanism and energy requirements for motility in *Flavobacterium* differ from that in *Escherichia coli* (Sections 3.13 and 3.14)?

17. In a few sentences, explain how a motile bacterium is able to sense the direction of an attractant and move toward it (Section 3.15).

18. In the experiment described in Figure 3.48, why is it essential to have a control (Section 3.15)?

Application Questions

1. Calculate the surface-to-volume ratio of a spherical cell 15 μm in diameter and of a cell 2 μm in diameter. What are the consequences of these differences in surface-to-volume ratio for cell function?

2. Assume you are given two cultures, one of a species of gram-negative *Bacteria* and one of a species of *Archaea*. Other than by phylogenetic analyses, discuss at least four different ways you could tell which culture was which.

3. Calculate the amount of time it would take a cell of *Escherichia coli* (1×2 μm) swimming at maximum speed (60 cell lengths per second) to travel all the way up a 3-cm-long capillary tube containing a chemical attractant.

4. Assume you are given two cultures of rod-shaped bacteria, one gram-positive and the other gram-negative. How could you differentiate them using (a) light microscopy; (b) electron microscopy; (c) chemical analyses of cell walls; and (d) phylogenetic analyses?

Need more practice? Test your understanding with Quantitative Questions; access additional study tools including tutorials, animations, and videos; and then test your knowledge with chapter quizzes and practice tests at **www.microbiologyplace.com**.

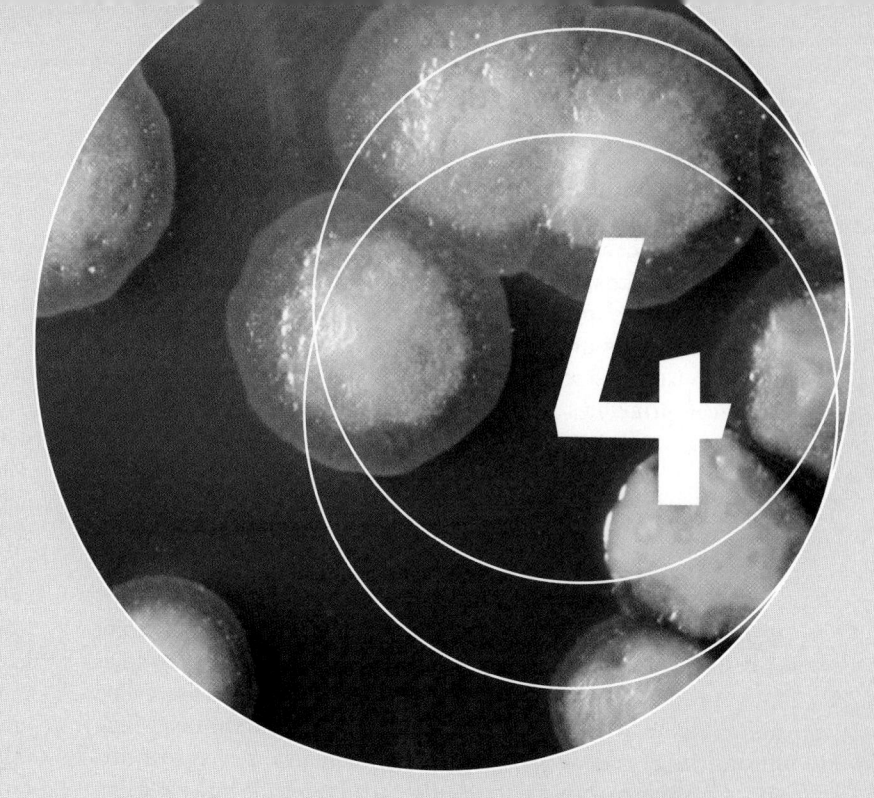

Nutrition, Culture, and Metabolism of Microorganisms

A microbial cell carries out a host of metabolic reactions to yield the energy necessary to divide and form two cells. Continued growth on a solid surface leads to visible masses of cells, called colonies.

Recall from Chapter 2 that all cells require energy to drive life processes. The requisite energy is obtained from organic chemicals by chemoorganotrophs, from inorganic chemicals by chemolithotrophs, and from light by phototrophs. In this chapter we explore how cells conserve and use their energy and nutrients. We assume that the reader has some background in cell chemistry and refer the reader who needs a refresher on the chemical principles of life to an overview of this topic at www.microbiologyplace.com

I   Nutrition and Culture of Microorganisms

Before a cell can replicate, it must coordinate many different chemical reactions and organize many different molecules into specific structures. Collectively, these reactions are called **metabolism**. Metabolic reactions are either **catabolic**, which means *energy releasing*, or **anabolic**, which means *energy requiring*. Catabolism breaks molecular structures down, releasing energy in the process, and anabolism uses energy to build larger molecules from smaller ones.

We examine some of the key catabolic and anabolic reactions of cells in this chapter. However, before we do, we consider how microorganisms are grown in the laboratory and the nutrients they need for growth. Indeed, most of what we know about the metabolism of microorganisms has emerged from the study of laboratory cultures. Our initial focus is on chemoorganotrophs; later in the chapter we consider chemolithotrophs and phototrophs.

4.1 Nutrition and Cell Chemistry

In this section we learn how to care for and feed microorganisms. Nutrition is the part of microbial physiology that deals with the nutrients required for growth. Different organisms need different complements of nutrients, and not all nutrients are required in the same amounts. Some nutrients, called *macronutrients*, are required in large amounts, while others, called *micronutrients*, are required in just trace amounts.

All microbial nutrients are compounds constructed from the chemical elements. However, just a handful of elements dominate living systems and are essential: hydrogen (H), oxygen (O), carbon (C), nitrogen (N), phosphorus (P), sulfur (S), and selenium (Se). In addition to these, at least 50 other elements, although not required, are metabolized in some way by microorganisms (**Figure 4.1**). An approximate chemical formula for a cell is $CH_2O_{0.5}N_{0.15}$, indicating that C, H, O, and N constitute the bulk of a living organism.

Besides water, which makes up 70–80% of the wet weight of a microbial cell (a single cell of *Escherichia coli* weighs just 10^{-12} g), cells consist primarily of macromolecules—proteins, nucleic acids, lipids, and polysaccharides. The essential elements make up the building blocks (monomers) of these macromolecules, the amino acids, nucleotides, fatty acids, and sugars. Proteins dominate the macromolecular composition of a cell, making up 55% of total cell dry weight. Moreover, the diversity of proteins exceeds that of all other macromolecules combined. Interestingly, as important as DNA is to a cell, it contributes a very small percentage of a cell's dry weight; RNA is far more abundant (Figure 4.1c).

The data shown in Figure 4.1 are from actual analyses of cells of *E. coli*; comparable data vary a bit from one microorganism to the next. But in any microbial cell, carbon and nitrogen are important macronutrients, and thus we begin our study of microbial nutrition with these key elements.

Carbon and Nitrogen

All cells require carbon, and most prokaryotes require *organic* (carbon-containing) compounds as their source of carbon. Heterotrophic bacteria assimilate organic compounds and use them to make new cell material. Amino acids, fatty acids, organic acids, sugars, nitrogen bases, aromatic compounds, and countless other organic compounds can be transported and catabolized by one or another bacterium. Autotrophic microorganisms build their cellular structures from carbon dioxide (CO_2) with energy obtained from light or inorganic chemicals.

A bacterial cell is about 13% nitrogen, which is present in proteins, nucleic acids, and several other cell constituents. The bulk of nitrogen available in nature is in inorganic form as ammonia (NH_3), nitrate (NO_3^-), or nitrogen gas (N_2). Virtually all prokaryotes can use NH_3 as their nitrogen source, and many can also use NO_3^-. By contrast, N_2 can only be used by nitrogen-fixing prokaryotes, discussed in detail in later chapters. Nitrogen in organic compounds, for example, in amino acids, may also be available to microorganisms; if organic N is available and is taken up, the compound can immediately enter the monomer pool for biosynthesis or be catabolized as an energy source.

Other Macronutrients: P, S, K, Mg, Ca, Na

In addition to C, N, O, and H, many other elements are needed by cells, but in smaller amounts (Figure 4.1b). Phosphorus is a key element in nucleic acids and phospholipids and is typically supplied to a cell as phosphate (PO_4^{2-}). Sulfur is present in the amino acids cysteine and methionine and also in several vitamins, including thiamine, biotin, and lipoic acid. Sulfur can be supplied to cells in several forms, including sulfide (HS^-) and sulfate (SO_4^{2-}). Potassium (K) is required for the activity of several enzymes, whereas magnesium (Mg) functions to stabilize ribosomes, membranes, and nucleic acids and is also required for the activity of many enzymes. Calcium (Ca) is not required by all cells but can play a role in helping to stabilize microbial cell walls, and it plays a key role in the heat stability of endospores. Sodium (Na) is required by some, but not all, microorganisms, and its requirement is typically a reflection of the habitat. For example, seawater contains relatively high levels of Na^+, and marine microorganisms typically require Na^+ for growth. By contrast, freshwater species are usually able to grow in the absence of Na^+. K, Mg, Ca, and Na are all supplied to cells as salts, typically as chloride or sulfate salts.

Micronutrients: Iron and Other Trace Metals

Microorganisms require several metals for growth (Figure 4.1a). Chief among these is iron (Fe), which plays a major role in cellular respiration. Iron is a key component of cytochromes and of iron–sulfur proteins involved in electron transport reactions

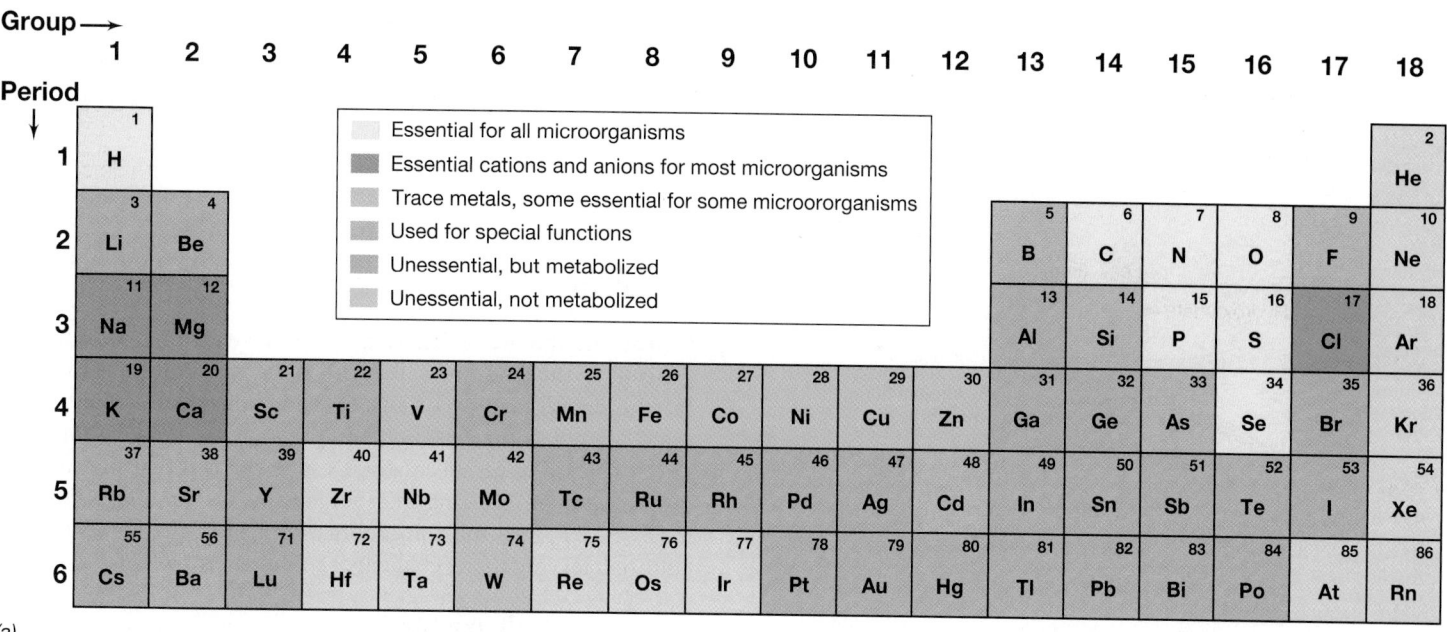

Group →

| Period | 1 | 2 | 3 | 4 | 5 | 6 | 7 | 8 | 9 | 10 | 11 | 12 | 13 | 14 | 15 | 16 | 17 | 18 |

- Essential for all microorganisms
- Essential cations and anions for most microorganisms
- Trace metals, some essential for some microorganisms
- Used for special functions
- Unessential, but metabolized
- Unessential, not metabolized

(a)

Essential elements as a percent of cell dry weight

C	50%	P	2.5%
O	17%	S	1.8%
N	13%	Se	<0.01%
H	8.2%		

(b)

Macromolecular composition of a cell

Macromolecule	Percent of dry weight
Protein	55
Lipid	9.1
Polysaccharide	5.0
Lipopolysaccharide	3.4
DNA	3.1
RNA	20.5

(c)

Figure 4.1 Elemental and macromolecular composition of a bacterial cell. *(a)* A microbial periodic table of the elements. With the exception of uranium, which can be metabolized by some prokaryotes, elements in period 7 or beyond in the complete periodic table of the elements are not known to be metabolized. *(b)* Contributions of the essential elements to cell dry weight. *(c)* Relative abundance of macromolecules in a bacterial cell. Data in *(b)* from *Aquat. Microb. Ecol. 10:* 15–27 (1996) and in *(c)* from Escherichia coli *and* Salmonella typhimurium: *Cellular and Molecular Biology.* ASM, Washington, DC (1996).

(Section 4.9). Under anoxic conditions, iron is generally in the ferrous (Fe^{2+}) form and soluble. However, under oxic conditions, iron is typically in the ferric (Fe^{3+}) form as part of insoluble minerals. To obtain Fe^{3+} from such minerals, cells produce iron-binding molecules called **siderophores** that function to bind Fe^{3+} and transport it into the cell. A major group of siderophores is the hydroxamic acids, organic molecules that chelate Fe^{3+} strongly. As **Figure 4.2** shows, after the iron–hydroxamate complex reaches the cytoplasm, the iron is released, and the hydroxamate is excreted and can be used again for iron transport.

Many other types of siderophores are known. Some bacteria produce phenolic siderophores (for example, the enterobactins) whereas others produce peptide siderophores (for example, aquachelin). Both classes of siderophore have extremely high binding affinities and easily bind iron at levels as low as 1 nanogram per liter. However, as important as iron is for most cells, some organisms can grow in the absence of iron. For example,

many lactic acid bacteria such as species of *Lactobacillus* do not contain detectable iron and grow normally in its absence. In these organisms, manganese (Mn^{2+}) often plays a role similar to that just described for iron. Many other metals are required or otherwise metabolized by microorganisms (Figure 4.1a). Like iron, these micronutrients are called *trace elements* or *trace metals.* Micronutrients typically play a role as cofactors for enzymes. **Table 4.1** lists the major micronutrients and examples of enzymes in which each plays a role in the cell.

Micronutrients: Growth Factors

Growth factors are *organic* compounds that, like trace metals, are required in only very small amounts. Growth factors are vitamins, amino acids, purines, pyrimidines, or various other organic molecules. Although most microorganisms are able to biosynthesize the growth factors they need, some must obtain one or more of them from the environment and thus must be supplied with these compounds when cultured in the laboratory.

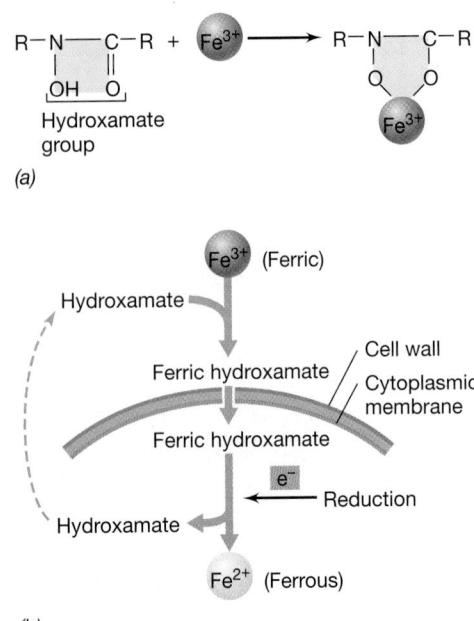

(a)

(b)

Figure 4.2 Mechanism of hydroxamate siderophores. *(a)* Iron is bound as Fe^{3+} and *(b)* transported and released inside the cell and reduced to Fe^{2+}. The hydroxamate then exits the cell and repeats the cycle.

Vitamins are the most commonly required growth factors. Most vitamins function as coenzymes, which are nonprotein components of enzymes. Vitamin requirements vary among microorganisms, ranging from none to several. Lactic acid bacteria, which include the genera *Streptococcus*, *Lactobacillus*, and *Leuconostoc* (↩ Section 18.1), are renowned for their many vitamin requirements, which are even more extensive than those of humans (see Table 4.2).

MiniQuiz

• Which four elements make up the bulk of a cell's dry weight?

• Which two classes of macromolecules contain most of a cell's nitrogen?

• What roles does iron play in cellular metabolism? How do cells sequester iron?

4.2 Culture Media

Culture media are the nutrient solutions used to grow microorganisms in the laboratory. Because laboratory culture is required for the detailed study of any microorganism, careful attention must be paid to the selection and preparation of media for laboratory culture to be successful. Despite the fact that microbiologists have been growing microorganisms in laboratory cultures for over 125 years, most microorganisms in nature have yet to be cultured, leaving many challenges to the microbiologist today.

Classes of Culture Media

Two broad classes of culture media are used in microbiology: defined media and complex media. **Defined media** are prepared by adding precise amounts of highly purified inorganic or organic chemicals to distilled water. Therefore, the *exact composition* of a defined medium (in both a qualitative and quantitative sense) is known. Of major importance in any culture medium is the carbon source because all cells need large amounts of carbon to make new cell material (Figure 4.1). The particular carbon source and its concentration depend on the organism to be cultured. **Table 4.2** lists recipes for four culture media. Some defined media, such as the one listed for *Escherichia coli*, are said to be "simple" because they contain only a single carbon source. In this medium, cells of *E. coli* make all organic molecules from this carbon source.

Table 4.1 *Micronutrients (trace elements) needed by microorganisms*[a]

Element	Cellular function or molecule of which a part
Boron (B)	Autoinducer for quorum sensing in bacteria; also found in some polyketide antibiotics
Chromium (Cr)	Possible but not proven component for glucose metabolism (necessary in mammals)
Cobalt (Co)	Vitamin B_{12}; transcarboxylase (only in propionic acid bacteria)
Copper (Cu)	In respiration, cytochrome *c* oxidase; in photosynthesis, plastocyanin, some superoxide dismutases
Iron (Fe)[b]	Cytochromes; catalases; peroxidases; iron–sulfur proteins; oxygenases; all nitrogenases
Manganese (Mn)	Activator of many enzymes; component of certain superoxide dismutases and of the water-splitting enzyme in oxygenic phototrophs (photosystem II)
Molybdenum (Mo)	Certain flavin-containing enzymes; some nitrogenases, nitrate reductases, sulfite oxidases, DMSO-TMAO reductases; some formate dehydrogenases
Nickel (Ni)	Most hydrogenases; coenzyme F_{430} of methanogens; carbon monoxide dehydrogenase; urease
Selenium (Se)	Formate dehydrogenase; some hydrogenases; the amino acid selenocysteine
Tungsten (W)	Some formate dehydrogenases; oxotransferases of hyperthermophiles
Vanadium (V)	Vanadium nitrogenase; bromoperoxidase
Zinc (Zn)	Carbonic anhydrase; alcohol dehydrogenase; RNA and DNA polymerases; and many DNA-binding proteins

[a]Not every micronutrient listed is required by all cells; some metals listed are found in enzymes or cofactors present in only specific microorganisms.
[b]Needed in greater amounts than other trace metals.

Table 4.2 *Examples of culture media for microorganisms with simple and demanding nutritional requirements*[a]

Defined culture medium for Escherichia coli	Defined culture medium for Leuconostoc mesenteroides	Complex culture medium for either E. coli or L. mesenteroides	Defined culture medium for Thiobacillus thioparus
K_2HPO_4 7 g KH_2PO_4 2 g $(NH_4)_2SO_4$ 1 g $MgSO_4$ 0.1 g $CaCl_2$ 0.02 g Glucose 4–10 g Trace elements (Fe, Co, Mn, Zn, Cu, Ni, Mo) 2–10 µg each Distilled water 1000 ml pH 7	K_2HPO_4 0.6 g KH_2PO_4 0.6 g NH_4Cl 3 g $MgSO_4$ 0.1 g Glucose 25 g Sodium acetate 25 g Amino acids (alanine, arginine, asparagine, aspartate, cysteine, glutamate, glutamine, glycine, histidine, isoleucine, leucine, lysine, methionine, phenylalanine, proline, serine, threonine, tryptophan, tyrosine, valine) 100–200 µg of each Purines and pyrimidines (adenine, guanine, uracil, xanthine) 10 mg of each Vitamins (biotin, folate, nicotinic acid, pyridoxal, pyridoxamine, pyridoxine, riboflavin, thiamine, pantothenate, *p*-aminobenzoic acid) 0.01–1 mg of each Trace elements (as in first column) 2–10 µg each Distilled water 1000 ml pH 7	Glucose 15 g Yeast extract 5 g Peptone 5 g KH_2PO_4 2 g Distilled water 1000 ml pH 7	KH_2PO_4 0.5 g NH_4Cl 0.5 g $MgSO_4$ 0.1 g $CaCl_2$ 0.05 g KCl 0.5 g $Na_2S_2O_3$ 2 g Trace elements (as in first column) Distilled water 1000 ml pH 7 Carbon source: CO_2 from air

(a)   (b)

[a]The photos are tubes of *(a)* the defined medium described, and *(b)* the complex medium described. Note how the complex medium is colored from the various organic extracts and digests that it contains. Photo credits: Cheryl L. Broadie and John Vercillo, Southern Illinois University at Carbondale.

For culturing many microorganisms, knowledge of the exact composition of a medium is not essential. In these instances complex media may suffice and may even be advantageous. **Complex media** employ digests of microbial, animal or plant products, such as casein (milk protein), beef (beef extract), soybeans (tryptic soy broth), yeast cells (yeast extract), or any of a number of other highly nutritious yet impure substances. These digests are commercially available in dehydrated form and can be easily prepared. However, the disadvantage of a complex medium is its imprecise nutritional composition. That is, although one may know approximately what is in the medium, its exact composition is unknown. An *enriched medium*, often used for the culture of otherwise difficult-to-grow nutritionally demanding (fastidious) microorganisms, starts with a complex base and is embellished with additional nutrients such as serum, blood, or other highly nutritious substances. Culture media are often made to be selective or differential (or both), especially media used in diagnostic microbiology. A *selective* medium contains compounds that inhibit the growth of some microorganisms but not others. For example, media are available for the selective isolation of pathogenic strains of *E. coli* from food products, such as ground beef, that could be contaminated with this organism. By contrast, a *differential* medium is one in which an indicator, typically a reactive dye, is added that reveals whether a particular chemical reaction has occurred during growth. Differential media are quite useful for distinguishing different species of bacteria and are therefore widely used in clinical diagnostics and systematic microbiology. Differential and selective media are further discussed in Chapter 31.

Nutritional Requirements and Biosynthetic Capacity

Of the four recipes in Table 4.2, three are defined and one is complex. The complex medium is easiest to prepare and supports growth of both of the chemoorganotrophs, *Escherichia coli* and *Leuconostoc mesenteroides*, the examples used in the table. However, the simple defined medium supports growth of *E. coli* but not of *L. mesenteroides*. Growth of the latter organism, a fastidious (nutritionally demanding) bacterium, in a defined medium requires the addition of several nutrients not needed by *E. coli*. By contrast, *E. coli* can synthesize everything it needs from a single carbon compound, in this case, glucose. The nutritional needs of *L. mesenteroides* can be satisfied by preparing either a highly supplemented defined medium, a rather laborious undertaking because of all the individual nutrients that need to be added (Table 4.2), or by preparing a complex medium, a much less demanding operation.

The fourth medium listed in Table 4.2 supports growth of the sulfur chemolithotroph *Thiobacillus thioparus*; this medium would not support growth of the chemoorganotrophs. *T. thioparus* is an autotroph and thus has no organic carbon requirements. *T. thioparus* derives all of its carbon from CO_2 and obtains its energy from the oxidation of the reduced sulfur compound thiosulfate ($Na_2S_2O_3$). Thus, *T. thioparus* has the greatest biosynthetic capacity of all the organisms listed in the table, surpassing even *E. coli* in this regard.

In a nutshell, what does Table 4.2 tell us? Simply put, it reveals the fact that different microorganisms can have vastly different nutritional requirements. Thus, for successful cultivation of any microorganism, it is necessary to understand its nutritional requirements and then supply it with the nutrients it needs in the proper form and in the proper amounts. If care is taken in preparing culture media, it is fairly easy to culture many different types of microorganisms in the laboratory. We discuss some procedures for doing this now.

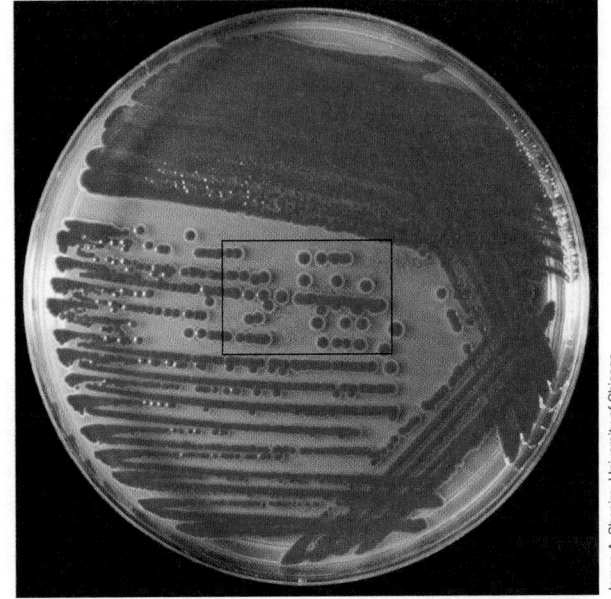

MiniQuiz

- Why would the routine culture of *Leuconostoc mesenteroides* be easier in a complex medium than in a chemically defined medium?

- In which medium shown in Table 4.2, defined or complex, do you think *E. coli* would grow the fastest? Why? *E. coli* will not grow in the medium described for *Thiobacillus thioparus*; why?

4.3 Laboratory Culture

Once a culture medium has been prepared and made sterile to render it free of all life forms, organisms can be inoculated and the culture can be incubated under conditions that will support growth. In a laboratory, inoculation will typically be with a **pure culture**, a culture containing only a single kind of microorganism.

It is essential to prevent other organisms from entering a pure culture. Such unwanted organisms, called *contaminants*, are ubiquitous (as Pasteur discovered over 125 years ago, ↩ Section 1.7), and microbiological techniques are designed to avoid contamination. A major method for obtaining pure cultures and for assessing the purity of a culture is the use of solid media, specifically, solid media prepared in the Petri plate, and we consider this now.

(a)

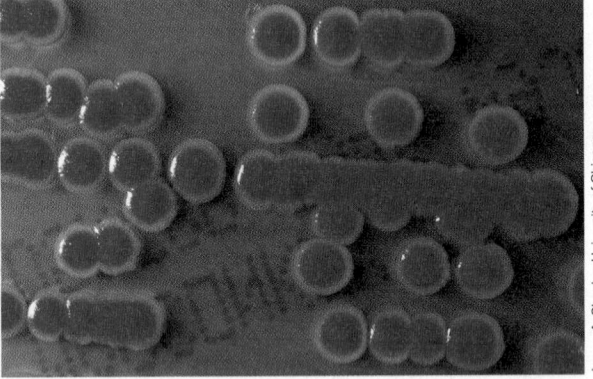

(b)

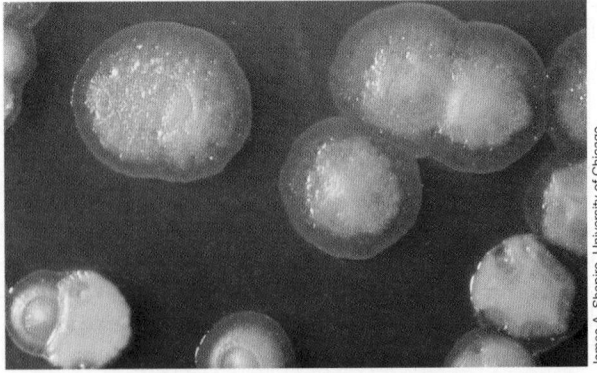

(c)

Solid and Liquid Culture Media

Liquid culture media are sometimes solidified by the addition of a gelling agent. Solid media immobilize cells, allowing them to grow and form visible, isolated masses called *colonies* (**Figure 4.3**). Microbial colonies are of various shapes and sizes depending on the organism, the culture conditions, the nutrient supply, and several other physiological parameters, and can contain several billion individual cells. Some microorganisms produce pigments that cause the colony to be colored (Figure 4.3). Colonies permit the microbiologist to visualize the composition and presumptive

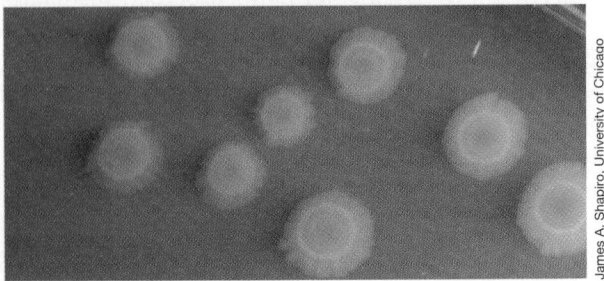

(d)

Figure 4.3 Bacterial colonies. Colonies are visible masses of cells formed from the division of one or a few cells and can contain over a billion (10^9) individual cells. *(a) Serratia marcescens*, grown on MacConkey agar. *(b)* Close-up of colonies outlined in part a. *(c) Pseudomonas aeruginosa*, grown on trypticase soy agar. *(d) Shigella flexneri*, grown on MacConkey agar.

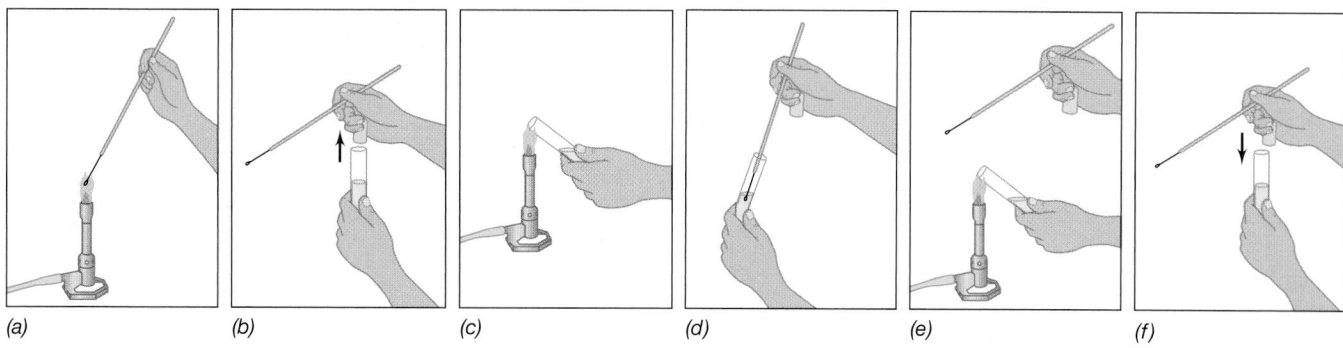

Figure 4.4 Aseptic transfer. *(a)* Loop is heated until red hot and cooled in air briefly. *(b)* Tube is uncapped. *(c)* Tip of tube is run through the flame. *(d)* Sample is removed on sterile loop for transfer to a sterile medium. *(e)* The tube is reflamed. *(f)* The tube is recapped. Loop is reheated before being taken out of service.

purity of the culture. Plates that contain more than one colony type are indicative of a contaminated culture. The appearance and uniformity of colonies on a Petri plate has been used as one criterion of culture purity for over 100 years (⮎ Section 1.8).

Solid media are prepared in the same way as liquid media except that before sterilization, *agar*, a gelling agent, is added to the medium, typically at a concentration of 1–2%. The agar melts during the sterilization process, and the molten medium is then poured into sterile glass or plastic plates and allowed to solidify before use (Figure 4.3). www.microbiologyplace.com **Online Tutorial 4.1: Aseptic Transfer and the Streak Plate Method**

Aseptic Technique

Because microorganisms are everywhere, culture media must be sterilized before use. Sterilization is typically achieved with moist heat in a large pressurized chamber called an *autoclave*. We discuss the operation and principles of the autoclave later, along with other methods of sterilization (⮎ Section 26.1).

Once a sterile culture medium has been prepared, it is ready to receive an inoculum to start the growth process. This manipulation requires **aseptic technique**, a series of steps to prevent contamination during manipulations of cultures and sterile culture media (**Figures 4.4** and **4.5**). A mastery of aseptic technique is

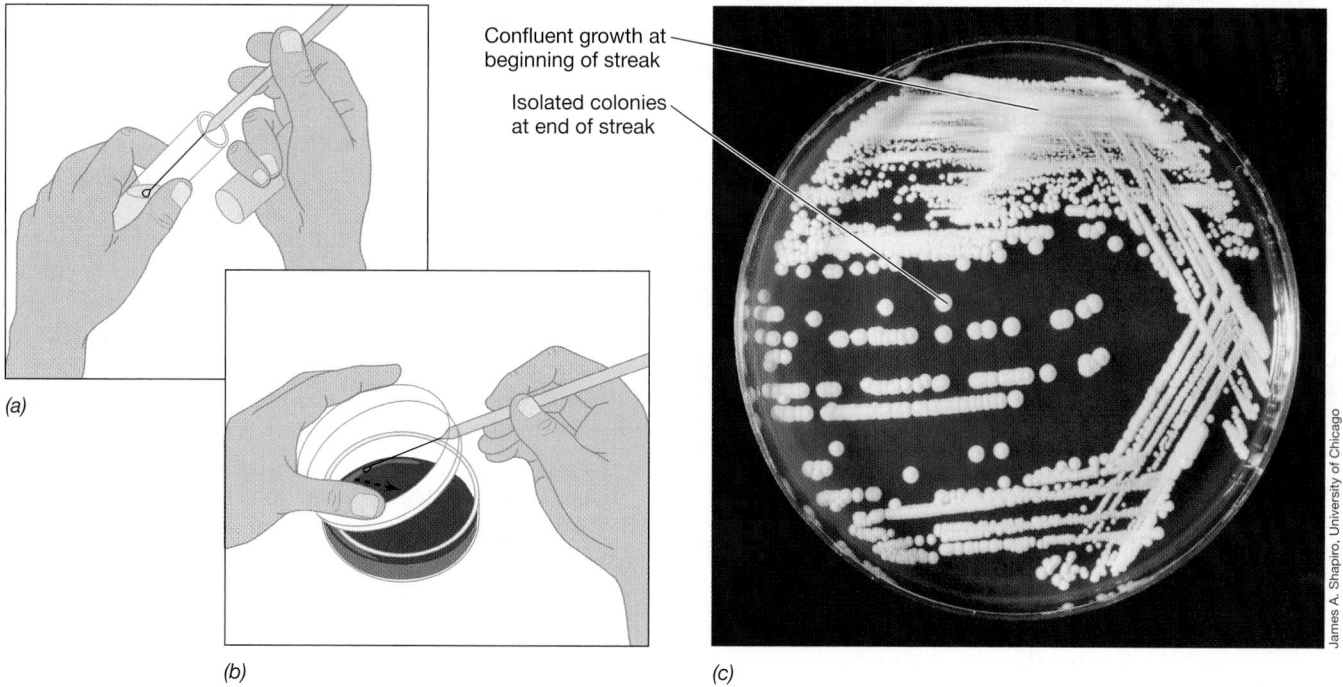

Confluent growth at beginning of streak

Isolated colonies at end of streak

James A. Shapiro, University of Chicago

Figure 4.5 Making a streak plate to obtain pure cultures *(a)* Loop is sterilized and a loopful of inoculum is removed from tube. *(b)* Streak is made and spread out on a sterile agar plate. Following the initial streak, subsequent streaks are made at angles to it, the loop being resterilized between streaks. *(c)* Appearance of a well-streaked plate after incubation, showing colonies of the bacterium *Micrococcus luteus* on a blood agar plate. It is from such well-isolated colonies that pure cultures can usually be obtained.

required for success in the microbiology laboratory, and it is one of the first methods learned by the novice microbiologist. Airborne contaminants are the most common problem because the dust in laboratory air contains microorganisms. When containers are opened, they must be handled in such a way that contaminant-laden air does not enter (Figures 4.4 and 4.5).

Aseptic transfer of a culture from one tube of medium to another is typically accomplished with an inoculating loop or needle that has previously been sterilized in a flame (Figure 4.4). Cells from liquid cultures can also be transferred to the surface of agar plates where colonies develop from the growth and division of single cells (Figure 4.5). Picking an isolated colony and restreaking it is the main method for obtaining pure cultures from samples containing several different organisms.

MiniQuiz

- What is meant by the word sterile? What would happen if freshly prepared culture media were not sterilized and then left at room temperature?
- Why is aseptic technique necessary for successful cultivation of pure cultures in the laboratory?

 Energetics and Enzymes

Regardless of how a microorganism makes a living—whether by chemoorganotrophy, chemolithotrophy, or phototrophy—it must be able to conserve some of the energy released in its energy-yielding reactions. Here we discuss the principles of energy conservation, using some simple laws of chemistry and physics to guide our understanding. We then consider enzymes, the cell's catalysts.

4.4 Bioenergetics

Energy is the ability to do work. In microbiology, energy is measured in kilojoules (kJ), a unit of heat energy. All chemical reactions in a cell are accompanied by *changes* in energy, energy either being required for or released during the reaction.

Basic Energetics

Although in any chemical reaction some energy is lost as heat, in microbiology we are interested in **free energy** (abbreviated **G**), which is the energy available to do work. The *change* in free energy during a reaction is expressed as $\Delta G^{0\prime}$, where the symbol Δ is read as "change in." The "0" and "prime" superscripts indicate that the free-energy value is for standard conditions: pH 7, 25°C, 1 atmosphere of pressure, and all reactants and products at molar concentrations.

Consider the reaction

$$A + B \rightarrow C + D$$

If $\Delta G^{0\prime}$ for this reaction is *negative* in arithmetic sign, then the reaction will proceed with the *release* of free energy, energy that the cell may conserve as ATP. Such energy-yielding reactions are called **exergonic**. However, if $\Delta G^{0\prime}$ is *positive*, the reaction *requires* energy in order to proceed. Such reactions are called

Table 4.3 *Free energy of formation for a few compounds of biological interest*

Compound	Free energy of formation $(G_f^0)^a$
Water (H_2O)	−237.2
Carbon dioxide (CO_2)	−394.4
Hydrogen gas (H_2)	0
Oxygen gas (O_2)	0
Ammonium (NH_4^+)	−79.4
Nitrous oxide (N_2O)	+104.2
Acetate ($C_2H_3O_2^-$)	−369.4
Glucose ($C_6H_{12}O_6$)	−917.3
Methane (CH_4)	−50.8
Methanol (CH_3OH)	−175.4

^{a}The free energy of formation values are in kJ/mol. See Table A1.1 in Appendix 1 for a more complete list of free energies of formation.

endergonic. Thus, exergonic reactions *release* energy whereas endergonic reactions *require* energy.

Free Energy of Formation and Calculating $\Delta G^{0\prime}$

To calculate the free-energy yield of a reaction, one first needs to know the free energy of its reactants and products. This is the free energy of formation (G_f^0), the energy released or required during the formation of a given molecule from the elements. **Table 4.3** gives a few examples of G_f^0. By convention, the free energy of formation of the elements in their elemental and electrically neutral form (for instance, C, H_2, N_2) is zero. The free energies of formation of compounds, however, are not zero. If the formation of a compound from its elements proceeds exergonically, then the G_f^0 of the compound is negative (energy is released). If the reaction is endergonic, then the G_f^0 of the compound is positive (energy is required).

For most compounds G_f^0 is negative. This reflects the fact that compounds tend to form spontaneously (that is, with energy being released) from their elements. However, the positive G_f^0 for nitrous oxide (N_2O) (+104.2 kJ/mol, Table 4.3) indicates that this compound does not form spontaneously. Instead, over time it decomposes spontaneously to yield N_2 and O_2. The free energies of formation of more compounds of microbiological interest are given in Appendix 1.

Using free energies of formation, it is possible to calculate $\Delta G^{0\prime}$ of a given reaction. For the reaction $A + B \rightarrow C + D$, $\Delta G^{0\prime}$ is calculated by subtracting the sum of the free energies of formation of the reactants (A + B) from that of the products (C + D). Thus

$$\Delta G^{0\prime} = G_f^0[C + D] - G_f^0[A + B]$$

The value obtained for $\Delta G^{0\prime}$ tells us whether the reaction is exergonic or endergonic. The phrase "products minus reactants" is a simple way to recall how to calculate changes in free energy during chemical reactions. However, before free-energy calculations can be made, it is first necessary to balance the reaction.

Appendix 1 details the steps in balancing reactions both electrically and atomically and calculating free energies for any hypothetical reaction.

$\Delta G^{0\prime}$ versus ΔG

Although calculations of $\Delta G^{0\prime}$ are usually reasonable estimates of actual free-energy changes, under some circumstances they are not. We will see later in this book that the actual concentrations of products and reactants in nature, which are rarely at molar levels, can alter the bioenergetics of reactions, sometimes in significant ways. Thus, what may be most relevant to a bioenergetic calculation is not $\Delta G^{0\prime}$, but ΔG, the free-energy change that occurs under the actual conditions in which the organism is growing. The equation for ΔG takes into account the actual concentrations of reactants and products in the reaction and is

$$\Delta G = \Delta G^{0\prime} + RT \ln K$$

where R and T are physical constants and K is the equilibrium constant for the reaction (Appendix 1). We distinguish between $\Delta G^{0\prime}$ and ΔG in important ways in Chapter 14, where we consider metabolic diversity in more detail, but for now, we only need to focus on the expression $\Delta G^{0\prime}$ and what it tells us about a chemical reaction catalyzed by a microorganism. Only reactions that are exergonic yield energy that can be conserved by the cell as ATP.

MiniQuiz

- What is free energy?

- Using the data in Table 4.3, calculate $\Delta G^{0\prime}$ for the reaction $CH_4 + \frac{1}{2} O_2 \rightarrow CH_3OH$. How does $\Delta G^{0\prime}$ differ from ΔG?

- Does glucose formation from the elements release or require energy?

4.5 Catalysis and Enzymes

Free-energy calculations reveal only whether energy is released or required in a given reaction. The value obtained says nothing about the *rate* of the reaction. Consider the formation of water from gaseous oxygen (O_2) and hydrogen (H_2). The energetics of this reaction are quite favorable: $H_2 + \frac{1}{2}O_2 \rightarrow H_2O$, $\Delta G^{0\prime} = -237$ kJ. However, if we were to mix O_2 and H_2 together in a sealed bottle and leave it for years, no measurable amount of water would form. This is because the bonding of oxygen and hydrogen atoms to form water requires that their chemical bonds first be broken. The breaking of these bonds requires some energy, and this energy is called **activation energy**.

Activation energy is the energy required to bring all molecules in a chemical reaction into the reactive state. For a reaction that proceeds with a net release of free energy (that is, an exergonic reaction), the situation is as diagrammed in **Figure 4.6**. Although the activation energy barrier is virtually insurmountable in the absence of a catalyst, in the presence of the proper catalyst, this barrier is greatly reduced.

Enzymes

The concept of activation energy leads us to consider catalysis and enzymes. A **catalyst** is a substance that lowers the activation

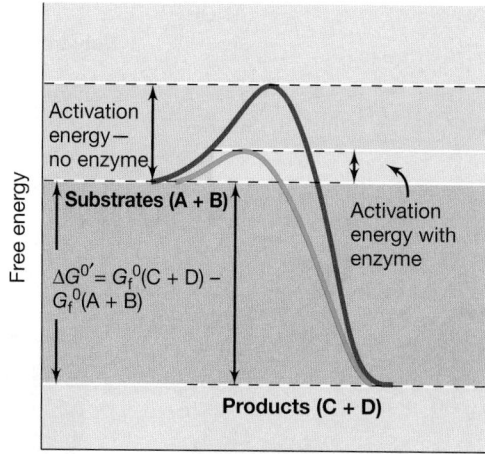

Figure 4.6 Activation energy and catalysis. Even chemical reactions that release energy may not proceed spontaneously, because the reactants must first be activated. Once they are activated, the reaction proceeds spontaneously. Catalysts such as enzymes lower the required activation energy.

energy of a reaction, thereby increasing the reaction rate. Catalysts facilitate reactions but are not consumed or transformed by them. Moreover, catalysts do not affect the energetics or the equilibrium of a reaction; catalysts affect only the *rate* at which reactions proceed.

Most cellular reactions do not proceed at useful rates without catalysis. Biological catalysts are called **enzymes**. Enzymes are proteins (or in a few cases, RNAs) that are highly specific for the reactions they catalyze. That is, each enzyme catalyzes only a single type of chemical reaction, or in the case of some enzymes, a single class of closely related reactions. This specificity is a function of the precise three-dimensional structure of the enzyme molecule.

In an enzyme-catalyzed reaction, the enzyme (E) combines with the reactant, called a *substrate* (S), forming an enzyme–substrate complex (E—S). Then, as the reaction proceeds, the *product* (P) is released and the enzyme is returned to its original state:

$$E + S \rightleftarrows E—S \rightleftarrows + P$$

The enzyme is generally much larger than the substrate(s), and the portion of the enzyme to which substrate binds is called the *active site*; the entire enzymatic reaction, from substrate binding to product release, may take only a few milliseconds.

Many enzymes contain small nonprotein molecules that participate in catalysis but are not themselves substrates. These small molecules can be divided into two classes based on the way they associate with the enzyme: *prosthetic groups* and *coenzymes*. Prosthetic groups bind very tightly to their enzymes, usually covalently and permanently. The heme group present in cytochromes (Section 4.9) is an example of a prosthetic group. **Coenzymes**, by contrast, are loosely bound to enzymes, and a single coenzyme molecule may associate with a number of different enzymes. Most coenzymes are derivatives of vitamins, and $NAD^+/NADH$, a derivative of the vitamin niacin, is a good example.

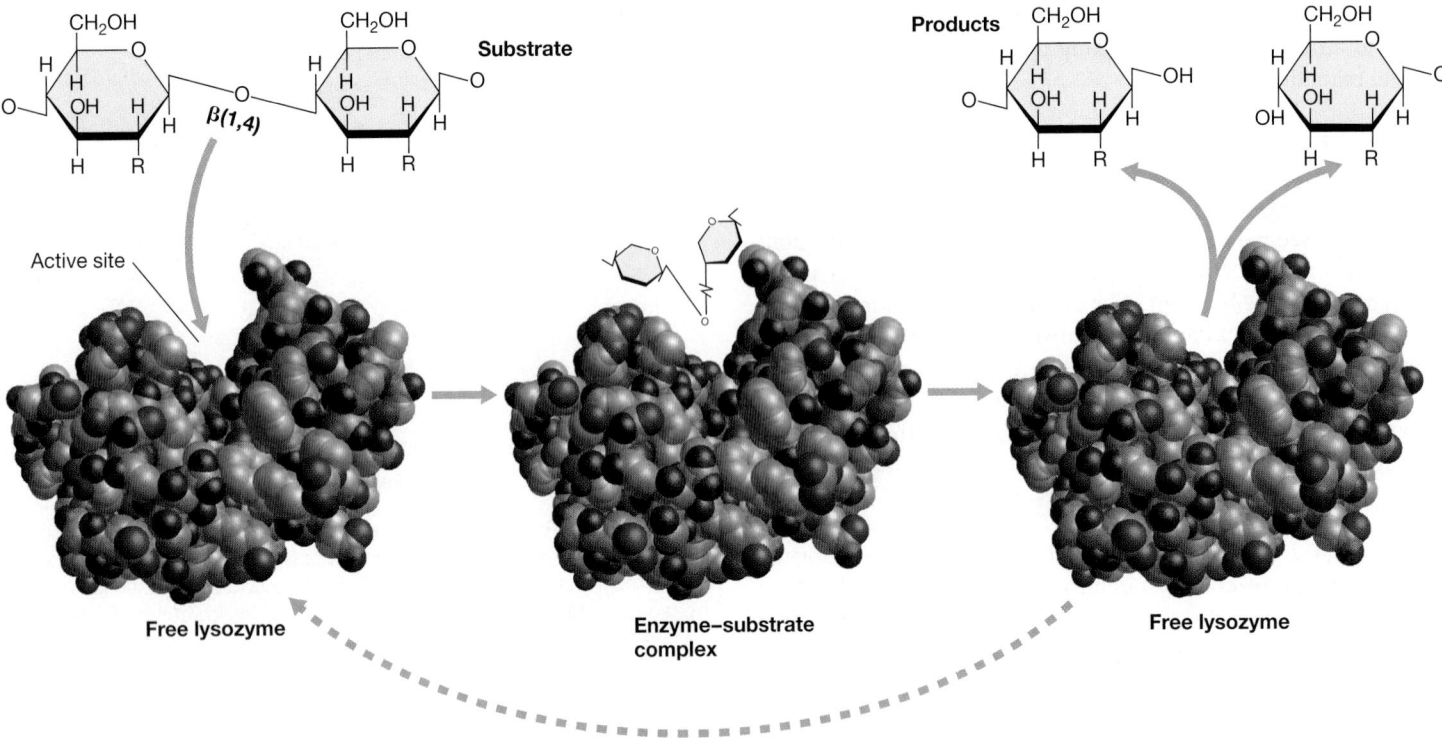

Figure 4.7 The catalytic cycle of an enzyme. The enzyme depicted here, lysozyme, catalyzes the cleavage of the β-1,4-glycosidic bond in the polysaccharide backbone of peptidoglycan. Following binding in the enzyme's active site, strain is placed on the bond, and this favors breakage. Space-filling model of lysozyme courtesy of Richard Feldmann.

Enzyme Catalysis

The catalytic power of enzymes is impressive. Enzymes increase the rate of chemical reactions anywhere from 10^8 to 10^{20} times over that which would occur spontaneously. To catalyze a specific reaction, an enzyme must do two things: (1) bind its substrate and (2) position the substrate relative to the catalytically active amino acids in the enzyme's active site. The enzyme–substrate complex (**Figure 4.7**) aligns reactive groups and places strain on specific bonds in the substrate(s). The net result is a reduction in the activation energy required to make the reaction proceed from substrate(s) to product(s) (Figure 4.6). These steps are shown in Figure 4.7 for the enzyme lysozyme, an enzyme whose substrate is the polysaccharide backbone of the bacterial cell wall polymer, peptidoglycan (Figure 3.16).

The reaction depicted in Figure 4.6 is exergonic because the free energy of formation of the substrates is greater than that of the products. Enzymes can also catalyze reactions that require energy, converting energy-poor substrates into energy-rich products. In these cases, however, not only must an activation energy barrier be overcome, but sufficient free energy must also be put into the reaction to raise the energy level of the substrates to that of the products. This is done by coupling the energy-*requiring* reaction to an energy-*yielding* one, such as the hydrolysis of ATP.

Theoretically, all enzymes are reversible in their activity. However, enzymes that catalyze highly exergonic or highly endergonic reactions typically act only unidirectionally. If a particularly exergonic or endergonic reaction needs to be reversed, a different enzyme usually catalyzes the reverse reaction.

MiniQuiz

- What is the function of a catalyst? What are enzymes made of?
- Where on an enzyme does the substrate bind?
- What is activation energy?

Ⅲ Oxidation–Reduction and Energy-Rich Compounds

The energy released in oxidation–reduction (redox) reactions is conserved in cells by the simultaneous synthesis of energy-rich compounds, such as ATP. Here we first consider oxidation–reduction reactions and the major electron carriers present in the cell. We then examine the compounds that actually conserve the energy released in oxidation–reduction reactions.

4.6 Electron Donors and Electron Acceptors

An *oxidation* is the removal of an electron or electrons from a substance, and a *reduction* is the addition of an electron or electrons to a substance. Oxidations and reductions are common in cellular biochemistry and can involve just electrons or an electron plus a proton (a hydrogen atom; H).

1. $H_2 \rightarrow 2\ e^- + 2\ H^+$

Electron-donating half reaction

2. $\frac{1}{2} O_2 + 2\ e^- \rightarrow O^{2-}$

Electron-accepting half reaction

3. $2\ H^+ + O^{2-} \rightarrow H_2O$

Formation of water

4. Electron donor, Electron acceptor
$H_2 + \frac{1}{2} O_2 \rightarrow H_2O$

Net reaction

Figure 4.8 **Example of an oxidation–reduction reaction.** The formation of H_2O by reaction of the electron donor H_2 and the electron acceptor O_2.

Redox Reactions

Redox reactions occur in pairs. For example, hydrogen gas (H_2) can release electrons and protons and become oxidized (**Figure 4.8**). However, electrons cannot exist alone in solution; they must be part of atoms or molecules. Thus, the equation as drawn does not itself represent an independent reaction. The reaction is only a *half reaction*, a term that implies the need for a second half reaction. This is because for any substance to be oxidized, another substance must be reduced.

The oxidation of H_2 can be coupled to the reduction of many different substances, including oxygen (O_2), in a second half reaction. This reduction half reaction, when coupled to the oxidation of H_2, yields the overall balanced reaction in step 4 of Figure 4.8. In reactions of this type, we refer to the substance *oxidized* (in this case, H_2) as the **electron donor**, and the substance *reduced* (in this case, O_2) as the **electron acceptor**. The concept of electron donors and electron acceptors is very important in microbiology and underlies virtually all aspects of energy metabolism.

Reduction Potentials and Redox Couples

Substances differ in their tendency to be electron donors or electron acceptors. This tendency is expressed as their **reduction potential** (E_0', standard conditions), measured in volts (V) in reference to that of a standard substance, H_2 (**Figure 4.9**). By convention, reduction potentials are given for half reactions written as *reductions*, with reactions at pH 7 because the cytoplasm of most cells is neutral, or nearly so.

A substance can be either an electron donor or an electron acceptor under different circumstances, depending on the substances with which it reacts. The constituents on each side of the arrow in half reactions are called a *redox couple*, such as $2\ H^+/H_2$, or $\frac{1}{2} O_2/H_2O$ (Figure 4.8). By convention, when writing a redox couple, the *oxidized* form of the couple is always placed on the left, before the forward slash, followed by the *reduced* form after the forward slash. In the example of Figure 4.8, the E_0' of the $2\ H^+/H_2$ couple is -0.42 V and that of the $\frac{1}{2} O_2/H_2O$ couple is $+0.82$ V. We will learn shortly that these values mean that O_2 is an excellent electron *acceptor* and H_2 is an excellent electron *donor*.

In redox reactions, the *reduced* substance of a redox couple whose E_0' is more negative donates electrons to the *oxidized* substance of a redox couple whose E_0' is more positive. Thus, in the couple $2\ H^+/H_2$, H_2 has a greater tendency to donate electrons than the tendency of $2\ H^+$ to accept them, and in the couple $\frac{1}{2} O_2/H_2O$, H_2O has a very weak tendency to donate electrons, whereas O_2 has a great tendency to accept them. It then follows

that in a reaction of H_2 and O_2, H_2 will be the electron donor and become oxidized, and O_2 will be the electron acceptor and become reduced (Figure 4.8).

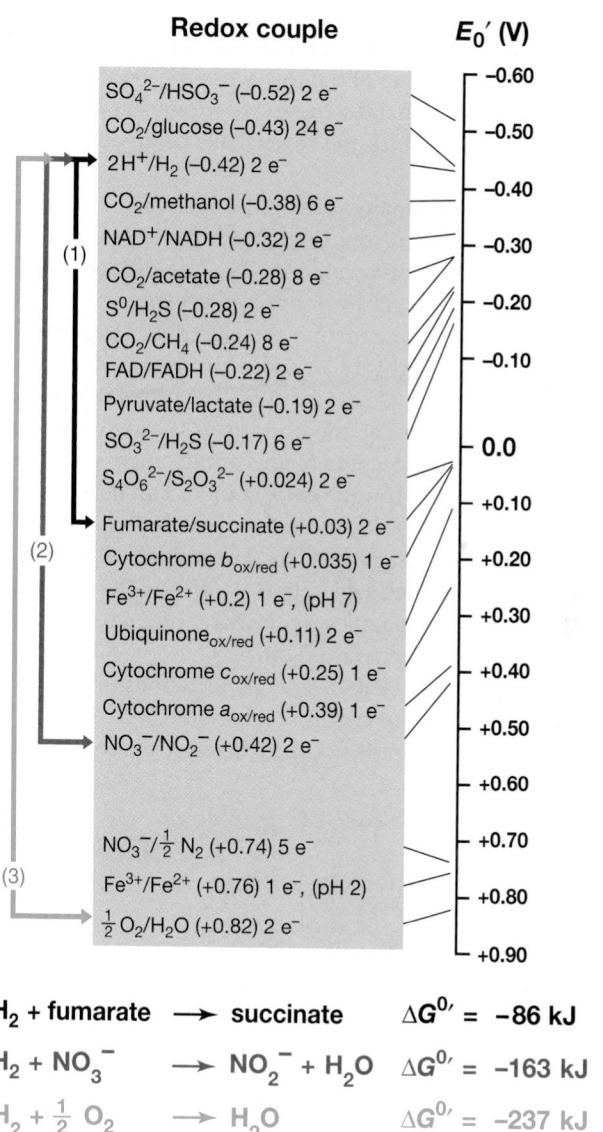

Redox couple	E_0' (V)
SO_4^{2-}/HSO_3^- (−0.52) 2 e^-	−0.60
CO_2/glucose (−0.43) 24 e^-	−0.50
$2H^+/H_2$ (−0.42) 2 e^-	
CO_2/methanol (−0.38) 6 e^-	−0.40
$NAD^+/NADH$ (−0.32) 2 e^-	−0.30
CO_2/acetate (−0.28) 8 e^-	
S^0/H_2S (−0.28) 2 e^-	−0.20
CO_2/CH_4 (−0.24) 8 e^-	
$FAD/FADH$ (−0.22) 2 e^-	−0.10
Pyruvate/lactate (−0.19) 2 e^-	
SO_3^{2-}/H_2S (−0.17) 6 e^-	0.0
$S_4O_6^{2-}/S_2O_3^{2-}$ (+0.024) 2 e^-	
Fumarate/succinate (+0.03) 2 e^-	+0.10
Cytochrome $b_{ox/red}$ (+0.035) 1 e^-	+0.20
Fe^{3+}/Fe^{2+} (+0.2) 1 e^-, (pH 7)	
Ubiquinone$_{ox/red}$ (+0.11) 2 e^-	+0.30
Cytochrome $c_{ox/red}$ (+0.25) 1 e^-	+0.40
Cytochrome $a_{ox/red}$ (+0.39) 1 e^-	
NO_3^-/NO_2^- (+0.42) 2 e^-	+0.50
	+0.60
$NO_3^-/\frac{1}{2} N_2$ (+0.74) 5 e^-	+0.70
Fe^{3+}/Fe^{2+} (+0.76) 1 e^-, (pH 2)	+0.80
$\frac{1}{2} O_2/H_2O$ (+0.82) 2 e^-	+0.90

(1) H_2 + fumarate $\longrightarrow$ succinate $\Delta G^{0'} = -86$ kJ

(2) $H_2 + NO_3^- \longrightarrow NO_2^- + H_2O$ $\Delta G^{0'} = -163$ kJ

(3) $H_2 + \frac{1}{2} O_2 \longrightarrow H_2O$ $\Delta G^{0'} = -237$ kJ

Figure 4.9 **The redox tower.** Redox couples are arranged from the strongest donors at the top to the strongest acceptors at the bottom. Electrons can be "caught" by acceptors at any intermediate level as long as the donor couple is more negative than the acceptor couple. The greater the difference in reduction potential between electron donor and electron acceptor, the more free energy is released. Note the differences in energy yield when H_2 reacts with three different electron acceptors, fumarate, nitrate, and oxygen.

UNIT 2

As previously mentioned, all half reactions are written as reductions. However, in an actual reaction between two redox couples, the half reaction with the more negative E_0' proceeds as an oxidation and is therefore written in the opposite direction. In the reaction between H_2 and O_2 shown in Figure 4.8, H_2 is thus oxidized and is written in the reverse direction from its formal half reaction.

The Redox Tower and Its Relationship to $\Delta G^{0\prime}$

A convenient way of viewing electron transfer reactions in biological systems is to imagine a vertical tower (Figure 4.9). The tower represents the range of reduction potentials possible for redox couples in nature, from those with the most negative E_0' on the top to those with the most positive E_0' at the bottom; thus, we can call the tower a *redox tower*. The *reduced* substance in the redox couple at the top of the tower has the greatest tendency to donate electrons, whereas the *oxidized* substance in the redox couple at the bottom of the tower has the greatest tendency to accept electrons.

Using the tower analogy, imagine electrons from an electron donor near the top of the tower falling and being "caught" by electron acceptors at various levels. The difference in reduction potential between the donor and acceptor redox couples is expressed as $\Delta E_0'$. The further the electrons drop from a donor before they are caught by an acceptor, the greater the amount of energy released. That is, $\Delta E_0'$ is *proportional* to $\Delta G_0'$ (Figure 4.9). Oxygen, at the bottom of the redox tower, is the strongest electron acceptor of any significance in nature. In the middle of the redox tower, redox couples can be either electron donors or acceptors depending on which redox couples they react with. For instance, the $2\,H^+/H_2$ couple (-0.42 V) can react with the fumarate/succinate ($+0.03$ V), NO_3^-/NO_2^- ($+0.42$ V), or $\frac{1}{2}O_2/H_2O(+8.82\,V)$ couples, with increasing amounts of energy being released, respectively (Figure 4.9).

Electron donors used in energy metabolism are also called *energy sources* because energy is released when they are oxidized (Figure 4.9). The point is not that the electron donor per se con-

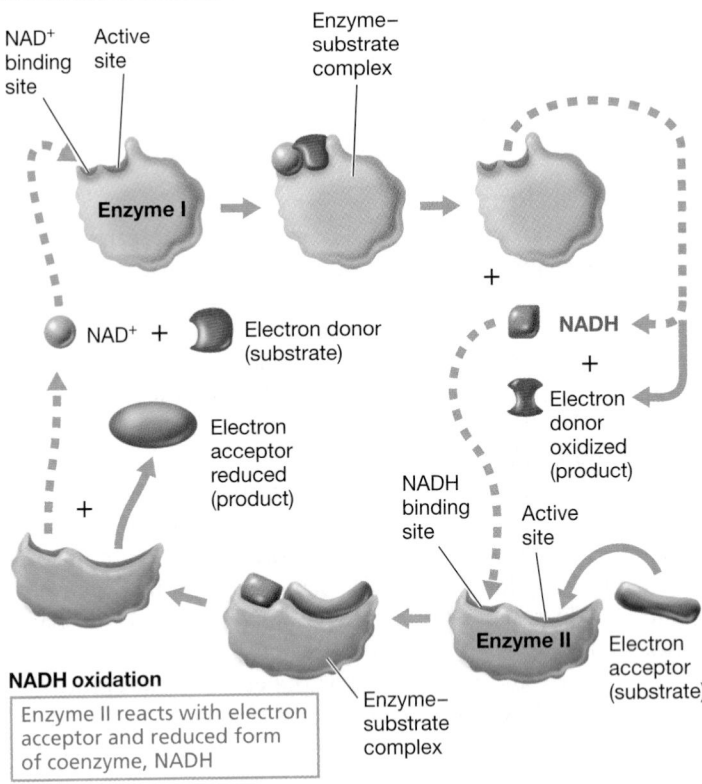

NAD+ reduction

Enzyme I reacts with electron donor and oxidized form of coenzyme, NAD+

NAD+ binding site

Active site

Enzyme–substrate complex

Enzyme I

NAD+ +

Electron donor (substrate)

NADH

+

Electron donor oxidized (product)

Electron acceptor reduced (product)

NADH binding site

Active site

Enzyme II

Electron acceptor (substrate)

NADH oxidation

Enzyme II reacts with electron acceptor and reduced form of coenzyme, NADH

Enzyme–substrate complex

Figure 4.11 NAD+/NADH cycling. A schematic example of redox reactions in two different enzymes linked by their use of either NAD+ or NADH.

tains energy but that the chemical reaction in which the electron donor participates releases energy. The presence of a suitable electron acceptor is just as important as the presence of a suitable electron donor. Lacking one or the other, the energy-releasing reaction cannot proceed. Many potential electron donors exist in nature, including a wide variety of organic and inorganic compounds.

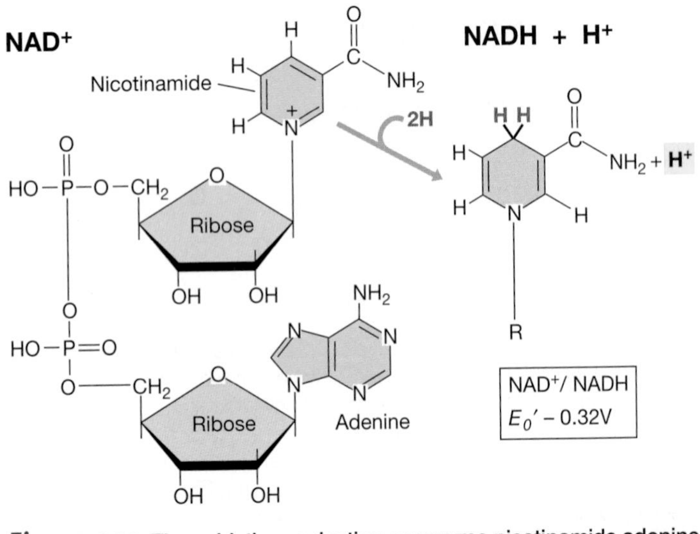

Electron Carriers and NAD/NADH Cycling

Redox reactions in microbial cells are typically mediated by one or more small molecules. A very common carrier is the coenzyme nicotinamide adenine dinucleotide (NAD^+) (**Figure 4.10**). NAD^+ is an electron plus proton carrier, transporting $2\,e^-$ and $2\,H^+$ at the same time.

The reduction potential of the $NAD^+/NADH$ couple is -0.32 V, which places it fairly high on the electron tower; that is, NADH is a good electron donor (Figure 4.10). Coenzymes such as NADH increase the diversity of redox reactions possible in a cell by allowing chemically dissimilar electron donors and acceptors to interact, with the coenzyme acting as the intermediary. For example, electrons removed from an electron donor can reduce NAD^+ to NADH, and the latter can be converted back to NAD^+ by donating electrons to the electron acceptor. **Figure 4.11** shows an example of such electron shuttling by $NAD^+/NADH$. In the reaction, NAD^+ and NADH facilitate the redox reaction without being

NAD+

Nicotinamide

NH_2

$HO-P-O-CH_2$

Ribose

OH   OH

$HO-P=O$

CH_2

Ribose

Adenine

NH_2

OH   OH

NADH + H+

2H

$NH_2 + $ **H+**

R

NAD+/ NADH
$E_0' - 0.32V$

Figure 4.10 The oxidation–reduction coenzyme nicotinamide adenine dinucleotide (NAD^+). NAD^+ undergoes oxidation–reduction as shown and is freely diffusible. "R" is the adenine dinucleotide portion of NAD^+.

consumed in the process. Recall that the cell requires large amounts of a primary electron donor (the substance that was oxidized to yield NADH) and a final electron acceptor (such as O_2). But the cell needs only a tiny amount of $NAD^+/NADH$ because they are constantly being recycled. All that is needed is an amount sufficient to service the redox enzymes in the cell that use these coenzymes in their reaction mechanisms (Figure 4.11).

$NADP^+/NADPH$ is a related redox coenzyme in which a phosphate group is added to $NAD^+/NADH$. $NADP^+/NADPH$ typically participate in redox reactions distinct from those that use $NAD^+/NADH$, most commonly in anabolic (biosynthetic) reactions in which oxidations and reductions occur.

MiniQuiz

• In the reaction $H_2 + \frac{1}{2}O_2 \rightarrow H_2O$, what is the electron donor and what is the electron acceptor?

• Why is nitrate (NO_3^-) a better electron acceptor than fumarate?

• Is NADH a better electron donor than H_2? Is NAD^+ a better acceptor than H^+? How do you determine this?

4.7 Energy-Rich Compounds and Energy Storage

Energy released from redox reactions must be conserved by the cell if it is to be used later to drive energy-requiring cell functions. In living organisms, chemical energy released in redox reactions is conserved primarily in phosphorylated compounds. The free energy released upon hydrolysis of the phosphate in these *energy-rich compounds* is significantly greater than that of

the average covalent bond in the cell, and it is this released energy that is conserved by the cell.

Phosphate can be bonded to organic compounds by either ester or anhydride bonds, as illustrated in **Figure 4.12**. However, not all phosphate bonds are energy-rich. As seen in the figure, the $\Delta G^{0\prime}$ of hydrolysis of the phosphate *ester* bond in glucose 6-phosphate is only -13.8 kJ/mol. By contrast, the $\Delta G^{0\prime}$ of hydrolysis of the phosphate *anhydride* bond in phosphoenolpyruvate is -51.6 kJ/mol, almost four times that of glucose 6-phosphate. Although either compound could be hydrolyzed to yield energy, cells typically use a small group of compounds whose $\Delta G^{0\prime}$ of hydrolysis is greater than -30 kJ/mol as energy "currencies" in the cell. Thus, phosphoenolpyruvate is energy-rich whereas glucose 6-phosphate is not. Notice in Figure 4.12 that ATP contains three phosphates, but only two of them have free energies of hydrolysis of >30 kJ. Also notice that the thioester bond between the C and S atoms of coenzyme A has a free energy of hydrolysis of >30 kJ.

Adenosine Triphosphate

The most important energy-rich phosphate compound in cells is **adenosine triphosphate (ATP)**. ATP consists of the ribonucleoside adenosine to which three phosphate molecules are bonded in series. ATP is the prime energy currency in all cells, being generated during exergonic reactions and consumed in endergonic reactions. From the structure of ATP (Figure 4.12), it can be seen that two of the phosphate bonds are phosphoanhydrides that have free energies of hydrolysis greater than 30 kJ. Thus, the reactions ATP → ADP + P_i and ADP → AMP + P_i each release roughly 32 kJ/mol of energy. By contrast, AMP is not energy-rich because its free energy of hydrolysis is only about half that of ADP or ATP (Figure 4.12).

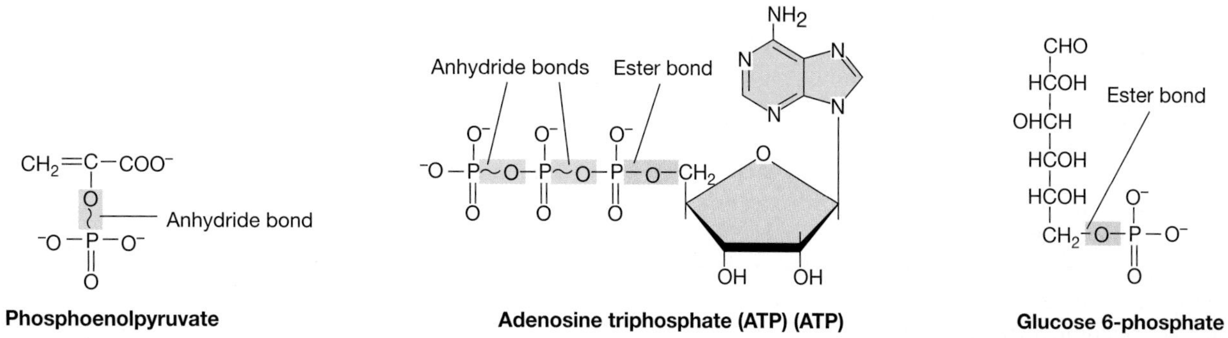

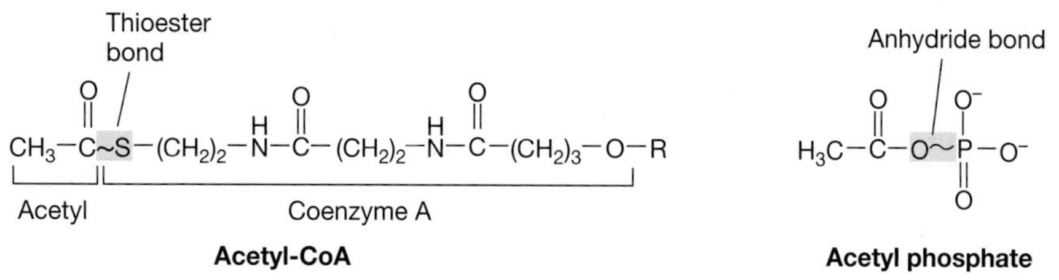

Compound	$G^{0\prime}$ kJ/mol
$\Delta G^{0\prime} > 30kJ$	
Phosphoenolpyruvate	−51.6
1,3-Bisphosphoglycerate	−52.0
Acetyl phosphate	−44.8
ATP	−31.8
ADP	−31.8
Acetyl-CoA	−35.7
$\Delta G^{0\prime} < 30kJ$	
AMP	−14.2
Glucose 6-phosphate	−13.8

Figure 4.12 Phosphate bonds in compounds that conserve energy in bacterial metabolism. Notice, by referring to the table, the range in free energy of hydrolysis of the phosphate bonds highlighted in the compounds. The "R" group of acetyl-CoA is a 3′ phospho ADP group.

Although the energy released in ATP hydrolysis is −32 kJ, a caveat must be introduced here to define more precisely the energy requirements for the synthesis of ATP. In an actively growing *Escherichia coli* cell, the ratio of ATP to ADP is about 7.5:1. This deviation from equilibrium affects the energy requirements for ATP synthesis. In such a cell, the actual energy expenditure (that is, the ΔG, Section 4.4) for the synthesis of 1 mole of ATP is on the order of −55 to −60 kJ. Nevertheless, for the purposes of learning and applying the basic principles of bioenergetics, we assume that reactions conform to "standard conditions" ($\Delta G^{0\prime}$), and thus we assume that the energy required for synthesis or hydrolysis of ATP is 32 kJ/mol.

Coenzyme A

Cells can use the free energy available in the hydrolysis of other energy-rich compounds as well as phosphorylated compounds. These include, in particular, derivatives of *coenzyme A* (for example, acetyl-CoA; see structure in Figure 4.12). Coenzyme A derivatives contain thioester bonds. Upon hydrolysis, these yield sufficient free energy to drive the synthesis of an energy-rich phosphate bond. For example, in the reaction

$$\text{acetyl-S-CoA} + H_2O + ADP + P_i \rightarrow$$
$$\text{acetate}^- + \text{HS-CoA} + ATP + H^+$$

the energy released in the hydrolysis of coenzyme A is conserved in the synthesis of ATP. Coenzyme A derivatives (acetyl-CoA is just one of many) are especially important to the energetics of anaerobic microorganisms, in particular those whose energy metabolism depends on fermentation. We return to the importance of coenzyme A derivatives many times in Chapter 14.

Energy Storage

ATP is a dynamic molecule in the cell; it is continuously being broken down to drive anabolic reactions and resynthesized at the expense of catabolic reactions. For longer-term energy storage, microorganisms produce insoluble polymers that can be catabolized later for the production of ATP.

Examples of energy storage polymers in prokaryotes include glycogen, poly-β-hydroxybutyrate and other polyhydroxyalkanoates, and elemental sulfur, stored from the oxidation of H_2S by sulfur chemolithotrophs. These polymers are deposited within the cell as large granules that can be seen with the light or electron microscope (Section 3.10). In eukaryotic microorganisms, polyglucose in the form of starch and lipids in the form of simple fats are the major reserve materials. In the absence of an external energy source, a cell can break down these polymers to make new cell material or to supply the very low amount of energy, called *maintenance energy*, needed to maintain cell integrity when it is in a nongrowing state.

MiniQuiz

- How much energy is released per mole of ATP converted to ADP + P_i under standard conditions? Per mole of AMP converted to adenosine and P_i?

- During periods of nutrient abundance, how can cells prepare for periods of nutrient starvation?

Essentials of Catabolism

Two series of reactions—fermentation and respiration—are linked to energy conservation in chemoorganotrophs: **Fermentation** is the form of anaerobic catabolism in which an organic compound is both an electron donor and an electron acceptor, and ATP is produced by substrate-level phosphorylation; and **respiration** is the catabolism in which a compound is oxidized with O_2 (or an O_2 substitute) as the terminal electron acceptor, usually accompanied by ATP production by oxidative phosphorylation. In both series of reactions, ATP synthesis is coupled to energy released in oxidation–reduction reactions.

One can look at fermentation and respiration as alternative metabolic choices available to some microorganisms. In organisms that can both ferment and respire, such as yeast, fermentation is necessary when conditions are anoxic and terminal electron acceptors are absent. When O_2 is available, respiration can take place. We will see that much more ATP is produced in respiration than in fermentation and thus respiration is the preferred choice (see the Microbial Sidebar, "Yeast Fermentation, the Pasteur Effect, and the Home Brewer"). But many microbial habitats lack O_2 or other electron acceptors that can substitute for O_2 in respiration (see Figure 4.22), and in such habitats, fermentation is the only option for energy conservation by chemoorganotrophs.

4.8 Glycolysis

In fermentation, ATP is produced by a mechanism called **substrate-level phosphorylation**. In this process, ATP is synthesized directly from energy-rich intermediates during steps in the catabolism of the fermentable substrate (**Figure 4.13a**). This

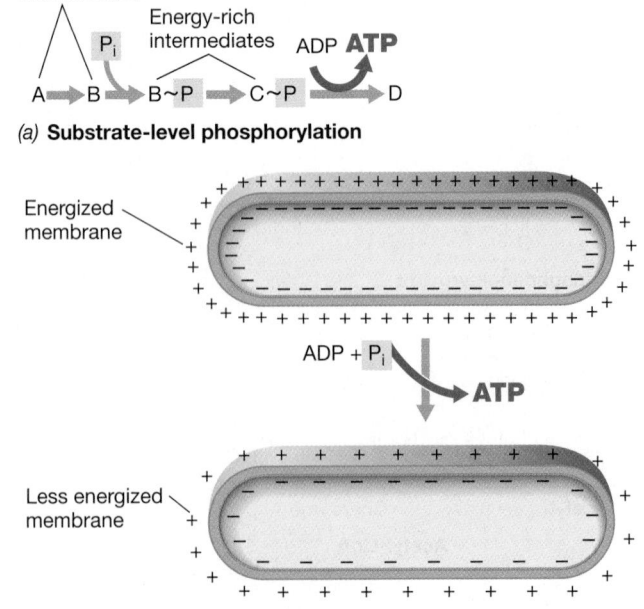

(a) **Substrate-level phosphorylation**

(b) **Oxidative phosphorylation**

Figure 4.13 Energy conservation in fermentation and respiration. *(a)* In fermentation, substrate-level phosphorylation produces ATP. *(b)* In respiration, the cytoplasmic membrane, energized by the proton motive force, dissipates energy to synthesize ATP from ADP + P_i by oxidative phosphorylation.

Yeast Fermentation, the Pasteur Effect, and the Home Brewer

Every home wine maker, brewer, and baker is an amateur microbiologist, perhaps without even realizing it. Indeed, anaerobic mechanisms of microbial energy generation are at the heart of some of the most commonly consumed fermented foods and beverages (Figure 1).

In the production of breads and most alcoholic beverages, the yeast *Saccharomyces cerevisiae* or a related species is exploited to produce ethanol (ethyl alcohol) and carbon dioxide (CO_2). Found in various sugar-rich environments such as fruit juices and nectar, yeasts can carry out the two opposing modes of chemoorganotrophic metabolism discussed in this chapter, *fermentation* and *respiration*. When oxygen (O_2) is present in high amounts, yeast grows efficiently on various sugars, making yeast cells and CO_2 (the latter from the citric acid cycle, Section 4.11) in the process. However, when conditions are anoxic, yeasts switch to fermentative metabolism using the glycolytic pathway. This reduces the production of new cells but yields significant amounts of the fermentation products ethanol and CO_2.

During his studies on fermentation, the early microbiologist Louis Pasteur (Section 1.7) recognized that yeast switch between aerobic and anaerobic metabolism. He showed that the ratio of glucose consumed by a yeast suspension to the weight of cells produced varied with the concentration of O_2 supplied; the ratio was maximal in the absence of O_2. In Pasteur's own words, "the ferment lost its fermentative abilities in proportion to the concentration of this gas." He referred to the yeast cells as "the ferment" because it had not yet been established that the yeast in the fermenting mixture were actually living cells! He described what has come to be known as the "Pasteur effect," a phenomenon that occurs in any organism (even humans) that can both ferment and respire glucose. The fermentation of glucose is maximal under anoxic conditions and is incrementally inhibited by O_2 because respiration yields much more energy per glucose than does fermentation. As a rule, cells carry out the metabolism that is most energetically beneficial to them.

The Pasteur effect occurs in alcoholic beverage fermentation. When grapes are squeezed to make juice, called *must*, small numbers of yeast cells present on the grapes are transferred to the must. During the first several days of the wine-making process, yeast grow primarily by respiration and consume O_2, making the juice anoxic. The yeast respire the glucose in the juice rather than fermenting it because more energy is available from the respiration of glucose than from its fermentation. However, as soon as the O_2 in the grape juice is depleted, fermentation begins along with alcohol formation. This switch from aerobic to anaerobic metabolism is crucial in wine making, and care must be taken to ensure that O_2 is kept out of the fermentation vessel. The vessel is thus sealed against the introduction of air. Laboratory studies of yeast have shown that the introduction of O_2 to a fermenting yeast culture triggers the expression of hundreds of genes necessary for respiration, and such events would interrupt ethanol formation and other desirable reactions in wine production.

Wine is only one of many alcoholic products made with yeast. Others include beer and distilled spirits such as brandy, whisky, vodka, and gin (Chapter 15). In distilled spirits, the ethanol, produced in relatively low amounts (10–15% by volume) by the yeast, is concentrated by distilling to make a beverage containing 40–70% alcohol. Even alcohol for motor fuel is made with yeast in parts of the world where sugar is plentiful but petroleum is in short supply (such as Brazil). In the United States, ethyl alcohol for use as an industrial solvent and motor fuel is produced using corn starch as a source of the fermentable substrate (glucose). Yeast also serves as the leavening agent in bread, although here it is not the alcohol that is important, but CO_2, the *other* product of the alcohol fermentation (see Figure 4.14). The CO_2 raises the dough, and the alcohol produced along with it is volatilized during the baking process. We discuss yeast and yeast products in Chapters 15 and 20.

The yeast cell, forced to carry out a fermentative lifestyle because the O_2 it needs for respiration is absent, has had a considerable impact on the lives of humans. Substances that from the physiological standpoint of the yeast cell are "waste products" of the glycolytic pathway—ethanol and CO_2—are, respectively, the foundation of the alcoholic beverage and baking industries.

Figure 1 Major food and beverage products of fermentation by the yeast *Saccharomyces cerevisiae*.

is in contrast to **oxidative phosphorylation**, typical of respiration, in which ATP is produced at the expense of the proton motive force (Figure 4.13*b*).

The fermentable substrate in a fermentation is both the electron donor and electron acceptor; not all compounds can be fermented, but sugars, especially hexoses such as glucose, are excellent fermentable substrates. A common pathway for the catabolism of glucose is **glycolysis**, which breaks down glucose into pyruvate. Glycolysis is also called the *Embden–Meyerhof–Parnas pathway* for its major discoverers. Whether glucose is fermented or respired, it travels through this pathway. Here we focus on the reactions of glycolysis and the reactions that follow under anoxic conditions.

Glycolysis can be divided into three stages, each involving a series of enzymatic reactions. Stage I comprises "preparatory" reactions; these are not redox reactions and do not release energy but instead lead to the production of a key intermediate of the

pathway. In Stage II, redox reactions occur, energy is conserved in the form of ATP, and two molecules of pyruvate are formed. The reactions of glycolysis are finished at this point. However, redox balance has not yet been achieved. So, in Stage III, redox reactions occur once again and fermentation products are formed (**Figure 4.14**).

Stage I: Preparatory Reactions

In Stage I glucose is phosphorylated by ATP, yielding glucose 6-phosphate; the latter is then isomerized to fructose 6-phosphate. A second phosphorylation leads to the production of fructose 1,6-bisphosphate. The enzyme aldolase then splits fructose 1,6-bisphosphate into two 3-carbon molecules, *glyceraldehyde 3-phosphate* and its isomer, *dihydroxyacetone phosphate*, which can be converted into glyceraldehyde 3-phosphate. To this point, all of the reactions, including the consumption of ATP, have proceeded without redox reactions.

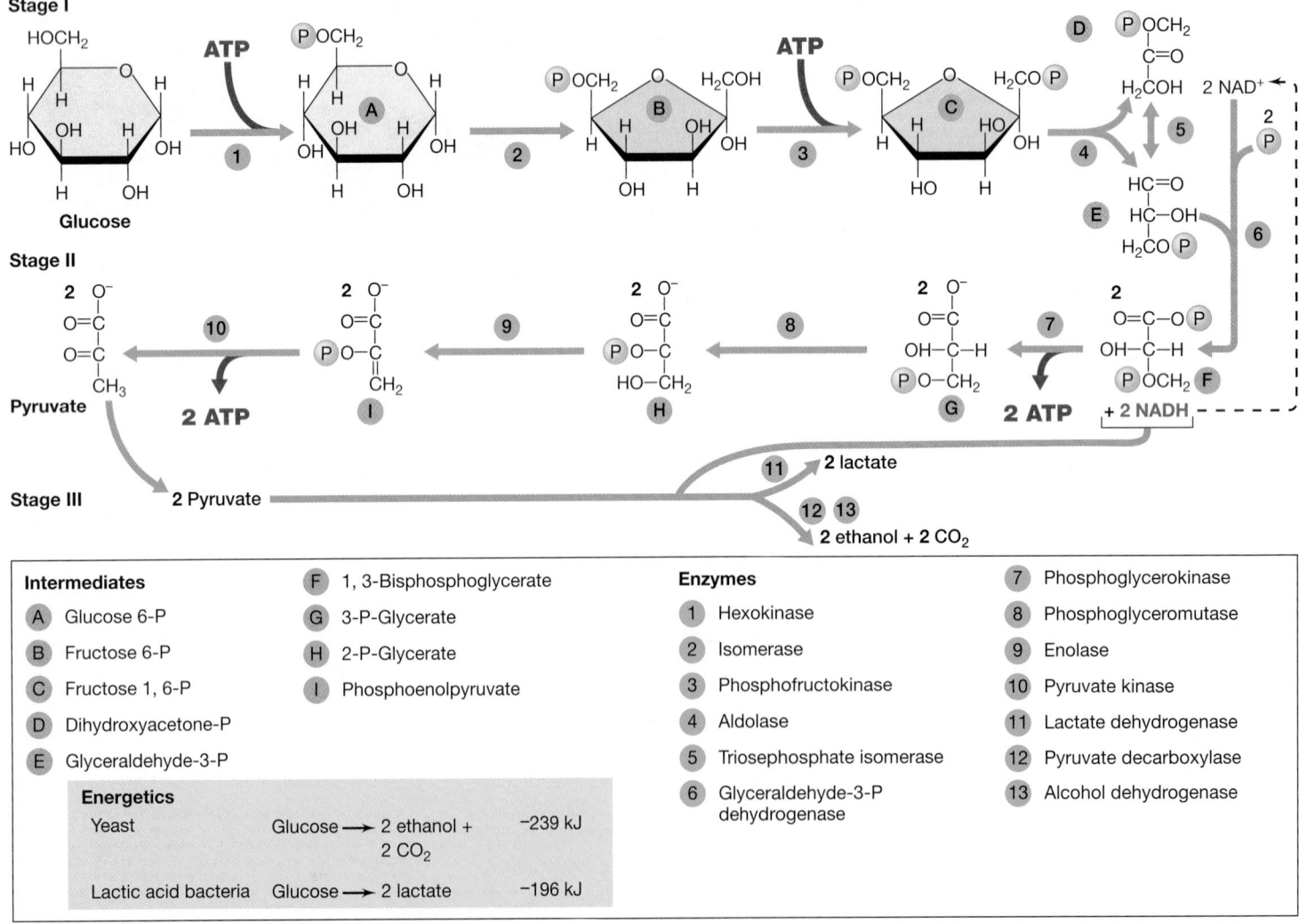

Figure 4.14 Embden–Meyerhof–Parnas pathway (glycolysis). The sequence of reactions in the catabolism of glucose to pyruvate and then on to fermentation products. Pyruvate is the end product of glycolysis, and fermentation products are made from it. The blue table at the bottom left lists the energy yields from the fermentation of glucose by yeast or lactic acid bacteria.

Stage II: Production of NADH, ATP, and Pyruvate

The first redox reaction of glycolysis occurs in Stage II during the oxidation of glyceraldehyde 3-phosphate to 1,3-bisphosphoglyceric acid. In this reaction (which occurs twice, once for each of the two molecules of glyceraldehyde 3-phosphate produced from glucose), the enzyme glyceraldehyde-3-phosphate dehydrogenase reduces its coenzyme NAD^+ to NADH. Simultaneously, each glyceraldehyde 3-phosphate molecule is phosphorylated by the addition of a molecule of inorganic phosphate. This reaction, in which inorganic phosphate is converted to organic form, sets the stage for energy conservation. ATP formation is possible because 1,3-bisphosphoglyceric acid is an energy-rich compound (Figure 4.12). ATP is then synthesized when (1) each molecule of 1,3-bisphosphoglyceric acid is converted to 3-phosphoglyceric acid, and (2) each molecule of phosphoenolpyruvate is converted to pyruvate (Figure 4.14).

During Stages I and II of glycolysis, *two* ATP molecules have been consumed and *four* ATP molecules have been synthesized (Figure 4.14). Thus, the net energy yield in glycolysis is *two molecules of ATP per molecule of glucose fermented.*

Stage III: Consumption of NADH and Production of Fermentation Products

During the formation of two molecules of 1,3-bisphosphoglyceric acid, two NAD^+ are reduced to NADH (Figure 4.14). However, as previously discussed (Section 4.6 and Figure 4.11), NAD^+ is only an electron shuttle, not a net (terminal) acceptor of electrons. Thus, the NADH produced in glycolysis must be oxidized back to NAD^+ in order for glycolysis to continue, and this is accomplished when pyruvate is reduced (by NADH) to fermentation products (Figure 4.14). For example, in fermentation by yeast, pyruvate is reduced to ethanol with the subsequent production of carbon dioxide (CO_2). By contrast, lactic acid bacteria reduce pyruvate to lactate. Many other possibilities for pyruvate reduction are possible depending on the organism (see sections on fermentative diversity in Chapter 14), but the net result is the same: NADH is reoxidized to NAD^+ during the production of fermentation products, allowing reactions of the pathway that depend on NAD^+ to continue.

Glucose Fermentation: Net and Practical Results

During glycolysis, glucose is consumed, two ATPs are made, and fermentation products are generated. For the organism the crucial product is ATP, which is used in energy-requiring reactions; fermentation products are merely waste products. However, fermentation products are not considered wastes by the distiller, the brewer, the cheese maker, or the baker (see the Microbial Sidebar). Thus, fermentation is more than just an energy-yielding process for a cell; it is also a means of making natural products useful to humans.

MiniQuiz

- Which reactions in glycolysis involve oxidations and reductions?
- What is the role of NAD^+/NADH in glycolysis?
- Why are fermentation products made during glycolysis?

4.9 Respiration and Electron Carriers

We have just seen that fermentation is an anaerobic process and releases only a small amount of energy. As a result, only a few ATP molecules are synthesized. Why is more energy not conserved in fermentation? The simple answer is that, although the fermentation products excreted still contain a large amount of potential energy, the organism cannot oxidize these further because O_2 is absent. By contrast, if O_2 (or other usable terminal acceptors, see Figure 4.22) are present, pyruvate can be oxidized to CO_2 instead of being reduced to fermentation products and excreted. When pyruvate is oxidized to CO_2, a far higher yield of ATP is possible. Oxidation using O_2 as the terminal electron acceptor is called *aerobic respiration*; oxidation using other acceptors under anoxic conditions is called *anaerobic respiration* (Section 4.12).

Our discussion of respiration covers both carbon transformations and redox reactions and focuses on two issues: (1) how electrons are transferred from the organic compound to the terminal electron acceptor and how this is coupled to energy conservation, and (2) the pathway by which organic carbon is oxidized into CO_2. During the former, ATP is synthesized at the expense of the proton motive force (Figure 4.13b); thus we begin with a consideration of electron transport, the series of reactions that lead to the proton motive force.

Electron transport is a membrane-mediated process and has two basic functions: (1) facilitating the transfer of electrons from primary donor to terminal acceptor and (2) participating in membrane events whose end result is energy conservation. Several types of oxidation–reduction enzymes participate in electron transport. These include *NADH dehydrogenases, flavoproteins* (**Figure 4.15**), *iron–sulfur proteins*, and *cytochromes*. Also participating are nonprotein electron carriers called *quinones*. The carriers are arranged in the membrane in order of increasingly more positive reduction potential, with NADH dehydrogenase first and the cytochromes last (see Figure 4.19).

NADH dehydrogenases are proteins bound to the inside surface of the cytoplasmic membrane. They have an active site that binds NADH and accepts two electrons plus two protons ($2 e^- + 2 H^+$) when NADH is oxidized to NAD^+ (Figures 4.10

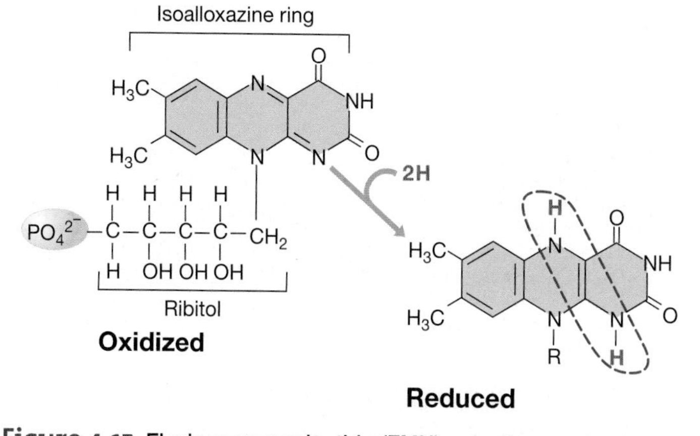

Figure 4.15 Flavin mononucleotide (FMN), a hydrogen atom carrier. The site of oxidation–reduction (dashed red circle) is the same in FMN and the related coenzyme flavin adenine dinucleotide (FAD, not shown). FAD contains an adenosine group bonded through the phosphate group on FMN.

and 4.11). The $2\,e^- + 2\,H^+$ are then transferred to a flavoprotein, the next carrier in the chain.

Flavoproteins contain a derivative of the vitamin riboflavin. The flavin portion, which is bound to a protein, is a prosthetic group that is reduced as it accepts $2\,e^- + 2\,H^+$ and oxidized when $2\,e^-$ are passed on to the next carrier in the chain. Note that flavoproteins *accept* $2\,e^- + 2\,H^+$ but *donate* only electrons. We will consider what happens to the $2\,H^+$ later. Two flavins are commonly found in cells, flavin mononucleotide (FMN) and flavin adenine dinucleotide (FAD). In the latter, FMN is bonded to ribose and adenine through a second phosphate. Riboflavin, also called vitamin B_2, is a source of the parent flavin molecule in flavoproteins and is a required growth factor for some organisms.

The cytochromes are proteins that contain heme prosthetic groups (**Figure 4.16**). Cytochromes undergo oxidation and reduction through loss or gain of a single electron by the iron atom in the heme of the cytochrome:

$$\text{Cytochrome—Fe}^{2+} \leftrightarrows \text{Cytochrome—Fe}^{3+} + e^-$$

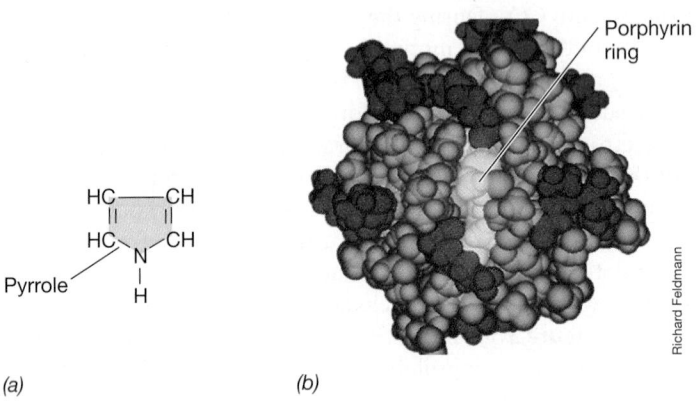

Porphyrin ring

Richard Feldmann

(a)   (b)

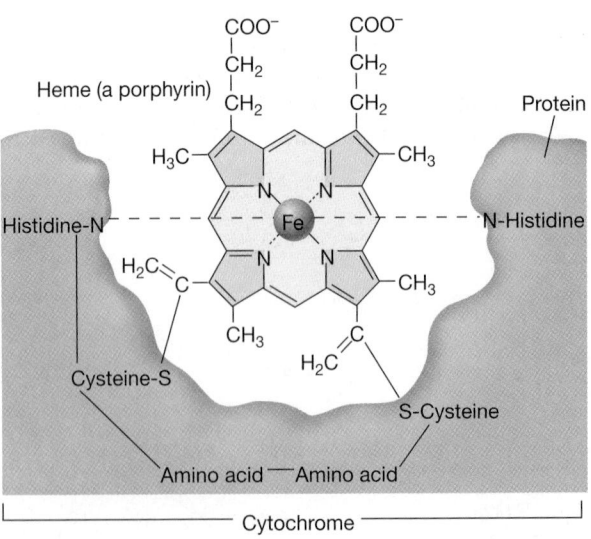

Heme (a porphyrin)

Protein

Histidine-N — — — — N-Histidine

Cysteine-S

S-Cysteine

Amino acid — Amino acid

Cytochrome

(c)

Figure 4.16 Cytochrome and its structure. *(a)* Structure of pyrrole, which is the building block of porphyrins such as heme in part c. *(b)* Space-filling model of cytochrome c; the porphyrin (light blue) is covalently linked via disulfide bridges to cysteine residues in the protein. *(c)* Schematic of cytochrome c model. Cytochromes carry electrons only; the redox site is the iron atom, which can alternate between the Fe^{2+} and Fe^{3+} oxidation states.

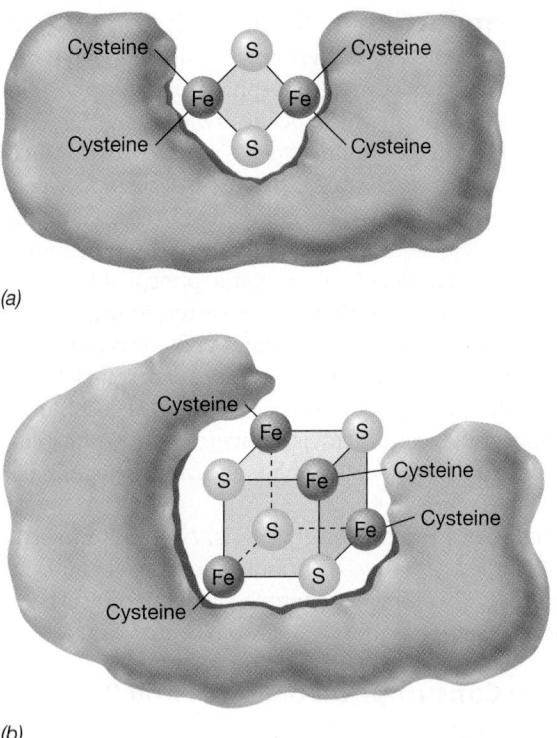

(a)

(b)

Figure 4.17 Arrangement of the iron–sulfur centers of nonheme iron–sulfur proteins. *(a)* Fe_2S_2 center. *(b)* Fe_4S_4 center. The cysteine linkages are from the protein portion of the molecule.

Several classes of cytochromes are known, differing widely in their reduction potentials (Figure 4.9). Different classes of cytochromes are designated by letters, such as cytochrome *a*, cytochrome *b*, cytochrome *c*, and so on, depending upon the type of heme they contain. The cytochromes of a given class in one organism may differ slightly from those of another, and so there are designations such as cytochromes a_1, a_2, a_3, and so on among cytochromes of the same class. Occasionally, cytochromes form complexes with other cytochromes or with iron–sulfur proteins. An important example is the cytochrome bc_1 complex, which contains two different *b*-type cytochromes and one *c*-type cytochrome. The cytochrome bc_1 complex plays an important role in energy metabolism, as we will see later.

In addition to the cytochromes, in which iron is bound to heme, one or more proteins with nonheme iron are typically present in electron transport chains. Centered in these proteins are clusters of iron and sulfur atoms, with Fe_2S_2 and Fe_4S_4 clusters being the most common (**Figure 4.17**). *Ferredoxin*, a common nonheme iron–sulfur protein, has an Fe_2S_2 configuration.

The reduction potentials of iron–sulfur proteins vary over a wide range depending on the number of iron and sulfur atoms present and how the iron centers are embedded in the protein. Thus, different iron–sulfur proteins can function at different locations in the electron transport chain. Like cytochromes, nonheme iron–sulfur proteins carry electrons only.

Quinones (**Figure 4.18**) are hydrophobic molecules that lack a protein component. Because they are small and hydrophobic, quinones are free to move about within the membrane. Like the

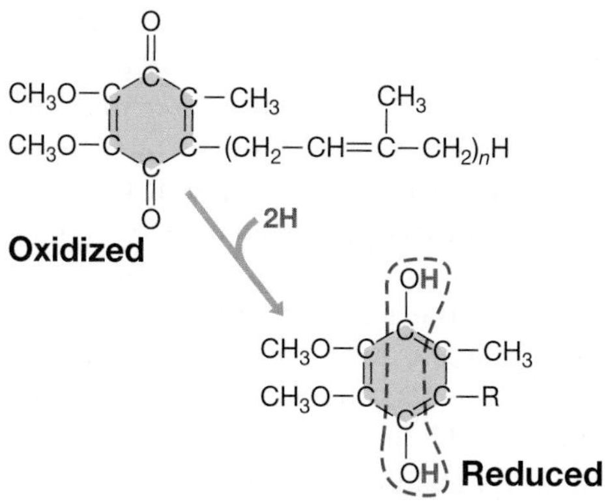

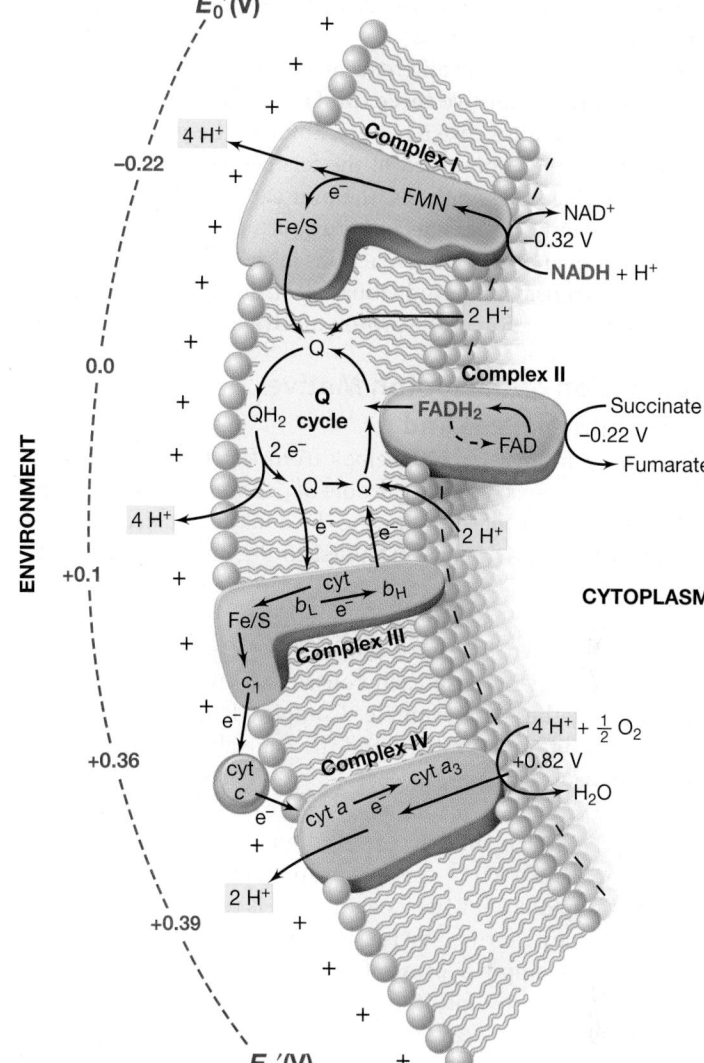

Figure 4.18 Structure of oxidized and reduced forms of coenzyme Q, a quinone. The five-carbon unit in the side chain (an isoprenoid) occurs in a number of multiples, typically 6–10. Oxidized quinone requires 2 e^- and 2 H^+ (2 H) to become fully reduced (dashed red circles).

flavoproteins, quinones accept 2 e^- + 2 H^+ but transfer only 2 e^- to the next carrier in the chain; quinones typically participate as links between iron–sulfur proteins and the first cytochromes in the electron transport chain.

MiniQuiz

- In what major way do quinones differ from other electron carriers in the membrane?

- Which electron carriers described in this section accept 2 e^- + 2 H^+? Which accept electrons only?

4.10 The Proton Motive Force

The conservation of energy by oxidative phosphorylation is linked to an energized state of the membrane (Figure 4.13b). This energized state is established by electron transport reactions between the electron carriers just discussed. To understand how electron transport is linked to ATP synthesis, we must first understand how the electron transport system is oriented in the cytoplasmic membrane. Electron transport carriers are oriented in the membrane in such a way that, as electrons are transported, protons are separated from electrons. Two electrons plus two protons enter the electron transport chain from NADH through NADH dehydrogenase to initiate the process. Carriers in the electron transport chain are arranged in the membrane in order of their increasingly positive reduction potential, with the final carrier in the chain donating the electrons plus protons to a terminal electron acceptor such as O_2 (**Figure 4.19**).

During electron transport, H^+ are extruded to the outer surface of the membrane. These H^+ originate from two sources: (1) NADH and (2) the dissociation of water (H_2O) into H^+ and OH^- in the cytoplasm. The extrusion of H^+ to the environment results in the accumulation of OH^- on the inside of the membrane. However, despite their small size, neither H^+ nor OH^- can diffuse through the membrane because they are charged (⮌ Section 3.4). As a result of the separation of H^+ and OH^-, the two sides of the membrane differ in both charge and pH.

Figure 4.19 Generation of the proton motive force during aerobic respiration. The orientation of electron carriers in the membrane of *Paracoccus denitrificans*, a model organism for studies of respiration. The + and – charges at the edges of the membrane represent H^+ and OH^-, respectively. E_0' values for the major carriers are shown. Note how when a hydrogen atom carrier (for example, FMN in Complex I) reduces an electron-accepting carrier (for example, the Fe/S protein in Complex I), protons are extruded to the outer surface of the membrane. Abbreviations: FMN, flavin mononucleotide; FAD, flavin adenine dinucleotide; Q, quinone; Fe/S, iron–sulfur–protein; cyt a, b, c, cytochromes (b_L and b_H, low- and high-potential b-type cytochromes, respectively). At the quinone site, electrons are recycled during the "Q cycle." This is because electrons from QH_2 can be split in the bc_1 complex (Complex III) between the Fe/S protein and the b-type cytochromes. Electrons that travel through the cytochromes reduce Q (in two, one-electron steps) back to QH_2, thus increasing the number of protons pumped at the Q-bc_1 site. Electrons that travel to Fe/S proceed to reduce cytochrome c_1, then cytochrome c, and then a-type cytochromes in Complex IV, eventually reducing O_2 to H_2O (2 electrons and 4 protons are required to reduce $\frac{1}{2} O_2$ to H_2O along with 2 H^+ extruded, and these come from electrons through cyt c and cytoplasmic protons, respectively). Complex II, the succinate dehydrogenase complex, bypasses Complex I and feeds electrons directly into the quinone pool at a more positive E_0' than NADH (see the electron tower in Figure 4.9).

The result of electron transport is thus the formation of an electrochemical potential across the membrane (Figure 4.19). This potential, along with the difference in pH across the membrane, is called the **proton motive force (pmf)** and causes the membrane to be energized much like a battery. Some of the potential energy in the pmf is then conserved in the formation of ATP. However, besides driving ATP synthesis, the pmf can also be tapped to do other forms of work, such as ion transport, flagellar rotation, and a few other energy-requiring reactions in the cell.

We now consider the individual electron transport reactions that lead to formation of the proton motive force.

Generation of the Proton Motive Force: Complexes I and II

The proton motive force develops from the activities of flavin enzymes, quinones, the cytochrome bc_1 complex, and the terminal cytochrome oxidase. Following the donation of NADH + H$^+$ to form FMNH$_2$, 4 H$^+$ are extruded to the outer surface of the membrane when FMNH$_2$ donates 2 e$^-$ to a series of nonheme iron proteins (Fe/S), forming the membrane protein section of *Complex I* (shown in Figure 4.19). These electron carriers are called *complexes* because each consists of several proteins that function together. For example, Complex I in *Escherichia coli* contains 14 different proteins and the equivalent complex in the mitochondrion contains at least 44 proteins. Complex I is also called *NADH:quinone oxidoreductase* because the reaction is one in which NADH is initially oxidized and quinone is ultimately reduced. Notably, 2 H$^+$ are taken up from the dissociation of H$_2$O in the cytoplasm when coenzyme Q is reduced at a catalytic site of Complex 1 formed by Fe/S centers (Figure 4.19).

Complex II simply bypasses Complex I and feeds e$^-$ and H$^+$ from FADH directly into the quinone pool. Complex II is also called the *succinate dehydrogenase complex* because of the specific substrate, succinate (a product of the citric acid cycle, Section 4.11), that it oxidizes. However, because Complex II bypasses Complex I, fewer H$^+$ are pumped per 2 e$^-$ that enter the electron transport chain here than for 2 e$^-$ that enter from NADH (Figure 4.19).

Complexes III and IV: bc_1 and *a*-Type Cytochromes

Reduced coenzyme Q passes electrons one at a time to the cytochrome bc_1 complex (*Complex III*, Figure 4.19). The cytochrome bc_1 complex consists of several proteins that contain hemes (Figure 4.16) or other metal cofactors. These include two *b*-type hemes (b_L and b_H), one *c*-type heme (c_1), and one iron–sulfur protein. The bc_1 complex is present in the electron transport chain of almost all organisms that can respire. It also plays a fundamental role in photosynthetic electron flow of phototrophic organisms (Sections 13.4 and 13.5).

The major function of the cytochrome bc_1 complex is to transfer e$^-$ from quinones to cytochrome *c*. Electrons travel from the bc_1 complex to a molecule of cytochrome *c*, located in the periplasm. Cytochrome *c* functions as a shuttle to transfer e$^-$ to the high-potential cytochromes *a* and a_3 (*Complex IV*, Figure 4.19). Complex IV is the terminal oxidase and reduces O$_2$ to H$_2$O in the final step of the electron transport chain. Complex IV also

pumps protons to the outer surface of the membrane, thereby increasing the strength of the proton motive force (Figure 4.19).

Besides transferring e$^-$ to cytochrome *c*, the cytochrome bc_1 complex can also interact with quinones in such a way that on average, two additional H$^+$ are pumped at the Q-bc_1 site. This happens in a series of electron exchanges between cytochrome bc_1 and Q, called the *Q cycle*. Because quinone and bc_1 have roughly the same E_0' (near 0 V, Figure 4.19), quinone molecules can alternately become oxidized and reduced using e$^-$ fed back to quinones from the bc_1 complex. This mechanism allows on average a total of 4 H$^+$ (instead of 2 H$^+$) to be pumped to the outer surface of the membrane at the Q-bc_1 site for every 2 e$^-$ that enter the chain in Complex I.

The electron transport chain shown in Figure 4.19 is one of many different sequences of electron carriers known from different organisms. However, three features are characteristic of all electron transport chains: (1) arrangement of carriers in order of increasingly more positive E_0', (2) alternation of electron-only and electron-plus-proton carriers in the chain, and (3) generation of a proton motive force.

As we will see now, it is this last characteristic, the proton motive force, that drives ATP synthesis.

ATP Synthase

How does the proton motive force generated by electron transport actually drive ATP synthesis? Interestingly, a strong parallel exists between the mechanism of ATP synthesis and the mechanism of the motor that drives rotation of the bacterial flagellum (Section 3.13). In analogy to how dissipation of the pmf applies torque that rotates the bacterial flagellum, the pmf also creates torque in a large protein complex that makes ATP. This complex is called **ATP synthase**, or **ATPase** for short.

ATPases consist of two components, a multiprotein cytoplasmic complex called F$_1$ that carries out the chemical function (ATP synthesis), connected to a membrane-integrated component called F$_o$ that carries out the ion-translocating function (**Figure 4.20**). ATPase catalyzes a reversible reaction between ATP and ADP + P$_i$ as shown in the figure. The structure of ATPase proteins is highly conserved throughout all the domains of life, suggesting that this mechanism of energy conservation was a very early evolutionary invention (Section 16.2).

F$_1$ and F$_o$ are actually two rotary motors. Pmf-driven H$^+$ movement through F$_o$ causes rotation of its *c* proteins. This generates a torque that is transmitted to F$_1$ via the coupled rotation of the γε subunits (Figure 4.20). The latter activity causes conformational changes in the β subunits that allows them to bind ADP + P$_i$. ATP is synthesized when the β subunits return to their original conformation, releasing the free energy needed to drive the synthesis.

ATPase-catalyzed ATP synthesis is called *oxidative phosphorylation* if the proton motive force originates from respiration reactions and *photophosphorylation* if it originates from photosynthetic reactions. Quantitative measures (stoichiometry) of H$^+$ consumed by ATPase per ATP produced yield a number between 3 and 4.

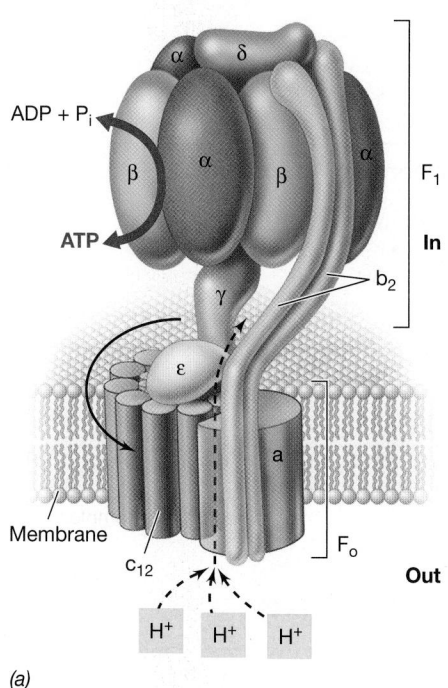

(a)

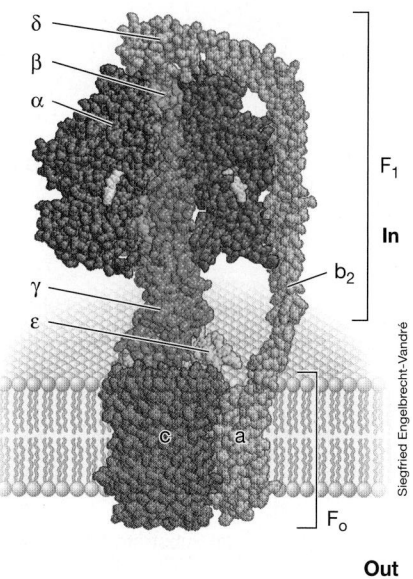

(b)

Siegfried Engelbrecht-Vandré

Figure 4.20 Structure and function of ATP synthase (ATPase) in *Escherichia coli*. *(a)* Schematic. F_1 consists of five different polypeptides forming an $\alpha_3\beta_3\gamma\epsilon\delta$ complex, the stator. F_1 is the catalytic complex responsible for the interconversion of ADP + P_i and ATP. F_o, the rotor, is integrated in the membrane and consists of three polypeptides in an ab_2c_{12} complex. As protons enter, the dissipation of the proton motive force drives ATP synthesis (3 H^+/ATP). ATPase is reversible in that ATP hydrolysis can drive formation of a proton motive force. *(b)* Space-filling model. The color-coding corresponds to the art in part *a*. Since proton translocation from outside the cell to inside the cell leads to ATP synthesis by ATPase, it follows that proton translocation from inside to outside in the electron transport chain (Figure 4.19) represents work done on the system and a source of potential energy.

Reversibility of ATPase

ATPase is reversible. The hydrolysis of ATP supplies torque for $\gamma\epsilon$ to rotate in the opposite direction from that in ATP synthesis, and this catalyzes the pumping of H^+ from the inside to the outside of the cell through F_o. The net result is *generation* instead of *dissipation* of the proton motive force. Reversibility of the ATPase explains why strictly fermentative organisms that lack electron transport chains and are unable to carry out oxidative phosphorylation still contain ATPases. As we have said, many important reactions in the cell, such as motility and transport, require energy from the pmf rather than from ATP. Thus, ATPase in organisms incapable of respiration, such as the strictly fermentative lactic acid bacteria, for example, functions unidirectionally to generate the pmf necessary to drive these important cell functions.

MiniQuiz

- How do electron transport reactions generate the proton motive force?
- What is the ratio of H^+ extruded per NADH oxidized through the electron transport chain of *Paracoccus* shown in Figure 4.19? At which sites in the chain is the proton motive force being established?
- What structure in the cell converts the proton motive force to ATP? How does it function?

4.11 The Citric Acid Cycle

Now that we have a grasp of how ATP is made in respiration, we need to consider the important reactions in carbon metabolism associated with formation of ATP. Our focus here is on the citric acid cycle, also called the Krebs cycle, a key pathway in virtually all cells.

Respiration of Glucose

The early biochemical steps in the respiration of glucose are the same as those of glycolysis; all steps from glucose to pyruvate (Figure 4.14) are the same. However, whereas in fermentation pyruvate is reduced and converted into products that are excreted, in respiration pyruvate is oxidized to CO_2. The pathway by which pyruvate is completely oxidized to CO_2 is called the **citric acid cycle** (CAC), summarized in **Figure 4.21**.

Pyruvate is first decarboxylated, leading to the production of CO_2, NADH, and the energy-rich substance *acetyl-CoA* (Figure 4.12). The acetyl group of acetyl-CoA then combines with the four-carbon compound oxalacetate, forming the six-carbon compound citric acid. A series of reactions follow, and two additional CO_2 molecules, three more NADH, and one FADH are formed. Ultimately, oxalacetate is regenerated to return as an acetyl acceptor, thus completing the cycle (Figure 4.21).

CO_2 Release and Fuel for Electron Transport

The oxidation of pyruvate to CO_2 requires the concerted activity of the citric acid cycle and the electron transport chain. For each pyruvate molecule oxidized through the citric acid cycle, three CO_2 molecules are released (Figure 4.21). Electrons released during the oxidation of intermediates in the citric acid cycle are transferred to NAD^+ to form NADH, or to FAD to form $FADH_2$. This is where respiration and fermentation differ in a major way. Instead of being used in the reduction of pyruvate as in fermentation (Figure 4.14), in respiration, electrons from NADH and $FADH_2$ are fuel for the electron transport chain, ultimately resulting in the reduction of an electron acceptor (O_2) to H_2O. This allows for the complete oxidation of glucose to CO_2 along with a much greater yield of energy. Whereas only *2 ATP* are produced per glucose fermented in alcoholic or lactic acid fermentations (Figure 4.14), a total of *38 ATP* can be made by aerobically respiring the same glucose molecule to CO_2 + H_2O (Figure 4.21*b*).

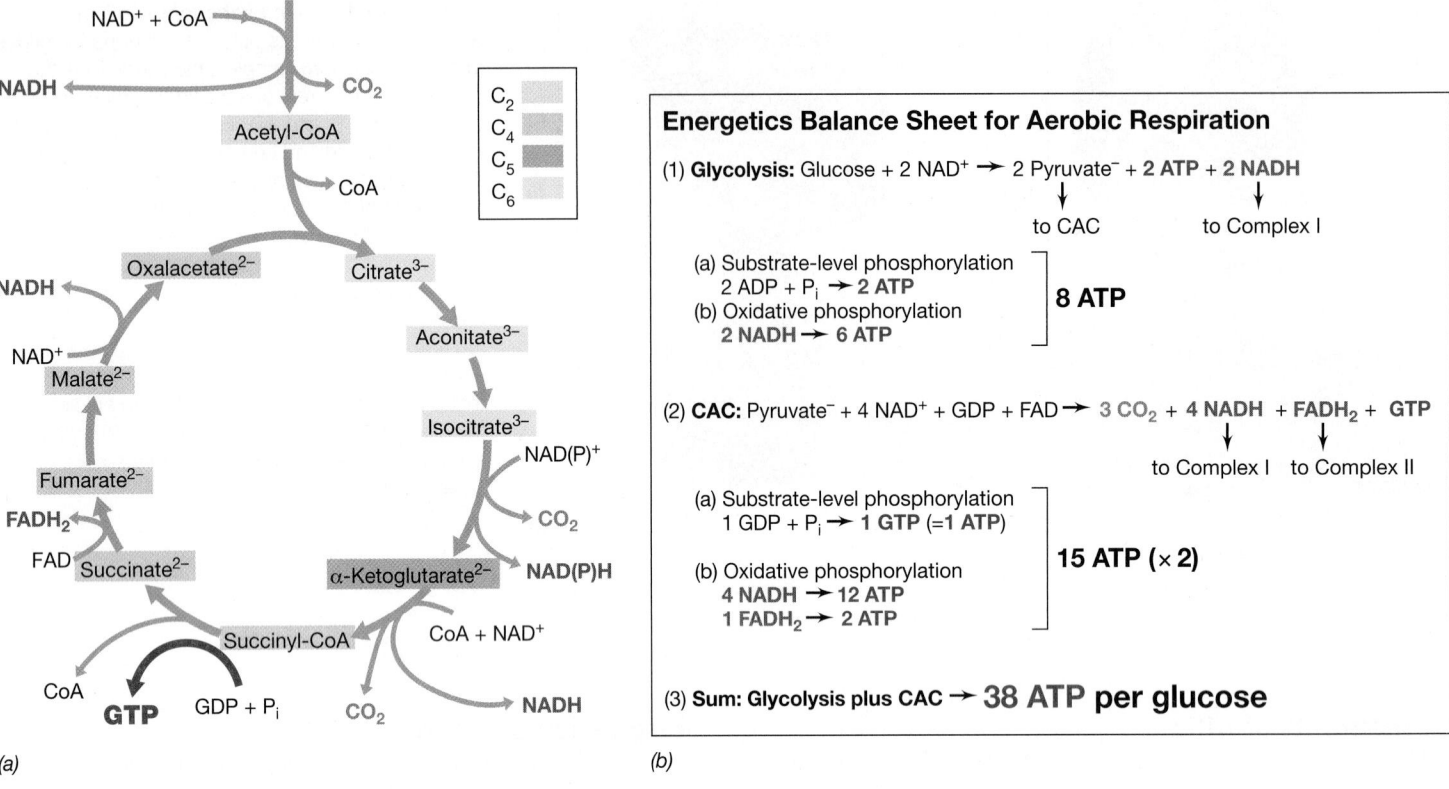

(a)

(b)

Figure 4.21 The citric acid cycle. *(a)* The citric acid cycle (CAC) begins when the two-carbon compound acetyl-CoA condenses with the four-carbon compound oxalacetate to form the six-carbon compound citrate. Through a series of oxidations and transformations, this six-carbon compound is ultimately converted back to the four-carbon compound oxalacetate, which then begins another cycle with addition of the next molecule of acetyl-CoA. *(b)* The overall balance sheet of fuel (NADH/FADH$_2$) for the electron transport chain and CO$_2$ generated in the citric acid cycle. NADH and FADH$_2$ feed into electron transport chain Complexes I and II, respectively (Figure 4.19).

Biosynthesis and the Citric Acid Cycle

Besides playing a key role in catabolism, the citric acid cycle plays another important role in the cell. The cycle generates several key compounds, small amounts of which can be drawn off for biosynthetic purposes when needed. Particularly important in this regard are α-ketoglutarate and oxalacetate, which are precursors of several amino acids (Section 4.14), and succinyl-CoA, needed to form cytochromes, chlorophyll, and several other tetrapyrrole compounds (Figure 4.16). Oxalacetate is also important because it can be converted to phosphoenolpyruvate, a precursor of glucose. In addition, acetate provides the starting material for fatty acid biosynthesis (Section 4.15, and see Figure 4.27). The citric acid cycle thus plays two major roles in the cell: *bioenergetic* and *biosynthetic*. Much the same can be said about the glycolytic pathway, as certain intermediates from this pathway are drawn off for various biosynthetic needs as well (Section 4.13).

MiniQuiz

• How many molecules of CO$_2$ and pairs of electrons are released per pyruvate oxidized in the citric acid cycle?

• What two major roles do the citric acid cycle and glycolysis have in common?

4.12 Catabolic Diversity

Thus far in this chapter we have dealt only with catabolism by chemoorganotrophs. We now briefly consider catabolic diversity, some of the alternatives to the use of organic compounds as electron donors, with emphases on both electron and carbon flow. **Figure 4.22** summarizes the mechanisms by which cells generate energy other than by fermentation and aerobic respiration. These include *anaerobic respiration*, *chemolithotrophy*, and *phototrophy*.

Anaerobic Respiration

Under anoxic conditions, electron acceptors other than oxygen can be used to support respiration in certain prokaryotes. These processes are called **anaerobic respiration**. Some of the electron acceptors used in anaerobic respiration include nitrate (NO$_3^-$, reduced to nitrite, NO$_2^-$, by *Escherichia coli* or to N$_2$ by *Pseudomonas* species), ferric iron (Fe^{3+}, reduced to Fe^{2+} by *Geobacter* species), sulfate (SO$_4^{2-}$, reduced to hydrogen sulfide, H$_2$S, by *Desulfovibrio* species), carbonate (CO$_3^{2-}$, reduced to methane, CH$_4$, by methanogens or to acetate by acetogens), and even certain organic compounds. Some of these acceptors, for example Fe^{3+}, are often only available in the form of insoluble

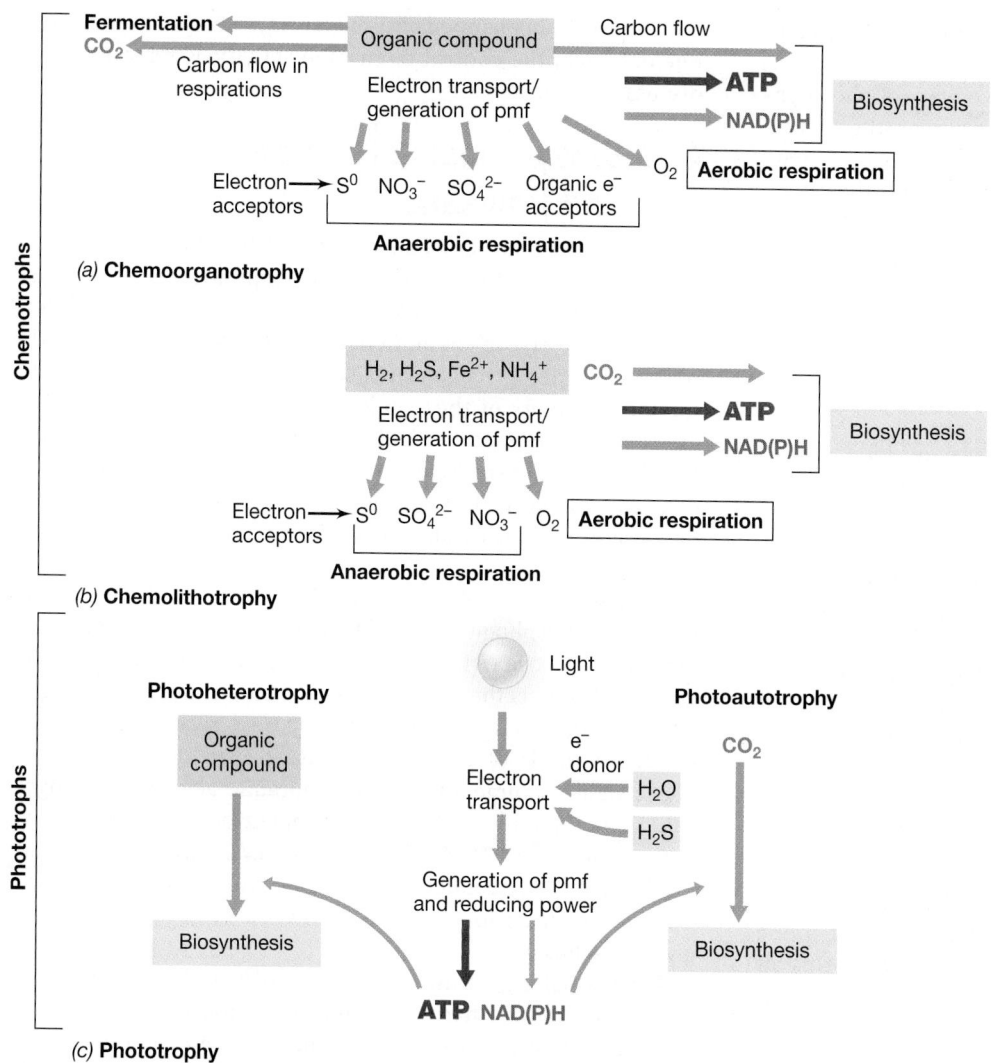

Figure 4.22 Catabolic diversity.
(a) Chemoorganotrophs. *(b)* Chemolithotrophs. *(c)* Phototrophs. Chemoorganotrophs differ from chemolithotrophs in two important ways: (1) The nature of the electron donor (organic versus inorganic compounds, respectively), and (2) The nature of the source of cellular carbon (organic compounds versus CO_2 respectively). However, note the importance of electron transport driving proton motive force formation in all forms of respiration and in photosynthesis.

minerals, such as metal oxides. These common minerals, widely distributed in nature, allow for anaerobic respiration in a wide variety of microbial habitats.

Because of the positions of these alternative electron acceptors on the redox tower (none has an E_0' as positive as the O_2/H_2O couple; Figure 4.9), less energy is released when they are reduced instead of oxygen (recall that $\Delta G^{0'}$ is proportional to $\Delta E_0'$; Section 4.6). Nevertheless, because O_2 is often limiting or absent in many microbial habitats, anaerobic respirations can be very important means of energy generation. As in aerobic respiration, anaerobic respirations involve electron transport, generation of a proton motive force, and the activity of ATPase.

Chemolithotrophy

Organisms able to use *inorganic* chemicals as electron donors are called **chemolithotrophs**. Examples of relevant inorganic electron donors include H_2S, hydrogen gas (H_2), Fe^{2+}, and NH_3.

Chemolithotrophic metabolism is typically aerobic and begins with the oxidation of the inorganic electron donor (Figure 4.22). Electrons from the inorganic donor enter an electron transport chain and a proton motive force is formed in

exactly the same way as for chemoorganotrophs (Figure 4.19). However, one important distinction between chemolithotrophs and chemoorganotrophs, besides their electron donors, is their source of carbon for biosynthesis. Chemoorganotrophs use organic compounds (glucose, acetate, and the like) as carbon sources. By contrast, chemolithotrophs use carbon dioxide (CO_2) as a carbon source and are therefore **autotrophs** (organisms capable of biosynthesizing all cell material from CO_2 as the sole carbon source). We consider many examples of chemolithotrophy in Chapter 13.

Phototrophy

Many microorganisms are **phototrophs**, using light as an energy source in the process of photosynthesis. The mechanisms by which light is used as an energy source are complex, but the end result is the same as in respiration: generation of a proton motive force that is used to drive ATP synthesis. Light-mediated ATP synthesis is called **photophosphorylation**. Most phototrophs use energy conserved in ATP for the assimilation of CO_2 as the carbon source for biosynthesis; they are called *photoautotrophs*. However, some phototrophs use organic compounds as carbon sources with light as the energy source; these are the *photo-heterotrophs* (Figure 4.22).

As we discussed in Chapter 2, there are two types of photosynthesis: *oxygenic* and *anoxygenic*. Oxygenic photosynthesis, carried out by cyanobacteria and their relatives and also by green plants, results in O_2 evolution. Anoxygenic photosynthesis is a simpler process used by purple and green bacteria that does not evolve O_2. The reactions leading to proton motive force formation in both forms of photosynthesis have strong parallels, as we see in Chapter 13.

The Proton Motive Force and Catabolic Diversity

Microorganisms show an amazing diversity of bioenergetic strategies. Thousands of organic compounds, many inorganic compounds, and light can be used by one or another microorganism as an energy source. With the exception of fermentations, in which substrate-level phosphorylation occurs (Section 4.8), energy conservation in respiration and photosynthesis is driven by the proton motive force.

Whether electrons come from the oxidation of organic or inorganic chemicals or from phototrophic processes, in all forms of respiration and photosynthesis, energy conservation is linked to the pmf through ATPase (Figure 4.20). Considered in this way, respiration and anaerobic respiration are simply metabolic variations employing different electron acceptors. Likewise, chemoorganotrophy, chemolithotrophy, and photosynthesis are simply metabolic variations upon a theme of different electron donors. Electron transport and the pmf link all of these processes, bringing these seemingly quite different forms of metabolism into a common focus. We pick up on this theme in Chapters 13 and 14.

MiniQuiz

- In terms of their electron donors, how do chemoorganotrophs differ from chemolithotrophs?
- What is the carbon source for autotrophic organisms?
- Why can it be said that the proton motive force is a unifying theme in most of bacterial metabolism?

Essentials of Anabolism

We close this chapter with a brief consideration of biosynthesis. Our focus here will be on biosynthesis of the building blocks of the four classes of macromolecules—sugars, amino acids, nucleotides, and fatty acids. Collectively, these biosyntheses are called *anabolism*. In Chapters 6 and 7 we consider synthesis of the macromolecules themselves, in particular, nucleic acids and proteins.

Many detailed biochemical pathways support the metabolic patterns we present here, but we will keep our focus on the essential principles. We finish with a glimpse at how the enzymes that drive these biosynthetic processes are controlled by the cell. For a cell to be competitive, it must regulate its metabolism. This happens in several ways and at several levels, one of which, the control of enzyme activity, is relevant to our discussion here.

4.13 Biosynthesis of Sugars and Polysaccharides

Polysaccharides are key constituents of the cell walls of many organisms, and in *Bacteria*, the peptidoglycan cell wall (⟳ Section 3.6) has a polysaccharide backbone. In addition, cells often store carbon and energy reserves in the form of the polysaccharides glycogen and starch (⟳ Section 3.10). The monomeric units of these polysaccharides are six-carbon sugars called *hexoses*, in particular, glucose or glucose derivatives. In addition to hexoses, five-carbon sugars called *pentoses* are common in the cell. Most notably, these include ribose and deoxyribose, present in the backbone of RNA and DNA, respectively.

In prokaryotes, polysaccharides are synthesized from either uridine diphosphoglucose (UDPG; **Figure 4.23**) or adenosine diphosphoglucose (ADPG), both of which are *activated* forms of glucose. ADPG is the precursor for the biosynthesis of glycogen. UDPG is the precursor of various glucose derivatives needed for the biosynthesis of other polysaccharides in the cell, such as *N*-acetylglucosamine and *N*-acetylmuramic acid in peptidoglycan or the lipopolysaccharide component of the gram-negative outer membrane (⟳ Sections 3.6 and 3.7). Polysaccharides are produced by adding glucose (from the activated form) to the preexisting polymer; for example, ADPG + glycogen → ADP + glycogen-glucose.

When a cell is growing on a hexose such as glucose, obtaining glucose for polysaccharide synthesis is obviously not a problem. But when the cell is growing on other carbon compounds, glucose must be synthesized. This process, called *gluconeogenesis*, uses phosphoenolpyruvate, one of the intermediates of glycolysis (Figure 4.14), as starting material. Phosphoenolpyruvate can be synthesized from oxalacetate, a citric acid cycle intermediate (Figure 4.21). An overview of gluconeogenesis is shown in Figure 4.23*b*.

Pentoses are formed by the removal of one carbon atom from a hexose, typically as CO_2. The pentoses needed for nucleic acid synthesis, ribose and deoxyribose, are formed as shown in Figure 4.23*c*. The enzyme ribonucleotide reductase converts ribose into deoxyribose by reduction of the hydroxyl (−OH) group on the 2′ carbon of the 5-carbon sugar ring. Interestingly, this reaction occurs after, not before, synthesis of nucleotides. Thus, *ribo*nucleotides are biosynthesized, and some of them are later reduced to *deoxy*ribonucleotides for use as precursors of DNA.

MiniQuiz

- How does anabolism differ from catabolism? Give an example of each.
- What form of activated glucose is used in the biosynthesis of glycogen by bacteria?
- What is gluconeogenesis?

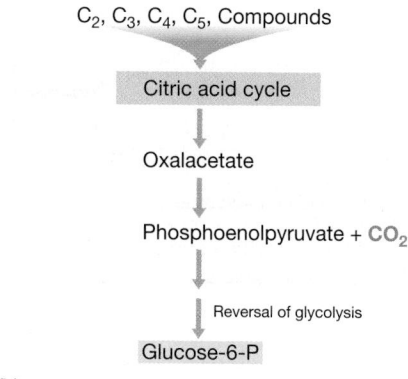

Uridine diphosphoglucose (UDPG)

(a)

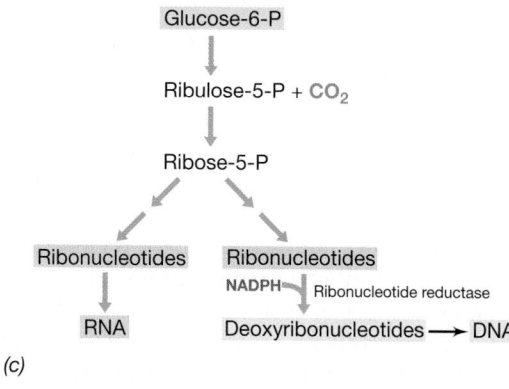

(b)

(c)

Figure 4.23 Sugar metabolism. *(a)* Polysaccharides are synthesized from activated forms of hexoses such as UDPG. Glucose is shown here in blue. *(b)* Gluconeogenesis. When glucose is needed, it can be biosynthesized from other carbon compounds, generally by the reversal of steps in glycolysis. *(c)* Pentoses for nucleic acid synthesis are formed by decarboxylation of hexoses such as glucose-6-phosphate. Note how the precursors of DNA are produced from the precursors of RNA by the enzyme ribonucleotide reductase. This enzyme reduces the 2′ hydroxyl group of the sugar, converting ribose to deoxyribose and reducing the hydroxyl group to water, and is active on all four ribonucleotides.

4.14 Biosynthesis of Amino Acids and Nucleotides

The monomers in proteins and nucleic acids are amino acids and nucleotides, respectively. Their biosyntheses are often long, multistep pathways and so we approach their biosyntheses here by identifying the key carbon skeletons needed to begin the biosynthetic pathways.

Monomers of Proteins: Amino Acids

Organisms that cannot obtain some or all of their amino acids preformed from the environment must synthesize them from other sources. Amino acids are grouped into structurally related *families* that share several biosynthetic steps. The carbon skeletons for amino acids come almost exclusively from intermediates of glycolysis (Figure 4.14) or the citric acid cycle (Figure 4.21; **Figure 4.24**).

The amino group of amino acids is typically derived from some inorganic nitrogen source in the environment, such as ammonia (NH_3). Ammonia is most often incorporated in formation of the amino acids glutamate or glutamine by the enzymes *glutamate dehydrogenase* and *glutamine synthetase*, respectively (**Figure 4.25**). When NH_3 is present at high levels, glutamate dehydrogenase or other amino acid dehydrogenases are used. However, when NH_3 is present at low levels, glutamine synthetase, with its energy-consuming reaction mechanism (Figure 4.25*b*) and high affinity for substrate, is employed. We discuss control of the activity of the important enzyme glutamine synthetase in Section 4.16.

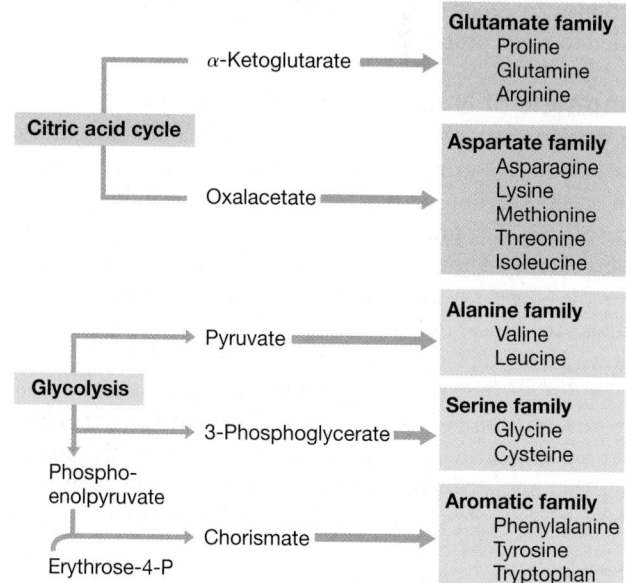

Figure 4.24 Amino acid families. The citric acid cycle and glycolysis provide the carbon skeletons for most amino acids. Synthesis of the various amino acids in a family may require many steps starting with the parent amino acid (shown in bold as the name of the family). Glycolysis is discussed in Section 4.8 (see Figure 4.14) and the citric acid cycle is discussed in Section 4.11 (see Figure 4.21).

UNIT 2

(a) α-Ketoglutarate + NH_3 $\xrightarrow[\text{Glutamate dehydrogenase}]{\text{NADH}}$ Glutamate

(b) Glutamate + NH_3 $\xrightarrow[\text{Glutamine synthetase}]{\text{ATP}}$ Glutamine

(c) Glutamate + Oxalacetate $\xrightarrow[\text{Transaminase}]{}$ α-Ketoglutarate + Aspartate

(d) Glutamine + α-Ketoglutarate $\xrightarrow[\text{Glutamate synthase}]{\text{NADH}}$ 2 Glutamate

Figure 4.25 Ammonia incorporation in bacteria. To emphasize the flow of nitrogen, both free ammonia (NH_3) and the amino groups of all amino acids are shown in green. Two major pathways for NH_3 assimilation in bacteria are those catalyzed by the enzymes *(a)* glutamate dehydrogenase and *(b)* glutamine synthetase. *(c)* Transaminase reactions transfer an amino group from an amino acid to an organic acid. *(d)* The enzyme glutamate synthase forms two glutamates from one glutamine and one α-ketoglutarate.

Once ammonia is incorporated into glutamate or glutamine, the amino group can be transferred to form other nitrogenous compounds. For example, glutamate can donate its amino group to oxalacetate in a transaminase reaction, producing α-ketoglutarate and aspartate (Figure 4.25c). Alternatively, glutamine can react with α-ketoglutarate to form two molecules of glutamate in an aminotransferase reaction (Figure 4.25d). The end result of these types of reactions is the shuttling of ammonia into various carbon skeletons from which further biosynthetic reactions can occur to form all 22 amino acids (Figure 6.29) needed to make proteins.

Monomers of Nucleic Acids: Nucleotides

The biochemistry behind purine and pyrimidine biosynthesis is quite complex. Purines are constructed literally atom by atom from several carbon and nitrogen sources, including even CO_2 (**Figure 4.26**). The first key purine, inosinic acid (Figure 4.26b), is the precursor of the purine nucleotides adenine and guanine. Once these are synthesized (in their triphosphate forms) and attached to ribose, they are ready to be incorporated into DNA (following ribonucleotide reductase activity) or RNA.

Like the purine ring, the pyrimidine ring is also constructed from several sources (Figure 4.26c). The first key pyrimidine is the compound uridylate (Figure 4.26d), and from this the pyrimidines thymine, cytosine, and uracil are derived. Structures of all of the purines and pyrimidines are shown in Figure 6.1.

MiniQuiz

- What is an amino acid family?
- List the steps required for the cell to incorporate NH_3 into amino acids.
- Which nitrogen bases are purines and which are pyrimidines?

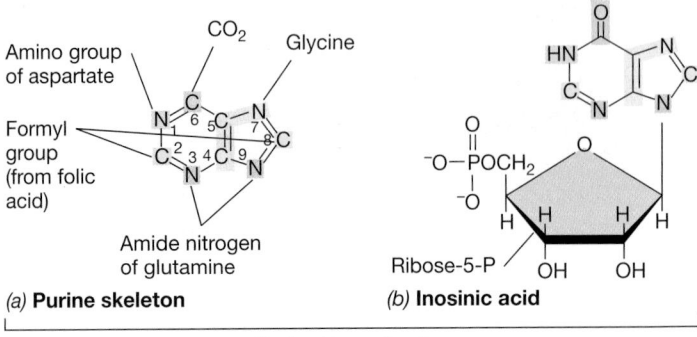

(a) **Purine skeleton** (b) **Inosinic acid**

Purine biosynthesis

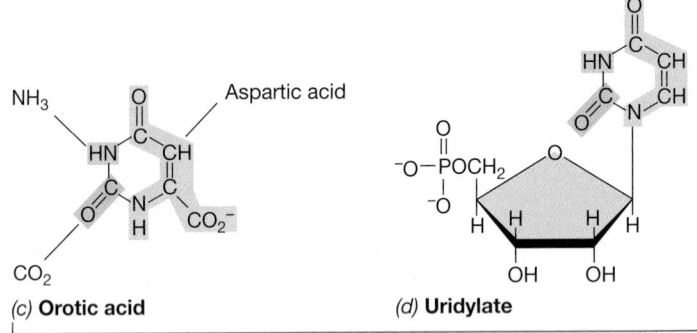

(c) **Orotic acid** (d) **Uridylate**

Pyrimidine biosynthesis

Figure 4.26 Composition of purines and pyrimidines. *(a)* Components of the purine skeleton. *(b)* Inosinic acid, the precursor of all purine nucleotides. *(c)* Components of the pyrimidine skeleton, orotic acid. *(d)* Uridylate, the precursor of all pyrimidine nucleotides. Uridylate is formed from orotate following a decarboxylation and the addition of ribose 5-phosphate.

4.15 Biosynthesis of Fatty Acids and Lipids

Lipids are important constituents of cells, as they are major structural components of membranes. Lipids can also be carbon and energy reserves. Other lipids function in and around the cell surface, including, in particular, the lipopolysaccharide layer of the outer membrane of gram-negative bacteria (Section 3.7). A cell can make many different types of lipids, some of which are produced only under certain conditions or have special functions in the cells. The biosynthesis of fatty acids is thus a major series of reactions in cells. Recall that *Archaea* do not contain fatty acids in their membrane lipids, but have instead branched side chains constructed of multiples of isoprene, a C_5 branched chained hydrocarbon (Figure 3.7).

Fatty Acid Biosynthesis

Fatty acids are biosynthesized two carbon atoms at a time with the help of a protein called *acyl carrier protein* (ACP). ACP holds the growing fatty acid as it is being synthesized and releases it once it has reached its final length (**Figure 4.27**). Although fatty acids are constructed *two* carbons at a time, each C_2 unit originates from the C_3 compound malonate, which is attached to the ACP to form malonyl-ACP. As each malonyl residue is donated, one molecule of CO_2 is released (Figure 4.27).

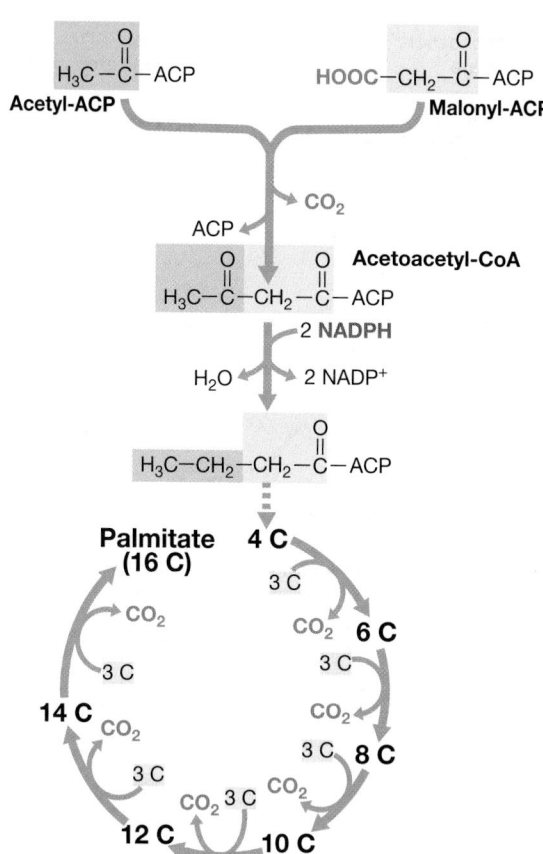

Figure 4.27 The biosynthesis of the C$_{16}$ fatty acid palmitate. The condensation of acetyl-ACP and malonyl-ACP forms acetoacetyl-CoA. Each successive addition of an acetyl unit comes from malonyl-ACP.

The fatty acid composition of cells varies from species to species and can also vary within a species due to differences in temperature. Growth at low temperatures promotes the biosynthesis and insertion in membrane lipids of shorter-chain fatty acids whereas growth at higher temperatures promotes longer-chain fatty acids. The most common fatty acids in lipids of *Bacteria* are those with chain lengths of C$_{12}$–C$_{20}$.

In addition to saturated, even-carbon-number fatty acids, fatty acids can also be unsaturated, branched, or have an odd number of carbon atoms. Unsaturated fatty acids contain one or more double bonds in the long hydrophobic portion of the molecule. The number and position of these double bonds is often species-specific or group-specific, and double bonds typically form by desaturation reactions after the saturated fatty acid has formed. Branched-chain fatty acids are biosynthesized using an initiating molecule that contains a branched-chain fatty acid, and odd-carbon-number fatty acids are biosynthesized using an initiating molecule that contains a propionyl (C$_3$) group.

Lipid Biosynthesis

In the assembly of lipids in cells of *Bacteria* and *Eukarya*, fatty acids are added to glycerol. For simple triglycerides (fats), all three glycerol carbons are esterified with fatty acids. In complex lipids, one of the carbon atoms in glycerol contains a molecule of phosphate, ethanolamine, carbohydrate, or some other polar

substance (Figure 3.4a). In *Archaea*, membrane lipids contain phytanyl (C$_{15}$) or biphytanyl (C$_{30}$) side chains (Figure 3.7) instead of fatty acids, and the biosynthesis of phytanyl is distinct from that described here for fatty acids. However, as for the lipids of *Bacteria* or *Eukarya*, the glycerol backbone of archaeal membrane lipids also contains a polar group (a sugar, phosphate, sulfate, or polar organic compound) that facilitates formation of the typical membrane architecture: a hydrophobic interior with hydrophilic surfaces (Figure 3.7).

MiniQuiz

- Explain why in fatty acid synthesis fatty acids are constructed two carbon atoms at a time even though the immediate donor for these carbons contains three carbon atoms.

4.16 Regulating the Activity of Biosynthetic Enzymes

We have just reviewed some of the key cellular biosyntheses. Anabolism requires hundreds of different enzymatic reactions, and many of the enzymes that catalyze these reactions are highly regulated. The advantage of regulation is clear: If the compound to be biosynthesized is available from the environment, neither carbon nor energy need be wasted in its biosynthesis.

There are two major modes of enzyme regulation in cells, one that controls the *amount* (or even the complete presence or absence) of an enzyme and another that controls the *activity* of an enzyme. In prokaryotic cells, the amount of a given enzyme is regulated at the gene level, and we reserve discussion of this until after we have considered some principles of molecular biology. Here we focus on what the cell can do to control the activity of enzymes already present in the cell.

Inhibition of an enzyme's activity is typically the result of either covalent or noncovalent changes in its structure. We begin with feedback inhibition and isoenzymes, both examples of noncovalent interactions, and end with the example of covalent modification of the enzyme glutamine synthetase.

Feedback Inhibition

A major means of controlling enzymatic activity is by **feedback inhibition**. This mechanism temporarily shuts off the reactions in an entire biosynthetic pathway. The reactions are shut off because an excess of the end product of the pathway inhibits activity of an early (typically the *first*) enzyme of the pathway. Inhibiting an early step effectively shuts down the entire pathway because no intermediates are generated for enzymes farther down the pathway (**Figure 4.28**). Feedback inhibition is reversible, however, because once levels of the end product become limiting, the pathway again becomes functional.

How can the end product of a pathway inhibit the activity of an enzyme whose substrate is quite unrelated to it? This occurs because the inhibited enzyme is an **allosteric enzyme**, an enzyme that has two binding sites, the *active site* (where substrate binds, Section 4.5), and the *allosteric site*, where the end product of the pathway binds. When the end product is in excess, it binds at the

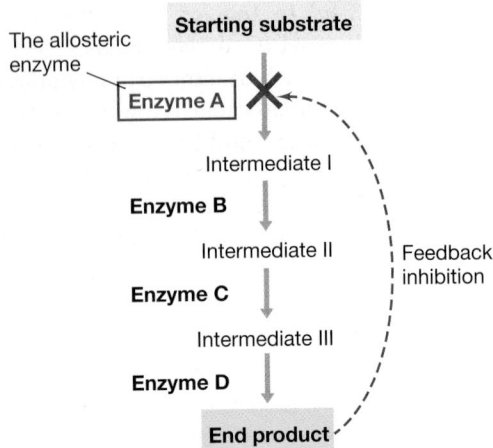

Figure 4.28 Feedback inhibition of enzyme activity. The activity of the first enzyme of the pathway is inhibited by the end product, thus shutting off the production of the three intermediates and the end product.

allosteric site, changing the conformation of the enzyme such that the substrate can no longer bind at the active site (**Figure 4.29**). When the concentration of the end product in the cell begins to fall, however, the end product no longer binds to the allosteric site, so the enzyme returns to its catalytic form and once again becomes active.

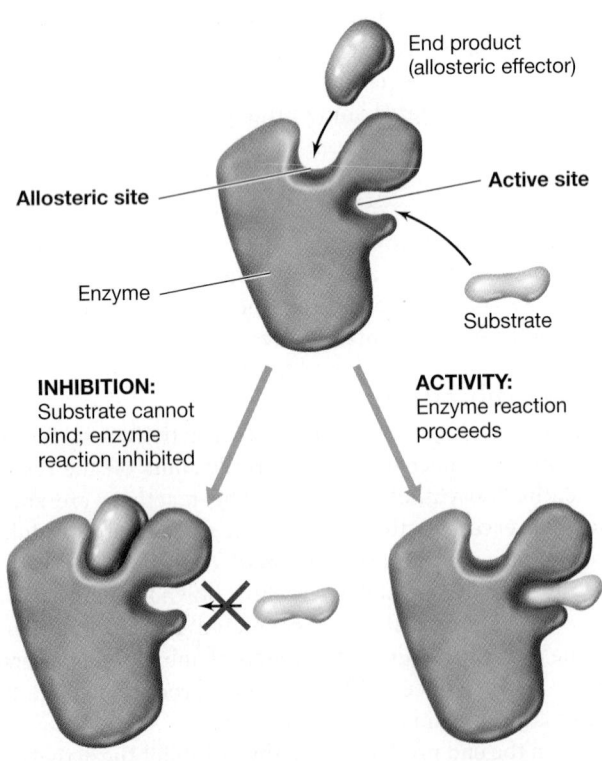

Figure 4.29 The mechanism of allosteric inhibition by the end product of a pathway. When the end product binds at the allosteric site, the conformation of the enzyme is so altered that the substrate can no longer bind to the active site. However, inhibition is reversible, and end product limitation will once again activate the enzyme.

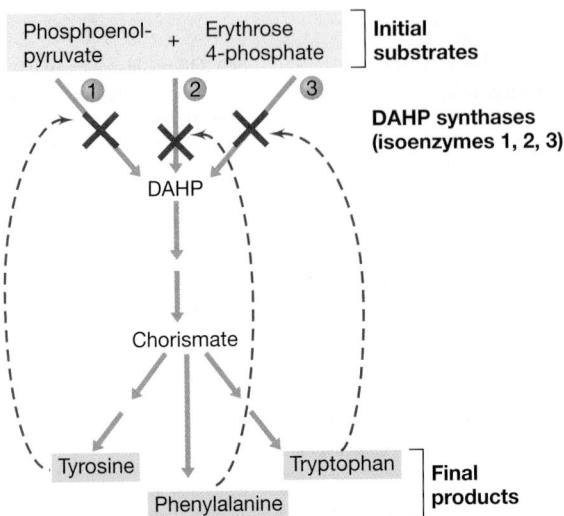

Figure 4.30 Isoenzymes and feedback inhibition. In *Escherichia coli*, the pathway leading to the synthesis of the aromatic amino acids contains three isoenzymes of DAHP synthase. Each of these enzymes is feedback-inhibited by one of the aromatic amino acids. However, note how an excess of all three amino acids is required to completely shut off the synthesis of DAHP. In addition to feedback inhibition at the DAHP site, each amino acid feedback inhibits its further metabolism at the chorismate step.

Isoenzymes

Some biosynthetic pathways controlled by feedback inhibition employ *isoenzymes* ("iso" means "same"). Isoenzymes are different enzymes that catalyze the same reaction but are subject to different regulatory controls. Examples are enzymes required for the synthesis of the aromatic amino acids tyrosine, tryptophan, and phenylalanine in *Escherichia coli*.

The enzyme 3-deoxy-D-arabino-heptulosonate 7-phosphate (DAHP) synthase plays a central role in aromatic amino acid biosynthesis. In *E. coli*, three DAHP synthase isoenzymes catalyze the first reaction in this pathway, each regulated independently by a different one of the end-product amino acids. However, unlike the example of feedback inhibition where an end product completely inhibits enzyme activity, enzyme activity is diminished incrementally; enzyme activity falls to zero only when *all three* end products are present in excess (**Figure 4.30**).

Enzyme Regulation by Covalent Modification

Some biosynthetic enzymes are regulated by covalent modification, typically the attachment or removal of some small molecule to the protein that affects its activity. Binding of the small molecule changes the conformation of the protein, inhibiting its catalytic activity. Removal of the molecule then returns the enzyme to an active state. Common modifiers include the nucleotides adenosine monophosphate (AMP) and adenosine diphosphate (ADP), inorganic phosphate (PO_4^{2-}), and methyl (CH_3) groups. We consider here the well-studied case of glutamine synthetase (GS), a key enzyme in ammonia (NH_3) assimilation, whose activity is modulated by the addition of AMP, a process called *adenylylation*.

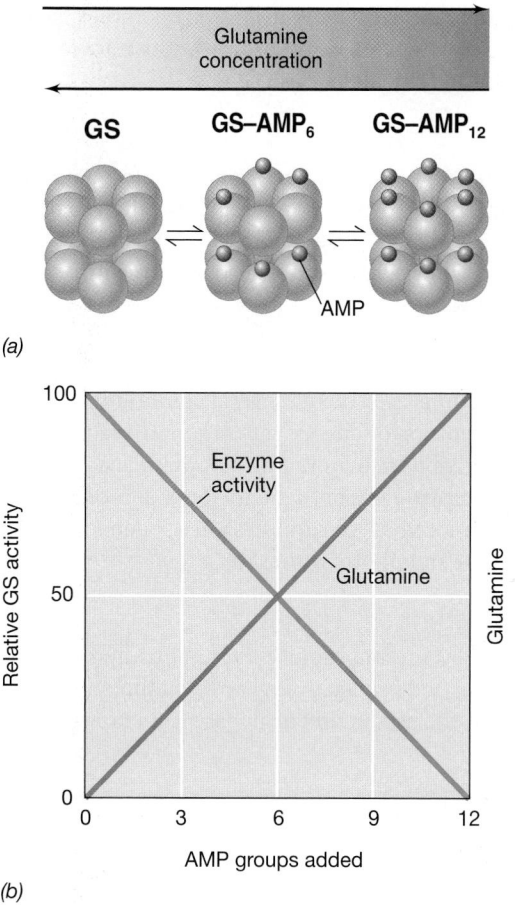

(a)

(b)

Figure 4.31 Regulation of glutamine synthetase by covalent modification. *(a)* When cells are grown with excess ammonia (NH_3), glutamine synthetase (GS) is covalently modified by adenylylation; as many as 12 AMP groups can be added. When cells are NH_3-limited, the groups are removed, forming ADP. *(b)* Adenylylated GS subunits are catalytically inactive, so the overall GS activity decreases progressively as more subunits are adenylylated. See Figure 4.25b for the reaction carried out by glutamine synthetase.

Each molecule of GS is composed of 12 identical subunits, and each subunit can be adenylylated. When the enzyme is fully adenylylated (that is, each molecule of GS contains 12 AMP groups), it is catalytically inactive. When it is partially adenylylated, it is partially active. As the glutamine pool in the cell increases, GS becomes more adenylylated, and its activity diminishes. As glutamine levels diminish, GS becomes less adenylylated and its activity increases (**Figure 4.31**). Other enzymes in the cell add and remove the AMP groups from GS, and these enzymes are themselves controlled, ultimately by levels of NH_3 in the cell.

Why should there be all of this elaborate regulation surrounding the enzyme GS? The activity of GS requires ATP (Figure 4.25*b*), and nitrogen assimilation is a major biosynthetic process in the cell. However, when NH_3 is present at high levels in the cell, it can be assimilated into amino acids by enzymes that do not consume ATP (Figure 4.25*a*); under these conditions, GS remains inactive. When NH_3 levels are very low, however, GS is forced to become catalytically active. By having GS active only when NH_3 is present at low levels, the cell conserves ATP that would be used unnecessarily if GS were active when NH_3 was present at high levels.

The modulation of GS activity in this very precise way stands in contrast to enzymes subject to feedback inhibition (Figures 4.29 and 4.30), whose activity is either "on" or "off", depending on the concentration of the effector molecule. This finer type of control allows GS to remain partially active until NH_3 is at such high levels that NH_3 assimilating systems that have a lower affinity for NH_3 than does GS and that do not require ATP, have sufficient NH_3 to be fully active.

MiniQuiz
- What is feedback inhibition?
- What is an allosteric enzyme?
- In glutamine synthetase, what does adenylylation do to enzyme activity?

Big Ideas

4.1
Cells are primarily composed of the elements H, O, C, N, P, and S. The various chemical compounds found in a cell are formed from nutrients present in the environment. Elements required in fairly large amounts are called macronutrients, whereas metals and organic compounds needed in very small amounts (micronutrients) are trace elements and growth factors, respectively.

4.2
Culture media that supply the nutritional needs of microorganisms are either defined or complex. "Selective," "differential," and "enriched" are terms that describe media used for the culture of particular species or for comparative studies of microorganisms.

4.3
Many microorganisms can be grown in the laboratory in liquid or solid culture media that contain the nutrients they require. Pure cultures of microorganisms can be cultured and maintained if aseptic technique is practiced.

4.4
Chemical reactions in the cell are accompanied by changes in energy, expressed in kilojoules. A chemical reaction may release free energy (may be exergonic) or may consume free energy (may be endergonic). $\Delta G^{0'}$ is a measure of the energy released or consumed in a reaction.

4.5

Enzymes are protein catalysts that speed up the rate of biochemical reactions by activating the substrates when they bind to their active site. Enzymes are highly specific in the reactions they catalyze, and this specificity resides in the three-dimensional structures of the polypeptides that make up the proteins.

4.6

Oxidation–reduction reactions require electron donors and electron acceptors. The tendency of a compound to accept or release electrons is expressed quantitatively by its reduction potential, E_0'. Redox reactions in a cell typically employ electron carriers such as $NAD^+/NADH$.

4.7

The energy released in redox reactions is conserved in compounds that contain energy-rich phosphate or sulfur bonds. The most common of these compounds is ATP, the prime energy carrier in the cell. Longer-term storage of energy is linked to the formation of polymers, which can be consumed to yield ATP.

4.8

Fermentation through the glycolytic pathway, which breaks glucose down to pyruvate, is a widespread mechanism of anaerobic catabolism. Glycolysis releases a small amount of ATP and makes fermentation products. For each molecule of glucose consumed in glycolysis, two ATPs are produced.

4.9

Electron transport systems consist of membrane-associated electron carriers that function in an integrated fashion to carry electrons from the primary electron donor to the terminal electron acceptor (oxygen in aerobic respiration).

4.10

When electrons are transported through an electron transport chain, protons are extruded to the outside of the membrane, forming the proton motive force. Key electron carriers include flavins, quinones, the cytochrome bc_1 complex, and other cytochromes. The cell uses the proton motive force to make ATP through the activity of ATPase.

4.11

Respiration completely oxidizes an organic compound to CO_2 with an energy yield that is much greater than that of fermentation. The citric acid cycle generates CO_2 and electrons for the electron transport chain and is also a source of key biosynthetic intermediates.

4.12

When conditions are anoxic, several compounds can be terminal electron acceptors for energy generation in anaerobic respiration. Chemolithotrophs use inorganic compounds as electron donors, whereas phototrophs use light to form a proton motive force. The proton motive force supports energy generation in all forms of respiration and photosynthesis.

4.13

Polysaccharides are important structural components of cells and are biosynthesized from activated forms of their monomers. Gluconeogenesis is the production of glucose from nonsugar precursors.

4.14

Amino acids are formed from carbon skeletons to which ammonia is added from either glutamate or glutamine. Nucleotides are biosynthesized using carbon from several sources.

4.15

Fatty acids are synthesized two carbons at a time and then attached to glycerol to form lipids. Only in *Bacteria* and *Eukarya* do lipids contain fatty acids.

4.16

Enzyme activity is regulated. In feedback inhibition, an excess of the final product of a biosynthetic pathway inhibits an allosteric enzyme at the beginning of the pathway. Enzyme activity can also be modulated by isoenzymes or by reversible covalent modification.

Review of Key Terms

Activation energy the energy required to bring the substrate of an enzyme to the reactive state

Adenosine triphosphate (ATP) a nucleotide that is the primary form in which chemical energy is conserved and utilized in cells

Allosteric enzyme an enzyme containing an active site for binding substrate and an allosteric site for binding an effector molecule such as the end product of a biochemical pathway

Anabolic reactions (Anabolism) the sum total of all biosynthetic reactions in the cell

Anaerobic respiration a form of respiration in which oxygen is absent and alternative electron acceptors are reduced

Aseptic technique manipulations to prevent contamination of sterile objects or microbial cultures during handling

ATPase (ATP synthase) a multiprotein enzyme complex embedded in the cytoplasmic membrane that catalyzes the synthesis of ATP coupled to dissipation of the proton motive force

Autotroph an organism capable of biosynthesizing all cell material from CO_2 as the sole carbon source

Catabolic reactions (Catabolism) biochemical reactions leading to energy conservation (usually as ATP) by the cell

Catalyst a substance that accelerates a chemical reaction but is not consumed in the reaction

Chemolithotroph an organism that can grow with inorganic compounds as electron donors in energy metabolism

Citric acid cycle a cyclical series of reactions resulting in the conversion of acetate to two molecules of CO_2

Coenzyme a small and loosely bound nonprotein molecule that participates in a reaction as part of an enzyme

Complex medium a culture medium composed of chemically undefined substances such as yeast and meat extracts

Culture medium an aqueous solution of various nutrients suitable for the growth of microorganisms

Defined medium a culture medium whose precise chemical composition is known

Electron acceptor a substance that can accept electrons from an electron donor, becoming reduced in the process

Electron donor a substance that can donate electrons to an electron acceptor, becoming oxidized in the process

Endergonic requires energy

Enzyme a protein that can speed up (catalyze) a specific chemical reaction

Exergonic releases energy

Feedback inhibition a process in which an excess of the end product of a multistep pathway inhibits activity of the first enzyme in the pathway

Fermentation anaerobic catabolism in which an organic compound is both an electron donor and an electron acceptor and ATP is produced by substrate-level phosphorylation

Free energy (G) energy available to do work; $G^{0\prime}$ is free energy under standard conditions

Glycolysis a biochemical pathway in which glucose is fermented, yielding ATP and various fermentation products; also called the Embden–Meyerhof–Parnas pathway

Metabolism the sum total of all the chemical reactions in a cell

Oxidative phosphorylation the production of ATP from a proton motive force formed by electron transport of electrons from organic or inorganic electron donors

Photophosphorylation the production of ATP from a proton motive force formed from light-driven electron transport

Phototrophs organisms that use light as their source of energy

Proton motive force a source of energy resulting from the separation of protons from hydroxyl ions across the cytoplasmic membrane, generating a membrane potential

Pure culture a culture that contains a single kind of microorganism

Reduction potential ($E_0\prime$) the inherent tendency, measured in volts under standard conditions, of a compound to donate electrons

Respiration the process in which a compound is oxidized with O_2 (or an O_2 substitute) as the terminal electron acceptor, usually accompanied by ATP production by oxidative phosphorylation

Siderophore an iron chelator that can bind iron present at very low concentrations

Substrate-level phosphorylation production of ATP by the direct transfer of an energy-rich phosphate molecule from a phosphorylated organic compound to ADP

Review Questions

1. What is the difference between a catabolic and an anabolic reaction (Section 4.1)?

2. What are siderophores and why are they necessary (Section 4.1)?

3. How do you distinguish a selective medium from a differential medium (Section 4.2)?

4. What is aseptic technique and why is it necessary (Section 4.3)?

5. Describe how you would calculate $\Delta G^{0\prime}$ for the reaction: glucose + 6 O_2 → 6 CO_2 + 6 H_2O. If you were told that this reaction is highly *exergonic*, what would be the arithmetic sign (negative or positive) of the $\Delta G^{0\prime}$ you would expect for this reaction (Section 4.4)?

6. Distinguish between $\Delta G^{0\prime}$, ΔG, and G_f^0 (Section 4.4).

7. What are enzymes? Why are they important to cells (Section 4.5)?

8. The following is a series of coupled electron donors and electron acceptors (written as donor/acceptor). Using just the data in Figure 4.9, order this series from most energy yielding to least energy yielding: H_2/Fe^{3+}, H_2S/O_2, methanol/NO_3^- (producing NO_2^-), H_2/O_2, Fe^{2+}/O_2, NO_2^-/Fe^{3+}, and H_2S/NO_3^- (Section 4.6).

9. What is the reduction potential of the $NAD^+/NADH$ couple (Section 4.7)?

10. Why is acetyl phosphate considered an energy-rich compound but glucose 6-phosphate is not (Section 4.7)?

11. How is ATP made in fermentation and in respiration (Section 4.8)?

12. Where in glycolysis is NADH produced? Where is NADH consumed (Section 4.8)?

13. List some of the important electron carriers found in electron transport chains (Section 4.9).

14. What is meant by the term proton motive force, and why is this concept so important in biology (Section 4.10)?

15. How is rotational energy in the ATPase used to produce ATP (Section 4.10)?

16. Work through the energy balance sheets for fermentation and respiration, and account for all sites of ATP synthesis. Organisms can obtain nearly 20 times more ATP when growing aerobically on glucose than by fermenting it. Write one sentence that accounts for this difference (Section 4.11).

17. Why can it be said that the citric acid cycle plays two major roles in the cell (Section 4.11)?

18. What are the differences in electron donor and carbon source used by *Escherichia coli* and *Thiobacillus thioparus* (a sulfur chemolithotroph) (Section 4.12 and Table 4.2)?

19. What two catabolic pathways supply carbon skeletons for sugar and amino acid biosyntheses (Sections 4.13 and 4.14)?

20. Describe the process by which a fatty acid such as palmitate (a C_{16} straight-chain saturated fatty acid) is synthesized in a cell (Section 4.15).

21. Why is feedback inhibition important (Section 4.16)?

Application Questions

1. Design a defined culture medium for an organism that can grow aerobically on acetate as a carbon and energy source. Make sure all the nutrient needs of the organism are accounted for and in the correct relative proportions.

2. *Desulfovibrio* can grow anaerobically with H_2 as electron donor and SO_4^{2-} as electron acceptor (which is reduced to H_2S). Based on this information and the data in Table A1.2 (Appendix 1), indicate which of the following components could not exist in the electron transport chain of this organism and why: cytochrome *c*, ubiquinone, cytochrome *c$_3$*, cytochrome *aa$_3$*, ferredoxin.

3. Again using the data in Table A1.2, predict the sequence of electron carriers in the membrane of an organism growing aerobically and producing the following electron carriers: ubiquinone, cytochrome *aa$_3$*, cytochrome *b*, NADH, cytochrome *c*, FAD.

4. Explain the following observation in light of the redox tower: Cells of *Escherichia coli* fermenting glucose grow faster when NO_3^- is supplied to the culture (NO_2^- is produced) and then grow even faster (and stop producing NO_2^-) when the culture is highly aerated.

Need more practice? Test your understanding with Quantitative Questions; access additional study tools including tutorials, animations, and videos; and then test your knowledge with chapter quizzes and practice tests at **www.microbiologyplace.com**.

5

Microbial Growth

The curved bacterium
Caulobacter has been a
model for studying the cell
division process, including
how shape-determining
proteins such as crescentin
(shown here stained red)
give cells their distinctive
shape.

Bacterial Cell Division

In the last two chapters we discussed cell structure and function (Chapter 3) and the principles of microbial nutrition and metabolism (Chapter 4). Before we begin our study of the biosynthesis of macromolecules in microorganisms (Chapters 6 and 7), we consider microbial growth. Growth is the ultimate process in the life of a cell—one cell becoming two.

5.1 Cell Growth and Binary Fission

In microbiology, **growth** is defined as *an increase in the number of cells*. Microbial cells have a finite life span, and a species is maintained only as a result of continued growth of its population. There are many reasons why understanding how microbial cells grow is important. For example, many practical situations call for the control of microbial growth, in particular, bacterial growth. Knowledge of how microbial populations can rapidly expand is useful for designing methods to control microbial growth, whether the methods are used to treat a life-threatening infectious disease or simply to disinfect a surface. We will study these control methods in Chapter 26. Knowledge of the events surrounding bacterial growth also allows us to see how these processes are related to cell division in higher organisms. As we will see, there are many parallels.

Bacterial cell growth depends upon a large number of cellular reactions of a wide variety of types. Some of these reactions transform energy. Others synthesize small molecules—the building blocks of macromolecules. Still others provide the various cofactors and coenzymes needed for enzymatic reactions. However, the key reactions of cell synthesis are polymerizations that make macromolecules from monomers. As macromolecules accumulate in the cytoplasm of a cell, they are assembled into new structures, such as the cell wall, cytoplasmic membrane, flagella, ribosomes, enzyme complexes, and so on, eventually leading to the process of cell division itself.

In a growing rod-shaped cell, elongation continues until the cell divides into two new cells. This process is called **binary fission** ("binary" to express the fact that two cells have arisen from one). In a growing culture of a rod-shaped bacterium such as *Escherichia coli*, cells elongate to approximately twice their original length and then form a partition that constricts the cell into two daughter cells (**Figure 5.1**). This partition is called a *septum* and results from the inward growth of the cytoplasmic membrane and cell wall from opposing directions; septum formation continues until the two daughter cells are pinched off. There are variations in this general pattern. In some bacteria, such as *Bacillus subtilis*, a septum forms without cell wall constriction, while in the budding bacterium *Caulobacter*, constriction occurs but no septum is formed. But in all cases, when one cell eventually separates to form two cells, we say that one *generation* has occurred, and the time required for this process is called the **generation time** (Figure 5.1 and see Figure 5.9).

During one generation, all cellular constituents increase proportionally; cells are thus said to be in *balanced growth*. Each daughter cell receives a chromosome and sufficient copies of ribosomes and all other macromolecular complexes, monomers,

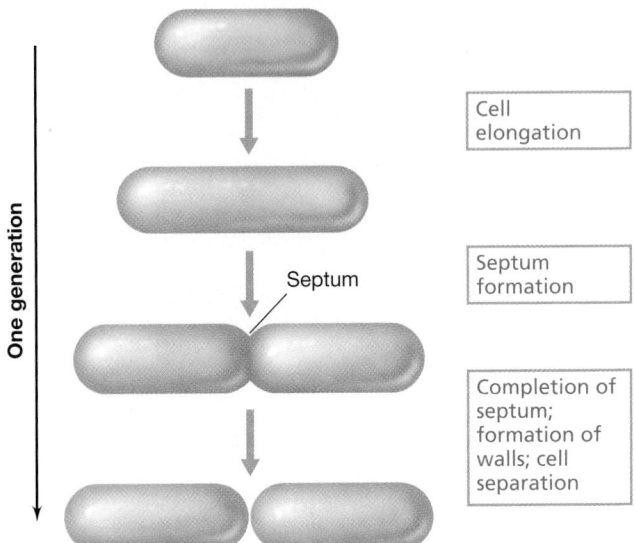

Figure 5.1 Binary fission in a rod-shaped prokaryote. Cell numbers double every generation.

and inorganic ions to exist as an independent cell. Partitioning of the replicated DNA molecule between the two daughter cells depends on the DNA remaining attached to the cytoplasmic membrane during division, with constriction leading to separation of the chromosomes, one to each daughter cell (see Figure 5.3).

The time required for a generation in a given bacterial species is highly variable and is dependent on nutritional and genetic factors, and temperature. Under the best nutritional conditions the generation time of a laboratory culture of *E. coli* is about 20 min. A few bacteria can grow even faster than this, but many grow much slower. In nature it is likely that microbial cells grow much slower than their maximum rate because rarely are all conditions and resources necessary for optimal growth present at the same time.

MiniQuiz

• Define the term generation. What is meant by the term generation time?

5.2 Fts Proteins and Cell Division

A series of proteins present in all *Bacteria*, called *Fts proteins*, are essential for cell division. The acronym *Fts* stands for *filamentous temperature sensitive*, which describes the properties of cells that have mutations in the genes that encode Fts proteins. Such cells do not divide normally, but instead form long filamentous cells that fail to divide. **FtsZ**, a key Fts protein, has been well studied in *Escherichia coli* and several other bacteria, and much is known concerning its important role in cell division.

FtsZ is found in all prokaryotes, including the *Archaea*; FtsZ-type proteins have even been found in mitochondria and chloroplasts, further emphasizing the evolutionary ties of these organelles to the *Bacteria*. Interestingly, the protein FtsZ is related to tubulin, the important cell-division protein in eukaryotes (⮎ Section 20.5). However, most other Fts proteins are found only in species of *Bacteria* and not in *Archaea*, so our discussion here will be restricted to the *Bacteria*. Among *Bacteria*, the gram-negative *E. coli* and the gram-positive *Bacillus subtilis* have been the model species.

Fts Proteins and Cell Division

Fts proteins interact to form a cell-division apparatus called the **divisome**. In rod-shaped cells, formation of the divisome begins with the attachment of molecules of FtsZ in a ring precisely around the center of the cell. This ring prescribes what will eventually become the cell-division plane. In a cell of *E. coli* about 10,000 FtsZ molecules polymerize to form the ring, and the ring attracts other divisome proteins, including *FtsA* and *ZipA* (**Figure 5.2**). ZipA is an anchor that connects the FtsZ ring to the cytoplasmic membrane and stabilizes it. FtsA, a protein related

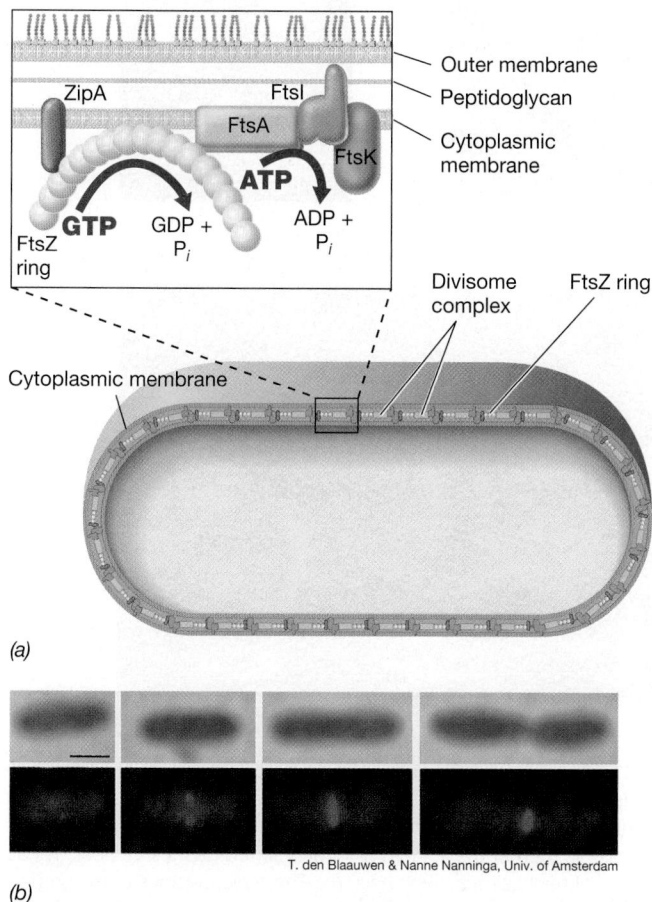

(a)

(b)

T. den Blaauwen & Nanne Nanninga, Univ. of Amsterdam

Figure 5.2 The FtsZ ring and cell division. *(a)* Cutaway view of a rod-shaped cell showing the ring of FtsZ molecules around the division plane. Blowup shows the arrangement of individual divisome proteins. ZipA is an FtsZ anchor, FtsI is a peptidoglycan biosynthesis protein, FtsK assists in chromosome separation, and FtsA is an ATPase. *(b)* Appearance and breakdown of the FtsZ ring during the cell cycle of *Escherichia coli*. Microscopy: upper row, phase-contrast; bottom row, cells stained with a specific reagent against FtsZ. Cell division events: first column, FtsZ ring not yet formed; second column, FtsZ ring appears as nucleoids start to segregate; third column, full FtsZ ring forms as cell elongates; fourth column, breakdown of the FtsZ ring and cell division. Marker bar in upper left photo, 1 μm.

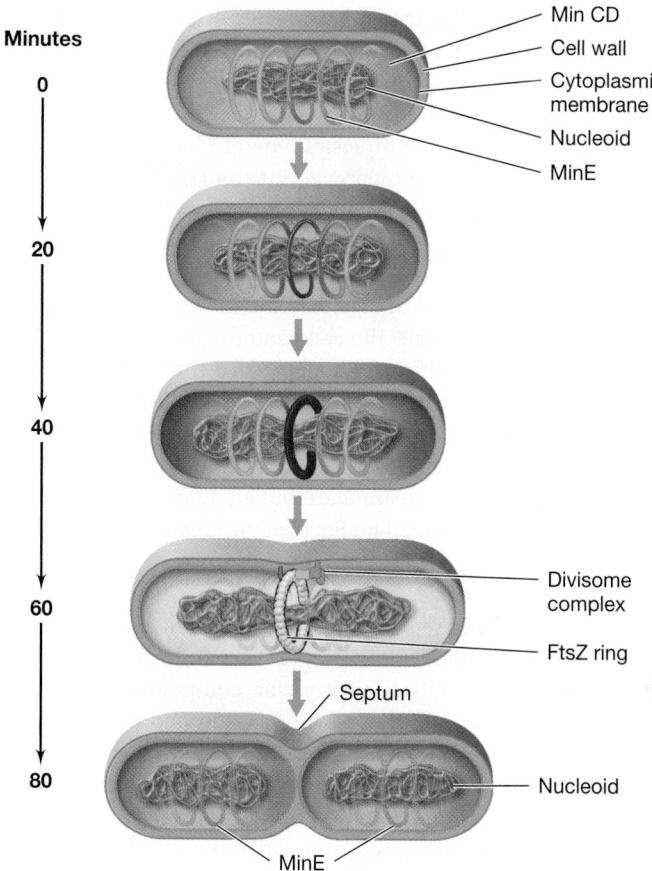

Figure 5.3 DNA replication and cell-division events. The protein MinE directs formation of the FtsZ ring and divisome complex at the cell-division plane. Shown is a schematic for cells of *Escherichia coli* growing with a doubling time of 80 min. MinC and MinD (not shown) are most abundant at the cell poles.

to actin, also helps to connect the FtsZ ring to the cytoplasmic membrane and has an additional role in recruiting other divisome proteins. The divisome forms well after elongation of a newborn cell has already begun. For example, in cells of *E. coli* the divisome forms about three-quarters of the way into cell division. However, before the divisome forms, the cell is already elongating and DNA is replicating (see Figure 5.3).

The divisome also contains Fts proteins needed for peptidoglycan synthesis, such as FtsI (Figure 5.2). FtsI is one of several *penicillin-binding proteins* present in the cell. Penicillin-binding proteins are so named because their activities are inhibited by the antibiotic penicillin (Section 5.4). The divisome orchestrates synthesis of new cytoplasmic membrane and cell wall material, called the *division septum*, at the center of a rod-shaped cell until it reaches twice its original length. Following this, the elongated cell divides, yielding two daughter cells (Figure 5.1).

DNA Replication, Min Proteins, and Cell Division

As we noted, DNA replicates before the FtsZ ring forms (**Figure 5.3**). The ring forms in the space between the duplicated nucleoids because, before the nucleoids segregate, they effectively block formation of the FtsZ ring. Location of the actual cell midpoint by FtsZ is facilitated by a series of proteins called *Min* proteins, especially MinC, MinD, and MinE. MinD forms a spiral structure on the inner surface of the cytoplasmic membrane and oscillates back and forth from pole to pole; MinD is also required to localize MinC to the cytoplasmic membrane. Together, MinC and D inhibit cell division by preventing the FtsZ ring from forming. MinE also oscillates from pole to pole, sweeping MinC and D aside as it moves along. Because MinC and MinD dwell longer at the poles than elsewhere in the cell during their oscillation cycle, on average the center of the cell has the lowest concentration of these proteins. As a result, the cell center is the most permissive site for FtsZ ring assembly, and the FtsZ ring thus defines the division plane. In this way, the Min proteins ensure that the divisome forms only at the *cell center* and not at the cell poles (Figure 5.3).

As cell elongation continues and septum formation begins, the two copies of the chromosome are pulled apart, each to its own daughter cell (Figure 5.3). The Fts protein *FtsK* and several other proteins assist in this process. As the cell constricts, the FtsZ ring begins to depolymerize, triggering the inward growth of wall materials to form the septum and seal off one daughter cell from the other. The enzymatic activity of FtsZ also hydrolyzes guanosine triphosphate (GTP, an energy-rich compound) to yield the energy necessary to fuel the polymerization and depolymerization of the FtsZ ring (Figures 5.2 and 5.3).

Properly functioning Fts proteins are essential for cell division. Much new information on cell division in *Bacteria* and *Archaea* has emerged in recent years, and genomic studies have confirmed that at least FtsZ is a key and universal cell-division protein. There is great practical interest in understanding bacterial cell division in great detail because such knowledge could lead to the development of new drugs that target specific steps in the growth of pathogenic bacteria. Like penicillin (a drug that targets bacterial cell wall synthesis), drugs that interfere with the function of specific Fts or other bacterial cell-division proteins could have broad applications in clinical medicine.

MiniQuiz
- When does the bacterial chromosome replicate in the binary fission process?
- How does FtsZ find the cell midpoint of a rod-shaped cell?

5.3 MreB and Determinants of Cell Morphology

Just as specific proteins direct cell *division* in prokaryotes, other specific proteins specify cell *shape*. Interestingly, these shape-determining proteins show significant homology to key cytoskeletal proteins in eukaryotic cells. As more is learned about these proteins, it has become clear that, like eukaryotes, prokaryotes also contain a cell cytoskeleton, one that is both dynamic and multifaceted.

Cell Shape and Actinlike Proteins in Prokaryotes

The major shape-determining factor in prokaryotes is a protein called *MreB*. MreB forms a simple cytoskeleton in cells of *Bacteria* and probably in *Archaea* as well. MreB forms

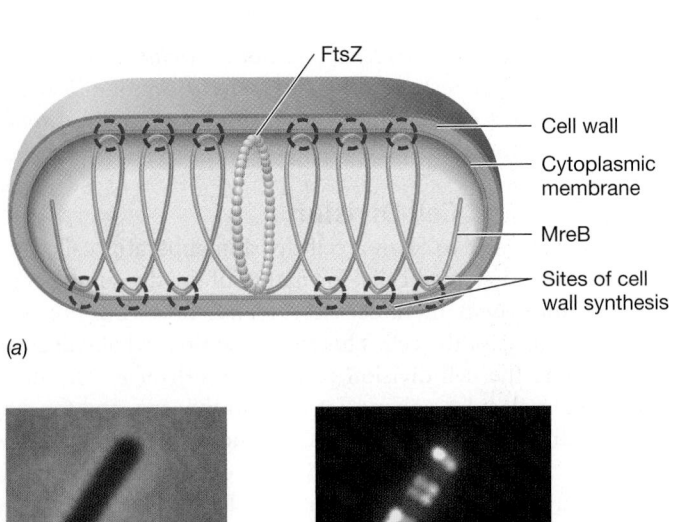

(a)

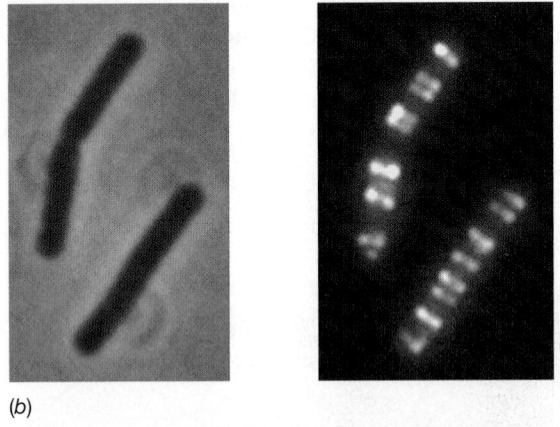

(b)

Alex Formstone

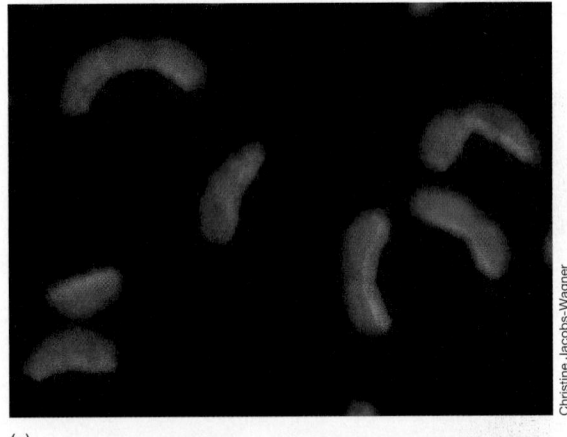

(c)

Christine Jacobs-Wagner

Figure 5.4 MreB and crescentin as determinants of cell morphology. *(a)* The cytoskeletal protein MreB is an actin analog that winds as a coil through the long axis of a rod-shaped cell, making contact with the cytoplasmic membrane in several locations (red dashed circles). These are sites of new cell wall synthesis. *(b)* Photomicrographs of the same cells of *Bacillus subtilis*. Left, phase-contrast; right, fluorescence. The cells contain a substance that makes the MreB protein fluoresce, shown here as bright white. *(c)* Cells of *Caulobacter crescentus*, a naturally curved (vibrio-shaped) cell. Cells are stained to show the shape-determining protein crescentin (red), which lies along the concave surface of the cell, and with DAPI, which stains DNA and thus the entire cell (blue).

spiral-shaped bands around the inside of the cell, just underneath the cytoplasmic membrane (**Figure 5.4**). Presumably, the MreB cytoskeleton defines cell shape by recruiting other proteins that orchestrate cell wall growth in a specific pattern. Inactivation of the gene encoding MreB in rod-shaped bacteria causes the cells to become coccoid (coccus-shaped). Interestingly, naturally coccoid bacteria lack the gene that encodes MreB and thus lack MreB. This indicates that the "default" shape for a bacterium is a sphere (coccus). Variations in the arrangement of MreB filaments in cells of nonspherical bacteria are probably responsible for the common morphologies of prokaryotic cells (Figure 3.1).

Besides cell shape, MreB plays other important roles in the bacterial cell; in particular, it assists in the segregation of the replicated chromosome such that one copy is distributed to each daughter cell. Other actinlike proteins also play a role in this regard. *Par proteins*, for example, are a series of proteins that function in an analogous fashion to the mitotic apparatus of eukaryotic cells, separating chromosomes and plasmids to the poles of the cell during the division process. Par proteins bind to the origin of replication of the bacterial chromosome. After the origin has been replicated, the Par proteins partition the two origins to opposite cell poles and then physically push or pull the two chromosomes apart.

Mechanism of MreB

How does MreB define a cell's shape? The answer is not entirely clear, but experiments on cell division and its link to cell wall synthesis have yielded two important clues. First, the helical structures formed by MreB (Figure 5.4) are not static, but instead can rotate within the cytoplasm of a growing cell. Second, newly synthesized peptidoglycan (Section 5.4) is associated with the MreB helices at points where the helices contact the cytoplasmic membrane (Figure 5.4). It thus appears that MreB functions to localize synthesis of new peptidoglycan and other cell wall components to specific locations along the cylinder of a rod-shaped cell during growth. This would explain the fact that new cell wall material in an elongated rod-shaped cell forms at several points along its long axis rather than from a single location at the FtsZ site outward, as in spherical bacteria (see Figure 5.5). By rotating within the cell cylinder and initiating cell wall synthesis where it contacts the cytoplasmic membrane, MreB would direct new wall synthesis in such a way that a rod-shaped cell would elongate only along its long axis.

Crescentin

Caulobacter crescentus, a vibrio-shaped species of *Proteobacteria* (Section 17.16), produces a shape-determining protein called *crescentin* in addition to MreB. Copies of crescentin protein organize into filaments about 10 nm wide that localize onto the concave face of the curved cell. The arrangement and localization of crescentin filaments are thought to somehow impart the characteristic curved morphology to the *Caulobacter* cell (Figure 5.4c). *Caulobacter* is an aquatic bacterium that undergoes a life cycle in which swimming cells, called *swarmers*, eventually form a stalk and attach to surfaces. Attached cells then undergo cell division to form new swarmer cells that are released to colonize new habitats. The steps in this life cycle are highly orchestrated at the genetic level, and *Caulobacter* has been used as a model system for the study of gene expression in cellular differentiation (Section 8.13). Although thus far crescentin has been found only in *Caulobacter*, proteins similar to crescentin have been found in other helically shaped cells, such as *Helicobacter*. This suggests that these proteins may be necessary for the formation of curved cells.

Archaeal Cell Morphology and the Evolution of Cell Division and Cell Shape

Although less is known about how cell morphology is controlled in *Archaea* than in *Bacteria*, the genomes of most *Archaea* contain genes that encode MreB-like proteins. Thus, it is likely that these function in *Archaea* as they do in *Bacteria*. Along with the finding that FtsZ also exists in *Archaea*, it appears that there are strong parallels in cell-division processes and morphological determinants in all prokaryotes.

How do the determinants of cell shape and cell division in prokaryotes compare with those in eukaryotes? Interestingly, the protein MreB is structurally related to the eukaryotic protein actin, and FtsZ is related to the eukaryotic protein tubulin. In eukaryotic cells actin assembles into structures called *microfilaments* that function as scaffolding in the cell cytoskeleton and in cytokinesis, whereas tubulin forms *microtubules* that are important in eukaryotic mitosis and other processes (Sections 7.1, 20.1, and 20.5). In addition, the shape-determining protein crescentin in *Caulobacter* is related to the keratin proteins that make up *intermediate filaments* in eukaryotic cells. Intermediate filaments are also part of the eukaryotic cytoskeleton and are fairly widespread among *Bacteria*. It thus appears that most of the proteins that control cell division and cell shape in eukaryotic cells have their evolutionary roots in prokaryotic cells, cells that preceded them on Earth by billions of years (Figure 1.6).

MiniQuiz

- What eukaryotic protein is related to MreB? What does this protein do in eukaryotic cells?
- What is crescentin and what does it do?

5.4 Peptidoglycan Synthesis and Cell Division

In the previous section we considered some of the key events in binary fission and learned that a major feature of the cell-division process is the production of new cell wall material. In most cocci, cell walls grow in opposite directions outward from the FtsZ ring (**Figure 5.5**), whereas the walls of rod-shaped cells grow at several locations along the length of the cell (Figure 5.4). However, in both cases preexisting peptidoglycan has to be severed to allow newly synthesized peptidoglycan to be inserted. How does this occur?

Beginning at the FtsZ ring (Figures 5.2 and 5.3), small gaps in the wall are made by enzymes called *autolysins*, enzymes that function like lysozyme (Section 3.6) to hydrolyze the β-1,4 glycosidic bonds that connect N-acetylglucosamine and N-acetylmuramic acid in the peptidoglycan backbone. New cell wall material is then added across the gaps (Figure 5.5a). The junction between new and old peptidoglycan forms a ridge on the cell surface of gram-positive bacteria called a *wall band*

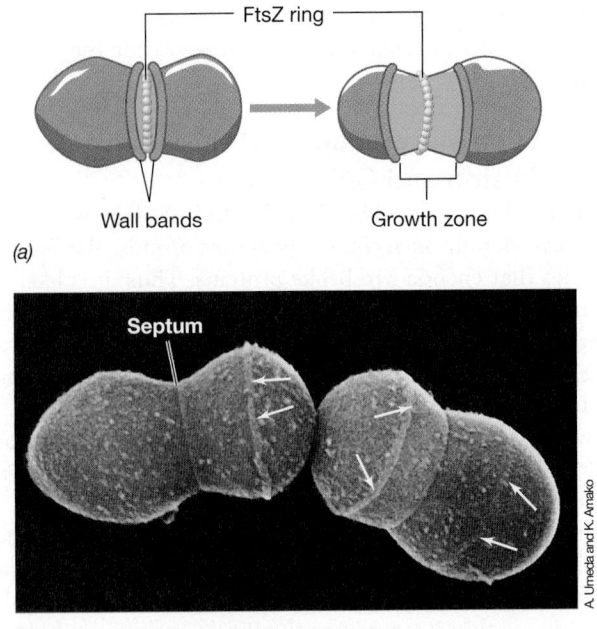

(a)

Wall bands   Growth zone

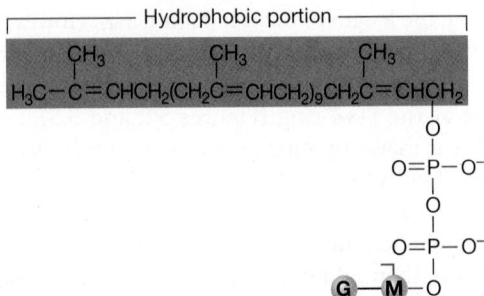

(b)

Figure 5.5 Cell wall synthesis in gram-positive *Bacteria*. *(a)* Localization of cell wall synthesis during cell division. In cocci, cell wall synthesis (shown in green) is localized at only one point (compare with Figure 5.4). *(b)* Scanning electron micrograph of cells of *Streptococcus hemolyticus* showing wall bands (arrows). A single cell is about 1 μm in diameter.

(Figure 5.5*b*), analogous to a scar. It is of course essential in peptidoglycan synthesis that new cell wall precursors (*N*-acetylmuramic acid/*N*-acetylglucosamine/tetrapeptide units, see Figure 5.7) be spliced into existing peptidoglycan in a coordinated and consistent manner in order to prevent a breach in peptidoglycan integrity at the splice point; a breach could cause spontaneous cell lysis, called *autolysis*.

Biosynthesis of Peptidoglycan

We discussed the general structure of peptidoglycan in Section 3.6. The peptidoglycan layer can be thought of as a stress-bearing fabric, much like a thin sheet of rubber. Synthesis of new peptidoglycan during growth requires the controlled cutting of preexisting peptidoglycan by autolysins along with the simultaneous insertion of peptidoglycan precursors. A lipid carrier molecule called *bactoprenol* (**Figure 5.6**) plays a major role in this process.

Hydrophobic portion

CH_3 CH_3 CH_3

$H_3C-C=CHCH_2(CH_2C=CHCH_2)_9CH_2C=CHCH_2$

O
|
O=P−O⁻
|
O
|
O=P−O⁻
|
G—M—O

Figure 5.6 Bactoprenol (undecaprenol diphosphate). This highly hydrophobic molecule carries cell wall peptidoglycan precursors through the cytoplasmic membrane.

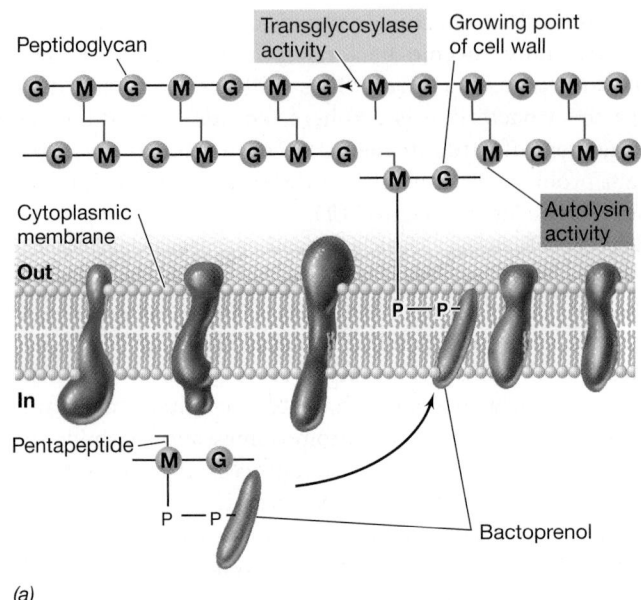

(a)

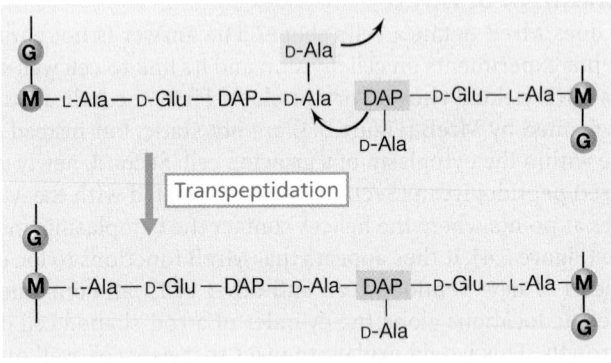

(b)

Figure 5.7 Peptidoglycan synthesis. *(a)* Transport of peptidoglycan precursors across the cytoplasmic membrane to the growing point of the cell wall. Autolysin breaks glycolytic bonds in preexisting peptidoglycan, while transglycosylase synthesizes them, linking old peptidoglycan with new. *(b)* The transpeptidation reaction that leads to the final cross-linking of two peptidoglycan chains. Penicillin inhibits this reaction.

Bactoprenol is a hydrophobic C_{55} alcohol that bonds to a *N*-acetylglucosamine/*N*-acetylmuramic acid/pentapeptide peptidoglycan precursor (**Figure 5.7*a***). Bactoprenol transports peptidoglycan precursors across the cytoplasmic membrane by rendering them sufficiently hydrophobic to pass through the membrane interior. Once in the periplasm, bactoprenol interacts with enzymes called *transglycosylases* that insert cell wall precursors into the growing point of the cell wall and catalyze glycosidic bond formation (Figure 5.7*b*).

Transpeptidation

The final step in cell wall synthesis is **transpeptidation**. Transpeptidation forms the peptide cross-links between muramic acid residues in adjacent glycan chains (Section 3.6 and Figures 3.16 and 3.17). In gram-negative bacteria such as *Escherichia coli*, cross-links form between diaminopimelic acid (DAP) on one peptide and D-alanine on the adjacent peptide (Figure 5.7*b*; see also ⇄ Figure 3.17). Initially, there are *two* D-alanine residues at the end of the peptidoglycan precursor, but only *one* remains in the final molecule as the other D-alanine molecule is removed during the transpeptidation reaction (Figure 5.7*b*). This reaction, which is exergonic, supplies the energy necessary to drive the reaction forward (transpeptidation occurs outside the cytoplasmic membrane, where ATP is unavailable). In *E. coli*, the protein FtsI (Figure 5.2*a*) is the key protein in transpeptidation at the division septum, while a separate transpeptidase enzyme cross-links peptidoglycan elsewhere in the growing cell. In gram-positive bacteria, where a glycine interbridge is common, cross-links occur across the interbridge, typically from an L-lysine of one peptide to a D-alanine on the other (⇄ Section 3.6 and Figure 3.17).

Transpeptidation and Penicillin

Transpeptidation is medically noteworthy because it is the reaction inhibited by the antibiotic penicillin. Several penicillin-binding proteins have been identified in bacteria, including the previously mentioned FtsI (Figure 5.2*a*). When penicillin is bound to penicillin-binding proteins the proteins lose their catalytic activity. In the absence of transpeptidation, the continued activity of autolysins (Figure 5.7) so weakens the cell wall that the cell eventually bursts.

Penicillin has been a successful drug in clinical medicine for at least two reasons. First, humans are *Eukarya* and therefore lack peptidoglycan; the drug can thus be administered in high doses and is typically nontoxic. And second, most pathogenic bacteria contain peptidoglycan and are thus potential targets of the drug. Nevertheless, the continual and excessive use of penicillin since it became commercially available following World War II has selected for resistant mutants of many common pathogens previously susceptible to this drug. Many of these are widespread in human and other animal populations because they make variants of penicillin-binding proteins that are catalytically active but no longer bind penicillin. In these cases, cell wall synthesis occurs uninterrupted in the presence of the drug, and other drugs need to be used to thwart the infection.

MiniQuiz
- What are autolysins and why are they necessary?
- What is the function of bactoprenol?
- What is transpeptidation and why is it important?

 Population Growth

As we mentioned earlier, microbial growth is defined as an increase in the *number* of cells in a population. So we now move on from considering the growth and division of an individual cell to consider the dynamics of growth in bacterial populations.

5.5 The Concept of Exponential Growth

During cell division one cell becomes two. During the time that it takes for this to occur (the generation time), both total cell *number* and *mass* double. Generation times vary widely among microorganisms. In general, most bacteria have shorter generation times than do most microbial eukaryotes. The generation time of a given organism in culture is dependent on the growth medium and the incubation conditions used. Many bacteria have minimum generation times of 0.5–6 h under the best of growth conditions, but a few very rapidly growing organisms are known whose doubling times are less than 20 min and a few slow-growing organisms whose doubling times are as long as several days or even weeks.

Exponential Growth

A growth experiment beginning with a single cell having a doubling time of 30 min is presented in **Figure 5.8**. This pattern of population increase, where the number of cells doubles during a constant time interval, is called **exponential growth**. When the cell number from such an experiment is graphed on arithmetic (linear) coordinates as a function of time, one obtains a curve with a continuously increasing slope (Figure 5.8*b*).

By contrast, when the number of cells is plotted on a logarithmic ($\log_{10}$) scale and time is plotted arithmetically (a *semilogarithmic* graph), as shown in Figure 5.8*b*, the points fall on a straight line. This straight-line function reflects the fact that the cells are growing exponentially and the population is doubling in a constant time interval. Semilogarithmic graphs are also convenient to use to estimate the generation time of a microbial culture from a set of growth data. Generation times may be read directly from

Time (h)	Total number of cells	Time (h)	Total number of cells
0	1	4	256 (2^8)
0.5	2	4.5	512 (2^9)
1	4	5	1,024 (2^{10})
1.5	8	5.5	2,048 (2^{11})
2	16	6	4,096 (2^{12})
2.5	32	.	.
3	64	.	.
3.5	128	10	1,048,576 (2^{19})

(a)

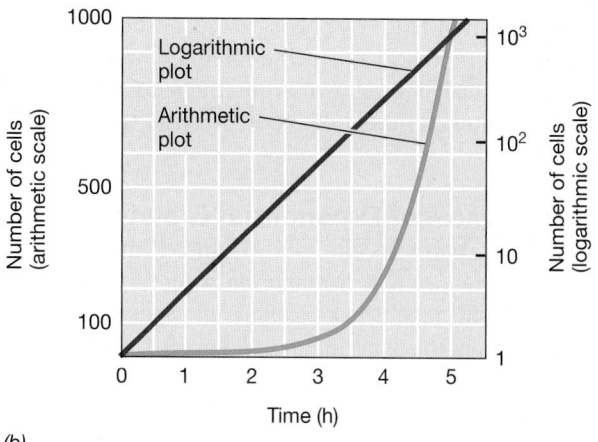

(b)

Figure 5.8 The rate of growth of a microbial culture. (a) Data for a population that doubles every 30 min. (b) Data plotted on arithmetic (left ordinate) and logarithmic (right ordinate) scales.

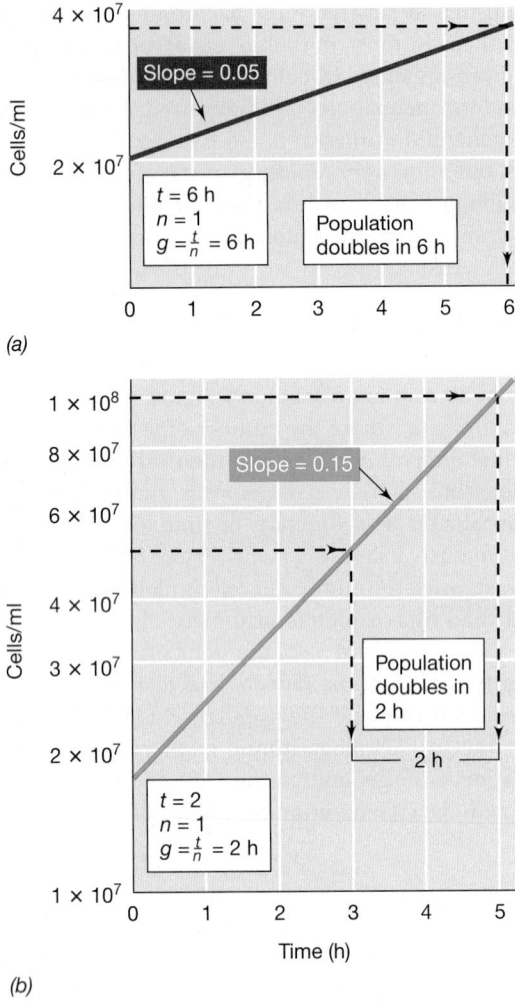

(a)

(b)

Time (h)

Figure 5.9 Calculating microbial growth parameters. Method of estimating the generation times (*g*) of exponentially growing populations with generation times of (*a*) 6 h and (*b*) 2 h from data plotted on semilogarithmic graphs. The slope of each line is equal to 0.301/*g*, and *n* is the number of generations in the time *t*. All numbers are expressed in scientific notation; that is, 10,000,000 is 1×10^7, 60,000,000 is 6×10^7, and so on.

the graph as shown in **Figure 5.9**. For example, when two points on the curve that represent one cell doubling on the Y axis are selected and vertical lines drawn from them to intersect the X axis, the time interval measured on the X axis is the generation time (Figure 5.9*b*).

The Consequences of Exponential Growth

During exponential growth, the increase in cell number is initially rather slow, but increases at an ever faster rate. In the later stages of growth, this results in an explosive increase in cell numbers. For example, in the experiment in Figure 5.8, the rate of cell production in the first 30 min of growth is 1 cell per 30 min. However, between 4 and 4.5 h of growth, the rate of cell production is considerably higher, 256 cells per 30 min, and between 5.5 and 6 h of growth it is 2048 cells per 30 min (Figure 5.8). Thus in an actively growing bacterial culture, cell numbers can get very large very quickly.

Consider the following practical implication of exponential growth. For a nonsterile and nutrient-rich food product such as milk to stand at room temperature for a few hours during the early stages of exponential growth, when total bacterial cell numbers are relatively low, is not detrimental. However, when cell numbers are initially much higher, standing for the same length of time leads to spoilage of the milk. The lactic acid bacteria responsible for milk spoilage contaminate milk during its collection. These harmless organisms grow only very slowly at refrigeration temperatures (~4°C), and only after several days of slow growth at this temperature are the effects of spoilage (rancid milk) noticeable. However, at room temperature or above, growth is greatly accelerated. Thus two bottles of milk that have expiration dates one week apart will contain considerably different bacterial cell numbers and have different outcomes if they are left at room temperature overnight; the fresher milk with still relatively low cell numbers may have no off taste while the older milk with much higher cell numbers is spoiled.

MiniQuiz

- Why does exponential growth lead to large cell populations in so short a period of time?

- What is a *semilogarithmic* plot and what information can we derive from it?

5.6 The Mathematics of Exponential Growth

The increase in cell number in an exponentially growing bacterial culture approaches a geometric progression of the number 2. As one cell divides to become two cells, we express this as $2^0 \rightarrow 2^1$. As two cells become four, we express this as $2^1 \rightarrow 2^2$, and so on (Figure 5.8*a*). A fixed relationship exists between the initial number of cells in a culture and the number present after a period of exponential growth, and this relationship can be expressed mathematically as

$$N = N_0 2^n$$

where N is the final cell number, N_0 is the initial cell number, and n is the number of generations during the period of exponential growth. The generation time (*g*) of the exponentially growing population is t/n, where t is the duration of exponential growth expressed in days, hours, or minutes. From a knowledge of the initial and final cell numbers in an exponentially growing cell population, it is possible to calculate n, and from n and knowledge of t, the generation time, *g*.

The Relationship of *N* and *N₀* to *n*

The equation $N = N_0 2^n$ can be expressed in terms of n as follows:

$$N = N_0 2^n$$

$$\log N = \log N_0 + n \log 2$$

$$\log N - \log N_0 = n \log 2$$

$$n = \frac{\log N - \log N_0}{\log 2} = \frac{\log N - \log N_0}{0.301}$$

$$= 3.3 \, (\log N - \log N_0)$$

With this simple formula, we can calculate generation times in terms of measurable quantities, N and N_0. As an example, consider actual growth data from the graph in Figure 5.9b, in which $N = 10^8$, $N_0 = 5 \times 10^7$, and $t = 2$:

$$n = 3.3 \, [\log(10^8) - \log(5 \times 10^7)] = 3.3(8 - 7.69) = 3.3(0.301) = 1$$

Thus, in this example, $g = t/n = 2/1 = 2$ h. If exponential growth continued for another 2 h, the cell number would be 2×10^8. Two hours later the cell number would be 4×10^8, and so on.

Other Growth Expressions

Besides determination of the generation time of an exponentially growing culture by inspection of graphical data (Figure 5.9b), g can be calculated from the slope of the straight-line function obtained in a semilogarithmic plot of exponential growth. The slope is equal to 0.301 n/t (or $0.301/g$). In the above example, the slope would thus be 0.301/2, or 0.15. Since g is equal to 0.301/slope, we arrive at the same value of 2 for g. The term $0.301/g$ is called the *specific growth rate*, abbreviated k.

Another useful growth expression is the reciprocal of the generation time, called the *division rate*, abbreviated v. The division rate is equal to $1/g$ and has units of reciprocal hours (h^{-1}). Whereas g is a measure of the *time* it takes for a population to double in cell number, v is a measure of the *number of generations* per unit of time in an exponentially growing culture. The slope of the line relating log cell number to time (Figure 5.9) is equal to $v/3.3$.

Armed with knowledge of n and t, one can calculate g, k, and v for different microorganisms growing under different culture conditions. This is often useful for optimizing culture conditions

for a particular organism and also for testing the positive or negative effect of some treatment on the bacterial culture. For example, compared with an unamended control, factors that stimulate or inhibit growth can be identified by measuring their effect on the various growth parameters discussed here.

MiniQuiz
- Distinguish between the terms specific growth rate and generation time.
- If in 8 h an exponentially growing cell population increases from 5×10^6 cells/ml to 5×10^8 cells/ml, calculate g, n, v, and k.

5.7 The Microbial Growth Cycle

The data presented in Figures 5.8 and 5.9 reflect only part of the growth cycle of a microbial population, the part called *exponential growth*. For several reasons, an organism growing in an enclosed vessel, such as a tube or a flask (a growth condition called a **batch culture**), cannot grow exponentially indefinitely. Instead, a typical *growth curve* for the population is obtained, as illustrated in **Figure 5.10**. The growth curve describes an entire growth cycle, and includes the lag phase, exponential phase, stationary phase, and death phase.

Lag Phase

When a microbial culture is inoculated into a fresh medium, growth usually begins only after a period of time called the *lag phase*. This interval may be brief or extended, depending on the history of the inoculum and the growth conditions. If an exponentially growing culture is transferred into the same medium under the same conditions of growth (temperature, aeration, and the like), there is no lag and exponential growth begins immediately. However, if the inoculum is taken from an old

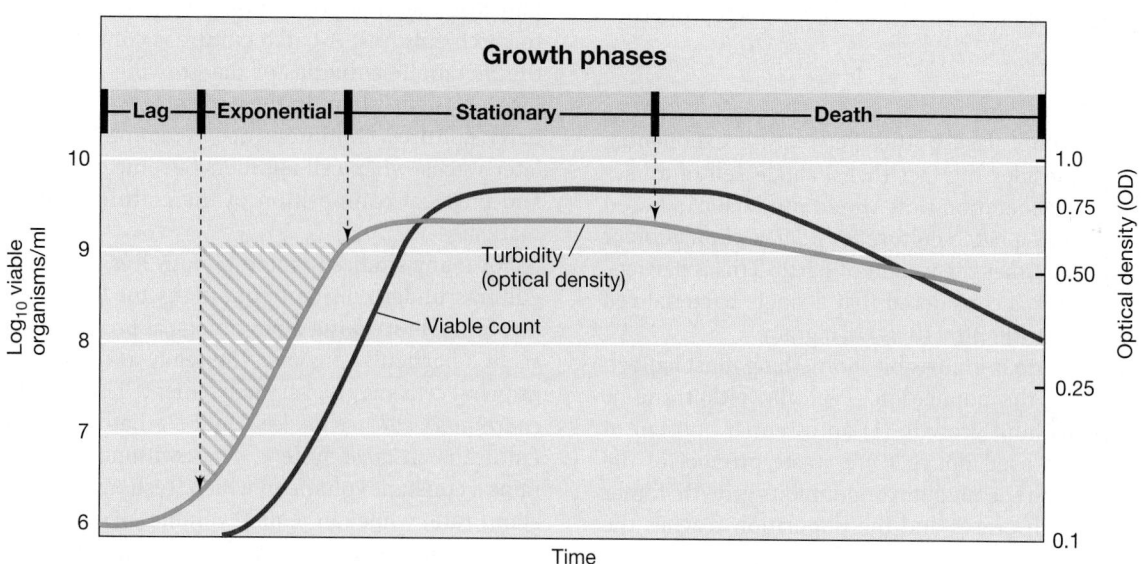

Figure 5.10 Typical growth curve for a bacterial population. A viable count measures the cells in the culture that are capable of reproducing. Optical density (turbidity), a quantitative measure of light scattering by a liquid culture, increases with the increase in cell number.

(stationary phase) culture and transferred into the same medium, there is usually a lag even if all the cells in the inoculum are alive. This is because the cells are depleted of various essential constituents and time is required for their biosynthesis. A lag also ensues when the inoculum consists of cells that have been damaged (but not killed) by significant temperature shifts, radiation, or toxic chemicals because of the time required for the cells to repair the damage.

A lag is also observed when a microbial population is transferred from a rich culture medium to a poorer one; for example, from a complex medium to a defined medium (⟳ Section 4.2). To grow in any culture medium the cells must have a complete complement of enzymes for synthesis of the essential metabolites not present in that medium. Hence, upon transfer to a medium where essential metabolites must be biosynthesized, time is needed for production of the new enzymes that will carry out these reactions.

Exponential Phase

As we saw in the previous section, during the *exponential phase* of growth each cell divides to form two cells, each of which also divides to form two more cells, and so on, for a brief or extended period, depending on the available resources and other factors. Cells in exponential growth are typically in their healthiest state and hence are most desirable for studies of their enzymes or other cell components.

Rates of exponential growth vary greatly. The rate of exponential growth is influenced by environmental conditions (temperature, composition of the culture medium), as well as by genetic characteristics of the organism itself. In general, prokaryotes grow faster than eukaryotic microorganisms, and small eukaryotes grow faster than large ones. This should remind us of the previously discussed concept of surface-to-volume ratio. Recall that small cells have an increased capacity for nutrient and waste exchange compared with larger cells, and this metabolic advantage can greatly affect their growth and other properties (⟳ Section 3.2).

Stationary Phase

In a batch culture (tube, flask bottle, Petri dish), exponential growth is limited. Consider the fact that a single cell of a bacterium with a 20-min generation time would produce, if allowed to grow exponentially in a batch culture for 48 h, a population of cells that weighed 4000 times the weight of Earth! This is particularly impressive when it is considered that a single bacterial cell weighs only about one-trillionth (10^{-12}) of a gram.

Obviously, this scenario is impossible. Something must happen to limit the growth of the population. Typically, either one or both of two situations limit growth: (1) an essential nutrient of the culture medium is used up, or (2) a waste product of the organism accumulates in the medium and inhibits growth. Either way, exponential growth ceases and the population reaches the *stationary phase.*

In the stationary phase, there is no net increase or decrease in cell number and thus the growth rate of the population is zero. Although the population may not grow during the stationary phase, many cell functions can continue, including energy metabolism and biosynthetic processes. Some cells may even divide during the stationary phase but no net increase in cell number occurs. This is because some cells in the population grow, whereas others die, the two processes balancing each other out. This is a phenomenon called *cryptic growth.*

Death Phase

If incubation continues after a population reaches the stationary phase, the cells may remain alive and continue to metabolize, but they will eventually die. When this occurs, the population enters the *death phase* of the growth cycle. In some cases death is accompanied by actual cell lysis. Figure 5.10 indicates that the death phase of the growth cycle is also an exponential function. Typically, however, the rate of cell death is much slower than the rate of exponential growth.

The phases of bacterial growth shown in Figure 5.10 are reflections of the events in a *population* of cells, not in individual cells. Thus the terms lag phase, exponential phase, stationary phase, and death phase have no meaning with respect to individual cells but only to cell populations. Growth of an individual cell is a necessary prerequisite for population growth. But it is population growth that is most relevant to the ecology of microorganisms, because measurable microbial activities require microbial populations, not just an individual microbial cell.

MiniQuiz
- In what phase of the growth curve in Figure 5.10 are cells dividing in a regular and orderly process?
- Under what conditions does a lag phase not occur?
- Why do cells enter stationary phase?

5.8 Continuous Culture: The Chemostat

Our discussion of population growth thus far has been confined to batch cultures. A batch culture is continually being altered by the metabolic activities of the growing organisms and is therefore a *closed* system. In the early stages of exponential growth in batch cultures, conditions may remain relatively constant. But in later stages, when cell numbers become quite large, the chemical and physical composition of the culture medium changes dramatically.

For many studies in microbiology it is useful to be able to keep cultures under constant conditions for long periods. For example, if one is studying a physiological process such as the synthesis of a particular enzyme, the ready availability of exponentially growing cells may be very convenient. This is only possible with a *continuous culture device.* Unlike a batch culture, a continuous culture is an *open* system. The continuous culture vessel maintains a constant volume to which fresh medium is added at a constant rate while an equal volume of spent culture medium (containing cells) is removed at the same rate. Once such a system is in equilibrium, the chemostat volume, cell number, and nutrient status remain constant, and the system is said to be in *steady state.*

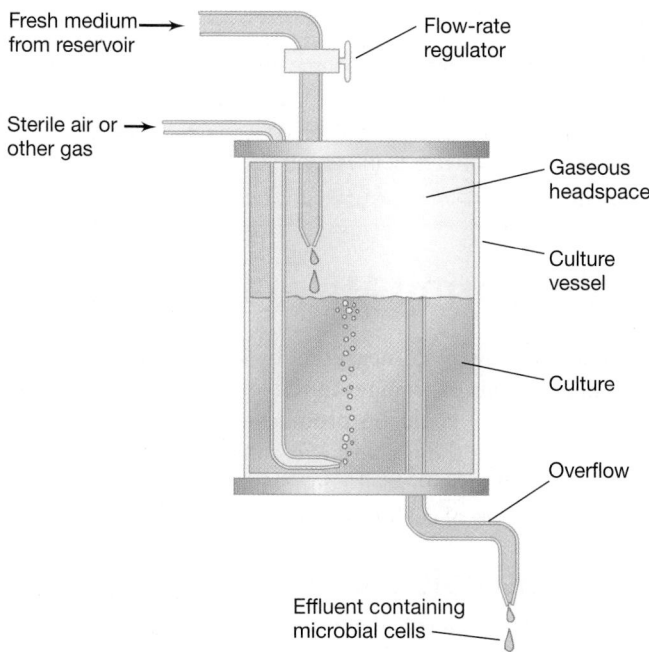

Figure 5.11 Schematic for a continuous culture device (chemostat). The population density is controlled by the concentration of limiting nutrient in the reservoir, and the growth rate is controlled by the flow rate. Both parameters can be set by the experimenter.

The Chemostat

The most common type of continuous culture device is the **chemostat** (**Figure 5.11**). In the chemostat, both growth rate and cell density of the culture can be controlled independently and simultaneously. Two factors govern growth rate and cell density respectively. These are: (1) the *dilution rate*, which is the rate at which fresh medium is pumped in and spent medium is removed; and (2) the *concentration of a limiting nutrient*, such as a carbon or nitrogen source, present in the sterile medium entering the chemostat vessel.

In a batch culture, the nutrient concentration can affect both growth rate and growth yield (**Figure 5.12**). At very low concentrations of a given nutrient, the growth rate is submaximal because the nutrient cannot be transported into the cell fast enough to satisfy metabolic demand. At moderate or higher nutrient levels, however, the growth rate plateaus, but the final cell density may continue to increase in proportion to the concentration of nutrients in the medium up to some fixed limit (Figure 5.12). In a chemostat, by contrast, growth rate and growth yield are controlled independently: The growth rate is set by the dilution rate, while the cell yield (number/milliliter) is controlled by the limiting nutrient. Independent control of these two growth parameters is impossible in a batch culture because it is a closed system where growth conditions are constantly changing with time.

Varying Chemostat Parameters

The effects on bacterial growth of varying the dilution rate and concentration of growth-limiting nutrient in a chemostat are shown in **Figure 5.13**. As seen, there are rather wide limits over

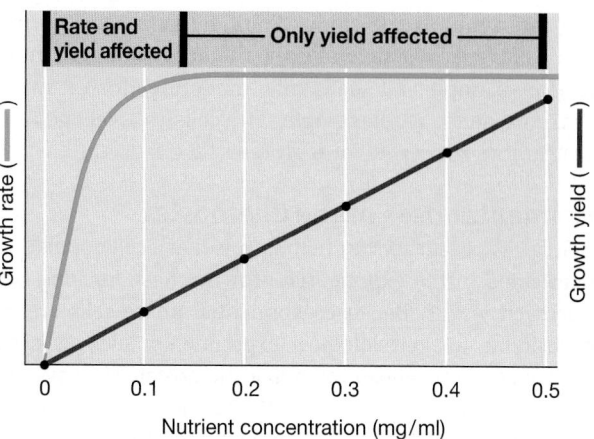

Figure 5.12 The effect of nutrients on growth. Relationship between nutrient concentration, growth rate (green curve), and growth yield (red curve) in a batch culture (closed system). Only at low nutrient concentrations are both growth rate and growth yield affected.

which the dilution rate controls growth rate, although at both very low and very high dilution rates the steady state breaks down. At too high a dilution rate, the organism cannot grow fast enough to keep up with its dilution and is washed out of the chemostat. By contrast, at too low a dilution rate, cells may die from starvation because the limiting nutrient is not being added fast enough to permit maintenance of cell metabolism. However, between these limits, different growth rates can be achieved by simply varying the dilution rate.

Cell density in a chemostat is controlled by a limiting nutrient, just as it is in a batch culture (Figure 5.12). If the concentration of this nutrient in the incoming medium is raised, with the dilution rate remaining constant, the cell density will increase while the

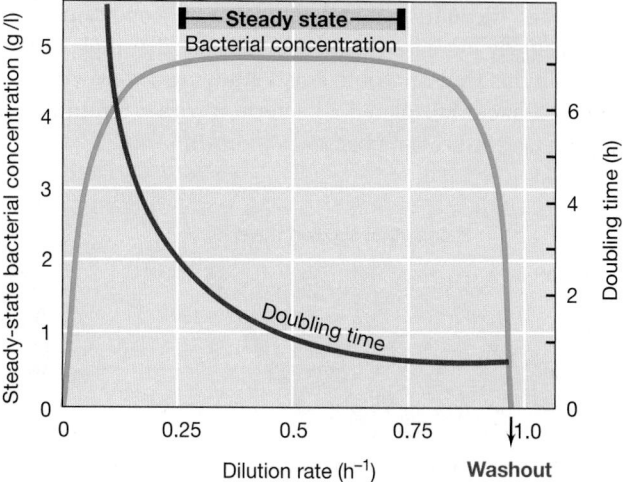

Figure 5.13 Steady-state relationships in the chemostat. The dilution rate is determined from the flow rate and the volume of the culture vessel. Thus, with a vessel of 1000 ml and a flow rate through the vessel of 500 ml/h, the dilution rate would be 0.5 h^{-1}. Note that at high dilution rates, growth cannot balance dilution, and the population washes out. Note also that although the population density remains constant during steady state, the growth rate (doubling time) can vary over a wide range.

growth rate remains the same. Thus, by adjusting the dilution rate and nutrient level accordingly, the experimenter can obtain dilute (for example, 10^5 cells/ml), moderate (for example, 10^7 cells/ml), or dense (for example, 10^9 cells/ml) cell populations growing at low, moderate, or high rates.

Experimental Uses of the Chemostat

A practical advantage to the chemostat is that a cell population may be maintained in the exponential growth phase for long periods, days or even weeks. Because exponential phase cells are usually most desirable for physiological experiments, such cells can be available at any time when grown in a chemostat. Moreover, repetition of experiments can be done with the knowledge that each time the cell population will be as close to being the same as possible. For some applications, such as the study of a particular enzyme, enzyme activities may be significantly lower in stationary phase cells than in exponential phase cells, and thus chemostat-grown cultures are ideal. In practice, after a sample is removed from the chemostat, a period of time is required for the vessel to return to its original volume and for steady state to be reachieved. Once this has occurred, the vessel is ready to be sampled once again.

The chemostat has been used in microbial ecology as well as in microbial physiology. For example, because the chemostat can easily mimic the low substrate concentrations that often prevail in nature, it is possible to prepare mixed or pure bacterial populations in a chemostat and study the competitiveness of different organisms at particular nutrient concentrations. Using these methods together with the powerful tools of phylogenetic stains and gene tracking (Chapters 16 and 23), experimenters can monitor changes in the microbial community in the chemostat as a function of different growth conditions. Such experiments often reveal interactions within the population that are not obvious from growth studies in batch culture.

Chemostats have also been used for enrichment and isolation of bacteria. From a natural sample, one can select a stable population under the nutrient and dilution-rate conditions chosen and then slowly increase the dilution rate until a single organism remains. In this way, microbiologists studying the growth rates of various soil bacteria isolated a bacterium with a 6-min doubling time—the fastest-growing bacterium known!

MiniQuiz

• How do microorganisms in a chemostat differ from microorganisms in a batch culture?

• What happens in a chemostat if the dilution rate exceeds the maximal growth rate of the organism?

• Do pure cultures have to be used in a chemostat?

Measuring Microbial Growth

Population growth is measured by tracking changes in the number of cells or changes in the level of some cellular component. The latter could be protein, nucleic acids, or the dry weight of the cells themselves. We consider here two common measures of cell growth: cell counts and turbidity, the latter of which is a measure of cell mass.

5.9 Microscopic Counts

A total count of microbial numbers can be achieved using a microscope to observe and enumerate the cells present in a culture or natural sample. The method is simple, but the results can be unreliable.

The most common total count method is the microscopic cell count. Microscopic counts can be done on either samples dried on slides or on samples in liquid. Dried samples can be stained to increase contrast between cells and their background (⮌ Section 2.2). With liquid samples, specially designed counting chambers are used. In such a counting chamber, a grid with squares of known area is marked on the surface of a glass slide (**Figure 5.14**). When the coverslip is placed on the chamber, each square on the grid has a precisely measured volume. The number of cells per unit area of grid can be counted under the microscope, giving a measure of the number of cells per small chamber volume. The number of cells per milliliter of suspension is calculated by employing a conversion factor based on the volume of the chamber sample (Figure 5.14).

A second method of enumerating cells in liquid samples is with a flow cytometer. This is a machine that employs a laser beam and complex electronics to count individual cells. Flow cytometry is

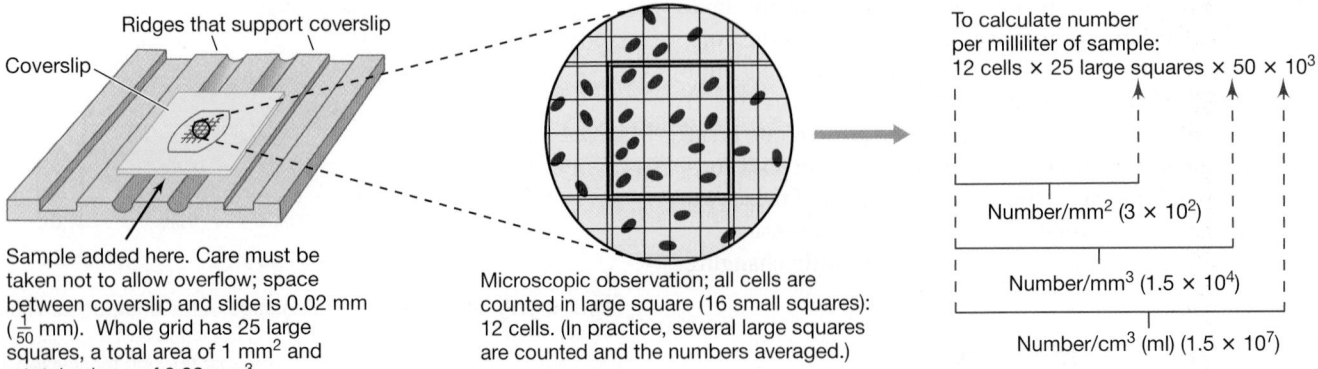

Coverslip

Ridges that support coverslip

Sample added here. Care must be taken not to allow overflow; space between coverslip and slide is 0.02 mm ($\frac{1}{50}$ mm). Whole grid has 25 large squares, a total area of 1 mm^2 and a total volume of 0.02 mm^3.

Microscopic observation; all cells are counted in large square (16 small squares): 12 cells. (In practice, several large squares are counted and the numbers averaged.)

To calculate number per milliliter of sample:
12 cells × 25 large squares × 50 × 10^3

Number/mm^2 (3 × 10^2)

Number/mm^3 (1.5 × 10^4)

Number/cm^3 (ml) (1.5 × 10^7)

Figure 5.14 Direct microscopic counting procedure using the Petroff–Hausser counting chamber. A phase-contrast microscope is typically used to count the cells to avoid the necessity for staining.

rarely used for the routine counting of microbial cells, but has applications in the medical field for counting and differentiating blood cells and other cell types from clinical samples. It has also been used in microbial ecology to separate different types of cells for isolation purposes.

Microscopic counting is a quick and easy way of estimating microbial cell number. However, it has several limitations: (1) Without special staining techniques (⌘ Section 22.3), dead cells cannot be distinguished from live cells. (2) Small cells are difficult to see under the microscope, and some cells are inevitably missed. (3) Precision is difficult to achieve. (4) A phase-contrast microscope is required if the sample is not stained. (5) Cell suspensions of low density (less than about 10^6 cells/milliliter) have few if any bacteria in the microscope field unless a sample is first concentrated and resuspended in a small volume. (6) Motile cells must be immobilized before counting. (7) Debris in the sample may be mistaken for microbial cells.

In microbial ecology, total cell counts are often performed on natural samples using stains to visualize the cells. The stain DAPI (⌘ Section 2.2 and Figure 2.6c) stains all cells in a sample because it reacts with DNA. By contrast, fluorescent stains that are highly specific for certain organisms or groups of related organisms can be prepared by attaching the fluorescent dyes to specific nucleic acid probes. For example, phylogenetic stains that stain only species of *Bacteria* or only species of *Archaea* can be used in combination with nonspecific stains to measure cell numbers of each domain in a given sample; the use of these stains will be discussed in Section 16.9. If cells are present at low densities, for example in a sample of open ocean water, this problem can be overcome by first concentrating cells on a filter and then counting them after staining. Because they are easy to do

and often yield useful information, microscopic cell counts are very common in microbial studies of natural environments. www.microbiologyplace.com **Online Tutorial 5.1: Direct Microscopic Counting Procedure**

MiniQuiz
- What are some of the problems that can arise when unstained preparations are used to make total cell counts of samples from natural environments?

5.10 Viable Counts

A **viable** cell is one that is able to divide and form offspring, and in most cell-counting situations, these are the cells we are most interested in. For these purposes, we can use a viable counting method. To do this, we typically determine the number of cells in a sample capable of forming colonies on a suitable agar medium. For this reason, the viable count is also called a **plate count**. The assumption made in the viable counting procedure is that each viable cell can grow and divide to yield one colony. Thus, colony numbers are a reflection of cell numbers.

There are at least two ways of performing a plate count: the *spread-plate method* and the *pour-plate method* (**Figure 5.15**). In the spread-plate method, a volume (usually 0.1 ml or less) of an appropriately diluted culture is spread over the surface of an agar plate using a sterile glass spreader. The plate is then incubated until colonies appear, and the number of colonies is counted. The surface of the plate must not be too moist because the added liquid must soak in so the cells remain stationary. Volumes greater than about 0.1 ml are avoided in this method because the excess

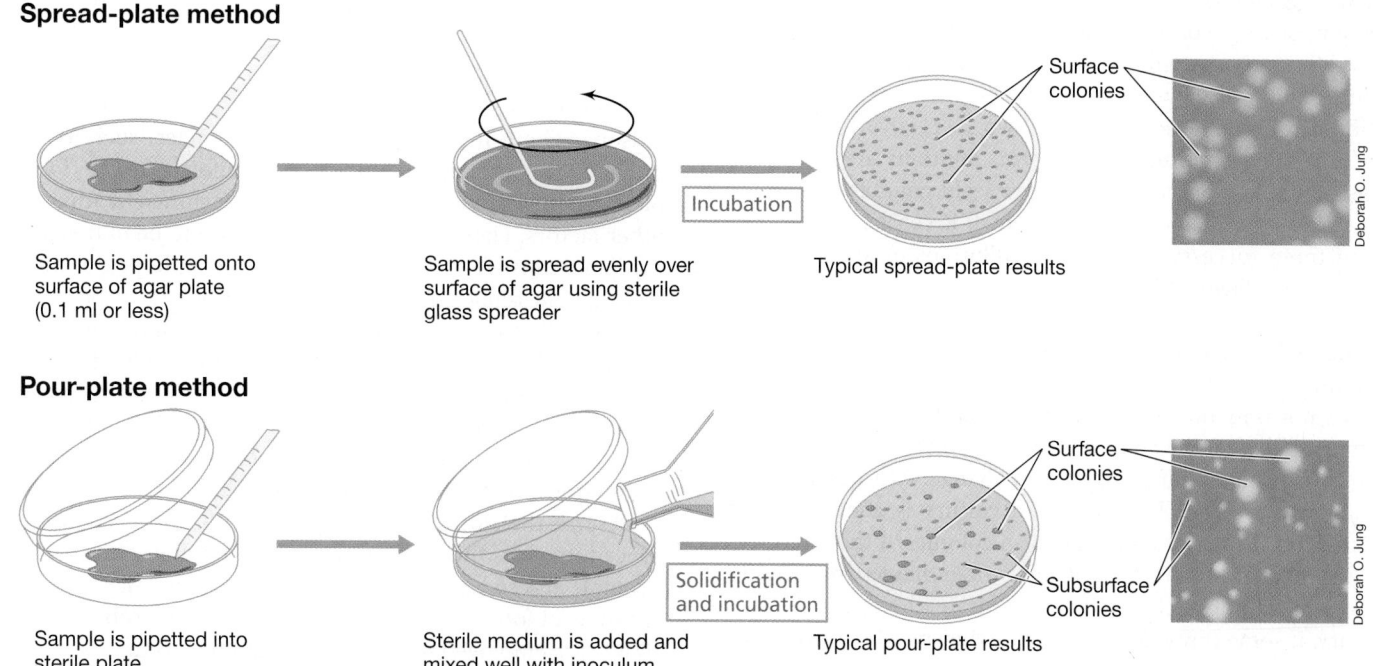

Figure 5.15 Two methods for the viable count. In the pour-plate method, colonies form within the agar as well as on the agar surface. On the far right are photos of colonies of *Escherichia coli* formed from cells plated by the spread-plate method (top) or the pour-plate method (bottom).

liquid does not soak in and may cause the colonies to coalesce as they form, making them difficult to count.

In the pour-plate method (Figure 5.15), a known volume (usually 0.1–1.0 ml) of culture is pipetted into a sterile Petri plate. Melted agar medium, tempered to just about gelling temperature, is then added and mixed well by gently swirling the plate on the benchtop. Because the sample is mixed with the molten agar medium, a larger volume can be used than with the spread plate. However, with this method the organism to be counted must be able to withstand brief exposure to the temperature of molten agar (~45–50°C). Here, colonies form throughout the medium and not just on the agar surface as in the spread-plate method. The plate must therefore be examined closely to make sure all colonies are counted. If the pour-plate method is used to enumerate cells from a natural sample, another problem may arise; any debris in the sample must be distinguishable from actual bacterial colonies or the count will be erroneous.

Diluting Cell Suspensions before Plating

With both the spread-plate and pour-plate methods, it is important that the number of colonies developing on or in the medium not be too many or too few. On crowded plates some cells may not form colonies, and some colonies may fuse, leading to erroneous measurements. If the number of colonies is too small, the statistical significance of the calculated count will be low. The usual practice, which is most valid statistically, is to count colonies only on plates that have between 30 and 300 colonies.

To obtain the appropriate colony number, the sample to be counted must almost always be diluted. Because one may not know the approximate viable count ahead of time, it is usually necessary to make more than one dilution. Several 10-fold dilutions of the sample are commonly used (**Figure 5.16**). To make a 10-fold (10^{-1}) dilution, one can mix 0.5 ml of sample with 4.5 ml of diluent, or 1.0 ml of sample with 9.0 ml of diluent. If a 100-fold (10^{-2}) dilution is needed, 0.05 ml can be mixed with 4.95 ml of diluent, or 0.1 ml with 9.9 ml of diluent. Alternatively, a 10^{-2} dilution can be achieved by making two successive 10-fold dilutions. With dense cultures, such *serial* dilutions are needed to reach a suitable dilution for plating to yield countable colonies. Thus, if a 10^{-6}($1/10^6$) dilution is needed, it can be achieved by making three successive 10^{-2}($1/10^2$) dilutions or six successive 10^{-1} dilutions (Figure 5.16).

Sources of Error in Plate Counting

The number of colonies obtained in a viable count experiment depends not only on the inoculum size and the viability of the culture, but also on the culture medium and the incubation conditions. The colony number can also change with the length of incubation. For example, if a mixed culture is used, the cells deposited on the plate will not all form colonies at the same rate; if a short incubation time is used, fewer than the maximum number of colonies will be obtained. Furthermore, the size of colonies may vary. If some tiny colonies develop, they may be missed during the counting. With pure cultures, colony development is a more synchronous process and uniform colony morphology is the norm.

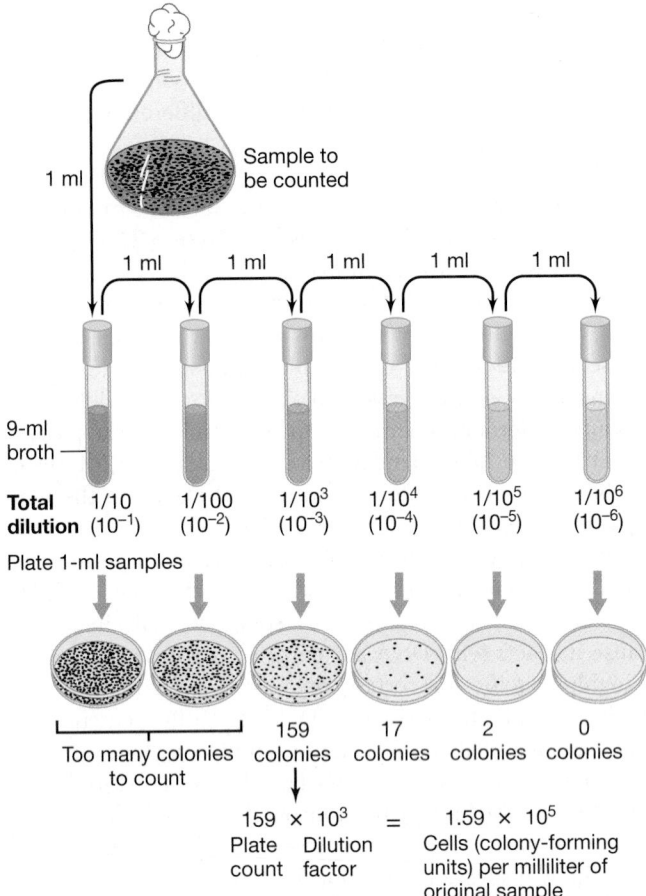

Figure 5.16 Procedure for viable counting using serial dilutions of the sample and the pour-plate method. The sterile liquid used for making dilutions can simply be water, but a solution of mineral salts or actual growth medium may yield a higher recovery. The dilution factor is the reciprocal of the dilution.

Viable counts can be subject to rather large errors for several reasons. These include plating inconsistencies, such as inaccurate pipetting of a liquid sample, a nonuniform sample (for example, a sample containing cell clumps), insufficient mixing, and other factors. Hence, if accurate counts are to be obtained, great care and consistency must be taken in sample preparation and plating, and replicate plates of key dilutions must be prepared. Note also that if two or more cells are in a clump, they will grow to form only a single colony. So if a sample contains many cell clumps, a viable count of that sample may be erroneously low. Data are often expressed as the number of *colony-forming units* obtained rather than the actual number of viable cells, because a colony-forming unit may contain one or more cells.

Despite the difficulties associated with viable counting, the procedure gives the best estimate of the number of viable cells in a sample and so is widely used in many areas of microbiology. For example, in food, dairy, medical, and aquatic microbiology, viable counts are employed routinely. The method has the virtue of high sensitivity, because as few as one viable cell per sample plated can be detected. This feature allows for the sensitive detection of microbial contamination of foods or other materials.

Targeted Plate Counts

The use of highly selective culture media and growth conditions in viable counting procedures allows one to target only particular species, or in some cases even a single species, in a mixed population of microorganisms present in the sample. For example, a complex medium containing 10% NaCl is very useful in isolating species of *Staphylococcus* from skin, because the salt inhibits growth of most other bacteria. In practical applications such as in the food industry, viable counting on both complex and selective media allows for both quantitative and qualitative assessments of the microorganisms present in a food product. That is, with a single sample one medium may be employed for a total count and a second medium used to target a particular organism, such as a specific pathogen. Targeted counting is common in wastewater and other water analyses. For instance, enteric bacteria originate from feces and are easy to target using selective media; if enteric bacteria are detected in a water sample from a swimming site, for example, their presence is a signal that the water is unsafe for human contact.

The Great Plate Count Anomaly

Direct microscopic counts of natural samples typically reveal far more organisms than are recoverable on plates of any single culture medium. Thus, although a very sensitive technique, plate counts can be highly unreliable when used to assess total cell numbers of natural samples, such as soil and water. Some microbiologists have referred to this as "the great plate count anomaly."

Why do plate counts show lower numbers of cells than direct microscopic counts? One obvious factor is that microscopic methods count dead cells, whereas viable methods by definition will not. More important, however, is the fact that different organisms, even those present in a very small natural sample, may have vastly different requirements for nutrients and growth conditions in laboratory culture (↩ Sections 4.1 and 4.2). Thus, one medium and set of growth conditions can at best be expected to support the growth of only a subset of the total microbial community. If this subset makes up, for example, 10^6 cells/g in a total viable community of 10^9 cells/g, the plate count will reveal only 0.1% of the viable cell population, a vast underestimation of the actual number.

Plate count results can thus carry a large caveat. Targeted plate counts using highly selective media, as in, for example, the microbial analysis of sewage or food, can often yield quite reliable data since the physiology of the targeted organisms are known. By contrast, "total" cell counts of the same samples using a single medium and set of growth conditions may be, and usually are, underestimates of actual cell numbers by one to several orders of magnitude.

MiniQuiz

- Why is a viable count more sensitive than a microscopic count? What major assumption is made in relating plate count results to cell number?

- Describe how you would dilute a bacterial culture by 10^{-7}.

- What is the "great plate count anomaly"?

5.11 Turbidimetric Methods

During exponential growth, all cellular components increase in proportion to the increase in cell numbers. Thus, instead of measuring changes in cell number over time, one could instead measure the increase in protein, DNA, or dry weight of a culture as a barometer of growth. However, since cells are actual objects instead of dissolved substances, cells scatter light, and a rapid and quite useful method of estimating cell numbers based on this property is *turbidity*.

A suspension of cells looks cloudy (turbid) to the eye because cells scatter light passing through the suspension. The more cells that are present, the more light is scattered, and hence the more turbid the suspension. What is actually assessed in a turbidimetric measurement is total cell mass. However, because cell mass is proportional to cell number, turbidity can be used as a measure of cell numbers and can also be used to follow an increase in cell numbers of a growing culture.

Optical Density

Turbidity is measured with a spectrophotometer, an instrument that passes light through a cell suspension and measures the unscattered light that emerges; the more cells that are present in the cell suspension, the more turbid it will be (**Figure 5.17**). A spectrophotometer employs a prism or diffraction grating to generate incident light of a specific wavelength (Figure 5.17a). Commonly used wavelengths for bacterial turbidity measurements include 480 nm (blue), 540 nm (green), 600 nm (orange), and 660 nm (red). Sensitivity is best at shorter wavelengths, but measurements of dense cell suspensions are more accurate at longer wavelengths. The unit of turbidity is called *optical density* (*OD*) at the wavelength specified, for example, OD_{540} for spectrophotometric measurements at 540 nm (Figure 5.17). The term *absorbance* (A), for example A_{540}, is also commonly used, but it should be understood that it is light scattering, not absorbance per se, that is being measured in turbidimetric measurements of microbial growth.

Relating OD to Cell Numbers

For unicellular organisms, OD is proportional, within certain limits, to cell number. Turbidity readings can therefore be used as a substitute for total or viable counting methods. However, before this can be done, a standard curve must be prepared that relates cell number (microscopic or viable count), dry weight, or protein content to turbidity. As can be seen in such a plot, proportionality only holds within limits (Figure 5.17c). Thus, at high cell concentrations, light scattered away from the spectrophotometer's photocell by one cell can be scattered back by another. To the photocell, this is as if light had never been scattered in the first place. At such high cell densities, the one-to-one correspondence between cell number and turbidity deviates from linearity, and OD measurements become less accurate. However, up to this limit, turbidity measurements can be accurate measures of cell number or dry weight. Also, because different organisms differ in size and shape, equal cell numbers of two different bacterial species will not necessarily yield the same OD. Thus, to relate OD to actual cell numbers, a standard curve relating these two

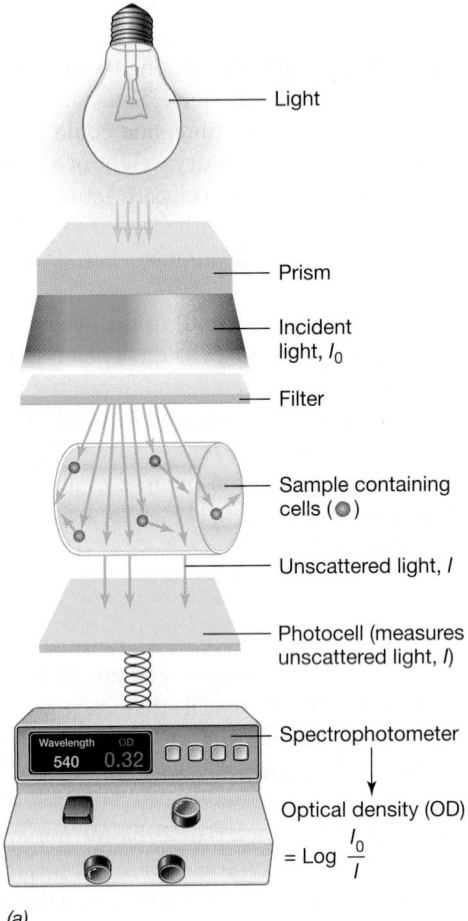

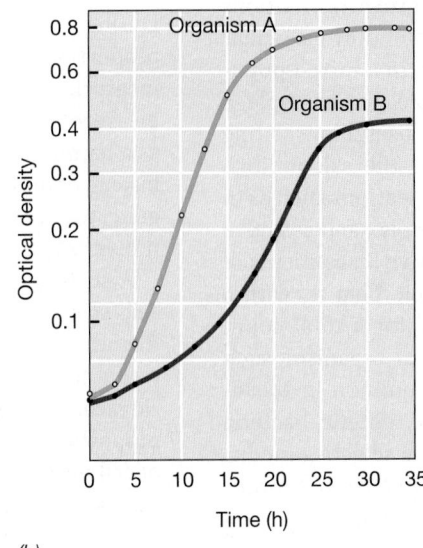

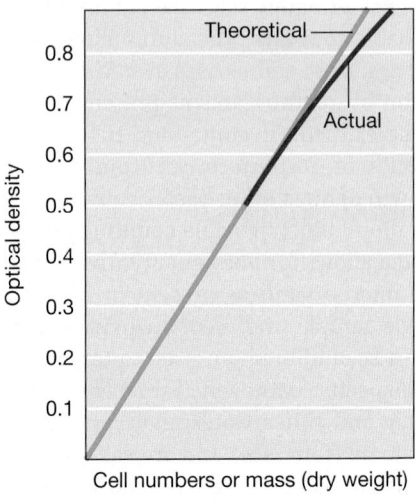

(b)

(c)

Figure 5.17 Turbidity measurements of microbial growth. *(a)* Measurements of turbidity are made in a spectrophotometer. The photocell measures incident light unscattered by cells in suspension and gives readings in optical density units. *(b)* Typical growth curve data for two organisms growing at different growth rates. For practice, calculate the generation time (*g*) of the two cultures using the formula $n = 3.3(\log N - \log N_0)$ where N and N_0 are two different OD readings with a time interval t between the two. Which organism is growing faster, A or B? *(c)* Relationship between cell number or dry weight and turbidity readings. Note that the one-to-one correspondence between these relationships breaks down at high turbidities.

parameters must be made for each different organism grown routinely in the laboratory.

Turbidity measurements have the virtue of being quick and easy to perform. Turbidity measurements can typically be made without destroying or significantly disturbing the sample. For these reasons, turbidity measurements are widely employed to monitor growth of microbial cultures. The same sample can be checked repeatedly and the measurements plotted on a semilogarithmic plot versus time. From these, it is easy to calculate the generation time and other parameters of the growing culture (Figure 5.17*b*).

Turbidity measurements are sometimes problematic. Although many microorganisms grow in even suspensions in liquid medium, many others do not. Some bacteria form small to large clumps, and in such instances, OD measurements may be quite inaccurate as a measure of total microbial mass. In addition, many bacteria grow in films on the sides of tubes or other growth vessels, mimicking in laboratory culture how they actually grow in nature. Thus for ODs to accurately reflect cell mass (and thus cell numbers) in a liquid culture, clumping and biofilms have to be minimized. This can often be accomplished by stirring, shaking, or in some way keeping the cells well mixed during the growth process to prevent the formation of cell aggregates and biofilms.

MiniQuiz
- List two advantages of using turbidity as a measure of cell growth.
- Describe how you could use a turbidity measurement to tell how many colonies you would expect from plating a culture of a given OD.

Ⓥ Temperature and Microbial Growth

The activities of microorganisms including growth are greatly affected by the chemical and physical state of their environment. Many environmental factors can be considered. However, four key factors control the growth of all microorganisms: temperature, pH, water availability, and oxygen; we consider each of these here. Some other factors can potentially affect the growth of microorganisms, such as pressure and radiation. These more specialized environmental factors will be considered later in this book when we encounter microbial habitats in which they play major roles. However, it is important to remember that for the successful culture of any microorganism, both medium and growth conditions must be suitable.

Microbial Growth in Aquatic Systems: Cyanobacterial Blooms

In this chapter we have discussed the effects of different environmental factors on microbial growth. In addition to physical factors such as temperature, nutritional factors—as well as the interaction of these two sets of factors—also influence microbial growth. In the fermentation industry, the positive effects of environmental factors on microbial growth are generally to optimize growth conditions so that a higher biomass concentration and higher productivity can be achieved. We will examine this in more detail in Chapter 15. In the environment, however, the dense growth of certain microorganisms can sometimes be harmful to aquatic life as well as humans, for example, in cases of dense growth of certain phototrophic microorganisms in aquatic areas.

Algae and cyanobacteria play a crucial role in the production of oxygen and in food web dynamics in aquatic systems. These organisms are diverse in their size, morphology, and nutritional requirements. Certain cyanobacteria can grow densely and form blooms in water as a response to eutrophication, the organic enrichment of an ecosystem through the input of organic nutrients in sewage, industrial wastes, and agricultural fertilizers, or high concentrations of key inorganic nutrients, such as phosphorus and nitrogen. Cyanobacterial blooms have been reported in many parts of the world, such as in Lake Victoria in Africa, Lake Taihu in China, and the Barwon-Darling River in Australia.

Cyanobacterial blooms are influenced by a number of environmental factors, such as temperature, pH, solar radiation, stratification, concentration of organic compounds, and water residence time. An elevated nutrient input, especially of phosphorus, is commonly associated with the formation of cyanobacterial blooms and the frequency of bloom events. Cyanobacterial blooms typically occur more densely in surface waters and can form a dark green paint-like layer floating on the water surface, or even a blue-green scum several inches thick. Cyanobacterial blooms increase the turbidity of aquatic ecosystems, alter the abundance of rooted aquatic plants, reduce the abundance of fish associated with weed beds, and decrease the recreational value of affected lakes and reservoirs. The bloom-forming cyanobacteria cannot maintain the abnormally high population for a long period and usually die within a few weeks. If environmental factors for their growth remain favorable, one species may replace another as nutrient conditions change following the first bloom.

The die-off of cyanobacteria is associated with oxygen depletion and fish death in bloom-forming areas. The odorous compounds—geosmin, 2-methylisoborneol, and β-cyclocitral—produced by some cyanobacteria can taint drinking water, making water treatment processes more expensive and challenging. Several species of cyanobacteria also produce toxins called *cyanotoxins,* which have neurotoxic or hepatotoxic effects on fish and may also cause serious illness in humans. The occurrence of cyanobacterial blooms in aquatic areas, therefore, is a threat to both public health and to water quality (Figure 1).

Figure 1 Cyanobacterial blooms in Lake Taihu in China.

5.12 Effect of Temperature on Growth

Temperature is probably *the* most important environmental factor affecting the growth and survival of microorganisms. At either too cold or too hot a temperature, microorganisms will not be able to grow and may even die. The minimum and maximum temperatures for growth vary greatly among different microorganisms and usually reflect the temperature range and average temperature of their habitats.

Cardinal Temperatures

Temperature affects microorganisms in two opposing ways. As temperatures rise, chemical and enzymatic reactions in the cell proceed at more rapid rates and growth becomes faster; however, above a certain temperature, cell components may be irreversibly damaged. Thus, as the temperature is increased within a given range, growth and metabolic function increase up to a point where denaturation reactions set in. Above this point, cell functions fall to zero. For every microorganism there is a *minimum* temperature below which growth is not possible, an *optimum* temperature at which growth is most rapid, and a *maximum* temperature above which growth is not possible (**Figure 5.18**). These three temperatures, called the **cardinal temperatures**, are characteristic for any given microorganism.

The cardinal temperatures of different microorganisms differ widely; some organisms have temperature optima as low as 4°C and some higher than 100°C. The temperature range throughout which microorganisms grow is even wider than this, from below freezing to well above the boiling point of water. However, no single organism can grow over this whole temperature range, as the range for any given organism is typically 25–40 degrees.

The maximum growth temperature of an organism reflects the temperature above which denaturation of one or more essential cell components, such as a key enzyme, occurs. The factors controlling an organism's minimum growth temperature are not as

clear. However, as previously discussed, the cytoplasmic membrane must be in a semifluid state for transport (⮎ Section 3.5) and other important functions to take place. An organism's minimum temperature may well be governed by membrane functioning; that is, if an organism's cytoplasmic membrane stiffens to the point that it no longer functions properly in transport or can no longer develop or consume a proton motive force, the organism cannot grow. The growth temperature *optimum* reflects a state in which all or most cellular components are functioning at their maximum rate and is typically closer to the maximum than to the minimum (see Figure 5.19).

Temperature Classes of Organisms

Although there is a continuum of organisms, from those with very low temperature optima to those with high temperature optima, it is possible to distinguish four classes of microorganisms in relation to their growth temperature optima: **psychrophiles**, with low temperature optima; **mesophiles**, with midrange temperature optima; **thermophiles**, with high temperature optima; and **hyperthermophiles**, with very high temperature optima (**Figure 5.19**).

Mesophiles are widespread in nature. They are found in warm-blooded animals and in terrestrial and aquatic environments in temperate and tropical latitudes. Psychrophiles and thermophiles are found in unusually cold and unusually hot environments, respectively. Hyperthermophiles are found in extremely hot habitats such as hot springs, geysers, and deep-sea hydrothermal vents.

Escherichia coli is a typical mesophile, and its cardinal temperatures have been precisely defined. The optimum temperature for most strains of *E. coli* is near 39°C, the maximum is 48°C, and the minimum is 8°C. Thus, the temperature *range* for *E. coli* is about 40 degrees, near the high end for prokaryotes (Figure 5.19).

We now turn to the interesting cases of microorganisms growing at very low or very high temperatures, some of the physiological problems they face, and some of the biochemical solutions they have evolved to survive under extreme conditions.

MiniQuiz

- How does a hyperthermophile differ from a psychrophile?
- What are the cardinal temperatures for *Escherichia coli*? To what temperature class does it belong?
- *E. coli* can grow at a higher temperature in a complex medium than in a defined medium. Why?

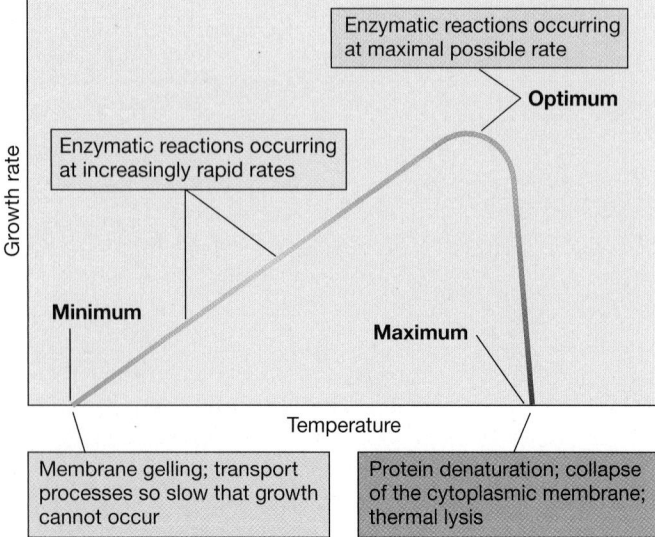

Figure 5.18 The cardinal temperatures: minimum, optimum, and maximum. The actual values may vary greatly for different organisms (see Figure 5.19).

5.13 Microbial Life in the Cold

Because humans live and work on the surface of Earth where temperatures are generally moderate, it is natural to consider very hot and very cold environments as "extreme." However, many microbial habitats are very hot or very cold. The organisms that live in these environments are therefore called **extremophiles** (⮎ Section 2.8 and Table 2.1). Interestingly, in most cases these organisms have evolved to grow *optimally* at their environmental temperature. We consider the biology of these fascinating organisms here and in the next section.

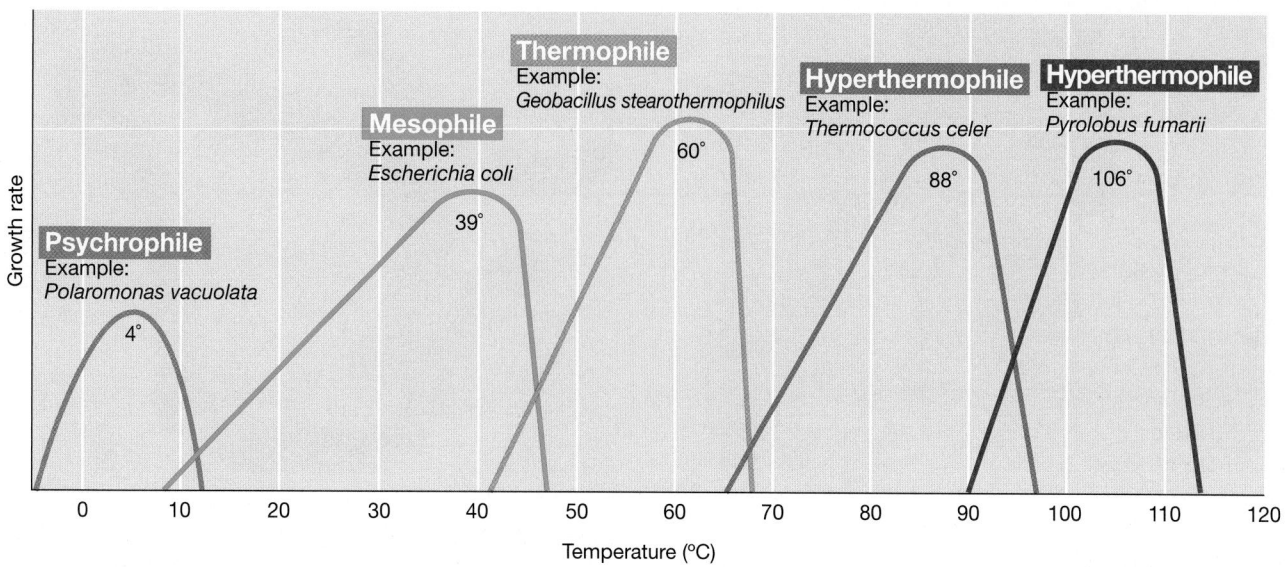

Figure 5.19 Temperature and growth response in different temperature classes of microorganisms. The temperature optimum of each example organism is shown on the graph.

Cold Environments

Much of Earth's surface is cold. The oceans, which make up over half of Earth's surface, have an average temperature of 5°C, and the depths of the open oceans have constant temperatures of 1–3°C. Vast land areas of the Arctic and Antarctic are permanently frozen or are unfrozen for only a few weeks in summer (**Figure 5.20**). These cold environments are not sterile, as viable microorganisms can be found growing at any low-temperature environment in which some liquid water remains. Salts and other solutes, for example, depress the freezing point of water and allow microbial growth to occur below the freezing point of pure water, 0°C. But even in frozen materials there are often small pockets of liquid water where solutes have concentrated and microorganisms can metabolize and grow. Within glaciers, for example, there exists a network of liquid water channels in which prokaryotes thrive and reproduce.

In considering cold environments, it is important to distinguish between environments that are *constantly* cold and those that are only *seasonally* cold. The latter, characteristic of temperate climates, may have summer temperatures as high as 40°C. A temperate lake, for example, may have a period of ice cover in the winter, but the time that the water is at 0°C is relatively brief. Such highly variable environments are less favorable habitats for cold-active microorganisms than are the constantly cold environments characteristic of polar regions, high altitudes, and the depths of the oceans. For example, lakes in the Antarctic McMurdo Dry Valleys contain a permanent ice cover several meters thick (Figure 5.20*d*). The water column below the ice in these lakes remains at 0°C or colder year round and is thus an ideal habitat for cold-active microorganisms.

Psychrophilic Microorganisms

As noted earlier, organisms with low temperature optima are called *psychrophiles.* A psychrophile is defined as an organism with an optimal growth temperature of 15°C or lower, a maximum

growth temperature below 20°C, and a minimal growth temperature at 0°C or lower. Organisms that grow at 0°C but have optima of 20–40°C are called **psychrotolerant**.

Psychrophiles are found in environments that are constantly cold and may be rapidly killed by warming, even to as little as 20°C. For this reason, their laboratory study requires that great care be taken to ensure that they never warm up during sampling, transport to the laboratory, isolation, or other manipulations. In open ocean waters, where temperatures remain constant at about 3°C, various cold-active *Bacteria* and *Archaea* are present, although only a relatively few have been isolated in laboratory culture. Temperate environments, which warm up in summer, cannot support the heat-sensitive psychrophiles because they cannot survive the warming.

Psychrophilic microbial communities containing algae and bacteria grow in dense masses within and under sea ice (frozen seawater that forms seasonally) in polar regions (Figures 5.20*a*, *b*), and are also often present on the surfaces of snowfields and glaciers at such densities that they impart a distinctive coloration to the surface (**Figure 5.21a**). The common snow alga *Chlamydomonas nivalis* is an example of this, its spores responsible for the brilliant red color of the snow surface (Figure 5.21*b*). This green alga grows within the snow as a green-pigmented vegetative cell and then sporulates. As the snow dissipates by melting, erosion, and ablation (evaporation and sublimation), the spores become concentrated on the surface. Related species of snow algae contain different carotenoid pigments, and thus fields of snow algae can also be green, orange, brown, or purple.

In addition to snow algae, several psychrophilic bacteria have been isolated, mostly from marine sediments or sea ice, or from Antarctica. Some of these, particularly isolates from sea ice such as *Polaromonas* (Figure 5.20*c*), have very low growth temperature optima and maxima (4°C and 12°C, respectively). A species of the sea ice bacterium *Psychromonas* grows at −12°C, the lowest temperature for any known bacterium. But even this is

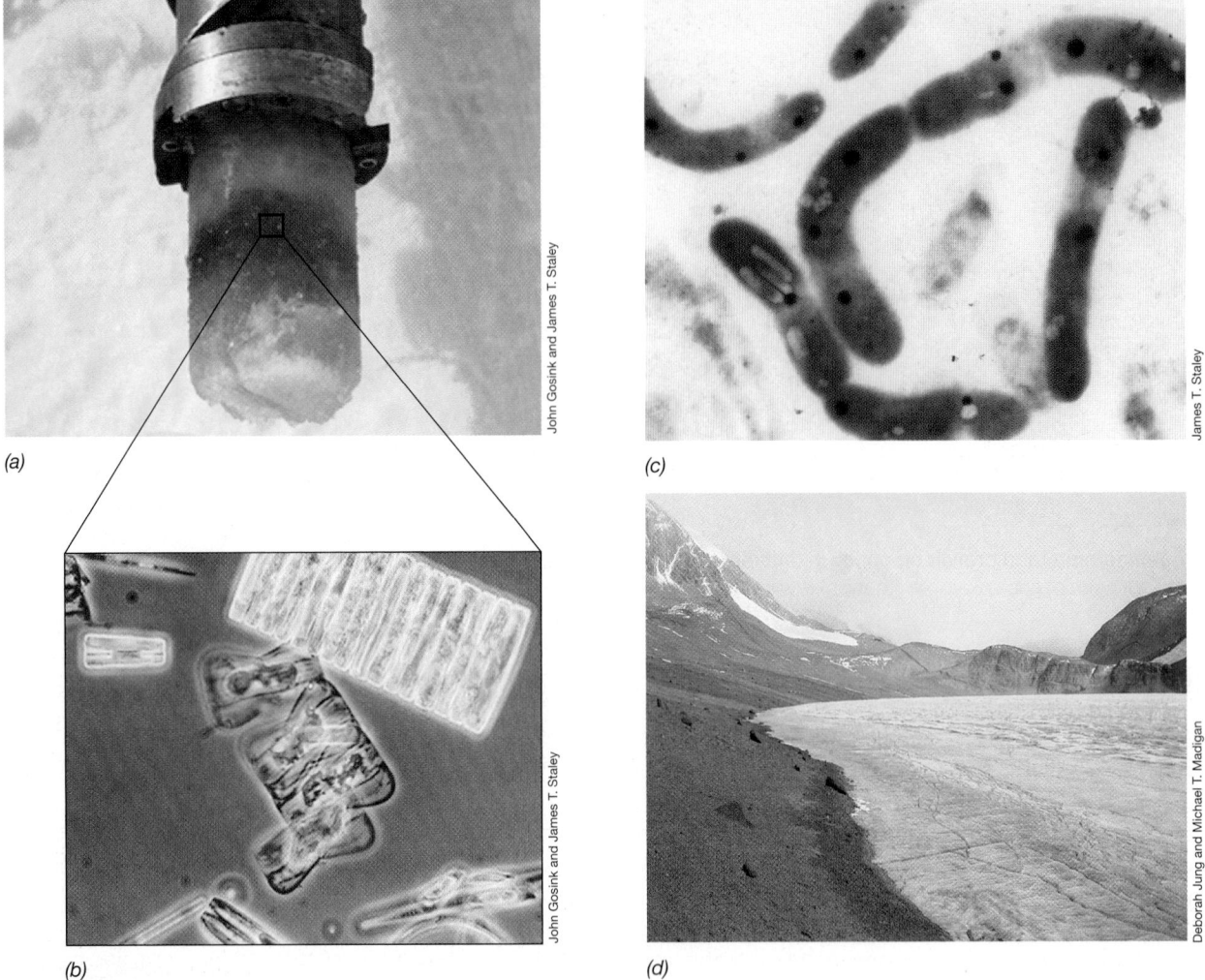

(a)

(b)

(c)

(d)

Figure 5.20 Antarctic microbial habitats and microorganisms. *(a)* A core of frozen seawater from McMurdo Sound, Antarctica. The core is about 8 cm wide. Note the dense coloration due to pigmented microorganisms. *(b)* Phase-contrast micrograph of phototrophic microorganisms from the core shown in part a. Most organisms are either diatoms or green algae (both eukaryotic phototrophs). *(c)* Transmission electron micrograph of *Polaromonas*, a gas vesiculate bacterium that lives in sea ice and grows optimally at 4°C. *(d)* Photo of the surface of Lake Bonney, McMurdo Dry Valleys, Antarctica. Like many other Antarctic lakes, Lake Bonney, which is about 40 m deep, remains permanently frozen and has an ice cover of about 5 m. The water column of Lake Bonney remains near 0°C and contains both oxic and anoxic zones; thus both aerobic and anaerobic microorganisms inhabit the lake. However, no higher eukaryotic organisms inhabit Dry Valley lakes, making them uniquely microbial ecosystems.

unlikely to be the lower temperature limit for bacterial growth, which is probably closer to −20°C. Pockets of liquid water can exist at −20°C, and studies have shown that enzymes from cold-active bacteria can still function under such conditions. Growth rates at such cold temperatures would likely be extremely low, with doubling times of months, or even years. But if an organism can grow, even if only at a very slow rate, it can remain competitive and maintain a population in its habitat.

Psychrotolerant Microorganisms

Psychrotolerant microorganisms are more widely distributed in nature than are psychrophiles and can be isolated from soils and water in temperate climates, as well as from meat, milk and other dairy products, cider, vegetables, and fruit stored at refrigeration temperatures (~4°C). As noted, psychrotolerant microorganisms

grow best at a temperature between 20 and 40°C. Moreover, although psychrotolerant microorganisms do grow at 0°C, most do not grow very well at that temperature, and one must often wait several weeks before visible growth is seen in laboratory cultures. Various *Bacteria*, *Archaea*, and microbial eukaryotes are psychrotolerant.

Molecular Adaptations to Psychrophily

Psychrophiles produce enzymes that function optimally in the cold and that may be denatured or otherwise inactivated at even very moderate temperatures. The molecular basis for this is not entirely understood, but is clearly linked to protein structure. For example, several cold-active enzymes show greater amounts of α-helix and lesser amounts of β-sheet secondary structure (↩ Section 6.21) than do enzymes that are inactive in the cold.

Because β-sheet secondary structures tend to be more rigid than α-helices, the greater α-helix content of cold-active enzymes allows these proteins greater flexibility for catalyzing their reactions at cold temperatures.

Cold-active enzymes also tend to have greater polar and lesser hydrophobic amino acid content than their mesophilic and thermophilic counterparts (↩ Figure 6.29 for structures of amino acids). Moreover, cold-active proteins tend to have lower numbers of weak bonds, such as hydrogen and ionic bonds, and fewer specific interactions between regions (domains) compared with proteins from organisms that grow best at higher temperatures. Collectively, these molecular features probably help these enzymes remain flexible and functional under cold conditions.

Another feature of psychrophiles is that compared to mesophiles, transport processes (↩ Section 3.5) function optimally at low temperature. This is an indication that the cytoplasmic membranes of psychrophiles are structurally modified in such a way that low temperatures do not inhibit membrane functions. Cytoplasmic membranes from psychrophiles tend to have a higher content of unsaturated and shorter-chain fatty acids. This helps the membrane remain in a semifluid state at low temperatures (membranes composed of predominantly saturated or long-chain fatty acids would become stiff and waxlike at low temperatures). In addition, the lipids of some psychrophilic bacteria contain polyunsaturated fatty acids, something very uncommon in prokaryotes. For example, the psychrophilic bacterium *Psychroflexus* contains fatty acids with up to five double bonds. These fatty acids remain more flexible at low temperatures than saturated or monounsaturated fatty acids.

Other molecular adaptations to cold include "cold-shock" proteins and cryoprotectants. Cold-shock proteins are a series of proteins that have several functions including helping the cell maintain other proteins in an active form under cold conditions or binding to specific mRNAs and facilitating their translation. These mRNAs include, in particular, those that encode other cold-functional proteins, most of which are not produced when the cell is growing near its temperature optimum. Cryoprotectants include dedicated antifreeze proteins or specific solutes, such as glycerol or certain sugars that are produced in large amounts at cold temperatures; these agents help prevent the formation of ice crystals that can puncture the cytoplasmic membrane.

Freezing

Although temperatures below −20°C prevent microbial growth, such temperatures, or even much colder ones, do not necessarily cause microbial death. Microbial cells can continue to metabolize at temperatures far beneath that which will support growth. For example, microbial respiration as measured by CO_2 production has been shown in tundra soils at temperatures as low as −39°C. Thus, enzymes continue to function at temperatures far below those that allow for cell growth.

The medium in which cells are suspended also affects their sensitivity to freezing. If cryoprotectants such as glycerol or dimethyl sulfoxide (DMSO) are added to a cell suspension, this depresses the freezing point and prevents ice crystal formation. To freeze cells for long-term preservation, cells are typically suspended in

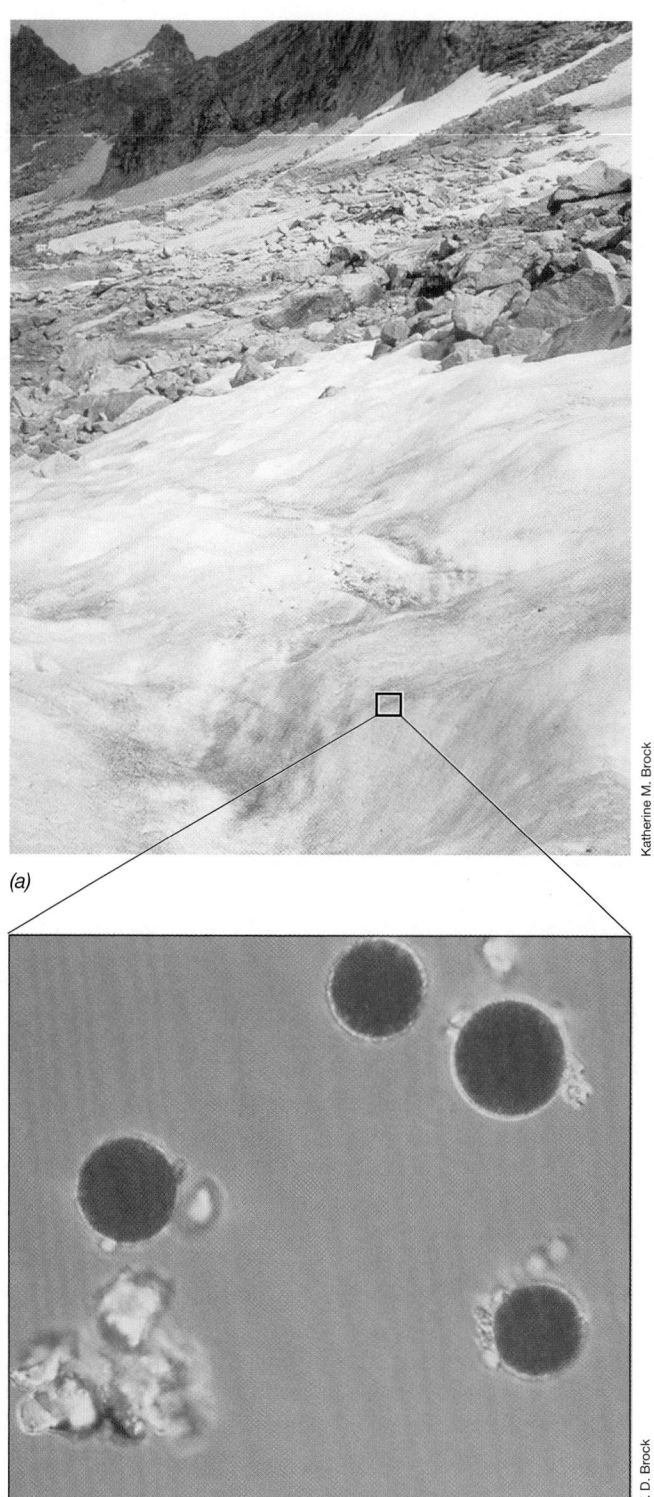

(a)

(b)

Figure 5.21 Snow algae. *(a)* Snow bank in the Sierra Nevada, California, with red coloration caused by the presence of snow algae. Pink snow such as this is common on summer snow banks at high altitudes throughout the world. *(b)* Photomicrograph of red-pigmented spores of the snow alga *Chlamydomonas nivalis*. The spores germinate to yield motile green algal cells. Some strains of snow algae are true psychrophiles but many are psychrotolerant, growing best at temperatures above 20°C. From a phylogenetic standpoint, *C. nivalis* is a green alga, and these organisms are covered in Section 20.20.

growth medium containing 10% DMSO or glycerol and quickly frozen at −80°C (ultracold-freezer temperature) or −196°C (liquid nitrogen temperature). Properly prepared frozen cells that do not thaw and refreeze can remain viable for decades or even longer.

We now travel to the other end of the thermometer and look at microorganisms growing at high temperatures.

MiniQuiz

- How do psychrotolerant organisms differ from psychrophilic organisms?
- What molecular adaptations to cold temperatures are seen in the cytoplasmic membrane of psychrophiles? Why are they necessary?

5.14 Microbial Life at High Temperatures

Microbial life flourishes in high-temperature environments, from sun-heated soils and pools of water to boiling hot springs, and the organisms present are typically highly adapted to their environmental temperature.

Thermal Environments

Organisms whose growth temperature optimum exceeds 45°C are called *thermophiles* and those whose optimum exceeds 80°C are called *hyperthermophiles* (Figure 5.19). Temperatures as high as these are found only in certain areas. For example, the surface of soils subject to full sunlight can be heated to above 50°C at midday, and some surface soils may become warmed to even 70°C. Fermenting materials such as compost piles and silage can also reach temperatures of 70°C. However, the most extensive and extreme high-temperature environments in nature are associated with volcanic phenomena. These include, in particular, hot springs.

Many hot springs have temperatures at or near boiling, and steam vents (fumaroles) may reach 150–500°C. Hydrothermal vents in the bottom of the ocean can have temperatures of 350°C or greater (↩ Section 23.12). Hot springs exist throughout the world, but they are especially abundant in the western United States, New Zealand, Iceland, Japan, Italy, Indonesia, Central America, and central Africa. The largest concentration of hot springs in the world is in Yellowstone National Park, Wyoming (USA).

Although some hot springs vary widely in temperature, many are nearly constant, varying less than 1–2°C over many years. In addition, different springs have different chemical compositions and pH values. Above 65°C, only prokaryotes are present (**Table 5.1**), but the diversity of *Bacteria* and *Archaea* may be extensive.

Hyperthermophiles in Hot Springs

In boiling hot springs (**Figure 5.22**), a variety of hyperthermophiles are typically present, including both chemoorganotrophic and chemolithotrophic species. Growth of natural populations of hyperthermophiles can be studied very simply by immersing a microscope slide into a spring and then retrieving it a few days later; microscopic examination reveals colonies of

Table 5.1 *Presently known upper temperature limits for growth of living organisms*

Group	Upper temperature limits (°C)
Macroorganisms	
Animals	
Fish and other aquatic vertebrates	38
Insects	45–50
Ostracods (crustaceans)	49–50
Plants	
Vascular plants	45 (60 for one species)
Mosses	50
Microorganisms	
Eukaryotic microorganisms	
Protozoa	56
Algae	55–60
Fungi	60–62
Prokaryotes	
Bacteria	
Cyanobacteria	73
Anoxygenic phototrophs	70–73
Chemoorganotrophs/chemolithotrophs	95
Archaea	
Chemoorganotrophs/chemolithotrophs	122

prokaryotes that have developed from single cells that attached to and grew on the glass surface (Figure 5.22b). Scrapings of cell material can then be used for molecular analyses. Such ecological studies of organisms living in boiling springs have shown that growth rates are often rapid; doubling times as short as 1 h have been recorded.

Cultures of many hyperthermophiles have been obtained, and a variety of morphological and physiological types of both *Bacteria* and *Archaea* are known. Phylogenetic studies using ribosomal RNA (rRNA) gene sequencing have shown great evolutionary diversity among these hyperthermophiles as well. Some hyperthermophilic *Archaea* have growth-temperature optima above 100°C, while no species of *Bacteria* are known to grow above 95°C. Growing laboratory cultures of organisms with optima above the boiling point requires pressurized vessels that permit temperatures in the growth medium to rise above 100°C. Such organisms typically originate from undersea hot springs (hydrothermal vents). The most heat-tolerant of all known *Archaea* is *Methanopyrus*, a methanogenic organism capable of growth at 122°C.

Thermophiles

Many thermophiles (optima 45–80°C) are also present in hot springs and other thermal environments. In hot springs, as boiling water overflows the edges of the spring and flows away from the source, it gradually cools, setting up a thermal gradient. Along this gradient, various microorganisms grow, with different species

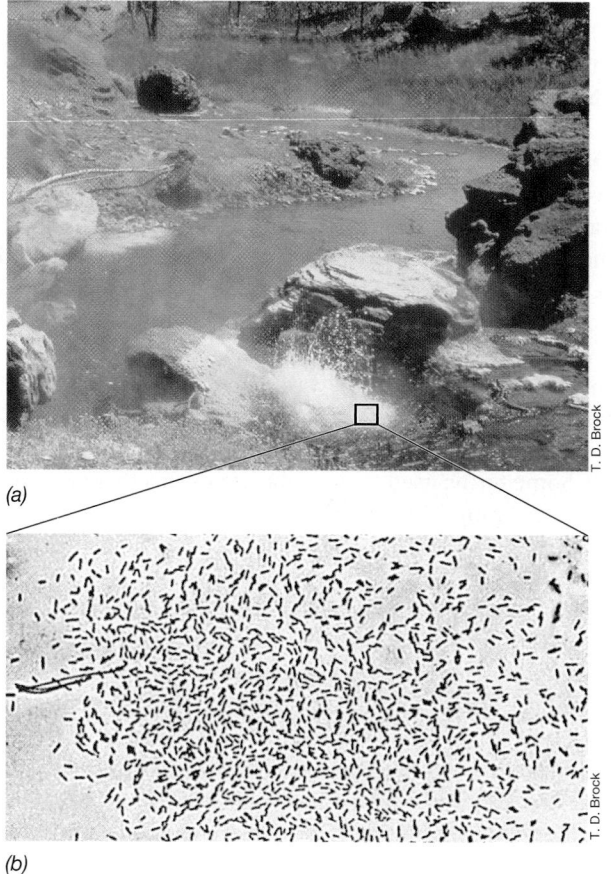

(a)

(b)

Figure 5.22 Growth of hyperthermophiles in boiling water.
(a) Boulder Spring, a small boiling spring in Yellowstone National Park. This spring is superheated, having a temperature 1–2°C above the boiling point. The mineral deposits around the spring consist mainly of silica and sulfur. *(b)* Photomicrograph of a microcolony of prokaryotes that developed on a microscope slide immersed in such a boiling spring.

Figure 5.23 Growth of thermophilic cyanobacteria in a hot spring in Yellowstone National Park. Characteristic V-shaped pattern (shown by the dashed white lines) formed by cyanobacteria at the upper temperature for phototrophic life, 70–74°C, in the thermal gradient formed from a boiling hot spring. The pattern develops because the water cools more rapidly at the edges than in the center of the channel. The spring flows from the back of the picture toward the foreground. The light-green color is from a high-temperature strain of the cyanobacterium *Synechococcus*. As water flows down the gradient, the density of cells increases, less thermophilic strains enter, and the color becomes more intensely green.

growing in the different temperature ranges (**Figure 5.23**). By studying the species distribution along such thermal gradients and by examining hot springs and other thermal habitats at different temperatures around the world, it has been possible to determine the upper temperature limits for each type of organism (Table 5.1). From this information we can conclude that (1) prokaryotic organisms are able to grow at far higher temperatures than are eukaryotes, (2) the most thermophilic of all prokaryotes are certain species of *Archaea*, and (3) nonphototrophic organisms can grow at higher temperatures than can phototrophic organisms.

Thermophilic prokaryotes have also been found in artificial thermal environments, such as hot water heaters. The domestic or industrial hot water heater has a temperature of 60–80°C and is therefore a favorable habitat for the growth of thermophilic prokaryotes. Organisms resembling *Thermus aquaticus*, a common hot spring thermophile, have been isolated from domestic and industrial hot water heaters. Electric power plants, hot water discharges, and other artificial thermal sources also provide sites where thermophiles can grow. Many of these organisms can be readily isolated using complex media incubated at the temperature of the habitat from which the sample originated.

Protein Stability at High Temperatures

How do thermophiles and hyperthermophiles survive at high temperature? First, their enzymes and other proteins are much more heat-stable than are those of mesophiles and actually function *optimally* at high temperatures. How is heat stability achieved? Amazingly, studies of several heat-stable enzymes have shown that they often differ very little in amino acid sequence from heat-sensitive forms of the enzymes that catalyze the same reaction in mesophiles. It appears that critical amino acid substitutions at only a few locations in the enzyme allow the protein to fold in such a way that it is heat-stable.

Heat stability of proteins in hyperthermophiles is also bolstered by an increased number of ionic bonds between basic and acidic amino acids and their often highly hydrophobic interiors; the latter property is a natural resistance to unfolding in an aqueous cytoplasm. Finally, solutes such as di-inositol phosphate, diglycerol phosphate, and mannosylglycerate are produced at high levels in certain hyperthermophiles, and these may also help stabilize their proteins against thermal degradation.

Membrane Stability at High Temperatures

In addition to enzymes and other macromolecules in the cell, the cytoplasmic membranes of thermophiles and hyperthermophiles must be heat-stable. We mentioned earlier that psychrophiles have membrane lipids rich in unsaturated fatty acids, making the membranes semifluid and functional at low temperatures. Conversely, thermophiles typically have lipids rich in saturated fatty acids. This feature allows the membranes to remain stable and functional at high temperatures. Saturated fatty acids form a stronger hydrophobic environment than do unsaturated fatty acids, which helps account for membrane stability.

Hyperthermophiles, most of which are *Archaea*, do not contain fatty acids in their membranes but instead have C_{40} hydrocarbons composed of repeating units of isoprene ( Figures 3.6*c* and 3.7*b*) bonded by ether linkage to glycerol phosphate. In addition, however, the architecture of the cytoplasmic membranes of hyperthermophiles takes a unique twist: The membrane forms a lipid *monolayer* rather than a lipid *bilayer* ( Figure 3.7*e*). This structure prevents the membrane from melting (peeling apart) at the high growth temperatures of hyperthermophiles. We consider other aspects of heat stability in hyperthermophiles, including that of DNA stability, in Chapter 19.

Thermophily and Biotechnology

Thermophiles and hyperthermophiles are interesting for more than just basic biological reasons. These organisms offer some major advantages for industrial and biotechnological processes, many of which can be run more rapidly and efficiently at high temperatures. For example, enzymes from thermophiles and hyperthermophiles are widely used in industrial microbiology. Such enzymes can catalyze biochemical reactions at high temperatures and are in general more stable than enzymes from mesophiles, thus prolonging the shelf life of purified enzyme preparations.

A classic example of a heat-stable enzyme of great importance to biology is the DNA polymerase isolated from *T. aquaticus*. *Taq polymerase*, as this enzyme is known, has been used to automate the repetitive steps in the polymerase chain reaction (PCR) technique ( Section 6.11), an extremely important tool for biology. Several other uses of heat-stable enzymes and other heat-stable cell products are also known or are being developed for industrial applications.

MiniQuiz

- Which domain of prokaryotes includes species with optima of >100°C? What special techniques are required to culture them?
- What is the structure of membranes of hyperthermophilic *Archaea*, and why might this structure be useful for growth at high temperature?
- What is *Taq* polymerase and why is it important?

(V) Other Environmental Factors Affecting Growth

Temperature has a major effect on the growth of microorganisms. But many other factors do as well, chief among these being pH, osmolarity, and oxygen.

5.15 Acidity and Alkalinity

Acidity or alkalinity of a solution is expressed by its **pH** on a scale on which neutrality is pH 7 (**Figure 5.24**). pH values less than 7 are *acidic* and those greater than 7 are *alkaline*. It is important to remember that pH is a logarithmic function—a change of one pH unit corresponds to a 10-fold change in hydrogen ion (H^+) concentration. Thus, vinegar (pH near 2) and household ammonia (pH near 11) differ in hydrogen ion concentration by a billionfold.

Every microorganism has a pH range within which growth is possible and typically shows a well-defined growth pH optimum. Most organisms show a growth range of 2–3 pH units. Most natural environments have a pH between 4 and 9, and organisms with optima in this range are most commonly encountered. Only a few species can grow at pH values of lower than 3 or greater than 9. Some terms used to describe organisms that grow best in particular pH ranges are shown in **Table 5.2**.

Acidophiles

Organisms that grow optimally at a pH value in the range termed *circumneutral* (pH 5.5 to 7.9) are called **neutrophiles** (Table 5.2). By contrast, organisms that grow best below pH 5.5 are called **acidophiles**. There are different classes of acidophiles, some growing best at moderately acidic pH and others at very low pH. Many fungi and bacteria grow best at pH 5 or even below, while a more restricted number grow best below pH 3, including in particular the genus *Acidithiobacillus*. An even more restricted group grow best below pH 2 and those with pH optima below 1 are

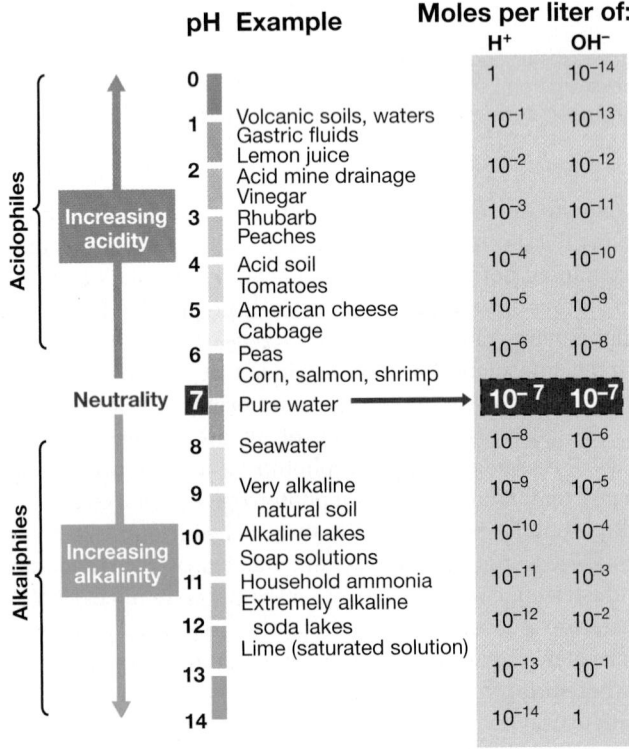

Figure 5.24 The pH scale. Although some microorganisms can live at very low or very high pH, the cell's internal pH remains near neutrality.

Table 5.2 *Relationships of microorganisms to pH*

Physiological class (optima range)	Approximate pH optimum for growth	Example organism[a]
Neutrophile (pH >5.5 and <8)	7	*Escherichia coli*
Acidophile (pH <5.5)	5	*Rhodopila globiformis*
	3	*Acidithiobacillus ferrooxidans*
	1	*Picrophilus oshimae*
Alkaliphile (pH ≥8)	8	*Chloroflexus aurantiacus*
	9	*Bacillus firmus*
	10	*Natronobacterium gregoryi*

[a] *Picrophilus* and *Natronobacterium* are *Archaea*; all others are *Bacteria*.

extremely rare. Most acidophiles cannot grow at pH 7 and many cannot grow at greater than two pH units above their optimum.

A critical factor governing acidophily is the stability of the cytoplasmic membrane. When the pH is raised to neutrality, the cytoplasmic membranes of strongly acidophilic bacteria are destroyed and the cells lyse. This indicates that these organisms are not just acid-*tolerant* but that high concentrations of hydrogen ions are actually *required* for membrane stability. For example, the most acidophilic prokaryote known is *Picrophilus oshimae*, a species of *Archaea* that grows optimally at pH 0.7 and 60°C (the organism is also a thermophile). Above pH 4, cells of *P. oshimae* spontaneously lyse. As one would expect, *P. oshimae* inhabits extremely acidic thermal soils associated with volcanic activity.

Alkaliphiles

A few extremophiles have very high pH optima for growth, sometimes as high as pH 10, and some of these can still grow at even higher pH. Microorganisms showing growth pH optima of 8 or higher are called **alkaliphiles**. Alkaliphilic microorganisms are typically found in highly alkaline habitats, such as soda lakes and high-carbonate soils. The most well-studied alkaliphilic prokaryotes are certain *Bacillus* species, such as *Bacillus firmus*. This organism is alkaliphilic, but has an unusually broad pH range for growth, from 7.5 to 11. Some extremely alkaliphilic bacteria are also halophilic (salt-loving), and most of these are *Archaea* (⟳ Section 19.2). Some phototrophic purple bacteria (⟳ Section 17.2) are strongly alkaliphilic. Certain alkaliphiles have industrial uses because they produce hydrolytic enzymes, such as proteases and lipases, which are excreted from the cell and thus function well at alkaline pH. These enzymes are produced on a large scale and added as supplements to laundry detergents.

Alkaliphiles are of basic interest for several reasons but particularly because of the bioenergetic problems they face living at such high pH. For example, imagine trying to generate a proton motive force (⟳ Section 4.10) when the external surface of your cytoplasmic membrane is so alkaline. Some strategies for this are known. In *B. firmus* a sodium (Na^+) motive force rather than a proton motive force drives transport reactions and motility. Remarkably, however, a proton motive force drives ATP synthesis in cells of *B. firmus*, even though the external membrane surface is

awash in hydroxyl ions (OH^-). It is thought that H^+ are in some way kept very near the outer surface of the cytoplasmic membrane such that they cannot combine with OH^- to form water.

Internal Cell pH

The optimal pH for growth of any organism is a measure of the pH of the *extracellular* environment only. The *intracellular* pH must remain relatively close to neutrality to prevent destruction of macromolecules in the cell. For the majority of microorganisms whose pH optimum for growth is between pH 6 and 8, organisms called *neutrophiles*, the cytoplasm remains neutral or very nearly so. However, in acidophiles and alkaliphiles the internal pH can vary from neutrality. For example, in the previously mentioned acidophile *P. oshimae*, the internal pH has been measured at pH 4.6, and in extreme alkaliphiles an intracellular pH as high as 9.5 has been measured. If these are not the lower and upper limits of cytoplasmic pH, respectively, they are extremely close to the limits. This is because DNA is acid-labile and RNA is alkaline-labile; if a cell cannot maintain these key macromolecules in a stable state, it obviously cannot survive.

Buffers

In batch cultures, the pH can change during growth as the result of metabolic reactions of microorganisms that consume or produce acidic or basic substances. Thus, *buffers* are frequently added to microbial culture media to keep the pH relatively constant. However, a given buffer works over only a narrow pH range. Hence, different buffers must be used at different pH values.

For near neutral pH ranges, potassium phosphate (KH_2PO_4) and calcium carbonate ($CaCO_3$) are good buffers. Many other buffers for use in microbial growth media or for the assay of enzymes extracted from microbial cells are available, and the best buffering system for one organism or enzyme may be considerably different from that for another. Thus, the optimal buffer for use in a particular situation must usually be determined empirically. For assaying enzymes *in vitro*, though, a buffer that works well in an assay of the enzyme from one organism will usually work well for assaying the same enzyme from other organisms.

MiniQuiz
- What is the increase in concentration of H^+ when going from pH 7 to pH 3?
- What terms are used to describe organisms whose growth pH optimum is either very high or very low?

5.16 Osmotic Effects

Water is the solvent of life, and water availability is an important factor affecting the growth of microorganisms. Water availability not only depends on the absolute water content of an environment, that is, how moist or dry it is, but it is also a function of the concentration of solutes such as salts, sugars, or other substances that are dissolved in the water. Dissolved substances have an affinity for water, which makes the water associated with solutes less available to organisms.

Table 5.3 *Water activity of several substances*

Water activity (a_w)	Material	Example organisms[a]
1.000	Pure water	*Caulobacter, Spirillum*
0.995	Human blood	*Streptococcus, Escherichia*
0.980	Seawater	*Pseudomonas, Vibrio*
0.950	Bread	Most gram-positive rods
0.900	Maple syrup, ham	Gram-positive cocci such as *Staphylococcus*
0.850	Salami	*Saccharomyces rouxii* (yeast)
0.800	Fruit cake, jams	*Saccharomyces bailii, Penicillium* (fungus)
0.750	Salt lakes, salted fish	*Halobacterium, Halococcus*
0.700	Cereals, candy, dried fruit	*Xeromyces bisporus* and other xerophilic fungi

[a]Selected examples of prokaryotes or fungi capable of growth in culture media adjusted to the stated water activity.

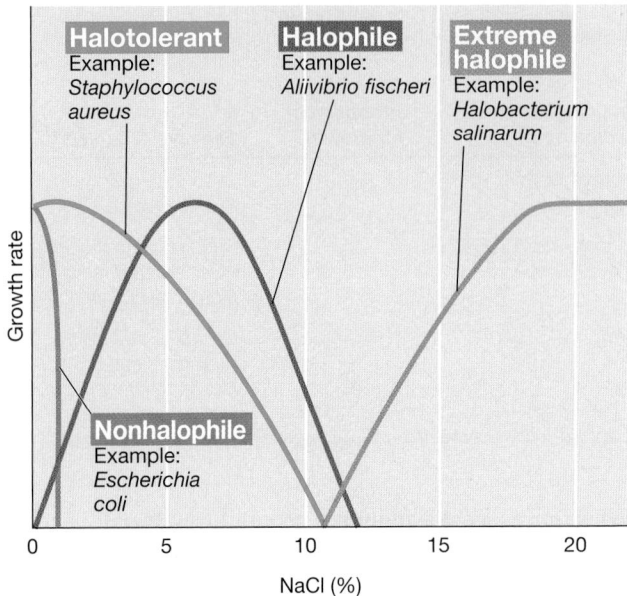

Figure 5.25 **Effect of sodium chloride (NaCl) concentration on growth of microorganisms of different salt tolerances or requirements.** The optimum NaCl concentration for marine microorganisms such as *Aliivibrio fischeri* is about 3%; for extreme halophiles, it is between 15 and 30%, depending on the organism.

Water Activity and Osmosis

Water availability is expressed in physical terms as **water activity**. Water activity, abbreviated a_w, is defined as the ratio of the vapor pressure of the air in equilibrium with a substance or solution to the vapor pressure of pure water. Thus, values of a_w vary between 0 and 1; some representative values are given in **Table 5.3**. Water activities in agricultural soils generally range between 0.90 and 1.

Water diffuses from regions of high water concentration (low solute concentration) to regions of lower water concentration (higher solute concentration) in the process of osmosis. The cytoplasm of a cell typically has a higher solute concentration than the environment, so water tends to diffuse into the cell. Under such conditions, the cell is said to be in *positive water balance*. However, when a cell finds itself in an environment where the solute concentration exceeds that of the cytoplasm, water will flow out of the cell. This can cause serious problems if a cell has no way to counteract it because a dehydrated cell cannot grow.

Halophiles and Related Organisms

In nature, osmotic effects are of interest mainly in habitats with high concentrations of salts. Seawater contains about 3% sodium chloride (NaCl) plus small amounts of many other minerals and elements. Marine microorganisms usually have a specific requirement for NaCl and grow optimally at the water activity of seawater (**Figure 5.25**). Such organisms are called **halophiles**. By definition, halophiles require at least some NaCl for growth, but the optimum varies with the organism and its habitat. For example, marine organisms typically grow best with 1–4% NaCl, organisms from hypersaline environments (environments that are more salty than seawater), 3–12%, and organisms from extremely hypersaline environments require even higher levels of NaCl. And the growth requirement for NaCl cannot be replaced by KCl, meaning that halophiles have an absolute requirement for Na+.

Most microorganisms are unable to cope with environments of very low water activity and either die or become dehydrated and dormant under such conditions. **Halotolerant** organisms can tolerate some reduction in the a_w of their environment, but grow best in the absence of the added solute (Figure 5.25). By contrast, some organisms thrive and indeed require low water activity for growth. These organisms are of interest not only from the standpoint of their adaptation to life under these conditions, but also from an applied standpoint, for example, in the food industry, where solutes such as salt and sucrose are commonly used as preservatives to inhibit microbial growth.

Organisms capable of growth in very salty environments are called **extreme halophiles** (Figure 5.25). These organisms require 15–30% NaCl, depending on the species, for optimum growth. Organisms able to live in environments high in sugar as a solute are called **osmophiles**, and those able to grow in very dry environments (made dry by lack of water rather than from dissolved solutes) are called **xerophiles**. Examples of these various classes of organisms are given in **Table 5.4**.

Compatible Solutes

When an organism grows in a medium with a low water activity, it can obtain water from its environment only by increasing its internal solute concentration and driving water in by osmosis. The internal solute concentration can be raised by either pumping solutes into the cell from the environment or by synthesizing a solute. Many organisms are known that employ one or the other of these strategies, and several examples are given in Table 5.4.

The solute used inside the cell for adjustment of cytoplasmic water activity must be noninhibitory to macromolecules within the cell. Such compounds are called **compatible solutes**, and

Table 5.4 *Compatible solutes of microorganisms*

Organism	Major cytoplasmic solute(s)	Minimum a_w for growth	
Nonphototrophic *Bacteria*/freshwater cyanobacteria	Amino acids (mainly glutamate or proline[a])/sucrose, trehalose[b]	0.98–0.90	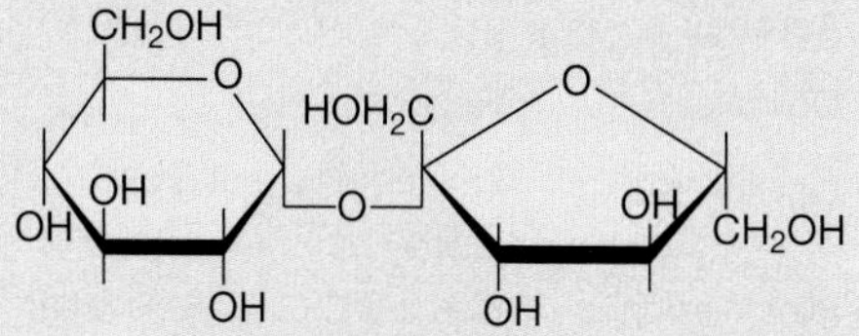
Marine cyanobacteria	α-Glucosylglycerol[b]	0.92	
Marine algae	Mannitol,[b] various glycosides, dimethylsulfoniopropionate	0.92	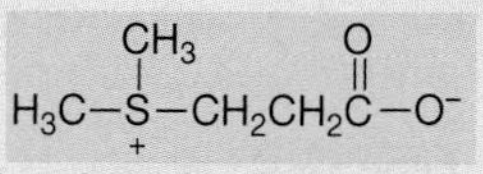
Salt lake cyanobacteria	Glycine betaine	0.90–0.75	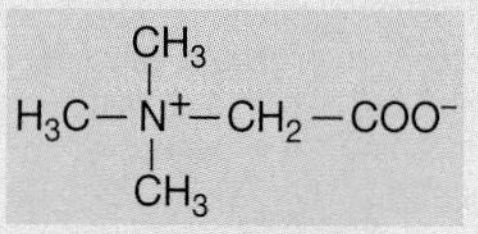
Halophilic anoxygenic phototrophic purple *Bacteria*	Glycine betaine, ectoine, trehalose[b]	0.90–0.75	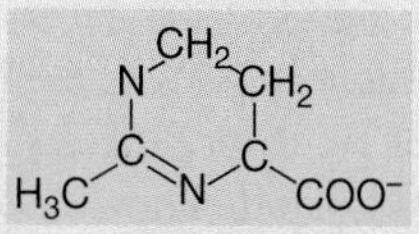
Extremely halophilic *Archaea* and some *Bacteria*	KCl	0.75	
Dunaliella (halophilic green alga)	Glycerol	0.75	
Xerophilic and osmophilic yeasts	Glycerol	0.83–0.62	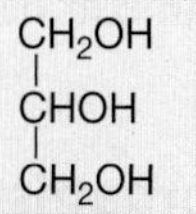
Xerophilic filamentous fungi	Glycerol	0.72–0.61	

Sucrose

Dimethylsulfoniopropionate

Glycine betaine

Ectoine

Glycerol

[a] See Figure 6.29 for the structures of amino acids.
[b] Structures not shown. Like sucrose, trehalose is a C_{12} disaccharide; glucosylglycerol is a C_9 alcohol; mannitol is a C_6 alcohol.

several such solutes are known. These substances are typically highly water-soluble molecules, such as sugars, alcohols, or amino acid derivatives (Table 5.4). The compatible solute of extremely halophilic *Archaea*, such as *Halobacterium*, and a very few extremely halophilic *Bacteria*, is KCl ($\rightleftarrows$ Section 19.2).

The concentration of compatible solute in a cell is a function of the level of solutes present in its environment; however, in any given organism the maximal amount of compatible solute is a genetically directed characteristic. As a result, different organisms can tolerate different ranges of water potential (Tables 5.3 and 5.4). Nonhalotolerant, halotolerant, halophilic, and extremely halophilic microorganisms (Figure 5.25) are to a major extent defined by their genetic capacity to produce or accumulate compatible solutes.

Gram-positive cocci of the genus *Staphylococcus* are notoriously halotolerant (in fact, a common isolation procedure for them is to use media containing 7.5–10% NaCl), and these organisms use the amino acid proline as a compatible solute. Glycine betaine is an analog of the amino acid glycine in which the hydrogen atoms on the amino group are replaced by methyl groups. This places a positive charge on the N atom and greatly increases

solubility. Glycine betaine is widely distributed as a compatible solute among halophilic phototrophic bacteria, as is ectoine, a cyclic derivative of the amino acid aspartate (Table 5.4). Other common compatible solutes include various sugars and dimethylsulfoniopropionate produced by marine algae, and glycerol produced by several organisms including xerophilic fungi that grow at the lowest water potential of all known organisms (Table 5.4).

MiniQuiz

• What is the a_w of pure water?

• What are compatible solutes, and when and why are they needed by the cell? What is the compatible solute of *Halobacterium*?

5.17 Oxygen and Microorganisms

Because animals require molecular oxygen (O_2), it is easy to assume that all organisms require O_2. However, this is not true; many microorganisms can, and some must, live in the total absence of oxygen.

Table 5.5 *Oxygen relationships of microorganisms*

Group	Relationship to O_2	Type of metabolism	Example[a]	Habitat[b]
Aerobes				
Obligate	Required	Aerobic respiration	*Micrococcus luteus* (B)	Skin, dust
Facultative	Not required, but growth better with O_2	Aerobic respiration, anaerobic respiration, fermentation	*Escherichia coli* (B)	Mammalian large intestine
Microaerophilic	Required but at levels lower than atmospheric	Aerobic respiration	*Spirillum volutans* (B)	Lake water
Anaerobes				
Aerotolerant	Not required, and growth no better when O_2 present	Fermentation	*Streptococcus pyogenes* (B)	Upper respiratory tract
Obligate	Harmful or lethal	Fermentation or anaerobic respiration	*Methanobacterium formicicum* (A)	Sewage sludge, anoxic lake sediments

[a]Letters in parentheses indicate phylogenetic status (B, *Bacteria*; A, *Archaea*). Representative of either domain of prokaryotes are known in each category. Most eukaryotes are obligate aerobes, but facultative aerobes (for example, yeast) and obligate anaerobes (for example, certain protozoa and fungi) are known.
[b]Listed are typical habitats of the example organism.

Oxygen is poorly soluble in water, and because of the constant respiratory activities of microorganisms in aquatic habitats, O_2 can quickly become exhausted. Thus, *anoxic* (O_2-free) microbial habitats are common in nature and include muds and other sediments, bogs, marshes, water-logged soils, intestinal tracts of animals, sewage sludge, the deep subsurface of Earth, and many other environments. In these anoxic habitats, microorganisms, particularly prokaryotes, thrive.

Oxygen Classes of Microorganisms

Microorganisms vary in their need for, or tolerance of, O_2. In fact, microorganisms can be grouped according to their relationship with O_2, as outlined in **Table 5.5**. **Aerobes** can grow at full oxygen tensions (air is 21% O_2) and respire O_2 in their metabolism. Many aerobes can even tolerate elevated concentrations of oxygen (hyperbaric oxygen). **Microaerophiles**, by contrast, are aerobes that can use O_2 only when it is present at levels reduced from that in air (microoxic conditions). This is because of their limited capacity to respire or because they contain some O_2-sensitive molecule such as an O_2-labile enzyme. Many aerobes are **facultative**, meaning that under the appropriate nutrient and culture conditions they can grow under either oxic or anoxic conditions.

Some organisms cannot respire oxygen; such organisms are called **anaerobes**. There are two kinds of anaerobes: **aerotolerant anaerobes**, which can tolerate O_2 and grow in its presence even though they cannot use it, and **obligate anaerobes**, which are inhibited or even killed by O_2 (Table 5.5). The reason obligate anaerobes are killed by O_2 is unknown, but it is likely because they are unable to detoxify some of the products of O_2 metabolism (Section 5.18).

So far as is known, obligate anaerobiosis is found in only three groups of microorganisms: a wide variety of *Bacteria* and *Archaea*, a few fungi, and a few protozoa. The best-known group of obligately anaerobic *Bacteria* belongs to the genus *Clostridium*, a group of gram-positive endospore-forming rods. Clostridia are widespread in soil, lake sediments, and the intestinal tracts of warm-blooded animals, and are often responsible for spoilage of canned foods. Other obligately anaerobic organisms are the methanogens and many other *Archaea*, the sulfate-reducing and acetogenic bacteria, and many of the bacteria that inhabit the animal gut and oral cavity. Among obligate anaerobes, however, the sensitivity to O_2 varies greatly. Some species can tolerate traces of O_2 or even full exposure to O_2, whereas others cannot.

Culture Techniques for Aerobes and Anaerobes

For the growth of many aerobes, it is necessary to provide extensive aeration. This is because the O_2 that is consumed by the organisms during growth is not replaced fast enough by simple diffusion from the air. Therefore, forced aeration of liquid cultures is needed and can be achieved by either vigorously shaking the flask or tube on a shaker or by bubbling sterilized air into the medium through a fine glass tube or porous glass disc. Aerobes typically grow better with forced aeration than with O_2 supplied only by diffusion.

For the culture of anaerobes, the problem is not to provide O_2, but to exclude it. Obligate anaerobes vary in their sensitivity to O_2, and procedures are available for reducing the O_2 content of cultures. Some of these techniques are simple and suitable mainly for less O_2-sensitive organisms; others are more complex, but necessary for growth of obligate anaerobes. Bottles or tubes filled completely to the top with culture medium and provided with tightly fitting stoppers provide suitably anoxic conditions for organisms that are not overly sensitive to small amounts of O_2. A chemical called a *reducing agent* may be added to culture media; the reducing agent reacts with oxygen and reduces it to water (H_2O). An example is thioglycolate, which is added to thioglycolate broth, a medium commonly used to test an organism's requirements for O_2 (**Figure 5.26**).

Thioglycolate broth is a complex medium containing a small amount of agar, making the medium viscous but still fluid. After thioglycolate reacts with O_2 throughout the tube, O_2 can penetrate

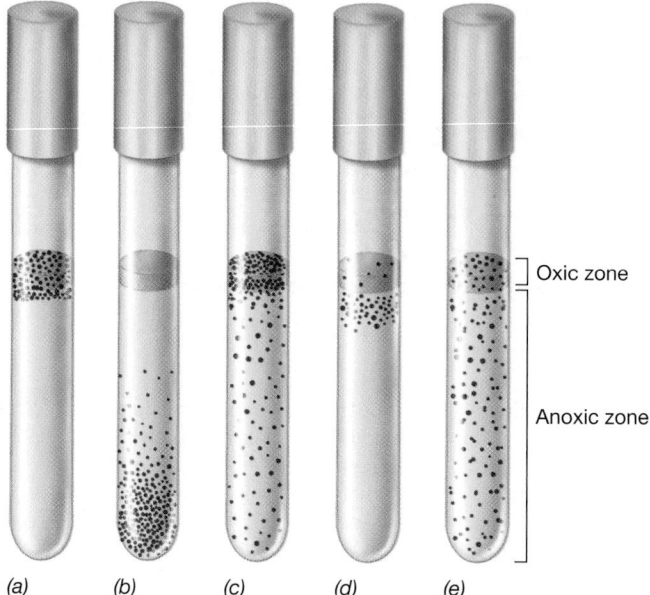

(a)   (b)   (c)   (d)   (e)

Oxic zone

Anoxic zone

Figure 5.26 Growth versus oxygen (O₂) concentration. From left to right, aerobic, anaerobic, facultative, microaerophilic, and aerotolerant anaerobe growth, as revealed by the position of microbial colonies (depicted here as black dots) within tubes of thioglycolate broth culture medium. A small amount of agar has been added to keep the liquid from becoming disturbed. The redox dye, resazurin, which is pink when oxidized and colorless when reduced, has been added as a redox indicator. (a) O₂ penetrates only a short distance into the tube, so obligate aerobes grow only close to the surface. (b) Anaerobes, being sensitive to O₂, grow only away from the surface. (c) Facultative aerobes are able to grow in either the presence or the absence of O₂ and thus grow throughout the tube. However, growth is better near the surface because these organisms can respire. (d) Microaerophiles grow away from the most oxic zone. (e) Aerotolerant anaerobes grow throughout the tube. Growth is not better near the surface because these organisms can only ferment.

only near the top of the tube where the medium contacts air. Obligate aerobes grow only at the top of such tubes. Facultative organisms grow throughout the tube but grow best near the top. Microaerophiles grow near the top but not right at the top. Anaerobes grow only near the bottom of the tube, where O_2 cannot penetrate. The redox indicator dye *resazurin* is added to the medium to differentiate oxic from anoxic regions; the dye is pink when oxidized and colorless when reduced and so gives a visual assessment of the degree of penetration of O_2 into the medium (Figure 5.26).

To remove all traces of O_2 for the culture of strict anaerobes, one can place an oxygen-consuming system in a jar holding the tubes or plates. One of the simplest devices for this is an anoxic jar, a glass or gas-impermeable plastic jar fitted with a gastight seal within which tubes, plates, or other containers are placed for incubation (**Figure 5.27a**). The air in the jar is replaced with a mixture of H_2 and CO_2, and in the presence of a palladium catalyst, the traces of O_2 left in the jar and culture medium are consumed in the formation of water ($H_2 + O_2 \rightarrow H_2O$), eventually leading to anoxic conditions.

For obligate anaerobes it is usually necessary to not only remove all traces of O_2, but also to carry out all manipulations of cultures in a completely anoxic atmosphere. Strict anaerobes can be killed by even a brief exposure to O_2. In these cases, a culture medium is first boiled to render it O_2-free, then a reducing agent such as H_2S is added, and the mixture is sealed under an O_2-free gas. All manipulations are carried out under a jet of sterile O_2-free H_2 or N_2 that is directed into the culture vessel when it is open, thus driving out any O_2 that might enter. For extensive research on anaerobes, special devices called *anoxic glove bags* permit work with open cultures in completely anoxic atmospheres (Figure 5.27b).

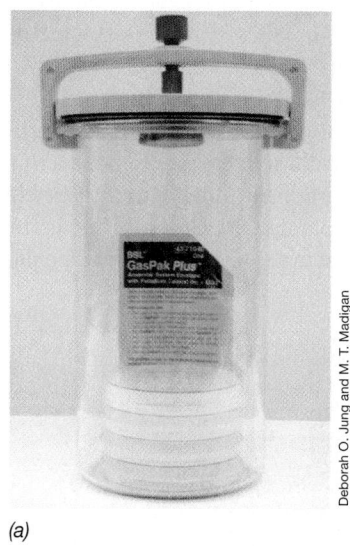

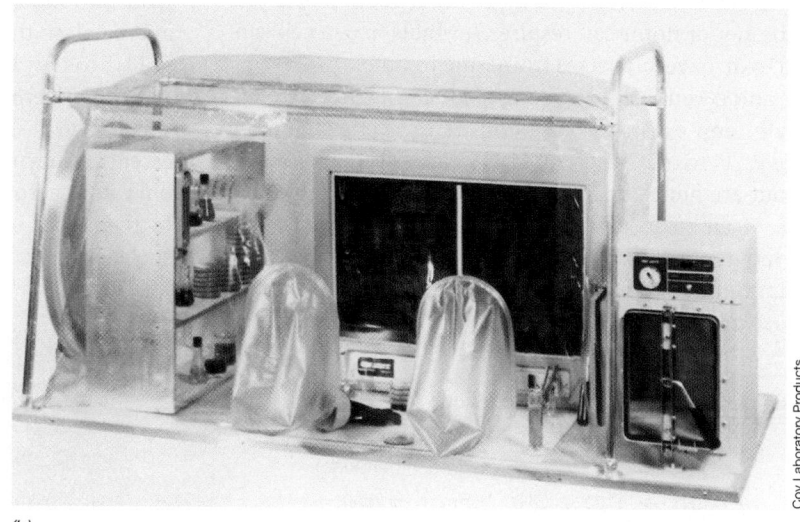

(a)   (b)

Deborah O. Jung and M. T. Madigan

Coy Laboratory Products

Figure 5.27 Incubation under anoxic conditions. (a) Anoxic jar. A chemical reaction in the envelope in the jar generates $H_2 + CO_2$. The H_2 reacts with O_2 in the jar on the surface of a palladium catalyst to yield H_2O; the final atmosphere contains N_2, H_2, and CO_2. (b) Anoxic glove bag for manipulating and incubating cultures under anoxic conditions. The airlock on the right, which can be evacuated and filled with O_2-free gas, serves as a port for adding and removing materials to and from the glove bag.

UNIT 2

MiniQuiz

- How does an obligate aerobe differ from a facultative aerobe?
- How does a reducing agent work? Give an example of a reducing agent.

5.18 Toxic Forms of Oxygen

O_2 is a powerful oxidant and the best electron acceptor for respiration. But O_2 can also be a poison to obligate anaerobes. Why? It turns out that O_2 itself is not poisonous, but instead it is toxic derivatives of oxygen that can damage cells that are not prepared to deal with them. We consider this topic here.

Oxygen Chemistry

Oxygen in its ground state is called *triplet* oxygen ($^3O_2$). However, other electronic configurations of oxygen are possible, and most are toxic to cells. One major form of toxic oxygen is *singlet* oxygen ($^1O_2$), a higher-energy form of oxygen in which outer shell electrons surrounding the nucleus become highly reactive and can carry out spontaneous and undesirable oxidations within the cell. Singlet oxygen is produced both photochemically and biochemically, the latter through the activity of various peroxidase enzymes. Organisms that frequently encounter singlet oxygen, such as airborne bacteria and phototrophic microorganisms, often contain colored pigments called *carotenoids*, which function to convert singlet oxygen to nontoxic forms.

Superoxide and Other Toxic Oxygen Species

Besides singlet oxygen, many other toxic forms of oxygen exist, including *superoxide anion* (O_2^-), *hydrogen peroxide* (H_2O_2), and *hydroxyl radical* (OH•). All of these are produced as by-products of the reduction of O_2 to H_2O in respiration (**Figure 5.28**). Flavoproteins, quinones, and iron–sulfur proteins (Section 4.9), found in virtually all cells, can also catalyze the reduction of O_2 to O_2^-. Thus, whether or not it can respire O_2 (Table 5.5), a cell can be exposed to toxic oxygen species from time to time.

Superoxide anion and OH• are strong oxidizing agents and can oxidize virtually any organic compound in the cell, including macromolecules. Peroxides such as H_2O_2 can also damage cell components but are not as toxic as O_2^- or OH•. The latter is the most reactive of all toxic oxygen species but is transient and quickly removed in other reactions. Later we will see that certain cells of the immune system make toxic oxygen species for the specific purpose of killing microbial invaders (Section 29.1).

Reactants	Products	
$O_2 + e^- \rightarrow$	O_2^-	(superoxide)
$O_2^- + e^- + 2\,H^+ \rightarrow$	H_2O_2	(hydrogen peroxide)
$H_2O_2 + e^- + H^+ \rightarrow$	H_2O + OH•	(hydroxyl radical)
OH• $+ e^- + H^+ \rightarrow$	H_2O	(water)

Outcome:
$O_2 + 4\,e^- + 4\,H^+ \rightarrow 2\,H_2O$

Figure 5.28 Four-electron reduction of O_2 to H_2O by stepwise addition of electrons. All the intermediates formed are reactive and toxic to cells, except for water, of course.

$H_2O_2 + H_2O_2 \rightarrow 2\,H_2O + O_2$

(a) **Catalase**

H_2O_2 + **NADH** $+ H^+ \rightarrow 2\,H_2O + NAD^+$

(b) **Peroxidase**

$O_2^- + O_2^- + 2\,H^+ \rightarrow H_2O_2 + O_2$

(c) **Superoxide dismutase**

$4\,O_2^- + 4\,H^+ \rightarrow 2\,H_2O + 3\,O_2$

(d) **Superoxide dismutase/catalase in combination**

$O_2^- + 2\,H^+ + \text{rubredoxin}_{\text{reduced}} \rightarrow H_2O_2 + \text{rubredoxin}_{\text{oxidized}}$

(e) **Superoxide reductase**

Figure 5.29 Enzymes that destroy toxic oxygen species. *(a)* Catalases and *(b)* peroxidases are porphyrin-containing proteins, although some flavoproteins may consume toxic oxygen species as well. *(c)* Superoxide dismutases are metal-containing proteins, the metals being copper and zinc, manganese, or iron. *(d)* Combined reaction of superoxide dismutase and catalase. *(e)* Superoxide reductase catalyzes the one-electron reduction of O_2^- to H_2O_2.

Superoxide Dismutase and Other Enzymes That Destroy Toxic Oxygen

With so many toxic oxygen derivatives to deal with, it is not surprising that organisms have evolved enzymes that destroy these compounds (**Figure 5.29**). Superoxide and H_2O_2 are the most common toxic oxygen species, so enzymes that destroy these compounds are widely distributed. The enzyme *catalase* attacks H_2O_2, forming O_2 and H_2O; its activity is illustrated in Figure 5.29*a* and **Figure 5.30**. Another enzyme that destroys H_2O_2 is *peroxidase* (Figure 5.29*b*), which differs from catalase in that it requires a reductant, usually NADH, for activity and produces only H_2O as a product. Superoxide is destroyed by the enzyme *superoxide dismutase*, an enzyme that generates H_2O_2 and O_2 from two molecules of O_2^- (Figure 5.29*c*). Superoxide dismutase and catalase thus work in concert to bring about the conversion of O_2^- to O_2 plus H_2O (Figure 5.29*d*).

Cells of aerobes and facultative aerobes typically contain both superoxide dismutase and catalase. Superoxide dismutase is an essential enzyme for aerobes, and the absence of this enzyme in obligate anaerobes was originally thought to explain why O_2 is toxic to them (but see next paragraph). Some aerotolerant anaerobes, such as the lactic acid bacteria, also lack superoxide dismutase, but

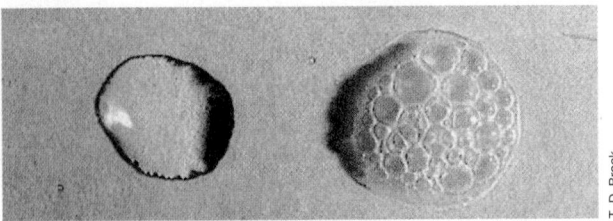

Figure 5.30 Method for testing a microbial culture for the presence of catalase. A heavy loopful of cells from an agar culture was mixed on a slide (right) with a drop of 30% hydrogen peroxide. The immediate appearance of bubbles is indicative of the presence of catalase. The bubbles are O_2 produced by the reaction $H_2O_2 + H_2O_2 \rightarrow 2\,H_2O + O_2$.

they use protein-free manganese (Mn^{2+}) complexes to carry out the dismutation of O_2^- to H_2O_2 and O_2. Such a system is not as efficient as superoxide dismutase, but may have functioned as a primitive form of this enzyme in ancient anaerobic organisms faced with O_2 for the first time when cyanobacteria first appeared on Earth.

Superoxide Reductase

Another means of superoxide disposal is present in certain obligately anaerobic *Archaea*. In the hyperthermophile *Pyrococcus furiosus*, for example, superoxide dismutase is absent, but a unique enzyme, *superoxide reductase*, is present and functions to remove O_2^-. However, unlike superoxide dismutase, superoxide reductase reduces O_2^- to H_2O_2 without the production of O_2 (Figure 5.29*e*), thus avoiding exposure of the organism to O_2. The electron donor for superoxide reductase activity is rubredoxin, an iron–sulfur protein with low reduction potential. *P. furiosus* also lacks catalase, an enzyme that, like superoxide dismutase, also generates O_2 (Figure 5.29*a*). Instead, the H_2O_2 produced by superoxide reductase is removed by the activity of peroxidase-like enzymes that yield H_2O as a final product (Figure 5.29*b*).

Superoxide reductases are present in many other obligate anaerobes as well, such as sulfate-reducing bacteria (*Bacteria*) and methanogens (*Archaea*), as well as in certain microaerophilic species of *Bacteria*, such as *Treponema*. Thus these organisms, previously thought to be O_2-sensitive because they lacked superoxide dismutase, can indeed consume superoxide. The sensitivity of these organisms to O_2 may therefore be for entirely different and as yet unknown reasons.

Many obligately anaerobic hyperthermophiles such as *Pyrococcus* inhabit deep-sea hydrothermal vents (Section 23.12) but are quite tolerant of cold, oxic conditions. Although they do not grow under these conditions, superoxide reductase presumably prevents their killing when they are exposed to O_2. It is thought that O_2 tolerance may be an important factor enabling transport of these organisms in fully oxic ocean water from one deep-sea hydrothermal system to another.

MiniQuiz

- How does superoxide dismutase protect a cell from toxic oxygen?
- How does the activity of superoxide dismutase differ from that of superoxide reductase?

Big Ideas

5.1

Microbial growth is defined as an increase in cell numbers and is the final result of the doubling of all cell components prior to the actual division event that yields two daughter cells. Most microorganisms grow by binary fission.

5.2

Cell division and chromosome replication are coordinately regulated, and the Fts proteins are keys to these processes. With the help of MinE, FtsZ defines the cell division plane and helps assemble the divisome, the protein complex that orchestrates cell division.

5.3

MreB protein helps define cell shape, and in rod-shaped cells, MreB forms a cytoskeletal coil that directs cell wall synthesis along the long axis of the cell. The protein crescentin plays an analogous role in *Caulobacter*, leading to formation of a curved cell. Shape and cell division proteins in eukaryotes have prokaryotic counterparts.

5.4

During bacterial growth new cell wall material is synthesized by the insertion of new glycan tetrapeptide units into preexisting wall material. Bactoprenol facilitates transport of these units through the cytoplasmic membrane. Transpeptidation completes the process of cell wall synthesis by cross-linking adjacent ribbons of peptidoglycan at muramic acid residues.

5.5

Microbial populations show a characteristic type of growth pattern called exponential growth. A plot of the logarithm of cell numbers versus time is called a semilogarithmic plot and can be used to derive the doubling time of the exponentially growing population.

5.6

From knowledge of the initial and final cell numbers and the time of exponential growth, the generation time and growth rate constant of a cell population can be calculated directly. Key parameters here are *n*, the number of generations; *t*, time; and *g*, generation time. The generation time is expressed as $g = t/n$.

5.7

Microorganisms show a characteristic growth pattern when inoculated into a fresh culture medium. There is usually a lag phase and then growth commences in an exponential fashion. As essential nutrients are depleted or toxic products build up, growth ceases and the population enters the stationary phase. If incubation continues, cells may begin to die.

5.8

The chemostat is an open system used to maintain cell populations in exponential growth for extended periods. In a chemostat, the rate at which a culture is diluted with fresh growth medium controls the doubling time of the population, while the cell density (cells/ml) is controlled by the concentration of a growth-limiting nutrient dissolved in the fresh medium.

5.9
Cell counts can be done under the microscope using special counting chambers. Microscopic counts measure the total number of cells in the sample and are very useful for assessing a microbial habitat for total cell numbers. Certain stains can be used to target specific cell populations in a sample.

5.10
Viable cell counts (plate counts) measure only the living population present in the sample with the assumption that each colony originates from the growth and division of a single cell. Depending on how they are used, plate counts can be fairly accurate assessments or highly unreliable.

5.11
Turbidity measurements are an indirect but very rapid and useful method of measuring microbial growth. However, in order to relate a turbidity value to a direct cell number, a standard curve plotting these two parameters against one another must first be established.

5.12
Temperature is a major environmental factor controlling microbial growth. An organism's cardinal temperatures describe the minimum, optimum, and maximum temperatures at which it grows and can differ dramatically from one organism to the next. Microorganisms can be grouped by their cardinal temperature as psychrophiles, mesophiles, thermophiles, and hyperthermophiles.

5.13
Organisms with cold temperature optima are called psychrophiles, and the most extreme representatives inhabit constantly cold environments. Psychrophiles have evolved macromolecules that function best at cold temperatures, but that can be unusually sensitive to warm temperatures.

5.14
Organisms with growth temperature optima between 45 and 80°C are called thermophiles and those with optima greater than 80°C are called hyperthermophiles. These organisms inhabit hot environments that can have temperatures even above 100°C. Thermophiles and hyperthermophiles produce heat-stable macromolecules.

5.15
The acidity or alkalinity of an environment can greatly affect microbial growth. Some organisms grow best at low or high pH (acidophiles and alkaliphiles, respectively), but most organisms grow best between pH 5.5 and 8. The internal pH of a cell must stay relatively close to neutral to prevent nucleic acid destruction.

5.16
The water activity of an aqueous environment is controlled by the dissolved solute concentration. To survive in high-solute environments, organisms produce or accumulate compatible solutes to maintain the cell in positive water balance. Some microorganisms grow best at reduced water potential and some even require high levels of salts for growth.

5.17
Aerobes require O_2 to live, whereas anaerobes do not and may even be killed by O_2. Facultative organisms can live with or without O_2. Special techniques are needed to grow aerobic and anaerobic microorganisms.

5.18
Several toxic forms of oxygen can form in the cell, but enzymes are present that neutralize most of them. Superoxide in particular seems to be a common toxic oxygen species.

Review of Key Terms

Acidophile an organism that grows best at low pH; typically below pH 5.5

Aerobe an organism that can use oxygen (O_2) in respiration; some require O_2

Aerotolerant anaerobe a microorganism unable to respire O_2 but whose growth is unaffected by oxygen

Alkaliphile an organism that has a growth pH optimum of 8 or higher

Anaerobe an organism that cannot use O_2 in respiration and whose growth is typically inhibited by O_2

Batch culture a closed-system microbial culture of fixed volume

Binary fission cell division following enlargement of a cell to twice its minimum size

Biofilm an attached polysaccharide matrix containing bacterial cells

Cardinal temperatures the minimum, maximum, and optimum growth temperatures for a given organism

Chemostat a device that allows for the continuous culture of microorganisms with independent control of both growth rate and cell number

Compatible solute a molecule that is accumulated in the cytoplasm of a cell for adjustment of water activity but that does not inhibit biochemical processes

Divisome a complex of proteins that directs cell division processes in prokaryotes

Exponential growth growth of a microbial population in which cell numbers double within a specific time interval

Extreme halophile a microorganism that requires very large amounts of salt (NaCl), usually greater than 10% and in some cases near to saturation, for growth

Extremophile an organism that grows optimally under one or more chemical or physical extremes, such as high or low temperature or pH

Facultative with respect to O_2, an organism that can grow in either its presence or absence

FtsZ a protein that forms a ring along the midcell division plane to initiate cell division

Generation time the time required for a population of microbial cells to double

Growth an increase in cell number

Halophile a microorganism that requires NaCl for growth

Halotolerant a microorganism that does not require NaCl for growth but can grow in the presence of NaCl, in some cases, substantial levels of NaCl

Hyperthermophile a prokaryote that has a growth temperature optimum of 80°C or greater

Mesophile an organism that grows best at temperatures between 20 and 45°C

Microaerophile an aerobic organism that can grow only when O_2 tensions are reduced from that present in air

Neutrophile an organism that grows best at neutral pH, between pH 5.5 and 8

Obligate anaerobe an organism that cannot grow in the presence of O_2

Osmophile an organism that grows best in the presence of high levels of solute, typically a sugar

pH the negative logarithm of the hydrogen ion (H^+) concentration of a solution

Psychrophile an organism with a growth temperature optimum of 15°C or lower and a maximum growth temperature below 20°C

Plate count a viable counting method where the number of colonies on a plate is used as a measure of cell numbers

Psychrotolerant capable of growing at low temperatures but having an optimum above 20°C

Thermophile an organism whose growth temperature optimum lies between 45 and 80°C

Transpeptidation formation of peptide cross-links between muramic acid residues in peptidoglycan synthesis

Viable capable of reproducing

Water activity the ratio of the vapor pressure of air in equilibrium with a solution to that of pure water

Xerophile an organism that is able to live, or that lives best, in very dry environments

Review Questions

1. What is binary fission? Briefly describe the process (Section 5.1).

2. Describe the role of proteins present at the divisome (Section 5.2).

3. In what way do derivatives of the rod-shaped bacterium *Escherichia coli* carrying mutations that inactivate the protein MreB look different microscopically from wild-type (unmutated) cells? What is the reason for this (Section 5.3)?

4. Why is penicillin effective against bacteria (Section 5.4)?

5. What is the difference between the specific growth rate (k) of an organism and its generation time (g) (Sections 5.5 and 5.6)?

6. Describe the growth cycle of a population of bacterial cells from the time this population is first inoculated into fresh medium (Section 5.7).

7. How can a chemostat regulate growth rate and cell numbers independently (Section 5.8)?

8. What is the difference between a total cell count and a viable cell count (Sections 5.9 and 5.10)?

9. Why is it advisable to dilute a sample prior to performing a plate count (Section 5.10)?

10. What are mesophiles? Thermophiles? Psychrophiles (Section 5.12)?

11. Describe a habitat where you would find a psychrophile; a hyperthermophile (Sections 5.13 and 5.14).

12. Concerning the pH of the environment and of the cell, in what ways are acidophiles and alkaliphiles different? In what ways are they similar (Section 5.15)?

13. Write an explanation in molecular terms for how a halophile is able to make water flow into the cell while growing in a solution high in NaCl (Section 5.16).

14. Contrast an aerotolerant and an obligate anaerobe in terms of sensitivity to O_2 and ability to grow in the presence of O_2. How does an aerotolerant anaerobe differ from a microaerophile (Section 5.17)?

15. Compare and contrast the enzymes catalase, superoxide dismutase, and superoxide reductase from the following points of view: substrates, oxygen products, organisms containing them, and role in oxygen tolerance of the cell (Section 5.18).

Application Questions

1. Calculate g and k in a growth experiment in which a medium was inoculated with 5×10^6 cells/ml of *Escherichia coli* cells and, following a 1-h lag, grew exponentially for 5 h, after which the population was 5.4×10^9 cells/ml.

2. *Escherichia coli* but not *Pyrolobus fumarii* will grow at 40°C, while *P. fumarii* but not *E. coli* will grow at 110°C. What is happening (or not happening) to prevent growth of each organism at the nonpermissive temperature?

3. In which direction (into or out of the cell) will water flow in cells of *Escherichia coli* (an organism found in your large intestine) suddenly suspended in a solution of 20% NaCl? What if the cells were suspended in distilled water? If growth nutrients were added to each cell suspension, which (if either) would support growth, and why?

Need more practice? Test your understanding with Quantitative Questions; access additional study tools including tutorials, animations, and videos; and then test your knowledge with chapter quizzes and practice tests at **www.microbiologyplace.com**.

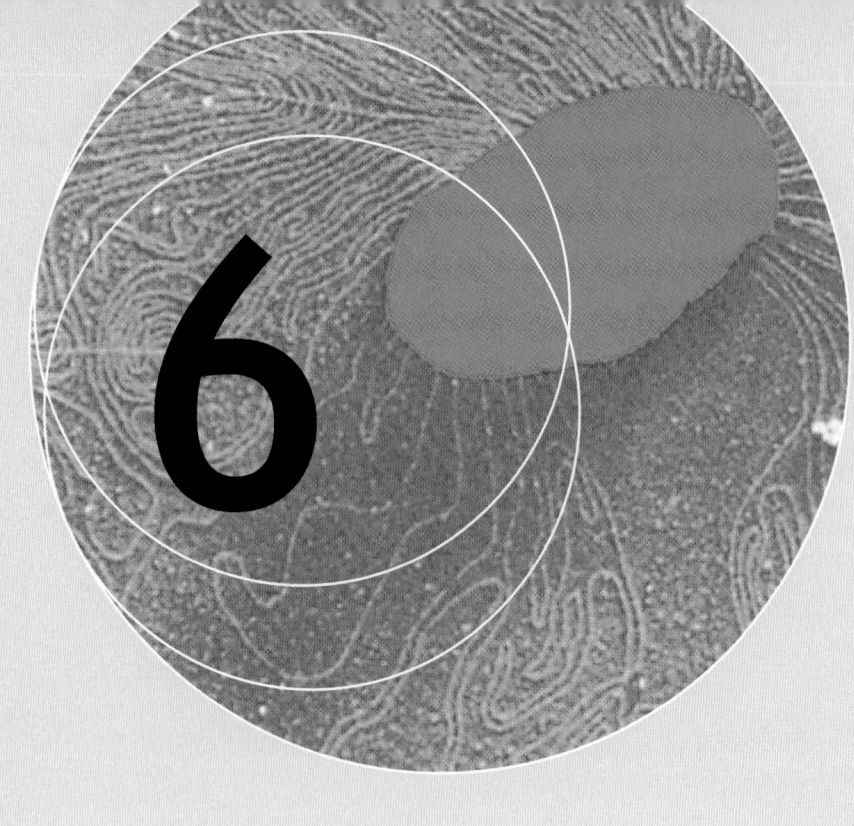

6

Molecular Biology of *Bacteria*

The essence of life is a cell's organization and the orderly replication of its DNA. Seen here, DNA is emerging from a bacterial cell treated to release its chromosome.

Cells may be regarded as chemical machines and coding devices. As chemical machines, cells transform their vast array of macromolecules into new cells. As coding devices, they store, process, and use genetic information. Genes and gene expression are the subject of molecular biology. In particular, the review of molecular biology in this chapter covers the chemical nature of genes, the structure and function of DNA and RNA, and the replication of DNA. We then consider the synthesis of proteins, macromolecules that play important roles in both the structure and the functioning of the cell. Our focus here is on these processes as they occur in *Bacteria*. In particular, *Escherichia coli*, a member of the *Bacteria*, is the model organism for molecular biology and is the main example used. Although *E. coli* was not the first bacterium to have its chromosome sequenced, this organism remains the best characterized of any organism, prokaryote or eukaryote.

DNA Structure and Genetic Information

6.1 Macromolecules and Genes

The functional unit of genetic information is the **gene**. All life forms, including microorganisms, contain genes. Physically, genes are located on chromosomes or other large molecules known collectively as **genetic elements**. Nowadays, in the "genomics era," biology tends to characterize cells in terms of their complement of genes. Thus, if we wish to understand how microorganisms function we must understand how genes encode information.

Chemically, genetic information is carried by the **nucleic acids** deoxyribonucleic acid, **DNA**, and ribonucleic acid, **RNA**. DNA carries the genetic blueprint for the cell and RNA is the intermediary molecule that converts this blueprint into defined amino acid sequences in proteins. Genetic information consists of the sequence of monomers in the nucleic acids. Thus, in contrast to polysaccharides and lipids, nucleic acids are **informational macromolecules**. Because the sequence of monomers in proteins is determined by the sequence of the nucleic acids that encode them, proteins are also informational macromolecules.

The monomers of nucleic acids are called **nucleotides**, consequently, DNA and RNA are **polynucleotides**. A nucleotide has three components: a pentose sugar, either ribose (in RNA) or deoxyribose (in DNA), a nitrogen base, and a molecule of phosphate, PO_4^{3-}. The general structure of nucleotides of both DNA and RNA is very similar (**Figure 6.1**). The nitrogen bases are either **purines** (*adenine* and *guanine*) which contain two fused heterocyclic rings or **pyrimidines** (*thymine, cytosine,* and *uracil*) which contain a single six-membered heterocyclic ring (Figure 6.1*a*). Guanine, adenine, and cytosine are present in both DNA and RNA. With minor exceptions, thymine is present only in DNA and uracil is present only in RNA.

The nitrogen bases are attached to the pentose sugar by a glycosidic linkage between carbon atom 1 of the sugar and a nitrogen atom in the base, either nitrogen 1 (in pyrimidine bases) or 9 (in purine bases). A nitrogen base attached to its sugar, but

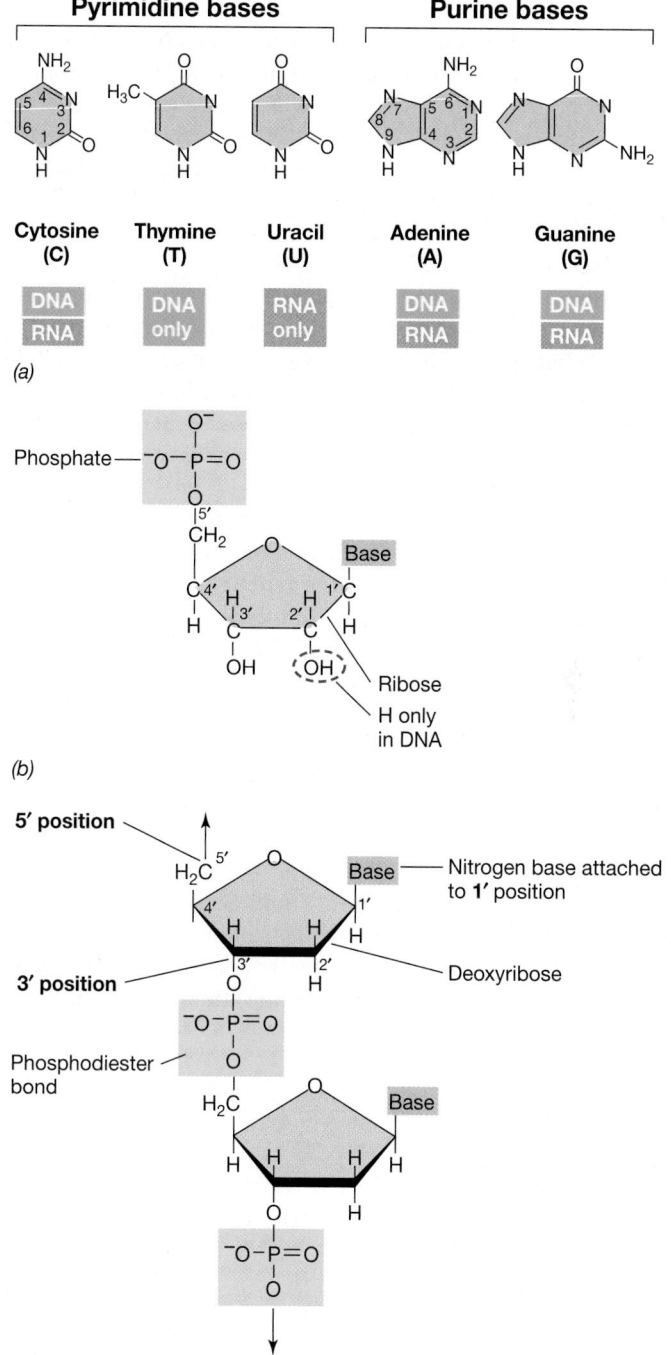

Figure 6.1 Components of the nucleic acids. *(a)* The nitrogen bases of DNA and RNA. Note the numbering system of the rings. In attaching itself to the 1′ carbon of the sugar phosphate, a pyrimidine base bonds through N-1 and a purine base bonds at N-9. *(b)* Nucleotide structure. The numbers on the sugar contain a prime (′) after them because the rings of the nitrogen bases are also numbered. In DNA a hydrogen is present on the 2′-carbon of the pentose sugar. In RNA, an OH group occupies this position. *(c)* Part of a DNA chain. The nucleotides are linked by a phosphodiester bond. In addition to the bases shown, transfer RNAs (tRNAs) contain unusual pyrimidines such as pseudouracil and dihydrouracil, and various modified purines not present in other RNAs (see Figure 6.33).

lacking phosphate, is called a **nucleoside**. Nucleotides are nucleosides plus one or more phosphates (Figure 6.1). Nucleotides play other roles in addition to comprising nucleic acids. Nucleotides, especially adenosine triphosphate (ATP) and guanosine triphosphate (GTP), carry chemical energy. Other nucleotides or derivatives function in redox reactions, as carriers of sugars in polysaccharide synthesis, or as regulatory molecules.

The Nucleic Acids, DNA and RNA

The nucleic acid backbone is a polymer of alternating sugar and phosphate molecules. The nucleotides are covalently bonded by phosphate between the 3'- (3 prime) carbon of one sugar and the 5'-carbon of the next sugar. [Numbers with prime marks refer to positions on the sugar ring; numbers without primes to positions on the rings of the bases.] The phosphate linkage is called a **phosphodiester bond** because the phosphate connects two sugar molecules by an ester linkage (Figure 6.1). The sequence of nucleotides in a DNA or RNA molecule is its **primary structure** and the sequence of bases forms the genetic information.

In the genome of cells, DNA is *double-stranded*. Each chromosome consists of two strands of DNA, with each strand containing hundreds of thousands to several million nucleotides linked by phosphodiester bonds. The strands are held together by hydrogen bonds that form between the bases in one strand and those of the other strand. When located next to one another, purine and pyrimidine bases can form hydrogen bonds (**Figure 6.2**). Hydrogen bonding is most stable when guanine (G) bonds with cytosine (C) and adenine (A) bonds with thymine (T). Specific base pairing, A with T and G with C, ensures that the two strands of DNA are *complementary* in base sequence; that is, wherever a G is found in one strand, a C is found in the other, and wherever a T is present in one strand, its complementary strand has an A.

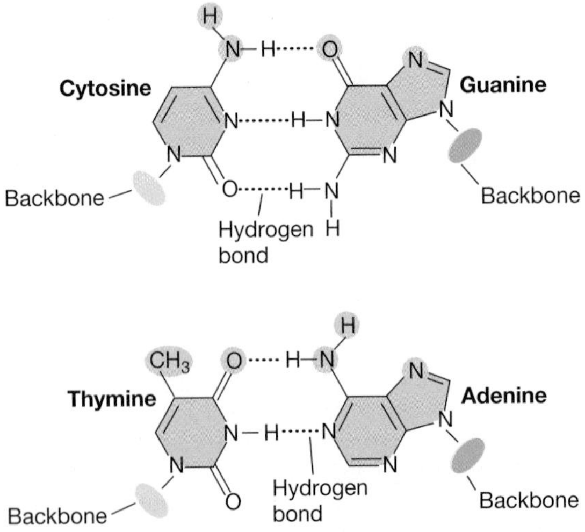

With a few exceptions, all RNA molecules are *single-stranded*. However, RNA molecules typically fold back upon themselves in regions where complementary base pairing is possible. The term **secondary structure** refers to this folding whereas primary structure refers to the nucleotide sequence. In certain large RNA molecules, such as ribosomal RNA (Section 6.19), some parts of the molecule are unfolded but other regions possess secondary structure. This leads to highly folded and twisted molecules whose biological function depends critically on their final three-dimensional shape.

Genes and the Steps in Information Flow

When genes are expressed, the information stored in DNA is transferred to ribonucleic acid (RNA). Several classes of RNA exist in cells. Three types of RNA take part in protein synthesis. **Messenger RNA** (mRNA) is a single-stranded molecule that carries the genetic information from DNA to the ribosome, the protein-synthesizing machine. **Transfer RNAs** (tRNAs) convert the genetic information on mRNA into the language of proteins. **Ribosomal RNAs** (rRNAs) are important catalytic and structural components of the ribosome. In addition to these, cells contain a variety of *small RNAs* that regulate the production or activity of proteins or other RNAs. The molecular processes of genetic information flow can be divided into three stages (**Figure 6.3**):

1. **Replication**. During replication, the DNA double helix is duplicated, producing two double helices.

2. **Transcription**. Transfer of information from DNA to RNA is called transcription.

3. **Translation**. Synthesis of a protein, using the information carried by mRNA, is known as translation.

Figure 6.2 Specific pairing between guanine (G) and cytosine (C) and between adenine (A) and thymine (T) via hydrogen bonds. These are the typical base pairs found in double-stranded DNA. Atoms that are found in the major groove of the double helix and that interact with proteins are highlighted in pink. The deoxyribose phosphate backbones of the two strands of DNA are also indicated. Note the different shades of green for the two strands of DNA, a convention used throughout this book.

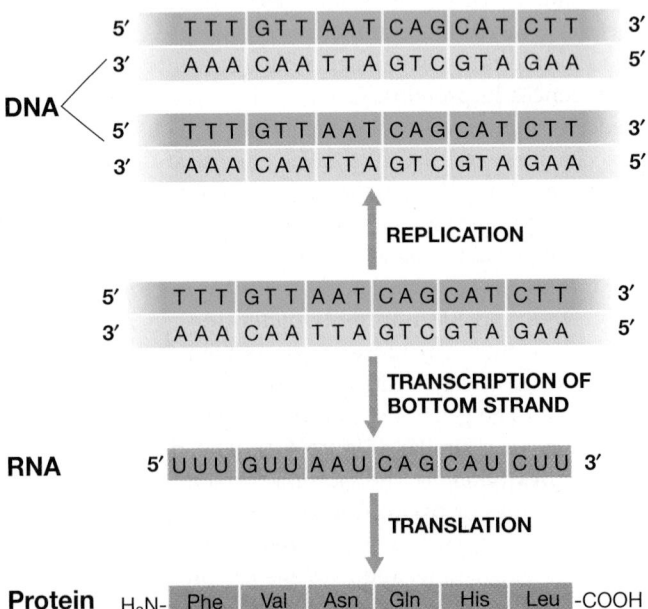

Figure 6.3 Synthesis of the three types of informational macromolecules. Note that for any particular gene only one of the two strands of the DNA double helix is transcribed.

There is a linear correspondence between the base sequence of a gene and the amino acid sequence of a polypeptide. Each group of three bases on an mRNA molecule encodes a single amino acid, and each such triplet of bases is called a **codon**. This genetic code is translated into protein by the ribosomes (which consist of proteins and rRNA), tRNA, and proteins known as translation factors.

The three steps shown in Figure 6.3 are used in all cells and constitute the central dogma of molecular biology (DNA → RNA → protein). Note that many different RNA molecules are each transcribed from a relatively short region of the long DNA molecule. In eukaryotes, each gene is transcribed to give a single mRNA (Chapter 7), whereas in prokaryotes a single mRNA may carry genetic information for several genes, that is, for several protein coding regions. Some viruses violate the central dogma (Chapter 9). Some viruses use RNA as the genetic material and must therefore replicate their RNA using RNA as template. In retroviruses such as HIV—the causative agent of AIDS—an RNA genome is converted to a DNA version by a process called reverse transcription.

MiniQuiz

- What components are found in a nucleotide?
- How does a nucleo*side* differ from a nucleo*tide*?
- Distinguish between the primary and secondary structure of RNA.
- What three informational macromolecules are involved in genetic information flow?
- In all cells there are three processes involved in genetic information flow. What are they?

6.2 The Double Helix

In all cells and many viruses, DNA exists as a double-stranded molecule with two polynucleotide strands whose base sequences are **complementary**. (As discussed in Chapter 9, the genomes of some DNA viruses are single-stranded.) The complementarity of DNA arises because of specific base pairing: adenine always pairs with thymine, and guanine always pairs with cytosine. The two strands of the double-stranded DNA molecule are arranged in an **antiparallel** fashion (**Figure 6.4**, distinguished as two shades of green). Thus, the strand on the left runs 5′ to 3′ from top to bottoms, whereas the other strand runs 5′ to 3′ from bottom to top.

The two strands of DNA are wrapped around each other to form a double helix (**Figure 6.5**) that forms two distinct grooves, the *major groove* and the *minor groove*. Most proteins that interact specifically with DNA bind in the major groove, where there is plenty of space. Because the double helix is a regular structure, some atoms of each base are always exposed in the major groove (and some in the minor groove). Key regions of nucleotides that are important in interactions with proteins are shown in Figure 6.2.

Several double-helical structures are possible for DNA. The Watson and Crick double helix is known as the B-form or *B-DNA* to distinguish it from the A- and Z-forms. The A-form is shorter and fatter than the B-form. It has 11 base pairs per turn, and the

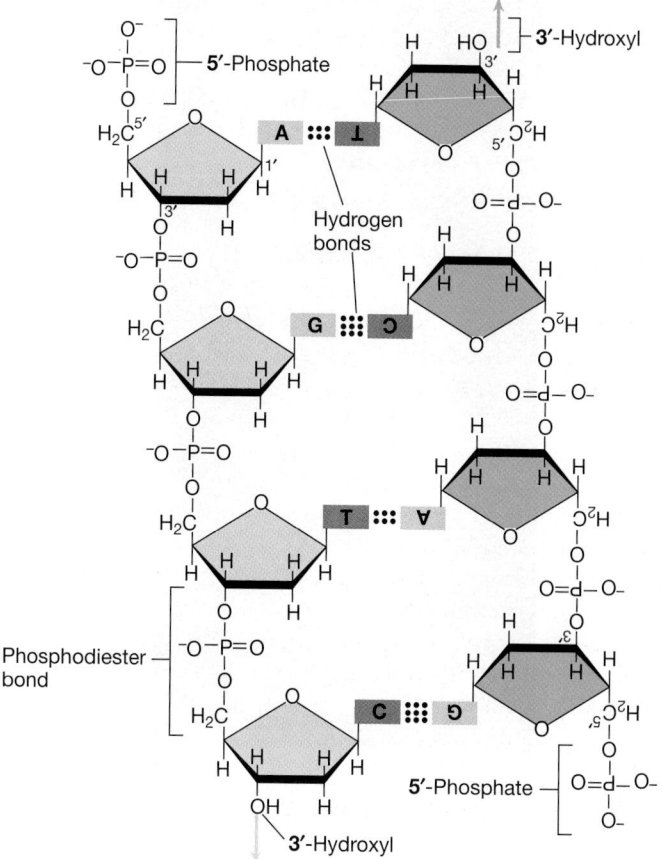

Figure 6.4 DNA structure. Complementary and antiparallel nature of DNA. Note that one chain ends in a 5′-phosphate group, whereas the other ends in a 3′-hydroxyl. The red bases represent the pyrimidines cytosine (C) and thymine (T), and the yellow bases represent the purines adenine (A) and guanine (G).

major groove is narrower and deeper. Double-stranded RNA or hybrids of one RNA plus one DNA strand often form the A-helix. The Z-DNA double helix has 12 base pairs per turn and is left-handed. Its sugar–phosphate backbone is a zigzag line rather than a smooth curve. Z-DNA is found in GC- or GT-rich regions, especially when negatively supercoiled. Occasional enzymes and regulatory proteins bind Z-DNA preferentially.

Size and Shape of DNA Molecules

The size of a DNA molecule is expressed as the number of nucleotide bases or base pairs per molecule. Thus, a DNA molecule with 1000 bases is 1 kilobase (kb) of DNA. If the DNA is a double helix, then *kilobase pairs* (kbp) is used. Thus, a double helix 5000 base pairs in size would be 5 kbp. The bacterium *Escherichia coli* has about 4640 kbp of DNA in its chromosome. When dealing with large genomes the term *megabase pair* (Mbp) for a million base pairs is used. The genome of *E. coli* is thus 4.64 Mbp.

Each base pair takes up 0.34 nanometer (nm) in length along the double helix, and each turn of the helix contains approximately 10 base pairs. Therefore, 1 kbp of DNA is 0.34 μm long with 100 helical turns. The *E. coli* genome is thus $4640 \times 0.34 = 1.58$ mm

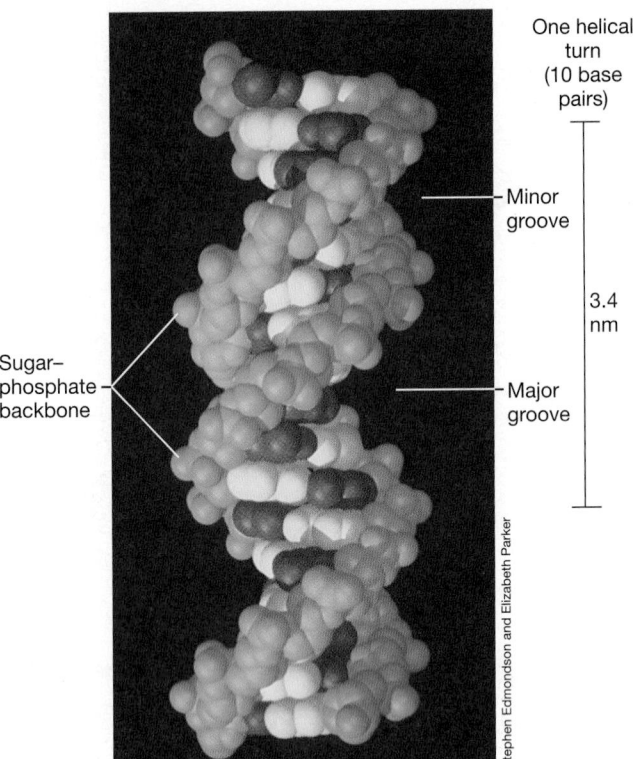

Figure 6.5 A computer model of a short segment of DNA showing the overall arrangement of the double helix. One of the sugar–phosphate backbones is shown in blue and the other in green. The pyrimidine bases are shown in red and the purines in yellow. Note the locations of the major and minor grooves (compare with Figure 6.2). One helical turn contains 10 base pairs.

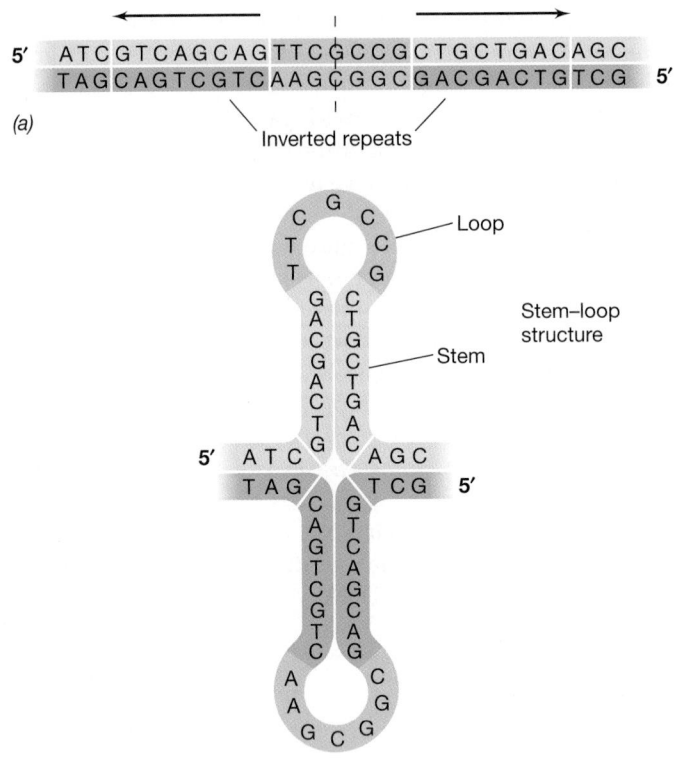

Figure 6.6 Inverted repeats and the formation of a stem–loop. *(a)* Nearby inverted repeats in DNA. The arrows indicate the symmetry around the imaginary axis (dashed line). *(b)* Formation of stem–loop structures by pairing of complementary bases on the same strand.

long. Since cells of *E. coli* are about 2 μm long, the chromosome is several hundred times longer than the cell itself!

Long DNA molecules are quite flexible, but stretches of DNA less than 100 base pairs are more rigid. Some short segments of DNA can be bent by proteins that bind them. However, certain base sequences themselves cause DNA to bend. Such sequences usually have several runs of five or six adenines, each separated by four or five other bases.

Inverted Repeats and Stem–Loop Structures

Short, repeated sequences are often found in DNA molecules. Many proteins bind to regions of DNA containing inverted repeat sequences (Chapter 8). As shown in **Figure 6.6**, nearby inverted repeats can form stem–loop structures. The stems are short double-helical regions with normal base pairing. The loop contains the unpaired bases between the two repeats.

The formation of stem–loop structures in DNA itself is relatively rare. However, the production of stem–loop structures in the RNA produced from DNA following transcription is common. Such secondary structures formed by base pairing within a single strand of RNA are found in transfer RNA (Section 6.18) and ribosomal RNA (Section 6.19). Even when a stem–loop does not form, inverted repeats in DNA are often binding sites for DNA-binding proteins that regulate transcription (Chapter 8) or for endonucleases that cut DNA (Section 11.1).

The Effect of Temperature on DNA Structure

Although individual hydrogen bonds are very weak, the large number of such bonds between the base pairs of a long DNA molecule hold the two strands together effectively. There may be millions or even hundreds of millions of hydrogen bonds in a long DNA molecule, depending on the number of base pairs. Recall that each adenine–thymine base pair has *two* hydrogen bonds, while each guanine–cytosine base pair has *three*. This makes GC pairs stronger than AT pairs.

When isolated from cells and kept near room temperature and at physiological salt concentrations, DNA remains double-stranded. However, if the temperature is raised, the hydrogen bonds will break but the covalent bonds holding a chain together will not, and so the two DNA strands will separate. This process is called denaturation (melting) and can be measured experimentally because single-stranded and double-stranded nucleic acids differ in their ability to absorb ultraviolet radiation at 260 nm (**Figure 6.7**).

DNA with a high percentage of GC pairs melts at a higher temperature than a similar-sized molecule with more AT pairs. If the heated DNA is allowed to cool slowly, the double-stranded DNA can re-form, a process called annealing. This can be used not only to re-form native DNA but also to form hybrid molecules whose two strands come from different sources. Hybridization, the artificial assembly of a double-stranded nucleic acid by complementary base pairing of two single strands, is a powerful technique in molecular biology (Section 11.2).

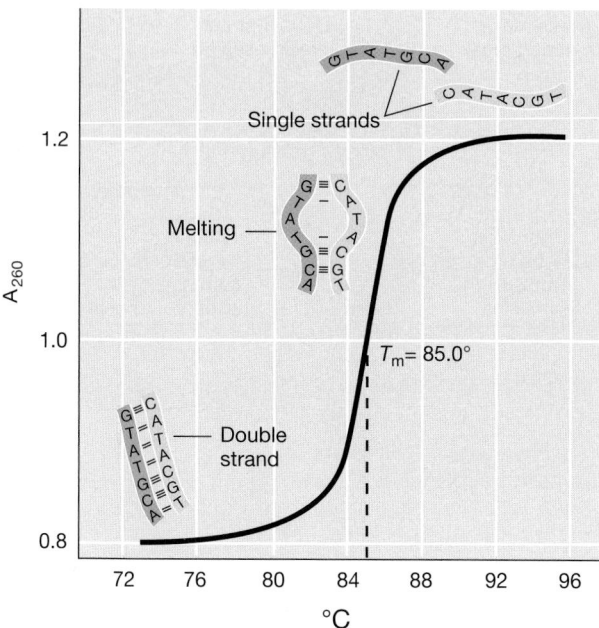

Figure 6.7 **Thermal denaturation of DNA.** DNA absorbs more ultraviolet radiation at 260 nm as the double helix is denatured. The transition is quite abrupt, and the temperature of the midpoint, T_m, is proportional to the GC content of the DNA. Although the denatured DNA can be renatured by slow cooling, the process does not follow a similar curve. Renaturation becomes progressively more complete at temperatures well below the T_m and then only after a considerable incubation time.

MiniQuiz

- What does antiparallel mean in terms of the structure of double-stranded DNA?
- Define the term complementary when used to refer to two strands of DNA.
- Define the terms denaturation, reannealing, and hybridization as they apply to nucleic acids.
- Why do GC-rich molecules of DNA melt at higher temperatures than AT-rich molecules?

6.3 Supercoiling

If linearized, the *Escherichia coli* chromosome would be over 1 mm in length, about 700 times longer than the *E. coli* cell itself. How is it possible to pack so much DNA into such a little space? The solution is the imposition of a "higher-order" structure on the DNA, in which the double-stranded DNA is further twisted in a process called *supercoiling.* **Figure 6.8** shows how supercoiling occurs in a circular DNA duplex. If a circular DNA molecule is linearized, any supercoiling is lost and the DNA becomes "relaxed." When relaxed, a DNA molecule has exactly the number of turns of the helix predicted from the number of base pairs.

Supercoiling puts the DNA molecule under torsion, much like the added tension to a rubber band that occurs when it is twisted. DNA can be supercoiled in either a positive or a negative manner. In positive supercoiling the double helix is overwound, whereas in negative supercoiling the double helix is underwound. Negative supercoiling results when the DNA is twisted about its

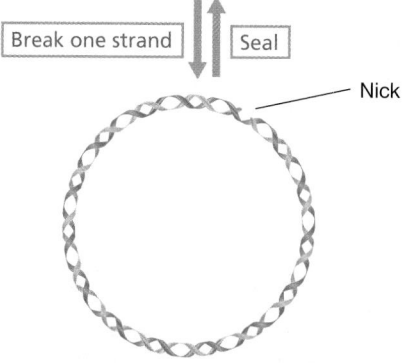

(a) **Relaxed, covalently closed circular DNA**

| Break one strand | Seal |

Nick

(b) **Relaxed, nicked circular DNA**

| Break one strand | Rotate one end of broken strand around helix and seal |

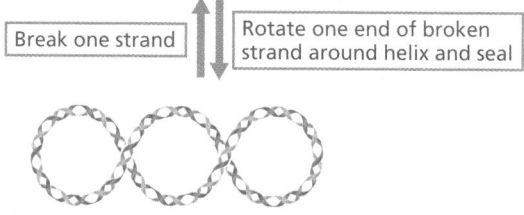

(c) **Supercoiled circular DNA**

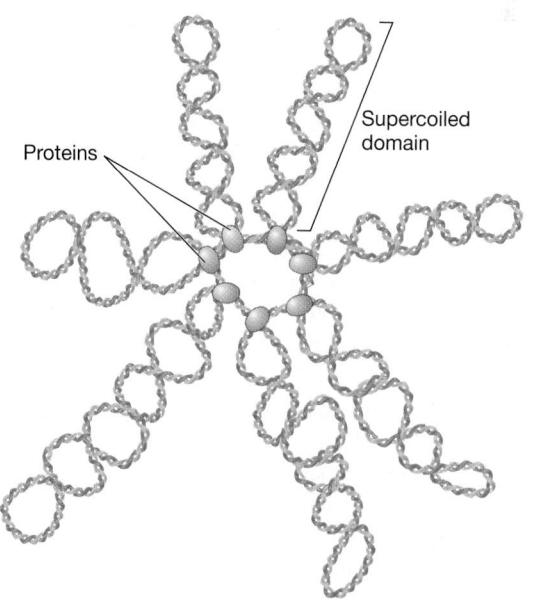

Supercoiled domain

Proteins

(d) **Chromosomal DNA with supercoiled domains**

Figure 6.8 **Supercoiled DNA.** *(a–c)* Relaxed, nicked, and supercoiled circular DNA. A nick is a break in a phosphodiester bond of one strand. *(d)* In fact, the double-stranded DNA in the bacterial chromosome is arranged not in one supercoil but in several supercoiled domains, as shown here.

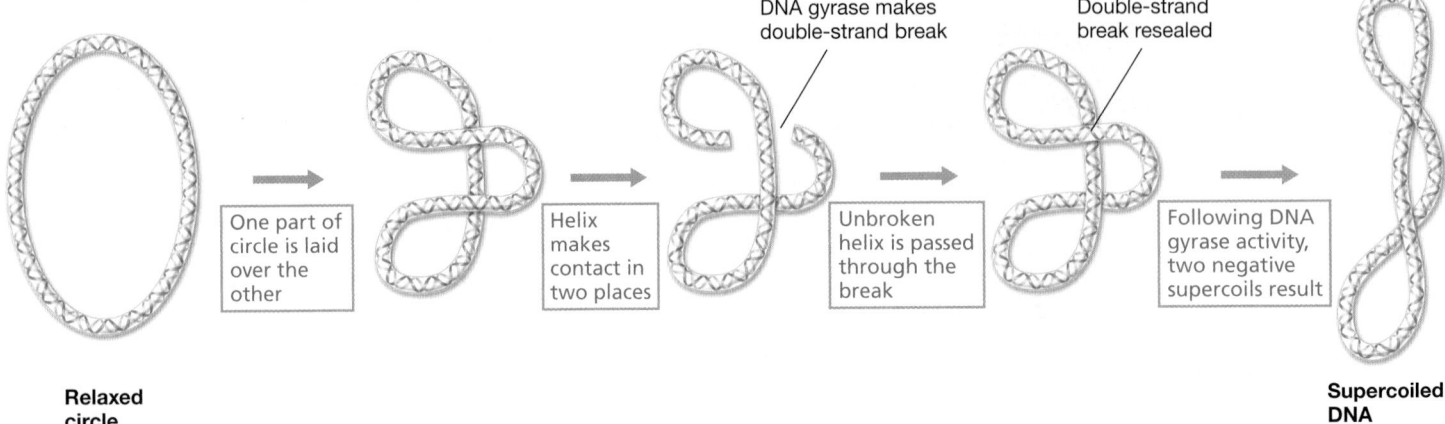

Figure 6.9 **DNA gyrase.** Introduction of negative supercoiling into circular DNA by the activity of DNA gyrase (topoisomerase II), which makes double-strand breaks.

axis in the opposite sense from the right-handed double helix. Negatively supercoiled DNA is the form predominantly found in nature. However, certain species of *Archaea* (Chapter 7) that grow at very high temperatures do contain positively supercoiled DNA. In *Escherichia coli* more than 100 supercoiled domains are thought to exist, each of which is stabilized by binding to specific proteins.

Topoisomerases: DNA Gyrase

Supercoils are inserted or removed by enzymes known as topoisomerases. Two major classes of topoisomerase exist with different mechanisms. Class I topoisomerases make a single-stranded break in the DNA that allows the rotation of one strand of the double helix around the other. Each rotation adds or removes a single supercoil. After this, the nick is resealed. For example, surplus supercoiling in DNA is generally removed by the class I enzyme, topoisomerase I. As shown in Figure 6.8, a break in the backbone (a nick) of either strand allows DNA to lose its supercoiling. However, to prevent the entire bacterial chromosome from becoming relaxed every time a nick is made, the chromosome contains supercoiled domains as shown in Figure 6.8*d*. A nick in the DNA of one domain does not relax DNA in the others. It is unclear precisely how these domains are formed, although specific DNA-binding proteins are involved.

Class II topoisomerases make double-stranded breaks, pass the double helix through the break, and reseal the break (**Figure 6.9**). Each such operation adds or removes two supercoils. Inserting supercoils into DNA requires energy from ATP, whereas releasing supercoils does not. In *Bacteria* and most *Archaea*, the class II topoisomerase, **DNA gyrase**, inserts negative supercoils into DNA. Some antibiotics inhibit the activity of DNA gyrase. These include the quinolones (such as nalidixic acid), the fluoroquinolones (such as ciprofloxacin), and novobiocin.

Through the activity of topoisomerases, a DNA molecule can be alternately supercoiled and relaxed. Supercoiling is necessary for packing the DNA into the cell and relaxation is necessary for DNA replication and transcription. In most prokaryotes, the level of negative supercoiling results from the balance between the activity of DNA gyrase and topoisomerase I. Supercoiling also affects gene expression. Certain genes are more actively transcribed when DNA is supercoiled, whereas transcription of other genes is inhibited by supercoiling.

MiniQuiz
- Why is supercoiling important?
- What mechanism is used by DNA gyrase?
- What function do topoisomerases serve inside cells?

6.4 Chromosomes and Other Genetic Elements

Structures containing genetic material (DNA in most organisms, but RNA in some viruses) are called *genetic elements*. The **genome** is the total complement of genes in a cell or virus. Although the main genetic element in prokaryotes is the **chromosome**, other genetic elements are found and play important roles in gene function in both prokaryotes and eukaryotes (**Table 6.1**). These include virus genomes, plasmids, organellar genomes, and transposable elements. A typical prokaryote has a single circular chromosome containing all (or most) of the genes found inside the cell. Although a single chromosome is the rule among prokaryotes, there are exceptions. A few prokaryotes contain two chromosomes. Eukaryotes have multiple chromosomes making up their genome (Section 7.5). Also, the DNA in all known eukaryotic chromosomes is linear in contrast to most prokaryotic chromosomes, which are circular DNA molecules.

Viruses and Plasmids

Viruses contain genomes, *either* of DNA or RNA, that control their own replication and their transfer from cell to cell. Both linear and circular viral genomes are known. In addition, the nucleic acid in viral genomes may be single-stranded or double-stranded. Viruses are of special interest because they often cause disease. We discuss viruses in Chapters 9 and 21 and a variety of viral diseases in later chapters.

Inside the figure:
DNA gyrase makes double-strand break
Double-strand break resealed

One part of circle is laid over the other

Helix makes contact in two places

Unbroken helix is passed through the break

Following DNA gyrase activity, two negative supercoils result

Relaxed circle

Supercoiled DNA

Table 6.1 *Kinds of genetic elements*

Organism	Element	Type of nucleic acid	Description
Prokaryote	Chromosome	Double-stranded DNA	Extremely long, usually circular
Eukaryote	Chromosome	Double-stranded DNA	Extremely long, linear
All organisms	Plasmid[a]	Double-stranded DNA	Relatively short circular or linear, extrachromosomal
All organisms	Transposable element	Double-stranded DNA	Always found inserted into another DNA molecule
Mitochondrion or chloroplast	Genome	Double-stranded DNA	Medium length, usually circular
Virus	Genome	Single- or double-stranded DNA or RNA	Relatively short, circular or linear

[a]Plasmids are uncommon in eukaryotes.

Plasmids are genetic elements that replicate separately from the chromosome. The great majority of plasmids are double-stranded DNA, and although most plasmids are circular, some are linear. Most plasmids are much smaller than chromosomes. Plasmids differ from viruses in two ways: (1) They do not cause cellular damage (generally they are beneficial), and (2) they do not have extracellular forms, whereas viruses do. Although only a few eukaryotes contain plasmids, one or more plasmids have been found in most prokaryotic species and can be of profound importance. Some plasmids contain genes whose protein products confer important properties on the host cell, such as resistance to antibiotics.

What is the difference, then, between a large plasmid and a chromosome? A chromosome is a genetic element that contains genes whose products are necessary for essential cellular functions. Such essential genes are called *housekeeping genes*. Some of these encode essential proteins, such as DNA and RNA polymerases, and others encode essential RNAs, such as ribosomal and transfer RNA. In contrast to the chromosome, plasmids are usually expendable and rarely contain genes required for growth under all conditions. There are many genes on a chromosome that are unessential as well, but the presence of *essential* genes is necessary for a genetic element to be classified as a chromosome.

Transposable Elements

Transposable elements are segments of DNA that can move from one site on a DNA molecule to another site, either on the same molecule or on a different DNA molecule. Transposable elements are not found as separate molecules of DNA but are inserted into other DNA molecules. Chromosomes, plasmids, virus genomes, and any other type of DNA molecule may act as host molecules for transposable elements. Transposable elements are found in both prokaryotes and eukaryotes and play important roles in genetic variation. In prokaryotes there are three main types of transposable elements: insertion sequences, transposons, and some special viruses. Insertion sequences are the simplest type of transposable element and carry no genetic information other than that required for them to move about the chromosome. Transposons are larger and contain other genes. We discuss both of these in more detail in Chapter 10. In Chapter 21 we discuss a bacterial virus, Mu, that is itself a transposable element. The unique feature common to all transposable elements is that they replicate as part of some other molecule of DNA.

MiniQuiz

- What is a genome?
- What are viruses and plasmids?
- What genetic material is found in all cellular chromosomes?
- What defines a chromosome in prokaryotes?

 Chromosomes and Plasmids

6.5 The *Escherichia coli* Chromosome

Today, many bacterial genomes, including that of *Escherichia coli*, have been completely sequenced, thus revealing the number and location of the genes they possess. However, the genes of *E. coli* were initially mapped long before sequencing was performed, using conjugation and transduction (⟳ Sections 10.8 and 10.9). The genetic map of *E. coli* strain K-12 is shown in **Figure 6.10**. Map distances are given in "minutes" of transfer that derive from conjugation experiments, with the entire chromosome containing 100 minutes (or centisomes). Zero is arbitrarily set at *thrABC* (the threonine operon), because the *thrABC* genes were the first shown to be transferred by conjugation in *E. coli*. The genetic map in Figure 6.10 shows only a few of the several thousand genes in the *E. coli* chromosome. The size of the chromosome is given in both minutes and in kilobase pairs of DNA.

The strain of *E. coli* whose chromosome was originally sequenced, strain MG1655, is a derivative of *E. coli* K-12, the traditional strain used for genetics. Wild-type *E. coli* K-12 has bacteriophage lambda integrated into its chromosome (⟳ Section 9.10) and also contains the F plasmid. However, strain MG1655 had both of these removed before sequencing (lambda by radiation and the F plasmid by acridine treatment). The chromosome of strain MG1655 contains 4,639,221 bp. Analysis revealed 4288 possible protein-encoding genes that account for about 88% of the genome. Approximately 1% of the genome consists of genes encoding tRNAs and rRNAs. Regulatory sequences—promoters, operators, origin and terminus of DNA replication, and so on—comprise around 10% of the genome. The remaining 0.5% consists of noncoding, repetitive sequences.

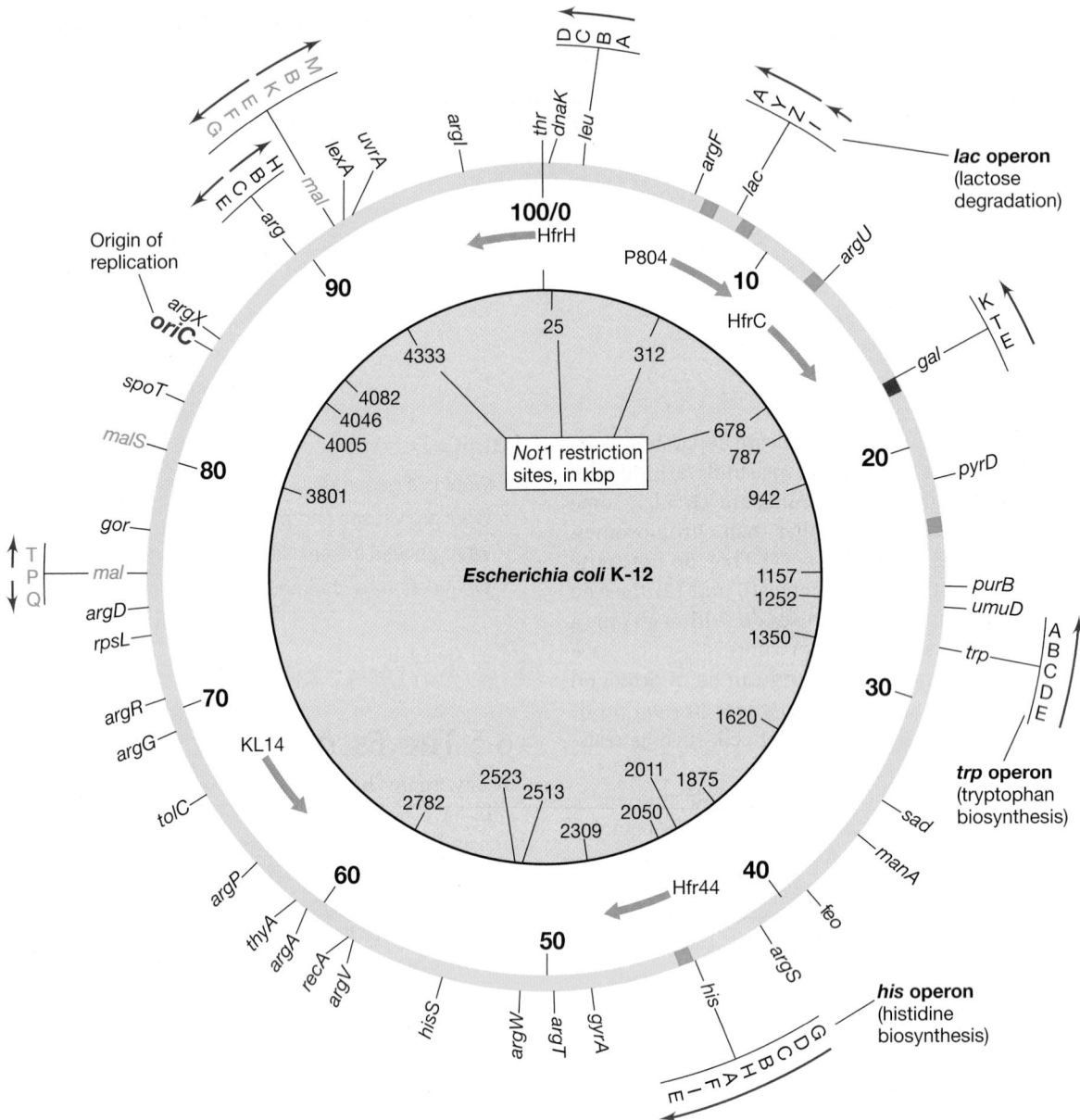

Figure 6.10 The chromosome of *Escherichia coli* strain K-12. The *E. coli* chromosome contains 4,639,221 base pairs and 4288 open reading frames (an indicator of genes; Section 6.17). On the outer edge of the map, the locations of a few genes are indicated. A few operons are also shown, with their directions of transcription. Around the inner edge, the numbers from 0 to 100 refer to map position in minutes. Note that 0 is located by convention at the *thr* locus. Replication proceeds bidirectionally from the origin of DNA replication, *oriC*, at 84.3 min. The inner circle shows the locations, in kilobase pairs, of the sites where the restriction enzyme *Not*I cuts. The origins and directions of transfer of a few Hfr strains are also shown (arrows). The locations of five copies of the transposable element IS3 found in a particular strain are shown in blue. The site where bacteriophage lambda integrates is shown in red. If lambda were present, it would add an extra 48.5 kbp (slightly over 1 min) to the map. The genes of the maltose regulon, which includes several operons, are indicated by green labels. The maltose genes are abbreviated *mal* except for *lamB*, which encodes an outer membrane protein for maltose uptake that is also the receptor for bacteriophage lambda. The gene *rpsL* (73 min) encodes a ribosomal protein. This gene was once called *str* because mutations in it lead to streptomycin resistance.

Arrangement of Genes on the *Escherichia coli* Chromosome

Genetic mapping of the genes that encode the enzymes of a single biochemical pathway in *E. coli* has shown that these genes are often clustered. On the genetic map in Figure 6.10, a few such clusters are shown. Notice, for instance, the *gal* gene cluster at 18 min, the *trp* gene cluster at about 28 min, and the *his* cluster at 44 min. Each of these gene clusters constitutes an *operon* that is transcribed as a single mRNA carrying multiple coding sequences, that is, a *polycistronic* mRNA (Section 6.15).

Genes for some other biochemical pathways in *E. coli* are not clustered. For example, genes for arginine biosynthesis (*arg* genes) are scattered throughout the chromosome. The early discovery of multigene operons and their use in studying gene regulation (for example, the *lac* operon; ↪ Section 8.5), often gives the impression that such operons are the rule in prokaryotes. However, sequence analysis of the *E. coli* chromosome has shown that over 70% of the 2584 predicted or known transcriptional units contain only a single gene. Only about 6% of the operons have four or more genes.

In *E. coli* the transcription of some genes proceeds clockwise around the chromosome, whereas transcription of others proceeds counterclockwise. This means that some coding sequences are on one strand of the chromosome whereas others are on the opposite strand. There are about equal numbers of genes on both strands. The direction of transcription of a few multigene operons is shown by the arrows in Figure 6.10. Many genes that are highly expressed in *E. coli* are oriented so that they are transcribed in the same direction that the DNA replication fork moves through them. The two replication forks start at the origin, *oriC* located at about 84 min, and move in opposite directions around the circular chromosome toward the terminus, which is located at approximately 34 min. All seven of the rRNA operons of *E. coli* and 53 of its 86 tRNA genes are transcribed in the same direction as replication. Presumably, this arrangement for highly expressed genes allows RNA polymerase to avoid collision with the replication fork, because this moves in the same direction as the RNA polymerase.

Almost 2000 *E. coli* proteins, or genes encoding proteins, were identified by classical genetic analyses before its chromosome was sequenced. Sequence analyses indicate that approximately 4225 different proteins may be encoded by the *E. coli* chromosome. Around 30% of these proteins are of unknown function or are hypothetical. The average *E. coli* protein contains slightly more than 300 amino acid residues, but many proteins are smaller and many are much larger. The largest gene in *E. coli* encodes a protein of 2383 amino acids that is still uncharacterized. This giant protein shows similarities to proteins found in pathogenic enteric bacteria closely related to *E. coli* and may thus play some role in infection.

Although sequence analysis yields much information, to understand the function, particularly of regulatory sequences, it is still necessary to isolate mutants, map the mutations, and use biochemical and physiological analyses to determine their effects on the organism. This is especially true of the 20–40% of genes that show up in all genomic analyses (↪ Section 12.3) as encoding proteins of unknown function. This huge repository of hypothetical proteins doubtless holds new biochemical secrets that will expand the known metabolic capabilities of prokaryotes. In addition, because many prokaryotic genes have homologs in eukaryotes including humans, understanding gene function in prokaryotes aids our understanding of human genetics.

Although *E. coli* has very few duplicate genes, computer analyses have shown that many of its protein-encoding genes arose by gene duplication during evolutionary history (↪ Section 12.10). The *E. coli* genome also contains some large gene families—groups of genes with related sequences encoding products with related functions. For example, there is a family of 70 genes that all encode membrane transport proteins. Gene families are common, both within a species and across broad taxonomic lines. Thus gene duplication plays a major role in evolution.

Insertions within the *Escherichia coli* Chromosome and Horizontal Gene Transfer

Several other genetic elements are inserted into the *E. coli* chromosome and are consequently replicated with it. There are multiple copies of several different insertion sequences (IS elements), including seven copies of IS2 and five of IS3. Both of these IS elements are also found on the F plasmid, and both take part in the formation of Hfr strains (↪ Section 10.10). There are several defective integrated viruses that vary from nearly complete virus genomes to small fragments. Three of these are related to bacteriophage lambda.

E. coli obtained part of its genome by *horizontal* (*lateral*) gene transfer from other organisms. Horizontal transfer contrasts with *vertical* gene transfer in which genes move from mother cell to daughter cell. In fact, it has been estimated that nearly 20% of the *E. coli* genome originated from horizontal transfers. Horizontally transferred segments of DNA can often be detected because they have significantly different GC ratios (the ratio of guanine–cytosine base pairs to adenine–thymine base pairs) or codon distributions (codon bias, Section 6.17) from those of the host organism.

Horizontal gene transfer may cause large-scale changes in a genome. For example, strains of *E. coli* are known that contain virulence genes located on large, unstable regions of the chromosome called *pathogenicity islands* that can be acquired by horizontal transfer (↪ Sections 12.12 and 12.13). Horizontal transfer does not necessarily result in an ever-larger genome size. Many genes acquired in this way provide no selective advantage and so are lost by deletion. This keeps the chromosome of a given species at roughly the same size over time. For example, comparisons of genome sizes of several strains of *E. coli* have shown them all to be about 4.5–5.5 Mbp, despite the fact that prokaryotic genomes can vary from under 0.5 to over 10 Mbp. Genome size is therefore a species-specific trait.

MiniQuiz

- Genetic maps of bacterial chromosomes are now typically made using only molecular cloning and DNA sequencing. Why were other methods also used for *E. coli*?
- How large is an average bacterial protein?
- Approximately how large is the *E. coli* genome in base pairs? How many genes does it contain?

6.6 Plasmids: General Principles

Many prokaryotic cells contain other genetic elements, in particular, **plasmids**, in addition to the chromosome. Plasmids are genetic elements that replicate independently of the host chromosome, in the sense of possessing their own origin of replication. However, they do rely on chromosomally encoded enzymes for their replication. Unlike viruses, plasmids do not have an extracellular form and exist inside cells as free, typically circular, DNA. Plasmids differ from chromosomes in carrying only nonessential (but often very helpful) genes. Essential genes reside on chromosomes. Thousands of different plasmids are known. Indeed, over 300 different naturally occurring plasmids have been isolated from strains of *Escherichia coli* alone. In this section we discuss their basic properties.

Plasmids have been widely exploited in genetic engineering. Countless new, artificial plasmids have been constructed in the laboratory. Genes from a wide variety of sources have been incorporated into such plasmids, thus allowing their transfer across any species barrier. The only requirements for artificial plasmids are that they carry genes controlling their own replication and are stably maintained in the host of choice. This topic is discussed further in Chapter 11.

Physical Nature and Replication of Plasmids

Almost all known plasmids consist of double-stranded DNA. Most are circular, but many linear plasmids are also known. Naturally occurring plasmids vary in size from approximately 1 kbp to more than 1 Mbp. Typical plasmids are circular double-stranded DNA molecules less than 5% the size of the chromosome (**Figure 6.11**). Most plasmid DNA isolated from cells is supercoiled, this being the most compact form that DNA takes within the cell (Figure 6.8). Some bacteria may contain several different types of plasmids. For example, *Borrelia burgdorferi* (the Lyme disease pathogen, ↩ Section 34.4) contains 17 different circular and linear plasmids!

The enzymes that replicate plasmids are normal cell enzymes. The genes carried by the plasmid itself are concerned primarily with controlling the initiation of replication and with partitioning replicated plasmids between daughter cells. Different plasmids

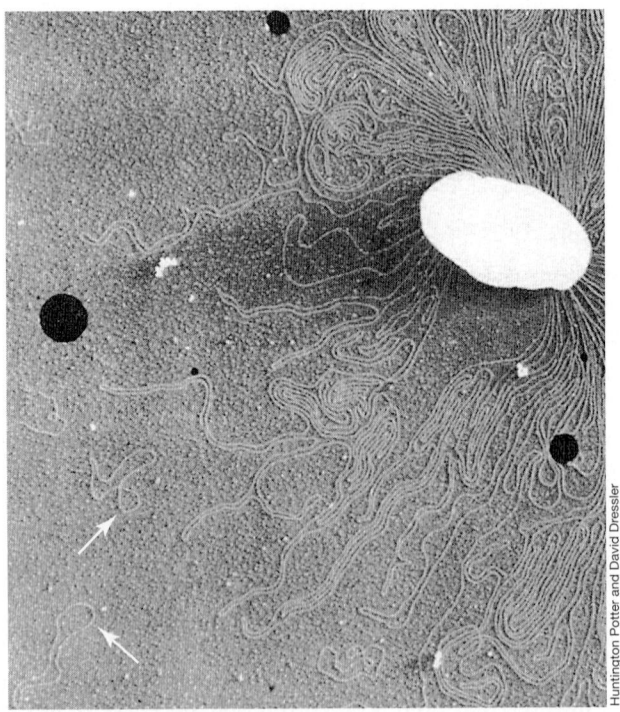

Figure 6.11 The bacterial chromosome and bacterial plasmids, as seen in the electron microscope. The plasmids (arrows) are the circular structures and are much smaller than the main chromosomal DNA. The cell (large, white structure) was broken gently so the DNA would remain intact.

Huntington Potter and David Dressler

are present in cells in different numbers; this is called the *copy number*. Some plasmids are present in the cell in only 1–3 copies, whereas others may be present in over 100 copies. Copy number is controlled by genes on the plasmid and by interactions between the host and the plasmid.

Most plasmids in gram-negative *Bacteria* replicate in a manner similar to that of the chromosome. This involves initiation at an origin of replication and bidirectional replication around the circle, giving a theta intermediate (Section 6.10). However, some small plasmids have unidirectional replication, with just a single replication fork. Because of the small size of plasmid DNA relative to the chromosome, plasmids replicate very quickly, perhaps in a tenth or less of the total time of the cell division cycle.

Most plasmids of gram-positive *Bacteria*, plus a few from gram-negative *Bacteria* and *Archaea*, replicate by a rolling circle mechanism similar to that used by bacteriophage ϕX174 (↩ Section 21.2). This mechanism proceeds via a single-stranded intermediate. Most linear plasmids replicate by using a protein bound to the 5′ end of each strand to prime DNA synthesis (↩ Section 7.7).

Plasmid Incompatibility and Plasmid Curing

Many bacterial cells contain multiple different plasmids. However, when two different plasmids are closely related genetically, they cannot both be maintained in the same cell. The two plasmids are then said to be *incompatible*. When a plasmid is transferred into a cell that already carries another related and incompatible plasmid, one or the other will be lost during subsequent cell replication. A number of incompatibility (Inc) groups exist. Plasmids belonging to the same Inc group exclude each other but can coexist with plasmids from other groups. Plasmids within each Inc group are related in sharing a common mechanism of regulating their replication. Therefore, although a bacterial cell may contain different kinds of plasmids, each is genetically distinct.

Some plasmids, called *episomes*, can integrate into the chromosome. Under such conditions their replication comes under control of the chromosome. This situation is analogous to that of several viruses whose genomes can integrate into the host genome (↩ Section 9.10).

Plasmids can sometimes be eliminated from host cells by various treatments. This removal, called *curing*, results from inhibition of plasmid replication without parallel inhibition of chromosome replication. As a result, the plasmid is diluted out during cell division. Curing may occur spontaneously, but is greatly increased by treatments with certain chemicals such as acridine dyes, which insert into DNA, or other treatments that interfere more with plasmid replication than with chromosome replication.

Cell-to-Cell Transfer of Plasmids

How do plasmids manage to infect new host cells? Some prokaryotic cells can take up free DNA from the environment (↩ Section 10.7). Consequently, plasmids released by the death and disintegration of their previous host cell may be taken up by a new host. However, few bacterial species have this ability, and it is unlikely to account for much plasmid transfer. The main mechanism of plasmid transfer is *conjugation*, a function encoded by some plasmids themselves that involves cell-to-cell contact (Section 10.9).

Plasmids capable of transferring themselves by cell-to-cell contact are called *conjugative*. Not all plasmids are conjugative. Transfer by conjugation is controlled by a set of genes on the plasmid called the *tra* (for transfer) region. These genes encode proteins that function in DNA transfer and replication and others that function in mating pair formation. If a plasmid possessing a *tra* region becomes integrated into the chromosome, the plasmid can then mobilize the chromosomal DNA, which may be transferred from one cell to another (Section 10.10).

Most conjugative plasmids can only move between closely related species of bacteria. However, some conjugative plasmids from *Pseudomonas* have a broad host range. This means that they are transferable to a wide variety of other gram-negative *Bacteria*. Such plasmids can transfer genes between distantly related organisms. Conjugative plasmids have been shown to transfer between gram-negative and gram-positive *Bacteria*, between *Bacteria* and plant cells, and between *Bacteria* and fungi. Even if the plasmid cannot replicate independently in the new host, transfer of the plasmid itself could have important evolutionary consequences if genes from the plasmid recombine with the genome of the new host.

MiniQuiz

- How does a plasmid differ from a virus?
- How can a large plasmid be differentiated from a small chromosome?
- What function do the *tra* genes of the F plasmid carry out?

6.7 The Biology of Plasmids

Clearly, all plasmids must carry genes that ensure their own replication. In addition, some plasmids also carry genes necessary for conjugation. Although plasmids do not carry genes that are essential to the host, plasmids may carry genes that profoundly influence host cell physiology. In some cases plasmids encode properties fundamental to the ecology of the bacterium. For example, the ability of *Rhizobium* to interact with plants and form nitrogen-fixing root nodules requires certain plasmid functions (Section 25.8). Other plasmids confer special metabolic properties on bacterial cells, such as the ability to degrade toxic pollutants. Indeed, plasmids are a major mechanism for conferring special properties on bacteria and for mobilizing these properties by horizontal gene flow. Some special properties conferred by plasmids are summarized in **Table 6.2**.

Resistance Plasmids

Among the most widespread and well-studied groups of plasmids are the resistance plasmids, usually just called *R plasmids*, which confer resistance to antibiotics and various other growth inhibitors. Several antibiotic resistance genes can be carried by a single R plasmid, or, alternatively, a cell may contain several R plasmids. In either case, the result is multiple resistance. R plasmids were first discovered in Japan in the 1950s in strains of enteric

Table 6.2 *Examples of phenotypes conferred by plasmids in prokaryotes*

Phenotype class	Organisms[a]
Antibiotic production	*Streptomyces*
Conjugation	Wide range of bacteria
Metabolic functions	
Degradation of octane, camphor, naphthalene	*Pseudomonas*
Degradation of herbicides	*Alcaligenes*
Formation of acetone and butanol	*Clostridium*
Lactose, sucrose, citrate, or urea utilization	Enteric bacteria
Pigment production	*Erwinia, Staphylococcus*
Gas vesicle production	*Halobacterium*
Resistance	
Antibiotic resistance	Wide range of bacteria
Resistance to toxic metals	Wide range of bacteria
Virulence	
Tumor production in plants	*Agrobacterium*
Nodulation and symbiotic nitrogen fixation	*Rhizobium*
Bacteriocin production and resistance	Wide range of bacteria
Animal cell invasion	*Salmonella, Shigella, Yersinia*
Coagulase, hemolysin, enterotoxin	*Staphylococcus*
Toxins and capsule	*Bacillus anthracis*
Enterotoxin, K antigen	*Escherichia coli*

bacteria that had acquired resistance to sulfonamide antibiotics. Since then they have been found throughout the world. The emergence of bacteria resistant to antibiotics is of considerable medical significance and is correlated with the increasing use of antibiotics for treating infectious diseases (Section 26.12). Soon after these resistant strains were isolated, it was shown that they could transfer resistance to sensitive strains via cell-to-cell contact. The infectious nature of conjugative R plasmids permitted their rapid spread through cell populations.

In general, resistance genes encode proteins that either inactivate the antibiotic or protect the cell by some other mechanism. Plasmid R100, for example, is a 94.3-kbp plasmid (**Figure 6.12**) that carries genes encoding resistance to sulfonamides, streptomycin, spectinomycin, fusidic acid, chloramphenicol, and tetracycline. Plasmid R100 also carries several genes conferring resistance to mercury. Plasmid R100 can be transferred between enteric bacteria of the genera *Escherichia, Klebsiella, Proteus, Salmonella*, and *Shigella*, but does not transfer to gram-negative bacteria outside the enteric group. Different R plasmids with genes for resistance to most antibiotics are known. Many drug-resistant modules on R plasmids, such as those on R100, are also transposable elements (Section 12.11), and this, combined with the fact that many of these plasmids are conjugative, have made them a serious threat to traditional antibiotic therapy.

Plasmids Encoding Virulence Characteristics

Pathogenic microorganisms possess a variety of characteristics that enable them to colonize hosts and establish infections. Here we note two major characteristics of the virulence (disease-causing

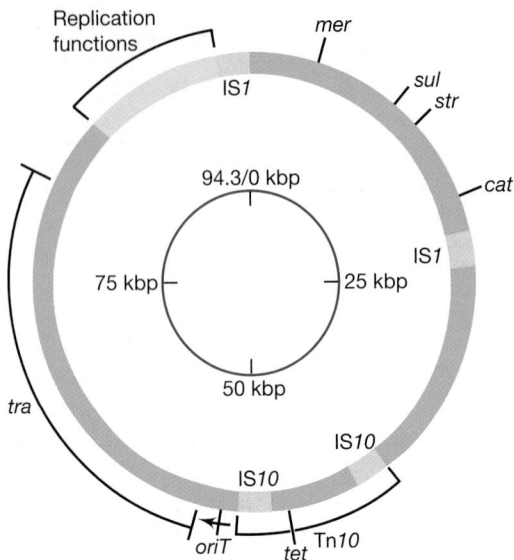

Figure 6.12 Genetic map of the resistance plasmid R100. The inner circle shows the size in kilobase pairs. The outer circle shows the location of major antibiotic resistance genes and other key functions: *mer*, mercuric ion resistance; *sul*, sulfonamide resistance; *str*, streptomycin resistance; *cat*, chloramphenicol resistance; *tet*, tetracycline resistance; *oriT*, origin of conjugative transfer; *tra*, transfer functions. The locations of insertion sequences (IS) and the transposon Tn*10* are also shown. Genes for plasmid replication are found in the region from 88 to 92 kbp.

ability) of pathogens that are often plasmid encoded: (1) the ability of the pathogen to attach to and colonize specific host tissue and (2) the production of toxins, enzymes, and other molecules that cause damage to the host.

Enteropathogenic strains of *Escherichia coli* are characterized by the ability to colonize the small intestine and to produce a toxin that causes diarrhea. Colonization requires a cell surface protein called *colonization factor antigen*, encoded by a plasmid. This protein confers on bacterial cells the ability to attach to epithelial cells of the intestine. At least two toxins in enteropathogenic *E. coli* are encoded by plasmids: the hemolysin, which lyses red blood cells, and the enterotoxin, which induces extensive secretion of water and salts into the bowel. It is the enterotoxin that is responsible for diarrhea (Section 27.11).

Some virulence factors are encoded on plasmids. Other virulence factors are encoded by other mobile genetic elements, such as transposons and bacteriophages. Some virulence factors are chromosomal. Several examples are known in which multiple virulence genes are present on different genetic elements within the same cell. For instance, the genes encoding the virulence determinants of Shiga toxin–producing strains of *E. coli* (Section 36.9) are distributed among the chromosome, a bacteriophage, and a plasmid.

Bacteriocins

Many bacteria produce proteins that inhibit or kill closely related species or even different strains of the same species. These agents are called **bacteriocins** to distinguish them from antibiotics. Bacteriocins have a narrower spectrum of activity than antibiotics. The genes encoding bacteriocins and the proteins needed for processing and transporting them and for conferring immunity on the producing organism are usually carried on plasmids or transposons. Bacteriocins are often named after the species of organism that produces them. Thus, *E. coli* produces *colicins*; *Yersinia pestis* produces *pesticins*, and so on.

The Col plasmids of *E. coli* encode various colicins. Col plasmids can be either conjugative or nonconjugative. Colicins released from the producer cell bind to specific receptors on the surface of susceptible cells. The receptors for colicins are typically proteins whose normal function is to transport growth factors or micronutrients across the outer membrane of the cell. Colicins kill cells by disrupting some critical cell function. Many colicins form channels in the cell membrane that allow potassium ions and protons to leak out, leading to loss of the ability to generate energy. Another major group of colicins are nucleases and degrade DNA or RNA. For example, colicin E2 is a DNA endonuclease that cleaves DNA, and colicin E3 is a ribonuclease that cuts at a specific site in 16S rRNA and therefore inactivates ribosomes.

The bacteriocins or bacteriocin-like agents of gram-positive bacteria are quite different from the colicins but are also often encoded by plasmids; some even have commercial value. For instance, lactic acid bacteria produce the bacteriocin nisin A, which strongly inhibits the growth of a wide range of gram-positive bacteria and is used as a preservative in the food industry.

MiniQuiz

- What properties does an R plasmid confer on its host cell?
- What properties does a virulence plasmid typically confer on its host cell?
- How do bacteriocins differ from antibiotics?

DNA Replication

DNA replication is necessary for cells to divide, whether to reproduce new organisms, as in unicellular microorganisms, or to produce new cells as part of a multicellular organism. DNA replication must be sufficiently accurate that the daughter cells are genetically identical to the mother cell (or almost so). This involves a host of special enzymes and processes.

6.8 Templates and Enzymes

DNA exists in cells as a double helix with complementary base pairing. If the double helix is opened up, a new strand can be synthesized as the complement of each parental strand. As shown in **Figure 6.13**, replication is **semiconservative**, meaning that the two resulting double helices consist of one new strand and one parental strand. The DNA strand that is used to make a complementary daughter strand is called the template, and in DNA replication each parental strand is a template for one newly synthesized strand (Figure 6.13).

The precursor of each new nucleotide in the DNA strand is a deoxynucleoside 5′-triphosphate. The two terminal phosphates are removed and the innermost phosphate is then attached

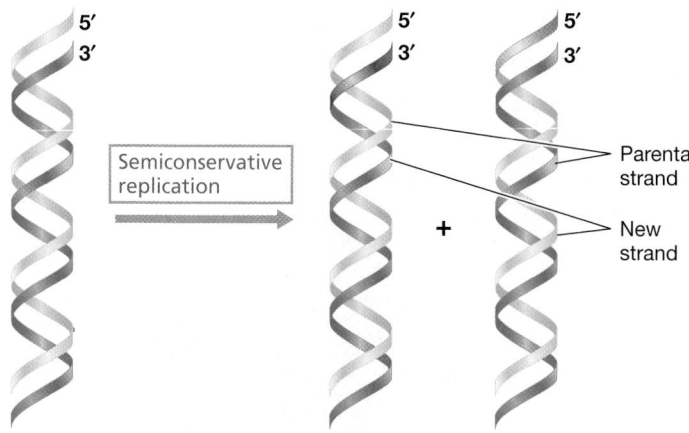

Figure 6.13 Overview of DNA replication. DNA replication is a semi-conservative process in all cells. Note that the new double helices each contain one new strand (shown topped in red) and one parental strand.

Figure 6.15 The RNA primer. Structure of the RNA–DNA hybrid formed during initiation of DNA synthesis.

covalently to a deoxyribose of the growing chain (**Figure 6.14**). This addition of the incoming nucleotide requires the presence of a free hydroxyl group, which is available only at the 3′ end of the molecule. This leads to the important principle that

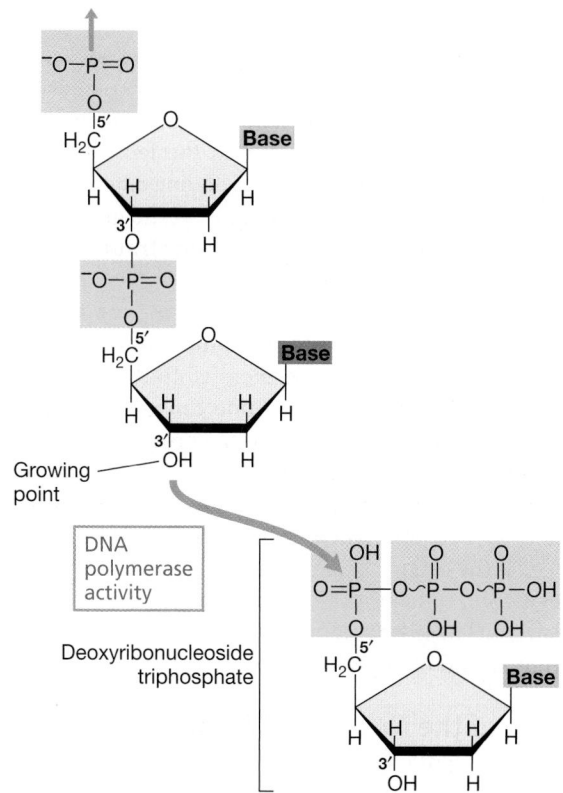

Figure 6.14 Extension of a DNA chain by adding a deoxyribonucleoside triphosphate at the 3′ end. Growth proceeds from the 5′-phosphate to the 3′-hydroxyl end. DNA polymerase catalyzes the reaction. The four precursors are deoxythymidine triphosphate (dTTP), deoxyadenosine triphosphate (dATP), deoxyguanosine triphosphate (dGTP), and deoxycytidine triphosphate (dCTP). Upon nucleotide insertion, the two terminal phosphates of the triphosphate are split off as pyrophosphate (PP$_i$). Thus, two energy-rich phosphate bonds are consumed when adding each nucleotide.

DNA replication always proceeds from the 5′ end to the 3′ end, the 5′-phosphate of the incoming nucleotide being attached to the 3′-hydroxyl of the previously added nucleotide.

Enzymes that catalyze the addition of deoxynucleotides are called **DNA polymerases**. Several such enzymes exist, each with a specific function. There are five different DNA polymerases in *Escherichia coli*, called DNA polymerases I, II, III, IV, and V. DNA polymerase III (Pol III) is the primary enzyme for replicating chromosomal DNA. DNA polymerase I (Pol I) is also involved in chromosomal replication, though to a lesser extent (see below). The other DNA polymerases help repair damaged DNA (Section 10.4).

All known DNA polymerases synthesize DNA in the 5′ → 3′ direction. However, no known DNA polymerase can initiate a new chain; all of these enzymes can only add a nucleotide onto a preexisting 3′-OH group. To start a new chain, a **primer**, a nucleic acid molecule to which DNA polymerase can attach the first nucleotide, is required. In most cases this primer is a short stretch of RNA.

When the double helix is opened at the beginning of replication, an RNA-polymerizing enzyme makes the RNA primer. This enzyme, called **primase**, synthesizes a short stretch of RNA of around 11–12 nucleotides that is complementary in base pairing to the template DNA. At the growing end of this RNA primer is a 3′-OH group to which DNA polymerase can add the first deoxyribonucleotide. Continued extension of the molecule thus occurs as DNA rather than RNA. The newly synthesized molecule has a structure like that shown in **Figure 6.15**. The primer will eventually be removed and replaced with DNA, as described later.

MiniQuiz

- To which end (5′ end or 3′ end) of a newly synthesized strand of DNA does polymerase add a base?
- Why is a primer required for DNA replication?

6.9 The Replication Fork

Much of our understanding of the details of DNA replication has been obtained from studying the bacterium *Escherichia coli*, and the following discussion deals primarily with this organism. However, DNA replication is probably quite similar in all *Bacteria*. By contrast, although most species of *Archaea* have circular chromosomes, many events in DNA replication resemble those in eukaryotic cells more than those in *Bacteria* (Chapter 7). This again reflects the phylogenetic affiliation between *Archaea* and *Eukarya*.

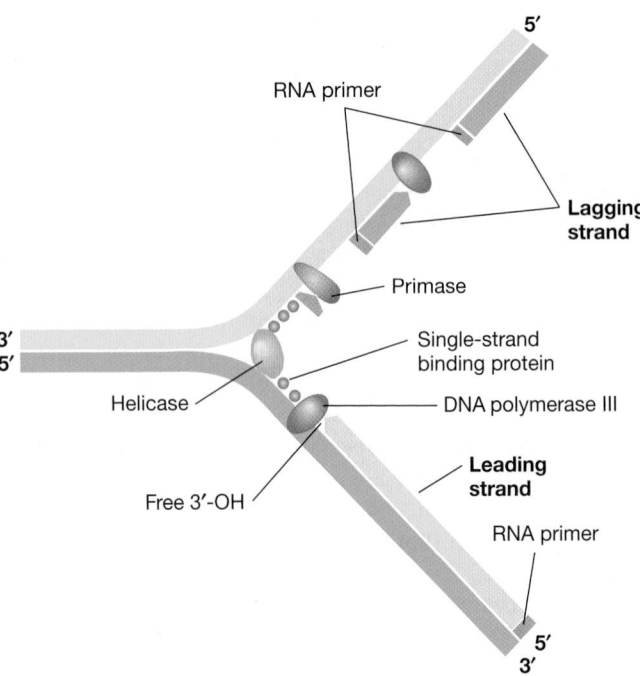

Figure 6.16 Events at the DNA replication fork. Note the polarity and antiparallel nature of the DNA strands.

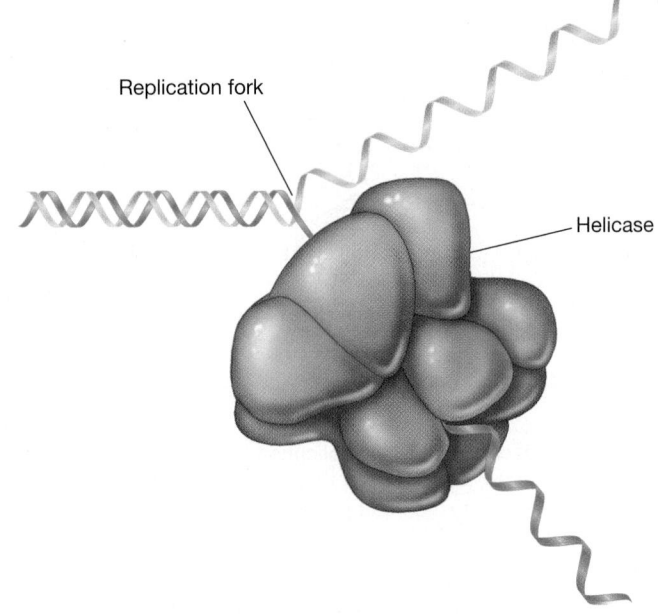

Figure 6.17 DNA helicase unwinding a double helix. In this figure, the protein and DNA molecules are drawn to scale. Simple diagrams often give the incorrect impression that most proteins are relatively small compared to DNA. Although DNA molecules are generally extremely long, they are relatively thin compared to many proteins.

Initiation of DNA Synthesis

Before DNA polymerase can synthesize new DNA, the double helix must be unwound to expose the template strands. The zone of unwound DNA where replication occurs is known as the **replication fork**. The enzyme DNA helicase unwinds the double helix, using energy from ATP, and exposes a short single-stranded region (**Figures 6.16** and **6.17**). Helicase moves along the DNA and separates the strands just in advance of the replication fork. The single-stranded region is covered by single-strand binding protein. This stabilizes the single-stranded DNA and prevents the double helix from re-forming.

Unwinding of the double helix by helicase generates positive supercoils ahead of the advancing replication fork. To counteract this, DNA gyrase travels along the DNA ahead of the replication fork and inserts negative supercoils to cancel out the positive supercoiling.

Bacteria possess a single location on the chromosome where DNA synthesis is initiated, the origin of replication (*oriC*). This consists of a specific DNA sequence of about 250 bases that is recognized by initiation proteins, in particular a protein called DnaA (**Table 6.3**), which binds to this region and opens up the double helix. Next to assemble is the helicase (known as DnaB), which is helped onto the DNA by the helicase loader protein (DnaC). Two helicases are loaded, one onto each strand, facing in opposite directions. Next, two primase and then two DNA polymerase enzymes are loaded onto the DNA behind the helicases. Initiation of DNA replication then begins on the two single strands. As replication proceeds, the replication fork appears to move along the DNA (Figure 6.16). www.microbiologyplace.com

Online Tutorial 6.1: DNA Replication

Leading and Lagging Strands

Figure 6.16 shows an important distinction in replication between the two DNA strands due to the fact that replication always proceeds from $5' \rightarrow 3'$ (always adding a new nucleotide to the 3'-OH of the growing chain). On the strand growing from the $5'$-PO_4^{2-} to the $3'$-OH, called the **leading strand**, DNA synthesis occurs continuously because there is always a free 3'-OH at the replication fork to which a new nucleotide can be added. But on the opposite strand, called the **lagging strand**, DNA synthesis occurs discontinuously because there is no 3'-OH at the replication fork to which a new nucleotide can attach. Where is the 3'-OH on this strand? It is located at the opposite end, away from the replication fork. Therefore, on the lagging strand, RNA primers must be synthesized by primase multiple times to provide free 3'-OH groups. By contrast, the leading strand is primed only once, at the origin. As a result, the lagging strand is made in short segments, called *Okazaki fragments*, after their discoverer, Reiji Okazaki. These fragments are joined together later to give a continuous strand of DNA.

Synthesis of the New DNA Strands

After synthesizing the RNA primer, primase is replaced by Pol III. This enzyme is a complex of several proteins (Table 6.3), including the polymerase core enzyme itself. Each polymerase is held on the DNA by a sliding clamp, which encircles and slides along the single template strands of DNA. Consequently, the replication fork contains two polymerase core enzymes and two sliding clamps, one set for each strand. However, there is only a single clamp-loader complex. This is needed to assemble the sliding clamps onto the DNA. After assembly on the lagging

Table 6.3 *Major enzymes involved in DNA replication in* Bacteria

Enzyme	Encoding genes	Function
DNA gyrase	*gyrAB*	Replaces supercoils ahead of replisome
Origin-binding protein	*dnaA*	Binds origin of replication to open double helix
Helicase loader	*dnaC*	Loads helicase at origin
Helicase	*dnaB*	Unwinds double helix at replication fork
Single-strand binding protein	*ssb*	Prevents single strands from annealing
Primase	*dnaG*	Primes new strands of DNA
DNA polymerase III		Main polymerizing enzyme
Sliding clamp	*dnaN*	Holds Pol III on DNA
Clamp loader	*holA–E*	Loads Pol III onto sliding clamp
Dimerization subunit (Tau)	*dnaX*	Holds together the two core enzymes for the leading and lagging strands
Polymerase subunit	*dnaE*	Strand elongation
Proofreading subunit	*dnaQ*	Proofreading
DNA polymerase I	*polA*	Excises RNA primer and fills in gaps
DNA ligase	*ligA, ligB*	Seals nicks in DNA
Tus protein	*tus*	Binds terminus and blocks progress of the replication fork
Topoisomerase IV	*parCE*	Unlinking of interlocked circles

strand, the elongation component of Pol III, DnaE, then adds deoxyribonucleotides until it reaches previously synthesized DNA (**Figure 6.18**). At this point, Pol III stops.

The next enzyme to take part, Pol I, has more than one enzymatic activity. Besides synthesizing DNA, Pol I has a $5' \rightarrow 3'$ exonuclease activity that removes the RNA primer preceding it (Figure 6.18). When the primer has been removed and replaced with DNA, Pol I is released. The last phosphodiester bond is made by an enzyme called **DNA ligase**. This enzyme seals nicks in DNAs that have an adjacent $5'$-PO_4^{2-} and $3'$-OH (something that Pol III is unable to do), and along with Pol I, it also participates in DNA repair. DNA ligase is also important for sealing genetically manipulated DNA during molecular cloning (↺ Section 11.3).

MiniQuiz

• Why are there leading and lagging strands?

• What recognizes the origin of replication?

• What enzymes take part in joining the fragments of the lagging strand?

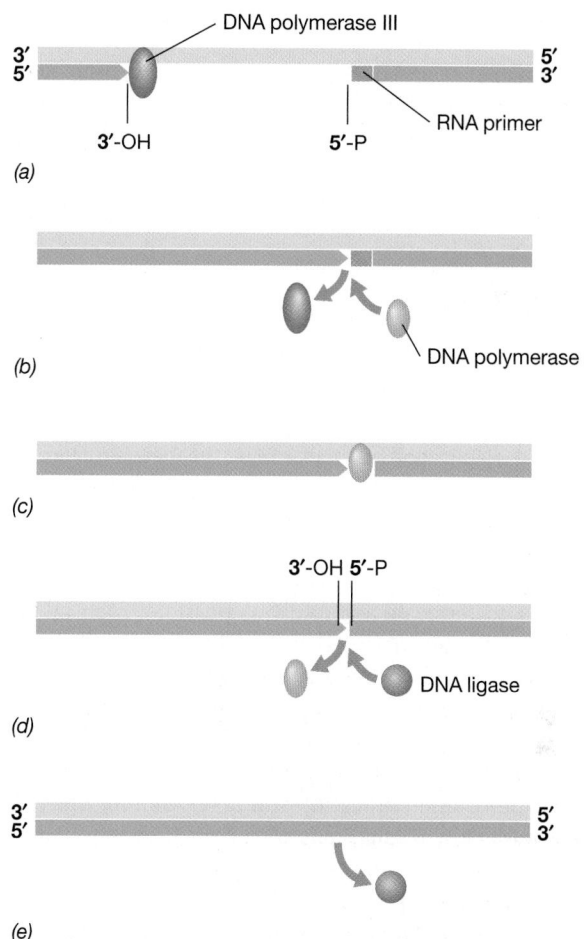

Figure 6.18 Sealing two fragments on the lagging strand. *(a)* DNA polymerase III is synthesizing DNA in the $5' \rightarrow 3'$ direction toward the RNA primer of a previously synthesized fragment on the lagging strand. *(b)* On reaching the fragment, DNA polymerase III leaves and is replaced by DNA polymerase I. *(c)* DNA polymerase I continues synthesizing DNA while removing the RNA primer from the previous fragment. *(d)* DNA ligase replaces DNA polymerase I after the primer has been removed. *(e)* DNA ligase seals the two fragments together.

6.10 Bidirectional Replication and the Replisome

The circular nature of the chromosome of *Escherichia coli* and most other prokaryotes creates an opportunity for speeding up replication. In *E. coli*, and probably in all prokaryotes with circular chromosomes, replication is *bidirectional* from the origin of replication, as shown in **Figures 6.19** and **6.20**. There are thus two replication forks on each chromosome moving in opposite directions. These are held together by the two Tau protein subunits. In circular DNA, bidirectional replication leads to the formation of characteristic shapes called theta structures (Figure 6.19). Most large DNA molecules, whether from prokaryotes or eukaryotes, have bidirectional replication from fixed origins. In fact, large eukaryotic chromosomes have multiple origins (↺ Section 7.7). During bidirectional replication, synthesis occurs in both a leading and lagging fashion on each template strand (Figure 6.20).

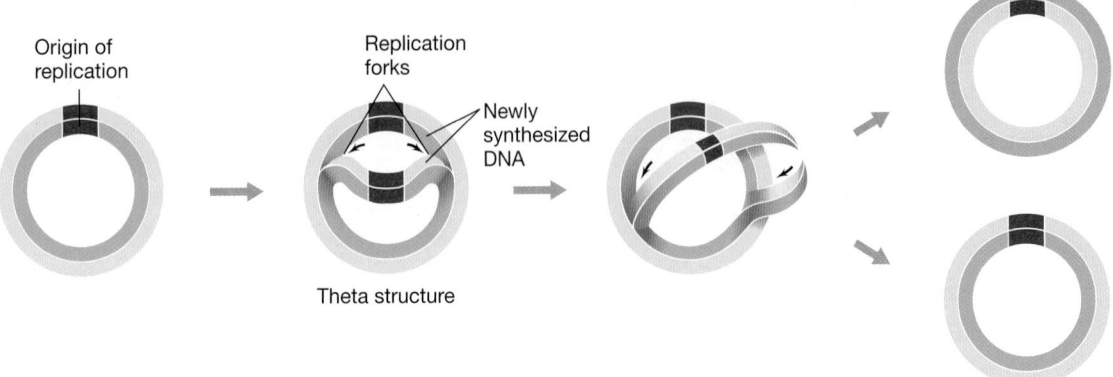

Figure 6.19 Replication of circular DNA: the theta structure. In circular DNA, bidirectional replication from an origin forms an intermediate structure resembling the Greek letter theta (θ).

Bidirectional DNA synthesis around a circular chromosome allows DNA to replicate as rapidly as possible. Even taking this into account and considering that Pol III can add nucleotides to a growing DNA strand at the rate of about 1000 per second, chromosome replication in *E. coli* still takes about 40 min. Interestingly, under the best growth conditions, *E. coli* can grow with a doubling time of about 20 min. However, even under these conditions, chromosome replication still takes 40 min. The solution to this conundrum is that cells of *E. coli* growing at doubling times shorter than 40 min contain multiple DNA replication forks. That is, a new round of DNA replication begins before the last round has been completed (**Figure 6.21**). Only in this way can a generation time shorter than the chromosome replication time be maintained.

The Replisome

Figure 6.16 shows the differences in replication of the leading and the lagging strands and the enzymes involved. From such a simplified drawing it would appear that each replication fork contains a host of different proteins all working independently. Actually, this is not so. These proteins aggregate to form a large replication complex called the *replisome* (**Figure 6.22**). The lagging strand of DNA loops out to allow the replisome to move smoothly along both strands, and the replisome literally pulls the DNA template through it as replication occurs. Therefore, it is

the DNA, rather than DNA polymerase, that moves during replication. Note also how helicase and primase form a subcomplex, called the *primosome*, which aids their working in close association during the replication process.

In summary, in addition to Pol III, the replisome contains several key replication proteins: (1) DNA gyrase, which removes supercoils; (2) DNA helicase and primase (the primosome), which unwind and prime the DNA; and (3) single-strand binding protein, which prevents the separated template strands from reforming a double helix (Figure 6.22). Table 6.3 summarizes the properties of proteins essential for DNA replication.

Fidelity of DNA Replication: Proofreading

DNA replicates with a remarkably low error rate. Nevertheless, when errors do occur, a backup mechanism exists to detect and correct them. Errors in DNA replication introduce mutations, changes in DNA sequence. Mutation rates in cells are remarkably low, between 10^{-8} and 10^{-11} errors per base pair inserted. This accuracy is possible partly because DNA polymerases get two chances to incorporate the correct base at a given site. The first chance comes when complementary bases are inserted opposite the bases on the template strand by Pol III according to the base-pairing rules, A with T and G with C. The second chance depends upon a second enzymatic activity of both Pol I and Pol III, called *proofreading* (**Figure 6.23**). In Pol III, a separate protein subunit,

Figure 6.20 Dual replication forks in the circular chromosome. At an origin of replication that directs bidirectional replication, two replication forks must start. Therefore, two leading strands must be primed, one in each direction. In *Escherichia coli*, the origin of replication is recognized by a specific protein, DnaA. Note that DNA synthesis is occurring in both a leading and a lagging manner on each of the new daughter strands. Compare this figure with the description of the replisome shown in Figure 6.22.

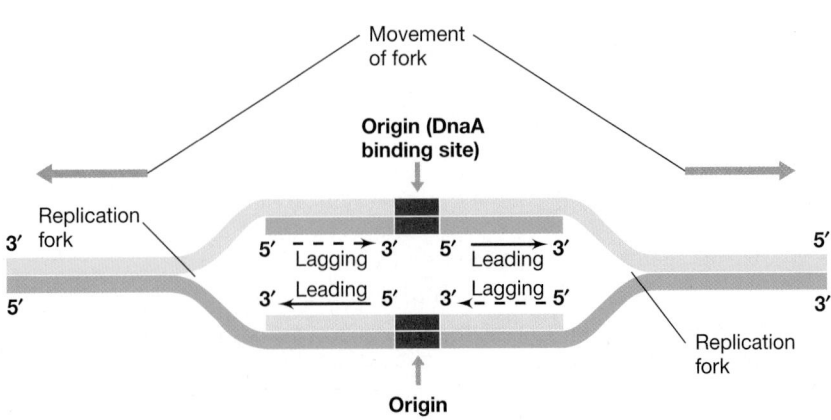

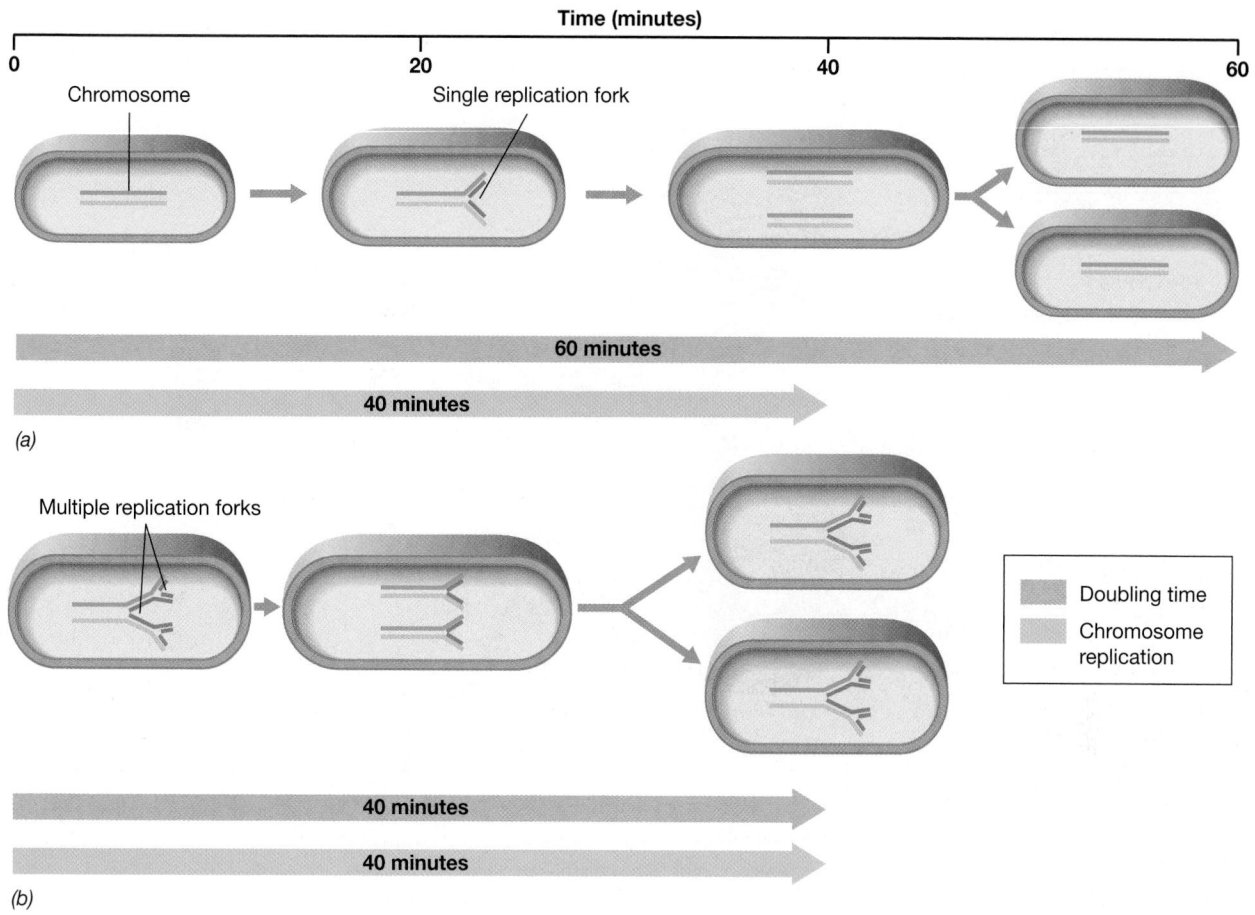

Figure 6.21 **Cell division versus chromosome duplication.** *(a)* Cells of *Escherichia coli* take approximately 40 min to replicate the chromosome and an additional 20 min for cell division. *(b)* When cells double in less than 60 min, a new round of chromosome replication must be initiated before the previous round is finished.

DnaQ, performs the proofreading, whereas in Pol I a single protein performs all functions.

Proofreading activity occurs if an incorrect base has been inserted because this creates a mismatch in base pairing. Both Pol I and Pol III possess a $3' \rightarrow 5'$ exonuclease activity that can remove such wrongly inserted nucleotides. The polymerase senses the mistake because incorrect base pairing causes a slight distortion in the double helix. After the removal of a mismatched nucleotide, the polymerase then gets a second chance to insert the correct nucleotide (Figure 6.23). The proofreading exonuclease activity is distinct from the $5' \rightarrow 3'$ exonuclease activity of Pol I that removes the RNA primer from both the leading and lagging strands. Only Pol I has this latter activity. Exonuclease proofreading occurs in prokaryotes, eukaryotes, and viral DNA replication systems. However, many organisms have additional mechanisms for reducing errors made during DNA replication, which operate after the replication fork has passed by. We will discuss some of these in Chapter 10.

Termination of Replication

Eventually the process of DNA replication is finished. How does the replisome know when to stop? On the opposite side of the circular chromosome from the origin is a site called the terminus of replication. Here the two replication forks collide as the new circles of DNA are completed. The details of termination are not fully known. However, in the terminus region there are several DNA sequences called *Ter* sites that are recognized by a protein called Tus, whose function is to block progress of the replication forks. When replication of the circular chromosome is complete, the two circular molecules are linked together, much like the links of a chain. They are unlinked by another enzyme, topoisomerase IV. Obviously, it is critical that, after DNA replication, the DNA is partitioned so that each daughter cell receives a copy of the chromosome. This process may be assisted by the important cell division protein FtsZ, which helps orchestrate several key events of cell division (↻ Section 5.2).

MiniQuiz

- What is the replisome and what are its components?
- How can *Escherichia coli* carry out cell division in less time than it takes to duplicate its chromosome?
- How is proofreading carried out during DNA replication?
- What brings the replication forks to a halt in the terminus region of the chromosome?

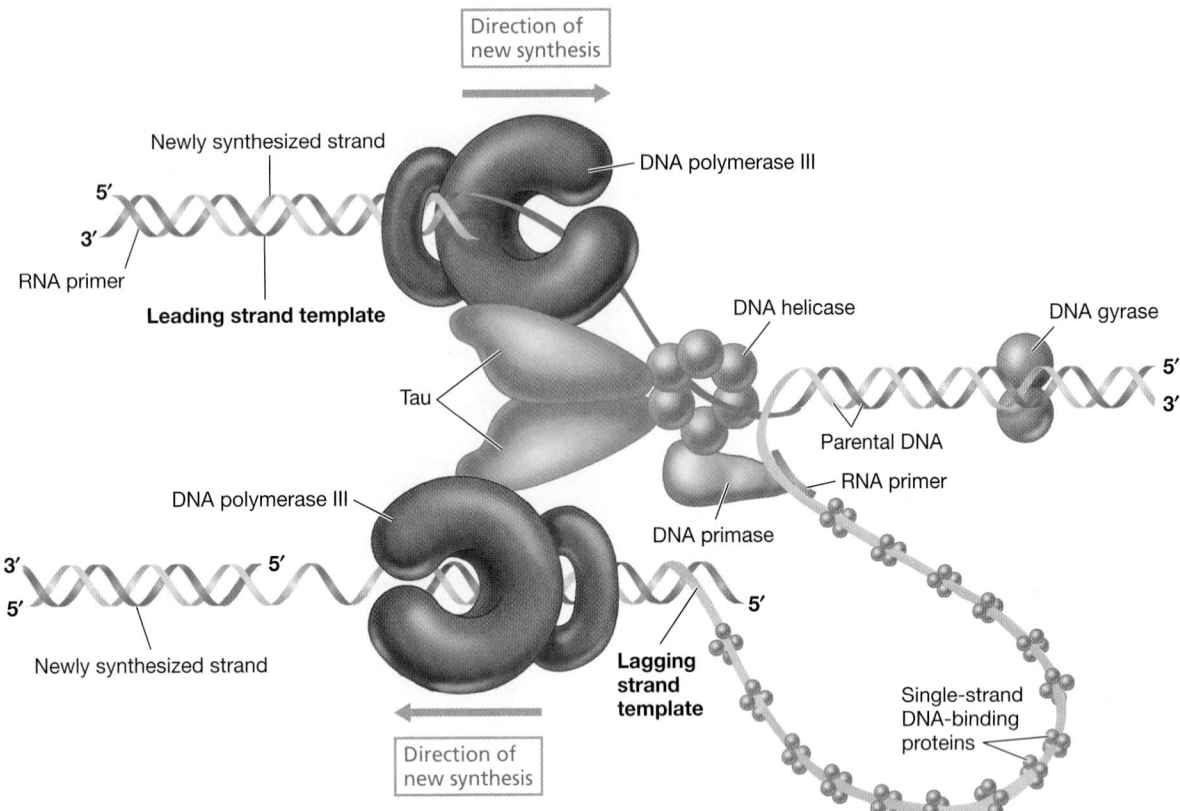

Figure 6.22 The replisome. The replisome consists of two copies of DNA polymerase III, plus helicase and primase (together forming the primosome), and many copies of single-strand DNA-binding protein. The tau subunits hold the two DNA polymerase assemblies and helicase together. Just upstream of the replisome, DNA gyrase removes supercoils in the DNA to be replicated. Note that the two polymerases are replicating the two individual strands of DNA in opposite directions. Consequently, the lagging-strand template loops around so that the whole replisome moves in the same direction along the chromosome.

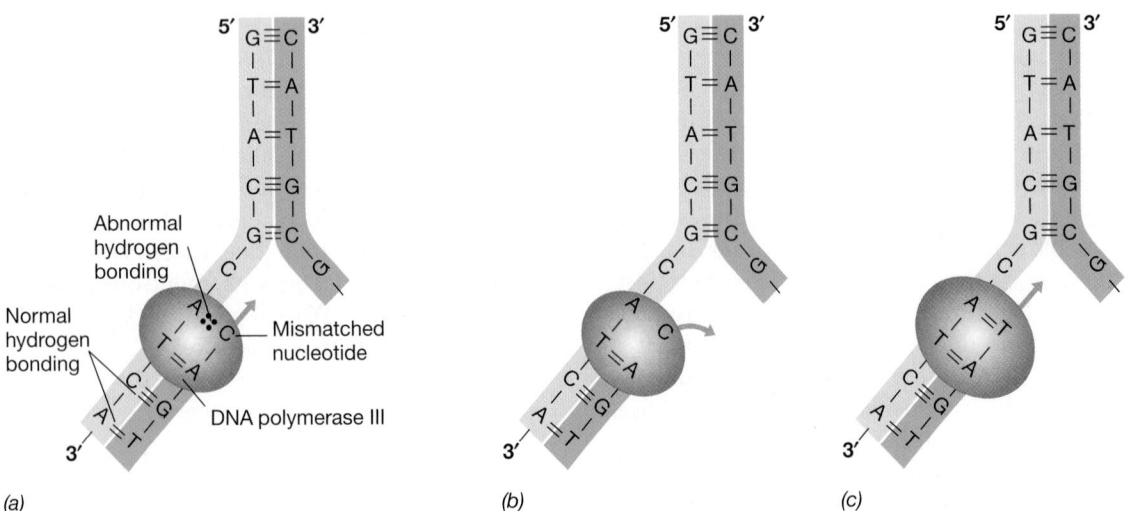

Figure 6.23 Proofreading by the 3′ → 5′ exonuclease activity of DNA polymerase III. (a) A mismatch in base pairing at the terminal base pair causes the polymerase to pause briefly. This signals the proofreading activity to (b) cut out the mismatched nucleotide, after which (c) the correct base is inserted by the polymerase activity.

6.11 The Polymerase Chain Reaction (PCR)

The **polymerase chain reaction (PCR)** is essentially DNA replication *in vitro*. The PCR can copy segments of DNA by up to a billionfold in the test tube, a process called *amplification*. This yields large amounts of specific genes or other DNA segments that may be used for a host of applications in molecular biology. PCR uses the enzyme DNA polymerase, which naturally copies DNA molecules (Section 6.8). Artificially synthesized primers (↩ Section 11.4) are used to initiate DNA synthesis, but are made of DNA (rather than RNA like the primers used by cells). PCR does not actually copy whole DNA molecules but amplifies stretches of up to a few thousand base pairs (the *target*) from within a larger DNA molecule (the *template*). PCR was conceived by Kary Mullis, who received a Nobel Prize for this achievement.

The steps in PCR amplification of DNA can be summarized as follows (**Figure 6.24**):

1. The template DNA is denatured by heating.

2. Two artificial DNA oligonucleotide primers flanking the target DNA are present in excess. This ensures that most template strands anneal to a primer, and not to each other, as the mixture cools (Figure 6.24*a*).

3. DNA polymerase then extends the primers using the original DNA as the template (Figure 6.24*b*).

4. After an appropriate incubation period, the mixture is heated again to separate the strands. The mixture is then cooled to allow the primers to hybridize with complementary regions of newly synthesized DNA, and the whole process is repeated (Figure 6.24*c*).

The power of PCR is that the products of one primer extension are templates for the next cycle. Consequently, each cycle doubles the amount of the original target DNA. In practice, 20–30 cycles are usually run, yielding a 10^6-fold to 10^9-fold increase in the target sequence (Figure 6.24*d*). Because the technique consists of several highly repetitive steps, PCR machines, called *thermocyclers*, are available that run through the heating and cooling cycles automatically. Because each cycle requires only about 5 min, the automated procedure gives large amplifications in only a few hours.

PCR at High Temperature

The original PCR technique employed the DNA polymerase *Escherichia coli* Pol III, but because of the high temperatures needed to denature the double-stranded copies of DNA, the enzyme was also denatured and had to be replenished every cycle. This problem was solved by employing a thermostable DNA polymerase isolated from the thermophilic hot spring bacterium *Thermus aquaticus*. DNA polymerase from *T. aquaticus*, called *Taq polymerase*, is stable to 95°C and thus is unaffected by the denaturation step employed in the PCR. The use of Taq DNA polymerase also increased the specificity of the PCR because the DNA is copied at 72°C rather than 37°C. At such high temperatures, nonspecific hybridization of primers to nontarget DNA is

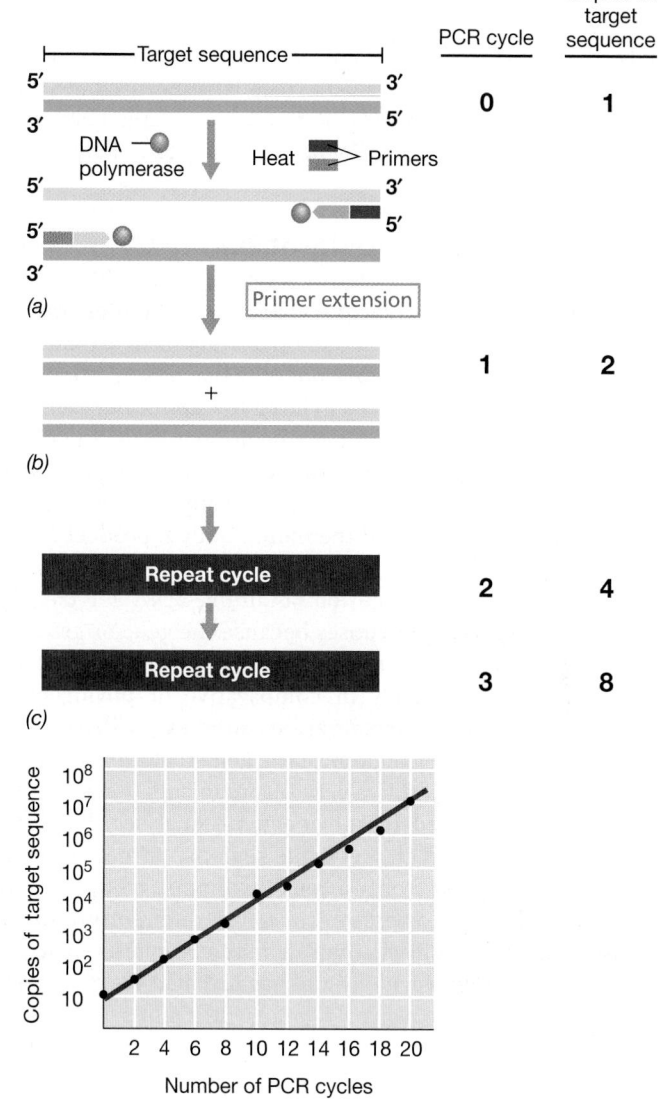

Figure 6.24 The polymerase chain reaction (PCR). The PCR amplifies specific DNA sequences. *(a)* Target DNA is heated to separate the strands, and a large excess of two oligonucleotide primers, one complementary to each strand, is added along with DNA polymerase. *(b)* Following primer annealing, primer extension yields a copy of the original double-stranded DNA. *(c)* Two additional PCR cycles yield four and eight copies, respectively, of the original DNA sequence. *(d)* Effect of running 20 PCR cycles on a DNA preparation originally containing ten copies of a target gene. Note that the plot is semilogarithmic.

rare, thus making the products of Taq PCR more homogeneous than those obtained using the *E. coli* enzyme. On the other hand, the primer hybridization step is often carried out at lower temperatures, which may allow some nonspecific binding.

DNA polymerase from *Pyrococcus furiosus*, a hyperthermophile with a growth temperature optimum of 100°C (↩ Section 19.5) is called *Pfu polymerase* and is even more thermostable than Taq polymerase. Moreover, unlike Taq polymerase, Pfu polymerase has proofreading activity (Section 6.10), making it especially useful when high accuracy is crucial. Thus, the error rate for Taq polymerase under standard conditions is 8.0×10^{-6} (per base duplicated), whereas for Pfu polymerase it is only 1.3×10^{-6}. To supply the commercial demand for thermostable DNA polymerases, the genes for these enzymes have been cloned into *E. coli*, allowing the enzymes to be produced in large quantities. www.microbiologyplace.com **Online Tutorial 6.2: Polymerase Chain Reaction (PCR)**

Applications and Sensitivity of PCR

PCR is a powerful tool. It is easy to perform, extremely sensitive and specific, and highly efficient. During each round of amplification the amount of product doubles, leading to an exponential increase in the DNA. This means not only that a large amount of amplified DNA can be produced in just a few hours, but that only a few molecules of target DNA need be present in the sample to start the reaction. The reaction is so specific that, with primers of 15 or so nucleotides and high annealing temperatures, there is almost no "false priming," and therefore the PCR product is virtually homogeneous.

PCR is extremely valuable for obtaining DNA for cloning genes or for sequencing purposes because the gene or genes of interest can easily be amplified if flanking sequences are known. PCR is also used routinely in comparative or phylogenetic studies to amplify genes from various sources. In these cases the primers are made for regions of the gene that are conserved in sequence across a wide variety of organisms. Because 16S rRNA, a molecule used for phylogenetic analyses, has both highly conserved and highly variable regions, primers specific for the 16S rRNA gene from various taxonomic groups can be synthesized. These may be used to survey different groups of organisms in any specific habitat. This technique is in widespread use in microbial ecology and has revealed the enormous diversity of the microbial world, much of it not yet cultured (⮔ Section 22.5).

Because it is so sensitive, PCR can be used to amplify very small quantities of DNA. For example, PCR has been used to amplify and clone DNA from sources as varied as mummified human remains and fossilized plants and animals. The ability of PCR to amplify and analyze DNA from cell mixtures has also made it a common tool of diagnostic microbiology. For example, if a clinical sample shows evidence of a gene specific to a particular pathogen, then it can be assumed that the pathogen was present in the sample. Treatment of the patient can then begin without the need to culture the organism, a time-consuming and often fruitless process. PCR has also been used in forensics to identify human individuals from very small samples of their DNA.

MiniQuiz

- Why is a primer needed at each end of the DNA segment being amplified by PCR?
- From which organisms are thermostable DNA polymerases obtained?
- How has PCR improved diagnostic clinical medicine?

 # RNA Synthesis: Transcription

Transcription is the synthesis of ribonucleic acid (RNA) using DNA as a template. There are three key differences in the chemistry of RNA and DNA: (1) RNA contains the sugar ribose instead of deoxyribose; (2) RNA contains the base uracil instead of thymine; and (3) except in certain viruses, RNA is not double-stranded. The change from deoxyribose to ribose affects the chemistry of a nucleic acid; enzymes that act on DNA usually have no effect on RNA, and vice versa. However, the change from thymine to uracil does not affect base pairing, as these two bases pair with adenine equally well.

RNA plays several important roles in the cell. Three major types of RNA are involved in protein synthesis: **messenger RNA (mRNA)**, **transfer RNA (tRNA)**, and **ribosomal RNA (rRNA)**. Several other types of RNA also occur that are mostly involved in regulation (Chapter 8). These RNA molecules all result from the transcription of DNA. It should be emphasized that RNA operates at two levels, genetic and functional. At the genetic level, mRNA carries genetic information from the genome to the ribosome. In contrast, rRNA has both a functional and a structural role in ribosomes and tRNA has an active role in carrying amino acids for protein synthesis. Indeed, some RNA molecules including rRNA have enzymatic activity (ribozymes, ⮔ Section 7.8). Here we focus on how RNA is synthesized in the *Bacteria*, using *Escherichia coli* as our model organism.

6.12 Overview of Transcription

Transcription is carried out by the enzyme **RNA polymerase**. Like DNA polymerase, RNA polymerase catalyzes the formation of phosphodiester bonds but between ribonucleotides rather than deoxyribonucleotides. RNA polymerase uses DNA as a template. The precursors of RNA are the ribonucleoside triphosphates ATP, GTP, UTP, and CTP. The mechanism of RNA synthesis is much like that of DNA synthesis. During elongation of an RNA chain, ribonucleoside triphosphates are added to the 3'-OH of the ribose of the preceding nucleotide. Polymerization is driven by the release of energy from the two energy-rich phosphate bonds of the incoming ribonucleoside triphosphates. In both DNA replication and RNA transcription the overall direction of chain growth is from the 5' end to the 3' end; thus the new strand is antiparallel to the template strand. Unlike DNA polymerase, however, RNA polymerase can initiate new strands of nucleotides on its own; consequently, no primer is necessary.

RNA Polymerases

The template for RNA polymerase is a double-stranded DNA molecule, but only one of the two strands is transcribed for any given gene. Nevertheless, genes are present on both strands of DNA and thus DNA sequences on both strands are transcribed, although at different locations. Although these principles are true for transcription in all organisms, there are significant differences among RNA polymerase from *Bacteria*, *Archaea*, and *Eukarya*. The following discussion deals only with RNA polymerase from *Bacteria*, which has the simplest structure and about which most is known (RNA polymerase in *Archaea* and *Eukarya* is discussed in Chapter 7).

RNA polymerase from *Bacteria* has five different subunits, designated β, β′, α, ω (omega), and σ (sigma), with α present in two copies. The β and β′ (beta prime) subunits are similar but not identical. The subunits interact to form the active enzyme, called the RNA polymerase holoenzyme, but the sigma factor is not as tightly bound as the others and easily dissociates, leading to the formation of the RNA polymerase core enzyme, $\alpha_2\beta\beta'\omega$. The core enzyme alone synthesizes RNA, whereas the sigma factor recognizes the appropriate site on the DNA for RNA synthesis to begin. The omega subunit is needed for assembly of the core enzyme but not for RNA synthesis. RNA synthesis is illustrated in **Figure 6.25**. www.microbiologyplace.com **Online Tutorial 6.3: Transcription**

Promoters

RNA polymerase is a large protein and makes contact with many bases of DNA simultaneously. Proteins such as RNA polymerase can interact specifically with DNA because portions of the bases are exposed in the major groove. However, in order to initiate RNA synthesis correctly, RNA polymerase must first recognize the initiation sites on the DNA. These sites, called **promoters**, are recognized by the sigma factor (**Figure 6.26**).

Once the RNA polymerase has bound to the promoter, transcription can proceed. In this process, the DNA double helix at the promoter is opened up by the RNA polymerase to form a transcription bubble. As the polymerase moves, it unwinds the DNA in short segments. This transient unwinding exposes the template strand and allows it to be copied into the RNA complement. Thus, promoters can be thought of as pointing RNA polymerase in one direction or the other along the DNA. If a region of DNA has two nearby promoters pointing in opposite directions, then transcription from one will proceed in one direction (on one of the DNA strands) while transcription from the other promoter will proceed in the opposite direction (on the other strand).

Once a short stretch of RNA has been formed, the sigma factor dissociates. Elongation of the RNA molecule is then carried out by the core enzyme alone (Figure 6.25). Sigma is only needed to form the initial RNA polymerase–DNA complex at the promoter. As the newly made RNA dissociates from the DNA, the opened DNA closes back into the original double helix. Transcription stops at specific sites called transcription terminators (Section 6.14).

Unlike DNA replication, which copies entire genomes, transcription copies much smaller units of DNA, often as little as a single gene. This system allows the cell to transcribe different genes at different frequencies, depending on the needs of the cell for different proteins. In other words, gene expression is regulated. As we shall see in Chapter 8, regulation of transcription is an important and elaborate process that uses many different mechanisms and is very efficient at controlling gene expression and conserving cell resources.

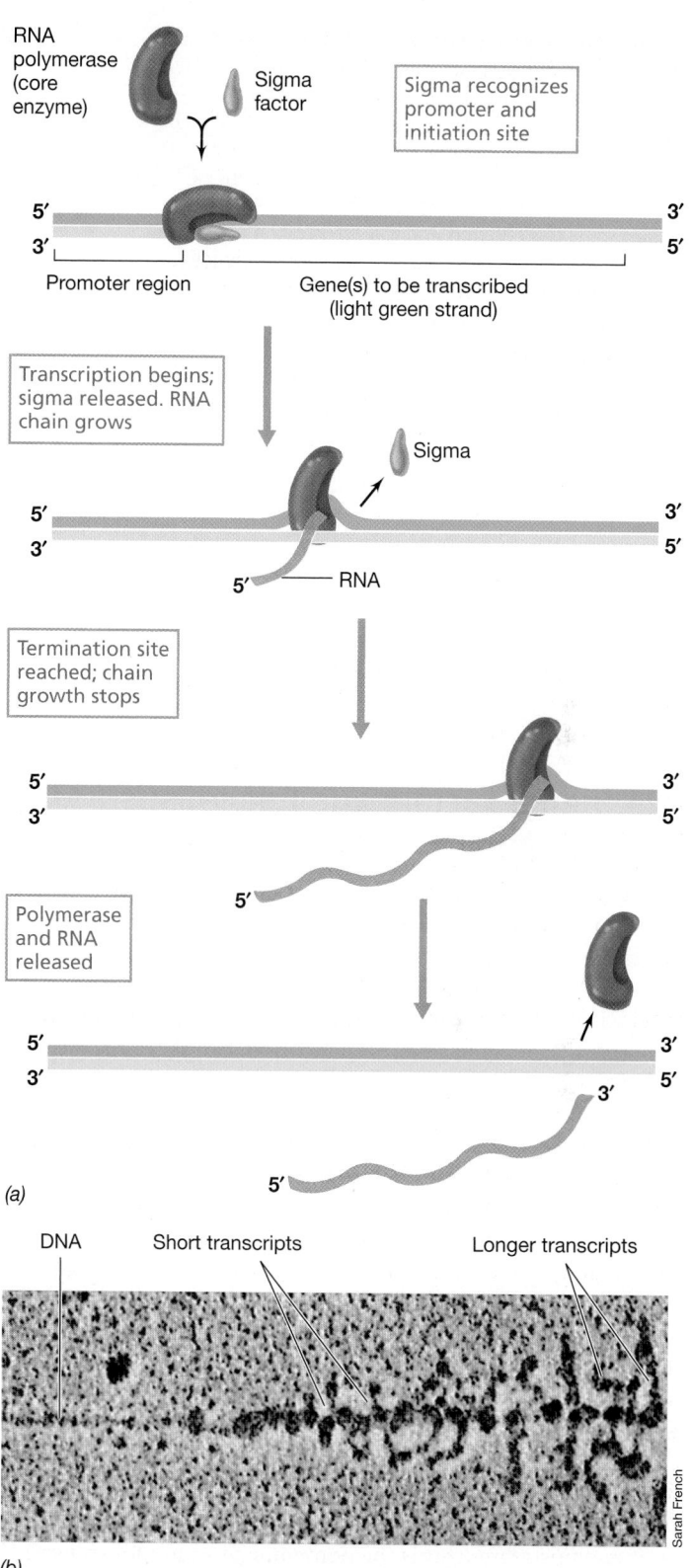

(a)

(b)

Figure 6.25 Transcription. *(a)* Steps in RNA synthesis. The initiation site (promoter) and termination site are specific nucleotide sequences on the DNA. RNA polymerase moves down the DNA chain, temporarily opening the double helix and transcribing one of the DNA strands. *(b)* Electron micrograph of transcription along a gene on the *Escherichia coli* chromosome. The region of active transcription is about 2 kb pairs of DNA. Transcription is proceeding from left to right, with the shorter transcripts on the left becoming longer as transcription proceeds.

MiniQuiz

- In which direction (5′ → 3′ or 3′ → 5′) along the template strand does transcription proceed?
- What is a promoter? What protein recognizes the promoters in *Escherichia coli*?
- What is the role of the omega subunit of RNA polymerase?

Figure 6.26 The interaction of RNA polymerase with the promoter. Shown below the RNA polymerase and DNA are six different promoter sequences identified in *Escherichia coli*, a species of *Bacteria*. The contacts of the RNA polymerase with the −35 sequence and the Pribnow box (−10 sequence) are shown. Transcription begins at a unique base just downstream from the Pribnow box. Below the actual sequences at the −35 and Pribnow box regions are consensus sequences derived from comparing many promoters. Note that although sigma recognizes the promoter sequences on the 5′ → 3′ (dark green) strand of DNA, the RNA polymerase core enzyme will actually transcribe the light green strand running 3′ → 5′ because core enzyme works only in a 5′ → 3′ direction.

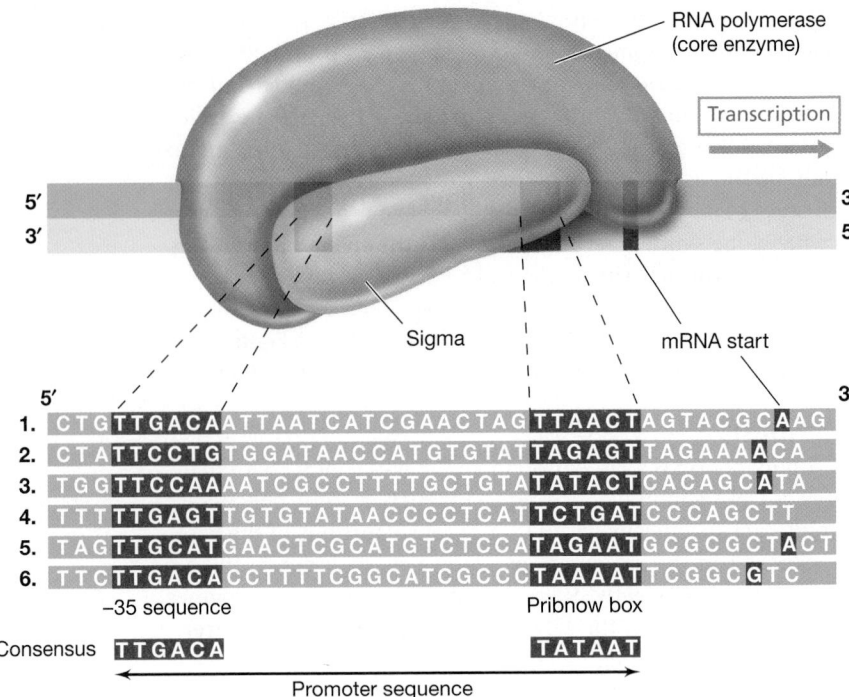

6.13 Sigma Factors and Consensus Sequences

Promoters are specific DNA sequences that bind RNA polymerase. Figure 6.26 shows the sequence of several promoters from *Escherichia coli*. All these sequences are recognized by the same sigma factor, the major sigma factor in *E. coli*, called σ^{70} (the superscript 70 indicates the size of this protein, 70 kilodaltons). Although these sequences are not identical, two shorter sequences within the promoter region are highly conserved, and it is these that sigma recognizes.

Both conserved sequences are upstream of the transcription start site. One is 10 bases before the transcription start, the −10 region, or *Pribnow box*. Although promoters differ slightly, most bases are the same within the −10 region. Comparison of many −10 regions gives the consensus sequence: TATAAT. In our example, each promoter matches from three to five of these bases. The second conserved region is about 35 bases from the start of transcription. The consensus sequence in the −35 region is TTGACA (Figure 6.26). Again, most promoters differ slightly, but are very close to consensus.

In Figure 6.26, six alternative sequences are shown for only one strand of the DNA. This is conventional "shorthand" for writing DNA sequences. By convention, the strand shown is the one with its 5′ end upstream (this is the nontemplate strand for transcription). In reality, RNA polymerase binds to double-stranded DNA and then unwinds it. A single strand of the unwound DNA is then used as template by the RNA polymerase. Although it binds to both DNA strands, sigma makes most of its contacts with the nontranscribed strand where it recognizes the specific sequences in the −10 and −35 regions.

Some sigma factors in other bacteria are much more specific in regard to binding sequences than σ^{70} of *E. coli*. In such cases,

very little leeway is allowed in the critical bases that are recognized. In *E. coli*, promoters that are most like the consensus sequence are usually more effective in binding RNA polymerase. Such promoters are called strong promoters and are very useful in genetic engineering, as discussed in Chapter 11.

Alternative Sigma Factors in *Escherichia coli*

Most genes in *E. coli* require the standard sigma factor, σ^{70} or RpoD, for transcription and have promoters like those in Figure 6.26. However, several alternative sigma factors are known that recognize different consensus sequences (**Table 6.4**). Each alternative sigma factor is specific for a group of genes required under special circumstances. Thus σ^{38}, also known as RpoS, recognizes a consensus sequence found in the promoters of genes expressed during stationary phase. Consequently, it is possible to control the expression of each family of genes by regulating the availability of the corresponding sigma factor. This may be done by changing either the rate of synthesis or the rate of degradation of the sigma factor. In addition, the activity of alternative sigma factors can be blocked by other proteins called *anti-sigma factors*. These may temporarily inactivate a particular sigma factor in response to environmental signals.

In total there are seven different sigma factors in *E. coli*, and each recognizes different consensus sequences (Table 6.4). Sigma factors were originally named according to their molecular weight. More recently, they have been named according to their roles, for example, RpoN stands for "*RNA polymerase—Nitrogen*." Most of these sigma factors have counterparts in other *Bacteria*. The endospore-forming bacterium *Bacillus subtilis* has 14 sigma factors, with 4 different sigma factors dedicated to the transcription of endospore-specific genes (⮂ Section 8.12).

Table 6.4 *Sigma factors in* Escherichia coli

Name[a]	Upstream recognition sequence[b]	Function
σ⁷⁰ RpoD	TTGACA	For most genes, major sigma factor for normal growth
σ⁵⁴ RpoN	TTGGCACA	Nitrogen assimilation
σ³⁸ RpoS	CCGGCG	Stationary phase, plus oxidative and osmotic stress
σ³² RpoH	TNTCNCCTTGAA[c]	Heat shock response
σ²⁸ FliA	TAAA	For genes involved in flagella synthesis
σ²⁴ RpoE	GAACTT	Response to misfolded proteins in periplasm
σ¹⁹ FecI	AAGGAAAAT	For certain genes in iron transport

[a]Superscript number indicates size of protein in kilodaltons. Many factors also have other names, for example, σ⁷⁰ is also called σ^D.
[b]N = any nucleotide.

MiniQuiz

- What is a consensus sequence?
- To what parts of the promoter region does sigma bind?
- How are families of genes required during specialized conditions controlled as a group using sigma factors?

6.14 Termination of Transcription

Only those genes that need to be expressed should be transcribed. Therefore it is important to terminate transcription at the correct position. **Termination** of RNA synthesis is governed by specific base sequences on the DNA. In *Bacteria* a common termination signal on the DNA is a GC-rich sequence containing an inverted repeat with a central nonrepeating segment (Section 6.2). When such a DNA sequence is transcribed, the RNA forms a stem–loop structure by intra-strand base pairing (**Figure 6.27**). Such stem–loop structures, followed by a run of adenosines in the DNA template and therefore a run of uridines in the mRNA, are effective transcription terminators. This is due to the formation of a stretch of U:A base pairs that holds the RNA and DNA template together. This structure is very weak as U:A base pairs have only two hydrogen bonds each. The RNA polymerase pauses at the stem–loop, and the DNA and RNA come apart at the run of uridines. This terminates transcription. Sequence patterns that terminate transcription without the intervention of any extra factors are referred to as intrinsic terminators.

The other mechanism for transcription termination uses a specific protein factor, known in *Escherichia coli* as Rho. Rho does not bind to RNA polymerase or to the DNA, but binds tightly to RNA and moves down the chain toward the RNA polymerase–DNA complex. Once RNA polymerase has paused at a Rho-dependent termination site (a specific sequence on the DNA template), Rho causes both the RNA and RNA polymerase to be released from the DNA, thus terminating transcription. Although

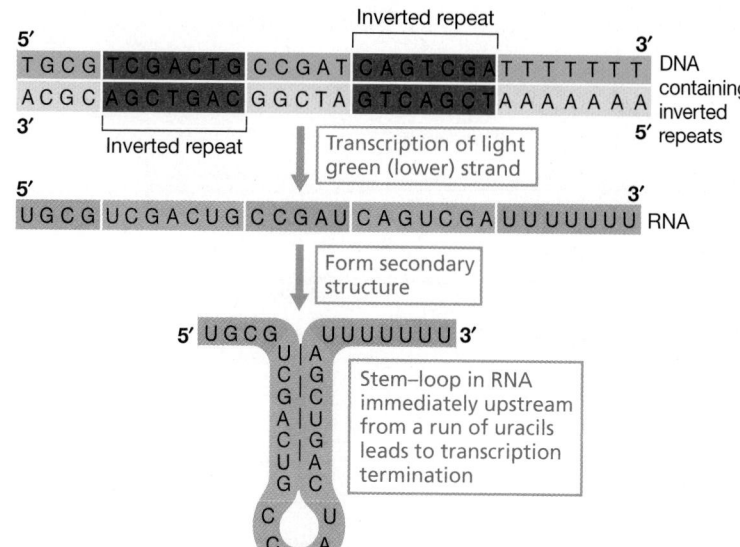

Figure 6.27 Inverted repeats and transcription termination. Inverted repeats in transcribed DNA form a stem–loop structure in the RNA that terminates transcription when followed by a run of uracils.

the termination sequences function at the level of RNA, remember that RNA is transcribed from DNA. Consequently, transcription termination is ultimately determined by specific nucleotide sequences on the DNA.

MiniQuiz

- What is a stem–loop structure?
- What is an intrinsic terminator?
- How does Rho protein terminate transcription?

6.15 The Unit of Transcription

Genetic information on chromosomes is organized into transcription units. These are segments of DNA that are transcribed into a single RNA molecule. Each transcription unit is bounded by sites where transcription is initiated and terminated. Some units of transcription include only a single gene. Others contain two or more genes. These genes are said to be cotranscribed, yielding a single RNA molecule.

Ribosomal and Transfer RNAs and RNA Longevity

Most genes encode proteins, but others encode nontranslated RNAs, such as ribosomal RNA or transfer RNA. There are several different types of rRNA in an organism. Prokaryotes have three types: 16S rRNA, 23S rRNA, and 5S rRNA (with a ribosome having one copy of each; Section 6.19). As shown in **Figure 6.28**, transcription units exist that contain one gene for each of these rRNAs, and these genes are therefore cotranscribed. The situation is similar in eukaryotes. Therefore, in all organisms the unit of transcription for most rRNA is longer than a single gene. In prokaryotes tRNA genes are often cotranscribed with each other or even, as shown in Figure 6.28, with genes for rRNA.

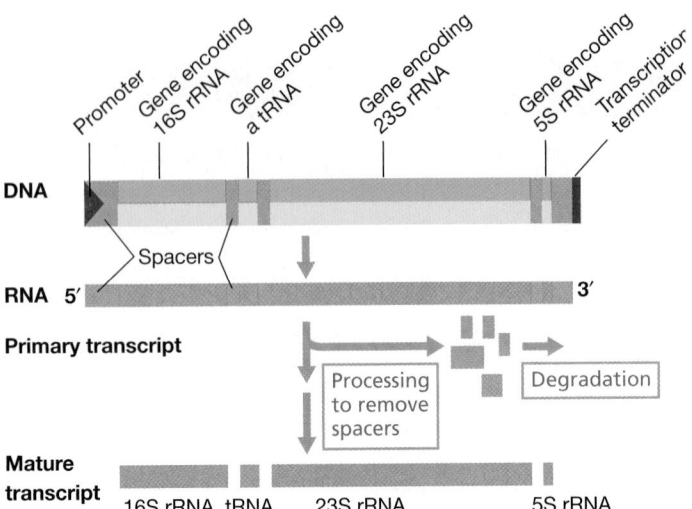

Figure 6.28 A ribosomal rRNA transcription unit from *Bacteria* and its subsequent processing. In *Bacteria* all rRNA transcription units have the genes in the order 16S rRNA, 23S rRNA, and 5S rRNA (shown approximately to scale). Note that in this particular transcription unit the spacer between the 16S and 23S rRNA genes contains a tRNA gene. In other transcription units this region may contain more than one tRNA gene. Often one or more tRNA genes also follow the 5S rRNA gene and are cotranscribed. *Escherichia coli* contains seven rRNA transcription units.

These cotranscribed transcripts must be processed by cutting into individual units to yield mature (functional) rRNAs or tRNAs. Overall, RNA processing is rare in prokaryotes but common in eukaryotes, as we will see later (Chapter 7).

In prokaryotes, most messenger RNAs have a short half-life (on the order of a few minutes), after which they are degraded by cellular ribonucleases. This is in contrast to rRNA and tRNA, which are stable RNAs. This stability is due to tRNAs and rRNAs forming highly folded structures that prevent them from being degraded by ribonucleases. By contrast, normal mRNA does not form such structures and is susceptible to ribonuclease attack. The rapid turnover of prokaryotic mRNAs permits the cell to quickly adapt to new environmental conditions and halt translation of messages whose products are no longer needed.

Polycistronic mRNA and the Operon

In prokaryotes, genes encoding related enzymes are often clustered together. RNA polymerase proceeds through such clusters and transcribes the whole group of genes into a single, long mRNA molecule. An mRNA encoding such a group of cotranscribed genes is called a *polycistronic mRNA*. When this is translated, several polypeptides are synthesized, one after another, by the same ribosome.

A group of related genes that are transcribed together to give a single polycistronic mRNA is known as an **operon**. Assembling genes for the same biochemical pathway or genes needed under the same conditions into an operon allows their expression to be coordinated. Despite this, eukaryotes do not have operons and polycistronic mRNA (Chapter 7). Often, transcription of an operon is controlled by a specific region of the DNA found just upstream of the protein-coding region of the operon. This is considered in more detail in Chapter 8.

MiniQuiz
- What is the role of messenger RNA (mRNA)?
- What is a transcription unit?
- What is a polycistronic mRNA?
- What are operons and why are they useful to prokaryotes?

Ⓥ Protein Structure and Synthesis

6.16 Polypeptides, Amino Acids, and the Peptide Bond

Proteins play major roles in cell function. Two major classes of proteins are *catalytic* proteins (enzymes) and *structural* proteins. **Enzymes** are the catalysts for chemical reactions that occur in cells. Structural proteins are integral parts of the major structures of the cell: membranes, walls, ribosomes, and so on. Regulatory proteins control most cell processes by a variety of mechanisms, including binding to DNA. However, all proteins show certain basic features in common.

Proteins are polymers of **amino acids**. All amino acids contain an amino group ($-NH_2$) and a carboxylic acid group ($-COOH$) that are attached to the α-carbon (**Figure 6.29a**). Linkages between the carboxyl carbon of one amino acid and the amino nitrogen of a second (with elimination of water) are known as **peptide bonds** (**Figure 6.30**). Two amino acids bonded by peptide linkage constitute a dipeptide; three amino acids, a tripeptide; and so on. When many amino acids are linked they form a **polypeptide**. A protein consists of one or more polypeptides. The number of amino acids differs greatly from one protein to another, from as few as 15 to as many as 10,000.

Each amino acid has a unique side chain (abbreviated R). These vary considerably, from as simple as a hydrogen atom in the amino acid glycine to aromatic rings in phenylalanine, tyrosine, and tryptophan (Figure 6.29b). Amino acids exist as pairs of **enantiomers**. These are optical isomers that have the same molecular and structural formulas, except that they are mirror images and are designated as either D or L, depending on whether a pure solution rotates light to the right or left, respectively. Natural proteins employ L-amino acids only. Nevertheless, D-amino acids are occasionally found in cells, most notably in the cell wall polymer peptidoglycan (🔗 Section 3.6) and in certain peptide antibiotics (🔗 Section 26.9). Cells can interconvert certain enantiomers by enzymes called *racemases*.

The chemical properties of an amino acid are governed by its side chain. Amino acids with similar chemical properties are grouped into related "families" (Figure 6.29b). For example, the side chain may contain a carboxylic acid group, as in aspartic acid or glutamic acid, rendering the amino acid acidic. Others contain additional amino groups, making them basic. Several amino

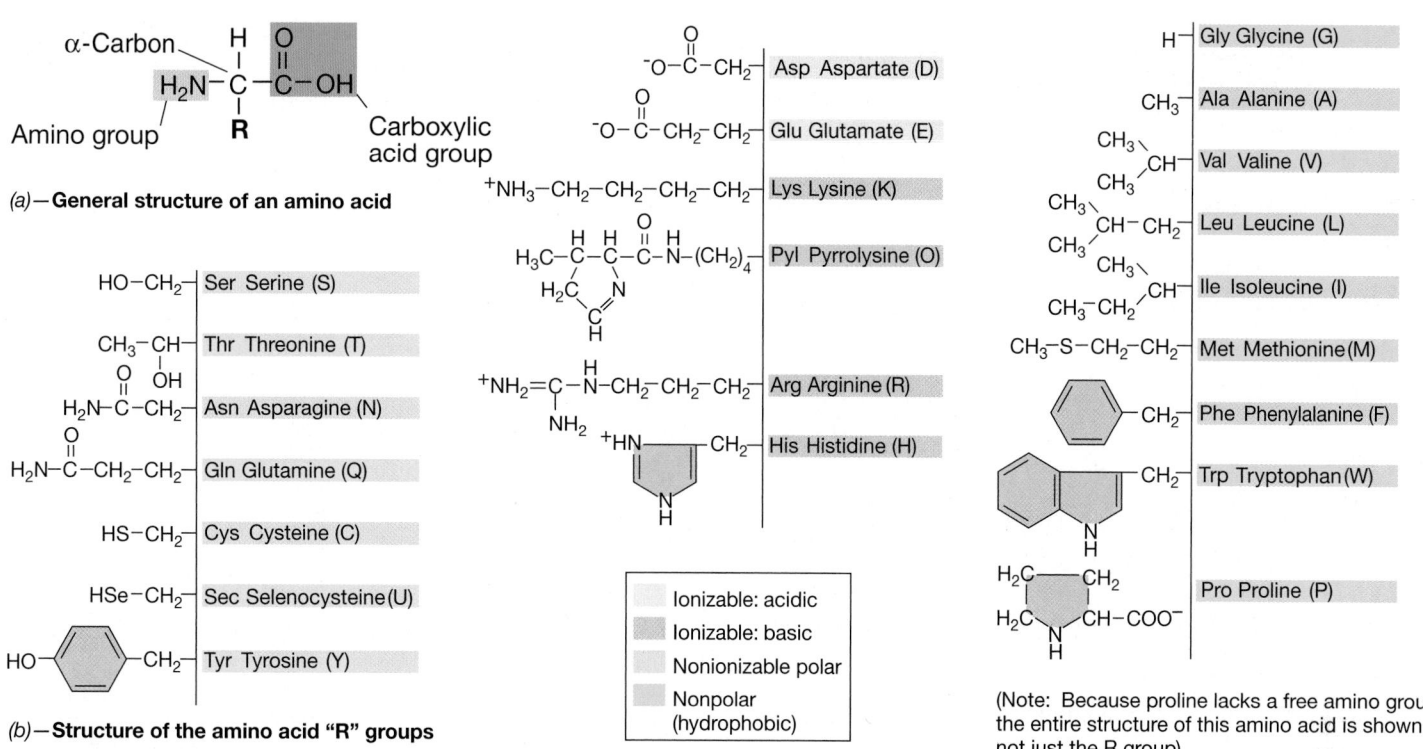

Figure 6.29 Structure of the 22 genetically encoded amino acids. *(a)* General structure. *(b)* R group structure. The three-letter codes for the amino acids are to the left of the names, and the one-letter codes are in parentheses to the right of the names. Pyrrolysine has thus far been found only in certain methanogenic *Archaea* (Section 19.3).

acids contain hydrophobic side chains and are known as nonpolar amino acids. Cysteine contains a sulfhydryl group (–SH). Using their sulfhydryl groups, two cysteines can form a disulfide linkage (R–S–S–R) that connects two polypeptide chains.

The diversity of chemically distinct amino acids makes possible an enormous number of unique proteins with widely different biochemical properties. If one assumes that an average polypeptide contains 300 amino acids, there are 22^{300} different polypep-

tide sequences that are theoretically possible. No cell has anywhere near this many different proteins. In practice, a cell of *Escherichia coli* contains around 2000 different kinds of proteins.

The linear sequence of amino acids in a polypeptide is the **primary structure**. This, ultimately, determines the further folding of the polypeptide, which in turn determines the biological activity. The two ends of a polypeptide are designated as the "C-terminus" and "N-terminus" depending on whether a free carboxylic acid group or a free amino group is found (Figure 6.30).

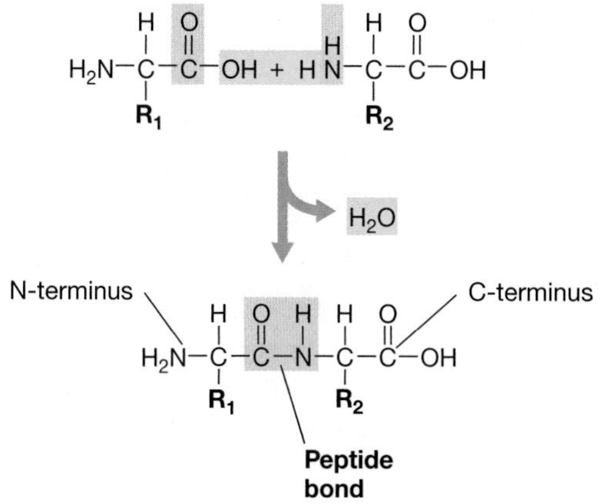

MiniQuiz

- What chemical groups do amino acids contain?
- Draw the structure of a dipeptide containing the amino acids alanine and tyrosine. Outline the peptide bond.
- Which enantiomeric form of amino acids is found in proteins?
- Glycine does not have two different enantiomers; why?

Figure 6.30 Peptide bond formation. R_1 and R_2 refer to the side chains of the amino acids. Note that, following peptide bond formation, a free OH group is present at the C-terminus for formation of the next peptide bond.

6.17 Translation and the Genetic Code

In the first two steps in biological information transfer, replication and transcription, nucleic acids are synthesized on nucleic acid templates. The last step, **translation**, also uses a nucleic acid as template, but in this case the product is a protein rather than a nucleic acid. The heart of biological information transfer is the correspondence between the nucleic acid template and the amino acid sequence of the polypeptide product. This is known

Table 6.5 *The genetic code as expressed by triplet base sequences of mRNA*

Codon	Amino acid	Codon	Amino acid	Codon	Amino acid	Codon	Amino acid
UUU	Phenylalanine	UCU	Serine	UAU	Tyrosine	UGU	Cysteine
UUC	Phenylalanine	UCC	Serine	UAC	Tyrosine	UGC	Cysteine
UUA	Leucine	UCA	Serine	UAA	None (stop signal)	UGA	None (stop signal)
UUG	Leucine	UCG	Serine	UAG	None (stop signal)	UGG	Tryptophan
CUU	Leucine	CCU	Proline	CAU	Histidine	CGU	Arginine
CUC	Leucine	CCC	Proline	CAC	Histidine	CGC	Arginine
CUA	Leucine	CCA	Proline	CAA	Glutamine	CGA	Arginine
CUG	Leucine	CCG	Proline	CAG	Glutamine	CGG	Arginine
AUU	Isoleucine	ACU	Threonine	AAU	Asparagine	AGU	Serine
AUC	Isoleucine	ACC	Threonine	AAC	Asparagine	AGC	Serine
AUA	Isoleucine	ACA	Threonine	AAA	Lysine	AGA	Arginine
AUG (start)[a]	Methionine	ACG	Threonine	AAG	Lysine	AGG	Arginine
GUU	Valine	GCU	Alanine	GAU	Aspartic acid	GGU	Glycine
GUC	Valine	GCC	Alanine	GAC	Aspartic acid	GGC	Glycine
GUA	Valine	GCA	Alanine	GAA	Glutamic acid	GGA	Glycine
GUG	Valine	GCG	Alanine	GAG	Glutamic acid	GGG	Glycine

[a]AUG encodes *N*–formylmethionine at the beginning of polypeptide chains of *Bacteria*.

as the **genetic code**. A triplet of three bases called a *codon* encodes each specific amino acid. The 64 possible codons (four bases taken three at a time $= 4^3$) of mRNA are shown in **Table 6.5**. The genetic code is written as RNA rather than as DNA because it is mRNA that is translated. Note that in addition to the codons for amino acids, there are also specific codons for starting and stopping translation.

Properties of the Genetic Code

There are 22 amino acids that are encoded by the genetic information carried on mRNA. (A variety of others are created by modification of these after translation.) Consequently, because there are 64 codons, many amino acids are encoded by more than one codon. Although knowing the codon at a given location unambiguously identifies the corresponding amino acid, the reverse is not true. Knowing the amino acid does not mean that the codon at that location is known. A code such as this that lacks one-to-one correspondence between "word" (that is, the amino acid) and code is called a degenerate code. However, knowing the DNA sequence and the correct reading frame, one can specify the amino acid sequence of a protein. This permits the determination of amino acid sequences from DNA base sequences and is at the heart of genomics (Chapter 12). In most cases where multiple codons encode the same amino acid, the multiple codons are closely related in base sequence (Table 6.5).

A codon is recognized by specific base pairing with a complementary sequence of three bases called the **anticodon**, which is found on tRNAs. If this base pairing were always the standard pairing of A with U and G with C, then at least one specific tRNA would be needed to recognize each codon. In some cases, this is true. For instance, there are six different tRNAs in *Escherichia coli* for the amino acid leucine, one for each codon (Table 6.5). By contrast, some tRNAs can recognize more than one codon. Thus, although there are two lysine codons in *E. coli*, there is only one lysyl tRNA whose anticodon can base-pair with either AAA or

AAG. In these special cases, tRNA molecules form standard base pairs at only the first two positions of the codon while tolerating irregular base pairing at the third position. This phenomenon is called **wobble** and is illustrated in **Figure 6.31**, where a pairing between G and U (rather than G with C) is illustrated at the wobble position.

Stop and Start Codons

A few codons do not encode any amino acid (Table 6.5). These codons (UAA, UAG, and UGA) are the **stop codons**, and they signal the termination of translation of a protein-coding sequence on the mRNA. Stop codons are also called **nonsense codons**, because they interrupt the "sense" of the growing polypeptide when they terminate translation.

A coding sequence on messenger RNA is translated beginning with the **start codon (AUG)**, which encodes a chemically modified methionine, *N*-formylmethionine. Although AUG at the beginning of a coding region encodes *N*-formylmethionine, AUG

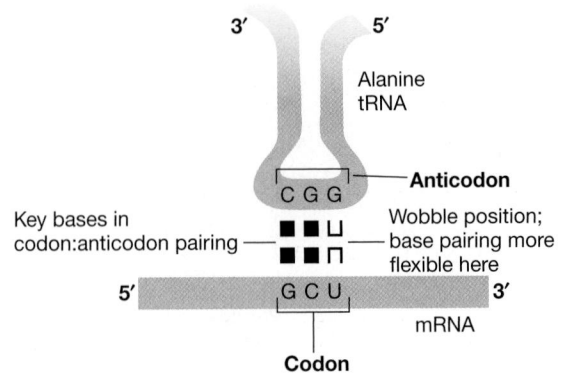

Figure 6.31 The wobble concept. Base pairing is more flexible for the third base of the codon than for the first two. Only a portion of the tRNA is shown here.

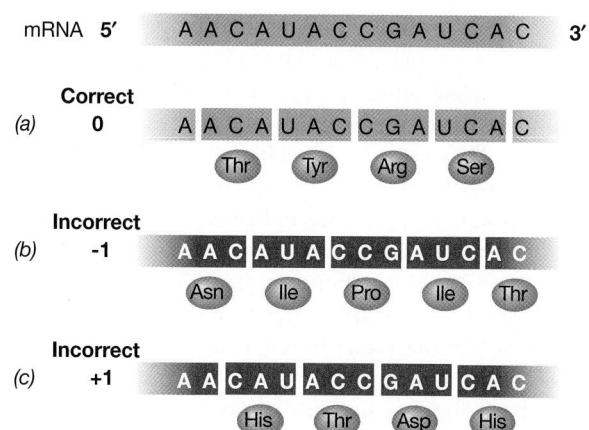

Figure 6.32 Possible reading frames in an mRNA. An interior sequence of an mRNA is shown. *(a)* The amino acids that would be encoded if the ribosome is in the correct reading frame (designated the "0" frame). *(b)* The amino acids that would be encoded by this region of the mRNA if the ribosome were in the −1 reading frame. *(c)* The amino acids that would be encoded if the ribosome were in the +1 reading frame.

within the coding region encodes methionine. Two different tRNAs are involved in this process (Section 6.19). With a triplet code it is critical for translation to begin at the correct nucleotide. If it does not, the whole reading frame of the mRNA will be shifted and thus an entirely different protein will be made. If the shift introduces a stop codon into the reading frame, the protein will terminate prematurely. By convention the reading frame that is translated to give the protein encoded by the gene is called the 0 frame. As can be seen in **Figure 6.32**, the other two possible reading frames (−1 and +1) do not encode the same amino acid sequence. Therefore it is essential that the ribosome finds the correct start codon to begin translation and, once it has, that it moves down the mRNA exactly three bases at a time. How is the correct reading frame ensured?

Reading frame fidelity is governed by interactions between mRNA and rRNA within the ribosome. Ribosomal RNA recognizes a specific AUG on the mRNA as a start codon with the aid of an upstream sequence in the mRNA called the Shine–Dalgarno sequence. This alignment requirement explains why occasional mRNA from *Bacteria* can use other start codons, such as GUG. However, even these unusual start codons direct the incorporation of *N*-formylmethionine as the initiator amino acid.

Open Reading Frames

One common method of identifying protein-encoding genes is to examine each strand of the DNA sequence for **open reading frames (ORFs)**. Remember that RNA is transcribed from DNA, so that if one knows the sequence of DNA, one also knows the sequence of mRNA that is transcribed from it. If an mRNA can be translated, it contains an open reading frame: a start codon (typically AUG) followed by a number of codons and then a stop codon in the same reading frame as the start codon. In practice, only ORFs long enough to encode a protein of realistic length are accepted as true coding sequences. Although most functional

proteins are at least 100 amino acids in length, a few protein hormones and regulatory peptides are much shorter. Consequently, it is not always possible to tell from sequence data alone whether a relatively short ORF is merely due to chance or encodes a genuine, albeit short, protein.

A computer can be programmed using the above guidelines to scan long DNA base sequences to look for open reading frames. In addition to looking for start and stop codons, the search may include promoters and Shine–Dalgarno ribosome-binding sequences as well. The search for ORFs is very important in genomics (Chapter 12). If an unknown piece of DNA has been sequenced, the presence of an ORF implies that it can encode protein.

Codon Bias

Several amino acids are encoded by multiple codons. One might assume that such multiple codons would be used at equal frequencies. However, this is not so, and sequence data show major **codon bias**. In other words, some codons are greatly preferred over others even though they encode the same amino acid. Moreover, this bias is organism-specific. In *E. coli*, for instance, only about 1 out of 20 isoleucine residues in proteins is encoded by the isoleucine codon AUA, the other 19 being encoded by the other isoleucine codons, AUU and AUC (Table 6.5). Codon bias is correlated with a corresponding bias in the concentration of different tRNA molecules. Thus a tRNA corresponding to a rarely used codon will be in relatively short supply.

The origin of codon bias is unclear, but it is easily recognized and may be taken into account in practical uses of gene sequence information. For example, a gene from one organism whose codon usage differs dramatically from that of another may not be translated efficiently if the gene is cloned into the latter using genetic engineering (Chapter 11). This is due to a shortage of the tRNA for codons that are rare in the host but frequent in the cloned gene. However, this problem can be corrected or at least compensated for by genetic manipulation.

Modifications to the Genetic Code

All cells appear to use the same genetic code. Therefore, the genetic code is a universal code. However, this view has been tempered a bit by the discovery that some organelles and a few cells use genetic codes that are slight variations of the "universal" genetic code.

Alternative genetic codes were first discovered in the genomes of animal mitochondria. These modified codes typically use nonsense codons as sense codons. For example, animal (but not plant) mitochondria use the codon UGA to encode tryptophan instead of using it as a stop codon (Table 6.5). Several organisms are known that also use slightly different genetic codes. For example, in the genus *Mycoplasma* (*Bacteria*) and the genus *Paramecium* (*Eukarya*), certain nonsense codons encode amino acids. These organisms simply have fewer nonsense codons because one or two of them are used as sense codons. In a few rare cases, nonsense codons encode unusual amino acids rather than one of the 20 common amino acids (Section 6.20).

MiniQuiz
- What are stop codons and start codons?
- Why is it important for the ribosome to read "in frame"?
- What is codon bias?
- If you were given a nucleotide sequence, how would you find ORFs?

6.18 Transfer RNA

A transfer RNA carries the anticodon that base-pairs with the codon on mRNA. In addition, each tRNA is specific for the amino acid that corresponds to its own anticodon (that is, the cognate amino acid). The tRNA and its specific amino acid are linked by specific enzymes called **aminoacyl-tRNA synthetases**. These ensure that a particular tRNA receives the correct amino acid and must thus recognize both the tRNA and its cognate amino acid.

General Structure of tRNA

There are about 60 different tRNAs in bacterial cells and 100–110 in mammalian cells. Transfer RNA molecules are short, single-stranded molecules that contain extensive secondary structure and have lengths of 73–93 nucleotides. Certain bases and secondary structures are constant for all tRNAs, whereas other parts are variable. Transfer RNA molecules also contain some purine and pyrimidine bases that differ somewhat from the standard bases found in RNA because they are chemically modified. These modifications are made to the bases after transcription. These unusual bases include pseudouridine (ψ), inosine,

dihydrouridine (D), ribothymidine, methyl guanosine, dimethyl guanosine, and methyl inosine. The mature and active tRNA also contains extensive double-stranded regions within the molecule. This secondary structure forms by internal base pairing when the single-stranded molecule folds back on itself (**Figure 6.33**).

The structure of a tRNA can be drawn in a cloverleaf fashion, as in Figure 6.33*a*. Some regions of tRNA secondary structure are named after the modified bases found there (the TψC and D loops) or after their functions (anticodon loop and acceptor stem). The three-dimensional structure of a tRNA is shown in Figure 6.33*b*. Note that bases that appear widely separated in the cloverleaf model may actually be much closer together when viewed in three dimensions. This allows some of the bases in one loop to pair with bases in another loop.

The Anticodon and the Amino Acid–Binding Site

One of the key variable parts of the tRNA molecule is the anticodon, the group of three bases that recognizes the codon on the mRNA. The anticodon is found in the anticodon loop (Figure 6.33). The three nucleotides of the anticodon recognize the codon by specifically pairing with its three bases. By contrast, other portions of the tRNA interact with both the rRNA and protein components of the ribosome, nonribosomal translation proteins, and the aminoacyl synthetase enzyme.

At the 3′ end, or acceptor stem, of all tRNAs are three unpaired nucleotides. The sequence of these three nucleotides is always cytosine-cytosine-adenine (CCA), and they are absolutely essential for function. Curiously, in most organisms these three nucleotides are not encoded by the tRNA genes on the chromosome. Instead they are added, one after another, by an enzyme

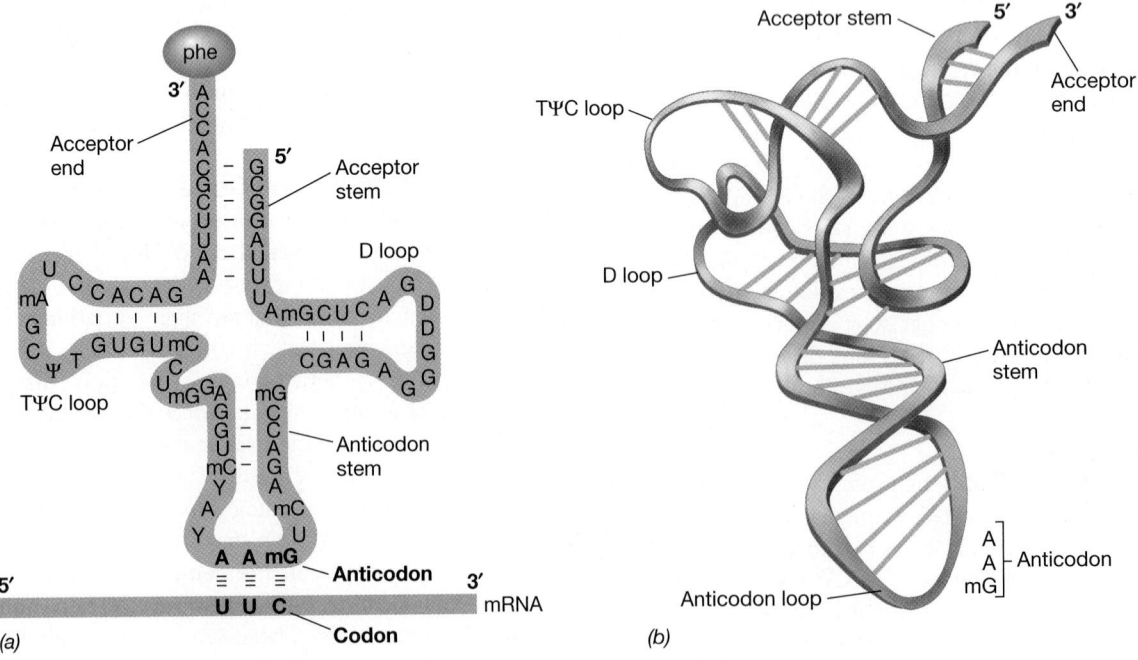

Figure 6.33 Structure of a transfer RNA. *(a)* The conventional cloverleaf structural drawing of yeast phenylalanine tRNA. The amino acid is attached to the ribose of the terminal A at the acceptor end. A, adenine; C, cytosine; U, uracil; G, guanine; T, thymine; ψ, pseudouracil; D, dihydrouracil; m, methyl; Y, a modified purine. *(b)* In fact, the tRNA molecule folds so that the D loop and TψC loops are close together and associate by hydrophobic interactions.

called the CCA-adding enzyme, using CTP and ATP as substrates. The cognate amino acid is covalently attached to the terminal adenosine of the CCA end by an ester linkage to the ribose sugar. As we shall see, from this location on the tRNA, the amino acid is incorporated into the growing polypeptide chain on the ribosome by a mechanism described in the next section.

Recognition, Activation, and Charging of tRNAs

Recognition of the correct tRNA by an aminoacyl-tRNA synthetase involves specific contacts between key regions of the tRNA and the synthetase (**Figure 6.34**). As might be expected because of its unique sequence, the anticodon of the tRNA is important in recognition by the synthetase. However, other contact sites between the tRNA and the synthetase are also important. Studies of tRNA binding to aminoacyl-tRNA synthetases, in which specific tRNA bases have been changed by mutation, have shown that only a small number of key nucleotides in tRNA are involved in recognition. These other key recognition nucleotides

are often part of the acceptor stem or D loop of the tRNA (Figure 6.33). It should be emphasized that the fidelity of this recognition process is crucial, for if the wrong amino acid is attached to the tRNA, it will be inserted into the growing polypeptide, likely leading to the synthesis of a faulty protein.

The specific reaction between amino acid and tRNA catalyzed by the aminoacyl-tRNA synthetase begins with activation of the amino acid by reaction with ATP:

$$\text{Amino acid} + \text{ATP} \longleftrightarrow \text{aminoacyl—AMP} + \text{P—P}$$

The aminoacyl-AMP intermediate formed normally remains bound to the enzyme until collision with the appropriate tRNA molecule. Then, as shown in Figure 6.34*a*, the activated amino acid is attached to the tRNA to form a charged tRNA:

$$\text{Aminoacyl—AMP} + \text{tRNA} \longleftrightarrow \text{aminoacyl—tRNA} + \text{AMP}$$

The pyrophosphate (PP$_i$) formed in the first reaction is split by a pyrophosphatase, giving two molecules of inorganic phosphate.

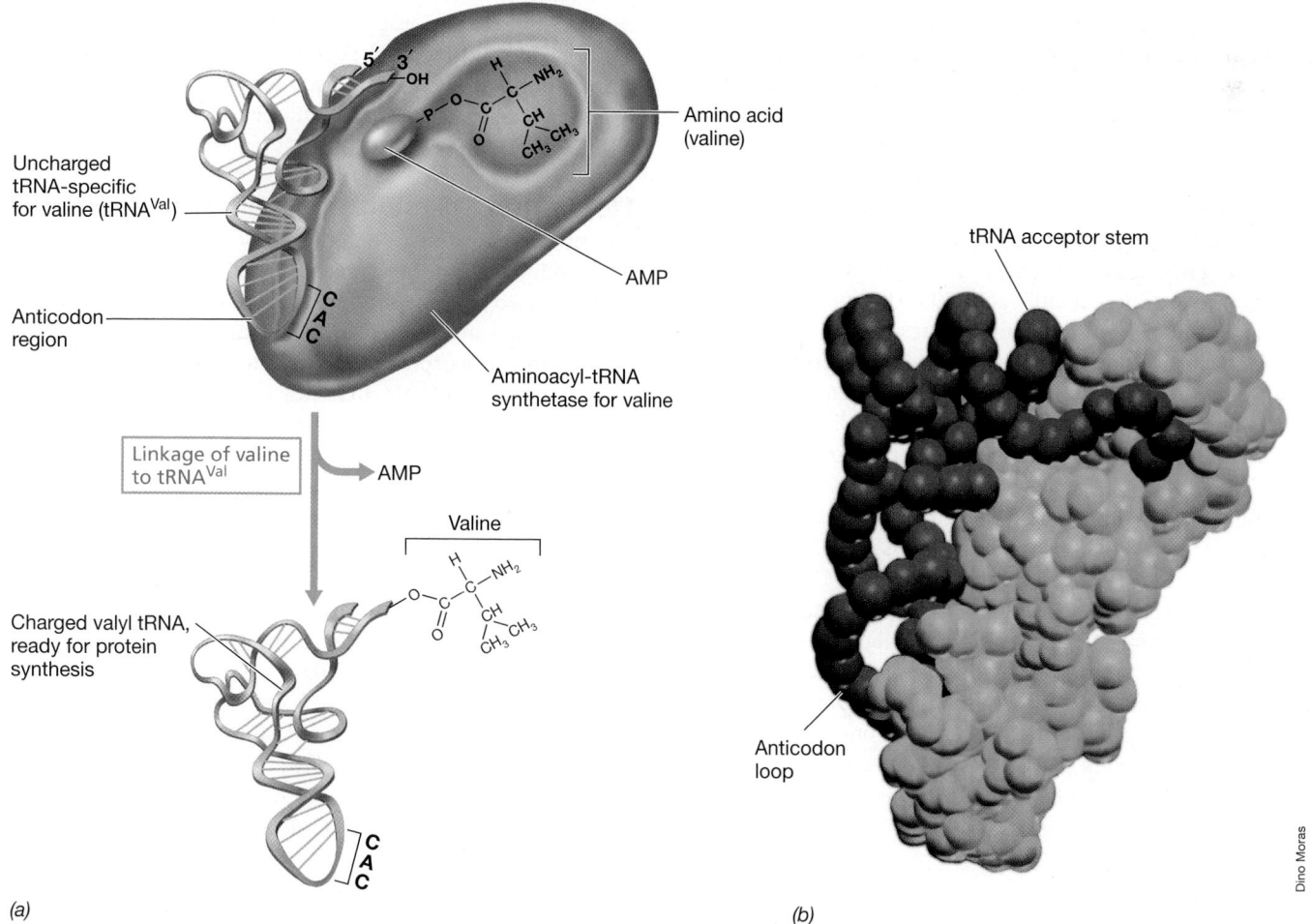

(a)

(b)

Dino Moras

Figure 6.34 Aminoacyl-tRNA synthetase. *(a)* Mode of activity of an aminoacyl-tRNA synthetase. Recognition of the correct tRNA by a particular synthetase involves contacts between specific nucleic acid sequences in the D loop and acceptor stem of the tRNA and specific amino acids of the synthetase. In this diagram, valyl-tRNA synthetase is shown catalyzing the final step of the reaction, where the valine in valyl-AMP is transferred to tRNA. *(b)* A computer model showing the interaction of glutaminyl-tRNA synthetase (blue) with its tRNA (red). Reprinted with permission from M. Ruff et al. 1991. *Science* 252: 1682–1689. © 1991, AAAS.

UNIT 3

Because ATP is used and AMP is formed in these reactions, a total of two energy-rich phosphate bonds are needed to charge a tRNA with its cognate amino acid. After activation and charging, the aminoacyl-tRNA leaves the synthetase and travels to the ribosome where the polypeptide is synthesized.

6.19 Steps in Protein Synthesis

It is vital for proper functioning of proteins that the correct amino acids are inserted at the proper locations in the polypeptide chain. This is the task of the protein-synthesizing machinery, the ribosome. Although protein synthesis is a continuous process, it can be broken down into a number of steps: initiation, elongation, and termination. In addition to mRNA, tRNA, and ribosomes, the process requires a number of proteins designated initiation, elongation, and termination factors. The energy-rich compound guanosine triphosphate (GTP) provides the necessary energy for the process. The key steps in protein synthesis are shown in Figure 6.35.

Ribosomes

Ribosomes are the sites of protein synthesis. A cell may have many thousand ribosomes, the number increasing at higher growth rates. Each ribosome consists of two subunits (**Figure 6.35**). Prokaryotes possess 30S and 50S ribosomal subunits, yielding intact 70S ribosomes. The S-values are Svedberg units, which refer to the sedimentation coefficients of ribosomal subunits (30S and 50S) or intact ribosomes (70S) when subjected to centrifugal force in an ultracentrifuge. (Although larger particles do have larger S-values, the relationship is not linear and S-values cannot be added together.)

Each ribosomal subunit contains specific ribosomal RNAs and ribosomal proteins. The 30S subunit contains 16S rRNA and 21 proteins, and the 50S subunit contains 5S and 23S rRNA and 31 proteins. Thus, in *Escherichia coli*, there are 52 distinct ribosomal proteins, most present at one copy per ribosome. The ribosome is a dynamic structure whose subunits alternately associate and dissociate and also interact with many other proteins. There are several proteins that are essential for ribosome function and interact with the ribosome at various stages of translation. These are regarded as "translation factors" rather than "ribosomal proteins" per se.

Initiation of Translation

In *Bacteria*, such as *E. coli*, initiation of protein synthesis begins with a free 30S ribosomal subunit. From this, an initiation complex forms consisting of the 30S subunit, plus mRNA, formylmethionine tRNA, and several initiation proteins called IF1, IF2, and IF3. GTP is also required for this step. Next, a 50S ribosomal subunit is added to the initiation complex to form the active 70S ribosome. At the end of the translation process, the ribosome separates again into 30S and 50S subunits.

Just preceding the start codon on the mRNA is a sequence of three to nine nucleotides called the Shine–Dalgarno sequence or ribosome-binding site that helps bind the mRNA to the ribosome. The ribosome-binding site is toward the 5′ end of the mRNA and is complementary to base sequences in the 3′ end of the 16S rRNA. Base pairing between these two molecules holds the ribosome–mRNA complex securely together in the correct reading frame. Polycistronic mRNA has multiple Shine–Dalgarno sequences, one upstream of each coding sequence. This allows bacterial ribosomes to translate several genes on the same mRNA because the ribosome can find each initiation site within a message by binding to its Shine–Dalgarno site.

Translational initiation always begins with a special initiator aminoacyl-tRNA binding to the start codon, AUG. In *Bacteria* this is formylmethionyl-tRNA. After polypeptide completion, the formyl group is removed. Consequently, the N-terminal amino acid of the completed protein will be methionine. However, in many proteins this methionine is removed by a specific protease. Because the Shine–Dalgarno sequences (and other possible interactions between the rRNA and the mRNA) direct the ribosome to the proper start site, prokaryotic mRNA can use a start codon other than AUG. The most common alternative start codon is GUG. When used in this context, however, GUG calls for formylmethionine initiator tRNA (and not valine, see Table 6.5).

Elongation, Translocation, and Termination

The mRNA threads through the ribosome primarily bound to the 30S subunit. The ribosome contains other sites where the tRNAs interact. Two of these sites are located primarily on the 50S subunit, and they are termed the A site and the P site (Figure 6.35). The A site, the acceptor site, is the site on the ribosome where the incoming charged tRNA first attaches. Loading of tRNA into the A site is assisted by the elongation factor EF-Tu.

The P site, the peptide site, is the site where the growing polypeptide chain is held by the previous tRNA. During peptide bond formation, the growing polypeptide chain moves to the tRNA at the A site as a new peptide bond is formed. Several nonribosomal proteins are required for elongation, especially the elongation factors, EF-Tu and EF-Ts, as well as more GTP (to simplify Figure 6.35, the elongation factors are omitted and only part of the ribosome is shown).

Following elongation, the tRNA holding the polypeptide is translocated (moved) from the A site to the P site, thus opening the A site for another charged tRNA (Figure 6.35). Translocation requires the elongation factor EF-G and one molecule of GTP for each translocation event. At each translocation step the ribosome advances three nucleotides, exposing a new codon at the A site. Translocation pushes the now empty tRNA to a third site, called the E site. It is from this exit site that the tRNA is actually released from the ribosome. The precision of the translocation step is critical to the accuracy of protein synthesis. The ribosome must move exactly one codon at each step. Although mRNA appears to be moving through the ribosome complex, in reality, the ribosome is moving along the mRNA. Thus, the three sites on the ribosome shown in Figure 6.35 are not static locations but are moving parts of a complex biomolecular machine.

TRANSLATION: Initiation

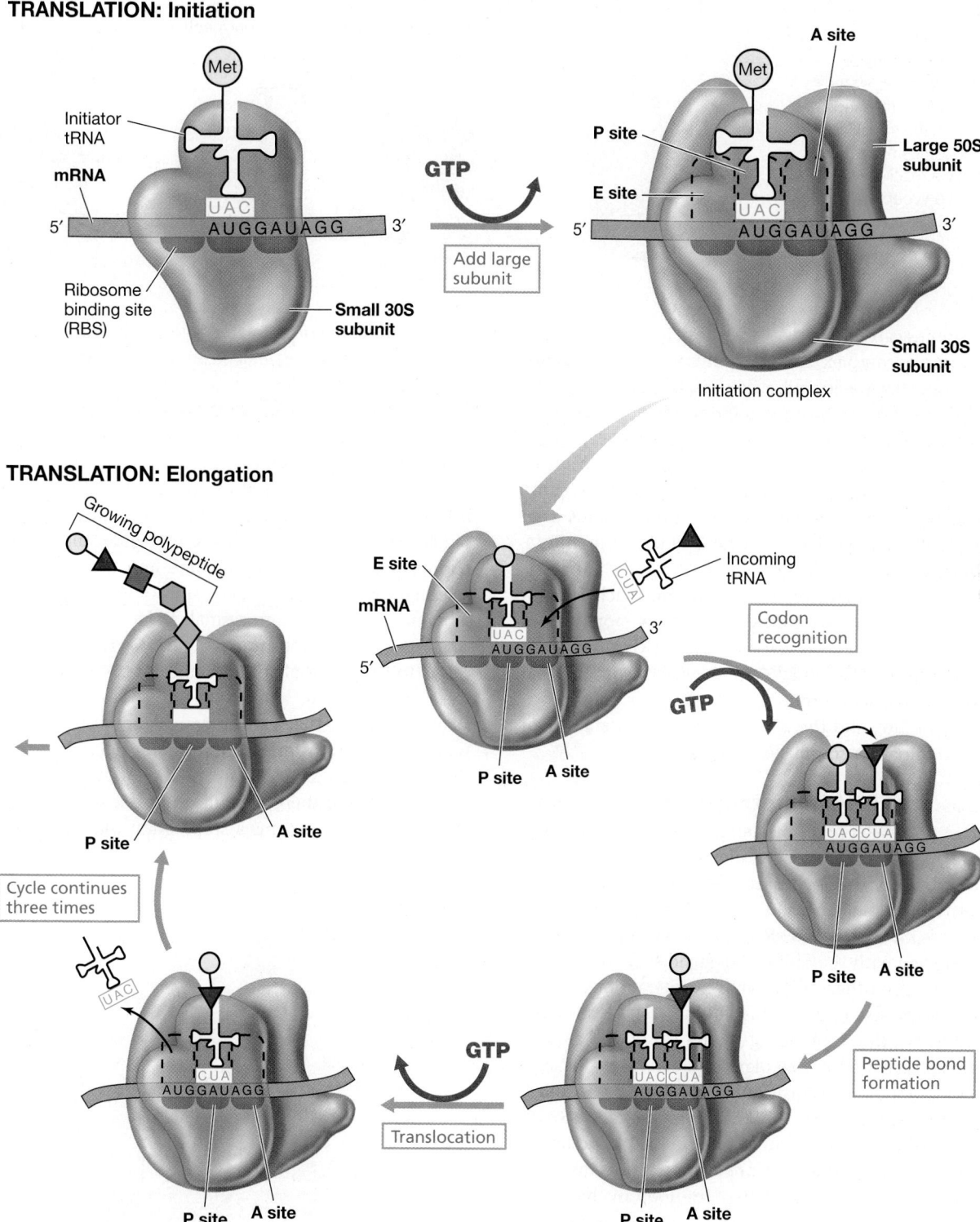

TRANSLATION: Elongation

Figure 6.35 The ribosome and protein synthesis. *Initiation* of protein synthesis. The mRNA and initiator tRNA, carrying N-formylme-thionine ("Met"), bind first to the small subunit of the ribosome. Initiation factors (not shown) use energy from GTP to promote the addition of the large ribosomal subunit. The initiator tRNA starts out in the P site. *Elongation* cycle of translation. Elongation factors (not shown) use GTP to install the incoming tRNA into the A site. Peptide bond formation is then catalyzed by the 23S rRNA. Translocation of the ribosome along the mRNA from one codon to the next requires hydrolysis of another GTP. The outgoing tRNA is released from the E site. The next charged tRNA binds to the A site and the cycle repeats. The genetic code, expressed in the language of mRNA, is shown in Table 6.5.

Figure 6.36 Polysomes.
Translation by several ribosomes on a single messenger RNA forms the polysome. Note how the ribosomes nearest the 5′ end of the message are at an earlier stage in the translation process than ribosomes nearer the 3′ end, and thus only a relatively short portion of the final polypeptide has been made.

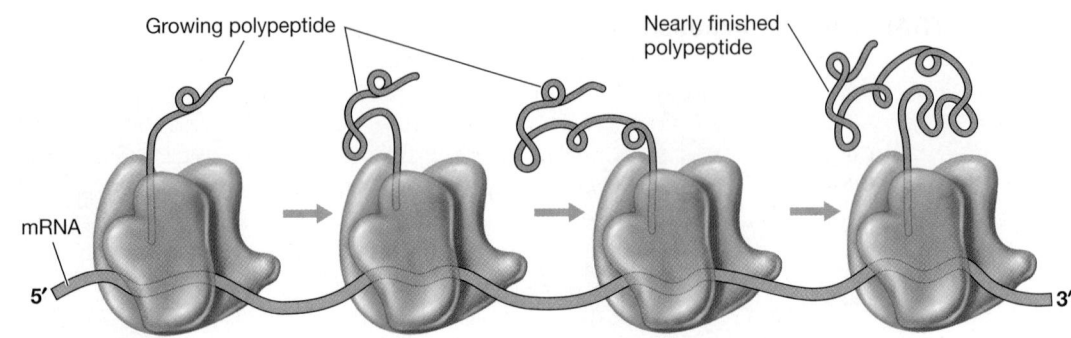

Several ribosomes can simultaneously translate a single mRNA molecule, forming a complex called a polysome (**Figure 6.36**). Polysomes increase the speed and efficiency of translation, and because the activity of each ribosome is independent of that of its neighbors, each ribosome in a polysome complex makes a complete polypeptide. Note in Figure 6.36 how ribosomes closest to the 5′ end (the beginning) of the mRNA molecule have short polypeptides attached to them because only a few codons have been read, while ribosomes closest to the 3′ end of the mRNA have nearly finished polypeptides.

Protein synthesis terminates when the ribosome reaches a stop codon (nonsense codon). No tRNA binds to a stop codon. Instead, specific proteins called *release factors* (RFs) recognize the stop codon and cleave the attached polypeptide from the final tRNA, releasing the finished product. Following this, the ribosomal subunits dissociate, and the 30S and 50S subunits are then free to form new initiation complexes and repeat the process.

Role of Ribosomal RNA in Protein Synthesis

Ribosomal RNA plays vital roles in all stages of protein synthesis, from initiation to termination. The role of the many proteins present in the ribosome, although less clear, may be to act as a scaffold to position key sequences in the ribosomal RNAs.

In *Bacteria* it is clear that 16S rRNA is involved in initiation through base pairing with the Shine–Dalgarno sequence on the mRNA. There are also other mRNA–rRNA interactions during elongation. On either side of the codons in the A and P sites, the mRNA is held in position by binding to 16S rRNA and ribosomal proteins. Ribosomal RNA also plays a role in ribosome subunit association, as well as in positioning tRNA in the A and P sites on the ribosome (Figure 6.35). Although charged tRNAs that enter the ribosome recognize the correct codon by codon–anticodon base pairing, they are also bound to the ribosome by interactions of the anticodon stem–loop of the tRNA with specific sequences within 16S rRNA. Moreover, the acceptor end of the tRNA (Figure 6.35) base-pairs with sequences in 23S rRNA.

In addition to all of this, the actual formation of peptide bonds is catalyzed by rRNA. The peptidyl transferase reaction happens on the 50S subunit of the ribosome and is catalyzed by the 23S rRNA itself, rather than by any of the ribosomal proteins. The 23S rRNA also plays a role in translocation, and the EF proteins are known to interact specifically with 23S rRNA. Thus, besides its role as the structural backbone of the ribosome, ribosomal

RNA plays a major catalytic role in the translation process as well. www.microbiologyplace.com **Online Tutorial 6.4: Translation**

Freeing Trapped Ribosomes

A defective mRNA that lacks a stop codon causes a problem in translation. Such a defect may arise, for example, from a mutation that removed the stop codon, defective synthesis of the mRNA, or partial degradation of the mRNA. If a ribosome reaches the end of an mRNA molecule and there is no stop codon, release factor cannot bind and the ribosome cannot be released from the mRNA. The ribosome is trapped.

Bacterial cells contain a small RNA molecule, called *tmRNA*, that frees stalled ribosomes (**Figure 6.37**). The "tm" in its name refers to the fact that tmRNA mimics both tRNA, in that it carries the amino acid alanine, and mRNA, in that it contains a short stretch of RNA that can be translated. When tmRNA collides with a stalled ribosome, it binds alongside the defective mRNA. Protein synthesis can then proceed, first by adding the alanine on the tmRNA and then by translating the short tmRNA message. Finally, tmRNA contains a stop codon that allows release factor

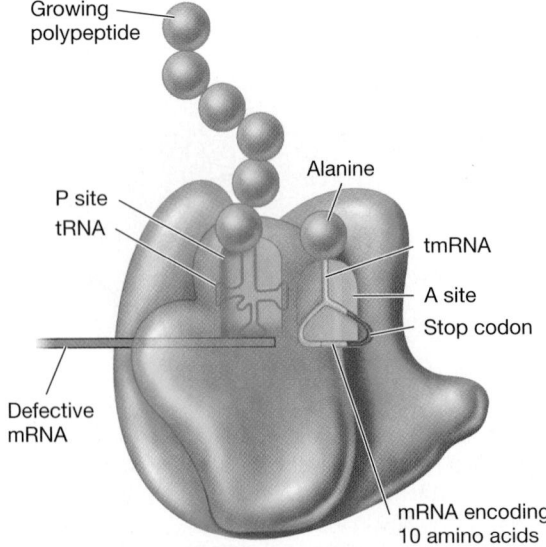

Figure 6.37 Freeing of a stalled ribosome by tmRNA. A defective mRNA lacking a stop codon stalls a ribosome that has a partly synthesized polypeptide attached to a tRNA (blue) in the P site. Binding of tmRNA (yellow) in the A site releases the polypeptide. Translation then continues up to the stop codon provided by the tmRNA.

to bind and disassemble the ribosome. The protein made as a result of this rescue operation is defective and is subsequently degraded. The short sequence of amino acids encoded by tmRNA and added to the end of the defective protein is a signal for a specific protease to degrade the protein. Thus, through the activity of tmRNA, stalled ribosomes are freed up to participate in protein synthesis once again.

Effect of Antibiotics on Protein Synthesis

A large number of antibiotics inhibit protein synthesis by interacting with the ribosome. These interactions are quite specific, and many involve rRNA. Some antibiotics are useful research tools because they are specific for different steps in protein synthesis. For instance, streptomycin inhibits initiation, whereas puromycin, chloramphenicol, cycloheximide, and tetracycline inhibit elongation. Several of these antibiotics are clinically useful. Many antibiotics specifically inhibit ribosomes of organisms from only one or two of the phylogenetic domains. For example, chloramphenicol and streptomycin are specific for the ribosomes of *Bacteria* and cycloheximide for ribosomes of *Eukarya*. The mode of action of these and other antibiotics will be discussed in Chapter 26.

MiniQuiz

- What are the components of a ribosome?
- What functional roles does rRNA play in protein synthesis?
- What roles do the initiation and elongation factors play in protein synthesis?
- How is a completed polypeptide chain released from the ribosome?
- How does tmRNA free stalled ribosomes?

6.20 The Incorporation of Selenocysteine and Pyrrolysine

The universal genetic code has codons for 20 amino acids (Table 6.5). However, many proteins contain other amino acids. In fact, more than 100 different amino acids have been found in various proteins. Most of these are made by modifying a standard amino acid after it is incorporated into a protein, a process called *posttranslational modification*. However, two nonstandard amino acids are genetically encoded, although in an unusual manner, and are thus inserted during protein synthesis itself. These exceptions are selenocysteine and pyrrolysine, the 21st and 22nd genetically encoded amino acids (Figure 6.29).

Selenocysteine has the same structure as cysteine except it contains selenium instead of sulfur. It is formed by modifying serine after it has been attached to selenocysteine tRNA. Pyrrolysine is a lysine derivative with an extra aromatic ring. Pyrrolysine is fully synthesized and only then attached to pyrrolysyl tRNA.

Both selenocysteine and pyrrolysine are encoded by stop codons (UGA and UAG, respectively). Both have their own tRNAs that contain anticodons that read these stop codons. Both selenocysteine and pyrrolysine also have specific aminoacyl-tRNA synthetases to charge the tRNA with the amino acids.

Most stop codons in organisms that use selenocysteine and pyrrolysine do indeed indicate stop. However, occasional stop codons are recognized as encoding selenocysteine or pyrrolysine. For selenocysteine this depends on a recognition sequence just downstream of the special UGA codon. This forms a stem–loop that binds the SelB protein. The SelB protein also binds charged selenocysteine tRNA and brings it to the ribosome when needed. Similarly, pyrrolysine incorporation relies on a recognition sequence just downstream of the pyrrolysine-encoding UAG codon.

Selenocysteine and pyrrolysine are both relatively rare. *Escherichia coli* makes only a handful of proteins with selenocysteine, including two different formate dehydrogenase enzymes. It was sequencing the genes for these enzymes that led to the discovery of selenocysteine. Most organisms, including plants and animals, have a few proteins that contain selenocysteine. Pyrrolysine is rarer still. It has been found in certain *Archaea* and *Bacteria* but was first discovered in species of methanogenic *Archaea*, organisms that generate methane (Section 19.3). In certain methanogens the enzyme methylamine methyltransferase contains a pyrrolysine residue. Whether there are yet other genetically encoded amino acids is unlikely but remains a possibility.

MiniQuiz

- Explain the term posttranslational modification.
- What specific components (apart from a ribosome and a stop codon) are needed for the insertion of selenocysteine into a growing polypeptide chain?

6.21 Folding and Secreting Proteins

For a protein to function properly it must be folded correctly and it must also end up in the correct location in the cell. Here we briefly discuss these two related processes.

Levels of Protein Structure

Once formed, a polypeptide does not remain linear; instead it folds to form a more stable structure. Hydrogen bonding, between the oxygen and nitrogen atoms of two peptide bonds, generates the **secondary structure** (**Figure 6.38a**). One common type of secondary structure is the *α-helix*. To envision an α-helix, imagine a linear polypeptide wound around a cylinder (Figure 6.38b). This positions peptide bonds close enough to allow hydrogen bonding. The large number of such hydrogen bonds gives the α-helix its inherent stability. In the β-sheet, the polypeptide chain folds back and forth upon itself instead of forming a helix. However, as in the α-helix, the folding in a β-sheet positions peptide bonds so that they can undergo hydrogen bonding (Figure 6.38c).

Many polypeptides contain regions of both α-helix and β-sheet secondary structure, the type of folding and its location in the molecule being determined by the primary structure and the available opportunities for hydrogen bonding. A typical protein is thus made up of many folded subdomains.

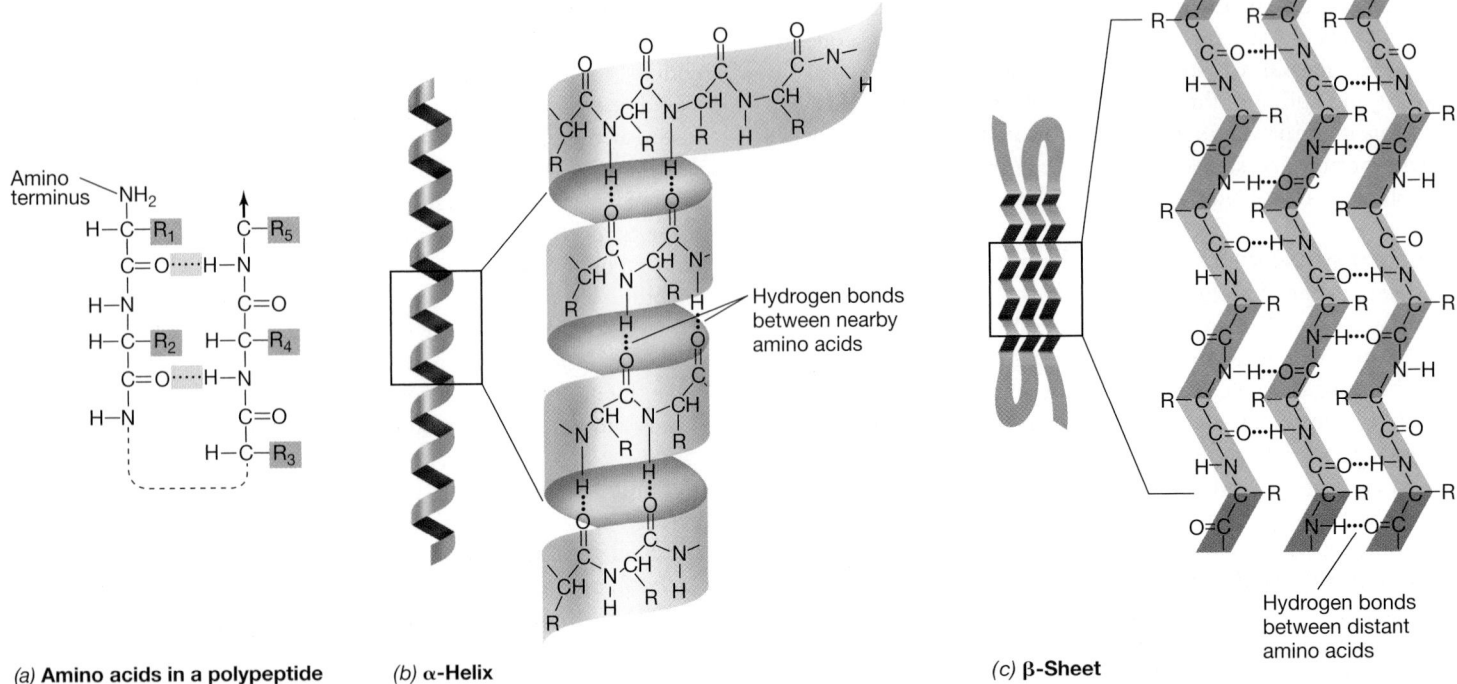

(a) **Amino acids in a polypeptide**   (b) **α-Helix**   (c) **β-Sheet**

Figure 6.38 **Secondary structure of polypeptides.** *(a)* Hydrogen bonding in protein secondary structure. R represents the side chain of the amino acid. *(b)* α-Helix secondary structure. *(c)* β-Sheet secondary structure. Note that the hydrogen bonding is between atoms in the peptide bonds and does not involve the R groups.

Interactions between the R groups of the amino acids in a polypeptide generate two further levels of structure. The **tertiary structure** depends largely on hydrophobic interactions, with lesser contributions from hydrogen bonds, ionic bonds, and disulfide bonds. The tertiary folding generates the overall three-dimensional shape of each polypeptide chain (**Figure 6.39**). Many proteins consist of two or more polypeptide chains. The **quaternary structure** refers to the number and type of polypeptides that form the final protein. In proteins with quaternary structure, each polypeptide is called a *subunit* and has its own primary, secondary, and tertiary structure. Some proteins have multiple copies of a single subunit. A protein with two identical subunits, for example, is called a *homodimer*. Other proteins may contain nonidentical subunits, each present in one or more copies (a *heterodimer*, for example, has one copy each of two different polypeptides). The subunits are held together by the same forces as for tertiary structure.

Both tertiary and quaternary structures may be stabilized by disulfide bonds between two adjacent sulfhydryl groups of appropriately positioned cysteine residues. If the two cysteine residues are located in different polypeptides, the disulfide bond covalently links the two molecules. Alternatively, a single polypeptide chain can fold and bond to itself if a disulfide bond can form within the molecule.

Chaperonins Assist Protein Folding

Most polypeptides fold spontaneously into their active form while they are being synthesized. However, some do not and require assistance from other proteins called **chaperonins** (also known as **molecular chaperones**) for proper folding or for assembly into larger complexes. The chaperonins themselves do not become part of the assembly but only assist in folding. Indeed, one important function of chaperonins is to prevent improper aggregation of proteins.

There are several different kinds of chaperonins. Some help newly synthesized proteins fold correctly. Other chaperonins are very abundant in the cell, especially under growth conditions that put protein stability at risk (for example, high temperatures).

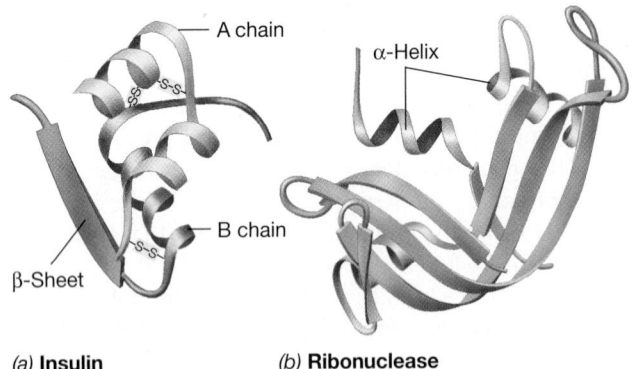

(a) **Insulin**   (b) **Ribonuclease**

Figure 6.39 **Tertiary structure of polypeptides.** *(a)* Insulin, a protein containing two polypeptide chains; note how the B chain contains both α-helix and β-sheet secondary structure and how disulfide linkages (S–S) help in dictating folding patterns (tertiary structure). *(b)* Ribonuclease, a large protein with several regions of α-helix and β-sheet secondary structure.

Chaperonins are widespread in all domains of life, and their sequences are highly conserved among all organisms.

Four key chaperonins in *Escherichia coli* are the proteins DnaK, DnaJ, GroEL, and GroES. DnaK and DnaJ are ATP-dependent enzymes that bind to newly formed polypeptides and keep them from folding too abruptly, a process that increases the risk of improper folding (**Figure 6.40**). Slower folding thus improves the chances of correct folding. If the DnaKJ complex is unable to fold the protein properly, it may transfer the partially folded protein to the two multi-subunit proteins GroEL and GroES. The protein first enters GroEL, a large barrel-shaped protein that uses the energy of ATP hydrolysis to fold the protein properly. GroES assists in this (Figure 6.40). It is estimated that only about 100 of the several thousand proteins of *E. coli* need help in folding from the GroEL–GroES complex, and of these approximately a dozen are essential for survival of the bacteria.

In addition to folding newly synthesized proteins, chaperonins can also refold proteins that have partially denatured in the cell. A protein may denature for many reasons, but often it is because the organism has temporarily experienced high temperatures. Chaperonins are thus one type of *heat shock protein*, and their synthesis is greatly accelerated when a cell is stressed by excessive heat (↩ Section 8.11). The heat shock response is an attempt by the cell to refold its partially denatured proteins for reuse before proteases recognize them as improperly folded and destroy them. Refolding is not always successful, and cells contain proteases whose function is to specifically target and destroy misfolded proteins, freeing their amino acids to make new proteins.

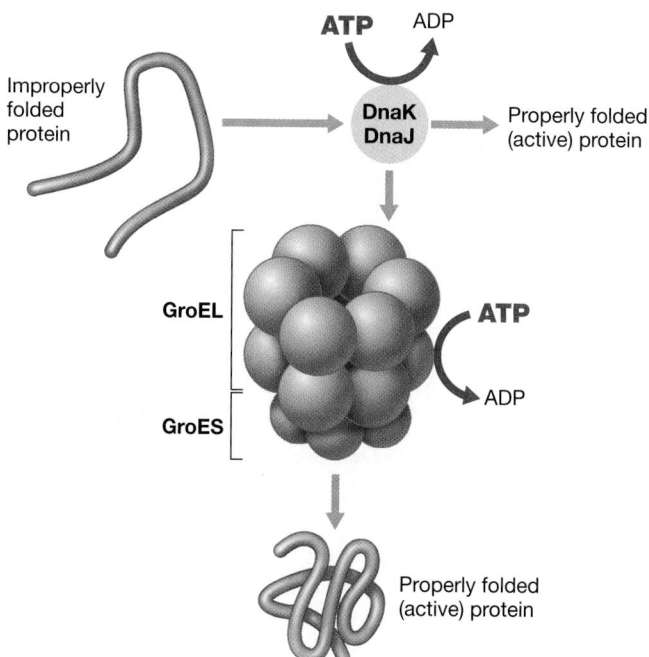

Figure 6.40 The activity of molecular chaperones. An improperly folded protein can be refolded by either the DnaKJ complex or by the GroEL–GroES complex. In both cases, energy for refolding comes from ATP.

Denaturation

When proteins are exposed to extremes of heat or pH or to certain chemicals that affect their folding, they may undergo **denaturation**. This results from the polypeptide chain unfolding, so destroying the higher-order (secondary, tertiary, and quaternary) structure of the protein. Depending on the severity of the denaturing conditions, the polypeptide may refold after the denaturant is removed. Typically, however, denatured proteins unfold to expose their hydrophobic regions. They then stick together to form protein aggregates that lack biological activity.

The biological properties of a protein are usually lost when it is denatured. Peptide bonds are not broken, so a denatured molecule retains its primary structure. This shows that biological activity is a function of the uniquely folded form of the protein as ultimately directed by primary structure. Denaturation of proteins is a major means of destroying microorganisms. Alcohols such as phenol and ethanol are effective disinfectants because they readily penetrate cells and irreversibly denature their proteins. Such chemical disinfectants have enormous practical value in household, hospital, and industrial applications. We discuss disinfectants, along with other agents used to destroy microorganisms, in Chapter 26.

Protein Secretion and the Signal Recognition Particle

Many proteins are located in the cytoplasmic membrane, in the periplasm of gram-negative cells, or even outside the cell proper. Such proteins must get from their site of synthesis on ribosomes into or through the cytoplasmic membrane. How is it possible for a cell to selectively transfer some proteins across a membrane while leaving most proteins in the cytoplasm?

Most proteins that must be transported into or through membranes are synthesized with an amino acid sequence of about 15–20 residues, called the **signal sequence**, at the beginning of the protein molecule. Signal sequences are quite variable, but typically they have a few positively charged residues at the beginning, a central region of hydrophobic residues, and then a more polar region. The signal sequence "signals" the cell's secretory system that this particular protein is to be exported and also helps prevent the protein from completely folding, a process that could interfere with its secretion. Because the signal sequence is the first part of the protein to be synthesized, the early steps in export may actually begin before the protein is completely synthesized (**Figure 6.41**).

Proteins to be exported are identified by their signal sequences either by the *SecA protein* or the *signal recognition particle* (*SRP*) (Figure 6.41). Generally, SecA binds proteins that are fully exported across the membrane into the periplasm whereas the SRP binds proteins that are inserted into the membrane but are not released on the other side. SRPs are found in all cells. In *Bacteria*, they contain a single protein and a small noncoding RNA molecule (4.5S RNA). Both SecA and the SRP deliver proteins to be secreted to the membrane secretion complex. In *Bacteria* this is normally the Sec system, whose channel consists of the three proteins SecYEG. The protein is exported across the cytoplasmic membrane through this channel. It may then either

Figure 6.41 Export of proteins via the major secretory system. The signal sequence is recognized either by SecA or by the signal recognition particle, which carries the protein to the membrane secretion system. The signal recognition particle binds proteins that are inserted into the membrane whereas SecA binds proteins that are secreted across the cytoplasmic membrane.

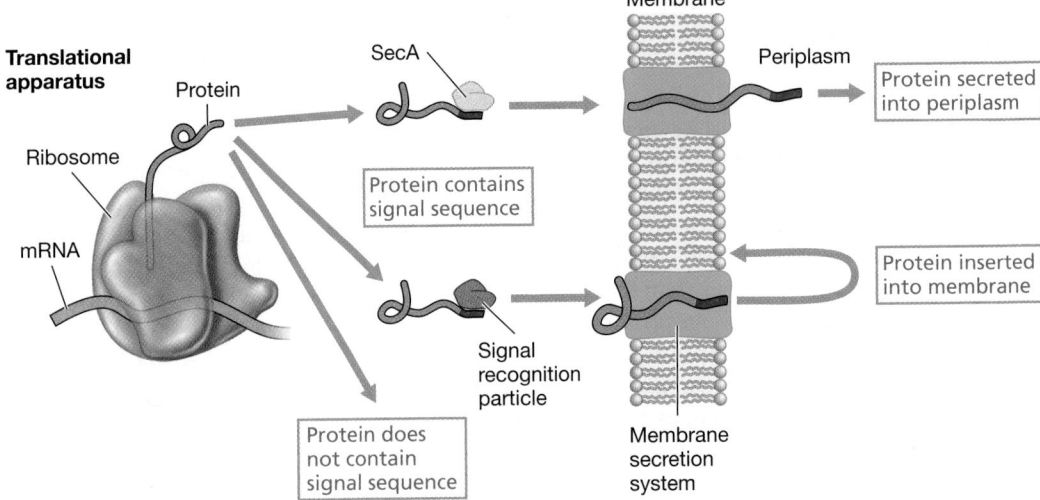

remain in the membrane or be released into the periplasm or the environment (Figure 6.41). After crossing the membrane, the signal sequence is removed by a protease.

Secretion of Folded Proteins: The Tat System

In the Sec system for protein export, the transported proteins are threaded through the cytoplasmic membrane in an unfolded state and only fold afterward (Figure 6.41). However, there are a few proteins that must be transported outside the cell after they have already folded. Usually this is because they contain small cofactors that must be inserted into the protein as it folds into its final form. Such proteins fold in the cytoplasm and then are exported by a transport system distinct from Sec, called the *Tat protein export system.*

The acronym Tat stands for "twin arginine translocase" because the transported proteins contain a short signal sequence containing a pair of arginine residues. This signal sequence on a folded protein is recognized by the TatBC proteins, which carry the protein to TatA, the membrane transporter. The energy required for transport is supplied by the proton motive force. A wide variety of proteins are transported by the Tat system, especially proteins required for energy metabolism that function in the periplasm. This includes iron–sulfur proteins and several

other redox-coupled proteins ($\mathcal{C}$ Section 4.9). In addition, the Tat pathway transports proteins needed for outer membrane biosynthesis and a few proteins that do not contain cofactors but can only fold properly within the cytoplasm.

MiniQuiz
- Define the terms primary, secondary, and tertiary structure with respect to proteins.
- How does a polypeptide differ from a protein?
- Describe the number and kinds of polypeptides present in a homotetrameric protein.
- What is a molecular chaperone?
- Why do some proteins have a signal sequence?
- What is a signal recognition particle?

In this chapter we have covered the essentials of the key molecular processes that occur in *Bacteria*. We next consider how archaeal and eukaryotic cells carry out the same processes. There are many similarities but also some major differences, in replication, transcription, and translation, among organisms in the three domains of life.

Big Ideas

6.1

The informational content of a nucleic acid is determined by the sequence of nitrogenous bases along the polynucleotide chain. Both RNA and DNA are informational macromolecules, as are the proteins they encode. RNA can fold into various configurations to generate secondary structure. The three key processes of macromolecular synthesis are: (1) DNA replication; (2) transcription (the synthesis of RNA from a DNA template); and

(3) translation (the synthesis of proteins using messenger RNA as template).

6.2

DNA is a double-stranded molecule that forms a helix. Its length is measured in terms of numbers of base pairs. The two strands in the double helix are antiparallel, but inverted repeats allow for the formation of secondary structure. The strands of a double-helical DNA molecule can be denatured by heat and allowed to reassociate following cooling.

6.3

Very long DNA molecules can be packaged into cells because they are supercoiled. In prokaryotes this supercoiling is brought about by enzymes called topoisomerases. DNA gyrase is a key enzyme in prokaryotes and introduces negative supercoils into the DNA.

6.4

In addition to the chromosome, a number of other genetic elements exist in cells. Plasmids are DNA molecules that exist separately from the chromosome of the cell. Viruses contain a genome, either DNA or RNA, that controls their own replication. Transposable elements exist as a part of other genetic elements.

6.5

The *Escherichia coli* chromosome has been mapped using conjugation, transduction, molecular cloning, and sequencing. *E. coli* has been a useful model organism, and a considerable amount of information has been obtained from it, not only about gene structure but also about gene function and regulation.

6.6

Plasmids are small circular or linear DNA molecules that carry nonessential genes. Although a cell can contain more than one plasmid, these cannot be closely related genetically. Although they have no extracellular form, plasmids can be transferred by the process of conjugation.

6.7

The genetic information that plasmids carry is not essential for cell function under all conditions but may confer a selective growth advantage under certain conditions. Examples include antibiotic resistance, enzymes for degradation of unusual organic compounds, and special metabolic pathways. Virulence factors of many pathogenic bacteria are plasmid encoded.

6.8

Both strands of the DNA helix serve as templates for the synthesis of two new strands (semiconservative replication). The two progeny double helices each contain one parental strand and one new strand. The new strands are elongated by addition of deoxyribonucleotides to the 3′ end. DNA polymerases require a primer made of RNA.

6.9

DNA synthesis begins at a unique location called the origin of replication. The double helix is unwound by helicase and is stabilized by single-strand binding protein. Extension of the DNA occurs continuously on the leading strand but discontinuously on the lagging strand. The fragments of the lagging strand are joined together later.

6.10

Starting from a single origin, DNA synthesis proceeds in both directions around circular chromosomes. Therefore, there are two replication forks in operation simultaneously. The proteins at the replication fork form a large complex known as the replisome. Most errors in base pairing that occur during replication are corrected by the proofreading functions of DNA polymerases. Incorrect nucleotides are removed and replaced. Finally, DNA replication terminates when the replication forks meet at a special terminus region on the chromosome.

6.11

The polymerase chain reaction is a procedure for amplifying DNA *in vitro* and employs heat-stable DNA polymerases. Heat is used to denature the DNA into two single-stranded molecules, each of which is copied by the polymerase. After each cycle, the newly formed DNA is denatured and a new round of copying proceeds. After each cycle, the amount of target DNA doubles.

6.12

The three major types of RNA are messenger RNA (mRNA), transfer RNA (tRNA), and ribosomal RNA (rRNA). Transcription of RNA from DNA is due to the enzyme RNA polymerase, which adds nucleotides onto 3′ ends of growing chains. Unlike DNA polymerase, RNA polymerase needs no primer and recognizes a specific start site on the DNA called the promoter.

6.13

In *Bacteria*, promoters are recognized by the sigma subunit of RNA polymerase. Regions of DNA recognized by a particular DNA-binding protein have very similar sequences. Alternative sigma factors allow joint regulation of large families of genes in response to growth conditions.

6.14

RNA polymerase stops transcription at specific sites called transcription terminators. Although encoded by DNA, these terminators function at the level of RNA. Some are intrinsic terminators and require no accessory proteins beyond RNA polymerase itself. In *Bacteria* these sequences are usually stem–loops followed by a run of uridines. Other terminators require proteins such as Rho.

6.15

The unit of transcription in prokaryotes often contains more than a single gene. Several genes are then transcribed into a single mRNA molecule that contains information for more than one polypeptide. A cluster of genes that are transcribed together from a single promoter constitute an operon. In all organisms, genes encoding rRNA are cotranscribed but then processed to form the final rRNA species.

6.16

Polypeptide chains contain 22 different genetically encoded amino acids that are linked via peptide bonds. Mirror-image (enantiomeric) forms of amino acids exist, but only the L-form is found in proteins. The primary structure of a protein is its amino acid sequence, but the folding (higher-order structure) of the polypeptide determines how the protein functions in the cell.

6.17

The genetic code is expressed as RNA, and a single amino acid may be encoded by several different but related codons. In addition to the nonsense codons, there is also a specific start codon that signals where the translation process should begin.

6.18

One or more tRNAs exist for each amino acid incorporated into proteins by the ribosome. Enzymes called aminoacyl-tRNA synthetases attach amino acids to their cognate tRNAs. Once the correct amino acid is attached to its tRNA, further specificity resides primarily in the codon–anticodon interaction.

6.19

The ribosome plays a key role in the translation process, bringing together mRNA and aminoacyl-tRNAs. There are three sites on the ribosome: the acceptor site, where the charged tRNA first combines; the peptide site, where the growing polypeptide chain is held; and an exit site. During each step of amino acid addition, the ribosome advances three nucleotides (one codon) along the mRNA, and the tRNA that is in the acceptor site moves to the peptide site. Protein synthesis terminates when a nonsense codon, which does not encode an amino acid, is reached.

6.20

Many nonstandard amino acids are found in proteins as a result of posttranslational modification. In contrast, the two rare amino acids selenocysteine and pyrrolysine are inserted into growing polypeptide chains during protein synthesis. They are both encoded by special stop codons that have a nearby recognition sequence that is specific for insertion of selenocysteine or pyrrolysine.

6.21

Proteins must be properly folded in order to function correctly. Folding may occur spontaneously but may also involve other proteins called molecular chaperones. Many proteins also need to be transported into or through membranes. Such proteins are synthesized with a signal sequence that is recognized by the cellular export apparatus and is removed either during or after export.

Review of Key Terms

Amino acid one of the 22 different monomers that make up proteins; chemically, contains a carboxylic acid, an amino group and a characteristic side chain all attached to the α-carbon

Aminoacyl-tRNA synthetase an enzyme that catalyzes attachment of an amino acid to its cognate tRNA

Anticodon a sequence of three bases in a tRNA molecule that base-pairs with a codon during protein synthesis

Antiparallel in reference to double-stranded DNA, the two strands run in opposite directions (one runs $5' \rightarrow 3'$ and the complementary strand $3' \rightarrow 5'$)

Bacteriocin a toxic protein secreted by bacteria that kills other, related bacteria

Chaperonin or molecular chaperone a protein that helps other proteins fold or refold from a partly denatured state

Chromosome a genetic element, usually circular in prokaryotes, carrying genes essential to cellular function

Codon a sequence of three bases in mRNA that encodes an amino acid

Codon bias nonrandom usage of multiple codons encoding the same amino acid

Complementary nucleic acid sequences that can base-pair with each other

Denaturation loss of the correct folding of a protein, leading (usually) to protein aggregation and loss of biological activity

DNA (deoxyribonucleic acid) a polymer of deoxyribonucleotides linked by phosphodiester bonds that carries genetic information

DNA gyrase an enzyme found in most prokaryotes that introduces negative supercoils in DNA

DNA ligase an enzyme that seals nicks in the backbone of DNA

DNA polymerase an enzyme that synthesizes a new strand of DNA in the $5' \rightarrow 3'$ direction using an antiparallel DNA strand as a template

Enantiomer a form of a molecule that is the mirror image of another form of the same molecule

Enzyme a protein or an RNA that catalyzes a specific chemical reaction in a cell

Gene a segment of DNA specifying a protein (via mRNA), a tRNA, an rRNA, or any other noncoding RNA

Genetic code the correspondence between nucleic acid sequence and amino acid sequence of proteins

Genetic element a structure that carries genetic information, such as a chromosome, a plasmid, or a virus genome

Genome the total complement of genes contained in a cell or virus

Informational macromolecule any large polymeric molecule that carries genetic information, including DNA, RNA, and protein

Lagging strand the new strand of DNA that is synthesized in short pieces and then joined together later

Leading strand the new strand of DNA that is synthesized continuously during DNA replication

Messenger RNA (mRNA) an RNA molecule that contains the genetic information to encode one or more polypeptides

Nonsense codon another name for a stop codon

Nucleic acid DNA or RNA

Nucleoside a nitrogenous base (adenine, guanine, cytosine, thymine, or uracil) plus a sugar (either ribose or deoxyribose) but lacking phosphate

Nucleotide a monomer of a nucleic acid containing a nitrogenous base (adenine, guanine, cytosine, thymine, or uracil), one or more molecules of phosphate, and a sugar, either ribose (in RNA) or deoxyribose (in DNA)

Open reading frame (ORF) a sequence of DNA or RNA that could be translated to give a polypeptide

Operon a cluster of genes that are cotranscribed as a single messenger RNA

Peptide bond a type of covalent bond linking amino acids in a polypeptide

Phosphodiester bond a type of covalent bond linking nucleotides together in a polynucleotide

Plasmid an extrachromosomal genetic element that has no extracellular form

Polymerase chain reaction (PCR) artificial amplification of a DNA sequence by repeated cycles of strand separation and replication

Polynucleotide a polymer of nucleotides bonded to one another by covalent bonds called phosphodiester bonds

Polypeptide a polymer of amino acids bonded to one another by peptide bonds

Primary structure the precise sequence of monomers in a macromolecule such as a polypeptide or a nucleic acid

Primase the enzyme that synthesizes the RNA primer used in DNA replication

Primer an oligonucleotide to which DNA polymerase attaches the first deoxyribonucleotide during DNA synthesis

Promoter a site on DNA to which RNA polymerase binds to commence transcription

Protein a polypeptide or group of polypeptides forming a molecule of specific biological function

Purine one of the nitrogenous bases of nucleic acids that contain two fused rings; adenine and guanine

Pyrimidine one of the nitrogenous bases of nucleic acids that contain a single ring; cytosine, thymine, and uracil

Quaternary structure in proteins, the number and types of individual polypeptides in the final protein molecule

Replication synthesis of DNA using DNA as a template

Replication fork the site on the chromosome where DNA replication occurs and where the enzymes replicating the DNA are bound to untwisted, single-stranded DNA

Ribosomal RNA (rRNA) types of RNA found in the ribosome; some participate actively in protein synthesis

Ribosome a cytoplasmic particle composed of ribosomal RNA and protein, whose function is to synthesize proteins

RNA (ribonucleic acid) a polymer of ribonucleotides linked by phosphodiester bonds that plays many roles in cells, in particular, during protein synthesis

RNA polymerase an enzyme that synthesizes RNA in the $5' \rightarrow 3'$ direction using a complementary and antiparallel DNA strand as a template

Secondary structure the initial pattern of folding of a polypeptide or a polynucleotide, usually dictated by opportunities for hydrogen bonding

Semiconservative replication DNA synthesis yielding two new double helices, each consisting of one parental and one progeny strand

Signal sequence a special N-terminal sequence of approximately 20 amino acids that signals

that a protein should be exported across the cytoplasmic membrane

Start codon a special codon, usually AUG, that signals the start of a protein

Stop codon a codon that signals the end of a protein

Termination stopping the elongation of an RNA molecule at a specific site

Tertiary structure the final folded structure of a polypeptide that has previously attained secondary structure

Transcription the synthesis of RNA using a DNA template

Transfer RNA (tRNA) a small RNA molecule used in translation that possesses an anticodon at one end and has the corresponding amino acid attached to the other end

Translation the synthesis of protein using the genetic information in RNA as a template

Wobble a less rigid form of base pairing allowed only in codon–anticodon pairing

Review Questions

1. In one sentence, describe the difference between DNA and RNA (Section 6.1).

2. Why is DNA negatively charged (Section 6.1)?

3. What do you understand "complementary base pairing" to mean in a DNA structure (Section 6.2)?

4. Is the sequence 5′-GCACGGCACG-3′ an inverted repeat? Explain your answer (Section 6.2).

5. DNA molecules that are AT-rich separate into two strands more easily when the temperature rises than do DNA molecules that are GC-rich. Explain this based on the properties of AT and GC base pairing (Section 6.2).

6. Describe how DNA, which is many times the length of a cell when linearized, fits into the cell (Section 6.3).

7. List the major genetic elements known in microorganisms (Section 6.4).

8. What is the size of the *Escherichia coli* chromosome? About how many proteins can it encode? How much noncoding DNA is present in the *E. coli* chromosome (Section 6.5)?

9. How do plasmids replicate and how does this differ from chromosomal replication (Section 6.6)?

10. What are R plasmids and why are they of medical concern (Section 6.7)?

11. A structure commonly seen in circular DNA during replication is the theta structure. Draw a diagram of the replication process and show how a theta structure could arise (Sections 6.9 and 6.10).

12. Why are errors in DNA replication so rare? What enzymatic activity, in addition to polymerization, is associated with DNA polymerase III and how does it reduce errors (Section 6.10)?

13. Describe the basic principles of gene amplification using the polymerase chain reaction (PCR). How have thermophilic and hyperthermophilic prokaryotes simplified the use of PCR (Section 6.11)?

14. Do genes for tRNAs have promoters? Do they have start codons? Explain (Sections 6.12 and 6.15).

15. The start and stop sites for mRNA synthesis (on the DNA) are different from the start and stop sites for protein synthesis (on the mRNA). Explain (Sections 6.15 and 6.19).

16. Why are amino acids so named? Write a general structure for an amino acid. What is the importance of the R group to final protein structure? Why does the amino acid cysteine have special significance for protein structure (Section 6.16)?

17. What is codon bias (Section 6.17)?

18. What are aminoacyl-tRNA synthetases and what types of reactions do they carry out? How does a synthetase recognize its correct substrates (Section 6.18)?

19. The enzyme activity that forms peptide bonds on the ribosome is called peptidyl transferase. Which molecule catalyzes this reaction (Section 6.19)?

20. Define the types of protein structure: primary, secondary, tertiary, and quaternary. Which of these structures are altered by denaturation (Section 6.21)?

21. Sometimes misfolded proteins can be correctly refolded, but sometimes they cannot and are destroyed. What kinds of proteins are involved in refolding misfolded proteins? What kinds of enzymes are involved in destroying misfolded proteins (Section 6.21)?

22. How does a cell know which of its proteins are designed to function outside of the cell (Section 6.21)?

Application Questions

1. The genome of the bacterium *Neisseria gonorrhoeae* consists of one double-stranded DNA molecule that contains 2220 kilobase pairs. Calculate the length of this DNA molecule in centimeters. If 85% of this DNA molecule is made up of the open reading frames of genes encoding proteins, and the average protein is 300 amino acids long, how many protein-encoding genes does *Neisseria* have? What kind of information do you think might be present in the other 15% of the DNA?

2. Compare and contrast the activity of DNA and RNA polymerases. What is the function of each? What are the substrates of each? What is the main difference in the behavior of the two polymerases?

3. What would be the result (in terms of protein synthesis) if RNA polymerase initiated transcription one base upstream of its normal starting point? Why? What would be the result (in terms of protein synthesis) if translation began one base downstream of its normal starting point? Why?

4. In Chapter 10 we will learn about mutations, inheritable changes in the sequence of nucleotides in the genome. By inspecting Table 6.5, discuss how the genetic code has evolved to help minimize the impact of mutations.

Need more practice? Test your understanding with Quantitative Questions; access additional study tools including tutorials, animations, and videos; and then test your knowledge with chapter quizzes and practice tests at **www.microbiologyplace.com**.

Archaeal and Eukaryotic Molecular Biology

The two domains of prokaryotes—*Bacteria* and *Archaea*—are phylogenetically distinct, despite their structural similarities. This distinction is especially obvious in the molecular biology of these organisms. In Chapter 6 we viewed molecular biology as illustrated by the model bacterium, *Escherichia coli*. In both *Archaea* and higher organisms, the overall flow of information from DNA to RNA to protein is the same as in *Bacteria*. However, some details differ, and in eukaryotic cells the presence of the nucleus as a separate compartment complicates the flow of genetic information.

Despite lacking a nucleus, in many of their properties *Archaea* resemble eukaryotes more closely than *Bacteria*. Indeed, *Archaea* are increasingly used as model organisms to investigate mechanisms they share with the eukaryotes that are more difficult to study in complex eukaryotic cells. Here we discuss key elements of the molecular biology of *Archaea*, in particular those aspects that show similarities to the eukaryotes.

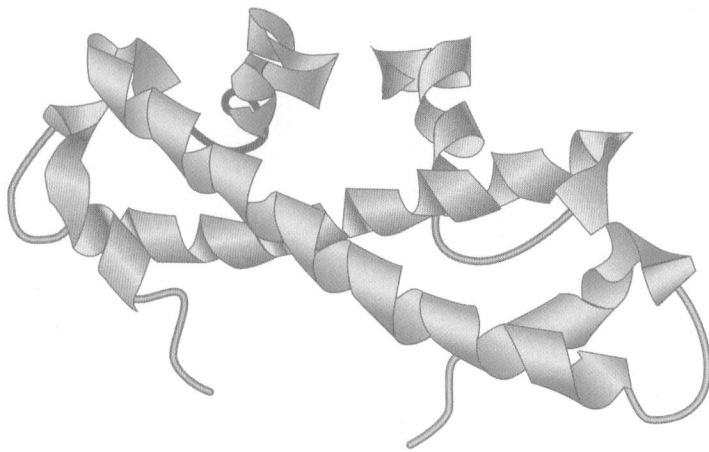

Figure 7.1 Histones in *Archaea*. Although archaeal histones are shorter than eukaryotic histones, they contain the same histone fold structure. Shown here is a dimer of the histone HPhA from the hyperthermophilic archaeon *Pyrococcus horikoshii*. Adapted from structure 1KU5 of the Protein Data Bank.

 # Molecular Biology of *Archaea*

The *Archaea* were originally regarded as aberrant members of the *Bacteria* because both groups share the same overall prokaryotic cell design in which there are no membrane-bound compartments and, in particular, no nucleus. Moreover, both groups typically contain a single circular chromosome and often transcribe several genes onto the same polycistronic messenger RNA (mRNA). However, comparative ribosomal RNA (rRNA) sequencing has revealed a closer relationship between *Archaea* and *Eukarya* than between *Bacteria* and *Archaea*. This is reflected in such areas as the use of histones for DNA packaging and the detailed mechanism of translation. We start by comparing the genes and chromosomes of *Archaea* with those of the other two domains.

7.1 Chromosomes and DNA Replication in *Archaea*

The chromosomes of *Archaea* resemble those of *Bacteria* in being circular and carrying from around 500 to a few thousand genes. However, DNA packaging and chromosome replication reveal greater similarities with the eukaryotes.

DNA Packaging in *Archaea*

In all organisms, long DNA molecules are packaged by supercoiling, although the mechanisms vary among the three domains. *Bacteria* use DNA gyrase to supercoil their DNA (๙ Section 6.3), whereas the eukaryotes wind their DNA around proteins known as histones (Section 7.5). Many *Archaea* possess both DNA gyrase and histones. Thus the DNA of *Archaea* is condensed by supercoiling mediated by DNA gyrase as in *Bacteria*, by binding to histones as in eukaryotes, or by a combination of these two mechanisms.

Histones (**Figure 7.1**) are present in all eukaryotes and in at least some members of all archaeal phyla, although not in every archaeal species. Histones are small, positively charged proteins that neutralize the negative charge of DNA (resulting from the

phosphate groups). The DNA is wound around clusters of histones, forming structures known as **nucleosomes**, which are spaced along the DNA double helix at regular intervals. Archaeal histones form clusters of four (sometimes referred to as tetrasomes) rather than the eight found in eukaryotic histones, and accommodate approximately 80 base pairs (bp) of DNA. Archaeal histones are shorter in length than eukaryotic histones, but are homologous in amino acid sequence and similar in their three-dimensional structure. Archaeal and eukaryotic histones share the so-called histone fold, the central region of the histone protein that is necessary for forming nucleosomes. Eukaryotic histones have extra N- and C-terminal domains that are not necessary for nucleosome assembly. Although *Bacteria* do not possess genuine histones, they do contain histone-like proteins that bind to DNA. These histone-like proteins are not homologous in sequence to true histones, nor do they form nucleosomes. Analogous histone-like proteins also occur in many *Archaea*.

Those *Archaea* that grow at extremely high temperatures (hyperthermophiles; ๙ Chapter 19) contain an enzyme called **reverse DNA gyrase**. This topoisomerase introduces positive supercoils into DNA. Reverse gyrase appears to play an important but undefined role in both *Bacteria* and *Archaea* that grow at extremely high temperatures. Although it is present in nearly all hyperthermophiles, it is not essential for cell viability. Because all cells require the genetic information in DNA to be accessible to the replication and transcription machinery, the roles of DNA gyrase or reverse DNA gyrase may be to ensure that the structure of DNA within the cell remains quite dynamic.

Replication of Chromosomes in *Archaea*

The circular archaeal chromosome is structurally similar to that of *Bacteria* and also undergoes bidirectional replication (๙ Section 6.10). Nonetheless, the protein machinery of chromosomal replication in *Archaea* shows greater similarity to that of eukaryotes.

Table 7.1 *Three-domain comparison of chromosomes and their replication*[a]

	Bacteria	Archaea	Eukarya
Chromosome topology	Circular	Circular	Linear
Chromosome number	One	One	Multiple
Origins per chromosome	One	One, two, or three	Many
Origin recognition	DnaA protein	Cdc6/Orc[b] protein	ORC[b]
Helicase	DnaB protein	MCM protein	MCM protein
Helicase loader	DnaC protein	Cdc6/Orc[b] protein	Cdc6 protein[b]
Sliding clamp	DnaN protein	PCNA protein	PCNA protein
Primase	DnaG protein	PriS and PriL	Pri1 and Pri2
Main DNA polymerase	PolC family	PolB family	PolB family

[a]Relatively uncommon exceptions, such as bacteria with linear chromosomes, are omitted from this table.
[b]The Cdc6/Orc protein of *Archaea* is similar to both the Cdc6 and some ORC (origin recognition complex) subunits of eukaryotes and may be bifunctional. Some *Archaea* have more than one Cdc6-type protein.

In contrast to bacteria, whose chromosomes have a single origin of replication, several *Archaea* are known whose single circular chromosome has multiple origins of replication (**Table 7.1**). For example, *Halobacterium* appears to have two replication origins on its chromosome and *Sulfolobus* has three.

The proteins of *Archaea* and *Eukarya* that recognize the origin of replication and help synthesize DNA show much greater similarity to each other than to functionally equivalent proteins of *Bacteria* (Table 7.1). In some cases, such as DNA helicase or the origin recognition complex (ORC), *Eukarya* have enzyme complexes consisting of multiple different (but related) protein subunits. In contrast, *Archaea* have only a single protein that forms equivalent complexes. For example, the eukaryotic helicase forms rings consisting of six different protein subunits, whereas the archaeal helicase forms rings of six identical subunits. In many respects, *Archaea* seem to have a simplified version of the eukaryotic replication apparatus.

All organisms possess multiple DNA polymerases specialized for different roles such as replication and DNA repair. There are three main structural families of DNA polymerase. Organisms in different domains use members of these families for different roles. For example, *Bacteria* use DNA polymerases of family C (for example, Pol III of *Escherichia coli*) as their main replicative enzymes, whereas DNA polymerases of family A and B are used mostly in DNA repair. In contrast, *Archaea* and *Eukarya* use DNA polymerases of family B as their main replicative enzymes (Table 7.1).

MiniQuiz

- What similarities and differences are there between the histones of *Archaea* and *Eukarya*?
- How do the activities of DNA gyrase and reverse gyrase differ?
- How many origins of replication are found on chromosomes from *Bacteria*, *Archaea*, and *Eukarya*?

7.2 Transcription and RNA Processing in *Archaea*

The fundamental phylogenetic relationship of the *Archaea* and the *Eukarya* was originally revealed by sequence comparisons of ribosomal RNA (↩ Section 2.7). Further comparisons have shown that transcription and translation in these two domains share many structural and mechanistic features, confirming that these two domains are more closely related to each other than either is to *Bacteria*. In this section we review transcription in *Archaea*.

Transcription in *Archaea*

As with their chromosome replication machinery, *Archaea* seem to have a simplified version of the eukaryotic transcription apparatus. Both the sequences of archaeal promoters and the structure and activity of RNA polymerase resemble those of eukaryotes. Conversely, the regulation of transcription in *Archaea* shares major similarities with *Bacteria*. Regulation of gene expression of both *Bacteria* and *Archaea* is covered in Chapter 8 (↩ Section 8.6 for *Archaea*).

Archaea contain only a single RNA polymerase, which most closely resembles eukaryotic RNA polymerase II (Section 7.9). The archaeal RNA polymerase typically has 11 or 12 subunits, while eukaryotic RNA polymerase II has 12 or more (**Figure 7.2**). RNA polymerase from *Bacteria* has only four different subunits (excluding the sigma recognition subunit) (↩ Section 6.12). The antibiotic rifampicin inhibits bacterial RNA polymerase, but does not inhibit either the eukaryotic or archaeal enzymes. Furthermore, the structure of archaeal promoters resembles that of eukaryotic promoters recognized by eukaryotic RNA polymerase II. Three main recognition sequences are part of the promoters in both domains, and these sequences are recognized by a series of proteins called *transcription factors* that are similar in *Eukarya* and *Archaea*.

The most important recognition sequence in archaeal and eukaryotic promoters is the 6- to 8-base-pair TATA box, located

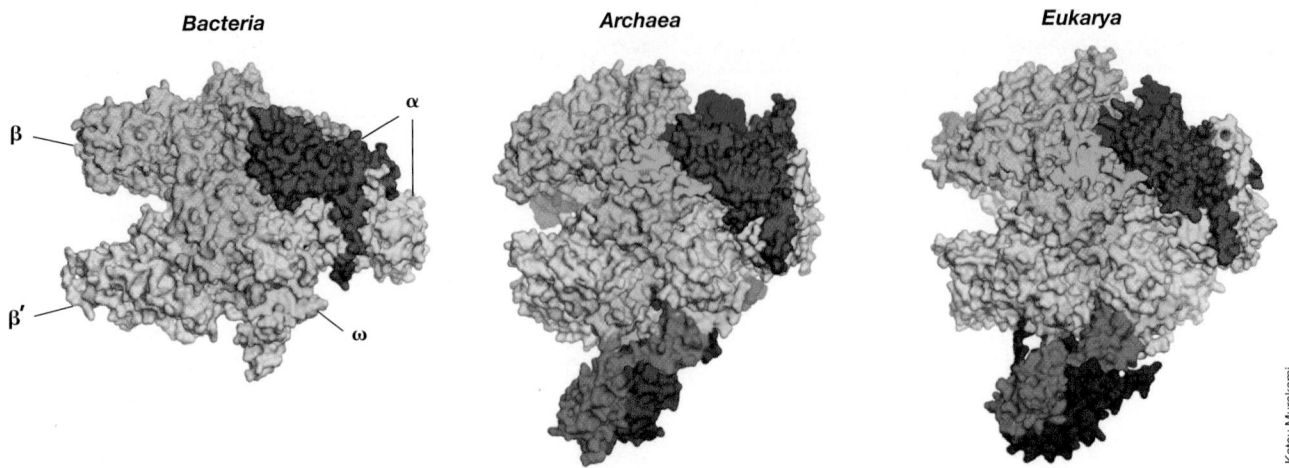

Figure 7.2 RNA polymerase from the three domains. Surface representation of multi-subunit cellular RNA polymerase structures from *Bacteria* (left, *Thermus aquaticus* core enzyme), *Archaea* (center, *Sulfolobus solfataricus*), and *Eukarya* (right, *Saccharomyces cerevisiae* RNA Pol II). Orthologous subunits are depicted by the same color. A unique subunit in the *S. solfataricus* RNA polymerase is not shown in this view.

18–27 nucleotides upstream of the transcriptional start site (**Figure 7.3**). This is recognized by the TATA-binding protein (TBP). Upstream of the TATA box is the B recognition element (BRE) sequence that is recognized by transcription factor B (TFB). In addition, the initiator element, located at the start of transcription, is also important. Once TBP has bound to the TATA box and TFB has bound to the BRE, then archaeal RNA polymerase

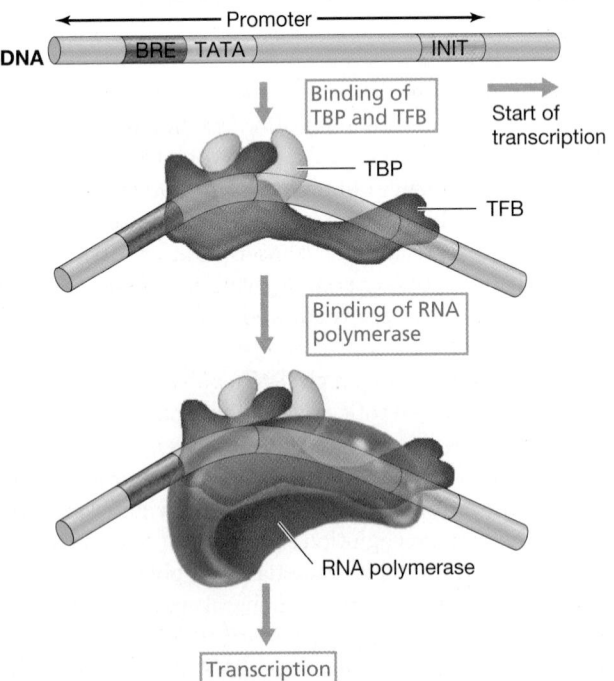

Figure 7.3 Promoter architecture and transcription in *Archaea*. Three promoter elements are critical for promoter recognition in *Archaea*: the initiator element (INIT), the TATA box, and the B recognition element (BRE). The TATA-binding protein (TBP) binds the TATA box; transcription factor B (TFB) binds to both BRE and INIT. Once both TBP and TFB are in place, RNA polymerase binds.

can bind and initiate transcription. This process is similar in eukaryotes except that more transcription factors are required (Section 7.9).

Less is known about the transcription termination signals in *Archaea*. Some archaeal genes have inverted repeats followed by an AT-rich sequence, sequences very similar to those found in many bacterial transcription terminators (Section 6.15). However, such termination sequences are not found in other archaeal genes. One other type of suspected transcription terminator lacks inverted repeats but contains repeated runs of thymines. In some way, this signals the archaeal termination machinery to terminate transcription. No Rho-like proteins have been found in *Archaea*. (Rho is needed for transcription termination of some genes in *Bacteria*, Section 6.14.)

Intervening Sequences in *Archaea*

As is the case in *Bacteria*, intervening sequences in genes that encode proteins are extremely rare in *Archaea*. This is in contrast to *Eukarya*, in which many such genes are split into two or more coding regions separated by noncoding regions (Section 7.8). The segments of coding sequence are called **exons**, and **introns** are the intervening noncoding regions. The term **primary transcript** refers to the RNA molecule that is originally transcribed before the introns are removed to generate the final mRNA, consisting solely of the exons.

However, several tRNA- and rRNA-encoding genes of *Archaea* do possess introns that must be removed after transcription to generate the mature tRNA or rRNA. These introns were named archaeal introns because they are processed by a different mechanism than are typical eukaryotic introns (Section 7.8). Instead, archaeal introns are excised by a specific endoribonuclease that recognizes exon–intron junctions (**Figure 7.4**). In occasional *Archaea*, tRNA molecules are even assembled by splicing together segments from two or three different primary transcripts.

Archaeal-style introns are also found in the nuclear tRNA genes of eukaryotes. Furthermore, the archaeal endoribonuclease

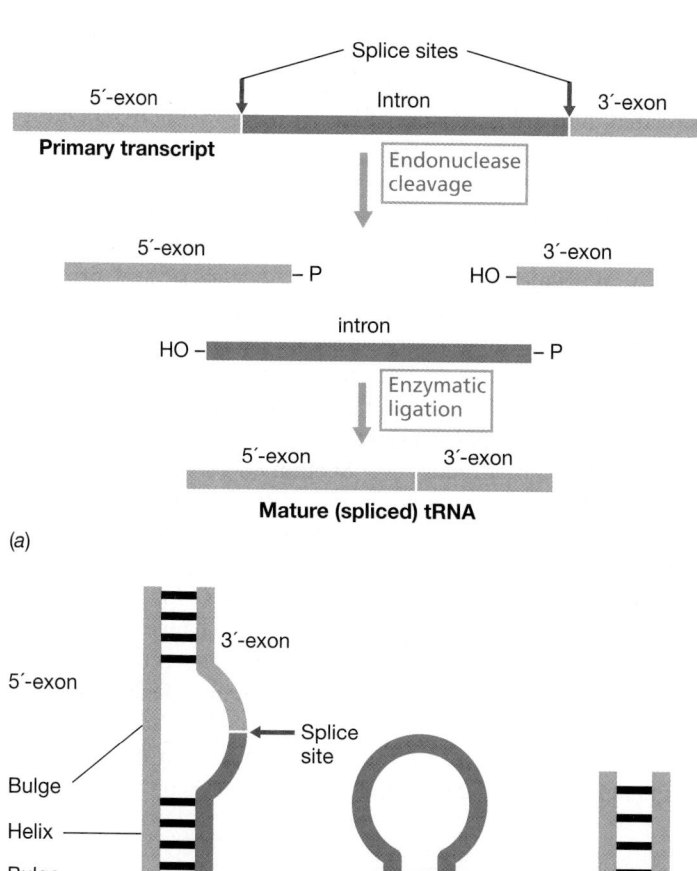

Figure 7.4 Splicing of archaeal introns. *(a)* Reaction scheme. Removal of archaeal introns is a two-step reaction. In the first step a specific endonuclease excises the intron. In the second step a ligase joins the 5′-exon to the 3′-exon, generating the mature, spliced tRNA. *(b)* Folding of the tRNA precursor. The two splice sites (red arrows) are recognized by their characteristic "bulge-helix-bulge" motifs. The products of the reaction are the tRNA and a circular intron.

that splices introns is homologous to two of the subunits of the enzyme complex that removes introns from eukaryotic nuclear tRNA. The overall frequency of intervening gene sequences is similar in *Bacteria* and *Archaea*, possibly because they both contain small, compact genomes. On the other hand, the splicing mechanism of archaeal introns is shared with a small subset of introns in the nucleus of eukaryotes.

MiniQuiz

• What three major components make up an archaeal promoter?

• What are archaeal introns?

7.3 Protein Synthesis in *Archaea*

As previously stated, the relatedness of *Archaea* and the eukaryotic nuclear genome was originally discovered by comparative sequencing of rRNA, which is a key component of the translation machinery (∂∂ Section 6.19). Analyses of most other components of the translation machinery also support the close relationship of *Archaea* to the eukaryotes.

The ribosome of eukaryotes contains several rRNA molecules plus 78 proteins. Of these proteins, 34 are common to all three domains of life, another 33 are shared by *Archaea* and *Eukarya*, and another 11 are unique to eukaryotes—none are common to only *Bacteria* and *Eukarya*. A similar situation is seen with the various translation factors. Some are common to all three domains, while others are shared by *Archaea* and *Eukarya*. In addition to the 11 uniquely eukaryotic ribosomal proteins, eukaryotes have one extra rRNA molecule, the 5.8S rRNA, which is not found in either *Bacteria* or *Archaea*.

Translation needs not only a functional ribosome, but also several translation factors (initiation, elongation, and release factors; ∂∂ Section 6.19). Eukaryotes and *Archaea* have approximately twice as many translation factors as *Bacteria*. Those of eukaryotes tend to have more subunits than those of *Archaea*, as is the case with many proteins involved in replication (Section 7.1). However, the translation factors of *Archaea* and *Eukarya* show a high degree of homology.

Overall, the components of the translation machinery are more closely related in *Archaea* and *Eukarya* than they are to those of *Bacteria*. Despite this, the mechanism of protein synthesis in *Archaea* shares some features with both *Bacteria* and *Eukarya*, and as a result, the situation is rather complicated. For example, the mRNAs in both *Bacteria* and *Archaea* have sequences complementary to the 3′ end of 16S rRNA (∂∂ Section 6.19) that are absent in *Eukarya*. Conversely, both *Archaea* and *Eukarya* insert methionine as the first amino acid, in contrast to *Bacteria*, which use *N*-formylmethionine (∂∂ Section 6.17). At present, few experimental data are available on the detailed mechanism of translation or the order in which the various components assemble for *Archaea*. Overall, sequence comparisons underscore the similarity between *Eukarya* and *Archaea* in the RNA and proteins that make up the translation machinery.

MiniQuiz

• How does translation in *Archaea* resemble that in *Bacteria*?

• How does translation in *Archaea* resemble that in *Eukarya*?

• Which components of the eukaryotic ribosome are missing in both *Bacteria* and *Archaea*?

7.4 Shared Features of *Bacteria* and *Archaea*

Despite the closer genetic relationship of *Archaea* to the eukaryotes, organisms in the two prokaryotic domains, *Bacteria* and *Archaea*, nonetheless share several fundamental properties that are absent from eukaryotes. **Figure 7.5** summarizes how the genetic features of the three domains are distributed. Both *Bacteria* and *Archaea* are typically single-celled microorganisms, most of which divide by binary fission, although some divide by budding. Neither *Bacteria* nor *Archaea* possess a nucleus or membrane-bound organelles. Consequently, *Bacteria* and *Archaea* share several features such as coupled transcription and translation (Section 2.5) that are impossible in eukaryotes because transcription occurs inside the nucleus. Other shared features, such as the near absence of intervening sequences in both *Bacteria* and *Archaea*, may be a result of a "prokaryotic lifestyle." In other words, relatively rapid growth as single cells creates pressure for a small, compact genome with little surplus, noncoding DNA.

Neither *Bacteria* nor *Archaea* possess the large, membrane-enclosed, eukaryotic type of flagellum that uses ATP as an energy source. Instead, both *Bacteria* and *Archaea* use flagella with rotary bases that are energized by the proton motive force and

whose shaft is a single helix of protein subunits (Section 3.13). However, the protein that makes up the flagellar shaft in *Archaea* is related to the pilus protein of *Bacteria* (Section 3.9), not to the bacterial flagellin protein.

Bacteria and *Archaea* share many other properties. Both possess a single circular chromosome and transcribe clusters of genes, controlled as a unit (an operon), into single polycistronic mRNA molecules that are translated to give several proteins. Furthermore, both *Bacteria* and *Archaea* have Shine–Dalgarno sequences and both lack the cap structure typical of eukaryotic mRNA (Section 7.8). In addition, regulation of transcription in *Archaea* is largely bacterial in nature, and relies on both repressor and activator DNA-binding proteins, which recognize sites within the promoter region (Section 8.7).

In eukaryotes, reproduction and genetic exchange are coupled during sexual reproduction, in which haploid gametes from two parents fuse to create a new diploid individual. Thus, eukaryotes alternate between haploid and diploid states. In *Bacteria* and *Archaea*, however, reproduction and gene exchange are two distinct processes. There is no diploid phase, and reproduction is equivalent to cell division. Several different mechanisms are responsible for genetic transfer in *Bacteria* and *Archaea*, but all are unidirectional and DNA is transferred from a donor organism to a recipient in processes that do not involve cell division (Chapter 10).

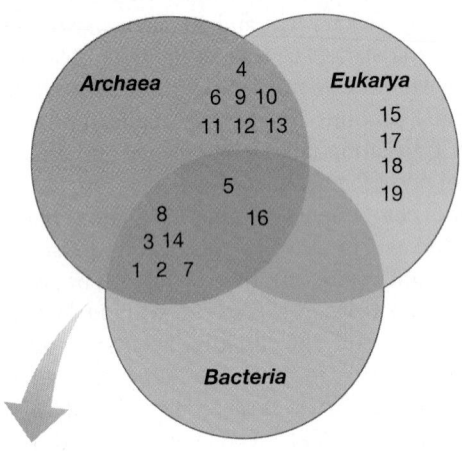

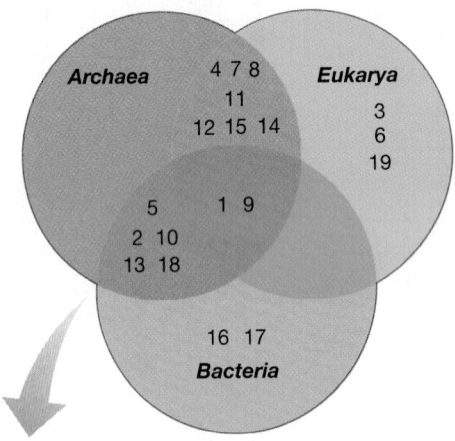

Genome

1 Chromosome circular versus linear
2 Single chromosome versus multiple chromosomes
3 Introns rare
4 Archaeal-type introns
5 Inteins
6 Histones
7 DNA gyrase
8 Reverse gyrase
9 Multiple chromosomal origins
10 Eukaryotic origin recognition complex
11 Eukaryotic-type helicase
12 B family DNA polymerase is major replicative enzyme
13 Eukaryotic-type sliding clamp
14 Restriction enzymes
15 RNAi
16 Genome of double-stranded DNA
17 Multiple retroelements in genome
18 Centromeres
19 Telomeres and telomerase

(a)

Transcription and Translation

1 RNA used as a genetic messenger
2 Polycistronic mRNA
3 Cap and tail on mRNA
4 TATA box and BRE sequence in promoter
5 Repressors binding directly to DNA in promoter
6 Multiple RNA polymerases
7 RNA polymerase II with 8 or more subunits
8 Multiple transcription factors needed
9 Ribosomes synthesize proteins
10 70S versus 80S ribosomes
11 Ribosomal RNA sequence homologies
12 Ribosomal protein sequence homologies
13 Shine–Dalgarno sequences
14 Multiple translation factors
15 Elongation factor sensitive to diphtheria toxin
16 *N*-Formylmethionine versus methionine
17 tmRNA rescues stalled ribosomes
18 16S and 23S rRNA
19 18S, 28S, and 5.8S rRNA

(b)

Figure 7.5 Molecular features of the three domains. Venn diagrams show which features are shared by the domains and which are unique. *(a)* Genomic features. *(b)* Features of transcription and translation.

RNA interference (RNAi), a major mechanism for protection against infection by RNA viruses, is found only in eukaryotes (Section 7.10), although proteins homologous to some of the RNAi components are found in *Archaea*. Conversely, both *Bacteria* and *Archaea* possess the CRISPR system of virus defense (⮡ Microbial Sidebar in Chapter 8, "The CRISPR Antiviral Defense System"). Although reminiscent of RNAi, the mechanism is distinct and CRISPR is targeted against DNA as well as RNA. *Bacteria* and *Archaea* also produce restriction endonucleases (⮡ Section 11.1) that function to destroy incoming foreign DNA, whereas eukaryotes lack restriction enzyme systems.

MiniQuiz

- What shared features of *Bacteria* and *Archaea* are likely due to the lack of a nucleus and membrane-bound organelles?
- What shared features of *Bacteria* and *Archaea* affect transcription and translation?

Eukaryotic Molecular Biology

The fact that eukaryotes have a nucleus has profound consequences. These include the need to disassemble the nucleus during cell division, as seen in mitosis and meiosis, and the physical separation of transcription (nuclear) and translation (cytoplasmic). Eukaryotes are also largely unique in the replication of linear (as opposed to circular) chromosomes and in the processing of mRNA—both of which occur inside the nucleus. A summary of molecular events in eukaryotes is given in **Figure 7.6**.

7.5 Genes and Chromosomes in *Eukarya*

In eukaryotes, protein-encoding genes are often split into two or more coding regions, known as exons, with noncoding regions, known as introns, separating them. Both introns and exons are transcribed into the primary RNA transcript. From this, the mature (functional) mRNA is formed by excision of the introns and splicing together the exons (Section 7.8). Only after this occurs does the mRNA exit the nucleus to be translated by the ribosomes in the cytoplasm. As we have seen (Section 7.2), a few genes in prokaryotes contain introns, but the vast majority do not.

Eukaryotes typically contain much more DNA than is needed to encode all the proteins required for cell function. For instance, in the human genome only about 3% of the total DNA encodes protein. In contrast, in prokaryotes this fraction often exceeds 90%. The "extra" DNA in eukaryotes is present as introns or repetitive sequences, some of which are repeated hundreds or thousands of times. Eukaryotic microorganisms have fewer introns than do higher eukaryotes. For example, about 70% of DNA in the yeast *Saccharomyces cerevisiae* encodes protein. Additionally, eukaryotes often have multiple copies of certain genes, such as those that encode tRNAs and rRNAs. The genes for rRNA are found repeated in some prokaryotes, but usually only a few copies are present.

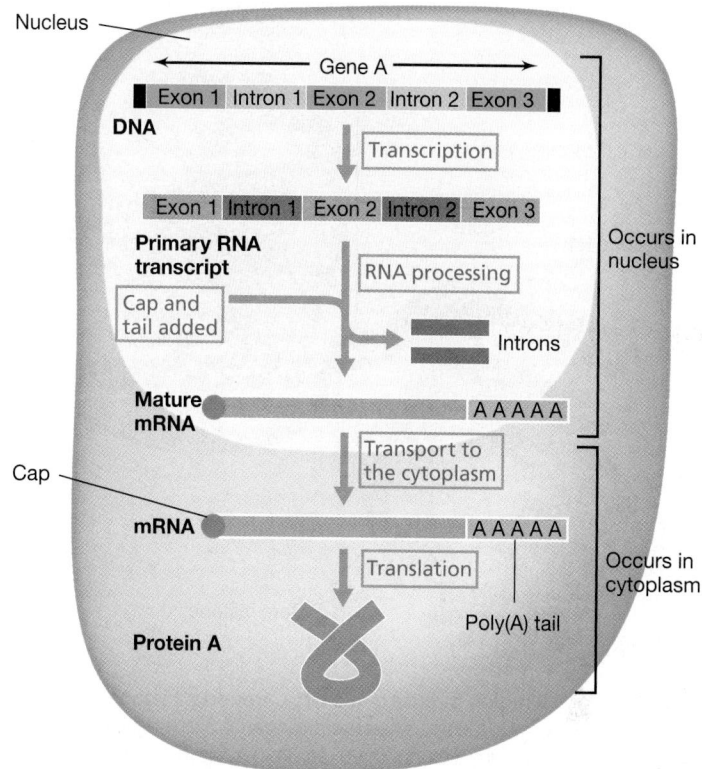

Figure 7.6 Information transfer in eukaryotes. Prior to translation, noncoding regions (introns) are removed from the primary RNA transcript, leaving exons to be joined to form the mature mRNA. In addition, a cap and a poly(A) tail are added before the mRNA is allowed to exit the nucleus.

In contrast to the single circular prokaryotic chromosome in the cytoplasm, a eukaryote has multiple linear chromosomes inside the nucleus. Because these chromosomes are in the nucleus and the ribosomes are in the cytoplasm, transcription and translation are spatially separated processes. Signals for gene expression that originate outside the cell or within the cytoplasm must be transmitted into the nucleus in order to take effect.

In eukaryotes, large amounts of protein are bound to the DNA in a very regular fashion. The linear DNA molecules that make up eukaryotic chromosomes are wound around proteins called histones to form structures called *nucleosomes* (**Figure 7.7**). In eukaryotic chromosomes, the formation of the nucleosome introduces negative supercoils into the DNA. The nucleosomes are spaced along the double helix at very regular intervals, and each nucleosome contains approximately 200 base pairs of DNA plus nine histone proteins, two each of the four core histones (H2A, H2B, H3, and H4) and one linker histone, H1 (Figure 7.7). As noted above (Section 7.1), histones are also present in *Archaea*. However, archaeal histones are shorter than eukaryotic histones and assemble into tetramers.

Histones are positively charged proteins that neutralize the negative charge of DNA caused by the presence of phosphate groups. Of all known proteins, eukaryotic core histones have been the most highly conserved during evolution. For example, only two amino acids out of 102 are different between histone H4 of cows and peas.

UNIT 3

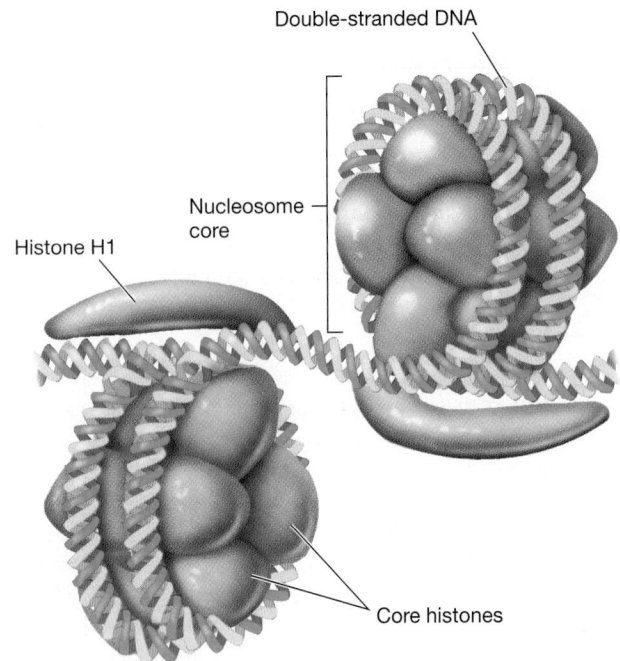

Figure 7.7 Packaging of eukaryotic DNA around a histone core to form a nucleosome. Nucleosomes are arranged along the DNA strand somewhat like beads on a string. In eukaryotes, nucleosomes consist of a core with eight proteins, two copies each of histones H2A, H2B, H3, and H4, plus a linker with one copy of histone H1.

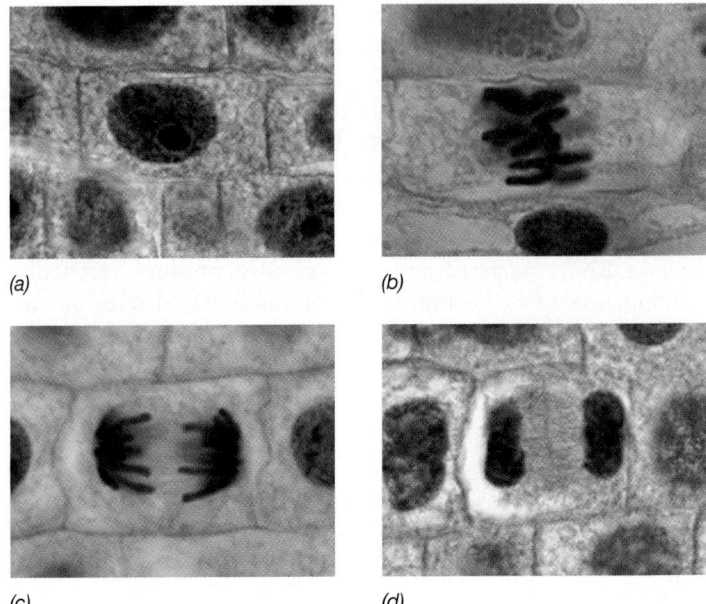

Figure 7.8 Mitosis, as seen in the light microscope. These are onion root tip cells that have been stained to reveal nucleic acid and chromosomes. *(a)* Interphase. *(b)* Metaphase. Chromosomes are paired in the center of the cell. *(c)* Anaphase. Chromosomes are separating. *(d)* Telophase.

This complex of DNA plus histones is called chromatin and can be further compacted by folding and looping to eventually form very dense structures. Highly condensed chromatin is known as heterochromatin and cannot be transcribed because it cannot be accessed by RNA polymerase. During eukaryotic cell division the DNA is condensed into heterochromatin and the chromosomes become much more compact. It is these compacted chromosomes that are visualized in eukaryotic cells during cell division.

MiniQuiz

- What effect does the presence of a nucleus have on genetic information flow in eukaryotes?
- What role do histones and nucleosomes play in organizing DNA in eukaryotes?

7.6 Overview of Eukaryotic Cell Division

Eukaryotic cells divide by a process in which the chromosomes are replicated, the nucleus is disassembled, the chromosomes are segregated into two sets, and a nucleus is finally reassembled in each daughter cell. Many eukaryotes also undergo sexual reproduction, during which male and female gametes are formed and fuse to form the zygote. This is fundamentally different from the unidirectional mating processes of prokaryotes (Chapter 10).

From a genetic standpoint, eukaryotic cells can exist in two forms: *haploid* or *diploid*. Diploid cells have two copies of each chromosome whereas haploid cells have only one. Cells of many single-celled eukaryotes, such as the brewer's yeast, *Saccharomyces cerevisiae*, can exist indefinitely in the haploid stage (containing 16 chromosomes) as well as in the diploid stage (32 chromosomes). Occasionally, two haploid yeast cells will fuse (mate) to yield a diploid cell (↩ Section 20.17). This process should not be confused with the situation in cells of multicellular plants and animals. In these organisms the diploid phase is present in the organism proper, and the haploid phase occurs only transiently, in the gametes.

Mitosis is the process that follows DNA replication in a eukaryotic cell. During mitosis, the chromosomes condense, divide, and are separated into two sets, one for each daughter cell (**Figure 7.8**). Diploid cells of both yeasts and higher eukaryotes undergo mitosis before each cell division to maintain the diploid number of chromosomes per cell. Similarly, haploid *S. cerevisiae* cells undergo mitosis before each cell division to maintain the haploid number of 16 chromosomes per cell.

Meiosis is the process of conversion from the diploid to the haploid stage. Meiosis consists of two cell divisions. In the first meiotic division, homologous chromosomes segregate into separate cells, changing the genetic state from diploid to haploid. The second meiotic division is essentially the same as mitosis, as the two haploid cells divide to form a total of four haploid gametes. In higher organisms these are the eggs and sperm; in eukaryotic microorganisms, they are spores or related structures.

MiniQuiz

- The diploid number of human chromosomes is 46. How many chromosomes are there in a human sperm cell?
- What are the important differences between mitosis and meiosis?

7.7 Replication of Linear DNA

Most prokaryotic chromosomes are circular, as are most plasmids, some virus genomes, and the genomes of most organelles (⟳ Section 12.4). In contrast to prokaryotes, the nucleus of eukaryotic cells contains linear DNA. Almost all the steps in DNA replication are the same whether the chromosome is linear or circular. However, there is a problem with the replication of linear genetic elements that is not an issue with circular ones. The problem is how to replace the RNA primer with DNA at the 5′ end of the strand.

To understand why this is a problem, first review Figure 6.15. Imagine that the left end of the DNA in this diagram is actually one end of a linear chromosome. Even if the RNA primer is very short and there is a special enzyme to remove it, no DNA polymerase can replace it with DNA because all known DNA polymerases require a primer. Therefore, if nothing was done about this problem, the DNA molecule would become shorter each time it was replicated. The replication of linear DNA thus requires special attention, and there are at least two solutions to this problem.

Replication of Linear DNA Using a Protein Primer

Viruses that contain linear DNA genomes (this includes many viruses that infect eukaryotes) and most linear plasmids solve the problem of replicating linear DNA by using a protein primer instead of an RNA primer. Although all DNA polymerases must add each nucleotide to a free hydroxyl (–OH) group, some DNA polymerases can add the first base onto an –OH group present on specific proteins that bind to the ends of linear chromosomes (**Figure 7.9**). These proteins are encoded by the plasmid or virus, and they recognize and bind the ends of

the chromosomes. These protein primers are not removed, so these plasmids and viruses have proteins permanently attached to the 5′ end of their DNA. Protein primers are also the means by which the occasional linear chromosomes of some *Bacteria* are replicated.

Telomeres and Telomerase

A special method is used to replicate the ends of eukaryotic chromosomes, which are called *telomeres*. Telomeres contain repetitive DNA—a short sequence (often 6 base pairs) tandemly repeated from 20 to several hundred times. The sequences from the telomeres of different eukaryotes are closely related, and one strand always has several guanines. During replication, this guanine-rich sequence is present on the leading strand of the DNA duplex and can base-pair with the 3′ end of a complementary RNA molecule present in the enzyme **telomerase** (**Figure 7.10**).

Telomerase becomes active once bound to the 3′ end of the guanine-rich overhang of telomeric DNA. Telomerase does not need a template to begin DNA synthesis because the enzyme itself contains a small RNA template as a cofactor. Technically, telomerase is an unusual type of reverse transcriptase (⟳ Section 9.12) that makes DNA using an RNA template. Telomerase works repetitively to make a long extension of the leading strand. Once this extension is made, the other strand (the lagging strand) can be primed with an RNA primer in the normal fashion and the DNA replicated (Figure 7.10*b*). Telomeres do not need to be a precise number of repeats long, but just long enough to ensure that no portion of a gene becomes lost during DNA replication.

Centromeres and Kinetochores

In addition to telomeres, eukaryotic chromosomes contain *centromeres*. These are regions that provide an attachment site for the spindle fibers that pull the pairs of homologous chromosomes apart during mitosis (**Figure 7.11**). Centromeres cannot be situated at the very ends of eukaryotic chromosomes and are often more or less centrally located. The term *kinetochore* refers to the complex assemblage of proteins that links the DNA of the centromere region to the spindle fibers.

Unlike telomeres, centromeres vary greatly in sequence and length from one eukaryotic organism to another. In higher animals such as humans, the centromere consists of multiple repeat sequences spread over about 1 Mbp, whereas in *Saccharomyces cerevisiae*, the total centromere sequence is 125 bp long. The short centromere sequence of yeast is fortunate from the viewpoint of genetic engineering because stable maintenance and replication of plasmids or other cloning vectors in eukaryotes requires the presence of a centromere sequence (⟳ Section 11.8). A short sequence reduces the space on the vector that must be devoted to the centromere.

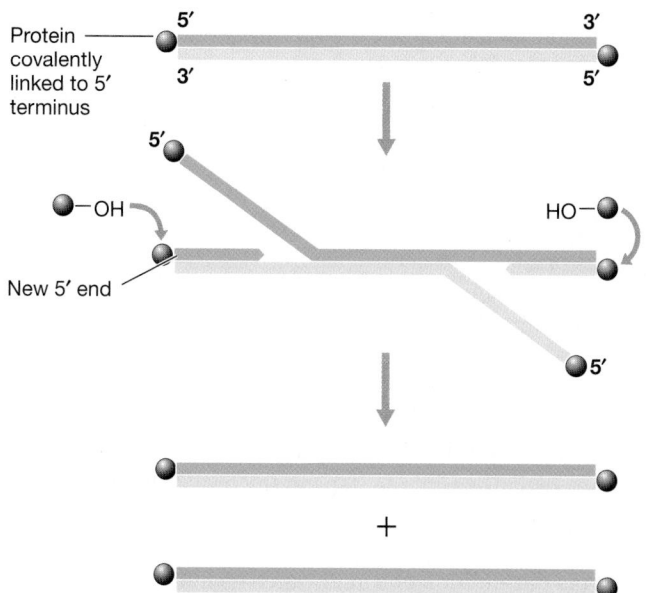

Figure 7.9 Replication of linear DNA using protein primers. New strands of DNA are primed by proteins covalently attached to their 5′ ends. Note the free –OH group on the protein. DNA polymerase III can add a nucleotide to this –OH group.

MiniQuiz

- How can a protein prime DNA for replication?
- What is telomerase and what is its function?
- Why are centromeres important?

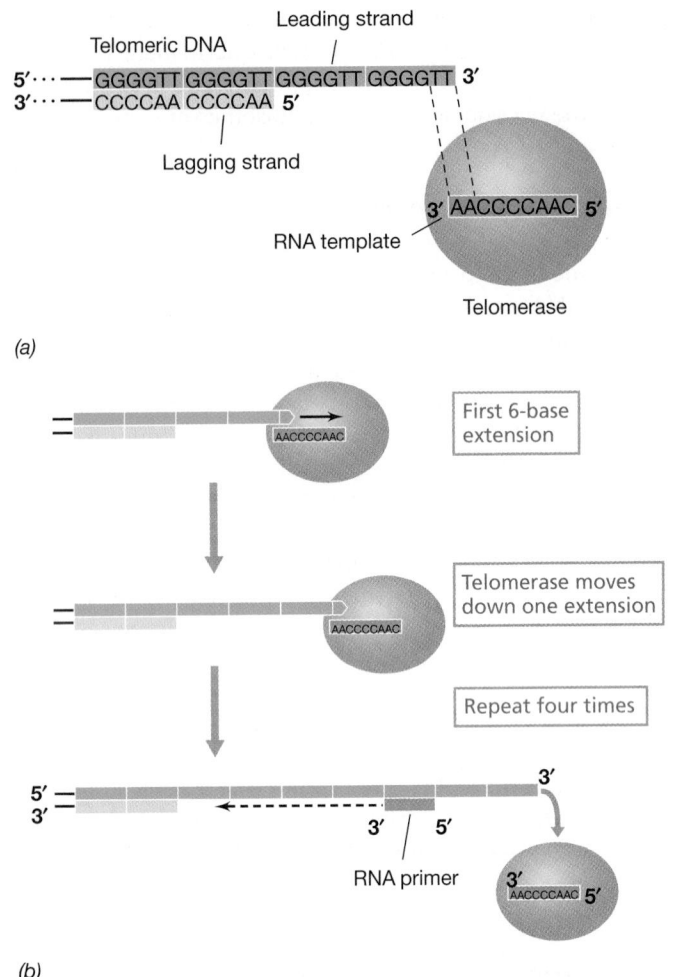

(a)

(b)

(c)

Figure 7.10 Model for the activity of telomerase at one end of a eukaryotic chromosome. *(a)* A diagram of the sequence of the end of the DNA in a telomere with four of the guanine-rich repeats and the enzyme telomerase, which contains a short RNA template. *(b)* Steps in elongation of the guanine-rich strand catalyzed by telomerase. After telomerase finishes, the lagging strand can be primed with an RNA primer by primase, followed by completion of the lagging strand by DNA polymerase and ligase. *(c)* A preparation of HeLa cell chromosomes stained with fluorescent dyes. The red dots are leading-strand telomeres and the green dots are lagging-strand telomeres.

7.8 RNA Processing

During transcription, RNA is formed from a DNA template. The initial RNA product of transcription is known as the primary transcript. However, many RNA molecules need alterations—known as **RNA processing**—before they are mature, that is, ready to carry out their role in the cell.

As we have seen, many genes in eukaryotes contain intervening sequences, the introns, between the protein-coding regions, the exons. These intervening sequences are removed from the primary transcript. The RNA is cleaved to remove the introns and the exons are joined to form a contiguous protein-coding sequence in the mature mRNA (Figure 7.6). The process by which introns are removed and exons are joined is called *splicing*. Occasional introns are found in prokaryotes, but the mechanism of removal is different. (See also the Microbial Sidebar, "Inteins and Protein Splicing.")

The Spliceosome

RNA splicing takes place in the nucleus. Splicing is done by a large macromolecular complex about the size of a ribosome, called the **spliceosome**. The spliceosome removes introns and joins adjacent exons to form mature mRNA (**Figure 7.12**). The spliceosome contains four large RNA–protein complexes, called small nuclear ribonucleoproteins (snRNPs), together with many protein factors;

indeed, over 100 proteins participate in its activity. The snRNPs each contain noncoding RNA molecules known as small nuclear RNA (snRNA) that recognize sequences on the primary transcript by base pairing. In particular, some of the snRNA molecules recognize conserved sequences at the splice junctions.

Once the splice junctions at both ends of an intron have been located by the snRNA, the proteins of the spliceosome cut out the intron and join the flanking exons together. The intron is removed as a lariat structure (Figure 7.12) that is then degraded, releasing free nucleotides. Many genes, especially in higher animals and plants, have multiple introns, so it is clearly important that they should all be recognized and removed by the spliceosome to generate the final mature mRNA.

RNA Capping and the Poly(A) Tail

There are two other unique steps in the processing of eukaryotic mRNA. Both steps take place in the nucleus prior to splicing. The first, called *capping*, occurs before transcription is complete. Capping is the addition of a methylated guanine nucleotide at the 5′-phosphate end of the mRNA. The cap nucleotide is added in reverse orientation relative to the rest of the mRNA molecule. Occasionally, other nucleotides next to the 5′ end of the eukaryotic mRNA are modified by methylation during capping. The guanosine cap is needed for translation and promotes the formation of

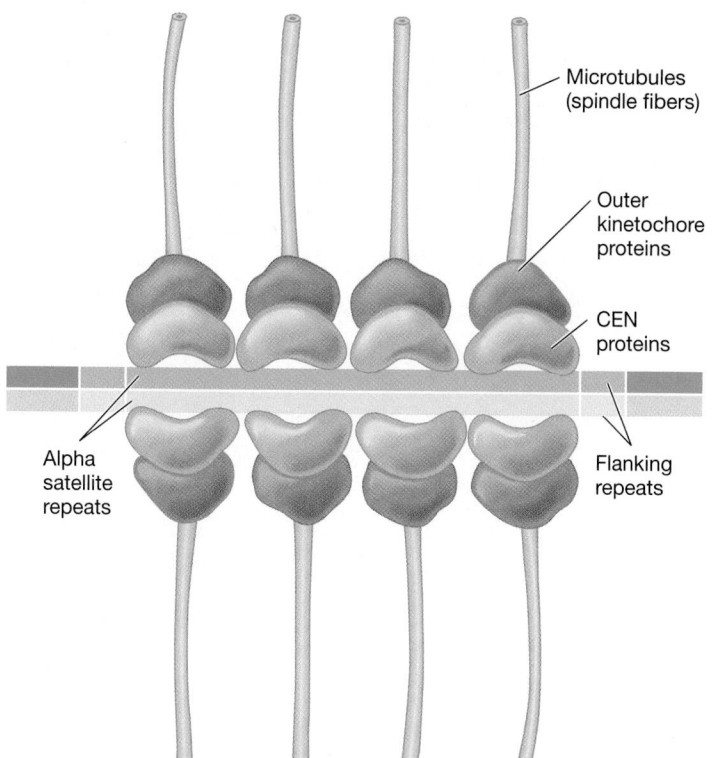

Figure 7.11 The eukaryotic centromere. DNA at the centromere of human chromosomes consists of multiple repeats of 171 bp flanked by other repeats, both of which are highly condensed into heterochromatin. The centromere (CEN) proteins bind directly to the centromere DNA, and other proteins forming the kinetochore complex assemble onto the CEN proteins. The microtubules that make up the spindle attach to the kinetochore.

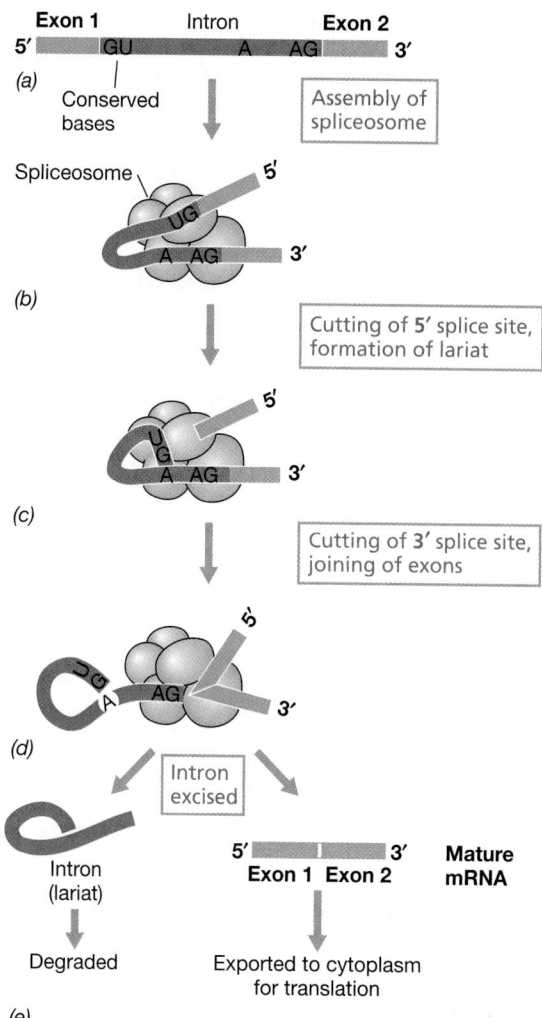

Figure 7.12 Activity of the spliceosome. Removal of an intron from the primary transcript of a protein-coding gene in a eukaryote. *(a)* A primary transcript containing a single intron. The sequence GU is conserved at the 5′ splice site, and AG is conserved at the 3′ splice site. There is also an interior A that serves as a branch point. *(b)* Several small ribonucleoprotein particles (shown in brown) assemble on the RNA to form a spliceosome. Each of these particles contains distinct small RNA molecules that take part in the splicing mechanism. *(c)* The 5′ splice site has been cut with the simultaneous formation of a branch point. *(d)* The 3′ splice site has been cut and the two exons have been joined. Note that overall, two phosphodiester bonds were broken, but two others were formed. *(e)* The final products are the joined exons (the mRNA) and the released intron.

the initiation complex between the mRNA and the ribosome through specific cap-binding proteins (Section 7.9).

The second processing step consists of trimming the 3′ end of the primary transcript and adding 100–200 adenylate residues as the *poly(A) tail*. The tail recognition sequence, AAUAAA, is located close to the 3′ end of the primary transcript and beyond the stop codon of the protein encoded by the mRNA. (Thus the poly(A) tail is not translated.) The poly(A) tail stabilizes mRNA and must be removed before the mRNA can be degraded. The poly(A) tail is also required for translation; it indicates to the translation machinery that the RNA is mRNA rather than some other form of RNA and that it is ready for translation. The three steps leading to the formation of mature eukaryotic mRNA are summarized in **Figure 7.13**. Only when all three are complete is the mature mRNA transported into the cytoplasm for translation.

Some bacterial mRNAs are also polyadenylated, although the tails are relatively short (10–40 bases) and the enzyme that adds the poly(A) tail is associated with the ribosomes. In addition, the role of the poly(A) tail on bacterial mRNA is quite different from that on eukaryotic mRNA, as the poly(A) tail triggers mRNA degradation. Similarly, adding a poly(A) tail to chloroplast mRNA promotes its degradation, which correlates with the prokaryotic ancestry of this organelle (Section 16.4).

Self-Splicing Introns

Certain RNAs have enzymatic activity. These catalytic RNAs, called **ribozymes**, participate in several cellular reactions, the most important being polypeptide synthesis. Ribozymes work like protein enzymes in possessing an "active site" that binds the substrate and catalyzes formation of a product (Section 4.5).

Self-splicing introns are introns that fold up to generate three-dimensional structures with ribozyme activity. This enzymatic activity allows them to excise themselves from an RNA molecule while joining adjacent exons together. Most self-splicing introns are found in the genes of mitochondria and chloroplasts. Others

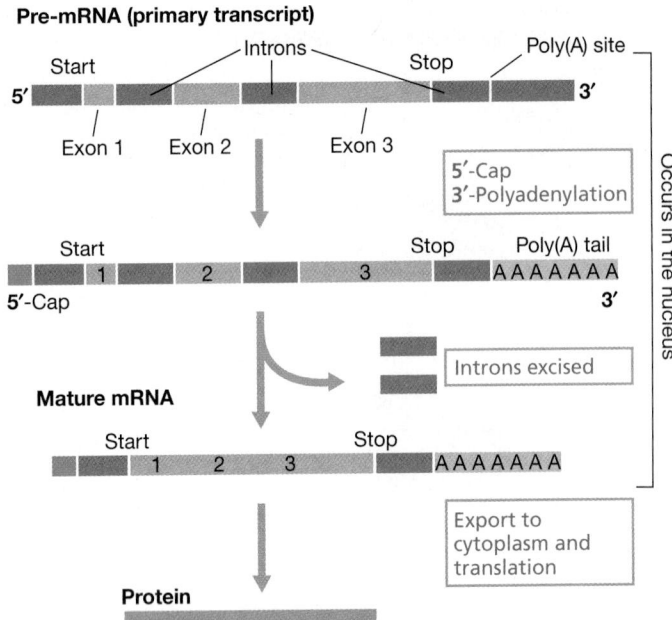

Figure 7.13 Processing of the primary transcript into mature mRNA in eukaryotes. The processing steps include adding a cap at the 5′ end, removing the introns, and clipping the 3′ end of the transcript while adding a poly(A) tail. All these steps are carried out in the nucleus. The location of the start and stop codons to be used during translation are indicated.

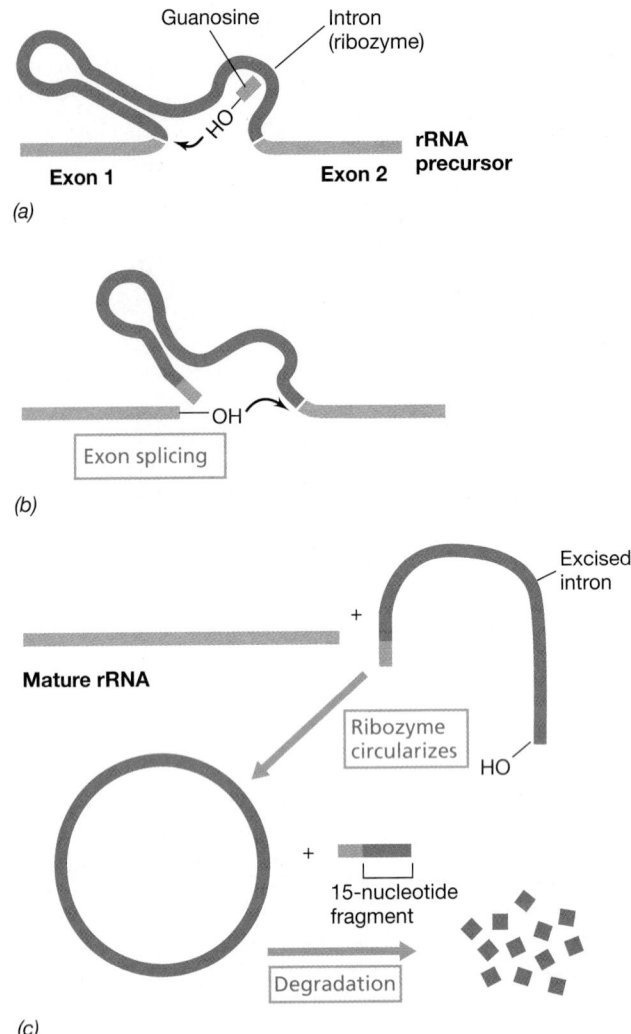

Figure 7.14 Self-splicing intron of the protozoan *Tetrahymena*. There is considerable secondary structure in such molecules, which is critical for the splicing reaction. *(a)* The rRNA precursor (primary transcript) contains a 413-nucleotide intron. *(b)* Following the addition of the nucleoside guanosine, the intron splices itself out and joins the two exons. *(c)* The intron is spliced out and then circularizes with the loss of a 15-nucleotide fragment.

are present in the rRNA (but not mRNA) of lower eukaryotes, such as the single-celled ciliate *Tetrahymena*. In this case, a 413-nucleotide intron splices itself out of the primary transcript and joins two adjacent exons to form the mature rRNA (**Figure 7.14**). This *Tetrahymena* ribozyme is thus a sequence-specific endoribonuclease that carries out a reaction analogous to that of the spliceosome. The excised intron circularizes after a short oligonucleotide fragment is removed from the intron (Figure 7.14). The *Tetrahymena* cell eventually degrades both fragments of the intron.

Two major classes of self-splicing introns are known. Those that require a free guanosine in the splicing reaction, such as the *Tetrahymena* intron in Figure 7.14, are group I introns. In contrast, group II introns use an internal adenosine residue to initiate splicing and generate a lariat product similar to that produced by spliceosomes (Figure 7.12).

Although they catalyze a specific reaction just as protein enzymes do, the ribozymes of self-splicing introns differ from protein enzymes in a key way. Unlike protein enzymes, a self-splicing ribozyme catalyzes its reaction *only once*. (Note that this is not true of ribozymes as a whole. Some ribozymes do indeed catalyze multiple reactions, just as protein enzymes do.)

Why do ribozymes exist in a world in which protein enzymes dominate? It has been proposed that ribozymes are the vestigial remains of a simpler form of life, "RNA life," which may have predated the era in which proteins are the cell's major catalysts. Indeed, it is possible that an RNA world existed long before cellular structures evolved. We discuss this concept further in

Chapter 16. Although proteins have replaced RNA enzymes in most areas of biocatalysis, a few reactions catalyzed by RNA remain. Either these are the last "holdouts" from the RNA world, or alternatively, they may be reactions that protein enzymes catalyze only very poorly and thus RNA catalysis is required to get the job done.

MiniQuiz

- What is splicing, where does it occur, and what is required for it to occur?
- What does a cap consist of and where is it located on eukaryotic mRNA?
- What small molecule do all group I introns require for enzymatic activity?

Inteins and Protein Splicing

A rather unusual type of processing removes and discards portions of a protein and then reconnects the active protein domains. The splicing out of noncoding intervening sequences that interrupt genes is normally done at the level of RNA, as discussed in Section 7.8. In this case, the introns are removed during processing of the primary transcript to yield the final mature mRNA, consisting of the exons. However, a few instances are known where intervening sequences are removed at the level of the protein instead of the RNA. This process is called **protein splicing**, the peptide removed is called an **intein**, and the final mature protein consists of the **exteins**.

Protein splicing is rare overall, but is found in proteins from *Archaea*, *Bacteria*, chloroplasts, and lower *Eukarya*. Relative to the number of described organisms, inteins are actually most frequent in *Archaea*. Usually there is just a single intein per protein, but occasional examples are known in which two inteins are inserted into the same host protein.

Figure 1 shows how protein splicing works in the production of DNA gyrase in the bacterium *Mycobacterium leprae*, the causative agent of leprosy (🔗 Section 33.4). The A subunit of the *M. leprae* DNA gyrase is the protein GyrA, which is encoded by the *gyrA* gene. Note that the flanking sequences in the

GyrA polypeptide, the exteins, are ligated together to form the final active gyrase protein, while the inteins in the precursor GyrA are discarded. Interestingly, inteins are *self-splicing* entities and thus have the enzymatic activity of a highly specific protease.

In addition to joining two exteins, the spliced-out intein polypeptide has a second enzyme activity; it acts as a site-specific deoxyribonuclease (DNase). Its function is to protect the existence of the intein. If a mutation occurred that deleted the intein DNA sequence from the middle of the host gene, the host cell would not be harmed, but the intein would be lost. However, the intein DNase will cut the DNA of such a host cell in the middle of the host gene. This is likely to kill any cell that deletes the useless intein. Only cells that keep the intein survive. This cutting occurs at the site where the intein is normally inserted; thus host genes with an intein are not cut because the nuclease recognition site is disrupted by the presence of the intein DNA sequence. Inteins are unnecessary for cell survival, and intein-encoding DNA may therefore be regarded as a form of selfish DNA (that is, DNA that promotes its own survival but confers no benefit on the host cell). The origin of inteins is obscure.

Inteins have found uses in biotechnology. Carrier proteins fused to a target protein are often used to facilitate protein purification

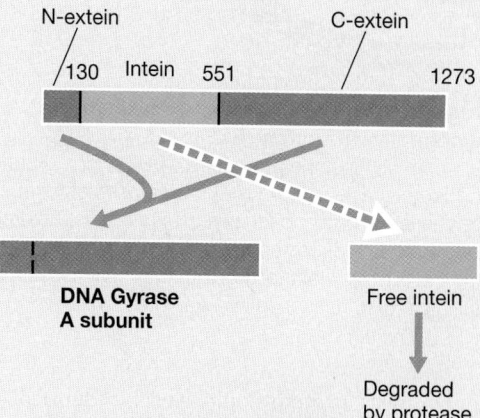

Figure 1 Protein splicing. *The protein synthesized from the gyrA mRNA in Mycobacterium leprae is 1273 amino acid residues in length. The N-extein is the amino terminal extein and the C-extein is the carboxyl terminal extein. Residues 131 to 550 make up an intein, which removes itself in a self-splicing reaction that generates the free intein and the DNA gyrase A subunit.*

(🔗 Section 15.10). However, after purification, the carrier must be removed. To accomplish this, inteins have been engineered as carrier proteins. Once the target protein has been purified, the intein is triggered to cut itself out. This releases the purified target protein without the need for additional and potentially complicated steps.

7.9 Transcription and Translation in *Eukarya*

The close genetic relationship between *Archaea* and *Eukarya* is most noticeable when comparing the cellular machinery for transcription and translation in the three domains of life. RNA polymerases from all sources contain some subunits that are evolutionarily conserved. Nonetheless, true to their phylogenetic roots, archaeal and eukaryotic RNA polymerases are more similar and structurally more complex than those of *Bacteria*. Similarly, protein synthesis in *Archaea* and *Eukarya* shares several features that contrast with *Bacteria* (Section 7.3).

Transcription in Eukaryotes

Eukaryotes have multiple RNA polymerases, unlike *Bacteria* and *Archaea*, which have just one. The eukaryotic nucleus contains

three RNA polymerases, RNA polymerases I, II, and III, that transcribe different categories of nuclear genes. The single RNA polymerase of *Archaea* most closely resembles eukaryotic RNA polymerase II. Whereas *Bacteria* contain a relatively simple (five-polypeptide) RNA polymerase, the RNA polymerases of *Archaea* and the eukaryotic nucleus have ten or more subunits (Figure 7.2). The mitochondria and chloroplasts also possess RNA polymerase, but this resembles the bacterial enzyme.

Eukaryotic RNA polymerase I transcribes the genes for the two large rRNA molecules, 18S and 28S. RNA polymerase III transcribes genes for tRNA, 5S rRNA, and other small RNA molecules. Protein-encoding genes are transcribed by RNA polymerase II, which is therefore responsible for making all of the cell's mRNA. Each RNA polymerase recognizes a distinct class of promoters. In contrast, in prokaryotes the promoter for a gene

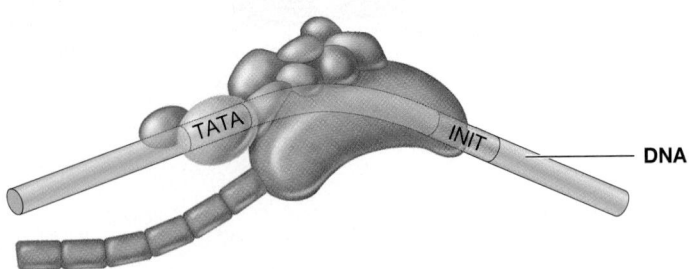

Figure 7.15 **The interaction of eukaryotic RNA polymerase II with a promoter.** The polymerase itself (brown) is positioned at the initiator element (INIT) of the promoter. A TATA-binding protein (yellow) is shown bound at the TATA box. The polymerase has a repetitive amino acid sequence at one end (shown as a tail-like structure) that can be phosphorylated, affecting activity of the polymerase. The other proteins shown in blue are a few of the large number of accessory factors required for initiation of transcription in eukaryotes.

encoding a protein could well be identical to a promoter for a gene encoding a tRNA.

RNA polymerase II is depicted binding to a promoter on the DNA in **Figure 7.15**. This promoter contains a TATA box, which is a conserved sequence resembling, in some respects, the Pribnow box in the promoters of *Bacteria* (⮌ Section 6.13). The promoter also contains an initiator element (INIT) very near the transcription start site. These two regions are key elements of eukaryotic promoters, although there may also be other sequences in any given promoter.

Eukaryotic RNA polymerases require transcription factors to recognize specific promoters just as they do in *Archaea*. However, these proteins (and those of *Archaea*) recognize the promoter independently, not as part of a polymerase holoenzyme as in *Bacteria* (⮌ Section 6.13). In both *Archaea* and *Eukarya*,

general transcription factors are needed for the functioning of all promoters recognized by a particular RNA polymerase. Specific transcription factors are needed for expression of certain genes under specific circumstances.

Translation in Eukaryotes

Protein synthesis by eukaryotic ribosomes is generally more complex than in *Bacteria*. The cytoplasmic ribosomes of eukaryotic cells (80S ribosomes) are larger than bacterial ribosomes (70S ribosomes) and contain more rRNA and protein molecules. In particular, the large ribosomal subunit contains three rRNA molecules, 5S, 5.8S, and 28S. The 5.8S rRNA is homologous to the 5′ end of bacterial 23S rRNA, and the 28S rRNA corresponds to the rest of bacterial 23S rRNA. Thus the eukaryotic 5.8S and 28S rRNAs taken together are equivalent to the 23S rRNA of bacteria. Eukaryotes also have more initiation factors and a more complex initiation procedure. (Note that eukaryotic cells also contain bacterial-type ribosomes in their mitochondria and chloroplasts. The term "eukaryotic ribosomes" refers to those found in the eukaryotic *cytoplasm* and whose components are encoded by nuclear genes.) These differences are summarized in **Table 7.2**.

In both *Bacteria* and *Archaea* mRNA is polycistronic and may be translated to give several proteins. In eukaryotes, mRNA carries only a single gene that is translated into a single protein. That is, eukaryotic mRNA is monocistronic (Figure 7.5).

Initiation of protein synthesis differs significantly among the three domains. All organisms have a special initiator tRNA that recognizes the start codon and inserts the first amino acid. *Bacteria* use *N*-formylmethionine as the first amino acid of all proteins, whereas eukaryotes and *Archaea* use methionine. However, unlike mRNA in both *Bacteria* and *Archaea*, eukaryotic mRNA has no ribosome-binding site (Shine–Dalgarno sequence).

Table 7.2 *Comparison of translation*

Bacteria	Archaea	Eukarya (cytoplasm)
Coupled transcription and translation	Coupled transcription and translation	No coupled transcription and translation for nuclear genes
Polycistronic mRNA	Polycistronic mRNA	Monocistronic mRNA
No cap on mRNA	No cap on mRNA	5′ end of mRNA recognized by cap
Start codon is next AUG after Shine–Dalgarno sequence	Start codon is often next AUG after Shine–Dalgarno sequence	First AUG in mRNA is used
First amino acid is *N*-formylmethionine	First methionine is unmodified	First methionine is unmodified
70S ribosomes with 30S and 50S subunits	70S ribosomes with 30S and 50S subunits	80S ribosomes with 40S and 60S subunits
Small subunit rRNA: 16S (1500 nucleotides)	Small subunit rRNA: 16S (1500 nucleotides)	Small subunit rRNA: 18S (2300 nucleotides)
Large subunit rRNA: 23S (2900 nucleotides) and 5S (120 nucleotides)	Large subunit rRNA: 23S (2900 nucleotides) and 5S (120 nucleotides)	Large subunit rRNA: 28S (4200 nucleotides), 5.8S (160 nucleotides), and 5S (120 nucleotides)
Protein homologies: bacterial	Protein homologies: eukaryotic	Protein homologies: eukaryotic
rRNA homologies: bacterial	rRNA homologies: eukaryotic	rRNA homologies: eukaryotic
Not inhibited by diphtheria toxin	Inhibited by diphtheria toxin	Inhibited by diphtheria toxin
Two elongation factors	Multiple elongation factors	Multiple elongation factors
Three initiation factors	Multiple initiation factors	Multiple initiation factors

Instead, the mRNA is recognized by its cap (Section 7.8). A specific protein, cap-binding protein, binds both the mRNA cap and the ribosome. The first AUG to be found on the mRNA is normally used as the start codon in eukaryotes. In addition, the order of assembly of the ribosomal initiation complex is different: In eukaryotes the initiator Met-tRNA binds to the small ribosomal subunit before the mRNA (the opposite is true in *Bacteria*).

Finally, translation is inhibited in eukaryotic cells by a protein toxin made by the bacterium *Corynebacterium diphtheriae*, the causative agent of the disease diphtheria. This toxin can actually kill cells and is a major factor in the pathology of the disease (ᶜᵉ Section 33.3). In line with the many similarities in the translational mechanisms of *Archaea* and eukaryotes, diphtheria toxin also inhibits translation in cells of *Archaea* but has no effect on the process in *Bacteria* (Table 7.2).

MiniQuiz

- Which type of eukaryotic RNA polymerase resembles most closely the RNA polymerase of *Archaea*?

- In eukaryotes, which type of genes are transcribed by RNA polymerases I and III?

- How is mRNA recognized by the ribosome in eukaryotes?

7.10 RNA Interference (RNAi)

Healthy cells contain double-stranded DNA (dsDNA) and single-stranded RNA (ssRNA) but not double-stranded RNA (dsRNA). The presence of dsRNA usually indicates the presence of an RNA virus genome within the cell. Note that even if the virion of an RNA virus contains single-stranded RNA, the virus genome must go through a double-stranded form (the replicative form) during replication. **RNA interference** (RNAi), a mechanism for defending against RNA viruses, both cleaves the dsRNA and destroys any ssRNA (usually mRNA) that corresponds to the targeted dsRNA sequence. RNAi is found only in eukaryotes, including protozoa, animals, and plants. RNAi is not found in *Bacteria* or *Archaea*; instead the CRISPR system plays a similar role in these cells but uses a different mechanism (ᶜᵉ Chapter 8 Microbial Sidebar, "The CRISPR Antiviral Defense System").

RNAi is triggered by dsRNA of greater than 20 base pairs in length. Longer molecules of dsRNA are cleaved into fragments of 21–23 bp by a dsRNA-specific nuclease known as Dicer. These short dsRNA fragments are known as **short interfering RNA** (siRNA) and are bound by the RNA-induced silencing complex (RISC). RISC recognizes and destroys ssRNA that corresponds in sequence to siRNA (**Figure 7.16**). In practice, such ssRNA is usually mRNA made by the infecting RNA virus. To achieve this, RISC separates the siRNA into its two strands. Any longer ssRNA in the cell that base-pairs with these fragments is then cleaved by the nuclease Slicer, which is part of RISC.

The RNAi effect can spread from cell to cell and may travel quite a long way through an organism. This is especially noticeable in plants. The spread is due to copying of siRNA by an enzyme known as RNA-dependent RNA polymerase (RdRP). The

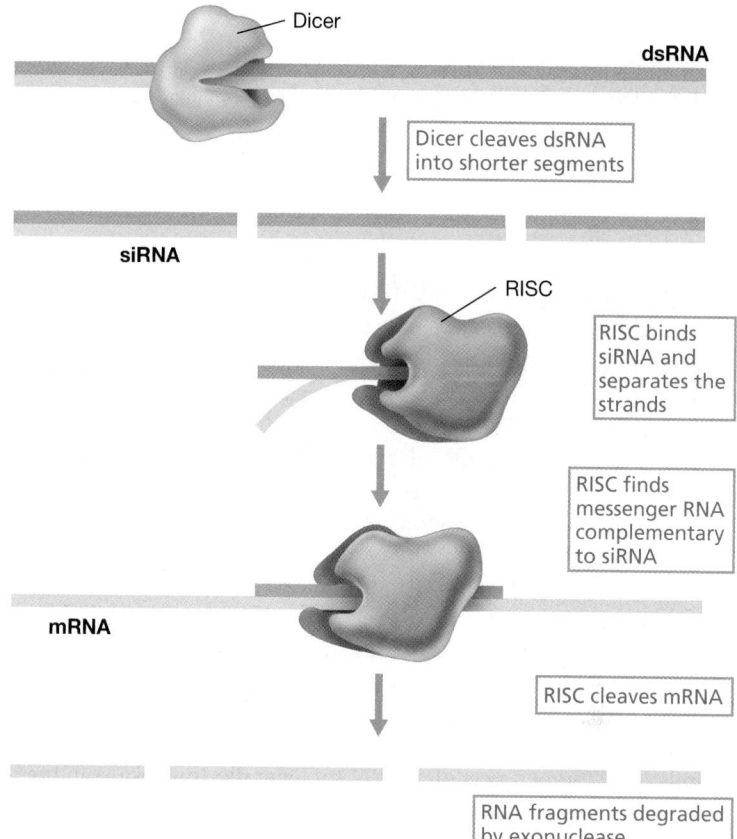

Figure 7.16 RNA interference. The nuclease Dicer cleaves double-stranded RNA into segments of 21–23 base pairs known as short interfering RNA (siRNA). This is recognized by RISC (RNA-induced silencing complex), which separates the strands of the siRNA. Finally, RISC cleaves target RNA that hybridizes to the siRNA.

siRNA then travels from cell to cell via intercellular connections called plasmodesmata. In lower animals, such as *Caenorhabditis elegans*, the RNAi effect can be "inherited" for several generations. However, mammals do not possess the RdRP needed to amplify RNAi, so the effect remains localized and cannot be transmitted.

RNAi is now widely used in research to prevent the expression of genes in animals and plants. If the sequence of a gene is known, as is often the case, it is possible to make synthetic siRNA corresponding to the target gene. Using RNAi to switch off a gene allows gene function to be tested without the need to make defective mutants. This is especially useful for eukaryotes, most of which are diploid and where classical genetic analysis requires introducing mutations into both copies of a gene. RNAi by its very nature prevents expression of all copies of the target gene and may even be used in organisms such as the protozoan *Paramecium* where multiple gene copies are often present. Experimentally, RNAi may be induced by providing long molecules of dsRNA that are cut into siRNA by Dicer. This generates dsRNA inside the cell. Alternatively, short dsRNA molecules of 21–23 nucleotides in length may be administered directly and will function as siRNA.

7.11 Regulation by MicroRNA

MicroRNAs (miRNA) are small dsRNA molecules that regulate translation in eukaryotic cells. The miRNA system is related to RNAi, but regulates expression of the cell's own genes rather than protecting against virus invasion.

Thus the miRNA molecules are encoded by the genome and are transcribed to give precursors (pre-miRNA) that fold to form double-stranded regions (**Figure 7.17**). Before leaving the nucleus, these are cut to release pre-miRNA by an enzyme known as Drosha that is related to Dicer. The pre-miRNA exits the nucleus and is trimmed by Dicer to form miRNA. This is then bound by a protein complex, miRISC, that is analogous to RISC. Here the strands are separated in a manner similar to what happens to siRNA in RISC.

The miRISC containing the single-stranded miRNA then binds to its target on cellular mRNA. Generally, this results in the blocking of translation, but the mRNA is not usually degraded. A single miRNA may repress translation of hundreds of proteins, but the effect is usually mild. Thus, miRNA tends to modulate the level of protein synthesis rather than switch it completely off.

A variety of other small RNA molecules that are involved in regulation have been discovered recently. In many cases their biological roles and mechanisms of action are still largely obscure. One important class is piRNA (Piwi-interacting RNA). These are encoded by the genome, but are not generated by Dicer/Drosha-style cleavage of dsRNA. Some of these protect eukaryotes against the spread of repeated DNA sequences and are especially active in the reproductive cells of higher animals.

Unlike most other small RNAs, the genes for miRNA are transcribed by RNA polymerase II. The genes are often found in clusters and are transcribed to give a precursor RNA containing several miRNA sequences. The miRNA precursor RNAs are capped and tailed like messenger RNA. Later, they are processed to release the separate miRNA molecules.

MicroRNAs are not only used by eukaryotic cells but also by some of the more complex viruses that infect them. Herpesviruses encode over 140 miRNAs. About two thirds of these control virus gene expression whereas the other third affect host cell gene expression. Not surprisingly, several of these target the host immune system. Fascinatingly, several herpesvirus miRNAs suppress the function of host cell miRNAs!

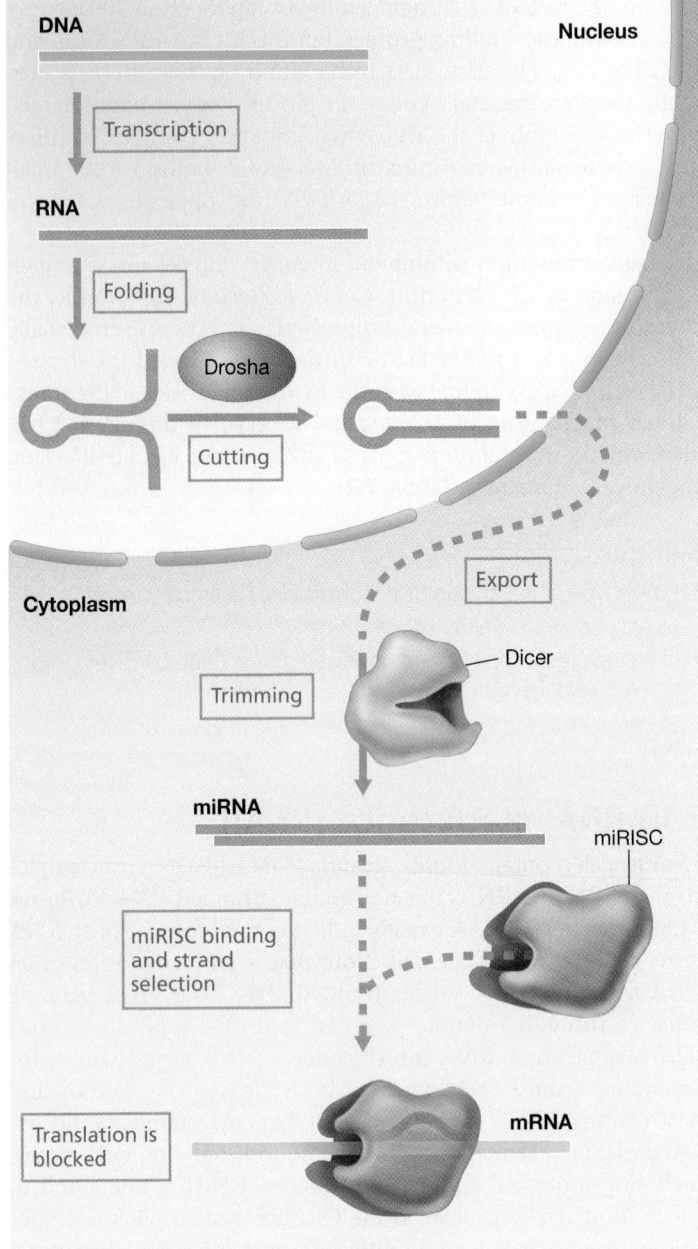

Figure 7.17 Mechanism of miRNA action. The primary transcript for miRNA is made in the nucleus. The miRNA is cut out from the primary transcript by Drosha and trimmed further by Dicer after moving from the nucleus into the cytoplasm. The miRNA is then bound by miRISC, which separates the strands. One strand is used to base-pair with a target mRNA. This prevents translation of the mRNA.

Big Ideas

7.1

The packaging of very long DNA molecules relies on some form of supercoiling. In *Bacteria* this is due to DNA gyrase, whereas in eukaryotes DNA is wound around nucleosomes that consist of proteins called histones. *Archaea* possess both DNA gyrase and histones. Some *Archaea* possess reverse gyrase, which introduces positive supercoiling. Despite resembling *Bacteria* in having single circular chromosomes, the DNA replication apparatus of *Archaea* is much more similar to that of eukaryotes.

7.2

The transcription apparatus and the promoter architecture of *Archaea* and *Eukarya* have many features in common, although the components are usually relatively more simple in *Archaea*. A few unusual intervening sequences are found in *Archaea* that are similar to those found in eukaryotic nuclear tRNA.

7.3

The translation apparatus of *Archaea* and *Eukarya* share many features, although the number of ribosomal components and translation factors is fewer in *Archaea*. However, *Archaea* show some similarities with *Bacteria*, such as the use of Shine–Dalgarno sequences.

7.4

Bacteria and *Archaea* share many properties related to their simple cell structure and relatively small genome size. In addition, they share several fundamental genetic features, including circular chromosomes, polycistronic mRNA, Shine–Dalgarno sequences, bacterial-style gene regulation, restriction enzymes, CRISPR (instead of RNAi), and flagella with rotary bases.

7.5

Due to the presence of a nucleus, the organization of genetic information is more complex in eukaryotes. Most eukaryotic genes have both coding regions (exons) and noncoding regions (introns).

7.6

Eukaryotic microorganisms can mate and exchange DNA during sexual reproduction. Mitosis ensures appropriate segregation of the chromosomes during asexual cell division. Meiosis generates haploid cells known as gametes that fuse to form a diploid zygote.

7.7

The ends of linear DNA present a problem to the replication machinery that circular DNA molecules do not. Some linear DNA molecules of prokaryotes and viruses solve this problem using a protein primer. Eukaryotes solve the problem using a special enzyme called telomerase to extend one strand of the DNA.

7.8

The processing of eukaryotic precursor mRNAs is unique and has three distinct steps: splicing, capping, and adding a poly(A) tail. Only after processing can eukaryotic mRNA exit the nucleus to be translated. Introns in some transcripts are self-splicing, and the RNA itself catalyzes the reaction.

7.9

In *Eukarya* the major classes of RNA are transcribed by different RNA polymerases, with RNA polymerase II responsible for transcribing mRNA. Translation in *Eukarya* involves larger ribosomes and more initiation factors. Eukaryotes begin each polypeptide chain with an unmodified methionine.

7.10

In *Eukarya* the mechanism of RNAi destroys viral mRNAs. RNAi is triggered by the presence of dsRNA and protects against infection by RNA viruses.

7.11

Several classes of small regulatory RNA are known that are specific to eukaryotes. The best known, miRNA, modulates gene expression by controlling the translation of mRNA.

Review of Key Terms

Exon the coding DNA sequences in a split gene (contrast with intron)

Extein the portion of a protein that remains and has biological activity after the splicing out of any inteins

Histones the basic proteins that protect and compact the DNA in eukaryotes and *Archaea*

Intein an intervening sequence in a protein; a segment of a protein that can splice itself out

Intron the intervening noncoding DNA sequences in a split gene (contrast with exon)

Meiosis the specialized form of nuclear division that halves the diploid number of chromosomes to the haploid number for gametes of eukaryotic cells

Mitosis the normal form of nuclear division in eukaryotic cells in which chromosomes are replicated and partitioned into two daughter nuclei

Nucleosome spherical complex of eukaryotic DNA plus histones

Primary transcript an unprocessed RNA molecule that is the direct product of transcription

Protein splicing the removal of intervening sequences from a protein

Reverse gyrase an enzyme that introduces positive supercoils into DNA

Ribozyme catalytic RNA

RNA interference (RNAi) a response that is triggered by the presence of double-stranded RNA and results in the degradation of single-stranded RNA homologous to the inducing dsRNA

RNA processing the conversion of a primary transcript RNA to its mature form

Self-splicing intron an intron that possesses ribozyme activity and splices itself out

Short interfering RNA (siRNA) short double-stranded RNA molecules that trigger RNA interference

Spliceosome a complex of ribonucleoproteins that catalyze the removal of introns from primary RNA transcripts

Telomerase an enzyme complex that replicates DNA at the end of eukaryotic chromosomes

Review Questions

1. List at least five features of *Archaea* that clearly differentiate the members of this domain from *Bacteria* (Sections 7.1–7.3).

2. What are nucleosomes (Section 7.1)?

3. What are exons and introns? In which organisms are they commonly found (Section 7.2)?

4. List at least five features of eukaryotic cells that clearly differentiate them from both types of prokaryotic cells (Section 7.4).

5. Why are most eukaryotic genes much longer than corresponding genes in *Bacteria* (Section 7.5)?

6. Compare and contrast the processes of mitosis and meiosis. Which process is absolutely necessary for growth of *Saccharomyces cerevisiae* and why (Section 7.6)?

7. Why would genes eventually be lost from eukaryotic DNA in the absence of telomerase activity (Section 7.7)?

8. Why do eukaryotic mRNAs have to be "processed," whereas most prokaryotic RNAs do not (Section 7.8)?

9. How many RNA polymerases does a eukaryotic cell have (Section 7.9)? Which eukaryotic RNA polymerase is similar to the RNA polymerase of *Archaea*?

10. Why is RNA interference useful (Section 7.10)?

11. What is the difference between siRNA and miRNA (Sections 7.10 and 7.11)?

Application Questions

1. *Archaea* and *Bacteria* both have single circular chromosomes. Suggest possible reasons for this similarity.

2. Eukaryotic cells usually have much larger genomes than prokaryotic cells. List three reasons why this is not surprising.

3. Eukaryotic cells often have interrupted genes that contain introns, whereas prokaryotes do not. Suggest possible reasons for this difference.

4. The siRNA and miRNA systems are very similar and probably evolutionarily related. Which do you think evolved first and why?

Need more practice? Test your understanding with Quantitative Questions; access additional study tools including tutorials, animations, and videos; and then test your knowledge with chapter quizzes and practice tests at **www.microbiologyplace.com**.

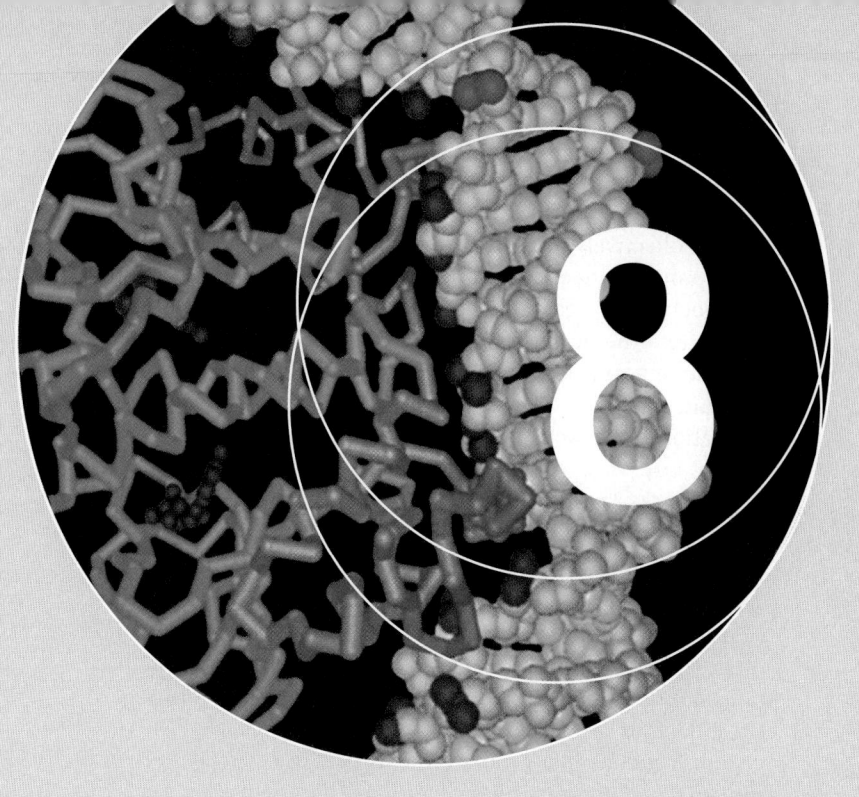

Regulation of Gene Expression

Regulation at the transcriptional level, a common mechanism for controlling gene expression in prokaryotes, is triggered by the attachment or release of DNA-binding proteins to specific genes on the DNA.

After the genetic information stored as DNA is transcribed into RNA, the information is translated to yield specific proteins. Collectively, these processes are called **gene expression**. Most proteins are enzymes that carry out the hundreds of different biochemical reactions needed for cell growth. To efficiently orchestrate the numerous reactions in a cell and to make maximal use of available resources, cells must *regulate* the kinds and amounts of proteins and other macromolecules they make. Such regulation is the focus of this chapter.

I Overview of Regulation

Some proteins and RNA molecules are needed in the cell at about the same level under all growth conditions. The expression of these molecules is said to be *constitutive*. However, more often a particular macromolecule is needed under some conditions but not others. For instance, enzymes required for using the sugar lactose are useful only if lactose is available. Microbial genomes encode many more proteins than are actually present in the cell under any particular condition. Thus, regulation is a major process in all cells and helps to conserve energy and resources.

Cells use two major approaches to regulating protein function. One controls the *activity* of an enzyme or other protein and the other controls the *amount* of an enzyme. The activity of a protein can be regulated only after it has been synthesized (that is, post-translationally). Regulating the activity of an enzyme in the cell is typically very rapid (taking seconds or less), whereas synthesizing an enzyme is relatively slow (taking several minutes). After synthesis of an enzyme begins, it takes some time before it is present in amounts sufficient to affect metabolism. Conversely, after synthesis of an enzyme stops, a considerable time may elapse before the enzyme is sufficiently diluted that it no longer affects metabolism. However, working together, regulation of enzyme activity and of enzyme synthesis efficiently controls cell metabolism.

8.1 Major Modes of Regulation

Most bacterial genes are transcribed into messenger RNA (mRNA), which in turn is translated into protein, as we discussed in Chapter 6. The components of a typical gene, with the corresponding mRNA and protein (the gene product), are summarized in **Figure 8.1**. The structural gene encodes the gene product and its expression is controlled by sequences in the upstream region. The amount of protein synthesized can be regulated at either the level of transcription, by varying the amount of mRNA made, or, less often, at the level of translation, by translating or not translating the mRNA. Occasionally the amount of protein may be regulated by degradation of the protein. Note that the sequences that determine the beginning and end of transcription are distinct from those that determine the beginning and end of translation. They are separated by small spacer regions, the 5′ and 3′ untranslated regions (5′-UTR and 3′-UTR).

Systems that control the level of expression of particular genes are varied, and genes are often regulated by more than one system. The processes that regulate the activity of enzymes have already been discussed (↩ Section 4.16). Here we consider how the synthesis of RNA and proteins is controlled.

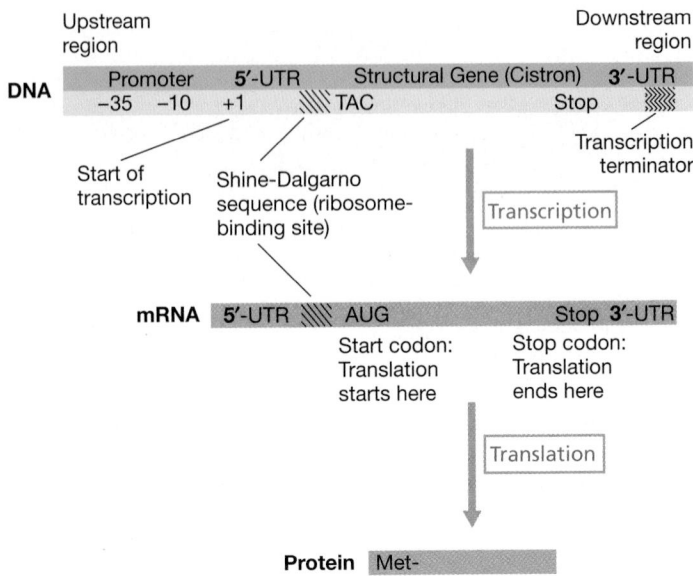

Figure 8.1 Components of a bacterial gene. The promoter, consisting of −35 and −10 regions, lies upstream of the gene. The 5′ untranslated region (5′-UTR) is a short region between the start of transcription and the start of translation. The 3′ untranslated region (3′-UTR) is a short region between the stop codon and the transcription terminator. The synthesis of the gene product (protein) may be regulated at the level of transcription or of translation or both.

MiniQuiz

- What steps in the synthesis of protein might be subject to regulation?
- Which is likely to be more rapid, the regulation of activity or the regulation of synthesis? Why?

II DNA-Binding Proteins and Regulation of Transcription

The amount of a protein present in a cell may be controlled at the level of transcription, at the level of translation, or, occasionally, by protein degradation. Our discussion begins with control at the level of transcription because this is the major means of regulation in prokaryotes.

The half-life of a typical mRNA in prokaryotes is short, only a few minutes at best. This allows prokaryotes to respond quickly to changing environmental parameters. Although there are energy costs in resynthesizing mRNAs that have been translated only a few times before being degraded, there are benefits to removing mRNAs rapidly when they are no longer needed, as this prevents the production of unneeded proteins. Thus, transcription and mRNA degradation coexist in the growing cell.

For a gene to be transcribed, RNA polymerase must recognize a specific promoter on the DNA and begin functioning (↩ Section 6.12). Regulation of transcription typically requires proteins that can bind to DNA. Thus, before discussing specific regulatory mechanisms, we must consider DNA-binding proteins.

8.2 DNA-Binding Proteins

Small molecules often take part in regulating transcription. However, they rarely do so directly. Instead, they typically influence the binding of certain proteins, called *regulatory proteins*, to specific sites on the DNA. It is these proteins that actually regulate transcription.

Interaction of Proteins with Nucleic Acids

Interactions between proteins and nucleic acids are central to replication, transcription, and translation, and also to the regulation of these processes. Protein–nucleic acid interactions may be nonspecific or specific, depending on whether the protein attaches anywhere along the nucleic acid or whether it recognizes a specific sequence. Histones (Section 7.5) are good examples of nonspecific binding proteins. Histones are universally present in *Eukarya* and are also present in many *Archaea*. Because they are positively charged, histones combine strongly and relatively nonspecifically with negatively charged DNA. If the DNA is covered with histones, RNA polymerase cannot bind and the DNA cannot be transcribed. However, removal of histones does not automatically lead to transcription, but simply leaves the DNA accessible to other proteins that control gene expression.

Most DNA-binding proteins interact with DNA in a sequence-specific manner. Specificity is provided by interactions between specific amino acid side chains of the proteins and specific chemical groups on the nitrogenous bases and the sugar–phosphate backbone of the DNA. Because of its size, the *major groove* of DNA is the main site of protein binding. Figure 6.2 identified atoms of the bases in the major groove that are known to interact with proteins. To achieve high specificity, the binding protein must interact simultaneously with several nucleotides. In practice, this means that a specific binding protein binds only to DNA containing a specific base sequence.

We have already described a structure in DNA called an *inverted repeat* (Figure 6.6). Such inverted repeats are frequently the locations at which regulatory proteins bind specifically to DNA (**Figure 8.2**). Note that this interaction does not involve the formation of stem–loop structures in the DNA. DNA-binding proteins are often homodimeric, meaning they are composed of two identical polypeptide subunits, each subdivided into **domains**; that is, regions of a protein with a specific structure and function. Each subunit has a domain that interacts specifically with a region of DNA in the major groove. When protein dimers interact with inverted repeats on DNA, each subunit binds to one of the inverted repeats. The dimer as a whole thus binds to both DNA strands (Figure 8.2). The DNA-binding protein recognizes base sequences by making a series of molecular contacts that are specific for that particular sequence.

Structure of DNA-Binding Proteins

DNA-binding proteins in both prokaryotes and eukaryotes possess several classes of protein domains that are critical for proper binding to DNA. One of the most common is the *helix-turn-helix* structure (**Figure 8.3**). This consists of two segments of polypeptide chain that have α-helix secondary structure connected by

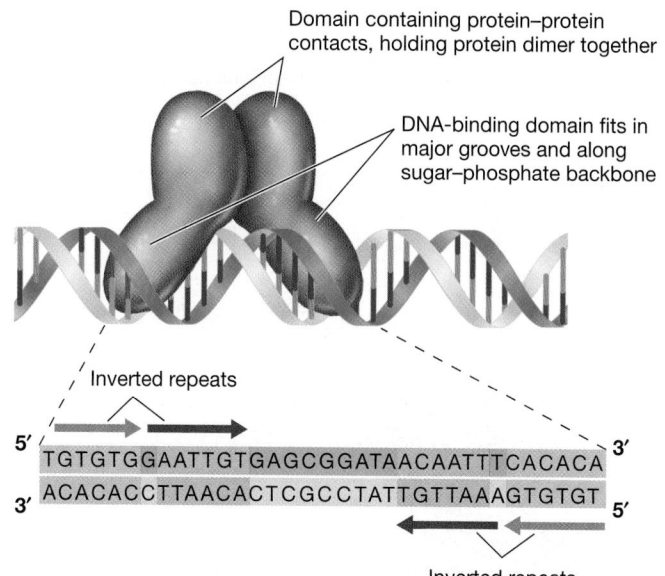

Domain containing protein–protein contacts, holding protein dimer together

DNA-binding domain fits in major grooves and along sugar–phosphate backbone

Inverted repeats

5′ TGTGTGGAATTGTGAGCGGATAACAATTTCACACA 3′
3′ ACACACCTTAACACTCGCCTATTGTTAAAGTGTGT 5′

Inverted repeats

Figure 8.2 DNA-binding proteins. Many DNA-binding proteins are dimers that combine specifically with two sites on the DNA. The specific DNA sequences that interact with the protein are inverted repeats. The nucleotide sequence of the operator gene of the lactose operon is shown, and the inverted repeats, which are sites at which the *lac* repressor makes contact with the DNA, are shown in shaded boxes.

a short sequence forming the "turn." The first helix is the *recognition helix* that interacts specifically with DNA. The second helix, the *stabilizing helix*, stabilizes the first helix by interacting hydrophobically with it. The turn linking the two helices consists of three amino acid residues, the first of which is typically a glycine. Sequences are recognized by noncovalent interactions, including hydrogen bonds and van der Waals contacts, between the recognition helix of the protein and specific chemical groups in the sequence of base pairs on the DNA.

Many different DNA-binding proteins from *Bacteria* contain the helix-turn-helix structure. These include many repressor proteins, such as the *lac* and *trp* repressors of *Escherichia coli* (Section 8.3), and some proteins of bacterial viruses, such as the bacteriophage lambda repressor (Figure 8.3b). Indeed, over 250 different known proteins with this motif bind to DNA to regulate transcription in *E. coli*.

Two other types of protein domains are commonly found in DNA-binding proteins. One of these, the *zinc finger*, is frequently found in regulatory proteins in eukaryotes (**Figure 8.4a**). The zinc finger is a protein structure that, as its name implies, binds a zinc ion. Part of the "finger" of amino acids that is created forms an α-helix, and this recognition helix interacts with DNA in the major groove. There are usually at least two or three zinc fingers on proteins that use them for DNA binding.

The other protein domain commonly found in DNA-binding proteins is the *leucine zipper* (Figure 8.4b). These are regions in which leucine residues are spaced every seven amino acids, somewhat resembling a zipper. Unlike the helix-turn-helix structure and the zinc finger, the leucine zipper does not interact with DNA itself but functions to hold two recognition helices in the correct orientation to bind DNA.

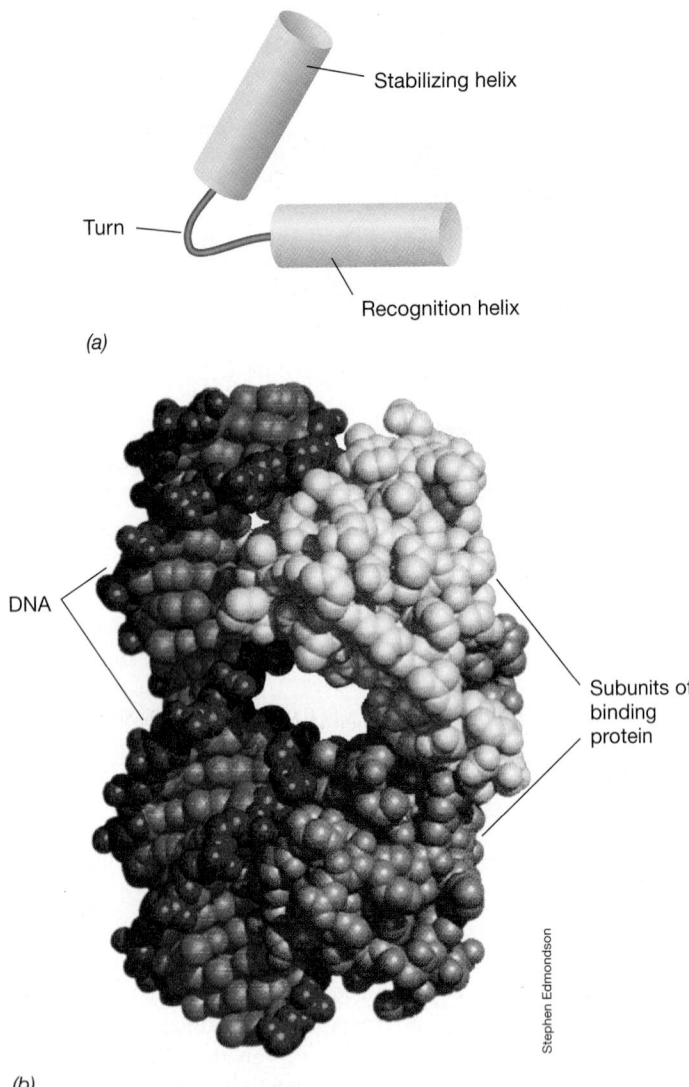

(a)

(b)

Stephen Edmondson

Figure 8.3 The helix-turn-helix structure of some DNA-binding proteins. *(a)* A simple model of the helix-turn-helix structure within a single protein subunit. *(b)* A computer model of both subunits of the bacteriophage lambda repressor, a typical helix-turn-helix protein, bound to its operator. The DNA is red and blue. One subunit of the dimeric repressor is shown in brown and the other in yellow. Each subunit contains a helix-turn-helix structure. The coordinates used to generate this image were downloaded from the Protein Data Base, Brookhaven, NY (http://www.pdb.org/pdb/home/).

Once a protein binds at a specific site on the DNA, various outcomes are possible. Some DNA-binding proteins are enzymes that catalyze a specific reaction on the DNA, such as transcription by RNA polymerase. In other cases, however, the binding event can either block transcription (negative regulation, Section 8.3) or activate it (positive regulation, Section 8.4).

MiniQuiz
- What is a protein domain?
- Why are some interactions of proteins with DNA specific to certain DNA sequences?

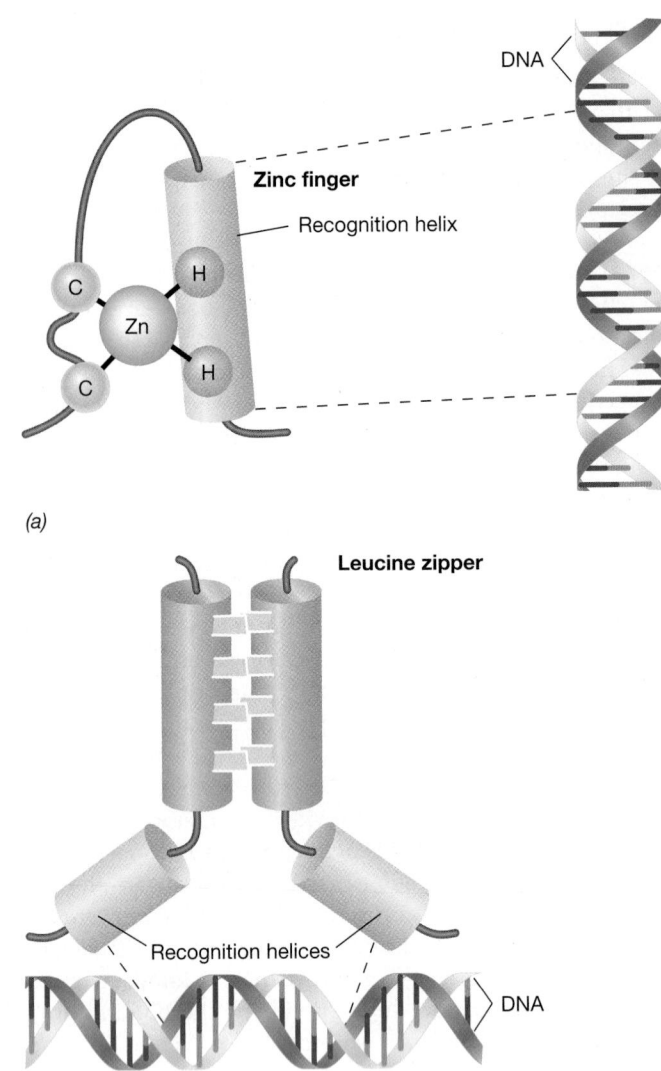

(a)

(b)

Figure 8.4 Simple models of protein substructures found in eukaryotic DNA-binding proteins. Cylinders represent α-helices. Recognition helices are the domains that bind DNA. *(a)* The zinc finger structure. The amino acids holding the Zn²⁺ ion always include at least two cysteine residues (C), with the other residues being histidine (H). *(b)* The leucine zipper structure. The leucine residues (yellow) are spaced exactly every seven amino acids. The interaction of the leucine side chains helps hold the two helices together.

8.3 Negative Control of Transcription: Repression and Induction

Transcription is the first step in biological information flow; because of this, it is simple and efficient to control gene expression at this point. If one gene is transcribed more frequently than another, there will be more of its mRNA available for translation and therefore a greater amount of its protein product in the cell. Several different mechanisms for controlling gene expression are known in bacteria, and all of them are greatly influenced by the environment in which the organism is growing, in particular by the presence or absence of specific small molecules. These molecules can interact with specific proteins such as the DNA-binding proteins just described. The result is the control of transcription or, more rarely, translation.

We begin by describing repression and induction, simple forms of regulation that govern gene expression at the level of transcription. In this section we deal with **negative control** of transcription, control that prevents transcription.

Enzyme Repression and Induction

Often the enzymes that catalyze the synthesis of a specific product are not made if the product is already present in the medium in sufficient amounts. For example, the enzymes needed to synthesize the amino acid arginine are made only when arginine is absent from the culture medium; an excess of arginine decreases the synthesis of these enzymes. This is called enzyme **repression**.

As can be seen in **Figure 8.5**, if arginine is added to a culture growing exponentially in a medium devoid of arginine, growth continues at the previous rate, but production of the enzymes for arginine synthesis stops. Note that this is a specific effect, as the synthesis of all other enzymes in the cell continues at the previous rate. This is because the enzymes affected by a particular repression event make up only a tiny fraction of the entire complement of proteins in the cell. Enzyme repression is widespread in bacteria as a means of controlling the synthesis of enzymes required for the production of amino acids and the nucleotide precursors purines and pyrimidines. In most cases, the final product of a particular biosynthetic pathway represses the enzymes of the pathway. This ensures that the organism does not waste energy and nutrients synthesizing unneeded enzymes.

Enzyme **induction** is conceptually the opposite of enzyme repression. In enzyme induction, an enzyme is made only when its substrate is present. Enzyme repression typically affects biosynthetic (anabolic) enzymes. In contrast, enzyme induction usually affects degradative (catabolic) enzymes.

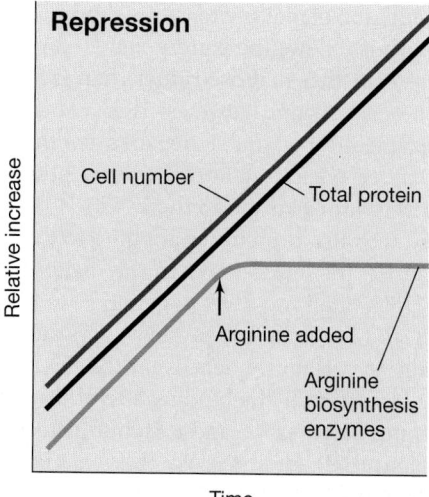

Figure 8.5 Enzyme repression. The addition of arginine to the medium specifically represses production of enzymes needed to make arginine. Net protein synthesis is unaffected.

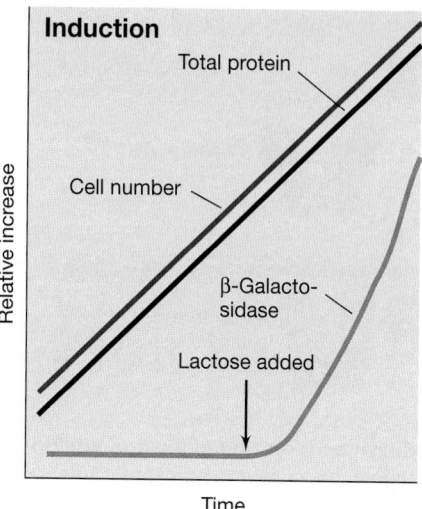

Figure 8.6 Enzyme induction. The addition of lactose to the medium specifically induces synthesis of the enzyme β-galactosidase. Net protein synthesis is unaffected.

Consider, for example, the utilization of the sugar lactose as a carbon and energy source by *Escherichia coli*. **Figure 8.6** shows the induction of β-galactosidase, the enzyme that cleaves lactose into glucose and galactose. This enzyme is required for *E. coli* to grow on lactose. If lactose is absent, the enzyme is not made, but synthesis begins almost immediately after lactose is added. The three genes in the *lac* operon encode three proteins, including β-galactosidase, that are induced simultaneously upon adding lactose. This type of control mechanism ensures that specific enzymes are synthesized only when needed.

Inducers and Corepressors

The substance that induces enzyme synthesis is called an *inducer* and a substance that represses enzyme synthesis is called a *corepressor*. These substances, which are normally small molecules, are collectively called *effectors*. Interestingly, not all inducers and corepressors are actual substrates or end products of the enzymes involved. For example, structural analogs may induce or repress even though they are not substrates of the enzyme. Isopropylthiogalactoside (IPTG), for instance, is an inducer of β-galactosidase even though IPTG cannot be hydrolyzed by this enzyme. In nature, however, inducers and corepressors are probably normal cell metabolites. Detailed studies of lactose utilization in *E. coli* have shown that the actual inducer of β-galactosidase is not lactose, but its isomer allolactose, which is made from lactose. www.microbiologyplace.com **Online Tutorial 8.1: Negative Control of Transcription and the *lac* Operon**

Mechanism of Repression and Induction

How can inducers and corepressors affect transcription in such a specific manner? They do this indirectly by binding to specific DNA-binding proteins, which, in turn, affect transcription. For an example of a repressible enzyme, we consider the arginine operon. **Figure 8.7a** shows transcription of the arginine genes, which proceeds when the cell needs arginine. When arginine is

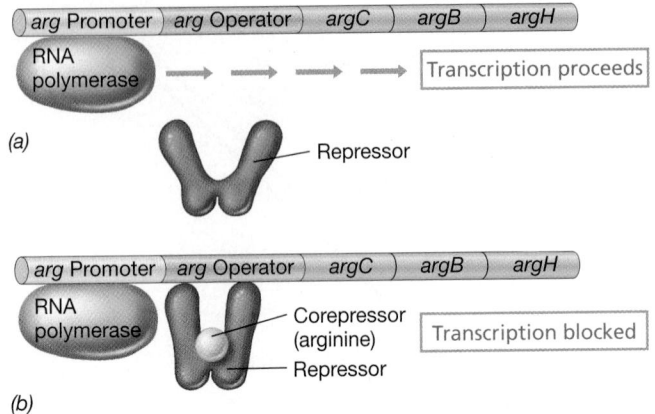

Figure 8.7 Enzyme repression in the arginine operon. *(a)* The operon is transcribed because the repressor is unable to bind to the operator. *(b)* After a corepressor (small molecule) binds to the repressor, the repressor binds to the operator and blocks transcription; mRNA and the proteins it encodes are not made. For the *argCBH* operon, the amino acid arginine is the corepressor that binds to the arginine repressor.

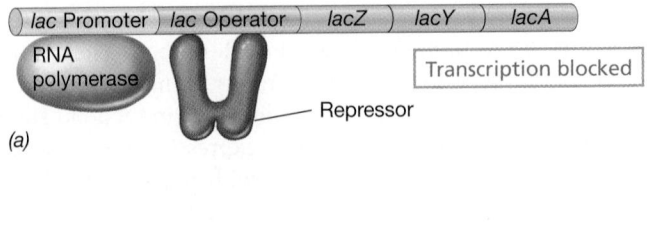

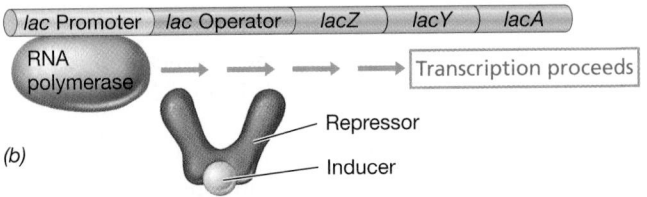

Figure 8.8 Enzyme induction in the lactose operon. *(a)* A repressor protein bound to the operator blocks the binding of RNA polymerase. *(b)* An inducer molecule binds to the repressor and inactivates it so that it no longer can bind to the operator. RNA polymerase then transcribes the DNA and makes an mRNA for that operon. For the *lac* operon, the sugar allolactose is the inducer that binds to the lactose repressor.

plentiful it acts as corepressor. As Figure 8.7*b* shows, arginine binds to a specific **repressor protein**, the arginine repressor, present in the cell. The repressor protein is allosteric (⮌ Section 4.16); that is, its conformation is altered when the corepressor binds to it.

By binding its effector, the repressor protein becomes active and can then bind to a specific region of the DNA near the promoter of the gene, known as the *operator*. This region gave its name to the **operon**, a cluster of genes arranged in a linear and consecutive fashion whose expression is under the control of a single operator (⮌ Section 6.5). All of the genes in an operon are transcribed as a single unit yielding a single mRNA. The operator is located downstream of the promoter where synthesis of mRNA is initiated (Figure 8.7). If the repressor binds to the operator, transcription is physically blocked because RNA polymerase can neither bind nor proceed. Hence, the polypeptides encoded by the genes in the operon cannot be synthesized. If the mRNA is polycistronic (⮌ Section 6.15), all the polypeptides encoded by this mRNA will be repressed.

Enzyme induction may also be controlled by a repressor. In this case, the repressor protein is active in the absence of the inducer, completely blocking transcription. When the inducer is added, it combines with the repressor protein and inactivates it; inhibition is overcome and transcription can proceed (**Figure 8.8**).

All regulatory systems employing repressors have the same underlying mechanism: inhibition of mRNA synthesis by the activity of specific repressor proteins that are themselves under the control of specific small effector molecules. And, as previously noted, because the repressor's role is inhibitory, regulation by repressors is called *negative control*. One point to note is that genes are not turned on and off completely like light switches. DNA-binding proteins vary in concentration and affinity and thus control is quantitative. Even when a gene is "fully repressed" there is often a very low level of basal transcription.

MiniQuiz
- Why is "negative control" so named?
- How does a repressor inhibit the synthesis of a specific mRNA?

8.4 Positive Control of Transcription

Negative control relies on a repressor protein to bring about repression of mRNA synthesis. By contrast, in **positive control** of transcription the regulatory protein is an **activator** that activates the binding of RNA polymerase to DNA. An excellent example of positive regulation is the catabolism of the sugar maltose in *Escherichia coli*.

Maltose Catabolism in *Escherichia coli*

The enzymes for maltose catabolism in *E. coli* are synthesized only after the addition of maltose to the medium. The expression of these enzymes thus follows the pattern shown for β-galactosidase in Figure 8.6 except that maltose rather than lactose is required to induce gene expression. However, the synthesis of maltose-degrading enzymes is not under negative control as in the *lac* operon, but under positive control; transcription requires the binding of an **activator protein** to the DNA.

The maltose activator protein cannot bind to the DNA unless it first binds maltose, the inducer. When the maltose activator protein binds to DNA, it allows RNA polymerase to begin transcription (**Figure 8.9**). Like repressor proteins, activator proteins bind specifically only to certain sequences on the DNA. However, the region on the DNA that is the binding site of the activator is not called an operator (Figures 8.7 and 8.8), but instead an *activator-binding site* (Figure 8.9). Nevertheless, the genes controlled by this activator-binding site are still called an operon.

Binding of Activator Proteins

The promoters of positively controlled operons have nucleotide sequences that bind RNA polymerase weakly and are poor

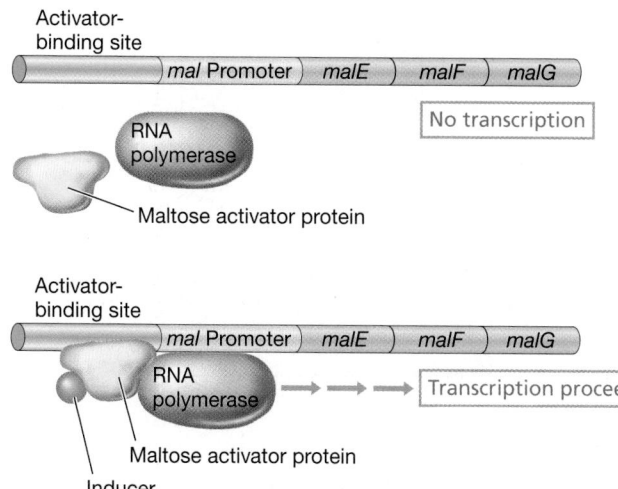

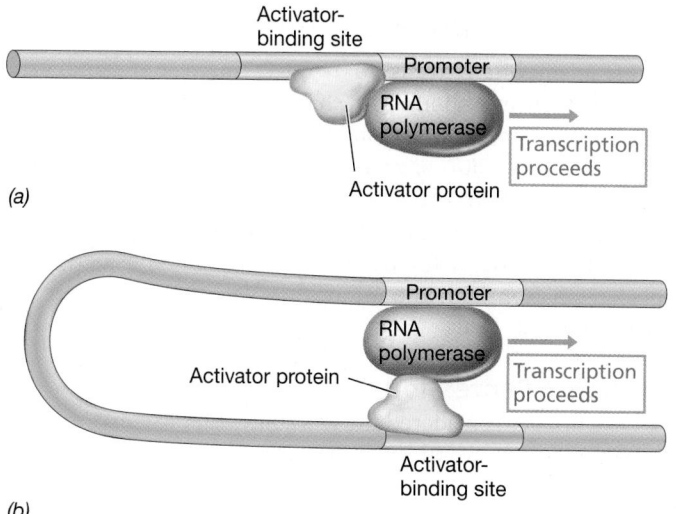

Figure 8.9 **Positive control of enzyme induction in the maltose operon.** *(a)* In the absence of an inducer, neither the activator protein nor the RNA polymerase can bind to the DNA. *(b)* An inducer molecule (for the *malEFG* operon it is the sugar maltose) binds to the activator protein (MalT), which in turn binds to the activator-binding site. This allows RNA polymerase to bind to the promoter and begin transcription.

Figure 8.11 **Activator protein interactions with RNA polymerase.** *(a)* The activator-binding site is near the promoter. *(b)* The activator-binding site is several hundred base pairs from the promoter. In this case, the DNA must be looped to allow the activator and the RNA polymerase to contact.

matches to the consensus sequence (⟳ Section 6.13). Thus, even with the correct sigma (σ) factor, the RNA polymerase has difficulty binding to these promoters. The role of the activator protein is to help the RNA polymerase recognize the promoter and begin transcription.

For example, the activator protein may modify the structure of the DNA by bending it (**Figure 8.10**), allowing the RNA polymerase to make the correct contacts with the promoter to begin transcription. Alternatively, the activator protein may interact directly with the RNA polymerase. This can happen either when the activator-binding site is close to the promoter (**Figure 8.11a**) or when it is several hundred base pairs away from the promoter,

a situation in which DNA looping is required to make the necessary contacts (Figure 8.11*b*).

Many genes in *E. coli* have promoters under positive control and many have promoters under negative control. In addition, many operons have promoters with multiple types of control and some have more than one promoter, each with its own control system! Thus, the simple picture outlined above is not typical of all operons. Multiple control features are common in the operons of virtually all prokaryotes, and thus their overall regulation can be very complex.

Operons versus Regulons

In *E. coli*, the genes required for maltose utilization are spread out over the chromosome in several operons, each of which has an activator-binding site to which a copy of the maltose activator protein can bind. Therefore, the maltose activator protein actually controls the transcription of more than one operon. When more than one operon is under the control of a single regulatory protein, these operons are collectively called a **regulon**. Therefore, the enzymes for maltose utilization are encoded by the maltose regulon.

Regulons are known for operons under negative control as well. For example, the arginine biosynthetic enzymes (Section 8.3) are encoded by the arginine regulon, whose operons are all under the control of the arginine repressor protein (only one of the arginine operons was shown in Figure 8.7). In regulon control a specific DNA-binding protein binds only at those operons it controls regardless of whether it is functioning as an activator or repressor; other operons are not affected.

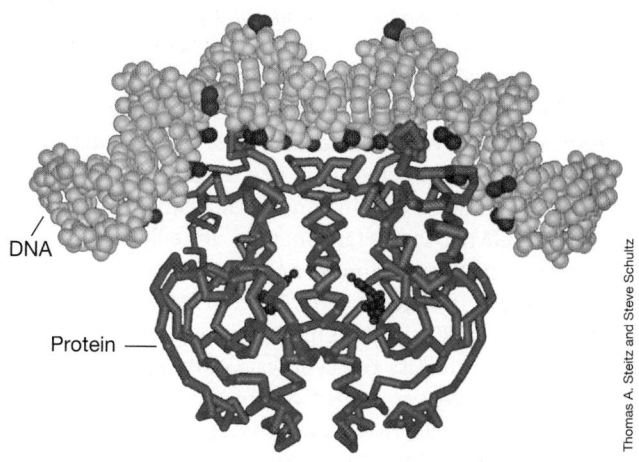

Thomas A. Steitz and Steve Schultz

Figure 8.10 **Computer model of a positive regulatory protein interacting with DNA.** This model shows the cyclic AMP receptor protein (CRP), a regulatory protein that controls several operons. The α-carbon backbone of this protein is shown in blue and purple. The protein is binding to a DNA double helix (green and light blue). Note that binding of the CRP protein to DNA has bent the DNA.

MiniQuiz

• Compare and contrast the activities of an activator protein and a repressor protein.

• Distinguish between an operon and a regulon.

8.5 Global Control and the *lac* Operon

An organism often needs to regulate many unrelated genes simultaneously in response to a change in its environment. Regulatory mechanisms that respond to environmental signals by regulating the expression of many different genes are called *global control systems*. Both the lactose operon and the maltose regulon respond to global controls in addition to their own controls discussed in Sections 8.3 and 8.4. We begin our consideration of global regulation with the *lac* operon and the choice between different sugars.

Catabolite Repression

We have not yet considered the possibility that bacteria might be confronted with several different carbon sources that could be used. For example, *Escherichia coli* can use many different sugars. When faced with several sugars, including glucose, do cells of *E. coli* use them simultaneously or one at a time? The answer is that glucose is used first. It would be wasteful to induce enzymes for using other sugars when glucose is available, because *E. coli* grows faster on glucose than on other carbon sources. **Catabolite repression** is a mechanism of global control that decides between different available carbon sources if more than one is present.

When cells of *E. coli* are grown in a medium that contains glucose, the synthesis of enzymes needed for the breakdown of other carbon sources (such as lactose or maltose) is repressed, even if those other carbon sources are present. Thus, the presence of a favored carbon source overrules the induction of pathways that catabolize other carbon sources. Catabolite repression is sometimes called the "glucose effect" because glucose was the first substance shown to cause this response. The key point is that the favored substrate is a better carbon and energy source for the organism. Thus, catabolite repression ensures that the organism uses the *best* available carbon and energy source first.

Why is catabolite repression called *global* control? In *E. coli* and other organisms for which glucose is the best energy source, catabolite repression prevents expression of most other catabolic operons as long as glucose is present. Dozens of catabolic operons are affected, including those for lactose, maltose, a host of other sugars, and most other commonly used carbon and energy sources for *E. coli*. In addition, genes for the synthesis of flagella are controlled by catabolite repression because if bacteria have a good carbon source available, there is no need to swim around in search of nutrients.

One consequence of catabolite repression is that it may lead to two exponential growth phases, a situation called *diauxic growth*. If two usable energy sources are available, the cells grow first on the better energy source. Growth stops when the better source is depleted, but then following a lag period, it resumes on the other energy source. Diauxic growth is illustrated in **Figure 8.12** for *E. coli* on a mixture of glucose and lactose. The cells grow more rapidly on glucose than on lactose. Although glucose and lactose are both excellent energy sources for *E. coli*, glucose is superior, and growth is faster.

The proteins of the *lac* operon, including the enzyme β-galactosidase, are required for using lactose and are induced in

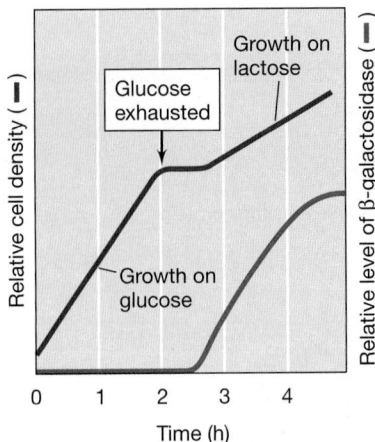

Figure 8.12 Diauxic growth of *Escherichia coli* on a mixture of glucose and lactose. The presence of glucose represses the synthesis of β-galactosidase, the enzyme that cleaves lactose into glucose and galactose. After glucose is depleted, there is a lag during which β-galactosidase is synthesized. Growth then resumes on lactose but at a slower rate.

its presence (Figures 8.6 and 8.8). But the synthesis of these proteins is also subject to catabolite repression. As long as glucose is present, the *lac* operon is not expressed and lactose is not used. However, when glucose is depleted, catabolite repression is abolished, the *lac* operon is expressed, and the cells grow on lactose.

Cyclic AMP and Cyclic AMP Receptor Protein

Despite its name, catabolite repression relies on an activator protein and is actually a form of positive control (Section 8.4). The activator protein is called the *cyclic AMP receptor protein* (*CRP*). A gene that encodes a catabolite-repressible enzyme is expressed only if CRP binds to DNA in the promoter region. This allows RNA polymerase to bind to the promoter. CRP is an allosteric protein and binds to DNA only if it has first bound a small molecule called *cyclic adenosine monophosphate* (*cyclic AMP* or *cAMP*) (**Figure 8.13**). Like many DNA-binding proteins (Section 8.2), CRP binds to DNA as a dimer.

Cyclic AMP is a key molecule in many metabolic control systems, both in prokaryotes and eukaryotes. Because it is derived from a nucleic acid precursor, it is a **regulatory nucleotide**. Other regulatory nucleotides include cyclic guanosine monophosphate (cyclic GMP; important mostly in eukaryotes), cyclic di-GMP (important in biofilm formation; ↩ Section 23.4), and guanosine tetraphosphate (ppGpp; Section 8.10). Cyclic AMP is synthesized from ATP by an enzyme called *adenylate cyclase*. However, glucose inhibits the synthesis of cyclic AMP and also stimulates cyclic AMP transport out of the cell. When glucose enters the cell, the cyclic AMP level is lowered, CRP protein cannot bind DNA, and RNA polymerase fails to bind to the promoters of operons subject to catabolite repression. Thus, catabolite repression is an indirect result of the presence of a better energy source (glucose). The direct cause of catabolite repression is a low level of cyclic AMP.

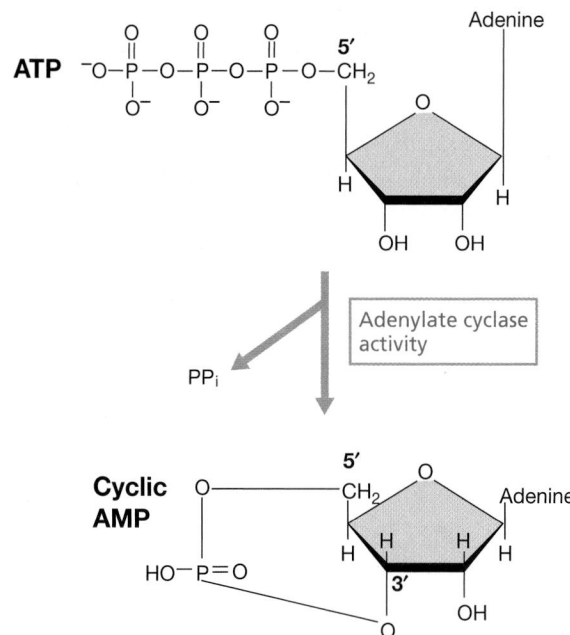

Figure 8.13 Cyclic AMP. Cyclic adenosine monophosphate (cyclic AMP) is made from ATP by the enzyme adenylate cyclase.

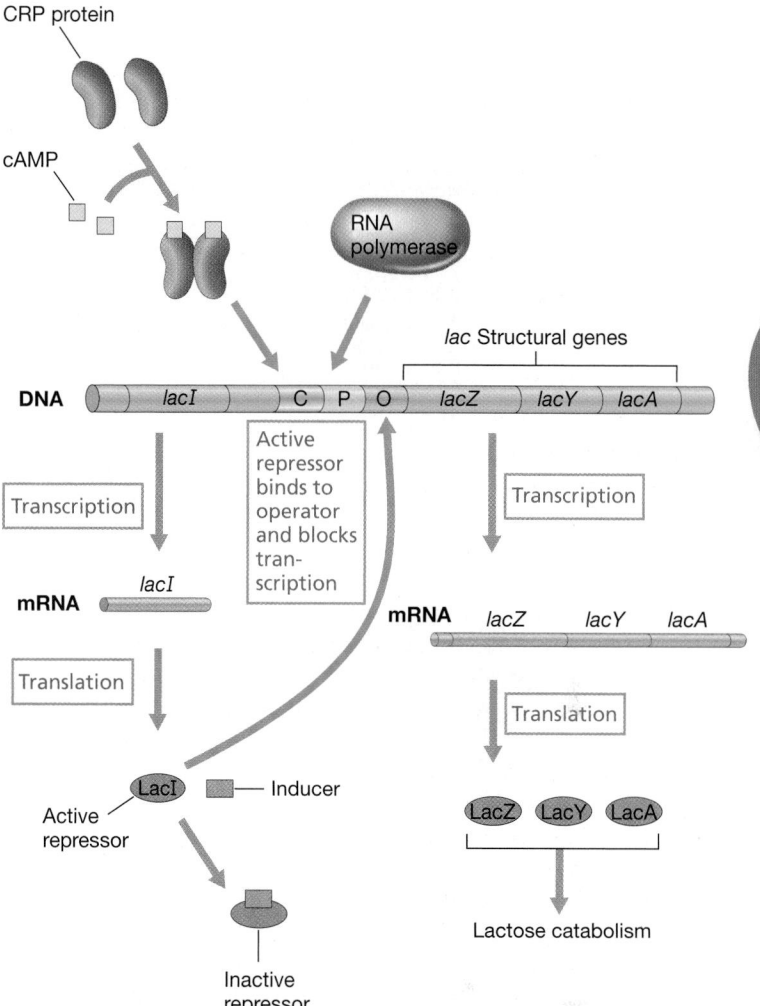

Figure 8.14 Overall regulation of the *lac* system. The *lac* operon consists of *lacZ*, encoding β-galactosidase, which breaks down lactose, plus two other genes, *lacY*, encoding lactose permease, and *lacA*, encoding lactose acetylase. The LacI repressor protein is encoded by a separate gene, *lacI*. LacI binds to the operator (O) unless the inducer is present. RNA polymerase binds to the promoter (P). CRP binds to the C site when activated by cyclic AMP. For the *lac* operon to be transcribed by RNA polymerase, the LacI repressor must be absent (that is, inducer must be present) and cyclic AMP levels must be high (due to the absence of glucose), thus allowing CRP to bind.

Let us return to the *lac* operon and include catabolite repression. The entire regulatory region of the *lac* operon is diagrammed in **Figure 8.14**. For *lac* genes to be transcribed, two requirements must be met: (1) The level of cyclic AMP must be high enough for the CRP protein to bind to the CRP-binding site (positive control), and (2) lactose or another molecule capable of acting as inducer must be present so that the lactose repressor (LacI protein) does not block transcription by binding to the operator (negative control). If these two conditions are met, the cell is signaled that glucose is absent and lactose is present; then and only then does transcription of the *lac* operon begin.

MiniQuiz

- Explain how catabolite repression depends on an activator protein.
- What role does cyclic AMP play in glucose regulation?
- Explain how the *lac* operon is both positively and negatively controlled.

8.6 Control of Transcription in *Archaea*

There are two alternative approaches to regulating the activity of RNA polymerase. One strategy, common in *Bacteria*, is to use DNA-binding proteins that either block RNA polymerase activity (repressor proteins) or stimulate RNA polymerase activity (activator proteins). The alternative, common in eukaryotes, is to transmit signals to the protein subunits of the RNA polymerase itself. Given the greater overall similarity between the mechanism of transcription in *Archaea* and *Eukarya* (Chapter 7), it is perhaps surprising that the regulation of transcription in *Archaea* more closely resembles that of *Bacteria*.

Few repressor or activator proteins from *Archaea* have yet been characterized in detail, but it is clear that *Archaea* have both types of regulatory proteins. Archaeal repressor proteins either block the binding of RNA polymerase itself or block the binding of TBP (TATA-binding protein) and TFB (transcription factor B), which are required for RNA polymerase to bind to the promoter in *Archaea* (⟳ Section 7.2). At least some archaeal activator proteins function in just the opposite way, by recruiting TBP to the promoter, thereby facilitating transcription.

A good example of an archaeal repressor is the NrpR protein from the methanogen *Methanococcus maripaludis;* this protein

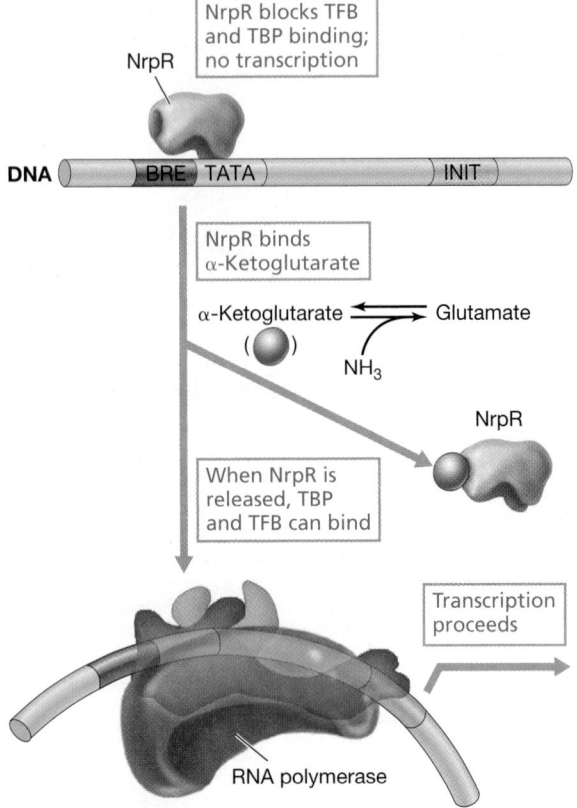

NrpR

NrpR blocks TFB
and TBP binding;
no transcription

DNA BRE TATA INIT

NrpR binds
α-Ketoglutarate

α-Ketoglutarate ⇌ Glutamate
() NH₃

NrpR

When NrpR is
released, TBP
and TFB can bind

Transcription
proceeds

RNA polymerase

Figure 8.15 Repression of genes for nitrogen metabolism in **Archaea.** The NrpR protein of *Methanococcus maripaludis* acts as a repressor. It blocks the binding of the TFB and TBP proteins, which are required for promoter recognition, to the BRE site and TATA box, respectively. If there is a shortage of ammonia, α-ketoglutarate is not converted to glutamate. The α-ketoglutarate accumulates and binds to NrpR, releasing it from the DNA. Now TBP and TFB can bind. This in turn allows RNA polymerase to bind and transcribe the operon.

represses genes active in nitrogen assimilation (**Figure 8.15**), such as those for nitrogen fixation (↩ Section 13.15) and glutamine synthesis (↩ Section 4.16). When organic nitrogen is plentiful in the *M. maripaludis* cell, NrpR represses nitrogen assimilation genes. However, if the level of nitrogen becomes limiting, α-ketoglutarate accumulates to high levels. This occurs because α-ketoglutarate, a citric acid cycle intermediate, is also a major acceptor of ammonia during nitrogen assimilation.

When levels of α-ketoglutarate rise, this signals that ammonia is limiting and that additional pathways need to be activated for obtaining ammonia, such as nitrogen fixation or the high-affinity nitrogen assimilation enzyme glutamine synthetase. Elevated levels of α-ketoglutarate function as an inducer by binding to the NrpR protein. In this state, NrpR loses its affinity for the promoter regions of its target genes and no longer blocks transcription from these promoters. In this respect, the NrpR protein resembles the LacI repressor and similar proteins of *Bacteria* (Section 8.3).

Other archaeal proteins regulate transcription in a positive manner. Thus their binding in the promoter region increases transcription. Some of these transcription activators are related to bacterial proteins whereas others appear to be unique to the *Archaea*. The SurR protein of *Pyrococcus furiosus* is an example of a regulatory protein that functions either as an activator or as a repressor, depending on the location of its binding site within the promoter region. SurR controls the response of *Pyrococcus furiosus* to elemental sulfur and its conversion to hydrogen sulfide.

MiniQuiz
- What is the major difference between transcriptional regulation in *Archaea* and eukaryotes?
- How do transcriptional activators in *Archaea* often differ in mechanism from those in *Bacteria*?

(III) Sensing and Signal Transduction

Prokaryotes regulate cell metabolism in response to environmental fluctuations, including temperature changes, changes in pH and oxygen availability, changes in the availability of nutrients, and even changes in the number of other cells present. Therefore, there must be mechanisms by which cells receive signals from the environment and transmit them to the specific target to be regulated.

Some signals are small molecules that enter the cell and function as effectors. However, in many cases the external signal is not transmitted directly to the regulatory protein but instead is detected by a sensor that transmits it to the rest of the regulatory machinery, a process called **signal transduction**.

8.7 Two-Component Regulatory Systems

Because most signal transduction systems contain two parts, they are called **two-component regulatory systems**. Characteristically, such systems consist of a specific **sensor kinase protein** usually located in the cytoplasmic membrane and a **response regulator protein** present in the cytoplasm.

A kinase is an enzyme that phosphorylates compounds, typically using phosphate from ATP. Sensor kinases detect a signal from the environment and phosphorylate themselves (a process called autophosphorylation) at a specific histidine residue (**Figure 8.16**). Sensor kinases thus belong to the class of enzymes called *histidine kinases*. The phosphate is then transferred to another protein inside the cell, the response regulator. This is typically a DNA-binding protein that regulates transcription, in either a positive or a negative fashion depending on the system. In the example shown in Figure 8.16, regulation is negative; the response regulator is a repressor that binds DNA, blocking transcription, until the transfer of the phosphate releases it, permitting transcription.

Although the term is rarely used, a *one-component regulatory system* consists of a single protein that both detects a signal and carries out a regulatory response. Examples include the LacI repressor, the MalT activator, and the Crp protein. All three bind a small molecule (the signal) and then bind to DNA to regulate transcription.

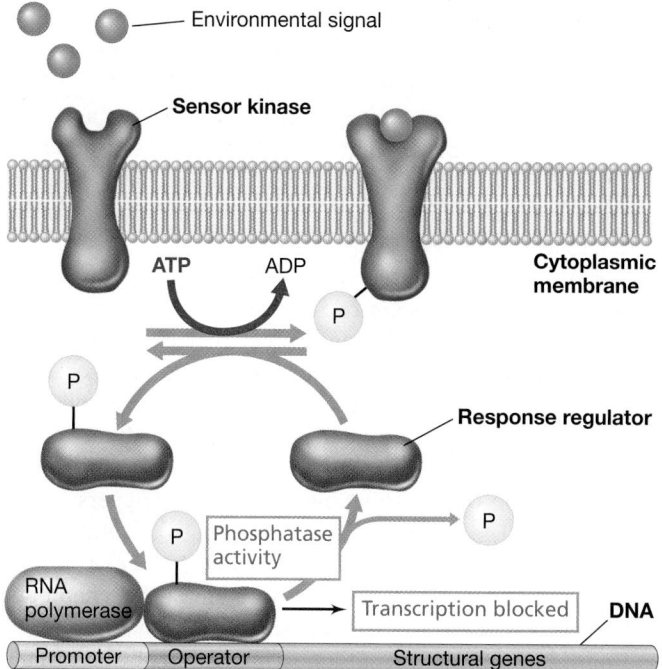

Figure 8.16 **The control of gene expression by a two-component regulatory system.** One component is a sensor kinase in the cytoplasmic membrane that phosphorylates itself in response to an environmental signal. The phosphoryl group is then transferred to the second component, a response regulator. The phosphorylated form of the response regulator then binds to DNA. In the system shown here, the phosphorylated response regulator is a repressor protein. The phosphatase activity of the response regulator slowly releases the phosphate from the response regulator and resets the system.

A balanced regulatory system must have a feedback loop, that is, a way to complete the regulatory circuit and terminate the response. This resets the system for another cycle. This feedback loop involves a phosphatase, an enzyme that removes the phosphate from the response regulator at a constant rate. This reaction is often carried out by the response regulator itself, although in some cases separate proteins are involved (Figure 8.16). Phosphatase activity is typically slower than phosphorylation. However, if phosphorylation ceases due to reduced sensor kinase activity, phosphatase activity eventually returns the response regulator to the fully nonphosphorylated state.

Examples of Two-Component Regulatory Systems

Two-component systems regulate a large number of genes in many different bacteria. Interestingly, two-component systems are rare or absent in *Archaea* and in *Bacteria* that live as parasites of higher organisms. A few key examples of two-component systems include those that respond to phosphate limitation, nitrogen limitation, and osmotic pressure. In *Escherichia coli* almost 50 different two-component systems are present, and several are listed in **Table 8.1**. For example, the osmolarity of the environment controls the relative levels of the proteins OmpC and OmpF in the *E. coli* outer membrane. OmpC and OmpF are porins, proteins that allow metabolites to cross the outer membrane of gram-negative bacteria (Section 3.7). If osmotic pressure is low, the synthesis of OmpF, a porin with a larger pore, increases; if osmotic pressure is higher, OmpC, a porin with a smaller pore, is made in larger amounts. The response regulator of this system is OmpR. When OmpR is phosphorylated, it activates transcription of the *ompC* gene and represses transcription of the *ompF* gene. The *ompF* gene in *E. coli* is also controlled by antisense RNA (Section 8.14).

Some signal transduction systems have multiple regulatory elements. For instance, in the Ntr regulatory system, which regulates nitrogen assimilation in many *Bacteria*, including *E. coli*, the response regulator is the activator protein nitrogen regulator I (NRI). NRI activates transcription from promoters recognized by RNA polymerase using the alternative sigma factor σ^{54} (RpoN) (Section 6.13). The sensor kinase in the Ntr system, nitrogen regulator II (NRII), fills a dual role as both kinase and phosphatase. The activity of NRII is in turn regulated by the addition or removal of uridine monophosphate groups from another protein, known as PII.

The Nar regulatory system (Table 8.1) controls a set of genes that allow the use of nitrate or nitrite or both as alternative electron acceptors during anaerobic respiration (Section 14.7). The Nar system contains two different sensor kinases and two different response regulators. In addition, all of the genes regulated by this system are also controlled by the FNR protein (fumarate nitrite regulator), a global regulator for genes of anaerobic respiration (see Table 8.3). This type of multiple regulation is common for systems of central importance to cellular metabolism.

Genomic analyses allow easy detection of genes encoding two-component regulatory systems because the histidine kinases show significant amino acid sequence conservation. Two-component

Table 8.1 *Examples of two-component systems that regulate transcription in* **Escherichia coli**

System	Environmental signal	Sensor kinase	Response regulator	Activity of response regulator[a]
Arc system	Oxygen	ArcB	ArcA	Repressor/activator
Nitrate and nitrite respiration (Nar)	Nitrate and nitrite	NarX	NarL	Activator/repressor
		NarQ	NarP	Activator/repressor
Nitrogen utilization (Ntr)	Shortage of organic nitrogen	NRII (= GlnL)	NRI (= GlnG)	Activator of promoters requiring RpoN/σ^{54}
Pho regulon	Inorganic phosphate	PhoR	PhoB	Activator
Porin regulation	Osmotic pressure	EnvZ	OmpR	Activator/repressor

[a]Note that many response regulator proteins act as both activators and repressors depending on the genes being regulated. Although ArcA can function as either an activator or a repressor, it functions as a repressor on most operons that it regulates.

systems closely related to those in *Bacteria* are also present in microbial eukaryotes, such as the yeast *Saccharomyces cerevisiae*, and even in plants. However, most eukaryotic signal transduction pathways rely on phosphorylation of serine, threonine, and tyrosine residues of proteins that are unrelated to those of bacterial two-component systems.

8.8 Regulation of Chemotaxis

We have previously seen that some prokaryotes can move toward attractants or away from repellents, a behavior called *chemotaxis* (∞ Section 3.15). We noted that prokaryotes are too small to sense spatial gradients of a chemical, but they can respond to temporal gradients. That is, they can sense the *change* in concentration of a chemical over time rather than the absolute concentration of the chemical stimulus. Prokaryotes use a modified two-component system to sense temporal changes in attractants or repellents and process this information to regulate flagellar rotation. Note that chemotaxis uses a two-component system to directly regulate the activity of preexisting flagella rather than the transcription of the genes encoding the flagella.

Step One: Response to Signal

The mechanism of chemotaxis is complex and depends upon multiple proteins. Several sensory proteins reside in the cytoplasmic membrane and sense the presence of attractants and repellents. These sensor proteins are not themselves sensor kinases but interact with cytoplasmic sensor kinases. These sensory proteins allow the cell to monitor the concentration of various substances over time.

The sensory proteins are called *methyl-accepting chemotaxis proteins* (*MCPs*). *Escherichia coli* possesses five different MCPs. Each MCP is a transmembrane protein that can sense certain compounds. For example, the Tar MCP of *E. coli* senses the attractants aspartate and maltose and the repellents cobalt and nickel. MCPs bind attractants or repellents directly or in some cases indirectly through interactions with periplasmic binding proteins. Binding of an attractant or repellent triggers interactions with cytoplasmic proteins that affect flagellar rotation.

MCPs make contact with the cytoplasmic proteins CheA and CheW (**Figure 8.17**). CheA is the sensor kinase for chemotaxis. When an MCP binds a chemical, it changes conformation and, with help from CheW, affects the autophosphorylation of CheA to form CheA-P. Attractants *decrease* the rate of autophosphorylation, whereas repellents *increase* this rate. CheA-P then passes the phosphate to CheY (forming CheY-P); this is the response regulator that controls flagellar rotation. CheA-P can also pass the phosphate to CheB (a response regulator active in step three). This phosphorylation is much slower than that of CheY, and is discussed later.

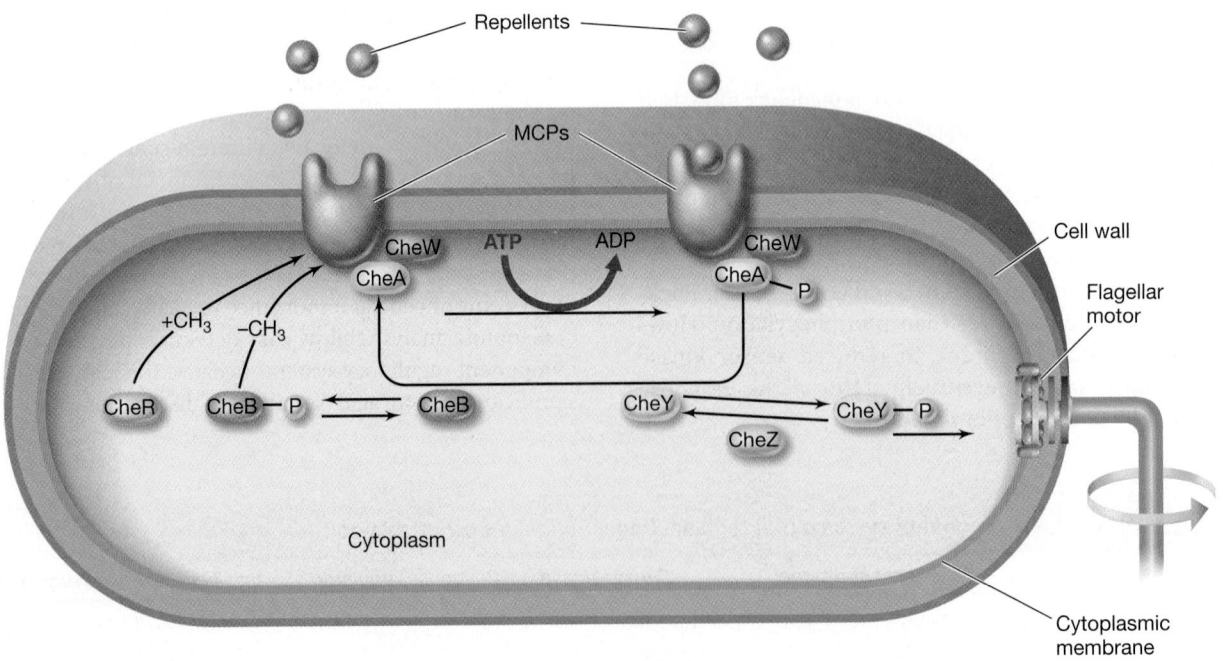

Figure 8.17 Interactions of MCPs, Che proteins, and the flagellar motor in bacterial chemotaxis. The methyl-accepting chemotaxis protein (MCP) forms a complex with the sensor kinase CheA and the coupling protein CheW. This combination triggers autophosphorylation of CheA to CheA-P. CheA-P can then phosphorylate the response regulators CheB and CheY. Phosphorylated CheY (CheY-P) binds to the flagellar motor switch. CheZ dephosphorylates CheY-P. CheR continually adds methyl groups to the MCP. CheB-P (but not CheB) removes them. The degree of methylation of the MCPs controls their ability to respond to attractants and repellents and leads to adaptation.

Step Two: Controlling Flagellar Rotation

CheY is a key protein in the system because it governs the direction of rotation of the flagellum. Recall that if rotation of the flagellum is counterclockwise, the cell will continue to move in a run, whereas if the flagellum rotates clockwise, the cell will tumble (Section 3.15). CheY-P interacts with the flagellar motor to induce clockwise flagellar rotation, which causes tumbling. When unphosphorylated, CheY cannot bind to the flagellar motor and the flagellum rotates counterclockwise; this causes the cell to run. Another protein, CheZ, dephosphorylates CheY, returning it to the form that allows runs instead of tumbles. Because repellents increase the level of CheY-P, they lead to tumbling, whereas attractants lead to a lower level of CheY-P and smooth swimming (runs).

Step Three: Adaptation

Once an organism has successfully responded to a stimulus, it must stop responding and reset the sensory system to await further signals. This is known as *adaptation*. During adaptation of the chemotaxis system, a feedback loop resets the system. This relies on the response regulator CheB, mentioned earlier.

As their name implies, MCPs can be methylated. The cytoplasmic protein CheR (Figure 8.17) continually adds methyl groups to the MCPs at a slow rate using *S*-adenosylmethionine as a methyl donor. The response regulator CheB is a demethylase that removes methyl groups from the MCPs. Phosphorylation of CheB greatly increases its rate of activity. The changes in methylation of the MCPs cause conformational changes similar to those due to binding of attractant or repellent. When MCPs are fully methylated they no longer respond to attractants, but are more sensitive to repellents. Conversely, when MCPs are unmethylated they respond highly to attractants, but are insensitive to repellents. Varying the methylation level thus allows adaptation to sensory signals.

If the level of an attractant remains high, CheY and CheB are not phosphorylated. Consequently, the cell swims smoothly. Methylation of the MCPs increases during this period because CheB-P is not present to rapidly demethylate them. However, MCPs no longer respond to the attractant when they become fully methylated. Therefore, if the level of attractant remains high but constant, the cell begins to tumble. Eventually, CheB becomes phosphorylated and CheB-P demethylates the MCPs. This resets the receptors and they can once again respond to further increases or decreases in level of attractants. Therefore the cell stops swimming if the attractant concentration is constant. It only continues to swim if even higher levels of attractant are encountered.

The course of events is just the opposite for repellents. Fully methylated MCPs respond best to an increasing gradient of repellents and send a signal for cell tumbling to begin. The cell then moves off in a random direction while MCPs are slowly demethylated. With this mechanism for adaptation, chemotaxis successfully achieves the ability to monitor small changes in the concentrations of both attractants and repellents over time.

Other Types of Taxis

In addition to chemotaxis, several other forms of taxis are known, for example, *phototaxis* (movement toward light) and *aerotaxis* (movement toward oxygen) (Section 3.15). Many of the cytoplasmic Che proteins that function in chemotaxis also play a role in these. In phototaxis, a light sensor protein replaces the MCPs of chemotaxis, and in aerotaxis, a redox protein monitors levels of oxygen. These sensors then interact with cytoplasmic Che proteins to direct runs or tumbles. Thus several different kinds of signals converge on the same flagellar control system.

MiniQuiz

- What are the primary response regulator and the primary sensor kinase for regulating chemotaxis?
- Why is adaptation during chemotaxis important?
- How does the response of the chemotaxis system to an attractant differ from its response to a repellent?

8.9 Quorum Sensing

Many prokaryotes respond to the presence in their surroundings of other cells of their species. Some prokaryotes have regulatory pathways that are controlled by the density of cells of their own kind. This is called **quorum sensing** (the word "quorum" in this sense means "sufficient numbers").

Mechanism of Quorum Sensing

Quorum sensing is a mechanism to assess population density. Many bacteria use this approach to ensure that sufficient cell numbers are present before starting activities that require a certain cell density to work effectively. For example, a pathogenic (disease-causing) bacterium that secretes a toxin will have no effect as a single cell; production of toxin by one cell alone would merely waste resources. However, if a sufficiently large population of cells is present, the coordinated expression of the toxin may successfully cause disease.

Quorum sensing is widespread among gram-negative bacteria but is also found in gram-positive bacteria. Each species that employs quorum sensing synthesizes a specific signal molecule called an **autoinducer**. This molecule diffuses freely across the cell envelope in either direction. Because of this, the autoinducer reaches high concentrations inside the cell only if there are many cells nearby, each making the same autoinducer. Inside the cell, the autoinducer binds to a specific activator protein and triggers transcription of specific genes (**Figure 8.18b**).

There are several different classes of autoinducers (**Table 8.2**). The first to be identified were the *acyl homoserine lactones* (AHLs) (Figure 8.18a). Several different AHLs, with acyl groups of different lengths, are found in different species of gram-negative bacteria. In addition, many gram-negative bacteria make autoinducer 2 (AI-2; a cyclic furan derivative). This is apparently used as a common autoinducer between many species of bacteria. Gram-positive bacteria generally use certain short peptides as autoinducers.

Quorum sensing was first discovered as the mechanism of regulating light emission in bioluminescent bacteria. Several bacterial species can emit light, including the marine bacterium *Aliivibrio fischeri* (Section 17.12). **Figure 8.19** shows bioluminescent colonies of *A. fischeri*. The light is generated by an

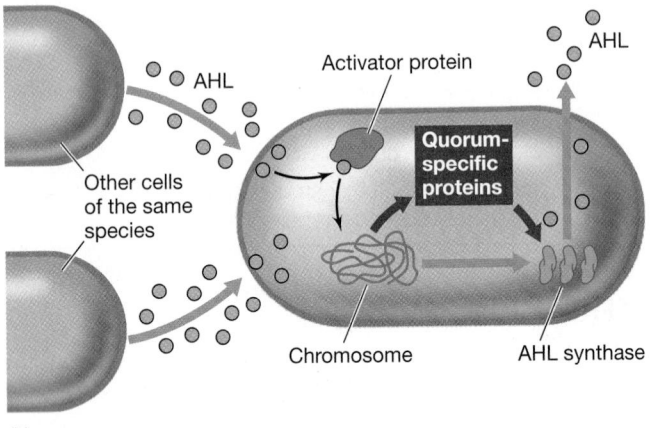

Acyl homoserine lactone (AHL)

(a)

(b)

Figure 8.18 Quorum sensing. *(a)* General structure of an acyl homoserine lactone (AHL). Different AHLs are variants of this parent structure. R = alkyl group (C_1–C_{17}); the carbon next to the R group is often modified to a keto group (C=O). *(b)* A cell capable of quorum sensing expresses AHL synthase at basal levels. This enzyme makes the cell's specific AHL. When cells of the same species reach a certain density, the concentration of AHL rises sufficiently to bind to the activator protein, which activates transcription of quorum-specific genes.

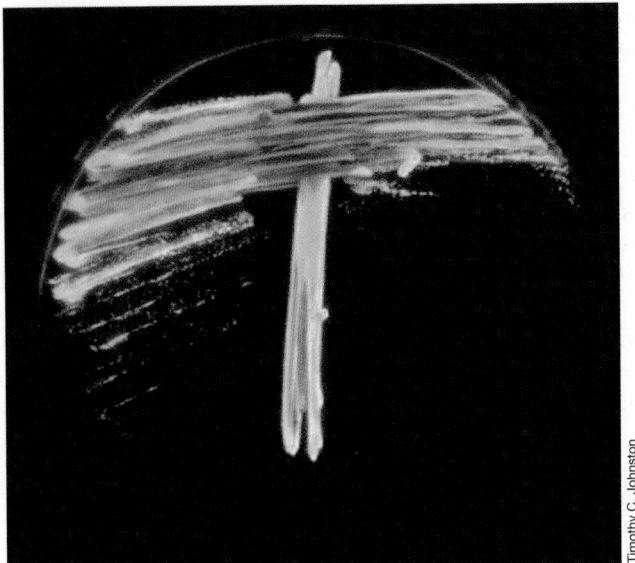

Figure 8.19 Bioluminescent bacteria producing the enzyme luciferase. Cells of the bacterium *Aliivibrio fischeri* were streaked on nutrient agar in a Petri dish and allowed to grow overnight. The photograph was taken in a darkened room using only the light generated by the bacteria.

enzyme called luciferase. The *lux* operons encode the proteins needed for bioluminescence. They are under control of the activator protein LuxR and are induced when the concentration of the specific *A. fischeri* AHL, *N*-3-oxohexanoyl homoserine lactone, becomes high enough. This AHL is synthesized by the enzyme encoded by the *luxI* gene.

Examples of Quorum Sensing

Various genes are controlled by quorum sensing, including some in pathogenic bacteria. For example, pseudomonads use 4-hydroxyalkyl quinolines as autoinducers to induce genes involved in virulence. In *Pseudomonas aeruginosa*, for instance, quorum

sensing triggers the expression of a large number of unrelated genes when the population density becomes sufficiently high. These genes assist cells of *P. aeruginosa* in the transition from growing freely suspended in liquid to growing in a semisolid matrix called a *biofilm* (↺ Section 23.4). The biofilm, formed by specific polysaccharides produced by *P. aeruginosa*, increases the pathogenicity of this organism and prevents the penetration of antibiotics.

The pathogenesis of *Staphylococcus aureus* (↺ Section 33.9) involves, among many other things, the production and secretion of small extracellular peptides that damage host cells or that interfere with the immune system. The genes encoding these virulence factors are under the control of a quorum-sensing system that uses a small peptide as autoinducer. The regulation of these virulence genes is quite complex and requires a regulatory RNA molecule as well as regulatory proteins that form a signal transduction system.

Quorum sensing also occurs in microbial eukaryotes. For example, in the yeast *Saccharomyces cerevisiae*, specific aromatic

Table 8.2 *Examples of quorum sensing and autoinducers*

Organism	Autoinducer	Receptor	Process regulated
Proteobacteria	Acyl homoserine lactones	LuxR protein	Diverse processes
Many diverse bacteria	AI-2 (furanone ± borate)[a]	LuxQ protein	Diverse processes
Pseudomonads	4-Hydroxyalkyl quinolines	PqsR protein	Virulence; biofilms
Streptomyces	Gamma-butyrolactones	ArpA repressor	Antibiotic synthesis; sporulation
Gram-positive bacteria	Oligopeptides (linear or cyclic)	Two-component systems	Diverse processes
Yeast	Aromatic alcohols	?	Filamentation

[a]The AI-2 autoinducer exists in several slightly different structures, some of which have an attached borate group.

alcohols are produced as autoinducers and control the transition between growth of *S. cerevisiae* as single cells and as elongated filaments. Similar transitions are seen in other fungi, some of which cause disease in humans. An example is *Candida*, whose quorum sensing is mediated by the long-chain alcohol farnesol.

Some eukaryotes produce molecules that interfere with bacterial quorum sensing. Most of those known so far are furanone derivatives with halogens attached. These mimic the AHLs or AI-2 and disrupt bacterial behavior that relies on quorum sensing. Quorum-sensing disruptors have been suggested to have possible future applications in dispersing bacterial biofilms and preventing the expression of virulence genes.

MiniQuiz

- What properties are required for a molecule to function as an autoinducer?
- How do the autoinducers used in quorum sensing by gram-negative bacteria differ from those used by gram-positive bacteria?

8.10 The Stringent Response

Nutrient levels in the natural environments of bacterial cells often change significantly, even if only briefly. Such changing conditions can easily be simulated in the laboratory, and much work has been done with *Escherichia coli* and other bacteria on the regulation of gene expression following a "shift down" or "shift up" in nutrient status. These include, in particular, the regulatory events triggered by starvation for amino acids or energy.

As a result of a shift down from amino acid excess to limitation, as occurs when a culture is transferred from a rich complex medium to a defined medium with a single carbon source, the synthesis of rRNA and tRNA ceases almost immediately (**Figure 8.20a**). No new ribosomes are produced. Protein and DNA synthesis is curtailed, but the biosynthesis of new amino acids is activated. Following such a shift, new proteins must be made to synthesize the amino acids no longer available in the environment; these are made by existing ribosomes. After a while, rRNA synthesis (and hence, the production of new ribosomes) begins again but at a new rate commensurate with the cell's reduced growth rate (Figure 8.20a). This course of events is called the **stringent response** (or stringent control) and is another example of global control.

The stringent response is triggered by a mixture of two regulatory nucleotides, *guanosine tetraphosphate* (ppGpp) and *guanosine pentaphosphate* (pppGpp); this mixture is often written as (p)ppGpp (Figure 8.20b). In *E. coli*, these nucleotides, which are also called *alarmones*, rapidly accumulate during a shift down from amino acid excess to amino acid starvation. Alarmones are synthesized by a specific protein, called RelA, using ATP as a phosphate donor (Figure 8.20b,c). RelA adds two phosphate groups from ATP to GTP or GDP, thus producing pppGpp or ppGpp, respectively. RelA is associated with the 50S subunit of the ribosome and is activated by a signal from the ribosome during amino acid limitation. When the growth of the cell is limited by a shortage of amino acids, the pool of *uncharged* tRNAs increases relative to *charged* tRNAs. Eventually, an uncharged

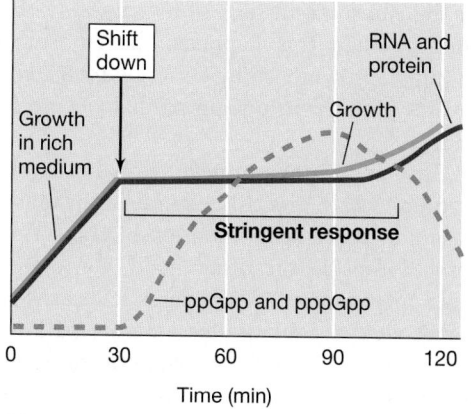

(a)

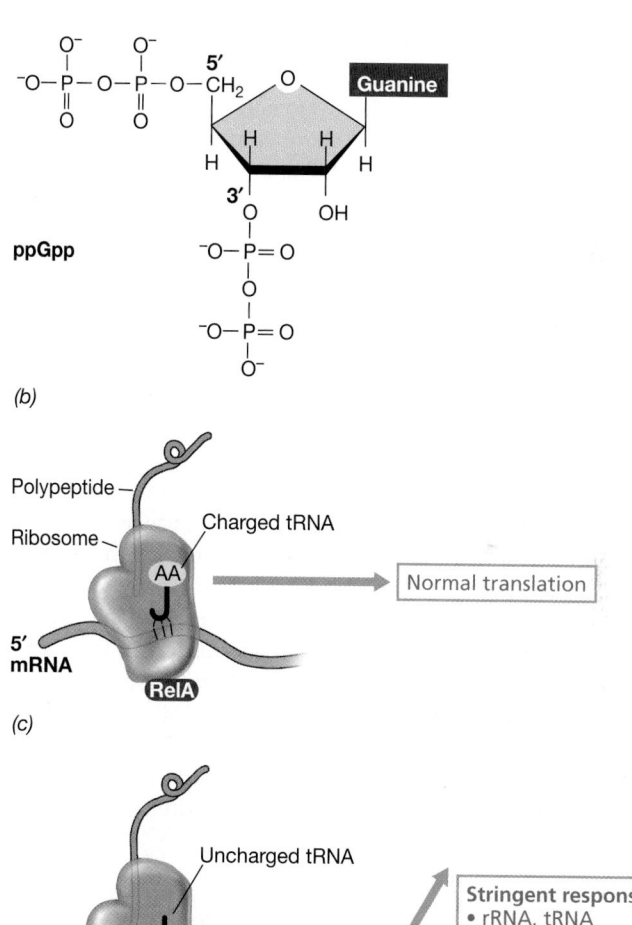

(b)

(c)

(d)

Figure 8.20 The stringent response. *(a)* Upon nutrient downshift, rRNA, tRNA, and protein syntheses temporarily cease. Sometime later, growth resumes at a new (decreased) rate. *(b)* Structure of guanosine tetraphosphate (ppGpp), a trigger of the stringent response. *(c)* Normal translation, which requires charged tRNAs. *(d)* Synthesis of ppGpp. When cells are starved for amino acids, an uncharged tRNA can bind to the ribosome, which stops ribosome activity. This event triggers the RelA protein to synthesize a mixture of pppGpp and ppGpp.

tRNA is inserted into the ribosome instead of a charged tRNA during protein synthesis. When this happens, the ribosome stalls, and this leads to (p)ppGpp synthesis by RelA (Figure 8.20*d*). The protein Gpp converts pppGpp to ppGpp so that ppGpp is the major overall product.

The alarmones ppGpp and pppGpp have global control effects. They strongly inhibit rRNA and tRNA synthesis by binding to RNA polymerase and preventing initiation of transcription of genes for these RNAs. On the other hand, alarmones activate the biosynthetic operons for certain amino acids as well as catabolic operons that yield precursors for amino acid synthesis. By contrast, operons that encode biosynthetic proteins whose amino acid products are present in sufficient amounts remain shut down. The stringent response also inhibits the initiation of new rounds of DNA synthesis and cell division and slows down the synthesis of cell envelope components, such as membrane lipids. Efficient binding of (p)ppGpp to RNA polymerase requires the protein DksA, which is needed to position the (p)ppGpp correctly in the channel that normally allows substrates (that is, nucleoside triphosphates) into the RNA polymerase active site.

In addition to RelA, another protein, SpoT, helps trigger the stringent response. The SpoT protein can either make (p)ppGpp or degrade it. Under most conditions, SpoT is responsible for degrading (p)ppGpp; however, SpoT synthesizes (p)ppGpp in response to certain stresses or when there is a shortage of energy. Thus the stringent response results not only from the absence of precursors for protein synthesis, but also from the lack of energy for biosynthesis.

The stringent response can be thought of as a mechanism for adjusting the cell's biosynthetic machinery to the availability of the required precursors and energy. By so doing, the cell achieves a new balance between anabolism and catabolism. In many natural environments, nutrients appear suddenly and are consumed rapidly. Thus a global mechanism such as the stringent response that balances the metabolic state of a cell with the availability of precursors and energy likely improves its ability to compete in nature.

The RelA/(p)ppGpp system is found only in *Bacteria* and in the chloroplasts of plants. *Archaea* and eukaryotes do not make (p)ppGpp in response to resource shortages. Although *Archaea* display an overall response similar to the stringent response of *Bacteria* when faced with carbon and energy shortages, they use regulatory mechanisms different from those described here to deal with these situations.

MiniQuiz

- Which genes are activated during the stringent response and why?
- Which genes are repressed during the stringent response and why?
- How are the alarmones ppGpp and pppGpp synthesized?

8.11 Other Global Control Networks

Catabolite repression and the stringent response are both examples of global control. There are several other global control systems in *Escherichia coli* (and probably in all prokaryotes), and a few of these are listed in **Table 8.3**. Global control systems regulate many genes comprising more than one regulon (Section 8.4). Global control networks may include activators, repressors, signal molecules, two-component regulatory systems (Section 8.7), regulatory RNA (Sections 8.14 and 8.15), and alternative sigma (σ) factors (⮰ Section 6.13).

An example of a global response that is widespread in all three domains of life is the response to high temperature. In many bacteria this **heat shock response** is largely controlled by alternative σ factors.

Heat Shock Proteins

Most proteins are relatively stable. Once made, they continue to perform their functions and are passed along at cell division. However, some proteins are less stable at elevated temperatures and tend to unfold. Improperly folded proteins are recognized by protease enzymes and are degraded. Consequently, cells that are heat stressed induce the synthesis of a set of proteins, the **heat shock proteins**, that help counteract the damage. Heat shock proteins assist the cell in recovering from stress. They are induced not only by heat, but also by several other stress factors that the cell can encounter. These include exposure to high levels of certain chemicals, such as ethanol, and exposure to high doses of ultraviolet (UV) radiation.

Table 8.3 *Examples of global control systems known in* **Escherichia coli**[a]

System	Signal	Primary activity of regulatory protein	Number of genes regulated
Aerobic respiration	Presence of O_2	Repressor (ArcA)	>50
Anaerobic respiration	Lack of O_2	Activator (FNR)	>70
Catabolite repression	Cyclic AMP level	Activator (CRP)	>300
Heat shock	Temperature	Alternative sigmas (RpoH and RpoE)	36
Nitrogen utilization	NH_3 limitation	Activator (NR$_I$)/alternative sigma RpoN	>12
Oxidative stress	Oxidizing agents	Activator (OxyR)	>30
SOS response	Damaged DNA	Repressor (LexA)	>20

[a]For many of the global control systems, regulation is complex. A single regulatory protein can play more than one role. For instance, the regulatory protein for aerobic respiration is a repressor for many promoters but an activator for others, whereas the regulatory protein for anaerobic respiration is an activator protein for many promoters but a repressor for others. Regulation can also be indirect or require more than one regulatory protein. Many genes are regulated by more than one global system.

In *E. coli* and in most prokaryotes examined, there are three major classes of heat shock protein, Hsp70, Hsp60, and Hsp10. We have encountered these proteins before, although not by these names (↪ Section 6.21). The Hsp70 protein of *E. coli* is DnaK, which prevents aggregation of newly synthesized proteins and stabilizes unfolded proteins. Major representatives of the Hsp60 and Hsp10 families in *E. coli* are the proteins GroEL and GroES, respectively. These are molecular chaperones that catalyze the correct refolding of misfolded proteins. Another class of heat shock proteins includes various proteases that degrade denatured or irreversibly aggregated proteins.

The heat shock proteins are very ancient and highly conserved. Molecular sequencing of heat shock proteins, especially Hsp70, has been used to help unravel the phylogeny of eukaryotes. Heat shock proteins are present in all cells, although the regulatory system that controls their expression varies greatly in different groups of organisms.

Heat Shock Response

In many bacteria, such as *E. coli*, the heat shock response is controlled by the alternative σ factors RpoH (σ^{32}) and RpoE (**Figure 8.21**). RpoH controls expression of heat shock proteins in the cytoplasm, and RpoE regulates the expression of a different set of

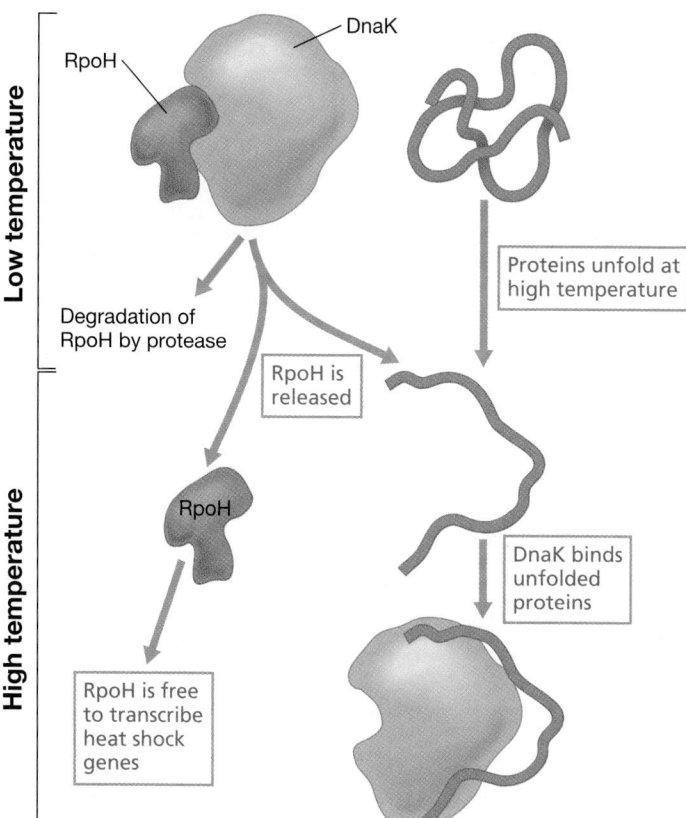

Figure 8.21 Control of heat shock in *Escherichia coli*. The RpoH alternative sigma factor is broken down rapidly by proteases at normal temperatures. This is stimulated by binding of the DnaK chaperonin to RpoH. At high temperatures, some proteins are denatured, and DnaK recognizes and binds to the unfolded polypeptide chains. This removes DnaK from RpoH, which slows the degradation rate. The level of RpoH rises, and the heat shock genes are transcribed.

heat shock proteins in the periplasm and cell envelope. RpoH is normally degraded within a minute or two of its synthesis. However, when cells suffer a heat shock, degradation of RpoH is inhibited and its level therefore increases. Consequently, transcription of those operons whose promoters are recognized by RpoH increases too. The rate of degradation of RpoH depends on the level of free DnaK protein, which inactivates RpoH. In unstressed cells the level of free DnaK is relatively high and the level of intact RpoH is correspondingly low. However, if heat stress unfolds proteins, DnaK binds preferentially to the unfolded proteins and so is no longer free to promote degradation of RpoH. Thus, the more denatured proteins there are, the lower the level of free DnaK and the higher the level of RpoH; the result is heat shock gene expression.

When the stress situation has passed, for example, upon a temperature downshift, RpoH is rapidly inactivated by DnaK and the synthesis of heat shock proteins is greatly reduced. Because heat shock proteins perform vital functions in the cell, there is always a low level of these proteins present, even under optimal conditions. However, the rapid synthesis of heat shock proteins in stressed cells emphasizes how important they are in surviving excessive heat, chemicals, or physical agents. Such stresses can generate large amounts of inactive proteins that need to be refolded (and in the process, reactivated) or degraded to release free amino acids for the synthesis of new proteins.

There is also a heat shock response in *Archaea*, even in species that grow best at very high temperatures. An analog of the bacterial Hsp70 is found in many *Archaea* and is structurally quite similar to those found in gram-positive species of *Bacteria*. Hsp70 is also present in eukaryotes. In addition, other types of heat shock proteins are present in *Archaea* that are unrelated to stress proteins of *Bacteria*.

One problem faced by all cells during cold shock is that RNA, including mRNA, tends to form stable secondary structures, especially stem–loop structures, that may interfere with translation. Cold shock proteins include several RNA-binding proteins. Some of these prevent secondary structure formation and others (RNA helicases) unwind base-paired regions in RNA.

MiniQuiz
- What triggers the heat shock response?
- Why do cells have more than one type of σ factor?
- Why might the proteins induced during heat shock not be needed during cold shock?

Regulation of Development in Model *Bacteria*

Differentiation and development are largely characteristics of multicellular organisms. Because most prokaryotic microorganisms grow as single cells, few show differentiation. Nonetheless, occasional examples among single-celled prokaryotes illustrate the basic principle of differentiation, namely that one cell gives rise to two genetically identical descendants that perform different roles and must therefore express different sets of genes. Here we discuss two well-studied examples, the formation of endospores in the gram-positive bacterium *Bacillus* and the formation of two cell types, motile and stationary, in the gram-negative bacterium *Caulobacter*.

Although forming just two different cell types may seem superficially simple, the regulatory systems that control these processes are highly complex. There are three major phases for the regulation of differentiation: (1) triggering the response, (2) asymmetric development of two sister cells, and (3) reciprocal communication between the two differentiating cells.

8.12 Sporulation in *Bacillus*

Many microorganisms, both prokaryotic and eukaryotic, respond to adverse conditions by forming spores (⮌ Section 3.12). Once favorable conditions return, the spore germinates and the microorganism returns to its normal lifestyle. Among the *Bacteria*, the genus *Bacillus* is well known for the formation of endospores, that is, spores formed inside a mother cell. Prior to endospore formation, the cell divides asymmetrically. The smaller cell develops into the endospore, which is surrounded by the larger mother cell. Once development is complete, the mother cell bursts, releasing the endospore.

Endospore formation in *Bacillus subtilis* is triggered by unfavorable conditions, such as starvation, desiccation, or growth-inhibitory temperatures. Multiple aspects of the environment are monitored by a group of five sensor kinases. These function via a phosphotransfer relay system whose mechanism resembles that of a two-component regulatory system (Section 8.7), but is

considerably more complex (**Figure 8.22**). The net result of multiple adverse conditions is the successive phosphorylation of several proteins called *sporulation factors*, culminating with sporulation factor Spo0A. When Spo0A is highly phosphorylated, sporulation proceeds. Spo0A controls the expression of several genes. The product of one of these, SpoIIE, is responsible for removing the phosphate from SpoIIAA. This allows SpoIIAA in turn to remove the anti-sigma factor, SpoIIAB, and liberate the σ factor, σF, as discussed below.

Once triggered, endospore development is controlled by four different σ factors, two of which, σF and σG, activate genes needed inside the developing endospore itself, and two of which, σE and σK, activate genes needed in the mother cell surrounding the endospore (Figure 8.22b). The sporulation signal, transmitted via Spo0A, activates σF in the smaller cell that is destined to become the endospore. σF is already present, but is inactive, as it is bound by an anti-σ factor. The signal from Spo0A activates a protein that binds to the anti-σ factor and inactivates it, so liberating σF. Once free, σF binds to RNA polymerase and promotes transcription (inside the spore) of genes whose products are needed for the next stage of sporulation. These include the gene for the sigma factor σG and the genes for proteins that cross into the mother cell and activate σE. Active σE is required for transcription inside the mother cell of yet more genes, including the gene for σK. The sigma factors σG (in the endospore) and σK (in the mother cell) are required for transcription of genes needed even later in the sporulation process.

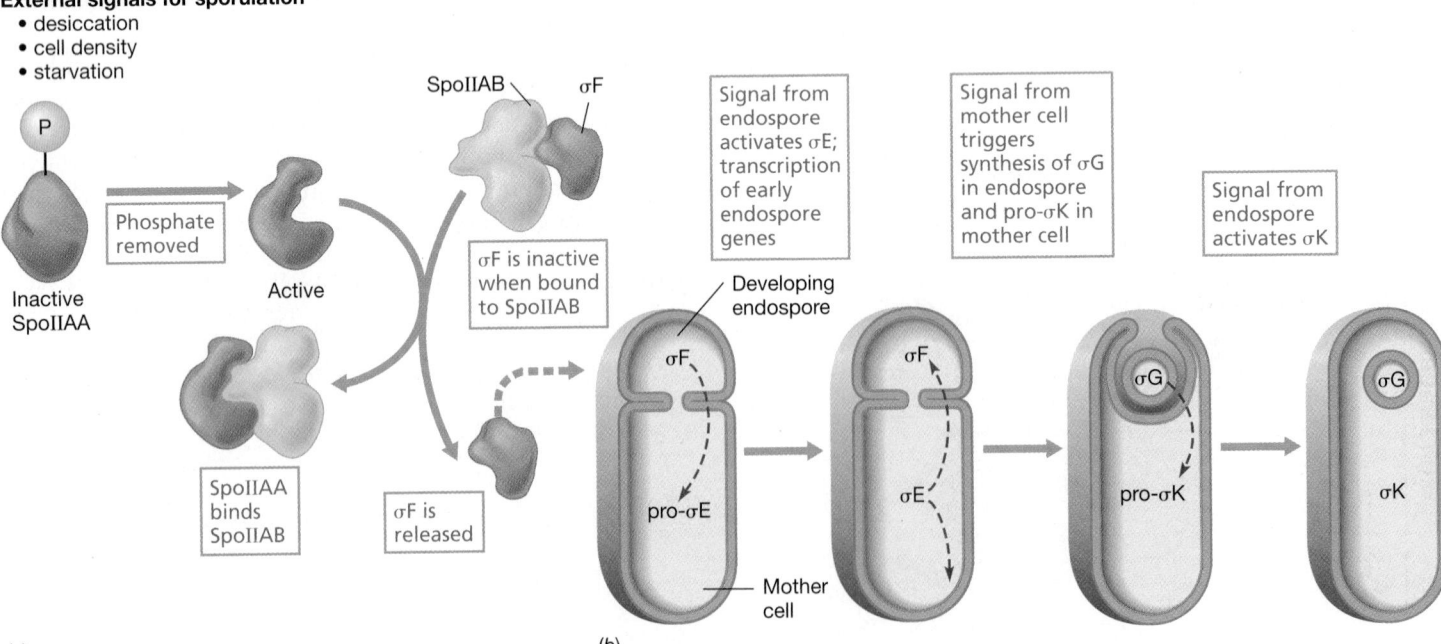

External signals for sporulation
- desiccation
- cell density
- starvation

Figure 8.22 Control of endospore formation in *Bacillus*. After an external signal is received, a cascade of sigma (σ) factors controls differentiation. *(a)* Active SpoIIAA binds the anti-σ factor SpoIIAB, thus liberating the first σ factor, σF. *(b)* σF initiates a cascade of sigma factors, some of which already exist and need to be activated, others of which are not yet present and whose genes must be expressed. These σ factors then promote transcription of genes needed for endospore development.

One fascinating aspect of endospore formation is that it is preceded by what is in effect cellular cannibalism. Those cells in which Spo0A has already become activated secrete a protein that lyses nearby cells of the same species whose Spo0A protein has not yet become activated. This toxic protein is accompanied by a second protein that delays sporulation of neighboring cells. Cells committed to sporulation also make an antitoxin protein to protect themselves against the effects of their own toxin. Their sacrificed sister cells are used as a source of nutrients for developing endospores. Shortages of certain nutrients, such as phosphate, increase the expression level of the toxin-encoding gene.

MiniQuiz

- How are different sets of genes expressed in the developing endospore and the mother cell?
- What is an anti-σ factor and how can its effect be overcome?

8.13 *Caulobacter* Differentiation

Caulobacter provides another example in which a cell divides into two genetically identical daughter cells that perform different roles and express different sets of genes. *Caulobacter* is a species of *Proteobacteria* that is common in aquatic environments, typically in waters that are nutrient-poor (↻ Section 17.16). In the *Caulobacter* life cycle, free-swimming (swarmer) cells alternate with cells that lack flagella and are attached to surfaces by a stalk with a holdfast at its end. The role of the swarmer cells is dispersal, as swarmers cannot divide or replicate their DNA. Conversely, the role of the stalked cell is reproduction.

The *Caulobacter* cell cycle is controlled by three major regulatory proteins whose concentrations oscillate in succession (**Figure 8.23**). Two of these are the transcriptional regulators, GcrA and CtrA. The third is DnaA, a protein that functions both in its normal role in initiating DNA replication and also as a transcriptional regulator. Each of these regulators is active at a specific stage in the cell cycle, and each controls many other genes that are needed at that particular stage in the cycle.

CtrA is activated by phosphorylation in response to external signals. Once phosphorylated, CtrA-P activates genes needed for the synthesis of the flagella and other functions in swarmer cells. Conversely, CtrA-P represses the synthesis of GcrA and also inhibits the initiation of DNA replication by binding to and blocking the origin of replication (Figure 8.23). As the cell cycle proceeds, CtrA is degraded by a specific protease; as a consequence, levels of DnaA rise. The absence of CtrA-P allows access to the chromosomal origin of replication, and, as in all *Bacteria*, DnaA binds to the origin and triggers the initiation of DNA replication (↻ Section 6.9). In addition, in *Caulobacter* DnaA activates several other genes needed for chromosomal replication. The level of DnaA then falls due to protease degradation, and the level of GcrA rises. The GcrA regulator promotes the elongation phase of chromosome replication, cell division, and the growth of the stalk on the immobile daughter cell. Eventually, GcrA levels fall and high levels of CtrA reappear (in the daughter cell destined to swim away) (Figure 8.23).

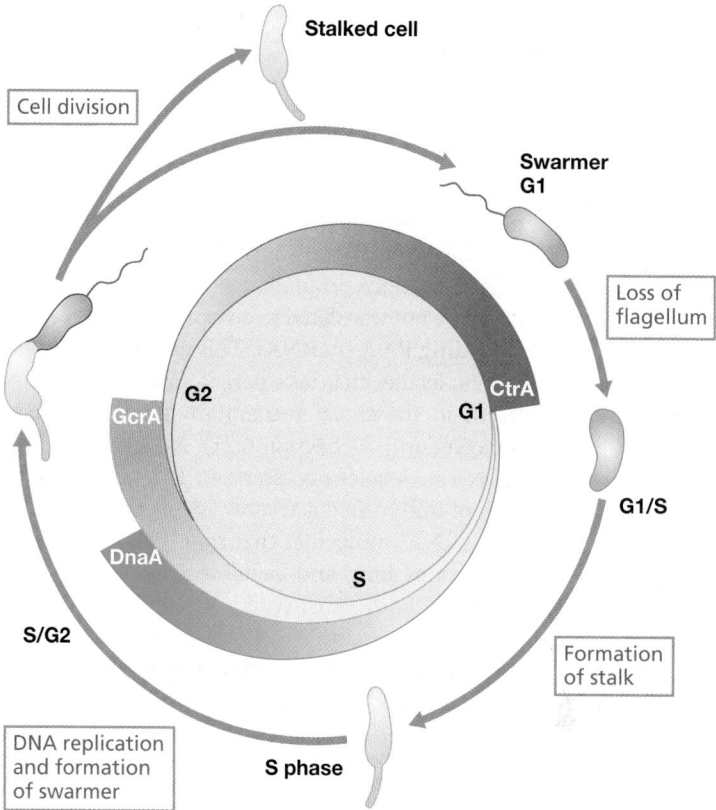

Figure 8.23 Cell cycle regulation in *Caulobacter*. Three global regulators, CtrA, DnaA, and GcrA, oscillate in levels through the cycle as shown. G1 and G2 are the two growth phases and S is the synthesis (of DNA) phase. In G1 swarmer cells, CtrA represses initiation of DNA replication and expression of GcrA. At the G1/S transition, CtrA is degraded and DnaA levels rise. DnaA binds to the origin of replication and initiates replication. GcrA also rises and activates genes for cell division and DNA synthesis. At the S/G2 transition, CtrA levels begin to rise again and shut down GcrA expression. GcrA levels slowly decline in the stalked cell but are rapidly degraded in the swarmer. CtrA is degraded in the stalked cell.

Many of the details of the regulation of the *Caulobacter* cell cycle are still uncertain. Both external stimuli and internal factors such as nutrient and metabolite levels affect the cycle, but how this information is integrated into the overall control system is only partly understood. However, since its genome has been sequenced and good genetic systems for gene transfer and analysis are available, differentiation in *Caulobacter* has been used as a model system for studying cell developmental processes.

MiniQuiz

- Why are the levels of DnaA protein controlled during the *Caulobacter* cell cycle?
- When do the regulators CtrA and GcrA carry out their main roles during the *Caulobacter* life cycle?

Ⓥ RNA-Based Regulation

Thus far we have focused on mechanisms in which regulatory proteins sense signals or bind to DNA. In some cases a single protein does both; in other cases separate proteins carry out these two activities. Nonetheless, all of these mechanisms rely on *regulatory proteins*. However, RNA itself may regulate gene expression, both at the level of transcription and the level of translation of mRNA to produce proteins.

RNA molecules that are not translated to give proteins are collectively known as **noncoding RNA (ncRNA)**. This category includes the rRNA and tRNA molecules that take part in protein synthesis and the RNA present in the signal recognition particle that is involved in protein secretion (⮌ Section 6.21). Noncoding RNA also includes small RNA molecules necessary for RNA processing, especially the splicing of mRNA in eukaryotes (⮌ Section 7.8). In addition, small RNA (sRNA) molecules that range from approximately 40–400 nucleotides long and regulate gene expression are widely distributed in both prokaryotes and eukaryotes. In *Escherichia coli*, for example, a number of sRNA molecules have been found to regulate various aspects of cell physiology by binding to other RNAs or in some cases even to other small molecules.

8.14 RNA Regulation and Antisense RNA

The most frequent way in which regulatory RNA molecules exert their effects is by base pairing with other RNA molecules, usually mRNA, that have regions of complementary sequence. These double-stranded regions tie up the mRNA and prevent its translation (**Figure 8.24**). Small RNAs (sRNAs) that show this activity are called *antisense RNA*, because the sRNA has a sequence complementary to the coding sense of the mRNA.

Antisense RNAs bind to their mRNA complements to form double-stranded RNAs that cannot be translated and are soon degraded by specific ribonucleases (Figure 8.24). This removes the mRNA and thus prevents the synthesis of new protein molecules from mRNAs already present in the cell but whose gene products are no longer needed because of a change in conditions.

Theoretically, antisense RNA could be made by transcribing the nontemplate strand of the same gene that yielded the target mRNA. Instead, a distinct "anti-gene" is used to form the antisense RNA. Only a relatively short piece of antisense RNA is needed to block transcription of mRNA, and therefore the "anti-gene" that encodes the antisense RNA is much shorter than the gene that encodes the original message. Typically, antisense RNAs

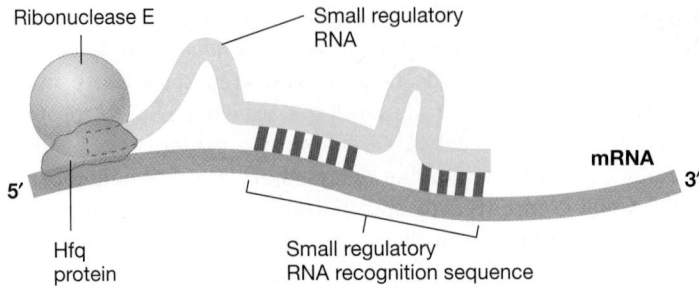

Figure 8.25 **The RNA chaperone Hfq holds RNAs together.** Binding of antisense RNA to mRNA often requires the Hfq protein. Antisense RNA molecules usually have several stem–loop structures. One consequence is that the complementary base sequence that recognizes the mRNA is noncontiguous. The antisense RNA blocks the ribosome-binding site on the mRNA and prevents its translation. Ribonuclease E, also bound by Hfq, then begins to degrade the mRNA.

are around 100 nucleotides long and bind to a target region approximately 30 nucleotides long. In addition, each antisense RNA can usually regulate several different mRNAs, all of which share the same target sequence for antisense RNA binding.

Transcription of antisense RNA is enhanced under conditions in which its target genes need to be turned off. For example, the RyhB antisense RNA of *Escherichia coli* is transcribed when iron is limiting for growth. RyhB antisense RNA binds to a dozen or more target mRNAs that encode proteins needed for iron metabolism or that use iron as cofactors. Binding of RyhB antisense RNA blocks translation of the mRNA. The base-paired RyhB/mRNA molecules are then degraded by ribonucleases, in particular, ribonuclease E. This forms part of the mechanism by which *E. coli* and related bacteria respond to a shortage of iron. Other responses to iron limitation in *E. coli* include transcriptional controls involving repressor and activator proteins (Sections 8.3 and 8.4) that increase the capacity of cells to take up iron and to tap into intracellular iron stores.

The binding of many antisense RNAs to their targets depends on a small protein, Hfq, that binds not only to both RNA molecules but also to ribonuclease E (**Figure 8.25**). Hfq protein forms hexameric rings with RNA-binding sites on both surfaces. Hfq and similar proteins are known as *RNA chaperones*, as they help small RNA molecules, including many antisense RNAs, maintain their correct structure.

Although antisense RNA usually blocks translation of mRNA, occasional examples are known in which antisense RNA does

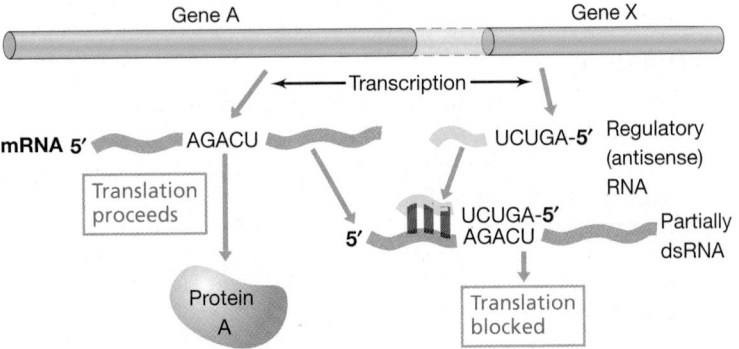

Figure 8.24 **Regulation by antisense RNA.** Gene A is transcribed from its promoter to yield an mRNA that can be translated to form protein A. Gene X is a small gene with a sequence identical to that of part of Gene A but with its promoter at the opposite end. Therefore, if it is transcribed, the resulting antisense RNA will be complementary to the mRNA of gene A. If these two RNAs base-pair, forming double-stranded RNA, translation will be blocked.

The CRISPR Antiviral Defense System

Ever since RNA interference (RNAi) was discovered in eukaryotes (Section 7.10), scientists have wondered whether bacteria have an equivalent system to protect themselves against virus attack. Recent discoveries have revealed that although bacteria do not have RNAi, they do have another RNA-based defense program to destroy invading virus genomes. In fact, the bacterial CRISPR system tackles both RNA and DNA viruses, unlike RNAi, which only works against RNA viruses.

CRISPR stands for Clustered Regularly Interspaced Short Palindromic Repeats. The CRISPR region on the bacterial chromosome is essentially a memory bank of hostile virus sequences. It consists of many different segments of virus sequence alternating with identical repeated sequences (Figure 1). The CRISPR system provides resistance to any viruses that contain the same or very closely related sequences.

The proteins of the CRISPR system (CRISPR associated proteins, or CAS proteins) perform two roles. Some use the stored sequence information to recognize intruding virus genomes and destroy them. Others are involved in obtaining and storing segments of virus sequence, a process that remains obscure. The CAS proteins are encoded by genes that lie upstream of the CRISPR sequences (Figure 1).

The CRISPR region is transcribed as a whole into a long RNA molecule that is then cleaved by CAS proteins in the middle of each of the repeated sequences. This converts it into individual virus-specific segments. If one of these segments base-pairs with the nucleic acid of an invading virus,

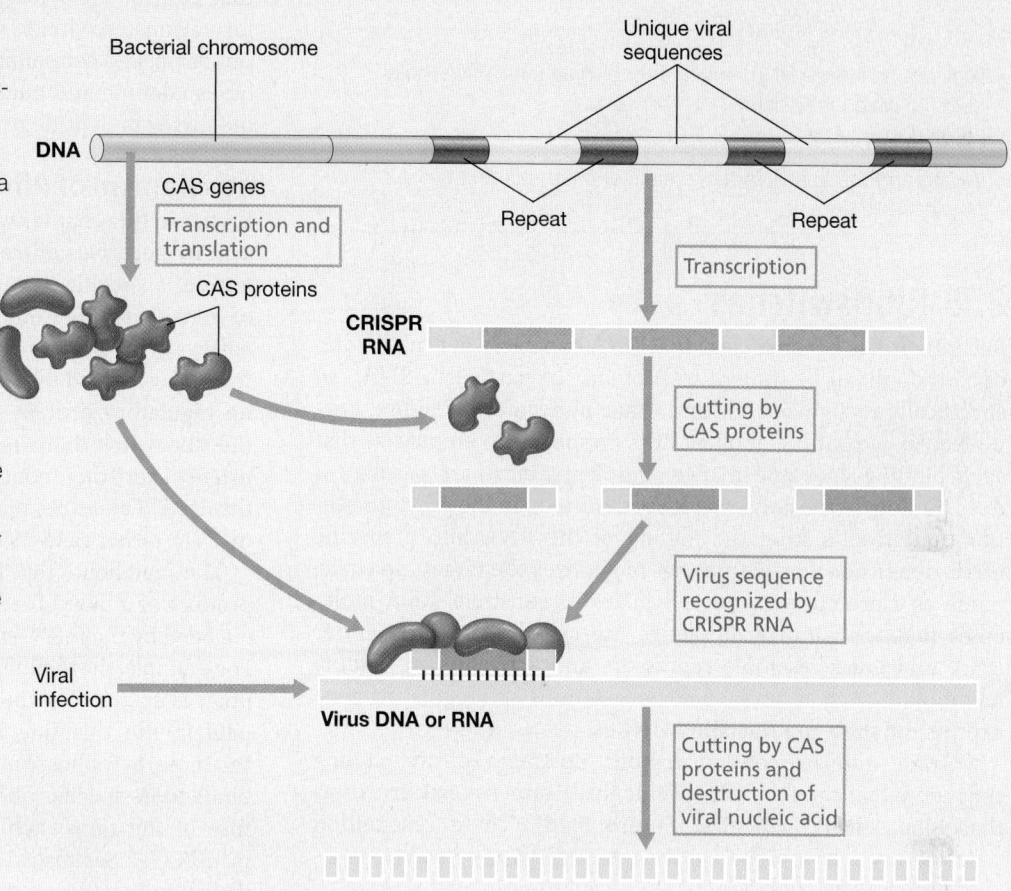

Figure 1 **Operation of the CRISPR system.** The CRISPR region on the bacterial chromosome is transcribed into a long RNA molecule that is then cut into segments by some of the CAS proteins. Each segment carries a single virus-specific sequence. If one of these short CRISPR RNA molecules recognizes a virus nucleic acid by base pairing, other CAS proteins destroy the virus DNA or RNA.

then the virus DNA or RNA is destroyed by other CAS proteins.

The CRISPR system is widely distributed in both *Archaea* and *Bacteria*. Approximately 90% of the sequenced genomes of *Archaea* and 70% of those of *Bacteria* possess the CRISPR system. However, many occurrences of the CRISPR system detected by genomic sequencing appear to be incomplete or defective.

just the opposite and actually enhances the translation of its target mRNA. It is hypothesized that in these cases the native mRNA forms a secondary structure that prevents translation. The antisense RNA is thought to bind to a short region of the mRNA and unfold it, thereby allowing access to the ribosome.

Antisense RNA does not always work via an effect on mRNA. For example, the replication of the high copy number plasmid ColE1 is regulated by an sRNA that primes DNA synthesis and its antisense partner that blocks initiation of DNA synthesis. The level of the antisense RNA determines how often replication is initiated.

Regulation by antisense RNA usually modulates the expression of genes that are also controlled by other systems, and many complex examples are known in higher organisms. For example, in the mold *Neurospora* the time of day controls growth via a complex mechanism that uses antisense RNA. The levels of sense and antisense transcripts for a biological-clock gene are found to cycle out of step with each other in response to day and night cycles.

Fragments of antisense RNA are also used to detect and destroy viral intruders in the CRISPR defense mechanism (⟳ Microbial Sidebar, "The CRISPR Antiviral Defense System").

MiniQuiz

- Why are antisense RNAs much shorter than the mRNA molecules to which they bind?
- How do cells synthesize antisense RNA molecules?
- What happens to mRNA molecules following binding of their antisense RNAs?

8.15 Riboswitches

Recently it has become clear that RNA can carry out many roles once thought to be limited to proteins. In particular, RNA can specifically recognize and bind other molecules, including low-molecular-weight metabolites. It is important to emphasize that such binding does not involve complementary base pairing (as does binding of the antisense RNA described in the previous section) but results from the folding of the RNA into a specific three-dimensional structure that recognizes the target molecule, much as a protein enzyme recognizes its substrate. RNA molecules that are catalytically active are called ribozymes. Other RNA molecules resemble repressors and activators in binding metabolites such as amino acids or vitamins and regulating gene expression; these are the **riboswitches**.

Certain mRNAs contain regions upstream of the coding sequences that can fold into specific three-dimensional structures that bind small molecules (**Figure 8.26**). These recognition

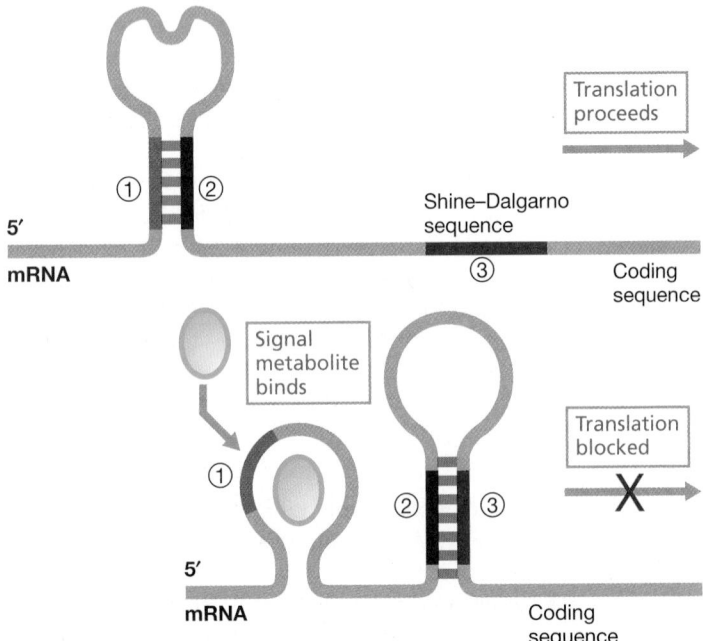

Figure 8.26 **Regulation by a riboswitch.** Binding of a specific metabolite alters the secondary structure of the riboswitch domain, which is located in the 5′ untranslated region of the mRNA, preventing translation. The Shine–Dalgarno site is where the ribosome binds the RNA.

domains are riboswitches and exist as two alternative structures, one with the small molecule bound and the other without. Alternation between the two forms of the riboswitch thus depends on the presence or absence of the small molecule, which in turn controls expression of the mRNA. Riboswitches have been found that control the synthesis of enzymes in biosynthetic pathways for various enzymatic cofactors, such as the vitamins thiamine, riboflavin, and cobalamin (B_{12}), for a few amino acids, for the purine bases adenine and guanine, and for glucosamine 6-phosphate, a precursor in peptidoglycan synthesis.

Mechanism of Riboswitches

Earlier in this chapter we discussed the regulation of gene expression by negative control of transcription (Section 8.3). The presence of a specific metabolite often shuts down the transcription of genes encoding enzymes for the corresponding biosynthetic pathway. In our example of the arginine biosynthetic pathway this is performed by a protein repressor. In a riboswitch, there is no regulatory protein. Instead, the metabolite binds directly to the riboswitch domain at the 5′ end of the mRNA. Riboswitches usually exert their control after the mRNA has already been synthesized. Therefore, most riboswitches control *translation* of the mRNA, rather than its *transcription*.

The metabolite that is bound by the riboswitch is typically the product of a biosynthetic pathway whose constituent enzymes are encoded by the mRNAs that carry the corresponding riboswitches. For example, the thiamine riboswitch that binds thiamine pyrophosphate is upstream of the coding sequences for enzymes that participate in the thiamine biosynthetic pathway. When the pool of thiamine pyrophosphate is sufficient in the cell, this metabolite binds to its specific riboswitch mRNA. The new secondary structure of the riboswitch blocks the ribosome-binding site on the mRNA (⟳ Section 6.19) and prevents the mRNA from binding to the ribosome; this prevents translation (Figure 8.26). If the concentration of thiamine pyrophosphate drops sufficiently low, this molecule can dissociate from its riboswitch mRNA. This unfolds the mRNA and exposes the ribosome-binding site, allowing the mRNA to bind to the ribosome and be translated.

The thiamine analog pyrithiamine blocks the synthesis of thiamine and, hence, inhibits bacterial growth. Until the discovery of riboswitches, the site of action of pyrithiamine remained mysterious. It now appears that pyrithiamine is converted by cells to pyrithiamine pyrophosphate, which then binds to the thiamine riboswitch. Thus the biosynthetic pathway is shut off even when no thiamine is available. Bacterial mutants selected for resistance to pyrithiamine have alterations in the sequence of the riboswitch that result in failure to bind both pyrithiamine pyrophosphate and thiamine pyrophosphate.

In *Bacillus subtilis*, where about 2% of the genes are under riboswitch control, the same riboswitch is present on several mRNAs that together encode the proteins for a particular pathway. For example, over a dozen genes in six operons are controlled by the thiamine riboswitch.

Despite being part of the mRNA, some riboswitches nevertheless do control transcription. The mechanism is similar to that seen in attenuation (Section 8.16) where a conformational change in the riboswitch causes premature termination of the synthesis of the mRNA that carries it.

Riboswitches and Evolution

How widespread are riboswitches and how did they evolve? Thus far riboswitches have been found only in some bacteria and a few plants and fungi. Some scientists believe that riboswitches are remnants of the RNA world, a period eons ago before cells, DNA, and protein, when it is hypothesized that catalytic RNAs were the only self-replicating life forms (Section 16.2). In such an environment, riboswitches may have been a primitive mechanism of metabolic control—a simple means by which RNA life forms could have controlled the synthesis of other RNAs. As proteins evolved, riboswitches might have been the first control mechanisms for their synthesis as well. If this is true, the riboswitches that remain today may be the last vestiges of this simple form of control because, as we have seen in this chapter, metabolic regulation is almost exclusively carried out by way of regulatory proteins.

MiniQuiz

- What happens when a riboswitch binds the small metabolite that regulates it?
- What are the major differences between using a repressor protein versus a riboswitch to control gene expression?

8.16 Attenuation

Attenuation is a form of transcriptional control that functions by premature termination of mRNA synthesis. That is, in attenuation, control is exerted *after* the initiation of transcription, but *before* its completion. Consequently, the number of completed transcripts from an operon is reduced, even though the number of initiated transcripts is not.

The basic principle of attenuation is that the first part of the mRNA to be made, called the *leader region*, can fold up into two alternative secondary structures. In this respect, the mechanism of attenuation resembles that of riboswitches. In attenuation, one mRNA secondary structure allows continued synthesis of the mRNA, whereas the other secondary structure causes premature termination. Folding of the mRNA depends either on events at the ribosome or on the activity of regulatory proteins, depending on the organism. The best examples of attenuation are the regulation of genes controlling the biosynthesis of certain amino acids in gram-negative *Bacteria*. The first to be described was in the tryptophan operon in *Escherichia coli*, and we focus on it here. Attenuation control has been documented in several other species of *Bacteria*, and genomic analyses of *Archaea* suggest that the mechanism is present in this domain as well. However, because the processes of transcription and translation are spatially separated in eukaryotes, attenuation control is absent from *Eukarya*.

Attenuation and the Tryptophan Operon

The tryptophan operon contains structural genes for five proteins of the tryptophan biosynthetic pathway plus the usual promoter and regulatory sequences at the beginning of the operon (**Figure 8.27**). Like many operons, the tryptophan operon has more than one type of regulation. The first enzyme in the pathway, anthranilate synthase (a multi-subunit enzyme encoded by *trpD* and *trpE*), is subject to feedback inhibition by tryptophan (Section 4.16). Transcription of the entire tryptophan operon

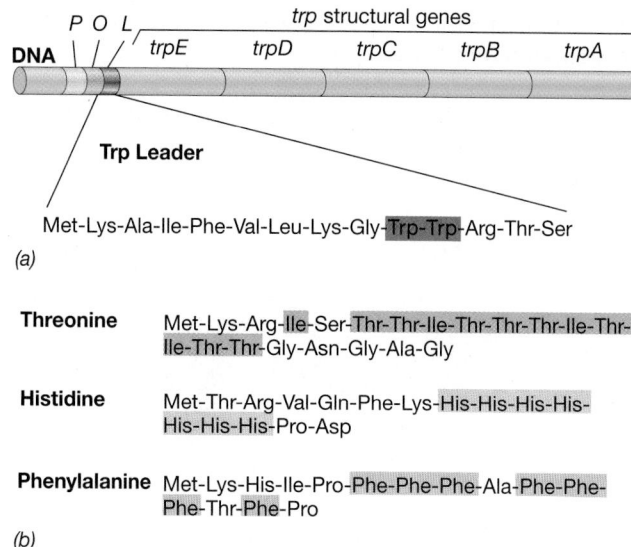

Figure 8.27 Attenuation and leader peptides in *Escherichia coli*. Structure of the tryptophan (*trp*) operon and of the tryptophan leader peptide and other leader peptides in *E. coli*. (a) Arrangement of the *trp* operon. Note that the leader (*L*) encodes a short peptide containing two tryptophan residues near its terminus (there is a stop codon following the Ser codon). The promoter is labeled *P*, and the operator is labeled *O*. The genes labeled *trpE* through *trpA* encode the enzymes needed for tryptophan synthesis. (b) Amino acid sequences of leader peptides of some other amino acid synthetic operons. Because isoleucine is made from threonine, it is an important constituent of the threonine leader peptide.

is also under negative control (Section 8.3). However, in addition to the promoter and operator regions needed for negative control, there is a sequence in the operon called the *leader sequence* that encodes a short polypeptide, the *leader peptide*. The leader sequence contains tandem tryptophan codons near its terminus and functions as an attenuator (Figure 8.27).

The basis of control of the tryptophan attenuator is as follows. If tryptophan is plentiful in the cell, there will be plenty of charged tryptophan tRNAs and the leader peptide will be synthesized. Synthesis of the leader peptide results in termination of transcription of the remainder of the *trp* operon, which includes the structural genes for the biosynthetic enzymes. On the other hand, if tryptophan is scarce, the tryptophan-rich leader peptide will not be synthesized. If synthesis of the leader peptide is halted by a lack of tryptophan, the rest of the operon is transcribed. www.microbiologyplace.com **Online Tutorial 8.2: Attenuation and the Tryptophan Operon**

Mechanism of Attenuation

How does translation of the leader peptide regulate transcription of the tryptophan genes downstream? Consider that in prokaryotic cells transcription and translation are simultaneous processes; as mRNA is released from the DNA, the ribosome binds to it and translation begins (Section 6.19). That is, while transcription of downstream DNA sequences is still proceeding, translation of already transcribed sequences is under way (**Figure 8.28**).

Transcription is attenuated because a portion of the newly formed mRNA folds into a unique stem–loop that inhibits RNA polymerase activity. The stem–loop structure forms in the mRNA

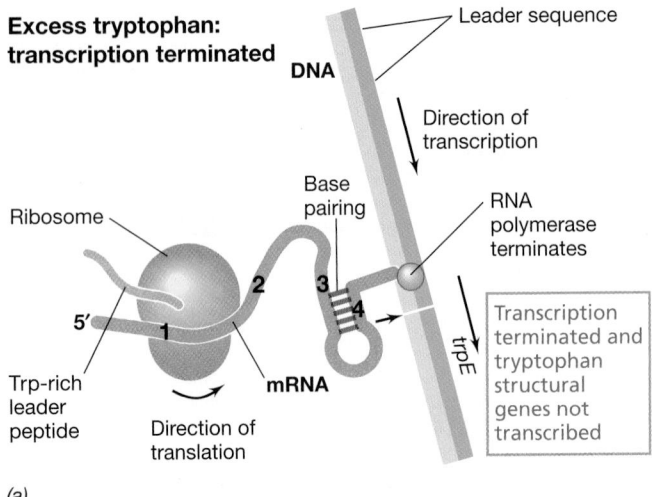

Excess tryptophan: transcription terminated

(a)

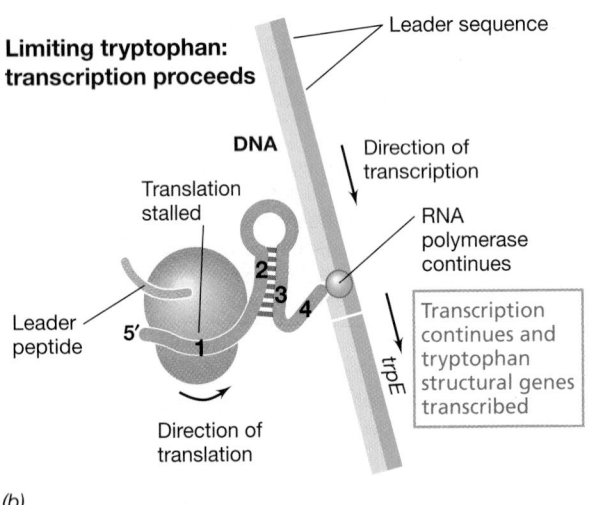

Limiting tryptophan: transcription proceeds

(b)

Figure 8.28 Mechanism of attenuation. Control of transcription of tryptophan (*trp*) operon structural genes by attenuation in *Escherichia coli*. The leader peptide is encoded by regions 1 and 2 of the mRNA. Two regions of the growing mRNA chain are able to form double-stranded loops, shown as 3:4 and 2:3. *(a)* When there is excess tryptophan, the ribosome translates the complete leader peptide, and so region 2 cannot pair with region 3. Regions 3 and 4 then pair to form a loop that terminates transcription. *(b)* If translation is stalled because of tryptophan starvation, a loop forms by pairing of region 2 with region 3, loop 3:4 does not form, and transcription proceeds past the leader sequence.

because two stretches of nucleotides near each other are complementary and can thus base-pair. If tryptophan is plentiful, the ribosome translates the leader sequence until it comes to the leader stop codon. The remainder of the leader sequence then forms a stem–loop, a transcription pause site, which is followed by a uracil-rich sequence that actually causes termination (Figure 8.28*a*).

If tryptophan is in short supply, transcription of genes encoding tryptophan biosynthetic enzymes is obviously desirable. During transcription of the leader, the ribosome pauses at a tryptophan codon because of a shortage of charged tryptophan tRNAs. The presence of the stalled ribosome at this position allows a stem–loop to form (sites 2 and 3 in Figure 8.28*b*) that differs from the terminator stem–loop. This alternative stem–loop is not a transcription termination signal. Instead, it prevents the terminator stem–loop

(sites 3 and 4 in Figure 8.28*a*) from forming. This allows RNA polymerase to move past the termination site and begin transcription of tryptophan structural genes. Thus, in attenuation control, the rate of transcription is influenced by the rate of translation.

Attenuation also occurs in *Escherichia coli* in the biosynthetic pathways for histidine, threonine–isoleucine, phenylalanine, and several other amino acids and essential metabolites. As shown in Figure 8.27*b*, the leader peptide for each of these amino acid biosynthetic operons is rich in that particular amino acid. The *his* operon is dramatic in this regard because its leader peptide contains seven histidines in a row near the end of the peptide (Figure 8.27*b*). This long stretch of histidines gives attenuation a major effect in regulation, which may compensate for the fact that unlike the *trp* operon, the *his* operon in *E. coli* is not also under negative control by a protein repressor.

Translation-Independent Attenuation Mechanisms

Gram-positive *Bacteria*, such as *Bacillus*, also use attenuation of transcription to regulate certain amino acid biosynthetic operons. And, as in gram-negative *Bacteria*, the mechanism relies on alternative mRNA secondary structures, which in one configuration lead to termination. However, the mechanism is independent of translation and requires an RNA-binding protein.

In the *Bacillus subtilis* tryptophan operon, the binding protein is called the *trp* attenuation protein. In the presence of sufficient amounts of the amino acid tryptophan, this regulatory protein binds to the leader sequence in the mRNA and causes transcription termination. By contrast, if tryptophan is limiting, the protein does not bind to the leader sequence. This allows the favorable secondary structure to form and transcription proceeds.

Attenuation also occurs with genes unrelated to amino acid biosynthesis. These mechanisms obviously do not rely on amino acid levels. Some of the operons for pyrimidine biosynthesis (the *pyr* operons) in *E. coli* are regulated by attenuation, and the same is true for *Bacillus*. The mechanisms in the two organisms are, however, quite different, although each employs a system to assess the level of pyrimidines in the cell. In *E. coli* the mechanism monitors the rate of transcription, not translation. If pyrimidines are plentiful, RNA polymerase moves along and transcribes the leader DNA at a normal rate; this allows a terminator stem–loop to form in the mRNA. By contrast, if pyrimidines are scarce, the RNA polymerase pauses at pyrimidine-rich sequences, which leads to formation of a nonterminator stem–loop that allows further transcription.

In *Bacillus*, a different mechanism is employed. For *pyr* attenuation, an RNA-binding protein controls the alternative stem–loop structures of the *pyr* mRNA, terminating transcription when pyrimidines are in excess. In this way the cell can maintain levels of pyrimidines, compounds that require significant cell resources to biosynthesize (⮌ Section 4.14), at levels needed to balance biosynthetic needs.

MiniQuiz

- Explain how the formation of one stem–loop in the RNA can block the formation of another.
- How does attenuation of the tryptophan operon differ between *Escherichia coli* and *Bacillus subtilis*?

Big Ideas

8.1

Most genes encode proteins and most proteins are enzymes. Expression of an enzyme-encoding gene is regulated by controlling the activity of the enzyme or controlling the amount of enzyme produced.

8.2

Certain proteins bind to DNA when specific domains of the proteins bind to specific regions of the DNA molecule. In most cases the interactions are sequence-specific. Proteins that bind to DNA are often regulatory proteins that affect gene expression.

8.3

The amount of a specific enzyme in the cell can be controlled by increasing (inducing) or decreasing (repressing) the amount of messenger RNA that encodes the enzyme. This transcriptional regulation is carried out by allosteric regulatory proteins that bind to DNA. In negative control of transcription, the regulatory protein is called a repressor and it functions by inhibiting mRNA synthesis.

8.4

Positive regulators of transcription are called activator proteins. They bind to activator-binding sites on the DNA and stimulate transcription. Inducers modify the activity of activating proteins. In positive control of enzyme induction, the inducer promotes the binding of the activator protein and thus stimulates transcription.

8.5

Global control systems regulate the expression of many genes simultaneously. Catabolite repression is a global control system that helps cells make the most efficient use of available carbon sources. The *lac* operon is under the control of catabolite repression as well as its own specific negative regulatory system.

8.6

Archaea resemble *Bacteria* in using DNA-binding activator and repressor proteins to regulate gene expression at the level of transcription.

8.7

Signal transduction systems transmit environmental signals to the cell. In prokaryotes, signal transduction is typically carried out by a two-component regulatory system that includes a membrane-integrated sensor kinase and a cytoplasmic response regulator. The activity of the response regulator depends on its state of phosphorylation.

8.8

Chemotactic behavior responds in a complex manner to attractants and repellents. The regulation of chemotaxis affects the activity of proteins rather than their synthesis. Adaptation by methylation allows the system to reset itself to the continued presence of a signal.

8.9

Quorum sensing allows cells to monitor their environment for cells of their own kind. Quorum sensing depends on the sharing of specific small molecules known as autoinducers. Once a sufficient concentration of the autoinducer is present, specific gene expression is triggered.

8.10

The stringent response is a global control mechanism triggered by amino acid starvation. The alarmones ppGpp and pppGpp are produced by RelA, a protein that monitors ribosome activity. Within the cell the stringent response achieves balance between protein production and amino acid requirements.

8.11

Cells can control sets of genes by employing alternative sigma factors. These recognize only certain promoters and thus allow transcription of a select category of genes that is appropriate under certain environmental conditions. Cells respond to both heat and cold by expressing sets of genes whose products help the cell overcome stress.

8.12

Sporulation in *Bacillus* during adverse conditions is triggered via a complex phosphotransfer relay system that monitors multiple aspects of the environment. The sporulation factor Spo0A then sets in motion a cascade of regulatory responses under the control of several alternative sigma factors.

8.13

Differentiation in *Caulobacter* consists of the alternation between motile cells and those that are attached to surfaces. Three major regulatory proteins—CtrA, GcrA, and DnaA—act in succession to control the three phases of the cell cycle. Each in turn controls many other genes needed at specific times in the cell cycle.

8.14

Cells can control genes in several ways by employing regulatory RNA molecules. One way is to take advantage of base pairing and use antisense RNA to form a double-stranded RNA that cannot be translated.

8.15

Riboswitches are RNA domains at the 5′ ends of mRNA that recognize small molecules and respond by changing their three-dimensional structure. This, in turn, affects the translation of the mRNA or, sometimes, premature termination of transcription. Riboswitches are mostly used to control biosynthetic pathways for amino acids, purines, and a few other metabolites.

8.16

Attenuation is a mechanism whereby transcription is controlled after initiation of mRNA synthesis. Attenuation mechanisms depend upon alternative stem–loop structures in the mRNA.

Review of Key Terms

Activator protein a regulatory protein that binds to specific sites on DNA and stimulates transcription; involved in positive control

Attenuation a mechanism for controlling gene expression that terminates transcription after initiation but before a full-length messenger RNA is produced

Autoinducer small signal molecule that takes part in quorum sensing

Catabolite repression the suppression of alternative catabolic pathways by a preferred source of carbon and energy

Cyclic AMP a regulatory nucleotide that participates in catabolite repression

Gene expression transcription of a gene followed by translation of the resulting mRNA into protein

Heat shock proteins proteins induced by high temperature (or certain other stresses) that protect against high temperature, especially by refolding partially denatured proteins or by degrading them

Heat shock response response to high temperature that includes the synthesis of heat shock proteins together with other changes in gene expression

Induction production of an enzyme in response to a signal (often the presence of the substrate for the enzyme)

Negative control a mechanism for regulating gene expression in which a repressor protein prevents transcription of genes

Noncoding RNA RNA that is not translated into protein; examples include ribosomal RNA, transfer RNA, and small regulatory RNAs

Operon one or more genes transcribed into a single RNA and under the control of a single regulatory site

Positive control a mechanism for regulating gene expression in which an activator protein functions to promote transcription of genes

Quorum sensing a regulatory system that monitors the population level and controls gene expression based on cell density

Regulatory nucleotide a nucleotide that functions as a signal rather than being incorporated into RNA or DNA

Regulon a series of operons controlled as a unit

Repression prevention of the synthesis of an enzyme in response to a signal

Repressor protein a regulatory protein that binds to specific sites on DNA and blocks transcription; involved in negative control

Response regulator protein one of the members of a two-component regulatory system; a protein that is phosphorylated by a sensor kinase and then acts as a regulator, often by binding to DNA

Riboswitch an RNA domain, usually in a messenger RNA molecule, that can bind a specific small molecule and alter its secondary structure; this, in turn, controls translation of the mRNA

Sensor kinase protein one of the members of a two-component regulatory system; a protein that phosphorylates itself in response to an external signal and then transfers the phosphoryl group to a response regulator protein

Signal transduction *see* two-component regulatory system

Stringent response a global regulatory control that is activated by amino acid starvation or energy deficiency

Two-component regulatory system a regulatory system consisting of two proteins: a sensor kinase and a response regulator

Review Questions

1. Describe why a protein that binds to a specific sequence of double-stranded DNA is unlikely to bind to the same sequence if the DNA is single-stranded (Section 8.2).

2. Name three major gene components involved in bacterial gene regulation (Section 8.1).

3. What is the difference between an operon and a regulon (Section 8.4)?

4. What is catabolite repression? Why is it also known as the "glucose effect" (Section 8.5)?

5. What are the two components that give their name to a signal transduction system in prokaryotes? What is the function of each of the components (Section 8.7)?

6. Adaptation allows the mechanism controlling flagellar rotation to be reset. How is this achieved (Section 8.8)?

7. How can quorum sensing be considered a regulatory mechanism for conserving cell resources (Section 8.9)?

8. What events trigger the stringent response? Why are the events in the stringent response a logical consequence of the trigger of the response (Section 8.10)?

9. What are the roles of heat shock proteins in a typical bacterial cell (Section 8.11)?

10. Explain how alternative sigma factors control sporulation in *Bacillus* (Section 8.12).

11. What role does the DnaA protein play in differentiation in *Caulobacter* (Section 8.13)?

12. Describe how antisense RNA is involved in RNA regulation (Section 8.14).

13. Describe how transcriptional attenuation works. What is actually being "attenuated" (Section 8.16)?

Application Questions

1. What would happen to regulation from a promoter under negative control if the region where the regulatory protein binds was deleted? What if the promoter was under positive control?

2. Promoters from *Escherichia coli* under positive control are not close matches to the promoter consensus sequence for *E. coli* (↻ Section 6.13). Why?

3. The attenuation control of some of the pyrimidine biosynthetic pathway genes in *Escherichia coli* actually involves coupled transcription and translation. Can you describe a mechanism whereby the cell could somehow make use of translation to help it measure the level of pyrimidine nucleotides?

4. Most of the regulatory systems described in this chapter involve regulatory proteins. However, regulatory RNA is also important. Describe how one could achieve negative control of the *lac* operon using either of two different types of regulatory RNA.

5. Many amino acid biosynthetic operons under attenuation control are also under negative control. Considering that the environment of a bacterium can be highly dynamic, what advantage could be conferred by having attenuation as a second layer of control?

6. How would you design a regulatory system to make *Escherichia coli* use succinic acid in preference to glucose? How could you modify it so that *E. coli* prefers to use succinic acid in the light but glucose in the dark?

Need more practice? Test your understanding with Quantitative Questions; access additional study tools including tutorials, animations, and videos; and then test your knowledge with chapter quizzes and practice tests at **www.microbiologyplace.com**.

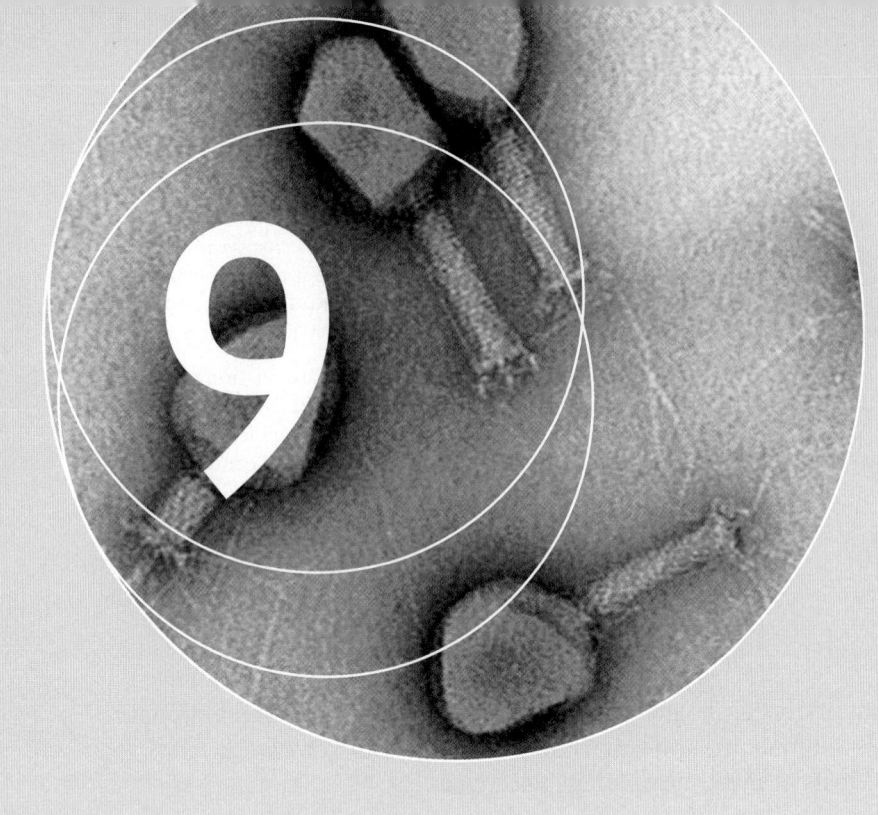

9

Viruses and Virology

Bacterial viruses such as the *Escherichia coli* bacterio-phage T4 have long been used as model systems for studying viral infection and replication processes.

Viruses are genetic elements that cannot replicate independently of a living cell, called the **host cell**. However, viruses do possess their own genetic information and are thus independent of the host cell's genome. Viruses rely on the host cell for energy, metabolic intermediates, and protein synthesis. Viruses are therefore obligate intracellular parasites that rely on entering a suitable living cell to carry out their replication cycle. However, unlike genetic elements such as plasmids (⮌ Section 6.6), viruses have an extracellular form, the virus particle, that enables them to exist outside the host and that facilitates transmission from one host cell to another. To multiply, viruses must enter a cell in which they can replicate, a process called *infection*.

Viruses can replicate in a way that is destructive to the host cell, and this accounts for the fact that some viruses are agents of disease. We cover a number of human diseases caused by viruses in Chapters 33 and 34. However, viruses may also inhabit a cell and replicate in step with the cell without destroying it. Like plasmids and transposable elements, viruses may confer important new properties on their host cells. These properties will be inherited when the host cell divides if each new cell also inherits the viral genome. These changes are not always harmful and may even be beneficial.

The study of viruses is called *virology*, and we introduce the essentials of the field here. There are four parts in this chapter. The first part introduces basic concepts of virus structure, infection of the host cell, and how viruses can be detected and quantified. The second part deals with the basic molecular biology of virus replication. The third part provides an overview of some key viruses that infect bacteria and animals; further coverage of viral diversity can be found in Chapter 21. The fourth part deals with subviral entities.

Viruses outnumber the living cells on our planet by at least 10-fold, and infect all types of cellular organisms. Therefore, they are interesting in their own right. However, scientists also study viruses for what they reveal about the genetics and biochemistry of cellular processes and, for many viruses, the development of disease. Furthermore, as we shall see in Chapters 10 and 11, viruses are also important in microbial genetics and genetic engineering.

 Virus Structure and Growth

9.1 General Properties of Viruses

Although viruses are not cells and thus are nonliving, they nonetheless possess a genome encoding the information they need in order to replicate. However, viruses rely on host cells to provide the energy and materials needed for replicating their genomes and synthesizing their proteins. Consequently, viruses cannot replicate unless the virus genome has gained entry into a suitable host cell.

Viruses can exist in either extracellular or intracellular forms. In its extracellular form, a virus is a microscopic particle containing nucleic acid surrounded by a protein coat and sometimes, depending on the specific virus, other macromolecules. The virus particle, or **virion**, is metabolically inert and cannot generate energy or carry out biosynthesis. The virus genome moves from the cell in which it was produced to another cell inside the virion. Once in the new cell, the intracellular state begins and the virus replicates. New copies of the virus genome are produced, and the components of the virus coat are synthesized. Certain animal viruses (such as polio and respiratory syncytial virus) may skip the extracellular stage when moving from cell to cell within the same organism. Instead, they mediate the fusion of infected cells with uninfected cells and transfer themselves in this way. However, when moving from one organism to another they are truly extracellular.

Viral genomes are usually very small, and they encode primarily proteins whose functions viruses cannot usurp from their hosts. Therefore, during replication inside a cell, viruses depend heavily on host cell structural and metabolic components. The virus redirects host metabolic functions to support virus replication and the assembly of new virions. Eventually, new viral particles are released, and the process can repeat itself.

Viral Genomes

All cells contain double-stranded DNA genomes. By contrast, viruses have either DNA or RNA genomes. (One group of viruses does use both DNA and RNA as their genetic material but at different stages of their replication cycle.) Virus genomes can be classified according to whether the nucleic acid in the virion is DNA or RNA and further subdivided according to whether the nucleic acid is single- or double-stranded, linear, or circular (**Figure 9.1**). Some viral genomes are circular, but most are linear.

Although those viruses whose genome consists of DNA follow the central dogma of molecular biology (DNA → RNA → protein, ⮌ Section 6.1), RNA viruses are exceptions to this rule. Nonetheless, genetic information still flows from nucleic acid to protein. Moreover, all viruses use the cell's translational machinery, and so regardless of the genome structure of the virus, messenger RNA (mRNA) must be generated that can be translated on the host cell ribosomes.

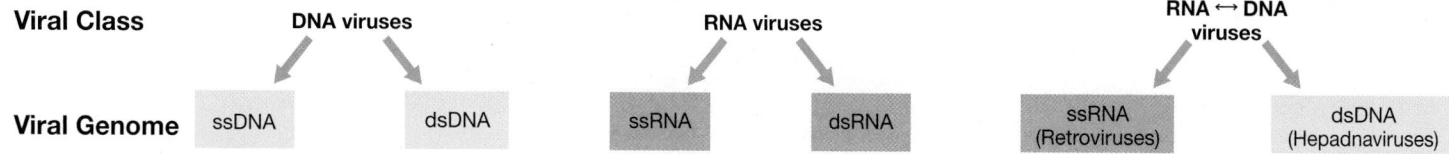

Figure 9.1 Viral genomes. The genomes of viruses can be either DNA or RNA, and some use both as their genomic material at different stages in their replication cycle. However, only one type of nucleic acid is found in the virion of any particular type of virus. This can be single-stranded (ss), double-stranded (ds), or in the hepadnaviruses, partially double-stranded. Some viral genomes are circular, but most are linear.

Viral Hosts and Taxonomy

Viruses can be classified on the basis of the hosts they infect as well as by their genomes. Thus, we have bacterial viruses, archaeal viruses, animal viruses, plant viruses, and viruses that infect other kinds of eukaryotic cells. Bacterial viruses, sometimes called **bacteriophages** (or phage for short; from the Greek *phagein*, meaning "to eat"), have been intensively studied as model systems for the molecular biology and genetics of virus replication. Species of both *Bacteria* and *Archaea* are infected by specific viruses. Indeed, many of the basic concepts of virology were first worked out with bacterial viruses and subsequently applied to viruses of higher organisms. Because of their frequent medical importance, animal viruses have been extensively studied, whereas plant viruses, although of enormous importance to modern agriculture, have been less well studied.

A formal system of viral classification exists that groups viruses into various taxa, such as orders, families, and even genus and species. The family taxon seems particularly useful. Members of a family of viruses all have a similar virion morphology, genome structure, and strategy of replication. Virus families have names that include the suffix -*viridae* (as in *Poxviridae*). We discuss a few of these in Chapter 21.

MiniQuiz

- How does a virus differ from a plasmid?
- How does a virion differ from a cell?
- What is a bacteriophage?
- Why does a virus need a host cell?

9.2 Nature of the Virion

Virions come in many sizes and shapes. Most viruses are smaller than prokaryotic cells, ranging in size from 0.02 to 0.3 μm (20–300 nm). A common unit of measure for viruses is the nanometer, which is one-thousandth of a micrometer. Smallpox virus, one of the largest viruses, is about 200 nm in diameter (about the size of the smallest cells of *Bacteria*). Poliovirus, one of the smallest viruses, is only 28 nm in diameter (about the size of a ribosome). Consequently, viruses could not be properly characterized until the invention of the electron microscope in the 1930s.

Viral genomes are smaller than those of most cells. Most bacterial genomes are between 1000 and 5000 kilobase pairs (kbp) of DNA, with the smallest known being about 500 kbp. (Interestingly, *Bacteria* with the smallest genomes are, like viruses, parasites that replicate in other cells; ↩ Table 12.1.) The largest known viral genome, that of *Mimivirus*, consists of 1.18 Mbp of double-stranded DNA. This virus, which infects protists such as *Amoeba*, is one of a few viruses currently known whose genome is larger than some cellular genomes. More typical virus genome sizes are listed in **Table 9.1**. Some viruses have genomes so small they contain fewer than five genes. Also, as can be seen in the table, the genome of some viruses, such as reovirus or influenza virus, is segmented into more than one molecule of nucleic acid.

Viral Structure

The structures of virions are quite diverse, varying widely in size, shape, and chemical composition. The nucleic acid of the virion is always located within the particle, surrounded by a protein shell called the **capsid**. This protein coat is composed of a number of

Table 9.1 *Some types of viral genomes*[a]

Virus	Host	DNA or RNA	Single- or double-stranded	Viral genome Structure	Viral genome Number of molecules	Size (bases or base pairs)[a]
H-1 parvovirus	Animals	DNA	Single-stranded	Linear	1	5,176
φX174	*Bacteria*	DNA	Single-stranded	Circular	1	5,386
Simian virus 40 (SV40)	Animals	DNA	Double-stranded	Circular	1	5,243
Poliovirus	Animals	RNA	Single-stranded	Linear	1	7,433
Cauliflower mosaic virus	Plants	DNA	Double-stranded	Circular	1	8,025
Cowpea mosaic virus	Plants	RNA	Single-stranded	Linear	2 different	9,370 (total)
Reovirus type 3	Animals	RNA	Double-stranded	Linear	10 different	23,549 (total)
Bacteriophage lambda	*Bacteria*	DNA	Double-stranded	Linear	1	48,514[b]
Herpes simplex virus type I	Animals	DNA	Double-stranded	Linear	1	152,260
Bacteriophage T4	*Bacteria*	DNA	Double-stranded	Linear	1	168,903
Human cytomegalovirus	Animals	DNA	Double-stranded	Linear	1	229,351

[a]The size is in bases or base pairs depending on whether the virus is single- or double-stranded. The sizes of the viral genomes chosen for this table are known accurately because they have been sequenced. However, this accuracy can be misleading because only a particular strain or isolate of a virus was sequenced. Therefore, the sequence and exact number of bases for other isolates may be slightly different. No attempt has been made to choose the largest and smallest viruses known, but rather to give a fairly representative sampling of the sizes and structures of the genomes of viruses containing both single- and double-stranded RNA and DNA.
[b]This includes single-stranded extensions of 12 nucleotides at either end of the linear form of the DNA (see Section 9.10).

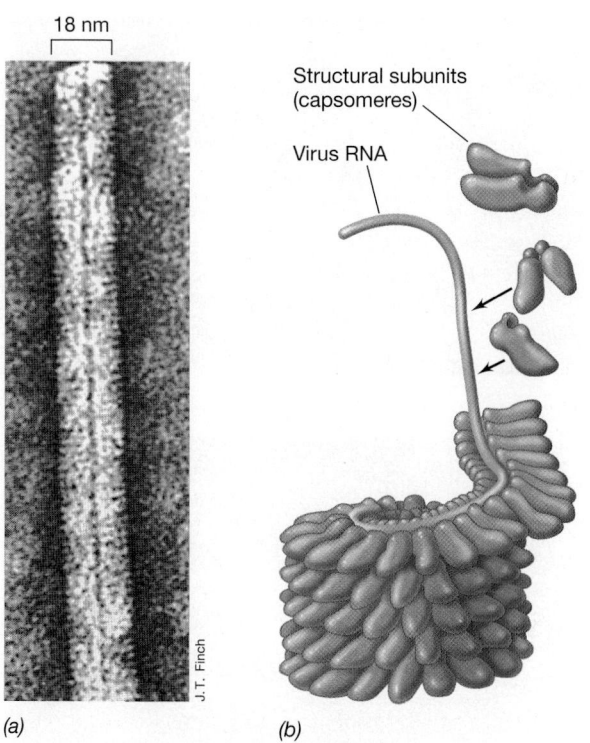

Figure 9.2 **The arrangement of nucleic acid and protein coat in a simple virus, tobacco mosaic virus.** *(a)* A high-resolution electron micrograph of a portion of the virus particle. *(b)* Assembly of the tobacco mosaic virus virion. The RNA assumes a helical configuration surrounded by the protein capsid. The center of the particle is hollow.

individual protein molecules, which are arranged in a precise and highly repetitive pattern around the nucleic acid (**Figure 9.2**).

The small genome size of most viruses restricts the number of different viral proteins that can be encoded. A few viruses have only a single kind of protein in their capsid, but most viruses have several distinct proteins that are associated in specific ways to form assemblies called **capsomeres** (Figure 9.2). The capsomere is the smallest morphological unit that can be seen with the electron microscope. A single virion can have a large number of capsomeres. The information for proper folding and assembly of the proteins into capsomeres is typically contained within the structure of the proteins themselves; hence, the overall process of virion assembly is called *self-assembly*. However, occasional virus proteins, such as the lambda capsid protein, require help from the chaperonin GroE (Section 6.21).

The complete complex of nucleic acid and protein packaged in the virion is called the virus **nucleocapsid**. Inside the virion are often one or more virus-specific enzymes. Such enzymes play a role during the infection and replication processes, as discussed later in this chapter. Some viruses are *naked*, whereas others possess lipid-containing layers around the nucleocapsid called an *envelope* (**Figure 9.3**).

Virus Symmetry

The nucleocapsids of viruses are constructed in highly symmetric ways. Symmetry refers to the way in which the capsomeres are arranged in the virus capsid. When a symmetric structure is

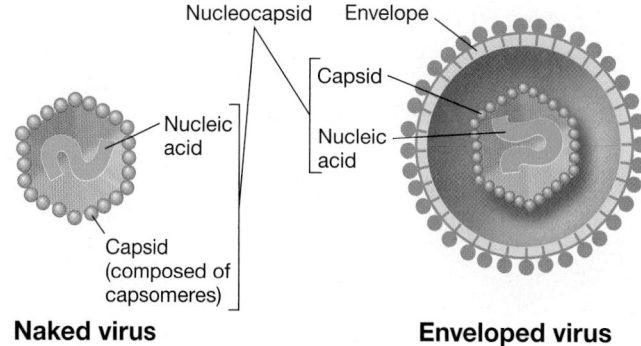

Figure 9.3 Comparison of naked and enveloped virus particles.

rotated around an axis, the same form is seen again after a certain number of degrees of rotation. Two kinds of symmetry are recognized in viruses, which correspond to the two primary shapes, rod and spherical. Rod-shaped viruses have *helical* symmetry, and spherical viruses have *icosahedral* symmetry. In all cases, the characteristic structure of the virus is determined by the structure of the capsomeres of which it is constructed.

A typical virus with helical symmetry is the tobacco mosaic virus (TMV) illustrated in Figure 9.2. It is an RNA virus in which the 2130 identical capsomeres are arranged in a helix. The overall dimensions of the TMV virion are 18×300 nm. The lengths of helical viruses are determined by the length of the nucleic acid, but the width of the helical virion is determined by the size and packaging of the capsomeres.

An **icosahedron** is a symmetric structure containing 20 triangular faces and 12 vertices and is roughly spherical in shape (**Figure 9.4**). Icosahedral symmetry is the most efficient arrangement of subunits in a closed shell because it uses the smallest number of capsomeres to build the shell. The simplest arrangement of capsomeres is three per face, for a total of 60 capsomeres per virion. Most viruses have more nucleic acid than can be packed into a shell made of just 60 capsomeres. The next possible structure that permits close packing contains 180 capsomeres, and many viruses have shells with this configuration. Other common configurations contain 240 or 420 capsomeres.

Figure 9.4*a* shows a model of an icosahedron. Figure 9.4*b* shows the same icosahedron viewed from three different angles to illustrate its complex 5-3-2 symmetry. The axes of symmetry divide the icosahedron into segments (5, 3, or 2) of identical size and shape. Figure 9.4*c* shows an electron micrograph of a typical icosahedral virus, human papillomavirus; this virus contains 360 capsomeres clustered into groups of five. Figure 9.4*d* shows a computer model of the same virus, where the five-capsomere clusters are more easily seen.

Enveloped Viruses

Enveloped viruses contain a membrane surrounding the nucleocapsid (**Figure 9.5*a***). Many viruses are enveloped, and most of these infect animal cells (for example, influenza virus), although occasional enveloped bacterial and plant viruses are also known. The viral envelope consists of a lipid bilayer with proteins, usually glycoproteins, embedded in it. The lipids of the viral membrane

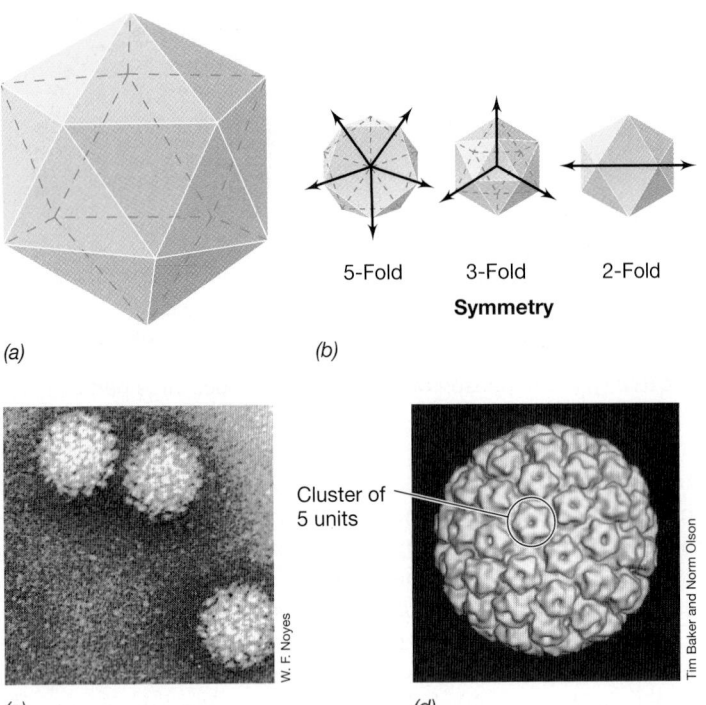

(a)

5-Fold   3-Fold   2-Fold
Symmetry

(b)

Cluster of 5 units

(c)   *(d)*

Figure 9.4 Icosahedral symmetry. *(a)* A model of an icosahedron. *(b)* Three views of an icosahedron showing the 5-3-2 symmetry. *(c)* Electron micrograph of human papillomavirus, a virus with icosahedral symmetry. The virion is about 55 nm in diameter. *(d)* Three-dimensional reconstruction of human papillomavirus calculated from images of frozen hydrated virions. The virus contains 360 units arranged in 72 clusters of 5 each.

are derived from the membranes of the host cell, but viral membrane proteins that are encoded by viral genes are also embedded in the membrane. The symmetry of enveloped viruses is not expressed by the virion as a whole, but by the nucleocapsid present inside the virus envelope.

Note that the envelope is the component of the virion that makes initial contact with the host cell. The specificity of virus infection and some aspects of virus penetration are thus controlled in part by characteristics of virus envelopes. The virus-specific envelope proteins are critical for attachment of the virion to the host cell during infection or for release of the virion from the host cell after replication.

Complex Viruses

Some virions are even more complex than anything discussed so far, being composed of several parts, each with separate shapes and symmetries. The most complicated viruses in terms of structure are some of the bacterial viruses, which possess icosahedral heads plus helical tails. In some bacterial viruses, such as bacteriophage T4 of *Escherichia coli* (Figure 9.5*b*), the tail itself has a complex structure. The complete T4 tail has almost 20 different proteins, and the T4 head has several more proteins. In such complex viruses, assembly is also quite involved. For instance, in T4 the complete tail is formed as a subassembly, and then the tail is added to the DNA-containing head. Finally, tail fibers formed from another protein are added to make the mature, infectious virion.

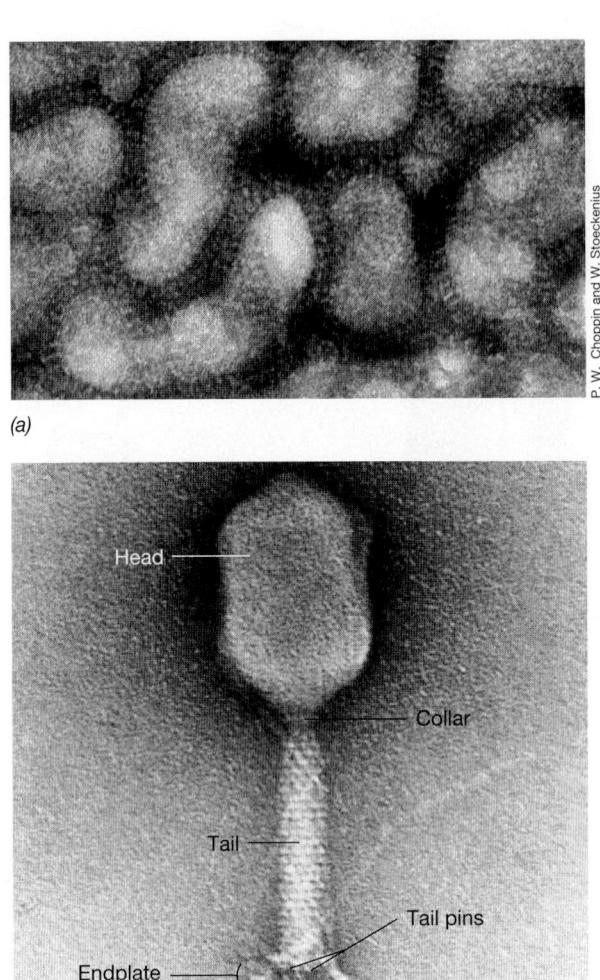

(a)

Head

Collar

Tail

Tail pins

Endplate

Tail fibers

(b)

Figure 9.5 Electron micrographs of animal and bacterial viruses. *(a)* Influenza virus, an enveloped virus. The virions are about 80 nm in diameter, but have no defined shape (⟲ Section 21.9). *(b)* Bacteriophage T4 of *Escherichia coli*. The tail components function in attachment of the virion to the host and injection of the nucleic acid (Figure 9.10). The head is about 85 nm in diameter.

Enzymes in Virions

Virions do not carry out metabolic processes and thus a virus is metabolically inert outside a host cell. However, some virions do contain enzymes that play important roles in infection. Some of these enzymes are required for very early events in the infection process. For example, some bacteriophages contain the enzyme lysozyme (⟲ Section 3.6), which they use to make a small hole in the bacterial cell wall. This allows the virus to inject its nucleic acid into the cytoplasm of the host cell. Lysozyme is again produced in large amounts in the later stages of infection, causing lysis of the bacterial cell and release of the new virions.

Many viruses contain their own nucleic acid polymerases for replication of the viral genome and for transcription of virus-specific RNA. For example, retroviruses are RNA viruses that replicate via DNA intermediates. These viruses possess an RNA-dependent DNA polymerase called *reverse transcriptase* that transcribes the viral RNA to form a DNA intermediate. Other viruses contain RNA genomes and require their own RNA polymerase. These virion enzymes are necessary because cells cannot make DNA or RNA from an RNA template (Sections 6.8 and 6.12).

Some viruses contain enzymes that aid in their release from the host. For example, certain animal viruses contain surface proteins called neuraminidases, enzymes that cleave glycosidic bonds in glycoproteins and glycolipids of animal cell connective tissue, thus liberating the virions. Although most virions lack their own enzymes, those that contain them do so for good reason: The host cell would not be able to produce virions in the absence of these extra enzymes.

MiniQuiz

- What is the difference between a naked virus and an enveloped virus?
- What kinds of enzymes can be found within the virions of specific viruses?

9.3 The Virus Host

Because viruses replicate only inside living cells, the cultivation of viruses requires the use of appropriate hosts. Viruses infecting prokaryotes are typically the easiest to grow in the laboratory. For the study of bacterial viruses, pure cultures are used either in liquid or on semisolid (agar) media. Most animal viruses and many plant viruses can be cultivated in tissue or cell cultures, and the use of such cultures has enormously facilitated research on these viruses. Plant viruses can be more difficult to work with, because their study sometimes requires use of the whole plant. This is a problem because plants grow much slower than bacteria, and plant viruses also often require a break in the thick plant cell wall in order to infect.

Animal cell cultures are derived from cells originally taken from an organ of an experimental animal. Unless blood cells are used, cell cultures are usually obtained by aseptically removing pieces of tissue and dissociating the cells by treatment with an enzyme that degrades the extracellular material that holds animal cells together. The resulting cell suspension is spread over a flat surface, such as the bottom of a culture flask or a Petri dish. The thin layer of cells adhering to the glass or plastic dish, called a *monolayer*, is overlaid with a suitable culture medium and incubated at a suitable temperature. The culture media used for cell cultures are typically quite complex, containing a number of amino acids and vitamins, salts, glucose, and a bicarbonate buffer system. To obtain the best growth, addition of a small amount of blood serum is usually necessary to provide vital nutrients, and several antibiotics are added to prevent bacterial contamination.

Some cell cultures prepared in this way can be subcultured and grown indefinitely as *permanent cell lines*. Cell lines are convenient for virus research because cell material is continuously available. In many cases, a culture will not grow indefinitely, but may

remain alive for a number of days. Such cultures, called *primary cell cultures*, may still be useful for growing a virus, although new cultures need to be prepared from fresh sources from time to time, an expensive and time-consuming process. In some cases, primary or permanent cell lines cannot be obtained, but whole organs or pieces of organs can successfully replicate the virus. Such organ cultures may still be useful in virus research because they permit growth of viruses under more or less controlled laboratory conditions.

MiniQuiz

- In virology, what is a host?
- Why is it helpful to use cell culture for viral research?

9.4 Quantification of Viruses

In virology it is often necessary to quantify the number of virions in a suspension. Although one can count virions using an electron microscope (Figures 9.4c and 9.5a), the number of virions in a suspension can be more easily quantified by measuring their effects on the host. Using such a method, we see that a virus infectious unit is the smallest unit that causes a detectable effect when added to a susceptible host. This can be as few as one virion, although a larger inoculum is more often required. By determining the number of infectious units per volume of fluid, a measure of virus quantity, called a *titer*, can be obtained.

Plaque Assay

When a virion initiates an infection on a layer of host cells growing on a flat surface, a zone of lysis may be seen as a clear area in the layer of growing host cells. This clearing is called a **plaque**, and it is assumed that each plaque originated from the replication of a single virion (**Figure 9.6**).

With bacteriophages, plaques may be obtained when virions are mixed into a small volume of melted agar containing host bacteria that is spread on the surface of an agar medium (Figure 9.6a). During incubation the bacteria grow and form a turbid layer that is visible to the naked eye. However, wherever a successful viral infection has been initiated, cells are lysed, forming a plaque (Figure 9.6b). By counting the number of *plaque-forming units*, one can calculate the titer, or number of virus infectious units, present in the virus sample.

The plaque assay also permits the isolation of pure virus strains. This is because if a plaque has arisen from a single virion, all the viruses in this plaque should be genetically identical. Some of the virions from this plaque can be picked and inoculated into a fresh bacterial culture to establish a pure virus line. The development of the plaque assay technique was as important for the advancement of virology as Koch's development of solid media (Section 1.8) was for pure culture microbiology.

Plaques may be obtained for animal viruses by using cultured animal cells as hosts. A monolayer of cultured animal cells is prepared on a plate or flat bottle and the virus suspension is overlaid.

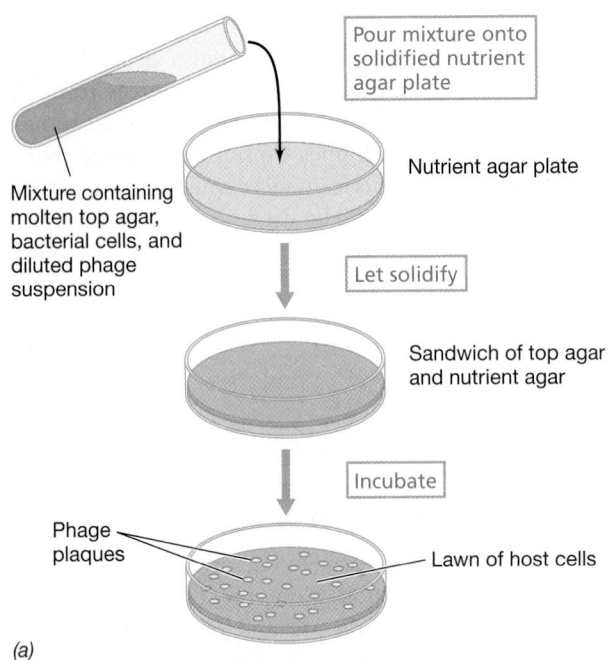

(a)

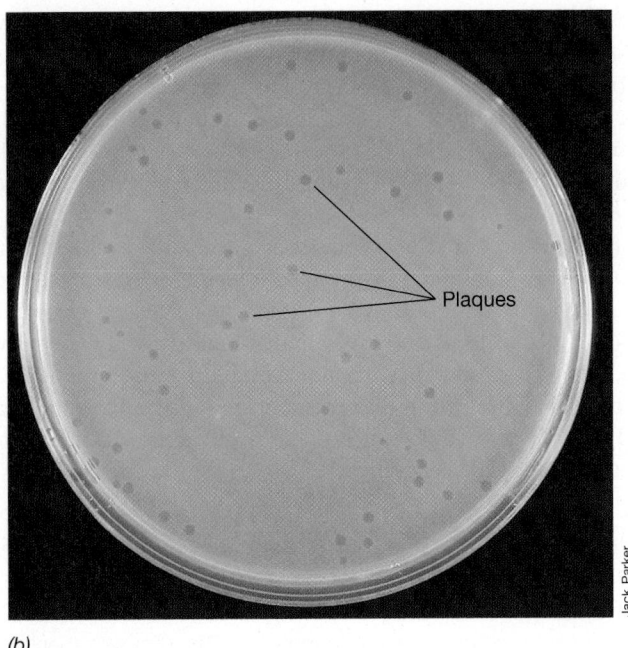

(b)

Figure 9.6 Quantification of bacterial virus by plaque assay using the agar overlay technique. *(a)* A dilution of a suspension containing the virus is mixed in a small amount of melted agar with the sensitive host bacteria. The mixture is poured on the surface of an agar plate of the appropriate medium. The host bacteria, which have been spread uniformly throughout the top agar layer, begin to grow, and after overnight incubation form a lawn of confluent growth. Virion-infected cells are lysed, forming plaques in the lawn. The size of the plaque depends on the virus, the host, and conditions of culture. *(b)* Photograph of a plate showing plaques formed by a bacteriophage on a lawn of sensitive bacteria. The plaques shown are about 1–2 mm in diameter.

Plaques are revealed by zones of destruction of the animal cells, and from the number of plaques produced, an estimation of the virus titer can be made (**Figure 9.7**).

Efficiency of Plating

The concept of *efficiency of plating* is important in quantitative virology. In any given viral system, the number of plaque-forming units is always lower than counts of the viral suspension made with an electron microscope. The efficiency with which virions

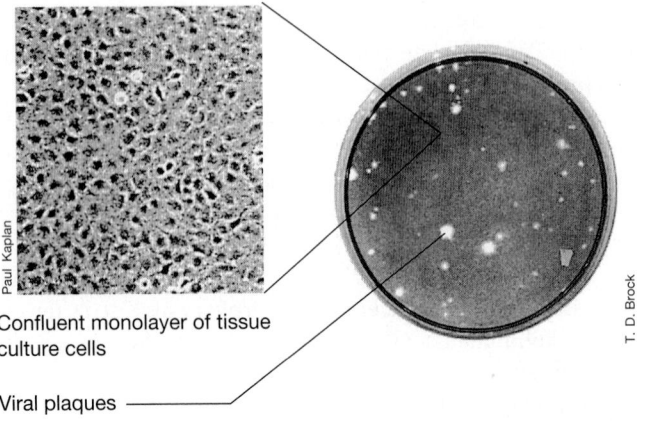

Confluent monolayer of tissue culture cells

Viral plaques

Figure 9.7 Cell cultures in monolayers grown on a Petri plate. Note the presence of plaques. Also shown is a photomicrograph of a cell culture.

infect host cells is thus rarely 100% and may often be considerably less. Virions that fail to cause infection are often inactive, although this is not always the case. Some viruses produce many incomplete virions during infection. In other cases, especially with RNA viruses, the viral mutation rate is so high that many virions contain defective genomes. However, sometimes a low efficiency of plating merely means that under the conditions used, some virions did not successfully infect cells. Although with bacterial viruses the efficiency of plating is often higher than 50%, with many animal viruses it may be much lower, 0.1% or 1%. Knowledge of plating efficiency is useful in cultivating viruses because it allows one to estimate how concentrated a viral suspension needs to be (that is, its titer) to yield a certain number of plaques.

Intact Animal Methods

Some viruses do not cause recognizable effects in cell cultures yet cause death in whole animals. In such cases, quantification can be done only by titration in infected animals. The general procedure is to carry out a serial dilution of the virus sample ($\rightleftarrows$ Section 5.10), generally at 10-fold dilutions, and to inject samples of each dilution into several sensitive animals. After a suitable incubation period, the fraction of dead and live animals at each dilution is tabulated and an end point dilution is calculated. This is the dilution at which, for example, half of the injected animals die (the lethal dose for 50% or LD_{50}, $\rightleftarrows$ Section 27.8).

Although using whole animals is much more cumbersome and much less accurate than cell culture methods, it may be essential for the study of certain types of viruses.

MiniQuiz

- Give a definition of efficiency of plating.
- What is a plaque-forming unit?

Viral Replication

9.5 General Features of Virus Replication

For a virus to replicate it must induce a living host cell to synthesize all the essential components needed to make more virions. These components must then be assembled into new virions that are released from the cell. The viral replication cycle can be divided into five steps (**Figure 9.8**).

1. *Attachment* (adsorption) of the virion to a susceptible host cell.
2. *Penetration* (entry, injection) of the virion or its nucleic acid into the host cell.
3. *Synthesis* of virus nucleic acid and protein by host cell metabolism as redirected by the virus.
4. *Assembly* of capsids (and membrane components in enveloped viruses) and *packaging* of viral genomes into new virions. This whole process is called *maturation*.
5. *Release* of mature virions from the cell.

The growth curve resulting from these stages of virus replication is illustrated in **Figure 9.9**. In the first few minutes after infection the virus is said to undergo an *eclipse*. During this period infectious particles cannot be detected in the culture medium. The eclipse begins as soon as infectious particles are removed from the environment by adsorbing to host cells. Once attached to host cells, the virions are no longer available to infect other cells. This is followed by the entry of viral nucleic acid (or intact virion) into the host cell. If the infected cell breaks open at this point, the virion no longer exists as an infectious entity since the viral genome is no longer inside its capsid. *Maturation* begins as the newly synthesized nucleic acid molecules become packaged inside protein coats. During the maturation phase, the titer of active virions inside the host cell rises dramatically. However, the new virus particles still cannot be detected in the culture medium unless the cells are artificially lysed to release them. Because newly synthesized virions have not yet appeared outside the cell, the eclipse and maturation periods together are called the *latent period*.

At the end of maturation, mature virions are released, either as a result of cell lysis or by budding or excretion, depending on the virus. The number of virions released, called the *burst size*, varies with the particular virus and the particular host cell, and can range from a few to a few thousand. The duration of the virus replication cycle varies from 20–60 min (in many bacterial viruses) to 8–40 h (in most animal viruses). Because the release of virions is more or less simultaneous, virus replication

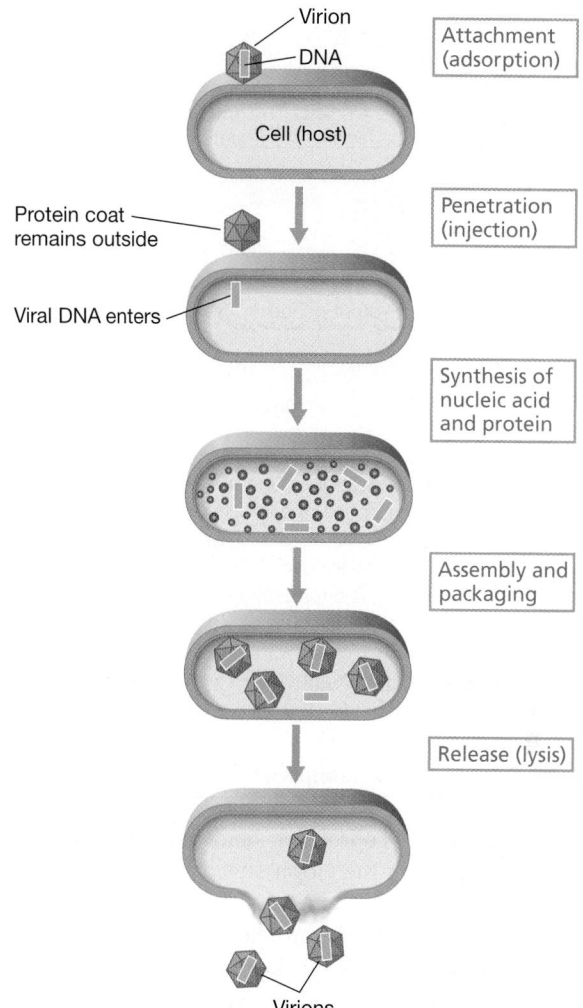

Figure 9.8 **The replication cycle of a bacterial virus.** Note that the viruses and cell are not drawn to scale.

is typically characterized by a *one-step growth curve* (Figure 9.9). In the next two sections we consider a few key steps of the virus replication cycle in more detail.

MiniQuiz

- What is packaged into the virions?
- Explain the term maturation.
- What events happen during the latent period of viral replication?

9.6 Viral Attachment and Penetration

In this section we focus on virus attachment and penetration, the first steps in the viral life cycle. In addition, we consider the mechanism by which some bacteria react to penetration by bacteriophage DNA.

Attachment

The most common basis for the host specificity of a virus depends upon attachment. The virion itself (whether naked or enveloped) has one or more proteins on its external surface that interact with

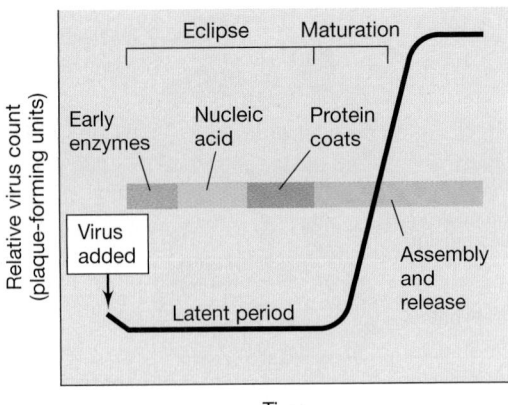

Figure 9.9 **The one-step growth curve of virus replication.** This graph displays the results of a single round of viral replication in a population of cells. Following adsorption, the infectivity of the virus particles disappears, a phenomenon called *eclipse*. This is due to the uncoating of the virus particles. During the latent period, viral nucleic acid replicates and protein synthesis occurs. The maturation period, when virus nucleic acid and protein are assembled into mature virus particles, follows. Finally, the virions are released, either with or without cell lysis. This general picture is amplified for bacteriophage T4 in Figure 9.15.

specific host cell surface components called receptors. These receptors are normal surface components of the host, such as proteins, carbohydrates, glycoproteins, lipids, lipoproteins, or complexes of these, to which the virion attaches. The receptors carry out normal functions for the cell. For example, the receptor for bacteriophage T1 is an iron-uptake protein and that for bacteriophage lambda is involved in maltose uptake. Animal virus receptors may include macromolecules needed for cell–cell contact or by the immune system. For example, the receptors for poliovirus and for HIV are normally used in interactions between human cells.

In the absence of its specific receptor, the virus cannot adsorb and hence cannot infect. Moreover, if the receptor is altered, for example, by mutation, the host may become resistant to virus infection. However, mutants of the virus can also arise that gain the ability to adsorb to previously resistant hosts. In addition, some animal viruses may be able to use more than one receptor, so the loss of one may not necessarily prevent attachment. Thus, the host range of a particular virus is, to some extent, determined by the availability of a suitable receptor that the virus can recognize. In multicellular organisms, cells in different tissues or organs often express different proteins on their cell surfaces. Consequently, viruses that infect animals often infect only cells of certain tissues. For example, many viruses that cause coughs and colds infect only cells of the upper respiratory tract.

Penetration

The attachment of a virus to its host cell results in changes to both the virus and the host cell surface that result in penetration. Viruses must replicate within cells. Therefore, at a minimum, the viral genome must enter the cell (Figure 9.8). Entry of the virus genome into a susceptible cell will not lead to virus replication if the information in the viral genome cannot be read. Consequently, as we mentioned (Section 9.2), for some viruses to replicate, certain viral proteins must also enter the host cell. A cell

that allows the complete replication cycle of a virus to take place is said to be *permissive* for that virus.

Different viruses have different strategies for penetration. Uncoating refers to the process in which the virions lose their outer coat and the viral genome is exposed. Some enveloped animal viruses are uncoated at the cytoplasmic membrane, releasing the virion contents into the cytoplasm. However, the entire virion of naked animal viruses and many enveloped animal viruses enters the cell via endocytosis. In such cases the virus must be uncoated inside the host cell so that the genome is exposed and replication can proceed. Some enveloped viruses are uncoated in the cytoplasm. Others (such as influenza) are uncoated at the nuclear membrane and the viral genome then enters the nucleus. In animal cells, wherever uncoating occurs, the viral genome must eventually enter the nucleus to be replicated, except in a few rare cases.

Tailed Bacteriophage Attachment and Penetration

Cells that have cell walls, such as most bacteria, are infected in a manner different from animal cells, which lack cell walls. The most complex penetration mechanisms have been found in viruses that infect bacteria. The bacteriophage T4, which infects *Escherichia coli*, is a good example.

The structure of bacteriophage T4 was shown in Figure 9.5b. The virion has a head, within which the viral linear double-stranded DNA is folded, and a long, fairly complex tail, at the end of which is a series of tail fibers and tail pins. The T4 virions first attach to *E. coli* cells by means of the tail fibers (**Figure 9.10**). The ends of the fibers interact specifically with polysaccharides that are part of the outer layer of the gram-negative cell envelope

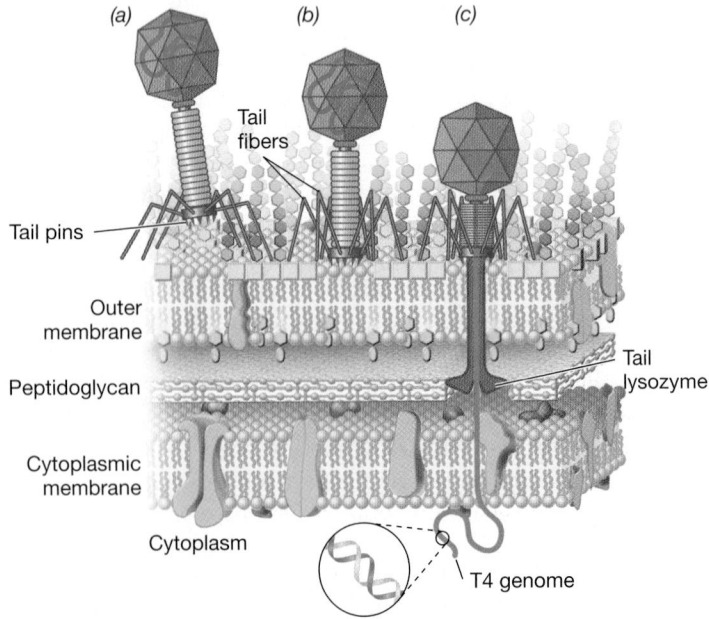

Figure 9.10 **Attachment of bacteriophage T4 to the cell wall of** *Escherichia coli* **and injection of DNA.** (a) Attachment of a T4 virion to the cell wall by the long tail fibers interacting with core lipopolysaccharide. (b) Contact of cell wall by the tail pins. (c) Contraction of the tail sheath and injection of the T4 genome. The tail tube penetrates the outer membrane, and the tail lysozyme digests a small opening through the peptidoglycan layer.

(Section 3.7). These tail fibers then retract, and the core of the tail makes contact with the cell wall of the bacterium through a series of fine tail pins at the end of the tail. The activity of a lysozyme-like enzyme forms a small pore in the peptidoglycan layer. The tail sheath then contracts, and the viral DNA passes into the cytoplasm of the host cell through a hole in the tip of the phage tail, with the majority of the coat protein remaining outside (Figure 9.10).

Virus Restriction and Modification by the Host

Animals can often eliminate invading viruses by immune defense mechanisms before the viral infection becomes widespread or sometimes even before the virus has penetrated target cells. In addition, eukaryotes, including animals and plants, possess an antiviral mechanism known as RNA interference (Section 7.10). Although they lack immune systems, both *Bacteria* and *Archaea* possess an antiviral mechanism similar to RNA interference, known as CRISPR (Chapter 8, Microbial Sidebar). In addition, prokaryotes destroy double-stranded viral DNA after it has been injected by using restriction endonucleases (Section 11.1), enzymes that cleave foreign DNA at specific sites, thus preventing its replication. This phenomenon is called *restriction* and is part of a general host mechanism to prevent the invasion of foreign nucleic acid. For such a system to be effective, the host must have a mechanism for protecting its own DNA. This is accomplished by specific modification of its DNA at the sites where the restriction enzymes cut (Section 11.1).

Restriction enzymes are specific for double-stranded DNA, and thus single-stranded DNA viruses and all RNA viruses are unaffected by restriction systems. Although host restriction systems confer significant protection, some DNA viruses have overcome host restriction by modifying their own DNA so that they are no longer subject to restriction enzyme attack. Two patterns of chemical modification of viral DNA are known: glucosylation and methylation. For instance, the T-even bacteriophages (T2, T4, and T6) have their DNA glucosylated to varying degrees, which prevents endonuclease attack. Many other viral DNAs can be modified by methylation. However, whether glucosylated or methylated, viral DNAs are modified after genomic replication has occurred by modification proteins encoded by the virus.

Other viruses, such as the bacteriophages T3 and T7, avoid destruction by host restriction enzymes by encoding proteins that inhibit the host restriction systems. To counter this, some bacteria have multiple restriction and methylation systems that help prevent infection by viruses that can circumvent only one of them. Bacteria also contain other DNA methylases in addition to those that protect them from their own restriction enzymes. Some of these methylases take part in DNA repair or in gene regulation, but others protect the host DNA from foreign endonucleases. This is necessary because some viruses encode restriction systems themselves that are designed to destroy host DNA! It is thus clear that viruses and hosts have responded to each other's defense mechanisms by continuing to evolve their own mechanisms to better their chances of infection or survival, respectively.

MiniQuiz
- How does the attachment process contribute to virus–host specificity?
- Why do some viruses need to be uncoated after penetration and others do not?

9.7 Production of Viral Nucleic Acid and Protein

Once a host has been infected, new copies of the viral genome must be made and virus-specific proteins must be synthesized in order for the virus to replicate. Typically, the production of at least some viral proteins begins very early after the viral genome has entered the cell. The synthesis of these proteins requires viral mRNA. For certain types of RNA viruses, the genome itself is the mRNA. For most viruses, however, the mRNA must first be transcribed from the DNA or RNA genome and then the genome must be replicated. We consider these important events here.

The Baltimore Classification Scheme and DNA Viruses

The virologist David Baltimore, who along with Howard Temin and Renato Dulbecco shared the Nobel Prize for Physiology or Medicine in 1975 for the discovery of retroviruses and reverse transcriptase, developed a classification scheme for viruses. The Baltimore classification scheme (**Table 9.2**) is based on the relationship of the viral genome to its mRNA and recognizes

Table 9.2 *The Baltimore classification system of viruses*

Class	Description of genome and replication strategy	Examples Bacterial viruses	Animal viruses
I	Double-stranded DNA genome	Lambda, T4	Herpesvirus, pox virus
II	Single-stranded DNA genome	φX174	Chicken anemia virus
III	Double-stranded RNA genome	φ6	Reoviruses (Section 21.10)
IV	Single-stranded RNA genome of plus configuration	MS2	Poliovirus
V	Single-stranded RNA genome of minus configuration		Influenza virus, rabies virus
VI	Single-stranded RNA genome that replicates with DNA intermediate		Retroviruses
VII	Double-stranded DNA genome that replicates with RNA intermediate		Hepatitis B virus

seven classes of viruses. Double-stranded (ds) DNA viruses are in class I. The mechanism of mRNA production and genome replication of class I viruses is the same as that used by the host cell genome, although different viruses use different strategies to ensure that viral mRNA is expressed in preference to host mRNA.

Class II viruses are single-stranded (ss) DNA viruses. Before mRNA can be produced from such viruses, a complementary DNA strand must be synthesized because RNA polymerase uses double-stranded DNA as a template (⮌ Section 6.8). These viruses form a dsDNA intermediate during replication that is also used for transcription (**Figure 9.11**). The synthesis of the dsDNA intermediate and its subsequent transcription can be carried out by cellular enzymes (although viral proteins may also be required). The dsDNA intermediate is also used to generate the viral genome; one strand becomes the genome while the other is discarded (Figure 9.11). Until recently, all known ssDNA viruses contained positive-strand DNA, which has the same sequence as their mRNA (see positive-strand viruses below). However, a novel virus is now known that contains circular ssDNA of negative polarity. Torque teno virus (TTV), as this virus is called, is widespread in humans and other animals but causes no obvious disease symptoms. The mode of replication of TTV has not yet been fully investigated.

Positive- and Negative-Strand RNA Viruses

The production of mRNA and genome replication is different for RNA viruses (classes III–VI). Recall that mRNA is complementary in base sequence to the template strand of DNA. By convention in virology, mRNA is of the plus (+) configuration. Its complement is thus of the minus (−) configuration. This convention is used to describe the genome of a single-stranded virus, whether its genome contains RNA or DNA (Figure 9.11). For example, a virus that has a ssRNA genome with the same orientation as its mRNA is a **positive-strand RNA virus**, while a virus whose ssRNA genome is complementary to its mRNA is a **negative-strand RNA virus**.

Cellular RNA polymerases do not normally catalyze the formation of RNA from an RNA template, but instead require a DNA template. Therefore, RNA viruses, whether positive, negative, or double-stranded, require a specific RNA-dependent RNA polymerase. The simplest case is the positive-strand RNA viruses (class IV) in which the viral genome is of the plus configuration and hence can function directly as mRNA (Figure 9.11). In addition to other required proteins, this mRNA encodes a virus-specific RNA-dependent RNA polymerase (also called RNA replicase). Once synthesized, this polymerase first makes complementary minus strands of RNA and then uses them as templates to make more plus strands. These plus strands can either be translated as mRNA or packaged as the genome in newly synthesized virions (Figure 9.11).

For negative-strand RNA viruses (class V), the situation is more awkward. The incoming RNA is the wrong polarity to serve as mRNA, and therefore mRNA must be synthesized first. Because cells do not have an RNA polymerase capable of this, these viruses must carry some of this enzyme in their virions, and the enzyme enters the cell along with the genomic RNA. The complementary plus strand of RNA is synthesized by this RNA-dependent RNA polymerase and is then used as mRNA. This plus-strand mRNA is also used as a template to make more negative-strand genomes (Figure 9.11). The dsRNA viruses (class III) face a similar problem. Although the virion does contain plus-strand RNA, this is part

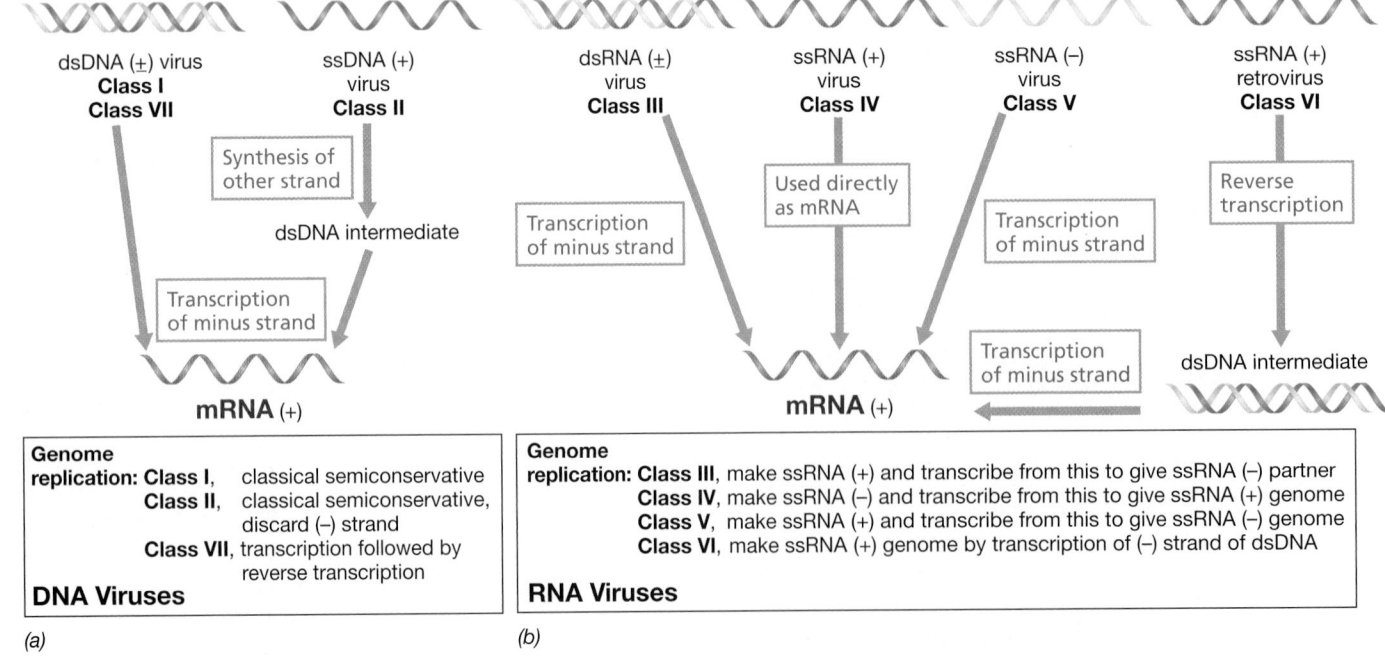

Figure 9.11 Formation of mRNA and new genomes in *(a)* DNA viruses and *(b)* RNA viruses. By convention, mRNA is always considered to be of the plus (+) orientation. Examples of each class of virus are given in Table 9.2.

of the dsRNA genome and cannot be released to act as mRNA. Consequently, the virions of dsRNA viruses must also contain RNA-dependent RNA polymerases that transcribe the dsRNA genome to produce plus-strand mRNA upon entry into the host cell.

Retroviruses

The **retroviruses** are animal viruses that are responsible for causing certain kinds of cancers and acquired immunodeficiency syndrome, AIDS. Retroviruses have ssRNA in their virions but replicate through a dsDNA intermediate (class VI). The process of copying the information found in RNA into DNA is called **reverse transcription**, and thus these viruses require an enzyme called **reverse transcriptase**. Although the incoming RNA of retroviruses is the plus strand, it is not used as mRNA, and therefore these viruses must carry reverse transcriptase in their virions. After infection, the virion ssRNA is converted to dsDNA via a hybrid RNA–DNA intermediate. The dsDNA is then the template for mRNA synthesis by normal cellular enzymes.

Finally, class VII viruses are those that have double-stranded DNA in their virions but replicate through an RNA intermediate. These unusual viruses also use reverse transcriptase. The strategy these viruses use to produce mRNA is the same as that of class I viruses (Figure 9.11), although their DNA replication is very unusual because, as we will see later, the genome is only partially double-stranded (Section 21.11).

While the Baltimore scheme covers most possibilities, there are exceptions. For example, ambiviruses contain a ssRNA genome, half of which is in the plus orientation (and can thus be used as mRNA) and half in the minus configuration (which cannot). A complementary strand must be synthesized from the latter half before the genes there can be translated. Evolution has clearly pushed viral genome diversity to the limits!

Viral Proteins

Once viral mRNA is made (Figure 9.11), viral proteins can be synthesized. These proteins can be grouped into two broad categories:

1. Proteins synthesized soon after infection, called **early proteins**, which are necessary for the replication of virus nucleic acid

2. Proteins synthesized later, called **late proteins**, which include the proteins of the virus capsid

Generally, both the timing and amount of virus proteins are highly regulated. Early proteins are typically enzymes that act catalytically and are therefore synthesized in smaller amounts. By contrast, late proteins are typically structural components of the virion and are made in much larger amounts.

Virus infection upsets the regulatory mechanisms of the host because there is a marked overproduction of viral nucleic acid and protein in the infected cell. In some cases, virus infection causes a complete shutdown of host macromolecular synthesis, whereas in other cases, host synthesis proceeds concurrently with virus synthesis. In either case, regulation of virus synthesis is under the control of the virus rather than the host. Several

aspects of this control resemble the regulatory mechanisms discussed in Chapter 8, but there are also some uniquely viral regulatory mechanisms. We discuss these regulatory mechanisms next when we consider some well-studied viruses.

MiniQuiz

- Why must some types of virus contain enzymes in the virion in order for mRNA to be produced?

- Distinguish between a positive-strand RNA virus and a negative-strand RNA virus.

- Both positive-strand RNA viruses and retroviruses contain plus configuration RNA genomes. Contrast mRNA production in these two classes of viruses.

III Viral Diversity

9.8 Overview of Bacterial Viruses

Bacteriophages are quite diverse, and examples of the various classes are illustrated in **Figure 9.12**. Most bacterial viruses that have been investigated in detail infect well-studied bacteria, such as *Escherichia coli* and *Salmonella enterica*. However, viruses are known that infect a wide range of *Bacteria* and *Archaea*.

Most known bacteriophages contain dsDNA genomes, and this type of bacteriophage is thought to be the most common in nature. However, many other kinds are known, including those with ssRNA genomes, dsRNA genomes, and ssDNA genomes (Figure 9.11). In fact, this remarkable diversity of genomes may have been an important factor in the evolution of nucleic acid function in cellular organisms (see the Microbial Sidebar, "Did Viruses Invent DNA?").

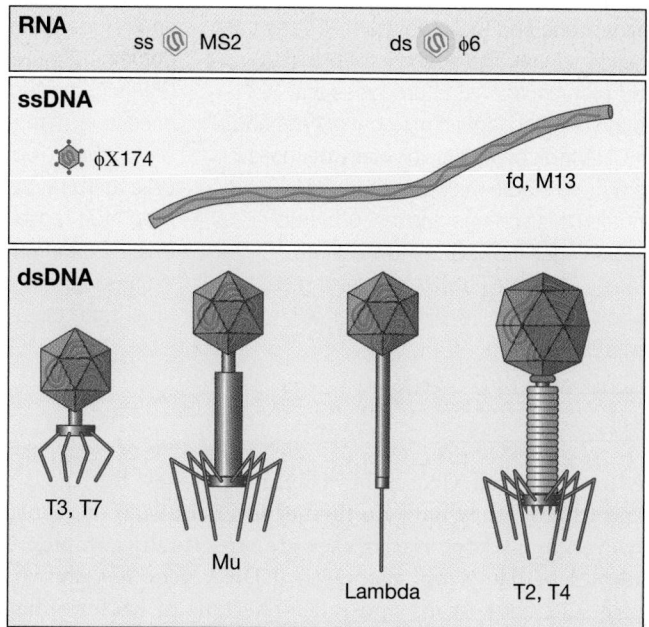

Figure 9.12 Schematic representations of the main types of bacterial viruses. Sizes are to approximate scale. The nucleocapsid of φ6 is surrounded by a membrane.

Did Viruses Invent DNA?

The three-domain theory of cellular evolution divides living cells into three lineages, *Bacteria*, *Archaea*, and *Eukarya*, based on the sequence of their ribosomal RNA (Section 16.9). In addition, molecular analyses of the cellular components required for transcription and translation support this scheme rather well. However, when molecular analyses of the components required for DNA replication, recombination, and repair are considered, the three-domain scheme does not hold up so well. For example, class II topoisomerases of the *Archaea* are more closely related to those of the *Bacteria* than to those of the *Eukarya*. In addition, viral DNA-processing enzymes show erratic relationships to those of cellular organisms. For example, the DNA polymerase of bacteriophage T4 is more closely related to the DNA polymerases of eukaryotes than to those of its bacterial hosts.

Another major problem in microbial evolution is where viruses fit into the universal tree of life. Did they emerge relatively late as rogue genetic elements escaping from cellular genomes, or were they around at the same time as the very earliest cells? A related issue is when DNA entered the evolutionary scene and took over from RNA as the genetic material. The scenario of the RNA world (Section 16.2) proposes that RNA was the original genetic material of cells and that DNA took over relatively early because it was a more stable molecule than RNA. Hence, in this scheme, the last universal common ancestor (LUCA) to the three domains of life was a DNA-containing cell. But how did the LUCA obtain its DNA?

Recently, Patrick Forterre of the Institut Pasteur has suggested a novel evolutionary scenario for how cells obtained DNA that also explains how the cellular machinery that deals with DNA originated in cells in the first place. Forterre argues that minor improvements in genetic stability would not have been sufficiently beneficial to select for the upheaval of converting an entire cellular genome from RNA to DNA. Instead, he suggests that viruses invented DNA as a modification mechanism to protect their genomes from host cell enzymes designed to destroy them (Figure 1). Viruses are known today that contain genomes of RNA, DNA, DNA containing uracil instead of thymine, and DNA containing hydroxymethylcytosine in place of cytosine (Figure 1*a*). Moreover, modern cells of all three domains contain systems designed to destroy incoming foreign DNA or RNA.

Forterre's hypothesis starts with an RNA world consisting of cells with RNA genomes plus viruses with RNA genomes. Viruses with DNA genomes were then selected because this protected them from degradation by cellular nucleases. This would have occurred before the LUCA (also containing an RNA genome) split into the three domains (Figure 1*b*). Then three nonvirulent DNA viruses ("founder viruses") infected the ancestors of the three domains. The three DNA viruses replicated inside their host cells as DNA plasmids, much as a P1 prophage replicates inside *Escherichia coli* today. Furthermore, two of the founder viruses were more closely related to each other (and these infected the ancestors of

today's *Archaea* and *Eukarya*) than to the third founder virus (which infected the ancestor of *Bacteria*). Gradually, cells converted their genes from RNA into DNA due to its greater stability. Reverse transcriptase is believed to be an enzyme of very ancient origin, and it is conceivable that it was involved in the conversion of RNA genes to DNA, as occurs in retroviruses today.

To recap the hypothesis, the LUCA diverged into the three cellular ancestors to the three domains of life, and this laid the groundwork for the transcription and translation machinery in cells—that is, those functions that involve RNA (but not DNA). However, the use of DNA as a storage system for genetic information—now a universal property of cells—was provided by a family of DNA viruses that infected cells eons ago. Because DNA is a more stable molecule than RNA, cells with RNA genomes that were not infected by DNA viruses never became DNA-based cells and eventually became extinct (Figure 1*b*).

The Forterre model explains the origin of DNA in cells and provides a mechanism for the gradual replacement of RNA genomes with DNA. And, importantly, it also explains the noncongruence of the DNA replication, recombination, and repair machinery of cells of the different domains as compared with the transcription and translation machinery. Although this hypothesis does not wholly explain the origin of viruses, it does explain their diversity of replication systems and the very ancient structural similarities between certain families of DNA and RNA viruses.

A few bacterial viruses have lipid envelopes, but most are naked (that is, they have no further layers outside the capsid). However, many bacterial viruses are structurally complex. All examples of bacteriophages with dsDNA genomes shown in Figure 9.12 have heads and tails. The tails of bacteriophages T2, T4, and Mu are contractile and function in DNA entry into the host (Figure 9.10). By contrast, the tail of phage lambda is flexible.

Although tailed bacterial viruses were first studied as model systems for understanding general features of virus replication, some of them are now used as convenient tools for genetic engineering. Understanding bacterial viruses is not only valuable as background for the discussion of animal viruses but is also essential for the material presented in the chapters on microbial genetics (Chapter 10) and genetic engineering (Chapter 11).

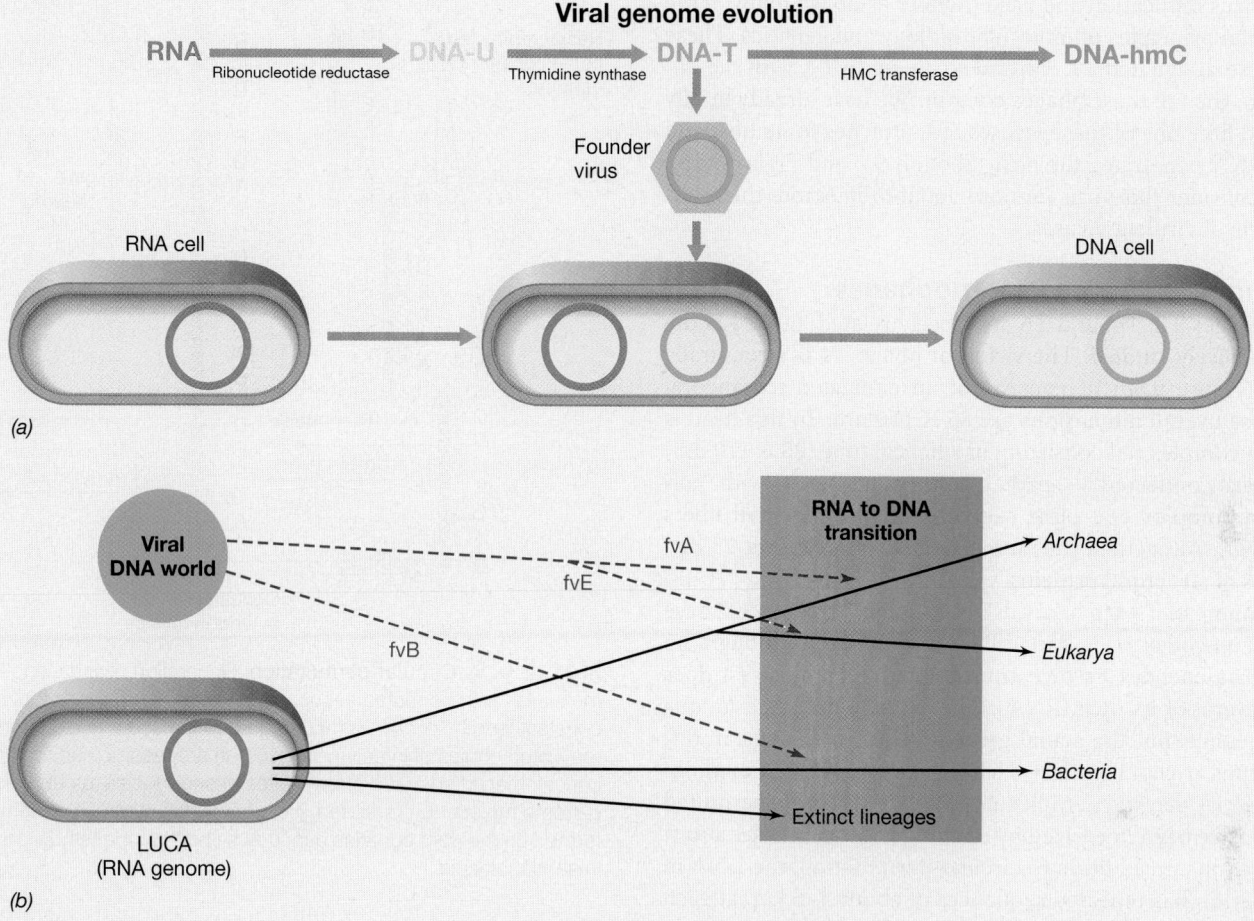

Viral genome evolution

RNA $\xrightarrow{\text{Ribonucleotide reductase}}$ DNA-U $\xrightarrow{\text{Thymidine synthase}}$ DNA-T $\xrightarrow{\text{HMC transferase}}$ DNA-hmC

Founder virus

RNA cell → → DNA cell

(a)

Viral DNA world

fvA

fvE

RNA to DNA transition

→ Archaea

→ Eukarya

fvB

→ Bacteria

→ Extinct lineages

LUCA (RNA genome)

(b)

Figure 1 Hypothesis of viral origin of DNA. *(a)* Several successive cycles of mutation and selection resulted in the appearance of viral nucleic acids more resistant to degradation by the host cell: DNA-U, DNA with uracil; DNA-T, DNA with thymine (i.e., normal DNA); DNA-hmC, DNA with 5-hydroxymethylcytosine. All four types of nucleic acid are found in present-day viruses, although DNA-U and DNA-hmC are rare. Conversion of RNA cellular genomes to DNA postulates lysogeny by a DNA "founder virus" followed by movement of host genes onto the DNA genome. *(b)* Three founder viruses, fvB, fvA, fvE, are hypothesized to have infected the ancestors of the *Bacteria*, *Archaea*, and Eukarya, respectively. Note that viruses fvA and fvE are more closely related to each other than to fvB. As a result of viral infection, the genomes of these three ancestral lines were eventually converted from RNA to DNA. Presumably, other cellular lineages derived from the last universal common ancestor (LUCA) that retained RNA genomes are extinct.

In the next two sections we examine two contrasting viral life cycles: *virulent* and *temperate*. In the virulent (or lytic) mode, viruses lyse or kill their hosts after infection, whereas in the temperate (or lysogenic) mode, viruses replicate their genomes in step with the host genome and without killing their hosts. A similar phenomenon is seen with viruses that infect higher organisms. When animal viruses divide in step with host cells, this is known as a "latent" infection.

MiniQuiz

- What type of nucleic acid is thought to be most common in bacteriophage genomes?
- What is the role of the contractile tails found in many bacteriophages?
- How do the virulent and temperate lifestyles of a bacteriophage differ?

9.9 Virulent Bacteriophages and T4

Virulent viruses kill their hosts after infection. The first such viruses to be studied in detail were bacteriophages with linear, dsDNA genomes that infect *Escherichia coli* and a number of related *Bacteria*. Virologists studied these viruses as model systems for virus replication and used them to establish many of the fundamental principles of molecular biology and genetics. These phages were designated T1, T2, and so on, up to T7, with the "T" referring to the tail these phages contain. We have already briefly mentioned how one of these viruses, T4, attaches to its host and how its DNA penetrates the host (Section 9.6 and Figure 9.10). Here we consider this virus in more detail to illustrate the replication cycle of virulent viruses.

The Genome of T-Even Bacteriophages

Bacteriophages T2, T4, and T6 are closely related, but T4 is the most extensively studied. The virion of phage T4 is structurally complex (Figure 9.5*b*). It consists of an elongated icosahedral head whose overall dimensions are 85×110 nm. To this head is attached a complex tail consisting of a helical tube (25×110 nm) to which are connected a sheath, a connecting "neck" with "collar," and a complex end plate carrying long, jointed tail fibers (Figure 9.5*b*). Altogether, the virus particle contains over 25 distinct types of structural proteins.

The genome of T4 is a linear dsDNA molecule of 168,903 base pairs that encodes over 250 different proteins. Although no known virus encodes its own translational apparatus, T4 does encode several of its own tRNAs. The T4 genome has a unique linear sequence, but the actual genomic DNA molecules in different virions are not identical. This is because the DNA of phage T4 is *circularly permuted*. Molecules that are circularly permuted appear to have been linearized by opening identical circles at different locations. In addition to circular permutation, the DNA in each T4 virion has repeated sequences of about 3–6 kbp at each end, called *terminal repeats*. Both of these factors affect genome packaging.

When T4 DNA enters a host cell, it is first replicated as a unit, and then several genomic units are recombined end to end to form a long DNA molecule called a *concatemer* (**Figure 9.13**). During the packaging of T4 DNA, the DNA is not cut at a specific sequence. Instead, a segment of DNA long enough to fill a phage head is cut from the concatemer. Because the T4 head holds slightly more than a genome length, this "headful mechanism" leads to circular permutation and terminal redundancy. T4 DNA contains the modified base *5-hydroxymethylcytosine* in place of cytosine (**Figure 9.14**). These residues are glucosylated (Section 9.6), and DNA with this modification is resistant to virtually all known restriction enzymes. Consequently, the incoming T4 DNA is protected from host defenses.

Events During T4 Infection

Things happen rapidly in a T4 infection. Early in infection T4 directs the synthesis of its own RNA and also begins to replicate its unique DNA. About 1 min after attachment and penetration of the host by T4 DNA, the synthesis of host DNA and RNA ceases and transcription of specific phage genes begins. Translation of

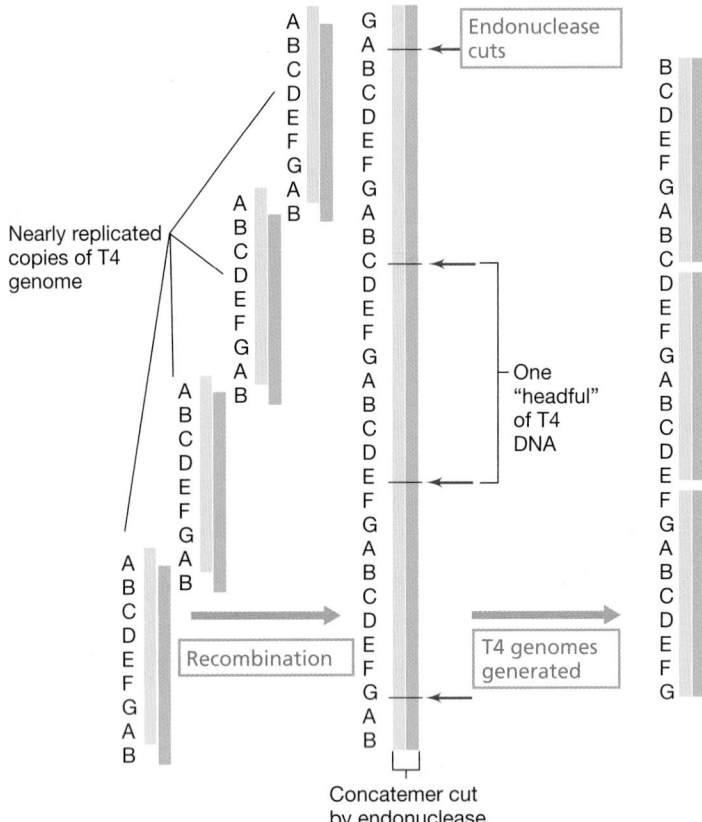

Figure 9.13 Circular permutation. Generation of virus-length T4 DNA molecules with permuted sequences by an endonuclease that cuts off constant lengths of DNA without regard to the sequence. Left: nearly replicated copies of infecting T4 genome are recombined to form a concatemer. Middle: red arrows, sites of endonuclease cuts. Right: genome molecules generated. Note how each of the T4 genomes formed on the right contains genes A–G, but that the termini are unique in each molecule.

viral mRNA begins soon after, and within 4 min of infection, phage DNA replication has begun.

The T4 genome can be divided into three parts, encoding *early proteins*, *middle proteins*, and *late proteins*, respectively (**Figure 9.15**). The early and middle proteins are primarily enzymes needed for DNA replication and transcription, whereas the late

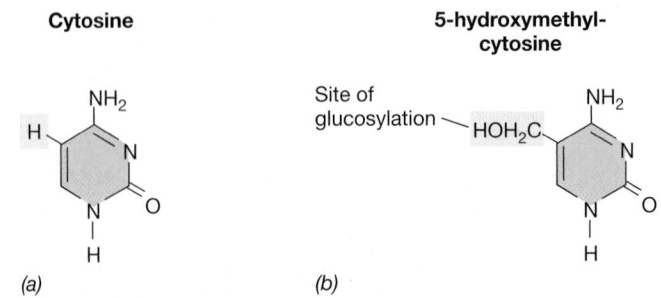

Figure 9.14 The unique base in the DNA of the T-even bacteriophages, 5-hydroxymethylcytosine. (*a*) Cytosine. (*b*) 5-Hydroxymethyl-cytosine. DNA containing glucosylated 5-hydroxymethylcytosine is resistant to cutting by restriction enzymes.

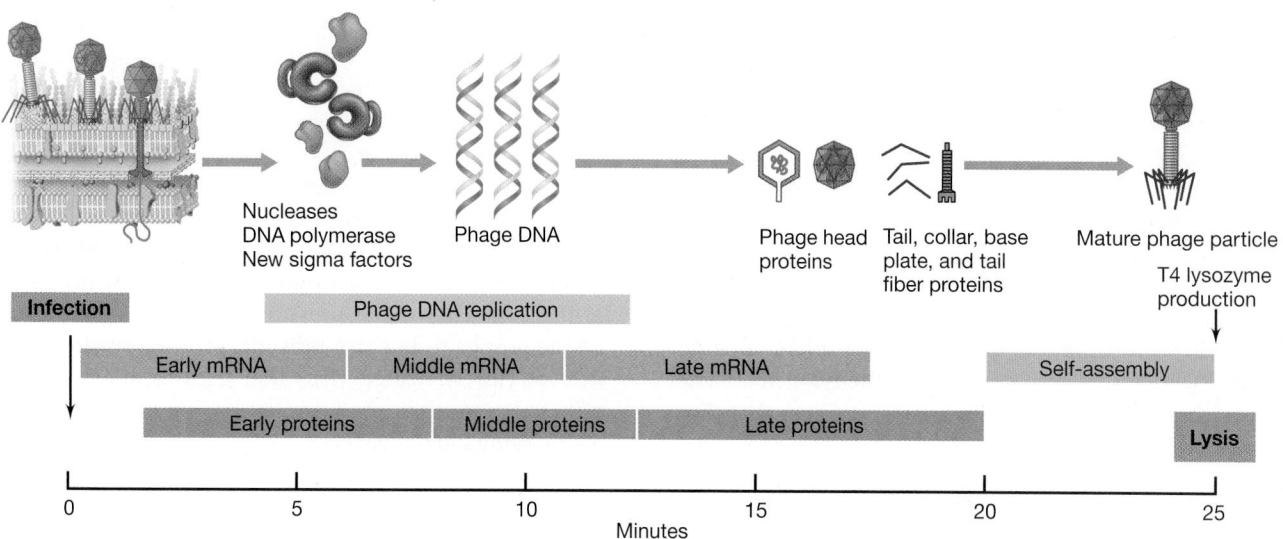

Figure 9.15 Time course of events in phage T4 infection. Following injection of DNA, early and middle mRNA is produced that codes for nucleases, DNA polymerase, new phage-specific sigma factors, and other proteins needed for DNA replication. Late mRNA codes for structural proteins of the phage virion and for T4 lysozyme, which is needed to lyse the cell and release new phage particles.

proteins are the head and tail proteins and the enzymes required to liberate the mature phage particles from the cell. The time course of events during T4 infection is shown in Figure 9.15.

Although T4 has a very large genome for a virus, it does not encode its own RNA polymerase. The control of T4 mRNA synthesis requires the production of proteins that sequentially modify the specificity of the host RNA polymerase so that it recognizes phage promoters. The early promoters are read directly by the host RNA polymerase and require the host sigma factor. Host transcription is shut down shortly after this by a phage-encoded anti-sigma factor that binds to host σ^{70} ($\rightleftarrows$ Section 6.13) and interferes with its recognition of host promoters.

Phage-specific proteins encoded by the early genes also covalently modify the host RNA polymerase α-subunits ($\rightleftarrows$ Section 6.12), and a few phage-encoded proteins also bind to the RNA polymerase. These modifications change the specificity of the host RNA polymerase so that it now recognizes T4 middle promoters. One of the T4 early proteins, MotA, recognizes a particular DNA sequence in T4 middle promoters and guides RNA polymerase to these sites. Transcription from the late promoters requires a new T4-encoded sigma factor. Sequential modification of host cell RNA polymerase as described here for phage T4 is used to regulate gene expression by many other bacteriophages as well.

T4 encodes over 20 new proteins that are synthesized early after infection. These include enzymes for the synthesis of the unusual base 5-hydroxymethylcytosine (Figure 9.14) and for its glucosylation, as well as an enzyme that degrades the normal DNA precursor deoxycytidine triphosphate. In addition, T4 encodes a number of enzymes that have functions similar to those of host enzymes in DNA replication, but that are formed in larger amounts, thus permitting faster synthesis of T4-specific DNA. Additional early proteins include those involved in the processing of newly replicated phage DNA (Figure 9.13).

Most late genes encode structural proteins for the virion, including those for the head and tail. The assembly of heads and tails is independent. The DNA is actively pumped into the head until the internal pressure reaches the required level, which is over ten times that of bottled champagne! The tail and tail fibers are added after the head has been filled (Figure 9.15). The phage encodes an enzyme, T4 lysozyme, which degrades the peptidoglycan layer of the host cell. The virus exits when the cell is lysed. After each replication cycle, which takes only about 25 min (Figure 9.15), over 100 new virions are released from each host cell, which itself has now been almost completely destroyed.

MiniQuiz

- What does it mean that the bacteriophage T4 genome is both circularly permuted and has terminal repeats?
- Explain how T4 ensures that its genes, rather than those of the host, are transcribed.

9.10 Temperate Bacteriophages, Lambda and P1

Bacteriophage T4 is virulent. However, some other viruses, although able to kill cells through a virulent cycle, also possess an alternative life cycle that results in a stable relationship with the host. Such viruses are called **temperate viruses**. Such viruses can enter into a state called **lysogeny**, where most virus genes are not expressed and the virus genome, called a **prophage**, is replicated in synchrony with the host chromosome. It is expression of the viral genome that harms the host cell, not the mere presence of viral DNA. Consequently, host cells can harbor viral genomes without harm, provided that the viral genes for lytic functions are not expressed. In cells that harbor a temperate virus, called **lysogens**, the phage genome is replicated in step with the host genome and, during cell division, is passed from one generation to the next. Under certain stressful conditions temperate viruses may revert to the **lytic pathway** and begin to produce virions.

The two best-characterized temperate phages are lambda and P1. Both have contributed significantly to the advance of molecular genetics and are used in bacterial genetics (phage P1, ⮌ Section 10.8) and molecular cloning (lambda, ⮌ Section 11.9). Lysogeny is also of ecological importance because most bacteria isolated from nature are lysogens for one or more bacteriophages. Lysogeny can confer new genetic properties on the bacterial host cell, and we will see several examples in later chapters of pathogenic bacteria whose virulence depends on the lysogenic bacteriophage they harbor. Many animal viruses persist in their host cells in ways that resemble lysogeny.

The Replication Cycle of a Temperate Phage

Temperate phages may enter the virulent mode after infecting a host cell or they may establish lysogeny. An overall view of the life cycle of a temperate bacteriophage is shown in **Figure 9.16**.

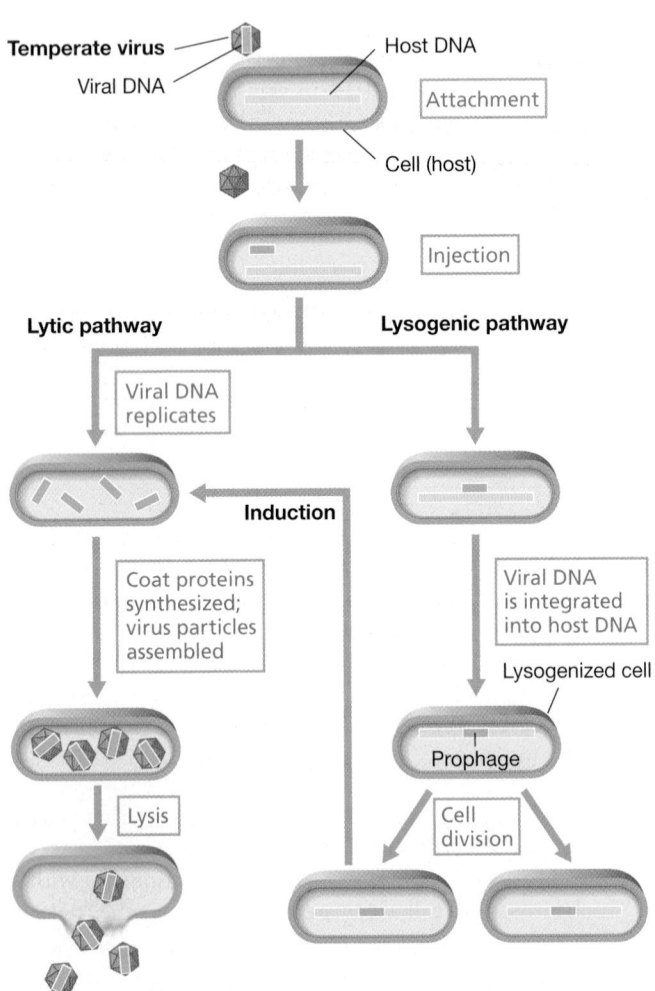

Figure 9.16 The consequences of infection by a temperate bacteriophage. The alternatives upon infection are replication and release of mature virus (lysis) or lysogeny, often by integration of the virus DNA into the host DNA, as shown here. The lysogen can be induced to produce mature virus and lyse.

During lysogeny, the temperate virus does not exist as a virus particle inside the cell. Instead, the virus genome is either integrated into the bacterial chromosome (e.g., bacteriophage lambda) or exists in the cytoplasm in plasmid form (e.g., bacteriophage P1). In either case, it replicates in step with the host cell as long as the genes activating its virulent pathway are not expressed. These forms of the virus are known as prophages. Typically, this control is due to a phage-encoded repressor protein (clearly, the gene encoding the repressor protein must be expressed). The virus repressor protein not only controls genes on the prophage, but also prevents gene expression by any identical or closely-related virus that tries to infect the same host cell. This results in the lysogens having immunity to infection by the same type of virus.

If the phage repressor is inactivated or if its synthesis is prevented, the prophage is induced (Figure 9.16). New virions are produced, and the host cell is lysed. Altered conditions, especially damage to the host cell DNA, induce the lytic pathway in some cases (e.g., in bacteriophage lambda). If the virus loses the ability to leave the host genome because of mutation, it becomes a *cryptic virus*. Genomic studies have shown that many bacterial chromosomes contain DNA sequences that were clearly once part of a viral genome. Thus, the establishment and breakdown of the lysogenic state is likely a dynamic process in prokaryotes. www.microbiologyplace.com **Online Tutorial 9.1: A Temperate Bacteriophage**

Bacteriophage Lambda

Bacteriophage lambda, which infects *Escherichia coli*, has been studied in great detail. As with other temperate viruses, both the virulent and the temperate pathways are possible (Figure 9.16). Lambda virions resemble those of other tailed bacteriophages, although no tail fibers are present in the commonly used laboratory strains (**Figure 9.17** and Figure 9.12). Wild-type lambda does have tail fibers. The lambda genome consists of linear dsDNA. However, at the 5′ terminus of each strand is a single-stranded region 12 nucleotides long. These single-stranded cohesive ends are complementary, and when lambda DNA enters the host cell

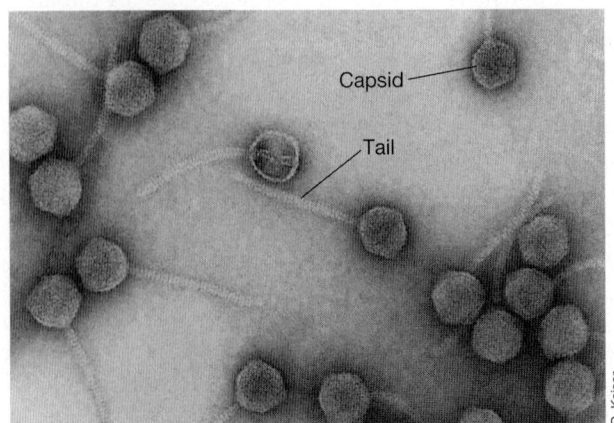

Figure 9.17 Bacteriophage lambda. Electron micrograph by negative staining of phage lambda virions. The head of each virion is about 65 nm in diameter and contains linear dsDNA.

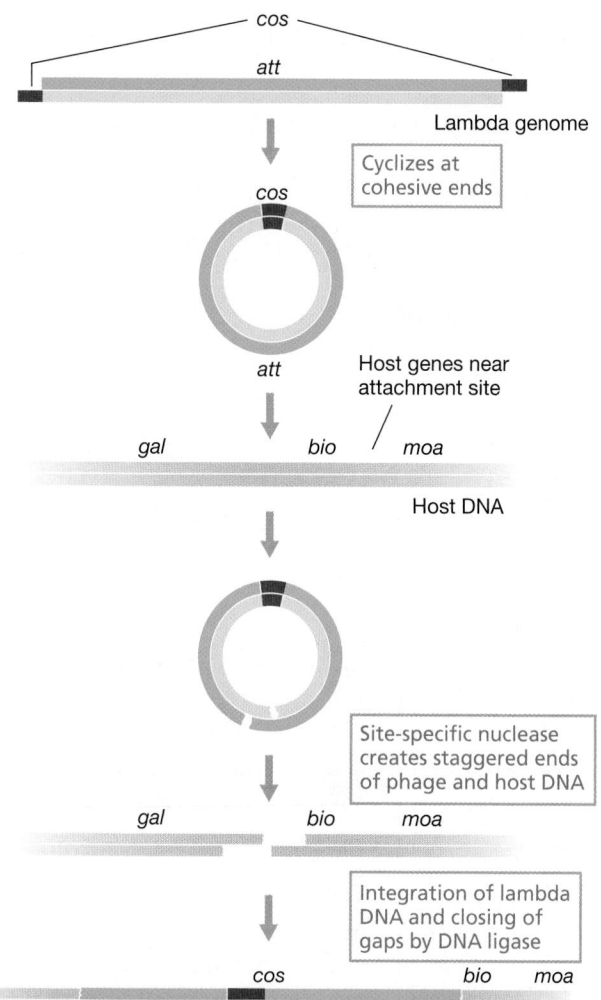

Figure 9.18 Integration of lambda DNA into the host. Integration always occurs at specific attachment sites (*att* sites) on both the host DNA and the phage. Some host genes near the attachment site are given: *gal* operon, galactose utilization; *bio* operon, biotin synthesis; *moa* operon, molybdenum cofactor synthesis. A site-specific enzyme (integrase) is required, and specific pairing of the complementary ends results in integration of phage DNA.

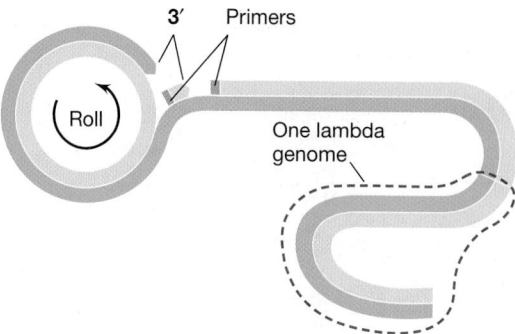

Figure 9.19 Rolling circle replication of the lambda genome. As the dark green strand rolls out, it is being replicated at its opposite end. Note that this synthesis is *asymmetric* because one of the parental strands continues to serve as a template and the other is used only once.

and circularizes, they base-pair, forming what is known as the *cos* site (**Figure 9.18**). The DNA is then ligated, forming a double-stranded circle.

When lambda is lysogenic, it integrates into the *E. coli* chromosome at a unique site known as the lambda attachment site, *att*λ. Integration requires the enzyme lambda integrase, which recognizes the phage and bacterial attachment sites (labeled *att* in Figure 9.18) and catalyzes integration. The integrated lambda DNA is then replicated along with the rest of the host genome and transmitted to progeny cells.

When lambda enters the virulent (lytic) pathway, it synthesizes long, linear concatemers of DNA by rolling circle replication (**Figure 9.19**). In contrast to semiconservative replication, this mechanism is asymmetrical and occurs in two stages. In the first stage, one strand of the circular lambda genome is nicked. Then a long single-stranded concatemer is made using the intact strand

as a template. In the next stage, a second strand is made using the single-stranded concatemer as a template. Finally, the double-stranded concatemer is cut into genome-sized lengths at the *cos* sites, resulting in cohesive ends. The linear genomes are packaged into phage heads and the tails are added; the host cell is then lysed by phage-encoded enzymes. Many DNA and RNA viruses and some plasmids use variants of rolling circle replication. In some cases, single-stranded concatemers are cut and packaged; in other cases, the complementary strand is made before packaging, as in lambda.

Lambda: Lysis or Lysogeny?

Whether lysis or lysogeny occurs during lambda infection depends on an exceedingly complex genetic switch. The key elements are two repressor proteins, the lambda repressor, or cI protein (Figure 9.20), and the repressor protein Cro. To establish lysogeny, two events must happen: (1) The production of late proteins must be prevented; and (2) a copy of the lambda genome must be integrated into the host chromosome. If cI is made, it represses the synthesis of all other lambda-encoded proteins and lysogeny is established. Conversely, Cro indirectly represses the expression of the lambda cII and cIII proteins, which are needed to maintain lysogeny, by inducing synthesis of the cI. Thus, when Cro is made in high amounts, lambda is committed to the lytic pathway. The degradation of cII by a host cell protease (FtsH protein) is also critical. The cIII protein protects cII against protease attack and stabilizes it. A summary of the steps controlling lambda lysis and lysogeny is shown in **Figure 9.20**. The final outcome is determined by whether Cro protein or cI dominates in a given infection. If Cro dominates regulatory events, the outcome is lysis, whereas if cI dominates, lysogeny will occur.

MiniQuiz
- What are the two pathways available to a temperate virus?
- What is a lysogen?
- What events need to happen for lambda to become a prophage?

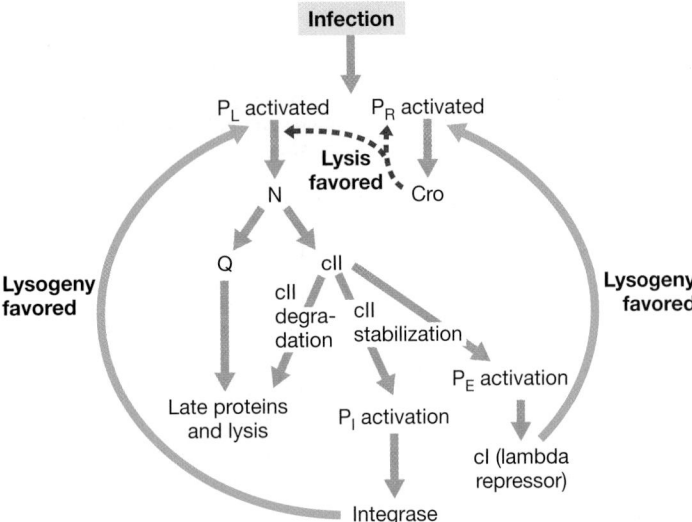

Figure 9.20 Summary of the steps in lambda infection. Lysis versus lysogeny is governed by whether or not the lambda repressor (cI) is made. High Cro activity prevents transcription (red dashed arrows) from the lambda leftward promoter, P_L, and the lambda rightward promoter, P_R. This prevents the synthesis of N protein, which in turn results in a decrease of both Q protein and cII protein. The lack of cII prevents synthesis of cI protein, and the result is lysis. The level of cII also depends on its degradation by host proteases versus its protection by lambda cIII protein (not shown). If sufficient cII is present, the promoters for cI (P_E) and integrase (P_I) are activated (green arrows) and both cI and integrase are made. This results in integration and lysogeny.

9.11 Overview of Animal Viruses

The first few sections of this chapter were devoted to general properties of viruses, and little was said about animal viruses. Here we consider animal viruses. It is important to remember that the bacteriophage host is a bacterial cell, whereas the host of an animal virus is a eukaryotic cell. We will expand on these important differences in Chapter 21, where we discuss several types of animal virus in more detail. However, the key points are that (1) unlike in prokaryotes, the entire virion typically enters the animal cell, and (2) eukaryotic cells contain a nucleus, where many animal viruses replicate.

Classification of Animal Viruses

Various types of animal viruses are illustrated in **Figure 9.21**. We discussed the principles of virus classification in Section 9.7. As for bacterial viruses, animal viruses are classified according to the Baltimore classification system (Table 9.2), which classifies viruses by genome type and reproductive strategy. Animal viruses are known in all replication categories, and an example of each is discussed in Chapter 21. Most animal viruses that have been studied in detail are those that can replicate in cell cultures (Section 9.3).

Note that there are many more kinds of enveloped animal viruses than enveloped bacterial viruses (Section 9.8). This relates to the differences in host cell exteriors. Unlike prokaryotic cells, animal cells lack a cell wall, and thus viruses are more easily released from the cell. Many animal viruses are enveloped and when these exit, they remove part of the animal cell's lipid bilayer as they pass through the membrane.

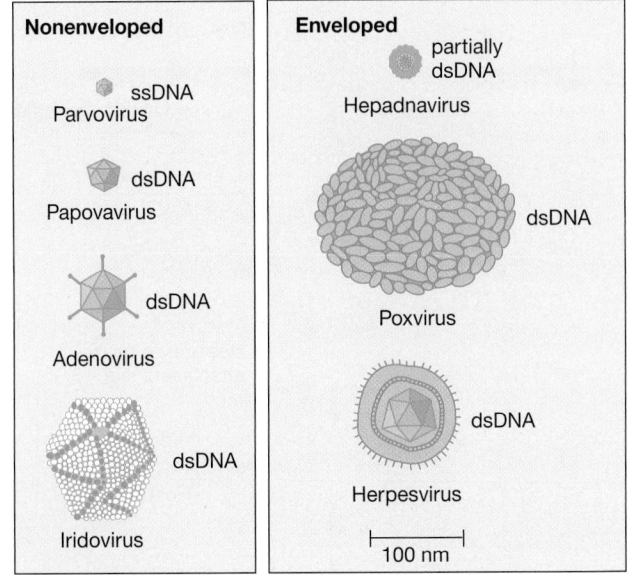

(a) **DNA viruses**

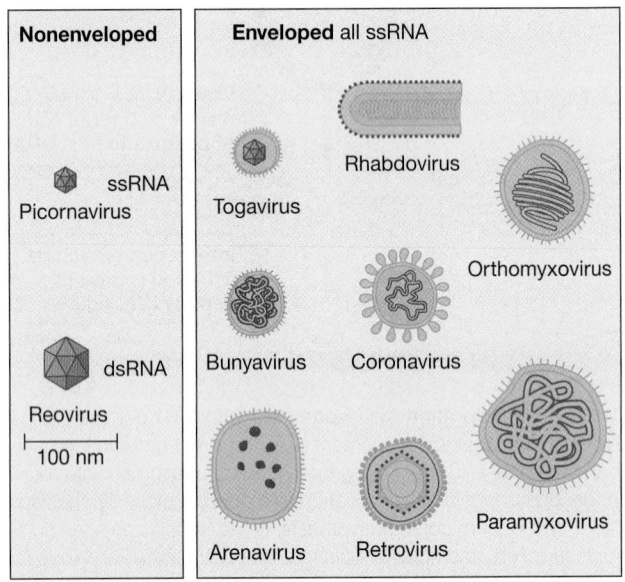

(b) **RNA viruses**

Figure 9.21 Diversity of animal viruses. The shapes and relative sizes of the major groups of vertebrate viruses. The hepadnavirus genome has one complete DNA strand and part of the complementary strand (↪ Section 21.11).

Consequences of Virus Infection in Animal Cells

Viruses can have several different effects on animal cells. Virulent infection results in the destruction of the host cell (**Figure 9.22**). With enveloped viruses, however, release of virions, which occurs by a kind of budding process, may be slow, and the host cell may not be lysed. The infected cell may therefore remain alive and continue to produce virus indefinitely. Such infections are called persistent infections (Figure 9.22).

Viruses may also cause latent infection of a host. In a latent infection, there is a delay between infection by the virus and host cell lysis. Fever blisters (cold sores), caused by the herpes simplex virus (↪ Section 21.14), are a typical example of a latent viral infection;

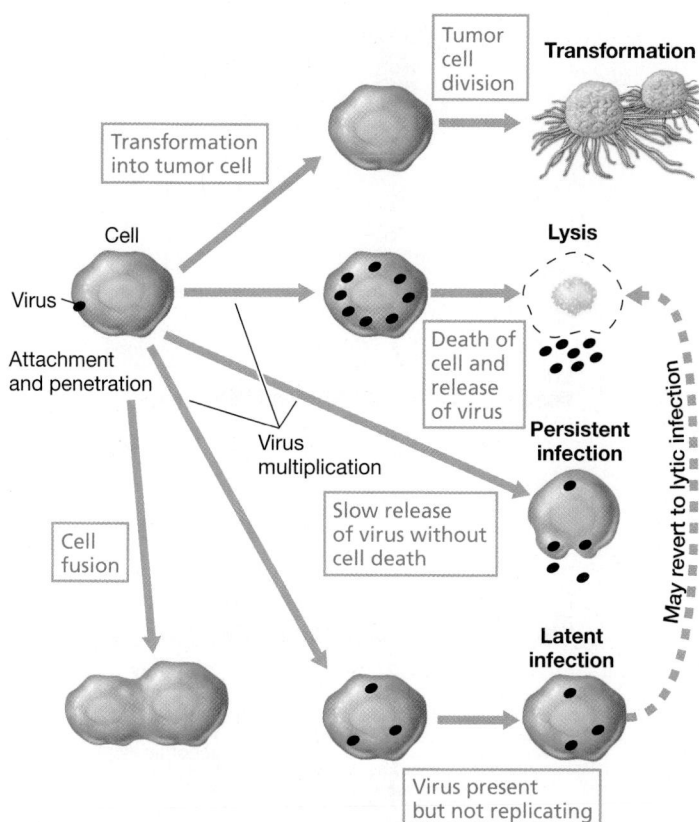

Figure 9.22 Possible effects that animal viruses may have on cells they infect. Most animal viruses are lytic, and only very few are known to cause cancer.

the symptoms (the result of lysed cells) reappear sporadically as the virus emerges from latency. The latent stage in viral infection of an animal cell is not usually due to integration of the viral genome into the host genome, as often happens with lysogenic infections by temperate bacteriophages. Instead, herpesviruses exist in a relatively inactive state within nerve cells. A low level of transcription continues, but the viral DNA does not replicate.

Some enveloped viruses promote fusion between multiple animal cells, creating giant cells with several nuclei (Figure 9.22). Not surprisingly, such fused cells fail to develop correctly and are short-lived. Cell fusion allows viruses to avoid exposure to the immune system by moving between host cell nuclei without emerging from the host cells. Finally, certain animal viruses can convert a normal cell into a tumor cell, a process called **transformation**. We discuss cancer-causing viruses in Sections 21.11 and 21.14.

Many different animal viruses are known. But of all the viruses listed in Figure 9.21, one group stands out as having an absolutely unique mode of replication. These are the retroviruses. We explore them next as an example of a complex and highly unusual animal virus with significant medical and evolutionary implications.

MiniQuiz

• Differentiate between a persistent and a latent viral infection.

• Contrast the ways in which animal viruses enter cells with those used by bacterial viruses.

9.12 Retroviruses

Retroviruses contain an RNA genome that is replicated via a DNA intermediate (Section 9.7 and Figure 9.11). The term *retro* means "backward," and the name retrovirus is derived from the fact that these viruses transfer information from RNA to DNA. Retroviruses employ the enzyme reverse transcriptase to carry out this interesting process. The use of reverse transcriptase is not restricted to the retroviruses. Hepatitis B virus (a human virus) and cauliflower mosaic virus (a plant virus) also use reverse transcription during their life cycles (ᴄ𝄐 Section 21.11). However, these other viruses carry the DNA version of their genome in the virion whereas retroviruses carry RNA.

Retroviruses are interesting for several other reasons. For example, they were the first viruses shown to cause cancer and have been studied for their carcinogenic characteristics. Also, one retrovirus, human immunodeficiency virus (HIV), causes acquired immunodeficiency syndrome (AIDS). This virus infects a specific kind of white blood cell (T-helper cell) in humans that is vital for proper functioning of the immune system. In later chapters we discuss the medical aspects of AIDS (ᴄ𝄐 Section 32.6).

Retroviruses are enveloped viruses (**Figure 9.23a**). There are several proteins in the virus envelope and typically seven internal proteins, four of which are structural and three of which are enzymatic. The enzymes found in the virion are reverse transcriptase, integrase, and a protease. The virion also contains specific cellular tRNA molecules used in replication (discussed later in this section).

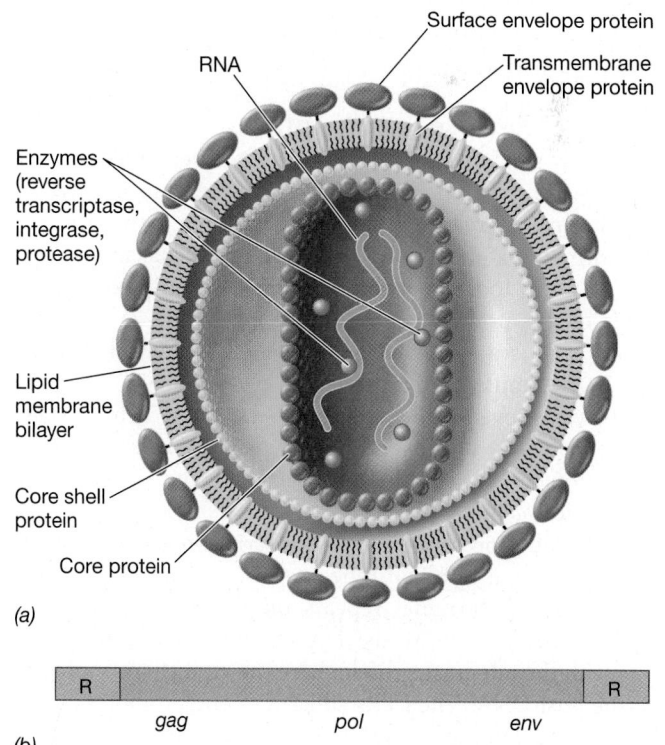

(a)

(b)

Figure 9.23 Retrovirus structure and function. *(a)* Structure of a retrovirus. *(b)* Genetic map of a typical retrovirus genome. Each end of the genomic RNA contains direct repeats (R).

Features of Retroviral Genomes and Replication

The genome of the retrovirus is unique. It consists of two identical single-stranded RNA molecules of the plus (+) orientation. A genetic map of a typical retrovirus genome is shown in Figure 9.23*b*. Although there are differences between the genetic maps of different retroviruses, all contain the following genes arranged in the same order: *gag*, encoding structural proteins; *pol*, encoding reverse transcriptase and integrase; and *env*, encoding envelope proteins. Some retroviruses, such as Rous sarcoma virus, carry a fourth gene downstream from *env* that is active in cellular transformation and cancer. The terminal repeats shown on the map are essential for viral replication.

The overall process of replication of a retrovirus can be summarized in the following steps (**Figure 9.24**):

1. Entry into the cell by fusion with the cytoplasmic membrane at sites of specific receptors
2. Removal of the virion envelope at the cytoplasmic membrane, but the genome and virus-specific enzymes remain in the virus core
3. Reverse transcription of one of the two identical genomic RNA molecules into a ssDNA that is subsequently converted by reverse transcriptase to a linear dsDNA molecule, which then enters the nucleus
4. Integration of retroviral DNA into the host genome
5. Transcription of retroviral DNA, leading to the formation of viral mRNAs and viral genomic RNA
6. Assembly and packaging of the two identical genomic RNA molecules into nucleocapsids in the cytoplasm
7. Budding of enveloped virions at the cytoplasmic membrane and release from the cell

Activity of Reverse Transcriptase

A very early step after the entry of the RNA genome into the cell is reverse transcription: conversion of RNA into a DNA copy using the enzyme reverse transcriptase present in the virion. The DNA formed is a linear double-stranded molecule and is synthesized in the cytoplasm within an uncoated viral core particle. Details of this process can be found in Section 21.11. Reverse transcriptase is a type of DNA polymerase and, like all DNA polymerases, must have a primer (⮌ Section 6.8). The primer for retrovirus reverse transcription is unusual in being a specific tRNA encoded by the host cell. The type of tRNA used as primer depends on the virus and is packaged into the virion from the previous host cell.

The overall process of reverse transcription generates a product that has long terminal repeats (LTRs, Figure 9.24) that are longer than the terminal repeats on the RNA genome itself (Figure 9.23). This entire dsDNA molecule enters the nucleus along with the integrase protein; here the viral DNA is integrated into the host DNA. The LTRs contain strong promoters of transcription and participate in the integration process. The integration of the retroviral DNA into the host genome is analogous to the integration of phage DNA into a bacterial genome to form a lysogen, except that the retrovirus cannot excise its DNA from the host genome. Thus, once integrated, the retroviral DNA, now

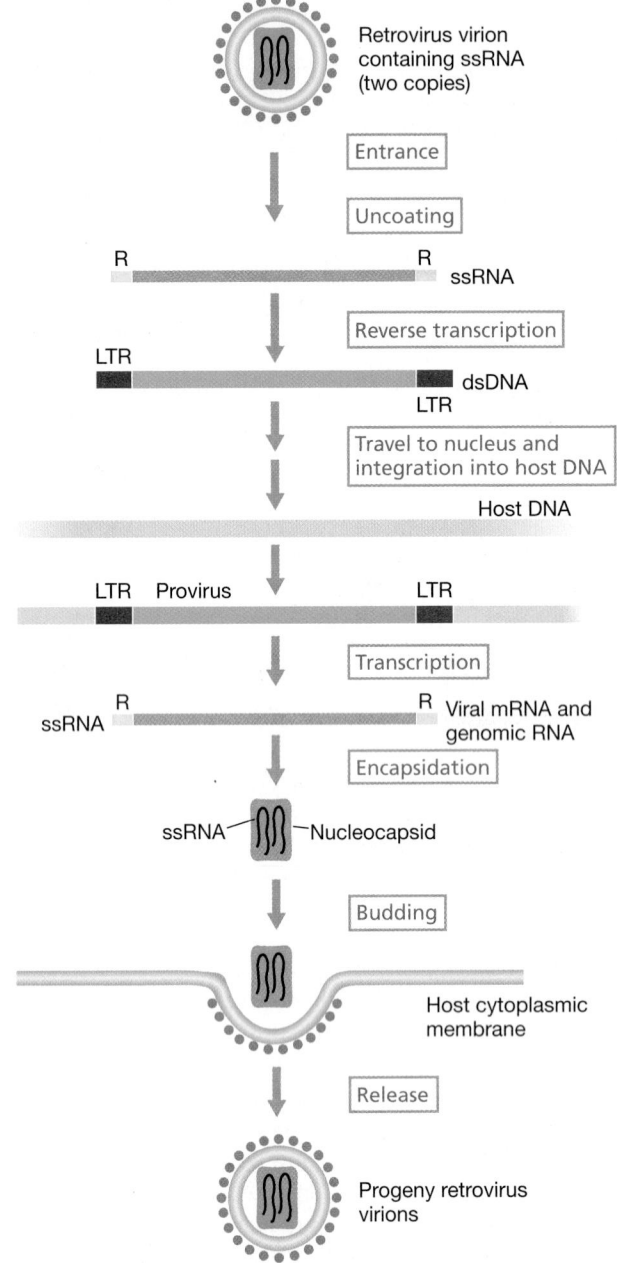

Figure 9.24 **Replication process of a retrovirus.** R, direct repeats; LTR, long terminal repeats. For more details on the conversion of RNA to DNA (reverse transcription, step 3), refer to Section 21.11.

called a **provirus**, becomes a permanent insertion into the host genome (Section 9.10). Viral DNA can be integrated anywhere in the host chromosomal DNA. Indeed, many higher eukaryotic genomes have high numbers of endogenous retroviral sequences. An estimated 8% of the sequences of the human genome are of retroviral origin.

If the promoters in the right-hand LTR are activated, the integrated proviral DNA is transcribed by a host cell RNA polymerase into RNA transcripts. These RNA transcripts may either be packaged into virus particles (as the genome) or may act as mRNA and be translated into virus proteins. Some virus proteins are made initially as a large primary *gag* protein that is split by proteolysis into the capsid proteins. Occasionally ribosomes fail to terminate at the *gag* stop codon (due either to inserting an amino acid at a stop codon or a shift in reading frame by the ribosome). This leads to the low-level translation of *pol*, the reverse transcriptase gene, which yields reverse transcriptase for insertion into virions. Note that reverse transcriptase is needed in much lower amounts than the retrovirus structural proteins.

When virus structural proteins have accumulated in sufficient amounts, nucleocapsids are assembled. Encapsidation of the RNA genome leads to the formation of mature nucleocapsids, which move to the cytoplasmic membrane for final assembly into the enveloped virions. As nucleocapsids bud through the cytoplasmic membrane they are sealed and then released and may infect neighboring cells (Figure 9.24).

MiniQuiz

- Why are some viruses known as retroviruses?
- How does the replication cycle of a temperate bacteriophage differ from that of a retrovirus?

IV Subviral Entities

We have defined a virus as a genetic element that subverts normal cellular processes for its own replication and that has an infectious extracellular form. There are several infectious agents that resemble viruses but whose properties are at odds with this definition, and are thus not considered viruses. Defective viruses are clearly derived from viruses but have become dependent on other, complete, viruses to supply certain gene products. In contrast, two of the most important subviral entities, viroids and prions, are not viruses at all, but differ in fundamental ways from viruses. They both illustrate the unusual ways that genetic elements can replicate and the unexpected ways they can subvert their host cells. However, prions stand out among all the entities we have considered in this chapter because the infectious transmissible agent lacks nucleic acid.

9.13 Defective Viruses

Some viruses cannot infect a host cell alone and rely on other viruses, known as **helper viruses**, to provide certain functions. Some of these so-called **defective viruses** merely rely on intact helper viruses of the same type to provide necessary functions. Far more interesting are those defective viruses, referred to as *satellite viruses*, for which no intact version of the same virus exists; these defective viruses rely on unrelated viruses as helpers.

Many defective viruses are known. For example, bacteriophage P4 of *Escherichia coli* can replicate, but its genome does not encode the major capsid protein. Instead, it relies on the related phage P2 as a helper to provide capsid proteins for the phage particle. However, P4 does encode an external scaffold protein that takes part in capsid assembly.

Satellite viruses are found in both animals and plants. For example, adeno-associated virus (AAV) is a satellite virus of humans that depends on adenovirus as a helper. AAV and adenovirus belong to two quite different virus families. Thus, AAV is not just a defective mutant of adenovirus, but is an unrelated virus that inhabits the same host cells. Because it causes little or no damage to the host, AAV is now being used as a eukaryotic cloning vector in gene therapy (↩ Section 15.17). In this system, AAV can be used to carry replacement genes to specific host tissues without causing disease itself.

MiniQuiz

- What is a helper virus?
- What is a satellite virus?

9.14 Viroids

Viroids are infectious RNA molecules that differ from viruses in lacking a capsid. Despite this lack, they have a reasonably stable extracellular form that travels from one host cell to another. Viroids are small, circular, single-stranded RNA molecules that are the smallest known pathogens. They range in size from 246 to 399 nucleotides and show a considerable degree of sequence homology to each other, suggesting that they have common evolutionary roots. Viroids cause a number of important plant diseases and can have a severe agricultural impact (**Figure 9.25**). A few well-studied viroids include coconut cadang-cadang viroid (246 nucleotides), citrus exocortis viroid (375 nucleotides), and potato spindle tuber viroid (359 nucleotides). No viroids are known that infect animals or prokaryotes.

Viroid Structure and Function

The extracellular form of the viroid is naked RNA; there is no protein capsid of any kind. Although the viroid RNA is a single-stranded, covalently closed circle, there is so much secondary

Figure 9.25 Viroids and plant diseases. Photograph of healthy tomato plant (left) and one infected with potato spindle tuber viroid (PSTV) (right). The host range of most viroids is quite restricted. However, PSTV infects tomatoes as well as potatoes, causing growth stunting, a flat top, and premature plant death.

Figure 9.26 Viroid structure. Viroids consist of single-stranded circular RNA that forms a seemingly double-stranded structure by intra-strand base pairing.

structure that it resembles a short double-stranded molecule with closed ends (**Figure 9.26**). This apparently makes the viroid sufficiently stable to exist outside the host cell. Because it lacks a capsid, the viroid does not use a receptor to enter the host cell. Instead, the viroid enters a plant cell through a wound, as from insect or other mechanical damage. Once inside, viroids move from cell to cell via the plasmodesmata, which are the thin strands of cytoplasm that link plant cells (**Figure 9.27**).

Even more curious, the viroid RNA molecule contains no protein-encoding genes, and therefore the viroid is almost totally dependent on host function for its replication. The viroid is replicated in the host cell nucleus or chloroplast by one of the plant RNA polymerases. The result is a multimeric RNA molecule consisting of many viroid units joined end to end. The viroid does contribute one function to its own replication; part of the viroid itself has ribozyme activity (⇄ Section 7.8). This ribozyme activity is used for self-cleavage of the multimeric RNA molecule, which releases individual viroids.

Viroid Disease

Viroid-infected plants can be symptomless or develop symptoms that range from mild to lethal, depending on the viroid (Figure 9.25). The mechanisms by which viroids cause plant diseases remain unclear. Most severe symptoms are growth

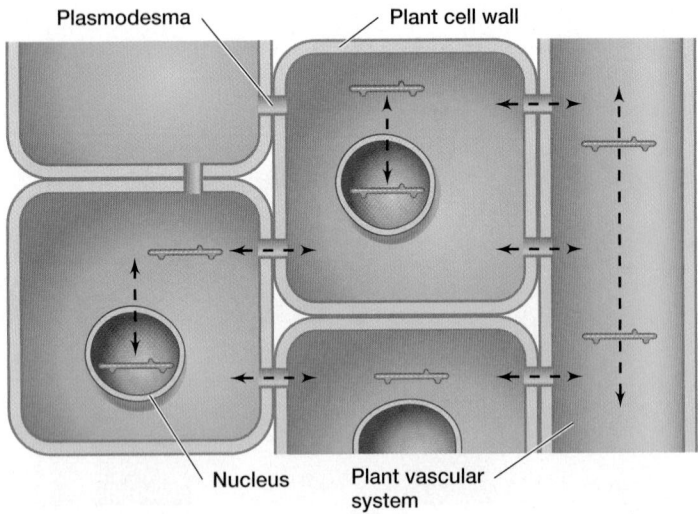

Figure 9.27 Viroid movement inside plants. After entry into a plant cell, viroids (orange) replicate either in the nucleus (shown here in purple) or in the chloroplast (not shown). Viroids can move between plant cells via the plasmodesmata (thin threads of cytoplasm that penetrate the cell walls and connect plant cells). In addition, on a larger scale, viroids can move around the plant via the plant vascular system.

related, suggesting that viroids mimic or interfere in some way with small regulatory RNAs (⇄ Section 7.11), examples of which are widely known in plants. Thus, viroids could themselves be derived from regulatory RNAs that have evolved away from carrying out beneficial roles in the cell to inducing destructive events. Recent data suggest that viroids give rise to siRNA (⇄ Section 7.10) as a side product during replication. It has been proposed that these siRNAs may then act via the RNA interference silencing pathway to suppress the expression of plant genes that show some homology to the viroid RNA. However, this is still unproven.

MiniQuiz

- If viroids are circular molecules, why are they usually drawn as compact rods?
- In what part of the host cell are viroids replicated?

9.15 Prions

Prions represent the other extreme from viroids. They have a distinct extracellular form, which consists entirely of protein. The prion particle contains neither DNA nor RNA. Nonetheless, it is infectious, and prions are known to cause diseases in animals, such as scrapie in sheep, bovine spongiform encephalopathy (BSE or "mad cow disease") in cattle, chronic wasting disease in deer and elk, and kuru and Creutzfeldt–Jakob disease (CJD) in humans. No prion diseases of plants are known, although prions have been found in yeast. Collectively, animal prion diseases are known as **transmissible spongiform encephalopathies (TSEs)**. In 1997 the American scientist Stanley B. Prusiner won the Nobel Prize for Physiology or Medicine for his pioneering work with these diseases and with the prion proteins.

In 1996 it became clear from disease tracking in England that the prion that causes BSE in cattle can also infect humans, resulting in a novel type of CJD called variant CJD (vCJD). Because transmission was from consumption of contaminated beef products, vCJD quickly became a worldwide health concern, with a major impact on the animal husbandry industry (⇄ Section 36.12). Most such instances of BSE occurred in the United Kingdom or other European Union (EU) countries and were linked to improper feeding practices in which protein supplements containing rendered cattle and sheep (including nervous tissues) were used to feed uninfected animals. Since 1994, this practice has been banned in all EU countries, and cases of BSE have dropped dramatically. Thus far, TSE transmission via other domesticated animals, such as swine, chicken, or fish, has not been found.

Forms of the Prion Protein

As prions lack nucleic acid, how is the protein they consist of encoded? The host cell contains a gene, *Prnp* (standing for "*Prion p*rotein") which encodes the native form of the prion protein, known as *PrP^C* (*Prion Protein Cellular*), that is primarily found in the neurons of healthy animals, especially in the brain. The pathogenic form of the prion protein is designated

PrP^Sc (prion protein *Scrapie*), because the first prion disease to be discovered was scrapie in sheep. PrP^Sc is identical in amino acid sequence to PrP^C from the same species, but has a different conformation. Prion proteins from different species of mammals are very similar, but are not identical in amino acid sequence. Susceptibility to infection depends on the protein sequence in a manner not fully understood. For example, PrP^Sc from cattle can infect humans, although at a very low frequency. However, PrP^Sc from sheep have never been observed to infect people. Native prions consist largely of α-helical segments, whereas pathogenic prions have less α-helix and more β-sheet regions instead. This causes the prion protein to lose its normal function, to become partially resistant to proteases, and to become insoluble, leading to aggregation within the neural cell (**Figure 9.28**). In this state, prion protein accumulates and neurological symptoms commence.

Prion Diseases and the Prion Infectious Cycle

When a pathogenic prion enters a host cell that is expressing native prion protein, it promotes the conversion of PrP^C protein into PrP^Sc. Thus the pathogenic prion does not subvert host enzymes or genes as a virus does; rather, it "replicates" by converting native prion proteins that already exist in the host cell into the pathogenic form. As the pathogenic prions accumulate, they form insoluble aggregates in the neural cells (Figure 9.28). This leads to disease symptoms that are invariably neurological and, in most cases, are due to destruction of brain or related

nervous tissue (Section 36.12). Whether the destruction of brain tissue is directly due to the accumulation of aggregated PrP^Sc is uncertain. PrP^C functions in the cell as a cytoplasmic membrane glycoprotein, and it has been shown that membrane attachment of pathogenic prions is necessary for disease symptoms to commence. Mutant versions of PrP^Sc that can no longer attach to nerve cell cytoplasmic membranes may still aggregate, but no longer cause disease.

Prion disease occurs by three distinct mechanisms, although all lead to the same result. In *infectious prion disease*, as described above, PrP^Sc is transmitted between animals or humans. In *sporadic prion disease*, random misfolding of a PrP^C molecule occurs in a normal, uninfected individual. This change is propagated as for infectious prion disease, and eventually PrP^Sc accumulates until symptoms appear. In humans this occurs in about one person in a million. In *inherited prion disease*, a mutation in the prion gene yields a prion protein that changes more often into PrP^Sc. Several different mutations are known whose symptoms vary slightly.

What happens if an incoming PrP^Sc protein finds no PrP^C to alter? The answer is that no disease results. This may seem surprising, but is logical given the mechanism of prion action. Mice that have been engineered with both copies of the *Prnp* gene disrupted and thus do not produce PrP^C are resistant to infection with pathogenic prions. Interestingly, such mice also live for a normal time and do not show any obvious behavioral abnormalities. This leaves wide open the puzzling question of what role PrP^C plays in brain cells.

Non-mammalian Prions

Are prions only found in mammals? Other vertebrates, including amphibians and fish, possess genes that are clearly homologous to the Prnp gene of mammals and that are expressed in nervous tissue. However, the proteins encoded by these genes do not have misfolded pathogenic versions and are therefore, by definition, not prions.

Curiously, proteins that fit the prion definition of an inherited self-perpetuating change in protein conformation are found in certain fungi, although they do not cause disease. Instead they adapt the fungal cells to altered conditions. In yeast, for example, the [URE3] prion is a transcription factor that regulates nitrogen metabolism. The normal, soluble form of this protein represses genes for using poor nitrogen sources. When the [URE3] prion accumulates, it forms insoluble aggregates, just as for mammalian prion protein. However, in yeast there is no pathogenic effect, instead the genes for nitrogen metabolism are derepressed.

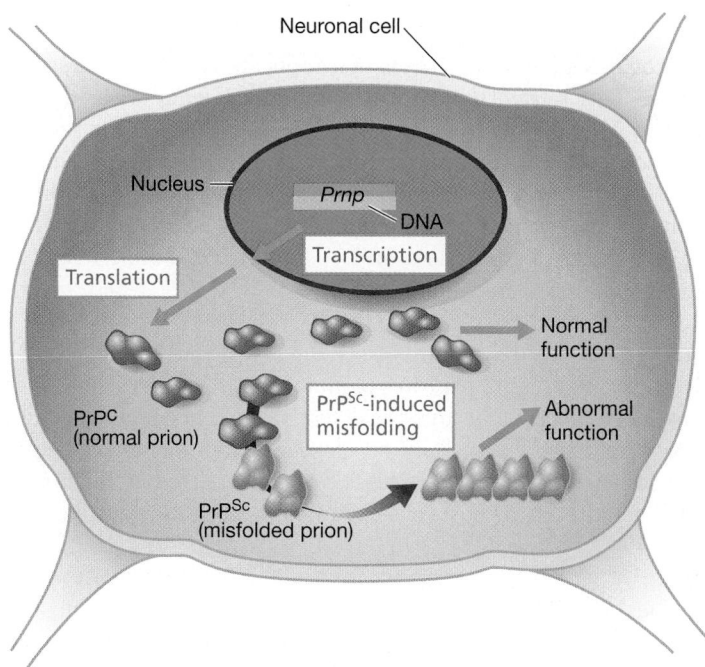

Figure 9.28 Mechanism of prion misfolding. Neuronal cells produce the native form of the prion protein. The pathogenic form of the prion protein catalyzes the refolding of native prions into the pathogenic form. The pathogenic form is protease resistant, insoluble, and forms aggregates in neural cells. This eventually leads to destruction of neural tissues and neurological symptoms.

MiniQuiz

- What is the difference between the native and pathogenic forms of the prion protein?
- How does sporadic prion disease differ from the transmitted form?
- How does a prion differ from a viroid?

Big Ideas

9.1

A virus is an obligate intracellular parasite that cannot replicate without a suitable host cell. A virion is the extracellular form of a virus and contains either an RNA or a DNA genome inside a protein shell. The virus genome may enter a new host cell by infection. The virus redirects the host metabolism to support virus replication. Viruses are classified by their nucleic acid and type of host.

9.2

In the virion of a naked virus, only nucleic acid and protein are present, with the nucleic acid on the inside; the whole unit is called the nucleocapsid. Enveloped viruses have one or more lipoprotein layers surrounding the nucleocapsid. The nucleocapsid is arranged in a symmetric fashion, with a precise number and arrangement of structural subunits surrounding the virus nucleic acid. Although virus particles are metabolically inert, one or more key enzymes are present within the virion in some viruses.

9.3

Viruses can replicate only in certain types of cells or in whole organisms. Bacterial viruses have proved useful as model systems because the host cells are easy to grow and manipulate in culture. Many animal and plant viruses can be grown in cultured cells.

9.4

Although only a single virion is required to initiate an infectious cycle, not all virions are equally infectious. One of the most accurate ways of measuring virus infectivity is by the plaque assay. Plaques are clear zones that develop on lawns of host cells. Theoretically, each plaque is due to infection by a single virus particle. The virus plaque is analogous to the bacterial colony.

9.5

The virus replication cycle can be divided into five stages: attachment (adsorption), penetration (injection), protein and nucleic acid synthesis, assembly and packaging, and virion release.

9.6

The attachment of a virion to a host cell is a highly specific process requiring complementary receptors on the surface of a susceptible host cell and its infecting virus. Resistance of the host to infection by the virus can involve restriction–modification systems that recognize and destroy foreign double-stranded DNA.

9.7

Before viral nucleic acid can replicate, new virus proteins are needed, and these are encoded by mRNA transcribed from the virus genome. In some RNA viruses, the viral genomic RNA is also the mRNA. In other viruses, the virus genome is a template for the formation of viral mRNA, and in certain cases, essential transcriptional enzymes are contained in the virion.

9.8

Bacterial viruses, or bacteriophages, are very diverse. The best-studied bacteriophages infect bacteria such as *Escherichia coli* and are structurally quite complex, containing heads, tails, and other components.

9.9

After a virion of T4 attaches to a host cell and the DNA penetrates into the cytoplasm, the expression of viral genes is regulated so as to redirect the host synthetic machinery to the production of viral nucleic acid and protein. New virions are then assembled and are released by lysis of the cell. T4 has a double-stranded DNA genome that is circularly permuted and terminally redundant.

9.10

Lysogeny is a state in which lytic events are repressed. Viruses capable of entering the lysogenic state are called temperate viruses. In lysogeny the virus genome becomes a prophage, either by integration into the host chromosome or by replicating like a plasmid in step with the host cell. However, lytic events can be induced by certain environmental stimuli.

9.11

There are animal viruses with all known modes of viral genome replication. Many animal viruses are enveloped, picking up portions of host membrane as they leave the cell. Not all infections of animal host cells result in cell lysis or death; latent or persistent infections are common, and a few animal viruses can cause cancer.

9.12

Retroviruses are RNA viruses that replicate via a DNA intermediate. The retrovirus human immunodeficiency virus (HIV) causes AIDS. The retrovirus particle contains an enzyme, reverse transcriptase, that copies the information from its RNA genome into DNA. The DNA is then integrated into the host chromosome in the manner of a temperate virus. The retrovirus DNA can be transcribed to yield mRNA (and new genomic RNA) or may remain in a latent state.

9.13

Defective viruses are parasites of intact helper viruses. The helper viruses supply proteins that the defective virus no longer encodes. Some defective viruses rely on closely related but intact helper viruses. However, satellite viruses rely on unrelated intact viruses that infect the same host cells to complete replication events.

9.14

Viroids are circular single-stranded RNA molecules that do not encode proteins and are dependent on host-encoded enzymes, except for the ribozyme activity of the viroid molecule itself. Viroids are the smallest known pathogens that contain nucleic acids.

9.15

Prions consist of protein, but have no nucleic acid. Prions exist in two conformations, the native cellular form and the pathogenic form. The pathogenic form "replicates" itself by converting native prion proteins into the pathogenic conformation.

Review of Key Terms

Bacteriophage a virus that infects prokaryotic cells

Capsid the protein shell that surrounds the genome of a virus particle

Capsomere the subunit of a capsid

Defective virus a virus that relies on another virus, the helper virus, to provide some of its components

Early protein a protein synthesized soon after virus infection and before replication of the virus genome

Helper virus a virus that provides some necessary components for a defective virus

Host cell a cell inside which a virus replicates

Icosahedron a three-dimensional figure with 20 triangular faces

Late protein a protein synthesized later in virus infection, after replication of the virus genome

Lysogen a bacterium containing a prophage

Lysogeny a state in which a viral genome is replicated as a prophage along with the genome of the host

Lytic pathway a series of steps after virus infection that leads to virus replication and destruction of the host cell

Negative-strand virus a virus with a single-stranded genome that has the opposite sense to the viral mRNA

Nucleocapsid the complex of nucleic acid and proteins of a virus

Plaque a zone of lysis or growth inhibition caused by virus infection of a lawn of sensitive host cells

Positive-strand virus a virus with a single-stranded genome that has the same complementarity as the viral mRNA

Prion an infectious protein whose extracellular form contains no nucleic acid

Prophage the lysogenic form of a bacterial virus

Provirus the genome of a temperate or latent virus when it is replicating in step with the host chromosome

Retrovirus a virus whose RNA genome is replicated via a DNA intermediate

Reverse transcriptase the enzyme that makes a DNA copy using RNA as template

Reverse transcription the process of copying information found in RNA into DNA

Temperate virus a virus whose genome can replicate along with that of its host without causing cell death, in a state called lysogeny

Transformation in eukaryotes, a process by which a normal cell becomes a cancer cell

Transmissible spongiform encephalopathy (TSE) a degenerative disease of the brain caused by prion infection

Virion the infectious virus particle; the viral genome surrounded by a protein coat and sometimes other layers

Virulent virus a virus that lyses or kills the host cell after infection; a nontemperate virus

Virus a genetic element containing either RNA or DNA surrounded by a protein capsid and that replicates only inside host cells

Viroid a small, circular, single-stranded RNA that causes certain plant diseases

Review Questions

1. Why are viruses classified as "nonliving" things (Section 9.1)?

2. Where does an enveloped virus acquire its envelope? Why is it an important component (Section 9.2)?

3. Define the term "host" as it relates to viruses (Section 9.3).

4. Describe the principle of plaque assay in virus quantification (Section 9.4).

5. Under some conditions, it is possible to obtain nucleic acid–free protein coats (capsids) of certain viruses. Under the electron microscope, these capsids look very similar to complete virions. What does this tell you about the role of the virus nucleic acid in the virus assembly process? Would you expect such particles to be infectious? Why (Section 9.5)?

6. Describe how a restriction endonuclease might play a role in resistance to bacteriophage infection. Why could a restriction endonuclease play such a role whereas a generalized DNase could not (Section 9.6)?

7. One can divide the replication process of a virus into five steps. Describe the events associated with each of these steps (Sections 9.6 and 9.7).

8. What is the difference between a positive-strand and a negative-strand RNA virus (Section 9.7)?

9. In terms of structure, how does the genome of bacteriophage T4 resemble and differ from that of *Escherichia coli* (Section 9.9)?

10. Many of the viruses we have considered have early genes and late genes. What is meant by these two classifications? What types of proteins tend to be encoded by early genes? What types of proteins by late genes? For bacteriophage T4 describe how expression of the late genes is controlled (Section 9.9).

11. Define the following: virulent, lysogeny, prophage (Section 9.10).

12. A strain of *Escherichia coli* that is missing the outer membrane protein responsible for maltose uptake is resistant to bacteriophage lambda infection. A lambda lysogen is immune to lambda infection. Describe the difference between resistance and immunity (Section 9.10).

13. Describe and differentiate the effects animal virus infection can have on an animal (Section 9.11).

14. Typically, tRNA is used in translation. However, it also plays a role in the replication of retroviral nucleic acid. Explain this role (Section 9.12).

15. What does a helper virus provide that allows a satellite virus to replicate (Section 9.13)?

16. What are the similarities and differences between viruses and viroids (Section 9.14)?

17. What are the similarities and differences between prions and viruses (Section 9.15)?

Application Questions

1. What causes the viral plaques that appear on a bacterial lawn to stop growing larger?

2. The promoters for mRNA encoding early proteins in viruses like T4 have a different sequence than the promoters for mRNA encoding late proteins in the same virus. Explain how this benefits the virus.

3. One characteristic of temperate bacteriophages is that they cause turbid rather than clear plaques on bacterial lawns. Can you think why this might be? (Remember the process by which a plaque develops in a lawn of bacteria.)

4. Suggest possible reasons why viroids infect only plants and not animals or bacteria.

5. Contrast the enzyme(s) present in the virions of a retrovirus and a positive-strand RNA bacteriophage. Why do they differ if each has plus configuration single-stranded RNA as their genome?

6. Since viral infection leads to more viral particles being formed, explain why the "growth curve" for viruses is stepped rather than smooth (as seen with bacterial multiplication).

7. What might be the advantage to bacterial host cells of carrying temperate viruses?

Need more practice? Test your understanding with Quantitative Questions; access additional study tools including tutorials, animations, and videos; and then test your knowledge with chapter quizzes and practice tests at **www.microbiologyplace.com**.

Genetics of *Bacteria* and *Archaea*

The Ames test uses bacteria to detect mutagens (chemical agents that cause mutations) and is used in the chemical and food industries to ensure that their products are safe for human use.

In this chapter we discuss the traditional principles and techniques of bacterial genetics. Many newer techniques are now routinely used to investigate the genomes of bacteria and other organisms. These newer approaches are discussed in Chapter 11, Genetic Engineering, and Chapter 12, Microbial Genomics. Here we first describe how alterations arise in the genetic material. Then we consider how genes can be transferred from one microorganism to another.

I Mutation

All organisms contain a specific sequence of nucleotide bases in their genome, their genetic blueprint. A **mutation** is a heritable change in the base sequence of that genome. Mutations can lead to changes—some good, some bad, but mostly neutral in effect—in an organism. Genetic alterations can also be brought about by recombination (Section 10.6), the physical exchange of DNA between genetic elements. Recombination creates new combinations of genes even in the absence of mutation. Whereas mutation usually brings about only a very small amount of genetic change in a cell, genetic recombination typically generates much larger changes. Entire genes, sets of genes, or even larger segments of DNA can be transferred between chromosomes or other genetic elements. Taken together, mutation and recombination fuel the evolutionary process.

Unlike most eukaryotes, prokaryotes do not reproduce sexually. However, prokaryotes possess mechanisms of horizontal genetic exchange that allow for both gene transfer and recombination. To detect genetic exchange between two prokaryotes, it is therefore necessary to employ genetic markers whose transfer can be detected. The term "marker" refers to any gene whose presence is monitored during a genetics experiment. If possible, markers are chosen that are relatively easy to detect. Genetically altered strains are used in gene transfer experiments, the alteration(s) being due to one or more mutations in their DNA. These mutations may involve changes in only one or a few base pairs or even the insertion or deletion of entire genes. Before discussing genetic exchange, we will therefore consider the molecular mechanism of mutation and the properties of mutant microorganisms.

10.1 Mutations and Mutants

A mutation is a heritable change in the base sequence of the nucleic acid in the genome of an organism or a virus or any other genetic entity. In all cells, the genome consists of double-stranded DNA. In viruses, by contrast, the genome may consist of single- or double-stranded DNA or RNA. A strain of any cell or virus carrying a change in nucleotide sequence is called a **mutant**. A mutant by definition differs from its parental strain in its **genotype**, the nucleotide sequence of the genome. In addition, the observable properties of the mutant—its **phenotype**—may also be altered relative to its parent. This altered phenotype is called a *mutant phenotype*. It is common to refer to a strain isolated from nature as a **wild-type strain**. The term wild-type may be used to refer to a whole organism or just to the status of a particular gene that is under investigation. Mutant derivatives can be obtained either directly from wild-type strains or from other strains previously derived from the wild type, for example, another mutant.

Genotype versus Phenotype

Depending on the mutation, a mutant strain may or may not differ in phenotype from its parent. By convention in bacterial genetics, the genotype of an organism is designated by three lowercase letters followed by a capital letter (all in italics) indicating a particular gene. For example, the *hisC* gene of *Escherichia coli* encodes a protein called HisC that functions in biosynthesis of the amino acid histidine. Mutations in the *hisC* gene would be designated as *hisC1*, *hisC2*, and so on, the numbers referring to the order of isolation of the mutant strains. Each *hisC* mutation would be different, and each *hisC* mutation might affect the HisC protein in different ways.

The phenotype of an organism is designated by a capital letter followed by two lowercase letters, with either a plus or minus superscript to indicate the presence or absence of that property. For example, a His^+ strain of *E. coli* is capable of making its own histidine, whereas a His^- strain is not. The His^- strain would require a histidine supplement for growth. A mutation in the *hisC* gene will lead to a His^- phenotype if it eliminates the function of the HisC protein.

Isolation of Mutants: Screening versus Selection

Virtually any characteristic of an organism can be changed by mutation. However, some mutations are selectable, conferring some type of advantage on organisms possessing them, whereas others are nonselectable, even though they may lead to a very clear change in the phenotype of an organism. A selectable mutation confers a clear advantage on the mutant strain under certain environmental conditions, so the progeny of the mutant cell are able to outgrow and replace the parent. A good example of a selectable mutation is drug resistance: An antibiotic-resistant mutant can grow in the presence of antibiotic concentrations that inhibit or kill the parent (**Figure 10.1*a***) and is thus selected for under these conditions. It is relatively easy to detect and isolate selectable mutants by choosing the appropriate environmental conditions. **Selection** is therefore an extremely powerful genetic tool, allowing the isolation of a single mutant from a population containing millions or even billions of parental organisms.

An example of a nonselectable mutation is color loss in a pigmented organism (Figure 10.1*b*). Nonpigmented cells usually have neither an advantage nor a disadvantage over the pigmented parent cells when grown on agar plates, although pigmented organisms may have a selective advantage in nature. We can detect such mutations only by examining large numbers of colonies and looking for the "different" ones, a process called **screening**.

Isolation of Nutritional Auxotrophs and Penicillin Selection

Although screening is more tedious than selection, methods are available for screening large numbers of colonies for certain types of mutations. For instance, nutritionally defective mutants can be detected by the technique of *replica plating* (**Figure 10.2**). An

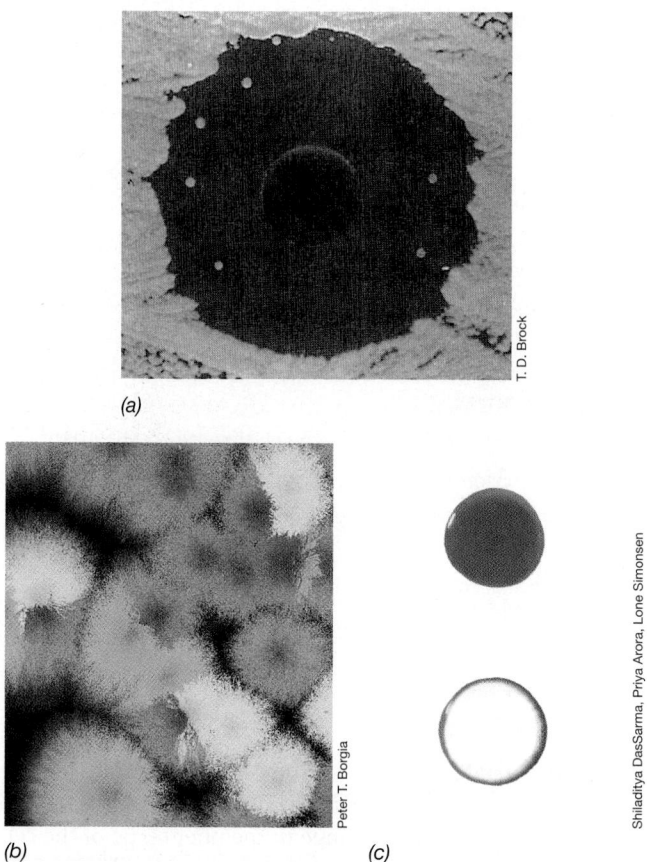

(a)

(b)   *(c)*

Figure 10.1 Selectable and nonselectable mutations. *(a)* Development of antibiotic-resistant mutants, a type of easily selectable mutation, within the inhibition zone of an antibiotic assay disc. *(b)* Nonselectable mutations. Spontaneous pigmented and nonpigmented mutants of the fungus *Aspergillus nidulans*. The wild type has a green pigment. The white or colorless mutants make no pigment, whereas the yellow mutants cannot convert the yellow pigment precursor to the normal (green) color. *(c)* Colonies of mutants of a species of *Halobacterium*, a member of the *Archaea*. The wild-type colonies are white. The orangish brown colonies are mutants that lack gas vesicles (⮩ Section 3.11). The gas vesicles scatter light and mask the color of the colony.

imprint of colonies from a master plate is made onto an agar plate lacking the nutrient by using sterile velveteen cloth or filter paper. Parental colonies will grow normally, whereas those of the mutant will not. Thus, the inability of a colony to grow on medium lacking the nutrient signals that it is a mutant. The colony on the master plate corresponding to the vacant spot on the replica plate can then be picked, purified, and characterized. A mutant with a nutritional requirement for growth is called an **auxotroph**, and the parent from which it was derived is called a *prototroph*. (A prototroph may or may not be the wild type. An auxotroph may be derived from the wild type or from a mutant derivative of the wild type.) For instance, mutants of *E. coli* with a His⁻ phenotype are histidine auxotrophs. Although of great utility, replica plating is nevertheless a screening process, and it can be laborious to isolate mutants by screening.

An ingenious method widely used to isolate auxotrophs is penicillin selection. Ordinarily, mutants that require specific nutrients are at a disadvantage in competition with the parent cells and so there is no direct way of isolating them. Moreover, auxotrophic mutants are rare in a mutagenized culture, and it is time consuming to obtain them by replica plating alone. However, penicillin selection can be used to enrich for auxotrophic mutants in a population of mutagenized cells, after which replica plating is much more effective. How does penicillin selection work?

Penicillin is an antibiotic that kills only growing cells. If penicillin is added to a population of cells growing in a medium lacking the nutrient required by the desired mutant, those cells capable of growth will be killed, whereas any nongrowing mutant cells will survive. After preliminary incubation in the absence of the nutrient in a penicillin-containing medium, the population is washed free of the penicillin and transferred to plates containing the nutrient. The colonies that appear include some wild-type cells that escaped penicillin killing, but also include a relatively increased proportion of the desired mutants. Penicillin selection is thus a kind of negative selection; the selection is not *for* the mutant, but instead *against* the wild type. Penicillin selection is

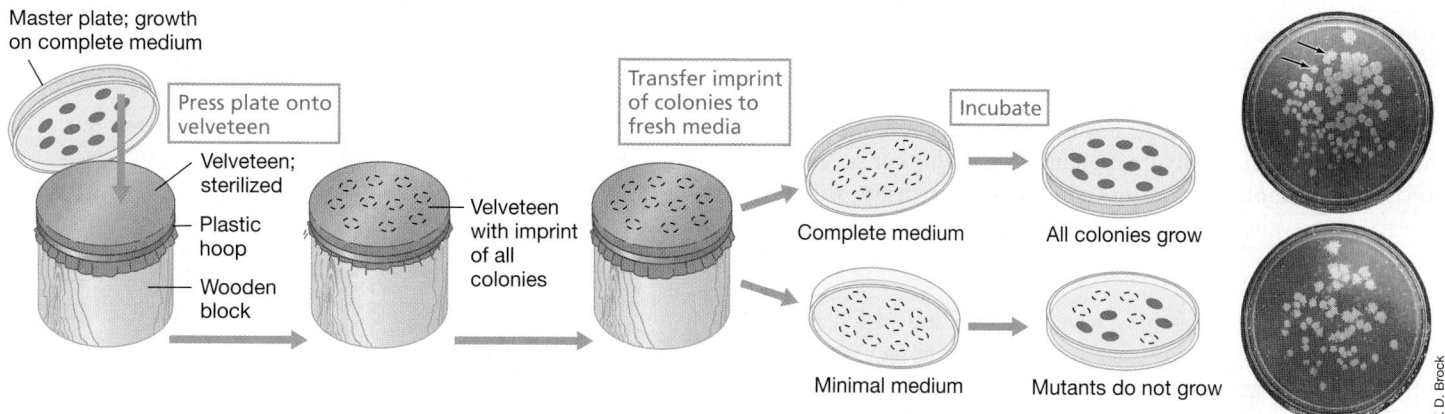

Figure 10.2 Screening for nutritional auxotrophs. The replica-plating method can be used for the detection of nutritional mutants. Photos: The photograph at the top right shows the master plate. Some of the colonies not appearing on the replica plate are indicated with arrows. The replica plate at bottom right lacked one nutrient (leucine) present in the master plate. Therefore, the colonies indicated with arrows on the master plate are leucine auxotrophs.

Table 10.1 *Kinds of mutants*

Phenotype	Nature of change	Detection of mutant
Auxotroph	Loss of enzyme in biosynthetic pathway	Inability to grow on medium lacking the nutrient
Temperature-sensitive	Alteration of an essential protein so it is more heat-sensitive	Inability to grow at a high temperature (for example, 40°C) that normally supports growth
Cold-sensitive	Alteration of an essential protein so it is inactivated at low temperature	Inability to grow at a low temperature (for example, 20°C) that normally supports growth
Drug-resistant	Detoxification of drug or alteration of drug target or permeability to drug	Growth on medium containing a normally inhibitory concentration of the drug
Rough colony	Loss or change in lipopolysaccharide layer	Granular, irregular colonies instead of smooth, glistening colonies
Nonencapsulated	Loss or modification of surface capsule	Small, rough colonies instead of larger, smooth colonies
Nonmotile	Loss of flagella or nonfunctional flagella	Compact instead of flat, spreading colonies
Pigmentless	Loss of enzyme in biosynthetic pathway leading to loss of one or more pigments	Presence of different color or lack of color
Sugar fermentation	Loss of enzyme in degradative pathway	Lack of color change on agar containing sugar and a pH indicator
Virus-resistant	Loss of virus receptor	Growth in presence of large amounts of virus

often used as a prelude to replica plating to increase the chances of obtaining auxotrophic mutants.

Examples of common classes of mutants and the means by which they are detected are listed in **Table 10.1**. www.microbiologyplace .com **Online Tutorial 10.1: Replica Plating**

MiniQuiz

- Distinguish between the words "mutation" and "mutant."
- Distinguish between the words "screening" and "selection."

10.2 Molecular Basis of Mutation

Mutations can be either spontaneous or induced. **Induced mutations** are those that are due to agents in the environment and include mutations made deliberately by humans. They can result from exposure to natural radiation (cosmic rays, and so on) that alters the structure of bases in the DNA. In addition, a variety of chemicals, including oxygen radicals (ᘰ Section 5.18), can chemically modify DNA. For example, oxygen radicals can convert guanine into 8-hydroxyguanine, and this causes mutations. **Spontaneous mutations** are those that occur without external intervention. The bulk of spontaneous mutations result from occasional errors in the pairing of bases during DNA replication.

Mutations that change only one base pair are called **point mutations**. Point mutations are caused by base-pair substitutions in the DNA or by the loss or gain of a single base pair. Most point mutations do not actually cause any phenotypic change, as discussed below. However, as for all mutations, any phenotypic change that does result from a point mutation depends on exactly where the mutation occurs and what the nucleotide change is.

Base-Pair Substitutions

If a point mutation is within the coding region of a gene that encodes a polypeptide, any change in the phenotype of the cell is most likely the result of a change in the amino acid sequence of the polypeptide. The error in the DNA is transcribed into mRNA, and the erroneous mRNA in turn is translated to yield a polypeptide. **Figure 10.3** shows the consequences of various

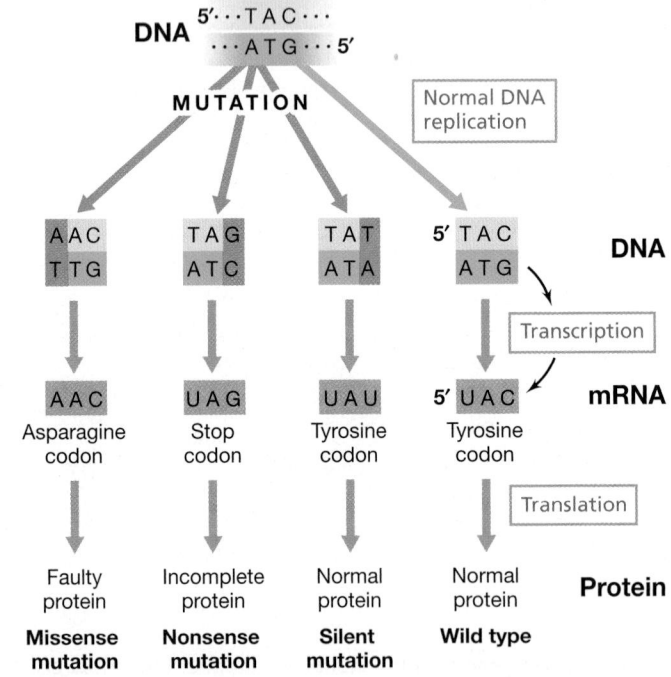

Figure 10.3 Possible effects of base-pair substitution in a gene encoding a protein. Three different protein products are possible from changes in the DNA for a single codon.

base-pair substitutions. (Occasionally, a base change that does not alter the amino acid sequence may nonetheless affect the cellular phenotype by changing the efficiency of translation of an mRNA molecule and thus altering protein levels. This is usually due to changes in secondary structure of the mRNA as a result of altered internal base pairing.)

In interpreting the results of a mutation, we must first recall that the genetic code is degenerate (Section 6.17 and Table 6.5). Consequently, not all mutations in the base sequence encoding a polypeptide will change the polypeptide. This is illustrated in Figure 10.3, which shows several possible results when the DNA that encodes a single tyrosine codon in a polypeptide is mutated. First, a change in the RNA from UAC to UA*U* would have no apparent effect because UAU is also a tyrosine codon. Although they do not affect the sequence of the encoded polypeptide, such changes in the DNA are indeed still mutations. They are one type of **silent mutation**, that is, a mutation that does not affect the phenotype of the cell. Note that silent mutations in coding regions are almost always in the third base of the codon (arginine and leucine can also have silent mutations in the first position).

Changes in the first or second base of the codon more often lead to significant changes in the polypeptide. For instance, a single base change from *U*AC to *A*AC (Figure 10.3) results in an amino acid change within the polypeptide from tyrosine to asparagine at a specific site. This is referred to as a **missense mutation** because the informational "sense" (precise sequence of amino acids) in the ensuing polypeptide has changed. If the change is at a critical location in the polypeptide chain, the protein could be inactive or have reduced activity. However, not all missense mutations necessarily lead to nonfunctional proteins. The outcome depends on where the substitution lies in the polypeptide chain and on how it affects protein folding and activity. For example, mutations in the active site of an enzyme are more likely to destroy activity than mutations in other regions of the protein.

Another possible outcome of a base-pair substitution is the formation of a nonsense (stop) codon. This results in premature termination of translation, leading to an incomplete polypeptide that would almost certainly not be functional (Figure 10.3). Mutations of this type are called **nonsense mutations** because the change is from a codon for an amino acid (sense codon) to a nonsense codon (Table 6.5). Unless the nonsense mutation is very near the end of the gene, the incomplete product will be completely inactive.

The terms "transition" and "transversion" are used to describe the type of base substitution in a point mutation. **Transitions** are mutations in which one purine base (A or G) is substituted for another purine, or one pyrimidine base (C or T) is substituted for another pyrimidine. **Transversions** are point mutations in which a purine base is substituted for a pyrimidine base or vice versa.

Frameshifts and Other Insertions or Deletions

Because the genetic code is read from one end of the nucleic acid in consecutive blocks of three bases (that is, as codons), any deletion or insertion of a single base pair results in a shift in the reading frame. These frameshift mutations often have

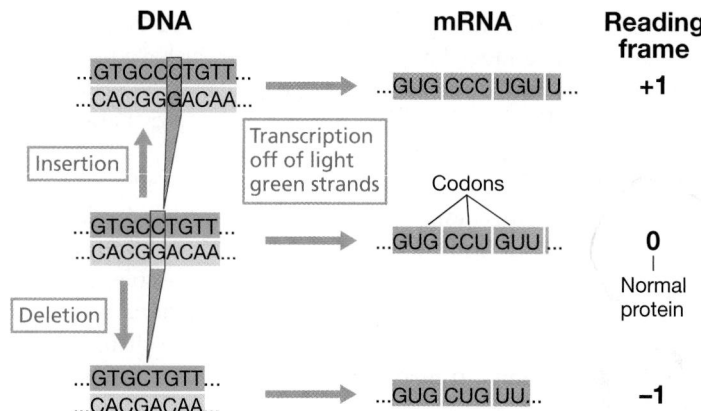

Figure 10.4 Shifts in the reading frame of mRNA caused by insertions or deletions. The reading frame in mRNA is established by the ribosome, which begins at the 5′ end (toward the left in the figure) and proceeds by units of three bases (codons). The normal reading frame is referred to as the 0 frame, that missing a base the −1 frame, and that with an extra base the +1 frame.

serious consequences. Single base insertions or deletions change the primary sequence of the encoded polypeptide, typically in a major way (**Figure 10.4**). Such microinsertions or microdeletions can result from replication errors. Insertion or deletion of two base pairs also causes a frameshift; however, insertion or deletion of three base pairs adds or removes a whole codon. This results in addition or deletion of a single amino acid in the polypeptide sequence. Although this may well be deleterious to protein function, it is usually not as bad as a frameshift, which scrambles the entire polypeptide sequence after the mutation point.

Insertions or deletions can also result in the gain or loss of hundreds or even thousands of base pairs. Such changes inevitably result in complete loss of gene function. Some deletions are so large that they may include several genes. If any of the deleted genes are essential, the mutation will be lethal. Such deletions cannot be restored through further mutations, but only through genetic recombination. Indeed, one way in which large deletions are distinguished from point mutations is that the latter are reversible through further mutations, whereas the former are not. Larger insertions and deletions may arise as a result of errors during genetic recombination. In addition, many insertion mutations are due to the insertion of specific identifiable DNA sequences 700–1400 base pairs (bp) in length called *insertion sequences*, a type of transposable element (Section 10.13). The effect of transposable elements on the evolution of bacterial genomes is discussed further in Section 12.12.

Other types of large-scale mutations are rearrangements brought about by errors in recombination. These include translocations, in which a large section of chromosomal DNA is moved to a new location (and in eukaryotes often to a different chromosome), and inversions, in which the orientation of a particular segment of DNA is reversed relative to the surrounding DNA. www.microbiologyplace.com **Online Tutorial 10.2: The Molecular Basis for Mutations**

Site-Directed Mutagenesis and Transposons

The mutations that we have considered thus far have been random, that is, not directed at any particular gene. However, recombinant DNA technology and the use of synthetic DNA make it possible to induce specific mutations in specific genes. The approach of generating mutations at specific sites is called *site-directed mutagenesis* and is discussed further in Chapter 11. Mutations can also be deliberately introduced by transposon mutagenesis (Section 10.13). If a transposable element inserts within a gene, loss of gene function generally results. Because transposable elements can insert into the chromosome at various locations, transposons are widely used to generate mutations.

Back Mutations or Reversions

Point mutations are typically reversible, a process known as **reversion**. A revertant is a strain in which the original phenotype that was changed in the mutant is restored. Revertants can be of two types. In *same-site revertants*, the mutation that restores activity is at the same site as the original mutation. If the back mutation is not only at the same site but also restores the original sequence, it is called a *true revertant.*

In second-site revertants, the mutation is at a different site in the DNA. Second-site mutations can restore a wild-type phenotype if they function as suppressor mutations—mutations that compensate for the effect of the original mutation. Several classes of suppressor mutations are known. These include (1) a mutation somewhere else in the same gene that restores enzyme function, such as a second frameshift mutation near the first that restores the original reading frame; (2) a mutation in another gene that restores the function of the original mutated gene; and (3) a mutation in another gene that results in the production of an enzyme that can replace the mutated one.

An interesting class of suppressor mutations are those due to alterations in tRNA. Nonsense mutations can be suppressed by changing the anticodon sequence of a tRNA molecule so that it now recognizes a stop codon. Such an altered tRNA is known as a suppressor tRNA and will insert the amino acid it carries at the stop codon that it now reads. Suppressor tRNA mutations would be lethal unless a cell has more than one tRNA for a particular codon. One tRNA may then be mutated into a suppressor, and the other performs the original function. Most cells have multiple tRNAs and so suppressor mutations are reasonably common, at least in microorganisms. Sometimes the amino acid inserted by the suppressor tRNA is identical to the original amino acid and the protein is fully restored. In other cases, a different amino acid is inserted and a partially active protein may be produced.

Unlike point mutations, large-scale deletions do not revert. By contrast, large-scale insertions can revert as the result of a subsequent deletion that removes the insertion. Typically, frameshift mutations of any magnitude are difficult to restore to the wild type, and mutants carrying frameshift mutations are therefore genetically quite stable. For this reason, geneticists often use them in genetic crosses to avoid accidental reversion of mutant strains during the course of a genetic study.

MiniQuiz
- What does it mean to say that point mutations can spontaneously revert?
- Do missense mutations occur in genes encoding tRNA? Why or why not?

10.3 Mutation Rates

The rates at which different kinds of mutations occur vary widely. Some types of mutations occur so rarely that they are almost impossible to detect, whereas others occur so frequently that they present difficulties for an experimenter trying to maintain a genetically stable stock culture. Furthermore, all organisms possess a variety of systems for DNA repair. Consequently, the observed mutation rate depends not only on the frequency of DNA alterations but also on the efficiency of DNA repair.

Spontaneous Mutation Frequencies

For most microorganisms, errors in DNA replication occur at a frequency of 10^{-6} to 10^{-7} per kilobase pair during a single round of replication. A typical gene has about 1000 base pairs. Therefore, the frequency of a mutation *in a given gene* is also in the range of 10^{-6} to 10^{-7} per generation. For instance, in a bacterial culture having 10^8 cells/ml, there are likely to be a number of different mutants for any given gene in each milliliter of culture. Eukaryotes with very large genomes tend to have replication error rates about 10-fold lower than typical bacteria, whereas DNA viruses, especially those with very small genomes, may have error rates 100-fold to 1000-fold higher than those of cellular organisms. RNA viruses have even higher error rates.

Single base errors during DNA replication are more likely to lead to missense mutations than to nonsense mutations because most single base substitutions yield codons that encode other amino acids (ᴄᴄ Table 6.5). The next most frequent type of codon change caused by a single base change leads to a silent mutation. This is because for the most part alternate codons for a given amino acid differ from each other by a single base change in the "silent" third position. A given codon can be changed to any of 27 other codons by a single base substitution, and on average, about two of these will be silent mutations, about one a nonsense mutation, and the rest will be missense mutations. There are also some DNA sequences, typically areas containing short repeats, that are hot spots for mutations because the error frequency of DNA polymerase is relatively high there. The error rate at hot spots is affected by the base sequence in the vicinity.

Unless a mutation can be selected for, its experimental detection is difficult, and much of the skill of the microbial geneticist involves increasing the efficiency of mutation detection. As we see in the next section, it is possible to greatly increase the mutation rate by mutagenic treatments. In addition, the mutation rate may change in certain situations, such as under high-stress conditions.

Mutations in RNA Genomes

Whereas all cells have DNA as their genetic material, some viruses have RNA genomes (ᴄᴄ Section 9.1). These genomes can also undergo mutation. Interestingly, the mutation rate in RNA genomes is about 1000-fold higher than in DNA genomes. Why should this be so?

Some RNA polymerases have proofreading activities like those of DNA polymerases (⇆ Section 6.10), thus limiting the total number of polymerase errors. However, although there are several repair systems for DNA that can correct changes before they become fixed in the genome as mutations (Section 10.4), comparable RNA repair mechanisms do not exist. This leads to heightened mutation rates for RNA genomes. This high mutation rate in RNA viruses has dramatic consequences. For example, the RNA genomes of viruses that cause disease can mutate very rapidly, presenting a constantly changing and evolving population of viruses. Such changes are one of many challenges to human medicine posed by the AIDS virus, HIV, which is an RNA virus with a notorious ability to undergo genetic changes that affect its virulence (⇆ Section 21.11).

MiniQuiz

• Which class of mutation, missense or nonsense, is more common, and why?

• Why are RNA viruses genetically highly variable?

10.4 Mutagenesis

The spontaneous rate of mutation is very low, but there are a variety of chemical, physical, and biological agents that can increase the mutation rate and are therefore said to induce mutations. These agents are called **mutagens**. We discuss some of the major categories of mutagens and their activities here.

Chemical Mutagens

An overview of some of the major chemical mutagens and their modes of action is given in **Table 10.2**. Several classes of chemical

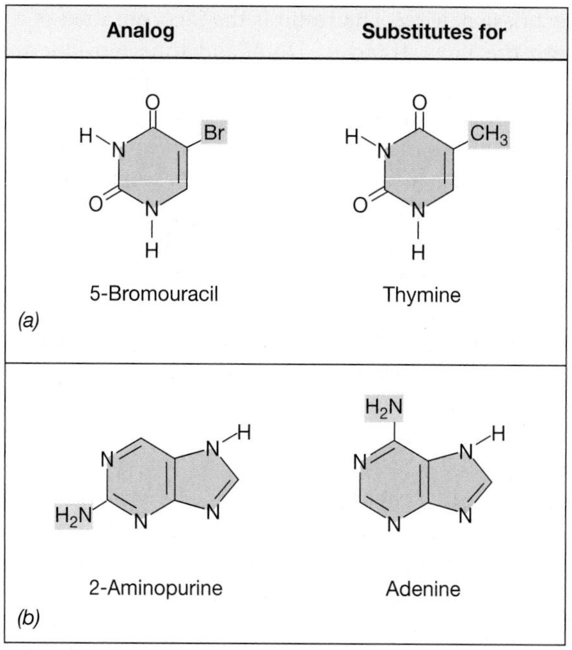

Figure 10.5 Nucleotide base analogs. Structure of two common nucleotide base analogs used to induce mutations and the normal nucleic acid bases for which they substitute. (a) 5-Bromouracil can base-pair with guanine, causing AT to GC substitutions. (b) 2-Aminopurine can base-pair with cytosine, causing AT to GC substitutions.

mutagens exist. The *nucleotide base analogs* are molecules that resemble the purine and pyrimidine bases of DNA in structure yet display faulty pairing properties (**Figure 10.5**). If a base analog is incorporated into DNA in place of the natural base, the DNA may replicate normally most of the time. However, DNA replication errors occur at higher frequencies at these sites due to

Table 10.2 *Chemical and physical mutagens and their modes of action*

Agent	Action	Result
Base analogs		
5-Bromouracil	Incorporated like T; occasional faulty pairing with G	AT → GC and occasionally GC → AT
2-Aminopurine	Incorporated like A; faulty pairing with C	AT → GC and occasionally GC → AT
Chemicals reacting with DNA		
Nitrous acid (HNO_2)	Deaminates A and C	AT → GC and GC → AT
Hydroxylamine (NH_2OH)	Reacts with C	GC → AT
Alkylating agents		
Monofunctional (for example, ethyl methanesulfonate)	Puts methyl on G; faulty pairing with T	GC → AT
Bifunctional (for example, mitomycin, nitrogen mustards, nitrosoguanidine)	Cross-links DNA strands; faulty region excised by DNase	Both point mutations and deletions
Intercalating dyes		
Acridines, ethidium bromide	Inserts between two base pairs	Microinsertions and microdeletions
Radiation		
Ultraviolet	Pyrimidine dimer formation	Repair may lead to error or deletion
Ionizing radiation (for example, X-rays)	Free-radical attack on DNA, breaking chain	Repair may lead to error or deletion

incorrect base pairing. The result is the incorporation of a wrong base into the new strand of DNA and thus introduction of a mutation. During subsequent segregation of this strand in cell division, the mutation is revealed.

Other chemical mutagens induce *chemical modifications* in one base or another, resulting in faulty base pairing or related changes (Table 10.2). For example, alkylating agents (chemicals that react with amino, carboxyl, and hydroxyl groups by substituting them with alkyl groups) such as nitrosoguanidine, are powerful mutagens and generally induce mutations at higher frequency than base analogs. Unlike base analogs, which have an effect only when incorporated during DNA replication, alkylating agents can introduce changes even in nonreplicating DNA. Both base analogs and alkylating agents tend to induce base-pair substitutions (Section 10.2).

Another group of chemical mutagens, the acridines, are planar molecules that function as *intercalating agents*. These mutagens become inserted between two DNA base pairs and push them apart. During replication, this abnormal conformation can lead to single base insertions or deletions in acridine-containing DNA. Thus, acridines typically induce frameshift mutations (Section 10.2). Ethidium bromide, which is often used to detect DNA in electrophoresis, is also an intercalating agent and therefore a mutagen.

Radiation

Several forms of radiation are highly mutagenic. We can divide mutagenic electromagnetic radiation into two main categories, nonionizing and ionizing (**Figure 10.6**). Although both kinds of radiation are used to generate mutations, nonionizing radiation such as ultraviolet (UV) radiation has the widest use.

The purine and pyrimidine bases of nucleic acids absorb UV radiation strongly, and the absorption maximum for DNA and

RNA is at 260 nm (Figure 6.7). Killing of cells by UV radiation is due primarily to its effect on DNA. Although several effects are known, one well-established effect is the production of pyrimidine dimers, in which two adjacent pyrimidine bases (cytosine or thymine) on the same strand of DNA become covalently bonded to one another. This results either in impeding DNA polymerase or in a greatly increased probability of DNA polymerase misreading the sequence at this point.

The UV radiation source most commonly used for mutagenesis is the germicidal lamp, which emits UV radiation in the 260-nm region. A dose of UV radiation is used that kills about 50–90% of the cell population, and mutants are then selected or screened for among the survivors. If much higher doses of radiation are used, the number of surviving cells is too low. If lower doses are used, damage to DNA is insufficient to generate enough mutations. When used at the correct dose, UV radiation is a very convenient tool for isolating mutants and avoids the need to handle toxic chemicals.

Ionizing Radiation

Ionizing radiation is a more powerful form of radiation than UV radiation and includes short-wavelength rays such as X-rays, cosmic rays, and gamma rays (Figure 10.6). These rays cause water and other substances to ionize, and mutagenic effects result indirectly from this ionization. Among the potent chemical species formed by ionizing radiation are chemical free radicals, the most important being the hydroxyl radical, OH• (Section 5.18).

Free radicals react with and damage macromolecules in the cell, including DNA. This causes double-stranded and single-stranded breaks that may lead to rearrangements or large deletions. At low doses of ionizing radiation only a few "hits" on DNA occur, but at higher doses, multiple hits cause fragmentation of DNA that sometimes cannot be repaired and thus leads to the death of the cell. In contrast to UV radiation, ionizing radiation penetrates readily through glass and other materials. Therefore, ionizing radiation is used frequently to induce mutations in animals and plants because its penetrating power makes it possible to reach the gamete-producing cells of these organisms. However, because ionizing radiation is more dangerous and is less readily available than UV radiation, it finds less use in microbial genetics.

DNA Repair Systems

By definition, a mutation is a *heritable* change in the genetic material. Therefore, if damaged DNA can be corrected before the cell divides, no mutation will occur. Most cells have a variety of different DNA repair processes to correct mistakes or repair damage. Most of these DNA repair systems are virtually error-free. However, some are error-prone and the repair process itself introduces the mutation. DNA repair processes may be grouped into three categories: direct reversal, repair of single-strand damage, and repair of double-strand damage.

Direct reversal applies to bases that have been chemically altered but whose identity is still recognizable. No base pairing (that is, no template strand) is needed. For example, some alkylated bases are repaired by direct chemical removal of the

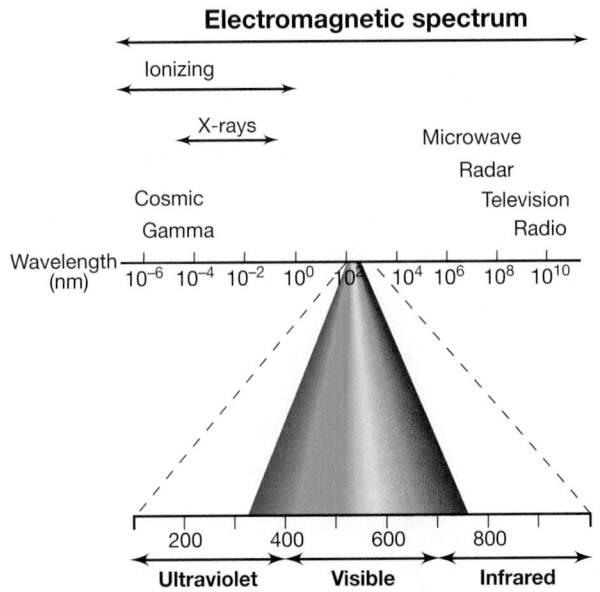

Figure 10.6 Wavelengths of radiation. Ultraviolet radiation consists of wavelengths just shorter than visible light. For any electromagnetic radiation, the shorter the wavelength, the higher the energy. DNA absorbs strongly at 260 nm.

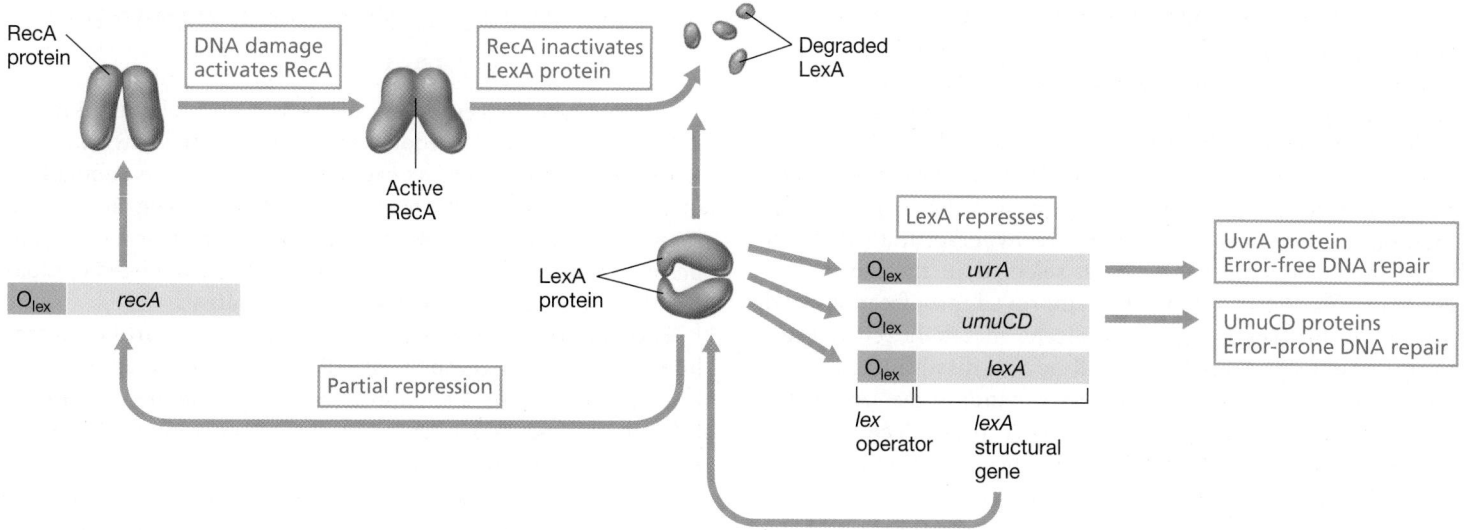

Figure 10.7 Mechanism of the SOS response. DNA damage activates RecA protein, which in turn activates the protease activity of LexA. The LexA protein then cleaves itself. LexA protein normally represses the activities of the *recA* gene and the DNA repair genes *uvrA* and *umuCD* (the UmuCD proteins are part of DNA polymerase V). However, repression is not complete. Some RecA protein is produced even in the presence of LexA protein. With LexA inactivated, these genes become highly active.

alkyl group. Another direct repair system is photoreactivation, which cleaves pyrimidine dimers generated by UV radiation. The enzyme photolyase absorbs blue light and uses the energy to drive the cleavage reaction.

Several systems exist that repair single-strand DNA damage. In these cases, the damaged DNA is removed from only one strand. Then the opposite (undamaged) strand is used as a template for replacing the missing nucleotides. In base excision repair, a single damaged base is removed and replaced. In nucleotide excision repair and mismatch repair, a short stretch of single-stranded DNA containing the damage is removed and replaced. Double-strand damage, including both cross-strand links and double-stranded breaks, is especially dangerous. These lesions are repaired by recombinational mechanisms and may require error-prone repair.

Mutations That Arise from DNA Repair: The SOS System

Some types of DNA damage, especially large-scale damage from highly mutagenic chemicals or large doses of radiation, may interfere with replication if such lesions are not removed before replication occurs. Lesions on the template DNA may lead to stalling of DNA replication, which is a lethal event. Stalled replication, as well as certain types of major DNA damage, activate the *SOS repair system*. The SOS system initiates a number of DNA repair processes, some of which are error-free. However, the SOS system also allows DNA repair to occur without a template, that is, without base pairing; as expected, this results in many errors and hence many mutations. This permits cell survival under conditions that are otherwise lethal.

In *Escherichia coli* the SOS repair system regulates the transcription of approximately 40 genes located throughout the chromosome that are involved in DNA damage tolerance and DNA

repair. In DNA damage tolerance, DNA lesions remain in the DNA, but are bypassed by specialized DNA polymerases that can move past DNA damage—a process known as translesion synthesis. Even if no template is available to allow insertion of the correct bases, it is less dangerous to fill the gap than let it remain. Consequently, translesion synthesis generates many errors. In *E. coli*, in which the process of mutagenesis has been studied in great detail, the two error-prone repair polymerases are DNA polymerase V, an enzyme encoded by the *umuCD* genes (**Figure 10.7**), and DNA polymerase IV, encoded by *dinB*. Both are induced as part of the SOS repair system.

The SOS system is a **regulon**, that is, a set of genes that are coordinately regulated although they are transcribed separately. The SOS system is regulated by two proteins, LexA and RecA. LexA is a repressor that normally prevents expression of the SOS regulon. The RecA protein, which normally functions in genetic recombination (Section 10.6), is activated by the presence of DNA damage, in particular by the single-stranded DNA that results when replication stalls (Figure 10.7). The activated form of RecA stimulates LexA to inactivate itself by self-cleavage. This leads to derepression of the SOS system and results in the coordinate expression of a number of proteins that take part in DNA repair. Because some of the DNA repair mechanisms of the SOS system are inherently error-prone, many mutations arise. Once the DNA damage has been repaired, the SOS regulon is repressed and further mutagenesis ceases.

Changes in Mutation Rate

High fidelity (low error frequency) in DNA replication is essential if organisms are to remain genetically stable. On the other hand, perfect fidelity is counterproductive because it would prevent evolution. Therefore, a mutation rate has evolved in cells that is very low, yet detectable. This allows organisms to balance the need for genetic stability with that for evolutionary improvement.

The fact that organisms as phylogenetically distant as *Archaea* and *E. coli* have about the same mutation rate might suggest that evolutionary pressure has selected organisms with the lowest possible mutation rates. However, this is not so. The mutation rate in an organism is subject to change. For example, mutants of some organisms that are hyperaccurate in DNA replication and repair have been selected in the laboratory. However, in these strains, the improved proofreading and repair mechanisms have a significant metabolic cost; thus, hyperaccurate mutants might well be at a disadvantage in the natural environment. On the other hand, some organisms seem to benefit from enhanced DNA repair systems that enable them to occupy particular niches in nature. A good example is the bacterium *Deinococcus radiodurans* (Section 18.17). This organism is 20 times more resistant to UV radiation and 200 times more resistant to ionizing radiation than is *E. coli*. This resistance, dependent in part upon redundant DNA repair systems and on a mechanism for exporting damaged nucleotides, allows the organism to survive in environments in which other organisms cannot, such as near concentrated sources of radiation or on the surfaces of dust particles exposed to intense sunlight.

In contrast to hyperaccuracy, some organisms actually benefit from an increased mutation rate. DNA repair systems are themselves genetically encoded and thus subject to mutation. For example, the protein subunit of DNA polymerase III involved in proofreading (Section 6.10) is encoded by the gene *dnaQ*. Certain mutations in *dnaQ* lead to mutants that are still viable but have an increased rate of mutation. These are known as hypermutable or **mutator strains**. Mutations leading to a mutator phenotype are known in several other DNA repair systems as well. The mutator phenotype is apparently selected for in complex and changing environments because strains of bacteria with mutator phenotypes appear to be more abundant under these conditions. Presumably, whatever disadvantage an increased mutation rate may have in such environments is offset by the ability to generate greater numbers of useful mutations. These mutations ultimately increase evolutionary fitness of the population and make the organism more successful in its ecological niche.

As indicated earlier, a mutator phenotype may be induced in wild-type strains by stressful situations. For instance, the SOS repair system includes error-prone repair. Therefore, when the SOS repair system is activated, the mutation rate increases. In some cases this is merely an inevitable by-product of DNA repair, but in other cases, the increased mutation rate may itself be of selective value to the organism for survival purposes.

MiniQuiz

- How do mutagens work?
- Why might a mutator phenotype be successful in an environment experiencing rapid changes?
- What is meant by "error prone" DNA repair?

10.5 Mutagenesis and Carcinogenesis: The Ames Test

The Ames test makes practical use of bacterial mutations to detect potentially hazardous chemicals in the environment. Because selectable mutants can be detected in large populations of bacteria with very high sensitivity, bacteria can be used to screen chemicals for potential mutagenicity. This is relevant because many mutagenic chemicals are also carcinogenic, capable of causing cancer in humans or other animals.

The variety of chemicals, both natural and artificial, that humans encounter through agricultural and industrial exposure is enormous. There is good evidence that some human cancers have environmental causes, most likely from various chemicals, making the detection of chemical carcinogens an important matter. Not every mutagen is also a carcinogen. The correlation, however, is high, and knowing that a compound is mutagenic to bacteria is a warning of possible danger. Bacterial tests for carcinogen screening were developed primarily by Bruce Ames and colleagues at the University of California in Berkeley and consequently, the mutagenicity test for carcinogens is known as the *Ames test* (**Figure 10.8**).

The standard way to test chemicals for mutagenesis is to look for an increase in the rate of back mutation (reversion) in auxotrophic strains of bacteria in the presence of the suspected mutagen. It is important that the auxotrophic strain carry a point mutation so that the reversion rate is measurable. Cells of such an auxotroph do not grow on a medium lacking the required nutrient (for example, an amino acid), and even very large populations of cells can be spread on the plate without formation of visible colonies. However, if back mutants (revertants) are present, those cells form colonies. Thus, if 10^8 cells are spread on the surface of a single plate, even as few as 10–20 revertants can be detected by the 10–20 colonies they form (Figure 10.8, left photo). However, if the reversion rate is increased by the presence of a chemical mutagen, the number of revertant colonies is

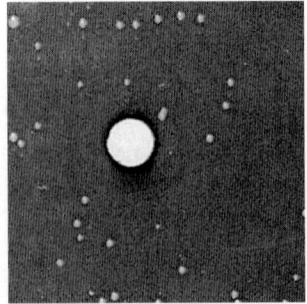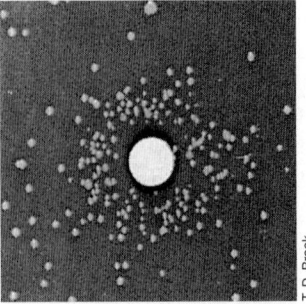

Figure 10.8 The Ames test for assessing the mutagenicity of a chemical. Two plates were inoculated with a culture of a histidine-requiring mutant of *Salmonella enterica*. The medium does not contain histidine, so only cells that revert back to wild type can grow. Spontaneous revertants appear on both plates, but the chemical on the filter-paper disc in the test plate (right) has caused an increase in the mutation rate, as shown by the large number of colonies surrounding the disc. Revertants are not seen very close to the test disc because the concentration of mutagen is lethally high there. The plate on the left was the negative control; its filter-paper disc had only water added.

even greater. Histidine auxotrophs of *Salmonella enterica* (Figure 10.8) and tryptophan auxotrophs of *Escherichia coli* have been the major tools of the Ames test.

Two additional elements have been introduced in the Ames test to make it much more powerful. The first of these is to use test strains that almost exclusively use error-prone pathways to repair DNA damage; normal repair mechanisms are thus thwarted (Section 10.4). The second important element in the Ames test is the addition of liver enzyme preparations to convert the chemicals to be tested into their active mutagenic (and potentially carcinogenic) forms. It has been well established that many carcinogens are not directly carcinogenic or mutagenic themselves, but undergo modifications in the human body that convert them into active substances. These changes take place primarily in the liver, where enzymes called mixed-function oxygenases, whose normal function is detoxification, generate activated forms of the compounds that are highly reactive (and thus mutagenic) toward DNA.

In the Ames test, a preparation of enzymes from rat liver is first used to activate the test compound. The activated complex is then soaked into a filter-paper disc, which is placed in the center of a plate on which the proper bacterial strain has been overlaid. After overnight incubation, the mutagenicity of the compound can be detected by looking for a halo of back mutations in the area around the paper disc (Figure 10.8). It is necessary to carry out this test with several different concentrations of the compound and with appropriate positive (known mutagens) and negative (no mutagen) controls, because compounds vary in their mutagenic activity and may be lethal at higher levels. A wide variety of chemicals have been subjected to the Ames test, and it has become one of the most useful screens for determining the potential carcinogenicity of a compound.

MiniQuiz

- Why does the Ames test measure the rate of back mutation rather than the rate of forward mutation?
- Of what significance is the detection of mutagens to the prevention of cancer?

 ## Gene Transfer

For genetic analyses, the microbial geneticist must cross strains of an organism that have different genotypes (and phenotypes) and look for recombinants. Three mechanisms of genetic exchange are known in prokaryotes: (1) *transformation*, in which free DNA released from one cell is taken up by another (Section 10.7); (2) *transduction*, in which DNA transfer is mediated by a virus (Section 10.8); and (3) *conjugation*, in which DNA transfer involves cell-to-cell contact and a conjugative plasmid in the donor cell (Sections 10.9 and 10.10). These processes are contrasted in **Figure 10.9**.

Before discussing the mechanisms of transfer, we must consider the fate of the transferred DNA. Whether it is transferred by transformation, transduction, or conjugation, the incoming DNA faces three possible fates: (1) It may be degraded by restriction

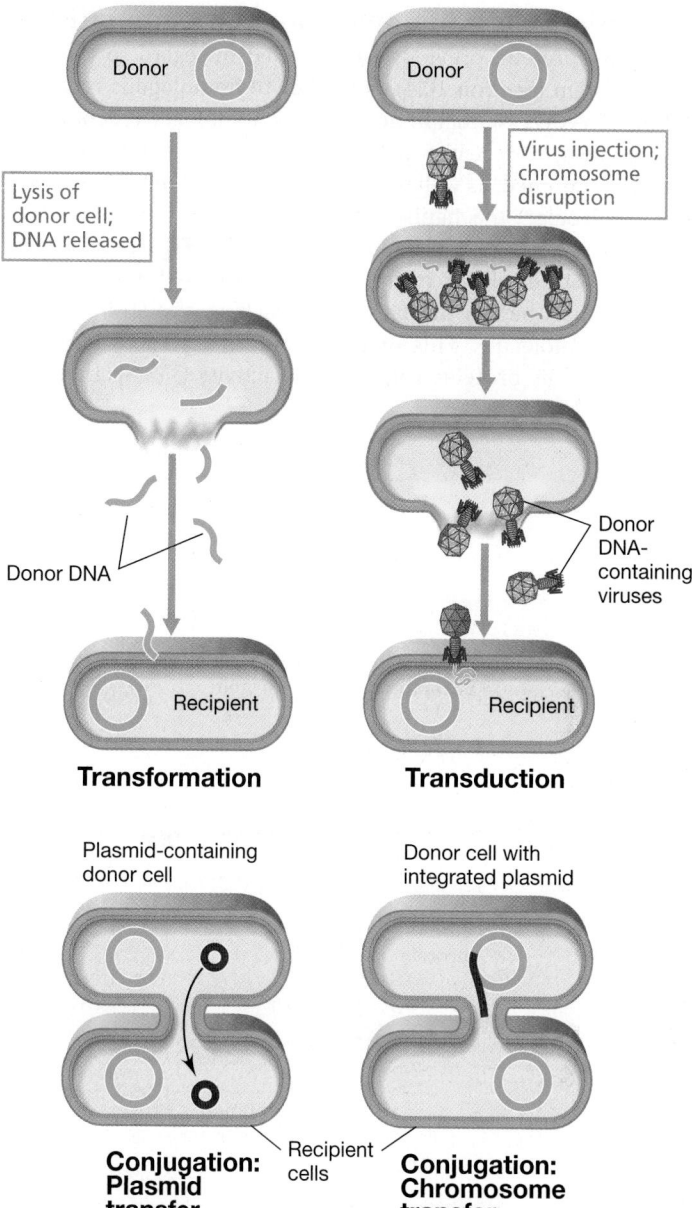

Figure 10.9 Processes by which DNA is transferred from donor to recipient bacterial cell. Just the initial steps in transfer are shown.

enzymes; (2) it may replicate by itself (but only if it possesses its own origin of replication such as a plasmid or phage genome); or (3) it may recombine with the host chromosome.

10.6 Genetic Recombination

Recombination is the physical exchange of DNA between genetic elements. In this section we focus on *homologous* recombination, a process that results in genetic exchange between homologous DNA sequences from two different sources. Homologous DNA sequences are those that have nearly the same sequence; therefore, bases can pair over an extended length of the two DNA molecules. This type of recombination is involved in the process referred to as "crossing over" in classical genetics.

Molecular Events in Homologous Recombination

The RecA protein, previously mentioned in regard to the SOS repair system (Section 10.4), is the key to homologous recombination. RecA is essential in nearly every homologous recombination pathway. RecA-like proteins have been identified in all bacteria examined, as well as in the *Archaea* and most *Eukarya*.

A molecular mechanism for homologous recombination between two DNA molecules is shown in **Figure 10.10**. An enzyme that cuts DNA in the middle of a strand, known as an endonuclease, begins the process by nicking one strand of the first DNA molecule. This nicked strand is separated from the other strand by proteins with helicase activity (Section 6.9).

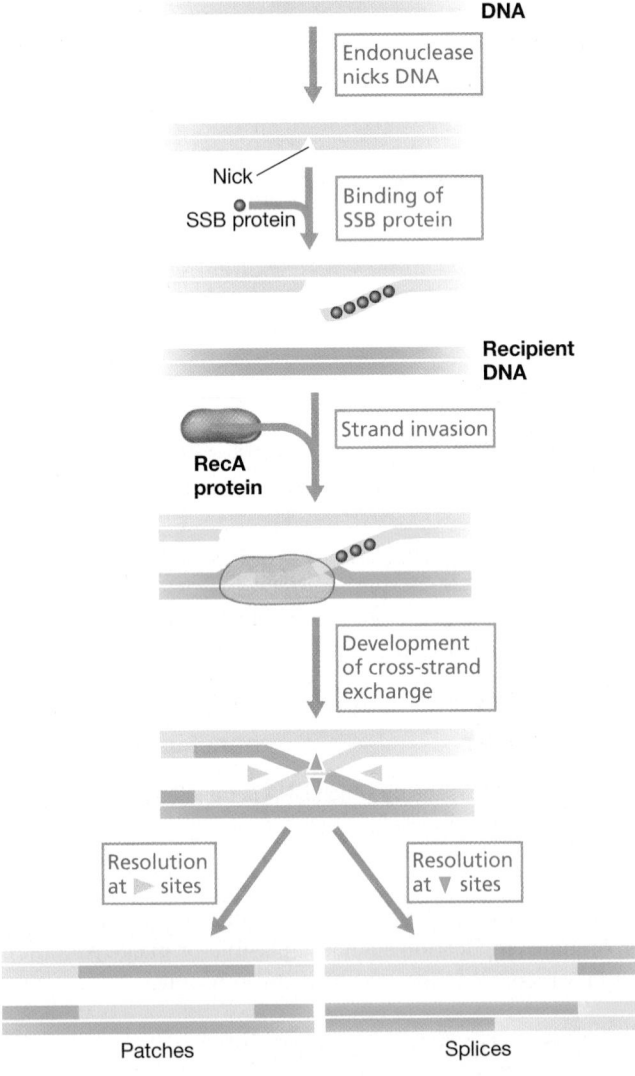

Figure 10.10 A simplified version of homologous recombination. Homologous DNA molecules pair and exchange DNA segments. The mechanism involves breakage and reunion of paired segments. Two of the proteins involved, single-strand binding (SSB) protein and the RecA protein, are shown. The other proteins involved are not shown. The diagram is not to scale: Pairing may occur over hundreds or thousands of bases. Resolution occurs by cutting and rejoining the cross-linked DNA molecules. Note that there are two possible outcomes, patches or splices, depending on where strands are cut during the resolution process.

In some recombination pathways specialized enzymes, such as the RecBCD enzyme of *Escherichia coli*, combine the endonuclease and helicase activities. Single-strand binding protein (Section 6.9) then binds to the resulting single-stranded segment. Next, the RecA protein binds to the single-stranded region. This results in a complex that promotes base pairing with the complementary sequence in the second DNA molecule. This in turn displaces the other strand of the second DNA molecule (Figure 10.10) and is therefore called *strand invasion*. The base pairing of one strand from each of the two DNA molecules over long stretches generates recombination intermediates containing long **heteroduplex** regions, where each strand has originated from a different chromosome. These structures are called *Holliday junctions* (after Robin Holliday, who proposed this model in 1964) and can migrate along the DNA; this migration is energized by a complex of several other proteins. Finally, the linked molecules are separated or "resolved" by resolvases that cut and rejoin the second (previously unbroken) strands of both original DNA molecules. In *E. coli*, the RecG and RuvC proteins both function as resolvases, and their activity generates two recombined DNA molecules. Depending on the orientation of the Holliday junction during resolution, two types of products, referred to as "patches" or "splices," are formed that differ in the conformation of the heteroduplex regions remaining after resolution (Figure 10.10).

Effect of Homologous Recombination on Genotype

For homologous recombination to generate new genotypes, the two homologous sequences must be related but genetically distinct. This is obviously the case in a diploid eukaryotic cell, which has two sets of chromosomes, one from each parent. In prokaryotes, genetically distinct but homologous DNA molecules are brought together in different ways, but the process of genetic recombination is equivalent. Genetic recombination in prokaryotes occurs after fragments of homologous DNA from a donor chromosome are transferred to a recipient cell by transformation, transduction, or conjugation. It is only after the transfer event, when the DNA fragment from the donor is in the recipient cell, that homologous recombination occurs. In prokaryotes, only part of a chromosome is transferred; therefore, if recombination does not occur, the DNA fragment will be lost because it cannot replicate independently. Thus, in prokaryotes, transfer is just the first step in generating recombinant organisms.

Detection of Recombination

To detect physical exchange of DNA segments, the cells resulting from recombination must be phenotypically different from both parents. Genetic crosses in bacteria usually depend on using recipient strains that lack some selectable character that the recombinants will gain. For instance, the recipient may be unable to grow on a particular medium, and genetic recombinants are selected that can. Various kinds of selectable markers, such as drug resistance and nutritional requirements, were discussed in Section 10.1.

The exceedingly great sensitivity of the selection process allows even a few recombinant cells to be detected in a large population of nonrecombinant cells (**Figure 10.11**). The only requirement for effective detection of recombination is that the back

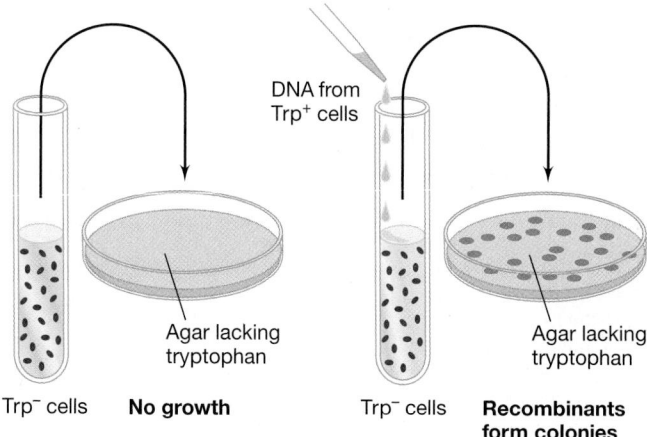

Figure 10.11 Using a selective medium to detect rare genetic recombinants. On the selective medium only the rare recombinants form colonies even though a very large population of bacteria was plated. Procedures such as this, which offer high resolution for genetic analyses, can ordinarily be used only with microorganisms. The type of genetic exchange being illustrated is transformation.

mutation rate for the selected characteristic should be low, because revertants will also form colonies. This problem can often be overcome by using double mutants—strains that carry two different mutations—in genetic crosses because it is very unlikely that two back mutations will occur in the same cell. Alternatively, frameshift mutants can be used, because their reversion rates are typically extremely low.

Much of the skill of the bacterial geneticist lies in the choice of proper mutants and selective media for efficient detection of genetic recombination. Because selection is so powerful and because crosses can be made using billions of individual cells, recombinational analysis following gene transfer is an important tool for the microbial geneticist.

MiniQuiz

- Which protein, found in all prokaryotes, facilitates the pairing required for homologous recombination?

- In eukaryotes, recombination involves entire chromosomes, but this is not true in prokaryotes. Explain.

10.7 Transformation

Transformation is a genetic transfer process by which free DNA is incorporated into a recipient cell and brings about genetic change. Several prokaryotes are naturally transformable, including certain species of both gram-negative and gram-positive *Bacteria* and also some species of *Archaea* (Section 10.12). Because the DNA of prokaryotes is present in the cell as a large single molecule, when the cell is gently lysed, the DNA pours out. Because of their extreme length (1700 μm in *Bacillus subtilis*, for example), bacterial chromosomes break easily. Even after gentle extraction, the *B. subtilis* chromosome of 4.2 megabase pairs (Mbp) is converted to fragments of about 10 kbp each. Because the DNA that corresponds to an average gene is about 1000

nucleotides, each of the fragments of *B. subtilis* DNA therefore contains about ten genes. This is a typical transformable size. A single cell usually incorporates only one or a few DNA fragments, so only a small proportion of the genes of one cell can be transferred to another by a single transformation event.

Transformation in the History of Molecular Biology

The discovery of transformation was one of the key events in biology, as it led to experiments demonstrating that DNA was the genetic material. This discovery became a cornerstone of molecular biology and modern genetics.

The British scientist Frederick Griffith obtained the first evidence of bacterial transformation in the late 1920s. Griffith was working with *Streptococcus pneumoniae* (pneumococcus), a bacterium that owes its ability to invade the body in part to the presence of a polysaccharide capsule (Section 3.9). Mutants can be isolated that lack this capsule and thus cannot cause disease. Such mutants are called *R strains* because their colonies appear rough on agar, in contrast to the smooth appearance of encapsulated strains, called *S strains*. A mouse infected with only a few cells of an S strain succumbs in a day or two to a massive pneumococcus infection. By contrast, even large numbers of R cells do not cause death when injected. Griffith showed that if heat-killed S cells were injected along with living R cells, the mouse developed a fatal infection and the bacteria isolated from the dead mouse were of the S type (**Figure 10.12**). Because the S cells isolated in such an experiment always had the capsule type of the heat-killed S cells, Griffith concluded that the R cells had been transformed into a new type. This process set the stage for the discovery of DNA.

Oswald T. Avery and his associates at the Rockefeller Institute in New York provided the molecular explanation for the transformation of pneumococcus in a series of studies during the 1930s and 1940s. Avery and his coworkers showed that transformation could be carried out in the test tube instead of the mouse and that a cell-free extract of heat-killed cells could induce transformation. In a series of painstaking biochemical experiments, the active fraction was purified from cell-free extracts and was shown to be DNA. The transforming activity of purified DNA preparations was very high, and only a very small amount of material was necessary. Subsequently, others showed that transformation in pneumococcus affected not only the capsule but also other genetic characteristics such as antibiotic resistance and sugar fermentation.

In 1953, James Watson and Francis Crick published their model of the structure of DNA, providing a theoretical framework for how DNA could serve as genetic material. Thus, three types of studies, the bacteriological ones of Griffith, the biochemical ones of Avery, and the structural ones of Watson and Crick, solidified the concept of DNA as the genetic material. In subsequent years, this work led to the whole field of molecular biology and molecular genetics.

Competence in Transformation

Even within transformable genera, only certain strains or species are transformable. A cell that is able to take up DNA and be

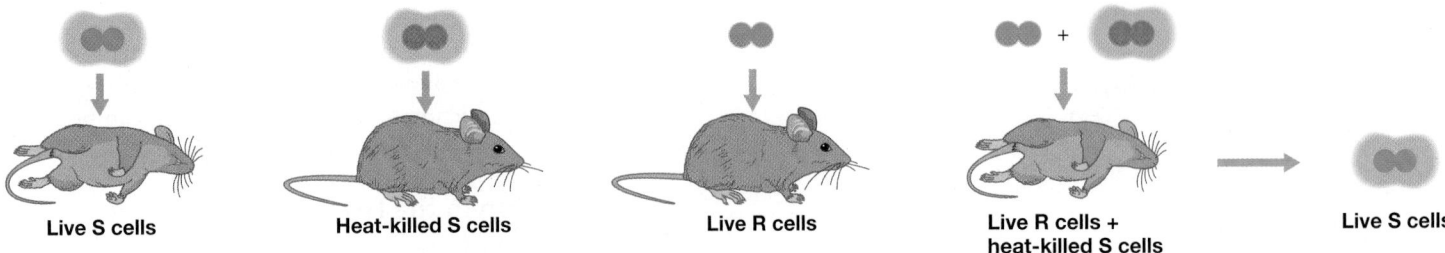

Live S cells   **Heat-killed S cells**   **Live R cells**   **Live R cells +**   **Live S cells**
heat-killed S cells

Figure 10.12 Griffith's experiments with pneumococcus. Live smooth (S) cells have a capsule and kill mice because immune cells cannot kill the encapsulated bacteria; the cells proliferate in the lung and cause a fatal pneumonia. Rough (R) cells have no capsule and are not pathogenic. But a combination of live R and dead S cells kill mice, and live S cells can be isolated from the animals. DNA carrying genes for capsule production is released from dead S cells and taken up by R cells, thus transforming them into S cells.

transformed is said to be *competent*, and this capacity is genetically determined.

Competence in most naturally transformable bacteria is regulated, and special proteins play a role in the uptake and processing of DNA. These competence-specific proteins include a membrane-associated DNA-binding protein, a cell wall autolysin, and various nucleases. One pathway of natural competence in *B. subtilis*—an easily transformed species—is regulated by quorum sensing (a regulatory system that responds to cell density; ∾ Section 8.9). Cells produce and excrete a small peptide during growth, and the accumulation of this peptide to high concentrations induces the cells to become competent. In *Bacillus*, roughly 20% of the cells in a culture become competent and stay that way for several hours. However, in *Streptococcus*, 100% of the cells can become competent, but only for a brief period during the growth cycle.

High-efficiency, natural transformation is rare among *Bacteria*. For example, *Acinetobacter, Bacillus, Streptococcus, Haemophilus, Neisseria*, and *Thermus* are naturally competent and easy to transform. By contrast, many *Bacteria* are poorly transformed, if at all, under natural conditions. *Escherichia coli* and many other gram-negative bacteria fall into this category. However, if cells of *E. coli* are treated with high concentrations of calcium ions and then chilled for several minutes, they become adequately competent. Cells of *E. coli* treated in this manner take up double-stranded DNA, and therefore transformation of this organism by plasmid DNA is relatively efficient. This is important because getting DNA into *E. coli*—the workhorse of genetic engineering—is critical for biotechnology, as we will see in Chapter 15.

Electroporation is a physical technique that is used to get DNA into organisms that are difficult to transform, especially those with thick cell walls. In electroporation, cells are mixed with DNA and then exposed to brief high-voltage electrical pulses. This makes the cell envelope permeable and allows entry of the DNA. Electroporation is a quick process and works for most types of cells, including *E. coli*, most other *Bacteria*, some members of the *Archaea*, and even yeast and certain plant cells.

Uptake of DNA in Transformation

During natural transformation, competent bacteria reversibly bind DNA. Soon, however, the binding becomes irreversible.

Competent cells bind much more DNA than do noncompetent cells—as much as 1000 times more. As noted earlier, the sizes of the transforming fragments are much smaller than that of the whole genome, and the fragments are further degraded during the uptake process. In *S. pneumoniae* each cell can bind only about ten molecules of double-stranded DNA of 10–15 kbp each. However, as these fragments are taken up, they are converted into single-stranded pieces of about 8 kb, with the complementary strand being degraded. The DNA fragments in the mixture compete with each other for uptake, and if excess DNA that does not contain the genetic marker under observation is added, the number of transformants decreases.

In preparations of transforming DNA, typically only about 1 out of 100–300 DNA fragments contains the genetic marker being studied. Thus, at high concentrations of DNA, the competition between DNA molecules results in saturation of the system, so even under the best of conditions it is impossible to transform all the cells in a population for a given marker. The maximum frequency of transformation that has so far been obtained is about 20% of the population; the usual values are between 0.1% and 1.0%. But when recipient population sizes are very high, even this low frequency is easy to detect. The minimum concentration of DNA yielding detectable transformants is about 0.01 ng/ml, which is so low that it is chemically undetectable.

Interestingly, transformation in *Haemophilus influenzae* requires the DNA fragment to have a particular 11-bp sequence for irreversible binding and uptake to occur. This sequence is found at an unexpectedly high frequency in the *Haemophilus* genome. Evidence such as this, and the fact that certain bacteria become competent in their natural environment, suggests that transformation is not a laboratory artifact but plays an important role in horizontal gene transfer in nature. By promoting new combinations of genes, naturally transformable bacteria increase the diversity and fitness of the microbial community as a whole.

Integration of Transforming DNA

Transforming DNA is bound at the cell surface by a DNA-binding protein. Next, either the entire double-stranded fragment is taken up, or a nuclease degrades one strand and the remaining strand is taken up, depending on the organism (**Figure 10.13**). After uptake, the DNA is bound by a competence-specific protein. This protects the DNA from nuclease attack until it reaches

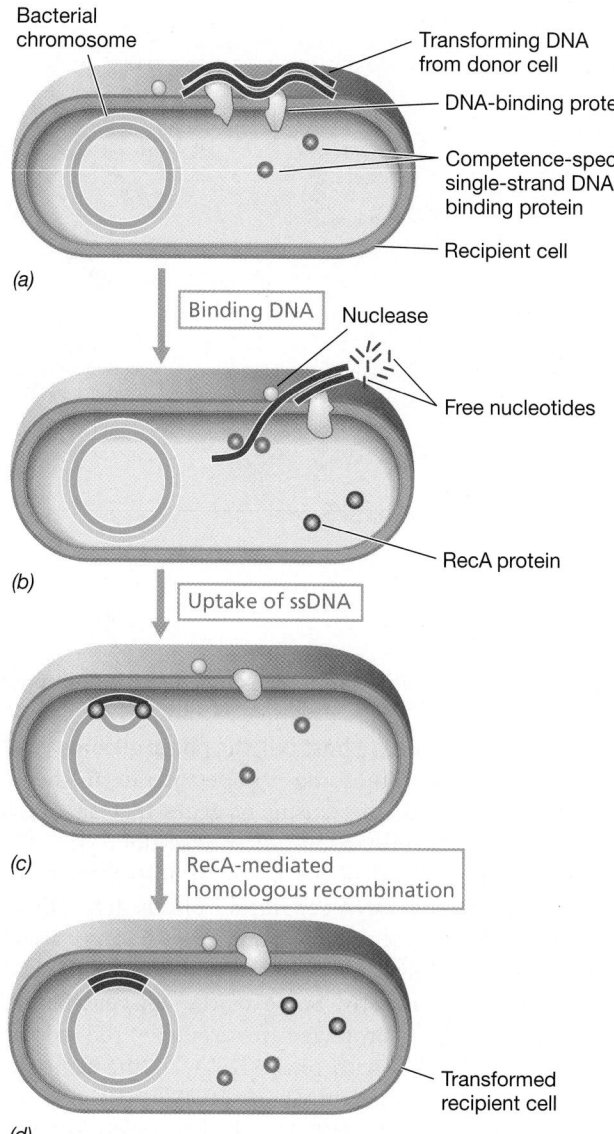

Bacterial
chromosome

Transforming DNA
from donor cell

DNA-binding protein

Competence-specific,
single-strand DNA-
binding protein

Recipient cell

(a)

Binding DNA   Nuclease

Free nucleotides

RecA protein

(b)

Uptake of ssDNA

(c)

RecA-mediated
homologous recombination

Transformed
recipient cell

(d)

Figure 10.13 Mechanism of transformation in a gram-positive bacterium. *(a)* Binding of double-stranded DNA by a membrane-bound DNA-binding protein. *(b)* Passage of one of the two strands into the cell while nuclease activity degrades the other strand. *(c)* The single strand in the cell is bound by specific proteins, and recombination with homologous regions of the bacterial chromosome is mediated by RecA protein. *(d)* Transformed cell.

the chromosome, where the RecA protein takes over. The DNA is integrated into the genome of the recipient by recombination (Figures 10.13 and 10.10). If single-stranded DNA is integrated, a heteroduplex DNA is formed. During the next round of chromosomal replication, one parental and one recombinant DNA molecule are generated. On segregation at cell division, the recombinant molecule is present in the transformed cell, which is now genetically altered compared to its parent. The preceding applies only to small pieces of linear DNA. Many naturally transformable *Bacteria* are transformed only poorly by plasmid DNA because the plasmid must remain double-stranded and circular in order to replicate.

Transfection

Bacteria can be transformed with DNA extracted from a bacterial virus rather than from another bacterium. This process is called *transfection*. If the DNA is from a lytic bacteriophage, transfection leads to virus production and can be measured by the standard phage plaque assay (Section 9.4). Transfection is useful for studying the mechanisms of transformation and recombination because the small size of phage genomes allows the isolation of a nearly homogeneous population of DNA molecules. By contrast, in conventional transformation the transforming DNA is typically a random assortment of chromosomal DNA fragments of various lengths, and this tends to complicate experiments designed to study the mechanism of transformation.

MiniQuiz

- The donor bacterial cell in a transformation is probably dead. Explain.
- Even in naturally transformable cells, competence is usually inducible. What does this mean?

10.8 Transduction

In **transduction**, a bacterial virus (bacteriophage) transfers DNA from one cell to another. Viruses can transfer host genes in two ways. In the first, called *generalized transduction*, DNA derived from virtually any portion of the host genome is packaged inside the mature virion in place of the virus genome. In the second, called *specialized transduction*, DNA from a specific region of the host chromosome is integrated directly into the virus genome—usually replacing some of the virus genes. This occurs only with certain temperate viruses (Section 9.10). The transducing bacteriophage in both generalized and specialized transduction is usually noninfectious because bacterial genes have replaced all or some necessary viral genes.

In generalized transduction, the donor genes cannot replicate independently and are not part of a viral genome. Unless the donor genes recombine with the recipient bacterial chromosome, they will be lost. In specialized transduction, homologous recombination may also occur. However, since the donor bacterial DNA is actually a part of a temperate phage genome, it may be integrated into the host chromosome during lysogeny (Section 9.10).

Transduction occurs in a variety of *Bacteria*, including the genera *Desulfovibrio*, *Escherichia*, *Pseudomonas*, *Rhodococcus*, *Rhodobacter*, *Salmonella*, *Staphylococcus*, and *Xanthobacter*, as well as *Methanothermobacter thermautotrophicus*, a species of *Archaea*. Not all phages can transduce, and not all bacteria are transducible, but the phenomenon is sufficiently widespread that it likely plays an important role in gene transfer in nature.

Generalized Transduction

In generalized transduction, virtually any gene on the donor chromosome can be transferred to the recipient. Generalized transduction was first discovered and extensively studied in the bacterium *Salmonella enterica* with phage P22 and has also been studied with phage P1 in *Escherichia coli*. An example of how

UNIT 4

Lytic cycle

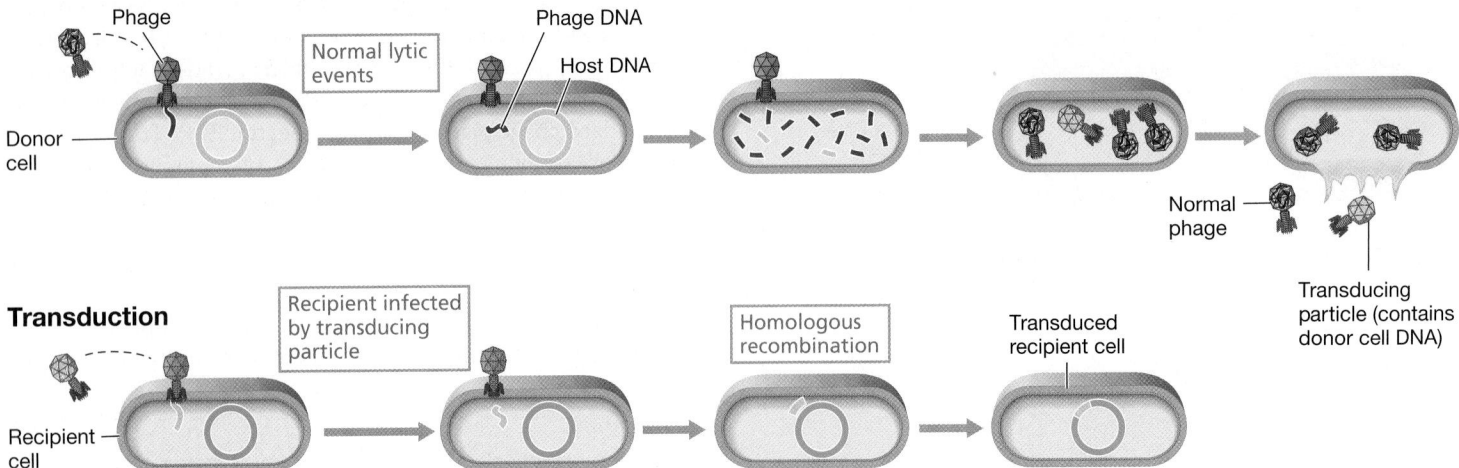

Figure 10.14 **Generalized transduction.** Note that "normal" virions contain phage genes, whereas a transducing particle contains host genes.

transducing particles are formed is given in **Figure 10.14**. When a bacterial cell is infected with a phage, the lytic cycle may occur. However, during lytic infection, the enzymes responsible for packaging viral DNA into the bacteriophage sometimes package host DNA accidentally. The result is called a *transducing particle*. These cannot lead to a viral infection because they contain no viral DNA, and are said to be *defective*. On lysis of the cell, the transducing particles are released along with normal virions that contain the virus genome. Consequently, the lysate contains a mixture of normal virions and transducing particles.

When this lysate is used to infect a population of recipient cells, most of the cells are infected with normal virus. However, a small proportion of the population receives transducing particles that inject the DNA they packaged from the previous host bacterium. Although this DNA cannot replicate, it can recombine with the DNA of the new host. Because only a small proportion of the particles in the lysate are defective, and each of these contains only a small fragment of donor DNA, the probability of a given transducing particle containing a particular gene is quite low. Typically, only about 1 cell in 10^6 to 10^8 is transduced for a given marker.

Phages that form transducing particles can be either temperate or virulent, the main requirements being that they have a DNA-packaging mechanism that accepts host DNA and that DNA packaging occurs before the host genome is completely degraded. Transduction is most likely when the ratio of input phage to recipient bacteria is low so that cells are infected with only a single phage particle; with multiple phage infection, the cell is likely to be killed by the normal virions in the lysate.

Phage Lambda and Specialized Transduction

Generalized transduction allows the transfer of any gene from one bacterium to another, but at a low frequency. In contrast, specialized transduction allows extremely efficient transfer, but is selective and transfers only a small region of the bacterial chromosome. In the first case of specialized transduction to be dis-

covered, the galactose genes were transduced by the temperate phage lambda of *E. coli*.

When lambda lysogenizes a host cell, the phage genome is integrated into the *E. coli* chromosome at a specific site (↩ Section 9.10). This site is next to the cluster of genes that encode the enzymes for galactose utilization. After insertion, viral DNA replication is under control of the bacterial host chromosome. Upon induction, the viral DNA separates from the host DNA by a process that is the reverse of integration (**Figure 10.15**). Usually, the lambda DNA is excised precisely, but occasionally, the phage genome is excised incorrectly. Some of the adjacent bacterial genes to one side of the prophage (for example, the galactose operon) are excised along with phage DNA. At the same time, some phage genes are left behind (Figure 10.15*b*).

One type of altered phage particle, called *lambda dgal* (λ*dgal*; *dgal* means "*d*efective *gal*actose"), is defective because of the lost phage genes. It will not make a mature phage in a subsequent infection. However, a viable lambda virion known as a helper phage can provide those functions missing in the defective particle. When cells are coinfected with λ*dgal* and the helper phage, the culture lysate contains a few λ*dgal* particles mixed in with a large number of normal lambda virions. When a galactose-negative (Gal⁻) bacterial culture is infected with such a lysate and Gal⁺ transductants selected, many are double lysogens carrying both lambda and λ*dgal*. When such a double lysogen is induced, the lysate contains large numbers of λ*dgal* virions and can transduce at high efficiency, although only for the restricted group of *gal* genes.

For a lambda virion to be viable, there is a limit to the amount of phage DNA that can be replaced with host DNA. Sufficient phage DNA must be retained to encode the phage protein coat and other phage proteins needed for lysis and lysogeny. However, if a helper phage is used together with a defective phage in a mixed infection, then far fewer phage-specific genes are needed in the defective phage. Only the *att* (attachment) region, the *cos* site (cohesive ends, for packaging), and the replication origin of the lambda genome are absolutely needed for production of a

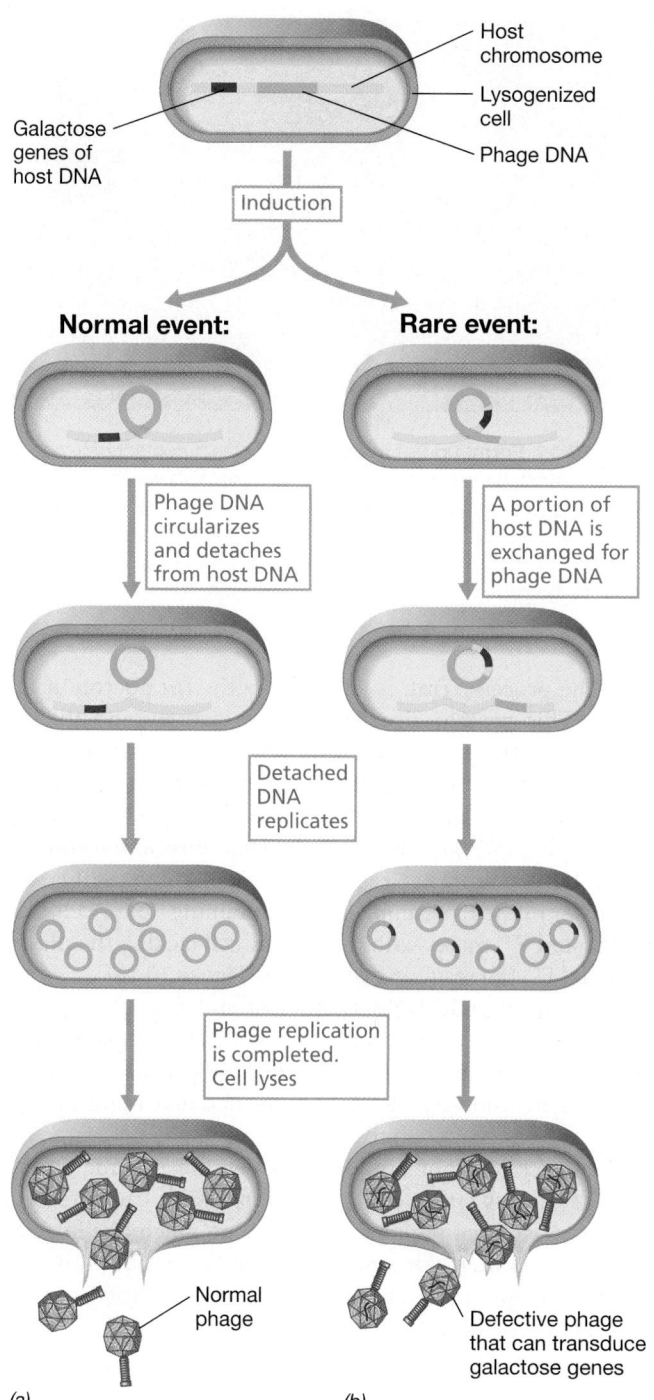

Figure 10.15 Specialized transduction. In an *Escherichia coli* cell containing a lambda prophage, *(a)* normal lytic events, and *(b)* the production of particles transducing the galactose genes. Only a short region of the circular host chromosome is shown in the figure.

transducing particle when a helper phage is used. By deleting the normal chromosomal *att* site and forcing lambda to integrate at other locations, specialized transducing phages covering many specific regions of the *E. coli* genome have been isolated. In addition, lambda transducing phages can be constructed by the techniques of genetic engineering to contain genes from any organism (↺ Section 11.9).

Phage Conversion

Alteration of the phenotype of a host cell by lysogenization is called *phage conversion*. When a normal (that is, nondefective) temperate phage lysogenizes a cell and becomes a prophage, the cell becomes immune to further infection by the same type of phage. Such immunity may itself be regarded as a change in phenotype. However, other phenotypic changes unrelated to phage immunity are often observed in lysogenized cells.

Two cases of phage conversion have been especially well studied. One involves a change in structure of a polysaccharide on the cell surface of *Salmonella anatum* on lysogenization with bacteriophage ε^{15}. The second involves the conversion of non-toxin-producing strains of *Corynebacterium diphtheriae* (the bacterium that causes the disease diphtheria) to toxin-producing (pathogenic) strains following lysogeny with phage β (↺ Section 33.3). In both cases, the genes responsible for the changes are an integral part of the phage genome and hence are automatically transferred upon infection by the phage and lysogenization.

Lysogeny probably carries a strong selective value for the host cell because it confers resistance to infection by viruses of the same type. Phage conversion may also be of considerable evolutionary significance because it results in efficient genetic alteration of host cells. Many bacteria isolated from nature are natural lysogens. It seems likely that lysogeny is often essential for survival of the host cells in nature.

MiniQuiz

- What is the major difference between generalized transduction and transformation?

- In specialized transduction, the donor DNA can replicate inside the recipient cell without homologous recombination taking place, but this is not true in generalized transduction. Explain.

10.9 Conjugation: Essential Features

Bacterial **conjugation** (mating) is a mechanism of genetic transfer that involves cell-to-cell contact. Conjugation is a plasmid-encoded mechanism. Conjugative plasmids use this mechanism to transfer copies of themselves to new host cells. Thus the process of conjugation involves a *donor* cell, which contains the conjugative plasmid, and a *recipient* cell, which does not. In addition, genetic elements that cannot transfer themselves can sometimes be mobilized during conjugation. These other genetic elements can be other plasmids or the host chromosome itself. Indeed, conjugation was discovered because the F plasmid of *Escherichia coli* can mobilize the host chromosome (see Figure 10.21). Transfer mechanisms may differ depending on the plasmid involved, but most plasmids in gram-negative *Bacteria* employ a mechanism similar to that used by the F plasmid.

F Plasmid

The F plasmid (F stands for "fertility") is a circular DNA molecule of 99,159 bp. **Figure 10.16** shows a genetic map of the F plasmid. One region of the plasmid contains genes that regulate DNA replication. It also contains a number of transposable elements

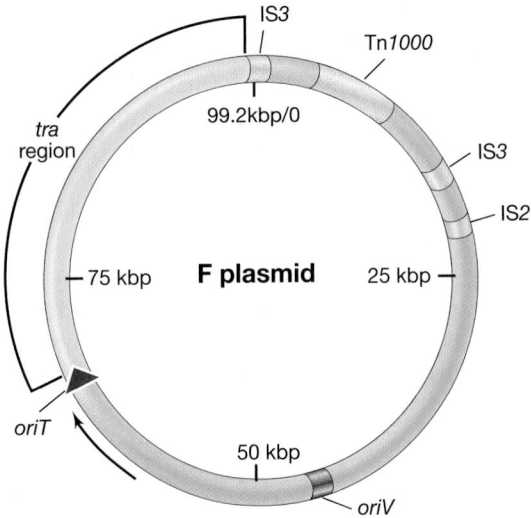

Figure 10.16 Genetic map of the F (fertility) plasmid of *Escherichia coli.* The numbers on the interior show the size in kilobase pairs (the exact size is 99,159 bp). The region in dark green at the bottom of the map contains genes primarily responsible for the replication and segregation of the F plasmid. The origin of vegetative replication is *oriV.* The light green *tra* region contains the genes needed for conjugative transfer. The origin of transfer during conjugation is *oriT.* The arrow indicates the direction of transfer (the *tra* region is transferred last). Insertion sequences are shown in yellow. These may recombine with identical elements on the bacterial chromosome, which leads to integration and the formation of different Hfr strains.

(Section 10.13) that allow the plasmid to integrate into the host chromosome. In addition, the F plasmid has a large region of DNA, the *tra* region, containing genes that encode transfer functions. Many genes in the *tra* region are involved in mating pair formation, and most of these have to do with the synthesis of a surface structure, the sex pilus (⮌ Section 3.9). Only donor cells produce these pili. Different conjugative plasmids may have slightly different *tra* regions, and the pili may vary somewhat in structure. The F plasmid and its relatives encode F pili.

Pili allow specific pairing to take place between the donor and recipient cells. All conjugation in gram-negative *Bacteria* is thought to depend on cell pairing brought about by pili. The pilus makes specific contact with a receptor on the recipient cell and then is retracted by disassembling its subunits. This pulls the two cells together (**Figure 10.17**). Following this process, donor and recipient cells remain in contact by binding proteins located in the outer membrane of each cell. DNA is then transferred from donor to recipient cell through this conjugation junction.

Mechanism of DNA Transfer During Conjugation

DNA synthesis is necessary for DNA transfer by conjugation. This DNA is synthesized not by normal bidirectional replication (⮌ Section 6.10), but by **rolling circle replication**, a mechanism also used by some viruses (⮌ Section 9.10) and shown in **Figure 10.18**. DNA transfer is triggered by cell-to-cell contact, at which time one strand of the circular plasmid DNA is nicked and is transferred to the recipient. The nicking enzyme required to

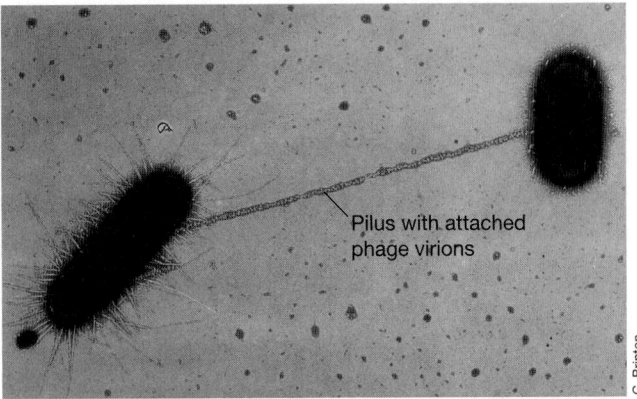

Figure 10.17 **Formation of a mating pair.** Direct contact between two conjugating bacteria is first made via a pilus. The cells are then drawn together to form a mating pair by retraction of the pilus, which is achieved by depolymerization. Certain small phages (F-specific bacteriophages; ⮌ Section 21.1) use the sex pilus as a receptor and can be seen here attached to the pilus.

initiate the process, TraI, is encoded by the *tra* operon of the F plasmid. This protein also has helicase activity and thus also unwinds the strand to be transferred. As this transfer occurs, DNA synthesis by the rolling circle mechanism replaces the transferred strand in the donor, while a complementary DNA strand is being made in the recipient. Therefore, at the end of the process, both donor and recipient possess complete plasmids. For transfer of the F plasmid, if an F-containing donor cell, which is designated F⁺, mates with a recipient cell lacking the plasmid, designated F⁻, the result is two F⁺ cells (Figure 10.18).

Transfer of plasmid DNA is efficient and rapid; under favorable conditions virtually every recipient cell that pairs with a donor acquires a plasmid. Transfer of the F plasmid, comprising approximately 100 kbp of DNA, takes about 5 minutes. If the plasmid genes can be expressed in the recipient, the recipient itself becomes a donor and can transfer the plasmid to other recipients. In this fashion, conjugative plasmids can spread rapidly among bacterial populations, behaving much like infectious agents. This is of major ecological significance because a few plasmid-containing cells introduced into a population of recipients can convert the entire population into plasmid-bearing (and thus donating) cells in a short time.

Plasmids can be lost from a cell by curing. This may happen spontaneously in natural populations when there is no selection pressure to maintain the plasmid. For example, plasmids conferring antibiotic resistance can be lost without affecting cell viability if there are no antibiotics in the cells' environment.

MiniQuiz
- In conjugation, how are donor and recipient cells brought into contact with each other?
- Explain how rolling circle DNA replication allows both donor and recipient to end up with a complete copy of plasmids transferred by conjugation.
- Why does F have two different origins of replication?

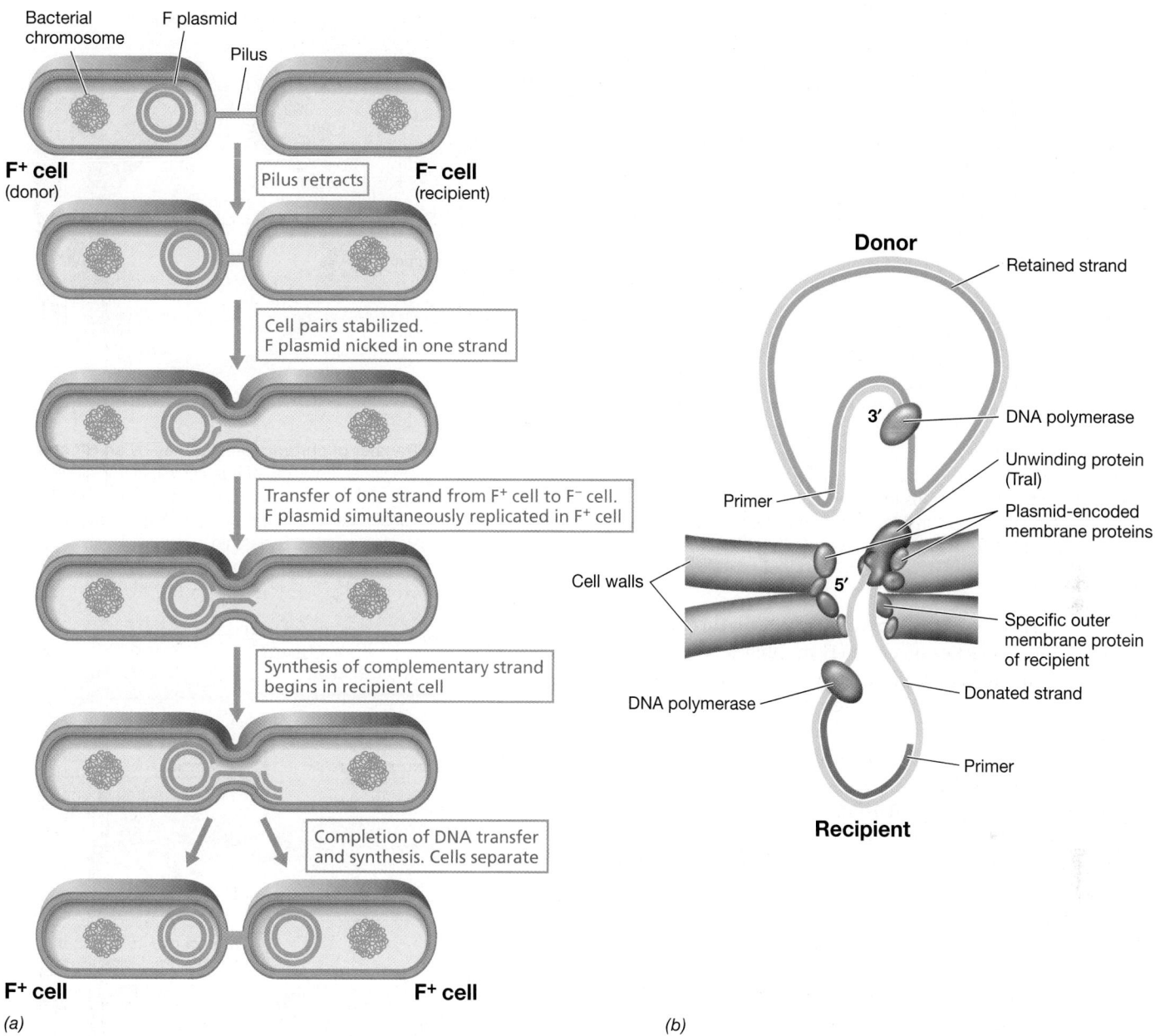

Figure 10.18 Transfer of plasmid DNA by conjugation. *(a)* The transfer of the F plasmid converts an F⁻ recipient cell into an F⁺ cell. Note the mechanism of rolling circle replication. *(b)* Details of the replication and transfer process.

10.10 The Formation of Hfr Strains and Chromosome Mobilization

Chromosomal genes can be transferred by plasmid-mediated conjugation. As mentioned above, the F plasmid of *Escherichia coli* can, under certain circumstances, mobilize the chromosome during cell-to-cell contact. The F plasmid is an *episome*, a plasmid that can integrate into the host chromosome. When the F plasmid is integrated, chromosomal genes can be transferred along with the plasmid. Following genetic recombination between donor and recipient DNA, horizontal transfer of chromosomal genes by this mechanism can be very extensive.

Cells possessing a nonintegrated F plasmid are called F⁺. Those with an F plasmid integrated into the chromosome are called **Hfr** (for *high frequency of recombination*) **cells**. This term refers to the high rates of genetic recombination between genes on the donor and recipient chromosomes. Both F⁺ and Hfr cells are donors, but unlike conjugation between an F⁺ and an F⁻, conjugation between an Hfr donor and an F⁻ leads to transfer of genes from the host chromosome. This is because the chromosome and plasmid now form a single molecule of DNA. Consequently, when rolling circle replication is initiated by the F plasmid, replication continues on into the chromosome. Thus, the chromosome is also replicated and transferred. Hence, integration of a conjugative plasmid provides a mechanism for *mobilizing* a cell's genome.

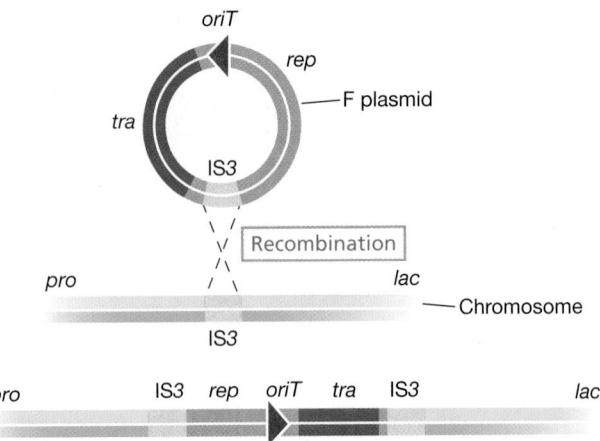

Figure 10.19 The formation of an Hfr strain. Integration of the F plasmid into the chromosome may occur at a variety of specific sites where IS elements are located. The example shown here is an IS3 located between the chromosomal genes *pro* and *lac*. Some of the genes on the F plasmid are shown. The arrow indicates the origin of transfer, *oriT*, with the arrow as the leading end. Thus, in this Hfr *pro* would be the first chromosomal gene to be transferred and *lac* would be among the last.

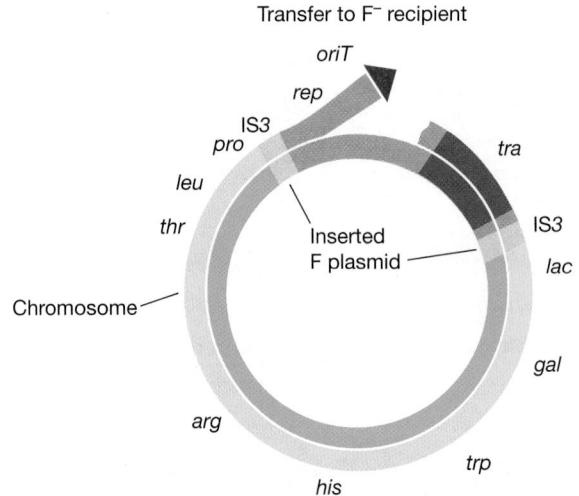

Figure 10.20 Transfer of chromosomal genes by an Hfr strain. The Hfr chromosome breaks at the origin of transfer within the integrated F plasmid. The transfer of DNA to the recipient begins at this point. DNA replicates during transfer as for a free F plasmid (Figure 10.18). This figure is not to scale; the inserted F plasmid is actually less than 3% of the size of the *Escherichia coli* chromosome.

Overall, the presence of the F plasmid results in three distinct alterations in the properties of a cell: (1) the ability to synthesize the F pilus, (2) the mobilization of DNA for transfer to another cell, and (3) the alteration of surface receptors so the cell can no longer act as a recipient in conjugation and is unable to take up a second copy of the F plasmid or genetically related plasmids.

Integration of F and Chromosome Mobilization

The F plasmid and the chromosome of *E. coli* both carry several copies of mobile elements called *insertion sequences* (IS; Section 10.13). These provide regions of sequence homology between chromosomal and F plasmid DNA. Consequently, homologous recombination between an IS on the F plasmid and a corresponding IS on the chromosome results in integration of the F plasmid into the host chromosome, as shown in **Figure 10.19**. Once integrated, the plasmid no longer replicates independently, but the *tra* operon still functions normally and the strain synthesizes pili. When a recipient is encountered, conjugation is triggered just as in an F⁺ cell, and DNA transfer is initiated at the *oriT* (origin of transfer) site. However, because the plasmid is now part of the chromosome, after part of the plasmid DNA is transferred, chromosomal genes begin to be transferred (**Figure 10.20**). As in the case of conjugation with just the F plasmid itself (Figure 10.18), chromosomal DNA transfer also involves replication.

Because the DNA strand typically breaks during transfer, only part of the donor chromosome is transferred. Consequently, the recipient does not become Hfr (or F⁺) because only part of the integrated F plasmid is transferred (**Figure 10.21**). However, after transfer, the Hfr strain remains Hfr because it retains a copy of the integrated F plasmid. Because a partial chromosome cannot replicate, for incoming donor DNA to survive, it must recombine with the recipient chromosome. Following recombination, the recipient cell may express a new phenotype due to incorporation of donor genes. Although Hfr strains transmit chromosomal

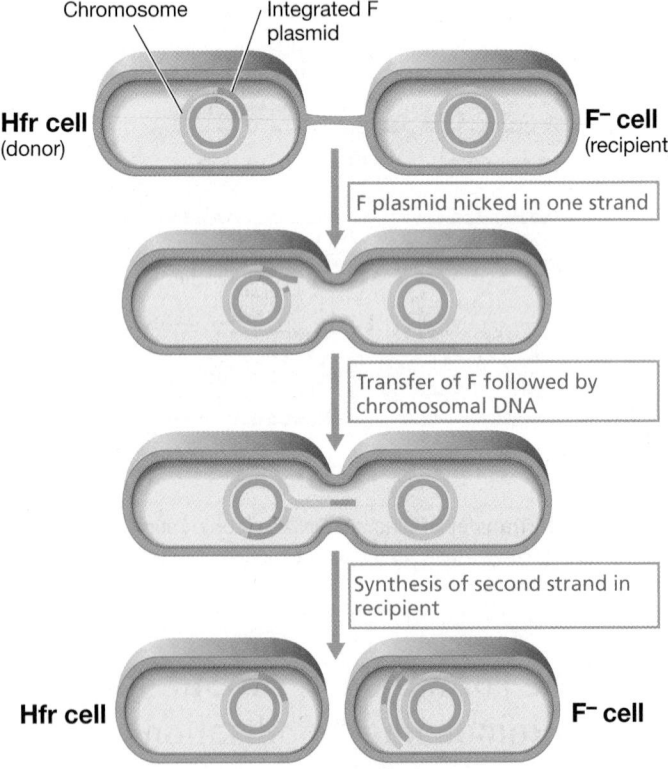

Figure 10.21 Transfer of chromosomal DNA by conjugation. Transfer of the integrated F plasmid from an Hfr strain results in the cotransfer of chromosomal DNA because this is linked to the plasmid. The steps in transfer are similar to those in Figure 10.18a. However, the recipient remains F⁻ and receives a linear fragment of donor chromosome attached to part of the F plasmid. For donor DNA to survive, it must be recombined into the recipient chromosome after transfer (not shown).

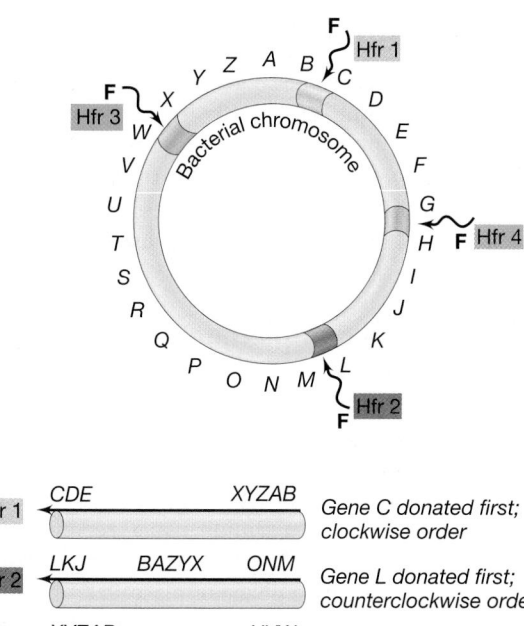

(a)

Hfr 1	CDE XYZAB	Gene C donated first; clockwise order
Hfr 2	LKJ BAZYX ONM	Gene L donated first; counterclockwise order
Hfr 3	XYZAB UVW	Gene X donated first; clockwise order
Hfr 4	GFE BAZYX JIH	Gene G donated first; counterclockwise order

(b)

Figure 10.22 Formation of different Hfr strains. Different Hfr strains donate genes in different orders and from different origins. *(a)* F plasmids can be inserted into various insertion sequences on the bacterial chromosome, forming different Hfr strains. *(b)* Order of gene transfer for different Hfr strains.

Use of Hfr Strains in Genetic Crosses

As for any system of bacterial gene transfer, the experimenter selects recombinants from conjugation. However, unlike transformation and transduction, both donor and recipient cells are viable during conjugation. It is thus necessary to choose selection

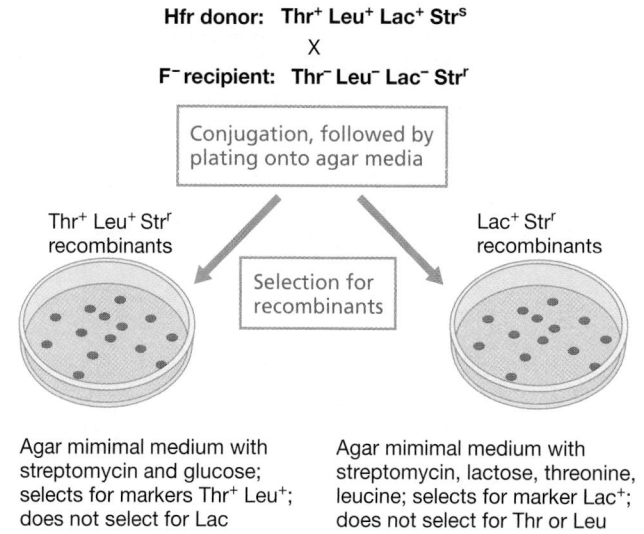

Figure 10.23 Example experiment for the detection of conjugation. Thr, threonine; Leu, leucine; Lac, lactose; Str, streptomycin. Note that each medium selects for specific classes of recombinants. The controls for the experiment are made by plating samples of the donor and the recipient before they are mixed. Neither should be able to grow on the selective media used.

conditions in which the desired recombinants can grow, but where neither of the parental strains can grow. Typically, a recipient is used that is resistant to an antibiotic, but is auxotrophic for some nutrient, and a donor is used that is sensitive to the antibiotic, but is prototrophic for the same nutrient. Thus, on minimal medium containing the antibiotic, only recombinant cells will grow following the mating.

For instance, in the experiment shown in **Figure 10.23**, an Hfr donor that is sensitive to streptomycin (Str^s) and is wild type for synthesis of the amino acids threonine and leucine (Thr^+ and Leu^+) and for utilization of lactose (Lac^+), is mated with a recipient cell that cannot make these amino acids or use lactose, but that is resistant to streptomycin (Str^r). The selective minimal medium contains streptomycin so that only recombinant cells can grow. The composition of each selective medium is varied depending on which genotypic characteristics are desired in the recombinant, as shown in Figure 10.23. The frequency of gene transfer is measured by counting the colonies grown on the selective medium.

The order of genes on the donor chromosome can also be determined by following the kinetics of transfer of individual markers. For example, in the process called *interrupted mating*, conjugating cells are separated by agitation in a mixer or blender. If mixtures of Hfr and F⁻ cells are agitated at various times after mixing and the genetic recombinants scored, it is found that the longer the time between pairing and agitation, the greater the number of genes from the Hfr that are found in the recombinant. As shown in **Figure 10.24**, genes located closer to the origin of transfer enter the recipient first and are present in a higher percentage of the recombinants than genes that are transferred later. In addition to showing that gene transfer from donor to recipient occurs sequentially, experiments of this kind allow the order of the genes on the bacterial chromosome to be determined.

genes at high frequency, they generally do not convert F⁻ cells to F⁺ or Hfr because the entire F plasmid is rarely transferred. Instead, an Hfr × F⁻ cross yields the original Hfr and an F⁻ cell that now has a new genotype. As in transformation and transduction, genetic recombination between Hfr genes and F⁻ genes involves homologous recombination in the recipient cell.

Because several distinct insertion sequences are present on the chromosome, a number of distinct Hfr strains are possible. A given Hfr strain always donates genes in the same order, beginning at the same position. However, Hfr strains that differ in the chromosomal integration site of the F plasmid transfer genes in different orders (**Figure 10.22**). At some insertion sites, the F plasmid is integrated with its origin pointing in one direction, whereas at other sites the origin points in the opposite direction. The orientation of the F plasmid determines which chromosomal genes enter the recipient first (Figure 10.22). By using various Hfr strains in mating experiments, it was possible to determine the arrangement and orientation of most of the genes in the *E. coli* chromosome long before it was sequenced.

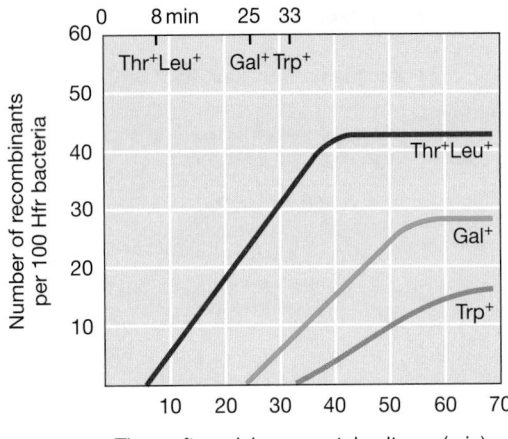

Figure 10.24 Time of gene entry in a mating culture. The rate of appearance of recombinants containing different genes after mating Hfr and F⁻ bacteria is shown. The location of the genes along the Hfr chromosome is shown at the upper left. Genes closest to the origin (0 min) are the first to be transferred. The experiment is done by mixing Hfr and F⁻ cells under conditions in which most Hfr cells find recipients. At various times, samples of the mixture are shaken violently to separate mating pairs and plated onto selective medium on which only recombinants can form colonies.

Transfer of Chromosomal Genes to the F Plasmid

Occasionally, integrated F plasmids may be excised from the chromosome. During excision, chromosomal genes may sometimes be incorporated into the liberated F plasmid. This can happen because both the F plasmid and the chromosome contain multiple identical insertion sequences where recombination can occur (Figure 10.20). F plasmids containing chromosomal genes are called *F′ plasmids*. When F′ plasmids promote conjugation, they transfer the chromosomal genes they carry at high frequency to the recipients. F′-mediated transfer resembles specialized transduction (Section 10.8) in that only a restricted group of chromosomal genes is transferred by any given F′ plasmid. Transferring a known F′ into a recipient allows one to establish diploids (two copies of each gene) for a limited region of the chromosome. Such partial diploids are important for complementation tests, as we will see in the next section.

Other Conjugation Systems

Although we have discussed conjugation almost exclusively as it occurs in *E. coli*, conjugative plasmids have been found in many other gram-negative *Bacteria*. Conjugative plasmids of the incompatibility group IncP can be maintained in virtually all gram-negative *Bacteria* and even transferred between different genera. Conjugative plasmids are also known in gram-positive *Bacteria* (for example, in *Streptococcus* and *Staphylococcus*). A process of genetic transfer similar to conjugation in *Bacteria* also occurs in some *Archaea* (Section 10.12).

MiniQuiz

- In conjugation involving the F plasmid of *Escherichia coli*, how is the host chromosome mobilized?

- Why does an Hfr × F⁻ mating not yield two Hfr cells?

- At which sites in the chromosome can the F plasmid integrate?

10.11 Complementation

In all three methods of bacterial gene transfer, only a portion of the donor chromosome enters the recipient cell. Therefore, unless recombination takes place with the recipient chromosome, the donor DNA will be lost because it cannot replicate independently in the recipient. Nonetheless, it is possible to stably maintain a state of partial diploidy for use in bacterial genetic analysis, and we consider this now.

Merodiploids and Complementation

A bacterial strain that carries two copies of any particular chromosomal segment is known as a partial diploid or *merodiploid*. In general, one copy is present on the chromosome itself and the second copy on another genetic element, such as a plasmid or a bacteriophage. Because it is possible to create specialized transducing phages or specific plasmids using recombinant DNA techniques (Chapter 11), it is possible to put any portion of the bacterial chromosome onto a phage or plasmid.

Consequently, if the chromosomal copy of a gene is defective due to a mutation, it is possible to supply a functional (wild-type) copy of the gene on a plasmid or phage. For example, if one of the genes for tryptophan biosynthesis has been inactivated, this will give a Trp⁻ phenotype. That is, the mutant strain will be a tryptophan auxotroph and will require the amino acid tryptophan for growth. However, if a copy of the wild-type gene is introduced into the same cell on a plasmid or viral genome, this gene will encode the necessary protein and restore the wild-type phenotype. This process is called *complementation* because the wild-type gene is said to complement the mutation, in this case converting the Trp⁻ cell into Trp⁺.

Complementation Tests and the Cistron

When two mutant strains are genetically crossed (whether by conjugation, transduction, or transformation), homologous recombination can yield wild-type recombinants unless both mutations affect exactly the same base pairs. For example, if two different Trp⁻ *Escherichia coli* mutants are crossed and Trp⁺ recombinants are obtained, it is obvious that the mutations in the two strains were not in the same base pairs. However, this kind of experiment cannot determine whether two mutations are in two different genes that both affect tryptophan synthesis or in different regions of the same gene. This can be determined by a complementation test.

To perform a complementation test, two copies of the region of DNA under investigation must be present and carried on two different molecules of DNA. One copy is normally present on the chromosome and the other is carried on a second DNA molecule, typically a plasmid. For example, if we are analyzing mutants in tryptophan biosynthesis, then two copies of the whole tryptophan operon must be present.

Suppose that we wish to know if two Trp⁻ strains have a mutation in the same gene. To do this we must arrange for one mutation to be present on the chromosome and the other on a plasmid. The mutations are then referred to as being in *trans* with respect to one another. If the two mutations are in the *same* gene, the recombinant cell will have two defective copies of the same gene and will

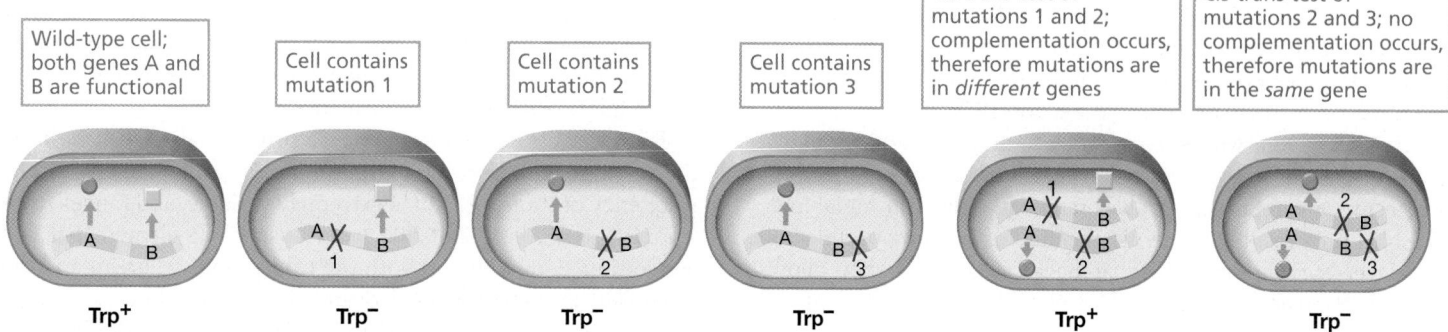

Figure 10.25 Complementation analysis. In this example, the proteins encoded by both genes A and B are required to synthesize tryptophan. Mutations 1, 2, and 3 each lead to the same phenotype, a requirement for tryptophan (Trp⁻). Complementation analysis indicates that mutations 2 and 3 are in the same gene but mutation 1 is in a separate gene.

display the negative phenotype. Conversely, if the two mutations are in *different* genes, the recipient cell will have one unmutated copy of each gene (one on the chromosome and the other on the plasmid) and be able to synthesize tryptophan. The possible combinations are shown diagrammatically in **Figure 10.25**. If one DNA molecule carries both mutations (that is, the mutations are in *cis*), a second DNA molecule can serve as a complement if it is wild type for both genes. Having the mutations in *cis* serves as a positive control in a complementation experiment. This type of complementation test is therefore called a *cis-trans* test.

A gene as defined by the *cis-trans* test is called a **cistron** and is equivalent to defining a structural gene as a segment of DNA that encodes a single polypeptide chain. If two mutations occur in genes encoding different enzymes, or even different protein subunits of the same enzyme, complementation of the two mutations is possible, and the mutations are therefore not in the same cistron (Figure 10.25). It is important to note that complementation does not rely on recombination; the two genes in question remain on separate genetic elements.

Although genetic crosses to test complementation are still done in bacterial genetics, it is often easier to sequence the gene in question to identify the nature and location of any mutations. This is especially true if the sequence of the wild-type gene is already known. The word "cistron" is now rarely used in microbial genetics except when describing whether an mRNA has the genetic information from one gene (*monocistronic* mRNA) or from more than one gene (*polycistronic* mRNA) (Section 6.15).

MiniQuiz

- What is a merodiploid?
- Complementation tests have been referred to as *cis-trans* tests. Explain.

10.12 Gene Transfer in *Archaea*

Although *Archaea* contain a single circular chromosome like most *Bacteria* (**Figure 10.26**) and the genomes of several species of *Archaea* have been entirely sequenced, the development of gene transfer systems lags far behind that for *Bacteria*. Practical

problems include the need to grow many *Archaea* under extreme conditions. Thus, the temperatures necessary to culture some hyperthermophiles will melt agar, and alternative materials are required to form solid media and obtain colonies.

Another problem is that most antibiotics do not affect *Archaea*. For example, penicillins do not affect *Archaea* because their cell walls lack peptidoglycan. The choice of selectable markers for genetic crosses is therefore often limited. However, novobiocin (a DNA gyrase inhibitor) and mevinolin (an inhibitor of isoprenoid biosynthesis) are used to inhibit extreme halophiles, and puromycin and neomycin (both protein synthesis inhibitors) inhibit methanogens.

No single species of *Archaea* has become a model organism for archaeal genetics, although more genetic work has probably been done on select species of extreme halophiles (*Halobacterium*, *Haloferax*, Section 19.2) than on any other *Archaea*. Instead, individual mechanisms for gene transfer have been found scattered among a range of *Archaea*. Examples of transformation, transduction, and conjugation are known. In addition, several plasmids have been isolated from *Archaea* and some have been used to construct cloning vectors, allowing genetic analysis through cloning and sequencing rather than traditional genetic crosses. Transposon

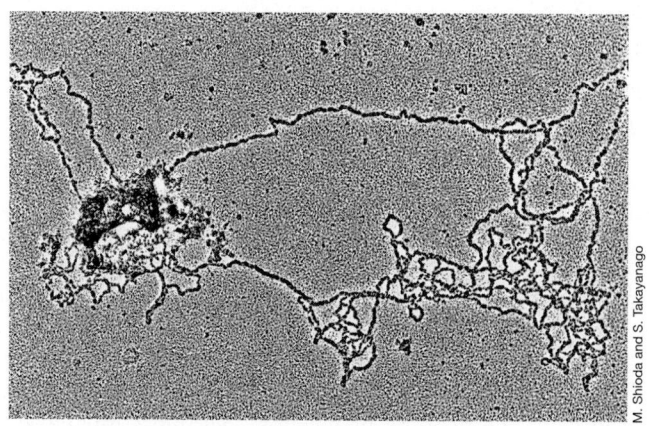

Figure 10.26 An archaeal chromosome, as shown in the electron microscope. The circular chromosome is from the hyperthermophile *Sulfolobus*, a member of the *Archaea*.

mutagenesis has been well developed in certain methanogen species including *Methanococcus* and *Methanosarcina*, and other tools such as shuttle vectors and other *in vitro* methods of genetic analysis have been developed for study of the highly unusual biochemistry of the methanogens (↵ Section 19.3).

Transformation works reasonably well in several *Archaea*. Transformation procedures vary in detail from organism to organism. One approach involves removal of divalent metal ions, which in turn results in the disassembly of the glycoprotein cell wall layer surrounding many archaeal cells and hence allows access by transforming DNA. However, *Archaea* with rigid cell walls have proven difficult to transform, although electroporation sometimes works. One exception is in *Methanosarcina* species, organisms with a thick cell wall, for which high-efficiency transformation systems have been developed that employ DNA-loaded lipid preparations (liposomes) to deliver DNA into the cell.

Although viruses that infect *Archaea* are plentiful, transduction is extremely rare. Only one archaeal virus, which infects the thermophilic methanogen *Methanothermobacter thermautotrophicus*, has been shown to transduce the genes of its host. Unfortunately the low burst size (about six phages liberated per cell) makes using this system for gene transfer impractical.

Two types of conjugation have been detected in *Archaea*. Some strains of *Sulfolobus solfataricus* (↵ Section 19.9) contain plasmids that promote conjugation between two cells in a manner similar to that seen in *Bacteria*. In this process, cell pairing occurs before plasmid transfer, and DNA transfer is unidirectional. However, most of the genes encoding these functions seem to have little similarity to those in gram-negative *Bacteria*. The exception is a gene similar to *traG* from the F plasmid, whose protein product is involved in stabilizing mating pairs. It thus seems likely that the actual mechanism of conjugation in *Archaea* is quite different from that in *Bacteria*.

Some halobacteria, in contrast, perform a novel form of conjugation. No fertility plasmids are involved, and DNA transfer is bidirectional. Cytoplasmic bridges form between the mating cells and appear to be used for DNA transfer. Neither type of conjugation has been developed to the point of being used for routine gene transfer or genetic analysis. However, these genetic resources will likely be useful for developing facile genetic systems in these organisms.

MiniQuiz
- Why is it usually more difficult to select recombinants with *Archaea* than with *Bacteria*?
- Why do penicillins not kill members of the *Archaea*?

10.13 Mobile DNA: Transposable Elements

As we have seen, molecules of DNA may move from one cell to another, but to a geneticist, "mobile DNA" has a specialized meaning. Mobile DNA refers to discrete segments of DNA that move as units from one location to another *within* other DNA molecules. Although the DNA of certain viruses can be inserted into and excised from the genome of the host cell, most mobile DNA consists of **transposable elements**. These are stretches of DNA that can move from one site to another. However, transposable

elements are always found inserted into another DNA molecule such as a plasmid, a chromosome, or a viral genome. Transposable elements do not possess their own origin of replication. Instead, they are replicated when the host DNA molecule into which they are inserted is replicated.

Transposable elements move by a process called *transposition* that is important both in evolution and in genetic analysis. The frequency of transposition is extremely variable, and ranges from 1 in 1000 to 1 in 10,000,000 per transposable element per cell generation, depending on both the transposable element and the organism. Transposition was originally observed in corn (maize) in the 1940s by Barbara McClintock before the DNA double helix was even discovered! She later received the Nobel Prize for this discovery. The molecular details of transposition were revealed using *Bacteria* due to the powerful genetic analyses possible in these organisms. Transposable elements are widespread in nature and can be found in the genomes of all three domains of life as well as in many viruses and plasmids.

Transposons and Insertion Sequences

The two major types of transposable elements in *Bacteria* are *insertion sequences* (IS) and *transposons*. Both elements have two important features in common: They carry genes encoding transposase, the enzyme necessary for transposition, and they have short inverted terminal repeats at their ends that are also needed for transposition. Note that the ends of transposable elements are not free but are continuous with the host DNA molecule into which the transposable element has inserted. **Figure 10.27** shows genetic maps of the insertion element IS2 and of the transposon Tn5.

Insertion sequences are the simplest type of transposable element. They are short DNA segments, about 1000 nucleotides

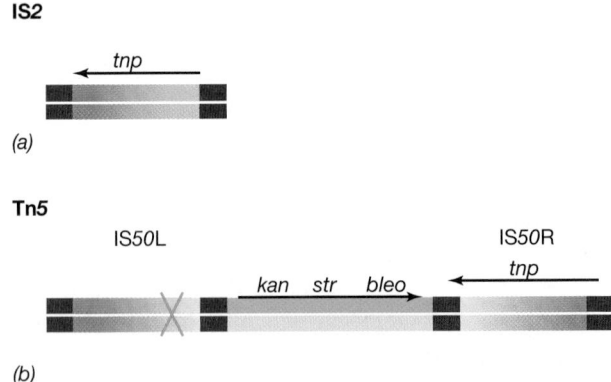

Figure 10.27 Maps of the transposable elements IS2 and Tn5. Inverted repeats are shown in red. The arrows above the maps show the direction of transcription of any genes on the elements. The gene encoding the transposase is *tnp*. *(a)* IS2 is an insertion sequence of 1327 bp with inverted repeats of 41 bp at its ends. *(b)* Tn5 is a composite transposon of 5.7 kbp containing the insertion sequences IS50L and IS50R at its left and right ends, respectively. IS50L is not capable of independent transposition because there is a nonsense mutation, marked by a blue cross, in its transposase gene. Otherwise, the two IS50 elements are almost identical. The genes *kan, str,* and *bleo* confer resistance to the antibiotics kanamycin (and neomycin), streptomycin, and bleomycin. Tn5 is commonly used to generate mutants in *Escherichia coli* and other gram-negative bacteria.

long, and typically contain inverted repeats of 10–50 bp. Each different IS has a specific number of base pairs in its terminal repeats. The only gene they possess is for the transposase. Several hundred distinct IS elements have been characterized. IS elements are found in the chromosomes and plasmids of both *Bacteria* and *Archaea*, as well as in certain bacteriophages. Individual strains of the same bacterial species vary in the number and location of the IS elements they harbor. For instance, one strain of *Escherichia coli* has five copies of IS2 and five copies of IS3. Many plasmids, such as the F plasmid, also carry IS elements. Indeed, integration of the F plasmid into the *E. coli* chromosome is due to homologous recombination between identical IS elements on the F plasmid and the chromosome (Section 10.10).

Transposons are larger than IS elements, but have the same two essential components: inverted repeats at both ends and a gene that encodes transposase. The transposase recognizes the inverted repeats and moves the segment of DNA flanked by them from one site to another. Consequently, any DNA that lies between the two inverted repeats is moved and is, in effect, part of the transposon. Genes included inside transposons vary widely. Some of these genes, such as antibiotic resistance genes, confer important new properties on the organism harboring the transposon. Because antibiotic resistance is both important and easy to detect, most highly investigated transposons have antibiotic resistance genes as selectable markers. Examples include transposon Tn5, which carries kanamycin resistance (Figure 10.27) and Tn*10*, with tetracycline resistance.

Because any genes lying between the inverted repeats become part of a transposon, it is possible to get hybrid transposons that display complex behavior. For example, conjugative transposons contain *tra* genes and can move between bacterial species by conjugation as well as transpose from place to place within a single bacterial genome. Even more complex is bacteriophage Mu, which is both a virus and a transposon (🔗 Section 21.4). In this case a complete virus genome is contained within a transposon. Other composite genetic elements consist of a segment of DNA lying between two identical IS elements. This whole structure can move as a unit and is called a *composite transposon*. The behavior of composite transposons indicates that novel transposons likely arise periodically in cells that contain IS elements located close to one another.

Mechanisms of Transposition

Both the inverted repeats found at the ends of transposable elements and transposase are essential for transposition. The transposase recognizes, cuts, and ligates the DNA during transposition. When a transposable element is inserted into target DNA, a short sequence in the target DNA at the site of integration is duplicated during the insertion process (**Figure 10.28**). The duplication arises because single-stranded DNA breaks are made by the transposase. The transposable element is then attached to the single-stranded ends that have been generated. Finally, enzymes of the host cell repair the single-strand portions, which results in the duplication.

Two mechanisms of transposition are known: *conservative* and *replicative* (**Figure 10.29**). In conservative transposition, as occurs with the transposon Tn5, the transposon is excised from one location and is reinserted at a second location. The copy number of a

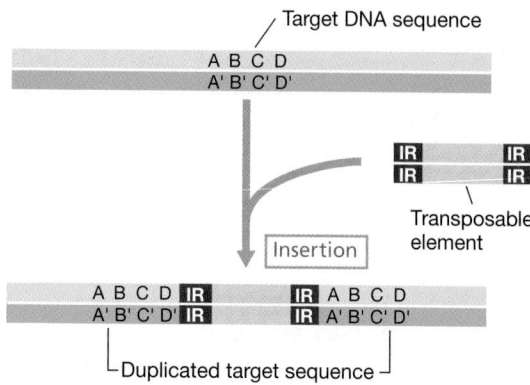

Figure 10.28 Transposition. Insertion of a transposable element generates a duplication of the target sequence. Note the presence of inverted repeats (IR) at the ends of the transposable element.

conservative transposon therefore remains at one. By contrast, during replicative transposition, a new copy is produced and is inserted at the second location. Thus, after a replicative transposition event, one copy of the transposon remains at the original site, and there is a second copy at the new site.

Transposition is a type of recombination called *site-specific* recombination, because specific DNA sequences (the inverted repeats and target sequence) are recognized by a protein (the transposase). This contrasts with *homologous* recombination (Section 10.6) in which homologous DNA sequences recognize each other by base pairing.

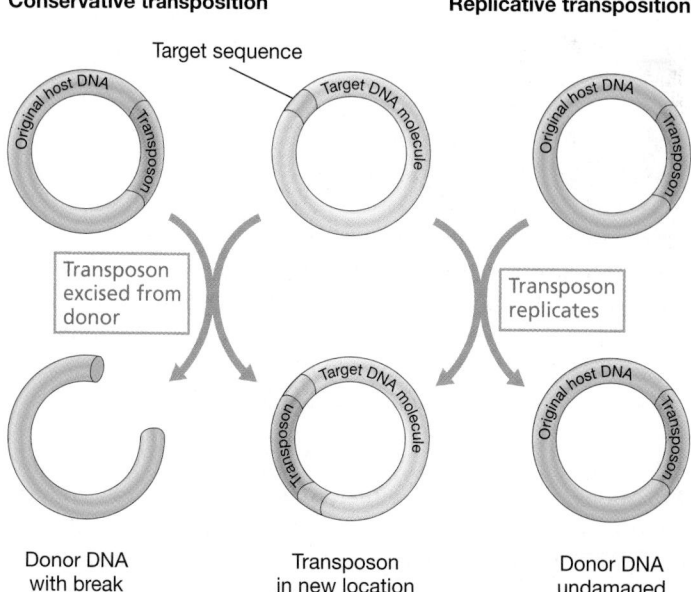

Figure 10.29 Two mechanisms of transposition. Donor DNA (carrying the transposon) is shown in green, and recipient DNA carrying the target sequence is shown in yellow. In both conservative and replicative transposition the transposase inserts the transposon (purple) into the target site (blue) on the recipient DNA. During this process, the target site is duplicated. In conservative transposition, the donor DNA is left with a double-stranded break at the previous location of the transposon. In contrast, after replicative transposition, both donor and recipient DNA possess a copy of the transposon.

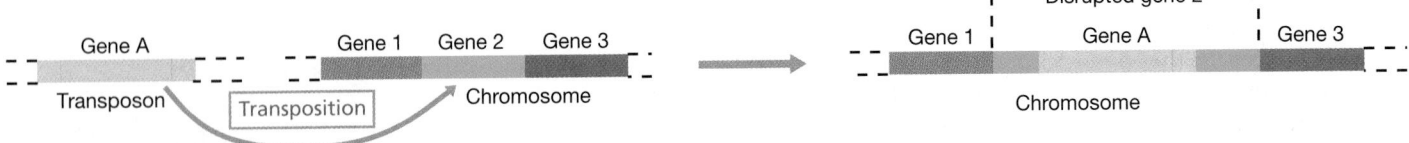

Figure 10.30 Transposon mutagenesis. The transposon moves into the middle of gene 2. Gene 2 is now disrupted by the transposon and is inactivated. Gene A in the transposon is expressed in both locations.

Mutagenesis with Transposons

When a transposon inserts itself within a gene, a mutation occurs in that particular gene (**Figure 10.30**). Mutations due to transposon insertion do occur naturally. However, deliberate use of transposons is a convenient way to create bacterial mutants in the laboratory. Typically, transposons carrying antibiotic resistance genes are used. The transposon is introduced into the target cell on a phage or plasmid that cannot replicate in that particular host. Consequently, antibiotic-resistant colonies will mostly be due to insertion of the transposon into the bacterial genome.

Because bacterial genomes contain relatively little noncoding DNA, most transposon insertions will occur in genes that encode proteins. If inserted into a gene encoding an essential protein, the mutation may be lethal under certain growth conditions and be suitable for genetic selection. For example, if transposon insertions are selected on rich medium on which all auxotrophs can grow, they can subsequently be screened on minimal medium supplemented with various nutrients to determine if a nutrient is required. Further analyses can then be performed to reveal which gene the transposon has disrupted. Auxotrophic mutations due to transposon insertions are very useful in bacterial genetics. Normally, auxotrophic recombinants cannot be isolated by positive selection. However, the presence of a transposon with an antibiotic resistance marker allows for positive selection.

Two transposons widely used for mutagenesis of *Escherichia coli* and related bacteria are Tn5 (Figure 10.27), which confers neomycin and kanamycin resistance, and Tn*10*, which confers tetracycline resistance. Many *Bacteria*, a few *Archaea*, and the yeast *Saccharomyces cerevisiae* have all been mutagenized using transposon mutagenesis. More recently, transposons have even been used to isolate mutations in animals, including mice.

MiniQuiz
- Which features do insertion sequences and transposons have in common?
- What is the significance of the terminal inverted repeats of transposons?

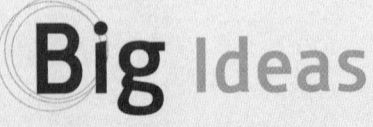

Big Ideas

10.1
Mutation is a heritable change in DNA sequence and may lead to a change in phenotype. Selectable mutations are those that give the mutant a growth advantage under certain environmental conditions and are especially useful in genetic research. If selection is not possible, mutants must be identified by screening.

10.2
Mutations, which can be either spontaneous or induced, arise because of changes in the base sequence of the nucleic acid of an organism's genome. A point mutation, which is due to a change in a single base pair, can lead to a single amino acid change in a polypeptide or to no change at all, depending on the particular codon. In a nonsense mutation, the codon becomes a stop codon and an incomplete polypeptide is made. Deletions and insertions cause more dramatic changes in the DNA, including frameshift mutations that often result in complete loss of gene function.

10.3
Different types of mutations occur at different frequencies. For a typical bacterium, mutation rates of 10^{-6} to 10^{-7} per kilobase pair are generally seen. Although RNA and DNA polymerases make errors at about the same rate, RNA genomes typically accumulate mutations at much higher frequencies than DNA genomes.

10.4
Mutagens are chemical, physical, or biological agents that increase the mutation rate. Mutagens can alter DNA in many different ways. However, alterations in DNA are not mutations unless they are inherited. Some DNA damage can lead to cell death if not repaired, and both error-prone and high-fidelity DNA repair systems exist.

10.5
The Ames test employs a sensitive bacterial assay system to identify chemical mutagens.

10.6

Homologous recombination occurs when closely related DNA sequences from two distinct genetic elements are combined together in a single element. Recombination is an important evolutionary process, and cells have specific mechanisms for ensuring that recombination takes place.

10.7

Certain prokaryotes exhibit competence, a state in which cells are able to take up free DNA released by other bacteria. Incorporation of donor DNA into a recipient cell requires the activity of single-strand binding protein, RecA protein, and several other enzymes. Only competent cells are transformable.

10.8

Transduction is the transfer of host genes from one bacterium to another by a bacterial virus. In generalized transduction, defective virus particles randomly incorporate fragments of the cell's chromosomal DNA, but the transducing efficiency is low. In specialized transduction, the DNA of a temperate virus excises incorrectly and takes adjacent host genes along with it; the transducing efficiency here may be very high.

10.9

Conjugation is a mechanism of DNA transfer in prokaryotes that requires cell-to-cell contact. Conjugation is controlled by genes carried by certain plasmids (such as the F plasmid) and involves transfer of the plasmid from a donor cell to a recipient cell. Plasmid DNA transfer involves replication via the rolling circle mechanism.

10.10

The donor cell chromosome can be mobilized for transfer to a recipient cell. This requires an F plasmid to integrate into the chromosome to form the Hfr phenotype. Transfer of the host chromosome is rarely complete but can be used to map the order of the genes on the chromosome. F′ plasmids are previously integrated F plasmids that have excised and captured some chromosomal genes.

10.11

A defective copy of a gene may be complemented by the presence of a second, unmutated copy of that gene. The construction of merodiploids carrying two copies of a specific gene or genes allows for complementation tests to determine if two mutations are in the same or different genes. This is necessary when mutations in different genes in the same pathway yield the same phenotype. Recombination does not occur in complementation tests.

10.12

Archaea lag behind *Bacteria* in the development of systems for gene transfer. Many antibiotics are ineffective against *Archaea*, making it difficult to select recombinants effectively. The unusual growth conditions needed by many *Archaea* also make genetic experimentation difficult. Nevertheless, the genetic transfer systems of *Bacteria*—transformation, transduction, and conjugation—are all known in *Archaea*.

10.13

Transposons and insertion sequences are genetic elements that can move from one location on a host DNA molecule to another by transposition, a type of site-specific recombination. Transposition can be either replicative or conservative. Transposons often carry genes encoding antibiotic resistance and can be used as biological mutagens.

Review of Key Terms

Auxotroph an organism that has developed a nutritional requirement, often as a result of mutation

Cistron a gene as defined by the *cis-trans* test; a segment of DNA (or RNA) that encodes a single polypeptide chain

Conjugation the transfer of genes from one prokaryotic cell to another by a mechanism involving cell-to-cell contact

Genotype the complete genetic makeup of an organism; the complete description of a cell's genetic information

Heteroduplex a DNA double helix composed of single strands from two different DNA molecules

Hfr cell a cell with the F plasmid integrated into the chromosome

Induced mutation a mutation caused by external agents such as mutagenic chemicals or radiation

Insertion sequence (IS) the simplest type of transposable element, which carries only genes involved in transposition

Missense mutation a mutation in which a single codon is altered so that one amino acid in a protein is replaced with a different amino acid

Mutagen an agent that causes mutation

Mutant an organism whose genome carries a mutation

Mutation a heritable change in the base sequence of the genome of an organism

Mutator strain a mutant strain in which the rate of mutation is increased

Nonsense mutation a mutation in which the codon for an amino acid is changed to a stop codon

Phenotype the observable characteristics of an organism

Plasmid an extrachromosomal genetic element that has no extracellular form

Point mutation a mutation that involves a single base pair

Recombination the process by which DNA molecules from two separate sources exchange sections or are brought together into a single DNA molecule

Regulon a set of genes or operons that are transcribed separately but are coordinately controlled by the same regulatory protein

Reversion an alteration in DNA that reverses the effects of a prior mutation

Rolling circle replication a mechanism of replicating double-stranded circular DNA that starts by nicking and unrolling one strand and using the other (still circular) strand as a template for DNA synthesis

Screening a procedure that permits the identification of organisms by phenotype or genotype, but does not inhibit or enhance the growth of particular phenotypes or genotypes

Selection placing organisms under conditions that favor or inhibit the growth of those with a particular phenotype or genotype

Silent mutation a change in DNA sequence that has no effect on the phenotype

Spontaneous mutation a mutation that occurs "naturally" without the help of mutagenic chemicals or radiation

Transduction the transfer of host cell genes from one cell to another by a virus

Transformation the transfer of bacterial genes involving free DNA (but see alternative usage in Chapter 9)

Transition a mutation in which a pyrimidine base is replaced by another pyrimidine or a purine is replaced by another purine

Transposable element a genetic element able to move (transpose) from one site to another on host DNA molecules

Transposon a type of transposable element that carries genes in addition to those involved in transposition

Transversion a mutation in which a pyrimidine base is replaced by a purine or vice versa

Wild-type strain a bacterial strain isolated from nature or one used as a parent in a genetics investigation

Review Questions

1. Write a one-sentence definition of the term "genotype." Do the same for "phenotype." Does the phenotype of an organism automatically change when a change in genotype occurs? Why or why not? Can phenotype change without a change in genotype? In both cases, give examples to support your answer (Section 10.1).

2. Explain why an *Escherichia coli* strain that is His⁻ is an auxotroph and one that is Lac⁻ is not. (*Hint:* Think about what *E. coli* does with histidine and lactose.) (Section 10.1)

3. Highlight the differences between base-pair substitution and frameshift mutation (Section 10.2).

4. Microinsertions that occur in promoters are not frameshift mutations. Define the terms microinsertion, frameshift, and mutation. Explain why this statement is true (Section 10.2).

5. Explain how it is possible for a frameshift mutation early in a gene to be corrected by another frameshift mutation farther along the gene (Section 10.2).

6. What are the implications of the high mutation rates of RNA viruses (Section 10.3)?

7. What are base analogs? Give an example of a base analog (Section 10.4).

8. How can the Ames test, an assay using bacteria, have any relevance to human cancer (Section 10.5)?

9. How does homologous recombination differ from site-specific recombination (Section 10.6)?

10. What is the difference between transformation and transduction (Sections 10.7 and 10.8)?

11. What is a sex pilus and which cell type, F⁻ or F⁺, would produce this structure (Section 10.9)?

12. What does an F⁺ cell need to do before it can transfer chromosomal genes (Section 10.10)?

13. What does it mean to complement a mutation "in *trans*" (Section 10.11)?

14. Explain why performing genetic selections is difficult when studying *Archaea*. Give examples of some selective agents that work well with *Archaea* (Section 10.12).

15. What are the major differences between insertion sequences and transposons (Section 10.13)?

16. The most useful transposons for isolating a variety of bacterial mutants are transposons containing antibiotic resistance genes. Why are such transposons so useful for this purpose (Section 10.13)?

Application Questions

1. A constitutive mutant is a strain that continuously makes a protein that is inducible in the wild type. Describe two ways in which a change in a DNA molecule could lead to the emergence of a constitutive mutant. How could these two types of constitutive mutants be distinguished genetically?

2. Although a large number of mutagenic chemicals are known, none is known that induces mutations in only a single gene (gene-specific mutagenesis). From what you know about mutagens, explain why it is unlikely that a gene-specific chemical mutagen will be found. How then is site-specific mutagenesis accomplished?

3. Why is it difficult in a single experiment to transfer a large number of genes to a recipient cell using transformation or transduction?

4. Transposable elements cause mutations when inserted within a gene. These elements disrupt the continuity of a gene. Introns also disrupt the continuity of a gene, yet the gene is still functional. Explain why the presence of an intron in a gene does not inactivate that gene but insertion of a transposable element does.

Need more practice? Test your understanding with Quantitative Questions; access additional study tools including tutorials, animations, and videos; and then test your knowledge with chapter quizzes and practice tests at **www.microbiologyplace.com**.

Genetic Engineering

Differentiating closely related strains of the same bacterial species is often a daunting task. Each of the 28 strains of *Escherichia coli* shown here stains a different color because cells of a given strain react with a unique nucleic acid probe that contains a specific fluorescent dye or combination of dyes.

In this chapter we discuss the basic techniques of genetic engineering, in particular those used to clone genes, alter genes, and to express them efficiently in host organisms. Performing genetics only *in vivo* (in living organisms) has many limitations that can be overcome by manipulating DNA *in vitro* (in a test tube). Some applications of genetic engineering are covered in Chapter 15 (Commercial Products and Biotechnology).

I Methods for Manipulating DNA

Genetic engineering refers to the use of *in vitro* techniques to alter genetic material in the laboratory. Such altered genetic material may be reinserted into the original source organism or into some other host organism. Genetic engineering depends upon our ability to cut DNA into specific fragments and to purify these for further manipulation. We begin by considering some of the basic tools of genetic engineering, including restriction enzymes, the separation of nucleic acids by electrophoresis, nucleic acid hybridization, and molecular cloning.

11.1 Restriction and Modification Enzymes

All cells contain enzymes that can chemically modify DNA in one way or another. One major class of such enzymes is the restriction endonucleases, or **restriction enzymes** for short. Restriction enzymes recognize specific base sequences (recognition sequences) within DNA and cut the DNA. Although they are widespread among prokaryotes (both *Bacteria* and *Archaea*), they are very rare in eukaryotes. *In vivo* restriction enzymes protect prokaryotes from hostile foreign DNA such as virus genomes. However, restriction enzymes are also essential for *in vitro* DNA manipulation, and their discovery gave birth to the field of genetic engineering.

Mechanism of Restriction Enzymes

Restriction endonucleases are divided into three major classes. The type I and III restriction enzymes bind to the DNA at their recognition sequences but cut the DNA a considerable distance away. In contrast, the type II restriction enzymes cleave the DNA within their recognition sequences, making this class of enzymes much more useful for the specific manipulation of DNA. **Figure 11.1** shows the 6-base-pair (bp) sequence that is recognized and cleaved by the restriction enzyme from *Escherichia coli* called *Eco*RI (this acronym stands for *Escherichia coli*, strain RY13, restriction enzyme I). The cleavage sites are indicated by arrows and the axis of symmetry by a dashed line. Note that the two strands of the recognition sequence have the same sequence if one is read from the left and the other from the right (or if both are read 5′ → 3′). Such inverted repeat sequences are called *palindromes* (the term palindrome is derived from the Greek, meaning "to run back again").

Most restriction enzymes are homodimeric proteins; that is, they are composed of two identical polypeptide subunits, and each subunit recognizes and cuts the DNA on one of its two

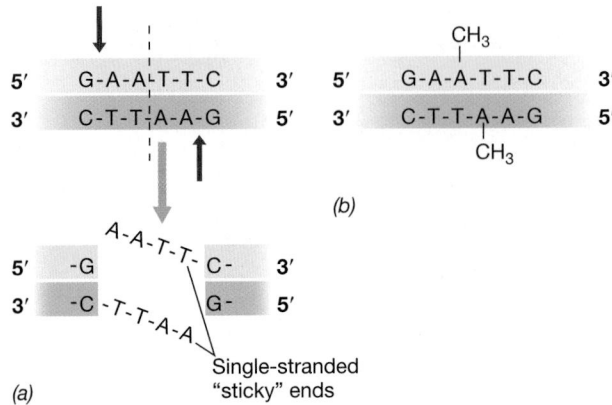

Figure 11.1 Restriction and modification of DNA. *(a)* (Top panel) The sequence of DNA recognized by the restriction endonuclease *Eco*RI. The red arrows indicate the bonds cleaved by the enzyme. The dashed line indicates the axis of symmetry of the sequence. (Bottom panel) Appearance of DNA after cutting with restriction enzyme *Eco*RI. Note the single-stranded "sticky ends." *(b)* The same sequence after modification by the *Eco*RI methylase. The methyl groups added by this enzyme are shown, and these protect the restriction site from cutting by *Eco*RI.

strands. This results in a double-stranded break. Many type II restriction enzymes make staggered cuts, leaving short, single-stranded overhangs known as "sticky ends" at the ends of the two fragments (Figure 11.1). As explained below, fragments with sticky ends have many uses, especially in molecular cloning. Other restriction enzymes cut both strands of the DNA directly opposite each other, yielding blunt ends.

Most of the DNA sequences recognized by type II restriction enzymes are short inverted repeats of from 4 to 8 bp. Consider again the enzyme *Eco*RI, which recognizes a specific 6-bp sequence (Figure 11.1). Any specific 6-base sequence should appear in a strand of DNA about once every 4096 nucleotides on average ($4096 = 4^6$; there are 4 possible bases at each of 6 positions). This assumes that all base pairs may occur at any given position with equal probability and that the DNA consists of 50% GC. Thus, several *Eco*RI cut sites should be present in any lengthy DNA molecule. The recognition sequences and cut sites for several restriction enzymes are given in **Table 11.1**. Several thousand restriction enzymes with several hundred different specificities are known; some leave "sticky ends" with a 5′ overhang and others with a 3′ overhang, whereas others generate blunt ends. Restriction enzymes are such important tools in modern molecular genetic research that they have become widely available commercially.

Modification: Protection from Restriction

The natural role of restriction enzymes is probably to protect the cell from invasion by foreign DNA, especially viral DNA. If foreign DNA enters the cell, the restriction enzymes will destroy it. However, a cell must protect its own DNA from inadvertent destruction by its own restriction enzymes. Such protection is conferred by **modification enzymes**. Each restriction enzyme is partnered with a corresponding modification enzyme that shares the same recognition sequence. The modification enzymes

Table 11.1 *Recognition sequences of a few restriction endonucleases*

Organism	Enzyme designation[a]	Recognition sequence[b]
Bacillus globigii	*Bgl*II	A↓GATCT
Bacillus subtilis	*Bsu*RI	GG↓C̆C
Brevibacterium albidum	*Bal*I	TGG↓C̆CA
Escherichia coli	*Eco*RI	G↓AĂTTC[c]
Haemophilus haemolyticus	*Hha*I	GC̆G↓C
Haemophilus influenzae	*Hind*II	GTPy↓PuAC̆
Haemophilus influenzae	*Hind*III	A↓AGCTT
Klebsiella pneumoniae	*Kpn*I	GGTAC↓C
Nocardia otitidiscaviarum	*Not*I	GC↓GGC̆CGC
Proteus vulgaris	*Pvu*I	CGAT↓CG
Serratia marcescens	*Sma*I	CCC↓GGG
Thermus aquaticus	*Taq*I	T↓CGĂ

[a]Nomenclature: The first letter of the three-letter abbreviation of a restriction endonuclease designates the genus from which the enzyme originates; the second two letters, the species. The roman numeral designates the order of discovery of enzymes in that particular organism, and any additional letters are strain designations.
[b]Arrows indicate the sites of enzymatic attack. Asterisks indicate the site of methylation (modification). G, guanine; C, cytosine; A, adenine; T, thymine; Pu, any purine; Py, any pyrimidine. Only the 5′ → 3′ sequence is shown.
[c]See Figure 11.1a.

chemically modify specific nucleotides in the restriction recognition sequences of the cell's own DNA. These modified sequences can no longer be cut by the corresponding restriction enzymes. Typically, modification consists of methylating specific bases within the recognition sequence, which prevents the restriction endonuclease from binding. For example, the sequence recognized by the *Eco*RI restriction enzyme (Figure 11.1a) can be modified by methylation of the two most interior adenines (Figure 11.1b). The enzyme that performs this modification is *Eco*RI methylase. If even a single strand is modified, the recognition sequence is no longer a substrate for the restriction enzyme *Eco*RI.

Gel Electrophoresis: Separation of DNA Molecules

Because the base sequences recognized by many restriction enzymes are four to six nucleotides long (Table 11.1), they cut DNA molecules into segments that range in length from a few hundred to a few thousand base pairs. After cleaving the DNA, the fragments generated can be separated from each other by **gel electrophoresis** and analyzed.

Electrophoresis is a procedure that separates charged molecules by migration in an electrical field. The rate of migration is determined by the charge on the molecule and by its size and shape. In gel electrophoresis (**Figure 11.2a**) the molecules are separated in a porous gel. Gels made of *agarose*, a polysaccharide, are used for separating DNA fragments. When an electrical current is applied, nucleic acids move through the gel toward the positive electrode due to their negatively charged phosphate

(a)

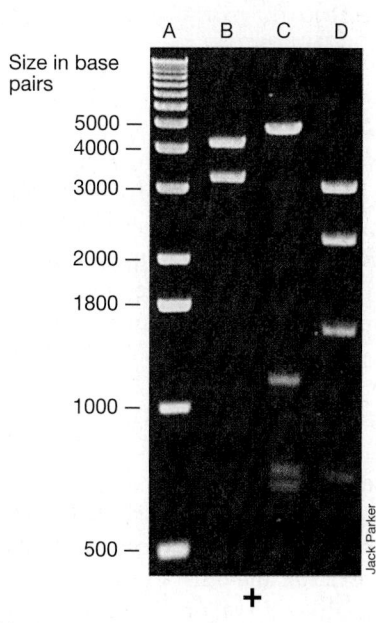

(b)

Figure 11.2 Agarose gel electrophoresis of DNA. *(a)* DNA samples are loaded into wells in a submerged agarose gel. *(b)* A photograph of a stained agarose gel. The DNA was loaded into wells toward the top of the gel (negative pole) as shown, and the positive electrode is at the bottom. The standard sample in lane A has fragments of known size that may be used to determine the sizes of the fragments in the other lanes. Bands stain less intensely at the bottom of the gel because the fragments are smaller, and thus there is less DNA to stain.

groups. The presence of the gel meshwork hinders the progress of the DNA, and small or compact molecules migrate more rapidly than large molecules. The higher the concentration of the gel, the more large molecules are hindered. Consequently, gels of different concentrations are used to separate molecules of different size ranges. After the gel has been run for sufficient time to separate the DNA molecules, it can be stained with a compound that binds to DNA, such as *ethidium bromide*, and the DNA will then fluoresce orange under ultraviolet light (Figure 11.2b). DNA fragments can be purified from gels and used for a variety of purposes.

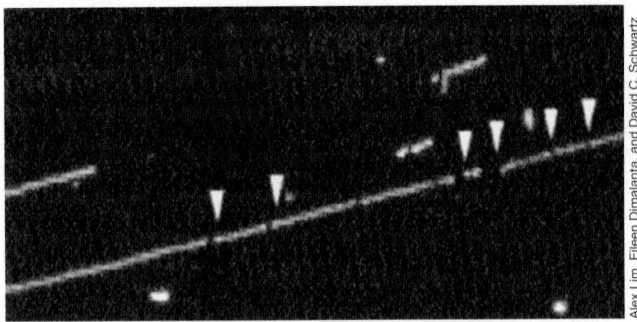

Figure 11.3 Optical mapping of restriction fragments. A digital fluorescence micrograph of a portion of the *Escherichia coli* chromosome digested with the restriction enzyme *Xho*I (isolated from *Xanthomonas holcicola*). Arrows indicate the sites of cutting by the restriction enzyme. Although the DNA strand itself is too small to be seen by light microscopy, fluorescence of the DNA is visible. The length of the DNA shown here is about 260 kbp.

Restriction Analyses of DNA

A typical agarose gel is shown in Figure 11.2*b*. Lane A contains a standard DNA sample consisting of DNA fragments of known sizes. The other lanes contain purified DNA (in this case a plasmid) cut by one or more restriction enzymes. Because a given restriction enzyme always cuts at the same site, the banding pattern of DNA molecules with the same sequence will always be the same. The size of the fragments can be determined by comparison with the standard sample. This approach can be used to generate a **restriction map** of the DNA. Combining restriction enzyme digestion of a single DNA molecule with fluorescent microscopy allows a restriction map to be made directly from the pattern of cuts along the DNA (**Figure 11.3**). This is called *optical mapping* and has been used to create restriction maps of the chromosomes of *E. coli* and other organisms.

Restriction analyses have many uses in molecular biology. They are useful in molecular cloning, where they provide a guide to which restriction enzyme and cut sites to use. Restriction analyses are also used to compare different but related DNA sequences for studies on classification of microorganisms. For example, the banding patterns generated from restriction analyses of either whole chromosomes or specific genes from a series of organisms can indicate their genetic relationships. Multiple restriction digests, using different enzymes either one at a time or in combination, can generate patterns that allow the relative relationships between different DNA molecules to be assessed.

MiniQuiz

- Why are restriction enzymes useful to the molecular biologist?
- What is the basis for separating molecules by electrophoresis?

11.2 Nucleic Acid Hybridization

When DNA is denatured (that is, the two strands are separated), the single strands can form hybrid double-stranded molecules with other single-stranded DNA (or RNA) molecules by complementary (or almost complementary) base pairing (⮎ Section 6.2).

This is called *nucleic acid hybridization,* or **hybridization** for short, and is widely used in detecting, characterizing, and identifying segments of DNA. Segments of single-stranded nucleic acids whose identity is already known and that are used in hybridization are called **nucleic acid probes** or, simply, probes. To allow detection, probes can be made radioactive or labeled with chemicals that are colored or yield fluorescent products. By varying the hybridization conditions, it is possible to adjust the "stringency" of the hybridization such that complementary base pairing must be nearly exact; this helps to avoid nonspecific pairing between sequences that are only partly complementary.

Hybridization can be very useful for finding related sequences in different chromosomes or other genetic elements or to find the location of a specific gene. In Southern blotting, named for its inventor, E.M. Southern, probes of known sequence are hybridized to target DNA fragments that have been separated by gel electrophoresis. The hybridization procedure in which DNA is the target sequence in the gel, and RNA or DNA is the probe, is called a **Southern blot**. By contrast, when RNA is the target sequence and DNA or RNA is the probe, the procedure is called a **Northern blot**.

In a Southern blot the DNA fragments in the gel are denatured to yield single strands and transferred to a synthetic membrane. The membrane is then exposed to a labeled probe. If the probe is complementary to any of the fragments, hybrids form, and the probe attaches to the membrane at the locations of the complementary fragments. Hybridization can be detected by monitoring the label that has bound to the membrane. **Figure 11.4** shows how a Southern blot can be used to identify fragments of DNA containing sequences that hybridize to the probe.

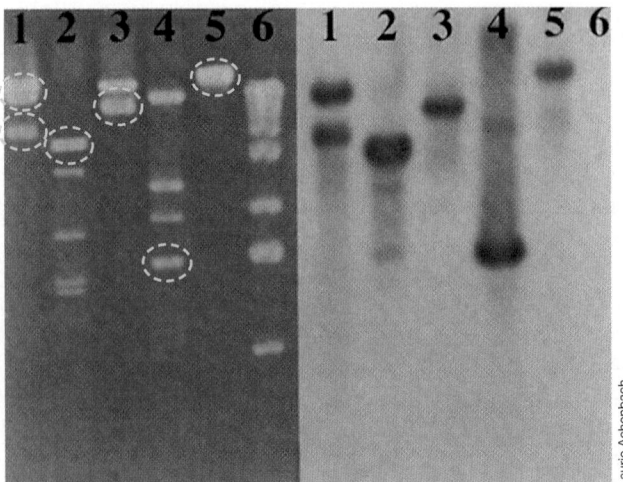

Figure 11.4 Southern blotting. (Left panel) Purified molecules of DNA from several different plasmids were treated with restriction enzymes and then subjected to agarose gel electrophoresis. (Right panel) Southern blot of the DNA gel shown to the left. After blotting, DNA in the gel was hybridized to a radioactive probe. The positions of the bands were visualized by X-ray autoradiography. Note that only some of the DNA fragments (circled in yellow) have sequences complementary to the labeled probe. Lane 6 contained DNA used as a size marker and none of the bands hybridized to the probe.

The procedure for Northern blots is analogous except that RNA molecules are separated on a gel and transferred to a synthetic membrane where they are probed. Northern blotting is often used to identify messenger RNA derived from specific genes. The intensity of a Northern blot gives a rough estimate of how much mRNA is present from the target gene and may therefore be used to monitor transcription.

Hybridization is often used to identify genes after cloning (Section 11.3). It is especially useful if the cloned gene is not expressed in the cloning host or if no assay is available to detect the gene product. In colony hybridization, a nucleic acid probe is used to detect recombinant DNA in colonies, as shown in Figure 11.6*b*. This procedure uses replica plating to produce a duplicate of the master plate on a membrane filter. (The same procedure can be carried out with virus vectors by blotting the plaques onto a membrane.) The cells on the filter are lysed in place to release their DNA, and the filter is treated to separate the DNA into single strands and fix them to the filter. This filter is then exposed to a labeled nucleic acid probe to allow hybridization, and unbound probe is washed away. The filter is then overlaid with X-ray film if a radioactive probe was used. After development, the X-ray film is examined for spots (Figure 11.6*b*). Colonies corresponding to these spots are then chosen and studied further.

MiniQuiz

- What is a Southern blot and what does it tell you?
- What is the difference between a Northern blot and a Southern blot?
- Does use of nucleic acid probes depend on gene expression? Explain.

11.3 Essentials of Molecular Cloning

In **molecular cloning** a fragment of DNA is isolated and replicated. The basic strategy of molecular cloning is to isolate the desired gene (or other segment of DNA) from its original location and move it to a small, simple genetic element, such as a plasmid or virus, which is called a **vector** (**Figure 11.5**). When the vector replicates, the cloned DNA that it carries is also replicated. Molecular cloning thus includes locating the gene of interest, obtaining and purifying a copy of the gene, and inserting it into a convenient vector. Once cloned, the gene of interest can be manipulated in various ways and may eventually be put back into a living cell. This approach provides the foundation for much of genetic engineering and has greatly helped the detailed analysis of genomes.

The first objective of gene cloning is to isolate copies of specific genes in pure form. Consider the problem. For a genetically "simple" organism such as *Escherichia coli*, an average gene is encoded by 1–2 kbp of DNA out of a genome of over 4600 kbp. An average *E. coli* gene is thus less than 0.05% of the total DNA in the cell. In human DNA the problem is even worse because the coding regions of average genes are not much larger than in *E. coli*, genes are typically split into pieces, and the genome is almost 1000 times larger! Nonetheless, our knowledge of DNA chemistry and enzymology allows us to break and rejoin and

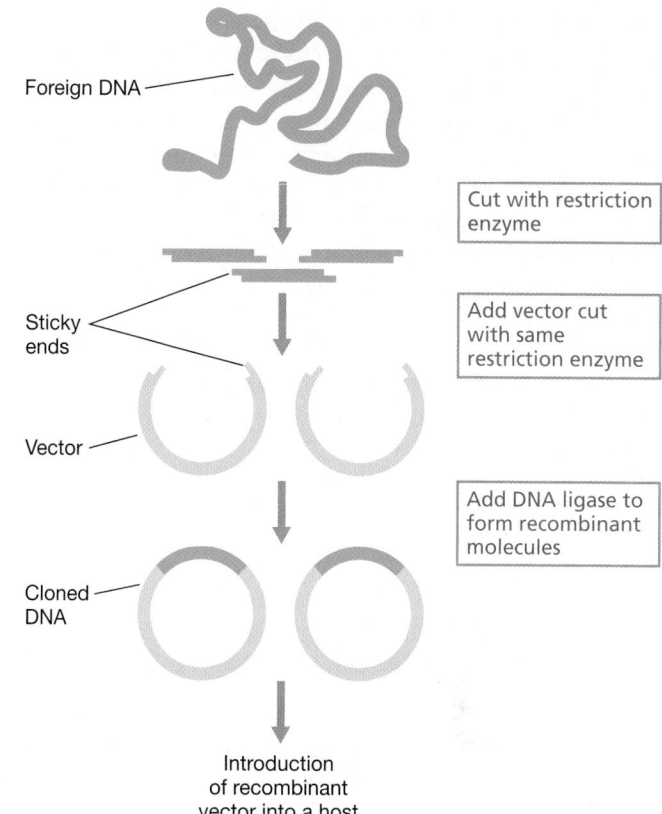

Figure 11.5 Major steps in gene cloning. The vector can be a plasmid or a viral genome. By cutting the foreign DNA and the vector DNA with the same restriction enzyme, complementary sticky ends are generated that allow foreign DNA to be inserted into the vector.

replicate DNA molecules *in vitro*. Restriction enzymes, DNA ligase, the polymerase chain reaction (PCR), and synthetic DNA are important tools for molecular cloning.

Steps in Gene Cloning: A Summary

1. **Isolation and fragmentation of the source DNA.** The source DNA can be total genomic DNA from an organism of interest, DNA synthesized from an RNA template by reverse transcriptase (⮕ Section 21.11), a gene or genes amplified by the polymerase chain reaction (⮕ Section 6.11), or even wholly synthetic DNA made *in vitro* (Section 11.4). If genomic DNA is the source, it is first cut with restriction enzymes (Section 11.1) to give a mixture of fragments of manageable size (Figure 11.5).

2. **Inserting the DNA fragment into a cloning vector.** Cloning vectors are small, independently replicating genetic elements used to carry and replicate cloned DNA segments. Most vectors are plasmids or viruses. Cloning vectors are typically designed to allow insertion of foreign DNA at a restriction site that cuts the vector without affecting its replication (Figure 11.5). If the source DNA and the vector are both cut with the same restriction enzyme that yields sticky ends, joining the two molecules is greatly assisted by annealing of the sticky ends. Blunt ends generated by some restriction enzymes can be joined by direct ligation or by using synthetic DNA linkers or adapters. In either

case, the strands are joined by DNA ligase, an enzyme that covalently links both strands of the vector and the inserted DNA.

3. **Introduction of the cloned DNA into a host organism.** Recombinant DNA molecules made in the test tube are introduced into suitable host organisms where they can replicate. Transformation (⮌ Section 10.7) is often used to get recombinant DNA into cells. In practice this often yields a mixture of recombinant constructs. Some cells contain the desired cloned gene, whereas other cells may contain other cloned genes from the same source DNA. Such a mixture is known as a **DNA (gene) library** because many different clones can be purified from the mixture, each containing different cloned DNA segments from the source organism. Making a gene library by cloning random fragments of a genome is called **shotgun cloning** and is widely used in genomic analyses.

Finding the Right Clone

Genetic engineering often begins by cloning a gene of interest. But first it is necessary to identify the host colony containing the correct clone. One can isolate host cells containing a plasmid vector by selecting for a marker such as antibiotic resistance, so that only these cells form colonies. When using a viral vector, one simply looks for plaques. These colonies or plaques can be screened for vectors carrying foreign DNA inserts by looking for the inactivation of a vector gene (Section 11.6). When cloning a single DNA fragment generated by PCR or purified by some other means, such simple selections or screenings are usually sufficient.

Another relatively simple case is when a mutant is available that is defective in the cloned gene. In this case, clones may be tested by complementation (⮌ Section 10.11). The vector plus cloned DNA is inserted into the mutant and colonies are screened for those that regain the wild-type phenotype.

However, a gene library may contain thousands or tens of thousands of clones, and often only one or a few may contain the genes of interest. Identifying cells carrying cloned DNA is only the first step. The biggest challenge remains finding the clone carrying the gene of interest. One must examine colonies of bacteria or plaques from viral-infected cells growing on agar plates and detect those few that contain the gene of interest. Sometimes the cloned gene is expressed and the protein is synthesized in the cloning host and may be detected. Often, however, the cloned gene is not expressed. Then the only option is to look for the DNA itself.

Nowadays, the correct clone is often located by DNA sequencing or by restriction digests performed on plasmids extracted from a large number of colonies. These procedures have been semiautomated and are much less labor intensive than they previously were. Another approach is to use hybridization as described in Section 11.2.

Detecting Proteins Expressed in the Cloning Host

If the foreign gene is expressed in the cloning host, the encoded protein can be screened for. Obviously, for this to work the host itself must not produce the protein being studied. Selection of cells containing cloned genes is relatively simple provided that the encoded protein can be assayed conveniently.

Antibodies can be used to detect a protein of interest. Antibodies are animal proteins that bind in a highly specific way to a target

molecule, the antigen (⮌ Section 28.4). In this case the protein encoded by the cloned gene is the antigen. Because the antibody combines specifically with the antigen, colonies that contain the antigen can be identified by observing the binding of the antibody. Because very little of the antigen is present in each colony, only a small amount of antibody is bound, and so a highly sensitive procedure for detecting bound antibody must be used. In practice, radioisotopes, fluorescent chemicals, or enzymes are used. Techniques for detecting antigens are discussed in Chapter 31.

The procedure using radioactive detection is outlined in **Figure 11.6a**. Replica plating (⮌ Figure 10.2) is used to make a duplicate of the master plate onto a synthetic membrane filter, and all further

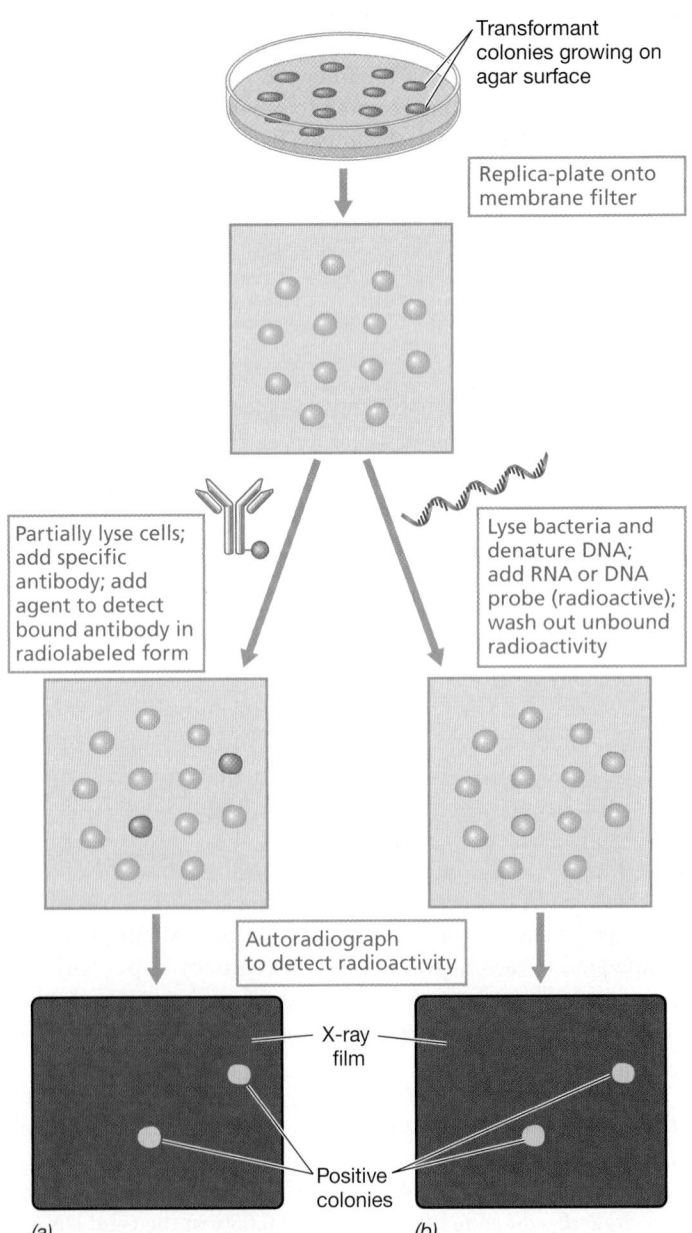

Figure 11.6 Finding the right clone. *(a)* Method for detecting production of protein by using a specific antibody. *(b)* Method for detecting recombinant clones by colony hybridization with a radioactive nucleic acid probe. Formation of a DNA duplex binds the DNA probe to a particular spot on the membrane.

manipulations are done with this filter. The duplicate colonies are lysed to release the antigen of interest. The antibody is then added and it binds the antigen. Unbound antibody is washed off and a radioactive reagent is added that is specific for the antibody. A sheet of X-ray film is placed over the filter and exposed. Radioactive colonies appear as spots on the X-ray film after it is developed (Figure 11.6*a*). The location of such spots corresponds to the location of a colony on the master plate that produces the protein. This colony is picked from the master plate and subcultured.

A major limitation of this procedure is that a specific antibody must be available for the antigen in question. Antibodies are made by injecting the antigen into an animal. But to be successful, the injected protein must be pure; otherwise, antibodies against multiple antigens will be formed. Thus, the protein of interest must be purified previously or false-positive reactions will make selection of clones difficult.

MiniQuiz

- What is the purpose of molecular cloning?
- What are the roles of a cloning vector, restriction enzymes, and DNA ligase in molecular cloning?
- How may cloned genes be identified?

11.4 Molecular Methods for Mutagenesis

As we have seen, conventional mutagens introduce mutations *at random* in the intact organism (Section 10.4). In contrast, *in vitro* mutagenesis, better known as **site-directed mutagenesis**, uses synthetic DNA plus DNA cloning techniques to introduce mutations into genes *at precisely determined sites*. In addition to changing just a few bases, mutations may also be engineered by inserting large segments of DNA at precisely determined locations.

Synthesizing DNA

Segments of DNA may be artificially synthesized and used as primers or probes or to provide altered versions of parts of genes or regulatory regions. When handling DNA segments it is often necessary to replicate the DNA to obtain sufficient quantities for experimentation. This is achieved by using the polymerase chain reaction (PCR). Automated systems for synthesizing DNA are available. Oligonucleotides of 30–35 bases are made routinely and oligonucleotides of over 100 bases in length can be made if necessary. Nowadays such oligonucleotides are commercially available. For the synthesis of longer polynucleotides, individual oligonucleotide fragments can be joined enzymatically using DNA ligase (Section 6.9).

DNA is synthesized *in vitro* in a solid-phase procedure in which the first nucleotide in the chain is fastened to an insoluble support (such as porous glass beads about 50 μm in size). Several steps are needed for the addition of each nucleotide, and the chemistry is intricate. After each step is completed, the reaction mixture is flushed out of the solid support and the series of reactions repeated for the addition of the next nucleotide. Once the oligonucleotide is the desired length, it is cleaved from the solid-phase support by a specific reagent and purified to eliminate by-products and contaminants.

Site-Directed Mutagenesis

Site-directed mutagenesis is a powerful tool, as it allows for a change to any base pair in a specific gene and thus has many uses in genetics. The basic procedure is to synthesize a short DNA oligonucleotide primer containing the desired base change (mutation) and to allow this to base-pair with single-stranded DNA containing the target gene. Pairing will be complete except for the short region of mismatch. Then the synthetic oligonucleotide is extended using DNA polymerase, thus copying the rest of the gene. The double-stranded molecule obtained is inserted into a host cell by transformation. Mutants are often selected by some sort of positive selection, such as antibiotic resistance; in this case, the modified DNA would also carry a nearby antibiotic resistance marker.

One procedure for site-directed mutagenesis is illustrated in **Figure 11.7**. The process starts with cloning the target gene into a single-stranded DNA vector. A widely used vector for this

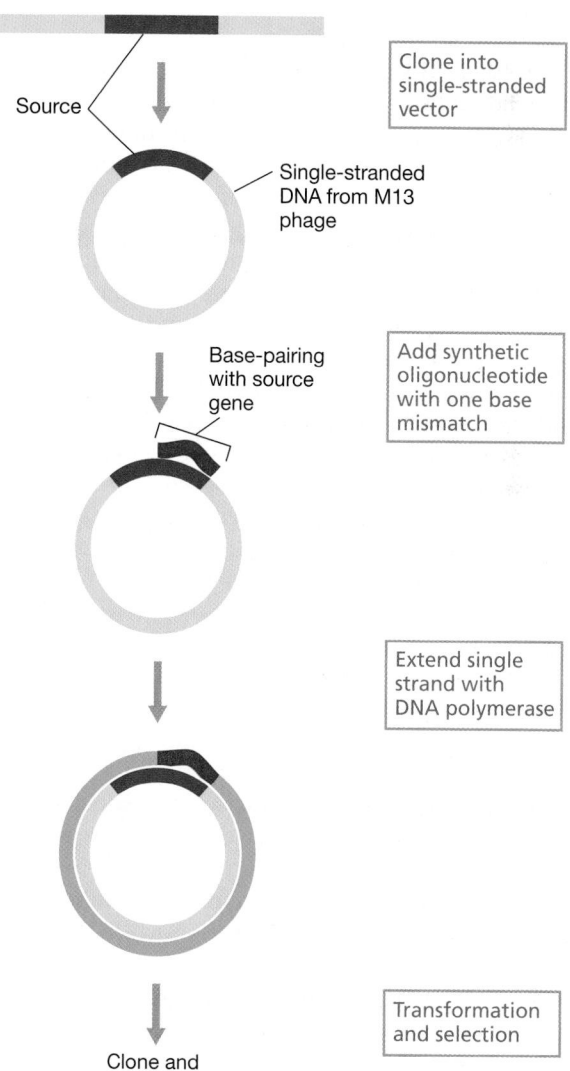

Source

Clone into single-stranded vector

Single-stranded DNA from M13 phage

Base-pairing with source gene

Add synthetic oligonucleotide with one base mismatch

Extend single strand with DNA polymerase

Transformation and selection

Clone and select mutant

Figure 11.7 Site-directed mutagenesis using synthetic DNA. Short synthetic oligonucleotides may be used to generate mutations. Cloning into the genome of bacteriophage M13 yields the single-stranded DNA needed for site-directed mutagenesis to work.

purpose is bacteriophage M13, a single-stranded DNA phage whose genome is easy to manipulate and that replicates in *Escherichia coli* (↩ Section 21.2). Several cloning methods are available based on M13 (Section 11.10). The mutagenized DNA can then bind by base pairing with the target gene (Figure 11.7). This gives a DNA molecule with a mismatch. After cell division and vector DNA replication, both daughter molecules will be fully base paired, but one daughter molecule will carry the mutation and the other will be wild type. Progeny bacteria are then screened for those with the mutation.

Site-directed mutagenesis may also be carried out using PCR. In this case, the short DNA oligonucleotide with the required mutation is used as a PCR primer. The mutation-carrying primer is designed to anneal to the target with the mismatch in the middle and must have enough matching nucleotides on both sides for binding to be stable during the PCR reaction. The mutant primer is paired with a normal primer. When the PCR reaction amplifies the target DNA, it incorporates the mutations in the mutant primer.

Applications of Site-Directed Mutagenesis

Site-directed mutagenesis can be used to investigate the activity of proteins with known amino acid substitutions. Suppose one was studying the active site of an enzyme. Site-directed mutagenesis could be used to change a specific amino acid in the active site, and the modified enzyme would then be assayed and compared to the wild-type enzyme. In such experiments, the vector encoding the mutant enzyme is inserted into a mutant host strain unable to make the original enzyme. Consequently, the activity measured is due to the mutant version of the enzyme alone.

Using site-directed mutagenesis, enzymologists can link virtually any aspect of an enzyme's activity, such as catalysis, resistance or susceptibility to chemical or physical agents, or interactions with other proteins, to specific amino acids in the protein. In particular, site-directed mutagenesis has given detailed information to structural biologists interested in which amino acids are critical for enzyme structure and function.

Cassette Mutagenesis and Gene Disruption

Because of the large number of restriction enzymes commercially available and therefore the large number of different DNA sequences that can be cut, it is usually possible to find several different restriction sites within any target gene. However, if sites for the appropriate restriction enzyme are not present at the required location, they can be inserted by site-directed mutagenesis as shown in Figure 11.7. If restriction sites are reasonably close together, the intervening DNA fragment can be excised and replaced by a synthetic DNA fragment in which one or more of the nucleotides have been changed. These synthetic fragments are called **DNA cassettes** (or cartridges), and the process is known as **cassette mutagenesis**. When using cassettes to replace sections of genes, the cassettes are typically the same size as the wild-type DNA fragments they replace.

Another type of cassette mutagenesis is called **gene disruption**. In this technique, cassettes are inserted into the middle of a gene, thus disrupting the coding sequence. Cassettes used for making insertion mutations can be almost any size and can even

carry an entire gene. To facilitate selection, cassettes that encode antibiotic resistance are commonly used. The process of gene disruption is illustrated in **Figure 11.8**. In this case, a DNA cassette carrying a gene conferring kanamycin resistance, the Kan cassette, is inserted at a restriction site in a cloned gene. The vector carrying the disrupted gene is then linearized by cutting with a different restriction enzyme. Finally, the linear DNA is transformed into the host, and kanamycin resistance is selected. The linearized plasmid cannot replicate, and so resistant cells arise mostly by homologous recombination (↩ Section 10.6) between the mutated gene on the plasmid and the wild-type gene on the chromosome.

Note that when a cassette is inserted, the cells not only gain antibiotic resistance but also lose the function of the gene into

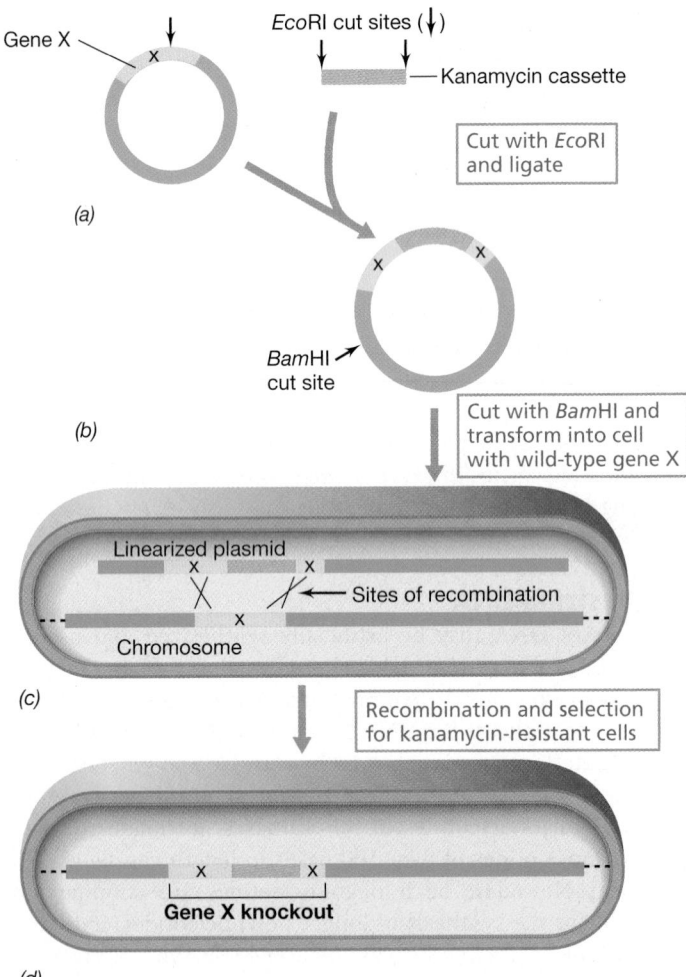

Figure 11.8 Gene disruption by cassette mutagenesis. *(a)* A cloned wild-type copy of gene X, carried on a plasmid, is cut with *Eco*RI and mixed with the kanamycin cassette. *(b)* The cut plasmid and the cassette are ligated, creating a plasmid with the kanamycin cassette as an insertion mutation within gene X. This new plasmid is cut with *Bam*HI and transformed into a cell. *(c)* The transformed cell contains the linearized plasmid with a disrupted gene X and its own chromosome with a wild-type copy of the gene. *(d)* In some cells, homologous recombination occurs between the wild-type and mutant forms of gene X. Cells that can grow in the presence of kanamycin have only a single, disrupted copy of gene X.

which the cassette is inserted. Such mutations are called *knockout mutations*. These are similar to insertion mutations made by transposons (Section 10.13), but here the experimenter chooses which gene will be mutated. Knockout mutations in haploid organisms (such as prokaryotes) yield viable cells only if the disrupted gene is nonessential. Indeed, the generation of gene knockouts may be used to investigate whether a given gene is essential.

MiniQuiz

- How can site-directed mutagenesis be useful to enzymologists?
- What are knockout mutations?
- Why is a solid support used during chemical synthesis of DNA?

11.5 Gene Fusions and Reporter Genes

DNA manipulation has revolutionized the study of gene regulation. A coding sequence from one source (the reporter) may be fused to a regulatory region from another source. Such gene fusions are often used in studying gene regulation (see below), especially where assaying the levels of the natural gene product is difficult or time consuming. They may also be used to increase expression of a desired gene product.

Reporter Genes

The key property of a **reporter gene** is that it encodes a protein that is easy to detect and assay. Reporter genes are used for a variety of purposes. They may be used to report on the presence or absence of a particular genetic element (such as a plasmid) or DNA inserted within a vector. They can also be fused to other genes or to the promoter of other genes so that gene expression can be studied.

The first gene to be used widely as a reporter was *lacZ*, which encodes the enzyme β-galactosidase (Section 8.3). Cells expressing β-galactosidase can be detected easily by their color on indicator plates that contain the artificial substrate Xgal (5-bromo-4-chloro-3-indolyl-β-D-galactopyranoside), which is cleaved by β-galactosidase to yield a blue color (Figure 11.12).

Another widely used reporter gene encodes luciferase. This enzyme makes cells expressing it luminescent (Section 8.9). Colonies containing this reporter system can be detected on agar plates by their luminescence against a large background of other colonies. However, the expression of luciferase depends on more than one gene because several accessory factors are needed. By contrast, the **green fluorescent protein** (GFP) needs no accessory factors and is widely used as a reporter (**Figure 11.9**). Although the gene for GFP was originally cloned from the jellyfish *Aequorea victoria*, the GFP protein may be expressed in most cells. It is stable and causes little or no disruption of host cell metabolism. If expression of a cloned gene is linked to that of GFP, expression of GFP signals that the cloned gene has also been expressed (Figure 11.9). Recent advances in fluorescent labeling now allow the simultaneous use of multiple fluorescent markers as discussed in the Microbial Sidebar, "Combinatorial Fluorescence Labeling."

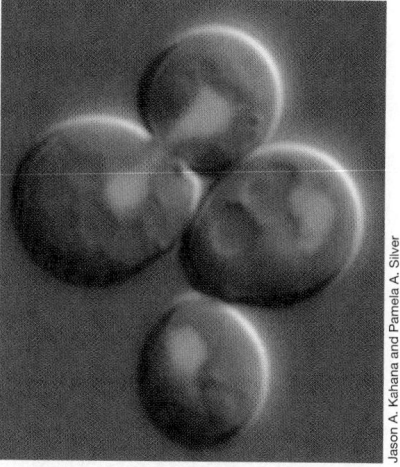

Figure 11.9 Green fluorescent protein (GFP). GFP can be used as a tag for protein localization *in vivo*. In this example, the gene encoding Pho2, a DNA-binding protein from the yeast *Saccharomyces cerevisiae*, was fused to the gene encoding GFP and photographed by fluorescence microscopy. The recombinant gene was transformed into budding yeast cells. These expressed the fluorescent fusion protein localized in the nucleus.

Gene Fusions

It is possible to engineer constructs that consist of segments from two different genes. Such constructs are known as **gene fusions**. If the promoter that controls a coding sequence is removed, the coding sequence can be fused to a different regulatory region to place the gene under the control of a different promoter. Alternatively, the promoter region can be fused to a gene whose product is easy to assay. There are two different types of gene fusions. In **operon fusions**, a coding sequence that retains its own translational start site and signals is fused to the transcriptional signals of another gene. In **protein fusions**, two coding sequences are fused with the result that they share the same transcriptional and translational start sites and signals.

Gene fusions are often used in studying gene regulation, especially if measuring the levels of the natural gene product is difficult or time consuming. The regulatory region of the gene of interest is fused to the coding sequence for a reporter gene, such as that for β-galactosidase or GFP. The reporter is then made under the conditions that would trigger expression of the target gene (**Figure 11.10**). The expression of the reporter is assayed under a variety of conditions to determine how the gene of interest is regulated. Operon fusions are used to assess transcriptional regulation, whereas protein fusions reveal translational control.

Gene fusions may also be used to test for the effects of regulatory genes. Mutations that affect regulatory genes are introduced into cells carrying gene fusions, and expression is measured and compared to cells lacking the regulatory mutations. This allows the rapid screening of multiple regulatory genes that are suspected of controlling the target gene.

MiniQuiz

- Why are gene fusions useful in studying gene regulation?
- What is a reporter gene?

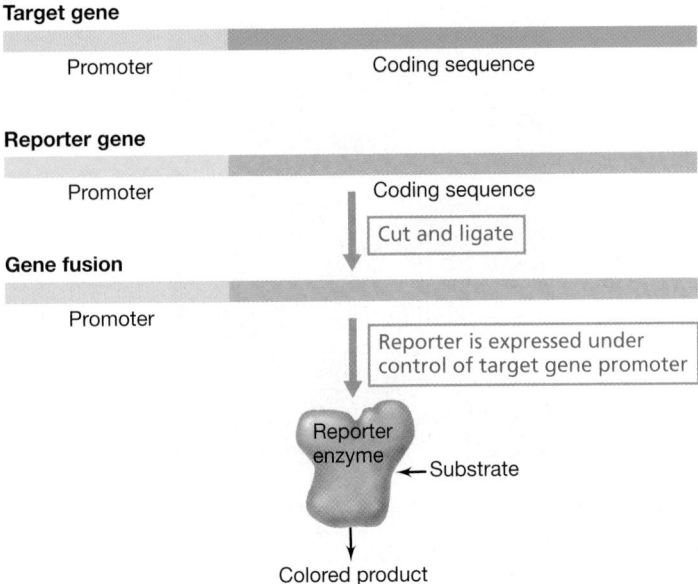

Figure 11.10 Construction and use of gene fusions. The promoter of the target gene is fused to the reporter coding sequence. Consequently, the reporter gene is expressed under those conditions where the target gene would normally be expressed. The reporter shown here is an enzyme (such as β-galactosidase) that converts a substrate to a colored product that is easy to detect. This approach greatly facilitates the investigation of regulatory mechanisms.

Ⅱ Gene Cloning

11.6 Plasmids as Cloning Vectors

The replication of plasmids in their host cell proceeds independently of direct chromosomal control. In addition to carrying genes required for their own replication, most plasmids are natural vectors because they often carry other genes that confer important properties on their hosts. In addition to independent replication, certain plasmids have other very useful properties as cloning vectors. These include (1) small size, which makes the DNA easy to isolate and manipulate; (2) multiple copy number, so many copies are present in each cell, thus giving both high yields of DNA and high-level expression of cloned genes; and (3) presence of selectable markers such as antibiotic resistance genes, which makes detection and selection of plasmid-borne clones easier.

Although conjugative plasmids are transferred by cell-to-cell contact in nature, most plasmid cloning vectors have been genetically modified to abolish conjugative transfer. This prevents unwanted movement of the vector into other organisms. However, vector transfer in the laboratory can be accomplished by chemically mediated transformation or electroporation (↺ Section 10.7). Depending on the host–plasmid system, replication of the plasmid may be under tight control, in which case only a few copies are made, or under relaxed control, in which case a large number of copies are made. Obtaining a high copy number is often important in gene cloning, and by proper selection of the host–plasmid system and manipulation of cellular macromolecule synthesis, plasmid copy numbers of several thousand per cell can be obtained.

An Example of a Cloning Vector: The Plasmid pUC19

The first plasmid cloning vectors used were natural isolates. In particular, the ColE plasmids of *Escherichia coli* that encode colicin E were used because they are relatively small and are naturally present in multiple copies, making DNA isolation easier. However, these were soon replaced by plasmids that were themselves the result of *in vitro* manipulations. A widely used plasmid cloning vector is pUC19 (**Figure 11.11**). This was derived in several steps from the ColE1 plasmid by removing the colicin genes and inserting genes for ampicillin resistance and for a blue–white color-screening system (see below). A segment of artificial DNA with cut sites for many restriction enzymes, called a *polylinker* or *multiple cloning site*, is inserted into the *lacZ* gene, the gene that encodes the lactose-degrading enzyme β-galactosidase (↺ Section 8.3). The presence of this short polylinker does not inactivate *lacZ*. Cut sites for restriction enzymes present in the polylinker are absent from the rest of the vector. Consequently, treatment with each of these enzymes opens the vector at a unique location, but does not cut the vector into multiple pieces. Plasmid pUC19 has a number of characteristics that make it suitable as a cloning vehicle:

1. It is relatively small, only 2686 base pairs.
2. It is stably maintained in its host (*E. coli*) in relatively high copy number, about 50 copies per cell.
3. It can be amplified to a very high number (1000–3000 copies per cell, about 40% of the cellular DNA) by inhibiting protein synthesis with the antibiotic chloramphenicol.

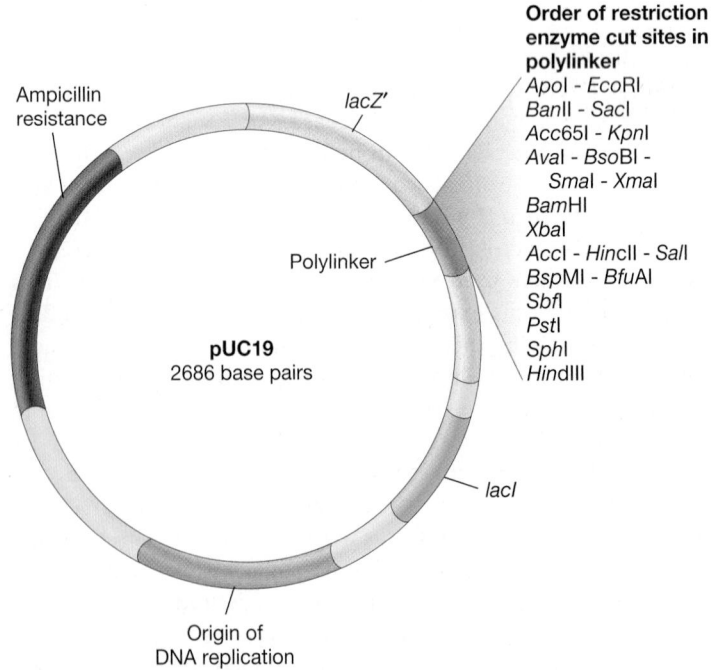

Figure 11.11 Cloning vector plasmid pUC19. Essential features include an ampicillin resistance marker and the polylinker with multiple restriction enzyme cut sites. Insertion of cloned DNA within the polylinker inactivates the truncated *lacZ'* gene that encodes part of β-galactosidase and allows for easy identification of transformants by blue–white screening.

Combinatorial Fluorescence Labeling

A wide range of fluorescent probes is now available. In addition to proteins such as GFP, many small fluorescent probes are available that can be covalently linked to macromolecules, including nucleic acids. In particular, fluorescently labeled oligonucleotides can be used as probes to find specific DNA target sequences. Such probes can thus be used to identify particular species or strains of bacteria by hybridizing to characteristic sequences in the genes for their 16S ribosomal RNA. This approach allows the identification of pathogens in clinical samples or bacteria of interest in environmental samples.

A major challenge is to use multiple probes simultaneously to identify several different target molecules (or different bacteria) within the same sample. The number of different fluorescent probes that could be used simultaneously was limited to three or four due to the overlap of their emission spectra. Recent advances in fluorescence microscopy and computerized signal analysis, referred to as "linear unmixing," have now overcome this problem.

A recent tour de force in multiple fluorescent labeling (Figure 1) demonstrated the simultaneous use of eight different oligonucleotide probes in binary combinations to distinguish between 28 different strains of *Escherichia coli*. These probes targeted 16S rRNA sequences that varied slightly from strain to strain. This technique should allow the simultaneous identification of multiple members of complex microbial associations.

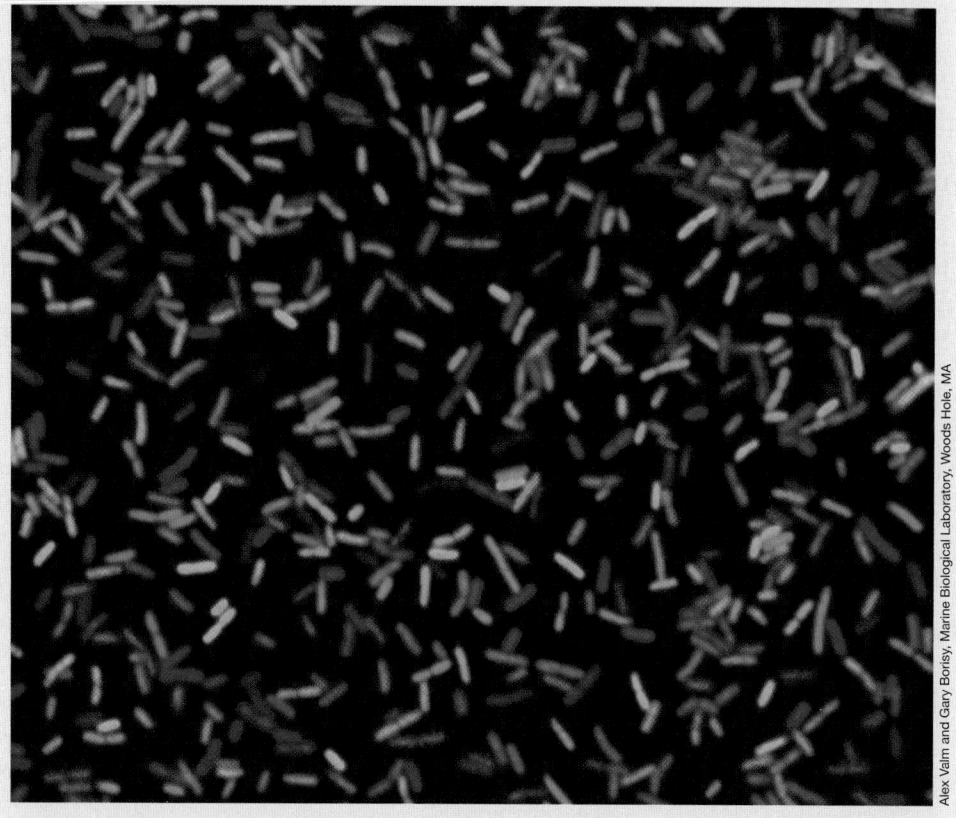

Alex Valm and Gary Borisy, Marine Biological Laboratory, Woods Hole, MA

Figure 1 Fluorescence spectral image of 28 differently labeled strains of *Escherichia coli*. Cells were labeled with combinations of fluorophore-conjugated oligonucleotides that are complementary to *E. coli* 16S rRNA.

Natural biofilms often contain many species of bacteria and are of great interest to microbial ecologists. Until now, tracking multiple strains of a species in a mixture has been impossible. However, multiple fluorescent labeling allows multiple strains to be monitored over time during the development of a mixed biofilm. Medical applications include differentiating strains of the same or closely related species, all of which may be present in a single clinical sample.

---

4. It is easy to isolate in the supercoiled form using routine techniques.

5. Moderate amounts of foreign DNA can be inserted, although inserts of more than 10 kbp lead to plasmid instability.

6. The complete base sequence of the plasmid is known, allowing identification of all restriction enzyme cut sites.

7. The polylinker contains single cut sites for a dozen restriction enzymes, such as *Eco*RI, *Sal*I, *Bam*HI, *Pst*I, and *Hin*dIII. Only a single cut site for each restriction enzyme used in cloning should occur in a cloning vector so that treatment with that enzyme linearizes the vector but does not cut it into pieces.

Single cut sites for multiple restriction enzymes increases the versatility of the vector. (One minor exception is that pairs of sites are sometimes used that allow replacement of the sequence between them.)

8. It has a gene conferring ampicillin resistance on its host. This permits ready selection of host cells containing the plasmid because such hosts gain resistance to the antibiotic.

9. It can be inserted into cells easily by transformation.

10. Insertion of foreign DNA into the polylinker can be detected by blue–white screening (see below) because of *lacZ*.

Cloning Genes into Plasmid Vectors

The use of plasmid vectors in gene cloning is shown in **Figure 11.12**. A suitable restriction enzyme with a cut site within the polylinker is chosen. Both the vector and the foreign DNA to be cloned are cut with this enzyme. The vector is linearized. Segments of the foreign DNA are inserted into the open cut site and ligated into position with DNA ligase. This disrupts the *lacZ* gene, a phenomenon called *insertional inactivation*. This may be used to detect the presence of foreign DNA within the vector. When the colorless reagent Xgal is added to the medium, β-galactosidase cleaves it, generating a blue product. Thus, cells containing the vector *without* cloned DNA form

blue colonies, whereas cells containing the vector *with* an insert of cloned DNA do not form β-galactosidase and are therefore white. After DNA ligation, the resulting plasmids are transformed into cells of *E. coli*. The colonies are selected on media containing both ampicillin, to select for the presence of the plasmid, plus Xgal, to test for β-galactosidase activity. Those colonies that are *blue* contain the plasmid without any inserted foreign DNA (i.e., the plasmid merely recyclized without picking up foreign DNA), whereas those colonies that are *white* contain plasmid with inserted foreign DNA and are picked for further analyses (see Figure 11.20 for a related example of the blue–white selection system).

Many subsequent vectors include features similar to those of pUC19 listed above. Sometimes insertional inactivation can be detected by selection, rather than by screening. For example, in some vectors, the gene carrying the polylinker normally produces a protein that is lethal to the host cell. Therefore, only cells containing a plasmid in which this gene has been inactivated can grow.

Cloning using plasmid vectors is versatile and widely used in genetic engineering, particularly when the fragment to be cloned is fairly small. Also, plasmids are often used as cloning vectors if expression of the cloned gene is desired, since regulatory genes can be engineered into the plasmid to obtain expression of the cloned genes under specific conditions (Section 11.8).

MiniQuiz

- Explain why in cloning it is necessary to use a restriction enzyme that cuts the vector in only one location.
- What is insertional inactivation?
- What is a polylinker?

11.7 Hosts for Cloning Vectors

To produce large amounts of cloned DNA, an ideal host should grow rapidly in an inexpensive culture medium. Ideally, the host should be nonpathogenic, be easy to transform with engineered DNA, be genetically stable in culture, and have the appropriate enzymes to allow replication of the vector. It is also helpful if considerable background information on the host and a wealth of tools for its genetic manipulation exist.

The most useful hosts for cloning are microorganisms that are easily grown and for which we have much information. These include the bacteria *Escherichia coli* and *Bacillus subtilis*, and the yeast *Saccharomyces cerevisiae* (**Figure 11.13**). Complete genome sequences are available for all of these organisms, and they are widely used as cloning hosts. However, in some cases other hosts and specialized vectors may be necessary to get the DNA properly cloned and expressed.

Prokaryotic Hosts

Although most molecular cloning has been done in *E. coli* (Figure 11.13), this host has a few disadvantages. *E. coli* is an excellent choice for initial cloning work, but is problematic as an expression host because it is found in the human intestine and some wild-type strains are potentially harmful. However, several modified *E. coli* strains have been developed for cloning

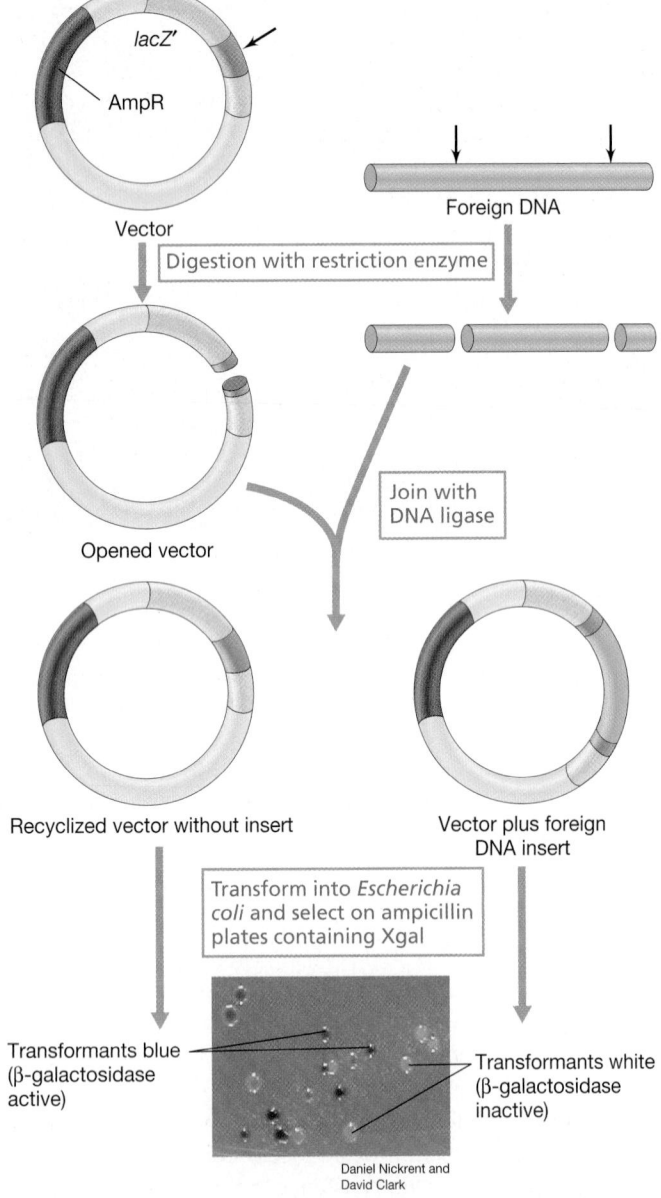

Vector

Foreign DNA

Digestion with restriction enzyme

Opened vector

Join with DNA ligase

Recyclized vector without insert

Vector plus foreign DNA insert

Transform into *Escherichia coli* and select on ampicillin plates containing Xgal

Transformants blue (β-galactosidase active)

Transformants white (β-galactosidase inactive)

Daniel Nickrent and David Clark

Figure 11.12 Cloning into the plasmid vector pUC19. The cloning vector is opened by cutting with a suitable restriction enzyme in the polylinker. Insertion of DNA inactivates β-galactosidase, allowing blue–white screening for the presence of the insert. The photo on the bottom shows colonies of *Escherichia coli* on an Xgal plate. The enzyme β-galactosidase can cleave the normally colorless Xgal to form a blue product.

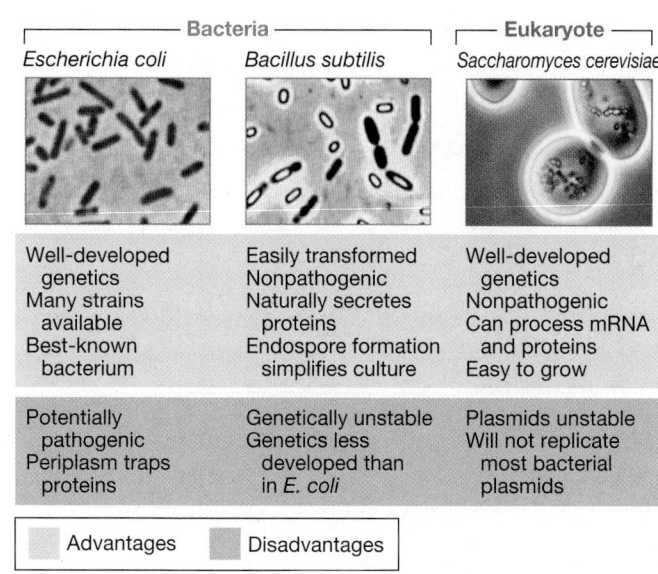

Bacteria		Eukaryote
Escherichia coli	*Bacillus subtilis*	*Saccharomyces cerevisiae*
Well-developed genetics Many strains available Best-known bacterium	Easily transformed Nonpathogenic Naturally secretes proteins Endospore formation simplifies culture	Well-developed genetics Nonpathogenic Can process mRNA and proteins Easy to grow
Potentially pathogenic Periplasm traps proteins	Genetically unstable Genetics less developed than in *E. coli*	Plasmids unstable Will not replicate most bacterial plasmids

☐ Advantages   ☐ Disadvantages

Figure 11.13 Hosts for molecular cloning. A summary of the advantages and disadvantages of some common cloning hosts.

purposes, and thus *E. coli* remains the organism of choice for most molecular cloning. A major problem with using any bacterial host, *E. coli* included, is the lack of systems to correctly modify eukaryotic proteins. This problem may be solved by using eukaryotic host cells, as discussed below.

Another problem with using *E. coli* is that, like all gram-negative bacteria, it has an outer membrane which hinders protein secretion. This issue may be resolved by using the gram-positive organism *B. subtilis* as a cloning host (Figure 11.13). Although the technology for cloning in *B. subtilis* is less advanced than for *E. coli*, several plasmids and phages suitable for cloning have been developed, and transformation is a well-developed procedure in *B. subtilis*. The main disadvantage of using *B. subtilis* as a cloning host is plasmid instability. It is often difficult to maintain plasmid replication over many subcultures of the organism. Also, foreign DNA is not as well maintained in *B. subtilis* as in *E. coli*; thus the cloned DNA is often unexpectedly lost.

Often host organisms for cloning must have specific genotypes to be effective. For instance, if the cloning vector uses the *lacZ* gene for screening, then the host must carry a mutation disabling this gene. Because the bacteriophage M13 infects only bacteria with F pili (Section 11.10), hosts used with M13-derived vectors must contain and express genes on the F plasmid. These types of considerations and others, such as the ease of selection of transformants, must be taken into account when choosing a cloning host.

Eukaryotic Hosts

Cloning in eukaryotic microorganisms has focused on the yeast *S. cerevisiae* (Figure 11.13). Plasmid vectors as well as artificial chromosomes (Section 11.10) have been developed for yeast. One important advantage of eukaryotic cells as hosts for cloning vectors is that they already possess the complex RNA and post-translational processing systems required for the production of eukaryotic proteins. Thus these systems do not have to be engineered into the vector or host cells as would be required if cloned eukaryotic DNA was to be expressed in a prokaryotic host.

For many applications, gene cloning in mammalian cells has been done. Cultured mammalian cells can be handled in some ways like microbial cultures, and are widely used in research on human genetics, cancer, infectious disease, and physiology. A disadvantage of using mammalian cells is that they are expensive and difficult to produce under large-scale conditions. Insect cell lines are simpler to grow, and vectors have been developed from an insect DNA virus, the baculovirus (Section 11.8). For some applications, in particular for plant agriculture, the cloning host can be a plant cell tissue culture line or even an entire plant. Indeed, genetic engineering has many applications in plant agriculture (⮌ Section 15.18). However, regardless of eukaryotic host type, it is necessary to get the vector DNA into the host cells.

Transfection of Eukaryotic Cells

The term "transfection" originally referred to transformation of viral nucleic acid into cells (both bacterial and eukaryotic) because this often resulted in viral infection. Many eukaryotic cells can take up DNA by transformation. However, because the word "transformation" is used to describe the conversion of mammalian cells to a cancerous state, the introduction of DNA into mammalian cells is usually called *transfection* even when no viral nucleic acid is involved.

Mammalian cells in culture may be transfected by adding DNA in combination with a variety of cationic carriers that bind to and protect DNA due to their positive charges. Natural or artificial polymers and cationic lipid vesicles (liposomes) have all been used. Transfection of cultured animal cells was originally accomplished by precipitating DNA in such a way that the cells would take it up by phagocytosis, which is possible because animal cells do not have cell walls. In animal cells, DNA can also be injected directly into the nucleus using micropipets, a technique called *microinjection*.

In yeast, which is an important organism for genetic engineering, transfection at low efficiencies can be performed by various methods. However, as with bacteria, electroporation is widely used for eukaryotic cells and can be used whether or not the cell wall of the organism is removed. Electroporation exposes host cells to pulsed electrical fields in the presence of cloned DNA. This treatment opens small pores in the cytoplasmic membrane that are not large enough to cause major cell damage or lysis, but are sufficient to allow cloned DNA to enter.

In addition to electroporation, a high-velocity microprojectile "gun" can be used to get DNA into cells. The particle gun or gene gun operates by using a propellant such as pressurized helium to fire DNA-coated particles through a small steel cylinder at target cells (**Figure 11.14**). The particles bombard the cell, piercing the cell wall and cytoplasmic membrane without actually killing the cells. The nucleic acid entering the cells can then recombine with host DNA. The particle gun has been used successfully to transfect yeast, algae, and higher plant cells. Moreover, unlike electroporation, the particle gun can be used to get DNA into intact tissues, such as plant seeds, and into mitochondria and chloroplasts.

MiniQuiz
- Why does molecular cloning require a host?
- Describe three mechanisms by which cells can take up DNA.

Before gas release

After gas release

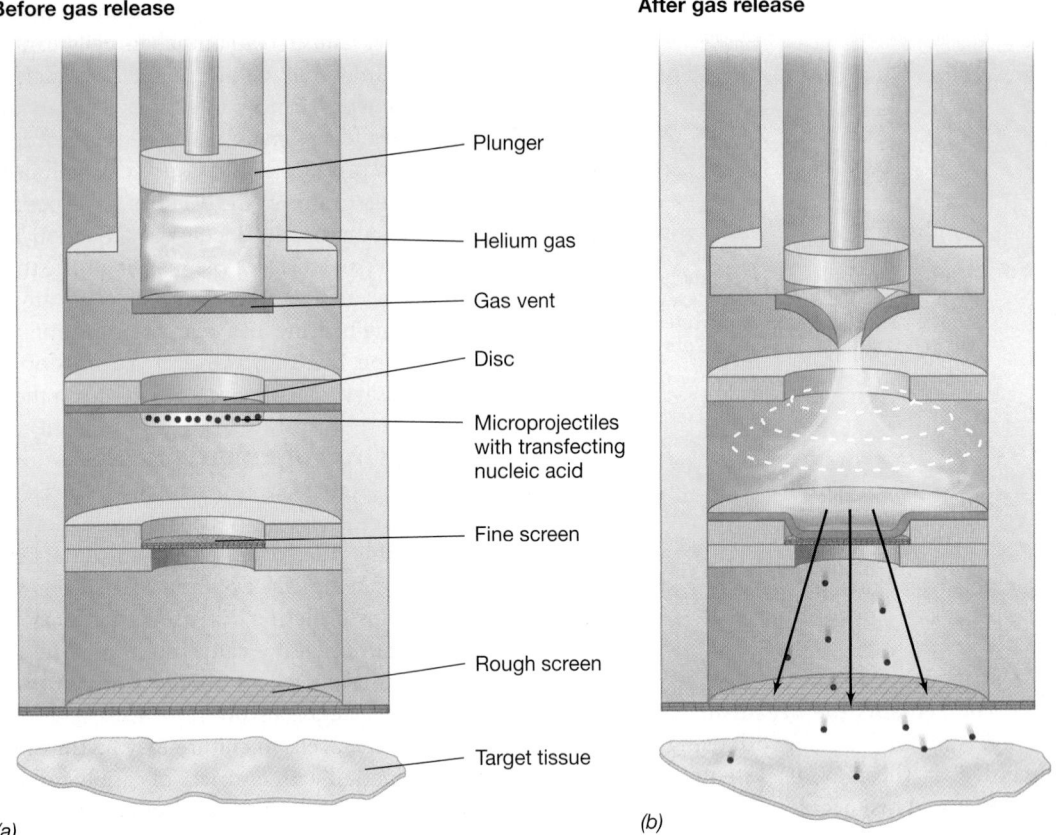

Plunger

Helium gas

Gas vent

Disc

Microprojectiles
with transfecting
nucleic acid

Fine screen

Rough screen

Target tissue

(a)

(b)

Figure 11.14 DNA gun for transfection of eukaryotic cells. The inner workings of the gun show how metal pellets coated with nucleic acids (microprojectiles) are projected at target cells. *(a)* Before firing and *(b)* after firing. A shock wave due to gas release throws the disc carrying the microprojectiles against the fine screen. The microprojectiles continue on into the target tissue.

11.8 Shuttle Vectors and Expression Vectors

Cloned genes are used for a variety of purposes. Specialized vectors have been engineered for use in different situations such as moving a cloned gene between organisms of different species or optimizing expression of the cloned gene. Two classes of vectors are involved, *shuttle* vectors and *expression* vectors.

Shuttle Vectors

Vectors that can replicate and are stably maintained in two (or more) unrelated host organisms are called **shuttle vectors**. Genes carried by a shuttle vector can thus be moved between unrelated organisms. Shuttle vectors have been developed that replicate in both *Escherichia coli* and *Bacillus subtilis*, *E. coli* and yeast, and *E. coli* and mammalian cells, as well as in many other pairs of organisms. The importance of a shuttle vector is that DNA cloned in one organism can be replicated in a second host without modifying the vector in any way to do so.

Many shuttle vectors have been designed to move genes between *E. coli* and yeast. Bacterial plasmid vectors were the starting point and were modified to function in yeast as well. Because bacterial origins of replication do not function in eukaryotes, it is necessary to provide a yeast replication origin. One bonus is that

DNA sequences of replication origins are similar in different eukaryotes, so the yeast origin functions in other higher organisms. When eukaryotic cells divide, the duplicated chromosomes are pulled apart by microtubules ("spindle fibers") attached to their centromeres (Section 7.6). Consequently, shuttle vectors for eukaryotes must contain a segment of DNA from the centromere in order to be properly distributed at cell division (**Figure 11.15**). Luckily, the yeast centromere recognition sequence, the CEN sequence, is relatively short and easy to insert into shuttle vectors.

Another requirement is a convenient marker to select for the plasmid in yeast. Unfortunately, yeast is not susceptible to most antibiotics that are effective against bacteria. In practice, yeast host strains that are defective in making a particular amino acid or purine or pyrimidine base are used. A functional copy of the biosynthetic gene that is defective in the host is inserted into the shuttle vector. For example, if the *URA3* gene, needed for synthesis of uracil, is used, the yeast will not grow in the absence of uracil unless it gains a copy of the shuttle vector.

Expression Vectors

Organisms have complex regulatory systems, and cloned genes are often expressed poorly or not at all in a foreign host cell. This problem is tackled by using **expression vectors** that are designed to allow the experimenter to control the expression of cloned

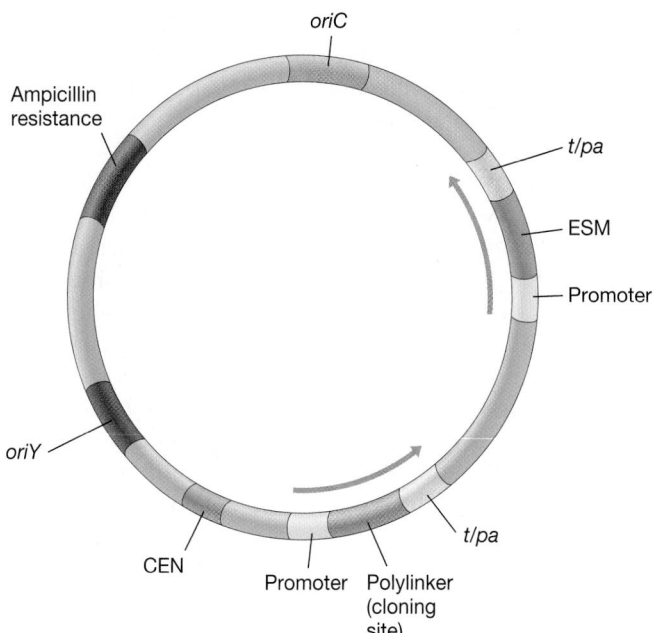

Figure 11.15 Genetic map of a shuttle vector used in yeast. The vector contains components that allow it to shuttle between *Escherichia coli* and yeast and be selected in each organism: *oriC*, origin of replication in *E. coli*; *oriY*, origin of replication in yeast; ESM, eukaryotic selectable marker; CEN, yeast centromere sequence; promoter; *t/pa*, transcription termination/polyadenylation signals. Arrows indicate the direction of transcription.

genes. Generally, the objective is to obtain high levels of expression, especially in biotechnological applications. However, when dealing with potentially toxic gene products, a low but strictly controlled level may be appropriate.

Expression vectors contain regulatory sequences that allow manipulation of gene expression. Usually the control is transcriptional because for high levels of expression it is essential to produce high levels of mRNA. In practice, high levels of transcription require strong promoters that bind RNA polymerase efficiently (ᴐᴐ Section 6.13). However, the native promoter of a cloned gene may work poorly in the new host. For example, promoters from eukaryotes or even from other bacteria function poorly or not at all in *E. coli*. Indeed, even some *E. coli* promoters function at low levels in *E. coli* because their sequences match the promoter consensus poorly and bind RNA polymerase inefficiently (ᴐᴐ Section 6.13).

For this reason, expression vectors must contain a promoter that functions efficiently in the host and one that is correctly positioned to drive transcription of the cloned gene. Promoters from *E. coli* that are used in expression vectors include *lac* (the *lac* operon promoter), *trp* (the *trp* operon promoter), *tac* and *trc* (synthetic hybrids of the *trp* and *lac* promoters), and lambda P$_L$ (the leftward lambda promoter; ᴐᴐ Section 9.10). These are all "strong" promoters in *E. coli* and in addition they can be specifically regulated.

Regulation of Transcription from Expression Vectors

Although producing very high levels of mRNA and having this translated into large amounts of protein is often useful, massive overproduction of foreign proteins often damages the host cell.

Therefore, it is important to regulate the expression of cloned genes. Often, in order to avoid damaging the host cells, the culture containing the expression vector is grown without expression of the foreign gene. Once a large population of healthy cells is obtained, expression of the cloned gene is then triggered by a genetic switch.

Regulating transcription by a repressor protein (ᴐᴐ Section 8.3) is a useful way to control a cloned gene. A strong repressor can completely block the synthesis of the proteins under its control by binding to the operator. When gene expression is required, the inducer is added. The repressor binds the inducer and is released from the DNA, thus allowing transcription of the regulated genes. The expression vector is designed such that the cloned gene is inserted just downstream from the chosen promoter and operator region. A strong ribosome-binding site is often included between the promoter and the cloned gene to give efficient translation. The overall result is control of the cloned gene by the chosen promoter together with efficient transcription and translation. The operator and promoter usually correspond to each other (for instance, the *lac* operator is used with the *lac* promoter), but this need not be so. For example, hybrid regulatory regions, such as fusing the *trp* promoter to the *lac* operator to form the *trc* regulatory element, are sometimes used.

Figure 11.16 shows an expression vector controlled by *trc*. This plasmid also contains a copy of the *lacI* gene that encodes the *lac* repressor. The level of repressor in a cell containing this plasmid

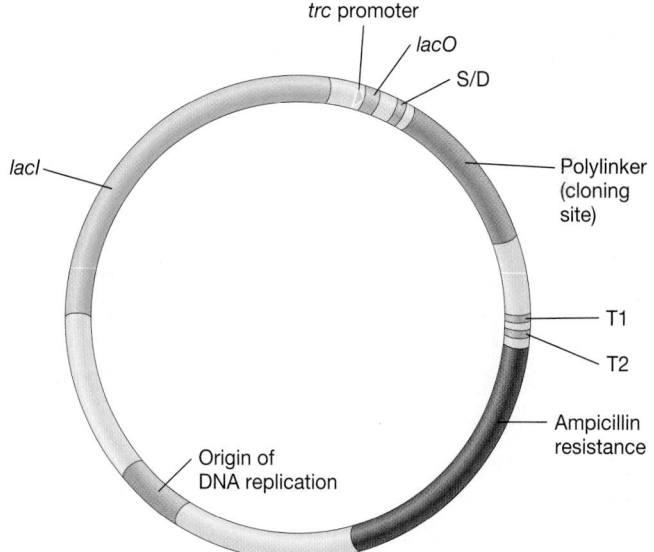

Figure 11.16 Genetic map of the expression vector pSE420. This vector was developed by Invitrogen Corp., a biotechnology company. The polylinker contains many different restriction enzyme recognition sequences to facilitate cloning. This region, plus the inserted cloned gene, are transcribed by the *trc* promoter, which is immediately upstream of the *lac* operator (*lacO*). Immediately upstream of the polylinker is a sequence that encodes a Shine–Dalgarno (S/D) ribosome-binding site on the resulting mRNA. Downstream of the polylinker are two transcription terminators (T1 and T2). The plasmid also contains the *lacI* gene, which encodes the *lac* repressor, and a gene conferring resistance to the antibiotic ampicillin. These two genes are under the control of their own promoters, which are not shown.

is sufficient to prevent transcription from the *trc* promoter until inducer is added. Addition of lactose or related *lac* inducers triggers transcription of the cloned DNA. In addition to a strong and easily regulated promoter, most expression vectors contain an effective transcription terminator (Section 6.14). This prevents transcription from the strong cloning promoter continuing on into other genes on the vector, which would interfere with vector stability. The expression vector shown in Figure 11.16 has strong transcription terminators to halt transcription immediately downstream from the cloned gene.

Regulating Expression with Bacteriophage T7 Control Elements

In some cases the transcriptional control system may not be a normal part of the host at all. An example of this is the use of the bacteriophage T7 promoter and T7 RNA polymerase to regulate expression. When T7 infects *E. coli*, it encodes its own RNA polymerase that recognizes only T7 promoters (Section 21.3). In T7 expression vectors, cloned genes are placed under control of the T7 promoter. To achieve this, the gene for T7 RNA polymerase must also be present in the cell under the control of an easily regulated system, such as *lac* (**Figure 11.17**). This is usually done by integrating the gene for T7 RNA polymerase with a *lac* promoter into the chromosome of a specialized host strain.

The BL21 series of host strains are especially designed to work with the pET series of T7 expression vectors. The cloned genes are expressed shortly after T7 RNA polymerase transcription has been switched on by a *lac* inducer, such as IPTG. Because it recognizes only T7 promoters, the T7 RNA polymerase transcribes only the cloned genes. The T7 RNA polymerase is so highly active that it uses most of the RNA precursors, thereby limiting transcription to the cloned genes. Consequently, host genes that require host RNA polymerase are for the most part not transcribed and thus the cells stop growing. Protein synthesis in such cells is then dominated by the protein of interest. Thus the T7 control system is very effective for generating extremely large amounts of a particular protein of interest.

Translation of the Cloned Gene

Expression vectors must also be designed to ensure that the mRNA produced is efficiently translated. To synthesize protein from an mRNA molecule, it is essential for the ribosomes to bind at the correct site and begin reading in the correct frame. In bacteria this is accomplished by having a ribosome-binding site (Shine–Dalgarno sequence, Section 6.19) and a nearby start codon on the mRNA. Bacterial ribosome-binding sites are not found in eukaryotic genes and must be engineered into the vector if high levels of expression of the eukaryotic gene are to be obtained. Once again, the vector shown in Figure 11.16 has such a site.

Often, adjustments have to be made to ensure high-efficiency translation after the gene has been cloned. For example, codon usage can be an obstacle. Codon usage is related to the concentration of the appropriate tRNA in the cell. Therefore, if a cloned gene has a very different codon usage from its expression host, it will probably be translated inefficiently in that host. Site-directed mutagenesis (Section 11.4) can be used to change selected codons in the gene, making it more amenable to the codon usage pattern of the host.

Finally, if the cloned gene contains introns, as eukaryotic genes typically do (Section 7.8), the correct protein product will not be made if the host is a prokaryote. This problem can also be corrected by using synthetic DNA. However, the usual method to create an intron-free gene is to obtain the mRNA (from which the introns have already been removed) and use reverse transcriptase (Section 21.11) to generate a complementary DNA (cDNA) copy.

Eukaryotic Vectors

It is often desirable to clone and express genes directly in eukaryotes, and vectors are available for cloning into yeast. Yeast is one of the few eukaryotes that naturally contains a plasmid, called the two-micron circle, and most yeast vectors are based on this. The use of yeast artificial chromosomes for cloning very large fragments of DNA is discussed in Section 11.10.

Virus vectors are commonly used in multicellular eukaryotes. For example, the DNA virus SV40 (Section 21.13), which causes tumors in primates, has been engineered as a cloning vector for cultured human cells. Derivatives of SV40 that do not induce tumors have been developed for cloning and expressing mammalian genes. Other mammalian vectors use adenovirus (Section 21.16) or vaccinia virus (Section 21.15). Vaccinia virus vectors in particular have been used to develop recombinant vaccines (Section 15.13). Vectors derived from baculovirus, a DNA virus that replicates in insect cells, can be used to make large quantities of the products of cloned genes.

Other expression vectors have been developed specifically to stably maintain and express cloned genes in an organism or tissue. These **integrating vectors** are maintained at low copy number

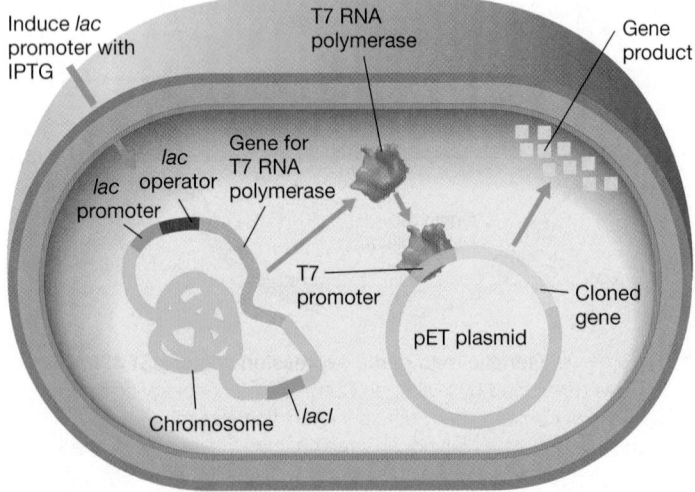

Figure 11.17 The T7 expression system. The gene for T7 RNA polymerase is in a gene fusion under control of the *lac* promoter and is inserted into the chromosome of a special host strain of *Escherichia coli*. Addition of IPTG induces the *lac* promoter, causing expression of T7 RNA polymerase. This transcribes the cloned gene, which is under control of the T7 promoter and is carried by the pET plasmid.

(typically one copy per genome) by integrating into the host chromosome. They have been developed in eukaryotes ranging from yeast to mammals, as well as in certain bacteria. Integrating vectors have uses in both basic science and in applications such as gene therapy (ᴒᴃ Section 15.17). In particular, retroviruses may be used to introduce genes into mammalian cells because these viruses replicate via a DNA form that is integrated into the host chromosome (ᴒᴃ Section 21.11).

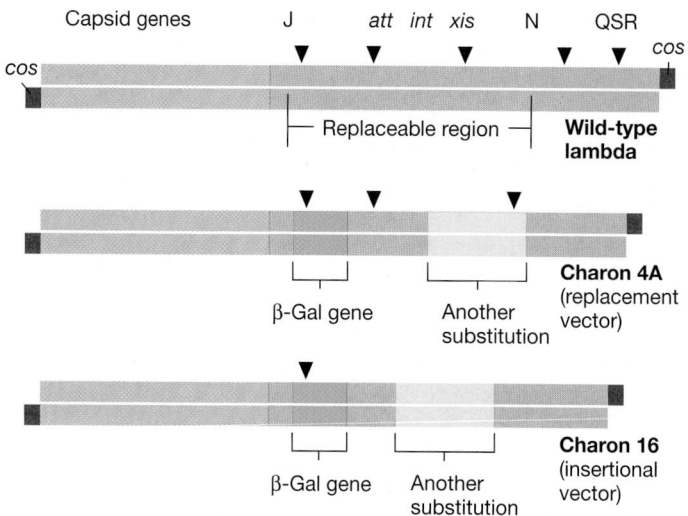

Figure 11.18 **Lambda cloning vectors.** Abbreviated genetic map of bacteriophage lambda showing the cohesive ends in red. Charon 4A and 16 are both derivatives of lambda with various substitutions and deletions in the nonessential region. Each has the *lacZ* gene, encoding the enzyme β-galactosidase, which permits detection of phage containing inserted clones. Whereas the wild-type lambda genome is 48.5 kbp, that for Charon 4A is 45.4 kbp and for Charon 16 is 41.7 kbp. The arrowheads above the maps of each phage indicate the sites recognized by the restriction enzyme *Eco*RI.

MiniQuiz

- Describe some of the components of an expression vector that improve expression of the cloned gene.
- Describe the components needed for an efficient shuttle vector.

11.9 Bacteriophage Lambda as a Cloning Vector

Recall that during the process of specialized transduction some host genes become incorporated into a bacteriophage genome, and that bacteriophage lambda is the most studied of the specialized transducing phages (ᴒᴃ Section 10.8). During specialized transduction, lambda acts as a vector, but recombination occurs in the cell, not in a test tube. Lambda can also be used as a cloning vector for *in vitro* recombination.

Lambda is a useful cloning vector because its biology is well understood, it can hold larger amounts of DNA than most plasmids, and DNA can be efficiently packaged into phage particles *in vitro*. These can be used to infect suitable host cells, and infection is much more efficient than transformation (transfection). Phage lambda has a large number of genes; however, the central third of the lambda genome, between genes J and N (**Figure 11.18**), is not essential for infectivity and can be replaced with foreign DNA. This allows relatively large DNA fragments, up to about 20 kilobase pairs (kbp), to be cloned into lambda. This is twice the cloning capacity of typical small plasmid vectors.

Modified Lambda Phages

Wild-type lambda is not suitable as a cloning vector because its genome has too many restriction enzyme sites. To avoid this difficulty, modified lambda phages have been constructed especially for cloning. In one set of modified lambda phages, called *Charon phages*, unwanted restriction enzyme sites have been removed by mutation. Foreign DNA can be inserted into variants that have only a single restriction site, such as Charon 16. By contrast, in variants with two sites, such as Charon 4A, foreign DNA can replace a specific segment of the lambda DNA (Figure 11.18). Such *replacement vectors* are especially useful in cloning large DNA fragments. When Charon 4A is used as a replacement vector, the two small interior fragments are cut out and discarded.

Steps in Cloning with Lambda

Cloning with lambda replacement vectors involves the following steps (**Figure 11.19**):

1. Isolating vector DNA from phage particles and cutting it into two fragments with the appropriate restriction enzyme.

2. Connecting the two lambda fragments to fragments of foreign DNA using DNA ligase. Conditions are chosen so molecules are formed of a length suitable for packaging into phage particles.

3. Packaging of the DNA by adding cell extracts containing the head and tail proteins and allowing the formation of viable phage particles to occur spontaneously.

4. Infecting *Escherichia coli* cells and isolating phage clones by picking plaques on a host strain.

5. Checking recombinant phage for the presence of the desired foreign DNA sequence using nucleic acid hybridization procedures, DNA sequencing, or observation of genetic properties.

Selection of recombinants is less of a problem with lambda replacement vectors (such as Charon 4A) than with plasmids because (1) the efficiency of transfer of recombinant DNA into the cell by lambda is very high and (2) lambda fragments that have not received new DNA are too small to be incorporated into phage particles. Thus, every viable phage virion should contain cloned DNA.

Both Charon vectors are also engineered to contain reporter genes, such as the gene for β-galactosidase previously discussed. When the vectors replicate in a lactose-negative (Lac⁻) strain of *E. coli*, β-galactosidase is synthesized from the phage gene, and the presence of lactose-positive (Lac⁺) plaques can be detected by using a color indicator agar (Section 11.5). However, if a foreign gene is inserted into the β-galactosidase gene, the Lac⁺ character is lost. Such Lac⁻ plaques can be readily detected as colorless plaques among a background of colored plaques (see Figure 11.20*b*).

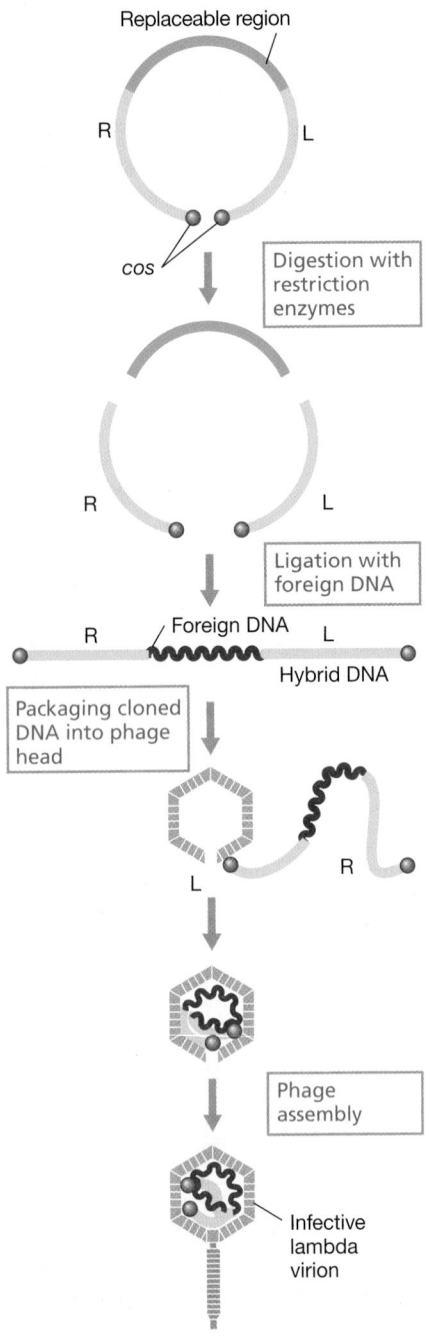

Figure 11.19 Bacteriophage lambda as a cloning vector. The maximum size of inserted DNA is about 20 kbp.

Cosmid Vectors

Like replacement vectors, cosmid vectors employ specific lambda genes and are packaged into lambda virions. *Cosmids* are plasmid vectors containing the *cos* site from the lambda genome, which yields cohesive ends when cut (🔁 Section 9.10). The *cos* site is required for packaging DNA into lambda virions. Cosmids are constructed from plasmids by ligating the lambda *cos* region to the plasmid DNA. Foreign DNA is then ligated into the vector. The modified plasmid, plus cloned DNA, can then be packaged into lambda virions *in vitro* as described previously and the phage particles used to transduce *E. coli*.

One major advantage of cosmids is that they can be used to clone large fragments of DNA, with inserts as large as 50 kbp accepted by the system. With big inserts, fewer clones are needed to cover a whole genome. Using cosmids also avoids the necessity of having to transform *E. coli*, which is especially inefficient with larger plasmids. Cosmids also permit storage of the DNA in phage particles instead of as plasmids. Phage particles are more stable than plasmids, so the recombinant DNA can be kept for long periods of time.

MiniQuiz

- What is a replacement vector?
- Why is the ability to package recombinant DNA in phage particles in a test tube useful?

11.10 Vectors for Genomic Cloning and Sequencing

Most of the techniques discussed in this section are variants of the *in vitro* techniques discussed above. The principles remain the same, but the emphasis is on the entire genome of an organism, rather than individual genes. Plasmid vectors and the Charon derivatives of bacteriophage lambda, for example, have been extensively used for cloning and sequencing, including for genomic analysis. Other, more specialized vectors for genome analysis include bacteriophage M13 and bacterial and yeast artificial chromosomes, and we focus on these here.

Vectors Derived from Bacteriophage M13

M13 is a filamentous bacteriophage with single-stranded DNA that replicates without killing its host (🔁 Section 21.2). Mature particles of M13 are released from host cells without lysing by a budding process, and infected cultures can provide continuous sources of phage DNA. Most of the genome of phage M13 contains genes essential for virus replication. However, a small region called the *intergenic sequence* can be used as a cloning site. Variable lengths of foreign DNA, up to about 5 kbp, can be cloned without affecting phage viability. As the genome gets larger, the virion simply grows longer.

Phage M13mp18 is a derivative of M13 in which the intergenic region has been modified to facilitate cloning (**Figure 11.20a**). One useful modification is the insertion of a functional fragment of *lacZ*, the *Escherichia coli* gene that encodes β-galactosidase. Cells infected with M13mp18 can be detected easily by their color on indicator plates. This is achieved using the artificial substrate, Xgal, which is cleaved by β-galactosidase to yield a blue color (Figure 11.20b). The *lacZ* gene has itself been modified to contain a 54-bp polylinker, which contains several restriction enzyme cut sites absent from the original M13 genome. The polylinker is inserted into the beginning of the coding portion of the *lacZ* gene, but does not affect enzyme activity. However, inserting additional cloned DNA into the polylinker inactivates the gene. Phages with such DNA inserts yield colorless plaques (no β-galactosidase activity), and can therefore be easily identified (Figure 11.20b).

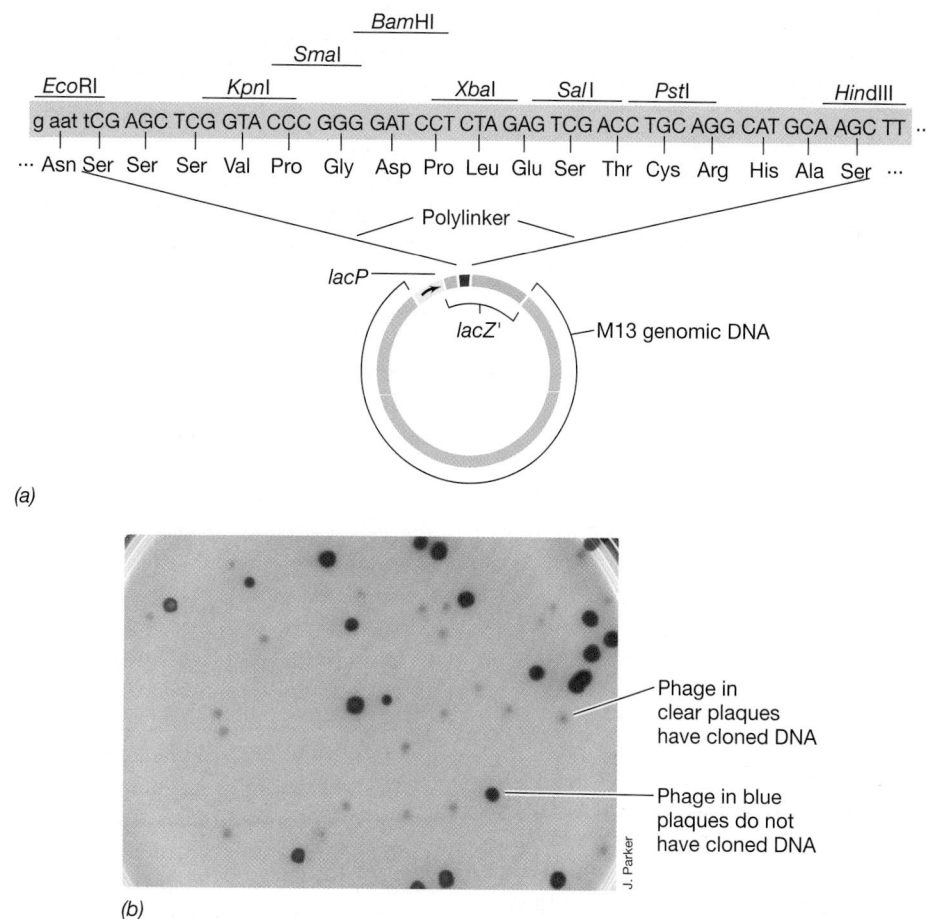

Figure 11.20 Cloning using bacteriophage M13. *(a)* A partial map of M13mp18, a derivative of M13 constructed for use as a cloning vector. The vector contains the *lac* promoter (*lacP*) and a truncated *lacZ'* gene, which encodes a functional part of β-galactosidase. At the beginning of this gene is a polylinker that contains several restriction sites but maintains the proper reading frame. The amino acids encoded by the polylinker are shown. Most DNA fragments cloned into the polylinker disrupt the *lacZ'* gene and abolish β-galactosidase activity. *(b)* Portion of an Xgal-containing agar plate showing white plaques formed by M13mp18 containing cloned DNA and blue plaques formed by phage lacking cloned DNA.

Use of M13 in Molecular Cloning

To clone DNA in M13 vectors, double-stranded replicative form DNA (⮌ Section 21.2) is isolated from the infected host and cut with a restriction enzyme. The foreign DNA is then treated with the same restriction enzyme. On ligation, double-stranded M13 molecules are obtained that contain the foreign DNA. When these molecules are introduced into the cell by transformation, they are replicated and produce single-stranded DNA bacteriophage particles containing the cloned DNA.

The single-stranded M13 DNA produced can be used directly in DNA sequencing. Because the base sequence next to the cut site where the foreign DNA is inserted is known, it is possible to construct an oligonucleotide primer complementary to this region and use this to determine the sequence of the DNA downstream from this point. In this way, M13 derivatives have proven extremely useful in sequencing foreign DNA, even rather long molecules, and have featured prominently in the sequencing of several genomes.

Artificial Chromosomes

Vectors that hold about 2–10 kbp of cloned DNA are adequate for making gene libraries for sequencing prokaryotic genomes. Bacteriophage lambda vectors, which hold 20 kbp or more, are also widely used in genomics projects. However, as the size of the genome increases, so does the number of clones needed to obtain a complete sequence. Therefore, for making libraries of DNA from eukaryotic microorganisms or from higher eukaryotes such as humans, it is useful to have vectors that can carry very large segments of DNA. This allows the size of the initial library to be manageable. Such vectors have been developed and are called **artificial chromosomes**.

Bacterial Artificial Chromosomes: BACs

Many bacteria contain large plasmids that are stably replicated within the cell, for example, the F plasmid of *E. coli* (⮌ Section 10.9). Naturally occurring derivatives of the F plasmid, called F' plasmids, are known that may carry large amounts of chromosomal DNA (⮌ Section 10.10). Because of these desirable properties, the

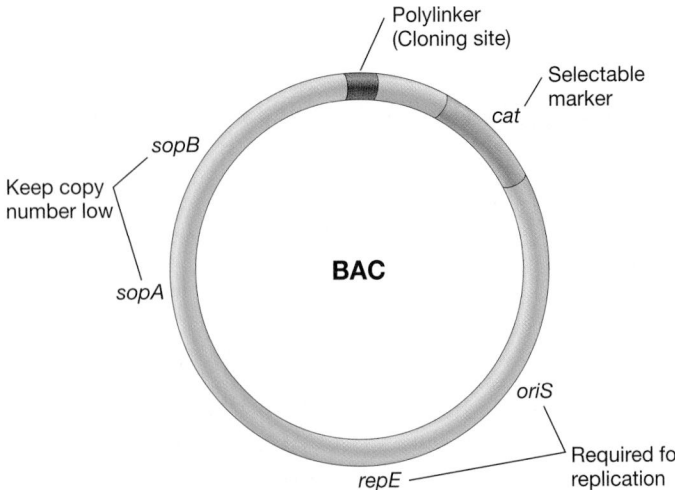

Figure 11.21 Genetic map of a bacterial artificial chromosome. The BAC shown is 6.7 kbp. The cloning region has several unique restriction enzyme sites. This BAC contains only a small fraction of the 99.2-kbp F plasmid.

F plasmid has been used to construct cloning vectors called **bacterial artificial chromosomes** (BACs).

Figure 11.21 shows the structure of a BAC based on the F plasmid. The vector is only 6.7 kbp compared to the 99.2-kbp of F itself. The BAC contains only a few genes from F, including *oriS* and *repE,* which are necessary for replication, and *sopA* and *sopB,* which keep the copy number very low. The plasmid also contains the *cat* gene, which confers chloramphenicol resistance on the host, and a polylinker that includes several restriction sites for cloning DNA. Foreign DNA of more than 300 kbp can be inserted and stably maintained in a BAC vector such as this. The host for a BAC is typically a mutant strain of *E. coli* that lacks the normal restriction and modification systems of the wild type (Section 11.1). This prevents the BAC from being destroyed. The host strain is usually also defective in recombination, which prevents recombination of the cloned DNA from the BAC into the host cell chromosome.

Eukaryotic Artificial Chromosomes

Historically, the first artificial chromosomes were **yeast artificial chromosomes** (YACs). These vectors replicate in yeast like normal chromosomes, but they have sites where very large fragments of DNA can be inserted. To function like normal eukaryotic chromosomes, YACs must have (1) an origin of DNA replication, (2) telomeres for replicating DNA at the ends of the chromosome (↩ Section 7.7), and (3) a centromere for segregation during mitosis (↩ Section 7.6). They must also contain a cloning site and a gene for selection following transformation into the host, which is typically the yeast *Saccharomyces cerevisiae.* **Figure 11.22** shows a diagram of a YAC vector into which foreign DNA has been cloned. YAC vectors are themselves only about 10 kbp, but they can have 200–800 kbp of cloned DNA inserted.

After it is confirmed that a particular fragment of DNA has been cloned into a BAC or a YAC, this segment can be subcloned into a plasmid or bacteriophage vector for more detailed analysis or sequencing. Although YACs can hold more DNA than BACs, there is a greater problem with recombination and rearrangement of the cloned DNA within yeast than within *E. coli.* For this reason BACs are now more widely used in genomic cloning than are YACs.

Human artificial chromosomes (HACs) have also been developed and are similar to YACs in overall structure. Oddly enough, either linear DNA molecules with telomeres or circular DNA will both replicate in cultured human cells, so most HACs are actually circular. They are constructed by adding segments of human DNA to BACs. The critical issue is to provide an efficient centromere. In mammals the centromere consists of long stretches of repeated sequences (↩ Section 7.6), and HACs must have long arrays of these repeats to be inherited stably. It is hoped that HACs will find future use in gene therapy to carry large human genes with their natural regulatory sequences.

MiniQuiz

- What do the acronyms BAC and YAC stand for?
- Compare the capacity for cloning foreign DNA in M13, lambda, BACs, and YACs.
- The yeast artificial chromosome behaves like a chromosome in a yeast cell. What makes this possible?

Figure 11.22 A yeast artificial chromosome containing foreign DNA. The foreign DNA was cloned into the vector at a *Not*I restriction site. The telomeres are labeled TEL and the centromere CEN. The origin of replication is labeled ARS (for autonomous *replication sequence*). The *URA3* gene is used for selection. The host into which the clone is transformed has a mutation in *URA3* and requires uracil for growth (Ura⁻). Host cells containing this YAC become Ura⁺. The diagram is not to scale; vector DNA is only 10 kbp whereas cloned DNA can be up to 800 kbp.

Big Ideas

11.1

Restriction enzymes recognize specific short sequences in DNA and make cuts in the DNA. The products of restriction enzyme digestion can be separated using gel electrophoresis.

11.2

Complementary nucleic acid sequences may be detected by hybridization. Probes composed of single-stranded DNA or RNA and labeled with radioactivity or a fluorescent dye are hybridized to target DNA or RNA sequences.

11.3

The isolation of a specific gene or region of a chromosome by molecular cloning is done using a plasmid or virus as the cloning vector. Restriction enzymes and DNA ligase are used *in vitro* to produce a chimeric DNA molecule composed of DNA from two or more sources. Once introduced into a suitable host, the cloned DNA can be produced in large amounts under the control of the cloning vector. Identification of cloned genes is performed by a range of molecular techniques.

11.4

Synthetic DNA molecules of desired sequence can be made *in vitro* and used to construct a mutated gene directly or to change specific base pairs within a gene by site-directed mutagenesis. Also, genes can be disrupted by inserting DNA fragments, called cassettes, into them, generating knockout mutants.

11.5

Reporter genes are genes whose products, such as β-galactosidase or GFP, are easy to assay or detect. They are used to simplify and increase the speed of genetic analysis. In gene fusions, segments from two different genes, one of which is usually a reporter gene, are spliced together.

11.6

Plasmids are useful cloning vectors because they are easy to isolate and purify and are often able to multiply to high copy numbers in bacterial cells. Antibiotic resistance genes on the plasmid are used to select bacterial cells containing the plasmid, and color-screening systems are used to identify colonies containing cloned DNA.

11.7

The choice of a cloning host depends on the final application. In many cases the host can be a prokaryote, but in others, it is essential that the host be a eukaryote. Any host must be able to take up DNA, and there are a variety of techniques by which this can be accomplished, both natural and artificial.

11.8

Many cloned genes are not expressed efficiently in a foreign host. Expression vectors have been developed for prokaryotic and eukaryotic hosts that contain genes or regulatory sequences that both increase transcription of the cloned gene and control the level of transcription. Signals to improve the efficiency of translation may also be present in the expression vector.

11.9

Bacteriophages such as lambda have been modified to make useful cloning vectors. Larger amounts of foreign DNA can be cloned with lambda than with many plasmids. In addition, the recombinant DNA can be packaged *in vitro* for efficient transfer to a host cell. Plasmid vectors containing the lambda *cos* sites are called cosmids, and they can carry large fragments of foreign DNA.

11.10

Specialized cloning vectors have been constructed for the sequencing and assembly of genomes. Some, like the M13 derivatives, are useful for both cloning and for direct DNA sequencing. Others, like artificial chromosomes, are useful for cloning very large fragments of DNA, fragments approaching a megabase in size.

Review of Key Terms

Artificial chromosome a single copy vector that can carry extremely long inserts of DNA and is widely used for cloning segments of large genomes

Bacterial artificial chromosome (BAC) a circular artificial chromosome with bacterial origin of replication

Cassette mutagenesis creating mutations by the insertion of a DNA cassette

DNA cassette an artificially designed segment of DNA that usually carries a gene for resistance to an antibiotic or some other convenient marker and is flanked by convenient restriction sites

DNA library (also called a gene library) a collection of cloned DNA segments that is big enough to contain at least one copy of every gene from a particular organism

Expression vector a cloning vector that contains the necessary regulatory sequences to allow transcription and translation of cloned genes

Gel electrophoresis a technique for separation of nucleic acid molecules by passing an electric current through a gel made of agarose or polyacrylamide

Gene disruption (also called gene knockout) the inactivation of a gene by insertion of a DNA fragment that interrupts the coding sequence

Gene fusion a structure created by joining together segments of two separate genes, in particular when the regulatory region of one gene is joined to the coding region of a reporter gene. Gene fusions include protein fusions and operon fusions

Genetic engineering the use of *in vitro* techniques in the isolation, alteration, and expression of DNA or RNA and in the development of genetically modified organisms

Green fluorescent protein (GFP) a protein that glows green and is widely used in genetic analysis

Human artificial chromosome (HAC) an artificial chromosome with human centromere sequence array

Hybridization the formation of a double helix by the base pairing of single strands of DNA or RNA from two different (but related) sources

Integrating vector a cloning vector that can be inserted into a host chromosome

Modification enzyme an enzyme that chemically alters bases within a restriction enzyme recognition site and thus prevents the site from being cut

Molecular cloning the isolation and incorporation of a fragment of DNA into a vector where it can be replicated

Northern blot a hybridization procedure where RNA is the target and DNA or RNA is the probe

Nucleic acid probe a strand of nucleic acid that can be labeled and used to hybridize to a complementary molecule from a mixture of other nucleic acids

Operon fusion a gene fusion in which a coding sequence that retains its own translational signals is fused to the transcriptional signals of another gene

Protein fusion a gene fusion in which two coding sequences are fused so that they share the same transcriptional and translational start sites

Reporter gene a gene used in genetic analysis because the product it encodes is easy to detect

Restriction enzyme an enzyme that recognizes a specific DNA sequence and then cuts the DNA; also known as a restriction endonuclease

Restriction map a map showing the location of restriction enzyme cut sites on a segment of DNA

Shotgun cloning making a gene library by random cloning of DNA fragments

Shuttle vector a cloning vector that can replicate in two or more dissimilar hosts

Site-directed mutagenesis construction of a gene with a specific mutation *in vitro*

Southern blot a hybridization procedure where DNA is the target and RNA or DNA is the probe

Vector (as in cloning vector) a self-replicating DNA molecule that is used to carry cloned genes or other DNA segments for genetic engineering

Yeast artificial chromosome (YAC) an artificial chromosome with yeast origin of replication and CEN sequence

Review Questions

1. What are restriction enzymes? Briefly describe the mechanism of restriction enzymes (Section 11.1).

2. How could you detect a colony containing a cloned gene if you already knew the sequence of the gene (Section 11.2)?

3. Genetic engineering depends on vectors. Describe the properties needed in a well-designed plasmid cloning vector (Section 11.3).

4. How could you detect a colony containing a cloned gene if you did not know the gene sequence but had available purified protein encoded by the gene (Section 11.3)?

5. What are the major uses for artificially synthesized DNA (Section 11.4)?

6. What is site-directed mutagenesis? Why is it a useful tool to study protein activity (Section 11.4)?

7. What is a reporter gene? Describe two widely used reporter genes (Section 11.5).

8. How are gene fusions used to investigate gene regulation (Section 11.5)?

9. How does the insertional inactivation of β-galactosidase allow the presence of foreign DNA in a plasmid vector such as pUC19 to be detected (Section 11.6)?

10. Describe two prokaryotic cloning hosts and the beneficial and detrimental features of each (Section 11.7).

11. What is the difference between a cloning vector and an expression vector (Sections 11.6 and 11.8)?

12. How has bacteriophage T7 been used in expressing foreign genes in *Escherichia coli,* and what desirable features does this regulatory system possess (Section 11.8)?

13. What advantages are there to using a lambda-based cloning vector rather than a plasmid vector (Section 11.9)?

14. What are the essential characteristics of an artificial chromosome? What is the difference between a BAC and a YAC? What characteristics of the F plasmid make it less useful *in vitro* (Section 11.10)?

Application Questions

1. Suppose you are given the task of constructing a plasmid expression vector suitable for molecular cloning in an organism of industrial interest. List the characteristics such a plasmid should have. List the steps you would use to create such a plasmid.

2. Suppose you have just determined the DNA base sequence for an especially strong promoter in *Escherichia coli* and you are interested in incorporating this sequence into an expression vector.

Describe the steps you would use. What precautions are necessary to be sure that this promoter actually works as expected in its new location?

3. Many genetic systems use the *lacZ* gene encoding β-galactosidase as a reporter. What advantages or problems would there be if (a) luciferase or (b) green fluorescent protein were used instead of β-galactosidase as reporters?

Need more practice? Test your understanding with Quantitative Questions; access additional study tools including tutorials, animations, and videos; and then test your knowledge with chapter quizzes and practice tests at **www.microbiologyplace.com**.

Microbial Genomics

The genomes of all strains of *Escherichia coli* are not identical; in addition to genes universally present in this common enteric bacterium, pathogenic strains contain genes that encode toxic proteins that are absent from harmless strains.

An organism's **genome** is its entire complement of genetic information, including the genes, their regulatory sequences, and noncoding DNA. Knowledge of the genome sequence of an organism reveals not only its genes, but also yields important clues to how the organism functions and its evolutionary history. Analyses of genomes make up the field of genomics and include the study of gene expression and the transcription and translation of the genetic information at a genome-wide level. The traditional approach to studying gene expression was to focus on a single gene or group of related genes. Today, the expression of all or most of an organism's genes can be examined in a single experiment.

Advances in genomics rely heavily on improvements in molecular technology. Major advances include shotgun cloning, the automation of DNA sequencing, and the development of powerful computational methods for analysis of DNA and protein sequences. Constant new advances have driven down the cost and increased the speed at which genomes are analyzed. Here we focus on microbial genomes, some techniques used to analyze these genomes, and what microbial genomics has revealed thus far.

I Genomes and Genomics

The word **genomics** refers to the discipline of mapping, sequencing, analyzing, and comparing genomes. At present sequencing of more than 2000 genomes of prokaryotes has either been completed or is in progress. Because new advances in DNA sequencing are introduced quite frequently, it is likely that the number of sequenced prokaryotic genomes will continue to grow rapidly. Here we explore what analysis of these genomes tells us.

12.1 Introduction to Genomics

The first genome sequenced was the 3569-nucleotide RNA genome of the virus MS2 (Section 21.1) in 1976. The first DNA genome sequenced was the 5386-nucleotide sequence of a small, single-stranded DNA virus, φX174 (Section 21.2), in 1977. This feat, accomplished by a group led by the British scientist Fred Sanger, introduced the dideoxy technique for DNA sequencing (Section 12.2). The first cellular genome sequenced was the 1,830,137-base-pair (bp) chromosome of *Haemophilus influenzae* that was published in 1995 by Hamilton O. Smith, J. Craig Venter, and their colleagues at The Institute for Genomic Research in Maryland.

In addition, we now have genome sequences of several animals, including the haploid human genome, which contains about 3 billion bp but only around 25,000 protein-coding genes. The largest genomes so far sequenced, in terms of the number of genes, are those of the plants rice and black cottonwood (a species of poplar) and of the protozoans *Paramecium* and *Trichomonas*, all of which have many more genes than humans. Information from genome sequences has provided new insight on topics as diverse as clinical medicine and microbial evolution.

The DNA sequences of many prokaryotic genomes are now available in public databases (for an up-to-date list of genome sequencing projects search http://www.genomesonline.org/),

and several thousand prokaryotic genomes have either been sequenced or are in progress. **Table 12.1** lists representative examples. These include many species of *Bacteria* and *Archaea* and both circular and linear genomes. Although rare, linear chromosomes are present in several *Bacteria*, including *Borrelia burgdorferi*, the causative agent of Lyme disease, and the important antibiotic-producing genus *Streptomyces*.

The list of prokaryotic genomes in Table 12.1 contains several pathogens. Naturally, such organisms have high priority for sequencing. The hyperthermophiles on the list may have important uses in biotechnology because the enzymes in these organisms are heat-stable. Indeed, the needs of the biomedical and biotechnology industries have greatly affected the choice of organisms to sequence. The list in Table 12.1 also includes organisms such as *Bacillus subtilis*, *Escherichia coli*, and *Pseudomonas aeruginosa* that are widely studied genetic model systems. Genome sequencing has now become so routine that sequencing projects are no longer driven so much by medical or biotechnological rationales. In several cases the genomes of several different strains of the same bacterium have been sequenced in order to reveal the extent of genetic variability within a species.

MiniQuiz
- How many genes are in the human genome?
- Name some organisms whose genomes are larger than the human genome.

12.2 Sequencing and Annotating Genomes

The term **sequencing** refers to determining the precise order of subunits in a polymer. In the case of DNA (or RNA) the sequence is the order of the nucleotides. In sequencing nucleic acids, short fragments of DNA with defined sequences, called *oligonucleotides* (typically 10–20 nucleotides), are artificially synthesized. These DNA molecules are used as **primers**. Primers are short segments of DNA or RNA that initiate the synthesis of new strands of nucleic acid. During DNA replication *in vivo*, RNA primers are used (Section 6.8), but in biotechnology DNA primers are used because they are more stable than RNA primers. Nucleic acid sequencing is a key tool for molecular biology, and we describe the essentials here.

DNA Sequencing: The Sanger Dideoxy Method

Much DNA sequencing today is still done by the dideoxy method invented by the British scientist Fred Sanger, who won a Nobel Prize for this important technique. This method generates DNA fragments of different lengths that are labeled with radioactivity or a fluorescent dye. There are fragments that terminate with each of the four bases—adenine, guanine, cytosine, and thymine. These fragments are separated by gel electrophoresis such that molecules that differ by only one nucleotide in length are clearly resolved. The procedure requires four separate reactions (and four separate gel lanes) for each sequence determination, one for fragments ending with each of the four bases. The positions of

Table 12.1 *Select prokaryotic genomes*[a]

Organism	Cell type[b]	Size (base pairs)	ORFs[c]	Comments
Bacteria				
Hodgkinia cicadicola	E	143,795	169	Degenerate aphid endosymbiont
Carsonella ruddii	E	159,662	182	Degenerate aphid endosymbiont
Buchnera aphidicola BCc	E	422,434	362	Primary aphid endosymbiont
Mycoplasma genitalium	P	580,070	470	Smallest nonsymbiotic bacterial genome
Borrelia burgdorferi	P	910,725	853	Spirochete, linear chromosome, causes Lyme disease
Chlamydia trachomatis	P	1,042,519	894	Obligate intracellular parasite, common human pathogen
Rickettsia prowazekii	P	1,111,523	834	Obligate intracellular parasite, causes epidemic typhus
Treponema pallidum	P	1,138,006	1041	Spirochete, causes syphilis
OM43 clade, strain HTCC2181	FL	1,304,428	1354	Marine methylotroph, smallest free-living genome
Pelagibacter ubique	FL	1,308,759	1354	Marine heterotroph
Aquifex aeolicus	FL	1,551,335	1544	Hyperthermophile, autotroph
Prochlorococcus marinus	FL	1,657,990	1716	Most abundant marine oxygenic phototroph
Streptococcus pyogenes	FL	1,852,442	1752	Causes strep throat and scarlet fever
Thermotoga maritima	FL	1,860,725	1877	Hyperthermophile
Chlorobaculum tepidum	FL	2,154,946	2288	Model green phototrophic bacterium
Neisseria gonorrhoeae NCCP11945	FL	2,232,025	2662	Causes gonorrhoea
Deinococcus radiodurans	FL	3,284,156	2185	Radiation resistant, multiple chromosomes
Synechocystis sp.	FL	3,573,470	3168	Model cyanobacterium
Bdellovibrio bacteriovorus	FL	3,782,950	3584	Predator of other prokaryotes
Caulobacter crescentus	FL	4,016,942	3767	Complex life cycle
Bacillus subtilis	FL	4,214,810	4100	Gram-positive genetic model
Mycobacterium tuberculosis	P	4,411,529	3924	Causes tuberculosis
Escherichia coli K-12	FL	4,639,221	4288	Gram-negative genetic model
Escherichia coli O157:H7	FL	5,594,477	5361	Enteropathogenic strain of *E. coli*
Bacillus anthracis	FL	5,227,293	5738	Pathogen, biowarfare agent
Rhodopseudomonas palustris	FL	5,459,213	4836	Metabolically versatile anoxygenic phototroph
Pseudomonas aeruginosa	FL	6,264,403	5570	Metabolically versatile opportunistic pathogen
Streptomyces coelicolor	FL	8,667,507	7825	Linear chromosome, produces antibiotics
Bradyrhizobium japonicum	FL	9,105,828	8317	Nitrogen fixation, nodulates soybeans
Sorangium cellulosum	FL	13,033,799	9367	Myxobacterium that forms multicellular fruiting bodies
Archaea				
Nanoarchaeum equitans	P	490,885	552	Smallest nonsymbiotic cellular genome
Thermoplasma acidophilum	FL	1,564,905	1509	Thermophile, acidophile
Methanocaldococcus jannaschii	FL	1,664,976	1738	Methanogen, hyperthermophile
Aeropyrum pernix	FL	1,669,695	1841	Hyperthermophile
Pyrococcus horikoshii	FL	1,738,505	2061	Hyperthermophile
Methanothermobacter thermautotrophicus	FL	1,751,377	1855	Methanogen
Archaeoglobus fulgidus	FL	2,178,400	2436	Hyperthermophile
Halobacterium salinarum	FL	2,571,010	2630	Extreme halophile, bacteriorhodopsin
Sulfolobus solfataricus	FL	2,992,245	2977	Hyperthermophile, sulfur chemolithotroph
Haloarcula marismortui	FL	4,274,642	4242	Extreme halophile, bacteriorhodopsin
Methanosarcina acetivorans	FL	5,751,000	4252	Acetotrophic methanogen

[a]Information on these and hundreds of other prokaryotic genomes can be found at **http://cmr.jcvi.org/cgi-bin/CMR/shared/ Genomes.cgi**, a website maintained by The J. Craig Venter Institute, Rockville, MD, a not-for-profit research institute, and at **http://www.genomesonline.org**. Links are listed there to other relevant websites.

[b]E, endosymbiont; P, parasite; FL, free-living. Parasitic species are shaded darker.

[c]Open reading frames. The purpose of reporting ORFs is to predict the total number of proteins that an organism might encode. Of course, genes encoding known proteins are included, as are all ORFs that could encode a protein greater than 100 amino acid residues. Smaller ORFs are typically not included unless they show similarity to a gene from another organism or unless the codon bias is typical of the organism being studied.

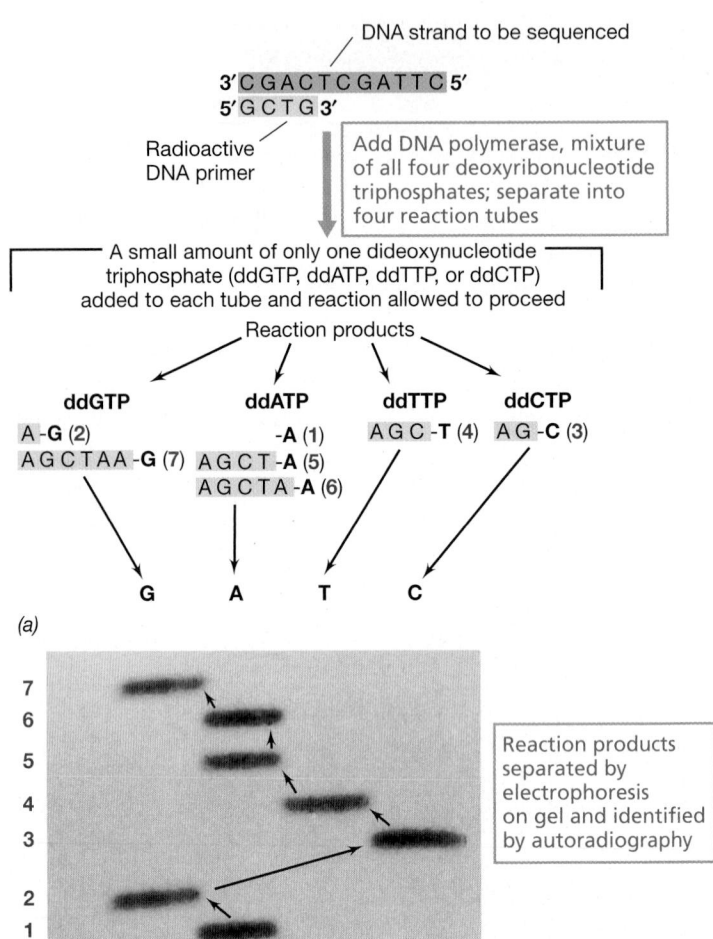

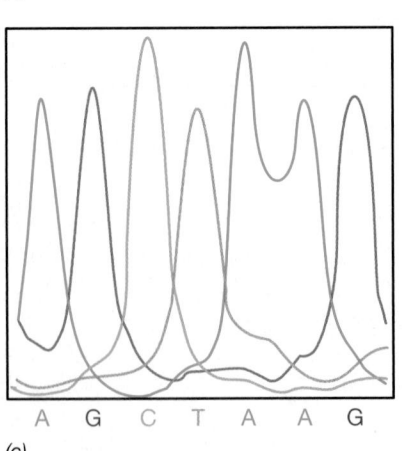

Figure 12.1 Dideoxynucleotides and Sanger sequencing. *(a)* A normal deoxyribonucleotide has a hydroxyl group on the 3′-carbon and a dideoxynucleotide does not. *(b)* Elongation of the chain terminates where a dideoxynucleotide is incorporated. Compare this figure with ↺↺ Figure 6.14.

the fragments are visualized by autoradiography (the darkening of photographic film by exposure to radiation) or fluorescence, and the sequences can then be read off the gel.

In the Sanger procedure the sequence is actually determined by making a copy of the single-stranded DNA using the enzyme DNA polymerase. As we have previously seen (↺↺ Section 6.9), this enzyme uses deoxyribonucleoside triphosphates as substrates and adds them to a growing chain. In each of the four incubation mixtures are small amounts of a different *dideoxy analog* of the deoxyribonucleoside triphosphates (**Figure 12.1**). Because the dideoxy sugar analog lacks the 3′-hydroxyl, elongation of the chain cannot continue after it has been inserted. The dideoxy analog thus acts as a specific chain-termination reagent. Oligonucleotide fragments of variable length are obtained, depending on the incubation conditions. Electrophoresis of these fragments is then carried out, and the positions of the bands are determined by exposing X-ray film or by fluorescence. By aligning the four dideoxynucleotide lanes and noting the vertical position of each fragment relative to its neighbor, the sequence of the DNA copy can be read directly from the gel (**Figure 12.2**).

The Sanger method can be used to sequence RNA as well as DNA. To sequence RNA, a single-stranded DNA copy is first made (using the RNA as the template) by the enzyme reverse transcriptase (↺↺ Section 9.12). The single-stranded DNA is then sequenced by the Sanger dideoxy method as described

Figure 12.2 DNA sequencing using the Sanger method. *(a)* Note that four different reactions must be run, one with each dideoxynucleotide. Because these reactions are run *in vitro*, the primer for DNA synthesis need not be RNA, and for convenience is DNA. *(b)* A portion of a gel containing the reaction products from part a. *(c)* Results of sequencing the same DNA as shown in parts a and b, but using an automated sequencer and fluorescent labels. Pyrosequencing (see text) is even faster than automated fluorescent sequencing and is used for most large-scale genomics sequencing.

above, and the RNA sequence is deduced by base-pairing rules from the DNA sequence.

Automated Sequencing

The demands of large-scale sequencing projects have led to the development of automated DNA-sequencing systems. With such systems, the sequencing reactions are still based on the dideoxy method, but fluorescent dye-labeled primers (or nucleotides) are used instead of radioactivity. The products are separated by automated electrophoresis and the bands detected by fluorescence spectroscopy. Each of the four different reactions uses a fluorescent label of a different color, and the lanes are scanned by a fluorescence-detecting laser. This allows all four reactions to be run on a single lane. The results are analyzed by computer and a sequence generated, with each of the four bases being distinguished by a separate color (Figure 12.2c).

454 Pyrosequencing

Recent technical advances have revolutionized DNA sequencing. In particular, the system developed by the 454 Life Sciences corporation can generate sequence data 100 times faster than previous methods. The 454 system relies on two major advances, *massively parallel liquid handling* and *pyrosequencing* instead of dideoxy sequencing.

In the 454 system the DNA sample is broken into single-stranded segments of about 100 bases each, and each fragment is attached to a microscopic bead. The DNA is amplified by the polymerase chain reaction (PCR; ↩ Section 6.11), resulting in each bead carrying several identical copies of the DNA strand. The beads are then put into a fiber-optic plate with over a million wells, each of which holds just one bead, and the four nucleotides are flowed sequentially over the plate in a fixed order; sequencing reactions then occur simultaneously across the entire plate.

Like Sanger sequencing, 454 sequencing involves synthesis of a complementary strand by DNA polymerase. However, instead of chain termination, each time a nucleotide is incorporated into the complementary strand in the 454 method, a molecule of pyrophosphate is released, which provides the energy needed for the release of light by the enzyme luciferase (↩ Section 11.5), also incorporated into the system. Each of the four possible deoxyribonucleoside triphosphates is tested in turn at each position in the growing chain. The one that yields a light pulse identifies which base was inserted. By base-pairing rules, the identity of the nucleotide in the sample DNA at this site is also revealed.

Although 454 sequencing is remarkably fast because the sample DNA in each well is being analyzed at the same time (thus the term "massively parallel"), it can only handle relatively short stretches of DNA. However, using computational analyses of sequence overlaps from different wells, entire genomes can be pieced together by the 454 system. It has been estimated that 454 sequencing will eventually be able to sequence an entire human genome in just a few days at costs that could make the procedure useful for medical diagnostics and related genetic analyses.

Several new "third generation" sequencing schemes that are even faster than the 454 system are presently in development. However, no single one of these has yet emerged as the standard method.

Assembling Genome Sequences

The analysis of a genome begins with the formation of a genomic library—the molecular cloning of DNA fragments that cover the entire genome (↩ Section 11.3). Virtually all genomic sequencing projects today employ **shotgun sequencing**. This technique relies on high-throughput sequencing, robotics, and powerful computational capacities. Shotgun sequencing has been used for sequencing genomes of both prokaryotes and eukaryotes, including the privately funded version of the human genome project.

In the whole genome shotgun approach, the entire genome, cleaved into fragments, is cloned. At this point the order and the orientation of the DNA fragments are unknown. The sequences are analyzed by a computer that searches for overlapping sequences. This allows the computer to assemble the sequenced fragments in the correct order. Much of the sequencing in the shotgun method is redundant. To ensure full coverage of a genome it is necessary to sequence a very large number of clones, many of which are identical or nearly identical. Typically, 7–10 replicate sequences are obtained for any given part of the genome. This greatly reduces errors because the redundancy in sequencing allows for a consensus nucleotide to be selected at any ambiguous point in the sequence.

For shotgun sequencing to work effectively, the cloning itself must be efficient (many clones are needed) and, as far as possible, the cloned DNA fragments should be randomly generated. Restriction sites are not random, but approximate randomization is achieved by using an enzyme that recognizes a short common sequence. Truly random fragments of DNA can be obtained by forcing the DNA through a nebulizer, which mechanically shears it. This device has a small opening nozzle that reduces the DNA solution to a spray; in the process, the DNA is sheared. The DNA fragments can be purified by size using gel electrophoresis (↩ Section 11.1) before cloning and sequencing.

Assembly and Annotation

Genome *assembly* consists of putting the fragments in the correct order and eliminating overlaps. Assembly generates a genome suitable for *annotation*, the process of identifying genes and other functional regions.

Sometimes shotgun sequencing and assembly does not yield a complete genome sequence and gaps are left in the sequence. In such situations, individual clones can be sought to cover the gap. One method of doing this is to perform PCR using primers complementary to the known sequences that flank a given gap. Note that these additional clones are targeted, not random, as in the shotgun method.

Some genome projects have the goal of obtaining a *closed genome*, meaning that the entire genome sequence is determined. Other projects stop at the *draft stage*, dispensing with sequencing the small gaps. Because shotgun sequencing and assembly are heavily automated procedures, but gap closure is not, a closed genome is much more expensive and time consuming to generate than a draft genome sequence.

After sequencing and assembly, the next step is genome *annotation*, the conversion of raw sequence data into a list of the genes present and other functional sequences in the genome.

Most genes encode proteins, and in most microbial genomes, the great majority of the genome consists of coding sequences. Because the genomes of microbial eukaryotes typically have fewer introns (⮌ Section 7.5) than plant and animal genomes, and prokaryotes have almost none, microbial genomes essentially consist of hundreds to thousands of open reading frames separated by short regulatory regions and transcriptional terminators. Recall that an **open reading frame (ORF)** is a sequence of DNA or RNA that could be translated to give a polypeptide.

How Does the Computer Find an ORF?

A *functional ORF* is one that actually encodes a protein. The simplest way to locate potential protein-encoding genes is to have a computer search the genome sequence for ORFs. Ribosomes establish a reading frame by initiating translation at a start codon, usually an AUG. The ribosome then proceeds until it reaches an in-frame stop codon (⮌ Section 6.17). Therefore, the first step in finding an ORF is to look for start and stop codons in the sequence. However, in-frame start and stop codons appear randomly with reasonable frequency. Thus, other clues are sought. One hint that an ORF is functional is its size. Most proteins contain 100 or more amino acids, so most functional ORFs are longer than 300 nucleotides (100 codons). However, simply programming the computer to ignore ORFs shorter than 100 codons will miss some genuine but short genes.

Thus, other factors such as **codon bias** (codon usage) must be considered. More than one codon exists for most of the 20 amino acids (⮌ Table 6.5), and some codons are used more frequently than others. Codon bias differs between organisms. If the codon bias in a given ORF is considerably different from the consensus codon bias, that ORF may not be functional or may be functional but obtained by horizontal gene transfer (Section 12.11). Furthermore, prokaryotic ribosomes start translation, not at the first possible start codon, but at one immediately downstream of a Shine–Dalgarno (ribosome-binding) sequence on the mRNA (⮌ Section 6.19). Therefore, searching the DNA sequence of a prokaryotic genome for potential Shine–Dalgarno sequences can help decide both whether an ORF is functional and which start codon is actually used.

Although any given gene is always transcribed from a single strand, in all but the smallest plasmid or viral genomes, both strands are transcribed in some part of the genome. Thus, computer inspection of both strands of DNA is required. In addition, some genes encode tRNAs and rRNAs that are not recognized by programs that search only for ORFs. However, genes for tRNAs and rRNAs can usually be located easily because the sequences of these RNAs are highly conserved (⮌ Section 16.9).

An ORF is more likely to be functional if its sequence is similar to the sequences of ORFs in the genomes of other organisms (regardless of whether they encode known proteins) or if some part of the ORF has a sequence known to encode a protein functional domain. This is because proteins with similar functions in different cells tend to be *homologous*; that is, they are related in an evolutionary sense and typically share sequence and structural features. Computers are used to search for such sequence similarities in databases such as GenBank. This database contains over 100 billion base pairs of information and can be found online at http://www.ncbi.nlm.nih.gov/Genbank/index.html. The most widely used database search tool is BLAST (Basic Local Alignment Search Tool), which has several variants depending on whether nucleic acid or protein sequences are used for searching. For example, the tool BLASTn searches nucleic acid databases using a nucleotide query, whereas BLASTp searches protein databases using a protein query.

It would be almost impossible to assemble even a very small genome and to locate genes and other important functional DNA sequences without the availability of sophisticated computational tools to handle large databases. Using computers to do analyses of this type is called working *in silico*, a term analogous to the terms *in vivo* and *in vitro* (*in silico* refers to the silicon processor chips present in the computer).

MiniQuiz

- Why are DNA primers necessary for DNA sequencing?
- Why is shotgun sequencing considered to give a very accurate genome sequence?
- What is done during genome assembly?
- What is an open reading frame (ORF)?
- How can protein homology assist in genome annotation?

12.3 Bioinformatic Analyses and Gene Distributions

Following assembly and annotation, genome analysis proceeds to the comparative stage. The issue here is how the genome of one organism compares with that of other organisms. These activities are a major part of the field of **bioinformatics**, the science that applies powerful computational tools to DNA and protein sequences for the purpose of analyzing, storing, and accessing the sequences for comparative purposes.

Size Range of Prokaryotic Genomes

Genomes of both *Bacteria* and *Archaea* show a strong correlation between genome size and open reading frame (ORF) content (**Figure 12.3**). Regardless of the organism, each megabase of prokaryotic DNA encodes about 1000 ORFs. As prokaryotic genomes increase in size, they also increase proportionally in gene number. This contrasts with eukaryotes, in which noncoding DNA may be a large fraction of the genome, especially in organisms with large genomes (see Table 12.3).

Analyzing genomic sequences can shed light on fundamental biological questions. For example, how many genes are necessary for life to exist? For free-living cells, the smallest genomes found so far encode approximately 1400 genes. Free-living *Bacteria* and *Archaea* are known that have genomes in this range (Table 12.1). These organisms are extremely efficient in their use of DNA. They have few or no introns, inteins, or transposons and have very short intergenic spaces. The largest prokaryotic genomes contain over 10,000 genes and are primarily from soil organisms, such as myxobacteria, with complex life cycles (⮌ Section 17.17).

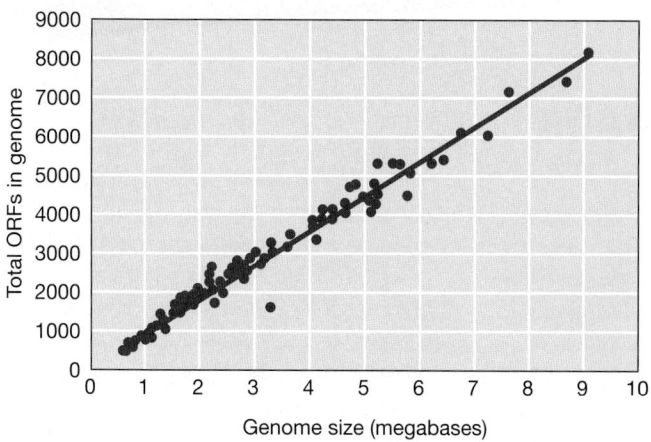

Figure 12.3 Correlation between genome size and ORF content in prokaryotes. These data are from analyses of 115 completed prokaryotic genomes and include species of both *Bacteria* and *Archaea*. Data from *Proc. Natl. Acad. Sci.* (*USA*) 101: 3160–3165 (2004).

Perhaps surprisingly, genomic analyses have shown that autotrophic organisms need only a few more genes than heterotrophs. For example, *Methanocaldococcus jannaschii* (*Archaea*) is an autotroph with only 1738 ORFs. This enables it to be not only free-living, but also to rely on carbon dioxide as its sole carbon source. *Aquifex aeolicus* (*Bacteria*) is also an autotroph and contains the smallest known genome of any autotroph at just 1.5 megabase pairs (Mbp, one million base pairs) (Table 12.1). Both *Methanocaldococcus* and *Aquifex* are also hyperthermophiles, growing optimally at temperatures above 80°C. Thus, a large genome is not necessary to support an extreme lifestyle, including that of autotrophs living in near-boiling water.

Small Genomes

The smallest cellular genomes belong to prokaryotes that are parasitic or endosymbiotic. A disproportionate number of such small genomes have been sequenced, partly because many are of medical importance and partly because smaller genomes are easier to sequence. Genome sizes for prokaryotes that are obligate parasites range from 490 kbp for *Nanoarchaeum equitans* (*Archaea*) to 4400 kbp for *Mycobacterium tuberculosis* (*Bacteria*). The genomes of *N. equitans* and several other bacteria, including *Mycoplasma*, *Chlamydia*, and *Rickettsia*, are smaller than the largest known viral genome, that of *Mimivirus* with 1.2 Mbp (⮑ Chapter 21, Microbial Sidebar).

Excluding endosymbionts, the smallest prokaryotic genome is that of the archaeon *N. equitans*, which is some 90 kbp smaller that that of *Mycoplasma genitalium* (Table 12.1). However, the genome of *N. equitans* actually contains more ORFs than the larger genome of *M. genitalium*. This is because the *N. equitans* genome is extremely compact with almost no noncoding DNA. *N. equitans* is a hyperthermophile and a parasite of a second hyperthermophile, the archaeon *Ignicoccus* (⮑ Section 19.7). Analyses of the gene content of *N. equitans* show it to be free of virtually all genes that encode proteins for anabolism or catabolism.

Using *Mycoplasma*, which has around 500 genes, as a starting point, several investigators have estimated that around 250–300 genes are the minimum for a viable cell. These estimates rely partly on comparisons with other small genomes. In addition, systematic mutagenesis has been performed to identify those *Mycoplasma* genes that are essential. However, these estimates are compromised by the fact that *Mycoplasma* is not free-living and cannot make many vital components, such as purines and pyrimidines. Systematic inactivation of genes in the free-living bacterium *Bacillus subtilis*, which has about 4100 genes, gave similar results. Approximately 270 genes were essential. However, the bacteria were provided with many nutrients, thus allowing them to survive without many genes needed for biosynthesis. Most of these essential genes are present in typical bacteria and approximately 70% are also found in *Archaea* and eukaryotes.

Smaller still than the genomes of prokaryotic parasites are those of some endosymbionts. Endosymbiotic bacteria live inside the cells of their hosts and are relatively common among insects.

Large Genomes

Some prokaryotes have very large genomes that are comparable in size to those of eukaryotic microorganisms. Because eukaryotes tend to have significant amounts of noncoding DNA and prokaryotes do not, some prokaryotic genomes actually have more genes than microbial eukaryotes, despite having less DNA. For example, *Bradyrhizobium japonicum*, which forms nitrogen-fixing root nodules on soybeans, has 9.1 Mbp of DNA and 8300 ORFs, whereas the yeast *Saccharomyces cerevisiae*, a eukaryote, has 12.1 Mbp of DNA but only 5800 ORFs (see Table 12.3). *Myxococcus xanthus* also has 9.1 Mbp of DNA, whereas its relatives in the *Deltaproteobacteria* have genomes approximately half this size (⮑ Section 17.17). It has been suggested that multiple duplication events of substantial segments of DNA happened in the evolutionary history of *M. xanthus*.

The largest prokaryotic genome known at present is that of *Sorangium cellulosum*. In contrast to *Bacteria*, the largest genomes found in *Archaea* thus far are around 5 Mbp (Table 12.1). Overall, prokaryotic genomes thus range in size from those of large viruses to those of eukaryotic microorganisms.

Gene Content of Prokaryotic Genomes

The complement of genes in a particular organism defines its biology. Conversely, genomes are molded by an organism's lifestyle. One might imagine, for instance, that obligate parasites such as *Treponema pallidum* (⮑ Section 18.16) would require relatively few genes for amino acid biosynthesis because the amino acids they need can be supplied by their hosts. This is indeed the case, as the *T. pallidum* genome lacks recognizable genes for amino acid biosynthesis, although genes are found encoding several proteases, enzymes that can convert peptides taken up from the host into free amino acids. In contrast, the free-living bacterium *Escherichia coli* has 131 genes for amino acid biosynthesis and metabolism and the soil bacterium *Bacillus subtilis* has over 200.

The Synthetic Cell: Assembly Details

How many and what parts need to be combined to build a cell? Despite the increasing availability of DNA sequence information, our understanding of the genome remains limited. It took 15 years to create a synthetic self-replicating cell, controlled by a computer-designed genome. In this quest to find the minimal cellular machinery required for life, groundbreaking tools were developed and the concept of the possibility of converting species was proven.

One approach to identifying the minimal genetic instructions required for life is to synthesize reduced genomes and monitor the viability of recipient cells. Such cell design was guided by protein networks modeled for *Mycoplasma pneumoniae,* and we also know that the smallest known genome of a non-endosymbiotic bacterium (*Mycoplasma genitalium* [⮌ Section 18.3], ~580 kbp) that lacks redundancy contains 43 RNA genes and encodes only 382 essential proteins (Figure 1).

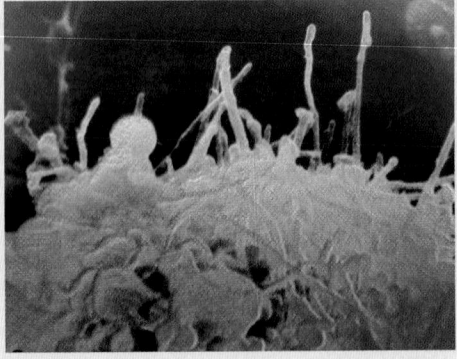

Figure 1 A scanning electron micrograph of *Mycoplasma* bacteria.

How was the "synthetic cell" created in practice?

The task was to synthesize and assemble 1.08 Mbp of a digitized *Mycoplasma mycoides* genome and resequence it to confirm its authenticity. "Watermarks" were included in the sequence to confirm the synthetic genome. For selection, a tetracycline resistance gene was inserted. The insert also disrupts a virulence gene, which makes the organism less of a health risk.

Genome synthesis required a series of assembly steps:

- Oligonucleotides were joined to produce overlapping cassettes of up to 5 kbp.
- Overlapping fragments of 100–144 kbp were assembled in three rounds of *in vitro* recombination, and each fragment was roughly a quarter of the *M. genitalium* genome.
- Fragments were then cloned into cells of *Escherichia coli* as bacterial artificial chromosome (BAC) plasmids (⮌ Section 11.10) and released by restriction.
- Fragments and a yeast centromeric plasmid were cotransformed into the yeast *Saccharomyces cerevisiae* to assemble the genome by homologous recombination (⮌ Section 10.6).

The yeast system has no restriction barriers—in fact, using it allows the combining of two different bacterial chromosomes in one yeast cell by mating. And gene toxicity is minimal, because bacterial genes are rarely expressed in yeast.

The synthetic *M. mycoides* genome was methylated (⮌ Section 11.1) and transplanted into restriction-minus *Mycoplasma capricolum* recipient cells. The recipient genome was lost and viable cells could only grow if they were controlled by the synthetic chromosome. Following this "species conversion," the cells have the *M. mycoides* proteome and phenotype. The next step would be minimizing a synthetic genome. However, the synthetic cell formed as described here was not truly "synthetic life" because the recipient cytoplasm was natural, not synthetic. But as these cells grow, they will at some point consist only of components synthesized using information on the synthetic genome. Creating true synthetic life would require synthesizing all components separately, including cytoplasm trapped within a synthetic cytoplasmic membrane, and then assembling all components of the cell and detecting growth.

Design, synthesis, assembly, transplantation, and propagation of genomes allow rapid progress in synthetic biology. The ethical implications of such dual use technology are reviewed routinely. Synthetic cells allow discovery and characterization of genomes of medically or industrially relevant organisms that are genetically intractable, not yet cultured, or problematic to culture in the laboratory for some reason. But it is clear that bacterial genomes can be engineered in yeast in an approach comparable to standard random mutagenesis but on a larger scale. Thus, more complex recombinant information can be introduced, opening avenues for vaccine development, pharmaceuticals, bioremediation, and other beneficial purposes.

Comparative analyses are useful in searching for genes that encode enzymes that almost certainly exist because of the known properties of an organism. *Thermotoga maritima* (*Bacteria*), for example, is a hyperthermophile found in hot marine sediments, and laboratory studies have shown that it can catabolize a large number of sugars. **Figure 12.4** summarizes some of the metabolic pathways and transport systems of *T. maritima* that have been deduced from analysis of its genome. In the *T. maritima* genome about 7% of the genes encode proteins for the metabolism of simple and complex sugars. As expected, its genome is also rich in

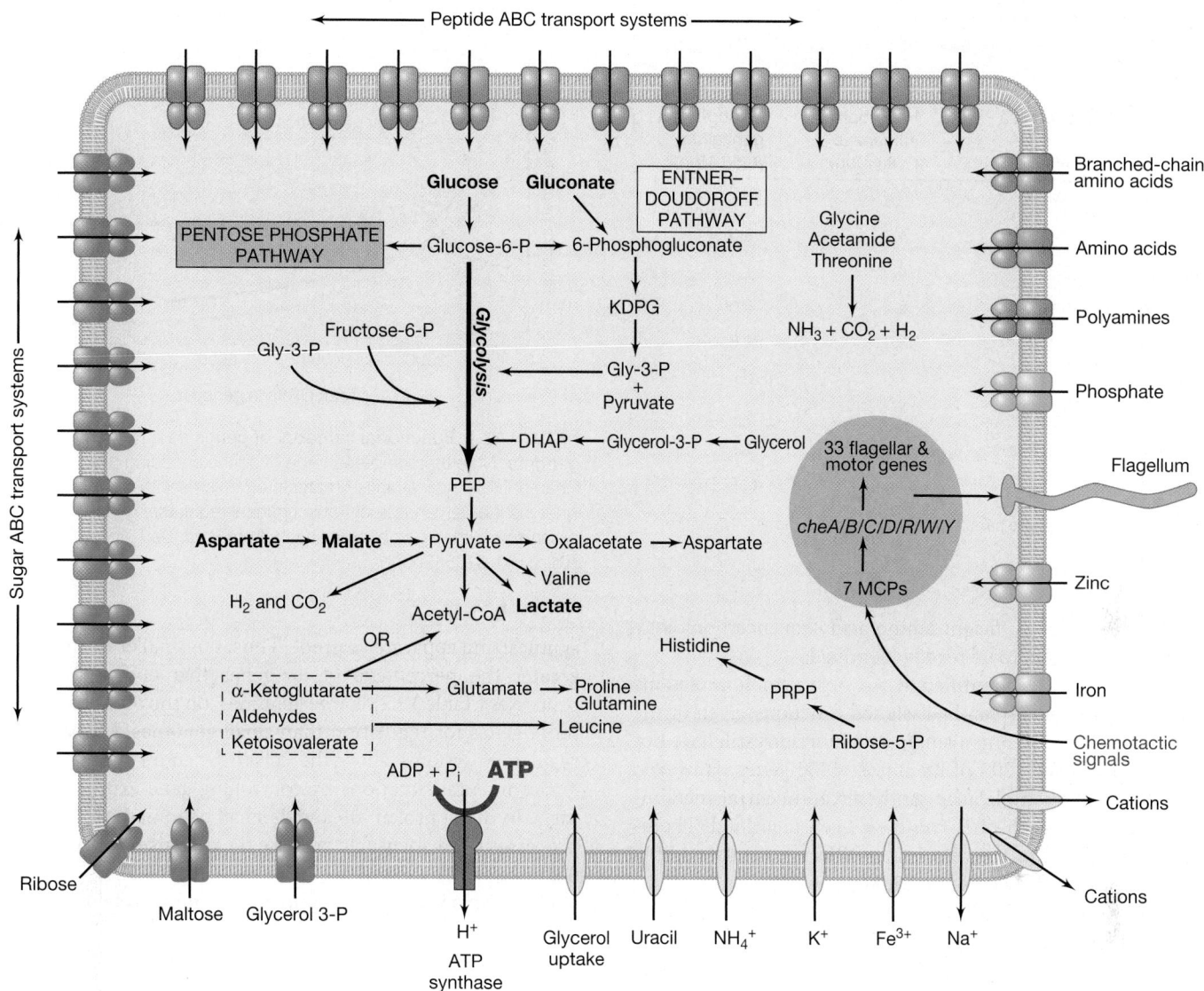

Figure 12.4 Overview of metabolism and transport in *Thermotoga maritima*. The figure outlines the metabolic capabilities of this organism. These include some of the pathways for energy production and the metabolism of organic compounds, including transport proteins that were identified from analysis of the genomic sequence. Gene names are not shown. The genome contains several ABC-type transport systems, 12 for carbohydrates, 14 for peptides and amino acids, and still others for ions. These are shown as multi-subunit structures in the figure. Other types of transport proteins have also been identified and are shown as simple ovals. The flagellum is shown, and this organism has seven transducers (MCPs) and several chemo-taxis (*che*) genes and genes required for flagellar assembly. A few aspects of sugar metabolism are also shown. This figure is adapted from one published by The Institute for Genomic Research.

genes for transport, particularly for carbohydrates and amino acids. All this suggests that *T. maritima* exists in an environment rich in organic material.

A functional analysis of genes and their activities in several prokaryotes is given in **Table 12.2**. Similar data are assembled when each new genome is published. Thus far a distinct pattern has emerged of gene distribution in prokaryotes. Metabolic genes are typically the most abundant class in prokaryotic genomes, although genes for protein synthesis overtake metabolic genes on a percentage basis as genome size decreases (Table 12.2, and see Figure 12.5). Interestingly, as vital as they are, genes for DNA replication and transcription make up only a minor fraction of the typical prokaryotic genome.

In addition to protein-encoding genes, most organisms have a substantial number of genes that encode nontranslated RNA. These include genes for rRNA, which are often present in multiple copies, tRNAs, and small regulatory RNAs (↩ Section 8.14).

Uncharacterized ORFs

Although there are differences among organisms, in most cases the number of genes whose role can be clearly identified in a given genome is 70% or less of the total number of ORFs detected. Uncharacterized ORFs are said to encode *hypothetical proteins*, which probably exist although their function is unknown. Uncharacterized ORFs possess an uninterrupted reading frame of reasonable length and start and stop codons. However, the

Table 12.2 *Gene function in bacterial genomes*

Functional categories	Percentage of genes		
	Escherichia coli (4.64 Mbp)[a]	Haemophilus influenzae (1.83 Mbp)[a]	Mycoplasma genitalium (0.58 Mbp)[a]
Metabolism	21.0	19.0	14.6
Structure	5.5	4.7	3.6
Transport	10.0	7.0	7.3
Regulation	8.5	6.6	6.0
Translation	4.5	8.0	21.6
Transcription	1.3	1.5	2.6
Replication	2.7	4.9	6.8
Other, known	8.5	5.2	5.8
Unknown	38.1	43.0	32.0

[a]Chromosome size, in megabase pairs. Each organism listed contains only a single circular chromosome.

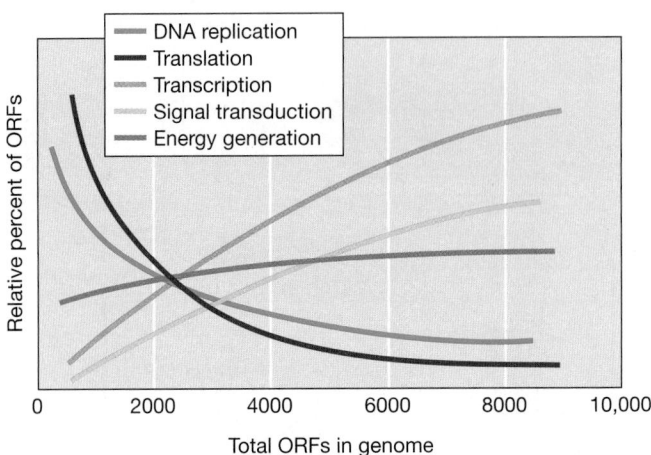

Figure 12.5 **Functional category of genes as a percentage of the genome.** Note that the percentage of genes encoding products for translation or DNA replication is greater in organisms with small genomes, whereas the percentage of transcriptional regulatory genes is greater in organisms with large genomes. Data from *Proc. Natl. Acad. Sci.* (*USA*) 101: 3160–3165 (2004).

encoded proteins lack sufficient amino acid sequence homology with any known protein to be readily identified.

As gene functions are identified in one organism, homologous ORFs in other organisms can be assigned functions. Even in the world's best-understood organism, *E. coli*, functions still have not been assigned to over 1000 of its almost 4300 genes. However, most genes for macromolecular syntheses and central metabolism essential for growth of *E. coli* have been identified. Therefore, as the functions of the remaining ORFs are identified, it is likely that most will encode nonessential proteins.

Many of the unidentified genes in *E. coli* are predicted to encode regulatory or redundant proteins; these might include proteins needed only under special conditions or as "backup" systems for key enzymes. However, it must be remembered that the precise function of many genes, even in well-studied organisms such as *E. coli*, are often unpredictable. Some gene identifications merely assign a given gene to a family or to a general function (such as "transporter"). By contrast, other genes are completely unknown and have only been predicted using bioinformatics. Moreover, some are actually incorrect; it has been estimated that as many as 10% of genes in databases are incorrectly annotated.

Gene Categories as a Function of Genome Size

As the data of Table 12.2 show, the percentage of an organism's genes devoted to one or another cell function is to some degree a function of genome size. This is summarized for a large number of bacterial genomes in **Figure 12.5**. Core cellular processes, such as protein synthesis, DNA replication, and energy production, show only minor variations in gene number with genome size (Figure 12.5). Consequently, the relative percentage of genes devoted to protein synthesis, for example, is large in organisms with small genomes. By contrast, genes that regulate transcription have high relative percentages in organisms with large genomes.

The data summarized in Figure 12.5 suggest that although many genes can be dispensed with, genes that encode the protein-

synthesizing apparatus cannot. Thus, the smaller the genome, the greater the percentage of its genes that encode translational processes (Table 12.2). Large genomes, on the other hand, contain more genes for regulation than small genomes. These additional regulatory systems allow the cell to be more flexible in diverse environmental situations by controlling gene expression accordingly. In many prokaryotes with small genomes, these regulatory processes are dispensable because the organisms are parasitic and obtain much of what they need from their hosts.

Organisms with large genomes can afford to encode many regulatory and specialized metabolic genes. This likely makes these organisms more competitive in their habitats, which, for many prokaryotes with very large genomes, is soil. In soil, carbon and energy sources are often scarce or available only intermittently, and vary greatly (↻ Section 23.6). A large genome with multiple metabolic options would thus be strongly selected for in such a habitat. All of the prokaryotes listed in Table 12.1 whose genomes are in excess of 6 Mbp inhabit soil.

Gene Distribution in *Bacteria* and *Archaea*

Analyses of gene categories have been done on several *Bacteria* and *Archaea* and the results are compared in **Figure 12.6**. Note that these data reflect the average gene content for several separate genomes. On average, species of *Archaea* devote a higher percentage of their genomes to energy and coenzyme production than do *Bacteria* (these results are undoubtedly skewed a bit due to the large number of novel coenzymes produced by methanogenic *Archaea*, ↻ Section 19.3). On the other hand, *Archaea* appear to contain fewer genes for carbohydrate metabolism or cytoplasmic membrane functions, such as transport and membrane biosynthesis, than do *Bacteria*. However, this finding may be skewed because the corresponding pathways are less studied in *Archaea* than in *Bacteria*, and many of the corresponding archaeal genes are probably still unknown.

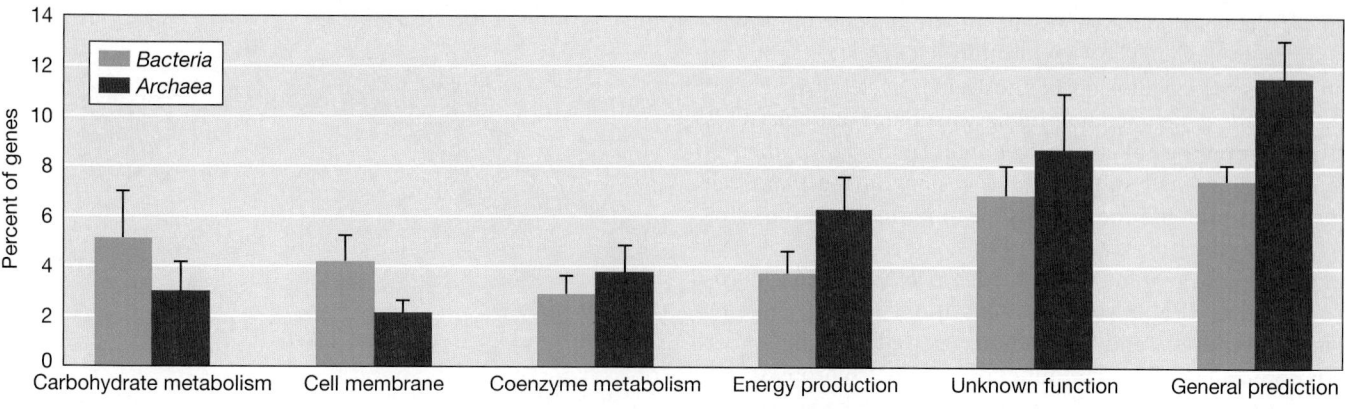

Figure 12.6 Variations in gene category in *Bacteria* and *Archaea*. Data are averages from 34 species of *Bacteria* and 12 species of *Archaea*. "Unknown function" represents genes known to encode proteins whose functions are unknown. Genes labeled as "general prediction" encode hypothetical proteins that may or may not exist. Data from *Proc. Natl. Acad. Sci.* (*USA*) 101: 3160–3165 (2004).

Both *Archaea* and *Bacteria* have relatively large numbers of genes whose functions are unknown or that encode only hypothetical proteins, although more uncertainty exists in both categories among the *Archaea* than the *Bacteria* (Figure 12.6). However, this may be an artifact due to the availability of fewer genome sequences from the *Archaea* than from the *Bacteria*.

MiniQuiz

- What lifestyle is typical of prokaryotic organisms that have genomes smaller than those of certain viruses?
- Approximately how many protein-encoding genes will a bacterial genome of 4 Mbp contain?
- Which organism is likely to have more genes, a prokaryote with 8 Mbp of DNA or a eukaryote with 10 Mbp of DNA? Explain.
- What is a hypothetical protein?
- What category of genes do prokaryotes contain the most of on a percentage basis?

12.4 The Genomes of Eukaryotic Organelles

Eukaryotic cells contain membrane-bound organelles in addition to the nucleus (Sections 20.1–20.5). Two of these organelles, the mitochondrion and the chloroplast, each contain a small genome. In addition, both contain the machinery necessary for protein synthesis, including ribosomes, transfer RNAs, and all the other components necessary for translation and formation of functional proteins. Indeed, these organelles share many traits in common with prokaryotic cells to which they are phylogenetically related (Section 20.4).

The Chloroplast Genome

Green plant cells contain chloroplasts, the organelles that perform photosynthesis. All known chloroplast genomes are circular DNA molecules. In each chloroplast are several identical copies of the genome. The typical chloroplast genome is about 120–160 kilobase pairs (kbp) with two inverted repeats of 6–76 kbp that each contain copies of the three rRNA genes (**Figure 12.7**). Several chloroplast genomes have been completely sequenced, and all are rather similar. The flagellated protozoan *Mesostigma viride* belongs to the earliest diverging green plant lineage. Its chloroplast genome contains more protein-encoding genes (92) and tRNA genes (37) than any other so far known.

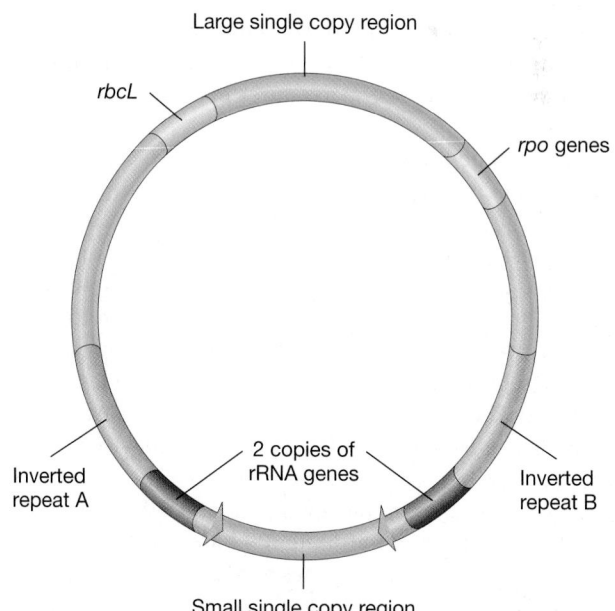

Figure 12.7 **Map of a typical chloroplast genome.** The genomes of chloroplasts are circular double-stranded DNA molecules. Most contain a large single copy region and two inverted repeat regions that flank a small single copy region. The inverted repeats each contain a copy of the three rRNA (5S, 16S, and 23S) genes. The large subunit of RubisCO is encoded by the *rbcL* gene in the large single copy region. The chloroplast also encodes its own RNA polymerase (*rpo* genes).

Many of the chloroplast genes encode proteins for photosynthesis and autotrophy. The enzyme RubisCO catalyzes the key step in fixing carbon dioxide in the Calvin cycle (Section 13.12). The *rbcL* gene encoding the large subunit of RubisCO is always present on the chloroplast genome (Figure 12.7), whereas the gene for the small subunit, *rbcS*, resides in the plant cell nucleus and its protein product must be imported from the cytoplasm into the chloroplast after synthesis.

The chloroplast genome also encodes rRNA used in chloroplast ribosomes, tRNA used in translation, several proteins used in transcription and translation, as well as some other proteins. Some proteins that function in the chloroplast are encoded by nuclear genes. These are thought to be genes that migrated to the nucleus as the chloroplast evolved from an endosymbiont into a photosynthetic organelle. Introns are common in chloroplast genes, and they are primarily of the self-splicing type (Section 7.8).

Analyses of chloroplast genomes firmly support the endosymbiotic hypothesis (Section 20.4). For example, chloroplast genomes contain genes that are homologs of genes in *Escherichia coli*, cyanobacteria, and other *Bacteria*. These include genes encoding proteins for cell division, suggesting that the mechanism of chloroplast division resembles that of bacterial cells. In addition, chloroplast genes for protein transport through membranes are highly related to those of *Bacteria*.

Mitochondrial Genomes

Mitochondria, which are the structures that produce energy by respiration, are found in most eukaryotic organisms. Mitochondrial genomes primarily encode proteins for oxidative phosphorylation and, as do chloroplast genomes, also encode proteins, rRNAs, and tRNAs for protein synthesis. However, most mitochondrial genomes encode many fewer proteins than do those of chloroplasts.

Several hundred mitochondrial genomes have been sequenced. The largest mitochondrial genome has 62 protein-encoding genes, but others encode as few as three proteins. The mitochondria of almost all mammals, including humans, encode only 13 proteins plus 22 tRNAs and 2 rRNAs. **Figure 12.8** shows a map of the 16,569-bp human mitochondrial genome. The mitochondrial genome of the yeast *Saccharomyces cerevisiae* is larger (85,779 bp), but has only 8 protein-encoding genes. Besides the genes encoding the RNA and proteins, the genome of yeast mitochondria contains large stretches of extremely adenine/thymine (AT)-rich DNA that has no apparent function. The mitochondrial genomes of plants are physically much larger than those of animal cells, and range from around 300 kbp to 2000 kbp. Despite this they carry only about 50 genes and contain large amounts of noncoding DNA.

Whereas chloroplasts use the "universal" genetic code, mitochondria use slightly different, simpler genetic codes (Section 6.17). These seem to have arisen from selection pressure for smaller genomes. For example, the 22 tRNAs produced in mitochondria are insufficient to read the universal genetic code, even with wobble pairing taken into consideration. Therefore, base pairing between the anticodon and the codon is even more flexible in mitochondria than it is in cells.

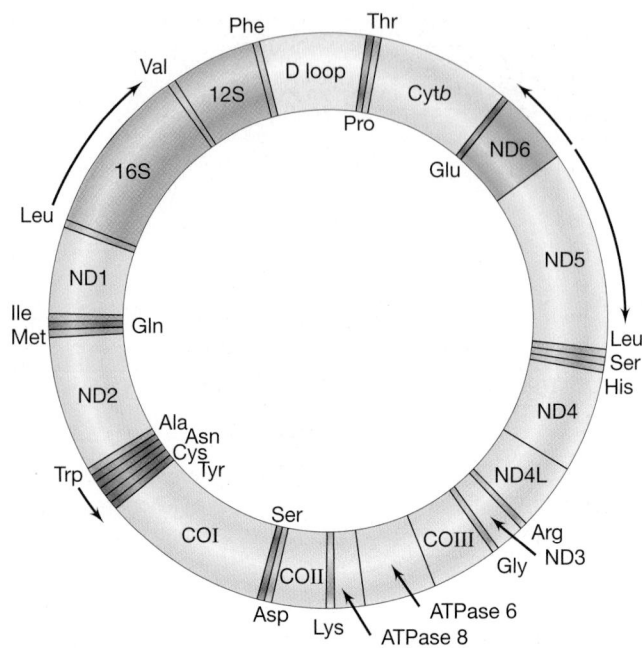

Figure 12.8 Map of the human mitochondrial genome. The circular genome of the human mitochondrion contains 16,569 bp. The genome encodes the 16S and 12S rRNA (corresponding to the prokaryotic 23S and 16S rRNAs) and 22 tRNAs. Genes that are transcribed counterclockwise (CCW) are in dark orange, and those transcribed clockwise (CW) in light orange. The amino acid designations for tRNA genes are on the outside for CCW-transcribed genes and on the inside of the map for CW genes. The 13 protein-encoding genes are shown in green (dark green, transcribed CCW; light green, transcribed CW). Cyt*b*, cytochrome *b*; ND1–6, components of the NADH dehydrogenase complex; COI–III, subunits of the cytochrome oxidase complex; ATPase 6 and 8, polypeptides of the mitochondrial ATPase complex. The two promoters are in the region called the D loop, which is also involved in DNA replication.

Unlike chloroplast genomes, which are all single, circular DNA molecules, the genomes of mitochondria are quite diverse. For example, some mitochondrial genomes are linear, including those of some species of algae, protozoans, and fungi. In other cases, such as in the yeast *S. cerevisiae*, although genetic analyses indicate that the mitochondrial genome is circular, it seems that the physical form is linear. (Recall that bacteriophage T4 has a genetically circular genome although it is physically linear, Section 9.9.) Finally, it should be noted that small plasmids exist in the mitochondria of several organisms, complicating mitochondrial genome analysis.

RNA Editing

RNA editing is the process of altering the base sequence of a messenger RNA *after* transcription. There are two forms of RNA editing. In one, nucleotides are either inserted or deleted. In the other, a base is chemically modified to change its identity. In either case, RNA editing can alter the coding sequence of an mRNA so that the amino acid sequence of the resulting polypeptide differs from that predicted from its gene sequence.

RNA editing is very rare in most organisms, especially animals. RNA editing is more common in the mitochondria and chloroplasts of plants. At specific sites in some mRNAs, a C is converted to a U by oxidative deamination. There are at least 25 sites of C to U conversion in the maize chloroplast. Depending on the location of the editing, a new codon may be formed, leading to an altered protein sequence.

Editing of mRNA by the insertion or deletion of nucleotides occurs in certain protozoa, notably the trypanosomes and their relatives, and more often in mitochondrial genes than in nuclear genes. Some mitochondrial transcripts are edited such that large numbers (hundreds in some cases) of uridines are added or, more rarely, deleted. RNA editing is precisely controlled by short sequences in the mRNA that "guide" editing enzymes. During insertional editing, the sequences in the mRNA are recognized by short guide RNA molecules that are complementary to the mRNA except that they have an extra A. The U residues are inserted into the mRNA opposite the extra A on the guide RNA. Obviously, this process must be very precisely controlled. Inserting too many or too few bases would generate frameshift errors that would yield dysfunctional proteins.

Organelles and the Nuclear Genome

Chloroplasts and mitochondria require many more proteins than they encode. For example, far more proteins are needed for translation in organelles than are encoded by the organelle genome. Thus, many organelle functions are encoded by nuclear genes.

It is estimated that the yeast mitochondrion contains over 400 different proteins; however, only eight of them are encoded by the yeast mitochondrial genome, the remaining proteins being encoded by nuclear genes. Although one might predict that proteins that function in specific processes in the eukaryotic nucleus and cytoplasm could be put to the same use in organelles, this is not the case. Although the genes for many organelle proteins are present in the nucleus, transcribed in the nucleus, and translated on the 80S ribosomes in the eukaryotic cytoplasm, the proteins are used specifically by the organelles and must be transported into them.

The nuclear-encoded proteins required for translation and energy generation in mitochondria are closely related to counterparts in the *Bacteria* rather than to those that function in the eukaryotic cytoplasm. Thus, it initially appeared that most genes encoding mitochondrial proteins had originally been in the genomes of the symbionts and had then been progressively transferred from the mitochondrion to the nucleus during the later stages of endosymbiosis. However, what was needed to confirm such a hypothesis was both the nuclear and mitochondrial genome sequences of a eukaryote, the genome sequence of a species of *Bacteria* phylogenetically closely related to the mitochondrial genome, and genome sequences of other *Bacteria* for comparative purposes. All these requirements were met for the yeast *S. cerevisiae* and certain *Bacteria*, and the analysis has been revealing.

Surprisingly, of the 400 nuclear genes encoding mitochondrial proteins, only about 50 were closely related to the phylogenetic lineage in *Bacteria* that led to mitochondria (the *Alphaproteobacteria*, Section 17.1). Another 150 were clearly related to proteins of *Bacteria*, but not necessarily *Alphaproteobacteria*. These *Bacteria*-like proteins were mostly needed for energy

conversions, translation, and biosynthesis. However, the remaining 200 or so mitochondrial proteins were encoded by genes that have no identifiable homologs among known genes of *Bacteria*. These proteins were mostly required for membranes, regulation, and transport. Thus, although the mitochondrion shows many signs of having originated from endosymbiotic events (Section 16.4), genomic analyses have shown that its genetic history is more complicated than previously thought.

MiniQuiz
- What is unusual about the genes that encode mitochondrial functions in yeast?
- How are genome size and gene content correlated in yeast and human mitochondria?
- What is RNA editing? How does it differ from RNA processing?

12.5 The Genomes of Eukaryotic Microorganisms

The genomes of several microbial and higher eukaryotes have now been sequenced (**Table 12.3**). The genomes of mammals, including human, mouse, and rat, have around 25,000 genes—about twice the number found in insects and four times that of yeast. However, the genomes of higher plants, such as rice and black poplar, contain even more genes, approaching twice that of humans. It is thought that sequencing of corn (maize) and other large plant genomes presently in progress will reveal even higher gene numbers.

Interestingly, certain single-celled protozoans, including *Paramecium* (40,000 genes) and *Trichomonas* (60,000 genes), have significantly more genes than humans do (Table 12.3). Indeed, *Trichomonas* presently holds the record for gene number of any organism. This is puzzling because *Trichomonas* is a human parasite, and as we have seen, such organisms typically have small genomes relative to comparable free-living organisms.

Of single-celled eukaryotes, the yeast *Saccharomyces cerevisiae* is most widely used as a model organism and is also extensively used in industry, and so we focus on it here.

The Yeast Genome

The haploid yeast genome contains 16 chromosomes ranging in size from 220 kbp to about 2352 kbp. The total yeast nuclear genome (excluding the mitochondria and some plasmid and virus-like genetic elements) is approximately 13,392 kbp. Why are the words "about" and "approximately" used to describe this genome when it has been completely sequenced? Yeast, like many other eukaryotes, has a large amount of repetitive DNA (Section 7.5). When the yeast genome was published in 1997, not all of the "identical" repeats had been sequenced. It is difficult to sequence a very long run of identical or nearly identical sequences and then assemble the data into a coherent framework. For example, yeast chromosome XII contains a stretch of approximately 1260 kbp containing 100–200 repeats of yeast rRNA genes. Another repeated sequence follows this long series of rRNA gene repeats. Because of such identical repeats, the sizes of eukaryotic genomes are inevitably only close approximations.

Table 12.3 *Some eukaryotic nuclear genomes*[a]

Organism	Comments	Organism/Cell type[b]	Genome size (Mbp)	Haploid chromosome number	Protein-encoding genes[c]
Nucleomorph of *Bigelowiella natans*	Degenerate endosymbiotic nucleus	E	0.37	3	331
Encephalitozoon cuniculi	Smallest known eukaryotic genome, human pathogen	P	2.9	11	2,000
Cryptosporidium parvum	Parasitic protozoan	P	9.1	8	3,800
Plasmodium falciparum	Malignant malaria	P	23	14	5,300
Saccharomyces cerevisiae	Yeast, a model eukaryote	FL	12.1	16	5,800
Ostreococcus tauri	Marine green alga, smallest free-living eukaryote	FL	12.6	20	8,200
Aspergillus nidulans	Filamentous fungus	FL	30	8	9,500
Giardia lamblia	Flagellated protozoan, causes acute gastroenteritis	P	12	5	9,700
Dictyostelium discoideum	Social amoeba	FL	34	6	12,500
Drosophila melanogaster	Fruit fly, model organism for genetic studies	FL	180	4	13,600
Caenorhabditis elegans	Roundworm, model organism for animal development	FL	97	6	19,100
Arabidopsis thaliana	Model plant for genetic studies	FL	125	5	26,000
Mus musculus	Mouse, a model mammal	FL	2,500	23	25,000
Homo sapiens	Human	FL	2,850	23	25,000
Oryza sativa	Rice, the world's most important crop plant	FL	390	12	38,000
Paramecium tetraurelia	Ciliated protozoan	FL	72	>50	40,000
Populus trichocarpa	Black poplar, a tree	FL	500	19	45,000
Trichomonas vaginalis	Flagellated protozoan, human pathogen	P	160	6	60,000

[a]All data are for the haploid nuclear genomes of these organisms.
[b]E, endosymbiont; P, parasite; FL, free-living.
[c]The number of protein-encoding genes is an estimate based on the number of known genes and sequences that seem likely to encode functional proteins.

In addition to having multiple copies of the rRNA genes, the yeast nuclear genome has approximately 300 genes for tRNAs (only a few are identical) and nearly 100 genes for other types of noncoding RNA. As with other eukaryotic genomes, the number of predicted ORFs in yeast has changed somewhat as sequence analysis is refined. As of 2006, approximately 5800 ORFs plus another 800 possible ORFs have been identified in the yeast genome; this is fewer than in some prokaryotic genomes (Tables 12.1 and 12.3). Of the yeast ORFs, about 3500 encode proteins whose functions are known. The wide variety of genetic and biochemical techniques available for studying this organism have resulted in significant advances in understanding the function of the remaining proteins as well (Section 12.8).

Minimal Gene Complement of Yeast

How many of the known yeast genes are actually essential? This question can be approached by systematically inactivating each gene in turn with knockout mutations (mutations that render a gene nonfunctional, ⮑ Section 11.4). Knockout mutations cannot normally be obtained in genes essential for cell viability in a haploid organism. However, yeast can be grown in both diploid

and haploid states (⮑ Section 20.17). By generating knockout mutations in diploid cells and then investigating whether they can also exist in haploid cells, it is possible to determine whether a particular gene is essential for cell viability.

Using knockout mutations, it has been shown that at least 877 yeast ORFs are essential, whereas 3121 clearly are not. Note that this number of essential genes is much greater than the approximately 300 genes (Section 12.3) predicted to be the minimal number required in prokaryotes. However, because eukaryotes are more complex than prokaryotes, a larger minimal gene complement would be expected.

Yeast Introns

Yeast is a eukaryote and contains introns (⮑ Section 7.5). However, the total number of introns in the protein-encoding genes of yeast is a mere 225. Most yeast genes with introns have only a single small intron near the 5′ end of the gene. This situation differs greatly from that seen in more complex eukaryotes. In the worm *Caenorhabditis elegans*, for example, the average gene has five introns, and in the fruit fly *Drosophila*, the average gene has four introns. Introns are also very common in the genes of plants.

Thus the plant *Arabidopsis* averages five introns per gene, and over 75% of *Arabidopsis* genes have introns. In humans almost all protein-encoding genes have introns, and it is not uncommon for a single gene to have 10 or more. Moreover, human introns are typically much larger than human exons. Indeed, exons make up only about 1% of the human genome, whereas introns account for 24%.

Other Eukaryotic Microorganisms

The genomes of several other eukaryotic microorganisms, mostly ones of medical importance, have been sequenced. The smallest eukaryotic cellular genome known belongs to *Encephalitozoon cuniculi*, an intracellular pathogen of humans and other animals that causes lung infections. *E. cuniculi* lacks mitochondria, and although its haploid genome contains 11 chromosomes, the genome size is only 2.9 Mbp with approximately 2000 genes (Table 12.3); this is smaller than many prokaryotic genomes (Table 12.1). As is also true in the prokaryotes, the smallest eukaryotic genome belongs to an endosymbiont (Table 12.3). Known as a *nucleomorph*, this is the degenerate remains of a eukaryotic endosymbiont found in certain green algae that have acquired photosynthesis by secondary endosymbiosis (Section 20.20). Nucleomorph genomes range from about 0.45 to 0.85 Mbp.

As previously mentioned, the largest eukaryotic genome belongs to *Trichomonas*, which has about 60,000 genes despite its parasitic existence (Table 12.3). The free-living ciliate *Paramecium* has about 40,000 genes, and the free-living social amoeba, *Dictyostelium*, has about 12,500 genes (but note that *Dictyostelium* has both single-celled and multicellular phases in its life cycle, Section 20.12). For comparison, the pathogenic amoeba *Entamoeba histolytica*, the causative agent of amebic dysentery, has approximately 10,000 genes.

Apart from the strange case of *Trichomonas*, parasitic eukaryotic microorganisms have genomes containing 10–30 Mbp of DNA and between 4000 and 11,000 genes. For example, the genome of the trypanosome *Trypanosoma brucei*, the agent of African sleeping sickness, has 11 chromosomes, 35 Mbp of DNA, and almost 11,000 genes. The most important eukaryotic parasite is *Plasmodium*, which causes malaria (Section 34.5). The 25-Mbp genome of *Plasmodium falciparum* consists of 14 chromosomes ranging in size from 0.7 to 3.4 Mbp. The estimated number of genes for *P. falciparum*, which infects humans, is 5300, and for the related species *Plasmodium yoelii*, which infects rodents, is 5900.

MiniQuiz

- How can you show whether a gene is essential?
- What is unusual about the genome of the eukaryote *Encephalitozoon*?

12.6 Metagenomics

Microbial communities contain many species of *Bacteria* and *Archaea*, most of which have never been cultured or formally identified. **Metagenomics**, also called environmental genomics, analyzes pooled DNA or RNA from an environmental sample containing organisms that have not been isolated and identified (Section 22.6). Just as the total gene content of an organism is its genome, so the total gene content of the organisms inhabiting an environment is known as its **metagenome**.

Several environments have been surveyed by large-scale metagenome sequencing projects. Extreme environments, such as acidic runoff waters from mines, tend to have low species diversity. From such environments it has been possible to isolate community DNA and assemble much of it into nearly complete individual genomes. Conversely, complex environments such as fertile soil yield too much sequence data to allow successful assembly at present.

In addition to metagenome analyses based on DNA sequencing, analyses based on RNA or proteins may be used to explore the patterns of gene expression in natural microbial communities. This topic is covered in more detail in Chapter 22 (see especially Section 22.10).

One curious recent revelation is that most DNA in natural habitats does not belong to living cells. About 50–60% of the DNA in the oceans is extracellular DNA found in deep-sea sediments. This is deposited when dead organisms from the upper layers of the ocean sink to the bottom and disintegrate. Because nucleic acids are repositories of phosphate, this DNA is a major contributor to the global phosphorus cycle.

MiniQuiz

- What is a metagenome?
- How is a metagenome analyzed?

Genome Function and Regulation

Despite the major effort required to generate an annotated genome sequence, in some ways the net result is simply a "list of parts." To understand how a cell functions, we need to know more than which genes are present. We must also investigate both gene expression (transcription) and the function of the final gene product. We focus here on gene expression. In analogy to the term "genome," the entire complement of RNA produced under a given set of conditions is known as the **transcriptome**.

12.7 Microarrays and the Transcriptome

Knowing the conditions under which a gene is transcribed may reveal a gene's function. We have already discussed how nucleic acid hybridization reveals the location of genes on specific fragments of DNA (Section 11.2). Hybridization techniques can also be used in conjunction with genomic sequence data to measure gene expression by hybridizing mRNA to specific DNA fragments. This technique has been radically enhanced with the development of microarrays.

Microarrays and the DNA Silica Chip

The **microarrays** used in genomics are small, solid supports to which genes or, more often, segments of genes are fixed and arrayed spatially in a known pattern; they are often called **gene chips**. The gene segments are synthesized by the polymerase chain reaction (PCR, ⮐ Section 6.11), or, alternatively, oligonucleotides are designed for each gene based on the genomic sequence. Once attached to the solid support, the DNA segments can be hybridized with mRNA from cells grown under specific conditions and scanned and analyzed by computer. Hybridization between a specific mRNA and a DNA segment on the chip indicates that the gene has been transcribed.

In practice, mRNA is present in amounts too low for direct use. Consequently, the mRNA sequences must first be amplified. Reverse transcriptase (RT) is used to generate complementary DNA (cDNA) from the mRNA. The cDNA may then be amplified by PCR (these two steps constitute the RT-PCR procedure). Alternatively, the cDNA may be used as a template by T7 RNA polymerase, which generates multiple RNA copies called *cRNA*. The cDNA or cRNA is then applied to the array.

A method for making and using microarrays is shown in **Figure 12.9**. Photolithography, a process used to produce computer chips, has been adapted to produce silica microarray chips 1 to 2 cm in size, each of which can hold thousands of different DNA fragments. In practice, each gene is often represented more than once in the array to provide increased reliability. Whole genome arrays contain DNA segments representing the entire genome of an organism. For example, there is a chip that contains the entire human genome (**Figure 12.10a**). This single chip can analyze over 47,000 human transcripts and has room for 6500 additional oligonucleotides for use in clinical diagnostics.

Figure 12.10b shows a part of a chip used to assay expression of the *Saccharomyces cerevisiae* genome. This chip easily holds the 5800 protein-encoding genes of *S. cerevisiae* (Section 12.5) so that global gene expression in this organism can be measured in a single experiment. To do this, the chip is hybridized with cRNA or cDNA derived from mRNA obtained from yeast cells grown under specific conditions. Any particular cRNA/cDNA binds only to the DNA on the chip that is complementary in sequence. To visualize binding, the cRNA/cDNA is tagged with a fluorescent dye, and the chip is scanned with a laser fluorescence detector. The signals are then analyzed by computer. A distinct pattern of hybridization is observed, depending upon which DNA sequences correspond to which mRNAs (Figures 12.9 and 12.10b). The intensity of the fluorescence gives a quantitative measure of gene expression (Figure 12.10b). This allows the computer to make a list of which genes were expressed and to what extent. Thus, using gene chips, the transcriptome of the organism of interest grown under specified conditions is revealed from the pattern and intensity of the fluorescent spots generated.

Applications of Gene Chips: Gene Expression

Gene chips may be used in several ways depending on the genes attached to the chip. Global gene expression is monitored by assembling an array of oligonucleotides complementary to each gene in the genome and then using the entire population of mRNA as the test sample (Figure 12.10b). Alternatively, one

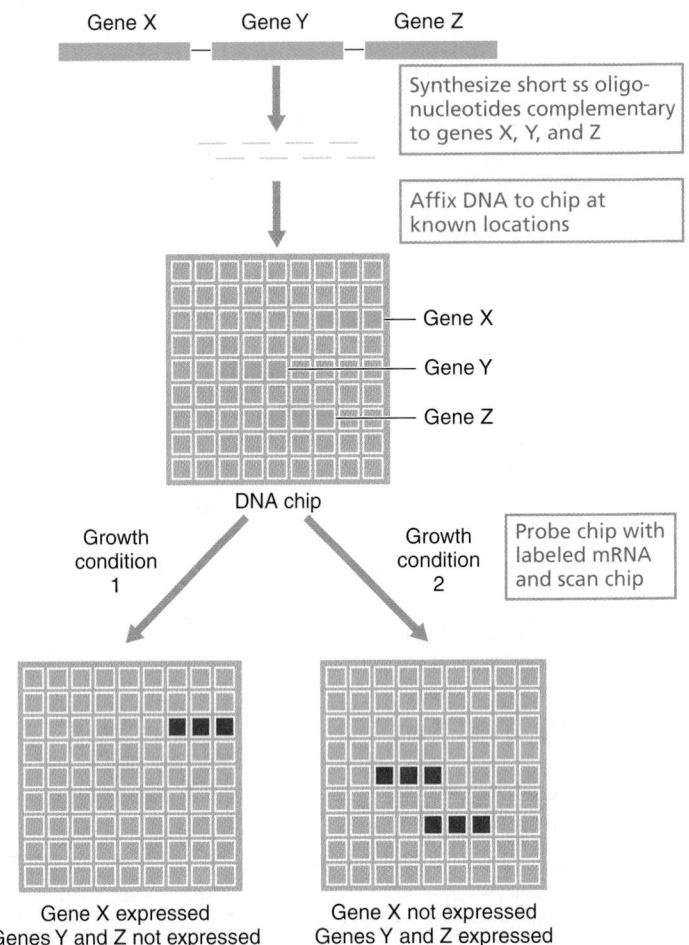

Figure 12.9 Making and using gene chips. Short single-stranded oligonucleotides corresponding to all the genes of an organism are synthesized individually and affixed at known locations to make a gene chip (microarray). The gene chip is assayed by hybridizing fluorescently labeled mRNA obtained from cells grown under a specific condition to the DNA probes on the chip and then scanning the chip with a laser.

can compare expression of specific groups of genes under different growth conditions. The ability to analyze the simultaneous expression of thousands of genes has tremendous potential for unraveling the complexities of metabolism and regulation. This is true both for organisms as "simple" as bacteria or as complex as higher eukaryotes.

The *S. cerevisiae* gene chip (Figure 12.10b) has been used to study metabolic control in this important industrial organism. Yeast can grow by fermentation and by respiration. Transcriptome analysis can reveal which genes are shut down and which are turned on when yeast cells are switched from fermentative (anaerobic) to respiratory (aerobic) metabolism or vice versa. Transcriptome analyses of such gene expression show that yeast undergoes a major metabolic "reprogramming" during the switch from anaerobic to aerobic growth. A number of genes that control production of ethanol (a key fermentation product) are strongly repressed, whereas citric acid cycle functions (needed for aerobic growth) are strongly activated by the switch. Overall, over 700 genes are turned on and over 1000 turned off during

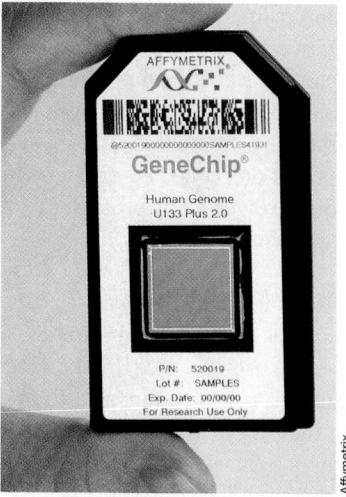

(a)

(b)

Figure 12.10 Using gene chips to assay gene expression. *(a)* The human genome chip containing over 40,000 gene fragments. *(b)* A hybridized yeast chip. The photo shows fragments from one-fourth of the genome of baker's yeast, *Saccharomyces cerevisiae*, affixed to a gene chip. Each gene is present in several copies and has been probed with fluorescently labeled cDNA derived from the mRNA extracted from yeast cells grown under a specific condition. The background of the chip is blue. Locations where the cDNA has hybridized are indicated by a gradation of colors up to a maximum number of hybridizations, which shows as white. Because the location of each gene on the chip is known, when the chip is scanned, it reveals which genes were expressed.

this metabolic transition. Moreover, by using a microarray, the expression pattern of genes of unknown function are also monitored during the fermentative to respiratory switch, yielding clues to their possible role. At present, no other available method can give as much information about gene expression as microarrays.

Another important use of DNA microarrays is the comparison of genes in closely related organisms. For example, this approach has been used to reveal how pathogenic bacteria evolved from their harmless relatives. In human medicine, microarrays are used to monitor gene loss or duplication in cancer cells. www.microbiologyplace.com **Online Tutorial 12.1: DNA Chips**

Applications in Identification

Besides probing gene expression, microarrays can be used to specifically identify microorganisms. In this case the array contains a set of characteristic DNA sequences from each of a variety of organisms or viruses. Such an approach can be used to differentiate between closely related strains by differences in their hybridization patterns. This allows very rapid identification of pathogenic viruses or bacteria from clinical samples or detection of these organisms in various other substances, such as food. For example, identification (ID) chips have been used in the food industry to detect particular pathogens, such as *Escherichia coli* O157:H7.

In environmental microbiology, microarrays have been used to assess microbial diversity. Phylochips, as they are called, contain oligonucleotides complementary to the 16S rRNA sequences of different bacterial species. After extracting bulk DNA or RNA from an environment, the presence or absence of each species can be assessed by the presence or absence of hybridization on the chip.

DNA chips are also available to identify higher organisms. A commercially available chip called the FoodExpert-ID contains 88,000 gene fragments from vertebrate animals and is used in the food industry to ensure food purity. For example, the chip can confirm the presence of the meat listed on a food label and can also detect foreign animal meats that may have been added as supplements to or substitutes for the official ingredients. The eventual goal is to have each meat product receive an "identity card" listing all the animal species whose tissues were detected in it. This is intended to give consumers more confidence in the wholesomeness of their food products. The FoodExpert-ID can also be used to detect vertebrate by-products in animal feed, a growing concern with the advent of prion-mediated diseases such as mad cow disease (ᘒᘒ Section 9.15).

MiniQuiz
- What do microarrays tell you that studying gene expression by assaying a particular enzyme cannot?
- Why is it useful to know how gene expression of the entire genome responds to a particular condition?

12.8 Proteomics and the Interactome

Which expressed genes actually yield protein products and what is the function of these proteins? The genome-wide study of the structure, function, and regulation of an organism's proteins is called **proteomics**.

The number and types of proteins present in a cell change in response to an organism's environment or other factors, such as developmental cycles. As a result, the term **proteome** has unfortunately become ambiguous. In its wider sense, a proteome refers to *all* the proteins encoded by an organism's genome. In its narrower sense, however, it refers to those proteins present in a cell *at any given time.*

Methods in Proteomics

The first major approach to proteomics was developed decades ago with the advent of two-dimensional (2D) polyacrylamide gel electrophoresis. This technique can separate, identify, and

M_r (kDa)

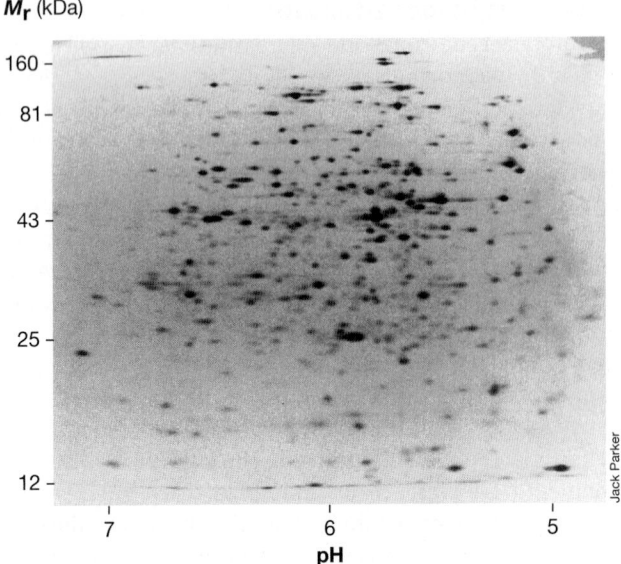

Figure 12.11 Two-dimensional polyacrylamide gel electrophoresis of proteins. Autoradiogram of the proteins of cells of *Escherichia coli*. Each spot on the gel is a different protein. The proteins are radioactively labeled to allow for visualization and quantification. The proteins were separated in the first dimension (X-direction) by isoelectric focusing under denaturing conditions. The second dimension (Y-direction) separates denatured proteins by their mass (M_r; in kilodaltons), with the largest proteins being toward the top of the gel.

measure all the proteins present in a cell sample. A 2D gel separation of proteins from *Escherichia coli* is shown in **Figure 12.11**. In the first dimension (the horizontal dimension in the figure), the proteins are separated by differences in their isoelectric points, the pH at which the net charge on each protein reaches zero. In the second dimension, the proteins are denatured in a way that gives each amino acid residue a fixed charge. The proteins are then separated by size (in much the same way as for DNA molecules; ↩ Section 11.1).

In studies of *E. coli* and a few other organisms, hundreds of proteins separated in 2D gels have been identified by biochemical or genetic means, and their regulation has been studied under various conditions. Using 2D gels, the presence of a particular protein under different growth conditions can be measured and related to environmental signals. One method of connecting an unknown protein with a particular gene using the 2D gel system is to elute the protein from the gel and sequence a portion of it, usually from its N-terminal end. More recently, eluted proteins have been identified by a technique called *mass spectrometry* (Section 12.9), usually after preliminary digestion to give a characteristic set of peptides. This sequence information may be sufficient to completely identify the protein. Alternatively, partial sequence data may allow the design of oligonucleotide probes or primers to locate the gene encoding the protein from genomic DNA by hybridization or PCR. Then, after sequencing of the DNA, the gene may be identified.

Today, liquid chromatography is increasingly used to separate protein mixtures. In high-pressure liquid chromatography (HPLC),

the sample is dissolved in a suitable liquid and forced under pressure through a column packed with a stationary phase material that separates proteins by variations in their chemical properties, such as size, ionic charge, or hydrophobicity. As the mixture travels through the column, it is separated by interaction of the proteins with the stationary phase. Fractions are collected at the column exit. The proteins in each fraction are digested by proteases and the peptides are identified by mass spectrometry.

Comparative Genomics and Proteomics

Although proteomics often requires intensive experimentation, *in silico* techniques can also be quite useful. Once the sequence of an organism's genome is obtained, it can be compared to that of other organisms to locate and identify genes that are similar to those already known. The sequence that is most important here is the amino acid sequence of the encoded proteins. Because the genetic code is degenerate (↩ Section 6.17), differences in DNA sequence may not necessarily lead to differences in the amino acid sequence (**Figure 12.12**).

Proteins with greater than 50% sequence identity frequently have similar functions. Proteins with identities above 70% are almost certain to have similar functions. Many proteins consist of distinct structural modules, called *protein domains*, each with characteristic functions. Such regions include metal-binding domains, nucleotide-binding domains, or domains for certain classes of enzyme activity, such as helicase or nuclease. Identification of domains of known function within a protein may reveal much about its role, even in the absence of complete sequence homology.

Structural proteomics refers to the proteome-wide determination of the three-dimensional (3D) structures of proteins. At present, it is not possible to predict the 3D structure of proteins directly from their amino acid sequences. However, structures of unknown proteins can often be modeled if the 3D structure is available for a protein with 30% or greater identity in amino acid sequence.

Coupling proteomics with genomics is yielding important clues about how gene expression in different organisms correlates with environmental stimuli. Not only does such information have important basic science benefits, but it also has potential

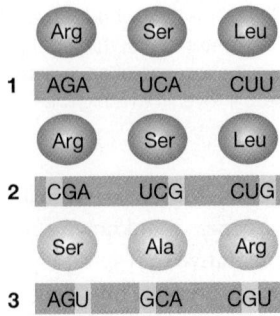

Figure 12.12 Comparison of nucleic acid and amino acid sequence similarities. Three different nucleotide sequences are shown (for convenience RNA is shown). Both sequence 2 and sequence 3 differ from sequence 1 in only three positions. However, the amino acid sequence encoded by 1 and 2 are identical, whereas that encoded by sequence 3 is unrelated to the other two.

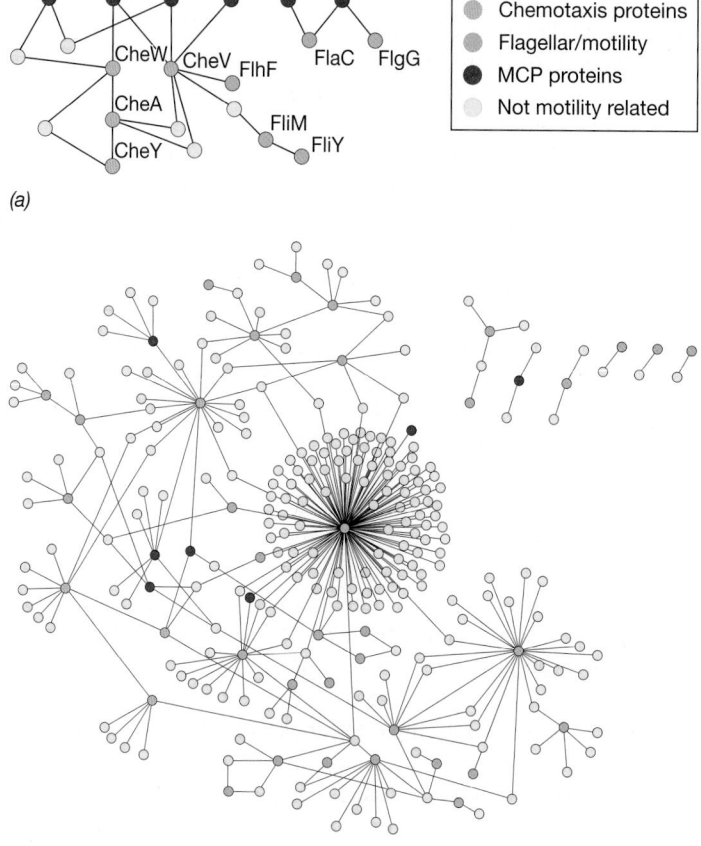

(a)

(b)

Figure 12.13 Motility protein interactome for *Campylobacter jejuni*. This network illustrates the way in which interactome data are depicted. (a) A subsection of the network highlighting the well-known proteins of the chemotaxis signal transduction pathway (CheW, CheA, and CheY) and their partners. MCP, methyl-accepting chemotaxis proteins (Section 8.8). (b) High-confidence interactions between all proteins known to have roles in motility. Note the six small networks that fall outside the single large network.

applications. These include advances in medicine, the environment, and agriculture. In all of these areas, understanding the link between the genome and the proteome and how it is regulated could give humans unprecedented control in fighting disease and pollution, as well as major benefits for agricultural productivity.

The Interactome

By analogy with the terms "genome" and "proteome," the **interactome** is the complete set of interactions among the macromolecules within a cell. Originally, the interactome applied to the interactions between proteins, many of which assemble into complexes. However, it is also possible to consider interactions between different classes of molecule, such as the protein–RNA interactome.

Interactome data are expressed in the form of network diagrams, with each node representing a protein and connecting lines representing the interactions. Diagrams of whole interactomes are extremely complex and difficult to interpret. More

useful are limited interactomes such as the motility protein network from *Campylobacter jejuni* (**Figure 12.13**). Figure 12.13*a* shows the core interactions between well-known components of the chemotaxis system (Section 8.8), and Figure 12.13*b* includes all other proteins that interact with these. At present, most larger interactomes are poorly validated, with different methods often giving conflicting results.

MiniQuiz

- Why is the term "proteome" ambiguous, whereas the term "genome" is not?
- What are the most common experimental methods used to survey the proteome?
- What is the interactome?

12.9 Metabolomics

The **metabolome** is the complete set of metabolic intermediates and other small molecules produced in an organism. Metabolomics has lagged behind other "omics" in large part due to the immense chemical diversity of small metabolites. This makes systematic screening technically challenging. Early attempts used nuclear magnetic resonance (NMR) analysis of extracts from cells labeled with ^{13}C-glucose. However, this method is limited in sensitivity, and the number of compounds that can be simultaneously identified in a mixture is too low for resolution of complete cell extracts.

The most promising approach to metabolomics is the use of newly developed variants of mass spectrometry. This approach is not limited to particular classes of molecules and can be extremely sensitive. The mass of carbon-12 is defined as exactly 12 molecular mass units (daltons). However, the masses of other atoms, such as nitrogen-14 or oxygen-16, are not exact integers. Mass spectrometry using extremely high mass resolution, which is now possible in special instruments, allows the unambiguous determination of the molecular formula of any small molecule. Clearly, isomers will have the same molecular formula, but they may be distinguished by their different fragmentation patterns during mass spectrometry. The same approach is used to identify the peptide fragments from digested proteins during proteome analyses (Section 12.8). In this case, identifying several oligopeptides allows the identity of the parent protein to be deduced provided that its amino acid sequence is known.

In the MALDI (matrix-assisted laser desorption ionization) version of mass spectrometry, the sample is ionized and vaporized by a laser (**Figure 12.14**). The ions generated are accelerated along the column toward the detector by an electric field. The time of flight (TOF) for each ion depends on its mass/charge ratio—the smaller this ratio, the faster the ion moves. The detector measures the TOF for each ion and the computer calculates the mass and hence the molecular formula. The combination of these two techniques is known as MALDI-TOF.

Metabolome analysis is especially useful in the study of plants, many of which produce several thousand different metabolites—more than most other types of organism. This is because plants make many *secondary metabolites*, such as scents, flavors,

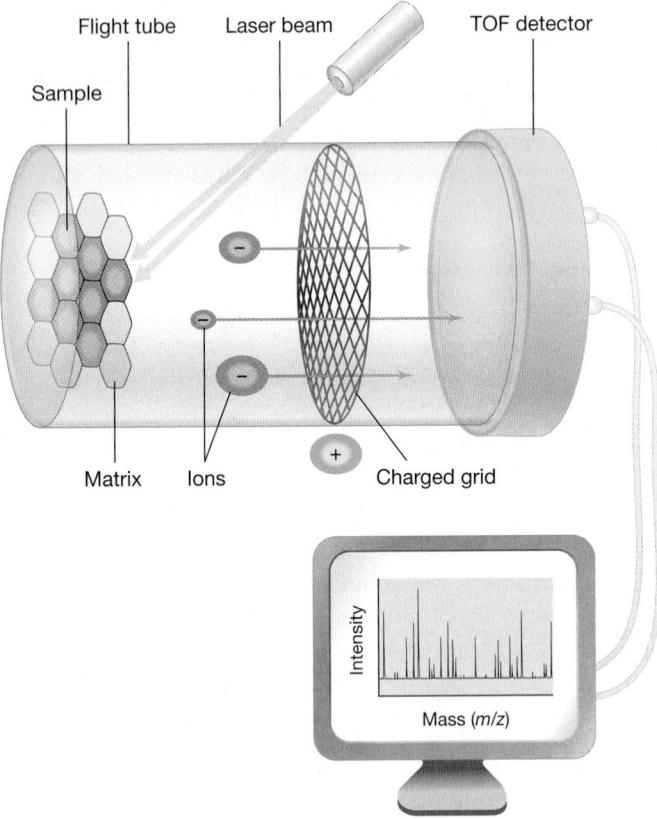

Figure 12.14 MALDI-TOF mass spectrometry. In matrix-assisted laser desorption ionization (MALDI) spectroscopy, the sample is ionized by a laser and the ions travel down the tube to the detector. The time of flight (TOF) depends on the mass/charge (*m/z*) ratio of the ion. The computer identifies the ions based on their time of flight, that is, the time to reach the detector.

alkaloids, and pigments, many of which are commercially important. Metabolomic investigations have monitored the levels of several hundred metabolites in the model plant *Arabidopsis*, and significant changes were observed in the levels of many of these metabolites in response to changes in temperature. Future directions for metabolomics presently under development include assessing the effect of disease on the metabolome of various human organs and tissues. Such results should greatly improve our understanding of how the human body fights off infectious and noninfectious disease.

MiniQuiz
- What techniques are used to monitor the metabolome?
- What is a secondary metabolite?

 The Evolution of Genomes

In addition to revealing how genes function and how organisms interact with the environment, comparative genomics can also illuminate evolutionary relationships between organisms. Reconstructing evolutionary relationships from genome sequences helps to distinguish between primitive and derived characteristics

and can resolve ambiguities in phylogenetic trees based on analyses of a single gene, such as small subunit rRNA (Section 16.9). Genomics is also a link to understanding early life forms and, eventually, may answer the most fundamental of all questions in biology: How did life first arise?

12.10 Gene Families, Duplications, and Deletions

Genomes from both prokaryotic and eukaryotic sources often contain multiple copies of genes that are related in sequence due to shared evolutionary ancestry; such genes are called **homologous** genes. Groups of gene homologs are called **gene families**. Not surprisingly, larger genomes tend to contain more individual members from a particular gene family.

Paralogs and Orthologs

Comparative genomics has shown that many genes have arisen by duplication of other genes. Such homologs may be subdivided, depending on their origins. Genes whose similarity is the result of gene duplication at some time in the evolution of an organism are called **paralogs**. Genes found in one organism that are similar to genes in another organism because of descent from a common ancestor are called **orthologs** (**Figure 12.15**). Orthologs are often not identical because of divergent evolution in lineages following speciation. An example of paralogous genes are those encoding several different lactate dehydrogenase (LDH) isoenzymes in humans. These enzymes are structurally distinct yet are all highly related and carry out the same enzymatic reaction. By contrast, the corresponding LDH from *Lactobacillus* is orthologous to all of the human LDH isoenzymes. Thus, gene families contain both paralogs and orthologs.

Gene Duplication

It is widely thought that gene duplication is the mechanism for the evolution of most new genes. If a segment of duplicated DNA is long enough to include an entire gene or group of genes, the organism with the duplication has multiple copies of these particular genes. After duplication, one of the duplicates is free to evolve while the other copy continues to supply the cell with the original function. In this way, evolution can "experiment" with one copy of the gene. Such gene duplication events, followed by diversification of one copy, are thought to be the major events that fuel microbial evolution. Genomic analyses have revealed numerous examples of protein-encoding genes that were clearly derived from gene duplication.

Duplications of genetic material may include just a handful of bases or even whole genomes. For example, comparison of the genomes of the yeast *Saccharomyces cerevisiae* and other fungi suggests that the ancestor of *Saccharomyces* duplicated its entire genome. This was followed by extensive deletions that eliminated much of the duplicated genetic material. Analysis of the genome of the model plant *Arabidopsis* suggests that there were one or more whole genome duplications in the ancestor of the flowering plants.

Did bacterial genomes evolve by whole genome duplication? The distribution of duplicated genes and gene families in the genomes of bacteria suggests that many frequent but relatively

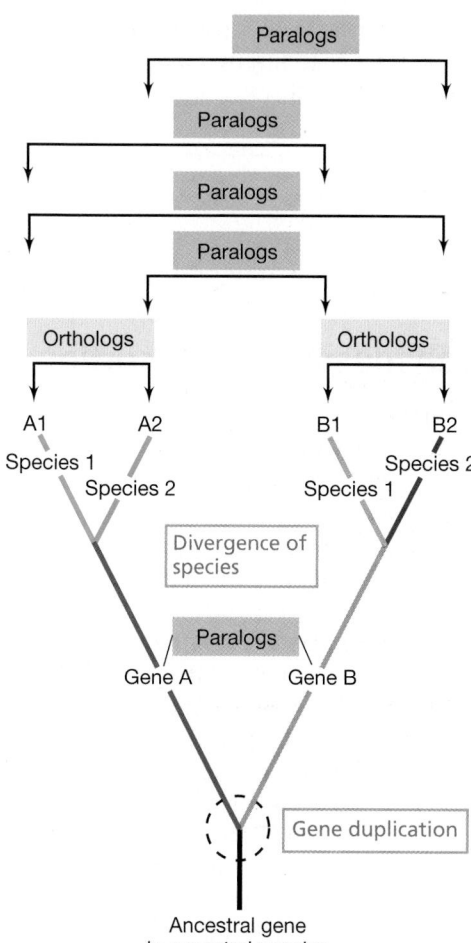

Figure 12.15 Orthologs and paralogs. This family tree depicts an ancestral gene that duplicated and diverged into two paralogous genes, A and B. Next, the ancestral species diverged into species 1 and species 2, both of which have genes for A and B (designated A1 and B1 and A2 and B2, respectively). Each such pair are paralogs. However, because species 1 and 2 are now separate species, A1 is an ortholog of A2 and B1 is an ortholog of B2.

small duplications have occurred. For example, among the *Deltaproteobacteria*, the soil bacterium *Myxococcus* has a genome of 9.1 Mbp. This is approximately twice that of the genomes of other typical *Deltaproteobacteria*, which range from 4 to 5 Mbp. Among the *Alphaproteobacteria*, genome sizes range from 1.1 to 1.5 Mbp for parasitic members to 4 Mbp for free-living *Caulobacter*, and up to 7–9 Mbp for plant-associated bacteria. However, in all of these cases gene distribution analysis points to frequent small-scale duplications rather than whole genome duplications. Conversely, in bacteria that are parasitic, frequent successive deletions have eliminated genes no longer needed for a parasitic lifestyle, leading to their unusually small genomes (Section 12.3).

Gene Analysis in Different Domains

The comparison of genes and gene families is a major task in comparative genomics. Because chromosomes from many different microorganisms have already been sequenced, such comparisons can be easily done, and the results are often surprising. For instance, genes in *Archaea* that are active in DNA replication, transcription, and translation are more similar to those in *Eukarya* than to those in *Bacteria*. Unexpectedly, however, many other genes in *Archaea*, for example those encoding metabolic functions other than information processing, are more similar to those in *Bacteria* than those in *Eukarya*. The powerful analytical tools of bioinformatics allow genetic relationships between any organisms to be deduced very quickly and at the single gene, gene group, or entire genome level. The results obtained thus far lend further support to the phylogenetic picture of life deduced originally by comparative rRNA sequence analysis (Section 16.9) and suggest that many genes in all organisms have common evolutionary roots. However, such analyses have also revealed instances of horizontal gene transfer, an important issue to which we now turn.

MiniQuiz
- What is a gene family?
- Contrast gene paralogs with gene orthologs.

12.11 Horizontal Gene Transfer and Genome Stability

Evolution is based on the transfer of genetic traits from one generation to the next. However, in prokaryotes, **horizontal gene transfer** (sometimes called lateral gene transfer) also occurs, and it can complicate evolutionary studies, especially those of entire genomes. Horizontal gene transfer refers to transfer from one cell to another by means other than the usual (vertical) inheritance process from mother cell to daughter cell. In prokaryotes, at least three mechanisms for horizontal gene transfer are known: *transformation*, *transduction*, and *conjugation* (Chapter 10).

Horizontal gene flow may be extensive in nature and may sometimes cross even phylogenetic domain boundaries. However, to be detectable by comparative genomics, the difference between the organisms must be rather large. For example, several genes with eukaryotic origins have been found in *Chlamydia* and *Rickettsia*, both human pathogens. In particular, two genes encoding histone H1-like proteins have been found in the *Chlamydia trachomatis* genome, suggesting horizontal transfer from a eukaryotic source, possibly even its human host. Note that this is opposite the situation in which genes from the ancestor of the mitochondrion were transferred to the eukaryotic nucleus (Section 12.4).

Detecting Horizontal Gene Flow

Horizontal gene transfers can be detected in genomes once the genes have been annotated. The presence of genes that encode proteins typically found only in distantly related species is one signal that the genes originated from horizontal transfer. However, another clue to horizontally transferred genes is the presence of a stretch of DNA whose guanosine/cytosine (GC) content or codon bias differs significantly from the rest of the genome (see Figure 12.17). With these clues, many likely examples of horizontal transfer have been documented in the genomes of various prokaryotes. A classic example exists with the organism *Thermotoga maritima*, a species of *Bacteria*, which was shown to contain over 400 genes (greater than 20% of its genome) of archaeal origin. Of these genes, 81 were found in discrete clusters. This strongly suggests that they were obtained by horizontal gene transfer, presumably from thermophilic *Archaea* that share the hot environments inhabited by *Thermotoga*.

Horizontally transferred genes typically encode metabolic functions other than the core molecular processes of DNA replication, transcription, and translation, and may account for the previously mentioned similarities of metabolic genes in *Archaea* and *Bacteria* (Section 12.10). In addition, there are several examples of virulence genes of pathogens having been transferred by horizontal means. It is obvious that prokaryotes are exchanging genes in nature, and the process likely functions to "fine-tune" an organism's genome to a particular situation or habitat.

It is necessary to be cautious when invoking horizontal gene transfer to explain the distribution of genes. When the human genome was first sequenced, a couple hundred genes were identified as being horizontal transfers from prokaryotes. However, when more eukaryotic genomes became available for examination, homologs were found for most of these genes in many eukaryotic lineages. Consequently, it now seems that most of these genes are in fact of eukaryotic origin. Only about a dozen human genes are now accepted as strong candidates for having relatively recent prokaryotic origins. The phrase "relatively recent" here refers to genes transferred from prokaryotes after separation of the major eukaryotic lineages, not to genes of possible ancient prokaryotic origin that are shared by eukaryotes as a whole.

Transplanting and Synthesizing Bacterial Genomes

Bacterial genomes have been artificially manipulated in a variety of ways on a small scale as described in previous chapters. However, it has recently become possible to artificially assemble and transfer whole bacterial genomes.

The whole genome (that is, the entire bacterial chromosome) was extracted and purified from one species of *Mycoplasma* that carried a gene for tetracycline resistance on its chromosome. The DNA was then transformed into another closely related species of *Mycoplasma*. The incoming donor genome was selected by resistance to tetracycline. The resident genome was lost and the cells that resulted from this genome transplantation were phenotypically identical to the donor strain of *Mycoplasma*.

It is also possible to completely synthesize a bacterial genome if the sequence is known. This has also been demonstrated using the approximately 580,000-bp *Mycoplasma* genome. Chemically synthesized oligonucleotides with overlapping sequences were assembled by recombination into segments of around 5–7 kb. These were successively assembled into segments of 24, 72, and 144 kb *in vitro*. The 144-kb "quarter chromosomes" were then cloned as bacterial artificial chromosomes in *Escherichia coli*. Finally, these were transformed into the yeast *Saccharomyces cerevisiae* and assembled by recombination. The fully assembled

bacterial chromosomes were isolated and checked by DNA sequencing, and were then inserted into a suitable *Mycoplasma* host cell.

MiniQuiz
• Which class of genes is rarely transferred horizontally? Why?
• List the major mechanisms by which horizontal gene transfer occurs in prokaryotes.

12.12 Transposons and Insertion Sequences

As described in Section 10.13, mobile DNA refers to segments of DNA that move from one location to another within host DNA molecules. Most mobile DNA consists of transposable elements, but integrated virus genomes and integrons are also found. All of these mobile elements play important roles in genome evolution.

Genome Evolution and Transposons

Transposons may move between different host DNA molecules, including chromosomes, plasmids, and viruses. In doing so they may pick up and horizontally transfer genes for various characteristics, including resistance to antibiotics and production of toxins. However, transposons may also mediate a variety of large-scale chromosomal changes. Bacteria that are undergoing rapid evolutionary change often contain relatively large numbers of mobile elements, especially insertion sequences. Recombination among identical elements generates chromosomal rearrangements such as deletions, inversions, or translocations. This is thought to provide a source of genome diversity upon which selection can act. Thus, chromosomal rearrangements that accumulate in bacteria during stressful growth conditions are often flanked by repeats or insertion sequences.

Conversely, once a species settles into a stable evolutionary niche, most mobile elements are apparently lost. For example, genomes of species of *Sulfolobus* (*Archaea*) have unusually high numbers of insertion sequences and show a high frequency of gene translocations. By contrast, *Pyrococcus* (*Archaea*) shows an almost complete lack of insertion sequences and a correspondingly low number of gene translocations. This suggests that for whatever reason, perhaps because of fluctuations in conditions in their habitats, the genome of *Sulfolobus* is more dynamic than the more stable genome of *Pyrococcus*.

Insertion Sequences

Chromosomal rearrangements due to insertion sequences have apparently contributed to the evolution of several bacterial pathogens. In *Bordetella*, *Yersinia*, and *Shigella*, the more highly pathogenic species show a much greater frequency of insertion sequences. For example, *Bordetella bronchiseptica* has a genome of 5.34 Mbp but carries no known insertion sequences. Its more pathogenic relative, *Bordetella pertussis*, has a smaller genome (4.1 Mb), but has more than 260 insertion sequences. Comparison of these genomes suggests that the insertion sequences are responsible for substantial genome rearrangement, including deletions responsible for the reduction of genome size in *B. pertussis*.

Insertion sequences also play a role in assembling genetic modules to generate novel plasmids. Thus 46% of the 220-kbp virulence megaplasmid of *Shigella flexneri* consists of insertion sequence DNA! In addition to full-length insertion sequences, there are many fragments in this plasmid that imply multiple ancestral rearrangements.

Integrons and Super-Integrons

Integrons are genetic elements that collect and express genes carried on mobile segments of DNA, called *cassettes*. Gene cassettes suitable for integration consist of a coding sequence lacking a promoter, but containing an integrase recognition site, the *attC* site. The integrons themselves contain a corresponding integration site, the *attI* site, into which gene cassettes may be integrated. The integron also possesses a gene encoding the **integrase**, the enzyme responsible for inserting cassettes. Integration occurs by recombination between the *attI* site and the *attC* site. Once a gene cassette has been inserted into an integron, the gene it carries may be expressed from a promoter that is provided by the integron.

Neither integrons nor gene cassettes are transposable elements (they do not have terminal inverted repeats, nor do they transpose). However, gene cassettes may exist transiently as free, non-replicating, circular DNA incapable of gene expression, or they may be found integrated into the *attI* site of an integron. Thus gene cassettes are a form of mobile DNA that may move from one integron to another. Most integrons are found on plasmids or in transposons and may collect multiple gene cassettes. A few integrons are found on bacterial chromosomes and may collect hundreds of gene cassettes, whereupon they are called *super-integrons*. For example, the second chromosome of *Vibrio cholerae* (causative agent of cholera) has a super-integron with approximately 200 genes, mostly of unknown function.

Most known integrons carry genes for antibiotic resistance. However, this is probably due to a bias in observation, because antibiotic resistance is of clinical importance. Over 40 different antibiotic resistance genes have been identified on integron gene cassettes, as have some genes associated with virulence in certain pathogenic bacteria. **Figure 12.16** shows the structure of two integrons from *Pseudomonas aeruginosa*, a potentially serious pathogen. Integrons have been found in various species of

Bacteria, often in clinical isolates, and their selection by horizontal gene transfer in such antibiotic-rich environments as hospitals and clinics is obvious. What is less obvious is the origin of the gene cassettes themselves. These are not simply random genes, as they must possess specific DNA sequences that are recognized by the integrase; they are incapable of expression until they become part of an integron and can be transcribed from the integron's promoter.

12.13 Evolution of Virulence: Pathogenicity Islands

Comparison of the genomes of pathogenic bacteria with those of their harmless close relatives often reveals extra blocks of genetic material that contain genes encoding *virulence factors*, special proteins or other molecules or structures that take part in causing disease (Section 27.9).

Some virulence genes are carried on plasmids or lysogenic bacteriophages (Section 9.10). However, many others are clustered in chromosomal regions called **pathogenicity islands**. For example, the identity and chromosomal location of most genes of pathogenic strains of *Escherichia coli* correspond to those of the harmless laboratory strain *E. coli* K-12, as would be expected. However, most pathogenic strains contain pathogenicity islands of considerable size that are absent from *E. coli* K-12 (**Figure 12.17**). Consequently, two strains of the same bacterial species may show significant differences in genome size. For example, as shown in Table 12.1, the enterohemorrhagic strain *E. coli* O157:H7 contains 20% more DNA and genes than *E. coli* K-12.

Pathogenicity islands are merely the best-known case of **chromosomal islands**. Such islands are presumed to have a "foreign" origin, based on several observations. First, these extra regions are often flanked by inverted repeats, implying that the whole region was inserted into the chromosome by transposition (Section 10.13) at some period in the recent evolutionary past. Second, the base composition and codon bias in chromosomal islands often differ significantly from that of the rest of the genome. Third, chromosomal islands are often found in some strains of a particular species but not in others.

Some chromosomal islands carry a gene for an integrase and are thought to move in a manner analogous to conjugative transposons (Section 10.13). Chromosomal islands are typically inserted into a gene for a tRNA; however, because the target site is duplicated upon insertion, an intact tRNA gene is regenerated during the insertion process. In a few cases, transfer of a whole chromosomal island between related bacteria has been demonstrated in the laboratory; transfer can presumably occur by any of the mechanisms of horizontal transfer previously discussed: transformation, transduction, and conjugation (Chapter 10). It is thought that after insertion into the genome of a new host cell,

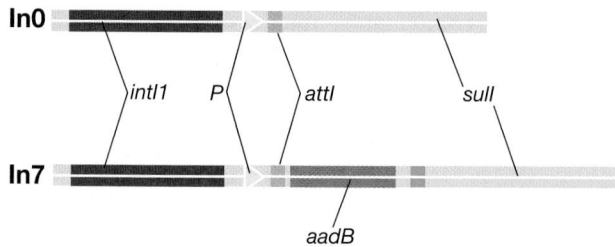

Figure 12.16 Structure of two naturally occurring integrons from *Pseudomonas*. Integron In0 has the basic set of genes: *intI1*, integrase; *attI*, the integration site; *P*, promoter; and *sulI*, a gene conferring sulfonamide resistance. Integron In7 contains all of these plus an integrated gene cassette. All cassettes contain a site (blue) for site-specific recombination. This cassette contains the *aadB* gene, which confers resistance to certain aminoglycoside antibiotics.

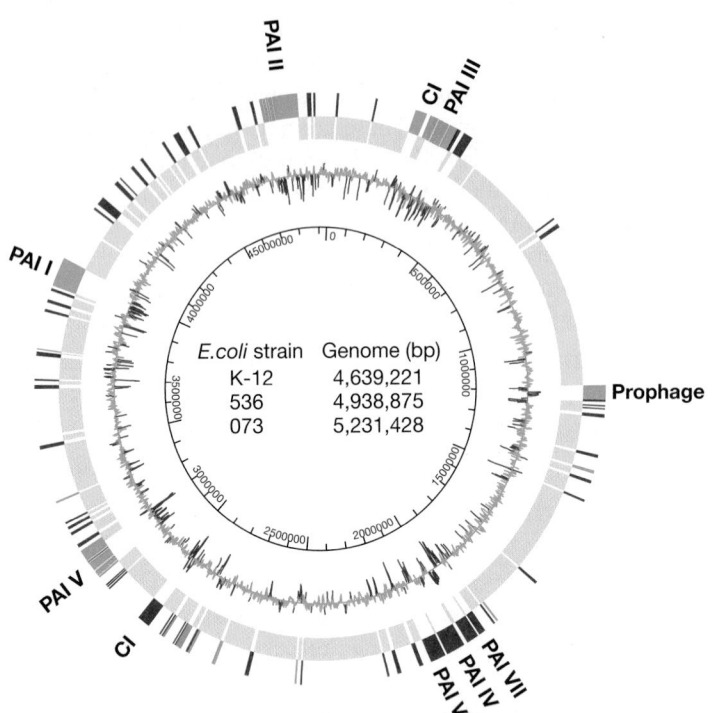

E.coli strain	Genome (bp)
K-12	4,639,221
536	4,938,875
073	5,231,428

Figure 12.17 Pathogenicity islands in *Escherichia coli*. Genetic map of *E. coli* strain 536, a urinary tract pathogen, compared with a second pathogenic strain (073) and the wild-type strain K-12. The pathogenic strains contain pathogenicity islands, and thus their chromosomes are larger than that of K-12. Inner circle, nucleotide base pairs. Jagged circle, DNA GC distribution; regions where GC content varies dramatically from the genome average are in red. Outermost circle, three-way genomic comparison: green, genes common to all strains; red, genes present in the pathogenic strains only; blue, genes found only in strain 536; orange, genes of strain 536 present in a different location in strain 073. Some very small inserts deleted for clarity. PAI, pathogenicity islands; CI, chromosomal island. Prophage, DNA from a temperate bacteriophage. Note the correlation between genomic islands and skewed GC content. Data adapted from *Proc. Natl. Acad. Sci.* (*USA*) 103: 12879–12884 (2006).

chromosomal islands gradually accumulate mutations—both point mutations and small deletions. Thus, over many generations, chromosomal islands tend to lose their ability to move.

Small pathogenicity islands that encode a series of virulence factors are present in certain strains of *Staphylococcus aureus* and can be moved between cells by temperate bacteriophages (Section 9.10). The islands are smaller than the phage genome, and when the islands excise from the chromosome and replicate, they induce the formation of defective phage particles that carry the genes for the islands but are too small to carry the phage genome. In this way, strains of *S. aureus* that lack the islands can quickly obtain them and become more effective pathogens.

Chromosomal islands contribute specialized functions that are not needed for simple survival. Not surprisingly, pathogenicity islands with their clinical relevance have drawn the most interest. However, chromosomal islands are also known that carry genes for the biodegradation of various substrates derived from human activity, such as aromatic hydrocarbons and herbicides. In addi-

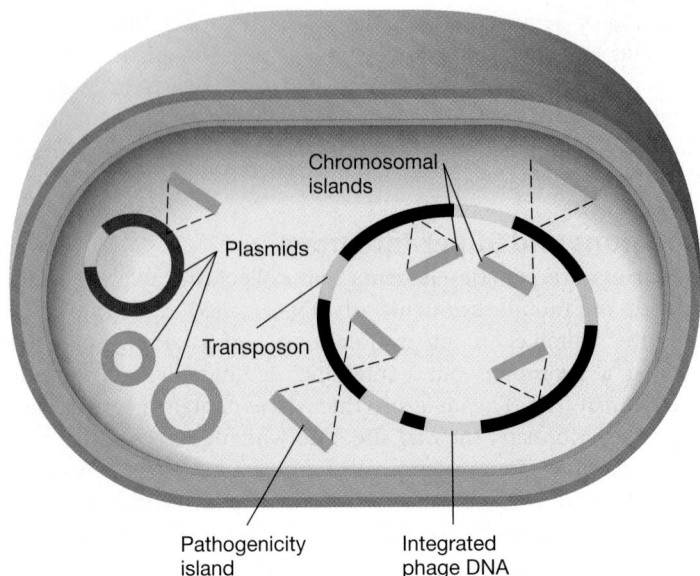

Figure 12.18 Pan genome versus core genome. The core genome is represented by the black regions of the chromosome and is present in all strains of a species. The pan genome includes elements that are present in one or more strains but not in all strains. Each colored wedge indicates a single insertion. Where two wedges emerge from the same location, they represent alternative islands that can insert at that site. However, only one insertion can be present at a given location. Plasmids, like the chromosome, may have insertions that are not present in all strains.

tion, many of the genes essential for the symbiotic relationship of rhizobia with plants in the root nodule symbiosis (Section 25.8) are carried in symbiosis islands inserted into the genome of these bacteria. Perhaps the most unique chromosomal island is the magnetosome island of the bacterium *Magnetospirillum*; this DNA fragment carries the genes needed for the formation of magnetosomes, intracellular magnetic particles used to orient the organism in a magnetic field and influence the direction of its motility (Section 3.10).

The presence or absence of chromosomal islands, transposable elements, integrated virus genomes, and plasmids means that there may be major differences in the total amount of DNA and the suite of accessory capabilities (virulence, symbiosis, or biodegradation) between strains of a single bacterial species. This has led to the concept that the genome of a bacterial species consists of two components, the *core* genome and the *pan* genome. The core genome is shared by all strains of the species, whereas the pan genome includes all of the optional extras present in some but not all strains of the species (**Figure 12.18**). In other words, one could say that the core genome is typical of the species as a whole, whereas the pan genome is unique to particular strains within a species.

MiniQuiz
- What is a chromosomal island?
- Why are chromosomal islands believed to be of foreign origin?

Big Ideas

12.1

Small viruses were the first organisms whose genomes were sequenced, but now many prokaryotic and eukaryotic cellular genomes have been sequenced.

12.2

DNA sequencing can be done using the Sanger chain termination method that employs dideoxynucleotides to block elongation of growing DNA chains. The use of radioactive labeling has largely given way to automated sequencing that relies on fluorescent labeling. Advances in technology have greatly increased the speed of DNA sequencing. Shotgun techniques employ random cloning and sequencing of small genome fragments followed by computer-generated assembly of the genome using overlaps as a guide. After sequencing is finished, computers search for ORFs and genes encoding protein homologs as part of the annotation process.

12.3

Sequenced prokaryotic genomes range in size from 0.16 Mbp to 13 Mbp, and even larger genomes are known. The smallest prokaryotic genomes are smaller than those of the largest viruses. The largest prokaryotic genomes have more genes than some eukaryotes. In prokaryotes gene content is typically proportional to genome size. Many genes can be identified by their sequence similarity to genes found in other organisms. However, a significant percentage of sequenced genes are of unknown function.

12.4

All eukaryotic cells (except for a few parasites) contain mitochondria. In addition, plant cells contain chloroplasts. Both organelles contain circular DNA genomes that encode rRNAs, tRNAs, and a few proteins needed for energy metabolism. Although the genomes of the organelles are independent of the nuclear genome, the organelles themselves are not. Many genes in the nucleus encode proteins required for organelle function.

12.5

The complete genomic sequence of the yeast *Saccharomyces cerevisiae* and that of many other microbial eukaryotes has been determined. Yeast may encode up to 5800 proteins, of which only about 900 appear essential. Relatively few of the protein-encoding genes of yeast contain introns. The number of genes in single-celled eukaryotes ranges from 2000 (less than many bacteria) to 60,000 (more than twice as many as humans).

12.6

Most microorganisms in the environment have never been cultured. Nonetheless, analysis of DNA samples has revealed colossal sequence diversity in most habitats. The concept of the metagenome embraces the total genetic content of all the organisms in a particular habitat.

12.7

Gene chips consist of genes or gene fragments attached to a solid support in a known pattern. mRNA is hybridized to these arrays and the gene chips are then analyzed to determine patterns of gene expression. The arrays are large enough and dense enough that the transcription pattern of an entire genome (the transcriptome) can be analyzed.

12.8

Proteomics is the analysis of all the proteins present in an organism. The ultimate aim of proteomics is to understand the structure, function, and regulation of these proteins. The interactome is the total set of interactions between macromolecules inside the cell.

12.9

The metabolome is the complete set of metabolic intermediates produced by an organism.

12.10

Genomics can be used to study the evolutionary history of an organism. Organisms contain gene families, genes with related sequences. If these arose because of gene duplication, the genes are said to be paralogs; if they arose by speciation, they are called orthologs.

12.11

Organisms may acquire genes from other organisms in their environment by horizontal gene transfer. Such transfer may cross the domain boundaries between *Bacteria*, *Archaea*, and *Eukarya*.

12.12

Mobile DNA elements, including transposons, integrons, and viruses, are important in genome evolution. Mobile DNA often carries genes encoding antibiotic resistance or virulence factors. Integrons are genetic structures that collect and express promoterless genes carried on gene cassettes.

12.13

Many bacteria contain relatively large chromosomal inserts of foreign origin known as chromosomal islands. These islands contain clusters of genes for specialized functions such as pathogenesis, biodegradation, or symbiosis.

Review of Key Terms

Bioinformatics the use of computational tools to acquire, analyze, store, and access DNA and protein sequences

Chromosomal island a bacterial chromosome region of foreign origin that contains clustered genes for some extra property such as virulence or symbiosis

Codon bias the relative proportions of different codons encoding the same amino acid; it varies in different organisms. Same as codon usage

Gene chip small solid-state supports to which genes or portions of genes are affixed and arrayed spatially in a known pattern (also called microarrays)

Gene family genes related in sequence to each other because of common evolutionary origin

Genome the total complement of genetic information of a cell or a virus

Genomics the discipline that maps, sequences, analyzes, and compares genomes

Homologs genes related in sequence to an extent that implies common genetic ancestry; includes both orthologs and paralogs

Horizontal gene transfer the transfer of genetic information between organisms as opposed to transfer from parent to offspring

Integrase the enzyme that inserts cassettes into an integron

Integron a genetic element that collects and expresses genes carried by cassettes

Interactome the total set of interactions between proteins (or other macromolecules) in an organism

Metabolome the total complement of small molecules and metabolic intermediates of a cell or organism

Metagenome the total genetic complement of all the cells present in a particular environment

Metagenomics the genomic analysis of pooled DNA or RNA from an environmental sample containing organisms that have not been isolated; same as environmental genomics

Microarray small, solid-state supports to which genes or portions of genes are affixed and arrayed spatially in a known pattern (also called gene chips)

Open reading frame (ORF) a sequence of DNA or RNA that could be translated to give a polypeptide

Ortholog a gene in one organism that is similar to a gene in another organism because of descent from a common ancestor (see also *Paralog*)

Paralog a gene whose similarity to one or more other genes in the same organism is the result of gene duplication (see also *Ortholog*)

Pathogenicity island a bacterial chromosome region of foreign origin that contains clustered genes for virulence

Primer an oligonucleotide to which DNA polymerase attaches the first deoxyribonucleotide during DNA synthesis

Proteome the total set of proteins encoded by a genome or the total protein complement of an organism

Proteomics the genome-wide study of the structure, function, and regulation of the proteins of an organism

RNA editing changing the coding sequence of an RNA molecule by altering, adding, or removing bases

Sequencing deducing the order of nucleotides in a DNA or RNA molecule by a series of chemical reactions

Shotgun sequencing sequencing of DNA from previously cloned small fragments of a genome in a random fashion, followed by computational methods to reconstruct the entire genome sequence.

Transcriptome the complement of all RNA produced in an organism under a specific set of conditions

Review Questions

1. Briefly describe the Sanger DNA sequencing method (Section 12.2).

2. What characteristics are used to identify open reading frames using sequence data (Section 12.2)?

3. What is the relationship between genome size and ORF content of prokaryotic genomes (Section 12.3)?

4. As a proportion of the total genome, which class of genes predominates in organisms with a small genome? In organisms with a large genome (Section 12.3)?

5. What is RNA editing? Briefly outline the two forms of RNA editing (Section 12.4).

6. How does your genome compare with that of yeast in overall size and gene number (Section 12.5)?

7. How can gene expression be measured in uncultured bacteria (Section 12.6)?

8. Most of the genetic information on our planet does not belong to cellular organisms. Discuss (Section 12.6).

9. Distinguish between the terms metagenomics, proteomics, and metabolomics (Sections 12.6, 12.8, and 12.9).

10. Why is the term proteome ambiguous (Section 12.8)?

11. What does a 2D protein gel show? How can the results of such a gel be correlated with protein function (Section 12.8)?

12. Why is investigation of the metabolome lagging behind that of the proteome (Section 12.9)?

13. What is the major difference in how duplications have contributed to the evolution of prokaryotic versus eukaryotic genomes (Section 12.10)?

14. Explain how horizontally transferred genes can be detected in a genome (Section 12.11).

15. Explain how transposons and insertion sequences promote the genome evolution of *Bacteria* (Section 12.12).

16. Explain how chromosomal islands might be expected to move between different bacterial hosts (Section 12.13).

Application Questions

1. Although the sequence of the yeast nuclear genome was published, the entire sequence was never actually completely determined. Describe the practical difficulties that were encountered in the sequencing.

2. Describe how one might determine which proteins in *Escherichia coli* are repressed when a culture is shifted from a minimal medium (containing only a single carbon source) to a rich medium containing many amino acids, bases, and vitamins. How might one study which genes are expressed during each growth condition?

3. In *Bacteria* and *Archaea* the acronym ORF is almost a synonym for the word "gene." However, in eukaryotes this is not, strictly speaking, true. Discuss, giving examples.

4. The gene encoding the beta subunit of RNA polymerase from *Escherichia coli* is said to be orthologous to the *rpoB* gene of *Bacillus subtilis*. What does that mean about the relationship between the two genes? What protein do you suppose the *rpoB* gene of *B. subtilis* encodes? The genes for the different sigma factors of *E. coli* are paralogous. What does that say about the relationship among these genes?

Need more practice? Test your understanding with Quantitative Questions; access additional study tools including tutorials, animations, and videos; and then test your knowledge with chapter quizzes and practice tests at **www.microbiologyplace.com**.

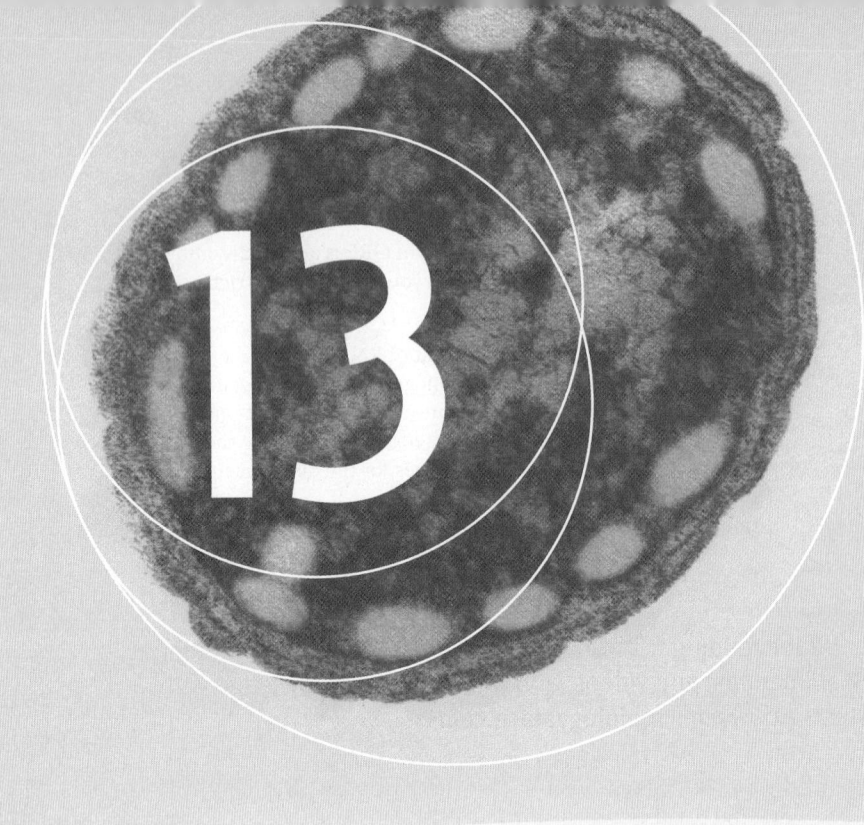

13

Phototrophy, Chemolithotrophy, and Major Biosyntheses

The enormous light-harvesting capacity of the chlorosomes of green bacteria (green structures in the cell periphery of this electron micrograph) allow these phototrophs to grow at the lowest light intensities for any known phototroph.

Amajor theme of microbiology is the great *phylogenetic* diversity of microbial life on Earth. We got a taste of this diversity in Chapter 2. In this unit we focus on *metabolic* diversity of microorganisms with special emphasis on the biochemical processes behind this diversity. We will then return to phylogenetic diversity with the necessary background to place it in the context of the great metabolic diversity characteristic of microbial life.

In this chapter we focus on the metabolic diversity of phototrophs and chemolithotrophs, organisms that use light or inorganic compounds, respectively, as their sources of energy. In addition, we discuss two important biosyntheses: *autotrophy*, the fixation of carbon dioxide (CO_2) into cell material, and *nitrogen fixation*, the reduction of atmospheric nitrogen (N_2) to ammonia (NH_3) to supply the cell's nitrogen requirements. In the next chapter we consider the metabolic diversity of chemoorganotrophs, especially the many forms of anaerobic metabolism in which prokaryotes excel.

 # Phototrophy

Phototrophy—the use of light energy—is widespread in the microbial world. In the first five sections we describe the major forms of phototrophy, including that which forms the very oxygen we breathe, and see how light energy is converted into the chemical energy of ATP.

13.1 Photosynthesis

The most important biological process on Earth is **photosynthesis**, the conversion of light energy to chemical energy. Organisms that carry out photosynthesis are called **phototrophs** (**Figure 13.1**). Most phototrophic organisms are also **autotrophs**, capable of growing with CO_2 as the sole carbon source. Energy from light is used in the reduction of CO_2 to organic compounds (*photoautotrophy*). However, some phototrophs use organic carbon as their carbon source; this lifestyle is called *photoheterotrophy* (Figure 13.1).

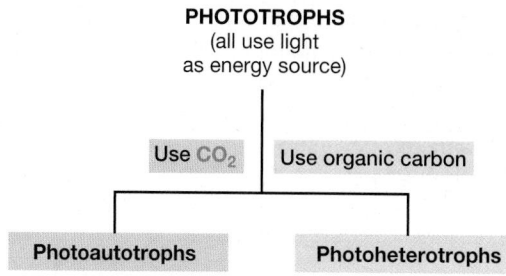

Figure 13.1 Classification of phototrophic organisms in terms of energy and carbon sources. Some photoautotrophs can also grow as photoheterotrophs when the opportunity arises.

Photosynthesis requires light-sensitive pigments, the *chlorophylls*, found in plants, algae, and several groups of prokaryotes. Sunlight reaches phototrophic organisms in packets of energy called *quanta*. Absorption of light energy by chlorophylls begins the process of photosynthetic energy conversion, and the net result is chemical energy, ATP.

Photoautotrophy (Figure 13.1) requires that two distinct sets of reactions operate in parallel: (1) ATP production and (2) CO_2 reduction to cell material. For autotrophic growth, energy is supplied from ATP, and electrons for the reduction of CO_2 come from NADH (or NADPH). The latter are produced by the reduction of NAD^+ (or $NADP^+$) by electrons originating from various electron donors. Some phototrophic bacteria obtain reducing power from electron donors in their environment, such as reduced sulfur sources, for example hydrogen sulfide (H_2S), or from hydrogen (H_2). By contrast, green plants, algae, and cyanobacteria use electrons from water (H_2O) as reducing power.

The oxidation of H_2O produces molecular oxygen (O_2) as a by-product. Because O_2 is produced, photosynthesis in these organisms is called **oxygenic photosynthesis**. However, in many phototrophic bacteria H_2O is not oxidized and O_2 is not produced, and thus the process is called **anoxygenic photosynthesis** (**Figure 13.2**). Oxygen, originally produced on Earth by the oxygenic photosynthesis of cyanobacteria (⮌ Figure 1.6),

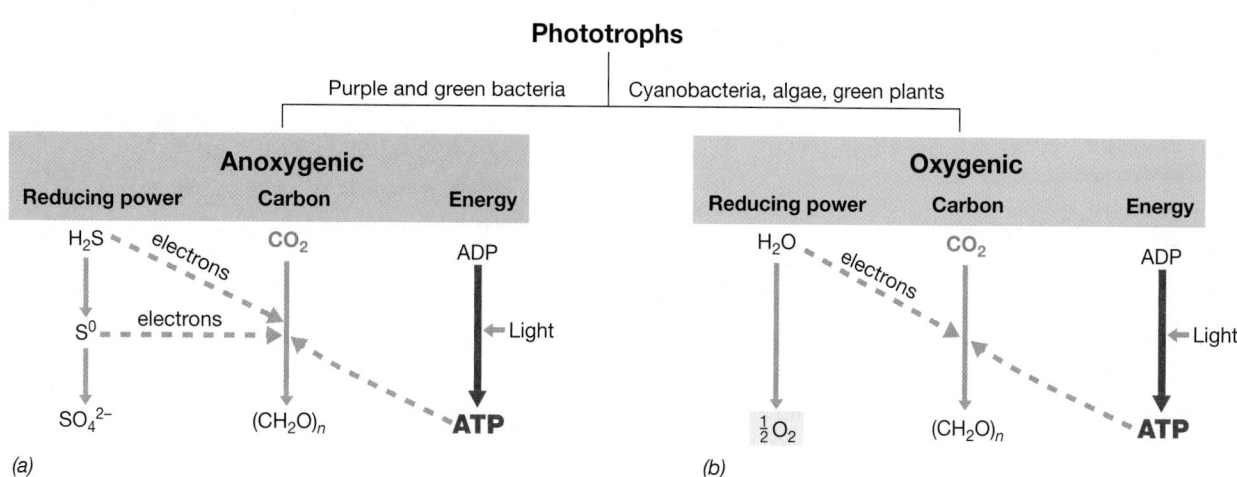

Figure 13.2 Patterns of photosynthesis. Energy and reducing power synthesis in *(a)* anoxygenic and *(b)* oxygenic phototrophs. Note that oxygenic phototrophs produce O_2, while anoxygenic phototrophs do not.

oxygenated the planet, converting it from an anoxic world able to support only anaerobic metabolisms to an oxic world where O_2 plays a key role in the biochemistry of many prokaryotes and virtually all eukaryotes. Oxygenation of Earth set the stage for an explosion of eukaryotic microbial diversity and eventually the rise of plants and animals.

13.2 Chlorophylls and Bacteriochlorophylls

Phototrophic organisms contain some form of **chlorophyll** (oxygenic phototrophs) or **bacteriochlorophyll** (anoxygenic phototrophs). Chlorophylls are related to porphyrins, tetrapyrroles that are the parent structure of the cytochromes (⭯ Section 4.9). But unlike cytochromes, chlorophylls contain magnesium instead of iron at the center of the ring. Chlorophylls also contain specific substituents on the rings as well as a hydrophobic alcohol that helps to anchor the chlorophyll into photosynthetic membranes.

The structure of chlorophyll *a*, the principal chlorophyll of higher plants, most algae, and the cyanobacteria, is shown in **Figure 13.3*a***. Chlorophyll *a* is green in color because it *absorbs* red and blue light preferentially and *transmits* green light. The spectral properties of any pigment can best be expressed by its *absorption spectrum*, a measure of the absorbance of the pigment at different wavelengths. The absorption spectrum of cells containing chlorophyll *a* shows strong absorption of red light

(maximum absorption at a wavelength of 680 nm) and blue light (maximum at 430 nm) (Figure 13.3*b*).

Chlorophyll Diversity

There are a number of different chlorophylls and bacteriochlorophylls, and each is distinguished by its unique absorption spectrum. Chlorophyll *b*, for instance, absorbs maximally at 660 nm rather than the 680-nm absorbance maximum of chlorophyll *a*. All plants contain chlorophylls *a* and *b*. Some prokaryotes contain chlorophyll *d*, while chlorophyll *c* is found only in certain eukaryotic phototrophs.

Among prokaryotes, cyanobacteria produce chlorophyll *a* and prochlorophytes produce chlorophylls *a* and *b*. Anoxygenic phototrophs, such as the phototrophic purple and green bacteria, produce one or more bacteriochlorophylls (**Figure 13.4**). Bacteriochlorophyll *a* (Figure 13.3), present in most purple bacteria (⭯ Section 17.2), absorbs maximally between 800 and 925 nm, depending on the species. Different species produce slightly different pigment-binding proteins, and the absorption maxima of bacteriochlorophyll *a* in any given organism depends to some degree on the nature of these proteins and how they are arranged to form photocomplexes in the photosynthetic membrane (see Figure 13.6). Other bacteriochlorophylls, whose distribution runs along phylogenetic lines, absorb in other regions of the visible and infrared spectrum (Figure 13.4).

Why do different phototrophs have different forms of chlorophyll or bacteriochlorophyll that absorb light of different wavelengths? This allows phototrophs to make better use of the available energy in the electromagnetic spectrum. Only light energy that is *absorbed* is useful for energy conservation. By having different pigments with different absorption properties, different phototrophs can coexist in the same habitat, each organism using wavelengths of light that others are not using.

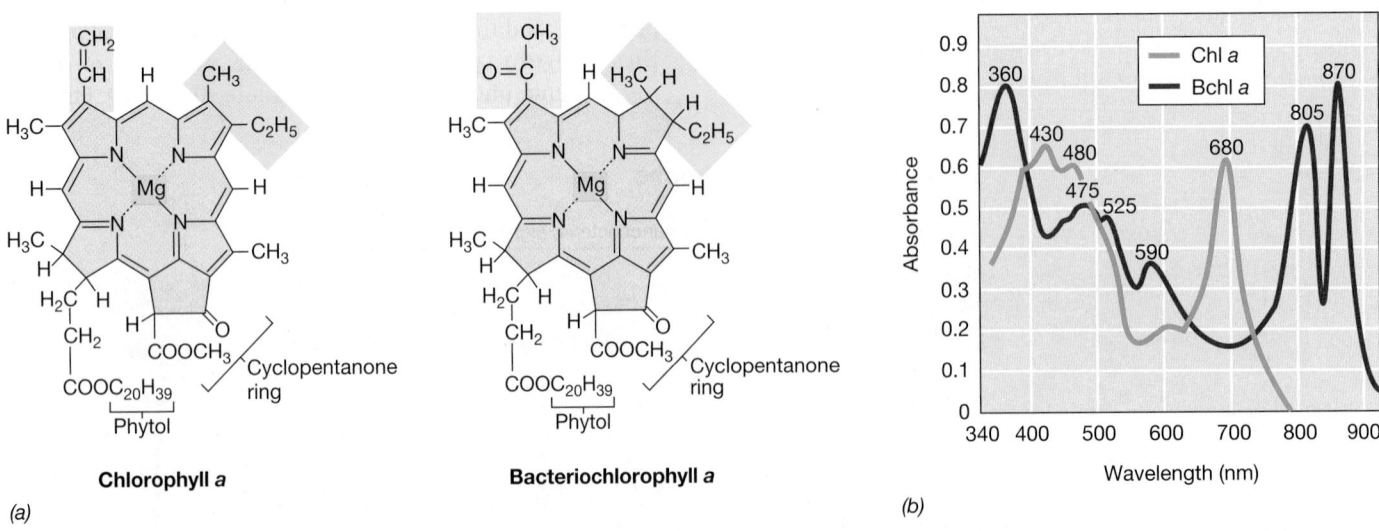

(a)   (b)

Figure 13.3 **Structures and spectra of chlorophyll a and bacteriochlorophyll a.** *(a)* The two molecules are identical except for those portions contrasted in yellow and green. *(b)* Absorption spectrum (green curve) of cells of the green alga *Chlamydomonas*. The peaks at 680 and 430 nm are due to chlorophyll *a*, and the peak at 480 nm is due to carotenoids. Absorption spectrum (red curve) of cells of the phototrophic purple bacterium *Rhodopseudomonas palustris*. Peaks at 870, 805, 590, and 360 nm are due to bacteriochlorophyll *a*, and peaks at 525 and 475 nm are due to carotenoids.

Pigment/Absorption maxima (*in vivo*)	R_1	R_2	R_3	R_4	R_5	R_6	R_7
Bchl *a* (purple bacteria)/ 805, 830–890 nm	—C(=O)—CH$_3$	—CH$_3$[a]	—CH$_2$—CH$_3$	—CH$_3$	—C(=O)—O—CH$_3$	P/Gg[b]	—H
Bchl *b* (purple bacteria)/ 835–850, 1020–1040 nm	—C(=O)—CH$_3$	—CH$_3$[c]	=C(H)—CH$_3$	—CH$_3$	—C(=O)—O—CH$_3$	P	—H
Bchl *c* (green sulfur bacteria)/745–755 nm	—C(H)(OH)—CH$_3$	—CH$_3$	—C$_2$H$_5$ / —C$_3$H$_7$[d] / —C$_4$H$_9$	—C$_2$H$_5$ / —CH$_3$	—H	F	—CH$_3$
Bchl *c$_s$* (green nonsulfur bacteria)/740 nm	—C(H)(OH)—CH$_3$	—CH$_3$	—C$_2$H$_5$	—CH$_3$	—H	S	—CH$_3$
Bchl *d* (green sulfur bacteria)/705–740 nm	—C(H)(OH)—CH$_3$	—CH$_3$	—C$_2$H$_5$ / —C$_3$H$_7$ / —C$_4$H$_9$	—C$_2$H$_5$ / —CH$_3$	—H	F	—H
Bchl *e* (green sulfur bacteria)/719–726 nm	—C(H)(OH)—CH$_3$	—C(H)(=O)	—C$_2$H$_5$ / —C$_3$H$_7$ / —C$_4$H$_9$	—C$_2$H$_5$	—H	F	—CH$_3$
Bchl *g* (heliobacteria)/ 670, 788 nm	—C(H)=CH$_2$	—CH$_3$[a]	—C$_2$H$_5$	—CH$_3$	—C(=O)—O—CH$_3$	F	—H

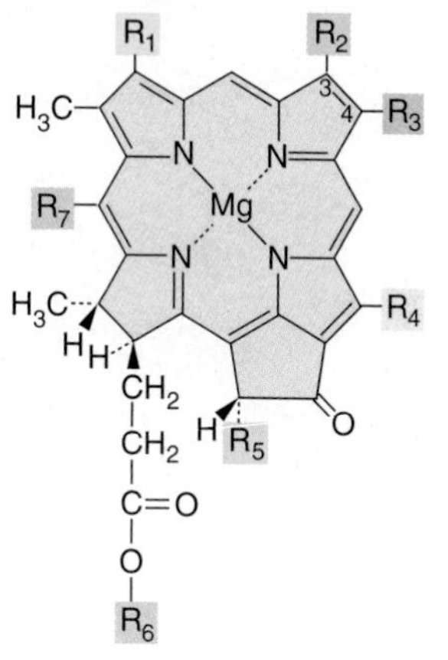

[a] No double bond between C$_3$ and C$_4$; additional H atoms are in positions C$_3$ and C$_4$.

[b] P, Phytyl ester (C$_{20}$H$_{39}$O—); F, farnesyl ester (C$_{15}$H$_{25}$O—); Gg, geranylgeraniol ester (C$_{10}$H$_{17}$O—); S, stearyl alcohol (C$_{18}$H$_{37}$O—).

[c] No double bond between C$_3$ and C$_4$; an additional H atom is in position C$_3$.

[d] Bacteriochlorophylls *c*, *d*, and *e* consist of isomeric mixtures with the different substituents on R$_3$ as shown.

Figure 13.4 Structure of all known bacteriochlorophylls (Bchl). The different substituents present in the positions R$_1$ to R$_7$ in the structure at the right are listed. Absorption properties can be determined by suspending intact cells of a phototroph in a viscous liquid such as 60% sucrose (this reduces light scattering and smooths out spectra) and running absorption spectra as shown in Figure 13.3*b*. *In vivo* absorption maxima are the physiologically relevant absorption peaks. The spectrum of bacteriochlorophylls extracted from cells and dissolved in organic solvents is often quite different.

Thus, pigment diversity has *ecological* significance for the successful coexistence of different phototrophs in the same habitat.

Photosynthetic Membranes and Chloroplasts

The chlorophyll pigments and all the other components of the light-gathering apparatus are present within membrane systems, the photosynthetic membranes. The location of the *photosynthetic membranes* differs between prokaryotic and eukaryotic microorganisms. In eukaryotic phototrophs, photosynthesis is associated with intracellular organelles, the chloroplasts, where the chlorophylls are attached to sheetlike membranes within the chloroplast (**Figure 13.5**). These photosynthetic membrane systems are called **thylakoids**; stacks of thylakoids form grana. The thylakoids are arranged so that the chloroplast is divided into two regions, the matrix space that surrounds the thylakoids and the inner space within the thylakoid array. This arrangement makes possible the generation of a light-driven proton motive force that is used to synthesize ATP, as will be described in Section 13.5.

Prokaryotes do not contain chloroplasts. Their photosynthetic pigments are integrated into internal membrane systems. These systems arise (1) from invagination of the cytoplasmic membrane (purple bacteria) (see Figure 13.12); (2) from the cytoplasmic membrane itself (heliobacteria; ⮌ Section 18.2); (3) in both the cytoplasmic membrane and specialized structures enclosed in a nonunit membrane, called chlorosomes (green bacteria; see Figure 13.7 and ⮌ Section 18.15); or (4) in thylakoid membranes (cyanobacteria, ⮌ Section 18.7).

Reaction Centers and Antenna Pigments

Chlorophyll or bacteriochlorophyll molecules within a photosynthetic membrane are attached to proteins to form photocomplexes consisting of anywhere from 50 to 300 molecules. Only a small number of these pigment molecules, called **reaction centers**,

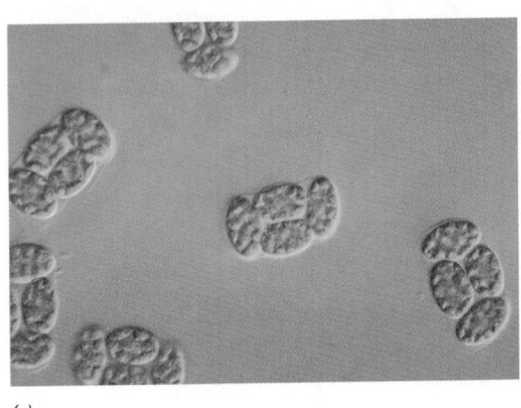

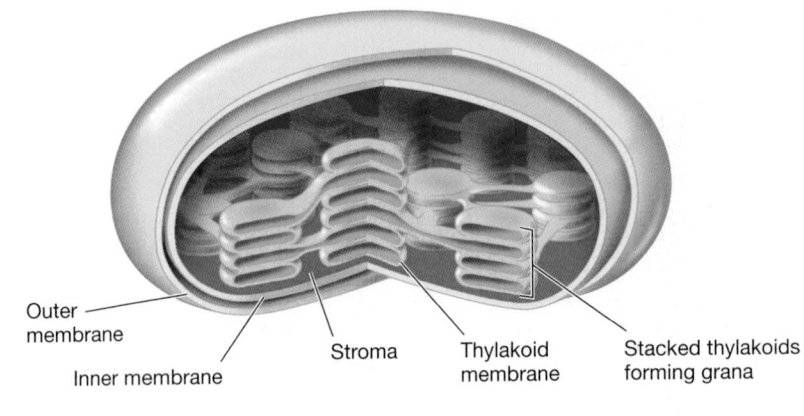

(a)   (b)

Figure 13.5 **The chloroplast.** *(a)* Photomicrograph of cells of the green alga *Makinoella*. Each of the four cells in a cluster contains several chloroplasts. *(b)* Details of chloroplast structure, showing how the convolutions of the thylakoid membranes define an inner space called the stroma and form membrane stacks called grana.

participate directly in reactions where light energy is converted into ATP (**Figure 13.6**). Reaction center chlorophylls or bacteriochlorophylls are surrounded by more numerous light-harvesting chlorophylls/bacteriochlorophylls. These **antenna pigments** (also called light-harvesting pigments) function to absorb light and funnel its energy to the reaction center (Figure 13.6). At the low light intensities that are often found in nature, this arrangement for concentrating energy allows reaction centers to collect light energy that would otherwise not be available to them.

Chlorosomes

The ultimate structure for capturing light at low intensity is the **chlorosome** (**Figure 13.7**). Chlorosomes are present in green sulfur bacteria (*Chlorobium*, ⮑ Section 18.15) and green nonsulfur bacteria (*Chloroflexus*, ⮑ Section 18.18). Chlorosomes are essentially giant antenna systems, but unlike the antenna pigments of other phototrophs, the bacteriochlorophyll molecules in the chlorosome are not attached to proteins. Chlorosomes contain bacteriochlorophyll *c*, *d*, or *e* (Figure 13.4) arranged in dense rodlike arrays running along the long axis of the structure. Light energy absorbed by these antenna pigments is transferred to bacteriochlorophyll *a* in the reaction center in the cytoplasmic membrane. This is accomplished through the Fenna–Matthews–Olson (FMO) protein, a protein that contains bacteriochlorophyll *a* and mediates energy transfer from the chlorosome to the reaction center (Figure 13.7).

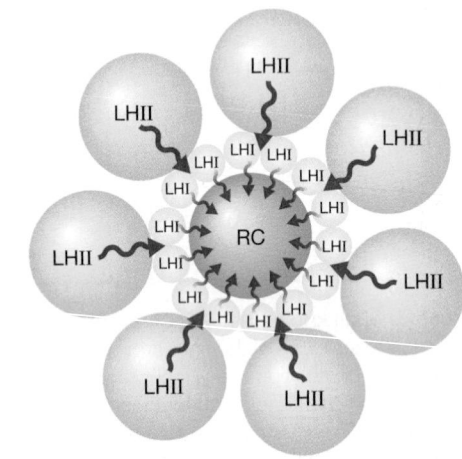

(a)

Figure 13.6 **Arrangement of light-harvesting chlorophylls/bacteriochlorophylls and reaction centers within a photosynthetic membrane.** *(a)* Light energy absorbed by light-harvesting (LH) molecules (light green) is transferred to the reaction centers (dark green, RC) where photosynthetic electron transport reactions begin. Pigment molecules are secured within the membrane by specific pigment-binding proteins. Compare this figure to Figures 13.13 and 13.15. *(b)* Atomic force micrograph of photocomplexes of the purple bacterium *Phaeospirillum molischianum*. This organism has two types of light-harvesting complexes, LHI and LHII. LHII complexes transfer energy to LHI complexes, and these transfer energy to the reaction center.

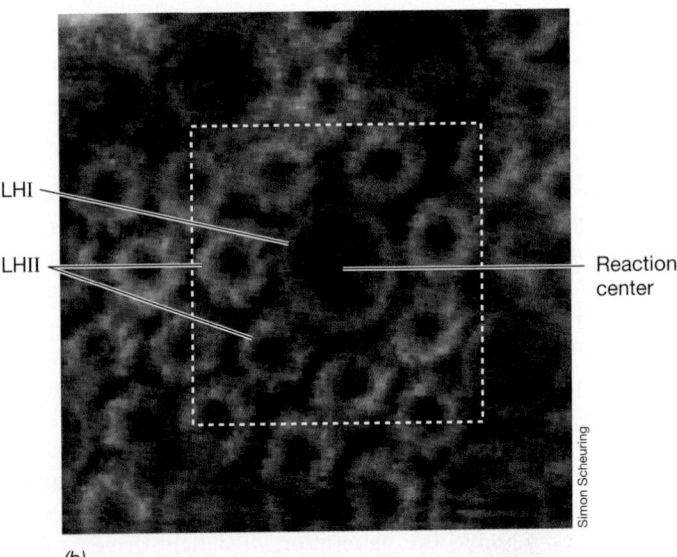

(b)

ety of organisms that form in hot springs and highly saline environments (⟿ Section 23.4). Microbial mats experience a steep light gradient, with light levels even a few millimeters into the mat approaching darkness. Thus, chlorosomes allow green nonsulfur bacteria in the mat to grow phototrophically with only the minimal light intensities available.

MiniQuiz

- What is the difference between the numbers of antenna and reaction center chlorophyll/bacteriochlorophyll molecules in a photosynthetic complex, and why?
- What pigments are found within the chlorosome?

13.3 Carotenoids and Phycobilins

Although chlorophyll or bacteriochlorophyll is required for photosynthesis, phototrophic organisms contain an assortment of accessory pigments as well. These include, in particular, the carotenoids and phycobilins. Carotenoids primarily play a photoprotective role in both anoxygenic and oxygenic phototrophs, while phycobilins function in energy metabolism as the major light-harvesting pigments in cyanobacteria.

Carotenoids

The most widespread accessory pigments in phototrophs are the **carotenoids**. Carotenoids are hydrophobic light-sensitive pigments that are firmly embedded in the photosynthetic membrane. **Figure 13.8** shows the structure of a common carotenoid, β-carotene. Carotenoids are typically yellow, red, brown, or green in color and absorb light in the blue region of the spectrum (Figure 13.3). The major carotenoids of anoxygenic phototrophs are shown in **Figure 13.9**. Because they tend to mask the color of bacteriochlorophylls, carotenoids are responsible for the brilliant colors of red, purple, pink, green, yellow, or brown that are observed in different species of anoxygenic phototrophs (see Figure 13.16 and ⟿ Figure 17.3).

Carotenoids are closely associated with chlorophyll or bacteriochlorophyll in photosynthetic complexes, and energy absorbed by carotenoids can be transferred to the reaction center. Nevertheless, carotenoids primarily function in phototrophic organisms as photoprotective agents. Bright light can be harmful to cells because it can catalyze photooxidation reactions that can lead to the production of toxic forms of oxygen, such as singlet oxygen ($^1O_2$) (⟿ Section 5.18). Singlet oxygen can oxidize components of the photosynthetic apparatus itself and render it nonfunctional. Carotenoids quench toxic oxygen species by absorbing much of this harmful light and prevent these dangerous photooxidations. Because phototrophic organisms must by their very nature live in

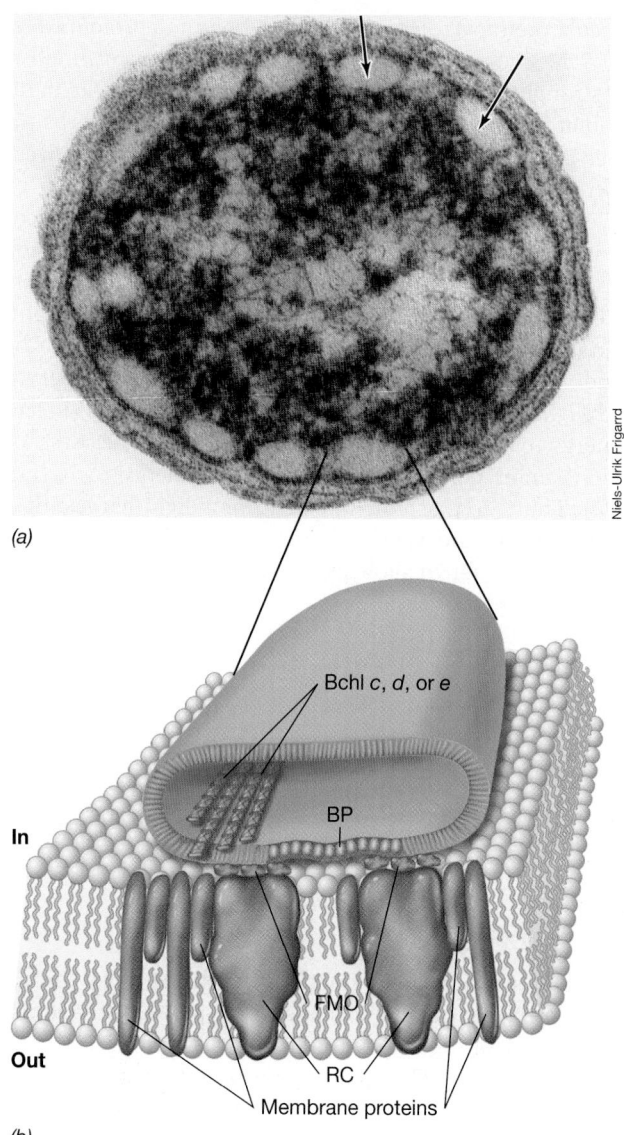

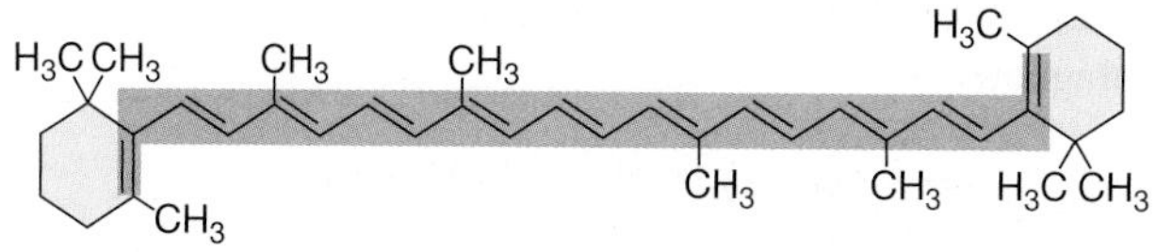

Figure 13.7 The chlorosome of green sulfur and green nonsulfur bacteria. *(a)* Transmission electron micrograph of a cross-section of a cell of the green sulfur bacterium *Chlorobaculum tepidum*. Note the chlorosomes (arrows). *(b)* Model of chlorosome structure. The chlorosome (green) lies appressed to the inside surface of the cytoplasmic membrane. Antenna bacteriochlorophyll (Bchl) molecules are arranged in tubelike arrays inside the chlorosome, and energy is transferred from these to reaction center (RC) Bchl *a* in the cytoplasmic membrane (blue) through a protein called FMO. Base plate (BP) proteins function as connectors between the chlorosome and the cytoplasmic membrane.

The arrangement of pigments in the chlorosome is remarkably efficient for absorbing light at low intensities. Light energy collected by the antenna pigments is forwarded to the reaction center where it is converted into chemical energy. Because they contain chlorosomes, green bacteria can grow at the lowest light intensities of all known phototrophs. Thus, green sulfur bacteria are typically found in the deepest waters of lakes, inland seas, and other anoxic aquatic habitats where other phototrophs cannot compete. Green nonsulfur bacteria are major components of microbial mats, thick biofilms containing a vari-

Figure 13.8 Structure of β-carotene, a typical carotenoid. The conjugated double-bond system is highlighted in orange.

I. Carotenes

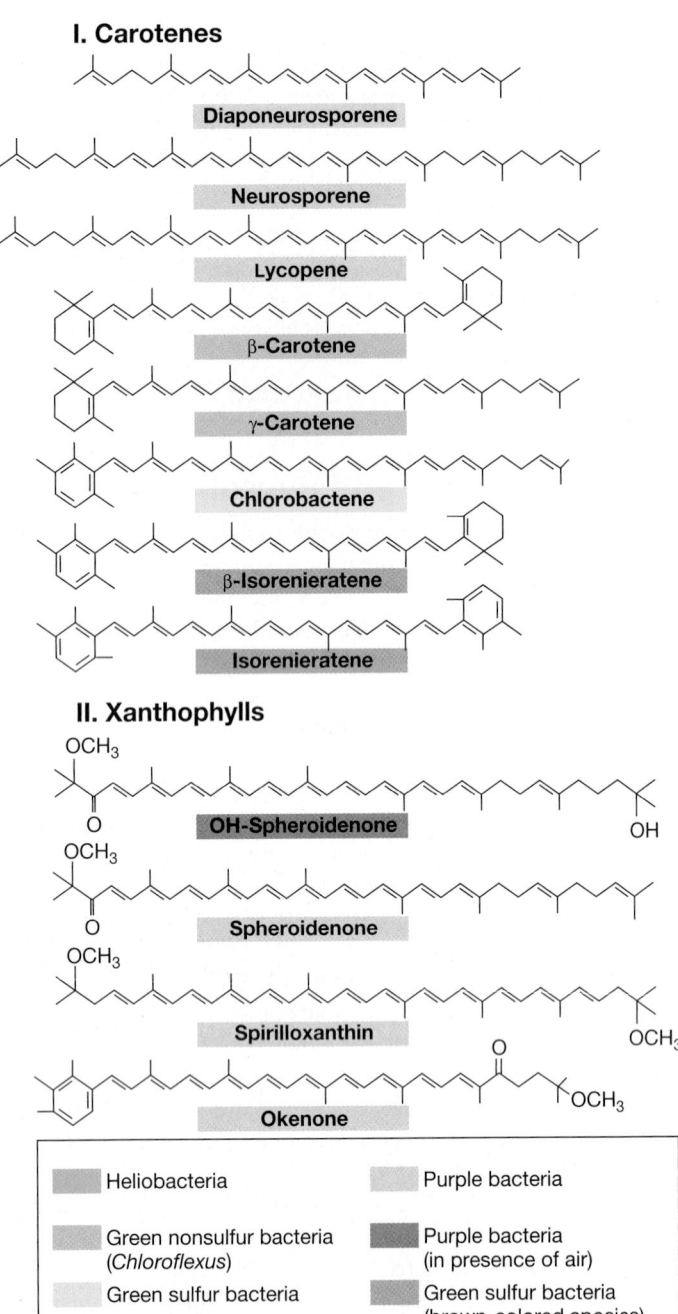

Diaponeurosporene

Neurosporene

Lycopene

β-Carotene

γ-Carotene

Chlorobactene

β-Isorenieratene

Isorenieratene

II. Xanthophylls

OH-Spheroidenone

Spheroidenone

Spirilloxanthin

Okenone

Heliobacteria	Purple bacteria
Green nonsulfur bacteria (*Chloroflexus*)	Purple bacteria (in presence of air)
Green sulfur bacteria	Green sulfur bacteria (brown-colored species)

Figure 13.9 Structures of some common carotenoids found in anoxygenic phototrophs. Carotenes are hydrocarbon carotenoids and xanthophylls are oxygenated carotenoids. Compare the structure of β-carotene shown in Figure 13.8 with how it is drawn here. For simplicity in the structures shown here, methyl (CH₃) groups are designated by bond only.

the light, the photoprotection conferred by carotenoids is clearly advantageous.

Phycobiliproteins and Phycobilisomes

Cyanobacteria and the chloroplasts of red algae contain **phycobiliproteins**, which are the main light-harvesting systems of these phototrophs. Phycobiliproteins consist of red or blue

open-chain tetrapyrroles, called *bilins*, bound to proteins (**Figure 13.10**). The red phycobiliprotein, called *phycoerythrin*, absorbs most strongly at wavelengths around 550 nm, whereas the blue phycobiliprotein, *phycocyanin* (Figure 13.10a), absorbs most strongly at 620 nm (**Figure 13.11**). A third phycobiliprotein, called *allophycocyanin*, absorbs at about 650 nm.

Phycobiliproteins assemble into aggregates called **phycobilisomes** that attach to thylakoids (Figure 13.10b,c). Phycobilisomes are arranged such that the allophycocyanin molecules are in direct contact with the photosynthetic membrane. Allophycocyanin is surrounded by phycocyanin or phycoerythrin (or both, depending on the organism). Phycocyanin and phycoerythrin absorb light of shorter wavelengths (higher energy) and transfer the energy to allophycocyanin, which is positioned closest to the reaction center chlorophyll and transfers energy to it (Figure 13.10b). Thus, in a fashion similar to how light-harvesting systems function in anoxygenic phototrophs, phycobilisomes facilitate energy transfer to allow cyanobacteria to grow at fairly low light intensities.

MiniQuiz

- In which phototrophs are carotenoids found? Phycobiliproteins?
- How does the structure of a phycobilin compare with that of a chlorophyll?
- Phycocyanin is blue-green. What are the wavelengths of light it is absorbing, and how do you know this?

13.4 Anoxygenic Photosynthesis

In the photosynthetic light reactions, electrons travel through a series of electron carriers arranged in a photosynthetic membrane in order of their increasingly more electropositive reduction potential (E_0'). This generates a proton motive force that drives ATP synthesis. Anoxygenic photosynthesis occurs in at least five phyla of *Bacteria*: the proteobacteria (purple bacteria); green sulfur bacteria; green nonsulfur bacteria; the gram-positive bacteria (heliobacteria); and the acidobacteria.

Photosynthetic Reaction Centers

The photosynthetic apparatus of purple bacteria has been best studied and is embedded in intracytoplasmic membrane systems of various morphologies. Membrane vesicles, sometimes called *chromatophores*, or membrane stacks called *lamellae* are common membrane morphologies (**Figure 13.12**). Reaction centers of purple bacteria consist of three polypeptides, designated L, M, and H. These proteins, along with a molecule of cytochrome *c*, are firmly embedded in the photosynthetic membrane and traverse the membrane several times (**Figure 13.13**).

The L, M, and H polypeptides bind pigments in the reaction center. The photocomplex consists of two molecules of bacteriochlorophyll *a*, called the *special pair*, two additional bacteriochlorophyll *a* molecules that function in photosynthetic electron flow, two molecules of bacteriopheophytin (bacteriochlorophyll *a* minus its magnesium atom), two molecules of quinone, and one carotenoid pigment (Figure 13.13a). All components of the reaction center are integrated in such a way that

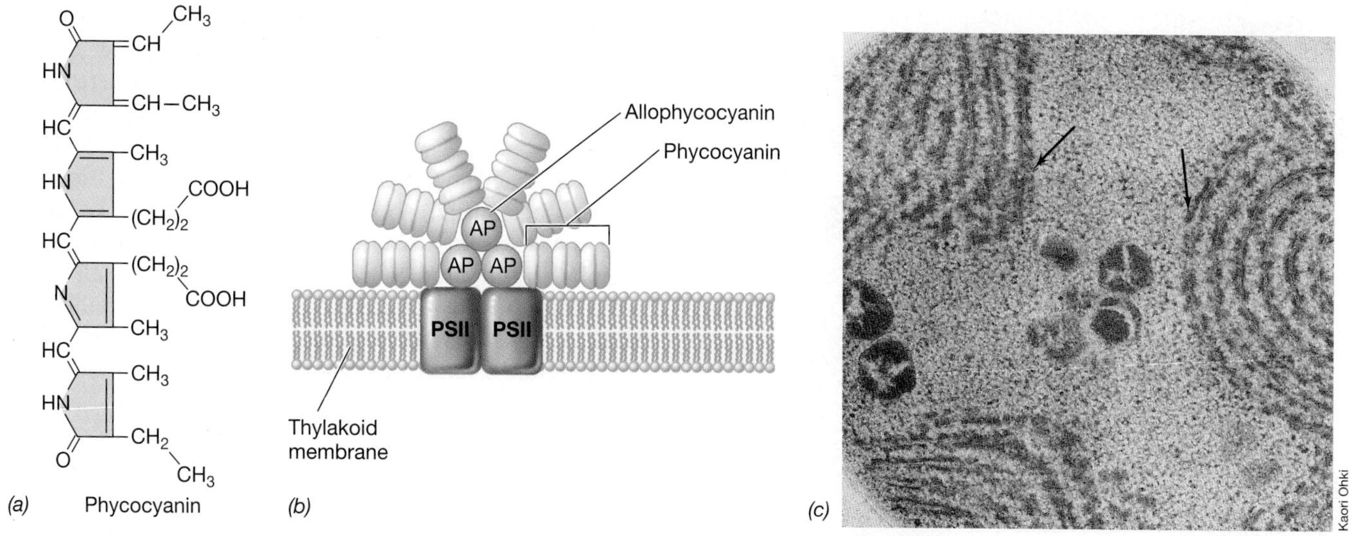

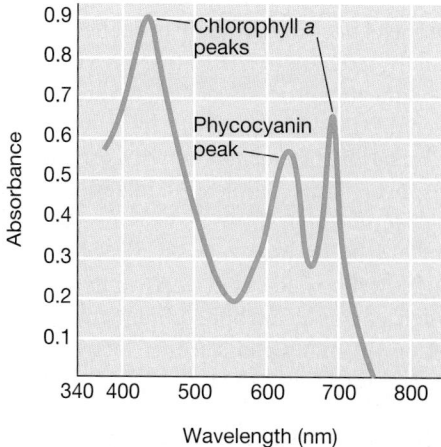

(a)   Phycocyanin   (b)   (c)

Figure 13.10 Phycobiliproteins and phycobilisomes. *(a)* Phycocyanin, a typical bilin. Phycocyanin is an open-chain tetrapyrrole derived biosynthetically from a closed porphyrin ring by loss of one carbon atom as carbon monoxide. The structure shown is the prosthetic group of phycocyanin, found in cyanobacteria (Section 18.7) and red algae (Section 20.19). *(b)* Structure of a phycobilisome. Phycocyanin absorbs at higher energies (shorter wavelengths) than allophycocyanin. Chlorophyll *a* absorbs at longer wavelengths (lower energies) than allophycocyanin. Energy flow is thus phycocyanin → allophycocyanin → chlorophyll *a* of PSII. *(c)* Electron micrograph of a thin section of the cyanobacterium *Synechocystis*. Note the darkly staining ball-like phycobilisomes (arrows) attached to the lamellar membranes.

they can interact in very fast electron transfer reactions in the early stages of photosynthetic energy conversion.

Photosynthetic Electron Flow in Purple Bacteria

Recall that photosynthetic reaction centers are surrounded by antenna pigments that function to funnel light energy to the reaction center (Figure 13.6). The energy of light is transferred from the antenna to the reaction center in packets called *excitons*, mobile forms of energy that migrate at high efficiency through the antenna pigments to the reaction center.

Figure 13.11 Absorption spectrum of a cyanobacterium that contains phycocyanin as an accessory pigment. Note how the presence of phycocyanin broadens the wavelengths of usable light energy (between 600 and 700 nm). Compare with Figure 13.3*b*.

The light reactions begin when exciton energy strikes the special pair of bacteriochlorophyll *a* molecules (Figure 13.13*a*). The absorption of energy excites the special pair, converting it from a relatively weak to a very strong electron donor (very electronegative reduction potential). Once this strong donor has been produced, the remaining steps in electron flow simply conserve the energy released when electrons flow through a membrane from carriers of low E_0' to those of high E_0', generating a proton motive force (**Figure 13.14**).

Before excitation, the purple bacterial reaction center, which is called *P870*, has an E_0' of about +0.5 V; after excitation, it has a potential of about −1.0 V (Figure 13.14). The excited electron within P870 proceeds to reduce a molecule of bacteriochlorophyll *a* within the reaction center (Figures 13.13*a* and 13.14). This transition takes place incredibly fast, taking only about three-trillionths (3×10^{-12}) of a second. Once reduced, bacteriochlorophyll *a* reduces bacteriopheophytin *a* and the latter reduces quinone molecules within the membrane. These transitions are also very fast, taking less than one-billionth of a second (Figure 13.14 and **Figure 13.15**). Relative to what has happened in the reaction center, further electron transport reactions proceed rather slowly, on the order of microseconds to milliseconds.

From the quinone, electrons are transported in the membrane through a series of iron–sulfur proteins and cytochromes (Figures 13.14 and 13.15), eventually returning to the reaction center. Key electron transport proteins include cytochrome bc_1 and cytochrome c_2 (Figure 13.14). Cytochrome c_2 is a periplasmic cytochrome (recall that the periplasm is the region between the cytoplasmic membrane and the outer membrane in gram-negative bacteria, Section 3.7) that functions as an electron

UNIT 5

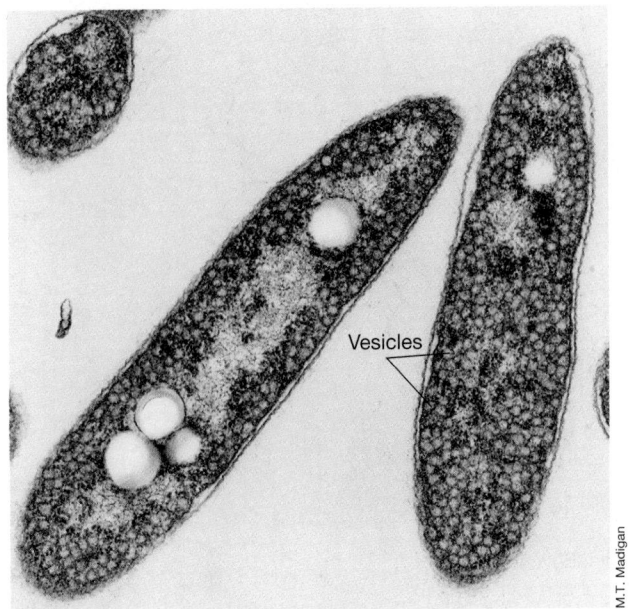

Vesicles

(a)

M.T. Madigan

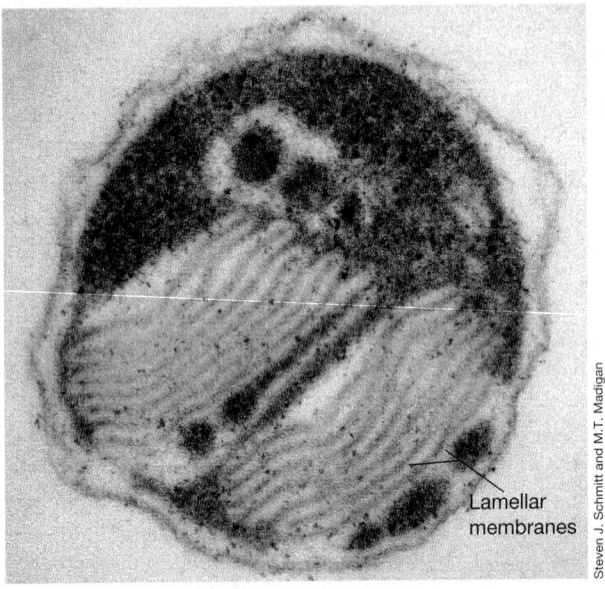

Lamellar
membranes

Steven J. Schmitt and M.T. Madigan

(b)

Figure 13.12 Membranes in anoxygenic phototrophs. *(a)* Chromatophores. Section through a cell of the purple bacterium *Rhodobacter* showing vesicular photosynthetic membranes. The vesicles are continuous with and arise by invagination of the cytoplasmic membrane. A cell is about 1 μm wide. *(b)* Lamellar membranes in the purple bacterium *Ectothiorhodospira*. A cell is about 1.5 μm wide. These membranes are also continuous with and arise from invagination of the cytoplasmic membrane, but instead of forming vesicles, they form membrane stacks.

shuttle between the membrane-bound bc_1 complex and the reaction center (Figures 13.14 and 13.15).

Photophosphorylation

ATP is synthesized during photosynthetic electron flow from the activity of ATPase that couples the proton motive force to ATP formation (⇄ Section 4.10). Electron flow is completed when

cytochrome c_2 donates an electron to the special pair (Figure 13.14); this returns these bacteriochlorophyll molecules to their original ground-state potential ($E_0' = +0.5$ V). The reaction center is then capable of absorbing new energy and repeating the process.

This mechanism of ATP synthesis is called **photophosphorylation**, specifically *cyclic photophosphorylation*, because electrons move within a closed loop. Cyclic photophosphorylation resembles respiration in that electron flow through the membrane establishes a proton motive force. However, unlike in respiration, in cyclic photophosphorylation there is no net input or consumption of electrons; electrons simply travel a circuitous route.

The spatial relationship of the electron transport components in the purple bacterial photosynthetic membrane is illustrated in Figure 13.15. Note that as in respiratory electron flow (⇄ Section 4.9), the cytochrome bc_1 complex interacts with the quinone pool during photosynthetic electron flow as a major means of establishing the proton motive force used to drive ATP synthesis (Figure 13.15).

Autotrophy in Purple Bacteria: Electron Donors and Reverse Electron Flow

For a purple bacterium to grow as a photoautotroph, the formation of ATP is not enough. Reducing power (NADH or NADPH) is also necessary so that CO_2 can be reduced to the redox level of cell material. As previously mentioned, reducing power for purple sulfur bacteria comes from hydrogen sulfide (H_2S), although sulfur (S^0), thiosulfate ($S_2O_3^{2-}$), ferrous iron (Fe^{2+}), nitrite (NO_2^-), and arsenite (AsO_3^{2-}) can also be used by one or another species. When H_2S is the electron donor in purple sulfur bacteria, globules of S^0 are stored inside the cells (**Figure 13.16a**).

Reduced substances used as photosynthetic electron donors are oxidized and electrons eventually end up in the "quinone pool" of the photosynthetic membrane (Figure 13.14). However, the E_0' of quinone (about 0 volts) is insufficiently electronegative to reduce NAD^+ (-0.32 V) directly. Instead, electrons from the quinone pool travel backwards against the thermodynamic gradient to eventually reduce $NAD(P)^+$ to $NAD(P)H$ (Figure 13.14). This energy-requiring process, called **reverse electron transport**, is driven by the energy of the proton motive force and involves a reversal of the normal activity of "Complex I" of the electron transport chain; Complex I oxidizes NADH and reduces quinone (⇄ Section 4.9). If NADPH is needed as a reductant instead of NADH, it can be produced from NADH by enzymes called *transhydrogenases*. Reverse electron flow is also the mechanism by which chemolithotrophs obtain their reducing power for CO_2 fixation, in many cases from electron donors of quite positive E_0' (Sections 13.6–13.11).

Photosynthesis in Other Anoxygenic Phototrophs

Our discussion of photosynthetic electron flow has thus far focused on the process as it occurs in purple bacteria. Although similar membrane reactions drive photophosphorylation in other

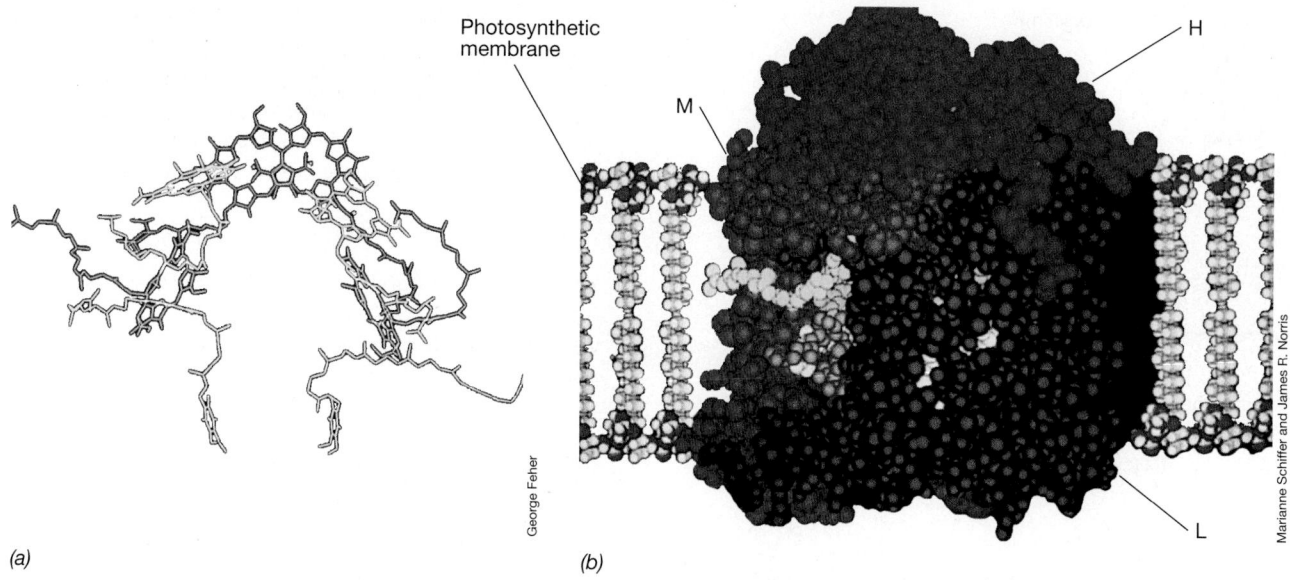

Photosynthetic membrane

M

H

L

George Feher

Marianne Schiffer and James R. Norris

(a)　　　　　(b)

Figure 13.13 Structure of the reaction center of purple phototrophic bacteria. (a) Arrangement of pigment molecules in the reaction center. The "special pair" of bacteriochlorophyll molecules overlap and are shown in orange at the top; quinones are in dark yellow and are at the bottom of the figure. The accessory bacteriochlorophylls are in lighter yellow near the special pair, and the bacteriopheophytin molecules are shown in blue. (b) Molecular model of the protein structure of the reaction center. The pigments described in part a are bound to membranes by protein H (blue), protein M (red), and protein L (green). The reaction center pigment–protein complex is integrated into the lipid bilayer.

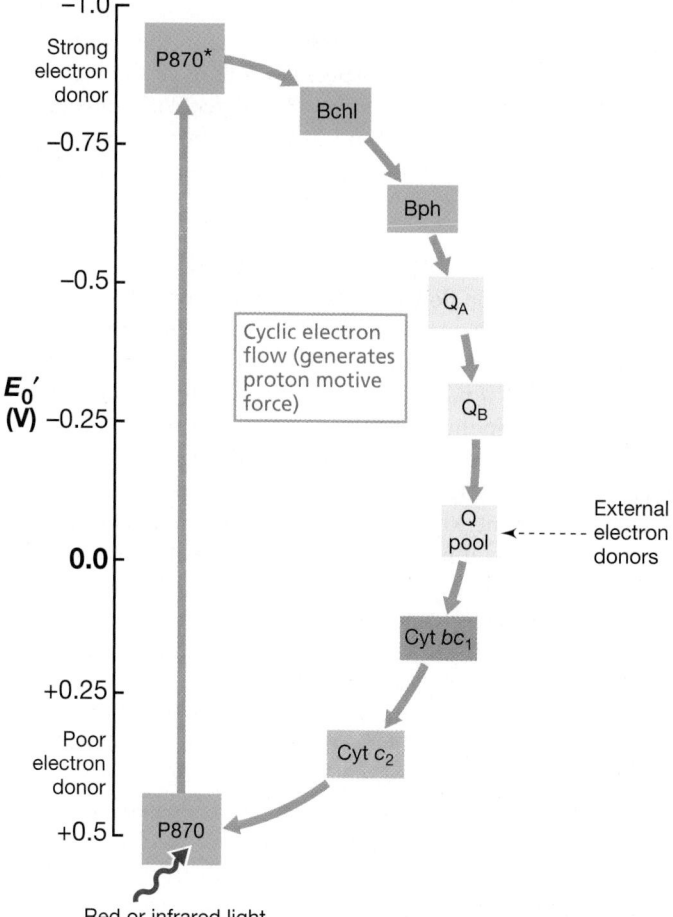

anoxygenic phototrophs, there are differences in certain details. The reaction centers of green nonsulfur bacteria and purple bacteria are structurally quite similar; however, the reaction centers of green sulfur bacteria and heliobacteria differ significantly from those of purple and green nonsulfur bacteria, and this affects some of the components in cyclic electron flow.

Figure 13.17 contrasts photosynthetic electron flow in purple and green bacteria and the heliobacteria. Note that in green bacteria and heliobacteria the excited state of the reaction center bacteriochlorophylls resides at a significantly more electronegative E_0' than in purple bacteria and that actual chlorophyll a (green bacteria) or a structurally modified form of chlorophyll a, called *hydroxychlorophyll a* (heliobacteria), is present in the reaction center. Thus, unlike in purple bacteria, where the first stable acceptor molecule (quinone) has an E_0' of about 0 V (Figure 13.14), the acceptors in green bacteria and heliobacteria iron–sulfur (FeS) proteins have a much more electronegative E_0' than does NADH. This has a major effect on reducing power synthesis in these organisms, as reverse electron flow, necessary in purple bacteria (Figure 13.14), is not required in green sulfur bacteria or heliobacteria.

Figure 13.14 Electron flow in anoxygenic photosynthesis in a purple bacterium. Only a single light reaction occurs. Note how light energy converts a weak electron donor, P870, into a very strong electron donor, P870*, and that following this event, the remaining steps in photosynthetic electron flow are much the same as those of respiratory electron flow. Bph, bacteriopheophytin; Q_A, Q_B, intermediate quinones; Q pool, quinone pool in membrane; Cyt, cytochrome. Compare this figure with Figures 13.15, 13.17, and 13.18.

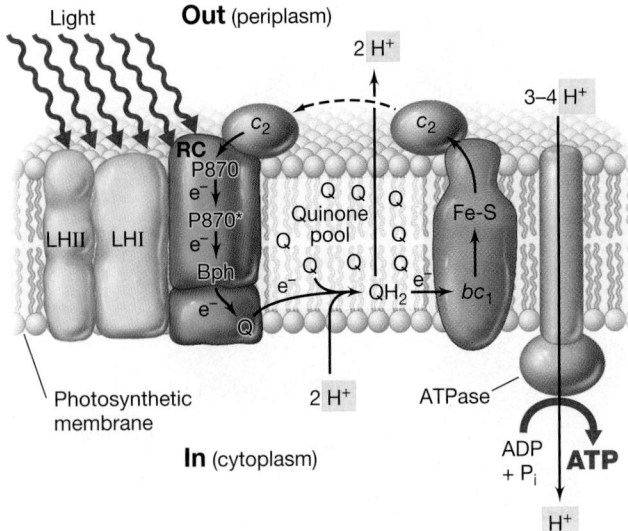

Figure 13.15 Arrangement of protein complexes in the purple bacterium reaction center. The light-generated proton gradient is used in the synthesis of ATP by the ATP synthase (ATPase). LH, light-harvesting bacteriochlorophyll complexes; RC, reaction center; Bph, bacteriopheophytin; Q, quinone; FeS, iron–sulfur protein; bc_1, cytochrome bc_1 complex; c_2, cytochrome c_2. For a description of ATPase function, see Section 4.10.

In green bacteria a protein called *ferredoxin* (reduced by the FeS protein, Figure 13.17) is the direct electron donor for CO_2 fixation. Thus, as in oxygenic phototrophs (to be discussed in the next section), in green bacteria both ATP and reducing power are direct products of the light reactions. A similar situation exists in the heliobacteria, but here the picture is complicated by the fact that autotrophic growth does not occur in heliobacteria (heliobacteria can grow phototrophically only as photoheterotrophs).

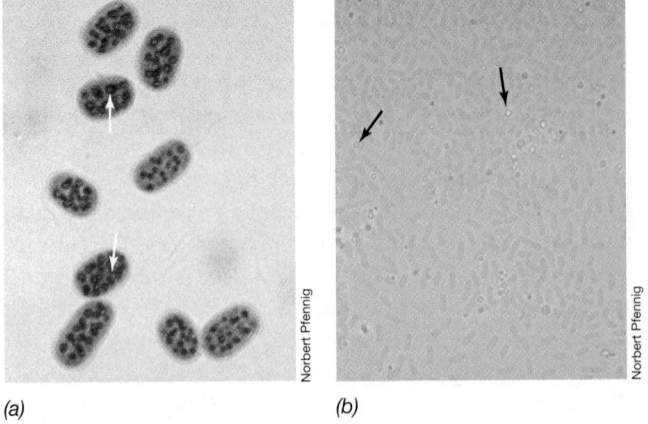

(a)   *(b)*

Figure 13.16 Phototrophic purple and green sulfur bacteria. *(a)* Purple bacterium, *Chromatium okenii*. Notice the sulfur granules deposited inside the cell (arrows). *(b)* Green bacterium, *Chlorobium limicola*. The refractile bodies are sulfur granules deposited outside the cell (arrows). In both cases the sulfur granules arise from the oxidation of H_2S to obtain reducing power. Cells of *C. okenii* are about 5 μm in diameter, and cells of *C. limicola* are about 0.9 μm in diameter. Both micrographs are bright-field images.

When H_2S is the electron donor for the synthesis of reducing power in green bacteria, globules of S^0 are produced as in purple bacteria but the globules are formed *outside* rather than *inside* the cells (Figure 13.16*b*). However, as in purple bacteria, the S^0 eventually disappears as it is oxidized to sulfate (SO_4^{2-}) to generate additional reducing power for CO_2 fixation.

MiniQuiz
- What parallels exist in the processes of photophosphorylation and oxidative phosphorylation?
- What is reverse electron flow and why is it necessary? Which phototrophs need to use reverse electron flow?

13.5 Oxygenic Photosynthesis

In contrast to electron flow in *anoxygenic* phototrophs, electron flow in *oxygenic* phototrophs proceeds through two distinct but interconnected series of light reactions. The two light systems are called *photosystem I* and *photosystem II*, each photosystem having a spectrally distinct form of reaction center chlorophyll *a*. Photosystem I (PSI) chlorophyll, called *P700*, absorbs light at long wavelengths (far red light), whereas PSII chlorophyll, called *P680*, absorbs light at shorter wavelengths (near red light).

As in anoxygenic photosynthesis, the oxygenic light reactions occur in membranes. In eukaryotic cells, these membranes are found in the chloroplast (Figure 13.5), whereas in cyanobacteria, photosynthetic membranes are arranged in stacks within the cytoplasm (Figure 13.10*c*). In both groups of phototrophs the membranes are arranged in a similar way, and the two forms of chlorophyll *a* are attached to specific proteins in the membrane and interact as shown in **Figure 13.18**. Oxygenic phototrophs use light to generate both ATP and NADPH, the electrons for the latter arising from the splitting of water into oxygen and electrons (Figure 13.2).

Electron Flow in Oxygenic Photosynthesis

The path of electron flow in oxygenic phototrophs resembles the letter Z turned on its side, and Figure 13.18 outlines this so-called "Z scheme" of photosynthesis. The reduction potential of the P680 chlorophyll *a* molecule in PSII is very electropositive, even more positive than that of the O_2/H_2O couple. This facilitates the first step in oxygenic electron flow, the splitting of water into oxygen and electrons (Figure 13.18). Light energy converts P680 into a strong reductant which reduces pheophytin *a* (chlorophyll *a* minus its magnesium atom), a molecule with an E_0' of about -0.5 V. An electron from water is then donated to the oxidized P680 molecule to return it to its ground-state reduction potential. From the pheophytin the electron travels through several membrane carriers of increasingly more positive E_0' that include quinones, cytochromes, and a copper-containing protein called *plastocyanin*; the latter donates the electron to PSI. The electron is accepted by the reaction center chlorophyll of PSI, P700, which has previously absorbed light energy and donated an electron that will eventually lead to the reduction of $NADP^+$. Electrons

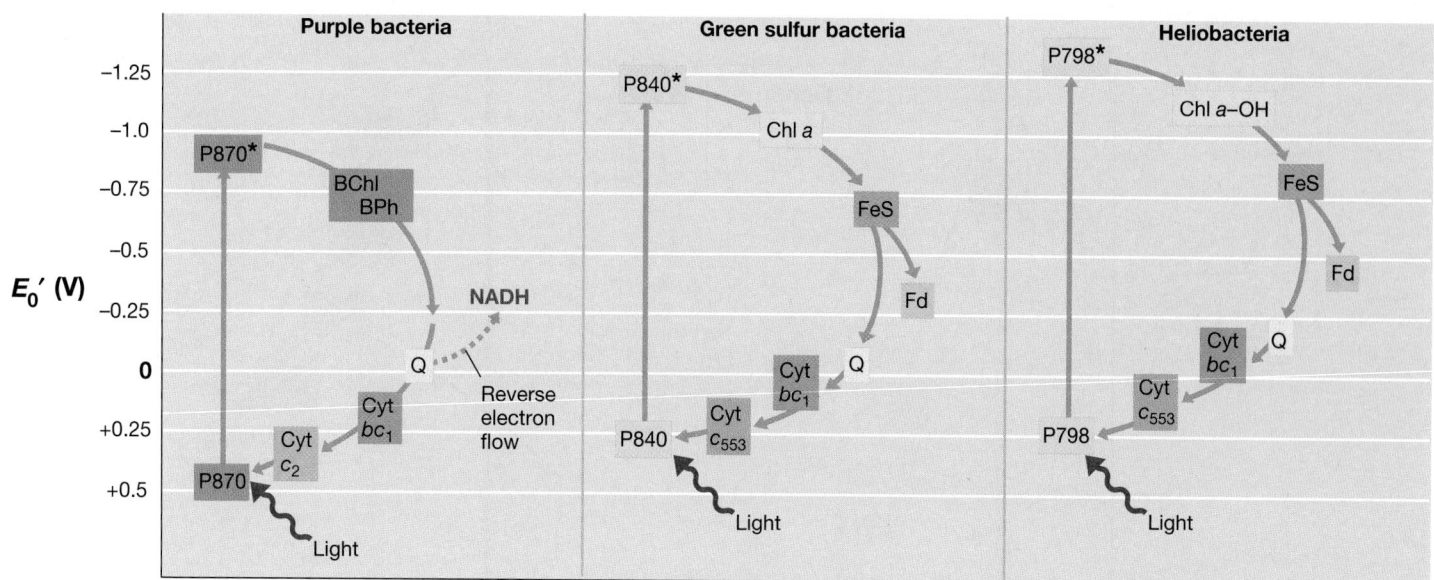

Figure 13.17 A comparison of electron flow in purple bacteria, green sulfur bacteria, and heliobacteria. Note that reverse electron flow in purple bacteria is necessary to produce NADH because the primary acceptor (quinone, Q) is more positive in potential than the $NAD^+/NADH$ couple. In green and heliobacteria, ferredoxin (Fd), whose E_0' is more negative than that of NADH, is produced by light-driven reactions for reducing power needs. Bchl, Bacteriochlorophyll; BPh, bacteriopheophytin. P870 and P840 are reaction centers of purple and green bacteria, respectively, and consist of Bchl a. The reaction center of heliobacteria (P798) contains Bchl g, and the reaction center of *Chloroflexus* is of the purple bacterial type. Note that forms of chlorophyll a are in the reaction centers of green bacteria and heliobacteria.

are transferred through several intermediates terminating with the reduction of $NADP^+$ to NADPH (Figure 13.18).

ATP Synthesis in Oxygenic Photosynthesis

Besides the net synthesis of reducing power (that is, NADPH), other important events take place while electrons flow in the photosynthetic membrane from PSII to PSI. Electron transport generates a proton motive force from which ATP can be produced by ATPase. This mechanism for ATP synthesis is called *noncyclic photophosphorylation* because electrons do not cycle back to reduce the oxidized P680, but instead are used in the reduction of $NADP^+$.

However, when reducing power is sufficient, ATP can also be produced in oxygenic phototrophs by *cyclic photophosphorylation*. This occurs when, instead of reducing $NADP^+$, electrons from ferredoxin are returned to travel the electron transport chain that connects PSII with PSI. In so doing, these electrons also generate a proton motive force that supports additional ATP synthesis (dashed line in Figure 13.18).

Anoxygenic Photosynthesis in Oxygenic Phototrophs and the Evolution of Photosynthesis

Photosystems I and II normally function in tandem in oxygenic photosynthesis. However, under certain conditions, for example if PSII activity is blocked, some algae and cyanobacteria can carry out cyclic photophosphorylation using only PSI, obtaining reducing power for CO_2 reduction from sources other than water. This is, in effect, anoxygenic photosynthesis (Figure 13.14).

Many cyanobacteria can use H_2S as an electron donor for anoxygenic photosynthesis, whereas many green algae can use H_2. When H_2S is used, it is oxidized to elemental sulfur (S^0), and sulfur granules similar to those produced by green sulfur bacteria (Figure 13.16b) are deposited outside the cyanobacterial cells. **Figure 13.19** shows an example of this with the filamentous cyanobacterium *Oscillatoria limnetica*. This organism lives in sulfide-rich saline ponds where it carries out anoxygenic photosynthesis along with photosynthetic green and purple bacteria and produces S^0 as an oxidation product of H_2S. In cultures of *O. limnetica*, electron flow from PSII is strongly inhibited by H_2S, necessitating anoxygenic photosynthesis if the organism is to survive.

From an evolutionary standpoint, the existence of cyclic photophosphorylation in both oxygenic and anoxygenic phototrophs is one of many indications of their close relationship. Oxygenic phototrophs such as *O. limnetica* contain PSII and hence have the ability to split water. However, *O. limnetica* retains the ability under certain conditions to use PSI alone, just as anoxygenic phototrophs do during phototrophic growth.

Further evidence of evolutionary relationships among phototrophs has been the discovery that the structure of the purple bacterial and green nonsulfur photosynthetic reaction center resembles that of PSII, whereas the structure of the reaction centers of green sulfur bacteria and heliobacteria resembles that of PSI. Because the evidence is strong that purple and green bacteria preceded cyanobacteria on Earth by perhaps as many as 0.5 billion years (🔗 Section 16.3), it is clear that anoxygenic photosynthesis was the first form of photosynthesis on Earth. Cyanobacteria appeared later by combining the two types of

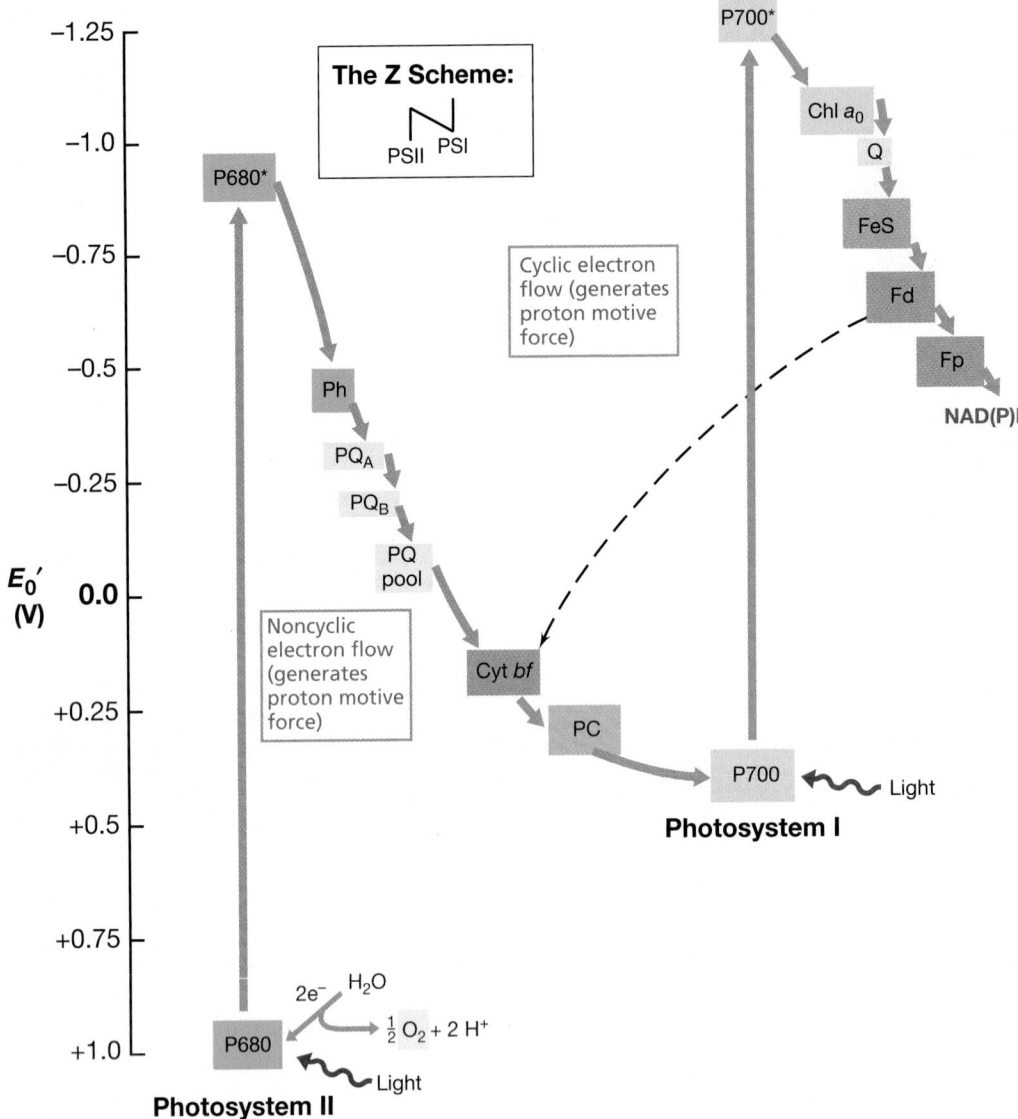

The Z Scheme:

PSII   PSI

Figure 13.18 Electron flow in oxygenic photosynthesis, the "Z" scheme. Electrons flow through two photosystems, PSI and PSII. Ph, pheophytin; Q, quinone; Chl, chlorophyll; Cyt, cytochrome; PC, plasto-cyanin; FeS, nonheme iron–sulfur protein; Fd, ferredoxin; Fp, flavoprotein; P680 and P700 are the reaction center chlorophylls of PSII and PSI, respectively. Compare with Figure 13.14.

anoxygenic photosynthetic reaction centers into one interconnected system along with evolving the key new process of using water as a photosynthetic electron donor.

MiniQuiz

- Why is the term noncyclic electron flow used in reference to oxygenic photosynthesis?
- Why has oxygenic photosynthesis been referred to as the Z scheme?
- What is the source of electrons for autotrophic CO_2 fixation in oxygenic phototrophs? In anoxygenic phototrophs?
- What evidence exists that anoxygenic photosynthesis and oxygenic photosynthesis are related processes?

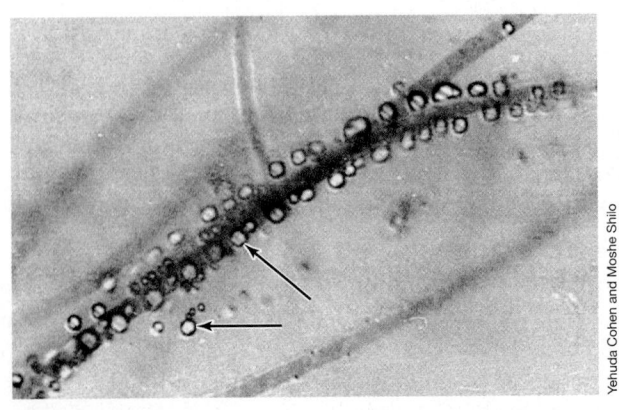

Figure 13.19 Oxidation of H_2S by *Oscillatoria limnetica*. Note the globules of S^0 (arrows), the oxidation product of H_2S, formed outside the cells. *O. limnetica* carries out oxygenic photosynthesis, but in the presence of H_2S, cells revert to the anoxygenic process.

Chemolithotrophy

We now turn our attention to the chemolithotrophs, high-lighting the strategies, problems, and advantages of a lifestyle of using inorganic chemicals as energy sources. Although chemolithotrophs lack photosynthetic pigments, they share with the phototrophs an ability to use CO_2 as their sole carbon source, and thus much of the biochemistry outside of that dealing with energy conservation is the same in these two groups.

13.6 The Energetics of Chemolithotrophy

Organisms that obtain energy from the oxidation of inorganic compounds are called **chemolithotrophs**. Most chemolithotrophic bacteria are also autotrophs. As we have noted, for growth on CO_2 as the sole carbon source an organism needs (1) ATP and (2) reducing power. Some chemolithotrophs grow as **mixotrophs**, meaning that although they conserve energy from the oxidation of an inorganic compound, they require an organic compound as their carbon source.

ATP generation in chemolithotrophs is similar to that in chemoorganotrophs, except that the electron donor is *inorganic* rather than *organic*. The electrons from the inorganic source undergo electron transport, and ATP synthesis occurs by way of ATPases. Reducing power in chemolithotrophs is obtained in either of two ways: directly from the inorganic compound, if it has a sufficiently low reduction potential, such as H_2, or by reverse electron transport reactions (as discussed in Section 13.4 for phototrophic purple bacteria), if the electron donor is more electropositive than NADH. As we will see, with most chemolithotrophs, reverse electron transport reactions are necessary.

Sources of Inorganic Electron Donors

Chemolithotrophs have many sources of inorganic electron donors. They may be geological, biological, or anthropogenic (the result of human activities). Volcanic activity is a major source of reduced sulfur compounds, primarily H_2S and S^0. Agricultural and mining operations add inorganic electron donors to the environment, especially reduced nitrogen and iron compounds, as does the burning of fossil fuels and the input of industrial wastes. Biological sources are also quite extensive, especially the production of H_2S, H_2, and NH_3. The ecological success and metabolic diversity of chemolithotrophs underscores the diversity of sources and abundance of inorganic electron donors available in nature.

Energetics of Chemolithotrophy

A review of reduction potentials listed in Table A1.2 reveals that the oxidation of a number of inorganic electron donors can provide sufficient energy for ATP synthesis. **Table 13.1** summarizes energy yields for some chemolithotrophic reactions. The organisms themselves are considered in Chapters 17–19.

Recall from Chapter 4 that the further apart two half reactions are in terms of the E_0' of their redox couples, the greater the amount of energy released. For instance, the difference in reduction potential between the $2\,H^+/H_2$ couple and the $\frac{1}{2}\,O_2/H_2O$ couple is -1.23 V, which is equivalent to a free-energy yield of -237 kJ/mol (Appendix 1 gives the calculations). On the other hand, the potential difference between the $2\,H^+/H_2$ couple and the NO_3^-/NO_2^- couple is less, -0.84 V, equivalent to a free-energy yield of -163 kJ/mol. This is still quite sufficient for the production of ATP (the energy-rich phosphate bond of ATP has a free energy of -31.8 kJ/mol). However, a similar calculation shows that there is insufficient energy available from, for example, the oxidation of H_2S to S^0 using CO_2 as the electron acceptor and forming CH_4 as the product (Appendix 1).

Energy calculations make it possible to predict the kinds of chemolithotrophs that should exist in nature. Because organisms must obey the laws of thermodynamics, only reactions that are thermodynamically favorable are potential energy-yielding reactions, and Table 13.1 lists all known classes of chemolithotrophs. We examine ecological aspects of chemolithotrophy in Chapter 24, where we will see that chemolithotrophic reactions form the heart of most nutrient cycles.

Table 13.1 *Energy yields from the oxidation of various inorganic electron donors*[a]

Electron donor	Chemolithotrophic reaction	Group of chemolithotrophs	E_0' of couple (V)	$\Delta G^{0'}$ (kJ/reaction)	Number of electrons/reaction	$\Delta G^{0'}$ (kJ/2e$^-$)
Phosphite[b]	$4\,HPO_3^{2-} + SO_4^{2-} + H^+ \rightarrow 4\,HPO_4^{2-} + HS^-$	Phosphite bacteria	-0.69	-91	2	-91
Hydrogen[b]	$H_2 + \frac{1}{2}O_2 \rightarrow H_2O$	Hydrogen bacteria	-0.42	-237.2	2	-237.2
Sulfide[b]	$HS^- + H^+ + \frac{1}{2}O_2 \rightarrow S^0 + H_2O$	Sulfur bacteria	-0.27	-209.4	2	-209.4
Sulfur[b]	$S^0 + 1\frac{1}{2}O_2 + H_2O \rightarrow SO_4^{2-} + 2\,H^+$	Sulfur bacteria	-0.20	-587.1	6	-195.7
Ammonium[c]	$NH_4^+ + 1\frac{1}{2}O_2 \rightarrow NO_2^- + 2\,H^+ + H_2O$	Nitrifying bacteria	$+0.34$	-274.7	6	-91.6
Nitrite[b]	$NO_2^- + \frac{1}{2}O_2 \rightarrow NO_3^-$	Nitrifying bacteria	$+0.43$	-74.1	2	-74.1
Ferrous iron[b]	$Fe^{2+} + H^+ + \frac{1}{4}O_2 \rightarrow Fe^{3+} + \frac{1}{2}H_2O$	Iron bacteria	$+0.77$	-32.9	1	-65.8

[a]Data calculated from E_0' values in Appendix 1; values for Fe^{2+} are for pH 2, and others are for pH 7. At pH 7 the value for the Fe^{3+}/Fe^{2+} couple is about $+0.2$ V.
[b]Except for phosphite, all reactions are shown coupled to O_2 as electron acceptor. The only known phosphite oxidizer couples to SO_4^{2-} as electron acceptor. H_2 and most sulfur compounds can be oxidized anaerobically using one or more electron acceptors, and Fe^{2+} can be oxidized at neutral pH with NO_3^- as electron acceptor.
[c]Ammonium can also be oxidized with NO_2^- as electron acceptor (anammox, Section 13.11).

UNIT 5

13.7 Hydrogen Oxidation

Hydrogen (H_2) is a common product of microbial metabolism, and a number of chemolithotrophs are able to use it as an electron donor in energy metabolism. Many anaerobic H_2-oxidizing *Bacteria* and *Archaea* are known, which differ in the electron acceptor they use (for example, nitrate, sulfate, ferric iron, and others), and these organisms are discussed in the next chapter. Here we consider the *aerobic* H_2-oxidizing bacteria, organisms that couple the oxidation of H_2 to the reduction of O_2, forming water.

Energetics of H_2 Oxidation

Synthesis of ATP during H_2 oxidation by O_2 is the result of electron transport reactions that generate a proton motive force. The overall reaction

$$H_2 + \tfrac{1}{2}O_2 \longrightarrow H_2O \qquad \Delta G^{0\prime} = -237 \text{ kJ}$$

is highly exergonic and can thus support the synthesis of ATP. In this reaction, which is catalyzed by the enzyme **hydrogenase**, the electrons from H_2 are initially transferred to a quinone acceptor. From there electrons pass through a series of cytochromes to eventually reduce O_2 to water (**Figure 13.20**).

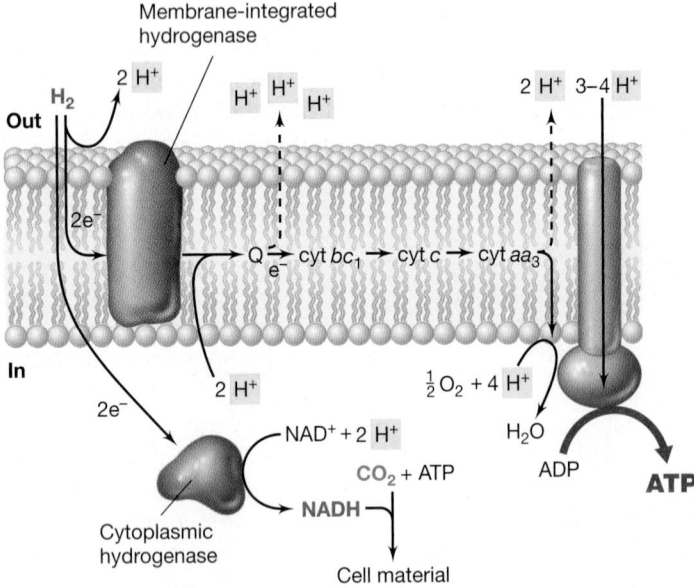

Figure 13.20 Bioenergetics and function of the two hydrogenases of aerobic hydrogen bacteria. In *Ralstonia eutropha* two hydrogenases are present; the membrane-bound hydrogenase participates in energetics, whereas the cytoplasmic hydrogenase makes NADH for the Calvin cycle. Some hydrogen bacteria have only the membrane-bound hydrogenase, and in these organisms reducing power is synthesized by reverse electron flow from Q back to NAD$^+$ to form NADH. Cyt, cytochrome; Q, quinone.

Some hydrogen bacteria synthesize two different hydrogenase enzymes, one cytoplasmic and one membrane-integrated. The latter participates in energetics, whereas the soluble hydrogenase binds H_2 and catalyzes the reduction of NAD$^+$ to NADH (the reduction potential of H_2, -0.42 V, is so negative that reverse electron flow reactions are unnecessary; Figure 13.20). The organism *Ralstonia eutropha* has been a model for studying aerobic H_2 oxidation, and we discuss some of the properties of this organism in Section 17.5.

Autotrophy in H_2 Bacteria

Although most hydrogen bacteria can also grow as chemoorganotrophs, when growing chemolithotrophically, they fix CO_2 by the Calvin cycle (Section 13.12). However, when readily usable organic compounds such as glucose are present, synthesis of Calvin cycle and hydrogenase enzymes by H_2 bacteria is typically repressed. Thus, H_2 bacteria can be considered *facultative chemolithotrophs*. In nature, H_2 levels in oxic environments are transient and low at best; most biological H_2 production is the result of fermentations, which are anoxic processes, and H_2 can be utilized by a number of different anaerobic prokaryotes. Thus aerobic hydrogen bacteria must closely regulate their catabolic enzymes and likely shift their metabolism between chemoorganotrophy and chemolithotrophy, depending on levels of usable organic compounds and H_2 in their habitats. Moreover, many aerobic H_2 bacteria grow best microaerobically and are probably most competitive as H_2 chemolithotrophs in oxic–anoxic interfaces where H_2 from fermentative metabolism is in greater and more continuous supply than in highly oxic habitats.

13.8 Oxidation of Reduced Sulfur Compounds

Many reduced sulfur compounds are used as electron donors by the colorless sulfur bacteria, called *colorless* to distinguish them from the bacteriochlorophyll-containing (pigmented) green and purple sulfur bacteria discussed earlier in this chapter (Figure 13.16). Historically, the concept of chemolithotrophy emerged from studies of the sulfur bacteria by the great Russian microbiologist Sergei Winogradsky (⮑ Section 1.9). From his observations of natural populations of the sulfur bacterium *Beggiatoa* (see Figure 13.21), Winogradsky first proposed the concept that an inorganic substance could provide a bacterium with energy, a process that we know today is a widespread means of supporting growth of prokaryotes.

Energetics of Sulfur Oxidation

The most common sulfur compounds used as electron donors are hydrogen sulfide (H_2S), elemental sulfur (S^0), and thiosulfate ($S_2O_3^{2-}$). In most cases the final product of oxidation is sulfate

(SO_4^{2-}), and the total number of electrons generated between H_2S (oxidation state of S, −2) and sulfate (oxidation state of S, +6) is eight. However, per electron consumed, roughly equivalent energy yields are obtained from all reduced sulfur compounds:

$$H_2S + 2\,O_2 \rightarrow SO_4^{2-} + 2\,H^+$$

(sulfide as substrate; final product, sulfate)
$$\Delta G^{0\prime} = -798.2 \text{ kJ/reaction } (-99.75 \text{ kJ/e}^-)$$

$$HS^- + \tfrac{1}{2}O_2 + H^+ \rightarrow S^0 + H_2O$$

(sulfide as substrate; final product, sulfur)
$$\Delta G^{0\prime} = -209.4 \text{ kJ/reaction } (-104.7 \text{ kJ/e}^-)$$

$$S^0 + H_2O + 1\tfrac{1}{2}O_2 \rightarrow SO_4^{2-} + 2\,H^+$$

(sulfur as substrate; final product, sulfate)
$$\Delta G^{0\prime} = -587.1 \text{ kJ/reaction } (-97.85 \text{ kJ/e}^-)$$

$$S_2O_3^{2-} + H_2O + 2\,O_2 \rightarrow 2\,SO_4^{2-} + 2\,H^+$$

(thiosulfate as substrate; final product, sulfate)
$$\Delta G^{0\prime} = -818.3 \text{ kJ/reaction } (-102 \text{ kJ/e}^-)$$

Sulfide oxidation occurs in stages, with the first oxidation step yielding elemental sulfur, S^0. Some sulfide-oxidizing bacteria, such as *Beggiatoa*, deposit this elemental sulfur inside the cell (**Figure 13.21**), where the sulfur exists as an energy reserve. When the supply of sulfide has been depleted, additional energy can then be obtained from the oxidation of sulfur to sulfate.

When elemental sulfur is provided externally as an electron donor, the organism must attach itself to the sulfur particle because elemental sulfur is rather insoluble (Figure 13.21b). By adhering to the particle, the organism can remove sulfur atoms

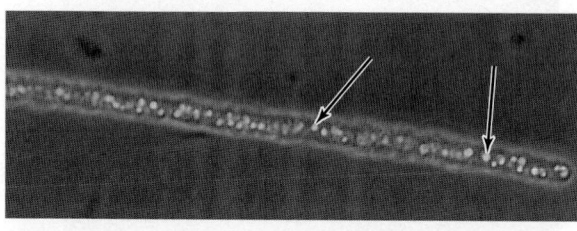

(a)

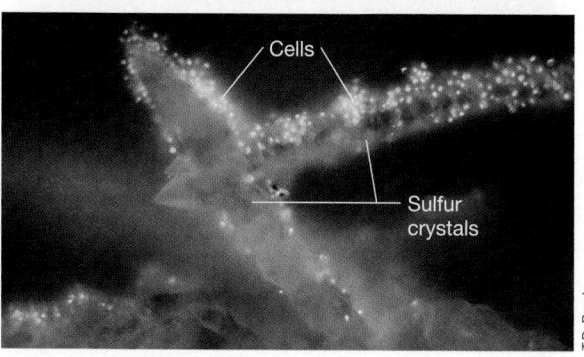

(b)

Figure 13.21 Sulfur bacteria. *(a)* Internal sulfur granules in *Beggiatoa* (arrows). *(b)* Attachment of cells of the sulfur-oxidizing archaeon *Sulfolobus acidocaldarius* to a crystal of elemental sulfur. Cells are visualized by fluorescence microscopy after being stained with the dye acridine orange. The sulfur crystal does not fluoresce.

for oxidation to sulfate. This occurs through the activity of membrane or periplasmic proteins that solubilize the sulfur by reduction of S^0 to HS^-, which is transported into the cell and enters chemolithotrophic metabolism (see Figure 13.22).

One product of the oxidation of reduced sulfur compounds is protons. Consequently, one result of the oxidation of reduced sulfur compounds by cultures of sulfur chemolithotrophs is the acidification of the medium. Because of this, many sulfur bacteria are acid-tolerant or even acidophilic. *Acidithiobacillus thiooxidans*, for example, grows best at a pH below 3.

Biochemistry of Sulfur Oxidation

The biochemical steps in the oxidation of various sulfur compounds are summarized in **Figure 13.22**. Several pathways for sulfur oxidation are known in sulfur chemolithotrophs. In two of the systems, the starting substrate, HS^-, $S_2O_3^{2-}$, or S^0, is first oxidized to sulfite (SO_3^{2-}); starting with sulfide, six electrons are released. Then the sulfite is oxidized to sulfate. This can occur in either of two ways. The most widespread system employs the enzyme *sulfite oxidase*. Sulfite oxidase transfers electrons from SO_3^{2-} directly to cytochrome *c*, and ATP is made from this during subsequent electron transport and proton motive force formation (Figure 13.22b). By contrast, some sulfur chemolithotrophs oxidize SO_3^{2-} to SO_4^{2-} via a reversal of the activity of the enzyme *adenosine phosphosulfate reductase*, an enzyme essential for the metabolism of sulfate-reducing bacteria (⇄ Section 14.8). This reaction, run in the direction of SO_4^{2-} production by sulfur chemolithotrophs, yields one energy-rich phosphate bond when AMP is converted to ADP (Figure 13.22a). When thiosulfate is the electron donor for sulfur chemolithotrophs, it is first split into S^0 and SO_3^{2-}, both of which are eventually oxidized to SO_4^{2-}.

Sox

A functionally distinct sulfide and thiosulfate oxidation system is present in *Paracoccus pantotrophus* and many other sulfur bacteria both chemolithotrophic and phototrophic. This system, called the *Sox* (for *s*ulfur *ox*idation) *system*, oxidizes reduced sulfur compounds directly to sulfate without the intermediate formation of sulfite (Figure 13.22a). The Sox system is encoded by over 15 genes for various cytochromes and other proteins necessary for the oxidation of reduced sulfur compounds directly to sulfate. The Sox system is present in several sulfur chemolithotrophs and is also present in some phototrophic sulfur bacteria that oxidize sulfide to obtain reducing power for CO_2 fixation rather than for energetic reasons. The fact that this biochemical system is distributed among prokaryotes that oxidize sulfide for different reasons is a good indication that the genes that encode Sox have been transferred between species by horizontal gene flow.

Energy from Sulfur Oxidation

Electrons from the oxidation of reduced sulfur compounds eventually reach the electron transport chain as shown in Figure 13.22b. Depending on the $E_0{\prime}$ of the electron donor couple, electrons enter at either the flavoprotein ($E_0{\prime} = -0.2$) or cytochrome *c* ($E_0{\prime} = +0.3$) level and are transported through the chain to O_2,

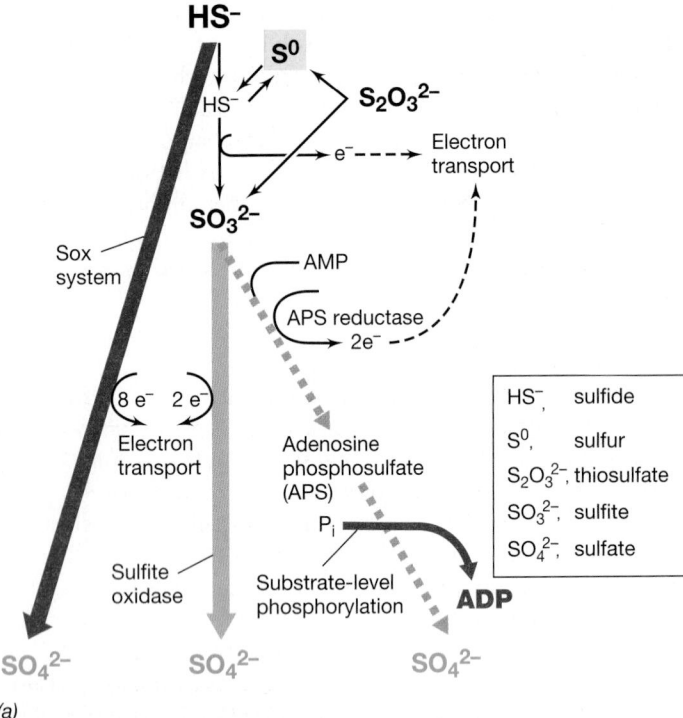

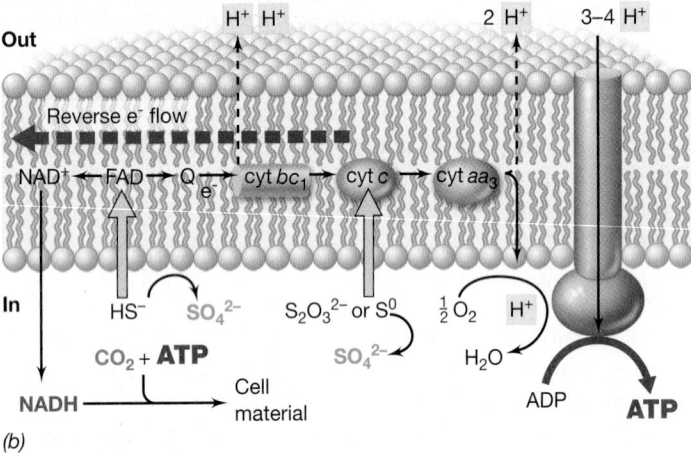

Figure 13.22 Oxidation of reduced sulfur compounds by sulfur chemolithotrophs. *(a)* Steps in the oxidation of different compounds. Three different pathways are known. *(b)* Electrons from sulfur compounds feed into the electron transport chain to drive a proton motive force; electrons from $S_2O_3^{2-}$ and S^0 enter at the level of cytochrome *c*. NADH is made by reverse electron flow. Cyt, cytochrome; FP, flavoprotein; Q, quinone. For the structure of APS, see Figure 14.14*a*.

generating a proton motive force that leads to ATP synthesis by ATPase. Electrons for autotrophic CO_2 fixation come from reverse electron flow (Section 13.4), eventually yielding NADH. Autotrophy is driven by reactions of the Calvin cycle or some other autotrophic pathway (Sections 13.12 and 13.13). Although the sulfur chemolithotrophs are primarily an aerobic group (↩ Section 17.4), some species can grow anaerobically using nitrate as an electron acceptor; the sulfur bacterium *Thiobacillus denitrificans* is a classic example of this, reducing nitrate to dinitrogen gas (the process of denitrification, ↩ Section 14.7).

MiniQuiz

- How many electrons are available from the oxidation of H_2S if S^0 is the final product? If SO_4^{2-} is the final product?
- How does the Sox system for oxidizing H_2S differ from other systems for oxidizing H_2S?

13.9 Iron Oxidation

The aerobic oxidation of ferrous (Fe^{2+}) to ferric (Fe^{3+}) iron supports growth of the chemolithotrophic "iron bacteria." At acidic pH, only a small amount of energy is available from this oxidation (Table 13.1), and for this reason the iron bacteria must oxidize large amounts of iron in order to grow. The ferric iron produced forms insoluble ferric hydroxide [$Fe(OH)_3$] and other iron precipitates in water that drive down the pH (**Figure 13.23**). This explains why most iron-oxidizing bacteria are obligately acidophilic.

Iron-Oxidizing Bacteria

The best-known iron bacteria, *Acidithiobacillus ferrooxidans* and *Leptospirillum ferrooxidans*, can both grow autotrophically using ferrous iron (Figure 13.23*b*) as electron donor at pH values below 1; growth is optimal at pH 2–3. These organisms are

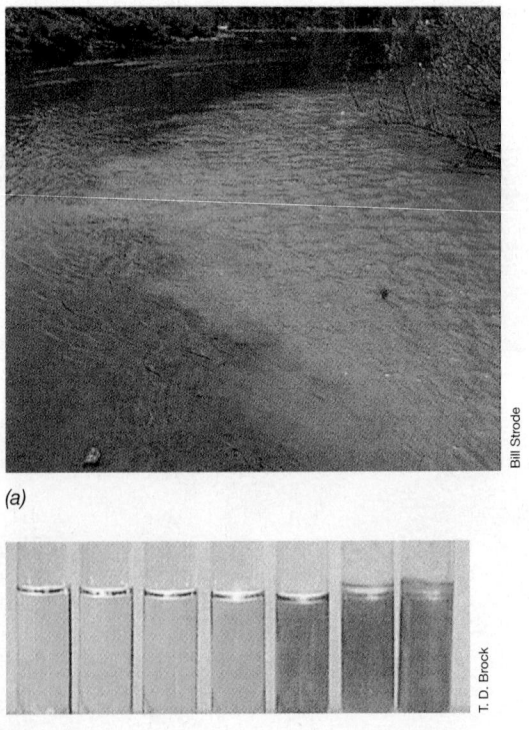

Figure 13.23 Iron-oxidizing bacteria. *(a)* Acid mine drainage, showing the confluence of a normal river and a creek draining a coal-mining area. The acidic creek is very high in Fe^{2+}. At low pH values, Fe^{2+} does not oxidize spontaneously in air, but *Acidithiobacillus ferrooxidans* carries out the oxidation; insoluble $Fe(OH)_3$ and complex ferric salts precipitate. *(b)* Cultures of *A. ferrooxidans*. Shown is a dilution series, with no growth in the tube on the left and increasing amounts of growth from left to right. Growth is evident from the production of Fe^{3+}, which readily complexes to form $Fe(OH)_3$ and protons.

common in acid-polluted environments such as coal-mining dumps (Figure 13.23*a*). *Ferroplasma*, a species of *Archaea*, is an extremely acidophilic iron oxidizer and can grow at pH values below 0 (↺ Section 19.4). We will discuss the role of all these organisms in acid-mine pollution and mineral oxidation in Sections 24.5 and 24.7.

At neutral pH, Fe^{2+} spontaneously oxidizes to Fe^{3+}, so opportunities for the iron bacteria are primarily in locations where Fe^{2+} is flowing from anoxic to oxic conditions. For example, anoxic groundwater often contains Fe^{2+}, and when it is released, as in an iron spring, it becomes exposed to O_2. At such interfaces, iron bacteria oxidize Fe^{2+} to Fe^{3+} before it oxidizes spontaneously. *Gallionella ferruginea*, *Sphaerotilus natans*, and *Leptothrix discophora* are examples of organisms that live at these interfaces. They are typically seen mixed in with the characteristic ferric iron deposits they form (↺ Figures 17.34 and 17.41).

Energy from Ferrous Iron Oxidation

The bioenergetics of iron oxidation by *Acidithiobacillus ferrooxidans* and other acidophilic iron oxidizers are remarkable because of the very electropositive reduction potential of the Fe^{3+}/Fe^{2+} couple at acidic pH (E_0' of +0.77 V at pH 2). The respiratory chain of *A. ferrooxidans* contains cytochromes of the *c* and aa_3 types and a periplasmic copper-containing protein called *rusticyanin* (**Figure 13.24**). There is also a key protein in iron oxidation located in the outer membrane of the cell. Because the reduction potential of the Fe^{3+}/Fe^{2+} couple is so high, the route

of electron transport to oxygen $\frac{1}{2}O_2/H_2O$, $E_0' = +0.82$ V) is by necessity very short.

Iron oxidation begins in the outer membrane where the organism contacts either soluble Fe^{2+} or insoluble ferrous iron minerals. Fe^{2+} is oxidized to Fe^{3+}, a one-electron transition, by an outer membrane cytochrome *c* that transfers electrons into the periplasm where rusticyanin ($E_0' = +0.68$ V) is the electron acceptor. This thermodynamically slightly unfavorable reaction is thought to be pulled forward by the removal of Fe^{3+} by $Fe(OH)_3$ formation (Figure 13.24). Rusticyanin then reduces a periplasmic cytochrome *c*, which transfers electrons to cytochrome aa_3, and it is the latter protein that reduces O_2 to H_2O; ATP is synthesized by ATPase in the usual fashion (Figure 13.24).

The proton motive force in *A. ferrooxidans* is of interest. In a highly acidic environment, a large gradient of protons already exists across the *A. ferrooxidans* membrane (the periplasm is pH 1–2, whereas the cytoplasm is pH 5.5–6, Figure 13.24). Although this unusual situation might make it appear at first as if *A. ferrooxidans* can make ATP "for free," this is not the case, as the organism cannot make ATP from this natural proton motive force in the absence of an electron donor. This is because H^+ ions that enter the cytoplasm via ATPase must be consumed in order to maintain the internal pH within acceptable limits. Proton consumption occurs during the reduction of O_2 in the electron transport chain and this reaction requires electrons; these come from the oxidation of Fe^{2+} to Fe^{3+} (Figure 13.24).

Autotrophy in *A. ferrooxidans* is supported by the Calvin cycle, and because of the high potential of the electron donor, Fe^{2+}, much energy must be consumed in reverse electron flow reactions to obtain the reducing power (NADH) necessary to drive CO_2 fixation. NADH is formed by reduction of NAD^+ by electrons obtained from Fe^{2+} that are pumped backwards through cytochrome bc_1 and the quinone pool (Figure 13.24). Overall, then, a relatively poor energetic yield coupled with large energetic demands for biosynthesis means that *A. ferrooxidans* must oxidize large amounts of Fe^{2+} to produce even a very small amount of cell material. Thus, in environments where acidophilic iron-oxidizing bacteria thrive, their presence is signaled not by the formation of much cell material but by the presence of large amounts of ferric iron precipitates (Figure 13.23; ↺ Figure 24.12). We consider the ecology of iron bacteria in Sections 24.5 and 24.7.

Ferrous Iron Oxidation under Anoxic Conditions

Fe^{2+} can be oxidized under anoxic conditions by certain chemolithotrophs and anoxygenic phototrophic bacteria (**Figure 13.25**). Fe^{2+} is used in this case as either an electron donor in energy metabolism and/or as a reductant for CO_2 fixation (autotrophy). At neutral pH where these organisms thrive, the E_0' of the Fe^{3+}/Fe^{2+} couple is 0.2 V, and thus, electrons from Fe^{2+} can reduce cytochrome *c* to initiate electron transport reactions. For chemolithotrophs, the electron acceptor is nitrate (NO_3^-), with either nitrite (NO_2^-) or dinitrogen gas (N_2) being the final product. Fe^{2+}-oxidizing phototrophs, which are certain species of phototrophic purple or green bacteria, can use either soluble Fe^{2+} or iron sulfide (FeS) as electron donor; with FeS, both Fe^{2+} and S^{2-} are oxidized, Fe^{2+} to Fe^{3+}, and HS^- to SO_4^{2-}.

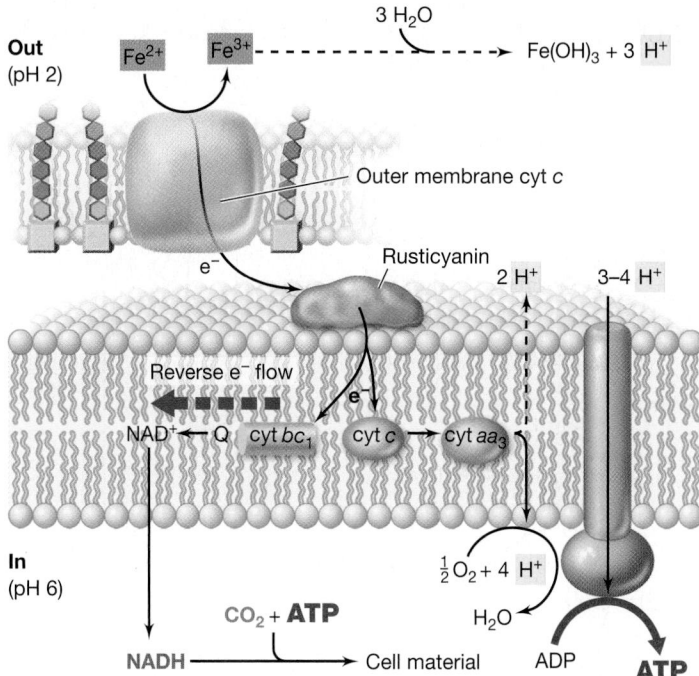

Figure 13.24 Electron flow during Fe^{2+} oxidation by the acidophile *Acidithiobacillus ferrooxidans*. The periplasmic copper-containing protein rusticyanin receives electrons from Fe^{2+} oxidized by a c-type cytochrome located in the outer membrane. From here, electrons travel a short electron transport chain resulting in the reduction of O_2 to H_2O. Reducing power comes from reverse electron flow. Note the steep pH gradient across the membrane.

UNIT 5

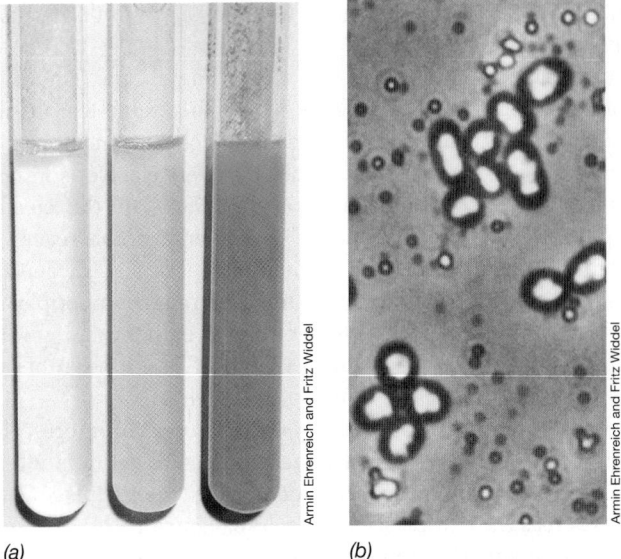

(a)   (b)

Figure 13.25 Fe^{2+} oxidation by anoxygenic phototrophic bacteria. (a) Oxidation in anoxic tube cultures. Left to right: Sterile medium, inoculated medium, a growing culture. The brown-red color is due to $Fe(OH)_3$. (b) Phase-contrast photomicrograph of an Fe^{2+}-oxidizing purple bacterium. The bright refractile areas within cells are gas vesicles (⇄ Section 3.11). The granules outside the cells are iron precipitates. This organism is phylogenetically related to the purple sulfur bacterium *Chromatium* (⇄ Section 17.2).

Anoxic Fe^{2+} oxidation may be the mechanism through which large deposits of iron minerals were laid down in ancient sediments on Earth. Fe^{3+} in ancient iron beds was always assumed to have formed from the abiotic or biological oxidation of Fe^{2+} by O_2 produced by oxygenic phototrophs. However, because of the age of these sediments, which in many cases predates the appearance of cyanobacteria on Earth (⇄ Section 16.3), it is more likely that the Fe^{3+} was formed by anoxygenic phototrophs or anaerobic chemolithotrophs oxidizing Fe^{2+} in iron-rich anoxic environments.

MiniQuiz

- Why is only a very small amount of energy available from the oxidation of Fe^{2+} to Fe^{3+} at acidic pH?
- What is the function of rusticyanin and where is it found in the cell?
- How can Fe^{2+} be oxidized under anoxic conditions?

13.10 Nitrification

The inorganic nitrogen compounds ammonia (NH_3) and nitrite (NO_2^-) are chemolithotrophic substrates and are oxidized aerobically by the "nitrifying bacteria" (⇄ Section 17.3) in the process of **nitrification**. Nitrifying bacteria are widely distributed in soils and water. One group (for example, *Nitrosomonas*) oxidizes NH_3 to nitrite (NO_2^-), and another group (for example, *Nitrobacter* and *Nitrospira*) oxidizes NO_2^- to NO_3^-. The complete oxidation of NH_3 to NO_3^-, an eight-electron transfer, is

thus carried out by the concerted activity of two groups of organisms, a process discovered by the Russian microbiologist Winogradsky at the end of the nineteenth century. Winogradsky also convincingly showed that the nitrifying bacteria were autotrophs, obtaining all of their carbon from CO_2. Along with his description of the process of chemolithotrophy, Winogradsky's discovery that nonphototrophic organisms could grow autotrophically was a truly revolutionary idea in biology in its day. We now know that nonphototototrophic autotrophy is widespread in the prokaryotic world.

Bioenergetics and Enzymology of Nitrification

The bioenergetics of nitrification is based on the same principles that govern other chemolithotrophic reactions: Electrons from reduced inorganic substrates (in this case, reduced nitrogen compounds) enter an electron transport chain, and electron flow establishes a proton motive force that drives ATP synthesis.

The electron donors for the nitrifying bacteria are not particularly strong. The E_0' of the NO_2^-/NH_3 couple (the first step in the oxidation of NH_3) is +0.34 V, and the E_0' of the NO_3^-/NO_2^- couple is even more positive, about +0.43 V. These reduction potentials force nitrifying bacteria to donate electrons to rather high-potential electron acceptors in their electron transport chains, which limits the energy available for energy conservation purposes.

Several key enzymes participate in the oxidation of reduced nitrogen compounds. In ammonia-oxidizing bacteria, NH_3 is oxidized by *ammonia monooxygenase* (monooxygenase enzymes are discussed in ⇄ Section 14.14), producing hydroxylamine (NH_2OH) and H_2O (**Figure 13.26**). A second key enzyme,

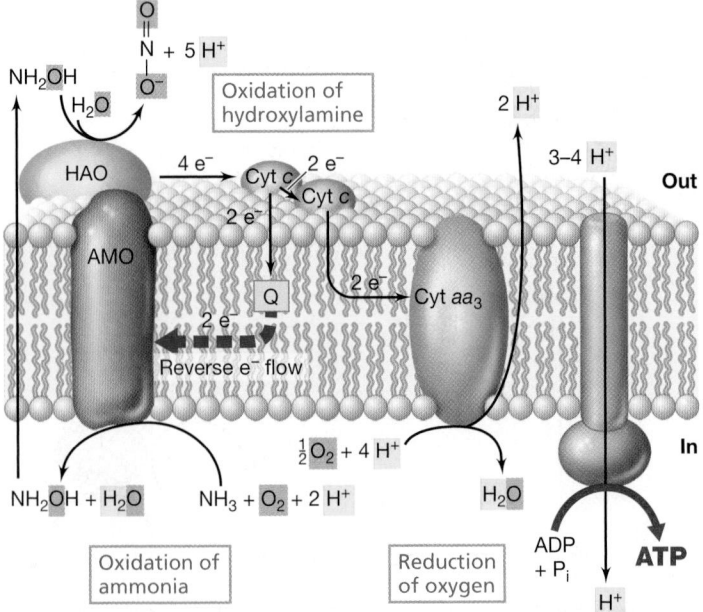

Figure 13.26 Oxidation of NH_3 and electron flow in ammonia-oxidizing bacteria. The reactants and the products of this reaction series are highlighted. The cytochrome *c* (Cyt *c*) in the periplasm is a different form of Cyt *c* than that in the membrane. AMO, ammonia monooxygenase; HAO, hydroxylamine oxidoreductase; Q, ubiquinone.

hydroxylamine oxidoreductase, then oxidizes NH_2OH to NO_2^-, removing four electrons in the process. Ammonia monooxygenase is an integral membrane protein, whereas hydroxylamine oxidoreductase is periplasmic (Figure 13.26). In the reaction carried out by ammonia monooxygenase,

$$NH_3 + O_2 + 2\,H^+ + 2\,e^- \rightarrow NH_2OH + H_2O$$

two electrons and protons are needed to reduce one atom of (O_2) to H_2O. These electrons originate from the oxidation of hydroxylamine and are supplied to ammonia monooxygenase from hydroxylamine oxidoreductase via cytochrome *c* and ubiquinone (Figure 13.26). Thus, for every *four* electrons generated from the oxidation of NH_3 to NO_2^-, only *two* actually reach cytochrome aa_3, the cytochrome that interacts with O_2 to form H_2O (Figure 13.26), and yield energy.

Nitrite-oxidizing bacteria employ the enzyme *nitrite oxidoreductase* to oxidize NO_2^- to NO_3^-, with electrons traveling a very short electron transport chain (because of the high potential of the NO_3^-/NO_2^- couple) to the terminal oxidase (**Figure 13.27**). Cytochromes of the *a* and *c* types are present in the electron transport chain of nitrite oxidizers, and the activity of cytochromes aa_3 generates a proton motive force (Figure 13.27). As in the iron oxidation reaction (Section 13.9), only small amounts of energy are available in this reaction. Thus, growth yields of nitrifying bacteria (grams of cells produced per mole of substrate oxidized) are low.

Other Nitrifying Prokaryotes

From a phylogenetic standpoint, all nitrifiers discussed thus far are *Bacteria*. However, at least one species of *Archaea* is a nitrifier. This organism, *Nitrosopumilus*, is an autotrophic ammonia-oxidizing chemolithotroph and member of the *Crenarchaeota* (⇄ Section 19.11). *Nitrosopumilus* contains genes related to those that encode ammonia monooxygenase in ammonia-oxidizing *Bacteria* such as *Nitrosomonas*, and thus it is likely that the physiology of NH_3 oxidation in *Bacteria* and *Archaea* is similar.

Thus far, nitrite-oxidizing *Archaea* are unknown. However, NO_2^- is an electron donor for certain species of anoxygenic purple phototrophic bacteria. In this case, however, NO_2^- is oxidized under *anoxic* conditions because anoxygenic photosynthesis occurs only under anaerobic conditions (Section 13.4). Moreover, the electrons derived from NO_2^- oxidation by these purple bacteria are not used to obtain energy as they are in nitrifying bacteria (Figure 13.27), but instead are used as a source of reducing power for autotrophic CO_2 fixation.

Carbon Metabolism and Ecology of Nitrifying Bacteria

Like sulfur- and iron-oxidizing chemolithotrophs, aerobic nitrifying bacteria employ the Calvin cycle for CO_2 fixation. The ATP and reducing power requirements of the Calvin cycle place additional burdens on an already relatively low-yielding energy-generating system (NADH to drive the Calvin cycle in nitrifiers is formed by reverse electron flow, Figures 13.26 and 13.27). The energetic constraints are particularly severe for NO_2^- oxidizers, and it is perhaps for this reason that most NO_2^- oxidizers have alternative energy-conserving mechanisms, being able to grow chemoorganotrophically on glucose and a few other organic substrates. By contrast, species of ammonia-oxidizing bacteria are either obligate chemolithotrophs or mixotrophs.

Nitrifying bacteria play key ecological roles in the nitrogen cycle, converting ammonia into nitrate, a key plant nutrient. Nitrifying bacteria are also important in sewage and wastewater treatment, removing toxic amines and ammonia and releasing less toxic nitrogen compounds. Nitrifying bacteria play a similar role in the water column of lakes, where ammonia produced in the sediments from the decomposition of organic nitrogenous compounds is oxidized to nitrate, a more favorable nitrogen source for algae and cyanobacteria.

MiniQuiz

- What is the inorganic electron donor for *Nitrosomonas*? For *Nitrospira*?
- What are the substrates for the enzyme ammonia monooxygenase?
- What do nitrifying bacteria use as their carbon source?

13.11 Anammox

Although the nitrifying bacteria just discussed are strict *aerobes*, NH_3 can also be oxidized under anoxic conditions. This process, known as **anammox** (for *anoxic ammonia oxidation*), is catalyzed by an unusual group of obligately anaerobic *Bacteria*.

Ammonia is oxidized in the anammox reaction with NO_2^- as the electron acceptor to yield N_2 as follows:

$$NH_4^+ + NO_2^- \rightarrow N_2 + 2\,H_2O \qquad \Delta G^{0\prime} = -357\text{ kJ}$$

The first anammox organism discovered, *Brocadia anammoxidans*, is a member of the *Planctomycetes* phylum of *Bacteria* (⇄ Section 18.10) (**Figure 13.28**). *Planctomycetes* are unusual *Bacteria*, lacking peptidoglycan and containing membrane-enclosed compartments of various types inside the cell. In cells of *B. anammoxidans* the

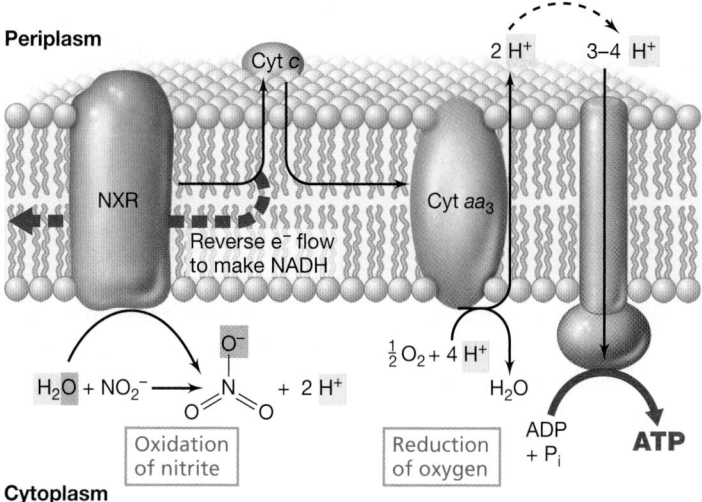

Figure 13.27 Oxidation of NO_2^- to NO_3^- by nitrifying bacteria. The reactants and products of this reaction series are highlighted to follow the reaction. NXR, nitrite oxidoreductase.

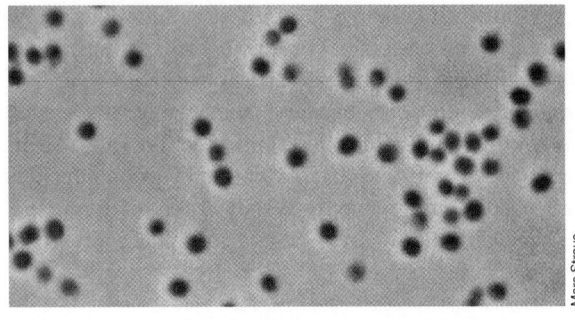

(a)

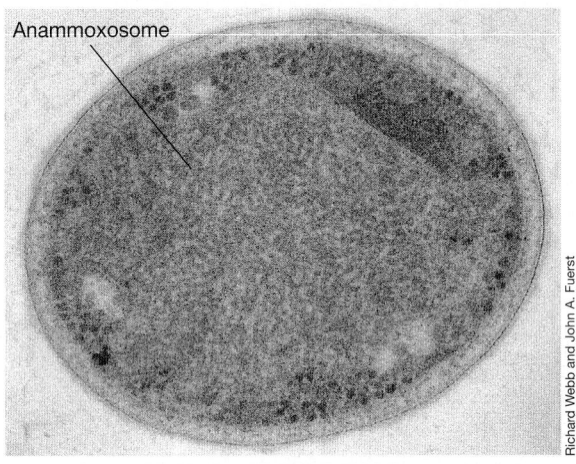

Anammoxosome

(b)

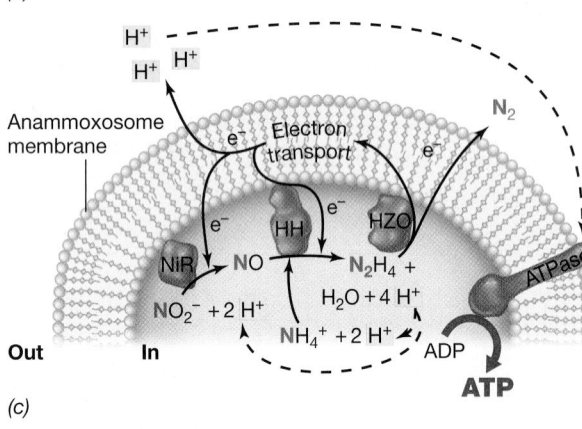

(c)

Figure 13.28 Anammox. (a) Phase-contrast photomicrograph of cells of *Brocadia anammoxidans*. A single cell is about 1 μm in diameter. (b) Transmission electron micrograph of a cell; note the membrane-enclosed compartments including the large fibrillar anammoxosome. (c) Reactions in the anammoxosome. NiR, nitrite reductase; HH, hydrazine hydrolase; HZO, hydrazine dehydrogenase.

The Anammoxosome

The anammoxosome is a unit membrane–enclosed structure (Figure 13.28b) and in this respect can be considered an organelle in the eukaryotic sense of the term (♻ Section 20.1).

compartment is the *anammoxosome*, and it is within this structure that the anammox reaction occurs (Figure 13.28c). In addition to *Brocadia*, several other genera of anammox bacteria are known, including *Kuenenia*, *Anammoxoglobus*, *Jettenia*, and *Scalindua*, all of which are related to *Brocadia* and contain anammoxosomes.

The anammoxosome takes up roughly half of the cell volume and does not contain normal cytoplasm, as the structure is devoid of ribosomes. Moreover, lipids in the anammoxosome membrane are not the typical lipids of *Bacteria* but instead consist of fatty acids that contain multiple cyclobutane (C_4) rings bonded to glycerol by both ester and ether bonds. These *ladderane lipids*, as they are called, aggregate in the membrane to form an unusually dense membrane structure that prevents diffusion of substances from the anammoxosome into the cytoplasm.

The strong anammoxosome membrane is required to protect the cell from the toxic intermediates produced during anammox reactions. These include, in particular, the compound *hydrazine* (N_2H_4), a very strong reductant. In the anammox reaction, NO_2^- is first reduced to nitric oxide (NO) by nitrite reductase, and then NO reacts with ammonium (NH_4^+) to yield N_2H_4 by activity of the enzyme hydrazine hydrolase (Figure 13.28c). N_2H_4 is then oxidized to N_2 plus electrons by the enzyme hydrazine dehydrogenase. Some of the electrons generated at this step enter the anammoxosome electron transport chain that yields a proton motive force and ATP by ATPase, while others feed back into the system to drive the electron-consuming earlier steps (Figure 13.28c).

Autotrophy in Anammox Bacteria

Like classical nitrifying bacteria, anammox bacteria are also autotrophs. Anammox organisms grow with CO_2 as their sole carbon source and use NO_2^- as an electron donor to produce cell material:

$$CO_2 + 2\,NO_2^- + H_2O \rightarrow CH_2O + 2\,NO_3^-$$

Although they are autotrophs, anammox bacteria lack Calvin cycle enzymes, and the mechanism of CO_2 fixation is instead the acetyl-CoA pathway, an autotrophic pathway widespread among obligately anaerobic bacteria (♻ Section 14.9). However, the high potential of the NO_2^-/NO_3^- couple (+0.42 V) precludes the use of NO_2^- directly as electron donor for CO_2 reduction because the acetyl-CoA pathway requires a very powerful reductant, either H_2 (−0.42 V) or the FeS redox protein ferredoxin (−0.4 V). Anammox organisms overcome this problem by using some of the N_2H_4, itself a very powerful electron donor (−0.5 V), to reduce ferredoxin, which serves as electron donor for autotrophic CO_2 fixation.

Ecology of Anammox

In nature the source of NO_2^- in the anammox reaction is presumably the aerobic ammonia-oxidizing bacteria. The two groups of ammonia oxidizers, aerobic and anaerobic, live together in ammonia-rich habitats such as sewage and other wastewaters. The abundant suspended particles in these habitats contain both oxic and anoxic zones in which ammonia oxidizers of different physiologies can coexist in close association. In mixed laboratory cultures, high levels of oxygen inhibit anammox and favor classic nitrification, and thus it is likely that in nature the fraction of ammonia oxidation catalyzed by anammox bacteria is governed by the concentration of O_2 in the system.

Before anammox was discovered, it was thought that NH_3 was stable under anoxic conditions and not oxidized biologically. We now know otherwise. From an environmental standpoint, anammox is a very beneficial process in the treatment of wastewaters. The anoxic removal of NH_3 and amines by the formation of N_2 (Figure 13.28c) helps reduce the input of fixed nitrogen from wastewater treatment facilities into rivers and streams, thereby maintaining higher water quality than would otherwise be possible. Ecological studies have shown that organisms in the genus *Scalindua* carry out anammox primarily in the marine environment. The activities of this anammox organism likely account for the significant fraction (>50%) of NH_3 known to disappear from marine sediments, the mechanism for which was previously unexplained. At least some ammonia-rich freshwater lake sediments also support anammox, and thus it appears that anammox can occur in any anoxic environment in which NH_3 and NO_2^- coexist.

MiniQuiz

- In what fundamental ways does anammox differ from ammonia oxidation by *Nitrosomonas*?
- Why are anammox reactions carried out in a special structure within the cell?
- What is the carbon source for anammox organisms?

III Major Biosyntheses: Autotrophy and Nitrogen Fixation

We have just discussed the energy conservation strategies of phototrophic and chemolithotrophic microorganisms. These characteristic mechanisms for using either light or inorganic chemicals as energy sources result in the synthesis of ATP. ATP then drives all of the energy-requiring reactions that phototrophs and chemolithotrophs carry out. At the top of this list is *autotrophy*, the process by which an energy-poor and highly oxidized form of carbon, CO_2, is reduced and assimilated into cell material. Many microorganisms are autotrophic, including virtually all phototrophs and chemolithotrophs. We focus here on autotrophy in phototrophs, of which at least three pathways are known. One pathway, the Calvin cycle, is widely distributed in chemolithotrophs as well. In Chapter 14 we will consider an additional autotrophic pathway, the acetyl-CoA pathway, widely distributed among obligately anaerobic *Bacteria* and *Archaea*.

We conclude this chapter by exploring the process of nitrogen fixation, the reduction of N_2 to NH_3. This energy-demanding process, widely distributed in the prokaryotic world, allows organisms inhabiting environments limiting in NH_3 or other forms of "fixed" nitrogen to use a source of nitrogen that is never limiting, N_2. The processes of autotrophy and nitrogen fixation thus bring into the cell the two chemical elements needed in highest amounts, C and N, respectively (Section 4.1), and are the most significant of the major biosyntheses carried out by microbial cells.

13.12 The Calvin Cycle

Several autotrophic pathways are known, but the **Calvin cycle**, named for its discoverer, Melvin Calvin, is the most widely distributed in nature. The Calvin cycle requires NAD(P)H, ATP, and several enzymes, two of which are unique to the cycle, *ribulose bisphosphate carboxylase* and *phosphoribulokinase*.

RubisCO and the Formation of PGA

The first step in the Calvin cycle is catalyzed by the enzyme ribulose bisphosphate carboxylase, **RubisCO** for short. The Calvin cycle is operative in purple bacteria, cyanobacteria, algae, green plants, most chemolithotrophic *Bacteria*, and even in a few *Archaea*. RubisCO catalyzes the formation of two molecules of 3-phosphoglyceric acid (PGA) from ribulose bisphosphate and CO_2 as shown in **Figure 13.29**. The PGA is then phosphorylated and reduced to a key intermediate of glycolysis, glyceraldehyde 3-phosphate. From here, glucose can be formed by reversal of the early steps in glycolysis (Section 4.8).

Stoichiometry of the Calvin Cycle

It is easiest to consider Calvin cycle reactions based on the incorporation of 6 molecules of CO_2, as this is what is required to make 1 molecule of glucose ($C_6H_{12}O_6$). For RubisCO to incorporate 6 molecules of CO_2, 6 molecules of ribulose bisphosphate (total, 30 carbons) are required; carboxylation of each yields 12 molecules of PGA (total, 36 carbon atoms) (**Figure 13.30**). These then form the carbon skeletons for 6 molecules of ribulose bisphosphate (total, 30 carbons) and one hexose (6 carbons), the latter to be used for cell biosynthesis.

A series of rearrangements between various sugars follow, eventually resulting in 6 molecules of ribulose 5-phosphate (total, 30 carbons). The final step in the Calvin cycle is the phosphorylation of ribulose 5-phosphate with ATP by the enzyme phosphoribulokinase to regenerate the acceptor molecule, ribulose bisphosphate.

In summary, the Calvin cycle consumes 12 molecules each of ATP and NADPH in the reduction of 12 PGAs to 12 glyceraldehyde 3-phosphates. Six more ATPs are required for the conversion of 6 ribulose phosphates to 6 ribulose bisphosphates. Thus, *12 NADPH and 18 ATP* are required to synthesize one C_6 sugar (hexose) from 6 CO_2 by the Calvin cycle. The hexose can then be polymerized into polysaccharide storage polymers such as glycogen or starch, or lipids such as poly-β-hydroxyalkanoates (Section 3.10) and used later to build new cell material. Alternatively, the hexose can be fed immediately into central metabolic pathways to form new cell material.

Carboxysomes

Several autotrophic prokaryotes that use the Calvin cycle produce polyhedral cell inclusions called **carboxysomes**. The inclusions, about 100 nm in diameter, are surrounded by a thin, protein membrane and consist of a crystalline array of RubisCO (**Figure 13.31**; Figure 17.9a), with about 250 RubisCO molecules present per carboxysome.

Carboxysomes appear to be a mechanism for concentrating CO_2 in the cell and making it readily available to RubisCO. Inorganic

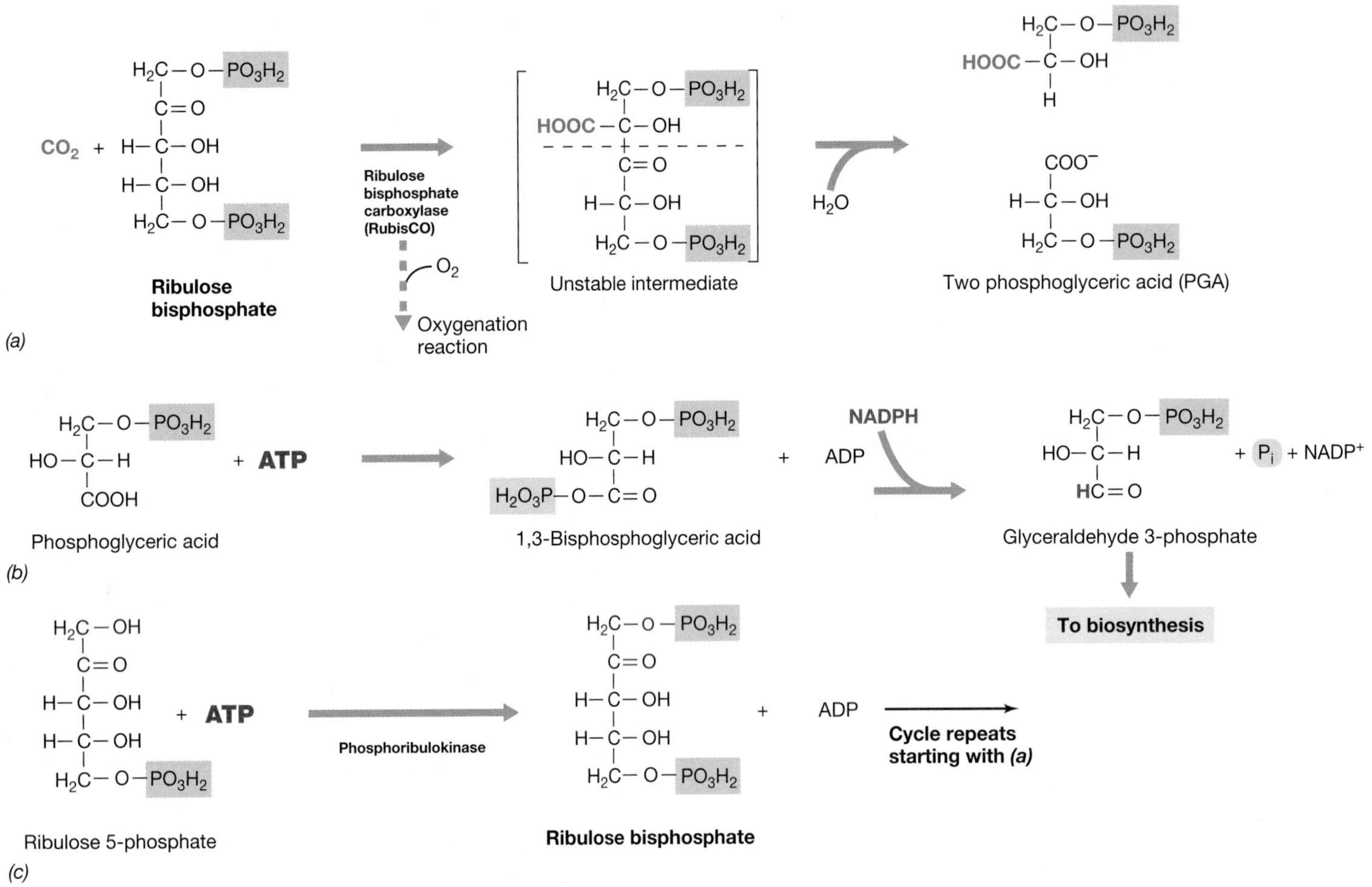

Figure 13.29 Key reactions of the Calvin cycle. *(a)* Reaction of the enzyme ribulose bisphosphate carboxylase. *(b)* Steps in the conversion of 3-phosphoglyceric acid (PGA) to glyceraldehyde 3-phosphate. Note that both ATP and NADPH are required. *(c)* Conversion of ribulose 5-phosphate to the CO_2 acceptor molecule ribulose 1,5-bisphosphate by the enzyme phosphoribulokinase.

carbon incorporated into the cell as bicarbonate (HCO_3^-) enters the carboxysome as CO_2 through the activity of a second carboxysome enzyme, carbonic anhydrase. CO_2 (rather than HCO_3^-) is the actual substrate for RubisCO, and once inside the carboxysome, the CO_2 is trapped and readily available for the first step of the Calvin cycle. The carboxysome also restricts access of RubisCO to O_2, an alternative substrate for this enzyme, and this ensures that RubisCO carboxylates rather than oxygenates ribulose bisphosphate (Figure 13.29). If ribulose 1,5-bisphosphate is oxygenated, more energy and reducing power are required to incorporate it into central metabolic pathways than if it is carboxylated.

MiniQuiz

- What reaction(s) does the enzyme RubisCO carry out?
- Why is reducing power needed for autotrophic growth?
- How much NADPH and ATP is required to make one hexose molecule by the Calvin Cycle?
- What is a carboxysome and what is its function?

13.13 Other Autotrophic Pathways in Phototrophs

Although they are autotrophs, the Calvin cycle does not operate in green sulfur and green nonsulfur bacteria. Instead, two novel autotrophic pathways are present, one in each group.

Autotrophy in Green Sulfur Bacteria

Green sulfur bacteria such as *Chlorobium* (Figure 13.16*b*) fix CO_2 by a reversal of steps in the citric acid cycle, a pathway called the **reverse citric acid cycle** (**Figure 13.32**). This pathway requires the activity of two ferredoxin-linked enzymes that catalyze the reductive fixation of CO_2; ferredoxin is produced in the light reactions of green sulfur bacteria (Figure 13.17).

   Ferredoxin is an electron donor with a very electronegative E_0', about -0.4 V. The two ferredoxin-linked reactions catalyze (1) the carboxylation of succinyl-CoA to α-ketoglutarate, and (2) the carboxylation of acetyl-CoA to pyruvate (Figure 13.32*a*). Most of the other reactions of the reverse citric acid cycle are catalyzed by enzymes working in reverse of the normal oxidative direction of the cycle. One exception is *citrate lyase*, an ATP-dependent

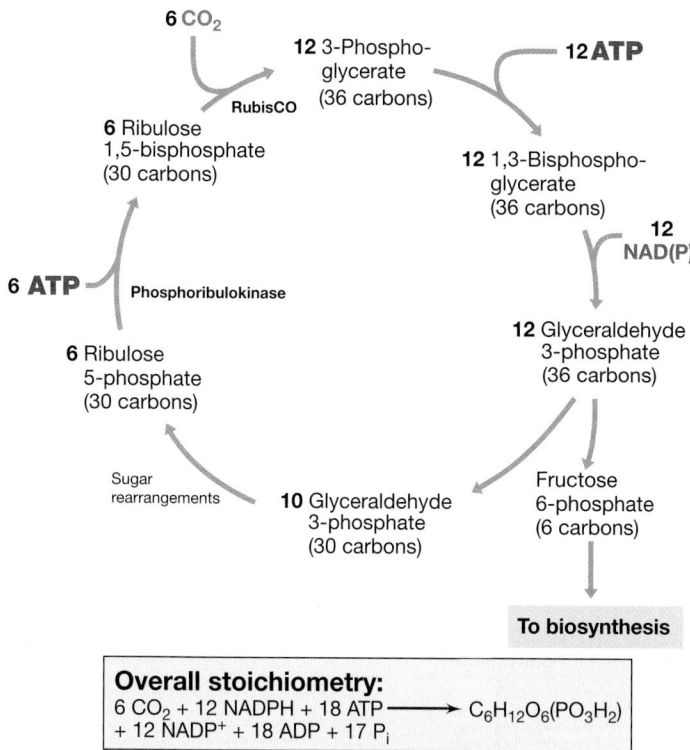

6 CO₂

12 3-Phospho-
glycerate
(36 carbons)

RubisCO

12 ATP

6 Ribulose
1,5-bisphosphate
(30 carbons)

12 1,3-Bisphospho-
glycerate
(36 carbons)

12
NAD(P)H

6 ATP — Phosphoribulokinase

6 Ribulose
5-phosphate
(30 carbons)

12 Glyceraldehyde
3-phosphate
(36 carbons)

Sugar
rearrangements

10 Glyceraldehyde
3-phosphate
(30 carbons)

Fructose
6-phosphate
(6 carbons)

To biosynthesis

Overall stoichiometry:
6 CO₂ + 12 NADPH + 18 ATP ⟶ C₆H₁₂O₆(PO₃H₂)
+ 12 NADP⁺ + 18 ADP + 17 Pᵢ

$$6\ CO_2 + 12\ NADPH + 18\ ATP \longrightarrow C_6H_{12}O_6(PO_3H_2) + 12\ NADP^+ + 18\ ADP + 17\ P_i$$

Figure 13.30 **The Calvin cycle.** Shown is the production of one hexose molecule from CO_2. For each six molecules of CO_2 incorporated, one fructose 6-phosphate is produced. In phototrophs, ATP comes from photophosphorylation and NAD(P)H from light or reverse electron flow. Use the color-coding here to follow the biochemical reactions in Figure 13.29.

enzyme that cleaves citrate into acetyl-CoA and oxalacetate (Figure 13.32*a*). In the oxidative direction of the cycle, this enzyme is absent and instead citrate is produced from these same precursors by the enzyme *citrate synthase.*

The reverse citric acid cycle also operates in certain nonphototrophic autotrophs as well. For example, the hyperthermophiles *Thermoproteus* and *Sulfolobus* (*Archaea*; ⟋⟍ Section 19.9) and *Aquifex* (*Bacteria*; ⟋⟍ Section 18.20) use the reverse citric acid cycle, as do certain sulfur chemolithotrophic bacteria, such as *Thiomicrospira.* Thus, this pathway, originally discovered in green sulfur bacteria and originally thought to be unique to

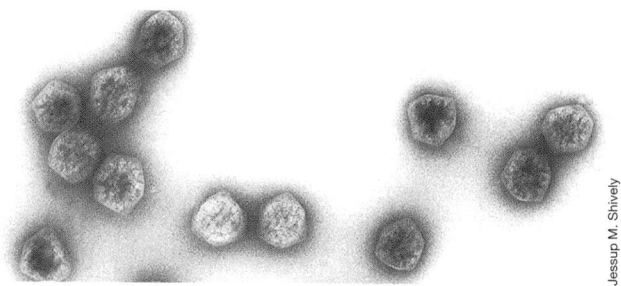

Jessup M. Shively

Figure 13.31 **Crystalline Calvin cycle enzymes: Carboxysomes.**
Electron micrograph of carboxysomes purified from the chemolithotrophic sulfur oxidizer *Halothiobacillus neapolitanus*. The structures are about 100 nm in diameter. Carboxysomes are present in a wide variety of obligately autotrophic aerobic prokaryotes.

these phototrophs, is likely distributed among several groups of autotrophic prokaryotes.

Autotrophy in *Chloroflexus*

The green nonsulfur phototroph *Chloroflexus* (⟋⟍ Section 18.18) grows autotrophically with either H_2 or H_2S as electron donor. However, neither the Calvin cycle nor the reverse citric acid cycle operates in this organism. Instead, two molecules of CO_2 are reduced to glyoxylate by the **hydroxypropionate pathway**. This pathway is so named because hydroxypropionate, a three-carbon compound, is a key intermediate (Figure 13.32*b*).

In phototrophic bacteria, the hydroxypropionate pathway has been confirmed only in *Chloroflexus*, the earliest branching anoxygenic phototroph on the phylogenetic tree of *Bacteria* (⟋⟍ Figure 17.1). This suggests that the hydroxypropionate pathway may have been one of the earliest, if not *the* earliest, mechanisms for autotrophy in anoxygenic phototrophs. But in addition to *Chloroflexus*, the hydroxypropionate pathway functions in several hyperthermophilic *Archaea*, including *Metallosphaera*, *Acidianus*, and *Sulfolobus* (⟋⟍ Section 19.9). These are all nonphototrophs that lie near the base of the phylogenetic tree of *Archaea*. The evolutionary roots of the hydroxypropionate pathway may thus be very deep. Indeed it is possible that this pathway was nature's first attempt at autotrophy in any prokaryotic organism.

MiniQuiz

- Contrast autotrophy in the following phototrophs: cyanobacteria; purple and green bacteria; *Chloroflexus*.

- Why might the hydroxypropionate pathway be of significant evolutionary importance?

13.14 Nitrogen Fixation and Nitrogenase

The biological utilization of dinitrogen (N_2) as cell nitrogen is called **nitrogen fixation**. The N_2 is reduced to ammonia (NH_3), a major form of fixed nitrogen, and then assimilated into organic forms, such as amino acids and nucleotides. The ability to fix nitrogen frees an organism from dependence on fixed nitrogen in its environment and confers a significant ecological advantage on it. The process of nitrogen fixation is also of enormous agricultural importance, supporting the nitrogen needs of key crops, such as soybeans.

Only certain prokaryotes can fix nitrogen, and an abbreviated list of nitrogen-fixing organisms is given in **Table 13.2**. Some nitrogen-fixing bacteria are *free-living* and require no host in order to carry out the process. By contrast, others are *symbiotic* and fix nitrogen only in association with certain plants (⟋⟍ Section 25.3). But it is the bacterium, not the plant, that fixes the N_2; no eukaryotic organisms are known that fix nitrogen. Many different physiological types of prokaryotes can fix nitrogen, including several that live in extreme environments. For example, nitrogen fixation has been recorded at temperatures below 0°C and as high as 92°C, and at pH 2 and pH 10, suggesting that few microbial environments would be off limits to nitrogen-fixing bacteria.

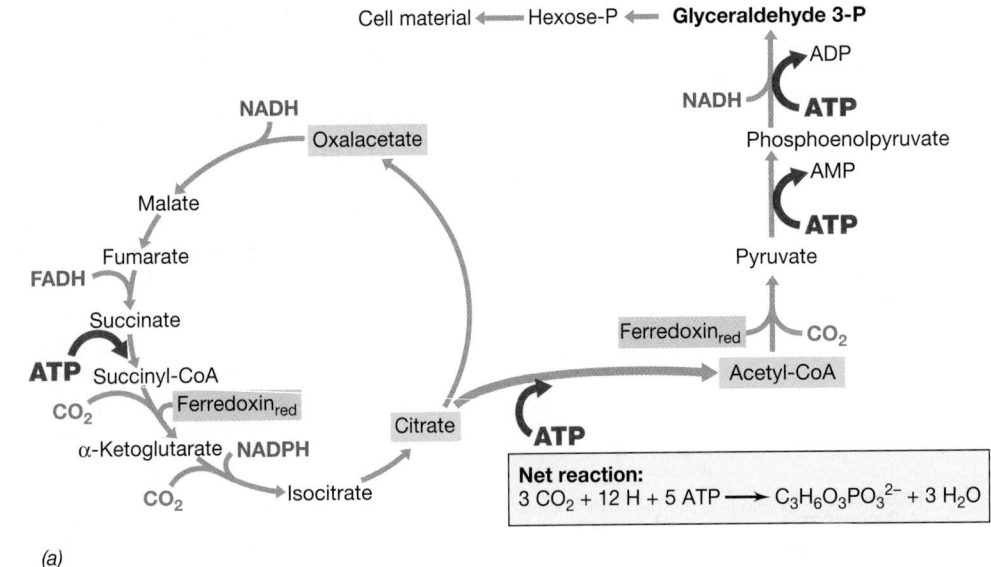

Figure 13.32 Unique autotrophic pathways in phototrophic green bacteria. *(a)* The reverse citric acid cycle is the mechanism of CO_2 fixation in green sulfur bacteria (Section 18.15). Ferredoxin$_{red}$ indicates carboxylation reactions requiring reduced ferredoxin (2 H each). Starting from oxalacetate, each turn of the cycle results in three molecules of CO_2 being incorporated and pyruvate as the product. The conversion of pyruvate to phosphoenolpyruvate consumes two energy-rich phosphate bond equivalents. NADH or FADH supply the other reducing power needs of the cycle. *(b)* The hydroxypropionate pathway is the autotrophic pathway in the green nonsulfur bacterium *Chloroflexus* (Section 18.18). Acetyl-CoA is carboxylated twice to yield methyl-malonyl-CoA. This intermediate is rearranged to yield a new acetyl-CoA acceptor molecule and a molecule of glyoxylate, which is converted to cell material. The source of reducing power in the hydroxypropionate pathway is NADPH.

Nitrogenase

Biological nitrogen fixation is catalyzed by a large enzyme complex called *nitrogenase*. Nitrogenase consists of two distinct proteins, *dinitrogenase* and *dinitrogenase reductase*. Both proteins contain iron, and dinitrogenase contains molybdenum as well. The iron and molybdenum in dinitrogenase are contained within a cofactor called the *iron–molybdenum cofactor* (*FeMo-co*, **Figure 13.33**), and actual reduction of N_2 occurs at this site. The composition of FeMo-co is MoFe$_7$S$_8$ • homo-citrate. Owing to the stability of the triple bond in N_2, it is extremely inert; its activation and reduction is therefore a very energy-demanding process. Six electrons must be transferred to reduce N_2 to NH_3; the three successive reduction steps occur directly on nitrogenase with no free intermediates accumulating (**Figure 13.34**).

Nitrogen fixation is inhibited by oxygen (O_2). This is because dinitrogenase reductase is rapidly and irreversibly inactivated by O_2, and this is true even if this enzyme is isolated from aerobic nitrogen fixers. In aerobic nitrogen-fixing bacteria, N_2 is fixed in intact cells but not in purified enzyme preparations. Nitrogenase in such organisms is protected from oxygen inactivation by one of several different mechanisms, including the rapid removal of

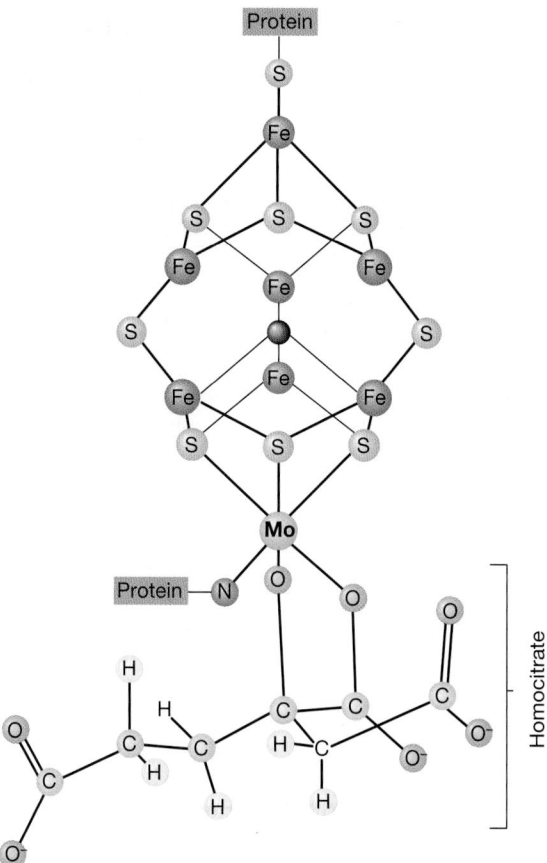

Figure 13.33 FeMo-co, the iron–molybdenum cofactor from nitrogenase. On the top is the Fe_7S_8 cube that binds to Mo along with O atoms from homocitrate (bottom, all O atoms shown in purple) and N and S atoms from dinitrogenase. The central atom shown in black is unknown, but could be C, O, or N.

Table 13.2 *Some nitrogen-fixing organisms*[a]		
Free-living aerobes/Facultative aerobes		
Chemoorganotrophs	*Phototrophs*	*Chemolithotrophs*
Azotobacter	Cyanobacteria	Alcaligenes
Azomonas		Thiobacillus
Agrobacterium		Acidithiobacillus
Klebsiella[b]		Streptomyces
Beijerinckia		thermoautotrophicus
Bacillus polymyxa		
Mycobacterium flavum		
Azospirillum lipoferum		
Citrobacter freundii		
Acetobacter diazotrophicus		
Methylomonas		
Methylococcus		
Methylosinus		
Free-living anaerobes		
Chemoorganotrophs	*Phototrophs*	*Chemolithotrophs*[c]
Clostridium	Chromatium	Methanosarcina
Desulfovibrio	Ectothiorhodospira	Methanococcus
Desulfobacter	Thiocapsa	Methanobacterium
Desulfotomaculum	Chlorobium	Methanospirillum
	Chlorobaculum	Methanolobus
	Rhodospirillum	Methanocaldococcus
	Rhodopseudomonas	
	Rhodomicrobium	
	Rhodopila	
	Rhodobacter	
	Heliobacterium	
	Heliobacillus	
	Heliophilum	
	Heliorestis	
Symbiotic		
Leguminous plants	*Nonleguminous plants*	
Soybeans, peas, clover, locust, alfalfa, and so on, in association with a bacterium of the genus *Rhizobium, Bradyrhizobium, Sinorhizobium,* or *Azorhizobium*	Alnus, Myrica, Ceanothus, Comptonia, Casuarina; in association with actinomycetes of the genus *Frankia*; *Anabaena* (a cyanobacterium) with the water fern *Azolla*	

[a]For some genera listed, nitrogen fixation occurs in only one or a few species.
[b]Nitrogen fixation occurs only under anoxic conditions.
[c]All are *Archaea*.

O_2 by respiration; the production of O_2-retarding slime layers (**Figure 13.35**); or, in certain cyanobacteria, by compartmentalization of nitrogenase in a special type of cell, the heterocyst (⮂ Section 18.7). In addition, although N_2 is not fixed in cell extracts exposed to oxygen, in aerobic nitrogen fixers such as *Azotobacter*, nitrogenase is protected from oxygen inactivation by complexing with a specific protein; this process is called *conformational protection* and is reversible. When oxygen is no longer present at inhibitory levels, the conformationally protected nitrogenase can once again become active.

Electron Flow in Nitrogen Fixation

The sequence of electron transfer in nitrogenase is as follows: electron donor → dinitrogenase reductase → dinitrogenase → N_2, with NH_3 being the final product. The electrons for N_2 reduction are transferred to dinitrogenase reductase from ferredoxin or flavodoxin, both of which are low-potential iron–sulfur proteins (⮂ Section 4.9); flavodoxin or ferredoxin is reduced by the oxidation of pyruvate (Figure 13.34). In addition to electrons, ATP is required for nitrogen fixation. In each cycle of electron transfer, dinitrogenase reductase is reduced and binds two molecules of ATP. ATP binding alters the con-

formation of dinitrogenase reductase and lowers its reduction potential, allowing it to interact with and reduce dinitrogenase. Upon electron transfer to dinitrogenase, the ATP is hydrolyzed and dinitrogenase reductase dissociates from dinitrogenase and begins another cycle of reduction and ATP binding (Figure 13.34). When fully reduced, dinitrogenase reduces N_2 to NH_3, with the actual reduction occurring at the FeMo-co center (Figure 13.33).

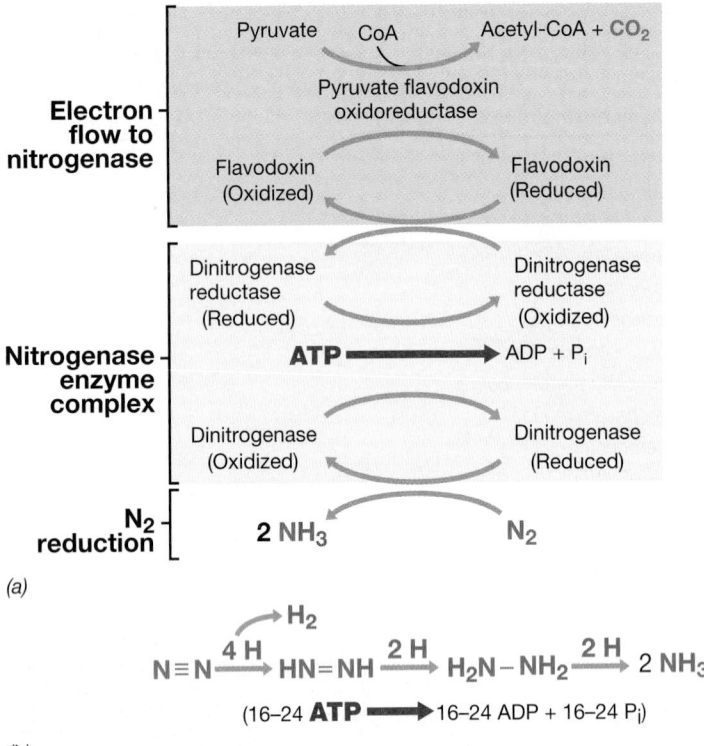

Electron flow to nitrogenase

Pyruvate   CoA   Acetyl-CoA + CO_2

Pyruvate flavodoxin oxidoreductase

Flavodoxin (Oxidized)   Flavodoxin (Reduced)

Nitrogenase enzyme complex

Dinitrogenase reductase (Reduced)   Dinitrogenase reductase (Oxidized)

ATP ➔ ADP + P_i

Dinitrogenase (Oxidized)   Dinitrogenase (Reduced)

N_2 reduction

2 NH_3 N_2

(a)

$$N \equiv N \xrightarrow{\text{4 H}} HN=NH \xrightarrow{\text{2 H}} H_2N-NH_2 \xrightarrow{\text{2 H}} 2\ NH_3$$

$$\xrightarrow{H_2}$$

(16–24 **ATP** ➔ 16–24 ADP + 16–24 P_i)

(b)

Figure 13.34 Nitrogenase function. *(a)* Shown are the steps in N_2 fixation starting from pyruvate as electron donor. Electrons are transferred from dinitrogenase reductase to dinitrogenase one at a time, and each electron supplied is associated with the hydrolysis of 2–3 ATPs. *(b)* Hypothetical steps in N_2 reduction showing the H_2 evolution step and a summary of nitrogenase activity.

Although only *six* electrons are necessary to reduce N_2 to two NH_3, *eight* electrons are actually consumed in the process, two electrons being lost as H_2 for each mole of N_2 reduced (Figure 13.34). The reason for this reducing power wastage is unknown, but it is clear that H_2 evolution is a necessary part of nitrogenase function and is an activity that cannot be eliminated by mutations in genes encoding nitrogenase proteins.

Alternative Nitrogenases

Some nitrogen-fixing *Bacteria* and *Archaea* produce nitrogenases that lack Mo under conditions in which Mo is either completely absent or severely limiting. These *alternative nitrogenases*, as they are called, are similar to the molybdenum nitrogenase but contain either vanadium (V) or Fe in place of Mo. Cofactors similar to FeMo-co are present in these alternative nitrogenases—FeVa-co in the vanadium nitrogenase, and an iron–sulfur cluster resembling FeMo-co and FeVa-co but lacking both Mo and V in the iron nitrogenase. Alternative nitrogenases are not synthesized when sufficient Mo is present. Molybdenum represses synthesis of alternative nitrogenases; thus the molybdenum nitrogenase is the main nitrogenase in the cell and the first to be synthesized when fixed nitrogen becomes limiting. Alternative nitrogenases function as a "backup system" to support nitrogen fixation when Mo is unavailable in the habitat.

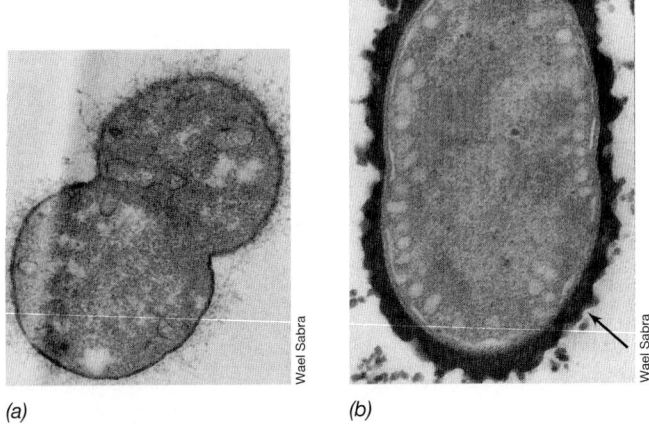

(a)   *(b)*

Wael Sabra

Figure 13.35 Induction of slime formation by O_2 in nitrogen-fixing cells of *Azotobacter vinelandii*. *(a)* Transmission electron micrograph of cells grown with 2.5% O_2; very little slime is evident. *(b)* Cells grown in air (21% O_2). Note the extensive darkly staining slime layer (arrow). The slime retards diffusion of O_2 into the cell, thus preventing nitrogenase inactivation by O_2. A single cell of *A. vinelandii* is about 2 μm in diameter.

A structurally and functionally novel molybdenum nitrogenase is produced by the gram-positive bacterium *Streptomyces thermoautotrophicus*. This organism is a thermophilic (optimum temperature 65°C), filamentous member of the *Actinobacteria* (♋ Section 18.6). *S. thermoautotrophicus* is a hydrogen bacterium that uses carbon monoxide (CO) as an electron donor in energy metabolism and is also an autotroph. The *S. thermoautotrophicus* nitrogenase contains Mo, but unlike the classic molybdenum nitrogenase, the *S. thermoautotrophicus* nitrogenase is insensitive to O_2. The dinitrogenase analog of the *S. thermoautotrophicus* nitrogenase, a protein called *Str1*, contains three different polypeptides that show slight structural similarity to dinitrogenase polypeptides from other nitrogen fixers. However, the dinitrogenase reductase analog, a protein called *Str2*, shows no similarity to other dinitrogenase reductases. Str2 is, however, related to the enzyme superoxide dismutase (♋ Section 5.18), and this activity plays a role in this unusual nitrogenase (**Figure 13.36**). Str2 supplies electrons to Str1 from superoxide (O_2^-), the O_2^- being formed from the reduction of O_2 by the oxidation of carbon monoxide (CO). Str1 then reduces N_2 to NH_3 (Figure 13.36).

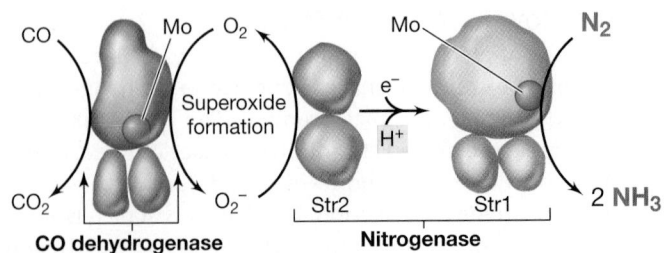

CO   Mo O_2

Superoxide formation

CO_2 O_2^-

e^-

H^+

Mo N_2

Str2   Str1   2 NH_3

CO dehydrogenase   **Nitrogenase**

Figure 13.36 Reactions of nitrogen fixation in *Streptomyces thermoautotrophicus*. Although Str2 and Str1 are distinct from dinitrogenase reductase and dinitrogenase proteins, they are functionally equivalent, respectively, to these proteins.

Other unique properties of the *S. thermoautotrophicus* nitrogenase include the fact that the enzyme requires less than half of the ATP of classical nitrogenases and is encoded by genes that are unrelated to "classical" nitrogenase genes. Moreover, the *S. thermoautotrophicus* nitrogenase does not reduce acetylene, a universal property of nitrogenases (see Figure 13.37). Thus any organism that contains an *S. thermoautotrophicus* nitrogenase or other unconventional nitrogenases may well go undetected in surveys for nitrogenase that employ acetylene reduction or genomic analyses as screening tools. It is therefore possible that many more prokaryotes fix nitrogen than microbiologists currently believe.

Assaying Nitrogenase: Acetylene Reduction

Classic nitrogenases are not entirely specific for N_2 because they also reduce other triply bonded compounds, such as cyanide (CN^-) and acetylene $(HC \equiv CH)$. The reduction of acetylene by nitrogenase is only a two-electron process, and *ethylene* $(H_2C \equiv CH_2)$ is produced. The reduction of acetylene to ethylene provides a simple and rapid method for measuring the activity of nitrogen-fixing systems by gas chromatography (**Figure 13.37**). This technique, known as the *acetylene reduction assay*, is widely used to detect and quantify nitrogen fixation.

Definitive proof for nitrogen fixation is obtained using an isotope of nitrogen, ^{15}N, as a tracer. (^{15}N is not a radioisotope but a stable isotope. It is detected with a mass spectrometer.) The gas phase of a culture is enriched with $^{15}N_2$, and after an incubation period, the cells and medium are digested to release NH_3 from all cellular nitrogenous compounds, and the NH_3 is assayed for its ^{15}N content. If there has been a significant production of $^{15}NH_3$, it is proof of nitrogen fixation.

Although $^{15}N_2$ assimilation is occasionally used to demonstrate nitrogen fixation, the acetylene reduction method is a more rapid and sensitive method for measuring this process. Therefore, biological acetylene reduction is most frequently used for studying nitrogen fixation, and is taken as strong evidence for the activity of nitrogenase. In an assay, the sample, which may be soil, water, a culture, or a cell extract, is incubated with $HC \equiv CH$, and the gas phase of the reaction mixture is later analyzed by gas chromatography for production of $H_2C \equiv CH_2$ (Figure 13.37). This method is far simpler and faster than other methods and can easily be adapted for use in ecological studies of nitrogen-fixing bacteria directly in their habitats.

MiniQuiz
- Write a balanced equation for the reaction catalyzed by nitrogenase.
- What is FeMo-co and what does it do?
- How is acetylene useful in studies of nitrogen fixation?

13.15 Genetics and Regulation of Nitrogen Fixation

Because the process of N_2 fixation is highly energy demanding, the synthesis and activity of nitrogenase and the many other enzymes required for N_2 fixation (Figure 13.34) are highly regulated.

Genetics of Nitrogen Fixation

The genes for the molybdenum nitrogenase of *Klebsiella pneumoniae*, a well-studied nitrogen fixer, are part of a regulon (several operons under common control, ↩ Section 8.4) called the *nif regulon*. The *K. pneumoniae nif* regulon spans 24 kilobase pairs of DNA and contains 20 genes arranged in operons such that genes whose products have similar functions are cotranscribed (**Figure 13.38**). In addition to nitrogenase structural genes, the genes for FeMo-co synthesis, genes controlling the electron transport proteins (Figure 13.34), and a number of regulatory genes are also present in the *nif* regulon.

Dinitrogenase is composed of two subunits, α (the product of *nifD*) and β (the product of *nifK*), each of which is present in two copies in the nitrogenase enzyme complex. Dinitrogenase reductase consists of two identical subunits, the product of *nifH*. FeMo-co is synthesized by enzymes encoded by several genes, including *nifN, V, Z, W, E, B*, and *Q*. The *nifA* gene encodes a positive regulatory protein that activates transcription of other *nif* genes (Figure 13.38).

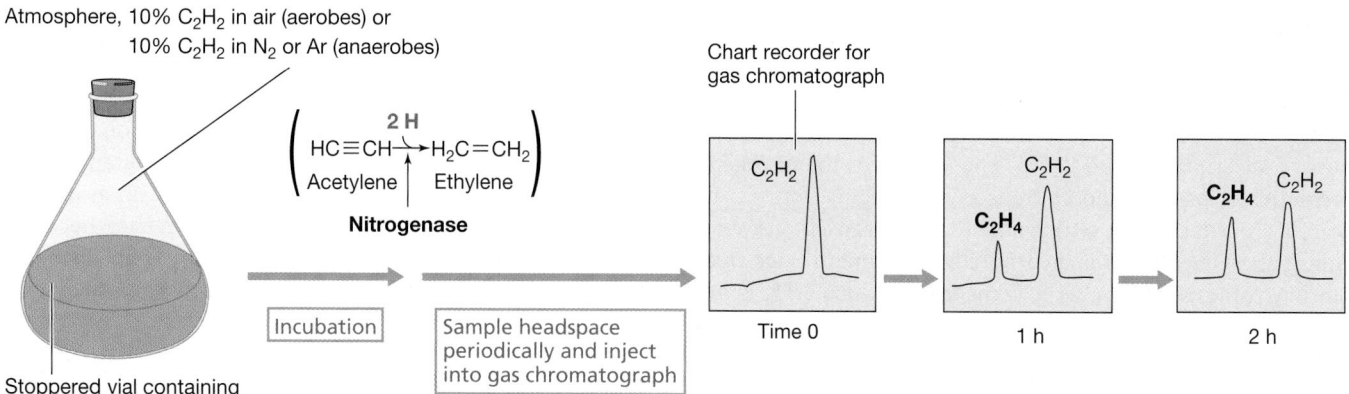

Figure 13.37 The acetylene reduction assay for nitrogenase activity. The results show no ethylene (C_2H_4) when the experiment begins (time 0), but increasing production of C_2H_4 as the assay proceeds. Note how as C_2H_4 is produced, C_2H_2 is consumed. If the vial contained an enzyme extract, conditions would need to be anoxic, even if nitrogenase was from an aerobic bacterium.

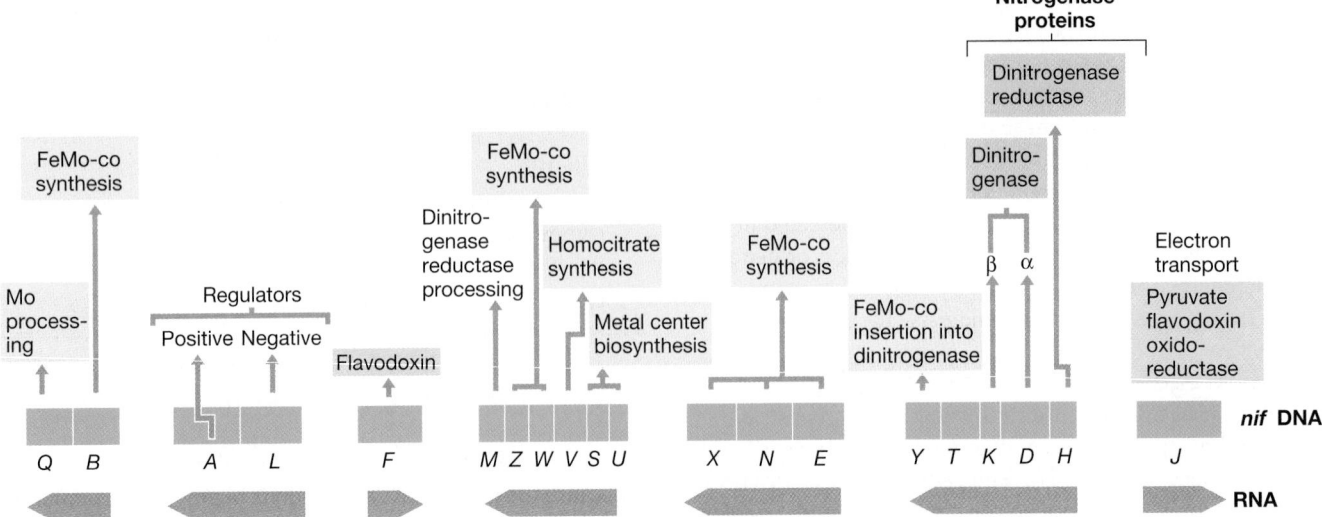

Figure 13.38 The *nif* regulon in *Klebsiella pneumoniae*, the best-studied nitrogen-fixing bacterium. The function of the *nifT* gene product is unknown. The mRNA transcripts are shown below the genes; arrows indicate the direction of transcription. Proteins that catalyze FeMo-co synthesis are shown in yellow.

Nitrogenase is a highly conserved protein, and the *nifHDK* genes that encode it have been used as probes to screen DNA from various prokaryotes for the presence of homologous genes, thus signaling their ability to fix N_2. Alternative nitrogenases (see Section 13.14) are encoded by their own structural genes, *vnfHDK* for the vanadium system and *anfHDK* for the iron-only system, but these genes show significant sequence similarity to *nifHDK*.

Regulation of Nitrogenase Synthesis

Nitrogenase is subject to strict regulatory controls. Nitrogen fixation is prevented by O_2 and by fixed forms of nitrogen, including NH_3, NO_3^-, and certain amino acids. A major part of this regulation occurs in the expression of *nif* structural genes, whose transcription is activated by the NifA protein (positive regulation, ⮂ Section 8.4). By contrast, NifL is a negative regulator of *nif* gene expression and contains a molecule of FAD (recall that FAD is a redox coenzyme for flavoproteins, ⮂ Section 4.9) that is involved in O_2 sensing. In the presence of sufficient O_2, NifL prevents synthesis of other *nif* genes, which in turn block synthesis of the oxygen-labile nitrogenase.

Ammonia prevents nitrogen fixation through a second protein, called NtrC, whose activity is regulated by the nitrogen status of the cell. When NH_3 is limiting, NtrC is active and promotes transcription of *nifA*. This encodes NifA, the nitrogen fixation activator protein, and *nif* transcription begins.

The NH_3 produced by nitrogenase does not itself prevent enzyme synthesis because it is incorporated into amino acids and used in biosynthesis as soon as it is made. But when NH_3 is in excess (as in natural environments or culture media high in NH_3), nitrogenase synthesis is prevented. In this way, ATP is not wasted in making ammonia when it is already available in ample amounts.

Regulation of Nitrogenase Activity

Besides regulating the *synthesis* of nitrogenase, the *activity* of nitrogenase already present in a cell can also be regulated by NH_3 in many nitrogen-fixing bacteria. Although there are likely different mechanisms for shutting off the activity of nitrogenase, they all seem to function by shutting down electron flow to dinitrogenase in response to excess NH_3, thus making nitrogenase inactive.

The mechanism of nitrogenase shut down by NH_3 is known in many nitrogen fixers and is called the *ammonia switch-off effect*. In this process, excess NH_3 causes a molecule of ADP to be added to dinitrogenase reductase, which results in a loss of enzyme activity. When NH_3 once again becomes limiting, the modified dinitrogenase reductase is converted back to its active form and N_2 fixation resumes. Ammonia switch-off is thus a rapid and reversible method of controlling ATP and reductant consumption by nitrogenase.

Ammonia switch-off has also been observed in nitrogen-fixing *Archaea*. In *Methanococcus* species, for example, NH_3 quickly inhibits the activity of nitrogenase. Here, however, covalent modification of dinitrogenase reductase is not involved. Instead, it appears that an NH_3-sensing protein exists in the cell that can bind to nitrogenase or in some other way inactivate it when NH_3 is in excess in the cell. Interestingly, the "ammonia-sensing system" in *Archaea* appears to function not by detecting NH_3 itself but by monitoring cytoplasmic levels of the carbon compound α-ketoglutarate, a precursor of the amino acid glutamate; a shortage of α-ketoglutarate is sensed by the organism as an excess of glutamate, and the latter inactivates nitrogenase because NH_3 triggers glutamate synthesis (⮂ Section 8.6). This indirect nitrogenase regulatory system controls nitrogenase activity in a few *Bacteria* as well.

MiniQuiz

- What chemical and physical factors affect the synthesis of nitrogenase?
- How can activity of preformed nitrogenase be regulated? How would a cell benefit by doing this?

Big Ideas

13.1

Phototrophs obtain their energy from light. In photosynthetic reactions ATP is generated from light and then is consumed in the reduction of CO_2 by NADH. Two forms of photosynthesis are known: oxygenic, where O_2 is produced, and anoxygenic, where it is not. Cyanobacteria and algae are oxygenic phototrophs, whereas purple bacteria, green bacteria, and heliobacteria are anoxygenic phototrophs.

13.2

Chlorophylls and bacteriochlorophylls reside in photosynthetic membranes where the light reactions of photosynthesis are carried out. Antenna chlorophylls harvest light energy and transfer it to reaction center chlorophylls. In green bacteria, chlorosomes function as a giant antenna system.

13.3

Accessory pigments such as carotenoids and phycobilins absorb light and transfer the energy to reaction center chlorophyll, thus broadening the wavelengths of light usable in photosynthesis. Carotenoids also play an important photoprotective role in preventing photooxidative damage to cells.

13.4

Electron transport reactions occur in the photosynthetic reaction center of anoxygenic phototrophs, resulting in the formation of a proton motive force and the synthesis of ATP. Reducing power for CO_2 fixation comes from substances such as H_2S, and NADH production in purple bacteria requires reverse electron transport.

13.5

In oxygenic photosynthesis, H_2O donates electrons to drive CO_2 fixation, and O_2 is a by-product. There are two separate photosystems, PSI and PSII, in oxygenic phototrophs, whereas anoxygenic phototrophs contain a single photosystem.

13.6

Chemolithotrophs oxidize inorganic electron donors to conserve energy and obtain reducing power. Most chemolithotrophs can also grow autotrophically.

13.7

The chemolithotrophic hydrogen bacteria use H_2 as an electron donor, reducing O_2 to H_2O. The enzyme hydrogenase is required to oxidize H_2, and H_2 also supplies reducing power for the fixation of CO_2 in these autotrophs.

13.8

Reduced sulfur compounds such as H_2S, $S_2O_3^{2-}$, and S^0 are electron donors for energy metabolism in sulfur chemolithotrophs. Electrons from these substances enter electron transport chains, yielding a proton motive force. Sulfur chemolithotrophs are also autotrophs and fix CO_2 by the Calvin cycle.

13.9

Chemolithotrophic iron bacteria oxidize Fe^{2+} as an electron donor. Most iron bacteria grow at acidic pH and are often associated with acidic pollution from mineral and coal mining. A few chemolithotrophic and phototrophic bacteria can oxidize Fe^{2+} to Fe^{3+} anaerobically.

13.10

Ammonia and nitrite are electron donors for the nitrifying bacteria. The ammonia-oxidizing bacteria produce nitrite, which is then oxidized by the nitrite-oxidizing bacteria to nitrate.

13.11

Anoxic ammonia oxidation (anammox) consumes both ammonia and nitrite, forming N_2. The anammox reaction occurs within a membrane-enclosed compartment within the cell, called the anammoxosome.

13.12

Autotrophy is supported in most phototrophic and chemolithotrophic bacteria by the Calvin cycle, in which the enzyme RubisCO plays a key role in converting CO_2 into sugar. Carboxysomes contain crystalline RubisCO and function to concentrate CO_2, the key substrate for this enzyme.

13.13

The reverse citric acid and hydroxypropionate cycles are autotrophic pathways in green sulfur and green nonsulfur bacteria, respectively, and are also found in a few nonphototrophic prokaryotes.

13.14

Nitrogen fixation is the reduction of N_2 to NH_3 and requires the enzyme nitrogenase. Most nitrogenases contain molybdenum or vanadium plus iron as metal cofactors, and the process of nitrogen fixation is highly energy demanding. Some other triply bonded compounds, such as acetylene, are also reduced by nitrogenase.

13.15

Nitrogenase and associated proteins are encoded by the *nif* regulon. Nitrogen fixation is highly regulated at both the transcriptional and enzyme activity levels, with O_2 and NH_3 being the two major regulatory effectors.

Review of Key Terms

Anammox anoxic ammonia oxidation

Anoxygenic photosynthesis photosynthesis in which O_2 is not produced

Antenna pigments light-harvesting chlorophylls or bacteriochlorophylls in photocomplexes that funnel energy to the reaction center

Autotroph an organism that uses CO_2 as its sole carbon source

Bacteriochlorophyll the chlorophyll pigment of anoxygenic phototrophs

Calvin cycle the biochemical pathway for CO_2 fixation in many autotrophic organisms

Carboxysomes crystalline inclusions of RubisCO

Carotenoid a hydrophobic accessory pigment present along with chlorophyll in photosynthetic membranes

Chemolithotroph a microorganism that oxidizes inorganic compounds as electron donors in energy metabolism

Chlorophyll a light-sensitive, Mg-containing porphyrin of phototrophic organisms that initiates the process of photophosphorylation

Chlorosome a cigar-shaped structure present in the periphery of cells of green sulfur and green nonsulfur bacteria and containing the antenna bacteriochlorophylls (*c*, *d*, or *e*)

Hydrogenase an enzyme, widely distributed in anaerobic microorganisms, capable of taking up or evolving H_2

Hydroxypropionate pathway an autotrophic pathway found in *Chloroflexus* and a few *Archaea*

Mixotroph an organism in which an inorganic compound serves as the electron donor in energy metabolism and organic compounds serve as the carbon source

Nitrification the microbial conversion of NH_3 to NO_3^-

Nitrogen fixation the biological reduction of N_2 to NH_3 by nitrogenase

Oxygenic photosynthesis photosynthesis carried out by cyanobacteria and green plants in which O_2 is evolved

Photophosphorylation the production of ATP in photosynthesis

Photosynthesis the series of reactions in which ATP is synthesized by light-driven reactions and CO_2 is fixed into cell material

Phototroph an organism that uses light as an energy source

Phycobiliprotein the antenna pigment complex in cyanobacteria that contains phycocyanin and allophycocyanin or phycoerythrin coupled to proteins

Phycobilisome an aggregate of phycobiliproteins

Reaction center a photosynthetic complex containing chlorophyll or bacteriochlorophyll and several other components, within which occurs the initial electron transfer reactions of photosynthetic electron flow

Reverse citric acid cycle a mechanism for autotrophy in green sulfur bacteria and a few other phototrophs

Reverse electron transport the energy-dependent movement of electrons against the thermodynamic gradient to form a strong reductant from a weaker electron donor

RubisCO the acronym for ribulose bisphosphate carboxylase, a key enzyme of the Calvin cycle

Thylakoids membrane stacks in cyanobacteria or in the chloroplast of eukaryotic phototrophs

Review Questions

1. In one sentence, highlight the difference between oxygenic and anoxygenic photosynthesis (Section 13.1).

2. What is the difference between chloroplasts and chlorosomes (Section 13.2)?

3. Where are the photosynthetic pigments located in a phototrophic purple bacterium? A cyanobacterium? A green alga? Considering the function of chlorophyll pigments, why are they not located elsewhere in the cell, for example, in the cytoplasm or in the cell wall (Section 13.2)?

4. Distinguish between the terms phycobiliproteins and phycobilisomes (Section 13.3).

5. How does light result in ATP production in an anoxygenic phototroph? In what ways are photosynthetic and respiratory electron flow similar? In what ways do they differ (Section 13.4)?

6. How is reducing power for autotrophic growth obtained in a purple bacterium? In a cyanobacterium (Section 13.4)?

7. How does the reduction potential of chlorophyll *a* in PSI and PSII differ? Why must the reduction potential of PSII chlorophyll be so highly electropositive (Section 13.5)?

8. Compare and contrast the utilization of H_2S by a purple phototrophic bacterium and by a colorless sulfur bacterium such as *Beggiatoa*. What role does H_2S play in the metabolism of each organism (Sections 13.4 and 13.8)?

9. Which inorganic electron donors are used by the organisms *Ralstonia* and *Thiobacillus* (Sections 13.6–13.8)?

10. Why can it be said that despite the chemistry of its environment, *Acidithiobacillus ferrooxidans* does not get an energetic "free lunch" (Section 13.9)?

11. Contrast classical nitrification with anammox in terms of oxygen requirements, organisms involved, and the need for monooxygenases (Sections 13.10 and 13.11).

12. Briefly describe the Calvin cycle (Section 13.12).

13. Which organisms employ the hydroxypropionate or reverse citric acid cycles as autotrophic pathways (Section 13.13)?

14. Write out the reaction catalyzed by the enzyme nitrogenase. How many electrons are required in this reaction? How many are actually used? Explain (Section 13.14).

15. How does the *Streptomyces thermoautotrophicus* nitrogenase differ from that of *Azotobacter* (Section 13.14)?

16. How is nitrogenase synthesis and activity controlled by NH_3 and O_2 (Section 13.15)?

Application Questions

1. Compare and contrast the absorption spectrum of chlorophyll *a* and bacteriochlorophyll *a*. Which wavelengths are preferentially absorbed by each pigment, and how do the absorption properties of these molecules compare with the regions of the spectrum visible to our eye? Why are most plants green?

2. The growth rate of the phototrophic purple bacterium *Rhodobacter* is about twice as fast when the organism is grown *phototrophically* in a medium containing malate as the carbon source as when it is grown with CO_2 as the carbon source (with H_2 as the electron donor). Discuss the reasons why this is true, and list the nutritional class in which we would place *Rhodobacter* when growing under each of the two different conditions.

3. Although physiologically distinct, chemolithotrophs and chemoorganotrophs share a number of features with respect to the production of ATP. Discuss these common features along with reasons why the growth yield (grams of cells per mole of substrate) of a chemoorganotroph respiring glucose is so much higher than for a chemolithotroph respiring sulfur.

4. Employing biotechnology, you would like to genetically engineer corn (maize) to fix nitrogen. Discuss what type of nitrogenase you would try to engineer into the corn plant and why this would be the most suitable enzyme for the purpose.

Need more practice? Test your understanding with Quantitative Questions; access additional study tools including tutorials, animations, and videos; and then test your knowledge with chapter quizzes and practice tests at **www.microbiologyplace.com**.

14

Catabolism of Organic Compounds

Methanogens produce natural gas (methane, CH_4) and are able to do so because they contain a series of unusual coenzymes, such as the green-fluorescing F_{420}, that participate in biochemical reactions unique to these organisms.

In Chapter 13 we considered phototrophy and chemolithotrophy, strategies for energy conservation that do not use organic compounds as electron donors. In this chapter we focus on organic compounds as electron donors and the many ways in which chemoorganotrophic microorganisms conserve energy. A major focus will be on *anaerobic* forms of metabolism, because novel strategies for anaerobic growth are a hallmark of prokaryotic diversity. We end the chapter with a consideration of the aerobic catabolism of key organic compounds, primarily monomers released from the degradation of macromolecules.

 # Fermentations

Two broad metabolic processes for the catabolism of organic compounds are *fermentation* and *respiration*. These processes differ fundamentally in terms of oxidation–reduction (redox) considerations and mechanism of ATP synthesis. In respiration, whether aerobic or anaerobic, exogenous electron acceptors are required to accept electrons generated from the oxidation of electron donors. In fermentation, this is not the case. Thus in respiration but not fermentation we will see a common theme of electron transport and the generation of a proton motive force.

We begin our exploration of organic catabolism with fermentations. Compared with respirations, fermentations are typically energetically marginal. However, we will see that a little free energy can go a long way and that bacterial fermentative diversity is both extensive and innovative.

14.1 Energetic and Redox Considerations

Many microbial habitats are **anoxic** (oxygen-free). In such environments, decomposition of organic material occurs anaerobically. If adequate supplies of electron acceptors such as sulfate (SO_4^{2-}), nitrate (NO_3^-), ferric iron (Fe^{3+}), and others to be considered later are unavailable in anoxic habitats, organic compounds are catabolized by **fermentation** (**Figure 14.1**). In Chapter 4 we discussed some key fermentations that yield alcohol or lactic acid as products by way of the glycolytic pathway. There we emphasized how fermentations are internally balanced redox processes in which the fermentable substrate becomes both oxidized and reduced.

An organism faces two major problems when it catabolizes organic compounds for the purpose of energy conservation: (1) ATP synthesis, and (2) redox balance. In fermentations, with rare exception ATP is synthesized by *substrate-level phosphorylation*. This is the mechanism in which energy-rich phosphate bonds from phosphorylated organic compounds are transferred directly to ADP to form ATP (♻ Section 4.7). The second problem, redox balance, is solved by the production and subsequent excretion of fermentation products generated from the original substrate (Figure 14.1).

Energy-Rich Compounds and Substrate-Level Phosphorylation

Energy can be conserved by substrate-level phosphorylation from many different compounds. However, central to an understanding of substrate-level phosphorylation is the concept of

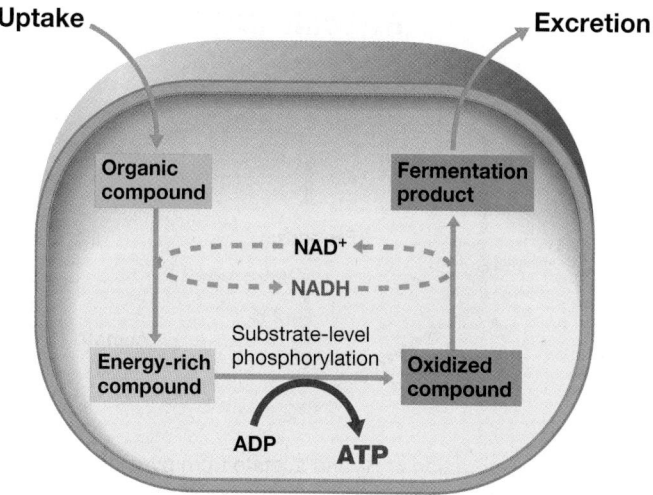

Figure 14.1 The essentials of fermentation. The fermentation product is excreted from the cell, and only a relatively small amount of the original organic compound is used for biosynthesis.

energy-rich compounds. These are organic compounds that contain an energy-rich phosphate bond or a molecule of coenzyme A; the hydrolysis of either of these is highly exergonic (♻ Figure 4.12). **Table 14.1** lists some energy-rich intermediates formed during biochemical processes. The hydrolysis of most of the compounds listed can be coupled to ATP synthesis ($\Delta G^{0\prime} = -31.8$ kJ/mol). In other words, if an organism can form one of

Table 14.1 *Energy-rich compounds involved in substrate-level phosphorylation*[a]

Compound	Free energy of hydrolysis, $\Delta G^{0\prime}$ (kJ/mol)[b]
Acetyl-CoA	−35.7
Propionyl-CoA	−35.6
Butyryl-CoA	−35.6
Caproyl-CoA	−35.6
Succinyl-CoA	−35.1
Acetyl phosphate	−44.8
Butyryl phosphate	−44.8
1,3-Bisphosphoglycerate	−51.9
Carbamyl phosphate	−39.3
Phosphoenolpyruvate	−51.6
Adenosine phosphosulfate (APS)	−88
N^{10}-Formyltetrahydrofolate	−23.4
Energy of hydrolysis of ATP (ATP → ADP + P_i)	−31.8

[a]Data from Thauer, R. K., K. Jungermann, and K. Decker, 1977. Energy conservation in chemotrophic anaerobic bacteria. *Bacteriol. Rev. 41*: 100–180.
[b]The $\Delta G^{0\prime}$ values shown here are for "standard conditions," which are not necessarily those of cells. Including heat loss, the energy costs of making an ATP are more like 60 kJ than 32 kJ, and the energy of hydrolysis of the energy-rich compounds shown here is thus likely higher. But for simplicity and comparative purposes, the values in this table will be taken as the actual energy released per reaction.

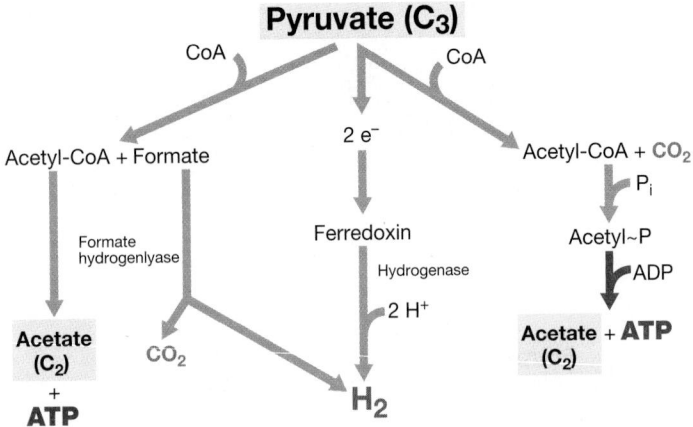

Pyruvate (C$_3$)

Figure 14.2 Production of H$_2$ and acetate from pyruvate. At least two mechanisms are known, one that produces H$_2$ directly and the other that makes formate as an intermediate. When acetate is produced, ATP synthesis is possible (Table 14.1).

Many anaerobic bacteria produce acetate as a major or minor fermentation product. The production of acetate and certain other fatty acids (Table 14.1) is energy conserving because it allows the organism to make ATP by substrate-level phosphorylation. The key intermediate generated in acetate production is acetyl-CoA (Table 14.1), an energy-rich compound. Acetyl-CoA can be converted to acetyl phosphate (Table 14.1) and the phosphate group of acetyl phosphate subsequently transferred to ADP, yielding ATP. One of the precursors of acetyl-CoA is pyruvate, a major product of glycolysis. The conversion of pyruvate to acetyl-CoA is a key oxidation reaction, and the electrons generated are used to form fermentation products or are released as H$_2$ (Figure 14.2).

MiniQuiz

- What is substrate-level phosphorylation?
- Why is acetate formation in fermentation energetically beneficial?

14.2 Lactic and Mixed-Acid Fermentations

these compounds during fermentative metabolism, it can make ATP by substrate-level phosphorylation.

Redox Balance, H$_2$, and Acetate Production

In any fermentation there must be atomic and redox balance; the total number of each type of atom and electrons in the products of the reaction must balance those in the substrates. This is obtained by the production and excretion from the cell of fermentation products (Figure 14.1). In several fermentations, redox balance is maintained by the production of molecular hydrogen, H$_2$. The production of H$_2$ is associated with the activity of the iron–sulfur protein ferredoxin, a very low-potential electron carrier, and is catalyzed by the enzyme **hydrogenase**, as illustrated in **Figure 14.2**. H$_2$ can also be produced from the C$_1$ fatty acid formate. Either way, the H$_2$ is then made available for use by other organisms.

Many fermentations are classified by either the substrate fermented or the products formed. **Table 14.2** lists some of the major fermentations classified on the basis of products formed. Note some of the broad categories, such as alcohol, lactic acid, propionic acid, mixed acid, butyric acid, and acetogenic. Some fermentations are described by the substrate fermented rather than the fermentation product. For instance, some endospore-forming anaerobic bacteria (genus *Clostridium*) ferment amino acids, whereas others ferment purines and pyrimidines. Other anaerobes ferment aromatic compounds (**Table 14.3**). Clearly, a wide variety of organic compounds can be fermented. Certain fermentations are carried out by only a very restricted group of anaerobes; in some cases this may be only a single known

Table 14.2 *Common bacterial fermentations and some of the organisms carrying them out*

Type	Reaction	Organisms
Alcoholic	Hexose → 2 ethanol + 2 CO$_2$	Yeast, *Zymomonas*
Homolactic	Hexose → 2 lactate$^-$ + 2 H$^+$	*Streptococcus*, some *Lactobacillus*
Heterolactic	Hexose → lactate$^-$ + ethanol + CO$_2$ + H$^+$	*Leuconostoc*, some *Lactobacillus*
Propionic acid	3 Lactate$^-$ → 2 propionate$^-$ + acetate$^-$ + CO$_2$ + H$_2$O	*Propionibacterium*, *Clostridium propionicum*
Mixed acid[a,b]	Hexose → ethanol + 2,3-butanediol + succinate^{2-} + lactate$^-$ + acetate$^-$ + formate$^-$ + H$_2$ + CO$_2$	Enteric bacteria including *Escherichia, Salmonella, Shigella, Klebsiella, Enterobacter*
Butyric acid[b]	Hexose → butyrate$^-$ + 2 H$_2$ + 2 CO$_2$ + H$^+$	*Clostridium butyricum*
Butanol[b]	2 Hexose → butanol + acetone + 5 CO$_2$ + 4 H$_2$	*Clostridium acetobutylicum*
Caproate/Butyrate	6 Ethanol + 3 acetate$^-$ → 3 butyrate$^-$ + caproate$^-$ + 2 H$_2$ + 4 H$_2$O + H$^+$	*Clostridium kluyveri*
Acetogenic	Fructose → 3 acetate$^-$ + 3 H$^+$	*Clostridium aceticum*

[a]Not all organisms produce all products. In particular, butanediol production is limited to only certain enteric bacteria. Reaction not balanced.
[b]Stoichiometry shows major products. Other products include some acetate and a small amount of ethanol (butanol fermentation only).

Table 14.3 *Some unusual bacterial fermentations*

Type	Reaction	Organisms
Acetylene	$2\ C_2H_2 + 3\ H_2O \rightarrow$ ethanol + acetate$^-$ + H$^+$	*Pelobacter acetylenicus*
Glycerol	4 Glycerol + $2\ HCO_3^- \rightarrow 7$ acetate$^-$ + $5\ H^+$ + $4\ H_2O$	*Acetobacterium* spp.
Resorcinol (aromatic)	$2\ C_6H_4(OH)_2 + 6\ H_2O \rightarrow 4$ acetate$^-$ + butyrate$^-$ + $5\ H^+$	*Clostridium* spp.
Phloroglucinol (aromatic)	$C_6H_6O_3 + 3\ H_2O \rightarrow 3$ acetate$^-$ + $3\ H^+$	*Pelobacter massiliensis* *Pelobacter acidigallici*
Putrescine	$10\ C_4H_{12}N_2 + 26\ H_2O \rightarrow 6$ acetate$^-$ + 7 butyrate$^-$ + $20\ NH_4^+$ + $16\ H_2$ + $13\ H^+$	Unclassified gram-positive nonsporulating anaerobes
Citrate	Citrate^{3-} + $2\ H_2O \rightarrow$ formate$^-$ + 2 acetate$^-$ + HCO_3^- + H^+	*Bacteroides* spp.
Aconitate	Aconitate^{3-} + H^+ + $2\ H_2O \rightarrow 2\ CO_2$ + 2 acetate$^-$ + H_2	*Acidaminococcus fermentans*
Glyoxylate	4 Glyoxylate$^-$ + $3\ H^+$ + $3\ H_2O \rightarrow 6\ CO_2$ + $5\ H_2$ + glycolate$^-$	Unclassified gram-negative bacterium
Benzoate	2 Benzoate$^- \rightarrow$ cyclohexane carboxylate$^-$ + 3 acetate$^-$ + HCO_3^- + $3\ H^+$	*Syntrophus aciditrophicus*

bacterium. A few examples are listed in Table 14.3. Many of these bacteria can be considered metabolic specialists, having evolved the capacity to catabolize a substrate not catabolized by other bacteria.

We begin with two very common fermentations of sugars in which lactic acid is a major product.

Lactic Acid Fermentation

The lactic acid bacteria are gram-positive organisms that produce lactic acid as a major or sole fermentation product (Section 18.1). Two fermentative patterns are observed. One, called **homofermentative**, yields a single fermentation product, lactic acid. The other, called **heterofermentative**, yields products in addition to lactate, mainly ethanol plus CO_2.

Figure 14.3 summarizes pathways for the fermentation of glucose by homofermentative and heterofermentative lactic acid bacteria. The differences observed can be traced to the presence or absence of the enzyme *aldolase*, a key enzyme of glycolysis (Figure 4.14). Homofermentative lactic acid bacteria contain aldolase and produce two molecules of lactate from glucose by the glycolytic pathway (Figure 14.3*a*). Heterofermenters lack aldolase and thus cannot break down fructose bisphosphate to triose phosphate. Instead, they oxidize glucose 6-phosphate to 6-phosphogluconate and then decarboxylate this to pentose phosphate. The pentose phosphate is then converted to triose phosphate and acetyl phosphate by the key enzyme *phosphoketolase* (Figure 14.3*b*). The early steps in catabolism by heterofermentative lactic acid bacteria are those of the pentose phosphate pathway (see Figure 14.38).

In heterofermenters, triose phosphate is converted to lactic acid with the production of ATP (Figure 14.3). However, to achieve redox balance the acetyl phosphate produced is used as an electron acceptor and is reduced by NADH (generated during the production of pentose phosphate) to ethanol. This occurs without ATP synthesis because the energy-rich CoA bond is lost during ethanol formation. Because of this, hetero-

fermenters produce only *one* ATP/glucose instead of the *two* ATP/glucose produced by homofermenters. In addition, because heterofermenters decarboxylate 6-phosphogluconate, they produce CO_2 as a fermentation product; homofermenters do not produce CO_2. Thus a simple way of differentiating a homofermenter from a heterofermenter is to observe for the production of CO_2 in laboratory cultures.

Entner–Doudoroff Pathway

A variant of the glycolytic pathway, called the *Entner–Doudoroff pathway*, is widely distributed in bacteria, especially among species of the pseudomonad group. In this pathway glucose 6-phosphate is oxidized to 6-phosphogluconic acid and NADPH; the 6-phosphogluconic acid is dehydrated and split into pyruvate and glyceraldehyde 3-phosphate (G-3-P), a key intermediate of the glycolytic pathway. G-3-P is then catabolized as in glycolysis, generating NADH and two ATP, and used as an electron acceptor to balance redox reactions (Figure 14.3*a*).

Because pyruvate is formed directly in the Entner–Doudoroff pathway and cannot yield ATP as can G-3-P (Figure 14.3), the Entner–Doudoroff pathway yields only half the ATP of the glycolytic pathway. Organisms using the Entner–Doudoroff pathway therefore share this physiological characteristic with heterofermentative lactic acid bacteria that also use a variant of the glycolytic pathway (Figure 14.3*b*). *Zymomonas*, an obligately fermentative pseudomonad, and *Pseudomonas*, a nonfermentative respiratory bacterium, are major genera that employ the Entner–Doudoroff pathway (Section 17.7).

Mixed-Acid Fermentations

In mixed-acid fermentations, characteristic of enteric bacteria (Section 17.11), three different acids are formed from the fermentation of glucose or other sugars—*acetic, lactic,* and *succinic*. Ethanol, CO_2, and H_2 are also formed. Glycolysis is the pathway used by mixed-acid fermenters, such as *Escherichia coli*, and we outlined the steps in that pathway in Figure 4.14.

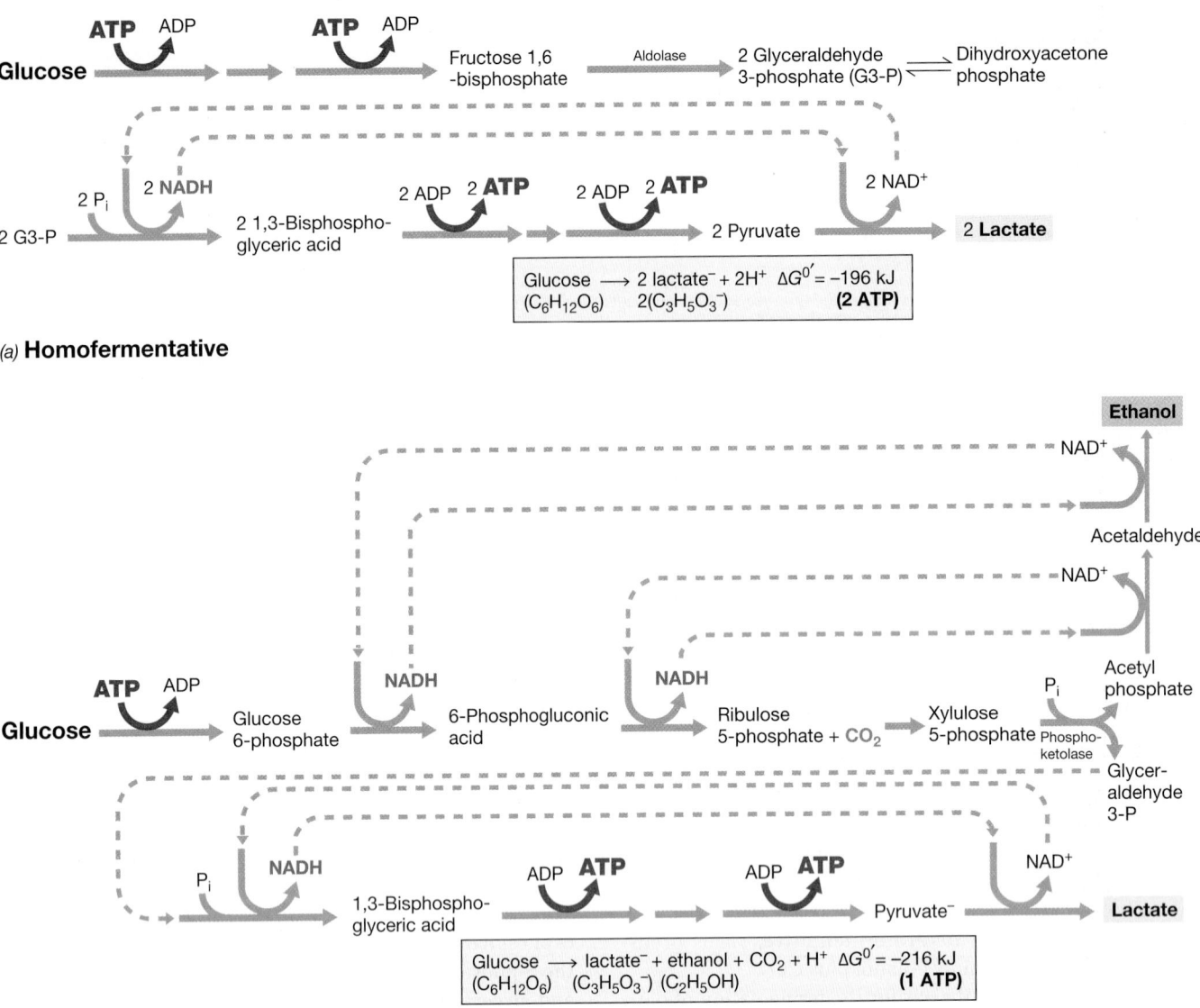

(a) **Homofermentative**

(b) **Heterofermentative**

Figure 14.3 The fermentation of glucose in *(a)* homofermentative and *(b)* heterofermentative lactic acid bacteria. Note that no ATP is made in reactions leading to ethanol formation in heterofermentative organisms.

Some enteric bacteria produce acidic products in lower amounts than *E. coli* and balance redox in their fermentations by producing larger amounts of neutral products. One key neutral product is the four-carbon alcohol *butanediol*. In this variation of the mixed-acid fermentation, butanediol, ethanol, CO_2, and H_2 are the main products observed (**Figure 14.4**). In the mixed-acid fermentation of *E. coli*, equal amounts of CO_2 and H_2 are produced, whereas in a butanediol fermentation, considerably more CO_2 than H_2 is produced. This is because mixed-acid fermenters produce CO_2 only from formic acid by means of the enzyme formate hydrogenlyase (Figure 14.2):

$$HCOOH \rightarrow H_2 + CO_2$$

By contrast, butanediol producers, such as *Enterobacter aerogenes*, produce CO_2 and H_2 from formic acid but also produce two additional molecules of CO_2 during the formation of each molecule of butanediol (Figure 14.4).

Because they produce fewer acidic products, butanediol fermenters do not acidify their environment as much as mixed-acid fermenters do, and this is presumably a reflection of differences in acid tolerance in the two groups that have significance for their competitive success in nature.

MiniQuiz

• How can homo- and heterofermentative lactic acid bacteria be differentiated in pure cultures?

• Butanediol production leads to greater ethanol production than in the mixed-acid fermentation of *Escherichia coli*. Why?

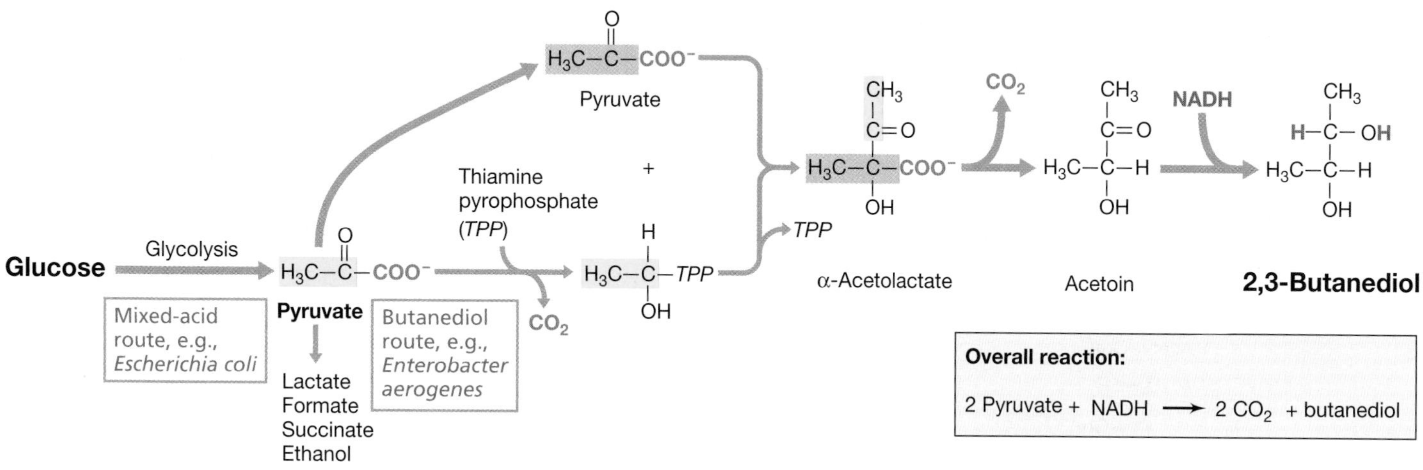

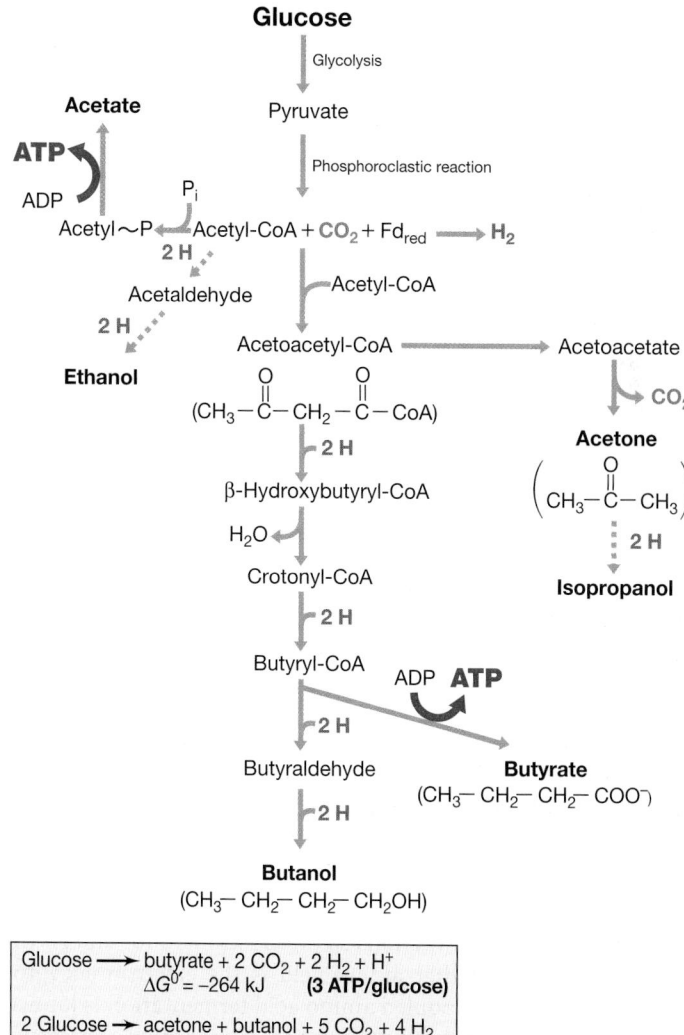

Figure 14.4 Butanediol production and mixed-acid fermentations. Note how only one NADH, but two molecules of pyruvate, are used to make one butanediol. This leads to redox imbalance and the production of more ethanol by butanediol producers than by mixed-acid fermenters.

14.3 Clostridial and Propionic Acid Fermentations

Species of the genus *Clostridium* are classical fermentative anaerobes (⮌ Section 18.2). Different clostridia ferment sugars, amino acids, purines and pyrimidines, and a few other compounds. In all cases ATP synthesis is linked to substrate-level phosphorylations either in the glycolytic pathway or from the hydrolysis of a CoA intermediate (Table 14.1). We begin with sugar-fermenting (saccharolytic) clostridia.

Sugar Fermentation by *Clostridium* Species

A number of clostridia ferment sugars, producing *butyric acid* as a major end product. Some species also produce the neutral products acetone and butanol, and *Clostridium acetobutylicum* is a classic example of this. The biochemical steps in the formation of butyric acid and neutral products from sugars are shown in **Figure 14.5**.

Glucose is converted to pyruvate via the glycolytic pathway, and pyruvate is split to yield acetyl-CoA, CO_2, and H_2 (through ferredoxin) by the phosphoroclastic reaction (Figure 14.2). Some of the acetyl-CoA is then reduced to butyrate or other fermentation products using NADH derived from glycolytic reactions as electron donor. The actual products observed are influenced by the duration and the conditions of the fermentation. During the early stages of the butyric fermentation, butyrate and a small amount of acetate are produced. But as the pH of the medium drops, synthesis of acids ceases and acetone and butanol begin to accumulate. However, if the pH of the medium is kept neutral by buffering, there is very little formation of neutral products and butyric acid production continues.

The accumulation of acidic products in the *C. acetobutylicum* fermentation lowers the pH, and this triggers derepression of genes responsible for solvent production. The production of butanol is actually a consequence of the production of acetone. For each acetone that is made, two NADH produced during glycolysis are not reoxidized as they would be if butyrate were produced (Figure 14.5). Because redox balance is necessary for any

Figure 14.5 The butyric acid and butanol/acetone fermentation. All fermentation products from glucose are shown in bold (dashed lines indicate minor products). Note how the production of acetate and butyrate lead to additional ATP by substrate-level phosphorylation. By contrast, formation of butanol and acetone reduces the ATP yield because the butyryl-CoA step is bypassed. 2 H, NADH; Fd$_{red}$, reduced ferredoxin.

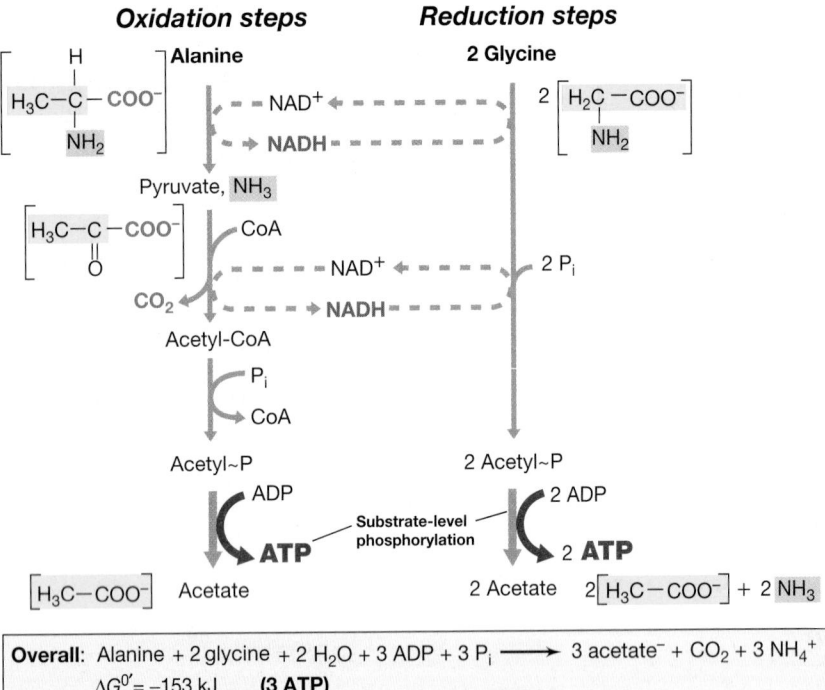

Amino acids participating in coupled fermentations (Stickland reaction)	
Amino acids oxidized:	**Amino acids reduced:**
Alanine	Glycine
Leucine	Proline
Isoleucine	Hydroxyproline
Valine	Tryptophan
Histidine	Arginine

Figure 14.6 The Stickland reaction. This example shows the cocatabolism of the amino acids alanine and glycine. The structures of key substrates, intermediates, and products are shown in brackets to allow the chemistry of the reaction to be followed. Note how in the reaction shown, alanine is the electron donor and glycine is the electron acceptor.

fermentation to proceed, the cell then uses butyrate as an electron acceptor. Butanol and acetone are therefore produced in equal amounts. Although neutral product formation helps the organism keep its environment from becoming too acidic, there is an energetic price to pay for this. In producing butanol, the cell loses the opportunity to convert butyryl-CoA to butyrate and thus ATP (Figure 14.5 and Table 14.1).

Amino Acid Fermentation by *Clostridium* Species and the Stickland Reaction

Some *Clostridium* species ferment amino acids. These are the "proteolytic" clostridia, organisms that degrade proteins released from dead organisms in nature. Some clostridia ferment individual amino acids, typically glutamate, glycine, alanine, cysteine, histidine, serine, or threonine. The biochemistry behind these fermentations is quite complex, but the metabolic strategy is quite simple. In virtually all cases, the amino acids are catabolized in such a way as to yield a fatty acid–CoA derivative, typically acetyl (C_2), butyryl (C_4), or caproyl (C_6). From these, ATP is produced by substrate-level phosphorylation (Table 14.1). Other products of amino acid fermentation include ammonia (NH_3) and CO_2.

Some clostridia ferment only an amino acid *pair*. In this situation one amino acid functions as the electron donor and is oxidized, whereas the other amino acid is the electron acceptor and is reduced. This coupled amino acid fermentation is known as a **Stickland reaction**. For instance, *Clostridium sporogenes* catabolizes a mixture of glycine and alanine; in this reaction alanine is the electron donor and glycine is the electron acceptor (**Figure 14.6**). Amino acids that can function as donors or acceptors in Stickland reactions are listed in Figure 14.6. The products of the Stickland reaction are NH_3, CO_2, and a car-

boxylic acid with one fewer carbons than the amino acid that was oxidized (Figure 14.6).

Many of the products of amino acid fermentation by clostridia are foul-smelling substances, and the odor that results from putrefaction is mainly a result of clostridial activity. In addition to fatty acids, other odoriferous compounds produced include hydrogen sulfide (H_2S), methylmercaptan (from sulfur amino acids), cadaverine (from lysine), putrescine (from ornithine), and NH_3. Purines and pyrimidines, released from the degradation of nucleic acids, lead to many of the same fermentation products and yield ATP from the hydrolysis of fatty acid–CoA derivatives (Table 14.1) produced in their respective fermentative pathway.

Clostridium kluyveri Fermentation

Another species of *Clostridium* also ferments a mixture of substrates in which one is the donor and one is the acceptor, as in the Stickland reaction. However, this organism, *C. kluyveri*, ferments not amino acids but instead ethanol plus acetate. In this fermentation, ethanol is the electron donor and acetate is the electron acceptor. The overall reaction is shown in Table 14.2.

The ATP yield in the caproate/butyrate fermentation is low, 1 ATP/6 ethanol fermented. However, *C. kluyveri* has a selective advantage over all other fermenters in its apparently unique ability to oxidize a highly reduced fermentation product (ethanol) and couple it to the reduction of another common fermentation product (acetate), reducing it to longer-chain fatty acids. The single ATP produced in this reaction comes from substrate-level phosphorylation during conversion of a fatty acid–CoA formed in the pathway to the free fatty acid. The fermentation of *C. kluyveri* is an example of a **secondary fermentation**, which is essentially a fermentation of fermentation products. We see another example of this now.

Propionic Acid Fermentation

The propionic acid bacterium *Propionibacterium* and some related bacteria produce *propionic acid* as a major fermentation product from either glucose or lactate. However, lactate, a fermentation product of the lactic acid bacteria, is probably the major substrate for propionic acid bacteria in nature, where these two groups live in close association. *Propionibacterium* is an important component in the ripening of Swiss (Emmentaler) cheese, to which the propionic and acetic acids produced give the unique bitter and nutty taste, and the CO_2 produced forms bubbles that leave the characteristic holes (eyes) in the cheese.

Figure 14.7 shows the reactions leading from lactate to propionate. When glucose is the starting substrate, it is first catabolized to pyruvate by the glycolytic pathway. Then pyruvate, produced either from glucose or from the oxidation of lactate, is carboxylated to form methylmalonyl-CoA, leading to the formation of oxalacetate and, eventually, propionyl-CoA (Figure 14.7). The latter reacts with succinate in a step catalyzed by the enzyme CoA transferase, producing succinyl-CoA and propionate. This results in a lost opportunity for ATP production from propionyl-CoA but avoids the energetic costs of having to activate succinate with ATP to form succinyl-CoA. The succinyl-CoA is then isomerized to methylmalonyl-CoA and the cycle is complete; propionate is formed and CO_2 regenerated (Figure 14.7).

NADH is oxidized in the steps between oxalacetate and succinate. Notably, the reaction in which fumarate is reduced to succinate is linked to electron transport and the formation of a proton

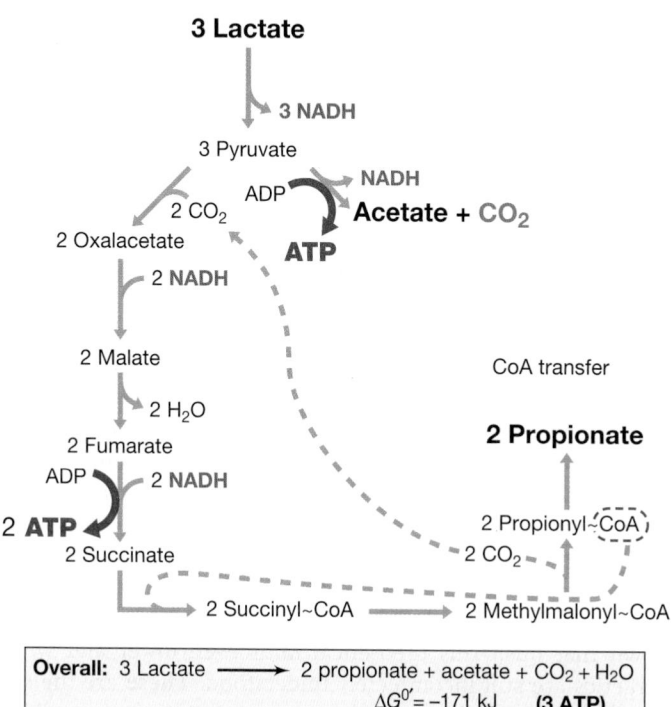

Figure 14.7 The propionic acid fermentation of *Propionibacterium*. Products are shown in bold. The four NADH made from the oxidation of three lactate are reoxidized in the reduction of oxalacetate and fumarate, and the CoA group from propionyl-CoA is exchanged with succinate during the formation of propionate.

motive force that yields ATP by oxidative phosphorylation. The propionate pathway also converts some lactate to acetate plus CO_2, which allows for additional ATP to be made (Figure 14.7). Thus, in the propionate fermentation both substrate-level *and* oxidative phosphorylation occur.

Propionate is also formed in the fermentation of succinate by the bacterium *Propionigenium*, but by a completely different mechanism than that described here for *Propionibacterium*. *Propionigenium*, to be considered next, is phylogenetically and ecologically unrelated to *Propionibacterium*, but energetic aspects of its metabolism are of considerable interest from the standpoint of bioenergetics.

MiniQuiz

- Compare the mechanisms for energy conservation in *Clostridium acetobutylicum* and *Propionibacterium*.
- What are the substrates for the *Clostridium kluyveri* fermentation? In nature, where do these come from?

14.4 Fermentations Lacking Substrate-Level Phosphorylation

Certain fermentations yield insufficient energy to synthesize ATP by substrate-level phosphorylation (that is, less than -32 kJ, Table 14.1), yet still support growth. In these cases, catabolism of the compound is linked to ion pumps that establish a proton motive force or sodium motive force across the cytoplasmic membrane. Examples of this include fermentation of succinate by *Propionigenium modestum* and the fermentation of oxalate by *Oxalobacter formigenes*.

Propionigenium modestum

Propionigenium modestum was first isolated in anoxic enrichment cultures lacking alternative electron acceptors and fed succinate as an electron donor. *Propionigenium* inhabits marine and freshwater sediments, and can also be isolated from the human oral cavity. The organism is a gram-negative short rod and, phylogenetically, is a species of *Actinobacteria* (⮂ Section 18.4). During studies of the physiology of *P. modestum*, it was shown to require sodium chloride (NaCl) for growth and to catabolize succinate under strictly anoxic conditions:

$$\text{Succinate}^{2-} + H_2O \rightarrow \text{propionate}^- + HCO_3^-$$
$$\Delta G^{0\prime} = -20.5 \text{ kJ}$$

This decarboxylation releases insufficient free energy to support ATP synthesis by substrate-level phosphorylation (Table 14.1) but sufficient free energy to pump a sodium ion (Na^+) across the cytoplasmic membrane from the cytoplasm to the periplasm. Energy conservation in *Propionigenium* is then linked to the sodium motive force that develops from Na^+ pumping; a sodium-translocating ATPase exists that uses the sodium motive force to drive ATP synthesis (**Figure 14.8a**).

In a related decarboxylation reaction, the bacterium *Malonomonas*, a species of *Deltaproteobacteria*, decarboxylates the C_3 dicarboxylic acid malonate, forming acetate plus CO_2. As for

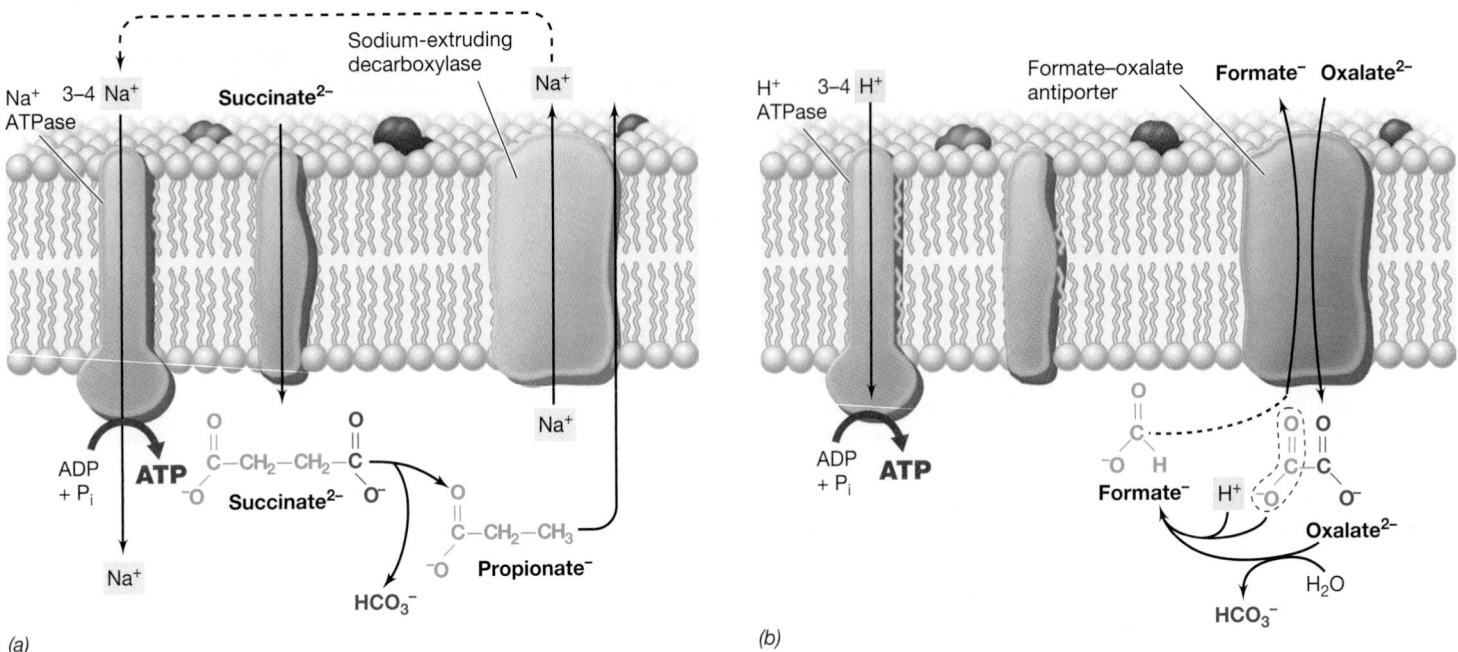

Figure 14.8 The unique fermentations of succinate and oxalate. *(a)* Succinate fermentation by *Propionigenium modestum*. Sodium export is linked to the energy released by succinate decarboxylation, and a sodium-translocating ATPase produces ATP. *(b)* Oxalate fermentation by *Oxalobacter formigenes*. Oxalate import and formate export by a formate–oxalate antiporter consume cytoplasmic protons. ATP synthesis is linked to a proton-driven ATPase. All substrates and products are shown in bold.

Propionigenium, energy metabolism in *Malonomonas* is linked to a sodium pump and sodium-driven ATPase. However, the mechanism of malonate decarboxylation is more complex than that of *Propionigenium* and involves many additional proteins. Interestingly, however, the energy yield of malonate fermentation by *Malonomonas* is even lower than that of *P. modestum*, −17.4 kJ. *Sporomusa*, an endospore-forming bacterium (⮌ Section 18.2) and also an acetogen (Section 14.9), is also capable of fermenting malonate, as are a few other *Bacteria*.

Oxalobacter formigenes

Oxalobacter formigenes is a bacterium present in the intestinal tract of animals, including humans. It catabolizes the C_2 dicarboxylic acid oxalate, producing formate plus CO_2. Oxalate degradation by *O. formigenes* is thought to be important in humans for preventing the accumulation of oxalate in the body, a substance that can accumulate to form calcium oxalate kidney stones. *O. formigenes* is a gram-negative strict anaerobe that is a species of *Betaproteobacteria*. *O. formigenes* carries out the following reaction:

$$\text{Oxalate}^{2-} + H_2O \rightarrow \text{formate}^- + HCO_3^- \qquad \Delta G^{0\prime} = -26.7 \text{ kJ}$$

As in the catabolism of succinate by *P. modestum*, insufficient energy is available from this reaction to drive ATP synthesis by substrate-level phosphorylation (Table 14.1). However, the reaction supports growth of the organism because the decarboxylation of oxalate is exergonic and forms formate, which is excreted from the cell. The internal consumption of protons during the oxidation of oxalate and production of formate is, in effect, a proton pump. That is, a divalent molecule (oxalate) enters the cell

while a univalent molecule (formate) is excreted. The continued exchange of oxalate for formate establishes a membrane potential that is coupled to ATP synthesis by the proton-translocating ATPase in the membrane (Figure 14.8*b*).

Energetics Lessons

The unique aspect of all of these decarboxylation-type fermentations is that ATP synthesis occurs *without substrate-level phosphorylation or electron transport*. Nevertheless, ATP synthesis can occur because the small amount of energy released is coupled to the pumping of an ion across the cytoplasmic membrane. Organisms such as *Propionigenium* or *Oxalobacter* thus teach us an important lesson in microbial bioenergetics: Energy conservation from reactions that yield less than −32 kJ is still possible if the reaction is coupled to an ion pump. However, a minimal requirement for an energy-conserving reaction is that it must yield sufficient free energy to pump a single ion. This is estimated to be about −12 kJ. Theoretically, reactions that release less energy than this should not be able to drive ion pumps and should therefore not be potential energy-conserving reactions. However, as we will see in the next section, there are bacteria known that push this theoretical limit even lower and whose energetics are still incompletely understood. These are the syntrophs, prokaryotes living on the energetic "edge of existence."

MiniQuiz
- Why does *Propionigenium modestum* require sodium for growth?
- Of what benefit is the organism *Oxalobacter* to human health?

14.5 Syntrophy

There are many examples in microbiology of **syntrophy**, a metabolic process in which two different organisms cooperate to degrade a substance—and conserve energy doing it—that neither can degrade alone. Most syntrophic reactions are secondary fermentations (Section 14.3) in which organisms ferment the fermentation products of other anaerobes. We will see in Section 24.2 how syntrophy is a key to the overall success of anoxic catabolism that leads to the production of methane (CH_4). Here we consider the microbiology and energetic aspects of syntrophy.

Table 14.4 lists some of the major groups of syntrophs and the compounds they degrade. Many organic compounds can be degraded syntrophically, including even aromatic and aliphatic hydrocarbons. But the major compounds of interest in freshwater syntrophic environments are fatty acids and alcohols.

Hydrogen Consumption in Syntrophic Reactions

The heart of syntrophic reactions is *interspecies H_2 transfer*, H_2 production by one partner linked to H_2 consumption by the other. The H_2 consumer can be any one of a number of physiologically distinct organisms: denitrifying bacteria, ferric iron–reducing bacteria, sulfate-reducing bacteria, acetogens, or methanogens, groups we will consider later in this chapter. Consider ethanol fermentation to acetate plus H_2 by a syntroph coupled to the production of methane (**Figure 14.9**). As can be seen, the syntroph carries out a reaction whose standard free-energy change ($\Delta G^{0\prime}$) is positive. However, the H_2 produced by the syntroph can be used as an electron donor by a methanogen in an exergonic reaction. When the two reactions are summed, the overall reaction is exergonic (Figure 14.9), and the free energy released is shared by both organisms.

Ethanol fermentation:

$$2\ CH_3CH_2OH + 2\ H_2O \rightarrow 4\ H_2 + 2\ CH_3COO^- + 2\ H^+$$

$$\Delta G^{0\prime} = +19.4\ \text{kJ/reaction}$$

Methanogenesis:

$$4\ H_2 + CO_2 \rightarrow CH_4 + 2\ H_2O$$

$$\Delta G^{0\prime} = -130.7\ \text{kJ/reaction}$$

Coupled reaction:

$$2\ CH_3CH_2OH + CO_2 \rightarrow CH_4 + 2\ CH_3COO^- + 2\ H^+$$

$$\Delta G^{0\prime} = -111.3\ \text{kJ/reaction}$$

(a) Reactions

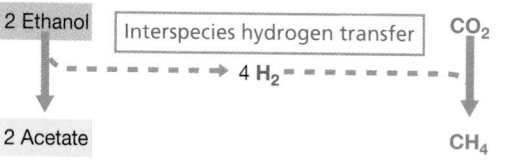

(b) Syntrophic transfer of H_2

Figure 14.9 Syntrophy: Interspecies H_2 transfer. Shown is the fermentation of ethanol to methane and acetate by syntrophic association of an ethanol-oxidizing syntroph and a H_2-consuming partner (in this case, a methanogen). *(a)* Reactions involved. The two organisms share the energy released in the coupled reaction. *(b)* Nature of the syntrophic transfer of H_2.

Another example of syntrophy is the oxidation of a fatty acid such as butyrate to acetate plus H_2 by the fatty acid–oxidizing syntroph *Syntrophomonas* (**Figure 14.10**):

$$\text{Butyrate}^- + 2\ H_2O \rightarrow 2\ \text{acetate}^- + H^+ + 2\ H_2$$

$$\Delta G^{0\prime} = +48.2\ \text{kJ}$$

The free-energy change of this reaction is highly unfavorable, and in pure culture *Syntrophomonas* will not grow on butyrate. However, if the H_2 produced by *Syntrophomonas* is consumed by a partner organism, *Syntrophomonas* grows on butyrate in coculture with the H_2 consumer. Why is this so?

Energetics of H_2 Transfer

Because it is such a powerful electron donor for anaerobic respirations, H_2 is quickly consumed in anoxic habitats. In a syntrophic relationship, the removal of H_2 by a partner organism pulls the reaction in the direction of product formation and thereby affects the energetics of the reaction. A review of the principles of free energy given in Appendix 1 indicates that the concentration of reactants and products in a reaction can have a major effect on energetics. This is usually not an issue for most fermentation products because they are not consumed to extremely low levels. H_2, by contrast, can be consumed to nearly undetectable levels, and at these tiny concentrations, the energetics of the reactions are dramatically affected.

For convenience, the $\Delta G^{0\prime}$ of a reaction is calculated on the basis of standard conditions—one molar concentration of

Table 14.4 *Properties of major syntrophic bacteria*[a]

Genus	Number of known species	Phylogeny[b]	Substrates fermented in coculture[c]
Syntrophobacter	4	Deltaproteobacteria	Propionate (C_3), lactate; some alcohols
Syntrophomonas	9	Firmicutes	C_4–C_{18} saturated/unsaturated fatty acids; some alcohols
Pelotomaculum	2	Firmicutes	Propionate, lactate, several alcohols; some aromatic compounds
Syntrophus	3	Deltaproteobacteria	Benzoate and several related aromatic compounds; some fatty acids and alcohols

[a]All syntrophs are obligate anaerobes.
[b]See Chapters 17 and 18.
[c]Not all species can use all substrates listed.

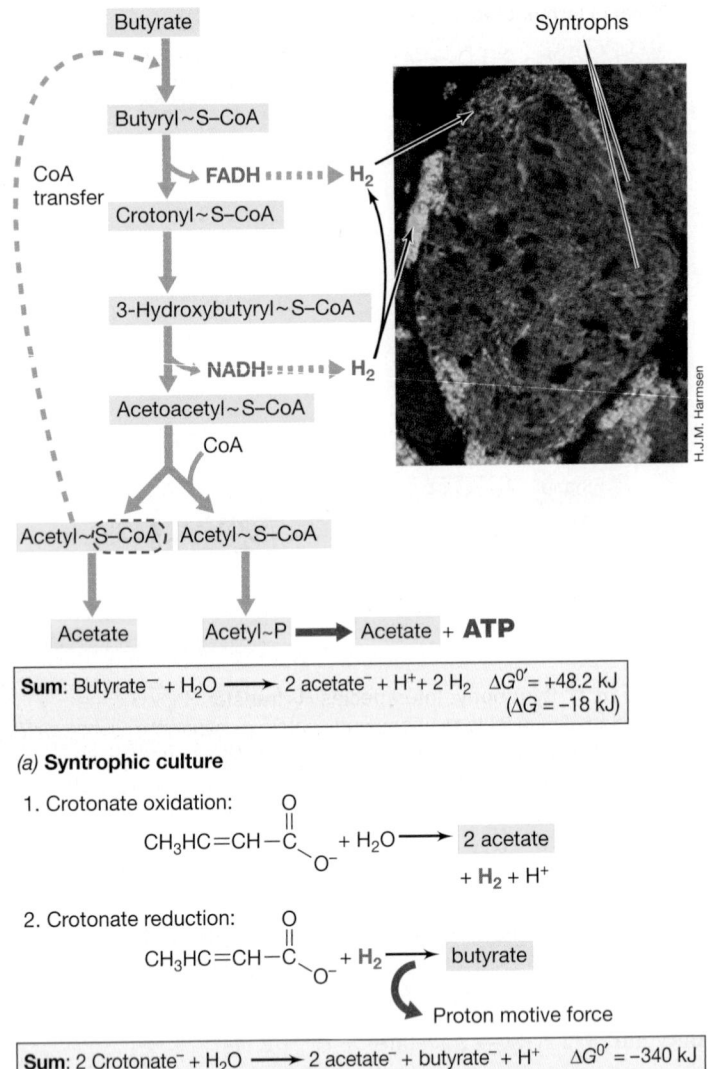

Sum: Butyrate⁻ + H_2O ⟶ 2 acetate⁻ + H⁺ + 2 H_2 $\Delta G^{0\prime} = +48.2$ kJ
($\Delta G = -18$ kJ)

(a) **Syntrophic culture**

1. Crotonate oxidation:

   $CH_3HC=CH-C(=O)O^- + H_2O$ ⟶ 2 acetate + H_2 + H⁺

2. Crotonate reduction:

   $CH_3HC=CH-C(=O)O^- + H_2$ ⟶ butyrate

   Proton motive force

Sum: 2 Crotonate⁻ + H_2O ⟶ 2 acetate⁻ + butyrate⁻ + H⁺ $\Delta G^{0\prime} = -340$ kJ

(b) **Pure culture**

Figure 14.10 Energetics of growth of *Syntrophomonas* in syntrophic culture and in pure culture. *(a)* In syntrophic culture, growth requires a H_2-consuming organism, such as a methanogen. H_2 production is driven by reverse electron flow because the $E_0\prime$ of the FADH and NADH couples are more electropositive than that of 2 H⁺/H_2. *(b)* In pure culture, energy conservation is linked to anaerobic respiration with crotonate reduction to butyrate. Inset: photomicrograph of FISH-stained cells (⮌ Section 16.9) of a fatty acid-degrading syntrophic bacterium in association with a methanogen.

products and reactants. By contrast, the related term ΔG is calculated on the basis of the actual concentrations of products and reactants present (Appendix 1 explains how to calculate ΔG). At very low levels of H_2, the energetics of the oxidation of ethanol or fatty acids to acetate plus H_2, a reaction that is endergonic under standard conditions, becomes exergonic. For example, if the concentration of H_2 is kept extremely low from consumption by the partner organism, ΔG for the oxidation of butyrate by *Syntrophomonas* yields −18 kJ (Figure 14.10*a*). As we learned in Section 14.4, this relatively low energy yield can still support growth of a bacterium.

Energetics in Syntrophs

Energy conservation in syntrophs is probably based on both substrate-level and oxidative phosphorylations. From biochemical studies of syntrophs, substrate-level phosphorylation has been shown to occur during the conversion of acetyl-CoA (generated by beta-oxidation of ethanol or the fatty acid) to acetate (Figure 14.10*a*), although the −18 kJ of energy released (ΔG) is in theory insufficient for this. However, the energy released is sufficient to produce a *fraction* of an ATP, so it is possible that two rounds of butyrate oxidation (Figure 14.10*a*) are necessary to couple to the production of one ATP by substrate-level phosphorylation.

Besides the syntrophic lifestyle, many syntrophs can carry out anaerobic respirations (Section 14.6) in pure culture by the disproportionation of unsaturated fatty acids (disproportionation is a process in which some molecules of a substrate are oxidized while some are reduced). For example, crotonate, an intermediate in syntrophic butyrate metabolism (Figure 14.10*a*), supports growth of *Syntrophomonas*. Under these conditions some of the crotonate is oxidized to acetate and some is reduced to butyrate (Figure 14.10*b*). Because crotonate reduction by *Syntrophomonas* is coupled to the formation of a proton motive force, as occurs in other anaerobic respirations that employ organic electron acceptors (such as fumarate reduction to succinate, Section 14.12), it is possible that some step or steps in syntrophic metabolism (Figure 14.10*a*) generate a proton motive force as well. Pumping protons or some other ion would almost certainly be required for benzoate- and propionate-fermenting syntrophs (Figure 14.10*a* inset) whose energy yield (ΔG) is only about −5 kJ or so.

Regardless of how ATP is made during syntrophic growth, an additional energetic burden occurs in syntrophy. During syntrophic metabolism, syntrophs produce H_2 ($E_0\prime$ −0.42 V) from more electropositive electron donors such as FADH ($E_0\prime$ −0.22 V) and NADH ($E_0\prime$ −0.32 V), generated during fatty acid oxidation reactions (Figure 14.10*a*); this cannot occur without an energy input. Thus, some fraction of the ATP generated by *Syntrophomonas* during syntrophic growth must be consumed to drive reverse electron flow reactions (⮌ Section 13.4), yielding H_2 for the H_2 consumer. When this energy drain is coupled to the inherently poor energetic yields of syntrophic reactions, it is clear that syntrophic bacteria are somehow making a living on a severely marginal energy economy. Even today syntrophs pose a significant challenge to our understanding of the minimal requirements for energy conservation in bacteria.

Ecology of Syntrophs

Ecologically, syntrophic bacteria are key links in the anoxic portions of the carbon cycle. Syntrophs consume highly reduced fermentation products and release a key product for anaerobic H_2 consumers. Without syntrophs, a bottleneck would develop in anoxic environments in which electron acceptors other than CO_2 were limiting (⮌ Section 24.2). By contrast, when conditions are oxic or alternative electron acceptors are abundant, syntrophic relationships are unnecessary. For example, if O_2 or NO_3^- is available as an electron acceptor, the energetics of the fermentation of a fatty acid or an alcohol is so favorable that syntrophic relationships are unnecessary. Thus, syntrophy is characteristic of anoxic catabolism in which methanogenesis or aceto-genesis are the terminal processes in the microbial ecosystem. Methanogenesis is a major process in anoxic wastewater biodegradation, and microbiological studies of sludge granules that form in such systems have shown the close physical relationship that develops between H_2 producer and H_2 consumer in such habitats (Figure 14.10*a* inset).

MiniQuiz

- Give an example of interspecies H_2 transfer. Why can it be said that both organisms benefit from this process?
- Predict how ATP is made during the syntrophic degradation of ethanol shown in Figure 14.9.

 Anaerobic Respiration

In the next several sections we survey the major forms of anaerobic respiration and see the many ways by which prokaryotes can conserve energy under anoxic conditions using electron acceptors *other* than oxygen (O_2).

14.6 Anaerobic Respiration: General Principles

We examined the process of aerobic respiration in Chapter 4. As we noted there, O_2 functions as a *terminal electron acceptor*, accepting electrons that have traveled through an electron transport chain. However, we also noted that other electron acceptors could be used instead of O_2, in which case the process is called **anaerobic respiration** (Section 4.12). Here we consider some of these processes.

Bacteria that carry out anaerobic respiration produce electron transport chains containing cytochromes, quinones, iron–sulfur proteins, and the other typical electron transport proteins that we have seen in aerobic respiration (Section 4.9) and in photosynthesis and chemolithotrophy (Chapter 13). In some organisms, such as the denitrifying bacteria, which are for the most part facultative aerobes (Section 5.17), anaerobic respiration competes with aerobic respiration. In such cases, if O_2 is present, the bacteria respire aerobically, and genes encoding proteins necessary for anaerobic respiration are repressed. However, when O_2 is depleted from the environment, the bacteria respire anaerobically, and the alternate electron acceptor is reduced. Many other organisms that carry out anaerobic respiration are obligate anaerobes and are unable to use O_2.

Alternative Electron Acceptors and the Redox Tower

The energy released from the oxidation of an electron donor using O_2 as electron acceptor is greater than if the same compound is oxidized with an alternate electron acceptor (Figure 4.9). These energy differences are apparent if the reduction potentials of each acceptor are examined (**Figure 14.11**). Because the O_2/H_2O couple is most electropositive, more energy is available when O_2 is used than when another electron acceptor is used.

This is why aerobic respiration is the dominant process and occurs to the exclusion of anaerobic respiration in an organism in which both processes are possible. Other electron acceptors that are near the O_2/H_2O couple are manganic ion (Mn^{4+}), ferric iron (Fe^{3+}), nitrate (NO_3^-), and nitrite (NO_2^-). Examples of more electronegative acceptors are sulfate (SO_4^{2-}), elemental sulfur (S^0), and carbon dioxide (CO_2), and organisms that use these acceptors are typically locked into an anaerobic lifestyle. A summary of the most common types of anaerobic respiration is given in Figure 14.11.

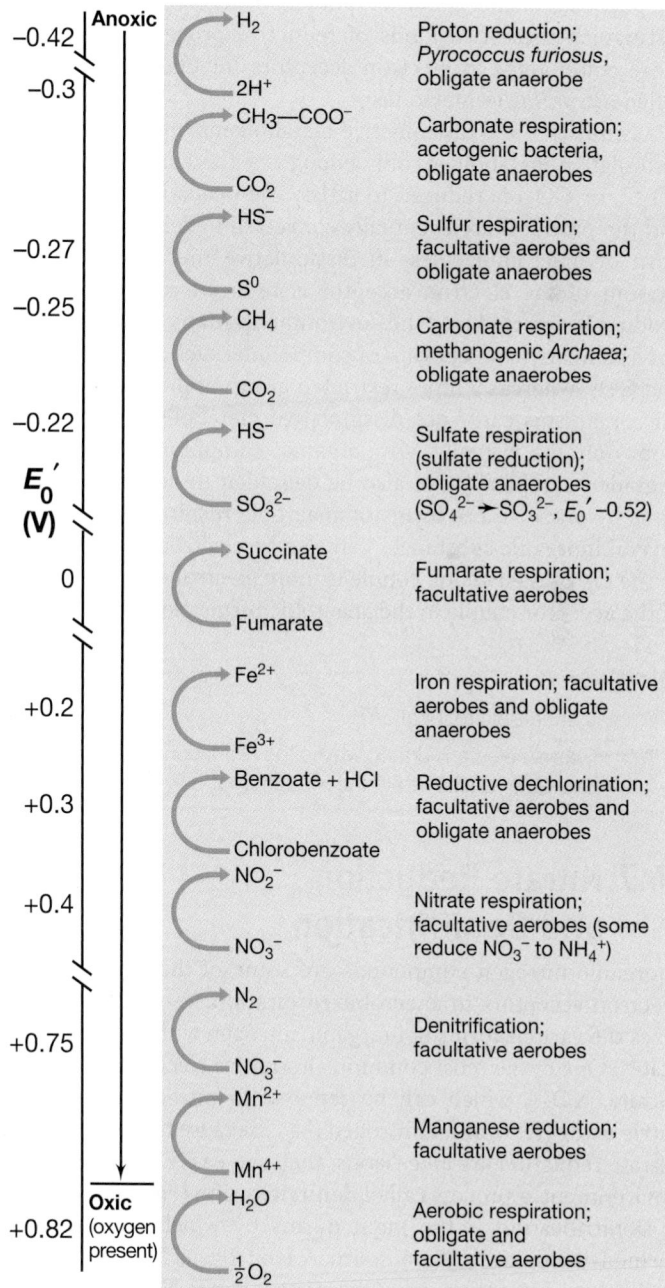

Figure 14.11 Major forms of anaerobic respiration. The redox couples are arranged in order from most electronegative E_0' (top) to most electropositive E_0' (bottom). See Figure 4.9 to compare how the energy yields of these anaerobic respirations vary.

Assimilative and Dissimilative Metabolism

Inorganic compounds such as NO_3^-, SO_4^{2-}, and CO_2 are reduced by many organisms as sources of cellular nitrogen, sulfur, and carbon, respectively. The end products of such reductions are amino groups (—NH_2), sulfhydryl groups (—SH), and organic carbon compounds, respectively. When an inorganic compound such as NO_3^-, SO_4^{2-}, or CO_2 is reduced for use in biosynthesis, it is said to be *assimilated*, and the reduction process is called *assimilative* metabolism. Assimilative metabolism of NO_3^-, SO_4^{2-}, and CO_2 is conceptually and physiologically quite different from the reduction of these electron acceptors for the purposes of energy conservation in anaerobic metabolism. To distinguish these two kinds of reductive processes, the use of these compounds as electron acceptors for energy purposes is called *dissimilative* metabolism.

Assimilative and dissimilative metabolisms differ markedly. In assimilative metabolism, only enough of the compound (NO_3^-, SO_4^{2-}, or CO_2) is reduced to satisfy the needs for biosynthesis, and the products are eventually converted to cell material in the form of macromolecules. In dissimilative metabolism, a large amount of the electron acceptor is reduced, and the reduced product is excreted into the environment. Many organisms carry out assimilative metabolism of compounds such as NO_3^-, SO_4^{2-}, and CO_2, whereas a more restricted group of primarily prokaryotic organisms carry out dissimilative metabolism. As for electron donors, virtually any organic compound that can be degraded aerobically can also be degraded under anoxic conditions by one or more forms of anaerobic respiration. Moreover, several inorganic substances can also be electron donors as long as the E_0' of their redox couple is more electronegative than that of the acceptor couple in the anaerobic respiration.

MiniQuiz

- What is anaerobic respiration?
- With H_2 as an electron donor, why is the reduction of NO_3^- a more favorable reaction than the reduction of S^0?

14.7 Nitrate Reduction and Denitrification

Inorganic nitrogen compounds are some of the most common electron acceptors in anaerobic respiration. **Table 14.5** summarizes the various forms of inorganic nitrogen with their oxidation states. One of the most common alternative electron acceptors is nitrate, NO_3^-, which can be reduced to nitrous oxide (N_2O), nitric oxide (NO), and dinitrogen (N_2). Because these products of nitrate reduction are all gaseous, they can easily be lost from the environment, a process called **denitrification** (**Figure 14.12**).

Denitrification is the main means by which gaseous N_2 is formed biologically. As a source of nitrogen, N_2 is much less available to plants and microorganisms than is NO_3^-, so for agricultural purposes, at least, denitrification is a detrimental process. For sewage treatment (⮌ Section 35.2), however, denitrification is beneficial because it converts NO_3^- to N_2. This transformation decreases the load of fixed nitrogen in the sewage treatment effluent that can stimulate algal growth in receiving waters, such as rivers and streams, or lakes (⮌ Section 24.2).

Table 14.5 *Oxidation states of key nitrogen compounds*

Compound	Oxidation state of N atom
Organic N (—NH_2)	−3
Ammonia (NH_3)	−3
Nitrogen gas (N_2)	0
Nitrous oxide (N_2O)	+1 (average per N)
Nitric oxide (NO)	+2
Nitrite (NO_2^-)	+3
Nitrogen dioxide (NO_2)	+4
Nitrate (NO_3^-)	+5

Biochemistry of Dissimilative Nitrate Reduction

The enzyme that catalyzes the first step of dissimilative nitrate reduction is *nitrate reductase*, a molybdenum-containing membrane-integrated enzyme whose synthesis is repressed by molecular oxygen. All subsequent enzymes of the pathway (**Figure 14.13**) are coordinately regulated and thus also repressed by O_2. But, in addition to anoxic conditions, NO_3^- must also be present before these enzymes are fully expressed.

The first product of nitrate reduction is nitrite (NO_2^-), and the enzyme nitrite reductase reduces it to NO (Figure 14.13c). Some organisms can reduce NO_2^- to ammonia (NH_3) in a dissimilative process, but the production of gaseous products—*denitrification*—is of greatest global significance. This is because denitrification consumes a fixed form of nitrogen (NO_3^-) and produces gaseous nitrogen compounds, some of which are of environmental significance. For example, N_2O can be converted

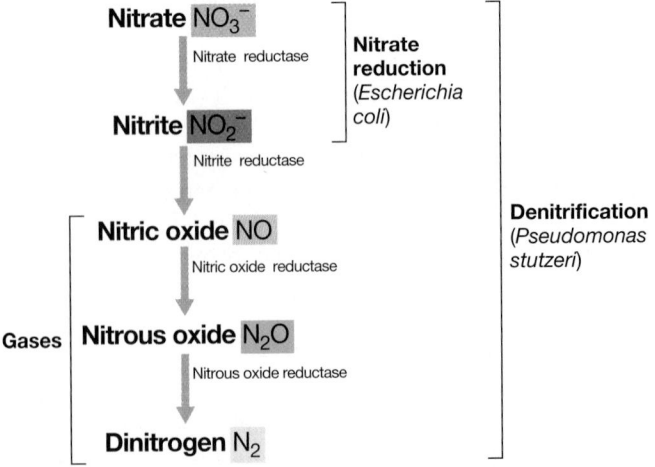

Figure 14.12 Steps in the dissimilative reduction of nitrate. Some organisms can carry out only the first step. All enzymes involved are derepressed by anoxic conditions. Also, some prokaryotes are known that can reduce NO_3^- to NH_4^+ in dissimilative metabolism. Note that colors used here match those used in Figure 14.13.

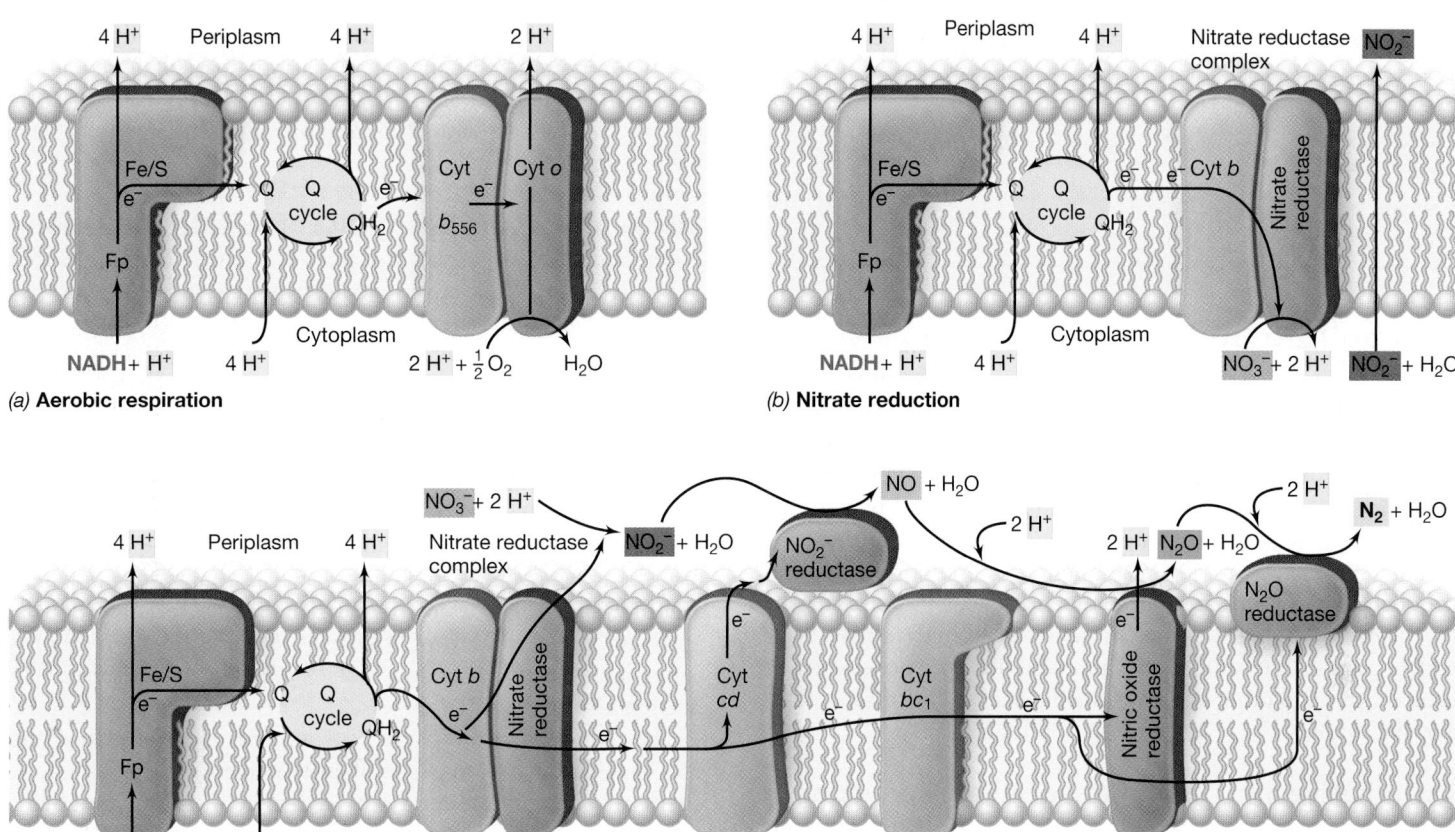

(a) **Aerobic respiration**

(b) **Nitrate reduction**

(c) **Denitrification**

Figure 14.13 **Respiration and nitrate-based anaerobic respiration.** Electron transport processes in the membrane of *Escherichia coli* when *(a)* O_2 or *(b)* NO_3^- is used as an electron acceptor and NADH is the electron donor. Fp, flavoprotein; Q, ubiquinone. Under high-oxygen conditions, the sequence of carriers is cyt $b_{562} \rightarrow$ cyt $o \rightarrow O_2$. However, under low-oxygen conditions (not shown), the sequence is cyt $b_{568} \rightarrow$ cyt $d \rightarrow O_2$. Note how more protons are translocated per two electrons oxidized aerobically during electron transport reactions than anaerobically with NO_3^- as electron acceptor, because the aerobic terminal oxidase (cyt *o*) pumps two protons. *(c)* Scheme for electron transport in membranes of *Pseudomonas stutzeri* during denitrification. Nitrate and nitric oxide reductases are integral membrane proteins, whereas nitrite and nitrous oxide reductases are periplasmic enzymes.

to NO by sunlight, and NO reacts with ozone (O_3) in the upper atmosphere to form NO_2^-. When it rains, NO_2^- returns to Earth as nitrous acid (HNO_2) in so-called acid rain. The remaining steps in denitrification are shown in Figure 14.13*c*.

The biochemistry of dissimilative nitrate reduction has been studied in detail in several organisms, including *Escherichia coli*, in which NO_3^- is reduced only to NO_2^-, and *Paracoccus denitrificans* and *Pseudomonas stutzeri*, in which denitrification occurs. The *E. coli* nitrate reductase accepts electrons from a *b*-type cytochrome, and a comparison of the electron transport chains in aerobic versus nitrate-respiring cells of *E. coli* is shown in Figure 14.13*a, b*.

Because of the reduction potential of the NO_3^-/NO_2^- couple (+0.43 V), fewer protons are pumped during nitrate reduction than in aerobic respiration (O_2/H_2O, +0.82 V). In *P. denitrificans* and *P. stutzeri*, nitrogen oxides are formed from NO_2^- by the enzymes nitrite reductase, nitric oxide reductase, and nitrous oxide reductase, as summarized in Figure 14.13*c*. During electron transport, a proton motive force is established, and ATPase functions to produce ATP in the usual fashion. Additional ATP is

available when NO_3^- is reduced to N_2 because the nitric oxide reductase is linked to proton extrusion (Figure 14.13*c*).

Other Properties of Denitrifying Prokaryotes

Most denitrifying prokaryotes are phylogenetically members of the *Proteobacteria* (Chapter 17) and, physiologically, facultative aerobes. Aerobic respiration occurs when O_2 is present, even if NO_3^- is also present in the medium. Many denitrifying bacteria also reduce other electron acceptors anaerobically, such as Fe^{3+} and certain organic electron acceptors (Section 14.12). In addition, some denitrifying bacteria can grow by fermentation and some are phototrophic purple bacteria (Section 13.4). Thus, denitrifying bacteria are quite metabolically diverse in alternative energy-generating mechanisms. Interestingly, at least one eukaryote has been shown to be a denitrifier. The protist *Globobulimina pseudospinescens*, a shelled amoeba (foraminifera, Section 20.11), can denitrify and likely employs this form of metabolism to survive in anoxic marine sediments where it resides.

- For *Escherichia coli*, why is more energy released in aerobic respiration than during NO_3^- reduction?

- How do the products of NO_3^- reduction differ between *E. coli* and *Pseudomonas*?

- Where is the dissimilative nitrate reductase found in the cell? What unusual metal does it contain?

14.8 Sulfate and Sulfur Reduction

Several inorganic sulfur compounds are important electron acceptors in anaerobic respiration. A summary of the oxidation states of key sulfur compounds is given in **Table 14.6**. Sulfate (SO_4^{2-}), the most oxidized form of sulfur, is a major anion in seawater and is reduced by the sulfate-reducing bacteria, a group that is widely distributed in nature. The end product of sulfate reduction is hydrogen sulfide, H_2S, an important natural product that participates in many biogeochemical processes (↩ Section 24.3). Species in the genus *Desulfovibrio* have been widely used for the study of sulfate reduction, and general properties of this and other sulfate-reducing bacteria are discussed in Section 17.18.

Assimilative and Dissimilative Sulfate Reduction

Again, as with nitrogen, it is necessary to distinguish between assimilative and dissimilative metabolism. Many organisms, including plants, algae, fungi, and most prokaryotes, use SO_4^{2-} as a source for biosynthetic sulfur needs. The ability to use SO_4^{2-} as an electron acceptor for energy-generating processes, however, involves the large-scale reduction of SO_4^{2-} and is restricted to the sulfate-reducing bacteria. In assimilative sulfate reduction, H_2S is formed on a very small scale and is assimilated into

organic form in sulfur-containing amino acids and other organic sulfur compounds. By contrast, in dissimilative sulfate reduction, H_2S can be produced on a very large scale and is excreted from the cell, free to react with other organisms or with metals to form metal sulfides.

Biochemistry and Energetics of Sulfate Reduction

As the reduction potentials in Table A1.2 and Figure 14.11 show, SO_4^{2-} is a much less favorable electron acceptor than is O_2 or NO_3^-. However, sufficient free energy to make ATP is available from sulfate reduction when an electron donor that yields NADH or FADH is oxidized. Table 14.6 lists some of the electron donors used by sulfate-reducing bacteria. Hydrogen (H_2) is used by virtually all species of sulfate-reducing bacteria, whereas use of the other donors is more restricted. For example, lactate and pyruvate are widely used by species found in freshwater anoxic environments, while acetate and longer-chain fatty acids are widely used by marine sulfate-reducing bacteria. Many morphological and physiological types of sulfate reducing bacteria are known, and with the exception of *Archaeoglobus* (↩ Section 19.6), a genus of *Archaea*, all known sulfate reducers are *Bacteria* (↩ Section 17.18).

The reduction of SO_4^{2-} to H_2S requires eight electrons and proceeds through a number of intermediate stages. Sulfate is chemically quite stable and cannot be reduced without first being activated; SO_4^{2-} is activated in a reaction requiring ATP. The enzyme ATP sulfurylase catalyzes the attachment of SO_4^{2-} to a phosphate of ATP, forming *adenosine phosphosulfate (APS)* as shown in **Figure 14.14**. In dissimilative sulfate reduction, the SO_4^{2-} in APS is reduced directly to sulfite (SO_3^{2-}) by the enzyme APS reductase with the release of AMP. In assimilative reduction, another phosphate is added to APS to form *phosphoadenosine phosphosulfate (PAPS)* (Figure 14.14*a*), and only then is the SO_4^{2-} reduced. However, in both cases the product of sulfate reduction is SO_3^{2-}. Once SO_3^{2-} is formed, H_2S is generated from the activity of the enzyme sulfite reductase (Figure 14.14*b*).

During dissimilative sulfate reduction, electron transport reactions lead to a proton motive force and this drives ATP synthesis by ATPase. A major electron carrier in this process is *cytochrome c_3*, a periplasmic low-potential cytochrome (**Figure 14.15**). Cytochrome c_3 accepts electrons from a periplasmically located hydrogenase and transfers these electrons to a membrane-associated protein complex. This complex, called *Hmc*, carries the electrons across the cytoplasmic membrane and makes them available to APS reductase and sulfite reductase, cytoplasmic enzymes that generate sulfite and sulfide, respectively (Figure 14.15).

The enzyme hydrogenase plays a central role in sulfate reduction whether *Desulfovibrio* is growing on H_2, per se, or on an organic compound such as lactate. This is because lactate is converted through pyruvate to acetate (the latter is for the most part excreted because *Desulfovibrio* is a non-acetate-oxidizing sulfate reducer; ↩ Section 17.18) with the production of H_2. The H_2 produced crosses the cytoplasmic membrane and is oxidized by the periplasmic hydrogenase to electrons, which are fed back into the system, and protons, which establish the proton motive force (Figure 14.15). Growth yields of sulfate-reducing bacteria suggest

Compound	Oxidation state of S atom
Oxidation states of key sulfur compounds	
Organic S (R—SH)	−2
Sulfide (H_2S)	−2
Elemental sulfur (S^0)	0
Thiosulfate (—S–SO_3^{2-})	−2/+6
Sulfur dioxide (SO_2)	+4
Sulfite (SO_3^{2-})	+4
Sulfate (SO_4^{2-})	+6
Some electron donors used for sulfate reduction	
H_2	Acetate
Lactate	Propionate
Pyruvate	Butyrate
Ethanol and other alcohols	Long-chain fatty acids
Fumarate	Benzoate
Malate	Indole
Choline	Various hydrocarbons

Table 14.6 *Sulfur compounds and electron donors for sulfate reduction*

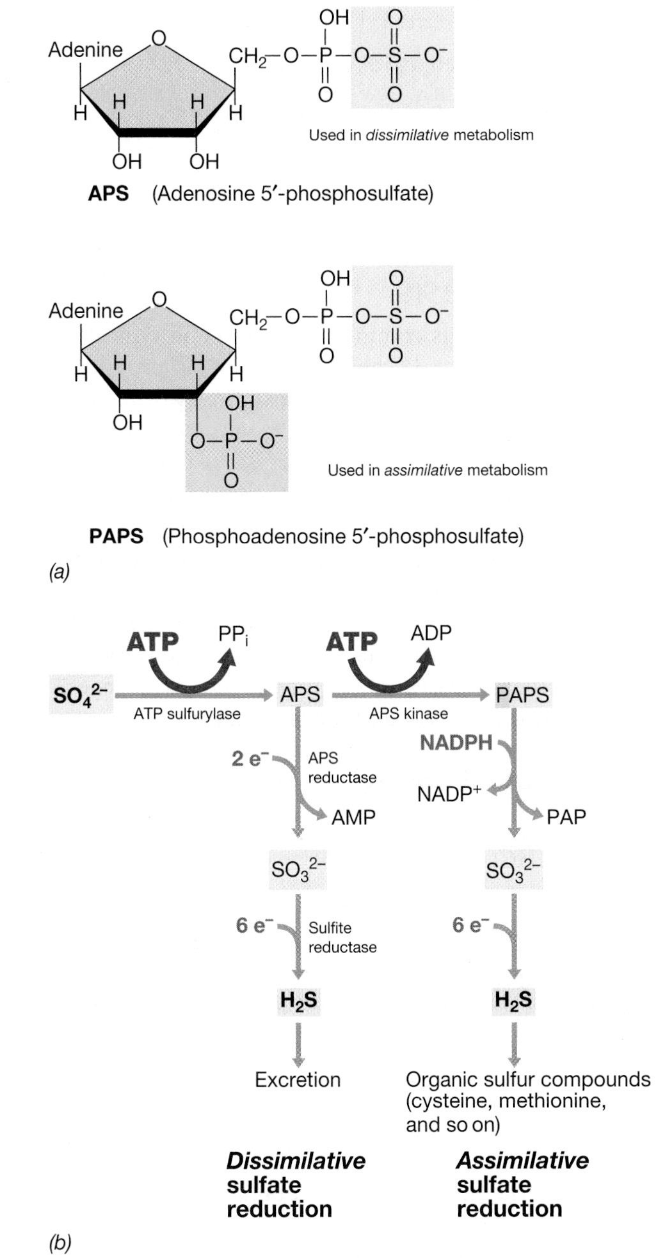

APS   (Adenosine 5′-phosphosulfate)

Used in *dissimilative* metabolism

PAPS   (Phosphoadenosine 5′-phosphosulfate)

(a)

(b)

Figure 14.14 Biochemistry of sulfate reduction: Activated sulfate. *(a)* Two forms of active sulfate can be made, adenosine 5′-phosphosulfate (APS) and phosphoadenosine 5′-phosphosulfate (PAPS). Both are derivatives of adenosine diphosphate (ADP), with the second phosphate of ADP being replaced by SO_4^{2-}. *(b)* Schemes of assimilative and dissimilative sulfate reduction.

that a net of one ATP is produced for each SO_4^{2-} reduced to HS^-. With H_2 as electron donor, the reaction is

$$4\,H_2 + SO_4^{2-} + H^+ \rightarrow HS^- + 4\,H_2O \qquad \Delta G^{0\prime} = -152\ \text{kJ}$$

When lactate or pyruvate is the electron donor, not only is ATP produced from the proton motive force, but additional ATP can be produced during the oxidation of pyruvate to acetate plus CO_2 via acetyl-CoA and acetyl phosphate (Table 14.1 and Figure 14.2).

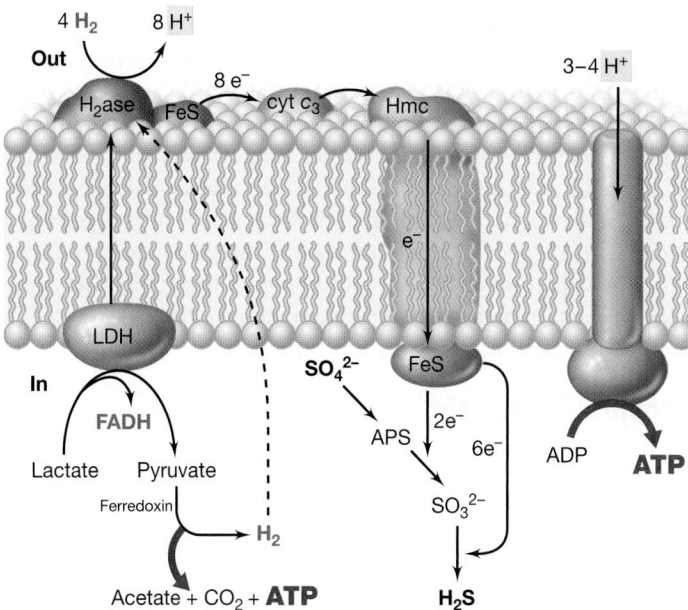

Figure 14.15 Electron transport and energy conservation in sulfate-reducing bacteria. In addition to external H_2, H_2 originating from the catabolism of organic compounds such as lactate and pyruvate can fuel hydrogenase. The enzymes hydrogenase (H_2ase), cytochrome (cyt) c_3, and a cytochrome complex (Hmc) are periplasmic proteins. A separate protein shuttles electrons across the cytoplasmic membrane from Hmc to a cytoplasmic iron–sulfur protein (FeS) that supplies electrons to APS reductase (forming SO_3^{2-}) and sulfite reductase (forming H_2S, Figure 14.14*b*). LDH, lactate dehydrogenase.

Acetate Use and Autotrophy

Many sulfate-reducing bacteria can oxidize acetate to CO_2 to obtain electrons for SO_4^{2-} reduction (Section 17.18):

$$CH_3COO^- + SO_4^{2-} + 3\,H^+ \rightarrow 2\,CO_2 + H_2S + 2\,H_2O$$
$$\Delta G^{0\prime} = -57.5\ \text{kJ}$$

The mechanism for acetate oxidation in most species is the *acetyl-CoA pathway*, a series of reversible reactions used by many anaerobes for acetate synthesis or acetate oxidation. This pathway employs the key enzyme *carbon monoxide dehydrogenase* (Section 14.9). A few sulfate-reducing bacteria can also grow autotrophically with H_2. When growing under these conditions, the organisms use the acetyl-CoA pathway for incorporating CO_2 into cell material. The acetate-oxidizing sulfate-reducing bacterium *Desulfobacter* lacks acetyl-CoA pathway enzymes and oxidizes acetate through the citric acid cycle (Figure 4.21), but this seems to be the exception rather than the rule.

Sulfur Disproportionation

Certain sulfate-reducing bacteria can disproportionate sulfur compounds of intermediate oxidation state. Disproportionation occurs when one molecule of a substance is oxidized while a second molecule is reduced, ultimately forming two different products. For example, *Desulfovibrio sulfodismutans* can disproportionate thiosulfate ($S_2O_3^{2-}$) as follows:

$$S_2O_3^{2-} + H_2O \rightarrow SO_4^{2-} + H_2S \qquad \Delta G^{0\prime} = -21.9\ \text{kJ/reaction}$$

Note that in this reaction one sulfur atom of $S_2O_3^{2-}$ becomes more oxidized (forming SO_4^{2-}), while the other becomes more reduced (forming H_2S). The oxidation of $S_2O_3^{2-}$ by *D. sulfodismutans* is coupled to proton pumping that is used by this organism to make ATP by ATPase. Other reduced sulfur compounds such as sulfite (SO_3^{2-}) and sulfur (S^0) can also be disproportionated. These forms of sulfur metabolism allow sulfate-reducing bacteria to recover energy from sulfur intermediates produced from the oxidation of H_2S by sulfur chemolithotrophs that coexist with them in nature and also from intermediates generated in their own metabolism during SO_4^{2-} reduction.

Phosphite Oxidation

At least one sulfate-reducing bacterium can couple phosphite (HPO_3^-) oxidation to SO_4^{2-} reduction. The reaction is chemolithotrophic, and the products are phosphate and sulfide:

$$4\,HPO_3^- + SO_4^{2-} + H^+ \rightarrow 4\,HPO_4^{2-} + HS^- \qquad \Delta G^{0'} = -364\,kJ$$

This bacterium, *Desulfotignum phosphitoxidans*, is an autotroph and a strict anaerobe, which by necessity it must be because phosphite spontaneously oxidizes in air. The natural sources of phosphite are likely to be organic phosphorous compounds called *phosphonates*, molecules generated from the anoxic degradation of organic phosphorous compounds. Along with sulfur disproportionation (also a chemolithotrophic process) and H_2 utilization, phosphite oxidation underscores the diversity of chemolithotrophic reactions carried out by sulfate-reducing bacteria.

Sulfur Reduction

Some organisms produce H_2S in anaerobic respiration, but are unable to reduce SO_4^{2-}; these are the elemental sulfur (S^0) reducers. Sulfur-reducing bacteria carry out the reaction

$$S^0 + 2\,H \rightarrow H_2S$$

The electrons for this process can come from H_2 or from various organic compounds. The first sulfur-reducing organism to be discovered was *Desulfuromonas acetoxidans* (⮌ Section 17.18). This organism oxidizes acetate, ethanol, and a few other compounds to CO_2, coupled with the reduction of S^0 to H_2S. Ferric iron (Fe^{3+}) also supports growth as an electron acceptor. The physiology of dissimilative sulfur-reducing bacteria is not as well understood as that of sulfate-reducing bacteria, but it is known that sulfur reducers lack the capacity to activate sulfate to APS (Figure 14.14), and presumably this prevents them from using SO_4^{2-} as an electron acceptor. *Desulfuromonas* contains high levels of several cytochromes, including an analog of cytochrome c_3, a key electron carrier in sulfate-reducing bacteria. Because the oxidation of acetate to CO_2 releases less energy than that needed to make an ATP by substrate-level phosphorylation, it is clear that oxidative phosphorylation plays a major role in the energetics of these organisms. A variety of other bacteria can use S^0 as an electron acceptor, including some species of the genera *Wolinella* and *Campylobacter*. In culture some sulfur reducers including *Desulfuromonas* can use Fe^{3+} as an electron acceptor, but S^0 is probably the major electron acceptor used in nature. It is the production of H_2S that connects the sulfur- and sulfate-reducing bacteria in an ecological sense.

MiniQuiz

- How is SO_4^{2-} converted to SO_3^{2-} during dissimilative sulfate reduction? Physiologically, how does *Desulfuromonas* differ from *Desulfovibrio*?
- Why is H_2 of importance to sulfate-reducing bacteria?
- Give an example of sulfur disproportionation.

14.9 Acetogenesis

Carbon dioxide, CO_2, is common in nature and typically abundant in anoxic habitats because it is a major product of the energy metabolisms of chemoorganotrophs. Two major groups of strictly anaerobic prokaryotes use CO_2 as an electron acceptor in energy metabolism. One of these groups is the *acetogens*, and we discuss them here. The other group, the *methanogens*, will be considered in the next section. H_2 is a major electron donor for both of these organisms, and an overview of their energy metabolism, **acetogenesis** and **methanogenesis**, is shown in **Figure 14.16**. Both processes are linked to ion pumps, of either protons (H^+) or sodium ions (Na^+), as the mechanism of energy conservation, and these pumps fuel ATPases in the membrane. Acetogenesis also conserves energy in a substrate-level phosphorylation reaction.

Organisms and Pathway

Acetogens carry out the reaction

$$4\,H_2 + H^+ + 2\,HCO_3^- \rightarrow CH_3COO^- + 4\,H_2O$$
$$\Delta G^{0'} = -105\,kJ$$

In addition to H_2, electron donors for acetogenesis include C_1 compounds, sugars, organic and amino acids, alcohols, and certain nitrogen bases, depending on the organism. Many acetogens can also reduce nitrate (NO_3^-) and thiosulfate ($S_2O_3^{2-}$). However, CO_2 reduction is probably the major reaction of ecological significance.

A major unifying thread among acetogens is the pathway of CO_2 reduction. Acetogens reduce CO_2 to acetate by the **acetyl-CoA pathway**, the major pathway in obligate anaerobes for the production or oxidation of acetate. **Table 14.7** lists the major groups of organisms that produce acetate or oxidize acetate via the acetyl-CoA pathway. Acetogens such as *Acetobacterium woodii* and *Clostridium aceticum* can grow either chemoorganotrophically by fermentation of sugars (reaction 1) or

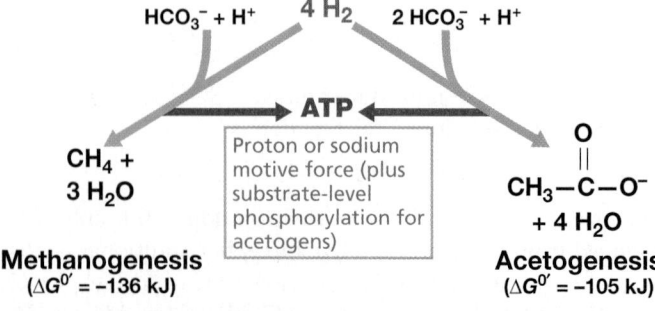

Figure 14.16 The contrasting processes of methanogenesis and acetogenesis. Note the difference in free energy released in the reactions.

Table 14.7 *Organisms employing the acetyl-CoA pathway*

I. **Pathway drives acetate synthesis for energy purposes**
   Acetoanaerobium noterae
   Acetobacterium woodii
   Acetobacterium wieringae
   Acetogenium kivui
   Acetitomaculum ruminis
   Clostridium aceticum
   Clostridium formicaceticum
   Clostridium ljungdahlii
   Moorella thermoacetica
   Desulfotomaculum orientis
   Sporomusa paucivorans
   Eubacterium limosum (also produces butyrate)
   Treponema primitia (from termite hindguts)

II. **Pathway drives acetate synthesis for cell biosynthesis**
   Acetogens
   Methanogens
   Sulfate-reducing bacteria

III. **Pathway drives acetate oxidation for energy purposes**
   Reaction: Acetate + 2 H_2O → 2 CO_2 + 8 H
      Group II sulfate reducers (other than *Desulfobacter*)
   Reaction: Acetate → CO_2 + CH_4
      Acetotrophic methanogens (*Methanosarcina, Methanosaeta*)

methanogens, most of which grow autotrophically on H_2 + CO_2 (⏎ Sections 19.3 and 14.10). By contrast, some bacteria employ the reactions of the acetyl-CoA pathway primarily in the reverse direction as a means of oxidizing acetate to CO_2. These include acetotrophic methanogens (⏎ Section 19.3) and sulfate-reducing bacteria (⏎ Sections 17.18 and 14.8).

Reactions of the Acetyl-CoA Pathway

Unlike other autotrophic pathways such as the Calvin cycle (⏎ Section 13.12), the reverse citric acid cycle, or the hydroxypropionate cycle (⏎ Section 13.13), the acetyl-CoA pathway of CO_2 fixation is not a cycle. Instead it catalyzes the reduction of CO_2 along two linear pathways; one molecule of CO_2 is reduced to the methyl group of acetate, and the other molecule of CO_2 is reduced to the carbonyl group. The two C_1 units are then combined at the end to form acetyl-CoA (**Figure 14.17**).

A key enzyme of the acetyl-CoA pathway is *carbon monoxide (CO) dehydrogenase*. CO dehydrogenase contains the metals Ni, Zn, and Fe as cofactors. CO dehydrogenase catalyzes the reaction

$$CO_2 + H_2 \rightarrow CO + H_2O$$

and the CO produced ends up as the *carbonyl* carbon of acetate (Figure 14.17). The methyl group of acetate originates from the reduction of CO_2 by a series of reactions requiring the coenzyme *tetrahydrofolate* (Figure 14.17). The methyl group is then

chemolithotrophically and autotrophically through the reduction of CO_2 to acetate with H_2 (reaction 2) as electron donor. In either case, the sole product is acetate:

$$(1)\ C_6H_{12}O_6 \rightarrow 3\ CH_3COO^- + 3\ H^+$$
$$(2)\ 2\ HCO_3^- + 4\ H_2 + H^+ \rightarrow CH_3COO^- + 4\ H_2O$$

Acetogens catabolize glucose by way of glycolysis, converting glucose to two molecules of pyruvate and two molecules of NADH (the equivalent of 4 H). From this point, two molecules of acetate are produced:

$$(3)\ 2\ Pyruvate^- \rightarrow 2\ acetate^- + 2\ CO_2 + 4\ H$$

The third acetate of reaction (1) comes from reaction (2), using the two molecules of CO_2 generated in reaction (3), plus the four H generated during glycolysis and the four H generated from the oxidation of two pyruvates to two acetates [reaction (3)]. Starting from pyruvate, then, the overall production of acetate can be written as

$$2\ Pyruvate^- + 4\ H \rightarrow 3\ acetate^- + H^+$$

Most acetogenic bacteria that produce and excrete acetate in energy metabolism are gram-positive *Bacteria*, and many are species of *Clostridium* or *Acetobacterium* (Table 14.7). A few other gram-positive and many different gram-negative *Bacteria* and *Archaea* use the acetyl-CoA pathway for autotrophic purposes, reducing CO_2 to acetate as a source of cell carbon.

The acetyl-CoA pathway functions in autotrophic growth for certain sulfate-reducing bacteria and is also used by the

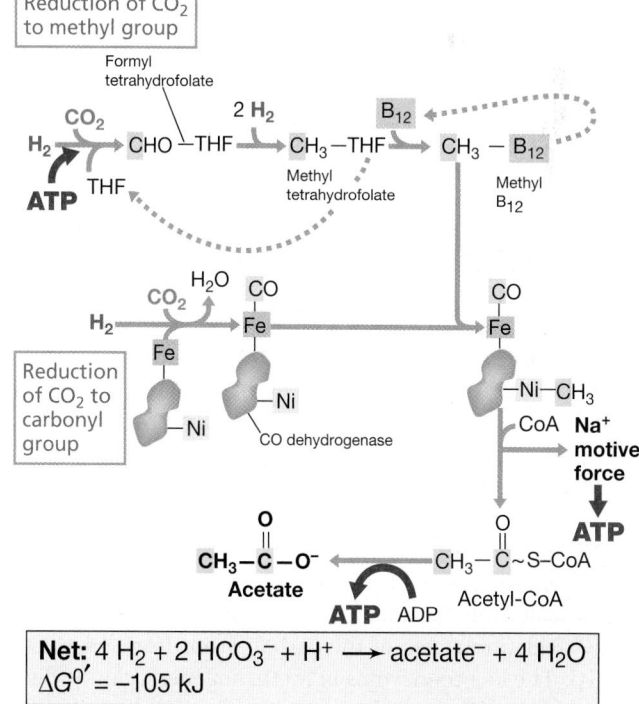

Figure 14.17 Reactions of the acetyl-CoA pathway. Carbon monoxide is bound to Fe and the CH_3 group to nickel in carbon monoxide dehydrogenase. Note that the formation of acetyl-CoA is coupled to the generation of a Na^+ motive force that drives ATP synthesis, and that ATP is also synthesized in the conversion of acetyl-CoA to acetate. THF, tetrahydrofolate; B_{12}, vitamin B_{12} in an enzyme-bound intermediate.

transferred from tetrahydrofolate to an enzyme that contains vitamin B_{12} as cofactor, and in the final step of the pathway, the methyl group is combined with CO by the enzyme CO dehydrogenase to form acetyl-CoA. Conversion of acetyl-CoA to acetate plus ATP completes the reaction series (Figure 14.17).

Energy Conservation in Acetogenesis

Energy conservation in acetogenesis is the result of substrate-level phosphorylation during the conversion of acetyl-CoA to acetate plus ATP (Section 14.1). There is also an energy-conserving step when a sodium motive force (analogous to a proton motive force) is established across the cytoplasmic membrane during the formation of acetyl-CoA. This energized state of the membrane allows for energy conservation from a Na^+-driven ATPase. Recall that we saw a similar situation in the succinate fermenter *Propionigenium*, where succinate decarboxylation was linked to Na^+ export and a Na^+-driven ATPase (Section 14.4). Acetogens need the ATP resulting from this reaction since the single ATP made by substrate-level phosphorylation is consumed in the first step of the acetyl-CoA pathway (Figure 14.17).

> **MiniQuiz**
> - Draw the structure of acetate and identify the carbonyl group and the methyl group. What key enzyme of the acetyl-CoA pathway produces the carbonyl group of acetate?
> - How do acetogens make ATP from the synthesis of acetate?
> - If fructose catabolism by glycolysis yields only two acetates, how does *Clostridium aceticum* produce three acetates from fructose?

14.10 Methanogenesis

The biological production of methane—*methanogenesis*—is carried out by a group of strictly anaerobic *Archaea* called the **methanogens**. The reduction of CO_2 by H_2 to form methane (CH_4) is a major pathway of methanogenesis and so we focus on this process and compare it with other forms of anaerobic respiration. We consider the basic properties, phylogeny, and taxonomy of the methanogens in Section 19.3; here we focus on their biochemistry and bioenergetics. Methanogenesis is a unique series of biochemical reactions that employs novel coenzymes. Because of this, we begin by considering these coenzymes and then move on to the actual pathway itself.

C_1 Carriers in Methanogenesis

Methanogenesis from CO_2 requires the input of eight electrons, and these electrons are added two at a time. This leads to intermediary oxidation states of the carbon atom from $+4$ (CO_2) to -4 (CH_4). The key coenzymes in methanogenesis can be divided into two classes: (1) those that carry the C_1 unit along its path of enzymatic reduction (C_1 carriers) and (2) those that donate electrons (redox coenzymes) (**Figure 14.18**, and see Figure 14.20). We consider the carriers first.

The coenzyme methanofuran is required for the first step of methanogenesis. Methanofuran contains the five-membered furan ring and an amino nitrogen atom that binds CO_2 (Figure 14.18a). Methanopterin (Figure 14.18b) is a methanogenic coenzyme that resembles the vitamin folic acid and plays a role analogous to that of tetrahydrofolate (a coenzyme that participates in C_1 transformations, see Figure 14.17) by carrying the C_1 unit in the intermediate steps of CO_2 reduction to CH_4. Coenzyme M (CoM) (Figure 14.18c) is a small molecule required for the terminal step of methanogenesis, the conversion of a methyl group (CH_3) to CH_4. Although not a C_1 carrier, the nickel (Ni^{2+})-containing tetrapyrrole coenzyme F_{430} (Figure 14.18d) is also needed for the terminal step of methanogenesis as part of the methyl reductase enzyme complex (discussed later).

Redox Coenzymes

The coenzymes F_{420} and 7-mercaptoheptanoylthreonine phosphate (also called coenzyme B, CoB), are electron donors in methanogenesis. Coenzyme F_{420} (Figure 14.18e) is a flavin derivative, structurally resembling the flavin coenzyme FMN ($\circlearrowleft$ Figure 4.15). F_{420} plays a role in methanogenesis as the electron donor in several steps of CO_2 reduction (see Figure 14.20). The oxidized form of F_{420} absorbs light at 420 nm and fluoresces blue-green. Such fluorescence is useful for the microscopic identification of a methanogen (**Figure 14.19**). CoB is required for the terminal step of methanogenesis catalyzed by the *methyl reductase enzyme complex*. As shown in Figure 14.18f, the structure of CoB resembles the vitamin pantothenic acid (which is part of acetyl-CoA) ($\circlearrowleft$ Figure 4.12).

Methanogenesis from $CO_2 + H_2$

Electrons for the reduction of CO_2 to CH_4 come primarily from H_2, but formate, carbon monoxide (CO), and even certain alcohols can also supply the electrons for CO_2 reduction in some methanogens. **Figure 14.20** shows the steps in CO_2 reduction by H_2:

1. CO_2 is activated by a methanofuran-containing enzyme and reduced to the formyl level. The immediate electron donor is the protein ferredoxin, a strong reductant with a reduction potential (E_0') near $-0.4V$.

2. The formyl group is transferred from methanofuran to an enzyme containing methanopterin (MP in Figure 14.20). It is subsequently dehydrated and reduced in two separate steps (total of 4 H) to the methylene and methyl levels. The immediate electron donor is reduced F_{420}.

3. The methyl group is transferred from methanopterin to an enzyme containing CoM by the enzyme methyl transferase. This reaction is highly exergonic and linked to the pumping of Na^+ across the membrane from inside to outside the cell.

4. Methyl-CoM is reduced to methane by methyl reductase; in this reaction, F_{430} and CoB are required. Coenzyme F_{430} removes the CH_3 group from CH_3—CoM, forming a Ni^{2+}–CH_3 complex. This complex is reduced by CoB, generating CH_4 and a disulfide complex of CoM and CoB (CoM-S—S-CoB).

5. Free CoM and CoB are regenerated by the reduction of CoM-S—S-CoB with H_2.

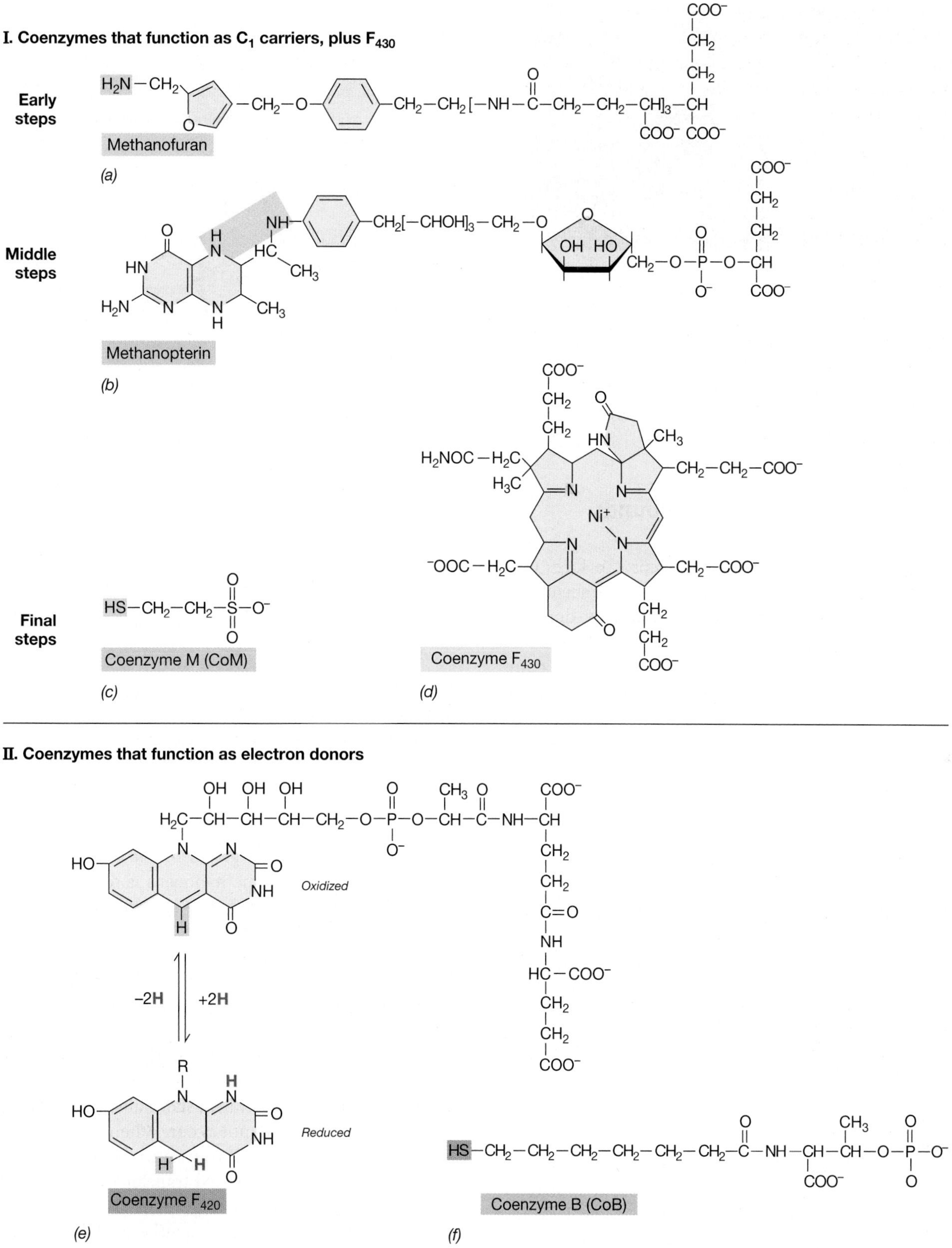

I. Coenzymes that function as C₁ carriers, plus F₄₃₀

Early steps

Methanofuran

(a)

Middle steps

Methanopterin

(b)

Final steps

Coenzyme M (CoM)

(c)

Coenzyme F₄₃₀

(d)

II. Coenzymes that function as electron donors

Oxidized

−2H   +2H

Reduced

Coenzyme F₄₂₀

(e)

Coenzyme B (CoB)

(f)

Figure 14.18 Coenzymes of methanogenesis. The atoms shaded in brown or yellow are the sites of oxidation–reduction reactions (brown in F_{420} and CoB) or the position to which the C_1 moiety is attached during the reduction of CO_2 to CH_4 (yellow in methanofuran, methanopterin, and coenzyme M). The colors used to highlight a particular coenzyme (CoB is orange, for example) are also in Figures 14.20–14.21 to follow the reactions in each figure.

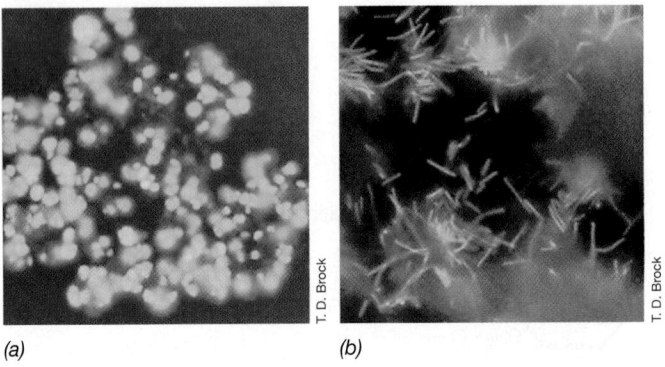

(a)   *(b)*

Figure 14.19 Fluorescence due to the methanogenic coenzyme F$_{420}$. *(a)* Autofluorescence in cells of the methanogen *Methanosarcina barkeri* due to the presence of the unique electron carrier F$_{420}$. A single cell is about 1.7 μm in diameter. The organisms were made visible by excitation with blue light in a fluorescence microscope. *(b)* F$_{420}$ fluorescence in cells of the methanogen *Methanobacterium formicicum*. A single cell is about 0.6 μm in diameter.

Methanogenesis from Methyl Compounds and Acetate

We will learn in Section 19.3 that methanogens can form CH$_4$ from methylated compounds such as methanol and acetate, as well as from H$_2$ + CO$_2$. Methanol is catabolized by donating methyl groups to a corrinoid protein to form CH$_3$–corrinoid (**Figure 14.21**). Corrinoids are the parent structures of compounds such as vitamin B$_{12}$ and contain a porphyrin-like corrin ring with a central cobalt atom (♺ Figure 15.8*a*). The CH$_3$–corrinoid complex then transfers the methyl group to CoM, yielding CH$_3$–CoM from which methane is formed in the same way as in the terminal step of CO$_2$ reduction (compare Figures 14.20 and 14.21*a*). If H$_2$ is unavailable to drive the terminal step, some of the methanol must be oxidized to CO$_2$ to yield electrons for this purpose. This occurs by reversal of steps in methanogenesis (Figures 14.20 and 14.21*a*).

When acetate is the substrate for methanogenesis, it is first activated to acetyl-CoA, which interacts with CO dehydrogenase from the acetyl-CoA pathway (Section 14.9). The methyl group of acetate is then transferred to the corrinoid enzyme to yield CH$_3$–corrinoid, and from there it goes through the CoM-mediated terminal step of methanogenesis. Simultaneously, the CO group is oxidized to yield CO$_2$ (Figure 14.21*b*).

Autotrophy

Autotrophy in methanogens occurs via the acetyl-CoA pathway (Section 14.9). As we have just seen, parts of this pathway are already integrated into the catabolism of methanol and acetate by methanogens (Figure 14.21). However, methanogens lack the tetrahydrofolate-driven series of reactions of the acetyl-CoA pathway that lead to the production of a methyl group (Figure 14.17). But this is not a problem because methanogens either derive methyl groups directly from their electron donors (Figure 14.21) or make methyl groups during methanogenesis from H$_2$ + CO$_2$ (Figure 14.20). Thus methanogens have abundant methyl groups, and the removal of some for biosynthesis is of little con-

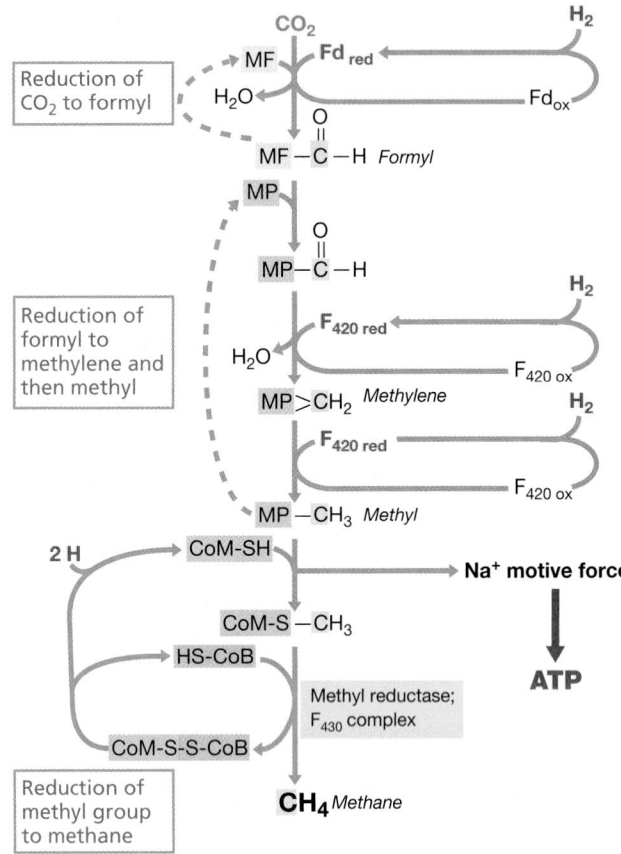

Figure 14.20 Methanogenesis from CO$_2$ plus H$_2$. The carbon atom reduced is shown in blue, and the source of electrons is highlighted in brown. See Figure 14.18 for the structures of the coenzymes. MF, Methanofuran; MP, methanopterin; CoM, coenzyme M; F$_{420red}$, reduced coenzyme F$_{420}$; F$_{430}$, coenzyme F$_{430}$; Fd, ferredoxin; CoB, coenzyme B.

sequence. The carbonyl group of the acetate produced during autotrophic growth of methanogens is derived from the activity of CO dehydrogenase, and the terminal step in acetate synthesis is as described for acetogens (Section 14.9 and Figure 14.17).

Energy Conservation in Methanogenesis

Under standard conditions the free energy from the reduction of CO$_2$ to CH$_4$ with H$_2$ is −131 kJ/mol, which is sufficient for the synthesis of at least one ATP. Energy conservation in methanogenesis occurs at the expense of a proton or sodium motive force, depending on the substrate used; substrate-level phosphorylation (Section 14.1) does not occur. When methanogenesis is supported by CO$_2$ + H$_2$, ATP is produced from the sodium motive force generated during methyl transfer from MP to CoM by the enzyme methyl transferase (Figure 14.20). This energized state of the membrane then drives the synthesis of ATP, probably by way of an H$^+$-linked ATPase following conversion of the sodium motive force into a proton motive force by exchange of Na$^+$ for H$^+$ across the membrane.

In some methanogens, such as *Methanosarcina*, a nutritionally versatile organism that can make methane from acetate or methanol as well as from CO$_2$, a different mechanism of energy

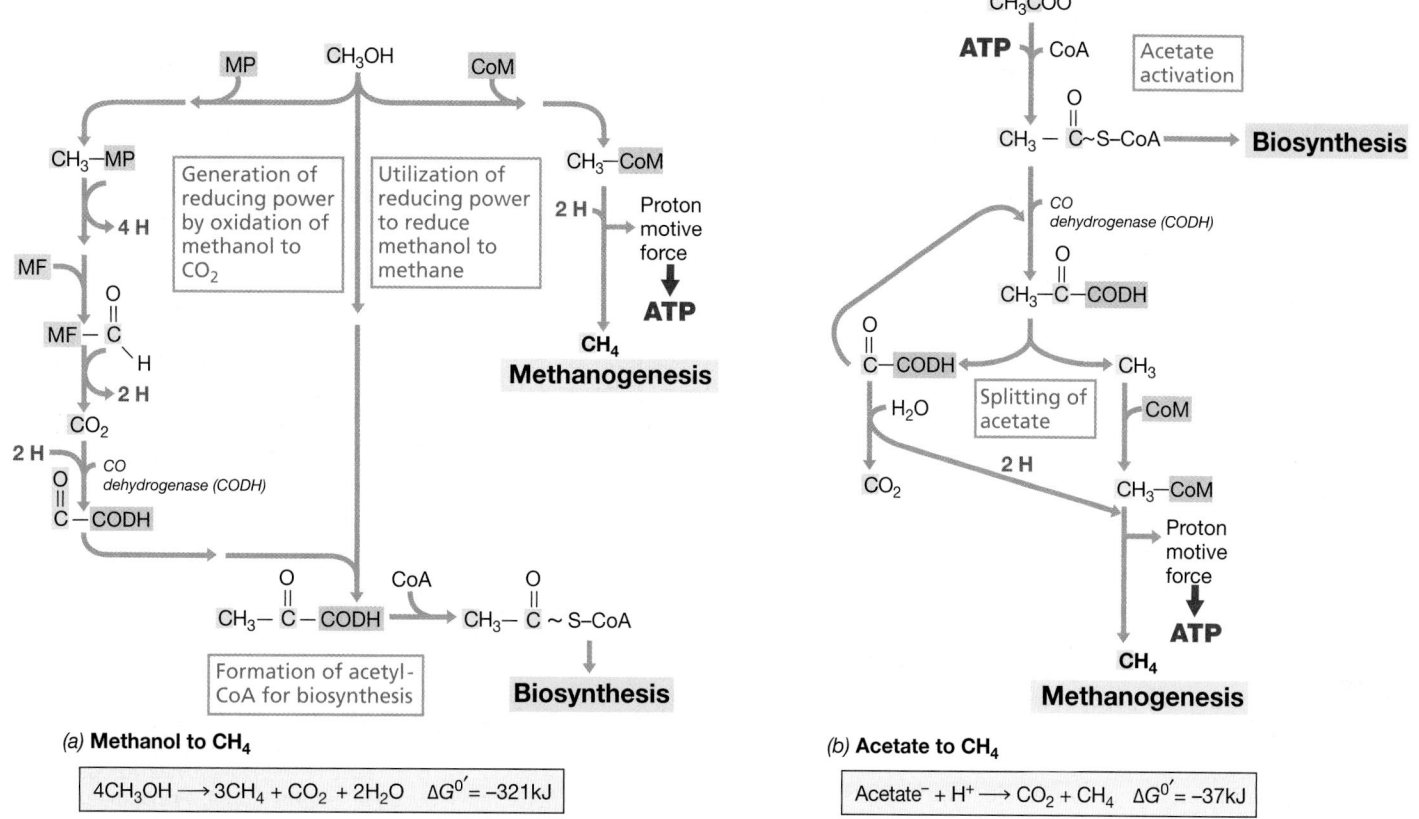

Figure 14.21 Methanogenesis from methanol and acetate. Both reaction series contain parts of the acetyl-CoA pathway. For growth on CH_3OH, most CH_3OH carbon is converted to CH_4, and a smaller amount is converted to either CO_2 or, via formation of acetyl-CoA, is assimilated into cell material. Abbreviations and color-coding are as in Figures 14.18 and 14.20; Corr, corrinoid-containing protein; CODH, carbon monoxide dehydrogenase.

conservation occurs in acetate- and methanol-grown cells, since the methyl transferase reaction cannot be coupled to the generation of a sodium motive force under these conditions. Instead, energy conservation in acetate- and methanol-grown cells is linked to the terminal step in methanogenesis, the methyl reductase step (Figure 14.20). In this reaction, the interaction of CoB with CH_3–CoM and methyl reductase forms CH_4 and a heterodisulfide, CoM-S—S-CoB. The latter is reduced by F_{420} to regenerate CoM-SH and CoB-SH (Figure 14.20). This reduction, carried out by the enzyme *heterodisulfide reductase*, is exergonic and is coupled to the pumping of H^+ across the membrane (**Figure 14.22**). Electrons from H_2 flow to the heterodisulfide reductase through a unique membrane-associated electron car-

Figure 14.22 Energy conservation in methanogenesis from methanol or acetate. (a) Structure of methanophenazine (MPH in part b), an electron carrier in the electron transport chain leading to ATP synthesis; the central ring of the molecule can be alternately reduced and oxidized. (b) Steps in electron transport. Electrons originating from H_2 reduce F_{420} and then methanophenazine. The latter, through a cytochrome of the *b* type, reduces heterodisulfide reductase with the extrusion of H^+ to the outside of the membrane. In the final step, heterodisulfide reductase reduces CoM-S—S-CoB to HS-CoM and HS-CoB. See Figure 14.18 for the structures of CoM and CoB.

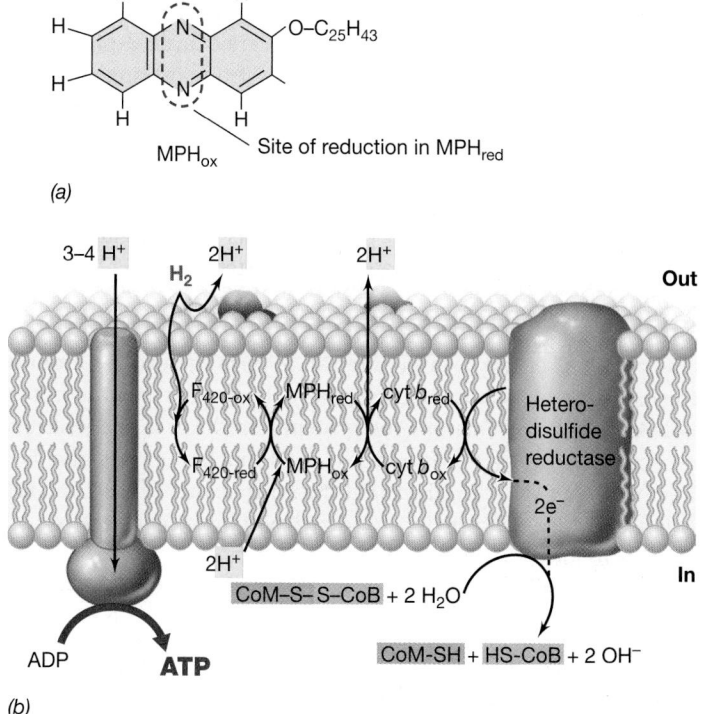

rier called *methanophenazine*. This compound is reduced by F_{420} and then oxidized by a *b*-type cytochrome, and the latter is the electron donor to the heterodisulfide reductase (Figure 14.22). Cytochromes and methanophenazine are lacking in methanogens that use only $H_2 + CO_2$ for methanogenesis.

In methanogens we thus see at least two mechanisms for energy conservation: (1) a proton motive force linked to the methylreductase reaction and used to drive ATP synthesis in acetate- or methanol-grown cells, and (2) a sodium motive force formed during methanogenesis from $H_2 + CO_2$.

MiniQuiz

- What coenzymes function as C_1 carriers in methanogenesis? As electron donors?

- In methanogens growing on $H_2 + CO_2$, how is carbon obtained for cell biosynthesis?

- How is ATP made in methanogenesis when the substrates are $H_2 + CO_2$? Acetate?

14.11 Proton Reduction

Perhaps the simplest of all anaerobic respirations is one carried out by the hyperthermophile *Pyrococcus furiosus*. *P. furiosus* is a species of *Archaea* and grows optimally at 100°C (⟳ Section 19.5) on sugars and small peptides as electron donors. *P. furiosus* was originally thought to use the glycolytic pathway because typical fermentation products such as acetate, CO_2, and H_2 were produced from glucose. However, analyses of sugar metabolism in this organism revealed an unusual and enigmatic situation.

During a key step of glycolysis, the oxidation of glyceraldehyde 3-phosphate forms 1,3-bisphosphoglyceric acid, an intermediate with two energy-rich phosphate bonds, each of which eventually yields ATP. In *P. furiosus*, this step is bypassed, yielding 3-phosphoglyceric acid directly from glyceraldehyde 3-phosphate (**Figure 14.23**). This prevents *P. furiosus* from making ATP by substrate-level phosphorylation at the 1,3-bisphosphoglyceric acid to 3-phosphoglyceric acid step, one of two sites of energy conservation in the glycolytic pathway (⟳ Figure 4.14). This yields *P. furiosus* a net of zero ATP from glycolytic steps that normally yield 2 ATP in other organisms. How can *P. furiosus* ferment glucose and ignore the most important energy-yielding steps?

Protons as Electron Acceptors

The riddle of energy conservation in *P. furiosus* revolves around the oxidation of 3-phosphoglyceric acid. In glycolysis this acceptor is NAD^+, but in *P. furiosus* the protein ferredoxin is the electron acceptor (Figure 14.23). Ferredoxin has a much more negative E_0' than that of $NAD^+/NADH$, about the same as that of the $2 H^+/H_2$ couple, –0.42 V. Ferredoxin is oxidized by transferring electrons to protons to form H_2 (Figure 14.23).

H_2 is typically produced during the oxidation of pyruvate to acetate plus CO_2 (Figure 14.2). This allows for ATP to be synthesized by substrate-level phosphorylation, and this also occurs in *P. furiosus* (Figure 14.23). But in addition, the H_2 released from ferredoxin is coupled to the pumping of protons (H^+) across the

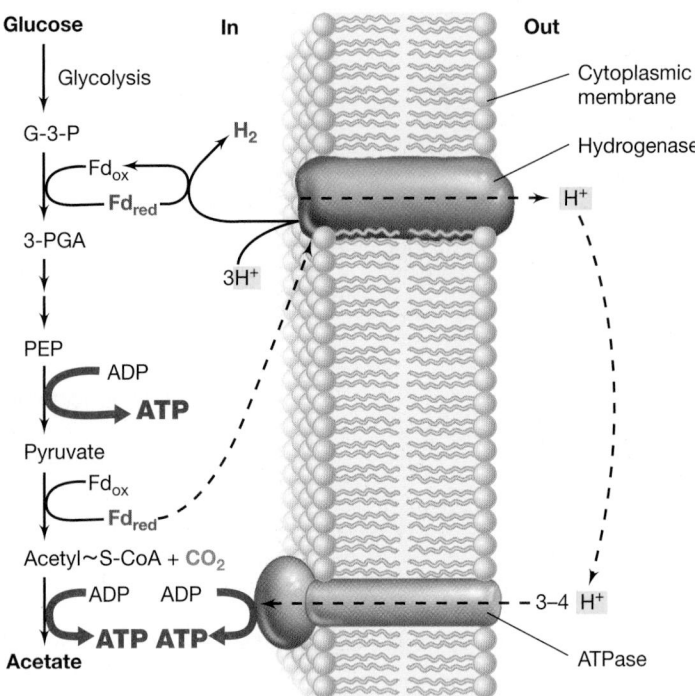

Figure 14.23 Modified glycolysis and proton reduction in anaerobic respiration in the hyperthermophile *Pyrococcus furiosus*. Hydrogen (H_2) production is linked to H^+ pumping by a hydrogenase that receives electrons from reduced ferredoxin (Fd_{red}). All intermediates from G-3-P downward in the pathway are present in two copies. Compare this figure with classical glycolysis in Figure 4.14. G-3-P, glyceraldehyde 3-phosphate; 3-PGA, 3-phosphoglycerate; PEP, phosphoenolpyruvate.

membrane by a membrane-integrated hydrogenase. This establishes a proton motive force that drives ATP synthesis by ATPase (Figure 14.23).

Although H^+ reduction by *P. furiosus* does not employ an electron transport chain per se, it can still be considered a form of anaerobic respiration because protons function as a net electron acceptor. The process differs from the proton pumping associated with decarboxylation reactions, such as those of *Oxalobacter*, where the free energy released during decarboxylation is coupled directly to H^+ translocation (Figure 14.8*b*). In *P. furiosus*, a proton is pumped during hydrogenase activity, analogous to how terminal electron carriers pump protons in aerobic or anaerobic respiratory processes (⟳ Figures 4.19 and 14.13).

Growth Yields and Evolution

Measurements of growth yields of *P. furiosus* on glucose indicate that, despite being unable to conserve energy from the main reactions in glycolysis, the organism actually synthesizes more ATP from glucose than most other glucose fermenters! Two ATP are produced by substrate-level phosphorylation during the conversion of two acetyl-CoA to acetate, and about one additional ATP is produced from H_2 production by hydrogenase (Figure 14.23).

Whether H^+ reduction by prokaryotes is more widespread than that in *P. furiosus* is unknown. However, the ancient phylogeny of *Pyrococcus* (⟳ Figure 19.1), coupled to its hot, anoxic habitat, similar to that of early Earth (⟳ Section 16.3), suggests that proton reduction, a bioenergetic mechanism that requires only a single membrane protein other than ATPase, might have been a very early form of anaerobic respiration, perhaps even nature's first proton pump.

MiniQuiz

• When fermenting glucose, how does *Pyrococcus furiosus* overcome the loss of most of the ATP produced by other glucose fermenters?

14.12 Other Electron Acceptors

In addition to the electron acceptors for anaerobic respiration discussed thus far, ferric iron (Fe^{3+}), manganic ion (Mn^{4+}), chlorate (ClO_3^-), perchlorate (ClO_4^-), and various organic compounds are important electron acceptors for bacteria in nature (**Figure 14.24**). Diverse bacteria are able to reduce these acceptors, especially Fe^{3+}, and many are able to reduce other acceptors, such as nitrate (NO_3^-) and elemental sulfur (S^0) (Sections 14.7 and 14.8), as well.

Ferric Iron Reduction

Ferric iron is an electron acceptor for energy metabolism in certain chemoorganotrophic and chemolithotrophic prokaryotes. Because Fe^{3+} is abundant in nature, its reduction supports a major form of anaerobic respiration. The reduction potential of the Fe^{3+}/Fe^{2+} couple is somewhat electropositive ($E_0' = +0.2$ V at pH 7), and thus, Fe^{3+} reduction can be coupled to the oxidation of several organic and inorganic electron donors. Electrons travel through an electron transport chain that generates a proton motive force and terminates in a ferric iron reductase system, reducing Fe^{3+} to ferrous iron (Fe^{2+}).

Much research on the energetics of Fe^{3+} reduction has been done with the gram-negative bacterium *Shewanella putrefaciens*, in which Fe^{3+}-dependent anaerobic growth occurs with various organic electron donors. Other important Fe^{3+} reducers include *Geobacter*, *Geospirillum*, and *Geovibrio*, and several hyperthermophilic *Archaea* (Chapters 17–19). *Geobacter metallireducens* has been a model for study of the physiology of Fe^{3+} reduction. *Geobacter* oxidizes acetate with Fe^{3+} as an acceptor in a highly exergonic reaction as follows:

$$Acetate^- + 8\ Fe^{3+} + 4\ H_2O \rightarrow 2\ HCO_3^- + 8\ Fe^{2+} + 9\ H^+$$
$$\Delta G^{0'} = -809\ kJ$$

Geobacter can also use H_2 or other organic electron donors, including the aromatic hydrocarbon toluene (see the Microbial Sidebar "Microbially Wired" in Chapter 24). This is of environmental significance because toluene from accidental spills or leakage from hydrocarbon storage tanks often contaminates iron-rich anoxic aquifers, and organisms such as *Geobacter* may be natural cleanup agents in such environments. Anoxic hydrocarbon metabolism is discussed in more detail shortly (Section 14.13).

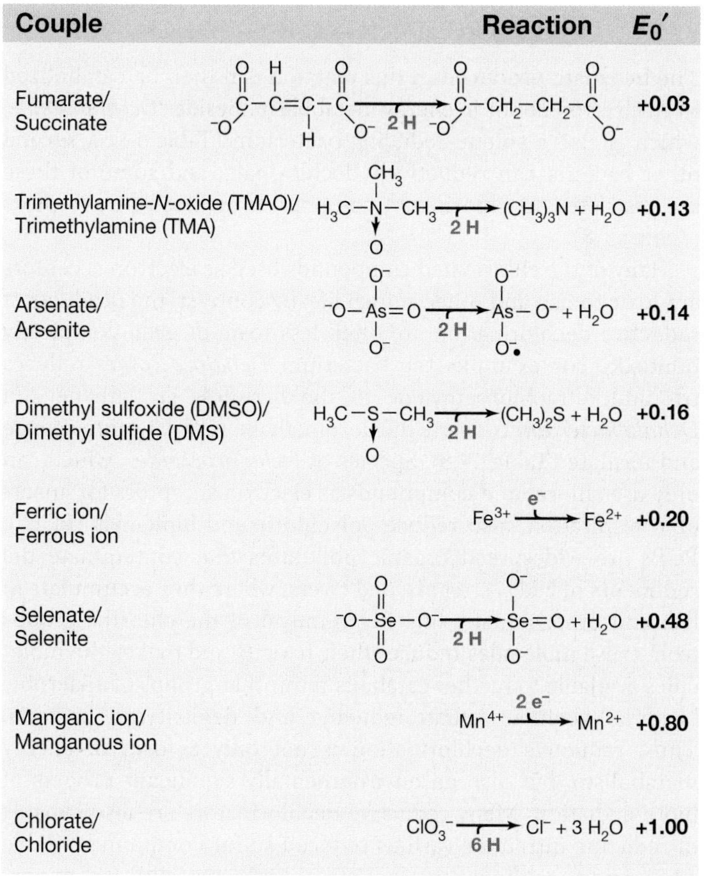

Reduction of Manganese and Other Inorganic Substances

Manganese has several oxidation states, of which manganic (Mn^{4+}) and manganous (Mn^{2+}) are the most relevant to microbial energetics. *S. putrefaciens* and a few other bacteria grow anaerobically on acetate or several other carbon sources with Mn^{4+} as electron acceptor. The reduction potential of the Mn^{4+}/Mn^{2+} couple is extremely high (Figure 14.24); thus, several compounds can donate electrons to Mn^{4+} reduction. This is also the case for chlorate (Figure 14.24). Several chlorate and perchlorate-reducing bacteria have been isolated, and most of them are facultative aerobes and thus also capable of aerobic growth.

Other inorganic substances can function as electron acceptors for anaerobic respiration. These include selenium and arsenic compounds (Figure 14.24). Although usually not abundant in natural systems, arsenic and selenium compounds are occasional pollutants and can support anoxic growth of various bacteria. The reduction of selenate (SeO_4^{2-}) to selenite (SeO_3^{2-}) and eventually to metallic selenium (Se^0) is an important method of selenium removal from water and has been used as a means of cleaning—a process called *bioremediation* (෴ Section 24.8)—selenium-contaminated soils. By contrast, the reduction of arsenate (AsO_4^{2-}) to arsenite (AsO_3^{2-}) can actually create a toxicity problem. Some groundwaters flow through rocks containing insoluble arsenate minerals. However, if the arsenate is reduced to arsenite by bacteria, the arsenite becomes more mobile and can contaminate groundwater. This has caused a serious problem of arsenic contamination of well water in some developing countries, such as Bangladesh, in recent years.

Couple	Reaction	E_0'
Fumarate/ Succinate	$^-O-\overset{O}{\overset{\|}{C}}-\overset{H}{\overset{\|}{C}}=C-\overset{O}{\overset{\|}{C}}-O^- \xrightarrow{2\ H} {}^-O-\overset{O}{\overset{\|}{C}}-CH_2-CH_2-\overset{O}{\overset{\|}{C}}-O^-$	+0.03
Trimethylamine-*N*-oxide (TMAO)/ Trimethylamine (TMA)	$H_3C-\overset{CH_3}{\overset{\|}{\underset{\|}{N}}}-CH_3 \xrightarrow{2\ H} (CH_3)_3N + H_2O$	+0.13
Arsenate/ Arsenite	$^-O-As=O \xrightarrow{2\ H} As-O^- + H_2O$	+0.14
Dimethyl sulfoxide (DMSO)/ Dimethyl sulfide (DMS)	$H_3C-\overset{\|}{\underset{O}{S}}-CH_3 \xrightarrow{2\ H} (CH_3)_2S + H_2O$	+0.16
Ferric ion/ Ferrous ion	$Fe^{3+} \xrightarrow{e^-} Fe^{2+}$	+0.20
Selenate/ Selenite	$^-O-\overset{O}{\underset{O}{Se}}-O^- \xrightarrow{2\ H} Se=O + H_2O$	+0.48
Manganic ion/ Manganous ion	$Mn^{4+} \xrightarrow{2\ e^-} Mn^{2+}$	+0.80
Chlorate/ Chloride	$ClO_3^- \xrightarrow{6\ H} Cl^- + 3\ H_2O$	+1.00

Figure 14.24 Some alternative electron acceptors for anaerobic respirations. Note the reaction and E_0' of each redox pair.

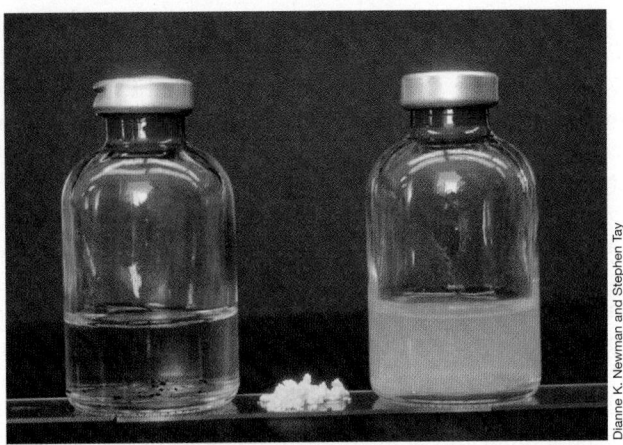

Figure 14.25 Biomineralization during arsenate reduction by the sulfate-reducing bacterium *Desulfotomaculum auripigmentum*. Left, appearance of culture bottle after inoculation. Right, following growth for two weeks and biomineralization of arsenic trisulfide, As_2S_3. Center, synthetic sample of As_2S_3.

Other forms of arsenate reduction are beneficial. For example, the sulfate-reducing bacterium *Desulfotomaculum* can reduce AsO_4^{3-} to AsO_3^{3-}, along with sulfate (SO_4^{2-}) to sulfide (HS^-). During this process a mineral containing arsenic and sulfide (As_2S_3, orpiment) precipitates spontaneously (**Figure 14.25**). The mineral is formed both intracellularly and extracellularly, and the process is an example of *biomineralization*, the formation of a mineral by bacterial activity. In this case As_2S_3 formation also functions as a means of detoxifying what would otherwise be a toxic compound (arsenic), and such microbial activities may have practical applications for the cleanup of arsenic-containing toxic wastes and groundwater.

Organic Electron Acceptors

Several organic compounds can be electron acceptors in anaerobic respirations. Of those listed in Figure 14.24, the compound that has been most extensively studied is *fumarate*, a citric acid cycle intermediate, which is reduced to succinate. The role of fumarate as an electron acceptor for anaerobic respiration derives from the fact that the fumarate–succinate couple has a reduction potential near 0 V (Figure 14.24), which allows coupling of fumarate reduction to the oxidation of NADH, FADH, or H_2. Bacteria able to use fumarate as an electron acceptor include *Wolinella succinogenes* (which can grow on H_2 as electron donor using fumarate as electron acceptor), *Desulfovibrio gigas* (a sulfate-reducing bacterium that can also grow under non-sulfate-reducing conditions), some clostridia, *Escherichia coli*, and many other bacteria.

Trimethylamine oxide (TMAO) (Figure 14.24) is an important organic electron acceptor. TMAO is a product of marine fish, where it functions as a means of excreting excess nitrogen. Various bacteria can reduce TMAO to trimethylamine (TMA), which has a strong odor and flavor (the odor of spoiled seafood is due primarily to TMA produced by bacterial action). Certain facultatively aerobic bacteria are able to use TMAO as an alternate elec-

tron acceptor. In addition, several phototrophic purple nonsulfur bacteria are able to use TMAO as an electron acceptor for anaerobic metabolism in darkness.

A compound similar to TMAO is dimethyl sulfoxide (DMSO), which is reduced by bacteria to dimethyl sulfide (DMS). DMSO is a common natural product and is found in both marine and freshwater environments. DMS has a strong, pungent odor, and bacterial reduction of DMSO to DMS is signaled by this characteristic odor. Bacteria, including *Campylobacter*, *Escherichia*, and many phototrophic purple bacteria, are able to use DMSO as an electron acceptor in energy generation.

The reduction potentials of the TMAO/TMA and DMSO/DMS couples are similar, near +0.15 V. This means that electron transport chains that terminate with the reduction of TMAO or DMSO must be rather short. As in fumarate reduction, in most instances of TMAO and DMSO reduction cytochromes of the *b* type (reduction potentials near 0 V) have been identified as terminal oxidases.

Halogenated Compounds as Electron Acceptors: Reductive Dechlorination

Several chlorinated compounds can function as electron acceptors for anaerobic respiration in the process called **reductive dechlorination** (also called *dehalorespiration*). For example, the bacterium *Desulfomonile* grows anaerobically with H_2 or organic compounds as electron donors and chlorobenzoate as an electron acceptor that is reduced to benzoate and hydrochloric acid (HCl):

$$C_7H_4O_2Cl^- + 2\,H \rightarrow C_7H_5O_2^- + HCl$$

The benzoate produced in this reaction can then be catabolized as an electron donor in energy metabolism. Besides *Desulfomonile*, which is also a sulfate-reducing bacterium (Table 14.6), several other bacteria can reductively dechlorinate, and some of these are restricted to chlorinated compounds as electron acceptors (**Table 14.8**).

Many of the chlorinated compounds used as electron acceptors are toxic to fish and other animal life; by contrast, the products of reductive dechlorination are often less toxic or even completely nontoxic. For example, the bacterium *Dehalococcoides* reduces tri- and tetrachloroethylene to the harmless gas ethene and *Dehalobacterium* converts dichloromethane (CH_2Cl_2) into acetate and formate (Table 14.8). Species of *Dehalococcoides*, which can only use chlorinated compounds as electron acceptors for anaerobic respiration, also reduce polychlorinated biphenyls (PCBs). PCBs are widespread organic pollutants that contaminate the sediments of lakes, streams, and rivers, where they accumulate in fish and other aquatic life. But removal of the chlorine groups from these molecules reduces their toxicity and makes the molecules available to further catabolism by other groups of anaerobic bacteria, such as sulfate-reducing and denitrifying bacteria. Thus, reductive dechlorination is not only a form of energy metabolism, but also an environmentally significant process of bioremediation. Many reductive dechlorinators are also capable of reducing nitrate or various reduced sulfur compounds (Table 14.8), and thus the group consists of both specialist and opportunist species.

Table 14.8 *Characteristics of some major genera of bacteria capable of reductive dechlorination*

Property	Genus				
	Dehalobacter	Dehalobacterium	Desulfitobacterium	Desulfomonile	Dehalococcoides
Electron donors	H_2	Dichloromethane (CH_2Cl_2) only	H_2, formate pyruvate, lactate	H_2, formate, pyruvate, lactate, benzoate	H_2, lactate
Electron acceptors	Trichloroethylene, tetrachloroethylene	Dichloromethane (CH_2Cl_2) only	Ortho-, meta-, or para-chlorophenols, NO_3^-, fumarate, SO_3^{2-}, $S_2O_3^{2-}$, S^0	Metachlorobenzoates, tetrachloroethylene, SO_4^{2-}, SO_3^{2-}, $S_2O_3^{2-}$	Trichloroethylene, tetrachloroethylene
Product of reduction of tetrachloroethylene	Dichloroethylene	Not applicable	Trichloroethylene	Dichloroethylene	Ethene
Other properties[a]	Contains cytochrome b	Grows only on CH_2Cl_2 and by disproportionation as follows: $CH_2Cl_2 \rightarrow$ formate + acetate + HCl ATP is formed by substrate-level phosphorylation	Can also grow by fermentation	Contains cytochrome c_3; requires organic carbon source; can grow by fermentation of pyruvate	Lacks peptidoglycan
Phylogeny[b]	Gram-positive *Bacteria*	Gram-positive *Bacteria*	Gram-positive *Bacteria*	*Deltaproteobacteria*	Green nonsulfur *Bacteria* (*Chloroflexi*)

[a]All organisms are obligate anaerobes.
[b]See Chapters 16–18.

UNIT 5

MiniQuiz

- With H_2 as electron donor, why is reduction of Fe^{3+} a more favorable reaction than reduction of fumarate?
- Give an example of biomineralization.
- What is reductive dechlorination and why is it environmentally relevant?

14.13 Anoxic Hydrocarbon Oxidation Linked to Anaerobic Respiration

Hydrocarbons are organic compounds that contain only carbon and hydrogen and are highly insoluble in water. We will see later in this chapter that *aerobic* hydrocarbon oxidation is a common microbial process in nature (Section 14.14). However, both aliphatic and aromatic hydrocarbons can be oxidized to CO_2 under anoxic conditions as well. Anoxic hydrocarbon oxidation occurs by way of various anaerobic respirations but has been best studied in denitrifying and sulfate-reducing bacteria.

Aliphatic Hydrocarbons

Aliphatic hydrocarbons are straight-chain saturated or unsaturated compounds, and many are substrates for denitrifying and sulfate-reducing bacteria. Saturated aliphatic hydrocarbons as long as C_{20} have been shown to support growth, although shorter-chain hydrocarbons are more soluble and readily catabolized. The mechanism of anoxic hydrocarbon degradation has been well studied for hexane (C_6H_{14}) metabolism in denitrifying bacteria. However, the mechanism appears to be the same for anoxic catabolism of longer-chain hydrocarbons and for hydrocarbon degradation by other anaerobic bacteria.

Hexane is a saturated aliphatic hydrocarbon. In anoxic hexane metabolism by *Azoarcus*, a species of *Proteobacteria*, hexane is attacked on carbon atom 2 by an *Azoarcus* enzyme that attaches a molecule of fumarate, an intermediate of the citric acid cycle (Section 4.11), forming the intermediate *1-methylpentylsuccinate* (**Figure 14.26a**). This compound now contains oxygen atoms and can be further catabolized anaerobically. Following the addition of coenzyme A, a series of reactions occurs that includes beta-oxidation (see Figure 14.42) and regeneration of fumarate. The electrons generated during beta-oxidation travel through an electron transport chain and generate a proton motive force. At the end of the chain, either nitrate (NO_3^-, in denitrifying bacteria) or sulfate (SO_4^{2-}, in sulfate-reducing bacteria) is reduced (Sections 14.7 and 14.8, respectively).

Aromatic Hydrocarbons

Aromatic hydrocarbons can be degraded anaerobically by some denitrifying, ferric iron-reducing, and sulfate-reducing bacteria. For anoxic catabolism of the aromatic hydrocarbon toluene, oxygen needs to be added to the compound to begin catabolism. Obviously this cannot come from O_2 if conditions are anoxic and occurs instead by the addition of fumarate, just as in aliphatic hydrocarbon catabolism (Figure 14.26). The reaction series eventually yields benzoyl-CoA, which is then further degraded by ring reduction (see Figure 14.27). Benzene (C_6H_6) can also be catabolized by nitrate-reducing bacteria, likely by a mechanism similar to that of toluene. Aromatic hydrocarbons containing multiple rings such as naphthalene ($C_{10}H_8$) can be degraded by certain sulfate-reducing and denitrifying bacteria. Growth on these substrates is very slow, and oxygenation of the hydrocarbon occurs by the addition of a molecule of CO_2 to the ring to form a carboxylic acid derivative.

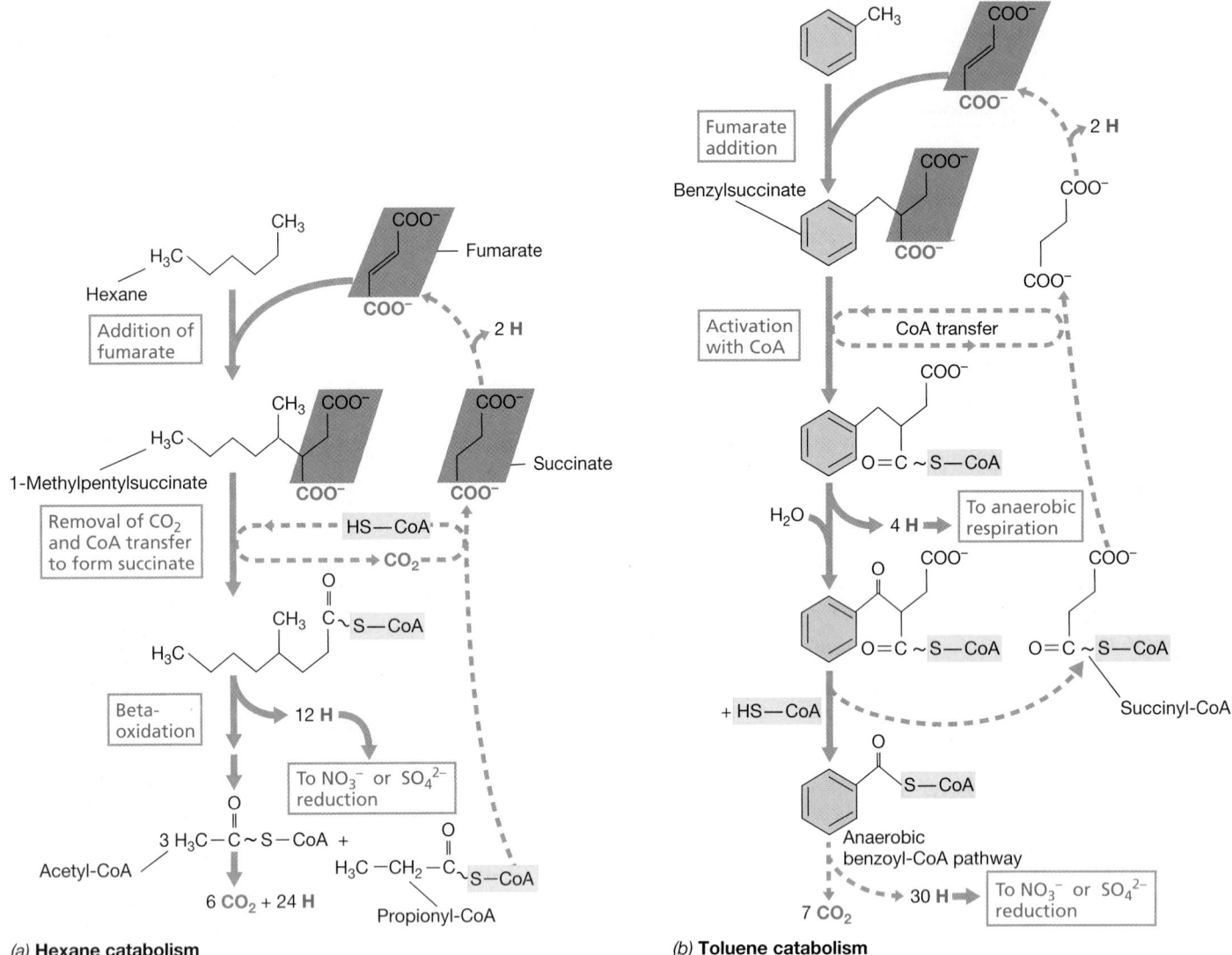

(a) **Hexane catabolism**

(b) **Toluene catabolism**

Figure 14.26 Anoxic catabolism of two hydrocarbons. *(a)* In anoxic catabolism of the aliphatic hydrocarbon hexane, the addition of fumarate provides the oxygen atoms necessary to form a fatty acid derivative that can be catabolized by beta-oxidation (see Figure 14.42) to yield acetyl-CoA. Electrons (H) generated from hexane catabolism are used to reduce sulfate or nitrate in anaerobic respirations. *(b)* Fumarate addition during the anoxic catabolism of the aromatic hydrocarbon toluene forms benzylsuccinate.

Besides the groups of anaerobes listed above, many other groups of bacteria can catabolize aromatic hydrocarbons anaerobically, including fermentative and phototrophic bacteria. However, except for toluene, only aromatic compounds that contain an O atom are degraded, and they are typically degraded by a common mechanism. When we examine the aerobic catabolism of aromatic compounds (Section 14.14), we will see that the biochemical mechanism occurs by way of ring *oxidation* (see Figure 14.30). By contrast, under anoxic conditions, the catabolism of aromatic compounds proceeds by ring *reduction*. Benzoate catabolism by the "benzoyl-CoA pathway" has been the focus of much of the work in this area, and the purple phototrophic bacterium *Rhodopseudomonas palustris*, an organism capable of catabolizing a wide variety of aromatic compounds, has been a model experimental organism (**Figure 14.27**). Benzoate catabolism in

this pathway begins by forming the coenzyme A derivative followed by ring cleavage to yield fatty or dicarboxylic acids that can be further catabolized to intermediates of the citric acid cycle.

Anoxic Oxidation of Methane

Methane (CH_4) is the simplest hydrocarbon. In freshwater ecosystems, methane is produced in anoxic sediments by methanogens and then oxidized to CO_2 by methanotrophs when it reaches oxic zones. These methanotrophs require O_2 for the catabolism of CH_4 because the first step in CH_4 oxidation employs a monooxygenase enzyme (see Section 14.14 and Figure 14.30). However, CH_4 can also be oxidized under *anoxic* conditions in marine and freshwater sediments.

In marine sediments, the anoxic oxidation of methane (AOM) is catalyzed by cell aggregates that contain both sulfate-reducing

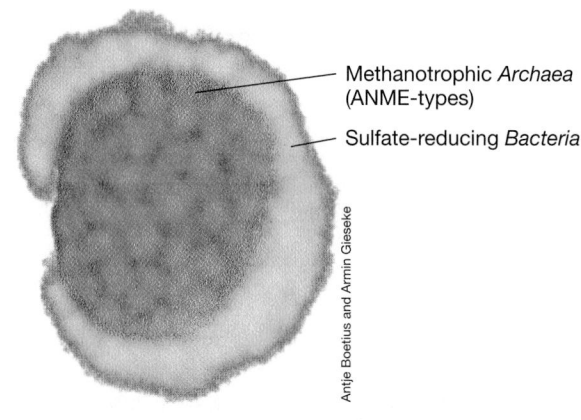

Figure 14.27 Anoxic degradation of benzoate by the benzoyl-CoA pathway. This pathway operates in the purple phototrophic bacterium *Rhodopseudomonas palustris* and many other facultative bacteria, both phototrophic and chemotrophic. Note that all intermediates of the pathway are bound to coenzyme A. The acetate produced is further catabolized in the citric acid cycle.

bacteria and *Archaea* phylogenetically related to methanogens (**Figure 14.28**). However, the archaeal component, called ANME (*an*oxic *me*thanotroph), of which there are several types, does not function in the consortium as a methano*gen*, but instead as a methano*troph*, oxidizing CH_4 as an electron donor. Electrons from methane oxidation are transferred to the sulfate reducer, which uses them to reduce SO_4^{2-} to H_2S (Figure 14.28*b*).

Details of the mechanism of AOM by the two organisms in the consortia are unclear, but it is thought that the methanotroph first activates CH_4 in some way and then oxidizes it to CO_2 by reversing the steps of methanogenesis, a series of reactions that would be highly endergonic (Section 14.10). Electrons are generated during the oxidative steps, but in what form the electrons are released to the sulfate reducer is unknown. Electrons are not released as H_2. Instead, electrons from the oxidation of CH_4 are shuttled from the methanotroph to the sulfate reducer in some organic intermediate, such as acetate, formate, or possibly as an organic sulfide (Figure 14.28*b*).

Regardless of mechanism, AOM yields only a very small amount of free energy:

$$CH_4 + SO_4^{2-} + H^+ \rightarrow CO_2 + HS^- + 2\,H_2O \qquad \Delta G^{0\prime} = -18\ \text{kJ}$$

How this energy is split between the methanotroph and the sulfate reducer is unknown. Substrate-level phosphorylation is unlikely, but as we have seen several times in this chapter, ion pumps can operate at these low energy yields and probably play a role in the energetics of AOM. In addition to oxidizing CH_4, the methanotrophic component of this consortium has been shown to fix nitrogen (⊘ Section 13.14), and it is possible that this fixed nitrogen supports the nitrogen needs of the entire consortium.

AOM is not limited to sulfate-reducing bacteria consortia. Methane-oxidizing denitrifying consortia are active in anoxic environments where CH_4 and NO_3^- coexist in significant amounts, such as certain freshwater sediments. In laboratory enrichments of these consortia some contain ANME-type methanotrophs while others are totally free of *Archaea*. AOM linked to ferric iron (Fe^{3+}) or manganic ion (Mn^{4+}) reduction also occurs. In both of these cases ANME-type methanotrophs have been identified, but in each system different ANME groups seem to predominate. Notably, however, the free-energy yield of AOM using Mn^{4+} or Fe^{3+} as electron acceptors is considerably more favorable than that of SO_4^{2-}, as would be expected from comparison of the reduction potentials of these different redox couples (Figure 14.11).

A newly discovered denitrifying bacterium employs a remarkable mechanism for anoxic methanotrophy not seen in any other methanotrophic system. The organism, provisionally named *Methylomirabilis oxyfera* because it is not yet in pure culture, oxidizes CH_4 with NO_3^- as an electron acceptor. During CH_4 oxidation, electrons reduce NO_3^- in steps we have previously seen in

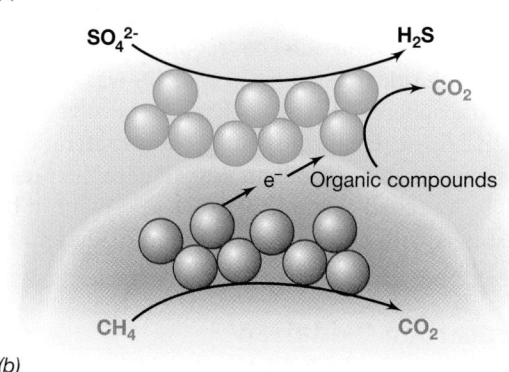

(a)

(b)

Figure 14.28 Anoxic methane oxidation. *(a)* Methane-oxidizing cell aggregates from marine sediments. The aggregates contain methanotrophic *Archaea* (red) surrounded by sulfate-reducing bacteria (green). Each cell type has been stained by a different FISH probe (⊘ Section 16.9). The aggregate is about 30 μm in diameter. *(b)* Mechanism for the cooperative degradation of CH_4. An organic compound or some other carrier of reducing power transfers electrons from methanotroph to sulfate reducer.

denitrifying bacteria such as *Pseudomonas* (Section 14.7). These steps include the reduction of NO_3^- to NO_2^-, and further on to N_2 (Figure 14.13c). But unlike *Pseudomonas*, in *M. oxyfera* NO_2^- is reduced to N_2 by way of nitric oxide (NO) without first producing nitrous oxide (N_2O) as an intermediate. Instead, *M. oxyfera* splits NO into N_2 and O_2 ($2\ NO \rightarrow N_2 + O_2$) using an enzyme called NO dismutase and then uses the O_2 produced as an electron acceptor for CH_4 oxidation. That is, the organism *produces its own* O_2 as an oxidant for electrons generated during the oxidation of CH_4 to CO_2.

The discovery of AOM by *Methylomirabilis oxyfera* has added a new twist to an already very intriguing story. The link between AOM and sulfate reduction was the first to be discovered and would naturally predominate in marine sediments because sulfate reduction is the dominant form of anaerobic respiration that occurs there (Section 24.3). But the list of alternative electron acceptors in anaerobic respiration is a very long one (Figure 14.11), and thus the discovery of AOM linked to oxidants other than SO_4^{2-}, NO_3^-, Mn^{4+}, or Fe^{3+} would not be surprising.

MiniQuiz

- Why is toluene a hydrocarbon whereas benzoate is not?
- How is hexane oxygenated during anoxic catabolism?
- What is AOM and which organisms participate in the process?

 # III Aerobic Chemoorganotrophic Processes

Many organic compounds are catabolized aerobically, and we survey some major aerobic processes here. We begin with a consideration of the oxygen requirements for some of these reactions.

14.14 Molecular Oxygen as a Reactant and Aerobic Hydrocarbon Oxidation

We previously discussed the role of molecular oxygen (O_2) as an *electron acceptor* in energy-generating reactions (Sections 4.9 and 4.10). Although this is by far the most important role of O_2 in cellular metabolism, O_2 plays an important role as a *reactant* in certain anabolic and catabolic processes as well.

Oxygenases

Oxygenases are enzymes that catalyze the incorporation of O_2 into organic compounds. There are two classes of oxygenases: *dioxygenases*, which catalyze the incorporation of both atoms of O_2 into the molecule, and *monooxygenases*, which catalyze the incorporation of one of the two oxygen atoms of O_2 into an organic compound; the second atom of O_2 is reduced to H_2O. For most monooxygenases, the required electron donor is NADH or NADPH (**Figure 14.29**). In the example of ammonia monooxygenase discussed previously (Section 13.10), the electron donor was cytochrome *c*, but this seems to be an exception.

Several types of reactions in living organisms require O_2 as a reactant. One of the best examples is O_2 in sterol biosynthesis.

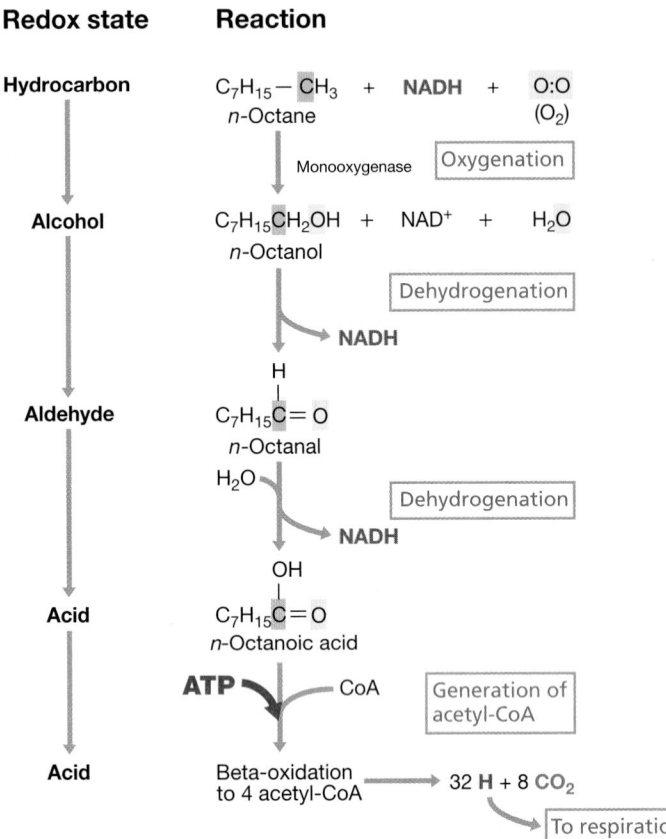

Figure 14.29 **Monooxygenase activity.** Steps in oxidation of an aliphatic hydrocarbon, the first of which is catalyzed by a monooxygenase. Some sulfate-reducing and denitrifying bacteria can degrade aliphatic hydrocarbons under anoxic conditions. For a description of beta-oxidation, see Figure 14.42.

Sterols are planar ring structures present in the membranes of eukaryotic cells and a few bacteria, and their biosynthesis requires O_2. Such a reaction obviously cannot take place under anoxic conditions, so organisms that grow anaerobically must either grow without sterols or obtain the needed sterols preformed from their environment. The requirement of O_2 in biosynthesis is of evolutionary significance, as O_2 was originally absent from the atmosphere of Earth when life first evolved. Oxygen became available on Earth only after the proliferation of cyanobacteria, approximately 2.7 billion years before the present (Section 16.3). A second example of O_2 as a reactant in biochemical processes is with aerobic hydrocarbon oxidation, and we consider this now.

Aerobic Hydrocarbon Oxidation

We saw in Section 14.13 how hydrocarbons could be catabolized under anoxic conditions; however, the aerobic oxidation of hydrocarbons is probably a much more extensive process in nature. Low-molecular-weight hydrocarbons are gases, whereas those of higher molecular weight are liquids or solids. Hydrocarbon consumption can be a natural process or can be a directed process for cleaning up spilled hydrocarbons from human activities (bioremediation, Section 24.7). Either way, the aerobic

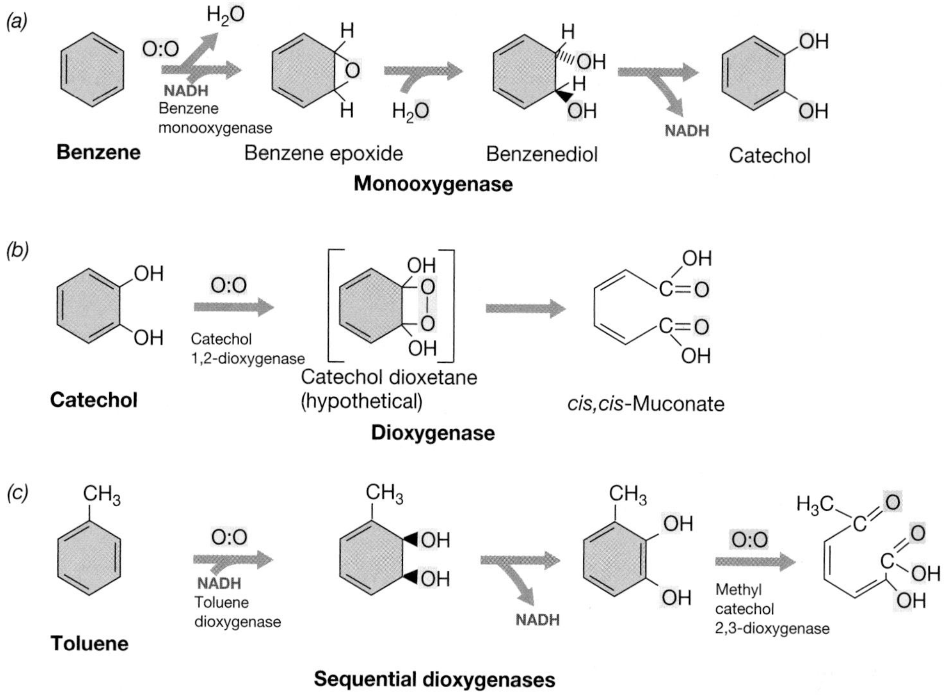

Figure 14.30 Roles of oxygenases in catabolism of aromatic compounds. Monooxygenases introduce one atom of oxygen from O_2 into a substrate, whereas diooxygenases introduce both atoms of oxygen. *(a)* Hydroxylation of benzene to catechol by a monooxygenase in which NADH is an electron donor. *(b)* Cleavage of catechol to *cis,cis*-muconate by an intradiol ring-cleavage dioxygenase. *(c)* The activities of a ring-hydroxylating dioxygenase and an extradiol ring-cleavage dioxygenase in the degradation of toluene. The oxygen atoms that each enzyme introduces are distinguished by different colors. Catechol and related compounds are common intermediates in aerobic aromatic catabolism. Compare aerobic toluene catabolism to anoxic toluene catabolism shown in Figure 14.26*b*.

catabolism of hydrocarbons can be very rapid owing to the metabolic advantage of having O_2 available as an electron acceptor compared with other acceptors of less positive reduction potential (Figure 14.11).

Several bacteria and fungi can use hydrocarbons as electron donors to support growth under aerobic conditions. The initial oxidation step of saturated aliphatic hydrocarbons by these organisms requires O_2 as a reactant, and one of the atoms of the oxygen molecule is incorporated into the oxidized hydrocarbon, typically at a terminal carbon atom. This reaction is carried out by a monooxygenase and a typical reaction sequence is shown in Figure 14.29. The end product of the reaction sequence known as beta-oxidation (see Figure 14.42) is acetyl-CoA, and this is oxidized in the citric acid cycle along with the production of electrons for the electron transport chain. The sequence is repeated to progressively degrade long hydrocarbon chains, and in most cases the hydrocarbon is oxidized completely to CO_2.

Aromatic Hydrocarbons

Many aromatic hydrocarbons can also be used as electron donors aerobically by microorganisms. The metabolism of these compounds, some of which contain several rings such as naphthalene or biphenyls, typically has as its initial stage the formation of catechol or a structurally related compound via catalysis by oxygenase enzymes, as shown in **Figure 14.30**. Once catechol is formed it can be further degraded and cleaved into compounds that can enter the citric acid cycle: succinate, acetyl-CoA, and pyruvate.

Several steps in the aerobic catabolism of aromatic hydrocarbons require oxygenases. Figure 14.30*a–c* shows four different oxygenase-catalyzed reactions, one using a monooxygenase, two using a ring-cleaving dioxygenase, and one using a ring-hydroxylating dioxygenase. As in aerobic aliphatic hydrocarbon catabolism, aromatic compounds, whether single or multiple

ringed, are typically oxidized completely to CO_2 and electrons enter an electron transport chain terminating with the reduction of O_2 to H_2O.

MiniQuiz

- How do monooxygenases differ in function from dioxygenases?
- What is the final product of catabolism of a hydrocarbon?
- What fundamental difference exists in the anaerobic degradation of an aromatic compound compared with its aerobic metabolism?

14.15 Methylotrophy and Methanotrophy

Methane (CH_4) and many other C_1 compounds can be catabolized aerobically by **methylotrophs**. Methylotrophs are organisms that use organic compounds that lack C—C bonds as electron donors and carbon sources (Section 17.6). The catabolism of compounds containing only a single carbon atom, such as methane and methanol (CH_3OH), have been the best studied of these substrates. We focus here on the physiology of methylotrophy, using CH_4 as an example.

Biochemistry of Methane Oxidation

The steps in CH_4 oxidation to CO_2 can be summarized as

$$CH_4 \rightarrow CH_3OH \rightarrow CH_2O \rightarrow HCOO^- \rightarrow CO_2$$

Methanotrophs are those methylotrophs that can use CH_4. Methanotrophs assimilate either all or one-half of their carbon (depending on the pathway used) at the oxidation state of formaldehyde (CH_2O). We will see later that this affords a major energy savings compared with the carbon assimilation of autotrophs, which also assimilate C_1 units, but exclusively from CO_2 rather than organic compounds.

Reactions and Bioenergetics of Aerobic Methanotrophy

The initial step in the aerobic oxidation of CH_4 is carried out by the enzyme *methane monooxygenase* (MMO). As we discussed in Section 14.14, monooxygenases catalyze the incorporation of oxygen atoms from O_2 into carbon compounds (and into some nitrogen compounds, ⮑ Section 13.10), thereby preparing them for further degradation. Methanotrophy has been especially well studied in the bacterium *Methylococcus capsulatus*. This organism contains two MMOs, one cytoplasmic and the other membrane-integrated. The electron donor for the cytoplasmic MMO is NADH, and NADH is probably the electron donor for the membrane-integrated MMO as well (**Figure 14.31**).

In the MMO reaction, an atom of oxygen is introduced into CH_4, and CH_3OH and H_2O are the products. Reducing power for the first step comes from later oxidative steps in the pathway. CH_3OH is oxidized by a periplasmic dehydrogenase, yielding formaldehyde and NADH (Figure 14.31). Once CH_2O is formed it is oxidized to CO_2 by either of two different pathways. One pathway uses enzymes that contain the coenzyme tetrahydrofolate, a coenzyme widely involved in C_1 transformations. The second and totally independent pathway employs the coenzyme methanopterin. Recall that methanopterin is a C_1 carrier in intermediate steps of the reduction of CO_2 to CH_4 by methanogenic *Archaea* (Section 14.10 and Figure 14.20). Methanotrophs use a methanopterin-containing reaction to drive the oxidation of CH_2O to formate plus NADH; formate is then oxidized to CO_2 by the enzyme formate dehydrogenase. However, regardless of the CH_2O oxidation pathway employed, electrons from the oxidation of CH_2O enter the electron transport chain, generating a proton motive force from which ATP is synthesized (Figure 14.31).

C_1 Assimilation into Cell Material

As will be discussed in Chapter 17, three phylogenetic groups of methanotrophs are known and at least two distinct pathways for C_1 incorporation into cell material exist. The **serine pathway**, utilized by type II methanotrophs, is outlined in **Figure 14.32**. In this pathway, a two-carbon unit, acetyl-CoA, is synthesized from one molecule of CH_2O (produced from the oxidation of CH_3OH, Figure 14.31) and one molecule of CO_2. The serine pathway requires reducing power and energy in the form of two molecules each of NADH and ATP, respectively, for each acetyl-CoA synthesized. The serine pathway employs a number of enzymes of the citric acid cycle and one enzyme, *serine transhydroxymethylase*, unique to the pathway (Figure 14.32).

The **ribulose monophosphate pathway**, used by type I methanotrophs, is outlined in **Figure 14.33**. This pathway is more efficient than the serine pathway because *all* of the carbon for cell material is derived from CH_2O. And, because CH_2O is at the same oxidation level as cell material, no reducing power is needed. The ribulose monophosphate pathway requires one molecule of ATP for each molecule of glyceraldehyde 3-phosphate (G-3-P) synthesized (Figure 14.33). Two G-3-Ps can be converted into glucose by the glycolytic pathway. Consistent with the lower

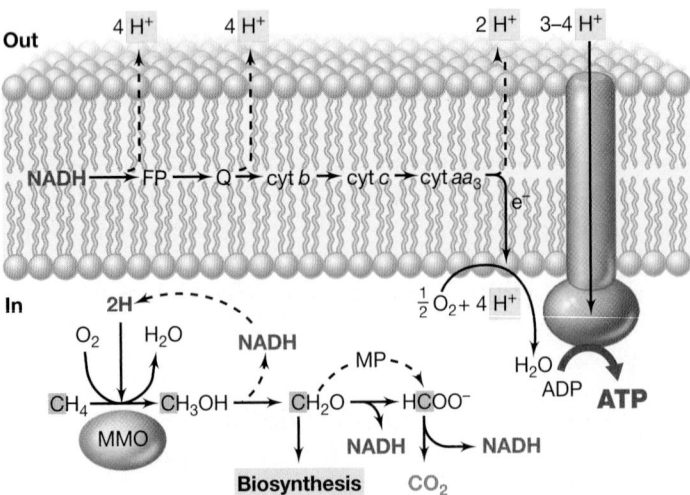

Figure 14.31 Oxidation of methane by methanotrophic bacteria. CH_4 is oxidized to CH_3OH by the enzyme methane monooxygenase (MMO). A proton motive force is established from electron flow in the membrane, and this fuels ATPase. Note how carbon for biosynthesis comes from CH_2O. Although not depicted as such, MMO is actually a membrane-associated enzyme and methanol dehydrogenase is periplasmic. FP, flavoprotein; cyt, cytochrome; Q, quinone; MP, methanopterin.

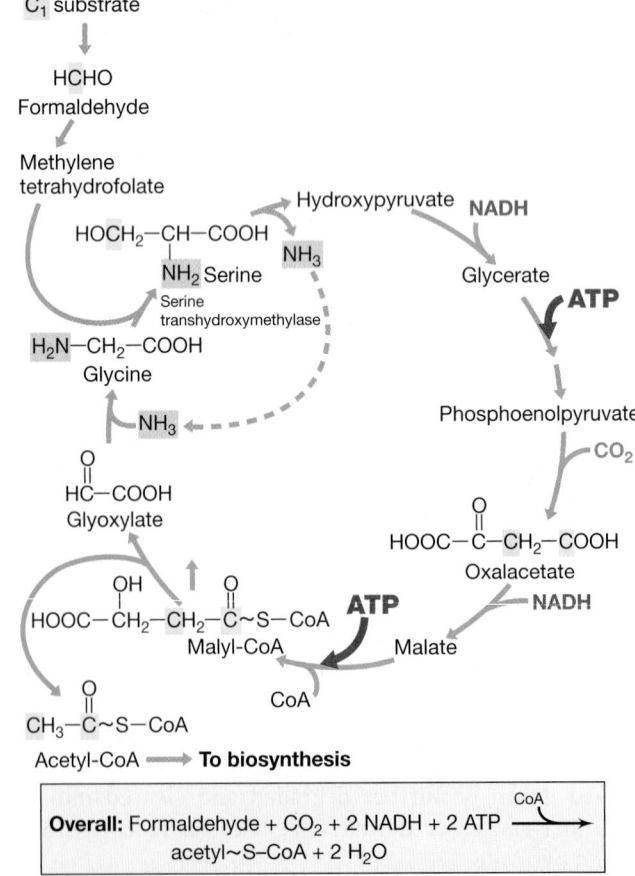

Figure 14.32 The serine pathway for the assimilation of C_1 units into cell material by methylotrophic bacteria. The product of the pathway, acetyl-CoA, is used as the starting point for making new cell material. The key enzyme of the pathway is serine transhydroxymethylase.

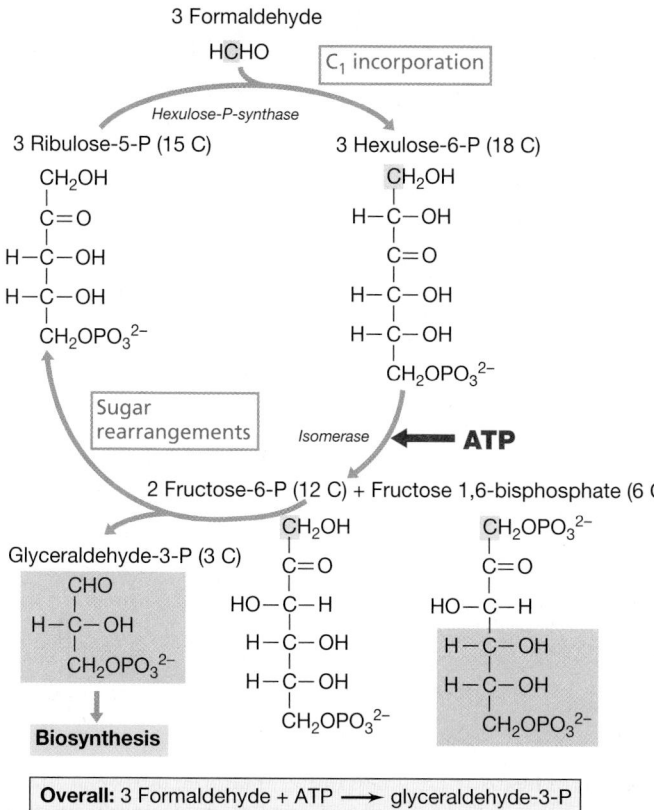

Figure 14.33 The ribulose monophosphate pathway for assimilation of C_1 units by methylotrophic bacteria. Three molecules of CH_2O are needed to complete the cycle, with the net result being one molecule of glyceraldehyde 3-phosphate. The key enzyme of this pathway is hexulose P-synthase. The sugar rearrangements require enzymes of the pentose phosphate pathway (Figure 14.38).

energy requirements of the ribulose monophosphate pathway, the cell yield (grams of cells produced per mole of CH_4 oxidized) of type I methanotrophs is higher than for type II methanotrophs.

The enzymes *hexulosephosphate synthase*, which condenses one molecule of formaldehyde with one molecule of ribulose 5-phosphate, and *hexulose 6-P isomerase* (Figure 14.33) are unique to the ribulose monophosphate pathway. The remaining enzymes of this pathway are widely distributed in bacteria. Finally, it should also be noted that the substrate for the initial reaction in this pathway, ribulose 5-phosphate, is very similar to the C_1 acceptor in the Calvin cycle, ribulose 1,5-bisphosphate (↩ Section 13.12), a signal that these two cycles likely share common evolutionary roots.

MiniQuiz

- Why are the energy and reducing power requirements for the ribulose monophosphate pathway different from those of the serine pathway?

- Why does the oxidation of CH_4 to CH_3OH require reducing power?

- Which pathway, the Calvin cycle or the ribulose monophosphate pathway, requires the greater energy input? Why?

14.16 Sugar and Polysaccharide Metabolism

Sugars and polysaccharides are common substrates for chemoorganotrophs, and we briefly consider their catabolism here.

Hexose and Polysaccharide Utilization

Sugars containing six carbon atoms, called *hexoses*, are the most important electron donors for many chemoorganotrophs and are also important structural components of microbial cell walls, capsules, slime layers, and storage products. The most common sources of hexose in nature are listed in **Table 14.9**, from which it can be seen that most are polysaccharides, although a few are disaccharides. Cellulose and starch are two of the most abundant natural polysaccharides.

Although both starch and cellulose are composed entirely of glucose, the glucose units are bonded differently (Table 14.9), and this profoundly affects their properties. Cellulose is more insoluble than starch and is usually less rapidly digested. Cellulose forms long fibrils, and organisms that digest cellulose are often found attached directly to these fibrils (**Figure 14.34**). In this way cellulase, the enzyme required to degrade cellulose, can contact its substrate and begin the digestive process. Many fungi are able to digest cellulose, and these are mainly responsible for the decomposition of plant materials on the forest floor. Among bacteria, however, cellulose digestion is restricted to relatively few groups, of which the gliding bacteria *Sporocytophaga* and *Cytophaga* (Figure 14.34 and **Figure 14.35**), clostridia, and actinomycetes are the most common.

Anoxic digestion of cellulose is carried out by a few *Clostridium* species, which are common in lake sediments, animal intestinal tracts, and systems for anoxic sewage digestion. Cellulose digestion is also a major process in the rumen of ruminant animals where *Fibrobacter* and *Ruminococcus* species actively degrade cellulose (↩ Section 25.9).

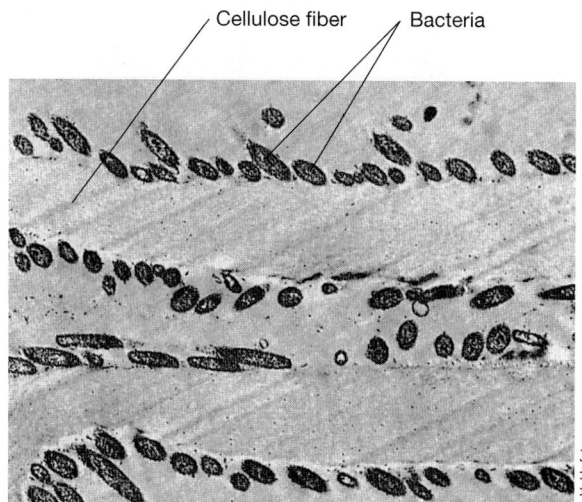

Figure 14.34 Cellulose digestion. Transmission electron micrograph showing attachment of the cellulose-digesting bacterium *Sporocytophaga myxococcoides* to cellulose fibers. Cells are about 0.5 μm in diameter.

Table 14.9 *Naturally occurring polysaccharides yielding hexose and pentose sugars*[a]

Substance	Composition	Sources	Catabolic enzymes
Cellulose	Glucose polymer (β-1,4-)	Plants (leaves, stems)	Cellulases (β, 1-4-glucanases)
Starch	Glucose polymer (α-1,4-)	Plants (leaves, seeds)	Amylase
Glycogen	Glucose polymer (α-1,4- and α-1,6-)	Animals (muscle) and microorganisms (granules)	Amylase, phosphorylase
Laminarin	Glucose polymer (β-1,3-)	Marine algae (*Phaeophyta*)	β-1,3-Glucanase (laminarinase)
Paramylon	Glucose polymer (β-1,3-)	Algae (*Euglenophyta* and *Xanthophyta*)	β-1,3-Glucanase
Agar	Galactose and galacturonic acid polymer	Marine algae (*Rhodophyta*)	Agarase
Chitin	*N*-Acetylglucosamine polymer (β-1,4-)	Fungi (cell walls) Insects (exoskeletons)	Chitinase
Pectin	Galacturonic acid polymer (from galactose)	Plants (leaves, seeds)	Pectinase (polygalacturonase)
Dextran	Glucose polymer	Capsules or slime layers of bacteria	Dextranase
Xylan	Heteropolymer of xylose and other sugars (β-1,4- and α-1,2 or α-1,3 side groups)	Plants	Xylanases
Sucrose	Glucose–fructose disaccharide	Plants (fruits, vegetables)	Invertase
Lactose	Glucose–galactose disaccharide	Milk	β-Galactosidase

[a]Each of these is subject to degradation by microorganisms.

Starch is digestible by many fungi and bacteria; this is illustrated for a laboratory culture in **Figure 14.36**. Starch-digesting enzymes, called *amylases*, are of considerable practical utility in industrial situations where starch must be digested, such as the textile, laundry, paper, and food industries, and fungi and bacteria are the commercial sources of these enzymes (⮌ Section 15.8).

All polysaccharides catabolized to support growth are first enzymatically hydrolyzed to monomeric or oligomeric units. In contrast, polysaccharides formed within cells as storage products are broken down not by *hydro*lysis, but by *phosphoro*lysis. This involves the addition of inorganic phosphate and results in the formation of hexose phosphate rather than free hexose. It may be summarized as follows for the degradation of starch, an α-1,4 polymer of glucose:

$$(C_6H_{12}O_6)_n + P_i \rightarrow (C_6H_{12}O_6)_{n-1} + \text{glucose 1-phosphate}$$

Because glucose 1-phosphate can be easily converted to glucose 6-phosphate—a key intermediate in glycolysis (⮌ Section 4.8)—with no energy expenditures, phosphorolysis represents a net energy savings to the cell.

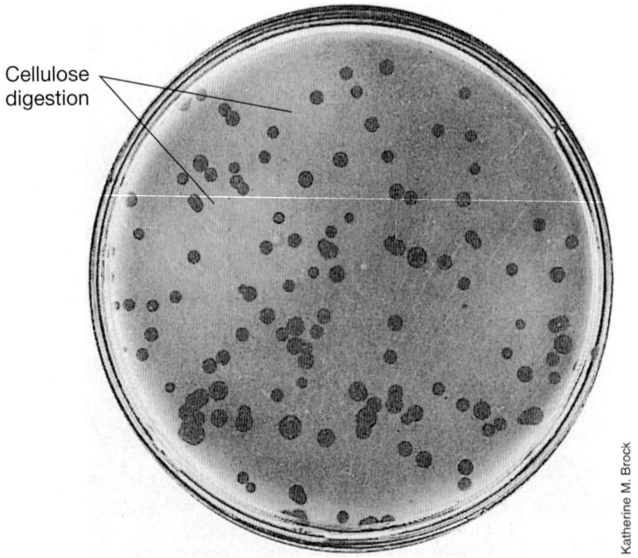

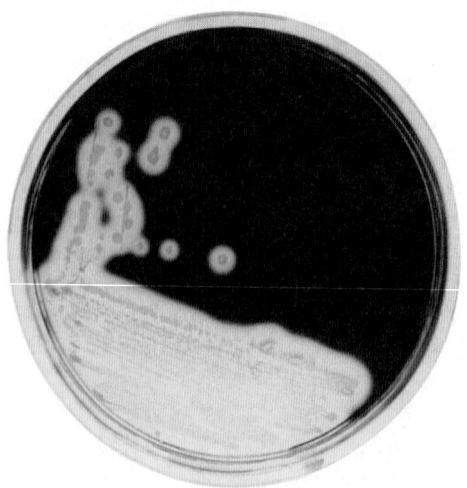

Figure 14.36 Hydrolysis of starch by *Bacillus subtilis*. After incubation, the starch–agar plate was flooded with Lugol's iodine solution. Where starch has been hydrolyzed, the characteristic purple-black color of the starch–iodine complex is absent. Starch hydrolysis extends some distance from the colonies because cells of *B. subtilis* produce the extracellular enzyme (exoenzyme) amylase, which diffuses into the surrounding medium.

Cellulose digestion

Katherine M. Brock

Figure 14.35 *Cytophaga hutchinsonii* colonies on a cellulose–agar plate. Areas where cellulose has been hydrolyzed are more translucent.

Disaccharides

Many microorganisms can use disaccharides for growth (Table 14.9). *Lactose* utilization by microorganisms is of considerable economic importance because milk-souring organisms produce lactic acid from lactose. *Sucrose*, the common disaccharide of higher plants, is usually first hydrolyzed to its component monosaccharides (glucose and fructose) by the enzyme invertase, and the monomers are then metabolized by the glycolytic pathway. *Cellobiose* (β-1,4-diglucose), a major product of cellulose digestion by cellulase, is degraded by cellulolytic bacteria but can also be degraded by many bacteria that are unable to degrade the cellulose polymer itself.

The microbial polysaccharide *dextran* is synthesized by some bacteria using the enzyme dextransucrase and sucrose as starting material:

$$n \text{ Sucrose} \rightarrow \underset{\text{dextran}}{(\text{glucose})_n} + n \text{ fructose}$$

Dextran is formed in this way by the bacterium *Leuconostoc mesenteroides* and a few others, and the polymer formed accumulates around the cells as a massive slime layer or capsule (**Figure 14.37**). Because sucrose is required for dextran formation, no dextran is formed when the bacterium is cultured on a medium containing glucose or fructose (⮌ Section 27.3). In nature, when cells that contain dextran or other polysaccharide capsules die, these materials once again become available for attack by fermentative or other chemoorganotrophic microorganisms.

The Pentose Phosphate Pathway

Pentose sugars are often available in nature. But if they are not available, they must be synthesized, because they form the backbone of the nucleic acids. Pentoses are made from hexose sugars, and the major pathway for this process is the **pentose phosphate pathway**.

Figure 14.38 summarizes the pentose phosphate pathway. Several important features should be noted. First, glucose can be oxidized to a pentose by loss of one carbon atom as CO_2. This generates NADPH and the key intermediate of the pathway, *ribulose 5-phosphate* (Figure 14.38). From the latter, ribose and from it deoxyribose are formed to supply the cell with nucleic acid precursors. Pentose sugars as electron donors can also feed

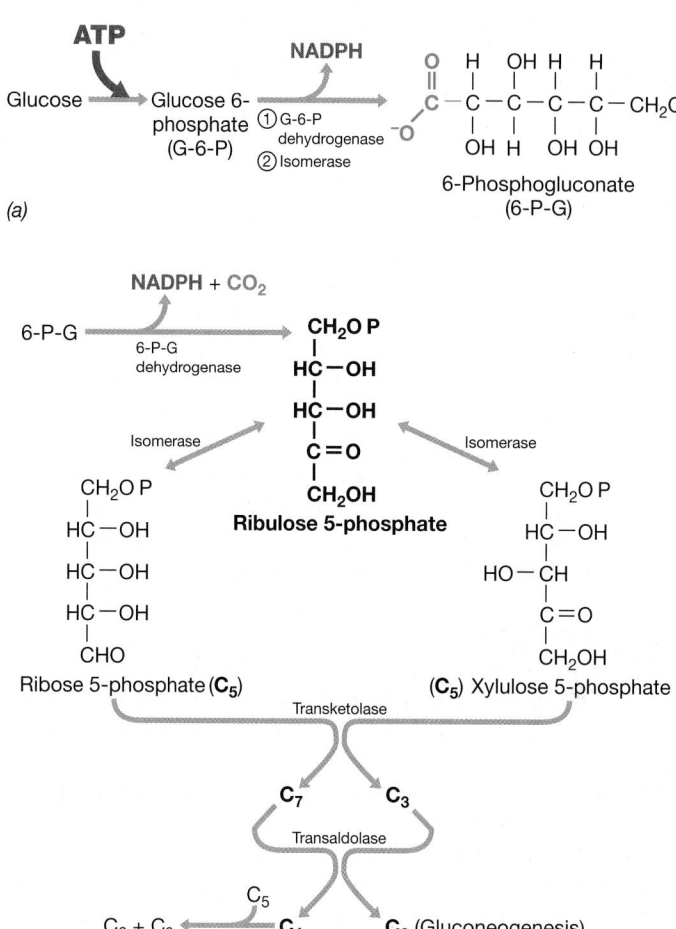

Figure 14.38 **The pentose phosphate pathway.** *(a)* The formation of 6-phosphogluconate. *(b)* The formation of pentoses from 6-phosphogluconate. The pathway is used to: (1) form pentoses from hexoses; (2) form hexoses from pentoses (gluconeogenesis); (3) catabolize pentoses as electron donors; and (4) generate NADPH. Some key enzymes of the pathway are indicated.

into the pentose phosphate pathway, typically becoming phosphorylated to form ribose phosphate or a related compound (Figure 14.38*b*) before being further catabolized.

A second important feature of the pentose phosphate pathway is the generation of sugar diversity. A variety of sugar derivatives, including C_4, C_5, C_6, and C_7, are formed in reactions of the pathway (Figure 14.38). This allows for pentose sugars to eventually yield hexoses for either catabolic purposes or for biosynthesis (gluconeogenesis, ⮌ Section 4.13).

A final important aspect of the pentose phosphate pathway is that it generates the redox coenzyme NADPH (Figure 14.38), and NADPH is used by the cell for many reductive biosyntheses; an important example would be ribonucleotide reductase, the enzyme that uses NADPH to convert ribonucleotides into deoxyribonucleotides (⮌ Section 4.14). Although most cells have an exchange mechanism for converting NADH into NADPH, the pentose phosphate pathway is the major means for direct synthesis of this important coenzyme.

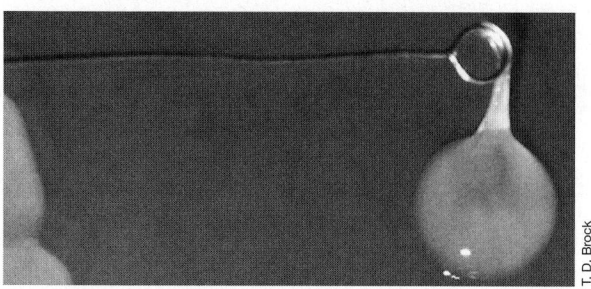

Figure 14.37 **Slime formation.** A slimy colony formed by the dextran-producing bacterium *Leuconostoc mesenteroides* growing on a sucrose-containing medium. When the same organism is grown on glucose, the colonies are small and not slimy because synthesis of dextran (a branched polysaccharide of glucose) specifically requires sucrose.

14.17 Organic Acid Metabolism

Various organic acids can be metabolized as carbon sources and electron donors by microorganisms. The intermediates of the citric acid cycle, *citrate, malate, fumarate,* and *succinate,* are common natural products formed by plants and are also fermentation products of microorganisms. Because the citric acid cycle has major biosynthetic as well as energetic functions (↩ Section 4.11), the complete cycle or major portions of it are nearly universal in microorganisms. Thus, it is not surprising that many microorganisms are able to use citric acid cycle intermediates as electron donors and carbon sources.

Glyoxylate Cycle

Unlike the utilization of organic acids containing four to six carbons, two- or three-carbon acids cannot be used as growth substrates by the citric acid cycle alone. The same is true for substrates such as hydrocarbons and lipids, degraded via beta-oxidation to acetyl-CoA (Section 14.18). The citric acid cycle can continue to operate only if the acceptor molecule, the four-carbon acid oxalacetate, is regenerated at each turn; any removal of carbon compounds for biosynthetic purposes would prevent completion of the cycle (↩ Figure 4.21).

When acetate is used, the oxalacetate needed to continue the cycle is produced through the **glyoxylate cycle** (**Figure 14.39**), so named because the C_2 compound glyoxylate is a key intermediate. This cycle is composed of citric acid cycle reactions plus two additional enzymes: *isocitrate lyase,* which splits isocitrate into succinate and glyoxylate, and *malate synthase,* which converts glyoxylate and acetyl-CoA to malate (Figure 14.39).

Biosynthesis through the glyoxylate cycle occurs as follows. The splitting of isocitrate into succinate and glyoxylate allows the succinate molecule (or another citric acid cycle intermediate derived from it) to be removed for biosynthesis because glyoxylate (C_2) combines with acetyl-CoA (C_2) to yield malate (C_4). Malate can be converted to oxalacetate to maintain the citric acid cycle after the C_4 intermediate (succinate) has been drawn off. Succinate is used in the production of porphyrins (needed for cytochromes, chlorophyll, and other tetrapyrroles). Succinate can also be oxidized to oxalacetate as a carbon skeleton for C_4 amino acids, or it can be converted (via oxalacetate and phosphoenolpyruvate) to glucose.

Pyruvate and C_3 Utilization

Three-carbon compounds such as pyruvate or compounds that can be converted to pyruvate (for example, lactate or carbohydrates) also cannot be catabolized through the citric acid cycle alone. Because some of the citric acid cycle intermediates are used for biosynthesis, the oxalacetate needed to keep the cycle going is synthesized from pyruvate or phosphoenolpyruvate by the addi-

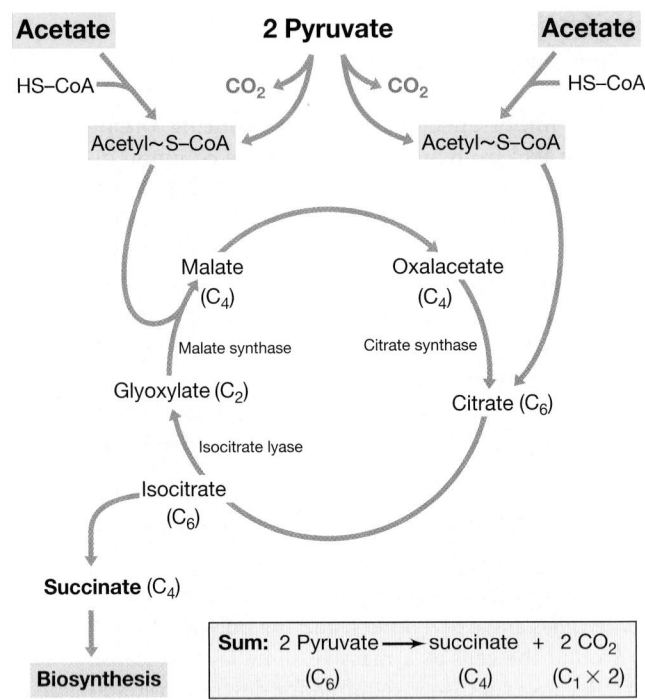

Figure 14.39 The glyoxylate cycle. Two unique enzymes, isocitrate lyase and malate synthase, operate along with most citric acid cycle enzymes. In addition to growth on pyruvate, the glyoxylate cycle also functions during growth on acetate.

tion of a carbon atom from CO_2. In some organisms this step is catalyzed by the enzyme *pyruvate carboxylase*:

$$\text{Pyruvate} + \text{ATP} + CO_2 \rightarrow \text{oxalacetate} + \text{ADP} + P_i$$

whereas in others it is catalyzed by *phosphoenolpyruvate carboxylase*:

$$\text{Phosphoenolpyruvate} + CO_2 \rightarrow \text{oxalacetate} + P_i$$

These reactions replace oxalacetate that is lost when intermediates of the citric acid cycle are removed for use in biosynthesis, and the cycle can continue to function.

14.18 Lipid Metabolism

Lipids are abundant in nature. The cytoplasmic membranes of all cells contain lipids, and many organisms produce lipid storage materials and contain lipids in their cell walls. These substances are biodegradable and are excellent substrates for microbial energy-yielding metabolism. When cells die, their lipids are thus catabolized, with CO_2 being the final product.

Fat and Phospholipid Hydrolysis

Fats are esters of glycerol and fatty acids and are readily available from the release of lipids from dead organisms. Microorganisms

Phospholipase activity: fatty acids released, leading to egg yolk precipitation

Clostridium perfringens

Inhibitor added: no phospholipase action, thus no precipitation of egg yolk

Figure 14.40 Phospholipase activity. Enzyme activity of phospholipase around a streak of *Clostridium perfringens* growing on an agar medium containing egg yolk. On half of the plate an inhibitor of phospholipase was added, preventing activity of the enzyme.

use fats only after hydrolysis of the ester bond, and extracellular enzymes called *lipases* are responsible for the reaction (**Figures 14.40** and **14.41**). Lipases attack fatty acids of various chain lengths. Phospholipids are hydrolyzed by enzymes called *phospholipases*, each of which is given a different letter designa-

Glycerol

H₂C — O — Fatty acid
|
HC — O — Fatty acid
|
H₂C — O — Fatty acid

Lipase

(a)

Phospholipase B

H₂C — O — Fatty acid
|
HC — O — Fatty acid
|
H₂C — O — P — O — (X)

Phospholipase A

Phospholipase C

Phospholipase D

(b)

Figure 14.41 Lipases. *(a)* Activity of lipases on a fat. *(b)* Phospholipase activity on phospholipid. The cleavage sites of the four distinct phospholipases A, B, C, and D are shown. X refers to a number of small organic molecules that may be at this position in different phospholipids.

tion depending on which ester bond it cleaves in the lipid (Figure 14.41). Phospholipases A and B cleave *fatty acid* esters, whereas phospholipases C and D cleave *phosphate* esters and hence are different classes of enzymes. The result of lipase activity is the release of free fatty acids and glycerol, and these substances can then be metabolized by chemoorganotrophic microorganisms.

Fatty Acid Oxidation

Fatty acids are oxidized by *beta-oxidation*, a series of reactions in which two carbons of the fatty acid are split off at a time (**Figure 14.42**). The fatty acid is first activated with coenzyme A; oxidation results in the release of acetyl-CoA by cleavage between the α and β carbons of the original fatty acid along with the formation of a new fatty acid two carbon atoms shorter (Figure 14.42). The process of beta-oxidation is then repeated, and another acetyl-CoA molecule is released.

There are two separate dehydrogenation reactions in beta-oxidation. In the first, electrons are transferred to flavin adenine dinucleotide (FAD), forming FADH, whereas in the second they are transferred to NAD⁺, forming NADH. Most

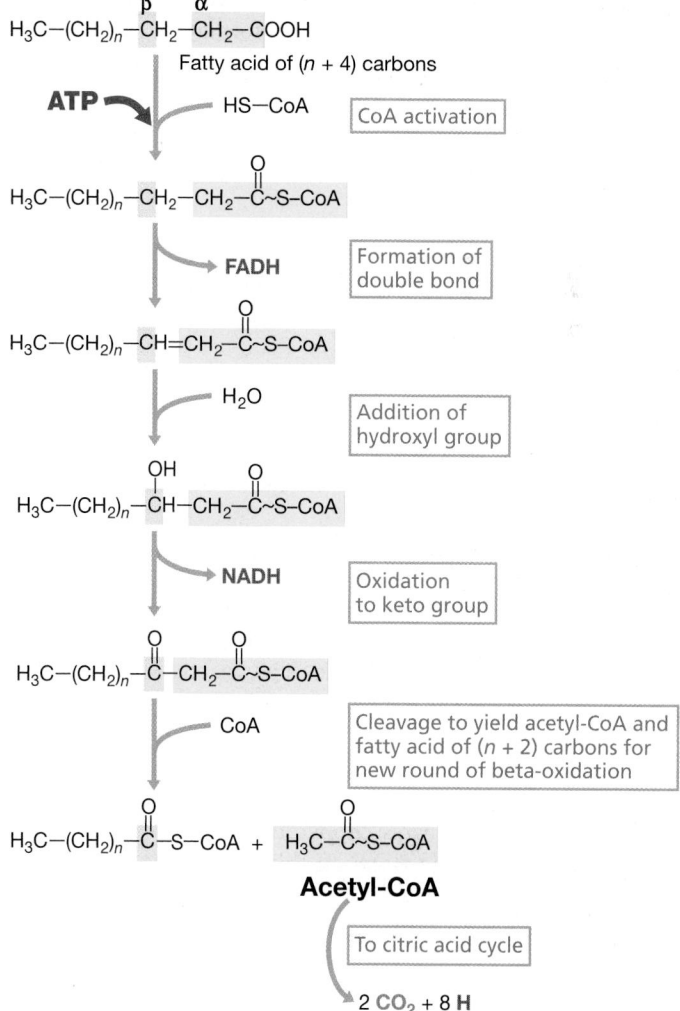

Figure 14.42 Beta-oxidation. Beta-oxidation of a fatty acid leading to the successive formation of acetyl-CoA.

fatty acids in a cell have an even number of carbon atoms, and complete oxidation yields acetyl-CoA. If odd-chain or branched-chain fatty acids are catabolized, propionyl-CoA or a branched-chain fatty acid–CoA remains after beta-oxidation, and these are either further metabolized to acetyl-CoA by ancillary reactions or excreted from the cell. The acetyl-CoA formed is then oxidized by the citric acid cycle or is converted to hexose and other cell constituents via the glyoxylate cycle (Figure 14.39).

Because they are highly reduced, fatty acids are excellent electron donors. For example, the oxidation of the 16-carbon saturated fatty acid palmitic acid can in theory generate 129 ATP molecules. These include oxidative phosphorylation from electrons generated during the formation of acetyl-CoA from beta-oxidations and from oxidation of the acetyl-CoA units themselves through the citric acid cycle.

MiniQuiz

- What are phospholipases and what do they do?
- How many electrons are released for every acetyl-CoA produced by beta oxidation of a fatty acid? For every acetyl-CoA oxidized to CO_2?

Big Ideas

14.1
In the absence of external electron acceptors, organic compounds can be catabolized anaerobically only by fermentation. A requirement for most fermentations is formation of an energy-rich organic compound that can yield ATP by substrate-level phosphorylation. Redox balance must also be achieved in fermentations, and H_2 production is a key means of disposing of excess electrons.

14.2
The lactic acid fermentation is carried out by homofermentative and heterofermentative species. The mixed-acid fermentation results in acids plus neutral products (ethanol, butanediol), depending on the organism.

14.3
Clostridia ferment sugars, amino acids, and other organic compounds. *Propionibacterium* produces propionate and acetate in a secondary fermentation of lactate.

14.4
The energy physiology of *Propionigenium*, *Oxalobacter*, and *Malonomonas* is linked to decarboxylation reactions that pump Na^+ or H^+ across the membrane. The reactions catalyzed by these organisms yield insufficient energy to make ATP by substrate-level phosphorylation.

14.5
In syntrophy two organisms cooperate to degrade a compound that neither can degrade alone. In this process H_2 produced by one organism is consumed by the partner. H_2 consumption affects the energetics of the reaction carried out by the H_2 producer, allowing it to make ATP where it otherwise could not.

14.6
Although O_2 is the most widely used electron acceptor in energy-yielding metabolism, certain other compounds can be used as electron acceptors. Anaerobic respiration yields less energy than aerobic respiration but can proceed in environments where O_2 is absent.

14.7
Nitrate is a common electron acceptor in anaerobic respiration. Nitrate reduction is catalyzed by the enzyme nitrate reductase, reducing NO_3^- to NO_2^-. Many bacteria that use NO_3^- in anaerobic respiration reduce it past NO_2^- to produce gaseous nitrogen compounds (denitrification).

14.8
Sulfate-reducing bacteria reduce SO_4^{2-} to H_2S. This process requires activation of SO_4^{2-} by ATP to form adenosine phosphosulfate (APS) and reduction by H_2 or organic electron donors. Disproportionation is an additional energy-yielding strategy for certain species. Some organisms, such as *Desulfuromonas*, cannot reduce SO_4^{2-} but produce H_2S from the reduction of S^0.

14.9
Acetogens are anaerobes that reduce CO_2 to acetate, usually with H_2 as electron donor. The mechanism of acetate formation is the acetyl-CoA pathway, a pathway widely distributed in obligate anaerobes for either autotrophic purposes or acetate catabolism.

14.10
Methanogenesis is the production of CH_4 from $CO_2 + H_2$ or from acetate or methanol by strictly anaerobic methanogenic *Archaea*. Several unique coenzymes are required for methanogenesis, and energy conservation is linked to either a proton or a sodium motive force.

14.11
The hyperthermophile *Pyrococcus furiosus* ferments glucose in an unusual fashion, reducing protons in an anaerobic respiration linked to ATPase activity.

14.12

Besides inorganic nitrogen and sulfur compounds and CO_2, several other substances can function as electron acceptors for anaerobic respiration. These include Fe^{3+}, Mn^{4+}, fumarate, and certain organic and chlorinated organic compounds.

14.13

Hydrocarbons can be oxidized under anoxic conditions, but oxygen must first be added to the molecule. This occurs by the addition of fumarate. Aromatic compounds are catabolized anaerobically by ring reduction and cleavage to form intermediates that can be catabolized in the citric acid cycle. Methane can be oxidized under anoxic conditions by consortia containing sulfate-reducing or denitrifying bacteria and methanotrophic *Archaea*.

14.14

In addition to its role as an electron acceptor, O_2 can also be a substrate; enzymes called oxygenases introduce atoms of oxygen from O_2 into a biochemical compound. Aerobic hydrocarbon oxidation is widespread in nature, and oxygenase enzymes are key to these catalyses. Unlike in anaerobic aromatic catabolism, the aerobic degradation of aromatic compounds proceeds by ring oxidation.

14.15

Methanotrophy is the use of CH_4 as both carbon source and electron donor, and the enzyme methane monooxygenase is a key enzyme in the catabolism of methane. In methanotrophs C_1 units are assimilated into cell material by either the ribulose monophosphate pathway or the serine pathway.

14.16

Polysaccharides are abundant in nature and can be broken down into hexose or pentose monomers and used as sources of both carbon and electrons. Starch and cellulose are common polysaccharides. The pentose phosphate pathway is the major means for generating pentose sugars for biosynthesis.

14.17

Organic acids are typically metabolized through the citric acid cycle or the glyoxylate cycle. Isocitrate lyase and malate synthase are the key enzymes of the glyoxylate cycle.

14.18

Fats are hydrolyzed by lipases or phospholipases to fatty acids plus glycerol. The fatty acids are oxidized by beta-oxidation reactions to acetyl-CoA, which is then oxidized to CO_2 by the citric acid cycle.

Review of Key Terms

Acetogenesis energy metabolism in which acetate is produced from either H_2 plus CO_2 or from organic compounds

Acetyl-CoA pathway a pathway of autotrophic CO_2 fixation and acetate oxidation widespread in obligate anaerobes including methanogens, acetogens, and sulfate-reducing bacteria

Anaerobic respiration respiration in which some substance, such as SO_4^{2-} or NO_3^-, is used as a terminal electron acceptor instead of O_2

Anoxic oxygen-free

Denitrification anaerobic respiration in which NO_3^- or NO_2^- is reduced to nitrogen gases, primarily N_2

Fermentation anaerobic catabolism of an organic compound in which the compound serves as both an electron donor and an electron acceptor and in which ATP is usually produced by substrate-level phosphorylation

Glyoxylate cycle a series of reactions including some citric acid cycle reactions that are used for aerobic growth on C_2 or C_3 organic acids

Heterofermentative producing a mixture of products, typically lactate, ethanol, and CO_2, from the fermentation of glucose

Homofermentative producing only lactic acid from the fermentation of glucose

Hydrogenase an enzyme, widely distributed in anaerobic microorganisms, capable of oxidizing or evolving H_2

Methanogen an organism that produces methane (CH_4)

Methanogenesis the biological production of CH_4

Methanotroph an organism that can oxidize CH_4

Methylotroph an organism capable of growth on compounds containing no C—C bonds; some methylotrophs are methanotrophic

Oxygenase an enzyme that catalyzes the incorporation of oxygen from O_2 into organic or inorganic compounds

Pentose phosphate pathway a major metabolic pathway for the production and catabolism of pentoses (C_5 sugars)

Reductive dechlorination (dehalorespiration) an anaerobic respiration in which a chlorinated organic compound is used as an electron acceptor, usually with the release of Cl^-

Ribulose monophosphate pathway a reaction series in certain methylotrophs in which formaldehyde is assimilated into cell material using ribulose monophosphate as the C_1 acceptor molecule

Secondary fermentation a fermentation in which the substrates are the fermentation products of other organisms

Serine pathway a reaction series in certain methylotrophs in which CH_2O plus CO_2 are assimilated into cell material by way of the amino acid serine

Stickland reaction the fermentation of an amino acid pair

Syntrophy a process whereby two or more microorganisms cooperate to degrade a substance neither can degrade alone

Review Questions

1. Define the term substrate-level phosphorylation. How does it differ from oxidative phosphorylation? Assuming an organism is facultative, what cultural conditions dictate whether the organism obtains energy from substrate-level rather than oxidative phosphorylation (Section 14.1)?

2. Briefly describe the differences between homofermentative and heterofermentative patterns (Section 14.2).

3. Give an example of a fermentation that does not employ substrate-level phosphorylation. How is energy conserved in this fermentation (Section 14.4)?

4. What is the Stickland reaction (Section 14.3)?

5. Why is NO_3^- a better electron acceptor for anaerobic respiration than is SO_4^{2-} (Section 14.6)?

6. In *Escherichia coli*, synthesis of the enzyme nitrate reductase is repressed by O_2. On the basis of bioenergetic arguments, why do you think this repression phenomenon might have evolved (Section 14.7)?

7. Why is hydrogenase a constitutive enzyme in *Desulfovibrio* (Section 14.8)?

8. Compare and contrast acetogens with methanogens in terms of (1) substrates and products of their energy metabolism, (2) ability to use organic compounds as electron donors in energy metabolism, and (3) phylogeny (Sections 14.9 and 14.10).

9. Why can it be said that in glycolysis in *Pyrococcus furiosus*, both fermentation and anaerobic respiration are occurring at the same time (Section 14.11)?

10. Compare and contrast ferric iron reduction with reductive dechlorination in terms of (1) product of the reduction and (2) environmental significance (Section 14.12).

11. How do denitrifying and sulfate-reducing bacteria degrade hydrocarbons without the participation of oxygenase enzymes (Sections 14.13–14.14)?

12. What are the functions of oxygenases? Are they important? Why (Section 14.14)?

13. Distinguish between the terms methylotrophy and methanotrophy (Section 14.15).

14. Compare and contrast the conversion of cellulose and intracellular starch to glucose units. What enzymes are involved and which process is the more energy efficient (Section 14.16)?

15. What is the major function of the glyoxalate cycle (Section 14.17)?

16. What is the product of the beta-oxidation of a fatty acid? How is this product oxidized to CO_2 (Section 14.18)?

Application Questions

1. When methane is made from CO_2 (plus H_2) or from methanol (in the absence of H_2), various steps in the pathway shown in Figures 14.20 and 14.21 are used. Compare and contrast methanogenesis from these two substrates and discuss why they must be metabolized in opposite directions.

2. Although dextran is a glucose polymer, glucose cannot be used to make dextran. Explain. How is dextran synthesis important in oral hygiene (⮌ Section 27.3)?

3. A fatty acid such as butyrate cannot be fermented in pure culture, although its anaerobic catabolism under other conditions occurs readily. How do these conditions differ, and why does the latter allow for butyrate catabolism? How then can butyrate be fermented in mixed culture?

Need more practice? Test your understanding with Quantitative Questions; access additional study tools including tutorials, animations, and videos; and then test your knowledge with chapter quizzes and practice tests at **www.microbiologyplace.com**.

15

Commercial Products and Biotechnology

The common baker's yeast is an important tool for many of the commercial processes of both industrial microbiology and biotechnology.

Many commercial products are produced on a large scale by microorganisms, and this is the field of **industrial microbiology**. These products include antibiotics, of course, but also a wide variety of other products. A common thread that unites these products is the *scale* of their production, which is usually very large, and the fact that they sell for a relatively low price. The products typically originate from enhancements of metabolic reactions that the microorganisms were already capable of carrying out, with the main goal being the *overproduction* of the product of interest.

Industrial microbiology contrasts with **biotechnology**, in which microorganisms are altered by **genetic engineering** to produce substances they would otherwise not be able to produce, for example, human hormones such as insulin. In addition, products of the biotech industry are typically made in relatively small amounts and have high intrinsic value. Thus while penicillin is produced by the ton, insulin is produced by the kilogram. In this chapter we see how both industrial microbiology and biotechnology are done and describe a few common products of each commercial enterprise.

I Putting Microorganisms to Work

Humans have been putting microorganisms to work for thousands of years. In the first half of this chapter, our discussion of industrial microbiology touches on the earliest human uses, which are still important today. In the second half, we explore the most recent uses, achieved through genetic engineering.

15.1 Industrial Products and the Microorganisms That Make Them

Major products of industrial microbiology include the microbial cells themselves—for example, yeast cultivated for food, baking, or brewing, and substances produced by microbial cells. Examples of substances produced by cells include enzymes, antibiotics, amino acids, vitamins, other food additives, commodity chemicals, and alcoholic beverages (**Table 15.1**).

The major organisms used in industrial microbiology are fungi (yeasts and molds) (Sections 20.13–20.18) and certain prokaryotes, in particular species of the genus *Streptomyces* (Section 18.6). Industrial microorganisms can be thought of as metabolic specialists, capable of synthesizing one or more products in high yield. Industrial microbiologists often use classical genetic methods to select for high-yielding mutant strains; their

goal is to increase the *yield* of the product to the point of being economically profitable. The genetics of the producing organism needs to be well understood. After selection, the metabolic behavior of the production strain may be far removed from that of the original wild-type strain.

A microorganism used in an industrial process must have other features in addition to being able to produce the substance of interest in high yield. First and foremost, the organism must be capable of growth and product formation in large-scale culture. Moreover, it should produce spores or some other reproductive cell so that it can be easily inoculated into the large vessels used to grow the producing organism on an industrial scale. It must also grow rapidly and produce the desired product in a relatively short period of time.

An industrially useful organism must also be able to grow in a liquid culture medium obtainable in bulk quantities at a low price. Many industrial microbiological processes use waste carbon from other industries as major or supplemental ingredients for large-scale culture media. These include *corn steep liquor* (a product of the corn wet-milling industry that is rich in nitrogen and growth factors) and *whey* (a waste liquid of the dairy industry containing lactose and minerals).

An industrial microorganism should not be pathogenic, especially to humans or economically important animals or plants. Because of the high cell densities in industrial microbial processes and the virtual impossibility of avoiding contamination of the environment outside the growth vessel, a pathogen would present potentially disastrous problems.

Finally, an industrial microorganism should be amenable to genetic analysis because the yields necessary to make an industrial process profitable typically demand the selection of high-yielding mutant derivatives of the original wild-type organism. Thus, an organism that can be genetically manipulated is a clear advantage for any potential industrial process.

MiniQuiz
- List three important products of industrial microbiology.
- List two desirable properties of an industrial microorganism.

15.2 Production and Scale

In Section 5.7 we considered microbial growth and described the various stages: *lag, exponential,* and *stationary*. Here we describe microbial growth and product formation in an industrial context. There are two types of microbial metabolites of interest to industrial microbiology, primary and secondary. A **primary metabolite** forms during the exponential growth phase of the microorganism. By contrast, a **secondary metabolite** forms near the end of growth, frequently at, near, or in the stationary phase of growth (**Figure 15.1**).

A typical primary metabolite is alcohol. Ethyl alcohol (ethanol) is a product of the fermentative metabolism of yeast and certain bacteria (Section 4.8) and is formed as part of energy metabolism. Because organisms can grow only if they produce energy, ethanol forms in parallel with growth (Figure 15.1a). By contrast, secondary metabolites not coupled directly to growth are some of the

Table 15.1 *Major products of industrial microbiology*

Product	Example
Antibiotics	Penicillin, tetracycline
Enzymes	Glucose isomerase, laundry proteases and lipases
Food additives	Vitamins, amino acids
Chemicals	Biofuels (alcohol and biodiesel), citric acid
Alcoholic beverages	Beer, wine, distilled spirits

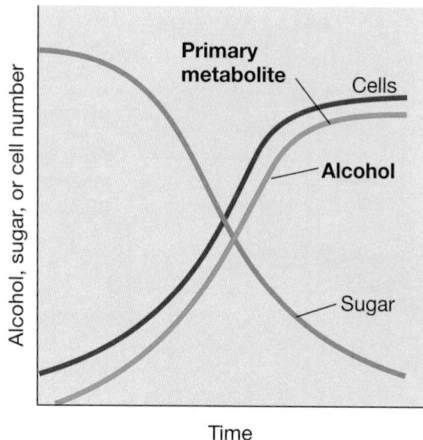

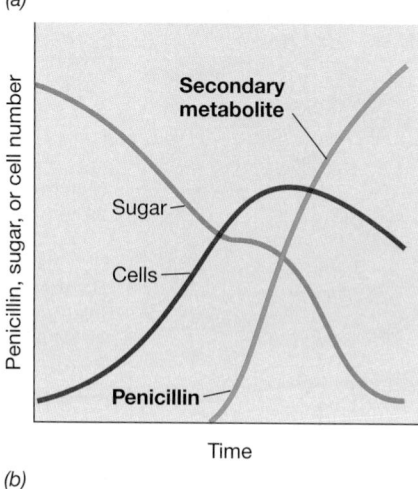

Figure 15.1 **Contrast between production of primary and secondary metabolites.** *(a)* Formation of alcohol by yeast—an example of a primary metabolite. *(b)* Penicillin production by the mold *Penicillium chrysogenum*—an example of a secondary metabolite. Note that penicillin is not made until after the exponential phase.

most complex and important metabolites of industrial interest (Figure 15.1*b*). Secondary metabolites typically share a number of characteristics. First, they are nonessential for growth and reproduction and their formation is highly dependent on growth conditions. Second, they are often produced as a group of closely related compounds and are often overproduced, sometimes in huge amounts. And finally, many secondary metabolites are the products of spore-forming microorganisms and production is linked to the sporulation process itself. Virtually all antibiotics, for example, are produced by either fungi or spore-forming prokaryotes.

Fermentors and the Characteristics of Large-Scale Fermentations

The vessel in which an industrial microbiology process is carried out is called a **fermentor**. In industrial microbiology, the term **fermentation** refers to *any* large-scale microbial process, whether or not it is, biochemically speaking, a fermentation. The size of fermentors varies from the small 5- to 10-liter laboratory scale to the enormous 500,000-liter industrial scale (**Figure 15.2**). The size of the fermentor used depends on the process and how

Table 15.2 *Fermentor sizes for various industrial fermentations*

Size of fermentor (liters)	Product
1,000–20,000	Diagnostic enzymes, substances for molecular biology
40,000–80,000	Some enzymes, antibiotics
100,000–150,000	Penicillin, aminoglycoside antibiotics, proteases, amylases, steroid transformations, amino acids, wine, beer
200,000–500,000	Amino acids (glutamic acid), wine, beer

it is operated. A summary of fermentor sizes for some common microbial fermentations is given in **Table 15.2**.

Large-scale industrial fermentors are almost always constructed of stainless steel. Such a fermentor is essentially a large cylinder, closed at the top and bottom, into which various pipes and valves have been fitted (Figure 15.2*b*). Because sterilization of the culture medium and removal of heat are vital for successful operation, the fermentor is fitted with an external cooling jacket through which steam (for sterilization) or water (for cooling) can be run. For very large fermentors, sufficient heat cannot be transferred through the jacket and so internal coils must be provided through which either steam (for sterilization) or cooling water (for growth) can be piped (Figure 15.2).

A critical part of the fermentor is the aeration system. With large-scale equipment, transfer of oxygen throughout the growth medium is critical, and elaborate precautions must be taken to ensure proper aeration. Oxygen is poorly soluble in water, and in a fermentor with a high density of microbial cells, there is a tremendous oxygen demand by the culture. Because of this, two different devices are used to ensure adequate aeration: an aerator, called a *sparger*, and a stirring device, called an *impeller* (Figure 15.2*b*). The sparger is typically just a series of holes through which filter-sterilized air can be passed into the fermentor. The air enters the fermentor as a series of tiny bubbles from which the oxygen passes by diffusion into the liquid. Stirring of the fermentor with an impeller (Figure 15.2*c*) accomplishes two things: It mixes the gas bubbles generated by the sparger and mixes the organisms through the liquid, ensuring that the microbial cells have uniform access to the nutrients.

During an actual production run, fermentors are monitored in real time for temperature, oxygen, pH, and the levels of key nutrients, such as ammonia and phosphate. This is done because it is often necessary to alter the conditions in the fermentor as the fermentation progresses. Computers are used to process environmental data as the fermentation proceeds and are programmed to respond by signaling for nutrient additions, increases in the rate of cooling water, impeller speed or sparger pressure, or changes in pH or other parameters, at just the right time to maintain high product yield.

Scale-Up from Laboratory to Commercial Fermentor

An important aspect of industrial microbiology is the transfer of a process from small-scale laboratory equipment to large-scale commercial equipment, a process called **scale-up**. An understanding

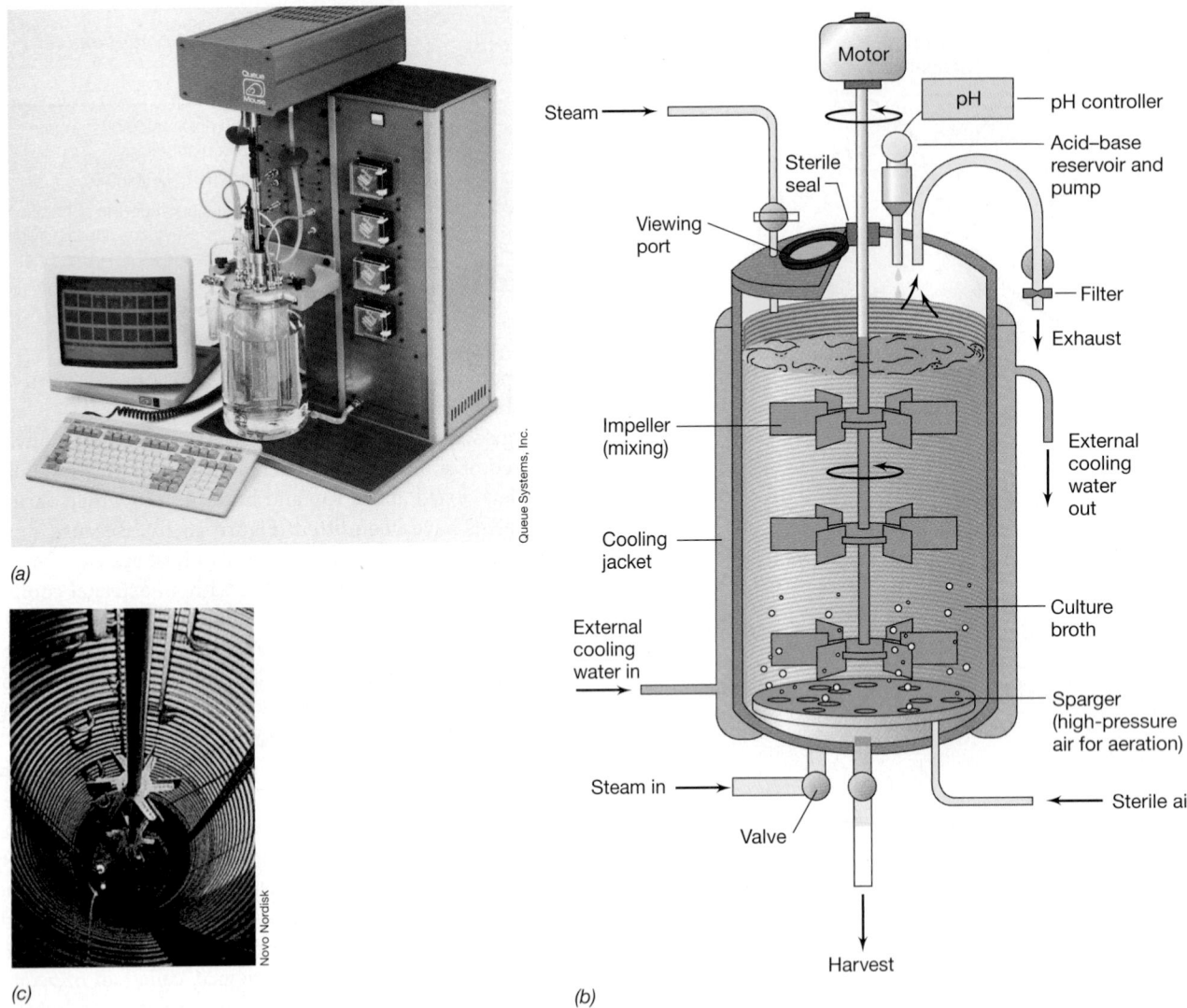

(a)

Queue Systems, Inc.

(c)

Novo Nordisk

(b)

Figure 15.2 Fermentors. *(a)* A small research fermentor with a volume of 5 liters. *(b)* Diagram of an industrial fermentor, illustrating construction and facilities for aeration and process control. *(c)* The inside of an industrial fermentor, showing the impeller and internal heating and cooling coils.

of scale-up is important because industrial processes rarely behave the same way in large-scale fermentors as in small-scale laboratory equipment (**Figure 15.3**). Many scale-up challenges arise from problems with aeration and mixing. Oxygen transfer is much more difficult to achieve in large fermentors than in small fermentors because the rich culture media used in industrial fermentations support high cell densities, and this leads to high oxygen demand. If oxygen levels become limiting, even for a short period, the culture may reduce—or even shut down—product formation.

In the development of an industrial process, everything begins in the laboratory flask. From here, a promising process is scaled-up to the laboratory fermentor, a small vessel, generally made of glass and 1 to 10 liters in size (Figures 15.2*a* and 15.3*a*). In the laboratory fermentor it is possible to test variations in culture media, temperature, pH, and other parameters, quickly and inexpensively. When these tests are successful, the process is scaled-up to the *pilot plant stage*, usually in fermentors of 300- to 3000-liter capacity. Here the conditions more closely approach

those of the actual commercial fermentor, but cost is not yet an issue. Finally, the process moves to the commercial fermentor itself, 10,000–500,000 liters in volume (Table 15.2, and Figure 15.2*b, c*). In all stages of scale-up, aeration is the key variable that is closely monitored; as scale-up proceeds, oxygen dynamics are carefully measured to determine how increases in volume affect oxygen demand in the fermentation.

MiniQuiz

- Is penicillin a primary or a secondary metabolite? How can you tell by looking at Figure 15.1?
- What are the size differences among a laboratory fermentor, a pilot plant fermentor, and a commercial fermentor? How is proper aeration ensured in a large-scale fermentation?
- What parameters in an industrial fermentation are typically monitored and why would adjustments need to be made in real time by automated systems?

(a)

(b)

Figure 15.3 Research and production fermentors. *(a)* A bank of small research fermentors used in process development. The fermentors are the glass vessels with the stainless steel tops. The small plastic bottles collect overflow. *(b)* A large bank of outdoor industrial-scale fermentors (each 240 m^3) used in commercial production of alcohol in Japan.

II Drugs, Other Chemicals, and Enzymes

We now consider some products of industrial microbiology, beginning with antibiotics and continuing with amino acids, vitamins, and enzymes. Of the microbial products manufactured commercially, the most important for the health industry are the antibiotics. Antibiotic production is a huge industry worldwide and one in which many important aspects of large-scale microbial culture were perfected.

15.3 Antibiotics: Isolation, Yield, and Purification

Antibiotics are substances produced by microorganisms that kill or inhibit the growth of other microorganisms and are typical secondary metabolites (Section 15.2). Most antibiotics used in human and veterinary medicine are produced by filamentous fungi or bacteria of the *Actinobacteria* group (Section 18.6). **Table 15.3** lists the most important antibiotics produced by large-scale industrial fermentations today.

Isolation of New Antibiotics

Modern drug discovery relies heavily on computer modeling of drug–target interactions (Section 26.13). However, in the past, and to a more limited extent today, laboratory screening programs are the route to discovery of new antibiotics. In this approach, possible antibiotic-producing microorganisms are obtained from nature in pure culture and are then tested for antibiotic production by assaying for diffusible materials that inhibit the growth of test bacteria (**Figure 15.4**). The test bacteria are selected to be either representative of or related to bacterial pathogens against which the antibiotics would actually be used.

Antibiotic production can be assayed by the *cross-streak method* (Figure 15.4*b*). Those isolates that show evidence of antibiotic production are then studied further to determine if the

Table 15.3 *Some antibiotics produced commercially*[a]

Antibiotic	Producing microorganism[b]
Bacitracin	*Bacillus licheniformis* (EFB)
Cephalosporin	*Cephalosporium* spp. (F)
Cycloheximide	*Streptomyces griseus* (A)
Cycloserine	*Streptomyces orchidaceus* (A)
Erythromycin	*Streptomyces erythreus* (A)
Griseofulvin	*Penicillium griseofulvum* (F)
Kanamycin	*Streptomyces kanamyceticus* (A)
Lincomycin	*Streptomyces lincolnensis* (A)
Neomycin	*Streptomyces fradiae* (A)
Nystatin	*Streptomyces noursei* (A)
Penicillin	*Penicillium chrysogenum* (F)
Polymyxin B	*Bacillus polymyxa* (EFB)
Streptomycin	*Streptomyces griseus* (A)
Tetracycline	*Streptomyces rimosus* (A)

[a]See Chapter 26 for structures and more discussion of these antibiotics.
[b]EFB, endospore-forming bacterium; F, fungus; A, actinomycete.

I. Isolation

II. Testing Activity Spectrum

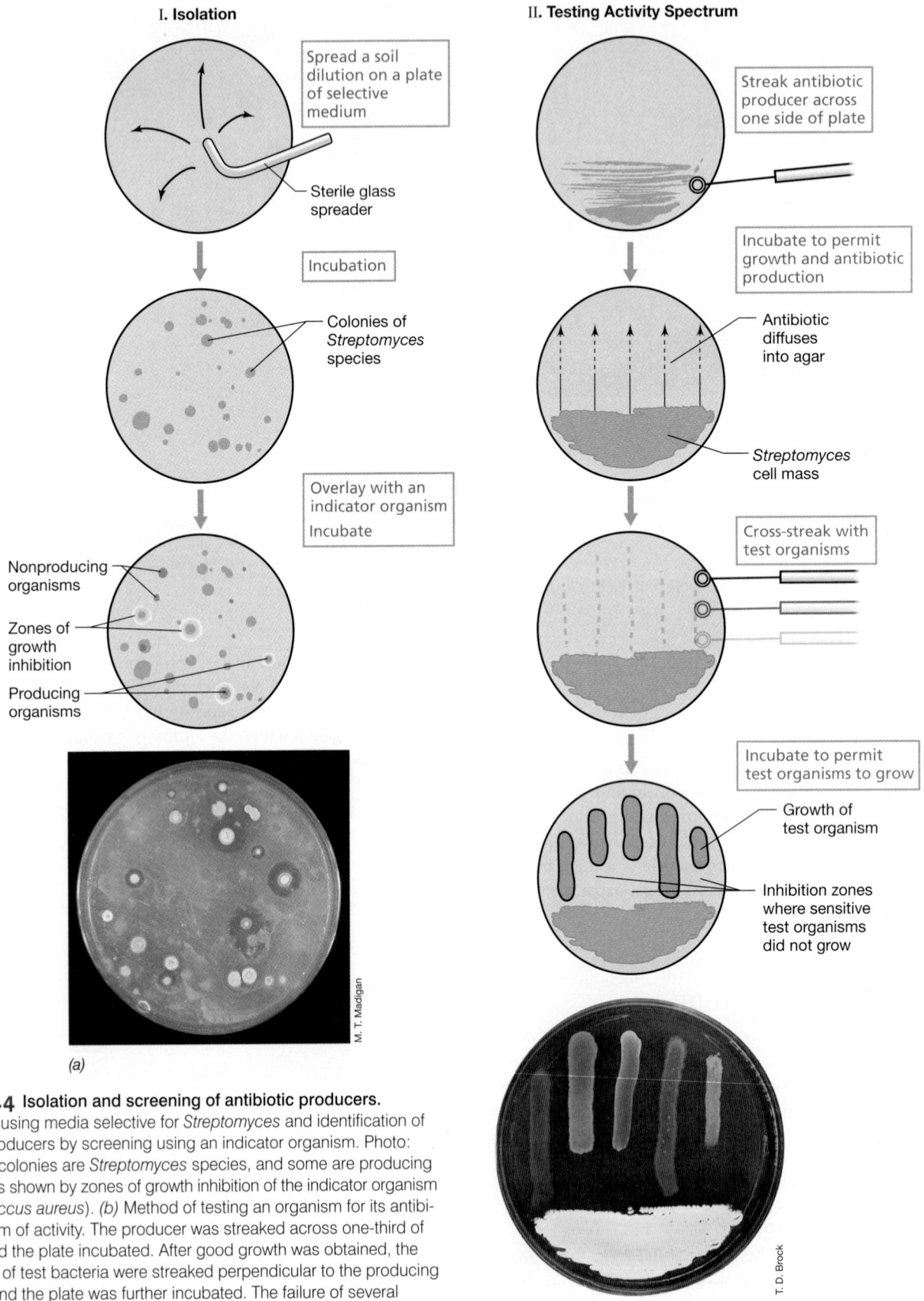

Spread a soil dilution on a plate of selective medium

Sterile glass spreader

Incubation

Colonies of *Streptomyces* species

Overlay with an indicator organism
Incubate

Nonproducing organisms

Zones of growth inhibition

Producing organisms

(a)

Streak antibiotic producer across one side of plate

Incubate to permit growth and antibiotic production

Antibiotic diffuses into agar

Streptomyces cell mass

Cross-streak with test organisms

Incubate to permit test organisms to grow

Growth of test organism

Inhibition zones where sensitive test organisms did not grow

(b)

M. T. Madigan

T. D. Brock

Figure 15.4 Isolation and screening of antibiotic producers.
(a) Isolation using media selective for *Streptomyces* and identification of antibiotic producers by screening using an indicator organism. Photo: Most of the colonies are *Streptomyces* species, and some are producing antibiotics as shown by zones of growth inhibition of the indicator organism (*Staphylococcus aureus*). *(b)* Method of testing an organism for its antibiotic spectrum of activity. The producer was streaked across one-third of the plate and the plate incubated. After good growth was obtained, the five species of test bacteria were streaked perpendicular to the producing organism, and the plate was further incubated. The failure of several species to grow near the producing organism indicates that it produced an antibiotic active against these bacteria. Photo: Test organisms streaked vertically (left to right) include *Escherichia coli*, *Bacillus subtilis*, *S. aureus*, *Klebsiella pneumoniae*, *Mycobacterium smegmatis*.

compounds they produce are new. Most of the isolates obtained produce known antibiotics, but when a new antibiotic is discovered, it is produced in sufficient amounts for structural analyses and then tested for toxicity and therapeutic activity in animals. Unfortunately, most new antibiotics fail these tests. However, a few prove to be medically useful and go on to be produced commercially. The time and costs in developing a new antibiotic, from discovery to clinical usage, average 15 years and 1 billion ($US). This includes many phases of clinical trials, which alone can take several years to complete, analyze, and submit for United States Food and Drug Administration (FDA) approval.

Yield and Purification

Rarely do antibiotic-producing strains just isolated from nature produce an antibiotic at sufficiently high concentration that commercial production can begin immediately. So one of the major tasks of the industrial microbiologist is to isolate *high-yielding strains*. Strain selection may involve mutagenizing the wild-type organism to obtain mutant derivatives that are so altered that they overproduce the antibiotic of interest. Product yield is a central issue with virtually all pharmaceuticals, and even after commercial production of an antibiotic has begun, research often continues to obtain higher-yielding strains or a more efficient process.

The next challenge is to purify the antibiotic specifically and efficiently, and elaborate methods for extraction and purification of the antibiotic are often necessary. The goal is to eventually obtain a crystalline product of high purity. Depending on the process, further purification steps may be necessary to remove traces of microbial cells or cell products if they co-purify with the antibiotic. These substances, called *pyrogens*, can cause severe or even fatal reactions in patients treated with the drug, and thus the purified product ready to ship must be pyrogen-free. www.microbiologyplace.com **Online Tutorial 15.1: Isolation and Screening of Antibiotic Producers**

MiniQuiz

- What are the major groups of microorganisms that produce antibiotics?
- What is meant by the word "screening" in the context of finding new antibiotics?

15.4 Industrial Production of Penicillins and Tetracyclines

Once a new antibiotic has been characterized and proven medically effective and nontoxic in tests on experimental animals, it is ready for clinical trials on humans. If the new drug proves clinically effective and passes toxicity and other tests, it is given FDA approval and is ready to be produced commercially. We focus here on the penicillins and tetracyclines, antibiotics that are produced by the ton for medical and veterinary use.

β-Lactam Antibiotics: Penicillin and Its Relatives

The penicillins, a class of **β-lactam antibiotics** characterized by the β-lactam ring (**Figure 15.5**), are produced by fungi of the genera *Penicillium* and *Aspergillus* and by certain prokaryotes. Commercial penicillin is produced in the United States using high-yielding strains of the mold *Penicillium chrysogenum*. Other important β-lactam antibiotics include the cephalosporins, produced commercially by the mold *Cephalosporium acremonium*. The penicillins and cephalosporins are covered in detail in Section 26.8.

Clinically useful penicillins are of several different types. The parent structure of all penicillins is the compound 6-aminopenicillanic acid (6-APA), which consists of a thiazolidine ring with a condensed β-lactam ring (Figure 15.5). The 6-APA carries a variable side chain in position 6. If the penicillin fermentation is carried out without addition of side-chain precursors, the **natural penicillins**, a group of related compounds, are produced. However, the final product can be specified by adding a side-chain precursor to the culture medium so that only one type of penicillin is produced in greatest amount. The product formed under these conditions is called **biosynthetic penicillin** (Figure 15.5).

To produce the most clinically useful penicillins, those with activity against gram-negative bacteria, researchers combined fermentation and chemical approaches, leading to the production of **semisynthetic penicillins**. To produce semisynthetic penicillins, a natural penicillin is treated to yield 6-APA that is then chemically modified by the addition of a side chain (Figure 15.5). Semisynthetic penicillins have many significant clinical advantages. These include in particular the fact that they are **broad-spectrum antibiotics**, meaning they are useful against a

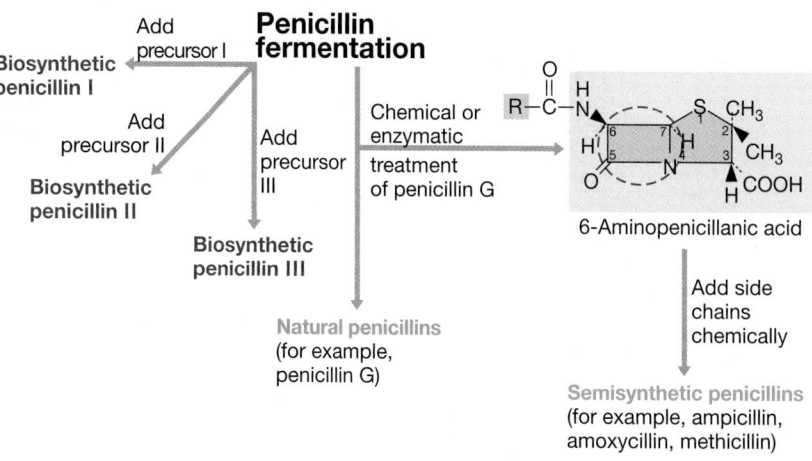

Figure 15.5 Industrial production of penicillins. The β-lactam ring is circled in red. The normal fermentation leads to the natural penicillins. If specific precursors are added during the fermentation, various biosynthetic penicillins are formed. Semisynthetic penicillins are produced by chemically adding a specific side chain to the 6-aminopenicillanic acid nucleus on the "R" group shown in purple. Semisynthetic penicillins are the most widely prescribed of all the penicillins today, primarily because of their broad spectrum of activity and ability to be taken orally.

UNIT 5

wide variety of bacterial pathogens, both gram-negative and gram-positive, and that most of them can be taken orally and thus do not require injection. The widely prescribed drug *ampicillin* is a good example of a semisynthetic penicillin. For these reasons, semisynthetic penicillins make up the bulk of the penicillin market today.

Production of Penicillins

Penicillin G is produced in fermentors of 40,000–200,000 liters. Penicillin production is a highly aerobic process, and efficient aeration is critical. Penicillin is a typical secondary metabolite. During the growth phase, very little penicillin is produced, but once the carbon source has been nearly exhausted, the penicillin production phase begins. By supplying additional carbon and nitrogen at just the right times, the production phase can be extended for several days (**Figure 15.6**).

A major ingredient of penicillin production media is corn steep liquor. This substance supplies the fungus with nitrogen and growth factors. High levels of glucose repress penicillin production, but high levels of lactose do not, so lactose (from whey) is added to the corn steep liquor in large amounts as a carbon source. As the lactose becomes limiting and cell densities in the fermentor become very high, "feedings" with low levels of glucose maximize penicillin yield (Figure 15.6). At the end of the production phase, the cells are removed by filtration and the pH is made acidic. The penicillin can then be extracted and concentrated into an organic solvent and, finally, crystallized.

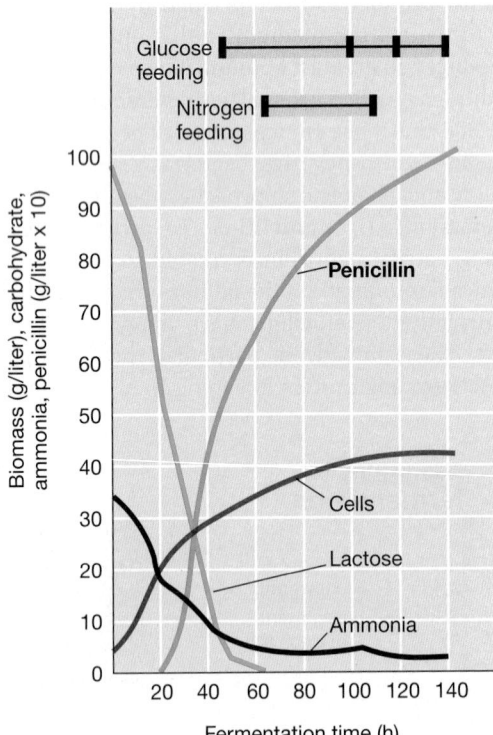

Figure 15.6 Kinetics of the penicillin fermentation with *Penicillium chrysogenum.* Note that penicillin is produced as cells are entering the stationary phase, when most of the carbon and nitrogen has been exhausted. Nutrient "feedings" keep penicillin production high over several days.

Production of Tetracyclines

The biosynthesis of **tetracyclines**, antibiotics containing the four-membered naphthacene ring, requires a large number of enzymatic steps. In the biosynthesis of chlortetracycline (**Figure 15.7**) for example, there are more than 72 intermediates. Genetic studies of *Streptomyces aureofaciens*, the producing organism in the chlortetracycline fermentation, have shown that a total of more than 300 genes are involved! With such a large number of genes, regulation of biosynthesis of this antibiotic is obviously quite complex. However, some key regulatory signals are known and are accounted for in the production scheme.

Chlortetracycline synthesis is repressed by both glucose and phosphate. Phosphate repression is especially significant, and so the medium used in commercial production contains relatively low phosphate concentrations. Figure 15.7 shows a tetracycline production scheme and the various stages in scale-up leading to the commercial fermentor. As in penicillin production, corn steep liquor is used, but sucrose rather than lactose is used as a carbon source. Glucose is avoided because glucose strongly represses antibiotic production through the transcriptional control mechanism known as catabolite repression (⮌ Section 8.5).

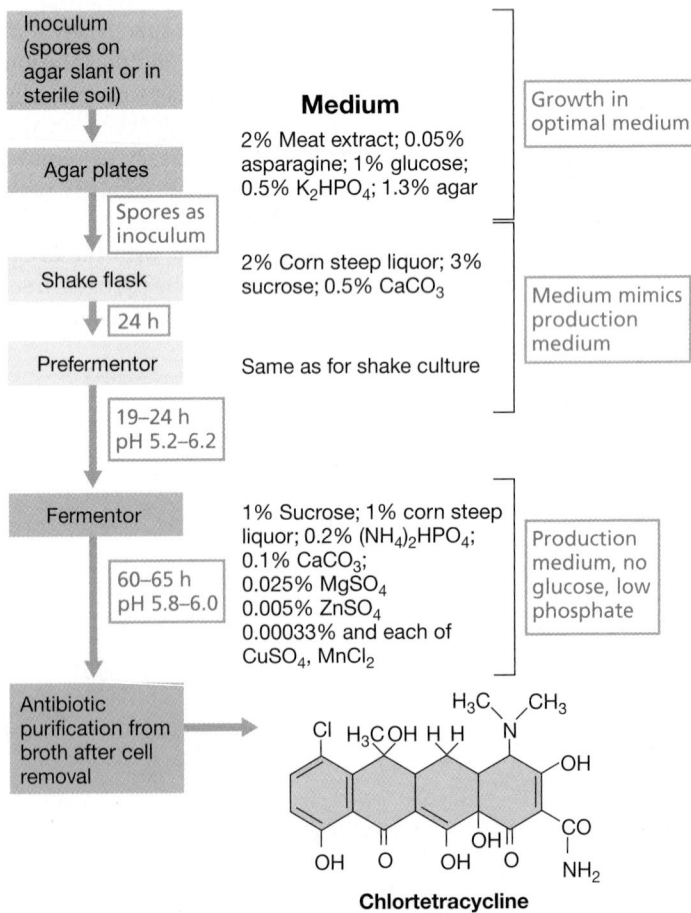

Figure 15.7 Production scheme for chlortetracycline using *Streptomyces aureofaciens.* The structure of chlortetracycline is shown on the bottom right. Glucose is used to grow the inoculum, but not for commercial production.

MiniQuiz

- What chemical structure is common to both the penicillins and the cephalosporins?

- In penicillin production, what is meant by the term semisynthetic? Biosynthetic?

- Why would corn syrup not be useful in the production of tetracyclines?

15.5 Vitamins and Amino Acids

Vitamins and amino acids are nutrients that are used in the pharmaceutical, nutraceutical (nutritional supplements), and food industries. Of these, several are produced on an industrial scale by microorganisms.

Vitamins

Vitamins are used as supplements for human food and animal feeds, and production of vitamins is second only to that of antibiotics in total sales of pharmaceuticals. Most vitamins are made commercially by chemical synthesis. However, a few are too complicated to be synthesized inexpensively but can be made in sufficient quantities relatively easily by microbial processes. Vitamin B_{12} and riboflavin are the most important of these vitamins. World production of B_{12} is on the order of 10,000 tons per year and of riboflavin about 1,000 tons per year.

Vitamin B_{12} (**Figure 15.8**) is synthesized in nature exclusively by microorganisms but is required as a growth factor by all animals. As a coenzyme, vitamin B_{12} plays an important role in microorganisms and animals in certain methyl transfers and related processes. In humans, a major deficiency of vitamin B_{12} leads to a debilitating condition called *pernicious anemia*, characterized by low production of red blood cells and nervous system disorders. For industrial production of vitamin B_{12}, microbial strains are employed that have been specifically selected for their high yields of the vitamin. Species of the bacteria *Propionibacterium* and *Pseudomonas* are the main commercial producers, especially *Propionibacterium freudenreichii*. The metal cobalt is present in vitamin B_{12} (Figure 15.8a), and commercial yields of the vitamin are greatly increased by addition of small amounts of cobalt to the culture medium.

Riboflavin (Figure 15.8b) is the parent compound of the flavins FAD and FMN, coenzymes that play important roles in enzymes for oxidation–reduction reactions (Section 4.9). Riboflavin is synthesized by many microorganisms, including bacteria, yeasts, and fungi. The fungus *Ashbya gossypii* naturally produces several grams per liter of riboflavin and is therefore the main organism used in microbial production. However, despite this good yield, there is significant economic competition between the microbiological process and strictly chemical synthesis.

Amino Acids

Amino acids are extensively used in the food and animal husbandry industries as additives, in the nutraceutical industry as nutritional supplements, and as starting materials in the chemical industry (**Table 15.4**). The most important commercial amino acid is glutamic acid, which is used as a flavor enhancer

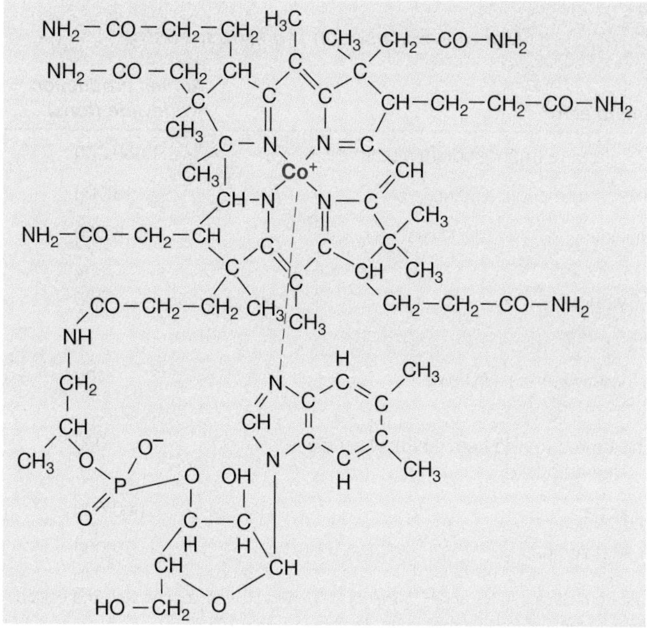

(a) **B₁₂**

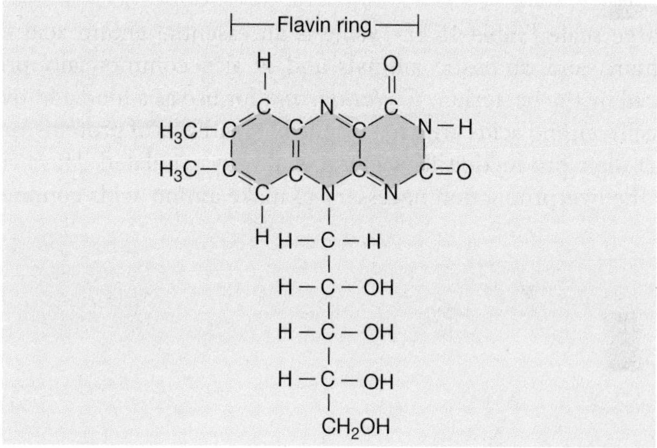

(b) **Riboflavin**

Figure 15.8 Vitamins produced by microorganisms on an industrial scale. (a) Vitamin B_{12}. Shown is the structure of cobalamin; note the central cobalt atom. The actual coenzyme form of vitamin B_{12} contains a deoxyadenosyl group attached to Co above the plane of the ring. (b) Riboflavin (vitamin B_2).

(monosodium glutamate, MSG). Over one million tons of this amino acid are produced annually by the gram-positive bacterium *Corynebacterium glutamicum*. To overproduce, the organism must be starved for the coenzyme biotin (Section 4.1). Biotin is important in the synthesis of fatty acids and thus the cytoplasmic membrane. Starving the organism for biotin weakens the membrane and makes it leaky and susceptible to glutamate excretion.

Two other important microbially produced amino acids, aspartic acid and phenylalanine, are used to synthesize the artificial sweetener **aspartame**, a nonnutritive and noncarbohydrate sweetener of diet soft drinks and other foods sold as low-calorie or sugar-free products. The amino acid lysine is also produced on

Table 15.4 *Amino acids used in the food industry[a]*

Amino acid[b]	Annual production worldwide (tons)	Uses	Purpose
L-Glutamate (monosodium glutamate, MSG)	1,000,000	Various foods	Flavor enhancer; meat tenderizer
L-Aspartate and L-alanine	13,000	Fruit juices	"Round off" taste
Glycine	6,000	Sweetened foods	Improves flavor; starting point for organic syntheses
L-Cysteine	700	Bread	Improves quality
		Fruit juices	Antioxidant
L-Tryptophan + L-Histidine	400	Various foods, dried milk	Antioxidant, prevent rancidity; nutritive additives
Aspartame (made from L-phenylalanine + L-aspartic acid)	7,000	Soft drinks, chewing gum, many other "sugar-free" products	Low-calorie sweetener
L-Lysine	800,000	Bread, cereal, and feed additives	Nutritive additive
DL-Methionine	70,000	Soy products, feed additives	Nutritive additive

[a]Data from Glazer, A. N., and H. Mikaido. 2007. *Microbial Biotechnology*, 2nd edition, W. H. Freeman, New York.
[b]The structures of these amino acids are shown in Figure 6.29.

a large scale (Table 15.4). Lysine is an essential amino acid for humans and domestic animals and is also commercially produced by the bacterium *C. glutamicum* for use as a food additive. In cells, amino acids are used for the biosynthesis of proteins, and thus their production in bacteria is strictly regulated. However, for the overproduction necessary to make amino acids commer-

cially from a microbial source, these regulatory mechanisms must be circumvented.

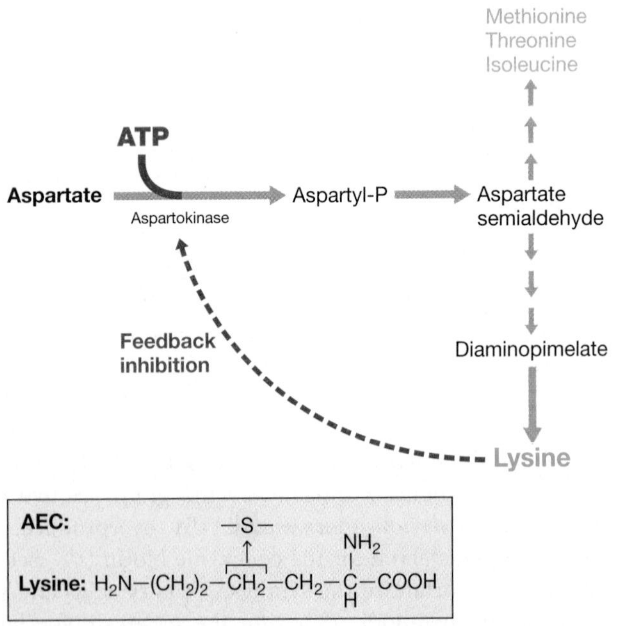

The production of lysine in *C. glutamicum* is controlled at the level of the enzyme aspartokinase; excess lysine feedback inhibits the activity of this enzyme (**Figure 15.9**; the phenomenon of feedback inhibition was described in Section 4.16). However, overproduction of lysine can be obtained by isolating mutants of *C. glutamicum* in which aspartokinase is no longer subject to feedback inhibition; this is done by isolating mutants resistant to the lysine analog *S*-aminoethylcysteine (AEC). AEC binds to the allosteric site of aspartokinase and inhibits activity of the enzyme (Figure 15.9). However, AEC-resistant mutants can be obtained easily and synthesize a modified form of aspartokinase whose allosteric site no longer recognizes AEC or lysine. In such mutants, feedback inhibition of this enzyme by lysine is nearly eliminated. For example, typical AEC-resistant mutants of *C. glutamicum* can produce over 60 g of lysine per liter in industrial fermentors, a concentration sufficiently high to make the process commercially viable. Once produced in the commercial fermentor, the amino acid must be purified and crystallized before it is ready to enter the market.

Figure 15.9 Industrial production of lysine using *Corynebacterium glutamicum*. Biochemical pathway leading from aspartate to lysine; note that lysine can feedback-inhibit activity of the enzyme aspartokinase, leading to cessation of lysine production. Shown also is the structure of lysine; the lysine analog *S*-aminoethylcysteine (AEC) is identical to lysine in structure except that a sulfur atom (S) replaces the CH₂ group shown. AEC normally inhibits growth, but AEC-resistant mutants of *C. glutamicum* have an altered allosteric site on their aspartokinase and grow and overproduce lysine because feedback inhibition no longer occurs.

MiniQuiz

• Which amino acid is commercially produced in the greatest amounts?

• Why is a mutant derivative of the bacterium *Corynebacterium glutamicum* required for commercial lysine production?

15.6 Enzymes as Industrial Products

Microorganisms produce many different enzymes, most of which are made in only small amounts and function within the cell. However, certain microbial enzymes are produced in much larger amounts and are excreted into the environment. These

Table 15.5 *Microbial enzymes and their applications*

Enzyme	Source	Application	Industry
Amylase (starch-digesting)	Fungi	Bread	Baking
	Bacteria	Starch coatings	Paper
	Fungi	Syrup and glucose manufacture	Food
	Bacteria	Cold-swelling laundry starch	Starch
	Fungi	Digestive aid	Pharmaceutical
	Bacteria	Removal of coatings (desizing)	Textile
	Bacteria	Removal of stains; detergents	Laundry
Protease (protein-digesting)	Fungi	Bread	Baking
	Bacteria	Spot removal	Dry cleaning
	Bacteria	Meat tenderizing	Meat
	Bacteria	Wound cleansing	Medicine
	Bacteria	Desizing	Textile
	Bacteria	Household detergent	Laundry
Invertase (sucrose-digesting)	Yeast	Soft-center candies	Candy
Glucose oxidase	Fungi	Glucose removal, oxygen removal	Food
		Test paper for diabetes	Pharmaceutical
Glucose isomerase	Bacteria	High-fructose corn syrup	Soft drink
Pectinase	Fungi	Pressing, clarification	Wine, fruit juice
Rennin	Fungi	Coagulation of milk	Cheese
Cellulase	Bacteria	Fabric softening, brightening; detergent	Laundry
Lipase	Fungi	Break down fat	Dairy, laundry
Lactase	Fungi	Breaks down lactose to glucose and galactose	Dairy, health foods
DNA polymerase	Bacteria, Archaea	DNA replication in polymerase chain reaction (PCR) technique (Section 6.11)	Biological research; forensics

extracellular enzymes, called **exoenzymes**, digest insoluble polymers such as cellulose, protein, lipids, and starch, and because of this, have commercial applications in the food and health industries and in the laundry and textile industries (**Table 15.5**).

Proteases, Amylases, and High-Fructose Syrup

Enzymes are produced industrially from fungi and bacteria. The microbial enzymes produced in the largest amounts on an industrial basis are the bacterial *proteases*, used as additives in laundry detergents. Most laundry detergents contain enzymes, usually proteases, but also amylases and lipases (Table 15.5). These enzymes help remove stains from food, blood, and other organic-rich substances by degrading the polymers into water-soluble components that wash away in the laundry cycle. Many laundry enzymes are isolated from alkaliphilic bacteria, organisms that grow best at alkaline pH (Section 5.15). The main producing organisms are species of *Bacillus*, such as *Bacillus licheniformis*. These enzymes, which have pH optima between 9 and 10, remain active at the alkaline pH of laundry detergent solutions.

Other important enzymes manufactured commercially are amylases and glucoamylases, which are used in the production of glucose from starch. The glucose is then converted by a second enzyme, glucose isomerase, to fructose, which is a much sweeter sugar than glucose. The final product is *high-fructose syrup* pro-

duced from glucose-rich starting materials, such as corn, wheat, or potatoes. High-fructose syrups are widely used in the food industry to sweeten soft drinks, juices, and many other products. Worldwide production of high-fructose syrups is over 10 billion kilograms per year.

Extremozymes: Enzymes with Unusual Stability

In many chapters in this book we consider prokaryotes able to grow at extremely high temperatures, the *hyperthermophiles*. These remarkable organisms can grow at such high temperatures because they synthesize heat-stable macromolecules, including enzymes. The term **extremozyme** has been coined to describe enzymes that function at some environmental extreme, such as high or low temperature or pH (**Figure 15.10**). The organisms that produce extremozymes are called *extremophiles* (Table 2.1).

Many industrial catalysts operate best at high temperatures, and so extremozymes from hyperthermophiles are widely used in both industry and research. Besides the *Taq* and *Pfu* DNA polymerases used in the polymerase chain reaction for amplifying specific DNA sequences (Section 6.11), thermostable proteases, amylases, cellulases, pullulanases (Figure 15.10b), and xylanases have been isolated and characterized from one or another species of hyperthermophile. However, it is not only thermostable enzymes that have found a market. Cold-active

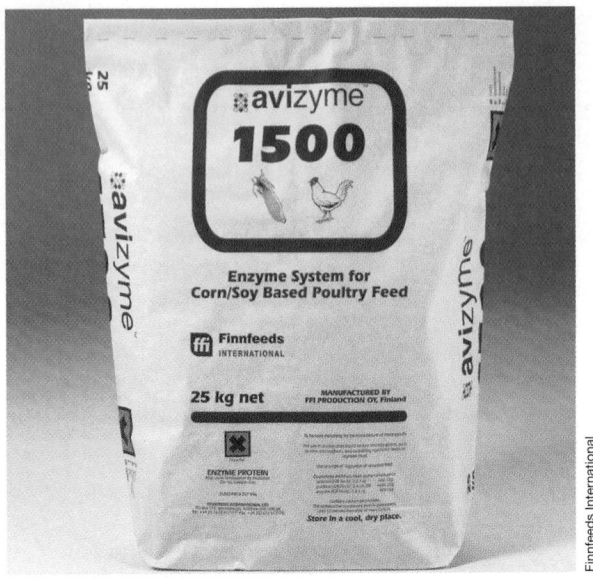

(a)

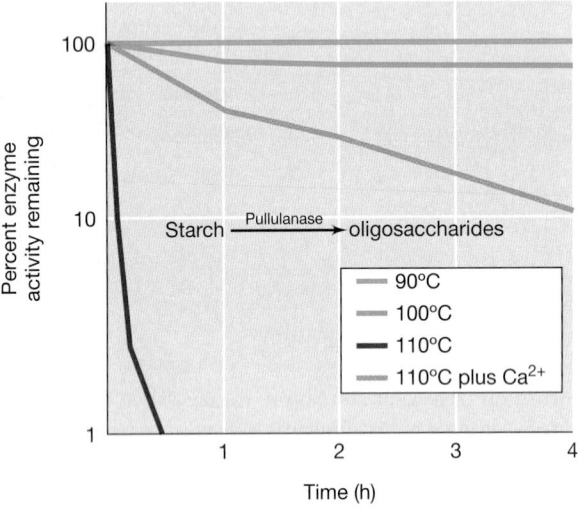

(b)

Figure 15.10 Examples of extremozymes, enzymes which function under environmentally extreme conditions. *(a)* An acid-tolerant enzyme mixture used as a feed supplement for poultry. The enzymes function in the bird's stomach to digest fibrous materials in the feed, thereby improving the nutritional value of the feed and promoting more rapid growth. *(b)* Thermostability of the enzyme pullulanase from *Pyrococcus woesei*, a hyperthermophile whose growth temperature optimum is 100°C. At 110°C the enzyme denatures, but calcium improves the heat stability of this enzyme dramatically.

enzymes (obtained from psychrophiles), enzymes that function at high salinity (obtained from halophiles), and enzymes active at high or low pH (obtained from alkaliphiles and acidophiles, respectively) (Figure 15.10*a*) have been applied commercially.

Immobilized Enzymes

For some industrial processes it is desirable to attach an enzyme to a solid surface to form an **immobilized enzyme**. Immobilization not only makes it easier to carry out the enzymatic reaction

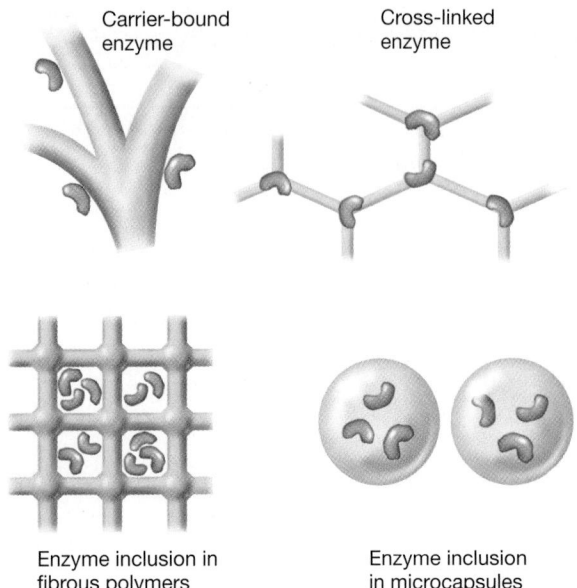

Figure 15.11 Procedures for the immobilization of enzymes. The procedure used varies with the enzyme, the product, and the production scale employed.

under large-scale continuous flow conditions, but also helps stabilize the enzyme to retard denaturation. Depending on the enzyme, the immobilized protein can remain active for up to several months.

A good example of the application of immobilized enzymes is in the starch-processing industry, mentioned previously. Starch is converted to high-fructose corn syrup by sequential treatment with amylase and glucose isomerase, but conventional treatment typically converts only about 50% of the glucose into fructose. However, this yield can be increased significantly by removing the fructose on a continuous basis and recycling the remaining glucose over an immobilized enzyme column of glucose isomerase.

Enzymes can be immobilized in three different ways (**Figure 15.11**). Enzymes can be bonded to a carrier made of cellulose, activated carbon, various minerals, or even glass beads through adsorption, ionic bonding, or covalent bonding. Enzyme molecules can also be linked to each other by chemical reaction with a cross-linking reagent such as dilute glutaraldehyde that reacts with amino acids in the enzyme and binds them together without affecting activity. And finally, enzymes can be enclosed in microcapsules, gels, semipermeable polymer membranes, or fibrous polymers such as cellulose acetate. Each of these methods for immobilizing enzymes has advantages and disadvantages, and the procedure used depends on the enzyme, the application, and the scale of the operation.

MiniQuiz

- How are enzymes of use in the laundry industry?
- What enzymes are needed to produce high-fructose corn syrup from starch?
- What is an extremozyme?

Ⅲ Alcoholic Beverages and Biofuels

Alcoholic beverages are a mainstay of human culture and have been produced on a large scale for centuries. Many different alcoholic beverages are known, with some having worldwide appeal and others more regional appeal. But all alcoholic beverages begin with a fermentation step in which some fermentable substance, typically a grain, vegetable, or fruit, is fermented by yeasts or bacteria to yield ethanol and carbon dioxide. The distinctive character of a given alcoholic beverage is the result of many factors, including natural flavors present in the fermentable substrate, chemicals other than alcohol produced during fermentation, and of course, the alcohol itself. We separate our coverage here into three sections, the first dealing with wine, the second with beer and distilled spirits, and the third with commodity alcohol and other biofuels.

15.7 Wine

Fruit juices undergo a natural fermentation by wild yeasts present in them. From these, particular strains of yeasts have been selected through the years for use in the wine industry. Wine production is a major industry worldwide and one that is growing rapidly with the influx of small specialty wineries, especially in the United States.

Wine Varieties

Most wine is made from grapes, and thus most wine is produced in parts of the world where quality grapes can be grown economically. These include the United States (**Figure 15.12**), New Zealand and Australia, South America, and many countries of the European Union—in particular, France, Spain, Italy, and Germany. Wine can also be made from many other fruits and from some nonfruit sugars, such as honey.

There are a great variety of wines, and their quality and character vary considerably. Dry wines are wines in which the sugars

(a)

(b)

(c)

(d)

Figure 15.12 Commercial wine making. *(a)* Equipment for transporting grapes to the winery for crushing. *(b)* Large tanks where the main wine fermentation takes place. *(c)* Large barrels used for aging wine in a large winery. *(d)* Smaller barrels used in a small French winery. Wine may be aged in these wooden casks for years. Red wines are almost always aged to some extent before being marketed, whereas white wines are rarely aged and can suffer in quality when aged significantly.

of the juice are almost completely fermented, whereas in sweet wines some of the sugar is left or additional sugar is added after the fermentation. A fortified wine is one to which brandy or some other alcoholic spirit is added after the fermentation; sherry and port are the best known fortified wines. A sparkling wine, such as champagne, is one in which considerable carbon dioxide (CO_2) is present, arising from a final fermentation by the yeast in the sealed bottle.

Wine Production

Wine production typically begins in the early fall with the harvesting of grapes. The grapes are crushed, and the juice, called *must*, is squeezed out. Depending on the grapes used and on how the must is prepared, either white or red wine may be produced (**Figure 15.13**). Typical varieties of white wine include Chablis, Rhine wine, sauterne, and chardonnay; typical red wines include burgundy, Chianti, claret, zinfandel, cabernet, and merlot. The yeasts that ferment wine are of two types: wild yeasts, which are present on the grapes as they are taken from the field and are transferred to the juice, and strains of the cultivated wine yeast, *Saccharomyces ellipsoideus*, which is added to the juice to begin the fermentation. Wild yeasts are less alcohol-tolerant than commercial wine yeasts and can also produce undesirable compounds affecting quality of the final product. Thus, it is the practice in most wineries to kill the wild yeasts present in the must by adding sodium metabisulfite ($Na_2S_2O_5$, labeled as "sulfites" on the bottle) at a level of 50–100 mg/l. Strains of *S. ellipsoideus* are resistant to this concentration of sulfite and are added to the must as a starter culture from a pure culture grown on sterilized grape juice.

The wine fermentation is carried out in fermentors of various sizes, from 200 to 200,000 liters, made of oak, cement, stone, or glass-lined metal (Figure 15.12). However, no matter what the construction, all fermentors must be designed so that the large amount of CO_2 produced during the fermentation can escape but air cannot enter, and this is accomplished by fitting the vessel with a special one-way valve.

Red and White Wines

A white wine is made either from white grapes or from the juice of red grapes from which the skins, containing the red coloring matter, have been removed. In the making of red wine, the skins, seeds, and pieces of stem, collectively called *pomace*, are left in during the fermentation. In addition to the color difference, red wine has a stronger flavor than white because of larger amounts of *tannins*, chemicals that are extracted into the juice from the grape skins during the fermentation.

In the production of a red wine, after about five days of fermentation, sufficient tannin and color have been extracted from the pomace that the wine can be drawn off for further fermentation in a new tank, usually for 1–2 weeks. The next step is called *racking*; the wine is separated from the sediment, which contains yeast cells and precipitate, and then stored at a lower temperature for aging, flavor development, and clarification. Clarification may be hastened by the addition of fining agents, materials such as casein, tannin, or bentonite clay that absorb particulates. Alternatively, the wine may be filtered through diatomaceous

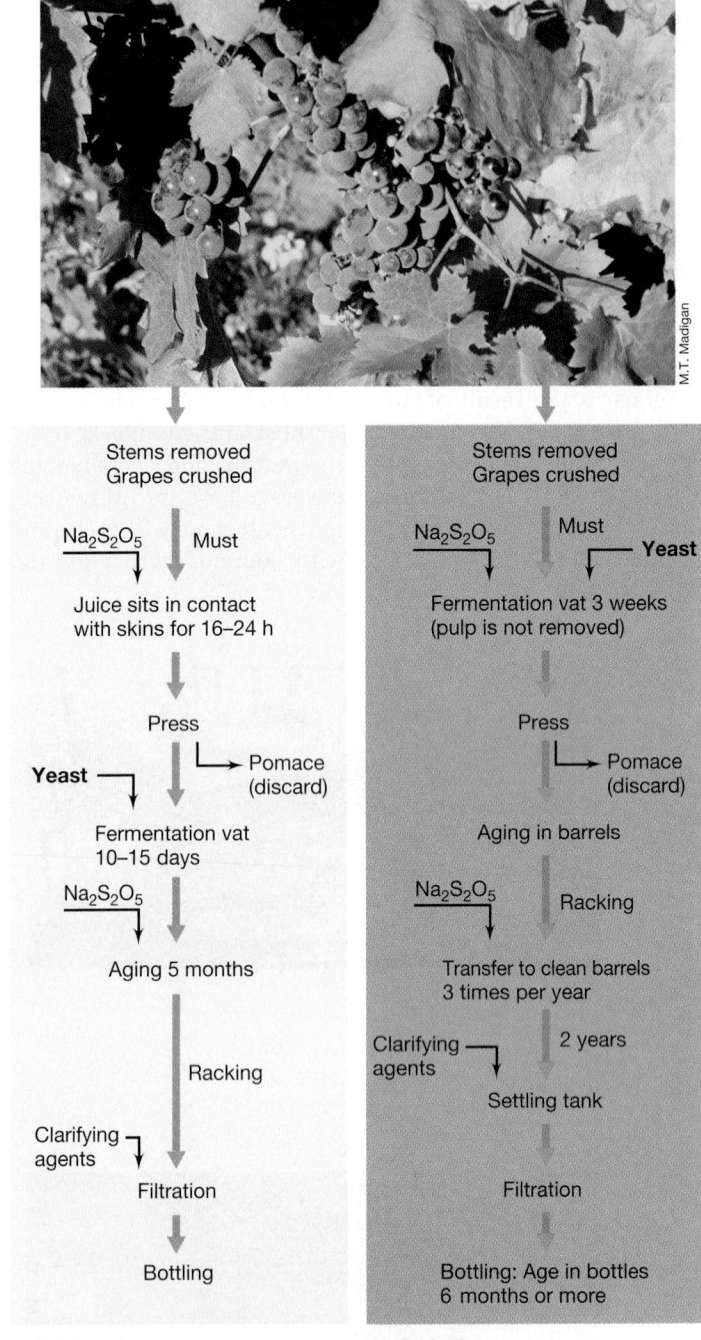

Figure 15.13 Wine production. *(a)* White wine. White wines vary from nearly colorless to straw-colored depending on the grapes used. *(b)* Red wine. Red wines vary in color from a faint red to a deep, rich burgundy. The background colors of parts a and b are those of chenin blanc, a typical white wine, and a rosé, a light red wine.

earth, asbestos, or membrane filters. The wine is then bottled and either stored for further aging, or sent to market.

Red wine is typically aged for months to several years (Figure 15.12*c, d*), but most white wine is sold without much aging. During aging, complex chemical changes occur, including the reduction of bitter components; this improves the flavor and odor, or *bouquet*, of the wine. The final alcohol content of wine varies from about 8% to 16% depending on the sugar content of the grapes, length of the fermentation, and strain of wine yeast used.

Malolactic Fermentation

Many high-quality dry red wines and a few white wines such as the chardonnays are subjected to a secondary fermentation following the primary fermentation by yeast. This is done before bottling and is called the *malolactic fermentation*. Full-bodied dry red wines are the typical candidates for malolactic fermentation.

In grape varieties used for dry wines a considerable amount of malic acid can be present in the grapes. The malic acid content of the grape varies locally due to climatic and soil conditions. Malic acid is a sharp and rather bitter acid. During the malolactic fermentation, malic acid is fermented to lactic acid, a softer, smoother acid, and this makes the wine less acidic and fruity but more complex and palatable. Many other constituents are produced during the malolactic fermentation, including diacetyl (2,3-butanedione), a major flavoring ingredient in butter; this also helps to impart a soft, smooth character to the wine.

The malolactic fermentation is catalyzed by species of lactic acid bacteria (↩ Section 18.1), including *Lactobacillus*, *Pediococcus*, and *Oenococcus*. These organisms are extremely acid-tolerant and can carry out the malolactic fermentation even if the initial pH of the wine is below pH 3.5. Commercial wineries typically use starter cultures of selected malolactic fermentation organisms and then store the wine in barrels especially for this purpose. Inocula for future rounds of malolactic fermentation come from lactobacilli that become attached to the insides of the barrel. The malolactic fermentation can take several weeks but is usually worth the wait, as the final product is often much smoother and far superior to the more sharp-tasting and bitter starting material.

MiniQuiz

- What production differences lead to red wine versus white wine?
- What occurs during the malolactic fermentation, and why is it carried out?

15.8 Brewing and Distilling

Beers and ales are popular alcoholic beverages produced worldwide from the fermentation of grains and other sources of starch. Although, like the wine industry, the brewing industry employs yeast to catalyze the fermentation itself, the amount of alcohol in brewed products is much lower than that in wine, and levels of CO_2 are typically much higher. Thus the two products—beer and wine—are quite different fermented beverages, and each has its own characteristic properties. And, like wine, the final brewed product can be greatly influenced by regional and cultural differences.

Making the Wort

Brewing is the production of alcoholic beverages from malted grains. Typical malt beverages include beer, ale, porter, and stout. *Malt* is prepared from germinated barley seeds, and it contains natural enzymes that digest the starch of grains and convert it to glucose. Because brewing yeasts are unable to digest starch, the malting process is essential for the generation of fermentable substrates.

The fermentable liquid for brewing is prepared by a process called *mashing*. The grain of the mash may consist only of malt, or other grains such as corn, rice, or wheat may be added. The mixture of ingredients in the mash is cooked and allowed to steep in a large mash tub at warm temperatures. During the heating period, enzymes from the malt cause digestion of the starches and liberate glucose, which will be fermented by the yeast. Proteins and amino acids are also liberated into the liquid, as are other nutrient ingredients necessary for the growth of yeast.

After mashing, the aqueous mixture, called *wort*, is separated by filtration. *Hops*, an herb derived from the female flowers of the hops plant, are added to the wort at this stage. Hops add flavors to the wort but also have antimicrobial properties, which help to prevent bacterial contamination in the subsequent fermentation. The wort is boiled for several hours, usually in large copper kettles (**Figure 15.14a,b**), during which time desired ingredients are extracted from the hops, undesirable proteins present in the wort are coagulated and removed, and the wort is sterilized. The wort is filtered again, cooled, and transferred to the fermentation vessel.

The Fermentation Process

Brewery yeast strains are of two major types: *top* fermenting and *bottom* fermenting. The main distinction between the two is that top-fermenting yeasts remain uniformly distributed in the fermenting wort and are carried to the top by the CO_2 gas generated during the fermentation, whereas bottom-fermenting yeasts settle to the bottom. Top yeasts are used in the brewing of *ales*, and bottom yeasts are used to make *lager beers*. Bottom yeasts have been given the species designation *Saccharomyces carlsbergensis*, whereas top yeasts are considered *Saccharomyces cerevisiae*. Top yeasts usually ferment at higher temperatures (14–23°C) than bottom yeasts (6–12°C) and thus complete the fermentation in a shorter time (5–7 days for top fermentation versus 8–14 days for bottom fermentation).

After bottom yeasts complete lager beer fermentation, the beer is pumped off into large tanks where it is stored in the cold (about −1°C) for several weeks (Figure 15.14c). Following this, the beer is filtered and placed in storage tanks (Figure 15.14d) from which it is packaged and sent to market. Top-fermented ale is stored for only short periods at a higher temperature (4–8°C), which assists in development of the characteristic ale flavor.

Home Brew

Amateur and small-scale commercial brewing has become popular in recent years, especially in the United States. Many styles of beer from English bitters and India pale ale to German bock and Russian Imperial stout can be made at home, and the character of a particular brew depends on many factors including the amounts and types of malt, sugar, hops, and grain used, the strain of yeast employed, the temperature and duration of the fermentation, and how the beer is aged. For home brewing only simple equipment is necessary such as a stainless steel container to prepare the wort, a 20-liter (5-gallon) fermentor fitted with a valve to

Figure 15.14 Brewing beer in a large commercial brewery. (a, b) The copper brew kettle is where the wort is mixed with hops and then boiled. From the brew kettle, the liquid passes to large fermentation tanks where yeast ferments glucose to ethanol plus CO_2. (c) If the beer is a lager, it is stored for several weeks at low temperature in tanks where particulate matter, including yeast cells, settles. (d) The beer is then filtered and placed in storage tanks from which it is packaged into kegs, bottles, or cans.

allow CO_2 to escape, and glass bottles to store the final product (**Figure 15.15**).

Home brewing is much the same as commercial brewing except that hop-flavored malt extract is often used directly as the wort instead of preparing the wort in the traditional way. Using the same basic equipment, various beers can be made, each with its own distinctive taste and character. Dark beers, which typically contain more alcohol than lighter beers, require more malt for their production and are usually brewed from a combination of different malts such as those obtained from darker varieties of grain or those that have been roasted to caramelize the sugars and yield a darker color. Most beers contain 3–6% alcohol. A typical American-style lager (Figure 15.15d) contains about 3.5% alcohol (by volume), a Munich-style dark about 4.5%, and bock beers about 5%. Some specialty beers and ales can contain upwards of 12% alcohol.

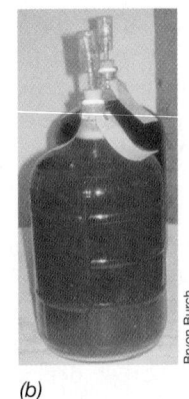

Figure 15.15 Home brewing. (a) A stainless steel pot to boil the wort. (b) The fermentation vessel. (c) Bottling and capping the beer. (d) Glasses of two common beers, a lager (pils) (left) and a common dark beer (bock) (right). The vessel is fitted with a fermentation lock that maintains anoxic conditions but allows CO_2 to escape.

The trend toward individuality in beer is evident not only by the growing number of home brewers, but also by the fact that many microbreweries are appearing. Total production by microbreweries in the United States is less than that of a major brewer, but the products often have their own distinctive character and local appeal. The particular properties of a given microbrew can be traced to the smaller scale on which the brewing is done, the unique sources of ingredients, water, and yeast strains employed, and to differences in times and temperatures used in the brewing process.

Distilled Alcoholic Beverages

Distilled alcoholic beverages are *distillates*, the products of heating fermented liquids to volatilize alcohol and other constituents. The distillate is condensed and collected, a process called *distilling*. A product much higher in alcohol content is obtained by distilling than is possible by fermentation alone. Virtually any alcoholic liquid can be distilled, and each yields a characteristic distilled beverage. The distillation of malt brews yields *whiskey*, distilled wine yields *brandy*, distilled fermented molasses yields *rum*, distilled fermented grain or potatoes yields *vodka*, and distilled fermented grain and juniper berries yields *gin* (**Figure 15.16**). Alcohol concentrations in distilled products vary from as little as 20% to as high as 95%. The "proof" rating, used primarily for labeling distilled spirits in the United States, is defined as twice the alcohol concentration. Thus a whiskey that is 80 proof contains 40% ethanol by volume.

The distillate contains not only alcohol but also other volatile products arising either from the yeast fermentation or released from the ingredients. Some of these other products add desirable flavor, whereas others are undesirable. To eliminate the latter, the distilled product is typically aged, usually in oak barrels. The aging removes undesirable products and allows desirable new flavors and aromatic ingredients to develop. The fresh distillate is typically colorless, whereas the aged product is often brown or yellow (Figure 15.16). The character of the final product is partly determined by the manner and length of aging; aging times of 5–10 years are common, but some very expensive distilled spirits are aged for 20 years or more.

MiniQuiz

- In brewing, why is the mashing process necessary?
- What are the major differences between a beer and an ale?
- How does whiskey differ from brandy?

15.9 Biofuels

Production of ethyl alcohol (ethanol) as a commodity chemical is a major industrial process, and today over 60 billion liters of ethanol are produced yearly worldwide from the fermentation of various feedstocks. In the United States most ethanol is obtained by yeast fermentation of glucose obtained from cornstarch. Ethanol is stripped from the fermentation broth by distillation (**Figure 15.17a**). Various yeasts have been used in commodity ethanol production, including species of *Saccharomyces*, *Kluyveromyces*, and *Candida*, but most ethanol in the United States is produced by *Saccharomyces*.

Figure 15.16 Distilled spirits. Aging in oak casks imparts a distinctive amber or yellow color to distilled spirits. Left to right, dark rum, brandy, whiskey. Gin and vodka (not shown) are not aged in oak and are colorless.

Ethanol as a Biofuel

Ethanol is currently the most important global **biofuel**, a term indicating that the fuel was made from the fermentation of recently grown plant material rather than being of ancient origin (that is, fossil fuel). Other major biofuels include biodiesel, made from vegetable oils, and algal fuels, alcohols and oils produced from green algae. The feedstock used for ethanol production has been a major issue in the debate over whether biofuels are the wave of the future. In the United States, for example, the increased demand for corn as a biofuel feedstock has driven up the price of human foods and livestock feeds. In other countries, for example Brazil, which is a major ethanol producer, not only corn but also sugar cane, whey, sugar beets, and even wood chips and waste paper are used as feedstocks for the fermentation. For cellulosic materials, the cellulose must first be treated to release glucose, which is then fermented to alcohol. Alternative feedstocks showing great promise for ethanol production are grasses such as switchgrass (Figure 15.17b), a rapidly growing and easily harvestable grass whose cellulosic cell walls can be degraded to glucose and fermented to ethanol.

In the United States *gasohol* is produced by adding ethanol to gasoline; at a final concentration of 10% ethanol, it can be used in virtually all gasoline engines. The combustion of gasohol produces lower amounts of carbon monoxide and nitrogen oxides than pure gasoline, and hence gasohol is a cleaner-burning fuel. The production of more ethanol-rich fuels such as E-85 (85% ethanol and 15% gasoline) is also growing in the United States, but this fuel can only be used in modified engines. However, E-85 fuel reduces emission of nitrogen oxides by nearly 90% and is thus a means for reducing important pollutants in the atmosphere and reducing dependence on conventional sources of oil. Many major cities concerned about air pollution are retrofitting their public transportation systems, especially buses, to burn E-85.

Total ethanol production to meet fuel demands in the United States is scheduled to top 30 billion liters by 2012. The major downside to ethanol production is that at present it takes about 25% more energy to produce a liter of ethanol than is present in the ethanol itself. However, because bioethanol is a product of recently fixed carbon rather than of buried fossil fuel, its use is considered a more sustainable and environmentally friendly way to supply the liquid fuel needs of the foreseeable future.

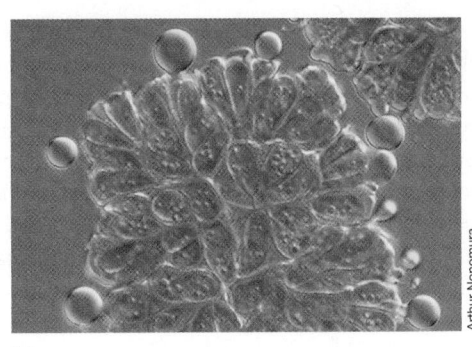

(a)   (b)   (c)

Figure 15.17 Biofuels. (a) A bioethanol production plant in Nebraska (USA). In the plant, glucose obtained from cornstarch is fermented by *Saccharomyces cerevisiae* to ethanol plus CO_2. The large tank in the left foreground is the ethanol storage tank, and the tanks and pipes in the background are for distilling ethanol from the fermentation broth. (b) Switchgrass, a promising feedstock for bioethanol production. The cellulose from this rapidly growing plant can be treated to yield glucose that can then be fermented to ethanol or butanol. (c) The petroleum-producing colonial green alga, *Botryococcus braunii*. Note the excreted oil droplets that appear as bubbles along the margin of the cells.

Petroleum Biofuels

In addition to bioethanol production, green energy initiatives have spurred research on many other biofuels. This includes the production of longer-chain alcohols, such as butanol, from fermentative processes, but also the direct synthesis of petroleum by green algae. For example, during growth the colonial green alga *Botryococcus braunii* excretes long-chain (C_{30}–C_{36}) hydrocarbons that have the consistency of crude oil (Figure 15.17c). In *B. braunii* about 30% of the cell dry weight consists of petroleum, and there has been heightened interest in using this and other oil-producing algae as renewable sources of petroleum. There is even evidence from biomarker studies that some known petroleum reserves originated from green algae that grew in lakebeds in ancient times rather than having been formed from the microbial degradation of plant materials. Although it is a promising source of oil from a "green" perspective, the major problem with algal petroleum is scale: In a world that currently uses about 90 million barrels of oil per day, the logistics of growing oil-producing algae that could contribute significantly to global oil demand are daunting.

MiniQuiz
- How can yeast help to solve global energy problems?
- What is the difference between gasohol and E-85?

 Products from Genetically Engineered Microorganisms

We now consider some products of the biotechnology industry, products synthesized by genetically engineered bacteria or other organisms. Compared with most of the products of industrial microbiology just considered, biotech drugs are inherently more valuable but are produced on a much smaller scale.

Before the era of biotechnology, diabetics relied on insulin extracted from animals to control their blood sugar levels. In most cases this worked well, but a small percentage of diabetics showed immune reactions to the foreign (porcine or bovine) version of insulin. Genetic engineering allowed genuine human insulin to be produced by bacteria. Indeed, human insulin was the first commercialized product from genetic engineering. Today several genetically engineered hormones and other human proteins are available for clinical use. These human proteins were originally produced by cloning human genes, inserting them into bacteria (typically *Escherichia coli*), and having the bacteria make the protein. However, several problems were encountered in this approach. Foreign proteins made in bacteria must be isolated by disrupting the bacterial cells and then purified. Traces of bacterial proteins that contaminate the desired protein may elicit an unwanted immune response. Moreover, traces of lipopolysaccharide from the gram-negative bacterial outer membrane are toxic (endotoxin, ↩ Section 27.12).

Other major problems revolved around the challenge of expressing eukaryotic genes in bacteria. These include the facts that (1) the genes must be placed under control of a bacterial promoter (↩ Section 11.8); (2) the introns (↩ Section 7.8) must be removed; (3) codon bias (↩ Section 6.17) affects the efficiency of translation; and (4) many mammalian proteins are modified after translation, and bacteria lack the ability to perform most such modifications. As a result, recent efforts in biotechnology have been to express mammalian proteins using genetically modified eukaryotic host cells. Both eukaryotic cells in culture and whole transgenic animals have been used. For example, transgenic goats have been used that secrete the protein of interest in their milk. Most recently, plants are being engineered to express mammalian proteins. We consider some major topics in biotechnology in more detail now.

15.10 Expressing Mammalian Genes in Bacteria

The procedures for cloning and manipulating genes were covered in Chapter 11. Here we are concerned with expressing cloned genes to manufacture a useful product. Expression vectors (↻ Section 11.8) are needed to express eukaryotic genes in bacteria. However, there are still obstacles to be faced, even if the mammalian gene has been cloned into an expression vector. One major issue is the presence of introns that disrupt the coding sequence of many eukaryotic genes, especially in higher organisms such as mammals (↻ Section 7.8). Introns must be spliced out for the genes to function; however, prokaryotic hosts lack the machinery to do so. To skirt this problem, introns are removed during the cloning process, before the genes are inserted into the host used for production. Typically, cloned mammalian genes no longer contain their original introns but consist of an uninterrupted coding sequence. The two major ways of achieving this are now described.

Cloning the Gene via mRNA

The standard way to obtain an intron-free eukaryotic gene is to clone it via its messenger RNA (mRNA). Because introns are removed during the processing of mRNA, the mature mRNA carries an uninterrupted coding sequence. Tissues expressing the gene of interest often contain large amounts of the corresponding mRNA, although other mRNAs are also present. In certain situations, however, a single mRNA dominates in a tissue type, and extraction of bulk mRNA from that tissue provides a useful starting point for gene cloning.

In a typical mammalian cell, about 80–85% of the RNA is ribosomal RNA, 10–15% is transfer RNA, and only 1–5% is mRNA. However, eukaryotic mRNA is unique because of the poly(A) tails found at the 3′ end (↻ Section 7.8), and this makes it easy to isolate, even though it is scarce. If a cell extract is passed over a chromatographic column containing strands of poly(T) linked to a cellulose support, most of the mRNA separates from other RNAs by the specific pairing of A and T bases. The RNA is released from the column by a low-salt buffer, which gives a preparation greatly enriched in mRNA.

Once mRNA has been isolated, the genetic information is converted into complementary DNA (cDNA). This is done by the enzyme reverse transcriptase. This enzyme, essential for retroviral replication (↻ Section 21.11), copies information from RNA into DNA, a process called **reverse transcription (Figure 15.18)**. Reverse transcriptase needs a primer to start DNA synthesis. When making DNA using mRNA as a template, a primer is used that is complementary to the poly(A) tail of the mRNA. This primer is hybridized to the mRNA, and reverse transcriptase is added.

Reverse transcriptase makes DNA that is complementary to the mRNA. As seen in Figure 15.18, the newly synthesized cDNA has a hairpin loop at its end. The loop forms because, after the enzyme completes copying the mRNA, it starts to copy the newly made DNA. This hairpin loop provides a convenient primer for synthesis of the second (complementary) strand of DNA by DNA polymerase I and is later removed by a single-strand-specific

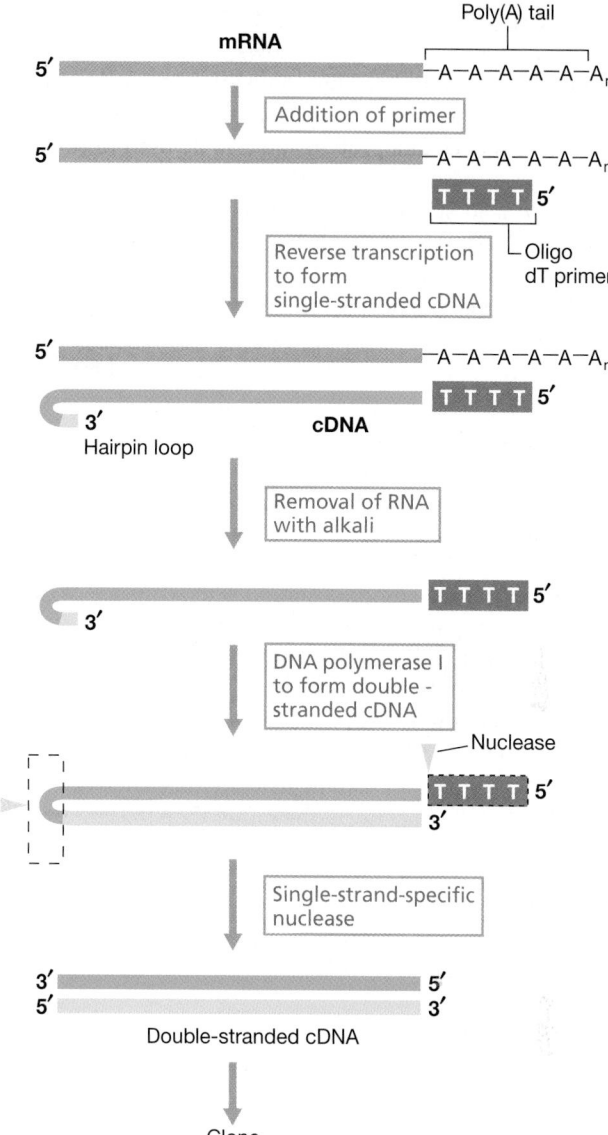

Figure 15.18 Complementary DNA (cDNA). Steps in the synthesis of cDNA from an isolated mRNA using the retroviral enzyme reverse transcriptase. The poly(A) tail is typical of eukaryotic mRNA.

nuclease. The result is a linear, double-stranded DNA molecule, one strand of which is complementary to the original mRNA (Figure 15.18).

This double-stranded cDNA contains the coding sequence but lacks introns. It can be inserted into a plasmid or other vector for cloning. However, because the cDNA corresponds to the mRNA, it lacks a promoter and other upstream regulatory sequences that are not transcribed into RNA. Special expression vectors with bacterial promoters and ribosome-binding sites are used to obtain high-level expression of genes cloned in this way (↻ Section 11.8).

A cDNA library is a gene library (↻ Section 11.3) consisting of cDNA versions of genes made from mRNA extracted from a eukaryotic cell. The library reflects only those genes expressed in the particular tissue under the existing conditions.

Finding the Gene via the Protein

Knowing the sequence of a gene allows the synthesis of a cDNA molecule to use as a probe. This can be used to find the gene by screening a gene library (Section 11.3). Knowledge of the amino acid sequence of a protein can also be used to construct a probe or even to synthesize a whole gene.

The amino acid sequence of a protein can be used to design and synthesize an oligonucleotide probe that encodes it. This process is illustrated in **Figure 15.19**. Unfortunately, degeneracy of the genetic code complicates this approach. Most amino acids are encoded by more than one codon (Table 6.5), and codon usage varies from organism to organism. Thus, the best region of a gene to synthesize as a probe is one that encodes part of the protein rich in amino acids specified by only a single codon (for example, methionine, AUG; tryptophan, UGG) or at most two codons (for example, phenylalanine, UUU, UUC; tyrosine, UAU, UAC; histidine, CAU, CAC). This strategy increases the chances that the probe will be nearly complementary to the mRNA or gene of interest. If the complete amino acid sequence of the protein is not known, partial sequence data may be used.

For certain small proteins there may be good reason to synthesize the entire gene. Many mammalian proteins (including high-value peptide hormones) are made by protease cleavage of larger precursors. Thus, to produce a short peptide hormone such as insulin, it may be more efficient to construct an artificial gene that encodes just the final hormone rather than the larger precursor protein from which it is derived naturally. Chemical synthesis also allows synthesis of modified genes that may make useful new proteins. Artificial synthesis of DNA is now routine, and it is possible to synthesize genes encoding proteins over 200 amino acid

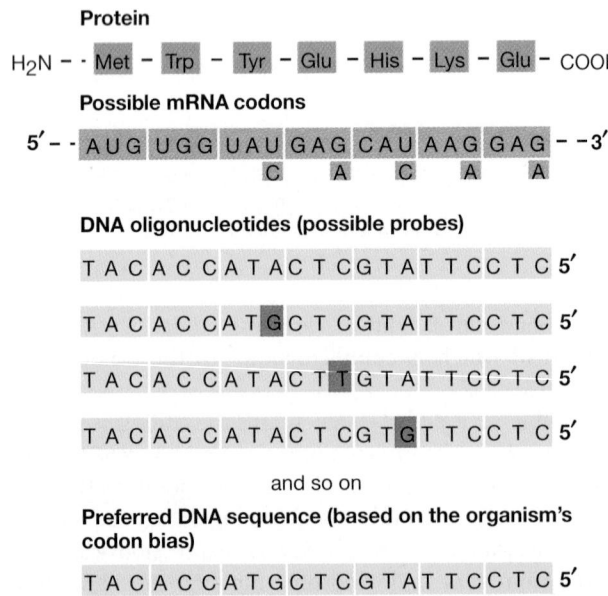

Figure 15.19 Deducing the best sequence of an oligonucleotide probe from the amino acid sequence of a protein. Because many amino acids are encoded by multiple codons, many nucleic acid probes are possible for a given polypeptide sequence. If codon usage by the target organism is known, a preferred sequence can be selected. Complete accuracy is not essential because a small amount of mismatch can be tolerated, especially with long probes.

residues in length (600 nucleotides). The synthetic approach was first used in a major way for production of the human hormone insulin in bacteria. Moreover, constructed genes are free of introns and thus the mRNA does not need processing. Also, promoters and other regulatory sequences can easily be built into the gene upstream of the coding sequences, and codon bias (Section 6.17) can be accounted for.

With these techniques many human and viral proteins have been expressed at high yield under the control of bacterial regulatory systems. These include insulin, somatostatin, viral capsid proteins, and interferon.

Protein Folding and Stability

The synthesis of a protein in a new host may bring additional problems. For example, some proteins are susceptible to degradation by intracellular proteases and may be destroyed before they can be isolated. Moreover, some eukaryotic proteins are toxic to prokaryotic hosts, and the host cell may be killed before a sufficient amount of the product is synthesized. Further engineering of either the host or the vector may eliminate these problems.

Sometimes when foreign proteins are massively overproduced, they form inclusion bodies inside the host. Inclusion bodies consist of aggregated insoluble protein that is often misfolded or partly denatured, and they are often toxic to the host cell. Although inclusion bodies are relatively easy to purify because of their size, the protein they contain is often difficult to solubilize and may be inactive. One possible solution to this problem is to use a host that overproduces molecular chaperones that aid in folding (Section 6.21).

Fusion Proteins for Improved Purification

Protein purification can often be made much simpler if the target protein is made as a **fusion protein** along with a carrier protein encoded by the vector. To do this, the two genes are fused to yield a single coding sequence. A short segment that is recognized and cleaved by a commercially available protease is included between them. After transcription and translation, a single protein is made. This is purified by methods designed for the carrier protein. The fusion protein is then cleaved by the protease to release the target protein from the carrier protein. Fusion proteins simplify purification of the target protein because the carrier protein can be chosen to have ideal properties for purification.

Several fusion vectors are available to generate fusion proteins. **Figure 15.20** shows an example of a fusion vector that is also an expression vector. In this example, the carrier protein is the *Escherichia coli* maltose-binding protein, and the fusion protein is easily purified by methods based on its affinity for maltose. Once purified, the two portions of the fusion protein are separated by a specific protease (factor Xa, a protease whose natural role is in blood clotting). In some cases the target protein is released from the carrier protein by specific chemical treatment, rather than by protease cleavage.

Fusion systems are also used for other purposes. One advantage of making a fusion protein is that the carrier protein can be chosen to contain the bacterial *signal sequence*, a peptide rich in hydrophobic amino acids that enables transport of the protein

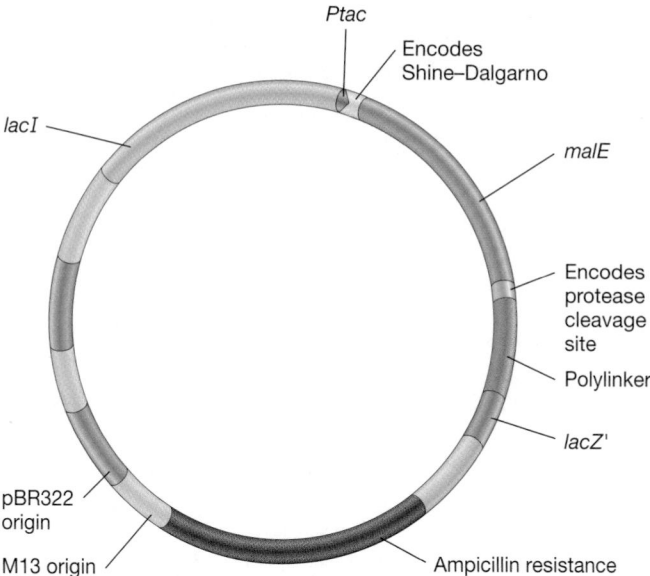

Figure 15.20 **An expression vector for fusions.** The gene to be cloned is inserted into the polylinker so it is in frame with the *malE* gene, which encodes maltose-binding protein. The insertion inactivates the gene for the alpha fragment of *lacZ*, which encodes β-galactosidase. The fused gene is under control of the hybrid *tac* promoter (*Ptac*). The plasmid also contains the *lacI* gene, which encodes the *lac* repressor. Therefore, an inducer must be added to turn on the *tac* promoter. The plasmid contains a gene conferring ampicillin resistance on its host. In addition to the plasmid origin of replication, there is a bacteriophage M13 origin. Thus, this vector is a phagemid and can be propagated either as a plasmid or as a phage. This vector was developed by New England Biolabs (Ipswich, MA).

across the cytoplasmic membrane (⇄ Section 6.21). This makes possible a bacterial expression system that not only makes mammalian proteins, but also secretes them. When the right strains and vectors are employed, the desired protein can make up as much as 40% of the protein molecules in a cell.

MiniQuiz

- What major advantage does cloning mammalian genes from mRNA or using synthetic genes have over PCR amplification and cloning of the native gene?

- How is a fusion protein made?

15.11 Production of Genetically Engineered Somatotropin

One of the most economically profitable areas of biotechnology today is the production of human proteins. Many mammalian proteins have high pharmaceutical value but are typically present in very low amounts in normal tissue, and it is therefore extremely costly to purify them. Even if the protein can be produced in cell culture, this is much more expensive and difficult than growing microbial cultures that produce the protein in high yield. Therefore, the biotechnology industry has genetically engineered microorganisms to produce many different mammalian

proteins. Although insulin was the first human protein to be produced in this manner, the procedure had several unusual complications, because insulin consists of two short polypeptides held together by disulfide bonds. A more typical example is somatotropin, and we focus on this here.

Growth hormone, or *somatotropin*, consists of a single polypeptide encoded by a single gene. Somatotropin from one mammalian species usually functions reasonably well in other species; indeed, transgenic animals have been made expressing foreign somatotropin genes, as discussed below. Lack of somatotropin results in hereditary dwarfism. Because the human somatotropin gene was successfully cloned and expressed in bacteria, children showing stunted growth can be treated with recombinant human somatotropin. However, dwarfism may also be caused by lack of the somatotropin receptor, and in this case administration of somatotropin has no effect. (People of the African Pygmy tribes have normal levels of human somatotropin, but most of them are no taller than 4 feet, 10 inches because they have defective growth hormone receptors.)

The somatotropin gene was cloned as complementary DNA (cDNA) from mRNA as described in Section 15.10 (**Figure 15.21**).

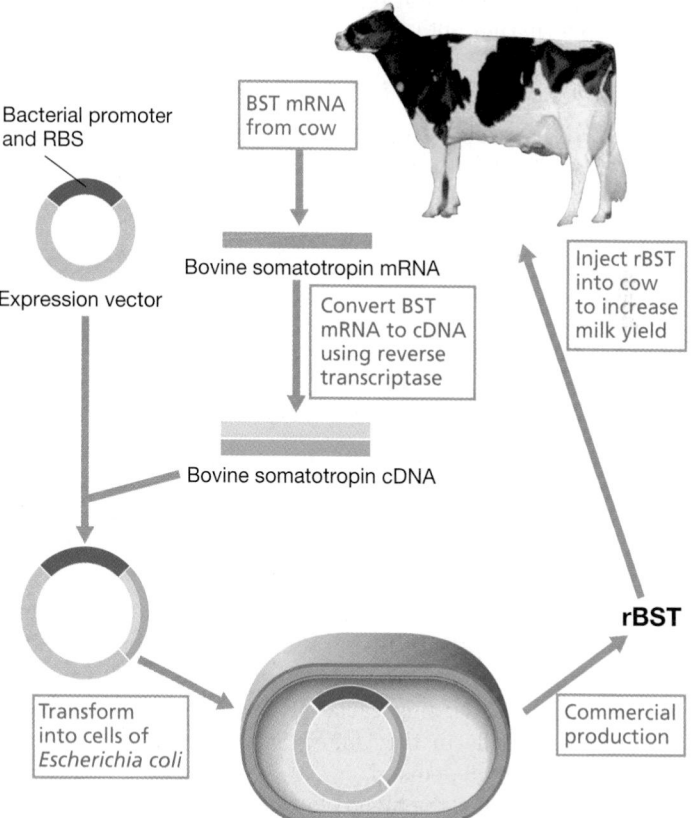

Figure 15.21 **Cloning and expression of bovine somatotropin.** The mRNA for bovine somatotropin (BST) is obtained from an animal. The mRNA is converted to cDNA by reverse transcriptase. The cDNA version of the somatotropin gene is then cloned into a bacterial expression vector that has a bacterial promoter and ribosome-binding site (RBS). The construct is transformed into cells of *Escherichia coli*, and recombinant bovine somatotropin (rBST) is produced. Milk production increases in cows treated with rBST.

The cDNA was then expressed in a bacterial expression vector. The main problem with producing relatively short polypeptide hormones such as somatotropin is their susceptibility to protease digestion. This problem can be countered by using bacterial host strains defective for several proteases.

Recombinant bovine somatotropin (rBST) is used in the dairy industry (Figure 15.21). Injection of rBST into cows does not make them grow larger but instead stimulates milk production. The reason for this is that somatotropin has two binding sites. One binds to the somatotropin receptor and stimulates growth, the other to the prolactin receptor and promotes milk production. However, excessive milk production by cows causes some health problems in the animals, including an increased frequency of infections of the udder and decreased reproductive capability.

When somatotropin is used to remedy human growth defects, it is desirable to avoid side effects from the hormone's prolactin activity (prolactin stimulates lactation). Site-directed mutagenesis (↺ Section 11.4) of the somatotropin gene was used to generate somatotropin that no longer binds the prolactin receptor. To accomplish this, several amino acids needed for binding to the prolactin receptor were altered by changing their coding sequences. Thus it is possible not merely to make genuine human hormones, but also to alter their specificity and activity to make them better pharmaceuticals.

MiniQuiz

- What is the advantage of using genetic engineering to make insulin?
- What are the major problems when manufacturing proteins in bacteria?
- How has biotechnology helped the dairy industry?

15.12 Other Mammalian Proteins and Products

Many other mammalian proteins are produced by genetic engineering (**Table 15.6**). These include, in particular, an assortment of hormones and proteins for blood clotting and other blood processes. For example, tissue plasminogen activator (TPA) is a blood protein that scavenges and dissolves blood clots that may form in the final stages of the healing process. TPA is primarily used in heart patients or others suffering from poor circulation to prevent the development of clots that can be life-threatening. Heart disease is a leading cause of death in many developed countries, especially in the United States, so microbially produced TPA is in high demand.

In contrast to TPA, the blood clotting factors VII, VIII, and IX are critically important for the *formation* of blood clots. Hemophiliacs suffer from a deficiency of one or more clotting factors and can therefore be treated with microbially produced clotting factors. In the past hemophiliacs have been treated with clotting factor extracts from pooled human blood, some of which was contaminated with viruses such as HIV and hepatitis C, putting hemophiliacs at high risk for contracting these diseases. Recombinant clotting factors have eliminated this problem.

Table 15.6 *A few therapeutic products made by genetic engineering*

Product	Function
Blood proteins	
Erythropoietin	Treats certain types of anemia
Factors VII, VIII, IX	Promotes blood clotting
Tissue plasminogen activator	Dissolves blood clots
Urokinase	Promotes blood clotting
Human hormones	
Epidermal growth factor	Wound healing
Follicle-stimulating hormone	Treatment of reproductive disorders
Insulin	Treatment of diabetes
Nerve growth factor	Treatment of degenerative neurological disorders and stroke
Relaxin	Facilitates childbirth
Somatotropin (growth hormone)	Treatment of some growth abnormalities
Immune modulators	
α-Interferon	Antiviral, antitumor agent
β-Interferon	Treatment of multiple sclerosis
Colony-stimulating factor	Treatment of infections and cancer
Interleukin-2	Treatment of certain cancers
Lysozyme	Anti-inflammatory
Tumor necrosis factor	Antitumor agent, potential treatment of arthritis
Replacement enzymes	
β-Glucocerebrosidase	Treatment of Gaucher disease, an inherited neurological disease
Therapeutic enzymes	
Human DNase I	Treatment of cystic fibrosis
Alginate lyase	Treatment of cystic fibrosis

Some mammalian proteins made by genetic engineering are enzymes rather than hormones (Table 15.6). For instance, *human DNase I* is used to treat the buildup of DNA-containing mucus in patients with cystic fibrosis. The mucus forms because cystic fibrosis is often accompanied by life-threatening lung infections by the bacterium *Pseudomonas aeruginosa*. The bacterial cells form biofilms (↺ Section 23.4) within the lungs that make drug treatment difficult. DNA is released when the bacteria lyse, and this fuels mucus formation. DNase digests the DNA and greatly decreases the viscosity of the mucus. There are more than 30,000 patients with cystic fibrosis in the United States alone. Treatment of cystic fibrosis with DNase was approved in 1994, and sales today of this life-saving enzyme exceed $100 million. A second enzyme, *alginate lyase*, also produced by genetic engineering, shows promise in treating cystic fibrosis because it degrades the polysaccharide produced by *P. aeruginosa* cells. Like DNA from lysed cells, this polymer also contributes to lung mucus, and thus its hydrolysis relieves respiratory symptoms.

Not all the enzymes produced by genetic engineering have therapeutic uses. Many commercial enzymes (Section 15.6) are now produced in this way. Sometimes the benefits of genetic engineering are quite unexpected. Rennet, which is an enzyme used to make cheese, is the product of slaughtered animals and thus cannot be consumed by strict vegetarians (vegans). However, "vegetarian cheese" containing recombinant rennet produced in a microorganism is being marketed and has found wide acceptance.

Further applications come from using site-directed mutagenesis (⮌ Section 11.4) on existing cloned genes to generate new products with new properties. Certain molecules, such as many antibiotics, are synthesized in cells by biochemical pathways that use a series of enzymes (Section 15.4). These enzymes can be modified by genetic engineering to produce modified forms of the antibiotics.

MiniQuiz

- Contrast the activity of TPA and blood factors VII, VIII, and IX.
- Explain how a DNA-degrading enzyme can be useful in treating a bacterial infection, such as that which occurs with cystic fibrosis.

15.13 Genetically Engineered Vaccines

Vaccines are substances that elicit immunity to a particular disease when injected into an animal (⮌ Section 28.7). Typically, vaccines are suspensions of killed or modified pathogenic microorganisms or viruses (or parts isolated from them). Often the part that elicits the immune response is a surface protein, for instance, a viral coat protein. Genetic engineering can be applied in many different ways to the production of vaccines.

Recombinant Vaccines

Recombinant DNA techniques can be used to modify the pathogen itself. For instance, one can delete pathogen genes that encode virulence factors but leave those whose products elicit an immune response. This yields a recombinant, live, attenuated vaccine. Conversely, one can add genes from a pathogenic virus to another, relatively harmless virus, referred to as a carrier virus. Such vaccines are called **vector vaccines**. This approach induces immunity to the pathogenic viral disease. Indeed, one can even combine the two approaches. For example, a recombinant vaccine is used to protect poultry against both fowlpox (a disease that reduces weight gain and egg production) and Newcastle disease (a viral disease that is often fatal). The fowlpox virus (a typical pox virus; ⮌ Section 21.15) was first modified by deleting genes that cause disease, but not ones that elicit immunity. Then immunity-inducing genes from the Newcastle virus were inserted. This resulted in a **polyvalent vaccine**, a vaccine that immunizes against two different diseases.

Vaccinia virus is widely used to prepare live recombinant vaccines for human use (⮌ Section 21.15). Vaccinia virus itself is generally not pathogenic for humans and has been used for over 100 years as a vaccine against the related smallpox virus. However, cloning genes into vaccinia virus requires a selective marker, which is provided by the gene for thymidine kinase. Vaccinia is unusual for a virus in carrying its own thymidine kinase, an enzyme that normally converts thymidine into thymidine triphosphate. However, this enzyme also converts the base analog

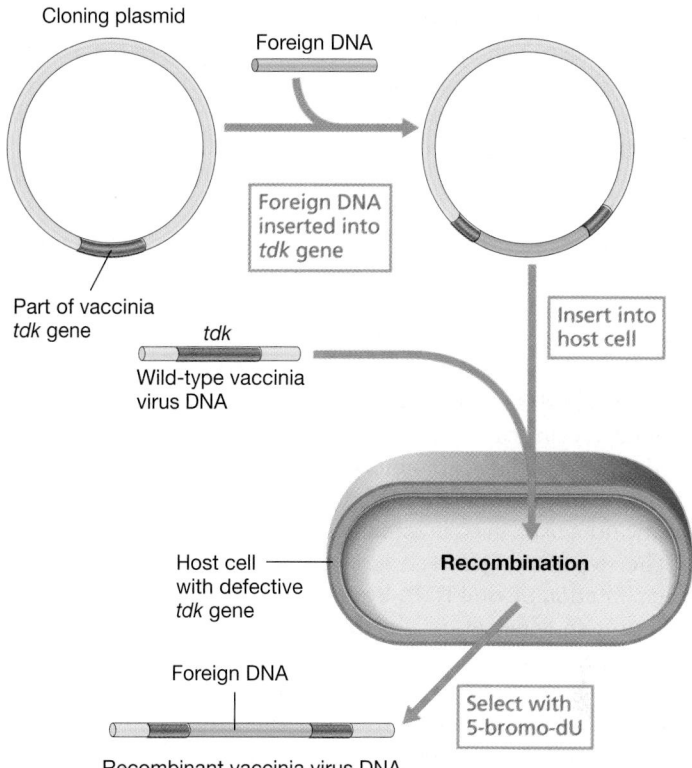

Figure 15.22 Production of recombinant vaccinia virus. Foreign DNA is inserted into a short segment of the thymidine kinase gene (*tdk*) from vaccinia virus carried on a plasmid. The plasmid with the insert and wild-type vaccinia virus are both put into the same host cell where they recombine. The cells are treated with 5-bromodeoxyuridine (5-bromo-dU), which kills cells with active thymidine kinase. Only recombinant vaccinia viruses whose *tdk* gene is inactivated by insertion of foreign DNA survive.

5-bromodeoxyuridine to a nucleotide that is incorporated into DNA, which is a lethal reaction. Therefore, cells that express thymidine kinase (whether from the host cell genome or from a virus genome) are killed by 5-bromodeoxyuridine.

Genes to be put into vaccinia virus are first inserted into an *Escherichia coli* plasmid that contains part of the vaccinia thymidine kinase (*tdk*) gene (**Figure 15.22**). The foreign DNA is inserted into the *tdk* gene, which is therefore disrupted. This recombinant plasmid is then transformed into animal cells whose own *tdk* genes have been inactivated. These cells are also infected with wild-type vaccinia virus. The two versions of the *tdk* gene— one on the plasmid and the other on the virus—recombine. Some viruses gain a disrupted *tdk* gene plus its foreign insert (Figure 15.22). Cells infected by wild-type virus, with active thymidine kinase, are killed by 5-bromodeoxyuridine. Cells infected by recombinant vaccinia virus with a disrupted *tdk* gene grow long enough to yield a new generation of viral particles (Figure 15.22). In other words, the protocol selects for viruses whose *tdk* gene contains a cloned insert of foreign DNA.

Vaccinia virus does not actually need thymidine kinase to survive. Consequently, recombinant vaccinia viruses can still infect human cells and express any foreign genes they carry. Indeed, vaccinia viruses can be engineered to carry genes from multiple viruses (that is, they are polyvalent vaccines). Currently, several vaccinia vector vaccines have been developed and licensed for veterinary use, including one for rabies. Many other vaccinia vaccines are at the clinical trial stage. Vaccinia vaccines are relatively benign, yet highly immunogenic in humans, and their use will likely increase in the coming years.

Subunit Vaccines

Recombinant vaccines need not include every protein from the pathogenic organism. Subunit vaccines may contain only a specific protein or two from a pathogenic organism. For a virus this protein is often the coat protein because coat proteins are typically highly immunogenic. The coat proteins are purified and used in high dosage to elicit a rapid and high level of immunity. Subunit vaccines are currently very popular because they can be used to produce large amounts of immunogenic proteins without the possibility that the purified products contain the entire pathogenic organism, even in minute amounts.

The steps in preparing a viral subunit vaccine are as follows: fragmentation of viral DNA by restriction enzymes; cloning viral coat protein genes into a suitable vector; providing proper promoters, reading frame, and ribosome-binding sites; and reinsertion and expression of the viral genes in a microorganism. Sometimes only certain portions of the protein are expressed rather than the entire protein, because immune cells and antibodies typically react with only small portions of the protein. (When this approach is used against an RNA virus, the viral genome must be converted to a cDNA copy first.)

When *E. coli* is used as the cloning host, viral subunit vaccines are often poorly immunogenic and fail to protect in experimental tests of infection. The problem is that the recombinant proteins produced by bacteria are nonglycosylated, and glycosylation is necessary for the proteins to be immunologically active. Glycosylation is accomplished in an animal cell infected by the virus when, in the course of viral replication, viral coat proteins are modified after translation by the addition of sugar residues (glycosylation). To solve the problem with the vaccine, a eukaryotic cloning host is used. For example, the first recombinant subunit vaccine approved for use in humans (against hepatitis B) was made in yeast. The gene encoding a surface protein from hepatitis B virus was cloned and expressed in yeast. The protein produced was glycosylated and formed aggregates very similar to those found in patients infected with the virus. These aggregates were purified and used to effectively vaccinate humans against infection by hepatitis B virus.

Subunit vaccines against many viruses and other pathogens are currently being developed. Cultured insect or mammalian cells are often used as hosts to prepare such recombinant vaccines. As noted, to obtain the correct glycosylation or other modifications of the immunogenic protein, it is often important to use a eukaryotic host. However, vaccines with correct glycosylation can often be produced in eukaryotic hosts relatively unrelated to humans, such as plants or insect cells. Recently, both yeast and insect cells have themselves been genetically engineered by the insertion of human genes that catalyze glycosylation. The resulting host cells add human-type glycosylation patterns to the proteins they produce.

The Future of Recombinant Vaccines

Genetically engineered recombinant vaccines will likely become increasingly common for several reasons. They are safer than normal attenuated or killed vaccines because it is impossible to transmit the disease in the vaccine. They are also more reproducible because their genetic makeup can be carefully monitored.

In addition, recombinant vaccines can usually be prepared much faster than those made by more traditional methods. For preparing vaccines for some diseases such as influenza, time is of the essence. Recombinant vaccines using cloned influenza virus hemagglutinin genes can be made in just 2 or 3 months. This contrasts with the 6–9 months needed to make an attenuated intact ("live") influenza virus vaccine. Preparation time is important in responding to an epidemic caused by a new strain of virus, a common situation with influenza outbreaks. Finally, recombinant vaccines are typically less expensive than those produced by traditional methods.

DNA Vaccines

Although vaccines have been extremely successful in the fight against infectious diseases, in some cases vaccines are difficult to produce. However, a conceptually new approach to vaccine production is possible—**DNA vaccines**—also known as *genetic vaccines*. DNA vaccines use the genome of the pathogen itself to immunize the individual. Defined fragments of the pathogen's genome or specific genes that encode immunogenic proteins are used. The key genes are cloned into a plasmid or viral vector and delivered by injection. When DNA is taken up by animal cells it may be degraded or it may be transcribed and translated. If it is translated and the protein produced is immunogenic, the animal will be effectively immunized against the pathogen. Thus, the immune response is made against the protein encoded by the vaccine DNA. The DNA itself is not immunogenic.

Several DNA vaccines, for example, vaccines against HIV, hepatitis B, and several cancers, have undergone clinical trials. For unknown reasons, the DNA vaccines so far tested have not proved potent enough to provide protective immunity to humans. It is hoped that future improvements will permit clinical use. Nonetheless, DNA vaccines have been licensed for use in animals (for example, a vaccine against West Nile virus for horses).

Unlike viral vaccines, DNA vaccines escape surveillance by the host immune system because nucleic acids themselves are poorly immunogenic. This prevents the animal from suffering autoimmune effects in which antibodies and immune cells attack host cells (ᴄᴄ Section 28.9). DNA vaccines have the advantage that they are both safe and inexpensive. In addition, DNA is more stable than live vaccines, which avoids the need for refrigeration—an important practical point for using vaccines in developing countries.

MiniQuiz

- Explain why recombinant vaccines might be safer than some vaccines produced by traditional methods.
- What are the important differences among a recombinant live attenuated vaccine, a vector vaccine, a subunit vaccine, and a DNA vaccine?

15.14 Mining Genomes

Just as the total genetic content of an organism is its *genome*, so the collective genomes of an environment is known as its *metagenome*. Complex environments, such as fertile soil, contain vast numbers of uncultured bacteria and other microorganisms together with the viruses that prey on them (⮰ Section 12.6). Taken together, these contain correspondingly vast numbers of novel genes. Indeed, most of the genetic information on Earth exists in microorganisms and their viruses that have not been cultured.

Environmental Gene Mining

Gene mining is the process of isolating potentially useful novel genes from the environment without culturing the organisms that carry them. Instead of being cultured, DNA (or RNA) is isolated directly from environmental samples and cloned into suitable vectors to construct a metagenomic library (**Figure 15.23**). The nucleic acid includes genes from uncultured organisms as well as DNA from dead organisms that has been released into the environment but has not yet been degraded. If RNA is isolated, it must be converted to a DNA copy by reverse transcriptase (Figure 15.18). However, isolating RNA is more time consuming and limits the metagenomic library to those genes that are transcribed and therefore active in the environment sampled.

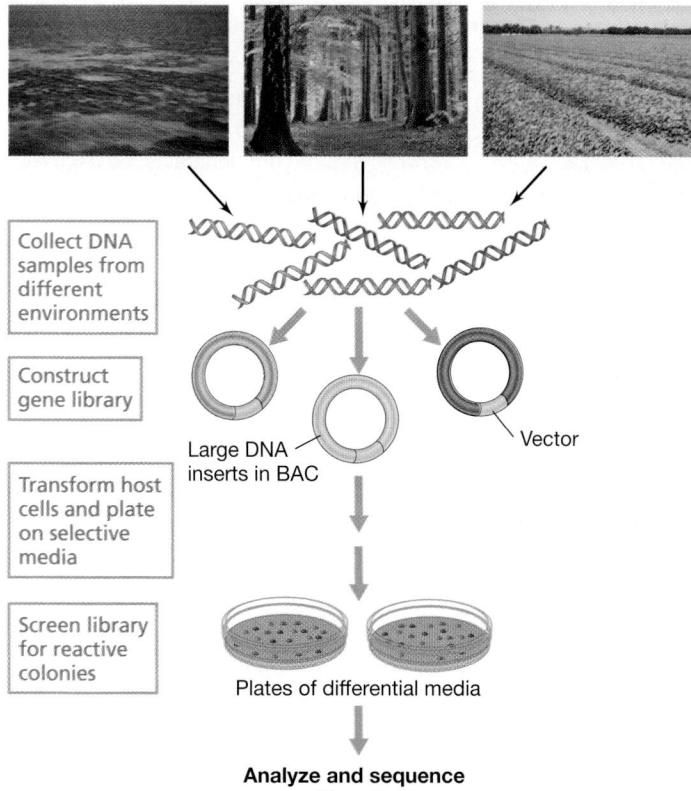

Collect DNA samples from different environments

Construct gene library

Large DNA inserts in BAC

Vector

Transform host cells and plate on selective media

Screen library for reactive colonies

Plates of differential media

Analyze and sequence positive clones

Figure 15.23 Metagenomic search for useful genes in the environment. DNA samples are obtained from different sites, such as seawater, forest soil, and agricultural soil. A clone library is constructed using bacterial artificial chromosomes (BACs) and screened for genes of interest. Possibly useful clones are analyzed further.

The metagenomic library is screened by the same techniques as any other clone library. Metagenomics has identified novel genes encoding enzymes that degrade pollutants and enzymes that make novel antibiotics. So far several lipases, chitinases, esterases, and other degradative enzymes with novel substrate ranges and other properties have been isolated by this approach. Such enzymes are used in industrial processes for various purposes (Section 15.6). Enzymes with improved resistance to industrial conditions, such as high temperature, high or low pH, and oxidizing conditions, are especially valuable and sought after.

Discovery of genes encoding entire metabolic pathways, such as for antibiotic synthesis, as opposed to single genes, requires vectors such as bacterial artificial chromosomes (BACs) that can carry large inserts of DNA (⮰ Section 11.10). BACs are especially useful for screening samples from rich environments, such as soil, where vast numbers of unknown genomes are present and correspondingly large numbers of genes are available to screen.

Targeted Gene Mining

Metagenomics can screen directly for enzymes with certain properties. Suppose one needed an enzyme or entire pathway capable of degrading a certain pollutant at a high temperature. The first step would be to find a hot environment polluted with the target compound. Assuming that microorganisms capable of degradation were present in the environment, a reasonable hypothesis, DNA from the environment would then be isolated and cloned. Host bacteria containing the clones would be screened for growth on the target compound. For convenience, this step is usually done in an *Escherichia coli* host, on the assumption that thermostable enzymes will still show some activity at 40°C (this is typically the case). Once suspects have been identified, enzyme extracts can be tested *in vitro* at high temperatures. Recently, thermophilic cloning systems have been developed that allow direct selection at high temperature. These rely on expression vectors that can replicate in both *E. coli* and the hot spring thermophile *Thermus thermophilus*.

MiniQuiz

- Explain why metagenomic cloning gives large numbers of novel genes.
- What are the advantages and disadvantages of isolating environmental RNA as opposed to DNA?

15.15 Engineering Metabolic Pathways

Although proteins are large molecules, expressing large amounts of a single protein that is encoded by a single gene is relatively simple. By contrast, small metabolites are typically made in biochemical pathways employing several enzymes. In these cases, not only are multiple genes needed, but their expression must be regulated in a coordinated manner as well.

Pathway engineering is the process of assembling a new or improved biochemical pathway using genes from one or more organisms. Most efforts so far have modified and improved existing pathways rather than creating entirely new ones (but see the Microbial Sidebar, "Synthetic Biology and Bacterial Photography").

Synthetic Biology and Bacterial Photography

The term "synthetic biology" refers to the use of genetic engineering to create novel biological systems out of available biological parts, often from several different organisms. An ultimate goal of synthetic biology is to synthesize a viable cell from component parts, a feat that will likely be accomplished in the near future. A major start in this direction was made in 2007 when a team of synthetic biologists transferred the entire chromosome of one species of bacterium into another species of bacterium. The latter species then took on all of the properties of the species whose genome it had acquired. In 2010, the team extended their work and placed a *laboratory-synthesized* chromosome of one bacterial species into the cell of a second bacterial species and got the synthetic chromosome to function normally and direct the activities of the recipient cell.

An interesting example of synthetic biology on a smaller scale is the use of genetically modified *Escherichia coli* cells to produce photographs. The engineered bacteria are grown as a lawn on agar plates. When an image is projected onto the lawn, bacteria in the dark make a dark pigment whereas bacteria in the light do not. The result is a primitive black-and-white photograph of the projected image.

Construction of the photographic *E. coli* required the engineering and insertion of three genetic modules: (1) a light detector and signaling module; (2) a pathway to convert heme (already present in *E. coli*) into the photoreceptor pigment phycocyanobilin; and (3) an enzyme encoded by a gene whose transcription can be switched on and off to make the dark pigment (Figure 1a). The light detector is a fusion protein. The outer half is the light-detecting part of the phytochrome protein from the cyanobacterium *Synechocystis*. This needs a special light-absorbing pigment, phycocyanobilin, which is not made by *E. coli*, hence the need to install the pathway to make phycocyanobilin.

The inner half of the light detector is the signal transmission domain of the EnvZ sensor protein from *E. coli*. EnvZ is part of a two-component regulatory system, its partner being OmpR (⟳ Section 8.7). Normally, EnvZ activates the DNA-binding protein OmpR. Activated OmpR in turn activates target genes by binding to the promoter. In the present case, the hybrid protein was designed to activate OmpR in the dark but not in the light. This is because phosphorylation of OmpR is required for activation, and red light converts the sensor to a state in which phosphorylation is inhibited. Consequently the target gene is off in the light and on in the dark. When a mask is placed over the Petri plate containing a lawn of the engineered *E. coli* cells (Figure 1b), cells in the dark make a pigment that cells in the light do not, and in this way a "photograph" of the masked image develops (Figure 1c).

The pigment made by the *E. coli* cells results from the activity of an enzyme naturally found in this organism that functions in lactose metabolism, β-galactosidase. The target gene, *lacZ*, encodes this enzyme. In the dark, the *lacZ* gene is expressed and β-galactosidase is made. The enzyme cleaves a lactose analog called *S-gal* present in the growth medium to release galactose and a colored dye. In the light, the *lacZ* gene is not expressed, no β-galactosidase is made, and so no black dye is released. The difference in contrast between cells producing the dye and cells that are not generates the bacterial photograph (Figure 1c).

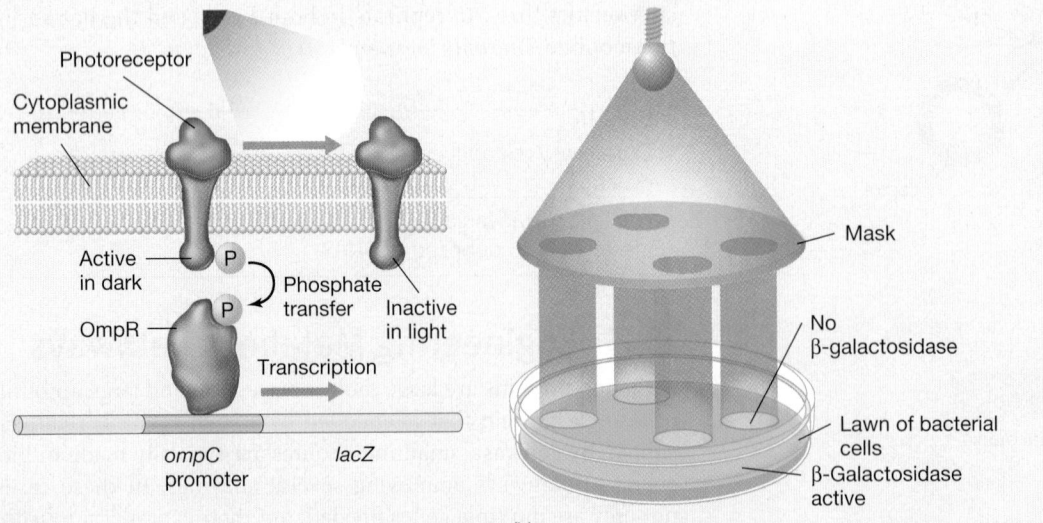

Figure 1 Bacterial photography. *(a)* Light-detecting *Escherichia coli* cells were genetically engineered using components from cyanobacteria and *E. coli* itself. Red light inhibits phosphate (P) transfer to the DNA-binding protein OmpR; phosphorylated OmpR is required to activate *lacZ* transcription (*lacZ* encodes β-galactosidase). *(b)* Set-up for making a bacterial photograph. The opaque portions of the mask correspond to zones where β-galactosidase is active and thus to the dark regions of the final image. *(c)* A bacterial photograph of a portrait of Charles Darwin.

Because genetic engineering of bacteria is simpler than that of higher organisms, most pathway engineering has been done with bacteria. Engineered microorganisms are used to make products, including alcohols, solvents, food additives, dyes, and antibiotics. They may also be used to degrade agricultural waste, pollutants, herbicides, and other toxic or undesirable materials.

An example of pathway engineering is the production of indigo by *Escherichia coli* (**Figure 15.24**). Indigo is an important dye used for treating wool and cotton. Blue jeans, for example, are made of cotton dyed with indigo. In ancient times indigo and related dyes were extracted from sea snails. More recently, indigo was extracted from plants, but today it is synthesized chemically. The demand for indigo by the textile industry has spawned new approaches for its synthesis, including a biotechnological one.

Because the structure of indigo is very similar to that of naphthalene, enzymes that oxygenate naphthalene also oxidize indole to its dihydroxy derivative, which then oxidizes spontaneously in air to yield indigo, a bright blue pigment. Enzymes for oxygenating naphthalene are present on several plasmids found in *Pseudomonas* and other soil bacteria. When genes from such plasmids were cloned into *E. coli*, the cells turned blue due to production of indigo; the blue cells had picked up the genes for the enzyme naphthalene oxygenase.

Although only the gene for naphthalene oxygenase was cloned during indigo pathway engineering, the indigo pathway consists of four steps, two enzymatic and two spontaneous (Figure 15.24). *E. coli* synthesizes the enzyme tryptophanase that carries out the first step, the conversion of tryptophan to indole. For indigo production, tryptophan must be supplied to the recombinant *E. coli* cells. This is accomplished by affixing the cells to a solid support in a bioreactor and trickling a tryptophan solution from waste protein or other sources over them. Recirculating the material over the cells several times, as is typically done in these types of immobilized cell industrial processes (Section 15.6), steadily increases indigo levels until the dye can be harvested.

MiniQuiz

- Explain why pathway engineering is more difficult than cloning and expressing a human hormone.
- How was *Escherichia coli* modified to produce indigo?

Ⓥ Transgenic Eukaryotes

A *transgene* is a gene from one organism that has been inserted into a different organism. Hence, a **transgenic organism** is one that contains a transgene. A related term is **genetically modified organism (GMO)**. Strictly speaking, this refers to genetically engineered organisms whether or not they contain foreign DNA. However, in common usage, especially in agriculture, "GMO" is often used interchangeably with "transgenic organism."

On the one hand, the genetic engineering of higher organisms is not truly microbiology. On the other hand, much of the DNA manipulation is carried out using bacteria and their plasmids long before the engineered transgene is finally inserted into the plant or animal. Furthermore, vectors based on viruses are widely used in the genetic engineering of higher organisms. Therefore we emphasize the microbial systems that have contributed to the genetic manipulation of plants and animals.

15.16 Genetic Engineering of Animals

Many foreign genes have been incorporated and expressed in laboratory research animals and in commercially important animals. The genetic engineering uses microinjection to deliver cloned genes to fertilized eggs; genetic recombination then incorporates the foreign DNA into the genomes of the eggs. The first transgenic animals were mice that were engineered as model systems for studying mammalian physiology. Genes for growth hormone from rats or humans were engineered for expression and inserted into eggs that developed into mice that expressed the growth hormone. The result, of course, was larger mice. More recently, farm animals have been genetically modified, not only to improve yields, but also so that their waste is less polluting.

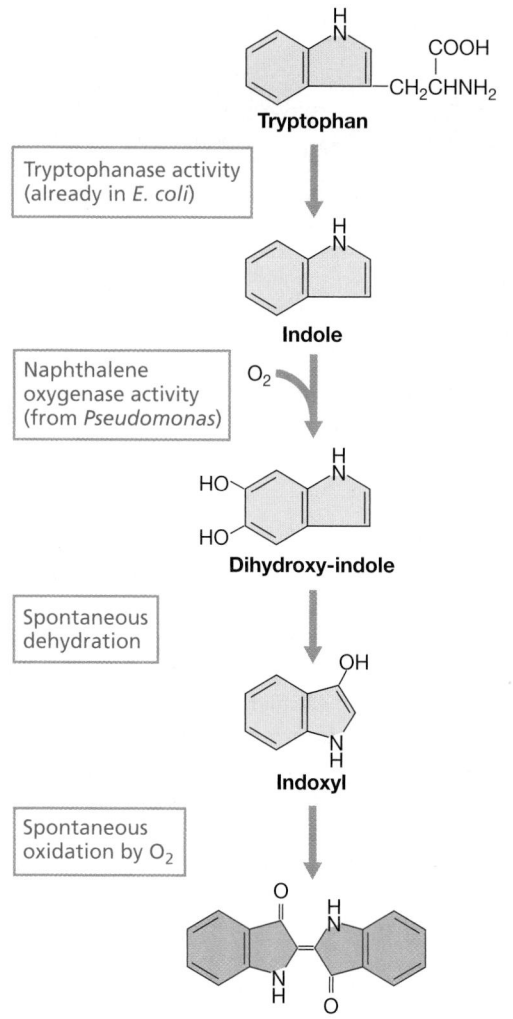

Figure 15.24 Engineered pathway for production of the dye indigo. *Escherichia coli* naturally expresses tryptophanase, which converts tryptophan into indole. Naphthalene oxygenase (originally from *Pseudomonas*) converts indole to dihydroxy-indole, which spontaneously dehydrates to indoxyl. Upon exposure to air, indoxyl dimerizes to form indigo, which is blue.

Transgenic Animals in Pharming

Transgenic animals can be used to produce proteins of pharmaceutical value—a process called *pharming*. Transgenic animals are particularly useful for producing human proteins that require specific posttranslational modifications for activity, such as blood-clotting enzymes. Proteins of this type are not made in an active form by microorganisms or by plants.

Some proteins have been genetically engineered to be secreted in high yield in animal milk. This is convenient for several reasons. First, this allows larger volumes of material to be made more simply and cheaply than by bacterial culture. Second, a milk-processing industry already exists, so little new technology is needed to purify the protein. Third, milk is a natural product that most humans can tolerate, so purification to remove possibly toxic bacterial proteins is unnecessary. Goats have proven useful for making several human proteins including tissue plasminogen activator, which is used to dissolve blood clots (Section 15.12).

Transgenic Animals in Medical Research

Transgenic animals have become increasingly important in basic biomedical research for studying gene regulation and developmental biology. For example, so-called "knockout" mice, which have had both copies of a particular gene inactivated by genetic engineering, are used to analyze genes active in animal physiology. For instance, knockout mice that lack both copies of the gene for myostatin, a protein that slows muscle growth, develop massive muscles. In contrast, transgenic mice that overproduce myostatin show reduced muscle mass. Many other strains of knockout mice have been developed for use in medical research, and a 2007 Nobel Prize was awarded for the development of this important genetic tool.

Improving Livestock and Other Food Animals

Livestock may be engineered to increase their productivity, nutritional value, and disease resistance. Occasionally, transgenic livestock are produced that do not have increased commercial value but that demonstrate the feasibility of certain genetic techniques. For example, pigs genetically engineered to express the reporter gene that encodes the green fluorescent protein (⟳ Section 11.5) are, as expected, green (**Figure 15.25a**).

One scheme to improve the nutrition of livestock is to insert entire metabolic pathways from bacteria into the animals. For example, genes that encode the enzymes of the metabolic pathway for making methionine, a required amino acid, could remove the need for this amino acid in the animals' diet. A notable technical success has been the insertion into pigs of a gene from *Escherichia coli* that helps degrade organic phosphate. The resulting Enviropig™ no longer needs phosphate supplements in its feed. However, most importantly, the manure from these animals is low in phosphate, and this prevents phosphate runoff from pig-manure waste ponds into freshwater; such an influx of inorganic nutrients can trigger algal blooms and fish die-offs (⟳ Section 23.8).

Pigs have also been genetically engineered to increase their levels of omega-3 fatty acids. These fatty acids reduce heart disease but are found in significant amounts only in cold-water fish,

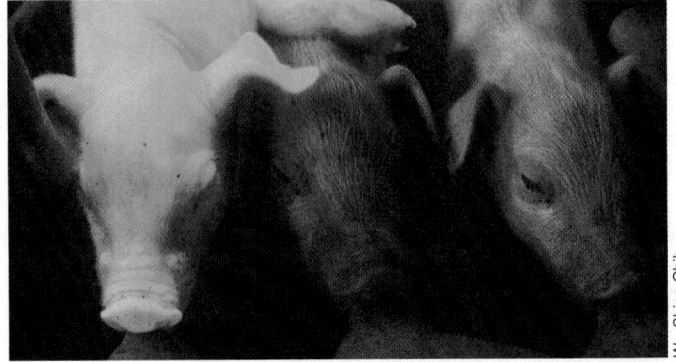
(a)
Wu-Shinn Chih

(b)
Aqua Bounty Technologies

Figure 15.25 **Transgenic animals.** *(a)* A piglet (left), seen under blue light, that has been genetically engineered to express the green fluorescent protein. Control piglets are shown in the center and right. *(b)* Fast-growing salmon. The *AquAdvantage*™ Salmon (top) was engineered by Aqua Bounty Technologies (St. Johns, Newfoundland, Canada). The transgenic and the control fish are 18 months old and weigh 4.5 kg and 1.2 kg, respectively.

such as salmon, and a few other rare foods. To create transgenic pigs with an altered fatty acid profile, a gene from the roundworm *Caenorhabditis elegans*, called *fat1*, was inserted into the pigs. The enzyme encoded by *fat1* converts the less healthy but more common omega-6 fatty acids into omega-3 fatty acids. Such animals should be healthier for consumers, especially those who have dietary restrictions on fat or who are at high risk for heart disease. It will be several years before the omega-3-enriched pigs reach the consumer market, assuming they receive government approval.

Another interesting practical example of a transgenic animal is the "fast-growing salmon" (Figure 15.25b). These transgenic salmon do not grow to be larger than normal salmon but simply reach market size much faster. The gene for growth hormone in natural salmon is activated by light. Consequently, salmon grow rapidly only during the summer months. In the genetically engineered salmon, the promoter for the growth hormone gene was replaced with the promoter from another fish that grows at a more or less constant rate all year round. The result was salmon that make growth hormone constantly and thus grow faster.

MiniQuiz

- What is pharming?
- Why are knockout mice useful in investigating human gene function?
- What environmental advantage does Enviropig™ have over normal pigs?

15.17 Gene Therapy in Humans

A large number of human genetic diseases are known. A list can be found in the OMIM (Online Mendelian Inheritance in Man) database, available online at http://www.ncbi.nlm.nih.gov/. Because conventional genetic experiments cannot be done with humans as they can with other animals, our understanding of human genetics has lagged behind that of many other organisms. However, by using recombinant DNA technology, coupled with conventional genetic studies (following family inheritance, and so on), it is possible to localize particular genetic defects to specific regions of particular chromosomes. Moreover, the human genome has been sequenced (↺ Section 12.1). Consequently, if the region encoding a presumed genetic defect is cloned and sequenced, the base sequence of the defective gene may be compared with that of the normal gene. Even without knowledge of the enzyme defect, it is possible to obtain information about the genetic changes causing many hereditary defects.

Human Hereditary Diseases

Many genes, including those for Huntington's disease, hemophilia types A and B, cystic fibrosis, Duchenne muscular dystrophy, multiple sclerosis, and breast cancer have been located with these techniques, and the mutations in the defective genes have been identified. With this in mind, how can genetic engineering be used to treat or cure these diseases?

The use of genetic engineering to treat human genetic diseases, including attacking cancer cells, is known as **gene therapy**. In the gene therapy procedure, a nonfunctional or dysfunctional gene in a person is "replaced" by a functional gene. Strictly speaking, it is not the defective *gene* that is replaced, but instead its *function*. The therapeutic wild-type gene is inserted elsewhere in the genome and its gene product corrects the genetic disorder. Major obstacles to this approach exist in targeting the correct cells for gene therapy and in successfully inserting the required gene into cell lines that will perpetuate the genetic alteration.

The first genetic disease for which the use of gene therapy was approved is a form of severe combined immune deficiency (SCID). This disease, caused by the absence of adenosine deaminase (ADA), an enzyme of purine metabolism in bone marrow cells, leads to a crippled immune system. The gene therapy approach uses a retrovirus as a vector to carry a wild-type copy of the ADA gene. T cells (part of the immune system; ↺ Section 28.1) are removed from the patient and infected with the retrovirus carrying the ADA gene. The retrovirus also carries a marker gene, encoding resistance to the antibiotic neomycin, so that T cells carrying the inserted retrovirus may be selected and identified. The engineered T cells are then placed back in the body. However, because T cells have a limited life span, the ther-

apy must be repeated every few months. Consequently, for newborn infants diagnosed with defects in the ADA gene, treatment protocols for SCID have been developed in which the ADA gene is inserted into stem cells obtained from the umbilical cord blood of the infants. The engineered stem cells are then returned to the infants. Because stem cells continue to divide and provide a fresh supply of new T cells, this effects a long-term cure.

Several other gene therapy treatments, some using other virus vectors, are currently being tested with various levels of success. After the first gene therapy experiment with ADA in 1990, there were no striking breakthroughs until 2000. Another form of SCID, caused by defects in a different gene, was successfully treated in several patients. It seems likely that this very rare form of the disease can now be successfully treated using gene therapy.

Technical Problems with Gene Therapy

Although gene therapy has tremendous potential, most applications remain distant prospects. Some current difficulties are related to the vectors being used. Although using retroviral vectors gives stable integration of the transgene, the site of insertion is unpredictable and expression of the cloned gene is often transient. The vectors also have limited infectivity and are rapidly inactivated in the host. Many nonretroviral vectors, such as the adenovirus vector, have similar problems, and adverse reactions to the vector itself can also be a severe problem. However, some promising new vectors for gene therapy have emerged, including human artificial chromosomes (↺ Section 11.10) and highly modified retroviral vectors.

It is important to recall that in the gene therapy protocols being tested, the defective copy of the gene is not actually replaced; rather, its *defective function* is replaced. The retrovirus (containing the good copy of the gene) simply integrates somewhere into the human genome of the target cells. Actual gene replacements in germ line cells (cells that give rise to gametes) can be accomplished in experimental animals, although the techniques of isolating individual animals with these changes cannot readily be applied to humans. Moreover, attempts to change the germ cells of humans would also raise ethical questions that will likely keep these types of procedures, even if they have great medical promise, only a very long-range possibility.

MiniQuiz

- Why is SCID such a serious disease, and how can gene therapy help someone afflicted with SCID?
- What problems arise from using a retrovirus as a vector in gene therapy?
- A person treated successfully by gene therapy will still have a defective copy of the gene. Explain.

15.18 Transgenic Plants in Agriculture

Genetic improvement of plants by traditional selection and breeding has a long history, but recombinant DNA technology has led to revolutionary changes. Plant DNA can be modified by genetic engineering and then transformed into plant cells by either electroporation or the particle gun (↺ Section 11.7). Alternatively, one can use plasmids from the bacterium *Agrobacterium tumefaciens*, which naturally transfers DNA directly into the cells of certain plants (Section 25.7).

Plants differ from animals in having no real separation of the germ line from the somatic cells. Consequently plants can often be regenerated from just a single cell. Moreover, it is possible to culture plant cells *in vitro*. Therefore plant genetic engineering is mostly done with plant cells growing in culture. After genetically altered clones have been selected, the cells are induced by treatment with plant hormones to grow into whole plants.

Many successes in plant genetic engineering have already been achieved, and several transgenic plants are in agricultural production. The public knows these plants as *genetically modified* (GM) plants. In this section we discuss how foreign genes are inserted into plant genomes and how transgenic plants may be used.

The Ti Plasmid and Transgenic Plants

The gram-negative plant pathogen *A. tumefaciens* contains a large plasmid, called the **Ti plasmid**, that is responsible for its virulence. This plasmid contains genes that mobilize DNA for transfer to the plant, which as a result contracts crown gall disease (Section 25.7). The segment of the Ti plasmid DNA that is actually transferred to the plant is called **T-DNA**. The sequences at the ends of the T-DNA are essential for transfer, and the DNA to be transferred must be included between these ends.

One common Ti-vector system that has been used for the transfer of genes to plants is a two-plasmid system called a *binary vector*, which consists of a cloning vector plus a helper plasmid. The cloning vector contains the two ends of the T-DNA flanking a multiple cloning site, two origins of replication so that it can replicate in both *Escherichia coli* (the host for cloning) and *A. tumefaciens*, and two antibiotic resistance markers, one for selection in plants and the other for selection in bacteria. The foreign DNA is inserted into the vector, which is transformed into *E. coli* and then moved to *A. tumefaciens* by conjugation (**Figure 15.26**).

This cloning vector lacks the genes needed to transfer T-DNA to a plant. However, when placed in an *A. tumefaciens* cell that contains a suitable helper plasmid, the T-DNA can be transferred to a plant. The "disarmed" helper plasmid, called *D-Ti*, contains the virulence (*vir*) region of the Ti plasmid but lacks the T-DNA. It lacks the genes that cause disease but supplies all the functions needed to transfer the T-DNA from the cloning vector. The cloned DNA and the kanamycin resistance marker of the vector are mobilized by D-Ti and transferred into a plant cell where they enter the nucleus (Figure 15.26d). Following integration into a plant chromosome, the foreign DNA can be expressed and confer new properties on the plant.

A number of transgenic plants have been produced using the Ti plasmid of *A. tumefaciens*. The Ti system works well with broadleaf plants (dicots), including crops such as tomato, potato, tobacco, soybean, alfalfa, and cotton. It has also been used to produce transgenic trees, such as walnut and apple. The Ti system does not work with plants from the grass family (monocots, including the important crop plant, corn), but other methods of introducing DNA, such as the particle gun (Section 11.7), have been used successfully for them.

Herbicide and Insect Resistance

Major areas targeted for genetic improvement in plants include herbicide, insect, and microbial disease resistance as well as improved product quality. The first GM crop to be grown commercially was tobacco grown in China in 1992 that was engineered for resistance to viruses. By the year 2005, the area planted worldwide with GM crops was estimated to exceed 1 billion acres (440 million hectares). The main GM crops today are soybeans, corn, cotton, and canola. Almost all the GM soybeans and canola planted were herbicide resistant, whereas the corn and cotton were herbicide resistant or insect resistant, or both. In 2005 the United States grew over half the world's total of GM crops. Argentina, Canada, Brazil, and China were the other major producers, with the rest of the world accounting for less than 5% of the total.

Herbicide resistance is genetically engineered into a crop plant to protect it from herbicides applied to kill weeds. Many herbicides inhibit a key plant enzyme or protein necessary for growth.

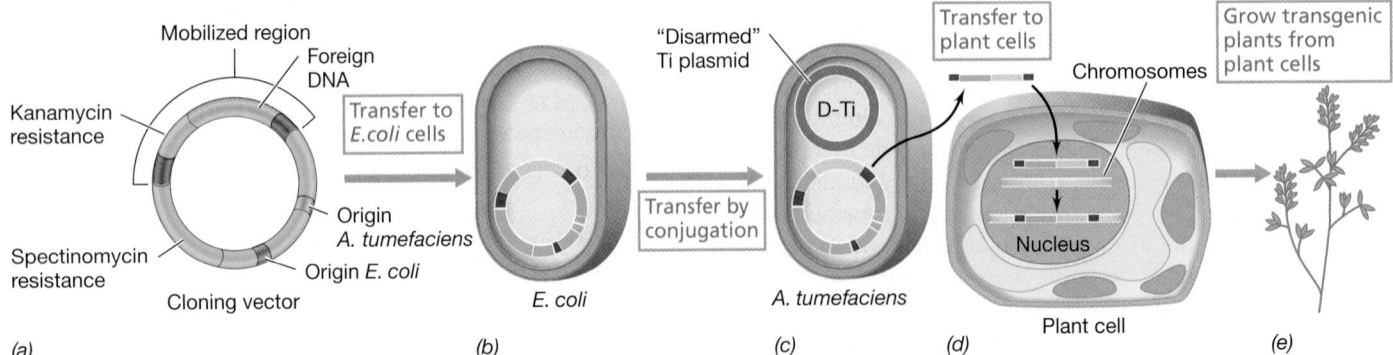

Figure 15.26 Production of transgenic plants using a binary vector system in *Agrobacterium tumefaciens*. *(a)* Plant cloning vector containing ends of T-DNA (red), foreign DNA, origins of replication, and resistance markers. *(b)* The vector is put into cells of *Escherichia coli* for cloning and then *(c)* transferred to *A. tumefaciens* by conjugation. The resident Ti plasmid (D-Ti) has been genetically engineered to remove key pathogenesis genes. *(d)* D-Ti can still mobilize the T-DNA region of the vector for transfer to plant cells grown in tissue culture. *(e)* From the recombinant plant cell, a whole plant can be grown.

Stephen R. Padgette, Monsanto Company

Figure 15.27 **Transgenic plants: herbicide resistance.** The photograph shows a portion of a field of soybeans that has been treated with Roundup™, a glyphosate-based herbicide manufactured by Monsanto Company (St. Louis, MO). The plants on the right are normal soybeans; those on the left have been genetically engineered to be glyphosate resistant.

For example, the herbicide glyphosate (Roundup™) kills plants by inhibiting an enzyme necessary for making aromatic amino acids. Some bacteria contain an equivalent enzyme and are also killed by glyphosate. However, mutant bacteria were selected that were resistant to glyphosate and contained a resistant form of the enzyme. The gene encoding this resistant enzyme from *A. tumefaciens* was cloned, modified for expression in plants, and transferred into important crop plants, such as soybeans. When sprayed with glyphosate, plants containing the bacterial gene are not killed (**Figure 15.27**). Thus glyphosate can be used to kill weeds that compete for water and nutrients with the growing crop plants. Herbicide-resistant soybeans are now widely planted in the United States.

Insect Resistance: Bt Toxin

Insect resistance has also been genetically introduced into plants. One widely used approach is based on introducing genes encoding the toxic proteins of *Bacillus thuringiensis* into plants. *B. thuringiensis* produces a crystalline protein called *Bt toxin* (↺ Section 18.2) that is toxic to moth and butterfly larvae. Many variants of Bt toxin exist that are specific for different insects. Certain strains of *B. thuringiensis* produce additional proteins toxic to beetle and fly larvae and mosquitoes.

Several different approaches are used to enhance the action of Bt toxin for pest control in plants. One approach is to use a single set of Bt toxins that is effective against many different insects. This is possible because the protein consists of separate structural regions (domains) that are responsible for specificity and toxicity. The toxic domain is highly conserved in all the various Bt toxins. Genetic engineers have made hybrid genes that encode the toxic domain and one of several different specificity domains to yield a suite of toxins, each best suited for a particular plant or pest situation.

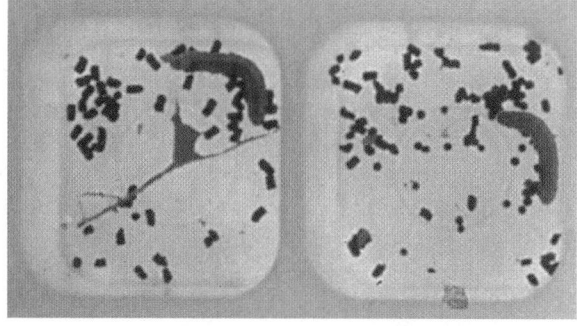

(a)

(b)

Kevin McBride, Calgene, Inc.

Figure 15.28 **Transgenic plants: insect resistance.** *(a)* The results of two different assays to determine the effect of beet armyworm larvae on tobacco leaves from normal plants. *(b)* The results of similar assays using tobacco leaves from transgenic plants that express Bt toxin in their chloroplasts.

The Bt transgene is normally inserted directly into the plant genome. For example, a natural Bt toxin gene was cloned into a plasmid vector under control of a chloroplast ribosomal RNA promoter and then transferred into tobacco plant chloroplasts by microprojectile bombardment (↺ Section 11.7). This yielded transgenic plants that expressed Bt toxin at levels that were extremely toxic to larvae from a number of insect species (**Figure 15.28**).

Although transgenic Bt toxin looked at first to be a great agricultural success, some problems have arisen, in particular, the selection of insects resistant to Bt toxin. Resistance to insecticides and herbicides is a common problem in agriculture, and the fact that a product has been produced by genetic engineering does not exempt it from this problem. In addition, Bt toxin often kills nontarget insects, some of which may be helpful. Many approaches must be used for pest control in agriculture, and Bt toxin is just one of many. Nevertheless, transgenic crops with Bt toxin are widely planted in the United States.

Bt toxin is harmless to mammals, including humans, for several reasons. First, cooking and food processing destroy the toxin. Second, any toxin that is ingested is digested and therefore inactivated in the mammalian gastrointestinal tract. Third, Bt toxin works by binding to specific receptors in the insect intestine that are absent from the intestines of other groups of organisms. Binding promotes a change in conformation of the toxin, which then generates pores in the intestinal lining of the insect that disrupt the insect digestive system and kill the insect.

UNIT 5

Other Uses of Plant Biotechnology

Not all genetic engineering is directed toward making plants disease resistant. Genetic engineering can also be used to develop GM plants that are more nutritious or that have more desirable consumer-oriented characteristics. For instance, the first GM food grown for sale in the United States market was a tomato in which spoilage was delayed, increasing the shelf life. In addition, transgenic plants can be genetically engineered to produce commercial or pharmaceutical products, as has been done with microorganisms and animals. For example, crop plants such as tobacco and tomatoes have been engineered to produce a number of products, such as the human protein *interferon*. Transgenic crop plants can also be used to produce human antibodies efficiently and inexpensively. These antibodies, called *plantibodies*, have potential as anticancer or antiviral drugs, and some are undergoing clinical trials. For example, transgenic tobacco plants have been used to make an antibody known as CaroRx that blocks bacteria that cause dental caries from attaching to teeth. CaroRx is made in high levels in tobacco leaves and is relatively easy to purify. So far clinical trials have shown it to be safe and effective. Plants are useful in producing these types of products because they typically modify proteins correctly and because crop plants can be efficiently grown and harvested in large amounts.

Crop plants are also being developed for the production of vaccines. For instance, a recombinant tobacco mosaic virus (♻ Section 21.7) has been engineered whose coat contains surface proteins of *Plasmodium vivax*, one of the microbial parasites that cause malaria (♻ Section 34.5). The *P. vivax* proteins elicit an immune response in humans. Hence, this recombinant virus could be used to produce a malaria vaccine cheaply in large amounts by simply harvesting infected tobacco or tomato plants and processing them for the immunogenic proteins. Another interesting approach is to produce a vaccine in an edible plant product. Such *edible vaccines* now under development could immunize humans against diseases caused by enteric bacteria, including cholera and diarrhea (♻ Section 35.5).

A rather different kind of transgenic plant is the *Amflora potato*, developed by BASF, a German chemical company. The Amflora potato is not intended for eating. Unlike normal potatoes, which produce two types of starch, amylopectin and amylose, the Amflora potato makes only amylopectin, a raw material in the paper and adhesives industries. Use of Amflora potatoes will avoid the expensive and energy-consuming purification that removes amylose from amylopectin. Approval of this transgenic crop is presently under consideration by the European Union.

Although public acceptance of GM crops remains high in the United States, there have been some concerns over the contamination of human food with GM corn, so far approved only for animal food. In some European Union countries there has been considerable public concern over GM organisms. Most concerns center around either the perception of adverse effects of foreign genes on humans or domesticated animals or the potential "escape" of transgenes from transgenic plants into native plants. At present, supporting evidence for either of these scenarios is not strong; however, there are indications that glyphosate resistance genes are beginning to spread into the weed plant population. Thus, concerns remain about GM plants and have served to control the rate at which new transgenic plants enter the marketplace.

MiniQuiz

- What is a transgenic plant?
- Give an example of a genetically modified plant and describe how its modification benefits agriculture.
- What advantages do plants have as vehicles for making antibodies?

Big Ideas

15.1

An industrial microorganism must synthesize a product in high yield, grow rapidly on inexpensive culture media available in bulk, be amenable to genetic analysis, and be nonpathogenic. Industrial products include both cells and substances made by cells.

15.2

Primary metabolites are produced during the exponential phase and secondary metabolites are produced later. Large-scale aerobic industrial fermentations require stirring and aeration, and the process must be continuously monitored to ensure high product yields. Scale-up is the process of converting an industrial fermentation from laboratory scale to production scale.

15.3

Industrial production of antibiotics begins with screening for antibiotic producers. Once new producers are identified, chemical analyses of the antimicrobial agent are performed. If the new antibiotic is biologically active and nontoxic in both experimental animals and humans, high-yielding strains are sought for more cost-efficient commercial production.

15.4

The β-lactam antibiotics penicillin and cephalosporin and the tetracyclines are major drugs of medical and veterinary importance. Biosynthetic and semisynthetic derivatives of natural penicillins are the most widely used penicillins today.

15.5

Vitamins produced by industrial microbiology include vitamin B_{12} and riboflavin, and the major amino acids produced commercially are glutamate and lysine. High yields of amino acids are obtained by modifying regulatory signals that control their synthesis.

15.6

Many microbial enzymes are used in the laundry industry to remove stains from clothing, and thermostable and alkali-stable enzymes have many advantages for this application. When an enzyme is used in a large-scale process, it is often immobilized by being bonded to an inert substrate.

15.7

Most wine is made by fermenting the juice of grapes. Complex chemical changes occur during the wine fermentation due to a suite of chemicals in addition to alcohol. The malolactic fermentation is used to remove bitterness and produce a smoother final product.

15.8

Brewed products are made from malted grains, and distilled spirits are made by the distillation of ethanol and other flavor ingredients from brews, other fermented products, and wines.

15.9

Ethanol for use in biofuels can be made by the fermentation of glucose from starch or cellulose. Gasohol is the major biofuel in the United States and usually consists of a blend of gasoline (90%) and ethanol (10%).

15.10

To achieve very high levels of expression of eukaryotic genes in prokaryotes, the expressed gene must be free of introns. This can be accomplished by synthesizing cDNA from the mature mRNA encoding the protein of interest or by making an entirely synthetic gene. Protein fusions are often used to stabilize or solubilize the cloned protein.

15.11

The first human protein made commercially using engineered bacteria was human insulin. Recombinant bovine somatotropin is widely used in the United States to increase milk yield in dairy cows.

15.12

Many proteins are extremely expensive to obtain by traditional purification methods because they are found in human or animal tissues in only small amounts. Many of these can now be made in large amounts from a cloned gene in a suitable expression system.

15.13

Many recombinant vaccines have been produced or are under development. These include live recombinant, vector, subunit, and DNA vaccines.

15.14

Genes for useful products may be cloned directly from DNA or RNA in environmental samples without first isolating the organisms that carry them.

15.15

In pathway engineering, genes that encode the enzymes for a metabolic pathway are assembled. These genes may come from one or more organisms, but the engineering must achieve regulation of the coordinated sequence of expression required in the pathway.

15.16

Genetic engineering can make transgenic animals that produce proteins of pharmaceutical value and animal models of human diseases for medical research. Most recently, attempts are being made to improve livestock for human consumption and to reduce harmful environmental effects of mass-produced livestock.

15.17

A major hope of genetic engineering is in human gene therapy, a process whereby functional copies of a gene are inserted into a person to treat a genetic disease.

15.18

Genetic engineering can make plants resistant to disease, improve product quality, and make crop plants a source of recombinant proteins and vaccines. The Ti plasmid of the bacterium *Agrobacterium tumefaciens* can transfer DNA into plant cells. Genetically engineered commercial plants are called genetically modified (GM) organisms.

Review of Key Terms

Aspartame a nonnutritive sweetener composed of the amino acids aspartate and phenylalanine, the latter as a methyl ester

β-Lactam antibiotic a member of a group of antibiotics including penicillin that contain the four-membered heterocyclic β-lactam ring

Biofuel a fuel made by microorganisms from the fermentation of carbon-rich feedstocks

Biosynthetic penicillin a form of penicillin produced by supplying the synthesizing microorganism with a specific side-chain precursor

Biotechnology the use of genetically engineered organisms in industrial, medical, or agricultural applications

Brewing the manufacture of alcoholic beverages such as beer from the fermentation of malted grains

Broad-spectrum antibiotic an antimicrobial drug useful in treating a wide variety of bacterial diseases caused by both gram-negative and gram-positive bacteria

Distilled alcoholic beverage a beverage containing alcohol concentrated by distillation

DNA vaccine a vaccine that uses the DNA of a pathogen to elicit an immune response

Exoenzyme an enzyme produced by a microorganism and then excreted into the environment

Extremozyme an enzyme able to function in one or more chemical or physical extremes, for example, high temperature or low pH

Fermentation in an industrial context, any large-scale microbial process, whether carried out aerobically or anaerobically

Fermentor a tank in which an industrial fermentation is carried out

Fusion protein a genetically engineered protein made by fusing two DNA sequences encoding different proteins together into a single gene

Gene therapy the treatment of a disease caused by a dysfunctional gene by introducing a functional copy of the gene

Genetically modified organism (GMO) an organism whose genome has been altered using genetic engineering; the abbreviation GM is also used in terms such as GM crops and GM foods

Genetic engineering the use of *in vitro* techniques in the isolation, manipulation, alteration, and expression of DNA (or RNA), and in the development of genetically modified organisms

Immobilized enzyme an enzyme attached to a solid support over which substrate is passed and converted to product

Industrial microbiology the large-scale use of microorganisms to make products of commercial value

Malolactic fermentation a secondary fermentation used to remove bitterness in the production of some wines by the conversion of malic acid to lactic acid

Natural penicillin the parent penicillin structure produced by cultures of *Penicillium* not supplemented with side-chain precursors

Pathway engineering the assembly of a new or improved biochemical pathway using genes from one or more organisms

Polyvalent vaccine a vaccine that immunizes against more than one disease

Primary metabolite a metabolite excreted during a microorganism's exponential growth phase

Protease an enzyme that degrades proteins by hydrolysis

Reverse transcription the conversion of an RNA sequence into the corresponding DNA sequence

Scale-up the conversion of an industrial process from a small laboratory setup to a large commercial fermentation

Secondary metabolite a metabolite excreted from a microorganism at the end of its exponential growth phase and into the stationary phase

Semisynthetic penicillin a penicillin produced using components derived from both microbial fermentation and chemical syntheses

T-DNA the segment of the *Agrobacterium tumefaciens* Ti plasmid that is transferred into plant cells

Tetracycline a member of a class of antibiotics containing the four-membered naphthacene ring

Ti plasmid a plasmid in *Agrobacterium tumefaciens* capable of transferring genes from bacteria to plants

Transgenic organism a plant or an animal with foreign DNA inserted into its genome

Vector vaccine a vaccine made by inserting genes from a pathogenic virus into a relatively harmless carrier virus

Review Questions

1. In what ways do industrial microorganisms differ from microorganisms in nature? In what ways are they similar (Section 15.1)?

2. List three major types of industrial products that can be obtained with microorganisms, and give two examples of each (Section 15.1).

3. Compare and contrast primary and secondary metabolites, and give an example of each. List at least two molecular explanations for why some metabolites are secondary rather than primary (Section 15.2).

4. How does an industrial fermentor differ from a laboratory culture vessel? How does a fermentor differ from a fermenter (Section 15.2)?

5. List three examples of antibiotics that are important products of industrial microbiology. For each of these antibiotics, list the producing organisms and the general chemical structure (Sections 15.3 and 15.4).

6. Outline the major steps in the production of tetracycline (Section 15.4).

7. What unusual characteristics must an organism have if it is to overproduce and excrete an amino acid such as lysine (Section 15.5)?

8. What is high-fructose syrup, how is it produced, and what is it used for in the food industry (Section 15.6)?

9. What is the advantage of using immobilized enzymes (Section 15.6)?

10. In what way is the manufacture of beer similar to the manufacture of wine? In what ways do these two processes differ? How does the production of distilled alcoholic beverages differ from that of beer and wine (Sections 15.7 and 15.8)?

11. What is the major liquid biofuel made worldwide? How is it currently being made in the United States? Why is it necessary for new feedstocks to be developed (Section 15.9)?

12. How do fusion proteins help to improve protein purification (Section 15.10)?

13. What classes of mammalian proteins are produced by biotechnology? How are the genes for such proteins obtained (Sections 15.11 and 15.12)?

14. Briefly describe the two different types of vaccines that are widely used currently (Section 15.13).

15. How has metagenomics been used to find novel useful products (Section 15.14)?

16. What is pathway engineering? Why is it more difficult to produce an antibiotic than to produce a single enzyme via genetic engineering (Section 15.15)?

17. What is a knockout mouse? Why are knockout mice important for the study of human physiology and hereditary defects (Section 15.16)?

18. How has genetic engineering benefited the treatment of SCID and cystic fibrosis (Sections 15.12 and 15.17)?

19. What is the Ti plasmid and how has it been of use in genetic engineering (Section 15.18)?

20. List several examples in which crop plants have been improved by genetic engineering. How have genetically engineered plants helped human medicine (Section 15.18)?

Application Questions

1. As a researcher in a pharmaceutical company you are assigned the task of finding and developing an antibiotic effective against a new bacterial pathogen. Outline a plan for this process, starting from isolation of the low-yield producing organism to high-yield industrial production of the new antibiotic.

2. You wish to produce high yields of the amino acid phenylalanine for use in production of the sweetener aspartame. The overproducing organism you wish to use is not subject to feedback inhibition by phenylalanine, but is subject to typical repression of phenylalanine biosynthesis enzymes by excess phenylalanine. Applying the principles of enzyme regulation studied in Chapter 8 and microbial genetics in Chapter 10, describe two classes of mutants you could isolate that would overcome this problem, and detail the genetic lesions each would have.

3. You have just discovered a protein in mice that may be an effective cure for cancer, but it is present only in tiny amounts. Describe the steps you would use to produce this protein in therapeutic amounts. Which host would you want to clone the gene into and why? Which host would you use to express the protein in and why?

4. Gene therapy is used to treat people who have a genetic disease and, if successful, it will cure them. However, such people will still be able to pass on the genetic disease to their offspring. Explain. Why do you believe this might be an area of research that is not attracting as much attention as treatment of the individual?

5. Compare the advantages and disadvantages of using transgenic crops (such as Bt corn) as a source of human food. Consider various viewpoints, including that of the farmer, the environmentalist, and the consumer.

Need more practice? Test your understanding with Quantitative Questions; access additional study tools including tutorials, animations, and videos; and then test your knowledge with chapter quizzes and practice tests at **www.microbiologyplace.com**.

16

Microbial Evolution and Systematics

Fluorescent dyes bound to specific nucleic acid probes can differentiate cells in natural samples that are morphologically similar but phylogenetically distinct.

A unifying theme in all of biology is **evolution**. By deploying its major tools of descent through modification and selection of the fittest, evolution has affected all life on Earth, from the first self-replicating entities, be they cells or otherwise, to the modern cells we see today. Since its origin, Earth has undergone a continuous process of physical and geological change, eventually establishing conditions conducive to the origin of life. After microbial life appeared, Earth continued to present it with new opportunities and challenges. As microbial metabolisms and physiologies evolved in response, microbial activities changed planet Earth in significant ways to yield the biosphere we see today.

This chapter focuses on the evolution of microbial life, from the origins of the earliest cells and metabolisms to the microbial diversity we see today. Methods for discerning evolutionary relationships among modern-day descendants of early microbial lineages are a major theme. Overall, the goal of this chapter is to provide an evolutionary and systematic framework for the diversity of contemporary microbial life that we will explore in the next four chapters.

Early Earth and the Origin and Diversification of Life

In these first few sections, we consider the possible conditions under which life arose, the processes that might have given rise to the first cellular life, its divergence into two evolutionary lineages, *Bacteria* and *Archaea*, and the later formation, through endosymbiosis, of a third lineage, the *Eukarya*. Although much about these events and processes remains speculative, geological and molecular evidence has combined to build a plausible scenario for how life might have arisen and diversified.

16.1 Formation and Early History of Earth

Before considering how life arose, we need to go back even farther, and ask how Earth itself formed.

Origin of Earth

Earth is thought to have formed about 4.5 billion years ago, based on analyses of slowly decaying radioactive isotopes. Our planet and the other planets of our solar system arose from materials making up a disc-shaped nebular cloud of dust and gases released by the supernova of a massive old star. As a new star—our sun—formed within this cloud, it began to compact, undergo nuclear fusion, and release large amounts of energy in the form of heat and light. Materials left in the nebular cloud began to clump and fuse due to collisions and gravitational attractions, forming tiny accretions that gradually grew larger to form clumps that eventually coalesced into planets. Energy released in this process heated the emerging Earth as it formed, as did energy released by radioactive decay within the condensing materials, forming a planet Earth of fiery hot magma. As Earth cooled over time, a metallic core, rocky mantle, and a thin lower-density surface crust formed.

The inhospitable conditions of early Earth, characterized by a molten surface under intense bombardment from space by asteroids and other objects, are thought to have persisted for over 500 million years. Water on Earth originated from innumerable collisions with icy comets and asteroids and from volcanic outgassing of the planet's interior. At this time, due to the heat, water would have been present only as water vapor. No rocks dating to the origin of planet Earth have yet been discovered, presumably because they have undergone geological metamorphosis. However, ancient sedimentary rocks, which formed under liquid water, have been found in several locations on Earth. Some of the oldest sedimentary rocks discovered thus far are in southwestern Greenland; these rocks date to about 3.86 billion years ago. The sedimentary composition of these rocks indicates by that time Earth had at least cooled sufficiently (<100°C) for the water vapor to have condensed and formed the early oceans.

Even more ancient materials, crystals of the mineral zircon ($ZrSiO_4$), however, have been discovered, and these materials give us a glimpse of even earlier conditions on Earth. Impurities trapped in the crystals and the mineral's isotopic ratios of oxygen (⮌ Section 22.8) indicate that Earth cooled much earlier than previously believed, with solid crust forming and water condensing into oceans perhaps as early as 4.3 billion years ago. The presence of liquid water implies that conditions could have been compatible with life within a couple of hundred million years after Earth was formed.

Evidence for Microbial Life on Early Earth

The fossilized remains of cells and the isotopically "light" carbon abundant in these rocks provide evidence for early microbial life (we discuss the use of isotopic analyses of carbon and sulfur as indications of living processes in Section 22.8). Some ancient rocks contain what appear to be bacteria-like microfossils, typically simple rods or cocci (**Figure 16.1**).

In rocks 3.5 billion years old or younger, microbial formations called **stromatolites** are common. Stromatolites are microbial mats consisting of layers of filamentous prokaryotes and trapped

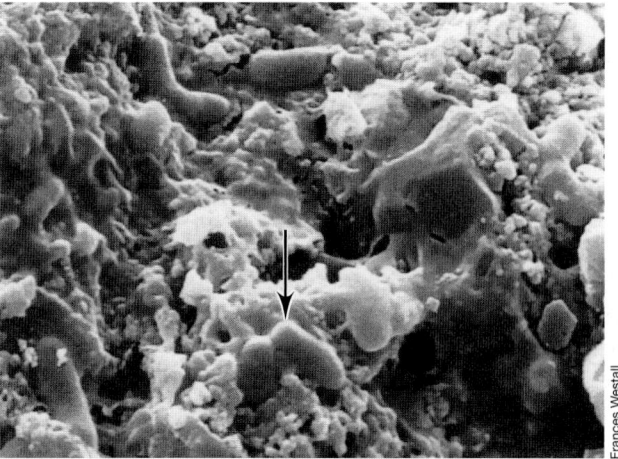

Figure 16.1 Ancient microbial life. Scanning electron micrograph of microfossil bacteria from 3.45 billion-year-old rocks of the Barberton Greenstone Belt, South Africa. Note the rod-shaped bacteria (arrow) attached to particles of mineral matter. The cells are about 0.7 μm in diameter.

(a)

(b)

(c)

(e)

(d)

Figure 16.2 Ancient and modern stromatolites. (a) The oldest known stromatolite, found in a rock about 3.5 billion years old, from the Warra-woona Group in Western Australia. Shown is a vertical section through the laminated structure preserved in the rock. Arrows point to the laminated layers. (b) Stromatolites of conical shape from 1.6 billion-year-old dolomite rock from northern Australia. (c) Modern stromatolites in Shark Bay, Western Australia. (d) Modern stromatolites composed of ther-mophilic cyanobacteria growing in a thermal pool in Yellowstone National Park. Each structure is about 2 cm high. (e) Another view of modern and very large stromatolites from Shark Bay. Individual structures are 0.5–1 m in diameter.

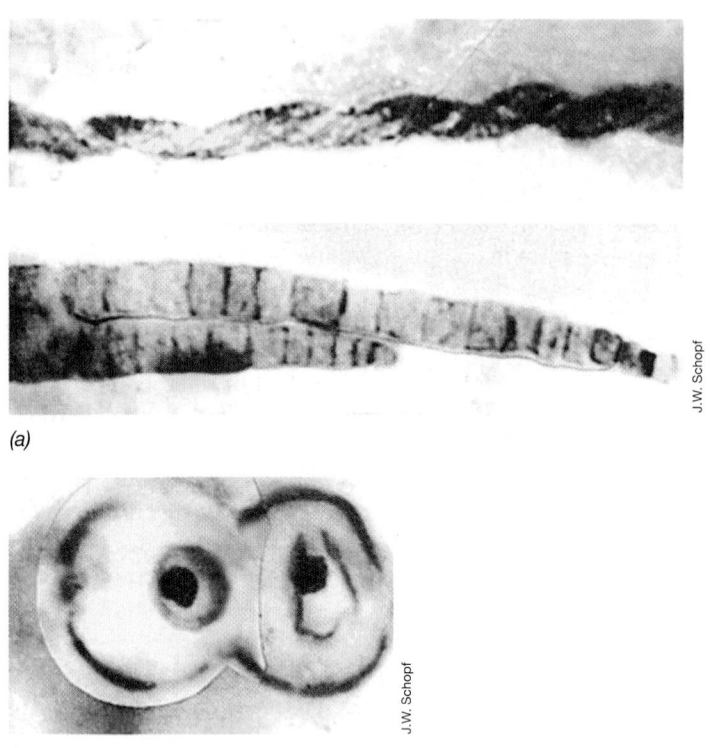

(a)

(b)

Figure 16.3 More recent fossil bacteria and eukaryotes. (a) One billion-year-old microfossils from central Australia that resemble modern filamentous cyanobacteria. Cell diameters, 5–7 μm. (b) Microfossils of eukaryotic cells from the same rock formation. The cellular structure is similar to that of certain modern green algae, such as *Chlorella* species. Cell diameter, about 15 μm. Color was added to make cell form more apparent.

mineral materials; they may become fossilized (**Figure 16.2a, b**) (we discuss microbial mats in Section 23.5). What kind of organisms were these ancient stromatolitic bacteria? By comparing ancient stromatolites with modern stromatolites growing in shallow marine basins (Figure 16.2c and e) or in hot springs (Figure 16.2d; ⊖⊃ Figure 23.9b), we can see it is likely that ancient stromatolites formed from filamentous phototrophic bacteria, such as ancestors of the green nonsulfur bacterium *Chloroflexus* (⊖⊃ Section 18.18). **Figure 16.3** shows photomicrographs of thin sections of more recent rocks containing microfossils that appear remarkably similar to modern species of cyanobacteria and green algae, both of which are oxygenic phototrophs (⊖⊃ Sections 18.7 and 20.20). The age of these microfossils, about 1 billion years, is well within the time frame that such organisms were thought to be present on Earth (⊖⊃ Figure 1.6).

In summary, microfossil evidence strongly suggests that microbial life was present within at least 1 billion years of the formation of Earth and probably somewhat earlier, and that by that time, microorganisms had already attained an impressive diver-

sity of morphological forms. We tackle the issue of how life first evolved from nonliving materials in the next section. But regardless of when self-replicating life forms first appeared, the process of evolution began at the same time, selecting for improvements that would eventually lead to microbial cells' inhabiting every ecosystem on Earth that was chemically and physically compatible with life.

MiniQuiz

• How did planet Earth form?

• What evidence is there that microbial life was present on Earth 3 billion years ago?

• What do crystals of the mineral zircon tell us about conditions for early life?

16.2 Origin of Cellular Life

Here we consider the issue of how living organisms might have originated, focusing on two questions: (1) How might the first cells have arisen? (2) What might those early cells have been like? But along the way, we will consider the likely possibility that self-replicating RNAs preceded cellular life and how these molecules may have paved the way for cellular life.

Surface Origin Hypothesis

One hypothesis for the origin of life holds that the first membrane-enclosed, self-replicating cells arose out of a primordial soup rich in organic and inorganic compounds in a "warm little pond," as Charles Darwin suggested in *On the Origin of Species*—in other words, life arose on Earth's surface. Although there is experimental evidence that organic precursors to living cells can form spontaneously under certain conditions, surface conditions on early Earth are thought to have been hostile to both life and its inorganic and organic precursors. The dramatic temperature fluctuations and mixing resulting from meteor impacts, dust clouds, and storms, along with intense ultraviolet radiation, make a surface origin for life unlikely.

Subsurface Origin Hypothesis

A more likely hypothesis is that life originated at hydrothermal springs on the ocean floor, well below Earth's surface, where conditions would have been much less hostile and much more stable. A steady and abundant supply of energy in the form of reduced inorganic compounds, for example, hydrogen (H_2) and hydrogen sulfide (H_2S), would have been available at these spring sites. When the very warm (90–100°C) hydrothermal water flowed up through the crust and mixed with cooler, iron-containing and more oxidized oceanic waters, precipitates of colloidal pyrite (FeS), silicates, carbonates, and magnesium-containing montmorillonite clays formed. These precipitates built up into structured mounds with gel-like adsorptive surfaces, semipermeable enclosures, and pores. *Serpentinization*, the abiotic process by which Fe/Mg silicates (serpentines) react with other minerals and H_2, was a likely source of the first organic compounds, such as hydrocarbons and fatty acids. These could then have reacted with iron and nickel sulfide minerals to eventually form amino acids, simple peptides, sugars, and nitrogenous bases (**Figure 16.4**).

With phosphate from seawater, nucleotides such as AMP and ATP could have been formed and polymerized into RNA by montmorillonite clay, a material known to catalyze such reactions. The flow of H_2 and H_2S from the crust provided steady sources of electrons for this prebiotic chemistry, and the process was perhaps powered by redox and pH gradients developed across semipermeable FeS membrane-like surfaces, providing a prebiotic proton motive force (↩ Section 4.10).

An important point to keep in mind here is that before life appeared on Earth, organic precursors of life would not have been consumed by organisms, as they would be today. So the possibility that millions of years ago organic matter accumulated to levels where self-replicating entities emerged, is not an unreasonable hypothesis.

An RNA World and Protein Synthesis

The synthesis and concentration of organic compounds by prebiotic chemistry set the stage for self-replicating systems, the precursors to cellular life. How might self-replicating systems have arisen? One possibility is that there was an early *RNA world,* in which the first self-replicating systems were molecules of RNA (Figure 16.4). Although fragile, RNA could have survived in the cooler temperatures where the gel-like precipitates formed at ocean floor warm springs. Because RNA can bind small mole-

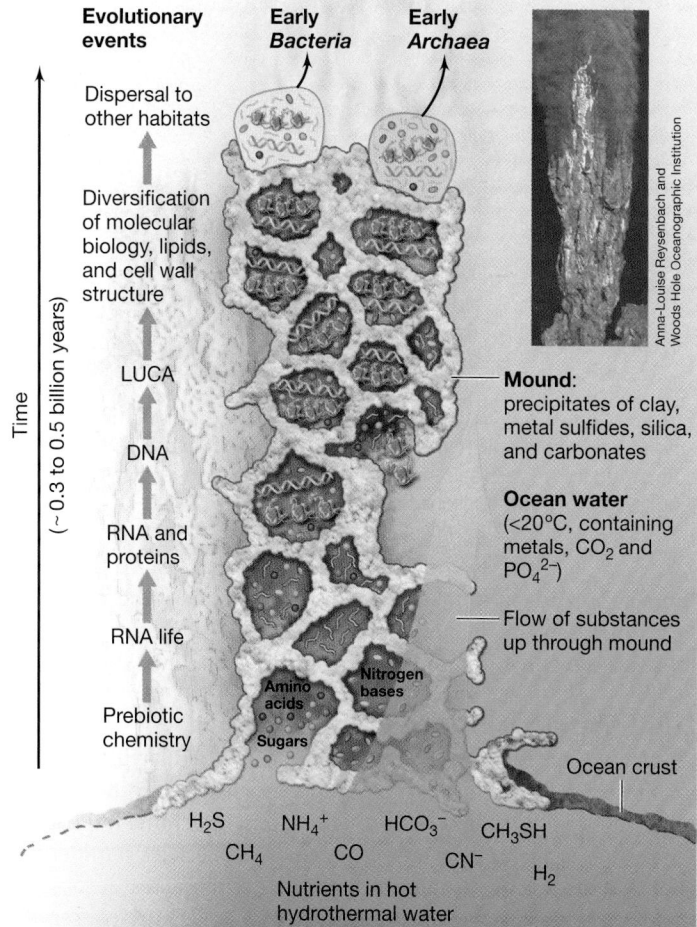

Figure 16.4 Submarine mounds and their possible link to the origin of life. Model of the interior of a hydrothermal mound with hypothesized transitions from prebiotic chemistry to cellular life depicted. Key milestones are self-replicating RNA, enzymatic activity of proteins, and DNA taking on a genetic coding function, leading to early cellular life. This was followed by diversification of molecular biology and biochemistry, eventually giving rise to early *Bacteria* and *Archaea*. LUCA, last universal common ancestor. Inset: photo of an actual hydrothermal mound. Hot mineral-rich hydrothermal fluid mixes with cooler, more oxidized, ocean water, forming precipitates. The mound is composed of precipitates of Fe and S compounds, clays, silicates, and carbonates.

cules, such as ATP and other nucleotides, and has catalytic activity (ribozymes, ↩ Section 7.8), RNA might have catalyzed its own synthesis from the available sugars, bases, and phosphate.

RNA also can bind other molecules, such as amino acids, catalyzing the synthesis of primitive proteins. As different proteins were made and then accumulated in the RNA world, they coated the inner surfaces of the hydrothermal mounds. Later, as different types of proteins emerged, some with catalytic abilities, proteins began to take over the catalytic role of RNAs (Figure 16.4). Eventually, DNA, a molecule more stable than RNA and therefore a better repository of genetic (coding) information, arose and assumed the template role for RNA synthesis. This three-part system—DNA, RNA, and protein—became fixed early on as the fittest solution to biological information processing. Following these steps, one can envision a time of intensive biochemical

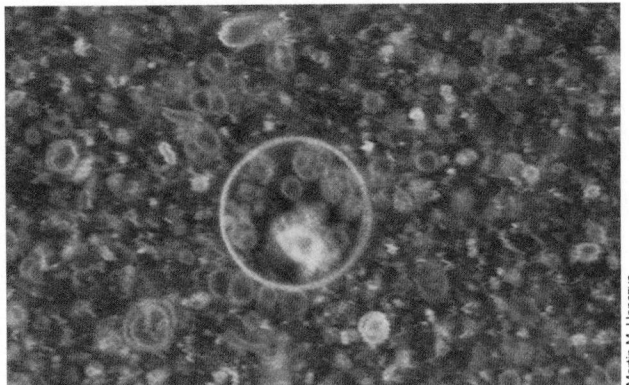

Figure 16.5 Lipid vesicles made in the laboratory from the fatty acid myristic acid and RNA. The vesicle itself stains green, and the RNA complexed inside the vesicle stains red. Vesicle synthesis is catalyzed by the surfaces of montmorillonite clay particles.

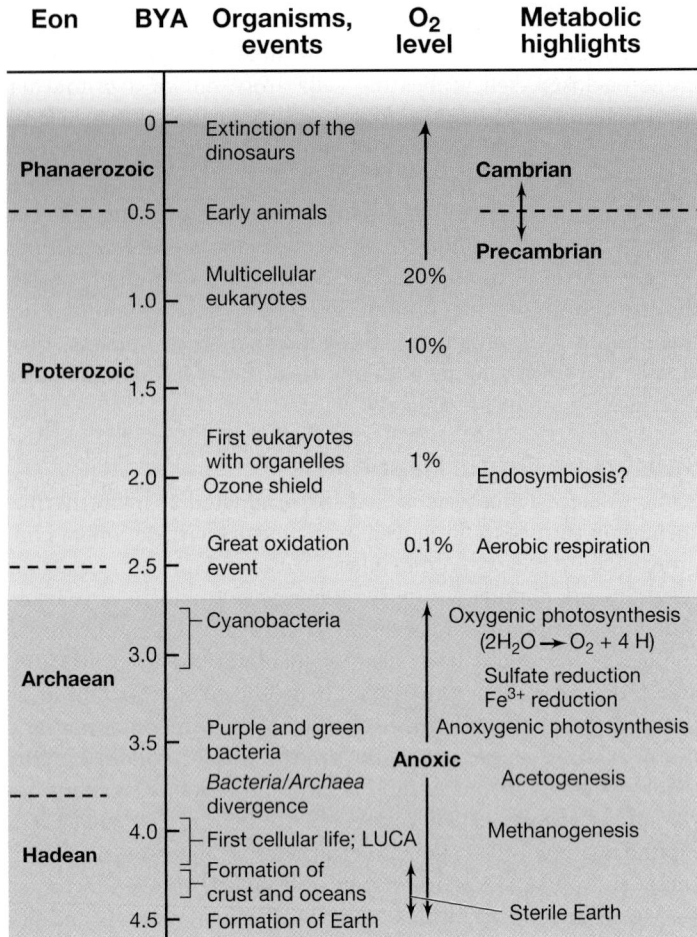

Figure 16.6 Major landmarks in biological evolution, Earth's changing geochemistry, and microbial metabolic diversification. The maximum time for the origin of life is fixed by the time of the origin of Earth, and the minimum time for the origin of oxygenic photosynthesis is fixed by the Great Oxidation Event, about 2.4 billion years ago (BYA). Note how the oxygenation of the atmosphere from cyanobacterial metabolism was a gradual process, occurring over a period of about 2 billion years. *Bacteria* respiring at low O_2 levels likely dominated Earth for a billion years or so before Earth's atmosphere reached current levels of oxygen. Compare this figure with the introduction to the antiquity of life on Earth shown in Figure 1.6.

innovation and experimentation in which much of the structural and functional machinery of these earliest self-replicating systems was invented and refined by natural selection.

Lipid Membranes and Cellular Life

Anther important step in the emergence of cellular life was the synthesis of phospholipid membrane vesicles that could enclose the evolving biochemical and replication machinery. Proteins embedded in the lipids would have made the vesicles semipermeable and thus able to shuttle nutrients and wastes across the membrane, setting the stage for the evolution of energy-conserving processes and ATP synthesis. By entrapping RNA and DNA, these lipoprotein vesicles, which may have been similar to montmorillonite clay vesicles that can be synthesized in the laboratory (**Figure 16.5**), may have enclosed the first self-replicating entities, partitioning the biochemical machinery in a unit not unlike the cells we know today.

From this population of structurally very simple early cells, referred to as the *last universal common ancestor (LUCA)*, cellular life began to evolve in two distinct directions, possibly in response to physiochemical differences in their most successful niches (Figure 16.4). These two populations of cells would have then undergone strong selection for improvements in transport, metabolism, motility, energy conservation, and the many other structural and functional aspects we associate with cells today. In the two lineages similar overarching processes evolved, but many of the underlying details differed. For example, the two populations evolved different lipids, cell walls, specialized metabolisms, and enzymatic machinery for nucleic acid replication and protein synthesis. As natural selection continued, these two prokaryotic lineages, the *Bacteria* and the *Archaea*, became ever more distinct, displaying the characteristic properties we associate with each lineage today (see Table 16.1).

Early Metabolism

From the time of formation, the early ocean and all of Earth was anoxic. Molecular oxygen (O_2) did not appear in any significant quantities until oxygenic photosynthesis by cyanobacteria

evolved (**Figure 16.6**). Thus, the energy-generating metabolism of primitive cells would have been exclusively anaerobic and would likely have had to be heat-stable because of the temperature of early Earth. Carbon metabolism may well have been autotrophic because consumption of abiotically formed organic compounds for cellular material probably would have exhausted these compounds relatively quickly. The possibility that the use of CO_2 as a carbon source (autotrophy) was an early physiological lifestyle is also supported by the metabolism of many of the earliest lineages on the phylogenetic tree of life (see Figure 16.16); for example, the genera *Aquifex* (*Bacteria*) and *Pyrolobus* (*Archaea*) are autotrophs and, not surprisingly, are also hyperthermophiles.

It is widely thought that H_2 was a major fuel for energy metabolism of early cells. This hypothesis is also supported by the tree of life, in that virtually all of the earliest branching organisms in the *Bacteria* and *Archaea* lineages use H_2 as an electron donor in energy metabolism. Abiotic reactions between iron sulfide minerals and hydrogen sulfide have been proposed as a source of the needed H_2:

$$FeS + H_2S \rightarrow FeS_2 + H_2 \qquad \Delta G^{0\prime} = -42 kJ$$

Also, ferrous iron (Fe^{2+}) can reduce protons to H_2 in the presence of ultraviolet radiation as an energy source. Regardless of the source, H_2 could have fueled a primitive ATPase in the cytoplasmic membrane of early cells to yield ATP (**Figure 16.7**). However, with H_2 as an electron donor, an electron acceptor would also have been required to form a redox pair (Section 4.6); this could have been elemental sulfur (S^0). As shown in Figure 16.7, the oxidation of H_2 with the reduction of S^0 to yield H_2S is exergonic and would likely have required few enzymes. Moreover, because of the abundance of H_2 and sulfur compounds on early Earth, this scheme would have provided cells with a nearly limitless supply of energy.

These early forms of chemolithotrophic metabolism driven by H_2 would likely have supported the production of large amounts of organic compounds from autotrophic CO_2 fixation. Over time, these organic materials would have accumulated and could have provided the environment needed for the appearance of new chemoorganotrophic bacteria with diverse metabolic strategies to conserve energy from organic compounds; metabolic diversity (Chapter 14) would have been off and running.

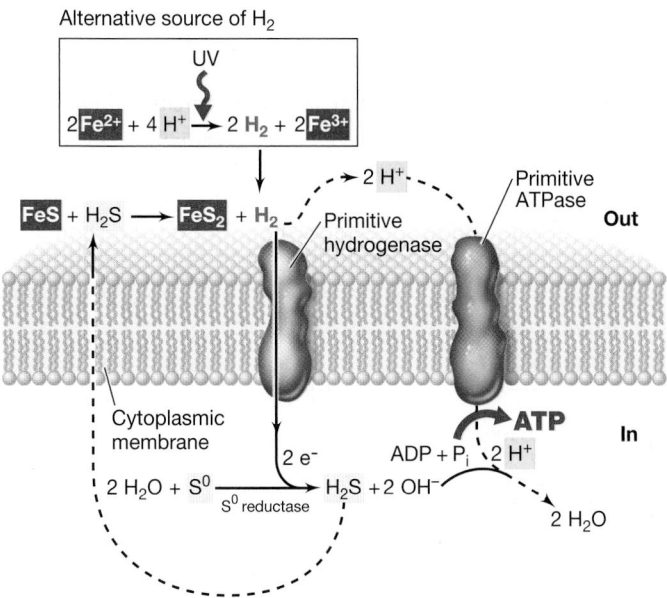

Figure 16.7 **A possible energy-generating scheme for primitive cells.** Formation of pyrite leads to H_2 production and S^0 reduction, which fuels a primitive ATPase. Note how H_2S plays only a catalytic role; the net substrates would be FeS and S^0. Also note how few different proteins would be required. $\Delta G^{0\prime} = -42$ kJ for the reaction $FeS + H_2S \rightarrow FeS_2 + H_2$. An alternative source of H_2 could have been the UV-catalyzed reduction of H^+ by Fe^{2+} as shown.

MiniQuiz

• What roles did the mounds of mineral-rich materials at warm hydrothermal springs play in the origin of life?

• What important cell structure was necessary for life to proceed from an RNA world to cellular life?

• How could cells have obtained energy from $FeS + H_2S$?

16.3 Microbial Diversification: Consequences for Earth's Biosphere

Following the origin of cells and the development of early forms of energy and carbon metabolism, microbial life underwent a long process of metabolic diversification, taking advantage of the various and abundant resources available on Earth. As particular resources were consumed and became limiting, evolution selected for more efficient and novel metabolisms. Also, microbial life, through its metabolic activity, altered the biosphere, depleting some resources and creating others through the production of waste products and cellular material. Here we examine the scope of metabolic diversification and focus on one key metabolic waste product in particular, molecular oxygen (O_2), a molecule that had a profound impact on the further evolution of life on Earth.

Metabolic Diversification

Geological and molecular data allow us to look back in time to gain insight into microbial diversification. Molecular evidence, in contrast to geological materials, which can be examined directly, is indirect; phylogenies are based on comparisons of DNA sequences that only estimate when the ancestors of modern bacteria first appeared. In Section 16.6 we describe molecular clocks and DNA sequence-based analysis, but here we use molecular information to estimate a timescale for the appearance of the major metabolic groups of bacteria.

LUCA, the last universal common ancestor, may have existed as early as 4.3 billion years ago (Figure 16.6). Molecular evidence suggests that ancestors of modern-day *Bacteria* and *Archaea* had already diverged by 3.8–3.7 billion years ago (Figure 1.6*b*). As these lineages diverged, they developed distinct metabolisms. Early *Bacteria* may have used H_2 and CO_2 to produce acetate, or ferrous iron (Fe^{2+}) compounds, for energy generation, as noted above. At the same time, early *Archaea* developed the ability to use H_2 and CO_2, or possibly acetate as it accumulated, as substrates for methanogenesis (the production of methane, CH_4), according to the following formulas:

$$4 H_2 + CO_2 \rightarrow CH_4 + 2 H_2O$$
$$H_3CCOO^- + H_2O \rightarrow CH_4 + HCO_3^-$$

Phototrophy (Sections 13.1–13.5) arose somewhat later, about 3.3 billion years ago, and apparently only in *Bacteria*. The ability to use solar radiation as an energy source allowed phototrophs to diversify extensively. With the exception of the early-branching hyperthermophilic *Bacteria* (genera such as *Aquifex* and *Thermotoga*), the common ancestor of all other *Bacteria* appears to be an anaerobic phototroph, possibly similar to *Chloroflexus*. About

2.7–3 billion years ago, the cyanobacteria lineage developed a photosystem that could use H_2O in place of H_2S for photosynthetic reduction of CO_2, releasing O_2 instead of elemental sulfur (S^0) as a waste product; the evolution of this process opened up many new metabolic possibilities, in particular aerobic respirations (Figure 16.6).

The Rise of Oxygen: Banded Iron Formations

Molecular and chemical evidence indicates that oxygenic photosynthesis first appeared on Earth about 300 million years before significant levels of O_2 appeared in the atmosphere. By 2.5 billion years ago, O_2 levels had risen to one part per million, a tiny amount by present-day standards, but enough to initiate what has come to be called the *Great Oxidation Event* (Figure 16.6). What delayed the buildup of O_2 for so long?

The O_2 that cyanobacteria produced did not begin to accumulate in the atmosphere until it first reacted and consumed the bulk of reduced materials, especially reduced iron minerals such as FeS and FeS_2, in the oceans; these materials oxidize slowly but spontaneously with O_2. The Fe^{3+} produced from the oxidation of these minerals became a prominent marker in the geological record. Much of the iron in rocks of Precambrian origin (>0.5 billion years ago, see Figure 16.6) exists in **banded iron formations** (**Figure 16.8**), laminated sedimentary rocks formed in deposits of iron- and silica-rich materials. The metabolism of cyanobacteria yielded O_2 that oxidized Fe^{2+} to Fe^{3+}. The Fe^{3+} formed various iron oxides that accumulated in layers as banded iron formations (Figure 16.8). Once the abundant Fe^{2+} on Earth was consumed, the stage was set for O_2 to accumulate in the atmosphere, but not until 800–900 million years ago did atmospheric O_2 accumulate to present-day levels (~21%, Figure 16.6).

New Metabolisms and the Ozone Shield

As O_2 accumulated on Earth, the atmosphere gradually changed from anoxic to oxic (Figure 16.6). Species of *Bacteria* and *Archaea* unable to adapt to this change were increasingly

Figure 16.8 **Banded iron formations.** An exposed cliff made of sedimentary rock about 10 m in height in Western Australia contains layers of iron oxides (arrows) interspersed with layers containing iron silicates and other silica materials. The iron oxides contain iron in the ferric (Fe^{3+}) form produced from ferrous iron (Fe^{2+}) primarily by the oxygen released by cyanobacterial photosynthesis.

restricted to anoxic habitats because of the toxicity of O_2 and because it oxidized the reduced substances upon which their metabolisms were dependent. However, the oxic atmosphere also created conditions for the evolution of various new metabolic pathways, such as sulfate reduction, nitrification, and the various other chemolithotrophic processes (Chapters 13 and 14). Prokaryotes that evolved the ability to respire O_2 gained a tremendous energetic advantage because of the high reduction potential of the O_2/H_2O couple (⟳ Section 4.6) and so were capable of producing larger cell populations from a given amount of resources than were anaerobic organisms. Larger and rapidly growing cell populations increased the chances for natural selection of new types of metabolic schemes.

As Earth became more oxic, organelle-containing eukaryotic microorganisms arose (Section 16.4), and the rise in O_2 spurred their rapid evolution. The oldest microfossils known to be eukaryotic because they have recognizable nuclei are about 2 billion years old. Multicellular and increasingly complex microfossils of algae are evident from 1.9 to 1.4 billion years ago. By 0.6 billion years ago, with O_2 near present-day levels, large multicellular organisms, the Ediacaran fauna, were present in the sea (Figure 16.6). In a relatively short time, multicellular eukaryotes diversified into the ancestors of modern-day algae, plants, fungi, and animals (⟳ Section 16.8).

An important consequence of O_2 for the evolution of life was the formation of *ozone* (O_3), a gas that provides a barrier preventing much of the intense ultraviolet (UV) radiation of the sun from reaching the Earth. When O_2 is subject to UV radiation, it is converted to O_3, which strongly absorbs wavelengths up to 300 nm. Until an ozone shield developed in Earth's upper atmosphere, evolution could have continued only beneath the ocean surface and in protected terrestrial environments where organisms were not exposed to the lethal DNA damage from the sun's intense UV radiation. However, as Earth developed an ozone shield, organisms could range over the surface of Earth, exploiting new habitats and evolving ever-greater diversity. Figure 16.6 summarizes some landmarks in biological evolution and Earth's geochemistry as Earth transitioned from an anoxic to a highly oxic planet.

MiniQuiz

- Why is the advent of cyanobacteria considered a critical step in evolution?
- In what oxidation state is iron present in banded iron formations?
- What role did ozone play in biological evolution, and how did cyanobacteria make the production of ozone possible?

16.4 Endosymbiotic Origin of Eukaryotes

Up to about 2 billion years ago, all cells apparently lacked a membrane-enclosed nucleus and organelles, the key characteristics of eukaryotic cells (domain *Eukarya*). Here we consider the origin of the *Eukarya* and show how eukaryotes are genetic chimeras containing genes from at least two different phylogenetic domains.

Endosymbiosis

The lineages that gave rise to modern-day *Bacteria* and *Archaea* had existed as the only life forms on our planet for about 2 billion years before eukaryotes appeared (Figure 16.6). This timing tells us that the origin of eukaryotes came after the rise in atmospheric O_2, the development of respiratory metabolism and photosynthesis in *Bacteria*, and the evolution of enzymes such as superoxide dismutase (Section 5.18) that could detoxify the oxygen radicals generated as a by-product of aerobic respiration. How might *Eukarya* have arisen and in what ways did the availability of oxygen influence evolution?

A well-supported explanation for the origin of the eukaryotic cell is the **endosymbiotic hypothesis**. The hypothesis posits that the mitochondria of modern-day eukaryotes arose from the stable incorporation of a respiring bacterium into other cells, and that chloroplasts similarly arose from the incorporation of a cyanobacterium-like organism that carried out oxygenic photosynthesis. Oxygen was almost certainly a driving force in endosymbiosis through its consumption in energy metabolism by the ancestor of the mitochondrion and its production in photosynthesis by the ancestor of the chloroplast. The greater amounts of energy released by aerobic respiration undoubtedly contributed to rapid evolution of eukaryotes, as did the ability to exploit sunlight for energy.

The overall physiology and metabolism of mitochondria and chloroplasts and the sequence and structures of their genomes support the endosymbiosis hypothesis. For example, both mitochondria and chloroplasts contain ribosomes of prokaryotic size (70S) and have 16S ribosomal RNA gene sequences (Section 16.6) characteristic of certain *Bacteria*. Moreover, the same antibiotics that inhibit ribosome function in free-living *Bacteria* inhibit ribosome function in these organelles. Mitochondria and chloroplasts also contain small amounts of DNA arranged in a covalently closed, circular form, typical of *Bacteria* (Section 2.6). Indeed, these and many other telltale signs of *Bacteria* are present in organelles from modern eukaryotic cells (Section 20.4).

There are, however, two other questions germane to how the eukaryotic cell arose: (1) What kind of cell was it that acquired endosymbionts? (2) How did the nuclear membrane form?

Formation of the Eukaryotic Cell

Two hypotheses have been put forward to explain the formation of the eukaryotic cell (**Figure 16.9**). In one, eukaryotes began as a nucleus-bearing cell that later acquired mitochondria and chloroplasts by endosymbiosis (Figure 16.9*a*). In this hypothesis, the nucleus-bearing cell line arose in a lineage of cells that split from the *Archaea*; the nucleus is thought to have arisen in this cell line during evolutionary experimentation with increasing cell and genome size, probably in response to oxic events that were transforming the geochemistry of Earth (Section 16.3). However, a major problem with this hypothesis is that it does not easily account for the fact that *Bacteria* and *Eukarya* have similar membrane lipids, in contrast to those of *Archaea* (Section 3.3).

The second hypothesis, called the *hydrogen hypothesis,* proposes that the eukaryotic cell arose from an association between a H_2-producing species of *Bacteria,* the symbiont, which eventually gave rise to the mitochondrion, and a species of H_2-consuming *Archaea,* the host (Figure 16.9*b*). In this

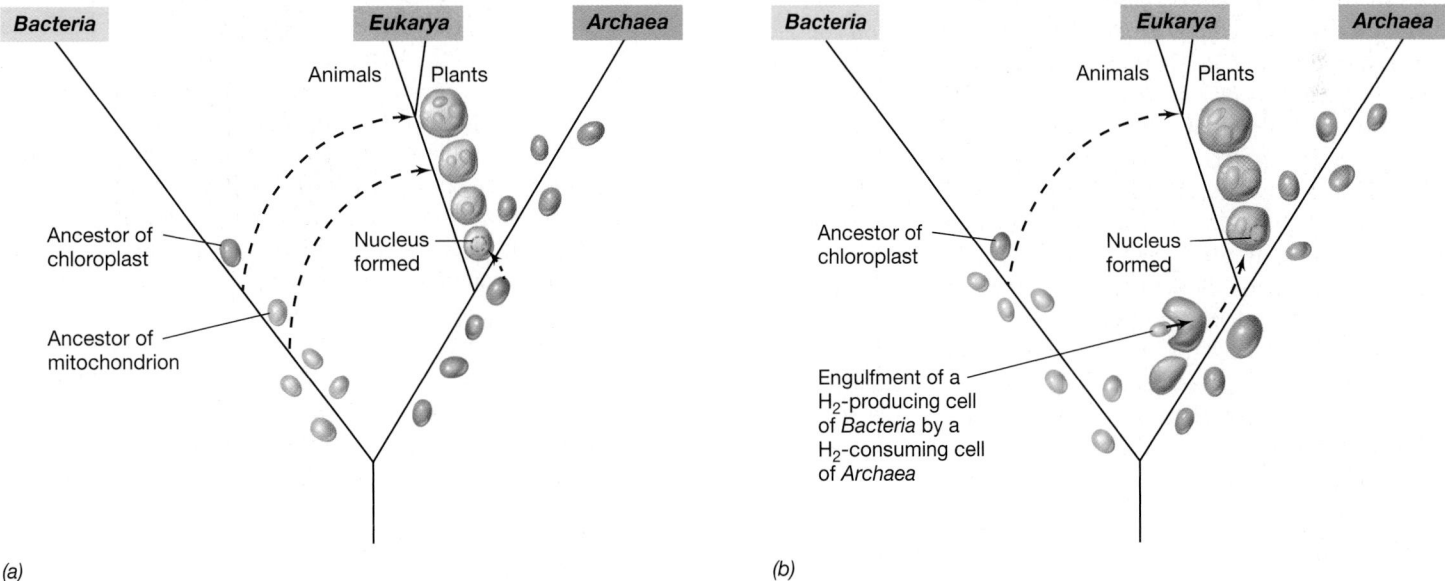

Figure 16.9 Endosymbiotic models for the origin of the eukaryotic cell. *(a)* The nucleated line diverged from the archaeal line and later acquired by endosymbiosis the bacterial ancestor of the mitochondrion and then the cyanobacterial ancestor of the chloroplast, at which point the nucleated line diverged into the lineages giving rise to plants and animals. *(b)* The hydrogen hypothesis. The bacterial ancestor of the mitochondrion was taken up endosymbiotically by a species of *Archaea* and the nucleus developed later followed by the endosymbiotic acquisition of the cyanobacterial ancestor of the chloroplast. Note the position of the mitochondrion and chloroplast on the universal phylogenetic tree shown in Figure 16.16.

hypothesis, the nucleus arose after genes for lipid synthesis were transferred from the symbiont to the host. This led to the synthesis of lipids containing fatty acids by the host, lipids that may have been more conducive to the formation of internal membranes, such as the nuclear membrane system (Section 20.1). The simultaneous increase in size of the host genome led to sequestering DNA within a membrane, which organized it and made replication and gene expression more efficient. Later, this mitochondrion-containing, nucleated cell line acquired chloroplasts by endosymbiosis, leading to the first phototrophic eukaryotes (Figure 16.9*b*).

Both hypotheses to explain the origin of eukaryotes point to the eukaryotic cell as a genetic chimera, a cell made up of genes from both *Bacteria* and *Archaea*. However, the hydrogen hypothesis nicely accounts for the observation that eukaryotic cells contain bacterial (rather than archaeal) lipids, yet share with *Archaea* many molecular features of transcription and translation (see Table 16.1 and Chapter 7). The hydrogen hypothesis predicts that aspects of energy metabolism—that is, ATP-producing pathways in mitochondria, hydrogenosomes (degenerate mitochondria, Section 20.2), and the cytoplasm, as well as glycolytic enzymes in the cytoplasm—should be more similar in eukaryotes and *Bacteria* than in eukaryotes and *Archaea*, and research has shown this to be true.

Consequences of the Evolution of the Modern Eukaryote

No matter how the eukaryotic cell arose, the appearance of eukaryotes was a major step in evolution, creating complex cells with new capabilities powered by a respiratory organelle and, in phototrophic cells, a photosynthetic organelle as well. Like the origin of the first cells from abiotic materials and the diversification of *Bacteria* and *Archaea*, evolution of the eukaryotic cell with its many individual components probably took long periods of time and had many dead ends. Like the explosion in diversity of *Bacteria* and *Archaea*, each step along the way to the modern eukaryotic cell created new opportunities for variation to arise and natural selection to work, with some functions being discarded while others were being refined, eventually producing a totally new model for cellular life, a model upon which complex multicellular organisms could be built. Indeed, the period from about 2 billion years ago to the present saw the rise and diversification of unicellular eukaryotic microorganisms, the origin of multicellularity, and the appearance of structurally complex plant, animal, and fungal life (Figures 16.6, 16.9, and see Figure 16.16).

MiniQuiz

- What evidence supports the idea that the eukaryotic mitochondrion and chloroplast were once free-living members of the domain *Bacteria*?

- Why does the hydrogen hypothesis for endosymbiosis best account for the properties of modern eukaryotes?

- In what ways are modern eukaryotes a combination of attributes of *Bacteria* and *Archaea*?

II ● Microbial Evolution

We begin here by reviewing the evolutionary process. We consider how scientists reconstruct the evolutionary history of life using methods of molecular genetics, and will see, as summarized in the universal tree of life, how microorganisms are related to each other and to other living things.

16.5 The Evolutionary Process

Evolution, the process by which organisms undergo descent with modification, is driven by mutation and selection. In this Darwinian view of life, all organisms are related through descent from an ancestor that lived in the past. We have outlined a hypothesis for the origin of the most distant of those ancestors, the last universal common ancestor (LUCA, Section 16.2). Since the time of LUCA, life has undergone an extensive process of change as new kinds of organisms arose from other kinds existing in the past. Evolution has also led to the loss of life forms, with organisms less able to compete becoming extinct over time. Evolution accounts not only for the tremendous diversity we see today, but also for the high level of complexity in modern organisms. Indeed, no organism living today is primitive. All extant life forms are modern organisms, well adapted to and successful in their ecological niches, having arisen by evolution under the pressure of natural selection.

Genomic Changes

DNA sequence variation can arise in the genome of an organism from mutations including the loss or gain of whole genes. Mutations, which arise from errors in replication and from certain external factors such as ultraviolet radiation, are essential for life to evolve through natural selection. Adaptive mutations are those that improve the **fitness** of an organism, increasing its survival capacity or reproductive success compared with that of competing organisms. By contrast, harmful mutations lower an organism's fitness. Most mutations, however, are neutral, neither benefiting nor causing harm to the organism, and over time these mutations can accumulate in an organism's genome.

Recall that prokaryotes are genetically haploid; this affects their evolution because mutations in prokaryotic cells are not "covered" by a second copy of the gene, as they are in diploid organisms, but are instead immediately expressed. However, the process of *gene duplication* (Section 12.10) can set the stage for the origin of new functions as mutations in the duplicated sequence encode proteins that differ in greater and greater ways from the original protein. Mutations can also lead to gene loss, which eliminates from the cell the gene product and any competitive benefit accruing from it. Extreme cases of gene loss are often part of the evolutionary history of obligate symbionts and parasites, organisms that receive their essential nutrients from their hosts (Section 25.9).

Another process can also bring about heritable changes in the sequence of an organism's genome: **Horizontal gene transfer.** This process can bring in genes from near or distantly related lineages as cells exchange genes by any of several mechanisms (Chapters 10 and 12).

Selection and the Rapidity of Evolution in Prokaryotes

Regardless of whether a mutation or other change in a genome is neutral, beneficial, or harmful, these changes provide the opportunity for selection of genetically new organisms whose genomes have greater or fewer capacities. As environmental changes create new habitats, cells are presented with new conditions under which they may either survive and successfully compete for nutrients or become extinct. The heritable variation present in a population of cells provides the raw material for natural selection (see Figure 16.25). That is, reproduction of those individuals bearing mutations beneficial under the new circumstances is favored. Moreover, because bacteria typically form large populations that can increase in number quite rapidly, evolutionary events in bacterial populations can also occur quite rapidly. A classic example of this can be seen in laboratory experiments with purple phototrophic bacteria such as *Rhodobacter*, organisms that can grow both chemotrophically and phototrophically. When cultured under anoxic conditions, the cells produce bacteriochlorophyll and carotenoids. In the light, these pigments allow for photosynthetic reactions that lead to ATP synthesis (♻ Section 13.5). However, when cultured under anoxic conditions in darkness, the cells still make pigments because anoxia is the signal that triggers their synthesis. Is there a selective advantage (or disadvantage) to this metabolic strategy?

In nature, if dark-growing cells of phototrophic purple bacteria do not see light right away, they may see light a bit later, and by synthesizing pigments in the dark they are prepared to begin photosynthesis immediately when light returns; thus there is a selective advantage to this strategy. But when serially subcultured in darkness in the laboratory, making pigments that cannot be used is a metabolic disadvantage, and mutants incapable of photosynthesis quickly take over the population (**Figure 16.10**). These mutants no longer carry the burden of making all (or in some mutants any) of the photosynthetic pigments of the wild-type organisms. Since in darkness such pigments would be useless to these cells anyway, the mutants grow faster than any remaining wild-type cells. Although these mutants have reduced phototrophic capacities or in some cases have completely lost the ability to grow phototrophically (see photo inset in Figure 16.10), in permanent darkness they quickly become the fittest organisms in the population and therefore enjoy the greatest reproductive success. Such mutations affecting photosynthesis occur at the same rate in the light as in the dark, but in the light the selection for phototrophy is so strong that such mutants are quickly lost from the population.

The transitions shown in the experiment of Figure 16.10, which occurred in a matter of a few days, remind us of how fast evolutionary pressures can shift even major properties (such as metabolic strategies) of a microbial cell population. In accordance with evolutionary theory, the environment (in this case darkness) selected the fittest organisms for further propagation; that is, those cells whose dark growth rate was maximal. Cells unable to maintain such rapid growth rates are replaced, and eventually, a homogeneous population exists of cells that grow best under the given set of conditions.

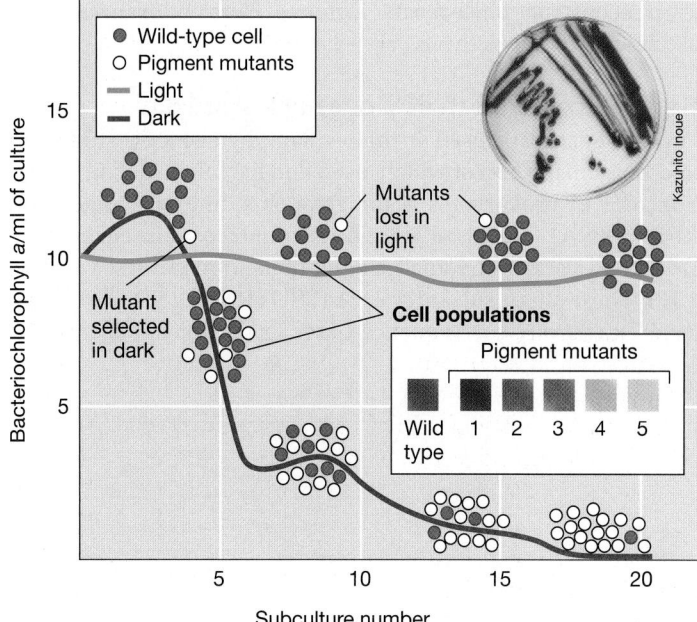

Figure 16.10 Survival of the fittest and natural selection in a population of phototrophic purple bacteria. Serial subculture of *Rhodobacter capsulatus* under anoxic dark conditions quickly selects for nonphototrophic mutants that outcompete and grow faster than cells still making bacteriochlorophyll and carotenoids. Photos: top, plate culture showing colonies of phototrophic cells of *R. capsulatus*; bottom, close-up photos of colonies of wild type and five pigment mutants (1–5) obtained during serial dark subculture. Wild-type cells are reddish-brown from their assortment of carotenoid pigments. The color of mutant colonies reflects the absence (or reduced synthesis) of one or more carotenoids. Mutant strain 5 lacked bacteriochlorophyll and was no longer able to grow phototrophically. Mutant strains 1–4 could grow phototrophically but at reduced growth rates from the wild type. Data adapted from Madigan, M.T., et. al. 1982. *J. Bacteriol.* **150**: 1422–1429.

MiniQuiz

- How can gene duplication assist the evolutionary process?
- How does the accumulation of mutations set the stage for selection?
- In the experiment of Figure 16.10, why did the dark cell population lose its pigments?

16.6 Evolutionary Analysis: Theoretical Aspects

The evolutionary history of a group of organisms is called its **phylogeny**, and a major goal of evolutionary analysis is to understand phylogenetic relationships. Because we do not have direct knowledge of the path of microbial evolution, phylogeny is inferred indirectly from nucleotide sequence data. Our premises are that (1) all organisms are related by descent, and (2) that the sequence of DNA in a cell's genome is a record of the organism's ancestry. Because evolution is a process of inherited nucleotide sequence change, comparative analyses of DNA sequences allow us to reconstruct phylogenetic histories. Here, we examine some of the ways in which this is carried out.

Genes Employed in Phylogenetic Analysis

Various genes are used in molecular phylogenetic studies of microorganisms. Most widely used and useful for defining relationships in prokaryotes is the gene encoding **16S ribosomal RNA (rRNA)** (Figure 16.11) and its counterpart in eukaryotes, 18S rRNA, parts of the small subunit of the ribosome (Section 6.19). These **small subunit rRNA (SSU rRNA)** genes have been used extensively for sequence-based evolutionary analysis as

pioneered by Carl Woese, an American scientist, in the 1970s. SSU rRNA genes are excellent candidates for phylogenetic analysis because they are (1) universally distributed, (2) functionally constant, (3) sufficiently conserved (that is, slowly changing), and (4) of adequate length to provide a deep view of evolutionary relationships. A large and constantly growing database of SSU rRNA gene sequences exists. For example, the **Ribosomal Database Project** (RDP; http://rdp.cme.msu.edu) contains a collection of such sequences, now numbering over 1.3 million, and provides computational programs for analytical purposes.

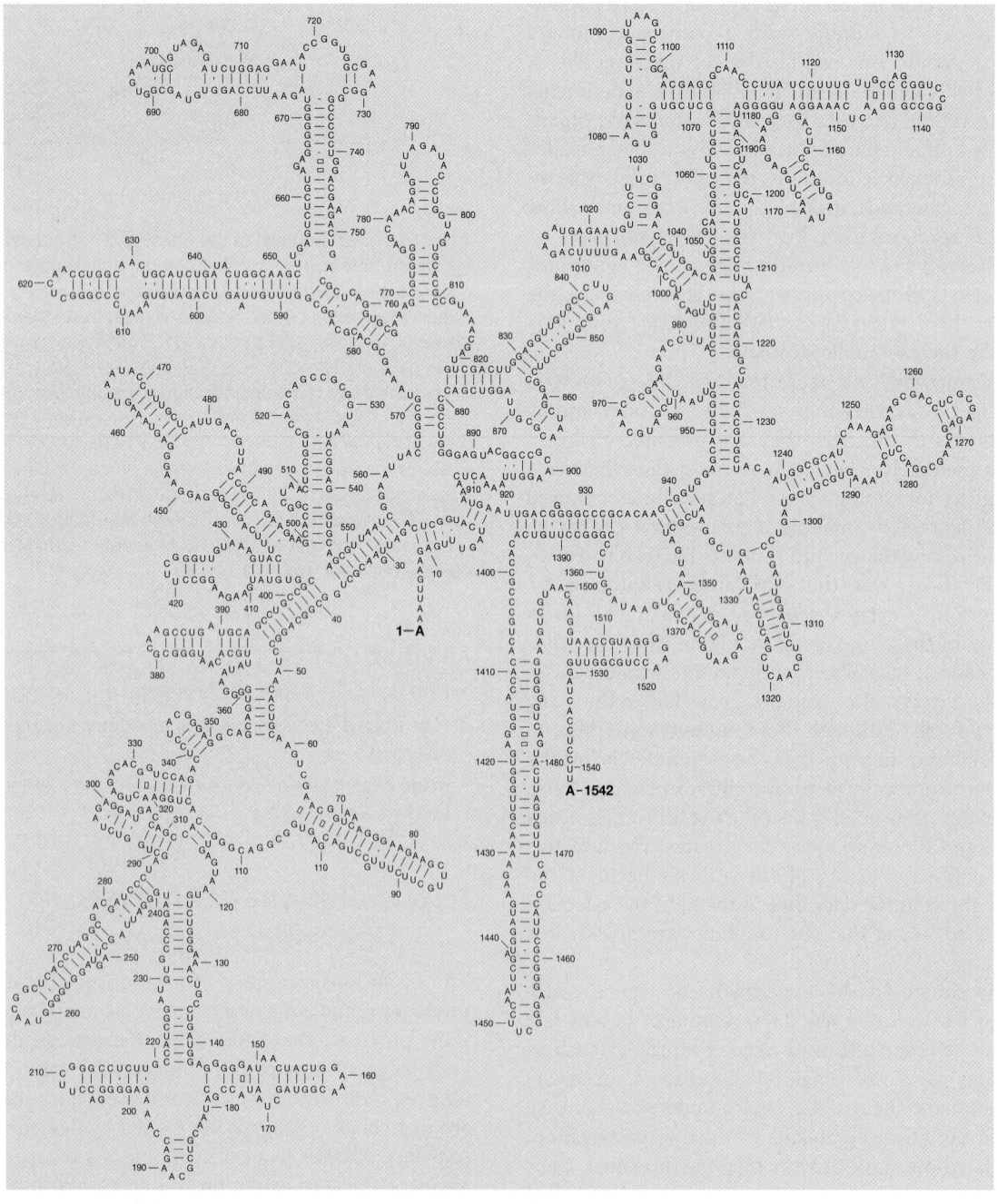

Figure 16.11 Ribosomal RNA (rRNA). Primary and secondary structure of 16S rRNA from *Escherichia coli* (*Bacteria*). The 16S rRNA from *Archaea* is similar in secondary structure (folding) but has numerous differences in primary structure (sequence).

Along with SSU genes, those for several highly conserved proteins have been used effectively in phylogenetic analysis, including genes encoding protein synthesis elongation factor EF-Tu (♻ Section 6.19), heat shock protein Hsp60 (♻ Section 8.11), and several transfer RNA (tRNA) synthetases (♻ Section 6.18). Although the highly conserved SSU genes are particularly useful for deep evolutionary analysis, the amount of variation present in SSU rRNA gene sequences is often insufficient to discriminate among closely related species. In Section 16.11 we discuss ways of bypassing this problem by using genes whose sequences have diverged more than the 16S rRNA gene, consequently revealing distinctions between closely related bacteria, and by using multiple genes simultaneously for evolutionary analyses.

Molecular Clocks

An unresolved question in phylogenetics is whether DNA (and protein) sequences change at a constant rate. The approach to answering the question focuses on pairs of homologous sequences—that is, sequences of shared evolutionary ancestry that encode functionally equivalent molecules. If sequences do change at a constant rate, such pairs would serve as an approximate **molecular clock**, allowing the time in the past when the two sequences diverged from a common ancestral sequence to be estimated. Major assumptions of the molecular clock approach are that nucleotide changes accumulate in a sequence in proportion to time, that such changes generally are neutral and do not interfere with gene function, and that they are random.

The molecular clock approach has been used to estimate the time of divergence of distantly related organisms, such as the domains *Archaea* and *Eukarya* (about 2.8 billion years ago, Figure 1.6*b*), as well as closely related organisms, such as the enteric bacteria *Escherichia coli* and *Salmonella typhimurium* (about 120–140 million years ago). These data have also been combined with evidence from the geological record on isotopes and specific biological markers to approximate when different metabolic patterns emerged in bacteria (Section 16.3 and Figure 16.6).

The main problem with the molecular clock approach, however, is that DNA sequences do change at different rates, which means that direct and reliable correlations to a timescale will be difficult to make. However, much of phylogenetic analysis is concerned with *relative* relationships among organisms, shown by their branching order on phylogenetic trees. These relationships are generally discernible from molecular sequence analyses regardless of whether different sequences change at similar rates, so the accuracy of the molecular clock approach is not a major concern.

MiniQuiz

- List three reasons that SSU rRNA genes are suitable for phylogenetic analyses.
- What information does the Ribosomal Database Project provide?
- What value do molecular clocks have in phylogenetic analysis?

16.7 Evolutionary Analysis: Analytical Methods

As we have seen, modern phylogenetics is based on nucleotide sequence comparisons, for which specific methods have been developed. We consider these methods here.

Obtaining DNA Sequences

Phylogenetic analysis using DNA sequences relies heavily on the polymerase chain reaction (PCR) to obtain sufficient copies of a gene for reliable sequencing (♻ Section 6.11). Specific oligonucleotide primers have been designed that bind to the ends of the gene of interest, or to DNA flanking the gene, allowing DNA polymerase to bind to and copy the gene. The source of DNA bearing a gene of interest typically is genomic DNA purified from particular bacterial strains, but could be DNA extracted from an environmental sample (Section 16.9). The PCR product is visualized by agarose gel electrophoresis, excised from the gel, extracted and purified from the agarose, and then sequenced, often using the same oligonucleotides as primers for the sequencing reactions. These steps are summarized in **Figure 16.12**.

An important aspect of PCR amplification is *primer design*, which is a matter of deciding which sequence to use to amplify a specific gene and then actually constructing the sequence. Standard primers exist for many highly conserved genes, such as the SSU rRNA genes (Figure 16.12). Primers are available that are domain-specific and can be used to amplify an SSU gene from any organism in a given domain. Other primers can be designed that are lineage-specific, or even more restrictive. At the other extreme, "universal" primers are available that will amplify SSU genes from *any* organism, prokaryote or eukaryote. Primer design is both an art and a science and often requires computational analyses, along with some trial and error, to construct primers that will effectively amplify the gene of interest.

Sequence Alignment

Phylogenetic analysis is based on *homology*, that is, analysis of DNA sequences that are related by common ancestry. Once the DNA sequence of a gene is obtained, the next step in phylogenetic analysis is to align that sequence with homologous sequences from other organisms. By doing this, nucleotide mismatches and insertions and deletions, some of which may be phylogenetically informative, can be pinpointed.

Figure 16.13 shows an example of sequence alignment. The web-based algorithm BLAST (*Basic Local Alignment Search Tool*) of the National Institutes of Health (http://www.ncbi.nlm.nih.gov/BLAST) aligns sequences automatically and can identify genes homologous to a specific sequence from among the many thousands already sequenced. Related sequences are then downloaded from GenBank (http://www.ncbi.nih.gov/Genbank), which is an annotated collection of all publicly available DNA sequences, and aligned. Proper sequence alignment is critical to phylogenetic analysis because the assignment of mismatches and gaps caused by deletions is in effect an explicit hypothesis of how the sequences have diverged from a common ancestral sequence. Genes that encode proteins usually are aligned with the aid of their inferred

UNIT 6

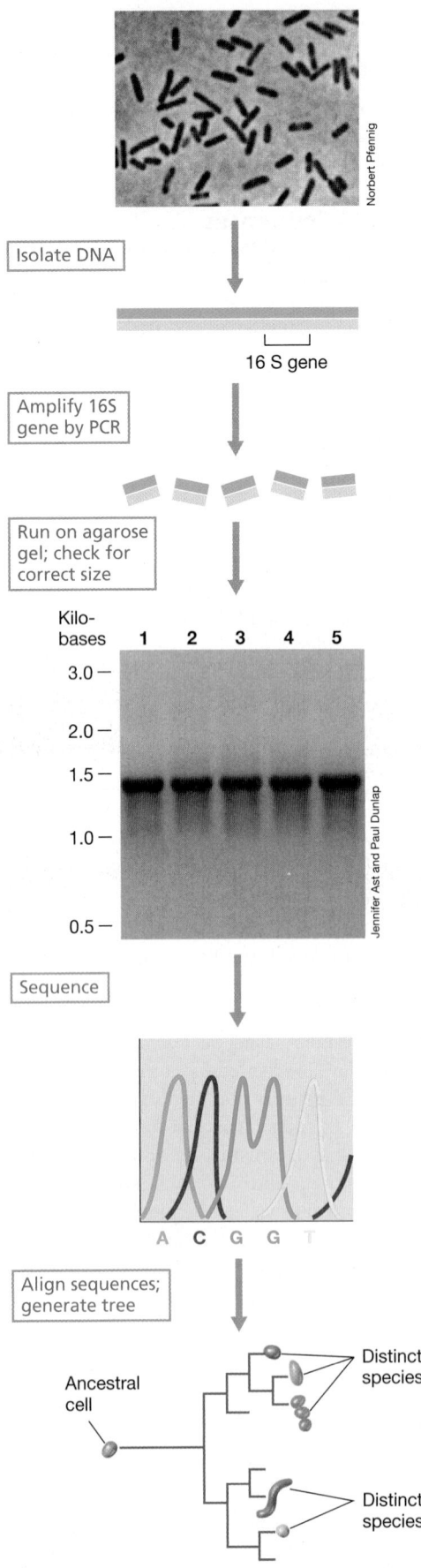

Isolate DNA

16 S gene

Amplify 16S
gene by PCR

Run on agarose
gel; check for
correct size

Kilo-
bases   1   2   3   4   5

3.0 —

2.0 —

1.5 —

1.0 —

0.5 —

Sequence

A   C   G G   T

Align sequences;
generate tree

Ancestral
cell

Distinct
species

Distinct
species

Figure 16.12 **PCR amplification of the 16S rRNA gene.** Following DNA isolation, primers complementary to the ends of the 16S rRNA (see Figure 16.11) are used to PCR-amplify the 16S rRNA gene from genomic DNA of five different unknown bacterial strains and the products are run on an agarose gel (photo). The bands of amplified DNA are approximately 1465 nucleotides in length. Positions of DNA kilobase size markers are indicated at the left. Excision from the gel and purification of these PCR products is followed by sequencing and analysis to identify the bacteria.

amino acid sequences. Other genes, such as those encoding 16S rRNA, can often be aligned by inspection or through the use of computer programs designed to minimize the number of mismatches and gaps. Secondary structure, the folding of the 16S rRNA (Figure 16.11), is also helpful in making accurate gene alignments because base mismatches that show up in the secondary structure of highly conserved regions of the molecule readily signal alignment errors.

Phylogenetic Trees

Reconstructing evolutionary history from observed nucleotide sequence differences includes construction of a phylogenetic tree, which is a graphic depiction of the relationships among sequences of the organisms under study, much like a family tree. A phylogenetic tree is composed of *nodes* and *branches* (**Figure 16.14**). The tips of the branches represent species that exist now and from which the sequence data were obtained. The nodes are points in evolution where an ancestor diverged into two new organisms, each of which then began to evolve along its separate pathway. The branches define both the order of descent and the ancestry of the nodes, whereas the branch length represents the number of changes that have occurred along that branch.

Phylogenetic trees can be constructed that are either *unrooted*, showing the relative relationships among the organisms under study but not the evolutionary path leading from an ancestor to a strain (Figure 16.14*a*), or *rooted*, in which case the unique path from an ancestor to each strain is defined (Figure 16.14*b*, *c*). Trees are rooted by the inclusion in the analysis of an *outgroup*, an organism that is less closely related to the organisms under study than the organisms are to each other, but that shares with them homologs of the gene under study.

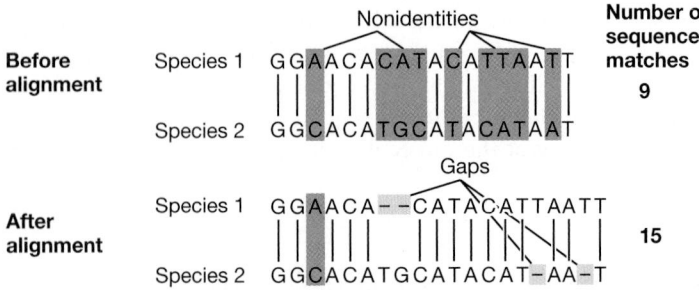

		Nonidentities	Number of sequence matches
Before alignment	Species 1	GGAACACATACATTAATT	
		∣∣ ∣∣∣ ∣∣∣ ∣∣	9
	Species 2	GGCACATGCATACATAAT	
		Gaps	
After alignment	Species 1	GGAACA‑‑CATACATTAATT	
		∣∣ ∣∣∣ ∣∣∣∣∣∣∣ ∣	15
	Species 2	GGCACATGCATACAT‑AA‑T	

Figure 16.13 **Alignment of DNA sequences.** Sequences for a hypothetical region of a gene are shown for two organisms, before alignment and after the insertion of gaps to improve the matchup of nucleotides, indicated by the vertical lines showing identical nucleotides in the two sequences. The insertion of gaps in the sequences substantially improves the alignment.

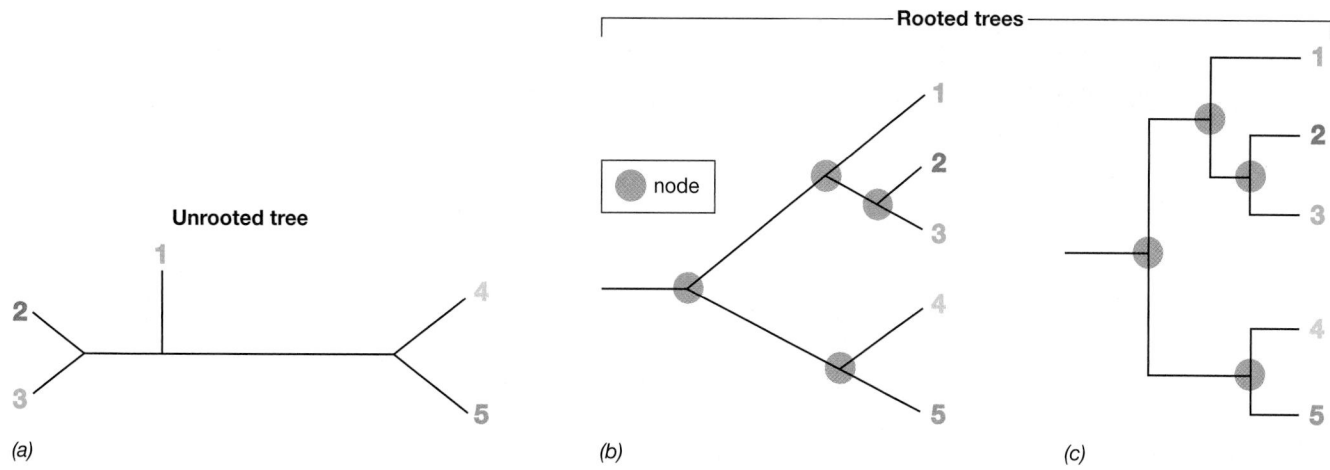

Figure 16.14 **Phylogenetic trees.** Unrooted *(a)* and rooted *(b, c)* forms of a phylogenetic tree are shown. The tips of the branches are species (or strains) and the nodes are ancestors. Ancestral relationships are revealed by the branching order in rooted trees.

In its most basic form, a phylogenetic tree is a depiction of lines of descent, and the relationship between two organisms therefore should be read in terms of common ancestry. That is, the more recently two species shared a common ancestor, the more closely related they are. The rooted trees in Figure 16.14*b* and *c* illustrate this point. Species 2 is more closely related to species 3 than it is to species 1 because 2 and 3 share a more recent common ancestor than do 2 and 1.

Tree Construction

Modern evolutionary analysis uses character-state methods, also called *cladistics*, for tree construction. Character-state methods define phylogenetic relationships by examining changes in nucleotides at particular positions in the sequence, using those characters that are *phylogenetically informative*. These are characters that define a **monophyletic** group; that is, a group that has descended from one ancestor. **Figure 16.15** describes how phylogenetically informative characters are recognized in aligned sequences. Computer-based analysis of these changes generates a phylogenetic tree, or *cladogram*.

A widely used cladistic method is *parsimony*, which is based on the assumption that evolution is most likely to have proceeded by the path requiring fewest changes. Computer algorithms based on parsimony provide a way of identifying the tree with the smallest number of character changes. Other cladistic

methods, *maximum likelihood* and *Bayesian analysis*, proceed like parsimony, but they differ by assuming a model of evolution, for example, that certain kinds of nucleotide changes occur more often than others. Inexpensive computer applications, such as PAUP (*Phylogenetic Analysis Under Parsimony*, and Other Methods), guidebooks, and web-accessible tutorials are available for learning the basic procedures of cladistic analysis and tree construction.

MiniQuiz

- How are DNA sequences obtained for phylogenetic analysis?
- What does a phylogenetic tree depict?
- Why is sequence alignment critical to phylogenetic analysis?

16.8 Microbial Phylogeny

Biologists previously grouped living organisms into five *kingdoms*: plants, animals, fungi, protists, and bacteria. DNA sequence-based phylogenetic analysis, on the other hand, has revealed that the five kingdoms do not represent five primary evolutionary lines. Instead, as previously outlined in Chapter 2, cellular life on Earth has evolved along *three* primary lineages, called **domains**. Two of these domains, the *Bacteria* and the *Archaea*, are exclusively composed of prokaryotic cells. The *Eukarya* contains the eukaryotes (**Figure 16.16**), including the plants, animals, fungi, and protists.

An SSU rRNA Gene–Based Phylogeny of Life

The **universal phylogenetic tree** based on small subunit rRNA genes (Figure 16.16) is a genealogy of all life on Earth. It depicts the evolutionary history of all cells and clearly reveals the three domains. The root of the universal tree represents a point in time when all extant life on Earth shared a common ancestor, the last universal common ancestor, LUCA (Figure 16.16).

The three-domain concept is also supported by sequence analysis of several other genes shared among all organisms. Analysis of

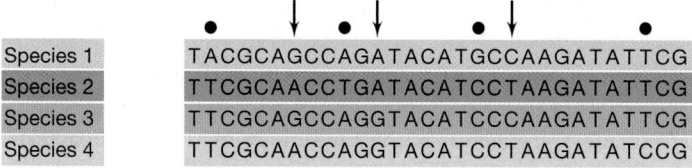

Figure 16.15 **Identification of phylogenetically informative sites.** Aligned sequences for four species are shown. Invariant sites are unmarked, and phylogenetically neutral sites are indicated by dots. Phylogenetically informative sites, varying in at least two of the sequences, are marked with arrows.

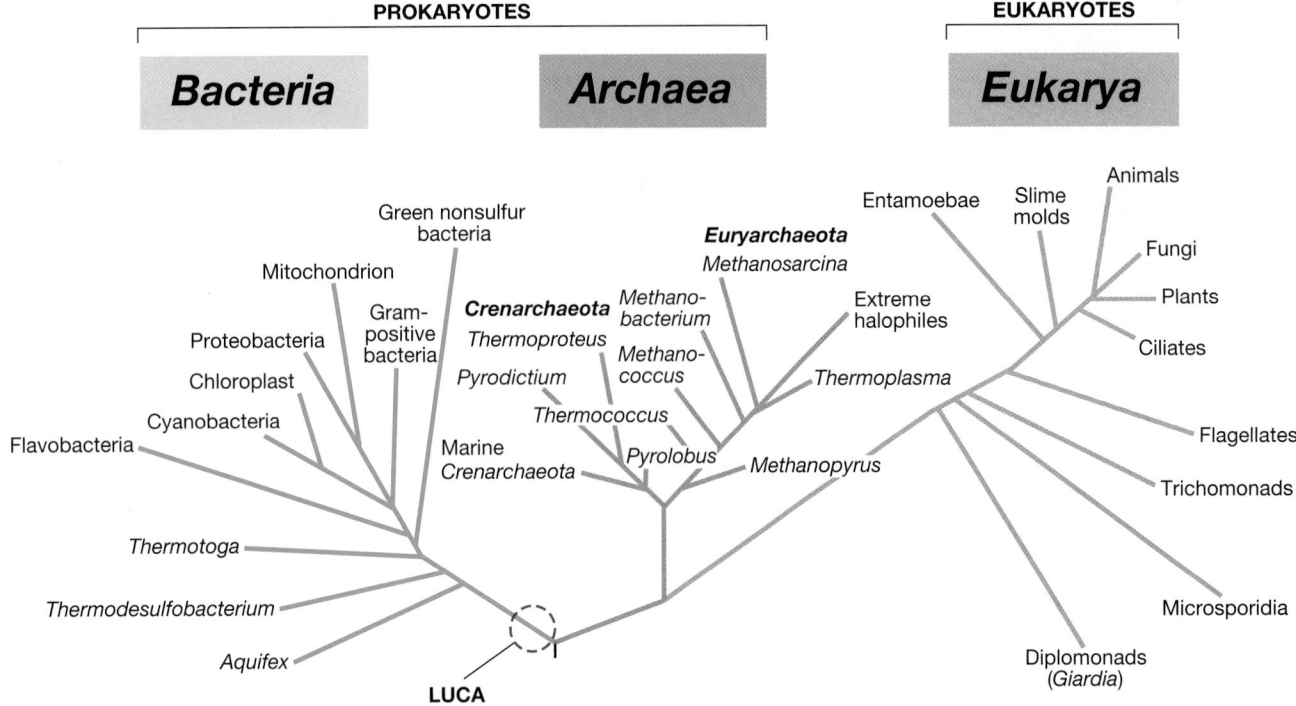

Figure 16.16 Universal phylogenetic tree as determined from comparative SSU rRNA gene sequence analysis. Only a few key organisms or lineages are shown in each domain. At least 80 lineages of *Bacteria* have now been identified although many of these have not yet been cultured. LUCA, last universal common ancestor.

over 30 genes present in nearly 200 species of *Bacteria, Archaea,* and *Eukarya* whose genomes have been completely sequenced confirms the distinct separation between these three lines of descent. Although branching orders and relationships among some lineages within the domains will likely be revised as more genetic data are obtained, analysis of multiple genes from genomic studies (Section 16.11) supports the basic structure of life proposed by Woese based on sequence analysis of SSU rRNA genes.

The presence of genes in common in *Bacteria, Archaea,* and *Eukarya,* of which there are many examples (ᐸᗒ Section 7.4 and Figure 7.5), raises an interesting question. If these lineages diverged from each other so long ago from a common ancestor, how is it they share so many genes? One hypothesis is that early in the history of life, before the primary domains had diverged, horizontal gene transfer (Chapters 10 and 12) was extensive. During this time, genes encoding proteins that conferred exceptional fitness, for example, genes for the core cellular functions of transcription and translation, were promiscuously transferred among a population of primitive organisms derived from a common ancestral cell. If true, this would explain why, as genome analyses have shown, all cells regardless of domain have many core functional genes in common, more than would be expected if all cells shared a primitive common ancestor (Figure 16.16).

But what about the unique genes present in each domain, of which there are several examples as well (ᐸᗒ Section 7.4)? It is hypothesized that over time barriers to unrestricted horizontal gene transfer evolved, perhaps from the selective colonization of habitats (thereby generating reproductive isolation) or as the

result of structural barriers that in some way prevented free genetic exchange. As a result, the previously genetically promiscuous population slowly began to sort out into the primary lines of evolutionary descent, the *Bacteria* and *Archaea* (Figure 16.16). As each lineage continued to evolve, certain unique biological traits became fixed within each group. Then, about 2.8 billion years ago, the *Archaea* and *Eukarya* diverged as distinct domains (ᐸᗒ Figure 1.6). Today, after a total of nearly 4 billion years of microbial evolution, we see the grand result: three domains of cellular life that on the one hand share many common features, but on the other hand, display distinctive evolutionary histories of their own. **Table 16.1** summarizes some major characteristics of the three domains.

Bacteria

Among *Bacteria,* at least 80 lineages (called phyla, singular **phylum,** or divisions) have been discovered thus far; only some key ones are shown in the universal tree in Figure 16.16. The *Bacteria* are discussed in detail in Chapters 17 and 18. Many lineages of *Bacteria* are known only from environmental sequences (phylotypes, Sections 16.9 and 22.5). Although some lineages are characterized by unique phenotypic traits, such as the morphology of the spirochetes or the physiology of the cyanobacteria, most major groups of *Bacteria* consist of species that, although specifically related from a phylogenetic standpoint, lack strong phenotypic cohesiveness. The largest group, the **Proteobacteria,** is a good example of this, as collectively this group shows all known forms of microbial physiology.

Table 16.1 *Major characteristics of Bacteria, Archaea, and Eukarya*[a]

Characteristic	Bacteria	Archaea	Eukarya
Morphological and genetic			
Prokaryotic cell structure	Yes	Yes	No
Cell wall	Peptidoglycan	No peptidoglycan	No peptidoglycan
Membrane lipids	Ester-linked	Ether-linked	Ester-linked
Membrane-enclosed nucleus	Absent	Absent	Present
DNA present in covalently closed and circular form	Yes	Yes	No
Histone proteins present	No	Yes	Yes
RNA polymerases (Figure 7.2)	One (4 subunits)	One (8–12 subunits)	Three (12–14 subunits each)
Ribosomes (mass)	70S	70S	80S
Initiator tRNA	Formylmethionine	Methionine	Methionine
Introns in most genes	No	No	Yes
Operons	Yes	Yes	No
Capping and poly(A) tailing of mRNA	No	No	Yes
Plasmids	Yes	Yes	Rare
Sensitivity to chloramphenicol, streptomycin, kanamycin, and penicillin	Yes	No	No
Physiological/special structures			
Dissimilative reduction of S^0 or SO_4^{2-} to H_2S, or Fe^{3+} to Fe^{2+}	Yes	Yes	No
Nitrification (ammonia oxidation)	Yes	Yes	No
Chlorophyll-based photosynthesis	Yes	No	Yes (in chloroplasts)
Denitrification	Yes	Yes	No
Nitrogen fixation	Yes	Yes	No
Rhodopsin-based energy metabolism	Yes	Yes	No
Chemolithotrophy (Fe, NH_3, S, H_2)	Yes	Yes	No
Endospores	Yes	No	No
Gas vesicles	Yes	Yes	No
Synthesis of carbon storage granules composed of poly-β-hydroxyalkanoates	Yes	Yes	No
Growth above 70°C	Yes	Yes	No
Growth above 100°C	No	Yes	No

[a]Note that for many features only particular representatives within a domain show the property.

The major eukaryotic organelles clearly originated from within the domain *Bacteria*, the mitochondrion from within the *Proteobacteria* and the chloroplast from within the cyanobacteria (Figure 16.16). As we discussed earlier, eukaryotic organelles originated from endosymbiotic events (Figure 16.9) that shaped the modern eukaryotic cell as a genetic chimera containing genes from two or more phylogenetic lineages.

Archaea

The domain *Archaea* consists of two major phyla, the *Crenarchaeota* and *Euryarchaeota* (Figure 16.16). We discuss *Archaea* in detail in Chapter 19. Branching close to the root of the universal tree are hyperthermophilic species of *Crenarchaeota*, such as *Pyrolobus* (Figure 16.16). These are followed by the

phylum *Euryarchaeota*, which includes the methane-producing (methanogenic) *Archaea* and the extreme halophiles and extreme acidophiles, such as *Thermoplasma* (Figure 16.16). As in the tree of *Bacteria*, there are some lineages of *Archaea* known only from the sampling of rRNA genes from the environment (Section 16.9). This list keeps expanding as more habitats are specifically sampled for archaeal diversity. It has become clear that cultured species of *Archaea*, primarily obtained from extreme environments such as hot springs, saline lakes, acidic soils, and the like, have many relatives in more moderate habitats such as freshwater lakes, streams, agricultural soils, and the oceans; at this point we have only limited knowledge of the activities and metabolic strategies of the *Archaea* that inhabit nonextreme environments.

Eukarya

Phylogenetic trees of species in the domain *Eukarya* have been constructed from comparative sequence analysis of the 18S rRNA gene, the functional equivalent of the 16S rRNA gene. In Chapter 20 where we consider microbial eukaryotes in detail, we will see that the SSU phylogenetic picture of eukaryotes is probably inaccurate. The 18S tree shows some "early-branching" microbial eukaryotes, such as the microsporidia and the diplomonads (Figure 16.16). By contrast, the position of these organisms on multigene phylogenetic trees (Section 16.11) is quite different, and shows them to have arisen during a burst of evolutionary radiation that led to most lineages of microbial eukaryotes (⟳ Figure 20.12). It is likely that this burst in eukaryotic evolution was triggered by the onset of oxic conditions on Earth and subsequent development of the ozone shield (Section 16.3). The latter would have greatly expanded the number of surface habitats available for colonization. Nevertheless, although 18S rRNA sequencing appears to give a skewed view of eukaryotic microbial evolution, it still clearly sorts the eukaryotes out as a distinct domain of life with evolutionary roots more closely tied to the *Archaea* than to the *Bacteria* (Figure 16.16).

16.9 Applications of SSU rRNA Phylogenetic Methods

Many research tools make use of small subunit (SSU) rRNA gene sequencing. These tools include rRNA probes, used in both microbial ecology and diagnostic medicine, and DNA fingerprinting.

Phylogenetic Probes and FISH

Recall that a probe is a strand of nucleic acid that can be labeled and used to hybridize to a complementary nucleic acid (⟳ Section 11.2). Probes can be general or specific. For example, universal SSU rRNA probes are available that bind by complementary base pairing to conserved sequences in the rRNA of all organisms, regardless of domain. By contrast, specific probes can be designed that react only with the ribosomes of species in a single domain. Such **phylogenetic probes** can also be designed to target lineages within a domain, such as members of particular families, genera, or even species.

The binding of probes to cellular ribosomes can be seen microscopically if a fluorescent dye is attached to the probes. When cells are treated with the appropriate reagents, their membranes become permeable and allow penetration of the probe–dye mixture. After hybridization of the probe directly to rRNA in ribosomes, the cells become uniformly fluorescent and

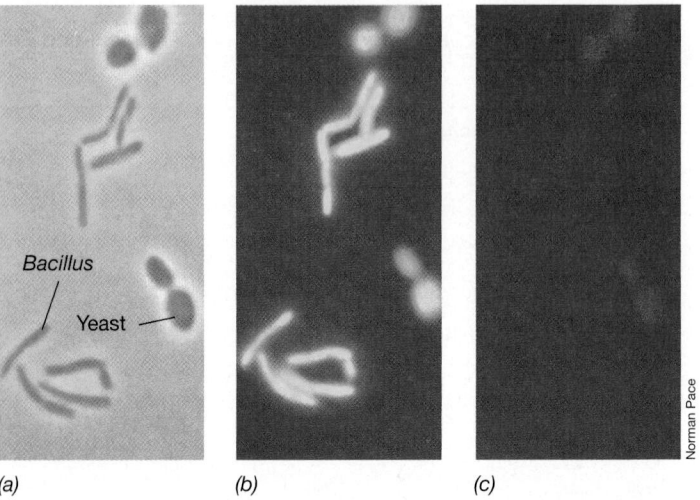

(a)   *(b)*   *(c)*

Figure 16.17 Fluorescently labeled rRNA probes: Phylogenetic stains. *(a)* Phase-contrast photomicrograph of cells of *Bacillus megaterium* (rod, *Bacteria*) and the yeast *Saccharomyces cerevisiae* (oval cells, *Eukarya*). *(b)* Same field; cells stained with a yellow-green universal rRNA probe (this probe reacts with species from any domain). *(c)* Same field; cells stained with a eukaryal probe (only cells of *S. cerevisiae* react). Cells of *B. megaterium* are about 1.5 μm in diameter and cells of *S. cerevisiae* are about 6 μm in diameter.

can be observed under a fluorescent microscope (**Figure 16.17**). This technique is called **FISH** (*f*luorescent *in situ h*ybridization) and can be applied to cells in culture or in a natural environment (the term *in situ* means "in the environment"). In essence, FISH is a *phylogenetic stain*.

FISH technology is widely used in microbial ecology and clinical diagnostics. In ecology, FISH can be used for the microscopic identification and tracking of organisms directly in the environment. FISH also offers a method for assessing the composition of microbial communities directly by microscopy (⟳ Section 22.4). In clinical diagnostics, FISH has been used for the rapid identification of specific pathogens from patient specimens. The technique circumvents the need to grow an organism in culture. Instead, microscopic examination of a specimen can confirm the presence of a specific pathogen, thus facilitating a rapid diagnosis and treatment. By contrast, isolation and identification of pathogens by classical means typically takes 24–48 hours and can take much longer.

Microbial Community Analysis

Polymerase chain reaction (PCR)-amplified rRNA genes (Figure 16.12) need not originate from a pure culture grown in the laboratory. Using methods described in detail in Chapter 22, a phylogenetic "snapshot" of a natural microbial community can be taken using PCR to amplify the genes encoding SSU rRNA from organisms in that community. Such genes can easily be sorted out, sequenced, and aligned. From these data, a phylogenetic tree can be constructed of sequences that depict the different rRNA genes present in the natural community. From this tree, the presence of specific organisms can be inferred even though none of them were actually cultivated or otherwise identified. Such *microbial community analyses*, a major tool of microbial ecology research today, have revealed many key features of microbial community structure and microbial interactions.

Ribotyping

Information from rRNA-based phylogenetic analyses also finds application in a technique for bacterial identification called **ribotyping**. Unlike comparative sequencing methods, however, ribotyping does not require sequencing. Instead, it generates a specific pattern of bands, a kind of *DNA fingerprint* called a *ribotype*, when DNA from an organism is digested by a restriction enzyme and the fragments are separated by gel electrophoresis and probed with an rRNA gene probe (**Figure 16.18**). Differences between organisms in the sequence of their 16S rRNA genes translate into the presence or absence of sites cut by different restriction endonucleases (Section 11.1). The ribotype of a particular organism may therefore be unique and diagnostic, allowing identification of different species and even different strains of a species if there are differences in their SSU rRNA gene sequences.

In ribotyping, following digestion and separation DNA fragments are transferred from the gel onto nylon membranes and hybridized with a labeled rRNA gene probe. The pattern of the fragments on the gel is then digitized, and compared with patterns of reference organisms in a computer database (Figure 16.18). Ribotyping is highly specific and rapid because it bypasses the PCR, sequencing, sequence alignment, and sequence analysis steps of SSU rRNA phylogenetic analysis (Figure 16.12). For these reasons, ribotyping has found many applications in clinical diagnostics and the microbial analyses of food, water, and beverages.

MiniQuiz

- How can oligonucleotide probes be made visible under the microscope? What is this technology called?
- What kinds of questions can be addressed using microbial community analysis?
- How is ribotyping able to distinguish between different bacteria?

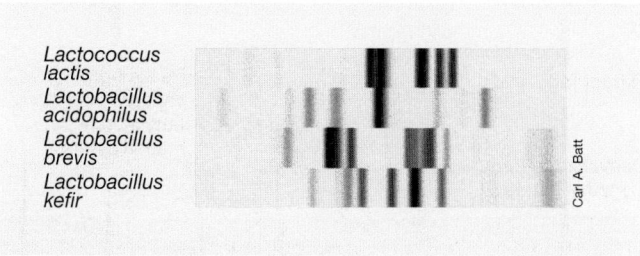

Figure 16.18 Ribotyping. Ribotype results for four different lactic acid bacteria. DNA was taken from a colony of each bacterium, digested into fragments by restriction enzymes, separated by gel electrophoresis, and then probed with a 16S rRNA gene probe. For each species the electrophoresis produced a unique pattern of bands. Variations in position and intensity of the bands are important in identification.

Microbial Systematics

Systematics is the study of the diversity of organisms and their relationships. It links together phylogeny, just discussed, with **taxonomy**, in which organisms are characterized, named, and placed into groups according to several defined criteria. Bacterial taxonomy traditionally has focused on practical aspects of identification and description, activities that have relied heavily on phenotypic comparisons. At present, the growing use of genetic information, especially DNA sequence data, is increasingly allowing taxonomy to reflect phylogenetic relationships as well.

Bacterial taxonomy has changed substantially in the past few decades, embracing a combination of methods for the identification of bacteria and description of new species. This *polyphasic approach* to taxonomy uses three kinds of methods—phenotypic, genotypic, and phylogenetic—for the identification and description of bacteria. Phenotypic analysis examines the morphological, metabolic, physiological, and chemical characteristics of the cell. Genotypic analysis considers characteristics of the genome. These two kinds of analysis group organisms based on similarities. They are complemented by phylogenetic analysis, which seeks to place organisms within an evolutionary framework.

16.10 Phenotypic Analysis: Fatty Acid Methyl Esters (FAME)

The observable characteristics—the phenotype—of a bacterium provide many traits that can be used to differentiate species. Typically, for either describing a new species or identifying a bacterium, several of these traits are determined for the organism of interest. The results are then compared with phenotypes of known organisms, either examined in parallel with the unknowns or from published information. The specific traits used depend on the kind of organism, and which traits are chosen for testing may arise from the investigator's purpose and from substantial prior knowledge of the bacterial group to which the new organism likely belongs. For example, in applied situations, such as in clinical diagnostic microbiology, where identification may be an end in itself and time is of the essence, a well-defined subset of traits is typically used that quickly discriminates between likely possibilities. **Table 16.2** lists general categories and examples of some phenotypic traits used in identifications and species descriptions, and we examine one of these traits here.

The types and proportions of fatty acids present in cytoplasmic membrane lipids and the outer membrane lipids of gram-negative bacteria are major phenotypic traits of interest. The technique for identifying these fatty acids has been nicknamed **FAME**, for *f*atty *a*cid *m*ethyl *e*ster, and is in widespread use in clinical, public health, and food and water-inspection laboratories where pathogens routinely must be identified. FAME analyses are also widely used in the characterization of new species of bacteria.

The fatty acid composition of *Bacteria* varies from species to species in chain length and in the presence or absence of double

Table 16.2 *Some phenotypic characteristics of taxonomic value*

Category	Characteristics
Morphology	Colony morphology; Gram reaction; cell size and shape; pattern of flagellation; presence of spores, inclusion bodies (e.g., PHB,[a] glycogen, or polyphosphate granules, gas vesicles, magnetosomes); capsules, S-layers or slime layers; stalks or appendages; fruiting-body formation
Motility	Nonmotile; gliding motility; swimming (flagellar) motility; swarming; motile by gas vesicles
Metabolism	Mechanism of energy conservation (phototroph, chemoorganotroph, chemolithotroph); utilization of individual carbon, nitrogen, or sulfur compounds; fermentation of sugars; nitrogen fixation; growth factor requirements
Physiology	Temperature, pH, and salt ranges for growth; response to oxygen (aerobic, facultative, anaerobic); presence of catalase or oxidase; production of extracellular enzymes
Cell lipid chemistry	Fatty acids[b]; polar lipids; respiratory quinones
Cell wall chemistry	Presence or absence of peptidoglycan; amino acid composition of cross-links; presence or absence of cross-link interbridge
Other traits	Pigments; luminescence; antibiotic sensitivity; serotype; production of unique compounds, for example, antibiotics

[a]PHB, poly-β-hydroxybutyric acid (⟳ Section 3.10).
[b]Figure 16.19

bonds, rings, branched chains, or hydroxy groups (**Figure 16.19***a*). Hence, a fatty acid profile can often identify a particular bacterial species. For the analyses, fatty acids extracted from cell hydrolysates of a culture grown under standardized conditions are chemically derivatized to form their corresponding methyl esters. These now volatile derivatives are then identified by gas chromatography. A chromatogram showing the types and amounts of fatty acids from the unknown bacterium is then compared with a database containing the fatty acid profiles of thousands of reference bacteria grown under the same conditions. The best matches to that of the unknown are then selected (Figure 16.19*b*).

As a phenotypic trait for species identification and description, FAME does have some drawbacks. In particular, FAME analyses require rigid standardization because fatty acid profiles of an organism, like many other phenotypic traits, can vary as a function of temperature, growth phase (exponential versus stationary), and to a lesser extent, growth medium. Thus, for consistent results, it is necessary to grow the unknown organism on a specific medium and at a specific temperature for comparison of its fatty acid profile with those of organisms from the database that have been grown in the same way. For many organisms this is impossible, of course, and thus FAME analyses are limited to those organisms that can be grown under the specified conditions. In addition, the extent of variation in FAME profiles among strains of a species, a necessary consideration in studies to discriminate between species, is not yet well documented.

Figure 16.19 Fatty acid methyl ester (FAME) analysis in bacterial identification. *(a)* Classes of fatty acids in *Bacteria*. Only a single example is given of each class, but in fact, more than 200 different fatty acids are known from bacterial sources. A methyl ester contains a methyl group (CH$_3$) in place of the proton on the carboxylic acid group (COOH) of the fatty acid. *(b)* Procedure. Each peak from the gas chromatograph is due to one particular fatty acid methyl ester, and the peak height is proportional to the amount.

Classes of Fatty Acids in *Bacteria*

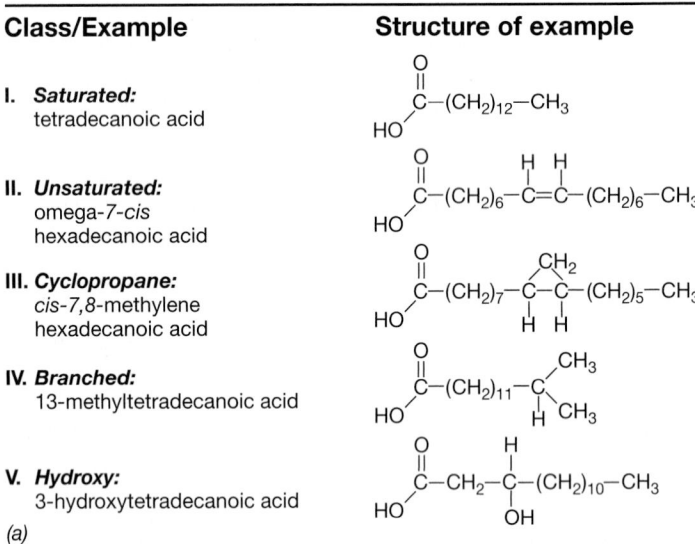

Class/Example

I. *Saturated:*
tetradecanoic acid

II. *Unsaturated:*
omega-7-*cis*
hexadecanoic acid

III. *Cyclopropane:*
cis-7,8-methylene
hexadecanoic acid

IV. *Branched:*
13-methyltetradecanoic acid

V. *Hydroxy:*
3-hydroxytetradecanoic acid

(a)

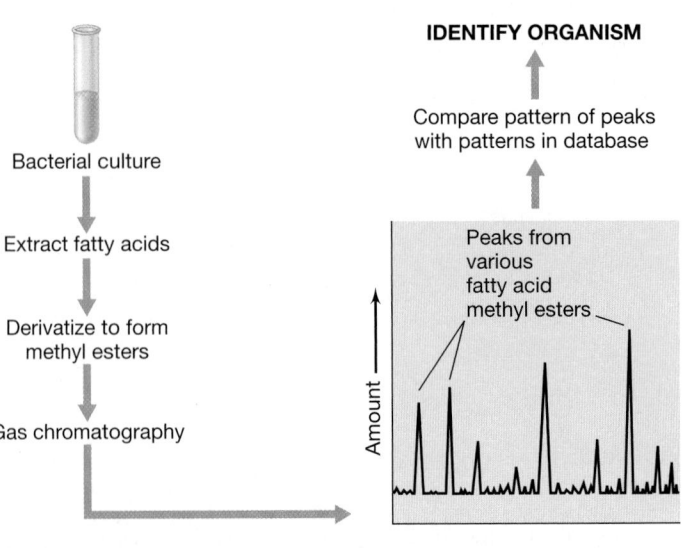

(b)

16.11 Genotypic Analyses

Comparative analysis of the genome provides many traits for discriminating between species of bacteria. Genotypic analysis has particular appeal in microbial taxonomy because of the insights it provides at the DNA level. The method of genotypic analysis used depends on the question(s) posed, with DNA–DNA hybridization and DNA profiling among the more commonly used in microbial taxonomy.

DNA–DNA Hybridization

When two organisms share many identical or highly similar genes, their DNAs are expected to hybridize in approximate proportion to the similarities in their DNA sequences. For this reason, measurement of **DNA–DNA hybridization** between the genomes of two organisms provides a rough index of their similarity to each other. DNA–DNA hybridization therefore is useful for differentiating between organisms as a complement to small subunit rRNA gene sequencing.

We discussed the theory and methodology of nucleic acid hybridization in Section 11.2. In a hybridization experiment, genomic DNA isolated from one organism is made radioactive with radioactive phosphorus (^{32}P) or tritium (^{3}H), sheared to a relatively small size, heated to separate the two strands, and mixed with an excess of unlabeled DNA prepared in the same way from a second organism (**Figure 16.20**). The DNA mixture is then cooled to allow the single strands to reanneal. The double-stranded DNA is separated from any remaining unhybridized DNA. Following this, the amount of radioactivity in the hybridized DNA is determined and compared with the control, which is taken as 100% (Figure 16.20). Several nonradioactive DNA labeling systems are also available (Section 11.2).

DNA–DNA hybridization is a sensitive method for revealing subtle differences in the genomes of two organisms and is therefore often useful for differentiating very similar organisms. Although there is no fixed convention as to how much hybridization between two DNAs is necessary to assign two organisms to the same taxonomic rank, hybridization values of 70% or greater are recommended as evidence that two isolates are the same species. Values of at least 25% are required to argue that two organisms are in the same genus (Figure 16.20c). DNAs from more distantly related organisms, for example, *Clostridium* (gram-positive) and *Salmonella* (gram-negative), would hybridize at only background levels, 10% or less.

GC Ratios

Another method that has been used to compare and describe bacteria is the **GC ratio** of their DNA. The GC ratio is the percentage of guanine (G) plus cytosine (C) in an organism's genomic DNA. GC ratios vary over a wide range, with values as low as 17% and as high as nearly 80% among species of *Bacteria* and *Archaea*, a range that is somewhat broader than for eukaryotes. It is typically the case that if two organisms' GC ratios differ

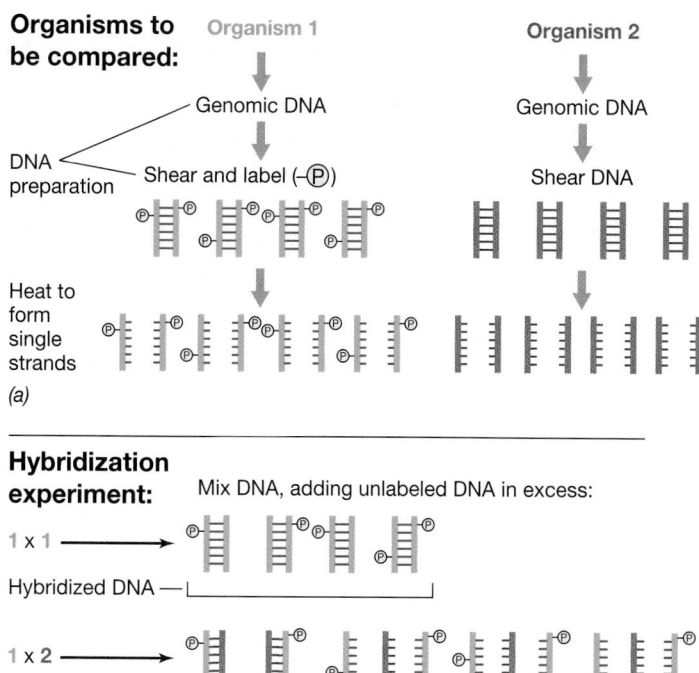

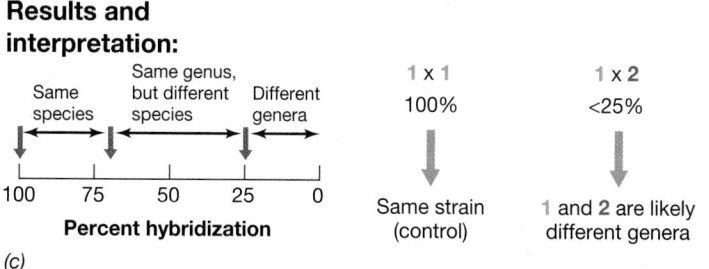

Figure 16.20 Genomic hybridization as a taxonomic tool.
(a) Genomic DNA is isolated from test organisms. One of the DNAs is labeled (shown here as radioactive phosphate in the DNA of Organism 1). (b) Excess unlabeled DNA is added to prevent labeled DNA from reannealing with itself. Following hybridization, hybridized DNA is separated from unhybridized DNA. Radioactivity in the hybridized DNA is measured. (c) Radioactivity in the control (Organism 1 DNA hybridizing to itself) is taken as the 100% hybridization value.

by more than about 5%, they have few DNA sequences in common and are therefore unlikely to be closely related. However, two organisms can have identical GC ratios and yet be unrelated because very different nucleotide sequences are possible from DNA of the same overall base composition. In this case, the identical GC ratios are taxonomically misleading. Because gene sequence data are increasingly easy to obtain, GC ratios are applied less commonly in bacterial taxonomy than in the past.

DNA Profiling Methods

There are several methods that generate DNA fragment patterns for analysis of genotypic similarity among bacterial strains. One of these DNA profiling methods, ribotyping, was described earlier. Other commonly used methods for rapid genotyping of bacteria

Table 16.3 *Some genotypic methods used in bacterial taxonomy*

Method	Description/application
DNA–DNA hybridization	Genome-wide comparison of sequence similarity. Useful for distinguishing species within a genus
DNA profiling	Ribotyping (Section 16.9), AFLP, rep-PCR (Figure 16.21). Rapid method to distinguish between species and strains within a species
Multilocus sequence typing	Strain typing using DNA sequences of multiple genes (Figure 16.22). High resolution, useful for distinguishing even very closely related strains within a species
GC ratio	Percentage of guanine–cytosine base pairs in the genome. If the GC ratio of two organisms differs by more than about 5%, they cannot be closely related, but organisms with similar or even identical GC ratios may be unrelated. Not much used now in taxonomy because of poor resolution
Multiple-gene or whole genome phylogenetic analyses	Application of cladistic methods to subsets of genes or to whole genomes from the organisms to be compared. Yields better phylogenetic picture than single-gene analyses

include *repetitive extragenic palindromic PCR* (*rep-PCR*) and *amplified fragment length polymorphism* (*AFLP*) (**Table 16.3**). In contrast to ribotyping, which focuses on a single gene, rep-PCR and AFLP assay for variations in DNA sequence throughout the genome.

The rep-PCR method is based on the presence of highly conserved repetitive DNA elements interspersed randomly around the bacterial chromosome. The number and positions of these elements differ between strains of a species that have diverged in genome sequence. Oligonucleotide primers designed to be complementary to these elements enable PCR amplification of elements from different genomic fragments that can be visualized by gel electrophoresis as patterns of bands. The patterns differ among different strains, giving what amounts to strain-specific DNA "fingerprints" (**Figure 16.21**).

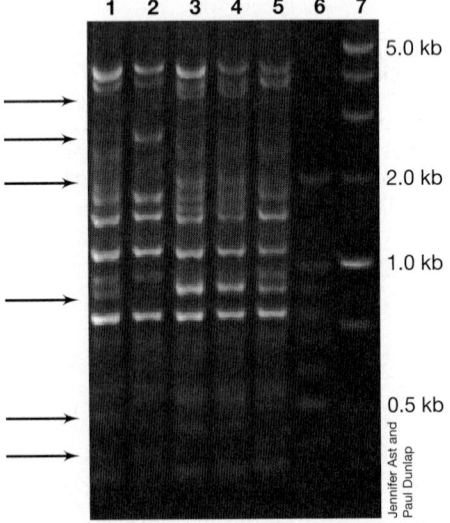

Figure 16.21 DNA fingerprinting with rep-PCR. Genomic DNAs from five strains (1–5) of a single species of bacteria were PCR-amplified using specific primers called *rep* (repetitive extragenic palindromic); the PCR products were separated in an agarose gel on the basis of size to generate DNA fingerprints. Arrows indicate some of the differing bands. Strains 3 and 4 have very similar DNA profiles. Lanes 6 and 7 are 100-bp and 1-kbp DNA size markers, respectively, used for estimating sizes of the DNA fragments.

AFLP is based on the digestion of genomic DNA with one or two restriction enzymes and selective PCR amplification of the resulting fragments, which are then separated by agarose gel electrophoresis. Strain-specific banding patterns similar to those of rep-PCR or other DNA fingerprinting methods are generated, with the large number of bands giving a high degree of discrimination between strains within a species. A technique similar to AFLP called T-RFLP (*terminal restriction fragment length polymorphism*) is widely used in phylogenetic analyses of natural microbial communities (⟲ Section 22.5).

Multilocus Sequence Typing

One of the limitations of both rRNA gene sequence analysis and ribotyping (but not of strain typing with rep-PCR or AFLP) is that these analyses focus on only a single gene, which may not provide sufficient information for unequivocal discrimination of bacterial strains. **Multilocus sequence typing (MLST)** circumvents this problem and is a powerful technique for characterizing strains within a species.

MLST consists of sequencing several different "housekeeping" genes from an organism and comparing their sequences with sequences of the same genes from different strains of the same organism. Housekeeping genes encode essential functions in cells and are located on the chromosome rather than on a plasmid. For each gene, an approximately 450-base-pair sequence is amplified using PCR and is then sequenced. Each nucleotide along the sequence is compared and differences are noted. Each difference, or sequence variant, is called an **allele** and is assigned a number. The strain being studied is then assigned a series of numbers as its allelic profile, or multilocus sequence type. In MLST, strains with identical sequences for a given gene have the same allele number for that gene, and two strains with identical sequences for all the genes have the same allelic profile (and would be considered identical by this method). The relatedness between each allelic profile is expressed in a dendrogram of linkage distances that vary from 0 (strains are identical) to 1 (strains are only distantly related, if at all) (**Figure 16.22**).

MLST has sufficient resolving power to distinguish among even very closely related strains. In practice, strains can be discriminated on the basis of a single nucleotide change in just one

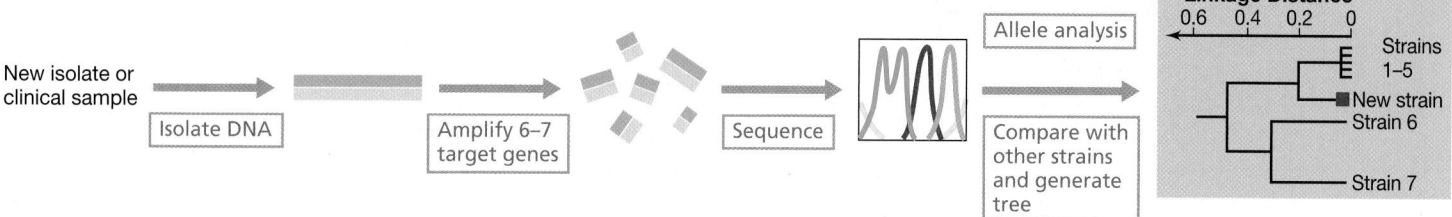

Figure 16.22 Multilocus sequence typing. Steps in MLST leading to a similarity phenogram are shown. Strains 1–5 are virtually identical, whereas strains 6 and 7 are distinct from one another and from strains 1–5.

of the analyzed genes. MLST is not useful, however, for comparing organisms above the species level; its resolution is too sensitive to yield meaningful information for grouping higher-order taxa such as genera and families.

MLST has found its greatest use in clinical microbiology, where it has been used to differentiate strains of a particular pathogen. This is important because some strains within a species—*Escherichia coli* K-12, for example—may be harmless, whereas others, such as strain O157:H7, can cause serious and even fatal infections (↩ Section 36.9). MLST is also widely used in epidemiological studies to track a virulent strain of a bacterial pathogen as it moves through a population, and in environmental studies to define the geographic distributions of strains.

Multigene and Whole Genome Analyses

Sequence comparisons of particular genes can provide valuable insight for taxonomy as well as phylogeny. The 16S rRNA gene, for example, the importance of which in microbial phylogeny was described in Section 16.6, has proven exceptionally useful in taxonomy as well, serving as a "gold standard" for the identification and description of new species. Other highly conserved genes, such as *recA,* which encodes a recombinase protein, and *gyrB,* which encodes a DNA gyrase protein, also can be useful for distinguishing bacteria at the species level. But for many reasons, including the facts that single-gene analyses give only a very limited genomic snapshot and that some genes may have been subject to horizontal gene flow that could lead to incorrect taxonomic conclusions, multigene and whole genomic analyses are becoming popular in microbial systematics.

The use of multiple genes for the identification and description of bacteria can avoid problems associated with reliance on individual genes. Multigene sequence analysis is similar to MLST (Figure 16.22), except that complete or nearly complete gene sequences are obtained and comparisons are made using cladistic methods (Section 16.7 and see Figure 16.24). By sequencing several functionally unrelated genes, one can obtain a more representative sampling of the genome than is possible with a single gene, and instances of horizontal gene transfer can be detected and those genes excluded from further consideration. Analyses of whole genome sequences provide an even greater depth of genotypic analysis. For example, differences between species in genome structure, including size and number of chromosomes, their GC content, and whether the chromosomes are linear or circular may have taxonomic significance. Comparative analysis

of gene content (presence or absence of genes) and the order of genes in the genome can also provide insights.

MiniQuiz

- What is DNA fingerprinting and how is it useful for distinguishing bacteria?
- Hybridization of 90% of two organisms' DNA indicates that they are _____?
- How do AFLP and MLST differ from ribotyping?
- What advantages do multigene and whole genome analyses have over single-gene analyses?

16.12 The Species Concept in Microbiology

At present, there is no universally accepted concept of **species** for prokaryotes. Microbial systematics combines phenotypic, genotypic, and sequence-based phylogenetic data within a framework of standards and guidelines for describing and identifying prokaryotes, but the issue of what actually constitutes a prokaryotic species remains controversial. Because species are the fundamental units of biological diversity, how the concept of species is defined in microbiology determines how we distinguish and classify the units of diversity that make up the microbial world.

Current Definition of Prokaryotic Species

A prokaryotic species is defined operationally as a group of strains sharing a high degree of similarity in several independent traits. Traits currently considered most important for grouping strains together as a species include 70% or greater genomic DNA–DNA hybridization and 97% or greater identity ($<3\%$ difference) in 16S rRNA gene sequence (Sections 16.6 and 16.11). Experimental data suggest that these two criteria are valid, reliable, and consistent in identifying new species of prokaryotes (**Figure 16.23**). Based on genotypic criteria such as these, over 7000 species of *Bacteria* and *Archaea* have been formally recognized. What criteria should be used to define a genus, the next highest taxon, is more a matter of judgment, but 16S rRNA gene sequence differences of more than 5% from all other organisms is considered good evidence that an organism constitutes its own genus. Above the level of genus to the family, order, and other ranks of higher taxa, no consensus ribosomal RNA sequence-based criteria exist for delineating

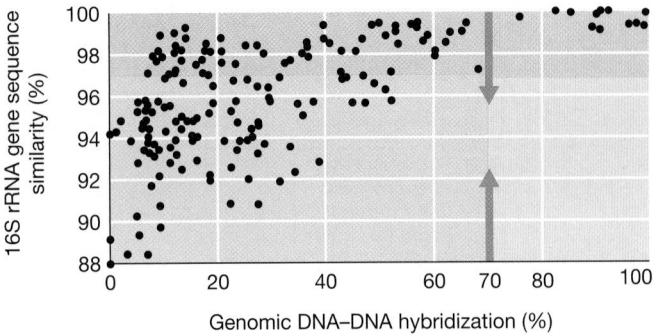

Figure 16.23 Relationship between 16S rRNA gene sequence similarity and genomic DNA–DNA hybridization for pairs of organisms. These data are the results from several independent experiments with various species of the domain *Bacteria*. Points in the darker tan region at the upper right represent pairs for which 16S rRNA gene sequence similarity and genomic hybridization were both very high; thus, in each case, the two organisms tested were clearly the same species. Points in the green box represent pairs that appear to be different species, and both methods show this. The blue box shows examples of pairs that seem to be different species as measured by genomic DNA–DNA hybridization, but not by 16S rRNA gene sequence. Note that above 70% DNA–DNA hybridization, no 16S rRNA gene similarities were found that were less than 97%. Data from Rosselló-Mora, R., and R. Amann. 2001. *FEMS Microbiol. Revs.* 25:39–67.

these ranks. **Table 16.4** gives an example of species definition in practice, listing relevant traits for the classification of the phototrophic purple bacterium *Allochromatium warmingii* from the domain down to the species level.

The Biological and Phylogenetic Species Concepts

The *biological species concept* posits that a species is an interbreeding population of organisms that is reproductively isolated from other interbreeding populations; it is widely accepted as effective for defining species of eukaryotic organisms. However, the biological species concept is not meaningful for *Bacteria* and *Archaea* because they are haploid organisms that do not reproduce sexually.

An alternative to the biological species concept suitable for haploid organisms is the *phylogenetic species concept*. This concept defines a prokaryotic species as a group of strains that cluster closely with each other and are distinct from other groups of strains based on multiple-gene cladistic analyses (Sections 16.7 and 16.11). An example of such analysis using six genes from three species of the bacterium *Photobacterium* is shown in **Figure 16.24**.

The DNA sequence of the 16S rRNA gene has diverged relatively little throughout evolutionary history and therefore provides good family- and genus-level resolution. But 16S rRNA

Table 16.4 *Taxonomic hierarchy for the purple sulfur bacterium* Allochromatium warmingii

Taxon	Name	Properties	Confirmed by	
Domain	*Bacteria*	Bacterial cells; rRNA gene sequences typical of *Bacteria*	Microscopy; 16S rRNA gene sequence analysis; presence of unique biomarkers, for example, peptidoglycan	Sulfur (S⁰) globules
Phylum	*Proteobacteria*	rRNA gene sequence typical of *Proteobacteria*	16S rRNA gene sequence analysis	
Class	*Gammaproteobacteria*	Gram-negative bacteria; rRNA sequence typical of *Gammaproteobacteria*	Gram-staining, microscopy	
Order	*Chromatiales*	Phototrophic purple bacteria	Characteristic pigments (Figure 13.3)	
Family	*Chromatiaceae*	Purple sulfur bacteria	Ability to oxidize H_2S and store S^0 within cells; microscopic observation of S^0 (see photo); 16S rRNA gene sequence	
Genus	*Allochromatium*	Rod-shaped purple sulfur bacteria; <95% 16S gene sequence identity with all other genera	Microscopy (see photo)	
Species	*warmingii*	Cells 3.5–4.0 μm × 5–11 μm; storage of sulfur mainly in poles of cell (see photo); <97% 16S gene sequence identity with all other species	Cell size measured microscopically with a micrometer; observation of polar position of S^0 globules in cells (see photo); 16S rRNA gene sequence	

Photomicrograph of cells of the purple sulfur bacterium *Allochromatium warmingii*.

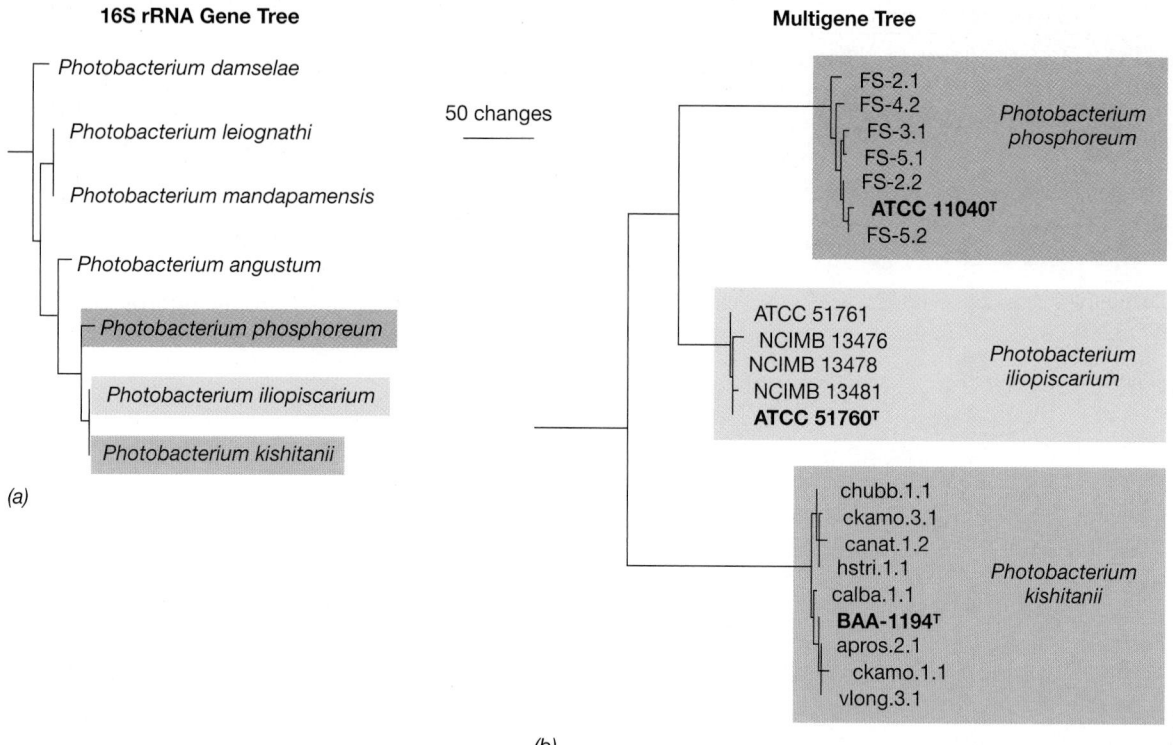

16S rRNA Gene Tree

Photobacterium damselae
Photobacterium leiognathi
Photobacterium mandapamensis
Photobacterium angustum
Photobacterium phosphoreum
Photobacterium iliopiscarium
Photobacterium kishitanii

50 changes

(a)

Multigene Tree

FS-2.1
FS-4.2
FS-3.1
FS-5.1
FS-2.2
ATCC 11040T
FS-5.2

Photobacterium phosphoreum

ATCC 51761
NCIMB 13476
NCIMB 13478
NCIMB 13481
ATCC 51760T

Photobacterium iliopiscarium

chubb.1.1
ckamo.3.1
canat.1.2
hstri.1.1
calba.1.1
BAA-1194T
apros.2.1
ckamo.1.1
vlong.3.1

Photobacterium kishitanii

(b)

Figure 16.24 Multigene phylogenetic analysis. A phylogeny is shown for species in the genus *Photobacterium* (*Gammaproteobacteria*). *(a)* 16S rRNA gene tree, showing the species to be poorly resolved. *(b)* Multigene analysis based on combined parsimony analysis of the 16S rRNA gene, *gyrB*, and *luxABFE* genes in three *Photobacterium* species. Because the *gyrB* and *luxABFE* sequences diverge more than the 16S rRNA sequence, multigene analysis clearly resolves the 21 different strains analyzed into three distinct clades (phylogenetic species), *P. phosphoreum* (7 strains), *P. iliopiscarium* (5 strains), and *P. kishitanii* (9 strains). The scale bar indicates the branch length equal to a total of 50 nucleotide changes. The type strain of each species is designated with a superscript T appended to the strain designation and shown in bold. Phylogenetic analyses courtesy of Tory Hendy and Paul V. Dunlap.

gene sequence analyses do not necessarily provide good species-level resolution when sequences differ very little, as is the case here (Figure 16.24*a*). For better species-level resolution *gyrB*, the gene encoding DNA gyrase subunit B, and *luxABFE*, a series of genes encoding luminescence enzymes (Figure 1.1), were used, in addition to the 16S rRNA gene. The *gyrB* and *luxABFE* genes are less functionally constrained than the 16S rRNA gene, meaning that their sequences can vary more without a loss of function of the proteins they encode. The multigene analysis clearly resolves the strains of *Photobacterium* into three distinct evolutionary clades, and each clade can be considered a phylogenetic species (Figure 16.24*b*). In this way, multigene phylogenetic analyses and the phylogenetic species concept can be used to distinguish bacterial species that cannot be resolved by rRNA gene sequence analyses alone.

Speciation in Prokaryotes

How do new prokaryotic species arise? A likely possibility is by the process of periodic purges and selection within cell populations. Imagine a population of bacteria that originated from a single cell and that occupies a particular niche in a habitat. In theory, these cells are genetically identical. If cells in this population share a particular resource (for example, a key nutrient), the population is considered an **ecotype**.

Different ecotypes can coexist in a habitat, but each is only most successful within its prime niche in the habitat. However, within each ecotype, genes mutate at random over time as the cells grow. Most of these mutations are neutral and have no effect. However, if there is a beneficial mutation (one that increases fitness) in a cell in one of the ecotypes, that cell will produce more progeny over time, and this will purge the population of the original, less well-adapted cells (see Figure 16.10). Repeated rounds of mutation and selection in this ecotype lead it to become more and more distinct genetically from the other ecotypes. Then, given enough time, cells in this lineage will carry a sufficiently large set of unique traits that they emerge as their own species (**Figure 16.25**). Selection of strains bearing beneficial mutations can proceed gradually, or it can occur quite suddenly due to rapid environmental change. Note that this series of events *within* an ecotype has no effect on *other* ecotypes, because different ecotypes do not compete for the same resources (Figure 16.25).

It is also possible that a new genetic capacity in an ecotype may arise from genes obtained from cells of another ecotype by horizontal gene transfer, rather than from mutation and selection. The extent of horizontal gene transfer among bacteria is variable. Genome sequence analyses have revealed examples in which horizontal transfer of genes has apparently been frequent and others in which it has been rare (Section 12.11). However,

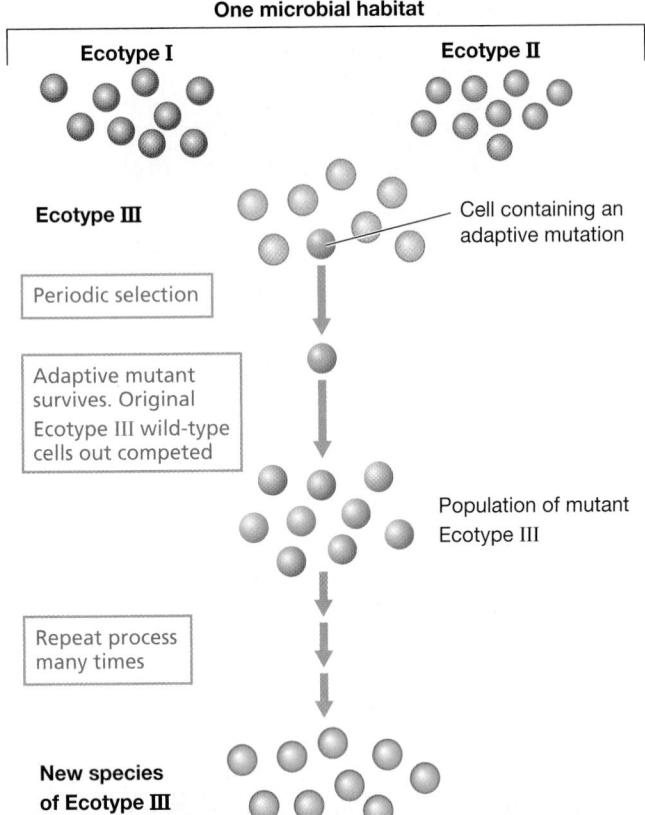

One microbial habitat

Ecotype I   **Ecotype II**

Ecotype III

Cell containing an adaptive mutation

Periodic selection

Adaptive mutant survives. Original Ecotype III wild-type cells out competed

Population of mutant Ecotype III

Repeat process many times

New species of Ecotype III

Figure 16.25 A model for bacterial speciation. Several ecotypes can coexist in a single microbial habitat, each occupying its own prime ecological niche. A cell within an ecotype that has a beneficial mutation may grow to become a population that eventually replaces the original ecotype. As this is repeated within a given ecotype, a genetically distinct population of cells arises that is a new species. Because other ecotypes do not compete for the same resources, they are unaffected by genetic and selection events outside their prime niche.

despite the potential impact of horizontal gene transfer for the instant acquisition of new metabolic capacities, speciation in *Bacteria* and *Archaea* is likely driven primarily by mutation and periodic selection (Figure 16.25). This is because horizontal gene transfer often confers upon the recipient cell only temporary benefits because the transferred genes will likely be lost if there is insufficient selective pressure to retain them.

How Many Prokaryotic Species Are There?

The result of nearly 4 billion years of evolution is the prokaryotic world we see today (Figure 16.16). Microbial taxonomists agree that no firm estimate of the number of prokaryotic species can be given at present, in part because of uncertainty about what defines a species. However, they also agree that in the final analysis, this number will be very large. Over 7000 species of *Bacteria* and *Archaea* are already known, based primarily on 16S rRNA gene sequencing, and thousands more, perhaps as many as 100,000–1,000,000, are thought likely to exist. If we were to attack this problem using multigene analyses, which provide better species-level resolution than 16S rRNA analyses alone (Figure

16.24), the species estimate would increase by one to two orders of magnitude.

Microbial community analyses (Sections 16.9, 22.5, and 22.6) indicate that we have only scratched the surface in our ability to culture the diversity of *Bacteria* and *Archaea* in nature. With the future application of more powerful tools— both molecular and cultural—for revealing diversity, it is likely that the already impressive list of species known will grow even larger. But the reality today is that an accurate estimate of prokaryotic species is simply out of reach of our current understanding and technology.

MiniQuiz

- How do the biological and phylogenetic species concepts differ? Which is suitable for prokaryotes and why?
- What is an ecotype?
- How many species of *Bacteria* and *Archaea* are already known? How many likely exist?

16.13 Classification and Nomenclature

We conclude this chapter with a brief description of how *Bacteria* and *Archaea* are classified and named; the science of *taxonomy*. Information is also presented on culture collections, which serve as repositories for scientific deposition of microbial cultures, on some key taxonomic resources available for microbiology, and the procedures for naming new species. The formal description of a new prokaryotic species and deposition of living cultures into a culture collection form an important foundation for prokaryotic systematics.

Taxa and Naming of Prokaryotes

Classification is the organization of organisms into progressively more inclusive groups on the basis of either phenotypic similarity or evolutionary relationship. The hierarchical nature of classification was shown in Table 16.4. A species is made up of one to several strains, and similar species are grouped into genera (singular, genus). Similar genera are grouped into families, families into orders, orders into classes, up to the domain, the highest-level taxon.

Nomenclature is the actual naming of organisms and follows the **binomial system** of nomenclature devised by the Swedish medical doctor and botanist, Carl Linnaeus, and used throughout biology; organisms are given genus names and species epithets. The names are Latin or Latinized Greek derivations, often descriptive of some key property of the organism, and are printed in *italics*. By classifying organisms into groups and naming them, we order the natural microbial world and make it possible to communicate effectively about all aspects of particular organisms, including their behavior, ecology, physiology, pathogenesis, and evolutionary relationships. The creation of new names must follow the rules described in *The International Code of Nomenclature of Bacteria*. This source presents the formal framework by which *Bacteria* and *Archaea* are to be officially named and the procedures by which existing names can be changed, for example, when new data warrants taxonomic rearrangements.

Table 16.5 *Some national microbial culture collections*

Collection	Name	Location	Web address
ATCC	American Type Culture Collection	Manassas, Virginia	**http://www.atcc.org**
BCCM/LMG	Belgium Coordinated Collection of Microorganisms	Ghent, Belgium	**http://bccm.belspo.be**
CIP	Collection de l'Institut Pasteur	Paris, France	**http://www.pasteur.fr**
DSMZ	Deutsche Sammlung von Mikroorganismen und Zellkulturen	Braunschweig, Germany	**http://www.dsmz.de**
JCM	Japan Collection of Microorganisms	Saitama, Japan	**http://www.jcm.riken.go.jp**
NCCB	Netherlands Culture Collection of Bacteria	Utrecht, The Netherlands	**http://www.cbs.knaw.nl/nccb**
NCIMB	National Collection of Industrial, Marine and Food Bacteria	Aberdeen, Scotland	**http://www.ncimb.com**

Bergey's Manual and *The Prokaryotes*

Because taxonomy is largely a matter of scientific judgment, there is no "official" classification of *Bacteria* and *Archaea*. Presently, the classification system most widely accepted by microbiologists is that of *Bergey's Manual of Systematic Bacteriology,* a major taxonomic treatment of *Bacteria* and *Archaea* (see Appendix 2 for a list of genera and higher-order taxa from *Bergey's Manual*). Widely used, *Bergey's Manual* has served the community of microbiologists since 1923 and is a compendium of information on all recognized prokaryotes. Each chapter, written by experts, contains tables, figures, and other systematic information useful for identification purposes.

A second major source in bacterial diversity is *The Prokaryotes*, a reference that provides detailed information on the enrichment, isolation, and culture of *Bacteria* and *Archaea*. This work is available online by subscription through university libraries. Collectively, *Bergey's Manual* and *The Prokaryotes* offer microbiologists both the concepts as well as the details of the biology of *Bacteria* and *Archaea* as we know it today; they are the primary resources for microbiologists characterizing newly isolated organisms.

Culture Collections

National microbial culture collections (**Table 16.5**) are an important foundation of microbial systematics. These permanent collections catalog and store microorganisms and provide them upon request, usually for a fee, to researchers in academia, medicine, and industry. The collections play an important role in protecting microbial biodiversity, just as museums do in preserving plant and animal specimens for future study. However, unlike museums, which maintain collections of chemically preserved or dried, dead specimens, microbial culture collections store microorganisms as *viable cultures*, typically frozen or in a freeze-dried state. These storage methods maintain the cells indefinitely in a living state.

A related and key role of culture collections is as repositories for *type strains*. When a new species of bacteria is described in a scientific journal, a strain is designated as the nomenclatural type of the taxon for future taxonomic comparison with other strains of that species. Deposition of this type strain in the national culture collections of at least two countries, thereby making the strain publicly available, is a prerequisite for validation of the new

species name. Some of the large national culture collections are listed in Table 16.5. Their websites contain searchable databases of strain holdings, together with information on the environmental sources of strains and publications on them.

Describing New Species

When a new prokaryote is isolated from nature and thought to be unique, a decision must be made as to whether it is sufficiently different from other prokaryotes to be described as a new taxon. To achieve formal validation of taxonomic standing as a new genus or species, a detailed description of the organism's characteristics and distinguishing traits, along with its proposed name, must be published, and, as just mentioned, viable cultures of the organism must be deposited in at least two international culture collections (Table 16.5). The manuscript describing and naming a new taxon undergoes peer review before publication. A major vehicle for the description of new taxa is the *International Journal of Systematic and Evolutionary Microbiology* (*IJSEM*), the official publication of record for the taxonomy and classification of *Bacteria* and *Archaea*. In each issue, the *IJSEM* publishes an approved list of newly validated names. By providing validation of newly proposed names, publication in *IJSEM* paves the way for their inclusion in *Bergey's Manual of Systematic Bacteriology*. Two websites provide listings of valid, approved bacterial names: List of Prokaryotic Names with Standing in Nomenclature (http://www.bacterio.cict.fr), and Bacterial Nomenclature Up-to-Date (http://www.dsmz.de/bactnom/bactname.htm).

The International Committee on Systematics of Prokaryotes (ICSP) is responsible for overseeing nomenclature and taxonomy of *Bacteria* and *Archaea*. The ICSP oversees the publication of *IJSEM* and the *International Code of Nomenclature of Bacteria*, and it gives guidance to several subcommittees that establish and revise standards for the description of new species in the different groups of prokaryotes.

MiniQuiz

- What roles do culture collections play in microbial systematics?
- What is the *IJSEM* and what taxonomic function does it fulfill?
- Why might viable cell cultures be of more use in microbial taxonomy than preserved specimens?

Big Ideas

16.1

Planet Earth is about 4.5 billion years old. The first evidence for microbial life can be found in rocks 3.86 billion years old. In rocks 3.5 billion years old or younger, microbial formations called stromatolites are abundant and show extensive microbial diversification.

16.2

Life may have first arisen at submerged hydrothermal springs, and the first self-replicating life forms may have been RNAs. Eventually, DNA evolved and the DNA plus RNA plus protein model for cellular life was fixed. Early microbial metabolism was anaerobic and likely chemolithotrophic, exploiting abundant abiotic sources of H_2, FeS, and H_2S. The earliest carbon metabolism may have been autotrophic.

16.3

Early *Bacteria* and *Archaea* diverged from a common ancestor as long as 4 billion years ago. Microbial metabolism diversified on early Earth with the evolution of methanogenesis and anoxygenic photosynthesis. Oxygenic photosynthesis eventually led to an oxic Earth, banded iron formations, and great bursts in metabolic and cellular evolution.

16.4

The eukaryotic cell developed from endosymbiotic events. In the most likely scenario, a H_2-producing species of *Bacteria* was incorporated as an endosymbiont into a H_2-consuming species of *Archaea* (the host). The modern eukaryotic cell is a chimera with genes and characteristics from both *Bacteria* and *Archaea*.

16.5

Evolution is descent with modification. Natural selection works by favoring the survival and reproductive success of organisms that by chance have mutations that confer high fitness under the existing environmental conditions.

16.6

Phylogeny, the evolutionary history of life, can be reconstructed through analysis of homologous DNA sequences. Genes encoding SSU rRNAs have been used as molecular clocks to construct a phylogeny of all organisms, prokaryotes as well as eukaryotes.

16.7

Analytical methods for evolutionary analysis include sequence alignment and construction of phylogenetic trees that, if rooted, indicate a path of evolution based on common ancestry. Character-state methods such as parsimony are commonly used for tree construction.

16.8

Life on Earth evolved in three major directions, forming the domains *Bacteria*, *Archaea*, and *Eukarya*. Each domain contains several major lineages. The universal tree of life shows that the two prokaryotic domains, *Bacteria* and *Archaea*, split from each other eons ago, and that *Eukarya* split from *Archaea* later in the history of life.

16.9

Phylogenetic analyses of SSU rRNA genes have led to the development of research tools useful in ecology and medicine. Key among these is FISH, which uses fluorescently labeled phylogenetic probes to identify organisms in a natural sample. Other key methods include microbial community analysis and ribotyping.

16.10

Systematics is the study of the diversity and relationships of living organisms. Polyphasic taxonomy is based on phenotypic, genotypic, and phylogenetic information. Phenotypic traits useful in taxonomy include morphology, motility, metabolism, and cell chemistry, especially lipid analyses.

16.11

Genotypic analysis examines traits of the genome. Bacterial species can be distinguished genotypically on the basis of DNA–DNA hybridization, DNA profiling, MLST, multigene or whole genome analyses, and by the GC content of their DNA.

16.12

At present a prokaryotic species is defined operationally based on shared genetic and phenotypic traits. The biological species concept is unsuitable for prokaryotes because their mode of reproduction is not sexual. New species of prokaryotes arise from periodic purging and selection within an ecotype, and the number of distinct species of prokaryotes in nature is surely enormous.

16.13

Formal recognition of a new prokaryotic species requires depositing a sample of the organism in culture collections and publishing the new species name and description. *Bergey's Manual of Systematic Bacteriology* and *The Prokaryotes* are major taxonomic compilations of *Bacteria* and *Archaea*.

Review of Key Terms

Allele a sequence variant of a given gene

Archaea phylogenetically related prokaryotes distinct from *Bacteria*

Bacteria phylogenetically related prokaryotes distinct from *Archaea*

Banded iron formation iron oxide–rich ancient sedimentary rocks containing zones of oxidized iron (Fe^{3+}) formed by oxidation of Fe^{2+} by O_2 produced by cyanobacteria

Binomial system the system devised by the Swedish scientist Carl Linnaeus for naming living organisms in which an organism is given a genus name and a species epithet

Cladistics phylogenetic methods that group organisms by their evolutionary relationships, not by their phenotypic similarities

Domain in a taxonomic sense, the highest level of biological classification

DNA–DNA hybridization the experimental determination of genomic similarity by measuring the extent of hybridization of DNA from one organism with that of another

Ecotype a population of genetically identical cells sharing a particular resource within an ecological niche

Endosymbiotic hypothesis the idea that a chemoorganotrophic bacterium and a cyanobacterium were stably incorporated into another cell type to give rise, respectively, to the mitochondria and chloroplasts of modern-day eukaryotes

Eukarya all eukaryotes: algae, protists, fungi, slime molds, plants, and animals

Evolution descent with modification; DNA sequence variation and the inheritance of that variation

FAME fatty acid methyl ester; a technique for identifying microorganisms from their fatty acids

FISH fluorescent *in situ* hybridization; a staining technique for phylogenetic studies

Fitness the capacity of an organism to survive and reproduce as compared to that of competing organisms

GC ratio in DNA from an organism, the percentage of the total nucleic acid that consists of guanine and cytosine bases

Horizontal gene transfer the transfer of DNA from one cell to another, possibly distantly related, cell

Molecular clock a DNA sequence, such as the gene for rRNA, that can be used as a comparative temporal measure of evolutionary divergence

Monophyletic in phylogeny, a group descended from one ancestor

Multilocus sequence typing (MLST) a taxonomic tool for classifying organisms from gene sequence variations in several housekeeping genes

Phylogenetic probe an oligonucleotide, sometimes made fluorescent by attachment of a dye, complementary in sequence to some sequence in rRNA

Phylogeny the evolutionary history of an organism

Phylum a major lineage of cells in one of the three domains of life

Proteobacteria a large group of phylogenetically related, gram-negative *Bacteria*

Ribosomal Database Project (RDP) a large database of small subunit (SSU) rRNA

sequences that can be retrieved electronically and used in comparative rRNA sequence studies

Ribotyping a means of identifying microorganisms from analysis of DNA fragments generated from restriction enzyme digestion of the genes encoding their 16S rRNA

16S rRNA a large polynucleotide (~1500 bases) that functions as part of the small subunit of the ribosome of *Bacteria* and *Archaea* and from whose gene sequence evolutionary information can be obtained; its eukaryotic counterpart is 18S rRNA

Small subunit (SSU) rRNA RNA from the 30S ribosomal subunit of *Bacteria* and *Archaea* or the 40S ribosomal subunit of eukaryotes; that is, 16S or 18S rRNA, respectively

Species defined in microbiology as a group of strains that all share the same major properties and differ in one or more significant properties from other groups of strains; defined phylogenetically as a monophyletic, exclusive group based on DNA sequence analyses

Stromatolite a laminated microbial mat, typically built from layers of filamentous *Bacteria* and other microorganisms, which can become fossilized

Systematics the study of the diversity of organisms and their relationships; includes taxonomy and phylogeny

Taxonomy the science of identification, classification, and nomenclature

Universal phylogenetic tree a tree that shows the positions of representatives of all domains of cells

Review Questions

1. What is the age of planet Earth? When did the oceans form, and what is the age of the earliest known microfossils (Section 16.1)?

2. Under what conditions did life likely originate? What were the steps leading from prebiotic chemistry to living cells (Section 16.2)?

3. What kind of energy and carbon metabolisms likely characterized early cellular life (Section 16.2)?

4. Why was the evolution of cyanobacteria of such importance to the further evolution of life on Earth? What component of the geological record is used to date the evolution of cyanobacteria (Section 16.3)?

5. Briefly describe the endosymbiotic origin of eukaryotic cells (Section 16.4).

6. What does the phrase "descent with modification" imply about natural relationships among living organisms (Section 16.5)?

7. Define phylogeny. Why are SSU rRNA genes good candidates for phylogenetic studies (Section 16.6)?

8. Describe the steps for determining an SSU phylogeny of three bacteria you have isolated from nature (Section 16.7).

9. What major evolutionary finding emerged from the study of rRNA sequences? How has this discovery supported the endosymbiotic theory of eukaryotic origins (Section 16.8)?

10. What major physiological and biochemical properties do *Archaea* share with *Eukarya* or with *Bacteria* (Section 16.8)?

11. What is FISH technology? Give an example of how it could be used (Section 16.9).

12. What is ribotyping? Why is it useful (Section 16.9)?

13. What is measured in FAME analyses (Section 16.10)?

14. Briefly describe the various methods of genotypic analysis. Name one advantage for each method (Section 16.11).

15. How is multigene phylogenetic analysis an improvement over analyses based on individual genes (Sections 16.11 and 16.12)?

16. How is it thought that new bacterial species arise? How many bacterial species are there? Why don't we know this number more precisely (Section 16.12)?

17. What roles do microbial culture collections play in microbial systematics (Section 16.13)?

Application Questions

1. Compare and contrast the physical and chemical conditions on Earth at the time life first arose with conditions today. From a physiological standpoint, discuss at least two reasons why *animals* could not have existed on early Earth.

2. In what ways has microbial metabolism altered Earth's biosphere? How might life on Earth be different if oxygenic photosynthesis had not evolved?

3. For the following sequences, identify the phylogenetically informative sites (assume that the sequences are properly aligned). Identify also the phylogenetically neutral sites and those that are invariant.

   Taxon 1: TCCGTACGTTA

   Taxon 2: TCCCCACGGTT

   Taxon 3: TCGGTACCGTA

   Taxon 4: TCGGTACCGTA

4. Imagine that you have been given several bacterial strains from various countries around the world and that all the strains are thought to cause the same gastrointestinal disease and to be genetically identical. Upon carrying out a DNA fingerprint analysis of the strains, you find that four different strain types are present. What methods could you use to test whether the different strains are actually members of the same species?

5. Imagine that you have discovered a new form of microbial life, one that appears to represent a fourth domain. How would you go about characterizing the new organism and determining if it actually is evolutionarily distinct from *Bacteria, Archaea,* and *Eukarya*?

Need more practice? Test your understanding with Quantitative Questions; access additional study tools including tutorials, animations, and videos; and then test your knowledge with chapter quizzes and practice tests at **www.microbiologyplace.com**.

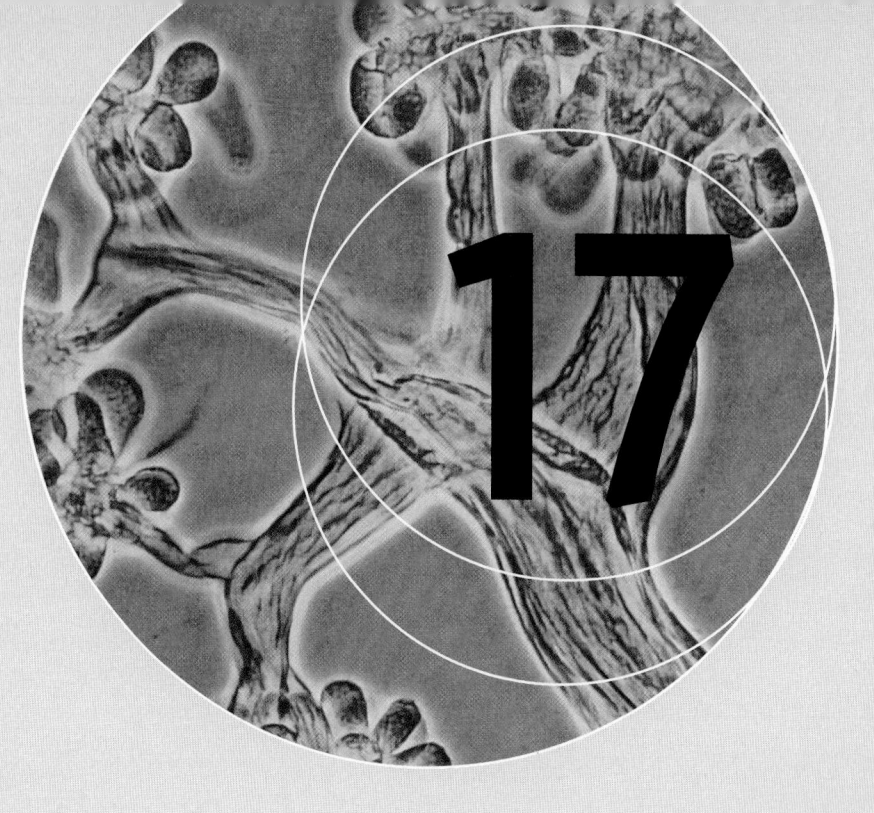

Bacteria: The *Proteobacteria*

The myxobacteria, a major group of *Proteobacteria*, the largest known phylum of *Bacteria*, are highly social bacteria that cooperate to form macroscopically visible masses of cells called fruiting bodies.

I The Phylogeny of *Bacteria*

In the last chapter we examined evolutionary relationships among microorganisms. In this and the next three chapters we expand on these concepts with a consideration of the properties and diversity of major microbial groups. We begin our tour with a focus on *Proteobacteria*, a major group within *Bacteria* that contains many of the most commonly encountered bacteria. Then, in the three chapters that follow, we examine representatives of other major lineages of *Bacteria*, of *Archaea*, and of microbial *Eukarya*, respectively.

With several thousand species of bacteria described, we obviously cannot consider them all. Therefore, using a phylogenetic tree to focus our discussion, we will explore some of the best-known species, particularly ones for which much phenotypic information is available. For more detailed information on prokaryotic diversity the reader is directed to the two major reference sources: *Bergey's Manual of Systematic Bacteriology* and *The Prokaryotes* (⮀ Section 16.13).

17.1 Phylogenetic Overview of *Bacteria*

Many major lineages, called *phyla*, of *Bacteria* are known from the study of laboratory cultures, and many others have been identified from the retrieval and sequencing of ribosomal RNA (rRNA) genes from microbial communities in natural habitats. **Figure 17.1** gives an overview of the phylogeny of major phyla of *Bacteria* for which laboratory cultures have been obtained. When one includes phyla of *Bacteria* known only from 16S rRNA sequences retrieved from the environment (⮀ Section 22.6), well over 80 phyla can be distinguished. However, since little phenotypic information is available on species in which cultures have not yet been obtained, we focus here on phyla with cultured species.

As Figure 17.1 clearly shows, the most phylogenetically ancient (least derived) phylum contains the genus *Aquifex* and relatives, all of which are hyperthermophilic H$_2$-oxidizing chemolithotrophs. Other "early" phyla such as *Thermodesulfobacterium*, *Thermotoga*, and the *Chloroflexus* group (green nonsulfur bacteria) also contain thermophilic species.

Continuing past the green nonsulfur bacteria, we see the deinococci and relatives, the morphologically unique spirochetes, the phototrophic green sulfur bacteria, the chemoorganotrophic *Flavobacterium* and *Cytophaga* groups, the budding *Planctomyces–Pirellula* and the *Verrucomicrobium* groups, the *Chlamydia*, and the genera *Nitrospira* and *Deferribacter* (Figure 17.1). Other major groups include the gram-positive bacteria and the cyanobacteria. The gram-positive bacteria are a large group of primarily chemoorganotrophic *Bacteria* and are discussed in detail in Chapter 18. They can be separated into two subgroups

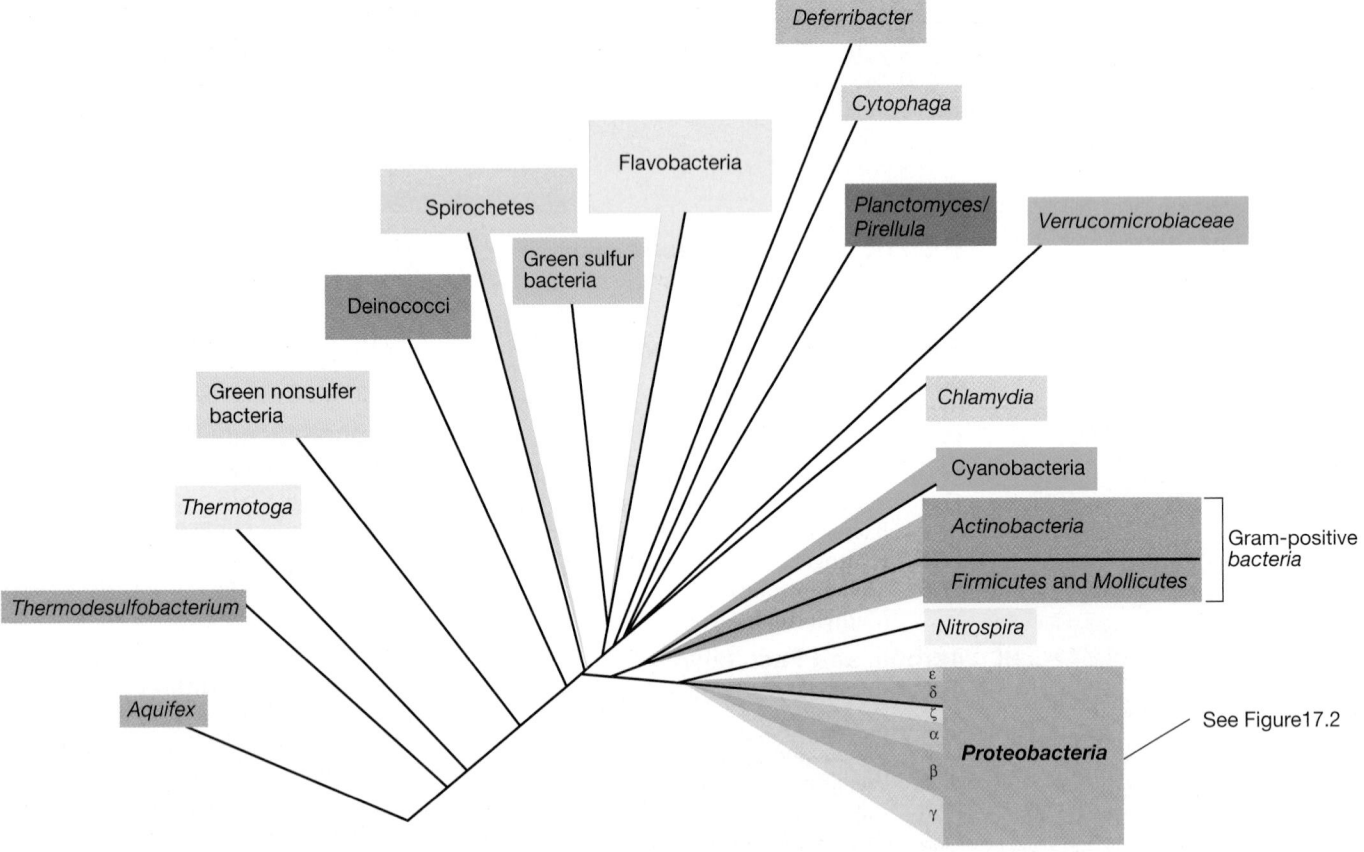

Figure 17.1 Some major phyla of *Bacteria* based on 16S ribosomal RNA gene sequence comparisons. Over 80 phyla of *Bacteria* are currently known, including many phyla known only from environmental sequences obtained in community sampling.

called the *Firmicutes* and the *Actinobacteria*. The cyanobacteria are oxygenic phototrophic bacteria (⇄ Section 13.5) with evolutionary roots near those of the gram-positive *Bacteria*; these organisms are covered in Chapter 18.

The remaining phylum of cultured *Bacteria*, the **Proteobacteria** (**Figure 17.2**), is by far the largest and most metabolically diverse of all *Bacteria*. *Proteobacteria* constitute the majority of known bacteria of medical, industrial, and agricultural significance. As a group, the *Proteobacteria* are all gram-negative bacteria. They

show an exceptionally wide diversity of energy-generating mechanisms, with chemolithotrophic, chemoorganotrophic, and phototrophic species (Figure 17.2). Indeed, we saw in Chapters 13 and 14 the great diversity of energy metabolisms used by various representatives of this group. The *Proteobacteria* are equally diverse in terms of their relationship to oxygen (O_2), with anaerobic, microaerophilic, and facultatively aerobic species known. Morphologically, they also exhibit a wide range of cell shapes, including straight and curved rods, cocci, spirilla, filamentous, budding, and appendaged forms.

Based on 16S rRNA gene sequences, the phylum *Proteobacteria* can be divided into six classes, *Alphaproteobacteria*, *Betaproteobacteria*, *Gammaproteobacteria*, *Deltaproteobacteria*, *Epsilonproteobacteria*, and *Zetaproteobacteria*, each containing many genera and species. The *Zeta* class is currently composed of only one organism, the marine iron-oxidizing bacterium *Mariprofundus*, but other relatives almost certainly exist. Despite the phylogenetic breadth of the *Proteobacteria*, species in different classes often have similar or even identical metabolisms. For example, phototrophy and methylotrophy occur in species of three different classes of *Proteobacteria*, and ammonia and nitrite-oxidizing (nitrifying) bacteria span four different classes of *Proteobacteria* (Figure 17.2) plus an additional genus that forms the heart of a separate phylum of *Bacteria*! This observation strongly suggests that gene sharing by horizontal gene flow (⇄ Section 12.11) has played a major role in shaping the metabolic diversity of the *Proteobacteria*. The sharing of metabolic traits in the different classes of *Proteobacteria* is also a good reminder that phenotype and phylogeny often give different views of prokaryotic diversity.

We now consider the major groups of *Proteobacteria*, grouping them along some common phenotypic themes.

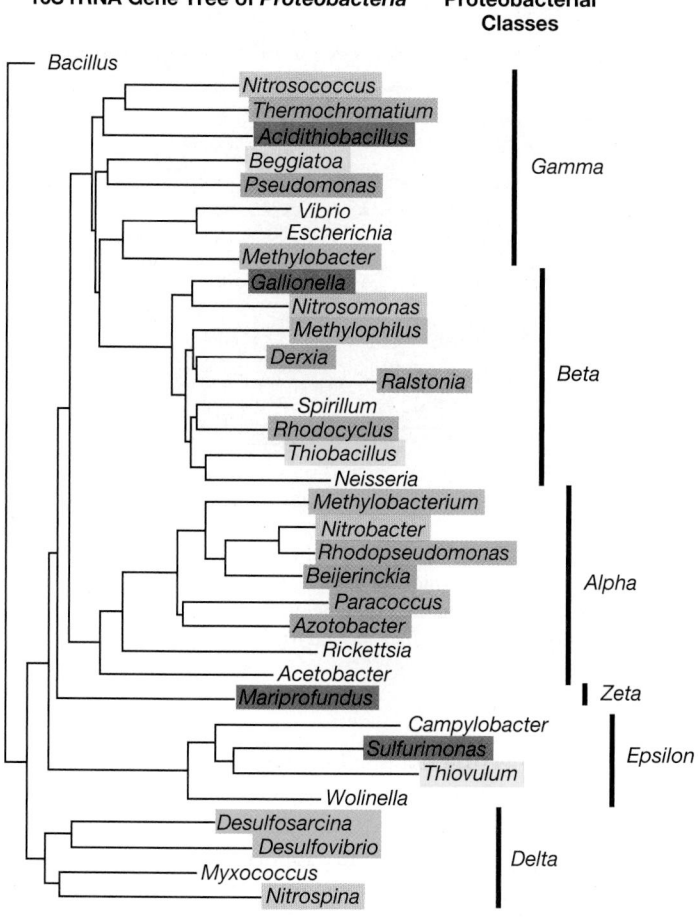

16S rRNA Gene Tree of *Proteobacteria*   **Proteobacterial Classes**

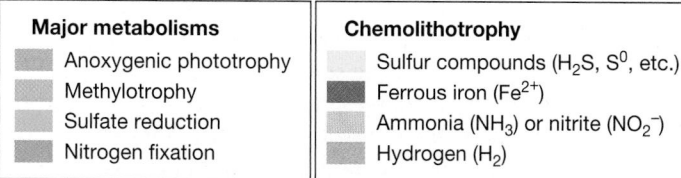

Major metabolisms	**Chemolithotrophy**
Anoxygenic phototrophy	Sulfur compounds (H_2S, S^0, etc.)
Methylotrophy	Ferrous iron (Fe^{2+})
Sulfate reduction	Ammonia (NH_3) or nitrite (NO_2^-)
Nitrogen fixation	Hydrogen (H_2)

Figure 17.2 Phylogenetic tree of some key genera of *Proteobacteria*. The tree was constructed by comparative 16S rRNA sequencing as described in Chapter 16. Note how identical metabolisms are often distributed in phylogenetically distinct genera, suggesting that horizontal gene flow has been extensive in the *Proteobacteria*. Some organisms listed may have multiple properties; for example, some sulfur chemolithotrophs are also iron or hydrogen chemolithotrophs, and several of the organisms listed can fix nitrogen. Phylogenetic analyses were performed and the phylogenetic tree constructed by Marie Asao.

MiniQuiz

• How many phyla of *Bacteria* are thought to exist? How many of these are actually in laboratory culture?

• How is it thought that a particular energy-yielding metabolism, such as phototrophy, arose in species of several different classes of *Proteobacteria*?

Phototrophic, Chemolithotrophic, and Methanotrophic *Proteobacteria*

The first groups of *Proteobacteria* we consider are those able to carry out anoxygenic photosynthesis, or sulfur-, iron-, hydrogen-, or nitrogen-dependent chemolithotrophy (all of these metabolisms are discussed in Chapter 13). We will also consider bacteria that oxidize methane (CH_4) (⇄ Section 14.15). We begin with purple bacteria, classic examples of the phototrophic lifestyle.

UNIT 6

17.2 Purple Phototrophic Bacteria

Key Genera: Chromatium, Ectothiorhodospira, Rhodobacter, Rhodospirillum

The purple phototrophic bacteria carry out *anoxygenic* photosynthesis. Thus, unlike the cyanobacteria (⮌ Section 18.7), which are *oxygenic* phototrophs, no O_2 is released. Purple bacteria are a morphologically diverse group, and the classification of these organisms has been established along phylogenetic, morphological, and physiological lines. Different genera fall within the *Alpha-, Beta-,* or *Gammaproteobacteria* (see Tables 17.1 and 17.2).

Purple bacteria contain bacteriochlorophylls and carotenoid pigments. Together, these pigments give purple bacteria their spectacular colors, usually purple, red, or orange (**Figure 17.3**). In Sections 13.2 and 13.3 we examined the structure of these pigments and learned how they function in light-mediated energy conservation (photophosphorylation).

Purple bacteria produce intracytoplasmic photosynthetic membrane systems into which their pigments are inserted. These membranes can be of various arrangements (**Figure 17.4**) but in all cases originate from invaginations of the cytoplasmic membrane. These internal membranes allow purple bacteria to increase the amount of pigment they contain and to thus better utilize the available light. When cells are grown at high light intensities, photosynthetic membranes are few and pigment contents are low. By contrast, at low light intensities, the cells are packed with membranes and pigments.

Purple Sulfur Bacteria

Purple bacteria that utilize hydrogen sulfide (H_2S) as an electron donor for CO_2 reduction in photosynthesis are called **purple sulfur bacteria** (**Table 17.1**). The H_2S is oxidized to elemental sulfur (S^0) that is stored in globules inside the cells (**Figure 17.5**); the sulfur later disappears as it is oxidized to sulfate (SO_4^{2-}). Many purple sulfur bacteria can also use other reduced sulfur compounds as

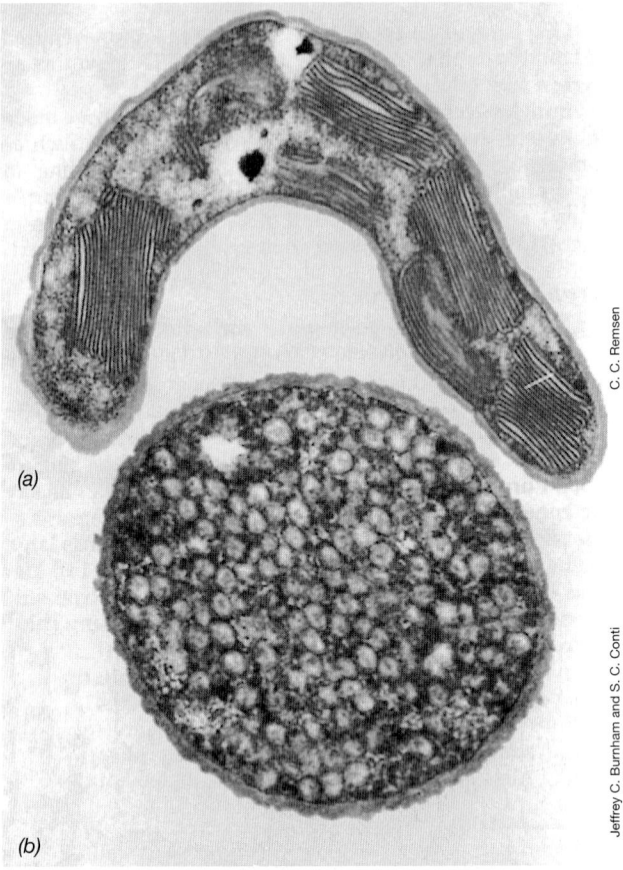

(a)

(b)

Figure 17.4 Membrane systems of phototrophic purple bacteria as revealed by the electron microscope. *(a) Ectothiorhodospira mobilis,* showing the photosynthetic membranes in flat sheets (lamellae). *(b) Allochromatium vinosum,* showing the membranes as individual, spherical vesicles.

Figure 17.3 Photograph of liquid cultures of phototrophic purple bacteria showing the color of species with various carotenoid pigments. The blue culture is a carotenoidless mutant derivative of *Rhodospirillum rubrum* showing that bacteriochlorophyll *a* is actually blue. The bottle on the far right (*Rhodobacter sphaeroides* strain G) lacks one of the carotenoids of the wild type and thus is greener.

photosynthetic electron donors; for example, thiosulfate ($S_2O_3^{2-}$) is commonly used to grow laboratory cultures. All purple sulfur bacteria discovered thus far are *Gammaproteobacteria* and use the Calvin cycle (⮌ Section 13.12) to support autotrophy.

Purple sulfur bacteria are generally found in illuminated anoxic zones of lakes and other aquatic habitats where H_2S accumulates, and also in "sulfur springs," where geochemically or biologically produced H_2S can trigger the formation of mass developments of cells of purple sulfur bacteria (**Figure 17.6**). The most favorable lakes for development of purple sulfur bacteria are meromictic (permanently stratified) lakes. Meromictic lakes stratify because they have denser (usually saline) water in the bottom and less dense (usually freshwater) nearer the surface. If sufficient sulfate is present to support sulfate reduction, the sulfide, produced in the sediments, diffuses upward into the anoxic bottom waters, and here purple sulfur bacteria can form dense cell masses, called *blooms,* usually in association with green phototrophic bacteria (Figure 17.6c).

Unlike other purple sulfur bacteria, the genera *Ectothiorhodospira* and *Halorhodospira* produce S^0 (from the oxidation of H_2S) *outside* rather than *inside* the cell (Figure 17.5d). These genera are also interesting because many species are extremely

Table 17.1 *Genera and characteristics of purple sulfur bacteria*[a]

Characteristics	Genus
Sulfur deposited externally	
Spirilla, polar flagella	*Ectothiorhodospira*
Spirilla, extreme alkaliphiles	*Thiorhodospira*
Spirilla, extreme halophiles	*Halorhodospira*
Sulfur deposited internally	
Do not contain gas vesicles	
Ovals or rods, polar flagella	*Chromatium*
	Allochromatium
	Halochromatium
	Rhabdochromatium
	Thermochromatium
	Isochromatium
	Marichromatium
Spheres, alkaliphilic	*Thioalkalicoccus*
Spheres, contain bacteriochlorophyll *b*	*Thioflavicoccus*
Spheres, diplococci, tetrads, nonmotile; cells 1.2–3 μm in diameter	*Thiocapsa*
Spheres or ovals, polar flagella; cells 2.5–3 μm in diameter	*Thiocystis*
Spheres, 1.5–2.5 μm in diameter	*Thiohalocapsa*
Spheres, 1–2 μm in diameter	*Thiorhodococcus*
Spheres, 1.2–1.5 μm in diameter	*Thiococcus*
Large spirilla, polar flagella	*Thiospirillum*
Small spirilla	*Thiorhodovibrio*
Contain gas vesicles	
Irregular spheres forming platelets of 4–16 cells	*Thiolamprovum*
Rods	*Lamprobacter*
Spheres, ovals, polar flagella	*Lamprocystis*
Rods, nonmotile; forming irregular network	*Thiodictyon*
Spheres, nonmotile; forming flat sheets of tetrads	*Thiopedia*

[a]From a phylogenetic standpoint, all are species of *Gammaproteobacteria*.

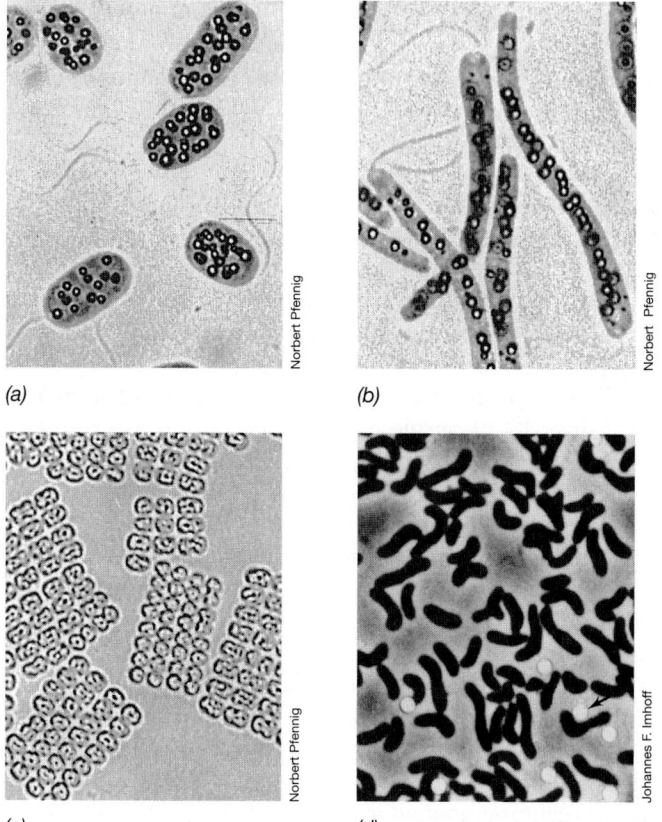

(a)   (b)
(c)   (d)

Figure 17.5 Bright-field and phase-contrast photomicrographs of purple sulfur bacteria. *(a) Chromatium okenii;* cells are about 5 μm wide. Note the globules of elemental sulfur inside the cells. *(b) Thiospirillum jenense,* a very large, polarly flagellated spiral; cells are about 30 μm long. Note the sulfur globules. *(c) Thiopedia rosea;* cells are about 1.5 μm wide. *(d)* Phase micrograph of cells of *Ectothiorhodospira mobilis.* Cells are about 0.8 μm wide. Note external sulfur globules (arrow). Compare the photo of *Chromatium okenii* with the drawings of purple sulfur bacteria made by the Russian microbiologist Sergei Winogradsky over 120 years ago, shown in Figure 1.22a.

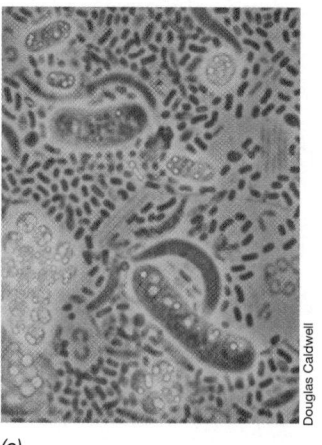

(a)   (b)   (c)

Figure 17.6 Blooms of purple sulfur bacteria. *(a) Lamprocystis roseopersicina,* in a sulfide spring. The bacteria grow near the bottom of the spring pool and float to the top (by virtue of their gas vesicles) when disturbed. The green color is from cells of the eukaryotic alga *Spirogyra* (Figure 20.41d). *(b)* Sample of water from a depth of 7 m in Lake Mahoney, British Columbia. The major organism is *Amoebobacter purpureus.* *(c)* Phase-contrast photomicrograph of layers of purple sulfur bacteria from a small, stratified lake in Michigan. The purple sulfur bacteria include *Chromatium* species (large rods) and *Thiocystis* (small cocci).

UNIT 6

halophilic (salt-loving) or alkaliphilic and are among the most extreme in these characteristics of all known *Bacteria*. These organisms are typically found in saline lakes, soda lakes, and salterns, where abundant levels of SO_4^{2-} support sulfate-reducing bacteria (Section 17.18), the organisms that produce H_2S.

Purple Nonsulfur Bacteria

Some purple bacteria are called **purple nonsulfur bacteria** because it was originally thought that they were unable to use H_2S as an electron donor for the reduction of CO_2 to cell material. In fact, H_2S can be used by most species in this group, although the levels ideal for purple sulfur bacteria (1–3 mM) are typically toxic to most purple nonsulfur bacteria. The morphological diversity of purple nonsulfur bacteria is as extensive as that of purple sulfur bacteria (**Table 17.2** and **Figure 17.7**), and all purple nonsulfur bacteria isolated thus far are either *Alpha-* or *Betaproteobacteria* (Table 17.2).

Some purple nonsulfur bacteria can also grow anaerobically in the dark using fermentative or anaerobic respiratory metabolism, and most can grow aerobically in darkness by respiration. Under the latter conditions, synthesis of the photosynthetic machinery is repressed by O_2, and the electron donor can be an organic compound or in some species even an inorganic compound, such as H_2. However, it is the capacity of this group for *photoheterotrophy* (a condition where light is the energy source and an organic compound is the carbon source) that likely spells their competitive success in nature. Purple nonsulfur bacteria are

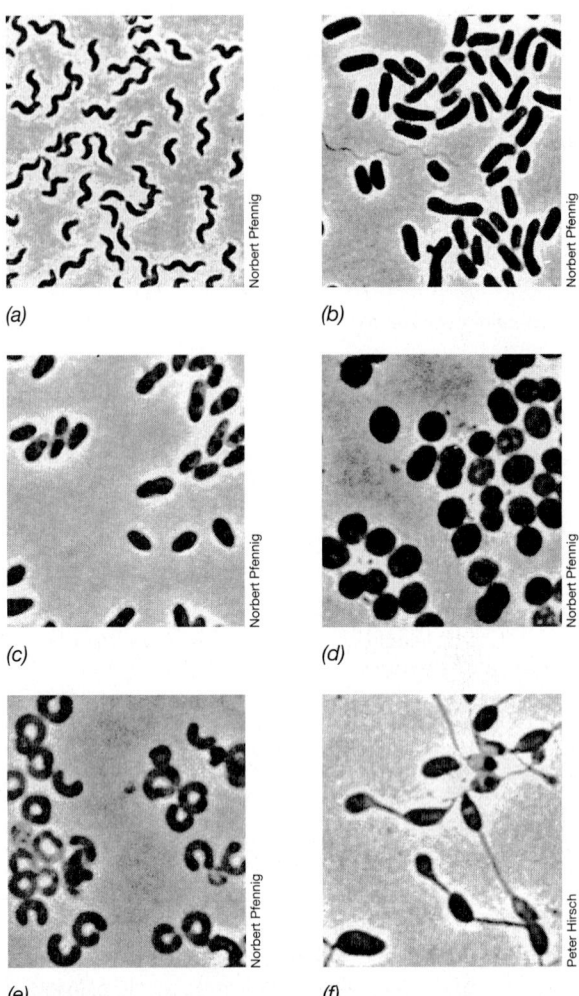

(a) (b) (c) (d) (e) (f)

Figure 17.7 **Representatives of several genera of purple nonsulfur bacteria.** *(a) Phaeospirillum fulvum*; cells are about 3 μm long. *(b) Rhodoblastus acidophilus*; cells are about 4 μm long. *(c) Rhodobacter sphaeroides*; cells are about 1.5 μm wide. *(d) Rhodopila globiformis*; cells are about 1.6 μm wide. *(e) Rhodocyclus purpureus*; cells are about 0.7 μm in diameter. *(f) Rhodomicrobium vannielii*; cells are about 1.2 μm wide. See also Table 17.2.

Table 17.2 Genera and characteristics of purple nonsulfur bacteria[a]

Characteristics	Genus
Alphaproteobacteria	
Spirilla, polarly flagellated	*Rhodospirillum*
	Phaeospirillum
	Rhodovibrio
	Rhodothalassium
	Roseospira
	Rhodospira
	Roseospirillum
Rods, polarly flagellated; divide by budding	*Rhodopseudomonas*
	Rhodoplanes
	Rhodobium
Rods; divide by binary fission	*Rhodobacter*
Ovoid to rod-shaped cells	*Rhodovulum*
Ovals, peritrichously flagellated; growth by budding and hypha formation	*Rhodomicrobium*
Large spheres, acidophilic (pH 5 optimum)	*Rhodopila*
Small spheres, alkaliphilic (pH 9 optimum)	*Rhodobaca*
Betaproteobacteria	
Ring-shaped or spirilla	*Rhodocyclus*
Curved rods	*Rubrivivax*
Curved rods	*Rhodoferax*

[a]All are members of the *Proteobacteria* (see Figure 17.2).

typically nutritionally diverse, using fatty, organic, or amino acids; sugars; alcohols; or even aromatic compounds like benzoate or toluene as carbon sources. Most species can also grow photoautotrophically with CO_2 and either H_2 or low levels of H_2S as reductant.

Enrichment and isolation of purple nonsulfur bacteria is easy using a mineral salts medium supplemented with an organic acid as carbon source. Such media, inoculated with a mud, lake water, or sewage sample and incubated anaerobically in the light, invariably select for purple nonsulfur bacteria. Enrichment cultures can be made even more selective by omitting fixed nitrogen sources (for example, NH_4^+) or organic nitrogen sources (for example, yeast extract or peptone) from the medium and supplying a gaseous headspace of N_2. Virtually all purple nonsulfur bacteria can fix N_2 (Section 13.14) and will thrive under such conditions, rapidly outcompeting other bacteria.

MiniQuiz
- What is meant by the term anoxygenic?
- Can any purple bacteria grow in the absence of light?

17.3 The Nitrifying Bacteria

Key Genera: *Nitrosomonas, Nitrobacter*

Many species of *Bacteria* are **chemolithotrophs**, organisms that can use *inorganic* electron donors as energy sources. Most chemolithotrophs are also capable of autotrophic growth and in this way share a major physiological trait with anoxygenic phototrophic bacteria and cyanobacteria. We focus here on the best-studied chemolithotrophs: those capable of oxidizing reduced nitrogen or sulfur compounds, ferrous iron (Fe^{2+}), or H_2.

Ammonia and Nitrite Oxidizers

Bacteria able to grow chemolithotrophically at the expense of reduced inorganic nitrogen compounds are called **nitrifying bacteria (Figure 17.8)**. Several genera are recognized on the basis of morphology and phylogeny as well as the particular steps in the oxidation sequences that they carry out (**Table 17.3**). Phylogenetically, the majority of nitrifying bacteria are scattered among four of the *Proteobacteria* classes: *Alpha, Beta, Gamma,* and *Delta*. The genus *Nitrospira* forms its own phylum of *Bacteria* (see Figure 17.1 and Section 18.21) and is related to other nitrifying bacteria in a metabolic sense only. Nevertheless, ecological studies of nitrification suggest that *Nitrospira* is the most abundant nitrifying bacterium in nature. In addition, certain *Archaea* oxidize ammonia (NH_3) as a chemolithotrophic substrate, and seem to be the dominant nitrifiers in the oceans, where ammonia levels are very low (Sections 13.10 and 19.11).

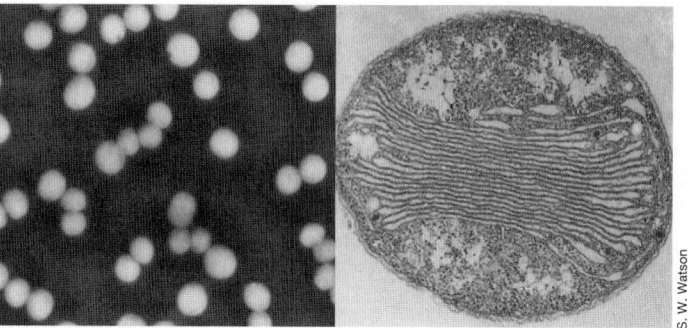

Reaction: $NH_3 + 1\frac{1}{2} O_2 \longrightarrow NO_2^- + H_2O$

(a)

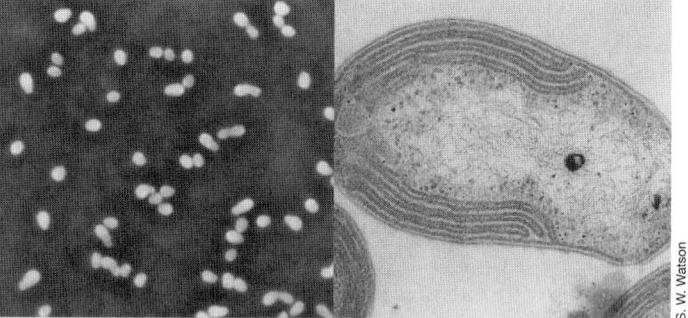

Reaction: $NO_2^- + \frac{1}{2} O_2 \longrightarrow NO_3^-$

(b)

Figure 17.8 Nitrifying bacteria. *(a)* Phase photomicrograph (left) and electron micrograph (right) of the ammonia-oxidizing bacterium *Nitrosococcus oceani*. A single cell is about 2 μm in diameter. *(b)* Phase-contrast photomicrograph (left) and electron micrograph (right) of the nitrite-oxidizing bacterium *Nitrobacter winogradskyi*. A cell is about 0.7 μm in diameter. Beneath each panel is shown the chemolithotrophic reaction that each organism catalyzes.

Table 17.3 *Characteristics of the nitrifying bacteria*

Characteristics	Genus	Phylogenetic group[a]	Primary habitats
Oxidize ammonia			
Gram-negative short to long rods, motile (polar flagella) or nonmotile; peripheral membrane systems	*Nitrosomonas*	Beta	Soil, sewage, freshwater, marine
Large cocci, motile; vesicular or peripheral membranes	*Nitrosococcus*	Gamma	Marine
Spirals, curved or lobed cells, motile (peritrichous flagella); no obvious membrane system	*Nitrosospira*	Beta	Soil, freshwater
Oxidize nitrite			
Short rods, reproduce by budding, occasionally motile (single subterminal flagellum); membrane system arranged as a polar cap	*Nitrobacter*	Alpha	Soil, freshwater, marine
Short rods forming cell aggregates; no internal membranes; psychrophile	*Nitrotoga*	Beta	Siberian permafrost
Long, slender rods, nonmotile; no internal membrane system	*Nitrospina*	Delta	Marine
Large cocci, motile (one or two subterminal flagella); membrane system randomly arranged in tubes	*Nitrococcus*	Gamma	Marine
Helical to vibrioid-shaped cells, nonmotile; no internal membranes	*Nitrospira*	*Nitrospira* group	Marine, sponges, soil, wastewater, hot springs

[a]Phylogenetically, all nitrifying bacteria thus far examined are *Proteobacteria*, except for *Nitrospira*, which forms its own phylogenetic lineage (Figure 17.1 and Section 18.21), or certain marine crenarchaeotes (Section 19.11).

No chemolithotroph is known that carries out the complete oxidation of NH_3 to nitrate (NO_3^-). Thus, nitrification results from the sequential activities of two physiological groups of organisms, the *ammonia-oxidizing bacteria* (which oxidize NH_3 to nitrite, NO_2^-) (Figure 17.8a), and the *nitrite-oxidizing bacteria*, the actual nitrate-producing bacteria, which oxidize NO_2^- to NO_3^- (Figure 17.8b). Ammonia-oxidizing bacteria typically have genus names beginning in *Nitroso-*, whereas genus names of nitrate producers begin with *Nitro-* (Table 17.3).

Many species of nitrifying bacteria have internal membrane stacks (Figure 17.8) that closely resemble the photosynthetic membranes found in their close phylogenetic relatives, the purple phototrophic bacteria (Section 17.2) and the methane-oxidizing (methanotrophic) bacteria (Section 17.6). The membranes are the location of key enzymes in nitrification: *ammonia monooxygenase*, which oxidizes NH_3 to hydroxylamine (NH_2OH), and *nitrite oxidoreductase*, which oxidizes NO_2^- to NO_3^- (♻ Section 13.10).

Ecology, Isolation, and Culture

The nitrifying bacteria are widespread in soil and water. They are present in highest numbers in habitats where NH_3 is abundant, such as sites with extensive protein decomposition (ammonification), and also in sewage treatment facilities (♻ Section 35.2). Nitrifying bacteria develop especially well in lakes and streams that receive inputs of sewage or other wastewaters because these are frequently high in NH_3 (♻ Figure 23.16).

Enrichment cultures of nitrifying bacteria can be achieved using mineral salts media containing NH_3 or NO_2^- as electron donors and bicarbonate (HCO_3^-) as the sole carbon source. Because these organisms produce very little ATP from their electron donors (♻ Section 13.10), visible turbidity may not develop in cultures even after extensive nitrification has occurred. An easy means of monitoring growth is thus to assay for the production of NO_2^- (with NH_3 as electron donor) or NO_3^- (with NO_2^- as electron donor). Most of the nitrifying bacteria are obligate chemolithotrophs and obligate aerobes. Species of *Nitrobacter* are an exception and are able to grow chemoorganotrophically on acetate or pyruvate as the sole carbon and energy source. One group, the *anammox* bacteria, is phylogenetically distinct from the nitrifiers considered here and oxidizes NH_3 anaerobically (♻ Section 13.11).

MiniQuiz

- What is the electron donor for *Nitrosomonas*? For *Nitrobacter*?
- What is the final product of nitrification?

17.4 Sulfur- and Iron-Oxidizing Bacteria

Key Genera: *Thiobacillus, Acidithiobacillus, Achromatium, Beggiatoa*

The ability to grow chemolithotrophically on reduced sulfur compounds is spread among organisms in four classes of *Proteobacteria* (**Table 17.4**). Two broad ecological classes of sulfur-oxidizing

---

Table 17.4 *Physiological characteristics of sulfur-oxidizing chemolithotrophic bacteria*

Genus and species	Inorganic electron donor	Range of pH for growth	Phylogenetic group[a]
Species growing poorly if at all in organic media			
Thiobacillus thioparus	H_2S, sulfides, S^0, $S_2O_3^{2-}$	6–8	Beta
Thiobacillus denitrificans[b]	H_2S, S^0, $S_2O_3^{2-}$	6–8	Beta
Halothiobacillus neapolitanus	S^0, $S_2O_3^{2-}$	6–8	Gamma
Acidithiobacillus thiooxidans	S^0	2–4	Gamma
Acidithiobacillus ferrooxidans	S^0, metal sulfides, Fe^{2+}	2–4	Gamma
Species growing well in organic media			
Starkeya novella	$S_2O_3^{2-}$	6–8	Alpha
Thiomonas intermedia	$S_2O_3^{2-}$	3–7	Beta
Filamentous sulfur chemolithotrophs			
Beggiatoa	H_2S, $S_2O_3^{2-}$	6–8	Gamma
Thiothrix	H_2S	6–8	Gamma
Thioploca[c]	H_2S, S^0	—	Gamma
Other genera			
Achromatium	H_2S	—	Gamma
Thiomicrospira	$S_2O_3^{2-}$, H_2S	6–8	Gamma
Thiosphaera	H_2S, $S_2O_3^{2-}$, H_2	6–8	Alpha
Thermothrix	H_2S, $S_2O_3^{2-}$, SO_3^{2-}	6.5–7.5	Beta
Thiovulum[d]	H_2S, S^0	6–8	Epsilon

[a]All are *Proteobacteria*.
[b]Facultative aerobes; use NO_3^- as electron acceptor anaerobically.
[c]Pure cultures not yet available.
[d]See Figure 17.49

bacteria exist, those living at neutral pH and those living at acidic pH. Some of the acidophiles also have the ability to grow chemolithotrophically using ferrous iron (Fe^{2+}) as an electron donor. We considered the biochemistry of these metabolisms in Chapter 13 and consider the ecology of the organisms themselves in Chapter 23.

Thiobacillus and *Achromatium*

The genus *Thiobacillus* and related genera contain several gram-negative, rod-shaped *Betaproteobacteria*, indistinguishable morphologically from most other gram-negative rods (**Figure 17.9a**); they are the best studied of the sulfur chemolithotrophs. The electron donors most commonly used by the sulfur chemolithotrophs are H_2S, S^0, and thiosulfate ($S_2O_3^{2-}$). The oxidation of these substrates generates sulfuric acid (H_2SO_4), and thus several thiobacilli are acidophilic. One highly acidophilic species, *Acidithiobacillus ferrooxidans*, can also grow chemolithotrophically by the oxidation of Fe^{2+} and is a major biological agent for the oxidation of this metal. Iron pyrite (FeS_2) is a major natural

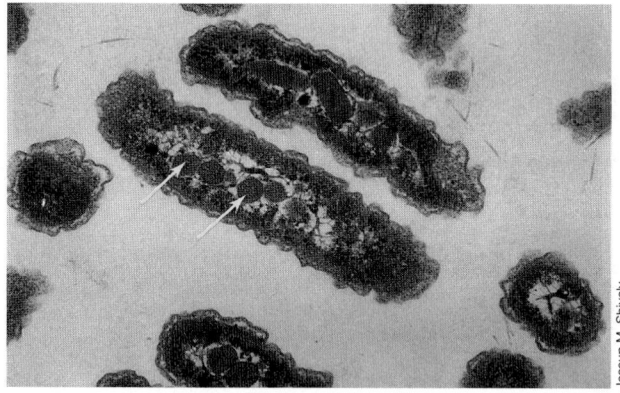

(a)

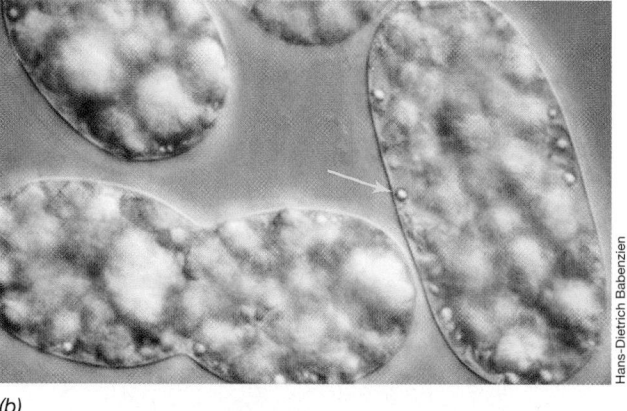

(b)

Figure 17.9 **Nonfilamentous sulfur chemolithotrophs.** *(a)* Transmission electron micrograph of cells of the chemolithotrophic sulfur oxidizer *Halothiobacillus neapolitanus*. A single cell is about 0.5 μm in diameter. Note the polyhedral bodies (carboxysomes) distributed throughout the cell (arrows). *(b) Achromatium.* Cells photographed by differential interference contrast microscopy. The small globular structures near the periphery of the cells (arrow) are elemental sulfur, and the large granules are calcium carbonate. A single *Achromatium* cell is about 25 μm in diameter.

source of Fe^{2+} as well as sulfide. The oxidation of FeS_2, especially in mining operations, can be both beneficial (because leaching of the ore releases the iron from the sulfide mineral) and ecologically disastrous (the environment can become acidic and contaminated with toxic metals such as aluminum, cadmium, and lead) (ⅇ Sections 24.5 and 24.7).

Achromatium is a spherical sulfur-oxidizing chemolithotroph that is common in freshwater sediments containing H_2S. Cells of *Achromatium* are large cocci that can have diameters of 10–100 μm (Figure 17.9b). Phylogenetic analyses of natural populations of *Achromatium* have shown that several species likely exist (probably each of distinct size), although pure cultures of this organism have not yet been achieved. *Achromatium* is a species of *Gammaproteobacteria* and is specifically related to purple sulfur bacteria, such as its phototrophic counterpart *Chromatium* (Section 17.2 and Figure 17.5a). Like *Chromatium*, cells of *Achromatium* store S^0 internally (Figure 17.9b); the granules later disappear as S^0 is oxidized to SO_4^{2-}. Cells of *Achromatium* also store large granules of calcite ($CaCO_3$) (Figure 17.9b), possibly as a carbon source (in the form of CO_2) for autotrophic growth.

Culture

Some sulfur chemolithotrophs are *obligate chemolithotrophs*, locked into a lifestyle of using inorganic instead of organic compounds as electron donors. When growing in this fashion, they are also autotrophs, converting CO_2 into cell material by reactions of the Calvin cycle (ⅇ Section 13.12). **Carboxysomes** are often present in cells of obligate chemolithotrophs (Figure 17.9a). These structures contain high levels of Calvin cycle enzymes and probably increase the rate at which these organisms fix CO_2.

Other sulfur chemolithotrophs are *facultative chemolithotrophs*, facultative in the sense that they can grow either chemolithotrophically (and thus, also as autotrophs) or chemoorganotrophically (Table 17.4). Most species of *Beggiatoa*, however, can obtain energy from the oxidation of inorganic sulfur compounds but lack enzymes of the Calvin cycle. They thus require organic compounds as carbon sources. Such a nutritional lifestyle is called **mixotrophy**.

Beggiatoa

Organisms of this genus are filamentous, gliding, sulfur-oxidizing *Gammaproteobacteri*a. Filaments of *Beggiatoa* are usually large in both diameter and length, consisting of many short cells attached end to end (**Figure 17.10**). Filaments then flex and twist so that many filaments may become intertwined to form a complex tuft.

Beggiatoa is found in nature primarily in habitats rich in H_2S, such as sulfur springs (Figure 17.10b), decaying seaweed beds, mud layers of lakes, and waters polluted with sewage. In such environments, filaments of *Beggiatoa* are typically filled with S^0 (Figure 17.10a). *Beggiatoa* are also common inhabitants of hydrothermal vents (underwater hot springs, ⅇ Section 23.12). Although a few strains of *Beggiatoa* are truly chemolithotrophic autotrophs, most grow best mixotrophically with reduced sulfur compounds as electron donors and organic compounds as carbon sources.

An interesting habitat of *Beggiatoa* is the rhizosphere of plants (rice, cattails, and other swamp plants) living in flooded, and hence anoxic, soils. Such plants pump O_2 down into their roots, so a sharply defined oxic–anoxic boundary develops between the

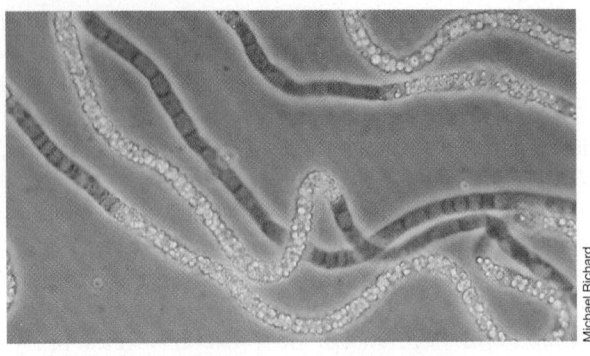

(a)

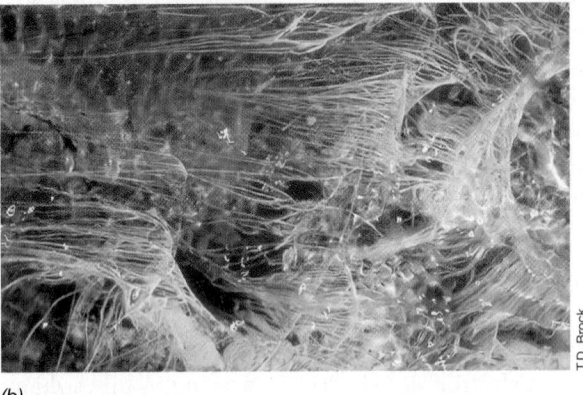

(b)

Figure 17.10 Filamentous sulfur-oxidizing bacteria. *(a)* Phase-contrast photomicrograph of a *Beggiatoa* species isolated from a sewage treatment plant. Note the abundant elemental sulfur granules in some of the cells. *(b)* Sulfur-oxidizing bacteria in the outflow of a small sulfide spring. The filamentous cells twist together to form thick streamers, and the white color is due to accumulated elemental sulfur.

root and the soil. *Beggiatoa* and other sulfur bacteria develop at this interface and play a beneficial role for the plant by oxidizing (and thus detoxifying) the H$_2$S.

Thioploca and *Thiothrix*

Other filamentous sulfur-oxidizing *Gammaproteobacteria* include *Thioploca* (**Figure 17.11**) and *Thiothrix* (**Figure 17.12**). *Thioploca* is a large, filamentous sulfur-oxidizing chemolithotroph that forms cell bundles surrounded by a common sheath (Figure 17.11). Thick mats of a marine *Thioploca* species have been found on the ocean floor off the coast of South America. Ecological studies of these organisms have shown that they carry out the anoxic oxidation of H$_2$S coupled to the reduction of nitrate (NO$_3^-$) to ammonium (NH$_4^+$). Cells of *Thioploca* can accumulate huge amounts of NO$_3^-$ intracellularly, and this NO$_3^-$ can then support extended periods of anaerobic respiration with H$_2$S as electron donor. It is thought that these *Thioploca* mats fix substantial amounts of CO$_2$ and also play a major role in sulfur and nitrogen cycling in the marine environment.

Thiothrix is a filamentous sulfur-oxidizing organism in which the filaments group together at their ends by way of a holdfast to form cell arrangements called rosettes (Figure 17.12*b*). Physiologically, *Thiothrix* is an obligately aerobic mixotroph, and in this and most other respects it resembles *Beggiatoa*.

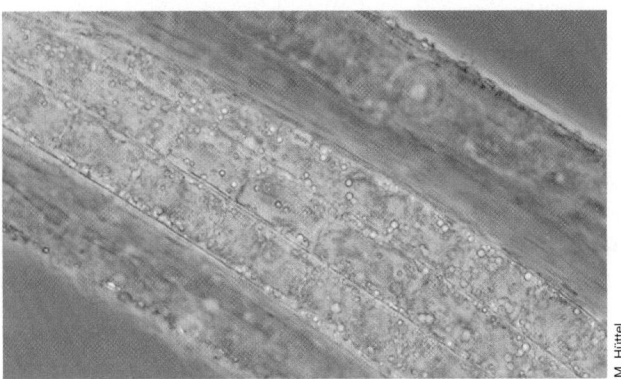

Figure 17.11 Cells of a large marine *Thioploca* species. Cells contain sulfur granules (yellow) and are about 40–50 μm wide.

MiniQuiz

- What is the final product of H$_2$S oxidation by sulfur chemolithotrophs?
- Name a common genus of rod-shaped sulfur chemolithotroph.

(a)

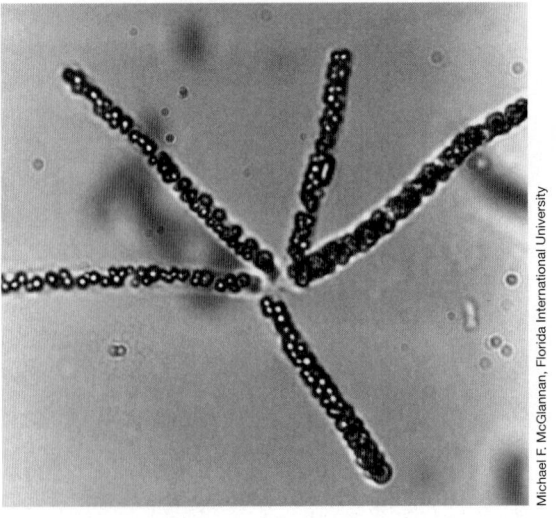

(b)

Figure 17.12 *Thiothrix*. *(a)* A sulfide-containing artesian spring in Florida (USA). The outside of the spring is coated with a mat of *Thiothrix*. The mat is about 1.5 m in diameter. *(b)* Phase-contrast photomicrograph of a rosette of cells of *Thiothrix* isolated from the spring. Note the internal sulfur globules produced from the oxidation of sulfide. Each filament is about 4 μm in diameter.

17.5 Hydrogen-Oxidizing Bacteria

Key Genera: *Ralstonia, Paracoccus*

Many bacteria can grow with H_2 as sole electron donor and O_2 as electron acceptor in their energy metabolism:

$$H_2 + \tfrac{1}{2} O_2 \rightarrow H_2O \qquad \Delta G^{0\prime} = -237 \text{ kJ}$$

Most of these organisms, known collectively as the "hydrogen bacteria," can also grow autotrophically (using reactions of the Calvin cycle to incorporate CO_2) and are grouped together here as the chemolithotrophic hydrogen-oxidizing bacteria. All hydrogen bacteria contain one or more hydrogenase enzymes that function to bind H_2 and use it either to produce ATP (♻ Section 13.7) or for reducing power for autotrophic growth (**Table 17.5**).

Different hydrogen-oxidizing *Proteobacteria* are scattered among the *Alpha, Beta*, and *Gamma* subclasses. These organisms should be distinguished from the many strictly anaerobic prokaryotes that oxidize H_2 in anaerobic respirations; for example, acetogens, methanogens, and sulfate-reducing bacteria. Both gram-positive and gram-negative hydrogen bacteria are known, with the best-studied representatives classified in the genera *Ralstonia* (**Figure 17.13**), *Pseudomonas*, and *Paracoccus* (Table 17.5). *Paracoccus denitrificans* can also oxidize H_2 anaerobically with nitrate (NO_3) as electron acceptor, forming N_2 (denitrification), and

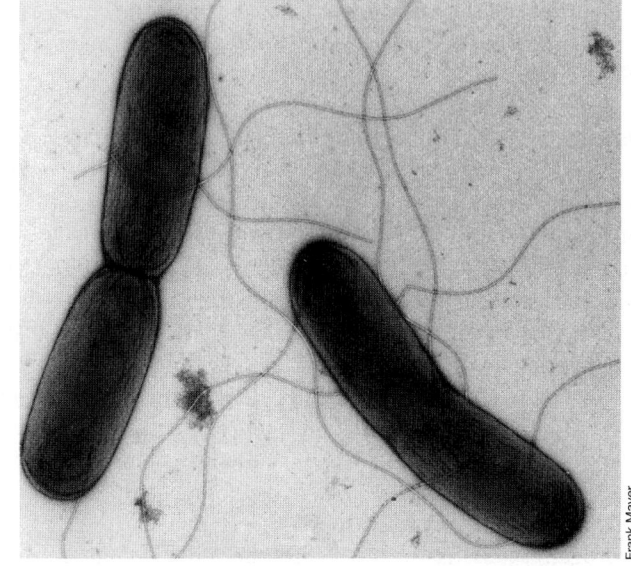

Frank Mayer

Figure 17.13 Hydrogen bacteria. Transmission electron micrograph of negatively stained cells of the hydrogen-oxidizing chemolithotroph *Ralstonia eutropha*. A cell is about 0.6 μm in diameter and contains several flagella.

Table 17.5 *Differential characteristics of a few common species of hydrogen-oxidizing bacteria*

Genus and species	Denitrification	Growth on fructose	Motility	Phylogenetic group[a]	Other characteristics
Gram-negative					
Acidovorax facilis	−	+	+	Beta	Membrane-bound hydrogenase
Ralstonia eutropha	+	+	+	Beta	Membrane-bound and cytoplasmic hydrogenases
Achromobacter xylosoxidans	−	+	+	Beta	Membrane-bound and cytoplasmic hydrogenases
Aquaspirillum autotrophicum	−	−	+	Beta	Membrane-bound hydrogenase
Pseudomonas carboxydovorans	−	−	+	Gamma	Membrane-bound hydrogenase; also oxidizes CO
Hydrogenophaga flava	−	+	+	Beta	Colonies are bright yellow
Paracoccus denitrificans	+	+	−	Alpha	Membrane-bound hydrogenase; strong denitrifier
Aquifex pyrophilus	+	−	+	Aquifex group[b]	Hyperthermophile, grows microaerophilically or anaerobically (with NO_3^-), obligate chemolithotroph; also uses S^0 or $S_2O_3^{2-}$ as electron donor
Hydrogenobacter thermophilus	−	−	−	Aquifex group[b]	As for *Aquifex*, but obligate aerobe (microaerophile)
Gram-positive					
Bacillus schlegelii	−	−	+	Firmicutes[c]	Produces endospores; thermophile; also uses CO or $S_2O_3^{2-}$ as electron donor
Arthrobacter sp.	−	+	−	Actinobacteria[d]	Membrane-bound hydrogenase
Mycobacterium gordonae	−	?	−	Actinobacteria[e]	Acid-fast; colonies yellow to orange

[a]Aerobic hydrogen bacteria are *Proteobacteria* except as indicated.
[b]See Section 18.20.
[c]See Section 18.2.
[d]See Section 18.4.
[e]See Section 18.5.

UNIT 6

has been particularly well studied for its bioenergetics of electron transport and generation of a proton motive force (Section 4.10 and Figure 4.19).

Physiology and Ecology of Hydrogen Bacteria

When growing chemolithotrophically on H_2, most hydrogen bacteria grow best under microaerophilic (5–10% O_2) conditions because hydrogenases are typically oxygen-sensitive. The element nickel (Ni^{2+}) must be present in the medium for chemolithotrophic growth of hydrogen bacteria because virtually all hydrogenases contain Ni^{2+} as a key metal cofactor. A few hydrogen bacteria also fix nitrogen (Section 13.14), making possible their culture in a mineral salts medium supplied with only H_2, O_2, CO_2, and N_2 as carbon, energy, and nitrogen sources! Virtually all hydrogen bacteria are facultative chemolithotrophs, meaning that they can also grow chemoorganotrophically with organic compounds as energy sources.

Hydrogen-oxidizing bacteria can be enriched if a small amount of mineral salts medium containing trace metals (especially Ni^{2+} and Fe^{2+}) is inoculated with soil or water and incubated in a large, sealed flask containing a headspace of 5% O_2, 10% CO_2, and 85% H_2. When the liquid becomes turbid, plates of the same medium are streaked and incubated in a glass jar containing the same gas mixture (one must exercise care here, as mixtures of O_2 and H_2 are potentially explosive).

CO Oxidation

Some hydrogen bacteria can grow aerobically on carbon monoxide (CO) as electron donor. CO-oxidizing bacteria, called *carboxydotrophic* bacteria, grow autotrophically using the Calvin cycle (Section 13.12) to fix CO_2 generated from the oxidation of CO. Electrons from the oxidation of CO to CO_2 by the enzyme carbon monoxide dehydrogenase travel through an electron transport chain that forms a proton motive force. Interestingly, CO is a potent inhibitor of many cytochromes, acting as a respiratory poison. However, carboxydotrophic bacteria get around this problem by synthesizing CO-resistant cytochromes and are thus immune to any toxic effects of CO. Like the hydrogen bacteria, virtually all carboxydotrophic bacteria also grow chemoorganotrophically by oxidizing organic compounds, a likely indication that CO levels are quite variable in nature and a backup means of energy metabolism is essential.

CO consumption by carboxydotrophic bacteria on a global basis is a significant ecological process. Although much CO is generated from human and other sources, CO levels in air have not risen significantly over many years. Because the most significant releases of CO (primarily from automobile exhaust, incomplete combustion of fossil fuels, and the catabolism of lignin, a plant product) are in oxic environments, carboxydotrophic bacteria in the upper layers of soil probably represent the most significant sink for CO in nature.

MiniQuiz

- What key enzyme is necessary for growth of chemolithotrophs on H_2 as electron donor?
- What is the product of CO oxidation?

17.6 Methanotrophs and Methylotrophs

Key Genera: *Methylomonas, Methylobacter*

Methane (CH_4) is found extensively in nature. It is produced in anoxic environments by methanogenic *Archaea* (Sections 14.10 and 19.3) and is a major gas of anoxic muds, marshes (Figure 1.11*a*), anoxic zones of lakes, the rumen, and the mammalian intestinal tract. Methane is the major constituent of "natural gas" widely used as a heating and industrial fuel, and is also present in many coal formations.

Methanotrophs oxidize methane and a few other one-carbon compounds as electron donors in energy metabolism and as carbon sources. We discussed the physiology of methanotrophy in Section 14.15. Methanotrophs grow aerobically and are widespread in soils and waters. They exhibit diverse morphologies but are related in terms of their phylogeny and ecology. In marine sediments and a few other anoxic environments, methane is oxidized under strictly anoxic conditions by a consortium of methanogenic *Archaea* and sulfate-reducing bacteria (Section 14.13), but we consider here only the aerobic methanotrophs.

C₁ Metabolism

A list of compounds catabolized by methanotrophs is given in **Table 17.6**. From a biochemical point of view, these compounds share a key characteristic, the absence of carbon–carbon bonds. Thus, in methanotrophs all organic compounds in the cell must be synthesized from C_1 precursors. Organisms that can grow using carbon compounds that lack C—C bonds are called **methylotrophs**. Many but not all methylotrophs are also methanotrophs. However, methanotrophs are unique in that they can grow not only on some of the more oxidized one-carbon compounds (Table 17.6), but also on methane.

Methanotrophs possess a key enzyme, *methane monooxygenase*, which catalyzes the incorporation of an atom of oxygen from O_2 into CH_4, forming methanol (CH_3OH, Section 14.15). The requirement for O_2 as a reactant in the initial oxygenation of CH_4 thus explains why these methanotrophs are obligate aerobes. Most methanotrophs are obligate C_1 utilizers, unable to use compounds

Table 17.6 *Substrates used by methylotrophic bacteria*[a]

I. Substrates used for growth	
Methane, CH_4[b]	Formate, $HCOO^-$
Methanol, CH_3OH	Formamide, $HCONH_2$
Methylamine, CH_3NH_2	Carbon monoxide, CO
Dimethylamine, $(CH_3)_2NH$	Dimethyl ether, $(CH_3)_2O$
Trimethylamine, $(CH_3)_3N$	Dimethyl carbonate, $CH_3OCOOCH_3$
Tetramethylammonium, $(CH_3)_4N^+$	Dimethyl sulfoxide, $(CH_3)_2SO$
Trimethylamine *N*-oxide, $(CH_3)_3NO$	Dimethyl sulfide, $(CH_3)_2S$
Trimethylsulfonium, $(CH_3)_3S^+$	Chloromethane, CH_3Cl

II. Substrates oxidized but not used for growth	
Ammonium, NH_4^+	Bromomethane, CH_3Br
Ethylene, $H_2C{=}CH_2$	Higher hydrocarbons (ethane, propane)

[a]A single isolate does not use all of the above, but at least one methylotrophic bacterium has been reported to oxidize each of the listed compounds.
[b]Methylotrophs able to oxidize methane are called *methanotrophs*.

Table 17.7 *Some characteristics of methanotrophic bacteria*

Organism	Morphology	Phylogenetic group[a]	Internal membranes[b]	Carbon assimilation pathway[c]	N_2 fixation
Methylomonas	Rod	*Gamma*	I	Ribulose monophosphate	No
Methylomicrobium	Rod	*Gamma*	I	Ribulose monophosphate	No
Methylobacter	Coccus to ellipsoid	*Gamma*	I	Ribulose monophosphate	No
Methylococcus	Coccus	*Gamma*	I	Ribulose monophosphate and Calvin cycle	Yes
Methylosinus	Rod or vibrioid	*Alpha*	II	Serine	Yes
Methylocystis	Rod	*Alpha*	II	Serine	Yes
Methylocella[d]	Rod	*Alpha*	II	Serine	Yes
Methylacidiphilum[d]	Rod	*Verrucomicrobiaceae*[d] (see Figure 17.1)	Membrane vesicles	Serine and Calvin cycle	Yes

[a]All except for *Methylacidiphilum* are *Proteobacteria*.

[b]Internal membranes: type I, bundles of disc-shaped vesicles distributed throughout the organism; type II, paired membranes running along the periphery of the cell. See Figure 17.14.

[c]See Figures 14.32 and 14.33.

[d]Acidophiles. For properties of *Verrucomicrobiaceae*, see Section 18.11.

containing carbon–carbon bonds. By contrast, most nonmethanotrophic methylotrophs can use organic acids, ethanol, and sugars.

Methane-oxidizing bacteria are virtually unique among bacteria in possessing relatively large amounts of sterols. Sterols are rigid planar molecules found in the cytoplasmic and other membranes of eukaryotes but are absent from most bacteria. Sterols may be an essential part of the complex internal membrane system for methane oxidation (see Figure 17.14). The only other group of bacteria in which sterols are widely distributed is in the cell wall–less mycoplasmas (↩ Section 18.3).

Classification of Methanotrophs

Table 17.7 gives a taxonomic overview of the methanotrophs. These bacteria were initially distinguished on the basis of morphology and formation of resting stages. However, now they are classified into two major groups based on their internal cell structure, phylogeny, and carbon assimilation pathway. *Type I methanotrophs* assimilate one-carbon compounds via the ribulose monophosphate cycle and are phylogenetically *Gammaproteobacteria*. By contrast, *type II methanotrophs* assimilate C_1 intermediates via the serine pathway and are phylogenetically *Alphaproteobacteria* (Table 17.7). We discussed the biochemical details of these pathways in Section 14.15.

Both groups of methanotrophs contain extensive internal membrane systems for methane oxidation. Membranes in type I methanotrophs are arranged as bundles of disc-shaped vesicles distributed throughout the cell (**Figure 17.14b**). Type II species possess paired membranes running along the periphery of the cell (Figure 17.14a). The key enzyme methane monooxygenase is located in these membranes.

The genus *Methylacidiphilum* contains a phylogenetically and physiologically unique methanotroph. Species in this genus are

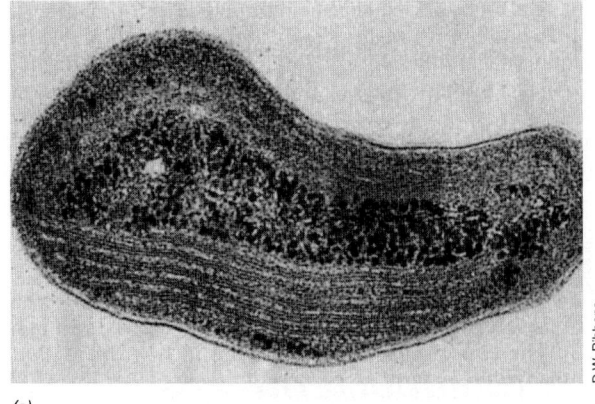

(a)

D.W. Ribbons

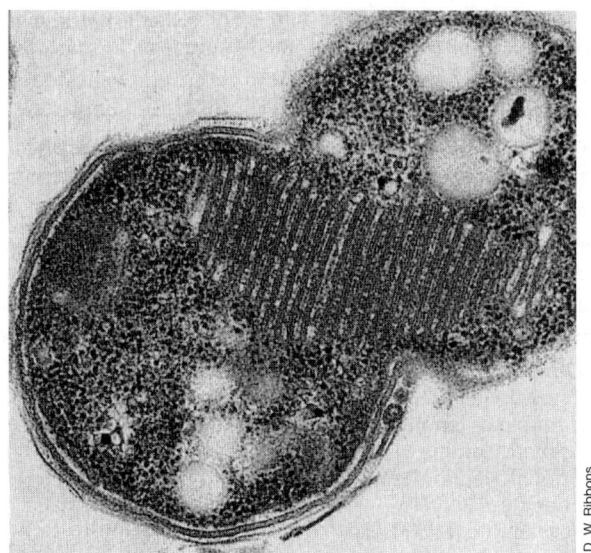

(b)

D. W. Ribbons

Figure 17.14 Methanotrophs. *(a)* Electron micrograph of a cell of *Methylosinus*, illustrating a type II membrane system. Cells are about 0.6 μm in diameter. *(b)* Electron micrograph of a cell of *Methylococcus capsulatus*, illustrating a type I membrane system. Cells are about 1 μm in diameter.

UNIT 6

thermophilic and extremely acidophilic, growing optimally at pH 2 and capable of growth below pH 1. *Methylacidiphilum* inhabits acidic geothermal environments where CH_4 is released in vented gas; at the interface of the anoxic geothermal gas and the atmosphere, cells have the two key substrates they need (CH_4 plus O_2) to grow as methanotrophs. *Methylacidiphilum* is related to species of the genus *Verrucomicrobium*, an organism that forms its own lineage of *Bacteria* (Figure 17.1), and grows in laboratory culture only on CH_4 or CH_3OH at acidic pH. *Methylacidiphilum* also fixes nitrogen, as do many other methanotrophs (Table 17.7), but at the acidic pH values at which *Methylacidiphilum* thrives, it is probably one of the most acidophilic of all known nitrogen-fixing bacteria.

Ecology and Isolation

Methanotrophs are widespread in aquatic and terrestrial environments, being found wherever stable sources of CH_4 are present. Methane produced in the anoxic regions of lakes rises through the water column, and methanotrophs are often concentrated in a narrow band at the zone where CH_4 and O_2 meet. Methane-oxidizing bacteria therefore play an important role in the carbon cycle, converting CH_4 derived from anoxic decomposition back into cell material and CO_2.

Methanotrophic bacteria and certain marine mussels and sponges have developed symbiotic relationships. Some marine mussels live in the vicinity of hydrocarbon seeps on the seafloor, places where CH_4 is released in substantial amounts. Isolated mussel gill tissues consume CH_4 at high rates in the presence of O_2. In these tissues, coccoid-shaped bacteria are present in high numbers (**Figure 17.15**). The bacterial symbionts contain intracytoplasmic membranes typical of methanotrophs (Figure 17.15*b*). The symbionts reside in vacuoles within animal cells near the gill surface, which probably ensures an effective gas exchange with seawater, and the fact that these symbionts are indeed methanotrophs has been shown by phylogenetic analyses. Assimilated CH_4 is distributed throughout the animal by the excretion of organic compounds by the methanotrophs. These methanotrophic symbioses are therefore conceptually quite similar to those that develop between sulfide-oxidizing chemolithotrophs and hydrothermal vent tube worms and giant clams ( Section 25.12).

All that is needed to enrich methanotrophs is a mineral salts medium containing a headspace about 50% each of CH_4 and air. Once good growth is obtained, purification can be achieved by repeated streaking on mineral salts agar plates incubated in a CH_4–air mixture. Colonies appearing on the plates are typically of two types: nonmethanotrophic chemoorganotrophs growing on traces of organic matter in the medium, which appear in 1–2 days, and methanotrophs, which appear after about a week. The colonies of some methanotrophs are pink from the presence of various carotenoid pigments and high levels of cytochromes in their membranes, and this feature can assist in identifying these organisms on plates.

Methanotrophs and Ammonia-Oxidizing Bacteria

Besides CH_4, methanotrophs can also oxidize ammonia (NH_3) (Table 17.6), and ammonia-oxidizing bacteria can also oxidize CH_4; however, neither group can actually *grow* using the other group's substrate as electron donor. It has been hypothesized that methanotrophic bacteria evolved from ammonia-oxidizing

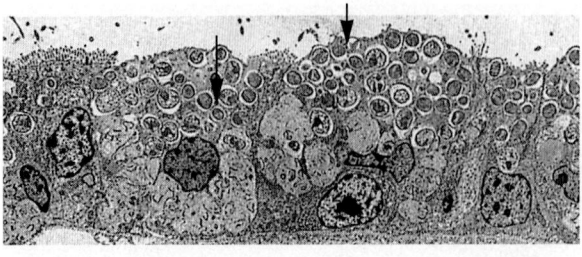

(a)

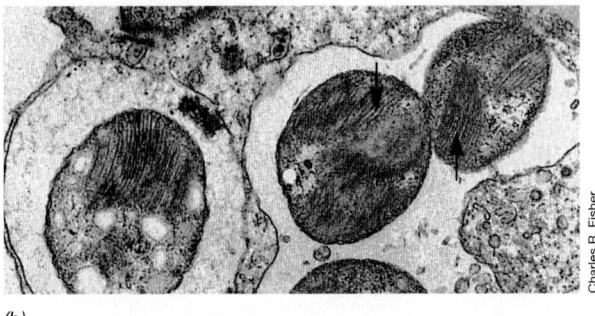

(b)

Figure 17.15 Methanotrophic symbionts of marine mussels.
(a) Electron micrograph of a thin section at low magnification of gill tissue from a marine mussel living near hydrocarbon seeps in the Gulf of Mexico. Note the symbiotic methanotrophs (arrows) in the tissues. *(b)* High-magnification view of gill tissue showing methanotrophs with type I membrane bundles (arrows). The methanotrophs are about 1 μm in diameter. Compare with Figure 17.14*b*.

bacteria via selection for the conversion of an ammonia monooxygenase into a methane monooxygenase. The fact that methanotrophs and nitrifiers have similar internal membrane systems (Section 17.3 and Figure 17.8) and are phylogenetically closely related (Figure 17.2) supports this theory. However, it has also been found that methanotrophic bacteria contain some of the same genes and make some of the same proteins as methanogenic (methane-producing) *Archaea*, and so the evolution of methanotrophy is still very much an open question. We saw how the contrasting processes of methanogenesis and methanotrophy are related in Sections 14.10 and 14.15.

MiniQuiz

- What is the difference between a methanotroph and a methylotroph?
- What features differentiate type I from type II methanotrophs?
- Why is *Methylacidiphilum* such a unique methanotroph?

Ⅲ Aerobic and Facultatively Aerobic Chemoorganotrophic *Proteobacteria*

The next few groups to be considered are the classic examples of chemoorganotrophic bacteria that carry out respiratory metabolisms. Here we will meet the pseudomonads, the enteric bacteria, the aerobic nitrogen-fixing bacteria, and many of their close relatives.

17.7 *Pseudomonas* and the Pseudomonads

Key Genera: *Pseudomonas, Burkholderia, Zymomonas, Xanthomonas*

All the genera in this group are straight or slightly curved gram-negative, chemoorganotrophic rods with *polar* flagella (**Figure 17.16**). Common genera are *Pseudomonas, Comamonas, Ralstonia,* and *Burkholderia,* discussed in some detail here. Other important genera include *Xanthomonas, Zoogloea,* and *Gluconobacter* (see Section 17.8). Phylogenetically, the pseudomonads scatter within the *Proteobacteria* (Figure 17.2 and **Table 17.8**).

Characteristics of Pseudomonads

Some major distinguishing characteristics of the pseudomonad group include an obligately respiratory metabolism, the absence of gas formation from glucose, and a positive oxidase test, all of which help to distinguish pseudomonads from enteric bacteria (Section 17.11). Although obligately respiratory, many pseudomonads can still grow under anoxic conditions with nitrate, fumarate, or many of the other electron acceptors that support anaerobic respiration (Chapter 14). Key species of the genus *Pseudomonas* and related genera are defined on the basis of phylogeny and various physiological and other phenotypic characteristics, as outlined in Table 17.8.

Pseudomonads typically have very simple nutritional requirements and one of their characteristic properties is the ability to use many different organic compounds as carbon and energy sources; some species utilize over 100 different compounds. On the other hand, pseudomonads generally lack the hydrolytic enzymes necessary to break down polymers into their component monomers. The genomes of these nutritionally versatile pseudomonads encode numerous inducible operons (⮌ Section 8.3), some of which encode a large number of enzymes to handle the catabolism of the many different organic substrates that are used.

The pseudomonads are ecologically important in soil and water and are probably responsible for the degradation of many low-molecular-weight compounds derived from the breakdown of plant and animal materials in oxic habitats. They are also capable of catabolizing many xenobiotic (not naturally occurring) compounds, such as pesticides and other toxic chemicals, and are thus important agents of bioremediation in the environment (⮌ Section 24.10).

Pathogenic Pseudomonads

A number of pseudomonads are pathogenic (**Table 17.9**). Among the fluorescent pseudomonads, the species *Pseudomonas aeruginosa* is frequently associated with infections of the urinary and respiratory tracts in humans. *P. aeruginosa* infections are also common in patients receiving treatment for severe burns or other traumatic skin damage and in people suffering from cystic fibrosis. *P. aeruginosa* is not an obligate pathogen. Instead, the organism is an opportunist, initiating infections in individuals with weakened immune systems. In addition to urinary tract infections, it can also cause systemic infections, usually in individuals who have experienced extensive skin damage.

P. aeruginosa is naturally resistant to many of the widely used antibiotics, so treatment of infections is often difficult. Resistance

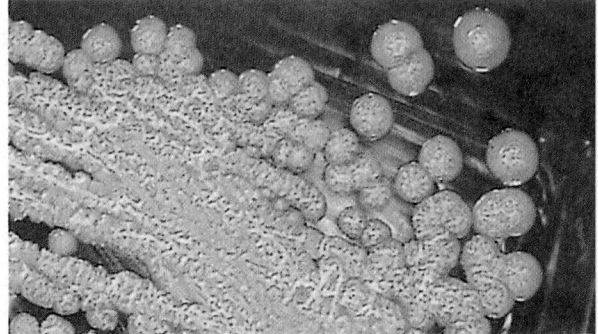

(a)

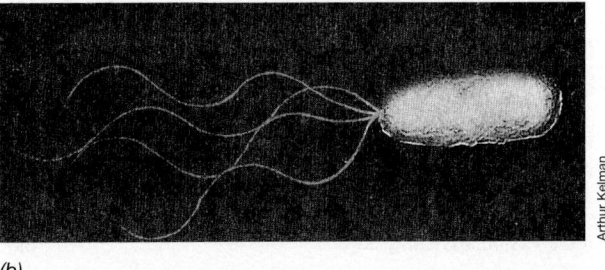

(b)

Figure 17.16 Typical pseudomonad colonies and cell morphology of pseudomonads. *(a)* Photograph of colonies of *Burkholderia cepacia* on an agar plate. *(b)* Shadow-cast transmission electron micrograph of a *Pseudomonas* cell. The cell measures about 1 μm in diameter.

is due to a resistance transfer plasmid (R plasmid) (⮌ Sections 10.9 and 26.12), which is a plasmid carrying genes encoding proteins that detoxify various antibiotics or pump them out of the cell. *P. aeruginosa* is commonly found in the hospital environment and can easily infect patients receiving treatment for other illnesses (healthcare-associated infections, ⮌ Section 32.7). Polymyxin, an antibiotic not ordinarily used in human therapy because of its toxicity, is effective against *P. aeruginosa* and is used in certain medical situations.

Certain species of *Pseudomonas, Ralstonia,* and *Burkholderia* and the genus *Xanthomonas* are well-known plant pathogens (phytopathogens) (Table 17.9). Phytopathogens frequently inhabit nonhost plants (in which disease symptoms are not apparent) and from there are transmitted to host plants and initiate infection. Disease symptoms vary considerably, depending on the particular phytopathogen and host plant. The pathogen releases plant toxins, lytic enzymes, plant growth factors, and other substances that destroy or distort plant tissue. In many cases the disease symptoms help identify the phytopathogen. Thus, *Pseudomonas syringae* is typically isolated from leaves showing chlorotic (yellowing) lesions, whereas *Pseudomonas marginalis,* a typical "soft-rot" pathogen, infects stems and shoots, but rarely leaves.

Zymomonas

The genus *Zymomonas* consists of large, gram-negative rods that carry out a vigorous fermentation of sugars to ethanol. Although distinct from other pseudomonads by its strictly fermentative metabolism, *Zymomonas* shows phylogenetic affiliation with

Table 17.8 *Subgroups and characteristics of pseudomonads*

Group	Phylogenetic group[a]	Characteristics
Fluorescent subgroup	*Gamma*	**Most produce water-soluble, yellow-green fluorescent pigments; do not form poly-β-hydroxybutyrate; single DNA homology group**
Pseudomonas aeruginosa		Pyocyanin production; growth at up to 43°C; single polar flagellum; capable of denitrification
Pseudomonas fluorescens		Does not produce pyocyanin or grow at 43°C; tuft of polar flagella
Pseudomonas putida		Similar to *P. fluorescens*, but does not liquefy gelatin and does grow on benzylamine
Pseudomonas syringae		Lacks arginine dihydrolase; oxidase-negative; pathogenic to plants
Pseudomonas stutzeri		Soil saprophyte; strong denitrifier and nonfluorescent
Acidovorans subgroup	*Beta*	**Nonpigmented; form poly-β-hydroxybutyrate; tuft of polar flagella; do not use carbohydrates; single DNA homology group**
Delftia acidovorans		Uses muconic acid as sole carbon source and electron donor
Comamonas testosteroni		Uses testosterone as sole carbon source
Pseudomallei-cepacia subgroup	*Beta*	**No fluorescent pigments; tuft of polar flagella; forms poly-β-hydroxybutyrate; single DNA homology group**
Burkholderia cepacia		Extreme nutritional versatility; some strains pathogenic to plants; human pathogen
Burkholderia pseudomallei		Causes melioidosis in animals; nutritionally versatile
Burkholderia mallei		Causes glanders in animals; nonmotile; nutritionally restricted
Diminuta-vesicularis subgroup	*Alpha*	**Single flagellum of very short wavelength; require vitamins (pantothenate, biotin, B$_{12}$)**
Brevundimonas diminuta		Nonpigmented; does not use sugars
Brevundimonas vesicularis		Carotenoid pigment; uses sugars
Ralstonia subgroup	*Beta*	
Ralstonia solanacearum		Plant pathogen
Pelomonas saccharophila		Grows chemolithotrophically with H$_2$; digests starch
Stenotrophomonas maltophilia	*Gamma*	Requires methionine; does not use NO$_3^-$ as N source; oxidase-negative
Others		**Not pathogenic; unusual ecology or physiology**
Zoogloea	*Beta*	Forms an extracellular fibrillar polymer that causes the cells to aggregate into distinctive flocs as a major component of activated sewage sludge (↝ Section 35.2).
Gluconobacter	*Alpha*	Incompletely oxidizes glucose to gluconic acid or ethanol to acetic acid; important in vinegar production
Zymomonas	*Alpha*	Obligately fermentative; ferments glucose to ethanol plus CO$_2$

[a]All pseudomonads are species of *Proteobacteria*.

Table 17.9 *Pathogenic pseudomonads*

Species	Relationship to disease
Animal pathogens	
Pseudomonas aeruginosa	Opportunistic pathogen, especially in hospitals; in patients with metabolic, hematologic, and malignant diseases; hospital-acquired (nosocomial) infections from catheterizations, tracheostomies, lumbar punctures, and intravenous infusions; in patients given prolonged treatment with immunosuppressive agents, corticosteroids, antibiotics, and radiation; may contaminate surgical wounds, abscesses, burns, ear infections, lungs of patients treated with antibiotics; lungs of those with cystic fibrosis; primarily a soil organism
Pseudomonas fluorescens	Rarely pathogenic, as it does not grow well at 37°C; may grow in and contaminate blood and blood products under refrigeration
Stenotrophomonas maltophilia	A ubiquitous, free-living organism that is a common nosocomial pathogen
Burkholderia cepacia	Isolated from humans and from environmental sources of medical importance; also plant pathogen, causes onion bulb rot
Burkholderia pseudomallei	Causes melioidosis, a disease endemic in animals and humans in Southeast Asia
Burkholderia mallei	Causes glanders, a disease of horses that is occasionally transmitted to humans
Pseudomonas stutzeri	Often isolated from humans and environmental sources; may live saprophytically in the body
Plant pathogens	
Ralstonia solanacearum	Causes wilts of many cultivated plants (for example, potato, tomato, tobacco, peanut)
Pseudomonas syringae	Attacks foliage, causing chlorosis and necrotic lesions on leaves; rarely found free in soil
Pseudomonas marginalis	Causes soft rot of various plants; active pectinolytic species
Xanthomonas campestris	Causes necrotic lesions on foliage, stems, fruits; also causes wilts and tissue rots; rarely found free in soil

these organisms (Table 17.8), and like other pseudomonads it employs the Entner–Doudoroff pathway for glucose catabolism (Section 14.2).

Zymomonas carries out alcoholic fermentations of various plant saps, and in many tropical areas of South and Central America, Africa, and Asia it is found in various fermented beverages made from plant saps, such as pulque (agave), palm sap, sugarcane juice, and honey. Although *Zymomonas* is rarely the sole organism responsible for these alcoholic fermentations, it is probably responsible for the production of most of the ethanol in these beverages where yeasts are present in only low numbers. *Zymomonas* is also responsible for spoilage of fruit juices, such as apple and pear ciders, and is also a frequent constituent of the bacterial flora of spoiled beer.

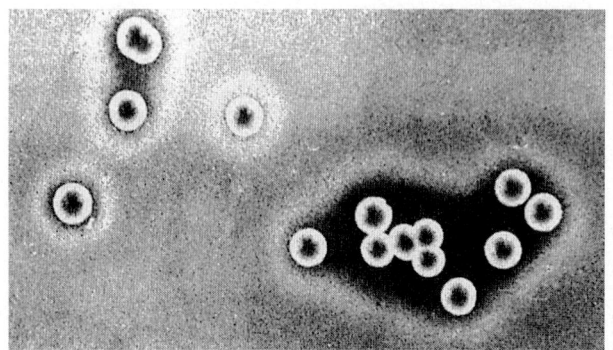

Figure 17.17 Colonies of *Acetobacter aceti* on calcium carbonate ($CaCO_3$) agar containing ethanol as electron donor. Note the clearing around the colonies due to the dissolution of $CaCO_3$ by the acetic acid produced.

MiniQuiz

- Other than Gram reaction and morphology, list two key characteristics of pseudomonads.
- If pseudomonads are "obligately respiratory," does this mean they cannot grow under anoxic conditions?

17.8 Acetic Acid Bacteria

Key Genera: *Acetobacter, Gluconobacter*

The acetic acid bacteria comprise gram-negative, obligately aerobic, motile rods that carry out the incomplete oxidation of alcohols and sugars, leading to the accumulation of organic acids as end products. With ethanol (C_2H_2OH) as a substrate, acetic acid ($C_2H_4O_2$) is produced, which gives the acetic acid bacteria their name. As one would expect, acetic acid bacteria are tolerant of acidic conditions; most strains can grow well at pH values lower than 5. The acetic acid bacteria are a heterogeneous assemblage of *Alphaproteobacteria*, comprising both peritrichously flagellated (*Acetobacter*) and polarly flagellated (*Gluconobacter*) organisms. In addition to flagellation, *Acetobacter* differs physiologically from *Gluconobacter* in being able to further oxidize the acetic acid it forms to CO_2; that is, with *Gluconobacter*, acetic acid is a "dead-end" product.

Ecology and Industrial Uses

The acetic acid bacteria are commonly found in fermenting fruit juices, such as hard cider or wine, or in beer. Colonies of acetic acid bacteria can be recognized on calcium carbonate ($CaCO_3$) agar plates containing ethanol, because the acetic acid produced dissolves and causes a clearing of the otherwise insoluble $CaCO_3$ (**Figure 17.17**). Cultures of acetic acid bacteria are used in the commercial production of vinegar (Section 36.3). In addition to ethanol, the acetic acid bacteria carry out an incomplete oxidation of some higher alcohols and sugars. For instance, glucose is oxidized to gluconic acid, galactose to galactonic acid, arabinose to arabonic acid, and so on. This property of "underoxidation" is exploited in the industrial manufacture of ascorbic acid (vitamin C). Ascorbic acid can be formed from sorbose, but sorbose is difficult to synthesize chemically. It is, however, conveniently obtainable from acetic acid bacteria, which oxidize sorbitol (a readily available sugar alcohol) to sorbose.

Another interesting property of some acetic acid bacteria is their ability to synthesize cellulose. The cellulose formed does not differ significantly from plant cellulose, with the exception that it is pure and not mixed in with other polymers like the hemicelluloses, pectin, or lignins of plants. Cellulose from acetic acid bacteria is formed as a matrix outside the cell wall and causes cells to become embedded in a tangled mass of cellulose microfibrils. When these species of acetic acid bacteria grow in an unshaken vessel, they form a surface pellicle of cellulose in which the bacteria develop. Because these bacteria are obligate aerobes, the ability to form such a pellicle may be a means by which the organisms remain at the surface of the liquid where oxygen is readily available.

MiniQuiz

- Why are acetic acid bacteria commonly found in alcoholic beverages?
- Which industrial processes use acetic acid bacteria?

17.9 Free-Living Aerobic Nitrogen-Fixing Bacteria

Key Genera: *Azotobacter, Azomonas, Beijerinckia*

A variety of free-living chemoorganotrophic bacteria inhabit soil and are capable of aerobic nitrogen (N_2) fixation (Section 13.14). The genus *Azotobacter* (**Figure 17.18**) was the first such organism and was discovered by the Dutch microbiologist Martinus Beijerinck, early in the twentieth century (Section 1.9). Beijerinck employed an aerobic enrichment culture devoid of a combined nitrogen source but exposed to air (Section 22.1). Phylogenetically, free-living nitrogen-fixing bacteria are *Alpha-, Beta-,* or *Gammaproteobacteria* (**Table 17.10** and Figure 17.2).

Taxonomy

The major free-living nitrogen-fixing bacteria that have been well studied include *Azotobacter, Azospirillum,* and *Beijerinckia*. *Azotobacter* cells are large rods or cocci, many isolates being almost the size of yeasts (eukaryotes), with diameters of 2–4 μm

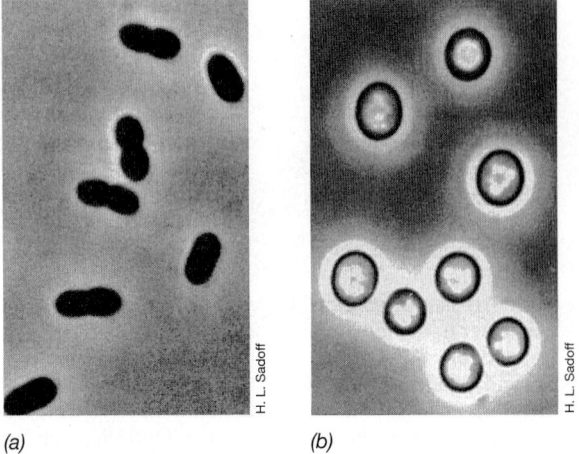

(a)   *(b)*

Figure 17.18 *Azotobacter vinelandii.* *(a)* Vegetative cells and *(b)* cysts visualized by phase-contrast microscopy. A cell measures about 2 μm in diameter and a cyst about 3 μm. Compare with Figure 1.21*b*.

or more, and some species are motile by peritrichous flagella. When they are growing on N_2 as a nitrogen source, extensive capsules or slime layers are typically produced by species of free-living nitrogen-fixing bacteria (**Figure 17.19** and see Figure 17.20; Figure 13.35). This layer helps protect the enzyme nitrogenase in the cytoplasm. Despite the fact that *Azotobacter* is an obligate aerobe, its nitrogenase, the enzyme that catalyzes N_2 fixation (↩ Section 13.14), is O_2-sensitive. It is thought that the high respiratory rate characteristic of *Azotobacter* cells and the abundant capsular slime they produce help protect nitrogenase from O_2. *Azotobacter* is able to grow on many different carbohydrates, alcohols, and organic acids, and metabolism is strictly oxidative. All species fix nitrogen, but can also grow on simple forms of combined nitrogen.

Azotobacter can form resting structures called *cysts* (Figure 17.18*b*). Like bacterial endospores, *Azotobacter* cysts show negli-

Table 17.10	Genera of free-living, aerobic nitrogen-fixing bacteria
Characteristics	**Genus**
Gammaproteobacteria	
Large rod; produces cysts; primarily found in neutral to alkaline soils	*Azotobacter*
Large rod; no cysts; primarily aquatic	*Azomonas*
Alphaproteobacteria	
Microaerophilic rod; associates with plants	*Azospirillum*
Pear-shaped rod with large lipid bodies at each end; produces extensive slime; inhabits acidic soils	*Beijerinckia*
Betaproteobacteria	
Small curved cells; no cysts	*Azoarcus*
Very thin curved cells; no cysts	*Azovibrio*
Very thin rods to vibrios; no cysts	*Azospira*
Cells form coils up to 50 μm long; no cysts	*Azonexus*
Rods; form coarse, wrinkled colonies	*Derxia*

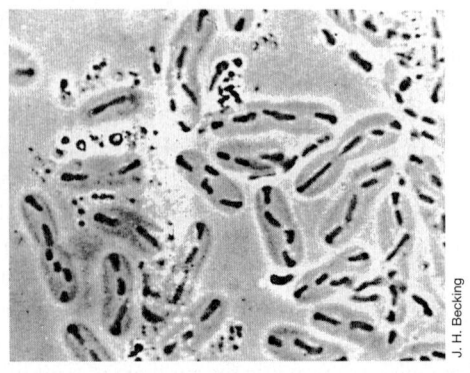

(a)

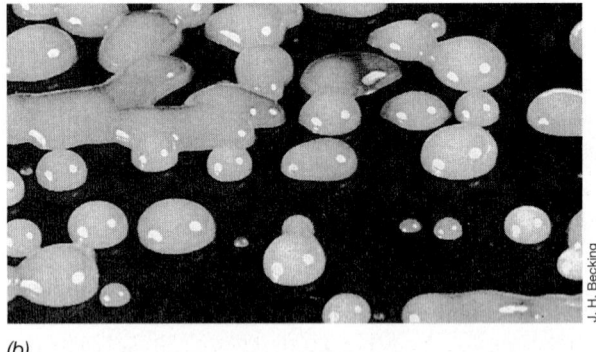

(b)

Figure 17.19 Examples of slime production by free-living N_2-fixing bacteria. *(a)* Cells of *Derxia gummosa* encased in slime. Cells are about 1–1.2 μm wide. *(b)* Colonies of *Beijerinckia* species growing on a carbohydrate-containing medium. Note the raised, glistening appearance of the colonies due to abundant capsular slime.

gible endogenous respiration and are resistant to desiccation, mechanical disintegration, and ultraviolet and ionizing radiation. In contrast to endospores, however, cysts are not very heat-resistant, and they are not completely dormant because they rapidly oxidize carbon sources if supplied.

The remaining major genera of free-living nitrogen fixers include *Azomonas*, a genus of large coccus to rod-shaped bacteria that resemble *Azotobacter*, except that they do not produce cysts and are primarily aquatic, and *Beijerinckia* and *Derxia* (**Figure 17.20**), two genera that grow well in acidic soils. *Azospirillum*, a rod- to spirillum-shaped nitrogen-fixing bacterium that forms nonspecific symbiotic associations with plants, in particular, corn (Table 17.10), rounds out this group.

Azotobacter and Alternative Nitrogenases

We considered the important process of biological N_2 fixation in Sections 13.14 and 13.15, and learned of the central importance of the metals molybdenum (Mo) and iron (Fe) to the enzyme nitrogenase. The species *Azotobacter chroococcum* was the first nitrogen-fixing bacterium shown capable of growth on N_2 in the absence of molybdenum. It was shown in *A. chroococcum* that either of two "alternative nitrogenases" are formed when Mo limitation prevents the normal Mo nitrogenase from being synthesized. These nitrogenases are less efficient than the Mo nitrogenase and contain either vanadium (V) or Fe in place of Mo. Subsequent investigations of other nitrogen-fixing bacteria have

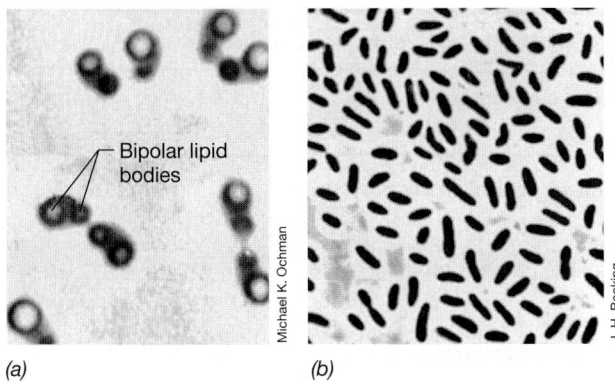

Figure 17.20 Phase-contrast photomicrographs of two genera of acid-tolerant, free-living N₂-fixing bacteria. *(a) Beijerinckia indica.* The cells are roughly pear-shaped, about 0.8 μm in diameter, and contain a large globule of poly-β-hydroxybutyrate at each end. *(b) Derxia gummosa.* Cells are about 1 μm in diameter.

Table 17.11	Characteristics of the genera of gram-negative cocci[a]	
Characteristics		**Genus**
I. Oxidase-positive, penicillin-sensitive		
Cocci; complex nutrition, utilize carbohydrates, obligate aerobes; *Betaproteobacteria* or *Gammaproteobacteria*		*Neisseria* *Moraxella*
Rods or cocci; generally no growth factor requirements, generally do not utilize carbohydrates; do not contain flagella, but some species exhibit twitching motility; many are commensals or pathogens of animals; *Betaproteobacteria*		*Branhamella* *Kingella*
II. Oxidase-negative, penicillin-resistant		
Some strains can utilize a restricted range of sugars, and some exhibit twitching motility; saprophytes in soil, water, and sewage; *Gammaproteobacteria*		*Acinetobacter*

[a]All are *Proteobacteria.*

shown that these genetically distinct "backup" nitrogenases are widely distributed among nitrogen-fixing bacteria, including *Archaea*, of which a few species fix nitrogen.

MiniQuiz

• What enzyme does *Azotobacter* need in order to fix N₂?

• Why do free-living nitrogen-fixing bacteria produce abundant slime?

17.10 *Neisseria, Chromobacterium, and Relatives*

Key Genera: *Neisseria, Chromobacterium*

This group of *Beta-* and *Gammaproteobacteria* comprises a diverse collection of organisms that are related phylogenetically as well as by Gram stain, morphology, lack of swimming motility, and aerobic metabolism. The genera *Neisseria, Moraxella, Branhamella, Kingella,* and *Acinetobacter* are distinguished as outlined in **Table 17.11**.

In the genus *Neisseria*, the cells are always cocci (**Figure 17.21c**), whereas cells of the other genera are rod-shaped, becoming coccoid only in the stationary phase of growth. This has led to designation of these organisms as *coccobacilli*. Organisms of the genera *Neisseria, Kingella,* and *Moraxella* are commonly isolated from animals, and some of them are pathogenic. We discuss the clinical microbiology of *Neisseria gonorrhoeae*, the causative agent of the disease gonorrhea, in Sections 31.1 and 31.2 and the pathogenesis of gonorrhea itself in Section 33.12. Some *Neisseria* are free-living saprophytes and reside in the oral cavity and other moist areas on the animal body, while others, such as *Neisseria meningitidis*, are serious pathogens that can cause a potentially fatal inflammation of the membranes lining the brain (meningitis, ↩ Section 33.5).

Species of *Acinetobacter* are common soil and water organisms, although they are occasionally found as parasites of animals and have been implicated in some nosocomial (healthcare-associated) infections. Some strains of *Moraxella* and *Acinetobacter*

possess the interesting property of *twitching motility*, exhibited as brief translocative movements or "jumps" covering distances of about 1–5 μm. Twitching bacteria contain special force-generating pili (↩ Section 3.9) that facilitate their movement.

Chromobacterium is a close phylogenetic relative of *Neisseria* but is rod-shaped in morphology, resembling the pseudomonads or enteric bacteria. The best-known *Chromobacterium* species is *C. violaceum*, a purple-pigmented organism (Figure 17.21a) found in soil and water and occasionally in pus-forming infections of humans and other animals. *C. violaceum* and a few other chromobacteria produce the purple pigment *violacein* (Figure 17.21b), a water-insoluble pigment with both antimicrobial and

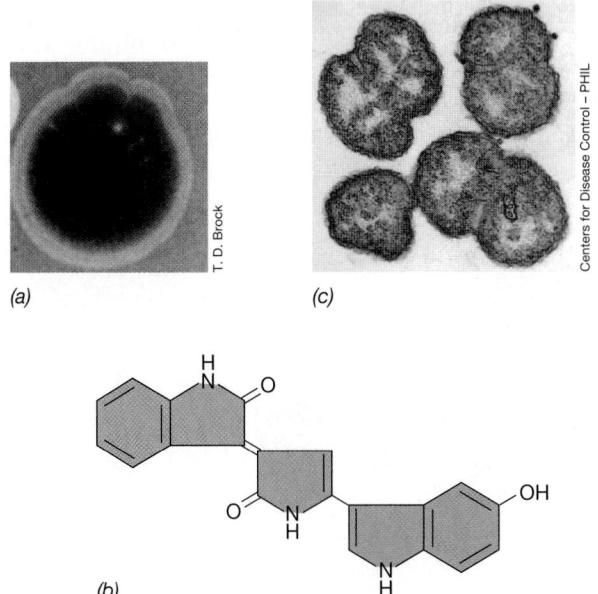

Figure 17.21 *Chromobacterium* and *Neisseria*. *(a)* A large colony of *Chromobacterium violaceum*. The purple pigment is an aromatic compound called violacein, the structural formula of which is shown in *(b)*. *(c)* Transmission electron micrograph of cells of *Neisseria gonorrhoeae* showing the typical diplococcus cell arrangements.

antioxidant properties that is produced only in media containing the amino acid tryptophan, the starting substrate for its synthesis. Like enteric bacteria, *Chromobacterium* is a facultative aerobe, growing fermentatively on sugars and aerobically on various carbon sources.

MiniQuiz

- What species causes the disease gonorrhea?
- What makes colonies of *Chromobacterium violaceum* pigmented?

17.11 Enteric Bacteria

Key Genera: *Escherichia, Salmonella, Proteus, Enterobacter*

The **enteric bacteria** comprise a relatively homogeneous phylogenetic group within the *Gammaproteobacteria* and consist of facultatively aerobic, gram-negative, nonsporulating rods that are either nonmotile or motile by peritrichous flagella (**Figure 17.22**). Enteric bacteria are also oxidase-negative, have relatively simple nutritional requirements, and ferment sugars to a variety of end products. The defining phenotypic characteristics that distinguish enteric bacteria from other bacteria of similar morphology and physiology are given in **Table 17.12**.

Among the enteric bacteria are many species pathogenic to humans, other animals, or plants, as well as other species of industrial importance. *Escherichia coli*, the best known of all microorganisms, is the classic example of an enteric bacterium. Because of the medical importance of many enteric bacteria, an extremely large number of isolates have been characterized, and numerous genera have been defined, largely for ease in identification purposes in clinical microbiology. However, because enteric bacteria are genetically very closely related, their positive identification often presents considerable difficulty. In clinical laboratories, identification is typically based on the combined analysis of a large number of diagnostic tests carried out using miniaturized rapid diagnostic media kits along with immunological and nucleic acid probes to identify signature proteins or genes of particular species (Chapter 31).

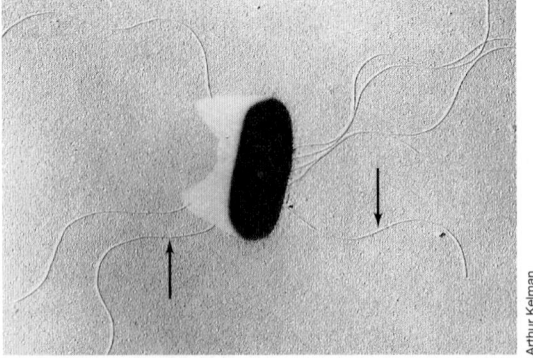

Figure 17.22 Butanediol producer. Electron micrograph of a shadow-cast preparation of cells of the butanediol-producing bacterium *Erwinia carotovora*. The cell is about 0.8 μm wide. Note the peritrichously arranged flagella (arrows), typical of enteric bacteria.

Table 17.12 *Defining characteristics of the enteric bacteria*
General characteristics
Gram-negative straight rods; motile by peritrichous flagella, or nonmotile; nonsporulating; facultative aerobes, producing acid from glucose; catalase-positive and oxidase-negative; usually reduce NO_3^- to NO_2^- but not to N_2 anaerobically; *Gammaproteobacteria* (Figure 17.2)
Some major genera
Mixed-acid fermenters: *Escherichia, Salmonella, Shigella, Citrobacter, Proteus, Yersinia* **Butanediol producers:** *Enterobacter, Klebsiella, Erwinia, Serratia*
Key biochemical tests to distinguish enteric bacteria from other bacteria of similar morphology[a]
Oxidase test: Enterics always negative—separates enterics from oxidase-positive bacteria of genera *Pseudomonas, Aeromonas, Vibrio, Alcaligenes, Achromobacter, Flavobacterium, Cardiobacterium*, which may have similar morphology
Nitrate reduced only to nitrite (assay for nitrite after growth)—distinguishes enteric bacteria from bacteria that reduce NO_3^- to N_2 (gas formation detected), such as *Pseudomonas* and many other oxidase-positive bacteria
Ability to ferment glucose—distinguishes enterics from obligately aerobic bacteria

[a]See Section 31.2.

Fermentation Patterns in Enteric Bacteria

One major taxonomic characteristic separating the various genera of enteric bacteria is the type and proportion of fermentation products generated from the fermentation of glucose. Two broad patterns are recognized, the *mixed-acid* fermentation and the *2,3-butanediol* fermentation (Table 17.12 and **Figure 17.23**).

In the mixed-acid fermentation, three acids are formed in significant amounts: acetic, lactic, and succinic; ethanol, CO_2, and H_2 are also formed, but not butanediol. In the butanediol fermentation, smaller amounts of acids are formed, and butanediol, ethanol, CO_2, and H_2 are the main products (⟳ Figure 14.4). As a result of mixed-acid fermentation, equal amounts of CO_2 and H_2 are produced, whereas in the butanediol fermentation, considerably more CO_2 than H_2 is produced. This is because mixed-acid fermenters produce CO_2 only from formic acid by means of the enzyme system formate hydrogen lyase:

$$HCOOH \rightarrow H_2 + CO_2$$

and this reaction results in equal amounts of CO_2 and H_2. The butanediol fermenters also produce CO_2 and H_2 from formic acid, but they produce two additional molecules of CO_2 during the formation of each molecule of butanediol (Figure 17.23*b*).

The Genus *Escherichia*

Species of *Escherichia* are almost universal inhabitants of the intestinal tract of humans and other warm-blooded animals, although they are by no means the dominant organisms in this habitat. *Escherichia* may play a nutritional role in the intestinal tract by synthesizing vitamins, particularly vitamin K. As a facultative aerobe, this organism probably also helps consume O_2, thus rendering the large intestine anoxic. Wild-type *Escherichia* strains

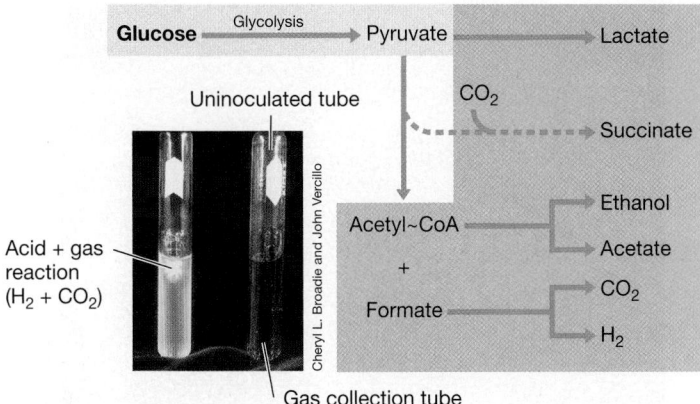

(a) **Mixed-acid fermentation** (for example, *Escherichia coli*)

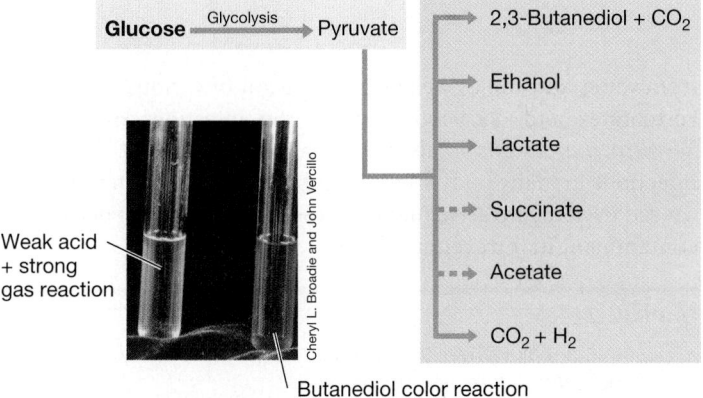

(b) **Butanediol fermentation** (for example, *Enterobacter aerogenes*)

Figure 17.23 **Enteric fermentations.** Distinction between *(a)* mixed-acid and *(b)* butanediol fermentation in enteric bacteria (ᚼᚼ Figure 14.4). The solid arrows indicate reactions leading to major products. Dashed arrows indicate minor products. The upper photo shows the production of acid (yellow) and gas (in the inverted Durham tube) in a culture of *Escherichia coli* carrying out a mixed-acid fermentation (purple tube was uninoculated). The bottom photo shows the pink-red color in the Voges–Proskauer (VP) test, which indicates butanediol production, following growth of *Enterobacter aerogenes*. The left (yellow) tube was not inoculated. Note that the mixed-acid fermentation produces less CO_2 but more acid products from glucose than does the butanediol fermentation.

rarely show any growth-factor requirements and are able to grow on a wide variety of carbon and energy sources such as sugars, amino acids, organic acids, and so on.

Some strains of *Escherichia* are pathogenic and have been implicated in diarrheal diseases, especially in infants, a major public health problem in developing countries. *Escherichia* is a major cause of urinary tract infections in women. Enteropathogenic *E. coli* (abbreviated as EPEC) are becoming more frequently implicated in gastrointestinal infections and generalized fevers. As noted in Sections 27.4, 27.11, and 36.9, certain of these strains form a surface structure called the *K antigen*, permitting attachment and colonization of the small intestine, and these strains produce enterotoxin, responsible for the signs and symptoms of diarrhea. Some strains, such as enterohemorrhagic *E. coli* (abbreviated as EHEC), an important representa-

tive of which is strain O157:H7, can cause sporadic outbreaks of severe foodborne disease. Infection occurs primarily through consumption of contaminated foods, such as raw or undercooked ground beef, unpasteurized milk, or contaminated water. In a small percentage of cases, *E. coli* O157:H7 causes a life-threatening complication related to its production of enterotoxin.

Salmonella, Shigella, and *Proteus*

Salmonella and *Escherichia* are quite closely related, the two genera showing about 50% genomic hybridization (ᚼᚼ Section 16.11). However, in contrast to most *Escherichia*, species of *Salmonella* are usually pathogenic, either to humans or to other warm-blooded animals; *Salmonella* also is found in the intestines of cold-blooded animals, such as turtles and lizards. In humans the most common diseases caused by salmonellas are typhoid fever and gastroenteritis (ᚼᚼ Sections 35.8 and 36.8). The salmonellas are characterized immunologically on the basis of three cell surface markers called *antigens*: the O, or cell wall (somatic) antigen; the H, or flagellar, antigen; and the Vi (outer polysaccharide layer) antigen, found primarily in strains of *Salmonella* causing typhoid fever. The O antigens are part of the lipopolysaccharides that constitute the outer membrane of gram-negative bacteria (ᚼᚼ Sections 3.7 and 27.12). Although there is little correlation between the antigenic type of *Salmonella* and the disease symptoms elicited, immunological typing permits tracking a single strain type in an epidemic.

The shigellas are also genetically very closely related to *Escherichia*. Tests for DNA hybridization show that strains of *Shigella* have 70% or even higher genomic hybridization with *E. coli* and therefore probably constitute a single species (ᚼᚼ Section 16.11). Moreover, genomic analyses strongly suggest that *Shigella* and *Escherichia* have exchanged a significant number of genes by horizontal gene flow. In contrast to most *Escherichia*, however, species of *Shigella* are typically pathogenic to humans, causing a rather severe gastroenteritis called *bacillary dysentery*. *Shigella dysenteriae*, transmitted by food- and waterborne routes, is a good example of this. The bacterium, which contains endotoxin, invades intestinal epithelial cells, where it excretes a neurotoxin that causes acute gastrointestinal distress.

Cells in the genus *Proteus* are typically highly motile (**Figure 17.24**) and produce the enzyme *urease*. By genomic DNA hybridization, *Proteus* shows only a distant relationship to *E. coli*. *Proteus* is a frequent cause of urinary tract infections in humans and probably benefits in this regard from its ready ability to degrade urea. Because of the rapid motility of *Proteus* cells, colonies growing on agar plates often exhibit a characteristic swarming phenotype (Figure 17.24*b*). Cells at the edge of the growing colony are more rapidly motile than those in the center of the colony. The former move a short distance away from the colony in a mass and then undergo a reduction in motility, settle down, and divide, forming a new population of motile cells that again swarm. As a result, the mature colony appears as a series of concentric rings, with higher concentrations of cells alternating with lower concentrations (Figure 17.24*b*).

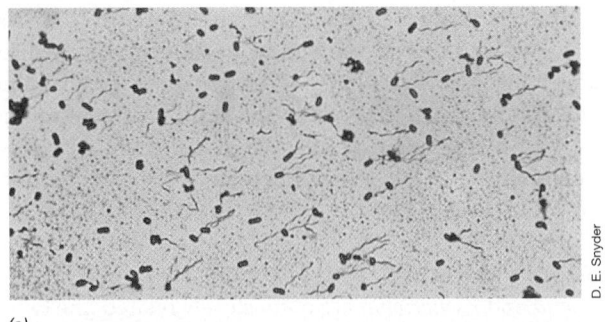

(a)

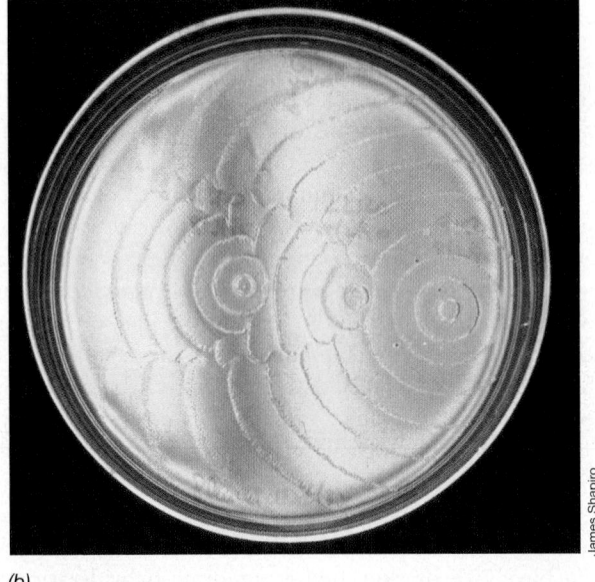

(b)

Figure 17.24 Swarming in *Proteus*. *(a)* Cells of *Proteus mirabilis* stained with a flagella stain; the peritrichous flagella of each cell form into a bundle to rotate in synchrony. *(b)* Photo of a swarming colony of *Proteus vulgaris*. Note the concentric rings.

Butanediol Fermenters: *Enterobacter*, *Klebsiella*, and *Serratia*

The butanediol fermenters are genetically more closely related to each other than to the mixed-acid fermenters, a finding that is in agreement with the observed physiological differences (Figure 17.23). *Enterobacter aerogenes* is a common species in water and sewage as well as the intestinal tract of warm-blooded animals and is an occasional cause of urinary tract infections. One species of *Klebsiella*, *K. pneumoniae*, occasionally causes pneumonia in humans, but klebsiellas are most commonly found in soil and water. Most *Klebsiella* strains also fix nitrogen (Section 13.14), a property unknown in other enteric bacteria.

The genus *Serratia* forms a series of red pyrrole-containing pigments called *prodigiosins* (**Figure 17.25**). Prodigiosin is produced in stationary phase as a secondary metabolite (Section 15.2) and is of interest because it contains the pyrrole ring also found in the pigments for energy transfer: porphyrins, chlorophylls, and phycobilins (Sections 13.1–13.3). However, it is unknown if prodigiosin plays any role in energy transfer, and its exact function is unknown. Species of *Serratia* can be isolated

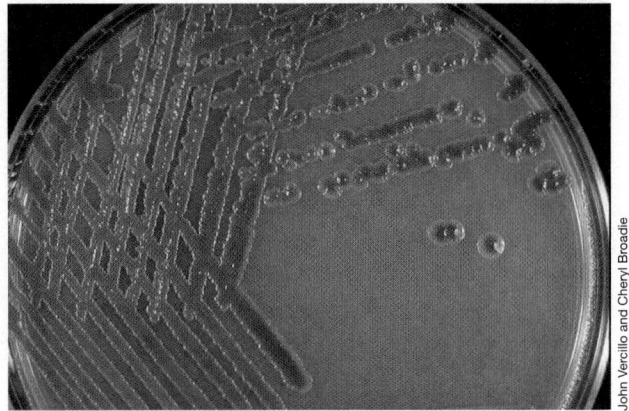

Figure 17.25 Colonies of *Serratia marcescens*. The orange-red pigmentation is due to the pyrrole-containing pigment prodigiosin.

from water and soil as well as from the gut of various insects and vertebrates and occasionally from the intestines of humans. *Serratia marcescens* is also a human pathogen that can cause infections in many body sites. It has been implicated in infections caused by some invasive medical procedures and is an occasional contaminant in intravenous fluids.

MiniQuiz

- Describe a major physiological difference between *Escherichia coli* and *Enterobacter aerogenes*.
- Which enteric bacteria are major pathogens and what kind of diseases do they cause?

17.12 *Vibrio, Aliivibrio, and Photobacterium*

Key Genera: *Vibrio, Aliivibrio, Photobacterium*

The *Vibrio* group contains gram-negative, facultatively aerobic rods and curved rods that employ a fermentative metabolism. Most species of *Vibrio* are polarly flagellated, although some are peritrichously flagellated. One key difference between the *Vibrio* group and enteric bacteria is that members of the former are oxidase-*positive*, a test for the presence of cytochrome *c* (Table 31.3), whereas members of the latter are oxidase-*negative*. Although *Pseudomonas* species are also polarly flagellated and oxidase-positive, they are not fermentative and so are clearly distinct from *Vibrio* species. The best-known genera in this group are *Vibrio, Aliivibrio,* and *Photobacterium*.

Most vibrios and related bacteria are aquatic, found in marine, brackish, or freshwater habitats. *Vibrio cholerae* is the specific cause of the disease cholera in humans (Sections 27.11 and 35.5); the organism does not normally cause disease in other hosts. Cholera is one of the most common human infectious diseases in developing countries and is transmitted almost exclusively via water.

Vibrio parahaemolyticus inhabits the marine environment and is a major cause of gastroenteritis in Japan, where raw fish is widely consumed; the organism has also been implicated in outbreaks of gastroenteritis in other parts of the world, including the United States. *V. parahaemolyticus* can be isolated from seawater itself or from shellfish and crustaceans, and its primary habitat is probably marine animals, with humans being an accidental host.

Bacterial Bioluminescence

Several species of bacteria can emit light, a process called **bioluminescence** (**Figure 17.26**). Most bioluminescent bacteria have been classified in the genera *Photobacterium, Aliivibrio*, and *Vibrio*, but a few species are found also in *Shewanella*, a genus of primarily marine bacteria, and in *Photorhabdus*, a genus of terrestrial bacteria. Most bioluminescent bacteria inhabit the marine environment and some species colonize specialized *light organs* of certain marine fishes and squids, producing light that the animal uses for signaling, avoiding predators, and attracting prey (Figure 17.26c–f and ↻ Section 25.11). When living symbiotically in light organs of fish and squids, or saprophytically, for example, on the skin of a dead fish or parasitically in the body of a crustacean, luminous bacteria can be recognized by the light they produce. Because some pathogenic strains of *Vibrio* are also luminous, such as certain strains of *V. cholerae* and *V. vulnificus*, care should always be taken when isolating and handling luminous bacteria.

Mechanism and Ecology of Bioluminescence

Although *Photobacterium, Aliivibrio*, and *Vibrio* isolates are facultative aerobes, they are bioluminescent only when O_2 is present. Luminescence in bacteria requires the genes *luxCDABE*

(↻ Section 16.12) and is catalyzed by the enzyme *luciferase*, which uses O_2, a long-chain aliphatic aldehyde such as tetradecanal, and reduced flavin mononucleotide ($FMNH_2$) as substrates:

$$FMNH_2 + O_2 + RCHO \xrightarrow{Luciferase} FMN + RCOOH + H_2O + light$$

The light-generating system constitutes a metabolic route for shunting electrons from $FMNH_2$ to O_2 directly, without employing other electron carriers such as quinones and cytochromes.

Luminescence in many luminous bacteria only occurs at high population density. The enzyme luciferase and other proteins of the bacterial luminescence system exhibit a population density–responsive induction, called **autoinduction**, in which transcription of the *luxCDABE* genes is controlled by a regulatory protein, LuxR, and an inducer molecule, acyl homoserine lactone (AHL, ↻ Section 8.9 and Figures 8.18 and 8.19). During growth, cells produce AHL, which can rapidly cross the cytoplasmic membrane in either direction, diffusing in and out of cells. Under conditions in which a high local population density of cells is attained, as in a test tube, a colony on a plate, or in the light organ of a fish or squid (↻ Section 25.11), AHL accumulates. Only when it reaches a certain concentration in the cell is it bound by LuxR, forming a complex that activates transcription of *luxCDABE*, and cells become luminous (Figure 17.26b; ↻ Figure 1.1). This gene regulatory mechanism is also called *quorum sensing* because of the population density–dependent nature of the phenomenon (↻ Section 8.9).

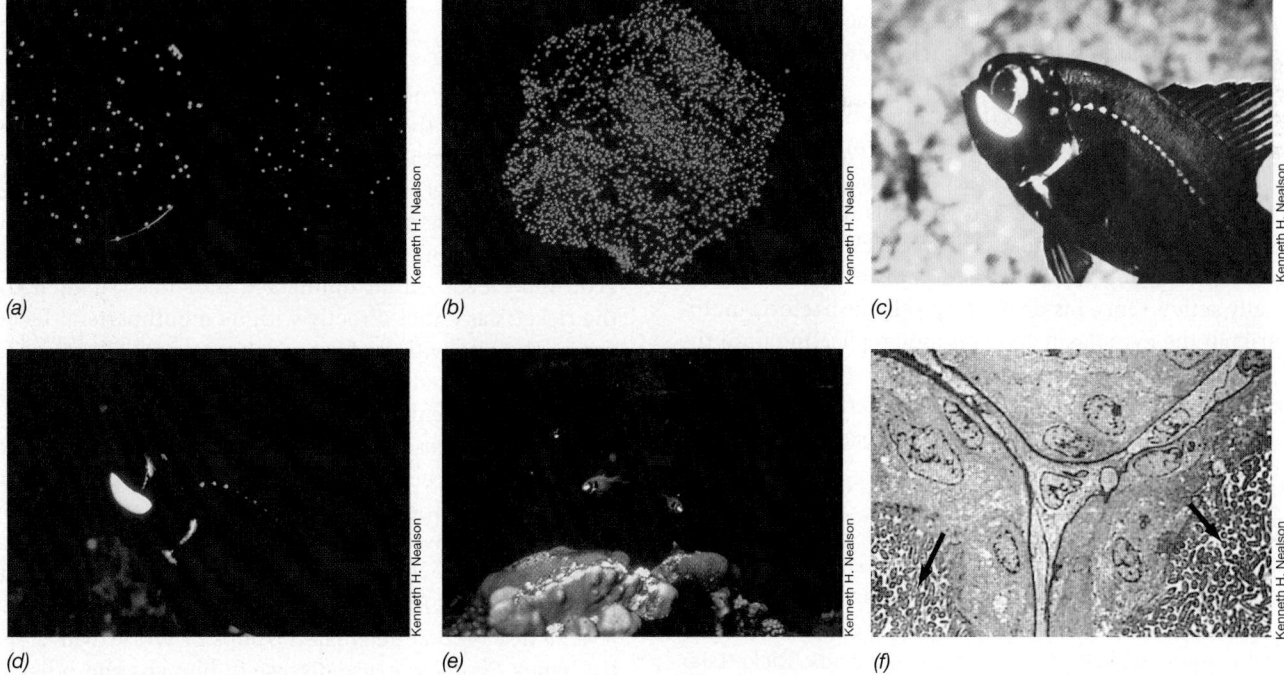

Figure 17.26 Bioluminescent bacteria and their role as light organ symbionts in the flashlight fish. *(a)* Two Petri plates of luminous bacteria photographed by their own light. Note the different colors. Left, *Aliivibrio fischeri* strain MJ-1, blue light, and right, strain Y-1, green light. *(b)* Colonies of *Photobacterium phosphoreum* photographed by their own light. *(c)* The flashlight fish *Photoblepharon palpebratus*; the bright area is the light organ containing bioluminescent bacteria. *(d)* Same fish photographed by its own light. *(e)* Underwater photograph taken at night of *P. palpebratus*. *(f)* Electron micrograph of a thin section through the light-emitting organ of *P. palpebratus* showing the dense array of bioluminescent bacteria (arrows).

In saprophytic, parasitic, and symbiotic habitats (Figure 17.26*c–f*), the rationale for population density–responsive induction of luminescence is to ensure that luminescence develops only when sufficiently high population densities are reached to allow the light produced to be visible to animals. The bacterial light can then attract animals to feed on the luminous material, thereby bringing the bacteria into the animal's nutrient-rich gut for further growth. Alternatively, the luminous material may function as a light source in symbiotic, light organ associations. Quorum sensing is a form of regulation that has also been found in many different nonluminous bacteria, including several animal and plant pathogens. Quorum sensing in these bacteria controls activities such as the production of extracellular enzymes and expression of virulence factors for which a high population density is beneficial if the bacteria are to have a biological effect.

MiniQuiz

- What substrates and enzyme are required for an organism such as *Aliivibrio* to emit visible light?

- What is quorum sensing and how does it control bioluminescence?

17.13 Rickettsias

Key Genera: *Rickettsia, Wolbachia*

The rickettsias are small, gram-negative, coccoid or rod-shaped *Alpha-* or *Gammaproteobacteria* in the size range of 0.3–0.7 × 1–2 μm. They are, with one exception, obligate intracellular parasites and have not yet been cultivated in the absence of host cells (**Figure 17.27**). Rickettsias are the causative agents of several human diseases, including typhus, Rocky Mountain spotted fever, and Q fever (🔗 Section 34.3).

Electron micrographs of thin sections of rickettsial cells show a typical prokaryotic morphology (Figure 17.27*b*); both cell wall and cytoplasmic membrane are clearly present. The rickettsial cell wall contains peptidoglycan, and the cells divide by binary fission. The penetration of a host cell by a rickettsial cell is an active process, requiring both host and parasite to be alive and metabolically active. Once inside the host cell, the bacteria multiply primarily in the cytoplasm and continue replicating until the host cell is loaded with parasites (Figure 17.27; 🔗 Figure 34.6). The host cell then bursts and liberates the bacterial cells. Several genera of rickettsias are known, and the properties of five key genera are shown in **Table 17.13**.

Metabolism and Pathogenesis

Most rickettsias (an exception is *Coxiella burnetii*, the causative agent of the disease Q fever) possess a highly specific energy metabolism: They oxidize only the amino acids glutamate or glutamine and cannot oxidize glucose or organic acids. Rickettsias synthesize a respiratory chain complete with cytochromes and are able to carry out electron transport phosphorylation using NADH as electron donor. They are also able to synthesize at least some of the small molecules needed for macromolecular synthesis and growth, but obtain the rest of their nutrients from the host cell. Thus, although parasites, rickettsias maintain a number of independent metabolic functions.

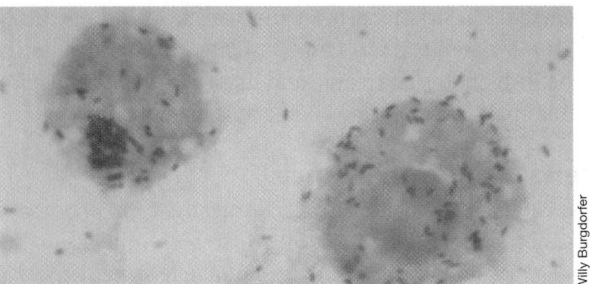

(a)

Willy Burgdorfer

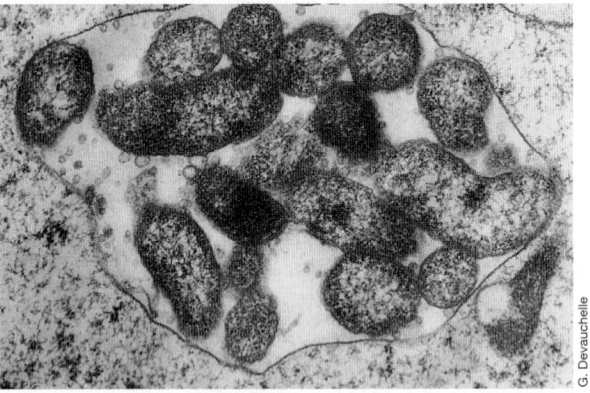

(b)

G. Devauchelle

Figure 17.27 Rickettsias growing within host cells. *(a) Rickettsia rickettsii* in tissue culture. Cells are about 0.3 μm in diameter. *(b)* Electron micrograph of cells of *Rickettsiella popilliae* within a blood cell of its host, the beetle *Melolontha melolontha*. The bacteria grow inside a vacuole within the host cell.

Rickettsias do not survive long outside their hosts, and this may explain why they must be transmitted from animal to animal by arthropod vectors. When the arthropod obtains a blood meal from an infected animal, rickettsias present in the blood are ingested and penetrate the epithelial cells of the gastrointestinal tract; there they multiply and appear later in the feces. When the arthropod feeds on an uninfected individual, it then transmits the rickettsias either directly with its mouthparts or by contaminating the bite with its feces. *C. burnetii* (🔗 Section 34.3) can also be transmitted person-to-person in infectious aerosols.

Other pathogenic rickettsias include the genera *Rochalimaea* and *Ehrlichia*. *Rochalimaea* is an atypical rickettsia because it can be grown in culture and is thus not an obligate intracellular parasite. In addition, when growing in tissue culture, cells of *Rochalimaea* grow on the outside surface of the eukaryotic host cells rather than within the cytoplasm or the nucleus as do other rickettsias. *Rochalimaea quintana* is the causative agent of *trench fever*, a disease that decimated troops in World War I. Species of the genus *Ehrlichia* cause disease in humans and other animals. Two of these diseases, ehrlichiosis in humans and Potomac fever in horses, can be quite debilitating.

Wolbachia

The genus *Wolbachia* contains species of rod-shaped *Alphaproteobacteria* that are intracellular parasites of several families of insects, a huge group that constitutes 70% of all known arthropod

Table 17.13 *Some characteristics of rickettsias*

Genus and species	Rickettsial group	Alternate host	Cellular location	Phylogenetic group[a]	DNA hybridization to R. rickettsii DNA (%)[b]
Rickettsia					
R. rickettsii	Spotted fever	Tick	Cytoplasm and nucleus	Alpha	100
R. prowazekii[c]	Typhus	Louse	Cytoplasm	Alpha	53
R. typhi	Typhus	Flea	Cytoplasm	Alpha	36
Rochalimaea					
R. quintana	Trench fever	Louse	Epicellular	Alpha	30
R. vinsonii	—	Vole	Epicellular	Alpha	30
Coxiella					
C. burnetii	Q fever	Tick	Vacuoles	Gamma	—
Ehrlichia					
E. chaffeensis	Ehrlichiosis (humans)	Tick or domestic animals	Mononuclear leukocytes	Alpha	—
E. equi	Potomac fever (horses)	Tick	Granulocyte	Alpha	—
Wolbachia[d]					
W. pipientis	—	Arthropods	Cytoplasm	Alpha	—

[a]All are *Proteobacteria*.
[b]For discussion of DNA–DNA hybridization, see Section 16.11.
[c]The genome of this organism has been sequenced and shows several similarities to the mitochondrial genome.
[d]Insect endosymbiont, not a pathogen of humans or other animals.

species (**Figure 17.28**). *Wolbachia* are phylogenetically related to the rickettsias and can have any of several effects on their insect hosts. These include inducing parthenogenesis (development of unfertilized eggs), the killing of males, and feminization (the conversion of male insects into females).

Wolbachia pipientis is the best-studied species in the genus. *W. pipientis* has a relatively small genome (about 1.5 Mbp), which is actually quite large by insect symbiont standards (⮰ Section 25.9). Cells of *W. pipientis* colonize the insect egg (Figure 17.28), where they multiply in vacuoles of host cells surrounded by a membrane of host origin. Cells of *W. pipientis* are passed from an infected female to her offspring through this egg infection. *Wolbachia*-induced parthenogenesis occurs in a number of species of wasps. In these insects, males normally arise from unfertilized eggs (which contain only one set of chromosomes), while females arise from fertilized eggs (which contain two sets of chromosomes). However, in unfertilized eggs infected with *Wolbachia*, the organism somehow triggers a doubling of the chromosome number, thus yielding only females. Predictably, if female insects are fed antibiotics that kill *Wolbachia*, parthenogenesis ceases.

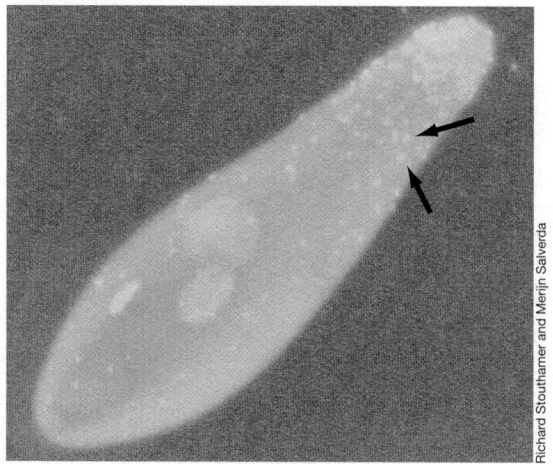

Figure 17.28 *Wolbachia.* Photomicrograph of a DAPI-stained egg of the parasitoid wasp *Trichogramma kaykai* infected with *Wolbachia pipientis*, which induces parthenogenesis. The *W. pipientis* cells are primarily located in the narrow end of the egg (arrows).

Richard Stouthamer and Merijn Salverda

MiniQuiz

- What is meant by the term "obligate intracellular parasite"? Name a disease caused by a *Rickettsia* species.
- What effects can *Wolbachia* have on its insect hosts?

IV Morphologically Unusual Proteobacteria

Some *Proteobacteria* have unusual morphologies or undergo fascinating life cycles. We consider some of these here, focusing on common curved and spiral-shaped bacteria, a few filamentous bacteria that encase their cells in a sheath, and the morphologically unusual prosthecate and stalked bacteria. In the latter group, we will focus on the important model bacterium for the study of cell differentiation, *Caulobacter*. Physiologically, all of these organisms are chemoorganotrophs, consuming organic matter in a wide variety of primarily aquatic habitats.

17.14 Spirilla

Key Genera: *Spirillum, Magnetospirillum, Bdellovibrio*

The **spirilla** are gram-negative, motile, spiral-shaped bacteria that show a wide variety of physiological attributes. Some key taxonomic criteria used are cell shape, size, and number of flagella, relation to oxygen (obligately aerobic, microaerophilic), relationship to higher organisms, and certain other physiological characteristics, such as nitrogen fixation and halophilism. The genera to be covered here are given in **Table 17.14**, where it can be seen that spirilla are found in each of the five classes of *Proteobacteria* (the genera *Campylobacter* and *Helicobacter* are *Epsilonproteobacteria* and covered in Section 17.19).

Spirillum, Aquaspirillum, Oceanospirillum, and Azospirillum

The spirilla, which are helically curved rods, are motile by means of polar flagella, usually tufts at both poles (**Figure 17.29**). The number of turns in the helix may vary from less than one complete turn (in which case the organism looks like a vibrio) to many turns. Spirilla with many turns can superficially resemble spirochetes (Section 18.16) but differ distinctly from the latter phylogenetically. In addition, spirilla do not have the outer sheath and endoflagella of spirochetes, but instead contain typical bacterial flagella (Section 3.13).

The cells of some spirilla are very large and are easily observed. For example, *Spirillum volutans*, a large spirillum, is common in aquatic environments and is microaerophilic, requiring O_2 but inhibited by O_2 at normal atmospheric levels. Another prominent characteristic of cells of *S. volutans* is the formation of intracellular inclusions called *volutin granules*, consisting of polyphosphate (Figure 17.29a; Section 3.10). *Azospirillum lipoferum* is a nitrogen-fixing organism and of considerable interest because it enters into a symbiotic relationship with tropical grasses and grain crops such as corn. Although not the intimate association that forms between root nodule bacteria of the genus *Rhizobium* and leguminous plants (Section 25.3), the *A. lipoferum*–corn association clearly benefits the corn plant by fixed nitrogen supplied by nitrogen fixation.

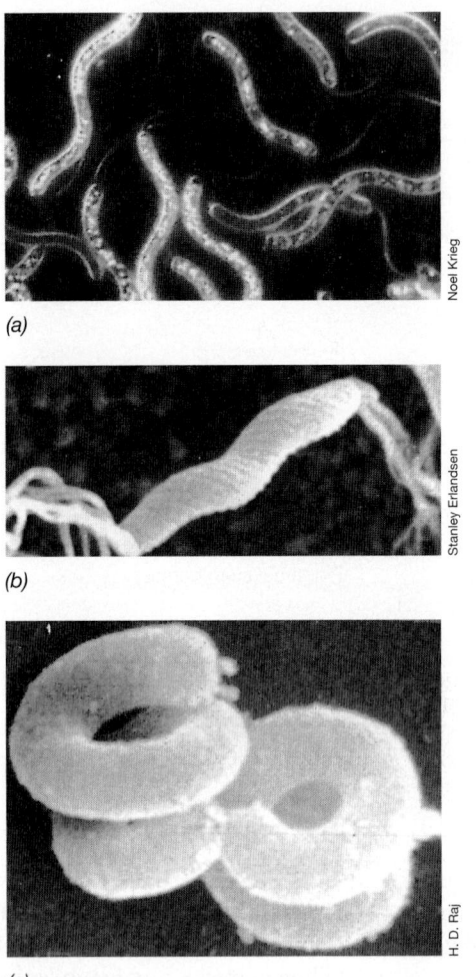

(a)

(b)

(c)

Figure 17.29 Spirilla. *(a) Spirillum volutans*, visualized by dark-field microscopy, showing flagellar bundles and volutin (polyphosphate) granules. Cells are about 1.5×25 μm. *(b)* Scanning electron micrograph of an intestinal spirillum. Note the polar flagellar tufts and the spiral structure of the cell surface. *(c)* Scanning electron micrograph of cells of *Ancylobacter aquaticus*. Cells are about 0.5 μm in diameter.

Table 17.14 *Characteristics of some major genera of spiral-shaped bacteria*[a]

Genus	Phylogenetic group[b]	Characteristics
Spirillum	*Beta*	Cell diameter 1.7 μm; microaerophilic; freshwater
Aquaspirillum	*Beta*	Cell diameter 0.2–1.5 μm; aerobic; freshwater
Magnetospirillum	*Alpha*	Vibrio to spirillum-shaped; cell diameter about 0.3 μm; contains magnetosomes; microaerophilic
Oceanospirillum	*Gamma*	Cell diameter 0.3–1.2 μm; aerobic; marine (require 3% NaCl)
Azospirillum	*Alpha*	Cell diameter 1 μm; microaerophilic; soil and rhizosphere; fixes N_2
Herbaspirillum	*Beta*	Cell diameter 0.6–0.7 μm; microaerophilic; soil and rhizosphere; fixes N_2
Bdellovibrio	*Delta*	Cell diameter 0.25–0.4 μm; aerobic; predatory on other bacteria; single polar sheathed flagellum
Ancylobacter	*Alpha*	Cell diameter 0.5 μm; curved rods forming rings; nonmotile, aerobic; sometimes gas vesiculate

[a]All are gram-negative and respiratory, but never fermentative.
[b]All are *Proteobacteria*. The spirilla *Campylobacter* and *Helicobacter* are covered in Section 17.19.

The small-diameter spirilla are fully aerobic and have been separated into two genera, *Aquaspirillum* and *Oceanospirillum*. The former includes freshwater species and the latter includes species that inhabit seawater and require sodium chloride (NaCl) for growth (Table 17.14). Numerous species of *Aquaspirillum* and *Oceanospirillum* have been described, and the various species are separated on physiological and phylogenetic grounds. These organisms undoubtedly play an important role in the recycling of organic matter in oxic aquatic environments.

Magnetotactic Spirilla

Highly motile, microaerophilic, magnetic spirilla have been isolated from freshwater habitats. These organisms demonstrate a dramatic directed movement in a magnetic field called *magnetotaxis*. The spirillum *Magnetospirillum magnetotacticum* (**Figure 17.30** and Table 17.14) is a major organism in this group. In an artificial magnetic field, magnetic spirilla quickly orient their long axis along the north–south magnetic moment of the field. Within the cells are chains of magnetic particles called *magnetosomes*, consisting of the iron minerals magnetite (Fe_3O_4) and greigite (Fe_3S_4) (↺ Section 3.10 and Figure 3.28).

Magnetic bacteria display one of two magnetic polarities depending on the orientation of magnetosomes within the cell. Cells in the Northern Hemisphere have the north-seeking pole of their magnetosomes forward with respect to their flagella and thus move in a northward direction. Cells in the Southern Hemisphere have the opposite polarity and move southward. Although the ecological role of bacterial magnets is unclear, the ability to orient in a magnetic field may be of selective advantage in maintaining these microaerophilic organisms in zones of low O_2 concentration near the oxic–anoxic interface.

Bdellovibrio

Bdellovibrio is a genus of small, highly motile and curved bacteria that prey on other bacteria, using the cytoplasmic constituents of their hosts as nutrients (*bdello* is a prefix meaning "leech"). After attachment of a *Bdellovibrio* cell to its prey, the predator penetrates the cell wall of the prey and replicates in the periplasmic space, eventually forming a spherical structure called a *bdelloplast*. Two stages of penetration are shown in electron micrographs in **Figure 17.31** and diagrammatically in **Figure 17.32**. A wide variety of gram-negative prey bacteria can be attacked by *Bdellovibrio*, but gram-positive cells are not attacked.

Bdellovibrio is an obligate aerobe, obtaining its energy from the oxidation of amino acids and acetate. In addition, *Bdellovibrio* assimilates nucleotides, fatty acids, peptides, and even some intact proteins directly from its host without first breaking them down. Prey-independent derivatives of predatory strains of *Bdellovibrio* can be isolated and grown on complex media, however, showing that predation is not obligatory. Phylogenetically, bdellovibrios are species of *Deltaproteobacteria* (Figure 17.1) and are widespread in soil and aquatic habitats. Procedures for their isolation are similar to those used to isolate bacterial viruses (↺ Section 9.4). Prey bacteria are spread on the surface of an agar plate to form a lawn and the surface is inoculated with a small amount of soil suspension that has been filtered through a membrane filter; the filter retains most bacteria, but allows the small *Bdellovibrio* cells to pass. On incubation of the agar plate, plaques analogous to bacteriophage plaques (↺ Figure 9.6) are formed at locations where *Bdellovibrio* cells are growing. Pure cultures of *Bdellovibrio* can then be isolated from

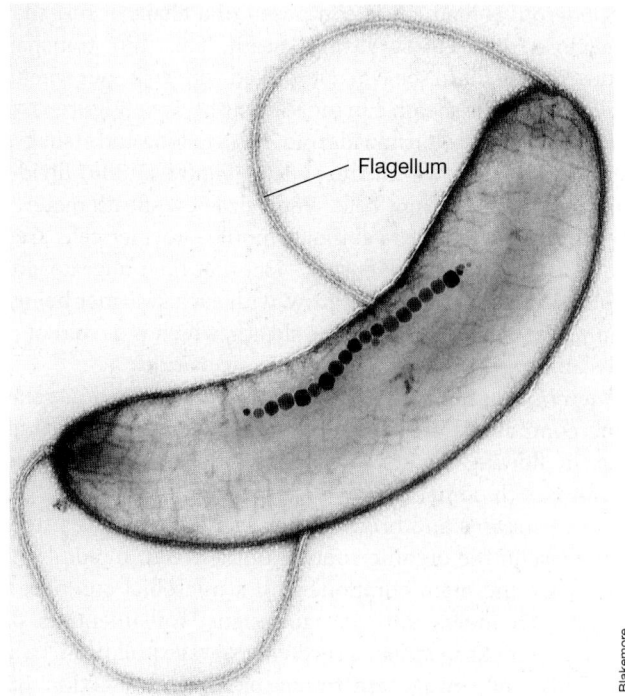

R. Blakemore

Figure 17.30 A magnetotactic spirillum. Electron micrograph of a single cell of *Magnetospirillum magnetotacticum*; a cell measures 0.3 × 2 μm. The cell contains particles of magnetosomes made of Fe_3O_4 arranged in a chain.

Flagellum

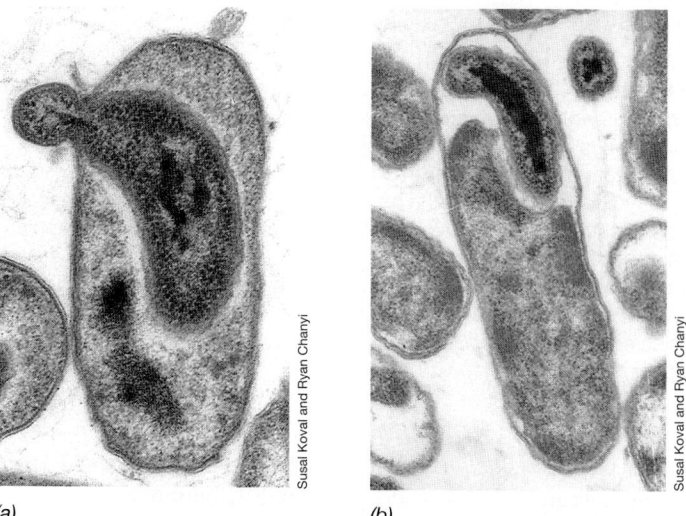

Susal Koval and Ryan Chanyi

(a)   *(b)*

Figure 17.31 Attack on a prey cell by *Bdellovibrio*. Thin-section electron micrographs of *Bdellovibrio* attacking a cell of *Delftia* (formerly *Comamonas*) *acidovorans*. *(a)* Entry of the predator cell. *(b)* *Bdellovibrio* cell inside the host. The *Bdellovibrio* cell is enclosed in the bdelloplast and replicates in the periplasmic space. A *Bdellovibrio* cell measures about 0.3 μm in diameter.

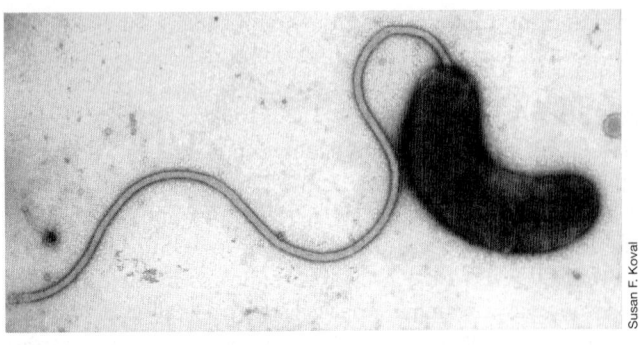

(a)

Susan F. Koval

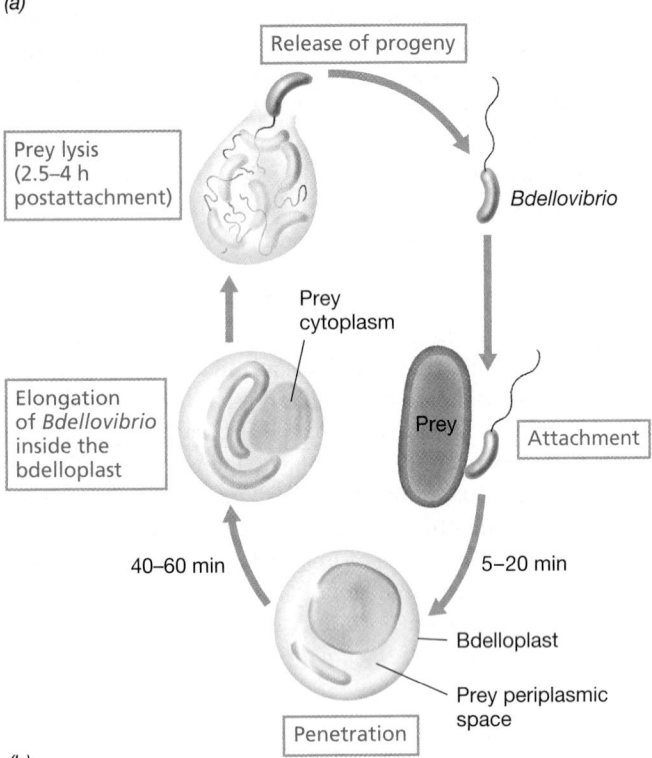

(b)

Figure 17.32   Developmental cycle of the bacterial predator *Bdellovibrio bacteriovorus.* *(a)* Electron micrograph of a cell of *Bdellovibrio bacteriovorus.* Note the very thick flagellum (compare with ♻ Figure 3.39). *(b)* Events in predation. Following primary contact with a gram-negative bacterium, the highly motile *Bdellovibrio* cell attaches to and penetrates into the prey periplasmic space. Once inside the periplasmic space, *Bdellovibrio* cells elongate and within 4 h progeny cells are released. The number of progeny cells released varies with the size of the prey bacterium. For example, 5–6 bdellovibrios are released from each infected *Escherichia coli* cell and 20–30 for a larger cell, such as a species of *Aquaspirillum.*

these plaques. *Bdellovibrio* cultures have been obtained from many soils and are thus widespread in distribution.

Ancylobacter

Species of *Ancylobacter* are ring-shaped, nonmotile, extremely diverse nutritionally, aerobic and chemoorganotrophic bacteria (Figure 17.29c). They resemble very tightly coiled vibrios and are widely distributed in aquatic environments. A phototrophic counterpart to *Ancylobacter* is the purple nonsulfur bacterium

Rhodocyclus purpureus, considered earlier (Section 17.2 and Figure 17.7e).

MiniQuiz
• What is a volutin granule?
• What is unique about the spirilla *Bdellovibrio* and *Magnetospirillum*?

17.15   Sheathed *Proteobacteria*: *Sphaerotilus* and *Leptothrix*

Key Genera: *Sphaerotilus, Leptothrix*

Sheathed bacteria are filamentous *Betaproteobacteria* (Figure 17.1) with a unique life cycle in which flagellated swarmer cells form within a long tube or sheath. Under unfavorable growth conditions, the swarmer cells move out and become dispersed to new environments, leaving behind the empty sheath. Under favorable conditions, the cells grow vegetatively within the sheath, leading to the formation of long, cell-packed sheaths.

Sheathed bacteria are common in freshwater habitats that are rich in organic matter, such as wastewaters and polluted streams. Because they are typically found in flowing waters, they are also abundant in trickling filters and activated sludge digesters in sewage treatment plants (♻ Section 35.2). In habitats in which reduced iron (Fe^{2+}) or manganese (Mn^{2+}) is present, the sheaths may become coated with ferric hydroxide $Fe(OH)_3$ or various manganese oxides from the oxidation of these metals.

Sphaerotilus

The *Sphaerotilus* filament is composed of a chain of rod-shaped cells enclosed in a closely fitting sheath. This thin, transparent structure is difficult to see when it is filled with cells, but when it is partially empty, the sheath can more easily be seen (**Figure 17.33a**). Individual cells are 1–2 μm wide and 3–8 μm long and stain gram-negatively. The cells within the sheath (Figure 17.33b) divide by binary fission, and the new cells synthesize new sheath material at the tips of the filaments. Eventually, motile swarmer cells are liberated from the sheaths (Figure 17.33c) that then migrate, attach to a solid surface, and begin to grow, with each swarmer being the forerunner of a new filament. The sheath, which is devoid of peptidoglycan, consists of protein and polysaccharide.

Sphaerotilus cultures are nutritionally versatile and use simple organic compounds as carbon and energy sources. Befitting its habitat in flowing waters, *Sphaerotilus* is an obligate aerobe. Large masses (blooms) of *Sphaerotilus* often occur in the fall of the year in streams and brooks when leaf litter causes a temporary increase in the organic content of the water. In addition, its filaments are the main component of a microbial complex that wastewater engineers call "sewage fungus," a filamentous slime found on the rocks in streams receiving sewage pollution. In activated sludge of sewage treatment plants (♻ Section 35.2), *Sphaerotilus* is often responsible for a condition called *bulking,* where the tangled masses of *Sphaerotilus* filaments so increase the bulk of the sludge that it remains suspended and does not settle as it should. This has a negative effect on the oxidation of

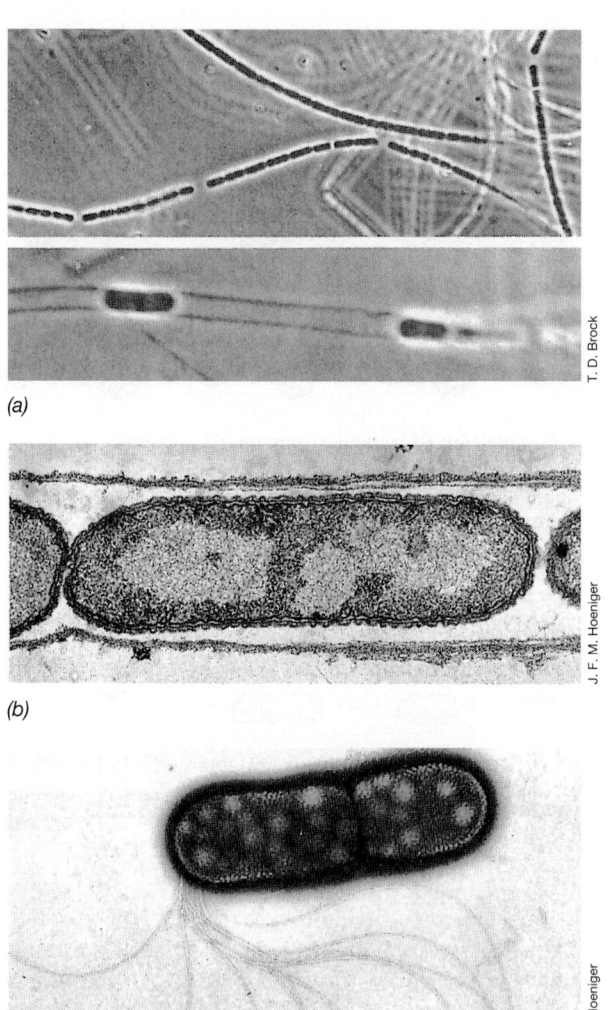

(a)

(b)

(c)

Figure 17.33 *Sphaerotilus natans*. A single cell is about 2 μm wide. *(a)* Phase-contrast photomicrographs of material collected from a polluted stream. Active growth stage (above) and swarmer cells leaving the sheath. *(b)* Electron micrograph of a thin section through a filament, clearly showing the sheath. *(c)* Electron micrograph of a negatively stained swarmer cell. Notice the polar flagellar tuft.

organic matter and the recycling of inorganic nutrients and leads to treatment plant discharges with high nitrogen and carbon loads.

Leptothrix

The ability of *Sphaerotilus* and *Leptothrix* to precipitate iron oxides on their sheaths is well established, and when sheaths become iron encrusted, as occurs in iron-rich waters, they can frequently be seen microscopically (**Figure 17.34**). Iron precipitates when ferrous iron (Fe^{2+}), chelated to organic materials such as humic or tannic acids, is oxidized to iron oxides. Fe^{2+} binds to the sheath, and the organic constituents are taken up and used as a carbon or energy source. Iron oxidation is fortuitous and the organism does not use Fe^{2+} as an electron donor in energy metabolism as many iron-oxidizing bacteria do (⮌ Section 13.9).

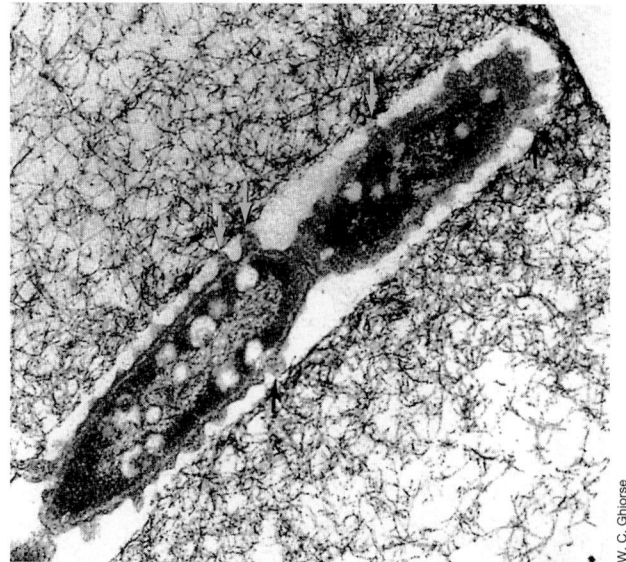

Figure 17.34 *Leptothrix* and iron precipitation. Transmission electron micrograph of a thin section of *Leptothrix* in a sample from a ferromanganese film in a swamp in Ithaca, New York. A single cell measures about 0.9 μm in diameter. Note the protuberances of the cell envelope that contact the sheath (arrows).

Besides Fe^{2+} oxidation, *Leptothrix* can also oxidize manganese, typically Mn^{2+}-containing minerals to Mn^{4+}-containing minerals, but as for Fe^{2+}, Mn^{2+} oxidation does not yield energy for the organism.

MiniQuiz

- Describe how a sheathed bacterium such as *Sphaerotilus* grows.
- List two metals that are oxidized by sheathed bacteria.

17.16 Budding and Prosthecate/ Stalked Bacteria

Key Genera: *Hyphomicrobium, Caulobacter*

This large and heterogeneous group of primarily *Alphaproteobacteria* contains organisms that form various kinds of cytoplasmic extrusions: *stalks, hyphae,* or *appendages* (**Table 17.15**, page 534). Extrusions of these kinds, which are smaller in diameter than the mature cell and contain cytoplasm and a cell wall, are collectively called **prosthecae** (**Figure 17.35**).

Budding Division

Budding bacteria divide as a result of unequal cell growth. In contrast to binary fission that forms two equivalent cells (⮌ Figure 5.1), cell division in stalked and budding bacteria forms a totally new daughter cell, with the mother cell retaining its original identity (**Figure 17.36**). A fundamental difference between these bacteria and bacteria that divide by binary fission is the formation of new cell wall material from a single point (polar growth) rather than throughout the whole cell (intercalary growth) as in binary fission (⮌ Sections 5.1–5.4). Several genera not

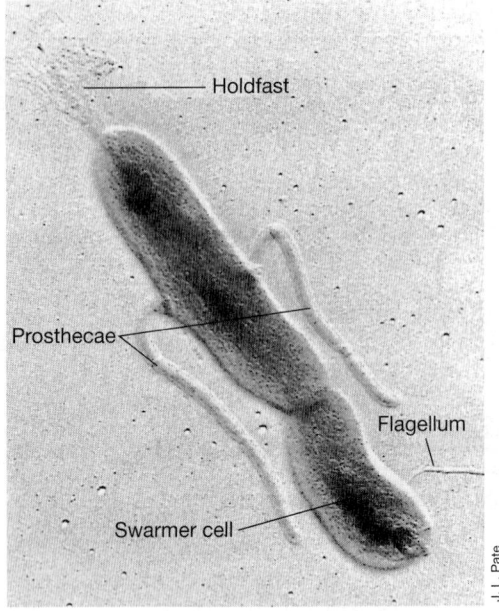

Holdfast

Prosthecae

Flagellum

Swarmer cell

J. L. Pate

(a)

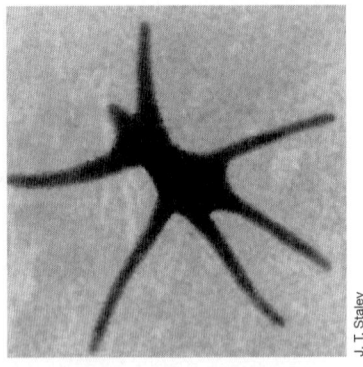

J. T. Staley

(b)

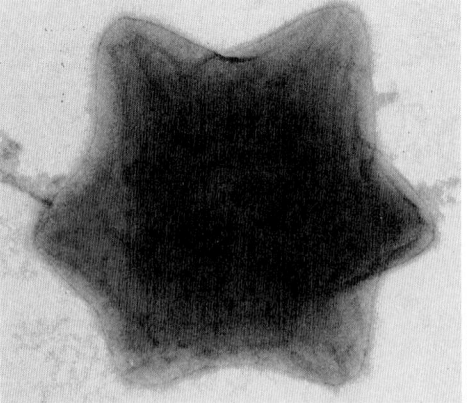

H. Schlesner

(c)

Figure 17.35 Prosthecate bacteria. *(a)* Electron micrograph of a shadow-cast preparation of *Asticcacaulis biprosthecum*, illustrating the location and arrangement of the prosthecae, the holdfast, and swarmer cell. The swarmer cell breaks away from the mother cell and begins a new cell cycle. Cells are about 0.6 μm wide. *(b)* Negatively stained electron micrograph of a cell of *Ancalomicrobium adetum*. The prosthecae are bounded by the cell wall, contain cytoplasm, and are about 0.2 μm in diameter. *(c)* Electron micrograph of the star-shaped bacterium *Stella*. Cells are about 0.8 μm in diameter.

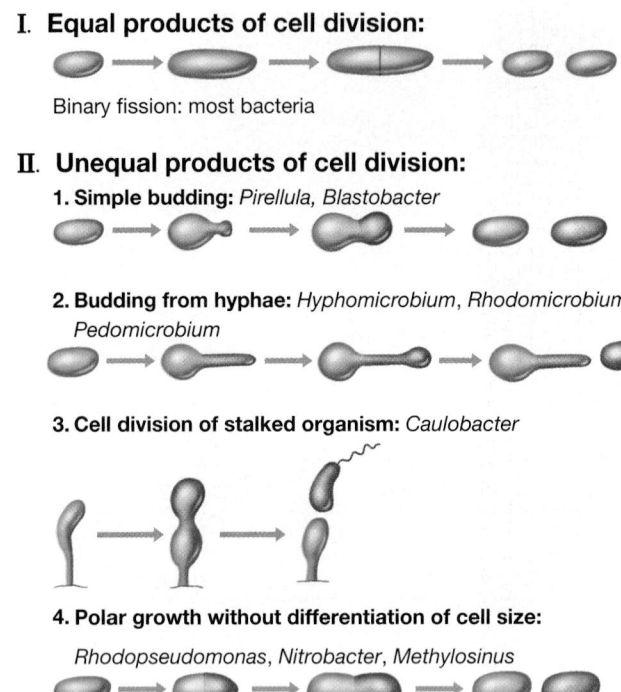

I. Equal products of cell division:

Binary fission: most bacteria

II. Unequal products of cell division:

1. Simple budding: *Pirellula, Blastobacter*

2. Budding from hyphae: *Hyphomicrobium, Rhodomicrobium, Pedomicrobium*

3. Cell division of stalked organism: *Caulobacter*

4. Polar growth without differentiation of cell size:

Rhodopseudomonas, Nitrobacter, Methylosinus

Figure 17.36 Cell division in different bacteria. Contrast between cell division in conventional bacteria and in various budding and stalked bacteria.

normally considered to be budding bacteria show polar growth without differentiation of cell size (Figure 17.36). An important consequence of polar growth is that internal structures, such as membrane complexes, are not partitioned in the cell division process and must be formed *de novo*. However, this has an advantage in that more complex internal structures can be formed in budding cells than in cells that divide by binary fission, since the latter cells would have to partition these structures between the two newly forming daughter cells. Not coincidentally, many budding bacteria, particularly phototrophic and chemolithotrophic species, contain extensive internal membrane systems.

Budding Bacteria: *Hyphomicrobium*

Two well-studied budding bacteria are closely related phylogenetically: *Hyphomicrobium*, which is chemoorganotrophic, and *Rhodomicrobium*, which is phototrophic. These organisms release buds from the ends of long, thin hyphae. The hypha is a direct cellular extension and contains cell wall, cytoplasmic membrane, and ribosomes, and can contain DNA.

Figure 17.37 shows the life cycle of *Hyphomicrobium*. The mother cell, which is often attached by its base to a solid substrate, forms a thin outgrowth that lengthens to become a hypha. At the end of the hypha, a bud forms. This bud enlarges, forms a flagellum, breaks loose from the mother cell, and swims away. Later, the daughter cell loses its flagellum and after a period of maturation forms a hypha and buds. More buds can also form at the hyphal tip of the mother cell, leading to arrays of cells connected by hyphae. In some cases, a bud begins to form directly

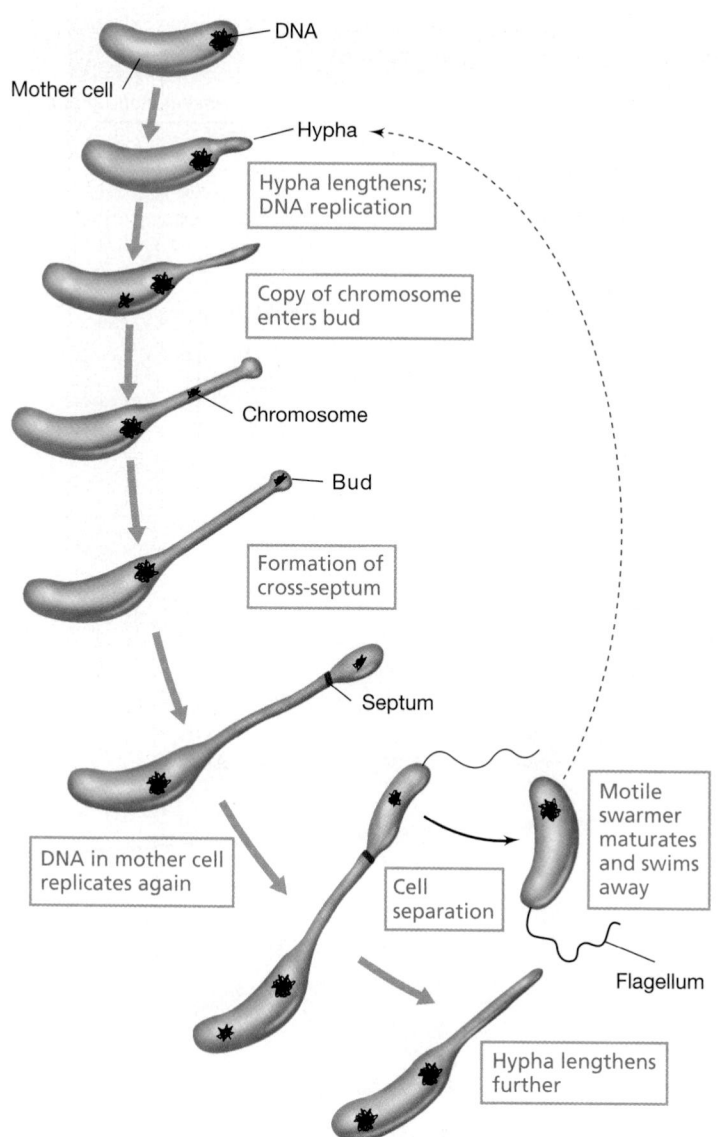

Figure 17.37 Stages in the *Hyphomicrobium* cell cycle. The single chromosome of *Hyphomicrobium* is circular.

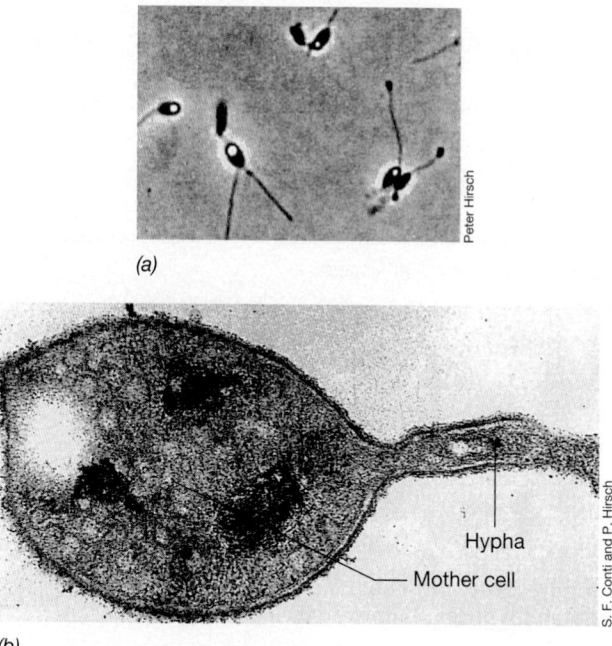

Figure 17.38 Morphology of *Hyphomicrobium*. *(a)* Phase-contrast micrograph of cells of *Hyphomicrobium*. Cells are about 0.7 μm wide. *(b)* Electron micrograph of a thin section of a single *Hyphomicrobium* cell. The hypha is about 0.2 μm wide.

from the mother cell without the intervening formation of a hypha, whereas in other cases a single cell forms hyphae from each end (**Figure 17.38**). Nucleoid replication events occur before the bud emerges, and then once a bud has formed, a copy of the chromosome moves down the hypha and into the bud. A cross-septum then forms, separating the still-developing bud from the hypha and mother cell (Figure 17.37).

Physiologically, *Hyphomicrobium* is a methylotrophic bacterium (Section 17.6), and it is widespread in freshwater, marine, and terrestrial habitats. Preferred carbon sources are C_1 compounds such as methanol (CH_3OH), methylamine (CH_3NH_2), formaldehyde (CH_2O), and formate ($HCOO^-$). A fairly specific enrichment procedure for *Hyphomicrobium* uses CH_3OH as electron donor with nitrate (NO_3^-) as electron acceptor in a dilute medium incubated under anoxic conditions. The only rapidly growing denitrifying bacterium known that uses CH_3OH as electron donor is *Hyphomicrobium* and so

this procedure can select this organism out of a wide variety of environments.

Prosthecate and Stalked Bacteria

Prosthecate (Figure 17.35) and stalked (**Figure 17.39**) bacteria are appendaged organisms that attach to particulate matter, plant material, or other microorganisms in aquatic habitats. Although a major function of these appendages is attachment, they also significantly increase the surface-to-volume ratio of the cells (prosthecae have large surface areas but almost no volume). Recall that the high surface-to-volume ratio of prokaryotic cells in general confers an increased ability to take up nutrients and expel wastes (Section 3.2). The unusual morphology of appendaged bacteria (Figure 17.35) carries this theme to an extreme, and may be an evolutionary adaptation to life in oligotrophic (nutrient-poor) waters where these organisms are most commonly found. Prosthecae may also function to reduce cell sinking. Because these organisms are typically strict aerobes, prosthecae may keep cells from sinking into anoxic zones in their aquatic environments where they would be unable to respire.

Caulobacter and *Gallionella*

Two common stalked bacteria are *Caulobacter* (Figure 17.39) and *Gallionella* (see Figure 17.41). The former is a chemoorganotroph that produces a cytoplasm-filled stalk, that is, a prostheca, while the latter is a chemolithotrophic iron-oxidizing bacterium whose stalk is composed of ferric hydroxide [$Fe(OH)_3$]. *Caulobacter* cells are often seen on surfaces in aquatic environments with the stalks of several cells attached to form

UNIT 6

Table 17.15 *Characteristics of major genera of stalked, appendaged (prosthecate), and budding bacteria*

Characteristics	Genus	Phylogenetic group[a]
Stalked bacteria		
Stalk an extension of the cytoplasm and involved in cell division	Caulobacter	Alpha
Stalked, fusiform-shaped cells	Prosthecobacter	Verrucomicrobiaceae[b]
Stalked, but stalk is an excretory product not containing cytoplasm:		
Stalk depositing iron, cell vibrioid	Gallionella	Beta
Laterally excreted gelatinous stalk not depositing iron	Nevskia	Gamma
Appendaged (prosthecate) bacteria		
Single or double prosthecae	Asticcacaulis	Alpha
Multiple prosthecae:		
Short prosthecae, multiply by fission, some with gas vesicles	Prosthecomicrobium	Alpha
Flat, star-shaped cells, some with gas vesicles	Stella	Alpha
Long prosthecae, multiply by budding, some with gas vesicles	Ancalomicrobium	Alpha
Budding bacteria		
Phototrophic, produce hyphae	Rhodomicrobium	Alpha
Phototrophic, budding without hyphae	Rhodopseudomonas	Alpha
Chemoorganotrophic, rod-shaped cells	Blastobacter	Alpha
Chemoorganotrophic, buds on tips of slender hyphae:		
Single hyphae from parent cell	Hyphomicrobium	Alpha
Multiple hyphae from parent cell	Pedomicrobium	Alpha

[a]All but *Prosthecobacter* are *Proteobacteria*.
[b]See Section 18.11.

rosettes (Figure 17.39*a*). At the end of the stalk is a structure called a *holdfast* by which the stalk anchors the cell to a surface.

The *Caulobacter* cell division cycle (**Figure 17.40**; ⟳ Figure 8.23) is unique because cells undergo unequal binary fission. A stalked cell of *Caulobacter* divides by elongation of the cell followed by binary fission, and a single flagellum forms at the pole opposite the stalk. The flagellated cell so formed, called a *swarmer*, separates from the nonflagellated mother cell and eventually attaches to a new surface, forming a new stalk at the flagellated pole; the flagellum is then lost. Stalk formation is a necessary precursor of cell division and is coordinated with DNA synthesis (Figure 17.40). The cell division cycle in *Caulobacter* is thus more complex than simple binary fission or budding division because the stalked and swarmer cells are structurally different and the growth cycle must include both forms.

Gallionella forms a twisted stalklike structure containing $Fe(OH)_3$ from the oxidation of ferrous iron (Fe^{2+}) (**Figure 17.41**). However, the stalk of *Gallionella* is not an integral part of the cell but is simply excreted from the cell surface. It contains an organic matrix on which the $Fe(OH)_3$ accumulates. *Gallionella* is common in the waters draining bogs, iron springs, and other habitats

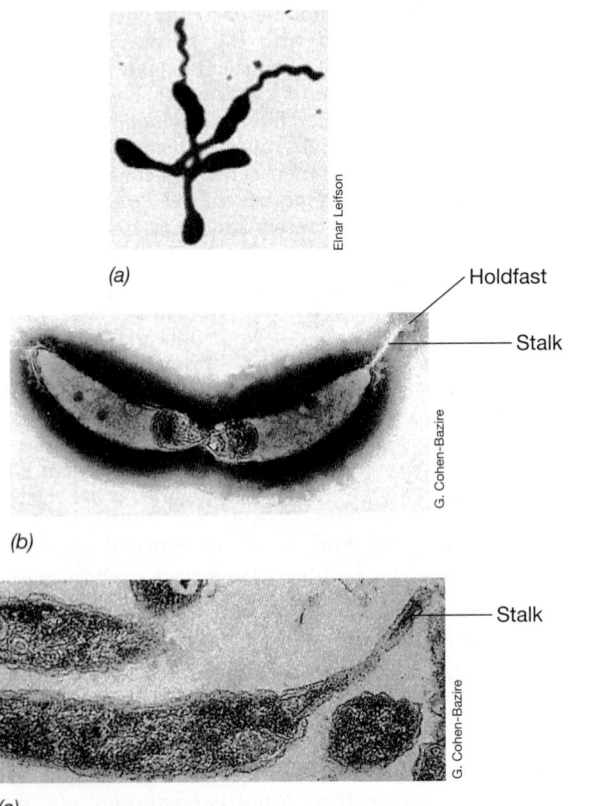

(a)

Holdfast

Stalk

(b)

Stalk

(c)

Figure 17.39 Stalked bacteria. *(a)* A *Caulobacter* rosette. A single cell is about 0.5 μm wide. The five cells are attached by their stalks, which are also prosthecae. Two of the cells have divided, and the daughter cells have formed flagella. *(b)* Negatively stained preparation of a *Caulobacter* cell in division. *(c)* A thin section of *Caulobacter* showing that cytoplasm is present in the stalk. Parts b and c are electron micrographs.

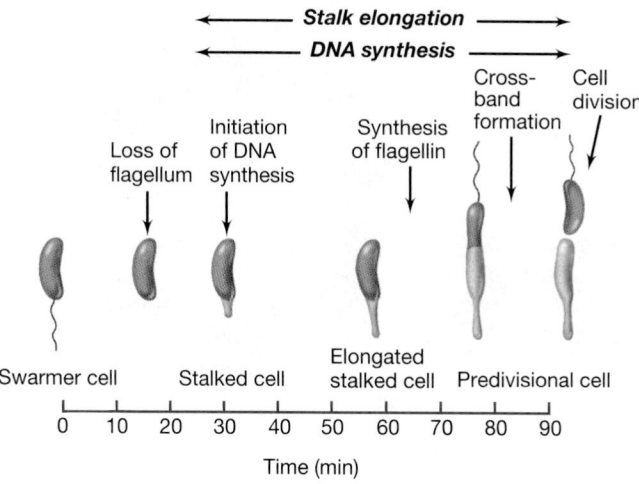

Figure 17.40 Growth of *Caulobacter*. Stages in the *Caulobacter* cell cycle, beginning with a swarmer cell. Compare with Figure 8.23.

where ferrous iron (Fe^{2+}) is present, usually in association with sheathed bacteria such as *Sphaerotilus*. *Gallionella* is an autotrophic chemolithotroph containing enzymes of the Calvin cycle ($\circlearrowleft$ Section 13.12) by which CO_2 is incorporated into cell material with Fe^{2+} as electron donor.

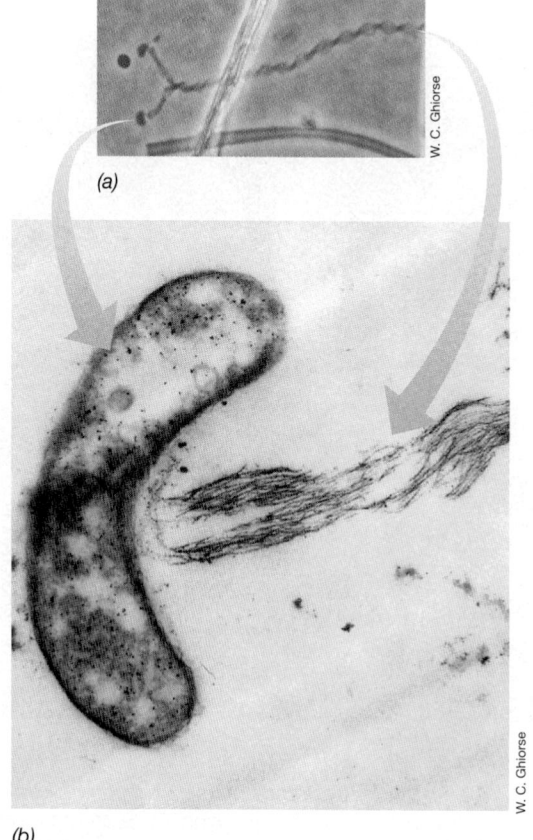

Figure 17.41 The neutrophilic ferrous iron oxidizer, *Gallionella ferruginea*, from an iron seep near Ithaca, New York. *(a)* Photomicrograph of two bean-shaped cells with stalks that combine to form one twisted mass. *(b)* Transmission electron micrograph of a thin section of a *Gallionella* cell with stalk. Cells are about 0.6 μm wide.

 Delta- and Epsilonproteobacteria

We round out the *Proteobacteria* with a consideration of the *Delta-* and *Epsilonproteobacteria*. As is typical of *Proteobacteria* in general, an assortment of metabolic patterns exists in species of these two proteobacterial classes, and several other phenotypic properties are unique to each group. We begin with the myxobacteria, socially active prokaryotes that interact to form macroscopically visible masses of cells.

17.17 Myxobacteria

Key Genera: *Myxococcus, Stigmatella*

Some bacteria exhibit a form of motility called *gliding* ($\circlearrowleft$ Section 3.14). Gliding bacteria, typically either long rods or filaments in morphology, lack flagella but can move when in contact with surfaces. One group of gliding bacteria, the myxobacteria, form multicellular structures called *fruiting bodies* and show life cycles involving intercellular communication. The fruiting myxobacteria are classified on morphological grounds using characteristics of the vegetative cells, the myxospores, and fruiting body structure (**Table 17.16**), and on phylogenetic grounds,

Table 17.16 *Classification of the fruiting myxobacteria[a]*

Characteristics	Genus
Vegetative cells tapered	
Spherical or oval myxospores, fruiting bodies usually soft and slimy without well-defined sporangia or stalks	*Myxococcus*
Rod-shaped myxospores:	*Archangium*
Myxospores not contained in sporangia, fruiting bodies without stalks	
Myxospores embedded in slime envelope:	
Fruiting bodies without stalks	*Cystobacter*
Stalked fruiting bodies, single sporangia	*Melittangium*
Stalked fruiting bodies, multiple sporangia	*Stigmatella*
Fruiting bodies are dark-brown clusters consisting of tiny spherical or disclike sporangia with an outer wall	*Angiococcus*
Vegetative cells not tapered (blunt, rounded ends); myxospores resemble vegetative cells; sporangia always produced	
Fruiting bodies without stalks; myxospores rod-shaped	*Polyangium*
Fruiting bodies without stalks; myxospores oval; highly cellulolytic	*Sorangium*
Fruiting bodies without stalks; myxospores coccoid	*Nannocystis*
Stalked fruiting bodies	*Chondromyces*

[a]Phylogenetically, all known myxobacteria are *Deltaproteobacteria*.

UNIT 6

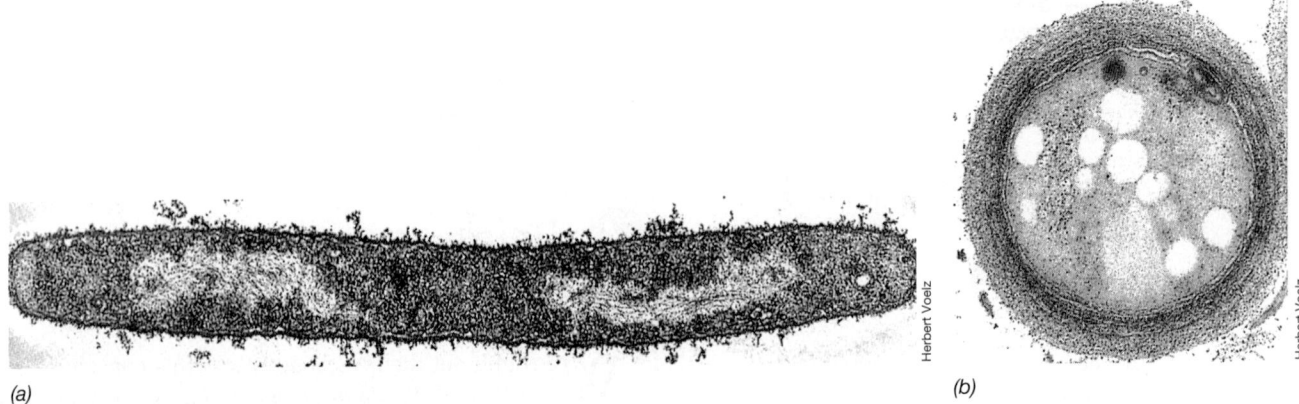

(a)   (b)

Herbert Voelz

Herbert Voelz

Figure 17.42 *Myxococcus.* *(a)* Electron micrograph of a thin section of a vegetative cell of *Myxococcus xanthus*. A cell measures about 0.75 μm wide. *(b)* Myxospore of *M. xanthus*, showing the multilayered outer wall. Myxospores measure about 2 μm in diameter.

using 16S rRNA gene sequence analyses to differentiate these different species of *Deltaproteobacteria*.

Fruiting Bodies

The fruiting myxobacteria exhibit the most complex behavioral patterns and life cycles of all known bacteria. To encode this complexity, the chromosome of some myxobacteria is very large. *Myxococcus xanthus*, for example, has a single circular chromosome of 9.2 megabase pairs, twice the size of the *Escherichia coli* chromosome (Table 12.1). The vegetative cells of the fruiting myxobacteria are simple, nonflagellated, gram-negative rods (**Figure 17.42**) that glide across surfaces and obtain their nutrients primarily by lysing other bacteria and utilizing the released nutrients. Under appropriate conditions, a swarm of vegetative cells aggregate and form fruiting bodies, within which some of the cells become converted to resting structures called *myxospores* (Figure 17.42*b*).

The fruiting bodies of the myxobacteria vary from simply masses of myxospores embedded in slime to complex forms with a fruiting body wall and a stalk (**Figure 17.43**). The fruiting bodies are often strikingly colored and morphologically elaborate (**Figure 17.44**), and can often be seen with the aid of a hand lens forming on moist pieces of decaying wood or plant material. Fruiting bodies of myxobacteria often develop on dung pellets (for example, rabbit pellets) after the pellets have been incubated for a few days in a moist chamber, and by using an inoculating loop or needle, one can transfer cells from a fruiting body to a plate for isolation. Many myxobacteria can be grown in the laboratory on complex media containing peptone or casein hydrolysate, which provides amino acids or small peptides as nutrients, and most are obligate aerobes.

Life Cycle of a Fruiting Myxobacterium

The life cycle of a typical fruiting myxobacterium is shown in **Figure 17.45**. A vegetative cell excretes slime, and as it moves across a solid surface, it leaves behind a slime trail (**Figure 17.46**). This slime trail is then used by other cells, such that a characteristic radiating pattern soon emerges with cells migrating along

(a)

Hans Reichenbach

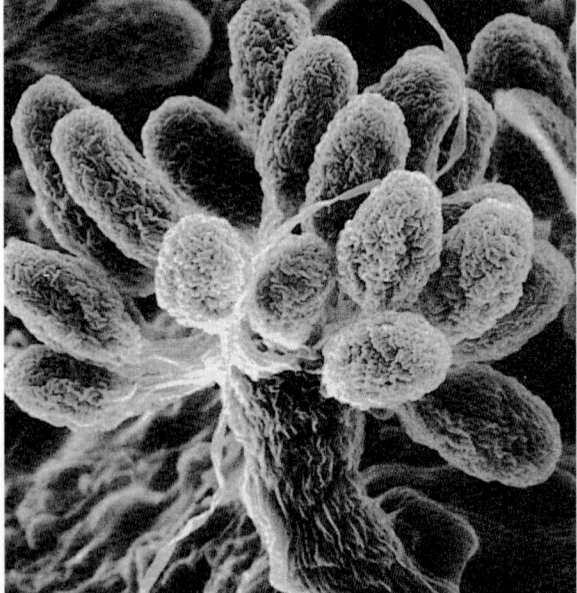

(b)

David White

Figure 17.43 *Stigmatella aurantiaca.* (a) Color photo of a single fruiting body. The structure is about 150 μm high. (b) Scanning electron micrograph of a fruiting body growing on a piece of wood. Note the individual cells visible in each fruit. The color of the fruiting body shown in part a is due to the production of structurally complex glucosylated carotenoid pigments.

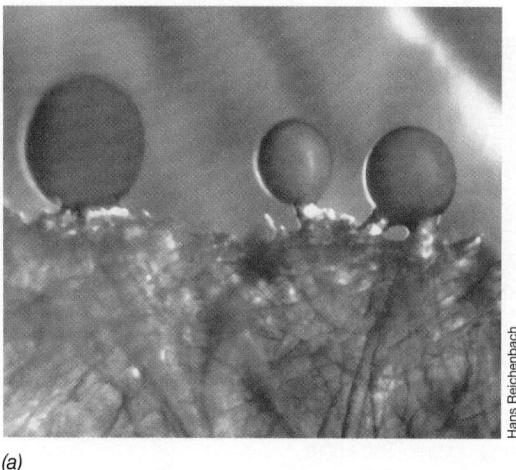
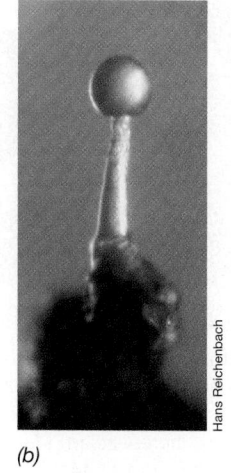
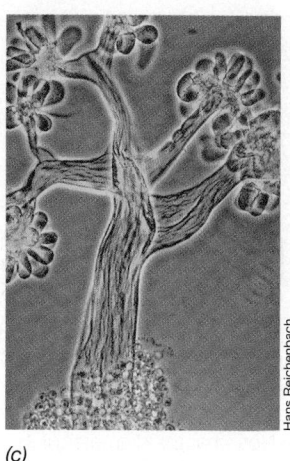

(a)   (b)   (c)

Figure 17.44 Fruiting bodies of three species of fruiting myxobacteria. *(a) Myxococcus fulvus* (125 μm high). *(b) Myxococcus stipitatus* (170 μm high). *(c) Chondromyces crocatus* (560 μm high).

established slime trails (Figure 17.46). The fruiting body ultimately formed (Figures 17.43 and 17.44) is a complex structure produced by cells that synthesize a stalk and those in the head that form myxospores.

Fruiting bodies do not form if adequate nutrients for vegetative growth are present, but upon nutrient exhaustion, the vegetative swarms begin to fruit. Cells aggregate, likely through chemotactic or quorum sensing responses (Sections 8.8 and 8.9), with the cells migrating toward each other and forming

mounds or heaps (**Figure 17.47**); a single fruiting body may have more than a billion cells. As the cell masses become higher, the differentiation of the fruiting body into stalk and head begins. The stalk is composed of slime within which a few cells are trapped. The majority of the cells migrate to the fruiting body head, where they undergo differentiation into myxospores (Figures 17.43–17.47).

Compared to vegetative cells, myxospores are more resistant to drying, ultraviolet radiation, and heat, but the degree of heat resistance is much less than that of the bacterial endospore (Section 3.12). The main function of the myxospore is probably to allow the organism to survive desiccation during dispersal or during intermittent drying of the habitat. Upon dissemination to a suitable habitat or restoration of adequate growth conditions, the myxospore eventually germinates to form a new vegetative cell.

Figure 17.45 Life cycle of *Myxococcus xanthus.* Aggregation assembles vegetative cells that then undergo fruiting body formation, within which some vegetative cells undergo morphogenesis to form resting cells called myxospores. The myxospores germinate under favorable nutritional and physical conditions to yield vegetative cells.

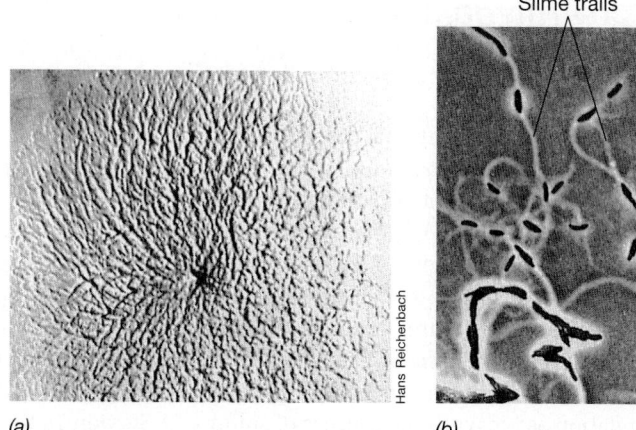

(a)   (b)

Figure 17.46 Swarming in *Myxococcus.* *(a)* Photomicrograph of a swarming colony (5-mm radius) of *Myxococcus xanthus* on agar. *(b)* Single cells of *Myxococcus fulvus* from an actively gliding culture, showing the characteristic slime trails on the agar. A cell of *M. fulvus* is about 0.8 μm in diameter.

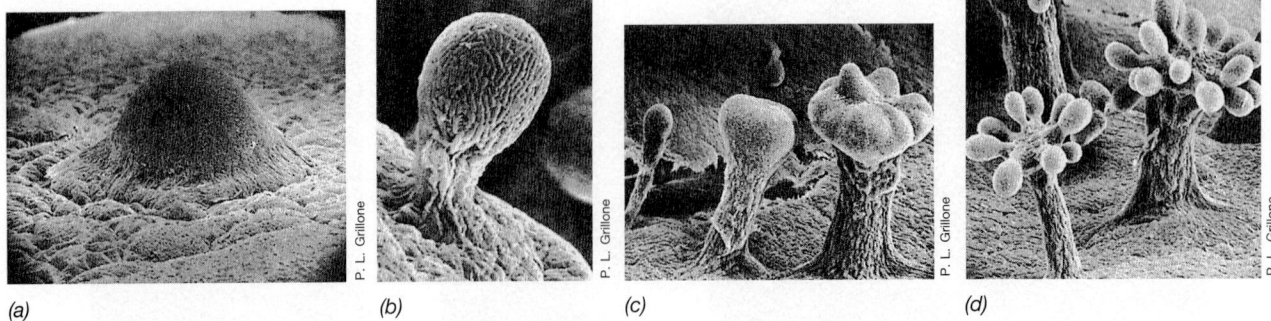

Figure 17.47 Scanning electron micrographs of fruiting body formation in *Chondromyces crocatus*. *(a)* Early stage, showing aggregation and mound formation. *(b)* Initial stage of stalk formation. Slime formation in the head has not yet begun and so the cells that compose the head are still visible. *(c)* Three stages in head formation. Note that the diameter of the stalk also increases. *(d)* Mature fruiting bodies. The entire fruiting structure is about 600 μm in height (compare with Figure 17.44c).

Many myxobacteria synthesize carotenoid pigments and thus their fruiting bodies are brightly colored (Figures 17.43*a* and 17.44). Pigment formation is promoted by light, and at least one function of these pigments is likely photoprotection, which would be beneficial in nature since the myxobacteria are typically exposed to sunlight. In the genus *Stigmatella* (Figure 17.43), light greatly stimulates fruiting body formation and catalyzes production of the compound 2,5,8-trimethyl-8-hydroxy-nonane-4-one. This substance, a type of pheromone, promotes cell aggregation, the initial step in fruiting body formation (Figure 17.45).

MiniQuiz

• What environmental conditions trigger fruiting body formation in myxobacteria?

• What is a myxospore, and how does it compare with an endospore?

17.18 Sulfate- and Sulfur-Reducing Proteobacteria

Key Genera: *Desulfovibrio, Desulfobacter, Desulfuromonas*

Sulfate (SO_4^{2-}) and sulfur (S^0) are electron acceptors for a large group of anaerobic *Deltaproteobacteria* that utilize organic compounds or H_2 as electron donors in anaerobic respirations. Hydrogen sulfide (H_2S) is the product of both SO_4^{2-} and S^0 reduction. Over 40 genera of these organisms, collectively called the dissimilative **sulfate-reducing bacteria** and **sulfur-reducing bacteria**, are known, and some of the key ones are shown in **Table 17.17**. The word *dissimilative* refers to the use of SO_4^{2-} or S^0 as electron acceptors in energy generation instead of their assimilation as biosynthetic sources of sulfur (Section 14.8).

General Properties

The genera of dissimilative sulfate-reducing bacteria form two physiological groups, those that can oxidize acetate and other fatty acids completely to CO_2, and those that cannot. The latter group includes the best studied of sulfate-reducing bacteria,

Desulfovibrio, along with *Desulfomonas, Desulfotomaculum*, and *Desulfobulbus* (**Figure 17.48**). These organisms utilize lactate, pyruvate, ethanol, or certain fatty acids as electron donors, reducing SO_4^{2-} to H_2S. The acetate oxidizers include *Desulfobacter* (Figure 17.48*d*), *Desulfococcus, Desulfosarcina* (Figure 17.48*e*), and *Desulfonema* (Figure 17.48*b*), among many others, and specialize in the complete oxidation of fatty acids, in particular acetate, reducing SO_4^{2-} to H_2S. The sulfate-reducing bacteria are, for the most part, obligate anaerobes, and strict anoxic techniques must be used in their cultivation (Figure 17.48*g*).

Sulfate-reducing bacteria are widespread in aquatic and terrestrial environments that contain some SO_4^{2-} and become anoxic as a result of microbial decomposition processes. *Desulfotomaculum*, phylogenetically a member of the *Firmicutes* (gram-positive *Bacteria*), consists of endospore-forming rods found primarily in soil. Growth and reduction of SO_4^{2-} by *Desulfotomaculum* in certain canned foods leads to a type of spoilage called *sulfide stinker*. The remaining genera of sulfate reducers are indigenous to anoxic freshwater or marine environments and can occasionally be isolated from the mammalian intestine.

Dissimilative Sulfur Reduction

The dissimilative sulfur-reducing bacteria can reduce S^0 to H_2S but are unable to reduce SO_4^{2-}. Species of *Desulfuromonas* (Figure 17.48*f*) grow anaerobically by coupling the oxidation of acetate or a few other organic compounds to the reduction of S^0. However, the ability to reduce S^0, as well as other sulfur compounds such as thiosulfate ($S_2O_3^{2-}$), sulfite (SO_3^{2-}), or dimethyl sulfoxide (DMSO), is widespread in a number of chemoorganotrophic, generally facultatively aerobic bacteria (for example, *Proteus, Campylobacter, Pseudomonas*, and *Salmonella*). *Desulfuromonas* differs from these aerobic bacteria in that it is an obligate anaerobe and utilizes only S^0 as an electron acceptor (Table 17.17).

Dissimilative sulfur-reducing bacteria reside in many of the same habitats as dissimilative sulfate-reducing bacteria and often form associations with bacteria that oxidize H_2S to S^0, such as green sulfur bacteria (Section 18.15). The S^0 produced from H_2S oxidation is then reduced back to H_2S during metabolism of the sulfur reducer, completing an anoxic sulfur cycle (Section 24.4).

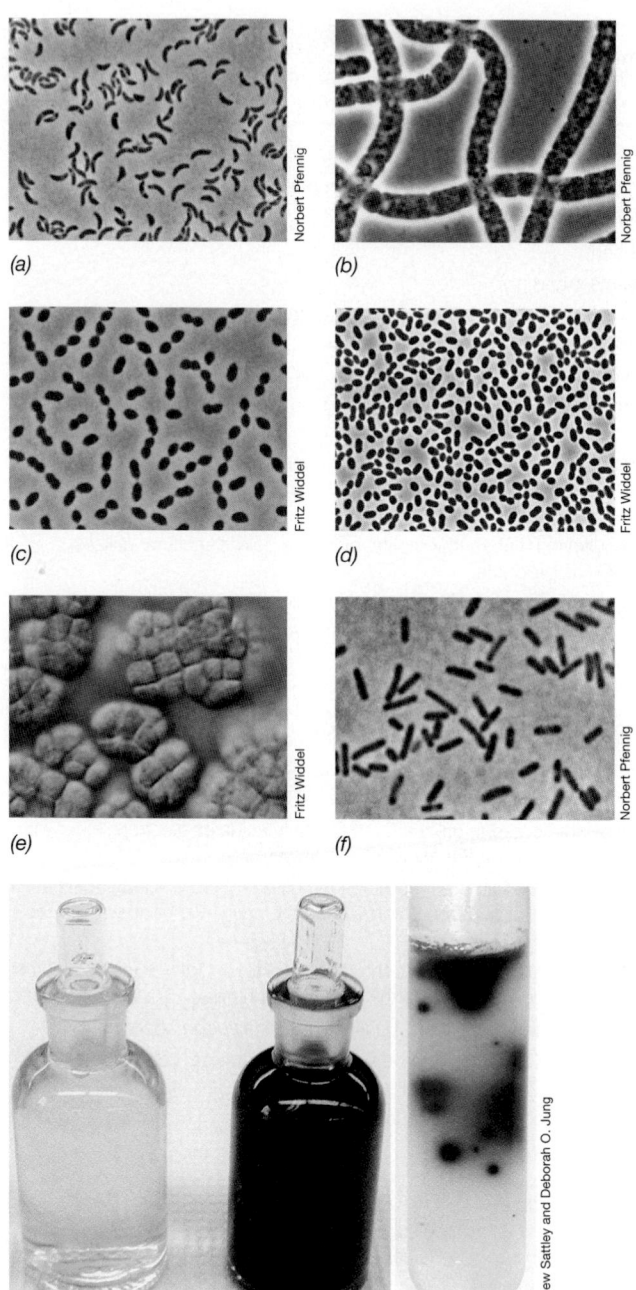

Figure 17.48 **Representative sulfate-reducing and sulfur-reducing bacteria.** *(a) Desulfovibrio desulfuricans*; cell diameter about 0.7 μm. *(b) Desulfonema limicola*; cell diameter 3 μm. *(c) Desulfobulbus propionicus*; cell diameter about 1.2 μm. *(d) Desulfobacter postgatei*; cell diameter about 1.5 μm. *(e) Desulfosarcina variabilis*; cell diameter about 1.25 μm. *(f) Desulfuromonas acetoxidans*; cell diameter about 0.6 μm. *(g)* Enrichment culture of sulfate-reducing bacteria. Left, sterile medium; center, a positive enrichment showing black FeS; right, colonies of sulfate-reducing bacteria in a dilution tube. Photos a–d and f are phase-contrast photomicrographs; part e is an interference contrast micrograph.

Physiology of Sulfate-Reducing Bacteria

The biochemistry of sulfate reduction was discussed in Section 14.8, so here we consider some of the more general physiological properties of this group. The range of electron donors used by sulfate-reducing bacteria is fairly broad. Hydrogen (H_2), lactate, and pyruvate are almost universally used and many species also oxidize certain alcohols (for example, ethanol, propanol, and butanol) as electron donors. Some strains of *Desulfotomaculum* utilize glucose, but this is rare among sulfate reducers. *Desulfovibrio* species typically oxidize lactate, pyruvate, or ethanol to acetate and then excrete this fatty acid as an end product. Depending on the species, fatty acid–oxidizing, sulfate-reducing bacteria oxidize acetate as well as longer-chain fatty acids (Table 17.17) completely to CO_2. Some species, such as *Desulfosarcina and Desulfonema*, grow chemolithotrophically and autotrophically with H_2 as an electron donor, SO_4^{2-} as an electron acceptor, and CO_2 as the sole carbon source. A few sulfate reducers can oxidize individual hydrocarbons and even crude oil itself as electron donors. A few species of sulfate-reducing bacteria also fix nitrogen, but this property is not common among sulfate-reducing bacteria. In addition to using SO_4^{2-} as an electron acceptor, many sulfate-reducing bacteria can use nitrate; sulfonates, such as isethionate ($HO—CH_2—CH_2—SO_3^-$); and S^0.

Certain organic compounds can also be fermented by sulfate-reducing bacteria. The most common of these is pyruvate, which is fermented by way of the phosphoroclastic reaction to acetate, CO_2, and H_2 (⮌ Figure 14.2). Moreover, although generally obligate anaerobes, a few sulfate-reducing bacteria, primarily strains isolated from microbial mats where they coexist with O_2-producing cyanobacteria, are quite O_2-tolerant and can respire with O_2 as the electron acceptor. At least one species, *Desulfovibrio oxyclinae*, can actually grow with O_2 as the electron acceptor under microaerophilic conditions.

Isolation

The enrichment of *Desulfovibrio* species is easy in an anoxic lactate–sulfate medium containing ferrous iron (Fe^{2+}). A reducing agent, such as thioglycolate or ascorbate, is required to achieve a low reduction potential (E_0') in the medium. When sulfate-reducing bacteria grow, the H_2S formed from SO_4^{2-} reduction combines with the ferrous iron to form black, insoluble ferrous sulfide (Figure 17.48g). This blackening not only indicates sulfate reduction, but the iron also binds and detoxifies the H_2S, making possible growth to higher cell densities. Purification can be accomplished by diluting the culture in molten agar tubes; a small amount of liquid from the original enrichment is added to a tube of molten agar growth medium, mixed thoroughly, and sequentially diluted through a series of molten agar tubes (⮌ Section 22.2 and Figure 22.3). Upon solidification, individual cells of sulfate-reducing bacteria become distributed throughout the agar and grow to form black colonies (Figure 17.48g) that can be removed aseptically to yield pure cultures.

MiniQuiz

- Physiologically, how does *Desulfobacter* differ from *Desulfovibrio*?

- Physiologically, how does *Desulfuromonas* differ from *Desulfovibrio*?

Table 17.17 *Characteristics of some key genera of sulfate- and sulfur-reducing bacteria*[a]

Genus	Characteristics
Species unable to oxidize acetate	
Desulfovibrio	Polarly flagellated, curved rods, gram-negative; contain desulfoviridin; one thermophilic
Desulfomicrobium	Motile rods, gram-negative; desulfoviridin absent
Desulfobotulus	Vibrios; gram-negative; motile; desulfoviridin absent
Desulfofustis	Motile rods, specialize in the degradation of glycolate and glyoxalate
Desulfotomaculum	Straight or curved rods; motile by peritrichous or polar flagellation; gram-negative; desulfoviridin absent; produce endospores; capable of utilizing acetate as energy source; related to *Firmicutes*
Desulfomonile	Rods; capable of reductive dechlorination of 3-chlorobenzoate to benzoate (Section 14.12)
Desulfobacula	Oval to coccoid cells, marine; can oxidize various aromatic compounds including the aromatic hydrocarbon toluene to CO_2
Archaeoglobus	Archaeon; hyperthermophile, temperature optimum 83°C; contains some unique coenzymes of methanogenic bacteria, makes small amount of methane during growth; H_2, formate, glucose, lactate, and pyruvate are electron donors; SO_4^{2-}, $S_2O_3^{2-}$, or SO_3^{2-} are electron acceptors (Section 19.6)
Desulfobulbus	Ovoid or lemon-shaped cells, gram-negative; desulfoviridin absent; if motile, by single polar flagellum; utilizes propionate as electron donor with acetate + CO_2 as products
Desulforhopalus	Curved rods, gas vacuolate, psychrophile; uses propionate, lactate, or alcohols as electron donor
Thermodesulfobacterium	Small, gram-negative rods; desulfoviridin present; thermophilic, optimum growth at 70°C; a member of the *Bacteria* but contains ether-linked lipids (Section 18.19)
Acetate-oxidizing species	
Desulfobacter	Rods, gram-negative; desulfoviridin absent; if motile, by single polar flagellum; utilizes only acetate as electron donor and oxidizes it to CO_2 via the citric acid cycle
Desulfobacterium	Rods, some with gas vesicles, marine; capable of autotrophic growth via the acetyl-CoA pathway
Desulfococcus	Spherical cells; nonmotile; gram-negative; desulfoviridin present; utilizes C_1 to C_{14} fatty acids as electron donor with complete oxidation to CO_2; capable of autotrophic growth via the acetyl-CoA pathway
Desulfonema	Large, filamentous gliding bacteria, gram-positive; desulfoviridin present or absent; utilizes C_2 to C_{12} fatty acids as electron donor with complete oxidation to CO_2; capable of autotrophic growth via the acetyl-CoA pathway (H_2 as electron donor)
Desulfosarcina	Cells in packets (sarcina arrangement), gram-negative; desulfoviridin absent; utilizes C_2 to C_{14} fatty acids as electron donor with complete oxidation to CO_2; capable of autotrophic growth via the acetyl-CoA pathway (H_2 as electron donor)
Desulfarculus	Vibrios; gram-negative; motile; desulfoviridin absent; utilizes only C_1 to C_{18} fatty acids as electron donor
Desulfacinum	Cocci to oval-shaped cells; gram-negative; utilizes C_1 to C_{18} fatty acids, very nutritionally diverse, capable of autotrophic growth; thermophilic
Desulforhabdus	Rods, gram-negative; nonmotile; utilizes fatty acids with complete oxidation to CO_2
Thermodesulforhabdus	Gram-negative motile rods; thermophilic; uses fatty acids up to C_{18}
Dissimilative sulfur reducers	
Desulfuromonas	Straight rods, single lateral flagellum, gram-negative; does not reduce sulfate; acetate, succinate, ethanol, or propanol used as electron donor; obligate anaerobe; one species is capable of the reductive dechlorination of trichloroethylene (Section 14.12)
Desulfurella	Motile short rods; gram-negative; requires acetate; thermophilic
Sulfurospirillum	Small vibrios, reduces S^0 with H_2 or formate as electron donors
Campylobacter	Curved, vibrio-shaped rods with polar flagella, gram-negative; unable to reduce sulfate but can reduce sulfur, sulfite, thiosulfate, nitrate, or fumarate anaerobically with acetate or a variety of other carbon or electron donor sources; facultative aerobe; microaerophilic

[a]Phylogenetically, most sulfate- and sulfur-reducing bacteria are species of *Deltaproteobacteria*. *Archaeoglobus* is a species of *Archaea* (Section 19.6); *Thermodesulfobacterium* is a deeply branching hyperthermophilic bacterium (Section 18.19); and *Sulfurospirillum* and *Campylobacter* are species of *Epsilonproteobacteria* (Section 17.19 and Table 17.18).

17.19 The *Epsilonproteobacteria*

The *Epsilon* class of *Proteobacteria* was initially defined by only a few pathogenic bacteria; in particular, by species of *Campylobacter* and *Helicobacter*. However, environmental studies of marine and terrestrial microbial habitats have shown that a diversity of *Epsilonproteobacteria* exist in nature, and their numbers and metabolic capabilities suggest they play important ecological roles.

Species of *Epsilonproteobacteria* are especially abundant at oxic–anoxic interfaces in sulfur-rich environments, such as those surrounding hydrothermal vents (Section 23.12); in these habitats they catalyze metabolic transformations of sulfur and associate with animals that live near the vents. Many of these bacteria are autotrophs and use H_2, formate, or reduced sulfur compounds, with nitrate, oxygen, or elemental sulfur as electron acceptors, depending on the species. Here we describe major

Table 17.18 *Characteristics of key genera of* Epsilonproteobacteria

Genus	Habitat	Descriptive characters	Physiology and metabolism
Campylobacter	Reproductive organs, oral cavity, and intestinal tract of humans and other animals; pathogenic	Slender, spirally curved rods; corkscrew-like motility by single polar flagellum	Microaerophilic; chemoorganotrophic
Arcobacter	Diverse habitats (freshwater, sewage, saline environments, animal reproductive tract, plants); some species pathogenic for humans and other animals	Slender, curved rods; motile by single polar flagellum	Microaerophilic; aerotolerant or aerobic; chemoorganotrophic; oxidation of sulfide to elemental sulfur (S^0) by some species; nitrogen fixation in one species
Helicobacter	Intestinal tract and oral cavity of humans and other animals; pathogenic	Rods to tightly spiral; some species with tightly coiled periplasmic fibers	Microaerophilic, chemoorganotrophic; produce high levels of urease (nitrogen assimilation)
Sulfurospirillum	Freshwater and marine habitats containing sulfur	Vibrioid to spiral-shaped cells; motile by polar flagella	Microaerophilic; reduces elemental sulfur (S^0)
Thiovulum	Freshwater and marine habitats containing sulfur; not yet in pure culture	Cells contain orthorhombic S^0 granules; rapid motility by peritrichous flagella	Microaerophilic; chemolithotrophic oxidizing H_2S
Wolinella	Bovine rumen	Rapidly motile by polar flagellum; single species known: *W. succinogenes*	Anaerobe; anaerobic respiration using fumarate, nitrate, or other compounds as terminal electron acceptor, and with H_2 or formate as electron donor

groups of cultured *Epsilonproteobacteria* (see Figure 17.2) and briefly consider the diversity of uncultured members of this class.

Campylobacter and *Helicobacter*

These two genera are key representatives of the *Epsilonproteobacteria* that share a number of characteristics. They are all gram-negative, motile spirilla, and most species are pathogenic to humans or other animals (**Table 17.18**). These organisms are also microaerophilic (⟲ Section 5.17) and must therefore be cultured from clinical specimens in culture media incubated at low (3–15%) O_2 and high (3–10%) CO_2.

Campylobacter species, over a dozen of which have been described, cause acute gastroenteritis that typically results in a bloody diarrhea. Pathogenesis is due to several factors, including an enterotoxin that is related to cholera toxin (⟲ Section 27.11). *Helicobacter pylori*, also a pathogen, causes both chronic and acute gastritis, leading to the formation of peptic ulcers. We consider these diseases, including their modes of transmission and clinical symptoms, in more detail in Sections 33.10 and 36.10.

Arcobacter

The genus *Arcobacter*, a relative of *Campylobacter*, is unusual among *Epsilonproteobacteria* in that its various species show an unusually wide diversity of habitats. Some species are pathogenic, infecting the reproductive and intestinal tracts of humans and other animals where they can cause reproductive failures, diarrhea-like diseases, and gastroenteritis and appendicitis. *Arcobacter* species have also been found in sewage and water reservoirs, so a fecal–oral route is likely the means by which gastrointestinal infections of *Arcobacter* are transmitted. One species, *Arcobacter nitrofigilis*, is associated with sediment and the roots of the salt marsh plant *Spartina* and can fix nitrogen.

Sulfurospirillum, *Thiovulum*, and *Wolinella*

Species of *Sulfurospirillum*, a *Campylobacter* relative, are nonpathogenic, free-living microaerophiles found in freshwater and marine habitats. These bacteria also carry out anaerobic respirations using elemental sulfur (S^0), selenate, or arsenate as electron acceptors (⟲ Section 14.12). *Thiovulum* also is microaerophilic and is found in freshwater and marine habitats in which sulfide-rich muds interface with oxic zones. At such interfaces populations of *Thiovulum* oxidize H_2S formed by sulfate-reducing bacteria and store the product, S^0, as globules inside the cell (**Figure 17.49**). When motile, *Thiovulum* cells, which are peritrichously flagellated, swim at exceptionally high speed, perhaps the fastest of all known bacteria (~0.6 mm/sec). *Thiovulum* cells are fairly large (10–20 μm) and secrete a slime that forms a veil-like film that helps cells attach to solid surfaces such as sand grains

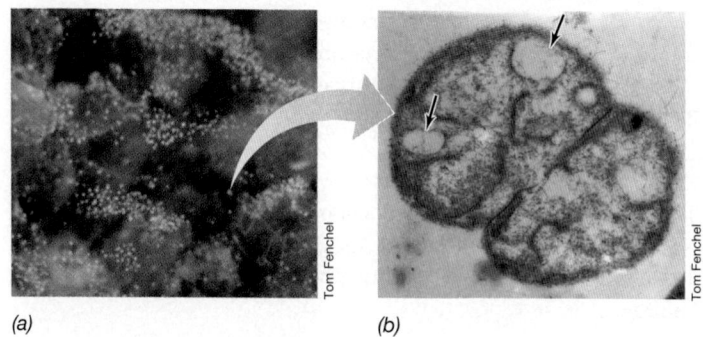

(a)   *(b)*

Figure 17.49 The sulfur-oxidizing epsilonproteobacterium *Thiovulum*. *(a)* Macrophotograph of cells of *Thiovulum* (yellow dots) that formed a thin veil in marine sand containing H_2S (large, irregular structures are sand grains). *(b)* Transmission electron micrograph of a dividing cell of *Thiovulum*. Sulfur (S^0) globules are shown with arrows. Single cells of *Thiovulum* are typically 10–20 μm in diameter.

(Figure 17.49*a*). *Thiovulum* is presumably a chemolithotrophic autotroph but pure cultures of the organism have yet to be achieved to rigorously test this hypothesis. However, if it is true, the physiology and ecology of *Thiovulum* would be very similar to other colorless sulfur bacteria such as *Achromatium*, *Beggiatoa*, and *Thiobacillus* (Section 17.4), sulfur chemolithotrophs that reside in other classes of *Proteobacteria* (Figure 17.2).

Wolinella is an anaerobic bacterium isolated from the bovine rumen (Table 17.18; ↩ Section 25.7). Unlike other *Epsilonproteobacteria*, the single known species, *W. succinogenes*, grows best as an anaerobe and can catalyze anaerobic respirations using fumarate or nitrate as electron acceptors with H_2 or formate as electron donors. Although *W. succinogenes* has thus far been found only in the rumen, its genome shows significant homologies to both the *Campylobacter* and *Helicobacter* genomes and contains additional genes that encode nitrogen fixation, extensive cell signaling mechanisms, and virtually complete metabolic pathways, absent from closely related genomes. This suggests that *Wolinella* inhabits diverse environments outside of the rumen.

Environmental *Epsilonproteobacteria*

In addition to cultured representatives of the genera mentioned above, and many additional species and genera not considered here, there are large groups within this class that are known only from 16S ribosomal RNA gene sequences obtained from the environment (↩ Section 22.6). Through environmental sequencing studies and ongoing cultivation efforts, species of *Epsilonproteobacteria* are now becoming recognized as ubiquitous in marine and terrestrial environments where sulfur-cycling activities are ongoing, particularly in deep-sea hydrothermal vent habitats where sulfide-rich and oxygenated waters mix. Also, living attached to the surface of animals such as the tube worm *Alvinella* and the shrimp *Rimicaris* that reside near hydrothermal vents, a large variety of as yet uncultured *Epsilonproteobacteria* may, through their sulfur metabolism, detoxify H_2S that would otherwise be deleterious to their animal hosts, allowing the animals to thrive in a chemically hostile environment (↩ Section 25.12). It is thus likely that further exploration of the phylogeny, metabolic activities, and ecological roles of *Epsilonproteobacteria* will generate exciting new aspects of prokaryotic diversity.

MiniQuiz

- Why is *Wolinella* physiologically unusual among the *Epsilonproteobacteria*?
- From what kinds of habitats might one likely find diverse species of *Epsilonproteobacteria*?

Big Ideas

17.1
Proteobacteria are a huge class of gram-negative bacteria characterized by a wide distribution in nature and broad metabolic diversity. Many bacteria commonly cultured from soil, water, and animal bodies are species of *Proteobacteria*.

17.2
Phototrophic purple bacteria are anoxygenic phototrophs that grow phototrophically and obtain carbon from CO_2 plus H_2S (purple sulfur bacteria) or organic compounds (purple nonsulfur bacteria). Purple nonsulfur bacteria are physiologically diverse, and most can also grow as chemoorganotrophs in darkness. Purple bacteria belong to the *Alpha-*, *Beta-*, and *Gammaproteobacteria*.

17.3
The nitrifying bacteria include chemolithotrophs that can oxidize NH_3 to NO_2^- (*Nitrosomonas*) and NO_2^- to NO_3^- (*Nitrobacter*). These organisms grow autotrophically in oxic nitrogen-rich habitats.

17.4
Sulfur chemolithotrophs oxidize H_2S and other reduced sulfur compounds for energy metabolism with O_2 or NO_3^- as electron acceptors and use either CO_2 or organic compounds as carbon sources.

17.5
The hydrogen bacteria oxidize H_2 with O_2 as the electron acceptor and fix CO_2 as the carbon source. Some hydrogen bacteria, the carboxydobacteria, oxidize carbon monoxide (CO). Most of these bacteria can also grow on organic compounds.

17.6
Methylotrophs are bacteria able to grow on carbon compounds that lack carbon–carbon bonds. Some methylotrophs are also methanotrophs, organisms able to catabolize methane. Most methanotrophs are *Proteobacteria* that contain extensive internal membranes and incorporate C_1 units by way of either the serine or ribulose monophosphate pathway. Methanotrophs reside in water and soil and are also symbionts of certain marine animals.

17.7
Pseudomonads include many gram-negative, chemoorganotrophic, aerobic rods, some of which are pathogenic. The genus *Pseudomonas* includes many species that are nutritionally diverse and widespread in nature in soil, water, and on the surfaces of plants and animals.

17.8
The acetic acid bacteria *Acetobacter* and *Gluconobacter* produce acetate from the oxidation of ethanol, and these acid-tolerant

bacteria are often found in the fermenting fluids of alcoholic beverages.

17.9
The free-living nitrogen-fixing bacteria are obligately aerobic bacteria that can use N_2 as their nitrogen source. A common genus is *Azotobacter*, an organism that produces slimy colonies in laboratory cultures.

17.10
Neisseria, *Chromobacterium*, and their relatives can be isolated from animals, and some species of this group are pathogenic, causing severe diseases in humans including gonorrhea and one form of meningitis.

17.11
Enteric bacteria are a major group of highly related, gram-negative, facultative bacteria of major medical importance. The different genera are distinguished primarily on phenotypic grounds, including metabolic properties, such as fermentation patterns.

17.12
Vibrio, *Aliivibrio*, and *Photobacterium* species are marine organisms, some of which are pathogenic and bioluminescent. Bioluminescence, catalyzed by the enzyme luciferase, is controlled by a quorum-sensing mechanism that ensures that light is not emitted until a large cell population has been attained.

17.13
Rickettsias are obligate intracellular parasites that are deficient in many metabolic functions and obtain key metabolites from their hosts. Species of the genus *Rickettsia* cause diseases such as typhus and Rocky Mountain spotted fever, while the genus *Wolbachia* contains common insect endosymbionts.

17.14
Spirilla are spiral-shaped, chemoorganotrophic bacteria that are widespread in the aquatic environment. Species are distributed among five classes of *Proteobacteria*. *Magnetospirillum* displays magnetotaxis and *Bdellovibrio* is a predatory bacterium, attacking and killing other gram-negative bacteria.

17.15
Sheathed bacteria are filamentous *Proteobacteria* in which individual cells form chains within an outer layer called the sheath. *Sphaerotilus* and *Leptothrix* are major genera of sheathed bacteria and can oxidize metals, such as Fe^{2+} and Mn^{2+}.

17.16
Budding and prosthecate bacteria are appendaged cells that form stalks or prosthecae used for attachment or nutrient absorption and are primarily aquatic. *Hyphomicrobium*, *Caulobacter*, and *Gallionella* are major genera.

17.17
The fruiting myxobacteria are rod-shaped gliding bacteria that aggregate to form complex masses of cells called fruiting bodies. Myxobacteria are mostly aerobic chemoorganotrophic soil bacteria that live by consuming dead organic matter or other bacterial cells.

17.18
Sulfate- and sulfur-reducing bacteria are a large group of *Deltaproteobacteria* unified physiologically by their ability to reduce SO_4^{2-} or S^0 to H_2S under anoxic conditions. Two physiological classes of sulfate-reducing bacteria are known, those that are capable of oxidizing acetate to CO_2 and those that are not.

17.19
Epsilonproteobacteria contain both pathogenic species, such as *Campylobacter* and *Helicobacter*, and nonpathogenic species, such as *Sulfospirillum* and *Thiovulum*. Many of the latter inhabit the oxic–anoxic interface of sulfide-rich environments.

Review of Key Terms

Autoinduction a gene regulatory mechanism involving small, diffusible signal molecules that are produced in larger amounts as population size increases

Bioluminescence the enzymatic production of visible light by living organisms

Carboxysome a polyhedral cellular inclusion of crystalline ribulose bisphosphate carboxylase (RubisCO), the key enzyme of the Calvin cycle

Chemolithotroph an organism able to oxidize inorganic compounds (such as H_2, Fe^{2+}, S^0, or NH_4^+) as energy sources (electron donors)

Enteric bacteria a large group of gram-negative rod-shaped *Bacteria* characterized by a facultatively aerobic metabolism and commonly found in the intestines of animals

Methanotroph an organism capable of oxidizing methane (CH_4) as an electron donor in energy metabolism

Methylotroph an organism capable of oxidizing organic compounds that do not contain carbon–carbon bonds; if able to oxidize CH_4, also a methanotroph

Mixotroph an organism that can conserve energy from the oxidation of inorganic compounds but requires organic compounds as a carbon source

Nitrifying bacteria chemolithotrophs capable of carrying out the transformation $NH_3 \rightarrow NO_2^-$, or $NO_2^- \rightarrow NO_3^-$

Prosthecae extrusions of cytoplasm, often forming distinct appendages, bounded by the cell wall

Proteobacteria a major lineage of *Bacteria* that contains a large number of gram-negative rods and cocci

Purple nonsulfur bacteria a group of phototrophic bacteria containing bacteriochlorophyll *a* or *b* and that grow best as photoheterotrophs and without H_2S

Purple sulfur bacteria a group of phototrophic bacteria containing bacteriochlorophylls *a* or *b* and that can oxidize H_2S and store elemental sulfur inside the cells (or in some species, outside the cell)

Spirilla spiral-shaped cells (singular, spirillum)

Sulfate-reducing and **sulfur-reducing bacteria** two groups of *Bacteria* that respire anaerobically with SO_4^{2-} or S^0, respectively, as electron acceptors, producing H_2S as final product

Review Questions

1. Compare and contrast the metabolism, morphology, and phylogeny of purple sulfur and purple nonsulfur bacteria (Section 17.2).

2. Compare and contrast the nitrogen metabolism of ammonia-oxidizing bacteria with that of nitrifying bacteria (Section 17.3).

3. What is CO oxidation (Section 17.5)?

4. Discuss the general properties of acetic acid bacteria (Section 17.8).

5. What is unusual about the growth requirements of most *Rickettsia* species (Section 17.13)?

6. List an electron donor for energy metabolism for each of the following *Proteobacteria* and state whether the organism is an aerobe or an anaerobe: *Thiobacillus, Nitrosomonas, Ralstonia eutropha, Methylomonas, Pseudomonas, Acetobacter,* and *Gallionella.*

7. How do you differentiate enteric bacteria from other bacteria of similar morphology (Section 17.11)?

8. Describe the mechanism of bioluminescence (Section 17.12).

9. Describe a key physiological feature of the following *Proteobacteria* that would differentiate each from the others: *Acetobacter, Methylococcus, Azotobacter, Photobacterium, Desulfovibrio,* and *Spirillum.*

10. What physiological trait unites the majority of cultured *Epsilonproteobacteria* (Section 17.19)?

Application Questions

1. Defend the following statement using phylogenetic, ecological, and physiological arguments: *Wolbachia pipientis* is a more highly evolved bacterium than *Escherichia coli.*

2. Defend or refute the following statements, using examples from this chapter: (a) Cell morphology has phylogenetic predictive value. (b) Major physiological differences between *Proteobacteria* correlate with the different classes in this phylum.

Need more practice? Test your understanding with Quantitative Questions; access additional study tools including tutorials, animations, and videos; and then test your knowledge with chapter quizzes and practice tests at **www.microbiologyplace.com**.

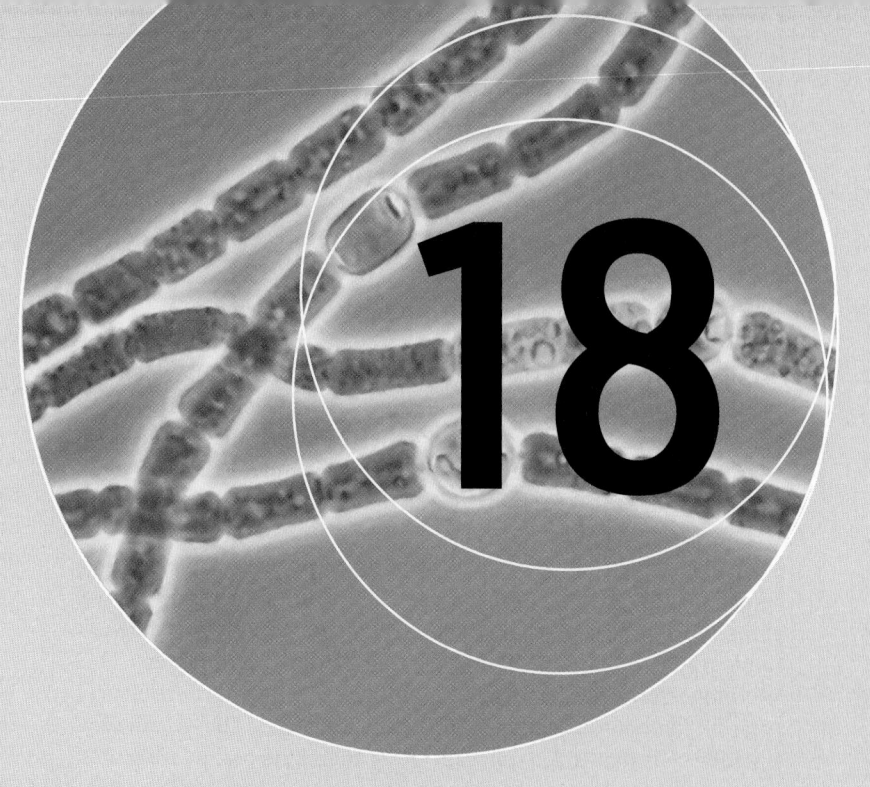

Other *Bacteria*

Cyanobacteria, such as *Anabaena* shown here, are oxygenic phototrophs that slowly converted the anoxic atmosphere of early Earth to the highly oxic atmosphere we see today.

Firmicutes, Mollicutes, and *Actinobacteria*

In Chapter 17 we examined the *Proteobacteria*, a large and diverse phylum of *Bacteria*. In this chapter, we focus on several of the many other phyla of *Bacteria* (➪ Figure 17.1), exploring well-known cultured species in the various evolutionarily distinct groups. Again we mention the resources to be found in *Bergey's Manual of Systematic Bacteriology* and *The Prokaryotes* (➪ Section 16.13) for more detailed information on these bacterial groups.

We start with an examination of gram-positive bacteria. The gram-positive bacteria are a large collection of bacteria that we present in three groups: the classes *Mollicutes* and *Actinobacteria*, and other major members of the phylum *Firmicutes*. The *Actinobacteria* include the actinomycetes, a huge group of primarily filamentous, sporulating, soil bacteria. The *Mollicutes* include cells that lack a cell wall. Other members of the *Firmicutes* include the endospore-forming bacteria, the lactic acid bacteria, and several other groups.

18.1 Nonsporulating *Firmicutes*

Key Genera: *Staphylococcus, Micrococcus, Streptococcus, Lactobacillus, Sarcina*

In this section we consider genera of lactic acid bacteria and their relatives, classical nonsporulating gram-positive rods and cocci, and some relatives. We also consider *Micrococcus* here, although it is actually a member of the *Actinobacteria*, because it is morphologically quite similar to *Staphylococcus* (**Table 18.1**).

Staphylococcus and *Micrococcus*

Staphylococcus (**Figure 18.1**) and *Micrococcus* are both aerobic organisms with a typical respiratory metabolism. They are catalase-positive, and this permits their distinction from *Streptococcus* and some other genera of gram-positive cocci. Gram-positive cocci are relatively resistant to reduced water potential and tolerate drying and high salt (NaCl) fairly well. Their ability to grow in media containing salt provides a selective means for isolation. For example, if an appropriate inoculum such as a skin swab, dry soil, or room dust is spread on a rich-medium agar plate containing 7.5% NaCl and the plate is incubated aerobically, gram-positive cocci often form the predominant colonies. Many species are pigmented, and this provides an additional aid in selecting gram-positive cocci.

The genera *Micrococcus* and *Staphylococcus* can easily be separated based on the oxidation–fermentation (OF) test (➪ Table 31.3). *Micrococcus* is an obligate aerobe and produces acid from glucose only under aerobic conditions, whereas *Staphylococcus* is a facultative aerobe and produces acid from glucose both aerobically and anaerobically. *Staphylococcus* also typically forms cell clusters (Figure 18.1a), whereas *Micrococcus* does not.

Staphylococci are common commensals and parasites of humans and animals, and they occasionally cause serious infections. In humans, there are two major species, *Staphylococcus epidermidis*, a nonpigmented, nonpathogenic organism usually found on the skin or mucous membranes, and *Staphylococcus aureus* (Figure 18.1), a yellow-pigmented species that is most commonly associated with pathological conditions including boils, pimples, pneumonia, osteomyelitis, meningitis, and arthritis. We discuss the pathogenesis of *S. aureus* in Section 27.2 and staphylococcal diseases in Sections 33.9 and 36.6.

Table 18.1 *Distinguishing features of major gram-positive cocci*

Genus	Motility	Arrangement of cells	Growth by fermentation	Phylogenetic group[a]	Other characteristics
Micrococcus	−	Clusters, tetrads	−	*Actinobacteria*	Strict aerobe
Staphylococcus	−	Clusters, pairs	+	*Firmicutes*	Only genus of this group to contain teichoic acid in cell wall
Stomatococcus	−	Clusters, pairs	+	*Actinobacteria*	Only genus of this group containing a capsule
Enterococcus	−	Singly, pairs, or short chains	+	*Firmicutes*	Strictly fermentative but aerotolerant; part of intestinal flora of most animals
Planococcus	+	Pairs, tetrads	−	*Firmicutes*	Primarily marine
Sarcina	−	Cuboidal packets of eight or more cells	+	*Firmicutes*	Extremely acid-tolerant; cellulose in cell wall
Ruminococcus	+	Pairs, chains	+	*Firmicutes*	Obligate anaerobe; inhabits rumen, cecum, and large intestine of many animals
Peptococcus	−	Clusters, pairs	+	*Firmicutes*	Obligate anaerobe; ferments peptone but not sugars
Peptostreptococcus	−	Clumps, short chains	+	*Firmicutes*	Obligate anaerobe; ferments peptone; common member of human normal flora, skin, intestine, vagina; also isolated from vaginal and purulent discharges

[a]See Figure 17.1.

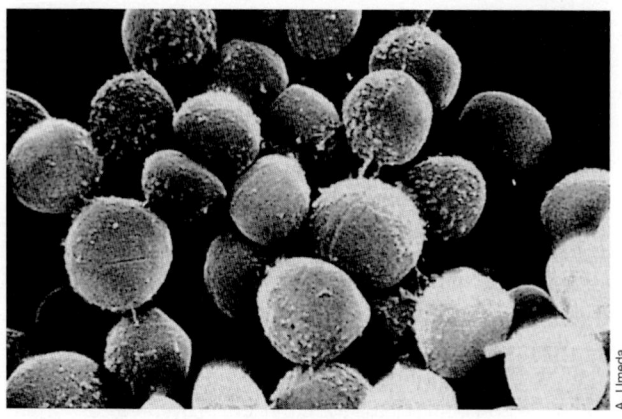

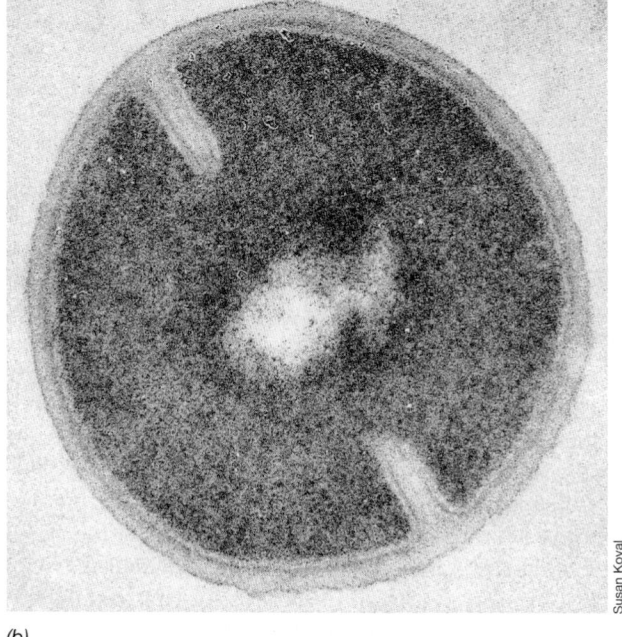

(a)

(b)

Figure 18.1 *Staphylococcus.* *(a)* Scanning electron micrograph of typical *Staphylococcus aureus* cells, showing the irregular arrangement of the cell clusters. Individual cells are about 0.8 μm in diameter. *(b)* Transmission electron micrograph of a dividing cell of *S. aureus*. Note the thick gram-positive cell wall.

Micrococcus species can also be isolated from skin but are much more common on surfaces of inanimate objects, on dust particles, and in soil.

Sarcina

The genus *Sarcina* contains species of bacteria that divide in three perpendicular planes to yield packets of eight or more cells (**Figure 18.2**). *Sarcina* are obligate anaerobes and are extremely acid-tolerant, being able to ferment sugars and grow in environments at a pH as low as 2. Cells of one species, *Sarcina ventriculi*, contain a thick fibrous layer of cellulose surrounding the cell wall (Figure 18.2*b*). The cellulose layers of adjacent cells become attached, and this functions as a cementing material to hold together packets of *S. ventriculi* cells.

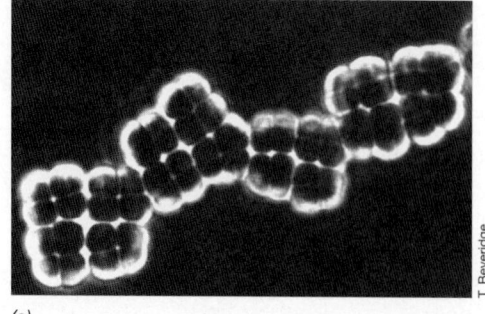

(a)

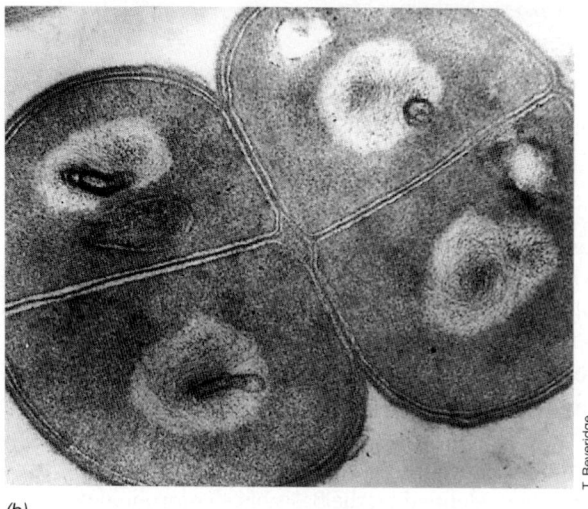

(b)

Figure 18.2 *Sarcina.* *(a)* Phase-contrast photomicrograph of cells of a typical gram-positive coccus *Sarcina*. A single cell is about 2 μm in diameter. *(b)* Electron micrograph of a thin section. The outermost layer of the cell consists of cellulose.

Sarcina species can be isolated from soil, mud, feces, and stomach contents. Because of its extreme acid tolerance, *S. ventriculi* is one of only a few bacteria that can inhabit and grow in the stomach of humans and other monogastric animals. Rapid growth of *S. ventriculi* is observed in the stomach of humans suffering from certain gastrointestinal disorders, such as pyloric ulcerations. These pathological conditions retard the flow of food to the intestine and often require surgery to correct.

Lactic Acid Bacteria and Lactic Acid Fermentations

The lactic acid bacteria are gram-positive rods and cocci that produce lactic acid as a major or sole fermentation product. Members of this group lack porphyrins and cytochromes, do not carry out oxidative phosphorylation, and hence obtain energy only by substrate-level phosphorylation. All lactic acid bacteria grow anaerobically. Unlike many anaerobes, however, most lactic acid bacteria are not sensitive to oxygen (O_2) and can grow in its presence; thus they are called *aerotolerant anaerobes*. Most lactic acid bacteria obtain energy only from the metabolism of sugars and therefore are usually restricted to habitats in which sugars are present. They typically have limited biosynthetic abilities, and their complex nutritional requirements include needs for amino acids, vitamins, purines, and pyrimidines (for example, ↺ Table 4.2 for *Leuconostoc mesenteroides*).

Table 18.2 *Differentiation of the principal genera of lactic acid bacteria*[a]

Cell form and arrangement	Genus
Cocci in chains or tetrads	
Homofermentative	*Streptococcus*
	Enterococcus
	Lactococcus
	Pediococcus
Heterofermentative	*Leuconostoc*
Rods, typically in chains	
Homofermentative	*Lactobacillus*
Heterofermentative	*Lactobacillus*

[a]The physiology of homofermentative and heterofermentative lactic acid bacteria is covered in Section 14.2.

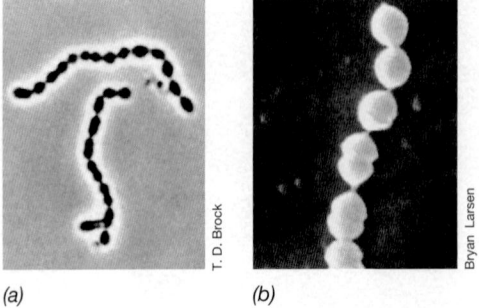

(a)   *(b)*

Figure 18.3 Gram-positive cocci. *(a) Lactococcus lactis*, phase-contrast micrograph. *(b) Streptococcus* sp., scanning electron micrograph. Cells in both photos are 0.5–1 μm in diameter.

One important difference between subgroups of the lactic acid bacteria lies in the pattern of products formed from the fermentation of sugars. One group, called **homofermentative**, produces a single fermentation product, *lactic acid*. The other group, called **heterofermentative**, produces other products, mainly ethanol and CO_2, as well as lactate (**Table 18.2**; Section 14.2 provides additional information on homofermentative and heterofermentative pathways in these bacteria). The various genera of lactic acid bacteria have been defined on the basis of cell morphology, phylogeny, and type of fermentative metabolism, as shown in Table 18.2.

Streptococcus and Other Cocci

The genus *Streptococcus* (**Figure 18.3**) contains homofermentative species with quite distinct habitats and activities that are of considerable practical importance to humans. Some species are pathogenic to humans and animals (Section 33.2). As producers of lactic acid, other streptococci play important roles in the production of buttermilk, silage, and other fermented products (Section 36.3), and certain species play a major role in the formation of dental caries (Section 27.3). The genus *Lactococcus* contains those streptococci of dairy significance, whereas the genus *Enterococcus* includes streptococci that are primarily of fecal origin. Species of the genera *Peptococcus* and *Peptostreptococcus* are obligate anaerobes that ferment proteins rather than sugars (Table 18.1).

Organisms in the genus *Streptococcus* have been divided into two groups of related species on the basis of characteristics enumerated in **Table 18.3**. Hemolysis on blood agar is of considerable importance in the subdivision of the genus into species. For example, species that produce streptolysin O or S form colonies surrounded by a large zone of complete red blood cell hemolysis when plated on blood agar, a condition called *β-hemolysis* (Figure 27.20*a*). Streptococci are also divided into immunological groups (designated by different letters), based on the presence of specific carbohydrate antigens (antigens are substances that elicit an immune response). Those β-hemolytic streptococci found in humans usually contain the group A antigen, whereas enterococci contain the group D antigen. Streptococci with group B antigen, usually found in association with animals, are a cause of mastitis (inflammation of the udder) in cows and have also been implicated in certain human infections. Lactococci are of antigen group N and are not pathogenic.

Heterofermentative lactococci are placed in the genus *Leuconostoc*. Strains of *Leuconostoc* also produce the flavoring ingredients diacetyl and acetoin from the catabolism of citrate; they have been used as starter cultures in dairy fermentations. Some strains of *Leuconostoc* produce large amounts of glucose or fructose polysaccharides, especially when cultured on sucrose as the carbon and energy source (Figure 14.37), and some of these polymers have found medical use as plasma extenders in blood transfusions.

Lactobacillus

Lactobacilli are typically rod-shaped, varying from long and slender to short, bent rods (**Figure 18.4**). Most species are homofermentative, but some are heterofermentative (Table 18.2). Lactobacilli are common in dairy products, and some strains are

Table 18.3 *Differential characteristics of streptococci, lactococci, and enterococci*

Group	Antigenic groups	Representative species	Type of hemolysis on blood agar	Habitat
Streptococci				
Pyogenes subgroup	A,B,C,F,G	*Streptococcus pyogenes*	Lysis (β)	Respiratory tract, systemic
Viridans subgroup	—	*Streptococcus mutans*	Greening (α)	Mouth, intestine
Enterococci	D	*Enterococcus faecalis*	Lysis (β), greening (α), or none	Intestine, vagina, plants
Lactococci	N	*Lactococcus lactis* (Figure 18.3a)	None	Plants, dairy products

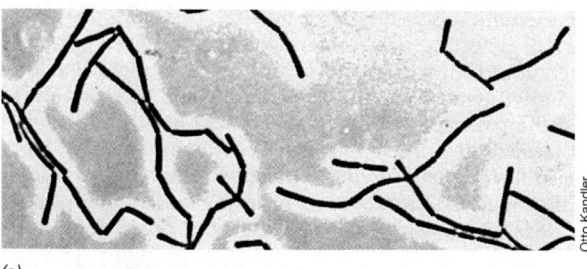

(a)

Otto Kandler

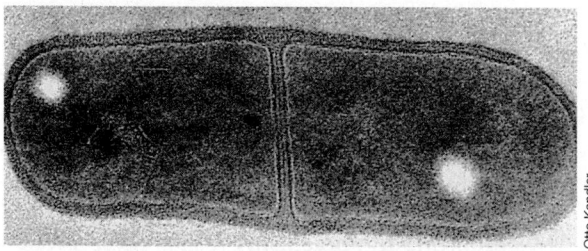

(b)

Otto Kandler

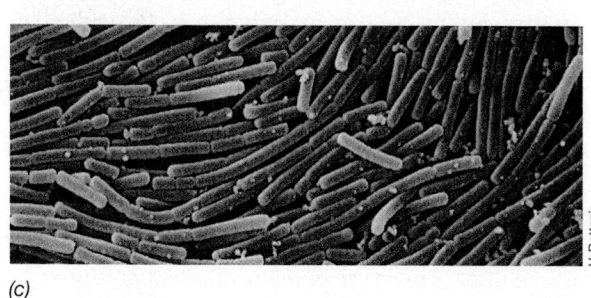

(c)

V. Bottazi

Figure 18.4 *Lactobacillus* **species.** *(a) Lactobacillus acidophilus*, phase-contrast. Cells are about 0.75 μm wide. *(b) Lactobacillus brevis*, transmission electron micrograph. Cells measure about 0.8 × 2 μm. *(c) Lactobacillus delbrueckii*, scanning electron micrograph. Cells are about 0.7 μm in diameter.

used in the preparation of fermented milk products. For instance, *Lactobacillus acidophilus* (Figure 18.4*a*) is used in the production of acidophilus milk; *Lactobacillus delbrueckii* (Figure 18.4*c*) is used in the preparation of yogurt; and other species are used in the production of sauerkraut, silage, and pickles (⊂⊃ Section 36.3). The lactobacilli are usually more resistant to acidic conditions than are the other lactic acid bacteria and are able to grow well at pH values as low as 4. Because of this, they can be selectively enriched from dairy products and fermenting plant material by use of acidic carbohydrate-containing media.

The acid resistance of the lactobacilli enables them to continue growing during natural lactic fermentations, even when the pH value has dropped too low for other lactic acid bacteria to grow. The lactobacilli are therefore typically responsible for the final stages of most lactic acid fermentations. They are rarely, if ever, pathogenic.

Listeria

Listeria are gram-positive coccobacilli that tend to form chains of three to five cells (⊂⊃ Figure 36.16). *Listeria* is phylogenetically related to *Lactobacillus* species, and like homofermentative lactic

acid bacteria, they produce acid but not gas from glucose. True lactic acid bacteria, however, are capable of growth under strictly anoxic conditions and lack the enzyme catalase. *Listeria*, in contrast, requires microoxic or fully oxic conditions for growth and produces catalase. Although several species of *Listeria* are known, the species *Listeria monocytogenes* is most noteworthy because it causes a major foodborne illness, *listeriosis* (⊂⊃ Section 36.11). The organism is transmitted in contaminated, usually ready-to-eat foods, such as cheese and sausages, and can cause anything from a mild illness to a fatal form of meningitis.

MiniQuiz

- How do heterofermentative and homofermentative bacteria differ physiologically?
- How can *Staphylococcus* and *Micrococcus* be selectively isolated from nature?

18.2 Endospore-Forming *Firmicutes*

Key Genera: *Bacillus, Clostridium, Sporosarcina, Heliobacterium*

Several genera of endospore-forming bacteria have been recognized (**Table 18.4**) and are distinguished on the basis of cell morphology, shape and cellular position of the endospore

Table 18.4 *Major genera of endospore-forming bacteria*

Characteristics	Genus
Rods	
Aerobic or facultative, catalase produced	*Bacillus* *Paenibacillus*
Microaerophilic, no catalase; homofermentative lactic acid producer	*Sporolactobacillus*
Anaerobic:	
Sulfate-reducing	*Desulfotomaculum*
Does not reduce sulfate, fermentative	*Clostridium*
Thermophilic, temperature optimum 65–70°C, fermentative	*Thermoanaerobacter*
Gram-negative; can grow as homoacetogen on $H_2 + CO_2$	*Sporomusa*
Halophile, isolated from the Dead Sea	*Sporohalobacter*
Produces up to five spores per cell; fixes N_2	*Anaerobacter*
Other rod-shaped endospore formers	
Acidophile, pH optimum 3	*Alicyclobacillus*
Alkaliphile, pH optimum 9	*Amphibacillus*
Phototrophic	*Heliobacterium, Heliophilum, Heliorestis*
Syntrophic, degrades fatty acids but only in coculture with a H_2-utilizing bacterium (⊂⊃ Section 14.5)	*Syntrophospora*
Reductively dechlorinates chlorophenols (⊂⊃ Section 14.12)	*Desulfitobacterium*
Cocci (usually arranged in tetrads or packets), aerobic	*Sporosarcina* (Figure 18.7)

UNIT 6

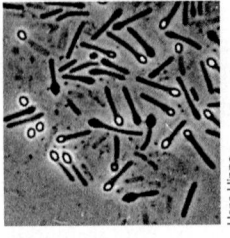
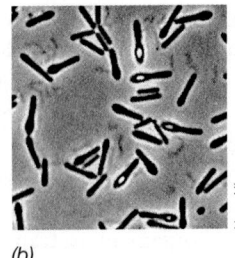
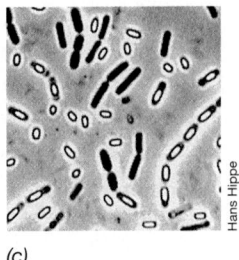

(a)   (b)   (c)

Figure 18.5 *Clostridium* **species and endospore location.** *(a) Clostridium cadaveris*, terminal endospores. Cells are about 0.9 μm wide. *(b) Clostridium sporogenes*, subterminal endospores. Cells are about 1 μm wide. *(c) Clostridium bifermentans*, central endospores. Cells are about 1.2 μm wide. All are phase-contrast micrographs.

(**Figure 18.5**), relationship to O_2, and energy metabolism. The structure and heat resistance of the bacterial endospore along with the process of endospore formation itself was discussed in Section 3.12. The two genera about which most is known are *Bacillus*, species of which are aerobic or facultatively aerobic, and *Clostridium*, which contains fermentative species. All endospore-forming bacteria are ecologically related because they are found in nature primarily in soil. Even those species that are pathogenic to humans or other animals are primarily saprophytic soil organisms and infect animals only incidentally. Indeed, the ability to produce endospores should be advantageous for a soil microorganism because soil is a highly variable environment in terms of nutrient levels, temperature, and water activity.

Endospore-forming bacteria can be selectively isolated from soil, food, dust, and other materials by heating the sample to 80°C for 10 min, a treatment that effectively kills vegetative cells while any endospores present remain viable. Streaking such heat-treated samples on plates of the appropriate medium and incubating either aerobically or anaerobically selectively yield species of *Bacillus* or *Clostridium*, respectively.

Bacillus and *Paenibacillus*

A list of representatives in the *Bacillus* group is shown in **Table 18.5**. Species of *Bacillus* and *Paenibacillus* grow well on defined media containing any of a number of carbon sources. Many bacilli produce extracellular hydrolytic enzymes that break down complex polymers such as polysaccharides, nucleic acids, and lipids, permitting the organisms to use these products as carbon sources and electron donors. Many bacilli produce antibiotics, including bacitracin, polymyxin, tyrocidine, gramicidin, and circulin. In most cases the antibiotics are released when the culture enters the stationary phase of growth and is committed to sporulation.

Several bacilli, most notably *Paenibacillus popilliae* and *Bacillus thuringiensis*, produce insect larvicides. *P. popilliae* causes a fatal condition called milky disease in Japanese beetle larvae and larvae of closely related beetles of the family *Scarabaeidae*. *B. thuringiensis* causes a fatal disease of larvae of many different groups of insects, although individual strains are specific as to the host affected. Both of these insect pathogens form a crystalline protein during sporulation called the *parasporal body*, which is deposited within the

Table 18.5 *Characteristics of representative species of bacilli*		
Characteristics	**Genus/Species**	**Endospore position**
I. Endospores oval or cylindrical, facultative aerobes, casein and starch hydrolyzed		
Sporangia not swollen, endospore wall thin		
Thermophiles and acidophiles	*Bacillus coagulans*	Central or terminal
	Alicyclobacillus acidocaldarius	Terminal
Mesophiles	*Bacillus licheniformis*	Central
	Bacillus cereus	Central
	Bacillus anthracis	Central
	Bacillus megaterium	Central
	Bacillus subtilis	Central
Insect pathogen	*Bacillus thuringiensis*	Central
Sporangia distinctly swollen, spore wall thick		
Thermophile	*Geobacillus stearothermophilus*	Terminal
Mesophiles	*Paenibacillus polymyxa*	Terminal
	Bacillus macerans	Terminal
	Bacillus circulans	Central or terminal
Insect pathogens	*Paenibacillus larvae*	Central or terminal
	Paenibacillus popilliae	Central
II. Endospores spherical, obligate aerobes, casein and starch not hydrolyzed		
Sporangia swollen	*Bacillus sphaericus*	Term
Sporangia not swollen	*Sporosarcina pasteurii*	Terminal

Endospore   Crystal

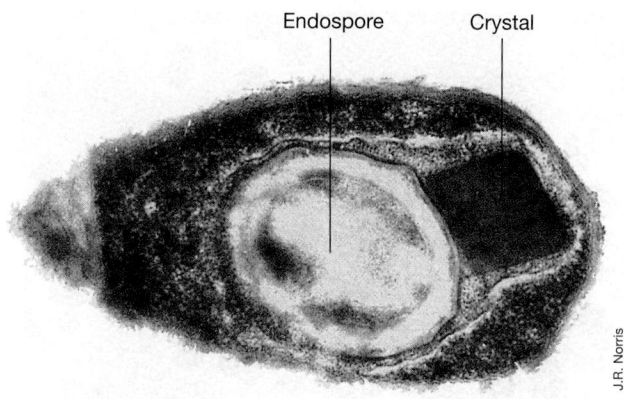

J.R. Norris

Figure 18.6 The toxic parasporal crystal in the insect pathogen *Bacillus thuringiensis*. Electron micrograph of a thin section of a sporulating cell. The crystalline protein (Bt toxin) is toxic to certain insects by causing lysis of their intestinal cells.

sporangium but outside the endospore proper (**Figure 18.6**). In *B. thuringiensis*, the parasporal body is a protoxin that is converted to a toxin by proteolytic cleavage in the larval gut. The toxin binds to intestinal epithelial cells and induces pore formation that causes leakage of the host cell cytoplasm followed by lysis. Endospore preparations derived from *B. thuringiensis* and *P. popilliae* are commercially available as biological insecticides.

Genes encoding crystal proteins from several *B. thuringiensis* strains have been isolated. The genes for the *B. thuringiensis* crystal protein (known commercially as "Bt toxin") have been introduced into plants to render the plants "naturally" resistant to insects. Genetically altered Bt toxins have also been developed by genetic engineering to help increase toxicity and reduce resistance (↶ Section 15.18).

Clostridium

The clostridia lack a respiratory chain; unlike *Bacillus* species, they obtain ATP only by substrate-level phosphorylation. Many anaerobic energy-yielding mechanisms are known in the clostridia (clostridial fermentations are discussed in Section 14.3). Indeed, the separation of the genus *Clostridium* into subgroups is based primarily on these properties and on the fermentable substrate used (**Table 18.6**). A number of clostridia are *saccharolytic* and ferment sugars, producing butyric acid as a major end product. Some of these also produce acetone and butanol, such as *Clostridium pasteurianum*, which is also a vigorous nitrogen-fixing bacterium. One group of clostridia ferments cellulose with the formation of acids and alcohols, and these are likely the major organisms decomposing cellulose anaerobically in soil.

Another group of clostridia are *proteolytic* and conserve energy from the fermentation of amino acids. Some species ferment individual amino acids, but others ferment only amino

Table 18.6 *Characteristics of some groups of clostridia*

Key characteristics	Other characteristics	Species
Ferment carbohydrates		
Ferment cellulose	Fermentation products: acetate, lactate, succinate, ethanol, CO_2, H_2	*C. cellobioparum*[a] *C. thermocellum*
Ferment sugars, starch, and pectin; some ferment cellulose	Fermentation products: acetone, butanol, ethanol, isopropanol, butyrate, acetate, propionate, succinate, CO_2, H_2; some fix N_2	*C. butyricum* *C. cellobioparum* *C. acetobutylicum* *C. pasteurianum* *C. perfringens*
Ferment sugars primarily to acetic acid	Total synthesis of acetate from CO_2; cytochromes present in some species	*C. aceticum* *Moorella thermoacetica* *C. formicaceticum*
Ferments only pentoses or methylpentoses	Ring-shaped cells form left-handed, helical chains; fermentation products: acetate, propionate, *n*-propanol, CO_2, H_2	*C. methylpentosum*
Ferment amino acids	Fermentation products: acetate, other fatty acids, NH_3, CO_2, sometimes H_2; some also ferment sugars to butyrate and acetate; may produce exotoxins; causative agents of serious or fatal diseases	*C. sporogenes* *C. histolyticum* *C. putrefaciens* *C. tetani* *C. botulinum* *C. tetanomorphum*
	Ferments three-carbon amino acids (for example, alanine) to propionate, acetate, and CO_2	*C. propionicum*
Ferments carbohydrates or amino acids	Fermentation products from glucose: acetate, formate, small amounts of isobutyrate and isovalerate	*C. bifermentans*
Purine fermenters	Ferments uric acid and other purines, forming acetate, CO_2, NH_3	*C. acidurici*
Ethanol fermentation to fatty acids	Produces butyrate, caproate, and H_2; requires acetate as electron acceptor; does not use sugars, amino acids, or purines	*C. kluyveri*

[a]All genus names beginning with a "*C.*" are species of the genus *Clostridium*.

UNIT 6

acid pairs. The products of amino acid fermentation are typically acetate, butyrate, CO_2, and H_2. The coupled catabolism of an amino acid pair is called a *Stickland reaction*; for example, *Clostridium sporogenes* ferments glycine plus alanine. In the Stickland reaction, one amino acid functions as the electron donor and is oxidized, whereas the other is the electron acceptor and is reduced (Figure 14.6). Many of the products of amino acid fermentation by clostridia are foul-smelling substances, and the odor that results from putrefaction is mainly the result of clostridial action. In addition to butyric acid, other odoriferous compounds produced are isobutyric acid, isovaleric acid, caproic acid, hydrogen sulfide, methylmercaptan (from sulfur amino acids), cadaverine (from lysine), putrescine (from ornithine), and ammonia.

The main habitat of clostridia is the soil, where they live primarily in "pockets" made anoxic by facultative or obligately aerobic bacteria. In addition, a number of clostridia inhabit the anoxic environment of the mammalian intestinal tract. Several clostridia are capable of causing severe diseases in humans, as will be discussed in Section 27.10. For example, botulism is caused by *Clostridium botulinum*, tetanus by *Clostridium tetani*, and gas gangrene by *Clostridium perfringens* and a number of other clostridia, both sugar and amino acid fermenters. These pathogenic clostridia seem in no way unusual metabolically but are distinct in that they produce specific toxins or, in those causing gas gangrene, a group of toxins. *C. perfringens* and related species can also cause gastroenteritis in humans and domestic animals (Section 36.7), and botulism outbreaks are not uncommon in birds such as ducks and a variety of other animals.

Sporosarcina

The genus *Sporosarcina* is unique among endospore formers because cells are cocci instead of rods. *Sporosarcina* consists of strictly aerobic spherical to oval cells that divide in two or three perpendicular planes to form tetrads or packets of eight or more cells (**Figure 18.7**). The major species is *Sporosarcina ureae*. This bacterium can be enriched from soil by plating dilutions of a pasteurized soil sample on alkaline nutrient agar supplemented with 8% urea and incubating in air. Most soil bacteria are strongly inhibited by as little as 2% urea. However, *S. ureae* tolerates this, catabolizing urea to CO_2 and ammonia (NH_3), which dramatically raises the pH. *S. ureae* is remarkably alkaline-tolerant and

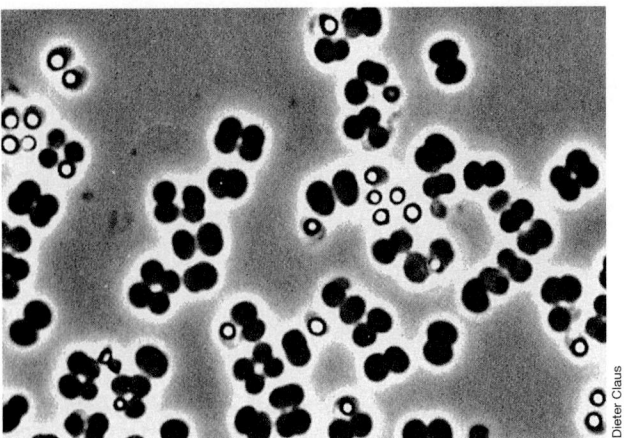

Figure 18.7 *Sporosarcina ureae*. Phase-contrast micrograph. A single cell is about 2 μm wide. Note bright refractile endospores. Most cell packets contain eight cells.

can be grown in media up to pH 10, and this feature can be used to advantage in its enrichment from soil.

Heliobacteria

Heliobacteria are phototrophic gram-positive *Bacteria*. **Heliobacteria** are anoxygenic phototrophs and produce a unique pigment, bacteriochlorophyll *g* (Figure 13.4). The group contains four genera: *Heliobacterium*, *Heliophilum*, *Heliorestis*, and *Heliobacillus*. All known heliobacteria produce rod-shaped or filamentous cells, but *Heliophilum* is morphologically unusual because its cells form into bundles (**Figure 18.8b**) that are motile as a unit.

Heliobacteria are strict anaerobes, but in addition to phototrophic growth they can grow chemotrophically in darkness by pyruvate fermentation (as can many clostridia, close relatives of the heliobacteria). Like the endospores of *Bacillus* or *Clostridium* species, the endospores of heliobacteria (Figure 18.8c) contain elevated calcium (Ca^{2+}) levels and the signature molecule of the endospore, dipicolinic acid (Section 3.12). Heliobacteria reside in soil, especially paddy (rice) field soils, where their nitrogen fixation activities may benefit rice productivity. A large diversity of heliobacteria have also been found in highly alkaline environments, such as soda lakes and surrounding alkaline soils.

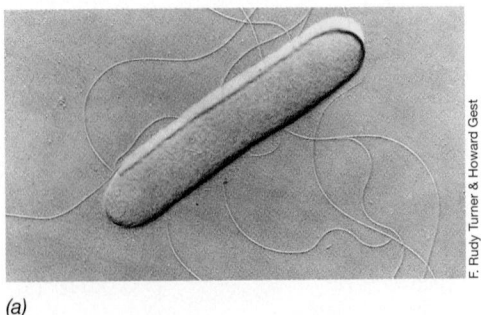

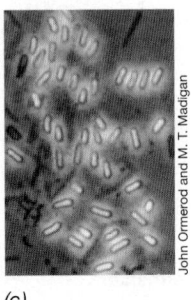

(a)   (b)   (c)

Figure 18.8 Cells and endospores of heliobacteria. *(a)* Electron micrograph of *Heliobacillus mobilis*, a peritrichously flagellated species. *(b)* *Heliophilum fasciatum* cell bundles as observed by electron microscopy. *(c)* Phase-contrast micrograph of endospores from *Heliobacterium gestii*.

18.3 *Mollicutes*: The Mycoplasmas

Key Genera: *Mycoplasma, Spiroplasma*

The *Mollicutes* (also called *Tenericutes*) are bacteria that lack cell walls (*mollis* is Latin for "soft") and are some of the smallest organisms known. Although they do not stain gram-positively (because they lack cell walls), the *Mollicutes*, often called the mycoplasmas after *Mycoplasma*, a key genus in the group, are phylogenetically related to the *Firmicutes*. They are organisms that likely once had cell walls but lost the need for them because of their special habitats, typically in or on the bodies of animal and plant hosts.

Properties of Mycoplasmas

The absence of cell walls in mycoplasmas has been confirmed by electron microscopy and chemical analyses, which show that peptidoglycan is absent. Mycoplasmas resemble protoplasts (bacteria treated to remove their cell walls), but they are more resistant to osmotic lysis and are able to survive conditions under which protoplasts lyse. This ability to resist osmotic lysis is at least partially determined by the presence of sterols, which make the cytoplasmic membranes of mycoplasmas more stable than that of other bacteria. Indeed, some mycoplasmas require sterols in their growth media, and this sterol requirement has been used to separate mycoplasmas into two physiological groups (**Table 18.7**).

In addition to sterols, certain mycoplasmas contain compounds called *lipoglycans* (Table 18.7). Lipoglycans are long-chain heteropolysaccharides covalently linked to membrane lipids and

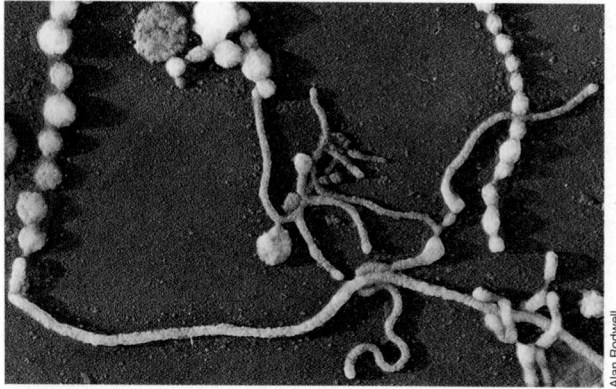

Figure 18.9 *Mycoplasma mycoides.* Metal-shadowed transmission electron micrograph. Note the coccoid and hyphalike elements. The average diameter of cells in chains is about 0.5 μm.

embedded in the cytoplasmic membrane of many mycoplasmas. Lipoglycans in some ways resemble the lipopolysaccharides in the outer membrane of gram-negative bacteria, except that they lack the lipid A backbone (Section 3.7). Lipoglycans function to help stabilize the cytoplasmic membrane and have also been identified as facilitating attachment of mycoplasmas to cell surface receptors of animal cells.

Growth of Mycoplasmas

Mycoplasmas can be grown in the laboratory and are small and pleomorphic cells. A single culture may exhibit small coccoid elements; larger, swollen forms; and filamentous forms, often highly branched (**Figure 18.9**). The small coccoid elements (0.2–0.3 μm in size) are probably the smallest of free-living cells (Section 3.2). The genomes of mycoplasmas are also smaller than those of most bacteria, between 500 and 1100 kilobase pairs of DNA in

Table 18.7 *Major characteristics of mycoplasmas*

Genus	Properties	Genome size (kilobase pairs)	Presence of lipoglycans
Require sterols			
Mycoplasma	Many pathogenic; facultative aerobes (Figure 18.9)	600–1350	+
Anaeroplasma	May or may not require sterols; obligate anaerobes; degrade starch, producing acetic, lactic, and formic acids plus ethanol and CO_2; found in the bovine and ovine rumen	1500–1600	+
Spiroplasma	Spiral to corkscrew-shaped cells; associated with various phytopathogenic (plant disease) conditions; facultative aerobe	940–2200	−
Ureaplasma	Coccoid cells; occasional clusters and short chains; growth optimal at pH 6; strong urease reaction; associated with certain urinary tract infections in humans; microaerophile	750	−
Entomoplasma	Facultative aerobe; associated with insects and plants	790–1140	Unknown
Do not require sterols			
Acholeplasma	Facultative aerobes	1500	+
Asteroleplasma	Obligate anaerobe; isolated from the bovine or ovine rumen	1500	+
Mesoplasma	Phylogenetically and ecologically related to *Entomoplasma*; facultative aerobes	870–1100	Unknown

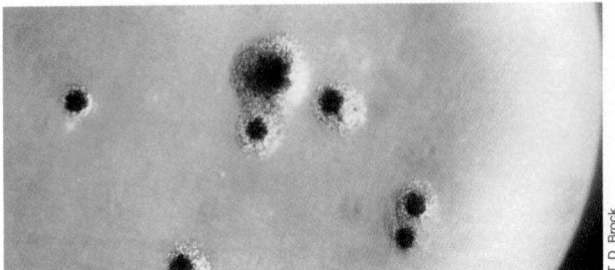

Figure 18.10 Colonies of a *Mycoplasma* species on agar. Note the typical "fried-egg" appearance. The colonies are about 0.5 mm in diameter.

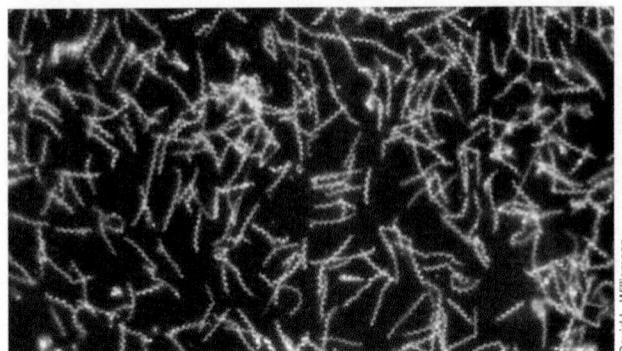

Figure 18.11 "Sex ratio" spiroplasma from the hemolymph of the fly *Drosophila pseudoobscura*. Dark-field micrograph. Female flies infected with the sex ratio spiroplasma bear only female progeny. Cells are about 0.15 μm in diameter.

most cases (Table 18.7). This is comparable to the genome size of the obligately parasitic chlamydia and rickettsia and about one-fifth to one-fourth that of *Escherichia coli*.

The mode of growth of mycoplasmas differs in liquid and agar cultures. On agar there is a tendency for the organisms to grow so that they become embedded in the medium. These colonies show a characteristic "fried-egg" appearance consisting of a dense central core that penetrates downward into the agar, surrounded by a circular spreading area that is lighter in color (**Figure 18.10**). As would be expected of cells lacking cell walls, growth of *Mollicutes* is not inhibited by antibiotics that inhibit cell wall synthesis. However, mycoplasmas are as sensitive as most *Bacteria* to antibiotics whose targets are other than the cell wall.

Media for the culture of mycoplasmas are typically quite complex. For many species, growth is poor or absent even in complex yeast extract–peptone–beef heart infusion media. Fresh serum or ascitic fluid (peritoneal fluids) is needed as well to provide unsaturated fatty acids and sterols. Some mycoplasmas can be cultivated on relatively simple culture media, however, and even defined media have been developed for some species. Most mycoplasmas use carbohydrates as carbon and energy sources and require vitamins, amino acids, purines, and pyrimidines as growth factors. The energy metabolism of mycoplasmas varies, with some species being strictly respiratory while others are facultative or even obligate anaerobes (Table 18.7).

Spiroplasma

The genus *Spiroplasma* consists of helical or spiral-shaped *Mollicutes*. Amazingly, although they lack a cell wall and flagella, spiroplasmas are motile by means of a rotary (screw) motion or a slow undulation. Intracellular fibrils that are thought to play a role in motility have been demonstrated. The organism has been isolated from ticks, the hemolymph (**Figure 18.11**) and gut of insects, vascular plant fluids and insects that feed on these fluids, and the surfaces of flowers and other plant parts. For example, *Spiroplasma citri* has been isolated from the leaves of citrus plants, where it causes a disease called citrus stubborn disease, and from corn plants suffering from corn stunt disease. A number of other mycoplasma-like organisms have been detected in diseased plants by electron microscopy, which indicates that a large group of plant-associated *Mollicutes* may exist. Some species of *Spiroplasma* are known that cause insect diseases, such as honey-bee spiroplasmosis and lethargy disease of the beetle *Melolontha*.

MiniQuiz

• Why do mycoplasmas need to have stronger cytoplasmic membranes than other bacteria?

• Motile spiroplasmas cannot contain a normal bacterial flagellum; why?

18.4 *Actinobacteria*: Coryneform and Propionic Acid Bacteria

Key Genera: *Corynebacterium, Arthrobacter, Propionibacterium*

A second major group of gram-positive bacteria is the *Actinobacteria*, which form their own subdivision within the gram-positive *Bacteria*. The *Actinobacteria* contain rod-shaped to filamentous and primarily aerobic bacteria that are common inhabitants of soil and plant materials. For the most part they are harmless commensals, species of *Mycobacterium* (for example, *Mycobacterium tuberculosis*) being notable exceptions. Some are of great economic value in either the production of antibiotics or certain fermented dairy products. We begin with the rod-shaped *Actinobacteria*.

Coryneform Bacteria

The coryneform bacteria are gram-positive, aerobic, nonmotile, rod-shaped organisms with the characteristic of forming irregular-shaped, club-shaped, or V-shaped cell arrangements during normal growth. V-shaped cells arise as a result of a snapping movement that occurs just after cell division, a process called *snapping division* (**Figure 18.12**). Snapping division occurs because the cell wall consists of two layers; only the inner layer participates in cross-wall formation, and so after the cross-wall is formed, the two daughter cells remain attached by the outer layer of the cell wall. Localized rupture of this outer layer on only one side of the cell results in a bending of the two cells away from the ruptured side (**Figure 18.13**) and thus development of V-shaped forms.

The main genera of coryneform bacteria are *Corynebacterium* and *Arthrobacter*. The genus *Corynebacterium* consists of an

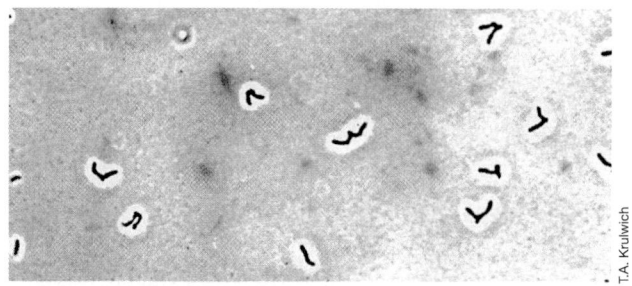

Figure 18.12 Snapping division in *Arthrobacter*. Phase-contrast micrograph of characteristic V-shaped cell groups in *Arthrobacter crystallopoietes* resulting from snapping division. Cells are about 0.9 μm in diameter.

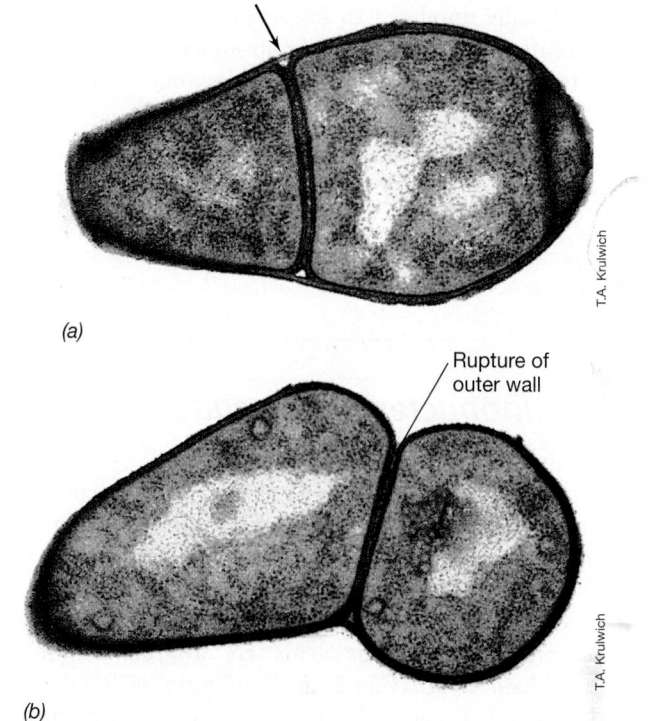

(a)

Rupture of outer wall

(b)

Figure 18.13 Cell division in *Arthrobacter*. Transmission electron micrograph of cell division in *Arthrobacter crystallopoietes*, illustrating how snapping division and V-shaped cell groups arise. *(a)* Before rupture of the outer cell wall layer (arrow). *(b)* After rupture of the outer layer on one side. Cells are 0.9–1 μm in diameter.

extremely diverse group of bacteria, including animal and plant pathogens and saprophytes. Some species, such as *Corynebacterium diphtheriae*, are pathogenic (diphtheria, Section 33.3). The genus *Arthrobacter*, consisting primarily of soil organisms, is distinguished from *Corynebacterium* on the basis of a developmental cycle involving conversion from rod to coccus and back to rod again (**Figure 18.14**). However, some coryneform bacteria are pleomorphic and form coccoid cells during growth, and so the distinction between the two genera on the basis of life cycle is not absolute. The *Corynebacterium* cell frequently has a swollen end, so it has a club-shaped appearance, whereas *Arthrobacter* species are less commonly club-shaped.

Along with the *Acidobacteria* (Section 18.13), species of *Arthrobacter* are among the most common of all soil bacteria. They are remarkably resistant to desiccation and starvation, despite the fact that they do not form spores or other resting cells. Arthrobacters are a heterogeneous group that have considerable nutritional versatility, and strains have been isolated that decompose herbicides, caffeine, nicotine, phenols, and other unusual organic compounds.

Propionic Acid Bacteria

The propionic acid bacteria (genus *Propionibacterium*) were first discovered in Swiss (Emmentaler) cheese, where their fermentative production of CO_2 produces the characteristic holes and the propionic acid they produce is at least partly responsible for the unique flavor of the cheese. The bacteria in this group are gram-positive anaerobes that ferment lactic acid, carbohydrates, and polyhydroxy alcohols, producing primarily propionic acid, acetic acid, and CO_2 (Section 14.3 and Figure 14.4).

The fermentation of lactate is of interest because lactate itself is an end product of fermentation for many bacteria (Section 18.1). The starter culture in Swiss cheese manufacture consists of a mixture of homofermentative streptococci and lactobacilli, plus propionic acid bacteria. The homofermentative organisms carry out the initial fermentation of lactose to lactic acid during formation of the curd (protein and fat). After the curd has been drained, the propionic acid bacteria develop rapidly. The eyes (or holes) characteristic of Swiss cheese are formed by the accumulation of CO_2, the gas diffusing through the curd and gathering at weak points. The propionic acid bacteria are thus able to obtain energy anaerobically from a product that other bacteria have produced by fermentation. This metabolic strategy is called a *secondary fermentation*.

Propionate is also formed in the fermentation of succinate by the bacterium *Propionigenium*. This organism is phylogenetically

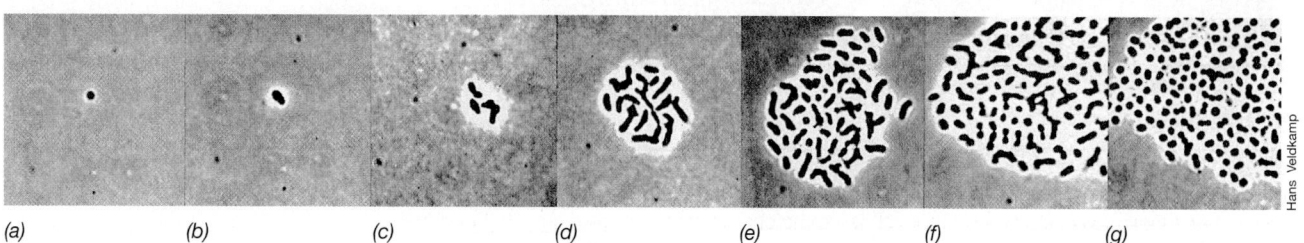

(a)   *(b)*   *(c)*   *(d)*   *(e)*   *(f)*   *(g)*

Figure 18.14 Stages in the life cycle of *Arthrobacter globiformis* as observed in slide culture. *(a)* Single coccoid element; *(b–e)* conversion to rod and growth of a microcolony consisting predominantly of rods; *(f–g)* conversion of rods to coccoid forms. Cells are about 0.9 μm in diameter.

and ecologically unrelated to *Propionibacterium*, but energetic aspects of its fermentation are of considerable interest. We discussed the mechanism of the *Propionigenium* fermentation in Section 14.4.

MiniQuiz

- What is snapping division and what organism exhibits it?
- What organism is involved in the production of Swiss cheese, and what products does it make that help to flavor the cheese and make the holes?

18.5 *Actinobacteria*: *Mycobacterium*

Key Genus: *Mycobacterium*

The genus *Mycobacterium* consists of rod-shaped organisms that at some stage of their growth cycle possess the distinctive staining property called **acid-fastness**. This property is due to the presence of unique lipids called *mycolic acids*, found only in species of the genus *Mycobacterium*, on the surface of the mycobacterial cell. Mycolic acids are a group of complex branched-chain hydroxylated lipids (**Figure 18.15**) covalently bound to peptidoglycan in the cell wall; the complex gives the cell surface a waxy, hydrophobic consistency.

Because of their waxy surface, mycobacteria do not stain well with Gram stain. A mixture of the red dye basic fuchsin and phenol is used in the acid-fast (Ziehl–Neelsen) stain. The stain is driven into the cells by slow heating and the role of the phenol is to enhance penetration of the fuchsin into the lipids. After washing in distilled water, the preparation is decolorized with acid alcohol and counterstained with methylene blue. Cells of acid-fast organisms stain red, whereas the background and non-acid-fast organisms appear blue (Figure 33.9).

Mycobacteria are somewhat pleomorphic and may undergo branching or even filamentous growth. However, in contrast to the filaments of the actinomycetes (Section 18.6), the filaments of

(a) **Mycolic acid; R_1 and R_2 are long-chain aliphatic hydrocarbons** *(b)* **Basic fuchsin**

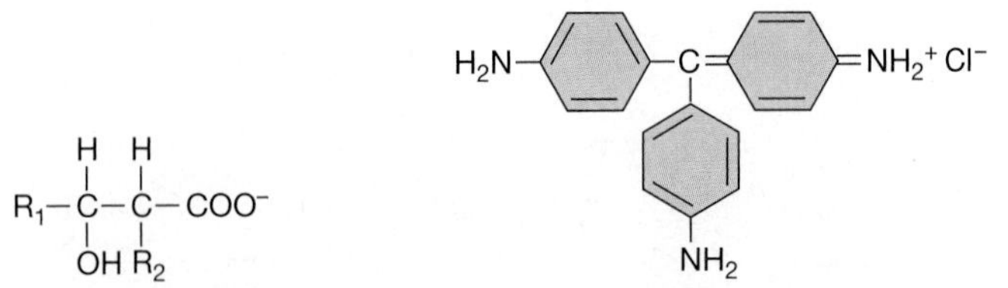

Figure 18.15 Acid-fast staining. Structure of *(a)* mycolic acid and *(b)* basic fuchsin, the dye used in the acid-fast stain. The fuchsin dye combines with mycolic acids in the cell wall via ionic bonds between COO^- and NH_2^+.

the mycobacteria become fragmented upon even the slight disturbance and a true mycelium is not formed. In general, mycobacteria can be separated into two major groups, slow growers and fast growers (**Table 18.8**). *Mycobacterium tuberculosis* is a typical slow grower, and visible colonies are produced from dilute inoculum only after days to weeks of incubation. When growing on solid media, mycobacteria form tight, compact, often wrinkled colonies (**Figure 18.16**). This colony morphology is probably due to the high lipid content and hydrophobic nature of the cell surface, which facilitate cells sticking together.

For the most part, mycobacteria have relatively simple nutritional requirements. Most species can grow aerobically in a simple mineral salts medium with ammonium as the nitrogen source and glycerol or acetate as the sole carbon source and electron donor. Growth of *M. tuberculosis* is more difficult and is stimulated by lipids and fatty acids. A glycerol–egg medium (Lowenstein–Jensen medium; eggs are a good source of lipids) is often used in primary isolation of *M. tuberculosis* from patient specimens. The virulence of *M. tuberculosis* cultures has been correlated with the formation of long, cordlike structures

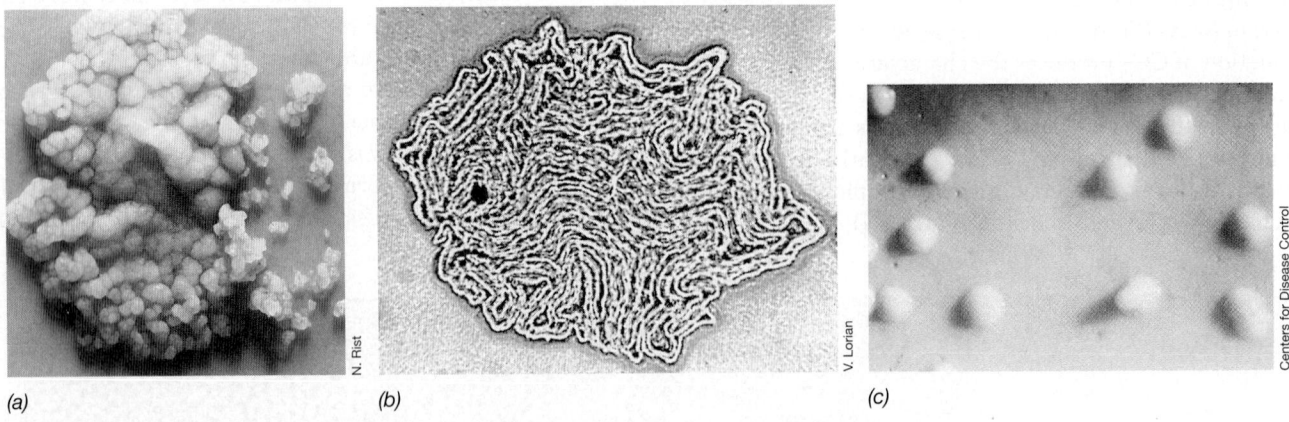

(a) *(b)* *(c)*

Figure 18.16 Characteristic colony morphology of mycobacteria. *(a) Mycobacterium tuberculosis*, showing the compact, wrinkled appearance of the colony. The colony is about 7 mm in diameter. *(b)* A colony of virulent *M. tuberculosis* at an early stage, showing the characteristic cordlike growth. Individual cells are about 0.5 μm in diameter. (See also the historic drawings of *M. tuberculosis* cells made by Robert Koch, Figure 1.20). *(c)* Colonies of *Mycobacterium avium* from a strain of this organism isolated as an opportunistic pathogen from an AIDS patient.

Table 18.8 *Some characteristics of representative mycobacteria*

Species	Human pathogen	Pigmentation
Slow-growing species		
Mycobacterium tuberculosis	+	None
Mycobacterium avium	+	Old colonies pigmented (see Figure 18.16c)
Mycobacterium bovis	+	None
Mycobacterium gordonae	+	Scotochromogenic
Fast-growing species		
Mycobacterium smegmatis	−	None
Mycobacterium phlei	−	Pigmented
Mycobacterium chelonae	+	None
Mycobacterium parafortuitum	−	Photochromogenic

(Figure 18.16*b*) that form due to side-to-side aggregation and intertwining of long chains of bacteria. Growth in cords reflects the presence of a characteristic glycolipid, the *cord factor*, on the cell surface (**Figure 18.17**). The pathogenesis of tuberculosis, along with the related mycobacterial disease leprosy, is discussed in Section 33.4.

Some mycobacteria produce yellow carotenoid pigments (Figure 18.16*c*), and based on pigmentation, the mycobacteria can be classified into three groups: (1) nonpigmented; (2) forming pigment only when cultured in light, a property called *photochromogenesis*; and (3) forming pigment even when cultured in the dark, a property called *scotochromogenesis* (Table 18.8). Photochromogenesis is triggered by blue wavelengths and is characterized by the photoinduction of one of the early enzymes in carotenoid biosynthesis. As with other carotenoid-containing bacteria, it is likely that carotenoids protect

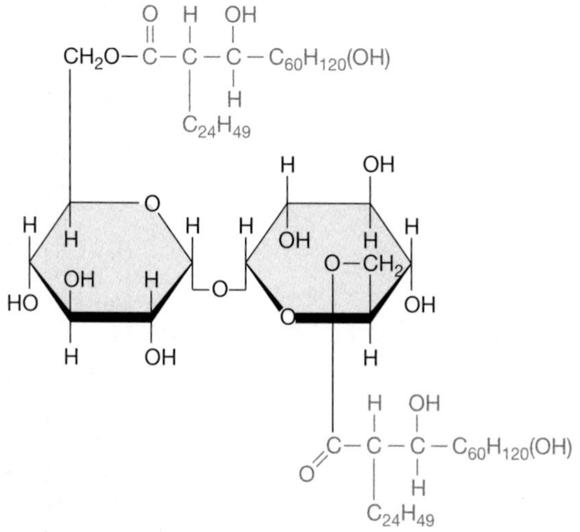

Figure 18.17 Structure of cord factor, a mycobacterial glycolipid: 6,6′-di-*O*-mycolyl trehalose. The two identical long-chain dialcohol groups are shown in purple.

mycobacteria against oxidative damage from singlet oxygen (Section 5.18).

MiniQuiz
- What is mycolic acid and what properties does this substance confer on mycobacteria?
- How does photochromogenesis differ from scotochromogenesis?

18.6 Filamentous *Actinobacteria*: *Streptomyces* and Relatives

Key Genera: *Streptomyces, Actinomyces, Nocardia*

The actinomycetes are a large group of phylogenetically related, filamentous and aerobic gram-positive *Bacteria*. As a result of successful growth and branching, a ramifying network of filaments called a *mycelium* is formed (**Figure 18.18**). Although it is of bacterial dimensions, the mycelium is analogous to the mycelium formed by filamentous fungi (Section 20.13). Most actinomycetes form spores; the manner of spore formation varies and is used in separating subgroups, as outlined in **Table 18.9**. We focus here on the genus *Streptomyces*, the most important genus in this group.

Streptomyces

Over 500 species of *Streptomyces* are recognized. *Streptomyces* filaments are typically 0.5–1.0 μm in diameter, are of indefinite length, and often lack cross-walls in the vegetative phase. *Streptomyces* grow at the tips of the filaments, often showing branching. Thus, the vegetative phase consists of a complex, tightly woven matrix, resulting in a compact, convoluted mycelium and subsequent colony. As the colony ages, characteristic aerial filaments called *sporophores* are formed, which project above the surface of the colony and give rise to spores (**Figure 18.19**).

Streptomyces spores, called *conidia*, are quite distinct from the endospores of *Bacillus* and *Clostridium*. Unlike the elaborate cellular differentiation that leads to the formation of an endospore, conidia are produced by the formation of cross-walls in the

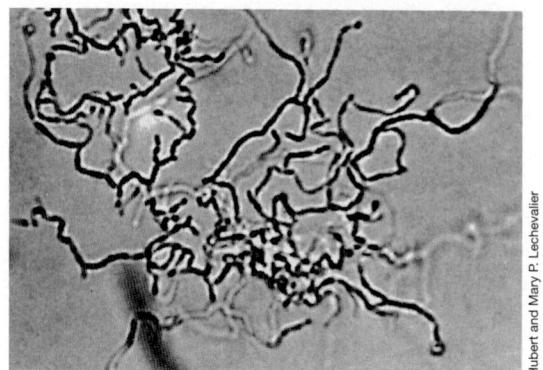

Figure 18.18 *Nocardia*. A young colony of an actinomycete of the genus *Nocardia*, showing typical filamentous cellular structure (mycelium). Each filament is about 0.8–1 μm in diameter.

Table 18.9 *Representative rod-shaped and filamentous actinomycetes and related genera*[a]

Coryneform group of bacteria: rods, often club-shaped, morphologically variable; not acid-fast or filamentous; snapping cell division

Corynebacterium: irregularly staining segments, sometimes granules; club-shaped swelling frequent; animal and plant pathogens, also soil saprophytes

Arthrobacter: coccus–rod morphogenesis; soil organisms

Cellulomonas: coryneform morphology; cellulose digested; facultative aerobe

Kurthia: rods with rounded ends occurring in chains; coccoid later

Brevibacterium: coccus–rod morphogenesis; cheese, skin

Propionic acid bacteria: anaerobic to aerotolerant; rods or filaments, branching

Propionibacterium: nonmotile; anaerobic to aerotolerant; produce propionic acid and acetic acid; dairy products (Swiss cheese); skin, may be pathogenic

Eubacterium: obligate anaerobes; produce mixture of organic acids, including butyric, acetic, formic, and lactic; intestine, infections of soft tissue, soil; may be pathogenic; probably the predominant member of the intestinal flora

Obligate anaerobes

Bifidobacterium: smooth microcolony, no filaments; coryneform cells common; found in intestinal tract of breast-fed infants

Acetobacterium: acetogen; sediments and sewage

Butyrivibrio: curved rods; rumen

Thermoanaerobacter: rods, thermophilic, found in hot springs

Actinomycetes: filamentous, often branching; highly diverse

Group I. Actinomycetes: not acid-fast; facultatively aerobic; mycelium not formed; branching filaments may be produced; rod, coccoid, or coryneform cells

Actinomyces: anaerobic to facultatively aerobic; filamentous microcolony, but filaments transitory and fragment into coryneform cells; may be pathogenic for humans or other animals; found in oral cavity

Other genera: *Arachnia, Bacterionema, Rothia, Agromyces*

Group II. Mycobacteria: acid-fast, filaments transitory

Mycobacterium: pathogens, saprophytes; obligate aerobes; lipid content of cells and cell walls high; waxes, mycolic acids; simple nutrition; growth slow; tuberculosis, leprosy, granulomas, avian tuberculosis; also soil organisms; hydrocarbon oxidizers

Group III. Nitrogen-fixing actinomycetes: nitrogen-fixing symbionts of plants; true mycelium produced

Frankia: forms nodules of two types on various plant roots; probably microaerophilic; grows slowly; fixes N_2

Group IV. Actinoplanes: true mycelium produced; spores formed, borne inside sporangia

Actinoplanes, Streptosporangium

Group V. Dermatophilus group: mycelial filaments divide transversely, and in at least two longitudinal planes, to form masses of motile, coccoid elements; aerial mycelium absent; occasionally responsible for epidermal infections

Dermatophilus, Geodermatophilus

Group VI. Nocardias: mycelial filaments commonly fragment to form coccoid or elongate elements; aerial spores occasionally produced; sometimes acid-fast; lipid content of cells and cell wall very high

Nocardia: common soil organisms; obligate aerobes; many hydrocarbon utilizers

Rhodococcus: soil saprophytes, also common in gut of various insects; utilize hydrocarbons

Group VII. Streptomycetes: mycelium remains intact, abundant aerial mycelium and long spore chains

Streptomyces: Nearly 500 recognized species, many produce antibiotics

Other genera (differentiated morphologically): *Streptoverticillium, Sporichthya, Kitasatoa*

Group VIII. Micromonosporas group: mycelium remains intact; spores formed singly, in pairs, or short chains; several thermophilic; saprophytes found in soil, rotting plant debris; one species produces endospores

Micromonospora, Microbispora, Thermobispora, Thermoactinomyces, Thermomonospora

[a]Phylogenetically, all species except for *Acetobacterium, Butyrivibrio*, and *Thermoanaerobacter* are *Actinobacteria*.

multinucleate sporophores followed by separation of the individual cells directly into spores (**Figure 18.20**). Differences in the shape and arrangement of aerial filaments and spore-bearing structures of various species are among the fundamental features used in classifying the *Streptomyces* species (**Figure 18.21**). The conidia and sporophores are often pigmented and contribute a characteristic color to the mature colony (**Figure 18.22**). The dusty appearance of the mature colony, its compact nature, and its color make detection of *Streptomyces* colonies on agar plates relatively easy (Figure 18.22*b*).

Ecology and Isolation of *Streptomyces*

Although a few streptomycetes are aquatic, they are primarily soil organisms. In fact, the characteristic earthy odor of soil is caused by the production of a series of complex metabolites all called *geosmin* by streptomycetes. Alkaline to neutral soils are more favorable for the development of *Streptomyces* than are acid soils. Moreover, higher numbers of *Streptomyces* are found in well-drained soils (such as sandy loams or soils covering limestone), where conditions are more likely to be aerobic, than in waterlogged soils, which quickly become anoxic.

(a)

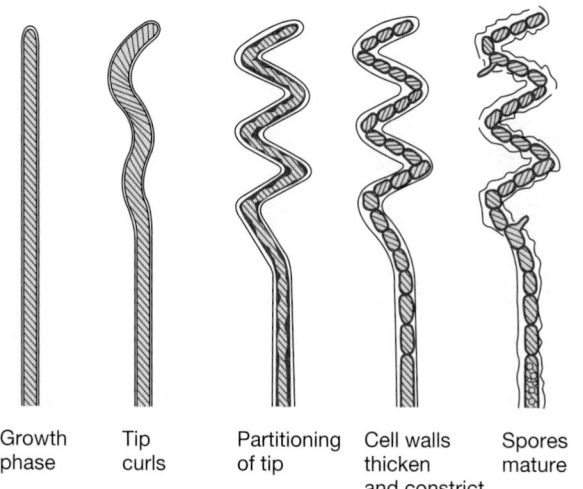

(b)

Figure 18.19 Spore-bearing structures of actinomycetes. Phase-contrast micrographs. *(a)* *Streptomyces*, a monoverticillate type. *(b)* *Streptomyces*, a closed spiral type. Filaments are about 0.8 μm wide in both types. Compare these photos with the art in Figure 18.21.

Isolation of *Streptomyces* from soil is relatively easy: A suspension of soil in sterile water is diluted and spread on selective agar medium, and the plates are incubated aerobically at 25°C (⮌ Figure 15.4). Media selective for *Streptomyces* contain mineral salts plus polymeric substances such as starch or casein as organic nutrients. Streptomycetes typically produce extracellular hydrolytic enzymes that permit utilization of polysaccharides (starch, cellulose, and hemicellulose), proteins, and fats, and

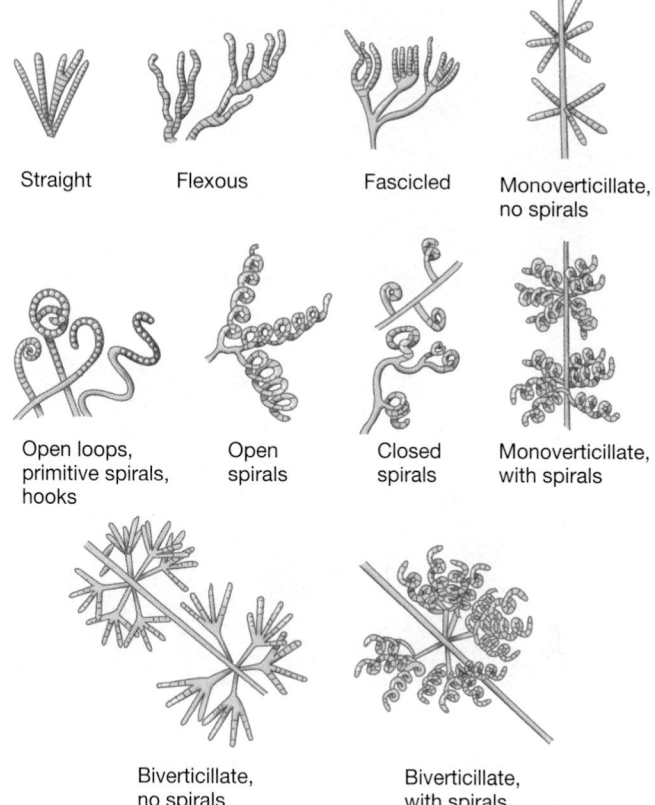

Figure 18.21 Morphologies of spore-bearing structures in the streptomycetes. A given species of *Streptomyces* produces only one morphological type of spore-bearing structure. The term "verticillate" means "whorls."

some strains can use hydrocarbons, lignin, tannin, and other polymers. After incubation for 5–7 days in air, the plates are examined for the presence of the characteristic *Streptomyces* colonies (Figure 18.22 and **Figure 18.23**), and spores from colonies can be restreaked to isolate pure cultures.

Antibiotics of *Streptomyces*

Perhaps the most striking physiological property of the streptomycetes is the extent to which they produce *antibiotics* (**Table 18.10** on page 561). Evidence for antibiotic production is often seen on the agar plates used in their initial isolation: Adjacent colonies of other bacteria show zones of inhibition (Figures 18.22*a* and 18.23*a*; ⮌ Figure 15.4).

About 50% of all *Streptomyces* isolated have been found to be antibiotic producers. Over 500 distinct antibiotics are produced by streptomycetes and many more are suspected (⮌ Sections 15.4 and 26.7); most of these have been identified chemically (Figure 18.23*b*). Some species produce more than one antibiotic, and often the several kinds produced by one organism are chemically unrelated. Although an antibiotic-producing organism is resistant to its own antibiotics, it usually remains sensitive to antibiotics produced by other streptomycetes. Many genes are required to encode the enzymes for antibiotic synthesis, and because of this, the genomes of *Streptomyces* species are typically quite large (8 megabase pairs and larger; ⮌ Table 12.1). More

Growth phase | Tip curls | Partitioning of tip | Cell walls thicken and constrict | Spores mature

Figure 18.20 Spore formation in *Streptomyces*. Diagram of stages in the conversion of an aerial hypha (sporophore) into spores (conidia).

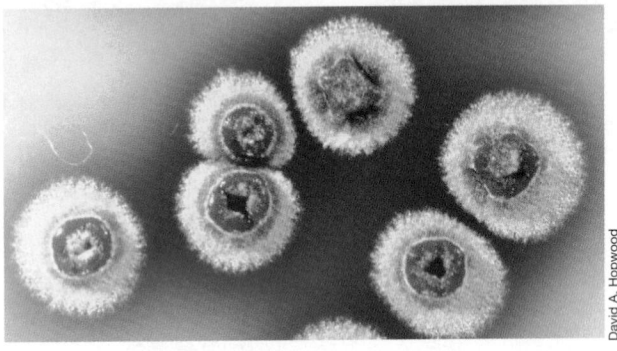

Figure 18.22 Streptomycetes. *(a)* Colonies of *Streptomyces* and other soil bacteria derived from spreading a soil dilution on a casein–starch agar plate. The *Streptomyces* colonies are of various colors (several black *Streptomyces* colonies are near the top of the plate) but can easily be identified by their opaque, rough, nonspreading morphology. *(b)* Close-up photo of colonies of *Streptomyces coelicolor*.

than 60 streptomycete antibiotics have been used in human and veterinary medicine. Some of the most common of these are listed in Table 18.10.

Ironically, despite the extensive research done on antibiotic-producing streptomycetes by the antibiotic industry and the fact that *Streptomyces* antibiotics are a multibillion-dollar-a-year industry, the ecology of *Streptomyces* remains poorly understood. The interactions of these organisms with other bacteria and the ecological rationale for antibiotic production remains an area about which we know relatively little. One hypothesis for why *Streptomyces* species produce antibiotics is that antibiotic production, which is linked to sporulation (a process itself triggered by nutrient depletion), might be a mechanism to inhibit the growth of other organisms competing with *Streptomyces* cells for limiting nutrients. This would allow the *Streptomyces* to complete the sporulation process and form a dormant structure that would have increased chances of survival.

MiniQuiz

- Contrast spores and sporulation in *Streptomyces* and *Bacillus* species.
- Name two clinically useful antibiotics produced by streptomycetes.
- Why might antibiotic production be of advantage to streptomycetes?

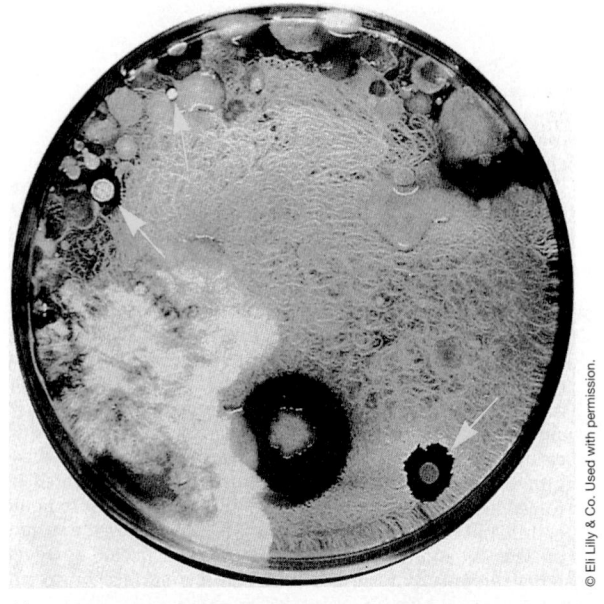

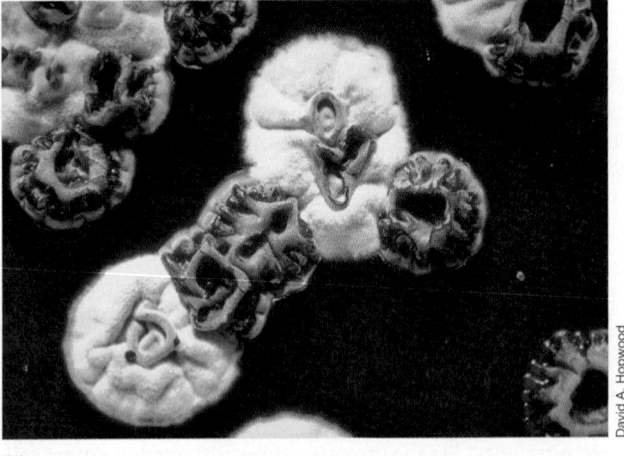

Figure 18.23 Antibiotics from *Streptomyces*. *(a)* Antibiotic action of soil microorganisms on a crowded plate. The smaller colonies surrounded by inhibition zones (arrows) are streptomycetes; the larger, spreading colonies are *Bacillus* species, some of which are also producing antibiotics. *(b)* The red-colored antibiotic undecylprodigiosin is being excreted by colonies of *S. coelicolor*.

Ⅱ Cyanobacteria and Prochlorophytes

18.7 Cyanobacteria

Key Genera: *Synechococcus, Oscillatoria, Anabaena*
Cyanobacteria comprise a large, morphologically and ecologically heterogeneous group of oxygenic, phototrophic *Bacteria*. Cyanobacteria represent one of the major phyla of *Bacteria* and show a distant relationship to gram-positive bacteria (⮂ Figure 17.1). As we saw in Section 16.3, these organisms were the first oxygen-evolving phototrophic organisms on Earth, and over

Table 18.10 *Some common antibiotics synthesized by species of* Streptomyces *and related* Actinobacteria

Chemical class	Common name	Produced by	Active against[a]
Aminoglycosides	Streptomycin	*S. griseus*[b]	Most gram-negative *Bacteria*
	Spectinomycin	*Streptomyces* spp.	*Mycobacterium tuberculosis*, penicillinase-producing *Neisseria gonorrhoeae*
	Neomycin	*S. fradiae*	Broad spectrum, usually used in topical applications because of toxicity
Tetracyclines	Tetracycline	*S. aureofaciens*	Broad spectrum, gram-positive and gram-negative *Bacteria*, rickettsias and chlamydias, *Mycoplasma*
	Chlortetracycline	*S. aureofaciens*	As for tetracycline
Macrolides	Erythromycin	*Saccharopolyspora erythraea*	Most gram-positive *Bacteria*, frequently used in place of penicillin; *Legionella*
	Clindamycin	*S. lincolnensis*	Effective against obligate anaerobes, especially *Bacteroides fragilis*, the major cause of anaerobic peritoneal infections
Polyenes	Nystatin	*S. noursei*	Fungi, especially *Candida* (a yeast) infections
	Amphotericin B	*S. nodosus*	Fungi
None	Chloramphenicol	*S. venezuelae*	Broad spectrum; drug of choice for typhoid fever

[a]Most antibiotics are effective against several different *Bacteria*. The entries in this column refer to the common clinical application of a given antibiotic. The structures and mode of action of many of these antibiotics are discussed in Sections 26.6–26.9.
[b]All genus names beginning with an "S." are species of *Streptomyces*.

billions of years converted the originally anoxic atmosphere of Earth to the highly oxic atmosphere we see today.

Structure and Classification of Cyanobacteria

The morphological diversity of the cyanobacteria is impressive (**Figure 18.24**). Both unicellular and filamentous forms are known, and there is considerable variation within these morphological types. Cyanobacteria can be divided into five morphological groups: (1) unicellular, dividing by binary fission (Figure 18.24*a*); (2) unicellular, dividing by multiple fission (colonial) (Figure 18.24*b*); (3) filamentous, containing differentiated cells called heterocysts that function in nitrogen fixation (Figure 18.24*d*, and see Figure 18.26); (4) filamentous nonheterocystous

forms (Figure 18.24*c*); and (5) branching filamentous species (Figure 18.24*e*). **Table 18.11** lists some major genera currently recognized in each group. Cyanobacterial cells range in size from 0.5–1 μm in diameter to cells as large as 40 μm in diameter. Phylogenetically, cyanobacteria group along morphological lines in most cases. Filamentous, heterocystous, and nonheterocystous species form distinct groups, as do the branching forms. However, unicellular cyanobacteria are highly diverse, with different representatives showing phylogenetic relationships to different morphological groups.

The cell wall of cyanobacteria is similar to that of gram-negative bacteria, and peptidoglycan is present in the walls. Many cyanobacteria produce extensive mucilaginous envelopes,

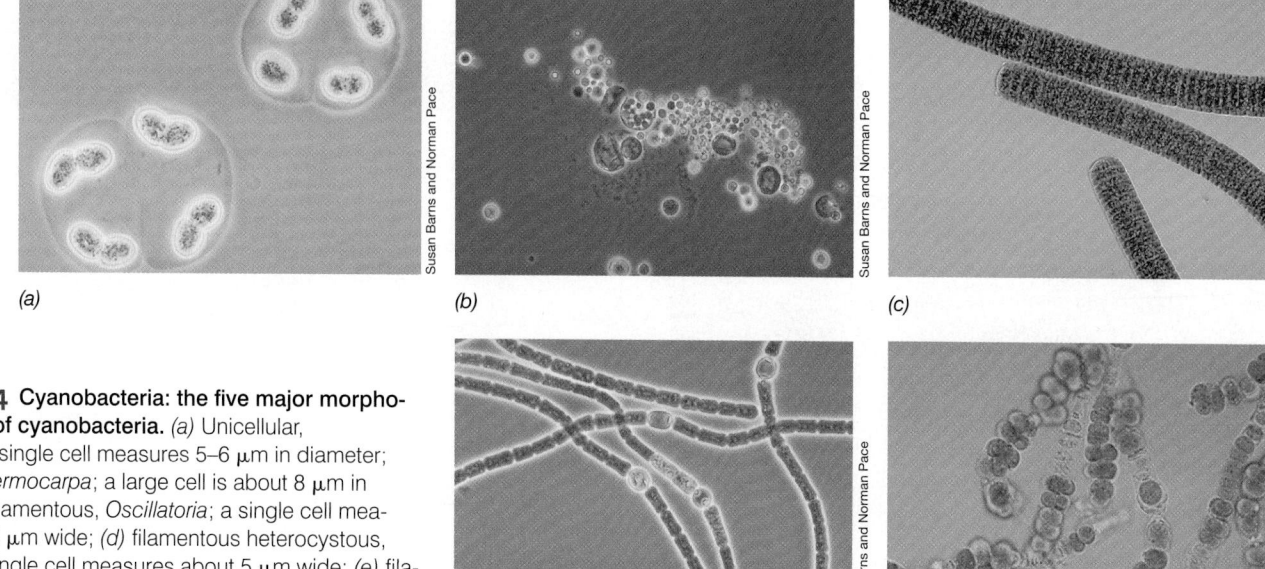

Figure 18.24 Cyanobacteria: the five major morphological types of cyanobacteria. *(a)* Unicellular, *Gloeothece*; a single cell measures 5–6 μm in diameter; *(b)* colonial, *Dermocarpa*; a large cell is about 8 μm in diameter; *(c)* filamentous, *Oscillatoria*; a single cell measures about 15 μm wide; *(d)* filamentous heterocystous, *Anabaena*; a single cell measures about 5 μm wide; *(e)* filamentous branching, *Fischerella*; a cell is about 10 μm wide. Micrographs in parts a, b, and d, phase contrast; parts c and e, bright field.

Table 18.11 *Genera and grouping of cyanobacteria*

Group	Genera
Group I. Unicellular: single cells or cell aggregates	*Gloeothece* (Figure 18.24a), *Gloeobacter, Synechococcus, Cyanothece, Gloeocapsa, Synechocystis, Chamaesiphon, Merismopedia*
Group II. Pleurocapsalean: reproduce by formation of small spherical cells called baeocytes produced through multiple fission	*Dermocarpa* (Figure 18.24b), *Xenococcus, Dermocarpella, Pleurocapsa, Myxosarcina, Chroococcidiopsis*
Group III. Oscillatorian: filamentous cells that divide by binary fission in a single plane	*Oscillatoria* (Figure 18.24c), *Spirulina, Arthrospira, Lyngbya, Microcoleus, Pseudanabaena*
Group IV. Nostocalean: filamentous cells that produce heterocysts	*Anabaena* (Figure 18.24d), *Nostoc, Calothrix, Nodularia, Cylindrospermum, Scytonema*
Group V. Branching: cells divide to form branches	*Fischerella* (Figure 18.24e), *Stigonema, Chlorogloeopsis, Hapalosiphon*

or sheaths, that bind groups of cells or filaments together (Figure 18.24*a*). The photosynthetic membrane system is often complex and multilayered (♷ Figure 13.10*c*), although the thylakoid membranes are regularly arranged in concentric circles around the periphery of the cytoplasm in some of the structurally simpler cyanobacteria (**Figure 18.25**).

Cyanobacteria produce chlorophyll *a*, and all of them also have characteristic biliprotein pigments, **phycobilins** (♷ Figure 13.10*a, b*), which function as accessory pigments in photosynthesis. One class of phycobilins, phycocyanins, are blue and, together with the green chlorophyll *a*, are responsible for the blue-green color of most cyanobacteria. However, some cyanobacteria produce phycoerythrin, a red phycobilin, and species producing phycoerythrin are red or brown.

Structural Variations: Gas Vesicles and Heterocysts

Among the cytoplasmic structures seen in many aquatic cyanobacteria are gas vesicles (♷ Section 3.11). The function of gas vesicles is to regulate cell buoyancy such that cells can remain in a position in the water column where light intensity is optimal for photosynthesis. Some filamentous cyanobacteria, such as *Anabaena*, form specialized cells called *heterocysts* distributed regularly at intervals along a filament, or sometimes located at one end of a filament (**Figure 18.26**). Heterocysts arise from differentiation of vegetative cells and are the

sites of nitrogen fixation (♷ Section 13.14) in heterocystous cyanobacteria.

Heterocysts have intercellular connections with adjacent vegetative cells that allow for mutual exchange of materials between these cells. The products of photosynthesis move from vegetative cells to heterocysts, and fixed nitrogen moves from heterocysts to vegetative cells (Figure 18.26*b*). Heterocysts lack photosystem II, the oxygen-evolving photosystem that generates reducing power from H_2O (♷ Section 13.5). Without photosystem II, heterocysts are unable to fix CO_2 and thus lack

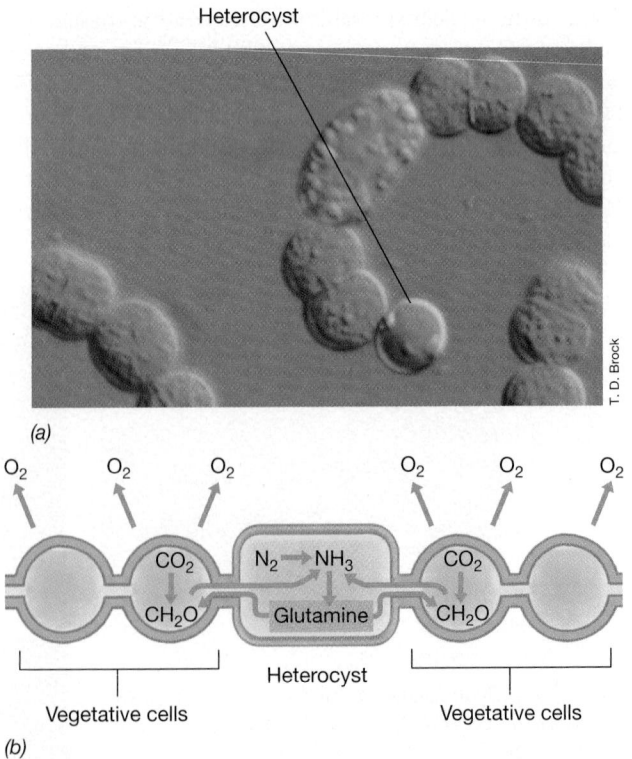

(a)

(b)

Figure 18.26 Heterocysts. *(a)* Heterocysts in the cyanobacterium *Anabaena. (b)* Heterocyst function. The heterocyst lacks oxygen-producing ability and obtains the needed reductant for nitrogen fixation from organic compounds produced by adjacent vegetative cells. Glutamine is the form of fixed nitrogen transported from heterocysts to vegetative cells.

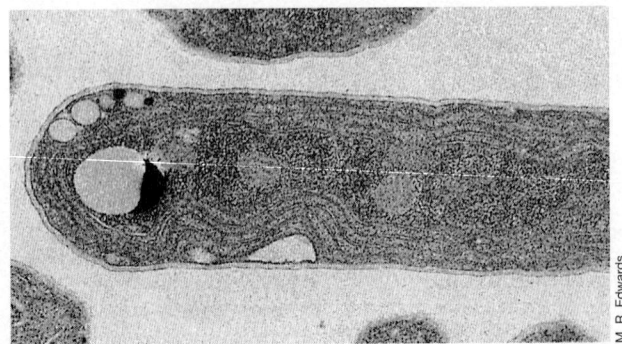

Figure 18.25 Thylakoids in cyanobacteria. Electron micrograph of a thin section of the cyanobacterium *Synechococcus lividus*. A cell is about 5 μm in diameter. Note thylakoid membranes running parallel to the cell wall.

the necessary electron donor (pyruvate) for nitrogen fixation. However, fixed carbon is imported by the heterocyst from adjacent vegetative cells (Figure 18.26*b*), and this is oxidized to yield electrons for nitrogen fixation. Along with the absence of photosystem II, heterocysts are surrounded by a thickened cell wall that slows the diffusion of O_2 into the cell. Because of the oxygen lability of the enzyme nitrogenase, the heterocyst thus maintains an anoxic environment, and by doing so stabilizes the nitrogen-fixing system in organisms that are not only aerobic but also oxygen producing.

Cyanophycin and Other Structures

A structure called *cyanophycin* can be seen in electron micrographs of many cyanobacteria. This structure is a copolymer of aspartic acid and arginine and can constitute up to 10% of the cell mass. Cyanophycin is a nitrogen storage product, and when nitrogen in the environment becomes deficient, this polymer is broken down and used as a cellular nitrogen source.

Many cyanobacteria exhibit gliding motility (♊ Section 3.14). Gliding occurs only when the cell or filament is in contact with a solid surface or with another cell or filament. In some cyanobacteria, gliding is not a simple translational movement but is accompanied by rotations, reversals, and flexings of filaments. Most gliding species exhibit directional movement toward light (phototaxis), and chemotaxis (♊ Section 3.15) may occur as well.

Among the filamentous cyanobacteria, the filaments can fragment to form small pieces called *hormogonia* (**Figure 18.27**); these break away from the filaments and glide off. In some species, resting structures called *akinetes* (Figure 18.27*c*) form, which protect the organism during periods of darkness, dessication, or cold. Akinetes are cells with thickened outer walls. When conditions are once again favorable, akinetes can germinate by breaking down the outer wall and growing out a new vegetative filament.

Physiology of Cyanobacteria

The nutrition of cyanobacteria is simple. Vitamins are not required, and nitrate or ammonia are used as nitrogen sources. Nitrogen-fixing species are common. Most species tested are obligate phototrophs, being unable to grow in the dark on organic compounds. However, some cyanobacteria can assimilate simple organic compounds such as glucose and acetate if light is present (photoheterotrophy). A few cyanobacteria, mainly filamentous species, can grow in the dark on glucose or sucrose, using the sugar as both carbon and energy source.

Several metabolic products of cyanobacteria are of considerable practical importance. Some cyanobacteria produce potent neurotoxins, and during water blooms when massive accumulations of cyanobacteria develop, animals ingesting such water may be killed. Many cyanobacteria are also responsible for the production of earthy odors and flavors in some freshwater, and if such waters are used as drinking water sources, aesthetic problems may arise. The major compound produced is geosmin, a substance also produced by many actinomycetes (Section 18.6).

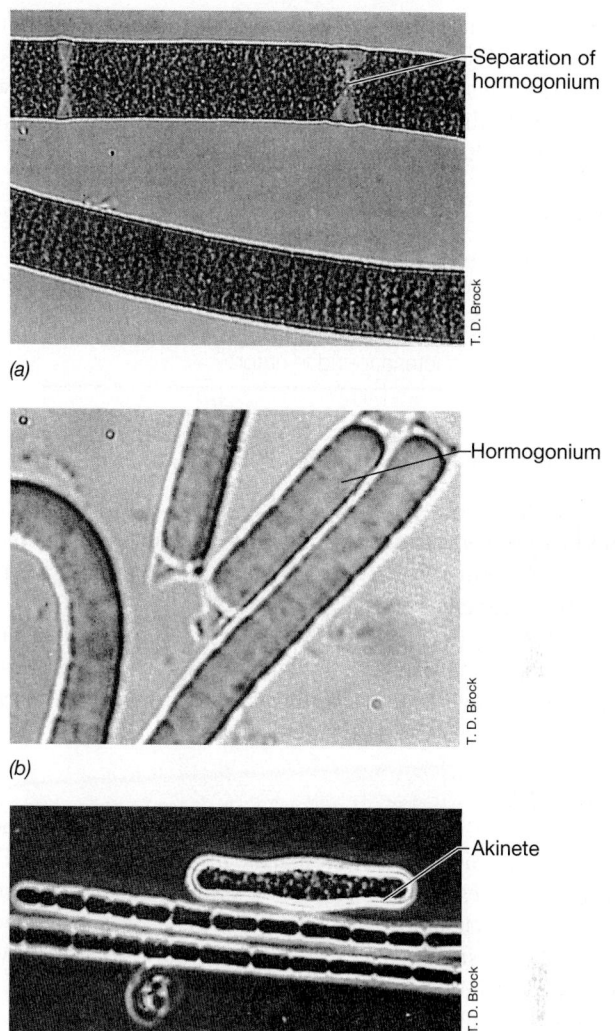

(a)

(b)

(c)

Figure 18.27 Structural differentiation in filamentous cyanobacteria. *(a)* Initial stage of hormogonium formation in *Oscillatoria*. Notice the empty spaces where the hormogonium is separating from the filament. *(b)* Hormogonium of a smaller *Oscillatoria* species. Notice that the cells at both ends are rounded. Differential interference contrast microscopy. *(c)* Akinete (resting spore) of *Anabaena* in a phase-contrast micrograph.

Ecology and Phylogeny of Cyanobacteria

Cyanobacteria are widely distributed in nature in terrestrial, freshwater, and marine habitats. In general, they are more tolerant of environmental extremes than are algae (eukaryotic cells) and are often the dominant or sole oxygenic phototrophic organisms in hot springs, saline lakes, desert soils, and other extreme environments. In some of these environments, cyanobacterial mats of variable thickness may form (♊ Figures 16.2 and 23.9). Freshwater lakes, especially those rich in inorganic nutrients, often develop blooms of cyanobacteria, especially in late summer when temperatures are warmest (♊ Figure 23.1). A few cyanobacteria are symbionts of liverworts, ferns, and cycads, and a number are phototrophic components of lichens, a symbiosis between a phototroph and a fungus (♊ Section 25.1). Small unicellular cyanobacteria, such as *Synechococcus* and the related organism *Prochlorococcus* (Section 18.8), are the most abundant phototrophs in the oceans, where they carry out photosynthesis that is responsible for a significant percentage of the CO_2 fixed globally. It is thus apparent that cyanobacteria not only prepared Earth for more diverse and higher life forms, but are key factors in maintaining the biosphere today.

MiniQuiz

- What is the major way in which cyanobacteria differ from phototrophic purple bacteria?
- What is a heterocyst and what is its function?
- Where are cyanobacteria found in nature?

18.8 Prochlorophytes

Key Genera: *Prochloron, Prochlorothrix, Prochlorococcus*

Prochlorophytes are oxygenic phototrophs that contain chlorophyll *a* and *b* but do not contain phycobilins. Prochlorophytes therefore resemble both cyanobacteria (because they are prokaryotes and have chlorophyll *a*) and the green plant/green alga chloroplast (because they contain chlorophyll *b* instead of phycobilins). Phylogenetically, prochlorophytes form a clade within the cyanobacteria phylum.

Prochloron

Prochloron was the first prochlorophyte discovered. It is found in nature as a symbiont of certain marine invertebrates, and cells of *Prochloron* expressed from cavities of animals are roughly spherical (**Figure 18.28**) and 8–10 μm in diameter. Electron micrographs of thin sections of *Prochloron* (Figure 18.28) show an extensive thylakoid membrane system similar to that of the chloroplast ($\wp$ Figure 13.5). Further evidence that *Prochloron* is phylogenetically a species of *Bacteria* is the presence of

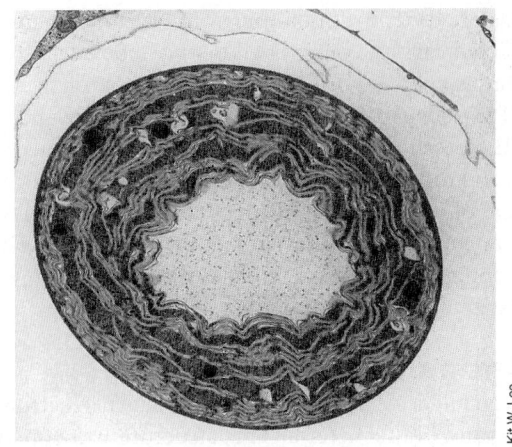

Figure 18.28 Electron micrograph of the prochlorophyte *Prochloron*. Note the extensive thylakoid membranes. Cells are about 10 μm in diameter.

N-acetylmuramic acid in the cell walls, indicating that peptidoglycan is present ($\wp$ Section 3.6). The pigments of *Prochloron* initially suggested that it was an ancestor of the green plant chloroplast, which also contain chlorophylls *a* and *b* and lack phycobilins. However, closer examination of the phylogeny of *Prochloron* has shown that prochlorophytes, along with cyanobacteria and the green plant chloroplast, all shared an ancestor in the distant past and that prochlorophytes are not themselves ancestral to green algae or the green plant chloroplast.

Prochlorothrix and *Prochlorococcus*

Prochlorothrix is a filamentous prochlorophyte (**Figure 18.29**) first isolated from a freshwater lake. Like *Prochloron*, *Prochlorothrix* contains chlorophylls *a* and *b* and lacks phycobilins, although the

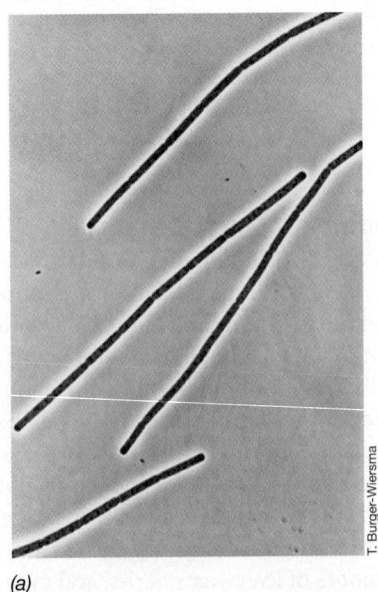
(a)

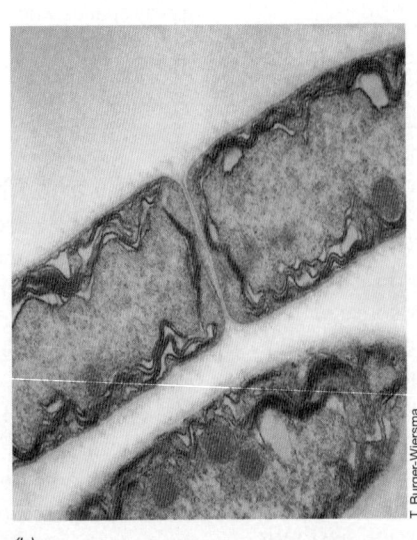
(b)

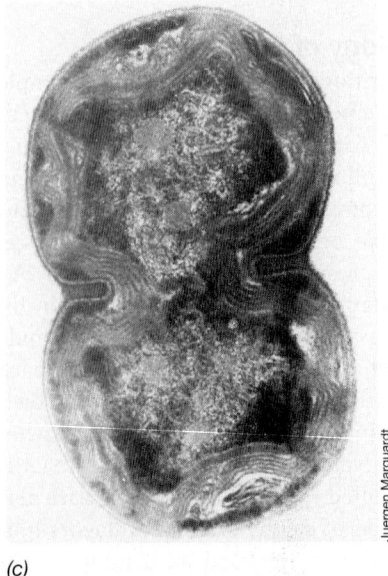
(c)

Figure 18.29 The prochlorophytes *Prochlorothrix* and *Acaryochloris*. *(a)* Phase-contrast micrograph. *(b)* Electron micrograph of thin section showing arrangement of membranes in the filamentous cells. The diameter of a cell is about 2 μm. *(c)* *Acaryochloris*, thin section. A cell is about 1.5 μm in diameter.

thylakoid membranes are less well developed than in *Prochloron* (compare Figures 18.28 and 18.29*b*). *Prochlorococcus*, a unicellular phototroph, inhabits the open oceans and is probably the most abundant and smallest oxygenic phototroph on Earth. Cells of *Prochlorococcus* are 0.5–0.8 μm in diameter (Figure 23.18), and like other prochlorophytes, contain chlorophyll *b*. However, *Prochlorococcus* lacks true chlorophyll *a* and produces instead a modified form of chlorophyll *a*, divinyl chlorophyll *a*.

Because its numbers in the oceans are very large, up to as many as 10^5 per milliliter of seawater, *Prochlorococcus* has considerable ecological significance as a primary producer. *Prochlorococcus* is found in temperate to tropical oligotrophic (nutrient-poor) open ocean waters. Two genetically distinct populations exist (Figure 22.17), one adapted to the lower light levels occurring deeper in the ocean's photic zone and the other adapted to higher light levels nearer the surface (euphotic zone). The adaptations relate to differences in the content and ratio of the chlorophylls in *Prochlorococcus*.

Other prochlorophytes have been isolated, including *Acaryochloris* (Figure 18.29*c*), which contains chlorophyll *d* as its major pigment. Chlorophyll *d* is common in red algae and absorbs near infrared light of 740 nm, light that is unavailable to oxygenic phototrophs containing other chlorophyll pigments.

MiniQuiz

- How are cyanobacteria, prochlorophytes, and the green plant chloroplast similar, and how do they differ?
- Of what ecological significance is *Prochlorococcus*?

 ## Chlamydia

We now consider the chlamydia, a group of very small gram-negative bacteria that cause some serious human and animal diseases.

18.9 The Chlamydia

Key Genera: *Chlamydia, Chlamydophila*

Organisms of the genera *Chlamydia* and *Chlamydophila* are obligate intracellular parasites with poor metabolic capacities; they constitute the phylum *Chlamydia* (Figure 17.1). Several species are recognized: *Chlamydophila psittaci*, the causative agent of the disease psittacosis; *Chlamydia trachomatis*, the causative agent of trachoma and a variety of other human diseases; and *Chlamydophila pneumoniae*, the cause of some respiratory syndromes (**Table 18.12**). Psittacosis is an epidemic disease of birds that is occasionally transmitted to humans and causes pneumonia-like symptoms. Trachoma is a debilitating disease of the eye characterized by vascularization and scarring of the cornea. Trachoma is the leading cause of blindness in humans. Other strains of *C. trachomatis* infect the genitourinary tract, and chlamydial infections are currently one of the leading sexually transmitted diseases (Section 33.13).

Molecular and Metabolic Properties

Besides being pathogens, chlamydias are intriguing because of the biological problems they pose. Biochemical studies show that the chlamydias have gram-negative-type cell walls, and they have both DNA and RNA; that is, they are clearly cellular. Electron microscopy of thin sections of mouse cells infected with chlamydias shows small bacterial cells dividing by binary fission (**Figure 18.30**). The biosynthetic capacities of the chlamydias are much more limited than even the rickettsias, the other group of obligate intracellular parasites known among the *Bacteria* (Section 17.13). Indeed, it was originally thought that chlamydias were "energy parasites," obtaining not only biosynthetic intermediates from their hosts, as do the rickettsias, but also ATP. However, this hypothesis has been questioned following the sequencing of the genome of *C. trachomatis*. The approximately 1-Mbp chromosome of *C. trachomatis* contains genes encoding proteins for ATP synthesis and even contains a complement of genes encoding peptidoglycan biosynthetic functions. This suggests that this organism may well contain peptidoglycan even though chemical analyses of chlamydial cells have been negative. Nevertheless, the chlamydias still probably have one of the simplest biochemical capacities of all known *Bacteria*, and this is summarized in **Table 18.13**.

Interestingly, the *C. trachomatis* genome lacks a gene encoding the protein FtsZ, a key protein in septum formation during cell division (Section 5.2). This protein was previously thought to be indispensable for growth of all prokaryotes. Moreover, genes are present in *C. trachomatis* that have a distinct "eukaryotic look" to them, suggesting that *C. trachomatis* has picked up some host genes that may encode functions that assist it in its pathogenic lifestyle (Table 18.12; Section 33.13).

Life Cycle of *Chlamydia*

The life cycle of a typical chlamydial species is shown in **Figure 18.31**. Two cellular types are seen in the life cycle: (1) a small, dense cell, called an *elementary body*, which is relatively resistant to drying and is the means of dispersal, and (2) a larger, less dense cell, called a *reticulate body*, which divides by binary fission and is the vegetative form.

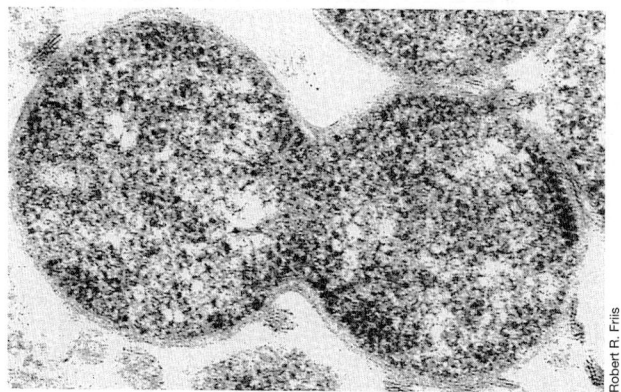

Figure 18.30 *Chlamydia.* Thin-section electron micrograph of a dividing reticulate body of *Chlamydia psittaci* within a mouse tissue-culture cell. A single chlamydial cell is about 1 μm in diameter.

UNIT 6

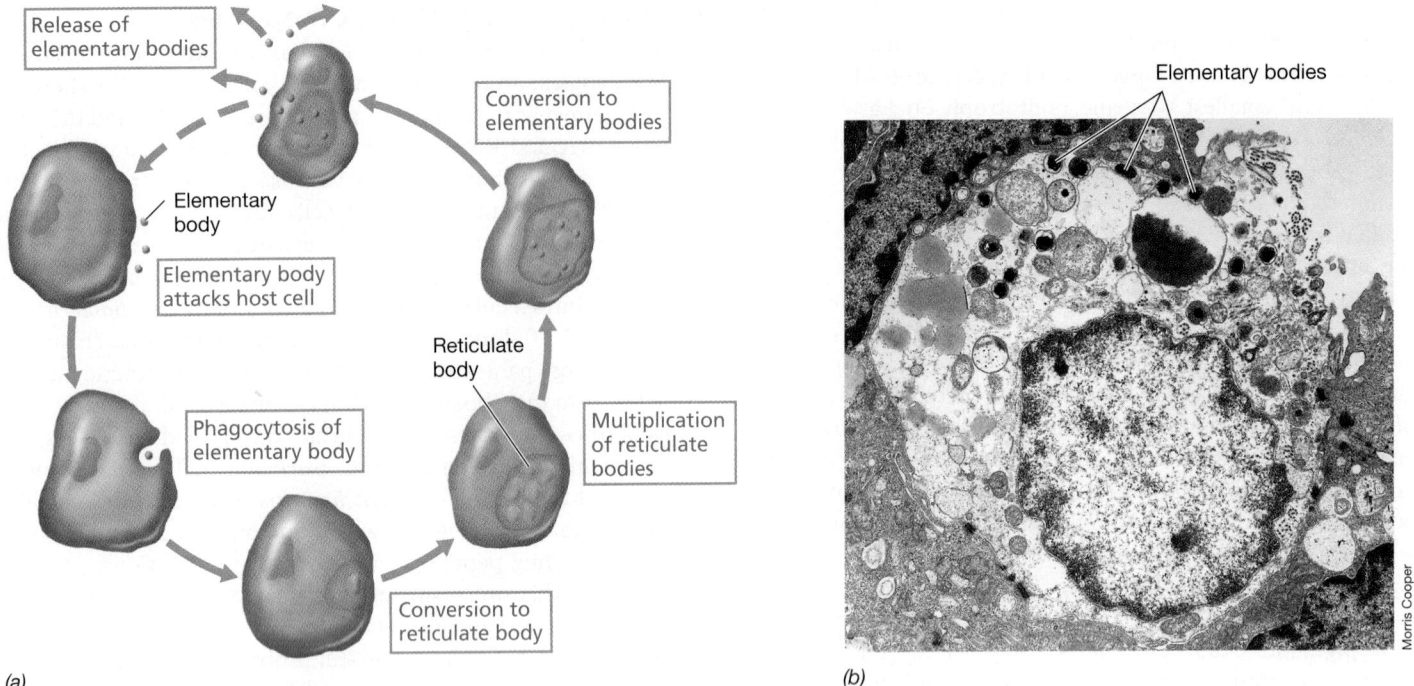

Figure 18.31 The infection cycle of a chlamydia. *(a)* Schematic diagram of the cycle: The entire cycle takes about 48 h. *(b)* Human chlamydial infection. Elementary bodies (~0.3 μm in diameter) are the infectious form and reticulate bodies (~1 μm in diameter), the multiplying form. An infected fallopian tube cell is bursting, releasing mature elementary bodies.

Elementary bodies are nonmultiplying cells specialized for infectious transmission. By contrast, reticulate bodies are noninfectious forms that function only to multiply inside host cells to form a large inoculum for transmission. Unlike the rickettsias, the chlamydias are not transmitted by arthropods but are primarily airborne invaders of the respiratory system—hence the significance of resistance to drying of the elementary bodies. A dividing reticulate body can be seen in Figure 18.30. After a number of cell divisions, these vegetative cells are converted into elementary bodies that are released when the host cell disintegrates (Figure 18.31*b*) and can then infect other nearby host cells. Generation times of 2–3 h have been measured for reticulate bodies, which are considerably faster than those found for the rickettsias.

Table 18.12 *Differential characteristics of species of the genera* Chlamydia *and* Chlamydophila

Characteristic	Chlamydia trachomatis	Chlamydophila psittaci	Chlamydophila pneumoniae
Hosts	Humans	Birds, mammals, occasionally humans	Humans
Usual site of infection	Mucous membrane	Multiple sites	Respiratory mucosa
Human-to-human transmission	Common	Rare	Probable
Percent homology to *C. trachomatis* DNA by DNA–DNA hybridization[a]	100	10	10
DNA, kilobase pairs/genome (*Escherichia coli* = 4639)	1000	550	~1000
Human diseases	Trachoma, otitis media, nongonococcal urethritis (males), urethral inflammation (females), lymphogranuloma venereum, cervicitis	Psittacosis	Respiratory syndromes
Domesticated animal diseases	—	Avian chlamydiosis (parrots, parakeets), pneumonia, synovial tissue arthritis, or conjunctivitis (kittens, lambs, calves, piglets, foals)	—

[a]For discussion of DNA–DNA hybridization, see Section 16.11.

Table 18.13 *The biology of obligate intracellular parasites: rickettsias, chlamydias, and viruses*

Property	Rickettsias	Chlamydias	Viruses
Structural			
Nucleic acid	RNA and DNA	RNA and DNA	Either RNA or DNA (single- or double-stranded), never both
Ribosomes	Present	Present	Absent
Cell wall	Peptidoglycan present	Peptidoglycan present[a]	No wall
Structural integrity during multiplication	Maintained	Maintained	Lost
Metabolic capacities			
Macromolecular synthesis	Carried out	Carried out	Only with use of host machinery
ATP-generating system	Present	Present[a]	Absent
Capable of oxidizing glutamate	Yes	No	No
Sensitivity to antibacterial antibiotics	Sensitive	Sensitive (except for penicillin)	Resistant
Phylogeny	*Alphaproteobacteria*	Chlamydial phylum	Not cells

[a]The genome of *Chlamydia trachomatis* (and several other chlamydial species) has been sequenced (⮌ Section 12.1), and genes for peptidoglycan synthesis and ATP synthesis are present. However, the lack of penicillin sensitivity of the chlamydia raises doubt whether peptidoglycan is synthesized.

In sum, the chlamydias appear to have evolved an efficient and effective survival strategy including parasitizing the resources of the host (Table 18.13) and the production of resistant cell forms for transmission. It is thus not surprising that chlamydias have been associated with so many different disease syndromes (Table 18.12; ⮌ Section 33.13).

MiniQuiz

- From data in Table 18.13, describe how chlamydias can be differentiated from rickettsias and viruses.
- What is the difference between an elementary body and a reticulate body?

 IV The *Planctomycetes*

We now take a look at a very unusual group of bacteria that are not only phylogenetically unique, but also stretch our definition of the word "prokaryote" to the limit.

18.10 *Planctomyces*: A Phylogenetically Unique Stalked Bacterium

Key Genera: *Planctomyces, Pirellula, Gemmata*

The *Planctomyces/Pirellula* group, commonly called the planctomycetes, contains a number of morphologically unique bacteria including the genera *Planctomyces, Pirellula, Gemmata*, and *Isosphaera*. The best studied of these is *Planctomyces* (**Figure 18.32**). In Section 17.16 we considered the stalked proteobacterium *Caulobacter*. *Planctomyces* is also a stalked bacterium. However, unlike *Caulobacter*, the stalk of *Planctomyces* consists of protein and does not contain a cell wall or cytoplasm (compare Figure 18.32 with Figure 17.39). The *Planctomyces* stalk presumably functions in attachment, but it is a much narrower and finer structure than the prosthecal stalk of *Caulobacter*.

Other Features of *Planctomyces*

Planctomyces and relatives are also of interest because they lack peptidoglycan and their cell walls are of an S-layer type (⮌ Section 3.8), consisting of protein containing large amounts of cysteine (as cystine) and proline. As would be expected of organisms lacking peptidoglycan, these organisms are resistant to antibiotics such as penicillin and cephalosporin that disrupt peptidoglycan synthesis.

Like *Caulobacter* (⮌ Figures 8.23 and 17.39), *Planctomyces* is a budding bacterium with a life cycle wherein motile swarmer cells attach to a surface, grow a stalk from the attachment point, and generate a new cell from the opposite pole by budding. This daughter cell produces a flagellum, breaks away from the attached mother cell, and begins the cycle anew. Physiologically, *Planctomyces* species are facultatively aerobic chemoorganotrophs, growing either by fermentation or respiration of sugars.

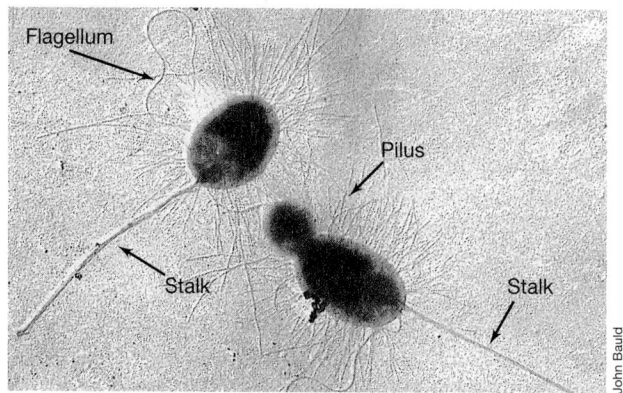

Figure 18.32 *Planctomyces maris.* Metal-shadowed transmission electron micrograph. A single cell is about 1–1.5 μm long. Note the fibrillar nature of the stalk. Pili are also abundant. Note also the flagella (curly appendages) on each cell and the bud that is developing from the nonstalked pole of one cell.

UNIT 6

The habitat of *Planctomyces* is primarily aquatic, both freshwater and marine, and the genus *Isosphaera* is a filamentous, gliding hot spring bacterium. The isolation of *Planctomyces* and relatives, like that of *Caulobacter*, requires dilute media, and because all known members of this group lack peptidoglycan, enrichments can be made even more selective by the addition of penicillin.

Compartmentalization in *Planctomycetes*

We learned in Section 2.5 of the major structural differences between prokaryotic and eukaryotic cells. In particular, eukaryotes have a membrane-enclosed nucleus whereas in prokaryotes, DNA supercoils and compacts to form the nucleoid present in the cytoplasm. However, *Planctomycetes* are unique among all known prokaryotes in that they show extensive cell compartmentalization, including in some cases a membrane-enclosed nuclear structure.

All *Planctomycetes* produce a structure enclosed by a nonunit membrane called a *pirellulosome*; this structure contains the nucleoid, ribosomes, and other necessary cytoplasmic components. But in some *Planctomycetes*, for example, in the bacterium *Gemmata* (**Figure 18.33**), the nucleoid itself is surrounded by a "nuclear envelope" consisting of a double membrane layer, analogous to the nuclear membrane in eukaryotes. DNA in *Gemmata* remains in a covalently closed, circular, and supercoiled form, typical of prokaryotes (↩ Section 6.3), but it is highly condensed and remains partitioned from the remaining cytoplasm by a true unit membrane (Figure 18.33). Another interesting compartment is the anammoxosome of *Brocadia anammoxidans*, a relative of *Planctomyces*. This bacterium carries out the anaerobic oxidation of ammonia (NH_3) within the enclosed anammoxosome structure. The structure protects cytoplasmic components from toxic substances produced in the reaction (↩ Section 13.11).

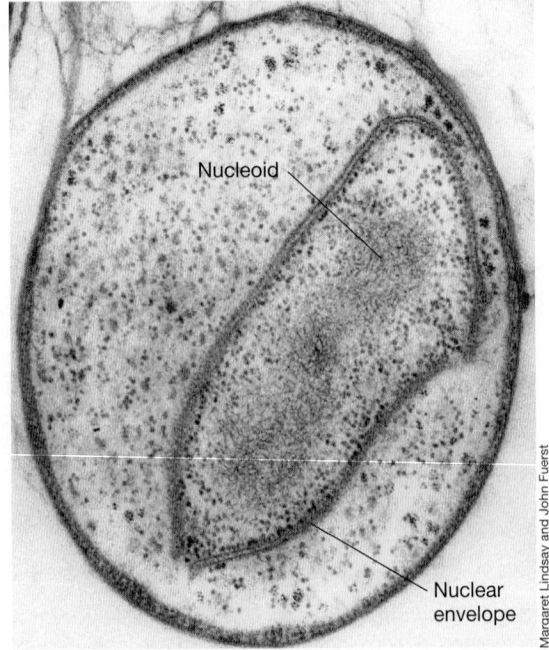

Figure 18.33 *Gemmata*: a nucleated bacterium. Thin-section electron micrograph of a cell of *Gemmata obscuriglobus* showing the nucleoid surrounded by a nuclear envelope. The cell is about 1.5 µm in diameter.

Thus far, all *Planctomycetes* have been found to contain internal cell compartments of one sort or another, and in no other known bacteria do internal compartments so closely resemble those of the eukaryotic cell. With this in mind, it would be easy to hypothesize that this lineage was nature's first experimentation with the eukaryotic cell plan. However, *Planctomycetes* occupy a phylogenetic position deep within the heart of the domain *Bacteria* (↩ Figure 17.1), and thus their structural resemblance to the eukaryotic cell is likely just coincidental.

MiniQuiz

- How does the stalk of *Planctomyces* differ from the stalk of *Caulobacter*?
- What is unusual about the bacterium *Gemmata*?

The *Verrucomicrobia*

With the *Verrucomicrobia* we find more morphologically unique bacteria, but a group phylogenetically distinct from the *Planctomycetes*.

18.11 *Verrucomicrobium* and *Prosthecobacter*

Key Genera: *Verrucomicrobium, Prosthecobacter*

This class of bacteria shares with prosthecate *Proteobacteria* (↩ Section 17.16) the formation of cytoplasmic appendages called *prosthecae*. The genera *Verrucomicrobium* and *Prosthecobacter* produce two to several prosthecae per cell (**Figure 18.34**). Also, unlike cells of *Caulobacter* (↩ Figures 8.23 and 17.39), which contain a single prostheca and produce flagellated and nonprosthecate swarmer cells, *Verrucomicrobium* and *Prosthecobacter* divide symmetrically, and both mother and daughter cells contain prosthecae at the time of cell division. The genus name *Verrucomicrobium* derives from Greek roots meaning "warty," which is an appropriate description of cells of

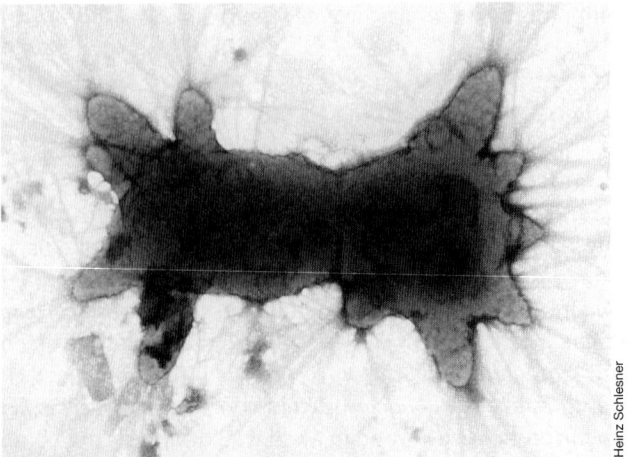

Figure 18.34 *Verrucomicrobium spinosum.* Negatively stained transmission electron micrograph. Note the wartlike prosthecae. A cell is about 1 µm in diameter.

Verrucomicrobium spinosum with their multiple projecting prosthecae (Figure 18.34).

Species of *Verrucomicrobia* share with other prosthecate bacteria the presence of peptidoglycan in their cell walls and are aerobic to facultatively aerobic bacteria capable of fermenting various sugars. *Verrucomicrobia* are widespread in nature, inhabiting freshwater and marine environments as well as forest and agricultural soils. From a phylogenetic standpoint, *Verrucomicrobia* are distinct from all known *Bacteria*. The group shows a loose phylogenetic affiliation with the *Planctomyces* and *Chlamydia* phyla (Figure 17.1), but is clearly sufficiently distinct from both of these groups to form its own separate lineage.

Species of the genus *Prosthecobacter* contain two genes that show significant homology to the genes that encode tubulin in eukaryotic cells. Tubulin is the key protein that makes up the cytoskeleton of eukaryotic cells (Section 20.5). Although the important cell division protein FtsZ (Section 5.2) is also a tubulin homolog, the *Prosthecobacter* proteins are structurally more similar to eukaryotic tubulin than is FtsZ. The role of the tubulin proteins in *Prosthecobacter* is unknown since a eukaryotic-like cytoskeleton has not been observed in these organisms. However, these genes likely signal that a specific relationship exists between *Verrucomicrobia* and eukaryotic cells, either in terms of shared ancestry or from horizontal gene transfer between cells of the two domains.

MiniQuiz
- Describe two ways that *Verrucomicrobia* differ from *Planctomycetes*.

VI The *Flavobacteria* and *Acidobacteria*

The *Flavobacteria* range from obligate aerobes to obligate anaerobes, but are unified by a common phylogenetic thread. The organisms inhabit many different types of environments, and we focus here on two main genera in the group. The *Acidobacteria* form a novel phylogenetic lineage of *Bacteria* and are ecologically important in the soil environment.

18.12 *Bacteroides* and *Flavobacterium*

Key Genera: *Bacteroides, Flavobacterium*

The genus *Bacteroides* contains obligately anaerobic, nonsporulating bacteria that are saccharolytic, fermenting sugars or proteins (depending on the species) to acetate and succinate as major fermentation products. *Bacteroides* are normally commensals, found in the intestinal tract of humans and other animals (Section 27.4). In fact, *Bacteroides* species are the numerically dominant bacteria in the human large intestine, where measurements have shown that 10^{10}–10^{11} prokaryotic cells are present per gram of feces (Section 25.6). However, species of *Bacteroides* can occasionally be pathogens and are the most important anaerobic bacteria associated with human infections such as bacteremia (bacteria in the blood).

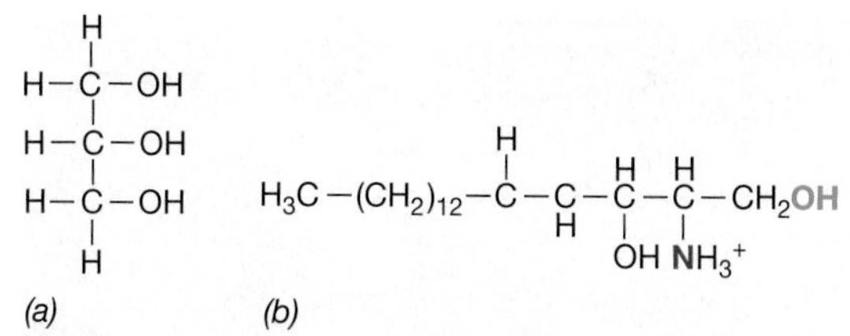

Figure at top right:

Figure 18.35 Sphingolipids. Comparison of *(a)* glycerol with *(b)* sphingosine. In sphingolipids, characteristic of *Bacteroides* species, sphingosine is the esterifying alcohol; a fatty acid is bonded by peptide linkage through the N atom (shown in red), and the terminal —OH group (shown in green) can be any of a number of compounds including phosphatidylcholine (sphingomyelin) or various sugars (cerebrosides and gangliosides).

Species of *Bacteroides* are unusual in that they are one of the few groups of bacteria to synthesize a special type of lipid called *sphingolipid,* a heterogeneous collection of lipids characterized by the long-chain amino alcohol sphingosine in place of glycerol in the lipid backbone (**Figure 18.35**). Sphingolipids such as sphingomyelin, cerebrosides, and gangliosides are common in mammalian tissues, especially in the brain and other nervous tissues, but rare in bacteria other than the *Bacteroides* group.

In contrast to *Bacteroides*, *Flavobacterium* species are primarily found in aquatic habitats, both freshwater and marine, as well as in foods and food-processing plants. Colonies of *Flavobacterium* are frequently yellow pigmented. Physiologically these organisms are aerobes and rather nutritionally restricted, using glucose as a carbon and energy source but very few other carbon compounds. *Flavobacteria* are rarely pathogenic; however, one species, *Flavobacteria meningosepticum,* has been associated with cases of infant meningitis, and several fish pathogens are also known.

Other important genera in this group are psychrophilic or psychrotolerant (Section 5.13). These include, in particular, the genera *Polaribacter* and *Psychroflexus,* organisms commonly isolated from cold environments, especially permanently cold environments such as polar waters and sea ice. Many other genera in the group are also capable of good growth below 20°C and can thus participate in food spoilage. None are pathogenic.

MiniQuiz
- How do *Bacteroides* and *Flavobacterium* differ in their physiology and ecology?

18.13 *Acidobacteria*

Key Genera: *Acidobacterium, Geothrix, Holophaga, Chloracidobacterium*

The *Acidobacteria* form the heart of a phylum of *Bacteria* common in soils and other habitats. The name *Acidobacteria* was originally coined to reflect the fact that these organisms are common in acidic soils, but further work has shown them to be abundant in virtually all soils. Three species of *Acidobacteria* have been well characterized, *Acidobacterium capsulatum, Geothrix fermentans,* and *Holophaga foetida;* all are gram-negative

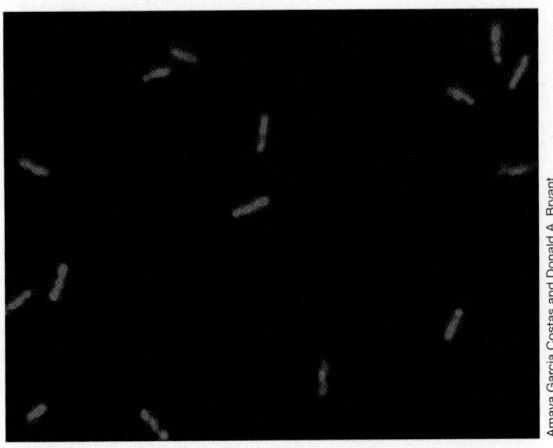

Amaya García Costas and Donald A. Bryant

Figure 18.36 A phototrophic acidobacterium. In the fluorescence photomicrograph of cells of *Chloracidobacterium thermophilum*, the red color is from the fluorescence of bacteriochlorophyll *c* present in chlorosomes. A cell of *C. thermophilum* is about 0.8 μm wide.

chemoorganotrophs. *A. capsulatum* is an acidophilic, encapsulated, aerobic bacterium isolated from acid mine drainage; it utilizes various sugars and organic acids. *G. fermentans*, a strict anaerobe, oxidizes simple organic acids (acetate, propionate, lactate, fumarate) to CO_2, with ferric iron (Fe^{3+}) as the electron acceptor, and can also ferment citrate to acetate plus succinate. *H. foetida*, also a strictly anaerobic bacterium, can degrade methylated aromatic compounds to acetate. Some *Acidobacteria* degrade polymers such as cellulose and chitin (Section 14.16), and at least one genus, *Chloracidobacterium*, is phototrophic (**Figure 18.36**).

Other *Acidobacteria* are known, but most are not yet cultured and are known only from environmental 16S ribosomal RNA gene sequences. Evidence for as many as 25 major subgroups within the *Acidobacteria* has been obtained based on 16S rRNA gene sequences, indicating substantial phylogenetic and metabolic diversity of the species in this phylum. Based on environmental 16S rRNA gene sequence data, *Acidobacteria* are abundant in soils, and in fact, may well be *the* most abundant of all soil bacteria. *Acidobacteria* also inhabit freshwater, hot spring microbial mats, wastewater treatment reactors, and sewage sludge. Their abundance, widespread distribution, and likely metabolic diversity indicate they play important ecological roles, especially in soil.

Chloracidobacterium (Figure 18.36) is a physiologically unique acidobacterium. This organism inhabits alkaline hot spring microbial mats and is an anoxygenic phototroph, containing much of the same photosynthetic machinery as green sulfur bacteria, including bacteriochlorophylls *a* and *c* and chlorosomes (Section 18.15 and Sections 13.2 and 13.4). However, *Chloracidobacterium* is an aerobe, in contrast to the strictly anaerobic green sulfur bacteria, and grows in nature in close association with cyanobacteria. The discovery of *Chloracidobacterium* brings to six the number of cultured phyla of *Bacteria* known to carry out photosynthesis (the others include the purple bacteria, green sulfur bacteria, green nonsulfur bacteria, heliobacteria, and cyanobacteria), underscoring the importance of photosynthesis in the evolutionary history of *Bacteria*.

MiniQuiz
- What is a major habitat of *Acidobacteria*?
- Among *Acidobacteria*, what is unusual about the organism *Chloracidobacterium*?

The *Cytophaga* Group

The *Cytophaga* group includes many common gram-negative terrestrial and aquatic bacteria.

18.14 *Cytophaga* and Relatives

Key Genera: *Cytophaga, Flexibacter, Rhodothermus, Salinibacter*

Organisms of the *Cytophaga* group are long, slender, gram-negative rods, often containing pointed ends, and move by gliding (**Figure 18.37**). The related genus, *Sporocytophaga*, is similar to *Cytophaga* in morphology and physiology, but the cells form resting spherical structures called *microcysts* (Figure 18.37*d*), similar to those produced by some fruiting myxobacteria (Section 17.17). Cytophagas are widespread in soil and water, often in great abundance.

Cytophaga and *Flexibacter*

Many cytophagas digest polysaccharides such as cellulose (Figure 18.37*c*), agar (Figure 18.37*a*), or chitin. The cellulose decomposers can easily be isolated by placing small crumbs of soil on pieces of cellulose filter paper laid on the surface of mineral salts agar. The bacteria attach to and digest the cellulose fibers, forming spreading colonies (Figure 18.37*c*). In pure culture *Cytophaga* can be grown on agar containing embedded cellulose fibers, and the presence of the organism is indicated by the clearing that occurs as the cellulose is digested (Figure 18.37*b* and Figures 14.34 and 14.35). Species of *Cytophaga* and *Sporocytophaga* are obligately aerobic and probably account for much of the cellulose digestion by bacteria in oxic environments in nature.

Several species of *Cytophaga* are fish pathogens and can cause serious problems in the cultivated fish industry. Two of the most important diseases are *columnaris disease*, caused by *Cytophaga columnaris*, and *cold-water disease*, caused by *Cytophaga psychrophila*. Both diseases preferentially affect stressed fish, such as those living in waters receiving pollutant discharges or living in high-density confinement situations such as fish hatcheries and aquaculture facilities. Infected fish show tissue destruction, frequently around the gills, probably from proteolytic activities of the *Cytophaga* pathogen.

Species of *Flexibacter* differ from the cytophagas in that they usually require complex media for good growth and are not cellulolytic. Cells of some *Flexibacter* species also undergo changes in cell morphology from long, gliding, threadlike filaments lacking cross-walls to short, nonmotile rods. Many species are pigmented due to carotenoids located in the cytoplasmic membrane, or related pigments called flexirubins, located in the gram-negative outer membrane. *Flexibacter* species are common soil and freshwater saprophytes, and none have been identified as pathogens.

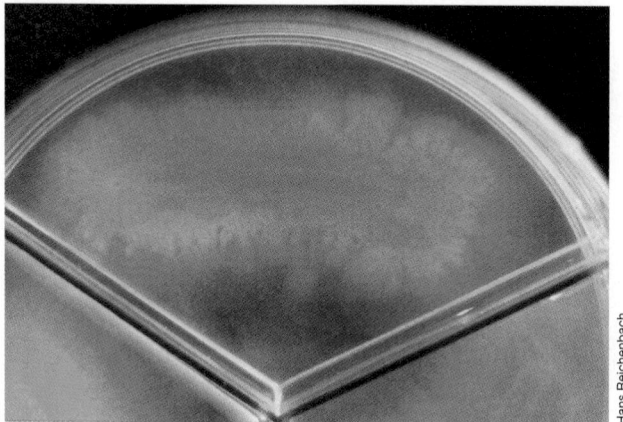

(a)

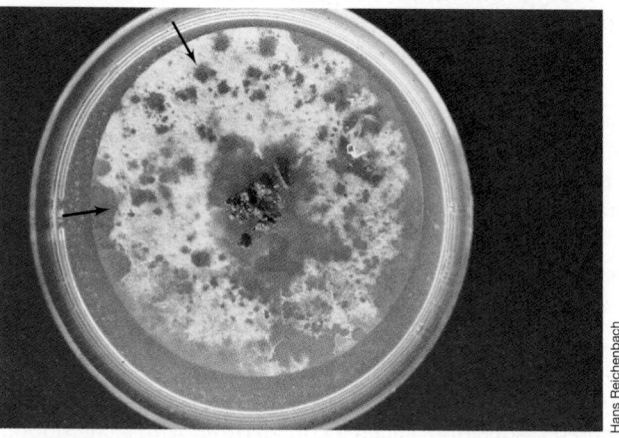

(b)

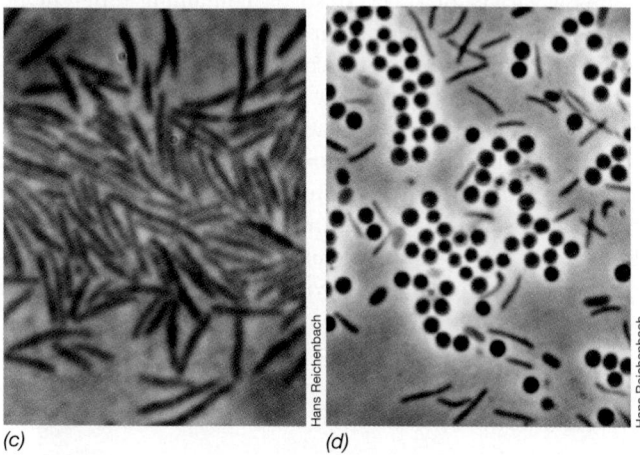

(c)   *(d)*

Figure 18.37 *Cytophaga* and *Sporocytophaga*. *(a)* Streak of an agarolytic marine *Cytophaga* hydrolyzing agar in a Petri dish. *(b)* Colonies of *Sporocytophaga* growing on cellulose. Note the clearing zones (arrows) where the cellulose has been degraded. *(c)* Phase-contrast micrograph of cells of *Cytophaga hutchinsonii* grown on cellulose filter paper (cells are about 1.5 μm in diameter). *(d)* Phase-contrast micrograph of the rod-shaped cells and spherical microcysts of *Sporocytophaga myxococcoides* (cells are about 0.5 μm and microcysts about 1.5 μm in diameter). Although *Sporocytophaga* microcysts are only slightly more heat tolerant than vegetative cells, they are extremely resistant to dessication and thus help the organism survive dry periods in soil.

Rhodothermus and *Salinibacter*

The genera *Rhodothermus* and *Salinibacter* are in the *Cytophaga* group, but are only distant relatives. *Rhodothermus* and *Salinibacter* are gram-negative, red- or yellow-pigmented, obligatory aerobic chemoorganotrophic bacteria. *Rhodothermus* is thermophilic, with a temperature optimum near 60°C. *Rhodothermus* grows best on sugars or on simple and complex polysaccharides. The organism inhabits shallow-water submarine hot springs and has also been detected in terrestrial hot springs. *Rhodothermus* produces heat-stable hydrolytic enzymes of biotechnological interest, including an amylase (that degrades starch), a cellulase (that degrades cellulose), and a xylanase (that degrades hemicelluloses abundant in plant cell walls), among many others.

Salinibacter is a genus of extremely halophilic red bacteria that is perhaps the most salt-tolerant and salt requiring of all *Bacteria*. In fact, its salt requirements rival those of extremely halophilic *Archaea*, such as *Halobacterium* (⟳ Section 19.2). *Salinibacter ruber*, the only known species, lives in saltern crystallization ponds and other highly saline environments. *Salinibacter* shares with *Halobacterium* the use of K^+ as a compatible solute, a property that is rarely found among halophiles of the domain *Bacteria* (most halophilic *Bacteria* either synthesize or accumulate organic solutes to maintain water balance in their salty environments, ⟳ Section 5.16). In contrast to the highly hydrolytic *Rhodothermus*, *Salinibacter* grows best with amino acids as electron donors, similar to *Halobacterium*. Thus, although they are phylogenetically unique, we see in *Halobacterium* and *Salinibacter* a likely case of convergent evolution, where physiological strategies have developed along the same pattern, probably because of the unique demands of the extreme environment that these organisms share.

MiniQuiz

- Describe a method for isolating *Cytophaga* species from nature.
- Contrast the habitat and physiology of *Rhodothermus* and *Salinibacter*.

Ⅷ Green Sulfur Bacteria

Here we see a phylogenetic lineage in which cultured relatives are all phototrophic; however, their mechanism of photosynthesis is quite different from that of purple bacteria, heliobacteria, or the cyanobacteria.

18.15 *Chlorobium* and Other Green Sulfur Bacteria

Key Genera: *Chlorobium, Chlorobaculum, "Chlorochromatium"*

Green sulfur bacteria are a phylogenetically distinct group of nonmotile anoxygenic and strictly anaerobic phototrophic bacteria. The group is morphologically restricted and includes short to long rods (**Table 18.14** and **Figure 18.38**). Like purple sulfur bacteria, green sulfur bacteria utilize hydrogen sulfide (H_2S) as an

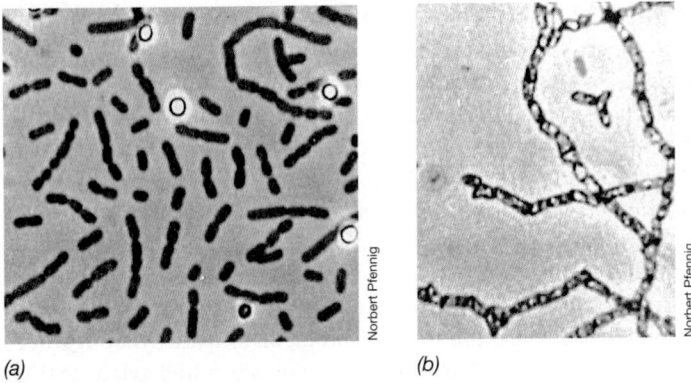

(a)

(b)

Figure 18.38 **Phototrophic green sulfur bacteria.** *(a) Chlorobium lim-icola*; cells are about 0.8 μm wide. Note the spherical sulfur granules deposited extracellularly. *(b) Chlorobium clathratiforme*, a bacterium forming a three-dimensional network; cells are about 0.8 μm wide.

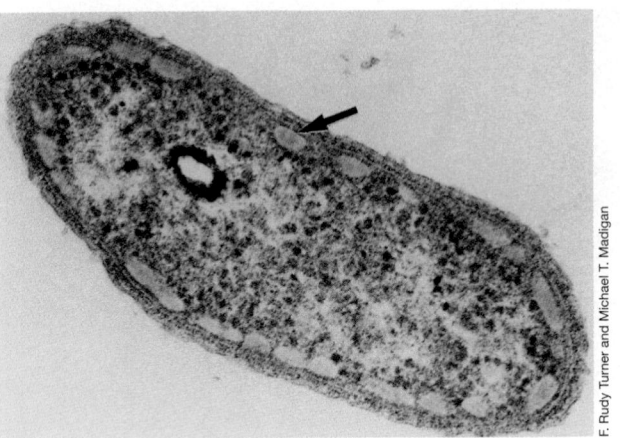

Figure 18.39 **The thermophilic green sulfur bacterium** *Chlorobaculum tepidum*. Thin-section electron micrograph. Note chlorosomes (arrow) in the cell periphery. A cell is about 0.7 μm wide.

electron donor, oxidizing it first to sulfur (S^0) and then to sulfate (SO_4^{2-}). But unlike purple sulfur bacteria, the S^0 produced by green sulfur bacteria is deposited outside the cell (Figure 18.38*a* and ↩ Figure 13.16*b*). Autotrophy is supported not by the reactions of the Calvin cycle, as in purple bacteria, but instead by a reversal of steps in the citric acid cycle (↩ Section 13.13), a unique means of autotrophy in phototrophic species.

Pigments and Ecology

The bacteriochlorophylls present in green sulfur bacteria include bacteriochlorophyll *a*, and either bacteriochlorophyll *c*, *d*, or *e*. The latter pigments function as an antenna (↩ Section 13.2) and are located in unique structures called **chlorosomes** (**Figure 18.39**). Chlorosomes are oblong bacteriochlorophyll-rich bodies bounded by a thin, nonunit membrane and attached to the cytoplasmic membrane in the periphery of the cell (Figure 18.39 and ↩ Figure 13.7). Their function is to funnel energy into the photosynthetic reaction center that eventually leads to ATP synthesis. Both green- and brown-colored species of green sulfur bacteria are known, the brown-colored species containing bacteriochlorophyll *e* and carotenoids that make cell suspensions brown (**Figure 18.40**; ↩ Figure 13.9).

Like purple sulfur bacteria (↩ Section 17.2), green sulfur bacteria live in anoxic, sulfidic aquatic environments, and because the chlorosome is such an efficient light-harvesting structure, they are typically found in lakes at the greatest depths of any phototrophic organism. One species, *Chlorobaculum tepidum*

(Figure 18.39), is thermophilic and forms dense microbial mats in high-sulfide hot springs. *C. tepidum* also grows rapidly and is amenable to genetic manipulation by both conjugation and transformation. Because of these features, *C. tepidum* has become the model organism for studying the molecular biology of green sulfur bacteria.

Green Sulfur Bacteria Consortia

Certain species of green sulfur bacteria form an intimate two-membered association, called a **consortium,** with a chemoorganotrophic bacterium. In the consortium, each organism benefits, and thus a variety of such consortia containing different phototrophic and chemotrophic components probably exist in nature. The phototrophic component, called the *epibiont*, is physically attached to the nonphototrophic central cell (**Figure 18.41**) and communicates with it in various ways (↩ Section 25.2).

The name *"Chlorochromatium aggregatum"* (not a formal name because this is a mixed culture) has been used to describe a commonly observed consortium that is green because the epibionts are green sulfur bacteria that contain green-colored carotenoids (Figure 18.41*b*). Evidence that the epibionts are

Table 18.14 *Genera and characteristics of phototrophic green sulfur bacteria*	
Characteristics	*Genus*
Straight or curved rods, some branching; nonmotile; color green or brown; some contain gas vesicles	*Chlorobium* *Chlorobaculum*
Spheres and ovals, nonmotile; forming prosthecae; green or brown	*Prosthecochloris*
Rods, motile by gliding; green	*Chloroherpeton*

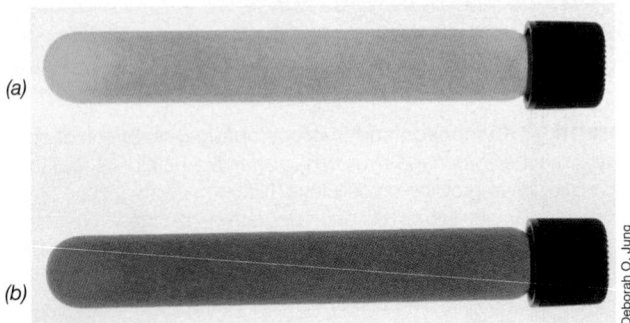

(a)

(b)

Figure 18.40 **Green and brown chlorobia.** Tube cultures of *(a) Chlorobaculum tepidum* and *(b) Chlorobaculum phaeobacteroides*. Cells of *C. tepidum* contain bacteriochlorophyll *c* and green carotenoids, and cells of *C. phaeobacteroides* contain bacteriochlorophyll *e* and isorenieratene, a brown carotenoid (↩ Figure 13.9).

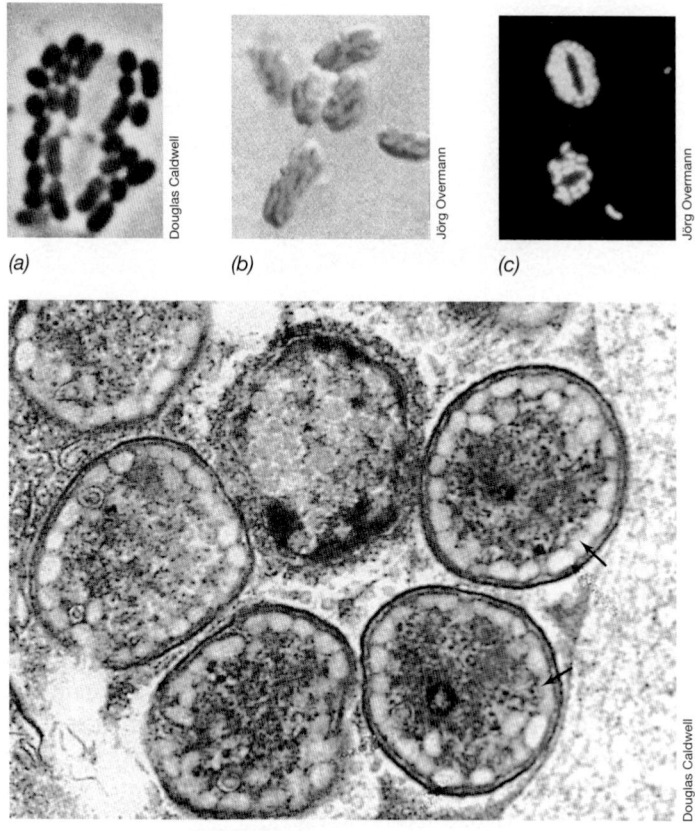

(a)   *(b)*   *(c)*

Douglas Caldwell

Jörg Overmann

Jörg Overmann

(d)

Douglas Caldwell

Figure 18.41 *"Chlorochromatium aggregatum."* Consortia of green sulfur bacteria and a chemoorganotroph. *(a)* In a phase-contrast micrograph, the nonphototrophic central organism is lighter in color than the pigmented phototrophic bacteria. *(b)* Green carotenoids lend their color to the phototrophs in a differential interference contrast micrograph. *(c)* A fluorescent micrograph shows the cells stained with a phylogenetic FISH probe specific for green sulfur bacteria. *(d)* Transmission electron micrograph of a cross section through a single consortium; note the chlorosomes (arrows) in the epibionts. The entire consortium is about 3 μm in diameter.

indeed green sulfur bacteria comes from pigment analyses, the presence of chlorosomes (Figure 18.41*d*), and from phylogenetic FISH staining (Figure 18.41*c*). A structurally similar organism called *"Pelochromatium roseum"* is brown because its epibionts produce brown-colored carotenoids (⟲ Figure 25.3). We examine the nature of the *Chlorochromatium* consortium—including how and why it forms—in more detail in Section 25.2.

MiniQuiz
- Which pigments are present in the chlorosome?
- What evidence exists that the epibionts of green bacterial consortia are truly green sulfur bacteria?

IX The Spirochetes

I n this section we find a morphologically unique group of *Bacteria* where morphology is a predictor of phylogeny.

18.16 Spirochetes

Key Genera: *Spirochaeta, Treponema, Cristispira, Leptospira, Borrelia*

Spirochetes are gram-negative, motile, tightly coiled *Bacteria*, typically slender and flexuous in shape (**Figure 18.42**). These morphologically unique bacteria form a major phylogenetic lineage of *Bacteria* (⟲ Figure 17.1). Spirochetes are widespread in aquatic environments and in animals. Some cause diseases, including syphilis, an important human sexually transmitted disease (⟲ Section 33.12).

The spirochete cell is made up of a protoplasmic cylinder, consisting of the regions enclosed by the cell wall and cytoplasmic membrane. Spirochete motility is conferred by flagella that emerge from each pole. However, unlike typical bacteria flagella (⟲ Section 3.13), spirochete flagella fold back from each pole onto the protoplasmic cylinder itself and remain in the periplasm of the cell; because of this, they have been called *endoflagella* (**Figure 18.43**). In addition, both endoflagella and the protoplasmic cylinder are surrounded by a flexible membrane called the *outer sheath* (Figure 18.43*b*).

Motility of Spirochetes

Spirochetes have an unusual mode of motility. Each endoflagellum is anchored at one end and extends about two-thirds of the length of the cell. Endoflagella rotate, as do typical bacterial flagella. However, when both endoflagella rotate in the same direction, the protoplasmic cylinder rotates in the opposite direction, placing torsion on the cell (Figure 18.43*b*). This causes the spirochete cell to move by flexing or lashing motions due to torque exerted at the ends of the protoplasmic cylinder by the rotating endoflagella (Figure 18.43*b*). Thus, despite the fact that endoflagella do not extend away from the cell, their

UNIT 6

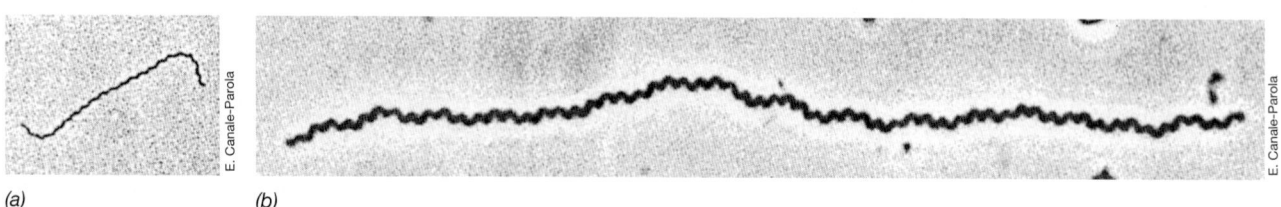

(a)   *(b)*

E. Canale-Parola

E. Canale-Parola

Figure 18.42 **Morphology of spirochetes.** Two spirochetes at the same magnification, showing the wide size range in the group. *(a) Spirochaeta stenostrepta*, by phase-contrast microscopy. A single cell is 0.25 μm in diameter. *(b) Spirochaeta plicatilis*. A single cell is 0.75 μm in diameter and can be up to 250 μm (0.25 mm) in length.

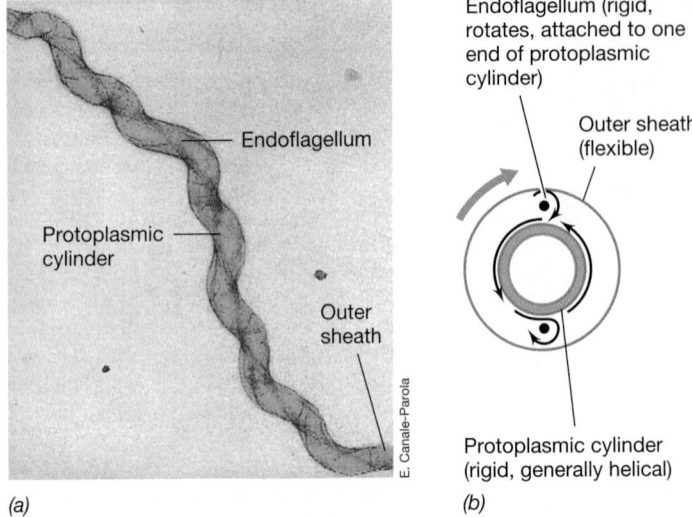

(a)

(b)

Figure 18.43 Motility in spirochetes. *(a)* Electron micrograph of a negatively stained cell of *Spirochaeta zuelzerae*, showing the position of the endoflagellum; the cell is about 0.3 μm in diameter. *(b)* Diagram of a spirochete cell, showing the arrangement of the protoplasmic cylinder, endoflagella, and external sheath, and how rotation of the endoflagellum generates rotation of both the protoplasmic cylinder and the external sheath.

rotation provides motility, albeit of a more irregular and jerky form than motility provided by the flagella of other bacteria. Spirochetes are classified into eight genera primarily on the basis of habitat, pathogenicity, phylogeny, and morphological and physiological characteristics. **Table 18.15** lists the major genera and their characteristics.

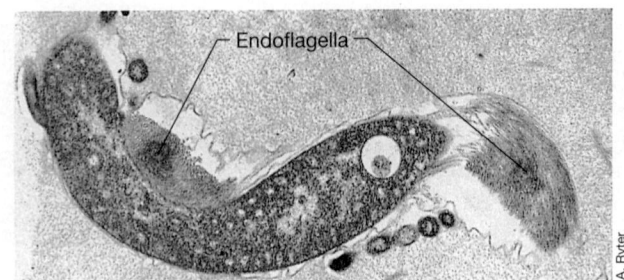

Figure 18.44 *Cristispira.* Electron micrograph of a thin section of a cell of *Cristispira*, a large spirochete. The cell measures about 2 μm in diameter. Notice the numerous endoflagella.

Spirochaeta and *Cristispira*

The genus *Spirochaeta* includes free-living, anaerobic, and facultatively aerobic spirochetes. These organisms, of which several species are known, are common in aquatic environments such as freshwater and sediments, and also in the oceans. *Spirochaeta plicatilis* (Figure 18.42b) is a large spirochete found in sulfidic freshwater and marine habitats. The 20 or so endoflagella inserted at each pole of *S. plicatilis* are arranged in a bundle that winds around the coiled protoplasmic cylinder. Another species, *Spirochaeta stenostrepta* (Figure 18.42a), is an obligate anaerobe commonly found in H$_2$S-rich black muds. It ferments sugars to ethanol, acetate, lactate, CO$_2$, and H$_2$.

Cristispira (**Figure 18.44**) is a unique spirochete found in nature only in the crystalline style of certain molluscs, such as clams and oysters. The crystalline style is a flexible, semisolid rod seated in a sac and rotated against a hard surface of the digestive

Table 18.15 *Genera of spirochetes and their characteristics*

Genus	Dimensions (μm)	General characteristics	Number of endoflagella	Habitat	Diseases
Cristispira	30–150 × 0.5–3.0	3–10 complete coils; bundle of endoflagella visible by phase-contrast microscopy	>100	Digestive tract of molluscs; has not been cultured	None known
Spirochaeta	5–250 × 0.2–0.75	Anaerobic or facultatively aerobic; tightly or loosely coiled	2–40	Aquatic, free-living, freshwater and marine	None known
Treponema	5–15 × 0.1–0.4	Microaerophilic or anaerobic; helical or flattened coil amplitude up to 0.5 μm	2–32	Commensal or parasitic in humans, other animals	Syphilis, yaws, swine dysentery, pinta
Borrelia	8–30 × 0.2–0.5	Microaerophilic; 5–7 coils of approximately 1 μm amplitude	7–20	Humans and other mammals, arthropods	Relapsing fever, Lyme disease, ovine and bovine borreliosis
Leptospira	6–20 × 0.1	Aerobic, tightly coiled, with bent or hooked ends; requires long-chain fatty acids	2	Free-living or parasitic in humans, other mammals	Leptospirosis
Leptonema	6–20 × 0.1	Aerobic; does not require long-chain fatty acids	2	Free-living	None known
Brachyspira	7–10 × 0.35–0.45	Anaerobe	8–28	Intestine of warm-blooded animals	Causes diarrhea in chickens and swine
Brevinema	4–5 × 0.2–0.3	Microaerophile, by 16S rRNA gene sequence analysis, forms deep branch in spirochete lineage (∂ Figure 17.1)	2	Blood and tissue of mice and shrews	Infectious for laboratory mice

tract, thereby mixing with and grinding the small particles of food taken in by the animal. Being large spirochetes, the cristispiras can easily be seen microscopically within the mollusc style as they rapidly rotate forward and backward in corkscrew fashion. *Cristispira* lives in both freshwater and marine molluscs, but not all species of molluscs possess them. Unfortunately, *Cristispira* has not been cultured, and so the physiological rationale for its restriction to this unique habitat is unknown.

Treponema and *Borrelia*

Anaerobic or microaerophilic host-associated spirochetes that are commensals or parasites of humans and animals reside in the genus *Treponema*. *T. pallidum*, the causal agent of syphilis (⊘ Section 33.12), is the best-known species of *Treponema*. It differs in morphology from other spirochetes; the cell is not helical but is flat and wavy. The *T. pallidum* cell is remarkably thin, measuring only 0.2 μm in diameter. Because of this, dark-field microscopy has long been used to examine exudates from suspected syphilitic lesions (⊘ Figure 33.31).

Although *T. pallidum* has not been cultured, other species of *Treponema* have, and are commensals found in both humans and other animals. For example, *Treponema denticola* is a major oral treponeme. It ferments amino acids such as cysteine and serine, forming acetate as the major fermentation acid, as well as CO_2, NH_3, and H_2S. Spirochetes are also common in the rumen, the digestive organ of ruminant animals (⊘ Section 25.7). For instance, *Treponema saccharophilum* (**Figure 18.45**) is a large, pectinolytic spirochete found in the bovine rumen. *T. saccha-*

rophilum is an obligate anaerobe that ferments pectin, starch, inulin, and other plant polysaccharides. This and other spirochetes likely play an important role in the rumen by fermenting plant polysaccharides to volatile fatty acids usable as energy sources by the ruminant (⊘ Figure 25.27).

Metabolically unusual species of the genus *Treponema* have been isolated from the hindgut of the termite. This environment is highly cellulolytic, as termites live mostly on wood and wood products. H_2 and CO_2 are produced from the fermentation of glucose released from the cellulose. *Treponema primitia* converts this H_2 plus CO_2 to acetate; that is, it is an acetogen (acetogenesis is covered in Section 14.9). This was the first instance of this form of energy metabolism to be discovered in a phylogenetic group of *Bacteria* outside of the clostridia and their relatives. The hindgut spirochete *Treponema azotonutricium* is capable of nitrogen fixation (⊘ Section 13.14), a property also not previously recognized in spirochetes.

The majority of species of *Borrelia* are animal or human pathogens. *Borrelia recurrentis* is the causative agent of relapsing fever in humans and is transmitted via an insect vector, usually by the human body louse. Relapsing fever is characterized by a high fever and generalized muscular pain that lasts for 3–7 days, followed by a recovery period of 7–9 days. Left untreated, the fever returns in two to three more cycles (hence the name, relapsing fever) and can cause death from hemorrhaging and organ failure. *Borrelia burgdorferi* (Figure 18.45b) is the causative agent of the tickborne disease called *Lyme disease*, which infects humans and animals. Lyme disease is discussed in Section 34.4. *B. burgdorferi* is also of interest because it is as yet one of the few known bacteria that has a linear (as opposed to a circular) chromosome (⊘ Section 12.1). Other species of *Borrelia* are primarily of veterinary importance, causing diseases in cattle, sheep, horses, and birds. In most cases, the bacterium is transmitted to the animal host from the bite of a tick.

Leptospira and *Leptonema*

The genera *Leptospira* and *Leptonema* contain strictly aerobic spirochetes that use long-chain fatty acids (for example, oleic acid) as electron donor and carbon sources. With few exceptions, these are the only substrates utilized for growth. Leptospiras are thin, finely coiled, and usually bent at each end into a semicircular hook. At present, several species are recognized in this group, some free-living and many parasitic. Two major species of *Leptospira* are *L. interrogans* (parasitic) and *L. biflexa* (free-living). Strains of *L. interrogans* are parasitic for humans and animals. Rodents are the natural hosts of most leptospiras, although dogs and pigs are also important carriers of certain strains.

In humans the most common leptospiral syndrome is *leptospirosis*, a disorder in which the organism localizes in the kidneys and can cause renal failure or even death. Leptospiras ordinarily enter the body through the mucous membranes or through breaks in the skin during contact with an infected animal. After a transient multiplication in various parts of the body, the organism localizes in the kidneys and liver, causing nephritis and jaundice. Domestic animals such as dogs are vaccinated against leptospirosis with a killed virulent strain in the combined

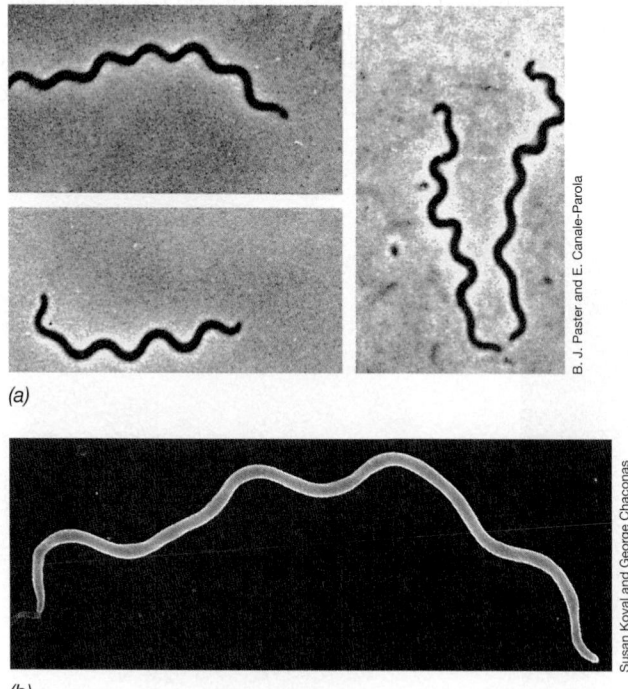

(a)

(b)

B. J. Paster and E. Canale-Parola

Susan Koval and George Chaconas

Figure 18.45 *Treponema and Borrelia. (a)* Phase-contrast micrographs of *Treponema saccharophilum*, a large pectinolytic spirochete from the bovine rumen. A cell measures about 0.4 μm in diameter. Left, regularly coiled cells; right, irregularly coiled cells. *(b)* Scanning electron micrograph of a cell of *Borrelia burgdorferi*, the causative agent of Lyme disease.

UNIT 6

distemper-leptospira-hepatitis vaccine. In humans, prevention of leptospirosis is effected primarily by elimination of the disease from animals, thereby eliminating the disease reservoir.

The Deinococci

In this lineage of *Bacteria* we see two genera that have revolutionized biology in two quite different ways: one because of its ability to withstand enormous doses of radiation and the other by virtue of a key heat-stable enzyme.

18.17 *Deinococcus* and *Thermus*

Key Genera: *Deinococcus, Thermus*

The deinococci group contains only a few genera, the best studied being *Deinococcus* and *Thermus*. The latter contains thermophilic chemoorganotrophic bacteria including *Thermus aquaticus*, the organism from which *Taq* DNA polymerase is obtained. Because it is so heat-stable, this enzyme is the major one used in the polymerase chain reaction (PCR) technique for amplifying DNA, as was described in Section 6.11.

Deinococci stain gram-positively and contain an unusual form of peptidoglycan in which ornithine is present in place of diaminopimelic acid in the *N*-acetylmuramic acid cross-bridges (⮌ Section 3.6). A number of species of *Thermus* and *Deinococcus* have been described, and all grow aerobically by catabolism of sugars, amino and organic acids, or various complex mixtures. We focus the rest of this discussion on *Deinococcus*.

The genus *Deinococcus* contains four species of gram-positive cocci; *Deinococcus radiodurans* is the best-studied species. The *D. radiodurans* cell wall is structurally complex and consists of several layers, including an outer membrane (**Figure 18.46**), normally present only in gram-negative bacteria (⮌ Section 3.7). However, unlike the outer membrane of gram-negative bacteria such as *Escherichia coli*, the outer membrane of *D. radiodurans* lacks lipid A. The organism primarily inhabits soil and can also be isolated from dust particles.

Radiation Resistance of *Deinococcus radiodurans*

Most deinococci are red or pink due to carotenoids, and many strains are highly resistant to ultraviolet (UV) radiation and to desiccation. Resistance to UV radiation can be used to advantage in isolating deinococci. These remarkable organisms can be selectively isolated from soil, ground meat, dust, and filtered air following exposure of the sample to intense UV (or even gamma) radiation and plating on a rich medium containing tryptone and yeast extract. Because many strains of *D. radiodurans* are even more resistant to radiation than are bacterial endospores, treatment of a sample with strong doses of radiation effectively sterilizes the sample except for cells of *D. radiodurans*, making isolation of deinococci relatively straightforward. For example, *D. radiodurans* cells can survive exposure to up to 15,000 grays (Gy) of ionizing radiation (1 Gy = 100 rad). This is sufficient to shatter the organism's chromosome into hundreds of fragments (by contrast, a human can be killed by exposure to less than 10 Gy).

In addition to impressive radiation resistance, *D. radiodurans* is resistant to the mutagenic effects of many other agents. The only chemical mutagens that seem to work on *D. radiodurans* are agents such as nitrosoguanidine, which induces deletions in DNA. Deletions are apparently not repaired as efficiently as point mutations in this organism, and mutants of *D. radiodurans* can be isolated in this way.

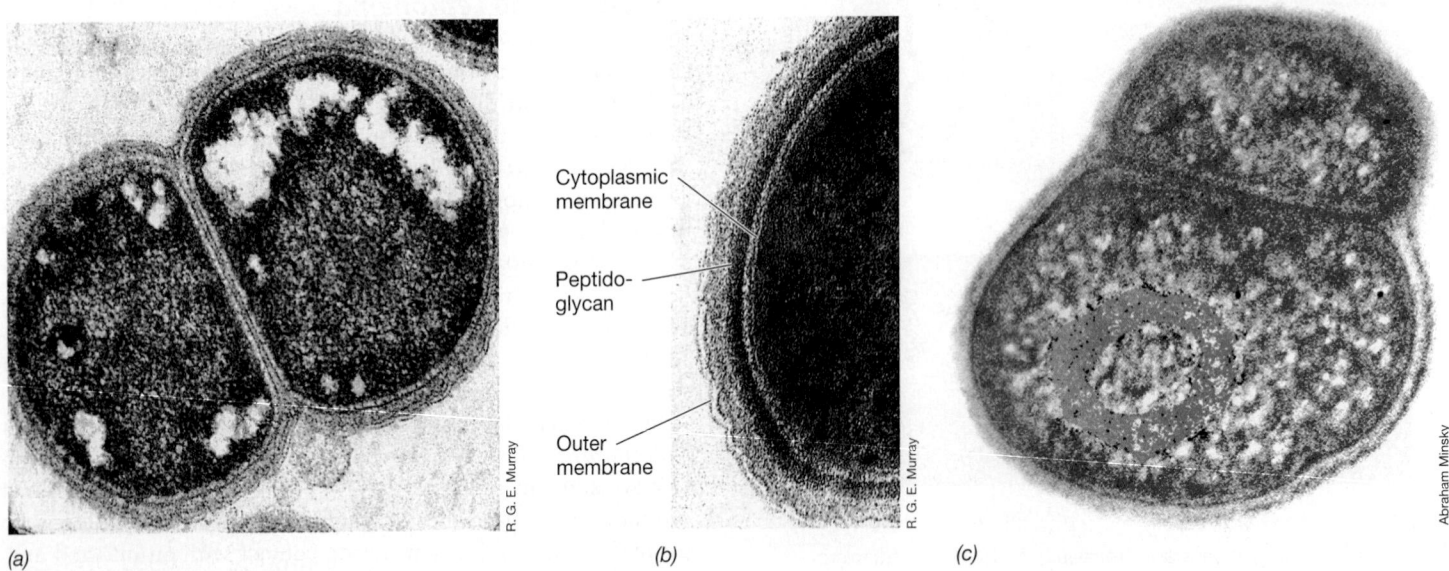

(a)   *(b)*   *(c)*

Cytoplasmic membrane
Peptido-glycan
Outer membrane

Figure 18.46 The radiation-resistant coccus *Deinococcus radiodurans*. An individual cell is about 2.5 μm in diameter. *(a)* Transmission electron micrograph of *D. radiodurans*. Note the outer membrane layer. *(b)* High-magnification micrograph of wall layer. *(c)* Transmission electron micrograph of cells of *D. radiodurans* colored to show the toroidal morphology of the nucleoid (green).

DNA Repair in *Deinococcus radiodurans*

Studies of *D. radiodurans* have shown that it is highly efficient in repairing damaged DNA. Several different DNA repair enzymes exist in *D. radiodurans*. In addition to the DNA repair enzyme RecA (Section 10.4), several RecA-independent DNA systems exist in *D. radiodurans* that can repair breaks in single- or double-stranded DNA, and excise and repair misincorporated bases. In fact, repair processes are so effective that the chromosome can even be reassembled from a fragmented state.

It is also thought that the unique arrangement of DNA in *D. radiodurans* cells plays a role in radiation resistance. Cells of *D. radiodurans* always exist as pairs or tetrads (Figure 18.46*a*). Instead of scattering DNA within the cell as in a typical nucleoid, DNA in *D. radiodurans* is ordered into a toroidal (coiled, or stack of rings) structure (Figure 18.46*c*). Repair is then facilitated by the fusion of nucleoids from adjacent compartments, because their toroidal structure provides a platform for homologous recombination. From this extensive recombination, a single repaired chromosome emerges, and the cell containing this chromosome can then grow and divide.

MiniQuiz
- Describe an unusual biological feature of *Deinococcus radiodurans* and a commercial application of *Thermus aquaticus*.

The Green Nonsulfur Bacteria: *Chloroflexi*

The green nonsulfur bacteria (phylum *Chloroflexi*) are phylogenetically distinct and contain just a few genera, the best known being the anoxygenic phototroph *Chloroflexus*. Most cultured *Chloroflexi* are thermophilic and many inhabit microbial mats in hot springs.

18.18 *Chloroflexus* and Relatives

Key Genera: *Chloroflexus, Heliothrix, Roseiflexus*

Chloroflexus and most other green nonsulfur bacteria (also called the filamentous anoxygenic phototrophs) are thermophilic, filamentous bacteria that, along with cyanobacteria, form thick microbial mats in neutral to alkaline hot springs (**Figure 18.47**; Figure 23.9*b*). Nonthermophilic *Chloroflexus*-like organisms have also been found in marine microbial mats. From a phylogenetic standpoint, the genus *Chloroflexus* is clearly the earliest known phototrophic bacterium (Figure 17.1).

Although an anoxygenic phototroph, *Chloroflexus* is a "hybrid" phototroph in the sense that its photosynthetic features resemble those of both green sulfur bacteria (Section 18.15) and purple phototrophic bacteria (Section 17.2). Like green sulfur bacteria, *Chloroflexus* contains bacteriochlorophyll *c* in chlorosomes (Figure 18.39); however, the photosynthetic reaction center of *Chloroflexus* is structurally more similar to that of the purple bacteria. Physiologically, *Chloroflexus* can grow both photoautotrophically and photoheterotrophically. The hydroxypropionate cycle, a pathway of CO_2 incorporation unique to only a

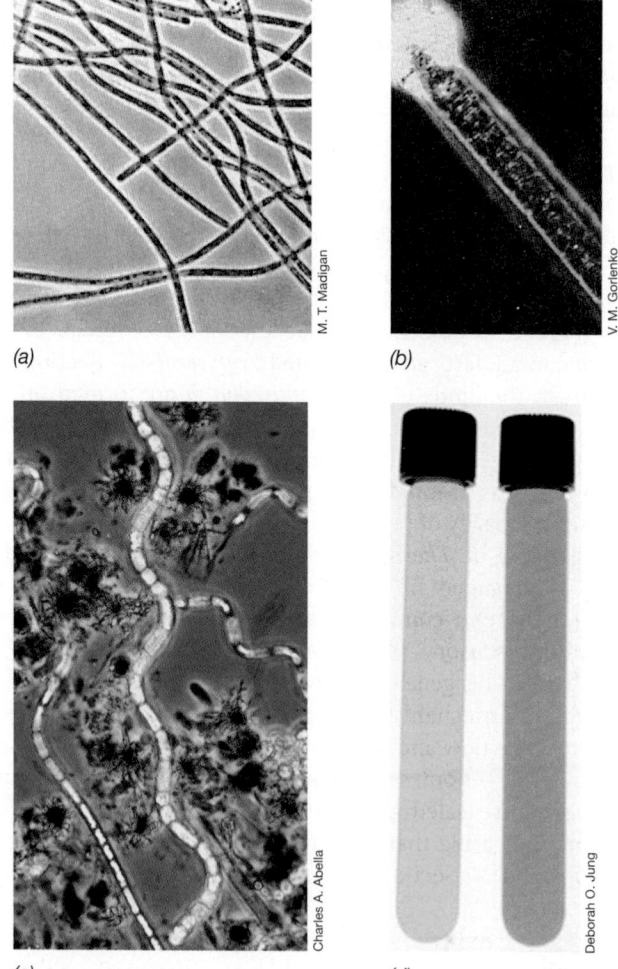

Figure 18.47 Green nonsulfur bacteria. *(a)* Phase-contrast micrograph of the anoxygenic phototroph *Chloroflexus aurantiacus*; cells are about 1 μm in diameter. *(b)* Phase-contrast micrograph of the large phototroph *Oscillochloris*; cells are about 5 μm wide. The brightly contrasting material on the top is a holdfast, used for attachment. *(c)* Phase-contrast micrograph of filaments of a *Chloronema* species; the cells are wavy filaments and about 2.5 μm in diameter. *(d)* Tube cultures of *C. aurantiacus* (right) and *Roseiflexus* (left). *Roseiflexus* is yellow because it lacks bacteriochlorophyll *c* and chlorosomes.

few *Bacteria* and *Archaea*, supports autotrophic growth of *Chloroflexus*. *Chloroflexus* also grows well in the dark on a wide variety of carbon sources by aerobic respiration.

Other Phototrophic Green Nonsulfur Bacteria

In addition to *Chloroflexus*, other phototrophic green nonsulfur bacteria include the thermophile *Heliothrix* and the large-celled mesophiles *Oscillochloris* (Figure 18.47*b*) and *Chloronema* (Figure 18.47*c*). *Oscillochloris* and *Chloronema* form rather large cells, 2–5 μm wide and up to several hundred micrometers long (Figure 18.47*c*). Species of both genera inhabit freshwater lakes containing low levels of H_2S. *Roseiflexus* and *Heliothrix* are similar to *Chloroflexus* in their filamentous morphology and thermophilic lifestyle, but differ in a major photosynthetic respect. *Roseiflexus* and *Heliothrix* lack bacteriochlorophyll *c* and chlorosomes and thus more closely resemble purple phototrophic bacteria (Section 17.2) than *Chloroflexus*. This can be seen in cultures of *Roseiflexus* that are yellow-orange from their extensive carotenoid pigments and lack of bacteriochlorophyll *c* (Figure 18.47*d*).

Thermomicrobium

Thermomicrobium is a chemotrophic genus of *Chloroflexi* and a strictly aerobic, gram-negative rod, growing optimally in complex media at 75°C. Besides its phylogenetic properties, *Thermomicrobium* is also of interest because of its membrane lipids (**Figure 18.48**). Recall that the lipids of *Bacteria* and *Eukarya* contain fatty acids esterified to *glycerol* (Section 3.3). By contrast, the lipids of *Thermomicrobium* are formed on *1,2-dialcohols* instead of glycerol, and have neither ester *nor* ether linkages (Figure 18.48). In addition, cells of *Thermomicrobium* contain only tiny amounts of peptidoglycan, and the cell wall is composed primarily of protein.

The genome of *Thermomicrobium roseum*, the best-studied species, is arranged in two circular chromosomes, with the smaller of the two containing all genes necessary for flagellar motility and sensory phenomena, such as chemotaxis (Section 3.15). This gene assignment, so far unique to this organism, suggests a mechanism for transferring motility functions by horizontal gene flow and may turn out to be more common once more bacterial genomes are sequenced. The *T. roseum* genome sequence also revealed genes encoding carbon-monoxide (CO) oxidation, indicating that this unusual bacterium may also be a methylotroph (Section 17.6).

MiniQuiz

- In what ways do *Chloroflexus* and *Roseiflexus* resemble *Chlorobium*? *Rhodobacter*?
- What is unique about *Thermomicrobium*?

Figure 18.48 The unusual lipids of *Thermomicrobium*. *(a)* Membrane lipids from *Thermomicrobium roseum* contain long-chain diols like the one shown here (13-methyl-1,2-nonadecanediol). Note that unlike the lipids of other *Bacteria* or of *Archaea*, neither ester- nor ether-linked side chains are present. *(b)* To form a bilayer membrane, dialcohol molecules oppose each other at the methyl groups, with the —OH groups being the inner and outer hydrophilic surfaces. Small amounts of the diols have fatty acids esterified to the secondary —OH group (shown in red), whereas the primary —OH group (shown in green) can bond a hydrophilic molecule like phosphate.

XII Hyperthermophilic Bacteria

Three groups of hyperthermophilic bacteria cluster deep in the phylogenetic tree of *Bacteria*, near to the root (Figure 17.1). Each group consists of one or two major genera, and a key physiological feature of most species is hyperthermophily—optimal growth at temperatures above 80°C (Section 5.14). We begin with *Thermotoga* and *Thermodesulfobacterium*, each representative of its own lineage.

18.19 Thermotoga and Thermodesulfobacterium

Key Genera: *Thermotoga, Thermodesulfobacterium*

Thermotoga is a rod-shaped hyperthermophile that forms a sheathlike envelope (called a *toga* and thus the genus name) (**Figure 18.49**), stain gram-negatively, and are nonsporulating. *Thermotoga* is an anaerobic, fermentative chemoorganotroph, catabolizing sugars or starch, and producing lactate, acetate, CO_2, and H_2 as fermentation products. The organism can also grow by anaerobic respiration using H_2 as an electron donor and ferric iron (Fe^{3+}) as an electron acceptor. Species of *Thermotoga* have been isolated from terrestrial hot springs as well as marine hydrothermal vents.

The genome of *Thermotoga* has been completely sequenced, and interestingly, this organism contains many genes that show strong homology to genes from hyperthermophilic *Archaea*. In fact, over 20% of the genes of *Thermotoga* probably originated from hyperthermophilic species of *Archaea* by horizontal gene transfers (Section 12.11). Although a few archaeal-like genes have been identified in the genomes of other *Bacteria* and vice

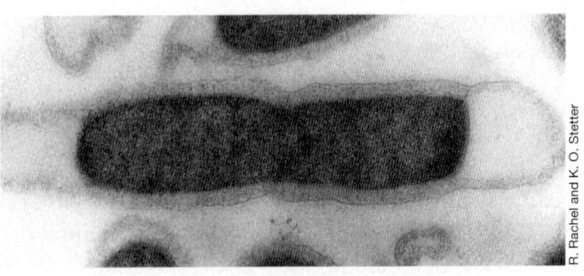

(a)

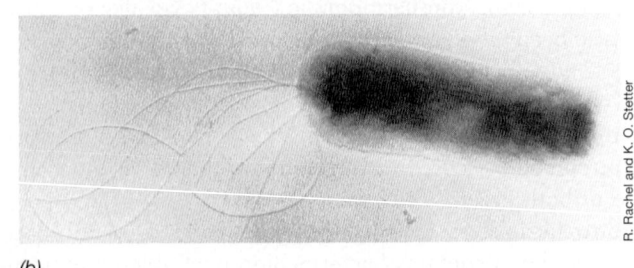

(b)

Figure 18.49 Hyperthermophilic *Bacteria*. Electron micrographs of two hyperthermophiles: *(a) Thermotoga maritima*—temperature optimum, 80°C. Note the outer covering, the toga. *(b) Aquifex pyrophilus*—temperature optimum, 85°C. Cells of *Thermotoga* measure 0.6 × 3.5 μm; cells of *Aquifex* measure 0.5 × 2.5 μm.

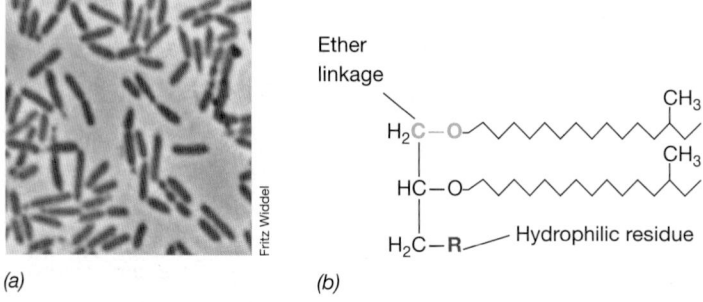

(a)   *(b)*

Figure 18.50 *Thermodesulfobacterium.* (a) Phase-contrast micrograph of cells of *Thermodesulfobacterium thermophilum.* (b) Structure of one of the lipids of *T. mobile.* Note that although the two hydrophobic side chains are ether-linked, they are not phytanyl units, as in *Archaea.* The designation "R" is for a hydrophilic residue, such as a phosphate group.

versa, only in *Thermotoga* has such large-scale horizontal transfer of genes between domains been detected thus far.

Thermodesulfobacterium (**Figure 18.50**) is a thermophilic sulfate-reducing bacterium, positioned on the phylogenetic tree as a separate phylum between *Thermotoga* and *Aquifex* (Figure 17.1). *Thermodesulfobacterium* is a strict anaerobe and cannot utilize acetate as an electron donor in its energy metabolism. Instead, *Thermodesulfobacterium* uses compounds such as lactate, pyruvate, and ethanol as electron donors, as do sulfate-reducing bacteria such as *Desulfovibrio* (Section 17.18), reducing SO_4^{2-} to H_2S.

An unusual biochemical feature of *Thermodesulfobacterium* is the production of *ether-linked lipids.* Recall that such lipids are a hallmark of the *Archaea* and that a polyisoprenoid C_{20} hydrocarbon (phytanyl) replaces fatty acids as the side chains in archaeal lipids (Section 3.3). However, the ether-linked lipids in *Thermodesulfobacterium* are unusual because the glycerol side chains are not phytanyl groups, as they are in *Archaea,* but instead are composed of a unique C_{17} hydrocarbon along with some fatty acids (Figure 18.50b). Thus we see in *Thermodesulfobacterium* both a deep phylogenetic lineage (Figure 17.1) and a lipid profile that combines features of both the *Archaea* and the *Bacteria.* However, a few other *Bacteria* have also been found to contain ether-linked lipids, and thus these lipids may be more common among *Bacteria* than previously thought.

MiniQuiz

- What is unique about the genome of *Thermotoga* and the lipids of *Thermodesulfobacterium*?

18.20 *Aquifex, Thermocrinis,* and Relatives

Key Genera: *Aquifex, Thermocrinis*

The genus *Aquifex* (Figure 18.49b) is an obligately chemolithotrophic and autotrophic hyperthermophile and is the most thermophilic of all known *Bacteria.* Various *Aquifex* species utilize H_2, sulfur (S^0), or thiosulfate ($S_2O_3^{2-}$) as electron donors and O_2 or nitrate (NO_3^-) as electron acceptors, and can grow at temperatures up to 95°C. *Aquifex* can tolerate only very low O_2 concentrations, and it remains one of the few aerobic (or more exactly, microaerophilic) hyperthermophiles known. Nutritional studies of *Aquifex* species have shown them totally unable to grow chemoorganotrophically on organic compounds, including complex mixtures like yeast or meat extract. *Hydrogenobacter,* a relative of *Aquifex,* shows most of the same properties as *Aquifex,* but is an obligate aerobe.

Aquifex and Autotrophy

Autotrophy in *Aquifex* occurs by way of the reverse citric acid cycle, a series of reactions previously found only in green sulfur bacteria (Section 18.15 and Section 13.13) within the domain *Bacteria.* The complete genome sequence of *Aquifex aeolicus* has been determined, and its entirely chemolithotrophic and autotrophic lifestyle is encoded by a very small genome of only 1.55 megabase pairs (one-third the size of the *Escherichia coli* genome). The discovery that so many hyperthermophilic species of *Archaea* and *Bacteria,* like *Aquifex,* are H_2 chemolithotrophs, coupled with the finding that they branch as very early lineages on their respective phylogenetic trees (Figure 17.1), suggests that H_2 was a key electron donor for energy metabolism in primitive organisms that appeared on the early Earth (Sections 16.2 and 19.14).

Thermocrinis

Thermocrinis (**Figure 18.51**) is a relative of *Aquifex* and *Hydrogenobacter.* This bacterium grows optimally at 80°C as a chemolithotroph oxidizing H_2, $S_2O_3^{2-}$, or S^0 as electron donors, with O_2 as electron acceptor. *Thermocrinis ruber,* the only known species, grows in the outflow of certain hot springs in Yellowstone National Park where it forms pink "streamers" consisting of a filamentous form of the cells attached to siliceous sinter (Figure 18.51a). In static culture, cells of *T. ruber* grow as individual rod-shaped cells (Figure 18.51b). However, when cultured in a flowing system in which growth medium is trickled over a solid glass surface to which cells can attach, *Thermocrinis* assumes the streamer morphology it forms in its constantly flowing habitat in nature.

T. ruber is of historical significance in microbiology because it was one of the organisms discovered in the 1960s by Thomas Brock, a pioneer in the field of thermal microbiology. The discovery by Brock that the pink streamers (Figure 18.51a) formed by *Thermocrinus* contained protein and nucleic acids clearly indicated that they were living organisms and not just mineral debris. Moreover, the presence of streamers in 80–90°C outflow waters but not those of lower temperatures supported Brock's hypothesis that hot spring microorganisms actually *required* heat for growth and were therefore likely to be present in even boiling water. Both of these conclusions were subsequently supported by the discovery of literally dozens of genera of hyperthermophilic bacteria inhabiting hot springs, hydrothermal vents, and other thermal environments by Brock and other microbiologists. More coverage of hyperthermophiles can be found in Sections 5.12, 5.14, and in Chapter 19.

UNIT 6

(a)

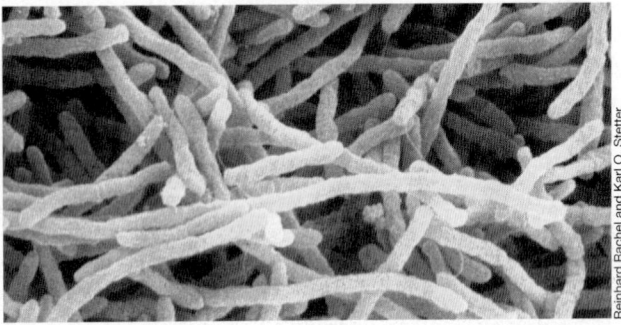

(b)

Figure 18.51 *Thermocrinis.* *(a)* Cells of *Thermocrinis ruber* growing as filamentous streamers (arrow) attached to siliceous sinter in the outflow (85°C) of Octopus Spring, Yellowstone National Park. The pink color is due to a carotenoid pigment. *(b)* Scanning electron micrograph of rod-shaped cells of *T. ruber* grown on a silicon-coated cover glass. The hair-like structures are silicon. A single cell of *T. ruber* is about 0.4 μm in diameter and from 1 to 3 μm long.

MiniQuiz

• What evolutionary significance is there to the fact that organisms in the *Aquifex* lineage are both hyperthermophilic and H_2 chemolithotrophs?

XIII *Nitrospira* and *Deferribacter*

We close this chapter by considering two genera of *Bacteria*, each of which forms the heart of its own phylogenetic lineage and each of which has its own special physiological twists.

18.21 *Nitrospira* and *Deferribacter*

Key Genera: *Nitrospira, Deferribacter*

In addition to what we have covered thus far in this chapter, several other phyla of *Bacteria* have been identified by 16S ribosomal RNA gene sequence analysis of cultures, although relatively little is known about them. Two such phyla contain the organisms *Nitrospira* and *Deferribacter* as key representatives (⭗ Figure 17.1). Physiologically, these organisms are either chemolithotrophs or chemoorganotrophs and range from mesophiles to thermophiles.

Nitrospira

Like nitrite-oxidizing *Proteobacteria* (the nitrifying bacteria; ⭗ Sections 13.10 and 17.3), *Nitrospira* oxidizes nitrite (NO_2^-) to nitrate (NO_3^-) and grows autotrophically (**Figure 18.52**). However, despite this close physiological resemblance to the classical nitrifying bacteria, *Nitrospira* is phylogenetically quite distinct from them (Figure 17.1). Also, *Nitrospira* lacks the extensive internal membranes found in species of nitrifying *Proteobacteria* (⭗ Figure 17.8). Nevertheless, *Nitrospira* inhabits many of the same environments as nitrite-oxidizing *Proteobacteria* such as *Nitrobacter*, so it has been suggested that its physiological capacities may have been transferred to it by horizontal gene flow from nitrifying *Proteobacteria* (or vice versa). As we know, this mechanism for acquiring physiological traits has been widely exploited in the bacterial world (⭗ Section 12.11). However, environmental surveys for the presence of nitrifying bacteria in nature have shown *Nitrospira* to be much more abundant than *Nitrobacter*; thus most of the NO_2^- oxidized in natural environments where nitrification is a significant process, such as in wastewater treatment plants and ammonia-rich soils, is probably due to the activities of *Nitrospira*.

Other genera in the *Nitrospira* group include *Leptospirillum*, an iron-oxidizing chemolithotroph responsible for much of the acid mine drainage associated with the mining of coal and iron (⭗ Section 24.7), and *Thermodesulfovibrio*, a thermophilic sulfate-reducing bacterium that inhabits hot spring microbial mats. Note that although similar in name and physiology, *Thermodesulfovibrio* and *Thermodesulfobacterium* (Section 18.19) are phylogenetically distinct organisms.

Deferribacter and *Geovibrio*

The genus *Deferribacter* forms its own distinct lineage (⭗ Figure 17.1) and is composed of species that specialize in anaerobic energy metabolism. Other genera in this group include *Geovibrio* and *Flexistipes*; the latter genus is an obligately anaerobic and fermentative bacterium.

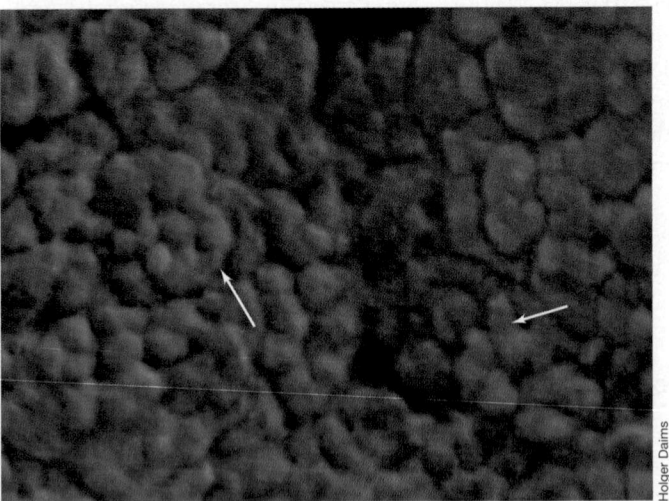

Figure 18.52 **The nitrifying bacterium *Nitrospira*.** An aggregate of *Nitrospira* cells enriched from activated sludge from a wastewater treatment facility. Individual cells are curved (arrows) and group into tetrads in the aggregate. A single cell of *Nitrospira* is about 0.3 × 1–2 μm.

Deferribacter and *Geovibrio* are extremely versatile at anaerobic respiration using a variety of electron acceptors, including nitrate, fumarate, and the metals ferric iron (Fe^{3+}) and manganic iron (Mn^{4+}). We discussed anaerobic respiration in Chapter 14 and showed that the process can be linked to many different terminal electron acceptors. Members of the *Deferribacter* group appear to be unusual in the vast number of alternative electron acceptors they can use and in the fact that they are obligate anaerobes. By contrast, most organisms capable of growing by anaerobic respiration with NO_3^- or metals as electron acceptors are facultative aerobes, organisms able to grow fully aerobically as well as by anaerobic respiration (⮂ Section 14.6). In addition to reducing Fe^{3+} and NO_3^-, *Geovibrio* can reduce elemental sulfur (S^0) to H_2S, and so ecologically, it resembles sulfur and sulfate-reducing bacteria (⮂ Section 17.18).

MiniQuiz
- Contrast the metabolic features of *Nitrospira* and *Deferribacter*. How do *Nitrospira* and *Deferribacter* differ?

Big Ideas

18.1
Lactic acid bacteria include organisms such as *Streptococcus*, *Lactobacillus*, *Staphylococcus*, and many others. Homofermentative or heterofermentative metabolisms dominate the group, and some species are pathogens of humans and other animals.

18.2
Production of endospores is a hallmark of the key genera *Bacillus* and *Clostridium*. Gram-positive bacteria are major agents for the degradation of organic matter in soil, and a few species are pathogenic.

18.3
The *Mollicutes* (mycoplasma group) contains organisms that lack cell walls and have a very small genome. Many species require sterols to strengthen their cytoplasmic membranes, and several are pathogenic for humans, other animals, and plants.

18.4
Corynebacterium and *Arthrobacter* are common gram-positive soil bacteria. *Propionibacterium* ferments lactate to propionate and is the key agent responsible for the unique flavor and texture of Swiss cheese.

18.5
Species of the genus *Mycobacterium* are mainly harmless soil saprophytes, but *Mycobacterium tuberculosis* causes the disease tuberculosis. Cells of *M. tuberculosis* have a lipid-rich, waxy outer surface layer that requires special staining procedures (the acid-fast stain) in order to observe the cells microscopically.

18.6
The streptomycetes are a large group of filamentous, gram-positive bacteria that form spores at the end of aerial filaments. Many clinically useful antibiotics such as tetracycline and neomycin have come from *Streptomyces* species.

18.7
Cyanobacteria are oxygenic phototrophic bacteria that oxygenated Earth's atmosphere. Several morphological/phylogenetic groups of cyanobacteria are known. Their habitats vary widely from aquatic to terrestrial, including many extreme environments.

18.8
Prochlorophytes are relatives of the cyanobacteria but differ most clearly from cyanobacteria in that they contain chlorophyll *b* or *d* and lack phycobilins. *Prochlorococcus* is thought to be the most abundant oxygenic phototrophic organism on Earth, inhabiting the photic zone of the open oceans.

18.9
Chlamydia are a group of small obligate intracellular parasites that cause various diseases in humans and other animals.

18.10
The *Planctomycetes* are a group of stalked, budding bacteria that form intracellular compartments, in some cases indistinguishable from the nucleus of eukaryotic cells.

18.11
Species of *Verrucomicrobia* are distinguished by their multiple prosthecate cells and their unique phylogeny.

18.12
Bacteroides and *Flavobacterium* are gram-negative bacteria that employ anaerobic and aerobic metabolisms, respectively, and that form their own phylogenetic lineage. Some species are pathogens but most are commensals in the intestine (*Bacteroides*) or aquatic environments and foods (*Flavobacterium*).

18.13
Acidobacteria are aerobic bacteria that are probably the most abundant of all soil bacteria. One organism, *Chloracidobacterium*, is phototrophic.

18.14
The *Cytophaga* group includes obligately aerobic chemoorganotrophic bacteria that live in soil, water, hot springs, saline, and thermal environments.

UNIT 6

18.15

Green sulfur bacteria are obligately anaerobic, anoxygenic phototrophs that can grow at very low light intensities because of their chlorosomes and oxidize H_2S to support autotrophy. Consortia containing green bacteria and a nonphototrophic central cell are common in sulfidic aquatic environments.

18.16

Spirochetes are tightly coiled, motile, helical bacteria that include both free-living and pathogenic species. Common genera include *Spirochaeta*, *Treponema*, *Borrelia*, and *Leptospira*. Spirochetes can be either aerobes or anaerobes and have an unusual mode of motility based on the rotation of endoflagella.

18.17

Deinococcus and *Thermus* are the major genera in a distinct lineage of *Bacteria*. *Thermus* is the source of the key enzyme in automated PCR, whereas *Deinococcus* is the most radiation-resistant bacterium known, exceeding even endospores in this regard.

18.18

The *Chloroflexi* include *Chloroflexus* and relatives, which are anoxygenic phototrophs showing properties characteristic of both phototrophic purple bacteria and green sulfur bacteria, and *Thermomicrobium*, a chemotrophic bacterium that produces unusual lipids.

18.19

Thermotoga and *Thermodesulfobacterium* form two deep lineages within the *Bacteria*. These hot spring bacteria have proven that extensive horizontal gene transfer has occurred from *Archaea* to *Bacteria* (*Thermotoga*) and that ether-linked lipids are not limited to the *Archaea* (*Thermodesulfobacterium*).

18.20

The *Aquifex* lineage contains a group of hyperthermophilic, H_2-oxidizing bacteria that form the earliest branch on the tree of the domain *Bacteria*.

18.21

Nitrospira and *Deferribacter* each form their own phylum of *Bacteria* but differ dramatically in their physiology. *Nitrospira* is an autotrophic nitrite-oxidizing bacterium, whereas *Deferribacter* and its relative *Geovibrio* specialize in various forms of anaerobic respiration.

Review of Key Terms

Acid-fastness a property of *Mycobacterium* species in which cells stained with the dye basic fuchsin resist decolorization with acidic alcohol

Chlorosome a cigar-shaped structure bounded by a nonunit membrane and containing the light-harvesting bacteriochlorophyll (*c*, *d*, or *e*) in green sulfur bacteria and *Chloroflexus*

Consortium a two- or more-membered association of bacteria, usually living in an intimate symbiotic fashion

Cyanobacteria prokaryotic oxygenic phototrophs that contain chlorophyll *a* and phycobilins, but not chlorophyll *b*

Green sulfur bacteria anoxygenic phototrophs containing chlorosomes and bacteriochlorophyll *c*, *d*, or *e* as light-harvesting chlorophyll

Heliobacteria anoxygenic phototrophs containing bacteriochlorophyll *g*

Heterofermentative in reference to lactic acid bacteria, capable of making more than one fermentation product

Homofermentative in reference to lactic acid bacteria, producing only lactic acid as a fermentation product

Phycobilin the light-capturing open chain tetrapyrrole component of phycobiliproteins

Prochlorophyte a bacterial oxygenic phototroph that contains chlorophylls *a* and *b* but lacks phycobilins

Spirochete a slender, tightly coiled gram-negative bacterium characterized by possession of endoflagella used for motility

Review Questions

1. Describe key physiological and morphological features that would differentiate *Staphylococcus* from *Micrococcus* (Section 18.1).

2. What physiological features distinguish *Bacillus* from *Clostridium* (Section 18.2)?

3. What is the key feature that differentiates *Mycobacterium* from other gram-positive bacteria such as *Bacillus* and *Propionibacterium* (Sections 18.4 and 18.5)?

4. What is the most significant characteristic of *Streptomyces* (Section 18.6)?

5. What is the most significant characteristic of *Chlamydia* (Section 18.9)?

6. Describe a key feature that would differentiate each of the following *Bacteria*: *Streptococcus*, *Planctomyces*, *Verrucomicrobium*, *Gemmata*, and *Spirochaeta* (Sections 18.1, 18.10, 18.11, 18.16).

7. Describe a key physiological feature of the following *Bacteria* that would differentiate each from the others: *Lactobacillus*, *Nitrospira*, and *Oscillatoria* (Sections 18.1, 18.7, 18.21).

8. What do species in the *Planctomycetes* have in common with *Archaea*? With *Eukarya* (Section 18.10)?

9. List a key feature that differentiates *Acidobacteria*, cytophagas, and *Bacteroides* (Sections 18.12–18.14).

10. Describe the possible benefits to each member of a consortium composed of a green sulfur bacterium and a chemoorganotroph (Section 18.15).

11. What major physiological property unites species of *Thermotoga*, *Aquifex*, and *Thermocrinis* (Sections 18.19, 18.20)?

Application Questions

1. Defend or refute the claim that convergent evolution accounts for the presence of phototrophic representatives in several different phyla of *Bacteria*.

2. Describe a major property of the earliest lineages on the phylogenetic tree of *Bacteria* and explain why this is consistent with our geochemical picture of early Earth. Would you expect organisms on these lineages to have large genomes or small genomes, and why?

Need more practice? Test your understanding with Quantitative Questions; access additional study tools including tutorials, animations, and videos; and then test your knowledge with chapter quizzes and practice tests at **www.microbiologyplace.com**.

19

Archaea

The tiny parasitic cells of *Nanoarchaeum* (red) contain the smallest genome of any species of *Archaea* and grow attached to cells of *Ignicoccus* in the near boiling waters of hot springs worldwide.

We now consider organisms in the domain *Archaea*. In Chapter 7 we discussed the major phenotypic similarities and differences between *Archaea* and *Bacteria*—the two domains of prokaryotic cells. In Chapter 16 we emphasized the profound phylogenetic differences between *Bacteria* and *Archaea*. Here we consider the organisms themselves. Some major characteristics of *Archaea* include the absence of peptidoglycan in cell walls and the presence of ether-linked lipids and structurally complex RNA polymerases. But beyond this, *Archaea* show enormous phenotypic diversity, and we focus on this diversity here.

Diversity

We begin this chapter with an overview of the phylogeny of *Archaea* and then proceed to a discussion of the *Euryarchaeota* and *Crenarchaeota*, the two phyla of *Archaea*.

19.1 Phylogenetic and Metabolic Diversity of *Archaea*

A phylogenetic tree of *Archaea* is shown in **Figure 19.1**. The tree, based on sequences of 16S ribosomal RNA genes, reveals a major evolutionary split of *Archaea* into two phyla, the *Crenarchaeota* and the *Euryarchaeota*. The separation of these groups is also supported by genomic analyses, which show that each group has its own pattern of genes but also that they share many genes in common.

Crenarchaeota

Among *Archaea* in laboratory culture, the **Crenarchaeota** contain mostly **hyperthermophiles**—organisms whose growth temperature optimum is greater than 80°C—including those able to grow at the highest temperatures of all known organisms. Interestingly however, several nonthermophilic *Crenarchaeota* related to hyperthermophilic species are known to inhabit aquatic and terrestrial environments. Many hyperthermophiles are chemolithotrophic autotrophs and because no phototrophs can survive such temperatures, these organisms are the sole primary producers in these habitats.

Hyperthermophilic species of *Crenarchaeota* tend to cluster closely together and occupy short branches on the phylogenetic tree (↺ Figures 16.16 and 19.1). These organisms are therefore thought to be more slowly evolving than other lineages in the domain. Hyperthermophilic *Crenarchaeota* are, therefore, our best available models of "early" *Archaea* and perhaps early life forms in general (↺ Section 16.2); we return to this theme at the end of this chapter (Section 19.14). By contrast, cold-dwelling relatives of hyperthermophilic crenarchaeotes have been identified

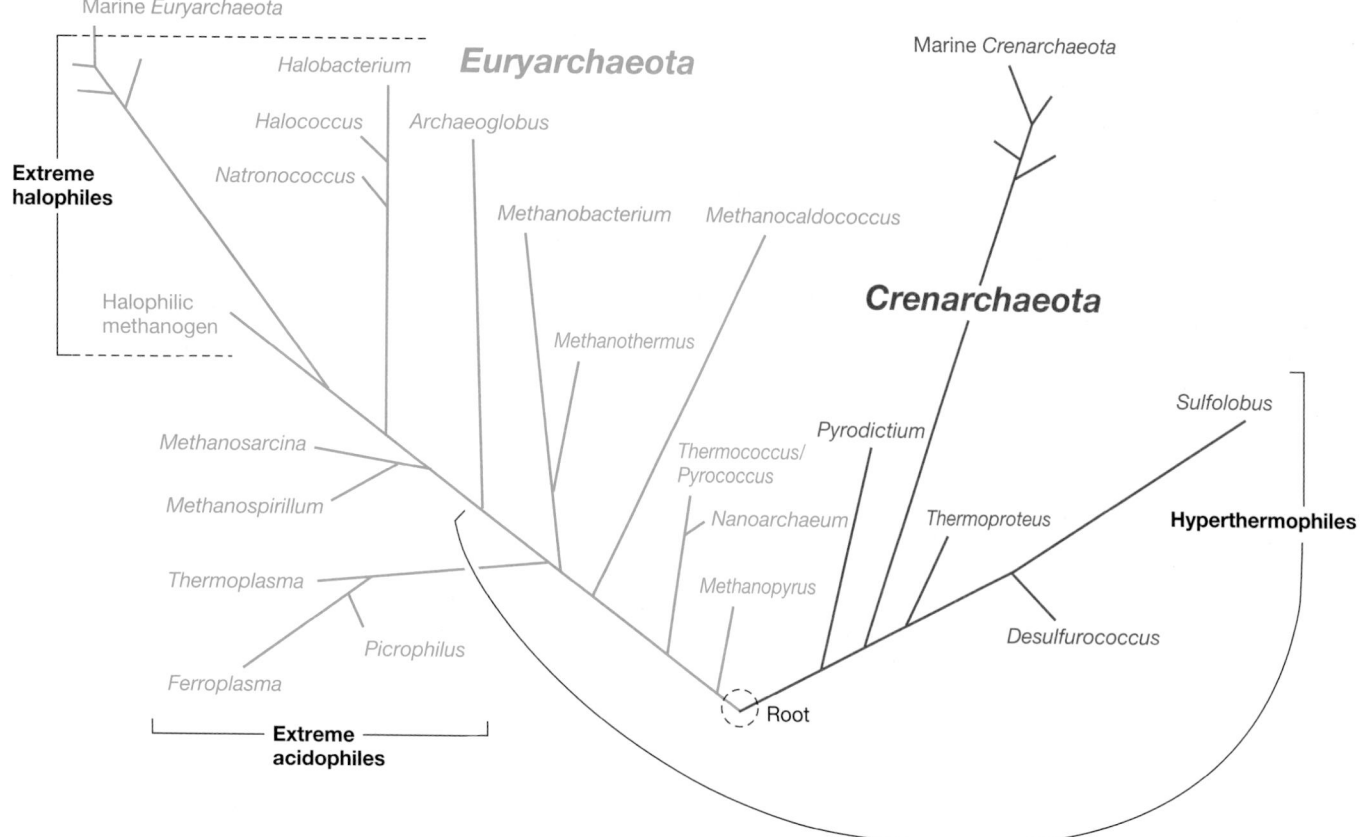

Figure 19.1 Detailed phylogenetic tree of the *Archaea* based on 16S rRNA gene sequence comparisons. Several members of the marine *Euryarchaeota* and marine *Crenarchaeota*, have been detected by community sampling of 16S rRNA genes, and at least one crenarchaeote has been cultured (see Section 19.11).

by community analysis of rRNA genes (🔗 Section 22.5) from *Archaea* in the oceans and various other temperate and even polar environments, and some of these organisms have been cultured. From a phylogenetic perspective, these species occupy longer branches on the tree and have therefore undergone rapid evolution, probably in the transition from hot to colder environments. We consider the *Crenarchaeota* in more detail in Sections 19.8–19.11.

Euryarchaeota

Euryarchaeota comprise a physiologically diverse group of *Archaea*. Like crenarchaeotes, many inhabit extreme environments of one kind or the other. This phylum includes methanogens— organisms whose metabolism produces methane (CH_4)—and several genera of extremely halophilic (salt-loving) *Archaea*, the "halobacteria" (Figure 19.1). As a study in physiological contrasts, these two groups are remarkable: Methanogens are the strictest of anaerobes while extreme halophiles are primarily obligate aerobes.

Other groups of euryarchaeotes include the hyperthermophiles *Thermococcus* and *Pyrococcus* and the methanogen *Methanopyrus*, all of which branch near the root of the archaeal tree (Figure 19.1), and the cell wall–less *Thermoplasma*, an organism phenotypically similar to the mycoplasmas (🔗 Section 18.3). Finally, in parallel to the *Crenarchaeota*, a large group of thus far uncultured euryarchaeotes inhabits marine environments and occupies long branches near the top of the archaeal tree (Figure 19.1). Many euryarchaeotes also inhabit freshwater and terrestrial habitats. We consider *Euryarchaeota* in more detail in Sections 19.2–19.7.

Metabolic Diversity of *Archaea*

Archaea include species that carry out chemoorganotrophic or chemolithotrophic metabolisms and we summarize many of these later (see Table 19.8). No true phototrophic species containing chlorophyll pigments are known, although a unique light-mediated form of energy generation does occur in some halophilic species. Energy metabolism in methanogens, a major group of *Archaea*, is unlike that of any other microbial group, *Bacteria* or *Archaea*. Methane (CH_4) is produced in either an anaerobic respiration where carbon dioxide (CO_2) is the electron acceptor and hydrogen (H_2) is the electron donor, or from the catabolism of a short list of organic compounds, acetate being the prime example. We consider the process of methanogenesis in Section 14.10.

Chemoorganotrophy in *Archaea* is supported by the oxidation of several different organic compounds, and the processes of fermentation and anaerobic respiration are well developed in certain species. Aerobic respiration occurs in the halophiles and some thermoacidophiles, but anaerobic respiration, especially forms employing elemental sulfur (S^0) as electron acceptor, is widespread, especially in *Crenarchaeota*. In these respirations electron transport reactions form a proton motive force that couples to ATP synthesis through membrane-bound ATPases, as in *Bacteria* (🔗 Section 4.12). Chemolithotrophy is also well established in the *Archaea*, with H_2 being a common electron donor

(Section 19.14). We examine chemolithotrophic metabolism of hyperthermophilic *Archaea* in Section 19.8.

Autotrophy is widespread in the *Archaea* and proceeds by several different pathways. These include, in particular, the *acetyl-CoA pathway* (🔗 Section 14.9) or some slight modification of it, but also the reverse (reductive) citric acid cycle and the 3-hydroxypropionate/4-hydroxypropionate cycle (🔗 Section 13.13). Enzymes of the Calvin cycle (🔗 Section 13.12), the most widespread autotrophic pathway in *Bacteria* and eukaryotes, have also been detected in a few hyperthermophilic *Archaea*.

We can conclude from this overview that many of the catabolic pathways in the *Archaea* are also those of the *Bacteria*, a good reminder that metabolism has a long evolutionary history. With this background and the phylogeny of *Archaea* (Figure 19.1) firmly in mind, we now consider the organismal diversity of this fascinating domain of life.

MiniQuiz
- Which phylum of *Archaea* contains organisms that make natural gas?
- What autotrophic pathways are found in *Archaea*?

 ## Euryarchaeota

Euryarchaeotes comprise a physiologically diverse group of *Archaea*, and many species inhabit extreme environments. We start with the extreme halophiles.

19.2 Extremely Halophilic *Archaea*

Key Genera: *Halobacterium, Haloferax, Natronobacterium*

Extremely halophilic *Archaea* (Figure 19.1), often called the "haloarchaea," are a diverse group that inhabits environments high in salt. These include naturally salty environments, such as solar salt evaporation ponds and salt lakes, and artificial saline habitats such as the surfaces of heavily salted foods, for example, certain fish and meats. Such salty habitats are called *hypersaline* (**Figure 19.2**). The term **extreme halophile** is used to indicate that these organisms are not only halophilic, but that their requirement for salt is very high, in some cases at levels near saturation.

An organism is considered an extreme halophile if it requires 1.5 M (about 9%) or more sodium chloride (NaCl) for growth. Most species of extreme halophiles require 2–4 M NaCl (12–23%) for optimal growth. Virtually all extreme halophiles can grow at 5.5 M NaCl (32%, the limit of saturation for NaCl), although some species grow very slowly at this salinity. Some phylogenetic relatives of extremely halophilic *Archaea*, for example species of *Haloferax* and *Natronobacterium*, are able to grow at much lower salinities, such as at or near that of seawater (about 2.5% NaCl); nevertheless, these organisms are grouped with other extreme halophiles. The isolation of these "low-salinity" haloarchaea highlights a theme emerging from environmental sampling and culture-based studies of *Archaea*, which is the presence of substantial previously unknown ecological, metabolic, and physiological diversity (Section 19.11).

(a)

(b)

(c)

(d)

Figure 19.2 Hypersaline habitats for halophilic *Archaea*. (a) The north arm of Great Salt Lake, Utah, a hypersaline lake in which the ratio of ions is similar to that in seawater, but in which absolute concentrations of ions are several times that of seawater. The green color is primarily from cells of cyanobacteria and green algae. (b) Aerial view near San Francisco Bay, California, of a series of seawater evaporating ponds where solar salt is prepared. The red-purple color is predominantly due to bacterioruberins and bacteriorhodopsin in cells of haloarchaea. (c) Lake Hamara, Wadi El Natroun, Egypt. A bloom of pigmented haloalkaliphiles is growing in this pH 10 soda lake. Note the deposits of trona ($NaHCO_3 \cdot Na_2CO_3 \cdot 2\ H_2O$) around the edge of the lake. (d) Scanning electron micrograph of halophilic bacteria including square *Archaea* present in a saltern in Spain.

Hypersaline Environments: Chemistry and Productivity

Hypersaline habitats are common throughout the world, but extremely hypersaline habitats are rare. Most such environments are in hot, dry areas of the world. Salt lakes can vary considerably in ionic composition. The predominant ions in a hypersaline lake depend on the surrounding topography, geology, and general climatic conditions.

Great Salt Lake in Utah (USA) (Figure 19.2a), for example, is essentially concentrated seawater; the relative proportions of the various ions are those of seawater, although the overall concentration of ions is much higher. Sodium (Na^+) is the predominant cation in Great Salt Lake, whereas chloride (Cl^-) is the predominant anion; significant levels of sulfate are also present at a slightly alkaline pH (**Table 19.1**). By contrast, another hypersaline basin, the Dead Sea, is relatively low in Na^+ but contains high levels of magnesium (Mg^{2+}) (Table 19.1).

Soda lakes are highly alkaline, hypersaline environments. The water chemistry of soda lakes resembles that of hypersaline lakes such as Great Salt Lake, but because high levels of carbonate minerals are also present in the surrounding strata, the pH of soda lakes is quite high. Waters of pH 10–12 are not uncommon in these environments (Table 19.1 and Figure 19.2c). In addition, calcium (Ca^{2+}) and Mg^{2+} are virtually absent from soda lakes because they precipitate out at high pH and carbonate concentrations (Table 19.1).

The diverse chemistries of hypersaline habitats have selected for a large diversity of halophilic microorganisms. Some organisms are known from one environment only while others are widespread in several habitats. Moreover, despite what may seem like rather harsh conditions, salt lakes can be highly productive ecosystems

Table 19.1 *Ionic composition of some hypersaline environments*[a]

	Concentration (g/l)		
Ion	Great Salt Lake[b]	Dead Sea	Lake Zugm[c]
Na^+	105	35.4	142
K^+	6.7	8.4	2.3
Mg^{2+}	11	47.5	<0.1
Ca^{2+}	0.3	0.47	<0.1
Cl^-	181	230	155
Br^-	0.2	6.4	—
SO_4^{2-}	27	0.4	23
HCO_3^-	0.7	0.2	67
pH	7.7	6.1	11

[a]For comparison, seawater contains (grams per liter): Na^+, 10.6; K^+, 0.38; Mg^{2+}, 1.27; Ca^{2+}, 0.4; Cl^-, 19; Br^-, 0.065; SO_4^{2-}, 2.65; HCO_3^-, 0.14; pH 7.8.
[b]See Figure 19.2a.
[c]Wadi El Natroun, Egypt (see Figure 19.2c).

Table 19.2 *Some genera of extremely halophilic* Archaea

Genus	Morphology	Habitat
Extreme halophiles		
Halobacterium	Rods	Salted fish; hides; hypersaline lakes; salterns
Halorubrum	Rods	Dead Sea; salterns
Halobaculum	Rods	Dead Sea
Haloferax	Flattened discs	Dead Sea; salterns
Haloarcula	Irregular discs	Salt pools, Death Valley, CA; marine salterns
Halococcus	Cocci	Salted fish; salterns
Halogeometricum	Pleomorphic flat cells	Solar salterns
Haloterrigena	Rods, ovals	Saline soil
Haloquadratum	Flat squares	Salterns
Haloalkaliphiles		
Natronobacterium	Rods	Highly saline soda lakes
Natrinema	Rods	Salted fish; hides
Natrialba	Rods	Soda lakes; beach sand
Natronomonas	Rods	Soda lakes
Natronococcus	Cocci	Soda lakes
Natronorubrum	Flattened cells	Soda lakes

(the word *productive* here means high levels of autotrophic CO_2 fixation). *Archaea* are not the only microorganisms present. The eukaryotic alga *Dunaliella* (⟲ Figure 20.41*a*) is the major, if not the sole, oxygenic phototroph in most salt lakes. In highly alkaline soda lakes where *Dunaliella* is absent, anoxygenic phototrophic purple bacteria of the genera *Ectothiorhodospira* and *Halorhodospira* (⟲ Section 17.2) predominate. Organic matter originating from primary production by oxygenic or anoxygenic phototrophs sets the stage for growth of haloarchaea, which are chemoorganotrophic organisms. In addition, a few extremely halophilic chemoorganotrophic *Bacteria*, such as *Halanaerobium*, *Halobacteroides*, and *Salinibacter*, thrive in such environments.

Marine salterns are also habitats for extreme halophiles. Marine salterns are enclosed basins filled with seawater that are left to evaporate, yielding solar sea salt (Figure 19.2*b, d*). As salterns approach the minimum salinity limits for haloarchaea, the waters turn a reddish purple color due to the massive growth—called a *bloom*—of cells (the red coloration apparent in Figure 19.2*b* and *c* is due to carotenoids and other pigments to be discussed later). Morphologically unusual *Archaea* are often present in salterns, including species with a square or cup-shaped morphology (Figure 19.2*d*). Extreme halophiles are also present in highly salted foods, such as certain types of sausages, marine fish, and salted pork.

Taxonomy and Physiology of Extremely Halophilic *Archaea*

Table 19.2 lists several of the currently recognized genera of extremely halophilic *Archaea*. Besides the term haloarchaea, these *Archaea* are commonly called "halobacteria," because the genus *Halobacterium* (**Figure 19.3**) was the first in this group to be described and is still the best-studied representative of the group. *Natronobacterium*, *Natronomonas*, and their relatives differ from other extreme halophiles in being extremely alkaliphilic as well as halophilic. As befits their soda lake habitat (Table 19.1

and Figure 19.2*c*), natronobacteria grow optimally at very low Mg^{2+} concentrations and high pH (9–11).

Haloarchaea stain gram-negatively, reproduce by binary fission, and do not form resting stages or spores. Cells of the various cultured genera are rod-shaped, cocci, or cup-shaped, but even cells that form squares are known (Figure 19.2*d*). A square isolate was recently obtained in pure culture and named

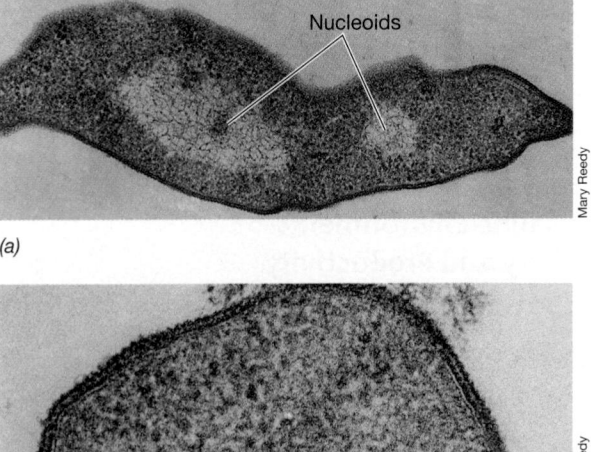

Figure 19.3 Electron micrographs of thin sections of the extreme halophile *Halobacterium salinarum*. A cell is about 0.8 μm in diameter. (*a*) Longitudinal section of a dividing cell showing the nucleoids. (*b*) High-magnification electron micrograph showing the glycoprotein subunit structure of the cell wall.

Haloquadratum to emphasize its unique morphology; the organism, which is only about 0.1 μm thick, also formed gas vesicles that allow it to float in its salty hypersaline habitat, perhaps as a means to be in contact with O_2 from air since most extreme halophiles are obligate aerobes. Most species of extreme halophiles lack flagella, but a few strains are weakly motile by flagella that rotate to propel the cell forward (Section 3.13). The genomes of *Halobacterium* and *Halococcus* are unusual in that large plasmids containing up to 30% of the total cellular DNA are present and the GC base ratio of these plasmids (near 60% GC) differs significantly from that of chromosomal DNA (66–68% GC). Plasmids from extreme halophiles are among the largest naturally occurring plasmids known (Section 6.6).

Most species of extremely halophilic *Archaea* are obligate aerobes. Most halobacteria use amino acids or organic acids as energy sources and require a number of growth factors (mainly vitamins; Section 4.1) for optimal growth. A few haloarchaea oxidize carbohydrates aerobically, but this capacity is rare; sugar fermentation does not occur. Electron transport chains containing cytochromes of the *a*, *b*, and *c* types are present in *Halobacterium*, and energy is conserved during aerobic growth via a proton motive force arising from electron transport. Some haloarchaea have been shown to grow anaerobically, as growth by anaerobic respiration (Section 14.6) linked to the reduction of nitrate or fumarate has been demonstrated in certain species.

Water Balance in Extreme Halophiles

Extremely halophilic *Archaea* require large amounts of Na^+ for growth, typically supplied as NaCl. Detailed salinity studies of *Halobacterium* have shown that the requirement for Na^+ cannot be satisfied by any other ion, even the chemically related ion potassium (K^+). However, cells of *Halobacterium* need *both* Na^+ and K^+ for growth, because each plays an important role in maintaining osmotic balance.

As we learned in Section 5.16, microbial cells must withstand the osmotic forces that accompany life. To do so in a high-solute environment such as the salt-rich habitats of *Halobacterium*, organisms must either accumulate or synthesize solutes intracellularly. These solutes are called **compatible solutes**. These compounds counteract the tendency of the cell to become dehydrated under conditions of high osmotic strength by placing the cell in positive water balance with its surroundings. Cells of *Halobacterium*, however, do not synthesize or accumulate organic compounds but instead pump large amounts of K^+ from the environment into the cytoplasm. This ensures that the concentration of K^+ *inside* the cell is even greater than the concentration of Na^+ *outside* the cell (**Table 19.3**). This ionic condition maintains positive water balance.

The *Halobacterium* cell wall (Figure 19.3*b*) is composed of glycoprotein and is stabilized by Na^+. Sodium ions bind to the outer surface of the *Halobacterium* wall and are absolutely essential for maintaining cellular integrity. When insufficient Na^+ is present, the cell wall breaks apart and the cell lyses. This is a consequence of the exceptionally high content of the *acidic* (negatively charged) amino acids aspartate and glutamate in the glycoprotein of the *Halobacterium* cell wall. The negative charge on the carboxyl group of these amino acids (Figure 6.29) is bound to

Table 19.3 *Concentration of ions in cells of Halobacterium salinarum*[a]

Ion	Concentration in medium (M)	Concentration in cells (M)
Na^+	4.0	1.4
K^+	0.032	4.6
Mg^{2+}	0.13	0.12
Cl^-	4.0	3.6

[a]Data from *Biochim. Biophys. Acta* 65: 506–508 (1962).

Na^+; when Na^+ is diluted away, the negatively charged parts of the proteins tend to repel each other, leading to cell lysis.

Halophilic Cytoplasmic Components

Like cell wall proteins, cytoplasmic proteins of *Halobacterium* are highly acidic, but it is K^+, not Na^+, that is required for activity. This makes sense, of course, because K^+ is the predominant cation in the cytoplasm of cells of *Halobacterium* (Table 19.3). Besides a high acidic amino acid composition, halobacterial cytoplasmic proteins typically contain lower levels of hydrophobic amino acids and lysine, a positively charged (basic) amino acid, than proteins of nonhalophiles. This is also to be expected because in a highly ionic cytoplasm, polar proteins would tend to remain in solution whereas nonpolar proteins would tend to cluster and perhaps lose activity. The ribosomes of *Halobacterium* also require high KCl levels for stability, whereas ribosomes of nonhalophiles have no KCl requirement.

Extremely halophilic *Archaea* are thus well adapted, both internally and externally, to life in a highly ionic environment. Cellular components exposed to the external environment require high Na^+ for stability, whereas internal components require high K^+. With the exception of a few extremely halophilic members of the *Bacteria* that also use KCl as a compatible solute, in no other group of bacteria do we find this unique requirement for such high amounts of specific cations.

Bacteriorhodopsin and Light-Mediated ATP Synthesis in Halobacteria

Certain species of haloarchaea can catalyze a light-driven synthesis of ATP. This occurs without chlorophyll pigments, so it is not photosynthesis. However, other light-sensitive pigments are present, including red and orange carotenoids—primarily C_{50} pigments called *bacterioruberins*—and inducible pigments involved in energy conservation; we discuss these pigments here.

Under conditions of low aeration, *Halobacterium salinarum* and some other haloarchaea synthesize a protein called **bacteriorhodopsin** and insert it into their cytoplasmic membranes. Bacteriorhodopsin is so named because of its structural and functional similarity to rhodopsin, the visual pigment of the eye. Conjugated to bacteriorhodopsin is a molecule of retinal, a carotenoid-like molecule that can absorb light energy and pump a proton across the cytoplasmic membrane. The retinal gives bacteriorhodopsin a purple hue. Thus cells of *Halobacterium* that are switched from growth under high-aeration conditions to

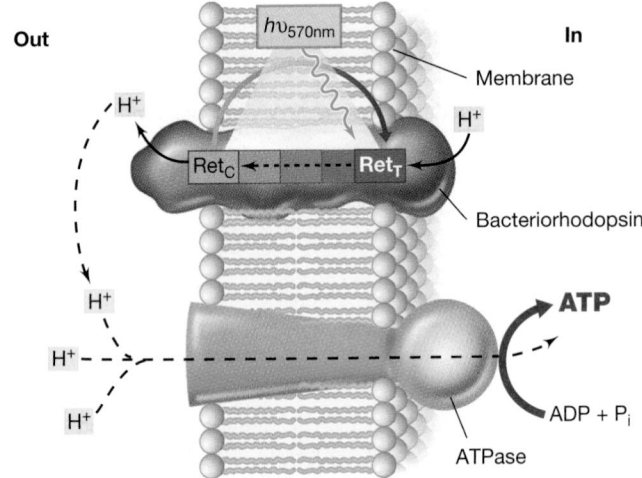

Figure 19.4 Model for the mechanism of bacteriorhodopsin. Light of 570 nm ($h\nu_{570nm}$) converts the protonated retinal of bacteriorhodopsin from the *trans* form (Ret$_T$) to the *cis* form (Ret$_C$), along with translocation of a proton to the outer surface of the cytoplasmic membrane, thus establishing a proton motive force. ATPase activity is driven by the proton motive force.

oxygen-limiting growth conditions (a trigger of bacteriorhodopsin synthesis) gradually change color from orange-red to purple-red as they synthesize bacteriorhodopsin and insert it into their cytoplasmic membranes.

Bacteriorhodopsin absorbs green light around 570 nm. Following absorption, the retinal of bacteriorhodopsin, which normally exists in a *trans* configuration (Ret$_T$), becomes excited and converts to the *cis* (Ret$_C$) form (**Figure 19.4**). This transformation is coupled to the translocation of a proton across the cytoplasmic membrane. The retinal molecule then decays to the *trans* isomer along with the uptake of a proton from the cytoplasm, and this completes the cycle. The proton pump is then ready to repeat the cycle (Figure 19.4). As protons accumulate on the outer surface of the membrane, a proton motive force is generated that is coupled to ATP synthesis through the activity of a proton-translocating ATPase (Section 4.10; Figure 19.4).

Bacteriorhodopsin-mediated ATP production in *H. salinarum* supports slow growth of this organism under anoxic conditions. The light-stimulated proton pump of *H. salinarum* also functions to pump Na$^+$ out of the cell by activity of a Na$^+$–H$^+$ antiport system and also drives the uptake of nutrients, including the K$^+$ needed for osmotic balance. Amino acid uptake by *H. salinarum* is indirectly driven by light because amino acids are cotransported into the cell with Na$^+$ by an amino acid–Na$^+$ symporter (Section 3.5); removal of Na$^+$ from the cell occurs by way of the light-driven Na$^+$–H$^+$ antiporter.

Other Rhodopsins

Besides bacteriorhodopsin, at least three other rhodopsins are present in the cytoplasmic membrane of *H. salinarum*. **Halorhodopsin** is a light-driven chloride (Cl$^-$) pump that brings Cl$^-$ into the cell as the anion for K$^+$. The retinal of halorhodopsin binds Cl$^-$ and transports it into the cell. Two other light sensors, called *sensory rhodopsins*, are present in *H. salinarum*. These light sensors control phototaxis (movement toward light, Section 3.15) by the

organism. Through the interaction of a cascade of proteins similar to those in chemotaxis (Section 8.8), sensory rhodopsins affect flagellar rotation, moving cells of *H. salinarum* toward light where bacteriorhodopsin can function to make ATP (Figure 19.4).

We will learn when we consider marine microbiology (Sections 23.9 and 23.10) that several *Bacteria* that contain bacteriorhodopsin-like proteins called *proteorhodopsins* inhabit the upper layers of the ocean. As far as is known, proteorhodopsin functions like bacteriorhodopsin except that several different spectral forms exist, each tuned to the absorption of different wavelengths of light. Proteorhodopsin as a mechanism for energy conservation in marine bacteria makes good ecological sense because levels of dissolved organic matter in the open oceans are typically very low, and thus a strictly chemoorganotrophic lifestyle would be difficult.

MiniQuiz

- What is the major physiological difference between *Halobacterium* and *Natronobacterium*?
- If cells of *Halobacterium* require high levels of Na$^+$ for growth, why is this not true for the organism's cytoplasmic enzymes?
- What benefit does bacteriorhodopsin confer on a cell of *Halobacterium salinarum*?

19.3 Methanogenic *Archaea*

Key Genera: *Methanobacterium, Methanocaldococcus, Methanosarcina*

Many *Euryarchaeota* are **methanogens**, microorganisms that produce methane (CH$_4$) as an integral part of their energy metabolism (methane production is called *methanogenesis*). In Section 14.10 we considered the biochemistry of methanogenesis. Later, we will learn how methanogenesis is the terminal step in the biodegradation of organic matter in many anoxic habitats in nature (Section 24.2). **Table 19.4** lists the major sources of biogenic methane in nature.

Diversity and Physiology of Methanogens

Methanogens show a variety of morphologies (**Figure 19.5** and **Table 19.5**). Their taxonomy is based on both phenotypic and phylogenetic analyses, with several taxonomic orders being

Table 19.4 *Habitats of methanogens*

I. Anoxic sediments: marsh, swamp, and lake sediments, paddy fields, moist landfills

II. Animal digestive tracts:
  (a) Rumen of ruminant animals such as cattle, sheep, elk, deer, and camels
  (b) Cecum of cecal animals such as horses and rabbits
  (c) Large intestine of monogastric animals such as humans, swine, and dogs
  (d) Hindgut of cellulolytic insects (for example, termites)

III. Geothermal sources of H$_2$ + CO$_2$: hydrothermal vents

IV. Artificial biodegradation facilities: sewage sludge digesters

V. Endosymbionts of various anaerobic protozoa

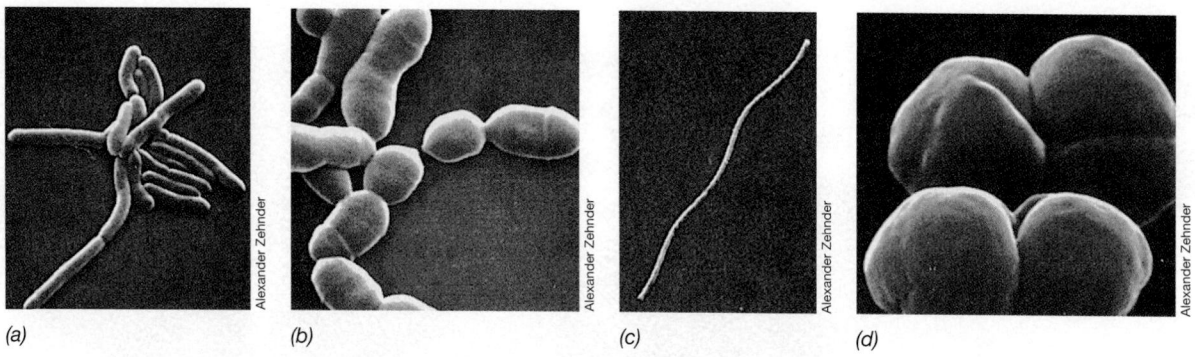

Figure 19.5 Scanning electron micrographs of cells of diverse species of methanogenic *Archaea*. *(a) Methanobrevibacter ruminantium*. A cell is about 0.7 μm in diameter. *(b) Methanobrevibacter arboriphilus*. A cell is about 1 μm in diameter. *(c) Methanospirillum hungatei*. A cell is about 0.4 μm in diameter. *(d) Methanosarcina barkeri*. A cell is about 1.7 μm wide.

Table 19.5 *Characteristics of some methanogenic* Archaea[a]

Order/Genus	Morphology	Substrates for methanogenesis
Methanobacteriales		
Methanobacterium	Long rods	$H_2 + CO_2$, formate
Methanobrevibacter	Short rods	$H_2 + CO_2$, formate
Methanosphaera	Cocci	Methanol + H_2 (both needed)
Methanothermus	Rods	$H_2 + CO_2$
Methanothermobacter	Rods	$H_2 + CO_2$, formate
Methanococcales		
Methanococcus	Irregular cocci	$H_2 + CO_2$, pyruvate + CO_2, formate
Methanothermococcus	Cocci	$H_2 + CO_2$, formate
Methanocaldococcus	Cocci	$H_2 + CO_2$
Methanotorris	Cocci	$H_2 + CO_2$
Methanomicrobiales		
Methanomicrobium	Short rods	$H_2 + CO_2$, formate
Methanogenium	Irregular cocci	$H_2 + CO_2$, formate
Methanospirillum	Spirilla	$H_2 + CO_2$, formate
Methanoplanus	Plate-shaped cells—occurring as thin plates with sharp edges	$H_2 + CO_2$, formate
Methanocorpusculum	Irregular cocci	$H_2 + CO_2$, formate, alcohols
Methanoculleus	Irregular cocci	$H_2 + CO_2$, alcohols, formate
Methanofollis	Irregular cocci	$H_2 + CO_2$, formate
Methanolacinia	Irregular rods	$H_2 + CO_2$, alcohols
Methanosarcinales		
Methanosarcina	Large, irregular cocci in packets	$H_2 + CO_2$, methanol, methylamines, acetate
Methanolobus	Irregular cocci in aggregates	Methanol, methylamines
Methanohalobium	Irregular cocci	Methanol, methylamines
Methanococcoides	Irregular cocci	Methanol, methylamines
Methanohalophilus	Irregular cocci	Methanol, methylamines, methyl sulfides
Methanosaeta	Long rods to filaments	Acetate
Methanosalsum	Irregular cocci	Methanol, methylamines, dimethylsulfide
Methanimicrococcus	Irregular cocci	Methanol, methylamines (H_2 needed with any methanogenic substrate)
Methanopyrales		
Methanopyrus	Rods in chains	$H_2 + CO_2$

[a]Taxonomic orders are listed in bold. An order is a taxonomic rank that consists of several families; families consist of several genera.

UNIT 6

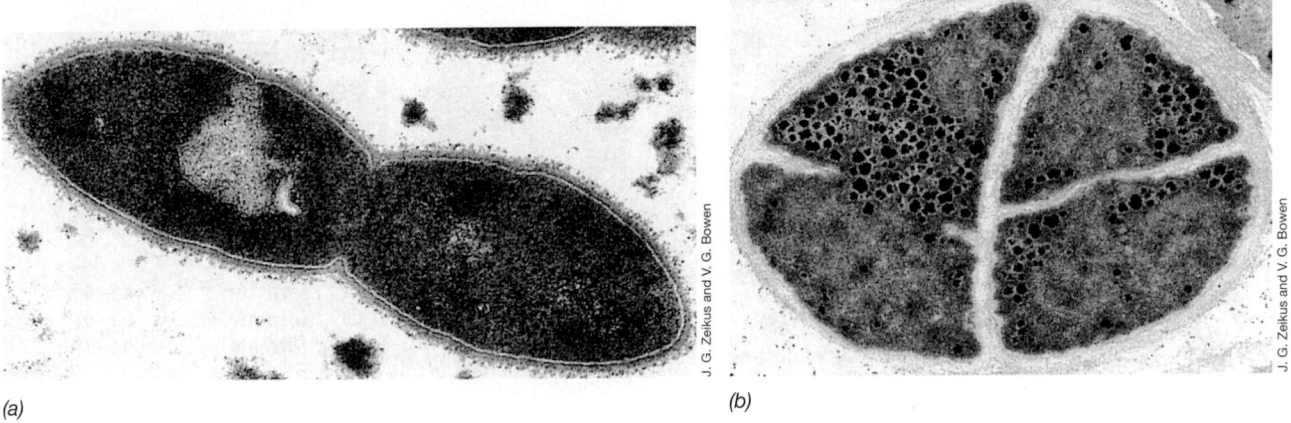

(a)

(b)

J. G. Zeikus and V. G. Bowen

Figure 19.6 Transmission electron micrographs of thin sections of methanogenic *Archaea*. *(a) Methanobrevibacter ruminantium*. A cell is 0.7 μm in diameter. *(b) Methanosarcina barkeri*, showing the thick cell wall and the manner of cell segmentation and cross-wall formation. A cell is 1.7 μm in diameter.

recognized (in taxonomy, an order contains groups of related families, each of which contains one or more genera; ↩ Section 16.13).

Methanogens show a diversity of cell wall chemistries. These include the pseudomurein walls of *Methanobacterium* species and relatives (**Figure 19.6a**), walls composed of methanochondroitin (so named because of its structural resemblance to chon-droitin, the connective tissue polymer of vertebrate animals) in *Methanosarcina* and relatives (Figure 19.6b), the protein or glycoprotein walls of *Methanocaldococcus* (**Figure 19.7a**) and *Methanoplanus* species, respectively, and the S-layer walls of *Methanospirillum* (Figure 19.5c; ↩ Section 3.8).

Physiologically, methanogens are obligate anaerobes, and strict anoxic techniques are necessary to culture them. Most

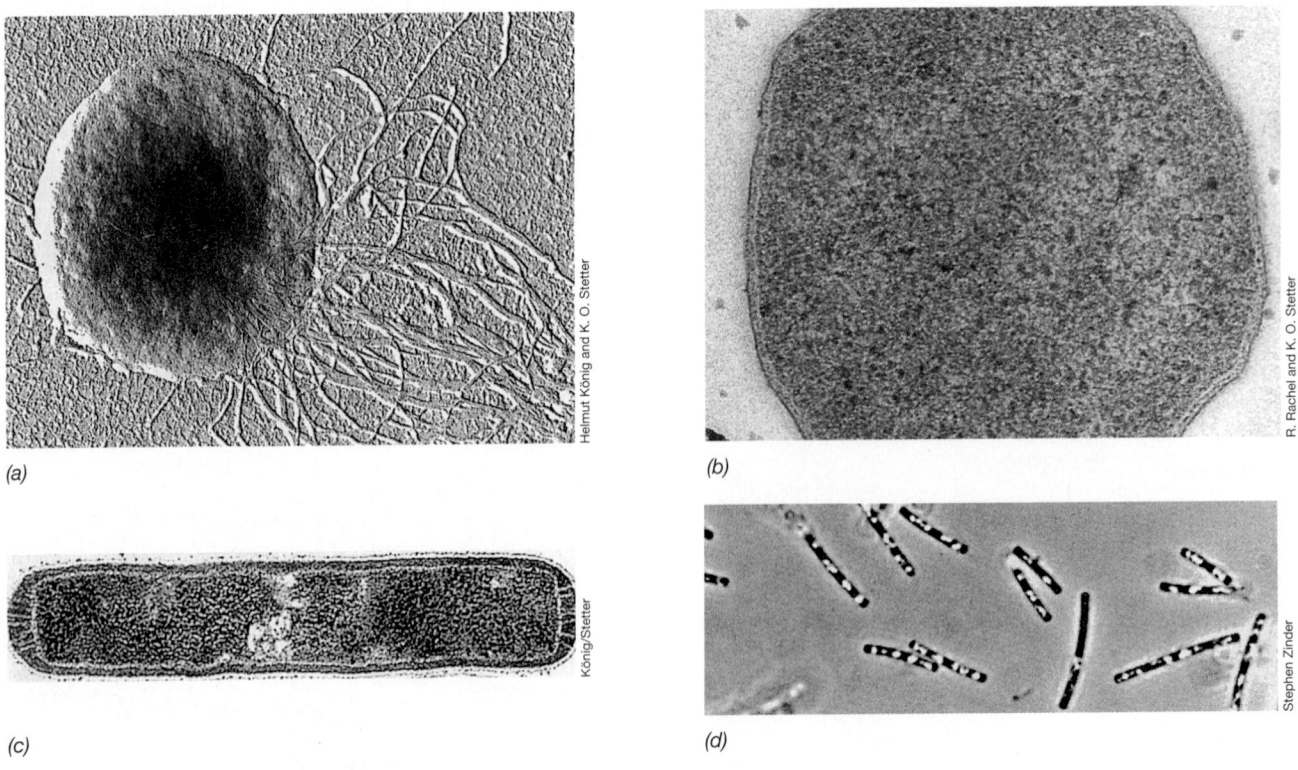

(a)

Helmut König and K. O. Stetter

(b)

R. Rachel and K. O. Stetter

(c)

König/Stetter

(d)

Stephen Zinder

Figure 19.7 Hyperthermophilic and thermophilic methanogens. *(a) Methanocaldococcus jannaschii* (temperature optimum, 85°C), shadowed preparation electron micrograph. A cell is about 1 μm in diameter. *(b) Methanotorris igneus* (temperature optimum, 88°C), thin section. A cell is about 1 μm in diameter. *(c) Methanothermus fervidus* (temperature optimum, 88°C), thin-sectioned electron micrograph. A cell is about 0.4 μm in diameter. *(d) Methanosaeta thermophila* (temperature optimum, 60°C), phase-contrast micrograph. A cell is about 1 μm in diameter. The refractile bodies inside the cells are gas vesicles.

methanogens are mesophilic and nonhalophilic, although "extremophilic" species growing optimally at very high (Figure 19.7) or very low temperatures, at very high salt concentrations, or at extremes of pH, have also been described. Several substrates can be converted to CH_4 by methanogens. Interestingly, these substrates do not include such common compounds as glucose and organic or fatty acids (other than acetate and pyruvate). Compounds such as glucose can be converted to CH_4, but only in reactions in which methanogens and other anaerobic bacteria cooperate. With the right mixture of organisms, virtually any organic compound, even hydrocarbons, can be converted to CH_4 plus CO_2 (⟲ Section 24.2).

Three classes of compounds make up the list of methanogenic substrates shown in **Table 19.6**. These are CO_2-*type* substrates, *methylated* substrates, and *acetate*. CO_2-type substrates include CO_2 itself, which is reduced to CH_4 using H_2 as the electron donor. Other substrates of this type include formate (which is $CO_2 + H_2$ in combined form) and CO, carbon monoxide. Methylated substrates include methanol (CH_3OH) and many others (Table 19.6). Methanol (CH_3OH) can be reduced using an external electron donor such as H_2, or, alternatively, in the absence of H_2, some CH_3OH can be oxidized to CO_2 to generate the electrons needed to reduce other molecules of CH_3OH to CH_4 (⟲ Figure 14.21). The final methanogenic process is the cleavage of acetate to CO_2 plus CH_4. Only a few methanogens are acetotrophic (Tables 19.5 and 19.6), although acetate is a major source of CH_4 in nature. The biochemistry of methanogenesis from each of these classes of substrates is considered in Section 14.10 along with how CH_4 formation is linked to energy conservation.

Methanocaldococcus jannaschii as a Model Methanogen

As described in Section 12.1, the genome of the hyperthermophilic methanogen *Methanocaldococcus jannaschii* (Figure 19.7a) and those of several other methanogens have been sequenced. The 1.66-Mbp circular genome of *M. jannaschii*, an

Table 19.6 *Substrates converted to methane by various methanogenic Archaea*

I. **CO_2-type substrates**
   Carbon dioxide, CO_2 (with electrons derived from H_2, certain alcohols, or pyruvate)
   Formate, $HCOO^-$
   Carbon monoxide, CO

II. **Methylated substrates**
   Methanol, CH_3OH
   Methylamine, $CH_3NH_3^+$
   Dimethylamine, $(CH_3)_2NH_2^+$
   Trimethylamine, $(CH_3)_3NH^+$
   Methylmercaptan, CH_3SH
   Dimethylsulfide, $(CH_3)_2S$

III. **Acetotrophic substrates**
   Acetate, CH_3COO^-
   Pyruvate, CH_3COCOO^-

organism that has been used as a model for the molecular study of methanogenesis, contains about 1700 genes, and genes encoding enzymes of methanogenesis and several other key cell functions have been identified. Interestingly, the majority of *M. jannaschii* genes encoding functions such as central metabolic pathways and cell division are similar to those in *Bacteria*. By contrast, most of the *M. jannaschii* genes encoding core molecular processes such as transcription and translation more closely resemble those of eukaryotes. These findings reflect the various traits shared by organisms in the three cellular domains and are consistent with our understanding of how the three domains evolved, as discussed in Chapter 16. However, analyses of the *M. jannaschii* genome also show that fully 40% of its genes have no counterparts in genes from either of the other domains. Some of these are genes that encode the enzymes needed for methanogenesis, of course, but many others likely encode novel cellular functions absent from cells in the other domains or may encode redundant functions carried out by classes of enzymes distinct from those found in *Bacteria* and *Eukarya*.

MiniQuiz
- What are the major substrates for methanogenesis?
- What is unusual about the *Methanocaldococcus jannaschii* genome?

19.4 *Thermoplasmatales*

Key Genera: *Thermoplasma, Picrophilus, Ferroplasma*

A phylogenetically distinct line of *Archaea* contains thermophilic and extremely acidophilic genera: *Thermoplasma, Ferroplasma*, and *Picrophilus* (Figure 19.1). These prokaryotes are among the most acidophilic of all known microorganisms, with *Picrophilus* being capable of growth even below pH 0. Most are also thermophilic as well. They also form their own taxonomic order within the *Euryarchaeota*, the *Thermoplasmatales*. We begin with a description of the mycoplasma-like organisms *Thermoplasma* and *Ferroplasma*.

Archaea Lacking Cell Walls

Thermoplasma and *Ferroplasma* lack cell walls, and in this respect they resemble the mycoplasmas (⟲ Section 18.3). *Thermoplasma* (**Figure 19.8**) is a chemoorganotroph that grows optimally at 55°C and pH 2 in complex media. Two species of *Thermoplasma* have been described, *Thermoplasma acidophilum* and *Thermoplasma volcanium*. Species of *Thermoplasma* are facultative aerobes, growing either aerobically or anaerobically by sulfur respiration (⟲ Section 14.8). Most strains of *T. acidophilum* have been obtained from self-heating coal refuse piles. Coal refuse contains coal fragments, pyrite (FeS_2), and other organic materials extracted from coal. When dumped into piles in surface-mining operations, coal refuse heats as a result of microbial metabolism bringing it to combustion temperature (**Figure 19.9**). This sets the stage for growth of *Thermoplasma*, which likely metabolizes organic compounds leached from the hot coal refuse. A second species, *T. volcanium*, has been isolated

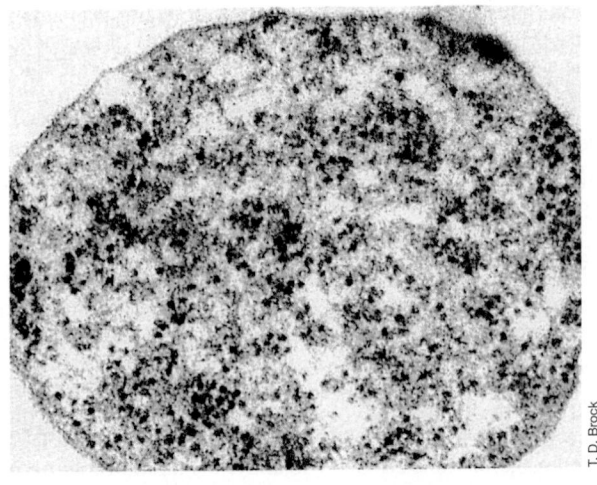

(a)

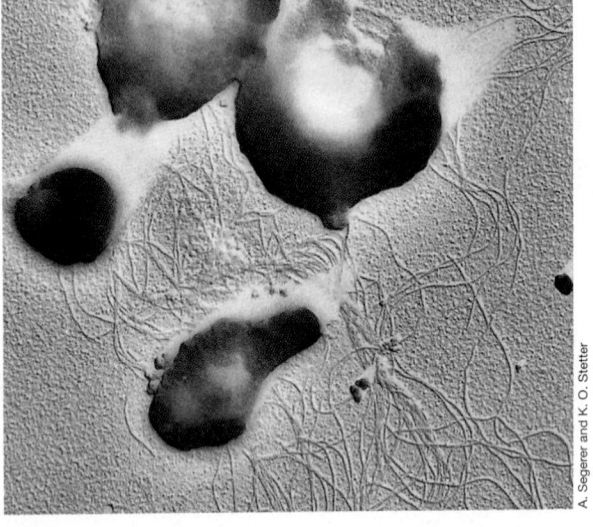

(b)

Figure 19.8 *Thermoplasma* **species.** *(a) Thermoplasma acidophilum,* an acidophilic and thermophilic mycoplasma-like archaeon; electron micrograph of a thin section. The diameter of cells varies from 0.2 to 5 μm. The cell shown is about 1 μm in diameter. *(b)* Shadowed preparation of cells of *Thermoplasma volcanium* isolated from hot springs. Cells are 1–2 μm in diameter. Notice the abundant flagella and irregular cell morphology.

in hot acidic soils throughout the world and is highly motile by multiple flagella (Figure 19.8*b*).

To survive the osmotic stresses of life without a cell wall and to withstand the dual environmental extremes of low pH and high temperature, *Thermoplasma* has evolved a unique cytoplasmic membrane structure. The membrane contains a lipopolysaccharide-like material called *lipoglycan*. This substance consists of a tetraether lipid monolayer membrane with mannose and glucose (**Figure 19.10**). This molecule constitutes a major fraction of the total lipid composition of *Thermoplasma*. The membrane also contains glycoproteins but not sterols. These molecules render the *Thermoplasma* membrane stable to hot, acidic conditions.

Like mycoplasmas (♻ Section 18.3), *Thermoplasma* contains a relatively small genome (1.5 Mbp). In addition, *Thermoplasma*

Figure 19.9 A typical self-heating coal refuse pile, habitat of *Thermoplasma*. The pile, containing coal debris, pyrite, and other microbial substrates, self-heats due to microbial metabolism.

DNA is complexed with a highly basic DNA-binding protein that organizes the DNA into globular particles resembling the nucleosomes of eukaryotic cells. This protein is homologous to the histone-like DNA-binding protein HU of *Bacteria*, which plays an important role in organization of the DNA in the cell. In contrast, several other *Euryarchaeota* contain basic proteins homologous to the DNA-binding histone proteins of eukaryotic cells (Section 19.13).

Ferroplasma

Ferroplasma is a chemolithotrophic relative of *Thermoplasma*. *Ferroplasma* is a strong acidophile; however, it is not a thermophile, as it grows optimally at 35°C. *Ferroplasma* oxidizes ferrous iron (Fe^{2+}) to ferric iron (Fe^{3+}) to obtain energy (this reaction generates acid, see Figure 19.16*d*) and uses CO_2 as its carbon source (autotrophy). *Ferroplasma* grows in mine tailings containing pyrite (FeS_2), which is its energy source. The extreme acidophily of *Ferroplasma* allows it to drive the pH of its habitat down to extremely acidic values. After moderate acidity is generated from Fe^{2+} oxidation by acidophilic organisms such as *Acidithiobacillus ferrooxidans* and *Leptospirillum ferrooxidans*, *Ferroplasma* becomes active and subsequently generates the very low pH values typical of acid mine drainage. Acidic waters at pH 0 can be generated by the activities of *Ferroplasma* (♻ Sections 24.5 and 24.7).

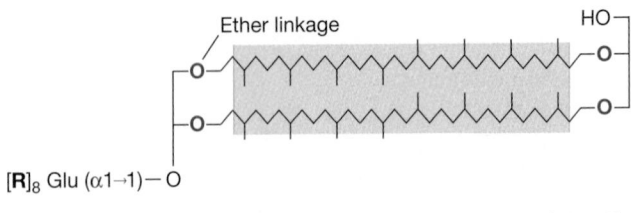

$[\mathbf{R}]_8$ Glu (α1→1)—O

$\mathbf{R}$ = Man (α1 → 2) Man (α1 → 4) Man (α1 → 3)

Figure 19.10 Structure of the tetraether lipoglycan of *Thermoplasma acidophilum*. Glu, Glucose; Man, mannose. Note the ether linkages and the fact that this lipid would form a monolayer rather than a bilayer membrane (compare the structure of lipoglycan with the membranes shown in Figure 3.7*e*).

Picrophilus

A phylogenetic relative of *Thermoplasma* and *Ferroplasma* is *Picrophilus*. Although *Thermoplasma* and *Ferroplasma* are extreme acidophiles, *Picrophilus* is even more so, growing optimally at pH 0.7 and capable of growth at pH values lower than 0. *Picrophilus* also has a cell wall (an S-layer; ⮌ Section 3.8) and a much lower DNA GC base ratio than does *Thermoplasma* or *Ferroplasma*. Although phylogenetically related, *Thermoplasma*, *Ferroplasma*, and *Picrophilus* have quite distinct genomes. Two species of *Picrophilus* have been isolated from acidic Japanese solfataras, and like *Thermoplasma*, both grow heterotrophically on complex media.

The physiology of *Picrophilus* is of interest as a model for extreme acid tolerance. Studies of its cytoplasmic membrane point to an unusual arrangement of lipids that forms a highly acid-impermeable membrane at very low pH. By contrast, at moderate acidities such as pH 4, the membranes of cells of *Picrophilus* become leaky and disintegrate. Obviously, this organism has evolved to survive only in highly acidic habitats.

MiniQuiz

- In what ways are *Thermoplasma* and *Picrophilus* similar? In what ways do they differ?
- How does *Thermoplasma* strengthen its cytoplasmic membrane to survive without a cell wall?
- How does *Ferroplasma* obtain energy for growth?

19.5 *Thermococcales* and *Methanopyrus*

Key Genera: *Thermococcus, Pyrococcus, Methanopyrus*

A few euryarchaeotes thrive in thermal environments and some are hyperthermophiles. We consider here three hyperthermophilic euryarchaeotes that branch on the archaeal tree very near the root (Figure 19.1). Two of these, *Thermococcus* and *Pyrococcus*, form a distinct taxonomic order: the *Thermococcales*. The third organism, *Methanopyrus*, is a methanogen that closely resembles other methanogens (Section 19.3 and Table 19.5) in its basic physiology but is unusual in its hyperthermophily, lipids, and phylogenetic position (Figure 19.1).

Thermococcus and Pyrococcus

Thermococcus is a spherical hyperthermophilic euryarchaeote indigenous to anoxic thermal waters in various locations throughout the world. The spherical cells contain a tuft of polar flagella and are thus highly motile (**Figure 19.11**). *Thermococcus* is an obligately anaerobic chemoorganotroph that metabolizes proteins and other complex organic mixtures (including some sugars) with elemental sulfur (S^0) as electron acceptor at temperatures from 55 to 95°C.

Pyrococcus is morphologically similar to *Thermococcus* (Figure 19.11b). *Pyrococcus* differs from *Thermococcus* primarily by its higher temperature requirements; *Pyrococcus* grows between 70 and 106°C with an optimum of 100°C. *Thermococcus* and *Pyrococcus* are also metabolically quite similar. Proteins, starch, or maltose are oxidized as electron donors, and S^0 is the terminal electron acceptor and is reduced to hydrogen sulfide (H_2S). Both *Thermococcus* and *Pyrococcus* form H_2S when S^0 is present, but form H_2 when S^0 is absent (see Table 19.8).

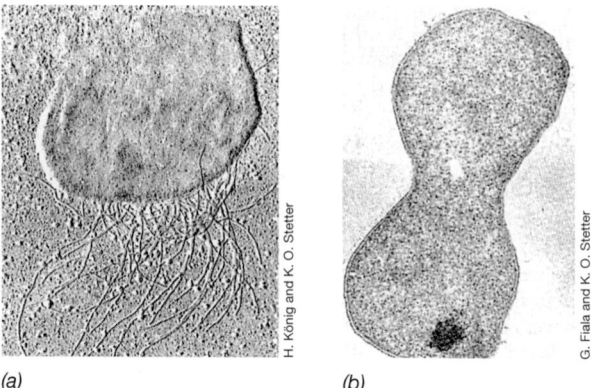

(a) (b)

Figure 19.11 Spherical hyperthermophilic *Euryarchaeota* from submarine volcanic areas. *(a)* *Thermococcus celer*. Electron micrograph of shadowed cells (note tuft of flagella). *(b)* Dividing cell of *Pyrococcus furiosus*. Electron micrograph of thin section. Cells of both organisms are about 0.8 μm in diameter.

Methanopyrus

Methanopyrus is a rod-shaped hyperthermophilic methanogen (**Figure 19.12**). *Methanopyrus* was isolated from hot sediments near submarine hydrothermal vents and from the walls of "black smoker" hydrothermal vent chimneys (Section 19.12; ⮌ Section 23.12). *Methanopyrus* lies near the base of the archaeal tree (Figure 19.1) and it shares phenotypic properties with both the hyperthermophiles and the methanogens. *Methanopyrus* produces CH_4 only from $H_2 + CO_2$ and grows rapidly for an autotrophic organism (generation time <1 h at 100°C). In special pressurized vessels, growth of one strain of *Methanopyrus* has been recorded at 122°C, the highest temperature yet shown to support microbial growth.

Methanopyrus is unusual because it contains membrane lipids found in no other known organism. Recall that in the lipids of

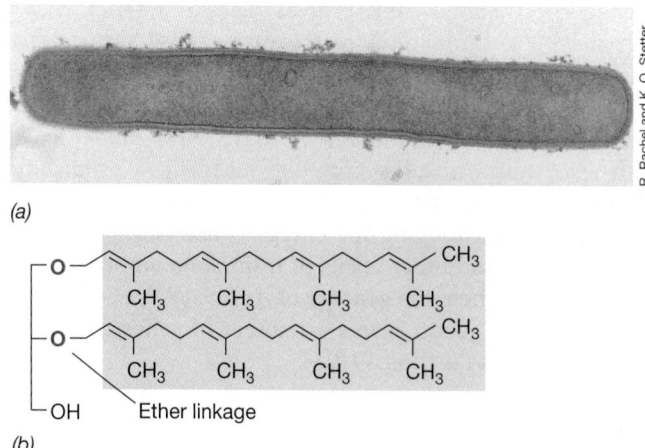

(a)

(b)

Ether linkage

Figure 19.12 *Methanopyrus*. *Methanopyrus* grows optimally at 100°C and can make CH_4 only from $CO_2 + H_2$. *(a)* Electron micrograph of a cell of *Methanopyrus kandleri*, the most thermophilic of all known organisms (upper temperature limit, 122°C). This cell measures 0.5 × 8 μm. *(b)* Structure of the novel lipid of *M. kandleri*. This is the normal ether-linked lipid of the *Archaea* with the exception that the side chains are an unsaturated form of phytanyl called geranylgeraniol.

Archaea, the glycerol side chains contain **phytanyl** rather than fatty acids bonded in ether linkage to the glycerol (↩ Section 3.3). In *Methanopyrus*, this ether-linked lipid is an unsaturated form of the otherwise saturated dibiphytanyl tetraethers found in all other hyperthermophilic *Archaea*; biochemically, the unsaturated lipid is considered a "primitive" characteristic. Along with its phylogenetic position, anaerobic metabolism, hyperthermophilic lifestyle, and primitive lipids, *Methanopyrus* may well be a modern descendant of one of Earth's earliest life forms (↩ Sections 16.2, 16.3, and 19.12–14).

MiniQuiz

- How do *Thermococcus* and *Pyrococcus* make ATP?
- In what way(s) does *Methanopyrus* display properties of key organisms in both the *Euryarchaeota* and *Crenarchaeota*, and what is unusual about its lipids?

19.6 Archaeoglobales

Key Genera: *Archaeoglobus, Ferroglobus*

We will see later that a number of hyperthermophilic *Crenarchaeota* catalyze anaerobic respirations in which elemental sulfur (S^0) is used as an electron acceptor, being reduced to H_2S (see Table 19.8). One hyperthermophilic euryarchaeote, *Archaeoglobus*, can reduce sulfate (SO_4^{2-}) and forms a phylogenetically distinct lineage within the *Euryarchaeota* (Figure 19.1).

Archaeoglobus

Archaeoglobus was isolated from hot marine sediments near hydrothermal vents. In its metabolism, *Archaeoglobus* couples the oxidation of H_2, lactate, pyruvate, glucose, or complex organic compounds to the reduction of SO_4^{2-} to H_2S. Cells of *Archaeoglobus* are irregular cocci (**Figure 19.13**) and grow optimally at 83°C.

Archaeoglobus and methanogens share some characteristics. We learned in Section 14.10 about the unique biochemistry of methanogenesis. Briefly, this process requires a series of novel coenzymes. With rare exceptions, these coenzymes have only been found in methanogens. Surprisingly, however, *Archaeoglobus* also contains many of these coenzymes and cultures of this organism actually produce small amounts of CH_4. Thus, *Archaeoglobus*, which also shows a rather close phylogenetic relationship to methanogens (Figure 19.1), may be a metabolically intermediate type of organism, bridging the energy-conserving processes of methanogenesis and other forms of respiration among *Archaea*. Not surprisingly then, the genome of *Archaeoglobus*, which contains about 2400 genes, shares a number of genes in common with methanogens (Section 19.3).

Ferroglobus

Ferroglobus (Figure 19.13*b*) is related to *Archaeoglobus* but is not a sulfate reducer. Instead, *Ferroglobus* is an iron-oxidizing chemolithotroph, conserving energy from the oxidation of Fe^{2+} to Fe^{3+} coupled to the reduction of nitrate (NO_3^-) to nitrite (NO_2^-) (see Table 19.8). *Ferroglobus* grows autotrophically and can also use H_2 or H_2S as electron donors in its energy metabo-

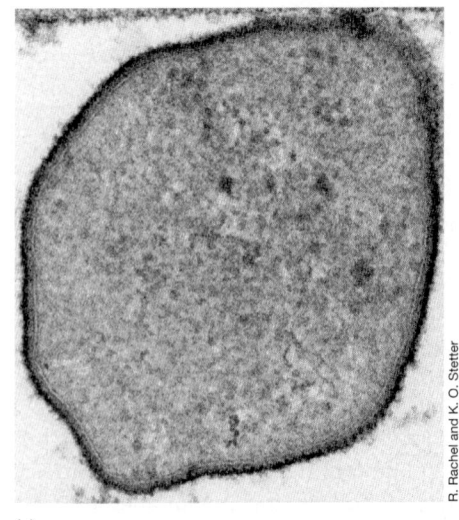

(a)

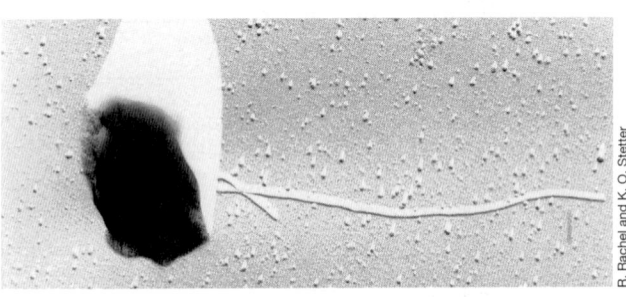

(b)

Figure 19.13 *Archaeoglobales.* *(a)* Transmission electron micrograph of the sulfate-reducing hyperthermophile *Archaeoglobus fulgidus*. The cell measures 0.7 μm in diameter. *(b)* Freeze-etched electron micrograph of *Ferroglobus placidus*, a ferrous iron–oxidizing, nitrate-reducing hyperthermophile. The cell measures about 0.8 μm in diameter.

lism. *Ferroglobus* was isolated from a shallow marine hydrothermal vent and grows optimally at 85°C.

Ferroglobus is interesting for several reasons, but especially for its ability to oxidize Fe^{2+} to Fe^{3+} under anoxic conditions. This process might help explain the origin of the abundant Fe^{3+} in ancient rocks such as the banded iron formations (↩ Section 16.3), rocks dated to before the predicted appearance of cyanobacteria on Earth. With organisms like *Ferroglobus*, it would have been possible for Fe^{2+} oxidation to proceed without the necessity for oxygen (O_2). The metabolism of *Ferroglobus* thus has implications for dating the origin of cyanobacteria and the subsequent oxygenation of Earth. Certain anoxygenic phototrophic bacteria can also oxidize Fe^{2+} under anoxic conditions (↩ Section 13.9), and so several anaerobic routes to ancient Fe^{3+} are possible. This makes it difficult to estimate when cyanobacteria first appeared on Earth and to what degree nonphototrophic organisms helped trigger the Great Oxidation Event (↩ Figure 16.6).

MiniQuiz

- Compare the energy-yielding metabolisms of *Archaeoglobus* and *Ferroglobus*.

19.7 *Nanoarchaeum* and *Aciduliprofundum*

Key Genera: *Nanoarchaeum* and *Aciduliprofundum*

Two *Euryarchaeota* deserve special mention here, *Nanoarchaeum* and *Aciduliprofundum*. *Nanoarchaeum equitans* is a very unusual prokaryote. As one of the smallest cellular organisms known and with the smallest genome among species of *Archaea* (0.49 Mb), *N. equitans* lives as an obligate symbiont of the crenarchaeote *Ignicoccus* (Section 19.10). The coccoid cells of *N. equitans* are very small, about 0.4 μm in diameter, having only about 1% of the volume of an *Escherichia coli* cell. They cannot grow in pure culture and replicate only when attached to the surface of *Ignicoccus* cells. *N. equitans* grows to 10 or more cells per *Ignicoccus* cell and lives an apparently parasitic lifestyle (**Figure 19.14**). The appearance of *N. equitans* cells is typical of *Archaea*, with a cell wall consisting of an S-layer (🔗 Section 3.8) that overlays what appears to be a periplasmic space (Figure 19.14*b*).

N. equitans and its host *Ignicoccus* were first isolated from a submarine hydrothermal vent (🔗 Section 23.12) off the coast of Iceland. However, environmental sampling of 16S rRNA genes (🔗 Section 22.5) indicates that organisms phylogenetically similar to *N. equitans* exist in other submarine hydrothermal vents and terrestrial hot springs, so *Archaea* of this kind are probably distributed worldwide in suitable hot habitats. Like its host *Ignicoccus*, *N. equitans* grows at temperatures from 70 to 98°C and optimally at 90°C.

The metabolism of *Nanoarchaeum* is not fully understood, but it appears to depend on its host for many metabolic functions. *Ignicoccus* is an autotroph that uses H_2 as an electron donor and S^0 as an electron acceptor and so likely supplies *N. equitans* with organic carbon. *N. equitans* is incapable of metabolizing H_2 and S^0 for energy, and whether it generates ATP from compounds obtained from *Ignicoccus* or obtains its ATP directly from its host is unknown.

Phylogeny and Genomics of *Nanoarchaeum equitans*

Although the sequence of its 16S rRNA gene clearly places *N. equitans* in the domain *Archaea*, the sequence differs at many sites from 16S rRNA gene sequences from other *Archaea*, even in regions of the molecule that are highly conserved among *Archaea*. These differences initially led to placing *N. equitans* in a separate, new phylum. However, more detailed phylogenetic analyses using genes for ribosomal proteins and other key cellular proteins suggest that *N. equitans* resides within the *Euryarchaeota* and is most closely related to the *Thermococcales* (Figure 19.1).

The sequence of the *N. equitans* genome provides insight into this organism's obligately symbiotic lifestyle. Its single, circular genome is only 490,885 nucleotides long, one of the smallest cellular genomes yet sequenced. Genes for several important metabolic functions are missing from the *N. equitans* genome, including those for the biosynthesis of amino acids, nucleotides, coenzymes, and lipids. Also missing are genes encoding proteins for key catabolic pathways, such as glycolysis. Presumably, all of these functions are carried out for *N. equitans* by its *Ignicoccus*

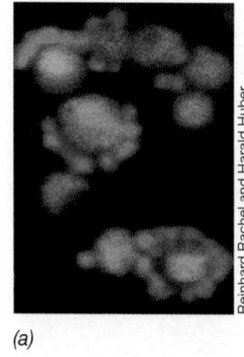
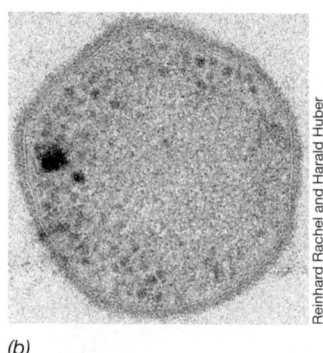

(a)   *(b)*

Figure 19.14 *Nanoarchaeum equitans.* *(a)* Fluorescence micrograph of cells of *Nanoarchaeum equitans* (red) attached to cells of *Ignicoccus* (green). Cells were stained by FISH (🔗 Sections 16.9 and 22.4) using specific nucleic acid probes targeted to each organism. *(b)* Transmission electron micrograph of a thin section of a cell of *N. equitans*. Note the distinct cell wall. Cells of *N. equitans* are about 0.4 μm in diameter.

host, with transfer of needed substances from *Ignicoccus* to the attached *N. equitans* cells. *N. equitans* also lacks some of the genes necessary to encode ATPase, and this indicates that it may not synthesize a functional ATPase—a first for cellular organisms. If no ATPase is present and substrate-level phosphorylation does not occur (due to the lack of glycolytic enzymes), then *N. equitans* would be dependent on *Ignicoccus* for ATP as well as carbon. With so many genes missing, which genes remain in the *N. equitans* genome? *N. equitans* contains genes encoding the key enzymes for DNA replication, transcription, and translation as well as genes for DNA repair enzymes. In addition to its small size, the genome of *N. equitans* is also among the most gene dense of any organism known; over 95% of the *N. equitans* chromosome encodes proteins. Clearly, then, *N. equitans* is living near the limits of life in terms of cell volume and genetic capacity.

Aciduliprofundum

Another remarkable genus of the *Euryarchaeota* is *Aciduliprofundum* (**Figure 19.15**). This motile anaerobic thermophile grows at 55–75°C. *Aciduliprofundum* is particularly notable because it is an acidophile, with best growth occurring about pH 4.5, and lives in sulfide deposits in hydrothermal vents (see Figure 19.24). Before its discovery, only neutrophilic or acid-tolerant microorganisms had been isolated from these deposits, which had been predicted to provide a suitable habitat for acidophiles. *Aciduliprofundum*, a chemoorganotroph that uses S^0 or Fe^{3+} as electron acceptors with complex organic compounds such as peptones and yeast extract as electron donors, constitutes upwards of 15% of the total archaeal cells present in these deposits and therefore is likely to play a major role in the biogeochemical cycling of sulfur and iron in deep-sea hydrothermal vents.

MiniQuiz

- Which aspects of the biology of *Nanoarchaeum equitans* make it especially interesting from an evolutionary point of view?
- Is *Aciduliprofundum* an aerobe or an anaerobe? How does it make ATP?

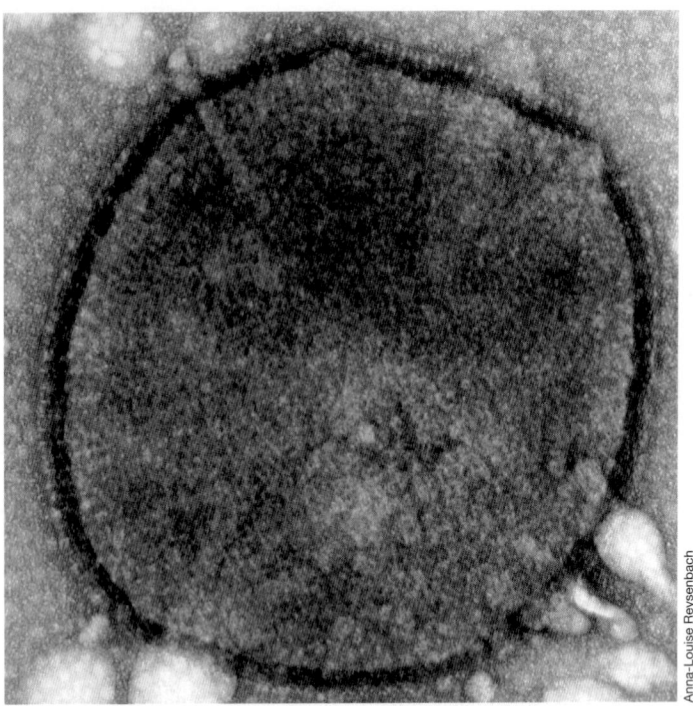

Anna-Louise Reysenbach

Figure 19.15 *Aciduliprofundum*. Transmission electron micrograph of a single cell, which measures about 0.7 μm in diameter.

Ⅲ *Crenarchaeota*

Crenarchaeotes are phylogenetically distinct from eury-archaeotes (Figure 19.1) and inhabit both ends of nature's temperature extremes: boiling and even superheated waters and freezing waters. Most cultured crenarchaeotes are hyperthermophiles and some actually have growth temperature optima above the boiling point of water. We start with an overview of the habitats and energy metabolism of crenarchaeotes and describe the properties of some key genera. We then conclude this section with a brief discussion of *Crenarchaeota* from temperate and cold environments, interesting new environments for a group once thought to be restricted to hot habitats.

19.8 Habitats and Energy Metabolism

A summary of the habitats of *Crenarchaeota* is shown in **Table 19.7**. Most hyperthermophilic *Archaea* have been isolated from geothermally heated soils or waters containing S^0 and H_2S, and most species metabolize sulfur in one way or another. In terrestrial environments, sulfur-rich springs, boiling mud, and soils may have temperatures up to 100°C and are mildly to extremely acidic owing to the production of sulfuric acid (H_2SO_4) from the biological oxidation of H_2S and S^0 (⟲ Sections 13.8 and 24.4). Such hot, sulfur-rich environments, called **solfataras**, are found throughout the world (**Figure 19.16**), including Italy, Iceland, New Zealand, and Yellowstone National Park in Wyoming (USA). Depending on the surrounding geology, solfataras can be mildly acidic to slightly alkaline (pH 5–8) or extremely acidic, with pH values below 1. Hyperthermophilic crenarchaeotes have been obtained from all of these environments, but the majority inhabit neutral or mildly acidic thermal habitats.

Hyperthermophilic *Crenarchaeota* also inhabit undersea hot springs called **hydrothermal vents**. We discuss the geology and microbiology of these habitats in Section 23.12. Here it is only necessary to note that submarine waters can be much hotter than surface waters because the water is under pressure. Indeed, all hyperthermophiles with growth temperature optima above 100°C have come from submarine sources. The latter include shallow (2–10 m depth) vents such as those off the coast of Vulcano, Italy, to deep (2000–4000 m depth) vents near ocean-spreading centers (see Figure 19.24). Deep hydrothermal vents are the hottest habitats so far known to yield prokaryotes.

With a few exceptions, hyperthermophilic *Crenarchaeota* are obligate anaerobes. Their energy-yielding metabolism is either chemoorganotrophic or chemolithotrophic (or both, for example, in *Sulfolobus*) and is dependent on diverse electron donors and acceptors. Fermentation is rare and most bioenergetic strategies involve anaerobic respirations (**Table 19.8**). Energy is conserved during these respiratory processes by the same general mechanism widespread in *Bacteria*: electron transfer within the cytoplasmic membrane leading to the formation of a proton motive force from which ATP is made by way of proton-translocating ATPases (⟲ Section 4.10).

Table 19.7 *Habitats of* Crenarchaeota

| Characteristic | Thermal area | | Nonthermal area |
	Terrestrial	Marine	
Locations	Solfataras (hot springs, fumaroles, mudpots, steam-heated soils); geothermal power plants; deep in Earth's crust	Submarine hot springs, hot sediments and vents ("black smokers"); deep oil reservoirs	Planktonic in oceans worldwide; near-shore and deep Antarctic waters; sea ice; symbionts of marine sponges
Temperature	Surface to 100°C; subsurface, above 100°C	Up to 400°C (smokers)	–2 to +4°C
Salinity/pH	Usually less than 1% NaCl; pH 0.5–9	Moderate, about 3% NaCl; pH 5–9	3–8% NaCl; pH 7–9
Gases and other nutrients	CO_2, CO, CH_4, H_2, H_2S, S^0, $S_2O_3^{2-}$, SO_4^{2-}, NH_4^+, N_2	Same as for terrestrial	CO_2, N_2, O_2; for chemolithotrophs, inorganic electron donors such as NH_4^+

(a) T. D. Brock

(b) T. D. Brock

(c) T. D. Brock

(d) T. D. Brock

Figure 19.16 Terrestrial habitats of hyperthermophilic *Archaea*: Yellowstone National Park **(Wyoming, USA).** *(a)* A typical solfatara; steam rich in H_2S rises to the surface. *(b)* Sulfur-rich hot spring, a habitat containing dense populations of *Sulfolobus*. The acidity in solfataras and sulfur springs comes from the oxidation of H_2S and S^0 to H_2SO_4 (sulfuric acid) by *Sulfolobus* and related prokaryotes. *(c)* A typical neutral pH boiling spring, Imperial Geyser. Many different species of hyperthermophilic *Archaea* may reside in such a habitat. *(d)* An acidic iron-rich geothermal spring, another *Sulfolobus* habitat; here the oxidation of Fe^{2+} to Fe^{3+} generates acidity.

Many hyperthermophilic crenarchaeotes can grow chemolithotrophically under anoxic conditions, with H_2 as the electron donor and S^0 or NO_3^- as the electron acceptor; a few can also oxidize H_2 aerobically (Table 19.8). H_2 respiration with ferric iron (Fe^{3+}) as electron acceptor is the process in several hyperthermophiles. Other chemolithotrophic lifestyles include the oxidation of S^0 and Fe^{2+} aerobically or Fe^{2+} anaerobically with NO_3^- as the acceptor (Table 19.8). Only one sulfate-reducing hyperthermophile is known (the euryarchaeote *Archaeoglobus*, Section 19.6). The only bioenergetic option apparently impossible is photosynthesis (the most thermophilic phototrophic microorganism known can grow up to only 73°C, see Figure 19.28).

MiniQuiz

- Why is it unlikely that hyperthermophiles with growth temperature optima >100°C would reside in terrestrial hot springs?
- What form of energy metabolism is widespread among hyperthermophiles?

19.9 *Crenarchaeota* from Terrestrial Volcanic Habitats

Key Genera: *Sulfolobus, Acidianus, Thermoproteus, Pyrobaculum*

Terrestrial volcanic habitats can have temperatures as high as 100°C and are thus suitable for hyperthermophilic *Archaea*. Two phylogenetically related organisms isolated from these environments are *Sulfolobus* and *Acidianus*. These genera form the heart of an order called the *Sulfolobales* (**Table 19.9**).

Sulfolobales

Sulfolobus grows in sulfur-rich acidic thermal areas (Figure 19.16) at temperatures up to 90°C and at pH values of 1–5. *Sulfolobus* is an aerobic chemolithotroph that oxidizes H_2S or S^0 to H_2SO_4 and fixes CO_2 as a carbon source. *Sulfolobus* can also grow chemoorganotrophically. Cells of *Sulfolobus* are more or less spherical but contain distinct lobes (**Figure 19.17**). Cells adhere

UNIT 6

Table 19.8 *Energy-yielding reactions of hyperthermophilic Archaea*

Nutritional class	Energy-yielding reaction	Metabolic[a] type	Example
Chemoorganotrophic	Organic compound + $S^0 \rightarrow H_2S + CO_2$	AnR	*Thermoproteus, Thermococcus, Desulfurococcus, Thermofilum, Pyrococcus*
	Organic compound + $SO_4^{2-} \rightarrow H_2S + CO_2$	AnR	*Archaeoglobus*
	Organic compound + $O_2 \rightarrow H_2O + CO_2$	AeR	*Sulfolobus*
	Organic compound $\rightarrow CO_2 + H_2$ + fatty acids	AnR	*Staphylothermus, Pyrodictium*
	Organic compound + $Fe^{3+} \rightarrow CO_2 + Fe^{2+}$	AnR	*Pyrodictium*
	Organic compound + $NO_3^- \rightarrow CO_2 + N_2$	AnR	*Pyrobaculum*
	Pyruvate $\rightarrow CO_2 + H_2$ + acetate	AnR	*Pyrococcus*
	Peptides ($H_2 + S^0 \rightarrow H_2S$ stimulates growth)	F	*Hyperthermus*
Chemolithotrophic	$H_2 + S^0 \rightarrow H_2S$	AnR	*Acidianus, Pyrodictium, Thermoproteus, Stygiolobus, Ignicoccus*
	$H_2 + NO_3^- \rightarrow NO_2^- + H_2O$ (NO_2^- is reduced to N_2 by some species)	AnR	*Pyrobaculum*
	$4 H_2 + NO_3^- + H^+ \rightarrow NH_4^+ + 2 H_2O + OH^-$	AnR	*Pyrolobus*
	$H_2 + 2 Fe^{3+} \rightarrow 2 Fe^{2+} + 2 H^+$	AnR	*Pyrobaculum, Pyrodictium, Archaeoglobus*
	$2 H_2 + O_2 \rightarrow 2 H_2O$	AeR	*Acidianus, Sulfolobus, Pyrobaculum*
	$2 S^0 + 3 O_2 + 2 H_2O \rightarrow 2 H_2SO_4$	AeR	*Sulfolobus, Acidianus*
	$2 FeS_2 + 7 O_2 + 2 H_2O \rightarrow 2 FeSO_4 + 2 H_2SO_4$	AeR	*Sulfolobus, Acidianus, Metallosphaera*
	$2 FeCO_3 + NO_3^- + 6 H_2O \rightarrow 2 Fe(OH)_3 + NO_2^- + 2 HCO_3^- + 2 H^+ + H_2O$	AnR	*Ferroglobus*
	$4 H_2 + SO_4^{2-} + 2 H^+ \rightarrow 4 H_2O + H_2S$	AnR	*Archaeoglobus*
	$4 H_2 + CO_2 \rightarrow CH_4 + 2 H_2O$	AnR	*Methanopyrus, Methanocaldococcus, Methanothermus*

[a]AnR, anaerobic respiration; AeR, aerobic respiration; F, fermentation.

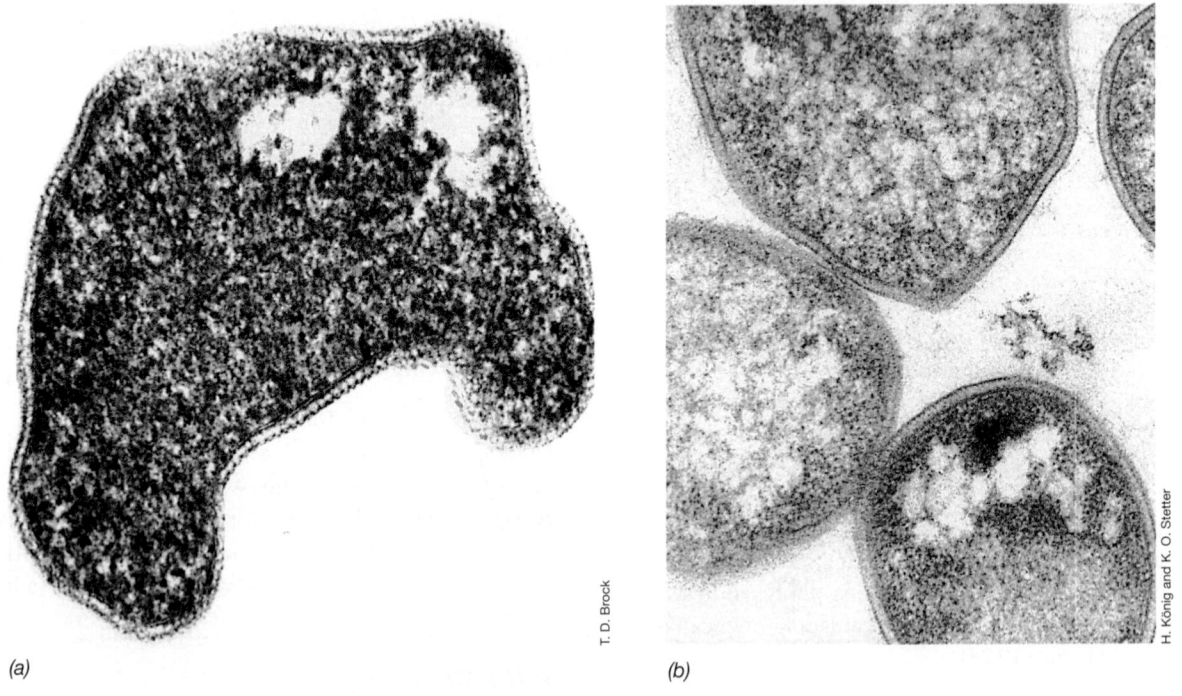

T. D. Brock

H. König and K. O. Stetter

(a)

(b)

Figure 19.17 Acidophilic hyperthermophilic *Archaea*, the *Sulfolobales*. *(a) Sulfolobus acidocaldarius*. Electron micrograph of a thin section. *(b) Acidianus infernus*. Electron micrograph of a thin section. Cells of both organisms vary from 0.8 to 2 μm in diameter.

Table 19.9 *Properties of some hyperthermophilic* Crenarchaeota

Order/Genus[a]	Morphology	Relationship to O_2[b]	Temperature Minimum	Optimum	Maximum	Optimum pH
Sulfolobales						
Sulfolobus	Lobed coccus	Ae	55	75	87	2–3
Acidianus	Coccus	Fac	60	88	95	2
Metallosphaera	Coccus	Ae	50	75	80	2
Stygiolobus	Lobed coccus	An	57	80	89	3
Sulfurisphaera	Coccus	Fac	63	84	92	2
Sulfurococcus	Coccus	Ae	40	75	85	2.5
Thermoproteales						
Thermoproteus	Rod	An	60	88	96	6
Thermofilum	Rod	An	70	88	95	5.5
Pyrobaculum	Rod	Fac	74	100	102	6
Caldivirga	Rod	An	60	85	92	4
Thermocladium	Rod	An	60	75	80	4.2
Desulfurococcales						
Desulfurococcus	Coccus	An	70	85	95	6
Aeropyrum	Coccus	Ae	70	95	100	7
Staphylothermus	Cocci in clusters	An	65	92	98	6–7
Pyrodictium	Disc-shaped with filaments	An	82	105	110	6
Pyrolobus	Lobed coccus	Fac	90	106	113	5.5
Thermodiscus	Disc-shaped	An	75	90	98	5.5
Ignicoccus	Irregular coccus	An	65	90	103	5
Hyperthermus	Irregular coccus	An	75	102	108	7
Stetteria	Coccus	An	68	95	102	6
Sulfophobococcus	Disc-shaped	An	70	85	95	7.5
Thermosphaera	Coccus	An	67	85	90	7
Strain 121	Coccus	An	85	106	121	7

[a]The group names ending in "ales" are order names.
[b]Ae, aerobe; An, anaerobe; Fac, facultative

tightly to sulfur crystals, where they can be seen with a microscope after preparation with fluorescent dyes (↝ Figure 13.21*b*). Besides the aerobic respiration of sulfur or organic compounds, *Sulfolobus* can also oxidize Fe^{2+} to Fe^{3+}, and this has been applied in the high-temperature leaching of iron and copper ores (↝ Section 24.7).

A facultative aerobe resembling *Sulfolobus* also lives in acidic solfataric springs. This organism, *Acidianus* (Figure 19.17*b*), differs from *Sulfolobus* most clearly by its ability to grow using S^0 both anaerobically as well as aerobically. Under aerobic conditions the organism uses S^0 as an electron *donor*, oxidizing S^0 to H_2SO_4, with O_2 as an electron acceptor. Anaerobically, *Acidianus* uses S^0 as an electron *acceptor* with H_2 as an electron donor, forming H_2S as the reduced product. Thus, the metabolic fate of S^0 in cultures of *Acidianus* depends on the presence or absence of O_2. Like *Sulfolobus*, *Acidianus* is roughly spherical in shape but is not as lobed (Figure 19.17*b*). It grows at temperatures from 65°C up to a maximum of 95°C, with an optimum of about 90°C. As a group, then, the *Sulfolobales* contain the most thermophilic of all highly acidophilic *Archaea*.

Thermoproteales

Key genera within the *Thermoproteales* are *Thermoproteus*, *Thermofilum*, and *Pyrobaculum*. The genera *Thermoproteus* and *Thermofilum* consist of rod-shaped cells that inhabit neutral or slightly acidic hot springs. Cells of *Thermoproteus* are rigid rods about 0.5 μm in diameter and highly variable in length, ranging from short cells of 1–2 μm (**Figure 19.18*a***) up to filaments 70–80 μm long. Filaments of *Thermofilum* are thinner, some 0.17–0.35 μm wide, with filament lengths ranging up to 100 μm (Figure 19.18*b*).

Both *Thermoproteus* and *Thermofilum* are strict anaerobes that carry out an S^0-based anaerobic respiration (Table 19.8). Most *Thermoproteus* isolates can grow chemolithotrophically on H_2 or chemoorganotrophically on complex carbon substrates such as yeast extract, small peptides, starch, glucose, ethanol, malate, fumarate, or formate (Table 19.8). *Pyrobaculum* (Figure 19.18*c*) is a rod-shaped hyperthermophile but is physiologically distinct from other *Thermoproteales* in that some species of *Pyrobaculum* can respire aerobically. However, *Pyrobaculum* can also grow by anaerobic respiration with NO_3^-,

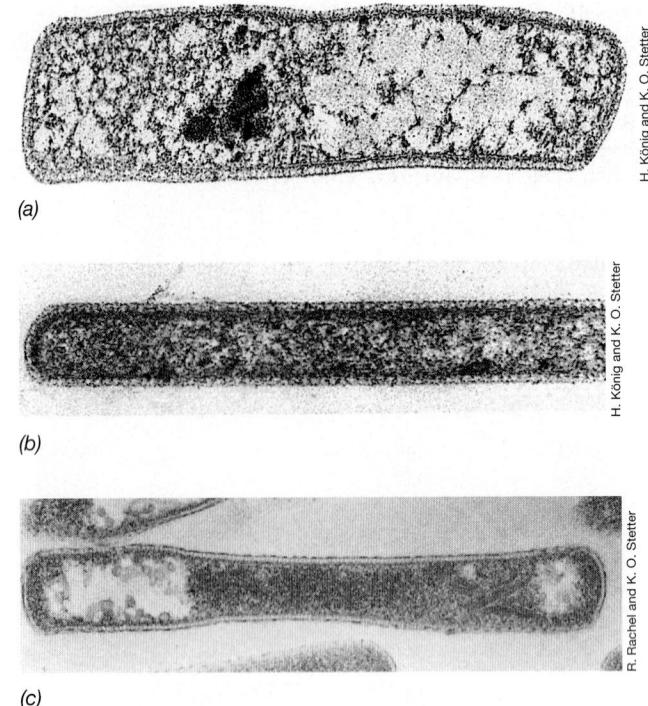

(a)

H. König and K. O. Stetter

(b)

H. König and K. O. Stetter

(c)

R. Rachel and K. O. Stetter

Figure 19.18 Rod-shaped hyperthermophilic *Archaea*, the *Thermoproteales*. (a) *Thermoproteus neutrophilus*. Electron micrograph of a thin section. A cell is about 0.5 μm in diameter. (b) *Thermofilum librum*. A cell is about 0.25 μm in diameter. (c) *Pyrobaculum aerophilum*. Transmission electron micrograph of a thin section; a cell measures 0.5 × 3.5 μm.

Fe^{3+}, or S^0 as electron acceptors and H_2 as an electron donor (that is, they can grow chemolithotrophically and autotrophically). Other species of *Pyrobaculum* can grow anaerobically on organic electron donors, reducing S^0 to H_2S. The growth temperature optimum of *Pyrobaculum* is 100°C, and species of this organism have been isolated from terrestrial hot springs and from hydrothermal vents.

MiniQuiz

• What are the major differences between *Sulfolobus* and *Pyrolobus*?

• Among *Thermoproteales*, what is unusual about the metabolism of *Pyrobaculum*?

19.10 *Crenarchaeota* from Submarine Volcanic Habitats

Key Genera: *Pyrodictium, Pyrolobus, Ignicoccus, Staphylothermus*

We now consider the microbiology of submarine volcanic habitats, homes to the most thermophilic of all known *Archaea*. These habitats include both shallow water thermal springs and deep-sea hydrothermal vents. We discuss the geology of these fascinating microbial habitats in Section 23.12 and the interesting animal communities that develop there in Section 25.12. The

organisms to be described here constitute an order of *Archaea* called the *Desulfurococcales* (Table 19.9).

Pyrodictium and *Pyrolobus*

Pyrodictium and *Pyrolobus* are examples of microorganisms whose growth temperature optimum lies above 100°C; the optimum for *Pyrodictium* is 105°C and for *Pyrolobus* is 106°C. Cells of *Pyrodictium* are irregularly disc-shaped and grow in culture in a mycelium-like layer attached to crystals of S^0. The cell mass consists of a network of fibers to which individual cells are attached (**Figure 19.19**). The fibers are hollow and consist of protein arranged in a fashion similar to that of bacterial flagella (Section 3.13). However, the filaments do not function in motility but instead as organs of attachment. The cell walls of *Pyrodictium* are composed of glycoprotein. Physiologically, *Pyrodictium* is a strict anaerobe that grows chemolithotrophically on H_2 with S^0 as an electron acceptor or chemoorganotrophically on complex mixtures of organic compounds (Table 19.8).

Pyrolobus fumarii (Figure 19.19c) is one of the most thermophilic of the hyperthermophiles. Its growth temperature maximum is 113°C (Table 19.9). *P. fumarii* lives in the walls of "black smoker" hydrothermal vent chimneys (Section 23.12 and Figure 23.32) where its autotrophic abilities contribute organic carbon to this otherwise inorganic environment. *P. fumarii* cells are coccoid-shaped (Figure 19.19c), and the cell wall is composed of protein. The organism is an obligate H_2 chemolithotroph, growing by the oxidation of H_2 coupled to the reduction of NO_3^- to ammonium (NH_4^+), thiosulfate ($S_2O_3^{2-}$) to H_2S, or very low concentrations of O_2 to H_2O. Besides its extremely thermophilic nature, *P. fumarii* can withstand temperatures substantially above its growth temperature maximum. For example, cultures of *P. fumarii* survive autoclaving (121°C) for 1 h, a condition that even bacterial endospores (Section 3.12) cannot withstand.

Another organism in this group shares with *Pyrolobus* a growth temperature optimum of 106°C. However, "Strain 121," as this organism has been called, actually shows weak growth at 121°C, and cells remain viable for 2 h at 130°C. Only *Methanopyrus*, a hyperthermophilic methanogen, can grow at a higher temperature (122°C, Section 19.5). Strain 121 consists of coccoid, flagellated cells (Figure 19.19d); the organism is also a strict anaerobe and grows chemolithotrophically and autotrophically with Fe^{3+} as electron acceptor and formate or H_2 as electron donors. It is thus clear that the *Pyrodictium/Pyrolobus* group collectively contains the most hyperthermophilic examples of all known prokaryotes.

Desulfurococcus and *Ignicoccus*

Other notable members of the *Desulfurococcales* include *Desulfurococcus*, the genus for which the order is named (**Figure 19.20**), and *Ignicoccus*. *Desulfurococcus* is a strictly anaerobic S^0-reducing organism like *Pyrodictium*, but differs from this organism in its phylogeny and the fact that it is much less thermophilic, growing optimally at about 85°C.

Ignicoccus grows optimally at 90°C, and its energy metabolism is based on H_2 as an electron donor and S^0 as an electron acceptor,

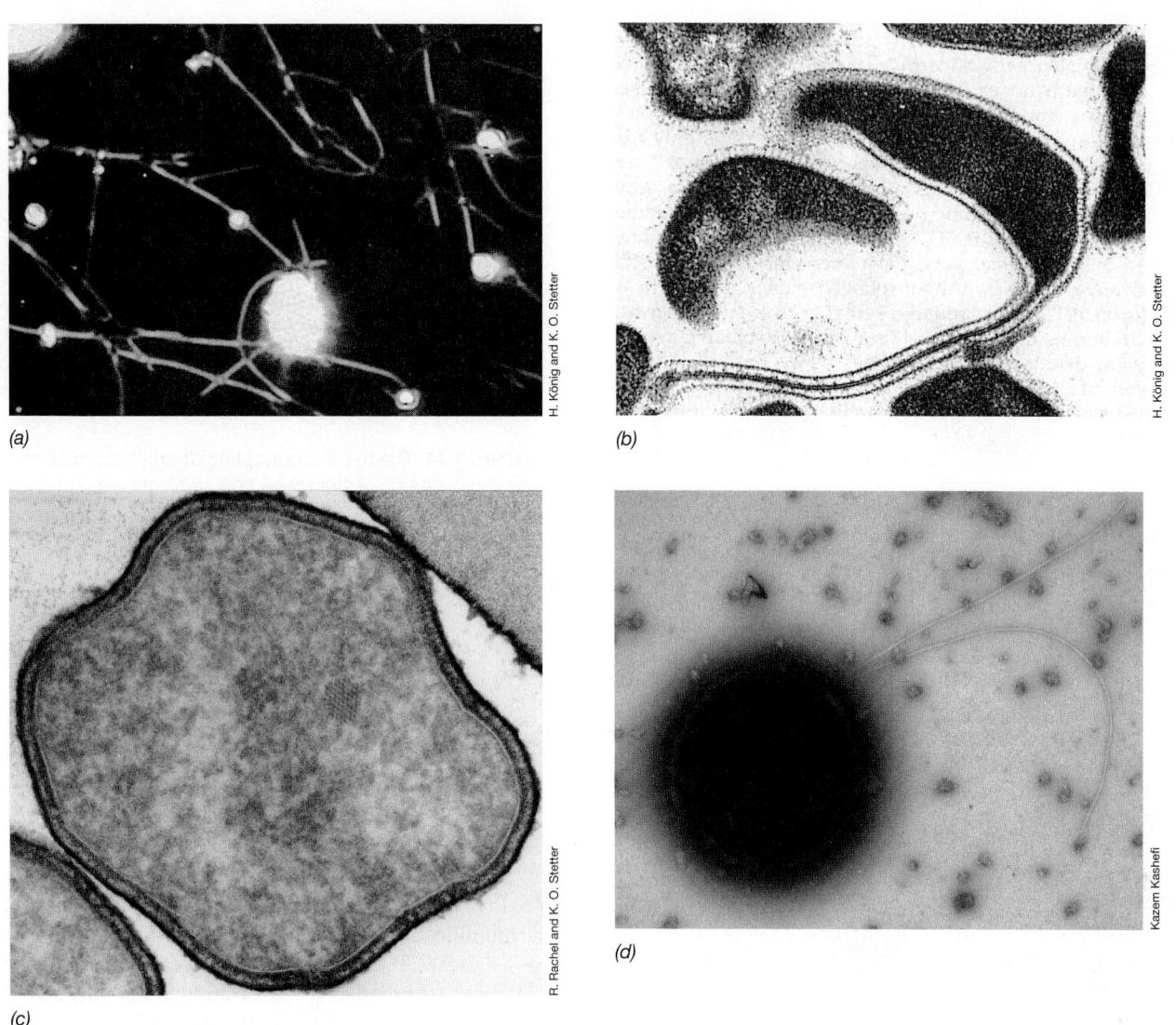

(a)

(b)

(c)

(d)

Figure 19.19 *Desulfurococcales* **with growth temperature optima >100°C.** *(a) Pyrodictium occultum* (growth temperature optimum, 105°C), dark-field micrograph. *(b)* Thin-section electron micrograph of *P. occultum*. Cells are highly variable in diameter from 0.3 to 2.5 μm. *(c)* Thin section of a cell of *Pyrolobus fumarii*, one of the most thermophilic of all known bacteria (growth temperature optimum, 106°C); a cell is about 1.4 μm in diameter. *(d)* Negative stain of a cell of "Strain 121," capable of growth at 121°C; a cell is about 1 μm wide.

as is that of so many hyperthermophilic *Archaea* (Table 19.8). *Ignicoccus* (Figure 19.20*b*) is a novel hyperthermophile because it contains an *outer membrane* similar to that of gram-negative *Bacteria* (⬅ Section 3.7). The outer membrane of *Ignicoccus* is unusual, however, in that it is present at some distance from the cytoplasm of the cell. This arrangement allows for an unusually large periplasm to form (Figure 19.20*b*). Indeed, the volume of the periplasm of *Ignicoccus* is some two to three times that of its cytoplasm, in contrast to that of gram-negative *Bacteria*, where periplasmic volume is about 25% that of the cytoplasm. The periplasm of *Ignicoccus* also contains membrane-bound vesicles (Figure 19.20*b*) that may function in exporting substances outside the cell. In addition, however, some *Ignicoccus* species are hosts to the small parasitic archaeon, *Nanoarchaeum equitans*, discussed in Section 19.7.

Staphylothermus

A morphologically unusual member of the order *Desulfurococcales* is the genus *Staphylothermus* (**Figure 19.21**). Cells of *Staphylothermus* are spherical, about 1 μm in diameter, and form aggregates of up to 100 cells, much like its morphological counterpart among the *Bacteria*, *Staphylococcus* (⬅ Figure 18.1). Unlike many hyperthermophiles, *Staphylothermus* is not a chemolithotroph, but instead a chemoorganotroph, growing optimally at 92°C. Energy is obtained from the fermentation of peptides, producing the fatty acids acetate and isovalerate as fermentation products (Table 19.8).

Isolates of *Staphylothermus* have been obtained from both shallow marine hydrothermal vents and very hot black smokers (see Figure 19.24; ⬅ Section 23.12). This organism is apparently

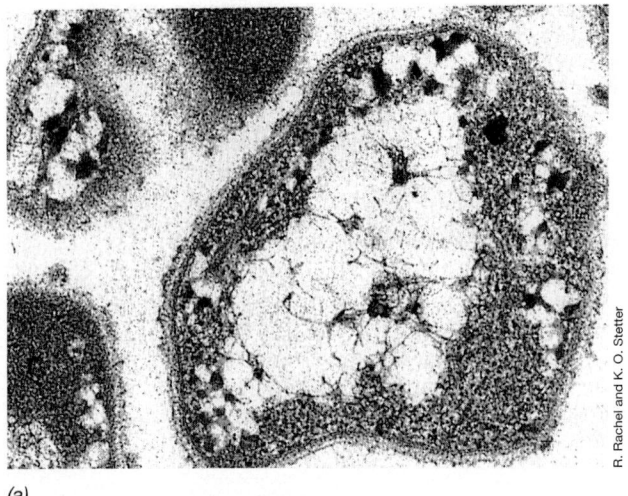

(a)

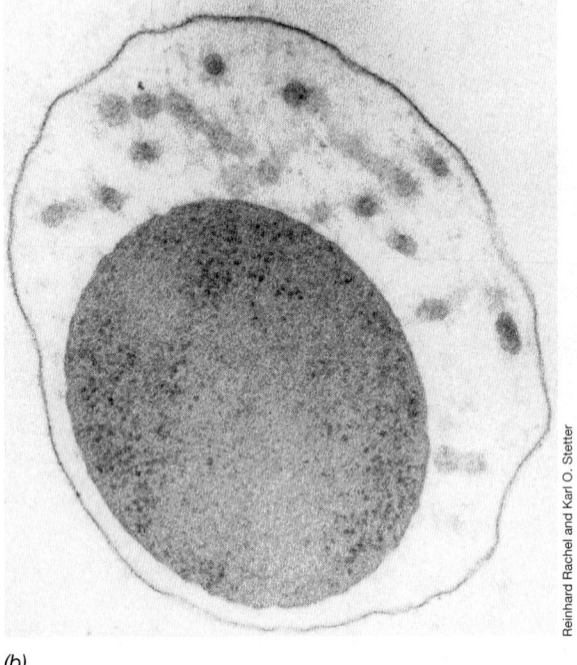

(b)

Figure 19.20 *Desulfurococcales* with growth temperature optima <100°C. *(a)* Thin section of a cell of *Desulfurococcus saccharovorans*; a cell is 0.7 μm in diameter. *(b)* Thin section of a cell of *Ignicoccus islandicus*. The cell proper is surrounded by an extremely large periplasm. The cell itself measures about 1 μm in diameter and the cell plus periplasm measures 1.4 μm.

widely distributed in submarine thermal areas, where it is likely to play a significant role in consuming proteins released from dead organisms.

MiniQuiz

• What can we conclude about the *Pyrodictium/Pyrolobus* group in terms of life at high temperature?

• What unusual structural features are present in *Ignicoccus* and *Staphylothermus*?

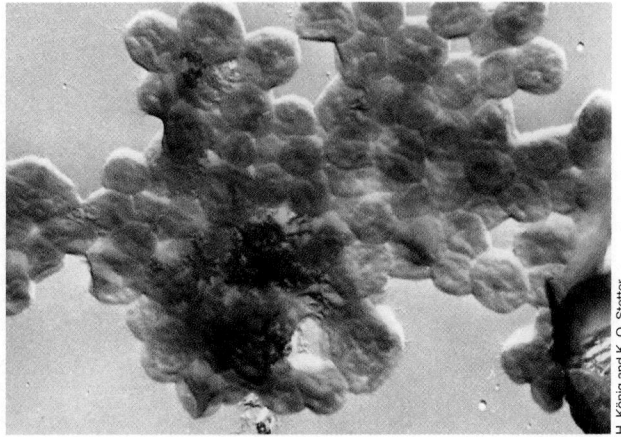

Figure 19.21 The hyperthermophile *Staphylothermus marinus*. Electron micrograph of shadowed cells. A single cell is about 1 μm in diameter.

19.11 *Crenarchaeota* from Nonthermal Habitats and Nitrification in *Archaea*

In contrast to hyperthermophiles, nonthermophilic crenarchaeotes (Table 19.7) have been identified from community sampling of ribosomal RNA genes from marine and terrestrial environments ranging in temperature from cool to frigid (Chapter 23). By using fluorescent phylogenetic probes (FISH, ⟳ Sections 16.9 and 22.4), crenarchaeotes have been detected in oxic marine waters worldwide. In stark contrast to the hyperthermophiles, marine crenarchaeotes thrive even in waters and sea ice near Antarctica (**Figure 19.22**). These organisms are planktonic (suspended freely or attached to suspended particles in the water column, Figure 19.22*b*) and present in significant numbers (~10^4/ml) in waters that are both nutrient-poor and very cold (0–4°C in seawater and below 0°C in sea ice).

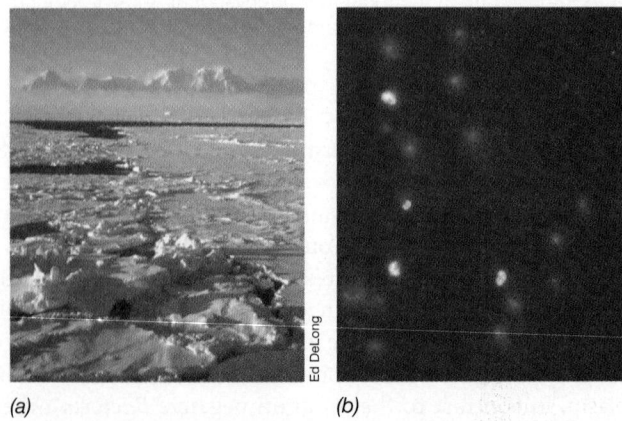

(a)   *(b)*

Figure 19.22 Cold-dwelling *Crenarchaeota*. *(a)* Photo of the Antarctic Peninsula taken from shipboard. The frigid waters that lie under the surface ice shown here are habitats for cold-dwelling crenarchaeotes. *(b)* Fluorescence photomicrograph of seawater treated with a FISH probe (⟳ Section 16.9) specific for species of *Crenarchaeota* (green cells). Blue cells are stained with DAPI, a DNA stain that stains all cells.

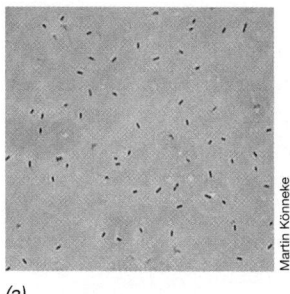
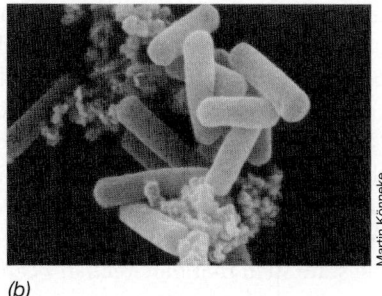

(a)   (b)

Figure 19.23 *Nitrosopumilus maritimus*, a nitrifying species of *Archaea*. This organism can oxidize NH_3 present at the very low amounts typical of marine environments. *(a)* Phase-contrast photomicrograph. *(b)* Scanning electron micrograph. A single cell of *N. maritimus* is about 0.2 μm in diameter.

Marine *Crenarchaeota* can account for up to 40% of the prokaryotes of deep ocean waters (💭 Section 23.10 and Figure 23.21). Lipid analyses of marine crenarchaeotes filtered from seawater have shown that they contain ether-linked lipids, the hallmark of the *Archaea* (💭 Section 3.3). Their high numbers, together with their ability to fix CO_2, suggest that marine *Crenarchaeota* play a major role in the global carbon cycle. Nonthermophilic *Euryarchaeota* are also present in marine environments, especially in temperate surface waters.

The physiology of marine crenarchaeotes remained a mystery until the discovery that at least some of these organisms are ammonia oxidizers. For example, *Nitrosopumilus maritimus* (**Figure 19.23**) has been isolated in laboratory culture and shown to grow chemolithotrophically by aerobically oxidizing ammonia (NH_3) to nitrite (NO_2^-), the first step in nitrification (💭 Sections 13.10, 17.3, and 24.3). This organism uses CO_2 as its sole carbon source (autotrophy), as do nitrifying *Bacteria* (💭 Section 17.3). However, unlike ammonia-oxidizing *Bacteria* such as *Nitrosomonas*, *N. maritimus* is adapted to life under extreme nutrient limitation, as would befit an organism indigenous to open ocean waters. In this regard, *N. maritimus* shows high NH_3 oxidation rates at the very low NH_3 concentrations found in open ocean waters. In addition, environmental analyses of genes that encode ammonia oxidation functions have revealed that nitrifying *Crenarchaeota* likely play a more important role than do nitrifying *Bacteria* in the marine nitrogen cycle.

MiniQuiz

- How does the organism *Nitrosopumilus maritimus* conserve energy and obtain carbon?

Ⅳ Evolution and Life at High Temperatures

Most of the hyperthermophiles discovered so far are species of *Archaea* and some grow near to what may be the upper temperature limit for life. Here we consider the major factors that likely define the upper temperature limit for life and the biologi-

cal adaptations of hyperthermophiles that permit them to exist at the exceptionally high temperatures of 100°C and higher. We end with a discussion of the importance of hydrogen (H_2) metabolism to the biology of hyperthermophiles.

19.12 An Upper Temperature Limit for Microbial Life

Habitats that contain liquid water—a prerequisite for cellular life—and that have temperatures higher than 100°C are only found where geothermally heated water flows out of vents or rifts in the ocean floor (💭 Figures 16.4, 23.30, and 23.31). The hydrostatic pressure that overlies the water keeps it from boiling, allowing it to reach temperatures of up to about 400°C in vents at several thousand meters' depth. In contrast, terrestrial hot springs can boil and therefore only attain temperatures near 100°C. It is therefore not surprising that hydrothermal vents have been rich sources of hyperthermophilic *Archaea* with growth temperature optima above 100°C (Table 19.9).

Black smokers emit hydrothermal vent fluid at 250–350°C or higher, forming metallic mounds or more upright structures called *chimneys* that form from the metal sulfides that precipitate out of the hot fluid as it mixes with the surrounding, much cooler seawater (**Figure 19.24**). As far as is known, the superheated vent water itself is sterile. However, hyperthermophiles thrive in mounds or smoker chimney walls where temperatures are compatible with their survival and growth (💭 Section 23.12 and Figure 23.32). By studying structures such as these, one can address the question, "What is the upper temperature for microbial (and presumably all forms of) life?"

What Is the Upper Temperature Limit for Life?

How high a temperature can hyperthermophiles withstand? Over the past several decades, the known upper temperature limit for life has been pushed higher and higher with the isolation

Figure 19.24 **Hydrothermal vents.** Hydrothermal mound from the Rainbow vent field, Mid-Atlantic Ridge hydrothermal system. The hydrothermal fluid emitting from the two short chimneys is >300°C.

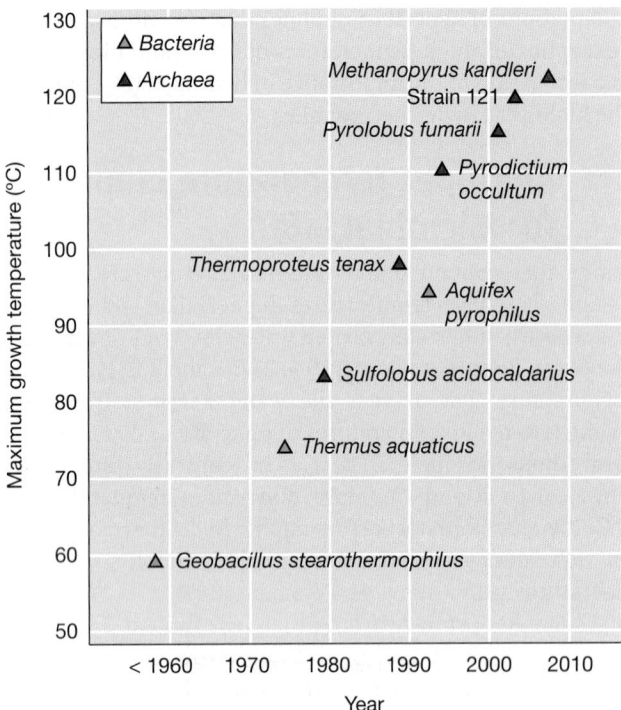

Figure 19.25 Thermophilic and hyperthermophilic prokaryotes. The graph gives the species that were, in turn, the record holders for growing at the highest temperature, from before 1960 to the present.

and characterization of new species of thermophiles and hyperthermophiles (**Figure 19.25**). Until recently, the record holder was *Pyrolobus fumarii*, with an upper temperature limit for growth of 113°C (Figure 19.19c). The current record holder, *Methanopyrus* (Section 19.5, Figure 19.12), however, has pushed the limit somewhat higher, with the ability to grow at 122°C and to survive substantial periods at even higher temperatures. Given the trend over the past several years (Figure 19.25), one can predict that *Archaea* even more hyperthermophilic than *Methanopyrus* may inhabit hydrothermal environments but have yet to be isolated. Indeed, many experts predict that the upper temperature limit for prokaryotic life is likely to exceed 140°C, perhaps even 150°C, and that the maximum temperature allowing survival but not growth is even hotter yet.

Biochemical Problems at Supercritical Temperatures

Whatever the upper temperature limit is for life, it is likely to be defined by one or more biochemical challenges that evolution has been unable to solve. There is obviously an upper limit, but we do not yet know what it is. Water samples taken directly from superheated (>250°C) hydrothermal vent discharges are devoid of measurable biochemical markers that would signal life as we know it (DNA, RNA, and protein), while vents emitting water at temperatures below about 150°C yield evidence for macromolecules. These results are consistent with laboratory experiments on the stability of key biomolecules. For example, ATP is degraded almost instantly at 150°C. Thus, above 150°C, organisms should be unable to overcome the heat lability of a molecule

that as far as we know is universally distributed in cells. As a caveat, however, the stability of small molecules such as ATP may be significantly greater under cytoplasmic conditions of high dissolved solutes than in pure solutions tested in the laboratory. Nevertheless, if life forms exist at temperatures above 150°C, they must be unique in many ways, either using a suite of novel small molecules absent from cells as we know them, or deploying special protection systems that maintain small molecules in a stable state such that biochemistry can proceed.

MiniQuiz
- Where are the hottest potential microbial habitats located on Earth?
- Why would it be impossible for organisms to grow at 200 or 300°C?

19.13 Molecular Adaptations to Life at High Temperature

Because all cellular structures and activities are affected by heat, hyperthermophiles are likely to exhibit multiple adaptations to the exceptionally high temperatures of their habitats. Here we briefly examine some adaptations employed by hyperthermophiles to protect their proteins and nucleic acids at high temperatures.

Protein Folding and Thermostability

Because most proteins denature at high temperatures, much research has been done to identify the properties of thermostable proteins. Protein thermostability derives from the folding of the molecule itself, not because of the presence of any special amino acids. Perhaps surprisingly, however, the amino acid composition of thermostable proteins is not particularly unusual except perhaps in their slight bias for increased levels of amino acids that promote alpha helical secondary structures. In fact, many enzymes from hyperthermophiles contain the same major structural features in both primary and higher-order structure (Section 6.21) as their heat-labile counterparts from organisms that grow best at much lower temperatures.

Thermostable proteins typically do display some structural features that likely improve their thermostability. These include having highly hydrophobic cores, which decrease the tendency of the protein to unfold in an ionic environment, and more ionic interactions on the protein surfaces, which also help hold the protein together and work against unfolding. Ultimately, it is the *folding* of the protein that most affects its heat stability, and noncovalent ionic bonds called *salt bridges* on a protein's surface likely play a major role in maintaining the biologically active structure. But, as previously stated, many of these changes are possible with only minimal changes in primary structure (amino acid sequence), as seen when thermostable and heat-labile forms of the same protein are compared.

Chaperonins: Assisting Proteins to Remain in Their Native State

Earlier we discussed a class of proteins called *chaperonins* (heat shock proteins; Sections 6.21 and 8.11) that function to refold partially denatured proteins. Hyperthermophilic *Archaea* have

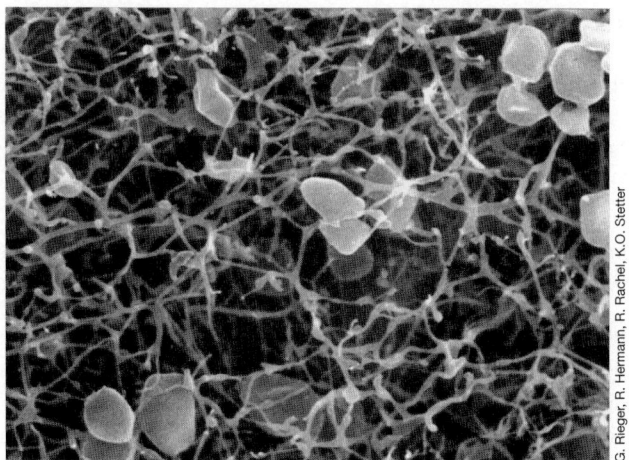

Figure 19.26 *Pyrodictium abyssi*, scanning electron micrograph. *Pyrodictium* has been studied as a model of macromolecular stability at high temperatures. Cells are enmeshed in a sticky glycoprotein matrix that binds them together.

special classes of chaperonins that function only at the highest growth temperatures. In cells of *Pyrodictium abyssi* (**Figure 19.26**), for example, a major chaperonin is the protein complex called the **thermosome**. This complex keeps other proteins properly folded and functional at high temperature, helping cells survive even at temperatures above their maximal growth temperature. Cells of *P. abyssi* grown near its maximum temperature (110°C) contain high levels of the thermosome. Possibly because of this, the cells can remain viable following a heat shock, such as a 1-h treatment in an autoclave (121°C). In cells experiencing such a treatment and then returned to the optimum temperature, the thermosome, which is itself quite heat resistant, is thought to refold sufficient copies of key denatured proteins that *P. abyssi* can once again begin to grow and divide. Thus, due to chaperonin activity, the upper temperature limit at which many hyperthermophiles can *survive* is higher than the upper temperature at which they can *grow*. The "safety net" of chaperonin activity probably ensures that cells in nature that briefly experience temperatures above their growth temperature maximum are not killed by the exposure.

DNA Stability at High Temperatures: Solutes, Reverse Gyrase, and DNA-Binding Proteins

What keeps DNA from melting at high temperatures? Various mechanisms are known to contribute. One such mechanism increases cellular solute levels, in particular potassium (K^+) or compatible organic compounds. For example, the cytoplasm of the hyperthermophilic methanogen *Methanopyrus* (Section 19.5) contains molar levels of potassium cyclic 2,3-diphosphoglycerate. This solute prevents chemical damage to DNA, such as depurination or depyrimidization (loss of a nucleotide base through hydrolysis of the glycosidic bond) from high temperatures, events that can lead to mutation (Section 10.2). This compound and other compatible solutes, such as potassium di-*myo*-inositol phosphate, which protects against osmotic stress, and the polyamines putrescine and spermidine, which stabilize both

ribosomes and nucleic acids at high temperature, help maintain key cellular macromolecules in hyperthermophiles in their active forms.

A unique protein found *only* in hyperthermophiles is responsible for DNA stability in these organisms. All hyperthermophiles produce a special DNA topoisomerase called **reverse DNA gyrase**. This enzyme introduces positive supercoils into the DNA of hyperthermophiles (in contrast to the negative supercoils introduced by DNA gyrase present in all other prokaryotes; Section 6.3). Positive supercoiling stabilizes DNA to heat and thereby prevents the DNA helix from spontaneously unwinding. The noticeable absence of reverse DNA gyrase in prokaryotes whose growth temperature optima lie below 80°C strongly suggests a specific role for this enzyme in DNA stability at high temperatures.

Species of *Euryarchaeota* also contain highly basic (positively charged) DNA-binding proteins that are remarkably similar in amino acid sequence and folding properties to the core histones of the *Eukarya* (Section 7.5). Archaeal histones from the hyperthermophilic methanogen *Methanothermus fervidus* (Figure 19.7c) have been particularly well studied. These proteins wind and compact DNA into nucleosome-like structures (**Figure 19.27**) (Figure 7.7) and maintain the DNA in a double-stranded form at very high temperatures. Archaeal histones are found in most *Euryarchaeota*, including extremely halophilic *Archaea*, such as *Halobacterium*. However, because the extreme halophiles are not thermophiles, archaeal histones may have other functions besides DNA stability, in particular in assisting in gene expression by opening the helix to allow for transcriptional proteins to bind.

Lipid and Ribosomal RNA Stability

How have the lipids and the protein-synthesizing machinery of hyperthermophiles adjusted to high temperatures? Virtually all hyperthermophilic *Archaea* synthesize lipids of the dibiphytanyl

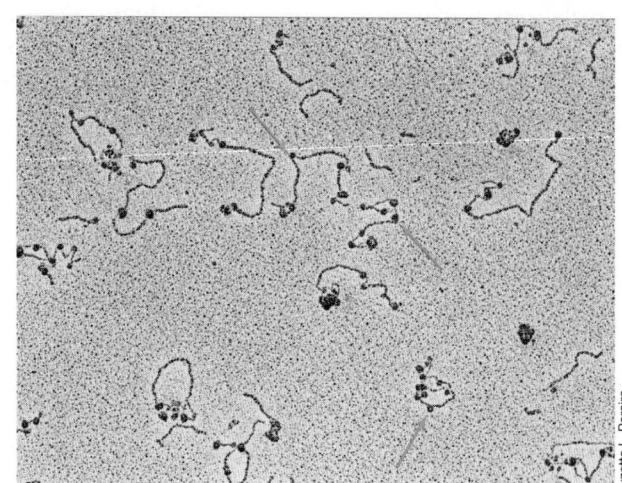

Figure 19.27 Archaeal histones and nucleosomes. Electron micrograph of linearized plasmid DNA wrapped around copies of archaeal histone Hmf (from the hyperthermophilic methanogen *Methanothermus fervidus*) to form the roughly spherical, darkly stained nucleosome structures (arrows). Compare this micrograph with an artist's depiction of the histones and nucleosomes of *Eukarya* shown in Figure 7.7.

UNIT 6

tetraether type ($\circlearrowright$ Section 3.3). These lipids are naturally heat resistant because the phytanyl units forming each half of the membrane structure are covalently bonded to one another; this yields a *lipid monolayer* membrane instead of the normal lipid bilayer ($\circlearrowright$ Figure 3.7). This structure resists the tendency of heat to pull apart a lipid bilayer constructed of fatty acid or phytanyl side chains that are not covalently bonded.

A final point on molecular adaptations to life at high temperatures is that of the base composition of ribosomal RNAs. Ribosomal RNAs (rRNAs) are key structural and function components of the ribosome, the cell's protein synthesizing apparatus ($\circlearrowright$ Section 6.19). Hyperthermophilic species of both *Bacteria* and *Archaea* show as much as a 15% greater proportion of GC base pairs in their small ribosomal subunit rRNAs compared with organisms that grow at lower temperatures. GC base pairs form three hydrogen bonds compared to the two of AU base pairs, and thus the higher GC content of the rRNAs should confer greater thermal stability on the ribosomes of these organisms and this should assist protein synthesis at high temperatures. By contrast to rRNAs, the GC content of genomic DNA of hyperthermophiles is often rather low, which suggests that the thermal stability of rRNA might be an especially significant factor for life under hyperthermophilic conditions.

MiniQuiz

- How do hyperthermophiles keep proteins and DNA from being destroyed by high heat?
- How are the lipids and ribosomes of hyperthermophiles protected from heat denaturation?

19.14 Hyperthermophilic *Archaea*, H_2, and Microbial Evolution

When cellular life first arose on Earth nearly 4 billion years ago, it is virtually certain that temperatures were far hotter than they are today. Thus, for hundreds of millions of years, Earth may have been suitable only for hyperthermophiles. Given the discussion above on the temperature limits to life, it has been hypothe-

sized that biological molecules, biochemical processes, and the first cells arose on Earth around hydrothermal springs and vents on the seafloor as they cooled to temperatures compatible with biological molecules ($\circlearrowright$ Section 16.2 and Figures 16.4 and 19.24). The phylogeny of modern hyperthermophiles (Figure 19.1), as well as the similarities in their habitats and metabolism to those of early cells on Earth, suggests that hyperthermophiles may be the closest remaining descendants of ancient cells and are a living window into the biology of ancient microbial life.

Phylogenetic Constraints on Hyperthermophiles

We have seen how sequencing of rRNA genes places hyperthermophilic species of *Archaea* and *Bacteria* on the deepest, shortest branches of their respective phylogenetic trees ($\circlearrowright$ Figures 17.1 and 19.1). This suggests that the rRNA genes of hyperthermophiles, and by extension all genes of hyperthermophiles, are subject to unusually strong constraints regarding sequence change. These constraints are likely related to the exceptionally high temperatures at which these organisms live, with only a limited amount of variation possible before heat stability and function become compromised. Indeed, the higher percentage of GC base pairs in the SSU rRNAs of hyperthermophiles (Section 19.13) may be an apt example of this. Thus it is possible that the allowable divergence in gene sequence of hyperthermophiles reached saturation early on in evolution and may thereafter have diverged relatively little from those of ancient cells. In other words, any mutations that compromised heat stability could be lethal and thus the biology of today's hyperthermophiles may differ very little from that of their ancient ancestors. If this hypothesis is true, what does it say about energy metabolism of ancient hyperthermophiles?

Hyperthermophilic Habitats and H_2 as an Energy Source

The oxidation of H_2 linked to the reduction of Fe^{3+}, S^0, NO_3^-, or, rarely, O_2 is a widespread form of energy metabolism in hyperthermophiles (Table 19.8 and **Figure 19.28**). This, coupled with the likelihood that these hyperthermophiles best characterize early Earth phenotypes, points to the important role H_2 has

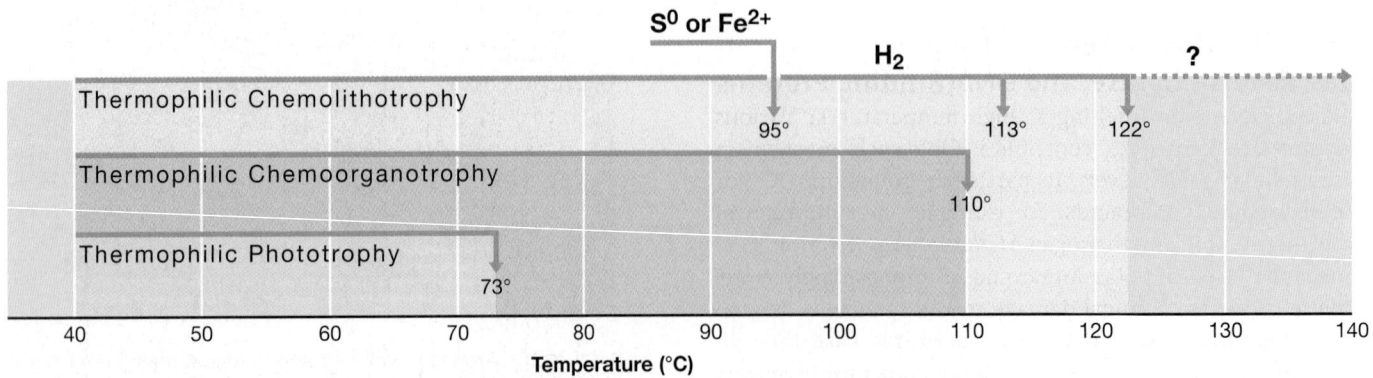

Figure 19.28 Upper temperature limits for energy metabolism. Phototrophy, *Synechococcus lividus* (*Bacteria*, cyanobacteria); chemoorganotrophy, *Pyrodictium occultum* (*Archaea*); chemolithotrophy with S^0 as electron donor, *Acidianus infernus* (*Archaea*); chemolithotrophy with Fe^{2+} as electron donor, *Ferroglobus placidus* (*Archaea*); chemolithotrophy with H_2 as electron donor, *Methanopyrus kandleri* (*Archaea*, 122°C).

played in the evolution of microbial life. Hydrogen metabolism may have evolved in primitive organisms because of the ready availability of H_2 and suitable inorganic electron acceptors in their primordial environments, but also because a H_2-based energy economy requires relatively few proteins ($\leftrightarrows$ Figure 16.7). As chemolithotrophs, these organisms may have obtained all of their carbon from CO_2 or might have assimilated available organic compounds directly for biosynthetic needs. Either way it is likely that the oxidation of H_2 was the energetic driving force for maintaining life processes.

If one compares microbial energy conservation mechanisms as a function of temperature from data of cultured prokaryotes, only chemolithotrophic organisms are known at the hottest temperatures (Figure 19.28). Chemoorganotrophy occurs up to at least 110°C, as this is the upper temperature limit for growth of *Pyrodictium occultum*, an organism that can conserve energy and grow by fermentation and by chemolithotrophic growth on H_2 with S^0 as electron acceptor (Table 19.9). Photosynthesis is the least heat-tolerant of all bioenergetic processes, with no hyperthermophilic representatives known and an apparent upper temperature limit of 73°C. This is consistent with the conclusion that anoxygenic photosynthesis first appeared on Earth some hundreds of millions of years after the first life forms are thought to have appeared ($\leftrightarrows$ Figure 16.6).

Comparisons of this kind (Figure 19.28) point to the H_2-oxidizing hyperthermophilic *Archaea* and *Bacteria* as the most likely extant examples of Earth's earliest cellular life forms. More so than any other prokaryotes, these organisms retain the metabolic and physiological traits one would predict to be necessary for existence on a hot early Earth.

MiniQuiz

- What phylogenetic and physiological evidence suggests that today's hyperthermophiles are the closest living links to Earth's earliest cells?

Big Ideas

19.1

The *Archaea* form two major phyla, the *Crenarchaeota*, which includes most hyperthermophilic species, and the *Euryarchaeota*, which includes the methanogens, extreme halophiles, and thermoacidophiles. With the exception of methanogenesis, metabolic patterns are rather similar in *Archaea* and *Bacteria*.

19.2

Extremely halophilic *Archaea* require large amounts of NaCl for growth and accumulate large levels of KCl in their cytoplasm as a compatible solute. These salts affect cell wall stability and enzyme activity. The light-mediated proton pump bacteriorhodopsin helps extreme halophiles make ATP.

19.3

Methanogenic *Archaea* are strict anaerobes whose metabolism is tied to the production of CH_4. Methane can be produced by CO_2 reduction by H_2, from methyl substrates such as CH_3OH, or from acetate.

19.4

Thermoplasma, *Ferroplasma*, and *Picrophilus* are extremely acidophilic thermophiles that form their own phylogenetic family of *Archaea*. Cells of *Thermoplasma* and *Ferroplasma* lack cell walls, resembling the mycoplasmas in this regard.

19.5

Methanopyrus is a hyperthermophilic methanogen that shows a set of characteristics predicted to have been present in early life forms and grows at the highest temperature.

19.6

Archaeoglobus and *Ferroglobus* are related anaerobic *Archaea* that carry out different anaerobic respirations. *Archaeoglobus* is a sulfate reducer and *Ferroglobus* is a nitrate reducer that oxidizes ferrous iron.

19.7

Nanoarchaeum equitans is a parasitic *Euryarchaeota* with a very small genome and thus depends on its host, *Ignicoccus*, for most of its cellular needs. *Aciduliprofundum* is a thermoacidophile that grows near sulfidic hydrothermal vents.

19.8

A wide variety of chemoorganotrophic and chemolithotrophic energy metabolisms have been found in hyperthermophilic *Crenarchaeota*, including fermentation and anaerobic respirations, and strictly autotrophic lifestyles are common.

19.9

Hyperthermophilic *Crenarchaeota* thrive in terrestrial hot springs of various chemistries. These include in particular organisms such as *Sulfolobus*, *Acidianus*, *Thermoproteus*, and *Pyrobaculum*.

19.10

In deep-sea hydrothermal systems, *Crenarchaeota* such as *Pyrolobus*, *Pyrodictium*, *Ignicoccus*, and *Staphylothermus* thrive. With the exception of the methanogen *Methanopyrus*, these genera contain species that grow at the highest temperatures currently known to support any life form.

19.11
Nonhyperthermophilic *Crenarchaeota* are widespread and abundant in cool and cold ocean waters. The organism *Nitrosopumilus* has been cultured and shown to be an ammonia-oxidizing bacterium.

19.12
Some hyperthermophiles can grow above the boiling point of water, but it is likely that life as we know it is limited to temperatures below 150°C.

19.13
Macromolecules in hyperthermophiles are protected from heat denaturation by their heat-stable folding patterns (proteins), solutes and binding proteins (DNA), unique monolayer membrane architecture (lipids), and the high GC content of their ribosomal RNAs.

19.14
Hydrogen metabolism is likely to have been the driving force behind the energetics of the earliest cells on Earth. Chemolithotrophic metabolisms based on H_2 as an electron donor are found in the most heat-tolerant of all known prokaryotes.

Review of Key Terms

Bacteriorhodopsin a protein containing retinal that is found in the membranes of certain extremely halophilic *Archaea* and that is involved in light-mediated ATP synthesis

Compatible solute an organic or inorganic substance accumulated in the cytoplasm of a halophilic organism that maintains osmotic pressure

Crenarchaeota a phylum of *Archaea* that contains both hyperthermophilic and cold-dwelling organisms

Euryarchaeota a phylum of *Archaea* that contains primarily methanogens, extreme halophiles, *Thermoplasma*, and some marine hyperthermophiles

Extreme halophile an organism whose growth is dependent on large concentrations (generally 9% or more) of NaCl

Halorhodopsin a light-driven chloride pump that accumulates Cl^- within the cytoplasm

Hydrothermal vent a deep-sea hot spring emitting warm (~20°C) to superheated (>300°C) water

Hyperthermophile an organism with a growth temperature optimum of 80°C or greater

Methanogen a CH_4-producing organism

Phytanyl a branched-chain hydrocarbon containing 20 carbon atoms and commonly found in the lipids of *Archaea*

Reverse DNA gyrase a protein universally present in hyperthermophiles that introduces positive supercoils into circular DNA

Solfatara a hot, sulfur-rich, generally acidic environment commonly inhabited by hyperthermophilic *Archaea*

Thermosome a heat shock (chaperonin) protein complex that functions to refold partially heat-denatured proteins in hyperthermophiles

Review Questions

1. What are some features that all *Archaea* have in common (Section 19.1)?

2. Which organism, *Pyrodictium, Thermoplasma*, or *Methanosarcina*, is the closest relative of the extreme halophile *Halobacterium* (*Hint*: The answer lies in Figure 19.1)? What form of energy metabolism does each of these organisms have (Sections 19.1–19.2)?

3. How can organisms such as *Halobacterium* survive in a high-salt environment, whereas an organism such as *Escherichia coli* cannot (Section 19.2)?

4. Contrast the roles of bacteriorhodopsin, halorhodopsin, and sensory rhodopsin in *Halobacterium salinarum* (Section 19.2).

5. What are methanogens (Section 19.3)?

6. How does *Ferroplasma* obtain its energy source for growth (Section 19.4)?

7. What is physiologically unique about *Methanopyrus* compared with another methanogen such as *Methanobacterium* (Section 19.5)? What is physiologically unique about *Archaeoglobus* (Section 19.6)?

8. How is *Nanoarchaeum* similar to other *Archaea*? How does it differ (Section 19.7)?

9. In one sentence, describe the habitat of *Crenarchaeota* (Section 19.8).

10. What is unusual about the metabolism of S^0 by *Acidianus* (Section 19.9)?

11. What is unusual about the organism *Pyrolobus fumarii* (Section 19.10)?

12. What is physiologically unusual about the marine crenarchaeote *Nitrosopumilus maritimus* (Section 19.11)?

13. How does protein folding help hyperthermophiles to adapt to high temperatures (Section 19.13)?

14. What is reverse DNA gyrase and why is it important to hyperthermophiles (Section 19.13)?

15. Why might H_2 metabolism have evolved as a mechanism for energy conservation in the earliest organisms on Earth (Section 19.14)?

Application Questions

1. Using the data in Figure 19.1 as a guide, discuss what indicates that bacteriorhodopsin may have been a late evolutionary invention.

2. Defend or refute the following statement: The upper temperature limit to life is unrelated to the stability of proteins or nucleic acids.

Need more practice? Test your understanding with Quantitative Questions; access additional study tools including tutorials, animations, and videos; and then test your knowledge with chapter quizzes and practice tests at **www.microbiologyplace.com**.

20

Eukaryotic Cell Biology and Eukaryotic Microorganisms

Diatoms are phototrophic microbial eukaryotes that produce tough, siliceous outer shells, called frustules, that can be buried for millions of years when the cells die and the frustules sink to the sediments.

In this chapter we consider the cell structure, phylogeny, and diversity of eukaryotic microorganisms. The unicellular eukaryotes are diverse, have an interesting evolutionary history, and are of major ecological importance. Several of them also directly impact humans, either favorably, for example through beneficial fermentations, or unfavorably, as human pathogens.

 # Eukaryotic Cell Structure and Function

The five sections that follow examine the structure of the eukaryotic cell and the ancestral endosymbiotic link between eukaryotic organelles and *Bacteria*.

20.1 Eukaryotic Cell Structure and the Nucleus

Compared with the prokaryotic cell, the eukaryotic cell is structurally more complex and typically much larger. Eukaryotic cells vary in the exact complement of organelles they contain, but a membrane-enclosed nucleus is universal and the hallmark of the eukaryotic cell.

General Structure

A schematic of a typical eukaryotic cell is shown in **Figure 20.1**. In contrast to prokaryotes, **eukaryotes** contain a membrane-enclosed nucleus and, depending on the organism, several other organelles. For example, mitochondria are nearly universal among eukaryotic cells, while pigmented chloroplasts are found only in phototrophic cells. Other internal structures typically include the Golgi complex, peroxisomes, lysosomes, endoplasmic reticula, and microtubules and microfilaments (Figure 20.1). These internal structures compartmentalize activities of the cell for efficient function. Some eukaryotic cells have flagella and cilia—organelles of motility—while others do not. Eukaryotic cells also may have extracellular components, such as a cell wall in fungi, algae, or plant cells (cell walls are not found in animal cells or most protists), or an extracellular matrix in animal cells. Also, many eukaryotes, even many eukaryotic microorganisms, form multicellular structures.

Nucleus

The **nucleus** contains the chromosomes of the eukaryotic cell (**Figure 20.2**). In eukaryotes, DNA within the nucleus is wound around basic (positively charged) proteins called **histones**, which help tightly pack the negatively charged DNA to form nucleosomes

UNIT 6

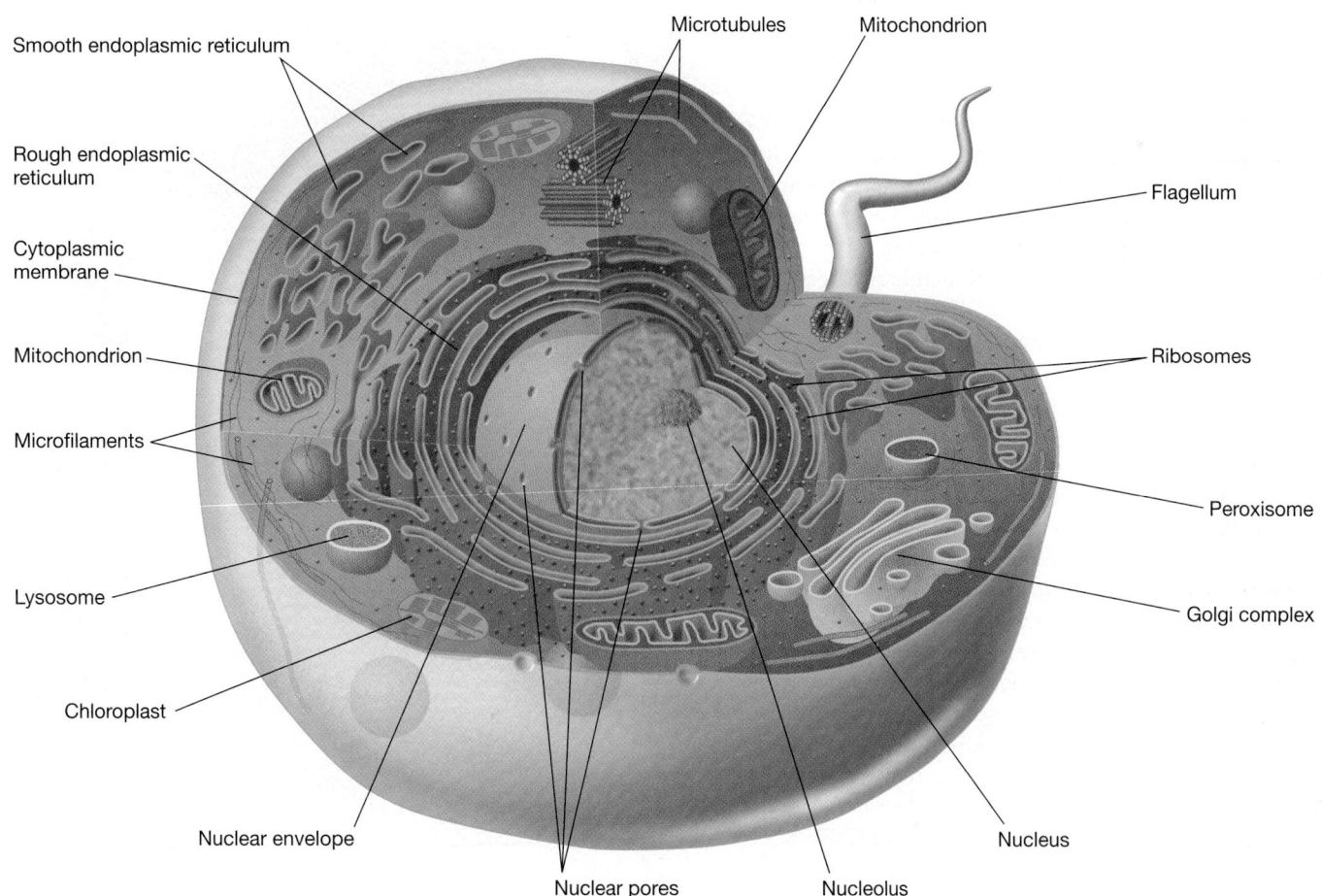

Figure 20.1 Cutaway schematic of a eukaryotic cell. Although all eukaryotic cells contain a nucleus, not all organelles and other structures shown are present in all eukaryotic cells. Not shown is the cell wall, found in fungi, algae, plants, and a few protists.

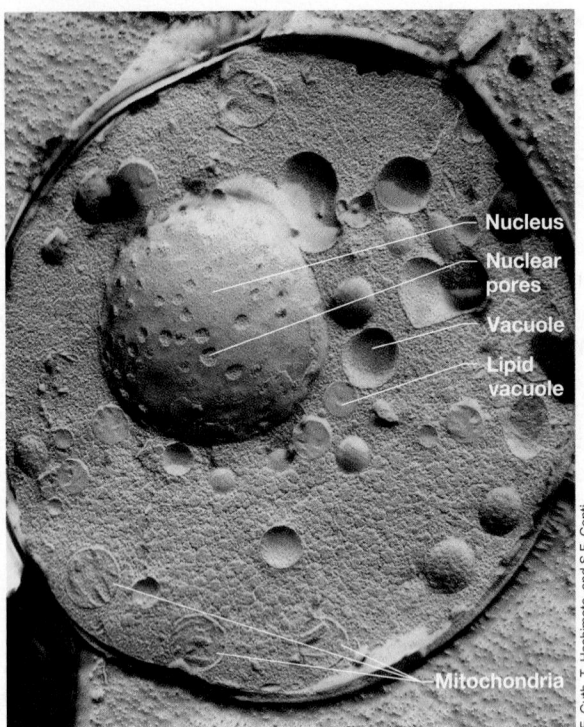

Figure 20.2 The nucleus. Electron micrograph of a yeast cell prepared by freeze-etching so as to reveal a surface view of the nucleus. The cell is about 8 μm wide and the nucleus about 1.5 μm wide.

and from them, chromosomes (Section 7.5, Figure 7.7). In many eukaryotic cells the nucleus is many micrometers in diameter and is easily visible with the light microscope, even without staining (Figure 2.5). In smaller eukaryotes, however, special staining procedures are often required to see the nucleus.

The nucleus is enclosed by a pair of membranes, each with its own function, separated by a space. The inner membrane is a simple sac; the outer membrane is in many places continuous with the endoplasmic reticulum. The inner and outer nuclear membranes specialize in interactions with the nucleoplasm and the cytoplasm, respectively. The nuclear membranes contain pores (Figure 20.2), formed from holes where the inner and outer membranes are joined. The pores allow a complex of proteins to import and export other proteins and nucleic acids into and out of the nucleus, a process called *nuclear transport*. Nuclear transport requires energy, and this comes from hydrolysis of the energy-rich compound guanosine triphosphate (GTP).

Within the nucleus is found the *nucleolus* (Figure 20.1), the site of ribosomal RNA (rRNA) synthesis. The nucleolus is rich in RNA, and ribosomal proteins synthesized in the cytoplasm are transported into the nucleolus and combine with rRNA to form the small and large subunits of eukaryotic ribosomes. These are then exported to the cytoplasm, where they associate to form the intact ribosome and function in protein synthesis.

MiniQuiz

• How is DNA arranged in eukaryotic chromosomes?

• What is the role of pores in the nuclear membrane?

20.2 The Mitochondrion and the Hydrogenosome

The mitochondrion and the hydrogenosome specialize in chemotrophic energy metabolism. Both organelles are enclosed by membranes, but have quite distinct functions.

Mitochondria

In aerobic eukaryotic cells, respiration and oxidative phosphorylation (a mechanism of ATP formation, Chapter 4) are localized in **mitochondria** (singular, **mitochondrion**). Mitochondria are of bacterial dimensions and can take on many shapes (**Figure 20.3**). A typical animal cell can contain over 1000 mitochondria, but the number per cell depends somewhat on the cell type and size. A yeast cell may have many fewer mitochondria per cell (Figure 20.2). The mitochondrion is enclosed by a double membrane system. The outer membrane, composed of an equal mixture of protein and lipid, is relatively permeable and contains numerous minute channels that allow passage of ions and small organic molecules. The inner membrane is more protein-rich than the outer membrane and is also less permeable.

Mitochondria also possess a series of folded internal membranes called **cristae**. These membranes, formed by invagination

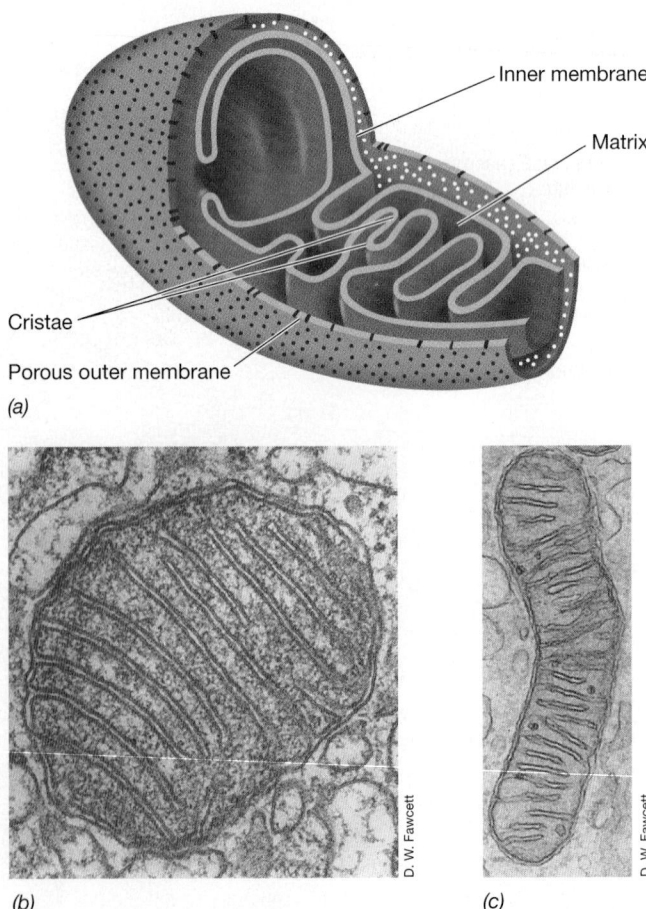

Figure 20.3 Structure of the mitochondrion. (a) Diagram showing the overall structure of the mitochondrion; note the inner and outer membranes. (b, c) Transmission electron micrographs of mitochondria from rat tissue showing the variability in morphology; note the cristae.

of the inner membrane, are the sites of enzymes for respiration and ATP production. Cristae also contain specific transport proteins that regulate the passage of metabolites, in particular ATP, into and out of the *matrix* of the mitochondrion (Figure 20.3*a*). The matrix contains enzymes for the oxidation of organic compounds—in particular, enzymes of the citric acid cycle (↺ Section 4.11).

The Hydrogenosome

Some anaerobic eukaryotic microorganisms lack mitochondria and instead contain **hydrogenosomes** (**Figure 20.4*a***). Although similar in size to a mitochondrion, the hydrogenosome lacks the citric acid cycle enzymes and usually also lacks cristae. Various microbial eukaryotes contain hydrogenosomes, and all are either obligate or aerotolerant anaerobes whose metabolism is strictly fermentative. Examples include human parasites such as the flagellate *Trichomonas* (Section 20.7 and Figure 20.13*b*) and various ciliated protists that inhabit the rumen of ruminant animals (↺ Section 25.7) or anoxic muds and sediments.

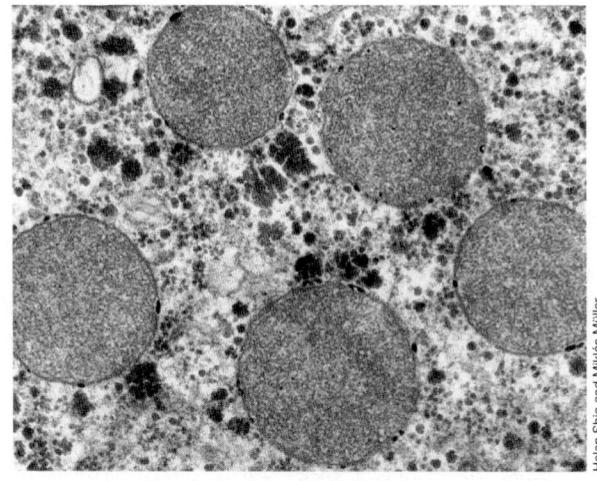

(a)

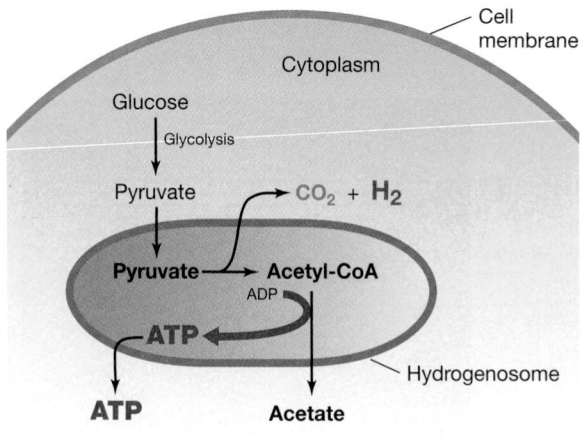

(b)

Figure 20.4 The hydrogenosome. *(a)* Electron micrograph of a thin section through a cell of the anaerobic parabasalid *Trichomonas vaginalis* showing five hydrogenosomes in cross section. Compare their internal structure with that of mitochondria in Figure 20.3. *(b)* Biochemistry of the hydrogenosome. Pyruvate is taken up by the hydrogenosome, and H_2, CO_2, acetate, and ATP are the products.

The major biochemical reactions in the hydrogenosome are the oxidation of pyruvate to H_2, CO_2, and acetate (Figure 20.4*b*). Pyruvate is oxidized to the energy-rich compound acetyl-CoA (from which additional ATP is made during the formation of acetate, ↺ Section 14.1), along with H_2 plus CO_2 (Figure 20.4*b*). The key enzymes in this process are *pyruvate:ferredoxin oxidoreductase* and *hydrogenase*. Some anaerobic eukaryotes have H_2-consuming endosymbiotic bacteria such as methanogens residing in their cytoplasm (↺ Figure 24.6*b, c*). The symbionts consume the H_2 and CO_2 produced by the hydrogenosome (Figure 20.4*b*), yielding methane (CH_4). Because hydrogenosomes lack an electron transport chain and enzymes of the citric acid cycle, they cannot oxidize the acetate produced from pyruvate catabolism as mitochondria do. Acetate is therefore excreted from the hydrogenosome into the cytoplasm of the host cell (Figure 20.4*b*).

MiniQuiz

- What key reactions occur in the mitochondrion, and what key product is made there?
- Compare and contrast the metabolic fate of pyruvate in the mitochondrion and the hydrogenosome.

20.3 The Chloroplast

Chloroplasts are chlorophyll-containing organelles found in phototrophic eukaryotes—plants, unicellular and multicellular algae, and certain unicellular organisms grouped as protists. Chloroplasts are relatively large and readily visible with the light microscope (**Figure 20.5**). The size, shape, and number of chloroplasts per cell vary markedly, and in contrast to mitochondria, chloroplasts are typically much larger than bacterial cells.

Like mitochondria, chloroplasts have a permeable outer membrane, a much less-permeable inner membrane, and an intermembrane space. The inner membrane surrounds the **stroma**, the membrane filling the lumen of the chloroplast, but the membrane

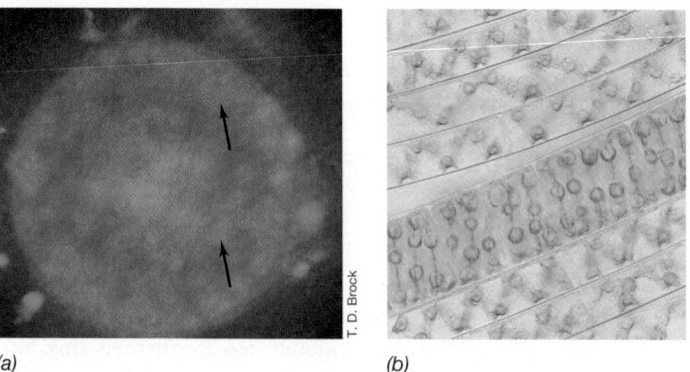

(a)   *(b)*

Figure 20.5 Photomicrographs of protist and green alga cells showing chloroplasts. *(a)* Fluorescence photomicrograph of the diatom *Stephanodiscus*. The chlorophyll in the chloroplasts (arrows) absorbs light and fluoresces red. The cell is about 40 μm wide. *(b)* Phase-contrast photomicrograph of the filamentous green alga *Spirogyra* showing the characteristic spiral-shaped chloroplasts (arrows) of this phototroph. A cell is about 20 μm wide.

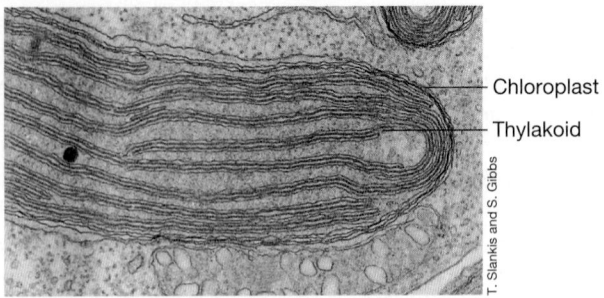

Figure 20.6 The chloroplast. Transmission electron micrograph showing a chloroplast of the stramenopile *Ochromonas danica*; note the thylakoids.

is not folded into cristae like the inner membrane of the mitochondrion (Figure 20.3*a*). Instead, chlorophyll and all other components needed for photosynthesis are located in a series of flattened membrane discs called **thylakoids** (**Figure 20.6**). The thylakoid membrane is highly impermeable to ions and other metabolites because its function is to establish the proton motive force necessary for ATP synthesis (↺ Section 4.10). In green algae and green plants, thylakoids are typically stacked into discrete structural units called *grana* (↺ Figure 13.5).

The chloroplast stroma contains large amounts of the enzyme *ribulose bisphosphate carboxylase* (RubisCO). RubisCO is a key catalyst of the **Calvin cycle**, the series of biosynthetic reactions by which many phototrophic organisms convert CO_2 to organic compounds (↺ Section 13.12). RubisCO makes up over 50% of the total chloroplast protein and catalyzes the formation of phosphoglyceric acid, a key compound for the biosynthesis of glucose (gluconeogenesis, ↺ Section 4.13). The permeability of the outer chloroplast membrane allows glucose and ATP produced during photosynthesis to diffuse into the cytoplasm where they can be used to build new cell material.

MiniQuiz
- Differentiate the stroma from thylakoids.
- What is RubisCO, what is its function, and where is it found?

20.4 Endosymbiosis: Relationships of Mitochondria and Chloroplasts to *Bacteria*

On the basis of their relative autonomy, size, and morphological resemblance to bacteria, it was hypothesized over 100 years ago that mitochondria and chloroplasts were descendants of ancient prokaryotic cells. Molecular evidence has confirmed this and clearly points to a free-living, facultatively aerobic species of *Alphaproteobacteria* as the ancestor of the mitochondrion, and a cyanobacterium-like organism acquired by a heterotrophic eukaryote some time after nucleated eukaryotic cells arose as the ancestor of the chloroplast. These two events are the major tenets of the **endosymbiotic hypothesis** (↺ Section 16.4). Through these associations, host cells obtained permanent part-

ners specializing in energy generation, whereas the endosymbionts received a stable and supportive growth environment.

Support for the Endosymbiotic Hypothesis

Several lines of molecular evidence support the endosymbiotic hypothesis, including the following:

1. **Mitochondria and chloroplasts contain DNA.** Although most of the proteins of the mitochondrion and the chloroplast are encoded by nuclear DNA, a few are encoded by a small genome residing within the organelle itself. These include certain proteins of the respiratory chain (mitochondrion) and photosynthetic apparatus (chloroplast), as well as ribosomal RNAs and transfer RNAs. Thus, nonphototrophic eukaryotic cells are genetic chimeras containing DNA from *two* different sources—the endosymbiont and the host cell nucleus. Phototrophic eukaryotes contain DNA from *three* different sources—the mitochondrial and chloroplast endosymbionts, and the nucleus. Most mitochondrial DNA and all chloroplast DNA is of a covalently closed circular form like that of most *Bacteria* (↺ Sections 2.5, 6.3). Mitochondrial DNA can be visualized in cells microscopically by using special staining methods (**Figure 20.7**). We discuss other features of organellar genomes in Section 12.4.

2. **The eukaryotic nucleus contains genes derived from *Bacteria*.** Genomic sequencing and other genetic studies have clearly shown that several nuclear genes encode properties unique to mitochondria and chloroplasts. Because the sequences of these genes more closely resemble those of *Bacteria* than those of *Archaea* or *Eukarya*, it is concluded that these genes were transferred to the nucleus from endosymbionts during the evolutionary transition from engulfed cell to organelle.

3. **Mitochondria and chloroplasts contain their own ribosomes.** Ribosomes, cell structures that function in protein synthesis, exist in either a large form (80S), typical of the cytoplasm of eukaryotic cells, or in a smaller form (70S), in

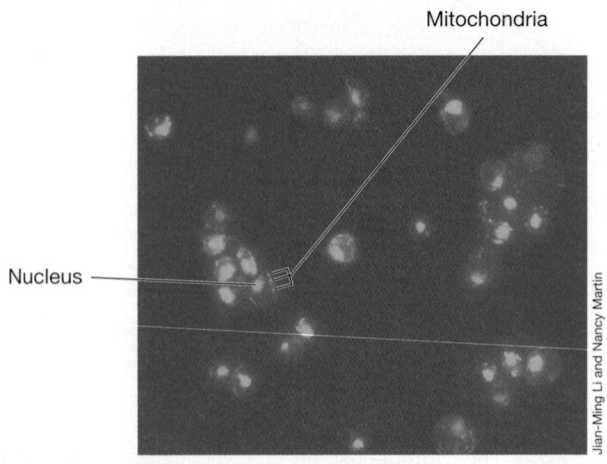

Figure 20.7 Cells of the ascomycete yeast *Saccharomyces cerevisiae*. The cells have been stained with the fluorescent dye DAPI that binds to DNA. Each mitochondrion has two to four circular chromosomes that stain blue with the dye.

Bacteria and *Archaea*. Mitochondria and chloroplasts also contain ribosomes and they are 70S in size, like those of *Bacteria* and *Archaea*.

4. **Antibiotic specificity.** Several antibiotics (streptomycin is one example) kill or inhibit *Bacteria* by disrupting 70S ribosome function. These same antibiotics also inhibit protein synthesis in mitochondria and chloroplasts.

5. **Molecular phylogeny.** Phylogenetic studies using comparative rRNA gene-sequencing methods (Chapter 16) and organellar genome studies (↩ Section 12.4) strongly support a model in which the chloroplast and mitochondrion originated from the domain *Bacteria* by the process of endosymbiosis.

6. **Hydrogenosomes.** Like mitochondria, hydrogenosomes contain DNA and ribosomes. Moreover, the nucleus of hydrogenosome-containing eukaryotes contains genes encoding proteins found in *Bacteria*. Hydrogenosomes are therefore thought to be metabolically degenerate mitochondria that exploit fermentation (rather than respiration) as a means of energy conservation in an anaerobic host. The mitochondrion and the hydrogenosome can thus be viewed as functionally related organelles that specialize in different metabolic strategies for making ATP. Other structures called *mitosomes* are present in some eukaryotic cells and are probably even more degenerate mitochondria, having lost virtually all energy-related functions.

Secondary Endosymbiosis

The mitochondrion, chloroplast, and hydrogenosome are thought to be the result of *primary endosymbiosis* events. Primary endosymbioses gave rise to the chloroplast in the common ancestor of green algae, red algae, and plants (see Figure 20.12). However, following this primary endosymbiosis event, several unrelated groups of nonphototrophic eukaryotes acquired chloroplasts by *secondary endosymbiosis*, the process of engulfing a green algal cell or a red algal cell, retaining its chloroplast and thereby becoming phototrophic. Secondary endosymbioses with green algae account for chloroplasts in euglenids and chlorarachniophytes (Sections 20.8 and 20.11; see Figure 20.12). Alveolates (ciliates, apicomplexans, and dinoflagellates; Section 20.9) and stramenopiles (Section 20.10) obtained their chloroplasts through secondary endosymbioses with red algae. The ancestral red algal chloroplasts were apparently lost from some lineages, such as the ciliates, or became greatly reduced in size in others, such as the apicomplexans. In some other organisms, such as the dinoflagellates, the red algal chloroplast was replaced with a chloroplast from a different alga, including green algae. These many examples of endosymbiotic events highlight the importance of endosymbiosis in the evolution and diversification of microbial eukaryotes (↩ Section 16.4 and Figure 16.9).

MiniQuiz

- Summarize the molecular evidence that supports the relationship of organelles to *Bacteria*.
- Distinguish between primary and secondary endosymbiosis.

20.5 Other Organelles and Eukaryotic Cell Structures

Other cytoplasmic structures are typically present in eukaryotic cells. These include the endoplasmic reticulum (ER), ribosomes, the Golgi complex, lysosomes, peroxisomes, and the organelles of motility—flagella and cilia. An extracellular matrix also can be observed surrounding animal cells. In contrast to mitochondria and chloroplasts, however, these additional cytoplasmic structures lack DNA and are not of endosymbiotic origin.

Endoplasmic Reticulum, Ribosomes, and the Golgi Complex

The endoplasmic reticulum (ER) is a network of membranes continuous with the nuclear membrane. Two types of endoplasmic reticulum exist: *rough*, which contains attached ribosomes, and *smooth*, which does not (Figure 20.1). Smooth ER participates in the synthesis of lipids and in some aspects of carbohydrate metabolism. Rough ER, through the activity of its ribosomes, is a major producer of glycoproteins and also produces new membrane material that is transported throughout the cell to enlarge the various membrane systems (Figure 20.1) before cell division.

The Golgi complex consists of a stack of membranes distinct from the ER (Figure 20.1 and **Figure 20.8**) but which functions in concert with the ER. In the Golgi complex, products of the ER are chemically modified and sorted into those destined for secretion—for example, hormones or digestive enzymes—and those that function in other membranous structures in the cell. Golgi arise from the division of preexisting Golgi and contain various enzymes that modify secretory and membrane proteins differently depending on their final destination in the cell. Many of the modifications are glycosylations (addition of sugar residues) that convert the proteins into specific glycoproteins.

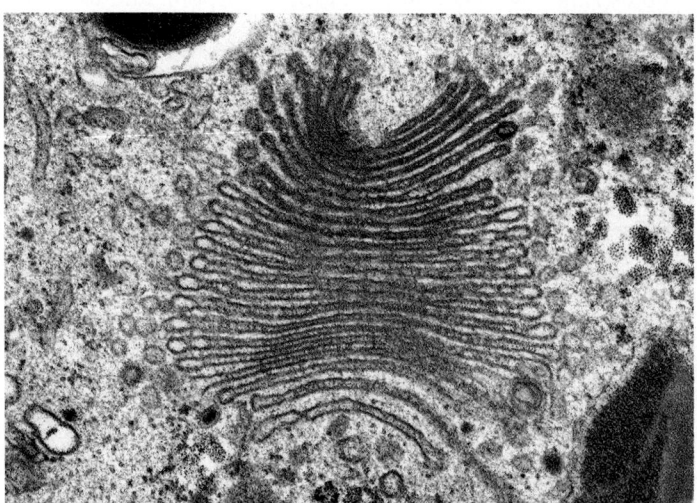

Figure 20.8 The Golgi complex. Transmission electron micrograph of a portion of a eukaryotic cell showing the Golgi complex (colored in gold). Note the multiple folded membranes of which the Golgi complex is composed (membrane stacks are 0.5–1.0 μm in diameter).

Lysosomes and Peroxisomes

Lysosomes (Figure 20.1) are membrane-enclosed compartments made from proteins and lipids transported from the Golgi complex. Lysosomes also receive proteins and lipids from the cytoplasmic membrane during the process of endocytosis and contain various digestive enzymes that hydrolyze macromolecules, such as proteins, fats, and polysaccharides, used in intracellular digestion. The lysosome fuses with food vacuoles, releasing its digestive enzymes, which break down these macromolecules for use in cellular biosynthesis and energy generation. Lysosomes also function in hydrolyzing damaged cellular components and recycling these materials for new biosyntheses. The internal pH of the lysosome is about 5 (two units lower than that of the cytoplasm), and the hydrolytic enzymes within the lysosome function optimally at this pH. These hydrolytic enzymes are nonspecific in their activity and could potentially destroy key cellular macromolecules if not contained. Thus, the lysosome allows the cell's lytic activities to be partitioned away from the cytoplasm proper. Following hydrolysis of macromolecules in the lysosome, the resulting monomers pass from the lysosome into the cytoplasm as nutrients for the cell.

The **peroxisome** is a specialized membrane-enclosed metabolic compartment (Figure 20.1). Peroxisomes originate in the cell by incorporating proteins and lipids from the cytoplasm, eventually becoming membrane-enclosed entities that can enlarge and divide in synchrony with the cell. The function of the peroxisome is to oxidize various compounds, such as alcohols and long-chain fatty acids, breaking them down into smaller molecules that are then used by the mitochondrion for energy generation. Peroxisomes also function to oxidize toxic compounds in the cell. The enzymes of the peroxisome transfer hydrogen from these compounds to O_2, producing hydrogen peroxide (H_2O_2) as a by-product. The H_2O_2 produced in the peroxisome is degraded to H_2O and O_2 by the enzyme catalase (Section 5.18). Peroxisomes play other roles as well, such as synthesizing bile salts that aid in the absorption and digestion of fats.

Microtubules, Microfilaments, and Intermediate Filaments

Just as buildings are supported by structural reinforcement, the large size of eukaryotic cells and their ability to move requires structural reinforcement. This internal structural network comes from proteins that form filamentous structures called *microtubules, microfilaments,* and *intermediate filaments*. Together, these structures form the cell **cytoskeleton** (Figure 20.1 and **Figure 20.9**).

Microtubules are tubes about 25 nm in diameter containing a hollow core and are composed of the proteins *α-tubulin* and *β-tubulin*. Microtubules function in maintaining cell shape, in cell motility by cilia and flagella (**Figure 20.10**), in chromosome movement during cell division, and in movement of organelles. **Microfilaments** are smaller filaments, about 7 nm in diameter, and are polymers of two intertwined strands of the protein actin. Microfilaments function in maintaining and changing cell shape, in cell motility by pseudopodia, and during cell division. **Intermediate**

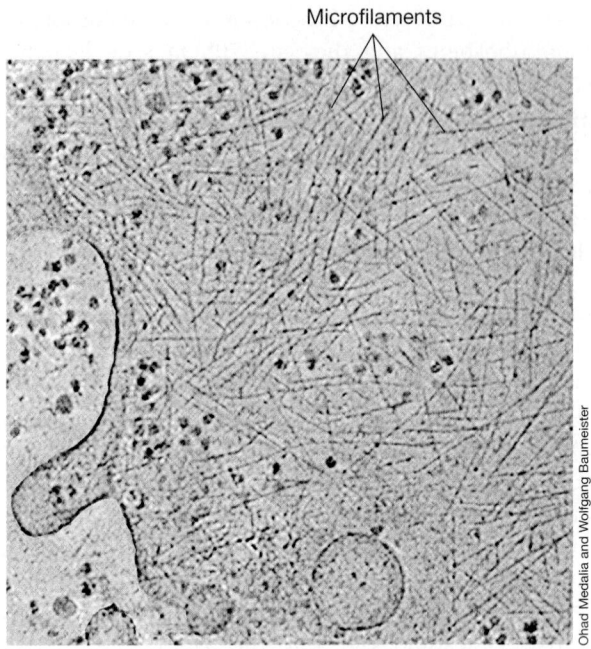

Microfilaments

Figure 20.9 Microfilaments and eukaryotic cell architecture. An electron tomographic image of a cell of the cellular slime mold *Dictyostelium discoideum* showing the network of actin microfilaments that along with microtubules functions as the cell cytoskeleton. Microfilaments are about 7 nm in diameter. Electron tomography allows a three-dimensional reconstruction of cells from a series of images taken with a transmission electron microscope.

filaments are fibrous keratin proteins supercoiled into thicker fibers 8–12 nm in diameter that function in maintaining cell shape and positioning organelles in the cell.

Flagella and Cilia

Flagella and cilia are present on many eukaryotic microorganisms and function as organelles of motility, allowing cells to move by swimming. Motility has survival value, as the ability to move allows motile organisms to move about their habitat and exploit new resources. *Cilia* are essentially short flagella that

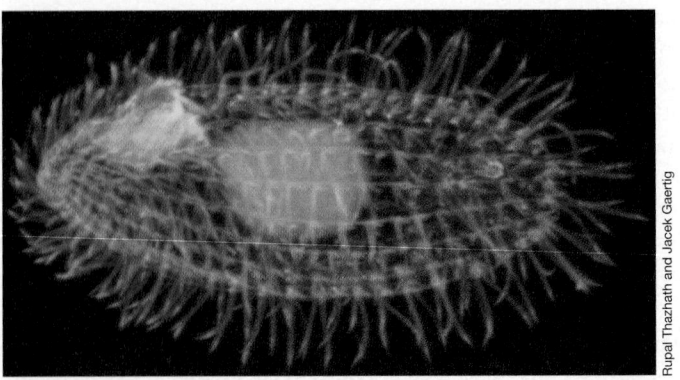

Figure 20.10 Tubulin of *Tetrahymena thermophila*. Fluorescence photomicrograph of a cell labeled with two types of antitubulin antibodies (red/green) and with DAPI, which stains DNA (blue, nucleus). A cell is about 10 μm wide.

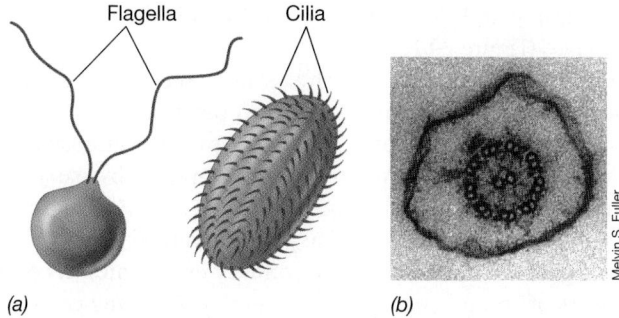

Flagella   Cilia

Melvin S. Fuller

(a)   (b)

Figure 20.11 Motility organelles in eukaryotic cells: Flagella and cilia. *(a)* Flagella can be present as single or multiple filaments. Cilia are structurally very similar to flagella but much shorter. Eukaryotic flagella move in a whiplike motion. *(b)* Cross section through a flagellum of the chytrid fungus *Blastocladiella emersonii* showing the outer sheath, the outer nine pairs of microtubules, and the central pair of microtubules.

beat in synchrony to propel the cell—usually quite rapidly—through the medium. *Flagella*, by contrast, are long appendages present singly or in groups that propel the cell along—typically more slowly than by cilia—through a whiplike motion (**Figure 20.11a**). The flagella of eukaryotic cells are structurally quite distinct from bacterial flagella and do not rotate as bacterial flagella do (🔗 Section 3.13).

In cross section, cilia and flagella are similar. Each contains a bundle of nine pairs of microtubules composed of tubulins, surrounding a central pair of microtubules called the *axoneme* (Figure 20.11b). A second protein, called *dynein*, is attached to the tubulin and functions as an ATPase, hydrolyzing ATP to yield the energy necessary to drive motility. Movement of flagella and cilia is similar. In both cases, movement involves the coordinated sliding of axonemal microtubules (Figure 20.11b) against one another in a direction toward or away from the base of the cell. This movement confers the whiplike action on the flagellum or cilium that results in cell propulsion.

A clear distinction can thus be made between the bacterial and eukaryotic flagellum. The filament of the bacterial flagellum is made from a helical array of a single protein, flagellin, and the structure itself is firmly anchored in a motor complex embedded in the cell wall and cytoplasmic membrane that rotates the filament 360° to gain propulsion at the expense of the proton motive force. The eukaryotic flagellum, by contrast, propels the cell through a whiplike motion of sliding microtubules driven by the energy of ATP.

Extracellular Components: The Cell Wall and Extracellular Matrix

Cell walls are present in fungi, algae, and plant cells, while animal cells and most protists lack cell walls. The cell wall functions to provide shape to the cell, protect it from the environment, and limit the uptake of water. In multicellular organisms such as plants, the structural support provided by the cell wall helps the plant to withstand gravity. The composition of the plant wall varies with the organism, but the polysaccharide cellulose, together with other polysaccharides and proteins, is used and forms a strong wall matrix ranging from 0.1 to a few millimeters

in thickness, much thicker than the cytoplasmic membrane that it surrounds.

In animal cells, some fungi, and some protists, an **extracellular matrix** (**ECM**) in which cells are embedded is present outside the cytoplasmic membrane. In animals, where the ECM has been best characterized, different glycoproteins make up the ECM. One of these proteins, called *fibronectin*, connects cells to the ECM by simultaneously attaching to other ECM glycoproteins such as proteoglycans and collagen, and also by way of *integrins*, which are proteins embedded in the cytoplasmic membrane. The ECM, through its connection to integrin proteins, provides a means for cells to integrate changes occurring outside and inside the cell and to coordinate the activities of adjacent cells. An ECM is also found in the fruiting bodies formed by some fungi (mushrooms) and slime molds (*Dictyostelium discoideum*), and in biofilms (🔗 Section 23.4) formed by some yeasts (*Candida albicans*).

MiniQuiz

- How does smooth endoplasmic reticulum differ from rough endoplasmic reticulum?
- Why are the activities in the lysosome best partitioned away from the cytoplasm proper?
- What functional roles do microtubules and the extracellular matrix of animal cells have?

 Eukaryotic Microbial Diversity

We now examine the diversity of microbial eukaryotes, cells that reside in the domain *Eukarya*. These include the *protists* and the *fungi*. **Protists** include both phototrophic and nonphototrophic unicellular eukaryotes other than fungi. We begin our tour with an overview of the phylogeny of microbial eukaryotes and then proceed to consider the individual groups. Microbial eukaryotes have a complex and intriguing evolutionary history and are remarkably diverse, varying in many aspects of their biology.

20.6 Phylogeny of the *Eukarya*

From the universal phylogenetic tree of life (🔗 Figure 16.16) we learned that *Eukarya* constitute a domain and that *Eukarya* are more closely related to *Archaea* than to *Bacteria*. The phylogeny of *Eukarya* was originally inferred from sequences of the 18S ribosomal RNA (rRNA) gene, which encodes the small subunit (SSU) RNA of the cytoplasmic ribosomes of eukaryotes. Although phylogenetic trees based on comparative sequencing of SSU rRNA genes are strongly believed to be the best trees available for *Bacteria* and *Archaea*, this does not appear to be true for microbial eukaryotes.

The SSU rRNA View and Other Views of Eukaryotic Evolution

The SSU rRNA view of eukaryotic phylogeny distinguishes certain organisms, such as the diplomonad *Giardia*, the microsporidian *Encephalitozoon*, and the parabasalid *Trichomonas*, as having diverged long ago, well before other eukaryotes, such as animals

UNIT 6

and plants (⮌ Figure 16.16). Supporting this was the fact that these "early-branching" eukaryotic groups initially appeared phenotypically primitive, lacking a mitochondrion for example, and to have therefore arisen before the primary endosymbiotic event(s) that led to the eukaryotic cell containing mitochondria that we know today (Section 20.4 and ⮌ Section 16.4 and Figure 16.9). However, we now know that "amitochondriate" eukaryotes contain structures analogous to the mitochondrion (hydrogenosomes or mitosomes, Sections 20.2 and 20.4), and therefore may not be as ancient as previously thought. It also appears from molecular analyses that for some reason these "early-branching" lineages have undergone one or more periods of rapid evolution that can lead to artifacts in 18S rRNA gene analyses, and so other tools have been brought to bear on the problem of eukaryotic microbial diversity.

Molecular sequencing of several other eukaryotic genes, and of genes encoding proteins such as tubulins, RNA polymerase, ATPase, and heat shock proteins, has been used to generate a new phylogenetic tree of *Eukarya*. Phylogenies based on these markers show several differences from the SSU rRNA gene-based tree of *Eukarya*. First, it appears that a major phylogenetic radiation took place as an early event in eukaryote evolution. This radiation included evolution of the ancestors of all, or essentially all, modern-day eukaryotic organisms. A tree representing this view (**Figure 20.12**) shows the diplomonads and parabasalids, amitochondriate organisms once thought to be *basal* (early evolving), instead as *derived* organisms, arising later in eukaryote evolution. Second, animals and fungi appear to be closely related (Figure 20.12). Third, the microsporidia, which branch very early in the SSU rRNA gene tree (⮌ Figure 16.16), are revealed to be a highly

derived group and close relatives of fungi, a highly derived group themselves (Figure 20.12).

The tree of *Eukarya* also shows how secondary endosymbioses account for the origin of chloroplasts in some unicellular phototrophic eukaryotes. Following primary endosymbiosis of the cyanobacterial ancestor of chloroplasts by early mitochondrion-containing eukaryotes (Figure 20.12; ⮌ Figure 16.9), these now phototrophic eukaryotes diverged into red and green algae. Then, in separate secondary endosymbiosis events, ancestors of certain euglenozoans (Section 20.8) and cercozoans (Section 20.11) engulfed green algae, and certain alveolates (Section 20.9) and stramenopiles (Section 20.10) engulfed red algae (Figure 20.12). These secondary endosymbioses help account for the great phylogenetic diversity of phototrophic eukaryotes and are likely to have occurred relatively recently in evolutionary time.

Eukaryotic Evolution: The Big Picture

What can we conclude from this emerging view of eukaryotic phylogeny? Although phylogenies based on SSU rRNA genes and on other genes and proteins all confirm the three domains of life, *Bacteria*, *Archaea*, and *Eukarya*, our overall view of eukaryotic evolution has changed dramatically with the new phylogenetic picture obtained from the study of other eukaryotic genes and proteins. Also, new aspects of eukaryotic biology, including the finding of hydrogenosomes or gene and protein remnants of mitochondria in organisms previously thought to have never contained them, coupled with recent morphological and sequence-based evidence of secondary endosymbiosis, have shifted our thinking. Taken together, these

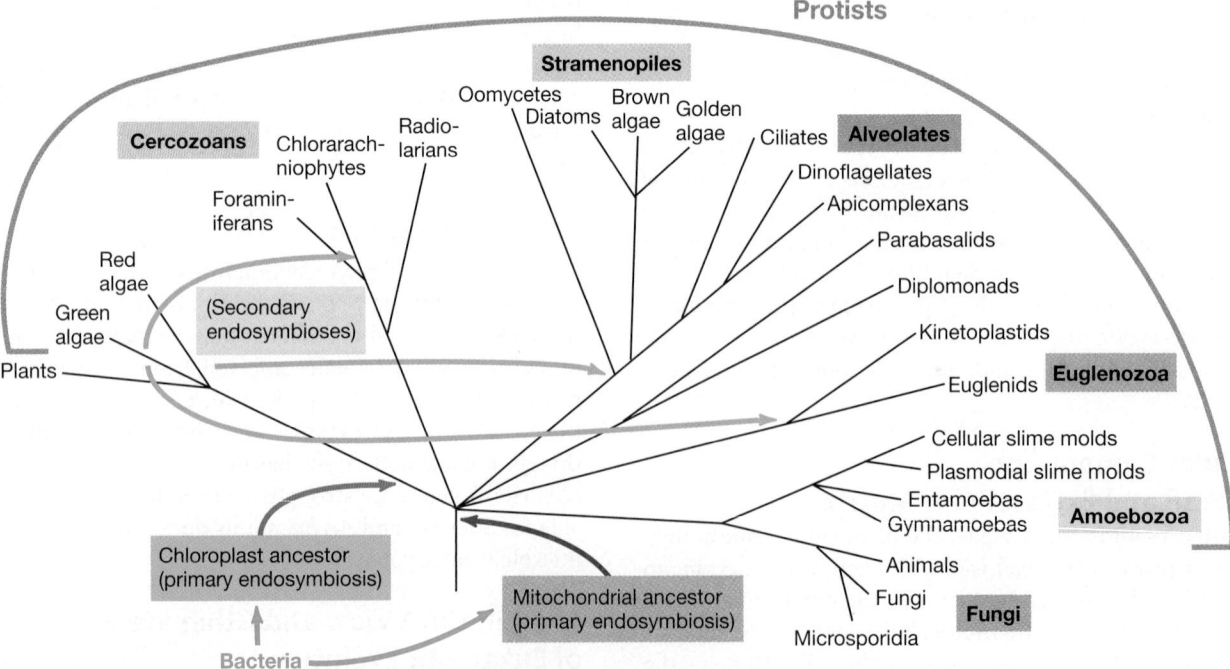

Figure 20.12 Phylogenetic tree of *Eukarya*. This composite tree is based on sequences of several genes and proteins. Dark green and red arrows indicate primary endosymbiotic events for the acquisition of the mitochondrion (red) and the chloroplast (green). Light green arrows indicate secondary endosymbiotic acquisition of chloroplasts from red and green algae by various protists.

data reveal that certain eukaryotic groups once thought to have arisen early in evolution probably arose within a more recent major evolutionary radiation. It follows that the origin of the mitochondrion may have predated this major radiation, as all extant *Eukarya* contain mitochondria or hydrogenosomes or at least some macromolecular traces of these structures, such as mitosomes.

The modern tree of *Eukarya* therefore points to the endosymbiotic acquisition of the ancestor of the mitochondrion, which likely provided the early eukaryotic cell with dramatic new metabolic capabilities, as the trigger for the evolutionary radiation of eukaryotic microorganisms. What triggered this endosymbiotic event is unknown, but was possibly the accumulation of O_2 in the atmosphere (Figure 16.6). Somewhat later in evolutionary time the ancestor of all phototrophic eukaryotes acquired the ancestor of the chloroplast in a primary endosymbiotic event. Eukaryotic phototrophic diversity unfolded later through secondary endosymbioses of chloroplast-containing red and green algae.

The tree shown in Figure 20.12 should not be considered the final word on eukaryotic evolution. As more results of comparative sequencing come to hand and as new studies reveal previously unsuspected aspects of eukaryotic biology, the true phylogeny of the eukaryotes will emerge. But it appears that a multigene tree, rather than the SSU rRNA tree, is the scaffold upon which the eukaryotic tree of life will rest.

MiniQuiz

- How do mitosomes differ from mitochondria?
- Explain why the absence of the mitochondrion could be false evidence that a eukaryotic organism evolved early.
- How does secondary endosymbiosis help explain the diversity of phototrophic eukaryotes?

 Protists

Now that we have the overall phylogeny of *Eukarya* in mind, we proceed to examine the major groups of eukaryotic microorganisms. We begin with the protists, a group that exhibits a wide range of morphologies, are widely distributed in nature, and represent a tremendous phylogenetic diversity. Indeed, protists represent much of the diversity found in the domain *Eukarya* (Figure 20.12).

20.7 Diplomonads and Parabasalids

Key Genera: *Giardia, Trichomonas*

Diplomonads and parabasalids are unicellular, flagellated protists that lack chloroplasts. They live in anoxic habitats, such as animal intestines, either symbiotically or as parasites, conserving energy from fermentation.

Diplomonads

Diplomonads, which characteristically contain two nuclei of equal size, also contain mitosomes, much reduced mitochondria lacking electron transport proteins and enzymes of the citric acid cycle (Section 20.6). The diplomonad *Giardia intestinalis*

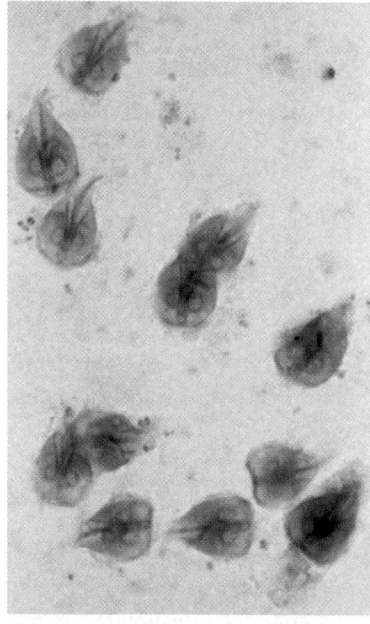
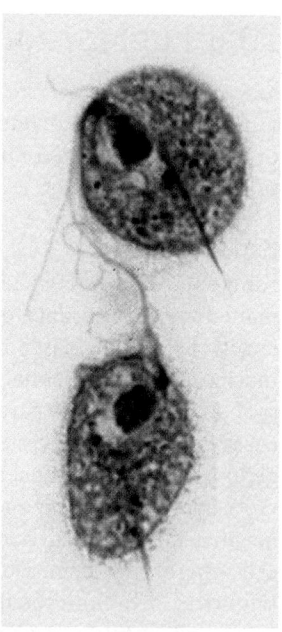

(a)   *(b)*

Figure 20.13 Diplomonads and parabasalids. *(a)* Light photomicrograph of cells of *Giardia lamblia*, a typical diplomonad. Note the dual nuclei and single flagellum. *(b)* Light photomicrograph of cells of the parabasalid *Trichomonas vaginalis*. Cells are about 6 μm wide.

(**Figure 20.13a**), also known as *Giardia lamblia*, causes giardiasis, one of the most common waterborne diarrheal diseases in the United States. We examine the disease giardiasis in Section 35.6.

Parabasalids

Parabasalids contain a *parabasal body* that, among other functions, gives structural support to the Golgi complex. They lack mitochondria but contain hydrogenosomes for anaerobic metabolism (Section 20.2). Parabasalids live in the intestinal and urogenital tract of vertebrates and invertebrates as parasites or as commensal symbionts (Chapter 25 and Section 33.13). The parabasalid *Trichomonas vaginalis* (Figure 20.13b) causes a sexually transmitted disease in humans.

The genomes of parabasalids are unique among eukaryotes in that most of them lack introns, the noncoding sequences characteristic of eukaryotic genes (Sections 7.5 and 12.5). In addition, the genome of *T. vaginalis* is huge for a parasitic organism, about 160 megabase pairs, and shows evidence of genes acquired from bacteria by horizontal gene transfer. Much of the genome of *T. vaginalis* contains repetitive sequences and transposable elements, which has made genomic analyses difficult. But *Trichomonas* is still thought to contain nearly 60,000 genes, about twice that of the human genome and near the upper limit observed thus far for eukaryotic genomes.

MiniQuiz

- How do diplomonads obtain energy?
- What is unusual about the *Trichomonas* genome?

20.8 Euglenozoans

Key Genera: *Trypanosoma, Euglena*

Euglenozoans are a diverse assemblage of unicellular, free-living or parasitic flagellated eukaryotes that includes the kinetoplastids and euglenids.

Kinetoplastids

Kinetoplastids are a well-studied group of euglenozoans and are named for the presence of the *kinetoplast*, a mass of DNA present in their single, large mitochondrion. Kinetoplastids live primarily in aquatic habitats, where they feed on bacteria. Some species, however, are parasites of animals and cause serious diseases in humans and vertebrate animals. Cells of *Trypanosoma*, a genus infecting humans, are small, about 20 μm long, thin, and crescent-shaped. Trypanosomes have a single flagellum that originates in a basal body and folds back laterally across the cell where it is enclosed by a flap of cytoplasmic membrane (**Figure 20.14**). Both the flagellum and the membrane participate in propelling the organism, making effective movement possible even in viscous liquids, such as blood.

Trypanosoma brucei (Figure 20.14) causes *African sleeping sickness*, a chronic and usually fatal human disease. The parasite lives and grows primarily in the bloodstream, but in the later stages of the disease it invades the central nervous system, causing an inflammation of the brain and spinal cord that is responsible for the characteristic neurological symptoms of the disease. The parasite is transmitted from host to host by the tsetse fly, *Glossina* spp., a bloodsucking fly found only in certain parts of Africa. After moving from the human to the fly via blood, the parasite proliferates in the intestinal tract of the fly and invades the insect's salivary glands and mouthparts, from which it is transferred to a new human host by a fly bite.

Euglenids

Another well-studied group of euglenozoans are the euglenids. Unlike the kinetoplasts, these organisms are nonpathogenic and phototrophic. They live exclusively in aquatic habitats and

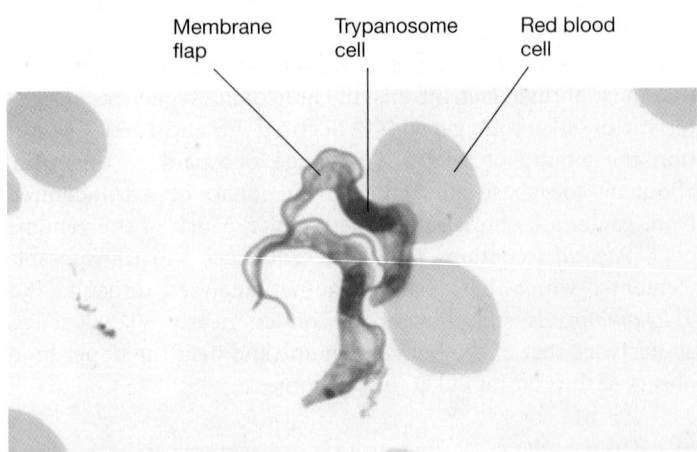

Membrane flap   Trypanosome cell   Red blood cell

Figure 20.14 Trypanosomes. Photomicrograph of the flagellated euglenozoan *Trypanosoma brucei*, the causative agent of African sleeping sickness. Blood smear preparation. A cell is about 3 μm wide.

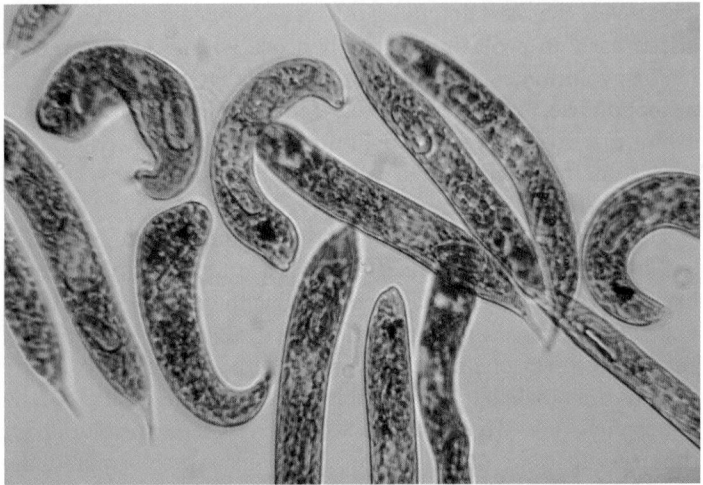

Figure 20.15 *Euglena*, a euglenozoan. This phototrophic protist, like other euglenozoans, is not pathogenic. A cell is about 15 μm wide.

contain chloroplasts, which allow for phototrophic growth (**Figure 20.15**). In darkness, however, cells of *Euglena*, a typical euglenid, can lose their chloroplasts and exist as chemoorganotrophs. Many euglenids can also feed on bacterial cells via **phagocytosis**, a process of surrounding a particle with a portion of their flexible cytoplasmic membrane to engulf the particle and bring it into the cell where it is digested.

MiniQuiz
- How are euglenozoans distinguished from other protists?
- How do cells of *Trypanosoma brucei* get from one human host to another?

20.9 Alveolates

Key Genera: *Gonyaulax, Plasmodium, Paramecium*

The alveolates as a group are characterized by the presence of *alveoli*, cytoplasmic sacs located just under the cytoplasmic membrane. Although the function of alveoli is unknown, they may help the cell maintain osmotic balance. Three phylogenetically distinct, although related, kinds of organisms make up the alveolates (Figure 20.12): the *ciliates*, which use cilia for motility; the *dinoflagellates*, which are motile by means of a flagellum; and the *apicomplexans*, which are animal parasites.

Ciliates

Ciliates possess *cilia* (**Figure 20.16**) at some stage of their life cycle. Cilia are structures that function in motility (Section 20.5) and may cover the cell or form tufts or rows, depending on the species. Probably the best-known and most widely distributed ciliates are those of the genus *Paramecium* (Figure 20.16). Like many other ciliates, *Paramecium* uses cilia not only for motility but also to obtain food by ingesting particulate materials such as bacterial cells through a distinctive funnel-shaped oral groove. Cilia that line the oral groove move material down the groove to the cell mouth (Figure 20.16*b*). There, it is enclosed in a vacuole

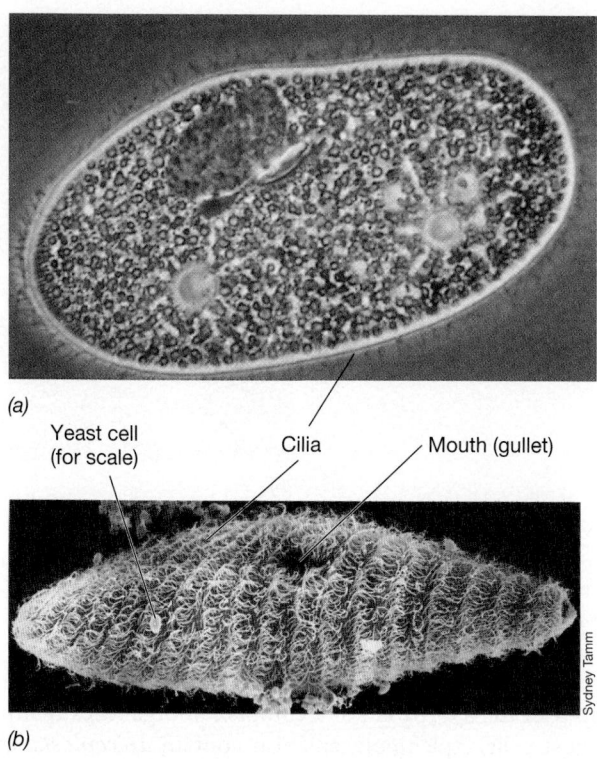

(a)

Yeast cell
(for scale)   Cilia   Mouth (gullet)

(b)

Figure 20.16 *Paramecium*, **a ciliated protist.** *(a)* Phase-contrast photomicrograph. *(b)* Scanning electron micrograph. Note the cilia in both micrographs. A single *Paramecium* cell is about 60 μm in diameter.

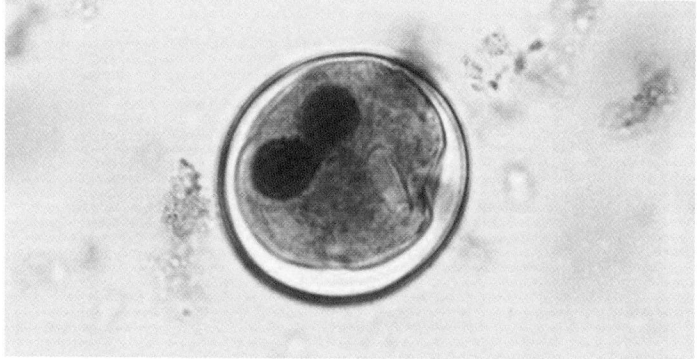

Figure 20.17 *Balantidium coli*, **a ciliated protist that causes a dysentery-like disease in humans.** The dark blue-stained structure in this *B. coli* cyst obtained from swine intestine is the macronucleus. The cell is about 50 μm wide.

by phagocytosis. Digestive enzymes secreted into the vacuole then break down the material as a source of nutrients. Ciliates are unique among protists in having two kinds of nuclei, *micronuclei* and *macronuclei*. Genes in the macronucleus regulate basic cellular functions, such as growth and feeding, whereas those of the micronucleus are involved in sexual reproduction, which occurs through a partial fusion of two *Paramecium* cells and exchange of micronuclei.

Many *Paramecium* species (as well as many other protists) are hosts for endosymbiotic prokaryotes that reside in the cytoplasm or sometimes in the macronucleus. These organisms may play a nutritional role, synthesizing vitamins or other growth factors used by the host cell. Ciliated protists that are commensal in the termite hindgut carry endosymbiotic methanogens (*Archaea*). These organisms metabolize H_2 produced from pyruvate oxidation in the hydrogenosome (Figure 20.4) to yield methane (CH_4), which is released to the atmosphere (Section 23.2). Moreover, ciliates themselves can be symbiotic, as obligately anaerobic ciliates are present in the rumen, the forestomach of ruminant animals. Rumen protists play a beneficial role in the digestive and fermentative processes of the animal (Section 25.7). In contrast to symbioses, some ciliates are animal parasites, although this lifestyle is less common in ciliates than in some other groups of protists. The species *Balantidium coli* (**Figure 20.17**), for example, is primarily an intestinal parasite of domestic animals, but occasionally infects the intestinal tract of humans, producing dysentery-like symptoms. Cells of *B. coli* form cysts (Figure 20.17) that can transmit the disease in infected food or water.

Dinoflagellates

Dinoflagellates are a diverse group of marine and freshwater phototrophs (**Figure 20.18**). Flagella encircling the cell impart spinning movements that give dinoflagellates their name (*dinos* is Greek for "whirling"). Dinoflagellates have two flagella of different lengths and with different points of insertion into the cell, transverse and longitudinal. The transverse flagellum is attached laterally, whereas the longitudinal flagellum originates from the lateral groove of the cell and extends lengthwise (see Figure 20.19*b*). Some dinoflagellates are free-living, whereas others live a symbiotic existence with animals that make up coral reefs, obtaining a sheltered and protected habitat in exchange for supplying phototrophically fixed carbon as a food source for the reef.

Several species of dinoflagellates are toxic. For example, dense suspensions of *Gonyaulax* cells, called "red tides" (**Figure 20.19***a*) due to the red-colored pigments of this organism, can form in warm and typically polluted, coastal waters. Such blooms are often associated with fish kills and poisoning in humans following consumption of mussels that have accumulated *Gonyaulax* through filter feeding. Toxicity results from a potent neurotoxin that can cause a condition called *paralytic shellfish poisoning*.

Figure 20.18 **The marine dinoflagellate** *Ornithocercus magnificus* **(an alveolate).** The cell proper is the globular central structure; the attached ornate structures are called *lists*. A cell is about 30 μm wide.

(a)

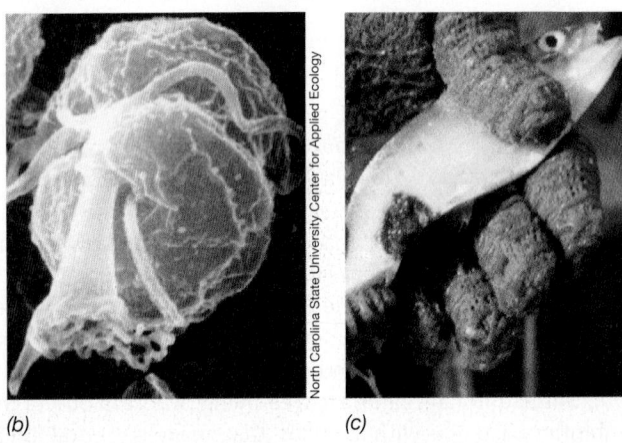

(b)   (c)

Figure 20.19 Toxic dinoflagellates (alveolates). *(a)* Photograph of a "red tide" caused by massive growth of toxin-producing dinoflagellates such as *Gonyaulax*. The toxin is excreted into the water and also accumulates in shellfish that feed on the dinoflagellates. *(b)* Scanning electron micrograph of a toxic spore of *Pfiesteria piscicida*; the structure is about 12 μm wide. *(c)* A fish killed by *P. piscicida*; note the lesions of decaying flesh.

Symptoms include numbness of the lips, dizziness, and difficulty breathing, and death can result from respiratory failure. *Pfiesteria* is another toxic dinoflagellate. Toxic spores of *Pfiesteria piscicida* (Figure 20.19*b*) infect fish and eventually kill them due to neurotoxins that affect movement and destroy skin. Lesions form on areas of the fish, allowing opportunistic bacterial pathogens to grow (Figure 20.19*c*). Human toxemia from *Pfiesteria* poisoning causes symptoms of skin rashes and respiratory problems.

Apicomplexans

Apicomplexans are obligate parasites that cause severe human diseases such as malaria (*Plasmodium* species) (**Figure 20.20*a***), toxoplasmosis (*Toxoplasma*) (Figure 20.20*b*), and coccidiosis (*Eimeria*). These organisms are characterized by nonmotile adult stages, and nutrients are taken up in soluble form across the cytoplasmic membrane as in bacteria and fungi.

Apicomplexans produce structures called *sporozoites* (Figure 20.20*b*), which function in transmission of the parasite to a new host, and the name apicomplexan derives from the presence at

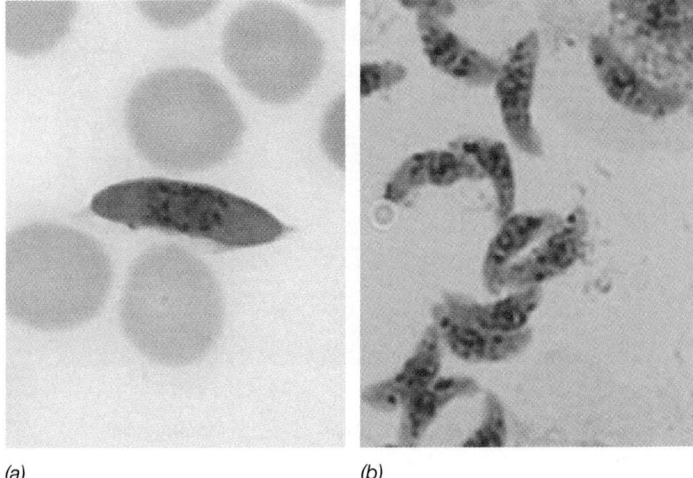

(a)   (b)

Figure 20.20 Apicomplexans. *(a)* A gametocyte of *Plasmodium falciparum* in a blood smear. The gametocyte is the stage in the malarial parasite life cycle that infects the mosquito vector. *(b)* Sporozoites of *Toxoplasma gondii*.

one apex of the sporozoite of a complex of organelles that penetrate host cells. Apicomplexans also contain *apicoplasts*. These are degenerate chloroplasts that lack pigments and phototrophic capacity, but contain a few of their own genes. Apicoplasts carry out fatty acid, isoprenoid, and heme biosyntheses, and export their products to the cytoplasm. It is hypothesized that apicoplasts are derived from red algal cells engulfed by apicomplexans in a secondary endosymbiosis (Figure 20.12). Over time, the chloroplast of the red algal cell degenerated to play a nonphototrophic role in the apicomplexan cell.

Both vertebrate and invertebrate hosts are known for apicomplexans. In some cases, an alternation of hosts takes place, with some stages of the life cycle linked to one host and some to another. Important apicomplexans are the coccidia, typically bird parasites, and species of *Plasmodium* (malaria parasites) (Figure 20.20*a*). We reserve detailed discussion of malaria—a disease that throughout history has killed more humans than any other disease—for Section 34.5.

MiniQuiz

• How does the organism *Paramecium* move?

• What health problem is associated with the organism *Gonyaulax*?

• What are apicoplasts, which organisms have them, and what functions do they carry out?

20.10 Stramenopiles

Key Genera: *Phytophthora, Nitzschia, Dinobryon*

The *stramenopiles* include both chemoorganotrophic and phototrophic organisms. Members of this group bear flagella with many short, hairlike extensions, and this morphological feature gives the group its name (from Latin *stramen* for "straw" and *pilus* for "hair"). The diatoms, oomycetes, and golden algae are the major stramenopiles (Figure 20.12).

Diatoms

Diatoms include over 200 genera of unicellular, phototrophic, microbial eukaryotes, and are major components of the planktonic (suspended) phytoplankton microbial community in marine and fresh waters. Diatoms characteristically produce a cell wall made of silica to which protein and polysaccharide are added. The wall, which protects the cell against predation, exhibits widely different shapes in different species and can be highly ornate (**Figure 20.21**). The external structure formed by this wall, called a *frustule*, often remains after the cell dies and the organic materials have disappeared. Diatom frustules typically show morphological symmetry, including *pinnate symmetry* (having similar parts arranged on opposite sides of an axis, as in the common diatom *Nitzschia*, Figure 20.21*b*), and *radial symmetry*, as in the marine diatoms *Thalassiosira* and *Asterolampra* (Figure 20.21*c*). Because the diatom frustules, which are composed mainly of silica (ↄ Figure 24.14) are resistant to decay, these structures can remain intact for long periods of time and often sink and remain in the sediments for millions of years. Diatom frustules constitute some of the best unicellular eukaryotic fossils known, and from dating this excellent fossil record, it can be seen that diatoms first appeared on Earth relatively recently, about 200 million years ago.

Oomycetes

The oomycetes ("egg fungi"), also called *water molds*, were previously grouped with fungi based on their filamentous growth and the presence of **coenocytic** (that is, multinucleate) hyphae, morphological traits characteristic of fungi. Their life cycle, unlike that

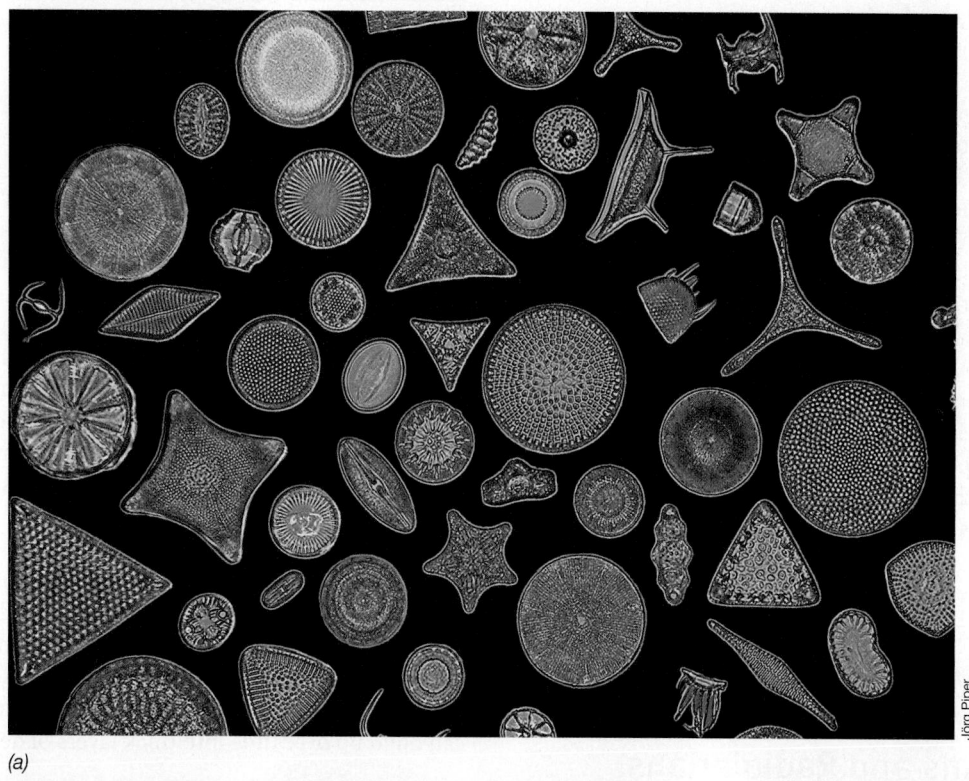

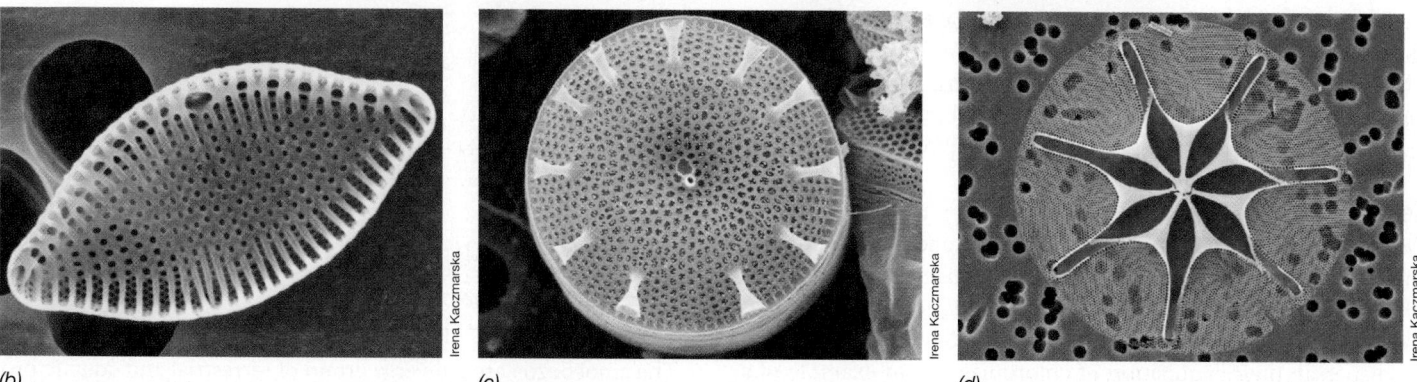

Figure 20.21 Diatom frustules. *(a)* Dark-field photomicrograph of a collage of frustules from different diatom species showing various forms of symmetry. *(b–d)* Scanning electron micrographs of diatom frustules showing pinnate (part b) or radial (parts c, d) symmetry. Diatoms vary considerably in size from very small species about 5 μm wide to larger species up to 200 μm wide.

of many fungi, includes a diploid, asexually reproducing phase and a diploid, sexually reproducing phase. Phylogenetically, however, the oomycetes are distant from fungi and are closely related to other stramenopiles (Figure 20.12). Oomycetes differ from fungi in other fundamental ways, as well. For example, the cell walls of oomycetes are typically made of cellulose, not chitin, as they are in fungi, and they have flagellated cells, which are lacking in all but a few fungi. Moreover, the diploid phase is dominant in oomycetes, whereas it is reduced in most fungi. Nonetheless, oomycetes are ecologically similar to fungi in that they grow as a mass of hyphae decomposing dead plant and animal material in aquatic habitats.

Oomycetes have had a major impact on human society. The oomycete *Phytophthora infestans*, which causes late blight disease of potatoes, contributed to massive famines in Ireland in the seventeenth century. The famines led to the death of a million Irish and to the migration of at least a million more to the United States, Australia, Canada, and other countries.

Golden Algae

Golden algae, also called *chrysophytes*, are primarily unicellular marine and freshwater phototrophs. Some species are chemoorganotrophs and feed by either phagocytosis or by transporting soluble organic compounds across the cytoplasmic membrane. Some golden algae, such as *Dinobryon*, found in freshwater, are colonial. However, most golden algae are unicellular and motile by the activity of two flagella of unequal length.

Golden algae are so named because of their golden-brown color. This is due to chloroplast pigments dominated by the brown-colored carotenoid fucoxanthin. Also, the major chlorophyll pigment in golden algae is chlorophyll *c* rather than chlorophyll *a*, and they lack the phycobiliproteins present in red algal chloroplasts (Section 20.19). Cells of the unicellular golden alga *Ochromonas*, the best-studied genus of this group, have only one or two chloroplasts.

MiniQuiz
- What structure of diatoms accounts for their excellent fossil record?
- In what ways do oomycetes differ from and resemble fungi?

20.11 Cercozoans and Radiolarians

Cercozoans and radiolarians are distinguished from other protists by their threadlike cytoplasmic extrusions (pseudopodia) by which they move and feed. Cercozoans were previously called *amoeba* because of their pseudopodia, but it is now known that many phylogenetically diverse organisms employ pseudopodia.

Cercozoans

Cercozoa include the chlorarachniophytes and foraminiferans, among other groups, and they are closely related to radiolarians, which also have threadlike pseudopodia. Chlorarachniophytes are phototrophic amoeba-like organisms that develop a flagellum for dispersal; their acquisition of chloroplasts is an example of a secondary symbiosis (Figure 20.12).

Foraminifera are exclusively marine cercozoa and live primarily in coastal waters. They form shell-like structures called *tests*, which have distinctive characteristics and are often quite ornate

Figure 20.22 A foraminiferan (a cercozoan). Note the ornate and multilobed test. The test is about 1 mm wide.

(**Figure 20.22**). Tests are typically made of organic materials reinforced with minerals such as calcium carbonate. The test is not firmly attached to the cell, and the amoeba-like cell may extend partway out of the test during feeding. However, because of the weight of the test, the cell usually sinks to the bottom of the water column, and it is thought that the organisms feed on particulate deposits in the sediments, primarily bacteria and the remains of dead organisms. Foraminiferan tests are relatively resistant to decay and hence are readily fossilized; the famous "White Cliffs of Dover" in England, for example, are composed of foraminiferan tests laid down in an ancient sea.

Radiolarians

Radiolarians are mostly marine, heterotrophic organisms, and like cercozoans, also have threadlike pseudopodia. The name "radiolarian" comes from the radial symmetry of their tests, which generally are made of silica and exist in one fused piece. The tests of marine radiolarians settle to the ocean floor when the cell dies and can build up over time into thick layers of decaying cell material.

MiniQuiz
- What structure distinguishes cercozoans and radiolarians from other protists?
- How are chlorarachniophytes thought to have acquired the ability to photosynthesize?

20.12 Amoebozoa

Key Genera: *Amoeba, Entamoeba, Physarum, Dictyostelium*

The amoebozoa are a diverse group of terrestrial and aquatic protists that use lobe-shaped pseudopodia for movement and feeding, in contrast to the threadlike pseudopodia of cercozoans and radiolarians. The major groups of amoebozoa are the *gymnamoebas*, the *entamoebas*, and the *plasmodial* and *cellular slime molds*.

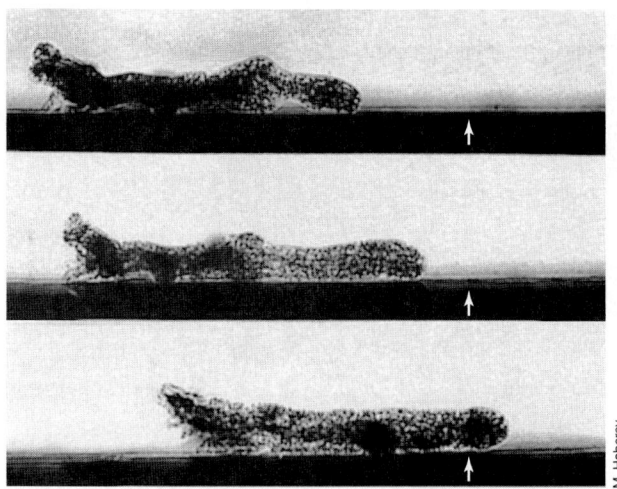

Figure 20.23 Time-lapse view of the motile amoebozoan *Amoeba proteus*. The time interval from top to bottom is about 6 sec. The arrows point to a fixed spot on the surface. A single cell is about 80 μm wide.

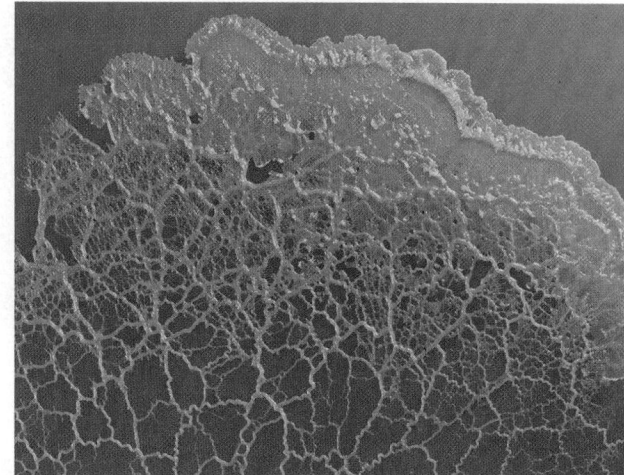

Figure 20.24 Slime mold. The plasmodial slime mold *Physarum* growing on an agar surface. The plasmodium is about 5 cm long and 3.5 cm wide.

Gymnamoebas and Entamoebas

The gymnamoebas are free-living protists that inhabit aquatic and soil environments. They use pseudopodia to move by a process called *amoeboid movement* (**Figure 20.23**) and feed by phagocytosis on bacteria, other protists, and organic materials. Amoeboid movement results from streaming of the cytoplasm as it flows forward at the less contracted and viscous cell tip, taking the path of least resistance. Cytoplasmic streaming is facilitated by microfilaments, which exist in a thin layer just beneath the cytoplasmic membrane (Section 20.5). *Amoeba* (Figure 20.23) is a common freshwater genus often present in pond water with species varying in size from 15 μm in diameter—clearly microsocopic—to over 750 μm—visible with the naked eye.

In contrast to gymnamoebas, the entamoebas are parasites of vertebrates and invertebrates. Their usual habitat is the oral cavity or intestinal tract of animals. *Entamoeba histolytica* (♻ Figure 35.19) is pathogenic in humans and can cause amebic dysentery, an ulceration of the intestinal tract that results in a bloody diarrhea. This parasite forms cysts that are transmitted from person to person by fecal contamination of water, food, and eating utensils. In Section 35.8 we discuss the etiology and pathogenesis of amebic dysentery, an important cause of death from intestinal parasites in humans.

Slime Molds

The **slime molds** were previously grouped with fungi since they undergo a similar life cycle (Section 20.13) and produce fruiting bodies with spores for dispersal. As protists, however, slime molds are motile and can move across a solid surface fairly quickly (see Figures 20.24–20.26). Phylogenetic analysis places the slime molds in the amoebozoa (Figure 20.12); they appear to be descended from the gymnamoebas.

The slime molds are divided into two groups, *plasmodial slime molds*, also called acellular slime molds, whose vegetative forms are masses of protoplasm of indefinite size and shape called plasmodia (**Figure 20.24**), and *cellular* slime molds, whose vegetative forms are single amoebae. Slime molds live primarily on decaying plant matter, such as leaf litter, logs, and soil. Their food consists mainly of other microorganisms, especially bacteria, which they ingest by phagocytosis. Slime molds can maintain themselves in a vegetative state for long periods but eventually form differentiated sporelike structures that can remain dormant and then germinate later to once again generate the active amoeboid state.

Plasmodial slime molds, such as *Physarum*, exist in the vegetative phase as an expanding single mass of protoplasm called the *plasmodium* that contains many diploid nuclei (Figures 20.24). The plasmodium is actively motile by amoeboid movement, the plasmodium flowing over the surface of the substratum, engulfing food particles as it moves. From the plasmodium, a sporangium containing haploid spores can be produced, and when conditions are favorable, the spores germinate to yield haploid flagellated swarm cells. The fusion of two swarm cells then regenerates a diploid plasmodium.

In contrast to plasmodial slime molds, cellular slime molds are haploid and form diploids only under certain conditions. In addition, instead of a single mass of protoplasm (Figures 20.24), cellular slime molds are individual, independent amoeboid cells, and when the available food is consumed, the cells aggregate to form a fruiting body.

The cellular slime mold *Dictyostelium discoideum* has been used as a model cellular slime mold. This organism undergoes a remarkable asexual life cycle in which vegetative cells aggregate, migrate as a cell mass, and eventually produce fruiting bodies in which cells differentiate and form spores (**Figures 20.25** and **20.26**). When cells of *Dictyostelium* are starved for nutrients, they aggregate and form a pseudoplasmodium; in this stage cells lose their individuality, but do not fuse. Aggregation is triggered by the production of cyclic adenosine monophosphate (cAMP); the first cells of *Dictyostelium* that produce this compound become centers for the attraction of neighboring amoeboid cells and trigger their aggregation into masses of moving cells called *slugs*. Fruiting body formation begins when the slug becomes stationary and vertically oriented (Figures 20.25 and 20.26). The emerging structure differentiates into a stalk and a head, with stalk cells forming cellulose, which provides the rigidity of the stalk, and the head cells differentiating into spores. Eventually, spores are released and dispersed, with each spore forming a new amoeba.

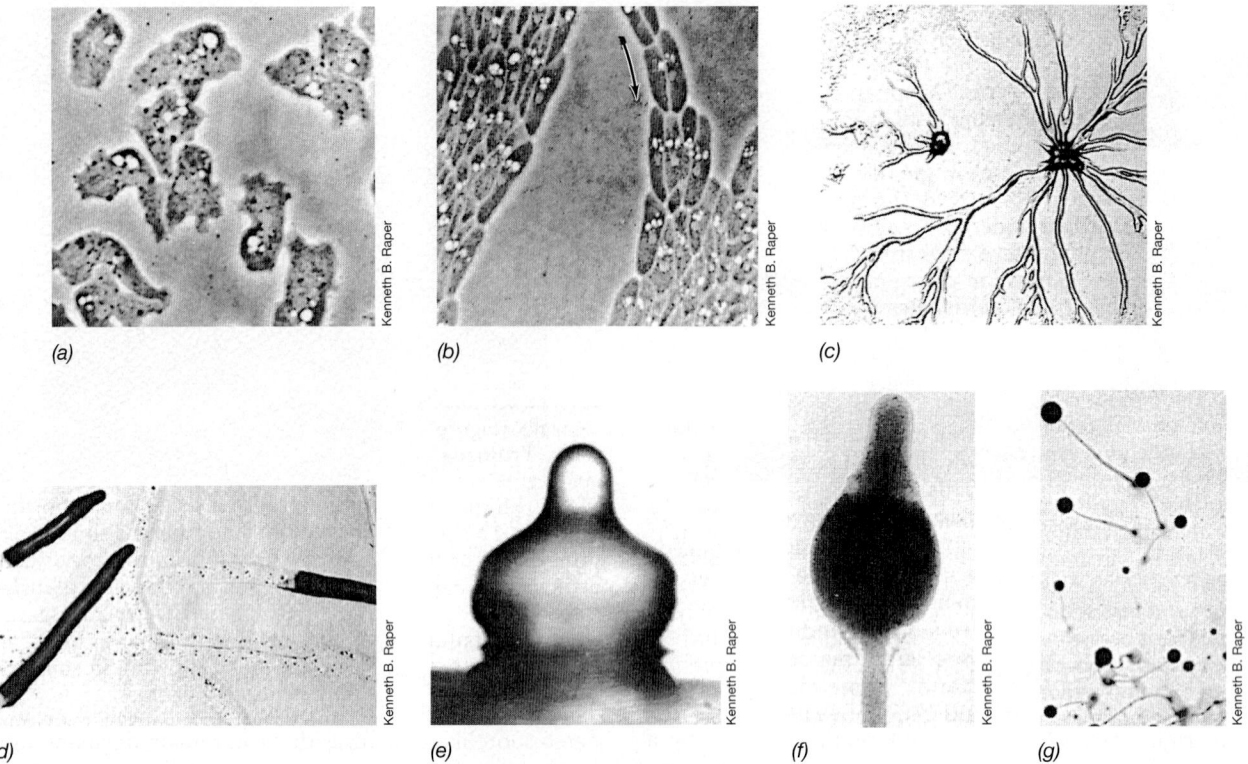

Figure 20.25 Photomicrographs of various stages in the life cycle of the cellular slime mold *Dictyostelium discoideum*. *(a)* Amoebae in preaggregation stage. *(b)* Aggregating amoebae. Amoebae are about 300 μm in diameter. *(c)* Low-power view of aggregating amoebae. *(d)* Migrating pseudoplasmodia (slugs) moving on an agar surface and leaving trails of slime behind. *(e, f)* Early stage of fruiting body. *(g)* Mature fruiting bodies. Figure 20.26 shows the sizes of these structures.

In addition to this asexual process, *Dictyostelium* can produce sexual spores. These form when two amoebae in an aggregate fuse to form a single large amoeba. A thick cellulose wall forms around this giant amoeba, forming a structure called the *macrocyst*, and this can remain dormant for long periods. Eventually, the diploid nucleus undergoes meiosis to form haploid nuclei that become integrated into new amoebae that can once again initiate the asexual cycle (Figures 20.25 and 20.26).

MiniQuiz

- How can amoebozoans be distinguished from cercozoans and radiolarians?
- Compare and contrast the lifestyles of gymnamoebas and entamoebas.
- Describe the major steps in the life cycle of *Dictyostelium discoideum*.

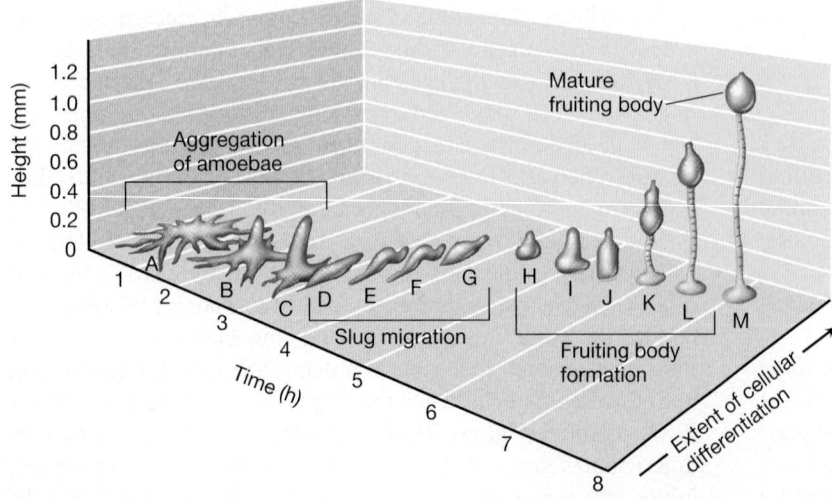

Figure 20.26 Stages in fruiting body formation in the cellular slime mold *Dictyostelium discoideum*. *(a–c)* Aggregation of amoebae. *(d–g)* Migration of the slug formed from aggregated amoebae. *(h–l)* Culmination of migration and formation of the fruiting body. *(m)* Mature fruiting body composed of stalk and head. Cells from the rear of the slug form the head and become spores. *Dictyostelium* also undergoes sexual reproduction (not shown) when two amoebae fuse to form a macrocyst; the fused nuclei in the macrocyst return to the haploid stage when meiosis forms new vegetative amoebae.

Ⓘ Fungi

Fungi are a large, diverse, and widespread group of organisms, consisting of the *molds*, *mushrooms*, and *yeasts*. Approximately 100,000 species of fungi have been described, and as many as 1.5 million species may exist. Fungi form a phylogenetic cluster distinct from other organisms and are most closely related to animals (Figure 20.12).

Most fungi are terrestrial. They inhabit soil or dead plant matter and play crucial roles in the mineralization of organic carbon. A large number of fungi are plant parasites. Indeed, fungi cause many of the economically significant diseases of crop plants. A few fungi cause disease in animals, including humans, although in general fungi are less important as animal pathogens than are other microorganisms. Fungi also establish symbiotic associations with many plants, facilitating the plant's acquisition of minerals from soil (↩ Section 25.5), and many fungi benefit humans through fermentation and the synthesis of antibiotics.

20.13 Fungal Physiology, Structure, and Symbioses

In this section we describe some general features of fungi, including their physiology, cell structure, and the symbiotic associations they develop with plants and animals. In the following section we examine fungal reproduction and phylogeny.

Nutrition and Physiology

Fungi are chemoorganotrophs, typically with simple nutritional requirements, and most are aerobes. Fungi feed by secreting extracellular enzymes that digest complex organic materials, such as polysaccharides or proteins, into sugars, peptides, amino acids, and so on, which are assimilated as sources of carbon and energy. As decomposers, fungi digest dead animal and plant materials. As parasites of plants or animals, fungi use the same mode of nutrition but take up nutrients from the living cells of the plants and animals they invade and infect rather than from dead organic materials.

A major ecological activity of fungi, especially basidiomycetes, is the decomposition of wood, paper, cloth, and other products derived from these natural sources. Lignin, a complex polymer in which the building blocks are phenolic compounds, is an important constituent of woody plants, and in association with cellulose, confers rigidity on them. Lignin is decomposed in nature almost exclusively through the activities of certain basidiomycetes called *wood-rotting fungi*. Two types of wood rot are known: *brown rot*, in which the cellulose is attacked preferentially and the lignin left unmetabolized, and *white rot*, in which both cellulose and lignin are decomposed. The white rot fungi are of major ecological importance because they play such a key role in decomposing woody material in forests.

Many fungi can grow at environmental extremes of low pH or high temperature (up to 62°C; ↩ Table 5.1) and this, coupled with the ease of dispersal of fungal spores, makes these organisms common contaminants of foods, microbial culture media, and surfaces of all sorts.

Fungal Morphology, Spores, and Cell Walls

Most fungi are multicellular, forming a network of filaments called *hyphae* (singular, hypha). Hyphae are tubular cell walls that surround the cytoplasmic membrane. Fungal hyphae are often septate, with cross-walls dividing each hypha into separate cells. In some cases, however, the vegetative cell of a fungal hypha contains more than one nucleus, and hundreds of nuclei can form due to repeated nuclear divisions without the formation of cross-walls, a condition called *coenocytic*. Each hyphal filament grows mainly at the tip by extension of the terminal cell (**Figure 20.27**).

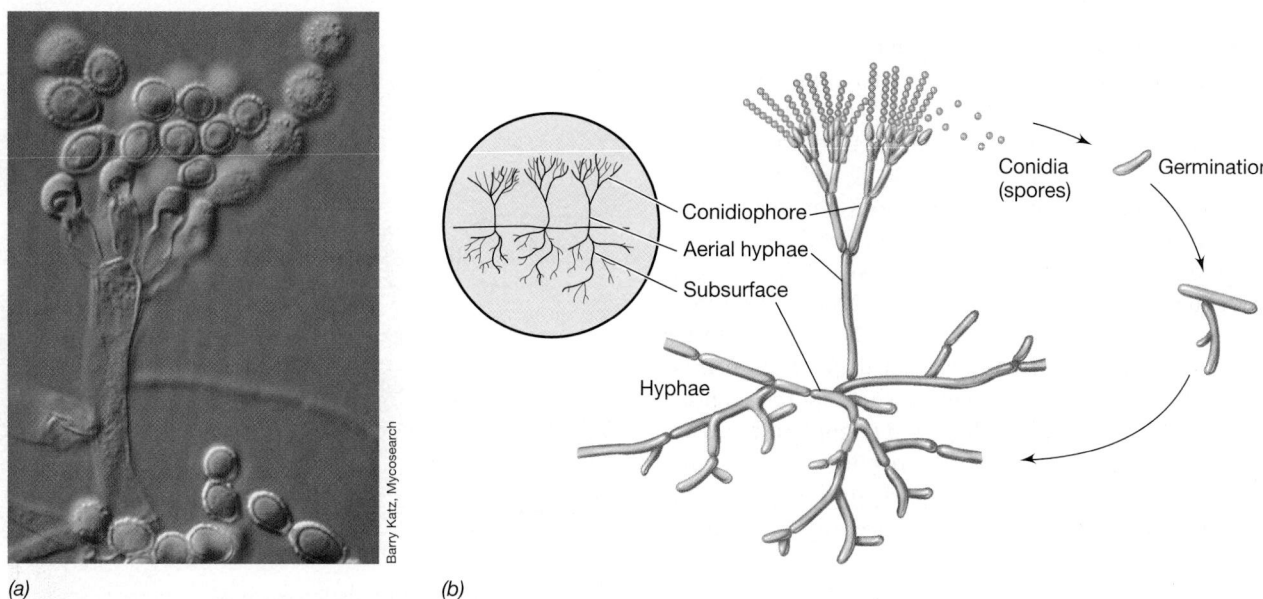

(a) Barry Katz, Mycosearch

(b)

Conidiophore
Aerial hyphae
Subsurface
Hyphae
Conidia (spores)
Germination

Figure 20.27 Fungal structure and growth. *(a)* Photomicrograph of a typical fungus. Spherical structures at the ends of aerial hyphae are asexual spores (conidia). *(b)* Diagram of a mold life cycle. The conidia can be dispersed by either wind or animals and are about 2 μm wide.

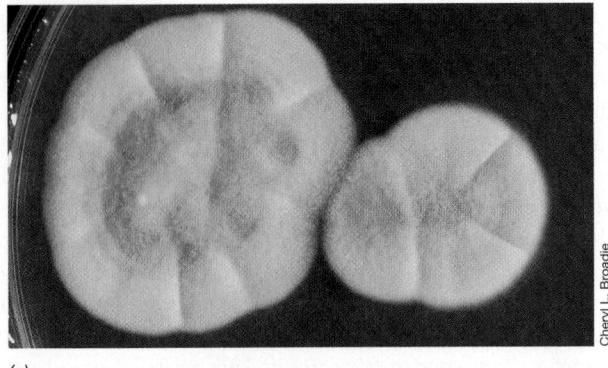

(a)

Cheryl L. Broadie

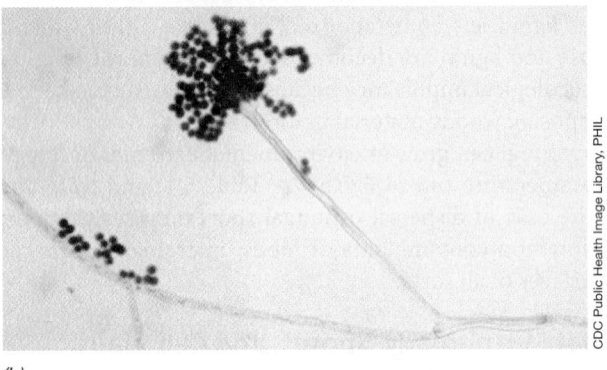

(b)

CDC Public Health Image Library, PHIL

Figure 20.28 Fungi. *(a)* Colonies of an *Aspergillus* species, an ascomycete, growing on an agar plate. Note the appearance of the masses of filamentous cells (the mycelium) and asexual spores that give the colonies a dusty, matted appearance. *(b)* Conidiophore and conidia of *Aspergillus fumigatus* (see Figure 20.27). The conidiophore is about 300 μm long and the conidia about 3 μm wide.

Hyphae typically grow together across a surface and form a compact, macroscopically visible tuft called a *mycelium* (**Figure 20.28**). The mycelium arises because the individual hyphae branch as they grow over and into the organic material on which the fungus is feeding, and these branches intertwine, forming a compact mat. From the mycelium, hyphal branches may reach up into the air above the surface, and spores called **conidia** are formed on these aerial branches (Figures 20.27 and 20.28). Conidia are asexual spores (their formation does not involve the fusion of gametes or meiosis), and they are often pigmented black, blue green, red, yellow, or brown (Figure 20.28). Conidia give the mycelium a dusty appearance (Figure 20.28*a*) and function to disperse the fungus to new habitats. Some fungi form macroscopic reproductive structures called *fruiting bodies* (**mushrooms** or puff balls, for example), in which spores are produced and from which they can be dispersed (**Figure 20.29**). The fruiting bodies can release millions of spores that are spread by wind, water, or animals to new habitats where the spores can then germinate. Some fungi grow as single-celled forms; these are the **yeasts**.

Most fungal cell walls consist of **chitin**, a polymer of the glucose derivative *N*-acetylglucosamine. Chitin is arranged in the walls in microfibrillar bundles, as is cellulose in plant cell walls, to form a thick, tough wall structure. Other polysaccharides such

as mannans and galactosans, or even cellulose itself, replace chitin in some fungal cell walls. Fungal cell walls are typically 80–90% polysaccharide, with proteins, lipids, polyphosphates, and inorganic ions making up the wall-cementing matrix.

Symbioses and Pathogenesis

Most plants are dependent on certain fungi to facilitate their uptake of minerals from soil. These fungi form symbiotic associations with the plant roots called *mycorrhizae* (the word means, literally, "fungus roots"). Mycorrhizal fungi establish close physical contact with the roots and help the plant obtain phosphate and other minerals and also water from the soil. In return, the fungi obtain nutrients such as sugars from the plant root.

There are two different kinds of mycorrhizal associations. One kind is *ectomycorrhizae*, typically formed between basidiomycetes and the roots of woody plants. In ectomycorrhizal associations the fungal hyphae form a sheath around the plant root but do not penetrate the root extensively. A second kind is *endomycorrhizae*, which form between glomeromycete fungi (Section 20.16) and many herbaceous (nonwoody) plants. In endomycorrhizae the fungal hyphae embed deeply in the plant root tissue, forming swollen vesicles or branching invaginations (arbuscules). Fungi also form symbiotic associations with cyanobacteria or green algae. These are called *lichens* and appear as colorful, crusty growths on the surfaces of trees and rocks (↺ Figure 25.1). Sections 25.1 and 25.5 consider mycorrhizal and lichen symbioses in more detail.

Fungi can invade and cause disease in plants and animals. Fungal plant pathogens cause widespread crop and plant damage worldwide, and fruit and grain crops in particular suffer significant yearly losses due to fungal infection. Many fungal plant pathogens form specialized hyphae, called *haustoria*, that penetrate the plant cell wall and consume cell cytoplasm. Human fungal diseases, called *mycoses*, range from relatively minor and

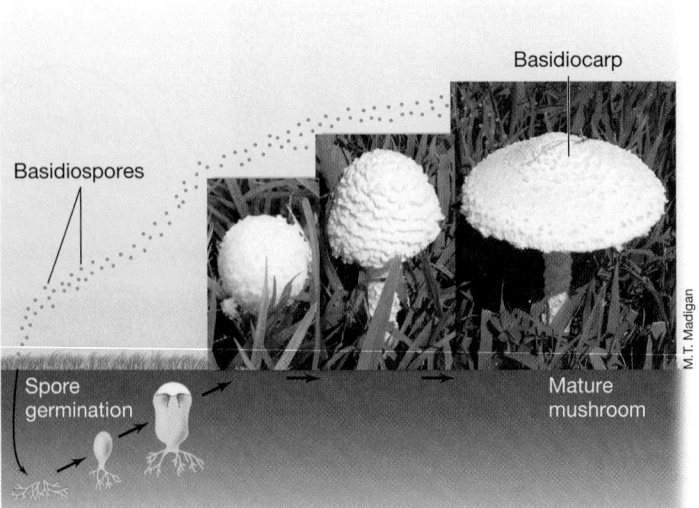

Figure 20.29 Mushroom life cycle. Mushrooms typically develop underground and then emerge on the surface rather suddenly (usually overnight), triggered by an influx of moisture. Photos of stages in formation of a common lawn mushroom (see also Section 20.18).

easily cured conditions, such as athlete's foot, to serious, life-threatening systemic mycoses, such as histoplasmosis. Section 34.8 describes some human diseases caused by fungi.

MiniQuiz

• What are conidia? How does a conidium differ from a hypha? A mycelium?

• What is chitin and where is it present in fungi?

• How do endomycorrhizal and ectomycorrhizal associations differ?

20.14 Fungal Reproduction and Phylogeny

Fungi reproduce by asexual means in one of three ways: (1) by the growth and spread of hyphal filaments; (2) by the asexual production of spores (conidia; Figures 20.27 and 20.28); or (3) by simple cell division, as in budding yeasts (**Figure 20.30**).

Sexual Spores of Fungi

Some fungi produce spores as a result of sexual reproduction. The spores develop from the fusion of either unicellular gametes or specialized hyphae called *gametangia*. Alternatively, sexual spores can originate from the fusion of two haploid cells to yield a diploid cell; this then undergoes meiosis and mitosis (⟳ Section 7.6) to yield individual haploid spores. Depending on the group, different types of sexual spores are produced. Spores formed within an enclosed sac (ascus) are called *ascospores*. Many yeasts produce ascospores, and we consider sporulation in the common baker's yeast *Saccharomyces cerevisiae* in Section 20.17 (see Figure 20.35). Sexual spores produced on the ends of a club-shaped structure (basidium) are *basidiospores* (Figure 20.29). *Zygospores*, produced by zygomycetous fungi, such as the common bread mold *Rhizopus* (Section 20.16 and see Figure 20.33), are macroscopically visible structures that result from the fusion of hyphae and genetic exchange. Eventually the zygospore matures and produces asexual spores that are dispersed by air and germinate to form new fungal mycelia.

Sexual spores of fungi are typically resistant to drying, heating, freezing, and some chemical agents. However, fungal sexual spores are not as resistant to heat as bacterial endospores (⟳ Section 3.12). Either an asexual or a sexual spore of a fungus can germinate and develop into a new hypha and mycelium.

The Phylogeny of Fungi

Fungi share a more recent common ancestor with animals than does any other group of eukaryotic organisms (Figure 20.12); fungi and animal cells are therefore phylogenetic *sister groups*. Fungi and animals are thought to have diverged approximately 1.5 billion years ago. The earliest fungal lineage is thought to be the chytridiomycetes, an unusual group in which cells produce flagellated spores (zoospores, Section 20.15). Thus the lack of flagella in most fungi indicates that motility is a characteristic that has been lost at various times in different fungal lineages.

A more detailed picture of fungal phylogeny than that shown in Figure 20.12 is depicted in the evolutionary tree in **Figure 20.31**. The phylogeny shown in this figure, based on comparative sequencing of small subunit ribosomal RNA (which can be used to resolve fairly close, but not distant, relationships among eukaryotes; see Section 20.6), defines several distinct fungal groups: the chytridiomycetes, zygomycetes, glomeromycetes,

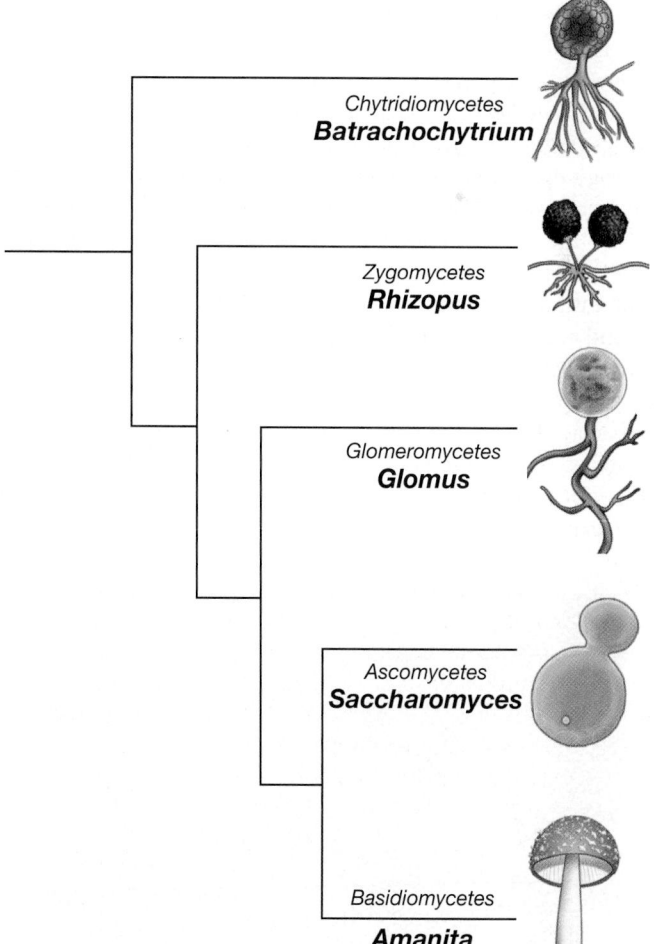

Chytridiomycetes
Batrachochytrium

Zygomycetes
Rhizopus

Glomeromycetes
Glomus

Ascomycetes
Saccharomyces

Basidiomycetes
Amanita

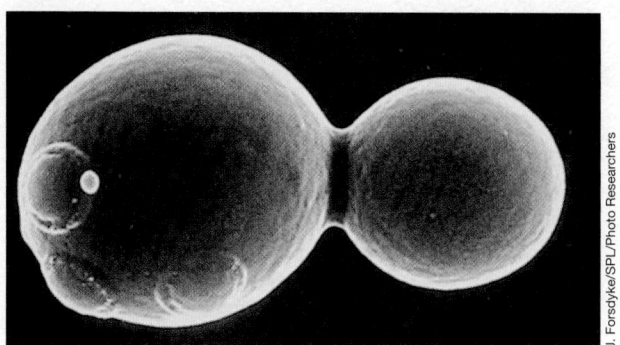

Figure 20.30 Scanning electron micrograph of the common baker's and brewer's yeast *Saccharomyces cerevisiae* **(ascomycetes).** Note the budding division and scars from previous budding. A single large cell is about 6 μm in diameter.

Figure 20.31 Phylogeny of fungi. This generalized phylogenetic tree based on 18S rRNA gene sequences depicts the relationships among the major groups (phyla) of fungi. A typical genus is listed for each group and depicted in the tree.

ascomycetes, and basidiomycetes. Figure 20.31 also supports the idea that the chytridiomycetes lie phylogenetically basal to all other fungal groups and that the most derived groups of fungi are the basidiomycetes and the ascomycetes.

MiniQuiz

- What are the major differences between ascospores and conidia?
- To what major group of macroorganisms are fungi most closely related?

20.15 Chytridiomycetes

Key Genera: *Allomyces, Batrachochytrium*

Chytridiomycetes, or *chytrids*, are the earliest diverging lineage of fungi (Figure 20.31), and their name refers to the structure of the fruiting body, which contains sexual spores called *zoospores*. These spores are unusual among fungal spores in being flagellated and motile, and are ideal for dispersal of these organisms in the aquatic environments, mostly freshwater and moist soils, where they are commonly found.

Many species of chytrids are known and some exist as single cells, whereas others form colonies with hyphae. They include both free-living forms that degrade organic material, such as *Allomyces*, and parasites of animals, plants, and protists. The chytrid *Batrachochytrium dendrobatidis* causes chytridiomycosis of frogs (**Figure 20.32**), a condition in which the organism infects the frog's epidermis and interferes with the ability of the frog to respire across the skin. Chytrids have been implicated in the massive die-off of amphibians worldwide, probably in response to increases in global temperatures that have stimulated chytrid proliferation and to increased animal susceptibility due to habitat loss and aquatic pollution.

Unresolved aspects of the phylogeny of chytrids suggest that this group is not monophyletic. That is, some organisms currently classified as chytrids may actually be more closely related to species of other fungal groups, such as the zygomycetes. As is true for protists, much about the evolution of fungi remains to be learned.

MiniQuiz

- What is one feature of chytrids that links them with other fungi?
- What is one feature of chytrids that distinguishes them from other fungi?

20.16 Zygomycetes and Glomeromycetes

Key Genera: *Rhizopus, Encephalitozoon*

We consider two groups of fungi here, the zygomycetes, known primarily for their role in food spoilage, and the glomeromycetes, important mycorrhizal fungi (Section 20.13). Zygomycetes are commonly found in soil and on decaying plant material, whereas glomeromycetes form symbiotic relationships with plant roots. All of these fungi are coenocytic (multinucleate), and a unifying characteristic is the formation of zygospores (Section 20.14).

Rhizopus, the Common Bread Mold

The black bread mold *Rhizopus stolonifer* (**Figure 20.33a**) is a common zygomycete. This organism undergoes a complex life cycle that includes both asexual and sexual reproduction. In the asexual phase the mycelia form sporangia within which haploid spores are produced. Once released, spores disperse and eventually germinate, giving rise to vegetatively growing mycelia. In the sexual phase, the gametangia (Section 20.14) of mycelia of different mating types (analogous to male and female plants or animals) yield a heterokaryotic (different nuclei) cell called a *zygosporangium*, which can remain dormant and resist dryness and other unfavorable conditions. When conditions are favorable, the different haploid nuclei fuse to form a diploid and then meiosis yields haploid spores. As in the asexual phase, the release of the spores, in this case genetically nonidentical spores, disperses the organism for vegetative hyphal growth.

Microsporidia

Closely related to the zygomycetes are the microsporidia, tiny (2–5 μm), unicellular, obligate parasites of animals and protists. Microsporidia are often one of the opportunistic pathogens that infect immune-compromised individuals, such as those who have AIDS (↩ Section 33.14). Based on small subunit ribosomal RNA

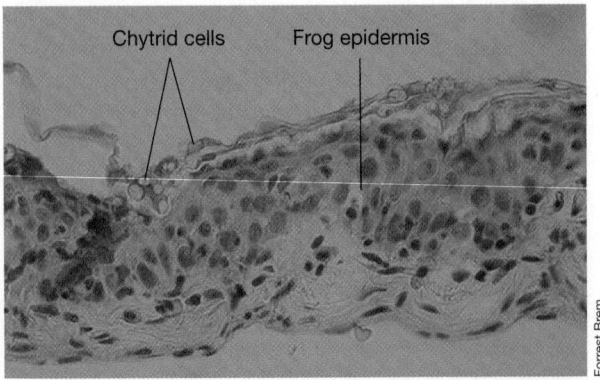

Figure 20.32 Chytridiomycetes. Cells of the chytrid *Batrachochytrium dendrobatidis* stained pink growing on the surface of the epidermis of a frog.

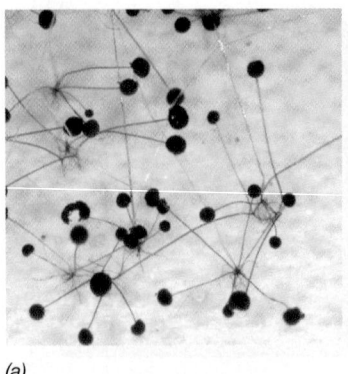
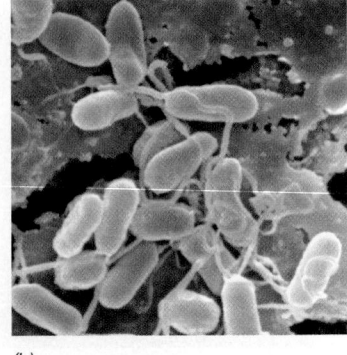

(a)   *(b)*

Figure 20.33 Zygomycetes and microsporidia. *(a)* Stained mycelium of the mold *Rhizopus* showing the spherical sporangia. *(b)* Scanning electron micrograph of cells of *Encephalitozoon intestinalis*.

gene sequencing and their lack of mitochondria, microsporidia were once thought to form an early-branching lineage of *Eukarya* (Section 20.6). However, the discovery in microsporidia of a few genes typical of *Bacteria*, the presence in some microsporidia of mitosome-like mitochondrial remnants, and the results of multigene phylogenetic analyses of microsporidia suggest that these organisms are not the phylogenetically atypical eukaryotes they were once thought to be, but instead have revealed them to be close relatives of zygomycetes.

Microsporidia have adapted to a parasitic lifestyle through the elimination or loss of many key aspects of eukaryotic biology; they are even more structurally stripped down than other amitochondriate eukaryotes. The microsporidium *Encephalitozoon* (Figure 20.33*b*), for example, lacks all organelles, including the Golgi complex and hydrogenosomes, and contains a very small genome. The genome of *Encephalitozoon* is only 2.9 Mbp and contains only about 2000 genes. This is 1.5 Mbp and 2600 genes smaller than that of the bacterium *Escherichia coli*! The *Encephalitozoon* genome lacks genes for seminal metabolic pathways, such as the citric acid cycle, meaning that this pathogen must be highly dependent on its host for even the most basic of metabolic processes.

Glomeromycetes and Arbuscular Mycorrhizae

The glomeromycetes are a relatively small and unique group of fungi in which all known species form endomycorrhizae, also called *arbuscular mycorrhizaes* (Section 20.13 and ⟳ Section 25.5), typically with the roots of herbaceous plants, but in some cases also with woody plants. As many as 80% or more of land plant species form endomycorrhizal associations in which the fungal hyphae enter the plant cell walls and produce swollen vesicles, or arbuscules. The increase in surface contact between the hyphae and the plant cell cytoplasm resulting from the arbuscule structure aids the plant's acquisition of minerals from the soil. In return, the fungus receives fixed carbon from the plant.

As plant symbionts, glomeromycetes are thought to have played an important role in the ability of early vascular plants to colonize land. So far, none of them have been grown independent of a plant. Glomeromycetes thus appear to be obligate symbionts. They reproduce only asexually and are mostly coenocytic in their morphology. Spores of *Glomus aggregatum*, a major arbuscular mycorrhizae (Figure 20.31), are collected from the roots of cultivated plants and used as an agricultural inoculant.

MiniQuiz

- What feature of zygomycetes gives this group its name?
- What is unusual about the microsporidia?
- How do arbuscular mycorrhizae aid the acquisition of nutrients by plants?

20.17 Ascomycetes

Key Genera: *Saccharomyces, Candida, Neurospora*

The ascomycetes are a large and highly diverse group of fungi. Ascomycetes range from primarily single-celled species, such as the baker's yeast *Saccharomyces* (Figure 20.30), to species that grow as filaments, such as the bread mold *Neurospora crassa*. The group ascomycetes, species of which are found in aquatic and terrestrial environments, takes its name from the production of *asci* (singular, ascus), cells in which two haploid nuclei from different mating types fuse, forming a diploid nucleus that then undergoes meiosis to form haploid ascospores. In some ascomycetes, the asci are formed within a fruiting body, called an *ascocarp*. Ascomycetes reproduce asexually by the production of conidia that form by mitosis at the tips of specialized hyphae called *conidiophores*. The ecological role of the ascomycetes is primarily that of decomposers of dead plant material, but many ascomycete species are ectomycorrhizal (Section 20.13 and ⟳ Section 25.5) with trees, and a large number are partners in lichen symbioses (Section 25.1) together with cyanobacteria or green algae. We focus here on the yeast *Saccharomyces* as a model ascomycete.

Saccharomyces cerevisiae

The cells of *Saccharomyces* and other single-celled ascomycetes are typically spherical, oval, or cylindrical, and cell division typically takes place by budding. In the budding process, a new cell forms as a small outgrowth of the old cell; the bud gradually enlarges and then separates from the parent cell (**Figure 20.34**, and see Figure 20.30). Although most ascomycete yeasts apparently reproduce only as single cells, some, including *Saccharomyces cerevisiae*, can form filaments in response to certain environmental conditions.

Yeast cells are typically much larger than bacterial cells and can be distinguished microscopically from bacteria by their larger size and by the obvious presence of internal cell structures, such as the nucleus or cytoplasmic vacuoles (Figure 20.34). Yeasts flourish in habitats where sugars are present, such as fruits, flowers, and the bark of trees. Yeasts are typically facultative aerobes, growing aerobically as well as by fermentation (⟳ Chapter 4 Microbial Sidebar). Several yeasts live symbiotically with animals, especially insects, and a few species are pathogenic for animals and humans (⟳ Section 34.8). The most important commercial yeasts are the baker's and brewer's yeasts, which are species of *Saccharomyces*. The yeast *S. cerevisiae* has been studied as a model eukaryote for many years and was the first eukaryote to have its genome completely sequenced (⟳ Section 12.5).

Sexual Reproduction in *Saccharomyces*

Saccharomyces can reproduce asexually or by sexual means in which two cells fuse. Within the fused cell, called a *zygote*, meiosis occurs and ascospores are eventually formed. The life cycle of

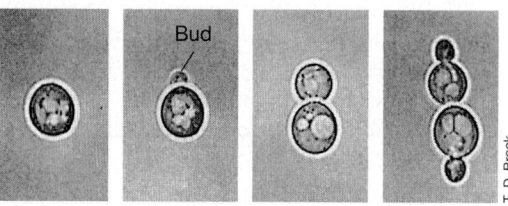

Figure 20.34 Growth by budding division in *Saccharomyces cerevisiae*. Shown is a time-lapse series of phase-contrast micrographs of the budding division process starting from a single cell. Note the pronounced nucleus. A single cell of *S. cerevisiae* is about 6 μm in diameter.

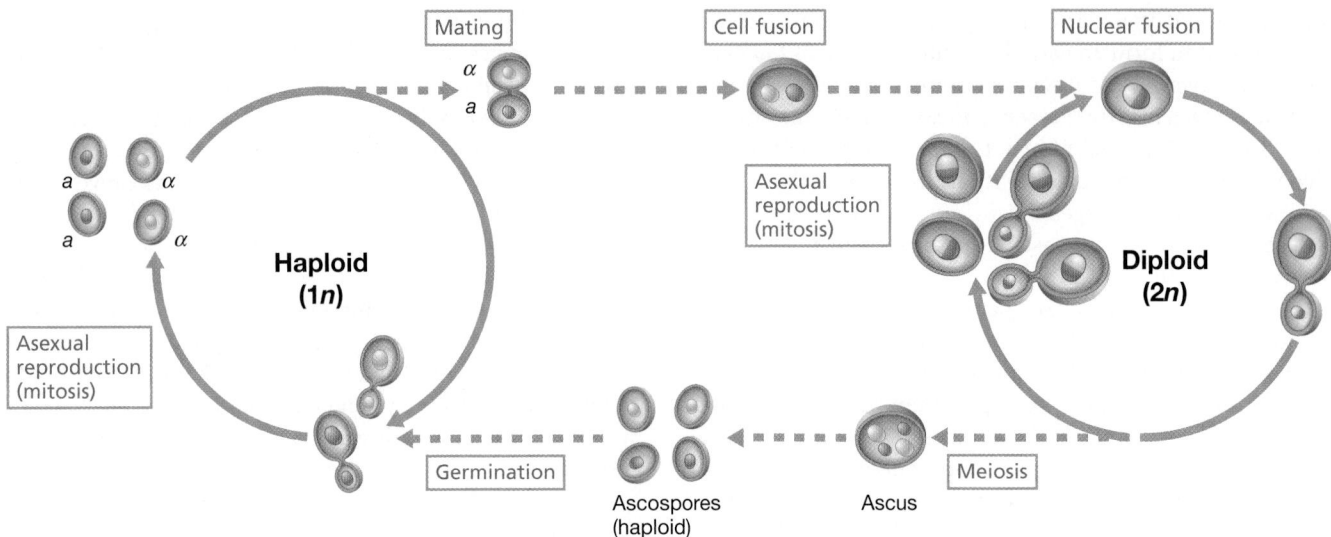

Figure 20.35 Life cycle of a typical ascomycete yeast, *Saccharomyces cerevisiae*. Cells can grow vegetatively for long periods as haploid cells or as diploid cells before life cycle events (dashed lines) generate the alternate genetic form.

the model yeast *S. cerevisiae*, including the important property of *mating types*, is described in **Figure 20.35**.

Cells of *S. cerevisiae* can grow vegetatively in either a haploid or diploid stage. *S. cerevisiae* forms two different types of haploid cells called mating types, and these can be considered analogous to male and female gametes. The mating of opposite mating types forms the diploid cell. When a diploid cell experiences nutrient-poor conditions, an ascus forms and four ascospores are produced, two of each mating type (Figure 20.35). The two mating types are designated α and *a*. Cells of type *a* mate only with cells of type α, and whether a cell is mating-type *a* or mating-type α is genetically determined, as will be discussed next.

Some haploid strains of *S. cerevisiae* remain *a* or α, but other strains are able to switch their mating type from one to the other (**Figure 20.36**). This switch in mating type occurs when the active mating gene is replaced with one of two otherwise silent genes. There is a single location on one of the *S. cerevisiae* chromosomes called the *MAT* (for *mating type*) locus, at which either gene *a* or gene α can be inserted. At this locus, the *MAT* promoter controls transcription of whichever gene is present. If gene *a* is at that locus, then the cell is mating-type *a*, whereas if gene α is at that locus, the cell is mating-type α. Elsewhere in the yeast genome are copies of genes *a* and α that are not expressed. These silent copies are the source of the inserted gene. In the switch (Figure 20.36), the appropriate gene, *a* or α, is copied from its silent site and inserted into the *MAT* location, replacing the gene already present. The old mating-type gene is excised and discarded, and the new gene is inserted, and whichever gene is inserted in the *MAT* locus is the one that will be transcribed.

The α and *a* genes of *S. cerevisiae* are regulatory genes. Among other things, they regulate the production of the peptide hormones α factor or *a* factor, which are excreted by yeast cells undergoing mating. These hormones bind to cells of the opposite mating type and bring about changes in their cell surfaces that enable the cells to fuse (**Figure 20.37**). Once mating has occurred, the nuclei fuse, forming a diploid zygote. The zygote grows vegetatively by budding (Figure 20.37*b*), but under starvation conditions will undergo meiosis and generate ascospores once again.

MiniQuiz

- In what ways does reproduction by budding differ from vegetative growth of hyphae?
- Explain how a *single* haploid cell of *Saccharomyces* can eventually yield a diploid cell.

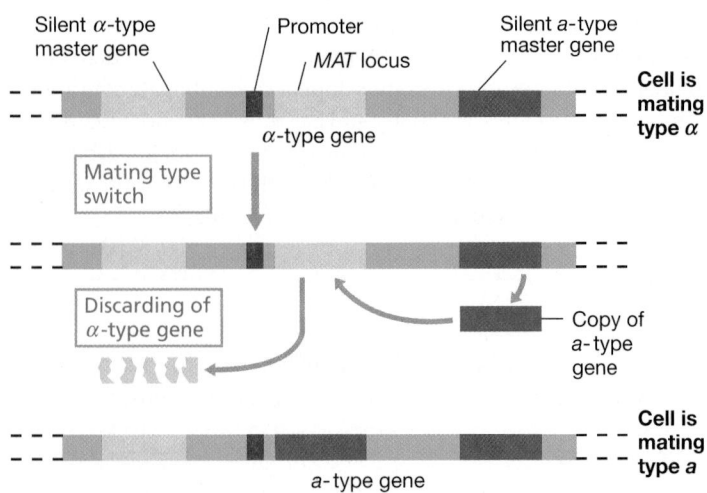

Figure 20.36 The cassette mechanism that switches an ascomycete yeast from mating type α to *a*. The cassette inserted at the *MAT* locus determines the mating type. The process shown is reversible, so type *a* can also revert to type α.

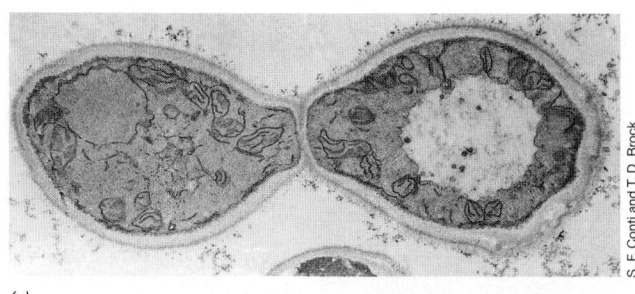

(a)

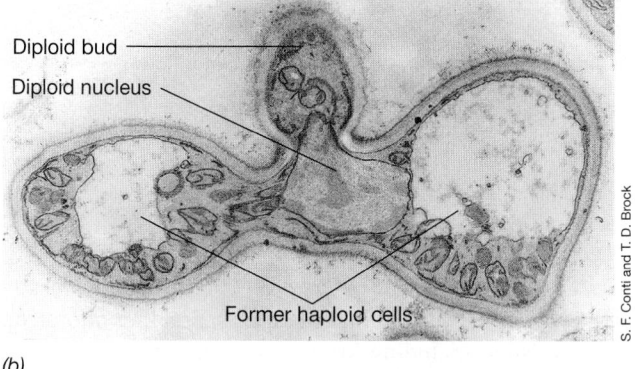

Diploid bud
Diploid nucleus

Former haploid cells

(b)

Figure 20.37 Electron micrographs of mating in the ascomycete yeast *Hansenula wingei*. *(a)* Two cells have fused at the point of contact. *(b)* Late stage of mating. The nuclei of the two cells have fused, and a diploid bud has formed at a right angle to the mating cells. This bud becomes the progenitor of a diploid cell line. A cell of *Hansenula* is about 10 μm in diameter.

20.18 Basidiomycetes and the Mushroom Life Cycle

Key Genera: *Agaricus, Amanita*

Basidiomycetes are a large group of fungi, with over 30,000 species described. Many are the commonly recognized mushrooms and toadstools, some of which are edible, such as the commercially grown mushroom *Agaricus* (⟳ Section 36.3), and others of which, such as *Amanita* (**Figure 20.38a**), are highly poisonous. Other basidiomycetes include puffballs, smuts, rusts, and an important human fungal pathogen, *Cryptococcus* (⟳ Section 34.8). The unifying characteristic of the basidiomycetes is the *basidium* (plural, basidia), a structure in which basidiospores are formed by meiosis. The basidium ("little pedestal") gives the group its name.

During most of its existence, a mushroom fungus lives as a simple haploid mycelium, growing vegetatively in soil, leaf litter, or decaying logs. It is the sexual reproductive phase of basidiomycetes that produces the visible mushroom structure (Figures 20.29 and 20.38). In this process, mycelia of different mating types (Section 20.17) fuse, and the faster growth of the dikaryotic (two nuclei per cell) mycelium formed from that fusion overgrows and crowds out the parental haploid mycelia. Then, when environmental conditions are favorable, usually following periods of wet and cool weather, the dikaryotic mycelium develops rapidly into the fruiting body.

(a)   *(b)*

Figure 20.38 Mushrooms. *(a)* *Amanita*, a highly poisonous mushroom. *(b)* Gills on the underside of the mushroom fruiting body contain the spore-bearing basidia.

A mushroom fruiting body, called a *basidiocarp*, begins as a mycelium that differentiates into a small button-shaped structure underground that then expands into the full-grown basidiocarp that we see aboveground, the mushroom (Figures 20.29 and 20.38). The dikaryotic basidia are borne on the underside of the basidiocarp on flat plates called gills, which are attached to the cap of the mushroom (Figure 20.38*b*). The basidia then undergo a fusion of the two nuclei, forming basidia with diploid nuclei. The two rounds of meiotic division generate four haploid nuclei in the basidia, and each of the nuclei becomes a basidiospore. The genetically different basidiospores can then be dispersed by wind to new habitats to begin the cycle again, germinating under favorable conditions and growing as haploid mycelia (Figure 20.29).

MiniQuiz
- Is the basidiocarp genetically haploid or diploid?

 ## Red and Green Algae

We conclude our tour of eukaryotic microbial diversity with the **algae**. As we have discussed, phototrophic eukaryotes originated from a primary endosymbiosis event, the engulfment and retention by a eukaryotic, heterotrophic cell of a cyanobacterium (Section 20.4 and Figure 20.12; ⟳ Section 16.4 and Figure 16.9). This early phototrophic protist diverged into the ancestors of the red and the green algae, and the lineage leading to the green algae later gave rise to the ancestor of plants. However, when various secondary endosymbioses occurred, in which cells of red and green algae were engulfed by other eukaryotic protists and their chloroplasts retained, a further diversification of phototrophic eukaryotes took place. Here we focus on the red and green algae, a large and diverse group of eukaryotic organisms that contain chlorophyll and carry out oxygenic photosynthesis.

20.19 Red Algae

Key Genus: *Cyanidium, Galdiera*

The red algae, also called *rhodophytes*, mainly inhabit the marine environment, but a few species are found in freshwater and terrestrial habitats. Both unicellular and multicellular (some macroscopic) species are known.

Basic Properties

Red algae are phototrophic and contain chlorophyll *a*; they are noteworthy among the algae in that their chloroplasts lack chlorophyll *b*, but contain phycobiliproteins, the major light-harvesting pigments of the cyanobacteria (⟲ Section 13.3). The reddish color (**Figure 20.39**) of many red algae results from phycoerythrin, an accessory pigment that masks the green color of chlorophyll. This pigment is present along with phycocyanin and allophycocyanin in structures called phycobilisomes, the light-harvesting (antenna) structure of cyanobacteria (⟲ Section 13.3). At greater depths in aquatic habitats, where less light penetrates, cells produce more phycoerythrin and are a darker red, whereas shallow-dwelling species often have less phycoerythrin and can be green in color (see Figure 20.40).

Most species of red algae are multicellular and lack flagella. Some are considered seaweeds and are the source of agar, the solidifying agent used in bacteriological media, (⟲ Microbial Sidebar Chapter 1) and carrageenans, thickening and stabilizing agents used throughout the food industry. Different species of red algae are filamentous, leafy, or, if they deposit calcium carbonate, *coralline* (coral-like) in morphology. Some coralline species play a role in the development of coral reefs (⟲ Section 25.14).

Cyanidium and Relatives

Unicellular species of red algae are also known. One such group, members of the *Cyanidiales* that includes the genera *Cyanidium*, *Cyanidioschyzon*, and *Galdieria* (**Figure 20.40**), live in acidic hot

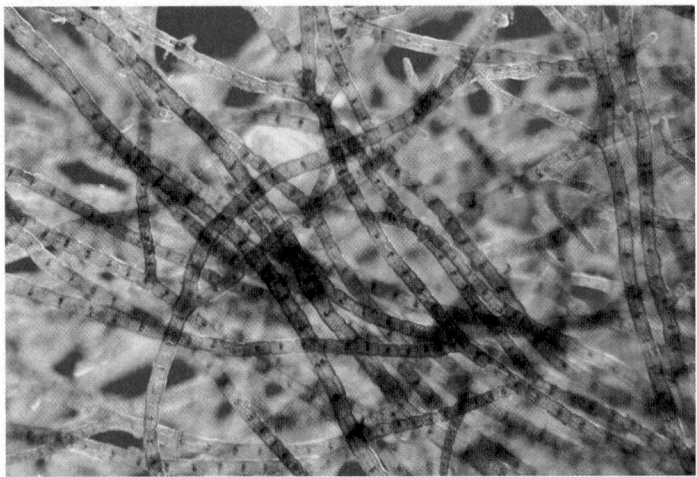

Figure 20.39 *Polysiphonia*, a marine red alga. Light micrograph. *Polysiphonia* grows attached to the surfaces of marine plants. Cells are about 150 μm wide.

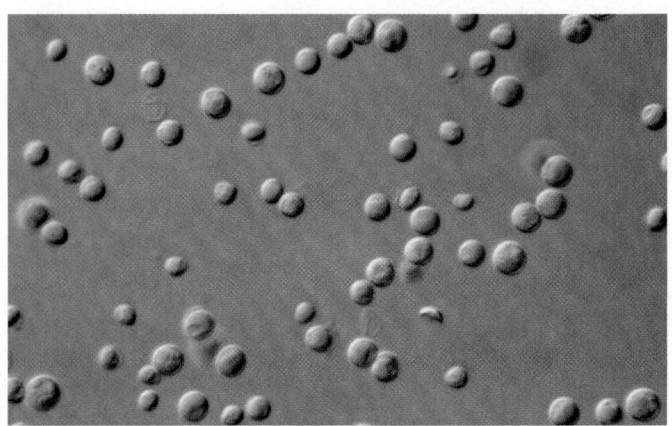

Figure 20.40 *Galdieria*, a red alga. This alga grows at low pH and high temperature in hot springs. The cells are about 25 μm in diameter and are green because *Galdieria* contains only low levels of phycoerythrin, a reddish accessory pigment.

springs at temperatures from 30 to 60°C and at pH values from 0.5 to 4.0; under these extreme conditions, no other phototrophic microorganisms, including anoxygenic phototrophs, can exist. The unicellular red algae are unusual in other ways as well. For example, cells of *Cyanidioschyzon merolae* are unusually small (1–2 μm in diameter) for eukaryotes, and the genome of this species, approximately 16.5 Mbp, is one of the smallest genomes known for a phototrophic eukaryote.

MiniQuiz

• What traits link cyanobacteria and red algae?

• What physiological properties would be necessary for *Galdieria* to live in its habitat?

20.20 Green Algae

Key Genera: *Chlamydomonas, Volvox*

The green algae, also called *chlorophytes*, bear chloroplasts containing chlorophylls *a* and *b*, which give them their characteristic green color, but lack phycobiliproteins. In the composition of their photosynthetic pigments, they are similar to plants and are closely related to plants phylogenetically. There are two main groups of green algae, the chlorophytes, examples of which are the unicellular *Chlamydomonas* and *Dunaliella* (**Figure 20.41a**), and the charophyceans, the algal group that is actually most closely related to land plants.

Most green algae inhabit freshwater while others are found in moist soil or growing in snow, to which they impart a pink color (⟲ Figure 5.21). Other green algae live as symbionts in lichens (⟲ Section 25.1). The morphology of chlorophytes ranges from unicellular (Figure 20.41a, c) to filamentous, with individual cells arranged end-to-end (Figure 20.41d) to colonial, as aggregates of cells (Figure 20.41b, e). Even multicellular species exist, an example of which is the seaweed *Ulva*. Most green algae have a complex life cycle, with both sexual and asexual reproductive stages.

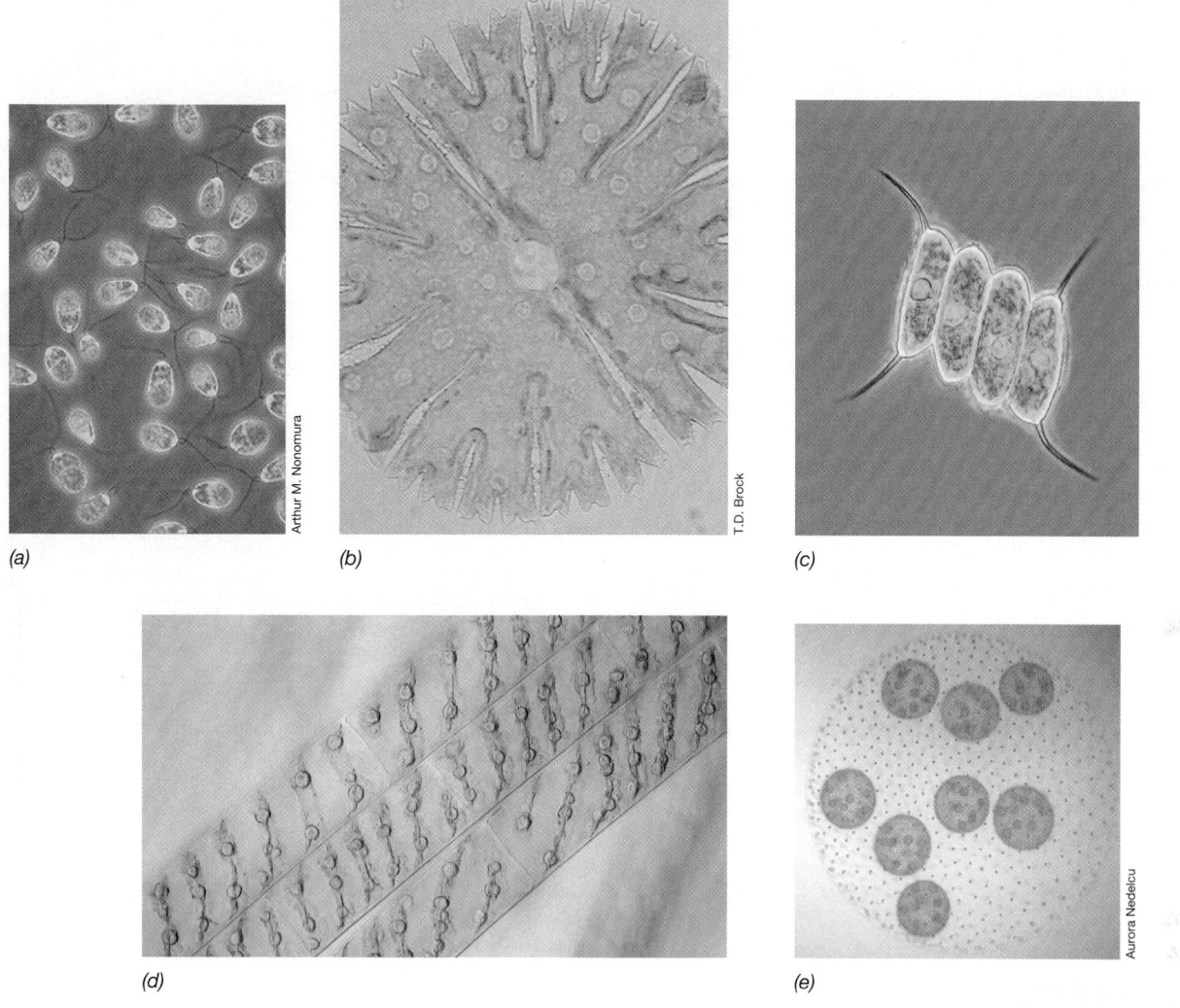

(a)   (b)   (c)

(d)   (e)

Figure 20.41 Light micrographs of repre-sentative green algae. *(a)* A single-celled, flag-ellated green alga, *Dunaliella*. A cell is about 5 μm wide. *(b) Micrasterias*. This single multilobed cell is about 100 μm wide. *(c) Scenedesmus*, showing packets of four cells each. *(d) Spirogyra*, a filamentous alga with cells about 20 μm wide. Note the green spiral-shaped chloroplasts. *(e) Volvox carteri* colony with eight daughter colonies. Daughter colonies are about 50 μm wide. The colony is made up of several hundred flagellated somatic cells (the small dots on the surface of the colony and on the surfaces of the smaller green daughter colonies). The daughter colonies contain reproductive cells called gonidia (the dark green structures within the daughter colonies).

Very Small Green Algae and Colonial Green Algae

One of the smallest eukaryotes known is the green alga *Ostreococcus tauri*, a common unicellular member of marine phytoplankton (♂♀ Section 23.9 and Figure 23.19*b*). Cells of *O. tauri* have a diameter of approximately 2 μm, and the organism contains the smallest genome of any known phototrophic eukaryote, approximately 12.6 Mbp. *Ostreococcus* thus provides a model organism for research into the evolution of genome reduction and specialization in eukaryotes.

At the **colonial** level of organization in green algae is *Volvox* (Figure 20.41*e*). This alga forms colonies composed of several hundred flagellated cells, some of which are motile and primarily carry out photosynthesis, while others specialize in repro-duction. Cells in a *Volvox* colony are interconnected by thin strands of cytoplasm that allow the entire colony to swim in a coordinated fashion. *Volvox* has been a long-term model for research on the genetic mechanisms controlling multicellularity and the distribution of functions among cells in multicellular organisms.

Endolithic Phototrophs

Some green algae grow inside rocks. These *endolithic* (*endo* means "inside") phototrophs inhabit porous rocks, such as those containing quartz, and are typically found in layers near the rock surface. Endolithic phototrophic communities are most common in dry environments such as deserts or cold dry environments such as the Antarctic. For example, in the Antarctic Dry Valleys, where temperatures and humidity are extremely low, life within a rock has its advantages. Rocks are heated by the sun, and water from snow melt can be absorbed and retained for relatively long periods, supplying moisture needed for growth. Moreover, water absorbed by a porous rock makes the rock more transparent, thus funneling more light to the algal layers.

A wide variety of phototrophs can form endolithic communities, including cyanobacteria and various green algae (**Figure 20.42**). In addition to being free-living phototrophs, green algae and cyanobacteria coexist with fungi in endolithic lichen communities (Section 25.1 for discussion of the lichen symbiosis). Metabolism and growth of these internal rock communities slowly weathers the rock, allowing gaps to develop where water can enter, freeze and thaw, and eventually crack the rock, producing new habitats for microbial colonization. The decomposing rock also forms a crude soil that can support development of plant and animal communities in environments where conditions (temperature, moisture, and so on) permit.

MiniQuiz

- What phototrophic properties link green algae and plants?
- What is unusual about the green algae *Ostreococcus* and *Volvox*?

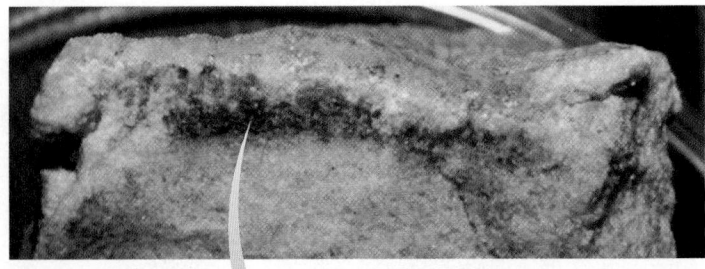

(a)

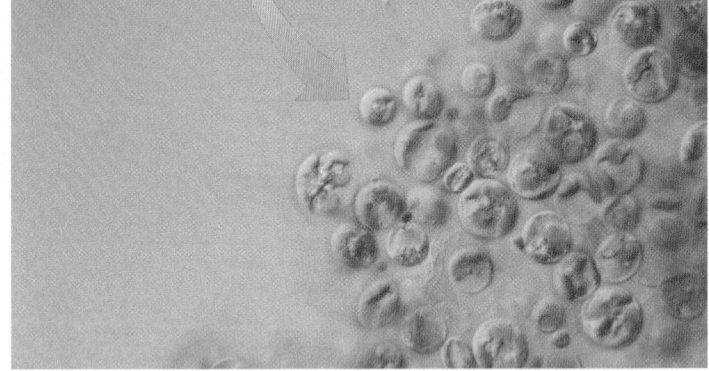

(b)

Figure 20.42 Endolithic phototrophs. *(a)* Photograph of a limestone rock from the Dry Valleys region of Antarctica broken open to show the layer of endolithic green algae. *(b)* Light micrograph of cells of the green alga *Trebouxia*, a widespread endolithic alga in Antarctica.

Big Ideas

20.1
The nucleus contains the cell's chromosomes and is surrounded by two membranes. The nucleolus is the site of ribosomal RNA synthesis.

20.2
The mitochondrion and hydrogenosome are energy-generating organelles of eukaryotic cells. Mitochondria carry out aerobic respiration, whereas hydrogenosomes ferment pyruvate to yield H_2, CO_2, acetate, and ATP.

20.3
Chloroplasts are the site of photosynthetic energy production and CO_2 fixation in eukaryotic phototrophs.

20.4
Key metabolic organelles of eukaryotes are the chloroplast, which functions in photosynthesis, and the mitochondrion or hydrogenosome, which function in respiration or fermentation. These organelles were originally *Bacteria* that established permanent residence inside other cells (endosymbiosis).

20.5
Endoplasmic reticula are membranous structures in eukaryotic cells that either contain ribosomes (rough) and are sites of protein synthesis, or do not (smooth) and are sites of lipid synthesis. Flagella and cilia are means of motility, while lysosomes are organelles that specialize in macromolecular hydrolyses. Structural proteins such as microtubules and microfilaments function as internal cell scaffolding.

20.6
SSU rRNA gene sequences do not seem to yield as reliable a phylogenetic tree of the *Eukarya* as do other genes and proteins. The modern, multigene tree of eukaryotes shows a major radiation of eukaryotic diversity emerging at some time following symbiotic events that led to the mitochondrion.

20.7
Diplomonads are unicellular, flagellated, nonphototrophic protists. Parabasalids such as *Trichomonas* are human pathogens and contain huge genomes that lack introns.

20.8

Euglenozoans are unicellular, flagellated protists. Some are phototrophic. This group includes some important human pathogens, such as *Trypanosoma*, and some well-studied non-pathogens, such as *Euglena*.

20.9

Three groups make up the alveolates: ciliates, dinoflagellates, and apicomplexans. Ciliates and dinoflagellates are free-living organisms, whereas apicomplexans are obligate parasites of animals.

20.10

Stramenopiles are protists that bear a flagellum with fine, hairlike extensions. They include oomycetes, diatoms, and golden algae.

20.11

Cercozoans and radiolarians are two related groups of protists. The cercozoans include the phototrophic chlorarachniophytes and foraminiferans, whereas the radiolarians are chemoorganotrophs.

20.12

Amoebozoa are protists that use pseudopodia for movement and feeding. Within amoebozoa are gymnamoebas, entamoebas, and slime molds. Plasmodial slime molds form masses of motile protoplasm, whereas cellular slime molds are individual cells that aggregate to form fruiting bodies from which spores are released.

20.13

Fungi include the molds, mushrooms, and yeasts. Other than phylogeny, fungi primarily differ from protists by their rigid cell wall, production of spores, and lack of motility.

20.14

A variety of sexual spores are produced by fungi, including ascospores, basidiospores, and zygospores. From a phylogenetic standpoint, fungi are the closest relatives of animals, and chytridiomycetes are the earliest lineage of fungi.

20.15

Chytrids are primarily aquatic fungi and are thought to be the most ancient of fungi, lying basal to all other known fungal groups on the 18S rRNA gene tree. Some chytrids are amphibian pathogens.

20.16

Zygomycetes form coenocytic hyphae and undergo both asexual and sexual reproduction, and the common bread mold *Rhizopus* is a good example. Microsporidia, once thought to be an early lineage of *Eukarya*, are closely related to the zygomycetes. Glomeromycetes are fungi that form arbuscular mycorrhizal associations with plants.

20.17

The ascomycetes are a large and diverse group of saprophytic fungi. Some, such as *Candida albicans*, can be pathogenic in humans. There are two mating types in the yeast *Saccharomyces cerevisiae*, and yeast cells can convert from one type to the other by a genetic switch mechanism.

20.18

Basidiomycetes include the mushrooms, puffballs, smuts, and rusts. Basidiomycetes undergo both vegetative reproduction as haploid mycelia and sexual reproduction via fusion of mating types and formation of haploid basidiospores.

20.19

Red algae are mostly marine and range from unicellular to multicellular. Their reddish color is due to the pigment phycoerythrin, a key cyanobacterial pigment, present in their chloroplast.

20.20

Green algae are common in aquatic environments and can be unicellular, filamentous, colonial, or multicellular. A unicellular green alga, *Ostreococcus*, has the smallest genome known for a phototrophic eukaryote, while the green alga *Volvox* is a model multicellular phototroph.

Review of Key Terms

Algae phototrophic eukaryotes, both microorganisms and macroorganisms

Calvin cycle the series of biosynthetic reactions by which most photosynthetic organisms convert CO_2 to organic compounds

Chitin a polymer of *N*-acetylglucosamine commonly found in the cell walls of fungi

Chloroplast the photosynthetic organelle of eukaryotic phototrophs

Ciliate any protist characterized in part by rapid motility driven by numerous short appendages called cilia

Coenocytic the presence of multiple nuclei in fungal hyphae without septa

Colonial the growth form of certain protists and green algae in which several cells live together and cooperate for feeding, motility, or reproduction; an early form of multicellularity

Conidia the asexual spores of fungi

Cristae the internal membranes of a mitochondrion

Cytoskeleton the cellular scaffolding typical of eukaryotic cells in which microtubules, microfilaments, and intermediate filaments define the cell's shape

Endosymbiotic hypothesis the idea that a chemoorganotrophic bacterium and a cyanobacterium were stably incorporated into another cell type to give rise, respectively, to the mitochondria and chloroplasts of modern-day eukaryotes

Eukarya eukaryotic organisms

Eukaryote a cell or organism having a unit membrane–enclosed nucleus and usually other organelles; a member of *Eukarya*

Extracellular matrix (ECM) proteins and polysaccharides that surround an animal cell and in which the cell is embedded

Fungi nonphototrophic eukaryotic microorganisms with rigid cell walls

Histones positively charged proteins that package eukaryotic DNA in nucleosomes

Hydrogenosome an organelle of endosymbiotic origin present in certain anaerobic eukaryotic microorganisms that functions to oxidize pyruvate to H_2, CO_2, and acetate along with the production of one molecule of ATP

Intermediate filament a filamentous polymer of fibrous keratin proteins, supercoiled into thicker fibers, that functions in maintaining cell shape and the positioning of certain organelles in the eukaryotic cell

Lysosome an organelle containing digestive enzymes for hydrolysis of proteins, fats, and polysaccharides

Microfilament a filamentous polymer of the protein actin that helps maintain the shape of a eukaryotic cell

Microtubule a filamentous polymer of the proteins α-tubulin and β-tubulin that functions in eukaryotic cell shape and motility

Mitochondrion the respiratory organelle of eukaryotic organisms

Mushroom the aboveground fruiting body, or basidiocarp, of basidiomycete fungi

Nucleus the organelle that contains the eukaryotic cell's chromosomes

Peroxisome an organelle that functions to rid the cell of toxic substances such as peroxides, alcohols, and fatty acids

Phagocytosis a mechanism for ingesting particulate material in which a portion of the cytoplasmic membrane surrounds the particle and brings it into the cell

Protist a unicellular eukaryotic microorganism; may be flagellate or aflagellate, phototrophic

or nonphototrophic, and most lack cell walls; includes algae and protozoa

Secondary endosymbiosis the acquisition by a mitochondrion-containing eukaryotic cell of a red or green algal cell

Slime mold a nonphototrophic protist that lacks cell walls and that aggregates to form fruiting structures (cellular slime molds) or masses of protoplasm (acellular slime molds)

Stroma the lumen of the chloroplast, surrounded by the inner membrane

Thylakoid a membrane layer containing the photosynthetic pigments in chloroplasts

Yeast the single-celled growth form of various fungi

Review Questions

1. List at least three features of eukaryotic cells that clearly differentiate them from prokaryotic cells (Section 20.1).

2. How are the mitochondrion and the hydrogenosome similar structurally? How do they differ? How do they differ metabolically (Section 20.2)?

3. What major physiological processes dealing with energy and carbon occur in the chloroplast (Section 20.3)?

4. Discuss some of the evidence that supports the endosymbiotic hypothesis (Section 20.4).

5. What are the functions of the following eukaryotic cell structures: endoplasmic reticulum, Golgi complex, lysosome, and peroxisome (Section 20.5)?

6. Distinguish between the functions of microtubules, microfilaments, and intermediate filaments (Section 20.5).

7. In what ways do diplomonads and parabasalids differ from each other (Section 20.7)?

8. Discuss the morphological differences between diplomonads, parabasalids, and euglenozoans (Sections 20.7 and 20.8).

9. What organism causes "red tides" and why is this organism toxic (Section 20.9)?

10. In what ways are oomycetes similar to and different from fungi (Section 20.10)?

11. What morphological trait links cercozoans and radiolarians and distinguishes them from other protists (Section 20.11)?

12. Why were slime molds once grouped with fungi (Section 20.12)?

13. Describe the nutritional requirements and physiology of fungi (Section 20.13).

14. What was the common ancestor of fungi and animals likely to have been like (Section 20.14)?

15. In what way do chytrids differ from other fungi (Section 20.15)?

16. Is the formation of zygospores in zygomycetes a sexual or an asexual process (Section 20.16)?

17. How is the mating type of a yeast cell determined (Section 20.17)?

18. What morphological feature unites the basidiomycetes, and where is this feature found (Section 20.18)?

19. In what kinds of habitats would one likely find red algae (Section 20.19)?

20. What traits link green algae and plants (Section 20.20)?

Application Questions

1. If the mitochondrion and chloroplast had originally been free-living eukaryotic cells, how would their molecular properties listed in Section 20.4 differ?

2. Explain why the process of endosymbiosis can be viewed as both an ancient event and a more recent event.

3. What does the presence of the apicoplast in apicomplexans reveal about their evolutionary history?

Viral Diversity

Rhabdoviruses, a family of single-stranded RNA viruses that are important pathogens of plants and animals, and which include the rabies virus, form characteristically bullet-shaped virions.

The previous chapters in this unit have explored the enormous diversity of microbial cells: *Bacteria, Archaea,* and *Eukarya.* Here we focus on the diversity of viruses and their genomes. This chapter extends Chapter 9, which introduced the principles of virology.

All cells contain double-stranded DNA as their genetic material. In contrast, viruses are known that have single-stranded RNA, double-stranded RNA, single-stranded DNA, or double-stranded DNA as their genetic material (↩ Section 9.1). This makes for interesting schemes of replication and gene expression. We illustrate viral diversity by exploring the replication processes of viruses in each of these groups. We separate our discussion by the type of host cell infected, because the differences in cell structure between the three domains of life place major constraints on the life cycles of the viruses that infect them.

I Viruses of *Bacteria* and *Archaea*

Many viruses are known that infect *Bacteria.* As mentioned in Chapter 9, most known bacterial viruses, or **bacteriophages** as they are called, have double-stranded DNA genomes (↩ Section 9.8). Even so, many bacteriophages have other types of genomes. **Table 21.1** lists several well-characterized bacteriophages that infect *Escherichia coli.* The simplest are those with RNA genomes, and we begin with these.

21.1 RNA Bacteriophages

Many bacteriophages contain RNA genomes of the plus (+) configuration. In such viruses the viral genome—**positive-strand RNA**—and the mRNA are of the same complementarity (↩ Section 9.7). RNA viruses of the enteric bacteria infect only bacterial cells that contain a type of plasmid, called a conjugative plasmid, that allows the bacterial cell to function as a donor in conjugation (↩ Section 10.9). This is because these RNA viruses infect bacteria by first attaching to pili (**Figure 21.1**) that are encoded by the plasmid and are present only on the donor cells.

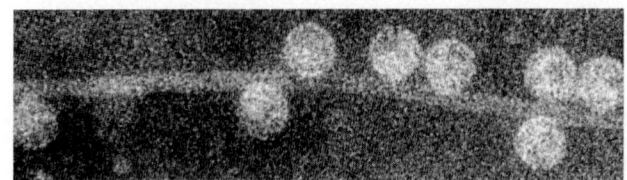

Figure 21.1 Small RNA bacteriophages. Electron micrograph of the pilus of a donor bacterial cell of *Escherichia coli* showing virions of a small RNA phage attached to the pilus.

Phage MS2

The bacterial RNA viruses are all quite small, about 25 nm in diameter, and they are all icosahedral (↩ Section 9.2) with 180 molecules of coat protein per virion. The complete nucleotide sequences of several RNA phage genomes are known. For example, the genome of the RNA phage MS2, which infects *Escherichia coli,* is 3569 nucleotides long.

The genetic map of phage MS2 is shown in **Figure 21.2a**, and the flow of events of MS2 replication is shown in Figure 21.2b. The small MS2 genome encodes only four proteins. These are the maturation protein (present in the mature virion as a single copy), coat protein, lysis protein (required for lysis of bacteria to release mature virions), and a subunit of **RNA replicase**, the enzyme that replicates the viral RNA. RNA replicase is a composite protein, consisting of one virus-encoded polypeptide and several host polypeptides. The maturation protein acts as a protease to process some viral proteins necessary for producing infective virions.

Because the genome of phage MS2 is positive-strand RNA, it is translated directly upon entry into the cell. After RNA replicase is made, it synthesizes negative-strand RNA using the genomic RNA as template (↩ Figure 9.11). After negative-strand RNA has been synthesized, more positive-strand RNA is made using the negative-strand RNA as a template. The newly made positive strands are translated for continued synthesis of the viral proteins.

Table 21.1 *Some bacteriophages of* Escherichia coli

Bacteriophage	Virion structure	DNA or RNA	Single- or Double-Stranded	Structure of genome	Size of genome[a]
MS2	Icosahedral	RNA	Single-stranded	Linear	3,569
φX174	Icosahedral	DNA	Single-stranded	Circular	5,386
M13, f1, and fd	Filamentous	DNA	Single-stranded	Circular	6,408
Lambda	Head & tail	DNA	Double-stranded	Linear	48,514[b]
T7 and T3	Head & tail	DNA	Double-stranded	Linear	39,936
T4	Head & tail	DNA	Double-stranded	Linear	168,903
Mu	Head & tail	DNA	Double-stranded	Linear	39,000[c]

[a]The sizes of some viral genomes are known accurately because they have been sequenced. However, the sequence and exact number of bases for related isolates may be slightly different. The size is in bases or base pairs depending on whether the virus is single- or double-stranded, respectively.
[b]This includes single-stranded extensions of 12 nucleotides at either end of the linear DNA.
[c]This includes 1800 bp of host DNA that are attached to the ends of Mu DNA in the virion.

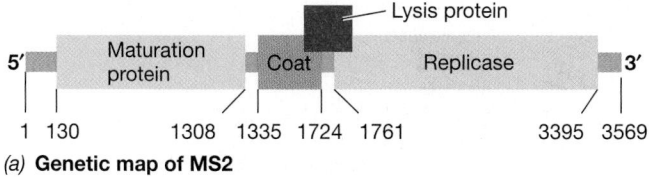

(a) **Genetic map of MS2**

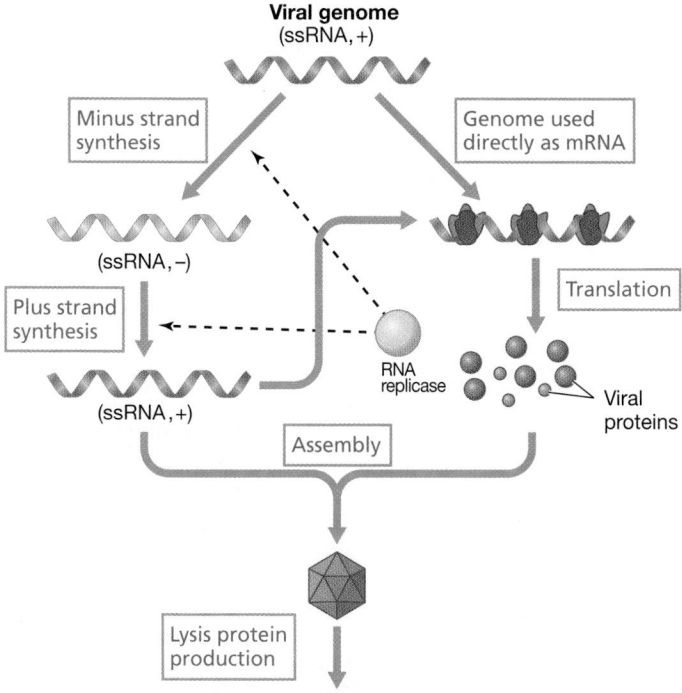

(b) **Flow of events during viral multiplication**

Figure 21.2 Bacteriophage MS2. *(a)* Note on the genetic map how the lysis protein gene overlaps with both the coat protein and replicase genes. The numbers refer to the nucleotide positions on the RNA. *(b)* Flow of events during multiplication.

Phage MS2 RNA is folded into a complex form with extensive secondary structure. Of the four AUG translational start sites on the mRNA, the most accessible to the translation machinery is that for the coat protein, and translation begins there very early in virus infection. The replicase mRNA is also translated early. As coat protein molecules increase in number in the cell, they bind to the RNA around the AUG start site for the replicase protein, effectively turning off synthesis of replicase. Although the gene for the maturation protein is at the 5′ end of the RNA, the extensive folding of the RNA limits access to the maturation protein start site. Consequently, only a few copies are synthesized. Thus, due to control of translation by RNA folding, the virus proteins are made in the relative amounts needed for virus assembly. In particular, the major virus protein synthesized is coat protein, which is needed in the largest amounts.

Overlapping Genes and Assembly of MS2

An interesting feature of bacteriophage MS2 is that the lysis protein is encoded by a gene that overlaps both the coat protein gene and the replicase gene (Figure 21.2*a*). This phenomenon of

overlapping genes is quite common in very small viral genomes (Section 21.2) and makes small genomes more efficient. The start codon of the MS2 lysis gene is not easily accessible to ribosomes because of the secondary structure of the RNA. However, when the ribosome terminates synthesis of the coat protein gene, the secondary structure in this region of the RNA is disrupted, and sometimes this disruption allows a ribosome to begin reading the lysis gene. This low level of translation prevents premature lysis of the cell. Only after sufficient copies of coat protein are available for the assembly of mature virions does cell lysis commence.

Ultimately, self-assembly of MS2 virions takes place, and virions are released from the cell as a result of cell lysis. The features of replication of small RNA viruses such as MS2 are thus fairly simple. The viral RNA itself functions as an mRNA, and regulation is based on controlling access of ribosomes to the appropriate translation start sites on the viral RNA.

Although several positive-strand RNA bacteriophages are known that are similar in their biology to that of MS2, no bacteriophage has yet been discovered that contains negative-strand RNA, a type of genome that is fairly common among eukaryotic viruses (Section 21.9). However, a few bacteriophages are known that have segmented, double-stranded RNA as their genetic material. The best studied of these is φ6, whose host is *Pseudomonas syringae*. This virus, which is **enveloped** (enclosed by a lipid membrane; *⟳* Figure 9.12), seems very closely related to the reoviruses (Section 21.10), which infect eukaryotes.

MiniQuiz

- What is the difference between the genomes of positive-strand RNA viruses and negative-strand RNA viruses?
- Describe what is meant by the term *overlapping genes*.

21.2 Single-Stranded DNA Bacteriophages

Some bacteriophages contain a positive single-stranded DNA (ssDNA) genome (there are no known negative-strand DNA bacteriophages). Before such a genome can be transcribed, a complementary strand of DNA must be synthesized, forming a double-stranded molecule, the **replicative form (RF)** (*⟳* Section 9.7 and Figure 9.11). Although these bacteriophages replicate this double-stranded DNA, they only package the positive strand of DNA into their virions. In this section we discuss two well-known single-stranded DNA phages that infect *Escherichia coli*, the icosahedral phage φX174 and the filamentous phage M13.

The Genome of Phage φX174

Bacteriophage φX174 contains a circular ssDNA genome inside an icosahedral virion. The φX174 genome consists of 5386 nucleotides and was the first DNA molecule to be completely sequenced, a remarkable achievement when it was accomplished by Frederick Sanger and colleagues in 1977. This virus is very small, about 25 nm in diameter, and the principal building block of the coat is a single protein present in 60 copies, the minimum number of protein subunits possible in an icosahedral virus. Attached at the vertices of the icosahedron are several other proteins that make up spike-like structures (Figure 9.12). These small DNA viruses possess only a few genes, and the host cell DNA replication machinery is used exclusively in the replication of viral DNA.

Phage φX174 is also of note because it was the first genetic element shown to have *overlapping genes,* a condition already described for the RNA phage MS2 (Section 21.1). In very small viruses such as φX174, there is insufficient DNA to encode all viral-specific proteins unless parts of the genome are read more than once in different reading frames. For example, in the φX174 genome, gene B resides within gene A and gene K resides within both genes A and C (**Figure 21.3**). Genes D and E also overlap, gene E being contained completely within gene D. Also, the termination codon of gene D overlaps the initiation codon of gene J (Figure 21.3*a*). In addition to overlapping genes, a small protein in φX174 called A* protein that functions in shutting down host DNA synthesis is synthesized by the reinitiation of *translation* (not transcription) within the mRNA for gene A. The A* protein is read from the same mRNA reading frame as A protein but has a different in-frame start codon and is thus a shorter protein.

DNA Replication by the Rolling Circle Mechanism

Several plasmids and viruses replicate by a mechanism different from the semiconservative replication of cellular DNA (Section 6.9). This alternative mechanism, **rolling circle replication**, is used by some viruses with ssDNA genomes, including φX174. Upon infection, their positive-strand viral DNA is separated from the protein coat. Entry into the cell is accompanied by the conversion of this ssDNA into the double-stranded replicative form (Figure 21.3*b*). Cell-encoded enzymes that help convert viral DNA to RF DNA include primase, DNA polymerase, ligase, and gyrase (Section 6.10).

In cells, replication of the lagging strand requires the synthesis of short RNA primers by primase (Section 6.8). These RNA primers are made at intervals along the lagging strand and are later removed and replaced with DNA by the enzyme DNA polymerase I. The situation with φX174 DNA is similar, except that the DNA in this case is a single-stranded closed circle. To begin replication of this DNA, primase synthesizes a short RNA primer at one or more specific initiation sites. DNA is then synthesized by DNA polymerase III, and the primer is removed and replaced by DNA using DNA polymerase I, exactly as in the case of lagging strand synthesis. This results in the formation of the complete, circular, double-stranded replicative form. Once the replicative form is completed, copies are made by conventional semiconservative replication with theta-form intermediates.

In the formation of single-stranded viral genomes, the rolling circle arises because the positive strand of the replicative form is cut and the 3′ end of the exposed DNA is used to prime

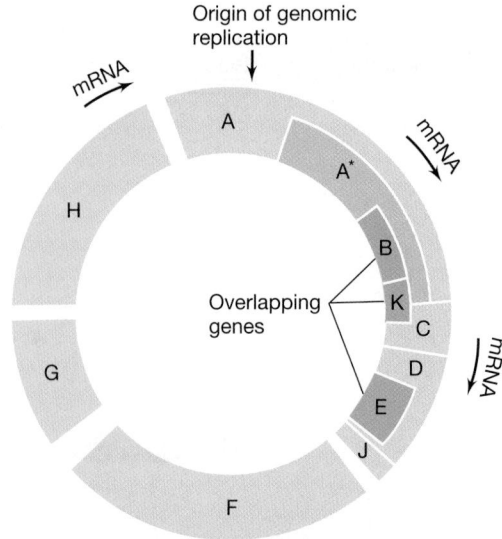

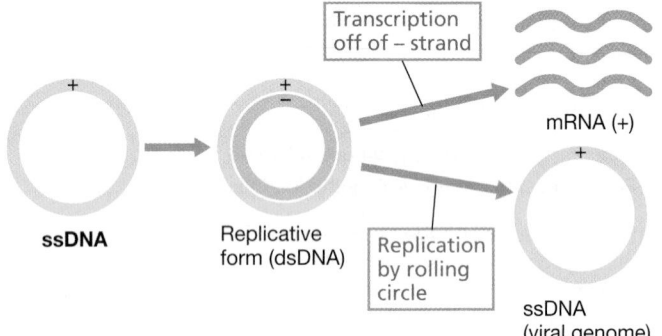

A	Replicative form DNA synthesis	E	Host cell lysis
A*	Shutoff of host DNA synthesis	F	Major capsid protein
B	Formation of capsid precursors	G	Major spike protein
C	DNA maturation	H	Minor spike protein
D	Capsid assembly	J	DNA packaging protein
		K	Function unknown

(a) **Genetic map of φX174**

(b) **Flow of events during φX174 replication**

Figure 21.3 Bacteriophage φX174, a single-stranded DNA phage. *(a)* Note the regions of gene overlap in the genetic map. Intergenic regions are not colored. Protein A* is formed using only part of the coding sequence of gene A by reinitiation of translation. The key indicates the functions of the proteins encoded by each gene. *(b)* Flow of events in φX174 replication. Progeny ssDNA is produced from replicative form dsDNA by rolling circle replication as shown in more detail in Figure 21.4.

synthesis of a new strand (**Figure 21.4**). Cutting of the plus strand is accomplished by the A protein. Continued rotation of the circle leads to the synthesis of a linear, single-stranded structure, the φX174 genome. Note that synthesis is asymmetric because only one of the strands (the negative strand) serves as a template. When the growing viral strand reaches unit length (5386 residues for φX174), the A protein cleaves and then ligates the two ends of the newly synthesized single strand to give an ssDNA circle.

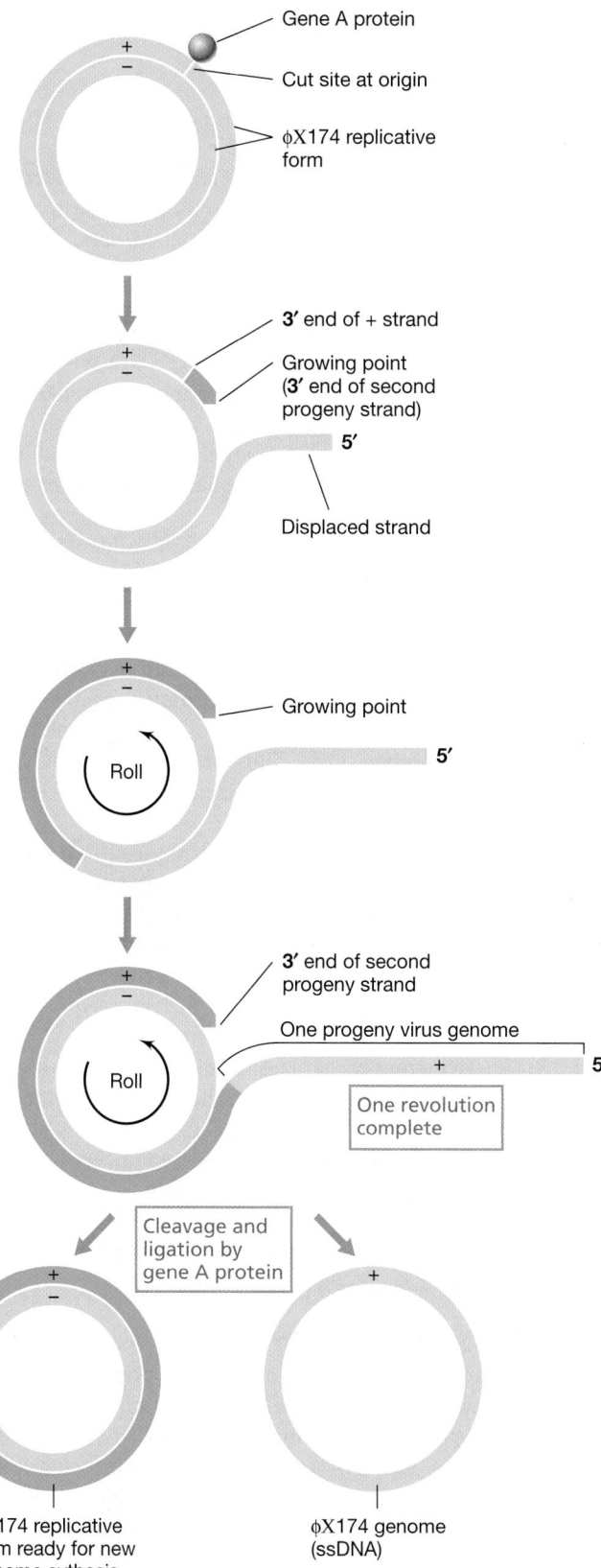

Gene A protein

Cut site at origin

φX174 replicative form

3′ end of + strand

Growing point (**3′** end of second progeny strand)

5′

Displaced strand

Growing point

Roll

5′

3′ end of second progeny strand

One progeny virus genome

Roll

5′

One revolution complete

Cleavage and ligation by gene A protein

φX174 replicative form ready for new genome sythesis

φX174 genome (ssDNA)

Figure 21.4 **Rolling circle replication in phage φX174.** Replication begins at the origin of the double-stranded replicative form with the cutting of the plus strand of DNA by gene A protein (both strands of DNA are shown in light green here to simplify the diagram.). After one new progeny strand has been synthesized (one revolution of the circle), the gene A protein cleaves the new strand and ligates its two ends.

Transcription and Translation in φX174

The replicative form is also used as a template for transcription of φX174 mRNA. Transcription of mRNA begins at several promoters and terminates at a number of sites. The polycistronic mRNA molecules are then translated into the various phage proteins. As already noted, several proteins are made from mRNA transcripts formed from different reading frames of the same DNA sequences (overlapping genes, Figure 21.3*a*). The efficiency with which such a small genome as that of φX174 can encode so many different proteins is impressive.

Ultimately, assembly of mature φX174 virions occurs. Release of virions from the cell takes place as a result of cell lysis. This is due to the E protein, which promotes lysis by inhibiting the activity of one of the enzymes involved in peptidoglycan synthesis in the cell wall (Section 5.4). Because of the resulting weakness in newly synthesized cell wall material, the cell eventually ruptures, releasing the viral particles.

Filamentous Single-Stranded DNA Bacteriophages

The filamentous DNA phages have helical rather than icosahedral symmetry. The most studied member of this group is phage M13, which infects *E. coli*, but related phages include f1 and fd. As with the small RNA bacteriophages, these filamentous DNA phages infect only conjugational donor cells, entering after attachment to the pilus (Figure 21.1). Although these phages are linear (filamentous) in shape, they possess circular single-stranded DNA.

Bacteriophage M13

M13 is the model filamentous bacteriophage. Its genome consists of circular ssDNA. It has found extensive use as a cloning and DNA-sequencing vector in genetic engineering (Section 11.10). The virion of phage M13 is only 6 nm in diameter but is 860 nm long. These filamentous DNA phages have the unusual property of being released without lysing the host cell. Thus, a cell infected with phage M13 can continue to grow, all the while releasing M13 virions. Virus infection causes a slowing of cell growth, but otherwise a cell is able to coexist with its virus. Typical plaques are thus not observed; instead, only areas of reduced turbidity are seen within the bacterial lawn.

Many aspects of DNA replication in filamentous phages are similar to those of φX174 (Figure 21.4); however, the virions are released without killing the cell by a process called *budding*. In this mechanism, the virus DNA is covered with the coat proteins as it crosses the cell envelope. Four minor coat proteins cover the ends of the virion, and the major coat protein (P8) covers the sides (**Figure 21.5**). Thus there is no intracellular accumulation of virions as with typical bacteriophages. Instead, virus assembly is coupled with the budding process.

Using M13 in Genetic Engineering

Several features of phage M13 make it useful as a cloning and DNA sequencing vehicle. First, it has ssDNA, which means that sequencing can easily be carried out by the Sanger dideoxynucleotide method (Section 12.2). Second, a double-stranded form of genomic DNA essential for cloning purposes is produced

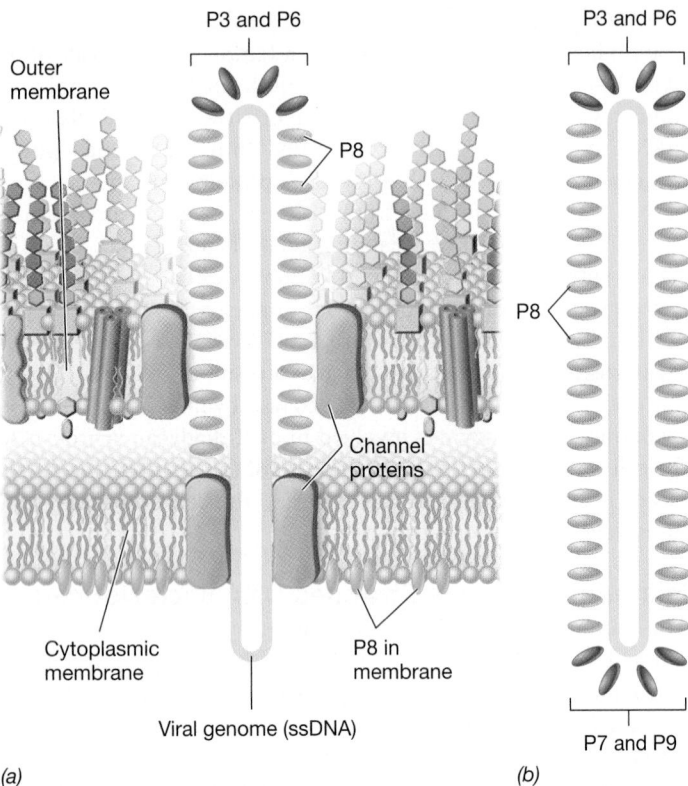

P3 and P6

Outer
membrane

P8

Channel
proteins

Cytoplasmic
membrane

P8 in
membrane

Viral genome (ssDNA)

(a)

P3 and P6

P8

P7 and P9

(b)

Figure 21.5 Release of filamentous phages. The virions of filamentous single-stranded phages (such as M13 or fd) exit from infected cells without lysis. *(a)* Budding. The virus DNA crosses the cell envelope through a channel constructed from virus-encoded proteins. As this occurs, the DNA is coated with phage proteins that have been embedded in the cytoplasmic membrane. *(b)* Complete virion. The two ends of the virion are covered with small numbers of the minor coat proteins P3 and P6 (front end) or P7 and P9 (rear end). The main part of the virion is covered by the major coat protein, P8.

naturally when the phage produces its replicative form. Third, as long as infected cells are kept growing, they can be maintained indefinitely, so a continuous source of the cloned DNA is available. And finally, like phage lambda (⟲ Section 11.9), there is an intergenic space in the genome of phage M13 that does not encode proteins and can be replaced by variable amounts of foreign DNA (⟲ Section 11.10). Consequently, phage M13 is an important part of the biotechnologist's toolbox.

MiniQuiz

- The genome of ϕX174, which is single-stranded and in the plus configuration, cannot be used directly as mRNA as in phage MS2. Why not?

- How does the replicative form of ϕX174 nucleic acid differ from the form found in the virion?

- Describe the genome of phage M13. How does the genome relate to the mRNA produced by this phage?

- How can M13 virions be released without killing the infected host cell?

21.3 Double-Stranded DNA Bacteriophages

The double-stranded DNA (dsDNA) bacteriophages are among the best studied of all viruses, and we have already discussed two important ones, T4 and lambda, in Chapter 9 (⟲ Sections 9.9 and 9.10). Because of their importance in molecular biology and gene regulation, we consider two more such viruses, T7 here and Mu in the next section.

Replication of Bacteriophage T7: Early Events

Bacteriophage T7 and its close relative T3 are relatively small DNA viruses that infect *Escherichia coli* and a few related enteric bacteria, notably *Shigella*. The virion has an icosahedral head and a very short tail. The T7 genome is a linear dsDNA molecule of 39,936 bp. The genetic map of T7 is shown in **Figure 21.6** and includes some overlapping genes.

The order of the genes on the T7 genome influences the regulation of virus replication. When the virion attaches to the bacterial cell, the DNA is injected with the genes at the "left end" of the genetic map entering the cell first. Several genes at this end of the T7 genome are quickly transcribed by host RNA polymerase and then translated. One of these early proteins inhibits the host restriction system, a mechanism for protecting the cell from foreign DNA (⟲ Section 11.1). This occurs very rapidly, as the anti-restriction protein is synthesized before the entire T7 genome even enters the cell!

Another early protein is T7 RNA polymerase. Two more early mRNA molecules encode proteins that inhibit host RNA polymerase, thus turning off the transcription of the early genes as well as the transcription of host genes. Host RNA polymerase is thus used just to transcribe the first few genes. T7 RNA polymerase then takes over and carries out the transcription of most phage genes.

T7 RNA polymerase recognizes only T7 promoters distributed along the T7 genome (Figure 21.6). These T7 promoters have a sequence unrelated to typical *E. coli* promoters. Because T7 RNA polymerase is extremely efficient, genetic engineers have used it to express cloned genes at very high levels (⟲ Section 11.8). Note that this strategy differs from that of phage T4. Phage T4 uses the host RNA polymerase but directs it only to phage genes using a T4-specific sigma factor (⟲ Section 9.9).

Genome Replication in T7

DNA replication in T7 begins at a single origin of replication (shown in Figure 21.6) and proceeds bidirectionally from this origin (**Figure 21.7**). Replicating molecules of T7 DNA can be recognized under the electron microscope by their characteristic structures. Because the origin of replication is near the left end, Y-shaped molecules are typically seen in transmission electron micrographs of replicating T7 DNA. Earlier in replication, bubble-shaped molecules can appear (Figure 21.7). Several virus-encoded proteins, including a T7-specific DNA polymerase, are required for T7 DNA replication. This differs from the use of host-encoded proteins by phage T4 (⟲ Section 9.9).

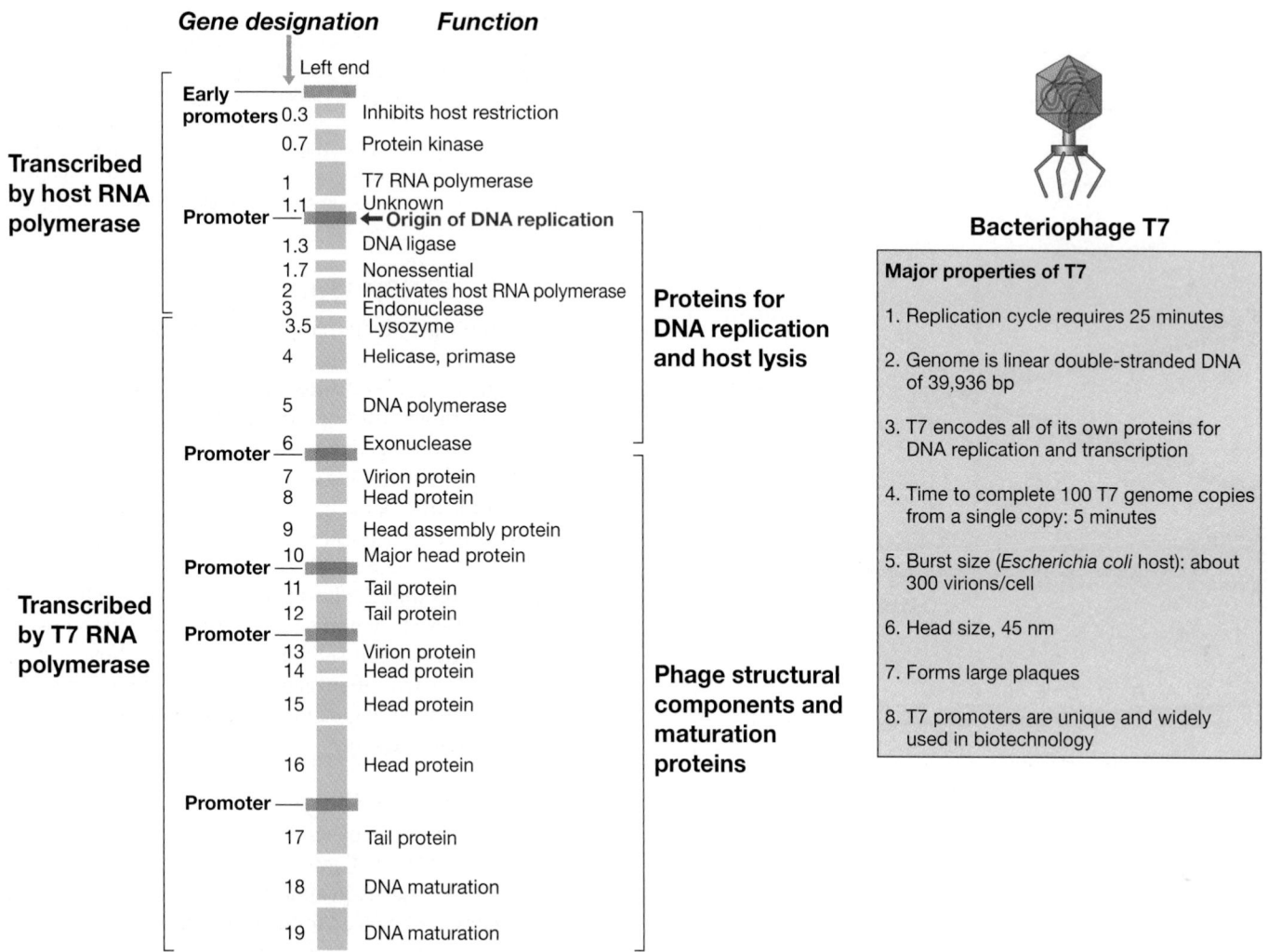

Figure 21.6 Genetic map of bacteriophage T7. The genes are designated by numbers. The map also shows their approximate sizes and the functions of gene products. Host RNA polymerase transcribes from the early promoters. T7 RNA polymerase transcribes from all other promoters.

Replication of linear DNA molecules must overcome the problem of the DNA shortening in each round of replication due to removal of the RNA primer used to initiate DNA synthesis. Different solutions to this problem exist (⮌ Section 7.7). The solution employed by T7 resembles that used by T4 and relies on repeated sequences (⮌ Section 9.9). T7 DNA has a direct terminal repeat of 160 bp at both ends of the molecule. To replicate DNA near the 5′ terminus, RNA primer molecules must be removed before replication is complete. This leaves an unreplicated segment of DNA at the 5′ terminus of each strand (lower part of Figure 21.7a). The opposite single 3′ strands on two separate DNA molecules, being complementary, can pair with these 5′ strands, forming a DNA molecule twice as long as the original T7 DNA (Figure 21.7b). The unreplicated portions of this structure are then completed through the activity of T7 DNA polymerase and ligase, resulting in a linear double molecule called a **concatemer.**

Continued replication and recombination can lead to concatemers of considerable length, but ultimately a phage-encoded endonuclease cuts each concatemer at a specific site, resulting in

the formation of virus-sized linear DNA molecules with terminal repeats (Figure 21.7c). Because T7 cuts the concatemer at specific sequences, the DNA sequence in each T7 virion is identical. Recall that this is not the case in phage T4, which processes DNA using a "headful mechanism" and whose DNA is therefore circularly permuted (⮌ Section 9.9).

Genomics has shown that some T7 genes have been shared with bacteria. For example, several DNA replication genes in mitochondria, the respiratory organelles of eukaryotic cells with evolutionary roots in the *Bacteria* (⮌ Section 20.4), were likely acquired from an ancestor of the T7 family of viruses.

MiniQuiz

- Of what significance is it that the T7 genome enters the cell in only one orientation?

- What is meant by terminal repeats?

- In what ways is DNA replication similar and in what ways is it different in phages T4 and T7?

UNIT 6

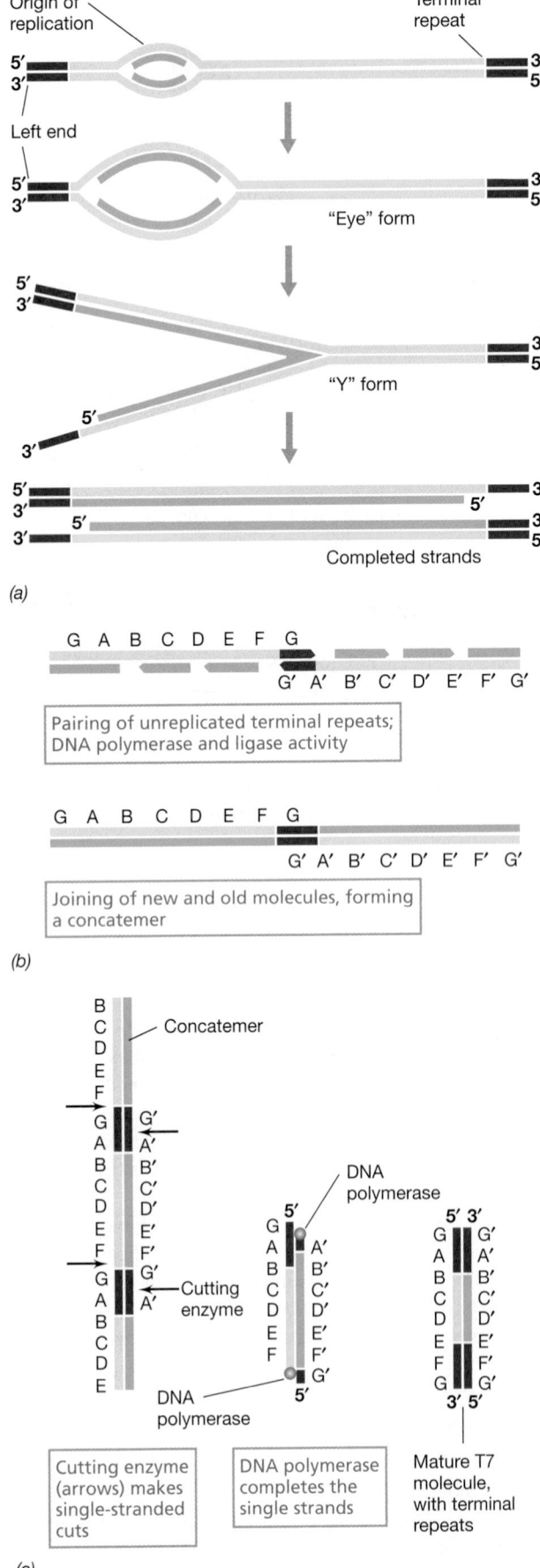

Origin of replication

Terminal repeat

Left end

"Eye" form

"Y" form

Completed strands

(a)

G A B C D E F G

G' A' B' C' D' E' F' G'

Pairing of unreplicated terminal repeats; DNA polymerase and ligase activity

G A B C D E F G

G' A' B' C' D' E' F' G'

Joining of new and old molecules, forming a concatemer

(b)

Concatemer

DNA polymerase

Cutting enzyme

DNA polymerase

Mature T7 molecule, with terminal repeats

Cutting enzyme (arrows) makes single-stranded cuts

DNA polymerase completes the single strands

(c)

Figure 21.7 Replication of the bacteriophage T7 genome. *(a)* The linear, double-stranded DNA undergoes bidirectional replication, giving rise to intermediate "eye" and "Y" forms (for simplicity, both template strands are shown in light green and both newly synthesized strands in dark green). *(b)* Formation of concatemers by joining DNA molecules at the unreplicated terminal ends. The designation of the genes is arbitrary. *(c)* Production of mature viral DNA molecules from T7 concatemers by activity of the cutting enzyme, an endonuclease.

21.4 The Transposable Phage Mu

The bacteriophage Mu is temperate, like lambda (⟲ Section 9.10), but has the unusual property of replicating by *transposition* (⟲ Section 10.13). Transposable elements are sequences of DNA that can move from one location on their host genome to another as discrete genetic units. They are found in both prokaryotes and eukaryotes and play important roles in genetic variation. Mu is both a bacteriophage and a very large transposable element that replicates its DNA by transposition.

This phage was named Mu because it generates *mu*tations when it integrates into the host cell chromosome. The mutations result because the Mu genome can be inserted within host genes, thus interrupting coding sequences. Hence, a host cell that is infected with Mu will often gain a mutant phenotype. Mu is a useful phage in bacterial genetics because it can be used to easily generate bacterial mutants.

Basic Properties of Phage Mu

Bacteriophage Mu is a large virus with an icosahedral head, a helical tail, and six tail fibers (**Figure 21.8**). The genome of Mu consists of linear dsDNA, and its genetic map is shown in **Figure 21.9a**. Most Mu genes are involved in the synthesis of head and tail proteins, and important genes at each end of the genome are involved in replication and host range.

The Mu virion contains approximately 39 kbp of DNA, but only 37.2 kbp constitute the actual Mu genome. The additional length is host DNA attached to the ends of the Mu genome, 50–150 bp at the left end and 1–2 kbp at the right end. These host DNA sequences are not unique but merely represent DNA adjacent to the location where Mu was inserted into the genome of its previous host. How does this happen?

When a Mu virion is formed, a length of DNA containing the Mu genome, just large enough to fill the phage head, is excised from the host. The DNA is packaged until the head is full, but the place at the right end where the DNA is cut varies from one virion to another. For that reason, as shown on the genetic map, there is a variable sequence of host DNA at the right-hand end of the phage (right of the *attR* site). Thus, each virion arising from a single infected cell is genetically unique in having different flanking host DNA.

The Invertible G Region of Mu

As shown on the genetic map (Figure 21.9a), a specific segment of the Mu genome called *G* is invertible, being present in the genome either in the orientation designated G^+ or in the inverted orientation G^-. The orientation of this segment determines the kind of tail fibers that are present on the phage. Because adsorption

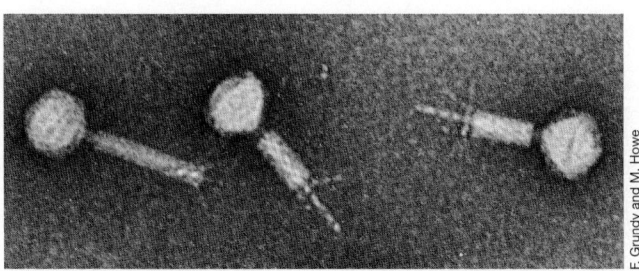

Figure 21.8 Bacteriophage Mu virions. Electron micrograph of virions of the double-stranded DNA phage Mu, the mutator phage.

to the host cell is controlled by molecular interactions between the tail fibers and the cell surface, the host range of Mu is determined by the orientation of the G segment in the phage. If the G segment is in the G$^+$ orientation, then the phage will make tail fibers that allow it to infect *Escherichia coli* strain K12. In contrast, if the G segment is in the G$^-$ orientation, then the phage will infect *E. coli* strain C or several other species of enteric bacteria. The two alternative tail fiber proteins are encoded on opposite strands within this small G segment.

Left of the G segment is a promoter that directs transcription into the G segment. In the orientation G$^+$, the promoter for genes S and U is active, whereas in the orientation G$^-$, a different promoter directs transcription of genes S' and U' on the opposite strand. Inversion of the G region is a rare event and is under the control of a gene adjacent to the G region. This inversion phenomenon is a simple mechanism the phage has evolved for attacking a variety of different host cells.

Replication of Mu

Note that the bacteriophages we have discussed that have double-stranded DNA genomes, T4 (↩ Section 9.9), T7 (Section 21.3), lambda (↩ Section 9.10), and Mu, all have *linear* genomes. Nonetheless, these viruses use three quite distinct schemes to replicate the ends of their genomes. Both T4 and T7 use terminal repeats to form concatemers, whereas lambda circularizes its genome after infection. Mu replicates in a completely different manner from all of these because its genome is replicated as part of a larger DNA molecule.

On infection of a host cell by Mu, the DNA is injected and is protected from host restriction by a modification system in which several adenine residues are modified by acetylation. In contrast with lambda, integration of Mu DNA into the host genome is essential for both lytic and lysogenic growth. Integration requires the activity of the gene A product, which is a **transposase** enzyme. A 5-bp duplication of host DNA is generated at the target site where the Mu DNA becomes integrated. As shown in Figure 21.9b, this host DNA duplication arises because staggered cuts are made at the point in the host genome where Mu is inserted. The resulting single-stranded segments are converted to the double-stranded form as part of the Mu integration process. Duplication of short stretches of host DNA is typical of transposable element insertion (↩ Section 10.13).

Mu can enter the lytic pathway either upon initial infection if the Mu repressor (the product of the *c* gene) is not made or by induc-

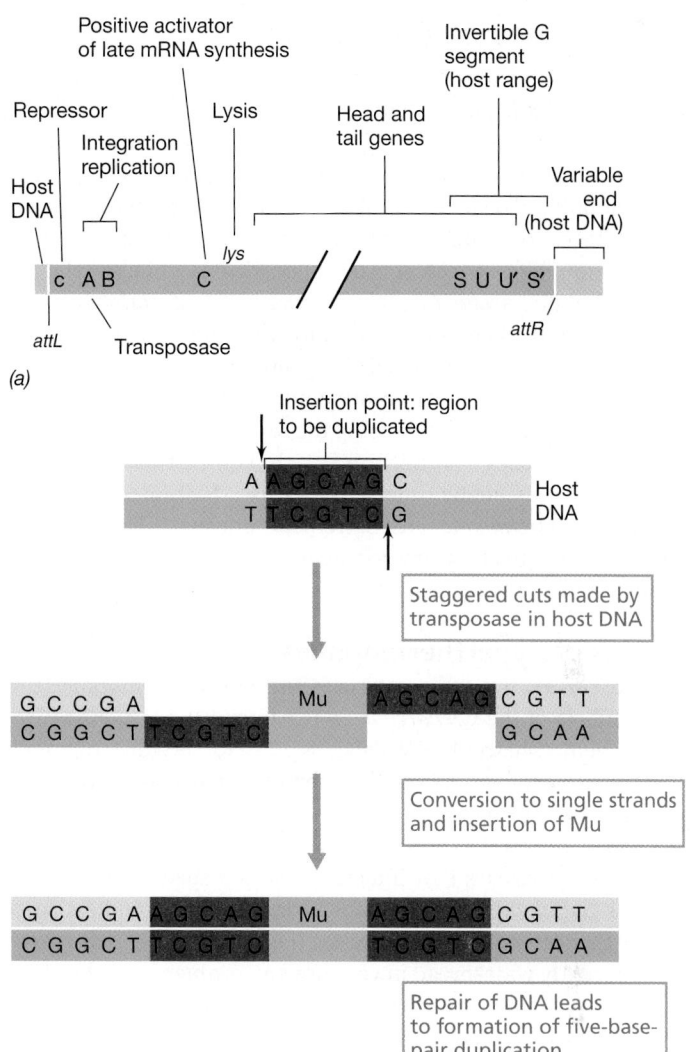

(b)

Figure 21.9 Bacteriophage Mu. *(a)* Genetic map of the Mu genome. Note that there is a lowercase *c* gene, which encodes a repressor protein, and an uppercase *C* gene, which encodes an activator protein. The region encoding the head and tail genes is not drawn to scale. *(b)* Integration of Mu into the host DNA, showing the generation of a 5-bp duplication of host DNA.

tion of a lysogen. In either case, Mu DNA is replicated by repeated transposition of Mu to multiple sites on the host genome. Initially, only the early genes of Mu are transcribed, but after the C protein is expressed (C is a positive activator of late transcription), the Mu head and tail proteins are synthesized. Eventually, the cell is lysed and mature phage particles are released. The lysogenic state in Mu requires the sufficient accumulation of repressor protein to prevent transcription of integrated Mu DNA.

MiniQuiz

- What mechanism is used to change the host range of Mu?
- What mechanism does Mu use to ensure that the ends of its linear genome are completely replicated?

21.5 Viruses of *Archaea*

Several DNA viruses have been discovered whose hosts are species of *Archaea*, including representatives of both the *Euryarchaeota* and *Crenarchaeota* (Chapter 19). Most viruses that infect species of *Euryarchaeota*, including both methanogenic and halophilic *Archaea*, are of the "head and tail" type, resembling phages that infect enteric bacteria, such as phage T4 (⇄ Section 9.9). In fact, certain tailed viruses of halophilic and haloalkaliphilic *Archaea*, such as phage φH of *Halobacterium salinarum* and phage φCh1 of *Natrialba magadii*, have linear double-stranded DNA (dsDNA) genomes that are circularly permuted and terminally redundant, as in phage T4. These viruses likely package DNA using the same "headful" mechanism that phage T4 uses (⇄ Section 9.9). In contrast to this familiar group of viruses, many other DNA viruses that infect members of the *Crenarchaeota* are extremely unusual. Thus far, no RNA viruses infecting any member of the *Archaea* have been found.

Viruses of Hyperthermophiles

The most distinctive archaeal viruses infect hyperthermophiles that are members of the *Crenarchaeota* (Chapter 19). For example, the sulfur chemolithotroph *Sulfolobus* is host to several structurally unusual viruses. Considering the habitat of *Sulfolobus*—hot, acidic soils and hot springs (⇄ Section 19.8)—these viruses must be remarkably resistant to heat and acid denaturation.

One of the viruses that infects *Sulfolobus* species, nicknamed SSV for *Sulfolobus* spindle-shaped *virus*, forms spindle-shaped virions that often cluster in rosettes (**Figure 21.10a**). Such viruses are apparently widespread in very hot environments, as they have been isolated from hot springs in Iceland and Japan, as well as Yellowstone National Park. Virions of SSV contain circular dsDNA about 15 kbp long, considerably smaller than the linear genome of a tailed bacteriophage like T4 (~168 kbp; no circular dsDNA viruses of *Bacteria* are known). A second morphological type of *Sulfolobus* bacteriophage is a rigid, helical rod (Figure 21.10b). Viruses in this class, nicknamed SIFV for *Sulfolobus islandicus* filamentous *virus*, contain linear dsDNA genomes of about 35 kbp. Many variations on the spindle- and rod-shaped patterns have been seen in viral isolation studies. These include very long rods similar in morphology to filamentous bacteriophages (Section 21.2) and spindle-shaped viruses that contain a large spindle-shaped center with appendages at each end (Figure 21.10d).

A spindle-shaped virus that infects *Acidianus convivator* (a species of *Archaea* closely related to *Sulfolobus*) shows an intriguing behavior. The virion is lemon-shaped when first released from the host cells. Following release from its host, the virion develops long thin tails, one at each end, and was thus named ATV for "*Acidianus* two-tailed virus" (Figure 21.10d). At 85–90°C, the optimum growth temperature for the host cell, the tails form in about an hour. When lemon-shaped ATV virions were stored at low temperature, tails did not develop. When the virions were returned to high temperatures, the tails formed. What is remarkable about ATV is that it provides the first example of virus development in the complete absence of host cell contact.

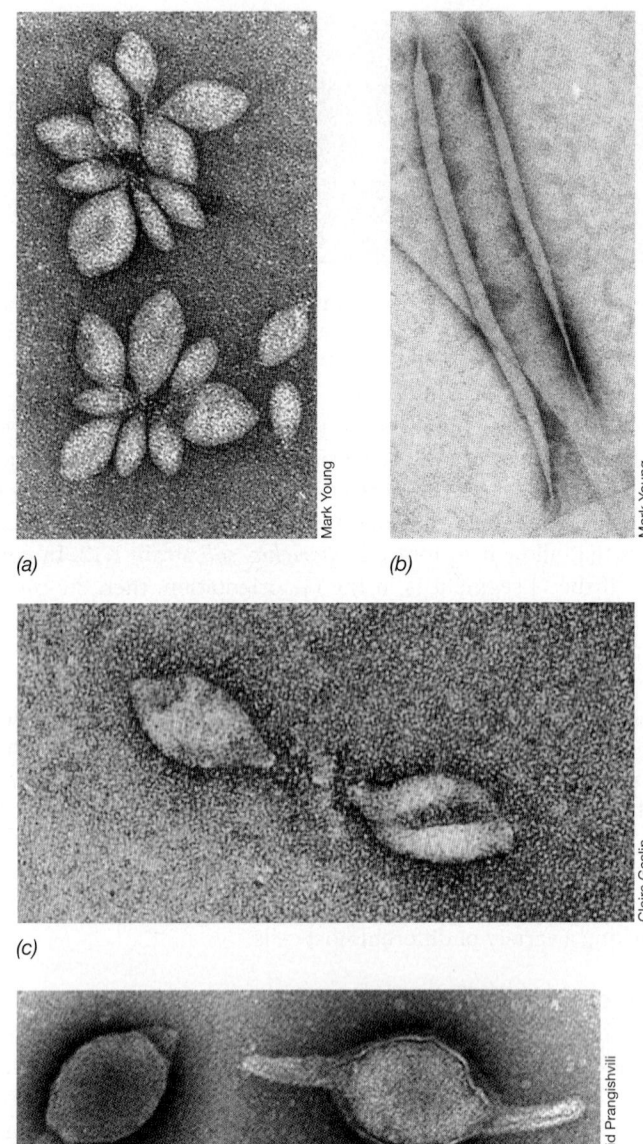

(a)   *(b)*

(c)

(d)

Figure 21.10 Archaeal viruses. Electron micrographs of viruses of *Crenarchaeota* (parts a, b, d), and a virus of a euryarchaeote (part c). *(a)* Spindle-shaped virus SSV1 that infects *Sulfolobus solfataricus*. The viral dimensions are 40 × 80 nm. *(b)* Filamentous virus SIFV that infects *S. solfataricus*. The viral dimensions are 50 × 900–1500 nm. *(c)* Spindle-shaped virus PAV1 that infects *Pyrococcus abyssi*. The viral dimensions are 80 × 120 nm. *(d)* ATV, the virus that infects the hyperthermophile *Acidianus convivator*. When released from the cell the virions are lemon-shaped (left virion), but they proceed to grow appendages on both ends (right virion) that can grow to over three times that shown here. The virions are about 100 nm in diameter.

ATV contains a circular dsDNA genome of about 68 kbp and is a lytic virus. It is thought that the extended tails of ATV help the virus in some way survive in the hot, acidic (pH 1.5) environment of *A. convivator*.

A spindle-shaped virus also infects *Pyrococcus*, a species in the archaeal phylum *Euryarchaeota* (⇄ Section 19.5). This virus, named PAV1 for *Pyrococcus abyssi virus 1*, resembles SSV but is larger and contains a very short tail (Figure 21.10c). Phage PAV1 has a circular dsDNA genome of about 18 kbp and, interestingly, is released from host cells without cell lysis, probably by a budding mechanism similar to that of the filamentous *Escherichia coli* phage M13 (Section 21.2). *Pyrococcus* is a hyperthermophile with a growth temperature optimum of about 100°C, meaning that PAV1 virions must be extremely heat-stable. Despite their similar morphologies, genomic comparisons of PAV1 and SSV-type viruses show little sequence similarity, indicating that the two types of viruses do not have common evolutionary roots.

Replication and Evolution of Archaeal Viruses

Replication studies of archaeal viruses still need to be done to define the major events in genome replication and virion assembly. However, considering that the genomes of these viruses are all dsDNA, it is unlikely that any major novel modes of replication will be found. However, many molecular details, such as the extent to which viral rather than host polymerases and related enzymes are used in replication, await further work on these novel viruses.

Spindle-shaped viruses are unknown among species of *Bacteria*. This, along with limited genomic similarity to viruses of *Bacteria*, suggests that these archaeal viruses do not share a common ancestor with known bacteriophages. By contrast, structural and genomic studies have shown that tailed viruses of *Archaea* may well have shared a common ancestry with tailed phages of *Bacteria*. This could be because these viruses share common evolutionary roots, or it could be the result of horizontal gene transfer (Section 12.12) of phage genes from *Bacteria* to *Archaea* (or vice versa). Both temperate viruses and transducing viruses (Section 10.8) exist within the *Archaea*, although examples are currently rare and have yet to be developed to the point of playing a major role in the laboratory genetics of these organisms.

MiniQuiz

- In which groups of *Archaea* have head- and tail-type phages been discovered? How are they related to the tailed phages of *Bacteria*?
- Compare the shapes of viruses that infect methanogenic and halophilic *Archaea* with those that grow at high temperatures.

21.6 Viral Genomes in Nature

The number of prokaryotic cells on Earth is far greater than the total number of eukaryotic cells; estimates of total prokaryotic cell numbers are on the order of 10^{30}. However, the number of viruses on Earth is even greater—an estimated 10^{31}. The best estimates of both cell and virus numbers in the environment come from the analysis of seawater, as this is much easier to analyze than soil or sediments. In the oceans, viruses account for about 95% of the total number of organisms present, but due to their small size, they constitute only about 5% of the total biomass.

There are millions of bacteria, and approximately ten times as many viruses, present in every milliliter of seawater (**Figure 21.11**). Not surprisingly, most of these viruses are bacteriophages, and these populations turn over rapidly. It has been estimated that every day 5–50% of the bacteria in seawater are killed by bacterio-

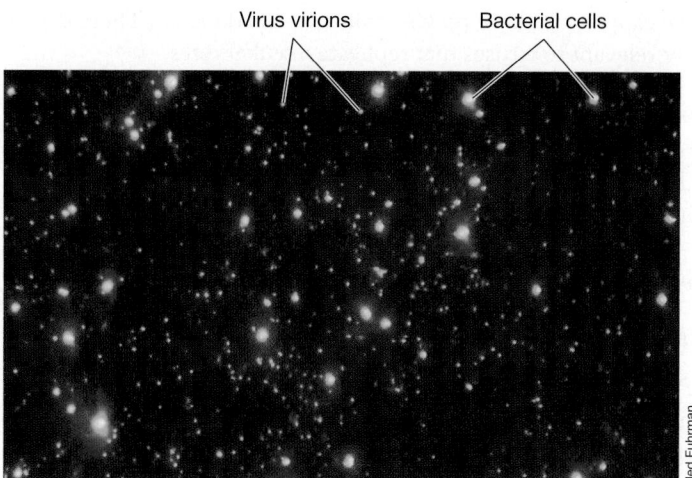

Figure 21.11 Viruses and bacteria in seawater. An epifluorescence photomicrograph of seawater stained with the dye SYBR Green to reveal prokaryotic cells and viruses. Although viruses are too small to be seen with the light microscope, fluorescence from a stained virus is visible.

phages, and most of the others are eaten by protozoa. By far the most common type of bacteriophages found in the oceans are head and tail phages containing dsDNA. RNA-containing bacteriophages are relatively rare.

Overall, most of the genetic diversity on Earth is thought to reside in viruses, mostly bacteriophages. Many bacteriophages can integrate into the genomes of their bacterial hosts (Section 9.10) and can transfer bacterial genes from one bacterium to another (Section 10.8). Thus, viruses have a massive influence on bacterial evolution, partly by culling the large bacterial populations on a daily basis and partly by catalyzing horizontal gene transfer.

The *viral metagenome* is the sum total of all the virus genes in a particular environment. Several viral metagenomic studies have been undertaken, and they invariably show that immense viral diversity exists on Earth. For example, approximately 75% of the gene sequences found in viral metagenomic studies show no similarity to any other genes presently in viral or cellular gene databases. By comparison, surveys of bacterial metagenomes typically reveal approximately 10% unknown genes. Thus, most viruses await discovery and most viral genes have unknown functions.

MiniQuiz

- How do viruses affect the evolution of bacterial genomes?
- What type of bacteriophages are most common in the oceans?

 RNA Viruses of Eukaryotes

The differences between prokaryotic and eukaryotic cells place some constraints on the mechanisms used by the viruses that infect them. For example, in prokaryotes transcription and translation can be coupled processes. In eukaryotes, in contrast, DNA is replicated and transcribed in the nucleus, whereas proteins are synthesized in the cytoplasm. In addition, mRNAs in eukaryotes are capped and have poly(A) tails (↩ Section 7.8). These details are relevant to viruses that replicate in eukaryotes.

We discussed some animal viruses in Chapter 9 (↩ Section 9.11). Our interest in animal viruses springs mostly from their role in human disease, and we mainly focus on animal viruses in the remainder of this chapter. However, plants also suffer from infectious diseases, some of which are due to viruses. Indeed, several virus diseases of crop plants are of major economic importance.

21.7 Plant RNA Viruses

Plant cell walls are extremely thick and strong. Yet a considerable number of viruses can infect plants and, in multicellular plants, spread from the infected cell to neighboring cells. Viruses often gain entry to plant cells when insects, such as aphids, attack the plant and damage the walls. The great majority of known plant viruses are positive-strand RNA viruses, perhaps because these small genomes can be transferred easily from cell to cell within the plant.

Tobacco Mosaic Virus: General Properties

In 1892, the Russian scientist Dmitri Ivanovsky showed that the causative agent of tobacco mosaic disease could pass through filters that retain bacteria. In 1898, the famous Dutch microbiologist Martinus Beijerinck (↩ Section 1.9) showed that this agent was not only filterable, but that it had many of the properties of a living organism. This agent, the first virus of any type to be recognized, was tobacco mosaic virus (TMV).

TMV has a rod-shaped virion with helical symmetry that contains 2130 copies of a coat protein plus a single copy of the positive-strand RNA genome. It was TMV that was first used to show that RNA could be the genetic material of viruses, just as DNA is in other viruses and in cells. TMV remains a serious agricultural problem because it infects tomato plants as well as tobacco. TMV infection of a plant requires damage to plant cell walls through which the virion enters. Uncoating takes place in the cell.

Genome and Replication of TMV

The RNA genome of TMV contains 6395 nucleotides, and a map of the genome is shown in **Figure 21.12**. Like the bacteriophage MS2 (Section 21.1), TMV encodes only four proteins. The genome has a 5′ cap (↩ Section 7.8), thus it can be used directly as an mRNA and translated in the plant cell. The 3′ end of the TMV genome folds into a transfer RNA–like structure (Figure

21.12). In bacteria, the genome of a positive-strand RNA virus may act as the viral mRNA (↩ Section 9.7). However, eukaryotes cannot translate polycistronic mRNA (↩ Section 7.9), so the expression of TMV genes differs from that of bacterial RNA viruses.

The first gene in TMV encodes a protein called MTH with two enzymatic activities, a methyltransferase that caps RNA and an RNA helicase. The next gene encodes the RNA-dependent RNA polymerase (RNA replicase) that the virus needs to make a negative-strand RNA copy from which it can then make more copies of the genomic RNA. This protein is synthesized as part of a *polyprotein* (Section 21.8) including the MTH domains and is only synthesized when a ribosome accidentally reads through the stop codon at the end of the MTH gene. Because this happens rather infrequently, this longer protein is made only at low levels.

The remaining two genes encode the movement protein and the coat protein. These two proteins are translated from small monocistronic mRNAs that are transcribed from the negative-strand RNA. Like coat proteins of other viruses, that of TMV is essential to the formation of the virion, which is essential for infecting new plants. However, it is the movement protein that enables TMV to infect neighboring cells in an already infected plant. Plant cells are interconnected by cytoplasmic strands called *plasmodesmata* that connect the cytoplasm of neighboring cells (↩ Section 9.14). Plasmodesmata have very narrow channels, so narrow in fact that neither the TMV virion nor free RNA can easily traverse these openings. The TMV movement protein binds to the new genomic positive-strand RNA and forms a complex that is extremely thin (about 2.5 nm) and that can move through the plasmodesmata and infect neighboring cells.

Many positive-strand RNA viruses are known. Some infect plants, others infect animals (Section 21.8), and yet others infect *Bacteria* (Section 21.1). Genomic analyses indicate that many of these viruses are closely related despite their quite different hosts and that all of their RNA replicases are related. Nonetheless, they have slightly different replication modes, usually related to differences in the host cells.

MiniQuiz

- Most plant RNA viruses encode a movement protein. What is its role in infection?
- Although the TMV genome is used as a messenger RNA, not all the proteins encoded by the virus can be translated from it. Explain.

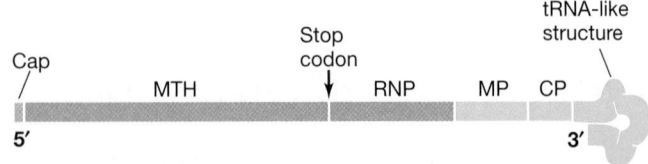

Figure 21.12 Genetic map of tobacco mosaic virus. The genome is a positive-strand RNA that is capped at its 5′ end and has a tRNA-like structure (drawn larger than scale) at its 3′ end. MTH, methyl transferase/helicase; RNP, RNA polymerase (replicase); MP, movement protein; CP, coat protein.

21.8 Positive-Strand RNA Animal Viruses

Several positive-strand RNA animal viruses cause disease in humans and other animals. These include poliovirus, the rhinoviruses that cause many cases of the common cold, the coronaviruses that cause respiratory syndromes, including severe acute respiratory syndrome (SARS), and the hepatitis A virus. The first animal virus discovered, foot-and-mouth disease virus, the causative agent of a debilitating and eventually fatal disease of cloven-hoofed (ruminant) animals, is also in this group. With the exception of the coronaviruses, which are much larger, these viruses are typically very small (about 30 nm in diameter) and contain single-stranded RNA. We focus here on poliovirus and the coronaviruses.

At one time, polio was a major infectious disease of humans, but the development of an effective vaccine has brought the disease almost completely under control. The World Health Organization (WHO) has a vaccination program intended to eradicate the disease worldwide, and at present there are reports of the virus in only a few countries in Africa and Asia. However, an unfortunate complication is slowing polio eradication in parts of Asia. For unknown reasons, possibly because of genetic variations in human populations, the effectiveness of the oral polio vaccine varies in different areas.

Poliovirus: General Features

The picornaviruses (*pico* means "small") are a family of very small, single-stranded RNA (ssRNA) viruses that includes poliovirus. The virion of poliovirus has an icosahedral structure with 60 morphological units per virion, each unit consisting of four distinct proteins (**Figure 21.13a**). The genome of poliovirus is a linear ssRNA molecule of 7433 bases. At the 5′ terminus of the viral RNA is a protein, called the VPg protein, that is attached covalently to the genomic RNA. At the 3′ terminus of the RNA is a poly(A) tail. The RNA genome of the virus is also the mRNA, even though the RNA is not capped, normally a prerequisite for translation in eukaryotes (∞ Section 7.8). Instead, the 5′ end of poliovirus RNA has a long sequence that can fold into a complex structure comprising multiple stem–loops. The VPg protein and the stem–loops mimic the cap-binding complex, and this permits binding of the poliovirus mRNA to the ribosome. In addition, poliovirus cleaves several proteins belonging to the host cell's translation initiation complex. This shuts down host cell translation and frees the ribosomes for use by the virus.

Although the viral RNA is monocistronic and has only a single start codon, it nonetheless encodes all the proteins of the virus. This trick is accomplished by making a single protein called a **polyprotein**. This giant protein (about 2200 amino acid residues) then undergoes self-cleavage into about 20 smaller proteins (including cleavage intermediates), among which are the four different structural proteins of the virion. Other proteins include the RNA-linked VPg protein, an RNA replicase responsible for synthesis of both minus-strand and plus-strand RNA, and at least one virus-encoded protease, which carries out the polyprotein cleavage. This process, called posttranslational cleavage, occurs in many animal viruses as well as in normal cell metabolism in animal cells. Translation of a positive-strand mRNA to yield a single polyprotein that is subsequently cleaved is a scheme also used by retroviruses (Section 21.11).

Replication of Poliovirus RNA

An overview of poliovirus replication is illustrated in Figure 21.13. The whole poliovirus replication process occurs in the cell cytoplasm. To initiate infection, the poliovirus virion attaches to a specific receptor on the surface of a sensitive cell and enters the cell. Once inside the cell, the virus particle is uncoated, and the free RNA associates with ribosomes. The viral RNA is then translated to yield the large polyprotein as described above.

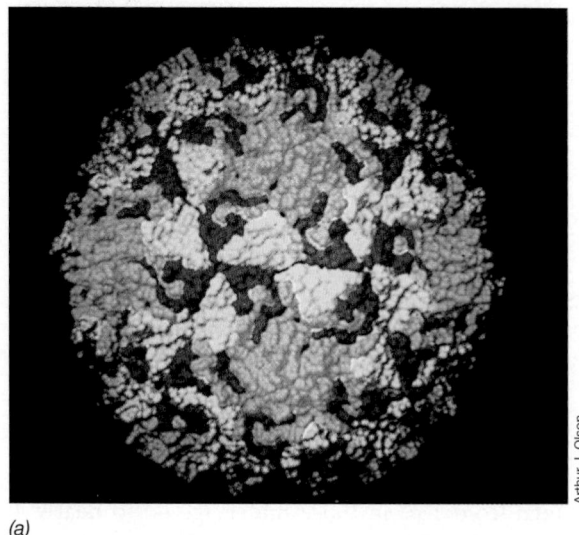

(a)

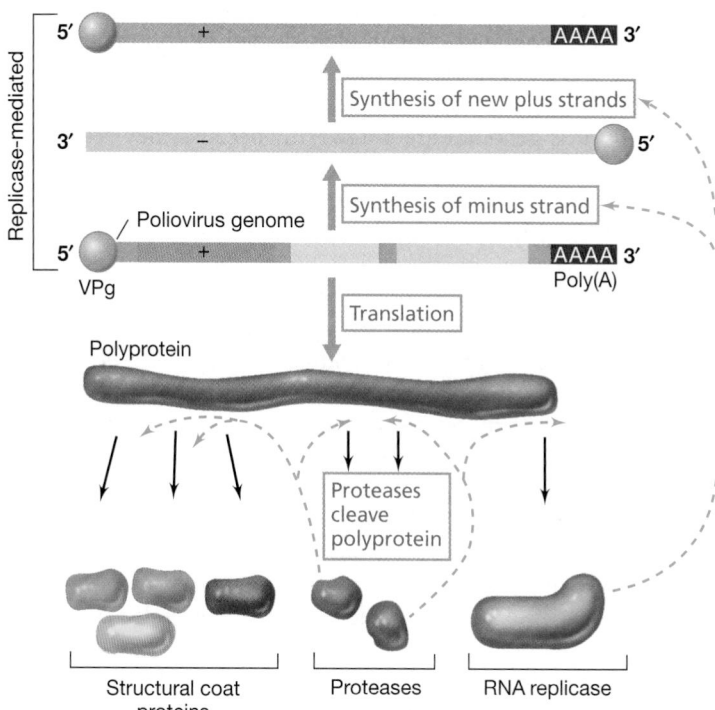

(b)

Figure 21.13 Poliovirus. *(a)* A computer model based on electron diffraction analysis of poliovirus virions. The various structural proteins are shown in distinct colors. *(b)* The replication and translation of poliovirus. Note the importance of the RNA replicase of poliovirus; this is needed to transcribe the poliovirus genome because animal cells cannot make RNA off of an RNA template.

Replication of viral RNA begins within a short time after infection and is catalyzed by the RNA replicase released by cleavage of the polyprotein. This RNA replicase uses the positive-strand viral RNA as a template to synthesize a complementary negative-strand RNA. The same virus-encoded RNA replicase uses this negative strand as the template for repeated synthesis of progeny positive strands. Some of the progeny positive strands may again be used to make more negative strands, and several thousand negative strands may eventually accumulate in the cell. From these, as many as a million positive-strand copies may ultimately be formed. Both the positive and negative strands become covalently linked to the tiny VPg protein (only 22 amino acids long), which functions as a primer for RNA synthesis.

Once poliovirus replication begins, host RNA and protein syntheses are inhibited. Host protein synthesis is inhibited as a result of the destruction of an important host protein, the cap-binding protein, required for translation of capped mRNAs (↺ Section 7.9). Poliovirus mRNA itself circumvents this limitation as previously mentioned through the extensive secondary structure of the 5′ end of its genome that allows binding to the ribosome. www.microbiologyplace.com **Online Tutorial 21.1: Replication of Poliovirus**

Coronaviruses and SARS

Coronaviruses are single-stranded plus-strand RNA viruses that, like poliovirus, replicate in the cytoplasm, but differ from poliovirus in their larger size and details of replication. Coronaviruses cause respiratory infections in humans and other animals, including about 15% of common colds (↺ Section 33.7). In 2003, a novel coronavirus caused several outbreaks of SARS, an occasionally fatal infection of the lower respiratory tract in humans.

Coronavirus virions are enveloped and more or less spherical, 60–220 nm in diameter, and contain club-shaped glycoprotein spikes on their surfaces (**Figure 21.14**). These give the virus the appearance of having a "crown" (the term *corona* is Latin for crown). Coronavirus genomes are noteworthy because they are the largest of any known RNA viruses (27–31 kb; strains of SARS virus average 29,700 nucleotides in length).

The coronavirus genome has a 5′ methylated cap and a 3′ poly(A) tail and so can function directly in the animal as mRNA. However, most viral proteins are not made from the translation of genomic RNA. Instead, upon infection, only a portion of the genome is translated, yielding a viral RNA replicase (Figure 21.14*b*). The replicase uses the genomic RNA as a template to produce a full-length negative RNA strand from which several monocistronic mRNAs are transcribed. The latter are then translated to produce viral proteins. Progeny plus-strand RNA genomes are also made by using the negative-strand RNA as template. The virions are assembled within the Golgi apparatus, a major secretory organelle in eukaryotic cells (↺ Section 20.5), with fully assembled virions being released at the cell surface.

Coronavirus thus differs from poliovirus in terms of virion size, genome size and capping, lack of the VPg protein, and the absence of polyprotein formation and cleavage. Nevertheless, like poliovirus, coronaviruses can cause severe disease, including a

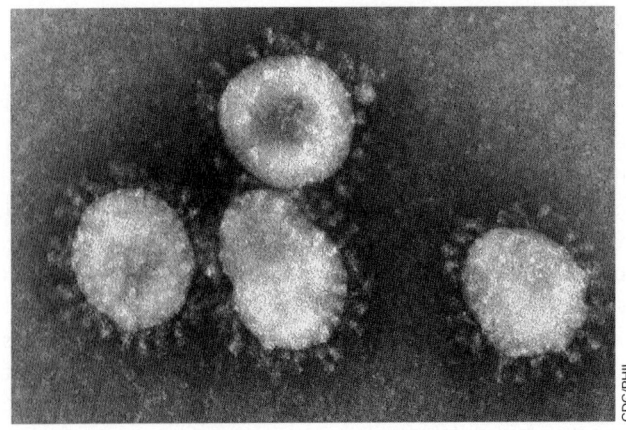

(a)

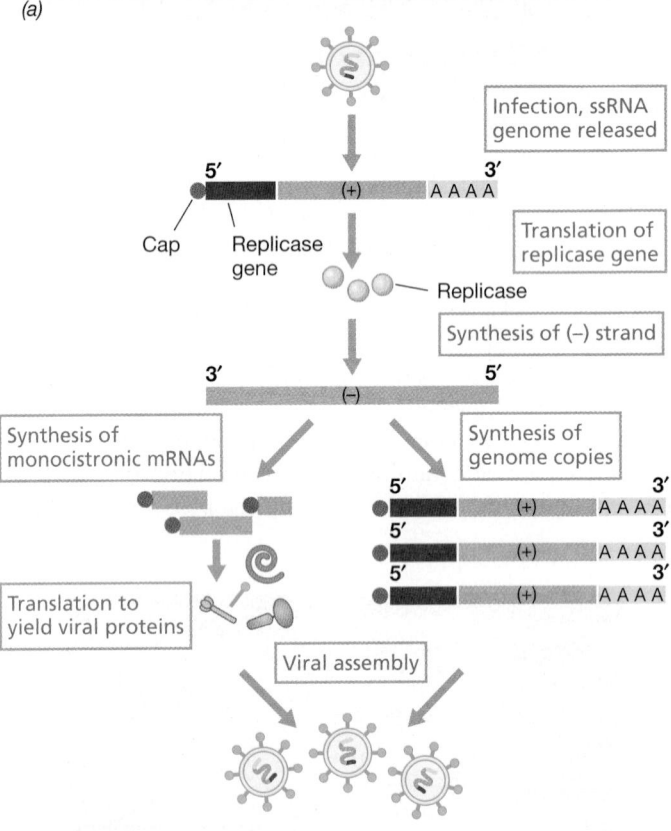

(b)

Figure 21.14 Coronaviruses. *(a)* Electron micrograph of a coronavirus. *(b)* Steps in coronavirus replication. The mRNA encoding viral proteins is transcribed from the negative strand made by the RNA replicase using the viral genome as template.

fatal acute infection that is rarely seen in polio. Fatality rates from SARS range from 13% in those under age 60 to nearly 45% in those above 60 (↺ Microbial Sidebar in Chapter 32, "SARS as a Model of Epidemiological Success").

MiniQuiz

- How can poliovirus RNA be synthesized in the cytoplasm whereas host RNA must be made in the nucleus?
- How are protein synthesis and genomic replication similar or different in poliovirus and the SARS virus?

21.9 Negative-Strand RNA Animal Viruses

Poliovirus and coronavirus replication requires conversion of the positive-strand genome into a negative-strand intermediate from which new positive strands are synthesized. However, in a number of RNA animal viruses the RNA genome itself is **negative-strand RNA**—that is, RNA that is *complementary* to the mRNA. These are thus called *negative-strand RNA viruses*. We discuss here two important examples: rhabdoviruses, including rabies virus, and orthomyxoviruses, including influenza virus. The Ebola virus, a human pathogen responsible for an emerging infectious disease, is also a negative-strand RNA virus. There are no known negative-strand RNA bacteriophages or archaeal viruses.

Rhabdoviruses: General Features

One of the most important negative-strand RNA viral pathogens is the rabies virus, which causes the disease rabies in animals and humans. Worldwide, there are over 30,000 (mostly fatal) cases of rabies each year in humans and countless more in domesticated and wild animals (↩ Section 34.1). Rabies virus is called a rhabdovirus, from *rhabdo* meaning "rod," which refers to the shape of the virus particle. Another rhabdovirus that has been extensively studied is vesicular stomatitis virus (VSV) (**Figure 21.15**), which causes the disease vesicular stomatitis in cattle, pigs, horses, and sometimes humans. Many rhabdoviruses, such as potato yellow dwarf virus, infect both insects and plants and can cause major losses of agricultural products.

The rhabdoviruses are enveloped viruses, with an extensive and complex lipid envelope surrounding the nucleocapsid. The virion is bullet-shaped, about 70 nm in diameter and 175 nm long (Figure 21.15). The nucleocapsid has helical symmetry and makes up only a small part of the virus particle weight (about 2–3% of the virion is RNA).

Replication of Rhabdoviruses

The rhabdovirus virion contains several enzymes that are essential for the infection process. One of these is an RNA-dependent RNA polymerase (RNA replicase). As previously discussed

(↩ Section 9.7), the presence of such an enzyme is essential because the genome of negative-strand viruses cannot be translated directly but must first be converted into the positive strand; host enzymes that transcribe RNA from an RNA template are not available.

The RNA of rhabdoviruses is transcribed in the cytoplasm into two distinct classes of RNAs (**Figure 21.16**). The first is a series of mRNAs encoding the structural genes of the virus (for example, VSV has five genes). Each mRNA is monocistronic, encoding a single protein. The second type of RNA is a positive-strand RNA that is a copy of the complete viral genome (the VSV genome is 11,162 nucleotides long). These full-length plus-strand RNAs are templates for the synthesis of negative-strand genomic RNA molecules for progeny virions. Once mRNA encoding the virus RNA polymerase is made in the primary transcription process, synthesis of more copies of the virus RNA polymerase occurs, leading to the formation of many plus-strand RNA molecules, both mRNAs and genomic RNA templates (Figure 21.16).

Assembly of Rhabdoviruses

Translation of viral mRNAs leads to the synthesis of viral coat proteins. Assembly of an enveloped virus is considerably more complex than assembly of a naked virion. Two kinds of coat proteins are formed, nucleocapsid proteins and envelope proteins. The nucleocapsid is formed first by association of the nucleocapsid protein molecules around the viral RNA.

The envelope proteins possess hydrophobic amino acid leader sequences at their amino-terminal ends (↩ Section 6.21). As these proteins are synthesized, sugar residues are added, leading to the formation of glycoproteins. Such glycoproteins are characteristic of

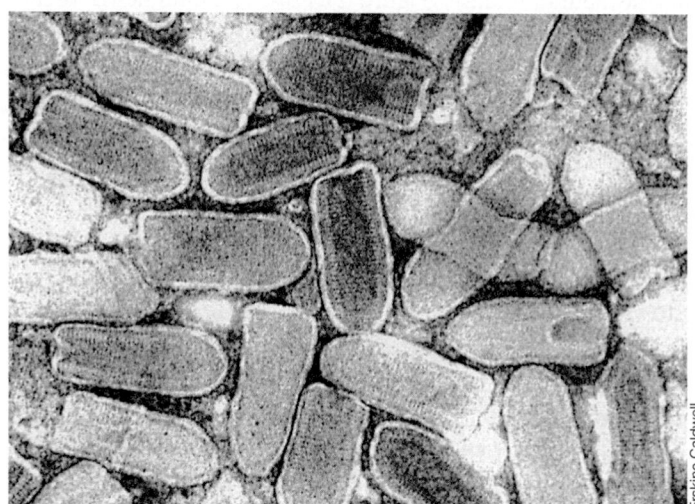

Figure 21.15 Electron micrograph of a rhabdovirus (vesicular stomatitis virus). A particle is about 65 nm in diameter.

Erskine Caldwell

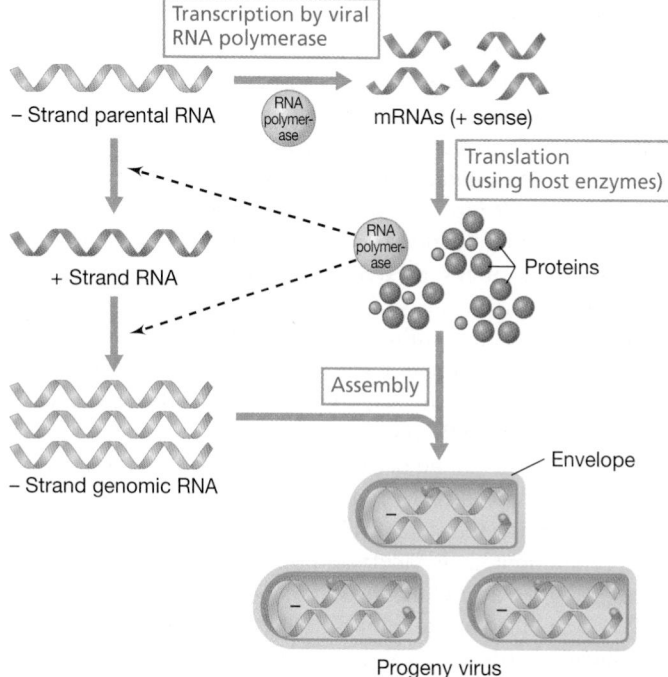

Figure 21.16 Flow of events during replication of a negative-strand RNA virus. Note the importance of the viral RNA polymerase, carried in the virion. This is critical because animal cells cannot make RNA using an RNA template.

eukaryotic membrane-associated proteins. Thus they migrate to the cytoplasmic membrane, where the leader sequences are removed and they are inserted into the membrane. Nucleocapsids then migrate to the areas on the cytoplasmic membrane where these virus-specific glycoproteins are located, recognizing the virus glycoproteins with great specificity. The nucleocapsids align with the glycoproteins and bud through them, becoming coated by the glycoproteins in the process.

The final result is an enveloped virion with an internal nucleocapsid surrounded by a membrane whose lipid is derived from the host cytoplasmic membrane but whose proteins are encoded by the virus. The budding process itself does not cause detectable damage to the cell, which may continue to release virions in this way for a considerable time. Host damage does eventually occur, but is brought about by factors other than just the production of virus.

Influenza and Other Orthomyxoviruses

Another group of negative-strand viruses is the orthomyxoviruses, which contain the important human pathogen, *influenza virus*. The term "myxo" refers to the mucus or slime of cell surfaces with which these viruses interact. In the case of influenza virus, this mucus is on the mucous membrane of the respiratory tract; thus influenza virus is primarily transmitted by the respiratory route (⟲ Section 33.8). The term "ortho" distinguishes this group from another family of negative-strand viruses, the paramyxovirus group. The paramyxoviruses, which include such important human pathogens as mumps and measles viruses, are quite similar to rhabdoviruses in their molecular biology.

The orthomyxoviruses have been extensively studied over many years, beginning with early work during the 1918 influenza pandemic that caused the deaths of millions of people worldwide. The orthomyxoviruses are enveloped viruses in which the viral RNA is present in the virion in a number of separate pieces. The genome of the orthomyxoviruses is thus segmented. In the case of influenza A virus, the genome is segmented into eight linear single-stranded molecules ranging in size from 890 to 2341 nucleotides. The influenza virus nucleocapsid is of helical symmetry, about 6–9 nm in diameter and about 60 nm long. This nucleocapsid is embedded in an envelope that has a number of virus-specific proteins as well as lipid derived from the host (**Figure 21.17**).

Because of the way influenza virus buds as it leaves the cell, the virion has no defined shape and is said to be pleomorphic (Figure 21.17*a*). Several proteins on the outside of the virion envelope interact with the host cell surface. One of these is hemagglutinin, so named because it causes agglutination (clumping) of red blood cells. The red blood cell is not the normal host for influenza virus but it has on its surface the same type of membrane component, sialic acid, as the mucous membrane cells of the respiratory tract. Thus, the red blood cell is a convenient cell type for assaying agglutination activity. An important feature of the influenza virus hemagglutinin is that antibodies against this hemagglutinin prevent the virus from infecting a cell. Consequently, antibody against the hemagglutinin neutralizes the virus, and this is the mechanism by which immunity to influenza is brought about by immunization (⟲ Section 33.8).

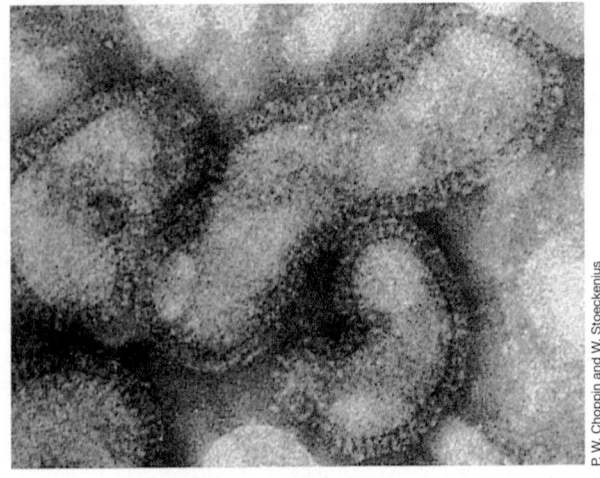

(a)

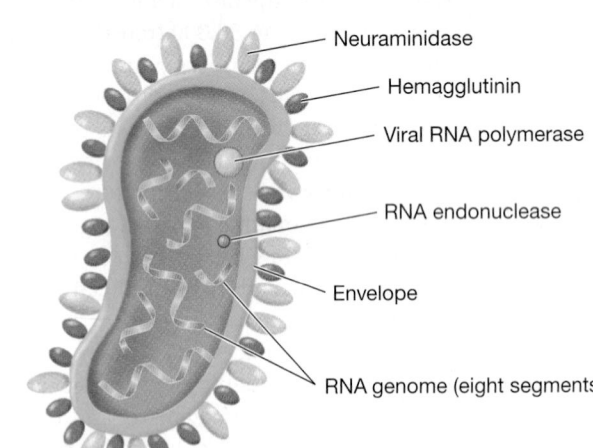

(b)

Figure 21.17 **Influenza virus.** *(a)* Electron micrograph of human influenza virions. *(b)* Diagram, showing some of the components including the segmented genome.

A second type of protein on the influenza virus surface is the enzyme neuraminidase (Figure 21.17*b*). Neuraminidase breaks down the sialic acid component of the host cytoplasmic membrane, which is a derivative of neuraminic acid. Neuraminidase functions primarily in virus assembly, destroying host membrane sialic acid that would otherwise block assembly or become incorporated into the mature virus particle.

Replication of Influenza Virus

In addition to neuraminidase, influenza virions possess two other key enzymes, an RNA-dependent RNA polymerase (RNA replicase), which converts the negative-strand genome into a positive strand (as already discussed for the rhabdoviruses), and an RNA endonuclease, which cuts a primer from the host's capped mRNA precursors.

After the virion enters the cell, the nucleocapsid separates from the envelope and migrates to the nucleus. The viral genome then replicates in the nucleus. Uncoating activates the virus RNA replicase. The mRNA molecules are then transcribed in the nucleus from the virus genomic RNA, using oligonucleotide primers cut from the 5′ ends of newly made, capped cellular mRNAs by the viral endonuclease. Thus, the viral mRNAs have 5′ caps. The poly(A) tails of the viral mRNAs are added, and the virus mRNA molecules move to the cytoplasm for translation. Thus, although influenza virus RNA replicates in the nucleus, influenza virus proteins, like all proteins, are synthesized in the cytoplasm.

Ten proteins are encoded by the eight segments of the influenza virus genome. The mRNAs transcribed from six segments each encode a single protein, whereas the other two segments encode two proteins each. The latter are not expressed by using true polycistronic mRNA as in prokaryotes because eukaryotic ribosomes recognize only the AUG closest to the 5′ end of the mRNA as a start codon (ᙂ Section 7.9). Therefore, they can make only one protein from a given RNA. The original full-length mRNAs transcribed from these two segments are each translated to give one protein. But in each case, an additional protein is translated from these messages following processing of the message by the host's RNA splicing machinery.

Some of the viral proteins are needed for influenza virus RNA replication, whereas others are structural proteins of the virion. The overall pattern of genomic RNA synthesis resembles that of the rhabdoviruses, with primary RNA synthesis resulting in the formation of positive-strand RNA that is then used as a template for making negative-strand RNA molecules. The complete enveloped virion forms by budding, as for the rhabdoviruses.

Antigenic Shift versus Antigenic Drift in Influenza

The segmented genome of the influenza virus has important practical consequences. Influenza virus and other viruses of this family exhibit a phenomenon called **antigenic shift** in which portions of the RNA genome from two genetically distinct strains of virus infecting the same cell are reassorted. This generates virions that express a set of surface proteins significantly different from that of both original viruses. Surface proteins are the major target of antibodies that form as a result of immunization by artificial means or through natural infection. However, after an antigenic shift, previous immunity to either original influenza virus is insufficient to ward off infection by the genetically novel viruses. Antigenic shift is thought to bring about major pandemics and epidemics of influenza because immunity to the new forms of the virus is essentially absent from the population.

Antigenic shift can be contrasted with **antigenic drift**. In the latter, the structure of the neuraminidase and hemagglutinin proteins on the surface of the influenza virus virion is altered, usually in a subtle way, by mutation in the genes encoding them. The alterations change the surface properties of influenza virus sufficiently that antibodies that recognized the virus previously no longer do so, or do so less effectively. Thus, for effective immunity to be achieved, new antibodies must be produced. This is a major reason why influenza vaccines rarely confer protection for more than 1 year (ᙂ Section 33.8). Genetic drift during the previous year yields slightly modified forms of influenza virus for which new vaccines must be made.

21.10 Double-Stranded RNA Viruses: Reoviruses

Reoviruses are an important family of animal viruses with double-stranded RNA (dsRNA) genomes. Rotavirus, a typical reovirus, is the most common cause of diarrhea in infants from 6 to 24 months of age and is responsible for massive mortality among infants in many developing nations. Other reoviruses cause respiratory infections and some infect plants; we previously mentioned that the bacteriophage φ6 seems closely related to the reoviruses (ᙂ Section 21.1). Reovirus virions consist of a nonenveloped nucleocapsid 60–80 nm in diameter, with a double shell of icosahedral symmetry (**Figure 21.18**). Predictably, the virions of these dsRNA viruses contain virus-encoded enzymes needed to synthesize mRNA and replicate the RNA genomes.

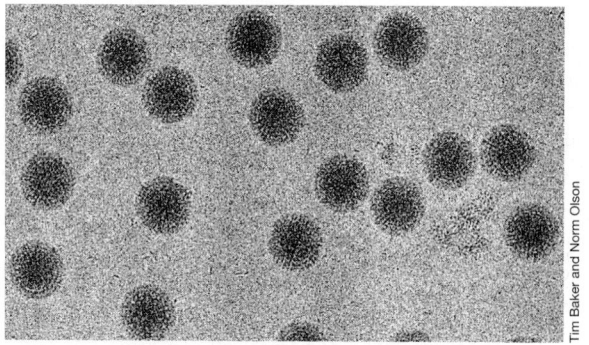
(a)

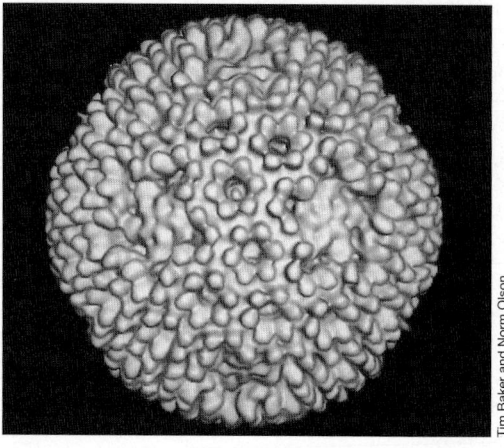
(b)

Figure 21.18 Double-stranded RNA viruses: The reoviruses. *(a)* An electron micrograph showing reovirus virions (each with a diameter of about 70 nm). *(b)* Three-dimensional reconstruction of a reovirus virion calculated from electron micrographs of frozen-hydrated virions.

The genome of reoviruses is segmented into 10–12 molecules of linear dsRNA. Replication occurs exclusively in the cytoplasm of the host. The dsRNA is inactive as mRNA, and the first step in reovirus replication is synthesis of plus-sense mRNA by a viral-encoded RNA-dependent RNA polymerase, using the minus strand as a template. The mRNA is capped and methylated by viral enzymes and then translated.

Generally, each molecule of RNA in the genome encodes a single protein, although in a few cases the protein formed is cleaved to yield the final products. However, one of the reovirus mRNAs actually encodes two proteins, but the RNA does not have to be processed in order to translate both of these. Instead, a ribosome sometimes "misses" the start codon for the first gene in this message and travels on to the start codon of the second gene. This is one of the occasional exceptions to the generalization that eukaryotic ribosomes initiate at the first AUG codon in an mRNA.

In the initial infection process, the reovirus virion binds to a cellular receptor protein. The attached virus then enters the cell and is transported into lysosomes (Section 20.5), where normally it would be destroyed. However, within the lysosome the outer shell of the virus particle is modified by removal of two proteins and cleavage of another by lysosomal enzymes. The viral core, surrounded only by the inner protein shell, is then released into the cytoplasm of the host cell.

This uncoating process activates the viral RNA polymerase and hence initiates virus replication. This occurs within the viral core, called the subviral particle, which remains intact in the cell. Each of the ten capped, single-stranded plus-strand RNAs serve as templates for the synthesis of complementary minus-strand RNA. This eventually yields progeny viral genomic dsRNA that is encapsidated. When enough viral capsid proteins are present, mature virions are assembled and released by cell lysis.

MiniQuiz

- What does the reovirus genome consist of?
- How does reovirus genome replication resemble that of influenza virus, and how does it differ?

21.11 Retroviruses and Hepadnaviruses

In Section 9.12 we discussed the **retroviruses**, a group of viruses that replicates through reverse transcription using the enzyme reverse transcriptase. There are two different types of viruses that use reverse transcriptase, and they differ in the type of nucleic acid in their genomes. The retroviruses have RNA genomes, whereas the hepadnaviruses have DNA genomes. In this section we deal with examples of both replication patterns.

Retroviruses: General Principles

Recall that retroviruses have enveloped virions that contain two copies of the RNA genome (Figure 9.24). The virion also contains several enzymes, including reverse transcriptase, and also a specific tRNA. Enzymes for retrovirus replication must be carried in the virion because although the retroviral genome is of the plus sense and is capped and tailed, it is not used directly as mRNA. Instead, one of the copies of the genome is converted to DNA by reverse transcriptase and is integrated into the host genome. The DNA that is eventually formed is a linear double-stranded molecule and is synthesized in the cytoplasm within an uncoated viral core particle. An outline of the steps in reverse transcription is given in **Figure 21.19**.

Activity of Reverse Transcriptase

Reverse transcriptase is essentially a DNA polymerase, but it actually possesses three enzymatic activities: (1) **reverse transcription** (the synthesis of DNA from an RNA template), (2) synthesis of DNA from a DNA template, and (3) ribonuclease H activity (an enzymatic activity that degrades the RNA strand of an RNA:DNA hybrid). Like all DNA polymerases, reverse transcriptase needs a primer for DNA synthesis. The primer for retrovirus reverse transcription is a specific cellular tRNA (Figure 21.19). This is packaged in the virion during its assembly in the previous host cell.

Using the tRNA primer, the 100 or so nucleotides at the 5′ terminus of the RNA are reverse-transcribed into DNA. Once reverse transcription reaches the 5′ end of the RNA, the process stops. To copy the remaining RNA, which is the bulk of the RNA of the virus, a different mechanism comes into play. First, terminally redundant RNA sequences at the 5′ end of the molecule are removed by the ribonuclease H activity of reverse transcriptase. This leads to the formation of a small, single-stranded DNA that is complementary to the RNA segment at the other end of the viral RNA. This short, single-stranded piece of DNA then hybridizes with the other end of the viral RNA molecule, where copying of the viral RNA sequences recommences.

As summarized in Figure 21.19, continued reverse transcription and ribonuclease H activities lead to the formation of a double-stranded DNA molecule with long terminal repeats (LTRs) at each end. These LTRs take part in the integration process and also contain strong transcriptional promoters. The integration of the viral DNA into the host genome is analogous to the integration of virus Mu DNA (Section 21.4). Integration can occur anywhere in the DNA of the host cell chromosomes, and once integrated, the retroviral DNA is a permanent component of the host chromosome.

Retroviral Gene Expression, Processing, and Virion Assembly

The integrated retroviral genome may be expressed, or it may remain in a latent state and not be expressed. If the promoters in the right LTR are activated, the retroviral DNA is transcribed by a cellular RNA polymerase into RNA transcripts that are capped and polyadenylated. These RNA transcripts may be either packaged into virions as genomic RNA or they may be used as mRNA and processed and translated to give virus proteins.

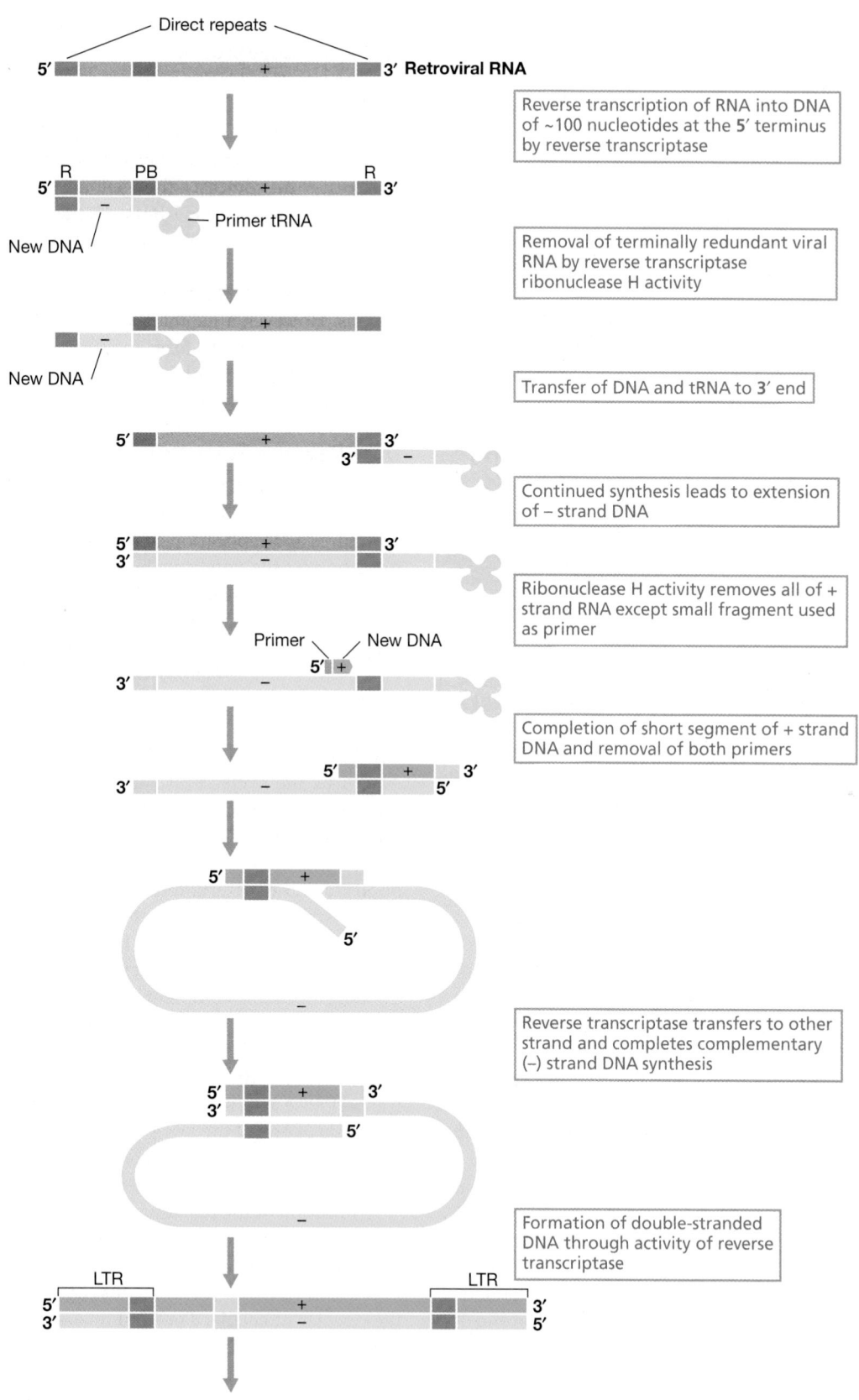

Integration into host chromosomal DNA to form provirus state

Figure 21.19 Formation of double-stranded DNA from retrovirus single-stranded RNA. The sequences labeled R on the RNA are direct repeats found at either end. The sequence labeled PB is where the primer (tRNA) binds. Note that DNA synthesis has yielded longer direct repeats on the DNA than were originally on the RNA. These are called long terminal repeats (LTRs).

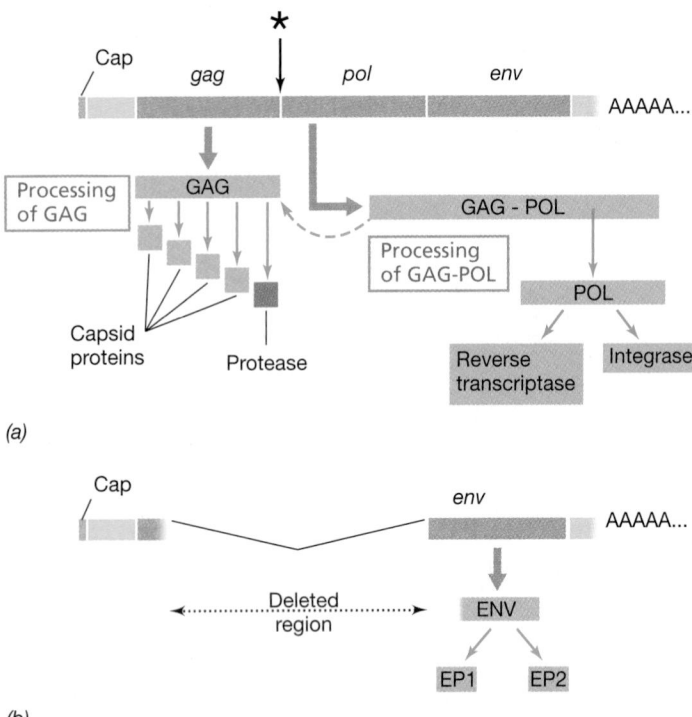

(a)

(b)

Figure 21.20 Translation of retrovirus mRNA and processing of the proteins. *(a)* The full-length mRNA with the three genes *gag*, *pol*, and *env* is shown at the top. The asterisk shows the site where a ribosome must read through a stop codon or do a precise shift of reading frame to synthesize the GAG-POL polyprotein. The thick blue arrows indicate translation, and the thin blue arrows indicate protein-processing events. One of the *gag* gene products is a protease. The POL product is processed to give reverse transcriptase (RT) and integrase (IN). *(b)* The mRNA has been processed to remove most of the *gag-pol* region. This shortened message is translated to give the ENV polyprotein, which is cleaved into two envelope proteins (EP), EP1 and EP2.

The translation and processing of such an mRNA from a retrovirus is shown in **Figure 21.20**. All retroviruses have the three genes *gag*, *pol*, and *env*, arranged in that order in the genome. The *gag* gene at the 5′ end of the mRNA actually encodes several small viral structural proteins. These are first synthesized as a polyprotein that is subsequently processed by a protease (which itself is part of the polyprotein). The structural proteins make up the capsid, and the protease is packaged in the virion.

Next, the *pol* gene is translated into a large polyprotein that also contains the *gag* proteins (Figure 21.20). Compared to structural proteins, the *pol* products are required in only small amounts. This is achieved because their synthesis requires the ribosome to make an error. To produce *pol* gene products, the ribosome must either read through a stop codon at the end of the *gag* gene or make a precise switch to a different reading frame in this region. These are rather rare events.

Once produced, the *pol* gene product is processed in two ways. First, it is removed from the *gag* proteins, and second, reverse transcriptase is cleaved from integrase, a protein required for DNA integration that is packaged in the virion along with reverse transcriptase (Figure 21.20). For the *env* gene to be translated, the full-length mRNA is first processed to remove the *gag* and *pol*

regions. The *env* product is made and then processed into two distinct envelope proteins.

HIV as a Retrovirus

Although the replication pattern of retroviruses described above is complex, this is actually the pattern typical of a rather "simple" retrovirus. In contrast, the genome of the retrovirus human immunodeficiency virus 1 (HIV-1), the causative agent of AIDS, is more complex than this and includes several additional small genes. Its expression not only requires extensive protein processing by proteases but also complex patterns of alternative splicing of introns. However, a hallmark of all retroviruses is the many protease steps required to form mature proteins. Consequently, specific protease inhibitors have been developed for treating retroviral infections, including HIV-1 infections (⟲⟳ Section 33.14).

DNA Reverse-Transcribing Viruses: Hepadnaviruses

The life cycles of viruses show a variety of unexpected genome structures and replication schemes. But none is stranger than the **hepadnaviruses**, such as human hepatitis B virus, a serious bloodborne pathogen (⟲⟳ Section 33.11). The term "hepadnavirus" comes from the fact that the virus infects the liver (thus "hepa") and the genome consists of DNA (thus "dna"). Hepatitis B virions are small, irregular, rod-shaped particles (**Figure 21.21a**).

The genomes of hepadnaviruses are unusual for several reasons, including the fact they are among the smallest known of any viruses, some 3–4 kb, and are only *partially* double-stranded. The virus life cycle is also unique and very complex. Like the retroviruses, hepadnaviruses use reverse transcriptase in their replication cycle. However, unlike retroviruses, the DNA genome of hepadnaviruses is replicated via an RNA intermediate, the opposite of the pattern in retroviruses.

One strand of the genome of hepadnaviruses is incomplete, and both strands have gaps (Figure 21.21b). Nevertheless, the two strands are held together in a circular form by hydrogen bonding between complementary base pairs. On entering the cytoplasm, a viral DNA polymerase carried in the virion completes the replication of this molecule. This polymerase is extremely versatile. It contains DNA polymerase plus reverse transcriptase activities and also functions as a protein primer for synthesis of one of the DNA strands! Despite the small size of the hepadnavirus genome, it encodes several proteins using overlapping genes, as is frequently seen in small viruses.

Replication of the hepadnavirus genome involves transcription by host RNA polymerase (in the nucleus), yielding a transcript with terminal repeats (Figure 21.21b). The repeats are made because the polymerase proceeds slightly more than once around the circular molecule. The viral reverse transcriptase then copies this into DNA, very much as in the replication of retroviruses, but in this case the DNA (rather than RNA as in retroviruses) becomes packaged into newly formed virions.

Hepadnaviruses are thus incredible examples of making the most of a small genome. The hepatitis B virus, for example, at 3.4 kb has a smaller genome than either the single-stranded RNA or DNA bacteriophages we explored at the beginning of this chapter

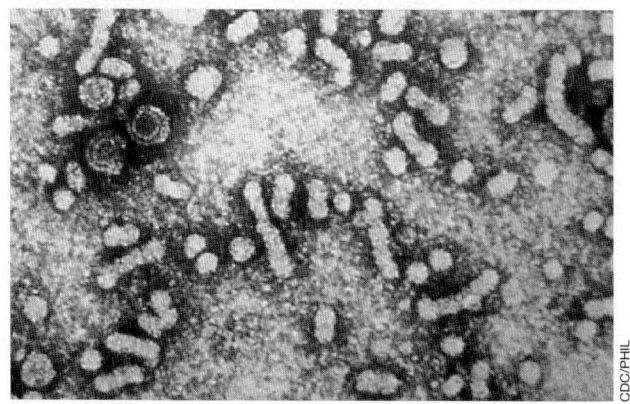

(a)

(b)

Figure 21.21 Hepadnaviruses. *(a)* Electron micrograph of hepatitis B virions. *(b)* Hepatitis B genome. The partially double-stranded genome is shown in green. Note that the positive strand is incomplete. The sizes of the transcripts are also shown. All of the genes in the hepatitis B virus overlap, and collectively, they cover every base in the genome. Reverse transcriptase produces the DNA genome from a single genome-length mRNA made by host RNA polymerase.

(Sections 21.1–21.3). Nevertheless, this small virus can cause a serious human disease (↩ Section 33.11).

MiniQuiz

- Why are protease inhibitors an effective treatment for human AIDS?

- How does the role of reverse transcriptase in the replication cycles of retroviruses and hepadnaviruses differ?

III DNA Viruses of Eukaryotes

DNA viruses of eukaryotes are fewer in number than RNA viruses. Both single-stranded and double-stranded DNA viruses are known for both animals and plants. The dsDNA viruses include some interesting large viruses of plants and protozoa. Many dsDNA viruses of animals are also known, including several human

pathogens. These include the polyoma- and papillomaviruses, the herpesviruses, the pox viruses, and the adenoviruses. The genomes of all of these replicate in the nucleus, except for the pox viruses, which replicate in the cytoplasm. In the following sections, we briefly discuss the replication of each of these groups of viruses.

21.12 Plant DNA Viruses

DNA viruses of plants are especially rare. However, some unusually large viruses are known that infect single-celled plants. Related viruses infect certain protozoa (Microbial Sidebar, "Mimivirus and Viral Evolution").

DNA Plant Viruses: *Chlorella* Viruses

Green algae are plants, and *Chlorella* is a widely distributed genus of microscopic, single-celled green alga. Most species of *Chlorella* are free-living, but some *Chlorella*-like algae are endosymbionts of freshwater or marine animals, including protozoa such as *Paramecium*. Many of these endosymbionts can be grown in the laboratory independently of the organism in which they reside, and some of these are hosts to viruses; the best studied of these is *Paramecium bursaria Chlorella* virus 1 (PBCV-1). PBCV-1 belongs to a large family of viruses known as phycodnaviruses, which are widespread in nature. Members of this family infect both *Chlorella* strains found as endosymbionts in other cells and also many free-living species of single-celled algae.

PBCV-1 has large icosahedral virions (**Figure 21.22**) and a linear dsDNA genome. The virions have a lipid component that is essential for infectivity but it is inside the capsid; therefore, the virion is not enveloped. The genomes of the *Chlorella* viruses are extremely large; all are over 300 kbp (for a comparison with other viruses see ↩ Table 9.1). In many cases the DNA is also extensively modified by methylation. The genome of PBCV-1 has been completely sequenced. This

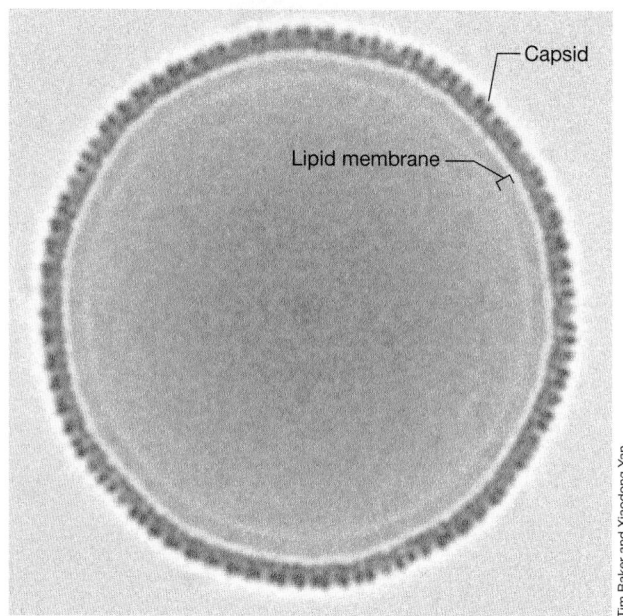

Figure 21.22 A cross section of a virion of the *Chlorella* virus PBCV-1. A lipid bilayer membrane is visible beneath the capsid shell. The virion has a diameter of approximately 170 nm. The image is reconstructed from several transmission electron micrographs.

Mimivirus and Viral Evolution

Mimivirus is the largest virus presently known, both in terms of its physical size and in genome size. When first discovered, it was misidentified as a gram-positive coccus because it stained with the Gram stain and is as large as some small bacteria. Mimivirus has an icosahedral capsid about 0.5 μm in diameter surrounded by filaments of 125 nm, giving a total diameter of about 0.75 μm (Figure 1 photomicrograph). The capsid shows three layers of dense matter under the electron microscope, probably consisting of two lipid membranes inside a protein shell.

Mimivirus contains 1.2 Mbp of double-stranded DNA that encodes an estimated 911 proteins. Its genome is thus over twice as large as that of the next largest known virus, bacteriophage G of *Bacillus subtilis* (497,513 bp) and three times that of the *Paramecium bursaria Chlorella* virus 1 (330,742 bp) discussed in Section 21.12. In addition, the Mimivirus genome is larger than that of several cellular organisms, notably the mycoplasmas. Of the 911 Mimivirus proteins, only about 300 have functions predicted by homology. However, of the 600 "unknown/unidentified" proteins with no close relatives in any sequence database, nearly 50 are found in the virus particle. Thus, at least these 50 are genuine functional proteins, not merely "genetic junk" as is sometimes suggested for the many unidentified genes and proteins found in viruses.

Mimivirus normally infects the amoeba *Acanthamoeba polyphaga*. It may also cause pneumonia in humans, although the evidence is largely indirect and consists of finding antibodies to Mimivirus in certain pneumonia patients. However, a laboratory technician working with Mimivirus developed pneumonia that was almost certainly due to infection by Mimivirus because virus particles were isolated from the technician's blood.

Mimivirus belongs to a group of large viruses that contain large genomes known as the *nucleocytoplasmic large DNA viruses* (NCLDV) (Figure 1 phylogenetic tree). The NCLDV comprises several virus families, including the pox viruses (Section 21.15) and phycodnaviruses (Section 21.12). These viruses all share a set of highly homologous proteins, mostly involved in DNA replication. All NCLDV also share an internal lipid layer surrounding the central core with a protein shell outside the lipid. Thus, these viruses are not truly enveloped. All of these large viruses contain not only certain enzymes but also mRNA inside the virion. In particular, they contain mRNA encoding virus DNA polymerase.

Database searches of DNA sequences present in environmental samples implies that unisolated large viruses related to Mimivirus are reasonably frequent in nature. In particular, genes related to those of Mimivirus have been found among DNA sequences from marine environments.

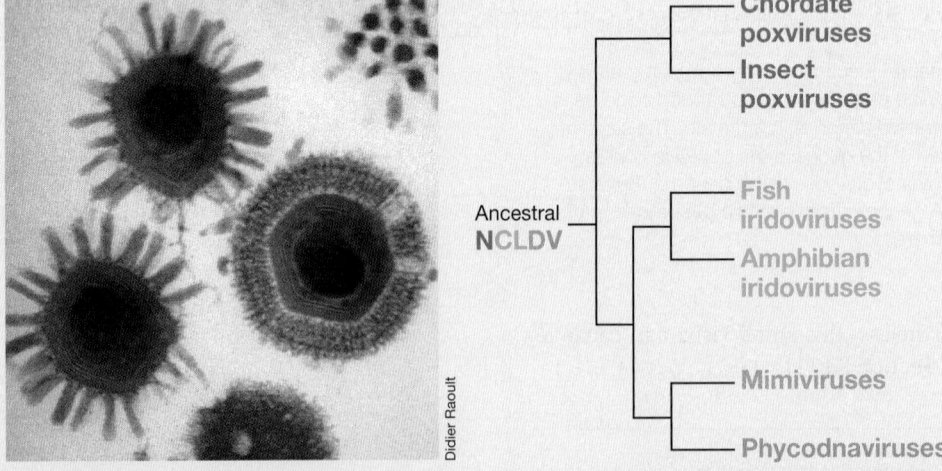

Didier Raoult

Figure 1 Mimivirus: The largest known virus. Mimivirus virions are shown in this thin-section transmission electron micrograph. Each virion is approximately 0.75 μm in diameter, roughly three-quarters the diameter of an *Escherichia coli* cell. Mimivirus is a member of the Nucleocytoplasmic large DNA viruses (NCLDV). The phylogenetic tree shows the genomic relationships between the poxviruses, iridoviruses (large viruses of fish and amphibians), phycodnaviruses (large double-stranded DNA viruses of plants and some green algae), and Mimivirus.

330,742-bp genome encodes over 370 different proteins and 10 tRNAs. The ends of the genome are incompletely base-paired hairpin loops very similar to those of pox viruses (Section 21.15).

Replication of *Chlorella* Viruses

PBCV-1 enters cells somewhat like bacteriophages do. The virion binds specifically to the cell wall of the host, and then at least five different enzymes carried by the virion digest away the cell wall at the point of contact. Viral DNA is then released into the cell, leaving the empty virion behind. As for most dsDNA viruses of eukaryotes, the DNA of PBCV-1 is replicated in the nucleus, and RNA is also synthesized there. PBCV-1 encodes several enzymes for DNA replication, including a DNA polymerase. However, although PBCV-1 encodes some transcription factors, it does not encode its own RNA polymerase. A few of the genes of PBCV-1 contain introns that must be removed (Section 7.8). The virus mRNA is capped and some of the enzymes involved are virus encoded. Early mRNA, but not late mRNA, has poly(A) tails. Like bacteriophage T4 and some other large DNA viruses, PBCV-1 encodes several of its own tRNAs.

There is one other extremely interesting fact to note about the *Chlorella* viruses: Their genomes encode several restriction and modification enzyme systems (Section 11.1). Indeed, they are the only source of restriction enzymes outside of the prokaryotes and a few bacteriophages. Thus we see features in *Chlorella* viruses typical of both prokaryotic and eukaryotic genetic elements.

MiniQuiz
- Which features of *Chlorella* viruses are prokaryote-like?
- Which features of *Chlorella* viruses are eukaryote-like?
- What is unusual about the Mimivirus genome?

21.13 Polyomaviruses: SV40

Some viruses of the polyomavirus family induce tumors in animals; indeed, the suffix "oma" means tumor. One of these DNA tumor viruses was called *simian virus 40* or *SV40* because it was first isolated from monkeys. SV40 was one of the first viruses to be studied by genetic engineering techniques and has been extensively used as a vector for moving genes into eukaryotic cells (Section 11.8).

The SV40 virion is a nonenveloped particle 45 nm in diameter with an icosahedral head containing 72 protein subunits. There are no enzymes in the virion. The genome of SV40 consists of one molecule of double-stranded DNA of 5243 bp. The DNA is circular (**Figure 21.23**) and exists in a supercoiled configuration within the virion. The complete base sequence of SV40 is known, and a genetic map is shown in **Figure 21.24**.

The SV40 genome is replicated in the nucleus, and the proteins are synthesized in the cytoplasm. The virion is assembled in the nucleus. The replication of SV40 can be divided into two distinct stages, early and late. During the early stage, the early region of the viral DNA is transcribed (Figure 21.24). A single RNA molecule—the primary transcript—is made by the cellular RNA polymerase, but it is then processed into two mRNAs, a large one and a small one, both of which are capped. Introns are present in

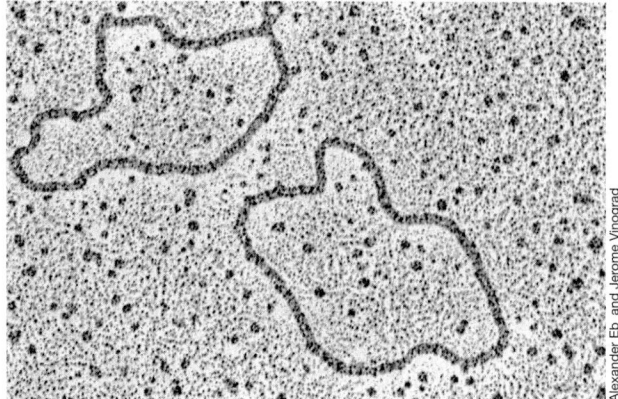

Figure 21.23 Polyomaviruses. Electron micrograph of relaxed (nonsupercoiled) circular DNA from a tumor virus. The contour length of each circle is about 1.5 μm.

the SV40 genome and are excised out of the primary RNA transcript. In the cytoplasm, the mRNAs are translated to yield two proteins. One of these proteins, the T antigen, binds to the site on the virus DNA that is the origin of replication; this initiates viral genome synthesis.

The genome of SV40 is too small to encode its own DNA polymerase, so host DNA polymerases are used. DNA is replicated in a bidirectional fashion from a single origin of replication. The process involves the same events that have already been described for host cell DNA replication (Sections 6.9 and 6.10).

Late SV40 mRNA molecules are synthesized using the strand complementary to that used for early mRNA synthesis. Transcription begins at a promoter near the origin of replication. This late RNA is then processed by splicing, capping, and polyadenylation

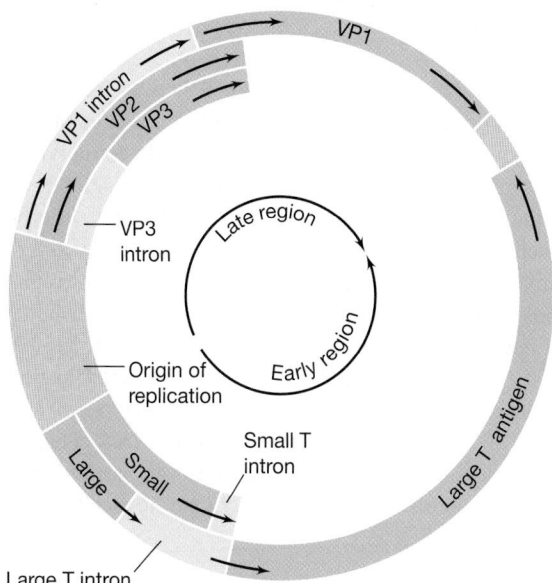

Figure 21.24 Genetic map of the polyomavirus SV40. VP1, VP2, and VP3 are the genes encoding the three proteins that make up the coat of SV40. The arrows show the direction of transcription. Note how genes encoding VP1, VP2, and VP3 overlap.

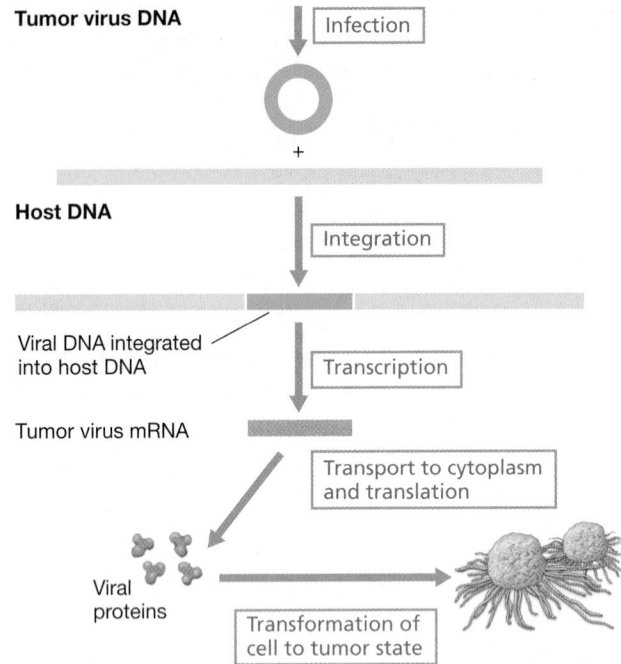

Tumor virus DNA

Infection

+

Host DNA

Integration

Viral DNA integrated into host DNA

Transcription

Tumor virus mRNA

Transport to cytoplasm and translation

Viral proteins

Transformation of cell to tumor state

Figure 21.25 Events in cell transformation by a polyomavirus such as SV40. Either all or portions of the viral DNA are incorporated into host cell DNA. The viral genes responsible for cell transformation are transcribed and processed to viral mRNA molecules, which are transported to the cytoplasm. Here they are translated to form proteins that transform host cells into tumor cells.

to yield mRNA corresponding to the three coat proteins: VP1, VP2, and VP3. The genes for these proteins overlap (Figure 21.24), a phenomenon seen in several other small viruses. SV40 coat protein mRNAs are transported to the cytoplasm and translated into the viral coat proteins. The latter are then transported back into the nucleus where virion assembly occurs. New SV40 virions are released by cell lysis.

Some polyomaviruses cause cancer. When a virus of the polyomavirus group infects a host cell, one of two modes of replication can occur, depending on the type of host cell. In some types of host cells, known as permissive cells, virus infection results in the usual formation of new virions and the lysis of the host cell. In other types of host cells, known as nonpermissive, efficient replication does not occur. Instead, the virus DNA becomes integrated into host DNA, analogous to a prophage (Section 9.10), genetically altering the cells in the process (**Figure 21.25**). Such cells can show loss of growth inhibition and become tumor cells, a process called transformation (Figure 9.22). As in certain tumor-causing retroviruses (Section 9.12), expression of specific polyomavirus genes converts cells to the transformed state (Figure 21.25). In particular, the large and small T proteins bind to and inactivate host cell proteins that control cell division.

MiniQuiz

- Why doesn't a virus like SV40 need to carry enzymes in the virion as influenza virus does?

- How can one transcript yield more than one mRNA in SV40?

21.14 Herpesviruses

The herpesviruses are a large group of double-stranded DNA viruses that cause diseases in humans and animals, including fever blisters (cold sores), venereal herpes, chicken pox, shingles, and infectious mononucleosis. Some of these diseases are discussed in Chapter 33. Herpesviruses are able to remain latent in the body for long periods of time, becoming active only under conditions of stress or when the immune system is compromised. The latent herpesvirus genomes replicate in step with the host cell genome. Both **herpes simplex**, the virus that causes fever blisters and genital herpes, and varicella-zoster virus, the cause of chicken pox (varicella) and shingles (zoster), are able to remain latent in the neurons of the sensory ganglia, from which they emerge periodically to cause infections of the skin.

An important group of herpesviruses are tumorigenic, causing clinical forms of cancer. For example, Epstein–Barr virus causes Burkitt's lymphoma, a tumor common among children in central Africa and New Guinea. Burkitt's lymphoma was among the first human cancers to be linked to virus infection. Epstein–Barr virus can also cause a nonmalignant general malaise called infectious mononucleosis in humans. Human herpesvirus 8 (HHV-8) is responsible for Kaposi's sarcoma, an atypical cancer frequently seen in AIDS patients (Section 33.14).

Herpesviruses: General Features

The herpesvirus virion is complex, consisting of four distinct structural layers. Herpes simplex 1 is an enveloped virus about 150 nm in diameter. The center of the virus, called the core, consists of linear dsDNA. The herpesvirus nucleocapsid is of icosahedral symmetry and consists of 162 capsomeres, each of which is composed of a number of distinct proteins. Outside the nucleocapsid is an amorphous layer called the *tegument*, a fibrous structure unique to herpesviruses. Surrounding the tegument is an envelope whose outer surface contains many small spikes.

A large number of separate proteins are present within the virion, but not all of them have been characterized. The genome of herpes simplex 1 virus consists of one large linear dsDNA molecule of 152,260 bp (about 30 times larger than the SV40 genome) that encodes at least 84 distinct polypeptides.

Herpesvirus Infection and Replication

Herpesvirus infection occurs following attachment of virions to specific cell receptors. Following fusion of the cytoplasmic membrane with the virus envelope, the nucleocapsid is released into the cell. The nucleocapsids are transported to the nucleus, where viral DNA is uncoated. Proteins in the virus particle inhibit macromolecular synthesis by the host.

Following infection, three classes of mRNA are produced: immediate early, which encodes five regulatory proteins; delayed early, which encodes DNA replication proteins including DNA polymerase; and late, which encodes structural proteins of the virus particle (**Figure 21.26**). During the immediate early stage,

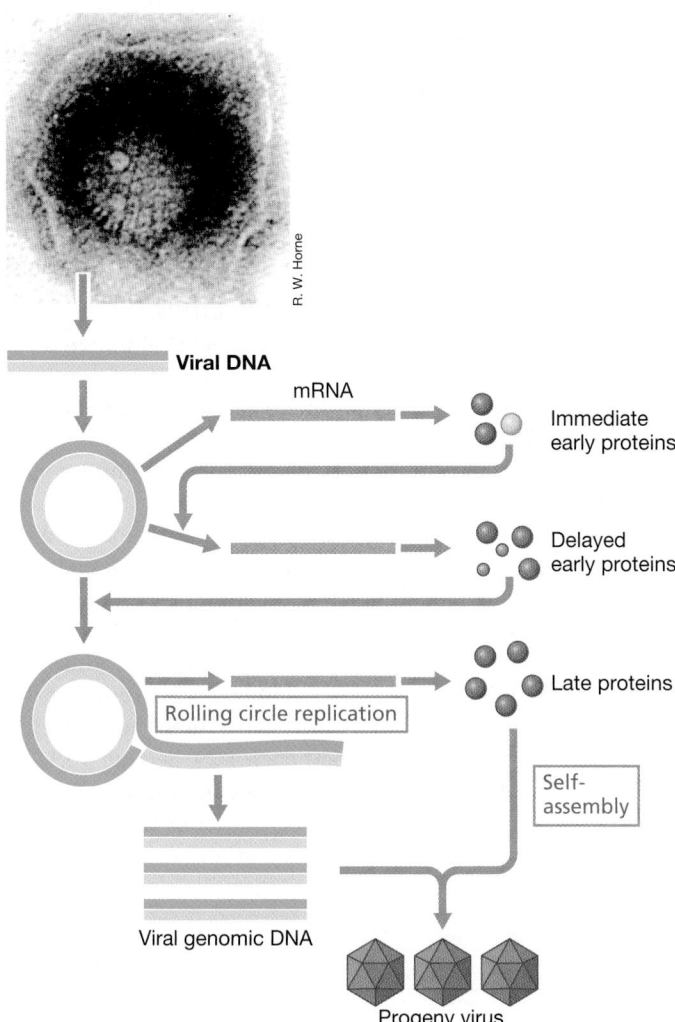

Figure 21.26 Herpesvirus. Flow of events in replication of herpes simplex virus starting from an electron micrograph of a herpes virion (diameter about 150 nm). Note in particular how the linear viral DNA circularizes during the replication process.

about one-third of the viral genome is transcribed by a host cell RNA polymerase. Early mRNA encodes certain regulatory proteins that stimulate the synthesis of the delayed early proteins. The delayed early proteins appear only after the immediate early proteins have been made. During this stage, about 40% of the viral genome is transcribed. Among the many key proteins synthesized during the delayed early stage are a viral-specific DNA polymerase, enzymes for synthesis of deoxyribonucleotides, and a DNA-binding protein. These enzymes are all needed for viral DNA replication.

Herpesvirus DNA synthesis itself takes place in the nucleus. After infection, the herpesvirus genome circularizes (remarkably like bacteriophage lambda; ↪ Section 9.10) and replicates by a rolling circle mechanism (↪ Figure 9.19). However, there seem to be three origins of replication. Long concatemers are formed that become processed into virus-length genomic DNA during the assembly process in a manner similar to that described for DNA bacteriophages (↪ Sections 9.9 and 21.3). Viral nucleocapsids are assembled in the cell

nucleus, and the viral envelope is added via a budding process through the inner membrane of the nucleus. Mature virions are subsequently released through the endoplasmic reticulum to the outside of the cell. Thus, the assembly of herpesvirus differs from that of the enveloped RNA viruses, which are assembled on the cytoplasmic membrane instead of the nuclear membrane.

Cytomegalovirus

A very widespread herpesvirus is cytomegalovirus (CMV). CMV replicates as a typical herpesvirus and is present in 50–85% of all adults in the United States by 40 years of age. For healthy individuals, infection with CMV comes with no apparent symptoms or long-term health consequences. However, CMV infection can cause serious disease in immune-compromised individuals, such as those receiving immunosuppressant drugs (for example, those who have received organ transplants and some cancer and dialysis patients) or those infected with the AIDS virus, HIV. In such individuals, CMV can cause pneumonia, retinitis (an eye condition), and gastrointestinal disease. These conditions can be serious and occasionally cause death.

In the healthy host the immune system keeps CMV in check. However, CMV remains dormant within cells of an infected individual and can become reactivated whenever the immune system is compromised, leading to symptoms whose severity depends on the degree to which the immune system is suppressed. Like the Epstein–Barr virus, CMV also causes some cases of infectious mononucleosis.

MiniQuiz

- Of what use is it to the herpesviruses to circularize their DNA prior to replication?
- Where is the herpesvirus nucleocapsid assembled and how does this affect the protein content of its envelope?
- Under what conditions is cytomegalovirus infection dangerous?

21.15 Pox Viruses

Pox viruses are among the most complex and largest animal viruses known (**Figure 21.27**). These viruses are also unique in being DNA viruses that replicate in the cytoplasm. Thus, a host cell infected with a pox virus exhibits DNA synthesis outside the nucleus, something that otherwise occurs in eukaryotic cells only in organelles.

General Properties of Pox Viruses

Pox viruses have been important historically as well as medically. Smallpox was the first virus to be studied in any detail and was the first virus for which a vaccine was developed (by Edward Jenner in 1798). By applying this vaccine worldwide, the disease smallpox has been eradicated in the wild, the first infectious disease to be eliminated in this fashion. Other pox viruses of importance are cowpox and rabbit myxomatosis virus, an important infectious agent of rabbits that was intentionally introduced

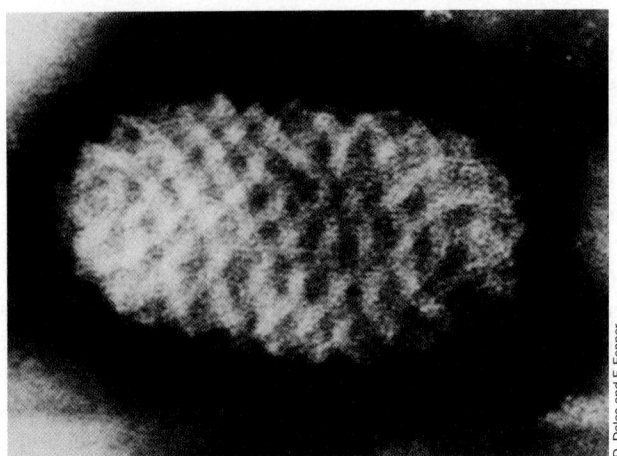

Figure 21.27 **Pox viruses.** Electron micrograph of a negatively stained vaccinia virus virion. The virion is approximately 400 nm (0.4 μm) long. Compare the size of pox virus with that of Mimivirus in the Microbial Sidebar.

in Australia to control the Australian rabbit population (♻ Section 32.5). Some pox viruses also cause tumors.

The pox viruses are very large, so large in fact, that they can be seen under the light microscope. Most research on pox viruses has been done with vaccinia virus, a close relative of smallpox virus and the virus used as a smallpox vaccine. The vaccinia virion is a brick-shaped structure about $400 \times 240 \times 200$ nm. The virion lacks an envelope but is covered on its outer surface with protein tubules arranged in a lattice-like pattern (Figure 21.27). Within the virion are two lateral bodies, composed mostly of protein, and a nucleocapsid, which contains DNA bounded by a layer of protein subunits.

The pox virus genome consists of linear double-stranded DNA. The vaccinia virus genome has about 185 kbp and about 180 genes. Pox virus DNA is unique because the two strands of the double helix are cross-linked at their termini as a result of phosphodiester bonds between adjacent strands. The ends of pox virus DNA are therefore very similar to those of the large *Chlorella* viruses discussed previously (Section 21.12).

Replication of Pox Viruses

Vaccinia virions are taken up into cells and the nucleocapsids liberated in the cytoplasm. The uncoating of the viral genome requires the activity of a viral protein that is synthesized after infection. The viral gene encoding this protein is transcribed by a viral-encoded RNA polymerase contained within the virion. In addition to this uncoating gene, a number of other viral genes are transcribed. The primary transcripts are turned into mRNAs by capping and polyadenylation while they are still inside the nucleocapsid. Copies of the genome are then made by a viral-encoded DNA polymerase.

Once the vaccinia DNA is fully uncoated, the formation of inclusion bodies within the cytoplasm begins. Within these inclusion bodies, the DNA is transcribed, replicated, and encapsidated into progeny virions. Each infecting virion initiates its own inclusion body, so the number of inclusions depends on the multiplicity of infection. Progeny DNA molecules form a pool from which individual molecules are incorporated into virions. Mature virions accumulate in the cytoplasm. The virions are released when the infected cell disintegrates, or they may be actively propelled up to and through the cell membrane by so-called "actin rockets" that use the contractile filaments of the host cell cytoskeleton.

Pox Viruses and Recombinant Vaccines

Vaccinia virus has been used as a host for genetically altered proteins of other viruses, permitting the construction of genetically engineered vaccines (♻ Section 15.13). As we will see in Section 28.7, a vaccine is a substance capable of eliciting an immune response in an animal and protects the animal from future infection with the same agent. Vaccinia virus causes no serious health effects in humans but is highly immunogenic. Therefore, as a carrier of proteins from pathogenic viruses, vaccinia virus is a relatively safe and effective tool for stimulating the immune response.

Molecular cloning methods (Chapter 11) have been used to express key proteins of various viral pathogens (including influenza virus, rabies virus, herpes simplex 1 virus, and hepatitis B virus) in vaccinia virions, which have then been used to develop a vaccine against that pathogen (♻ Section 15.13). A similar vaccine delivery system using adenovirus (discussed in the next section) as a vehicle has been developed, because, like vaccinia virus, adenoviruses are also of minor health consequence to humans.

MiniQuiz

- Why is it notable that pox viruses replicate their DNA in the cytoplasm?
- How are poxviruses of use to genetic engineering?

21.16 Adenoviruses

The adenoviruses are a major group of icosahedral linear double-stranded DNA viruses. The term *adeno* is derived from the Latin for "gland" and refers to the fact that these viruses were first isolated from the tonsils and adenoid glands of humans. Adenoviruses cause mild respiratory infections in humans, and a number of such viruses can be isolated, even from healthy individuals.

The genomes of adenoviruses consist of linear dsDNA of about 36 kbp. Attached covalently to the 5′ end of the DNA is a protein called the terminal protein, essential for replication of the DNA. The DNA also has inverted terminal repeats of 100–1800 bp (the number varies with the virus strain) that are important in the replication process. Adenoviruses are now widely used as gene therapy vectors (♻ Section 15.17).

Replication of Adenoviruses: Early Events

Adenoviral DNA replicates in the nucleus. After the virus particle has been transported to the nucleus, the nucleocapsid is released and converted to a viral DNA–histone complex. Early transcription is carried out by a host RNA polymerase, and a number of primary transcripts are made. The transcripts contain introns and so must first be spliced to yield mature transcripts that are capped and polyadenylated before translation. Adenovirus primary transcripts were the original system used to study RNA splicing in the eukaryotic nucleus because they are present at such high levels.

As is typical for many viruses, early adenoviral proteins regulate DNA replication and later proteins are structural. Viral DNA replication uses the terminal protein as a primer and another virus-encoded protein as DNA polymerase. The terminal protein contains a covalently bound cytosine residue from which DNA replication proceeds.

Replication of Adenoviruses: Genome Replication

Replication of the adenoviral genome begins at either end, the two strands being replicated asynchronously (**Figure 21.28**). The products of a round of replication are a double-stranded and a single-stranded molecule. However, then a unique replication mechanism proceeds. The single strand cyclizes by means of its inverted terminal repeats, and a new complementary strand is synthesized beginning from the 5' end (Figure 21.28). This mechanism of replication is noteworthy because it does not require the formation of a lagging strand as conventional DNA replication does (↩ Section 6.10). Instead, DNA synthesis proceeds in leading fashion on both newly synthesized DNA strands.

MiniQuiz

- How does adenovirus replicate its double-stranded DNA genome without synthesizing a lagging strand?
- Describe the structure of an adenovirus particle.

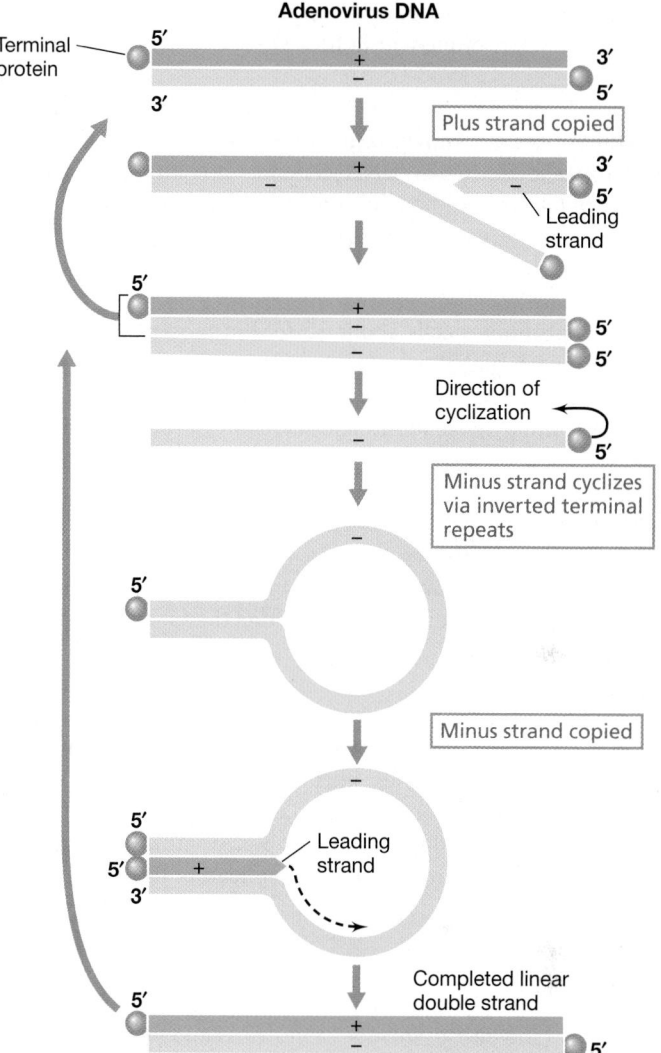

Figure 21.28 Replication of adenovirus DNA. Because of loop formation (cyclization), there is no lagging strand; DNA synthesis is leading on both strands.

Big Ideas

21.1
Many RNA viruses that infect bacteria are known. The small RNA genome of these bacterial viruses is translated directly and encodes only a few proteins.

21.2
The single-stranded DNA genome of the virus φX174 is so small that genes overlap to encode all of its essential proteins. The viral DNA replicates by a rolling circle mechanism. Some ssDNA viruses, such as M13, have filamentous virions that are released without cell lysis. These viruses are useful tools for DNA sequencing and genetic engineering.

21.3
The bacteriophage T7 double-stranded DNA genome always enters the host cell in the same orientation. The late genes in T7 are transcribed by a virus-encoded RNA polymerase. Replication of the T7 genome employs T7 DNA polymerase and involves terminal repeats and the formation of concatemers.

21.4
Bacteriophage Mu is a temperate virus that is also a transposable element. In either the lytic or lysogenic pathway, its genome is integrated into the host chromosome by the activity of a transposase. Even in the lytic pathway, its genome is replicated as part

of a larger DNA molecule. The genome is packaged into the virion with short sequences of host DNA at either end.

21.5

Several viruses are known to infect *Archaea*. Many of these have double-stranded circular DNA genomes not known in bacteriophages of *Bacteria*. Although head-and-tail-type viruses are known, many archaeal viruses have an unusual spindle-shaped morphology.

21.6

Viruses outnumber cells about 10-fold in most habitats. Consequently, most global genetic information is carried by virus genomes. Most virus genes are presently uncharacterized and many are unlike those found in cells of any of the domains of life.

21.7

Most plant viruses have positive-strand RNA genomes, and one example is tobacco mosaic virus (TMV), the first virus discovered. The genomes of these viruses can move within the plant through intercellular connections that span the cell walls.

21.8

In small positive-strand RNA viruses such as poliovirus, the viral RNA is translated directly, producing a long polyprotein that is broken down by enzymes into the numerous small proteins necessary for nucleic acid replication and virus assembly. Coronavirus is a large, single-stranded RNA virus that resembles poliovirus in some but not all of its replication features.

21.9

In negative-strand viruses the virus RNA is not the mRNA, but it is copied into mRNA by an enzyme present in the virion. Important negative-strand viruses include rabies virus and influenza virus.

21.10

Reoviruses contain segmented linear double-stranded RNA genomes. Like negative-strand RNA viruses, reoviruses contain an RNA-dependent RNA polymerase within the virion.

21.11

The retroviruses contain RNA genomes and use reverse transcriptase to make a DNA copy during their life cycle. The hepadnaviruses contain DNA genomes and use reverse transcriptase to make genomic DNA from an RNA copy. Both types of viruses have complex patterns of gene expression.

21.12

Some of the largest known viruses infect single-celled algae. The *Chlorella* viruses have very large double-stranded DNA genomes that encode several hundred proteins.

21.13

Most double-stranded DNA animal viruses, such as SV40, replicate in the nucleus. SV40 has a tiny genome and employs the strategy of overlapping genes to boost its genetic-coding potential. Some of these viruses cause cancer.

21.14

Herpesviruses are large double-stranded DNA viruses. The viral DNA circularizes and is replicated by a rolling circle mechanism. Herpesviruses cause a variety of disease syndromes and can maintain themselves in a latent state in the host indefinitely, initiating viral replication periodically.

21.15

The pox viruses are very large viruses that, unlike other DNA viruses, replicate entirely in the cytoplasm. These viruses are responsible for several human diseases, but a vaccination campaign has eradicated the smallpox virus in the wild.

21.16

Different double-stranded DNA animal viruses have different genome replication mechanisms. Adenovirus replication occurs within the nucleus. It requires protein primers and has a mechanism that avoids the synthesis of a lagging strand.

Review of Key Terms

Antigenic drift in influenza virus, minor changes in viral proteins (antigens) due to gene mutation

Antigenic shift in influenza virus, major changes in viral proteins (antigens) due to gene reassortment

Bacteriophage a virus that infects prokaryotic cells

Concatemer two or more identical linear nucleic acid molecules joined covalently in tandem

Enveloped in reference to a virus, having a lipoprotein membrane surrounding the virion

Hepadnavirus a virus whose DNA genome replicates by way of an RNA intermediate

Herpes simplex the virus that causes both genital herpes and cold sores

Negative strand a nucleic acid strand that has the opposite sense to (is complementary to) the mRNA

Overlapping genes two or more genes in which part or all of one gene is embedded in the other

Positive strand a nucleic acid strand that has the same sense as the mRNA

Polyprotein a large protein expressed from a single gene and subsequently cleaved to form several individual proteins

Replicative form a double-stranded DNA or RNA molecule that is an intermediate in the replication of viruses with single-stranded genomes

Retrovirus a virus whose RNA genome has a DNA intermediate as part of its replication cycle

Reverse transcription the process of copying genetic information found in RNA into DNA

RNA replicase an enzyme that can produce RNA from an RNA template

Rolling circle replication a mechanism used by some plasmids and viruses of replicating circular DNA, which starts by nicking and unrolling one strand and using the other, still circular strand as a template for DNA synthesis

Transposase an enzyme that catalyzes the insertion of DNA segments into other DNA molecules

Review Questions

1. What are bacteriophages? Name two examples of bacteriophages that infect *Escherichia coli* (Section 21.1).

2. What are overlapping genes? Give examples of viruses that have overlapping genes (Sections 21.1, 21.2, 21.3, 21.11, and 21.14).

3. Describe the rolling circle mechanism of DNA replication (Section 21.2).

4. Positive-strand RNA viruses are known for *Bacteria*, for plants, and for animals. What are the important distinctions between gene expression in these viruses, particularly between those that infect *Bacteria* and those that infect eukaryotes (Sections 21.1 and 21.8)?

5. One end of the T7 genome always enters the cell first during infection. Explain why this is necessary for the infection process to proceed (Section 21.3).

6. Why is bacteriophage Mu mutagenic? What features are necessary for Mu to insert into DNA (Section 21.4)?

7. What is unusual about the genome of archaeal viruses such as PAV1 compared to the genomes of double-stranded DNA viruses of bacteriophages (Section 21.5)?

8. Define the viral metagenome (Section 21.6).

9. What is the function of the VPg protein of poliovirus, and how, since their genome is also plus-sense RNA, can coronaviruses replicate without a VPg protein (Section 21.8)?

10. What is the difference between antigenic drift and antigenic shift (Section 21.9)?

11. The reovirus genome is unique in all of biology. Explain (Section 21.10).

12. Several types of animal virus can set up latent infections. Describe any two of these (Sections 21.11 and 21.14).

13. What makes retrovirus replication sensitive to protease inhibitors (Section 21.11)?

14. What three types of reaction does the reverse transcriptase of retroviruses carry out? What molecule does it use as a primer (Section 21.11)?

15. Although the genomes of herpesviruses and adenoviruses both contain double-stranded DNA, their replication features are quite distinct. Explain how they differ (Sections 21.14 and 21.16).

16. Of all the double-stranded DNA animal viruses, pox viruses stand out concerning one unique aspect of their DNA replication process. What is this unique aspect and how can this be accomplished without special enzymes being packaged in the virion (Section 21.15)?

Application Questions

1. Not all proteins are made from the RNA genome of bacteriophage MS2 in the same amounts. Can you explain why? One of the proteins functions very much like a repressor, but it functions at the translational level. Which protein is it and how does it function?

2. Give a mechanistic explanation of why the genomes of double-stranded RNA viruses might be segmented.

3. The mechanism of replication of both strands of DNA in some viruses, such as adenoviruses, is continuous (leading). Show how this can be without violating the "rule" that DNA synthesis always occurs in the overall direction of $5' \rightarrow 3'$.

4. Most genetic elements that express reverse transcriptase make it in small amounts and/or as part of a polyprotein. Can you think of any reason(s) why this might be so?

 Need more practice? Test your understanding with Quantitative Questions; access additional study tools including tutorials, animations, and videos; and then test your knowledge with chapter quizzes and practice tests at **www.microbiologyplace.com.**

Methods in Microbial Ecology

Fluorescent *in situ* hybridization (FISH) is a microscopic method that employs fluorescent dyes attached to specific nucleic acid probes to reveal the phylogeny of microorganisms in natural samples.

We now begin a new unit devoted to microorganisms in their natural habitats. We learned in Chapter 1 that *microbial communities* consist of various cell populations living in association with other populations in nature. The science of **microbial ecology** is focused on how microbial populations assemble to form communities and how these communities interact with each other and their environment.

The major components of microbial ecology are *biodiversity* and *microbial activity*. To study biodiversity, microbial ecologists must identify and quantify microorganisms in their habitats. Knowing how to do this is often helpful for isolating organisms of interest as well, another goal of microbial ecology. To study microbial activity, microbial ecologists must measure the metabolic processes that microorganisms carry out in their habitats. In this chapter we consider modern methods for assessing microbial diversity and activity. Chapter 23 will outline the basic principles of microbial ecology and examine the types of environments that microorganisms inhabit. Chapters 24 and 25 will complete our coverage of microbial ecology with a consideration of nutrient cycles, applied microbiology, and the role that microorganisms play in symbiotic associations with higher life forms.

We begin with the microbial ecologist's toolbox, which includes a collection of very powerful tools for dissecting the structure and function of microbial communities in relation to their natural habitats.

 I

Culture-Dependent Analyses of Microbial Communities

The vast majority of microorganisms, well over 99% of all species, have never been grown in laboratory cultures. Recognition of this fact, based on molecular surveys (Sections 22.3–22.7) of microbial habitats, has stimulated the development of new methods for separating out particular microbial species—that is, isolating them—to establish pure cultures. Culturing a microorganism remains the only way to fully characterize its properties and predict its impact on an environment.

In the first part of this chapter we cover the enrichment approach, a time-honored and useful method for isolating microorganisms from nature but one that is not without limitations. Enrichment is based on culturing in a selective growth medium, and the tools used in this approach are referred to, collectively, as *culture-dependent* analyses. In the second and third parts of this chapter we consider *culture-independent* analyses, techniques that can tell us much about the structure and function of microbial communities in the absence of actual laboratory cultures.

22.1 Enrichment

For an **enrichment culture**, a medium and a set of incubation conditions are established that are selective for the desired organism and counterselective for undesired organisms. Effective enrichment cultures duplicate as closely as possible the resources and conditions of a particular ecological niche.

Literally hundreds of different enrichment strategies have been devised, and **Tables 22.1** and **22.2** provide an overview of some successful ones.

Inocula

Successful enrichment cultures require an appropriate inoculum containing the organism of interest. Thus, the making of an enrichment culture begins with collecting a sample from the appropriate habitat to serve as the inoculum (Tables 22.1 and 22.2). Enrichment cultures are established by placing the inoculum into selective media and incubating under specific conditions. In this way, many common prokaryotes can be isolated. For example, the great Dutch microbiologist Martinus Beijerinck, who conceptualized the enrichment culture technique (Section 1.9), used enrichment cultures to isolate for the first time the nitrogen-fixing bacterium *Azotobacter* (**Figure 22.1**). Because *Azotobacter* is a rapidly growing bacterium capable of N_2 fixation in air, enrichment using media devoid of fixed nitrogen and incubation in air selects strongly for this bacterium and its close relatives. Non-nitrogen-fixing bacteria and anaerobic nitrogen-fixing bacteria are counterselected in this technique.

Enrichment Culture Outcomes

For success with enrichment cultures, attention to both the culture medium and the incubation conditions are important. That is, *resources* (nutrients) and *conditions* (temperature, pH, osmotic considerations, and the like) must mimic those of the habitat to give the best chance of obtaining the organism of

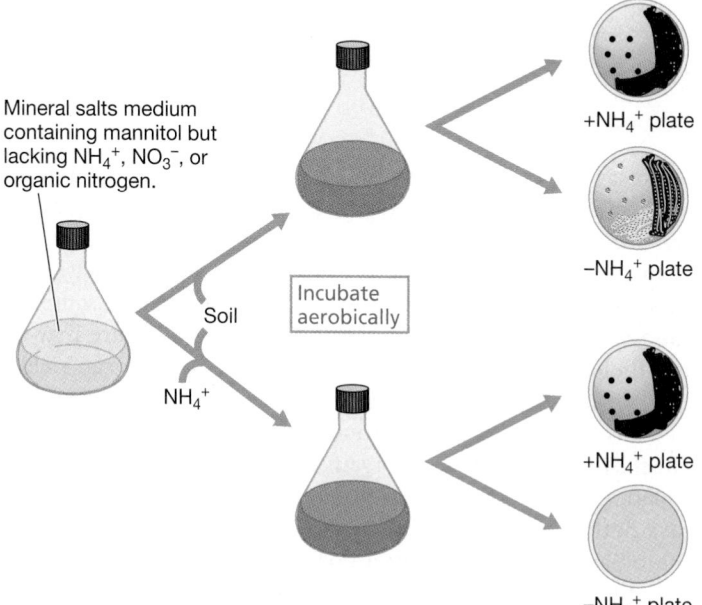

Figure 22.1 The isolation of *Azotobacter*. Selection for aerobic N_2-fixing bacteria usually results in the isolation of *Azotobacter* or its relatives. By contrast, enrichment with fixed forms of nitrogen such as NH_4^+ rarely results in isolating nitrogen-fixing bacteria because there is no selective pressure for nitrogen fixation. See Section 1.9 and Figure 1.21 for more on the historical importance of the bacterium *Azotobacter*.

Table 22.1 *Some enrichment culture methods for phototrophic and chemolithotrophic bacteria*

Light-phototrophic bacteria: main C source, CO_2

Incubation condition	Organisms enriched	Inoculum
Incubation in air		
N_2 as nitrogen source	Cyanobacteria	Pond or lake water; sulfide-rich muds; stagnant water; raw sewage; moist, decomposing leaf litter; moist soil exposed to light
NO_3^- as nitrogen source, 55°C	Thermophilic cyanobacteria	Hot spring microbial mat
Anoxic incubation		
H_2 or organic acids; N_2 as sole nitrogen source	Purple nonsulfur bacteria, heliobacteria	Same as above plus hypolimnetic lake water; pasteurized soil (heliobacteria); microbial mats for thermophilic species
H_2S as electron donor	Purple and green sulfur bacteria	
Fe^{2+}, NO_2^- as electron donor	Purple bacteria	

Dark-chemolithotrophic bacteria: main C source, CO_2 (medium must lack organic C)

Electron donor	Electron acceptor	Organisms enriched	Inoculum
Incubation in air: aerobic respiration			
NH_4^+	O_2	Ammonia-oxidizing bacteria (*Nitrosomonas*)	Soil, mud; sewage effluent
NO_2^-	O_2	Nitrite-oxidizing bacteria (*Nitrobacter, Nitrospira*)	
H_2	O_2	Hydrogen bacteria (various genera)	
H_2S, S^0, $S_2O_3^{2-}$	O_2	*Thiobacillus* spp.	
Fe^{2+}, low pH	O_2	*Acidithiobacillus ferrooxidans*	
Anoxic incubation			
S^0, $S_2O_3^{2-}$	NO_3^-	*Thiobacillus denitrificans*	Mud, lake sediments, soil
H_2	NO_3^-	*Paracoccus denitrificans*	
Fe^{2+}, neutral pH	NO_3^-	*Acidovorax* and various other gram-negative autotrophic bacteria	

interest (↺ Table 23.1). Some enrichment cultures yield nothing. This may be because the organism capable of growing under the enrichment conditions specified is absent from the habitat (that is, not every organism is everywhere). Alternatively, even though the organism of interest exists in the habitat sampled, the resources and conditions of the laboratory culture may be insufficient for its growth. Thus enrichment cultures can yield a firm positive conclusion (that an organism with certain capacities exists in a particular environment), but never a firm negative conclusion (that such an organism does not). Moreover, the isolation of the desired organism from an enrichment culture says nothing about the ecological importance or abundance of the organism in its habitat; a positive enrichment proves only that the organism was present in the sample and in theory requires that only a single viable cell be present.

The Winogradsky Column

The **Winogradsky column** is an artificial microbial ecosystem that serves as a long-term source of bacteria for enrichment cultures. Winogradsky columns have been used to isolate purple and green phototrophic bacteria, sulfate-reducing bacteria, and many other anaerobes. Named for the famous Russian microbiologist

Sergei Winogradsky (↺ Section 1.9), the column was first used by Winogradsky in the late nineteenth century to study soil microorganisms.

A Winogradsky column is prepared by filling a glass cylinder about half full with organically rich, preferably sulfide-containing mud into which carbon substrates have been mixed. The substrates determine which organisms are enriched, and fermentative substrates (such as glucose) that can lead to acidic conditions and excessive gas formation, which can create air pockets in the enrichment, are avoided. The mud is supplemented with small amounts of calcium carbonate ($CaCO_3$) as a buffer and gypsum ($CaSO_4$) as a source of sulfate. The mud is packed tightly in the cylinder (taking care to avoid trapping air) and then covered with lake, pond, or ditch water (or seawater if it is a marine column). The top of the cylinder is covered to prevent evaporation, and the container is placed near a window that receives diffuse sunlight for a period of months.

In a typical Winogradsky column a diverse community of organisms develops (**Figure 22.2a**). Algae and cyanobacteria appear quickly in the upper portions of the water column; by producing O_2 these organisms help to keep this zone of the column oxic. Decomposition processes in the mud lead to the

Table 22.2 *Some enrichment culture methods for chemoorganotrophic and strictly anaerobic bacteria*[a]

Electron donor (and nitrogen source)	Electron acceptor	Typical organisms enriched	Inoculum
Incubation in air: aerobic respiration			
Lactate + NH_4^+	O_2	*Pseudomonas fluorescens*	Soil, mud; lake sediments; decaying
Benzoate + NH_4^+	O_2	*Pseudomonas fluorescens*	vegetation; pasteurize inoculum
Starch + NH_4^+	O_2	*Bacillus polymyxa*, other *Bacillus* spp.	(80 °C for 15 min) for all *Bacillus* enrichments
Ethanol (4%) + 1% yeast extract, pH 6.0	O_2	*Acetobacter, Gluconobacter*	
Urea (5%) + 1% yeast extract	O_2	*Sporosarcina ureae*	
Hydrocarbons (e.g., mineral oil, gasoline, toluene) +NH_4^+	O_2	*Mycobacterium, Nocardia, Pseudomonas*	
Cellulose + NH_4^+	O_2	*Cytophaga, Sporocytophaga*	
Mannitol or benzoate, N_2 as N source	O_2	*Azotobacter*	
CH_4 + NO_3^-	O_2	*Methylobacter, Methylomicrobium*	Lake sediments, thermocline of stratified lake
Anoxic incubation: anaerobic respiration			
Organic acids	NO_3^-	*Pseudomonas* (denitrifying species)	Soil, mud; lake sediments
Yeast extract	NO_3^-	*Bacillus* (denitrifying species)	
Organic acids	SO_4^{2-}	*Desulfovibrio, Desulfotomaculum*	
Acetate, propionate, butyrate	SO_4^{2-}	Fatty acid–oxidizing sulfate reducers	As above; or sewage digester sludge; rumen contents;
Acetate, ethanol	S^0	*Desulfuromonas*	marine sediments
Acetate	Fe^{3+}	*Geobacter, Geospirillum*	
Acetate	ClO_3^-	Various chlorate-reducing bacteria	
H_2	CO_2	Methanogens (chemolithotrophic species only), homoacetogens	Mud, sediments, sewage sludge
CH_3OH	CO_2	*Methanosarcina barkeri*	
CH_3NH_2 or CH_3OH	NO_3^-	*Hyphomicrobium*	
Hydrocarbons	SO_4^{2-} or NO_3^-	Anoxic hydrocarbon-degrading bacteria	Freshwater or marine sediments
Acetate + H_2 + NH_4^+	*Tetrachloroethene* (PCE)	*Dehalococcoides* spp.	PCE polluted groundwater
Anoxic incubation: fermentation			
Glutamate or histidine	No exogenous electron acceptors added	*Clostridium tetanomorphum* or other proteolytic *Clostridium* species	Mud, lake sediments; rotting plant or animal material; dairy products (lactic and propionic acid bacteria); rumen or intestinal contents (enteric bacteria); sewage sludge; soil; pasteurize inoculum for *Clostridium* enrichments
Starch + NH_4^+	None	*Clostridium* spp.	
Starch + N_2 as N source	None	*Clostridium pasteurianum*	
Lactate + yeast extract	None	*Veillonella* spp.	
Glucose or lactose + NH_4^+	None	*Escherichia, Enterobacter*, other fermentative organisms	
Glucose + yeast extract (pH 5)	None	Lactic acid bacteria (*Lactobacillus*)	
Lactate + yeast extract	None	Propionic acid bacteria	
Succinate + NaCl	None	*Propionigenium*	
Oxalate	None	*Oxalobacter*	
Acetylene	None	*Pelobacter* and other acetylene fermenters	

[a]All media must contain an assortment of mineral salts including N, P, S, Mg^{2+}, Mn^{2+}, Fe^{2+}, Ca^{2+}, and other trace elements (⮌ Sections 4.1–4.3). Certain organisms may have requirements for vitamins or other growth factors. This table is meant as an overview of enrichment methods and does not speak to the effect incubation temperature might have in isolating thermophilic (high temperature), hyperthermophilic (very high temperature), and psychrophilic (low temperature) species, or the effect that extremes of pH or salinity might have, assuming an appropriate inoculum was available.

UNIT 7

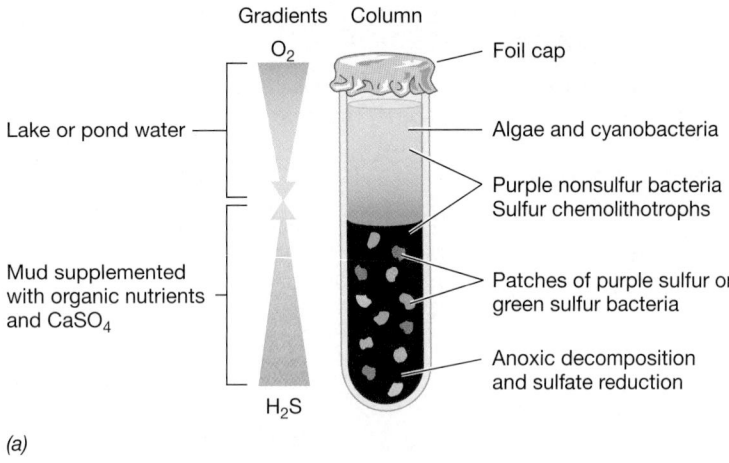

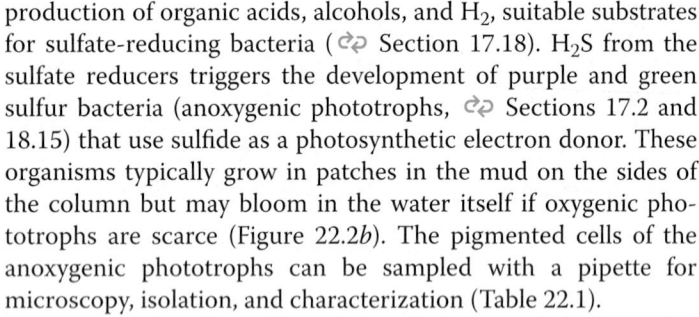

(a)

(b)

Figure 22.2 The Winogradsky column. (a) Schematic view of a typical column. The column is incubated in a location that receives subdued sunlight. Anoxic decomposition leading to sulfate reduction creates the gradient of H_2S. (b) Photo of Winogradsky columns that have remained anoxic up to the top; each column had a bloom of a different phototrophic bacterium. Left to right: *Thiospirillum jenense, Chromatium okenii,* (both purple sulfur bacteria), and *Chlorobium limicola* (green sulfur bacteria).

production of organic acids, alcohols, and H_2, suitable substrates for sulfate-reducing bacteria (Section 17.18). H_2S from the sulfate reducers triggers the development of purple and green sulfur bacteria (anoxygenic phototrophs, Sections 17.2 and 18.15) that use sulfide as a photosynthetic electron donor. These organisms typically grow in patches in the mud on the sides of the column but may bloom in the water itself if oxygenic phototrophs are scarce (Figure 22.2b). The pigmented cells of the anoxygenic phototrophs can be sampled with a pipette for microscopy, isolation, and characterization (Table 22.1).

Winogradsky columns have been used to enrich both aerobic and anaerobic prokaryotes. Besides supplying a ready source of inocula for enrichment cultures, columns can also be supplemented with a specific compound to test the hypothesis that an organism in the inoculum can degrade it. Once a crude enrichment has been established in the column, culture media can be inoculated for the isolation of pure cultures, as discussed in the next section. www.microbiologyplace.com **Online Tutorial 22.1: Enrichment Cultures**

Enrichment Bias

Although the enrichment culture technique is powerful, there exists a bias, and sometimes very severe bias, in the outcome of enrichments. This bias is typically most profound in liquid enrichment cultures where the most rapidly growing organism(s) for the chosen set of conditions dominate. However, using molecular techniques to be described later, we now know that often the most rapidly growing organisms in laboratory cultures are only minor components of the microbial community rather than the most abundant and ecologically relevant organisms carrying out the process of interest. This could be for several reasons including the fact that the levels of resources available

in laboratory cultures are typically much higher than those in nature, and the conditions in the natural habitat, including both the types and proportions of different organisms present as well as the physical and chemical conditions, are nearly impossible to reproduce and maintain for long periods in laboratory cultures.

This problem of **enrichment bias** can be demonstrated by comparing the results obtained in dilution cultures (Section 22.2) with classical liquid enrichment. Dilution of an inoculum followed by liquid enrichment or plating often yields different organisms than liquid enrichments established with the same but undiluted inocula. It is thought that dilution of the inoculum eliminates quantitatively insignificant but rapidly growing "weed" species, allowing development of organisms that are more abundant in the community but slower growing. Dilution of the inoculum is thus a common practice in enrichment culture microbiology today. As discussed below, the problem of overgrowth by "weed" species can also be circumvented by physical isolation of the desired organism before introducing it into a growth medium. This is partly accomplished by use of the dilution method. However, more recently, sophisticated methods have been developed to physically isolate single cells of interest (or single cell types) and place them in a growth medium that is free of undesired cells. We consider these techniques in the next section.

MiniQuiz

- Describe the enrichment strategy behind Beijerinck's isolation of *Azotobacter*.
- Why is sulfate (SO_4^{2-}) added to a Winogradsky column?
- What is enrichment bias?

22.2 Isolation

A pure culture—one containing a single kind of microorganism—can be isolated from an enrichment culture in many ways. Common isolation procedures include the streak plate, the agar shake, and liquid dilution. For organisms that form colonies on agar plates, the streak plate is quick, easy, and the method of choice (**Figure 22.3**); if a well-isolated colony is selected and restreaked several successive times, a pure culture can usually be obtained. With proper incubation facilities (for example, anoxic jars or anoxic chambers for anaerobes, ⮂ Section 5.17), it is possible to purify both aerobes and anaerobes on agar plates by the streak plate method.

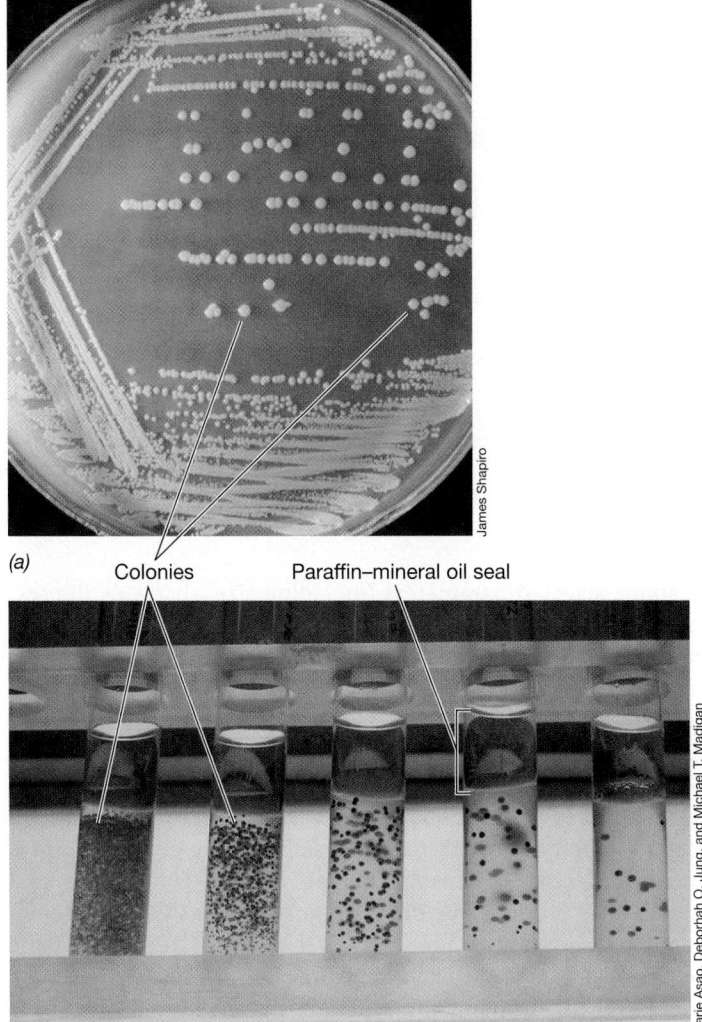

(a) Colonies Paraffin–mineral oil seal

(b)

Figure 22.3 Pure culture methods. *(a)* Organisms that form distinct colonies on plates are usually easy to purify. *(b)* Colonies of phototrophic purple bacteria in agar dilution tubes; the molten agar was cooled to approximately 45°C before inoculation. A dilution series was established from left to right, eventually yielding well-isolated colonies. The tubes were sealed with a 1:1 mixture of sterile paraffin and mineral oil to maintain anaerobiosis.

Agar Dilution Tubes and the Most-Probable-Number Technique

In the agar dilution tube method, a mixed culture is diluted in tubes of molten agar, resulting in colonies embedded in the agar. This agar shake method is useful for purifying anaerobic organisms such as phototrophic sulfur bacteria and sulfate-reducing bacteria from samples taken from Winogradsky columns. A culture is purified by successive dilutions of cell suspensions in tubes of molten agar medium (Figure 22.3*b*). Repeating this procedure using a colony from the highest-dilution tube as inoculum for a new set of dilutions eventually gives pure cultures. A related procedure called the *roll tube method* uses tubes containing a thin layer of agar on their inner surface. The agar can then be streaked for isolated colonies. Because the tubes can be flushed with an oxygen-free gas during streaking, the roll tube method is primarily used for the isolation of anaerobic prokaryotes.

Another purification procedure is the serial dilution of an inoculum in a liquid medium until the final tube in the series shows no growth. When a 10-fold serial dilution is used, for example, the last tube showing growth should have originated from ten or fewer cells. Besides being a method for obtaining pure cultures, serial dilution techniques are widely used to estimate viable cell numbers in the **most-probable-number (MPN) technique** (**Figure 22.4**). MPN methods have been used for estimating the numbers of microorganisms in foods, wastewater, and other samples in which cell numbers need to be assessed routinely. An MPN count can be done using highly selective media and incubation conditions to target one or a small group of organisms, such as a particular pathogen, or a count can be done using complex media to get a general estimate of viable cell numbers (but see the caveat

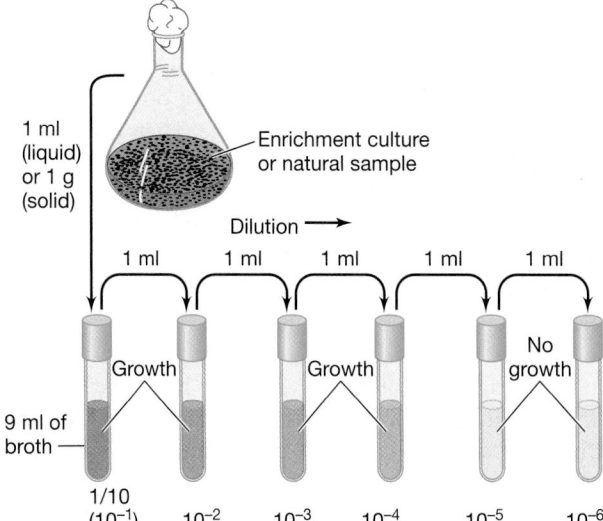

1 ml (liquid) or 1 g (solid)

Enrichment culture or natural sample

Dilution →

1 ml   1 ml   1 ml   1 ml   1 ml

Growth   Growth   No growth

9 ml of broth

1/10 (10^{-1}) 10^{-2} 10^{-3} 10^{-4} 10^{-5} 10^{-6}

Figure 22.4 Procedure for a most-probable-number (MPN) analysis. An existing enrichment culture or a natural sample of water and soil is added to a selective culture medium, and serial dilutions are made. The last tube showing growth (in the example shown it is the 10^{-4} dilution) should have developed from 10 or fewer cells. Since particle-attached microorganisms can skew numbers significantly, gentle methods to dis-associate microorganisms from particles are often used prior to dilution.

that applies to such estimates in Section 5.10). Use of several replicate tubes at each dilution improves accuracy of the final MPN obtained. www.microbiologyplace.com **Online Tutorial 22.2: Serial Dilutions and a Most-Probable-Number Analysis**

Criteria for Purity

Regardless of the methods used to purify a culture, once a putative pure culture has been obtained, it is essential to verify its purity. This should be done through a combination of (1) microscopy, (2) observation of colony characteristics on plates or in shake tubes, and (3) tests of the culture for growth in other media. In the latter, it is important to test the culture for growth in media in which it is predicted that the desired organism will grow poorly or not at all, but that contaminants will grow vigorously. In the final analysis, the microscopic observation of a single morphological type of cell that displays uniform staining characteristics (for example, in a Gram stain) coupled with uniform colony characteristics and the absence of contamination in growth tests with various culture media is good evidence that a culture is pure. The molecular methods described in the following sections for characterizing environmental populations can also be applied to verification of culture purity. However, these techniques are generally complementary and do not substitute for the more basic observations of culture characteristics and cellular morphology.

Selective Single-Cell Isolation: The Laser Tweezers and Flow Cytometry

In addition to the more classical methods just described, technological advances have yielded new tools for obtaining single cells that may grow into pure cultures, including the use of laser tweezers and flow cytometry.

Laser tweezers consist of an inverted light microscope equipped with a strongly focused infrared laser and a micromanipulation device. Trapping a single cell is possible because the laser beam creates a force that pushes down on a microbial cell (or other small object) and holds it in place. Then when the laser beam is moved, the trapped cell moves along with it (**Figure 22.5a**). If a mixed sample is in a capillary tube, a single cell can be optically trapped and moved away from contaminating organisms. The cell can then be isolated by breaking the tube at a point between the cell and the contaminants and flushing the cell into a small tube of sterile medium (Figure 22.5b).

Laser tweezers are especially useful for isolating slow-growing bacteria that would be overgrown by "weed" species in enrichment cultures or for organisms present in such low numbers that they would be missed using dilution-based enrichment methods. Laser tweezers, when coupled with staining techniques that identify particular organisms in a microscope field (Sections 22.3 and 22.4), can be used to select organisms of interest from a mixture for purification and further laboratory study.

A second method for selective isolation of single cells employs **flow cytometry**, a technique for counting and examining microscopic particles by suspending them in a stream of fluid and passing them through an electronic detector. Flow cytometers

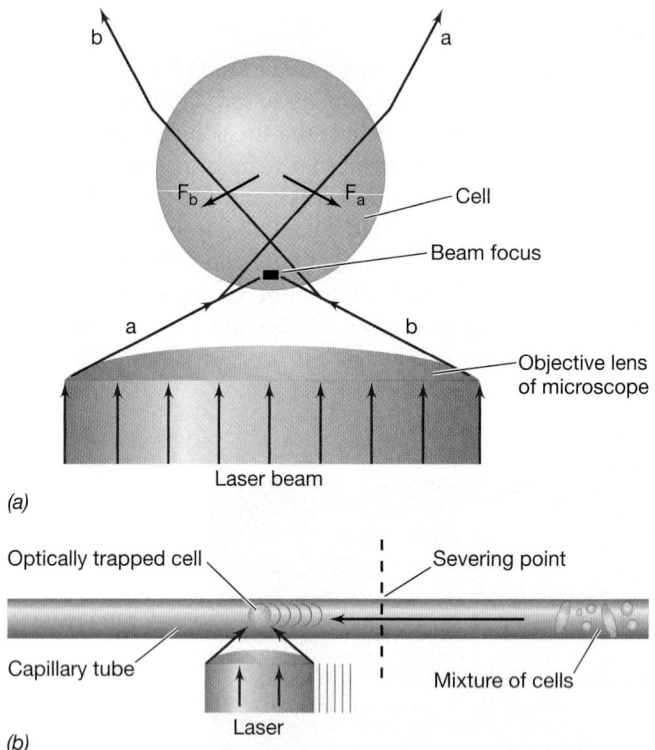

(a)

(b)

Figure 22.5 The laser tweezers for the isolation of single cells. *(a)* Principle of the laser tweezers. A laser beam strongly focused on a very small object (O) such as a cell creates downward radiation forces (F_a, F_b) that allow the cell to be dragged in any direction. *(b)* Isolation. The laser beam can lock onto a single cell present in a mixture in a capillary tube and drag the optically trapped cell away from the other cells. Once the desired cell is far enough away from the other cells, the capillary is severed and the cell is flushed into a tube of sterile medium.

examine selected parameters (including size, shape, or fluorescent properties) of single cells as they pass through a detector at rates of many thousands of cells per second. Specialized flow cytometers also sort individual cells based on selected parameters (see Section 22.10 and Figure 22.26). Single-cell sorting capacity has been used to arrange single cells on the surface of a solid growth medium or deposit them into individual wells of a multiwelled (microtiter) plate, where each well contains a slightly different growth medium. Because the growth requirements of some organisms include organic compounds and metabolites produced by other organisms that share their environment, addition of filter-sterilized source water (for aquatic organisms) or soil water extract (for soil organisms) has been successful for bringing some previously uncultured organisms into laboratory culture. Continuing developments of these and other methods for screening of enrichment cultures have spawned a new field of *high-throughput technology* for culturing previously uncultured microorganisms. High-throughput methods typically employ robotic systems to quickly test hundreds or thousands of combinations of nutrients for growth or to assay hundreds or thousands of different wells for DNA sequences that will identify the organisms being enriched, all run simultaneously to yield rapid results.

II Culture-Independent Analyses of Microbial Communities

Microbial ecologists quantify cells in a microbial habitat to estimate relative abundances of different species. Cell stains are necessary to obtain these types of data, and we detail these methods here. Organisms in natural environments can also be detected by assaying their genes. Genes encoding either ribosomal RNA (rRNA, ⮌ Sections 16.6–16.9) or enzymes that support a specific physiology are the usual targets in these studies. Environmental genomics (Section 22.7) is a method for assessing the entire gene complement of a habitat, revealing both the biodiversity and metabolic capabilities of the microbial community at the same time.

22.3 General Staining Methods

Several staining methods are suitable for quantifying microorganisms in natural samples. Although these methods do not reveal the physiology or phylogeny of the cells, they are nonetheless reliable and widely used by microbial ecologists for measuring total cell numbers. One method described in this section also allows for an assessment of cell viability.

Fluorescent Staining with Dyes That Bind Nucleic Acids

Fluorescent dyes can be used to stain microorganisms from virtually any microbial habitat. **DAPI** (4′,6-diamidino-2-phenylindole) has long been a popular stain for this purpose, as has the dye

acridine orange. There is also increasing use of *SYBR Green I*, a dye that confers very bright fluorescence to both bacteria and virus particles. These stains bind to DNA and are strongly fluorescent when exposed to ultraviolet (UV) radiation (DAPI absorption maximum, 400 nm; acridine orange absorption maximum, 500 nm; SYBR Green I absorption maximum, 497 nm), making the microbial cells in the sample readily visible and easy to enumerate. Cells stained with DAPI fluoresce blue, cells stained with acridine orange fluoresce orange or greenish-orange, and cells stained with SYBR Green I fluoresce green (**Figure 22.6**).

Dyes that stain DNA are widely used for the enumeration of microorganisms in environmental, food, and clinical samples. Depending on the sample, background staining is occasionally a problem with fluorescent stains, but because these dyes stain nucleic acids, they are for the most part nonreactive with inert matter. Thus, for many samples, from soil as well as aquatic sources, they can give a reasonable estimate of the cell numbers present. Staining with the brightly fluorescent SYBR Green I also provides excellent enumeration of aquatic virus populations (⮌ Section 23.10). For dilute aquatic samples, cells can be stained after filtering.

DNA staining is nonspecific; *all* microorganisms in a sample are stained. Although this may at first seem desirable, it is not necessarily so. For example, DAPI and acridine orange fail to differentiate between living and dead cells or between different species of microorganisms, so they cannot be used to assess cell viability or to track species of microorganisms in an environment.

Viability Staining

Viability staining differentiates live cells from dead ones. Viability stains thus both enumerate and yield viability data at the same time. The basis of differentiating between live and dead cells lies with whether a cell's cytoplasmic membrane is intact. Two dyes that fluoresce green and red are added to a sample; the green fluorescing dye penetrates all cells, viable or not, whereas the red dye, which contains the chemical propidium iodide, penetrates only those cells whose cytoplasmic membrane is no longer intact and that are therefore dead. Thus, when viewed microscopically,

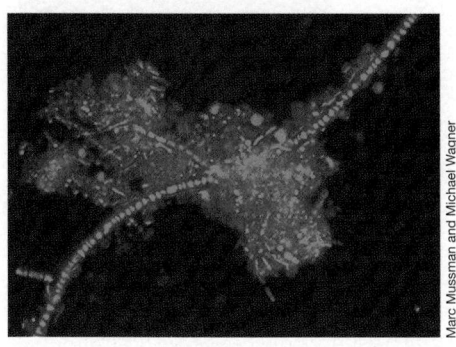
(a)

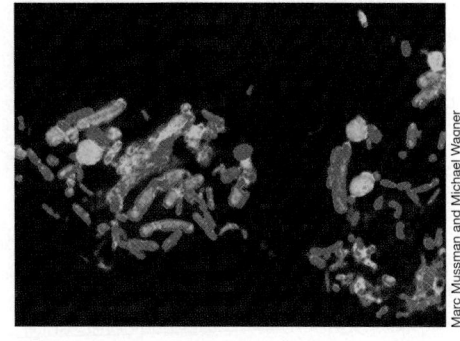
(b)

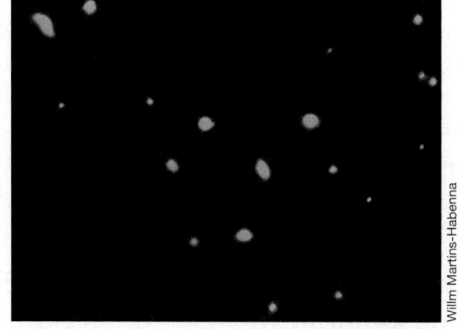
(c)

Figure 22.6 Nonspecific fluorescent stains. *(a)* DAPI and *(b)* acridine orange staining showing microbial communities inhabiting activated sludge in a municipal wastewater treatment plant. With acridine orange, cells containing low RNA levels stain green. *(c)* SYBR Green–stained sample of Puget Sound (Washington, USA) surface water showing green-fluorescing bacterial cells. The large cells near the center of the field are 0.8–1.0 μm in diameter.

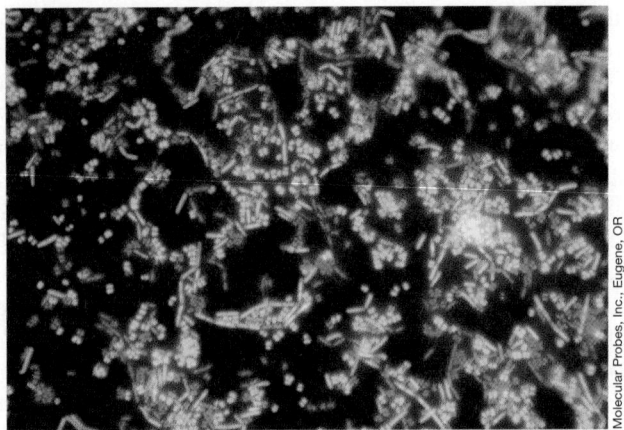

Molecular Probes, Inc., Eugene, OR

Figure 22.7 Viability staining. Live (green) and dead (red) cells of *Micrococcus luteus* (cocci) and *Bacillus cereus* (rods) stained by the LIVE/DEAD BacLight Bacterial Viability Stain.

green cells are scored as alive and red cells as dead, yielding an instant assessment of both abundance and viability (**Figure 22.7**).

Although useful for research that uses laboratory cultures, the live/dead staining method is not suitable for use in the direct microscopic examination of samples from many natural habitats because of problems with nonspecific staining of background materials. However, procedures have been developed for use of the live/dead stain in viability analyses of aquatic environments; a water sample is filtered and the filters are stained with the live/dead stain and examined microscopically. Thus in aquatic microbiology, live/dead staining is often used to measure the viability of cell populations in the water column of lakes or oceans, or in the flowing waters of streams, rivers, and other aquatic environments.

Fluorescent Antibodies

Staining techniques can be even more specific if fluorescent antibodies are used. Antibodies are specific biological reagents, and we discuss the use of fluorescent antibodies in Section 31.9. The great specificity of antibodies against cell surface constituents can be exploited as a means of identifying or tracking a specific organism in a habitat that contains a mixture of many organisms, such as in soil or clinical samples. The method requires the preparation of specific antibodies against the organism of interest, often a time-consuming and laborious procedure, and in most cases requires that a pure culture of the desired organism be available. For pathogenic microorganisms, however, highly specific antibodies can be purchased commercially and are widely used in clinical microbiology laboratories for the microscopic diagnosis of infectious diseases.

Green Fluorescent Protein as a Cell Tag and Reporter Gene

Bacterial cells can be altered by genetic engineering to make them autofluorescent. As discussed earlier (⬅ Section 11.5), a gene encoding a protein called the **green fluorescent protein (GFP)** can be inserted into the genome of virtually any cultured bacterium. When the GFP gene (*gfp*) is expressed, cells fluoresce

green when observed with ultraviolet microscopy. Although not useful for the study of natural populations of microorganisms (because these cells lack the GFP gene), GFP-tagged cells can be introduced into an environment, such as plant roots, and then tracked over time by microscopy. Using this method, microbial ecologists can study competition between the native microflora and a GFP-tagged introduced strain and can assess the effect of perturbations of an environment on the survivability of the introduced strain. GFP tagging is also used extensively in the study of microbial symbiotic associations with plants and animals (Chapter 25). However, GFP requires oxygen (O_2) to become fluorescent, and thus the GFP method is not suitable for tracking cells introduced into strictly anoxic habitats.

The gene *gfp* has also been used extensively in laboratory cultures of various bacteria and in controlled environments as a *reporter gene*. When this gene is fused with an operon under the control of a specific regulatory protein, transcription can be studied by using fluorescence as the indicator (a "reporter") of activity. That is, when the genes containing the fused GFP gene are transcribed, *gfp* is also transcribed, GFP is made, and cells fluoresce green (⬅ Section 11.5 and Figure 11.9). For example, expression of *gfp* was used to demonstrate that colonization of alfalfa roots by *Sinorhizobium meliloti* (legume—root nodule symbiosis, ⬅ Section 25.3) is promoted by sugars and dicarboxylic acids released by the plant (**Figure 22.8**).

Limitations of Microscopy

The microscope is an essential tool for exploring microbial diversity and for enumerating and identifying microorganisms in natural samples. However, microscopy alone does not

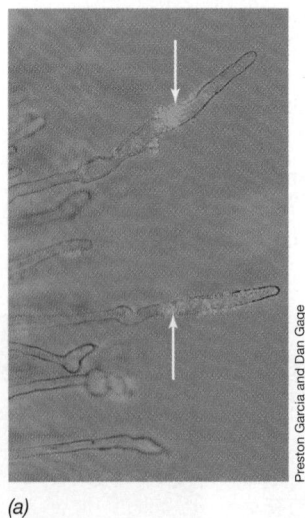 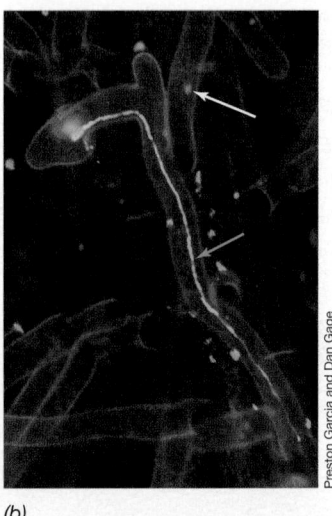

Preston Garcia and Dan Gage

(a)   *(b)*

Figure 22.8 The green fluorescent protein (GFP). *(a)* Cells of *Sinorhizobium meliloti* (arrows) carrying a plasmid with an alpha-galactoside inducible promoter fused to GFP on clover seedling roots. Green fluorescence indicates that alpha-galactosides are released and available to support the growth of this bacterium. *(b)* *S. meliloti* cells (white arrow) carrying a plasmid with a succinate-inducible promoter fused to GFP; green fluorescence indicates succinate or other C_4 dicarboxylic acids secreted by the plant root hairs. This image also shows an infection thread (blue arrow; ⬅ Section 25.3) that developed within a root hair to deliver *S. meliloti* cells to the root interior.

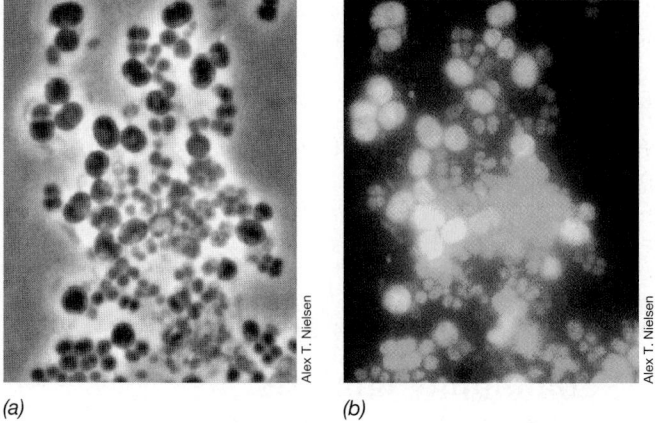

(a)   *(b)*

Figure 22.9 Morphology and genetic diversity. The photomicrographs shown here, produced by *(a)* phase-contrast and *(b)* a technique called phylogenetic FISH (Section 22.4), are of the same field of cells. Although the large oval cells are of a rather unusual morphology and size for prokaryotic cells and all look similar in phase-contrast microscopy, the phylogenetic stains reveal that there are two genetically distinct types (one stains yellow and one stains blue). The oval cells stained in blue or yellow are about 2.25 μm in diameter. The green cells in pairs or clusters are about 1 μm in diameter.

suffice for the study of microbial diversity. Prokaryotes vary greatly in size (Section 3.2 and Table 3.1). Very small cells can be a major problem and can go totally unnoticed, and some cells are near the limits of resolution of the light microscope. Such cells can easily be overlooked in the examination of natural samples, especially if the sample contains high levels of particulate matter or high numbers of large cells. Also, it is often difficult to differentiate live cells from dead cells or cells in general from certain inert materials in natural samples. However, the biggest limitation with the microscopic methods we have discussed thus far is that none of them reveal the phylogenetic diversity of the microorganisms in the habitat under study.

We will see in the next section and get a preview here (**Figure 22.9**) of powerful staining methods that can reveal the phylogeny of organisms observed in a natural sample. These methods have revolutionized microbial ecology. They have helped microbiologists overcome the major limitation of the light microscope in microbial ecology; that is, *identifying* observed organisms from a phylogenetic perspective. These methods have also taught microbial ecologists an important lesson: When observing unstained or nonspecifically stained natural populations of microorganisms under the microscope, one must remember that the sample almost certainly contains a genetically diverse community, even if many cells "look" the same (Figure 22.9). The simple shapes of bacteria conceal their remarkable diversity.

MiniQuiz

- Why is it incorrect to say that the GFP is a "staining" method?
- How do stains like DAPI differ from fluorescent antibodies in terms of specificity?

22.4 Fluorescent *In Situ* Hybridization (FISH)

Because of their great specificity, nucleic acid probes are powerful tools for identifying and quantifying microorganisms. Recall that a **nucleic acid probe** is a DNA or RNA oligonucleotide complementary to a sequence in a target gene or RNA; when the probe and the target come together, they hybridize (Section 11.2). As discussed in Section 16.9, nucleic acid probes can be made fluorescent by attaching fluorescent dyes to them. The fluorescent probes can be used to identify organisms that contain a nucleic acid sequence complementary to the probe. This technique is called **fluorescent *in situ* hybridization (FISH)**, and different applications are described here, including methods that target phylogeny or gene expression.

Phylogenetic Staining Using FISH

Phylogenetic FISH stains are fluorescing oligonucleotides complementary in base sequence to sequences in ribosomal RNA (16S or 23S rRNA in prokaryotes or 18S or 28S rRNA in eukaryotes) (Section 16.9). Phylogenetic stains penetrate the cell without lysing it and hybridize with rRNA directly in the ribosomes. Because ribosomes are scattered throughout the cell in prokaryotes, the entire cell becomes fluorescent (Figure 22.9).

Phylogenetic stains can be designed to be very specific and react with only one species or a handful of related microbial species, or they can be made more general and react with, for example, all cells of a given phylogenetic domain. Using FISH, an investigator can identify or track an organism of interest or a domain of interest in a natural sample. For example, if one wishes to determine the percentage of a given microbial population that are *Archaea*, an archaea-specific phylogenetic stain may be used in combination with DAPI (Section 22.3) to assess *Archaea* and total numbers, respectively, and a percentage could then be derived by calculation.

FISH technology can also employ multiple phylogenetic probes. With a suite of probes, each designed to react with a particular organism or group and each containing its own fluorescent dye, FISH can characterize the phylogenetic breadth of a habitat in a single experiment (**Figure 22.10**). If FISH is combined with confocal microscopy (Section 2.3), it is possible to explore microbial populations with depth, as, for example, in a biofilm (Section 23.4). In addition to microbial ecology, FISH is also an important tool in the food industry and in clinical diagnostics for the microscopic detection of specific pathogens.

CARD-FISH

Besides characterizing the phylogenetic diversity of a habitat, FISH can be used to measure *gene expression* in organisms in a natural sample. Because the target in this case is messenger RNA (mRNA), a form of RNA that is much less abundant than rRNA, standard FISH techniques cannot be applied. Instead, the signal (fluorescence) must be amplified. A FISH method that enhances the signal is called *catalyzed reporter deposition FISH* (*CARD-FISH*). In CARD-FISH the specific nucleic acid probe contains a molecule of the enzyme peroxidase conjugated to it instead of a fluorescent dye. After there has been time for hybridization, the preparation is treated with a highly fluorescent dye called

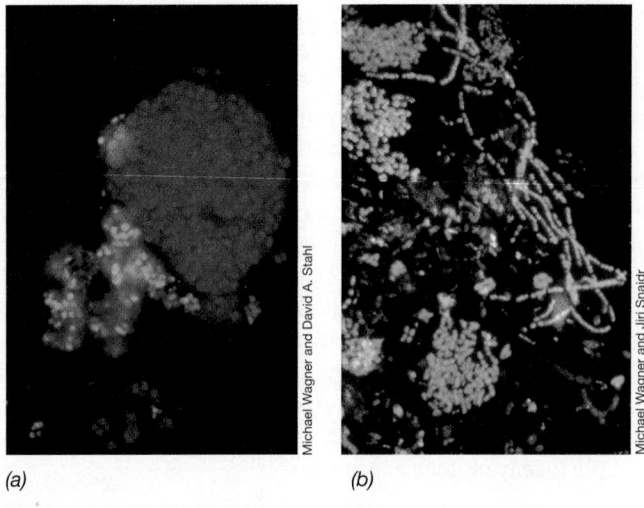

(a)   *(b)*

Figure 22.10 **FISH analysis of sewage sludge.** *(a)* Nitrifying bacteria. Red, ammonia-oxidizing bacteria; green, nitrite-oxidizing bacteria. *(b)* Confocal laser scanning micrograph of a sewage sludge sample. The sample was treated with three phylogenetic FISH probes, each containing a fluorescent dye (green, red, or purple) that identifies a particular group of *Proteobacteria*. Green-, red-, or purple-stained cells reacted with only a single probe; other cells reacted with multiple probes to give blue or yellow.

tyramide. The tyramide is converted by the activity of peroxidase into a very reactive intermediate that binds to adjacent protein, and this amplifies the signal sufficiently to be detected by fluorescence microscopy. Each molecule of peroxidase activates many molecules of tyramide so that even mRNA present in very low amounts can be visualized.

Besides detecting mRNA, CARD-FISH is also useful in phylogenetic studies of prokaryotes that may be growing very slowly, for example organisms inhabiting the open oceans where cold temperatures and low nutrient concentrations limit growth rates (**Figure 22.11**). Because such cells have few ribosomes compared

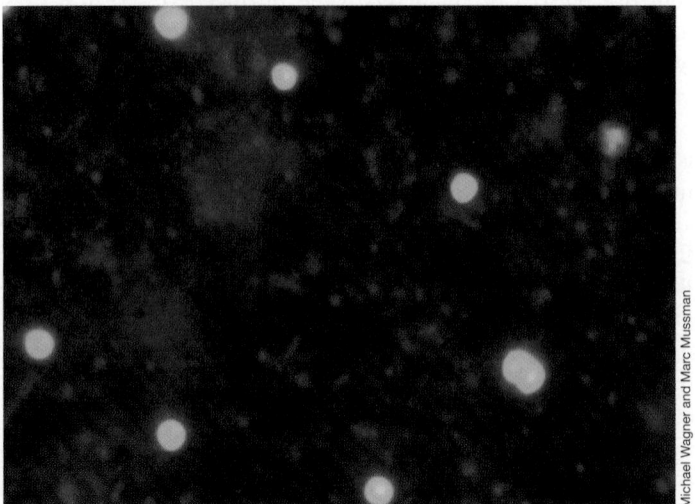

Figure 22.11 Catalyzed reporter deposition FISH (CARD-FISH) labeling of *Archaea*. Archaeal cells in this preparation fluoresce intensely (green) relative to DAPI-stained cells (blue).

with more actively growing cells, standard FISH often yields only a weak signal.

22.5 PCR Methods of Microbial Community Analysis

Many microbial biodiversity studies forgo isolating organisms or even quantifying or identifying them microscopically using the stains described in the previous sections. Instead, *specific genes* are used as measures of biodiversity and metabolic capacity. Some genes are unique to particular organisms. Detection of such a gene in an environmental sample implies that the organism is present. The major techniques employed in this type of microbial community analysis are the polymerase chain reaction (PCR), DNA fragment analysis by gel electrophoresis (DGGE, T-RFLP, ARISA) or molecular cloning, and DNA sequencing and analysis. In addition, as we will see in Section 22.7, entire genomes of environmental samples can also be analyzed as a measure of the biodiversity of microbial communities.

PCR and Microbial Community Analysis

We discussed the principle of PCR in Section 6.11. Recall the major steps in PCR: (1) Two nucleic acid primers are hybridized to a complementary sequence in a target gene, (2) DNA polymerase copies the target gene, and (3) multiple copies of the target gene are made by repeated melting of complementary strands, hybridization of primers, and new synthesis (🔗 Figure 6.24). From a single copy of a gene, several million copies can be made.

Which genes are suitable as target genes for microbial community analyses? Because genes encoding the small subunit ribosomal RNAs (rRNAs) are phylogenetically informative and techniques for their analysis well developed (🔗 Sections 16.5–16.9), they are widely used in community analyses. Moreover, because rRNA genes are universal and contain several regions of high sequence conservation, it is possible to amplify them from all organisms using only a few different PCR primers, even though the organisms may be phylogenetically distantly related. In addition to rRNA genes, genes that encode enzymes for metabolic functions unique to a specific organism or group of related organisms can be the target genes (**Table 22.3**).

Genes such as those encoding ribosomal RNAs that have changed in sequence over time as species have diverged are called *orthologs* (🔗 Section 12.10). Organisms that share the same or very closely related orthologous genes are called a **phylotype**. In microbial ecology, the phylotype concept is primarily used to provide a natural (phylogenetic) framework for describing the microbial diversity of a given habitat, regardless of whether the identified phylotypes are cultured organisms or not. Thus, the term phylotype is widely used to describe the microbial diversity of a habitat based solely on nucleic acid sequences. It is only

Table 22.3 *Genes commonly used for evaluating specific microbial processes in the environment using PCR*

Metabolic process[a]	Target gene	Encoded enzyme
Denitrification	narG	Nitrate reductase
	nirK, nirS	Nitrite reductase
	norB	Nitric oxide reductase
	nosZ	Nitrous oxide reductase
Nitrogen fixation	nifH	Nitrogenase
Nitrification	amoA	Ammonia monooxygenase
Methane oxidation	pmoA	Methane monooxygenase
Sulfate reduction	apsA	Adenosine phosphosulfate reductase
	dsrAB	Sulfite reductase
Methane production	mcrA	Methyl coenzyme M reductase
Degradation of petroleum compounds	nahA	Naphthalene dioxygenase
	alkB	Alkane hydroxylase
Anoxygenic photosynthesis	pufM	M subunit of photosynthetic reaction center

[a]All of these metabolic processes are discussed in Chapters 13 and 14.

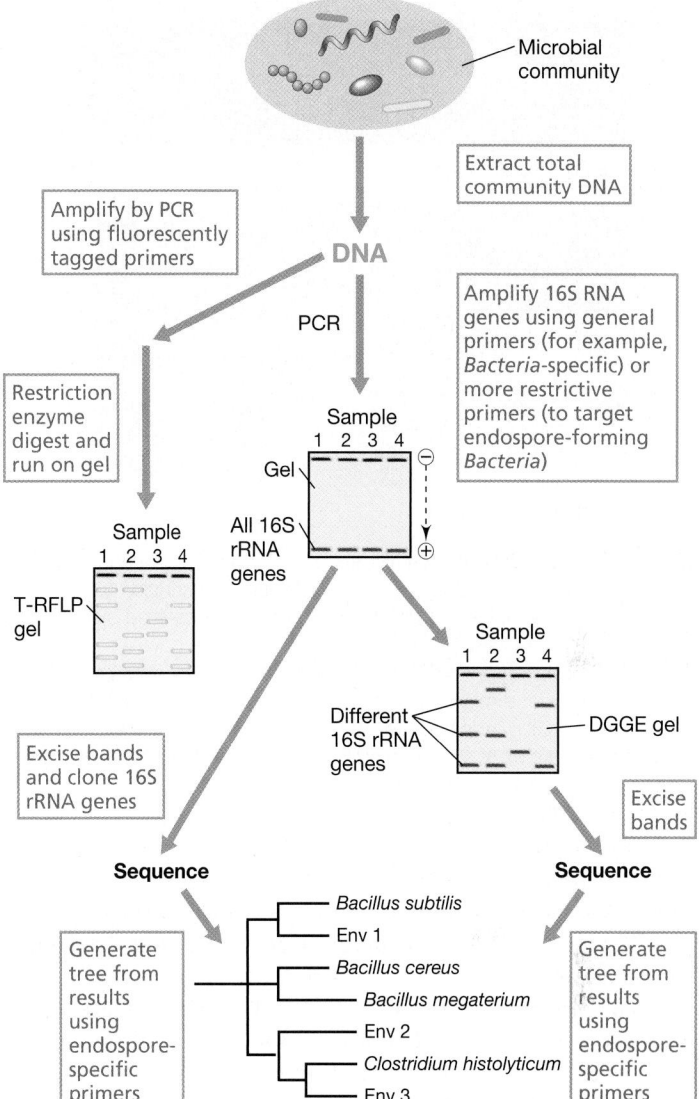

Figure 22.12 **Steps in single-gene biodiversity analysis of a microbial community.** From total community DNA, 16S rRNA genes are amplified using, in the DGGE example, primers that target only *Firmicutes*, a group of gram-positive *Bacteria*. The PCR bands are excised and the different 16S rRNA genes separated by either cloning or DGGE. Following sequencing, a phylogenetic tree is generated. "Env" indicates an environmental sequence (phylotype). In T-RFLP analyses, the number of bands indicates the number of phylotypes.

when additional physiological and genetic information becomes available, typically after the organism is brought into laboratory culture, that it becomes possible to propose a genus and species name for a phylotype.

In a typical community analysis experiment, total DNA is isolated from a microbial habitat (**Figure 22.12**). Commercially available kits that yield high-purity DNA from soil and other complex habitats are available for this purpose. The DNA obtained is a mixture of genomic DNA from all of the microorganisms that were in the sample from the habitat (Figure 22.13). From this mixture, PCR is used to amplify the target gene and make multiple copies of each variant of the target gene. If RNA is isolated instead of DNA (to detect those genes being transcribed), the RNA can be converted into complementary DNA by the enzyme reverse transcriptase (⮌ Section 9.12) and this subjected to PCR as for isolated DNA. However, regardless of whether DNA or RNA is originally isolated, following the PCR step the different variants of the target gene need to be sorted out before they can be sequenced, and we consider this problem next.

Denaturing Gradient Gel Electrophoresis: Separating Very Similar Genes

The results of a community sampling experiment may be visualized as bands on a gel that indicate the size of the target gene. **Figure 22.13a** shows such a gel on which only a single band

appears; however, because the amplified target gene came from a mixture of different cells, the phylotypes need to be sorted out before they are sequenced. This can be accomplished by molecular cloning (Figure 22.12, ⮌ Section 11.3) and sequencing each cloned variant or by using one of the recently developed high-throughput (Section 22.2) sequencing systems that do not require cloning for sequence determination. One can also use **denaturing gradient gel electrophoresis (DGGE)**, a method that separates genes of the same *size* that differ in their melting (denaturing) profile because of differences in their *base sequence* (Figure 22.13b).

UNIT 7

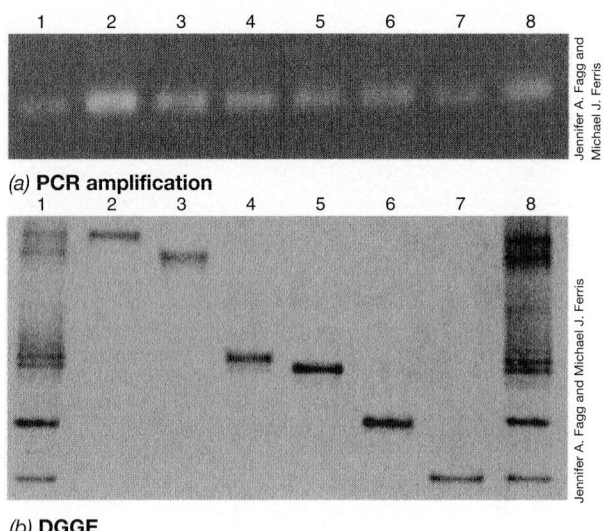

(a) **PCR amplification**

(b) **DGGE**

Jennifer A. Fagg and Michael J. Ferris

Figure 22.13 PCR and DGGE gels. Bulk DNA was isolated from a microbial community and amplified by PCR using primers for 16S rRNA genes of *Bacteria* (a, lanes 1 and 8). Six bands later resolved by DGGE (b, lanes 2–7) were excised and reamplified and each gave a single band at the same location on the PCR gel (a, lanes 2–7). However, by DGGE analysis, each band migrated to a different location on the gel (b, lanes 2–7). Note that all bands migrate to the same location in the PCR gel because they are all of the same size, but to different locations on the DGGE gel because they have different sequences.

DGGE employs a gradient of a DNA denaturant, such as a mixture of urea and formamide. When a double-stranded DNA fragment moving through the gel reaches a region containing sufficient denaturant, the strands begin to "melt"; at this point, their migration stops (Figures 22.12 and 22.13b). Differences in base sequence cause differences in the melting properties of DNA. Thus, the different bands observed in a DGGE gel are phylotypes that can differ in base sequence by as little as a single base change.

Once DGGE has been performed, the individual bands are excised and sequenced (Figure 22.12). With 16S rRNA as the target gene, for example, the DGGE pattern immediately reveals the number of phylotypes (distinct 16S rRNA genes) present in a habitat (Figure 22.13b). Following sequencing of each DGGE band, the actual species present in the community can be determined by phylogenetic analyses (↺ Sections 16.6 and 16.7; Figure 22.12). If PCR primers specific for genes other than 16S rRNA are used, such as a metabolic gene (Table 22.3), the variants of this specific gene that exist in the sample can also be assessed. Thus, although the number of bands on a DGGE gel is an overview of the biodiversity in a habitat, sequence analysis is required for identification and inferring phylogenetic relationship.

T-RFLP and ARISA

Another method for rapid microbial community analysis is *terminal restriction fragment length polymorphism* (*T-RFLP*). In this method a target gene (usually an rRNA gene) is amplified by PCR from community DNA using a primer set in which one of the primers is end-labeled with a fluorescent dye. The PCR

products are then treated with a restriction enzyme (↺ Section 11.1) that cuts the DNA at specific sequences. Restriction enzymes with recognition sites of only four base pairs are commonly used because they cut frequently within a relatively short PCR product. This generates a series of DNA fragments of varying length, the number of which depends on how many restriction cut sites are present in the DNA. The fluorescently labeled terminal fragments are then separated by gel electrophoresis. The digestion products are generally separated and sized on an automated DNA sequencer that detects fragments based on dye-conferred fluorescence. Therefore, only the terminal dye-labeled fragments are detected. The pattern obtained shows the rRNA sequence variation in the community sampled (Figure 22.12).

DGGE and T-RFLP both measure single-gene diversity, but in different ways. The pattern of bands on a DGGE gel reflects the number of same-length sequence variants of a single gene (Figure 22.13), whereas the pattern of bands on a T-RFLP gel reflects variants differing in DNA sequence of a single gene as measured by differences in restriction enzyme cut sites. The information obtained from a T-RFLP analysis, in addition to providing insight into the diversity and population abundances of a microbial community, can also be used to infer phylogeny. Diagnostic information for each fragment includes knowledge of sequences near both ends (primer sequence and restriction enzyme cut site), knowledge that a second restriction site does not exist within the fragment, and fragment length. Using specialized software, this information can be used to search for matching 16S rRNA sequences in public databases. Although this is of some predictive value, often there are many matching unique sequences. Thus, T-RFLP generally underestimates the diversity within a microbial community.

A technique related to T-RFLP that provides more detailed analysis of communities is *automated ribosomal intergenic spacer analysis* (*ARISA*), which takes advantage of the proximity of the 16S rRNA and 23S rRNA genes in prokaryotes. The DNA separating these two genes, called the *internal transcribed spacer* (*ITS*) region, differs in length among species and often also differs in length among the multiple rRNA operons of a single species (**Figure 22.14a**). The PCR primers for ARISA are complementary to conserved sequences in the 16S and 23S rRNA genes that flank the spacer region. Amplification (Figure 22.14b) and analysis (Figure 22.14c) is conducted as described for T-RFLP, resulting in a complex pattern of bands that can be used for community analysis. The technique differs from T-RFLP in that it does not require a restriction enzyme digestion following PCR amplification. The word "automated" refers to the use of a DNA sequencer that automatically identifies and assigns sizes to each dye-labeled fragment, as can also be done in T-RFLP analyses (Figure 22.14c). ARISA has received greatest application in the study of microbial community dynamics by monitoring, for example, changes in the presence and relative abundance of a specific community member through time and space.

Results of PCR Phylogenetic Analyses

Phylogenetic analyses of microbial communities have yielded surprising results. For example, using the gene encoding the 16S rRNA as the target, analyses of natural microbial communities

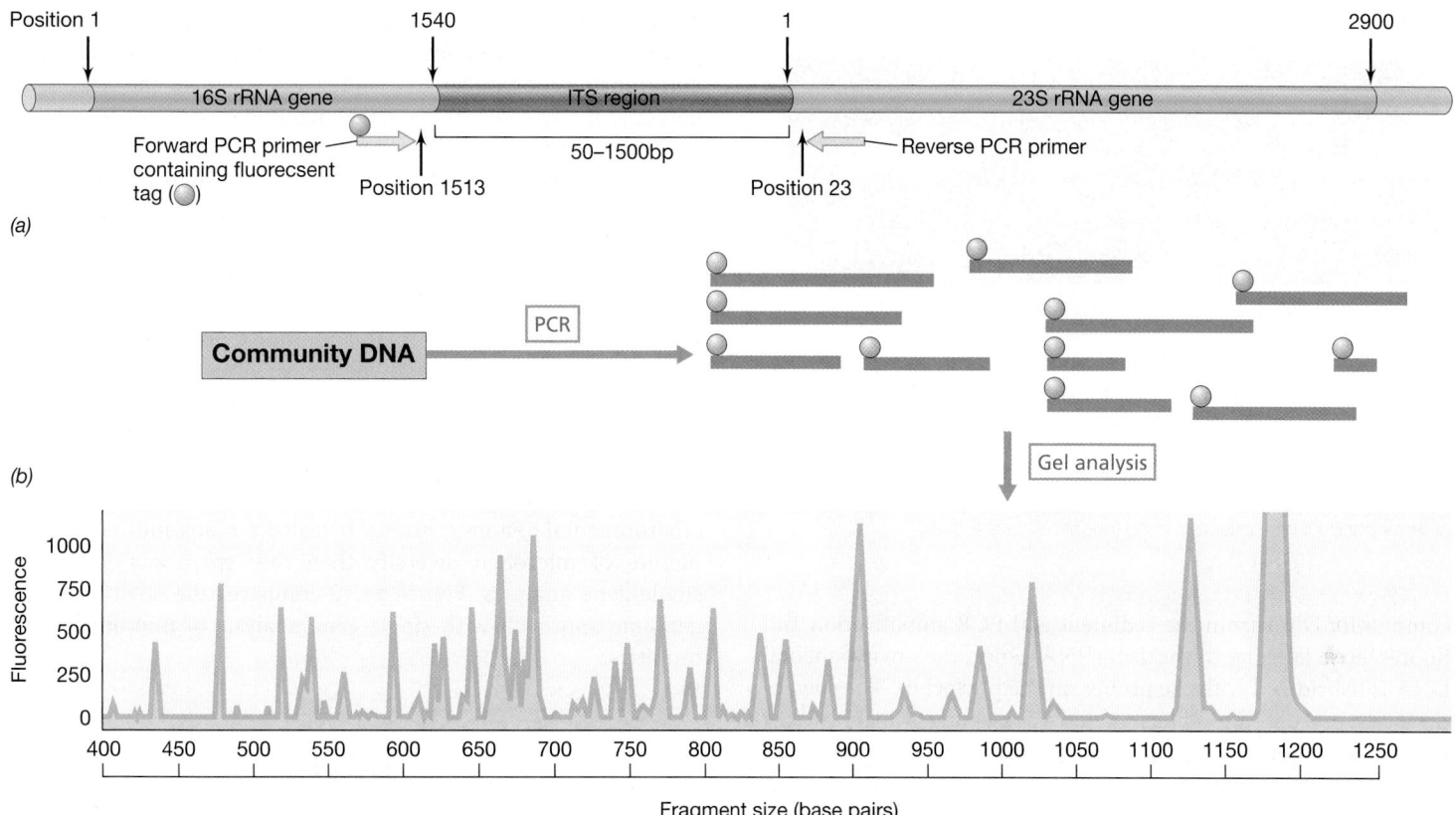

Figure 22.14 Automated ribosomal intergenic spacer analysis (ARISA). *(a)* Structure of rRNA operon spanning the 16S rRNA gene (positions 1–1540), an internal transcribed spacer (ITS) region of variable length, and the 23S rRNA gene (positions 1–2900). The PCR primers, one labeled with a fluorescent dye, are complementary to conserved sequences near the ITS region. *(b)* Amplified DNA fragments of different lengths, each corresponding to a community member. *(c)* Fragment analysis determined by an automated DNA sequencer. Peaks, corresponding to different ITS regions, can be identified by cloning and sequencing the amplified products.

typically show that many phylogenetically distinct prokaryotes (phylotypes) are present whose rRNA gene sequences differ from those of all known laboratory cultures (Figure 22.12). Moreover, using additional methods that allow a quantitative assessment of each phylotype, it has been found that with few exceptions the most abundant phylotypes in a natural microbial community are ones that have thus far defied laboratory culture. These sobering results make it clear that our knowledge of microbial diversity from enrichment cultures is very incomplete and that enrichment bias (Section 22.1) is a serious problem in culture-dependent biodiversity studies. In fact, microbial ecologists estimate that less than 0.1% of the phylotypes revealed by molecular community analyses have ever been grown in laboratory cultures.

MiniQuiz

- What could you conclude from PCR/DGGE analysis of a sample that yielded one band by PCR and one band by DGGE? One band by PCR and four bands by DGGE?

- What surprising finding has come out of many molecular studies of natural habitats using 16S rRNA as the target gene?

22.6 Microarrays and Microbial Diversity: Phylochips

We previously considered the use of DNA chips, a type of *microarray*, for assessing overall gene expression in microorganisms (Section 12.7 and Figures 12.9 and 12.10). More targeted microarrays can be constructed for rapid analyses in biodiversity studies. These phylogenetic microarrays, called *phylochips*, have been developed for screening microbial communities for specific groups of prokaryotes.

Phylochips are constructed by affixing rRNA probes or rRNA gene–targeted oligonucleotide probes to the chip surface in a known pattern. Each phylochip can be made as specific or general as required for the study by adjusting the specificity of the probes, and several thousand different probes can be added to a single phylochip. As an example, consider a phylochip designed to assess the diversity of sulfate-reducing bacteria in a sulfidic environment, such as a marine sediment. Arranged in a known pattern on the phylochip are oligonucleotides complementary to specific sequences in the 16S rRNA genes of all known sulfate-reducing bacteria (over 100 species). Then, following the isolation of total

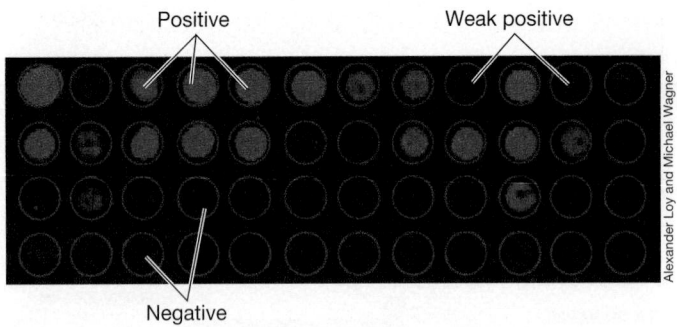

Positive Weak positive

Negative

Figure 22.15 Phylochip analysis of sulfate-reducing bacteria diversity. Each spot on the microarray shown has an oligonucleotide complementary to a sequence in the 16S rRNA of a different species of sulfate-reducing bacteria. After the microarray is hybridized with 16S rRNA genes PCR-amplified from a microbial community and then fluorescently labeled, the presence or absence of each species is signaled by fluorescence (positive or weak positive) or nonfluorescence (negative), respectively.

community DNA from the sediment and PCR amplification and fluorescence labeling of the 16S rRNA genes, the environmental DNA is hybridized with the probes on the phylochip. The species that are present are determined by assessing which probes hybridized sample DNA (**Figure 22.15**). Alternatively, rRNA might be extracted directly from the microbial community, labeled with a fluorescent dye, and hybridized directly to the phylochip without an amplification step.

Phylochips circumvent many of the time-consuming steps—PCR, DGGE, cloning, and sequencing—that are done in the microbial community analyses considered earlier (Figure 22.12). Also, instead of using 16S rRNA as the target, chips can be made to carry probes targeting genes that encode proteins necessary for a key metabolic function, such as nitrogen fixation or ammonia oxidation (Table 22.3), to address whether nitrogen-fixing or nitrifying bacteria, respectively, are present in the sample, and if so, which ones. Thus, the phylochip is another important tool for the culture-independent assessment of microbial biodiversity.

MiniQuiz

• What is a phylochip and what can it tell you?

22.7 Environmental Genomics and Related Methods

A more encompassing approach to the molecular study of microbial communities is **environmental genomics**, also called **metagenomics**, which uses the sequencing and analysis of all microbial genomes in a particular environment as a means of characterizing the entire genetic content of that environment. Metagenomics was made possible by the development of extremely high-throughput DNA sequencing technology. Before the metagenomics era, microbial community analyses typically focused on the diversity of a *single* gene. By contrast, in environmental genomics, *all* genes in a given microbial community can be sampled, and the information obtained can support a much deeper understanding of the structure and function of the community than can single-gene analyses.

It is not the goal of environmental genomics to generate complete and finished genome sequences, as has been done for many cultured microorganisms (Chapter 12). Instead, the idea is to detect as many genes as possible encoding recognizable proteins and then, if possible, to determine the phylogeny of the organism(s) to which the genes belong. The latter can be accomplished by sequencing larger fragments of environmental DNA that contain both genes for metabolic function and for assigning phylogeny, such as 16S rRNA genes. First-generation metagenomics employed random shotgun sequencing (↩ Section 12.2) of total DNA cloned from a microbial community or community DNA previously cloned into large vectors such as bacterial artificial chromosomes (BACs, ↩ Section 11.10) to characterize the gene content of that community. Although the amount of DNA that could be analyzed in a single study was rather limited compared with current metagenomic sequencing capacities, these early environmental genomic studies revealed a much more extensive picture of microbial diversity than the "snapshots" taken by single-gene analyses. **Figure 22.16** compares the environmental genomic approach with single-gene analysis of microbial communities.

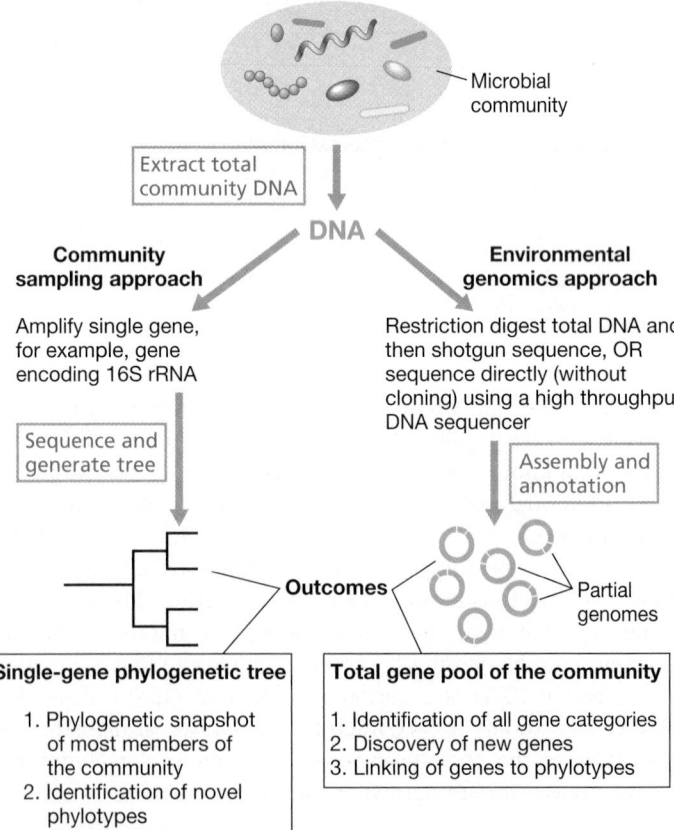

Figure 22.16 Single-gene versus environmental genomic approaches to microbial community analysis. Note that in the environmental genomic approach, all community DNA can be sequenced. Genomes are assembled and annotated but complete "finished" genomes are not obtained. Thus, not every gene is accounted for and the genomes contain some gaps. Total gene recovery is variable and depends, among other factors, on the complexity of the habitat; recovery is typically better in low-diversity habitats.

New Metagenomic Technologies

An early metagenomic study of prokaryotes in the Sargasso Sea (a low-nutrient region of the Atlantic Ocean near Bermuda) revealed remarkable diversity. This study was based on analysis of about 1 billion base pairs of sequence data from a random shotgun library of DNA recovered from surface water. The results suggested that at least 1800 bacterial and archaeal species were present, including 148 previously unknown phylotypes and many novel genes. Many of these species had previously been missed by rRNA-based community analyses. This is because not all of the 16S rRNA genes that were present in the Sargasso Sea microbial community could be amplified with the primers used for PCR amplification, and genes that fail to amplify, of course, remain undetected in community analyses. Metagenomics sidesteps this problem by sequencing DNA *without first amplifying it* by PCR. Genes and genomes are thus sequenced whether they can be amplified or not.

Although 1 billion base pairs of sequence is an enormous single data set, it was insufficient to fully describe the microbial species diversity in the Sargasso Sea sample! More recent metagenomic studies have used new DNA sequencers that do not require the construction of a gene library and have much higher sequencing capacity. These sequencers (such as that manufactured by 454 Life Sciences, ⮎ Section 12.2) use novel technologies that can sequence in excess of a billion base pairs *in a single sequencing run*! This astonishing sequencing capacity allows for metagenomic analyses of both the abundant as well as the rare components in a given habitat but is also placing unprecedented demands on the computational capacity required for the analyses. Indeed, major leaps in computational capacities will be needed to keep pace with the volume of metagenomic data of the future.

Some Examples of Environmental Genomics

Environmental genomics can detect both new genes in known organisms and known genes in new organisms. In the Sargasso Sea study mentioned previously, for example, genes encoding proteins that function in known metabolisms were occasionally found embedded within the genomes of organisms not previously known to carry out such metabolisms. For instance, the discovery of genes related to those encoding ammonia monooxygenase, a key enzyme of ammonia-oxidizing *Bacteria* (Table 22.3; ⮎ Sections 13.10 and 17.3), on a DNA fragment containing archaeal genes suggested the possible existence of ammonia-oxidizing *Archaea*. This was later established when microbiologists were successful in isolating nitrifying *Archaea* from the marine environment (*Nitrosopumilus maritimus*, ⮎ Section 19.11).

In a second example from the Sargasso Sea study, genes encoding proteorhodopsin, the light-mediated proton pump known from certain *Proteobacteria* and related to bacteriorhodopsin of extreme halophiles (⮎ Section 19.2), were found within the genomes of several new phylogenetic lineages of *Bacteria* and *Archaea*. Because these organisms had not yet been cultured, the presence of proteorhodopsin in these groups had not been previously suspected. This discovery pointed to light as being important to the physiology and ecology of these organisms and suggested new strategies for how to enrich and isolate them in laboratory culture.

Genomic approaches have also revealed variations in genes associated with a single phylotype; that is, in strains that contain identical, or nearly identical, rRNA genes. For example, in studies of *Prochlorococcus*, the most abundant cyanobacterium (oxygenic phototroph) in the ocean (⮎ Section 23.9), comparison of the genome sequences of cultured strains with *Prochlorococcus* genes obtained from metagenomic analyses of ocean water identified extensive regions shared between the cultured and environmental populations (**Figure 22.17**). This high level of gene conservation

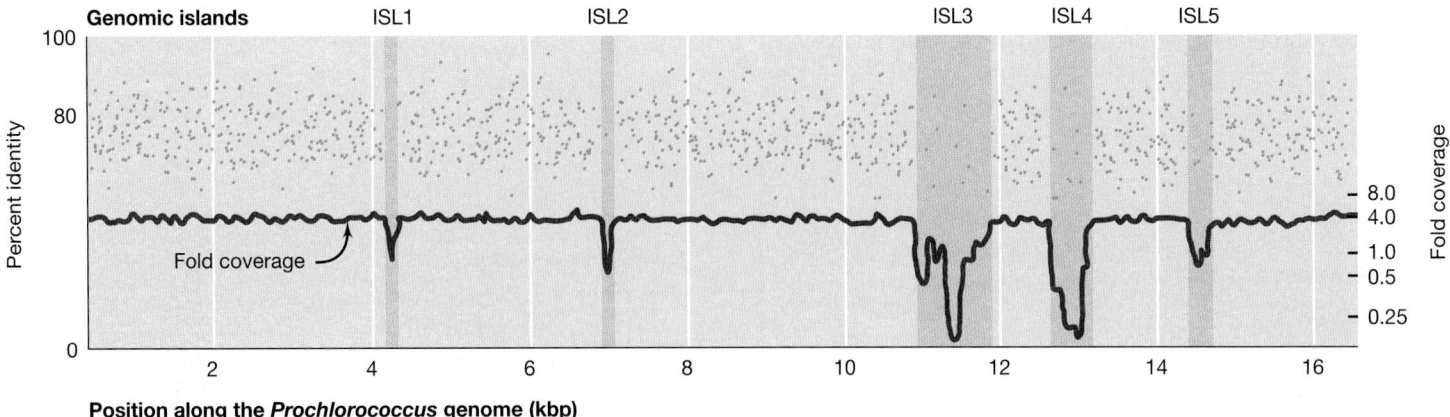

Figure 22.17 Metagenomic analysis. Sequences (represented as green dots) from the Sargasso Sea metagenome that align to the genome sequence of a cultured *Prochlorococcus*, showing regions where the cultured strain has genes of high similarity (high % identity) with sequences in the metagenome and other regions (shaded) where it lacks genes in common (genomic islands, ISL1–ISL5). Since the DNA sequence contained within the genomic islands is thought to encode niche-specific functions, the cultured strain would likely not exhibit the same environmental distribution as strains containing all island genes. Fold coverage is a measure of how completely the various regions in the *Prochlorococcus* genome are accounted for by similar sequences in the metagenome.

confirms that the organisms in culture are generally representative of environmental populations. However, these analyses also identified several highly variable regions in which the genomes of cultured strains differed significantly from environmental populations. These variable regions were clustered in the genome as *genomic islands* (Section 12.13) and likely encode functions that control the growth response of particular *Prochlorococcus* strains to environmental variables such as temperature or light quality and intensity.

Metatranscriptomics and Metaproteomics

The application of genomic methods has spawned two related techniques, metatranscriptomics and metaproteomics. *Metatranscriptomics* is analogous to metagenomics, but analyzes the sequences of total community *RNA*, rather than *DNA*. The isolated RNA is converted into DNA by reverse transcription (Section 9.12) before sequencing. Although meta*genomics* describes the functional capacities of the community (for example, the presence or absence of specific genes), meta*transcriptomics* reveals which genes in the community are being expressed at a specific time and place. Because the expression of most genes in prokaryotes is controlled at the level of transcription (Section 8.1), mRNA abundance can be considered a census of individual gene expression levels.

Metaproteomics, the measure of the diversity and abundance of different *proteins* in a community, is an even more direct measure of cell function than is metatranscriptomics. This is because different mRNAs have different half-lives and thus will not all yield the same number of protein copies. Metaproteomics is much more of a technical challenge than is either metagenomics or metatranscriptomics because protein identification requires that the proteins be physically separated before they can be identified. As a consequence, metaproteomics has thus far been restricted to simple microbial communities, such as those in some extreme environments, or to the characterization of only very abundant proteins in more complex communities. We discussed how proteins are identified in proteomic analyses in Section 12.8.

MiniQuiz

- What is a metagenome?
- How do environmental genomic approaches differ from microbial community analyses, such as that based on 16S rRNA gene analysis?
- How can the most metabolically active cell populations in a community be identified using environmental genomic methods?

Ⅲ Measuring Microbial Activities in Nature

S o far in this chapter our discussion has focused on measuring microbial *diversity*. We now turn to how microbial ecologists measure microbial *activity;* that is, what microorganisms are actually *doing* in their environment. The techniques to be

described include the use of gene expression, radioisotopes, microelectrodes, stable isotopes, and several genomic methods.

Activity measurements in a natural sample are *collective* estimates of the physiological reactions occurring in the entire microbial community, although two techniques to be discussed later, FISH-MAR and NanoSIMS, allow for a more targeted assessment of physiological activity. Activity measurements reveal both the types and rates of major metabolic reactions in a habitat, and the various techniques can be used alone or in combination in microbial community analysis. In conjunction with biodiversity estimates and gene expression analyses, these help define the structure and function of the microbial ecosystem, the ultimate goal of microbial ecology. Activity measurements can also provide valuable information for the design of enrichment cultures.

22.8 Chemical Assays, Radioisotopic Methods, and Microelectrodes

In many studies, direct chemical measurements of microbial reactions are sufficient for assessing microbial activity in an environment. For example, the fate of lactate oxidation by sulfate-reducing bacteria in a sediment sample can be tracked easily. If sulfate-reducing bacteria are present and active in a sediment sample, then lactate added to the sediment will be consumed and sulfate (SO_4^{2-}) will be reduced to hydrogen sulfide (H_2S). Since lactate, sulfate, and sulfide can all be measured with fairly high sensitivity using simple chemical assays, the transformations of these substances in a sample can be easily followed (**Figure 22.18a**).

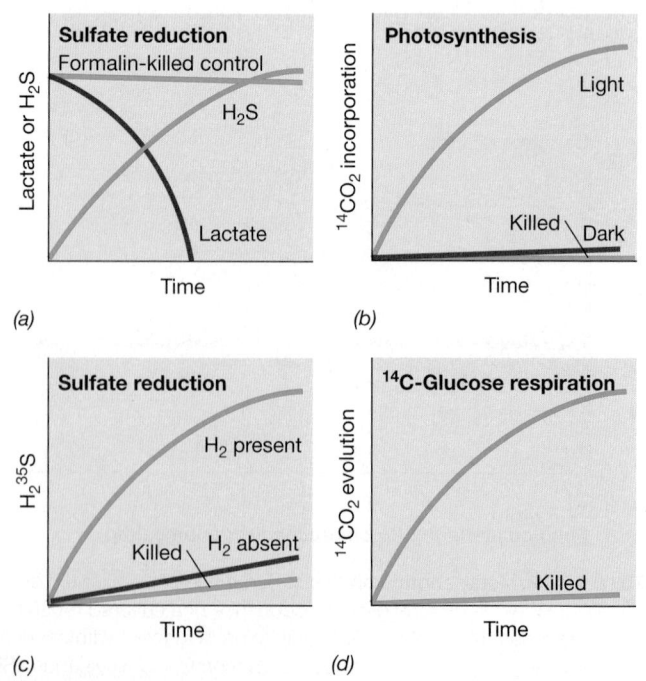

Figure 22.18 Microbial activity measurements. *(a)* Chemical measurements of lactate and H_2S transformations during sulfate reduction. Radioisotopic measurements: *(b)* photosynthesis measured with $^{14}CO_2$; *(c)* sulfate reduction measured with $^{35}SO_4^{2-}$; *(d)* production of $^{14}CO_2$ from ^{14}C-glucose.

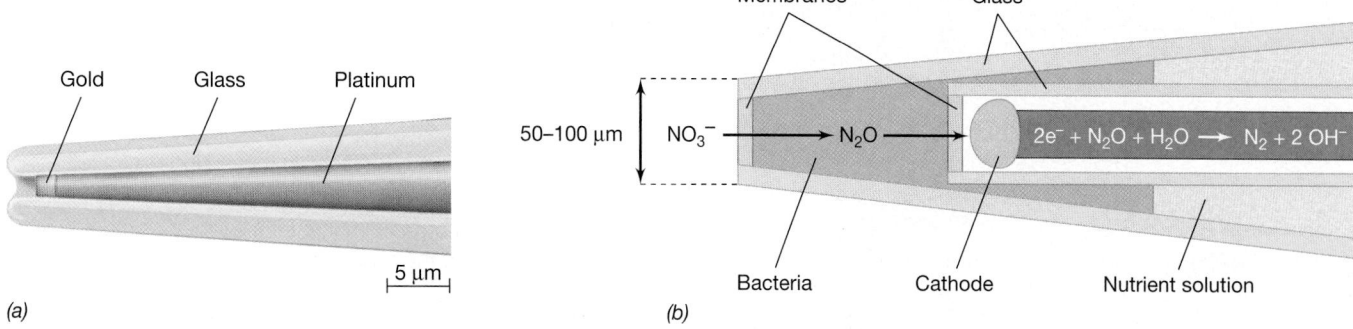

Figure 22.19 Microelectrodes. *(a)* Schematic drawing of an oxygen microelectrode. The platinum rod functions as a cathode, and when voltage is applied, O_2 is reduced to H_2O, generating a current. The current resulting from the reduction of O_2 at the gold surface of the cathode is proportional to the O_2 concentration in the sample. Note the scale of the electrode. *(b)* Biological microsensor for the detection of nitrate (NO_3^-). Bacteria immobilized at the sensor tip denitrify NO_3^- (or NO_2^-) to N_2O, which is detected by reduction to N_2 at the cathode. (Based on drawings by Niels Peter Revsbech)

Radioisotopes

When very high sensitivity is required, turnover rates need to be determined, or the fate of portions of a molecule is to be followed, *radioisotopes* are more useful than strictly chemical assays. For instance, if measuring photoautotrophy is the goal, the light-dependent uptake of radioactive carbon dioxide ($^{14}CO_2$) into microbial cells can be measured (Figure 22.18*b*). If sulfate reduction is of interest, the rate of conversion of $^{35}SO_4^{2-}$ to $H_2{}^{35}S$ can be assessed (Figure 22.18*c*). Heterotrophic activities can be measured by tracking the release of $^{14}CO_2$ from ^{14}C-labeled organic compounds (Figure 22.18*d*), and so on.

Isotopic methods are widely used in microbial ecology. To be valid, however, these must employ proper controls because some isotopic transformations might be due to abiotic processes. The *killed cell control* is the key control in such experiments. That is, it is essential to show that the transformation being measured stops when chemical agents or heat treatments that kill organisms are applied to the sample. Formalin at a final concentration of 4% is commonly used as a chemical sterilant in microbial ecology studies. This kills all cells, and transformations of radiolabeled materials in the presence of 4% formalin can be ascribed to abiotic processes (Figure 22.18).

Microelectrodes

Small glass electrodes called **microelectrodes** can be used to study the activity of microorganisms in nature. Microelectrodes have been constructed that measure many chemical species including pH, O_2, NO_2^-, NO_3^-, N_2O, CO_2, H_2, and H_2S. As the name *micro*electrode implies, these devices are very small, their tips ranging in diameter from 2 to 100 μm (**Figure 22.19a**). The electrodes are carefully inserted into the habitat in small increments to follow microbial activities over very short distances.

Microelectrodes have many applications. For example, O_2 concentrations in microbial mats (⮂ Figure 23.9), aquatic sediments, or soil particles (⮂ Figure 23.3) can be very accurately measured over extremely fine intervals using microelectrodes. A

micromanipulator is used to insert the electrodes gradually through the sample such that measurements can be taken every 50–100 μm (**Figure 22.20**). Using a bank of microelectrodes, each sensitive to a different chemical, simultaneous measurements of several transformations in a habitat can be made.

Microbial processes in the sea are extensively studied because they have a profound impact on nutrient cycles and the overall health of the planet. As it is difficult to reproduce the

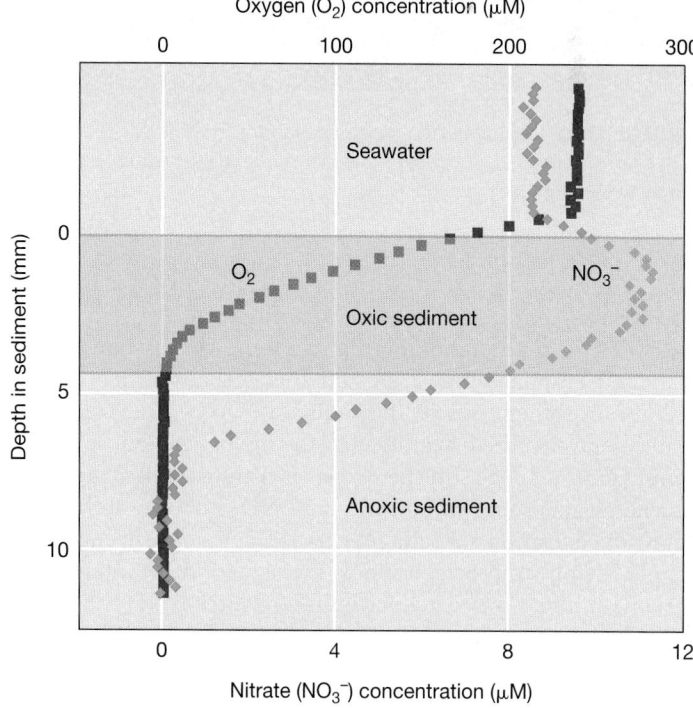

Figure 22.20 Depth profiles of oxygen and nitrate. Data obtained using the lander (see Figure 22.21) equipped with microelectrode sensors for remote chemical characterization of deep-sea sediments. (Based on data and drawings by Niels Peter Revsbech)

Figure 22.21 Deployment of deep-sea lander. The lander is equipped with a bank of microelectrodes (arrow) to measure distribution of chemicals in marine sediments.

Niels Peter Revsbech

conditions found at great depths in the laboratory, it is useful to use microelectrodes on robotic devices to analyze microbial activities on the seafloor. **Figure 22.21** shows deployment of a "lander" equipped with various microelectrodes so that the distribution of chemicals in the sediment can be analyzed at all water depths in the ocean. One of the biologically most important chemical species in the oceans is nitrate (NO_3^-), but electrochemical sensors cannot measure NO_3^- in seawater, as the high concentrations of salts interfere. To circumvent this problem, a "living" microelectrode was designed that contains bacteria within its tip that reduce NO_3^- (or nitrite, NO_2^-) to N_2O. The N_2O produced by the bacteria is then detected following its reduction to N_2 at the cathode of the microelectrode (Figure 22.19b); this provides an electrical impulse signaling the presence of NO_3^-. In the oxic layer of marine sediments, nitrate is produced from the oxidation of ammonium (nitrification, ↪ Section 13.10), so there is often a peak of NO_3^- in the sediment surface layer (Figure 22.20). In the deeper, anoxic layers of the sediment, NO_3^- is consumed by denitrification and dissimila-

tive reduction to ammonium (NH_4^+) (↪ Section 14.7), and NO_3^- therefore disappears a few millimeters below the oxic–anoxic interface (Figure 22.20).

MiniQuiz

- Why are radioisotopes so useful in measuring microbial activities?
- If a large pulse of organic matter entered the sediment, how would that change the profiles of nitrate and oxygen shown in Figure 22.20?

22.9 Stable Isotopes

For many of the chemical elements different isotopes exist, varying in their number of neutrons. Certain isotopes are unstable and break down as a result of radioactive decay. Others, called *stable isotopes*, are not radioactive, but are metabolized differently by microorganisms and can be used to study microbial transformations in nature. There are two methods in which stable isotopes can yield information on microbial activities. We describe *isotopic fractionation* in this section and *stable isotope probing* in the next section.

Isotopic Fractionation

The two elements most useful for stable isotope studies in microbial ecology are carbon and sulfur. Carbon (C) exists in nature primarily as ^{12}C, but about 5% exists as ^{13}C. Likewise, sulfur (S) with its four stable isotopes exists primarily as ^{32}S. Some sulfur is found as ^{34}S and very small amounts as ^{33}S and ^{36}S. The relative abundance of these isotopes changes when C or S is metabolized by microorganisms because enzymes typically favor the *lighter* isotope. That is, relative to the lighter isotope, the heavier isotope is discriminated against when both are metabolized by an enzyme (**Figure 22.22**). For example, when CO_2 is fixed into cell material by an autotrophic organism, the cellular carbon becomes *enriched* in ^{12}C and *depleted* in ^{13}C, relative to an inorganic carbon standard of known composition. Likewise, the sulfur atom in H_2S produced from the bacterial reduction of sulfate (SO_4^{2-}) is isotopically lighter than H_2S that has formed geochemically. These

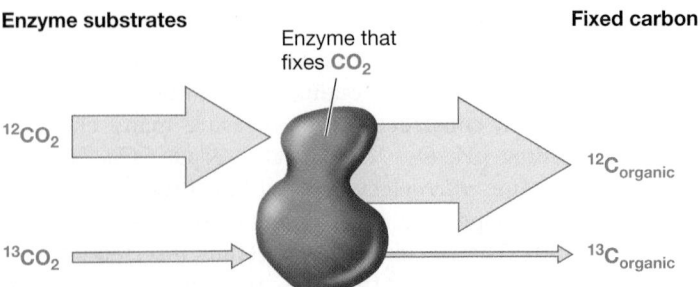

Enzyme substrates   **Fixed carbon**

Figure 22.22 Mechanism of isotopic fractionation with carbon as an example. Enzymes that fix CO_2 preferentially fix the lighter isotope (^{12}C). This results in fixed carbon being enriched in ^{12}C and depleted in ^{13}C relative to the starting substrate. The size of the arrows indicates the relative abundance of each isotope of carbon.

discriminations are called **isotopic fractionations** (Figure 22.22) and are typically the result of biological activities. They can therefore be used as a measure of whether or not a particular transformation has been catalyzed by microorganisms.

The isotopic fractionation of carbon in a sample is calculated as the extent of ^{13}C depletion relative to a standard having an isotopic composition of geological origin. The standard for carbon isotope analysis is a Cretaceous limestone formation (the PeeDee belemnite). Because the magnitude of fractionation is usually very small, depletion is calculated as "per mil" (‰, parts per thousand) and reported as the δ^{13}C (pronounced "delta C 13") of a sample using the following formula:

$$\delta^{13}C = \frac{(^{13}C/^{12}C \text{ sample}) - (^{13}C/^{12}C \text{ standard})}{(^{13}C/^{12}C \text{ standard})} \times 1000\text{ ‰}$$

The same formula form is used to calculate the fractionation of sulfur isotopes, in this case using iron sulfide mineral from the Canyon Diablo meteorite as the standard:

$$\delta^{34}C = \frac{(^{34}S/^{32}S \text{ sample}) - (^{34}S/^{32}S \text{ standard})}{(^{34}S/^{32}S \text{ standard})} \times 1000\text{ ‰}$$

Use of Isotopic Fractionation in Microbial Ecology

The isotopic composition of a material can reveal its past biological or geological origin. For example, plant material and petroleum (which is derived from plant material) have similar isotopic compositions (**Figure 22.23**). Carbon from both plants and petroleum is isotopically lighter than the CO_2 from which it was formed because the biochemical pathway used to fix CO_2 discriminated against $^{13}CO_2$ (Figures 22.22 and 22.23). Moreover,

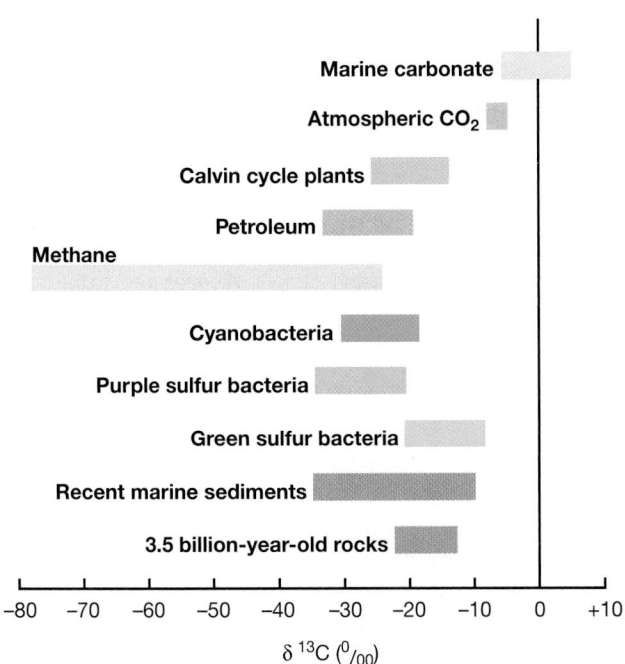

Figure 22.23 Isotopic geochemistry of ^{13}C and ^{12}C. Note that carbon fixed by autotrophic organisms is enriched in ^{12}C and depleted in ^{13}C. Methane shows extreme isotopic fractionation.

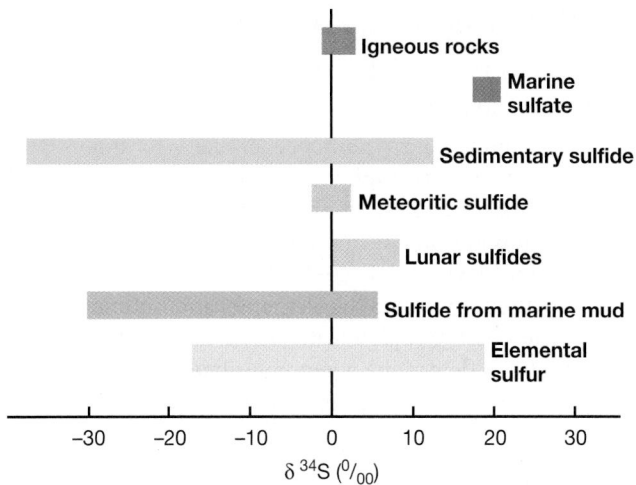

Figure 22.24 Isotopic geochemistry of ^{34}S and ^{32}S. Note that H_2S and S^0 of biogenic origin are enriched in ^{32}S and depleted in ^{34}S.

methane (CH_4) produced by methanogenic *Archaea* (↺ Section 19.3) is isotopically extremely light, indicating that methanogens discriminate strongly against $^{13}CO_2$ when they reduce CO_2 to CH_4 (↺ Section 14.10). By contrast, carbon in isotopically heavier marine carbonates is clearly of geological origin (Figure 22.23).

Because of the differences in the proportion of ^{12}C and ^{13}C in carbon of biological versus geological origin, the ^{13}C/^{12}C ratio of rocks of different ages has been used as evidence for or against past biological activity in Earth's ancient environments. Organic carbon in rocks as old as 3.5 billion years shows evidence of isotopic fractionation (Figure 22.23), supporting the idea that autotrophic life existed at this time. Indeed, we now believe that the first life on Earth appeared somewhat before this, about 3.8–3.9 billion years ago (↺ Sections 1.4 and 16.2).

The activity of sulfate-reducing bacteria is easy to recognize from their fractionation of stable sulfur isotopes in sulfides (**Figure 22.24**). As compared with an H_2S standard, sedimentary H_2S is highly enriched in ^{32}S (Figure 22.24). Fractionation during sulfate reduction allows one to identify biologically produced sulfur and has been widely used to trace the activities of sulfur-cycling prokaryotes through geological time. Sulfur isotopic analyses have also been used as evidence for the lack of life on the Moon. For example, the data in Figure 22.24 show that the isotopic composition of sulfides in lunar rocks closely approximates that of the H_2S standard, which represents primordial Earth, and differs from that of biogenic H_2S.

MiniQuiz

- How can the ^{13}C/^{12}C composition of a substance reveal its biological or geological origin?
- What is the simplest explanation for why lunar sulfides are isotopically similar to those of the primordial Earth?
- What is the expected isotopic composition of carbon in methanotrophs?

22.10 Linking Specific Genes and Functions to Specific Organisms

The isotopic methods introduced in the previous section used samples containing large numbers of cells to infer specific processes such as autotrophy or nitrogen fixation occurring within a community. These methods give an overview of community activities but do not reveal the contribution of individual cells. To do this, new isotopic methods have been developed that can measure the activity and the elemental and isotopic composition of single cells. Coupled with advanced DNA sequencing methods that can determine a genome sequence from the DNA contained in a single cell, these techniques are at the cutting edge of microbial ecology today.

Analysis of Single Cells by Secondary Ion Mass Spectrometry

Secondary ion mass spectrometry (SIMS) is based on the detection of ions released from a sample placed under a focused high-energy primary ion beam, for example, of cesium (Cs^+); from the data generated, the elemental and isotopic composition of released materials can be obtained. When the primary ion beam impacts the sample, most chemical bonds are broken and atoms or polyatomic fragments are ejected from a very thin layer of the surface (1–2 nm) either as neutral or charged particles (secondary ions), a process called *sputtering*. These secondary ions are directed to a mass spectrometer, an instrument that can determine their mass-to-charge ratio. The instrument also records where on the specimen the ion beam is directed such that a two-dimensional image of the distribution of specific ions on the sample surface is obtained. In addition, by focusing the ion beam on the same spot during repeated cycles of sputtering, material can be slowly burned away to expose deeper regions of the sample. The latest SIMS instruments are equipped with microprobes that can be focused on areas of less than a micrometer, measuring the distribution of different isotopes and elements within a single cell. This high-resolution SIMS analysis is called *NanoSIMS*. NanoSIMS instruments have multiple detectors that provide for the simultaneous analysis of ions of different mass to charge ratios originating from the same sample location (**Figure 22.25**).

When combined with FISH (Section 22.4), SIMS and NanoSIMS can be used to track the incorporation of different elements, natural isotopes, or isotope-labeled substrates into individual cells of different populations. An initial application of NanoSIMS technology was to characterize the composition of carbon isotopes in structured aggregates of anaerobic methane-oxidizing prokaryotes. A form of anaerobic methane oxidation widespread in marine sediments is the result of a syntrophic association of sulfate-reducing *Bacteria* and methane-oxidizing (methanotrophic) *Archaea* that form aggregates, with the sulfate reducers surrounding the methanotrophs; oxidation of CH_4 by the aggregates is accompanied by the transfer of metabolites from the methanotrophs to the sulfate reducers (↩ Section 14.13). To confirm which organism was actually oxidizing the methane, NanoSIMS technology was used. Because biogenic CH_4 is highly depleted in ^{13}C (Figure 22.23), it

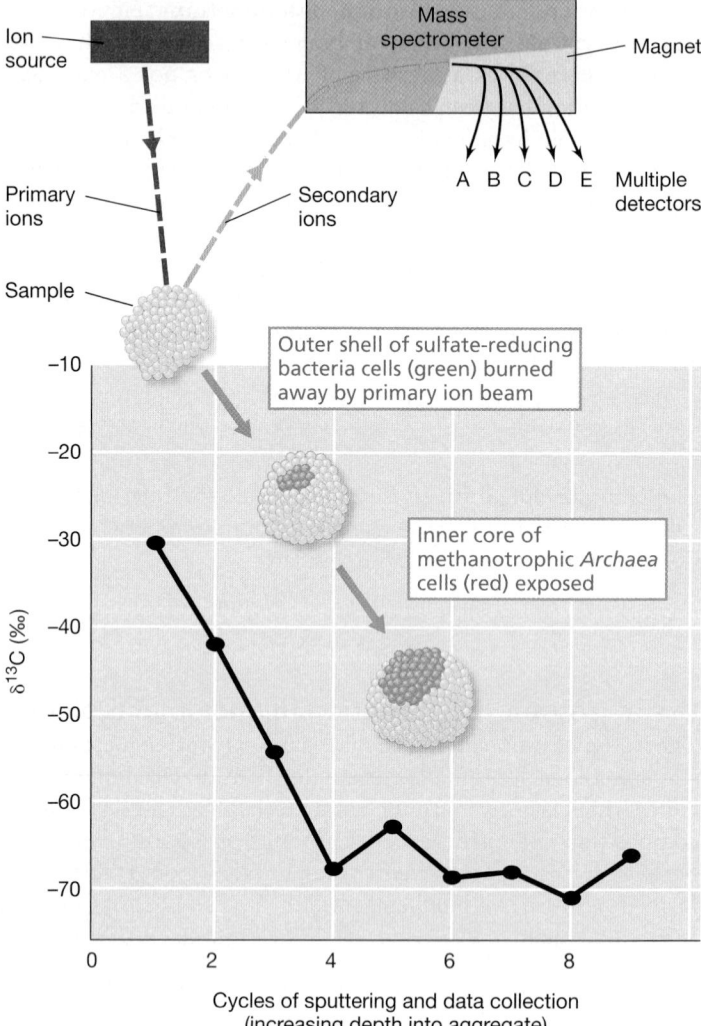

Figure 22.25 SIMS technology. Top, simplified diagram of a NanoSIMS instrument showing the beams of primary (red) and secondary (blue) ions and five different detectors, each of which identifies ions of a different mass to charge ratio. The graph is a depth profile of $\delta^{13}C$ (‰) from a structured aggregate of sulfate-reducing bacteria and methane-oxidizing *Archaea*. The isotopic signature of the carbon becomes increasingly light as the ion beam of the SIMS instrument burns away the outer layer of sulfate-reducing bacteria (green), exposing the methanotrophic *Archaea* (red) to the ion beam. The different cell populations were originally identified by FISH (see Figure 22.10). For more coverage of anoxic methanotrophy, see Section 14.15 and Figure 14.28.

is a natural tracer that can show the consumption (and incorporation) of CH_4 by the methanotroph. A depth profile of carbon isotope composition was created as the ion microprobe progressively burned through the cell aggregate (Figure 22.25). The $\delta^{13}C$ values approached −70‰ as the ion beam neared the interior of the aggregate, clearly showing that it was the *Archaea* that were oxidizing and incorporating the carbon from CH_4. NanoSIMS has also been used to track the assimilation of labeled substrates (for example, ^{15}N-labeled dinitrogen, $^{15}NH_4$, and ^{13}C-labeled CO_2) into single cells to identify which organism in a mixture is carrying out a specific metabolism.

Flow Cytometry and Multiparametric Analyses

Because of the large population sizes of natural microbial communities—typically well in excess of 1 million cells per milliliter of water or per gram of soil—methods that rely on microscopy can examine only a very small part of a whole community. Although image analysis software can help automate the process, most microscopic analyses still rely on the practiced eye of the investigator. It is particularly difficult to assess cell numbers by counting cells one by one, and this problem is compounded for populations present in low numbers. However, the technique known as *flow cytometry* offers an alternative to these constraints imposed by microscopy.

Flow cytometers can examine specific cell parameters such as size, shape, or fluorescent properties as the cells pass through a detector at rates of many thousands of cells per second (**Figure 22.26**; ⇄ Section 31.9). Fluorescence may be intrinsic (for example, chlorophyll fluorescence of phototrophic microorganisms; ⇄

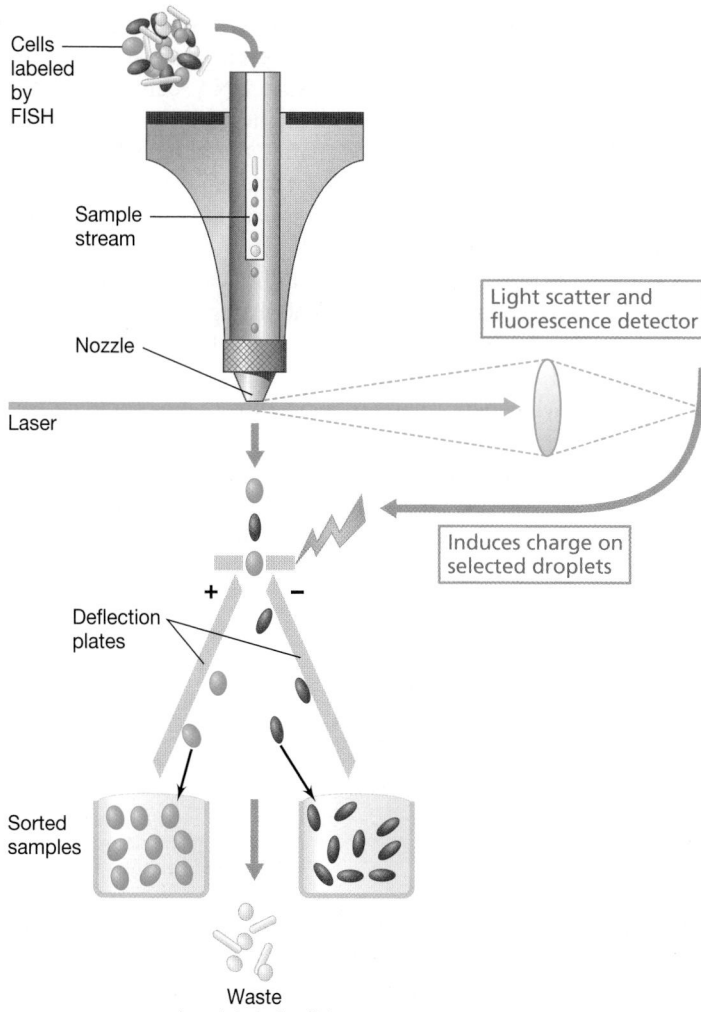

Figure 22.26 Flow cytometric cell sorting. As the fluid stream exits the nozzle, it is broken into droplets containing no more than a single cell. Droplets containing desired cell types (detected by fluorescence or light scatter) are charged and collected by redirection into collection tubes by positively or negatively charged deflection plates.

Figure 2.6*b*) or conferred by DNA staining, differential staining of live versus dead cells (vital stains), fluorescent DNA probes (FISH), or fluorescent antibodies, all methods discussed in this chapter. A major advantage of flow cytometry is the capacity to combine multiple parameters in analyzing a microbiological sample or finding a specific population. A remarkable example of this was the discovery in the late 1980s of a novel and abundant community of marine cyanobacteria, all species of the genus *Prochlorococcus*. *Prochlorococcus* cells are smaller and have different fluorescent properties than another common marine cyanobacterium, *Synechococcus*. Based on differences in size and fluorescence, flow cytometry resolved these two populations and *Prochlorococcus* was subsequently shown to be the predominant oxygenic phototroph in ocean waters between 40°S and 40°N latitude, reaching concentrations greater than 10^5 cell/ml. Based on this finding, it can be said that *Prochlorococcus* is the most abundant phototrophic organism on Earth (⇄ Section 23.9).

Radioisotopes in Combination with FISH: FISH-MAR

Radioisotopes are used as measures of microbial activity in a microscopic technique called **microautoradiography (MAR)**. In this method, cells from a microbial community are exposed to a substrate containing a radioisotope, such as an organic compound or CO_2. Heterotrophs take up the radioactive organic compounds and autotrophs take up the radioactive CO_2. Following incubation in the substrate, cells are affixed to a slide and the slide is dipped in photographic emulsion. While the slide is left in darkness for a period, radioactive decay from the incorporated substrate induces formation of silver grains in the emulsion; these appear as black dots above and around the cells. **Figure 22.27*a*** shows a MAR experiment in which an autotrophic cell has taken up $^{14}CO_2$.

Microautoradiography can be done simultaneously with FISH (Section 22.4) in *FISH-MAR*, a powerful technique that combines identification with activity measurements. FISH-MAR allows a microbial ecologist to determine which organisms in a natural sample are metabolizing a particular radiolabeled substance (by MAR) and also to identify these organisms (by FISH) (Figure 22.27*b*). FISH-MAR thus goes a step beyond phylogenetic identification; the technique reveals physiological information on the organisms as well. Such data are useful not only for understanding the activity of the microbial ecosystem but also for guiding enrichment cultures. For example, knowledge of the phylogeny and morphology of an organism metabolizing a particular substrate in a natural sample can be used to design an enrichment protocol to isolate the organism. In addition FISH-MAR can be used to detect quantitatively the amount of substrate consumed by single cells, allowing the activity distribution in a community to be described.

Stable Isotope Probing

We have seen how the combination of FISH with MAR or FISH with NanoSIMS allows for analyses of both microbial diversity and activity. These are powerful methods for linking specific

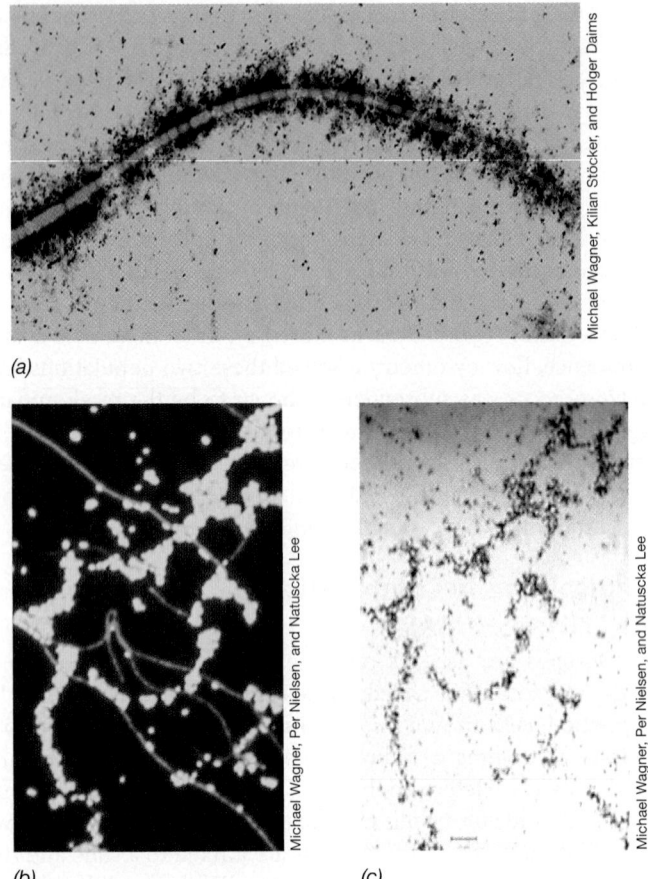

(a)

(b)   (c)

Figure 22.27 FISH-MAR. Fluorescent *in situ* hybridization (FISH) combined with microautoradiography (MAR). *(a)* An uncultured filamentous cell belonging to the *Gammaproteobacteria* (as revealed by FISH) is shown to be an autotroph (as revealed by MAR-measured uptake of $^{14}CO_2$). *(b)* Uptake of ^{14}C-glucose by a mixed culture of *Escherichia coli* (yellow cells) and *Herpetosiphon aurantiacus* (filamentous, green cells). *(c)* MAR of the same field of cells shown in part b. Incorporated radioactivity exposes the film and shows that glucose was assimilated mainly by cells of *E. coli*.

microbial populations with an activity or niche, but the organism's phylogeny must be known for the FISH probe to be developed (Section 22.4). An alternative method of coupling diversity to activity is **stable isotope probing (SIP)**. SIP, which employs substrates that contain a heavy (nonradioactive) isotope, is typically used to reveal the diversity behind specific metabolic transformations in the environment. Most of the SIP studies conducted thus far have used ^{13}C, a heavy isotope of carbon; however, stable isotope probing using either ^{15}N or ^{18}O have been successful as well. SIP reveals microbial diversity by yielding isotope-labeled DNA that can be used to analyze specific genes or the entire genome of the organism(s) that consumed the labeled substrate.

How is an SIP experiment done? Let's say the goal of a research project was to characterize organisms capable of catabolizing aromatic compounds in lake sediment. Using benzoate as a model aromatic compound, ^{13}C-enriched benzoate would be added to a sediment sample, the sample incubated for an appropriate period, and then total DNA extracted from the sample (**Figure 22.28**). As shown in Figure 22.12, such DNA originates from *all* of the organisms in the microbial community. However, organisms that incorporate ^{13}C-benzoate will synthesize DNA containing ^{13}C. ^{13}C-DNA is heavier, albeit only slightly heavier, than ^{12}C-DNA, but the difference is sufficient to separate the heavier DNA from the lighter DNA by a special type of centrifugation technique (Figure 22.28). Once the ^{13}C-DNA is isolated, it can be analyzed for the genes of interest.

Returning to the benzoate example, if the goal was to characterize the phylogeny of the organisms catabolizing the benzoate, PCR amplification of 16S ribosomal RNA genes from the ^{13}C-DNA could be used to do so (Figures 22.12 and 22.13). However, in addition to phylogenetic analyses, many other genes could be targeted once the ^{13}C-DNA is obtained. For example, SIP has been employed in studies of the phylogeny and metabolic pathways of methylotrophs, organisms that specialize in the catabolism of C_1 compounds (Section 14.15). In these studies, $^{13}CH_4$ or

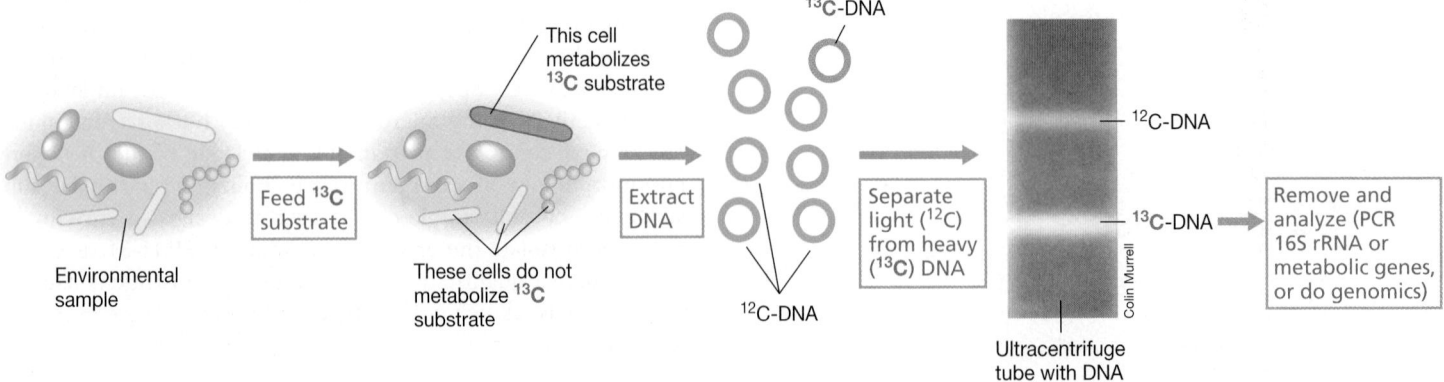

Figure 22.28 Stable isotope probing. The microbial community in an environmental sample is fed a specific ^{13}C-substrate. Organisms that can metabolize the substrate produce ^{13}C-DNA as they grow and divide; ^{13}C-DNA can be separated from lighter ^{12}C-DNA by density gradient centrifugation (photo). The isolated DNA is then subjected to specific gene analysis or entire genomic analysis.

$^{13}CH_3OH$ was used to label the methylotrophs followed by PCR amplification of 16S rRNA genes and genes encoding specific methane oxidation functions (Table 22.3) from the ^{13}C-DNA. Whole genome analyses are also possible using SIP. For example, in another methylotroph study SIP was used in combination with metagenomic analyses (Section 22.7) and pointed to a previously unsuspected methylotroph as being important in C_1 catabolism in that particular environment.

Stable isotope probing can also employ isotopes of nitrogen (N). In this case, the isotopically heavy isotope of N, ^{15}N, competes with the more abundant and lighter isotope, ^{14}N. To study nitrogen fixation, for example, a sample would be supplied with $^{15}N_2$, and those organisms that can fix N_2 (⮌ Section 13.14) will incorporate some of the $^{15}N_2$. The ^{15}N will end up in their DNA, making it isotopically "heavy"; such DNA can be separated from isotopically lighter DNA by ultracentrifugation (Figure 22.8) and analyzed for specific genes.

Single-Cell Genomics

A major stumbling block in any PCR-based gene recovery method is the requirement that a specific gene be identified prior to analysis that will react with the primers used in the amplification. Newer methods of DNA amplification now provide an alternative way to associate genes with a particular organism without the biases associated with PCR.

Multiple displacement amplification (MDA) (**Figure 22.29**) is a method to amplify chromosomal DNA from a single organism isolated from a natural environment using a cell sorting technique, such as flow cytometry. MDA uses a bacteriophage DNA polymerase to initiate replication of cell DNA at random points in the chromosome, displacing the complementary strand as each polymerase molecule synthesizes new DNA. The number of DNA copies produced is sufficient to determine the complete, or nearly complete, genome sequence of the organism. In this way, metabolic functions inferred from the genome sequence can be associated with the single cell from which the chromosomal DNA was derived, and PCR is not required.

MDA requires stringent control over purity to eliminate contaminating DNA, but when combined with high-throughput DNA sequencing methods, MDA provides a powerful tool for linking specific metabolic functions to individual cells that have never been grown in laboratory culture. Information about the metabolic capacities of these uncultured organisms can then be used to guide the development of strategies to recover them by enrichment culture methods.

MiniQuiz

- What are the advantages and disadvantages of flow cytometry, relative to microscopy, for characterizing a microbial community?
- How can stable isotope probing reveal the identity of an organism that carries out a particular process?
- What is the ecological value of single-cell genome sequencing?

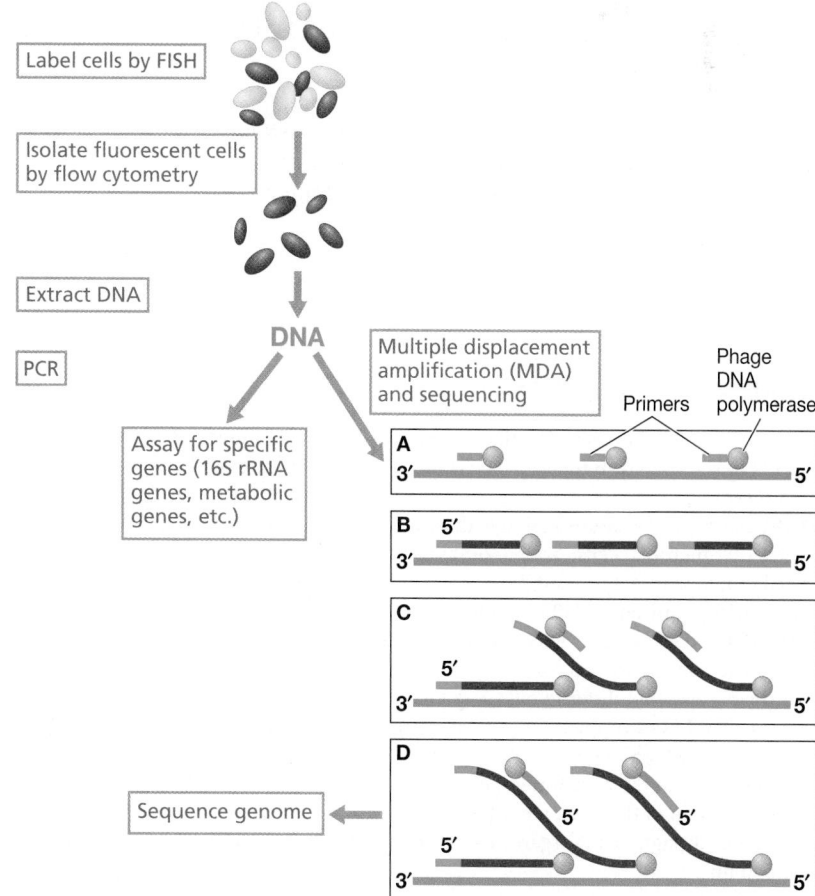

Figure 22.29 Genetic analysis of sorted cells. DNA is recovered from a specific population of cells following FISH labeling and flow cytometric sorting (Figure 22.28). DNA is characterized by PCR amplification and sequencing of specific genes, or by amplification of the entire genome by multiple displacement amplification (MDA) followed by sequencing. For MDA, an amount of DNA sufficient for full genome sequence determination is produced using short DNAs of random sequence as primers (A) to initiate genome replication by a bacteriophage DNA polymerase. The phage polymerase copies DNA from multiple points in the genome and also displaces newly synthesized DNA (B,C), thereby freeing additional DNA for primer annealing and (D) initiation of polymerization.

Big Ideas

22.1

The enrichment culture technique is a means of obtaining microorganisms from natural samples. Successful enrichment and isolation proves that an organism of a specific metabolic type was present in the sample, but does not indicate its ecological importance or abundance.

22.2

Once a successful enrichment culture has been established, pure cultures can often be obtained by conventional microbiological procedures, including streak plates, agar shakes, and dilution methods. Laser tweezers and flow cytometry allow one to isolate a cell from a microscope field and move it away from contaminants.

22.3

DAPI, acridine orange, and SYBR Green are general stains for quantifying microorganisms in natural samples. Some stains can differentiate live versus dead cells. Fluorescent antibodies that are specific for one or a small group of related organisms can make stains more specific. The green fluorescent protein makes cells autofluorescent and is a means for tracking cells introduced into the environment and reporting gene expression. In natural samples, morphologically identical cells may actually be genetically distinct.

22.4

FISH methods have combined the power of nucleic acid probes with fluorescent dyes and are thus highly specific in their staining properties. FISH methods include phylogenetic stains and CARD-FISH.

22.5

PCR can be used to amplify specific target genes such as ribosomal rRNA genes or key metabolic genes. DGGE can identify the different variants of these genes present in different species in a community.

22.6

Phylochips combine microarray and phylogenetic technologies and are used to screen microbial communities for specific groups of prokaryotes.

22.7

Environmental genomics (metagenomics) is based on cloning, sequencing, and analysis of the collective genomes of the organisms present in a microbial community. Metatranscriptomics and metaproteomics are offshoots of metagenomics whose focus is mRNA and proteins, respectively.

22.8

The activity of microorganisms in natural samples can be assessed very sensitively using radioisotopes or microelectrodes, or both. The measurements obtained give the net activity of the microbial community.

22.9

Isotopic compositions can reveal the biological origin and/or biochemical mechanisms involved in the formation of various substances. Isotopic fractionation is a result of the activity of enzymes that discriminate against the heavier form of an element when binding their substrates.

22.10

A variety of advanced technologies now offer tools to examine the activity and metabolic potential of microorganisms at multiple scales—from single cells, to populations, to communities. New technologies such as NanoSIMS, FISH-MAR, and SIP make it possible to examine metabolic activity, gene content, and gene expression in natural microbial communities in powerful ways.

Review of Key Terms

Acridine orange a nonspecific fluorescent dye used to stain DNA in microbial cells in a natural sample

DAPI a nonspecific fluorescent dye that stains DNA in microbial cells; used to obtain total cell numbers in natural samples

Denaturing gradient gel electrophoresis (DGGE) an electrophoretic technique capable of separating nucleic acid fragments of the same size that differ in base sequence

Enrichment bias a problem with enrichment cultures in which "weed" species tend to dominate in the enrichment, often to the exclusion of the most abundant or ecologically significant organisms in the inoculum

Enrichment culture highly selective laboratory culture methods for obtaining microorganisms from natural samples

Environmental genomics (metagenomics) the use of genomic methods (sequencing and analyzing genomes) to characterize natural microbial communities

FISH-MAR a technique that combines identification of microorganisms with measurement of metabolic activities

Flow cytometry a technique for counting and examining microscopic particles by suspending them in a stream of fluid and passing them by an electronic detection device

Fluorescent *in situ* hybridization (FISH) a method employing a fluorescent dye covalently bonded to a specific nucleic acid probe for identifying or tracking organisms in the environment

Green fluorescent protein (GFP) a fluorescing protein for tracking genetically modified organisms and determining conditions that induce the expression of specific genes.

High-throughput technology the employment of robotic systems and/or highly parallel reaction chemistries to run hundreds to thousands of procedures very quickly

Isotopic fractionation the discrimination by enzymes against the heavier isotope of the various isotopes of carbon or sulfur, leading to enrichment of the lighter isotopes

Laser tweezers a device for obtaining pure cultures by optically trapping a single cell with a laser beam and moving it away from surrounding cells into sterile growth medium

Metatranscriptomics the measurement of whole-community gene expression using RNA sequencing

Metaproteomics the measurement of whole-community protein expression using mass spectrometry to assign peptides to unique genes

Microautoradiography (MAR) the measurement of the uptake of radioactive substrates by visually observing the cells in an exposed photographic emulsion

Microbial ecology the study of the interaction of microorganisms with each other and their environment

Microelectrode a small glass electrode for measuring pH or specific compounds such as O_2 or H_2S that can be immersed into a microbial habitat at microscale intervals

Most-probable-number (MPN) technique the serial dilution of a natural sample to determine the highest dilution yielding growth

Multiple displacement amplification (MDA) a method to generate multiple copies of chromosomal DNA from a single organism

Nucleic acid probe an oligonucleotide, usually 10–20 bases in length, complementary in base sequence to a nucleic acid sequence in a target gene or RNA

Phylotype one or more organisms with the same or related sequences of a phylogenetic marker gene.

Stable isotope probing (SIP) a method for characterizing an organism that incorporates a particular substrate by supplying the substrate in ^{13}C form and then isolating ^{13}C-enriched DNA and analyzing the genes

Winogradsky column a glass column packed with mud and overlaid with water to mimic an aquatic environment, in which various bacteria develop over a period of months

Review Questions

1. What is the basis of the enrichment culture technique? Why is an enrichment medium usually suitable for the enrichment of only a certain group or groups of organisms (Section 22.1)?

2. What is the principle of the Winogradsky column, and what types of organisms does it serve to enrich? How might a Winogradsky column be used to enrich organisms present in an extreme environment, like a hot spring microbial mat (Section 22.1)?

3. Describe the principle of MPN for enumerating bacteria from a natural sample (Section 22.2).

4. Why would the laser tweezers be a method superior to dilution and liquid enrichment for obtaining an organism present in a sample in low numbers (Section 22.2)?

5. Compare and contrast the staining methods of using fluorescent dyes and fluorescent antibodies (Section 22.3).

6. Highlight the advantages of using FISH (Section 22.4).

7. What is the green fluorescent protein? In what ways does a green fluorescing cell differ from a cell fluorescing from, for example, phylogenetic staining (Sections 22.3 and 22.4)?

8. How can a phylogenetic picture of a microbial community be obtained without culturing its inhabitants (Section 22.5)?

9. What is the difference between DGGE and RFLP (Section 22.5)?

10. Why is a microarray not suitable for characterizing community-wide transcription (Sections 22.6 and 22.7)?

11. Why is metagenomics useful in characterizing microbial communities in an environment (Section 22.7)?

12. Why is environmental proteomics limited by natural abundance of microbial populations, whereas environmental genomics and metatranscriptomics are not so limited (Section 22.7)?

13. What are the major advantages of radioisotopic methods in the study of microbial ecology? What type of controls (discuss at least two) would you include in a radioisotopic experiment to show $^{14}CO_2$ incorporation by phototrophic bacteria or to show $^{35}SO_4^{2-}$ reduction by sulfate-reducing bacteria (Section 22.8)?

14. What can FISH-MAR tell you that FISH alone cannot (Section 22.10)?

15. Will autotrophic organisms contain more or less ^{12}C in their organic compounds than was present in the CO_2 that fed them (Section 22.9)?

16. What is the advantage of having multiple detectors on a NanoSIMS instrument (Section 22.10)?

17. How might you combine SIP and NanoSIMS to identify novel methane-consuming cells in a natural community (Section 22.10)?

Application Questions

1. Design an experiment for measuring the activity of sulfur-oxidizing bacteria in soil. If only certain species of the sulfur oxidizers present were metabolically active, how could you tell this? How would you prove that your activity measurement was due to biological activity?

2. You wish to know whether *Archaea* exist in a lake water sample but are unsuccessful in culturing any. Using techniques described in this chapter, how could you determine whether *Archaea* existed in the sample, and if they did, what proportion of the cells in the lake water were *Archaea*?

3. Design an experiment to solve the following problem. Determine the rate of methanogenesis (CO_2 + $4H_2$ → CH_4 + $2H_2O$) in anoxic lake sediments and whether or not it is H_2-limited. Also, determine the morphology of the dominant methanogen (recall that these are *Archaea*, ⮌ Section 19.3). Finally, calculate what percentage the dominant methanogen is of the total archaeal and total prokaryotic populations in the sediments. Do not forget to specify necessary controls.

4. Design a SIP experiment that would allow you to determine which organisms in a lake water sample were capable of oxidizing the hydrocarbon hexane (C_6H_{14}). Assume that four different species could do this. How would you combine SIP with other molecular analyses to identify these four species?

Need more practice? Test your understanding with quantitative questions; access additional study tools including tutorials, animations, and videos; and then test your knowledge with chapter quizzes and practice tests at **www.microbiologyplace.com**.

Major Microbial Habitats and Diversity

Besides "normal" habitats, such as lake or ocean water, prokaryotes inhabit a host of extreme environments, such as this superheated hydro-thermal vent chimney wall containing cells of *Archaea* (red) and *Bacteria* (green).

Microorganisms do not live alone in nature but instead interact with other organisms and with their environment. In so doing, microorganisms carry out many essential activities that support all life on Earth. In this chapter we explore some of the major habitats of microorganisms; these include soil, freshwater, and the oceans. But in addition to these, microorganisms have also established more specific, and often very intimate, associations with plants and animals. We examine a few examples of such microbial partnerships and symbioses in Chapter 25.

I Microbial Ecology

We begin with a broad overview of the science of microbial ecology, including ways that organisms interact with each other and their environments and the difference between species diversity and abundance. These basic ecological concepts pervade this and the next two chapters.

23.1 General Ecological Concepts

The distribution of microorganisms in nature resembles that of macroorganisms in the sense that a given species resides in certain places but not others; that is, everything is not everywhere. Also environments differ in their abilities to support diverse microbial populations. We examine these concepts here.

Ecosystems and Habitats

An **ecosystem** can be considered a dynamic complex of plant, animal, and microbial communities and their nonliving surroundings, which interact as a functional unit. An ecosystem contains many different **habitats**, parts of the ecosystem best suited to one or a few populations. Although microorganisms are present in any habitat containing plants and animals, many microbial habitats are unsuitable for plants and animals. For example, microorganisms are ubiquitous on Earth's surface and even deep within it; they inhabit boiling hot springs and solid ice, acidic environments near pH 0, saturated brines, environments contaminated with radionuclides and heavy metals, and the interior of porous rocks that contain little water. So some ecosystems are mostly or even exclusively microbial.

Collectively, microorganisms show great metabolic diversity and are the primary catalysts of nutrient cycles in nature (↺ Chapter 24). The *types* of microbial activities possible in an ecosystem are a function of the species present, their population sizes, and the physiological state of the microorganisms in each habitat. By contrast, the *rates* of microbial activities in an ecosystem are controlled by the nutrients and growth conditions that prevail. Depending on several factors, microbial activities in an ecosystem can have minimal or profound impacts and can diminish or enhance the activities of both the microorganisms themselves and the macroorganisms that may coexist with them.

Species Diversity in Microbial Habitats

A group of microorganisms of the same species that reside in the same place at the same time constitutes a microbial **population**. A microbial population is often descended from a single cell. As noted in earlier chapters, a microbial **community** consists of populations of one species living in association with populations of one or more other species. The species that inhabit a certain habitat are those best adapted to grow with the nutrients and conditions that prevail there.

The diversity of microbial species in a community can be expressed in two ways. One is **species richness**, the total *number* of different species present. Identifying cells is, of course, basic to determining microbial species richness, but this need not require their isolation and culture. Species richness may also be expressed in molecular terms by the diversity of phylotypes (for example ribosomal RNA genes, ↺ Section 22.5) observed in a given community. **Species abundance**, by contrast, is the *proportion* of each species in the community. Species richness and abundance can change quickly over a short time as shown in **Figure 23.1**. One goal of microbial ecology is to understand species richness and abundance in microbial communities along

(a)

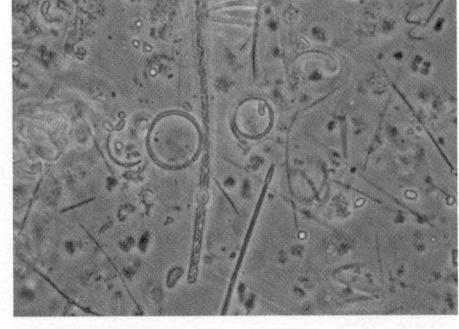

(b)

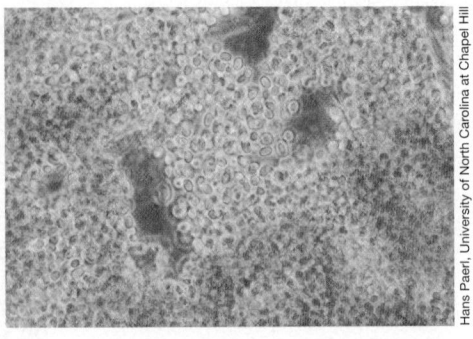

(c)

Hans Paerl, University of North Carolina at Chapel Hill

Figure 23.1 Microbial species diversity: Richness versus abundance. (a) Collecting samples from Lake Taihu, China, following a bloom of the cyanobacterium *Microcystis*. (b) High species richness in St. John's River, Florida, shown by microscopy of planktonic microorganisms including cyanobacteria, diatoms, green algae, flagellates, and bacteria. (c) Shift of St. John's River community to low richness but high abundance following a bloom of the cyanobacterium *Microcystis*.

Table 23.1 *Resources and conditions that determine microbial growth in nature*

Resources

Carbon (organic, CO_2)
Nitrogen (organic, inorganic)
Other macronutrients (S, P, K, Mg)
Micronutrients (Fe, Mn, Co, Cu, Zn, Mn, Ni)
O_2 and other electron acceptors (NO_3^-, SO_4^{2-}, Fe^{3+})
Inorganic electron donors (H_2, H_2S, Fe^{2+}, NH_4^+, NO_2^-)

Conditions

Temperature: cold → warm → hot
Water potential: dry → moist → wet
pH: 0 → 7 → 14
O_2: oxic → microoxic → anoxic
Light: bright light → dim light → dark
Osmotic conditions: freshwater → marine → hypersaline

with the community's associated activities and the nonliving environment. Once all of these factors are known, microbial ecologists can model the ecosystem by perturbing it in some way and observing whether predicted changes match experimental results.

The microbial species richness and abundance of a community are functions of the conditions and the kinds and amounts of nutrients available in the habitat. **Table 23.1** lists common nutrients and conditions relevant to microbial growth. In some microbial habitats, such as undisturbed organic-rich soils, high species richness is common (see Figure 23.13), with most species present at only moderate abundance. Nutrients in such a habitat are of many different types, and this helps select for high species richness. In other habitats, such as some extreme environments, species richness is often very low and abundance of one or a few species very high. This is because the conditions in the environment exclude all but a handful of species, and key nutrients are present at such high levels that the highly adapted species can grow to high cell densities. Bacteria that catalyze acid mine runoff from the oxidation of iron are a good example here. These organisms thrive in highly acidic, iron-rich but organic-poor waters, where acidic pH and the dearth of organic carbon limit species richness. However, the elevated levels of ferrous iron (Fe^{2+}) present, which is oxidized to Fe^{3+} in energy-yielding reactions (↩ Section 13.9), fuel high species abundance. We examine the activities of iron-oxidizing organisms in acidic environments in Sections 24.5 and 24.7.

MiniQuiz

- What is the difference between species richness and species abundance?
- How does an ecosystem differ from a habitat?
- What are the characteristics of a microbial population?

23.2 Ecosystem Service: Biogeochemistry and Nutrient Cycles

In any habitable ecosystem whose resources and growth conditions are suitable, individual microbial cells present there will grow to form populations. Metabolically similar microbial populations that exploit the same resources in a similar way are called **guilds**. The habitat that is shared by a guild and that supplies the nutrients and conditions the cells require for growth is called a **niche**. Sets of guilds form microbial communities (**Figure 23.2**). Microbial communities interact with macroorganisms and abiotic factors in the ecosystem in a way that defines the workings of that ecosystem.

Energy Inputs to the Ecosystem

Energy enters ecosystems as sunlight, organic carbon, and reduced inorganic substances. Light is used by phototrophs to make ATP and synthesize new organic matter (Figure 23.2). In addition to carbon (C), new organic matter contains nitrogen (N), sulfur (S), phosphorus (P), iron (Fe), and the other elements of life (↩ Section 4.1). This newly synthesized organic material along with organic matter that enters the ecosystem from the outside (called *allochthonous* organic matter) fuels the catabolic activities of heterotrophic organisms. These activities oxidize the organic matter to CO_2 by respiration or ferment it to various

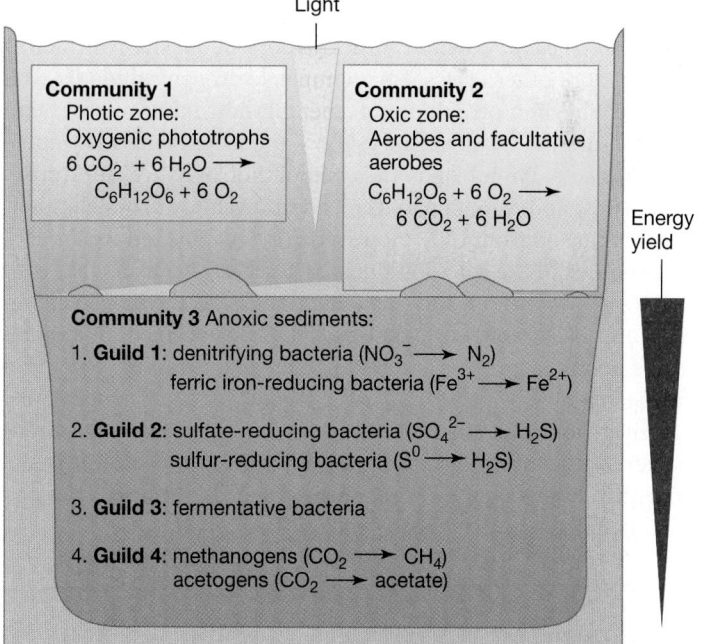

Figure 23.2 Populations, guilds, and communities. Microbial communities consist of populations of cells of different species. A freshwater lake ecosystem, for example, would likely have the communities shown here. The reduction of CO_2, SO_4^{2-}, S^0, NO_3^-, and Fe^{3+} are examples of anaerobic respirations. The region of greatest activity for each of the different respiratory processes would differ with depth in the sediment. As more energetically favorable electron acceptors are depleted by microbial activity near the surface, less favorable reactions occur deeper in the sediment.

reduced substances. If chemolithotrophs are present and metabolically active in the ecosystem, they obtain their energy from inorganic electron donors, such as H_2, Fe^{2+}, S^0, or NH_3 (Chapter 13) and contribute to the synthesis of new organic matter through their autotrophic activities (Figure 23.2).

Biogeochemical Cycling

Microorganisms play an essential role in cycling elements, in particular C, N, S, and Fe, between their different chemical forms. The study of these transformations is part of the science of **biogeochemistry**, an interdisciplinary science that includes biology, geology, and chemistry. Figure 23.2 shows how the activities of different guilds of microorganisms influence the chemistry of one environment, a lake ecosystem. The sequence of changing chemistry with increasing depth in the sediments corresponds to the layers of different microbial guilds. Where each guild resides is determined primarily by the available energy, which decreases with increasing depth in the sediments.

A *biogeochemical cycle* defines the transformations of an element that are catalyzed by either biological or chemical agents (or both). Many different microorganisms are involved in biogeochemical cycling reactions, and in many cases, microorganisms are the *only* biological agents capable of regenerating forms of the elements needed by other organisms, particularly plants. Thus, biogeochemical cycles are often also nutrient cycles, reactions that generate important nutrients for other organisms.

Most biogeochemical cycles proceed by oxidation–reduction reactions as the element moves through the ecosystem and are often tightly *coupled*, with transformations in one cycle impacting one or more other cycles. For example, hydrogen sulfide (H_2S) is oxidized by phototrophic and chemolithotrophic microorganisms to sulfur (S^0) and sulfate (SO_4^{2-}), the latter being a key nutrient for plants. Phototrophs and chemolithotrophs are autotrophic organisms, and thus impact the carbon cycle by producing new organic carbon from CO_2. However, SO_4^{2-} can be reduced to H_2S by activities of the sulfate-reducing bacteria, organisms that consume organic carbon, and this reduction closes the biogeochemical sulfur cycle while regenerating CO_2. The cycling of nitrogen is also a microbial process and is key to the regeneration of forms of nitrogen usable by plants and other organisms. The nitrogen cycle is driven by both chemolithotrophic and chemoorganotrophic bacteria, organisms that produce and consume organic carbon, respectively. We pick up the theme of biogeochemical cycles and their coupled nature in more detail in Chapter 24.

MiniQuiz

- How does a microbial guild differ from a microbial community?
- What is a biogeochemical cycle? Given an example using sulfur. Why are biogeochemical cycles also called nutrient cycles?

 II **The Microbial Environment**

Microorganisms define the limits of life throughout aquatic and terrestrial environments on our planet. Specific conditions required by a particular organism or group of organisms may be subject to rapid change due to inputs to and outputs from

their habitat and to microbial activities or physical disturbance. Thus, within one environment there can be multiple habitats, some of which are relatively stable and others that change rapidly over time and space.

23.3 Environments and Microenvironments

Besides living in the common habitats of soil and water, microorganisms thrive in extreme environments and also reside on and within the cells of other organisms. The intimate associations developed between microorganisms and other organisms will be presented in Chapter 25. Here we focus on terrestrial and aquatic microbial habitats.

The Microorganism, Niches, and the Microenvironment

The habitat in which a microbial community resides is governed by physiochemical conditions that are determined in part by the metabolic activities of the community. For example, the organic material used by one species may have been a metabolic by-product of a second species. Oxygen (O_2) can become limiting if biological consumption exceeds the rate at which it is supplied. Because microorganisms are very small, they directly experience only a tiny local environment; this small space is called their **microenvironment**. For example, for a typical 3-μm rod-shaped bacterium, a distance of 3 mm is equivalent to that which a human would experience over a distance of 2 km! As a consequence of the smallness of microorganisms, the variable metabolic activities of nearby microorganisms, and the changes in physiochemical conditions over short intervals of time and distance, numerous microenvironments can exist within a given habitat. The conditions supporting growth within a microenvironment correspond to the general requirements for growth we considered in Chapter 5.

Ecological theory states that for every organism there exists at least one niche, the *prime niche*, where it will be most successful. The organism dominates the prime niche but may also inhabit other niches; in other niches it is less ecologically successful than in its prime niche but it may still be able to compete. The full range of environmental conditions under which an organism can exist is called its *fundamental niche*. The word "niche" should not be confused with the word "microenvironment" because the microenvironment describes conditions at a specific location within a niche and can change rapidly. In other words, the general conditions that describe a specific niche may be transient at many places in a microenvironment.

Another important consequence for microorganisms of being small is that diffusion often determines the availability of resources. Consider, for example, the distribution of an important microbial nutrient such as O_2 in a soil particle. Microelectrodes (Chapter 22.8) can be used to measure oxygen concentrations throughout small soil particles. As shown in the data from an actual microelectrode experiment (**Figure 23.3**), soil particles are not homogeneous in terms of their O_2 content but instead contain many adjacent microenvironments. The outer layer of the soil particle may be fully oxic while the center, only a

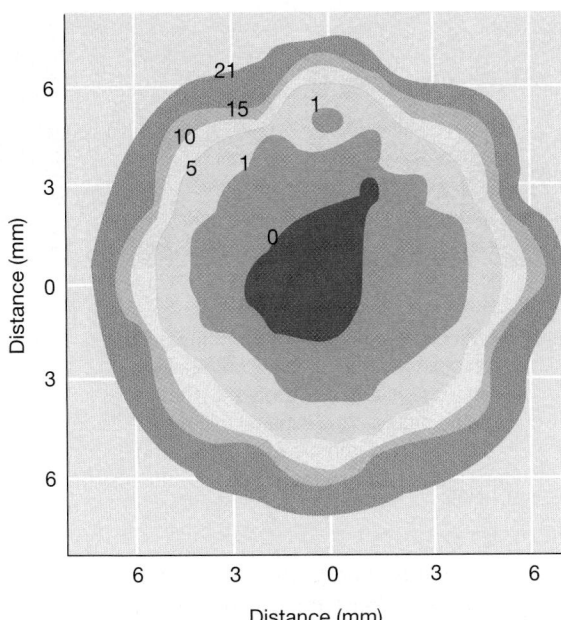

Figure 23.3 Oxygen microenvironments. Contour map of O_2 concentrations in a small soil particle as determined by a microelectrode (Section 22.8). The axes show the dimensions of the particle. The numbers on the contours are percentages of O_2 concentration (air is 21% O_2). Each zone can be considered a different microenvironment.

very short distance away (in human terms, but of course a great distance from a microbial standpoint), is anoxic (O_2-free). The microorganisms near the outer edges consume all of the O_2 before it can diffuse to the center of the particle. Thus, anaerobic organisms could thrive near the center of the particle, microaerophiles (aerobes that require very low oxygen levels) farther out, and obligately aerobic organisms in the outermost, fully oxic region of the particle. Facultatively aerobic bacteria could be distributed throughout the particle. Nutrient transfer is particularly important in thick assemblages of cells, such as biofilms and microbial mats, discussed in the next section.

Physiochemical conditions in a microenvironment are subject to rapid change in both time and space. For example, the O_2 concentrations shown in the soil particle in Figure 23.3 represent "instantaneous" values. Measurements taken in the same particle following a period of intense microbial respiration or disturbance due to wind, rain, or disruption by soil animals could differ dramatically from those shown. During such events certain populations may temporarily dominate the activities in the soil particle and grow to high numbers, while others remain dormant or nearly so. However, if the microenvironments shown in Figure 23.3 are eventually reestablished, the various microbial activities characteristic of the soil particle will eventually return as well.

Nutrient Levels and Growth Rates

Resources (Table 23.1) typically enter an ecosystem intermittently. A large pulse of nutrients—for example, an input of leaf litter or the carcass of a dead animal—may be followed by a period of nutrient deprivation. Because of this, microorganisms in nature often face a "feast-or-famine" existence. It is thus common for them to produce storage polymers as reserve

materials when resources are abundant and draw upon them in periods of starvation. Examples of storage materials are poly-β-hydroxyalkanoates, polysaccharides, and polyphosphate (Section 3.10).

Extended periods of exponential microbial growth in nature are probably rare. Microorganisms typically grow in spurts, linked closely to the availability and nature of resources. Because all relevant physiochemical conditions in nature are rarely optimal for microbial growth at the same time, growth rates of microorganisms in nature are usually well below the maximum growth rates recorded in the laboratory. For instance, the generation time of *Escherichia coli* in the intestinal tract of a healthy adult eating at regular intervals is about 12 h (two doublings per day), whereas in pure culture it can grow much faster, with a minimum generation time of about 20 min under the best conditions. Research-based estimates indicate that most cultured soil bacteria typically grow in nature at less than 1% of the maximal growth rate measured in the laboratory. These slow growth rates reflect the facts that (1) resources and growth conditions (Table 23.1) are frequently suboptimal; (2) the distribution of nutrients throughout the microbial habitat is not uniform; and (3) except for rare instances, microorganisms in nature grow in mixed populations rather than pure culture. An organism that grows rapidly in pure culture may grow much slower in a natural environment where it must compete with other organisms that may be better suited to the resources and growth conditions available.

Microbial Competition and Cooperation

Competition among microorganisms for resources in a habitat may be intense, with the outcome dependent on several factors, including rates of nutrient uptake, inherent metabolic rates, and ultimately, growth rates. A typical habitat contains a mixture of different species (Figures 23.1 and 23.2), with the density of each population dependent on how closely its niche resembles its prime niche.

Some microorganisms work together to carry out transformations that neither can accomplish alone. These microbial partnerships are particularly important for anoxic carbon cycling (Section 24.2). Metabolic cooperation can also be seen in the activities of organisms that carry out complementary metabolisms. For example, we have previously considered metabolic transformations that are carried out by two distinct groups of organisms, such as those of the nitrifying bacteria (Sections 13.10 and 17.3). Together, the nitrifying bacteria oxidize ammonia (NH_3) to nitrate (NO_3^-), although neither the ammonia oxidizers nor the nitrite oxidizers are capable of doing this alone. Because nitrite (NO_2^-), the product of the ammonia-oxidizing bacteria, is the substrate for the nitrite-oxidizing bacteria, the two groups of organisms often live in nature in tight association within their habitats (Figure 22.10).

MiniQuiz
- What characteristics define the prime niche of a particular microorganism?
- Why can many different physiological groups of organisms live in a single habitat?

UNIT 7

23.4 Surfaces and Biofilms

Surfaces are important microbial habitats, typically offering greater access to nutrients, protection from predation and physiochemical disturbances, and a means for cells to remain in a favorable habitat and not be washed away. Flow across a colonized surface increases transport of nutrients to the surface, providing more resources than are available to planktonic cells (cells that live a floating existence) in the same environment. A surface may also be provided by another organism or by a nutrient such as a particle of organic matter. For example, plant roots become heavily colonized by soil bacteria living on organic exudates from the plant, as revealed when fluorescent stains are used (**Figure 23.4a**).

Virtually any natural or artificial surface exposed to microorganisms will be colonized. For example, microscope slides have been used as experimental surfaces to which organisms can attach and grow. A slide can be immersed in a microbial habitat, left for a period of time, and then retrieved and examined microscopically (Figure 23.4b). Clusters of a few cells that develop from a single colonizing cell, called *microcolonies*, form readily on such surfaces, much as they do on natural surfaces in nature. In fact, periodic microscopic examination of immersed microscope slides has been used to measure growth rates of attached organisms in nature. Surface colonization may be sparse, consisting only of microcolonies and not visible to the eye, or may consist of so many cells that microbial accumulation becomes visible as, for example, in a stagnant toilet bowl. Surface growth can be particularly problematic in the hospital setting where microbial colonization of indwelling devices such as catheters and IV lines can cause serious infection. In a few extreme environments that lack small animal grazers (for example, hot springs), microbial accumulation on a surface can be many centimeters in thickness. Called *microbial mats*, such accumulations often contain highly complex yet very stable assemblages of phototrophic, autotrophic, and heterotrophic microorganisms (Section 23.5).

Biofilms

As bacterial cells grow on surfaces they commonly form **biofilms**—assemblages of bacterial cells attached to a surface and enclosed in an adhesive matrix secreted by the cells (**Figure 23.5**). The matrix is typically a mixture of polysaccharides, but can contain proteins and even nucleic acids. Biofilms trap nutrients for microbial growth and help prevent the detachment of cells on dynamic surfaces, such as in flowing systems (Figure 23.5c).

Biofilms typically contain several porous layers, and the cells in each layer can be examined by scanning laser confocal microscopy (↩ Section 2.3; Figure 23.5b). Biofilms may contain only one or two species or, more commonly, many species of bacteria. The biofilm that forms on a tooth surface, for example, contains several hundred different phylotypes, including species of both *Bacteria* and *Archaea*. Biofilms are thus functional and growing microbial communities and not just cells trapped in a sticky matrix.

Wherever submerged surfaces are present in natural environments, biofilm growth is almost always more extensive and

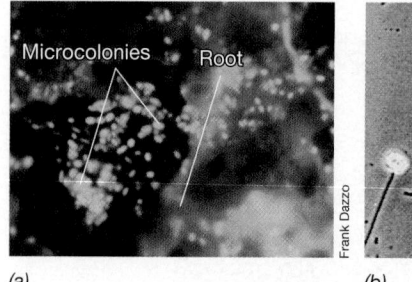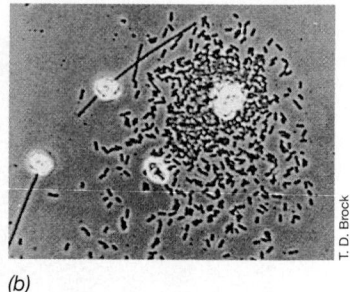

(a) *(b)*

Figure 23.4 Microorganisms on surfaces. *(a)* Fluorescence photomicrograph of a natural microbial community living on plant roots in soil. Note microcolony development. The preparation has been stained with acridine orange. *(b)* Bacterial microcolonies developing on a microscope slide that was immersed in a river. The bright particles are mineral matter. The short, rod-shaped cells are about 3 μm long.

diverse than the planktonic growth in the liquid that surrounds the surface. Biofilms differ from planktonic communities in supporting critical transport and transfer processes, which generally control growth in biofilm environments. For example, if consumption of O_2 by populations near the surface exceeds diffusion of O_2 into deeper regions of the biofilm, the deeper regions will

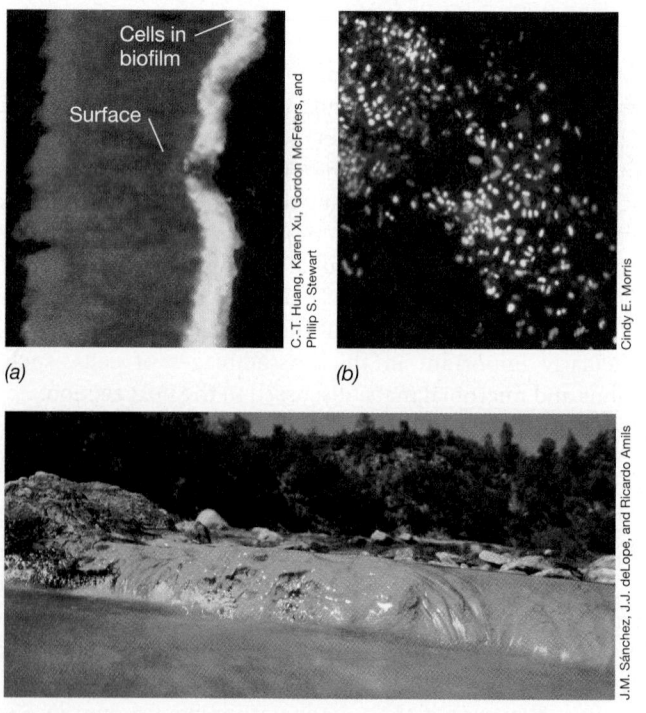

(a) *(b)*

(c)

Figure 23.5 Examples of microbial biofilms. *(a)* A cross-sectional view of an experimental biofilm made up of cells of *Pseudomonas aeruginosa*. The yellow layer (about 15 μm in depth) contains cells and is stained by a reaction showing activity of the enzyme alkaline phosphatase. *(b)* Confocal laser scanning microscopy of a natural biofilm (top view) on a leaf surface. The color of the cells indicates their depth in the biofilm: red, surface cells; green, 9-μm depth; blue, 18-μm depth. *(c)* A biofilm of iron-oxidizing bacteria attached to rocks in the iron-rich Rio Tinto, Spain. As Fe^{2+}-rich water passes over and through the biofilm, the organisms oxidize Fe^{2+} to Fe^{3+}.

become anoxic, opening up new niches for colonization by obligate anaerobes or facultative aerobes. This is similar to the depletion of O_2 in the interior of a soil particle that was depicted in Figure 23.3.

One of the most clinically and industrially relevant properties of biofilm microbial communities is their inherent tolerance to antibiotics and other antimicrobial stressors. A given species growing in a biofilm can be up to 1000 times more tolerant of an antimicrobial substance than planktonic cells of the same species. Reasons for the greater tolerance include slower growth rates in biofilms, reduced penetration of antimicrobial substances through the extracellular matrix, and different patterns of gene expression. The tolerance to antimicrobial substances may explain why biofilms are responsible for many untreatable or difficult-to-treat chronic infections and are also hard to eradicate in industrial systems where surface growth (fouling) impairs important processes.

Biofilm Formation

How do biofilms form? Random collision of cells with a surface accounts for initial cell attachment, with adhesion promoted by interaction between one or more cellular structures and the surface. Cellular structures include protein appendages (pili, flagella), cell surface proteins (for example, the large adhesion protein of *Pseudomonas fluorescens*), and polysaccharides. Attachment of a cell to a surface is a signal for the expression of biofilm-specific genes. These include genes encoding proteins that synthesize intercellular signaling molecules and initiate matrix formation (**Figure 23.6a**). Once committed to biofilm formation, a previously planktonic cell typically loses its flagella and becomes nonmotile.

Although the mechanism is yet to be discovered, bacteria somehow "sense" a suitable surface and this coordinates events that lead to the biofilm growth mode. How surface sensing takes place is an area of active research, but the actual switch from planktonic to biofilm growth is known to be triggered by the production of cyclic dimeric guanosine monophosphate (c-di-GMP), formed from two molecules of the nucleotide guanosine triphosphate (**Figure 23.7**). Most *Bacteria* use c-di-GMP as a *second messenger*, a communications molecule. Second messengers are intracellular regulatory molecules that transmit signals from the environment (first messenger) to the cellular machinery that generates the appropriate response, including motility, virulence, and biofilm formation. During the transition between planktonic and sessile growth states, c-di-GMP binds to proteins that modulate the activity of the flagellar motor and to enzymes that make the extracellular matrix of the biofilm. Studies of biofilm formation have revealed that part of the c-di-GMP signaling process is controlled by riboswitches (⮎ Section 8.15), regulatory messenger RNAs that interact directly with c-di-GMP and control transcription or translation of specific genes.

Pseudomonas aeruginosa and Biofilms

Besides the *intra*cellular activities triggered by c-di-GMP, *inter*cellular communication is necessary for the development and maintenance of bacterial biofilms. For example, in *Pseudomonas aeruginosa*, a notorious biofilm former (**Figure 23.8**), the major

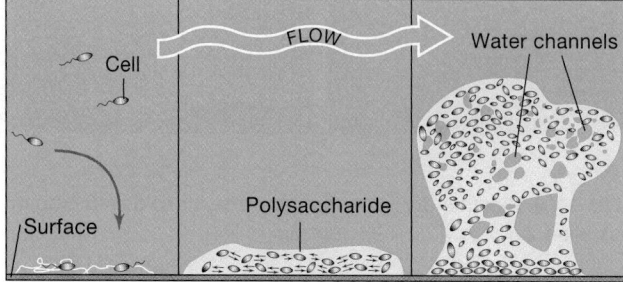

(a)

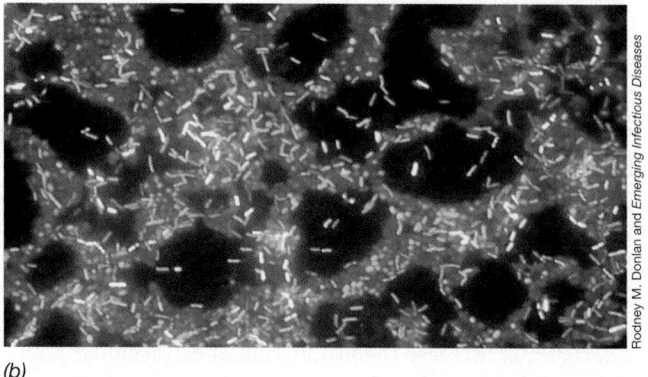

(b)

Figure 23.6 Biofilm formation. (a) Biofilms begin with the attachment of a few cells that then grow and communicate with other cells. The matrix is formed and becomes more extensive as the biofilm grows. (b) Photomicrograph of a DAPI-stained biofilm that developed on a stainless steel pipe. Note the water channels.

intercellular signaling molecules are *acylated homoserine lactones*. As these lactones accumulate, they signal adjacent *P. aeruginosa* cells that the population of this species is enlarging (quorum sensing, ⮎ Section 8.9). The signaling lactones then control expression of genes that contribute to biofilm formation.

c-di-GMP

Figure 23.7 Molecular structure of the second messenger cyclic dimeric guanosine monophosphate. This is used as an intracellular signaling molecule by many bacteria to control specific physiological processes.

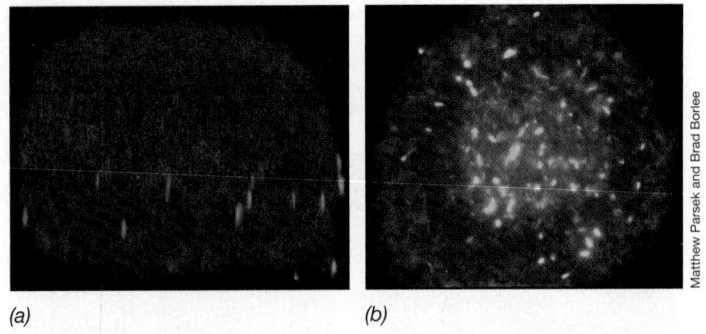

(a)   *(b)*

Matthew Parsek and Brad Borlee

Figure 23.8 Biofilms of *Pseudomonas aeruginosa* developing on a glass slide. *(a)* Side view of the mushroom-like microcolony (compare with Figure 23.5*a*). *(b)* View from the bottom of the same glass slide, looking up through the microcolony. Cells have been genetically modified to express the green fluorescent protein, and the adhesive matrix holding the biofilm together is stained red.

One of the genes turned on at this time encodes the biosynthesis of the second messenger c-di-GMP. In both *P. aeruginosa* and *P. fluorescens*, a related biofilm-forming organism, increases in c-di-GMP promote biofilm formation. However, the biofilm machinery regulated by c-di-GMP is very different in the two organisms. In *P. fluorescens*, changes in c-di-GMP impact secretion and cell surface localization of a protein called *adhesin* that sticks the cell to surfaces. By contrast, elevated c-di-GMP levels in *P. aeruginosa* increase the production of extracellular polysaccharide and decrease flagellar function. Over time, *P. aeruginosa* cells gather into large aggregates called "mushrooms" that can be over 0.1 mm high and contain billions of cells enmeshed in a sticky polysaccharide matrix (Figure 23.8). The final architecture of the biofilm is determined by multiple factors in addition to signaling molecules, including nutritional factors and local flow environment.

P. aeruginosa biofilms form in human lungs in patients with the genetic disease *cystic fibrosis*. In the biofilm state, *P. aeruginosa* is difficult to treat with antibiotics and the biofilm appears to help the bacteria persist in individuals with this disease. Like most biofilms, the biofilm found in the lungs of cystic fibrosis patients contains more than one bacterial species. So, in addition to *intraspecies* signaling, *interspecies* signaling is probably also occurring in the events that initiate and maintain biofilms containing more than one species.

Why Bacteria Form Biofilms

At least four reasons have been proposed for the formation of biofilms. First, biofilms are a means of microbial self-defense that increase survival. Biofilms resist physical forces that could otherwise remove cells only weakly attached to a surface. Biofilms also resist phagocytosis by cells of the immune system and the penetration of toxic molecules such as antibiotics. These advantages improve the chances for survival of cells in the biofilm. Second, biofilm formation allows cells to remain in a favorable niche. Biofilms attached to nutrient-rich surfaces, such as animal tissues, or to surfaces in flowing systems (Figure 23.5*c*) fix bacterial cells in locations where nutrients are more abundant or are con-

stantly replenished. Third, biofilms form because they allow bacterial cells to live in close association with each other. As we have already seen for *P. aeruginosa* and the biofilm that forms in cystic fibrosis patients, this facilitates cell-to-cell communication and increases chances for survival. Moreover, when cells are in close proximity to one another, there are more opportunities for nutrient and genetic exchange. Finally, biofilms seem to be the typical way bacterial cells grow in nature. The biofilm may be the "default" mode of growth for prokaryotes in natural environments, the latter of which differ dramatically in nutrient levels from the rich liquid culture media used in the laboratory. Planktonic growth may be the norm only for those bacteria adapted to life at extremely low nutrient concentrations (discussed in Section 23.9).

Biofilm Control

Biofilms have significant implications in human medicine and commerce. In the body, bacterial cells within a biofilm are protected from attack by the immune system, and antibiotics and other antimicrobial agents often fail to penetrate the biofilm. Besides cystic fibrosis, biofilms have been implicated in several medical and dental conditions, including periodontal disease, kidney stones, tuberculosis, Legionnaires' disease, and *Staphylococcus* infections. Medical implants are ideal surfaces for biofilm development. These include both short-term devices, such as a urinary catheter, as well as long-term implants, such as artificial joints. It is estimated that 10 million people a year in the United States experience biofilm infections from implants or intrusive medical procedures. Biofilms explain why routine oral hygiene is so important for maintaining dental health. Dental plaque is a typical biofilm and contains acid-producing bacteria responsible for dental caries (Section 27.3).

In industrial situations biofilms can slow the flow of water, oil, or other liquids through pipelines and can accelerate corrosion of the pipes themselves. Biofilms also initiate the degradation of submerged objects, such as structural components of offshore oil platforms, boats, and shoreline installations. The safety of drinking water may be compromised by biofilms that develop in water distribution pipes, many of which in the United States are nearly 100 years old (Figure 23.6*b*). Water-pipe biofilms mostly contain harmless microorganisms, but if pathogens successfully colonize a biofilm, standard chlorination practices may fail to kill them. Periodic releases of pathogenic cells can then lead to outbreaks of disease. There is concern that *Vibrio cholerae*, the causative agent of cholera (Section 35.5), may be propagated in this manner.

Biofilm control is big business, and thus far, only a limited number of tools exist to fight biofilms. Collectively, industries commit huge financial resources to treating pipes and other surfaces to keep them free of biofilms. New antimicrobial agents that can penetrate biofilms, as well as drugs that prevent biofilm formation by interfering with intercellular communication, are being developed. A class of chemicals called furanones, for example, has shown promise as biofilm preventatives on abiotic surfaces. Furanones are stable and some are relatively nontoxic, so they may have applications as antibiofilm agents in human medicine as well.

MiniQuiz

- Why might a biofilm be a good habitat for bacterial cells living in a flowing system?
- Give an example of a medically relevant biofilm that forms in virtually all healthy humans.

23.5 Microbial Mats

Among the most visibly conspicuous of microbial communities, *microbial mats* may be considered extremely thick biofilms. Built by phototrophic or chemolithotrophic bacteria, these layered communities can be many centimeters thick (**Figure 23.9**). The layers are composed of species of different microbial guilds whose activities are governed by light availability and other resources (Table 23.1). The combination of microbial metabolism and nutrient transport controlled by diffusion results in steep concentration gradients of different microbial nutrients and metabolites, creating unique niches at different depth intervals in the mats. The most abundant and versatile phototrophic mat builders are filamentous cyanobacteria, oxygenic phototrophs many of which grow under extreme environmental conditions. For example, some species of cyanobacteria grow in waters as hot as 73°C or as cold as 0°C and others tolerate salinities in excess of 12% and pH values as high as 10.

Cyanobacterial Mats

Cyanobacterial mats are complete microbial ecosystems, containing primary producers (the cyanobacteria) that along with populations of consumers mediate all key nutrient cycles in these ecosystems. Although this type of microbial ecosystem has existed for over 3.5 billion years, the evolution of metazoan grazers and competition with macrophytes (aquatic plants) triggered their decline about a billion years ago. Today, microbial mats develop only in aquatic environments where specific environmental stresses restrict grazing and competition, conditions most commonly found in hypersaline or geothermal habitats. Well-studied microbial mats are found in hypersaline solar evaporation basins, either formed naturally, such as Solar Lake (Sinai, Egypt), or those constructed for the recovery of sea salt (Figure 23.9*a*). Because microbial mats are restricted to extreme environments, most are found in remote locations and many are not readily accessible to study. In contrast, however, the cyanobacterial mats that colonize the outflow channels of hot springs in Yellowstone National Park (USA) and many other thermal regions in the world are easily accessible to scientific research (Figure 23.9*b, c*).

The chemical and biological structure of a microbial mat can change dramatically during a 24-h period (called a *diel cycle*) as a consequence of changing light intensity. Using microelectrodes (⟳ Section 22.8) it is possible to measure pH, H_2S, and O_2 repeatedly over a diel cycle in zones in the mat separated vertically by only a few micrometers. During the day, there is intense oxygen production in the photic surface layer of microbial mats and active sulfate reduction throughout the lower regions. Near the zone where O_2 and H_2S begin to mix, intense metabolic activity by phototrophic and chemolithotrophic sulfur bacteria may consume these substrates rapidly over very short vertical distances. Detecting the rate of these changes reveals the zones of greatest microbial activity (Figure 23.9*c*). These gradients disappear at night when the entire mat turns anoxic and H_2S accumulates. Some mat organisms rely on motility to follow the shifting chemical gradients. For example, sulfur-oxidizing filamentous phototrophic bacteria such as *Chloroflexus* and *Roseiflexus* follow the up-and-down movement of the O_2–H_2S interface on a diel basis.

Figure 23.9 Microbial mats. (*a*) Mat specimen collected from the bottom of a hypersaline pond at Guerrero Negro, Baja California (Mexico). Most of the bottom of this shallow pond is covered with mats built by the major primary producer, the filamentous cyanobacterium *Microcoleus chthonoplastes*. (*b*) Hot spring microbial mat core from an alkaline Yellowstone National Park (USA) hot spring. The upper (green) layer contains mainly cyanobacteria, while the reddish layers contain anoxygenic phototrophic bacteria. (*c*) Oxygen (O_2), H_2S, and pH profiles through a hot spring mat core such as that shown in part b.

Chemolithotrophic Mats

The most common types of chemolithotrophic mats are composed of filamentous sulfur-oxidizing bacteria, such as *Beggiatoa* and *Thioploca* species, which grow on marine sediment surfaces at the interface between O_2 supplied from the overlying water and H_2S produced by sulfate-reducing bacteria living in the sediment. In these habitats the bacteria oxidize H_2S to support energy conservation and autotrophic reactions (⟲ Sections 13.8 and 17.4).

Chemolithotrophic mats composed of sulfur-oxidizing *Thioploca* species (⟲ Figure 17.11) on sediments of the Chilean and Peruvian continental shelf are thought to be the most extensive microbial mats on Earth. *Thioploca* has developed a remarkable strategy to bridge spatially separated resources. These chemolithotrophic mat organisms contain large internal vacuoles that store high concentrations of nitrate (NO_3^-) for anaerobic respiration. Much like a scuba diver filling tanks with oxygen to dive into the water, cells of *Thioploca* migrate up to the sediment surface to charge internal vacuoles with NO_3^- from the water column. They then return ("dive") to the anoxic depths of the sediment (gliding at speeds of 3–5 mm per hour) to use their stored NO_3^- as an electron acceptor for H_2S oxidation.

The physical and biological structures of both biofilms and microbial mats are determined by metabolic interactions among microorganisms and the diffusion of nutrients. Thus, as biofilms form on a surface they become increasingly more complex, and in so doing generate new niches for organisms of differing physiologies. This diversity reaches its maximum in mature microbial mats (Figure 23.9), as these structures have been shown to be among the most complex microbial communities characterized thus far by molecular community sampling (⟲ Section 22.5).

MiniQuiz

• What is a microbial mat?

• How would aerobic bacteria respond to changing O_2 concentrations over a diel cycle?

 Terrestrial Environments

Extensive microbial habitats on Earth are in two terrestrial environments that are similar in lacking sunlight, being periodically or permanently anoxic, and having other physiochemical conditions in common. The two terrestrial environments are soils and water enclosed in soils and bedrock. In each of the sections we begin with the abiotic part of the environment and conclude with discussion of microbial life.

23.6 Soils

The word *soil* refers to the loose outer material of Earth's surface, a layer distinct from the bedrock that lies underneath (**Figure 23.10**). Soil develops over long periods of time through complex interactions among the parent materials (rock, sand, glacial drift materials, and so on), the topography, climate, and living organisms. Soils can be divided into two broad groups: *mineral soils* are derived from the weathering of rock and other inorganic materials, and *organic soils* are derived from sedimentation in bogs and marshes. Most soils are a mixture of these two basic types. Although mineral soils, which are the primary focus of this section, predominate in most terrestrial environments, there is increasing interest in the role that organic soils play in carbon storage. A detailed understanding of carbon storage (sinks) and sources (such as release of CO_2) is of great relevance to the science of climate change. The carbon cycle is considered in Chapter 24.

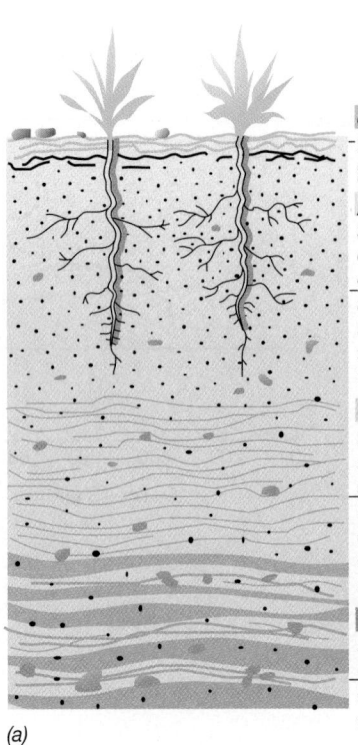

O horizon
Layer of undecomposed plant materials

A horizon
Surface soil (high in organic matter, dark in color, is tilled for agriculture; plants and large numbers of microorganisms grow here; microbial activity high)

B horizon
Subsoil (minerals, humus, and so on, leached from soil surface accumulate here; little organic matter; microbial activity detectable but lower than at A horizon)

C horizon
Soil base (develops directly from underlying bedrock; microbial activity generally very low)

(a)

(b)

Michael T. Madigan

Figure 23.10 Soil. *(a)* Profile of a mature soil. The soil horizons are zones defined by soil scientists. *(b)* Photo of a soil profile, showing O, A, and B horizons. This soil from Carbondale, Illinois (USA) is rich in clay and is very compact. Such soils are not as well drained as those that are rich in sand.

Soil Composition and Formation

Soils are composed of at least four components. These include (1) inorganic mineral matter, typically 40% or so of the soil volume; (2) organic matter, usually about 5%; (3) air and water, roughly 50%; and (4) microorganisms and macroorganisms, about 5%. Particles of various sizes are present in soil. Soil scientists classify soil particles on the basis of size: Those in the range of 0.1–2 mm in diameter are called *sand*, those between 0.002 and 0.1 mm *silt*, and those less than 0.002 mm *clay*. Different textural classes of soil are then given names such as "sandy clay" or "silty clay" based on the percentages of sand, silt, and clay they contain. A soil in which no one particle size dominates is called a *loam*.

Physical, chemical, and biological processes all contribute to the formation of soil. An examination of almost any exposed rock reveals the presence of algae, lichens, or mosses. These organisms are phototrophic and produce organic matter, which supports the growth of chemoorganotrophic bacteria and fungi. More complex chemoorganotrophic communities composed of *Bacteria*, *Archaea*, and eukaryotes then develop as the extent of the earlier colonizing organisms increases. Carbon dioxide produced during respiration becomes dissolved in water to form carbonic acid (H_2CO_3), which slowly dissolves the rock, especially rocks containing limestone ($CaCO_3$). In addition, many chemoorganotrophs excrete organic acids, which also promote the dissolution of rock into smaller particles.

Freezing, thawing, and other physical processes assist in soil formation by forming cracks in the rocks. As the particles generated combine with organic matter, a crude soil forms in these crevices, providing sites needed for pioneering plants to become established. The plant roots penetrate farther into the crevices, further fragmenting the rock; the excretions of the roots promote development in the **rhizosphere** (the soil that surrounds plant roots and receives plant secretions) of high microbial cell abundance (Figure 23.4a). When the plants die, their remains are added to the soil and become nutrients for more extensive microbial development. Minerals are rendered soluble, and as water percolates, it carries some of these substances deeper into the soil.

As weathering proceeds, the soil increases in depth and becomes able to support the development of larger plants and small trees. Soil animals such as earthworms colonize the soil and play an important role in keeping the upper layers of the soil mixed and aerated. Eventually, the movement of materials downward results

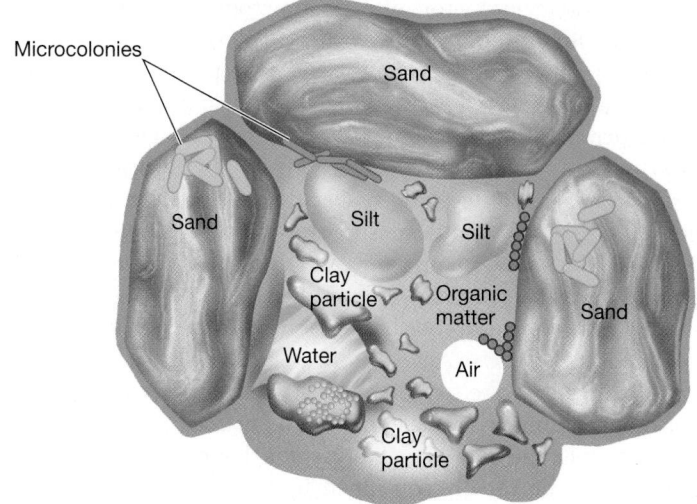

Figure 23.11 A soil microbial habitat. Very few microorganisms are free in the soil solution; most of them reside in microcolonies attached to the soil particles. Note the relative size differences among sand, clay, and silt particles.

in the formation of soil layers, called a soil profile (Figure 23.10). The rate of development of a typical soil profile depends on climatic and other factors, but it can take hundreds to thousands of years.

Soil as a Microbial Habitat

The most extensive microbial growth takes place on the surfaces of soil particles (**Figure 23.11**) and is highly promoted within, but is not limited to, the rhizosphere. As we have seen in Figure 23.3, even a single soil particle can contain many different microenvironments and can thus support the growth of several physiological types of microorganisms. To examine soil particles directly for microorganisms, fluorescence microscopes are often used, the organisms in the soil having been previously stained with a fluorescent dye. To visualize a specific microorganism in a soil particle, fluorescent antibody staining or gene probes (Sections 22.3, 22.4) can also be used. Microorganisms can also be observed on soil surfaces directly by scanning electron microscopy (**Figure 23.12**).

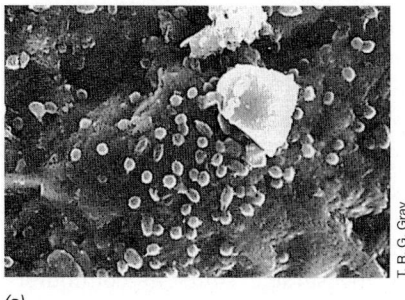
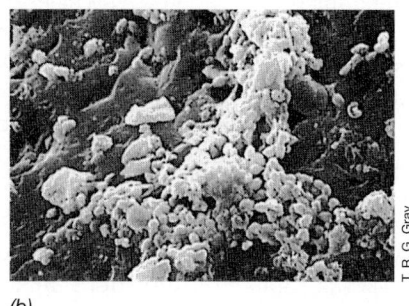
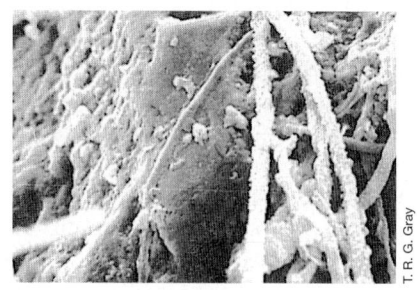

(a)   *(b)*   *(c)*

Figure 23.12 Scanning electron microscopy of microorganisms on the surface of soil particles.
(a) A microcolony of coccobacilli. *(b)* Actinomycete spores. The cells in part a and the spores in part b are about 1–2 μm wide. *(c)* Fungal hyphae. The hyphae are about 4 μm wide and are coated with mineral matter.

UNIT 7

One of the major factors affecting microbial activity in soil is the availability of water, and we have previously emphasized the importance of water for microbial growth (⮌ Section 5.16). Water is a highly variable component of soil, and soil water content depends on soil composition, rainfall, drainage, and plant cover. Water is held in the soil in two ways—by adsorption onto surfaces or as free water in thin sheets or films between soil particles (Figure 23.11). There is also water in the larger channels in soil, where bulk flow is important for rapid transport of microorganisms and their substrates and products.

The water present in soils has materials dissolved in it, and the mixture is called the *soil solution*. In well-drained soils, air penetrates readily, and the oxygen concentration of the soil solution can be high, similar to that of the soil surface. In waterlogged soils, however, the only oxygen present is that dissolved in water, and this can be rapidly consumed by the resident microflora. Such soils then become anoxic, and, as described for freshwater environments (Section 23.8), show profound changes in their biological activities.

The other major factor affecting microbial activity in soils is the extent of the resources present. The greatest microbial activity is in the organic-rich soil surface layers, especially in and around the rhizosphere. The numbers and activity of soil microorganisms depend to a great extent on the kinds and amounts of nutrients present. The limiting nutrients in soils are often inorganic nutrients such as phosphorus and nitrogen, key components of several classes of macromolecules.

A Phylogenetic Snapshot of Soil Prokaryotic Diversity

We learned in Chapter 22 that sequence analyses of 16S ribosomal RNA (rRNA) genes obtained from the environment can be used as a measure of prokaryotic diversity (⮌ Section 22.5). As yet, no natural communities have been so thoroughly characterized by these techniques that all resident species have been identified. However, within limits the method is widely considered to be a valid measure of microbial diversity and avoids the more serious problems of enrichment bias that plague culture-dependent diversity studies (⮌ Section 22.1). So here and in later sections of this chapter we present a "phylogenetic snapshot" of specific microbial habitats, with the goal of emphasizing trends and patterns rather than absolute details.

Molecular community sampling of surface soil prokaryotic diversity has shown typically *thousands* of different microbial species in a single gram of soil, likely reflecting the numerous microenvironments present there. A "species" is defined here operationally as a 16S rRNA gene sequence obtained from a microbial community that differs from all other sequences by more than 3% (⮌ Section 16.12). Such an environmental sequence is called a *phylotype*. Besides very large species numbers, soil microbial diversity studies have also showed that diversity varies with soil type and geographical location. For example, analysis of an Alaska forest soil, an Oklahoma prairie soil, and a Minnesota farm soil (all sites in the USA) revealed approximately 5000, 3700, and 2000 different phylotypes, respectively. The Alaska and Minnesota soils showed similar distributions at the phylum level of taxonomy (for example, *Proteobacteria*, *Acidobacteria*, *Bacteroidetes*, *Actinobacteria*, *Verrucomicrobia*, and *Planctomycetes*) but shared only about 20% of their species in common. This indicates that although the *proportions* of the dominant phyla in different soils are relatively constant, the *actual species present* within a phylum may vary considerably between different soils. In addition, lower bacterial diversity was observed in the farm soil than the Alaska soil, probably because modern intensive agricultural practices rely heavily on fertilization, low plant diversity, and the chemical suppression of unwanted plants and animals.

Figure 23.13 shows the general composition of soil microbial communities based on pooled 16S rRNA sequence data taken from several soils. As can be seen, *Proteobacteria* (Chapter 17) make up nearly half of the total phylotypes recovered, with all major subgroups except for *Epsilonproteobacteria* well represented. *Acidobacteria* and *Bacteroidetes* are also abundant groups; *Actinobacteria* and *Firmicutes* are less so. Also note that a major proportion of soil phylotypes are unclassified species or members of minor bacterial groups. This underscores the high bacterial diversity typical of soil ecosystems. In contrast to *Bacteria*, the diversity of *Archaea* in soil is minimal, with relatively few sequences within each major phylum of *Archaea* (*Euryarchaeota* and *Crenarchaeota*) represented.

A similar study to that shown in Figure 23.13 but performed on hydrocarbon-polluted soil showed that the general taxonomic makeup of polluted and unpolluted soils is similar: *Proteobacteria* comprise the largest fraction in both soil types, followed by significant representation of *Acidobacteria*, *Bacteroidetes*, *Actinobacteria*, and *Firmicutes*. However, there was a significant shift in fractional representation of these taxa in the two soils. Polluted soils are enriched in *Actinobacteria* and *Euryarchaeota* but diminished in *Bacteroidetes*, *Acidobacteria*, and unclassified *Bacteria* relative to nonpolluted soils. Notably, *Crenarchaeota* are absent from all surveys of hydrocarbon-polluted soils, suggesting that hydrocarbon pollutants eliminate this group, which includes the ammonia-oxidizing crenarchaeotes (*Archaea*, ⮌ Section 19.11). The impact of hydrocarbon pollution on *Bacteria* was most evident at lower taxonomic ranks, where polluted soils had a greater proportion of *Gammaproteobacteria* and only a single *Bacteroidetes* phylotype dominated. By contrast, unpolluted soils contained several phylotypes of *Bacteroidetes* (Figure 23.13). The diversity of *Acidobacteria* is also significantly reduced in polluted soils.

Although the *functional* significance of the observed diversity of microbial communities in polluted versus unpolluted soils is unknown, the shifts observed signal that the two soils will likely differ in their capacity to process carbon and nitrogen and to carry out other important nutrient cycling events. However, despite this lack of a functional connection, different 16S rRNA gene surveys of soils agree on two things: (1) undisturbed, unpolluted soils support very high prokaryotic diversity, and (2) perturbations in a soil trigger measurable shifts in community composition—presumably toward species that are more competitive in the disturbed soil environment—and an overall reduction in prokaryotic diversity.

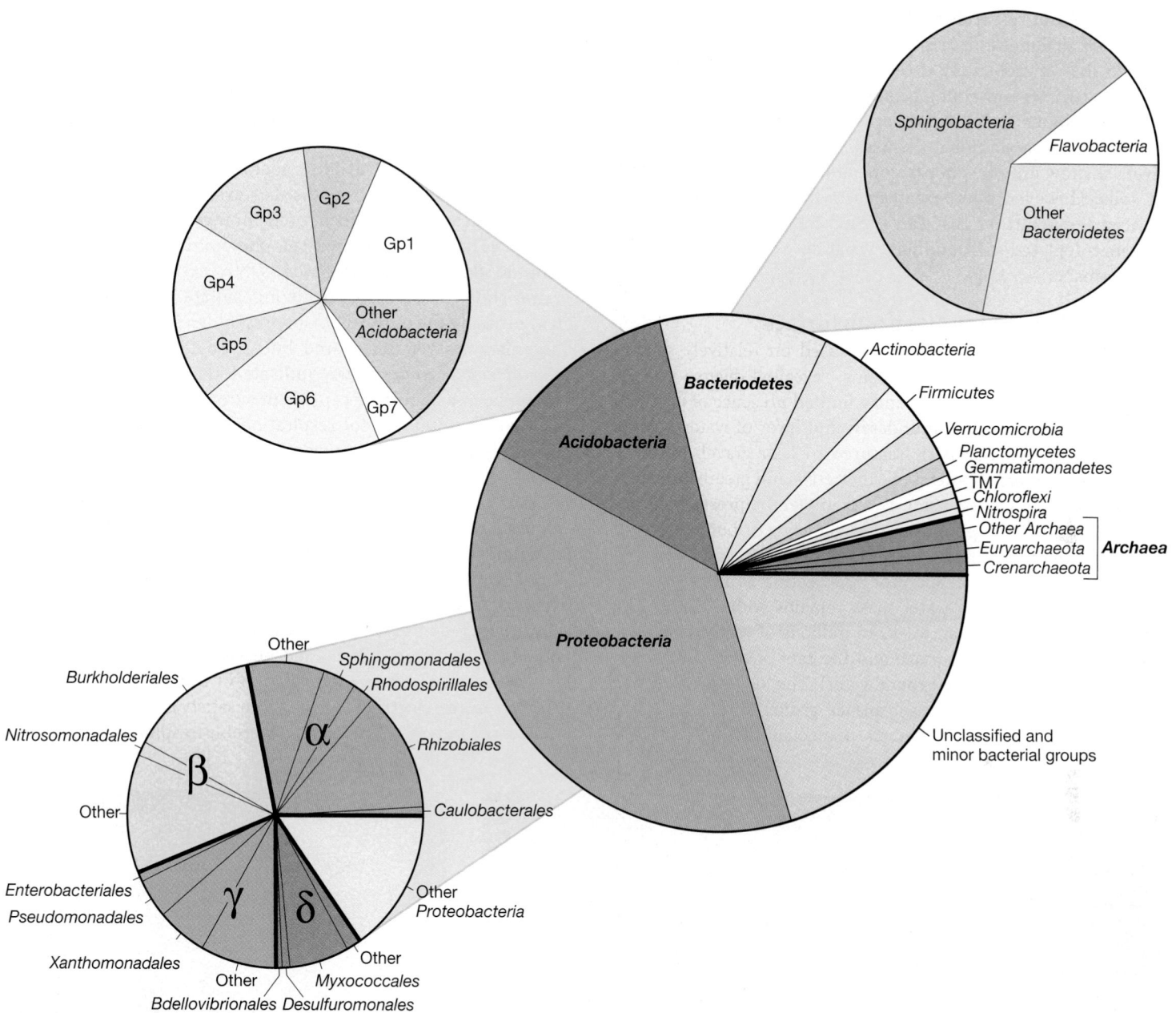

Figure 23.13 Soil prokaryotic diversity. The results are pooled analyses of 287,933 sequences from several studies of the 16S rRNA gene content of soil environments. Many of these groups are covered in Chapters 17 and 18 (*Bacteria*) or 19 (*Archaea*). For *Proteobacteria*, *Acidobacteria*, and *Bacteroidetes*, major subgroups are indicated (Gp, group). Note high species richness as indicated by the large proportion of the total community composed of unclassified and minor bacterial groups. Also note the relatively low proportion of the total prokaryotic soil community that consists of *Archaea* and that many soil *Archaea* are not clearly related to known species of *Euryarchaeota* or *Crenarchaeota*. Data assembled and analyzed by Nicolas Pinel.

MiniQuiz

- Which phylum of *Bacteria* dominates soil bacterial diversity?
- What factors govern the extent and type of microbial activity in soils?
- Which region of soil is the most microbially active?

23.7 The Subsurface

In the soils and rocks of Earth's subsurface, there is water. This underground water, called *groundwater,* is a vast but little-explored microbial habitat. As recently as three decades ago most microbiologists were of the opinion that significant microbial numbers were limited to the top 100 m or so of Earth's crust.

However, from research made possible by the development of improved drilling and aseptic sampling technology, it is now known that microbial life extends down at least 3 km into the Earth in regions containing trapped water. In fact, microorganisms in sediments and deeper crustal regions may account for as much as 40% of global biomass. The microbiology of relatively shallow groundwater is quite similar to the microbiology of soils. However, microorganisms in deep subsurface waters exist at temperatures that can exceed 50°C and in anoxic and nutrient-depleted surroundings. What do we know about these organisms?

Microbiology of the Deep Subsurface

Subsurface microbiology initially focused on relatively shallow and easily accessible aquifer systems, revealing diverse populations of *Archaea* and *Bacteria* and a limited presence of protozoa and fungi. An *aquifer* is an underground layer of water-bearing permeable material, such as fractured rock or gravel. Microorganisms in aquifers are metabolically active and greatly influence the chemistry of groundwater. For example, the presence of ferrous iron (Fe^{+2}) in groundwater is largely attributable to the activity of microorganisms such as *Geobacter* that reduce ferric iron (Fe^{3+}) as an electron acceptor (Section 14.12).

The period of time a water mass remains within a region of the subsurface varies from weeks to millions of years, depending on its proximity to the surface and the rate of recharge (movement of surface water into groundwater). The long-term isolation of microorganisms in deep subsurface groundwater that is not recharged has been suggested as a mechanism for *allopatric*

speciation (the emergence of new microbial species as a consequence of geographic isolation). However, the microbial diversity discovered in the subsurface thus far using culture-independent techniques (Chapter 22) has been unremarkable; the organisms closely resemble surface or near-surface species.

Research on the deep microbial biosphere has been facilitated by mining and drilling operations that expose water in fractured rock at great depths. For example, samples collected from a nearly 3-km-deep gold-mining operation in South Africa (**Figure 23.14**) revealed chemolithotrophic and autotrophic *Bacteria* and *Archaea*. DNA extracted from fissure water showed that an H_2-oxidizing, sulfate-reducing bacterium was virtually the only organism present. Genome analysis of the organism, as yet uncultured but given the provisional name *Desulforudis audaxviator*, indicated that it should be thermophilic and should be capable of autotrophic growth using H_2 as the electron donor for respiration and CO_2 fixation. In addition, the organism contained genes encoding nitrogen fixation proteins (Section 13.14), meaning that it could live on a diet of a few minerals, CO_2, SO_4^{2-}, N_2, and H_2. Such an organism would be well suited to long-term isolation in the deep subsurface and could be a model for the types of physiologies one would expect in such a nutrient-deficient environment.

Possible sources of H_2 for chemolithotrophs in the deep subsurface include the radiolysis of water by uranium, thorium, and other radioactive elements, and geochemical processes such as the release of H_2 from the oxidation of iron silicate minerals in aquifers. As an electron donor, H_2 can satisfy the needs of bacteria that carry out many different anaerobic respirations, including

(a)

(b)

Figure 23.14 Sampling the deep subsurface. *(a)* Sampling hot (55°C) fissure water from a depth of 3000 m in the Tau Tona South African gold mine. *(b)* Drilling to 600 m in Allendale, SC (USA), for the U.S. Department of Energy (DOE) Deep Subsurface Microbiology Program.

sulfate reduction, methanogenesis, acetogenesis, and ferric iron reduction, and examples of all these physiologies have been identified in various subsurface microbiology research projects. Thus the current consensus is that these types of chemolithotrophs likely dominate the deep subsurface.

Growth Rates and the Future of Subsurface Microbiology

Cell numbers in uncontaminated groundwater vary by several orders of magnitude (10^2–10^8 per ml), reflecting primarily nutrient availability, mostly in the form of dissolved organic carbon. Measured and estimated generation times for deep subsurface bacteria vary by many orders of magnitude, from days to centuries, as determined by the physiochemical environment, the physiology of the resident populations, and nutrient availability. However, relevant data in this regard are scarce and this is a question that will be greatly advanced by emerging technologies for direct characterization of single cells in the environment ($\curvearrowright$ Section 22.10). For example, microorganisms appear to be attached to surfaces or within biofilms in the nutrient-depleted subsurface, but it is unknown whether these are genetically or physiologically distinct from microorganisms in planktonic populations.

These many unanswered questions in subsurface microbiology have galvanized support for the establishment of permanent science laboratories at great depths in the Earth. For example, the U.S. National Science Foundation is constructing physics, geology, and microbiology research facilities at a depth of 2400 m in the Homestake Gold Mine in South Dakota (USA). Moreover, the Integrated Ocean Drilling Program, an international effort, has probed for microbial populations at great depths below the seafloor. Results thus far have shown *Archaea* and *Bacteria* as far down as 1600 m below the seafloor in rocks more than 100 million years old. Although this may sound ancient, such ages are relatively young compared with the viable bacteria that have been recovered from salt crystals nearly a half billion years old.

MiniQuiz

- Why could allopatric speciation be possible in the deep subsurface?
- What environmental factors determine the abundance and type of cells in the deep subsurface?

Aquatic Environments

Freshwater and marine environments differ in many ways including salinity, average temperature, depth, and nutrient content, but both provide many excellent habitats for microorganisms. In this unit we focus first on freshwater microbial habitats. We then consider two marine environments: (1) coastal and ocean waters, and (2) the deep sea. Much new information is emerging about marine microorganisms from studies using the molecular tools of microbial ecology, especially genetic stains, and microbial community sampling and metagenomics (Chapter 22).

23.8 Freshwaters

Freshwater environments are highly variable in the resources and conditions (Table 23.1) available for microbial growth. Both oxygen-producing and oxygen-consuming organisms are present in aquatic environments, and the balance between photosynthesis and respiration (Figure 23.2) controls the natural cycles of oxygen, carbon, and other nutrients (nitrogen, phosphorus, metals).

Among microorganisms, oxygenic phototrophs include the algae and cyanobacteria. These can either be *planktonic* (floating) and distributed throughout the water columns of lakes, sometimes accumulating in large numbers at a particular depth, or *benthic*, meaning they are attached to the bottom or sides of a lake or stream. Because oxygenic phototrophs obtain their energy from light and use water as an electron donor to reduce CO_2 to organic matter (Chapter 13), they are called **primary producers**.

The activity and diversity of heterotrophic aquatic microbial communities depend to a major extent on primary production, in particular its rates and temporal and spatial distributions. Oxygenic phototrophs produce new organic material as well as O_2. If primary production rates are very high, the resultant excessive organic matter production can lead to bottom-water O_2 depletion from respiration and the development of anoxic conditions. This in turn stimulates anaerobic metabolisms, including anaerobic respiration and fermentation. Like oxygenic phototrophs, anoxygenic phototrophs can also fix CO_2 into organic material. But these organisms use reduced substances other than water, such as H_2S or H_2, as electron donors in photosynthesis ($\curvearrowright$ Section 13.5). Organic matter produced by anoxygenic phototrophs can also support and enhance respiration, accelerating the spread of anoxia.

Oxygen Relationships in Freshwater Environments

The biological and nutrient structure of lakes is greatly influenced by seasonal changes in physical gradients of temperature and salinity. In many lakes in temperate climates the water column becomes *stratified*, separated into layers of differing physical and chemical characteristics. During the summer, warmer and less dense surface layers, called the **epilimnion**, are separated from the colder and denser bottom layers (the **hypolimnion**). The *thermocline* is the transition zone from epilimnion to hypolimnion (**Figure 23.15**). In the late fall and early winter, the surface waters become colder and thus more dense than the bottom layers. This, combined with wind-driven mixing, causes the cooled surface water to sink and the lake to "turn over," mixing surface and bottom waters. The separation of a relatively well-mixed surface layer from a relatively static bottom layer limits the transfer of nutrients between surface and bottom waters until fall turnover once again mixes the water layers.

During periods of stratification, transfer between surface and bottom waters is controlled not by mixing but by the much slower process of diffusion. As a result, bottom waters can experience seasonal periods of either low or no dissolved O_2. Although O_2 is one of the most plentiful gases in the atmosphere (21% of air), it has relatively limited solubility in water, and in a large body of water its exchange with the atmosphere is slow.

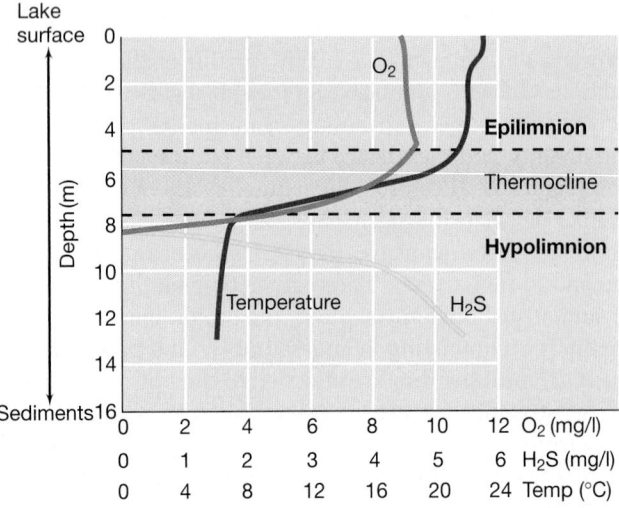

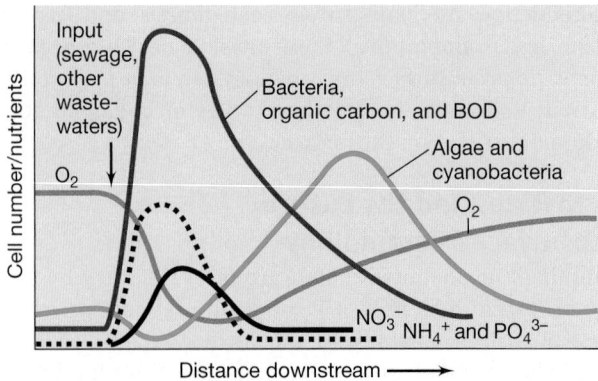

(a)

(b)

Figure 23.15 Development of anoxic conditions in a temperate lake due to summer stratification. The colder bottom waters are more dense and contain H_2S from bacterial sulfate reduction. The thermocline is the zone of rapid temperature change. As surface waters cool in the fall and early winter, they reach the temperature and density of hypolimnetic waters and sink, displacing bottom waters and effecting lake turnover. Data from a small freshwater lake in northern Wisconsin (USA).

Figure 23.16 Effect of the input of organic-rich wastewaters into aquatic systems. (a) In a river, bacterial numbers increase and O_2 levels decrease with a spike of organic matter. The rise in numbers of algae and cyanobacteria is primarily a response to inorganic nutrients, especially PO_4^{3-}. (b) Photo of a eutrophic (nutrient-rich) lake, Lake Mendota, Madison, Wisconsin (USA), showing algae, cyanobacteria, and aquatic plants that bloom in response to nutrient pollution from agricultural runoff.

Whether a body of water actually becomes O_2-depleted depends on several factors, including the amount of organic matter present and the degree of mixing of the water column. Organic matter that is not consumed in surface layers sinks to the depths and is decomposed by anaerobes (Figure 23.2). Lakes may contain high levels of dissolved organic matter because inorganic nutrients that run off the surrounding land can trigger algal and cyanobacterial blooms and these organisms typically excrete various organic compounds as well as die and decay. The combination of water body stratification during early summer, high organic loading, and limited O_2 transfer results in O_2 depletion of the bottom waters (Figure 23.15), making them unsuitable for aerobic organisms such as plants and animals.

This annual cycle allows the bottom waters to pass from oxic to anoxic and back to oxic. Microbial activity and community composition is altered with these changes in oxygen content, but other factors that accompany fall turnover of the water column, especially changes in temperature and nutrient levels, govern microbial diversity and activity as well. If organic matter is sparse, as it is in pristine lakes or in the open ocean, there may be insufficient substrate available for heterotrophs to consume all the oxygen. The microorganisms that dominate such environments are typically oligotrophs, organisms adapted to growth under very dilute conditions (discussed in Section 23.9). Alternatively, where currents are strong or there is turbulence because of wind mixing, the water column may be well mixed, and consequently oxygen may be transferred to the deeper layers.

Oxygen levels in rivers and streams are also of interest, especially those that receive inputs of organic matter from urban, agricultural, or industrial pollution. Even in a river well mixed by rapid water flow and turbulence, large organic inputs can lead to

a marked oxygen deficit from bacterial respiration (**Figure 23.16a**). As the water moves away from a point source input, for example, from an input of sewage, organic matter is gradually consumed, and the oxygen content returns to previous levels. As in lakes, nutrient inputs to rivers and streams from sewage or other pollutants can trigger massive blooms of cyanobacteria and algae (Figure 23.1) and aquatic plants (Figure 23.16b), thereby diminishing overall water quality and growth conditions for aquatic animals.

Biochemical Oxygen Demand

The microbial oxygen-consuming capacity of a body of water is called its **biochemical oxygen demand** (**BOD**). The BOD of water is determined by taking a sample, aerating it well to saturate the water with dissolved O_2, placing it in a sealed bottle, incubating it in the dark (usually for 5 days at 20°C), and determining the residual oxygen in the water at the end of incubation. A BOD determination gives a measure of the amount of organic material in the water that can be oxidized by the microorganisms present in the water. As a lake or river recovers from an input of organic matter or from excessive primary production, the initially high BOD becomes lower and is accompanied by a corresponding increase in dissolved oxygen in the ecosystem (Figure 23.16a). Another related measure of the organic material in a body of water is the *chemical oxygen demand* (*COD*). This determination uses a strong oxidizing agent, such as acidic potassium dichromate, to oxidize the organic matter to CO_2; the amount of organic matter present is proportional to the amount of dichromate consumed. COD is often used as a rapid measure of water quality and of its potential BOD.

We thus see that in freshwaters the oxygen and carbon cycles are linked, with the levels of organic carbon and oxygen being inversely related. Although photosynthesis produces O_2, the corresponding production of organic matter leads to O_2 deficiencies. Anoxic aquatic environments, which are typically rich in organic material, are the end result of respiratory processes that remove dissolved oxygen from the ecosystem, leaving the remaining organic material to be mineralized by organisms employing the anaerobic energy metabolisms we discussed in Chapter 14. It is also important to recognize the importance of storms, floods, and droughts in determining delivery, transport, and cycling of organic matter and inorganic nutrients in freshwater systems, including streams, rivers, lakes, and reservoirs. These less predictable changes also affect microbial productivity, diversity, distribution, and interactions in freshwater systems.

MiniQuiz

- What is a primary producer?
- In a freshwater lake, where is the epilimnion and where is the hypolimnion?
- Will addition of organic matter to a water sample increase or decrease its BOD?

23.9 Coastal and Ocean Waters: Phototrophic Microorganisms

Nutrient levels in the open ocean (the *pelagic zone*) are often very low compared with many freshwater environments. This is especially true of key inorganic nutrients for phototrophic organisms, such as nitrogen, phosphorus, and iron. In addition, water temperatures in the oceans are cooler and more constant seasonally than those of most freshwater lakes. The activity of marine phototrophs is limited by these factors, and thus overall microbial cell numbers are typically lower ($\sim 10^6$/ml) in the oceans than in freshwater environments ($\sim 10^7$/ml or higher).

Many different prokaryotes and eukaryotes inhabit ocean waters, but most are very small cells, a typical characteristic of organisms living in nutrient-poor environments. Smallness is an adaptive feature for nutrient-limited microorganisms in that it requires less energy for cellular maintenance. But the trade-off is that a greater number of transport enzymes relative to cell volume are needed for organisms to acquire nutrients from very dilute (oligotrophic) than from nutrient-rich (eutrophic) aquatic environments. For example, ammonia-oxidizing *Archaea* (*Nitrosopumilus*, ⮌ Section 19.11) are the dominant chemolithotrophs in pelagic waters and have very high-affinity transport systems for acquiring the ammonia they need as an electron donor.

In pelagic waters there is a lower return of nutrients from the bottom waters than in freshwater lakes, and thus lower average

Figure 23.17 **Distribution of chlorophyll in the western North Atlantic Ocean as recorded by satellite.** The east coast of the United States from the Carolinas to northern Maine is shown in dotted outline. Areas rich in phototrophic plankton are shown in red (>1 mg chlorophyll/m^3); blue and purple areas have lower chlorophyll concentrations (<0.01 mg/m^3). Note the high primary productivity of coastal areas and the Great Lakes.

primary productivity. However, because the oceans are so large, the collective carbon dioxide sequestration and oxygen production from oxygenic photosynthesis in the oceans are major factors in Earth's carbon balance. Salinity is more or less constant in the pelagic zone but is more variable in coastal areas. Terrestrial inputs, retention of nutrients, and upwelling of nutrient-rich waters combine to support higher populations of phototrophic microorganisms in near-shore waters than in pelagic waters (**Figure 23.17**); the more productive near-shore waters in turn support higher densities of heterotrophic bacteria and aquatic animals, such as fish and shellfish. In shallow marine waters such as marine bays and inlets, nutrient inputs can actually lead to the waters becoming intermittently anoxic from the removal of O_2 by respiration and the production of H_2S by sulfate-reducing bacteria.

Primary Productivity: *Prochlorococcus*

Much of the primary productivity in the open oceans, even at significant depths, comes from photosynthesis by prochlorophytes, tiny prokaryotic phototrophs that are phylogenetically related to cyanobacteria (⮌ Section 18.8); **prochlorophytes** contain chlorophylls *a* and *b* or chlorophylls *a* and *d*. The organism *Prochlorococcus* is a particularly important primary producer in the marine environment (**Figure 23.18**). Because *Prochlorococcus* lacks phycobilins, the accessory pigments of the cyanobacteria (⮌ Section 13.3), dense suspensions of *Prochlorococcus* cells are olive green (as are green algae) rather than the blue-green color of cyanobacteria (compare Figures 23.1*c* and 23.18).

A. Z. Worden and M. E. Breitbart

Penny Chisholm

Figure 23.18 *Prochlorococcus*, the most abundant oxygenic phototroph in the oceans. A bottle of *Prochlorococcus* showing the olive green color of the chlorophyll *a*- and *b*-containing cells. Inset: FISH-stained cells of *Prochlorococcus* in a marine water sample.

Prochlorococcus accounts for up to half of the photosynthetic biomass and production in the tropical and subtropical regions of the world's oceans, reaching cell densities of 10^5/ml. The prochlorophyte *Acaryochloris* contains chlorophylls *a* and *d* and is present in both marine waters and inland hypersaline lakes. At least four strains of *Prochlorococcus* have been identified, and each inhabits its own depth range in pelagic waters. The different *Prochlorococcus* strains are considered distinct *ecotypes*, genetic variants of a species that differ physiologically and therefore occupy slightly different niches (Section 16.12). The different *Prochlorococcus* ecotypes photosynthesize at different light intensities and use different inorganic and organic nitrogen and phosphorus sources. *Prochlorococcus* is thus distributed in both surface waters and deeper waters to depths of 200 m, which is near the bottom of the photic zone where light intensities are very low (see Figure 23.21). Genome sequences of several *Prochlorococcus* ecotypes have been determined and comparisons have revealed that although each ecotype contains about 2000 genes, only about 1100 are shared by all ecotypes. Each ecotype contains approximately 200 unique genes, which presumably have adaptive significance for growth in its prime niche. We illustrated this in Chapter 22 where we compared the genome of a single cultured *Prochlorococcus* ecotype to metagenome sequences obtained from pelagic waters (Section 22.7 and Figure 22.17).

Other Pelagic Oxygenic Phototrophs

In tropical and subtropical oceans, the planktonic filamentous marine cyanobacterium *Trichodesmium* (**Figure 23.19a**) is a widespread and occasionally abundant phototroph. Cells of *Trichodesmium* form puffs (colonies) of filaments. Each puff can contain many hundreds of individual filaments, each filament

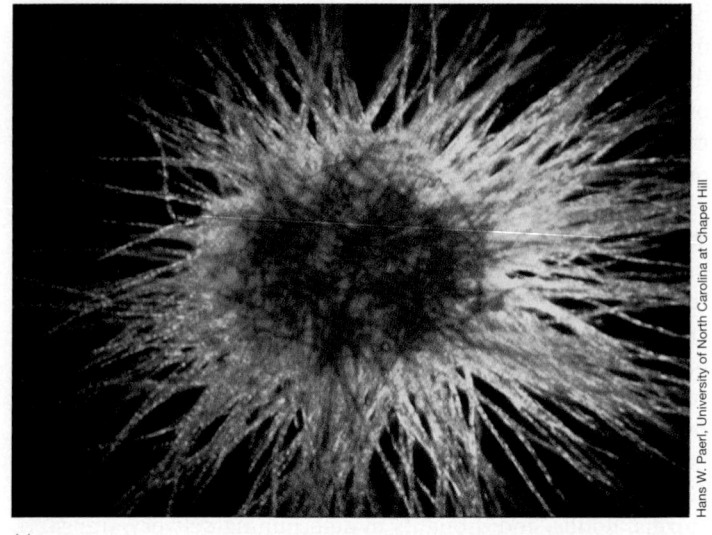

(a)

Hans W. Paerl, University of North Carolina at Chapel Hill

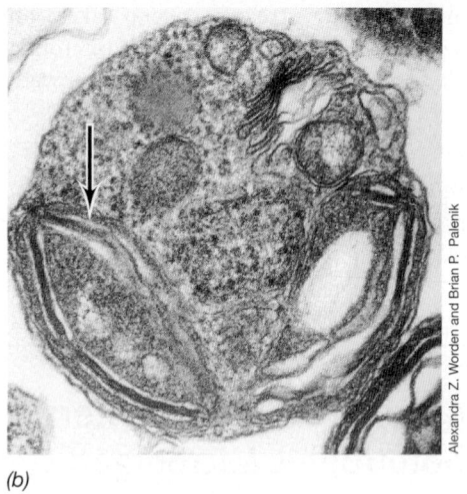

(b)

Alexandra Z. Worden and Brian P. Palenik

Figure 23.19 *Trichodesmium* and *Ostreococcus*. *(a)* Light photomicrograph of a puff of cells of the cyanobacterium *Trichodesmium*. This phototroph fixes nitrogen in tropical marine waters worldwide. The filaments in the puff are chains of cells. A cell is about 6 μm in diameter. *(b)* Transmission electron micrograph of a cell of *Ostreococcus*, a very small alga (eukaryote), found in substantial numbers in marine coastal waters. The arrow points to the chloroplast. An *Ostreococcus* cell is about 0.7 μm in diameter.

composed of 20–200 cells. In the Caribbean Sea, colonies of *Trichodesmium* can approach 100/m³. *Trichodesmium* is a nitrogen-fixing cyanobacterium, and the production of fixed nitrogen by this organism is thought to be an important link in the marine nitrogen cycle. *Trichodesmium* contains phycobilins, absent from prochlorophytes, and thus differs in its absorption properties from these organisms (Section 13.3).

Very small phototrophic eukaryotes also inhabit coastal and pelagic waters, and some of these are among the smallest eukaryotic cells known. *Ostreococcus*, for example, is a very small species of *Prasinophyceae*, a family of green algae that diverged early from other lineages (Section 20.20). Cells of *Ostreococcus* are cocci that measure only about 0.7 μm in diameter (Figure 23.19b); this is even smaller than a cell of *Escherichia coli*! Although cells of *Ostreococcus* and *Prochlorococcus* are roughly of the same dimensions and they are both oxygenic phototrophs, the genome of *Ostreococcus* is 12.6 Mbp (distributed over 20 chromosomes), which is more than seven times the size of the *Prochlorococcus* genome. In many marine waters small eukaryotic cells are present at about 10^4/ml. Although many of these are *Ostreococcus* or relatives, some are heterotrophs and some are phototrophs unrelated to *Ostreococcus* that incorporate small amounts of organic matter to supplement their primarily photosynthetic lifestyle.

Aerobic Anoxygenic Phototrophs

Besides *oxygenic* phototrophs, *anoxygenic* phototrophs are also present in pelagic waters. Like purple anoxygenic phototrophs, these organisms contain bacteriochlorophyll *a* (Sections 13.2, 13.4, and 17.2). However, unlike the classical purple bacteria that carry out photosynthesis only under anoxic conditions, these anoxygenic phototrophs carry out photosynthetic light reactions only when O_2 is available.

Aerobic anoxygenic phototrophs include bacteria such as *Erythrobacter*, *Roseobacter*, and *Citromicrobium* (**Figure 23.20**), all genera of *Alphaproteobacteria*. Aerobic anoxygenic phototrophs synthesize ATP by photophosphorylation when oxygen is present, which is all of the time in oxic pelagic waters, but they are unable to grow autotrophically and thus rely on organic carbon for their carbon sources. Aerobic anoxygenic phototrophs thus use the ATP produced by photophosphorylation to supplement their otherwise chemoorganotrophic metabolism. Surveys have shown that a great diversity of aerobic anoxygenic phototrophs exist in marine waters, especially near-shore waters. Oligotrophic and highly oxic freshwater lakes are also habitats for these interesting phototrophic bacteria. The physiology of the aerobic phototrophs is thus ideal for their illuminated and highly oxic habitats.

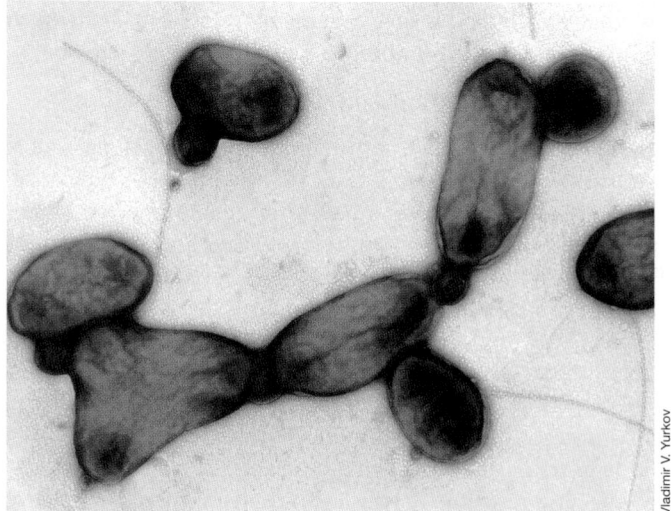

Figure 23.20 Aerobic anoxygenic phototrophic bacteria. Shown is a transmission electron micrograph of negatively stained cells of *Citromicrobium*. Cells of this marine, aerobic anoxygenic phototroph produce bacteriochlorophyll *a* only under oxic conditions and divide by both budding and binary fission, yielding morphologically unusual and irregular cells.

Vladimir V. Yurkov

MiniQuiz

- How does the organism *Prochlorococcus* contribute to both the carbon and oxygen cycles in the oceans?
- How does an organism like *Roseobacter* differ physiologically from *Prochlorococcus*?

23.10 Pelagic *Bacteria*, *Archaea*, and Viruses

Despite vanishingly low nutrient levels, significant numbers of prokaryotes live a planktonic existence in pelagic waters. These include species of both *Bacteria* and *Archaea*, and one organism in particular has garnered significant attention, a bacterium named *Pelagibacter*.

Distribution and Activity of *Archaea* and *Bacteria* in Pelagic Waters

The abundance of prokaryotic cells in the open oceans decreases with depth. In surface waters, cell numbers average about 10^6/ml. Below 1000 m, however, total cell numbers fall to between 10^3 and 10^5/ml. The distribution of *Bacteria* and *Archaea* with depth has been tracked in pelagic waters using fluorescent *in situ* hybridization (FISH) technology (Sections 16.9 and 22.4).

Species of *Bacteria* tend to predominate in waters above 1000 m, although cells of *Bacteria* and *Archaea* are found in near-equal abundance in deeper waters (**Figure 23.21**). Deep-water *Archaea* are almost exclusively species of *Crenarchaeota*, and many or perhaps even most are ammonia-oxidizing chemolithotrophs (Section 13.10); these organisms play an important role in coupling the marine carbon and nitrogen cycles (Section 24.1). Extrapolating from the data in Figure 23.21, it is estimated that 1.3×10^{28} and 3.1×10^{28} cells of *Archaea* and *Bacteria*, respectively, exist in the world's oceans. This means that the oceans contain the largest microbial biomass on the surface of the Earth. As we saw early on in this book, surface cell numbers pale by comparison to the huge number of prokaryotes in Earth's terrestrial and marine subsurfaces (Table 1.1).

Pelagic *Bacteria* and *Archaea* are ecologically important because they consume dissolved organic carbon in the oceans, one of the largest pools of unstable organic carbon on Earth. These small and free-living planktonic prokaryotes consume about half the total oceanic organic carbon produced from photosynthesis and are responsible for about half of all marine respiration and nutrient regeneration. Planktonic marine prokaryotes thus return organic matter to the marine food web that would otherwise be lost because of the inability of larger marine organisms to take up such diluted organic nutrients. This so-called "secondary production" is balanced by cell losses from bacterial grazing protists and from virus attack, leading to a near-steady state in which bacterial abundance in the open ocean remains roughly constant over time. But importantly, secondary production both recycles nutrients and allows some of the dissolved organic carbon in seawater to reach larger organisms, including fish, because protists are passed up the food web by the feeding activities of larger organisms.

UNIT 7

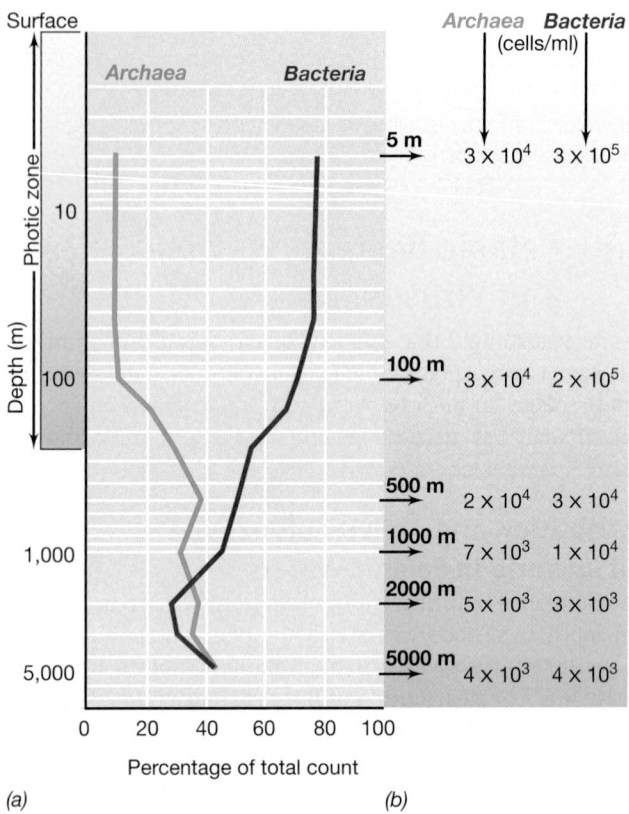

Figure 23.21 Percentage of total prokaryotes belonging to *Archaea* and *Bacteria* in North Pacific Ocean water. *(a)* Distribution of *Archaea* and *Bacteria* with depth. *(b)* Absolute numbers of *Archaea* and *Bacteria* with depth (per milliliter).

Pelagibacter

Very small planktonic heterotrophic bacteria inhabit pelagic marine waters in numbers of 10^5–10^6 cells/ml. The most abundant of these is *Pelagibacter*, a genus of *Alphaproteobacteria*. Cells of *Pelagibacter* are small rods that measure only 0.2– 0.5 μm, near the limits of resolution of the light microscope (**Figure 23.22**). What makes these organisms so successful in the open oceans?

Pelagibacter is an **oligotroph**, as are most pelagic prokaryotes. An oligotroph is an organism that grows best at very low concentrations of nutrients. *Pelagibacter* is a chemoorganotroph and grows in laboratory culture only up to the densities found in nature. However, in addition to respiring organic matter, *Pelagibacter* has genes that encode a form of the visual pigment rhodopsin that can convert light energy into ATP. In Section 19.2 we discussed the now well-studied molecule *bacteriorhodopsin*, a light-activated protein complex present in the extreme halophile *Halobacterium* (*Archaea*); bacteriorhodopsin functions in ATP synthesis as a simple light-driven proton pump (♻ Figure 19.4). The form of rhodopsin in *Pelagibacter* and other pelagic prokaryotes is structurally similar to bacteriorhodopsin and has been called **proteorhodopsin** ("proteo" referring to *Proteobacteria*). Although proteorhodopsin was first discovered in species of *Proteobacteria*, it is actually fairly widely distributed in *Bacteria*, including many *Gamma-* and *Alphaproteobacteria*,

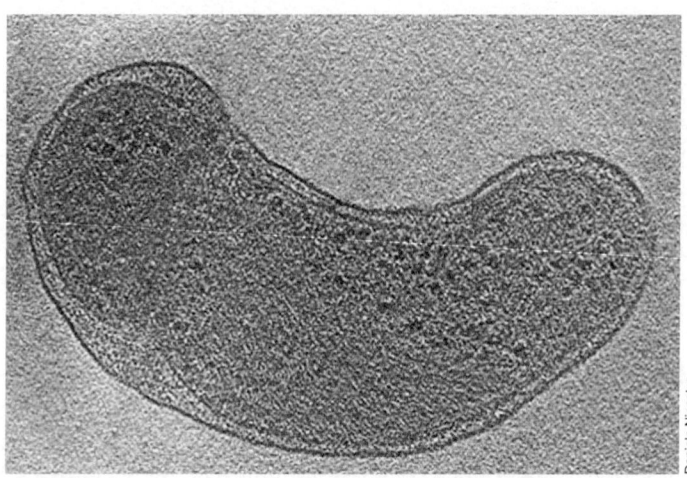

Figure 23.22 *Pelagibacter*, the most abundant prokaryote in the ocean. Electron micrograph taken by electron tomography, a technique for introducing a three-dimensional effect onto the image. A single cell of *Pelagibacter* is about 0.2 μm in diameter.

Bacteroidetes, and *Actinobacteria*, and has also been found in nonhalophilic species of *Archaea*, such as the *Thermoplasma* group (♻ Section 19.4). It is thought that proteorhodopsin supplements the energy metabolism of these organisms so that they do not have to rely solely on scarce organic carbon for their carbon and energy needs.

The genome of *Pelagibacter* is very small, only 1.3 Mbp. This is the smallest known genome for a free-living bacterium (Chapter 12). The genome encodes an unusually high number of ABC-type transport systems—transporters that have an extremely high affinity for their substrates (♻ Section 3.5)—and other enzymes useful for an oligotrophic organism. Environmental genomic studies (♻ Sections 12.6 and 22.7) have revealed a great abundance of *Pelagibacter* genes in pelagic waters, which correlates well with cell counts done using FISH (♻ Section 22.4) that clearly show that organisms related to *Pelagibacter* dominate prokaryotic numbers in pelagic waters worldwide.

Marine Viruses

In the oceans, viruses are more abundant than cellular microorganisms, often numbering over 10^7 virions/ml in typical seawater. In coastal waters, where bacterial cell numbers are higher than in the oceans, viral numbers are also higher, as many as 10^8 virions/ml. Most of the viruses are bacteriophages, which infect species of *Bacteria*, and archaeal viruses, which infect species of *Archaea*. The number of virions in seawater is about 10-fold greater than average prokaryotic cell numbers, suggesting that viruses are actively infecting their hosts, replicating, and being released into seawater (**Figure 23.23**). Only a small fraction of released viruses (an average of one per burst) successfully infects a new host, and most are inactivated or destroyed by sunlight and hydrolytic enzymes. In these ways, the entire viral population is replaced in periods of only a few days or weeks. We considered bacterial and archaeal viral diversity in Chapter 21.

Figure 23.23 Photomicrograph of abundant marine viruses. A seawater specimen was collected on a 0.2-μm filter, stained with SYBR Green, and then viewed by epifluorescence microscopy. The smallest green dots are viruses while the larger, brighter green dots are cells of *Bacteria* and *Archaea*. The average diameter of the prokaryotic cells is about 0.5 μm. Note the relative ratio of viruses to cells.

Along with feeding by protists, marine viruses probably help to maintain prokaryotic numbers at the levels that are observed, but viruses may also have other important ecosystem functions. These include facilitating genetic exchange between prokaryotic cells and allowing for lysogeny, the state in which a virus genome integrates within the cellular genome; lysogeny can confer new genetic properties on the cell (Section 9.10). For example, the discovery that some of the viruses that infect *Prochlorococcus*, the most abundant oxygenic phototroph in the oceans (Figure 23.18 and earlier discussion), contain genes that encode proteins needed for photosynthesis, indicates that even key metabolic properties may be subject to transfer by viral shuttles. Although the genetic diversity of marine viruses is just now being recognized, it is thought that the diversity of marine viral genomes could surpass even that of all prokaryotic cells, making the oceans a hotbed of genetic diversity.

A Phylogenetic Snapshot of Marine Prokaryotic Diversity

Several studies have attempted to characterize the diversity of planktonic marine prokaryotes by analysis of 16S rRNA genes obtained from seawater. The existence of abundant alphaproteobacterial populations to which *Pelagibacter* is affiliated was first revealed by such 16S rRNA sequence analysis. Mesophilic *Archaea* related to *Nitrosopumilus maritimus* (Section 19.11) were discovered using similar methods.

Major bacterial groups now recognized as abundant in the open ocean include *Alpha-* and *Gammaproteobacteria*, cyanobacteria, *Bacteroidetes*, and to a lesser extent, *Betaproteobacteria* and *Actinobacteria*; *Firmicutes* are only minor components (**Figure 23.24**). As for soil, a large proportion of unclassified and minor bacterial groups are also present in seawater. A major group of marine *Gammaproteobacteria* is the yet to be cultured "SAR 86 group," which accounts for approximately 10% of the total prokaryotic community in the ocean surface layer. Representing the *Archaea* in pelagic waters is a rather restricted diversity of *Euryarchaeota* and *Crenarchaeota,* most of which have never been brought into laboratory culture.

With the exception of the cyanobacteria, most marine *Bacteria* are thought to be heterotrophs adapted to extremely low nutrient availability, some augmenting energy conservation through proteorhodopsin or aerobic anoxygenic phototrophy (Section 23.9). With the discovery of the chemolithotroph *Nitrosopumilus*, it is possible that many marine *Archaea* specialize in ammonia oxidation, although heterotrophic species likely exist as well. "Dilution culture" methods employing very dilute culture media have been successful in bringing some pelagic prokaryotes into culture. It appears that most of these organisms

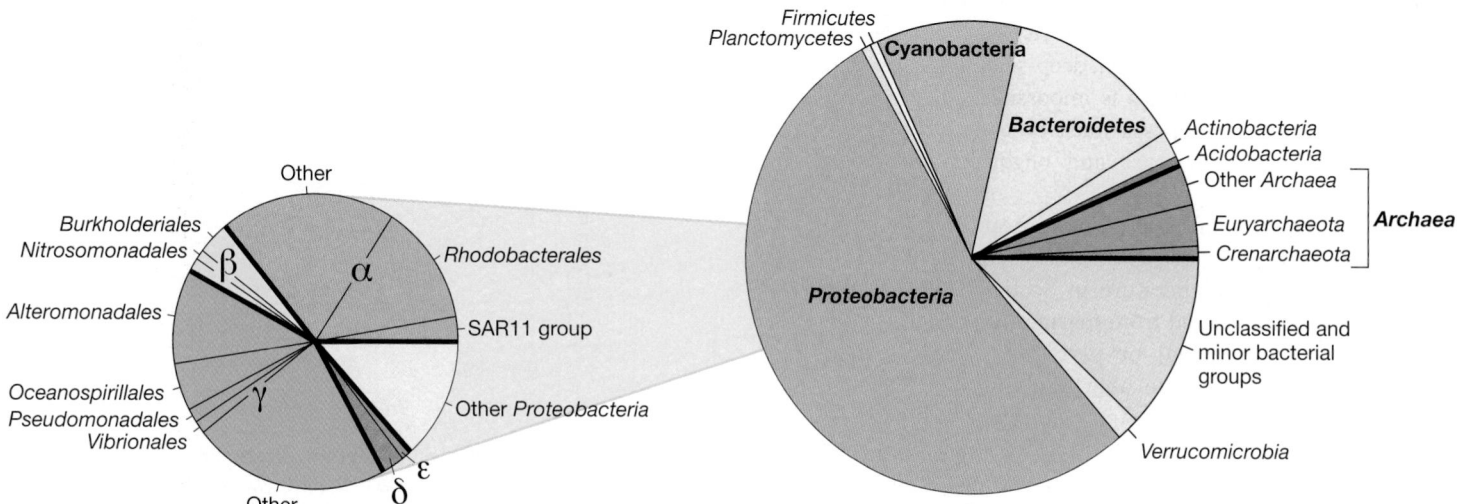

Figure 23.24 Ocean prokaryotic diversity. The results are pooled analyses of 25,975 sequences from several studies of the 16S rRNA gene content of pelagic ocean waters. Many of these groups are covered in Chapters 17 and 18 (*Bacteria*) or 19 (*Archaea*). For *Proteobacteria*, major subgroups are indicated. Note the high proportion of cyanobacterial and *Gammaproteobacteria* sequences. Data assembled and analyzed by Nicolas Pinel.

have evolved to grow only at very low nutrient concentrations and this makes mass culturing of them difficult. Cell densities of marine oligotrophs in laboratory cultures are similar to those in their natural environments (10^5–10^6/ml) and this renders many of the common tools for measuring cell growth (turbidity, microscopic counts) essentially useless on samples that are not first concentrated. Nevertheless, there have been notable successes with dilution culturing of marine bacteria and the aforementioned *Pelagibacter* is a good example.

MiniQuiz

- What is proteorhodopsin and why is it so named? Why might proteorhodopsin make a bacterium such as *Pelagibacter* more competitive in its habitat?
- How do numbers of pelagic prokaryotes and viruses compare?
- Which phylum and subgroups of *Bacteria* dominate pelagic marine waters?

23.11 The Deep Sea and Deep-Sea Sediments

Light penetrates no farther than about 300 m in pelagic waters; as has been mentioned, this illuminated region is called the *photic zone* (Figure 23.21). Beneath the photic zone, down to a depth of about 1000 m, there is still considerable biological activity. However, water at depths greater than 1000 m is, by comparison, much less biologically active and is known as the *deep sea*. Greater than 75% of all ocean water is deep-sea water, lying primarily at depths between 1000 and 6000 m. The deepest waters in the oceans lie below 10,000 m. However, because holes this deep are very rare, the waters in them make up only a very small proportion of all pelagic waters.

Conditions in the Deep Sea

Organisms that inhabit the deep sea face three major environmental extremes: (1) low temperature, (2) high pressure, and (3) low nutrient levels. In addition, deep-sea waters are completely dark such that photosynthesis is impossible. Thus, microorganisms that inhabit the deep sea must be chemotrophic and able to grow under high pressure and oligotrophic conditions in the cold.

Below depths of about 100 m, ocean water temperatures stay constant at 2–3°C. We discussed the responses of microorganisms to changes in temperature in Section 5.13. As would be expected, bacteria isolated from marine waters below 100 m are psychrophilic (cold-loving) or psychrotolerant. Deep-sea microorganisms must also be able to withstand the enormous hydrostatic pressures associated with great depths. Pressure increases by 1 atm for every 10 m of depth in a water column. Thus, an organism growing at a depth of 5000 m must be able to withstand pressures of 500 atm. We will see that microorganisms are remarkably tolerant of high hydrostatic pressures; many species can withstand pressures of 500 atm, and some species can withstand far more than this.

Piezotolerant and Piezophilic *Bacteria* and *Archaea*

Different physiological responses to pressure are observed in different deep-sea microorganisms. Some organisms simply tolerate high hydrostatic pressure, but do not grow optimally under such pressure; these organisms are **piezotolerant**. By contrast, others actually *grow best* under elevated hydrostatic pressure; these are called **piezophiles**. Organisms isolated from surface waters down to about 3000 m are typically piezotolerant. In piezotolerant organisms, higher metabolic rates are observed at 1 atm than at 300 atm, although growth rates at the two pressures may be similar (**Figure 23.25**). However, piezotolerant isolates typically do not grow at pressures greater than about 500 atm.

By contrast, cultures derived from samples taken at greater depths, 4000–6000 m, are typically piezophilic, growing optimally at pressures of around 300–400 atm. However, although piezophiles grow best under high pressure, they can still grow at 1 atm (Figure 23.25). In even deeper waters (for example, 10,000 m), **extreme piezophiles** are present. These organisms require very high pressure for growth. The extreme piezophile *Moritella*, isolated from the Mariana Trench (Pacific Ocean, >10,000-m depth) (**Figure 23.26**), grows optimally at a pressure of 700–800 atm and nearly as well at 1035 atm, the pressure it experiences in its natural habitat.

Unlike piezotolerant or piezophilic prokaryotes, extreme piezophiles, including species within the genus *Colwellia* as well as *Moritella*, do not grow at pressures of less than about 400 atm (Figure 23.25 and **Figure 23.27**). Interestingly, *Moritella* can tolerate moderate periods of decompression. However, viability is lost when a culture of the organism is left for several hours in a decompressed state. *Moritella* is also temperature sensitive. Its optimal growth temperature is its environmental temperature (2°C), and

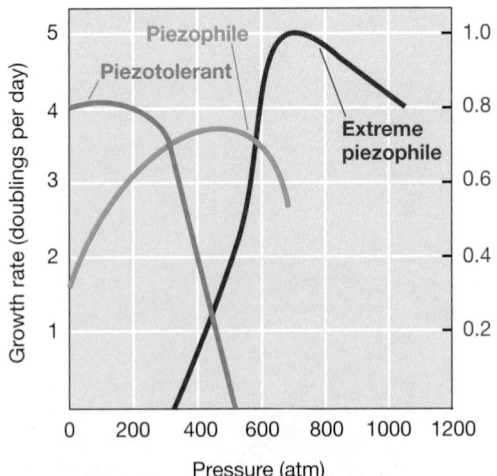

Figure 23.25 Growth of piezotolerant, piezophilic, and extremely piezophilic bacteria. The extreme piezophile (*Moritella*) was isolated from the Mariana Trench, off the Philippines, Pacific Ocean (Figure 23.26). Compare the slower growth rate of the extreme piezophile (right ordinate) with the growth rate of the piezotolerant and piezophilic bacteria (left ordinate), and note the inability of the extreme piezophile to grow at low pressures.

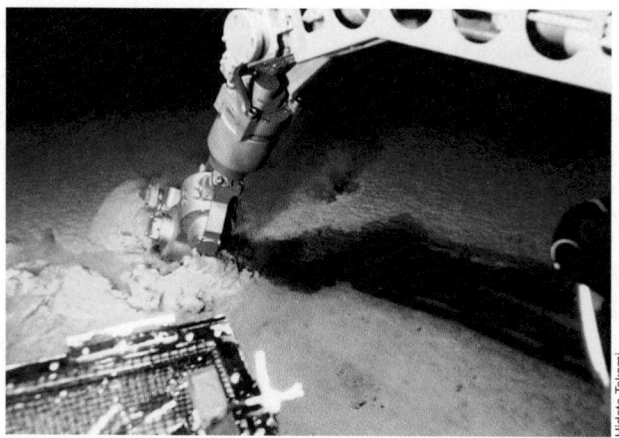

Figure 23.26 Sampling the deep sea. The unmanned submersible *Kaiko* collecting a sediment sample on the seafloor of the Mariana Trench at a depth of 10,897 m. The tubes of sediment are used for enrichment and isolation of piezophilic bacteria.

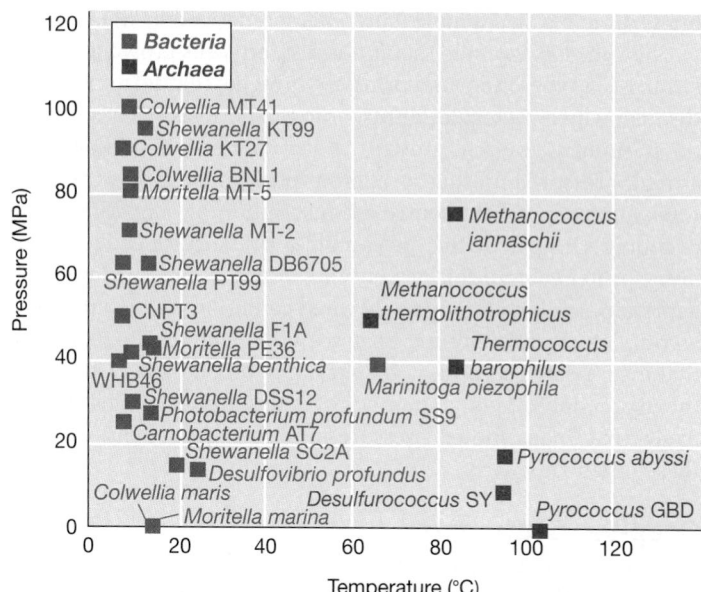

Figure 23.27 **Pressure and temperature optima for bacterial and archaeal piezophiles now available in culture.** Pressure is in pascals (Pa), the international system of units for pressure. One megapascal (Mpa) corresponds to approximately 10 atm. Note that different species of the same genus can have vastly different optima. Data assembled by Doug Bartlett.

temperatures above about 10°C significantly affect viability. A few piezophilic *Archaea* are also known, but thus far these have all been hyperthermophilic species that inhabit deep-sea hot springs (hydrothermal vents) (Figure 23.27).

Molecular Effects of High Pressure

High pressure affects cellular physiology and biochemistry in many ways. In general, pressure decreases the ability of the sub-units of multi-subunit proteins to interact. Thus, large protein complexes in extreme piezophiles must interact in such a way as to minimize pressure-related effects. Protein synthesis, DNA synthesis, and nutrient transport are sensitive to high pressure. Piezophilic bacteria grown under high pressure have a higher proportion of unsaturated fatty acids in their cytoplasmic membranes than when grown at 1 atm. Unsaturated fatty acids allow membranes to remain functional and keep from gelling at high pressures or at low temperatures. The rather slow growth rates of extreme piezophiles such as *Moritella* compared with other

marine bacteria (Figure 23.25) are likely due to the combined effects of pressure and low temperature; low temperature slows down the reaction rates of enzymes and this has a direct effect on cell growth (⮌ Section 5.12).

Besides enzymes, some other structural features accompany a piezophilic lifestyle. For example, for a gram-negative piezophile capable of growth at both 1 atm and 500–600 atm (Figure 23.25), it has been shown that growth at high pressure is accompanied by changes in the protein composition of the organism's outer membrane (⮌ Section 3.7). The studies that revealed these changes required special pressurized incubation devices (**Figure 23.28**). When grown under high pressure, a specific outer mem-

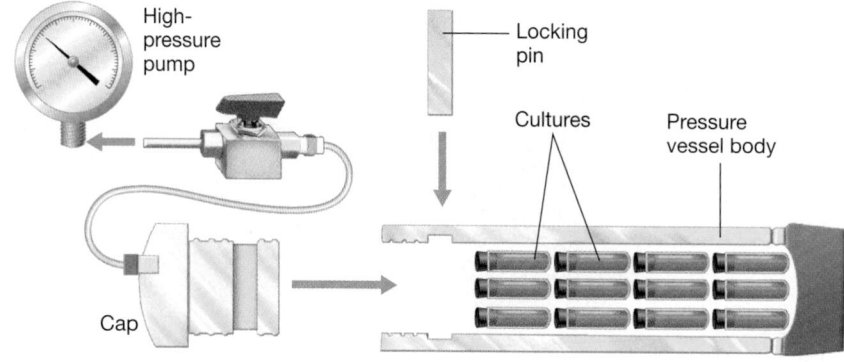

(a)   *(b)*

Figure 23.28 Pressure cells for growing piezophiles under elevated pressure. *(a)* Photo of several pressure cells incubating in a cold room (4°C). *(b)* Schematic design of a pressure cell. These vessels are designed to maintain pressures of 1000 atm.

brane protein called OmpH (*outer membrane protein* H) is present in cells of the piezophile that is absent from cells grown at 1 atm. OmpH is a type of porin. Porins are proteins that form channels through which molecules diffuse into the periplasm. Presumably, the porin made by cells grown at 1 atm cannot function properly at high pressure and thus a different porin must be synthesized. Interestingly, pressure controls transcription of *ompH*, the gene encoding OmpH. In this piezophile a pressure-sensitive membrane protein complex is present that monitors pressure and triggers transcription of *ompH* only when conditions of high pressure warrant it. Transcriptomic analyses (👁 Section 12.7) indicate that even relatively modest changes in hydrostatic pressure alter the expression of a large number of genes in piezophiles, so it is likely that many other pressure-monitoring proteins exist in these organisms.

Deep-Sea Sediments

Another vast and mostly unexplored microbial ecosystem exists deep below the seafloor. Deep drilling expeditions to explore the depths below the ocean seafloor have revealed both archaeal and bacterial populations as deep as 1600 m. Cell numbers typically decrease from about 10^9 cells/cm^3 of surface sediment to about 10^6 cells/cm^3 at 1000 m below the seafloor. Together, the subseafloor ecosystems are estimated to contain about 90 petagrams (1 petagram is 10^{15} g) of microbial cellular carbon and about 2.5% of the total organic carbon in the Earth's crust. Sequencing the 16S rRNA genes present in DNA extracted from drilling cores has identified relatively few sequences related to the classical sulfate-reducing bacteria (👁 Section 17.18) or methanogenic and methane-oxidizing *Archaea* (👁 Sections 14.10, 14.13, and 19.3) common in surface sediments. Remarkably, below about 1 m in depth *Archaea* predominate, comprising nearly 90% of total microbial biomass. Most of these *Archaea* have been identified only by their 16S rRNA sequences and are mainly species of *Euryarchaeota*, although most are unrelated to phylogenetic groups for which cultured organisms are available (Chapter 19). Thus, novel and uncultured archaeal lineages of unknown physiology populate the deep biosphere below the seafloor. How these organisms make a living in this extremely low-nutrient habitat has yet to be determined.

A Phylogenetic Snapshot of Marine Sediment Prokaryotic Diversity

Marine sediment communities have been explored only to a limited extent, given the great difficulty and expense of obtaining uncontaminated drilling cores from great depth. Analyses of available 16S rRNA sequences obtained from deep coring samples show these communities to be very distinct from open-ocean and soil communities. Most notably, *Archaea* of unknown affiliation make up a large fraction of the diversity (**Figure 23.29**). By contrast, in shallow marine sediments *Proteobacteria* dominate, as they do in all of the other habitats explored by culture-independent techniques (Figures 23.13 and 23.24, and see Figure 23.34). Within marine sediment *Proteobacteria*, phylotypes associated with sulfate-reducing bacteria such as the *Desulfobacterales* are quite common (Figure 23.29); sulfate reduction is the major form of anaerobic respiration in marine sediments (👁 Sections 14.8 and 17.18). *Bacteroidetes* and the unclassified/minor groups are also well represented in shallow marine sediments.

Although major players in marine waters, cyanobacteria make up just a tiny proportion of the total cell population in the permanently dark and anoxic sediments and probably represent cells that have reached the sediments following attachment to a particle or dead animal that eventually sank. What makes the deep marine sediments really stand out from the more shallow sediments is

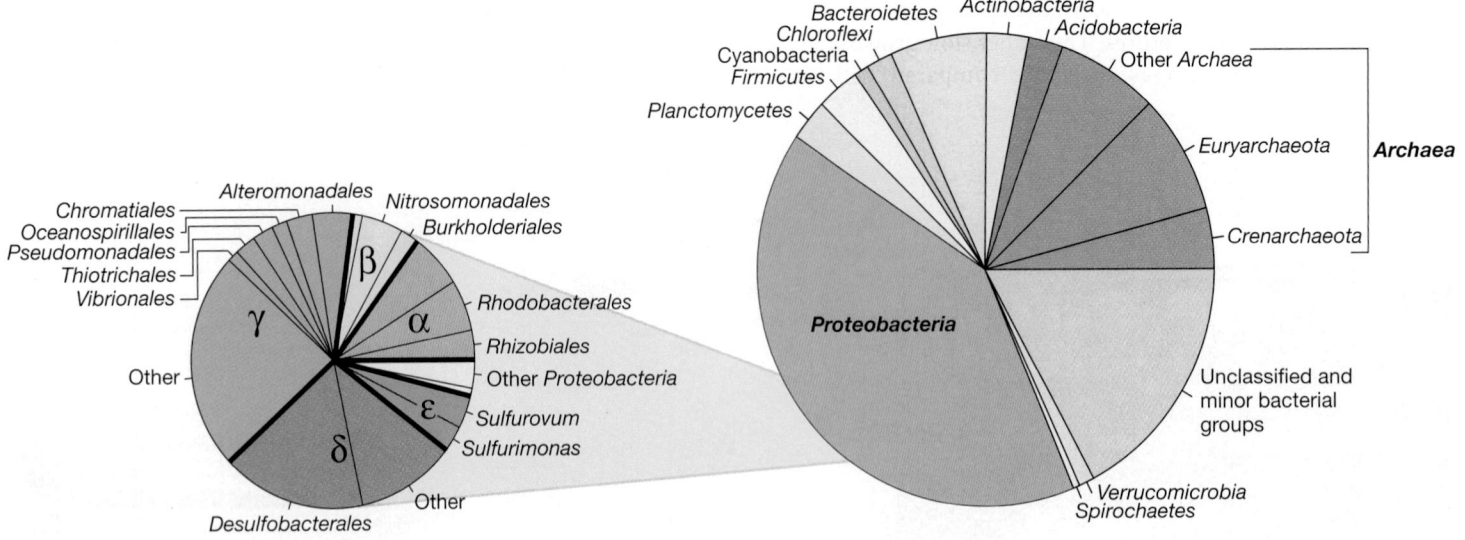

Figure 23.29 Marine sediment prokaryotic diversity. The results are pooled analyses of 13,360 sequences from several studies of the 16S rRNA gene content of shallow and deep marine sediments. Many of the groups indicated are covered in Chapters 17 and 18 (*Bacteria*) or 19 (*Archaea*). For *Proteobacteria*, major subgroups are indicated. Note the high proportion of archaeal sequences and of *Gamma-*, *Delta-*, and *Epsilonproteobacteria*. Data assembled and analyzed by Nicolas Pinel.

their high percentage of unusual *Archaea* that are absent from shallow sediments. How these organisms survive in the nutrient-depleted depths far below the seafloor is unclear, another mystery to emerge from culture-independent surveys of microbial distribution and diversity.

MiniQuiz

- How does pressure change with depth in a water column?
- What molecular adaptations are found in piezophiles that allow them to grow optimally under high pressure?
- Considering their metabolism, why are *Desulfobacterales* common in marine sediments?

23.12 Hydrothermal Vents

Although we have thus far described the deep sea as a remote, low-temperature, high-pressure environment suitable only for slow-growing piezotolerant and piezophilic microorganisms, there are some amazing exceptions. Thriving animal and microbial communities are found clustered in and around thermal springs in deep-sea waters throughout the world. These springs are located at depths from less than 1000 m to greater than 4000 m from the ocean surface in regions of the seafloor where volcanic magma and hot rock have caused the floor to rift apart at crustal spreading centers (**Figure 23.30**), or where iron and magnesium minerals associated with ancient rocks react with seawater and

generate heat. Seawater seeping into these dynamic cracking regions of the crust reacts with hot rock, resulting in hot springs saturated with chemical elements and dissolved gases. Collectively, these types of underwater hot springs are called **hydrothermal vents**. We will discuss several remarkable symbiotic associations between hydrothermal vent–associated animals and microorganisms in Chapter 25. Here we consider the vent environment as a habitat for free-living microorganisms.

Types of Vents

Volcanic hydrothermal systems are typically either warm (~5 to >50°C), diffuse vents or very hot vents that emit hydrothermal fluids at 270 to >400°C. The gently flowing, warm, diffuse fluids are emitted from cracks in the seafloor and the exterior walls of hydrothermal chimneys. The fluids originate from the mixing of cold seawater with hot hydrothermal fluids in subsurface regions of the sediments. Hot vents, called *black smokers*, form upright sulfide edifices called *chimneys* that can be less than 1 m to over 30 m in height. Chimneys form when acidic hydrothermal fluids rich in dissolved metals and magmatic gases are suddenly mixed with cold, oxygenated seawater. The rapid mixing causes fine-grained metal sulfide minerals such as pyrite and sphalerite to precipitate out, forming dark, buoyant plumes that rise above the seafloor (**Figure 23.31**).

A quite different type of hydrothermal vent environment is the "Lost City" formation located in the mid-Atlantic Ocean. Lost City is formed from the exposure of minerals associated with ocean crust 1–2 million years old that was once deep beneath the seafloor. Geological faults in these slow-spreading systems

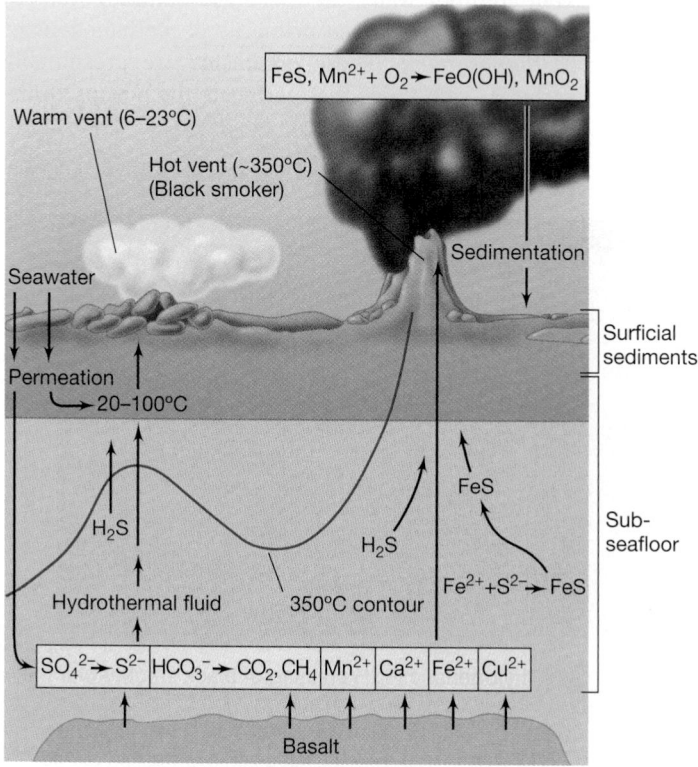

Figure 23.30 Hydrothermal vents. Schematic showing geological formations and major chemical species at warm vents and black smokers. In warm vents, the hot hydrothermal fluid is cooled by cold 2–3°C seawater permeating the sediments. In black smokers, hot hydrothermal fluid near 350°C reaches the seafloor directly.

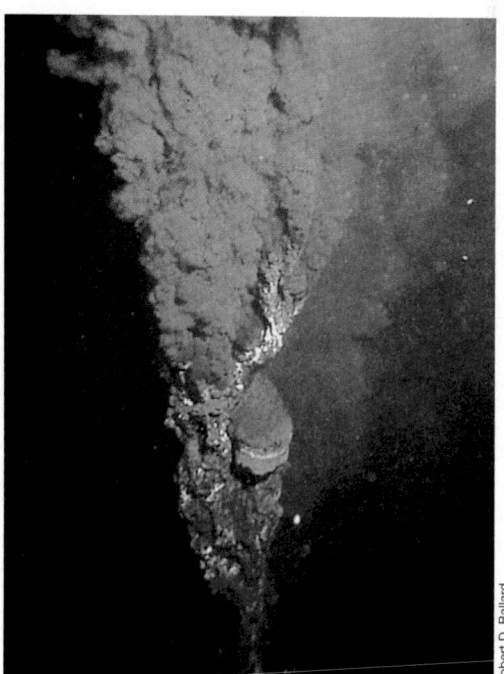

Figure 23.31 A hydrothermal vent black smoker emitting sulfide- and mineral-rich water at temperatures of 350°C. The walls of the black smoker chimneys display a steep temperature gradient and contain several types of prokaryotes.

Figure 23.32 **Massive carbonate chimney formation at Lost City peridotite-hosted vent system.** Microbial colonization of freshly exposed mineral surfaces was studied by placing sterile mineral fragments in the green-topped device placed over an actively venting area of the chimney. The diameter of the cylindrical collection device is approximately 10 cm.

exposed magnesium and iron-rich rocks called *peridotites* at the seafloor. Chemical reactions of seawater and newly exposed peridotite are highly exothermic, generating heat and also driving the pH up to as high as pH 11. Extremely high levels of H_2, CH_4, and other low-molecular-weight hydrocarbons are also present in the hot (200°C) hydrothermal fluids. In contrast to the acidic volcanic black smoker systems (Figure 23.30), which are relatively transient, mixing of these alkaline fluids with seawater results in the formation of calcium carbonate (limestone) chimneys that can reach up to 60 m in height and be active for 100,000 years or more (**Figure 23.32**).

Prokaryotes in Hydrothermal Vents

Bacteria displaying chemolithotrophic metabolisms dominate hydrothermal vent microbial ecosystems. Sulfidic vents support sulfur bacteria, whereas vents that emit other inorganic electron donors support nitrifying, hydrogen-oxidizing, iron- and manganese-oxidizing, or methylotrophic bacteria, the latter presumably growing on the CH_4 and carbon monoxide (CO) emitted from the vents. **Table 23.2** summarizes the inorganic electron donors and electron acceptors that are thought to play a role in hydrothermal vent chemolithotrophic metabolisms.

Although prokaryotes cannot survive in the superheated hydrothermal fluids of black smokers, thermophilic and hyper-

Table 23.2 *Chemolithotrophic prokaryotes present in the vicinity of deep-sea hydrothermal vents*[a]

Chemolithotroph	Electron donor	Electron acceptor	Product from donor
Sulfur-oxidizing	HS^-, S^0, $S_2O_3^{2-}$	O_2, NO_3^-	S^0, SO_4^{2-}
Nitrifying	NH_4^+, NO_2^-	O_2	NO_2^-, NO_3^-
Sulfate-reducing	H_2	S^0, SO_4^{2-}	H_2S
Methanogenic	H_2	CO_2	CH_4
Hydrogen-oxidizing	H_2	O_2, NO_3^-	H_2O
Iron- and manganese-oxidizing	Fe^{2+}, Mn^{2+}	O_2	Fe^{3+}, Mn^{4+}
Methylotrophic	CH_4, CO	O_2	CO_2

[a]See Chapters 13 and 14 for detailed discussions of these metabolisms.

thermophilic organisms do thrive in the *gradients* that form as the superheated water mixes with cold seawater. For example, the walls of smoker chimneys are teeming with hyperthermophiles such as *Methanopyrus*, a species of *Archaea* that oxidizes H_2 and makes CH_4 (⮌ Section 19.5). Phylogenetic FISH staining (⮌ Section 22.4) has detected cells of both *Bacteria* and *Archaea* in smoker chimney walls (**Figure 23.33**). The most thermophilic of all known sulfur-reducing prokaryotes, species of *Pyrolobus* and *Pyrodictium* (Chapter 19), were isolated from black smoker chimney walls. In contrast to the significant microbial diversity in volcanic vent chimney walls, the carbonate chimney walls of the Lost City vents are comprised primarily of methanogens of the genus *Methanosarcina*. These organisms are presumably nourished by the H_2-rich fluids that permeate the porous chimney walls.

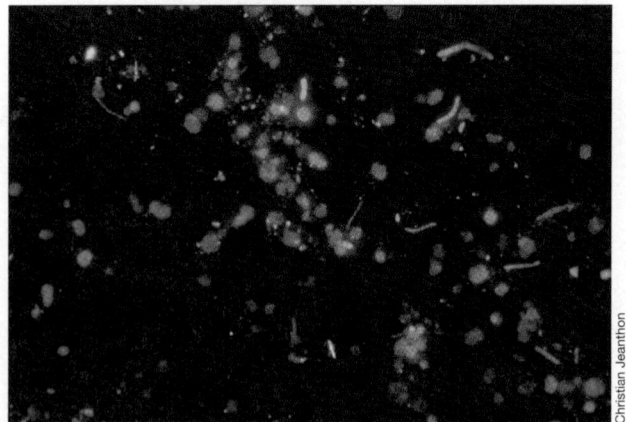

Figure 23.33 **Phylogenetic FISH staining of black smoker chimney material.** Snake Pit vent field, Mid-Atlantic Ridge (a tectonic Atlantic ocean-floor spreading center 3500 m deep and equidistant from the continents on either side of it). A green fluorescing dye was conjugated to a probe that reacts with the 16S rRNA of all *Bacteria* and a red dye to a 16S rRNA probe for *Archaea*. The hydrothermal fluid going through the center of this chimney is 300°C.

When smokers plug up from mineral debris, hyperthermophiles presumably drift away to colonize active smokers and somehow become integrated into the growing chimney wall. Surprisingly, although requiring very high temperatures for growth, hyperthermophiles are remarkably tolerant of cold temperatures and oxygen. Thus, transport of cells from one vent site to another in cold oxic seawater apparently is not a problem.

A Phylogenetic Snapshot of Hydrothermal Vent Prokaryotic Diversity

Using the powerful tools developed for microbial community sampling (⟲ Section 22.5), studies of prokaryotic diversity near volcanic hydrothermal vents have revealed an enormous diversity of *Bacteria*. These 16S rRNA gene sequence surveys include both warm and hot vents. Vent microbial communities are dominated by *Proteobacteria*, in particular *Epsilonproteobacteria* (⟲ Section 17.19; **Figure 23.34**). *Alpha-, Delta-,* and *Gammaproteobacteria* are also abundant, whereas *Betaproteobacteria* are much less so. Many *Epsilon-* and *Gammaproteobacteria* oxidize sulfide and sulfur as electron donors with either O_2 or nitrate (NO_3^-) as electron acceptors. As shown in the smaller pie diagram in Figure 23.34, vent *Epsilonproteobacteria* phylotypes most closely match those of chemolithotrophic sulfur bacteria such as *Sulfurimonas, Arcobacter, Sulfurovum,* and *Sulfurospirillum*. These bacteria oxidize reduced sulfur compounds as electron donors (⟲ Section 13.8), and such a physiology is consistent with their association near vent fluids charged with sulfur and sulfide. In addition, most *Deltaproteobacteria* specialize in anaerobic metabolisms using oxidized sulfur

compounds as electron acceptors. These include organisms such as *Desulfovibrio*, a sulfate-reducing bacterium that reduces sulfate (SO_4^{2-}) to sulfide, with lactate, pyruvate, or H_2 as electron donors (⟲ Section 14.8). However, in anoxic marine environments acetate-oxidizing sulfate-reducing bacteria dominate; *Desulfovibrio* cannot oxidize acetate. Acetate-oxidizing sulfate reducers include genera such as *Desulfobacter, Desulfococcus, Desulfosarcina,* and their relatives (⟲ Section 17.18).

In contrast to *Bacteria*, the diversity of volcanic hydrothermal vent *Archaea* is quite limited. Estimates of the number of unique phylotypes indicate that the diversity of *Bacteria* near hydrothermal vents is about ten times that of *Archaea*. However, *Archaea* are prevalent in samples recovered from the walls of hot vent chimneys (Figure 23.33). Most of the *Archaea* detected near hydrothermal vents are either methanogens (⟲ Section 19.3) or species of marine *Crenarchaeota* and *Euryarchaeota* (⟲ Figure 19.1). With the exception of the ammonia-oxidizing crenarchaeote *Nitrosopumilus* (⟲ Section 19.11), organisms in these groups remain uncultured and their physiologies poorly understood.

MiniQuiz

- How does a warm hydrothermal vent differ from a black smoker, both chemically and physically?
- Why is 350°C water emitted from a black smoker not boiling?
- Which phylum of *Bacteria* and which subgroups of this phylum dominate hydrothermal vent ecosystems, and why?

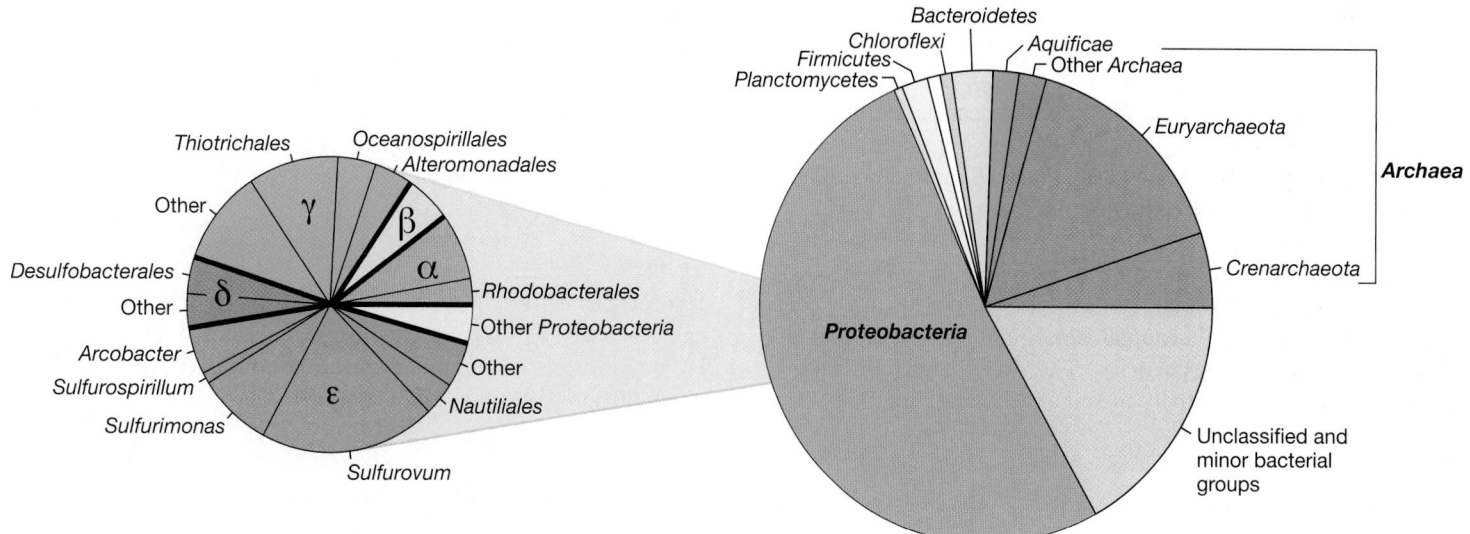

Figure 23.34 Hydrothermal vent prokaryotic diversity. The results are pooled analyses of 14,293 sequences from several studies of the 16S rRNA gene content of warm and hot hydrothermal vents. Many of these groups are covered in Chapters 17 and 18 (*Bacteria*) or 19 (*Archaea*). For *Proteobacteria*, major subgroups are indicated. Note high proportion of *Archaea* and of *Epsilonproteobacteria*. Data assembled and analyzed by Nicolas Pinel. The physiology of many of these organisms is summarized in Table 23.2.

Big Ideas

23.1

Ecosystems consist of organisms, their environments, and all of the interactions among the organisms and environments. The organisms are members of populations and communities and are adapted to habitats. Species richness and abundance are aspects of species diversity in a community and an ecosystem.

23.2

Microbial communities consist of guilds of metabolically related organisms. Microorganisms play major roles in energy transformations and biogeochemical processes that result in the recycling of elements essential to living systems.

23.3

The niche for a microorganism consists of the specific assortment of biotic and abiotic factors within a microenvironment in which that microorganism can be competitive. Microorganisms in nature often live a feast-or-famine existence such that only the best-adapted species reach high population density in a given niche. Cooperation among microorganisms is also important in many microbial interrelationships.

23.4

When surfaces are available, bacteria grow in attached masses of cells called biofilms. Biofilm formation involves both intra- as well as intercellular communication and confers several protective advantages on cells. Biofilms can have significant medical and economic impacts on humans when unwanted biofilms develop on inert as well as living surfaces.

23.5

Microbial mats are extremely thick biofilms consisting of microbial cells and trapped particulate materials. Microbial mats are widespread in hypersaline or thermal waters where grazing animals are prevented from feeding on the mat cells.

23.6

Soils are complex microbial habitats with numerous microenvironments and niches. Microorganisms are present in the soil primarily attached to soil particles. The most important factors influencing microbial activity in soil are the availability of water and nutrients.

23.7

The deep subsurface is a significant microbial habitat, most likely sustaining chemolithotrophic populations that can live on a diet of a few minerals, CO_2, SO_4^{2-}, N_2, and H_2. Hydrogen is thought to be continually produced by interaction of water with iron minerals or by the radiolysis of water.

23.8

In freshwater aquatic ecosystems, phototrophic microorganisms are the main primary producers. Most of the organic matter produced is consumed by bacteria, which can lead to depletion of oxygen in the environment. The BOD of a body of water indicates its relative content of organic matter that can be biologically oxidized.

23.9

Pelagic marine waters are more nutrient deficient than most freshwaters, yet substantial numbers of prokaryotes inhabit the oceans. The major phototrophs include *Prochlorococcus* (oxygenic) and the aerobic phototrophic bacteria (anoxygenic).

23.10

Species of *Bacteria* tend to predominate in marine surface waters, whereas in deeper waters *Archaea* comprise a larger fraction of the microbial community. Many pelagic *Bacteria* use light to make ATP by rhodopsin-driven proton pumps. Viruses outnumber prokaryotes by several orders of magnitude in marine waters.

23.11

The deep sea is a cold, dark habitat where hydrostatic pressure is high and nutrient levels are low. Piezophiles grow best under pressure but do not require pressure, whereas extreme piezophiles require high pressure, typically several hundred atmospheres, for growth.

23.12

Hydrothermal vents are deep-sea hot springs where either volcanic activity or unusual chemistry generates fluids containing large amounts of inorganic electron donors that can be used by chemolithotrophic bacteria.

Review of Key Terms

Biochemical oxygen demand (BOD) the microbial oxygen-consuming properties of a water sample

Biofilm colonies of microbial cells encased in a porous organic matrix and attached to a surface

Biogeochemistry the study of biologically mediated chemical transformations in the environment

Community two or more cell populations coexisting in a certain area at a given time

Ecosystem a dynamic complex of organisms and their physical environment interacting as a functional unit

Epilimnion the warmer and less dense surface waters of a stratified lake

Extreme piezophile a piezophilic organism unable to grow at a hydrostatic pressure of 1 atm and typically requiring several hundred atmospheres of pressure for growth

Guild a population of metabolically related microorganisms

Habitat an environment within an ecosystem where a microbial community could reside

Hydrothermal vents warm or hot water–emitting springs associated with crustal spreading centers on the seafloor.

Hypolimnion the colder, denser, and often anoxic bottom waters of a stratified lake

Microbial mat a thick, layered, diverse community in which cyanobacteria are essential to the formation of terrestrial mats that grow in a hypersaline or extremely hot aquatic environment

Microenvironment a micrometer-scale space surrounding a microbial cell or group of cells

Niche in ecological theory, the biotic and abiotic characteristics of the microenvironment

that contribute to an organism's competitive success

Oligotroph an organism that grows only or grows best at very low levels of nutrients

Piezophile an organism that grows best under a hydrostatic pressure greater than 1 atm

Piezotolerant able to grow under elevated hydrostatic pressures but growing best at 1 atm

Population a group of organisms of the same species in the same place at the same time

Primary producer an organism that uses light to synthesize new organic material from CO_2

Prochlorophyte a prokaryotic phototroph that

contains chlorophylls *a* and either *b* or *d*, and lacks phycobiliproteins

Proteorhodopsin a light-sensitive protein present in some open-ocean *Bacteria* that indirectly catalyzes ATP formation

Rhizosphere the region immediately adjacent to plant roots

Species abundance the proportion of each species in a community

Species richness the total number of different species present in a community

Stratified water column a body of water separated into layers having different physical and chemical characteristics

Review Questions

1. What is biogeochemical cycling? Why is it important in the microbial world (Section 23.2)?

2. Explain why both obligately anaerobic and obligately aerobic bacteria can often be isolated from the same soil sample (Section 23.3).

3. Outline some of the key uses of biofilms in industries (Section 23.4).

4. How can biofilms complicate treatment of infectious diseases (Section 23.4)?

5. How do microbial mats compare with biofilms in terms of dimensions and microbial diversity (Section 23.5)?

6. In what soil horizon are microbial numbers and activities the highest, and why (Section 23.6)?

7. How are nutrients for microbial growth replenished in the deep subsurface as opposed to the near subsurface (Section 23.7)?

8. What is BOD? What is the BOD of microorganisms in an aquatic environment? Why (Section 23.8)?

9. What organisms are the major phototrophs in the oceans (Section 23.9)?

10. List some major prokaryotes found in the open oceans and describe how they make ATP (Section 23.10).

11. Describe the physiological features of piezotolerant and piezophilic bacteria (Section 23.11).

12. Why are chemolithotrophic bacteria so prevalent at hydrothermal vents (Section 23.12)?

Application Questions

1. Imagine a sewage plant that is releasing sewage containing high levels of ammonia and phosphate and very low levels of organic carbon. Which types of microbial blooms might be triggered by this sewage? How would the graphs of oxygen near and beyond the plant's release point differ from the graph shown in Figure 23.16*a*?

2. Review the data of Figure 23.21. Keeping in mind that the open-ocean waters are highly oxic, predict the possible metabolic lifestyles of open-ocean *Archaea* and *Bacteria*. Why might proteorhodopsin be more abundant in one group of organisms than in the other?

 Need more practice? Test your understanding with Quantitative Questions; access additional study tools including tutorials, animations, and videos; and then test your knowledge with chapter quizzes and practice tests at **www.microbiologyplace.com.**

24

Nutrient Cycles, Biodegradation, and Bioremediation

Coccolithophores are marine algae that play a major role in the carbon cycle by consuming CO_2 and calcium in the oceans and helping to maintain ocean pH.

In the previous chapter we examined a variety of microbial habitats in order to set the stage for the consideration of some major microbial activities in this chapter. This chapter has two main themes: (1) *nutrient cycles* and (2) *biodegradation and bioremediation*. In both cases we focus on the biogeochemical activities of microorganisms and see how these activities interrelate.

 Nutrient Cycles

The key nutrients for life are cycled by both microorganisms and macroorganisms, but for any given nutrient, it is microbial activities that dominate. Understanding how microbial nutrient cycles work is important because the cycles and their many feedback loops are essential for plant agriculture and the overall health of sustainable planet life.

We begin our coverage of nutrient cycles with the carbon cycle. Major areas of interest here are the magnitude of carbon reservoirs on Earth, the rates of carbon cycling within and between reservoirs, and the coupling of the carbon cycle to other nutrient cycles. We particularly emphasize the compounds *carbon dioxide* (CO_2) and *methane* (CH_4) as major components of the carbon cycle.

24.1 The Carbon Cycle

On a global basis, carbon (C) cycles as CO_2 through all of Earth's major carbon reservoirs: the atmosphere, the land, the oceans, freshwaters, sediments and rocks, and biomass (**Figure 24.1**). As we have already seen for freshwater environments, the carbon and oxygen cycles are intimately linked (↺ Section 23.8). All nutrient cycles link in some way to the carbon cycle, but the nitrogen (N) cycle links particularly strongly because, other than water (H_2O), C and N make up the bulk of living organisms (↺ Section 4.1 and see Figure 24.4).

Carbon Reservoirs

The amount of C in reservoirs of Earth needs to be kept in balance with the amount that is cycling. By far the largest carbon reservoir on Earth is the sediments and rocks of Earth's crust (Figure 24.1), but the rate at which sediments and rocks decompose and carbon cycles out as CO_2 is so slow that flux out of this reservoir is insignificant on a human time scale. A large amount of C is found in land plants. This is the organic C of forests, grasslands, and agricultural crops—the major sites of phototrophic CO_2 fixation. However, more C is present in dead organic material, called **humus**, than in living organisms. Humus is a complex mixture of organic materials that have resisted rapid decomposition and is derived primarily from dead plants and microorganisms. Some humic substances are quite recalcitrant, with a decomposition time of several decades, but certain other humic components decompose much more rapidly.

The most rapid means of transfer of C is via the atmosphere. Carbon dioxide is removed from the atmosphere primarily by photosynthesis of land plants and marine microorganisms and is returned to the atmosphere by respiration of animals and chemoorganotrophic microorganisms (Figure 24.1). The single most important contribution of CO_2 to the atmosphere is by microbial decomposition of dead organic material, including humus. However, in the past 50 years human activities have increased atmospheric CO_2 levels by nearly 20%, primarily from the burning of fossil fuels. This rise in CO_2, a major *greenhouse gas*, has triggered a period of steadily increasing global temperatures called *global warming*. Although the consequences of global warming on microbial nutrient cycling are currently unpredictable, everything we know about the biology of microorganisms tells us that microbial activities in nature will change in response to higher temperatures. Whether these responses will be favorable or unfavorable to other organisms remains to be seen.

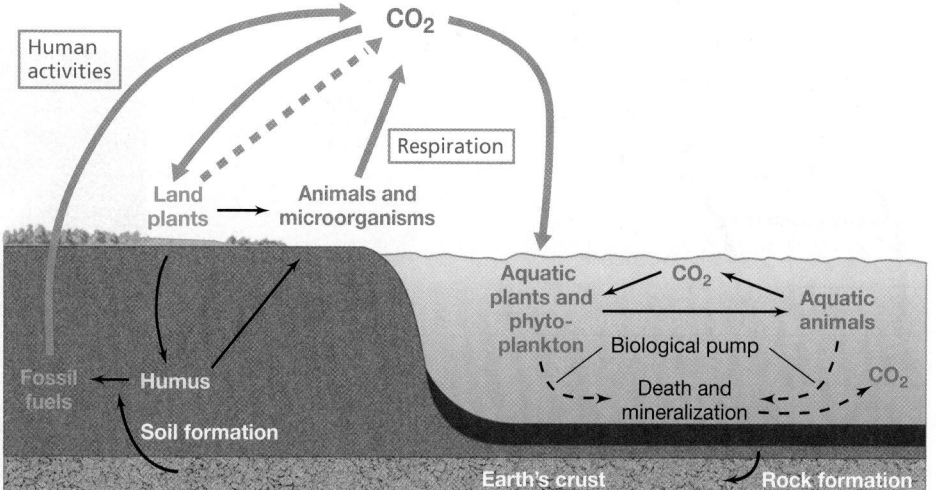

Major Carbon Reservoirs on Earth	
Reservoir	**Percent of Total [a]**
Rocks and sediments	99.5 (80% inorganic)
Oceans	0.05
Methane hydrates	0.014
Fossil fuels	0.006
Terrestrial biosphere	0.003
Aquatic biosphere	0.000002

[a] Total carbon, 76×10^{15} tons

Figure 24.1 The carbon cycle. The carbon and oxygen cycles are closely connected, as oxygenic photosynthesis both removes CO_2 and produces O_2, and respiratory processes both produce CO_2 and remove O_2. As the accompanying table shows, by far the greatest reservoir of carbon on Earth is in rocks and sediments, and most of this is in inorganic form as carbonates.

UNIT 7

Photosynthesis and Decomposition

New organic compounds are biologically synthesized on Earth only by CO_2 fixation by phototrophs and chemolithotrophs. Most organic compounds originate in photosynthesis and thus phototrophic organisms are the foundation of the carbon cycle (Figure 24.1). However, phototrophic organisms are abundant in nature only in habitats where light is available. The deep sea, deep terrestrial subsurface, and other permanently dark habitats are devoid of indigenous phototrophs. There are two groups of oxygenic phototrophic organisms: *plants* and *microorganisms*. Plants are the dominant phototrophic organisms of terrestrial environments, whereas phototrophic microorganisms dominate in aquatic environments.

The redox cycle for C (**Figure 24.2**) begins with photosynthetic CO_2 fixation, driven by the energy of light:

$$CO_2 + H_2O \longrightarrow (CH_2O) + O_2$$

CH_2O represents organic matter at the oxidation–reduction level of cell material. Phototrophic organisms also carry out respiration, both in the light and the dark. The overall equation for respiration is the reverse of oxygenic photosynthesis:

$$(CH_2O) + O_2 \longrightarrow CO_2 + H_2O$$

For organic compounds to accumulate, the rate of photosynthesis must exceed the rate of respiration. In this way, autotrophic organisms build biomass from CO_2, and then this biomass in one way or another supplies the carbon heterotrophic organisms

need. Anoxygenic phototrophs and chemolithotrophs also produce excess organic compounds, but in most environments the contributions of these organisms to the accumulation of organic matter are trivial compared to that of oxygenic phototrophs.

Organic compounds are degraded biologically to CH_4 and CO_2 (Figure 24.2). Carbon dioxide, most of which is of microbial origin, is produced by aerobic and other forms of respiration (↩ Section 14.6). Methane is produced in anoxic environments by *methanogens* from the reduction of CO_2 with hydrogen (H_2) or from the splitting of acetate. However, virtually any organic compound can eventually be converted to CH_4 from the cooperative activities of methanogens and fermentative bacteria, as we will see in the next section. Methane produced in anoxic habitats is insoluble and diffuses to oxic environments, where it is either released to the atmosphere or oxidized to CO_2 by *methanotrophs* (Figure 24.2). Hence, most of the carbon in organic compounds eventually returns to CO_2, and the links in the carbon cycle are closed.

Methane Hydrates

Although an even more minor component of the atmosphere than is CO_2, CH_4 is a potent greenhouse gas that is over 20 times more effective in trapping heat. Some CH_4 enters the atmosphere from methanogenic production, but not all biologically produced CH_4 is consumed or released to the atmosphere. Huge amounts of CH_4 derived primarily from microbial activities are trapped as *methane hydrates*, molecules of frozen methane. Methane hydrates form when sufficient CH_4 is present in environments of high pressure and low temperature such as beneath the permafrost in the Arctic and in marine sediments. These deposits can be up to several hundred meters thick and are estimated to contain 700–10,000 petagrams (1 petagram = 10^{15} g) of methane. This exceeds other known methane reserves on Earth by several orders of magnitude.

Methane hydrates are highly dynamic, absorbing and releasing CH_4 in response to changes in pressure, temperature (**Figure 24.3**), and fluid movement. The hydrates also fuel deep-water

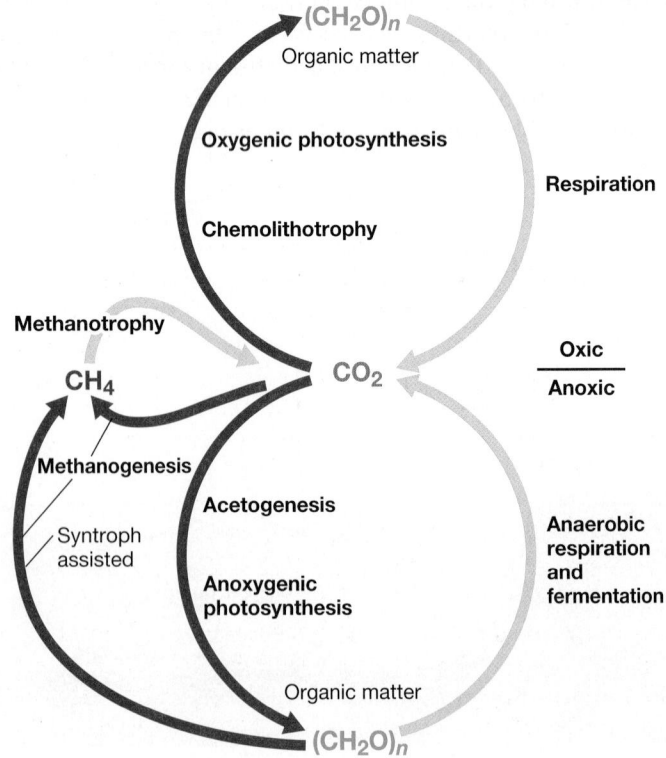

Figure 24.2 Redox cycle for carbon. The diagram contrasts autotrophic processes ($CO_2 \rightarrow$ organic compounds) and heterotrophic processes (organic compounds $\rightarrow CO_2$). Yellow arrows indicate oxidations; red arrows indicate reductions.

Figure 24.3 Burning methane hydrate. Frozen methane ice retrieved from marine sediments is ignited.

ecosystems, called *cold seeps*. Here, the slow release of CH_4 from seafloor hydrates nourishes not only anaerobic methane-oxidizing bacteria (Section 14.13), but also animal communities that contain methane-oxidizing endosymbionts that oxidize CH_4 and release organic matter to the animals (Section 17.6). Climate scientists now fear that global warming could catalyze a catastrophic release of CH_4 from methane hydrates, an event that would rapidly affect Earth's climate. In fact, the sudden release of large amounts of CH_4 from methane hydrates may have triggered the Permian–Triassic extinctions some 250 million years ago. These extinctions, the worst in Earth's history, wiped out virtually all marine animals and over 70% of all terrestrial plant and animal species.

Carbon Balances and Coupled Cycles

The amount of carbon in reservoirs on Earth needs to be kept in balance with the amount that is cycling if life is to continue as it has been for billions of years. Major reasons why significant environmental concern exists about global warming include not only the human-driven inputs of CO_2 from the combustion of fossil fuels (whose carbon has been buried and thus removed from the carbon cycle for millions of years), which is a relatively slow process, but the even more frightening possibility of warmer temperatures triggering the melting of methane hydrates. This would release enormous amounts of CH_4 that would greatly accelerate climate change and could affect microbial nutrient cycling in ways that would have serious downstream effects on all life forms.

Awareness of how the various nutrient cycles feed back upon one another and are interconnected is important. Although it is convenient to consider carbon cycling as a series of reactions separate from those in other nutrient cycles, in reality, all nutrient cycles are *coupled cycles*; major changes in one cycle affect the functioning of others. For example, consider the C and nitrogen cycles (**Figure 24.4**). The rate of primary productivity (CO_2 fixation) is controlled by several factors, in particular by the magnitude of photosynthetic biomass and by available nitrogen. Thus, large-scale reductions in biomass by, for instance, widespread deforestation, reduce rates of primary productivity and increase levels of CO_2. High levels of organic carbon stimulate

nitrogen fixation and this in turn adds more fixed N to the pool for primary producers; low levels of organic carbon have just the opposite effect. High levels of ammonia stimulate primary production and nitrification, but inhibit N_2 fixation. High levels of nitrate, an excellent N source for plants and aquatic phototrophs, stimulate primary production but also increase the rate of denitrification; the latter removes fixed forms of N from the environment and feeds back in a negative way on primary production (Figure 24.4).

This simple example illustrates how nutrient cycles are anything but isolated entities; they are coupled systems that maintain a delicate balance of inputs and outputs. Thus, one could expect these cycles to respond to large inputs in specific links (for example, through inputs of CO_2 or nitrogen fertilizers) in ways that are not always beneficial to the biosphere. This is particularly true of the C and N cycles because next to H_2O, C and N are the most abundant elements in living organisms and their cycles interact with each other in such major ways (Figure 24.4).

MiniQuiz

- How is new organic matter made in nature?
- In what ways are oxygenic photosynthesis and respiration related?
- What is a methane hydrate?

24.2 Syntrophy and Methanogenesis

Most organic compounds are oxidized in nature by *aerobic* microbial processes. However, because oxygen (O_2) is a poorly soluble gas and is actively consumed when available, much organic carbon still ends up in anoxic environments. Methanogenesis, the biological production of CH_4, is a major process in anoxic habitats and is catalyzed by a large group of *Archaea*, the *methanogens*, which are strict anaerobes. We discussed the biochemistry of methanogenesis in Section 14.10 and methanogens themselves in Section 19.3. Most methanogens can use CO_2 as a terminal electron acceptor in anaerobic respiration, reducing it to CH_4 with H_2 as electron donor. Only a very few other substrates, chiefly acetate, are directly converted to CH_4 by methanogens. To convert most organic compounds to CH_4, methanogens must team up with partner organisms that can supply them with precursors for methanogenesis. This is the job of the syntrophs.

Anoxic Decomposition and Syntrophy

In Section 14.5 we discussed the biochemistry of **syntrophy**, a process in which two or more organisms cooperate in the anaerobic degradation of organic compounds. Here we consider the interactions of syntrophic bacteria with their partner organisms and their significance for the C cycle. Our focus is on anoxic freshwater sediments and anoxic wastewater treatment, both of which are major sources of CH_4.

Polysaccharides, proteins, lipids, and nucleic acids from organic compounds find their way into anoxic habitats, where they are catabolized. Released by hydrolysis, the monomers become major electron donors for energy metabolism. For the

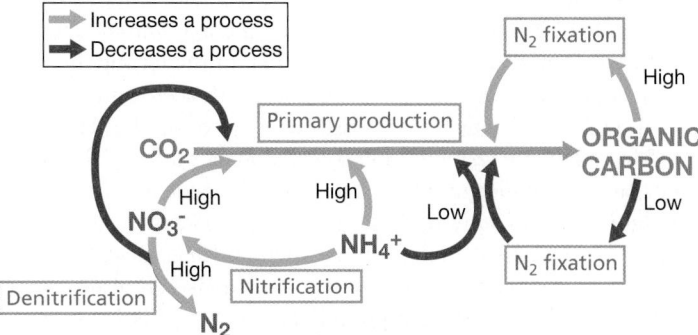

Figure 24.4 Coupled cycles. All nutrient cycles are interconnected, but the carbon and nitrogen cycles are extremely closely coupled. In the carbon cycle, CO_2 supplies the C for carbon compounds. The N cycle, shown in more detail in Figure 24.7, supplies N for many of the compounds.

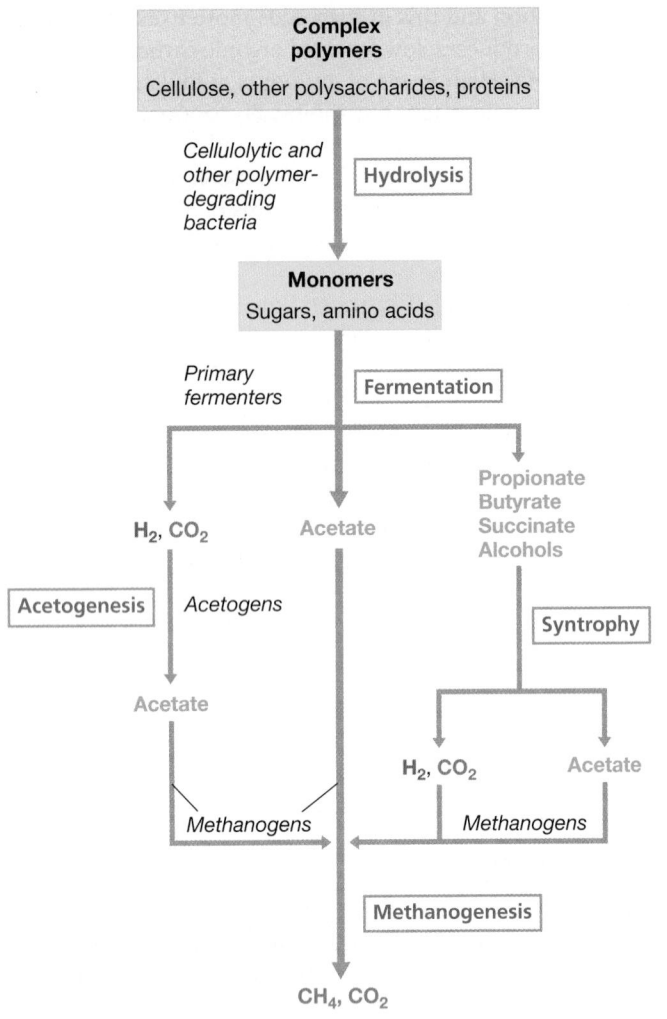

Figure 24.5 Anoxic decomposition. In anoxic decomposition various groups of fermentative anaerobes cooperate in the conversion of complex organic materials to CH_4 and CO_2. This picture holds for environments in which sulfate-reducing bacteria play only a minor role, for example, in freshwater lake sediments, sewage sludge bioreactors, or the rumen.

and Figure 24.5). Other species of *Syntrophomonas* use fatty acids up to C_{18} in length, including some unsaturated fatty acids. *Syntrophobacter wolinii* specializes in propionate (C_3) fermentation, generating acetate, CO_2, and H_2, and *Syntrophus gentianae* degrades aromatic compounds such as benzoate to acetate, H_2, and CO_2 (Table 24.1). However, syntrophs are unable to carry out these reactions in pure culture; their growth requires a H_2-consuming partner organism, and this requirement is directly connected to the energetics of syntrophic processes.

As described in Section 14.5, H_2 consumption by a partner organism is absolutely essential for growth of the syntrophs. When the reactions in Table 24.1 are written with all reactants at standard conditions (solutes, 1 M; gases, 1 atm, 25°C), the reactions yield free-energy changes ($\Delta G^{0\prime}$, ↩ Section 4.4) that are positive in arithmetic sign; that is, the reactions *require* rather than *release* energy. But the consumption of H_2 dramatically affects the energetics, making the reaction favorable and allowing energy to be conserved. This can be seen in Table 24.1, where the ΔG values (free-energy change measured under actual conditions in the habitat) are negative in arithmetic sign if H_2 concentrations are kept near zero through consumption by a partner organism.

The final products of the syntrophic partnership are thus CO_2 and CH_4 (Figure 24.5), and virtually any organic compound that enters a methanogenic habitat will eventually be converted to these products. This includes even complex aromatic and aliphatic hydrocarbons. Additional organisms other than those shown in Figure 24.5 may be involved in such degradations, but eventually fatty acids and alcohols will be generated, and they will be converted to methanogenic substrates by the syntrophs. Acetate produced by syntrophs (as well as by the activities of acetogenic bacteria, ↩ Section 14.9) is a direct methanogenic substrate and is converted to CO_2 and CH_4 by certain species of methanogens.

Methanogenic Symbionts and Acetogens in Termites

A variety of anaerobic protists that thrive under strictly anoxic conditions, including ciliates and flagellates, are known and play a major role in the carbon cycle. Methanogenic *Archaea* live within some of these protist cells as H_2-consuming endosymbionts. For example, methanogens are present *within* cells of trichomonal protists inhabiting the termite hindgut (**Figure 24.6**), where methanogenesis and acetogenesis are major processes. Methanogenic symbionts of protists are species of the genera *Methanobacterium* or *Methanobrevibacter* (↩ Section 19.3). In the termite hindgut, these endosymbiotic methanogens along with acetogenic bacteria are thought to benefit their protist hosts by consuming H_2 generated from glucose fermentation by cellulolytic protists. The acetogens are not endosymbionts but instead reside in the termite hindgut itself, consuming H_2 from primary fermenters and reducing CO_2 to make acetate. Unlike methanogens, acetogens can ferment glucose directly to acetate. Acetogens can also ferment methoxylated aromatic compounds to acetate. This is especially important in the termite hindgut because termites live on wood, which contains lignin, a complex polymer of methoxylated aromatic compounds; the acetate produced by acetogens in the termite hindgut is consumed by the animal as its primary energy source. Microbial symbioses in the termite hindgut are discussed in more detail in Section 25.10.

breakdown of a typical polysaccharide such as cellulose, the process begins with *cellulolytic* bacteria (**Figure 24.5**). These organisms hydrolyze cellulose into glucose, which is catabolized by fermentative organisms to short-chain fatty acids (acetate, propionate, and butyrate), alcohols such as ethanol and butanol, H_2, and CO_2. H_2 and acetate are consumed by methanogens directly, but the bulk of the carbon remains in the form of fatty acids and alcohols; these cannot be directly catabolized by methanogens and require the activities of syntrophic bacteria (↩ Section 14.5; Figure 24.5).

Role of the Syntrophs

The key bacteria in the conversion of organic compounds to CH_4 are the syntrophs, the bacteria that participate in syntrophy (**Table 24.1**; ↩ Table 14.4). Syntrophs are *secondary* fermenters because they ferment the products of the primary fermenters, yielding H_2, CO_2, and acetate as products. For example, *Syntrophomonas wolfei* oxidizes C_4 to C_8 fatty acids, yielding acetate, CO_2 (if the fatty acid was C_5 or C_7), and H_2 (Table 24.1

Table 24.1 *Major reactions in the anoxic conversion of organic compounds to methane[a]*

Reaction type	Reaction	Free-energy change (kJ/reaction)	
		$\Delta G^{0\prime b}$	ΔG^{c}
Fermentation of glucose to acetate, H_2, and CO_2	Glucose + $4 H_2O \rightarrow 2\,$acetate$^-$ + $2 HCO_3^- + 4 H^+ + 4 H_2$	−207	−319
Fermentation of glucose to butyrate, CO_2, and H_2	Glucose + $2 H_2O \rightarrow$ butyrate$^-$ + $2 HCO_3^- + 2 H_2 + 3 H^+$	−135	−284
Fermentation of butyrate to acetate and H_2	Butyrate$^-$ + $2 H_2O \rightarrow 2\,$acetate$^-$ + $H^+ + 2 H_2$	+48.2	−17.6
Fermentation of propionate to acetate, CO_2, and H_2	Propionate$^-$ + $3 H_2O \rightarrow$ acetate$^-$ + $HCO_3^- + H^+ + H_2$	+76.2	−5.5
Fermentation of ethanol to acetate and H_2	2 Ethanol + $2 H_2O \rightarrow 2\,$acetate$^-$ + $4 H_2 + 2 H^+$	+19.4	−37
Fermentation of benzoate to acetate, CO_2, and H_2	Benzoate$^-$ + $7 H_2O \rightarrow 3\,$acetate$^-$ + $3 H^+ + HCO_3^- + 3 H_2$	+70.1	−18
Methanogenesis from $H_2 + CO_2$	$4 H_2 + HCO_3^- + H^+ \rightarrow CH_4 + 3 H_2O$	−136	−3.2
Methanogenesis from acetate	Acetate$^-$ + $H_2O \rightarrow CH_4 + HCO_3^-$	−31	−24.7
Acetogenesis from $H_2 + CO_2$	$4 H_2 + 2 HCO_3^- + H^+ \rightarrow$ acetate$^-$ + $4 H_2O$	−105	−7.1

[a]Data adapted from Zinder, S. 1984. Microbiology of anaerobic conversion of organic wastes to methane: Recent developments. *Am. Soc. Microbiol.* 50:294–298.
[b]Standard conditions: solutes, 1 M; gases, 1 atm, 25°C.
[c]Concentrations of reactants in typical anoxic freshwater ecosystems: fatty acids, 1 mM; HCO_3^-, 20 mM; glucose, 10 μM; CH_4, 0.6 atm; H_2, 10^{-4} atm. For calculating ΔG from $\Delta G^{0\prime}$, refer to Appendix 1.

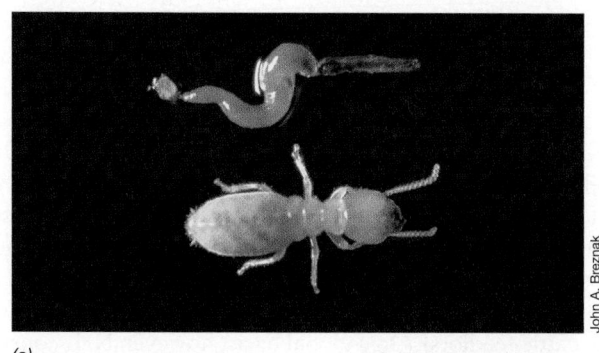

(a)

John A. Breznak

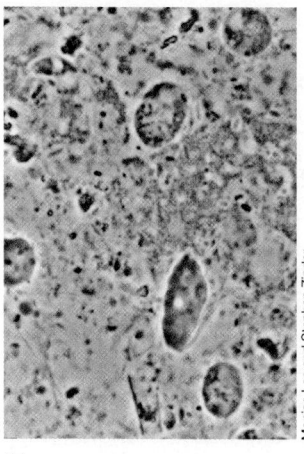

(b)

Monica Lee and Stephen Zinder

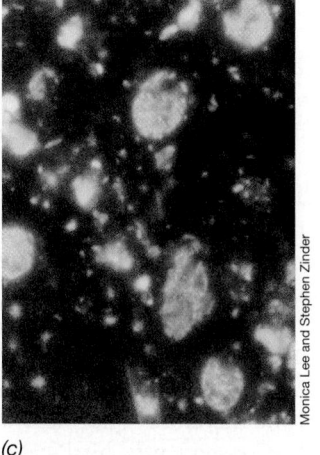

(c)

Monica Lee and Stephen Zinder

Figure 24.6 **Termites and their carbon metabolism.** *(a)* A subterranean termite worker larva shown beneath a hindgut dissected from another worker. The animal is about 0.5 cm long. Two views of the same microscopic field show termite hindgut protists photographed by *(b)* phase-contrast and *(c)* epifluorescence. Endosymbiotic methanogens in the protist cells fluoresce blue-green due to the methanogenic coenzyme F_{420} (compare with ⮌ Figure 14.19). The average diameter of the protist cells is 15–20 μm.

MiniQuiz

- Why does *Syntrophomonas* need a partner organism to ferment fatty acids or alcohols?
- What kinds of organisms are used in coculture with *Syntrophomonas*?
- What is the final product of acetogenesis?

24.3 The Nitrogen Cycle

Nitrogen (N) is an essential element for life (⮌ Section 4.1) and exists in a number of oxidation states. We have discussed four major microbial N transformations thus far: nitrification, denitrification, anammox, and nitrogen fixation (Chapters 13 and 14). These and other key N transformations are summarized in the redox cycle shown in **Figure 24.7**.

Nitrogen Fixation and Denitrification

Nitrogen gas (N_2) is the most stable form of N and is a major reservoir for N on Earth. However, only a relatively small number of prokaryotes are able to use N_2 as a cellular N source by *nitrogen fixation* ($N_2 + 8 H \rightarrow 2 NH_3 + H_2$) (⮌ Section 13.14). The N recycled on Earth is mostly already "fixed N"; that is, N in combination with other elements, such as in ammonia (NH_3) or nitrate (NO_3^-). In many environments, however, the short supply of fixed N puts a premium on biological nitrogen fixation, and in these habitats, nitrogen-fixing bacteria flourish.

We discussed the role of NO_3^- as an alternative electron acceptor in anaerobic respiration in Section 14.7. Under most conditions, the end product of NO_3^- reduction is N_2, NO, or N_2O. The reduction of NO_3^- to these gaseous nitrogen compounds, called **denitrification** (Figure 24.7), is the main means by which N_2 and N_2O is formed biologically. On the one hand, denitrification is a detrimental process. For example, if agricultural fields fertilized

UNIT 7

Key Processes and Prokaryotes in the Nitrogen Cycle

Processes	Example organisms
Nitrification ($NH_4^+ \rightarrow NO_3^-$)	
$NH_4^+ \rightarrow NO_2^-$	*Nitrosomonas*
$NO_2^- \rightarrow NO_3^-$	*Nitrobacter*
Denitrification ($NO_3^- \rightarrow N_2$)	*Bacillus, Paracoccus,*
	Pseudomonas
N_2 Fixation ($N_2 + 8 H \rightarrow NH_3 + H_2$)	
Free-living	
Aerobic	*Azotobacter*
	Cyanobacteria
Anaerobic	*Clostridium,* purple and
	green phototrophic bacteria
	Methanobacterium (Archaea)
Symbiotic	*Rhizobium*
	Bradyrhizobium
	Frankia
Ammonification (organic-N $\rightarrow NH_4^+$)	
	Many organisms can do this
Anammox ($NO_2^- + NH_3 \rightarrow 2 N_2$)	*Brocadia*

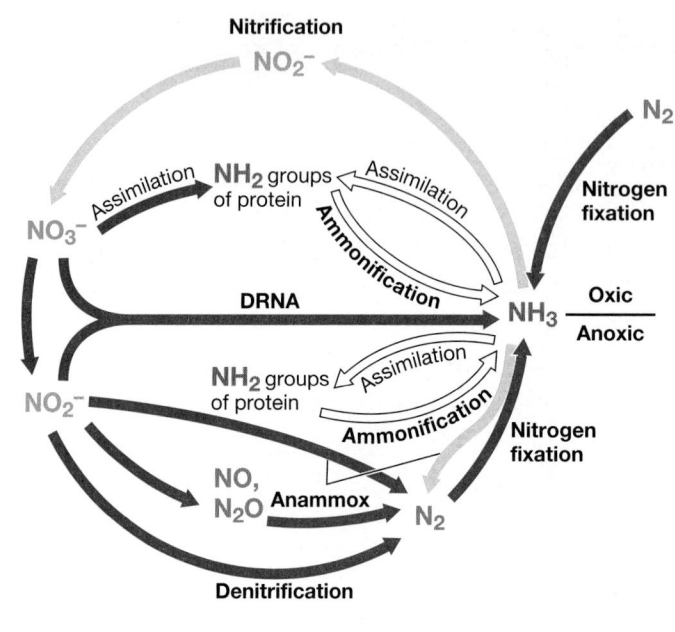

Figure 24.7 Redox cycle for nitrogen. Oxidation reactions are shown by yellow arrows and reductions by red arrows. Reactions without redox change are in white. The anammox reaction is $NH_3 + NO_2^- + H^+ \rightarrow N_2 + 2 H_2O$ (↺ Figure 13.11). DRNA, dissimilative reduction of nitrate to ammonia.

with nitrate fertilizer become waterlogged following heavy rains, anoxic conditions can develop and denitrification can be extensive; this removes fixed nitrogen from the soil. On the other hand, denitrification can aid in wastewater treatment (↺ Section 35.2). By removing NO_3^- as volatile forms of N, denitrification minimizes fixed N and thus algal growth when the treated sewage is discharged into lakes and streams.

The production of N_2O and NO by denitrification can have other environmental consequences. N_2O can be photochemically oxidized to NO in the atmosphere. NO reacts with ozone (O_3) in the upper atmosphere to form nitrite (NO_2^-), and this returns to Earth as nitric acid (HNO_2). Thus, denitrification contributes both to O_3 destruction, which increases passage of ultraviolet radiation to the surface of Earth, and to acid rain, which increases acidity of soils. Increases in soil acidity can change microbial community structure and function and, ultimately, soil fertility, impacting both plant diversity and agricultural yields of crop plants.

Ammonification and Ammonia Fluxes

Ammonia is released during the decomposition of organic nitrogen compounds such as amino acids and nucleotides, a process called *ammonification* (Figure 24.7). Another process contributing to the generation of NH_3 is the respiratory reduction of NO_3^- to NH_3, called *dissimilative reduction of nitrate to ammonia* (DRNA, Figure 24.7). DRNA dominates NO_3^- and nitrite (NO_2^-) reduction in reductant-rich anoxic environments, such as highly organic marine sediments and the human gastrointestinal tract. It is thought that nitrate-reducing bacteria exploit this pathway primarily when NO_3^- is limiting because DRNA consumes 8 electrons compared with the 4 and 5 consumed when NO_3^- is reduced only as far as N_2O or N_2, respectively.

At neutral pH, NH_3 exists as ammonium (NH_4^+). Much of the NH_4^+ released by aerobic decomposition in soils is rapidly recycled and converted to amino acids in plants and microorganisms. However, because NH_3 is volatile, some of it can be lost from alkaline soils by vaporization, and there are major losses of NH_3 to the atmosphere in areas with dense animal populations (for example, cattle feedlots). On a global basis, however, NH_3 constitutes only about 15% of the N released to the atmosphere, the rest being primarily N_2 or N_2O from denitrification.

Nitrification and Anammox

Nitrification, the oxidation of NH_3 to NO_3^-, is a major process in well-drained oxic soils at neutral pH, and is carried out by the nitrifying bacteria (Figure 24.7). Whereas denitrification *consumes* NO_3^-, nitrification *produces* NO_3^-. If materials high in NH_3, such as manure or sewage, are added to soils, the rate of nitrification increases. Nitrification is a two-step aerobic process in which some species oxidize NH_3 to NO_2^- and then other species oxidize NO_2^- to NO_3^-. Many species of *Bacteria* and at least one species of *Archaea* are nitrifiers (↺ Sections 13.10, 17.3, 18.21, 19.11).

Although NO_3^- is readily assimilated by plants, it is very soluble, and therefore rapidly leached or denitrified from waterlogged soils. Consequently, nitrification is not beneficial for plant agriculture. Ammonium, on the other hand, is positively charged and strongly adsorbed to negatively charged soils. Anhydrous NH_3 is therefore used extensively as an agricultural fertilizer, but to prevent its conversion to NO_3^-, chemicals are added to the NH_3 to inhibit nitrification. One common inhibitor is a pyridine compound called *nitrapyrin* (2-chloro-6-trichloromethylpyridine). Nitrapyrin specifically inhibits the *first* step in nitrification, the oxidation of NH_3 to NO_2^-. However, this effectively inhibits both steps in nitrification because the second step, $NO_2^- \rightarrow NO_3^-$, depends on the first (⮌ Section 13.10). The addition of nitrapyrin to anhydrous NH_3 has greatly increased the efficiency of crop fertilization and has helped prevent pollution of waterways by NO_3^- leached from nitrified soils.

Ammonia can be oxidized under anoxic conditions by the bacterium *Brocadia* in the process called *anammox*. In this reaction, NH_3 is oxidized anaerobically with NO_2^- as the electron acceptor, forming N_2 as the final product (Figure 24.7), which is released to the atmosphere. Although a major process in sewage and marine sediments, anammox is not significant in well-drained (oxic) soils. The microbiology and biochemistry of anammox was discussed in Section 13.11.

MiniQuiz

- What is nitrogen fixation and why is it important to the nitrogen cycle?
- How do the processes of nitrification and denitrification differ? How do nitrification and anammox differ?
- How does the compound nitrapyrin benefit both agriculture and the environment?

24.4 The Sulfur Cycle

Microbial transformations of sulfur (S) are even more complex than those of nitrogen because of the large number of oxidation states of S and the fact that several transformations of S also occur abiotically. Sulfate reduction and chemolithotrophic sulfur oxidation were covered in Sections 14.8 and 13.8, respectively. The redox cycle for microbial S transformations is shown in **Figure 24.8**.

Although a number of oxidation states of S are possible, only three are significant in nature, -2 (sulfhydryl, R–SH, and sulfide, HS^-), 0 (elemental sulfur, S^0), and $+6$ (sulfate, SO_4^{2-}). The bulk of

Earth's S is in sediments and rocks in the form of sulfate minerals, primarily gypsum ($CaSO_4$) and sulfide minerals (pyrite, FeS_2), but the oceans constitute the most significant reservoir of SO_4^{2-} in the biosphere. A significant amount of S, in particular sulfur dioxide (SO_2, a gas), enters the S cycle from human activities, primarily the burning of fossil fuels.

Hydrogen Sulfide and Sulfate Reduction

A major volatile S gas is hydrogen sulfide (H_2S). Hydrogen sulfide is produced from bacterial sulfate reduction ($SO_4^{2-} + 8 H^+ \rightarrow H_2S + 2 H_2O + 2 OH^-$) (Figure 24.8) or is emitted from sulfide springs and volcanoes. Although H_2S is volatile, different forms exist depending on pH: H_2S predominates below pH 7 and the nonvolatile HS^- and S^{2-} predominate above pH 7. Collectively, H_2S, HS^-, and S^{2-} are referred to as "sulfide."

Sulfate-reducing bacteria are a large and highly diverse group (⮌ Section 17.18) and are widespread in nature. However, in anoxic habitats such as freshwater sediments and many soils, sulfate reduction is SO_4^{2-}-limited. Moreover, because organic electron donors (or H_2, which is a product of the fermentation of organic compounds) are needed to support sulfate reduction, it only occurs where significant amounts of organic material are present.

In marine sediments, the rate of SO_4^{2-} reduction is typically C limited and can be greatly increased by an influx of organic matter. This is important because the disposal of sewage or garbage in the oceans or coastal regions can trigger sulfate reduction. Hydrogen sulfide is toxic to many plants and animals and therefore its formation is potentially detrimental (sulfide is toxic because it combines with the iron of cytochromes and blocks respiration). Sulfide is commonly detoxified in nature by combination with iron, forming the insoluble minerals FeS (pyrrhotite) and FeS_2 (pyrite). The black color of sulfidic sediments or sulfate-reducing bacterial cultures is due to these metal sulfide minerals (⮌ Figure 17.48).

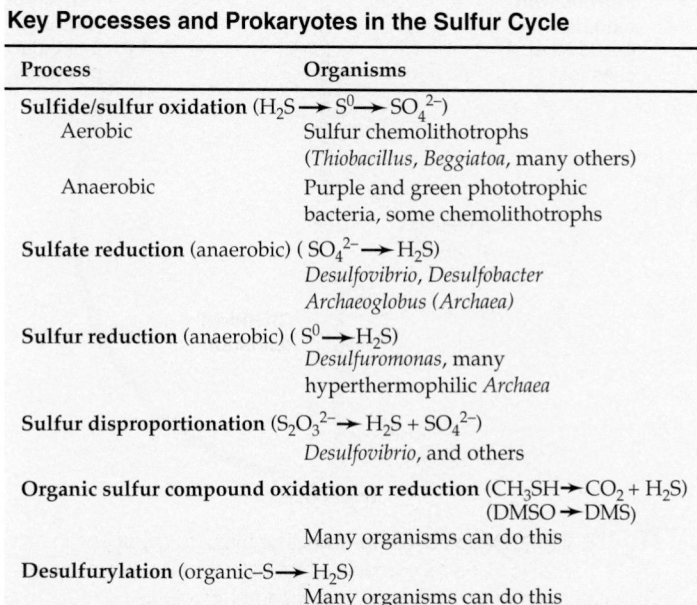

Key Processes and Prokaryotes in the Sulfur Cycle

Process	Organisms
Sulfide/sulfur oxidation ($H_2S \rightarrow S^0 \rightarrow SO_4^{2-}$)	
Aerobic	Sulfur chemolithotrophs (*Thiobacillus, Beggiatoa*, many others)
Anaerobic	Purple and green phototrophic bacteria, some chemolithotrophs
Sulfate reduction (anaerobic) ($SO_4^{2-} \rightarrow H_2S$)	*Desulfovibrio, Desulfobacter Archaeoglobus* (Archaea)
Sulfur reduction (anaerobic) ($S^0 \rightarrow H_2S$)	*Desulfuromonas*, many hyperthermophilic *Archaea*
Sulfur disproportionation ($S_2O_3^{2-} \rightarrow H_2S + SO_4^{2-}$)	*Desulfovibrio*, and others
Organic sulfur compound oxidation or reduction ($CH_3SH \rightarrow CO_2 + H_2S$) ($DMSO \rightarrow DMS$)	Many organisms can do this
Desulfurylation (organic–$S \rightarrow H_2S$)	Many organisms can do this

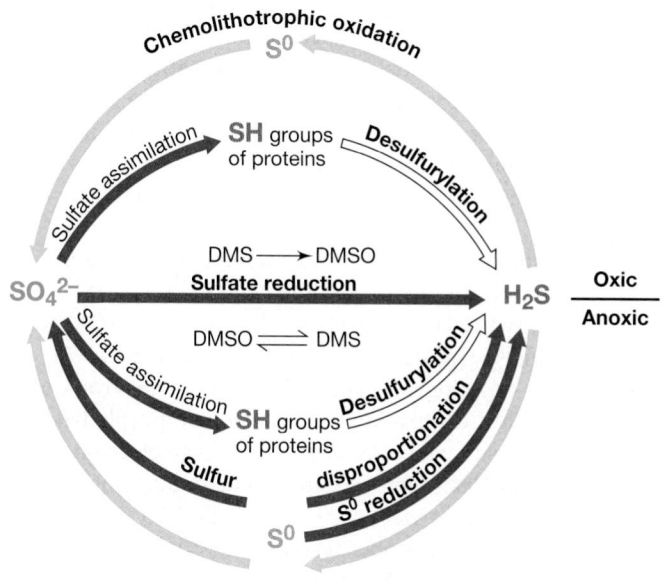

Figure 24.8 Redox cycle for sulfur. Oxidations are indicated by yellow arrows and reductions by red arrows. Reactions without redox changes are in white. DMS, dimethyl sulfide; DMSO, dimethyl sulfoxide.

Sulfide and Elemental Sulfur Oxidation–Reduction

Under oxic conditions, sulfide rapidly oxidizes spontaneously at neutral pH. Sulfur-oxidizing chemolithotrophic bacteria, most of which are aerobes (Sections 13.8 and 17.4), can catalyze the oxidation of sulfide. However, because of the rather rapid spontaneous reaction, microbial sulfide oxidation is significant only in areas where H_2S emerging from anoxic environments meets air. Where light is available, there can be anoxic oxidation of sulfide, catalyzed by the phototrophic purple and green sulfur bacteria (Sections 17.2 and 18.15).

Elemental sulfur is chemically stable but is readily oxidized by sulfur-oxidizing chemolithotrophic bacteria such as *Thiobacillus* and *Acidithiobacillus*. Because S^0 is insoluble, the bacteria that oxidize it must attach to the S^0 crystals to obtain their substrate (Figure 13.21*b*). The oxidation of S^0 forms sulfuric acid (H_2SO_4), and thus S^0 oxidation characteristically lowers the pH in the environment, sometimes drastically (Section 24.5). For this reason, S^0 is sometimes added to alkaline soils as an inexpensive and natural way to lower the pH, relying on the ubiquitous sulfur chemolithotrophs to carry out the acidification process.

Elemental sulfur can be reduced as well as oxidized. The reduction of S^0 to sulfide (a form of anaerobic respiration) is a major ecological process, especially among hyperthermophilic *Archaea* (Chapter 19). Although sulfate-reducing bacteria can also reduce S^0, in sulfidic habitats most S^0 is reduced by the physiologically specialized sulfur reducers, organisms that are incapable of SO_4^{2-} reduction (Section 17.18). The habitats of the sulfur reducers are generally those of the sulfate reducers, so from an ecological standpoint, the two groups form a metabolic guild unified by their formation of H_2S.

Organic Sulfur Compounds

In addition to *inorganic* forms of S, several *organic* S compounds are also cycled in nature. Many of these foul-smelling compounds are highly volatile and can thus enter the atmosphere. The most abundant organic S compound in nature is *dimethyl sulfide* (CH_3—S—CH_3); it is produced primarily in marine environments as a degradation product of dimethylsulfoniopropionate, a major osmoregulatory solute in marine algae (Section 5.16). This compound can be used as a carbon source and electron donor by microorganisms and is catabolized to dimethyl sulfide and acrylate. The latter, a derivative of the fatty acid propionate, is used to support growth.

Dimethyl sulfide released to the atmosphere undergoes photochemical oxidation to methanesulfonate (CH_3SO_3), SO_2, and SO_4^{2-}. By contrast, CH_3—S—CH_3 produced in anoxic habitats can be microbially transformed in at least three ways: (1) by methanogenesis (yielding CH_4 and H_2S), (2) as an electron donor for photosynthetic CO_2 fixation in phototrophic purple bacteria (yielding dimethyl sulfoxide, DMSO), and (3) as an electron donor in energy metabolism in certain chemoorganotrophs and chemolithotrophs (also yielding DMSO). DMSO can be an electron acceptor for anaerobic respiration (Section 14.12), producing CH_3—S—CH_3. Many other organic S compounds affect the global sulfur cycle, including methanethiol (CH_3SH),

dimethyl disulfide (H_3C—S—S—CH_3), and carbon disulfide (CS_2), but on a global basis, CH_3—S—CH_3 is the most significant.

MiniQuiz

- Is H_2S a substrate or a product of the sulfate-reducing bacteria? Of the chemolithotrophic sulfur bacteria?
- Why does the bacterial oxidation of sulfur result in a pH drop?
- What organic sulfur compound is most abundant in nature?

24.5 The Iron Cycle

Iron (Fe) is one of the most abundant elements in Earth's crust. On the surface of Earth, Fe exists naturally in two oxidation states, ferrous (Fe^{2+}) and ferric (Fe^{3+}). A third oxidation state, Fe^0, is abundant in Earth's core and is also a major product of human activities from the smelting of iron ores to form cast iron. In nature, Fe cycles primarily between the Fe^{2+} and Fe^{3+} forms. The redox reactions in the Fe cycle include both oxidations and reductions. Ferric iron is reduced both chemically and as a form of anaerobic respiration, and Fe^{2+} is oxidized both chemically and as a form of chemolithotrophic metabolism (**Figure 24.9**).

Bacterial Iron Reduction

Some *Bacteria* and *Archaea* can use Fe^{3+} as an electron acceptor in anaerobic respiration (Section 14.12). Ferric iron reduction is common in waterlogged soils, bogs, and anoxic lake sediments. Movement of groundwater from anoxic bogs or waterlogged soils may also move large amounts of Fe^{2+}. When this Fe^{2+}-laden water reaches oxic regions, the Fe^{2+} is oxidized

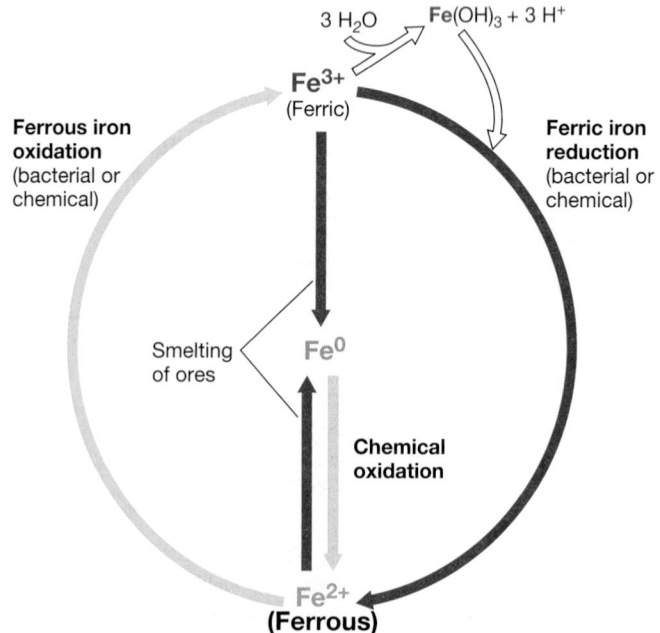

Figure 24.9 Redox cycle for iron. The major forms of iron in nature are Fe^{2+} and Fe^{3+}; Fe^0 is primarily a product of smelting of iron ores. Oxidations are shown by yellow arrows and reductions by red arrows. Fe^{3+} forms various minerals such as ferric hydroxide, $Fe(OH)_3$.

Microbially Wired

Regardless of the electron acceptor they use, when bacteria respire, they generate electricity. They do this when they oxidize an organic or inorganic electron donor and separate electrons from protons during electron transport reactions that generate the proton motive force.

In any form of respiration, electron disposal is necessary for energy conservation. When the electron acceptor is oxygen (O_2), nitrate (NO_3^-), or many of the other soluble substances used by bacteria as electron acceptors (👁 Section 14.6), the final product diffuses away from the cell. Many bacteria reduce ferric iron (Fe^{3+}) as an electron acceptor under anoxic conditions, including the bacterium *Geobacter sulfurreducens* (Figure 1). However, in contrast to soluble electron acceptors, Fe^{3+} is typically present in nature as an insoluble mineral, such as an iron oxide, and thus the reduction of Fe^{3+} occurs outside the cell where the mineral binds to outer cell surface structures. Under such conditions, the ferric iron functions as an electrical anode, and the bacterial cell facilitates transfer of electrons from the electron donor to the anode.[1] Research has shown that the iron oxide coatings on the *Geobacter* cell surface function as electrical "nanowires," much as copper wire does in a household electrical circuit. Being conductive structures, nanowires can transfer electrons to other electron acceptors or to nanowires on adjacent bacterial cells. In this way, electrons obtained by *Geobacter* from the oxidation of organic compounds or from hydrogen (H_2) can be shuttled from cell to cell within a microenvironment and, by this process, travel from one region of the habitat to another. Humic substances and other metals such as manganese, both of which can be electron donors or electron acceptors under the appropriate conditions (👁 Section 14.12), can facilitate these transfers by functioning as electron shuttles.

Surprisingly, electron shuttling by bacterial nanowires can occur over rather large spatial distances. In studies of hydrogen sulfide (H_2S) oxidation in anoxic marine sediments (sulfide is the product of sulfate-reducing bacteria), the oxidation of H_2S deep in the sediments released electrons that were shown to reduce O_2 at the sediment water interface some 20 cm away.[2]. Keeping in mind the typically small size of bacterial cells, electrons from H_2S must therefore have traveled through the nanowires of a huge network of bacterial cells to eventually reach the oxic zone.

In nature, electrical communication between bacterial cells may be a major way by which electrons generated from microbial metabolism in anoxic habitats are shuttled to oxic regions. Moreover, research on the microbiology of the process indicates that microbial electricity could be harnessed in the form of microbial "fuel cells" that could oxidize toxic and waste carbon compounds in anoxic environments, with the resulting electrons coupled to power generation. In such a scheme, bacteria would be exploited by functioning as the catalysts for diverting electrons from electron donors directly to artificial anodes, with the resulting electric current being siphoned off to supply a portion of human power needs.

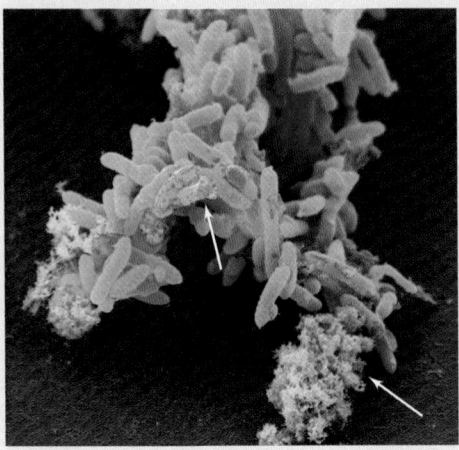

Figure 1 Cells of *Geobacter* attached to ferric iron precipitates (arrows) reduce Fe^{3+} to Fe^{2+}.

[1]Lovley, D.R. 2006. Bug juice: Harvesting electricity with microorganisms. *Nat. Rev. Microbiol. 4:* 497–508.
[2]Nielsen, L.P., N. Risgaard-Petersen, H. Fossing, P.B. Christensen, and M. Sayama. 2010. Electric currents couple spatially separated biogeochemical processes in marine sediment. *Nature 463:* 1071–1074.

---

chemically or by the iron bacteria. Fe^{3+} compounds then precipitate, leading to the formation of iron oxides, such as ferric hydroxide:

$$Fe^{2+} + \tfrac{1}{4}O_2 + 2\tfrac{1}{2}H_2O \longrightarrow Fe(OH)_3 + 2H^+$$

The $Fe(OH)_3$ precipitate can interact spontaneously with humic substances (Section 24.1) to reduce Fe^{3+} back to Fe^{2+} (Figure 24.9). Ferric iron can form complexes with various organic constituents. In this way Fe^{3+} becomes solubilized and once again available to ferric-reducing bacteria as an electron acceptor. In recent years it has been recognized that ferric precipitates on the surfaces of cells of bacteria such as *Geobacter* function as nanowires to move electrons around microbial habitats. This movement of electrons is a form of electricity, and the process may eventually have commercial applications for power generation (see the Microbial Sidebar, "Microbially Wired").

Ferrous Iron and Pyrite Oxidation at Acidic pH

The only electron acceptor able to oxidize Fe^{2+} abiotically is O_2. If anoxic, Fe^{2+}-rich groundwaters are exposed to air, Fe^{2+} is oxidized at the interface of these two zones by iron-oxidizing bacteria such as *Gallionella* and *Leptothrix* (👁 Sections 17.15 and 17.16). The most extensive bacterial Fe^{2+} oxidation, however, occurs at *acidic* pH, at which Fe^{2+} is not oxidized spontaneously. In extremely

Ricardo Amils

Figure 24.10 Oxidation of ferrous iron (Fe²⁺). A microbial mat growing in the Rio Tinto, Spain. The mat consists of acidophilic green algae (eukaryotes) and various iron-oxidizing chemolithotrophic prokaryotes. The Rio Tinto has a pH of about 2 and contains high levels of dissolved metals, in particular Fe^{2+}. The red-brown precipitates contain $Fe(OH)_3$ and other ferric minerals.

acidic, iron-rich habitats, the acidophilic chemolithotrophs *Acidithiobacillus ferrooxidans* and *Leptospirillum ferrooxidans* oxidize Fe^{2+} to Fe^{3+} (**Figure 24.10**). Very little energy is generated in the oxidation of Fe^{2+} to Fe^{3+} (⟳ Section 13.9) and so these bacteria must oxidize large amounts of Fe^{2+} in order to grow; consequently, even a relatively small population of cells can precipitate a large amount of iron minerals.

One of the most common forms of iron in nature is **pyrite** (FeS_2), which is often present in bituminous coals and in metal ores (**Figure 24.11**). Bacterial oxidation of FeS_2 contributes to the microbial leaching of ores described in Section 24.7. During coal-mining operations, acidic conditions develop as bacteria oxidize the FeS_2. The oxidation of FeS_2 is a combination of chemically and bacterially catalyzed reactions, and two electron acceptors participate in the process: O_2 and Fe^{3+}. When FeS_2 is first exposed in a coal-mining operation (Figure 24.11b), a slow chemical reaction with O_2 begins (Figure 24.11c). This reaction, called the *initiator reaction,* leads to the oxidation of HS^- to SO_4^{2-} and the development of acidic conditions as Fe^{2+} is released. *A. ferrooxidans* and *L. ferrooxidans* then oxidize Fe^{2+} to Fe^{3+}, and the Fe^{3+} formed under these acidic conditions, being soluble, reacts spontaneously with more FeS_2 and oxidizes the HS^- to sulfuric acid (H_2SO_4), which immediately dissociates into SO_4^{2-} and H^+:

$$FeS_2 + 14\ Fe^{3+} + 8\ H_2O \longrightarrow 15\ Fe^{2+} + 2\ SO_4^{2-} + 16\ H^+$$

Again, the bacteria oxidize Fe^{2+} to Fe^{3+}, and this Fe^{3+} reacts with more FeS_2. Thus, there is a progressive, rapidly increasing rate at which FeS_2 is oxidized, called the *propagation cycle* (Figure 24.11c). Under natural conditions some of the Fe^{2+} generated by the bacteria leaches away and is subsequently carried by anoxic groundwater into surrounding streams. However, bacterial oxidation of Fe^{2+} then takes place in the aerated streams, and, because O_2 is present, the insoluble $Fe(OH)_3$ is formed.

Ravin Donald

T. D. Brock

(a)   (b)

Propagation cycle

Initiator reaction

$$FeS_2\ +\ 3\tfrac{1}{2}\ O_2\ +\ H_2O \longrightarrow Fe^{2+}\ +\ 2\ SO_4^{2-}\ +\ H^+$$

Acidification

Spontaneous (Fe^{3+} is oxidant for propagation cycle) Fe^{3+} Bacteria or spontaneous

(c)

Figure 24.11 Coal and pyrite. *(a)* Coal from the Black Mesa formation in northern Arizona (USA); the gold-colored spherical discs (about 1 mm in diameter) are particles of pyrite (FeS_2). *(b)* A coal seam in a surface coal-mining operation. Exposing the coal to oxygen and moisture stimulates the activities of iron-oxidizing bacteria growing on the pyrite in the coal. *(c)* Reactions in pyrite degradation. The primarily abiotic initiator reaction sets the stage for the primarily bacterial oxidation of Fe^{2+} to Fe^{3+}. The Fe^{3+} attacks and oxidizes FeS_2 abiotically in the propagation cycle.

Acid Mine Drainage

Bacterial and spontaneous oxidation of sulfide minerals is the major cause of **acid mine drainage**, an environmental problem worldwide caused by surface coal-mining operations. As we have seen (Figure 24.11c), the breakdown of FeS_2 ultimately leads to the formation of H_2SO_4 and Fe^{2+}; in waters in which these products have formed, pH values can be lower than 1. Mixing of acidic mine waters into rivers (**Figure 24.12**) and lakes seriously degrades water quality because both the acid and the dissolved metals (iron, aluminum, and heavy metals such as cadmium and lead) are toxic to aquatic organisms. The O_2 requirement for the oxidation of Fe^{2+} to Fe^{3+} explains how acid mine drainage develops. As long as the coal is not mined, FeS_2 cannot be oxidized because O_2, water, and the bacteria cannot reach it. However, when a coal seam is exposed (Figure 24.11b), O_2 and water are introduced, making both spontaneous and bacterial oxidation of FeS_2 possible. The acid formed can then leach into surrounding aquatic systems (Figure 24.12).

Where acid mine drainage is extensive and Fe^{2+} levels high, a strongly acidophilic species of *Archaea, Ferroplasma,* is often present. This aerobic iron-oxidizing organism is capable of growth

Figure 24.12 Acid mine drainage from a surface coal-mining operation. The yellowish-red color is due to the precipitated iron oxides in the drainage (see Figure 24.11c for the reactions in acid mine drainage).

at pH 0 and at temperatures up to 50°C. Cells of *Ferroplasma* lack a cell wall and are phylogenetically related to *Thermoplasma*, also a cell-wall-lacking and strongly acidophilic (but chemoorganotrophic) member of the *Archaea* (⟳ Section 19.4).

MiniQuiz

- In what oxidation state is iron in the mineral $Fe(OH)_3$? In FeS? How is $Fe(OH)_3$ formed?
- Why does biological Fe^{2+} oxidation under oxic conditions occur mainly at acidic pH?

24.6 The Phosphorus, Calcium, and Silica Cycles

Many other chemical elements undergo microbial cycling and we focus on three key ones here—phosphorus (P), calcium (Ca), and silica (Si). The cycling of these elements is important in aquatic environments, particularly in the oceans, which are major reservoirs of Ca and Si. In ocean waters, huge amounts of Ca and Si are incorporated into the exoskeletons of certain microorganisms. Unlike the C, N, and S cycles, in the P, Ca, and Si cycles there are no redox changes or gaseous forms that can escape and alter Earth's atmospheric chemistry. Nevertheless, as we will see, keeping these cycles in balance, especially that of Ca, is important for maintaining sustainable planet life.

Phosphorus

Phosphorus is found in nature in the form of organic and inorganic phosphates. Its reservoirs are phosphate-containing minerals in rocks, dissolved phosphates in freshwaters and marine waters, and the nucleic acids and phospholipids of living organisms. Although P has multiple oxidation states, most environmental phosphates are at the +5 oxidation state (for example, inorganic phosphate, HPO_4^-). Phosphorus cycles through living organisms (cellular P), waters and soils (free organic P), and the Earth's crust (inorganic

P). P is typically the limiting nutrient for photosynthesis in freshwaters, which receive it from the weathering of rocks. In marine systems, a fraction of dissolved P is organic, in the form of phosphate esters and *phosphonates*. Phosphonates are organophosphate compounds that contain a P—C bond. Phosphonates are produced by certain microorganisms and comprise about a quarter of the organic P pool in nature; however, for many organisms phosphonates are a less available source of P than is phosphate because of the enzymes required to degrade phosphonates. Organisms lacking these enzymes can be P-limited even when sufficient P is present as phosphonates. Moreover, the degradation of methylphosphonate (CH_5O_3P) by some marine microorganisms, a process that liberates methane (CH_4), may explain the previously inexplicable observation of high CH_4 concentrations in highly oxygenated surface waters of the ocean (by contrast, methanogenic *Archaea* are strict anaerobes; ⟳ Section 19.3).

Calcium

The major global reservoirs of Ca are calcareous rocks and the oceans. In the oceans, where dissolved Ca exists as Ca^{2+}, Ca^{2+} cycling is a highly dynamic process, although the concentration of Ca^{2+} in seawater remains constant at about 10 mM. Several

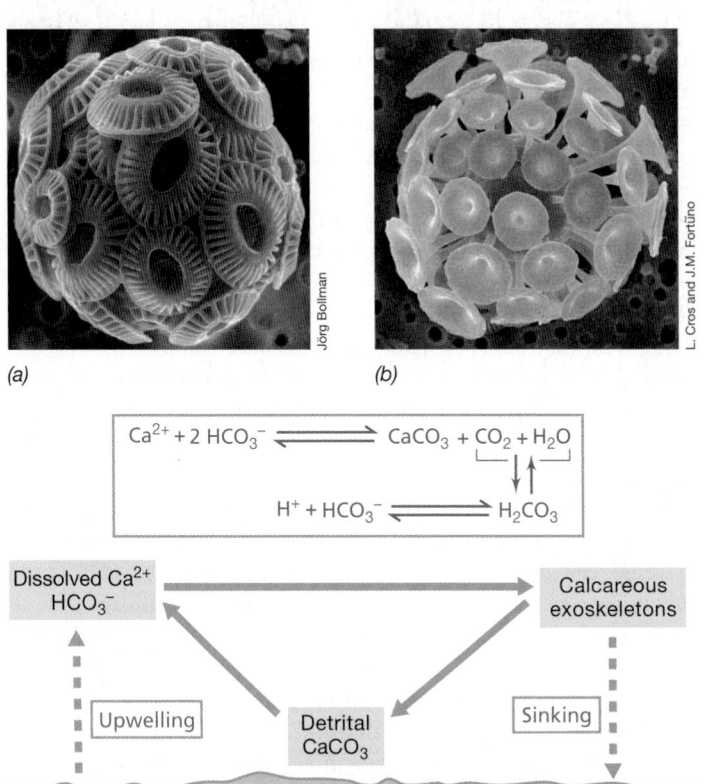

Figure 24.13 The marine calcium (Ca) cycle. Scanning electron micrographs of cells of the calcareous phytoplankton *(a) Emiliania huxleyi* and *(b) Discophaera tubifera*. The exoskeletons of these coccolithophores are made of calcium carbonate $(CaCO_3)$. A cell of *Emiliana* is about 8 μm wide and a cell of *Discophaera* is about 12 μm wide. *(c)* The marine calcium cycle; dynamic pools of Ca^{2+} are shaded in green. Detrital $CaCO_3$ is that in fecal pellets and other organic matter from dead organisms. Note how H_2CO_3 formation decreases ocean pH.

marine eukaryotic phototrophic microorganisms take up Ca^{2+} to form their calcareous exoskeletons; these include the coccolithophores and foraminifera (**Figure 24.13**; ↩ Section 20.11). The calcium-cycling activities of these planktonic phototrophs are also tightly coupled with inorganic components of the carbon cycle.

The precipitation of calcium carbonate ($CaCO_3$) to form the shells of calcareous phytoplankton controls both CO_2 flux into ocean surface water and inorganic carbon transport into deep ocean water and into the sediments. The formation of $CaCO_3$ both depletes surface dissolved bicarbonate (HCO_3^-) and increases the level of dissolved CO_2 (Figure 24.13c); the latter reduces the influx of atmospheric CO_2 into the surface ocean and this helps maintain the slightly alkaline pH of ocean waters. When these calcareous organisms die and sink toward the sediments, inorganic and organic C and Ca^{2+} are transported to the deep ocean from which they are slowly released over long periods.

The formation of $CaCO_3$ exoskeletons brings into play a delicate balance between Ca^{2+} and C and is sensitive to changes in CO_2 levels in the atmosphere. This is because increased levels of atmospheric CO_2 increase the formation of carbonic acid (H_2CO_3), and as this dissociates to form HCO_3^- and H^+, $CaCO_3$ dissolves and seawater pH decreases (Figure 24.13c). The more acidic oceans that will result from rising atmospheric CO_2 are expected to reduce the rate of formation of calcareous shells, which will likely have effects on other microbial nutrient cycles and plant and animal communities. For example, within a few decades it is predicted that the tropical oceans may be too acidic to sustain the growth of coral reefs, a major component of the marine biosphere (↩ Section 25.14), and that large parts of the polar oceans may become too corrosive for coccolithophores, important organisms in marine food webs. It is not clear at present what immediate effects these changes will have on Earth's biosphere in general. However, because nutrient cycles are closely coupled (Section 24.1), it can safely be predicted that any significant change in the C cycle will trigger changes in other nutrient cycles, some of which could have negative consequences for higher organisms.

Silica

The marine Si cycle is controlled primarily by unicellular eukaryotes (diatoms, silicoflagellates, and radiolarians) that build ornate external cell skeletons called *frustules* (**Figure 24.14a**). These structures are not of $CaCO_3$ as in the coccolithophores, but of opal (SiO_2), whose formation begins with the uptake by the cell of dissolved silicic acid (Figure 24.14b). Diatoms (↩ Section 20.10) are rapidly growing phototrophic eukaryotes and often dominate blooms of phytoplankton in coastal and open ocean waters. However, unlike other major phytoplankton groups, diatoms require Si and can become silica-limited when blooms develop. Also, because of their large size, diatom cells tend to sink faster than other organic particles; in this way, they contribute significantly to the return of Si and C to deeper ocean waters. The transport of organic material produced through primary production in near-surface waters to deeper ocean waters, primarily by sinking particles, is called the *biological pump* and is an important aspect of the carbon cycle in carbon burial and mineralization in marine environments (Figure 24.1).

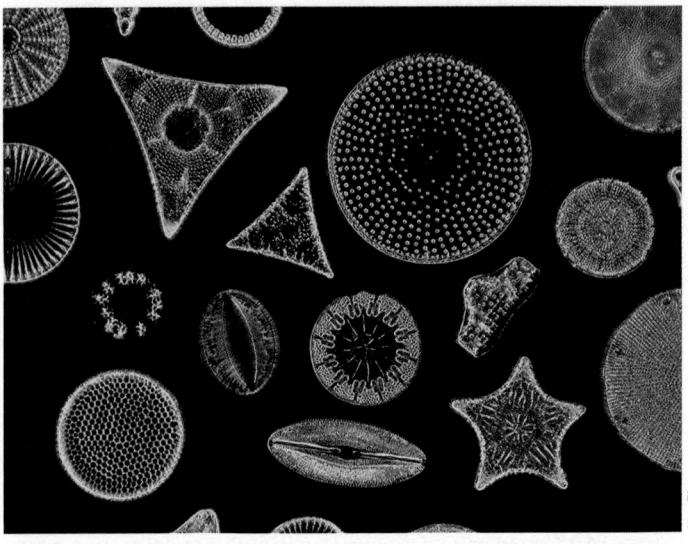

(a)

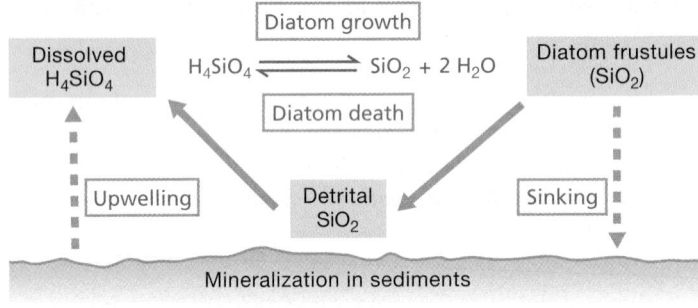

(b)

Figure 24.14 **The marine silica cycle.** (a) Dark-field photomicrograph of a collection of diatom shells (frustules). The frustules are made of SiO_2. (b) The marine silica cycle dynamic pools of Si are shaded in green.

In addition to the major nutrient requirements of any phototrophic organism (CO_2, N, P, Fe), diatoms require sufficient dissolved Si, and in nature this is primarily Si released from the skeletons of dead diatoms (Figure 24.14b). Although Si is released relatively rapidly following cell death, during periods of high diatom production in relatively shallow waters a significant fraction of dissolved Si can be buried in sediments and remain there for millions of years. This has consequences for continued diatom growth and their phototrophic consumption of dissolved CO_2 from ocean waters. The flux of CO_2 into and out of ocean water affects its pH (Figure 24.13c), and through this link, the Si and C cycles are coupled in a similar way as for the Ca and C cycles.

MiniQuiz

- How does the formation of $CaCO_3$ skeletons by calcareous phytoplankton retard CO_2 uptake and help maintain ocean water pH?
- How might Si depletion in the photic zone influence the biological pump?

Ⅱ Biodegradation and Bioremediation

The biogeochemical capacities of microorganisms seem almost limitless, and it is often said that microorganisms are "Earth's greatest chemists." The activities of these great little chemists have been exploited in many ways. Here we consider how microbial activities help extract valuable metals from low-grade ores and clean up environmental pollution.

24.7 Microbial Leaching

The acid production and dissolution of pyrite (FeS_2) by acidophilic bacteria discussed in Section 24.5 can be put to use in the mining of metal ores. Sulfide (HS^-) forms insoluble minerals with many metals, and many ores mined as sources of these metals are sulfide ores. If the concentration of metal in the ore is low, it may be economically feasible to mine the ore only if the metals of interest are first concentrated by **microbial leaching**. Leaching is especially useful for copper ores because copper sulfate ($CuSO_4$), formed during the oxidation of copper sulfide ores, is very water-soluble. Indeed, approximately a quarter of all copper mined worldwide is obtained by microbial leaching.

The Leaching Process

We have seen how *Acidithiobacillus ferrooxidans* and other metal-oxidizing chemolithotrophic bacteria can catalyze the oxidation of sulfide minerals, thus aiding in solubilization of the metal (Figure 24.11). The susceptibility to oxidation varies among minerals, and those minerals that are most readily oxidized are most amenable to microbial leaching. Thus, iron and copper sulfide ores such as pyrrhotite (FeS) and covellite (CuS) are readily leached, whereas lead and molybdenum ores are much less so. In microbial leaching, low-grade ore is dumped in a large pile called the *leach dump* and a dilute sulfuric acid solution at pH 2 is percolated down through the pile (**Figure 24.15**). The liquid emerging from the bottom of the pile (Figure 24.15b) is rich in dissolved metals and is transported to a precipitation plant (Figure 24.15c) where the desired metal is precipitated and purified (Figure 24.15d). The liquid is then pumped back to the top of the pile and the cycle repeated. As needed, acid is added to maintain an acidic pH.

We illustrate microbial leaching of copper with the common copper ore CuS, in which copper exists as Cu^{2+}. *A. ferrooxidans* oxidizes the sulfide in CuS to SO_4^{2-}, releasing Cu^{2+} as shown in **Figure 24.16**. However, this reaction can also occur spontaneously. Indeed, the key reaction in copper leaching is actually not the bacterial oxidation of sulfide in CuS but the spontaneous oxidation of sulfide by ferric iron (Fe^{3+}) generated from the bacterial oxidation of ferrous iron (Fe^{2+}) (Figure 24.16). In any copper ore, FeS_2 is also present, and its oxidation by bacteria leads to the formation of Fe^{3+} (Figures 24.11c and 24.16). The spontaneous reaction of CuS with Fe^{3+} proceeds in the absence of O_2 and forms Cu^{2+} plus Fe^{2+}; importantly for efficiency of the leaching process, this reaction can take place deep in the leach dump where conditions are anoxic.

(a)

(b)

(c)

(d)

Figure 24.15 The leaching of low-grade copper ores using iron-oxidizing bacteria. *(a)* A typical leaching dump. The low-grade ore has been crushed and dumped in such a way that the surface area exposed is as high as possible. Pipes distribute the acidic leach water over the surface of the pile. The acidic water slowly percolates through the pile and exits at the bottom. *(b)* Effluent from a copper leaching dump. The acidic water is very rich in Cu^{2+}. *(c)* Recovery of copper as metallic copper (Cu^0) by passage of the Cu^{2+}-rich water over metallic iron in a long flume. *(d)* A small pile of metallic copper removed from the flume, ready for further purification.

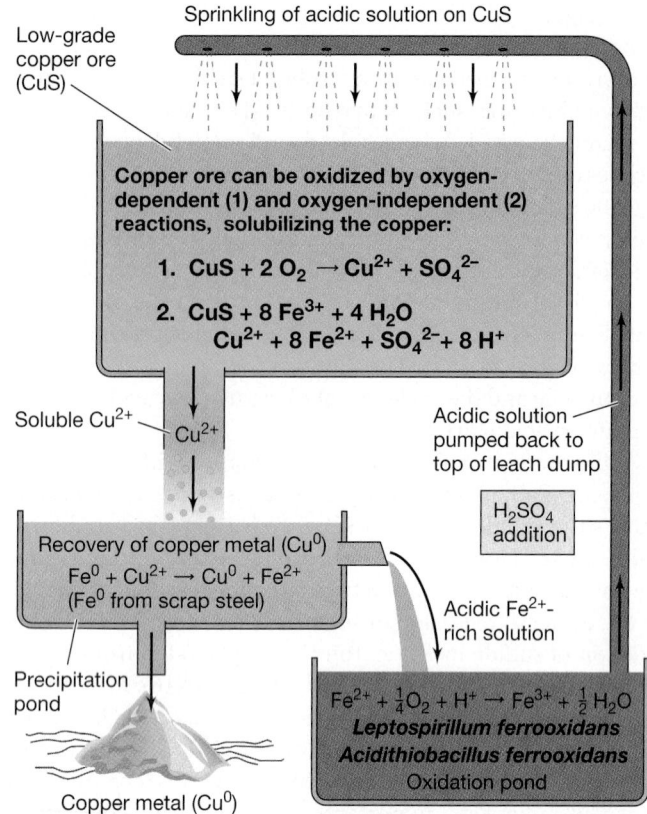

Figure 24.16 Arrangement of a leaching pile and reactions in the microbial leaching of copper sulfide minerals to yield metallic copper. Reaction 1 occurs both biologically and chemically. Reaction 2 is strictly chemical and is the most important reaction in copper-leaching processes. For reaction 2 to proceed, it is essential that the Fe^{2+} produced from the oxidation of sulfide in CuS to sulfate be oxidized back to Fe^{3+} by iron chemolithotrophs (see chemistry in the oxidation pond).

Metal Recovery

The precipitation plant is where the Cu^{2+} from the leaching solution is recovered (Figure 24.15c, d). Shredded scrap iron (a source of Fe^0) is added to the precipitation pond to recover copper from the leach liquid by the chemical reaction shown in the lower part of Figure 24.16. This results in a Fe^{2+}-rich liquid that is pumped to a shallow oxidation pond where iron-oxidizing chemolithotrophs oxidize the Fe^{2+} to Fe^{3+}. This now ferric iron–rich acidic liquid is pumped to the top of the pile and the Fe^{3+} is used to oxidize more CuS (Figure 24.16). The entire CuS leaching operation is thus driven by the oxidation of Fe^{2+} to Fe^{3+} by iron-oxidizing bacteria.

Temperatures rise in a leaching dump and this leads to shifts in the iron-oxidizing microbial populations. *A. ferrooxidans* is a mesophile, and when heat generated by microbial activities raises temperatures above about 30°C inside a leach dump, this bacterium is outcompeted by mildly thermophilic iron-oxidizing chemolithotrophs such as *Leptospirillum ferrooxidans* and *Sulfobacillus*. At even higher temperatures (60–80°C), hyperthermophilic *Archaea* such as *Sulfolobus* (🔗 Section 19.9) predominate in the leach dump.

Other Microbial Leaching Processes: Uranium and Gold

Bacteria are also used in the leaching of uranium (U) and gold (Au) ores. In uranium leaching, *A. ferrooxidans* oxidizes U^{4+} to U^{6+} with O_2 as an electron acceptor. However, U leaching depends more on the abiotic oxidation of U^{4+} by Fe^{3+} with *A. ferrooxidans* contributing to the process mainly through the reoxidation of Fe^{2+} to Fe^{3+}, as in copper leaching (Figure 24.16). The reaction observed is as follows:

$$UO_2 + Fe_2(SO_4)_3 \longrightarrow UO_2SO_4 + 2FeSO_4$$
$$(U^{4+}) \quad (Fe^{3+}) \qquad\qquad (U^{6+}) \qquad (Fe^{2+})$$

Unlike UO_2, the uranyl sulfate (UO_2SO_4) formed is highly soluble and is concentrated by other processes.

Gold is typically present in nature in deposits associated with minerals containing arsenic (As) and FeS_2. *A. ferrooxidans* and related bacteria can leach the arsenopyrite minerals, releasing the trapped Au:

$$2\,FeAsS[Au] + 7\,O_2 + 2\,H_2O + H_2SO_4 \longrightarrow$$
$$Fe_2(SO_4)_3 + 2\,H_3AsO_4 + [Au]$$

The Au is then complexed with cyanide (CN^-) by traditional gold-mining methods. Unlike copper leaching, which is done in a huge dump (Figure 24.15a), gold leaching is done in small bioreactor tanks (**Figure 24.17**), where more than 95% of the trapped Au can be released. Moreover, the potentially toxic As and CN^- residues from the mining process are removed in the gold-leaching bioreactor. Arsenic is removed as a ferric precipitate, and CN^- is removed by its bacterial oxidation to CO_2 plus urea in later stages of the Au recovery process. Small-scale microbial-bioreactor leaching has thus become popular as an alternative to the environmentally devastating gold-mining techniques that leave a toxic trail of As and CN^- at the extraction site. Pilot processes are also being developed for bioreactor leaching of zinc, lead, and nickel ores.

Bioremediation of Uranium-Contaminated Environments

Uranium contamination of groundwater at sites where uranium ores have been processed is a legacy from the nuclear weapons and power industries, and the movement of radioactive materials

Figure 24.17 Gold bioleaching. Gold leaching tanks in Ghana (Africa). Within the tanks, a mixture of *Acidithiobacillus ferrooxidans*, *Acidithiobacillus thiooxidans*, and *Leptospirillum ferrooxidans* solubilizes the pyrite/arsenic mineral containing trapped gold, which releases the gold.

offsite via groundwater is a threat to environmental and human health. Because the contamination is often widespread, making mechanical methods of recovery very expensive, there is great interest in the development of biological treatments that would exploit the ability of some bacteria to reduce U^{6+} to U^{4+}, a form of *bioremediation* (Section 24.9). Uranium as U^{4+} forms an immobile uranium mineral, *uraninite*, thus limiting the movement of U into groundwater and potential contact with humans and other animals.

Bacteria, including metal-reducing *Shewanella* and *Geobacter* species and sulfate-reducing *Desulfovibrio* species, couple the oxidation of organic matter and H_2 to the reduction of U^{6+} to U^{4+}. Field studies in which organic compounds have been injected into uranium-contaminated aquifers to stimulate U^{6+} reduction have shown that this approach can lower U levels to below the U.S. Environmental Protection Agency's drinking water standard of 0.126 μM. However, even though uraninite is stable under reducing conditions, if conditions become oxic, it reoxidizes. Thus, much ongoing uranium bioremediation research is focused on the questions of whether microbially reduced uranium is stable if the composition of the microbial community changes or if oxidants, such as O_2, NO_3^-, and Fe^{3+}, are introduced via groundwater.

MiniQuiz

- What is required to oxidize CuS under anoxic conditions?
- Which reaction, oxidation or reduction, is key to uranium leaching? Uranium bioremediation?

24.8 Mercury Transformations

Metals are typically present in rocks, soils, waters, and the atmosphere; however, some of these metals are toxic, including mercury (Hg), lead (Pb), arsenic (As), cadmium (Cd), and selenium (Se). Because of environmental concern and significant microbial involvement, we focus our discussion here on Hg. Mercury is not a biological nutrient but microbial transformations of various Hg species help to detoxify some of its most toxic forms.

Global Cycling of Mercury and Methylmercury

Mercury is a widely used industrial product, especially in the electronics industry, an active ingredient of many pesticides, a pollutant from the chemical industry and from the combustion of fossil fuels and municipal wastes, and a common contaminant of aquatic ecosystems and wetlands. Because of its propensity to concentrate in living tissues, Hg is of considerable environmental importance. The major form of Hg in the atmosphere is elemental mercury (Hg^0), which is volatile and is oxidized to mercuric ion (Hg^{2+}) photochemically. Most mercury thus enters aquatic environments as Hg^{2+} (**Figure 24.18**).

Microbial Redox Cycle for Mercury

Mercuric ion readily adsorbs to particulate matter and can be metabolized from there by microorganisms. Microbial activity methylates Hg, yielding *methylmercury*, CH_3Hg^+ (Figure 24.18).

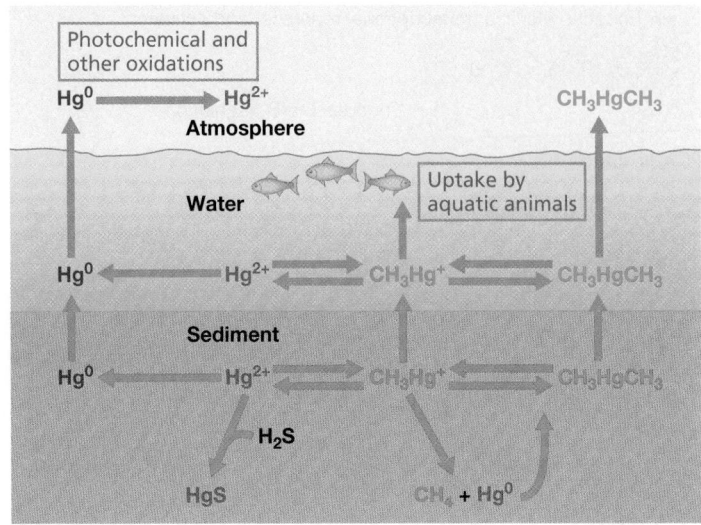

Figure 24.18 Biogeochemical cycling of mercury. The major reservoirs of mercury are water and sediments. Mercury in water can be concentrated in animal tissues; it can be precipitated as HgS from sediments. The forms of mercury commonly found in aquatic environments are each shown in a different color.

Methylmercury is extremely toxic to animals because it can be readily absorbed through the skin and is a potent neurotoxin. But in addition, CH_3Hg^+ is soluble and can be concentrated in the food chain, primarily in fish, or can be further methylated by microorganisms to yield the volatile compound *dimethylmercury* (CH_3—Hg—CH_3). Both CH_3Hg^+ and CH_3—Hg—CH_3 accumulate in animals, especially in muscle tissues. Methylmercury is about 100 times more toxic than Hg^0 or Hg^{2+}, and its accumulation in the aquatic food chain seems to be particularly acute in freshwater lakes and marine coastal waters where enhanced levels of CH_3Hg^+ have been detected in fish caught for human consumption. Mercuric compounds can cause liver and kidney damage in humans and other animals.

Several other microbial Hg transformations occur, including reactions catalyzed by sulfate-reducing bacteria ($H_2S + Hg^{2+} \rightarrow HgS$) and methanogens ($CH_3Hg^+ \rightarrow CH_4 + Hg^0$) (Figure 24.18). The solubility of mercuric sulfide (HgS) is very low, so in anoxic sulfate-reducing sediments, most Hg is present as HgS. But upon aeration, HgS can be oxidized to Hg^{2+} and SO_4^{2-} by metal-oxidizing bacteria, and the Hg^{2+} is eventually converted to CH_3Hg^+. Note, however, that it is not the Hg in HgS that is oxidized here, but instead the sulfide, probably by organisms related to *Acidithiobacillus*.

Mercury Resistance

At sufficiently high concentrations, Hg^{2+} and CH_3Hg^+ can be toxic to microorganisms as well as macroorganisms. However, several gram-positive and gram-negative bacteria convert toxic forms of Hg to nontoxic or less toxic forms. In mercury-resistant bacteria the enzyme *organomercury lyase* degrades the highly toxic CH_3Hg^+ to Hg^{2+} and methane (CH_4), and the NADPH (or NADH)-linked enzyme *mercuric reductase* reduces Hg^{2+} to Hg^0, which is volatile and thus mobile (**Figure 24.19**).

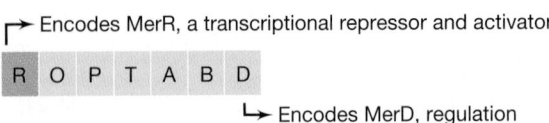

Encodes MerR, a transcriptional repressor and activator

R O P T A B D

Encodes MerD, regulation

(a) **mer operon**

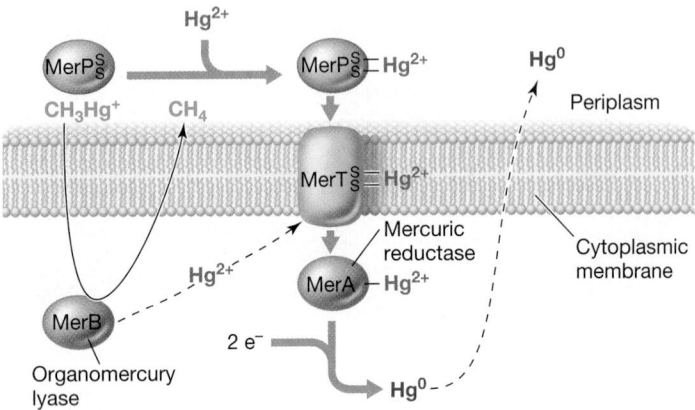

(b) **Mercury metabolism**

Figure 24.19 Mechanism of mercury transformations and resistance. *(a)* The *mer* operon. MerR can function as either a repressor (in the absence of Hg^{2+}) or transcriptional activator (in the presence of Hg^{2+}). *(b)* Transport and reduction of Hg^{2+} and CH_3Hg^+; the Hg^{2+} is bound by cysteine residues in the MerP and MerT proteins. MerA is the enzyme mercuric reductase and MerB is organomercurial lyase.

In many mercury-resistant bacteria, genes encoding Hg resistance reside on plasmids or transposons (Sections 6.7 and 12.12). These *mer* genes are arranged in an operon under control of the regulatory protein MerR, which can function as either a repressor or an activator of transcription (Sections 8.3 and 8.4), depending on Hg availability. In the absence of Hg^{2+}, MerR functions as a *repressor* and binds to the operator region of the *mer* operon, thus preventing transcription of the structural genes, *merTPCABD*. However, when Hg^{2+} is present, it forms a complex with MerR, which then binds to the *mer* operon and functions as an *activator* of transcription of *mer* structural genes.

The protein MerP is a periplasmic Hg^{2+}-binding protein. MerP binds Hg^{2+} and transfers it to the membrane transport protein MerT, which associates with mercuric reductase (MerA) to reduce Hg^{2+} to Hg^0 (Figure 24.19*b*). Thus, Hg^{2+} is not released into the cytoplasm and the final result is the release of Hg^0 from the cell. Mercuric ion produced from the activity of MerB is trapped by MerT and reduced by MerA, again releasing Hg^0 (Figure 24.19*b*). In this way, Hg^{2+} and CH_3Hg^+ are converted to the relatively nontoxic Hg^0.

MiniQuiz

- What forms of mercury are most toxic to organisms?
- How is mercury detoxified by bacteria?

24.9 Petroleum Biodegradation and Bioremediation

Petroleum is a rich source of organic matter, and because of this, microorganisms readily attack hydrocarbons when petroleum is pumped to Earth's surface and comes into contact with air and moisture. Under some circumstances, such as in bulk petroleum storage tanks, microbial growth is undesirable. However, in oil spills, biodegradation is desirable and can be promoted by the addition of inorganic nutrients to balance the huge influx of organic carbon from the oil. The term **bioremediation** refers to the microbial cleanup of oil, toxic chemicals, or other environmental pollutants, usually by stimulating the microorganisms' activities in some way. Although bioremediation of many toxic substances has been proposed, most successes have been in cleaning up spills of crude oil (**Figure 24.20**) or leakage of hydrocarbons from bulk storage tanks.

The biochemistry of hydrocarbon catabolism was covered in Sections 14.13 and 14.14. Both anoxic and oxic biodegradation is

(a)

(b)

(c)

Figure 24.20 Environmental consequences of large oil spills and the effect of bioremediation. *(a)* A contaminated beach along the coast of Alaska containing oil from the *Exxon Valdez* spill of 1989. *(b)* The rectangular plot (arrow) was treated with inorganic nutrients to stimulate bioremediation of spilled oil by microorganisms, whereas areas above and to the left were untreated. *(c)* Oil spilled into the Mediterranean Sea from the Jiyyeh (Lebanon) power plant that flowed to the port of Byblos during the 2006 war in Lebanon.

possible. We emphasized that under oxic conditions oxygenase enzymes play an important role in introducing oxygen atoms into the hydrocarbon. Our discussion here will focus on *aerobic* processes, because it is only when O_2 is present that oxygenase enzymes can function and hydrocarbon bioremediation can be effective in a relatively short time.

Hydrocarbon Decomposition

Diverse bacteria, fungi, and a few green algae can oxidize petroleum products aerobically. Small-scale oil pollution of aquatic and terrestrial ecosystems from human as well as natural activities is common. Oil-oxidizing microorganisms develop rapidly on oil films and slicks, and hydrocarbon oxidation is most extensive if the temperature is warm enough and supplies of inorganic nutrients (primarily N and P) are sufficient.

Because oil is insoluble in water and is less dense, it floats to the surface and forms slicks. There, hydrocarbon-degrading bacteria attach to the oil droplets (**Figure 24.21**) and eventually decompose the oil and disperse the slick. Certain oil-degrading bacteria are specialist species; for example, the bacterium *Alcanivorax borkumensis* grows only on hydrocarbons, fatty acids, or pyruvate. This organism produces surfactant chemicals that help break up the oil and solubilize it. Once solubilized, the oil can be incorporated more readily and catabolized as an electron donor and carbon source.

In large oil spills, such as those shown in Figure 24.20 or the more recent Gulf of Mexico spill off the coast of Louisiana (USA), volatile hydrocarbons, both aliphatic and aromatic, evaporate quickly without bioremediation, leaving nonvolatile components for cleanup crews and microorganisms to tackle. Microorganisms consume oil by oxidizing it to CO_2. When bioremediation activities are promoted by inorganic nutrient application, oil-oxidizing bacteria typically develop quickly after an oil spill (Figure 24.20*b*), and under ideal conditions, 80% or more of the nonvolatile oil components can be oxidized within one year. However, certain oil fractions, such as those containing branched-chain and polycyclic hydrocarbons, are not preferred microbial substrates and remain in the environment much longer. Spilled oil that finds its way into sediments is even more slowly degraded and can have a significant long-term impact on fisheries that depend on unpolluted waters for productive yields.

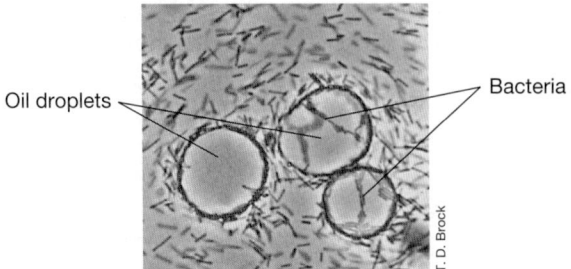

Figure 24.21 Hydrocarbon-oxidizing bacteria in association with oil droplets. The bacteria are concentrated in large numbers at the oil–water interface, but are actually not within the droplet itself.

Figure 24.22 Bulk petroleum storage tanks. Fuel tanks often support microbial growth at oil–water interfaces.

Degradation of Stored Hydrocarbons

Interfaces where oil and water meet often form on a large scale. Besides water that separates from crude petroleum during storage and transport, moisture can condense inside bulk fuel storage tanks (**Figure 24.22**) where there are leaks. This water eventually accumulates in a layer beneath the petroleum. Gasoline and crude oil storage tanks are thus potential habitats for hydrocarbon-oxidizing microorganisms. If sufficient sulfate (SO_4^{2-}) is present in the oil, as it often is in crude oils, sulfate-reducing bacteria can grow in the tanks, consuming hydrocarbons under anoxic conditions ($\hookleftarrow$ Sections 14.13 and 17.18). The sulfide (H_2S) produced is highly corrosive and causes pitting and subsequent leakage of the tanks along with souring of the fuel. Aerobic degradation of fuel components is less of a problem because the storage tanks are sealed and the fuel itself contains little dissolved O_2.

MiniQuiz

- What is bioremediation?
- Why might the addition of inorganic nutrients stimulate oil degradation whereas the addition of glucose would not?

24.10 Xenobiotics Biodegradation and Bioremediation

A **xenobiotic** is a synthetic chemical not produced by organisms in nature. Xenobiotics include pesticides, polychlorinated biphenyls (PCBs), munitions, dyes, and chlorinated solvents, among many other chemicals. Some xenobiotics differ chemically in such major ways from anything organisms have experienced in nature that they biodegrade extremely slowly, if at all. Other xenobiotics are structurally related to one or more natural compounds and can sometimes be degraded slowly by enzymes that normally degrade the structurally related natural compounds. We focus here on pesticides as examples of the potential of microorganisms to degrade xenobiotics.

UNIT 7

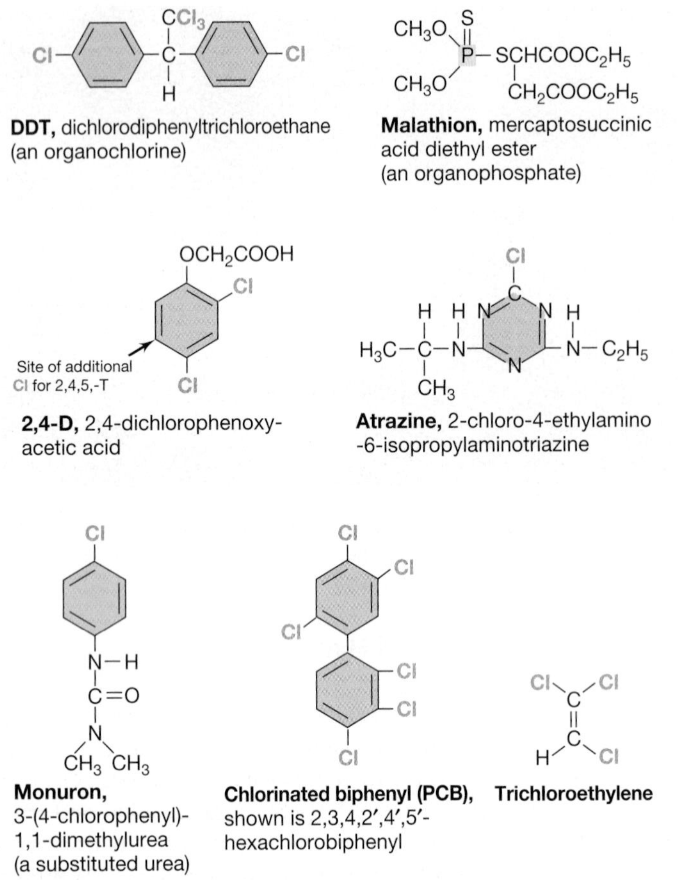

DDT, dichlorodiphenyltrichloroethane
(an organochlorine)

Malathion, mercaptosuccinic
acid diethyl ester
(an organophosphate)

2,4-D, 2,4-dichlorophenoxy-
acetic acid

Site of additional
Cl for 2,4,5,-T

Atrazine, 2-chloro-4-ethylamino
-6-isopropylaminotriazine

Monuron,
3-(4-chlorophenyl)-
1,1-dimethylurea
(a substituted urea)

Chlorinated biphenyl (PCB),
shown is 2,3,4,2′,4′,5′-
hexachlorobiphenyl

Trichloroethylene

Figure 24.23 Examples of xenobiotic compounds. Although none of these compounds exist naturally, microorganisms exist that can break them down.

Pesticide Catabolism

Over 1000 pesticides have been marketed worldwide for pest control purposes. Pesticides include *herbicides, insecticides,* and *fungicides.* Pesticides display a wide variety of chemistries, and include chlorinated, aromatic, and nitrogen- and phosphorus-containing compounds (**Figure 24.23**). Some of these substances can be used as carbon and energy sources by microorganisms, whereas others are utilized only poorly or not at all. Highly chlorinated compounds are typically the pesticides most resistant to microbial attack. However, related compounds may differ remarkably in their degradability. For example, chlorinated compounds such as DDT persist relatively unaltered for years in soils, whereas chlorinated compounds such as 2,4-D are significantly degraded in just a few weeks.

Environmental factors, such as temperature, pH, aeration, and organic content of the soil, influence the rate of pesticide decomposition, and some pesticides can disappear from soils nonbiologically by volatilization, leaching, or spontaneous chemical breakdown. In addition, some pesticides are degraded only when other organic material is present that can be used as the primary energy source, a phenomenon called *cometabolism.* In most cases, pesticides that are cometabolized are only partially degraded, generating new xenobiotic compounds that may be even more toxic or difficult to degrade than the original compound. Thus, from an environmental standpoint, cometabolism of a pesticide is not always good.

Dechlorination

Many xenobiotics are chlorinated compounds and their degradation proceeds through *dechlorination.* For example, the bacterium *Burkholderia* dechlorinates the pesticide 2,4,5-T (Figure 24.23) aerobically, releasing chloride ion (Cl^-) in the process (**Figure 24.24**); this reaction is catalyzed by oxygenase enzymes (Section 14.14). Following dechlorination, a dioxygenase enzyme breaks the aromatic ring to yield compounds that can enter the citric acid cycle and yield energy.

Although the aerobic breakdown of chlorinated xenobiotics is undoubtedly ecologically important, **reductive dechlorination** is probably more so because of the rapidity with which anoxic conditions develop in microbial habitats polluted with chlorinated compounds. We previously described reductive dechlorination as a form of anaerobic respiration in which chlorinated organic compounds such as chlorobenzoate ($C_7H_4O_2Cl^-$) are terminal electron acceptors (Section 14.12). Many compounds can be reductively dechlorinated including dichloro-, trichloro-, and tetrachloro- (perchloro-) ethylene, chloroform, dichloromethane, and polychlorinated biphenyls (Figure 24.23). In addition, several brominated and fluorinated organic compounds can be dehalogenated in analogous fashion. Many of these chlorinated or halogenated compounds are highly toxic and some have been linked to cancer (particularly trichloroethylene). Some of these compounds, such as PCBs, have been widely used as insulators in electrical transformers and enter anoxic environments from slow leakage of the transformer or from leaking storage containers. Eventually these compounds end up in groundwater, where they are the most common groundwater contaminants detected in the United States. There is therefore great interest in reductive dechlorination as a bioremediation strategy for their removal from anoxic environments.

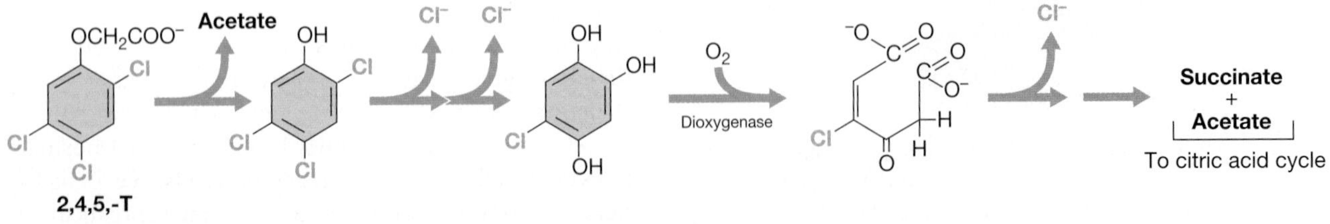

Figure 24.24 Biodegradation of the herbicide 2,4,5-T. Pathway of aerobic 2,4,5-T biodegradation; note the importance of oxygenase enzymes (Section 14.14) in the degradation process.

Plastics

Plastics are classic examples of xenobiotics, and the plastics industry worldwide produces over 40 million tons of plastic per year, almost half of which are discarded rather than recycled. Plastics are polymers of various chemistries (**Figure 24.25a**). Many plastics remain essentially unaltered for long periods in landfills, refuse dumps, and as litter in the environment. This problem has fueled the search for biodegradable alternatives called **microbial plastics** as replacements for some synthetic plastics.

Polyhydroxyalkanoates (PHAs) are a common bacterial storage polymer (♻ Section 3.10), and these readily biodegradable polymers have many of the desirable properties of xenobiotic plastics. PHAs can be biosynthesized in various chemical forms, each with its own unique physical properties (stiffness, shear and impact strength, and the like). A PHA *copolymer* containing equal amounts of poly-β-hydroxybutyrate and poly-β-hydroxyvalerate (Figure 24.25b) has been marketed in Europe as a container for personal care products and has had the greatest success as a plastic substitute thus far (Figure 24.25c). However, because synthetic plastics are currently less expensive than microbial plastics, synthetic petroleum-based plastics make up virtually the entire plastics market today.

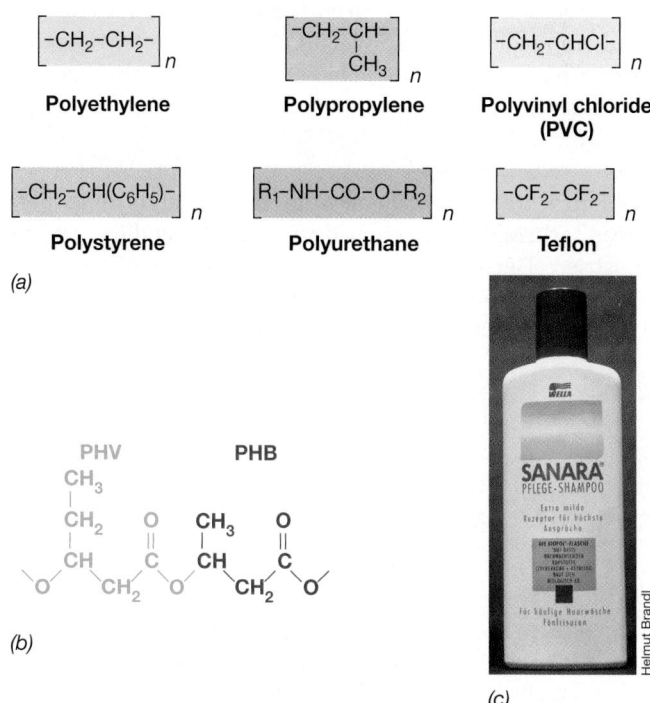

(a)

(b)

(c)

Figure 24.25 Synthetic and microbial plastics. *(a)* The monomeric structure of several synthetic plastics. *(b)* Structure of the copolymer of poly-β-hydroxybutyrate (PHB) and poly-β-hydroxyvalerate (PHV). *(c)* A brand of shampoo previously marketed in Germany and packaged in a bottle made of the PHB/PHV copolymer.

The bacterium *Ralstonia eutropha* has been used as a model organism for the commercial production of PHAs. This genetically manipulable and metabolically diverse bacterium (♻ Section 17.5) produces PHAs in high yield, and specific copolymers can be obtained by simple nutritional modifications. Nevertheless, the microbial plastics industry is burdened by the reality that the best substrates for PHA biosyntheses are glucose and related organic compounds, substances obtained from corn or other crops. And even at today's prices for oil, plant products cannot compete with oil as feedstocks for the plastics industry.

Contaminants of Emerging Concern

Until recently, studies of the environmental fate of chemicals have focused primarily on priority pollutants, including heavily used agricultural products and chemicals that demonstrate acute toxicity or carcinogenicity (Figure 24.23). However, it is now clear that new bioactive pollutants are entering the environment and will likely pose new challenges for microbial bioremediation. These pollutants include pharmaceuticals, active ingredients in personal care products, fragrances, household products, sunscreens, and many other unusual or xenobiotic molecules.

Unlike pesticides, these "new" pollutants are more or less continuously discharged to the environment primarily through release of treated or untreated sewage, and because of this, they do not need to persist to have environmental effects. For example, it is known that synthetic estrogen compounds, excreted in the urine of women taking birth control pills and eventually discharged from wastewater treatment plants, can activate estrogen response genes in aquatic animals such as fish and contribute to the feminization of males.

Wastewater treatment plants (♻ Section 35.2) were originally designed to handle natural materials, primarily human and industrial wastes, but now there is a growing interest in carefully researching the design of future treatment facilities to stimulate bioremediation of these emerging contaminants. Because these contaminants are often present in very low concentrations and are often new classes of xenobiotic chemicals, they may not actually support microbial growth but be degraded only by cometabolism or by highly specialized species. We can therefore expect that the bioremediation of emerging contaminants will be an active area of microbiological research and public policy in coming years.

MiniQuiz

- Which chemical class of pesticides is the most recalcitrant to microbial attack?
- What is reductive dechlorination and how does it differ from the reactions shown in Figure 24.24?
- What main advantage do microbial plastics have over synthetic plastics?
- Give an example of an "emerging" contaminant.

UNIT 7

Big Ideas

24.1
The oxygen and carbon cycles are interconnected through the complementary activities of autotrophic and heterotrophic organisms. Microbial decomposition is the single largest source of CO_2 released to the atmosphere.

24.2
Under anoxic conditions, organic matter is degraded to CH_4 and CO_2. Methane is formed primarily from the reduction of CO_2 by H_2 and from acetate, both supplied by syntrophic bacteria; these organisms depend on H_2 consumption as the basis of their energetics. On a global basis, biogenic CH_4 is a much larger source than abiogenic CH_4.

24.3
The principal form of nitrogen on Earth is N_2, which can be used as a N source only by nitrogen-fixing bacteria. Ammonia produced by nitrogen fixation or by ammonification can be assimilated into organic matter or oxidized to nitrate. Denitrification and anammox cause major losses of fixed nitrogen from the biosphere.

24.4
Bacteria play major roles in both the oxidative and reductive sides of the sulfur cycle. Sulfur- and sulfide-oxidizing bacteria produce SO_4^{2-}, whereas sulfate-reducing bacteria consume SO_4^{2-}, producing H_2S. Because sulfide is toxic and reacts with various metals, sulfate reduction is an important biogeochemical process. Dimethyl sulfide is the major organic sulfur compound of ecological significance in nature.

24.5
Iron exists naturally in two oxidation states, Fe^{2+} and Fe^{3+}. Bacteria reduce ferric iron in anoxic environments and oxidize Fe^{2+} aerobically at acidic pH. Ferrous iron oxidation is common in coal-mining regions, where it causes a type of pollution called acid mine drainage.

24.6
P, Ca, and Si are elements cycled by microbial activities, primarily in aquatic environments. Calcium and silica play important roles in the biogeochemistry of the oceans as components of the exoskeletons of coccolithophores and diatoms, respectively.

24.7
Bacterial solubilization of copper is a process called microbial leaching. Leaching is important in the recovery of copper, uranium, and gold from low-grade ores. Bacterial oxidation of Fe^{2+} to Fe^{3+} is the key reaction in most microbial leaching processes because Fe^{3+} can oxidize extractable metals in the ores under either oxic or anoxic conditions.

24.8
A major toxic form of Hg in nature is CH_3Hg^+, which can yield Hg^{2+}, which is reduced by bacteria to Hg^0. Genes conferring resistance to the toxicity of Hg, such as those that encode enzymes that can detoxify or pump out the metal, often reside on plasmids or transposons.

24.9
Hydrocarbons are excellent carbon and energy sources for bacteria and are readily oxidized when O_2 is available. Hydrocarbon-oxidizing bacteria bioremediate spilled oil, and their activities can be assisted by addition of inorganic nutrients.

24.10
Xenobiotics are chemicals new to nature, and some persist whereas others are readily degraded, depending on their chemistries. Dechlorination is a major means of detoxifying xenobiotics, but the accumulation of synthetic plastics is probably the major source of environmental concern in this area.

Review of Key Terms

Acid mine drainage acidic water containing H_2SO_4 derived from the microbial oxidation of iron sulfide minerals released by coal mining

Bioremediation the cleanup of oil, toxic chemicals, and other pollutants by microorganisms

Denitrification the biological reduction of NO_3^- to gaseous N compounds

Humus dead organic matter

Microbial leaching the extraction of valuable metals such as copper from sulfide ores by microbial activities

Microbial plastics polymers consisting of microbially produced (and thus biodegradable) substances, such as polyhydroxyalkanoates

Pyrite a common iron-containing ore, FeS_2

Reductive dechlorination the removal of chlorine as Cl^- from an organic compound by reduction of the carbon atom from C–Cl to C–H

Syntrophy the cooperation of two or more microorganisms to degrade anaerobically a substance neither can degrade alone

Xenobiotic a synthetic compound not produced by organisms in nature

Review Questions

1. Outline the redox cycle for carbon (Section 24.1).

2. How can organisms such as *Syntrophobacter* and *Syntrophomonas* grow when their metabolism is based on thermodynamically unfavorable reactions? How does coculture of these syntrophs with certain other bacteria allow them to grow (Section 24.2)?

3. Compare and contrast the processes of nitrification, ammonification, and denitrification (Section 24.3).

4. Which group of bacteria cycle sulfur compounds under anoxic conditions? If sulfur chemolithotrophs had never evolved, would there be a problem in the microbial cycling of sulfur compounds? Which organic sulfur compounds are most abundant in nature (Section 24.4)?

5. Discuss the implications of acid mine drainage (Section 24.5).

6. In what ways are Ca and Si cycling in ocean waters similar, and in what ways do they differ? How do the Ca and Si cycles couple to the carbon cycle (Section 24.6)?

7. Briefly describe the process of microbial leaching (Section 24.7).

8. How are Hg^{2+} and CH_3Hg^+ detoxified by the *mer* system (Section 24.8)?

9. What physical and chemical conditions are necessary for the rapid microbial degradation of oil in aquatic environments? Design an experiment that would allow you to test which conditions optimized the oil oxidation process (Section 24.9).

10. What are xenobiotic compounds and why might microorganisms have difficulty catabolizing them (Section 24.10)?

Application Questions

1. Compare and contrast the carbon, sulfur, and nitrogen cycles in terms of the physiologies of the organisms that participate in the cycle. Which physiologies are part of one cycle but not another?

2. ^{14}C-labeled cellulose is added to a vial containing a small amount of sewage sludge and sealed under anoxic conditions. A few hours later, $^{14}CH_4$ appears in the vial. Discuss what has happened to yield such a result.

3. Acid mine drainage is in part a chemical process and in part a biological process. Discuss the chemistry and microbiology that lead up to acid mine drainage and point out the key reactions that are biological. What ways can you think of to prevent acid mine drainage?

Need more practice? Test your understanding with Quantitative Questions; access additional study tools including tutorials, animations, and videos; and then test your knowledge with chapter quizzes and practice tests at **www.microbiologyplace.com**.

25

Microbial Symbioses

All animals harbor specific bacterial symbionts. In the bladder of the medicinal leech *Hirudo verbana*, several species of bacteria are present. Each phylogenetic group of species stains a different color— green, pink, or blue—in this FISH stain of bladder tissue.

In this chapter we consider relationships of microorganisms with other microorganisms or with macroorganisms—prolonged and intimate relationships of a type called **symbioses**, a word that means "living together." Microorganisms living within or on plants and animals can be categorized based on their effect on their hosts as *parasitic* (the microorganism benefits at some expense to the host), *pathogenic* (the microorganism actually causes a disease in the host), *commensal* (the microorganism has no discernible impact on the host), and *mutualistic* (the microorganism is beneficial to the host). In one way or another, all microbial symbioses benefit the microorganism.

Pathogenic and parasitic associations will be addressed in Chapter 27 and in following chapters covering specific diseases. Here we focus on a type of symbiosis called **mutualism**, a relationship in which both partners benefit. We view the microorganisms as intimate evolutionary partners that influence both the evolution and physiology of their hosts. Many mutualistic symbioses of microorganisms with plants and animals have origins many millions of years in the past. A mutualism that persists over evolutionary time beneficially modifies the physiology of both partners. This process of reciprocal change is called **coevolution** and, over time, the changes may be so extensive that the symbiosis becomes obligate—either the microorganism or the host (or both) cannot survive independent of the other.

I Symbioses between Microorganisms

Many microbial species—both prokaryotes and eukaryotes—have intimate and beneficial associations with other microbial species. Direct microscopic observations of samples from nature show that many microorganisms are not solitary entities, but are associated with other microorganisms on surfaces or as suspended aggregates of cells. In most cases the advantages conferred by an association are not known. Because microbial ecologists have recognized that *communities* of interacting microbial populations—not individual organisms—control critical environmental processes, research to discover the advantages of strictly microbial symbioses has increased. We present in Part I two types of microbial mutualisms where the advantages to both partners are clear.

25.1 Lichens

Lichens are the visible evidence of leafy or encrusting microbial symbioses often found growing on bare rocks, tree trunks, house roofs, and bare soils—surfaces where other organisms do not typically grow (**Figure 25.1**). A lichen is a mutualistic association between two microorganisms, a fungus and either an alga or a cyanobacterium. The alga or cyanobacterium is the phototrophic partner and produces organic matter, which then feeds the fungus. The fungus, unable to carry out photosynthesis, provides a firm anchor within which the phototrophic partner can grow, protected from erosion by rain or wind. Cells of the phototroph are embedded in defined layers or clumps among cells of the

(a)

(b)

Figure 25.1 Lichens. *(a)* A lichen growing on a branch of a dead tree. *(b)* Lichens coating the surface of a large rock.

fungus (**Figure 25.2**). The morphology of a lichen is primarily determined by the fungus, and many fungi are able to form lichen associations. Diversity among the phototrophs is much lower, and many different kinds of lichens can have the same phototrophic partner. Many cyanobacteria that partner with lichens are nitrogen-fixing species, organisms such as *Anabaena* or *Nostoc* (↺ Sections 13.14 and 18.7).

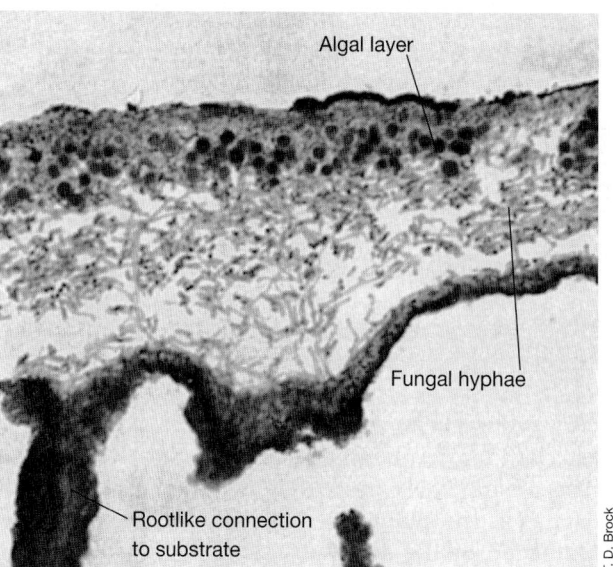

Algal layer

Fungal hyphae

Rootlike connection to substrate

Figure 25.2 Lichen structure. Photomicrograph of a cross section through a lichen. The algal layer is positioned within the lichen structure so as to receive the most sunlight.

The fungus clearly benefits from associating with the phototroph in the lichen symbiosis, but how does the phototroph benefit? *Lichen acids*, complex organic compounds excreted by the fungus, promote the dissolution and chelation of inorganic nutrients from the rock or other surface that are needed by the phototroph. Another role of the fungus is to protect the phototroph from drying; most of the habitats in which lichens live are dry, and fungi are, in general, better able to tolerate dry conditions than are the phototrophs. The fungus actually facilitates the uptake of water and sequesters some for the phototroph.

Lichens typically grow quite slowly. For example, a lichen 2 cm in diameter growing on the surface of a rock may be several years old. Lichen growth varies from 1 mm or less per year to over 3 cm per year, depending on the organisms composing the symbiosis and the amount of rainfall and sunlight received.

MiniQuiz

- What are the benefits to phototroph and fungus in the lichen mutualism?
- Besides organic compounds, of what benefit to the fungus is a mutualism with *Anabaena*?

25.2 "*Chlorochromatium aggregatum*"

In freshwater environments there are microbial mutualisms called **consortia**. A common consortium is between nonmotile, phototrophic, green sulfur bacteria, which may be colored either green or brown, and motile, nonphototrophic bacteria. These consortia are found worldwide in stratified freshwater lakes, and can account for up to 90% of the green sulfur bacteria and 67% of the bacterial biomass in these lakes. The basis of the mutualism of these consortia is in the photosynthetic production of organic matter by the green sulfur bacteria and the motility of the partner species. Each consortium has been given a genus and species name, but as these names do not denote true species (because they are not a single organism), the names are enclosed in quotation marks. We examined the general biology of these consortia in Section 18.15.

Nature of the Consortium

The morphology of a green sulfur bacterial consortium depends upon the species composition. The consortium generally consists of 13–69 green sulfur bacteria, called *epibionts*, surrounding and attached to a central, colorless, flagellated, rod-shaped bacterium (**Figure 25.3**). Several distinct motile phototrophic consortia have been recognized based on the color, morphology, and presence or absence of gas vesicles of the epibionts. For example, in "*Chlorochromatium aggregatum*" the central bacterium is surrounded by rod-shaped green bacteria. In "*Pelochromatium roseum*" the epibiont is brown. The consortium "*Chlorochromatium glebulum*" is bent and includes gas-vacuolated, green epibionts (Figure 25.3).

Green sulfur bacteria are obligately anaerobic phototrophs that form a distinct phylum (*Chlorobiaceae*). The green and brown species differ in the types of bacteriochlorophyll and

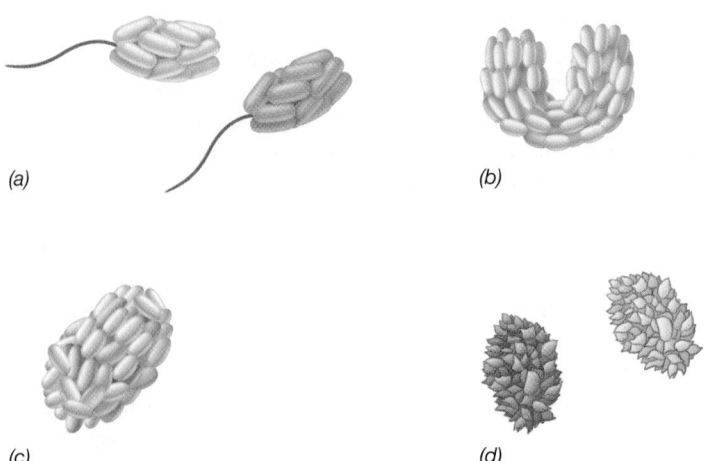

(a)

(b)

(c)

(d)

Figure 25.3 Drawings of some motile phototrophic consortia found in freshwater lakes. Green epibionts: *(a)* "*Chlorochromatium aggregatum*," *(b)* "*C. glebulum*," *(c)* "*C. magnum*," *(d)* "*C. lunatum*." Brown epibionts: *(a)* "*Pelochromatium roseum*," *(d)* "*P. selenoides*." The epibionts are 0.5–0.6 μm in diameter. Adapted from Overmann, J., and H. van Gemerden. 2000. *FEMS Microbiol. Rev. 24*: 591–599.

carotenoids they contain (⟳ Section 18.15). Both green and brown species are found in stratified lakes where light penetrates to depths at which the water contains hydrogen sulfide (H₂S). In the stratified lakes, the motile consortia reposition rapidly to remain where conditions are favorable for photosynthesis in the constantly changing gradients of light, oxygen, and sulfide (**Figure 25.4**). The consortia show dark aversion (scotophobotaxis, ⟳ Section 3.15) and a positive chemotaxis toward sulfide. Some free-living green sulfur bacteria, such as *Pelodictyon phaeoclathratiforme*, have gas vesicles that regulate buoyancy and vertical position in the water column. However,

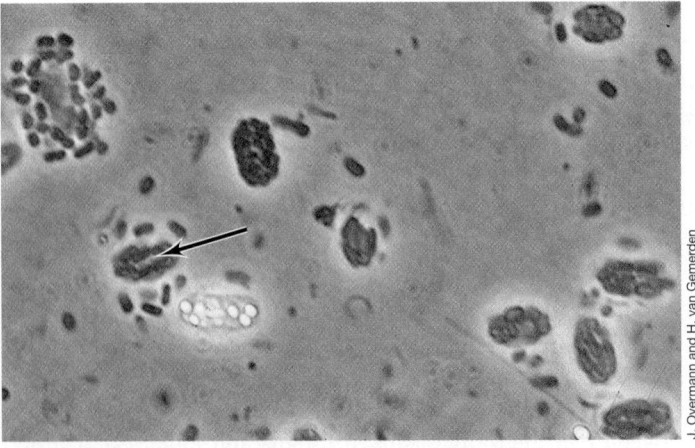

Figure 25.4 Phase-contrast micrograph of "*Pelochromatium roseum*" from Lake Dagow (Brandenburg, Germany). The preparation was compressed between a coverslip and microscope slide to reveal the central rod-shaped bacterium (arrow). A single consortium is about 3.5 μm in diameter. Used with permission from J. Overmann and H. van Gemerden. 2000. *FEMS Microbiol. Rev. 24*: 591–599.

the time they require for repositioning is from one to several days, which is not fast enough for tracking the more rapidly changing gradients. By contrast, motile consortia move up and down in the water column fast enough to follow the gradients of light and sulfide as they change on a diel basis.

Although green bacterial consortia were discovered almost a century ago, only with the advent of molecular techniques and newer culture methods has it become possible to study certain aspects of these remarkable associations. Sequencing of 16S ribosomal RNA (rRNA) genes revealed a significant *biogeography* of epibionts in lakes of Europe and the United States. Biogeography is the study of the geographic distribution of organisms, in this case, the genetically distinct phototrophic consortia in different lakes. Epibionts in neighboring lakes have identical 16S rRNA gene sequences, whereas the sequences of morphologically similar epibionts in widely separated lakes differ. Phylogenetic analysis has shown that the mechanisms of cell–cell recognition responsible for stable morphology have evolved between particular epibionts and their central bacterium.

Phylogeny of a Consortium

The epibiont of "*Chlorochromatium aggregatum*" has been isolated and grown in pure culture. Although this green sulfur bacterium, named *Chlorobium chlorochromatii*, can be grown in pure culture, no naturally free-living variant has been observed, supporting the view that in nature, a symbiotic lifestyle is obligate for epibionts. The central bacterium of "*Chlorochromatium aggregatum*" belongs to the *Betaproteobacteria*. Interestingly, this bacterium requires α-ketoglutarate, an intermediate of the citric acid cycle (Section 4.11), and this is presumably supplied by the epibiont. However, the central cell only assimilates fixed carbon in the presence of light and sulfide conditions in which the epibionts are active and can transfer nutrients to the central bacterium. Scanning electron microscopy of the consortium (**Figure 25.5**) has revealed that tubular extensions of the central bacterium's periplasm (Section 3.7) cover much of its surface and appear to fuse with the periplasm of the epibiont. If the two bacterial partners actually share a common periplasmic space, this would facilitate the transfer of nutrients from phototroph to chemotroph.

MiniQuiz
- What is the evidence that "*Chlorochromatium aggregatum*" is a stable product of evolution?
- What advantage does motility offer a phototrophic consortium?
- How might nutrients be shuttled between phototroph and chemotroph in the consortium?

 ## Plants as Microbial Habitats

Plants interact closely with microorganisms through their roots and leaf surfaces and even more intimately within their vascular tissue and cells. Most mutualisms between plants and microorganisms increase nutrient availability to the

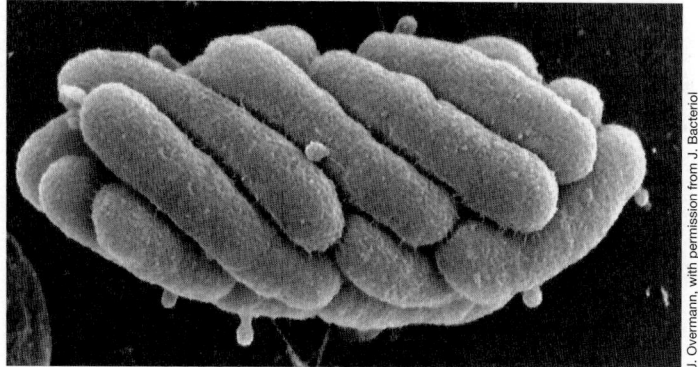
(a)

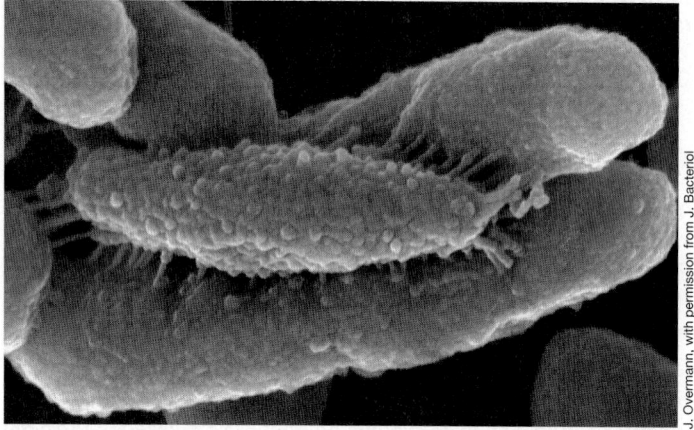
(b)

Figure 25.5 Scanning electron micrographs of "*Chlorochromatium aggregatum*." *(a) Chlorobium chlorochromatii* epibionts tightly clustered around a flagellated central bacterium. *(b)* The central bacterium exhibits numerous protrusions of its outer membrane that make intimate contact with the epibionts, possibly fusing the periplasms of the two organisms. Cells of the epibiont are about 0.6 μm in diameter. Used with permission from G. Wanner et al. 2008. *J. Bacteriol. 190:* 3721–3730.

plants or defend them against pathogens. We consider three examples in the following sections: (1) a mutualism (root nodules, Section 25.3), (2) a symbiosis that is harmful to the plant (crown gall disease, Section 25.4), and (3) a mutualism in which plants expand and interconnect their root system through association with a fungus (mycorrhizae, Section 25.5).

25.3 The Legume–Root Nodule Symbiosis

A plant–bacterial mutualism of great importance to humans is that of leguminous plants and nitrogen-fixing bacteria. *Legumes* are plants that bear their seeds in pods. This third largest family of flowering plants includes such agriculturally important species as soybeans, clover, alfalfa, beans, and peas. These plants are key commodities for the food and agricultural industries, and the ability of legumes to grow without nitrogen fertilizer saves farmers millions of dollars in fertilizer costs yearly.

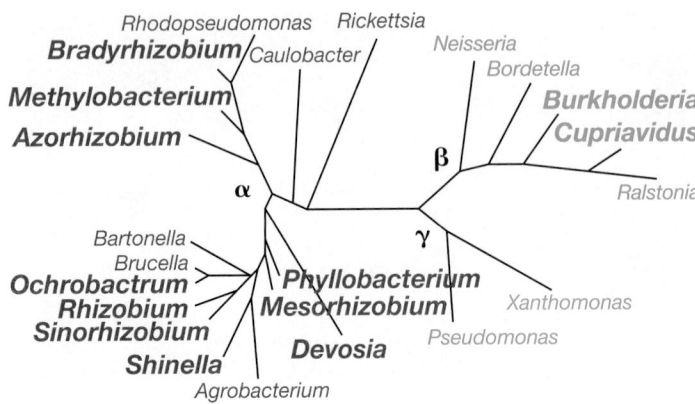

Figure 25.6 Phylogeny of rhizobial (names in boldface) and related genera inferred from analysis of 16S rRNA gene sequences. There are rhizobia in 12 genera and more than 70 species of *Alpha-* and *Betaproteobacteria*.

Figure 25.8 Effect of nodulation on plant growth. A field of unnodulated (left) and nodulated (right) soybean plants growing in nitrogen-poor soil. The yellow color is typical of chlorosis, the result of nutrient (in this case N) starvation.

The partners in a symbiosis are called *symbionts*, and most nitrogen-fixing bacterial symbionts of plants are collectively called *rhizobia*, derived from the name of a major genus, *Rhizobium*. Rhizobia are species of *Alpha-* or *Betaproteobacteria* (**Figure 25.6**) that can grow freely in soil or can infect leguminous plants and establish a symbiotic relationship. The same genus (or even species) can contain both rhizobial and nonrhizobial strains. Infection of legume roots by rhizobia leads to the formation of **root nodules** (**Figure 25.7**) in which the bacteria fix gaseous nitrogen (N_2) (⮌ Section 13.14). Nitrogen fixation in root nodules accounts for a fourth of the N_2 fixed annually on Earth and is of enormous agricultural importance, as it increases the fixed nitro-gen content of soil. Nodulated legumes can grow well on unfertil-ized bare soils that are nitrogen deficient, while other plants grow only poorly on them (**Figure 25.8**).

Leghemoglobin and Cross-Inoculation Groups

In the absence of its bacterial symbiont, a legume cannot fix N_2. Rhizobia, on the other hand, can fix N_2 when grown in pure culture under microaerophilic conditions (a low-oxygen environ-ment is necessary because nitrogenases are inactivated by high levels of O_2, ⮌ Section 13.14). In the nodule O_2 levels are precisely controlled by the O_2-binding protein **leghemoglobin**. Production of this iron-containing protein in healthy N_2-fixing nodules (**Figure 25.9**) is induced through the interaction of the plant and bacterial partners. Leghemoglobin functions as an "oxygen buffer," cycling between the oxidized (Fe^{3+}) and reduced (Fe^{2+}) forms of iron to keep unbound O_2 within the nodule low.

Figure 25.7 Soybean root nodules. The nodules developed from infection by *Bradyrhizobium japonicum*. The main stem of this soybean plant is about 0.5 cm in diameter.

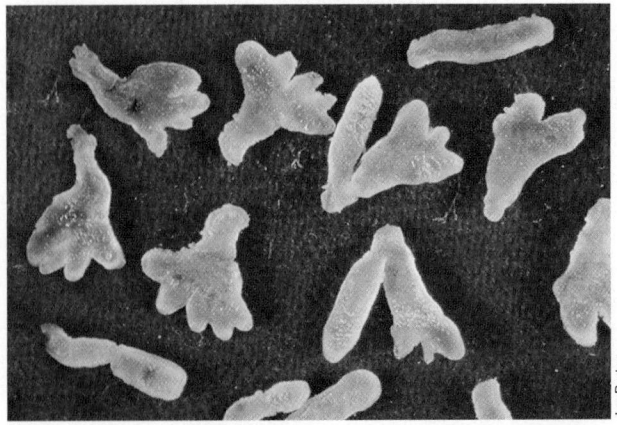

Figure 25.9 Root nodule structure. Sections of root nodules from the legume *Coronilla varia*, showing the reddish pigment leghemoglobin.

The ratio of leghemoglobin-bound O_2 to free O_2 in the root nodule is on the order of 10,000:1.

There is a marked specificity between the species of legume and rhizobium that can establish a symbiosis. A particular rhizobial species is able to infect certain species of legumes but not others. A group of related legumes that can be infected by a particular rhizobial species is called a *cross-inoculation group*—there is, for example, a clover group, a bean group, an alfalfa group, and so on (**Table 25.1**). If legumes are inoculated with the correct rhizobial strain, leghemoglobin-rich, N_2-fixing nodules develop on their roots (Figure 25.9).

Steps in Root Nodule Formation

How root nodules form is well understood for most rhizobia (**Figure 25.10**). The steps are as follows:

1. Recognition of the correct partner by both plant and bacterium and attachment of the bacterium to the root hairs
2. Secretion of oligosaccharide signaling molecules (nod factors) by the bacterium
3. Bacterial invasion of the root hair
4. Movement of bacteria to the main root by way of the infection thread
5. Formation of modified bacterial cells (bacteroids) within the plant cells and development of the N_2-fixing state
6. Continued plant and bacterial cell division, forming the mature root nodule

Another mechanism of nodule formation that does not require nod factors is used by some species of phototrophic rhizobia. This mechanism has yet to be elucidated, but appears to require bacterial production of *cytokinins*. Cytokinins are plant hormones, derived from adenine or phenylurea, necessary for cell growth and differentiation.

Host plant	Nodulated by
Pea	Rhizobium leguminosarum biovar viciae[a]
Bean	Rhizobium leguminosarum biovar phaseoli[a]
Bean	Rhizobium tropici
Lotus	Mesorhizobium loti
Clover	Rhizobium leguminosarum biovar trifolii[a]
Alfalfa	Sinorhizobium meliloti
Soybean	Bradyrhizobium japonicum
Soybean	Bradyrhizobium elkanii
Soybean	Sinorhizobium fredii
Sesbania rostrata (a tropical legume)	Azorhizobium caulinodans

Table 25.1 *Major cross-inoculation groups of leguminous plants*

[a]Several varieties (biovars) of *Rhizobium leguminosarum* exist, each capable of nodulating a different legume.

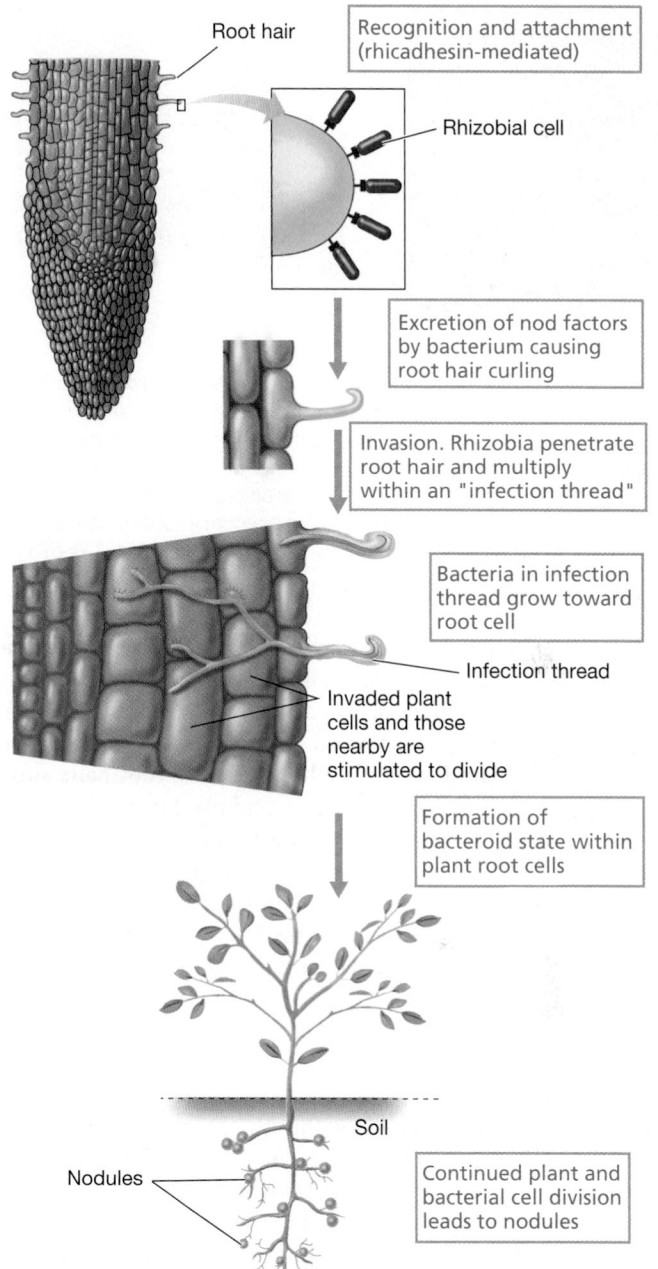

Figure 25.10 Steps in the formation of a root nodule in a legume infected by *Rhizobium*. Formation of the bacteroid state is a prerequisite for nitrogen fixation. See Figure 25.14 for physiological activities in the nodule.

Labels in figure:
- Root hair
- Recognition and attachment (rhicadhesin-mediated)
- Rhizobial cell
- Excretion of nod factors by bacterium causing root hair curling
- Invasion. Rhizobia penetrate root hair and multiply within an "infection thread"
- Bacteria in infection thread grow toward root cell
- Infection thread
- Invaded plant cells and those nearby are stimulated to divide
- Formation of bacteroid state within plant root cells
- Soil
- Nodules
- Continued plant and bacterial cell division leads to nodules

Attachment and Infection

The roots of leguminous plants secrete organic compounds that stimulate the growth of a diverse rhizosphere microbial community. If rhizobia of the correct cross-inoculation group are in the soil, they will form large populations and eventually attach to the root hairs that extend from the roots of the plant (Figure 25.10). An adhesion protein called *rhicadhesin* is present on the cell surfaces of rhizobia. Other substances, such as carbohydrate-containing proteins called *lectins* and specific receptors in the plant cytoplasmic membrane, also play roles in plant–bacterium attachment.

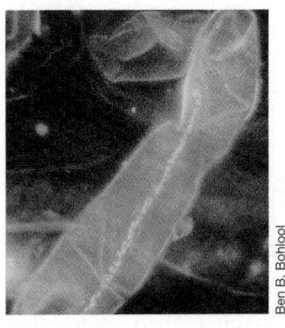
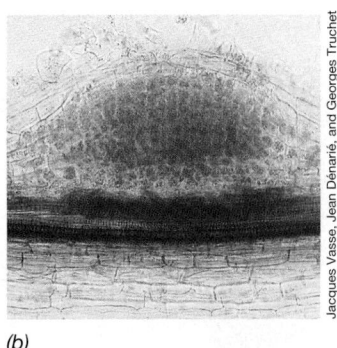
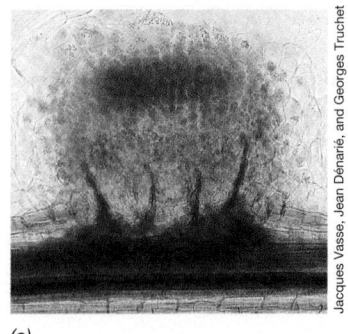
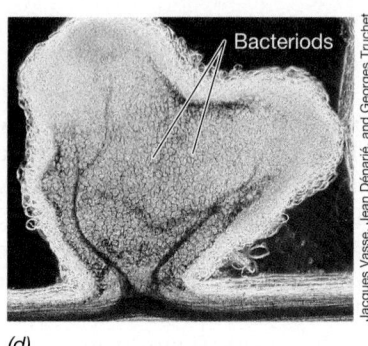

(a)   *(b)*   *(c)*   *(d)*

Figure 25.11 The infection thread and formation of root nodules. *(a)* An infection thread formed by cells of *Rhizobium leguminosarum* biovar *trifolii* on a root hair of white clover (*Trifolium repens*). The infection thread consists of a cellulosic tube through which bacteria move to root cells. *(b–d)* Nodules from alfalfa roots infected with cells of *Sinorhizobium meliloti* shown at different stages of development. Cells of both *R. leguminosarum* biovar *trifolii* and *S. meliloti* are about 2 mm long. The time course of nodulation events from infection to effective nodule is about 1 month in both soybean and alfalfa. Bacteroids are about 2 μm long. Photos b–d reprinted with permission from *Nature 351*:670–673 (1991), © Macmillan Magazines Ltd.

After attaching, a rhizobial cell penetrates into the root hair, which curls in response to substances excreted by the bacterium. The bacterium then induces formation by the plant of a cellulosic tube, called the **infection thread** (**Figure 25.11***a*), which spreads down the root hair. Root cells adjacent to the root hairs subsequently become infected by rhizobia, and plant cells divide. Continued plant cell division forms the tumorlike nodule (Figure 25.11*b–d*). A different mechanism of infection is used by some rhizobia adapted to aquatic or semiaquatic tropical legumes. These rhizobia enter the plant at the loose cellular junctions of roots emerging perpendicular from an established root (*lateral roots*). Following entry into the plant, some of the rhizobia develop infection threads, whereas others do not.

Bacteroids

The rhizobia multiply rapidly within the plant cells and become transformed into swollen, misshapen, and branched cells called **bacteroids**. A microcolony of bacteroids becomes surrounded by portions of the plant cytoplasmic membrane to form a structure called the *symbiosome* (Figure 25.11*d*), and only after the symbiosome forms does N_2 fixation begin. Nitrogen-fixing nodules can be detected experimentally by the reduction of acetylene to ethylene (⌘ Section 13.14). When the plant dies, the nodule deteriorates, releasing bacteroids into the soil. Although bacteroids are incapable of division, a small number of dormant rhizobial cells are always present in the nodule. These now proliferate, using some of the products of the deteriorating nodule as nutrients. The bacteria can then initiate infection the next growing season or maintain a free-living existence in the soil.

Nodule Formation: Nod Genes, Nod Proteins, and Nod Factors

Rhizobial genes that direct the steps in nodulation of a legume are called *nod genes*. It is thought that the ability to form nodules has independently emerged multiple times through the horizontal transfer of such genes as *nod* and *nif* that are located on plasmids or transferable regions of chromosomal DNA. In *Rhizobium leguminosarum* biovar *viciae*, which nodulates peas, ten *nod* genes have been identified. The *nodABC* genes encode proteins that produce oligosaccharides called **nod factors**; these induce root hair curling and trigger cell division in the pea plant, eventually leading to formation of the nodule (see Figure 25.14 for a description of root nodule biochemistry).

Nod factors consist of a backbone of *N*-acetylglucosamine to which various substituents are bonded (**Figure 25.12**). Which plants a given rhizobial species can infect is in part determined by the structure of the nod factor it produces. Besides the *nodABC* genes, which are universal and whose products synthesize

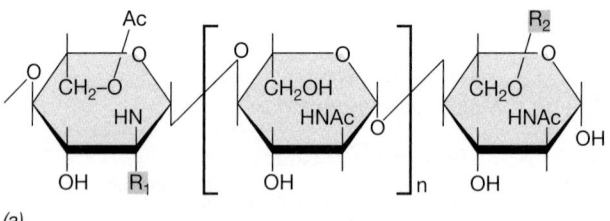

(a)

Species	R_1	R_2
Sinorhizobium meliloti (alfalfa)	C16:2 or C16:3	SO_4^{2-}
Rhizobium leguminosarum biovar *viciae* (pea)	C18:1 or C18:4	H or Ac

(b)

Figure 25.12 Nod factors. *(a)* General structure of the nod factors produced by *Sinorhizobium meliloti* and *Rhizobium leguminosarum* biovar *viciae* and *(b)* a table of the structural differences (R_1, R_2) that define the precise nod factor of each species. The central hexose unit can repeat up to three times. C16:2, palmitic acid with two double bonds; C16:3, palmitic acid with three double bonds; C18:1, oleic acid with one double bond; C18:4, oleic acid with four double bonds; Ac, acetyl.

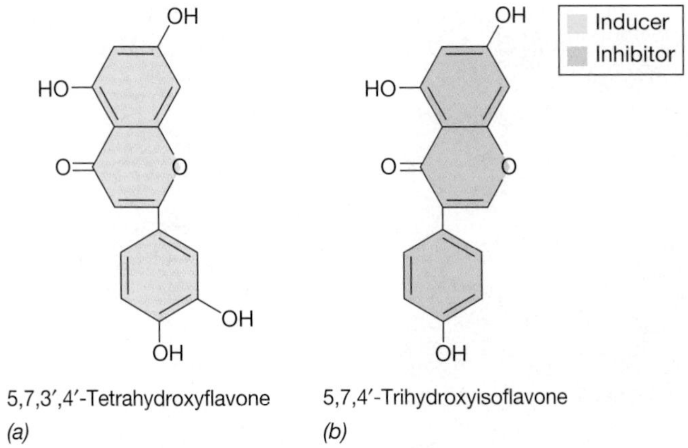

HO

OH

O

OH

OH

5,7,3′,4′-Tetrahydroxyflavone
(a)

HO

OH

O

OH

5,7,4′-Trihydroxyisoflavone
(b)

☐ Inducer
■ Inhibitor

Figure 25.13 Plant flavonoids and nodulation. Structures of flavonoid molecules that are *(a)* an inducer of *nod* gene expression and *(b)* an inhibitor of *nod* gene expression in *Rhizobium leguminosarum* biovar *viciae*. Note the similarities in the structures of the two molecules. The common name of the structure shown in part a is *luteolin*, and it is a flavone derivative. The structure in part b is called *genistein*, and it is an isoflavone derivative.

the nod backbone, each cross-inoculation group contains *nod* genes that encode proteins that chemically modify the nod factor backbone to form its species-specific molecule (Figure 25.12). In *R. leguminosarum* biovar *viciae*, *nodD* encodes the regulatory protein NodD, which controls transcription of other *nod* genes. After interacting with inducer molecules, NodD promotes transcription and is thus a positive regulatory protein (꩜ Section 8.4). NodD inducers are plant flavonoids, organic molecules that are widely excreted by plants (**Figure 25.13**). Some flavonoids that are structurally very closely related to *nodD* inducers in *R. leguminosarum* biovar *viciae* inhibit *nod* gene expression in other rhizobial species (Figure 25.13). This indicates that part of the specificity observed between plant and bacterium in the rhizobia–legume symbioses lies in the chemistry of the flavonoids excreted by each species of legume.

Biochemistry of Root Nodules

As discussed in Section 13.14, N_2 fixation requires the enzyme nitrogenase. Nitrogenase from bacteroids shows the same biochemical properties as the enzyme from free-living N_2-fixing bacteria, including O_2 sensitivity and the ability to reduce acetylene as well as N_2. Bacteroids are dependent on the plant for the electron donor for N_2 fixation. The major organic compounds transported across the symbiosome membrane and into the bacteroid proper are citric-acid-cycle intermediates—in particular, the C_4 organic acids *succinate, malate,* and *fumarate* (**Figure 25.14**). These are used as electron donors for ATP production and, following conversion to pyruvate, as the ultimate source of electrons for the reduction of N_2.

The product of N_2 fixation is ammonia (NH_3), and the plant assimilates most of this NH_3 by forming organic nitrogen compounds. The NH_3-assimilating enzyme glutamine synthetase is present in high levels in the plant cell cytoplasm and can convert glutamate and NH_3 into glutamine (꩜ Section 4.14). This and a

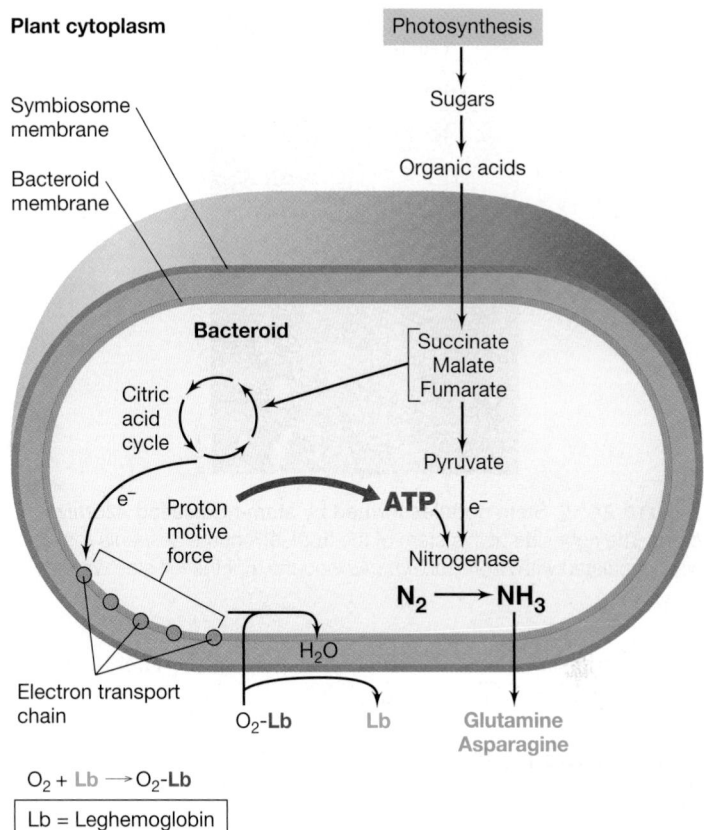

Figure 25.14 The root nodule bacteroid. Schematic diagram of major metabolic reactions and nutrient exchanges in the bacteroid. The symbiosome is a collection of bacteroids surrounded by a membrane originating from the plant.

few other organic nitrogen compounds transport bacterially fixed nitrogen throughout the plant. www.microbiologyplace.com **Online Tutorial 25.1: Root Nodule Bacteria and Symbiosis with Legumes**

Stem-Nodulating Rhizobia

Although most leguminous plants form N_2-fixing nodules on their *roots*, a few legume species bear nodules on their *stems*. Stem-nodulated leguminous plants are widespread in tropical regions where soils are often nitrogen deficient because of leaching and intense biological activity. The best-studied system is the tropical aquatic legume *Sesbania*, which is nodulated by the bacterium *Azorhizobium caulinodans* (**Figure 25.15**). Stem nodules typically form in the submerged portion of the stems or just above the water level. The general sequence of events by which stem nodules form in *Sesbania* resembles that of root nodules: attachment, formation of an infection thread, and bacteroid formation.

Some stem-nodulating rhizobia produce bacteriochlorophyll *a* and thus have the potential to carry out anoxygenic photosynthesis (꩜ Section 13.4). Bacteriochlorophyll-containing rhizobia, called photosynthetic *Bradyrhizobium*, are widespread in nature, particularly in association with tropical legumes. In these species, light energy converted to chemical energy (ATP) in

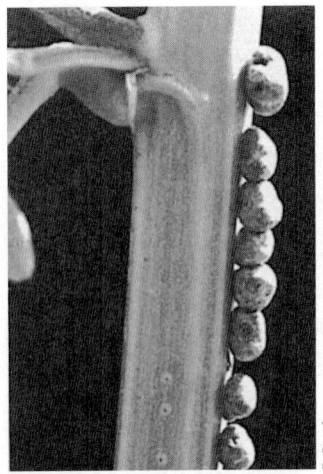

Figure 25.15 Stem nodules formed by stem-nodulating *Azorhizo-bium*. The right side of this stem of the tropical legume *Sesbania rostrata* was inoculated with *Azorhizobium caulinodans*, but the left side was not.

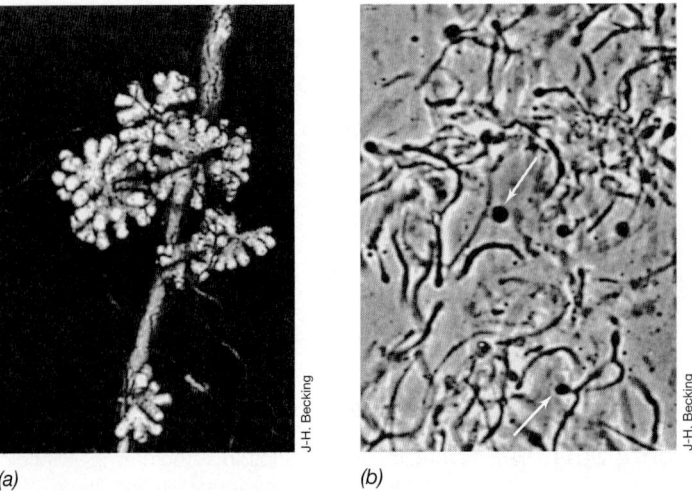

(a)   *(b)*

Figure 25.17 *Frankia* nodules and *Frankia* cells. *(a)* Root nodules of the common alder *Alnus glutinosa*. *(b) Frankia* culture purified from nodules of *Comptonia peregrina*. Note vesicles (arrows) on the tips of hyphal filaments.

photosynthesis is likely to be at least part of the energy source needed by the bacterium to support N₂ fixation.

Nonlegume N₂-Fixing Symbioses: *Azolla–Anabaena* and *Alnus–Frankia*

Various nonleguminous plants form N₂-fixing symbioses with bacteria other than rhizobia. For example, the water fern *Azolla* harbors within small pores of its fronds a species of heterocystous N₂-fixing cyanobacteria called *Anabaena azollae* (**Figure 25.16**). *Azolla* has been used for centuries to enrich rice paddies with fixed nitrogen. Before planting rice, the farmer allows the surface of the rice paddy to become densely covered with *Azolla*.

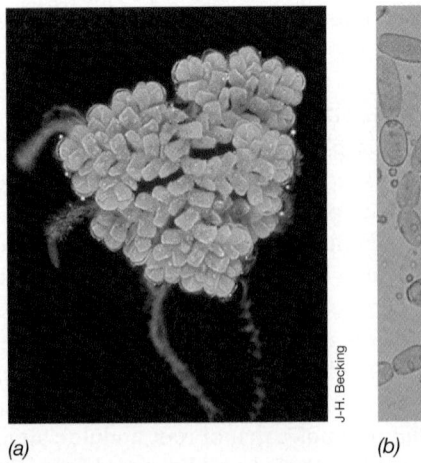 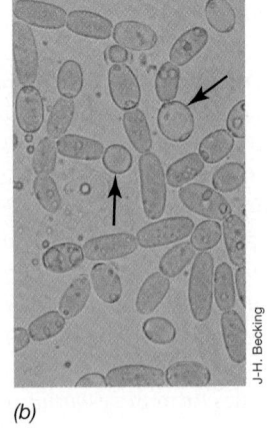

(a)   *(b)*

Figure 25.16 *Azolla–Anabaena* symbiosis. *(a)* Intact association showing a single plant of *Azolla pinnata*. The diameter of the plant is approximately 1 cm. *(b)* Cyanobacterial symbiont *Anabaena azollae* as observed in crushed leaves of *A. pinnata*. Single cells of *A. azollae* are about 5 μm wide. Vegetative cells are oblong; the spherical heterocysts (lighter color, arrows) are differentiated for nitrogen fixation.

As the rice plants grow, they eventually crowd out the *Azolla*, causing its death and the release of its nitrogen, which is assimilated by the rice plants. By repeating this process each growing season, rice farmers can obtain high yields of rice without applying nitrogenous fertilizers.

The alder tree (genus *Alnus*) has N₂-fixing root nodules (**Figure 25.17a**) that harbor filamentous, N₂-fixing actinomycetes of the genus *Frankia*. When assayed in cell extracts the nitrogenase of *Frankia* is sensitive to O₂, but cells of *Frankia* fix N₂ at full oxygen tensions. This is because *Frankia* protects its nitrogenase from O₂ by localizing the enzyme in terminal swellings on the cells called *vesicles* (Figure 25.17b). The vesicles contain thick walls that retard O₂ diffusion, thus maintaining the O₂ tension within vesicles at levels compatible with nitrogenase activity. In this regard, *Frankia* vesicles resemble the heterocysts produced by some filamentous cyanobacteria as localized sites of N₂ fixation (Section 13.14).

Alder is a characteristic pioneer tree able to colonize nutrient-poor soils, probably because of its ability to enter into a symbiotic N₂-fixing relationship with *Frankia*. A number of other small or bushy, woody plants are nodulated by *Frankia*. However, unlike the rhizobial symbionts of legumes, a single strain of *Frankia* can form nodules on several different species of plants, suggesting that the *Frankia*–root nodule symbiosis is less specific than that of leguminous plants.

MiniQuiz
- How do rhizobial root nodules benefit a plant?
- What are nod factors and what do they do?
- What is a bacteroid and what occurs within it? What is the function of leghemoglobin?
- What are the major similarities and differences between rhizobia and *Frankia*?

25.4 *Agrobacterium* and Crown Gall Disease

Some microorganisms develop parasitic symbioses with plants. The genus *Agrobacterium*, a relative of the root nodule bacterium *Rhizobium* (Figure 25.6), is such an organism, causing the formation of tumorous growths on diverse plants. The two species of *Agrobacterium* most widely studied are *Agrobacterium tumefaciens*, which causes *crown gall* disease, and *Agrobacterium rhizogenes*, which causes *hairy root* disease.

The Ti Plasmid

Although plants often form a benign accumulation of tissue called a *callus* when wounded, the growth in crown gall disease (**Figure 25.18**) is different in that it is uncontrolled growth, resembling an animal tumor. *A. tumefaciens* cells induce tumor formation only if they contain a large plasmid called the **Ti** (*t*umor *i*nducing) **plasmid**. In *A. rhizogenes*, a similar plasmid called the *Ri plasmid* is necessary for induction of hairy root disease. Following infection, a part of the Ti plasmid called the *transferred DNA* (T-DNA) is integrated into the plant's genome. T-DNA carries the genes for tumor formation and also for the production of a number of modified amino acids called *opines*. Octopine [N^2-(1,3-dicarboxyethyl)-L-arginine] and nopaline [N^2-(1,3-dicarboxypropyl)-L-arginine] are two common opines. Opines are produced by plant cells transformed by T-DNA and are a source of carbon and nitrogen, and sometimes phosphate, for the parasitic *A. tumefaciens* cells. These nutrients are the benefits for the bacterial symbiont.

Figure 25.18 Crown gall. Photograph of a crown gall tumor (arrow) on a tobacco plant caused by the crown gall bacterium *Agrobacterium tumefaciens*. The disease usually does not kill the plant but may weaken it and make it more susceptible to drought and diseases.

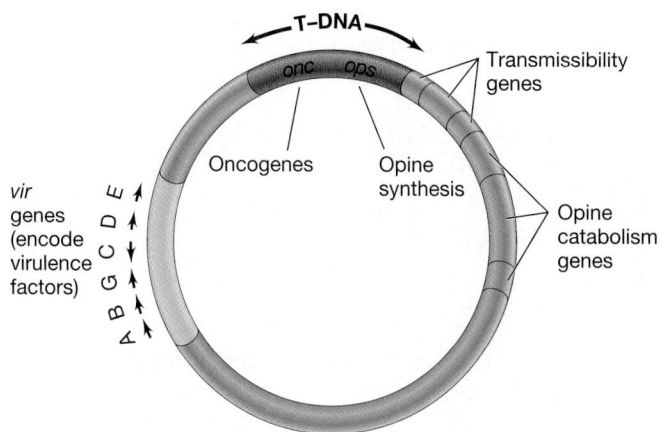

Figure 25.19 Structure of the Ti plasmid of *Agrobacterium tumefaciens*. T-DNA is the region transferred to the plant. Arrows indicate the direction of transcription of each gene. The entire Ti plasmid is about 200 kbp of DNA and the T-DNA is about 20 kbp.

Recognition and T-DNA Transfer

To initiate the tumorous state, *A. tumefaciens* cells attach to a wound site on the plant. Following attachment, the synthesis of cellulose microfibrils by the bacteria helps anchor them to the wound site, and bacterial aggregates form on the plant cell surface. This sets the stage for plasmid transfer from bacterium to plant.

The general structure of the Ti plasmid is shown in **Figure 25.19**. Only the T-DNA is actually transferred to the plant. The T-DNA contains genes that induce tumorigenesis. The *vir* genes on the Ti plasmid encode proteins that are essential for T-DNA transfer. Transcription of *vir* is induced by metabolites synthesized by wounded plant tissues. Examples of inducers include the phenolic compounds acetosyringone and ferulate. The transmissibility genes on the Ti plasmid (Figure 25.19) allow the plasmid to be transferred by conjugation from one bacterial cell to another.

The *vir* genes are the key to T-DNA transfer. The *virA* gene encodes a protein kinase (VirA) that interacts with inducer molecules and then phosphorylates the product of the *virG* gene (**Figure 25.20**). VirG is activated by phosphorylation and functions to activate other *vir* genes. The product of the *virD* gene (VirD) has endonuclease activity and nicks DNA in the Ti plasmid in a region adjacent to the T-DNA. The product of the *virE* gene is a DNA-binding protein that binds the single strand of T-DNA in the plant cell to protect it from destruction by nucleases. It is transferred into the plant cell independent of T-DNA. The *virB* operon encodes eleven different proteins that form a type IV secretion system for single-strand T-DNA and protein transfer between bacterium and plant (Figure 25.20) and thus resembles bacterial conjugation (⮂ Section 10.9). Laboratory studies of *A. tumefaciens* have shown that it can transfer T-DNA into many types of eukaryotic cells, including fungi, algae, protists, and even human cell lines.

Once inside the plant cell, T-DNA then becomes inserted into the genome of the plant. Tumorigenesis (*onc*) genes on the Ti

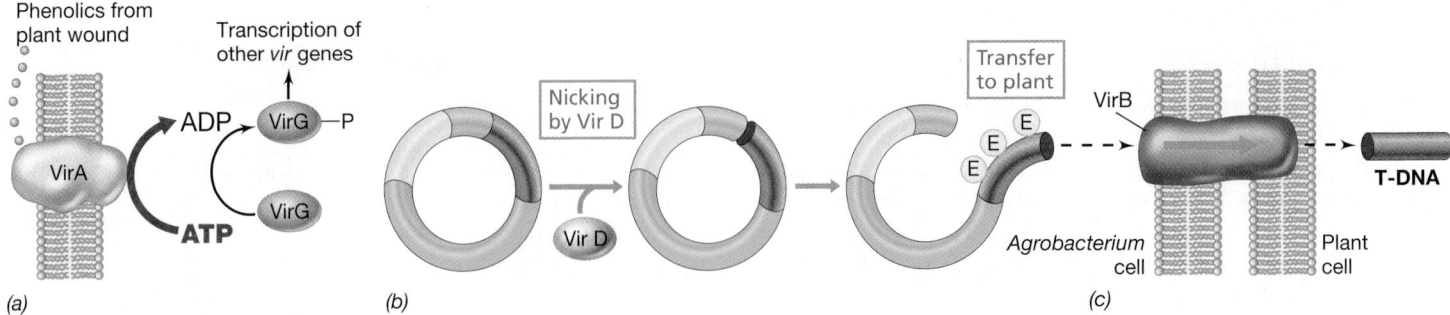

Figure 25.20 Mechanism of transfer of T-DNA to the plant cell by *Agrobacterium tumefaciens*. *(a)* VirA activates VirG by phosphorylation and VirG activates transcription of other *vir* genes. *(b)* VirD is an endonuclease that nicks the Ti plasmid, exposing the T-DNA. *(c)* VirB functions as a conjugation bridge between the *A. tumefaciens* cell and the plant cell, and VirE is a single-strand binding protein that assists in T-DNA transfer. Plant DNA polymerase produces the complementary strand to the transferred single strand of T-DNA.

plasmid (Figure 25.19) encode enzymes for plant hormone production and at least one key enzyme of opine biosynthesis. Expression of these genes leads to tumor formation and opine production. The Ri plasmid responsible for hairy root disease also contains *onc* genes. However, in this case the genes confer increased auxin responsiveness to the plant, and this promotes overproduction of root tissue and the symptoms of the disease. The Ri plasmid also encodes several opine biosynthetic enzymes.

Genetic Engineering with the Ti Plasmid

From the standpoint of microbiology and plant pathology, crown gall disease and hairy root disease both require intimate interactions that lead to genetic exchange from bacterium to plant. In other words, tumor induction in these diseases is the result of a natural plant-transformation system. Thus, in recent years interest in the Ti–crown gall system has shifted away from the disease itself toward applications of this natural genetic exchange process in plant biotechnology.

Several modified Ti plasmids that lack disease genes but that can still transfer DNA to plants have been developed by genetic engineering. These have been used for the construction of genetically modified (transgenic) plants. Many transgenic plants have been constructed thus far, including crop plants carrying genes for resistance to herbicides, insect attack, and drought. We discuss the use of the Ti plasmid as a vector in plant biotechnology in Section 15.18.

We discuss the use of the Ti plasmid as a vector in plant biotechnology in Section 15.18.

MiniQuiz

• What are opines and whom do they benefit?

• How do the *vir* genes differ from T-DNA in the Ti plasmid?

• How has an understanding of crown gall disease benefited plant agriculture?

25.5 Mycorrhizae

Mycorrhizae are mutualisms between plant roots and fungi in which nutrients are transferred in both directions. The fungus provides nutrients such as phosphorus from the soil to the plant, and the plant in turn transfers carbohydrates to the fungus.

These mutualisms are harnessed in agricultural applications. From fungal spores produced in culture or from root scrapings of infected plants, soil inoculants are produced that enhance plant growth.

Kinds of Mycorrhizae

There are two kinds of mycorrhizae. In *ectomycorrhizae*, fungal cells form an extensive sheath around the outside of the root with only a slight penetration into the root tissue itself (**Figure 25.21**). In *endomycorrhizae*, a part of the fungus becomes deeply embedded within the root tissue. Ectomycorrhizae are found mainly on the roots of forest trees, especially conifers, beeches, and oaks, and are most highly developed in boreal and temperate forests. In such forests, almost every root of every tree is mycorrhizal. The root system of a mycorrhizal tree such as a pine (genus *Pinus*) is composed of both long and short roots. The short roots, which are characteristically dichotomously branched in *Pinus* (Figure 25.21*a*), show typical fungal colonization, and long roots are also frequently colonized. Most mycorrhizal fungi do not catabolize cellulose and other leaf litter polymers. Instead, they catabolize simple carbohydrates and typically have one or more vitamin requirements. They obtain their carbon from root secretions and obtain inorganic minerals from the soil. Mycorrhizal fungi are rarely found in nature except in association with roots, and many are probably obligate symbionts.

Despite the close symbiotic association between fungus and root, a single species of tree can form multiple mycorrhizal associations. One pine species can associate with over 40 species of fungi. This relative lack of host specificity allows ectomycorrhizal mycelia to interconnect trees, providing linkages for transfer of carbon and other nutrients between trees of the same or different species. Nutrient transfer from well-illuminated overstory plants to shaded trees is thought to help equalize resource availability, subsidizing young trees and increasing biodiversity by promoting the coexistence of different species.

Arbuscular Mycorrhizae

Although ectomycorrhizal fungi have a significant impact on the ecology of forests, there is a greater diversity of endomycorrhizae. Most are *arbuscular mycorrhizae* (AM) that comprise a

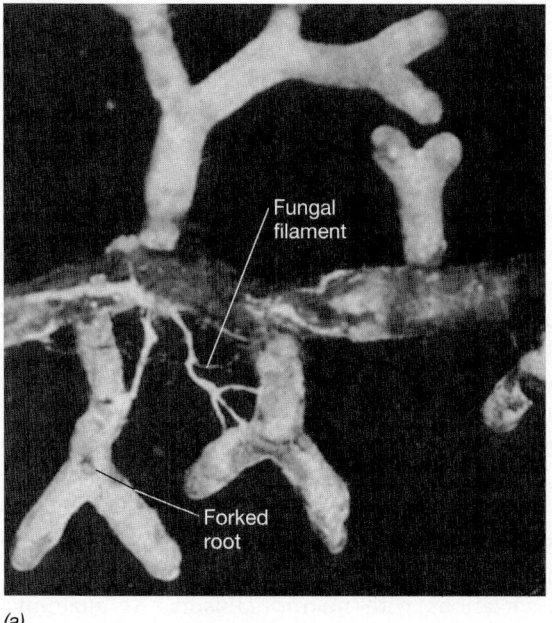

(a)   *(b)*

Figure 25.21 Mycorrhizae. *(a)* Typical ectomycorrhizal root of the pine *Pinus rigida* with filaments of the fungus *Thelophora terrestris*. *(b)* Seedling of *Pinus contorta* (lodgepole pine), showing extensive development of the absorptive mycelium of its fungal associate *Suillus bovinus*. This grows in a fanlike formation from the ectomycorrhizal roots to capture nutrients from the soil. The seedling is about 12 cm high.

phylogenetically distinct fungal division, the *Glomeromycota* (↺ Section 20.16), of which all or most species are obligate plant mutualists (the word "arbuscular" means "little tree"). AM colonize more than 85% of all terrestrial plants, including most grassland species and many crop species. The association between plants and the *Glomeromycota* is thought to be the ancestral type of mycorrhizae, established 400–460 million years ago and an important evolutionary step in the successful invasion of dry land by terrestrial plants.

AM fungi produce plant growth substances that induce morphological alterations in the roots, stimulating formation of the mycorrhizal state. Root colonization by an AM fungus begins with germination of a soil-borne spore, producing a short germination mycelium that recognizes the host plant through chemical signaling and then forms a contact structure with root epidermal cells called the *hyphopodium* (**Figure 25.22**). Penetrating hyphae extend into the plant from each hyphopodium, usually taking an intracellular path through epidermal and outer cortical cell layers of the root before forming dichotomously branched or coiled hyphal structures (the *arbuscules*) within cells of the inner cortex near the plant's vascular tissue. However, the arbuscular hyphae remain separated from plant protoplasm by an extensive plant cell membrane, which functions to increase the surface area of contact between plant and fungus.

Benefits for the Plant

The beneficial effect of the mycorrhizal fungus on the plant is best observed in poor soils where plants that are mycorrhizal thrive, but nonmycorrhizal ones do not. For example, if trees planted in prairie soils, which ordinarily lack a suitable fungal

inoculum, are artificially inoculated at the time of planting, they grow much more rapidly than uninoculated trees (**Figure 25.23**). The mycorrhizal plant can absorb nutrients from its environment more efficiently and thus has a competitive advantage. This improved nutrient absorption is due to the greater surface area provided by the fungal mycelium. For example, in the pine

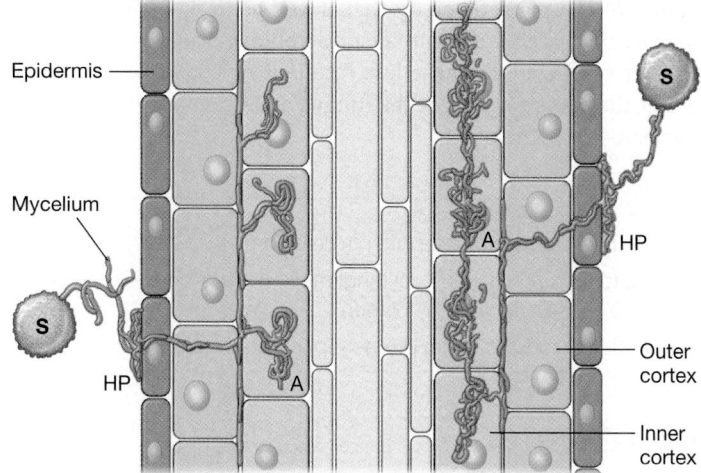

Figure 25.22 Arbuscular mycorrhizae root colonization. A spore (S) near a tree root generates a short mycelium that is attracted to the root by chemical signaling, forming an attachment structure called the hyphopodium (HP). The mycelium then enters the inner cortex region of the root by penetrating epidermal cells and cells of the outer cortex. Arbuscules (dichotomously branched invaginations, A) are formed by mycelia spreading either intercellularly (left) or intracellularly (right).

S. A. Wilde

Figure 25.23 Effect of mycorrhizal fungi on plant growth. Six-month-old seedlings of Monterey pine (*Pinus radiata*) growing in pots containing prairie soil: left, nonmycorrhizal; right, mycorrhizal.

seedling shown in Figure 25.21*b*, the ectomycorrhizal fungal mycelium makes up the overwhelming part of the absorptive capacity of the plant root system. The mycorrhizal plant is better able to function physiologically and compete successfully in a species-rich plant community, and the fungus benefits from a steady supply of organic nutrients.

In addition to helping plants absorb nutrients, mycorrhizae also play a significant role in supporting plant diversity. Field experiments have clearly shown a positive correlation between the abundance and diversity of mycorrhizae in a soil and the extent of the plant diversity that develops in it. Although most mycorrhizae are a true mutualistic symbiosis, there are also parasitic mycorrhizae. In these less frequent mycorrhizal symbioses either the plant parasitizes the fungus or, less commonly, the fungus parasitizes the plant.

MiniQuiz
- How do endomycorrhizae differ from ectomycorrhizae?
- What features of mycorrhizal fungi might have assisted in colonization of dry land by plants?
- How do mycorrhizal fungi promote plant diversity?

 # Mammals as Microbial Habitats

The evolution of animals has been shaped in part by a long history of symbiotic associations with microorganisms. To narrow our focus and look in depth at some details of these symbioses, we consider only mammals here. Microorganisms inhabit all sites on mammalian bodies, but the greatest diversity and density of microorganisms are found in the mammalian gut, and we center our discussion there. And finally, of the many mammals on Earth, we restrict our attention to ruminants and humans, the best-studied animals in terms of their gut microflora.

25.6 The Mammalian Gut

Some mammals are *herbivores*, consuming only plant materials, whereas others are *carnivores*, eating primarily the flesh of other animals. *Omnivores* eat both plants and animals. As **Figure 25.24** indicates, closely related mammals have evolved adaptations for differing diets. Notice that mammals of different lineages independently evolved the herbivorous lifestyle, mostly during the Jurassic period, an era in Earth's history of roughly 60 million years beginning about 200 million years ago. The massive evolutionary radiation of mammals during the Jurassic led to the evolution of several feeding strategies. Most mammalian species evolved gut structures that foster mutualistic associations with microorganisms. As anatomical differences evolved, microbial fermentation remained important or essential in mammalian digestion. *Monogastric* mammals, such as humans, have a single compartment, the stomach, positioned before the intestine. Such animals may get a substantial part of their energy requirement from microbial fermentation of otherwise indigestible foods, but herbivores are totally dependent on such fermentations.

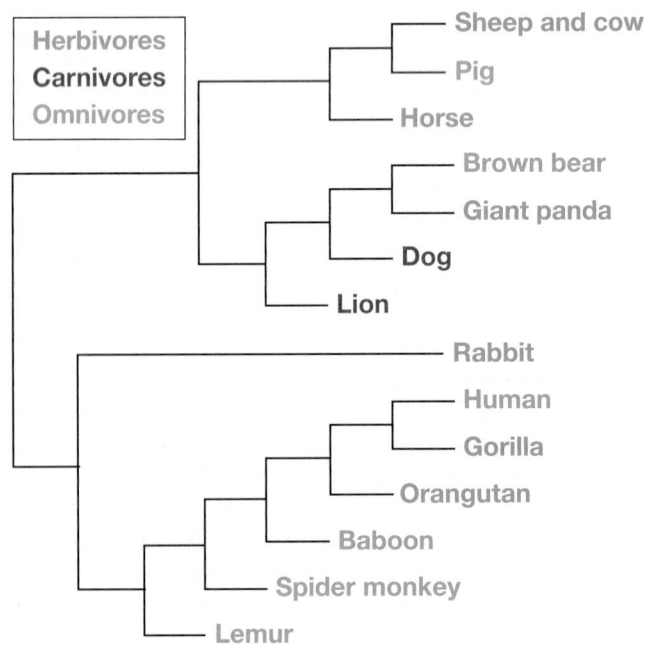

Figure 25.24 Phylogenetic tree showing multiple origins of herbivory among mammals. Some of the herbivores listed are foregut fermenters, while others are hindgut fermenters (Figure 25.25). Instead of animal flesh, some mammalian carnivores eat only insects (the insectivores, such as bats), or fish (the piscivores, such as the river otter).

Plant Substrates

Microbial associations with various mammalian species led to the capacity to catabolize plant fiber, the structural component of plant cell walls. Fiber is composed primarily of insoluble polysaccharides of which cellulose is the most abundant component. Mammals—and indeed almost all animals—lack the enzymes necessary to digest cellulose and certain other plant polysaccharides. Only microorganisms have genes encoding the glycoside hydrolases and polysaccharide lyases required to decompose these polysaccharides. As the most abundant organic compound on Earth and one composed exclusively of glucose, cellulose offers a rich source of carbon and energy for animals that can catabolize it. The two primary traits that evolved to support herbivory are (1) an enlarged anoxic fermentation chamber for holding ingested plant material, and (2) an extended retention time—the time that ingested material remains in the gut. A longer retention time allows for a longer association of microorganisms with the ingested material and thus a more complete degradation of the plant polymers.

Foregut versus Hindgut Fermenters

Two digestive plans have evolved in herbivorous mammals. In herbivores with a *foregut* fermentation, the microbial fermentation chamber *precedes* the small intestine. This gut architecture originated independently in ruminants, colobine monkeys, sloths, and macropod marsupials (**Figure 25.25**). These all share the common feature that ingested nutrients are degraded by the gut microbiota *before* reaching the acidic stomach and small intestine. We examine the digestive processes of ruminants, as examples of foregut fermenters, in the next section.

Horses and rabbits are herbivorous mammals, but they are not foregut fermenters. Instead, these animals are *hindgut* fermenters. They have only one stomach, but use an organ called the *cecum*, a digestive organ located between the small and large intestines, as their fermentation vessel. The cecum contains fiber- and cellulose-digesting (cellulolytic) microorganisms. Mammals, such as the rabbit, that rely primarily on microbial breakdown of plant fiber in the cecum are called *cecal fermenters*. In other hindgut

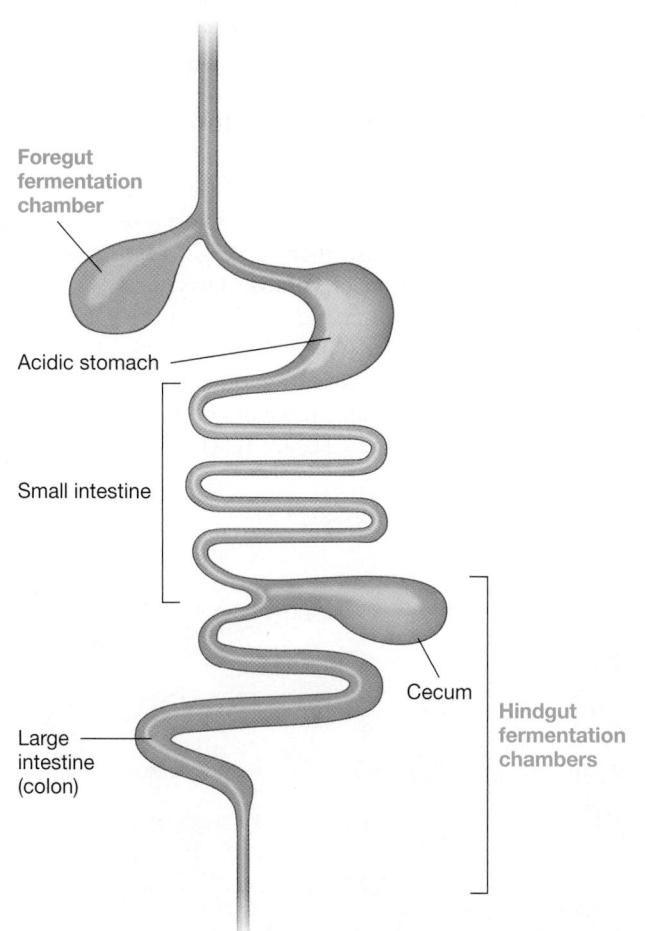

Foregut fermenters Examples: Ruminants (photo 1), colobine monkeys, macropod marsupials, hoatzin (photo 2)

Hindgut fermenters Examples: Cecal animals (photos 3 and 4), primates, some rodents, some reptiles

Figure 25.25 Variations on vertebrate gut architecture. All vertebrates have a small intestine, but vary in other gut structures. Most host absorption of dietary nutrients occurs in the small intestine, whereas microbial fermentation can occur in the forestomach, cecum, or large intestine (colon). Foregut fermentation is found in four major clades of mammals and one avian species (the hoatzin). Hindgut fermentation, either in the cecum or large intestine/colon, is common to many clades of mammals (including humans), birds, and reptiles. Compare with Figure 25.24.

fermenters, both the cecum and colon are major sites of fiber breakdown by microorganisms.

Anatomical differences among monogastric mammals, foregut fermenters, and hindgut fermenters are summarized in Figure 25.25. Nutritionally, foregut fermenters have an advantage over hindgut fermenters in that the cellulolytic microbial community of the foregut eventually passes through an acidic stomach. As this occurs, most microbial cells are killed by the acidity and become a protein source for the animal. By contrast, in animals such as horses and rabbits, the remains of the cellulolytic community pass out of the animal in the feces because of its position posterior to the acidic stomach.

MiniQuiz

- How do animals with foregut and hindgut fermentation differ in recovery of nutrients from plants?
- How does retention time affect microbial digestion of food in a gut compartment?

25.7 The Rumen and Ruminant Animals

A very successful group of foregut fermenters are *ruminants*, herbivorous mammals that possess a special digestive organ, the **rumen**, within which cellulose and other plant polysaccharides are digested by microorganisms. Some of the most important domesticated animals—cows, sheep, and goats—are ruminants. Camels, buffalo, deer, reindeer, caribou, and elk are also ruminants. Indeed, ruminants are Earth's dominant herbivores. Because the human food economy depends to a great extent on ruminant animals, rumen microbiology is of considerable economic significance and importance.

Rumen Anatomy and Activity

Unique features of the rumen as a site of cellulose digestion are its relatively large size (capable of holding 100–150 liters in a cow, 6 liters in a sheep) and its position in the gastrointestinal system before the acidic stomach. The rumen's warm and constant temperature (39°C), narrow pH range (5.5–7, depending on when the animal was last fed), and anoxic environment are also important factors in overall rumen function.

Figure 25.26a shows the relationship of the rumen to other parts of the ruminant digestive system. The digestive processes and microbiology of the rumen have been well studied, in part because it is possible to implant a sampling port, called a *fistula*, into the rumen of a cow (Figure 25.26*b*) or a sheep and remove samples for analysis.

After a cow swallows its food, it enters the first chamber of the four-compartment stomach, the reticulum. Digesta flow freely between the rumen and reticulum, sometimes referred to together as the reticulo-rumen. The main function of the reticulum is to collect smaller food particles and move them to the omasum. Larger food particles (called cud) are regurgitated, chewed, mixed with saliva containing bicarbonate, and returned to the reticulo-rumen, where they are digested by ruminal bacteria. Solids may remain in the rumen for more than a day during digestion. Eventually, small and more thoroughly digested food particles are passed to the omasum and from there to the abomasum, an organ similar to a true, acidic stomach. In the abomasum, chemical digestive processes begin that continue in the small and large intestine.

Microbial Fermentation in the Rumen

Food remains in the rumen for 20–50 hours depending on the feeding schedule and other factors. During this relatively long retention time, cellulolytic microorganisms hydrolyze cellulose, which frees glucose. The glucose then undergoes bacterial fermentation with the production of **volatile fatty acids (VFAs)**, primarily *acetic, propionic*, and *butyric* acids, and the gases carbon dioxide (CO_2) and methane (CH_4) (**Figure 25.27**). The VFAs pass through the rumen wall into the bloodstream and are oxidized by the animal as its main source of energy. The gaseous fermentation products CO_2 and CH_4 are released by eructation (belching).

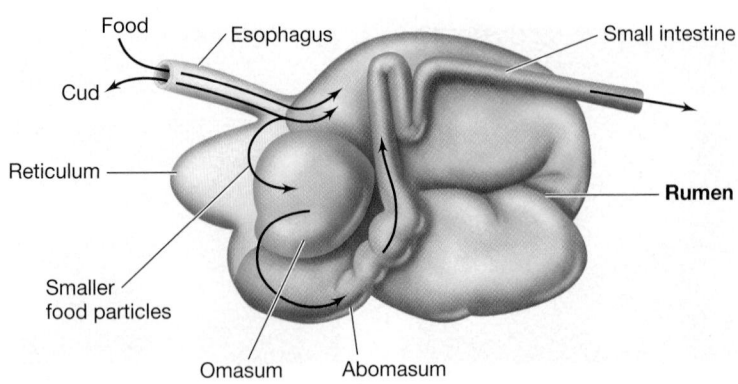

(a)

(b)

Figure 25.26 The rumen. (a) Schematic diagram of the rumen and gastrointestinal system of a cow. Food travels from the esophagus into the reticulo-rumen, consisting of the reticulum and rumen. Cud is regurgitated and chewed until food particles are small enough to pass from the reticulum into the omasum, abomasum, and intestines, in that order. The abomasum is an acidic vessel, analogous to the stomach of monogastric animals like pigs and humans. (b) Photo of a fistulated Holstein cow. The fistula, shown unplugged, is a sampling port that allows access to the rumen.

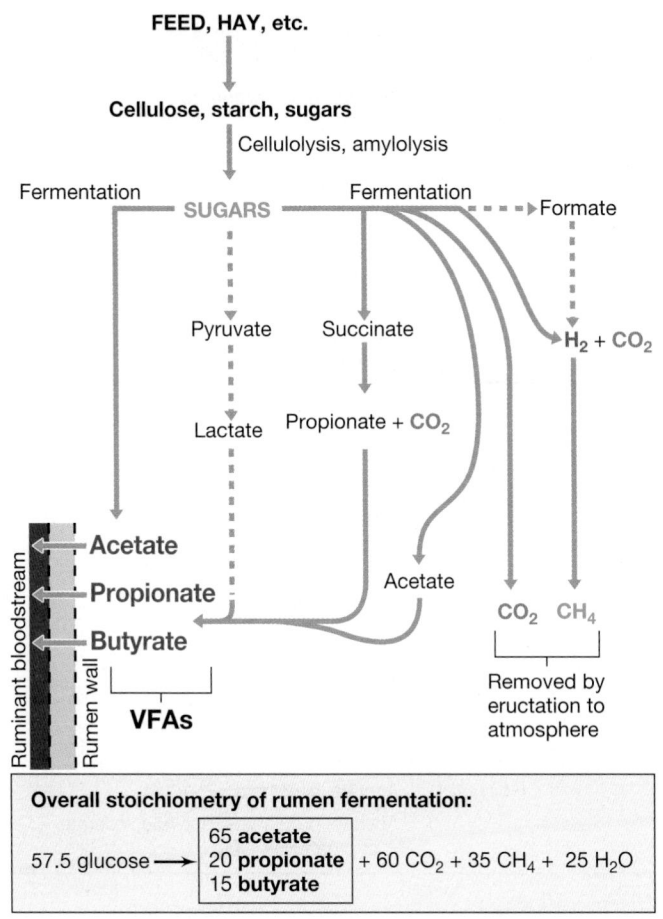

Figure 25.27 Biochemical reactions in the rumen. The major pathways are solid lines; dashed lines indicate minor pathways. Approximate steady-state rumen levels of volatile fatty acids (VFAs) are acetate, 60 mM; propionate, 20 mM; butyrate, 10 mM.

The rumen contains enormous numbers of bacteria (10^{10}–10^{11} cells/g of rumen contents). Most of the bacteria adhere tightly to food particles. These particles proceed through the gastrointestinal tract of the animal where they undergo further digestive processes similar to those of nonruminant animals. Bacterial cells that digested plant fiber in the rumen are themselves digested in the acidic abomasum. Because bacteria living in the rumen biosynthesize amino acids and vitamins, the digested bacterial cells are a major source of protein and vitamins for the animal.

Rumen Bacteria

Although some microbial eukaryotes are present, anaerobic bacteria dominate in the rumen because it is a strictly anoxic compartment. Cellulose is converted to fatty acids, CO_2, and CH_4 in a multistep microbial food chain, with several different anaerobes participating in the process. Recent estimates of ruminal microbial diversity from analysis of 16S rRNA gene sequences suggest that the typical rumen contains 300–400 bacterial "species" (defined as "operational taxonomic units" sharing less than 97% sequence identity, ↺ Section 16.12) (**Figure 25.28**). This is more than 10 times higher than culture-based diversity estimates. Molecular surveys show that species of *Firmicutes* and *Bacteroidetes* dominate the *Bacteria* in the rumen, while methanogens make up virtually the entire archaeal population.

A number of rumen anaerobes have been cultured and their physiology characterized (**Table 25.2**). Several different rumen bacteria hydrolyze cellulose to sugars and ferment the sugars to VFAs. *Fibrobacter succinogenes* and *Ruminococcus albus* are the two most abundant cellulolytic rumen anaerobes. Although both organisms produce cellulases, *Fibrobacter*, a gram-negative bacterium, produces enzymes localized to the outer membrane.

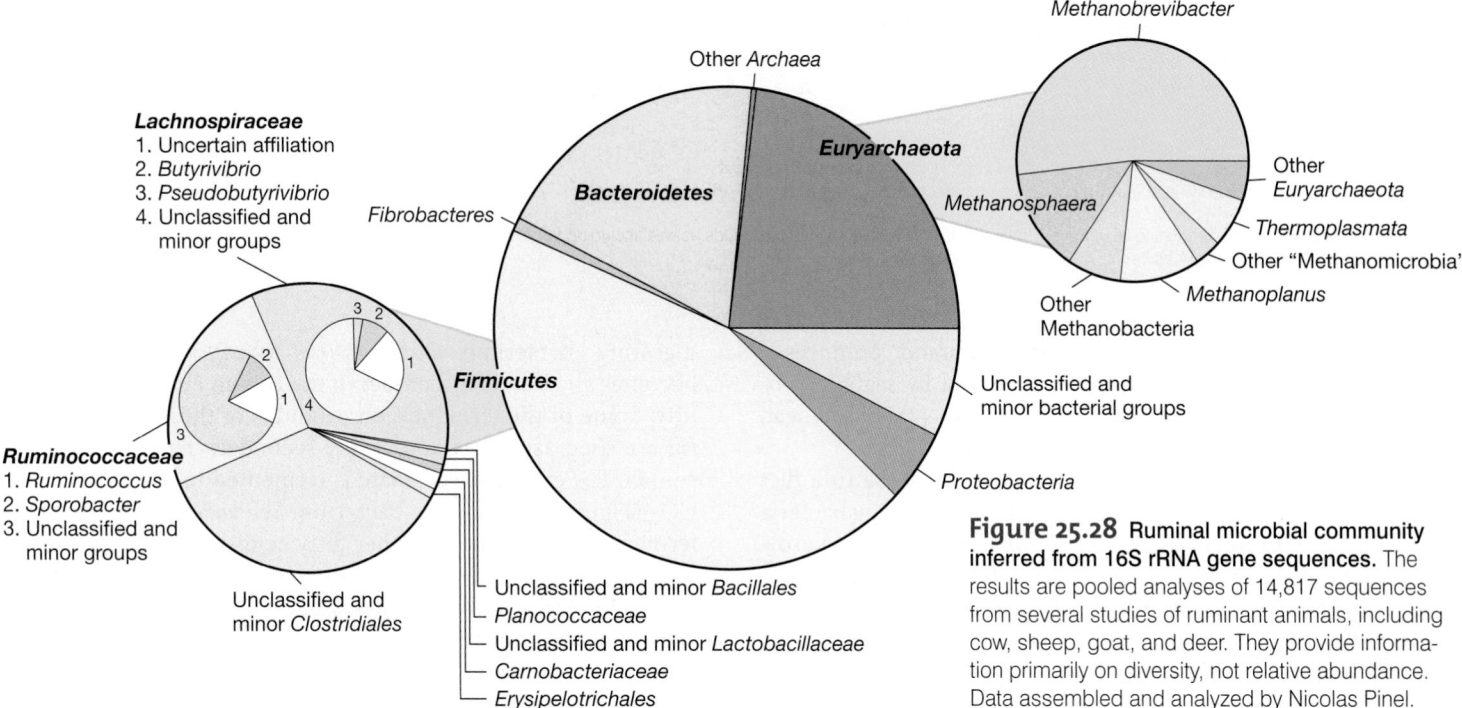

Figure 25.28 Ruminal microbial community inferred from 16S rRNA gene sequences. The results are pooled analyses of 14,817 sequences from several studies of ruminant animals, including cow, sheep, goat, and deer. They provide information primarily on diversity, not relative abundance. Data assembled and analyzed by Nicolas Pinel.

Table 25.2 *Characteristics of some rumen prokaryotes*

Organism[a]	Morphology	Fermentation products
Cellulose decomposers		
Gram-negative		
Fibrobacter succinogenes[b]	Rod	Succinate, acetate, formate
Butyrivibrio fibrisolvens[c]	Curved rod	Acetate, formate, lactate, butyrate, H_2, CO_2
Gram-positive		
Ruminococcus albus[c]	Coccus	Acetate, formate, H_2, CO_2
Clostridium lochheadii	Rod (endospores)	Acetate, formate, butyrate, H_2, CO_2
Starch decomposers		
Gram-negative		
Prevotella ruminicola[d]	Rod	Formate, acetate, succinate
Ruminobacter amylophilus	Rod	Formate, acetate, succinate
Selenomonas ruminantium	Curved rod	Acetate, propionate, lactate
Succinomonas amylolytica	Oval	Acetate, propionate, succinate
Gram-positive		
Streptococcus bovis	Coccus	Lactate
Lactate decomposers		
Gram-negative		
Selenomonas ruminantium subsp. *lactilytica*	Curved rod	Acetate, succinate
Megasphaera elsdenii	Coccus	Acetate, propionate, butyrate, valerate, caproate, H_2, CO_2
Succinate decomposer		
Gram-negative		
Schwartzia succinovorans	Rod	Propionate, CO_2
Pectin decomposer		
Gram-positive		
Lachnospira multipara	Curved rod	Acetate, formate, lactate, H_2, CO_2
Methanogens		
Methanobrevibacter ruminantium	Rod	CH_4 (from H_2 + CO_2 or formate)
Methanomicrobium mobile	Rod	CH_4 (from H_2 + CO_2 or formate)

[a] Except for the methanogens, which are *Archaea*, all organisms listed are species of *Bacteria*.
[b] These species also degrade xylan, a major plant cell wall polysaccharide (⮌ Section 14.16).
[c] Also degrades starch.
[d] Also ferments amino acids, producing NH_3. Several other rumen bacteria ferment amino acids as well, including *Peptostreptococcus anaerobius* and *Clostridium sticklandii*.

Ruminococcus, which lacks an outer membrane, produces a cellulose-degrading protein complex stabilized by scaffold proteins and bound to the cell wall. Both organisms therefore need to bind to cellulose particles in order to degrade them.

If a ruminant is gradually switched from cellulose to a diet high in starch (grain, for instance), the starch-digesting bacteria *Ruminobacter amylophilus* and *Succinomonas amylolytica* grow to high numbers in the rumen. On a low-starch diet these organisms are typically minor constituents. If an animal is fed legume hay, which is high in pectin, a complex polysaccharide containing both hexose and pentose sugars, then the pectin-digesting bacterium *Lachnospira multipara* (Table 25.2) becomes an abundant member of the rumen microbial community. Some of the fermentation products of these rumen bacteria are used as energy sources by secondary fermenters in the rumen. For example, succinate is fermented to propionate plus CO_2 (Figure 25.27) by the bacterium *Schwartzia*, and lactate is fermented to acetate and other fatty acids by *Selenomonas* and *Megasphaera* (Table 25.2). Hydrogen (H_2) produced in the rumen by fermentative processes never accumulates because it is quickly consumed by methanogens for the reduction of CO_2 to CH_4.

Dangerous Changes in the Rumen Microbial Community

Significant changes in the microbial composition of the rumen can cause illness or even death of the animal. For example, if a cow is changed abruptly from forage to a grain diet, the gram-positive bacterium *Streptococcus bovis* grows rapidly in the rumen. The normal level of *S. bovis*, about 10^7 cells/g, is an insignificant fraction of total rumen bacterial numbers. But if large amounts of grain are fed abruptly, numbers of *S. bovis* can quickly rise to dominate the rumen microbial community to over 10^{10} cells/g. This occurs because grasses contain mainly cellulose, which does not support growth of *S. bovis*, while grain contains high levels of starch, on which *S. bovis* grows rapidly.

Because *S. bovis* is a lactic acid bacterium (Sections 14.2 and 18.1), large populations are capable of producing large amounts of lactic acid. Lactic acid is a much stronger acid than the VFAs produced during normal rumen function. Lactate production thus acidifies the rumen below its lower functional limit of about pH 5.5, thereby disrupting the activities of normal rumen bacteria. Rumen acidification, a condition called *acidosis*, causes inflammation of the rumen epithelium, and severe acidosis can cause hemorrhaging in the rumen, acidification of the blood, and death of the animal. Despite the activities of *S. bovis*, ruminants such as cattle can be fed a diet exclusively of grain. However, to avoid acidosis, they must be switched from forage to grain *gradually* over a period of a few days. The slow introduction of starch selects for VFA-producing, starch-degrading bacteria (Table 25.2) instead of *S. bovis*, and thus normal rumen functions continue and the animal remains healthy.

Protective Changes in the Rumen Microbial Community

The overgrowth of *S. bovis* is an example of how a single microbial species can have a deleterious effect on animal health. There is also at least one well-studied example of how a single bacterial species can *enhance* the health of ruminant animals; in this case, animals fed the tropical legume, *Leucaena leucocephala*. This plant has a very high nutritional value, but contains an amino acid–like compound called *mimosine* that is converted to toxic 3-hydroxy-4(1H)-pyridone and 2,3-dihydroxypyridine (DHP) by rumen microorganisms (**Figure 25.29**). The observation that ruminants in Hawaii, but not Australia, could feed on *Leucaena* without toxic effect led investigators to hypothesize that further metabolism of DHP by bacteria present in Hawaiian ruminants alleviated DHP toxicity. This was subsequently confirmed by the isolation of the bacterium *Synergistes jonesii*, a unique anaerobe related to the *Deferribacter* group (Section 18.21) and not closely related to any other rumen bacteria. Inoculation of Australian ruminants with cells of *S. jonesii* conferred resistance to mimosine by-products, allowing them to feed on *Leucaena* without ill effect.

The success of this single-organism modification of the rumen microbial community has encouraged further studies of this sort, including genetic engineering of bacteria to improve their ability to utilize available nutrients or to detoxify toxic substances. A notable success has been inoculation of the rumen of sheep with

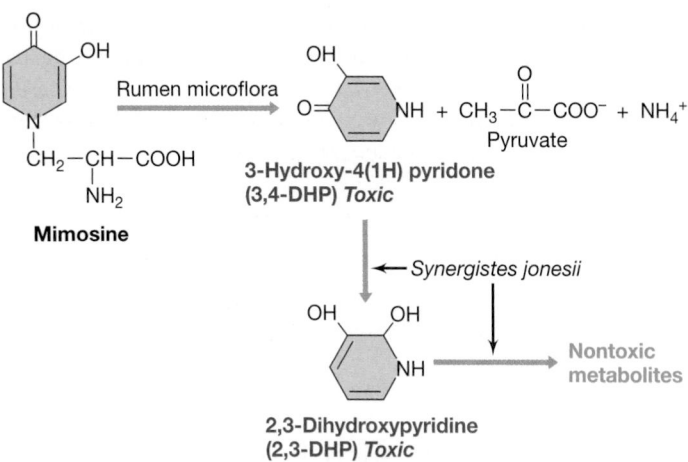

Figure 25.29 Conversion of mimosine to toxic pyridine and pyridone metabolites by ruminal microorganisms. Mimosine is converted to toxic 3,4-DHP by normal ruminal microbiota. *Synergistes jonesii* converts 3,4-DHP to nontoxic metabolites through a 2,3-DHP intermediate, preventing buildup of toxic metabolites of mimosine.

genetically engineered cells of *Butyrivibrio fibrisolvens* (Table 25.2) containing a gene encoding the enzyme fluoroacetate dehalogenase; this successfully prevented fluoroacetate poisoning of sheep fed plants containing high levels of this highly toxic inhibitor of the citric acid cycle.

Rumen Protists and Fungi

In addition to prokaryotes, the rumen has characteristic populations of ciliated protists (Chapter 20) present at a density of about 10^6 cells/ml. Many of these protists are obligate anaerobes, a property that is rare among eukaryotes. Although these protists are not essential for rumen fermentation, they contribute to the overall process. In fact, some protists are able to hydrolyze cellulose and starch and ferment glucose with the production of the same VFAs formed by cellulose-fermenting bacteria (Figure 25.27 and Table 25.2). Rumen protists also consume rumen bacteria and smaller rumen protists as food and are likely to play a role in controlling bacterial densities in the rumen. An interesting commensal interaction has been observed between rumen protists that produce VFAs and H_2 as products and methanogenic bacteria that consume the H_2, producing CH_4. Because they autofluoresce (Section 14.10), methanogens are easily observed in rumen fluid bound to the surface of H_2-producing protists.

Anaerobic fungi also inhabit the rumen and play a role in its digestive processes. Rumen fungi are typically species that alternate between a flagellated and a thallus form, and studies with pure cultures have shown that they can ferment cellulose to VFAs. *Neocallimastix*, for example, is an obligately anaerobic fungus that ferments glucose to formate, acetate, lactate, ethanol, CO_2, and H_2. Although a eukaryote, this fungus lacks mitochondria and cytochromes and thus lives an obligately fermentative existence. However, *Neocallimastix* cells contain a redox organelle called the *hydrogenosome*; this mitochondrial analog evolves H_2 and has thus far been found only in certain anaerobic protists (Section

20.2). Rumen fungi play an important role in the degradation of polysaccharides other than cellulose as well, including a partial solubilization of lignin (the strengthening agent in the cell walls of woody plants), hemicellulose (a derivative of cellulose that contains pentoses and other sugars), and pectin.

MiniQuiz

- What physical and chemical conditions prevail in the rumen?
- What are VFAs and of what value are they to the ruminant?
- Why is the metabolism of *Streptococcus bovis* of special concern to ruminant nutrition?

25.8 The Human Microbiome

The human *microbiome* encompasses all sites of the human body inhabited by microorganisms. These sites include the mouth, nasal cavities, throat, stomach, intestines, urogenital tracts, and skin (⮌ Sections 27.1–27.5). It is estimated that the number of microorganisms in the human microbiome is approximately 10^{14}, which is ten times more than the total number of human cells in a single person.

Importance to Human Health

The microbial community in the healthy human was once considered to consist of microorganisms that were merely commensals, but we now know that this microbial community is important in early development and overall health and predisposition to disease. Recognition that these microorganisms function as mutualists that play a central role in human health has prompted formation of an international research program called the *Human Microbiome Project* (HMP). Some of the major questions posed by the project include: (1) Do individuals share a core human microbiome? (2) Is there a correlation between microbial population structure and host genotype? (3) Do differences in the human microbiome correlate with differences in human health? (4) Are differences in the relative abundance of different bacteria important?

Human microbiome studies based on surveys of individuals using 16S rRNA gene sequencing and metagenomic analyses indicate that the diversity among individuals is so great that no one microbial species is found at high abundance in all individuals. Similarities between individuals are more evident at higher bacterial taxonomic levels such as phyla and in the distribution of genes of similar function in the gut community. The possible outcomes of these analyses for clinical medicine include the development of biomarkers for predicting an individual's predisposition to disease, the design of drugs targeting selected members of the intestinal microbial community, and personalized drug therapies.

The Human Gut Microbial Community

Humans are monogastric and omnivorous animals (Figure 25.25). In the human duodenum, ingested food passed down from the stomach is blended with bile, bicarbonate, and digestive enzymes. About 1–4 h after ingestion, food reaches the gut (large intestine) and by this time it is near neutral pH, and bacterial numbers have increased from about 10^4–10^8/g (**Figure 25.30**). Both the host and gut microorganisms share the easily digestible nutrients. The large intestine is the most heavily colonized area of the gastrointestinal tract and contains 10^{11}–10^{12} bacterial cells per g.

Colonization of an initially sterile gut begins immediately after birth; a succession of microbial populations replaces each other in turn until a stable, adult microbial community is established. The source of early colonizers is not clear, although some species are clearly transmitted from mother to infant. As is now recognized for most microbial communities, early descriptions of diversity based on culturing microorganisms greatly underestimated true diversity. For example, although we think of *Escherichia coli* as a significant gut bacterium, the entire phylum *Gammaproteobacteria* (to which *E. coli* belongs; ⮌ Section 17.1) makes up less than 1% of all gut bacteria. *E. coli* simply grows extremely well in laboratory culture and can thus be readily detected even when present in low numbers.

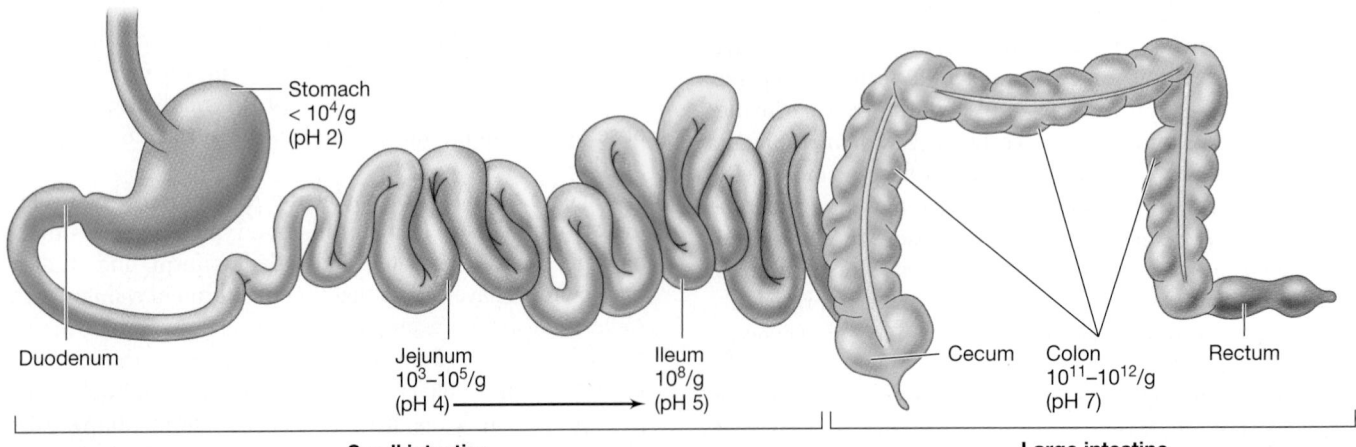

Figure 25.30 Numbers of bacteria in the monogastric human gastrointestinal tract. The small intestine is composed of the duodenum, jejunum, and ileum. Numbers in the individual sections are estimates of bacteria per gram of intestinal contents in healthy humans.

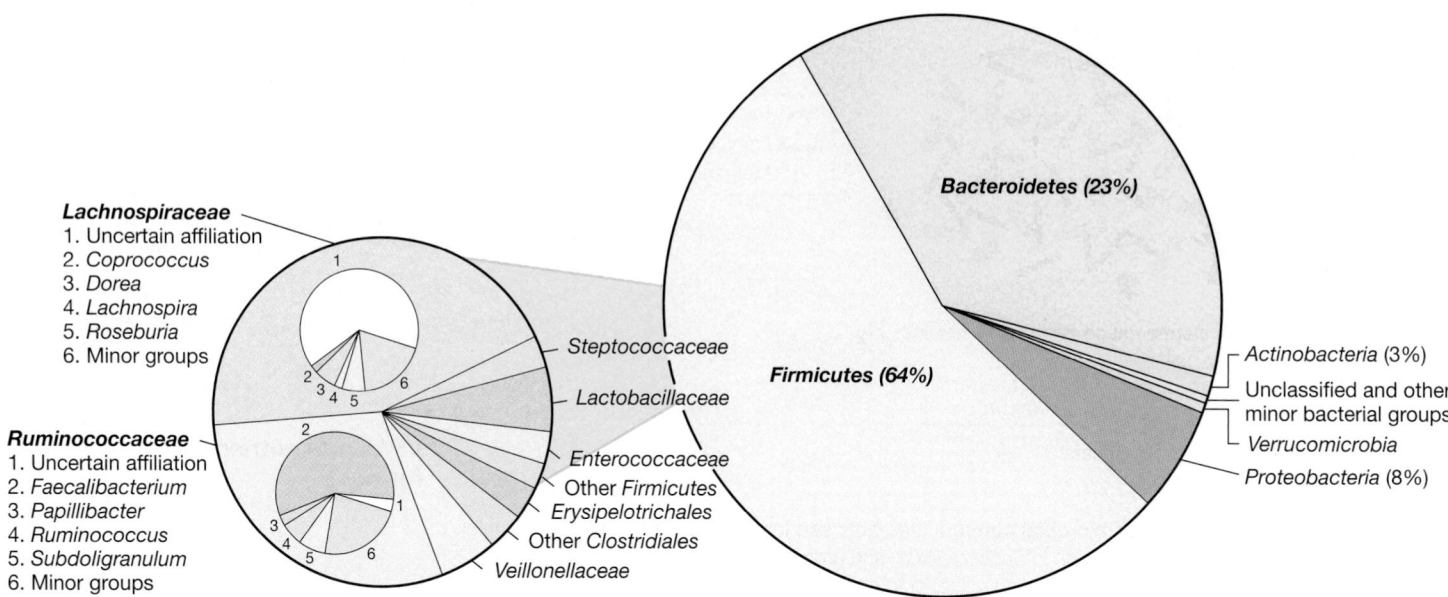

Lachnospiraceae
1. Uncertain affiliation
2. *Coprococcus*
3. *Dorea*
4. *Lachnospira*
5. *Roseburia*
6. Minor groups

Ruminococcaceae
1. Uncertain affiliation
2. *Faecalibacterium*
3. *Papillibacter*
4. *Ruminococcus*
5. *Subdoligranulum*
6. Minor groups

Bacteroidetes (23%)

Firmicutes (64%)

Steptococcaceae
Lactobacillaceae
Enterococcaceae
Other *Firmicutes*
Erysipelotrichales
Other *Clostridiales*
Veillonellaceae

Actinobacteria (3%)
Unclassified and other minor bacterial groups
Verrucomicrobia
Proteobacteria (8%)

Figure 25.31 Microbial composition of the human colon inferred from 16S rRNA gene sequences. The results are pooled analyses of 17,242 sequences mostly obtained from the distal colon (fecal samples) of several individuals. The data provide information primarily of diversity, not relative abundance. Data assembled and analyzed by Nicolas Pinel.

Somewhat surprisingly, mammalian gut communities are composed of only a few phyla and show a species composition distinct from that of any free-living microbial communities (Chapter 23). The vast majority (98%) of all human gut phylotypes fall into one of four bacterial groups: *Firmicutes* (64%), *Bacteroidetes* (23%), *Proteobacteria* (8%), and *Actinobacteria* (3%) (**Figure 25.31**). In contrast to the limited phylum-level diversity, the diversity of genera and species in the mammalian gut is enormous. A recent census of human intestinal diversity based on more than 50,000 bacterial 16S rRNA gene sequences identified at least 1800 genera, 16,000 species, and more than 36,000 strains. *Archaea* (represented by a phylotype closely related to the methanogen *Methanobrevibacter smithii*), yeasts, fungi, and protists make up only a minor part of the community.

Interestingly, there is high variability from person to person in microbial species abundance of the gut community. Studies of healthy men and women 27–94 years of age have revealed that each person harbors but a few hundred to no more than one thousand of the many thousands of different human gut microbial species detected and that a person's species composition is relatively stable over long periods. Comparative studies have also shown that humans share more genera with each other than with other species of mammals, and that there appears to be a core group of bacterial species shared by most healthy humans. This suggests that the precise mammalian gut microflora is "fine-tuned" to each mammalian species.

Contribution of Gut Microorganisms to Human Metabolism

Human gut microorganisms synthesize a large variety of enzymes that allow for the processing of complex dietary carbohydrates into monosaccharides and the production of VFAs. *Bacteroides* strains common in human adults have many genes whose products help catabolize polysaccharides, consistent with these bacteria being adapted to a gut environment rich in polysaccharides. Gut microorganisms also function in nitrogen metabolism. Of the 20 amino acids we require, 10 are said to be essential nutrients because we cannot synthesize them in adequate amounts. Although we obtain essential amino acids, such as lysine, from food, these nutrients may also be produced and excreted by certain gut microorganisms.

Gut microorganisms are also known to contribute to the "maturing" of the gastrointestinal tract. This includes triggering the expression of genes involved in nutrient uptake and metabolism in gut epithelial cells, priming the immune system early in life to recognize the gut microflora as nonforeign, and the development of a mucosal barrier to colonization by foreign bacteria. Studies of experimental colonization of germ-free mice with individual microbial species or microbial communities have demonstrated that colonization triggers the expression of genes for glucose uptake and lipid absorption and transport in the ileum. This also indicates that there may be a link between gut microbial composition and the ability of the host to harvest energy from its diet, contributing to nutritional abnormalities such as obesity, and we focus on this now.

Role of Gut Microorganisms in Obesity

Obesity is a significant health risk that contributes to high blood pressure, cardiovascular disease, and diabetes. Gut microorganisms may play a part in human obesity, although mechanisms remain unknown. Initial evidence relating gut microorganisms to host fat accumulation came from studies using germ-free mice. In these experiments, normal mice had 40% more total body fat than those raised under germ-free conditions, although both mouse populations were fed the same rations. After germ-free mice were inoculated with cecal material from a normal mouse, they developed

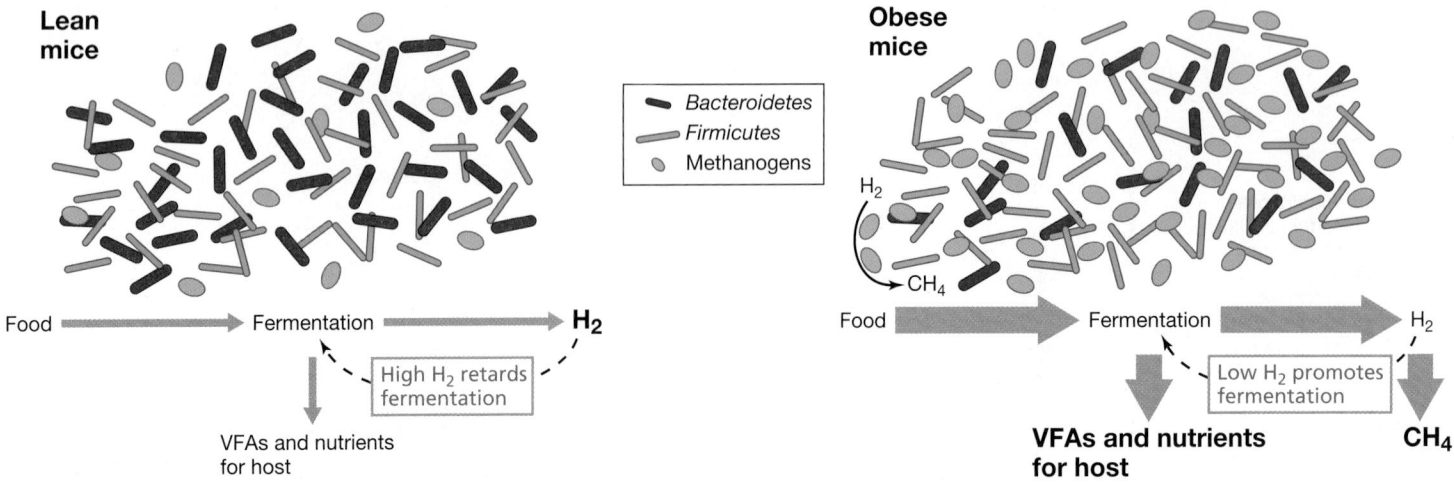

Figure 25.32 Differences in gut microbial communities between lean and obese mice. Obese mice have more methanogens, a 50% reduction in *Bacteroidetes*, and a proportional phylum-wide increase in *Firmicutes*. Nutrient production from fermentation is higher in obese mice due to removal of H_2 by methanogens.

a gut microflora and their total body fat increased although there had been no changes in food intake or energy expenditure. Mice that are genetically obese have microbial gut communities that differ from those of normal mice, with 50% fewer *Bacteroidetes*, a proportional increase in *Firmicutes*, and a greater number of methanogenic *Archaea* (**Figure 25.32**). Methanogens are thought to increase the efficiency of microbial conversion of fermentable substrates by removing hydrogen (H_2), as mentioned for fermentation in the rumen (Section 25.7). Hydrogen removal should stimulate fermentation, making more nutrients available for absorption by the host and thus contributing to obesity.

A tendency to accumulate fat associated with a particular gut microbial community is transmissible. A metagenomic analysis of 154 individuals—adult identical and fraternal twins and their mothers—examined correlations between the gut communities and host genotypes, host fat levels, and environment. As in the mouse models, obesity was associated with phylum-level differences in gut communities and lower bacterial diversity. Obese individuals had lower proportions of *Bacteroidetes* and a higher proportion of *Actinobacteria* (phylum *Firmicutes*). This study also indicated that inheritance of the microbiome from the mother was a more significant factor in obesity than host genotype. Although the exact role of the gut microflora in obesity remains unknown, it is likely that the microbial gut community governs the ability of the host to harvest organic nutrients from its diet. Moreover, the discovery that the gut microflora can affect obesity offers at least one nongenetic explanation for why obesity often "runs in families."

Microbial Communities in the Human Mouth

Besides the gut, the mouth and skin are also sites heavily colonized by microorganisms. Molecular characterization of the species diversity in these sites has not progressed as rapidly as for the gut communities, but available data have revealed similar patterns of high diversity, variability among individuals, and specific associations with both health and diseases.

As for all microbial communities reexamined by molecular methods, 16S rRNA-based sequence surveys of the oral cavity have shown that culture-based methods provided a very incomplete census of diversity. At least 750 species of aerobic and anaerobic microorganisms, including a minor representation of methanogenic *Archaea* and yeast, are known to reside in the oral cavity, distributed among teeth, tissue surfaces, and saliva. Because of the high species diversity, current research is focused on those genera having the largest representation in healthy adults. The oral cavity provides a variety of habitats, each colonized by species that are present primarily as biofilms (↩ Section 23.4). The primary colonizers of clean tooth surfaces are species of *Streptococcus*; obligate anaerobes such as *Veillonella* and *Fusobacterium* colonize habitats below the gum line. Most of these colonizers contribute to the health of the host by keeping pathogenic species in check and not allowing them to adhere to mucosal surfaces. Tooth decay, gum inflammation, and periodontal disease are among the most visible manifestations of a breakdown in these generally stable mutualisms. We discuss the normal microbial community of the oral cavity in more detail in Section 27.3.

Skin Microbial Communities

The skin is a critical human organ functioning primarily to prevent loss of moisture and restrict the entry of pathogens. Skin is also part of the human microbiome. Although total microbial numbers are typically low relative to the oral and gut communities, molecular analyses have shown that the skin harbors a rich and diverse microbial community of bacteria and fungi (primarily yeast) that vary significantly with location on the body. A 16S rRNA sequencing comparison of 20 diverse skin sites, loosely categorized as moist, dry, or oily, revealed tremendous diversity and variation among sites and individuals, but also showed some common patterns. All together, 19 bacterial phyla were detected. Most sequences affiliated with four groups: *Actinobacteria* (52%), *Firmicutes* (24%), *Proteobacteria* (16%), and *Bacteroidetes* (6%). A total of over 200 genera and many more species were detected, far exceeding culture-based estimates of diversity. About 60% of the sequences affiliated with only three genera: *Corynebacterium* (23%), *Propionibacterium* (23%), and *Staphylococcus* (17%), all gram-positive *Bacteria* (Chapter 18). *Propionibacterium* and *Staphylococcus* species predominated in oily sites, *Corynebacterium* species predominated in moist sites, and a mixed population of bacteria resided in dry sites, with *Betaproteobacteria* and *Flavobacteria* being common.

A comprehensive census of the skin microbiome is expected to contribute to the development of new therapies for skin disorders caused by specific organisms. With a comprehensive understanding of the microbial ecology, therapeutic intervention may include promoting the growth of protective symbiotic bacteria as well as inhibiting the growth of pathogenic bacteria. More specific coverage of the normal microbial community of human skin can be found in Section 27.2.

MiniQuiz

- Which major phyla of *Bacteria* dominate the human gut?
- How might increased numbers of methanogens in the gut contribute to obesity?
- What are expected practical outcomes of characterizing the human microbiome?

 IV Insects as Microbial Habitats

Insects are the most abundant class of animals living today, with over 1 million species known. As many as 20% of all insects are thought to support symbiotic microorganisms in a mutually beneficial way. The symbioses contribute to the insects' ecological success by providing them either nutritional advantages or protection. Some symbionts are found on insects' outer surfaces or in their digestive tracts. *Endosymbionts* are intracellular bacteria and are typically localized to specialized organs within the insect.

25.9 Heritable Symbionts of Insects

How symbionts are transferred from one generation to the next determines how a mutualism functions and how stable it is. Microbial symbionts can either be acquired by a host from an environmental reservoir (horizontal transmission) or be transferred directly from the parent to the next generation (*heritable* or *vertical* transmission). The mode of symbiont transmission is related to the specificity and persistence of an association. In general, less specificity is associated with horizontal transmission. In this section we focus only on mutualisms in which the microbial symbiont has no free-living form; that is, the symbionts are transmitted in a vertical fashion.

Types of Heritable Symbionts

All known heritable symbionts of insects lack a free-living replicative stage. Thus, they are *obligate* symbionts. However, although these bacteria require the host for replication, not all hosts are dependent upon the symbiont. Relative to host dependence, heritable symbionts are either primary symbionts or secondary symbionts. *Primary* symbionts are required for host reproduction. They are restricted to a specialized region called

the **bacteriome** present in several insect groups; within the bacteriome the bacterial cells reside in specialized cells called **bacteriocytes**. *Secondary* symbionts are not required for host reproduction. Unlike primary symbionts, secondary symbionts are not always present in every individual of a species and are not restricted to particular host tissues.

Secondary symbionts are broadly distributed among insect groups. Like pathogens, they invade different cell types and may live extracellularly within the insect's *hemolymph* (the fluid bathing the body cavity). In insects with bacteriomes, secondary symbionts can invade the bacteriocytes, co-residing with or sometimes displacing the primary symbionts (**Figure 25.33**). However, in order to persist in the insect host, the secondary symbiont must confer some benefit. Known benefits include nutritional advantages and protection from environmental stresses such as heat. Secondary symbionts may also provide protection against invasion by pathogens or predators. In most cases the basis for protection is unknown, but in one case a toxin encoded by a lysogenic bacteriophage (Section 9.10) carried by the symbiont is known to confer protection on the insect from infection by a parasitic wasp.

There are heritable parasitic symbionts that manipulate the host's reproductive system, increasing the frequency of female progeny (sex-ratio skewing, Section 17.13). Because most heritable symbionts are transmitted maternally, the suppression of male progeny serves to expand the number of infected individuals. An important by-product of improved basic understanding of insect symbionts is their increased use as biocontrol agents for pest management. For example, symbiotic *Wolbachia* (Section 17.13), which are reproductive manipulators, are widely distributed among insect species (possibly infecting as many as 60–70% of all insect species). The sperm of *Wolbachia*-infected males can sterilize uninfected females. Although the mechanism for sterilization is not fully understood, the phenomenon is being

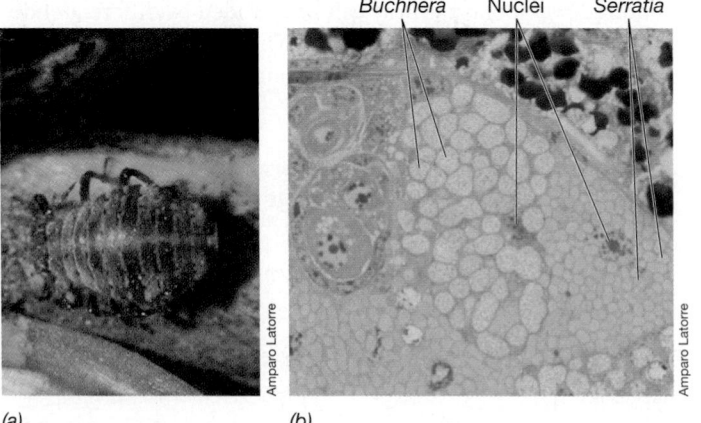

(a)   (b)

Figure 25.33 Primary and secondary symbionts of an aphid. *(a)* The cedar aphid *Cinara cedri*, a model organism for studies of symbioses. *(b)* Transmission electron micrograph of the bacteriome of *C. cedri* showing two bacteriocytes. Packed within each bacteriocyte are cells of *Buchnera aphidicola* (the primary symbiont) or *Serratia symbiotica*, the smaller, secondary symbiont. Arrows identify the nucleus of each bacteriocyte. The bacteriocyte containing *Buchnera* cells is about 40 μm wide.

UNIT 7

examined for possible pest management of the Mediterranean fruit fly (medfly), a serious pest of fruit crops. For example, an experimental release of *Wolbachia*-infected medfly males killed all embryos of uninfected females.

Functional Significance of Obligate Intracellular Symbionts of Insects

The association of bacterium and insect has allowed many insects to use food resources that are rich in some nutrients, but poor in others. To achieve adequate nutrition, some insects exploit the metabolic potential of their symbionts. For instance, aphids feed on the carbohydrate-rich but nutrient-poor sap of phloem vessels in plants. Early on it was suspected that obligate symbionts might benefit the insect by providing nutrients not provided by their primary diet. Molecular analyses have shown that most families of aphids harbor the bacterium *Buchnera* in their bacteriomes. The role of *Buchnera* in host nutrition was first indicated by experiments using defined diets to examine the nutrient requirements of aphids. Compared with infected controls, symbiont-free aphids required a diet containing all amino acids that are either lacking or rare in phloem sap. Subsequent genomic studies documented the presence in *Buchnera* of genes encoding the biosynthesis of nine amino acids missing from the sap. There are also examples of synergy between host and symbiont where the synthesis of certain amino acids becomes a joint venture. For example, *Buchnera* lacks the enzyme needed for the last step in leucine biosynthesis, but the necessary gene is present in the aphid's genome. Presumably, this enzyme is made by the aphid and participates in the leucine biosynthetic pathway along with the bacterial enzymes.

A secondary symbiont can also contribute to a joint venture. For example, the *Buchnera* symbiont of the cedar aphid is unable to supply tryptophan to the aphid. Two genes in the tryptophan biosynthetic pathway are present in *Buchnera*, but the remaining genes for the pathway are located on the chromosome of a secondary endosymbiont (Figure 25.33). Thus, different parts of a required metabolic pathway can be encoded by different endosymbionts present in the same insect. The fungus-cultivating ants provide yet another example of a complex symbiosis that has

formed between an insect and multiple microorganisms (see the Microbial Sidebar, "The Multiple Microbial Symbionts of Fungus-Cultivating Ants").

Genome Reduction and Gene Transfer Events

Common features of primary symbionts are extreme genome reduction (ᴄᴄ Table 12.1), high A+T content, and accelerated rates of mutation. Genomes of insect symbionts fall within a range from 160 to 800 kbp and 16.5 to 33% G+C (**Table 25.3**). The 160-kbp genome of *Carsonella* is the smallest genome known for any cell. In contrast, the genomes of related free-living bacteria range from 2 to 8 Mbp with a base composition closer to 50% G+C. Two common types of spontaneous mutation, cytosine deamination and the oxidation of guanosine, if not repaired, change a GC pair to an AT pair. Symbionts with reduced genomes have fewer DNA repair enzymes (ᴄᴄ Section 10.4) and this likely facilitates a shift over time to genomes of lower G+C content.

The streamlined genomes of insect symbionts have lost genes from most functional categories and have retained only genes required for host fitness and essential molecular processes, such as translation, replication, and transcription. Genome reduction implies that the symbionts are reliant on the host for many functions no longer encoded in the symbiont genome. For example, in many cases genes needed for cell wall components are missing, including lipid A and peptidoglycan, suggesting that the host supplies these functions or that the structures are not required to form stable cells within the bacteriocyte.

There is an interesting genomic contrast between primary symbionts and typical disease-causing bacteria (pathogens). While primary symbionts tend to lose genes encoding proteins required in *catabolic* pathways, pathogenic bacteria typically retain these, but lose genes for *anabolic* pathways. This reflects their differing relationships with their hosts; the insect symbiont provides the host with essential biosynthetic nutrients while the pathogen obtains important biosynthetic nutrients from the host.

Because genome sequences for both host and symbiont are now available, microbiologists can study gene transfer between them. Horizontal gene transfer is the movement of genetic information across normal mating barriers (Chapter 10). The

Table 25.3 *Genome features of some endosymbionts of animals*

Host	Symbiont (genus)	Symbiont genome size (Mbp)	G+C (%)	Gene number
Sharpshooters	Heterotroph (*Sulcia*)	0.25	22	227
Aphid	Heterotroph (*Buchnera*)	0.42–0.62	20–26	362–574
Tsetse fly	Heterotroph (*Wigglesworthia*)	0.70	22	617
Carpenter ant	Heterotroph (*Blochmannia*)	0.71–0.79	27–30	583–610
Clam (*Calyptogena okutanii*)	Sulfur chemolithotroph (unnamed)	1.0	32	975
Clam (*Calyptogena magnifica*)	Sulfur chemolithotroph (*Ruthia*)	1.2	34	1248
Tube worm (*Riftia pachyptila*)	Sulfur chemolithotroph (unnamed)	3.3[a]	58	Unknown

[a]The free-living sulfur chemolithotroph, *Thiomicrospira crunogena*, has a genome significantly smaller (2.4 Mbp) than this symbiont. All listed symbionts are obligately associated with their hosts, with the exception of the symbiont of *Riftia*, which is also free-living.

The Multiple Microbial Symbionts of Fungus-Cultivating Ants

The attine ants serve as an example of an elaborate symbiotic association between multiple microbial species and insect. These ants have established an obligate mutualism with a fungus they cultivate in fungal gardens for food, using small leaf fragments to mulch these gardens. A close symbiotic relationship between ant and fungus was first indicated by the observation that one specific fungus was cultivated by each ant lineage. The ants and their mutualistic fungi can be divided into five agricultural systems, each involving distinct lineages of ants and fungi. Ants grouped in the "lower attine agriculture" system form associations with specific groups of fungi they capture from the environment. The "higher attine agriculture" group cultivates fungi that apparently are no longer capable of existing apart from the ant mutualism. In addition to the close mutualistic relationship between ant species and the specific fungus they cultivate, this symbiosis is now known to include four other microbial symbionts: a small fungus that is parasitic on the garden fungus, nitrogen-fixing bacteria associated with the garden fungus, an actinobacterium that antagonizes the parasitic fungus, and a black yeast that interferes with the actino-bacterium.

The fungus is vertically transmitted between ant generations by colony-founding queens. The queen collects a pellet of fungus prior to her mating flight, storing it in a pouch in the oral cavity. After mating, she uses the fungus pellet to establish a new nest and fungus garden (Figure 1a). Nitrogen-fixing *Klebsiella* and *Pantoea* species associated with the fungus enrich the nutritional quality of the garden by adding new nitrogen to the nitrogen-poor leaf growth substrate. A single leaf-cutter ant colony may contribute as much as 1.8 kg of fixed nitrogen per year. This new nitrogen benefits the ant colony and also results in higher overall plant diversity near leaf-cutter colonies.

However, the garden is at risk of being destroyed by a parasitic ascomycete micro-fungus (genus *Escovopsis*). To repel the parasitic microfungus, the ant has formed another symbiotic association with an actinobacterium (genus *Pseudonocardia*) that appears as a "waxy bloom" growing on the cuticle of the ant (Figure 1b). These bacteria, housed in specialized cuticular modifications on the ant's body, secrete secondary metabolites that inhibit the growth of *Escovopsis*. The *Pseudonocardia* likely receive nourishment from the ant from glandular excretions through pores localized in regions of cuticular modification. Comparative sequencing has revealed good congruence between the phylogenies of the ants, fungal cultivars, *Escovopsis*, and *Pseudonocardia*, pointing to very specific interactions among microorganisms and ants in this complex symbiosis.

The fourth microbial participant identified recently in this symbiosis is an ascomycete yeast that grows in the same cuticular regions colonized by *Pseudonocardia*. This black yeast interferes with chemical protection of the garden by stealing nutrients from the *Pseudonocardia*, thereby indirectly reducing its ability to suppress *Escovopsis* growth.

(a)

(b)

Michael Poulsen and Cameron Currie

Figure 1 Attine ants. *(a)* Queen and worker ants in fungal gardens. *(b)* Mutualism with *Actinobacteria* can cover much of the exoskeleton of workers (white areas).

evidence is now clear that there can be extensive horizontal gene transfer between a symbiont and its host. For example, extensive stretches of the DNA of *Wolbachia* (⟳ Section 17.13) have been transferred from these bacteria to the nuclear genomes of their insect and nematode hosts. Amazingly, a complete copy of the *Wolbachia* genome has been found in the genome of the fruitfly, and several of the transferred genes have been shown to be transcribed.

MiniQuiz

- What factors stabilize the presence of a secondary insect symbiont?
- What are the consequences of symbiont genome reduction?
- How could you determine if a symbiont and host have experienced a long period of coevolution?

25.10 Termites

Microorganisms are primarily responsible for the degradation of wood and cellulose in natural environments. However, the activities of free-living microbial species have been exploited by certain groups of insects that have established symbiotic associations with protists and bacteria that can digest lignocellulosic materials. Like the rumen of herbivorous animals, the insect gut provides a protective niche for microbial symbionts, and in return, the insect gains access to nutrients derived from an otherwise indigestible carbon source. Termites are among the most abundant representatives of this type of symbiotic alliance.

Termite Natural History and Biochemistry

Enabled by the microbial communities in their guts, termites decompose the greater part of cellulose (74–99%) and hemicellulose (65–87%) in the plant material they ingest. In contrast to the insect examples previously discussed, most termites do not harbor intracellular bacteria. Termite diets include lignocellulosic plant materials (either intact or at various stages of decay), dung, and soil organic matter (humus). About two-thirds of the terrestrial environment supports one or more termite species, with greatest representation in tropical and subtropical regions, where termites may constitute as much as 10% of all animal biomass and 95% of soil insect biomass. In savannas, their numbers sometimes exceed $4000/m^2$, and their biomass density ($1–10$ g/m^2) may be higher than that of grazing mammalian herbivores.

Termites are categorized as higher or lower based on their phylogeny, and this classification correlates with different symbiotic strategies. The posterior alimentary tract of *higher* termites (family *Termitidae*, comprising about three-fourths of termite species) contains a dense and diverse community of mostly anaerobic bacteria, including cellulolytic species. In contrast, the *lower* termites harbor diverse populations of both anaerobic

bacteria and cellulolytic protists. Bacteria of lower termites participate little or not at all in cellulose digestion; only the protists phagocytize and degrade the wood particles ingested by the termites. The termite itself produces cellulases in the salivary glands or the midgut epithelium, but the relative contributions of microbial and termite enzymes to lignocellulosic breakdown is unknown.

The termite gut consists of a foregut (including the crop and muscular gizzard), a tubular midgut (site of secretion of digestive enzymes and absorption of soluble nutrients), and a relatively large hindgut of about one microliter volume (**Figure 25.34**). In lower termites the hindgut consists primarily of a single chamber, the *paunch* (Figure 25.34a). The hindgut of most higher termites is more complex, being divided into several compartments (Figure 25.34b). For both higher and lower termites, the hindgut harbors a dense and diverse microbial community and is a major site of nutrient absorption. Acetate and other organic acids are produced during microbial fermentation of carbohydrate in the hindgut, and these products are primary carbon and energy sources for the termite (Figure 25.34c). High O_2 consumption by bacteria near the gut wall keeps the interior of the hindgut anoxic. However, microelectrode measurements (Section 22.8) have shown that O_2 can penetrate up to 200 μm into the gut before it is completely removed by microbial respiratory activity. Thus, this tiny gut compartment offers distinct microbial niches with respect to O_2 and can support diverse microbial activities.

Bacterial Diversity and Lignocellulose Digestion in Higher Termites

In termites of different genera, the microbial gut communities differ significantly. Analysis of 16S rRNA gene sequences from hindgut contents of species of the higher termite genus *Nasutitermes* revealed a high diversity of microbial species from

Figure 25.34 Termite gut anatomy and function. Gut architecture of lower (a) and higher (b) termites, showing the foregut, midgut, and differing complexity of the hindgut compartments. (c) Photo of workers, gut architecture, and biochemical activities of the lower termite *Coptotermes formosanus*. Acetate and other products of microbial fermentations are assimilated by the termite. Hydrogen produced by fermentation is consumed primarily by CO_2-reducing acetogens, with a smaller amount going to hydrogenotrophic methanogens.

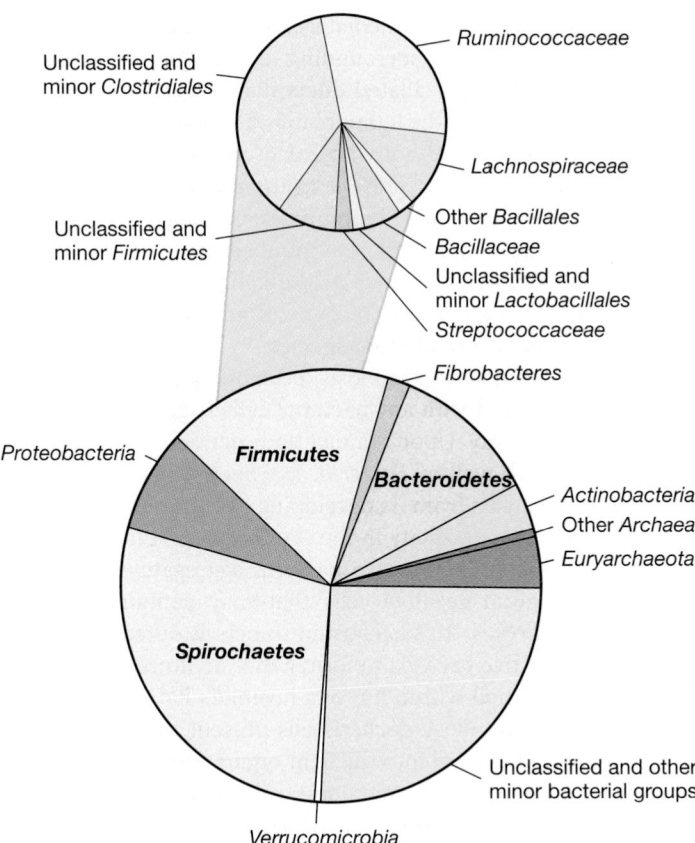

Figure 25.35 Microbial composition of termite hindgut inferred from 16S rRNA sequences. The results are pooled analyses of 5075 sequences from amplified or metagenomic sequencing studies of three genera of wood-feeding higher termites, *Nasutitermes*, *Reticulitermes*, and *Microcerotermes*. The data provide information primarily of diversity, not relative abundance. Data assembled and analyzed by Nicolas Pinel.

12 phyla of *Bacteria*, but few *Archaea* (**Figure 25.35**). Spirochetes of the genus *Treponema* (Section 18.16) dominated, with a lesser contribution from thus far uncultured organisms distantly related to the phylum *Fibrobacteres*, a group present in the rumen (Figure 25.28). Metagenomic analysis of the *Nasutitermes* hindgut microbial community revealed bacterial genes encoding glycosyl hydrolases that hydrolyze cellulose and hemicelluloses, and thus, although the corresponding cellulolytic bacteria have not yet been isolated from the higher termites, the metagenomic data clearly implicate spirochetes and *Fibrobacteres* in the digestion of lignocellulose (Figure 25.35). At every molting of an individual termite, gut symbionts are lost, yet there is good conservation of the gut community within each termite species. Stable horizontal transmission of gut symbionts likely occurs due to the intimate social behavior and close contact characteristic of termites.

Acetogenesis and Nitrogen Fixation in the Termite Gut

Genes encoding enzymes of the acetyl-CoA pathway (Section 14.9) are highly represented in the spirochetes of the *Nasutitermes* hindgut, consistent with their function as the major CO_2-

reducing acetogens. The termite gut microbial communities have long been recognized as important to host nitrogen metabolism, providing new fixed nitrogen through nitrogen fixation and helping to conserve nitrogen by recycling excretory nitrogen back to the insect for biosynthesis. Consistent with this, metagenomic analyses reveal that many bacteria, including *Fibrobacteres* and treponeme spirochetes, contain genes encoding nitrogenase (Section 13.14).

From a simple energetic viewpoint, methanogenesis from H_2 and CO_2 is more favorable than acetogenesis from the same substrates (-34 kJ/mol of H_2 versus -26 kJ/mol of H_2, respectively), and thus methanogens should have a competitive advantage in all habitats in which the two processes compete (Sections 14.9–14.10). However, in termites they do not. There are at least two reasons for this. First, unlike methanogens, acetogens are able to use other substrates such as sugars or methyl groups of lignin degradation products as electron donors for energy metabolism. Second, termite acetogens (which seem to consist mostly of spirochetes) can for some reason better colonize the H_2-rich termite gut center, whereas methanogens are largely restricted to the gut wall. On the wall, methanogens are located downstream of the H_2 gradient and thus receive only a fraction of the H_2 flux. In addition, the wall likely contains higher O_2 tensions, which may negatively affect the physiology of methanogens. So, despite the fact that termites are methanogenic, producing up to 150 terragrams of CH_4 per year on a global basis (1 terragram = 10^{12} grams), carbon and electron flow favor acetogenesis in this anoxic habitat.

MiniQuiz
- How are anoxic conditions maintained in the termite hindgut?
- Why does reductive acetogenesis predominate over methanogenesis in many termites?
- Which group of morphologically unusual bacteria, absent from molecular surveys of prokaryotes in the rumen, seem to dominate activities in the termite hindgut?

(V) Aquatic Invertebrates as Microbial Habitats

Thus far in this chapter we have discussed how certain macroorganisms that live in terrestrial environments provide habitats for microbial symbionts. Aquatic environments—especially marine environments—impose different constraints on symbioses and offer different opportunities for the evolution of symbioses between macroorganisms and microorganisms. Nevertheless, microbial symbioses with marine animals, especially with invertebrates, are common. By finding habitats in marine invertebrates, microorganisms establish a safe residence in a nutritionally rich environment. And the invertebrates benefit, too, as we will see with two well-studied examples: the squid and the hydrothermal vent animal symbioses. These microbial–animal associations are thus true symbioses, with both partners benefitting from the relationship.

UNIT 7

25.11 Hawaiian Bobtail Squid

The Hawaiian bobtail squid, *Euprymna scolopes*, is a small marine invertebrate (**Figure 25.36a**) that sequesters large populations of the bioluminescent gram-negative bacterium *Aliivibrio fischeri* (↩ Section 17.12) in a light organ located on its ventral side. Squid and bacterium are partners in a mutualism. The bacteria emit light that resembles moonlight penetrating marine waters, and this is thought to camouflage the squid from predators that strike from beneath. Several other species of *Euprymna* inhabit marine waters near Japan and Australia and in the Mediterranean, and these contain *Aliivibrio* symbionts as well.

The Squid–*Aliivibrio* System as a Model Symbiosis

Many features of the *E. scolopes*–*A. fischeri* symbiosis have made it an important model for studies of animal–bacterial symbioses. These include the facts that the animals can be grown in the laboratory and that there is only a single bacterial species in the symbiosis in contrast to the huge number in symbioses such as those of the rumen (Figure 25.28) or the mammalian large intestine (Figure 25.31). In addition, the symbiosis is not an essential one; both the squid and its bacterial partner can be cultured apart from each other in the laboratory. This allows juvenile squid to be grown without bacterial symbionts and then experimentally colonized. Experiments can be done to study specificity in the symbiosis, the number of bacterial cells needed to initiate an infection, the capacity of genetically defined mutants of *A. fischeri* to initiate infection of the squid, and many other aspects of the relationship. Moreover, because the genome of *A. fischeri* has been sequenced, the powerful techniques of microbial genomics may be employed.

Establishing the Squid–*Aliivibrio* Symbiosis

Juvenile squid just hatched from eggs do not contain cells of *A. fischeri*. Thus, transmission of bacterial cells to juvenile squid is a horizontal (environmental) rather than a vertical (parent–offspring) event. Almost immediately after juveniles emerge from eggs, cells of *A. fischeri* in surrounding seawater begin to colonize them, entering through ciliated ducts that end in the immature light organ. Amazingly, the light organ becomes colonized specifically with *A. fischeri* and not with any of the many other species of gram-negative bacteria present in the seawater. Even if large numbers of other species of bioluminescent bacteria are offered to juvenile squid along with low numbers of *A. fischeri*, only *A. fischeri* establishes residence in the light organ. This implies that the animal in some way recognizes and accepts *A. fischeri* cells and excludes those of other species.

The squid–*Aliivibrio* symbiosis develops in several stages. Contact of the squid with any bacterial cells triggers recognition in a very general way. Upon contact with peptidoglycan (a component of the cell wall of *Bacteria*, ↩ Section 3.6), the young squid secretes mucus from its developing light organ. The mucus is the first layer of specificity in the symbiosis, as it makes gram-negative but not gram-positive bacteria aggregate. Within the aggregates of gram-negative cells that may contain only low numbers of *A. fischeri*, this bacterium somehow outcompetes the other gram-negative bacteria to form a monoculture. The monoculture is established within 2 h of a juvenile's hatching from an egg. The highly motile *A. fischeri* cells present in the aggregate migrate up the ducts and into the light organ tissues. Once there, they lose their flagella, become nonmotile, divide to form dense populations (Figure 25.36b), and trigger developmental events that lead to maturation of the host light organ. The light organ in a mature *E. scolopes* contains between 10^8 and 10^9 *A. fischeri* cells.

Colonization of *A. fischeri* by the squid is assisted by the gas nitric oxide (NO). Nitric oxide is a well-known defense response of animal cells to attack by bacterial pathogens; the gas is a strong oxidant and causes sufficient oxidative damage to bacterial cells to kill them (↩ Section 29.1). Nitric oxide produced by the squid is incorporated into the mucus aggregates and is present in the light organ itself. As *A. fischeri* colonizes the light organ, NO levels diminish rapidly. It appears that cells of *A. fischeri* can tolerate exposure to NO and consume it through the activity of NO-inactivating enzymes. The inability of other gram-negative

(a)

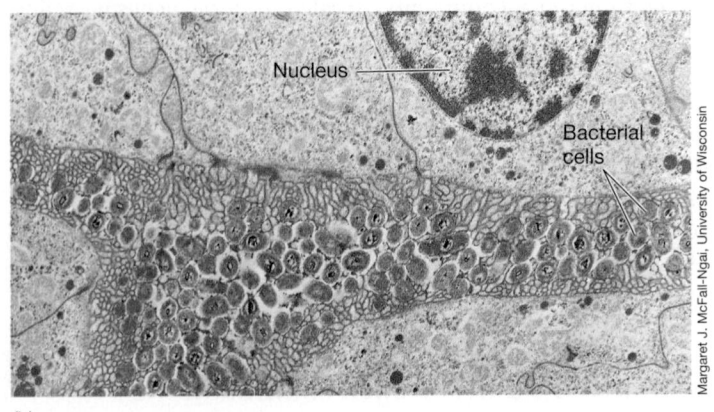

Nucleus

Bacterial cells

(b)

Figure 25.36 Squid–*Aliivibrio* symbiosis. *(a)* An adult Hawaiian bobtail squid, *Euprymna scolopes*, is about 4 cm long. *(b)* Thin-sectioned transmission electron micrograph through the *E. scolopes* light organ shows a dense population of bioluminescent *Aliivibrio fischeri* cells.

bacteria in the mucus aggregates to detoxify NO helps explain the sudden enrichment of *A. fischeri* in the ducts even before the actual colonization of the light organ. Then, after establishment, continued production of NO in the light organ prevents colonization by other bacterial species.

Propagating the Symbiosis

The squid matures into an adult in about 2 months and then lives a strictly nocturnal existence in which it feeds mostly on small crustaceans. During the day, the animal buries itself and remains quiescent in the sand. Each morning the squid nearly empties its light organ of *A. fischeri* cells and begins to grow a new population of the bacterium. The bacterial cells grow rapidly in the light organ; by midafternoon, the structure contains the dense populations of *A. fischeri* cells required for the production of visible light. The actual emission of light requires a certain density of cells and is controlled by the regulatory mechanism called *quorum sensing* (♻ Section 8.9). The daily expulsion of bacterial cells is thought to be a mechanism for seeding the environment with cells of the bacterial symbionts. This, of course, increases the chances that the next generation of juvenile squid will be colonized.

A. fischeri grows much faster in the light organ than in the open ocean, presumably because it is supplied with nutrients by the squid. Thus *A. fischeri* benefits from the symbiosis by having an alternative habitat to seawater in which rapid growth and dense populations are possible. Isolation studies have shown that *A. fischeri* is not a particularly abundant marine bacterium. Daily expulsion of *A. fischeri* cells from the light organ increases the bacterium's numbers in the microbial community. Thus, the symbiotic relationship of the bacterium with the squid probably helps maintain larger *A. fischeri* populations than would exist if all cells were free-living. Because the competitive success of a microbial species is to some degree a function of population size (♻ Section 23.1), this boost in cell numbers may confer an important ecological advantage on *A. fischeri* in its marine habitat.

MiniQuiz

- Of what value is the squid–*Aliivibrio* symbiosis to the squid? To the bacterium?
- What features of the squid–*Aliivibrio* symbiosis make it an ideal model for studying animal–bacterial symbioses?

25.12 Marine Invertebrates at Hydrothermal Vents and Gas Seeps

Diverse invertebrate communities develop near undersea hot springs called *hydrothermal vents*. We covered the geochemistry and microbiology of hydrothermal vents in Section 23.12. Here we focus on hydrothermal vent animals and their microbial symbionts.

Macroinvertebrates, including tube worms over 2 m in length and large clams and mussels, are present near these vents (**Figure 25.37**). Photosynthesis cannot support these invertebrate communities because they exist below the photic zone. However, hydrothermal fluids contain large amounts of reduced inorganic materials, including H_2S, Mn^{2+}, H_2, and CO (carbon monoxide),

(a)

(b)

Figure 25.37 Invertebrates living near deep-sea thermal vents. *(a)* Tube worms (family *Pogonophora*), showing the sheath (white) and plume (red) of the worm bodies. *(b)* Mussel bed in vicinity of a warm vent. Note yellow deposition of elemental sulfur from the oxidation of H_2S emitted from the vents.

and some vents contain high levels of ammonium (NH_4^+) instead of H_2S. All of these are good electron donors for chemolithotrophic prokaryotes, bacteria that use inorganic compounds as electron donors and fix CO_2 as their carbon source (Chapter 13). Thus, these hydrothermal vent invertebrates can exist in permanent darkness because they are nourished through a symbiotic association with these autotrophic bacteria.

Tube Worms and Giant Clams

Hydrothermal vent–associated animals either feed directly on free-living chemolithotrophic bacteria or have formed tight symbiotic associations with the bacteria. Mutualistic bacteria are either tightly attached to the animal surface (that is, as *epibionts*) or actually live within the animal tissues, supplying organic compounds to the animals in exchange for a safe residence and ready access to the electron donors needed for their energy metabolism. For example, the 2-m-long tube worms (Figure 25.37a) lack a mouth, gut, and anus, but contain an organ consisting primarily of spongy tissue called the *trophosome*. This structure, which constitutes half the worm's weight, is filled with sulfur granules

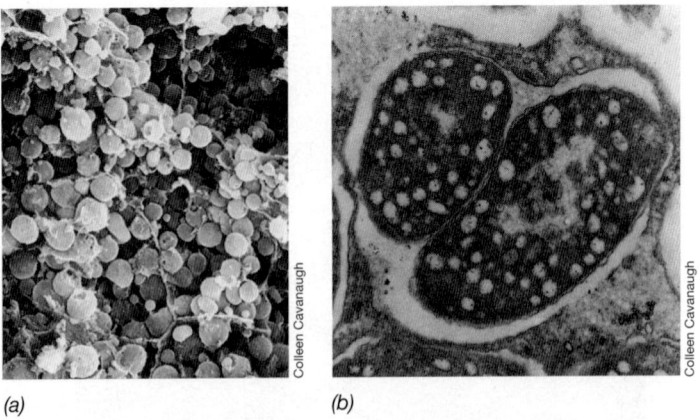

(a)   *(b)*

Figure 25.38 Chemolithotrophic sulfur-oxidizing bacteria associated with the trophosome tissue of tube worms from hydrothermal vents. *(a)* Scanning electron micrograph of trophosome tissue showing spherical chemolithotrophic sulfur-oxidizing bacteria. Cells are 3–5 μm in diameter. *(b)* Transmission electron micrograph of bacteria in sectioned trophosome tissue. The cells are frequently enclosed in pairs by an outer membrane of unknown origin. Reprinted with permission from *Science* 213: 340–342 (1981), © AAAS.

and large populations of spherical sulfur-oxidizing prokaryotes (**Figure 25.38**). Bacterial cells taken from trophosome tissue show activity of enzymes of the Calvin cycle, a major pathway for autotrophy (Section 13.12), but interestingly, contain enzymes of the reverse citric acid cycle, a second autotrophic pathway (Section 13.13), as well. In addition, they show a suite of sulfur-oxidizing enzymes necessary to obtain energy from reduced sulfur compounds (Section 13.8). The tube worms are thus nourished by organic compounds produced from CO_2 and excreted by the sulfur chemolithotrophs.

Along with tube worms, giant clams and mussels (Figure 25.37*b*) are also common near hydrothermal vents, and sulfur-oxidizing bacterial symbionts have been found in the gill tissues of these animals. Phylogenetic analyses have shown that each individual animal harbors a different strain of bacterial symbiont and that more different species of bacterial symbionts inhabit different species of vent animal. Although fairly closely related to free-living sulfur chemolithotrophs (Sections 13.8, 17.4, and

17.19), and the tube worm symbiont is known to have a free-living stage, none of the bacterial symbionts of hydrothermal vent animals have yet been obtained in laboratory culture.

The red plume of the tube worm (Figure 25.37*a*) is rich in blood vessels and is used to trap and transport inorganic substrates to the bacterial symbionts. The tube worms contain unusual hemoglobins that bind H_2S and O_2; these are then transported to the trophosome where they are released to the bacterial symbionts. The CO_2 content of tube-worm blood is also high, about 25 mM, and presumably this is released in the trophosome as a carbon source for the symbionts. In addition, stable isotope analyses (Section 22.9) of elemental sulfur from the trophosome have shown that its $^{34}S/^{32}S$ composition is the same as that of the sulfide emitted from the vent. This ratio is distinct from that of seawater sulfate and is further proof that geothermal sulfide is actually entering the worm in large amounts.

Other marine invertebrates have coevolved bacterial symbioses that supply their nutrition as well (**Table 25.4**). For example, methanotrophic (CH_4-consuming) symbionts are present in giant clams that live near natural gas seeps at relatively shallow depths in the Gulf of Mexico. Although not autotrophs (CH_4 is an organic compound), the methanotrophs do provide nutrition to the clams; the methanotrophs use CH_4 as their electron donor and carbon source and excrete organic carbon to the clams.

Genomics and Hydrothermal Vent Symbioses

Genome sequencing is revealing additional features of the metabolic interaction and coevolution of marine invertebrates and their prokaryotic symbionts. The genome sequence of the gill endosymbiont of the giant vent clam *Calyptogena magnifica* offers direct evidence for carbon fixation via the Calvin cycle; the genome encodes the key enzymes of the Calvin cycle, ribulose bisphosphate carboxylase (RubisCO) and phosphoribulokinase (Section 13.12), and genes encoding key sulfur oxidation processes. The genome of this symbiont also encodes the biosynthesis of most vitamins and cofactors and all 20 amino acids needed to support the host. However, because few substrate-specific transporters are encoded by the symbiont genome, it is suspected that the clam actually digests symbiont cells for nutrition, as do mussels (Table 25.4).

Like the obligate symbionts of insects, most symbionts of marine invertebrates have small genomes (Table 25.3), indicating

---

Table 25.4 *Marine animals with chemolithotrophic or methanotrophic endosymbiotic bacteria*

Host (genus or class)	Common name	Habitat	Symbiont type
Porifera (*Demospongiae*)	Sponge	Seeps	Methanotrophs
Platyhelminthes (*Catenulida*)	Flatworm	Shallow water	Sulfur chemolithotrophs
Nematoda (*Monhysterida*)	Mouthless nematode	Shallow water	Sulfur chemolithotrophs
Mollusca (*Solemya, Lucina*)	Clam	Vents, seeps, shallow water	Sulfur chemolithotrophs
Mollusca (*Calyptogena*)	Clam	Vents, seeps, whale falls[a]	Sulfur chemolithotrophs
Mollusca (*Bathymodiolus*)	Mussel	Vents, seeps, whale and wood falls[a]	Sulfur chemolithotrophs, methanotrophs
Mollusca (*Alviniconcha*)	Snail	Vents	Sulfur chemolithotrophs
Annelida (*Riftia*)	Tube worm	Vents, seeps, whale and wood falls[a]	Sulfur chemolithotrophs

[a]Whale and wood falls are sunken whale carcasses and wood, respectively.

reduced function and an obligate association with their host. The symbiont of the giant tube worm *Riftia pachyptila* is an exception, having a genome larger than some free-living sulfur-oxidizing chemolithotrophs (Table 25.3). The *R. pachyptila* symbiont is acquired by uninfected juvenile animals from the environment (horizontal transmission), and its larger genome is likely important for survival as a free-living bacterium.

MiniQuiz

- How do giant tube worms receive their nutrition?
- What are the similarities of the obligate symbioses of insects and thermal vent invertebrates?
- What factors determine the genome size of the symbionts of marine invertebrates?

25.13 Leeches

Leeches are parasitic annelids (segmented worms). Leeches are related to earthworms and share several properties with them. Some leeches live in marine environments, but our example here, the medicinal leech *Hirudo verbana* (**Figure 25.39a**), lives in freshwater.

Parasitic Lifestyle of Leeches

Like many animals that depend on a microbial partner, medicinal leeches have a restricted diet. They feed exclusively on vertebrate blood and secrete powerful anticoagulants and vasodilators that stimulate blood flow. In a single feeding, *H. verbana* can consume

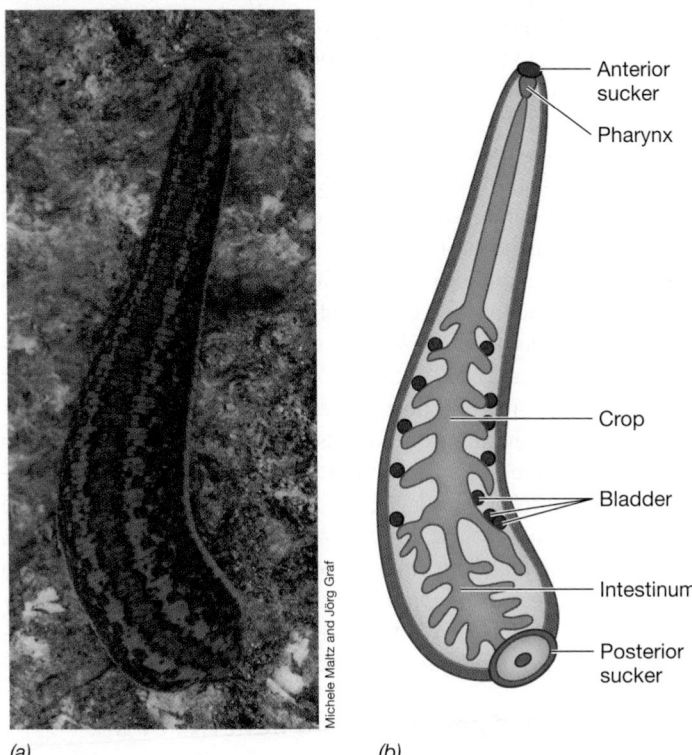

(a)   (b)

Figure 25.39 Medicinal leech *Hirudo verbana*. *(a)* An animal of about 6 cm in length. *(b)* Anatomy of *H. verbana*, showing the crop, intestinum, and bladder pairs.

over five times its body weight in blood. The blood meal is stored in the *crop*, which is the largest compartment of the leech digestive tract (Figure 25.39b). During feeding, water and salts are absorbed from the crop content until most water is removed and the fluid is in osmotic balance with the leech hemolymph. Excess water and nitrogenous waste are secreted through several pairs of bladders. Both the digestive tract and the bladder house microbial communities. It is thought that one function of the symbionts is to provide essential nutrients, such as vitamin B_{12}, absent or in low amounts in the blood meal.

This amazing ability of medicinal leeches to remove blood and secrete pharmacologically active compounds has been used for ages for the medical practice of *bloodletting*, and in recent times most commonly in plastic and reconstructive surgery. A challenge for medical replants and transplants is the connection of the veins. If, after transplant surgery, the number of functional veins exiting from the surgically introduced tissue is insufficient, the flow of fresh oxygenated blood into the tissue is stopped. The lack of oxygen can result in failure of the transplant. Leeches applied to the area remove blood, letting fresh blood enter the introduced tissue, and this procedure increases the transplant success rate.

The Leech Microbial Community

The leech digestive tract has two major compartments that house microbial communities, the digestive tract (the large crop and the smaller intestinum), where the digestion of the erythrocytes and absorption of nutrients are thought to occur, and the bladders (Figure 25.39b). The microbial community of the crop is surprisingly simple. Culture-independent studies using a combination of 16S rRNA gene analyses and fluorescent *in situ* hybridization (FISH, ↩ Section 16.9) revealed that the microbial community inside the crop is dominated by two species, *Aeromonas veronii* (*Gammaproteobacteria*) and a *Rikenella*-like (*Bacteroidetes*) bacterium. Farther along the alimentary canal toward the intestinum (Figure 25.39b) the complexity of the microbial community increases. In the intestinum various *Alpha-* and *Gammaproteobacteria*, along with *Bacteroidetes* and *Firmicutes*, prevail.

The unusually simple microbial community inside the crop suggests that there are mechanisms that prevent other microorganisms from colonizing. Specificity of symbiotic associations can be affected by the mode of transmission and molecular mechanisms that interfere with colonization or maintenance of microorganisms that enter the gut habitat. For example, leech hemocytes, invertebrate macrophage-like cells (↩ Section 28.1), patrol the gut and phagocytose bacteria. *A. veronii* is able to prevent phagocytosis and colonize the leech gut by injecting toxins directly into the hemocytes, using a bacterial secretion system that functions like a molecular syringe (↩ Section 27.11).

The bladders of leeches (Figure 25.39) house an interesting ensemble of microorganisms. The epithelial cells lining the lumen of the bladder are tightly packed with an *Ochrobactrum* species. These bacteria are related to beneficial and pathogenic alphaproteobacterial symbionts such as *Sinorhizobium meliloti* (Section 25.3) and *Brucella abortus* (↩ Section 27.8). The microbial community in the lumen of the bladder displays a distinct stratification: Two species of *Bacteroidetes* colonize the

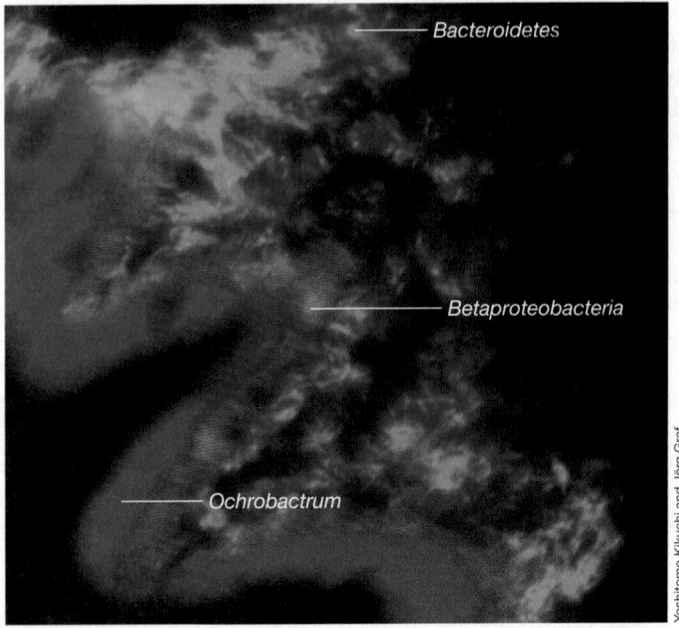

Yoshitomo Kikuchi and Jörg Graf

Figure 25.40 Micrograph of a FISH-stained microbial community in the bladder of *Hirudo verbana*. A probe (red) targeted at the 16S rRNA of *Betaproteobacteria* and a probe (green) targeted at the 16S rRNA of *Bacteroidetes* reveal distinct layers of different bacteria in the lumen of the bladder. Staining with DAPI (blue), which binds to DNA, reveals the intracellular alphaproteobacterium *Ochrobactrum* and host nuclei.

Host	Common name	Symbionts
Porifera	Sponge	Cyanobacteria, *Chlorella*, *Symbiodinium*
Cnidaria	Coral, sea anemone	*Symbiodinium*, *Chlorella*
Platyhelminthes	Flatworm	Diatoms, primitive chlorophytes
Mollusca	Snail, clam	*Symbiodinium*, *Chlorella*
Ascidia	Sea squirt	Cyanobacteria

Table 25.5 *Symbioses between animals and phototrophic symbionts*

epithelial side and two species of *Betaproteobacteria* colonize the luminal side of the biofilm-like structure that coats the bladder wall (**Figure 25.40**).

Symbiotic relationships require transmission of the microbial partners between host generations. Many gut symbionts are horizontally (environmentally) transmitted, but *A. veronii* appears to be vertically transmitted from the parent to the offspring via cocoons in which the embryos develop; juvenile leeches removed from cocoons are already infected with cells of *A. veronii*. Similarly, most of the bladder symbionts have been detected in juveniles taken from cocoons. Such vertical transmission of the symbionts ensures their safe transfer to the next host generation.

MiniQuiz

- How do leeches transmit symbionts to their progeny?
- In what way does the *A. veronii* symbiont of the leech resemble a pathogenic bacterium?

25.14 Reef-Building Corals

Coral reef ecosystems are the products of mutualistic associations between algae and simple marine animals. The extensive ecosystems associated with the worldwide distribution of these mutualisms support tens of thousands of species.

Phototrophic Symbioses with Animals

We saw in the beginning of this chapter that a lichen is a mutualism between a fungus and a phototrophic partner—an alga or cyanobacterium. Like the fungi, some animals establish mutualistic associations with photosynthetic algae or cyanobacteria

(**Table 25.5**). The animals in most of these associations are in phyla that display very simple body plans; for example, the Porifera (sponges) and Cnidaria (corals, sea anemones, and hydroids). These mutualistic animal–bacterial associations live in clear tropical waters where nutrients for the animals are scarce, and the animal body typically has a large surface area relative to its volume and is thus well suited for capturing light.

The coral skeleton is an extremely efficient light-gathering structure that greatly enhances light harvesting. There are only a few instances of algae forming associations with more complex animals, such as those in the phyla Platyhelminthes (flatworms), Mollusca (snails and clams), and Urochordata (sea squirts). In these cases either the animal has a suitable surface-to-volume ratio or has evolved specific light-gathering surfaces. The unicellular phototrophic symbionts are phylogenetically diverse and include cyanobacteria, rhodophytes, chlorophytes, diatoms, and dinoflagellates (◌ Section 18.7 and Chapter 20). Most common are the green algae *Chlorella* (associating with sponges and freshwater hydras), cyanobacteria (associating with marine sponges), and species of the dinoflagellate genus *Symbiodinium*.

The most spectacular and ecologically significant of these mutualisms is between the cnidarian stony corals (order Scleractinia) and the dinoflagellate *Symbiodinium* (**Figure 25.41**). Together the corals and dinoflagellates form the trophic and structural foundation of the coral reef ecosystem. The cnidarians possess a very simple two-tissue-layer body plan (ectoderm and gastroderm) and harbor the dinoflagellate symbiont intracellularly in vacuoles called *symbiosomes* within cells of the inner (gastrodermal) tissue layer (Figure 25.41c). The algae receive key inorganic nutrients from host metabolism and pass photosynthetically produced organic compounds to the corals. This mutualism has allowed coral reefs to develop in large expanses of nutrient-poor ocean waters. Dinoflagellates and other alveolates comprise eight genera and around 2000 extant species (◌ Section 20.9). Although dinoflagellate mutualisms are common, most are between species of *Symbiodinium* and marine invertebrates or protists (Figure 25.41). We focus here on the symbiotic association between *Symbiodinium* and the stony coral cnidarians.

Transmission, Specificity, and Benefits of the *Symbiodinium*–Coral Association

Reef-building corals reproduce sexually by releasing gametes into the seawater (broadcast spawning). A male and a female gamete fuse to form a free-swimming larva that later settles on a surface,

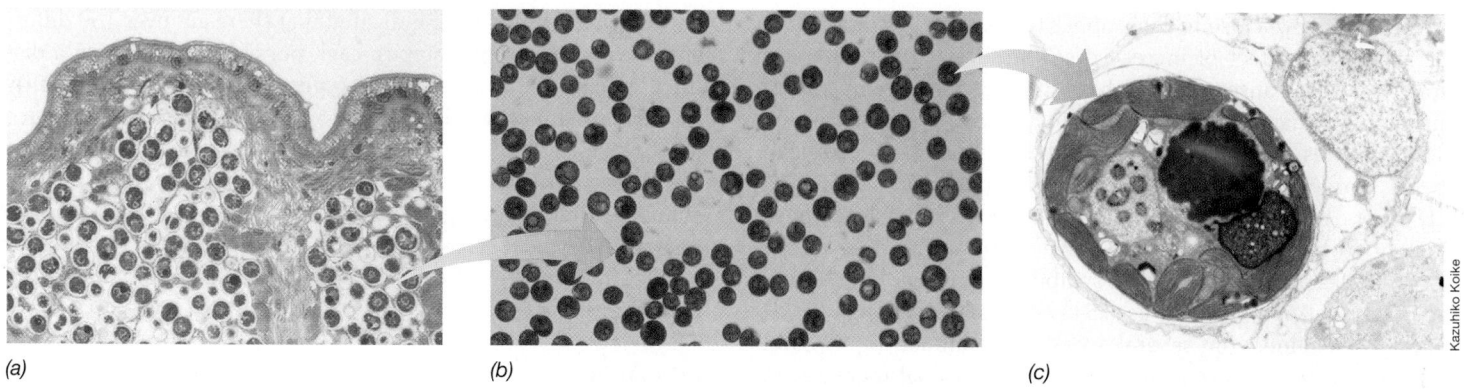

(a) (b) (c)

Kazuhiko Koike

Figure 25.41 *Symbiodinium* **symbiont of marine invertebrates.** *(a)* Thin-section micrograph of *Symbiodinium* in the mantle tissue of a giant clam. *(b) Symbiodinium* cells recovered from a soft coral. *(c)* Transmission electron micrograph of a *Symbiodinium* cell within a vacuole of a cell of the stony coral *Ctenactis echinata*. The *Symbiodinium* cell is about 10 μm in diameter.

where it may initiate a new coral colony. Algal symbionts are typically present in the egg before it is released from the parent (vertical transmission), although free-living *Symbiodinium* cells can also be ingested by juvenile corals (horizontal transmission). A developing coral that ingests dinoflagellates digests all of them except the particular *Symbiodinium* of its mutualism. After establishing an association, the coral controls the growth of *Symbiodinium* via chemical signaling and, following each cell division, each *Symbiodinium* daughter cell is allocated to a new symbiosome.

Both partners in the cnidarian–dinoflagellate mutualism have evolved adaptations for nutritional exchange. The dinoflagellates donate most of their photosynthetically fixed carbon (in the form of small molecules such as sugars, glycerol, and amino acids) to the cnidarian in exchange for inorganic nitrogen, phosphorus, and inorganic carbon from the host. Moreover, in addition to providing protection and inorganic nutrients, the calcium carbonate skeleton of corals is one of the most efficient collectors of solar radiation in nature, amplifying the incident light field for the symbionts by as much as fivefold; this benefits the symbiont in carrying out photosynthesis under a light-absorbing water column.

Coral Bleaching—the Risk of Harboring a Phototrophic Symbiont in a Changing World

The extensive coral reef systems in the oceans worldwide are now threatened with extinction, primarily as a consequence of human activities. Ongoing loss of these beautiful and productive ecosystems is thought to be the result of elevated atmospheric CO_2; namely, increased sea surface temperature, rising sea levels, and ocean acidification. These environmental changes are contributing to both bleaching and loss of coral structure from reduced calcification. Healthy corals harbor millions of cells of *Symbiodinium* per square centimeter of tissue. Coral bleaching is the loss of color from host tissues caused by the lysis of these symbionts, revealing the underlying white limestone skeleton (**Figure 25.42**).

UNIT 7

(a)

Ernesto Weil

(b)

Ernesto Weil

Figure 25.42 Coral bleaching. *(a)* Two colonies of the brain coral *Colpophyllia natans*. The coral on the left is a healthy brown color, whereas the coral on the right is fully bleached. *(b)* A large colony of partially bleached *Montastraea faveolata*.

Coral reefs live close to their upper temperature limit and it is the synergistic effect of increased sea surface temperature and irradiance that causes massive bleaching. Elevated temperature and high irradiance impair the photosynthetic apparatus of the dinoflagellates, resulting in the production of reactive oxygen species (for example, singlet oxygen and superoxide, ↩ Section 5.18) that cause damage to both host and symbiont. Bleaching is thought to be caused by a protective immune response of the host that destroys compromised symbionts. Increases in sea surface temperatures as small as 0.5–1.5°C above the local maximum, if sustained for several weeks, can induce rapid coral bleaching. Thermal stress, accentuated by seasonal increases in electromagnetic radiation of ultraviolet and some visible wavelengths, has resulted in bleaching of huge expanses of coral reefs. From projected increases in sea temperature owing to climate change, a collapse of Indian Ocean coral reef systems within only a few years is predicted, with global collapse of coral reefs occurring by the middle of this century. The ecological implications of bleaching are far-reaching, as corals are the foundation of coral reef ecosystems that support thousands of other animal species, including fish and other aquatic animals.

Molecular results have indicated that there are over 150 different *Symbiodinium* phylotypes, each possibly representing a distinct species. Although specific types appear to be consistently associated with particular species of coral, there is also evidence for host switching. Because the type of symbiont influences the ability of the coral to adapt to stresses associated with climate change, understanding alternative mechanisms of adaptive response, including possible symbiont switching, is essential to predicting the future health of corals, their symbionts, and the reefs they build.

MiniQuiz

- What gives corals their spectacular colors?
- What are the two mechanisms of *Symbiodinium* transfer to developing corals?
- What are the major environmental factors contributing to coral bleaching?

Big Ideas

25.1
Lichens are a mutualistic association between a fungus and an oxygenic phototroph.

25.2
The consortium "*Chlorochromatium aggregatum*" is a mutualism between a phototrophic green sulfur bacterium and a motile heterotroph. Mutual benefit is based on the phototroph supplying organic matter to the heterotroph in exchange for motility that permits rapid repositioning in stratified lakes to obtain optimal light and nutrients.

25.3
One of the most agriculturally important plant–microbial symbioses is that between legumes and nitrogen-fixing bacteria. The bacteria induce the formation of root nodules within which nitrogen fixation occurs. The plant provides the energy needed by the root nodule bacteria, and the bacteria provide fixed nitrogen for the plant.

25.4
The crown gall bacterium *Agrobacterium* enters into a unique relationship with plants. Part of the Ti plasmid in the bacterium can be transferred into the genome of the plant, initiating crown gall disease. The Ti plasmid has also been used for the genetic engineering of crop plants.

25.5
Mycorrhizae are mutualistic associations between the roots of plants and fungi that allow the plant to extend its root system via intimate interaction with an extensive network of fungal mycelia. The mycelia network provides the plant with essential nutrients such as phosphorus, and the plant, in turn, supplies carbohydrates to the fungus.

25.6
Microbial fermentation is important for digestion in all mammals. Several microbial mutualisms have evolved in different mammals that allow for the digestion of different types of food. Herbivores derive almost all of their carbon and energy from plant fiber.

25.7
The rumen, the digestive organ of ruminant animals, specializes in cellulose digestion, which is carried out by microorganisms. Bacteria, protists, and fungi in the rumen produce volatile fatty acids that provide energy for the ruminant. Rumen microorganisms synthesize vitamins and amino acids and are also a major source of protein—all used by the ruminant.

25.8
The human microbiome encompasses all sites of the human body inhabited by microorganisms. The microorganisms are

critical to early development, health, and predisposition to disease. The human gut microbial community is unique when compared with that of other mammals. The gut microflora affects energy recovery from food, and a shift in gut community structure is thought to contribute to obesity.

25.9

A large proportion of insects have established obligate mutualisms with bacteria, the basis of the mutualism often being bacterial biosynthesis of nutrients such as amino acids that are absent from the food the insect feeds on. Long-established obligate mutualisms are marked by extreme genome reduction of the symbiont, with retention of only those genes essential for the mutualism.

25.10

Termites associate symbiotically with bacteria and protists capable of digesting plant cell walls. The unique termite gut configuration and the hindgut microbial community composed largely of cellulolytic bacteria and protists and acetogenic bacteria result in high levels of acetate, the primary source of carbon and energy for the termite.

25.11

A light-emitting organ on the underside of the Hawaiian bobtail squid provides a habitat for bioluminescent cells of *Aliivibrio*

fischeri. From the mutualism in the light organ, the squid gains protection from predators while the bacterium benefits from a habitat in which it grows quickly and contributes cells to its free-living population.

25.12

Most invertebrates living on the seafloor near regions receiving hydrothermal fluids have established obligate mutualisms with chemolithotrophic bacteria. These mutualisms are nutritional, allowing the invertebrates to thrive in an environment enriched in reduced inorganic materials, such as H_2S, that are abundant in vent fluids. The invertebrates provide the symbionts an ideal nutritional environment in exchange for organic nutrients.

25.13

Leeches and particular bacterial species form symbioses in regions of the host body that are important for host nutrition and nitrogen retention. The existence of mechanisms for vertical transmission of the symbionts indicates that these mutualisms are highly evolved and functionally important.

25.14

The mutualism between the dinoflagellate *Symbiodinium* and the stony corals produces the extensive worldwide coral reef ecosystems that sustain a tremendous diversity of marine life. Coral bleaching caused by climate change threatens these ecosystems.

Review of Key Terms

Bacteriocyte a specialized insect cell in which bacterial symbionts reside

Bacteriome a specialized region in several insect groups that contains insect bacteriocyte cells packed with bacterial symbionts

Bacteroid the morphologically misshapen cells of rhizobia inside a leguminous plant root nodule; can fix N_2

Biogeography the study of the geographical distribution of organisms

Coevolution evolution that proceeds jointly in a pair of intimately associated species owing to the effects each has on the other

Consortium a mutualism between bacteria, for example, a phototrophic green sulfur bacteria and a motile nonphototrophic bacterium

Infection thread in the formation of root nodules, a cellulosic tube through which *Rhizobium* cells can travel to reach and infect root cells

Leghemoglobin an O_2-binding protein found in root nodules

Lichen a fungus and an alga (or cyanobacterium) living in symbiotic association

Mutualism a symbiosis in which both partners benefit

Mycorrhizae a symbiotic association between a fungus and the roots of a plant

Nod factors oligosaccharides produced by root nodule bacteria that help initiate the plant–bacterial symbiosis

Root nodule a tumorlike growth on plant roots that contains symbiotic nitrogen-fixing bacteria

Rumen the first vessel in the multichambered stomach of ruminant animals in which cellulose digestion occurs

Symbiosis an intimate relationship between two organisms, often developed through prolonged association and coevolution

Ti plasmid a conjugative plasmid present in the bacterium *Agrobacterium tumefaciens* that can transfer genes into plants

Volatile fatty acids (VFAs) the major fatty acids (acetate, propionate, and butyrate) produced during fermentation in the rumen

Review Questions

1. Briefly discuss the importance of the legume–root nodule symbiosis (Section 25.3).

2. What is Ti plasmid? How does acquiring it affect *Agrobacterium* (Section 25.4)?

3. How do mycorrhizae improve the growth of trees? How do they promote plant diversity (Section 25.5)?

4. What is a rumen and how do the digestive processes operate in the ruminant digestive tract? What are the major benefits and the disadvantages of a rumen system? How does a cecal animal compare with a ruminant (Section 25.7)?

5. What are VFAs? How important are they to animals such as cows (Section 25.7)?

6. What is a possible mechanism by which the microbial community of the human gut increases energy recovery, thereby contributing to obesity (Section 25.8)?

7. Why was *Escherichia coli* long thought to be a dominant member of the human gut microbial community (Section 25.8)?

8. How is it possible for aphids to feed on the carbohydrate-rich but nutrient-poor sap of phloem vessels in plants (Section 25.9)?

9. Why do symbionts that are transmitted horizontally show less genome reduction, as opposed to the significant genome reduction observed in heritable symbionts (Section 25.9)?

10. How do the microbial communities of guts of higher and lower termites differ in composition and degradation of cellulose (Section 25.10)?

11. Why is the squid–*Aliivibrio* symbiotic system widely used as a study model (Section 25.11)?

12. How does a tube worm obtain nutrients if it lacks a mouth, gut, and anus (Section 25.12)?

13. Compare the microbial communities in the medicinal leech crop, intestinum, and bladder (Section 25.13).

14. How does the body plan of corals influence their ability to symbiotically associate with *Symbiodinium* (Section 25.14)?

Application Question

1. Imagine that you have discovered a new animal that consumes only grass in its diet. You suspect it to be a ruminant and have available a specimen for anatomical inspection. If this animal is a ruminant, describe the position and basic components of the digestive tract you would expect to find and any key microorganisms and substances you might look for.

Need more practice? Test your understanding with Quantitative Questions; access additional study tools including tutorials, animations, and videos; and then test your knowledge with chapter quizzes and practice tests at **www.microbiologyplace.com**.

Microbial Growth Control

Filtration of an aqueous liquid through the tiny pores of a membrane filter traps any microbial cells that were present in the liquid and renders it sterile.

Ｗith this chapter we begin to study the relationships between microorganisms and humans. We start with the agents and methods used for control of microbial growth. The goal is to either reduce or eliminate the microbial load and limit microbial effects.

A few agents eliminate microbial growth entirely by **sterilization**—the killing or removal of all viable organisms from a growth medium or surface. In certain circumstances, however, sterility is not attainable or practical, as in fresh foods. Microorganisms can be effectively controlled by limiting or inhibiting their growth. For example, we wash fresh produce to remove most existing bacteria, limiting their growth. Likewise, we inhibit microbial growth on body surfaces by washing. Neither of these processes, however, kills or removes all microbes.

Methods for inhibiting rapid microbial growth include decontamination and disinfection. **Decontamination** is the treatment of an object or surface to make it safe to handle. For example, simply wiping a table after a meal removes contaminating microorganisms and their potential nutrients. **Disinfection**, in contrast, directly targets pathogens, although it may not eliminate all microorganisms. Specialized chemical or physical agents called *disinfectants* can kill microorganisms or inhibit microbial growth. Bleach (sodium hypochlorite) solution, for example, is a disinfectant used to clean and disinfect food preparation areas.

Under certain circumstances, it may be necessary to destroy all microorganisms. Such measures are necessary, for instance, when making microbiological media or preparing surgical instruments. Sterilization completely eliminates all microorganisms, including endospores, and also eliminates all viruses. Microbial control *in vivo* is much more difficult: Clinically useful bacteriocidal (bacteria killing) agents or bacteriostatic (bacteria inhibiting) agents must selectively prevent or reduce bacterial growth, while causing no harm to the host.

In this chapter we first examine methods of microbial control that are used *in vitro*. We then discuss antimicrobial drugs used in humans and animals.

Ⅰ Physical Antimicrobial Control

Ｐhysical methods are used in industry, medicine, and in the home to achieve microbial decontamination, disinfection, and sterilization. Heat, radiation, and filtration are commonly used to destroy or remove microorganisms. These methods prevent microbial growth or decontaminate areas or materials harboring microorganisms. Here we discuss physical control mechanisms and present some practical examples.

26.1 Heat Sterilization

Perhaps the most widespread method used for controlling microbial growth is the use of heat as a sterilization method. Factors that affect a microorganism's susceptibility to heat include the temperature and duration of the heat treatment and whether the heat is moist or dry.

Measuring Heat Sterilization

All microorganisms have a maximum growth temperature beyond which viability decreases ($\circlearrowleft$ Section 5.12). Microorganisms lose viability at very high temperatures because most

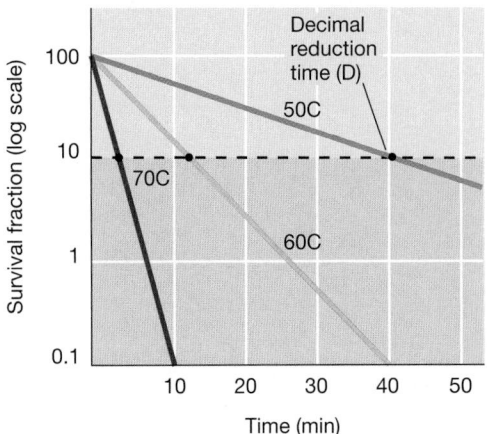

Figure 26.1 **The effect of temperature over time on the viability of a mesophilic bacterium.** The decimal reduction time, D, is the time at which only 10% of the original population of organisms remains viable at a given temperature. For 70°C, $D = 3$ min; for 60°C, $D = 12$ min; for 50°C, $D = 42$ min.

macromolecules lose structure and function, a process called *denaturation*. The effectiveness of heat as a sterilant is measured by the time required for a 10-fold reduction in the viability of a microbial population at a given temperature. This is the *decimal reduction time* or D. For example, over the range of temperatures usually used in food preparation (cooking and canning), the relationship between D and temperature is exponential; the logarithm of D plotted against temperature yields a straight line (**Figure 26.1**). The graph can be used to calculate processing times to achieve sterilization, for instance in a canning operation. The slope of the line indicates the sensitivity of the organism to heat under the conditions employed ($\circlearrowleft$ Section 36.2). Death from heating is an exponential (first-order) function, proceeding more rapidly as the temperature rises, as shown in **Figure 26.2**.

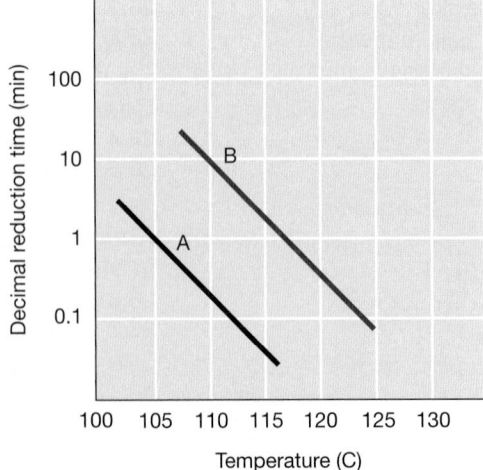

Figure 26.2 **The relationship between temperature and the rate of killing in mesophiles and thermophiles.** Data were obtained for decimal reduction times, D, at several different temperatures, as in Figure 26.1. For organism A, a typical mesophile, exposure to 110°C for less than 20 sec resulted in a decimal reduction, while for organism B, a thermophile, 10 min was required to achieve a decimal reduction.

The time necessary to kill a defined fraction (for example, 90%) of viable cells is independent of the initial cell concentration. As a result, sterilization of a microbial population takes longer at lower temperatures than at higher temperatures. The time and temperature, therefore, must be adjusted to achieve sterilization for each specific set of conditions. The type of heat is also important: Moist heat has better penetrating power than dry heat and, at a given temperature, produces a faster reduction in the number of living organisms.

Determination of a decimal reduction time requires a large number of viable count measurements (Section 5.10). An easier way to characterize the heat sensitivity of an organism is to measure the *thermal death time*, the time it takes to kill all cells at a given temperature. To determine the thermal death time, samples of a cell suspension are heated for different times, mixed with culture medium, and incubated. If all the cells have been killed, no growth is observed in the incubated samples. The thermal death time depends on the size of the population tested; a longer time is required to kill all cells in a large population than in a small one. When the number of cells is standardized, it is possible to compare the heat sensitivities of different organisms by comparing their thermal death times at a given temperature.

Endospores and Heat Sterilization

Some bacteria produce highly resistant cells called *endospores* (Section 3.12). The heat resistance of vegetative cells and endospores from the same organism differs considerably. For instance, in the autoclave (see below) a temperature of 121°C is normally reached. Under these conditions, endospores may require 4–5 minutes for a decimal reduction, whereas vegetative cells may require only 0.1–0.5 min at 65°C. To ensure adequate decontamination of any material, heat sterilization procedures must be designed to destroy endospores.

Endospores can survive heat that would rapidly kill vegetative cells of the same species. A major factor in heat resistance is the amount and state of water within the endospore. During endospore formation, the protoplasm is reduced to a minimum volume as a result of the accumulation of calcium (Ca^{2+})–dipicolinic acid complexes and small acid-soluble spore proteins (SASPs). This mixture forms a cytoplasmic gel, and a thick cortex then forms around the developing endospore. Contraction of the cortex results in a shrunken, dehydrated cell containing only 10–30% of the water of a vegetative cell (Section 3.12).

The water content of the endospore coupled with the concentration of SASPs determines its heat resistance. If endospores have a low concentration of SASPs and high water content, they exhibit low heat resistance. Conversely, if they have a high concentration of SASPs and low water content, they show high heat resistance. Water moves freely in and out of endospores, so it is not the impermeability of the endospore coat that excludes water, but the gel-like material in the endospore protoplast.

The medium in which heating takes place also influences the killing of both vegetative cells and endospores. Microbial death is more rapid at acidic pH, and acid foods such as tomatoes, fruits, and pickles are much easier to sterilize than neutral pH foods such as corn and beans. High concentrations of sugars, proteins, and fats decrease heat penetration and usually increase the resistance of organisms to heat, whereas high salt concentrations may either increase or decrease heat resistance, depending on the organism. Dry cells and endospores are more heat resistant than moist ones; consequently, heat sterilization of dry objects such as endospores always requires higher temperatures and longer heat application times than sterilization of wet objects such as liquid bacterial cultures.

The Autoclave

The **autoclave** is a sealed heating device that uses steam under pressure to kill microorganisms (**Figure 26.3a**). Killing of heat-resistant endospores requires heating at temperatures above 100°C, the boiling point of water at normal atmospheric pressure. The autoclave uses steam under 1.1 kilograms/square centimeter (kg/cm^2) [15 pounds/square inch (lb/in^2)] pressure, which yields a temperature of 121°C. At 121°C, the time to achieve sterilization of endospore-containing material is generally 10–15 minutes (Figure 26.3b).

If an object being sterilized is bulky, heat transfer to the interior is retarded, and the total heating time must be extended to ensure that the entire object is at 121°C for 10–15 minutes. Extended times are also required when large volumes of liquids are being autoclaved because large volumes take longer to reach sterilization temperatures. Note that it is not the *pressure* inside the autoclave that kills the microorganisms but the high *temperature* that can be achieved when steam is applied under pressure.

Pasteurization

Pasteurization uses precisely controlled heat to reduce the number of microorganisms found in milk and other heat-sensitive liquids. The process, named for Louis Pasteur (Section 1.7), was first used for controlling the spoilage of wine. Pasteurization does not kill all organisms and is therefore not a method of sterilization. Pasteurization does, however, reduce the *microbial load*, the number of viable microorganisms in a sample. At temperatures and times used for pasteurization of food products such as milk, pathogenic bacteria, especially the organisms causing tuberculosis, brucellosis, Q fever, and typhoid fever, are killed. These pathogens are no longer common in raw foods in developed countries, but pasteurization also controls commonly encountered pathogens such as *Listeria monocytogenes*, *Campylobacter* species, *Salmonella*, and *Escherichia coli* O157:H7; these pathogenic bacteria can be found in foods such as dairy products and juices (Sections 36.8–36.12). In addition, by decreasing the overall microbial load, pasteurization retards the growth of spoilage organisms, increasing the shelf life of perishable liquids (Sections 36.1 and 36.2).

Pasteurization of milk is usually achieved by passing the milk through a heat exchanger. The milk is pumped through tubing that is in contact with a heat source. Careful control of the milk flow rate and the size and temperature of the heat source raises the temperature of the milk to 71°C for 15 seconds. The milk is then rapidly cooled. This process is aptly called flash pasteurization.

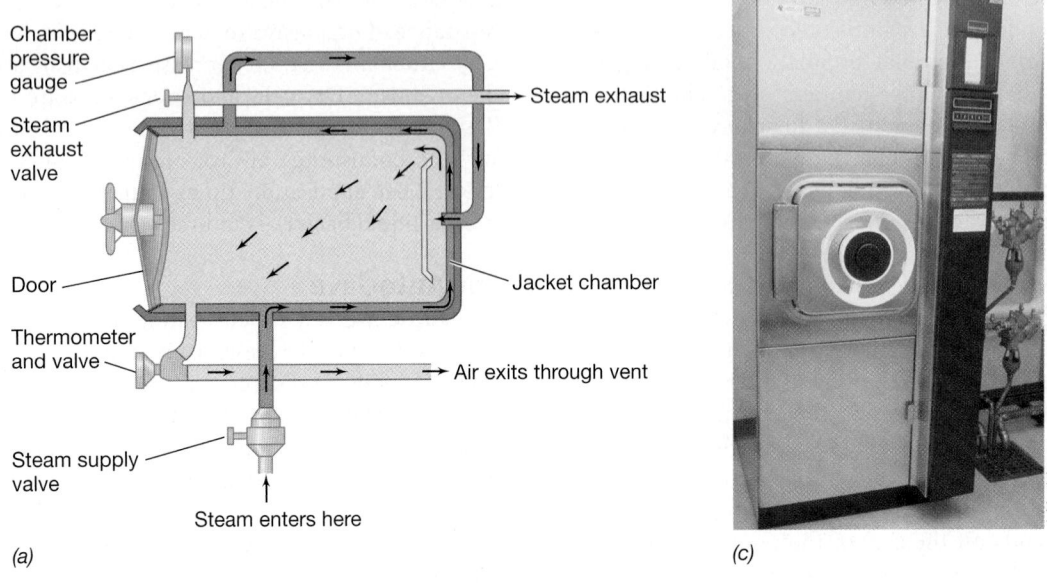

(a)

(c)

J. Martinko

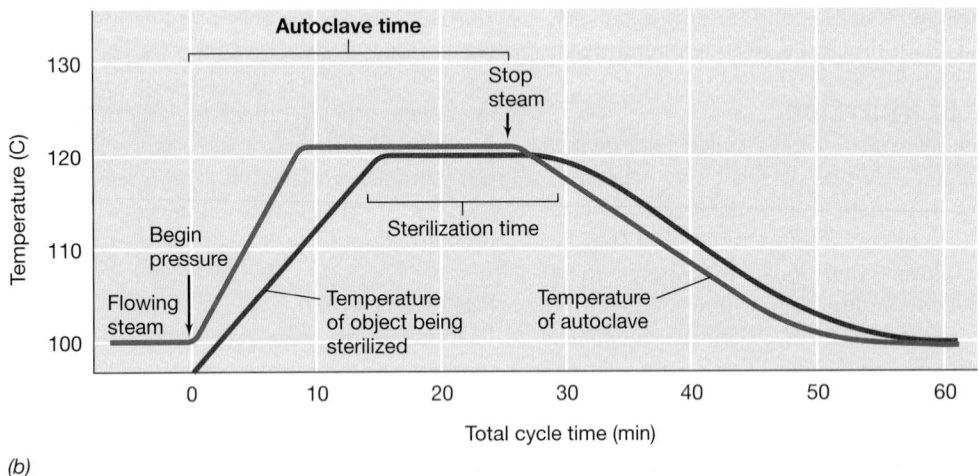

(b)

Figure 26.3 The autoclave and moist heat sterilization. *(a)* The flow of steam through an autoclave. *(b)* A typical autoclave cycle. The temporal heating profile of a fairly bulky object is shown. The temperature of the object rises and falls more slowly than the temperature of the autoclave. The temperature of the object must reach the target temperature and be held for 10–15 minutes to ensure sterility, regardless of the temperature and time recorded in the autoclave. *(c)* A modern research autoclave. Note the pressure-lock door and the automatic cycle controls on the right panel. The steam inlet and exhaust fittings are on the right side of the autoclave.

Milk can also be pasteurized in large quantities by heating in large vats to 63–66°C for 30 minutes. However, this bulk pasteurization method is less satisfactory because the milk heats and cools slowly and must be held at high temperatures for longer times. This slower heating and cooling of the milk alters the taste of the final product, rendering it generally less palatable for the consumer. Flash pasteurization, sometimes done at even higher temperatures and shorter times, alters the flavor less, kills heat-resistant organisms more effectively, and can be done on a continuous-flow basis, making it more adaptable to large dairy operations.

MiniQuiz

• Why is heat an effective sterilizing agent?

• Why is moist heat more effective than dry heat for sterilization?

• What steps are necessary to ensure the sterility of material contaminated with bacterial endospores?

• Distinguish between the sterilization of microbiological media and the pasteurization of dairy products.

26.2 Radiation Sterilization

Heat is just one form of energy that can sterilize or reduce microbial load. Microwaves, ultraviolet (UV) radiation, X-rays, gamma rays (γ-rays), and electrons can also effectively reduce microbial growth if applied in the proper dose and time. However, each type of energy has a different mode of action. For example, the antimicrobial effects of microwaves are due, at least in part, to thermal effects. Other forms of energy cause other modifications that lead to death or inactivation of microorganisms.

Ultraviolet Radiation

Ultraviolet radiation between 220 and 300 nm in wavelength has enough energy to cause modifications or actual breaks in DNA, sometimes leading to disruption of DNA and death of the exposed organism (Section 10.4). This "near-visible" UV light is useful for disinfecting surfaces, air, and materials such as water that do not absorb the UV waves. For example, laboratory laminar flow hoods, designed to maintain clean work areas, are equipped with a "germicidal" UV light to decontaminate the work surface after use (**Figure 26.4**). UV radiation, however, cannot penetrate solid, opaque, or light-absorbing surfaces, limiting its use to disinfection of exposed surfaces.

Ionizing Radiation

Ionizing radiation is electromagnetic radiation of sufficient energy to produce ions and other reactive molecular species from molecules with which the radiation particles collide. Ionizing radiation generates electrons, e⁻; hydroxyl radicals, OH· (Section 5.18), and hydride radicals, H·. Each of these highly reactive molecules is capable of altering and disrupting macromolecules such as DNA, lipids, and protein. The ionization and

Table 26.1 *Radiation sensitivity of microorganisms and biological functions*

Species or function	Type of microorganism	D10[a] (Gy)
Clostridium botulinum	Gram-positive, anaerobic, sporulating Bacteria	3300
Clostridium tetani	Gram-positive, anaerobic, sporulating Bacteria	2400
Bacillus subtilis	Gram-positive, aerobic, sporulating Bacteria	600
Escherichia coli O157:H7	Gram-negative Bacteria	300
Salmonella typhimurium	Gram-negative Bacteria	200
Lactobacillus brevis	Gram-positive Bacteria	1200
Deinococcus radiodurans	Gram-negative, radiation-resistant Bacteria	2200
Aspergillus niger	Mold	500
Saccharomyces cerevisiae	Yeast	500
Foot-and-mouth	Virus	13,000
Coxsackie	Virus	4500
Enzyme inactivation		20,000–50,000
Insect deinfestation		1000–5000

[a]D10 is the amount of radiation necessary to reduce the initial population or activity level 10-fold (1 logarithm). Gy = grays. 1 gray = 100 rads. The lethal dose for humans is 10 Gy.

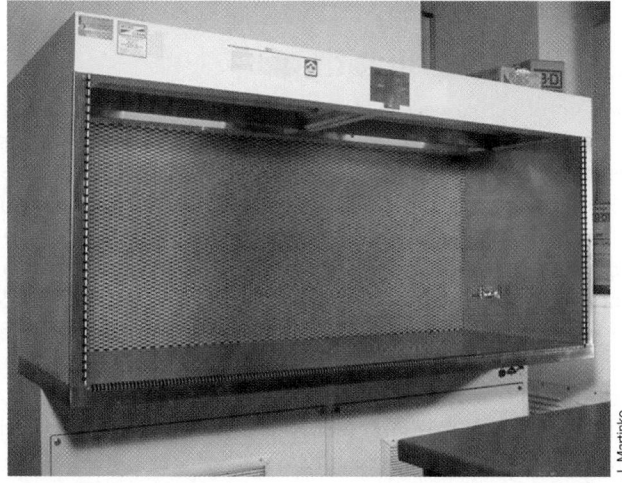

Figure 26.4 A laminar flow hood. An ultraviolet light source prevents contamination of the hood when it is not in use. When in use, air is drawn into the cabinet through a HEPA filter. The filtered air inside the cabinet is exhausted out of the cabinet, preventing contamination of the inside of the hood. The cabinet provides a contaminant-free workspace for microbial and tissue culture manipulations.

subsequent degradation of these biologically important molecules leads to the death of irradiated cells.

The unit of radiation is the *roentgen,* which is a measure of the energy output from a radiation source. The standard for biological applications such as sterilization is the absorbed radiation dose, measured in *rads* (100 erg/g) or *grays* (1 Gy = 100 rad). Some microorganisms are much more resistant to radiation than others. **Table 26.1** shows the dose of radiation necessary for a 10-fold (one log) reduction in the numbers of selected microorganisms or biological functions. For example, the amount of energy necessary to achieve a 10-fold reduction (*D*) of a radiation-sensitive bacterium such as *Escherichia coli* O157:H7 is 300 Gy. The *D* value is analogous to the decimal reduction time for heat sterilization: The relationship of the survival fraction plotted on a logarithmic scale versus the radiation dose in grays is essentially linear (**Figure 26.5** and compare to Figure 26.1).

In practice, this means that at a radiation dose of 300 Gy, 90% of *E. coli* O157:H7 in a given sample would be killed. A dose of 2 *D,* or 600 Gy, would kill 99% of the organism, and so on. A standard killing dose for radiation sterilization is 12 *D.* A killing dose of radiation-resistant endospores of a bacterium such as *Clostridium botulinum* for example, would be 3300 Gy × 12, or 39,600 Gy (Table 26.1). By contrast, the killing dose for *E. coli* O157:H7 is

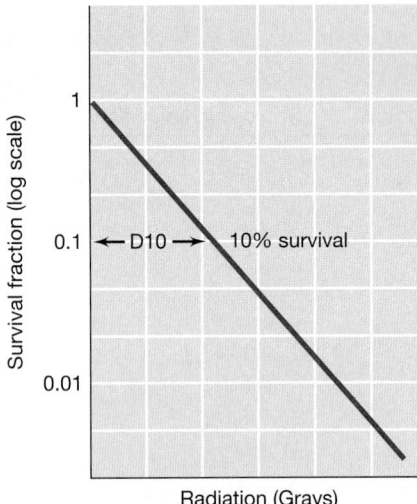

Figure 26.5 Relationship between the survival fraction and the radiation dose of a microorganism. The *D*10, or decimal reduction dose, can be interpolated from the data as shown.

only 3600 Gy. In general, microorganisms are much more resistant to ionizing radiation than are multicellular organisms. For example, the lethal radiation dose for humans can be as low as 10 Gy if delivered over a short time (several minutes)!

Radiation Practices

Several radiation sources are useful for sterilization. Common sources of ionizing radiation include cathode ray tubes that generate electron beams, X-ray machines, and radioactive nuclides ^{60}Co and ^{137}Cs, which are relatively inexpensive by-products of nuclear fission. These sources produce electrons (e^-), X-rays, or γ-rays, respectively, all of which have sufficient energy to efficiently kill microorganisms. In addition, X-rays and γ-rays penetrate solids and liquids, making them ideal for treatment of bulk items such as ground beef or cereal grains.

Radiation is currently used for sterilization and decontamination in the medical supplies and food industries. In the United States, the Food and Drug Administration has approved the use of radiation for sterilization of such diverse items as surgical supplies, disposable labware, drugs, and even tissue grafts (**Table 26.2**). However, because of the required specialized equipment,

Table 26.2 *Medical and laboratory products sterilized by radiation*

Tissue grafts	Drugs	Medical and laboratory supplies
Cartilage	Chloramphenicol	Disposable labware
Tendon	Ampicillin	Culture media
Skin	Tetracycline	Syringes
Heart valve	Atropine	Surgical equipment
	Vaccines	Sutures
	Ointments	

Table 26.3 *Recommended radiation dose for decontamination of selected foods*

Food type	kiloGrays
Fruit	1
Poultry	3
Spices, seasonings	30

costs, and hazards associated with radiation techniques, this type of sterilization is limited to large industrial applications or specialized facilities.

Certain foods and food products are also routinely irradiated to ensure sterilization, pasteurization, or insect deinfestation. Radiation is approved by the World Health Organization and can be used in the United States for decontamination of foods particularly susceptible to microbial contamination such as fresh produce, meat products, chicken, and spices (**Table 26.3** and ⮑ Section 36.2). The use of radiation for these purposes is an established and accepted technology in many countries. However, the practice has not been readily accepted in some countries such as the United States because of fears of possible radioactive contamination, alteration in nutritional value, production of toxic or carcinogenic products, and perceived "off" tastes in irradiated food.

MiniQuiz

- Define the decimal reduction dose and the killing dose for radiation treatment of microorganisms.
- Why is ionizing radiation more effective than UV radiation for sterilization of food products?

26.3 Filter Sterilization

Heat is an effective way to decontaminate most liquids and can even be used to treat gases. Heat-sensitive liquids and gases, however, must be treated by other methods. Filtration accomplishes decontamination and even sterilization without exposure to denaturing heat. The liquid or gas is passed through a filter, a device with pores too small for the passage of microorganisms, but large enough to allow the passage of the liquid or gas. The selection of filters for sterilization must account for the size range of the contaminants to be excluded. Some microbial cells are greater than 10 μm in diameter, while the smallest bacteria are less than 0.3 μm in diameter. Historically, selective filtration methods were used to define and isolate viruses, most of which range from 25 nm to 200 nm (0.2 μm) in diameter. **Figure 26.6** illustrates major types of filters.

Depth Filters

A depth filter is a fibrous sheet or mat made from a random array of overlapping paper or borosilicate (glass) fibers (Figure 26.6*a*). The depth filter traps particles in the network of fibers in the

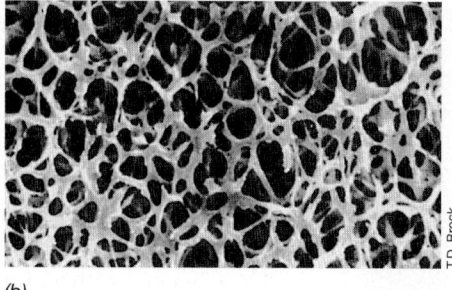
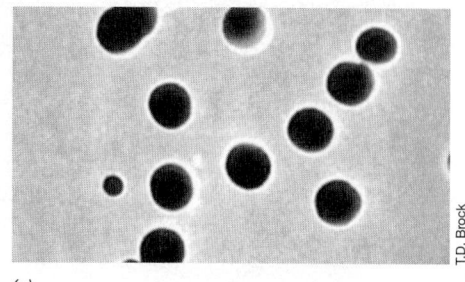

(a)   (b)   (c)

Figure 26.6 **Microbiological filters.** Scanning electron micrograph showing the structure of (a) a depth filter, (b) a conventional membrane filter, and (c) a nucleopore filter.

structure. Because the filtration material is arranged randomly in a thick layer, depth filters resist clogging and are often used as prefilters to remove larger particles from liquid suspensions so that the final filter in the sterilization process is not clogged. Depth filters are also used for the filter sterilization of air in industrial processes. In the home, the filter used in forced air heating and cooling systems is a simple depth filter designed to trap particulate matter such as dust, spores, and allergens.

Depth filters are important for biosafety applications. For example, manipulations of cell cultures, microbial cultures, and growth media require that contamination of both the operator and the experimental materials are minimized. These operations can be efficiently performed in a biological safety cabinet with airflow, both in and out of the cabinet, directed through a depth filter called a **HEPA filter**, or *h*igh-*e*fficiency *p*articulate *a*ir filter (Figure 26.4). A typical HEPA filter is a single sheet of borosilicate glass fibers that has been treated with a water-repellent binder. The filter, pleated to increase the overall surface area, is mounted inside a rigid, supportive frame. HEPA filters come in various shapes and sizes, from several square centimeters for vacuum cleaners, to several square meters for biological containment hoods and room air systems. Control of airborne particulate materials with HEPA filters allows the construction of "clean rooms" and isolation rooms for quarantine, as well as specialized biological safety laboratories (Section 31.4). HEPA filters typically remove 0.3-µm test particles with an efficiency of at least 99.97%; they remove both small and large particles, including most microorganisms, from the airstream.

Membrane Filters

Membrane filters are the most common type of filters used for liquid sterilization in the microbiology laboratory (Figure 26.6*b*). Membrane filters are composed of high tensile strength polymers such as cellulose acetate, cellulose nitrate, or polysulfone, manufactured to contain a large number of tiny holes, or pores. By adjusting the polymerization conditions during manufacture, the size of the holes in the membrane (and thus the size of the molecules that can pass through) can be precisely controlled. The membrane filter differs from the depth filter, functioning more like a sieve and trapping particles on the filter surface. About 80–85% of the membrane surface area consists of open pores. The porosity provides for a relatively high fluid flow rate.

Membrane filters for the sterilization of a liquid are illustrated in **Figure 26.7**. Presterilized membrane filter assemblies for sterilization of small to medium volumes of liquids such as growth media are routinely used in research and clinical laboratories. Filtration is accomplished by using a syringe, pump, or vacuum to force the liquid through the filtration apparatus into a sterile collection vessel.

Another type of membrane filter in common use is the nucleation track (nucleopore) filter. To make these filters, very thin polycarbonate film (10 µm) is treated with nuclear radiation and then etched with a chemical. The radiation causes local damage to the film, and the etching chemical enlarges these damaged locations into holes. The size of the holes can be controlled by varying the strength of the etching solution and the etching time. A typical nucleation track filter therefore has very uniform holes (Figure 26.6*c*). Nucleopore filters are commonly used to isolate specimens for scanning electron microscopy. Microorganisms are removed from liquid and concentrated in a single plane on the filter, where they can be observed with the microscope (**Figure 26.8**). Commonly used filter pore sizes for filter sterilization of small volumes, such as laboratory solutions, are 0.45 µm and 0.2 µm.

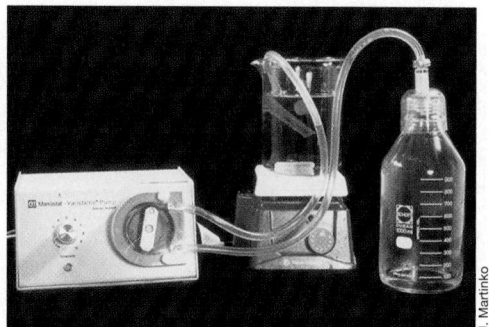

Figure 26.7 **Membrane filters.** Disposable, presterilized, and assembled membrane filter units. Left: a filter system designed for small volumes. Right: a filter system designed for larger volumes.

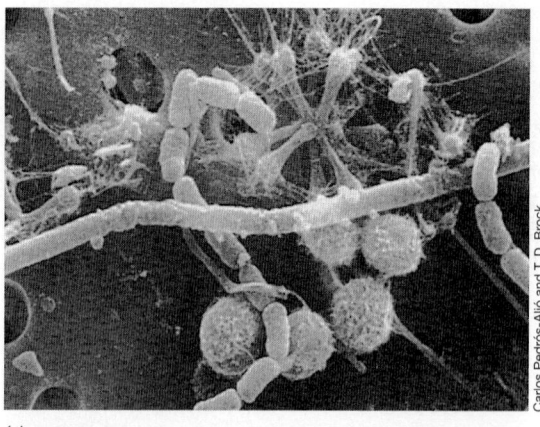

(a)

Carlos Pedrós-Alió and T. D. Brock

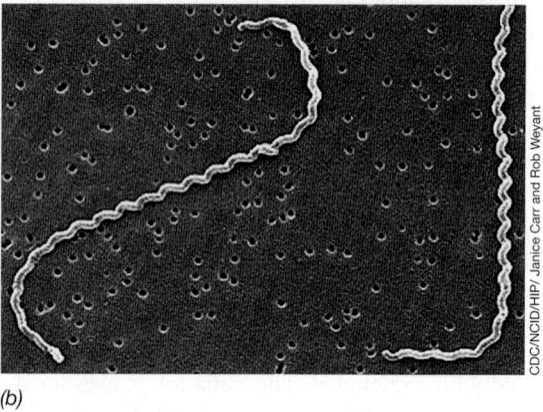

(b)

CDC/NCIDI/HIP/ Janice Carr and Rob Weyant

Figure 26.8 Scanning electron micrographs of bacteria trapped on nucleopore membrane filters. *(a)* Aquatic bacteria and algae. The pore size is 5 μm. *(b)* *Leptospira interrogans*. The bacterium is about 0.1 μm in diameter and up to 20 μm in length. The pore size of the filter is 0.2 μm.

MiniQuiz

• Why are filters used for sterilization of heat-sensitive liquids?

• Describe the use of depth filters for maintaining clean air in hospitals, laboratories, and the home.

II Chemical Antimicrobial Control

In the home, workplace, and laboratory, chemicals are routinely used to control microbial growth. An **antimicrobial agent** is a natural or synthetic chemical that kills or inhibits the growth of microorganisms. Agents that kill organisms are called *-cidal* agents, with a prefix indicating the type of microorganism killed. Thus, they are called **bacteriocidal**, **fungicidal**, and **viricidal** agents because they kill bacteria, fungi, and viruses, respectively. Agents that do not kill but only inhibit growth are called *-static* agents. These include **bacteriostatic**, **fungistatic**, and **viristatic** compounds.

26.4 Chemical Growth Control

Antimicrobial agents can differ in their selective toxicity. Nonselective agents have similar effects on all cells. Selective agents are more toxic for microorganisms than for animal tissues. Antimicrobial agents with selective toxicity are especially useful for treating infectious diseases because they kill selected microorganisms *in vivo* without harming the host. They are described later in this chapter. Here we discuss chemical agents that have relatively broad toxicity and are widely used for limiting microbial growth *in vitro*.

Effect of Antimicrobial Agents on Growth

Antibacterial agents can be classified as bacteriostatic, bacteriocidal, and bacteriolytic by observing their effects on bacterial cultures (**Figure 26.9**). Viable cells are measured by plate counts. The number of viable cells for a given organism is proportional to culture turbidity during the log phase of growth. Bacteriostatic agents are frequently inhibitors of protein synthesis and act by binding to ribosomes. If the concentration of the agent is lowered, the agent is released from the ribosome and growth resumes (Figure 26.9a). Many antibiotics work by this mechanism, and they will be discussed in Sections 26.6–26.9. Bacteriocidal

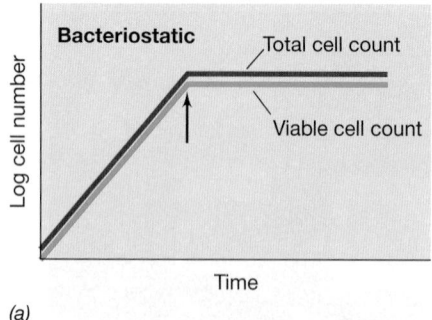

(a)

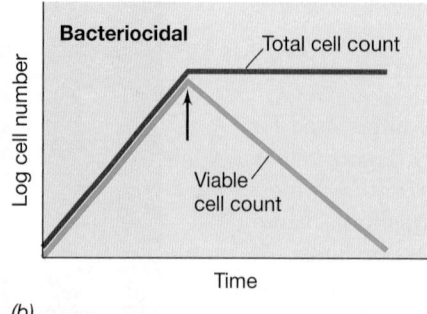

(b)

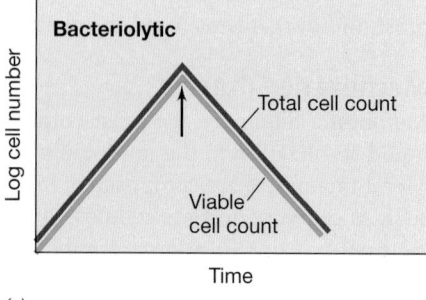

(c)

Figure 26.9 Bacteriostatic, bacteriocidal, and bacteriolytic antimicrobial agents. At the time indicated by the arrow, a growth-inhibitory concentration of each antimicrobial agent was added to an exponentially growing culture. The turbidity of each culture, coupled with viable plate counts, establishes the relationship between viable and total cell counts.

agents bind tightly to their cellular targets, are not removed by dilution, and kill the cell. The dead cells, however, are not destroyed, and total cell numbers, reflected by the turbidity of the culture, remain constant (Figure 26.9*b*). Some *-cidal* agents are also *-lytic* agents, killing by cell lysis and release of cytoplasmic contents. Lysis decreases the viable cell number and also the total cell number, shown by a decrease in culture turbidity (Figure 26.9*c*). Bacteriolytic agents include antibiotics that inhibit cell wall synthesis, such as penicillin, and chemicals such as detergents that rupture the cytoplasmic membrane.

Measuring Antimicrobial Activity

Antimicrobial activity is measured by determining the smallest amount of agent needed to inhibit the growth of a test organism, a value called the **minimum inhibitory concentration (MIC)**. To determine the MIC for a given agent against a given organism, a series of culture tubes is prepared and inoculated with the same number of microorganisms. Each tube contains medium with an increasing concentration of the agent. After incubation, the tubes are checked for visible growth (turbidity). The MIC is the lowest concentration of agent that completely inhibits the growth of the test organism (**Figure 26.10**). This is called the *tube dilution technique*.

The MIC is not a constant for a given agent; it varies with the test organism, the inoculum size, the composition of the culture medium, the incubation time, and the conditions of incubation, such as temperature, pH, and aeration. When culture conditions are standardized, however, different antimicrobial agents can be compared to determine which is most effective against a given organism.

Another common assay for antimicrobial activity is the *disc diffusion technique* (**Figure 26.11**). A Petri plate containing an agar medium is inoculated with a culture of the test organism. Known amounts of an antimicrobial agent are added to filter-

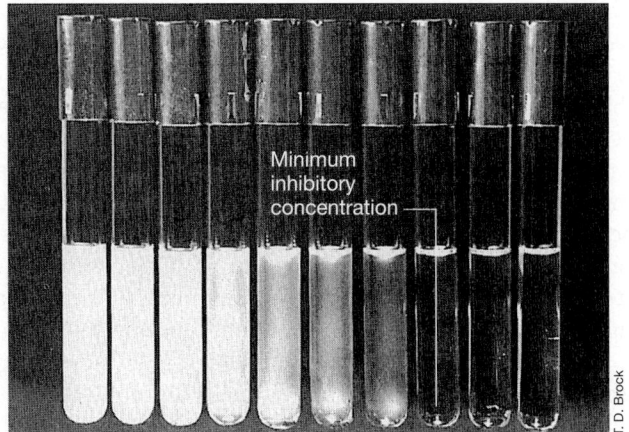

Figure 26.10 Antimicrobial agent susceptibility assay using dilution methods. The assay defines the minimum inhibitory concentration (MIC). A series of increasing concentrations of antimicrobial agent is prepared in the culture medium. Each tube is inoculated with a specific concentration of a test organism, followed by a defined incubation period. Growth, measured as turbidity, occurs in those tubes with antimicrobial agent concentrations below the MIC.

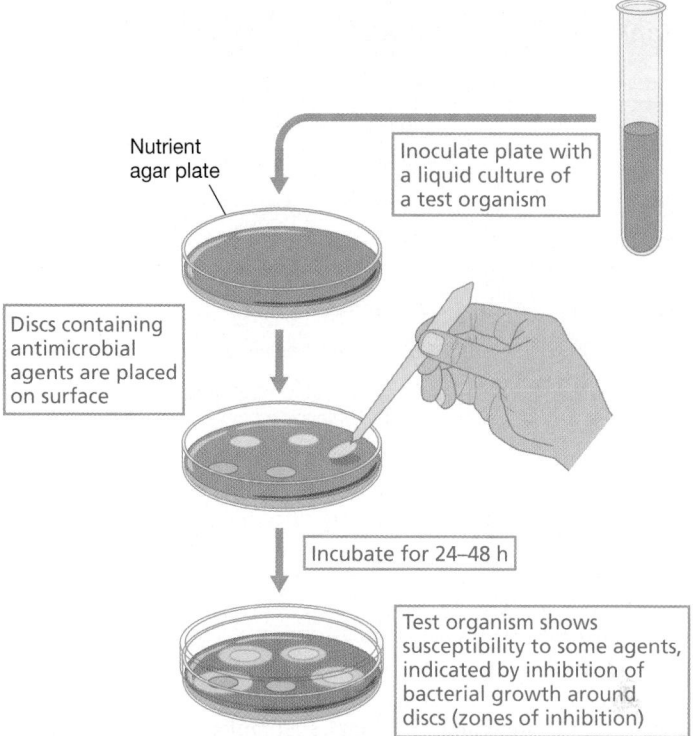

Figure 26.11 Antimicrobial agent susceptibility assay using diffusion methods. The antimicrobial agent diffuses from paper disks into the surrounding agar, inhibiting growth of susceptible microorganisms.

paper discs, which are then placed on the surface of the agar. During incubation, the agent diffuses from the disc into the agar, establishing a gradient; the farther the chemical diffuses away from the filter paper, the lower is the concentration of the agent. At some distance from the disc, the effective MIC is reached. Beyond this point the microorganism grows, but closer to the disc, growth is absent. A *zone of inhibition* is created with a diameter proportional to the amount of antimicrobial agent added to the disc, the solubility of the agent, the diffusion coefficient, and the overall effectiveness of the agent. The disc diffusion technique and other growth-dependent methods are routinely used to test pathogens for antibiotic susceptibility (Section 31.3).

MiniQuiz
- For antimicrobial agents, distinguish between the effects of -static, -cidal, and -lytic agents.
- Describe how the minimum inhibitory concentration of an antibacterial agent is determined.

26.5 Chemical Antimicrobial Agents for External Use

Chemical antimicrobial agents are divided into two categories. The first category contains antimicrobial products used to control microorganisms in industrial and commercial environments. These include chemicals used in foods, air-conditioning cooling

Table 26.4 *Industrial uses of antimicrobial chemicals*

Industry	Chemicals	Use
Paper	Organic mercurials, phenols,[a] methylisothiazolinone	To prevent microbial growth during manufacture
Leather	Heavy metals, phenols[a]	Antimicrobial agents present in the final product inhibit growth
Plastic	Cationic detergents	To prevent growth of bacteria on aqueous dispersions of plastics
Textile	Heavy metals, phenols[a]	To prevent microbial deterioration of fabrics, such as awnings and tents, that are exposed in the environment
Wood	Metal salts, phenols[a]	To prevent deterioration of wooden structures
Metal working	Cationic detergents	To prevent growth of bacteria in aqueous cutting emulsions
Petroleum	Mercurics, phenols,[a] cationic detergents, methylisothiazolinone	To prevent growth of bacteria during recovery and storage of petroleum and petroleum products
Air conditioning	Chlorine, phenols,[a] methylisothiazolinone	To prevent growth of bacteria (for example, *Legionella*) in cooling towers
Electrical power	Chlorine	To prevent growth of bacteria in condensers and cooling towers
Nuclear	Chlorine	To prevent growth of radiation-resistant bacteria in nuclear reactors

[a]Metallic (mercury, arsenic, and copper) compounds and phenolic compounds may produce environmentally hazardous waste products and create health hazards.

towers, textile and paper products, fuel tanks, and so on; some of these chemicals are so toxic that exposure can affect human health. **Table 26.4** provides examples of industrial applications for chemicals used to control microbial growth.

The second category of chemical antimicrobial agents contains products designed to prevent growth of human pathogens in inanimate environments and on external body surfaces. This category is subdivided into sterilants, disinfectants, sanitizers, and antiseptics.

Sterilants

Chemical **sterilants**, also called **sterilizers** or **sporicides**, destroy all forms of microbial life, including endospores. Chemical sterilants are used in situations where it is impractical to use heat (Section 26.1) or radiation (Section 26.2) for decontamination or sterilization. Hospitals and laboratories, for example, must be able to decontaminate and sterilize heat-sensitive materials, such as thermometers, lensed instruments, polyethylene tubing, catheters, and reusable medical equipment such as respirometers. Some form of cold sterilization is usually used for these purposes. Cold sterilization is performed in enclosed devices that resemble autoclaves, but which employ a gaseous chemical agent such as ethylene oxide, formaldehyde, peroxyacetic acid, or hydrogen peroxide. Liquid sterilants such as a sodium hypochlorite (bleach) solution or amylphenol are used for instruments that cannot withstand high temperatures or gas (**Table 26.5**).

Disinfectants and Sanitizers

Disinfectants are chemicals that kill microorganisms, but not necessarily endospores, and are used on inanimate objects. For example, disinfectants such as ethanol and cationic detergents are used to disinfect floors, tables, bench tops, walls, and so on.

These agents are important for infection control in, for example, hospitals and other medical settings. General disinfectants are used in households, swimming pools, and water purification systems (Table 26.5).

Sanitizers are agents that reduce, but may not eliminate, microbial numbers to levels considered to be safe. Food contact sanitizers are widely used in the food industry to treat surfaces such as mixing and cooking equipment, dishes, and utensils. Non–food contact sanitizers are used to treat surfaces such as counters, floors, walls, carpets, and laundry (Table 26.5).

Antiseptics and Germicides

Antiseptics and **germicides** are chemical agents that kill or inhibit growth of microorganisms and that are nontoxic enough to be applied to living tissues. Most of the compounds in this category are used for handwashing or for treating surface wounds (Table 26.5). Under some conditions, certain antiseptics are also effective disinfectants; they are effective antimicrobial agents when applied to inanimate surfaces. Ethanol, for example, is categorized as an antiseptic, but can also be a disinfectant. This depends on the concentration of ethanol used and the exposure time, with disinfection generally requiring higher ethanol concentrations and exposure times of several minutes. The Food and Drug Administration in the United States regulates the formulation, manufacture, and use of antiseptics and germicides because these agents involve direct human exposure and contact.

Antimicrobial Efficacy

Several factors affect the efficacy of chemical antimicrobial agents. For example, many disinfectants are neutralized by organic material. These materials reduce effective disinfectant concentrations and microbial killing capacity. Furthermore,

Table 26.5 *Antiseptics, sterilants, disinfectants, and sanitizers*

Agent	Use	Mode of action
Antiseptics		
Alcohol (60–85% ethanol or isopropanol in water)[a]	Topical antiseptic	Lipid solvent and protein denaturant
Phenol-containing compounds (hexachlorophene, triclosan, chloroxylenol, chlorhexidine)[b]	Soaps, lotions, cosmetics, body deodorants, topical disinfectants	Disrupts cytoplasmic membrane
Cationic detergents, especially quaternary ammonium compounds (benzalkonium chloride)	Soaps, lotion, topical disinfectants	Interact with phospholipids of cytoplasmic membrane
Hydrogen peroxide[a] (3% solution)	Topical antiseptic	Oxidizing agent
Iodine-containing iodophor compounds in solution[a] (Betadine®)	Topical antiseptic	Iodinates tyrosine residues of proteins; oxidizing agent
Octenidine	Topical antiseptic	Disrupts cytoplasmic membrane
Sterilants, disinfectants, and sanitizers[c]		
Alcohol (60–85% ethanol or isopropanol in water)[a]	Disinfectant for medical instruments and laboratory surfaces	Lipid solvent and protein denaturant
Cationic detergents (quaternary ammonium compounds, Lysol® and many related disinfectants)	Disinfectant and sanitizer for medical instruments, food and dairy equipment	Interact with phospholipids
Chlorine gas	Disinfectant for purification of water supplies	Oxidizing agent
Chlorine compounds (chloramines, sodium hypochlorite, sodium chlorite, chlorine dioxide)	Disinfectant and sanitizer for dairy and food industry equipment, and water supplies	Oxidizing agent
Copper sulfate	Algicide disinfectant in swimming pools and water supplies	Protein precipitant
Ethylene oxide (gas)	Sterilant for temperature-sensitive materials such as plastics and lensed instruments	Alkylating agent
Formaldehyde	3–8% solution used as surface disinfectant, 37% (formalin) or vapor used as sterilant	Alkylating agent
Glutaraldehyde	2% solution used as high-level disinfectant or sterilant, commonly used fixative in electron microscopy	Alkylating agent
Hydrogen peroxide[a]	Vapor used as sterilant	Oxidizing agent
Iodine-containing iodophor compounds in solution[a] (Wescodyne®)	Disinfectant for medical instruments and laboratory surfaces	Iodinates tyrosine residues
Mercuric dichloride[b]	Disinfectant for laboratory surfaces	Combines with –SH groups
OPA (ortho-phthalaldehyde)	High-level disinfectant for medical instruments	Alkylating agent
Ozone	Disinfectant for drinking water	Strong oxidizing agent
Peroxyacetic acid	Solution used as high-level disinfectant or sterilant	Strong oxidizing agent
Phenolic compounds[b]	Disinfectant for laboratory surfaces	Protein denaturant
Pine oils (Pine-Sol®) (contains phenolics and other detergents)	General disinfectant for household surfaces	Protein denaturant

[a]Alcohols, hydrogen peroxide, and iodine-containing iodophor compounds can act as antiseptics, disinfectants, sanitizers, or sterilants depending on concentration, length of exposure, and form of delivery.
[b]Use of heavy metal (mercury) compounds and phenolic compounds may produce environmentally hazardous waste products and may create health hazards.
[c]Many water-soluble antimicrobial compounds, with the exception of those containing heavy metals, can be used as sanitizers for food and dairy equipment and preparation areas, provided their use is followed by adequate draining before food contact.

Multi-Drug- and Extensively Drug-Resistant Tuberculosis

Tuberculosis (TB) is an infectious disease caused by *Myobacterium tuberculosis*, a pathogenic bacterium responsible for approximately 3 million deaths each year worldwide. The organism was first isolated in 1882 by Robert Koch during his pioneering work on tuberculosis (Section 1.8). Individuals infected with TB typically undergo an extensive course of antibiotic therapy lasting for 6–24 months, with most patients making a full recovery.[1] However, each year approximately 450,000 people are diagnosed with multi-drug-resistant TB (MDR TB), a form of TB that does not respond to treatment from two of the most effective first-line anti-TB drugs, isoniazid and rifampicin.[2]

MDR TB was first identified in the early 1990s, and arises from either primary infection with resistant TB bacilli or from the mismanagement or noncompliance with drug therapy of a patient being treated for a "standard" TB infection. MDR TB treatment requires the use of second-line drugs, which are less effective and more costly and cause unwelcomed side effects.[3]

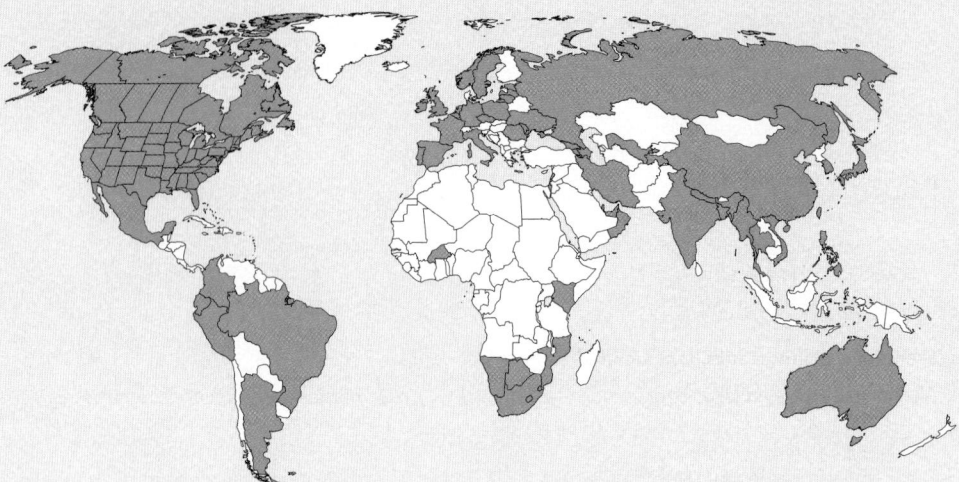

Figure 1 Shown in color are the countries and territories reporting at least one case of XDR TB as of January 2010.[4]

The World Health Organization estimates that 5.4% of MDR TB cases are in the form of extensively drug-resistant TB (XDR TB).[4] These strains are emerging at an alarming rate and exhibit additional resistance to many of the second-line anti-TB drugs.[5] XDR TB patients have limited treatment options, many of which are expensive and complicated to administer, especially in developing nations.[3] The true global incidence of XDR TB is

pathogens are often encased in particles or grow in large numbers as biofilms, covering surfaces of tissue or medical devices with several layers of microbial cells. Biofilms may slow or even completely prevent penetration of antimicrobial agents, reducing or negating their effectiveness.

Only sterilants are effective against bacterial endospores. Endospores are much more resistant to other agents than are vegetative cells because of their low water availability and reduced metabolism (Section 26.1). Some bacteria, such as *Mycobacterium tuberculosis*, the causal agent of tuberculosis, are resistant to the action of common disinfectants because of the waxy nature of their cell wall (Sections 18.5 and 33.4). Thus, the efficacy of antiseptics, disinfectants, sterilants, and other antimicrobial compounds used *in vitro* and *in vivo* for antimicrobial treatment must be empirically determined under the actual conditions of use.

MiniQuiz

- Distinguish between a sterilant, a disinfectant, a sanitizer, and an antiseptic.
- What disinfectants are routinely used for sterilization of water? Why are these disinfectants not harmful to humans?

unknown, largely due to the fact that health officials in many countries lack the necessary technologies to diagnose these rapidly emerging strains. Nevertheless, cases of XDR TB have been reported in several countries throughout the world (Figure 1).[4]

From this survey of disease incidence (Figure 1) and the fact that cases of XDR TB are rising rapidly in some cohorts, for example, in individuals with HIV-AIDS, it is clear that new cost-effective treatment regimens and new anti-TB drugs are necessary to

combat these highly drug-resistant forms of one of the oldest pathogens known to clinical medicine.

[1]Frieden, T. R., and S. S. Munsiff. 2005. The DOTS strategy for controlling the global tuberculosis epidemic. *Clin. Chest Med. 26*: 197–205.

[2]World Health Organization. 1997. Anti-tuberculosis drug resistance in the world: report No. 1 (The WHO/IUATLD Global Project on Anti-tuberculosis Drug Resistance Surveillance, 1994–1997, Geneva). WHO/TB.1997.229.

[3]Caminero, J. A., G. Sotgiu, A. Zumla, and G. B. Migliori. 2010. Best drug treatment for multidrug-resistant and extensively drug-resistant tuberculosis. *Lancet Infect. Dis. 10*: 621–629.

[4]World Health Organization. 2010. Multidrug and extensively drug-resistant TB (M/XDR-TB): 2010 global report on surveillance and response. WHO/HTM/TB/2010.3.

[5]Centers for Disease Control and Prevention. 2006. Emergence of *Mycobacterium tuberculosis* with extensive resistance to second-line drugs—worldwide, 2000–2004. *Morbid. Mortal. Weekly Rep. 55*: 301–305.

III Antimicrobial Agents Used *In Vivo*

Up to this point, we have considered the effects of physical and chemical agents used to inhibit microbial growth outside the human body. Most of the physical methods are too harsh and most chemicals mentioned are too toxic to be used inside the body; even relatively mild antiseptics can be used only on the skin. For control of infectious disease, chemical compounds that can be used internally are required. Discovery and development of antimicrobial drugs has played a major role in clinical and veterinary medicine, as well as in agriculture.

Antimicrobial drugs are classified based on their molecular structure, mechanism of action (**Figure 26.12**), and spectrum of antimicrobial activity (**Figure 26.13**). Worldwide, probably more than 10,000 metric tons of various antimicrobial drugs are manufactured and used annually (**Figure 26.14**). Antimicrobial agents fall into two broad categories, *synthetic agents* and *antibiotics*. We first concentrate on synthetic antimicrobial compounds. We then discuss naturally produced antibiotics.

26.6 Synthetic Antimicrobial Drugs

Systematic work on antimicrobial drugs was first initiated by the German scientist Paul Ehrlich. In the early 1900s, Ehrlich developed the concept of **selective toxicity**, the ability of a chemical

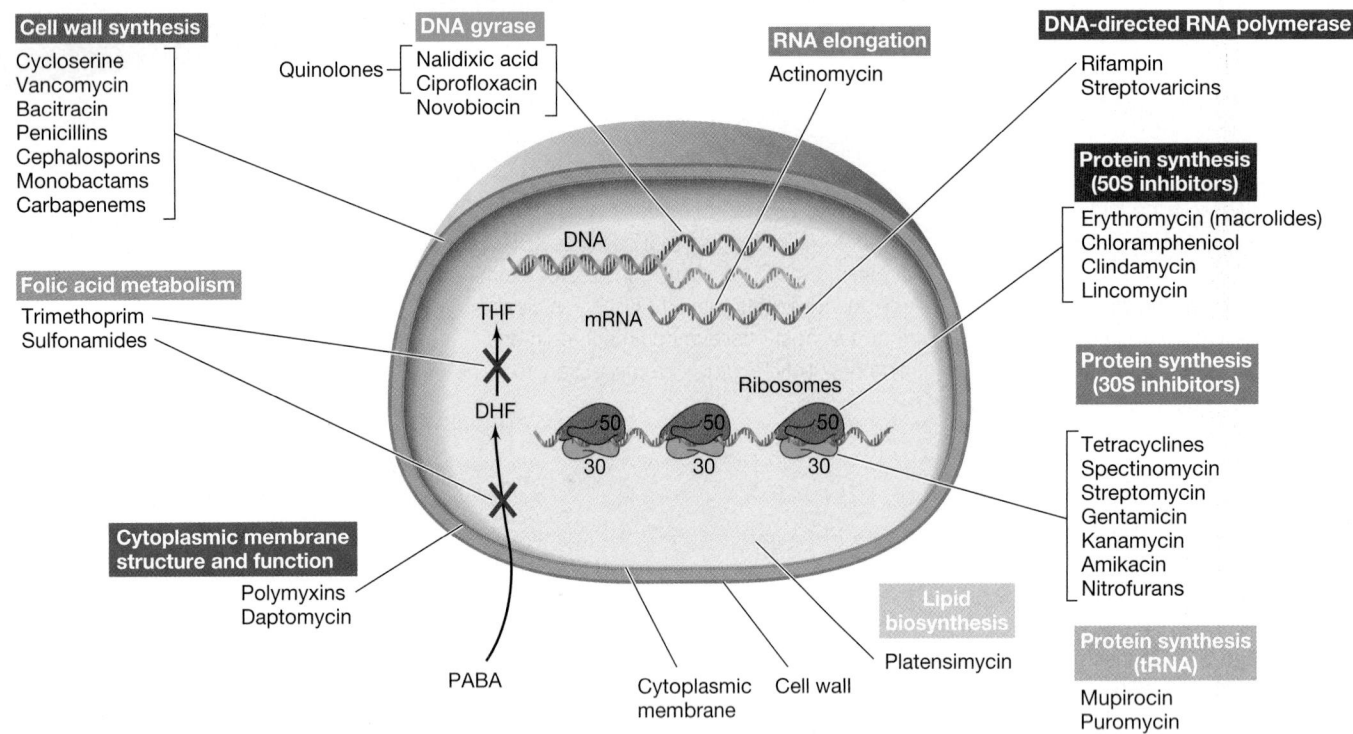

Figure 26.12 Mode of action of some major antimicrobial agents. Agents are classified according to their target structures in the bacterial cell. THF, tetrahydrofolate; DHF, dihydrofolate; mRNA, messenger RNA.

agent to inhibit or kill pathogenic microorganisms without adversely affecting the host. In his search for a "magic bullet" that would kill only pathogens, Ehrlich tested large numbers of chemical dyes for selective toxicity and discovered the first effective antimicrobial drugs, of which Salvarsan, an arsenic-containing compound used for the treatment of syphilis, was the most successful (**Figure 26.15**).

Growth Factor Analogs

We previously defined growth factors as specific chemical substances required in the medium because the organisms cannot synthesize them (↩ Section 4.1). A **growth factor analog** is a synthetic compound that is structurally similar to a growth factor, but subtle structural differences between the analog and

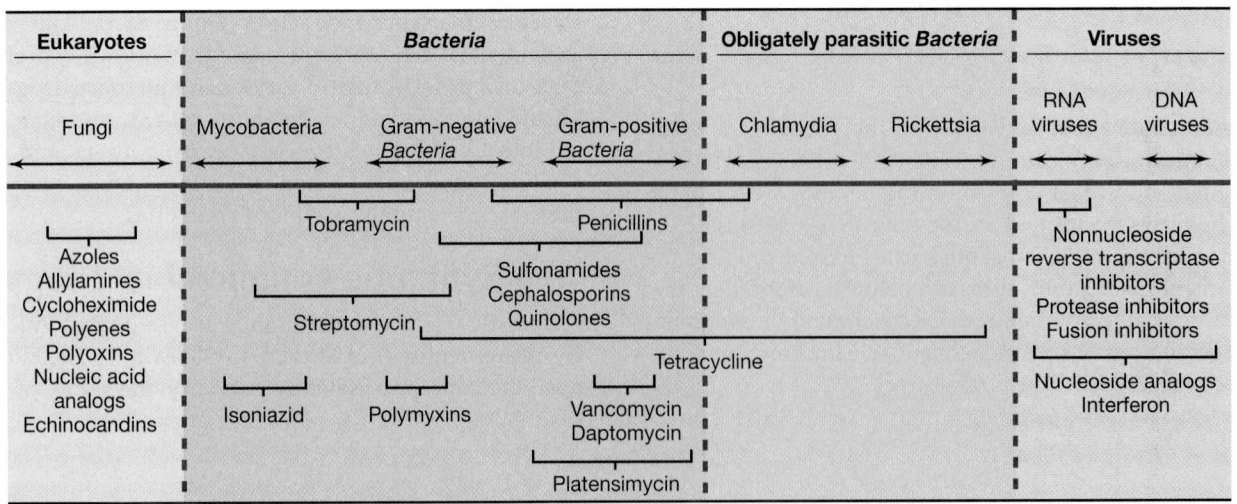

Figure 26.13 Antimicrobial spectrum of activity. Each antimicrobial agent affects a limited and well-defined group of microorganisms. A few agents are very specific and affect the growth of only a single genus. For example, isoniazid affects only organisms in the genus *Mycobacterium*.

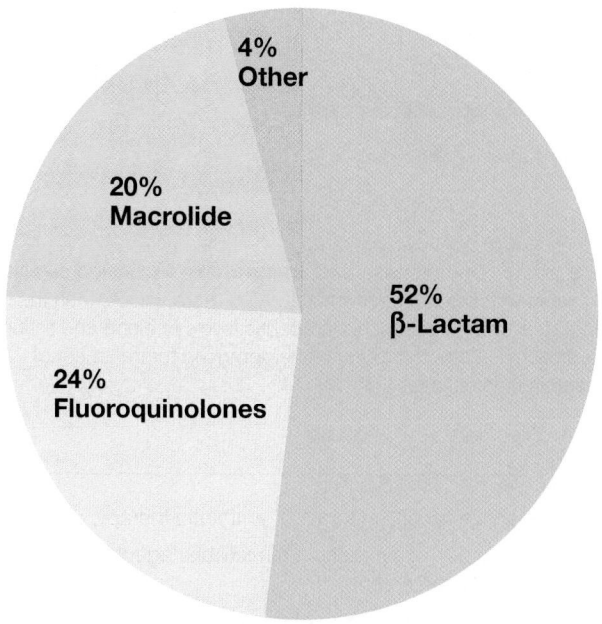

(a) Sulfanilamide   (b) p-Aminobenzoic acid

(c) Folic acid

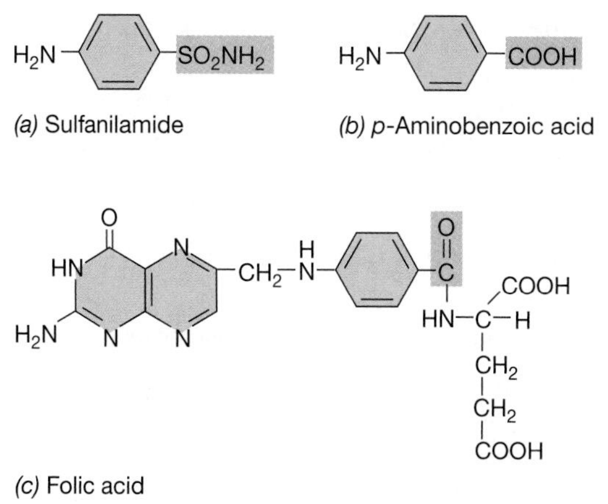

Figure 26.16 Sulfa drugs. (a) The simplest sulfa drug, sulfanilamide. (b) Sulfanilamide is an analog of p-aminobenzoic acid, a precursor of (c) folic acid, a growth factor.

Figure 26.14 Annual worldwide production and use of antibiotics. Each year an estimated 10,000 metric tons of antimicrobial agents are manufactured worldwide. The β-lactam antibiotics include cephalosporins (30%), penicillins (7%), and other β-lactams (15%). "Others" includes tetracyclines, aminoglycosides, and all other antimicrobial drugs.

the authentic growth factor prevent the analog from functioning in the cell, disrupting cell metabolism. Analogs are known for many important biomolecules, including vitamins, amino acids, purines and pyrimidines, and other compounds. We begin by considering antibacterial growth factor analogs. Growth factor analogs effective for the treatment of viral and fungal infections will be discussed in Sections 26.10 and 26.11.

Sulfa Drugs

Sulfa drugs, discovered by Gerhard Domagk in the 1930s, were the first widely used growth factor analogs that specifically inhibited the growth of bacteria. The discovery of the first sulfa drug

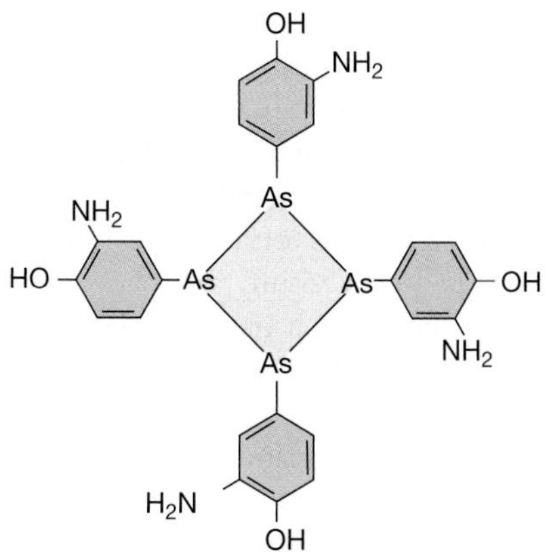

Figure 26.15 Salvarsan. This arsenic-containing compound was one of the first useful antimicrobial agents. It was used to treat syphilis.

resulted from the large-scale screening of chemicals for activity against streptococcal infections in experimental animals.

Sulfanilamide, the simplest sulfa drug, is an analog of *p*-aminobenzoic acid, which is itself a part of the vitamin folic acid, a nucleic acid precursor (**Figure 26.16**). Sulfanilamide blocks the synthesis of folic acid, thereby inhibiting nucleic acid synthesis. Sulfanilamide is selectively toxic in bacteria because bacteria synthesize their own folic acid, whereas most animals obtain folic acid from their diet. Initially, sulfa drugs were widely used for treatment of streptococcal infections (∅ Section 33.2). However, resistance to sulfonamides has been increasing because many formerly susceptible pathogens have developed an ability to take up folic acid from their environment. Antimicrobial therapy with sulfamethoxazole (a sulfa drug) plus trimethoprim, a related folic acid synthesis competitor, is still effective in many instances because the drug combination produces sequential blocking of the folic acid synthesis pathway. Resistance to this drug combination requires that two mutations in genes of the same pathway occur, a relatively rare event.

Isoniazid

Isoniazid (∅ Figure 33.11) is an important growth factor analog with a very narrow spectrum of activity (Figure 26.13). Effective only against *Mycobacterium*, isoniazid interferes with the synthesis of mycolic acid, a mycobacterial cell wall component. A nicotinamide (vitamin) analog, isoniazid is the most effective single drug used for control and treatment of tuberculosis (∅ Section 33.4).

Nucleic Acid Base Analogs

Analogs of nucleic acid bases formed by addition of a fluorine or bromine atom are shown in **Figure 26.17**. Fluorine is a relatively small atom and does not alter the overall shape of the nucleic acid base, but changes the chemical properties such that the compound does not function in cell metabolism, thereby blocking nucleic acid synthesis. Examples include fluorouracil, an analog of uracil, and bromouracil, an analog of thymine. Growth

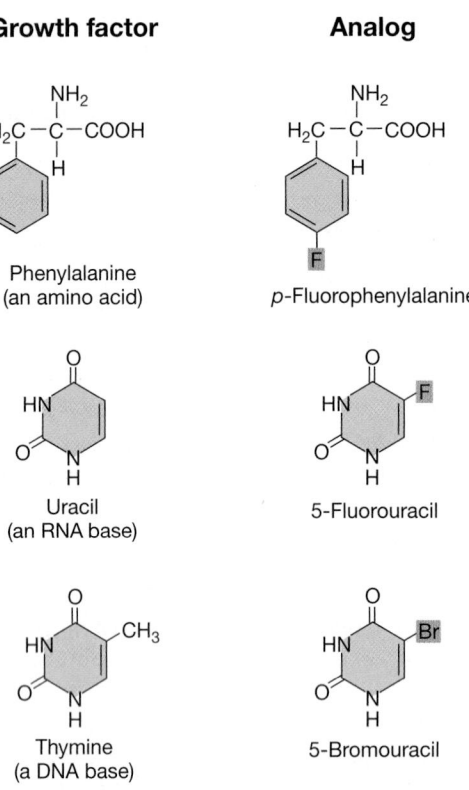

Growth factor   **Analog**

Phenylalanine
(an amino acid)

p-Fluorophenylalanine

Uracil
(an RNA base)

5-Fluorouracil

Thymine
(a DNA base)

5-Bromouracil

Figure 26.17 Growth factors and antimicrobial analogs. Structurally similar growth factors and their biologically active analogs are shown for comparison. The nutritional aspects of growth factors are discussed in Section 4.1(↺ Table 4.2).

factor analogs of nucleic acids are used in the treatment of viral and fungal infections and are also used as mutagens (Sections 26.10 and 26.11).

Quinolones

The **quinolones** are antibacterial compounds that interfere with bacterial DNA gyrase, preventing the supercoiling of DNA, a required step for packaging DNA in the bacterial cell (Figure 26.12; ↺ Section 6.3). Because DNA gyrase is found in all *Bacteria*, the fluoroquinolones are effective for treating both gram-positive and gram-negative bacterial infections (Figure 26.13). Fluoroquinolones such as ciprofloxacin (**Figure 26.18a**) are routinely used to treat urinary tract infections in humans. Ciprofloxacin is also the drug of choice for treating anthrax because some strains of *Bacillus anthracis*, the causative agent of anthrax (↺ Section 32.12), are resistant to penicillin. Moxifloxacin, a new fluoroquinolone, has been approved for treatment of tuberculosis, one of the few new drugs proven effective against *Mycobacterium tuberculosis* infections (**Figure 26.18b**). This drug, in combination with other anti-tuberculosis drugs (↺ Section 33.4), may significantly shorten the time necessary for treatment. Fluoroquinolones have also been widely used in the beef and poultry industries for prevention and treatment of respiratory diseases in animals.

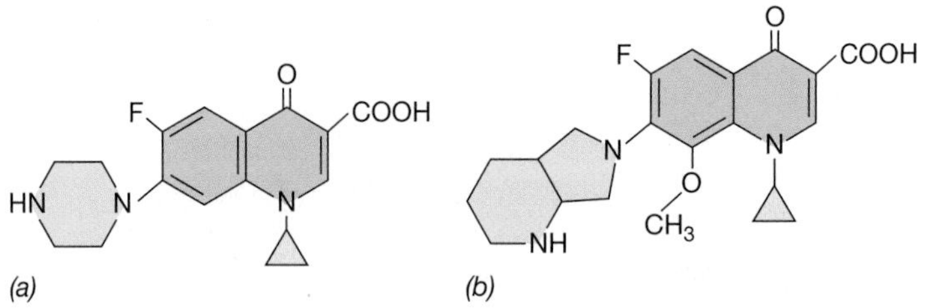

Figure 26.18 Quinolones. (a) Ciprofloxacin, a fluorinated derivative of nalidixic acid with broad-spectrum activity, is more soluble than the parent compound, allowing it to reach therapeutic levels in blood and tissues. (b) Moxifloxacin, a new fluoroquinolone approved for treatment of *Mycobacterium* infections.

MiniQuiz

- Explain selective toxicity in terms of antibiotic therapy.
- Distinguish the use of synthetic antimicrobial agents from antiseptics and disinfectants.
- Describe the action of any one of the synthetic antimicrobial drugs.

26.7 Naturally Occurring Antimicrobial Drugs: Antibiotics

Antibiotics are antimicrobial agents produced by microorganisms. Antibiotics are produced by a variety of bacteria and fungi and apparently have the sole function of inhibiting or killing other microorganisms. Although thousands of antibiotics are known, less than 1% are clinically useful, often because of host toxicity or lack of uptake by host cells. However, the clinically useful antibiotics have had a dramatic impact on the treatment of infectious diseases. *Natural* antibiotics can often be artificially modified to enhance their efficacy. These are said to be *semisynthetic* antibiotics. The isolation, characterization, and industrial production of antibiotics were discussed in Sections 15.3 and 15.4.

Antibiotics and Selective Antimicrobial Toxicity

The susceptibility of individual microorganisms to individual antimicrobial agents varies significantly (Figure 26.13). For example, gram-positive *Bacteria* and gram-negative *Bacteria* differ in their susceptibility to an individual antibiotic such as penicillin; gram-positive *Bacteria* are generally affected, whereas most gram-negative *Bacteria* are naturally resistant. Certain **broad-spectrum antibiotics** such as tetracycline, however, are effective against both groups. As a result, a broad-spectrum antibiotic finds wider medical use than a narrow-spectrum antibiotic. An antibiotic with a limited spectrum of activity may, however, be quite valuable for the control of pathogens that fail to respond to other antibiotics. A good example is vancomycin, a narrow-spectrum glycopeptide antibiotic that is a highly effective bacteriocidal agent for gram-positive, penicillin-resistant *Bacteria* from the genera *Staphylococcus*, *Bacillus*, and *Clostridium* (Figures 26.12 and 26.13).

Important targets of antibiotics in *Bacteria* are ribosomes, the cell wall, the cytoplasmic membrane, lipid biosynthesis enzymes, and DNA replication and transcription elements (Figure 26.12).

Antibiotics Affecting Protein Synthesis

Many antibiotics inhibit protein synthesis by interacting with the ribosome and disrupting translation (Figure 26.12). These interactions are quite specific and many involve binding to ribosomal RNA (rRNA). Several of these antibiotics are medically useful, and several are also effective research tools because they block defined steps in protein synthesis (Section 6.19). For instance, streptomycin inhibits protein chain initiation, whereas puromycin, chloramphenicol, cycloheximide, and tetracycline inhibit protein chain elongation.

Even when two antibiotics inhibit the same step in protein synthesis, the mechanisms of inhibition can be quite different. For example, puromycin binds to the A site on the ribosome, and the growing polypeptide chain is transferred to puromycin instead of the aminoacyl–transfer RNA (aminoacyl-tRNA) complex. The puromycin–peptide complex is then released from the ribosome, prematurely halting elongation. By contrast, chloramphenicol inhibits elongation by blocking formation of the peptide bond (Section 6.19).

Many antibiotics specifically inhibit ribosomes of organisms from only one phylogenetic domain. For example, chloramphenicol and streptomycin specifically target the ribosomes of *Bacteria*, whereas cycloheximide only affects the cytoplasmic ribosomes of *Eukarya*. Since the major organelles (mitochondria and chloroplasts) in *Eukarya* also have ribosomes that are similar to those of *Bacteria* (that is, 70S ribosomes), antibiotics that inhibit protein synthesis in *Bacteria* also inhibit protein synthesis in these organelles. For example, tetracycline antibiotics inhibit 70S ribosomes, but are still medically useful because eukaryotic mitochondria are affected only at higher concentrations than are used for antimicrobial therapy.

Antibiotics Affecting Transcription

A number of antibiotics specifically inhibit transcription by inhibiting RNA synthesis (Figure 26.12). For example, rifampin and the streptovaricins inhibit RNA synthesis by binding to the β-subunit of RNA polymerase. These antibiotics have specificity for *Bacteria*, chloroplasts, and mitochondria. Actinomycin inhibits RNA synthesis by combining with DNA and blocking RNA elongation. This agent binds most strongly to DNA at guanine–cytosine base pairs, fitting into the major groove in the double strand where RNA is synthesized.

Some of the most useful antibiotics are directed against unique structural features of *Bacteria*, such as their cell walls. We discuss these antibiotics and their targets in the next section.

MiniQuiz
- Distinguish antibiotics from growth factor analogs.
- What is a broad-spectrum antibiotic?
- Identify the potential target sites for the antibiotics that inhibit protein synthesis and transcription.

26.8 β-Lactam Antibiotics: Penicillins and Cephalosporins

One of the most important groups of antibiotics, both historically and medically, is the β-lactam group. **β-lactam antibiotics** include the medically important penicillins, cephalosporins, and cephamycins. These antibiotics share a characteristic structural component, the *β-lactam ring* (**Figure 26.19**). Together, the β-lactam antibiotics account for over one-half of all of the antibiotics produced and used worldwide (Figure 26.14).

Penicillins

In 1929, the British scientist Alexander Fleming characterized the first antibiotic, an antibacterial compound called *penicillin* because it was isolated from the fungus *Penicillium chrysogenum*

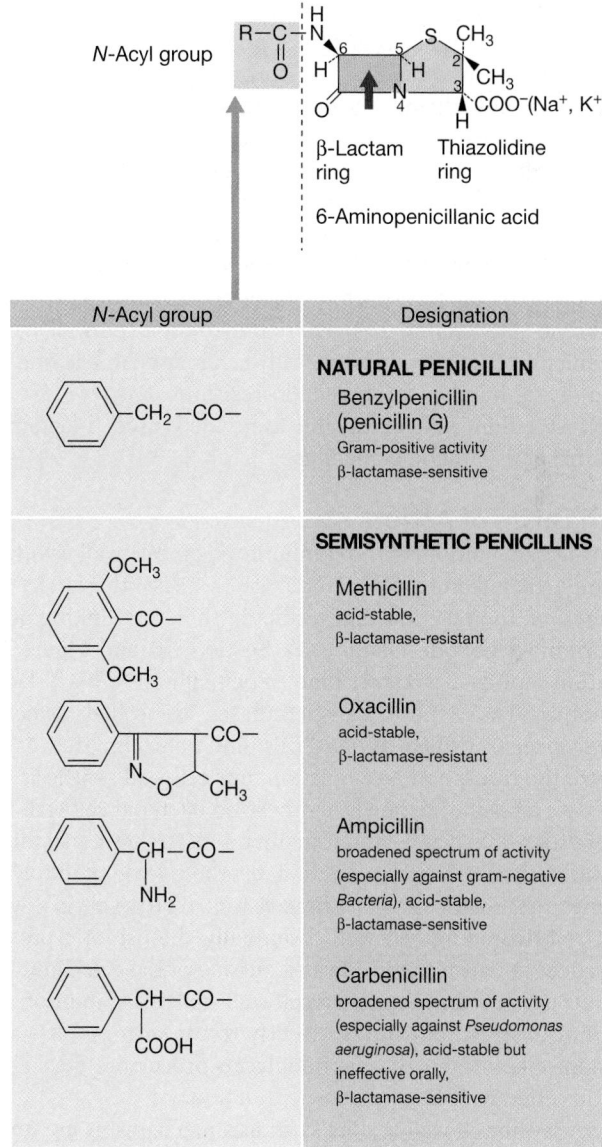

Figure 26.19 **Penicillins.** The red arrow (top panel) is the site of activity of most β-lactamase enzymes.

(Figure 26.19). The antibiotic, however, was not immediately recognized as a potentially important clinical drug. Even though sulfa drugs were widely available in the 1930s, their efficacy was mostly limited to the treatment of infections by gram-positive organisms such as *Streptococcus*; most other bacterial diseases were uncontrollable. However, in 1939, Howard Florey and his colleagues, motivated by the impending world war, developed a process for the large-scale production of **penicillin**. Penicillin G was the first clinically useful antibiotic. This new β-lactam antibiotic was dramatically effective in controlling staphylococcal and pneumococcal infections among military personnel and was more effective for treating streptococcal infections than sulfa drugs. By the end of World War II in 1945, penicillin became available for general use and pharmaceutical companies began to look for and develop other antibiotics, leading to drugs that revolutionized the treatment of infectious diseases.

Penicillin G is active primarily against gram-positive *Bacteria* because gram-negative *Bacteria* are impermeable to the antibiotic. Chemical modification of the penicillin G structure, however, significantly changes the properties of the resulting antibiotic. Many chemically modified semisynthetic penicillins are quite effective against gram-negative *Bacteria*. Figure 26.19 shows the structures of some of the penicillins. For example, ampicillin and carbenicillin, semisynthetic penicillins, are effective against some gram-negative *Bacteria*. The structural differences in the *N*-acyl groups of these semisynthetic penicillins allow them to be transported inside the gram-negative outer membrane (꒐ Section 3.7), where they inhibit cell wall synthesis. Penicillin G is also sensitive to β-lactamase, an enzyme produced by a number of penicillin-resistant *Bacteria* (Section 26.12). Oxacillin and methicillin are widely used β-lactamase-resistant semisynthetic penicillins.

Mechanism of Action

The β-lactam antibiotics are inhibitors of cell wall synthesis. An important feature of bacterial cell wall synthesis is *transpeptidation*, the reaction that results in the cross-linking of two glycan-linked peptide chains (꒐ Section 5.4 and Figure 5.7). The transpeptidase enzymes bind to penicillin or other β-lactam antibiotics. Thus, these transpeptidases are called *penicillin-binding proteins* (PBPs). When PBPs bind penicillin, they cannot catalyze the transpeptidase reaction, but cell wall synthesis continues. As a result, the newly synthesized bacterial cell wall is no longer cross-linked and cannot maintain its strength. In addition, the antibiotic–PBP complex stimulates the release of autolysins, enzymes that digest the existing cell wall. The result is a weakened, self-degrading cell wall. Eventually the osmotic pressure differences between the inside and outside of the cell cause lysis. By contrast, vancomycin, also a cell wall synthesis inhibitor, does not bind to PBPs, but binds directly to the terminal D-alanyl-D-alanine peptide on the peptidoglycan precursors (꒐ Figure 5.7); this effectively blocks transpeptidation.

Because the cell wall and its synthesis mechanisms are unique to *Bacteria*, the β-lactam antibiotics are highly selective and are not toxic to host cells. However, some individuals develop allergies to β-lactam antibiotics after repeated courses of antibiotic therapy.

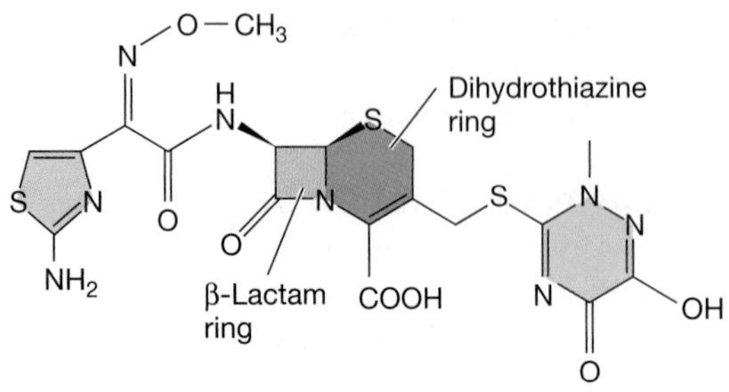

Figure 26.20 Ceftriaxone. Ceftriaxone is a β-lactam antibiotic that is resistant to most β-lactamases due to the adjacent six-member dihydrothiazine ring. Compare this structure to the five-member thiazolidine ring of the β-lactamase-sensitive penicillins (Figure 26.19).

Cephalosporins

The cephalosporins are another group of clinically important β-lactam antibiotics. Cephalosporins, produced by the fungus *Cephalosporium* sp., differ structurally from the penicillins. They retain the β-lactam ring but have a six-member dihydrothiazine ring instead of the five-member thiazolidine ring. The cephalosporins have the same mode of action as the penicillins; they bind irreversibly to PBPs and prevent the cross-linking of peptidoglycan. Clinically important cephalosporins are semisynthetic antibiotics with a broader spectrum of antibiotic activity than the penicillins. In addition, cephalosporins are typically more resistant to the enzymes that destroy β-lactam rings, the β-lactamases. For example, ceftriaxone (**Figure 26.20**) is highly resistant to β-lactamases and has replaced penicillin for treatment of *Neisseria gonorrhoeae* (gonorrhea) infections because many *N. gonorrhoeae* strains are now resistant to penicillin (Section 26.12, ꒐ Section 33.12).

MiniQuiz

- Draw the structure of the β-lactam ring and indicate the site of β-lactamase activity.
- How do the β-lactam antibiotics function?
- Of what clinical value are semisynthetic penicillins over natural penicillin?

26.9 Antibiotics from Prokaryotes

Many antibiotics active against *Bacteria* are also produced by *Bacteria*. These include many antibiotics that have major clinical applications, and we discuss their general properties here.

Aminoglycosides

Antibiotics that contain amino sugars bonded by glycosidic linkage are called **aminoglycosides**. Clinically useful aminoglycosides include streptomycin (produced by *Streptomyces griseus*) and its relatives, kanamycin (**Figure 26.21**), neomycin, gentamicin, tobramycin, netilmicin, spectinomycin, and amikacin. The aminoglycosides target the 30S subunit of the ribosome, inhibiting protein synthesis (Figure 26.12), and are clinically useful against gram-negative *Bacteria* (Figure 26.13).

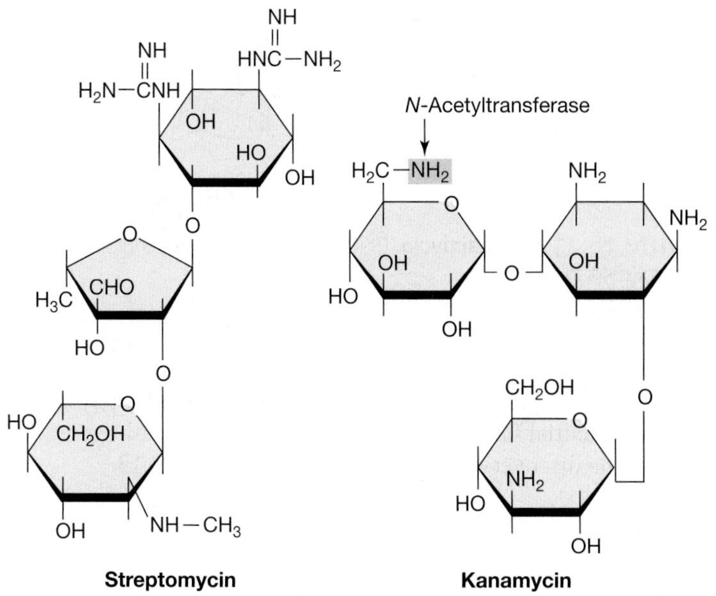

Figure 26.21 Aminoglycoside antibiotics: streptomycin and kanamycin. The amino sugars are in yellow. At the position indicated, kanamycin can be modified by a resistance plasmid that encodes *N*-acetyltransferase. Following acetylation, the antibiotic is inactive. Both kanamycin and streptomycin are synthesized by *Streptomyces* species.

Streptomycin was the first effective antibiotic used for the treatment of tuberculosis. The aminoglycoside antibiotics, however, are not widely used today, and together the aminoglycosides account for less than 4% of the total of all antibiotics produced and used. Because of serious side effects such as neurotoxicity and nephrotoxicity (kidney toxicity), streptomycin has been replaced by several synthetic antimicrobials for tuberculosis treatment. Bacterial resistance to aminoglycosides also develops readily. The use of aminoglycosides for treatment of gram-negative infections has decreased since the development of the semisynthetic penicillins (Section 26.8) and the tetracyclines (discussed later in this section). Aminoglycoside antibiotics are now considered reserve antibiotics used primarily when other antibiotics fail.

Macrolides

Macrolide antibiotics contain lactone rings bonded to sugars (**Figure 26.22**). Variations in both the lactone ring and the sugars result in a large number of macrolide antibiotics. The best-known macrolide is erythromycin (produced by *Streptomyces erythreus*). Other clinically useful macrolides include dirithromycin, clarithromycin, and azithromycin. The macrolides account for about 20% of the total world production and use of antibiotics (Figure 26.14). Erythromycin is a broad-spectrum antibiotic that targets the 50S subunit of the bacterial ribosome, inhibiting protein synthesis (Figure 26.12). Often used clinically in place of penicillin in patients allergic to penicillin or other β-lactam antibiotics, erythromycin is particularly useful for treating legionellosis (↺ Section 35.7).

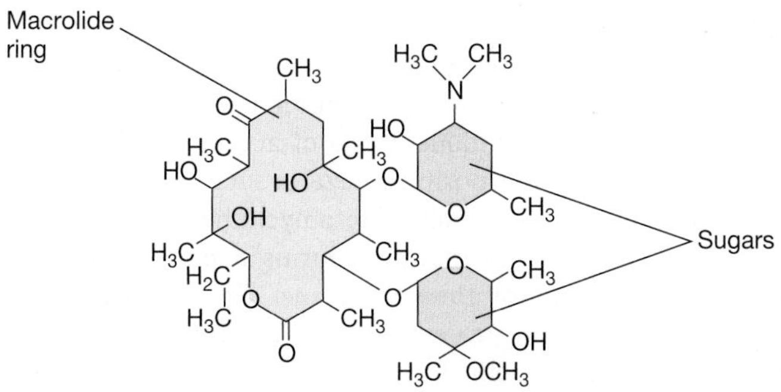

Figure 26.22 Erythromycin, a macrolide antibiotic. Erythromycin is a widely used broad-spectrum antibiotic.

Tetracyclines

The **tetracyclines**, produced by several species of *Streptomyces*, are an important group of antibiotics that find widespread medical use in humans (↺ Section 15.4). They were some of the first broad-spectrum antibiotics, inhibiting almost all gram-positive and gram-negative *Bacteria*. The basic structure of the tetracyclines consists of a naphthacene ring system (**Figure 26.23**). Substitutions to the basic naphthacene ring occur naturally and form new tetracycline analogs. Semisynthetic tetracyclines having substitutions in the naphthacene ring system have also been developed. Like erythromycin and the aminoglycoside antibiotics, tetracycline is a protein synthesis inhibitor, interfering with bacterial 30S ribosome subunit function (Figure 26.12).

The tetracyclines and the β-lactam antibiotics comprise the two most important groups of antibiotics in the medical field. The tetracyclines are also widely used in veterinary medicine and in some countries are used as nutritional supplements for poultry and swine. Because extensive nonmedical uses of medically important antibiotics have contributed to widespread antibiotic resistance, this use is now discouraged.

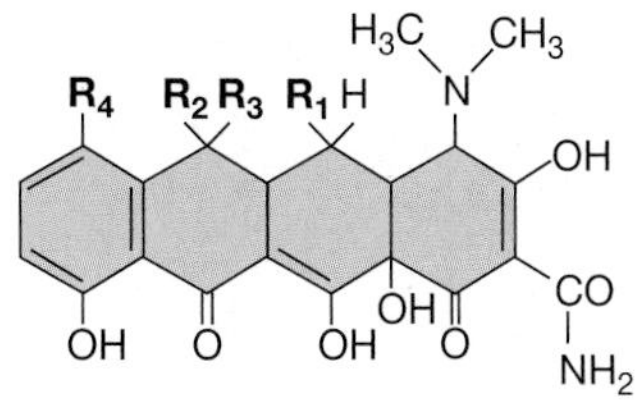

Tetracycline analog	R$_1$	R$_2$	R$_3$	R$_4$
Tetracycline	H	OH	CH$_3$	H
7-Chlortetracycline (aureomycin)	H	OH	CH$_3$	Cl
5-Oxytetracycline (terramycin)	OH	OH	CH$_3$	H

Figure 26.23 Tetracycline. The structure of tetracycline and its semisynthetic analogs.

Daptomycin

Daptomycin is another antibiotic produced by a member of the *Streptomyces* genus. This novel antibiotic is a cyclic lipopeptide (**Figure 26.24**) with a unique mode of action. Used mainly to treat infections by gram-positive *Bacteria* such as the pathogenic staphylococci and streptococci, daptomycin binds specifically to bacterial cytoplasmic membranes, forms a pore, and induces rapid depolarization of the membrane. The depolarized cell quickly loses its ability to synthesize macromolecules such as nucleic acids and proteins, resulting in cell death. Alterations in cytoplasmic membrane structure may account for rare instances of resistance.

Platensimycin

Platensimycin is the first member of a new structural class of antibiotics. Produced by *Streptomyces platensis*, this antibiotic (**Figure 26.25**) selectively inhibits a bacterial enzyme central to fatty acid biosynthesis, thus disrupting lipid biosynthesis. Platensimycin is effective against a broad range of gram-positive *Bacteria*, including nearly untreatable infections caused by methicillin-resistant *Staphylococcus aureus* and vancomycin-resistant enterococci. Already shown to be effective in eradicat-

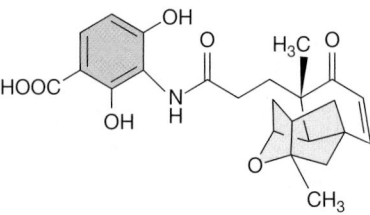

Figure 26.25 Platensimycin. Platensimycin selectively inhibits lipid biosynthesis in *Bacteria*.

ing *S. aureus* infections in mice, this antibiotic shows no toxicity. Platensimycin has a unique mode of action, and there is no known potential for development of resistance by pathogens. We discuss the discovery of platensimycin in Section 26.13.

MiniQuiz
- What are the biological sources of aminoglycosides, tetracyclines, macrolides, daptomycin, and platensimycin?
- How does the activity of each antibiotic class lead to death of the affected cell?

Ⓘ Ⅳ Control of Viruses and Eukaryotic Pathogens

Drugs that control growth of viruses and eukaryotic pathogens such as fungi and parasites are available, but they often affect eukaryotic host cells as well. As a result, selective toxicity for eukaryotic pathogens is very difficult to attain; only agents that preferentially affect pathogen-specific metabolic pathways or structural components are useful. There are a limited number of these drugs, and we discuss some important ones that affect viruses and fungi here. Drugs specific for treatment of parasitic diseases are discussed with the diseases themselves because they are extremely specific for individual parasites.

26.10 Antiviral Drugs

Because viruses use their eukaryotic hosts to reproduce and perform metabolic functions, most antiviral drugs also target host structures, resulting in host toxicity. However, several compounds are more toxic for viruses than for the host, and a few agents specifically target viruses. Largely because of efforts to find effective measures to control infections with the human immunodeficiency virus (HIV), the cause of AIDS (⟳ Section 33.14), significant achievements have been made in the development and use of antiviral agents.

Antiviral Agents

The most successful and commonly used agents for antiviral chemotherapy are the nucleoside analogs (**Table 26.6**). The first compound to gain universal acceptance in this category was zidovudine, or azidothymidine (AZT) (⟳ Figure 33.43). AZT

Figure 26.24 Daptomycin. Daptomycin is a cyclic lipopeptide that depolarizes cytoplasmic membranes in gram-positive *Bacteria*.

Table 26.6 *Antiviral compounds*

Category/drug	Mechanism of action	Virus affected
Fusion inhibitor		
Enfuvirtide	Blocks HIV–T lymphocyte membrane fusion	HIV[a]
Interferons		
α-Interferon	Induces proteins that inhibit viral replication	Broad spectrum (host-specific)
β-Interferon		
γ-Interferon		
Neuraminidase inhibitors		
Oseltamivir (Tamiflu®)	Block active site of influenza neuraminidase	Influenza A and B
Zanamivir (Relenza®)		Influenza A and B
Nonnucleoside reverse transcriptase inhibitor (NNRTI)		
Nevirapine	Reverse transcriptase inhibitor	HIV
Nucleoside analogs		
Acyclovir	Viral polymerase inhibitors	Herpes viruses, *Varicella zoster*
Ganciclovir		Cytomegalovirus
Trifluridine		Herpesvirus
Valacyclovir		Herpesvirus
Vidarabine		Herpesvirus, vaccinia, hepatitis B virus
Abacavir (ABC)	Reverse transcriptase inhibitors	HIV
Didanosine (dideoxyinosine or ddl)		HIV
Emtricitabine (FTC)		HIV
Lamivudine (3TC)		HIV, hepatitis B virus
Stavudine (d4T)		HIV
Zalcitabine (ddC)		HIV
Zidovudine (AZT) (♻ Figure 33.43)		HIV
Ribavirin	Blocks capping of viral RNA	Respiratory syncytial virus, influenza A and B, Lassa fever
Nucleotide analogs		
Cidofovir	Viral polymerase inhibitor	Cytomegalovirus, herpesviruses
Tenofovir (TDF)	Reverse transcriptase inhibitor	HIV
Protease inhibitors		
Amprenavir	Viral protease inhibitor	HIV
Indinavir (Figure 26.31)		HIV
Lopinavir		HIV
Nelfinavir		HIV
Saquinavir (Figure 26.31)		HIV
Pyrophosphate analog		
Phosphonoformic acid (foscarnet)	Viral polymerase inhibitor	Herpesviruses, HIV, hepatitis B virus
RNA polymerase inhibitor		
Rifamycin	RNA polymerase inhibitor	Vaccinia, pox viruses
Synthetic amines		
Amantadine	Viral uncoating blocker	Influenza A
Rimantadine		Influenza A

[a]Human immunodeficiency virus

inhibits retroviruses such as HIV (⮐ Sections 33.14 and 9.12). Azidothymidine is chemically related to thymidine but is a dideoxy derivative, lacking the 3′-hydroxyl group. AZT inhibits multiplication of retroviruses by blocking reverse transcription and production of the virally encoded DNA intermediate. This inhibits multiplication of HIV. A number of other nucleoside analogs having analogous mechanisms have been developed for the treatment of HIV and other viruses.

Nearly all nucleoside analogs, or **nucleoside reverse transcriptase inhibitors (NRTI)**, work by the same mechanism, inhibiting elongation of the viral nucleic acid chain by a nucleic acid polymerase. The nucleotide analog cidofovir works in the same way (Table 26.6). Because the normal cell function of nucleic acid replication is targeted, these drugs usually induce some host toxicity. Many NRTIs also lose their antiviral potency with time due to the emergence of drug-resistant viruses (⮐ Section 33.14).

Several other antiviral agents target the key enzyme of retroviruses, reverse transcriptase. Nevirapine, a **nonnucleoside reverse transcriptase inhibitor (NNRTI)**, binds directly to reverse transcriptase and inhibits reverse transcription. Phosphonoformic acid, an analog of inorganic pyrophosphate, inhibits normal internucleotide linkages, preventing synthesis of viral nucleic acids. As with the NRTIs, the NNRTIs generally induce some level of host toxicity because their action also affects normal host cell nucleic acid synthesis.

Protease inhibitors are another class of antiviral drugs that are effective for treatment of HIV (Table 26.6 and see Figure 26.31). These drugs prevent viral replication by binding the active site of HIV protease, inhibiting this enzyme from processing large viral proteins into individual viral components, thus preventing virus maturation (⮐ Sections 21.11, 26.13, and 33.14).

A final category of anti-HIV drugs is represented by a single drug, enfuvirtide, a **fusion inhibitor** composed of a 36-amino acid synthetic peptide that binds to the gp41 membrane protein of HIV (Table 26.6 and ⮐ Section 33.14). Binding of the gp41 protein by enfuvirtide stops the conformational changes necessary for the fusion of HIV and T lymphocyte membranes, thus preventing infection of cells by HIV.

Influenza Antiviral Agents

Two categories of drugs effectively limit influenza infection. The adamantanes amantadine and rimantadine are synthetic amines that interfere with an influenza A ion transport protein, inhibiting virus uncoating and subsequent replication. The neuraminidase inhibitors oseltamivir (brand name Tamiflu) and zanamivir (Relenza) block the active site of neuraminidase in influenza A and B viruses, inhibiting virus release from infected cells. Zanamivir is used only for treatment of influenza, whereas oseltamivir is used for both treatment and prophylaxis. The adamantanes are less useful than the neuraminidase inhibitors because resistance to adamantanes develops rapidly in strains of influenza virus (⮐ Section 33.8).

Interferons

Virus interference is a phenomenon in which infection with one virus interferes with subsequent infection by another virus. Several small proteins are the cause of interference; the proteins are called interferons. **Interferons** are small proteins in the cytokine family (⮐ Section 30.10) that prevent viral replication by stimulating the production of antiviral proteins in uninfected cells. Interferons are formed in response to live virus, inactivated virus, and viral nucleic acids. Interferon is produced in large amounts by cells infected with viruses of low virulence, but little is produced against highly virulent viruses. Highly virulent viruses inhibit cell protein synthesis before interferon can be produced. Interferons are also induced by natural and synthetic double-stranded RNA (dsRNA) molecules. In nature, dsRNA exists only in virus-infected cells as the replicative form of RNA viruses such as rhinoviruses (cold viruses) (⮐ Section 33.7); the dsRNA from the infecting virus signals the animal cell to produce interferon.

Interferons from virus-infected cells interact with receptors on uninfected cells, promoting the synthesis of antiviral proteins that function to prevent further virus infection. Interferons are produced in three molecular forms: *IFN-α* is produced by leukocytes, *IFN-β* is produced by fibroblasts, and *IFN-γ* is produced by immune lymphocytes.

Interferon activity is *host*-specific rather than *virus*-specific. That is, interferon produced by a member of one species can only activate receptors on cells from the same species. As a result, interferon produced by cells of an animal in response to, for example, a rhinovirus, could also inhibit multiplication of, for example, influenza viruses in cells within the same species, but has no effect on the multiplication of any virus in cells from other animal species.

Interferons produced *in vitro* have potential as possible antiviral and anticancer agents. Several approved recombinant interferons are available. However, the use of interferons as antiviral agents is not widespread because interferon must be delivered locally in high concentrations to stimulate the production of antiviral proteins in uninfected host cells. Thus, the clinical utility of these antiviral agents depends on our ability to deliver interferon to local areas in the host through injections or aerosols. Alternatively, appropriate interferon-stimulating signals such as viral nucleotides, nonvirulent viruses, or even synthetic nucleotides, if given to host cells prior to viral infection, might stimulate natural production of interferon.

MiniQuiz

- Why are there relatively few effective antiviral agents? Such agents are not used to treat common viral illnesses such as colds; why not?
- What steps in the viral maturation process are inhibited by nucleoside analogs? By protease inhibitors? By interferons?

26.11 Antifungal Drugs

Fungi, like viruses, pose special problems for the development of chemotherapy. Because fungi are *Eukarya*, much of their cellular machinery is the same as that of animals and humans; antifungal agents that act on metabolic pathways in fungi often affect corresponding pathways in host cells, making the drugs toxic. As a result, many antifungal drugs can be used only for topical (surface) applications. However, a few drugs are selectively toxic for fungi because they target unique fungal structures or metabolic processes. Fungus-specific drugs are becoming increasingly important as fungus infections in immunocompromised individuals become more prevalent (↩ Sections 33.14 and 34.8). We examine here the selective action and targets of several effective antifungal agents.

Ergosterol Inhibitors

Ergosterol in fungal cytoplasmic membranes replaces the cholesterol found in animal cytoplasmic membranes. Two types of antifungal compounds work by interacting with ergosterol or inhibiting its synthesis (**Table 26.7**). These include the polyenes, a group of antibiotics produced by species of *Streptomyces*. Polyenes bind to ergosterol, disrupting membrane function, causing membrane permeability and cell death (**Figure 26.26**). A second major type of antifungal compound includes the azoles and allylamines, synthetic agents that selectively inhibit ergosterol biosynthesis and therefore have broad antifungal activity. Treatment with azoles results in abnormal fungus cytoplasmic membranes, leading to membrane damage and alteration of critical membrane transport activities. Allylamines also inhibit ergosterol biosynthesis, but are restricted to topical use because they are not readily taken up by animal tissues.

Echinocandins

Echinocandins act by inhibiting 1,3-β-D-glucan synthase, the enzyme that forms glucan polymers in the fungal cell wall (Figure 26.26 and Table 26.7). Because mammalian cells do not have 1,3-β-D-glucan synthase (or cell walls), the action of these

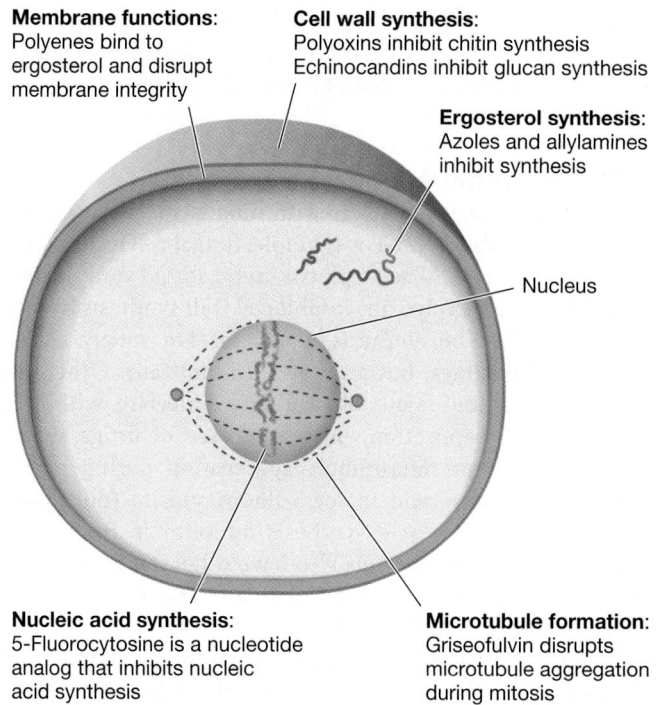

Membrane functions: Polyenes bind to ergosterol and disrupt membrane integrity

Cell wall synthesis: Polyoxins inhibit chitin synthesis Echinocandins inhibit glucan synthesis

Ergosterol synthesis: Azoles and allylamines inhibit synthesis

Nucleus

Nucleic acid synthesis: 5-Fluorocytosine is a nucleotide analog that inhibits nucleic acid synthesis

Microtubule formation: Griseofulvin disrupts microtubule aggregation during mitosis

Figure 26.26 Action of some antifungal agents. Traditional antibacterial agents are generally ineffective because fungi are eukaryotic cells. The cytoplasmic membrane and cell wall targets shown here are unique structures not present in vertebrate host cells.

Category	Target	Examples	Use
Allylamines	Ergosterol synthesis	Terbinafine	Oral, topical
Aromatic antibiotic	Mitosis inhibitor	Griseofulvin	Oral
Azoles	Ergosterol synthesis	Clotrimazole	Topical
		Fluconazole	Oral
		Itraconazole	Oral
		Ketoconazole	Oral
		Miconazole	Topical
		Posaconazole	Experimental
		Ravuconazole	Experimental
		Voriconazole	Oral
Chitin synthesis inhibitor	Chitin synthesis	Nikkomycin Z	Experimental
Echinocandins	Cell wall synthesis	Caspofungin	Intravenous
Nucleic acid analogs	DNA synthesis	5-Fluorocytosine	Oral
Polyenes	Ergosterol synthesis	Amphotericin B	Oral, intravenous
		Nystatin	Oral, topical
Polyoxins	Chitin synthesis	Polyoxin A	Agricultural
		Polyoxin B	Agricultural

Table 26.7 *Antifungal agents*

UNIT 8

agents is specific, resulting in selective fungal cell death. These agents are used to treat infections with fungi such as *Candida* and some fungi that are resistant to other agents (⟳ Sections 33.14 and 34.8).

Other Antifungal Agents

Other antifungal drugs interfere with fungus-specific structures and functions (Table 26.7). For example, fungal cell walls contain chitin, a polymer of *N*-acetylglucosamine found only in fungi and insects. Several polyoxins inhibit cell wall synthesis by interfering with chitin biosynthesis. Polyoxins are widely used as agricultural fungicides, but are not used clinically. Other antifungal drugs inhibit folate biosynthesis, interfere with DNA topology during replication, or, in the case of drugs such as griseofulvin, disrupt microtubule aggregation during mitosis. Moreover, the nucleic acid analog 5-fluorocytosine (flucytosine) is an effective nucleic acid synthesis inhibitor in fungi. Some very effective antifungal drugs also have other applications. For example, vincristine and vinblastine are effective antifungal agents and also have anticancer properties.

Predictably, the use of antifungal drugs has resulted in the emergence of populations of resistant fungi and the emergence of opportunistic fungal pathogens. For example, *Candida* species, which are normally not pathogenic, are known to produce disease in immunocompromised individuals. In addition, drug-resistant pathogenic *Candida* strains have developed in individuals who have been treated with antifungal drugs, and some are now resistant to all of the currently used antifungal agents (see Figure 26.29).

MiniQuiz
- Why are there very few clinically effective antifungal agents?
- What factors are contributing to an increased incidence of fungal infections?

 V # Antimicrobial Drug Resistance and Drug Discovery

Antimicrobial drug resistance is a major problem when dealing with many pathogenic microorganisms, especially in healthcare settings. Here we explore some of the mechanisms for drug resistance in microorganisms and present strategies for developing new antimicrobial agents.

26.12 Antimicrobial Drug Resistance

Antimicrobial drug resistance is the acquired ability of a microorganism to resist the effects of an antimicrobial agent to which it is normally susceptible. No single antimicrobial agent inhibits all microorganisms, and some form of antimicrobial drug resistance is an inherent property of virtually all microorganisms. As we have discussed, antibiotic producers are microorganisms.

In order to survive, the antibiotic-producing microorganism itself must be able to neutralize or destroy its own antibiotic. Thus, genes encoding antibiotic resistance must be present in virtually every organism that makes an antibiotic. Widespread antimicrobial drug resistance can then occur by horizontal transfer of resistance genes between and among microorganisms.

Resistance Mechanisms

For any of at least six different reasons, some microorganisms are naturally resistant to certain antibiotics.

1. The organism may lack the structure an antibiotic inhibits. For instance, some bacteria, such as the mycoplasmas, lack a bacterial cell wall and are therefore naturally resistant to penicillins.

2. The organism may be impermeable to the antibiotic. For example, most gram-negative *Bacteria* are impermeable to penicillin G and platensimycin.

3. The organism may be able to alter the antibiotic to an inactive form. Many staphylococci contain β-lactamases, an enzyme that cleaves the β-lactam ring of most penicillins (**Figure 26.27**).

4. The organism may modify the target of the antibiotic. In the laboratory, for example, antibiotic-resistant cells can be isolated from cultures that were grown from strains uniformly susceptible to the selecting antibiotic. The resistance of these isolates is usually due to mutations in chromosomal genes. In most cases, antibiotic resistance mediated by chromosomal genes arises because of a modification of the *target* of antibiotic activity (for example, a ribosome).

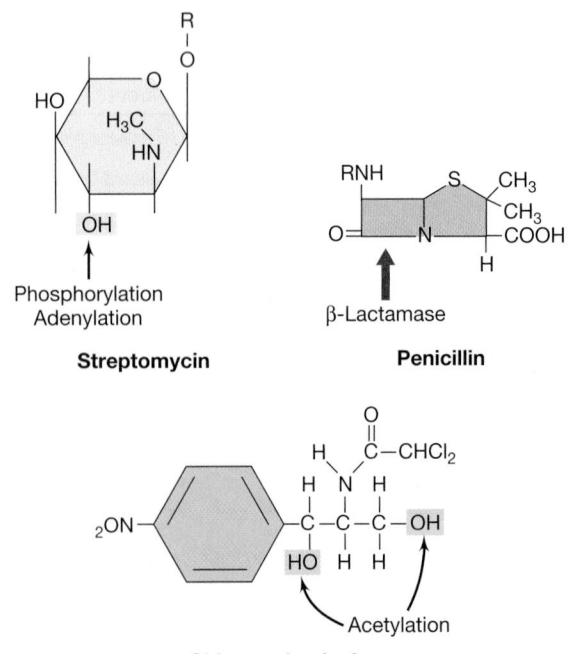

Figure 26.27 Sites at which antibiotics are attacked by enzymes encoded by R plasmid genes. Antibiotics may be selectively inactivated by chemical modification or cleavage. For the complete structure of streptomycin, see Figure 26.21 and for penicillin, Figure 26.19.

5. The organism may develop a resistant biochemical pathway. For example, many pathogens develop resistance to sulfa drugs that inhibit the production of folic acid in *Bacteria* (Section 26.6 and Figure 26.16). Resistant bacteria modify their metabolism to take up preformed folic acid from the environment, avoiding the need for the pathway blocked by the sulfa drugs.

6. The organism may be able to pump out an antibiotic entering the cell, a process called *efflux*.

Some specific examples of bacterial resistance to antibiotics are shown in **Table 26.8**.

Antibiotic resistance can be genetically encoded by the microorganism on either the bacterial chromosome or on a plasmid called an *R* (for *resistance*) *plasmid* (Sections 6.6 and 6.7) (Table 26.8). Because of widespread existing antibiotic resistance and continual emergence of new resistance, bacteria isolated from clinical specimens must be tested for antibiotic susceptibility using the MIC method or an agar diffusion method (Section 26.4). Details of the antibiotic susceptibility testing of clinical isolates are described in Section 31.3.

Mechanism of Resistance Mediated by R Plasmids

Most drug-resistant bacteria isolated from patients contain drug-resistance genes located on R plasmids rather than on the chromosome. Resistance is typically due to genes on the R plasmid that encode enzymes that modify and inactivate the drug (Figure 26.27) or genes that encode enzymes that prevent uptake of the drug or actively pump it out. For instance, the aminoglycoside antibiotics streptomycin, neomycin, kanamycin, and spectinomycin have similar chemical structures. Strains carrying R plasmids that encode resistance to these drugs make enzymes that phosphorylate, acetylate, or adenylate the drug. The modified drug then lacks antibiotic activity.

For the penicillins, R plasmids encode the enzyme *penicillinase* (a β-lactamase that splits the β-lactam ring, inactivating the antibiotic). Chloramphenicol resistance is due to an R plasmid–encoded enzyme that acetylates the antibiotic. Many R plasmids contain several different resistance genes and can confer multiple antibiotic resistance on a cell previously sensitive to each individual antibiotic.

Origin of Resistance Plasmids

R plasmids predated the widespread artificial use of antibiotics. A strain of *Escherichia coli* that was freeze-dried in 1946 contained a plasmid with genes conferring resistance to both tetracycline and streptomycin, even though neither of these antibiotics was used clinically until several years later. Similarly, R plasmid genes for resistance to semisynthetic penicillins existed before the semisynthetic penicillins had been synthesized.

Of perhaps even more ecological significance, R plasmids with antibiotic resistance genes are found in some nonpathogenic gram-negative soil bacteria. In the soil, R plasmids may confer selective advantages because major antibiotic-producing organisms (*Streptomyces* and *Penicillium*) are also soil organisms. R plasmids probably arose long before antibiotics were discovered, but, as we shall see, these naturally occurring plasmids have been propagated and spread as antibiotics were increasingly used in medicine and agriculture.

Table 26.8 *Bacterial resistance to antibiotics*

Resistance mechanism	Antibiotic example	Genetic basis of resistance	Mechanism present in:
Reduced permeability	Penicillins	Chromosomal	*Pseudomonas aeruginosa*
			Enteric *Bacteria*
Inactivation of antibiotic (for example, penicillinase; modifying enzymes such as methylases, acetylases, phosphorylases, and others)	Penicillins	Plasmid and chromosomal	*Staphylococcus aureus*
			Enteric *Bacteria*
	Chloramphenicol	Plasmid and chromosomal	*Neisseria gonorrhoeae*
			Staphylococcus aureus
	Aminoglycosides	Plasmid	Enteric *Bacteria*
			Staphylococcus aureus
Alteration of target (for example, RNA polymerase, rifamycin; ribosome, erythromycin and streptomycin; DNA gyrase, quinolones)	Erythromycin	Chromosomal	*Staphylococcus aureus*
	Rifamycin		Enteric *Bacteria*
	Streptomycin		Enteric *Bacteria*
	Norfloxacin		Enteric *Bacteria*
			Staphylococcus aureus
Development of resistant biochemical pathway	Sulfonamides	Chromosomal	Enteric *Bacteria*
			Staphylococcus aureus
Efflux (pumping out of cell)	Tetracyclines	Plasmid	Enteric *Bacteria*
	Chloramphenicol	Chromosomal	*Staphylococcus aureus*
			Bacillus subtilis
	Erythromycin	Chromosomal	*Staphylococcus* spp.

Spread of Antimicrobial Drug Resistance

The widespread use of antibiotics in medicine, veterinary medicine, and agriculture provides highly selective conditions for the spread of R plasmids. The resistance genes on R plasmids confer an immediate selective advantage and thus antibiotic resistance due to R plasmids is a predictable outcome of natural selection. The R plasmids and other sources of resistance genes pose significant limits on the long-term use of any single antibiotic as an effective antimicrobial agent.

Inappropriate use of antimicrobial drugs is the leading cause of rapid development of drug-specific resistance in disease-causing microorganisms. The discovery and clinical use of the many known antibiotics have been paralleled by the emergence of bacteria that resist them. **Figure 26.28** shows a correlation between the amounts of antibiotics used and the numbers of bacteria resistant to each antibiotic.

Overuse of antibiotics results in development of resistance. Increasingly, the antimicrobial agent prescribed for treatment of a particular infection must be changed because of increased resistance of the microorganism causing the disease. A classic example is the development of resistance to penicillin and other antimicrobial drugs in *Neisseria gonorrhoeae*, the bacterium that causes the sexually transmitted disease gonorrhea (Figure 26.28*b*). Prior to 1980, penicillin had been in continuous use for treatment for gonorrhea since it became available in the 1940s. However, penicillin is no longer a first-line treatment of gonorrhea because a significant percentage of clinical *N. gonorrhoeae*

isolates now produce β-lactamase, conferring penicillin resistance. Virtually all of these resistant isolates have developed since 1980; by 1990, penicillin-resistant strains were so common that fluoroquinolones such as ciprofloxacin replaced penicillin as the drug of choice for treatment. Soon after, however, the growing prevalence of fluoroquinolone-resistant *N. gonorrhoeae* strains isolated from Asia, Hawaii, and California in men who have sex with men again prompted a change in first-line drug recommendations for treatment of gonorrhea from the fluoroquinolone ciprofloxacin to ceftriaxone, a penicillinase-resistant β-lactam antibiotic (Figure 26.28*c*). Treatment guidelines are updated nearly every year to control continually emerging drug resistance in this organism (♻ Section 33.12).

Antibiotics are still used in clinical practice far more often than necessary. Antibiotic treatment is warranted in about 20% of individuals who are seen for infectious disease, but antibiotics are prescribed up to 80% of the time. Furthermore, in up to 50% of cases, prescribed doses or duration of treatments are not correct. This is compounded by patient noncompliance: Many patients stop taking medications, particularly antibiotics, as soon as they feel better. For example, the emergence of isoniazid-resistant tuberculosis correlates with a patient's failure to take the oral medication daily for the full course of 6–9 months (♻ Section 33.4). Exposure of virulent pathogens to sublethal doses of antibiotics for inadequate periods of time may select for drug-resistant strains.

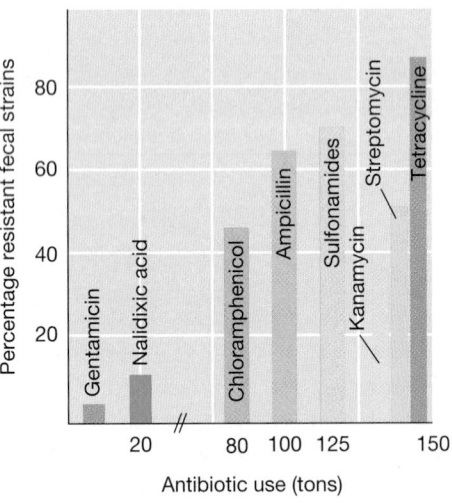

(a)

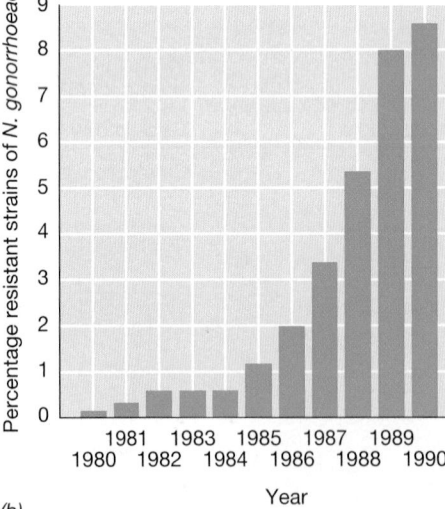

(b)

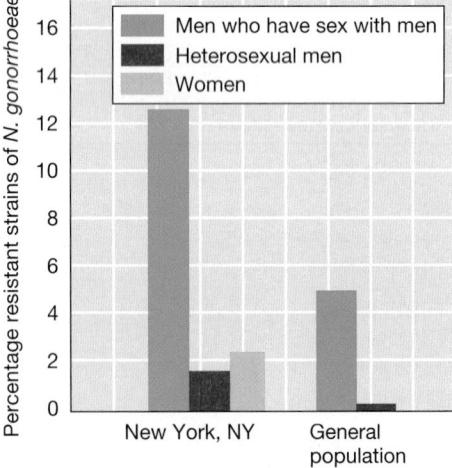

(c)

Figure 26.28 Patterns of drug resistance in pathogens. *(a)* The relationship between antibiotic use and the percentage of antibiotic-resistant bacteria isolated from diarrheal patients. Those agents that have been used in the largest amounts, as indicated by the amount produced commercially, are those for which drug-resistant strains are most frequent. *(b)* Percentage of

reported cases of gonorrhea caused by drug-resistant strains. The actual number of reported drug-resistant cases in 1985 was 9,000. This number rose to 59,000 in 1990. Greater than 95% of reported drug-resistant cases were due to penicillinase-producing strains of *Neisseria gonorrhoeae*. Since 1990, penicillin has not been recommended for treatment of gonorrhea because

of the emerging drug resistance. *(c)* The prevalence of fluoroquinolone-resistant *N. gonorrhoeae* in certain populations in the United States in 2003. Ciprofloxacin, a fluoroquinolone, is no longer recommended as a primary choice for treatment of *N. gonorrhoeae* infections. *Source*: Centers for Disease Control and Prevention, Atlanta, Georgia, USA.

Other recent studies, however, indicate that this trend is changing in the United States. Physicians prescribe about one-third fewer antibiotics for treatment of childhood infections than they did 10 years ago. This reduction has been done largely through efforts aimed at educating physicians, healthcare providers, and patients concerning the proper use of antibiotic therapy.

Indiscriminant, nonmedical use of antibiotics has also contributed to the emergence of resistant strains. In addition to their traditional use as a treatment for infections, antibiotics are used in agriculture as supplements to animal feeds both as growth-promoting substances and as prophylactic additives to prevent the occurrence of disease. Worldwide, about 50% of all antibiotics made are used in animal agriculture applications. Antibiotics are also extensively used in aquaculture (fish farming) and even in fruit production! Antibiotics used in the food supply far too frequently, over extended periods of time, and in high doses are a proven source of food infection outbreaks due to the selection of antibiotic-resistant pathogens. For example, fluoroquinolones, a group of broad-spectrum antibiotics that include clinically important therapeutic drugs such as ciprofloxacin, have been extensively used for over 20 years as growth-promoting and prophylactic agents in agriculture. As a result, fluoroquinolone-resistant *Campylobacter jejuni* has already emerged as a foodborne pathogen (Section 36.10), presumably because of the routine treatment of poultry flocks with fluoroquinolones to prevent respiratory diseases. Voluntary guidelines used by both poultry and drug producers are in place to monitor and reduce the use of fluoroquinolones. These measures may prevent development of resistance to new fluoroquinolone antibiotics.

Antibiotic-Resistant Pathogens

Largely as a result of failures to properly use antibiotics and monitor resistance, almost all pathogenic microorganisms have developed resistance to some antimicrobial agents since widespread use of antimicrobial drugs began in the 1950s (**Figure 26.29**). Penicillin and sulfa drugs, the first widely used antimicrobial agents, are not used as extensively today because many pathogens have acquired some resistance. Even the organisms that are still uniformly sensitive to penicillin, such as *Streptococcus pyogenes* (the bacterium that causes strep throat, scarlet fever, and rheumatic fever), now require larger doses of penicillin for successful treatment as compared to a decade ago.

A few pathogens have developed resistance to all known antimicrobial agents (Figure 26.29). Among these are several isolates of methicillin-resistant *Staphylococcus aureus* (MRSA) (methicillin is a semisynthetic penicillin; Section 26.8). Although MRSA is usually associated with healthcare settings, it also causes a significant number of community-associated infections. An increasing number of independently derived MRSA strains have developed reduced susceptibility to even vancomycin and are termed "vancomycin intermediate *Staphylococcus aureus*" (VISA) strains (Section 33.9). Vancomycin-resistant *Enterococcus faecium* (VRE) and some isolates of *Mycobacterium tuberculosis* and *Candida albicans* have also developed resistance to all known antimicrobial drugs. Antibiotic resistance can

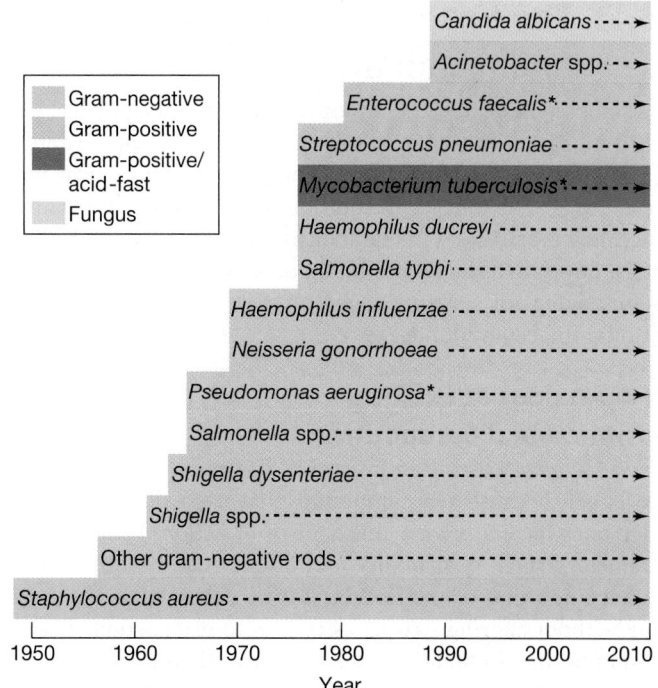

Figure 26.29 The appearance of antimicrobial drug resistance in some human pathogens. The asterisks indicate that some strains of these pathogens are now untreatable with known antimicrobial drugs.

be minimized if drugs are used only for treatment of susceptible diseases and are given in sufficiently high doses and for sufficient lengths of time to reduce the microbial population before resistant mutants can develop. Combining two unrelated antimicrobial agents may also reduce resistance; it is less likely that a mutant strain resistant to one antibiotic will also be resistant to the second antibiotic. However, certain common R plasmids confer multiple drug resistance and make multiple antibiotic therapy less useful as a clinical treatment strategy.

We now know that if the use of a particular antibiotic is stopped, the resistance to that antibiotic can be reversed over the course of several years. On the other hand, antibiotic-resistant organisms may persist in the gut for some time. This information implies that the efficacy of some antibiotics may be reestablished by withdrawing the antibiotic from use, but only by following a carefully monitored plan of prudent use upon reintroduction. Finally, as we discuss below, new antimicrobial agents are actively being developed using various strategies for drug design and discovery.

MiniQuiz

- Identify the six basic mechanisms of antibiotic resistance among bacteria.
- Identify the primary sources of antibiotic resistance genes.
- What practices encourage the development of antibiotic-resistant pathogens?

UNIT 8

26.13 The Search for New Antimicrobial Drugs

Resistance will develop to all known antimicrobial drugs, given sufficient drug exposure and time. Conservative, appropriate use of antibiotics is necessary to prolong the useful clinical life of these drugs. The long-term solution to antimicrobial drug resistance, however, resides in our ability to develop new antimicrobial drugs. Several strategies are being used to identify and produce useful analogs of existing agents or to design or discover novel antimicrobial compounds.

New Analogs of Existing Antimicrobial Compounds

New analogs of existing antimicrobial compounds are often effective, largely because new compounds that are structural mimics of older ones have a proven mechanism of action. In many cases, parameters such as solubility and affinity can be optimized by introducing minor modifications to the chemical structure of a drug without altering structures critical to drug action. The new compound may actually be more effective than the parent compound and, because resistance is based on structural recognition, the new compound may not be recognized by resistance factors. For example, Figure 26.23 shows the structure of tetracycline and two bioactive derivatives. Using authentic tetracycline as the lead compound, systematic chemical substitutions at the four R group sites can generate an almost endless series of tetracycline analogs. Using this basic strategy, new tetracycline-related compounds (Section 26.9), new β-lactam antibiotics (Section 26.8), and new analogs of vancomycin (**Figure 26.30**) have been synthesized.

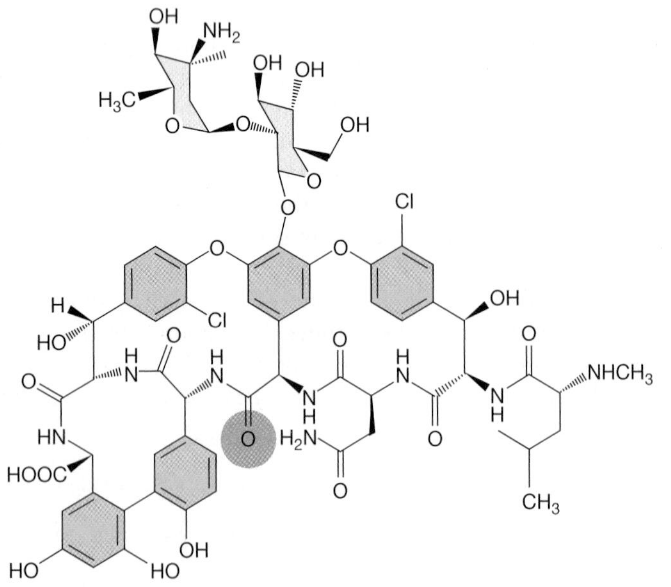

Vancomycin

Figure 26.30 Vancomycin. Intermediate drug resistance to the parent structure of vancomycin has developed in recent years. However, modification at the position shown in red by substitution of a methylene (=CH$_2$) group for the carbonyl oxygen restores much of the lost activity.

Some of these derivatives are as much as 100 times more potent than the parent compound.

The application of automated chemistry methods to drug discovery has dramatically increased our ability to rapidly generate new antimicrobial compounds. These automated methods, called *combinatorial chemistry*, initiate systematic modifications of a known antimicrobial product to yield large numbers of new analogs. For instance, using automated combinatorial chemistry and starting with pure tetracycline (Figure 26.23), five different reagents might be used to introduce substitutions at the four different tetracycline R groups. The substituted sites would yield $5 \times 5 \times 5 \times 5$ (five derivatives at each of four sites), or 625 different tetracycline derivatives from only five different reagents, all in a few hours' time! These compounds would then be assayed for *in vitro* biological activity on different test organisms using automated testing methods for antibiotic susceptibility. The automated synthesis and screening processes dramatically shorten drug discovery time and increase the number of new candidate drugs by a factor of 10 or more each year.

According to pharmaceutical industry estimates, about 7 million candidate compounds must be screened to yield a single useful clinical drug. Drugs effective in the laboratory must then be tested for efficacy and toxicity in animals and finally in clinical trials in humans. Animal testing requires multiple trials over several years to ensure that the candidate drug is both effective and safe. Clinical trials in humans to check efficacy and safety take additional years for each drug. Each year, the pharmaceutical industry spends up to $4 billion on new antimicrobial drug development. Discovery and development for each drug typically takes 10–25 years before it is approved for clinical use. The cost of discovery and development, from the laboratory through clinical trials, is estimated at over $500 million for each new drug approved for human use. This is a major reason why pharmaceutical drugs are so expensive.

Computer Drug Design

Novel antimicrobial compounds are much more difficult to identify than analogs of existing drugs because new antimicrobial compounds must work at unique sites in metabolism and biosynthesis or be structurally dissimilar to existing compounds to avoid inducing known resistance mechanisms. Computer technology and structural biology methods make it possible to design a drug to interact with specific microbial structures. Thus, drug discovery can now begin at the computer, where new drugs can be rapidly synthesized and tested for binding and efficacy in the computer environment at relatively low cost.

One of the most dramatic successes in computer-directed drug design is the development of *saquinavir*, a protease inhibitor that is used to slow the growth of the human immunodeficiency virus (HIV) in infected individuals (**Figure 26.31**). Designed by computer, saquinavir binds the active site of the HIV protease enzyme. The structure of saquinavir was based on the known three-dimensional structure of the protease–substrate complex. The HIV protease normally cleaves a virus-encoded precursor protein to produce the mature viral core and activate

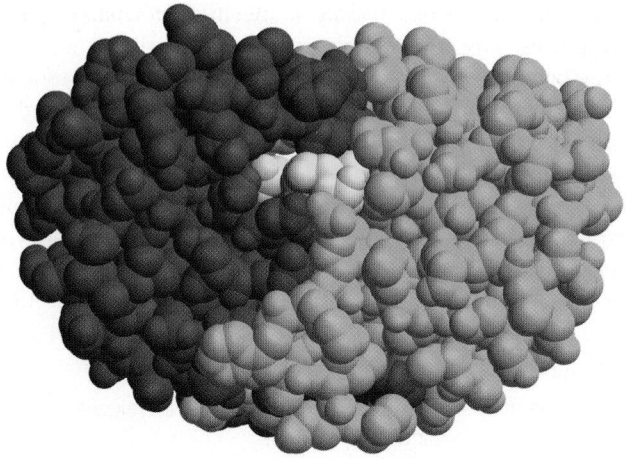

(a) **HIV protease**

Saquinavir

Indinavir

(b)

Figure 26.31 Computer-generated anti-HIV drugs. *(a)* The HIV protease homodimer. Individual polypeptide chains are shown in green and blue. A peptide (yellow) is bound in the active site. HIV protease cleaves an HIV precursor protein, a necessary step in virus maturation (⮌ Section 21.11). Blocking of the protease site by the bound peptide inhibits precursor processing and HIV maturation. This structure is derived from information in the Protein Data Bank. *(b)* These anti-HIV drugs are peptide analogs called protease inhibitors that were designed by computer to block the active site of HIV protease. The areas highlighted in orange show the regions analogous to peptide bonds in proteins.

the reverse transcriptase enzyme necessary for replication (⮌ Section 21.11). Saquinavir is a high-affinity peptide analog of the HIV precursor protein that displaces the authentic protein substrate, inhibiting virus maturation and growth in the human host. A number of other computer-designed protease inhibitors are in use as antiviral drugs for the treatment of AIDS (Table 26.6, Figure 26.31, and ⮌ Figure 33.43). Computer design and testing based on structural and biochemical modeling is a practical, rapid, and cost-effective method for designing antimicrobial drugs.

Natural Products as Antibiotics

As the first antibiotics were discovered and brought into clinical use in the 1930s and 1940s, researchers developed standard methods to isolate more new antibiotics. Candidate drugs were routinely isolated from natural sources such as *Streptomyces* or *Penicillium* cultures and systematically screened for antimicrobial activity using standard MIC or agar diffusion methods to find new antimicrobial compounds. As time passed, the yield from these traditional methods decreased, supplanted by higher yields from the combinatorial chemistry and computer design methods discussed above. Most of the effective natural antibiotics produced at reasonable levels by antibiotic-producing microorganisms had already been isolated. Remaining effective antibiotics, presumably present in concentrations so low that they were ineffective against test organisms, could not be identified.

Platensimycin (Figure 26.25), however, is an exception to this rule. This antibiotic was discovered using a modification of direct methods for screening natural products. Platensimycin represents a new class of antimicrobials that selectively inhibits bacterial lipid biosynthesis and is especially active against gram-positive pathogens, including MRSA, VISA, and VRE (Section 26.12). A key feature in the discovery of platensimycin was its selection using a novel method that may have broad applications for targeted drug discovery. To select an agent for a defined target, in this case an enzyme in the lipid synthesis pathway of gram-positive bacteria, scientists introduced a defect in the β-ketoacyl-(acyl-carrier protein) synthase I/II (FabF/B) gene in *Staphylococcus aureus* by using a strain expressing antisense FabF RNA (⮌ Section 8.14). The gene-specific antisense RNA decreased expression of FabF, reducing fatty acid synthesis and increasing the sensitivity of the crippled *S. aureus* strain to antibiotics that inhibit fatty acid synthesis. By screening 250,000 natural product extracts from 83,000 strains of potential antibiotic producers, the scientists were able to identify and isolate platensimycin from a soil microorganism, *Streptomyces platensis*. Although the screening of large numbers of strains is a huge task, the method identifies target-specific antibiotics present in low concentrations. This strategy is applicable to virtually any target for which the gene sequence (and, hence, the corresponding antisense RNA sequence) is known.

Drug Combinations

The efficacy of some antibiotics can be retained if they are given with compounds that inhibit antibiotic resistance. Several β-lactam antibiotics can be combined with β-lactamase inhibitors to preserve antibiotic activity in β-lactamase-resistant microorganisms. For example, the broad-spectrum β-lactam antibiotic ampicillin (Figure 26.19) can be mixed with sulbactam, a β-lactamase inhibitor. The inhibitor binds β-lactamase irreversibly, preventing degradation of the ampicillin and permitting it to disrupt cell wall formation in the affected cell. This combination preserves the effectiveness of the β-lactamase-sensitive ampicillin against β-lactamase producers such as staphylococci and certain gram-negative pathogens. Likewise, we have already mentioned the use of sulfamethoxazole–trimethoprim, a mixture of two folic acid synthesis inhibitors, to prevent the loss of efficacy through mutation and selection for resistance (Section 26.6).

Drug combination therapy approaches have revolutionized treatment of HIV infections. Currently, a combination therapy consisting of nucleoside analogs and a protease inhibitor is recommended. This drug treatment protocol is termed HAART, for highly active anti-retroviral therapy. As with antimicrobial combination regimens, HAART is designed to target two independent viral functions; the nucleoside analogs target virus replication and the protease inhibitors target virus maturation. Because the probability of a single virus developing resistance to multiple drugs is less than the probability of developing resistance to a single drug, HAART-resistant strains are relatively uncommon (♻ Section 33.14).

Bacteriophages

Bacteriophages are viruses that infect bacteria (♻ Sections 9.8–9.10). *Bacteriophage therapy* has been used on a limited basis for over 80 years to treat infections in animals and, in a few instances, in humans. Phages interact with individual bacterial cell surface components and show specificity for particular bacterial species. The attached phage enters the cell, replicates, and kills the bacterial host in the process. The efficacy and efficiency of these agents for human treatment is largely untested and somewhat controversial, although clinical trials of several products are ongoing. However, because bacteria can acquire resistance to a phage infection through mutations that alter receptors or reduce the susceptibility of the cell wall to phage enzymes, bacteriophage therapy will likely be susceptible to resistance, just as for most chemical antimicrobial agents.

MiniQuiz

- Explain the advantages and disadvantages of developing new drugs based on existing drug analogs.
- How can computer drug design aid in the search for new drugs?
- Explain the use of antisense RNA for drug discovery.

Big Ideas

26.1

Sterilization is the killing of all organisms and viruses, and heat is the most widely used method of sterilization. The temperature employed must eliminate the most heat-resistant organisms, usually bacterial endospores. An autoclave permits applications of steam heat under pressure, achieving temperatures above the boiling point of water, which kills endospores. Pasteurization does not sterilize liquids, but reduces microbial load, kills most pathogens, and inhibits the growth of spoilage microorganisms.

26.2

Controlled doses of electromagnetic radiation effectively inhibit microbial growth. Ultraviolet radiation is used for decontaminating surfaces and materials that do not absorb light, such as air and water. Ionizing radiation that can penetrate solid or light-absorbing materials is used for sterilization and decontamination in the medical and food industries.

26.3

Filters remove microorganisms from air or liquids. Depth filters, including HEPA filters, are used to remove microorganisms and other contaminants from liquids or air. Membrane filters are used for sterilization of heat-sensitive liquids, and nucleation track filters are used to isolate specimens for electron microscopy.

26.4

Chemicals are often used to control microbial growth. Chemicals that kill organisms are called -cidal agents; those that inhibit growth are called -static agents; those that lyse organisms are called -lytic agents. Antimicrobial agents are tested for efficacy by determining their ability to inhibit growth in vitro.

26.5

Sterilants, disinfectants, and sanitizers are used to decontaminate nonliving material. Antiseptics and germicides are used to reduce microbial growth on living tissues. Antimicrobial compounds have commercial, healthcare, and industrial applications.

26.6

Synthetic antimicrobial agents are selectively toxic for *Bacteria*, viruses, and fungi. Synthetic growth factor analogs such as sulfa drugs, isoniazid, and nucleic acid analogs are metabolic inhibitors. Quinolones inhibit the action of DNA gyrase in *Bacteria*.

26.7

Antibiotics are a chemically diverse group of antimicrobial compounds that are produced by microorganisms. Although many antibiotics are known, only a few are clinically effective. Each antibiotic works by inhibiting a specific cellular process or function in the target microorganisms.

26.8

The β-lactam compounds, including the penicillins and the cephalosporins, are the most important single class of clinical antibiotics. These antibiotics and their semisynthetic derivatives target cell wall synthesis in *Bacteria*. They have low host toxicity and collectively have a broad spectrum of activity.

26.9

The aminoglycosides, macrolides, and tetracycline antibiotics are structurally complex molecules produced by *Bacteria* and are active against other *Bacteria*. These antibiotics selectively interfere with protein synthesis in *Bacteria*. Daptomycin and platensimycin are structurally novel antibiotics that target cytoplasmic membrane functions and lipid biosynthesis, respectively.

26.10

Effective antiviral agents selectively target virus-specific enzymes and processes. Clinically useful antiviral agents include nucleoside analogs and other drugs that inhibit nucleic acid polymerases and viral genome replication. Agents such as protease inhibitors interfere with viral maturation steps. Host cells also produce the antiviral interferon proteins that stop viral replication.

26.11

Antifungal agents fall into many chemical categories. Because fungi are *Eukarya*, selective toxicity is hard to achieve, but some effective antifungal agents are available. Treatment of fungal infections is an emerging human health issue.

26.12

The use of antimicrobial drugs inevitably leads to resistance in the targeted microorganisms. The development of resistance can be accelerated by the indiscriminate use of antimicrobial drugs. A few pathogens have developed resistance to all known antimicrobial drugs.

26.13

New antimicrobial compounds are constantly being discovered and developed to deal with drug-resistant pathogens and to enhance our ability to treat infectious diseases. Analogs of existing drugs are often synthesized and used as next-generation antimicrobial compounds. Computer drug design is an important tool for drug discovery.

Review of Key Terms

Aminoglycoside an antibiotic such as streptomycin, containing amino sugars linked by glycosidic bonds

Antibiotic a chemical substance produced by a microorganism that kills or inhibits the growth of another microorganism

Antimicrobial drug resistance the acquired ability of a microorganism to grow in the presence of an antimicrobial drug to which the microorganism is usually susceptible

Antimicrobial agent a chemical compound that kills or inhibits the growth of microorganisms

Antiseptic (germicide) a chemical agent that kills or inhibits growth of microorganisms and is sufficiently nontoxic to be applied to living tissues

Autoclave a sealed heating device that destroys microorganisms with temperature and steam under pressure

Bacteriocidal agent an agent that kills bacteria

Bacteriostatic agent an agent that inhibits bacterial growth

Beta (β)-lactam antibiotic an antibiotic, including penicillin, that contains the four-membered heterocyclic β-lactam ring

Broad-spectrum antibiotic an antibiotic that acts on both gram-positive and gram-negative *Bacteria*

Decontamination a treatment that renders an object or inanimate surface safe to handle

Disinfectant an antimicrobial agent used only on inanimate objects

Disinfection the elimination of pathogens from inanimate objects or surfaces

Fungicidal agent an agent that kills fungi

Fungistatic agent an agent that inhibits fungal growth

Fusion inhibitor a peptide that blocks the fusion of viral and target cytoplasmic membranes

Germicide (antiseptic) a chemical agent that kills or inhibits growth of microorganisms and is sufficiently nontoxic to be applied to living tissues

Growth factor analog a chemical agent that is related to and blocks the uptake of a growth factor

HEPA filter a *h*igh-*e*fficiency *p*articulate *a*ir filter that removes particles, including microorganisms, from intake or exhaust air flow

Interferon a cytokine protein produced by virus-infected cells that induces signal transduction in nearby cells, resulting in transcription of antiviral genes and expression of antiviral proteins

Minimum inhibitory concentration (MIC) the minimum concentration of a substance necessary to prevent microbial growth

Nonnucleoside reverse transcriptase inhibitor (NNRTI) a nonnucleoside analog used to inhibit viral reverse transcriptase

Nucleoside reverse transcriptase inhibitor (NRTI) a nucleoside analog used to inhibit viral reverse transcriptase

Pasteurization the use of controlled heat to reduce the microbial load, including disease-producing microorganisms and spoilage microorganisms, in heat-sensitive liquids

Penicillin a class of antibiotics that inhibit bacterial cell wall synthesis, characterized by a β-lactam ring

Protease inhibitor an inhibitor of a viral protease

Quinolone a synthetic antibacterial compound that interacts with DNA gyrase and prevents supercoiling of bacterial DNA

Sanitizer an agent that reduces microorganisms to a safe level, but may not eliminate them

Selective toxicity the ability of a compound to inhibit or kill pathogenic microorganisms without adversely affecting the host

Sterilant (sterilizer, sporicide) a chemical agent that destroys all forms of microbial life

Sterilization the killing or removal of all living organisms and viruses from a growth medium

Tetracycline an antibiotic characterized by the four-ring naphthacene structure

Viricidal agent an agent that stops viral replication and activity

Viristatic agent an agent that inhibits viral replication

Review Questions

1. Highlight the differences between sterilization, decontamination, and disinfection (Section 26.1).

2. Describe the effects of a lethal dose of ionizing radiation at the molecular level (Section 26.2).

3. Compare and contrast sterilization by depth filters and membrane filters (Section 26.3).

4. Describe the principle of the disc diffusion test for antimicrobial susceptibility (Section 26.4).

5. Contrast the action of disinfectants and antiseptics. Disinfectants normally cannot be used on living tissue; why not (Section 26.5)?

6. Growth factor analogs are generally distinguished from antibiotics by a single important criterion. Explain (Section 26.6).

7. Identify common sources for naturally occurring antimicrobial drugs (Section 26.7).

8. Describe the mode of action that characterizes a β-lactam antibiotic. Why are these antibiotics generally more effective against gram-positive bacteria than against gram-negative bacteria (Section 26.8)?

9. Distinguish between the modes of action of at least three of the protein synthesis–inhibiting antibiotics (Section 26.9).

10. Discuss the different modes of action of two different antiviral agents (Section 26.10).

11. Identify the targets that allow selective toxicity of antifungal agents (Section 26.11).

12. Identify six mechanisms responsible for antibiotic resistance (Section 26.12).

13. Explain how application of antisense RNA methods can extend traditional natural product selection methods for antibiotic discovery (Section 26.13).

Application Questions

1. What are some potential drawbacks to the use of radiation in food preservation? Do you think these drawbacks could be manifested as health hazards? Why or why not? How would you distinguish between radiation-damaged and radiation-contaminated food?

2. Filtration is an acceptable means of pasteurization for some liquids. Design a filtration system for pasteurization of a heat-sensitive liquid. For a liquid of your choice, identify the advantages and disadvantages of a filtration system over a heat pasteurization system. Explain in terms of product quality, shelf life, and price.

3. Although growth factor analogs may inhibit microbial metabolism, only a few of these agents are useful in practice. Many growth factor analogs, including some in wide use, such as azidothymidine, exhibit significant host cell toxicity. Describe a growth factor analog that is effective and has low toxicity for host cells. Why is the toxicity low for the agent you chose? Also describe a growth factor analog that is effective against an infectious disease, but exhibits toxicity for host cells. Why might a toxic agent such as AZT still be used in certain situations to treat infectious diseases? What precautions would you take to limit the toxic effects of such a drug while maximizing the therapeutic activity? Explain your answer.

4. Although many antibiotics demonstrate clear selective toxicity for *Bacteria*, many groups of *Bacteria* are innately resistant to their effects. Indicate why gram-negative *Bacteria* are resistant to the effects of many, but not all, antibiotics. Further explain why some antibiotics are effective against these organisms.

5. List the features of an ideal antiviral drug, especially with regard to selective toxicity. Do such drugs exist? What factors might limit the use of such a drug?

6. Like viruses, fungi present special problems for drug therapy. Explain the problems inherent in drug treatment of both groups and explain whether or not you agree with the preceding statement. Give specific examples and suggest at least one group of agents that might target both types of infectious agents.

7. Explain the genetic basis of acquired resistance to β-lactam antibiotics in *Staphylococcus aureus*. Design experiments to reverse resistance to the β-lactam antibiotics. Do you think this can be done in the laboratory? Can your experiment be applied "in the field" to promote deselection of antibiotic-resistant organisms?

8. Design experiments to examine microorganisms for production of novel antibiotics. Which group or groups of microorganisms would you choose to screen for antibiotic production? Where could you find and isolate these organisms in a natural environment? What advantage, if any, would the production of an antibiotic provide for these organisms in nature? What *in vitro* methods would you use to test the efficacy of your potential new antibiotics? How might you increase the sensitivity of assays for natural products? Why are members of the genus *Streptomyces* still productive sources of novel antibiotics?

Need more practice? Test your understanding with Quantitative Questions; access additional study tools including tutorials, animations, and videos; and then test your knowledge with chapter quizzes and practice tests at **www.microbiologyplace.com**.

Microbial Interactions with Humans

Hemolysis, the lysis of red blood cells by proteins called hemolysins, is one of the many weapons pathogenic bacteria have evolved for destroying host tissues and releasing usable nutrients.

Humans have an extensive population of microorganisms, primarily bacteria, on the skin and the mucous membranes lining the mouth, gut, and excretory and reproductive systems. The human body contains 10^{13}–10^{14} cells, but about 10^{14}–10^{15} microorganisms live on or in the body. Most of these microorganisms are beneficial and some are necessary to maintain good health.

A few microorganisms called *pathogens* colonize, invade, and damage the human body through direct and indirect means. This is the process of infectious disease. Pathogens use a number of mechanisms to gain access to host tissues. These include the production of specialized attachment structures, unique growth factors, invasive enzymes, and potent biological toxins. These factors often lead to damage and occasionally death of the host.

In Chapter 26, we discussed physical and chemical mechanisms used to destroy or inhibit growth of microorganisms. Here we introduce normal microflora of the human body. We then look at pathogens and some of their disease-producing mechanisms. We conclude by introducing the natural nonspecific physical, anatomical, and biochemical defense mechanisms our bodies use to suppress or destroy most microbial pathogens and make microbial infectious disease a relatively infrequent event.

Beneficial Microbial Interactions with Humans

After a brief overview of human–microbial interactions, we will discuss microorganisms that inhabit the human body and contribute to overall good health under normal circumstances. In Part I of this chapter our focus will be on the microflora of the human oral cavity, skin, and colon. In Sections 25.6–25.8 we looked at the microbiology of the mammalian gut using the tools of molecular biology to assess microbial diversity.

27.1 Overview of Human– Microbial Interactions

Through normal everyday activities, the human body is exposed to countless microorganisms in the environment. In addition, hundreds of species and countless individual microbial cells, collectively referred to as the **normal microflora**, grow on or in the human body. Most, but not all, microorganisms are benign; a few contribute directly to our health, and even fewer pose direct threats to health.

Colonization by Microorganisms

Mammals *in utero* develop in a sterile environment and have no exposure to microorganisms. **Colonization**, the growth of a microorganism after it has gained access to host tissues, begins as animals are exposed to microorganisms in the birth process. The skin surfaces are readily colonized by many species. Likewise, the oral cavity and gastrointestinal tract acquire microorganisms through feeding and exposure to the mother's body, which, along with other environmental sources, initiates colo-

nization of the skin, oral cavity, upper respiratory tract, and gastrointestinal tract. Different populations of microorganisms colonize individuals in different localities and at different times. For example, *Escherichia coli*, a normal inhabitant of the human and animal gut, colonizes the guts of infants in developing countries within several days after birth. Infants in developed countries, however, typically do not acquire *E. coli* for several months; the first microorganisms to colonize the gut of these infants would more typically be *Staphylococcus aureus* and other microorganisms associated with the skin. Genetic factors also play a role. Thus, the normal microflora is highly dependent on the conditions to which an individual is exposed. The normal microflora is highly diverse in each individual and may differ significantly between individuals, even in a given population, but we will point out patterns of colonization by particular groups of organisms that inhabit specific niches, presumably because of their ability to access nutritional and metabolic support at particular body sites.

Pathogens

A **host** is an organism that harbors a **pathogen**, another organism that lives on or in the host and causes disease. The outcome of a host–pathogen relationship depends on **pathogenicity**, the ability of a pathogen to inflict damage on the host. Pathogenicity differs considerably among potential pathogens, as does the resistance or susceptibility of the host to the pathogen. An **opportunistic pathogen** causes disease only in the absence of normal host resistance.

Pathogenicity varies markedly for individual pathogens. The quantitative measure of pathogenicity is called **virulence**, the relative ability of a pathogen to cause disease. Virulence can be expressed quantitatively as the cell number that elicits disease in a host within a given time period. The host–pathogen interaction is a dynamic relationship between the two organisms, influenced by changing conditions in the pathogen, the host, and the environment. As a result, neither the virulence of the pathogen nor the relative resistance of the host is a constant factor.

Infection and Disease

Infection refers to any situation in which a microorganism is established and growing in a host, whether or not the host is harmed. **Disease** is damage or injury to the host that impairs host function. Infection is not synonymous with disease because growth of a microorganism on a host does not always cause host damage. Thus, species of the normal microflora have infected the host, but seldom cause disease. However, the normal microflora sometimes cause disease if host resistance is compromised, as may happen in diseases such as cancer and acquired immune deficiency syndrome (AIDS) (∞ Section 33.14).

Host–Pathogen Interactions

Animal hosts provide favorable environments for the growth of many microorganisms. They are rich in the organic nutrients and growth factors required by chemoorganotrophs, and provide conditions of controlled pH, osmotic pressure, and temperature. However, the animal body is not a uniform environment. Each region or organ differs chemically and physically from others and thus provides a selective environment where the growth of certain microorganisms is favored. The skin, respiratory tract, and gastrointestinal tract provide selective chemical and physical environments that support the growth of a highly diverse microflora. The relatively dry environment of the skin favors the growth of organisms that resist dehydration, such as the gram-positive bacterium *Staphylococcus aureus* ( ) Section 33.9); the highly oxygenated environment of the lungs favors the growth of the obligately aerobic *Mycobacterium tuberculosis* ( ) Section 33.4); and the anoxic environment of the large intestine supports growth of obligately anaerobic bacteria such as *Clostridium* and *Bacteroides* ( ) Sections 18.2 and 18.12). Animals also possess defense mechanisms that collectively prevent or inhibit microbial invasion and growth. The microorganisms that successfully colonize the host must circumvent these defense mechanisms.

The Infection Process

Infections frequently begin at sites in the animal's **mucous membranes**. Mucous membranes consist of single or multiple layers of *epithelial cells*, tightly packed cells that interface with the external environment. They are found throughout the body, lining the urogenital, respiratory, and gastrointestinal tracts. Mucous membranes are frequently coated with a protective liquid called **mucus** secreted by the epithelial cells. Mucus is a liquid secretion that contains water-soluble glycoproteins and proteins that retain moisture and aid in resistance to microbial invasion on mucosal surfaces. Microorganisms that contact host tissues at mucous membranes may associate loosely with the mucosal surface and are usually swept away by physical processes. Microorganisms may also adhere more strongly to the epithelial surface as a result of specific cell–cell recognition between pathogen and host. Tissue infection may follow, breaching

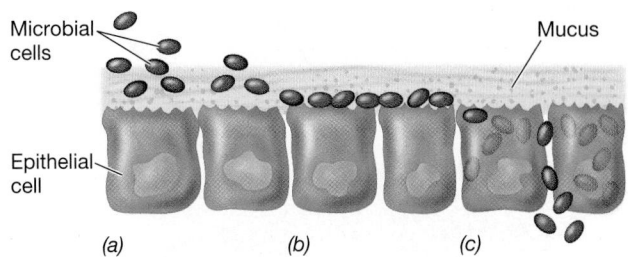

Figure 27.1 Bacterial interactions with mucous membranes. *(a)* Loose association. *(b)* Adhesion. *(c)* Invasion into submucosal epithelial cells.

the mucosal barrier and allowing the microorganism to invade deeper into submucosal tissues (**Figure 27.1**).

Microorganisms are almost always found on surfaces of the body, such as the skin, that are exposed to the environment. The mucosal surfaces of the oral cavity, respiratory tract, intestinal tract, and urogenital tract are also colonized with normal microflora. They are not normally found on or in the internal organs or in the blood, lymph, or nervous systems of the body. The growth of microorganisms in these normally sterile environments indicates serious infectious disease.

Table 27.1 shows some of the major types of microorganisms normally found in association with body surfaces. Mucosal surfaces have a diverse microflora because they offer a sheltered, moist environment and a large overall surface area. For example, a mucosal organ such as the small intestine has a surface area of about 400 m^2 available for nutrient transport, and this entire surface is a potential site for microbial growth.

MiniQuiz

- Distinguish between infection and disease.
- Why might one area of the body be more suitable for microbial growth than another?

Table 27.1 *Representative normal microflora of humans*	
Anatomical site	**Genera or major groups**[a]
Skin	*Acinetobacter, Corynebacterium, Enterobacter, Klebsiella, Malassezia (f), Micrococcus, Pityrosporum (f), Propionibacterium, Proteus, Pseudomonas, Staphylococcus, Streptococcus*
Mouth	*Streptococcus, Lactobacillus, Fusobacterium, Veillonella, Corynebacterium, Neisseria, Actinomyces, Geotrichum (f), Candida (f), Capnocytophaga, Eikenella, Prevotella,* spirochetes (several genera)
Respiratory tract	*Streptococcus, Staphylococcus, Corynebacterium, Neisseria, Haemophilus*
Gastrointestinal tract[b]	*Lactobacillus, Streptococcus, Bacteroides, Bifidobacterium, Eubacterium, Peptococcus, Peptostreptococcus, Ruminococcus, Clostridium, Escherichia, Klebsiella, Proteus, Enterococcus, Staphylococcus, Methanobrevibacter,* gram-positive bacteria, *Proteobacteria, Actinobacteria, Fusobacteria*
Urogenital tract	*Escherichia, Klebsiella, Proteus, Neisseria, Lactobacillus, Corynebacterium, Staphylococcus, Candida (f), Prevotella, Clostridium, Peptostreptococcus, Ureaplasma, Mycoplasma, Mycobacterium, Streptococcus, Torulopsis (f)*

[a]This list is not meant to be exhaustive, and not all of these organisms are found in every individual. Some organisms are more prevalent at certain ages (adults vs. children). Distribution may also vary between sexes. Many of these organisms can be opportunistic pathogens under certain conditions. Several genera are commonly found in more than one body area. (f), fungi.
[b]For a molecular picture of the prokaryotic diversity of the human large intestine, see Section 25.8 and Figure 25.31.

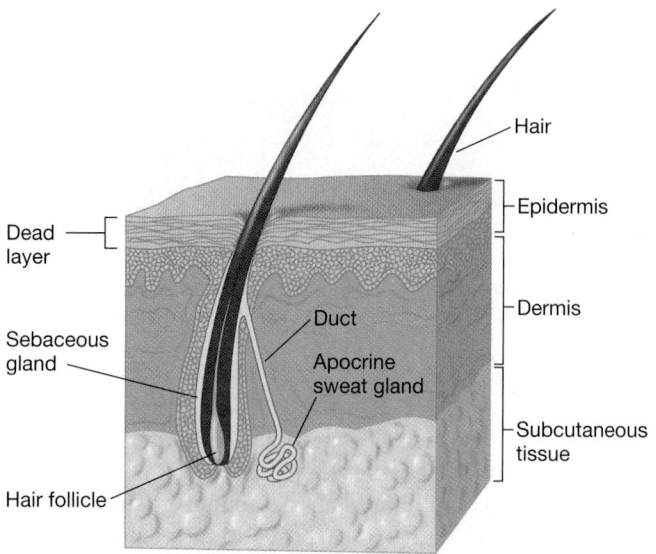

Figure 27.2 The human skin. Microorganisms are associated primarily with the sweat ducts and the hair follicles.

27.2 Normal Microflora of the Skin

An average adult human has about two square meters (2 m^2) of skin surface that varies greatly in chemical composition and moisture content. **Figure 27.2** shows the anatomy of the skin. Several distinct microenvironments are present in this organ. One distinct microenvironment includes moist skin areas such as the inside of the nostril, the armpit, and the umbilicus. This is separated by only a few centimeters from the dry microenvironment of the forearms and the palms of the hands. A third microenvironment consists of areas that have high concentrations of sebaceous glands. These glands produce an oily sub-

stance called sebum. The areas with sebaceous glands include the alar crease (by the side of the nose), the back of the scalp, the upper chest, and back.

To define the microbial population of the skin, standard culture methods have been used, but recently molecular methods using a metagenomic approach looked at 16S rRNA gene sequences from samples collected from 10 healthy volunteers at 20 environmentally and spatially diverse skin sites.

The skin sites sampled were divided according to the three general microenvironments. Nineteen different bacterial phyla were detected, but four phyla predominated, with the *Actinobacteria*, *Firmicutes*, *Proteobacteria*, and *Bacteroidetes* accounting for almost all of the sequences obtained (**Figure 27.3a**). Over 200 different genera were identified multiple times, but members of three genera, corynebacteria (*Actinobacteria*), propionibacteria (*Actinobacteria*), and staphylococci (*Firmicutes*) comprised more than 60% of the sequences (Figure 27.3b). Each microenvironment, however, had its own characteristic microbiota. Sebaceous areas have predominantly propionibacteria and staphylococci. The moist sites were dominated by corynebacteria and staphylococci. The dry environments have a mixed population dominated by *Betaproteobacteria*, corynebacteria, and *Flavobacteriales*. Significant trends in the composition of the microbiota that comprise each microenvironment are evident when data from all subjects are considered together (Figure 27.3b), but individuals showed large variations from the composite patterns.

Malassezia spp. are the most common fungi found on skin. At least five species of this yeast are typically found in healthy individuals. The lipophilic yeast *Pityrosporum ovale* is occasionally found on the scalp. In the absence of host resistance, as in patients with AIDS or in the absence of normal microflora, yeasts such as *Candida* and other fungi sometimes colonize the skin and cause serious infections.

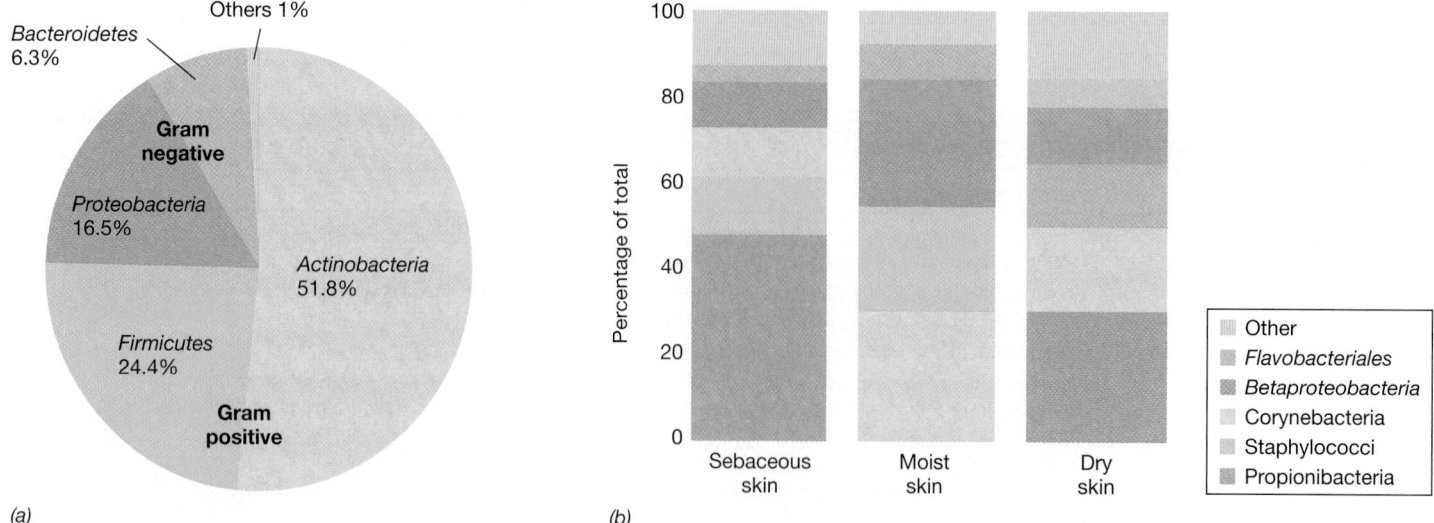

Figure 27.3 Normal skin microflora. *(a)* Analysis of the skin microbiome from 10 healthy human volunteers detected 19 bacterial phyla. Four phyla were predominant. *(b)* Composite populations of *Bacteria* from the same volunteers, divided according to sebaceous, moist, and dry skin microenvironments. Data are adapted from Grice et al., 2009, *Science 324:* 1190.

Although the resident microflora remains more or less constant, various environmental and host factors may influence its composition. (1) The *weather* may cause an increase in skin temperature and moisture, which increases the density of the skin microflora. (2) The *age* of the host has an effect; young children have a more varied microflora and carry more potentially pathogenic gram-negative *Bacteria* than do adults. (3) *Personal hygiene* influences the resident microflora; individuals with poor hygiene usually have higher microbial population densities on their skin. Organisms that do not survive on the skin generally succumb due to either the low moisture content or high organic acid content (low pH).

MiniQuiz

- Compare the populations of microorganisms in the three major skin microenvironments.
- Describe the properties of microorganisms that grow well on the skin.

27.3 Normal Microflora of the Oral Cavity

The oral cavity is a complex, heterogeneous microbial habitat. Saliva contains microbial nutrients, but it is not an especially good growth medium because the nutrients are present in low concentration and saliva contains antibacterial substances. For example, saliva contains lysozyme, an enzyme that cleaves glycosidic linkages in peptidoglycan of the bacterial cell wall, weakening the wall and causing cell lysis (Section 3.6). Lactoperoxidase, an enzyme in both milk and saliva, kills bacteria by a reaction in which singlet oxygen is generated (Section 5.18). Despite the activity of these antibacterial substances, food particles and cell debris provide high concentrations of nutrients near surfaces such as teeth and gums, creating favorable conditions for extensive local microbial growth, tissue damage, and disease.

The Teeth and Oral Microflora

The tooth consists of a mineral matrix of calcium phosphate crystals (enamel) surrounding living tooth tissue (dentin and pulp) (**Figure 27.4**). Bacteria found in the mouth during the first year of life (when teeth are absent) are predominantly aerotolerant anaerobes such as streptococci and lactobacilli. However, other bacteria, including some aerobes, are present. When the teeth appear, the balance of the microflora shift toward anaerobes that are specifically adapted to growth on surfaces of the teeth and in the gingival crevices.

Metagenomic analysis of the oral microflora indicates a complex population of microbes. Using samples acquired from a number of subjects, over 600 species have been identified. Among the most prevalent taxa are *Actinomyces* (4%), *Bacteroidetes* (2.4%), *Capnocytophaga* (2.6%), *Lachnospiraceae* (2.4%), *Lactobacillus* (3.4%), *Leptotrichia* (3.2%), *Neisseria* (3.2%), *Prevotella* (8.9%), *Selenomonas* (3.6%), *Streptococcus* (6.6%), and *Treponema* (7.9%). While most of these microorganisms have facultatively aerobic metabolisms, a few, such as *Bacteroidetes*, are obligately anaerobic and others, such as *Neisseria*, have an aerobic metabolism. Not all

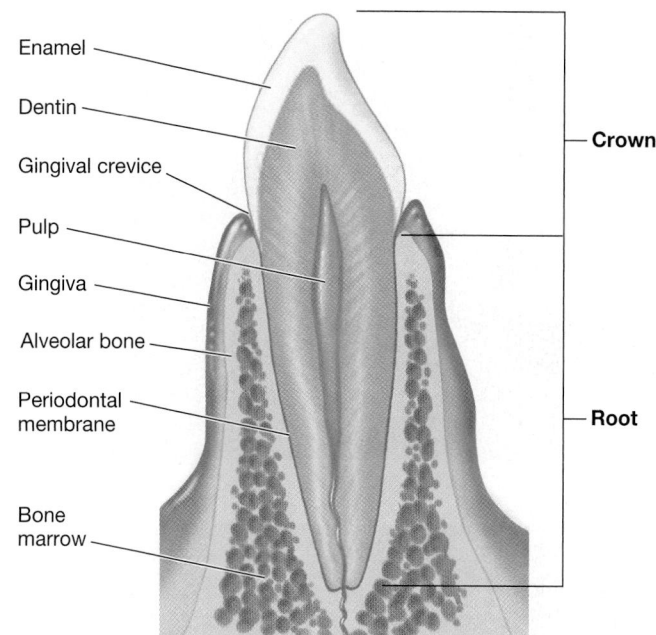

Figure 27.4 Section through a tooth. The diagram shows the tooth architecture and the surrounding tissues that anchor the tooth in the gum.

genera are similarly distributed in all subjects. This list of the most prevalent genera accounts for only about 48% of all species identified. Many other genera are present in even lower percentages, reflecting the highly complex microenvironments in the oral cavity.

Dental Plaque

Bacterial colonization of tooth surfaces begins with the attachment of single bacterial cells. Even on a freshly cleaned tooth surface, acidic glycoproteins from the saliva form a thin organic film several micrometers thick. This film provides an attachment site for bacterial microcolonies (**Figure 27.5**). Streptococci (primarily *Streptococcus sanguinis, S. sobrinus, S. mutans,* and *S. mitis*) can then colonize the glycoprotein film. Extensive growth of these organisms results in a thick bacterial layer called **dental plaque** (**Figures 27.6** and **27.7**).

If plaque continues to form, filamentous anaerobes such as *Fusobacterium* species begin to grow. The filamentous bacteria embed in the matrix formed by the streptococci and extend perpendicular to the tooth surface, making an ever-thicker bacterial layer. Associated with the filamentous bacteria are spirochetes such as *Borrelia* species, gram-positive rods, and gram-negative cocci. In heavy plaque, filamentous obligately anaerobic organisms such as *Actinomyces* may predominate. Thus, dental plaque is a mixed-culture biofilm (Section 23.4), consisting of a relatively thick layer of bacteria from several different genera as well as accumulated bacterial products.

The anaerobic nature of the oral microflora may seem surprising considering the intake of oxygen through the mouth. However, anoxia develops due to the metabolic activities of facultative bacteria growing on organic materials at the tooth surface. The plaque buildup produces a dense matrix that decreases oxygen

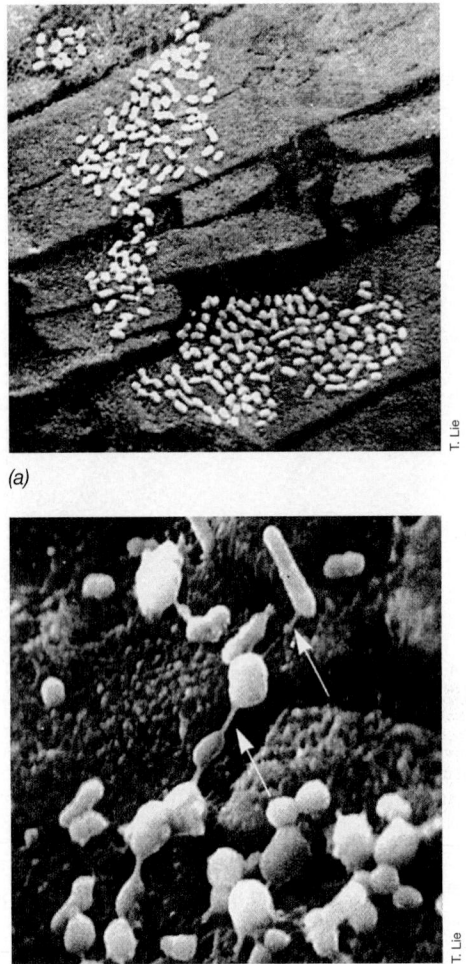

(a)

(b)

Figure 27.5 Microcolonies of bacteria. *(a)* The colonies are growing on a model tooth surface inserted into the mouth for 6 h. *(b)* Higher magnification of the preparation in part a. Note the diverse morphology of the organisms present and the slime layer (arrows) holding the organisms together.

diffusion to the tooth surface, forming an anoxic microenvironment. The microbial populations within dental plaque exist in a microenvironment of their own making and maintain themselves in the face of wide variations in the macroenvironmental conditions of the oral cavity.

Dental Caries

As dental plaque accumulates, the resident microflora produce locally high concentrations of organic acids that cause decalcification of the tooth enamel (Figure 27.4), resulting in **dental caries** (tooth decay). Tooth enamel is calcified tissue, and the ability of microorganisms to invade this tissue plays a role in the extent of dental caries. Thus, dental caries is an infectious disease.

The smooth, calcified surfaces of the teeth are relatively easy to clean and thus resist decay. The tooth surfaces in and near the gingival crevice, however, can retain food particles and are the sites where dental caries typically begins.

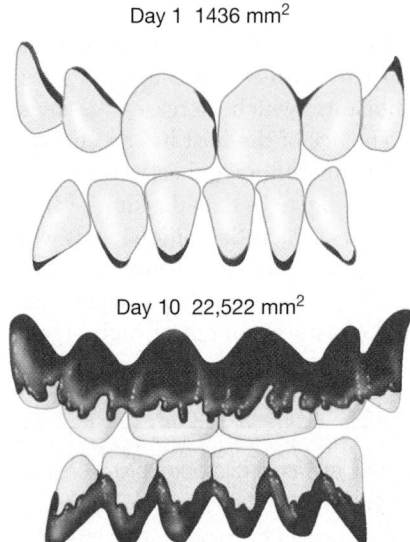

Day 1 1436 mm^2

Day 10 22,522 mm^2

Figure 27.6 Distribution of dental plaque. Plaque is revealed by use of a disclosing agent on unbrushed teeth after 1 day (top) and 10 days (bottom). The stained areas indicate plaque. Plaque buildup starts near the gum line, beginning directly adjacent to the mucous membranes of the gingiva.

Diets high in sucrose (table sugar) promote dental caries. Lactic acid bacteria ferment sugars to lactic acid. The lactic acid dissolves some of the calcium phosphate in localized areas, and proteolysis of the supporting matrix occurs through the action of bacterial proteolytic enzymes. Bacterial cells slowly penetrate further into the decomposing matrix.

Two bacteria implicated in dental caries are *Streptococcus sobrinus* and *Streptococcus mutans*, both lactic acid bacteria. *S. sobrinus* is probably the primary organism causing decay of smooth surfaces because of its specific affinity for salivary glycoproteins secreted onto smooth tooth surfaces (Figure 27.7). *S. mutans*, found predominantly in crevices and small fissures, produces dextran, a strongly adhesive polysaccharide that it uses to attach to tooth surfaces (**Figure 27.8**). *S. mutans* produces dextran through the activity of the enzyme dextransucrase but only in the presence of sucrose, the substrate for this enzyme:

$$n \text{ sucrose} \xrightarrow{\text{dextransucrase}} \text{dextran } (n \text{ glucose}) + n \text{ fructose}$$

Susceptibility to tooth decay varies and is affected by genetic traits in the individual as well as by diet and other extraneous factors. For example, sucrose, highly cariogenic because it is a substrate for dextransucrase, is part of the diet of most individuals in developed countries. Studies of the distribution of the cariogenic oral streptococci show a direct correlation between the presence of *S. mutans* and *S. sobrinus* and the extent of dental caries. In the United States and Western Europe, 80–90% of all individuals are infected by *S. mutans*, and dental caries is nearly universal. By contrast, *S. mutans* is absent from the plaque of Tanzanian children, and dental caries does not occur, presumably because sucrose is almost completely absent from their diets.

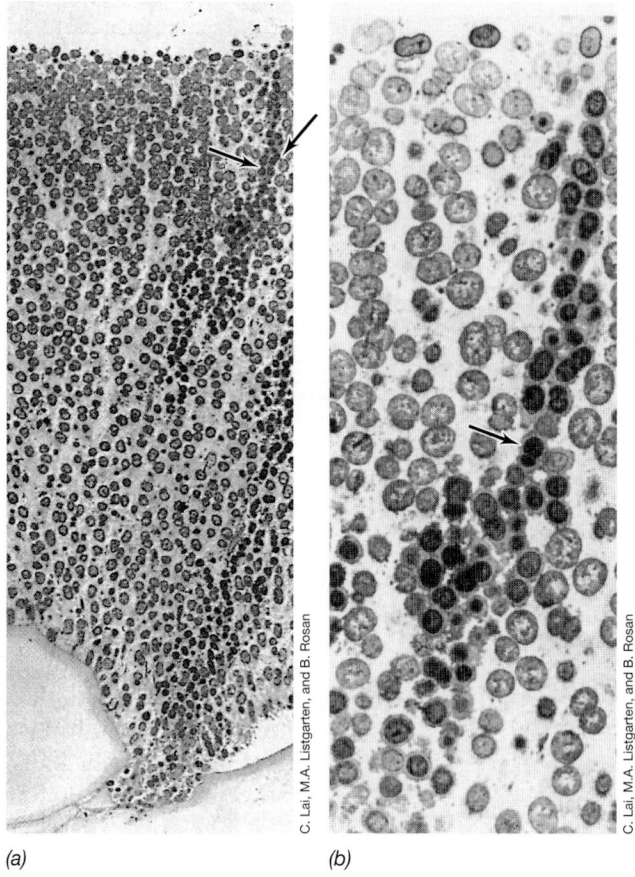

(a)   *(b)*

Figure 27.7 Dental plaque. The bottom of the photograph is the base of the plaque; the top is the portion exposed to the oral cavity. The thin-section scanning electron micrograph of the plaque layer is about 50 μm in depth. *(a)* Low-magnification electron micrograph. Organisms are predominantly streptococci. *Streptococcus sobrinus,* labeled by an antibody-microchemical technique, appears darker than the rest. *S. sobrinus* cells are seen as two distinct chains (arrows). *(b)* Higher-magnification electron micrograph showing the region with *S. sobrinus* cells (dark, arrow). Note the extensive slime layer surrounding the *S. sobrinus* cells. Individual cells are about 1 μm in diameter.

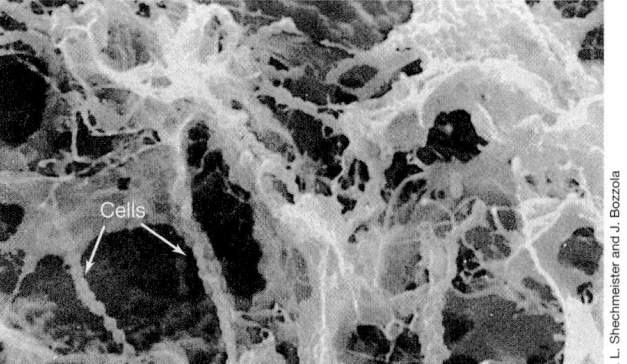

Figure 27.8 The cariogenic bacterium *Streptococcus mutans.* The sticky dextran material holds the cells together as filaments. The scanning electron micrograph shows individual cells about 1 μm in diameter.

Several strategies help control tooth decay in developed countries. For example, periodically brushing one's teeth removes the salivary glycoproteins from the smooth surfaces, inhibiting colonization by caries-producing bacteria. Another strategy is to include fluoride salts in municipal drinking water and toothpastes. The fluoride incorporated into the calcium phosphate crystals that make up the calcified matrix increases resistance to acid decalcification by colonizing bacteria.

Microorganisms in the mouth can also cause other infections. The areas along the periodontal membrane at or below the gingival crevice (periodontal pockets) (Figure 27.4) can be infected with microorganisms, causing inflammation of the gum tissues (gingivitis) leading to tissue- and bone-destroying periodontal disease. Some of the genera involved include fusiform bacteria (long, thin, gram-negative rods with tapering ends) such as the facultative aerobe *Capnocytophaga.* The aerobe *Rothia* and even strictly anaerobic methanogens such as *Methanobrevibacter* (*Archaea*) may also be present.

MiniQuiz

- Identify the potential microbial microenvironments in the oral cavity and the microorganisms that predominate in each.
- Is dental caries an infectious disease? Give at least one reason for your answer.
- Identify the contribution of the lactic acid bacteria to tooth decay.

27.4 Normal Microflora of the Gastrointestinal Tract

The human gastrointestinal tract consists of the stomach, small intestine, and large intestine (**Figure 27.9**). The gastrointestinal tract is responsible for digestion of food, absorption of nutrients, and the production of nutrients by the indigenous microflora. Starting with the stomach, the digestive tract is a column of nutrients mixed with microorganisms. The nutrients move one way through the column, encountering ever-changing populations of microorganisms. Here we examine the organisms as well as their functions and special properties.

Overall, about 10^{13} to 10^{14} microbial cells are present in the entire gastrointestinal tract. Our current view of the diversity and numbers of microorganisms that reside here has come from both standard culture methods and culture-independent molecular methods, such as microbial community analyses and metagenomics (Sections 12.6 and 22.6). In Section 25.8 we examine the microbial diversity of the human large intestine using the tools of molecular biology.

The Stomach

Because stomach fluids are highly acidic (about pH 2), the stomach is a chemical barrier to the entry of microorganisms into the gastrointestinal tract. However, microorganisms do populate this seemingly hostile environment. Studies using 16S rRNA sequences obtained from human stomach biopsies indicate that the stomach microbial population consists of several different phyla and a large

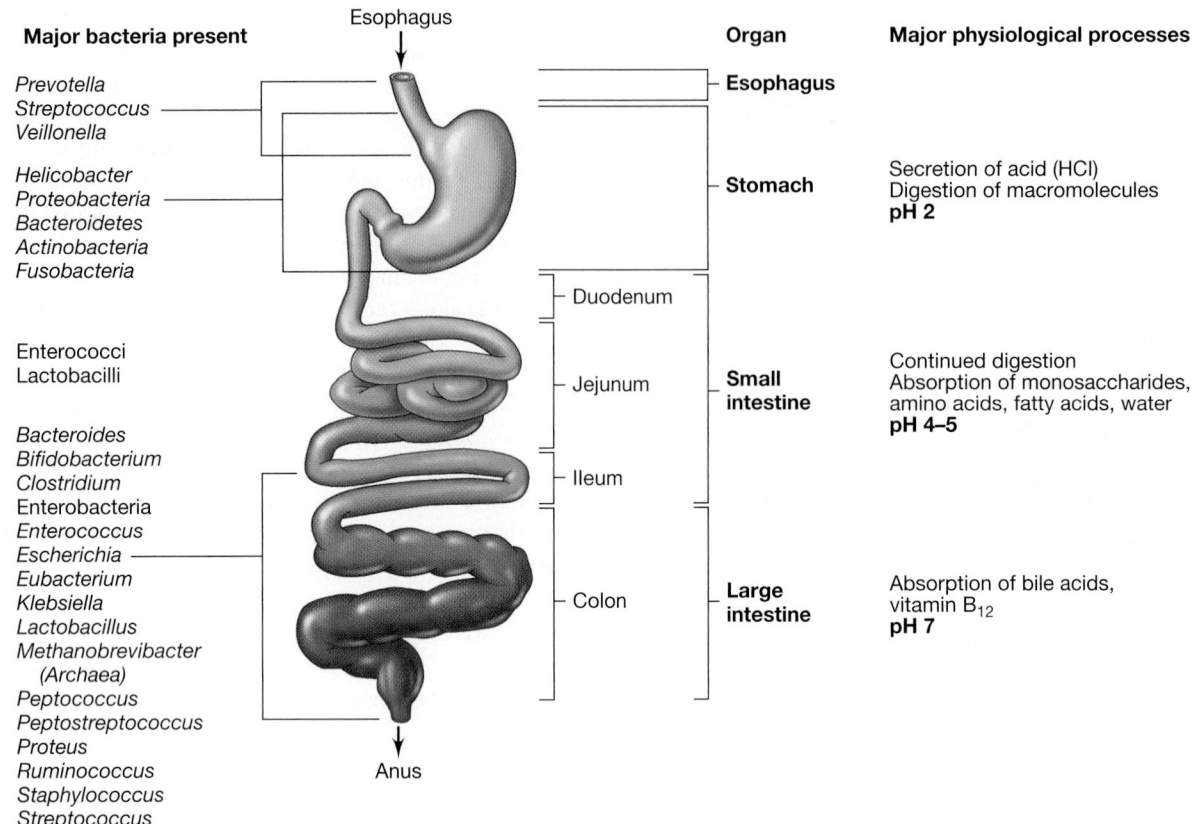

Major bacteria present		Organ	Major physiological processes
	Esophagus	Esophagus	
Prevotella *Streptococcus* *Veillonella*			
Helicobacter *Proteobacteria* *Bacteroidetes* *Actinobacteria* *Fusobacteria*		Stomach	Secretion of acid (HCl) Digestion of macromolecules **pH 2**
	Duodenum		
Enterococci Lactobacilli	Jejunum	Small intestine	Continued digestion Absorption of monosaccharides, amino acids, fatty acids, water **pH 4–5**
Bacteroides *Bifidobacterium* *Clostridium* Enterobacteria *Enterococcus* *Escherichia* *Eubacterium* *Klebsiella* *Lactobacillus* *Methanobrevibacter* (Archaea) *Peptococcus* *Peptostreptococcus* *Proteus* *Ruminococcus* *Staphylococcus* *Streptococcus*	Ileum Colon Anus	Large intestine	Absorption of bile acids, vitamin B$_{12}$ **pH 7**

Figure 27.9 The human gastrointestinal tract. The distribution of representative microorganisms often found in healthy adults. Not every individual harbors all of these microorganisms at any one time.

number of bacterial taxa. Individuals clearly have very different populations, but all contain several species of gram-positive bacteria as well as species of *Proteobacteria*, *Bacteroidetes*, *Actinobacteria*, and *Fusobacteria* (Figure 27.9). *Helicobacter pylori*, the most common single organism found, colonizes the stomach wall in many, but not all, individuals and can cause ulcers in susceptible hosts (↺ Section 33.10). Some of the bacteria that populate the stomach consist of organisms found in the oral cavity, introduced with the passage of food.

Distal to the stomach, the intestinal tract consists of the *small* intestine and the *large* intestine, each of which is divided into different anatomical segments. The composition of the intestinal microflora in humans varies considerably and is somewhat dependent on diet. For example, persons who consume a considerable amount of meat show higher numbers of *Bacteroides* and lower numbers of coliforms and lactic acid bacteria than do individuals with a vegetarian diet. Representative microorganisms found in the gastrointestinal tract are shown in Figure 27.9.

The Small Intestine

The small intestine has two distinct environments, in the *duodenum* and the *ileum*, which are connected by the *jejunum*. The duodenum, adjacent to the stomach, is fairly acidic and its normal microflora resembles that of the stomach. From the duodenum to the ileum, the pH gradually becomes less acidic and bacterial numbers increase. In the lower ileum, cell numbers of

10^5–10^7/gram of intestinal contents are common, even though the environment becomes progressively more anoxic. Fusiform anaerobic bacteria are typically present, attached to the intestinal wall at one end (**Figure 27.10**).

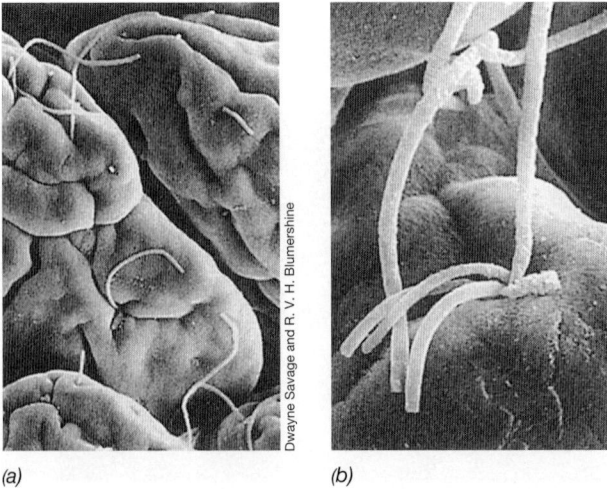

(a) *(b)*

Dwayne Savage and R. V. H. Blumershine

Figure 27.10 Scanning electron micrographs of the microbial community on the surface of the epithelial cells in the mouse ileum. *(a)* An overview at low magnification. Long, filamentous fusiform bacteria are apparent on the surface. *(b)* Higher magnification, showing several filaments attached at a single depression. The attachment is at the end of the filaments only. Individual cells are 10–15 μm long.

The Large Intestine

The ileum empties into the *cecum*, the connecting portion of the large intestine. The *colon* makes up the rest of the large intestine. In the colon, bacteria are present in enormous numbers. The colon is a fermentation vessel, and many bacteria live here, using nutrients derived from the digestion of food (Figure 27.9). Facultative aerobes such as *Escherichia coli* are present but in smaller numbers than other bacteria; total counts of facultative aerobes are less than 10^7/gram of intestinal contents. The facultative aerobes consume any remaining oxygen, making the large intestine strictly anoxic. This condition promotes growth of obligate anaerobes, including species of *Clostridium* and *Bacteroides*. The total number of obligate anaerobes in the colon is enormous. Bacterial counts of 10^{10} to 10^{11} cells/gram in distal gut and fecal contents are normal, with *Bacteroidetes* and gram-positive species accounting for greater than 99% of all bacteria. The archaeal methanogen *Methanobrevibacter smithii* can also be present in significant numbers (Figure 25.32). Protists are not found in the gastrointestinal tract of healthy humans, although various protists can cause opportunistic infections if ingested in contaminated food or water (Sections 35.6 and 36.12).

See Section 25.8 for a molecular snapshot of bacterial diversity in the human large intestine.

Functions and Products of Intestinal Microflora

Intestinal microorganisms carry out a wide variety of essential metabolic reactions that produce various compounds (**Table 27.2**). The composition of the intestinal microflora and the diet influence the type and amount of compounds produced. Among these products are vitamins B_{12} and K. These essential vitamins are not synthesized by humans, but are made by the intestinal microflora and absorbed from the gut. In addition, steroids, produced in the liver and released into the intestine from the gallbladder as bile acids, are modified in the intestine by the microflora; the modified bioactive steroid compounds are then absorbed from the gut.

Other products generated by the activities of fermentative bacteria and methanogens include gas (flatus) and the odor-producing substances listed in Table 27.2. Normal adults expel several hundred milliliters of gas from the intestines each day, of which about half is nitrogen (N_2) from swallowed air. Some foods metabolized by fermentative bacteria in the intestines result in the production of hydrogen (H_2) and carbon dioxide (CO_2). Methanogens, found in the intestines of over one-third of normal adults, convert H_2 and CO_2 produced by fermentative bacteria to methane (CH_4). The methanogens in the rumen of cattle produce significant amounts of methane, up to a quarter of the total global production (Section 25.7).

During the passage of food through the gastrointestinal tract, water is absorbed from the digested material, which gradually becomes more concentrated and is converted to feces. Bacteria make up about one-third of the weight of fecal matter. Organisms living in the lumen of the large intestine are continuously displaced downward by the flow of material, and bacteria that are lost are continuously replaced by new growth. Thus, the large intestine resembles the continuous culture properties of a chemostat (Section 5.8). The time needed for passage of material through the complete gastrointestinal tract is about 24 h in humans; the growth rate of bacteria in the lumen is one to two doublings per day. In humans, about 10^{13} bacterial cells are shed per day in feces.

Changing the Normal Microflora

When an antibiotic is taken orally, it inhibits the growth of the normal flora as well as pathogens, leading to the loss of antibiotic-susceptible bacteria in the intestinal tract. This is often signaled by loose feces or diarrhea. In the absence of the full complement of normal flora, opportunistic microorganisms such as antibiotic-resistant *Staphylococcus*, *Proteus*, *Clostridium difficile*, or the yeast *Candida albicans* can become established. The establishment of these opportunistic pathogens can lead to a harmful alteration in digestive function or even to disease. For example, antibiotic treatment allows microorganisms such as *C. difficile* that are less susceptible to antibiotics to grow without competition from the normal flora, causing infection and colitis. When antibiotic therapy is ended, however, the normal intestinal flora is quickly reestablished in adults. To speed the establishment of a competitive flora, recolonization of the gut by desired species can be accomplished by administration of **probiotics**, live cultures of intestinal bacteria that, when administered to a host, may confer a health benefit. Rapid recolonization of the gut may reestablish a competitive local flora and provide desirable microbial metabolic products.

Table 27.2 *Biochemical/metabolic contributions of intestinal microorganisms*

Process	Product
Vitamin synthesis	Thiamine, riboflavin, pyridoxine, B_{12}, K
Gas production	CO_2, CH_4, H_2
Odor production	H_2S, NH_3, amines, indole, skatole, butyric acid
Organic acid production	Acetic, propionic, butyric acids
Glycosidase reactions	β-Glucuronidase, β-galactosidase, β-glucosidase, α-glucosidase, α-galactosidase
Steroid metabolism (bile acids)	Esterified, dehydroxylated, oxidized, or reduced steroids

MiniQuiz

- How does the human digestive tract differ from that of a ruminant, such as a cow?
- Why might the small intestine be more suitable for growth of facultative aerobes than the large intestine?
- Identify several essential compounds made by indigenous intestinal microorganisms. What would happen if all microorganisms were completely eliminated from the intestine by the use of antibiotics?

Bacteriophages and Infectious Diseases

Bacterial pathogens remain a major threat to human health, even in the twenty-first century. Yet some or most of the harm caused by certain pathogens is due to features encoded not by the genome of the pathogen itself but instead by a virus infecting the pathogen.

Several pathogenic bacteria produce toxins as virulence factors. If these virulence factors are proteins, as many toxins are, they can be encoded by chromosomal DNA (e.g., as pathogenicity islands, ⮌ Section 12.13), extrachromosomal elements (e.g., plasmids), or bacteriophages (⮌ Sections 9.9 and 9.10; Figure 9.16). In the latter case, bacterial DNA from a pathogen that contains a functional virulence gene (i.e., encoding a virulence factor) can be packed into a bacteriophage during the lytic cycle. The resulting entity, called a transducing particle, can then infect a strain of the pathogen lacking the virulence gene during the transduction event (⮌ Section 10.8; Figure 10.14).

If these genes are then expressed, the process is called lysogenic (or phage) conversion, because it changes the phenotype of the bacterial cell. In this way, bacteriophages can turn nonpathogenic organisms into dangerous pathogens by simply infecting the pathogen and making available to the pathogen genes encoding new properties.

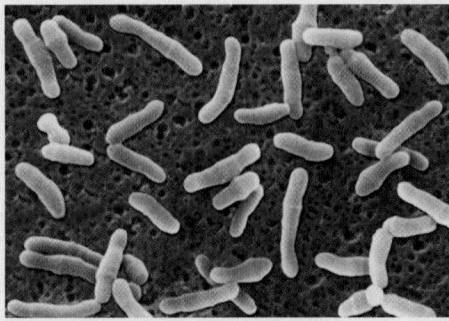

Figure 1 Scanning electron micrograph of *Corynebacterium diphtheriae*.

Several common infectious diseases have been linked to phage conversion. The nature of the virulence factors, their mode of action, and the resulting symptoms of disease vary widely. Here are a few examples:

- Pathogenic strains of *Corynebacterium diphtheriae* (Figure 1; ⮌ Section 33.3) produce diphtheria toxin (Figure 27.21), which eventually leads to death due to suffocation.
- The cholera toxin (Figure 27.24) produced by the bacterium *Vibrio cholerae* (⮌ Section 35.5) causes the loss of large amounts of fluid from intestinal tissues, which in turn can cause death from dehydration and loss of electrolytes.
- *Streptococcus pyogenes* (⮌ Section 33.2) produces certain exotoxins upon phage conversion. These exotoxins direct the course of disease symptoms toward either streptococcal pharyngitis or scarlet fever.
- *Staphylococcus aureus* produces heat-stable enterotoxins (⮌ Section 36.6) that cause a rather severe and common food poisoning.
- *Clostridium botulinum* (⮌ Section 36.7) produces an extremely poisonous neurotoxin, usually transmitted in contaminated foods, that causes death by suffocation.

From an ecological and survival standpoint, lysogenic conversion is advantageous for both the pathogen and its bacteriophage. In most cases the host's immune response can kill the pathogen, which would prove a dead end for the bacteriophage. But if the pathogen's virulence is increased by the conversion event, the survival of the pathogen as host to the phage is more probable.

Knowledge of the biology of virulence factors contributes to the development of treatments and even prevention of infectious diseases. A long-established example is the diphtheria vaccine (⮌ Table 28.4), which is directed against the diphtheria toxin and not against components of the bacterium. Therefore, vaccinated individuals are optimally protected against the bacterium itself. Pathogenic strains of *Corynebacterium diphtheriae* can still initiate an infection, but the onset of full-blown disease is prevented.

27.5 Normal Microflora of Other Body Regions

Each individual mucous membrane supports the growth of a specialized group of microorganisms. These organisms are part of the normal local environment and are characteristic of healthy tissue. In many cases, pathogenic microorganisms cannot colonize mucous membranes because of the competitive effects of the normal flora. Here we discuss two mucosal environments and their resident microorganisms.

Respiratory Tract

The anatomy of the respiratory tract is shown in **Figure 27.11**. In the **upper respiratory tract** (nasopharynx, oral cavity, larynx, and pharynx), microorganisms live in areas bathed with the secretions of the mucous membranes. Bacteria continually enter the upper respiratory tract from the air during breathing, but most are trapped in the mucus of the nasal and oral passages and expelled with the nasal secretions, or swallowed. A restricted group of microorganisms, however, colonizes respiratory mucosal surfaces in all individuals. The microorganisms most commonly found are staphylococci, streptococci, diphtheroid bacilli, and gram-negative cocci. Even potential pathogens such as *Staphylococcus aureus* and *Streptococcus pneumoniae* are often part of the normal flora of the nasopharynx of healthy individuals (Table 27.1). These individuals are *carriers* of the pathogens but do not normally develop disease, presumably

because the other resident microorganisms compete successfully for nutritional and metabolic resources and limit pathogen activities. The innate immune system (↩ Section 28.2) and components of the adaptive immune system such as IgA antibodies (↩ Section 29.7) are particularly active at mucosal surfaces and may also inhibit growth and invasion by the resident pathogens.

The **lower respiratory tract** (trachea, bronchi, and lungs) has no resident microflora in healthy adults, despite the large number of organisms potentially able to reach this region during normal breathing. Dust particles, which are fairly large, settle out in the upper respiratory tract. As the air passes into the lower respiratory tract, the flow rate decreases markedly, and organisms settle onto the walls of the respiratory passages. The walls of the entire respiratory tract are lined with ciliated epithelial cells, and the cilia, beating upward, push bacteria and other particulate matter toward the upper respiratory tract where they are then expelled in the saliva and nasal secretions or are swallowed. Only particles smaller than about 10 μm in diameter reach the lungs. Nevertheless, some pathogens can reach these locations and cause disease, most notably pneumonias caused by certain bacteria or viruses (↩ Sections 33.2 and 33.6).

Urogenital Tract

In healthy male and female urogenital tracts (**Figure 27.12**), the kidneys and bladder itself are sterile, but the epithelial cells lining the distal urethra are colonized by facultatively aerobic gram-negative rods and cocci (Table 27.1). Potential pathogens such as *Escherichia coli* and *Proteus mirabilis*, normally present in small numbers in the body or local environment, can multiply in the urethra and become pathogenic under altered conditions such as changes in pH. Such organisms are a frequent cause of urinary tract infections, especially in women.

The vagina of the adult female is weakly acidic (pH <5) and contains significant amounts of glycogen. *Lactobacillus acidophilus*, a resident organism in the vagina, ferments the polysaccharide glycogen, producing lactic acid that maintains a local acidic environment (Figure 27.12b). Other organisms, such as yeasts (*Torulopsis* and *Candida* species), streptococci, and *E. coli*, may also be present. Before puberty, the female vagina is neutral and does not produce glycogen, *L. acidophilus* is absent, and the flora consists predominantly of staphylococci, streptococci, diphtheroids, and *E. coli*. After menopause, glycogen production ceases, the pH rises, and the flora again resembles that found before puberty.

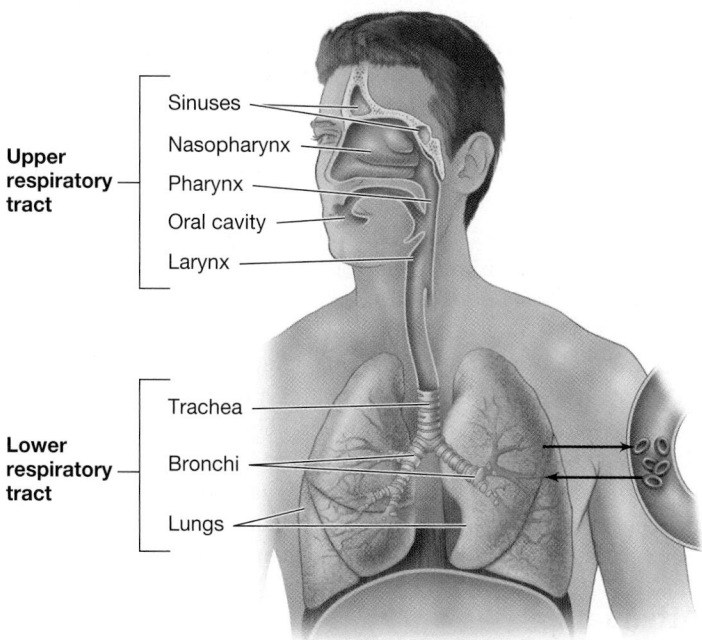

Figure 27.11 The respiratory tract. In healthy individuals the upper respiratory tract has a large variety and number of microorganisms. By contrast, the lower respiratory tract in a healthy person has few if any microorganisms. Additional aspects of the human respiratory tract are shown in Figure 27.26.

Upper respiratory tract: Sinuses, Nasopharynx, Pharynx, Oral cavity, Larynx

Lower respiratory tract: Trachea, Bronchi, Lungs

MiniQuiz

- Potential pathogens are often found in the normal flora of the upper respiratory tract. Why do they not cause disease in most cases?

- Why are upper respiratory tract infections much more common than lower respiratory tract infections?

- What is the importance of *Lactobacillus* found in the urogenital tract of healthy adult women?

Female

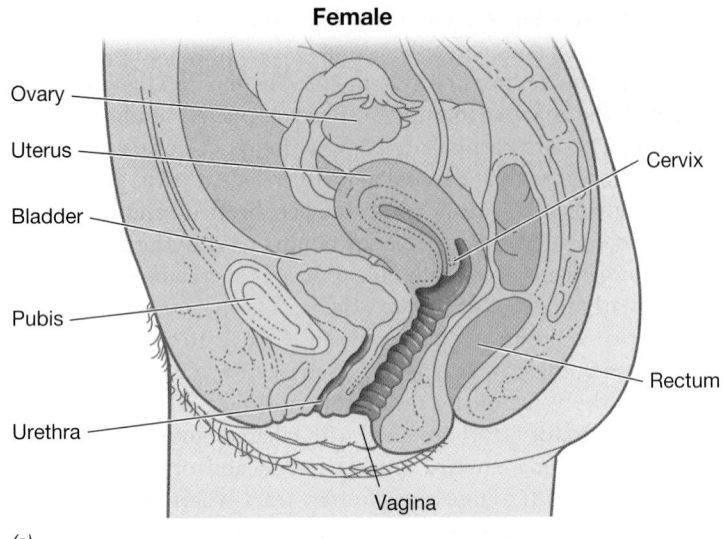

(a)

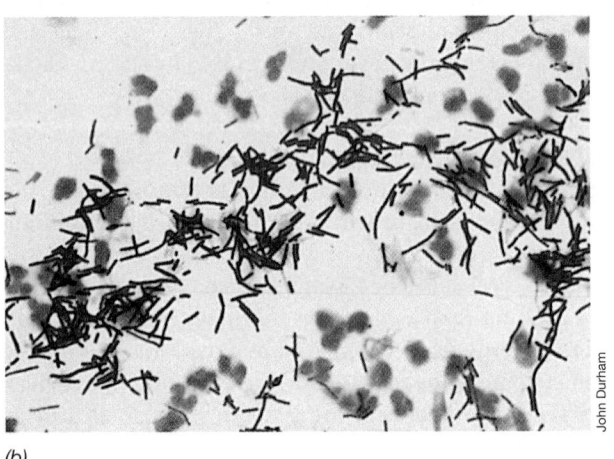

(b)

John Durham

Figure 27.12 Microbial growth in the genitourinary tract. *(a)* The genitourinary tracts of the human female and male, showing regions (red) where microorganisms often grow. The upper regions of the genitourinary tracts of both males and females are sterile in healthy individuals. *(b)* Gram stain of *Lactobacillus acidophilus*, the predominant organism in the vagina of women between the onset of puberty and the end of menopause. Individual rod-shaped cells are 3–4 μm long.

Male

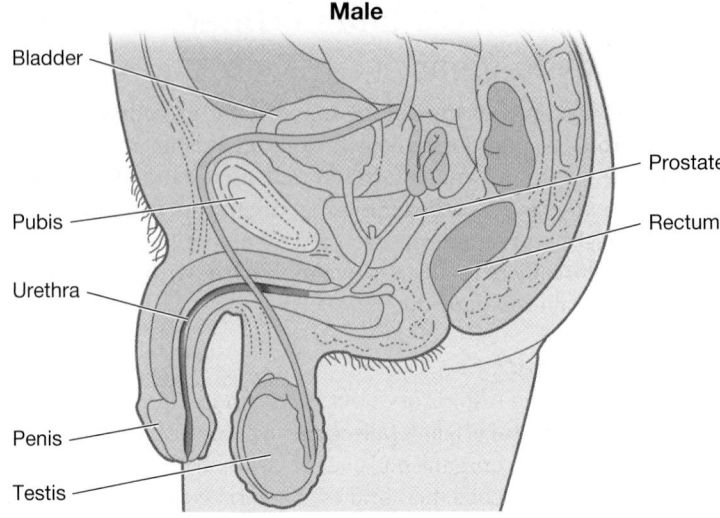

Ⅱ Microbial Virulence and Pathogenesis

Pathogenic microorganisms use several strategies to establish *virulence*, the relative ability of a pathogen to harm the host. Here we examine mechanisms of *microbial pathogenesis*, the process by which microorganisms cause disease. Microbial pathogenesis begins with exposure and adherence of the microorganisms to host cells, followed by invasion, colonization, and infection, or growth (**Figure 27.13**). Unchecked growth of the pathogen can result in host damage and disease.

27.6 Measuring Virulence

Here we discuss an objective method used to measure virulence. In the following sections, we provide examples of microbial virulence, highlighting the invasion and colonization factors that contribute to the infections and diseases caused by particular pathogens.

Virulence

The virulence of a pathogen can be estimated from experimental studies of the LD_{50} (lethal dose$_{50}$), the dose of an agent that kills 50% of the animals in a test group. Highly virulent pathogens fre-

quently show little difference in the number of cells required to kill 100% of the test group as compared with the number required to kill 50%. This is illustrated in **Figure 27.14** for experimental infections in mice. Only a few cells of virulent strains of *Streptococcus pneumoniae* are required to establish a fatal infection and kill all mice in a test population. As a result, the LD_{50} for *S. pneumoniae* in mice is hard to determine. By contrast, the LD_{50} for *Salmonella enterica* serovar Typhimurium, a much less virulent pathogen, is much higher. The number of cells of *S. enterica* ser. Typhimurium required to kill 100% of the population is more than 100 times greater than the number of cells needed to reach the LD_{50} and is proportionally related to the number of *Salmonella* cells introduced into the test mice.

Attenuation

Attenuation is the decrease or loss of virulence of a pathogen. When pathogens are kept in laboratory culture rather than isolated from diseased animals, their virulence is often decreased or even completely lost. Such organisms are said to be *attenuated*. Attenuation probably occurs because nonvirulent or weakly virulent mutants grow faster than virulent strains in laboratory media; after successive transfers to fresh media, such mutants are

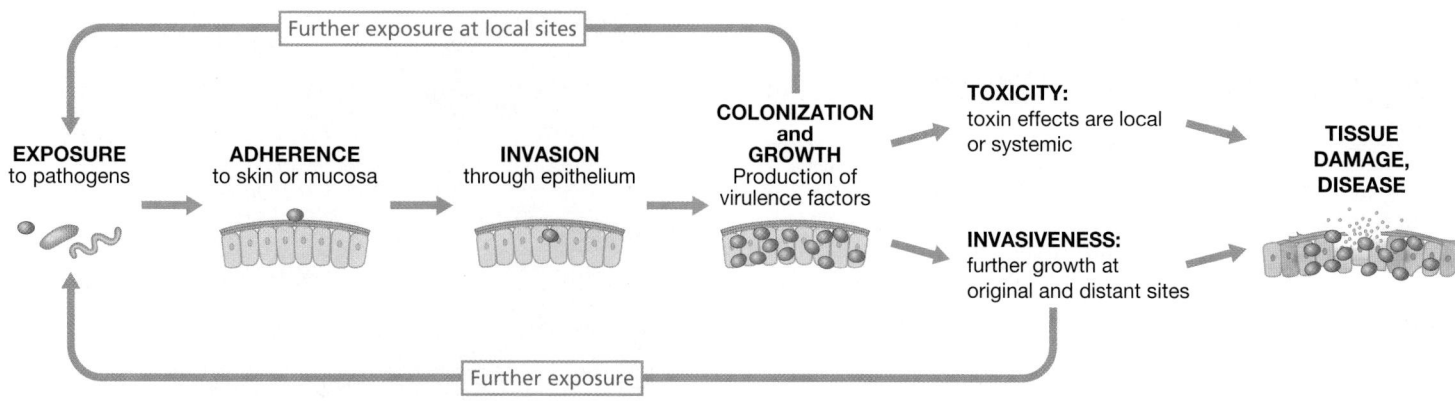

Figure 27.13 Microorganisms and mechanisms of pathogenesis. Following exposure to a pathogenic microorganism, subsequent pathogen-directed events can culminate in disease.

selectively favored. If an attenuated culture is reinoculated into an animal, the organism may regain its original virulence, especially with continued *in vivo* passage, but in many cases the loss of virulence is permanent. Attenuated strains are often used for production of vaccines, especially viral vaccines. For example, measles, mumps, and rubella vaccines, and animal vaccines for rabies, consist of attenuated viruses. An attenuated strain of *Mycobacterium bovis* is used as a vaccine for tuberculosis (⟲ Section 28.7).

MiniQuiz

- How can the LD_{50} test be used to define virulence of a pathogen?
- What circumstances can contribute to attenuation of a pathogen?

27.7 Entry of the Pathogen into the Host—Adherence

A pathogen must usually gain access to host tissues and multiply to cause disease. In most cases, this requires that the organisms penetrate the skin or mucous membranes, surfaces that are normally microbial barriers.

Most microbial infections begin at breaks or wounds in the skin or on the mucous membranes of the respiratory, digestive, or genitourinary tract. Bacteria or viruses able to initiate infection often adhere to epithelial cells through specific interactions between molecules on the pathogen and molecules on the host cell (**Figure 27.15**). In addition, pathogens often adhere to each other, forming biofilms.

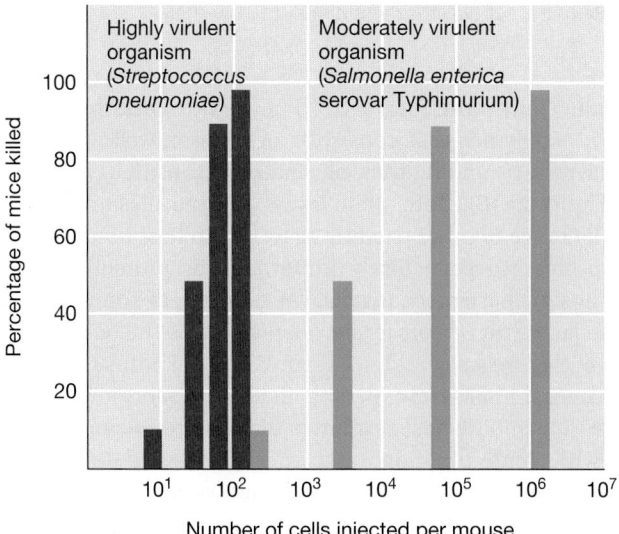

Figure 27.14 Microbial virulence. Differences in microbial virulence demonstrated by the number of cells of *Streptococcus pneumoniae* and *Salmonella enterica* serovar Typhimurium required to kill mice.

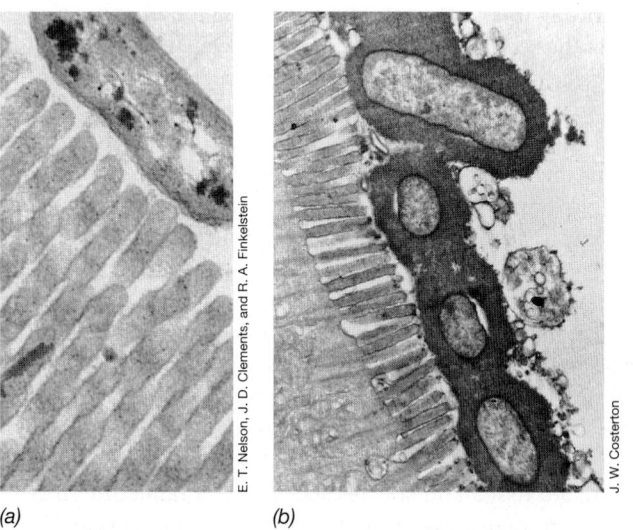

Figure 27.15 Adherence of pathogens to animal tissues. *(a)* Transmission electron micrograph of a thin section of *Vibrio cholerae* adhering to the brush border of rabbit microvilli in the intestine. This organism has no capsule. *(b)* Enteropathogenic *Escherichia coli* in a fatal model of infection in the newborn calf. The bacterial cells are attached to the brush border of calf intestinal microvilli through their distinct capsule. The rods are about 0.5 μm in diameter.

Table 27.3 *Major adherence factors used to facilitate attachment of microbial pathogens to host tissues*[a]

Factor	Example
Capsule/slime layer (Figures 27.5, 27.15, 27.16, 27.17)	Pathogenic *Escherichia coli*—capsule promotes adherence to the brush border of intestinal microvilli
	Streptococcus mutans—dextran slime layer promotes binding to tooth surfaces
Adherence proteins	*Streptococcus pyogenes*—M protein on the cell binds to receptors on respiratory mucosa
	Neisseria gonorrhoeae—Opa protein on the cell binds to CD66 receptors on epithelium
Lipoteichoic acid (🔗 Figure 3.18)	*Streptococcus pyogenes*—lipoteichoic acid facilitates binding to respiratory mucosal receptor (along with M protein)
Fimbriae (pili) (Figure 27.18)	*Neisseria gonorrhoeae*—pili facilitate binding to epithelium
	Salmonella species—type I fimbriae facilitate binding to epithelium of small intestine
	Pathogenic *Escherichia coli*—fimbrial colonization factor antigens (CFAs) facilitate binding to epithelium of small intestine

[a]Most receptor sites on host tissues are glycoproteins or complex lipids such as gangliosides or globosides.

Most pathogens selectively adhere to particular types of cells localized in a particular region of the body. For example, *Neisseria gonorrhoeae*, the pathogen that causes the sexually transmitted disease gonorrhea, adheres to mucosal epithelial cells in the genitourinary tract, eye, rectum, and throat. *N. gonorrhoeae* has a surface protein called Opa (*o*pacity *a*ssociated *p*rotein) that binds specifically to a host protein called CD66 found only on the surface of these cells (**Table 27.3**). Thus *N. gonorrhoeae* interacts with host cells by binding a specific cell surface protein.

Streptococcus pyogenes utilizes two cell-wall-associated molecules, the M protein and lipoteichoic acid, to form microfibrils that facilitate attachment to host cells (Table 27.3). M protein is also responsible for resistance to phagocytosis by neutrophils, cells important in antibacterial resistance (🔗 Sections 28.2 and 33.2).

Influenza virus occurs in nature as an avian pathogen, targeting the lung mucosal cells. Most influenza viruses remain in birds, but occasionally the glycoprotein responsible for adherence to cells, the hemagglutinin, mutates, allowing the virus to adhere to respiratory mucosal cells in other species such as pigs or humans, sometimes causing widespread infections. We will investigate tissue and species specificity of influenza virus more fully when we discuss respiratory diseases (🔗 Section 33.8).

Some macromolecules responsible for bacterial adherence are not covalently attached to the bacteria. These surface molecules are collectively known as a **glycocalyx**, a polymer secreted by a microorganism that coats the surface of the microorganism. These are usually polysaccharides, or, in the case of *Bacillus anthracis*, a polymer of D-glutamic acid, synthesized and secreted by the bacteria (🔗 Section 3.9). A loose network of polymers

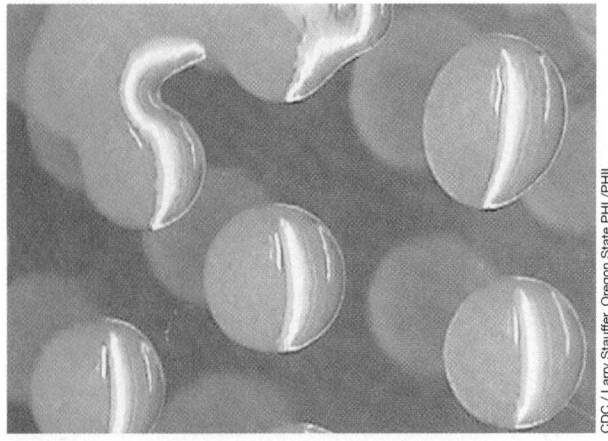

(a)

CDC / Larry Stauffer, Oregon State PHL/PHIL

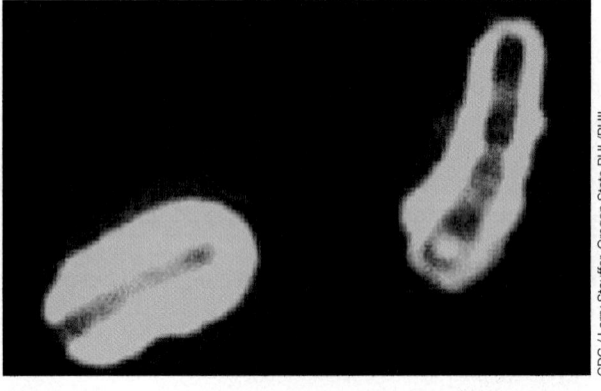

(b)

Figure 27.16 *Bacillus anthracis* **capsules.** *(a)* Capsules of *B. anthracis* on bicarbonate agar media. Encapsulated colonies are typically very large and mucoid in appearance. The individual encapsulated colonies are 0.5 cm in diameter. *(b)* Direct immunofluorescent stain of *B. anthracis* capsules. Antibodies coupled to fluorescein isothiocyanate (FITC) stain the capsule bright green, indicating that the capsule extends up to 1 μm from the cell, which is about 0.5 μm in diameter.

CDC / Larry Stauffer, Oregon State PHL/PHIL

extending outward from a cell is called a **slime layer** (Figure 27.5b). A polymer coat consisting of a dense, well-defined polymer layer surrounding the cell is called a **capsule** (Figure 27.15b and **Figure 27.16**). Both slime layers and capsules are important for adherence to other bacteria as well as to host tissues.

Capsules are particularly important for protecting bacteria from host defense mechanisms. For example, the only known virulence factor for *Streptococcus pneumoniae* is the polysaccharide capsule (**Figure 27.17** and 🔗 Figure 33.3b). Encapsulated strains of *S. pneumoniae* grow in lung tissues in enormous numbers, where they initiate host responses that lead to pneumonia, interfere with lung function, and cause extensive host damage. Nonencapsulated strains are less pathogenic; they are quickly and efficiently ingested and destroyed by phagocytes, white blood cells that ingest and kill bacteria by a process called *phagocytosis*. Thus, *S. pneumoniae* capsules are essential for pathogenicity; the capsules defeat a major defense mechanism used by the host to prevent invasion (🔗 Section 28.2).

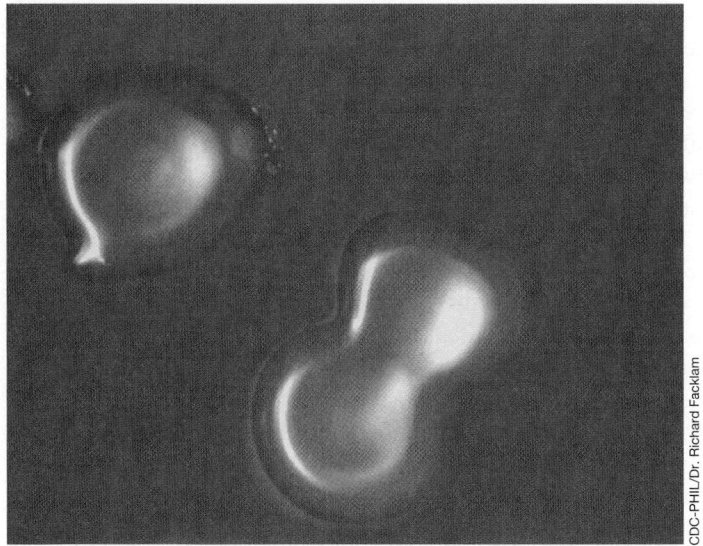

Figure 27.17 Capsule in *Streptococcus pneumoniae*. Colonies of *S. pneumoniae* strains with capsules show a mucoid morphology with a sunken center when grown on rich media such as the blood agar shown here. The colonies are about 2–3 mm in diameter. The mucoid appearance is due to the polysaccharide capsule.

Fimbriae and pili (Section 3.9) are bacterial cell surface protein structures that may function in the attachment process. For instance, the pili of *Neisseria gonorrhoeae* play a key role in attachment to the urogenital epithelium, and fimbriated strains of *Escherichia coli* (**Figure 27.18**) are more frequent causes of urinary tract infections than strains lacking fimbriae.

Among the best-characterized fimbriae are the type I fimbriae of enteric bacteria (*Escherichia, Klebsiella, Salmonella,* and *Shigella*). Type I fimbriae are uniformly distributed on the surface of cells. Pili are typically longer than fimbriae, with fewer pili found on the cell surface. Both pili and fimbriae function by binding host cell surface glycoproteins, initiating attachment. Flagella can also increase adherence to host cells.

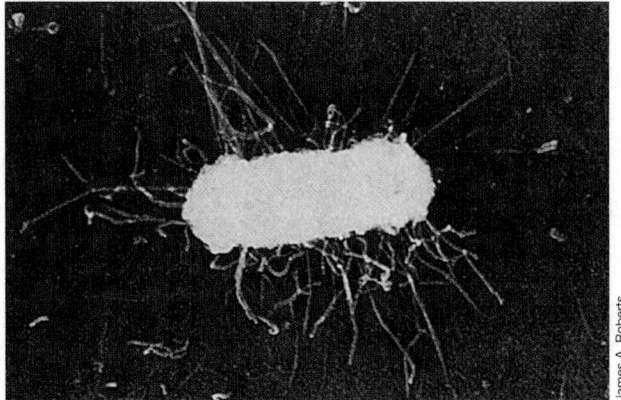

Figure 27.18 Fimbriae. Shadow-cast electron micrograph of the bacterium *Escherichia coli* showing type P fimbriae, which resemble type I fimbriae but are somewhat longer. The cell is about 0.5 μm in diameter.

Studies of diarrhea caused by enterotoxic strains of *E. coli* provide evidence for specific interactions between the mucosal epithelium and pathogens. Most strains of *E. coli* are normal, nonpathogenic inhabitants of the cecum and the colon (Figure 27.9). Several strains of *E. coli* are usually present in the body at the same time, and large numbers of these nonpathogens routinely pass through the body and are eliminated in feces. However, enterotoxic strains of *E. coli* contain genes encoding fimbrial CFA (colonization factor antigens); these proteins adhere specifically to cells in the small intestine. From here, they colonize and produce enterotoxins that cause diarrhea as well as other illnesses (Section 27.10). Nonpathogenic strains of *E. coli* seldom express CFA proteins. Some major factors important in microbial adherence are shown in Table 27.3.

MiniQuiz

- What is the difference between a slime layer and a capsule?
- How do Opa proteins on *Neisseria gonorrhoeae* and fimbrial CFA proteins on *Escherichia coli* influence adherence to mucosal tissues?

27.8 Colonization and Infection

The initial inoculum of a pathogen is usually too small to cause host damage even if a pathogen gains access to tissues. First the pathogen must multiply and colonize in the tissue. To do so the pathogen must find appropriate nutrients and environmental conditions to grow and cause infection in the host. The availability of microbial nutrients is most important, but temperature, pH, and the presence or absence of oxygen also affect pathogen growth.

Nutrient Availability

Not all vitamins and growth factors are in adequate supply in all tissues at all times, even in a vertebrate host. Soluble nutrients such as sugars, amino acids, organic acids, and growth factors are limited, favoring organisms able to use host-specific nutrients. *Brucella abortus*, for example, grows very slowly in most tissues of infected cattle, but grows very rapidly in the placenta. The placenta is the only tissue that contains high concentrations of erythritol, a sugar that is readily metabolized by *B. abortus*. The erythritol enhances *B. abortus* growth, causing abortion in cattle (see Table 27.6).

Trace elements may also be in short supply in host tissues and their availability can influence establishment of the pathogen. For example, iron is a growth-limiting micronutrient that influences microbial growth (Section 4.1). In the host, iron-specific host proteins called *transferrin* and *lactoferrin* have a very high affinity for iron and essentially sequester all iron. Because they limit the free iron available in host tissues, iron deficiency limits infection by many pathogens; a dietary iron supplement given to an infected animal greatly increases the virulence of some pathogens.

As we noted in Section 4.1, many bacteria produce iron-chelating compounds called *siderophores* that help them obtain iron from the environment. Siderophores from some pathogens are so efficient that they remove iron from transferrin and lactoferrin. For example, *aerobactin*, a plasmid-encoded siderophore produced by certain strains of *Escherichia coli*, readily removes iron bound to transferrin. Likewise, *Neisseria* species produce a transferrin-specific receptor that binds to and removes iron from transferrin. *Salmonella* species also have siderophores in their arsenal of virulence factors.

Localization in the Body

Some pathogens remain localized after initial entry, multiplying and producing a discrete focus of infection such as the boil that may arise from *Staphylococcus* skin infections (Section 33.9). Other pathogens may enter the lymphatic vessels and move to the lymph nodes, where they may be contained by the immune system. If bacterial growth in tissues occurs, some of the organisms may be shed into the bloodstream and be distributed to distant parts of the body, a condition called **bacteremia**. Bacteremia is the presence of bacteria in the bloodstream. Spread of the pathogen through the blood and lymph systems can also result in a bloodborne systemic infection called **septicemia**, and the organism may spread to other tissues. Septicemia may lead to massive inflammation, culminating in septic shock and rapid death, as we discuss in Section 28.5. Bacteremia and septicemia almost always start as a local infection in a specific organ such as the intestine, kidney, or lung.

MiniQuiz

- Why are colonization and growth necessary for the success of most pathogens?
- Identify host factors that limit or accelerate colonization and growth of a microorganism at selected local sites.

27.9 Invasion

The virulence of a pathogen is due to the toxicity and invasiveness of the pathogen, resulting in host damage. **Invasion** is the ability of a pathogen to enter into host cells or tissues, spread, and cause disease. Most pathogens must penetrate the epithelium to initiate disease. Growth may also begin on intact mucosal surfaces, especially if the normal flora is altered or eliminated, for example, by antibiotic therapy. Pathogen growth may also be established at sites distant from the original point of entry. Access to distant, usually interior, sites is through the blood or lymphatic circulatory system.

Toxins and Virulence Factors

A microorganism may be able to produce disease through a variety of toxins and virulence factors. **Toxicity** is the ability of an organism to cause disease by means of a preformed toxin that inhibits host cell function or kills host cells. For example, the dis-

ease tetanus is caused by an exotoxin produced by *Clostridium tetani* (exotoxins are proteins, Section 27.10). *C. tetani* cells rarely leave the wound where they were first introduced, growing relatively slowly at the wound site. Yet *C. tetani* can cause serious disease because tetanus toxin moves to distant parts of the body, initiating irreversible muscle contraction and often death of the host.

In addition to toxins, which we will discuss in detail in the following sections, many pathogens produce virulence factors that indirectly or directly enhance invasiveness by promoting pathogen colonization and growth. Many of these virulence factors are enzymes. For example, streptococci, staphylococci, and certain clostridia produce hyaluronidase (**Table 27.4**), an enzyme that promotes spreading of organisms in tissues by breaking down the polysaccharide hyaluronic acid, an intercellular cement in animals. Hyaluronidase digests the intercellular matrix, enabling these organisms to spread from an initial infection site. Similarly, the clostridia that cause gas gangrene produce collagenase, or κ-toxin (Table 27.4), which breaks down the tissue-supporting collagen network, enabling these organisms to spread through the body. Many pathogenic streptococci and staphylococci also produce proteases, nucleases, and lipases that degrade host proteins, nucleic acids, and lipids, respectively.

Fibrin, Clots, and Virulence

Fibrin is the insoluble blood protein that forms clots. Fibrin clots are often formed at a site of microbial invasion. The clotting mechanism, triggered by tissue injury, isolates the pathogens, limiting infection to a local region. Some pathogens counter this process by producing fibrinolytic enzymes that dissolve the fibrin clots and make further invasion possible. One fibrinolytic substance produced by *Streptococcus pyogenes* is called *streptokinase* (Table 27.4).

By contrast, other pathogens produce enzymes that promote the formation of fibrin clots. These clots localize and protect the organism. The best-studied microbial fibrin-clotting enzyme is *coagulase* (Table 27.4), produced by pathogenic *Staphylococcus aureus*. Coagulase causes insoluble fibrin to be deposited on *S. aureus* cells, protecting them from attack by host cells. The fibrin matrix produced as a result of coagulase activity may account for the extremely localized nature of many staphylococcal infections, as in boils and pimples (Figure 33.25). Coagulase-positive *S. aureus* strains are typically more virulent than coagulase-negative strains.

MiniQuiz

- Distinguish between *toxicity* and *invasiveness*. Give an example of a microorganism that relies almost exclusively on toxicity to promote virulence.
- Identify two virulence factors and relate them to the invasive properties of the pathogen.
- How can fibrin clots both promote and prevent bacterial infections?

Table 27.4 *Exotoxins and extracellular virulence factors produced by human pathogens*

Organism	Disease	Toxin or factor[a]	Action
Bacillus anthracis	Anthrax	Lethal factor (LF) Edema factor (EF) Protective antigen (PA) (AB)	PA is the cell-binding B component, EF causes edema, LF causes cell death
Bacillus cereus	Food poisoning	Enterotoxin complex	Induces fluid loss from intestinal cells
Bordetella pertussis	Whooping cough	Pertussis toxin (AB)	Blocks G protein signal transduction, kills cells
Clostridium botulinum	Botulism	Neurotoxin (AB)	Flaccid paralysis (Figure 27.22)
Clostridium tetani	Tetanus	Neurotoxin (AB)	Spastic paralysis (Figure 27.23)
Clostridium perfringens	Gas gangrene, food poisoning	α-Toxin (CT)	Hemolysis (lecithinase, Figure 27.19*b*)
		β-Toxin (CT)	Hemolysis
		γ-Toxin (CT)	Hemolysis
		δ-Toxin (CT)	Hemolysis (cardiotoxin)
		κ-Toxin (E)	Collagenase
		λ-Toxin (E)	Protease
		Enterotoxin (CT)	Alters permeability of intestinal epithelium
Corynebacterium diphtheriae	Diphtheria	Diphtheria toxin (AB)	Inhibits protein synthesis in eukaryotes (Figure 27.21)
Escherichia coli (enterotoxigenic strains only)	Gastroenteritis	Enterotoxin (Shiga-like toxin) (AB)	Inhibits protein synthesis, induces bloody diarrhea and hemolytic uremic syndrome
Haemophilus ducreyi	Chancroid	Cytolethal distending toxin[b] (AB)	Genotoxin (DNA lesions cause apoptosis in host cells)
Pseudomonas aeruginosa	*P. aeruginosa* infections	Exotoxin A (AB)	Inhibits protein synthesis
Salmonella spp.	Salmonellosis, typhoid fever, paratyphoid fever	Enterotoxin (AB)	Inhibits protein synthesis, lyses host cells
		Cytotoxin (CT)	Induces fluid loss from intestinal cells
Shigella dysenteriae	Bacterial dysentery	Shiga toxin (AB)	Inhibits protein synthesis, induces bloody diarrhea and hemolytic uremic syndrome
Staphylococcus aureus	Pyogenic (pus-forming) infections (boils and so on), respiratory infections, food poisoning, toxic shock syndrome, scalded skin syndrome	α-Toxin (CT)	Hemolysis
		Toxic shock syndrome toxin (SA)	Systemic shock
		Exfoliating toxin A and B (SA)	Peeling of skin, shock
		Leukocidin (CT)	Destroys leukocytes
		β-Toxin (CT)	Hemolysis
		γ-Toxin (CT)	Kills cells
		δ-Toxin (CT)	Hemolysis, leukolysis
		Enterotoxin A, B, C, D, and E (SA)	Induce vomiting, diarrhea, shock
		Coagulase (E)	Induces fibrin clotting
Streptococcus pyogenes	Pyogenic infections, tonsillitis, scarlet fever	Streptolysin O (CT)	Hemolysis
		Streptolysin S (CT)	Hemolysis (Figure 27.19*a*)
		Erythrogenic toxin (SA)	Causes scarlet fever
		Streptokinase (E)	Dissolves fibrin clots
		Hyaluronidase (E)	Dissolves hyaluronic acid in connective tissue
Vibrio cholerae	Cholera	Enterotoxin (AB)	Induces fluid loss from intestinal cells (Figure 27.24)

[a]AB, AB toxin; CT, cytolytic toxin; E, enzymatic virulence factor; SA, superantigen toxin; see Section 28.10.
[b]Cytolethal distending toxin is found in other gram-negative pathogens including *Actinobacillus actinomycetemocomitans*, *Campylobacter* sp., *Escherichia coli*, *Helicobacter* sp., *Salmonella enterica* serovar Typhi, and *Shigella dysenteriae*.

27.10 Exotoxins

Exotoxins are toxic proteins released from the pathogen cell as it grows. These toxins travel from a site of infection and cause damage at distant sites. Table 27.4 provides a summary of the properties and actions of some of the known bacterial exotoxins as well as other extracellular virulence factors.

Exotoxins fall into three categories: the *cytolytic toxins*, the *AB toxins*, and the *superantigen toxins*. The cytolytic toxins work by degrading cytoplasmic membrane integrity, causing lysis. The AB toxins consist of two subunits, A and B. The B component binds to a host cell surface receptor, facilitating the transfer of the A subunit across the targeted cytoplasmic membrane, where it damages the cell. The superantigens work by stimulating large numbers of immune cells, resulting in extensive inflammation and tissue damage, as we will discuss later (Section 28.10).

A subset of the exotoxins are the **enterotoxins**, exotoxins whose activity affects the small intestine, generally causing secretion of fluid into the intestinal lumen resulting in vomiting and diarrhea. Usually acquired by ingestion of contaminated food or water, enterotoxins are produced by a variety of bacteria, including the food-poisoning organisms *Staphylococcus aureus, Clostridium perfringens,* and *Bacillus cereus,* and the intestinal pathogens *Vibrio cholerae, Escherichia coli,* and *Salmonella enterica* serovar Typhimurium. As with the other exotoxins, enterotoxins may be cytolytic toxins, AB toxins, or superantigens. Here we concentrate on the cytotoxins and AB toxins.

Cytolytic Toxins

Cytolytic toxins are secreted, soluble, extracellular proteins produced by a variety of pathogens. Cytolytic toxins damage the host cytoplasmic membrane, causing cell lysis and death. Because the activity of these toxins is most easily observed with assays involving the lysis of red blood cells (erythrocytes), the toxins are often called *hemolysins* (Table 27.4). However, they also lyse cells other than erythrocytes. The production of hemolysin is demonstrated in the laboratory by streaking the pathogen on a blood agar plate (a rich medium containing 5% whole blood). During growth of the colonies, hemolysin is released and lyses the surrounding red blood cells, releasing hemoglobin and creating a clear area, called a zone of *hemolysis,* around the growing colonies (**Figure 27.19**).

Some hemolysins attack the phospholipid of the host cytoplasmic membrane. Because the phospholipid lecithin (phosphatidylcholine) is often used as a substrate, these enzymes are called *lecithinases* or *phospholipases.* An example is the α-toxin of *Clostridium perfringens,* a lecithinase that dissolves membrane lipids, resulting in cell lysis (Table 27.4, Figure 27.19*b*). Because the cytoplasmic membranes of all organisms contain phospholipids, phospholipases sometimes destroy bacterial as well as animal cytoplasmic membranes.

Some hemolysins, however, are not phospholipases. Streptolysin O, a hemolysin produced by streptococci, affects the sterols of the host cytoplasmic membrane. *Leukocidins* (Table 27.4) lyse white blood cells and may decrease host resistance (Section 28.2).

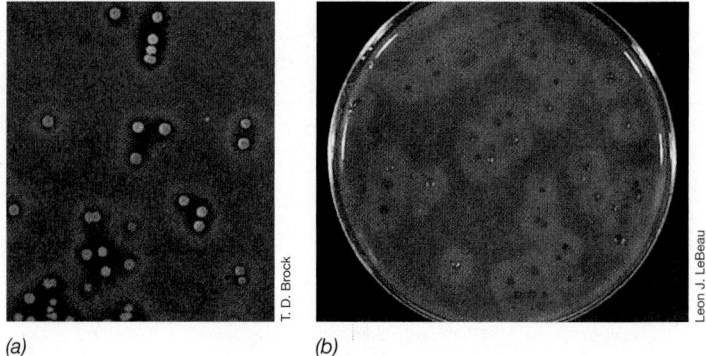

(a) (b)

Figure 27.19 Hemolysis. *(a)* Zones of hemolysis around colonies of *Streptococcus pyogenes* growing on a blood agar plate. *(b)* Action of lecithinase, a phospholipase, around colonies of *Clostridium perfringens* growing on an agar medium containing egg yolk, a source of lecithin. Lecithinase dissolves the cytoplasmic membranes of red blood cells, producing the cloudy zones of hemolysis around each colony.

Staphylococcal α-toxin (**Figure 27.20** and Table 27.4) kills nucleated cells and lyses erythrocytes. Toxin subunits first bind to the phospholipid bilayer. The subunits then oligomerize into nonlytic heptamers, now associated with the membrane. Following oligomerization, each heptamer undergoes conformational changes to produce a membrane-spanning pore, releasing the cell contents and allowing influx of extracellular material, disrupting cell function and causing cell death.

AB Toxins

Several pathogens produce AB exotoxins that inhibit protein synthesis. The diphtheria toxin produced by *Corynebacterium diphtheriae* is an AB toxin and an important virulence factor (Section 33.3). Rats and mice are relatively resistant to diphtheria toxin, but human, rabbit, guinea pig, and bird cells are very

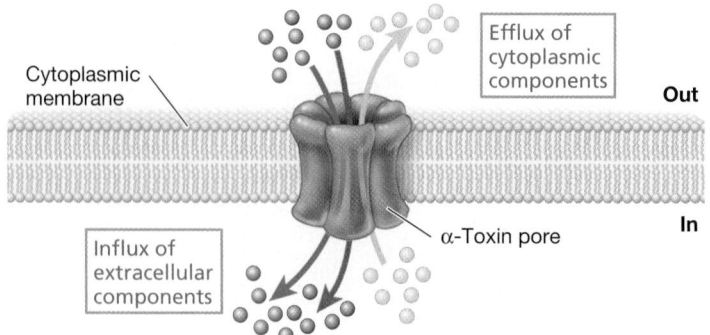

Figure 27.20 Staphylococcal α-toxin. Staphylococcal α-toxin is a pore-forming cytotoxin that is produced by growing *Staphylococcus* cells. Released as a monomer, seven identical protein subunits oligomerize in the cytoplasmic membrane of target cells. The oligomer forms a pore, releasing the contents of the cell and allowing the influx of extracellular material and the efflux of intracellular material. Eukaryotic cells swell and lyse. In erythrocytes, hemolysis occurs, visually indicating cell lysis.

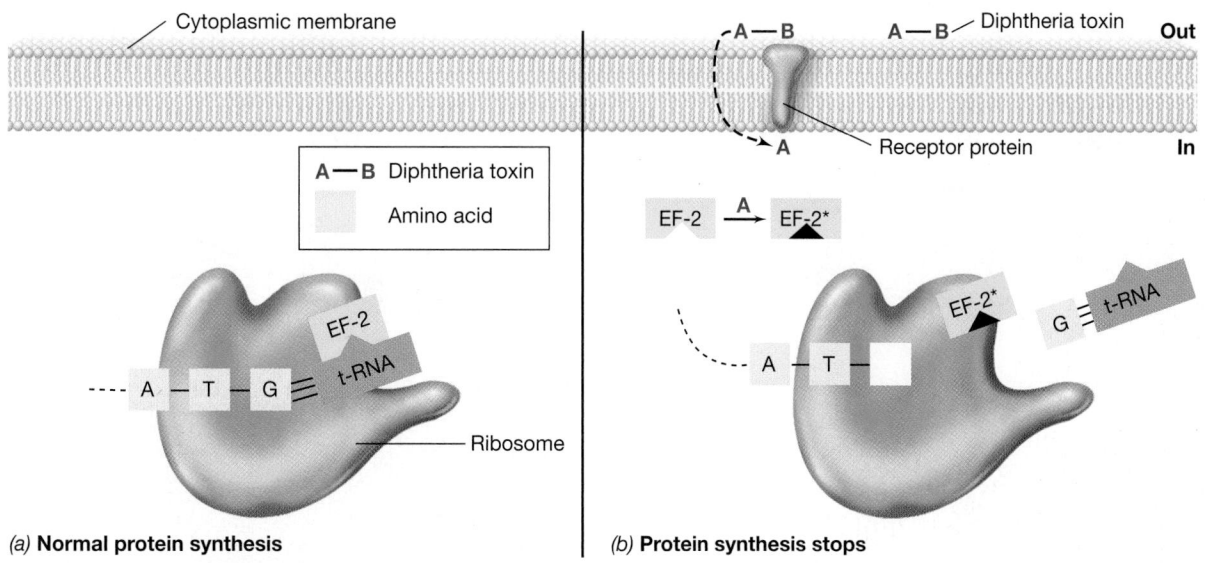

Cytoplasmic membrane

A—B Diphtheria toxin

Amino acid

EF-2

t-RNA

Ribosome

(a) **Normal protein synthesis**

A—B   A—B Diphtheria toxin   **Out**

A   Receptor protein   **In**

EF-2 →A→ EF-2*

EF-2*   G⫴ t-RNA

A—T

(b) **Protein synthesis stops**

Figure 27.21 The action of diphtheria toxin from *Corynebacterium diphtheriae*. *(a)* In a normal eukaryotic cell, elongation factor 2 (EF-2) binds to the ribosome, bringing an amino acid–charged tRNA to the ribosome, causing protein elongation. *(b)* In a cell affected by the diphtheria AB toxin, the toxin binds to the cytoplasmic membrane receptor protein via the B portion. Cleavage between the A and B toxin components occurs, and the A peptide is internalized. The A peptide catalyzes the ADP-ribosylation of elongation factor 2 (EF-2*). This modified elongation factor no longer binds the ribosome and cannot aid transfer of amino acids to the growing polypeptide chain, resulting in cessation of protein synthesis and death of the cell.

susceptible, with only a single toxin molecule required to kill each cell. Diphtheria toxin is secreted by *C. diphtheriae* as a single polypeptide. Fragment B specifically binds to a host cell receptor present on many eukaryotic cells, the heparin-binding epidermal growth factor (**Figure 27.21**). After binding, proteolytic cleavage between fragment A and B allows entry of fragment A into the host cytoplasm. Here fragment A disrupts protein synthesis by blocking transfer of an amino acid from a tRNA to the growing polypeptide chain (🔾 Section 6.19). The toxin specifically inactivates elongation factor 2 (EF-2), a protein involved in growth of the polypeptide chain, by catalyzing the attachment of adenosine diphosphate (ADP) ribose from NAD$^+$. Following ADP-ribosylation, the activity of the modified EF-2 decreases dramatically and protein synthesis stops.

Diphtheria toxin is encoded by the *tox* gene in a lysogenic bacteriophage called phage β. Toxigenic, pathogenic strains of *C. diphtheriae* are infected with phage β and encode the toxin. Nontoxigenic, nonpathogenic strains of *C. diphtheriae* can be converted to pathogenic strains by infection with phage β, a process called *phage conversion* (🔾 Section 10.8). www.microbiologyplace.com **Online Tutorial 27.1: Diphtheria and Cholera Toxin**

Exotoxin A of *Pseudomonas aeruginosa* functions similarly to diphtheria toxin, also modifying EF-2 by ADP-ribosylation (Table 27.4). The enterotoxin produced by *Shigella dysenteriae*, called *Shiga toxin*, and the Shiga-like toxin produced by enteropathogenic *E. coli* O157:H7 are also protein synthesis–inhibiting AB toxins (Table 27.4). The Shiga-like toxins target the small intestine cells near where the pathogen has colonized, shutting down protein synthesis and leading to bloody diarrhea and hemolytic

uremic syndrome, a kidney disease that may result in kidney failure, especially in children.

Tetanus and Botulinum Toxins

Clostridium tetani and *Clostridium botulinum* are endospore-forming bacteria commonly found in soil. These organisms occasionally cause disease in animals through potent AB exotoxins that are *neurotoxins*—they affect nervous tissue. Neither species is very invasive, and virtually all pathogenic effects are due to neurotoxicity. *C. botulinum* sometimes grows directly in the body, causing infant or wound botulism, and also grows and produces toxin in improperly preserved foods (🔾 Section 36.7). Death from botulism is usually from respiratory failure due to flaccid muscle paralysis. *C. tetani* grows in the body in deep wounds that become anoxic, such as punctures. Although *C. tetani* does not invade the body from the initial site of infection, the toxin can spread via the neural cells and cause spastic paralysis, the hallmark of tetanus, often leading to death (🔾 Section 34.9).

Botulinum toxins, the most potent biological toxins known, are seven related AB toxins. One milligram of botulinum toxin is enough to kill more than 1 million guinea pigs. Of the seven distinct botulinum toxins known, at least two are encoded on lysogenic bacteriophages specific for *C. botulinum*. The major toxin is a protein that forms complexes with nontoxic botulinum proteins to yield a bioactive protein complex. The complex then binds to presynaptic membranes on the termini of the stimulatory motor neurons at the neuromuscular junction, blocking the release of acetylcholine. Normal transmission of a nerve impulse to a muscle cell requires acetylcholine interaction with a

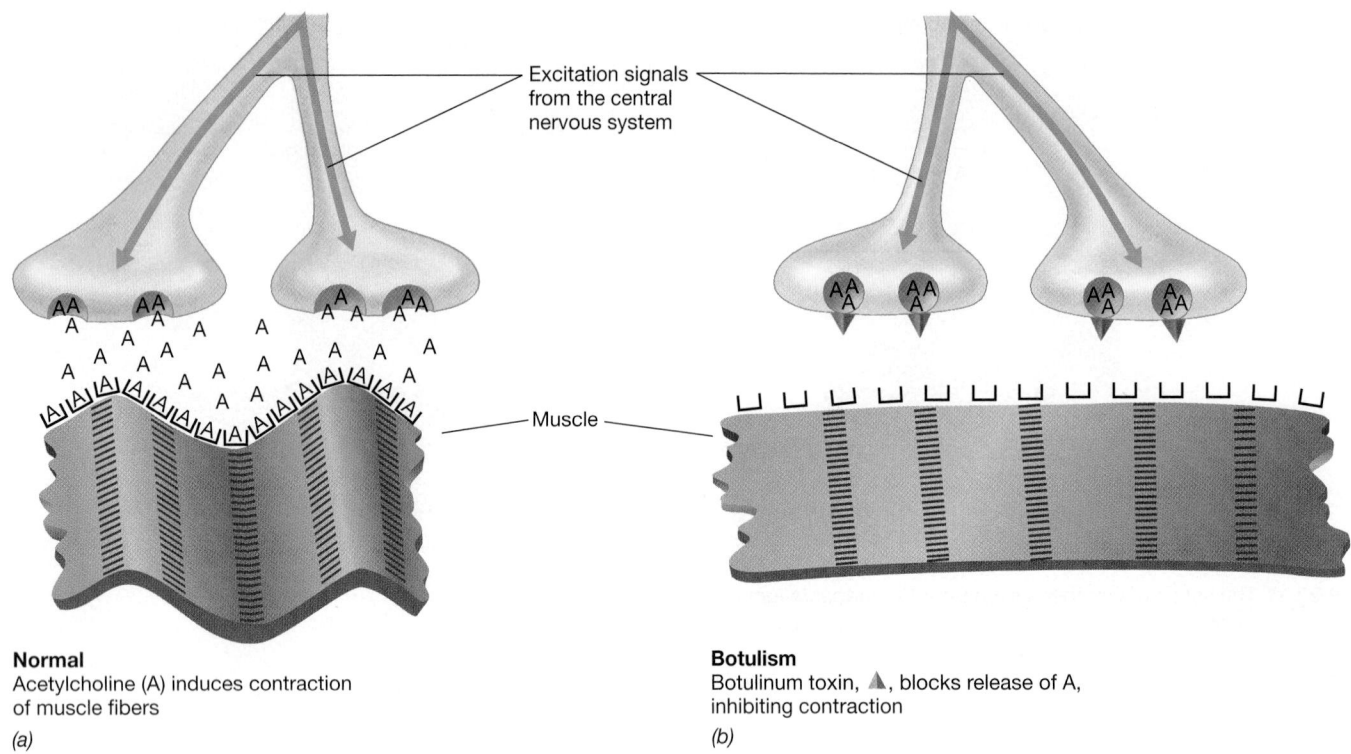

Normal
Acetylcholine (A) induces contraction
of muscle fibers

(a)

Botulism
Botulinum toxin, ▲, blocks release of A,
inhibiting contraction

(b)

Excitation signals
from the central
nervous system

Muscle

Figure 27.22 The action of botulinum toxin from *Clostridium botulinum.* *(a)* Upon stimulation of peripheral and cranial nerves, acetylcholine (A) is normally released from vesicles at the neural side of the motor end plate. Acetylcholine then binds to specific receptors on the muscle, inducing contraction. *(b)* Botulinum toxin acts at the motor end plate to prevent release of acetylcholine (A) from vesicles, resulting in a lack of stimulus to the muscle fibers, irreversible relaxation of the muscles, and flaccid paralysis.

muscle receptor; botulinum toxin prevents the poisoned muscle from receiving the excitatory acetylcholine signal (**Figure 27.22**). This prevents muscle contraction and leads to flaccid paralysis and death by suffocation, the outcome of botulism.

Tetanus toxin is also an AB protein neurotoxin. On contact with the central nervous system, this toxin is transported through the motor neurons to the spinal cord, where it binds specifically to ganglioside lipids at the termini of the inhibitory interneurons. The inhibitory interneurons normally work by releasing an inhibitory neurotransmitter, typically the amino acid glycine, which binds to receptors on the motor neurons. Glycine from the inhibitory interneurons then stops the release of acetylcholine by the motor neurons and inhibits muscle contraction, allowing relaxation of the muscle fibers. However, if tetanus toxin blocks glycine release, the motor neurons cannot be inhibited, resulting in tetanus, continual release of acetylcholine, and uncontrolled contraction of the poisoned muscles (**Figure 27.23**). The outcome is a spastic, twitching paralysis, and affected muscles are constantly contracted. If the muscles of the mouth are involved, the prolonged contractions restrict the mouth's movement, resulting in a condition called *lockjaw* (trismus). If respiratory muscles are involved, prolonged contraction may result in death due to asphyxiation (ᘒ Figure 34.24).

Tetanus toxin and botulinum toxin both block release of neurotransmitters involved in muscle control, but the symptoms are quite different and depend on the particular neurotransmitters involved.

Cholera Toxin

Cholera toxin, an enterotoxin produced by *V. cholerae*, causes cholera (ᘒ Section 35.5). Cholera is characterized by massive fluid loss from the intestines, resulting in severe diarrhea, life-threatening dehydration, and electrolyte depletion (**Figure 27.24**). The disease starts by ingestion of *V. cholerae* in contaminated food or water. The organism travels to the intestine, where it colonizes and secretes the cholera AB toxin. In the gut, the B subunit binds specifically to GM1 ganglioside, a complex glycolipid found in the cytoplasmic membrane of intestinal epithelial cells. The B subunit targets the toxin specifically to the intestinal epithelium but has no role in alteration of membrane permeability; the toxic action is a function of the A chain, which crosses the cytoplasmic membrane and activates adenylate cyclase, the enzyme that converts ATP to cyclic adenosine monophosphate (cAMP).

The cAMP molecule is a cyclic nucleotide that mediates many different regulatory systems in cells, including ion balance. The increased cAMP levels induced by the cholera enterotoxin induce secretion of chloride and bicarbonate ions from the epithelial cells into the intestinal lumen. This change in ion concentrations leads to the secretion of large amounts of water into the intestinal lumen. In acute cholera, the rate of water loss into the small intestine is greater than the possible reabsorption of water by the large intestine, resulting in a large net fluid loss. Cholera treatment is by oral fluid replacement with solutions containing electrolytes and other solutes to offset the dehydration-coupled ion imbalance.

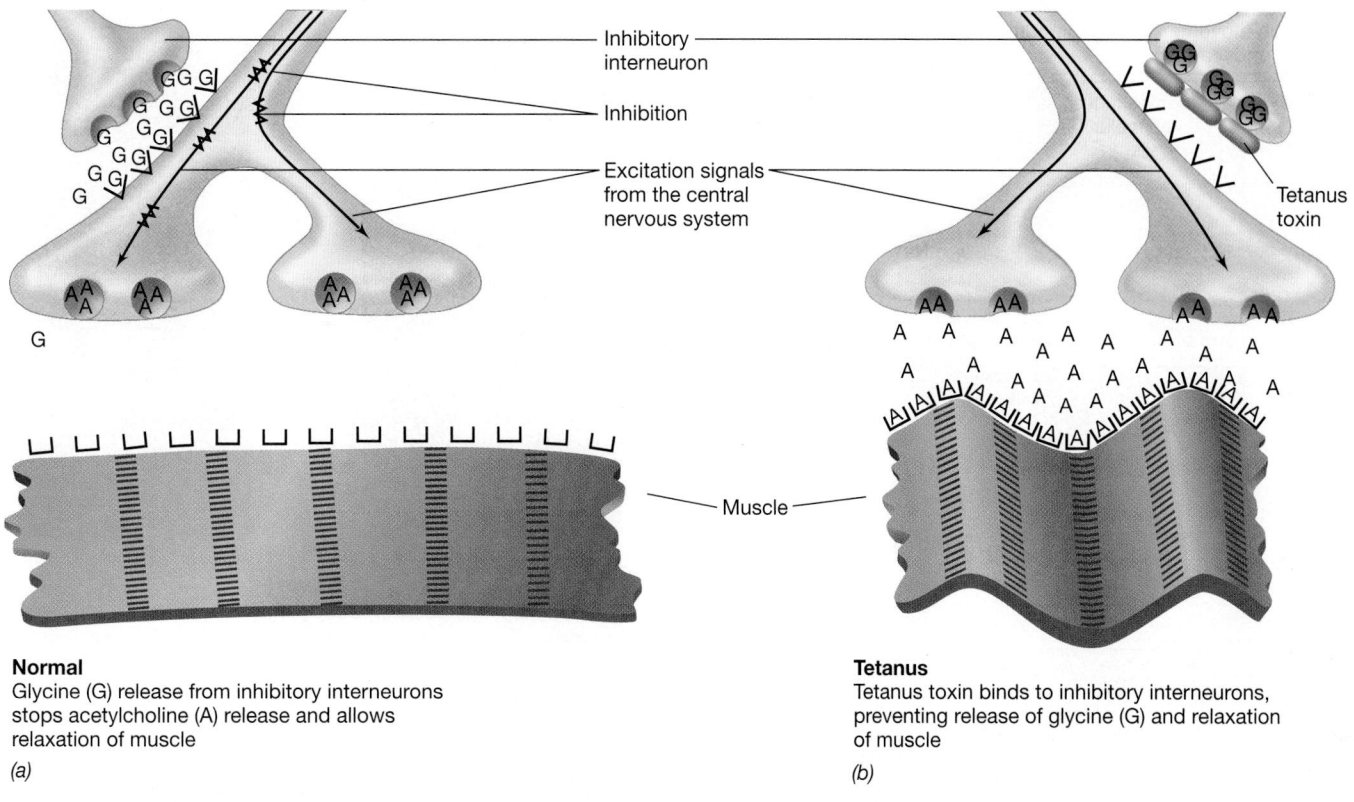

Figure 27.23 The action of tetanus toxin from *Clostridium tetani*. *(a)* Muscle relaxation is normally induced by glycine (G) release from inhibitory interneurons. Glycine acts on the motor neurons to block excitation and release of acetylcholine (A) at the motor end plate. *(b)* Tetanus toxin binds to the interneuron to prevent release of glycine from vesicles, resulting in a lack of inhibitory signals to the motor neurons, constant release of acetylcholine to the muscle fibers, irreversible contraction of the muscles, and spastic paralysis. For convenience, the inhibitory interneuron is shown near the motor end plate, but it is actually in the spinal cord.

Expression of cholera enterotoxin genes *ctxA* and *ctxB* is controlled by *toxR*. The *toxR* gene product is a transmembrane protein that controls cholera A and B chain production as well as other virulence factors, such as the outer membrane proteins and pili required for successful attachment and colonization of *V. cholerae* in the small intestine.

MiniQuiz

- What key features are shared by the AB exotoxins?
- Are bacterial growth and infection in the host necessary for the production of toxins? Explain and cite examples for your answer.

27.11 Endotoxins

Most gram-negative *Bacteria* produce toxic lipopolysaccharides as part of the outer layer of their cell envelope (💭 Section 3.7). These lipopolysaccharides are called **endotoxins**. In contrast to exotoxins, which are the secreted products of living cells, endotoxins are cell bound and released in large amounts only when the cells lyse. Endotoxins have been studied primarily in *Escherichia*, *Shigella*, and especially *Salmonella*, where they are another of the many virulence factors that contribute to patho-

genesis (see the Microbial Sidebar, "Virulence in *Salmonella*"). The properties of exotoxins and endotoxins are compared in **Table 27.5**.

Endotoxin Structure and Function

The structure of lipopolysaccharide (LPS) was diagrammed in Figures 3.19 and 3.20. LPS consists of three covalently linked subunits; the membrane-distal *O*-polysaccharide, lipid A, and a membrane-proximal core polysaccharide.

Endotoxins cause a variety of physiological effects. Fever is an almost universal result of endotoxin exposure because endotoxin stimulates host cells to release cytokines, soluble proteins secreted by phagocytes and other cells, that act as *endogenous pyrogens*, proteins that affect the temperature-controlling center of the brain, causing fever. Cytokines released due to endotoxin exposure can also cause diarrhea, a rapid decrease in the numbers of lymphocytes and platelets, and generalized inflammation (💭 Section 28.5). Large doses of endotoxin can cause death from hemorrhagic shock and tissue necrosis. Endotoxins are, however, generally less toxic than most exotoxins. For instance, in mice the LD_{50} for endotoxin is 200–400 μg per animal, whereas the LD_{50} for botulinum toxin is about 25 picograms (pg), about 10 million times less!

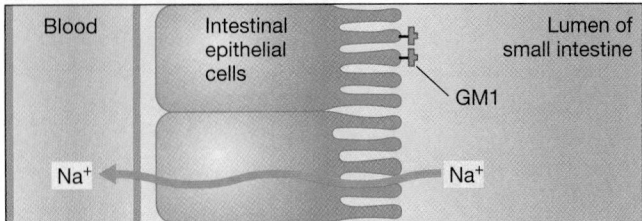

Normal ion movement, Na⁺ from lumen to blood, no net Cl⁻ movement

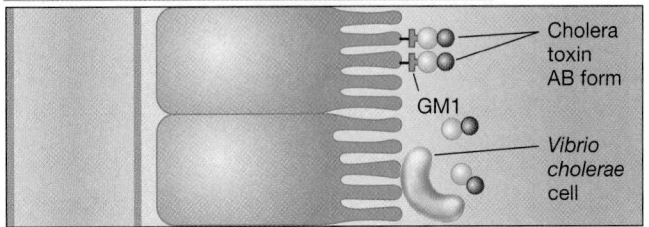

Colonization and toxin production by *V.cholerae*

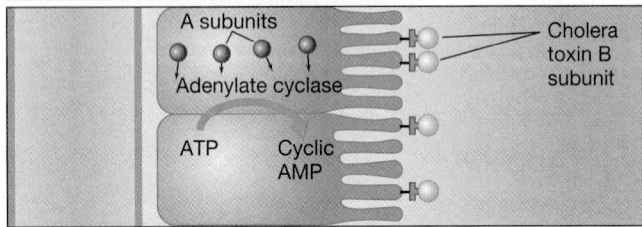

Activation of epithelial adenylate cyclase by cholera toxin

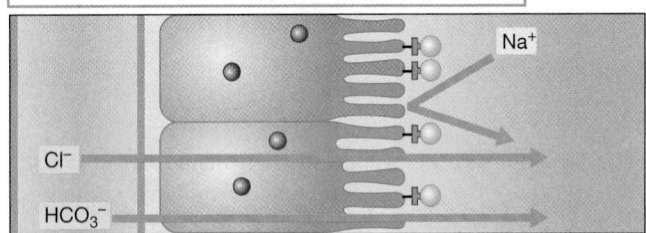

Na⁺ movement blocked, net Cl⁻ movement to lumen

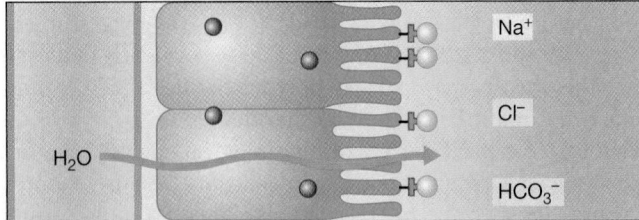

Massive water movement to the lumen; cholera symptoms

Figure 27.24 **The action of cholera enterotoxin.** Cholera toxin is a heat-stable AB enterotoxin that activates a second messenger pathway, disrupting normal ion flow in the intestine. Antibiotic treatment may shorten the course of the disease by limiting *Vibrio cholerae* growth, but does not affect the action of toxin that has already been produced.

The lipid A portion of LPS is responsible for toxicity, and the polysaccharide fraction makes the complex water-soluble and immunogenic. Animal studies indicate that both the lipid and polysaccharide fractions are necessary to induce toxic effects.

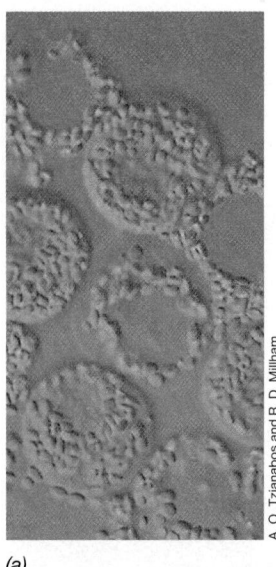
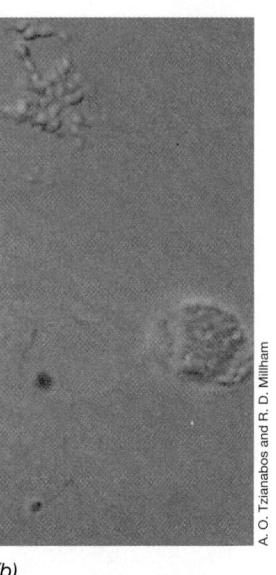

Figure 27.25 *Limulus* **amoebocytes.** *(a)* Normal amoebocytes from the horseshoe crab, *Limulus polyphemus*. *(b)* Amoebocytes following exposure to bacterial lipopolysaccharide (LPS). LPS contained in test samples induces degranulation and lysis of the cells.

Limulus Amoebocyte Lysate Assay for Endotoxin

Because endotoxins induce fever, pharmaceuticals such as antibiotics and intravenous solutions must be free of endotoxin. An endotoxin assay of very high sensitivity has been developed using lysates of amoebocytes from the horseshoe crab, *Limulus polyphemus*. Endotoxin specifically causes lysis of the amoebocytes (**Figure 27.25**). In the standard *Limulus* amoebocyte lysate (LAL) assay, *Limulus* amoebocyte extracts are mixed with the solution to be tested. If endotoxin is present, the amoebocyte extract forms a gel and precipitates, causing a change in turbidity. This reaction is measured quantitatively with a spectrophotometer and can detect as little as 10 pg/ml of LPS. The LAL is used to detect endotoxin in clinical samples such as serum or cerebrospinal fluid. A positive test is presumptive evidence for infection by gram-negative bacteria. Drinking water, water used for formulation of injectable drugs, and injectable aqueous solutions are routinely tested using the LAL to identify and eliminate endotoxin contamination from gram-negative organisms.

MiniQuiz
- Why do gram-positive bacteria not produce endotoxins?
- Why is it necessary to test water used for injectable drug preparations for endotoxin?

 Host Factors in Infection

Host factors influence the pathogenicity of a microorganism. Certain risk factors related to diet, stress, and pathogen exposure are controllable. Other host risk factors defined by, for example, age or genetics cannot be controlled. We conclude this

Table 27.5 *Properties of exotoxins and endotoxins*

Property	Exotoxins	Endotoxins
Chemical properties	Proteins, excreted by certain gram-positive or gram-negative *Bacteria*; generally heat-labile	Lipopolysaccharide–lipoprotein complexes, released on cell lysis as part of the outer membrane of gram-negative *Bacteria*; extremely heat-stable
Mode of action; symptoms	Specific; usually binds to specific cell receptors or structures; either cytotoxin, enterotoxin, or neurotoxin with defined, specific action on cells or tissues	General; fever, diarrhea, vomiting
Toxicity	Often highly toxic, sometimes fatal	Weakly toxic, rarely fatal
Immunogenicity response	Highly immunogenic; stimulate the production of neutralizing antibody (antitoxin)	Relatively poor immunogen; immune response not sufficient to neutralize toxin
Toxoid potential	Treatment of toxin with formaldehyde will destroy toxicity, but treated toxin (toxoid) remains immunogenic	None
Fever potential	Does not produce fever in host	Pyrogenic, often induces fever in host

chapter with a discussion of the passive physical and chemical barriers in humans that limit infection and colonization. In the following chapter we deal with active host responses to pathogen contact.

27.12 Host Risk Factors for Infection

A number of factors contribute to the susceptibility of the host to infection and disease. Here we introduce some of the host factors that may result in resistance to pathogens and explain how alterations in these factors may facilitate invasion by pathogens and lead to infectious disease.

Age as a Risk Factor

Age is an important factor for determining susceptibility to infectious disease. Infectious diseases are more common in the very young and in the very old. In the infant, for example, an intestinal microflora develops quickly, but the normal flora of an infant is not the same as that of an adult. Before the development of an adult flora, and especially in the days immediately following birth, pathogens have a greater opportunity to become established and produce disease. Thus, infants under one year of age often acquire diarrhea caused by enteropathogenic strains of *Escherichia coli* or viruses such as rotavirus (⟳ Sections 36.9 and 36.12).

Infant botulism results from an intestinal infection with *Clostridium botulinum* (⟳ Section 36.7). As the pathogen colonizes and grows, it secretes botulinum toxin, leading to flaccid paralysis (Section 27.10). Infant botulism, contracted after ingestion of *C. botulinum* from soil, air, or foods, occurs almost exclusively in infants under one year of age, presumably because establishment of the normal intestinal flora in older children and adults inhibits colonization by *C. botulinum*.

In individuals over 65 years of age, infectious diseases are much more common than in younger adults. For example, the elderly are much more susceptible to respiratory infections such as those caused by influenza virus (⟳ Section 33.8), probably because of a declining ability to make an effective immune response to respiratory pathogens.

Anatomical changes associated with age may also encourage infection. For example, enlargement of the prostate gland, a common condition in men over the age of 50, frequently leads to a decreased urinary flow rate, allowing pathogens to colonize the male urinary tract more readily and cause urinary tract infections (Figure 27.12).

Stress and Diet as Risk Factors

Stress can predispose a healthy individual to disease. In studies with rats and mice, physiological stressors such as fatigue, exertion, poor diet, dehydration, or drastic climate changes increase the incidence and severity of infectious diseases. For example, rats subjected to intense physical activity for long periods of time show a higher mortality rate from experimental *Salmonella* infections compared with rested control animals. Hormones that are produced under stress can inhibit normal immune responses and may play a role in stress-mediated disease. For example, cortisol, a hormone produced at high levels in the body in times of stress, is an anti-inflammatory agent that inhibits the activation of phagocytes and the immune response.

Diet plays a role in host susceptibility to infection. Inadequate diets low in protein and calories alter the normal flora, allowing opportunistic pathogens a better chance to multiply and increasing susceptibility of the host to known pathogens. For example, the number of *Vibrio cholerae* cells necessary to produce cholera in an exposed individual is drastically reduced if the individual is malnourished. The consumption of pathogen-contaminated food is an obvious way to acquire infections, and ingestion of pathogens with food can sometimes enhance the ability of the pathogen to cause disease. The number of organisms necessary

Virulence in *Salmonella*

The *Salmonella* genus includes at least 1400 different organisms that cause disease in humans (Figure 1). The pathogen produces intracellular infection in the gut that can lead to diarrheal diseases, including the severe and sometimes deadly typhoid fever. *Salmonella* employs a mixture of toxins and other virulence factors to promote invasiveness and enhance pathogenicity (Figure 2). At least three toxins, *enterotoxin*, *endotoxin*, and a *cytotoxin* that kills host cells by inhibiting protein synthesis and allowing calcium ions (Ca^{2+}) to escape, contribute to the virulence of these pathogens.

Several other virulence factors contribute to invasiveness and virulence in *Salmonella*. *Salmonella* produces iron-chelating siderophores, sequestering iron to aid in growth. Three separate structural entities, the cell surface polysaccharide O antigen, the flagellar H antigen, and fimbriae, enhance adherence. The capsular Vi polysaccharide interferes with interaction with host immunity by inhibiting complement binding and antibody-mediated killing.

The genes that initiate the invasion process are contained on the chromosomal *Salmonella* pathogenicity island 1 (SPI1). SPI1 is a collection of virulence genes flanked by sequences suggesting a transposable genetic element. Among these virulence genes, the *inv* (invasion) genes of *Salmonella* encode at least

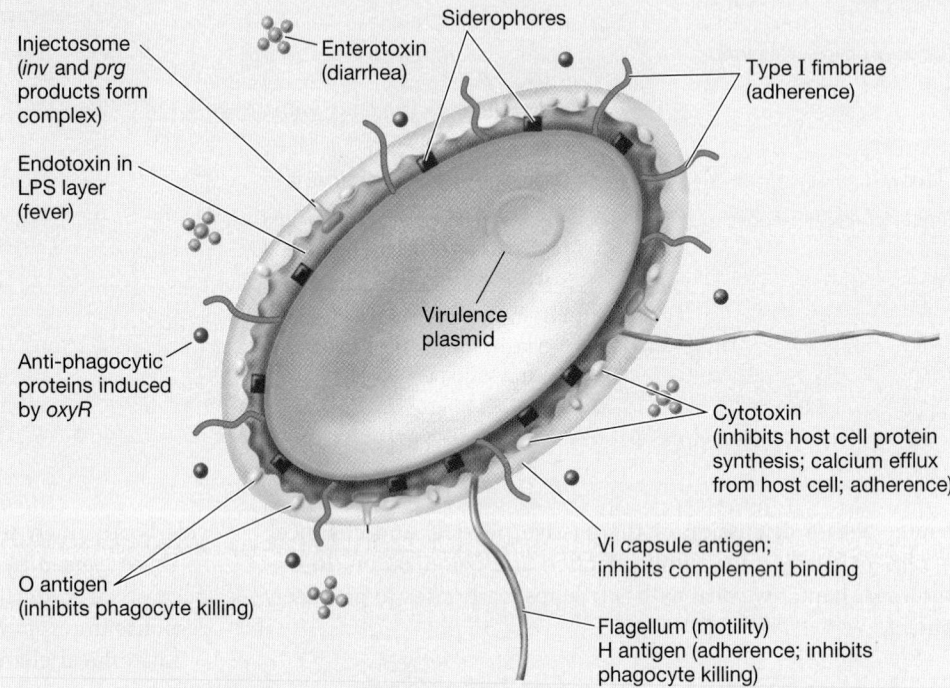

Figure 2 Virulence factors in *Salmonella pathogenesis.* Factors important for virulence and the development of pathogenesis are shown.

Siderophores

Injectosome (*inv* and *prg* products form complex)

Enterotoxin (diarrhea)

Type I fimbriae (adherence)

Endotoxin in LPS layer (fever)

Virulence plasmid

Anti-phagocytic proteins induced by *oxyR*

Cytotoxin (inhibits host cell protein synthesis; calcium efflux from host cell; adherence)

Vi capsule antigen; inhibits complement binding

O antigen (inhibits phagocyte killing)

Flagellum (motility) H antigen (adherence; inhibits phagocyte killing)

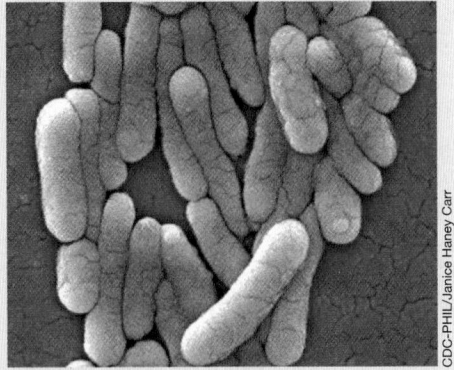

Figure 1 Scanning electron micrograph of cells of *Salmonella enterica* serovar **Typhimurium.** Each organism is about 3–5 μm long and up to 1 μm wide.

CDC-PHIL/Janice Haney Carr

ten different proteins that promote invasion. For example, *invH* encodes a surface adhesion protein. Other *inv* genes encode proteins important for trafficking of virulence proteins. The InvJ regulator protein controls assembly of structural proteins InvG, PrgH, PrgI, PrgJ, and PrgK that form a type III secretion system called the *injectosome*, an organelle in the bacterial envelope that allows direct transfer of virulence proteins into host cells through a needlelike assembly. Another *Salmonella* pathogenicity island, SPI2, contains genes that are responsible for systemic disease and resistance to host cell defenses (Section 12.13).

Salmonella species readily establish infections through intracellular parasitism. The infections start with ingestion and passage of the bacterial cells through the stomach to the intestine. *Salmonella* then invades and replicates inside intestinal epithelial cells called M cells. From here, *Salmonella* can invade the local phagocytes called

macrophages. *Salmonella* pathogens have virulence factors that target these cells. For example, the *Salmonella oxyR* gene encodes proteins that neutralize the toxic oxygen products produced by host macrophages as an antibacterial defense. In addition, the products of the *Salmonella phoP* and *phoQ* genes neutralize macrophage-produced antibacterial molecules called *defensins*. Thus, the *oxy* and *pho* gene products of *Salmonella* enhance pathogenicity-neutralizing host defenses that normally inhibit intracellular bacterial growth. Finally, several plasmid-borne virulence factors such as antibiotic resistance genes encoded on R plasmids can be spread between most *Salmonella* species as well as other enteric bacteria (Section 6.7).

In the final analysis, *Salmonella* and many other successful pathogens employ multiple strategies employing toxins, invasive factors, and other mechanisms to establish virulence and pathogenesis.

to induce cholera, for example, is greatly reduced when the *V. cholerae* is ingested in food, presumably because the food neutralizes stomach acids that would normally destroy the pathogen on its way to colonizing the small intestine.

In some cases, absence of a particular dietary substance may prevent disease by depriving a pathogen of critical nutrients. The best example here is the effect sucrose has on the development of dental caries. As we saw in Section 27.3, dietary restriction of sucrose, along with good oral hygiene, can virtually eliminate tooth decay. Without dietary sucrose, the highly cariogenic *Streptococcus mutans* and *Streptococcus sobrinus* are unable to synthesize the dextran layer needed to keep the bacterial cells attached to the teeth.

The Compromised Host

A compromised host is one in whom one or more resistance mechanisms are inactive and in whom the probability of infection is therefore increased. Many hospital patients with noninfectious diseases (for example, cancer and heart disease) acquire microbial infections more readily because they are compromised hosts (Section 32.7). Such **healthcare-associated infections**, sometimes called **nosocomial infections**, affect up to 2 million individuals each year in the United States, causing up to 100,000 deaths. Invasive healthcare procedures such as catheterization, hypodermic injection, spinal puncture, biopsy, and surgery may unintentionally introduce microorganisms into the patient. The stress of surgery and the anti-inflammatory drugs given to reduce pain and swelling can also reduce host resistance. For example, organ transplant patients are treated with immunosuppressive drugs to prevent immune rejection of the transplant, but suppressed immunity also reduces the ability of the patient to resist infection.

Some factors can compromise host resistance even outside the hospital. Smoking, excess consumption of alcohol, intravenous drug use, lack of sleep, poor nutrition, and acute or chronic infection with another agent are conditions that can reduce host resistance. For example, infection with the human immunodeficiency virus (HIV) predisposes a patient to infections from microorganisms that are not pathogens in uninfected individuals. HIV causes AIDS by destroying one type of immune cell, the CD4 T lymphocytes, involved in the immune response. The reduction in CD4 T cells reduces immunity, and an opportunistic pathogen, a microorganism that does not cause disease in a healthy, uninfected host, can then cause serious disease or even death (Sections 32.6 and 33.14).

Finally, certain genetic conditions compromise the host. For example, genetic diseases that eliminate important parts of the immune system predispose individuals to infections. Individuals with such conditions frequently die at an early age, not from the genetic condition itself, but from microbial infection.

MiniQuiz

- Identify age-related factors that influence susceptibility to infectious disease in infants and adults.
- Identify factors that influence susceptibility to infection and can be controlled by the host.

27.13 Innate Resistance to Infection

Hosts have innate resistance to most pathogens. Natural host resistance to pathogens due to a lack of pathogen receptors or targets is common. Eukaryotic hosts also possess specialized cells called phagocytes. Phagocytes have dedicated receptors that interact with broad classes of pathogens. We will discuss phagocytes and their pathogen-targeting receptors in Chapter 28 in the context of the innate immune response. Here we concentrate on several physical and chemical factors common to vertebrate hosts. These factors nonspecifically inhibit invasion by most pathogens (**Figure 27.26**).

Natural Host Resistance

Under certain circumstances, closely related species, or even members of the same species, may have different susceptibilities to a particular pathogen. The ability of a particular pathogen to cause disease in an individual animal species is highly variable. In rabies, for instance, death usually occurs in all species of mammals once symptoms of the disease develop. Nevertheless, certain animal

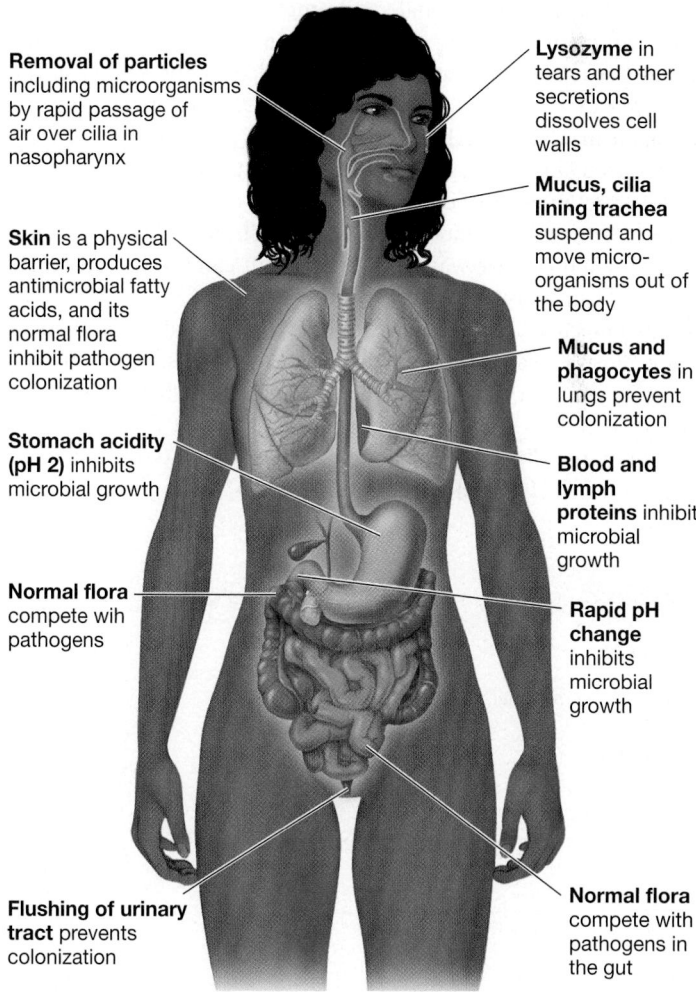

Removal of particles including microorganisms by rapid passage of air over cilia in nasopharynx

Skin is a physical barrier, produces antimicrobial fatty acids, and its normal flora inhibit pathogen colonization

Stomach acidity (pH 2) inhibits microbial growth

Normal flora compete wih pathogens

Flushing of urinary tract prevents colonization

Lysozyme in tears and other secretions dissolves cell walls

Mucus, cilia lining trachea suspend and move microorganisms out of the body

Mucus and phagocytes in lungs prevent colonization

Blood and lymph proteins inhibit microbial growth

Rapid pH change inhibits microbial growth

Normal flora compete with pathogens in the gut

Figure 27.26 Physical, chemical, and anatomical barriers to infection. These barriers provide natural resistance to colonization and infection by pathogens.

species are much more susceptible to rabies than others. Raccoons and skunks are extremely susceptible to rabies infection as compared with opossums, which rarely develop the disease (⊝⊅ Section 34.1). Anthrax infects many species of animals, causing disease symptoms varying from fatal blood poisoning in cattle to the mild pustules of human cutaneous anthrax. Introduction of the same pathogen by other routes, however, may challenge the resistance of the host. For example, pulmonary or airborne anthrax, such as that induced by weaponized strains used for bioterrorism (⊝⊅ Section 32.12), is almost universally fatal in humans. Anthrax causes a localized infection when acquired through the skin, but a lethal, systemic infection when acquired through the mucous membranes of the lungs. As another example of innate host resistance, diseases of warm-blooded animals are rarely transmitted to cold-blooded species, and vice versa. Presumably, the metabolic features of one group are not compatible with pathogens that infect the other.

Tissue Specificity

Most pathogens must adhere and colonize at the site of exposure to initiate infection. Even if pathogens adhere to an exposure site, the organisms cannot colonize if the site is not compatible with the pathogen's nutritional and metabolic needs. Thus, if *Clostridium tetani* were ingested, tetanus would not normally result because the pathogen is either killed by the acidity of the stomach or cannot compete with the well-developed intestinal flora. If, on the other hand, *C. tetani* cells or endospores were introduced into a deep wound, the organism would grow and produce tetanus toxin in the anoxic zones created by local tissue death. Conversely, enteric bacteria such as *Salmonella* and *Shigella* do not cause wound infections but can successfully colonize and cause disease in the intestinal tract.

In some cases, pathogens interact exclusively with members of a few closely related host species because the hosts share tissue-specific receptors. Human immunodeficiency virus (HIV), for instance, infects only higher primates including the great apes and humans. This is because a protein called CXCR4 found on human T cells (cells in the immune system) and a protein called CCR5 found on human macrophages (a phagocyte found in many human tissues) are also expressed in great apes. These proteins, the only cell surface receptors for HIV, bind the gp120 protein of HIV. Other animals, even most primates, lack the CXCR4 and CCR5 proteins, cannot bind HIV, and are therefore not susceptible to HIV infection (⊝⊅ Section 33.14). **Table 27.6** presents several examples of pathogen specficity for host tissue.

Physical and Chemical Barriers

The structural integrity of tissue surfaces poses a barrier to penetration by microorganisms. In the skin and mucosal tissues, potential pathogens must first adhere to tissue surfaces and then grow at these sites before traveling elsewhere in the body. Resistance to colonization and invasion is due to the production of host defense substances and to various anatomical mechanisms.

The skin is an effective barrier to the penetration of microorganisms. Sebaceous glands in the skin (Figure 27.2) secrete fatty acids and lactic acid, lowering the acidity of the skin to pH 5 and inhibiting colonization of many pathogenic bacteria (blood and internal organs are about pH 7.4). Microorganisms inhaled through the nose or mouth are removed by ciliated epithelial cells on the mucosal surfaces of the nasopharynx and trachea. Potential pathogens ingested in food or water must survive the strong acidity in the stomach (pH 2) and then must compete with the abundant resident microflora present in the small and large intestines. Finally, the lumen of the kidney, the eye, the respiratory system, and the cervical mucosa are constantly bathed with secretions such as tears and mucus containing lysozyme, an enzyme that can digest the cell wall and kill bacteria (⊝⊅ Section 3.6).

MiniQuiz

- Identify physical and chemical barriers to pathogens. How might these barriers be compromised?
- How might preexisting infection compromise an otherwise healthy host?

Table 27.6 *Tissue specificity in infectious disease*

Disease	Tissue infected	Organism
Acquired immunodeficiency syndrome (AIDS)	T helper lymphocytes	Human immunodeficiency virus (HIV)
Botulism	Motor end plate	*Clostridium botulinum*
Cholera	Small intestine epithelium	*Vibrio cholerae*
Dental caries	Oral epithelium	*Streptococcus mutans, S. sobrinus, S. sanguiins, S. mitis*
Diphtheria	Throat epithelium	*Corynebacterium diphtheriae*
Gonorrhea	Mucosal epithelium	*Neisseria gonorrhoeae*
Influenza	Respiratory epithelium	Influenza A and influenza B virus
Malaria	Blood (erythrocytes)	*Plasmodium* spp.
Pyelonephritis	Kidney medulla	*Proteus* spp.
Spontaneous abortion (cattle)	Placenta	*Brucella abortus*
Tetanus	Inhibitory interneuron	*Clostridium tetani*

Big Ideas

27.1

The animal body is a favorable environment for the growth of microorganisms, most of which do no harm. Microorganisms that cause harm are called pathogens. Pathogen growth initiated on host surfaces such as mucous membranes may result in infection and disease. The ability of a microorganism to cause or prevent disease is influenced by complex interactions between the microorganism and the host.

27.2

The skin has at least three different microenvironments, sebaceous, moist, and dry, that harbor distinctly different populations of microorganisms. Environmental and host factors influence the quantity and makeup of the normal skin microflora.

27.3

The complex microflora in the oral cavity can produce adherent substances and growth on tooth surfaces, typically resulting in mixed-culture biofilms called plaque. Acid produced by microorganisms in plaque damages tooth surfaces, resulting in dental caries. Further infection can result in periodontal disease.

27.4

The stomach and the intestinal tract support a diverse population of microorganisms in a variety of nutritional and environmental conditions. The populations of microorganisms are influenced by the diet of the individual and by the unique physical conditions in each distinct anatomical area.

27.5

A robust population of normal nonpathogenic microorganisms in the respiratory and urogenital tracts is essential for optimal organ function in normal individuals. The normal microflora help prevent the colonization of pathogens.

27.6

Virulence is determined by the invasiveness and toxicity of a pathogen. Pathogens use a wide variety of mechanisms and factors to establish virulence and pathogenicity.

27.7

Pathogens gain access to host tissues by adherence at mucosal surfaces through interactions between pathogen and host macro-molecules. Pathogen invasion starts at the site of adherence and may spread throughout the host via the circulatory or lymphatic systems.

27.8

A pathogen must gain access to nutrients and appropriate growth conditions before it can colonize and grow in substantial numbers in host tissue. Pathogens may grow locally at the site of invasion or may spread through the body.

27.9

Pathogens produce enzymes and other factors that enhance their ability to invade host tissue. These factors contribute to virulence by breaking down or altering host tissue to provide access and nutrients, and enhance colonization, infection, and pathogenesis.

27.10

Exotoxins contribute to the virulence of pathogens. Cytotoxins and AB toxins are potent exotoxins produced by microorganisms. Each exotoxin affects a specific host cell function. Enterotoxins are exotoxins that affect the small intestine. Bacterial exotoxins include some of the most potent biological toxins known.

27.11

Endotoxins are lipopolysaccharides derived from the outer membrane of gram-negative bacteria. Released upon cell lysis, endotoxins cause fever and other systemic toxic effects in the host. Endotoxins are generally less toxic than exotoxins. The presence of endotoxin indicates contamination by gram-negative bacteria.

27.12

Age, general health, genetic makeup, lifestyle factors such as stress and diet, and prior, concurrent, or chronic disease can contribute to susceptibility to infectious disease.

27.13

Innate resistance factors, as well as physical, anatomical, and chemical barriers, prevent colonization of the host by most pathogens. Breakdown of these passive defenses may result in susceptibility to infection and disease.

Review of Key Terms

Attenuation a decrease or loss of virulence

Bacteremia the presence of microorganisms in the blood

Capsule a dense, well-defined polysaccharide or protein layer closely surrounding a cell

Colonization the growth of a microorganism after it has gained access to host tissues

Dental caries tooth decay resulting from bacterial infection

Dental plaque bacterial cells encased in a matrix of extracellular polymers and salivary products, found on the teeth

Disease an injury to a host organism, caused by a pathogen or other factor, that affects the host organism's function

Endotoxin the lipopolysaccharide portion of the cell envelope of most gram-negative *Bacteria*, which is a toxin when solubilized

Enterotoxin a protein released extracellularly by a microorganism as it grows that produces immediate damage to the small intestine of the host

Exotoxin a protein released extracellularly by a microorganism as it grows that produces immediate host cell damage

Glycocalyx polymers secreted by a microorganism that coat the surface of the microorganism

Healthcare-associated infection (nosocomial infection) an infection contracted in a healthcare-associated setting

Host an organism that can harbor a pathogen

Infection the growth of organisms in the host

Invasion the ability of a pathogen to enter into host cells or tissues, spread, and cause disease

Lower respiratory tract the trachea, bronchi, and lungs

Mucous membrane layers of epithelial cells that interact with the external environment

Mucus a liquid secretion that contains water-soluble glycoproteins and proteins that retain moisture and aid in resistance to microbial invasion on mucosal surfaces

Normal microflora microorganisms that are usually found associated with healthy body tissue

Nosocomial infection (healthcare-associated infection) an infection contracted in a healthcare-associated setting

Opportunistic pathogen an organism that causes disease in the absence of normal host resistance

Pathogen an organism, usually a microorganism, that grows in or on a host and causes disease

Pathogenicity the ability of a pathogen to cause disease

Probiotic a live microorganism that, when administered to a host, may confer a health benefit

Septicemia a bloodborne systemic infection

Slime layer a diffuse layer of polymer fibers, typically polysaccharides, that forms an outer surface layer on the cell

Toxicity the ability of an organism to cause disease by means of a preformed toxin that inhibits host cell function or kills host cells

Upper respiratory tract the nasopharynx, oral cavity, and throat

Virulence the relative ability of a pathogen to cause disease

Review Questions

1. Define (a) opportunistic pathogens, (b) normal microflora, (c) infection, (d) disease (Section 27.1).

2. Identify the most common resident microorganisms on the skin. How were these resident microorganisms identified experimentally (Section 27.2)?

3. What contributes to dental plaques? Name at least two bacteria that cause dental plaques (Section 27.3).

4. Discuss the importance of intestinal microflora (Section 27.4).

5. Describe the relationship between *Lactobacillus acidophilus* and glycogen in the vaginal tract. What factors influence the differences between the normal vaginal flora of adult females as compared to that of prepubescent juvenile females (Section 27.5)?

6. Define virulence and identify parameters to distinguish between highly virulent and moderately virulent pathogens (Section 27.6).

7. Identify the role of the capsule and the fimbriae of bacteria in microbial adherence (Section 27.7).

8. Explain the role of the availability of nutritional factors in infection by microorganisms in the body (Section 27.8).

9. Identify the role of coagulase and streptokinase in the invasiveness of *Staphylococcus* and *Streptococcus*, respectively (Section 27.9).

10. What is the difference between exotoxin and endotoxin? Name one example for each type of toxin (Sections 27.10 and 27.11).

11. Describe the structure of a typical endotoxin. How does endotoxin induce fever? What microorganisms produce endotoxin (Section 27.11)?

12. Identify common factors that lead to host compromise. Indicate which factors are controllable by the host. Indicate which factors are not controllable by the host (Section 27.12).

13. In which body locations might pH values differ from standard body conditions? Which organisms might benefit or be inhibited by differences in body pH (Section 27.13)?

Application Questions

1. Mucous membranes are barriers against colonization and growth of microorganisms. However, mucous membranes, for example in the throat and the gut, are colonized with a variety of different microorganisms, some of which are potential pathogens. Explain how these potential pathogens are controlled under normal circumstances. Then describe at least one set of circumstances that might encourage pathogenicity.

2. Antibiotic therapy can significantly reduce the number of microorganisms residing in the gastrointestinal tract. What physiological symptoms might the reduction of normal flora produce in the host? Infection by opportunistic pathogens often follows long-term antimicrobial therapy. Many of these post-therapeutic infections are caused by the same microorganisms that produce opportunistic infections in individuals with AIDS. What pathogens might be involved? Why are individuals who have undergone antibiotic therapy particularly susceptible to these pathogens?

3. Design an experiment to increase the virulence and pathogenicity of *Streptococcus pneumoniae* (Hint: *S. pneumoniae* that is transferred for several passages *in vitro* loses its capsule and virulence for mice.) Would an increase in virulence confer a selective advantage for the organism? Be sure to consider the natural habitat.

4. Coagulase is a virulence factor for *Staphylococcus aureus* that acts by causing clot formation at the site of *S. aureus* growth. Streptokinase is a virulence factor for *Streptococcus pyogenes* that acts by dissolving clots at the site of *S. pyogenes* growth. Reconcile these opposing strategies for enhancing pathogenicity.

5. Although mutants incapable of producing exotoxins are relatively easy to isolate, mutants incapable of producing endotoxins are much harder to isolate. From what you know of the structure and function of these types of toxins, explain the differences in mutant recovery.

6. Identify the potential for infectious disease problems in the case of burns to the body. What microorganisms are likely to be involved in burn infections? Why does the normal local microflora fail to protect burn victims from microbial infections?

Need more practice? Test your understanding with Quantitative Questions; access additional study tools including tutorials, animations, and videos, and then test your knowledge with chapter quizzes and practice tests at **www.microbiologyplace.com**.

28

Immunity and Host Defense

The bacterium *Bordetella pertussis* is the causative agent of pertussis (whooping cough). Pertussis can be controlled by the DTaP (diphtheria, tetanus, acellular pertussis) vaccine, which contains a mixture of proteins from cells of *B. pertussis* along with inactivated toxins (toxoids) from the bacteria that cause diphtheria and tetanus.

We discussed passive protection against pathogen invasion, infection, and disease in Chapter 27. In the next three chapters, we shift our focus away from the microbiology of pathogens and passive protection toward the active mechanisms used by vertebrates to resist infection and disease.

The active ability to resist disease is called **immunity**. In this chapter, we begin with an overview of immune mechanisms and their importance in pathogen resistance. Multicellular organisms use certain cells and their products to kill or neutralize pathogens. The body has a built-in, or innate, immune system that targets and destroys most common pathogens regardless of their identity. A second tier of immunity, the adaptive immune system, targets specific strains of bacteria or viruses to neutralize their pathogenic effects.

We first look at how the innate immune system deals with most pathogens. We will then investigate the more complex mechanisms of the adaptive immune system. These immune mechanisms have evolved to protect animals from dangerous nonself pathogens; our survival is dependent on the functioning of these systems.

How does immunity prevent infectious disease? We address this question by considering natural and artificial immunity. We discuss planned vaccinations, a practical tool used to artificially recruit the adaptive immune response for protection against future pathogen challenges. We conclude by describing immune mechanisms that can themselves cause disease.

 # Immunity

In the course of evolution, the immune response was selected for because it recognizes and destroys dangerous pathogens. We start with **innate immunity**, the body's built-in ability to recognize and destroy pathogens or their products. Innate immunity is largely a function of **phagocytes**, cells that can engulf foreign particles, and can ingest, kill, and digest most bacterial pathogens. The phagocytes recognize structural features shared by many pathogens. Initial interactions with pathogens recruit large numbers of phagocytes to the site of infection. Here the phagocytes activate defense genes, leading to the transcription, translation, and expression of proteins that destroy the pathogen. This innate immune response develops within hours after contact with a pathogen.

Some pathogens, however, are so virulent that innate immune responses are not completely effective and infections sometimes still persist. When this happens, the phagocytes in the innate response can activate another defense mechanism called **adaptive immunity**, or specific immunity, to deal with these infections. Adaptive immunity is the acquired ability to recognize and destroy a specific pathogen or its products; it is activated by exposure of the immune system to the pathogen. Adaptive responses are directed at unique pathogen molecules called **antigens**. Phagocytes present antigen molecules to lymphocytes, key cells in the adaptive response. The antigens interact with specific receptors on the lymphocyte. The antigen–lymphocyte interactions activate the lymphocyte to transcribe and translate genes that produce pathogen-specific proteins. These proteins then interact with the specific pathogen, marking it for destruction. A protective adaptive response usually takes several days to develop because only a few lymphocytes are initially available to interact with each antigen; the strength of the adaptive response increases as the numbers of antigen-reactive lymphocytes multiply.

Here we introduce the cells active in the innate and adaptive host responses to pathogens and other foreign substances. We begin with the cells and organs common to the entire immune system and then consider the cells and mechanisms active in innate immunity. We finish with an overview of adaptive immunity, the focus of the rest of the chapter.

28.1 Cells and Organs of the Immune System

The cells active in both innate and adaptive immunity develop from common pluripotent precursors called **stem cells**. Immunity results from the actions of cells that circulate throughout the body, primarily through the blood and **lymph**, a fluid similar to blood that contains nucleated cells and proteins, but lacks red blood cells. Blood and lymph interact directly or indirectly with every major organ system, and some immune cells can move back and forth from organ interstitial spaces to the blood or lymph.

Stem Cells, Blood, and Lymph

Pluripotent stem cells are the progenitors of all blood cells, including the cells active in innate and adaptive immunity (**Figure 28.1**). Stem cells are produced and develop in the bone marrow where they differentiate to produce mature cells under the influence of soluble **cytokines**, proteins that influence many aspects of immunity, including growth of stem cells. After developing in the bone marrow, the differentiated cells travel through the blood and lymph to reach other parts of the body.

Blood consists of cellular and noncellular components, including many cells and molecules active in the immune response. The most numerous cells in human blood are erythrocytes (red blood cells), nonnucleated cells that function to carry oxygen from the lungs to the tissues (**Table 28.1**). About 0.1% of the cells in blood, however, are nucleated cells called **leukocytes** or white blood cells. Leukocytes include cells such as the phagocytes of the innate response and **lymphocytes**, the cells active in the adaptive response.

Whole blood is composed of suspended cells and **plasma**, a liquid containing proteins and other solutes. Outside the body, whole blood or plasma quickly forms an insoluble fibrin clot, remaining liquid only when an anticoagulant is added. Anticoagulants such as potassium citrate or heparin prevent fibrin formation, stopping the clotting process. When blood clots, the insoluble proteins trap the cells in a large, insoluble mass. The remaining fluid, called **serum**, contains no cells or clotting proteins. Serum does, however, contain a high concentration of other proteins, including soluble immune proteins called antibodies, which makes it useful for immunological investigations. The use of serum antibodies to detect antigens is called *serology*.

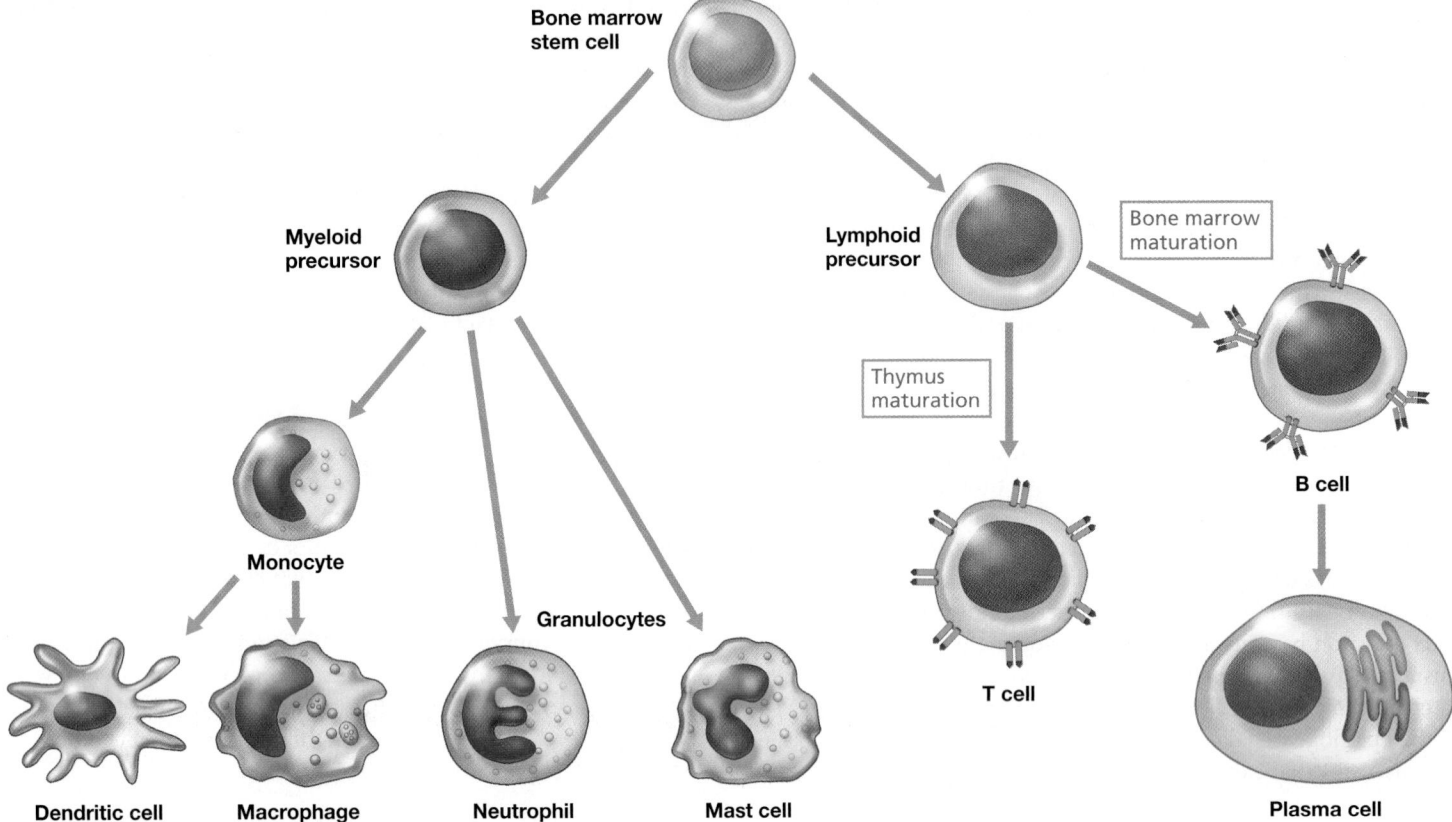

Figure 28.1 The origins of immune response cells. Immune response cells develop from pluripotent stem cells in the bone marrow into one or the other of two immune cell precursors. Myeloid precursors generate monocytes and granulocytes. In turn, monocytes develop into macrophages or dendritic cells, both phagocytic and active in antigen uptake and presentation. Granulocytes include phagocytic neutrophils, also called polymorphonuclear leukocytes or PMNs, and granule-releasing mast cells. Lymphoid precursors generate T and B cells, the lymphocytes that participate directly in the adaptive immune response. Plasma cells derived from B cells produce antibodies.

Blood and Lymph Circulation

Blood is pumped by the heart through arteries and capillaries throughout the body and is returned through the veins (**Figure 28.2b**). In the capillary beds, leukocytes and solutes pass to and from the blood into the lymphatic system, a separate circulatory system containing lymph, a fluid similar to blood that contains leukocytes, but lacks red blood cells (Figure 28.2a–c).

Lymph drains from extravascular tissues into lymphatic capillaries, lymph ducts, and then into lymph nodes throughout the lymph system (Figure 28.2d). Lymph nodes contain lymphocytes and phagocytes that are arranged to encounter microorganisms and antigens as they enter the nodes. The mucosa-associated lymphoid tissue (MALT), another part of the lymphatic system, interacts with antigens and microorganisms from the gut, the bronchial mucosal tissues, and other mucous membranes. The MALT also contains phagocytes and lymphocytes. Lymph fluid with antibodies and immune cells empties into the blood circulatory system via the thoracic lymph duct.

The white pulp in the spleen also consists of organized concentrations of lymphocytes and phagocytes, arranged to filter the blood. Collectively, the lymph nodes, MALT, and spleen are called *secondary lymphoid organs*. The secondary lymphoid organs are the sites where antigens interact with antigen-presenting phagocytes and lymphocytes to generate an adaptive immune response (Figure 28.2a).

Leukocytes

Leukocytes are nucleated white blood cells found in the blood and the lymph. Several distinct leukocytes (Table 28.1 and Figure 28.1) participate in innate or adaptive immunity.

Table 28.1 *Major cells found in normal human blood*

Cell type	Cells per milliliter
Erythrocytes	$4.2–6.2 \times 10^9$
Leukocytes[a]	$4.5–11 \times 10^6$
Lymphocytes	$1.0–4.8 \times 10^6$
Myeloid cells	Up to 7.0×10^6

[a]Leukocytes include all nucleated blood cells. They include lymphocytes and myeloid cells (monocytes and neutrophils).

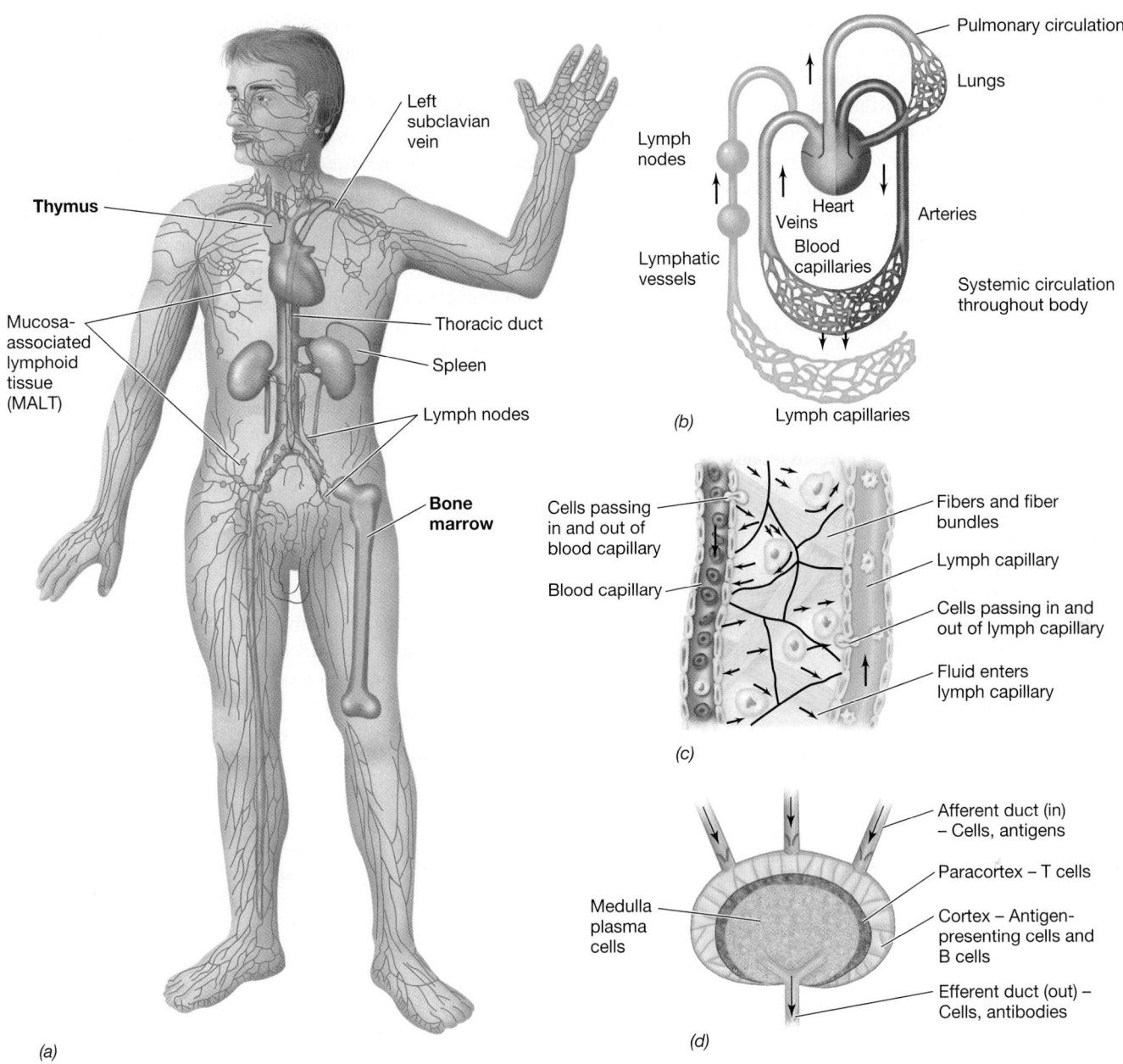

Figure 28.2 The blood and lymph systems.
(a) The lymphatic system. The major lymphatic organs and vessels are shown in green. The primary lymphoid organs are the bone marrow and thymus. The secondary lymphoid organs are the lymph nodes, spleen, and MALT. *(b)* Connections between the lymph and blood systems. Blood flows from the veins to the heart, to the lungs, and then through the arteries to the tissues. Lymph drains from the thoracic duct into the left subclavian vein of the blood circulatory system. *(c)* The exchange of cells between the blood and lymph systems is shown microscopically. Both blood and lymph capillaries are closed vessels, but cells pass from blood capillaries to lymph capillaries and back by a process known as extravasation. *(d)* A secondary lymphoid organ, the lymph node. The diagram identifies the node's major anatomic areas and the immune cells present in each area. The anatomy of the MALT and the spleen is analogous to that of the lymph nodes.

Myeloid cells, active in innate immunity, are derived from myeloid precursor cells. Mature myeloid cells can be divided into two lineages, the monocytes and the granulocytes (Figure 28.1). The monocyte lineage develops into specialized phagocytic cells, the **antigen-presenting cells** (**APCs**). These cells, in addition to the B cells we discuss below, engulf, process, and present antigens to lymphocytes. APCs include *macrophages* and *dendritic cells*. Immature cells called monocytes are circulating precursors of macrophages and dendritic cells. **Macrophages** are generally the first defense cells that interact with a pathogen. They are abundant in many tissues, especially spleen, lymph nodes, and MALT. **Dendritic cells** are also phagocytes with antigen-presenting properties.

Granulocytes are the second lineage of cells derived from myeloid precursors. Granulocytes contain cytoplasmic inclusions, or granules, that can be visualized by staining. These granules contain toxins or enzymes that are released to kill target cells. The phagocytic activity of one granulocyte, the **neutrophil**, also called a *polymorphonuclear leukocyte*, or a *PMN*, is central to innate immunity. Release of granules, a process called degranulation, from a granulocyte called a mast cell can cause allergy symptoms and inflammation.

Lymphocytes are specialized leukocytes involved exclusively in the adaptive immune response. Mature lymphocytes circulate through the blood and lymph system, but are concentrated in the lymph nodes and spleen where they interact with antigens. There are two types of lymphocytes, B cells (B lymphocytes) and T cells (T lymphocytes) (Figure 28.1). **B cells** originate and mature in the **bone marrow**. They are specialized APCs and the precursors of antibody-producing **plasma cells**. **Antibodies**, also called **immunoglobulins**, are soluble proteins produced by B cells and plasma cells. Antibodies interact with particular antigens. **T cells**, which interact with antigen, begin their development in the bone marrow, but travel to the **thymus** to mature. The bone marrow and thymus in mammals are called **primary lymphoid organs** because they are the sites where the lymphoid stem cells develop into functional antigen-reactive lymphocytes (Figure 28.2*a*).

All leukocytes actively move throughout the body and pass from blood to interstitial spaces, then to lymphatic vessels, and back to the blood circulatory system, a process called extravasation (Figure 28.2*c*).

MiniQuiz

- Trace the development of B cells, T cells, and macrophages from the common stem cell.

- Describe the circulation of a leukocyte from the blood to the lymph and back to the blood.

28.2 Innate Immunity

Eukaryotes from plants to vertebrates have developed molecular recognition mechanisms that lead to rapid and effective host defense. These evolutionarily related and conserved mechanisms are collectively called innate immunity, or in-built immunity: the noninducible, preexisting ability to recognize and destroy a pathogen or its products. Most importantly, innate immunity does not require previous exposure to a pathogen or its products. The innate immune response is mediated by phagocytes.

Pathogen-Associated Molecular Patterns

The macromolecules inside and on the surface of pathogens display **pathogen-associated molecular patterns** (**PAMPs**), consisting of repeating subunits. An example of a PAMP is the lipopolysaccharide (LPS) common to all gram-negative bacterial outer membranes (⟲ Section 3.7). Other PAMPs include bacterial flagellin, the double-stranded RNA (dsRNA) of certain viruses, and the lipoteichoic acids of gram-positive bacteria. All of these macromolecules contain repeating structural units.

PAMPs are shared among related pathogens and are highly conserved within that pathogen group.

Pattern Recognition Receptors

Phagocytes such as macrophages and neutrophils are generally the first line of defense against pathogens, especially those that the body has never before encountered. Phagocytes can interact speedily and effectively with pathogens because they have evolved specialized molecules that interact directly with PAMPs. The molecules are preformed receptors called **pattern recognition receptors** (**PRRs**) (**Figure 28.3**). Each PRR interacts with a particular PAMP to activate the phagocyte. One PRR found on all phagocytes, for example, interacts with the LPS on most gram-negative bacteria, including all pathogenic strains of *Salmonella* spp., *Escherichia coli*, and *Shigella* spp. Another phagocyte PRR interacts with the peptidoglycan on gram-positive cells. And still other PRRs interact with conserved pathogen features such as the dsRNA found in some viruses and flagellin on certain motile bacteria.

The interaction of a PAMP with a PRR activates the phagocyte to ingest and destroy the targeted pathogen by phagocytosis. All phagocytes have a number of preformed PRRs that are instantly available to interact with invasive pathogens. Innate immunity is an ancient response to infection. We know that the PRRs present in vertebrates, for example, have structural and evolutionary homologs in phylogenetic groups as distant as the insect *Drosophila* (fruit fly). Functionally similar phagocyte recognition and destruction systems are found in all multicellular organisms.

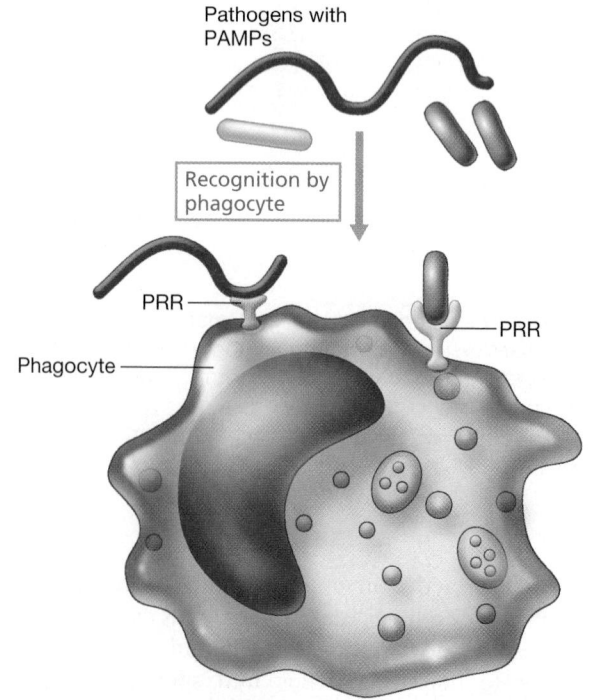

Figure 28.3 Innate immunity. Phagocytes interact with pathogens by recognizing pathogen-associated molecular patterns (PAMPs) with preformed pattern recognition receptors (PRRs). The interactions activate the phagocyte to ingest and destroy the pathogen and to produce cytokines that attract and activate other cells.

MiniQuiz
- Identify a pathogen-associated molecular pattern shared by a group of microorganisms.
- Identify the organisms and the cells that use pattern recognition receptors to provide innate immunity to pathogens.

28.3 Adaptive Immunity

The phagocytes responsible for innate immunity also initiate *adaptive* immunity in vertebrate animals. Adaptive immunity is the acquired ability to recognize and destroy an individual pathogen. In adaptive immunity, pathogen-specific receptors are produced in large numbers only after exposure to the pathogen or its products. Because adaptive immunity is directed toward an antigen, a molecular component of the pathogen, it is sometimes called antigen-specific immunity.

After the first exposure to an antigen, a **primary adaptive immune response** stimulates growth and multiplication of antigen-reactive cells, creating **clones**, large numbers of identical antigen-reactive cells. These clones may persist for years and confer long-term specific immunity.

Antigen-reactive lymphocytes are divided into two populations, T cells and B cells. Both lymphocyte populations produce unique proteins that interact with a single antigen and thus have **specificity** for that antigen. The unique antigen-reactive proteins of T cells are the **T cell receptors** (**TCRs**) and those of B cells are antibodies or immunoglobulins (Igs).

As compared to innate immunity, the adaptive response is inducible only when triggered by a unique antigen on a pathogen. For example, single polysaccharide antigens from a particular gram-negative organism's LPS molecule are unique for the genus and sometimes for the species. An individual lymphocyte clone that interacts with an LPS constituent on *Salmonella* will not interact with the LPS on other bacteria. The terminal sugars that constitute the antigen on the polysaccharides of *Salmonella* spp. are unique for the genus and are not shared by other bacteria, even other gram-negative organisms such as *Escherichia coli* or *Shigella* spp.

A second exposure to the same antigen activates the clones of antigen-reactive cells and generates a faster, stronger **secondary adaptive immune response** that peaks within several days (**Figure 28.4**). The products of this secondary immune response quickly target the pathogen for destruction. This rapid increase in adaptive immunity after a second antigen exposure is called **memory** (**immune memory**). Finally, the adaptive immune system exhibits **tolerance**, the acquired *inability* to make an immune response directed against self antigens. Tolerance ensures that adaptive immunity is directed to outside agents that pose genuine threats to the host, and not to host proteins.

T Cells and Antigen Presentation

Adaptive immunity begins with the interactions of immune T lymphocytes with antigens on infected cells. The infected cells that are first recognized by T cells may include the same phagocytes that were involved in the innate immune response. The T cell, with its TCR, can recognize antigen only when the antigens are complexed with self proteins that are known as **major**

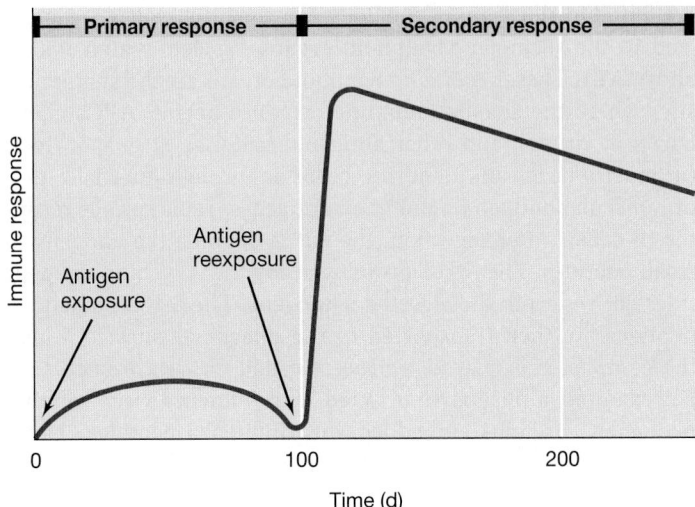

Figure 28.4 Primary and secondary immune responses. The antigens given at day 0 and day 100 must be identical to induce a secondary response. The secondary response may be more than 10-fold greater than the primary response.

histocompatibility complex (**MHC**) proteins and found on host cell surfaces (**Figure 28.5**). All host cells display MHC I proteins, and APCs (macrophages, dendritic cells, and B cells) also display an additional antigen-presenting protein, MHC II. Macrophages

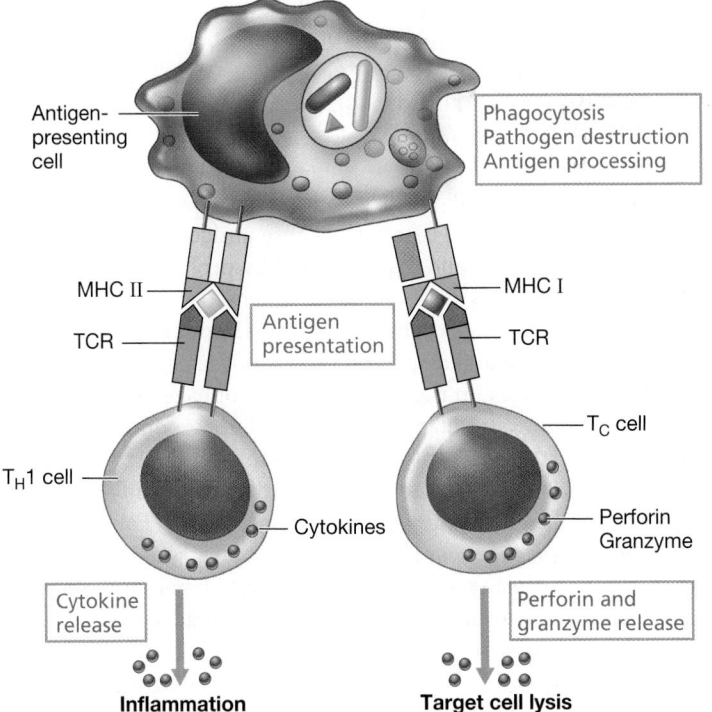

Figure 28.5 T cell immunity. Antigen-presenting cells such as the phagocytes in innate immunity ingest, degrade, and process antigens. They then present antigens to T cells that secrete protein cytokines that activate the adaptive immune response. Antigen-reactive T cells include inflammatory T-helper (T_H1) cells that make cytokines that activate other cells, causing inflammation. T-cytotoxic (T_C) cells produce perforins and granzymes, proteins that enter and lyse nearby target cells.

are found in all organs of the body, but the other APCs are localized in the secondary lymphoid organs—spleen, lymph nodes, and MALT. These secondary lymphoid organs are the anatomical sites where the adaptive immune response begins. APCs ingest bacteria, viruses, and other antigenic material by phagocytosis (in macrophages and dendritic cells) or through internalization of molecular antigen bound to an antigen-specific surface receptor (B cells). After ingestion, the APCs degrade the antigens to small peptides. The MHC proteins inside the APC bind the peptides derived from the digested pathogens. The MHC-embedded peptides are then transported to the phagocyte surface, where the complex is displayed, a process called *antigen presentation*. For example, a phagocyte infected with influenza virus will display MHC proteins embedded with influenza peptides. These MHC–peptide complexes are the targets for T cells.

T Lymphocyte Subsets

T cells interact with the peptide–MHC complex using the cell surface T cell receptor (TCR). Each T cell expresses a TCR that is specific for a single peptide–MHC complex. The antigen-specific T cells are found in the spleen, lymph nodes, and MALT closely associated with the APCs. The T cells constantly sample surrounding APC cells for peptide–MHC complexes. Peptide–MHC complexes that interact with the TCR send a signal to the T cell to grow and divide. The immune T cells produce antigen-reactive clones. These antigen-reactive T cells consist of three different T cell subsets, based on their functional properties. These T cell subsets interact with other cells to initiate immune reactions.

T-cytotoxic (T_C) *cells* recognize the antigen presented by an MHC I protein on an infected cell. When T_C cells interact with the infected cell, they secrete proteins that kill the antigen-bearing infected cell (Figure 28.5).

T-helper (T_H) *cells* interact with peptide–MHC II complexes on the surface of antigen-presenting cells. This interaction causes differentiation of the T_H cells, resulting in two subsets that indirectly mediate immune reactions. These antigen-activated T cell subsets, termed T_H1 and T_H2, respond by proliferating and producing soluble cytokines. Cytokines interact with receptors on other cells and activate them to initiate an immune response.

Differentiated antigen-specific T_H1 cells interact with peptide–MHC II complexes on the surface of macrophages (Figure 28.5). This interaction stimulates the T_H1 cell to produce cytokines that activate the macrophages, enhancing phagocytosis of any cells displaying the target antigen and causing inflammatory reactions that limit the spread of infections. For example, *Mycobacterium tuberculosis* infects macrophages and other cells in the lung, causing tuberculosis. Activated macrophages kill *M. tuberculosis* inside the cell, limiting spread to other cells. An inflammatory reaction associated with *M. tuberculosis* is termed the *tuberculin reaction* and is used as a diagnostic test for *M. tuberculosis* exposure. This test uses *tuberculin*, an extract from *M. tuberculosis*, to attract immune T_H1 cells that then produce cytokines, activating macrophages and causing a localized red, hot, hardened, and swollen area that typifies inflammation and effective immunity (**Figure 28.6**). Differentiated T_H2 cells, the other T_H subset, use cytokines to stimulate ("help") antigen-reactive B cells to produce antibodies, as we discuss below.

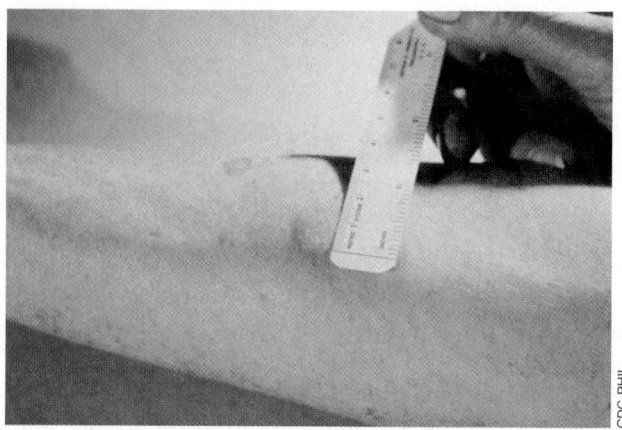

Figure 28.6 T_H1 **cells and macrophage activation.** This positive tuberculin test shows a reaction typical for inflammation due to T_H1 activation of macrophages. The raised area of inflammation on the forearm is about 1.5 cm in diameter. The action of macrophages activated by antigen-specific T_H1 cells caused the localized reaction to a tuberculosis antigen, tuberculin, at the site of injection.

MiniQuiz

- Explain the process of antigen presentation to T cells.
- Define the role of T_C and T_H1 cells in adaptive immunity.

28.4 Antibodies

Antibodies or immunoglobulins (Igs) are soluble proteins made by B cells and plasma cells in response to exposure to nonself antigens. Each antibody binds specifically to a single antigen. Antibody-mediated immunity controls the spread of infection by recognizing pathogens and their products in extracellular environments such as blood and mucus secretions.

B cells are specialized lymphocytes that have preformed antibodies on their surface; each B cell displays multiple copies of a single antibody that is specific for a single antigen. To make antibodies, B cells must first bind antigens through interactions with the surface antibodies. The surface antibody–antigen interaction induces the B cell to ingest the antigen-containing pathogen by phagocytosis. The B cell then kills and digests the pathogen, producing a battery of pathogen-derived peptide antigens. These antigens are then displayed, or presented, on the surface of the B cell to the antigen-specific T_H2 cell (**Figure 28.7**).

T_H2 cells do not interact directly with the pathogen, but stimulate ("help") other cells, in this case, the antigen-reactive B cells. T_H2 cells produce cytokines that stimulate antigen-reactive B cells, which, in turn, respond by growing and dividing, establishing clones of the original antigen-reactive B cell. These activated B cells then differentiate into plasma cells that produce antibodies (Figure 28.7). This initial antibody response, a primary adaptive response, is detectable within about five days, and antibodies reach peak quantities within several weeks. Subsequent exposure to the same antigen, for example by reinfection with the same pathogen, induces immune memory and a secondary

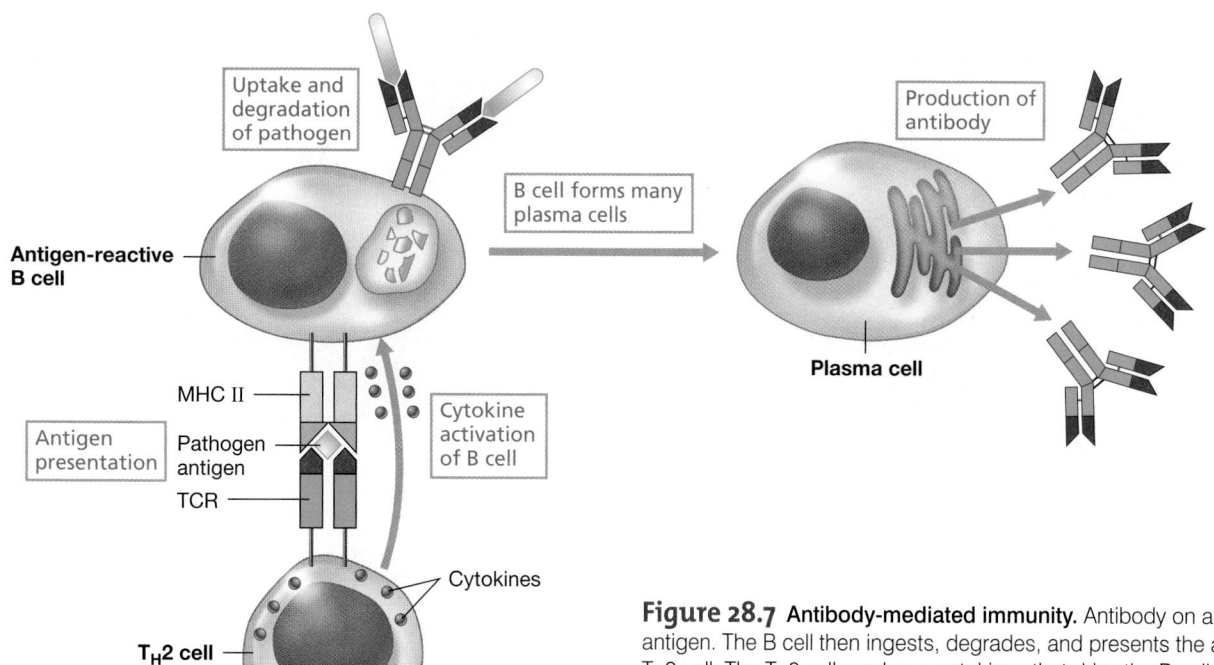

Figure 28.7 Antibody-mediated immunity. Antibody on a B cell binds antigen. The B cell then ingests, degrades, and presents the antigen to a T$_H$2 cell. The T$_H$2 cell produces cytokines that drive the B cell to form plasma cells, each producing antibodies. Antibody proteins each have two identical binding sites (red).

adaptive response, characterized by a faster development of higher quantities of antibodies (Figure 28.4).

Several different classes of antibodies are distinguished from one another by their primary amino acid sequence. Each antibody class has a specific general function. IgM and IgG are found in blood. The primary antibody response is characterized by production of large amounts of IgM antibodies, whereas the secondary antibody response is characterized by production of even larger amounts IgG. IgA is found in blood, and is also found in high concentrations in mucous membrane secretions, such as in the lungs and gut. IgE is found attached to the mast cells involved in parasite immunity and allergies (**Table 28.2**). IgD is found primarily as a surface immunoglobulin on B cells.

Antibodies released from the plasma cells interact with antigen on the pathogens. The antibody may have one or more effects on the pathogen. First, antibody interaction does not directly kill the pathogen, but may mark it for destruction by phagocytosis. Phagocytes have general antibody receptors called *Fc receptors* (*FcR*) that bind to any antibody attached to an antigen. This interaction results in enhanced phagocytosis of the antibody-coated cells, a process known as *opsonization*.

Antibody-mediated destruction of pathogens may also involve a group of proteins known collectively as *complement*. The complement proteins attach to pathogen surfaces, attracted by IgM or IgG antibodies bound to the pathogen. The complement proteins, concentrated at the cell surface by the antibody, have two possible effects on the pathogen. First, complement proteins can form a pore in the pathogen cytoplasmic membrane, directly lysing the pathogen cell. This complement–antibody interaction affects only those pathogen cells with bound antibodies. For example, antibodies specific for cell surface proteins of *Salmonella* interact only with *Salmonella*. Complement causes lysis of only the antibody-coated *Salmonella* cell, but not of a nearby *Escherichia coli* cell that is not coated with antibodies. Many pathogens, such as the thick-walled gram-positive *Streptococcus* spp., are relatively resistant to complement-mediated lysis because the cell wall makes the cytoplasmic membrane less accessible to complement proteins. However, antibodies to the external cell wall components can attract complement proteins to the pathogen surface. Here the complement proteins are bound by

Table 28.2 *Major soluble antibody classes*

Antibody class	Location	Functions
IgA	Serum and mucus secretions	Major effector of mucosal immunity
IgE	Bound to mast cells	Immediate hypersensitivity allergies, parasite immunity
IgG	Serum	Secondary serum antibody, highest concentration in blood (13.5 mg/ml)
IgM	Serum	Primary serum antibody, 1.5 mg/ml

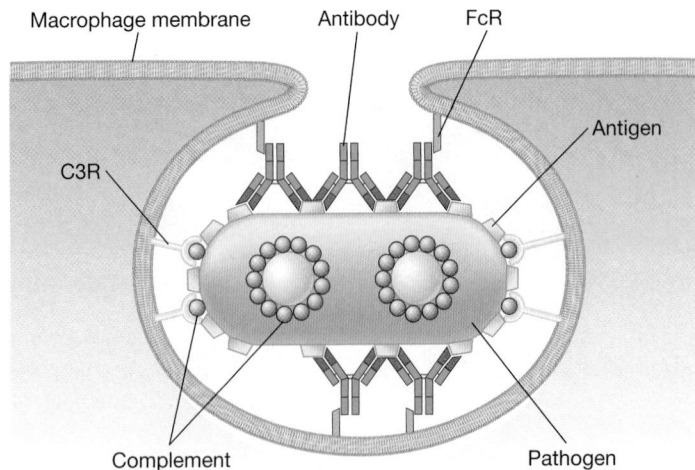

Figure 28.8 Antibodies, complement, and opsonization. Pathogen cells targeted by antibodies can be destroyed by several mechanisms. After antibody binds to antigen, complement proteins (red) are attracted to the cell. Complement may form pores and directly lyse the cell. In addition, phagocytes such as neutrophils and macrophages have receptors that bind antibodies (FcR; green) or complement (C3R; yellow). Interactions with these receptors enhance phagocytosis, a phenomenon called opsonization.

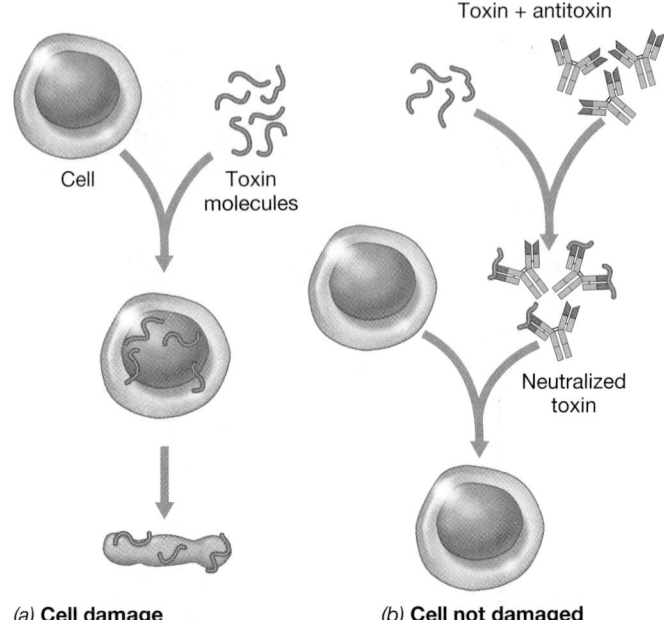

(a) Cell damage **(b) Cell not damaged**

Figure 28.9 Neutralization of an exotoxin by an antitoxin antibody. (a) Untreated toxin results in cell destruction. (b) Antitoxin antibody blocks toxin binding, neutralizing the toxin and preventing cell destruction.

complement receptors called *C3 receptors* (C3R) found on the surface of phagocytes such as neutrophils and macrophages. This interaction results in enhanced opsonization and phagocytosis of the antibody-complement sensitized cells (**Figure 28.8**).

Finally, antibodies may also block interactions between pathogens or their products and host cells. For example, IgA antibodies present in mucosal secretions and directed against influenza virus may interact with influenza virus antigens that bind to host cells, blocking attachment of the influenza virus to the host cell. Specific serum antibodies can also bind toxins such as tetanus toxin, again blocking the binding of toxin to host cell receptors. This process is called *neutralization* (**Figure 28.9**).

The antibody response is highly specific for the eliciting antigen. Antibodies interact with antigens, triggering lysis and phagocytosis of pathogens through opsonization and complement binding.

MiniQuiz

- Explain the process of antibody production starting with pathogen interaction with a B cell.
- Define the role of antibody and complement in pathogen destruction.

28.5 Inflammation

Inflammation is a general, nonspecific reaction to noxious stimuli such as toxins and pathogens. Inflammation is characterized by redness (erythema), swelling (edema), pain, and heat, usually

localized at the site of infection (Figure 28.6 and **Figure 28.10**). The molecular mediators of inflammation include a group of cell activators and chemoattractants called **cytokines** and **chemokines**. These proteins are produced by various cells, but especially by phagocytes and lymphocytes.

Infection recognition through either innate or adaptive immune response mechanisms can cause inflammation; both immune

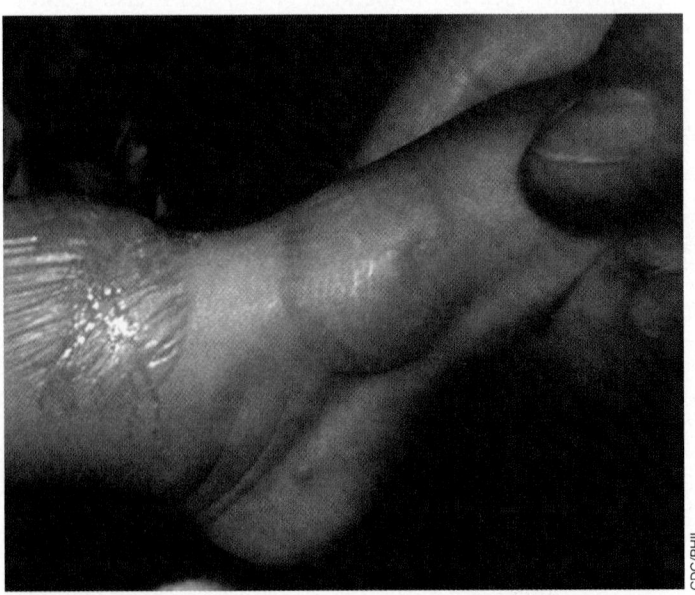

Figure 28.10 Inflammation. The swelling in this child's foot is due to infection with vaccinia virus and the resulting inflammation.

recognition systems induce inflammatory mediators such as cytokines and complement proteins to interact with or recruit and activate effector cells such as macrophages. Effective immunity stimulates the inflammatory response to isolate and limit tissue damage, destroying pathogen invaders and damaged cells. In some cases, however, inflammation can result in considerable damage to healthy host tissue.

Inflammatory Cells and Local Inflammation

Immune-mediated inflammation begins at the site of pathogen entry into the body. The innate PRRs on macrophages, the first immune cells to the site of infection, engage the pathogen PAMPs (Figure 28.3). This action activates the macrophage to produce and release cytokines and chemokines that interact with cytokine and chemokine receptors on other cells, such as neutrophils. For example, activated macrophages secrete a chemokine called CXCL8. Neutrophils, through a CXCL8 receptor, are activated and migrate along the chemokine gradient to the pathogens, where they begin to ingest and kill the pathogen. The neutrophils, in turn, secrete chemokines that attract other neutrophils, amplifying the response and destroying the pathogens.

The chemokine and cytokine mediators released by injured cells and phagocytes contribute to inflammation. For example, the macrophage produces *proinflammatory cytokines* including interleukin-1 (IL-1), IL-6, and tumor necrosis factor α (TNF-α). These cytokines increase vascular permeability, causing the swelling (edema), reddening (erythema), and local heating associated with inflammation. The edema stimulates local neurons, causing pain (**Figure 28.11a**).

The usual outcome of the inflammatory response is a rapid localization and destruction of the pathogen by macrophages and recruited neutrophils. As the pathogens are removed, the inflammatory cells are no longer stimulated, their numbers at the site are reduced, cytokine production decreases, chemoattraction stops, and inflammation subsides.

Systemic Inflammation and Septic Shock

In some cases, the inflammatory response fails to localize the pathogens and the reaction becomes systemic. Then, inflammatory cells and mediators contribute to inflammation on a larger scale. An inflammatory response that spreads inflammatory cells and mediators through the entire circulatory and lymphatic systems can lead to septic shock, a life-threatening condition. A common cause of septic shock is systemic infection by gram-negative enteric bacteria such as *Salmonella* or *Escherichia coli*, often caused by a ruptured or leaking bowel that releases the gram-negative organisms into the intraperitoneal cavity or the bloodstream. The primary infection is often cleared by the phagocytes or is treated successfully with antibiotics. However, the endotoxic outer membrane lipopolysaccharide (LPS) from these organisms interacts with a PRR on phagocytes, stimulating production of inflammatory cytokines, which are released into the systemic circulation. The cytokines then induce systemic responses that parallel the localized inflammatory response. In the case of systemic inflammation, however, many organs and

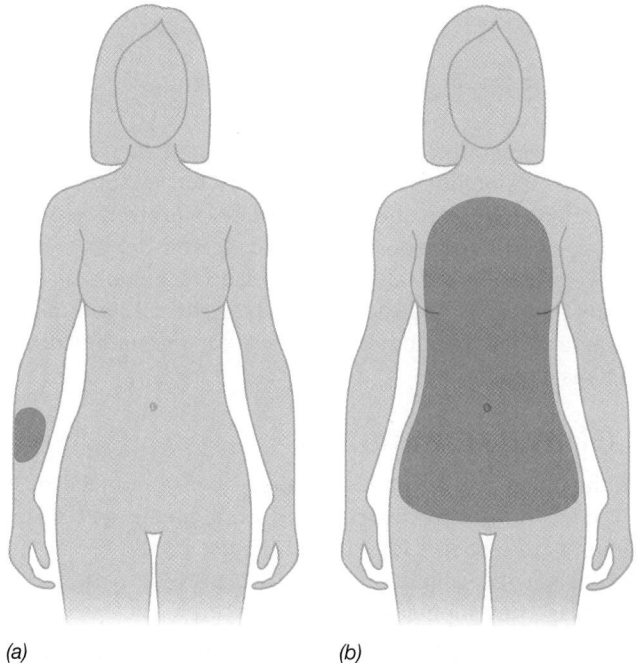

(a)   *(b)*

Figure 28.11 Local and systemic inflammation. *(a)* Inflammation at a site of local infection is mediated by the release of proinflammatory cytokines from the local surrounding macrophages. The result is a discrete area of inflammation that subsides as the infection is cleared. *(b)* Inflammation due to a systemic infection causes systemic release of proinflammatory cytokines, resulting in widespread systemic inflammatory symptoms including severe edema, fever, and shock. Severe systemic inflammation, known as septic shock, persists due to the cytokines released throughout the body even if the infection is controlled.

systems can be involved, leading to an extensive whole-body inflammatory event with potentially disastrous results. For example, the inflammatory cytokines IL-1, IL-6, and TNF-α are *endogenous pyrogens*, producing fever by stimulating release of prostaglandins in the brain. The systemic release of large quantities of endogenous pyrogens, instead of producing localized heating, induces uncontrollable high fever. When large amounts of inflammatory mediators are released systemically, the mechanism that causes local edema due to vasodilation and increased vascular permeability causes massive efflux of fluids from the central vascular tissue. The outcome is loss of systemic blood pressure and severe edema. The resulting condition, termed septic shock, is characterized by loss of blood volume as well as high fever and causes death in up to 30% of affected individuals (Figure 28.11b). Uncontrolled systemic inflammation can be more dangerous than the original infection.

MiniQuiz

- Identify the major symptoms of localized inflammation and of septic shock.
- Identify the molecular mediators of inflammation and define their individual roles.

UNIT 9

Ⅱ Prevention of Infectious Diseases

Immunity generated by natural exposure to pathogens is a very effective, if potentially dangerous, way to develop resistance to infections. We can, however, initiate a protective immune response by artificial exposure to nondangerous forms of a pathogen; we routinely and safely induce adaptive immune responses to many pathogens and their products. Artificial immunization remains our best public health defense for prevention of many infectious diseases.

28.6 Natural Immunity

Both the innate and adaptive immune responses protect the host from infections by pathogens, and both innate and adaptive immunity are essential for survival (**Figure 28.12**). For example, individuals with genetic defects that prevent neutrophil or macrophage development fail to produce phagocytes and thus lack innate immunity. Individuals with such defects cannot live without extraordinary intervention such as isolation from all environmental exposure. In a normal environment, they develop recurrent infections from bacteria, viruses, and fungi and die at an early age. Individuals who lack adaptive immunity have the same outcome, as we discuss below.

Active and Passive Immunity

Animals normally develop *natural active immunity* by acquiring a natural infection that initiates an adaptive immune response. Natural active immunity is the outcome of exposure to antigens through infection and usually results in protective immunity conferred by antibodies and T cells. For example, virtually all human adults have acquired active, protective immunity to many strains of influenza and cold viruses through immune responses to natural infections.

Natural passive immunity is a nonimmune person's acquisition of preformed immune cells or antibodies via natural transfer of cells or antibodies from an immune person. For example, for several months after birth, newborns have maternal IgG antibodies, transferred across the placenta before birth, in their blood. Also, IgA antibodies are transferred in breast milk. The antibodies that are protective for the infant were made in the mother, thus their designation as passively acquired antibodies. These preformed antibodies provide disease protection while the immune system of the newborn matures. Active and passive immunity are contrasted in **Table 28.3**.

Immune Deficiencies

The importance of active immunity in disease resistance is shown dramatically in individuals who are immunocompromised due to genetic defects or infection. For example, *agammaglobulinemia* is a disease in which patients cannot produce antibodies because of genetic defects in their B cells, and we know that antibody-mediated immunity is essential for protection from extracellular pathogens, especially bacteria. Therefore, persons with aggammaglobulinemia do not make protective antibodies and suffer from

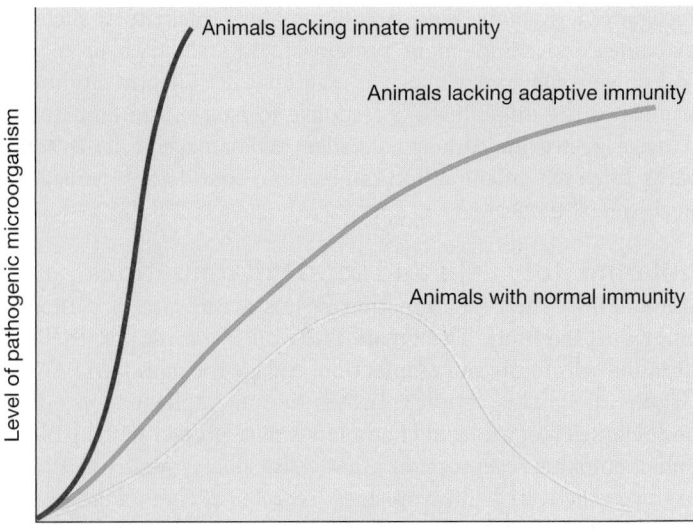

Figure 28.12 Infection and pathogen clearance in normal and immune-deficient animals. Animals with genetic defects that prevent development of the phagocytes critical for innate immunity have recurrent, incurable, lethal infections. Genetic defects that prevent development of mature, antigen-reactive B and T cells critical for adaptive immunity also allow recurrent infections, but the innate response controls these infections for a longer time and these animals live longer than the animals lacking innate immunity. Animals with normal innate and adaptive immunity rapidly clear most infections.

recurrent, life-threatening bacterial infections, but develop normal immune responses to viruses.

Individuals with *DiGeorge syndrome*, a developmental defect that prevents maturation of the thymus and inhibits production of mature T cells, suffer from serious recurrent infections with viruses and other intracellular pathogens. Thus, the T cell immune-deficiency problems seen in DiGeorge syndrome define the essential protective role for T cell immunity: protection from intracellular pathogens.

Table 28.3 *Active and passive immunity*

Active immunity	Passive immunity
Exposure to antigen; immunity achieved by injecting antigen or through infection	No exposure to antigen; immunity achieved by injecting antibodies or antigen-reactive T cells
Specific response made by individual achieving immunity	Specific immune response made by the donor of antibodies or T cells
Immune system activated by antigen; immune memory in effect	No immune system activation; no immune memory
Immune response can be maintained via stimulation of memory cells (i.e., booster immunization)	Immunity cannot be maintained and decays rapidly
Immune state develops over a period of weeks	Immunity develops immediately

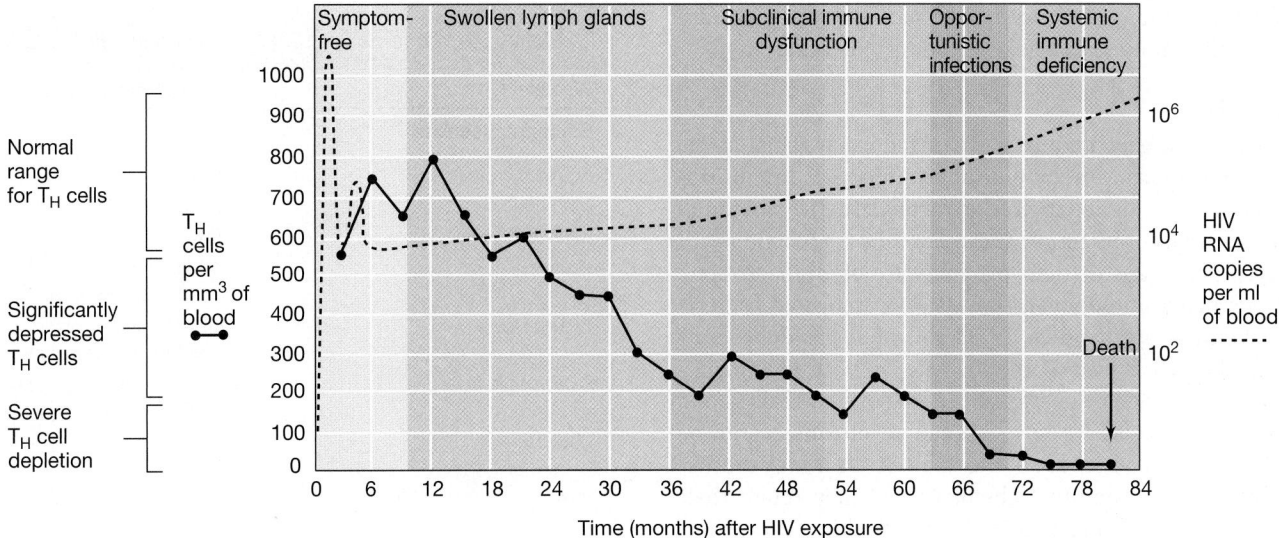

Figure 28.13 Decline of T-helper (T$_H$) lymphocytes and progress of HIV infection. During progression of untreated HIV infection, AIDS develops. There is a gradual loss in the number and functional ability of the T$_H$ cells, while the viral load, measured as HIV-specific RNA copies per milliliter of blood, gradually increases after an initial decline. The lack of an effective immune response leads to an increase in life-threatening infections.

Individuals with *severe combined immune deficiency syndrome* (*SCID*) have a genetic defect that prevents proper formation and expression of immunoglobulins or TCRs. As a result, they have no effective adaptive immunity.

The loss of the adaptive immune response is the defining characteristic of individuals with *acquired immunodeficiency syndrome* (*AIDS*). In AIDS patients, infection with the human immunodeficiency virus (HIV), if not controlled, causes nearly total depletion of T$_H$ cells, resulting in a lack of effective T cell immunity and antibodies (**Figure 28.13**). Such individuals suffer from recurrent, severe infections due to viral, bacterial, and fungal pathogens. Death from AIDS is characteristically due to secondary infections by one or more opportunistic pathogens (↺ Section 33.14). Other forms of acquired immunodeficiency can be caused by toxic reactions to drugs or environmental contaminants.

MiniQuiz

- Provide examples of natural active immunity and natural passive immunity.
- Describe the effects of the lack of B cell– or T cell–mediated immunity.

28.7 Artificial Immunity and Immunization

Immunization, the purposeful artificial induction of immunity to particular infectious diseases, is a major weapon for the prevention and treatment of these diseases. There are two ways by which artificial immunity can be induced. An individual may be purposefully exposed to a controlled dose of harmless antigen to induce formation of antibodies, a type of immunity called *artificial active immunity* because the recipient produces the antibodies. This process is commonly known as **vaccination**. Alternatively, an individual may receive injections of an *antiserum* (serum containing antibodies from the blood of an immune individual) or purified antibodies (immunoglobulin) derived from an immune individual. This is *artificial passive immunity* because the individual receiving the antibodies played no active part in antibody production.

In active immunity, introduction of antigen induces changes in the host: The immune system produces large quantities of antibodies and, more importantly, a population of immune memory cells in the primary response. A second ("booster") dose of the same antigen results in a faster response yielding much higher levels of antibodies and effector T cells due to this secondary, or memory, immune response. Active immunity often remains throughout life as a result of immune memory.

A passively immunized individual never has more antibodies than are received in the injection, and these antibodies gradually disappear from the body. Moreover, a later exposure to the antigen does not elicit a secondary response. Artificial passive immunity is usually therapeutic. Cells or antibodies from an immune individual are transferred to a nonimmune individual to prevent or cure active disease. For example, tetanus antiserum may be administered to passively immunize an individual suspected of being exposed to *Clostridium tetani* due to an acute injury such as a car accident. Such an individual needs immediate immune protection against the acute disease and cannot wait days to weeks for immunization to produce active immunity.

Artificial active immunity, as we discuss next, is often used as a prophylactic measure to protect a person against future attack by a pathogen. For example, immunization protects individuals against future encounters with *C. tetani* exotoxin, but is not an effective

UNIT 9

therapy for the trauma victim in the car accident above because effective adaptive immunity takes several days to develop.

Immunization

The antigen or antigen mixture used to induce artificial active immunity is known as a **vaccine** or an immunogen. Immunization with a vaccine designed to produce artificial active immunity may introduce risks of infection and other adverse reactions. To reduce risks, pathogens or their products are often inactivated. For example, many immunogens consist of pathogenic bacteria killed by chemical agents such as phenol or formaldehyde, or physical agents such as heat. Formaldehyde is also used to inactivate viruses for vaccines, such as in the inactivated (Salk) polio vaccine. Likewise, the active form of many exotoxins cannot be used as an immunogen because of the toxic effects. Many exotoxins, however, can be modified chemically so they retain their antigenicity but are no longer toxic. Such a modified exotoxin is called a **toxoid**. Toxoids such as the one that is the vaccine for *C. tetani* exotoxin can be given safely in doses large enough to induce protective immunity against the exotoxin.

Immunization with live cells or virus is usually more effective than immunization with dead or inactivated material. It is often possible to isolate a mutant strain of a pathogen that has lost its virulence but still retains the immunizing antigens; strains of this type are called *attenuated strains* (⮌ Section 27.6). However, because attenuated strains of pathogens are still viable, some individuals, especially those who are immunocompromised, may acquire active disease caused by the live, attenuated immunizing pathogen. Serious cases of disease have been caused by vaccine-acquired infections in immunocompromised individuals, for example, from attenuated poliovirus vaccines and smallpox virus vaccines.

A summary of vaccines available for use in humans is given in **Table 28.4**. Many effective viral vaccines are live attenuated vaccines. Attenuated vaccines tend to provide long-lasting T cell–mediated immunity, as well as a vigorous antibody response and a strong secondary response upon reimmunization. However, attenuated vaccine strains are difficult to select, standardize, and maintain. Because they are alive, attenuated vaccines usually have a limited shelf life and require refrigeration for storage. Killed virus vaccines, on the other hand, tend to provide short-lived immune responses without the development of a long-term memory response, but are relatively easy to store and maintain their potency for long periods of time.

Bacterial vaccines are nearly always provided in an inactivated form. Inactivated bacterial vaccines induce long-term antibody-mediated protection without exposing recipients to the risk of infection.

Immunization Practices

Infants acquire natural passive immunity from maternal antibodies transferred across the placenta or in breast milk. As a result, infants are relatively immune to common infectious diseases during the first 6 months of life. However, infants should be immunized to prevent key infectious diseases as soon as possible so that their own active immunity can replace the maternal passive immunity. As discussed in Section 28.3, a single exposure to

Table 28.4 *Vaccines for infectious diseases in humans*

Disease	Type of vaccine used
Bacterial diseases	
Anthrax	Toxoid
Diphtheria	Toxoid
Tetanus	Toxoid
Pertussis	Killed bacteria (*Bordetella pertussis*) or acellular proteins
Typhoid fever	Killed bacteria (*Salmonella enterica* serovar Typhi)
Paratyphoid fever	Killed bacteria (*Salmonella enterica* serovar Paratyphi)
Cholera	Killed cells or cell extract (*Vibrio cholerae*)
Plague	Killed cells or cell extract (*Yersinia pestis*)
Tuberculosis	Attenuated strain of *Mycobacterium tuberculosis* (BCG)
Meningitis	Purified polysaccharide from *Neisseria meningitidis*
Bacterial pneumonia	Purified polysaccharide from *Streptococcus pneumoniae*
Typhus fever	Killed bacteria (*Rickettsia prowazekii*)
Haemophilus influenzae meningitis	Conjugated vaccine (polysaccharide of *Haemophilus influenzae* conjugated to protein)
Viral diseases	
Influenza	Inactivated virus
Hepatitis A	Recombinant DNA vaccine
Hepatitis B	Recombinant DNA vaccine or inactivated virus
Human papillomavirus (HPV)	Recombinant DNA vaccine
Measles	Attenuated virus
Mumps	Attenuated virus
Rubella	Attenuated virus
Polio	Attenuated virus (Sabin) or inactivated virus (Salk)
Rabies	Inactivated virus (human) or attenuated virus (dogs and other animals)
Rotavirus	Attenuated virus
Smallpox	Cross-reacting virus (vaccinia)
Varicella (chicken pox)	Attenuated virus
Yellow fever	Attenuated virus

antigen does not lead to a high antibody *titer*, or antibody quantity. After an initial immunization, a series of secondary or "booster" immunizations are given to produce a secondary response and a high antibody titer. Current vaccine recommendations for children and adults in the United States are shown in **Figure 28.14**. Many vaccines require a series of immunizations to establish protective immunity; periodic reimmunization is often necessary to maintain immunity.

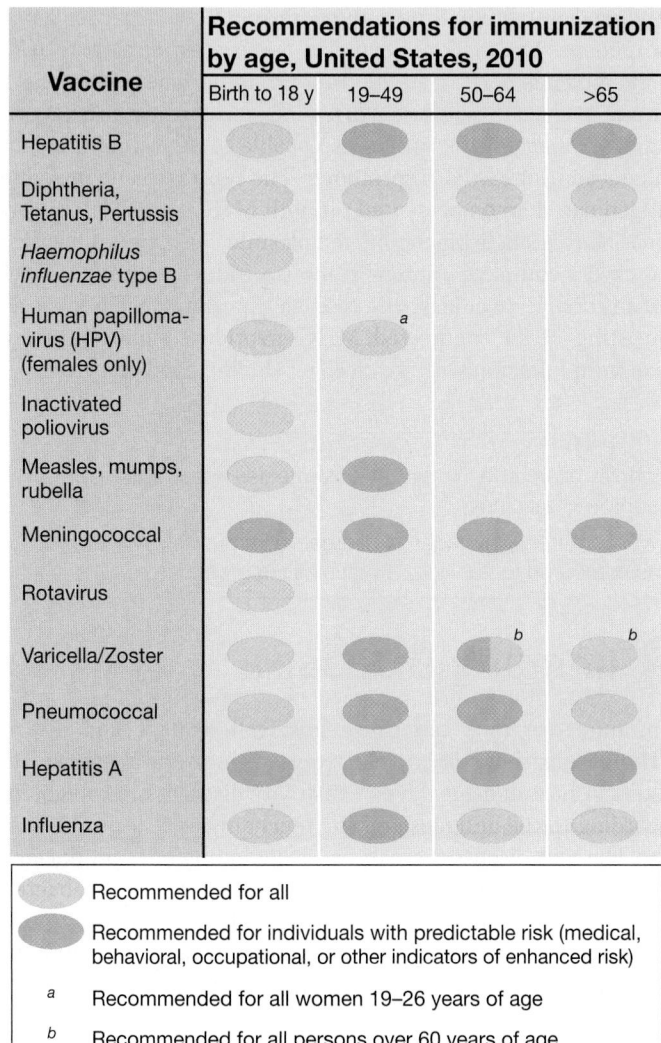

Vaccine	Recommendations for immunization by age, United States, 2010			
	Birth to 18 y	19–49	50–64	>65
Hepatitis B				
Diphtheria, Tetanus, Pertussis				
Haemophilus influenzae type B				
Human papilloma-virus (HPV) (females only)		*a*		
Inactivated poliovirus				
Measles, mumps, rubella				
Meningococcal				
Rotavirus				
Varicella/Zoster			*b*	*b*
Pneumococcal				
Hepatitis A				
Influenza				

- Recommended for all
- Recommended for individuals with predictable risk (medical, behavioral, occupational, or other indicators of enhanced risk)
- *a* Recommended for all women 19–26 years of age
- *b* Recommended for all persons over 60 years of age

Figure 28.14 Recommended immunizations for children and adults in the United States. This general course of immunizations is specified by the Centers for Disease Control and Prevention, Atlanta, Georgia, as of 2010. The CDC National Immunization Program website (http://www.cdc.gov/vaccines/) has specific immunization recommendations for timing and dose of immunizations for all age groups. In addition, the website has specific vaccine recommendations for international travelers, women of child-bearing age, and persons with medical conditions such as immunodeficiencies and chronic diseases.

The importance of immunization in controlling infectious diseases is well established. For example, introduction of an effective vaccine into a population has reduced the incidence of formerly epidemic childhood diseases such as measles, mumps, and rubella (ᐸ᠈ Figure 33.15) and has eliminated smallpox altogether (ᐸ᠈ Section 32.11). The degree of immunity obtained by vaccination, however, varies greatly with the individual as well as with the quality and quantity of the vaccine. Lifelong immunity is rarely achieved by means of a single injection, or even a series of injections, and the immune cells and antibodies induced by immunization gradually disappear from the body. On the other hand, natural infections may stimulate immune memory or *booster* responses. In the complete absence of antigenic stimulation, the length of effec-

tive immunity varies considerably with different antigens. For example, protective immunity to tetanus from toxoid immunization may last many years. As a result, current recommendations call for reimmunization in adults only every 10 years to maintain protective immunity. Immunity induced by a particular influenza virus vaccine, however, disappears within a year or two without reimmunization through active infection or vaccination.

As we noted above, passive immunity is introduced by injecting preformed antibodies. The antibody-containing preparation is known as an *antiserum*, or an *antitoxin* if the antibodies are directed against a toxin. Antisera are obtained from immunized animals, such as horses, or from humans with high antibody titers. These individuals are said to be *hyperimmune*. The antiserum or antitoxin is standardized to contain a known antibody titer; a sufficient number of units of antiserum must be injected to neutralize any antigen that might be present in the body. The immunoglobulin fraction separated from serum pooled from a number of individuals is also used for passive immunization. Pooled sera contain antibodies induced by artificial or natural exposure to various antigens.

Immunizations, whether active or passive, benefit the individual. Immunization is a major tool for public health disease-control programs because infections spread poorly in populations with a large proportion of immune individuals.

MiniQuiz
- Provide an example of artificial passive immunity. How does artificial passive immunity benefit the immunized individual?
- Review the immunization recommendations for individuals in your age group. How do these artificial active immunizations benefit the immunized individual?

28.8 New Immunization Strategies

Many vaccines are derived from whole organisms or toxoids, as described in the previous section. However, there are several other methods for producing antigens that are suitable for vaccines.

Synthetic and Genetically Engineered Vaccines

The simplest alternate approach to vaccine development is the use of *synthetic peptides*. To make a vaccine, a genetic engineer can synthesize a peptide that corresponds to an antigen of an infectious agent. For example, the structure of the toxin from the foot-and-mouth virus, an important animal pathogen, is known. Because the whole protein is toxic, it cannot be used as a vaccine. However, a peptide of 20 amino acids constitutes an important protective antigen in the protein. Because an antigen must be at least 100 amino acids long to be effective, a synthetic version of the peptide is not an effective vaccine by itself. Genetic engineers, however, attached the small peptide to a large, innocuous protein that acts as a carrier molecule. The synthetic vaccine produced a protective response to foot-and-mouth virus infection. This strategy has great promise for creating vaccines directed to a number of pathogens, but the entire sequence of the disease-causing protein must be known and the part that is the antigen must be identified before an effective vaccine can be engineered. The entire genomic sequences of many pathogens are now known, however, providing the information necessary to identify the antigenic part of each.

Molecular techniques can be used to make synthetic vaccines using information derived from pathogen genomics. For example, genes that encode antigens from virtually any virus can be cloned into the vaccinia virus genome and expressed. Inoculation with the genetically engineered vaccinia virus can then be used to induce immunity to the product of the cloned gene. Such a preparation is called a *recombinant-vector vaccine*. This method depends on the identification and cloning of the gene that encodes the antigen and also on the ability of the vaccinia virus to express the cloned gene as an antigenic protein. An effective recombinant vaccinia–rabies vaccine has been developed for use in animals. Recombinant DNA methods for vaccine development were discussed in Section 15.13.

Another immunization strategy involves the use of recombinant DNA proteins as immunogens. First, a pathogen gene must be cloned in a suitable microbial host that expresses the protein encoded by the cloned gene. The pathogen protein can then be harvested and used as a vaccine; such a vaccine is called a *recombinant-antigen vaccine*. For example, the current hepatitis B virus vaccine is a major hepatitis surface protein antigen (HbsAg) expressed by genetically modified yeast cells. A vaccine that is effective against human papillomavirus (HPV) is also a recombinant-antigen vaccine made in yeast cells (Microbial Sidebar, "The Promise of New Vaccines").

DNA Vaccines

A novel method for immunization is based on expression of cloned genes in host cells. *DNA vaccines* are bacterial plasmids that contain cloned DNA with the antigen of interest. Typically, the vaccine is injected intramuscularly into a host animal. Taken up by host cells, the DNA is transcribed and translated to produce immunogenic proteins, triggering a conventional immune response including T_C cells, T_H1 cells, and antibodies directed to the protein encoded by the cloned DNA.

DNA vaccine strategies may provide considerable advantages over conventional immunizations. For instance, because only a single pathogen gene is cloned and injected, there is no chance of an infection as there might be with an attenuated vaccine. Second, genes for individual antigens such as a tumor-specific antigen can

be cloned, targeting the immune response to a particular cell component. The response can also be targeted directly to APCs by including an MHC class II promoter in the gene construct. The promoter ensures selective expression in dendritic cells, B cells, and macrophages, the only cells capable of activating the genes influenced by the MHC II promoter. The expressed and processed antigen can then be presented on both MHC I and MHC II proteins. Thus, a single bioengineered plasmid can encode an antigen and elicit a complete immune response, inducing immune T cells and antibodies. In at least one case, an experimental DNA vaccine consisting of an engineered MHC–peptide complex protected mice from infection with a cancer-producing papillomavirus.

MiniQuiz
- Identify alternative immunization strategies already used for approved vaccines.
- What are the advantages of alternative immunization strategies as compared to traditional immunization procedures?

 Immune Diseases

Immune reactions can cause host cell damage and disease. **Hypersensitivity** is an inappropriate immune response that results in host damage. Hypersensitivity diseases are categorized according to the antigens and the mechanisms that produce disease. Here we discuss these diseases, including ones produced by superantigens, which are proteins produced by certain bacteria and viruses that cause widespread stimulation of immune cells, resulting in host damage by activating massive inflammatory responses.

28.9 Allergy, Hypersensitivity, and Autoimmunity

Antibody-mediated **immediate hypersensitivity** is commonly called *allergy*. Cell-mediated reactions also cause disease in the form of **delayed hypersensitivity**. *Autoimmune diseases* result from immune reactions directed against self antigens. These diseases are categorized as type I, II, III, or IV hypersensitivities based on symptoms, antigens, and immune effectors (**Table 28.5**).

Table 28.5 *Hypersensitivity*

Classification	Description	Immune mechanism	Time of latency	Examples
Type I	Immediate	IgE sensitization of mast cells	Minutes	Reaction to bee venom (sting) Hay fever
Type II	Cytotoxic[a]	IgG interaction with cell surface antigen	Hours	Drug reactions (penicillin)
Type III	Immune complex	IgG interaction with soluble or circulating antigen	Hours	Systemic lupus erythematosus (SLE)
Type IV	Delayed type	T_H1 inflammatory cell activation of macrophages	Days (24–48 h)	Poison ivy Tuberculin test

[a]Autoimmune diseases may be caused by type II, type III, or type IV reactions.

The Promise of New Vaccines

A vaccine has been licensed for protection against an infectious disease that is a major cause of cancer in women. This vaccine, called *Gardasil*, protects against infection by human papillomavirus (HPV). Gardasil has been licensed and recommended for females entering puberty and older. Various strains of HPV infect up to 75% of sexually active people and cause genital warts and vulvar, vaginal, and cervical cancers in infected women. The vaccine is targeted to HPV types 6, 11, 16, and 18. Together, these strains account for 70% of cervical cancers and 90% of genital warts (Figure 1).

The HPV vaccine is a preparation of virus-like particles of the major L1 capsid proteins from these viruses. The L1 proteins have been genetically engineered to be expressed by the yeast *Saccharomyces cerevisiae* and are released by disruption of the recombinant yeast cells as self-assembled virus-like particles. After purification, particles are adsorbed onto a chemical adjuvant. The adjuvant immobilizes particles, enhancing their ability to be taken up by phagocytes after injection.

The HPV vaccine is highly protective against viral infection, including genital warts (Figure 1) and all forms of cancers caused by the targeted HPVs. Because HPV is responsible for so much cervical cancer, risk in the immunized population will be significantly reduced. However, perhaps as important, the herd immunity (🔗 Section 32.5) resulting from immunization of a large proportion of the population will stop the spread of these viruses, providing protection even for individuals who are not immunized.

This vaccine is currently recommended for girls and women ranging in age from pre-puberty to the end of their reproductive years. While immunization of these susceptible individuals will certainly contribute to disease prevention, herd immunity can be raised to meaningful levels by immunizing all potential HPV sources such as the sexual partners of the women: The immunization of boys and men could provide the herd immunity necessary to eliminate transmission of this preventable sexually transmitted infection.

Another newly released vaccine is protective for infection by rotavirus, a common enterovirus that causes severe diarrhea, resulting in dehydration and even death in children worldwide. This intestinal disease can now be prevented by immunization with multiple doses of an oral, attenuated rotavirus. An earlier rotavirus vaccine was recalled due to postimmunization complications.

The HPV and rotavirus vaccines are two examples of effective vaccine development and implementation, but a number of important infectious diseases still cannot be prevented by vaccination. This list includes tuberculosis, malaria, and HIV, the three most important infectious diseases worldwide in terms of total disease and death. Over 1 million people die each year from each of these diseases. The current tuberculosis vaccine, BCG (🔗 Section 33.4), is considered inadequate and is not administered in many countries. No vaccine exists for the other diseases, although several vaccines are in development, and there are clinical trials for each of these. Finally, Gardasil is the only vaccine that is effective against any sexually transmitted infection in humans.

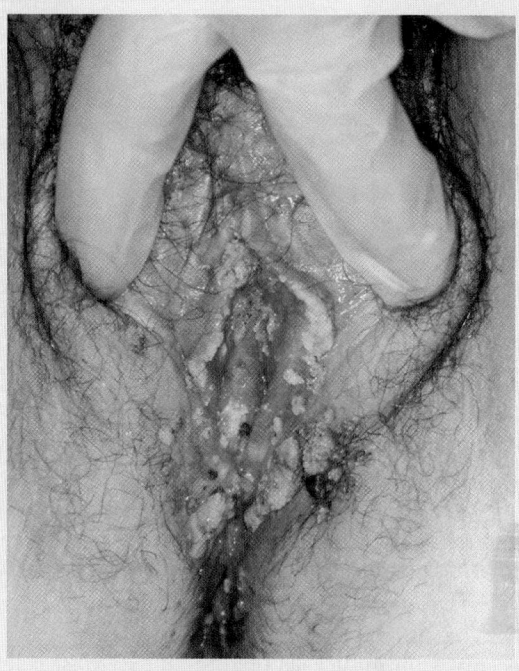

Figure 1 Genital warts caused by infection with human papillomavirus.

Even some effective vaccines have serious limitations. For example, influenza vaccines are only useful for one year because they are designed to target the strain-specific H and N antigens currently in circulation (🔗 Section 33.8). Development of a universal influenza vaccine that targets a common influenza virus antigen, M1, has been proposed, as in theory it should induce immunity to all influenza strains with a single vaccine. The immunogenicity and protection against influenza provided by M1 and other common antigens is, however, unproven.

Immediate Hypersensitivity

Immediate hypersensitivity, or type I hypersensitivity, is caused by release of vasoactive products from mast cells coated with IgE (**Figure 28.15**). Immediate hypersensitivity reactions occur within minutes after exposure to an *allergen*, the antigen that caused the type I hypersensitivity. Depending on the individual and the allergen, immediate hypersensitivity reactions can be very mild or can cause a life-threatening reaction called *anaphylaxis*.

About 20% of the population suffers from immediate hypersensitivity allergies to pollens, molds, animal dander, certain foods, insect venoms, and other agents (**Table 28.6**). Almost all allergens enter the body at the surface of mucous membranes such as the lungs or the gut. Initial exposure to allergens stimulates mucosa-associated T_H2 cells to produce cytokines that induce B cells to make IgE antibodies. Rather than circulating like IgG or IgM, the allergen-specific IgE antibodies bind to IgE

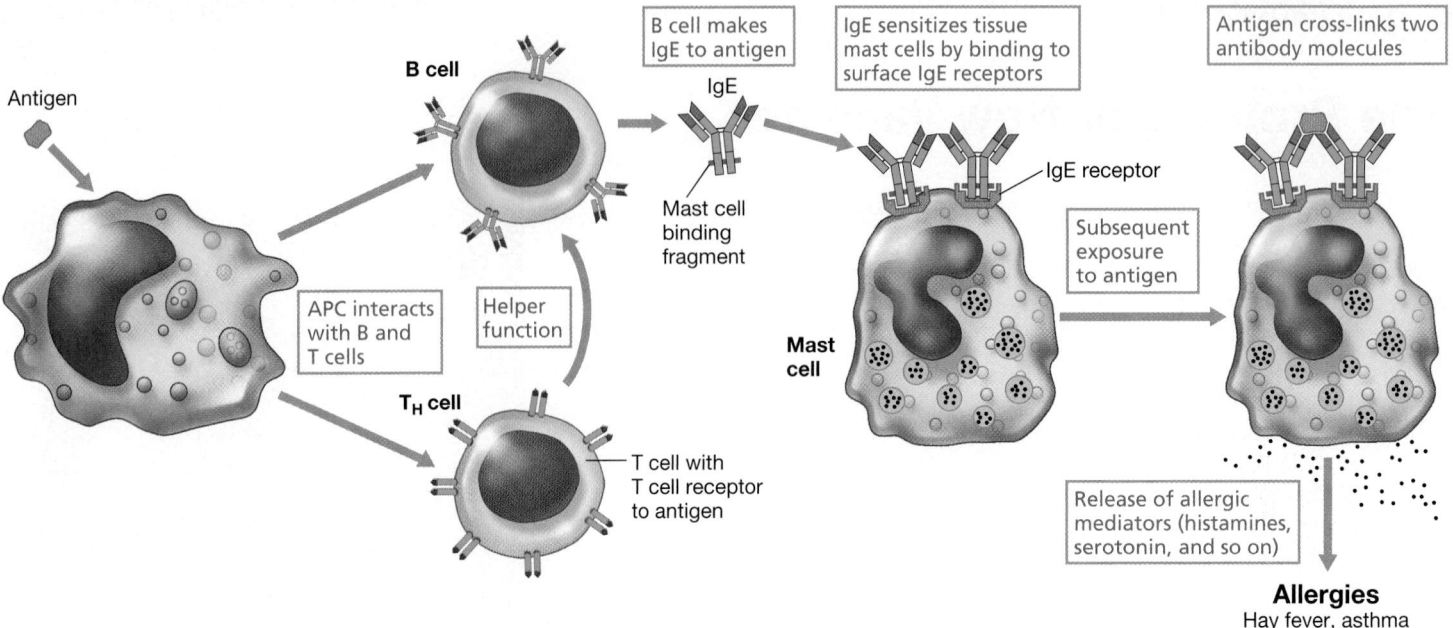

Figure 28.15 Immediate hypersensitivity. Certain antigens such as pollens stimulate IgE production. IgE binds to mast cells by means of a high-affinity surface receptor; binding sensitizes the mast cell. Antigen cross-links surface IgE, causing release of soluble mediators such as histamine. These mediators produce symptoms ranging from mild allergic symptoms to life-threatening anaphylaxis.

receptors on mast cells (Figure 28.15). Mast cells are nonmotile granulocytes (Section 28.1) associated with the connective tissue adjacent to capillaries throughout the body. With any subsequent exposure to the immunizing allergen, the mast cell–bound IgE molecules bind the antigen. Cross-linking of two or more IgEs by an antigen triggers the release of soluble allergic mediators from the mast cells, a process called *degranulation*. These mediators cause allergic symptoms within minutes of antigen exposure. In general, these symptoms are relatively short-lived. After initial sensitization by an allergen, the allergic individual responds to each subsequent reexposure to the antigen.

The primary chemical mediators released from mast cells are histamine and serotonin, modified amino acids that cause rapid dilation of blood vessels and contraction of smooth muscle, initiating the symptoms of systemic anaphylaxis. These symptoms include vasodilation (causing a sharp drop in blood pressure), severe respiratory distress, flushed skin, mucus production, sneezing, and itchy, watery eyes. If severe cases of anaphylaxis are not treated immediately with epinephrine to counter smooth muscle contraction, increase blood pressure, and promote breathing, the person can die from *anaphylactic shock*. Fortu-

nately, most allergic reactions are limited to mild local anaphylaxis with symptoms such as itchy, watery eyes. Less serious allergic symptoms are treated with drugs called *antihistamines* that neutralize the histamine mediators. Treatment for more serious symptoms may also include anti-inflammatory drugs such as steroids. Finally, immunization with increasing doses of the allergen may shift antibody production from IgE to IgG and IgA. The IgG and IgA interact with antigens and sequester them from sensitized mast cells, which prevents interactions with the IgE, stopping allergic symptoms and inhibiting production of more IgE. This procedure is called *desensitization*.

Delayed-Type Hypersensitivity

Delayed-type hypersensitivity (DTH), or type IV hypersensitivity, is cell-mediated hypersensitivity characterized by tissue damage due to inflammatory responses produced by T_H1 inflammatory cells (Table 28.5). Delayed-type hypersensitivity symptoms appear several hours after secondary exposure to the eliciting antigen, with a maximal response usually occurring in 24 to 48 hours. Typical antigens include components of certain microorganisms such as *Mycobacterium tuberculosis*. In addition, chemicals that covalently bind to skin proteins can create new antigens and elicit a DTH response. Hypersensitivity to these newly created antigens is known as *contact dermatitis* and results in, for example, skin reactions to poison ivy (**Figure 28.16**), jewelry, cosmetics, latex, and other chemicals. Several hours after exposure to the agent, the skin feels itchy at the site of contact. Reddening and swelling appear, often with localized tissue destruction in the form of blistering, and reach a maximum in one to three days. The delayed onset and the progress of the inflammatory response are typical

Table 28.6 *Immediate hypersensitivity allergens*
Pollen and fungal spores (hay fever)
Insect venoms (bee sting)
Certain foods (strawberries, nuts, shellfish, etc.)
Animal dander
Mites in house dust

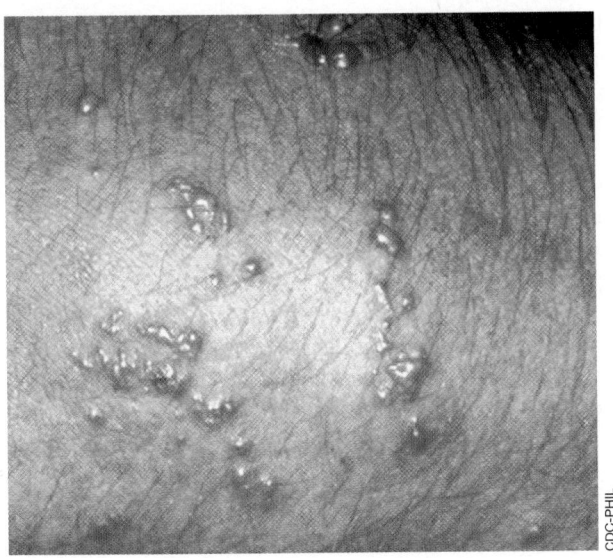

Figure 28.16 Delayed hypersensitivity. Poison ivy blisters on an arm. The raised rash appears about 24–48 hours after exposure to plans of the genus *Rhus* due to macrophage activation by T$_H$1 cells sensitized to *Rhus* antigens.

for a DTH reaction. As discussed below, certain self antigens may also elicit DTH responses, resulting in autoimmune disease.

Another example of delayed-type hypersensitivity is the development of protective immunity to the causal agent of tuberculosis, *M. tuberculosis*. This cellular immune response was discovered by Robert Koch in his classic studies on tuberculosis (Section 1.8). When antigens derived from the bacterium are injected subcuta-

neously into an animal previously infected with *M. tuberculosis*, a characteristic skin reaction develops. This is called the tuberculin test. A positive tuberculin reaction develops fully only after a period of 24–48 hours (Figure 28.6). (By contrast, skin reactions due to IgE-mediated immediate hypersensitivity develop within minutes after antigen injection.) T$_H$1 cells stimulated by the antigen release cytokines in the region of the introduced antigen that attract and activate large numbers of macrophages, which in turn produce a characteristic local inflammation, including induration, edema, erythema, pain, and heating of the skin. The activated macrophages then ingest and destroy the invading antigen. This DTH reaction is the basis for the tuberculin test used to determine previous exposure to *M. tuberculosis*.

A number of other microbial infections elicit DTH reactions. These include leprosy, brucellosis, psittacosis (all bacterial diseases); mumps (viral); and coccidioidomycosis, histoplasmosis, and blastomycosis (fungal). In all of these cellular immune reactions, visible antigen-specific skin responses resembling the tuberculin reaction occur after injection of antigens derived from the pathogens, indicating previous exposure to the pathogen.

Autoimmune Diseases

T and B cells destined to react with self antigens are normally eliminated during the process of lymphocyte maturation. In some individuals, however, T and B cells can be activated to produce immune reactions against self proteins, leading to autoimmune diseases (**Table 28.7**). For example, T$_H$1-mediated DTH can cause autoimmune responses directed against self antigens, as is the case for allergic encephalitis. In type 1 (juvenile) diabetes mellitus, T$_H$1 cells cause inflammatory reactions that destroy the

Table 28.7 *Autoimmune diseases of humans*

Disease	Organ, cell, or molecule affected	Mechanism (hypersensitivity type)[a]
Juvenile diabetes (insulin-dependent diabetes mellitus)	Pancreas	Cell-mediated immunity and autoantibodies against surface and cytoplasmic antigens of beta cells of pancreatic islets (II and IV)
Myasthenia gravis	Skeletal muscle	Autoantibodies against acetylcholine receptors on skeletal muscle (II)
Goodpasture's syndrome	Kidney	Autoantibodies against basement membrane of kidney glomeruli (II)
Rheumatoid arthritis	Cartilage	Autoantibodies against self IgG antibodies, which form complexes deposited in joint tissue, causing inflammation and cartilage destruction (III)
Hashimoto's disease (hypothyroidism)	Thyroid	Autoantibodies to thyroid surface antigens (II)
Male infertility (some cases)	Sperm cells	Autoantibodies agglutinate host sperm cells (II)
Pernicious anemia	Intrinsic factor	Autoantibodies prevent absorption of vitamin B$_{12}$ (III)
Systemic lupus erythematosus (SLE)	DNA, cardiolipin, nucleoprotein, blood clotting proteins	Autoantibody response to various cellular constituents results in immune complex formation (III)
Addison's disease	Adrenal glands	Autoantibodies to adrenal cell antigens (II)
Allergic encephalitis	Brain	Cell-mediated response against brain tissue (IV)
Multiple sclerosis	Brain	Cell-mediated and autoantibody response against central nervous system (II and IV)

[a]See Table 28.5.

beta cells of the pancreas. Many autoimmune diseases, however, are antibody mediated, as we now discuss.

Some autoimmune diseases are caused by **autoantibodies**, antibodies that interact with self antigens. In many cases, autoantibodies interact with organ-specific antigens. For example, in *Hashimoto's disease*, autoantibodies are made against thyroglobulin, a product of the thyroid gland. The disease affects thyroid function and is classified as type II hypersensitivity: Antibodies interact with antigens on the surface of host cells and initiate destruction of the tissue (Table 28.5). In this case, antibodies to thyroglobulin bind complement proteins, leading to local inflammation and destruction of the thyroid tissue. In juvenile diabetes, autoantibodies against the insulin-producing cells in the pancreas are observed, but tissue destruction occurs primarily through inflammatory reactions mediated by T_H1 cells.

Systemic lupus erythematosus (*SLE*) is an example of a disease caused by type III hypersensitivity. This disease and others like it are caused by autoantibodies directed against soluble, circulating self antigens. In SLE, the antigens include nucleoproteins and DNA. Antibodies bind to soluble proteins, producing insoluble immune complexes. Disease results when circulating antigen–antibody complexes deposit in different body tissues such as the kidney, lungs, and spleen. Here the antibodies bind complement, resulting in inflammation and local, often severe, cell damage. Thus, type III hypersensitivity is an immune complex disorder (Table 28.5).

Organ-specific autoimmune diseases are sometimes more easily controlled clinically than diseases that affect multiple organs. For example, the product of organ function, such as thyroxine in autoimmune hypothyroidism or insulin in juvenile diabetes, can often be supplied in pure form from another source. SLE, rheumatoid arthritis, and other autoimmune diseases that affect multiple organs and sites can often be controlled only by general immunosuppressive therapy, such as the use of steroid drugs. General immunosuppression, however, significantly increases chances of opportunistic infections.

Heredity influences the incidence, type, and severity of autoimmune diseases. Many autoimmune diseases correlate strongly with the presence of certain major histocompatibility complex (MHC) antigens (꘎ Section 29.4). Studies of model autoimmune diseases in mice support such a genetic link, but the precise conditions necessary for developing autoimmunity may also depend on other factors, such as prior infections, gender, age, and health status. Women, for example, are about 20 times more likely to develop SLE than are men.

MiniQuiz

- Discriminate between immediate hypersensitivity and delayed hypersensitivity with respect to antigens and immune effectors.
- Identify the two main categories of autoimmune disease with respect to antigens and immune effectors.

28.10 Superantigens:
Overactivation of T Cells

We discussed the mechanisms of action for several different categories of bacterial toxins in Chapter 27. Most toxins interact directly with host cells to cause tissue damage. Endotoxins, for example, interact directly with many cell types, causing release of endogenous pyrogens and other soluble mediators and producing fever and general inflammation (Section 28.5). Most exotoxins also interact directly with cells to cause cell damage. However, certain exotoxins, the superantigens, act indirectly on host cells, subverting the immune system so that T cells and their cytokine products extensively damage host cells.

Superantigens are proteins capable of eliciting a very strong response because they activate more T cells than a normal immune response. Superantigens, which interact with TCRs, are produced by many viruses and bacteria. Streptococci and staphylococci, for example, produce several different and very potent superantigens (꘎ Table 27.4).

Superantigen interaction with TCRs differs from conventional antigen–TCR binding. Conventional foreign antigens bind to a TCR at a defined antigen-binding site. However, superantigens bind to a site on the TCR that is outside the antigen-specific TCR binding site. A superantigen binds to all TCRs with a shared common structure, and many different TCRs share the same structure outside the antigen-binding site. In some cases, superantigens can bind 5–25% of all T cells, whereas less than 0.01% of all available T cells interact with a conventional foreign antigen in a typical immune response. The superantigens also bind to class II MHC molecules on APCs, again at a site outside the normal peptide-binding site (**Figure 28.17**). These cell surface interactions mimic conventional antigen presentation (Figure 28.5) and stimulate large numbers of T cells to grow and divide. As in normal responses, the activated T cells produce cytokines that stimulate other cells, such as macrophages and other phagocytes. The extensive cytokine production by the large proportion of superantigen-activated T cells

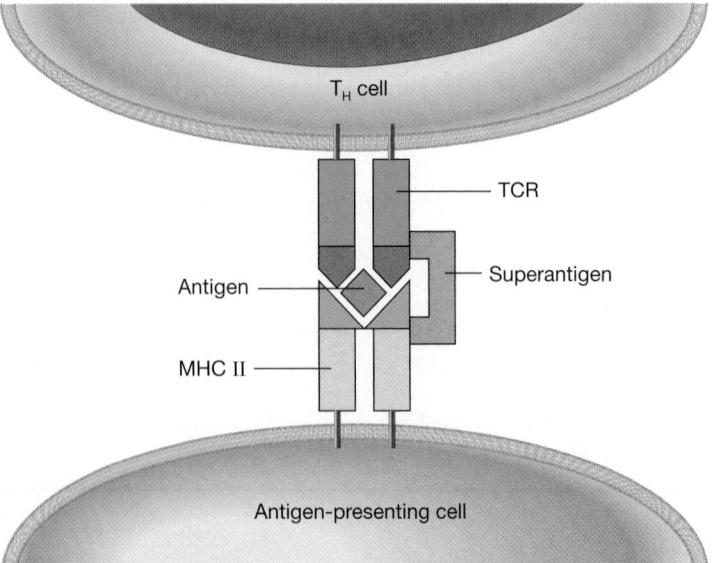

Figure 28.17 Superantigens. Superantigens act by binding to both the MHC protein and the TCR at positions outside the normal binding site. The superantigen binds to conserved regions of MHC and TCR proteins and can interact with large numbers of cells, causing massive T cell activation, cytokine release, and systemic inflammation.

triggers a widespread cell-mediated response characterized by systemic inflammatory reactions. The resulting fever, diarrhea, vomiting, mucus production, and even systemic shock may be fatal in extreme cases. Superantigen shock is virtually indistinguishable from septic shock (Section 28.5).

A very common superantigen disease is *Staphylococcus aureus* food poisoning, characterized by fever, vomiting, and diarrhea, and caused by one of several superantigen staphylococcal enterotoxins. *S. aureus* also produces the superantigen responsible for *toxic shock syndrome. Streptococcus pyogenes* produces erythrogenic toxin, the superantigen responsible for scarlet fever (⮌ Section 33.2).

MiniQuiz

- Discriminate between normal and superantigen activation of T cells.
- Identify the binding site for superantigens on T cells and APCs.

Big Ideas

28.1

Cells involved in innate and adaptive immunity originate from bone marrow stem cells. The blood and lymph systems circulate cells and proteins that are important components of the immune response. Leukocytes participate in immune responses in all parts of the body.

28.2

Innate immunity is a natural protective response to infection characterized by recognition of common pathogen-associated molecular patterns on pathogens. Phagocytes recognize these patterns through preformed pathogen recognition receptors. Interaction of pathogen-associated molecular patterns with pattern recognition receptors stimulates phagocytes to destroy the pathogens.

28.3

Adaptive immunity is triggered by the specific interactions of T cells with antigens presented on APCs. Peptide antigens embedded in MHC proteins are presented to T cells. T_C cells kill antigen-bearing target cells directly. T_H cells act through cytokines to promote immune reactions. T_H1 cells initiate inflammation and immunity by activating macrophages.

28.4

T_H2 cells stimulate B cells that have been exposed to antigen to differentiate into plasma cells. Plasma cells then produce antibodies. Antibodies are soluble, antigen-specific proteins that interact with antigens. Antibodies provide targets for interaction with proteins of the complement system, resulting in destruction of antigens through lysis or opsonization.

28.5

Inflammation, characterized by pain, swelling (edema), redness (erythema), and heat, is a normal and generally desirable outcome due to activation of nonspecific immune response effectors. Uncontrolled systemic inflammation, called septic shock, can lead to serious illness and death.

28.6

Adaptive immunity develops naturally and actively through immune responses to infections, or naturally and passively through antibody transfer across the placenta or in breast milk. A lack of innate or adaptive immunity results in death due to recurrent, uncontrollable infections.

28.7

Artificial immunity to infectious disease can be generated by passive or active means. Immunization with antigen induces artificial active immunity and is widely used to prevent infectious diseases. Artificial passive immunity involves transfer of antibodies or immune cells from an immune individual to a nonimmune individual. Vaccines are either attenuated or inactivated pathogens or pathogen products.

28.8

Immunization strategies using bioengineered molecules eliminate exposure to microorganisms and, in some cases, even to protein antigen. Application of these strategies is providing safer vaccines targeted to individual pathogen antigens.

28.9

Hypersensitivity is the induction by foreign antigens of cellular or antibody immune responses that damage host tissue. In autoimmunity, the immune response is directed against self antigens. Damage to host tissue is caused by the inflammation produced by immune mechanisms.

28.10

Superantigens are components of bacterial and viral pathogens that bind and activate large numbers of T cells. Superantigen-activated T cells may produce diseases characterized by systemic inflammatory reactions.

Review of Key Terms

Adaptive immunity the acquired ability to recognize and destroy a particular pathogen or its products, dependent on previous exposure to the pathogen or its products; also called specific immunity and antigen-specific immunity

Antibody a soluble protein produced by B cells and plasma cells that interacts with antigen; also called immunoglobulin

Antigen a molecule that interacts with specific components of the immune system

Antigen-presenting cell (APC) a macrophage, dendritic cell, or B cell that takes up and processes antigen and presents it to T-helper cells

Autoantibody an antibody that reacts to self antigens

B cell a lymphocyte with immunoglobulin surface receptors that produces immunoglobulin and may present antigens to T cells

Bone marrow the primary lymphoid organ containing the pluripotent precursor cells for all blood and immune cells

Chemokine a soluble protein that modulates an immune response

Clone a copy of an antigen-reactive lymphocyte, usually in large numbers

Cytokine a soluble protein produced by a leukocyte that modulates an immune response

Delayed hypersensitivity an inflammatory allergic response mediated by T_H1 lymphocytes

Dendritic cell a phagocytic antigen-presenting cell found in various body tissues; transports antigen to secondary lymphoid organs

Hypersensitivity an immune response leading to damage to host tissues

Immediate hypersensitivity an allergic response mediated by vasoactive products released from IgE-sensitized mast cells

Immunity the ability of an organism to resist infection

Immunization (vaccination) the inoculation of a host with inactive or weakened pathogens or pathogen products to stimulate protective immunity

Immunoglobulin (Ig) a soluble protein produced by B cells and plasma cells that interacts with antigen; also called antibody

Inflammation nonspecific reaction to noxious stimuli such as toxins and pathogens, characterized by redness (erythema), swelling (edema), pain, and heat (fever), usually localized at the site of infection

Innate immunity the noninducible ability to recognize and destroy an individual pathogen or its products that does not rely on previous exposure to a pathogen or its products; also called nonspecific immunity

Leukocyte a nucleated cell in blood; also called a white blood cell

Lymph a fluid that circulates through the lymphatic system, like blood but lacking red blood cells

Lymphocyte a subset of nucleated cells in blood involved in the adaptive immune response

Macrophage a large leukocyte found in tissues that has phagocytic and antigen-presenting capabilities

Major histocompatibility complex (MHC) a genetic region that encodes several proteins important for antigen processing and presentation. MHC I proteins are expressed on all cells. MHC II proteins are expressed only on antigen-presenting cells

Memory (immune memory) the ability to rapidly produce large quantities of specific immune cells or antibodies after subsequent exposure to a previously encountered antigen

Neutrophil a leukocyte exhibiting phagocytic properties, a granular cytoplasm (granulocyte), and a multilobed nucleus; also called polymorphonuclear leukocyte or PMN

Pathogen-associated molecular pattern (PAMP) a repeating structural component of a microorganism or virus recognized by a pattern recognition receptor (PRR)

Pattern recognition receptor (PRR) a protein in a phagocyte membrane that recognizes a pathogen-associated molecular pattern (PAMP)

Phagocyte a cell that engulfs foreign particles, and can ingest, kill, and digest most pathogens

Plasma the liquid portion of the blood containing proteins and other solutes

Plasma cell a differentiated B cell that produces antibodies

Primary adaptive immune response the production of antibodies or immune T cells on first exposure to antigen; the antibodies are mostly of the IgM class

Primary lymphoid organ an organ in which antigen-reactive lymphocytes develop and become functional; the bone marrow is the primary lymphoid organ for B cells; the thymus is the primary lymphoid organ for T cells

Secondary adaptive immune response the enhanced production of antibodies or immune T cells on second and subsequent exposures to antigen; the antibodies are mostly of the IgG class

Serum the liquid portion of the blood with clotting proteins removed

Specificity the ability of the immune response to interact with particular antigens

Stem cell a pluripotent cell that can develop into other cell types

Superantigen a pathogen product capable of eliciting an inappropriately strong immune response by stimulating greater than normal numbers of T cells

T cell a lymphocyte that interacts with antigens through a T cell receptor for antigen; T cells are divided into functional subsets including T_C (T-cytotoxic) cells and T_H (T-helper) cells. T_H cells are further subdivided into T_H1 (inflammatory) cells and T_H2 helper cells, which aid B cells in antibody formation

T cell receptor (TCR) an antigen-specific receptor protein on the surface of T cells

Thymus the primary lymphoid organ in which T cells develop

Tolerance the acquired inability to produce an immune response to particular antigens

Toxoid an attenuated form of a toxin that retains antigenicity but has lost toxicity

Vaccination (immunization) the inoculation of a host with inactive or weakened pathogens or pathogen products to stimulate protective immunity

Vaccine an inactivated or weakened pathogen or innocuous pathogen product used to stimulate protective immunity

Review Questions

1. Compare and contrast innate and adaptive immunity (Sections 28.1–28.3).

2. Name two roles of MHC in adaptive immunity (Section 28.3).

3. Identify the lymphocytes and the antigen-specific receptors involved in cell-mediated adaptive immunity (Section 28.3).

4. Identify the lymphocytes and the antigen-specific receptors involved in antibody-mediated adaptive immunity (Section 28.4).

5. Identify the cells that initiate inflammation and the cells that are activated by inflammatory signals (Section 28.5).

6. What is the difference between active and passive immunity (Sections 28.6 and 28.7)?

7. Compare and contrast attenuated vaccine and killed-virus vaccine (Section 28.7).

8. Describe a biotechnology-based immunization strategy that has been adapted for an approved vaccine. Did this vaccine replace an existing vaccine? If so, what advantage does the biotechnology-based vaccine have over the conventional vaccine (Section 28.8)?

9. Define the differences between immediate and delayed-type hypersensitivity in terms of immune effectors, target tissues, antigens, and clinical outcome (Section 28.9).

10. Describe the general mechanism used by superantigens to activate T cells. How does superantigen activation differ from T cell activation by conventional antigens (Section 28.10)?

Application Questions

1. Describe the relative importance of innate immunity compared to adaptive immunity. Is one more important than the other? Can we survive in a normal environment without immunity?

2. Inflammation is the hallmark of an activated immune response. Explain how inflammation is triggered by both innate and adaptive immune mechanisms. Are the inflammatory cells the same for both methods of activation? Why does inflammation subside as an infection is controlled?

3. Many infectious diseases have no effective vaccines. Pick several of these diseases (for example, AIDS, malaria, the common cold) and explain why current vaccine strategies have not been effective. Prepare some alternate strategies for immunization against the diseases you have chosen.

4. Are superantigen reactions desirable for the host? Do they confer protection for the host or do they benefit the pathogen?

Need more practice? Test your understanding with Quantitative Questions; access additional study tools including tutorials, animations, and videos; and then test your knowledge with chapter quizzes and practice tests at **www.microbiologyplace.com**.

29

Immune Mechanisms

Lymphocytes, different subsets of which are shown here stained different colors, are the key immune cells in all forms of adaptive immunity.

We discussed the basic features and most important outcomes of innate and adaptive immunity—protection against pathogen infection and disease—in Chapter 28. Our immune system is based on recognizing and interacting with antigens, the molecular components of pathogens. Here we concentrate on the specific mechanisms used to achieve effective immunity. We look closely at the organs, cells, and molecules active in innate and adaptive immunity.

 Overview of Immunity

We begin by expanding upon our discussions of innate and adaptive immunity, looking in depth at some of the mechanisms that are important for pathogen recognition and destruction.

Innate immunity is primarily a function of phagocytes. Innate responses recognize common structural features found on and in pathogens. Interactions with pathogens activate genes in the phagocytes that control the transcription, translation, and expression of proteins that destroy the pathogens. Innate immunity develops immediately when a phagocyte contacts a pathogen. **Adaptive immunity** is the acquired ability to recognize and destroy a particular pathogen or its products; adaptive immunity requires exposure to that particular pathogen.

Innate immune responses are not always effective, and dangerous infections sometimes still occur. However, certain phagocytes also activate adaptive immunity to deal with these infections. These phagocytes pass antigens to receptors on lymphocytes. The antigen–receptor interaction activates the lymphocytes to transcribe and translate genes that produce pathogen-specific proteins—antibodies and T cell receptors—that are the agents of adaptive immunity. An adaptive response takes several days to develop because only a few lymphocytes are initially available; the strength of the adaptive response increases as the pathogen-reactive lymphocytes multiply.

29.1 Innate Response Mechanisms

We presented an overview of innate immunity in Section 28.2 and Figure 28.3. Pathogens sometimes breach host barriers, leading to host infection. When infection starts, the immune system is mobilized to protect the host from further damage; innate immunity is the first line of defense after anatomical, physical, and chemical barriers have failed. Innate immunity, which begins immediately when a phagocyte contacts a pathogen or a pathogen product, is the most important for host protection for about four days after infection is initiated. Phagocytes engulf and destroy pathogens, often initiating complex host-mediated inflammatory reactions (Section 28.5). Here we introduce the innate mechanisms responsible for pathogen recognition, ingestion, and destruction by phagocytes.

Phagocytes

The first cell type active in the innate response is usually a **phagocyte** (literally, a cell that eats). The primary function of a phagocyte is to engulf and destroy pathogens. As just mentioned, some phagocytes also process and display the pathogen antigens that initiate the adaptive immune response (Section 28.1).

Phagocytes include macrophages, monocytes, neutrophils, and dendritic cells (**Figure 29.1**). Found in tissues and fluids throughout the body, phagocytes are usually motile, moving by amoeboid action. Most have inclusions called lysosomes, which contain bactericidal substances such as hydrogen peroxide, lysozyme, proteases, phosphatases, nucleases, and lipases. Phagocytes trap pathogens on surfaces such as blood vessel walls or fibrin clots. The membrane surrounding the pathogen pinches

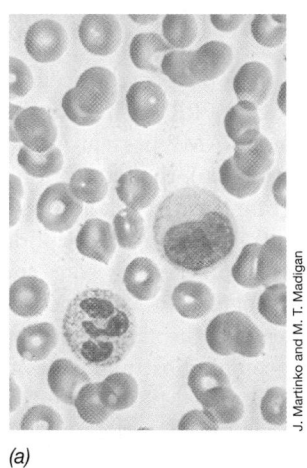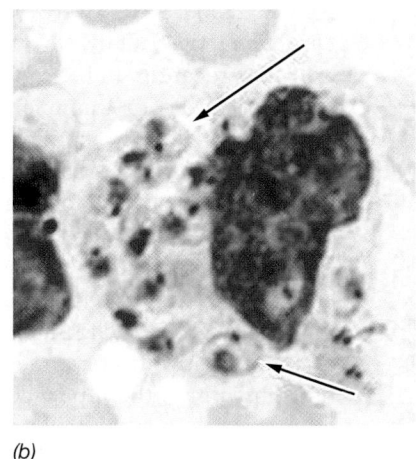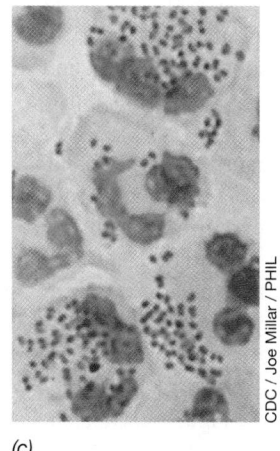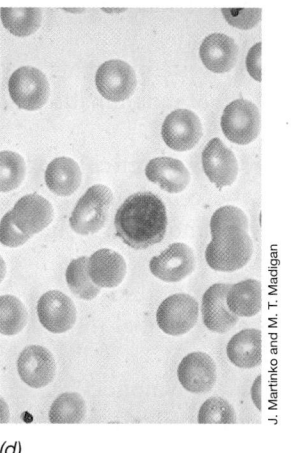

(a)   *(b)*   *(c)*   *(d)*

Figure 29.1 Major immune cell types.
(a) The nucleated cell in the lower left center is a neutrophil (PMN), characterized by a segmented nucleus (violet stain) and granular cytoplasm. The nucleated cell to the right and slightly above the neutrophil is a monocyte.

These phagocytes are 12–15 μm in diameter. The nonnucleated red blood cells are about 6 μm in diameter. *(b)* A skin macrophage that has ingested numerous *Leishmania* (arrows), a protozoan. *(c)* Neutrophils that have ingested *Neisseria gonorrhoeae*. The neutrophils are about

12–15 μm in diameter. Note the multilobed nucleus in each cell. Not all neutrophils have ingested bacteria. *(d)* The nucleated cell is a circulating lymphocyte. The lymphocyte is about 10 μm in diameter and has almost no visible cytoplasm.

UNIT 9

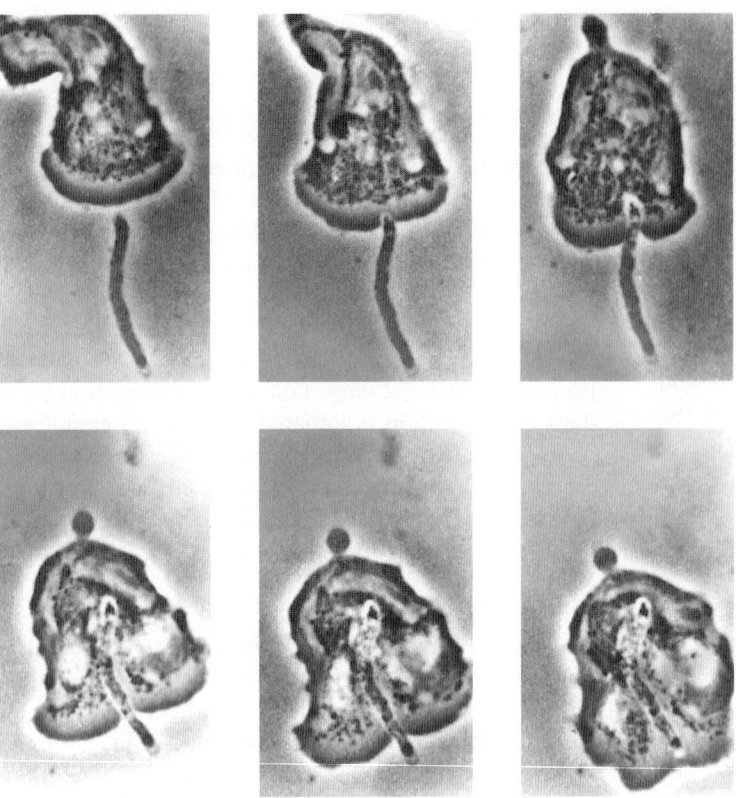

Figure 29.2 Phagocytosis. Time-lapse phase-contrast micrographs of the phagocytosis and digestion of a chain of *Bacillus megaterium* cells by a human macrophage. The bacterial chain is about 20 μm long. The macrophage is one of a group of cells that ingests and degrades pathogens and pathogen products.

off and forms a phagosome. The phagosome, now containing the engulfed pathogen, then moves inside the phagocyte and fuses with a lysosome to form a phagolysosome. The toxic substances and enzymes inside the phagolysosome usually kill and digest the engulfed microbial cell (**Figure 29.2**).

One phagocyte, the **neutrophil**, is an actively motile granulocyte containing large numbers of lysosomes (Figure 29.1*a, c*). Derived from myeloid stem cells (Figure 28.1), neutrophils are found predominantly in the bloodstream and bone marrow, from which they migrate to sites of active infection in tissues. Neutrophils present in higher than normal numbers in the blood or at a site of inflammation indicate an active response to a current infection.

Monocytes are circulating precursors of macrophages, a major phagocytic cell type (Figure 29.1). Macrophages are large cells found in almost all tissues, where they may constitute up to 10–15% of the total cells. Because they ingest and destroy most pathogens and foreign molecules that invade the body, macrophages are essential to the innate response. They are also critically important for initiating adaptive immunity by presenting antigens to T lymphocytes.

Dendritic cells (dendrocytes) (Figure 28.1) also have the dual function of phagocytosis and antigen presentation. Derived from the same monocyte progenitors as macrophages, immature dendritic cells are found throughout the body tissues, where they function as very active phagocytes. When the dendritic cells ingest antigen, they migrate to the lymph nodes, where they present antigen to T lymphocytes. The specialized antigen-presenting properties of macrophages and dendritic cells are examined in Section 29.4.

Pathogen Recognition by Phagocytes

Phagocytes have a pathogen-recognition system that triggers a timely and appropriate response, generally leading to recognition and either containment or destruction of the pathogen. This system employs evolutionarily conserved **pattern recognition receptors (PRRs)**. PRRs are membrane-bound phagocyte proteins that recognize a **pathogen-associated molecular pattern (PAMP)**, a structural component on a microbial cell or virus (Figure 28.3). PRRs were first observed in phagocytes in *Drosophila*, the fruit fly, where they are called *Toll receptors*. Each **Toll-like receptor (TLR)** on a human phagocyte recognizes a specific PAMP. For example, TLR-2, a PRR on human phagocytes, interacts with peptidoglycan, a PAMP present in the cell wall of nearly all bacteria (Section 3.6); the interaction activates the phagocytes, targeting gram-positive pathogens with exposed peptidoglycan (**Figure 29.3**). Access to the peptidoglycan of gram-negative cell walls is blocked by the surface lipopolysaccharides. Other TLRs recognize PAMPS such as the unmethylated CpG oligonucleotides and the lipopolysaccharide of gram-negative bacteria. Several soluble host molecules function similarly to these phagocyte-associated PRRs. Later in this chapter, we discuss the soluble PRRs in the context of their ability to activate proteins that enhance phagocytosis and destruction of pathogens (Section 29.9). For a more extensive discussion of TLRs, other PRRs, and PAMPs, see Section 30.1.

The PAMP–PRR interaction triggers a transmembrane signal that results in production of important defense proteins, such as some that produce toxic oxygen compounds that can cause pathogen death.

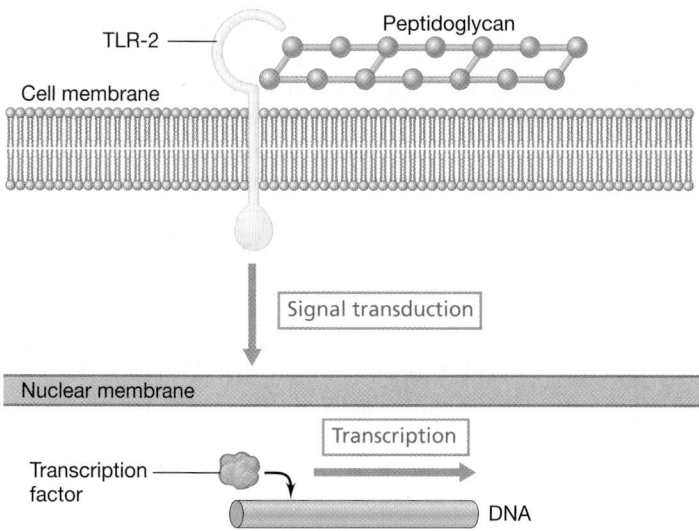

Figure 29.3 A Toll-like receptor. Membrane-spanning TLR-2 interacts with peptidoglycan from gram-positive pathogens. This interaction stimulates signal transduction, activating transcription factors in the nucleus. The result is translation of proteins that induce inflammation and other phagocyte activities. All Toll-like receptors have analogous mechanisms for activating innate immunity.

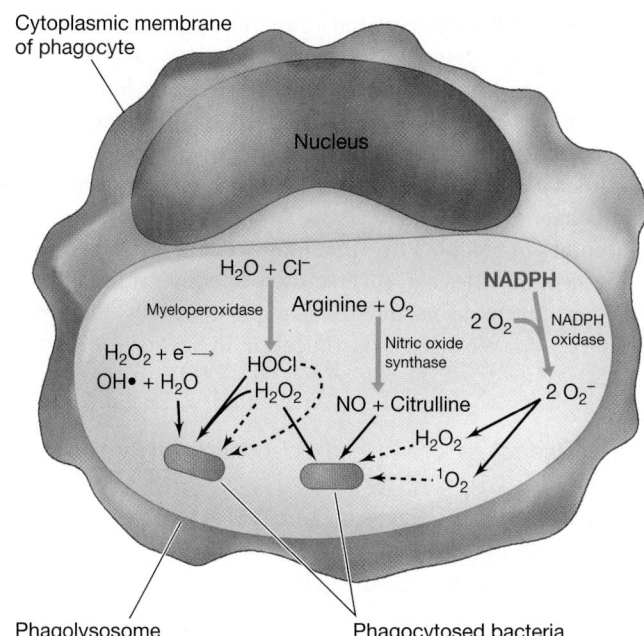

Figure 29.4 Action of phagocyte enzymes in generating toxic oxygen compounds. These compounds include hydrogen peroxide (H_2O_2), the hydroxyl radical ($OH\bullet$), hypochlorous acid (HOCl), the superoxide anion (O_2^-), singlet oxygen ($^1O_2$), and nitric oxide (NO). Formation of these toxic compounds requires a substantial increase in the uptake and utilization of molecular oxygen, O_2. This increase in oxygen uptake and consumption by activated phagocytes is known as the respiratory burst.

Oxygen-Dependent Killing in Phagocytes

Genes that control the production of oxygen compounds toxic to pathogens are up-regulated in activated phagocytes. These toxic compounds include hydrogen peroxide (H_2O_2), superoxide anions (O_2^-), hydroxyl radicals ($OH\bullet$), singlet oxygen ($^1O_2$), hypochlorous acid (HOCl), and nitric oxide (NO) (Section 5.18). The acidic conditions in the phagolysosome favor the production of these highly reactive compounds. Phagocytic cells use toxic oxygen compounds to kill ingested bacterial cells by oxidizing key cellular constituents. The reactions occur within the phagocyte, which is not damaged by the toxic oxygen products. Oxygen-mediated killing by phagocytes is summarized in **Figure 29.4**. Activated phagocytes take up and use larger than normal quantities of O_2 over a short time to produce toxic oxygen compounds. This increased rate of O_2 uptake by activated phagocytes is called the *respiratory burst*.

Phagocytes and Inflammation

Inflammation is a reaction to noxious stimuli such as toxins and pathogens. Inflammation causes redness (erythema), swelling (edema), pain, and heat, usually localized at the site of infection. The molecular mediators of inflammation include proteins called *cytokines* and *chemokines*. These proteins are produced by various immune cells, including phagocytes.

Inflammation is the usual outcome of either an innate or an adaptive immune response; both responses induce inflammatory mediators that recruit and activate the same phagocytes, especially macrophages and neutrophils. An effective inflammatory response isolates and limits tissue damage, destroying pathogen invaders and damaged cells at infection sites. In some cases, however, inflammation can damage healthy host tissue (Section 28.5).

Inhibiting Phagocytes

Some pathogens have developed mechanisms for neutralizing toxic phagocyte products, for killing the phagocytes, or for avoiding phagocytosis. For example, *Staphylococcus aureus* produces pigmented compounds called carotenoids that neutralize singlet oxygen and prevent killing (Section 33.9). Intracellular pathogens such as *Mycobacterium tuberculosis* (the cause of tuberculosis) grow and persist within phagocytic cells (Section 33.4). *M. tuberculosis* uses its cell wall glycolipids to absorb hydroxyl radicals and superoxide anions, the most lethal toxic oxygen species produced by phagocytes.

Some intracellular pathogens produce phagocyte-killing proteins called *leukocidins*. In such cases, the pathogen is ingested as usual, but the leukocidin kills the phagocyte, releasing the pathogen. Dead phagocytes make up much of the material of *pus*; organisms such as *Streptococcus pyogenes* and *S. aureus*, major leukocidin producers, are called *pyogenic* (pus-forming) pathogens. Localized infections by pyogenic bacteria often form boils or abscesses.

Another important pathogen defense against phagocytosis is the bacterial capsule (Section 3.9). Encapsulated bacteria are often highly resistant to phagocytosis, apparently because the capsule prevents adherence of the phagocyte to the bacterial cell. The clearest case of the importance of a capsule that prevents phagocytosis is that of *Streptococcus pneumoniae*. Fewer

UNIT 9

than ten cells of an encapsulated strain of *S. pneumoniae* can kill a mouse within a few days after injection (⮑ Figure 27.14). Nonencapsulated strains are completely avirulent. Surface components other than capsules can also inhibit phagocytosis. For instance, pathogenic *S. pyogenes* produces M protein, a substance that alters the surface of the pathogen in a way that inhibits phagocytosis.

Antibodies or soluble PRRs that interact with capsules or other cell surface molecules can reverse the protective effect of bacterial defense mechanisms and enhance phagocytosis, the process of opsonization. We discuss opsonization in Section 29.9.

MiniQuiz

- Describe the cellular location and molecular specificity of PAMPs and PRRs.
- Identify toxic oxygen products and explain the respiratory burst that takes place in activated phagocytes.
- Identify at least one mechanism used by pathogens to inhibit phagocytosis.

29.2 Adaptive Response Mechanisms

Adaptive immunity is the acquired ability to recognize and destroy a particular pathogen or its products that is dependent on exposure to that pathogen. The effector mechanisms for adaptive immunity require activation by pathogen exposure. We discussed the basic features of the adaptive immune response in Section 28.3. The adaptive response, as compared to the innate response, takes longer to provide effective immunity, becoming evident about 4 days after the onset of infection or antigen exposure. While the innate response is keeping the pathogen in check, some of the innate-response phagocytes—the macrophages and dendritic cells—and also B lymphocytes take up and digest pathogens. All of these **antigen-presenting cells (APCs)** process pathogen components into smaller pieces called *antigens*. The APCs then present the antigens to T lymphocytes. The T lymphocytes, called T cells, recognize the antigens with their cell surface **T cell receptors (TCRs)**; all the TCRs on a particular T cell recognize only one antigen. Some T cells, the T-cytotoxic (T_C) cells, directly attack and destroy antigen-bearing cells. Other T cells, the T-helper 1 (T_H1) cells, act indirectly by secreting cytokines that activate cells such as macrophages to destroy antigen-bearing cells. This **cell-mediated immunity** recognizes pathogen antigens on infected host cells and kills the cells.

Another subset of T cells, the T-helper 2 (T_H2) cells, interacts with B lymphocytes called **B cells**, stimulating them to make **antibodies (immunoglobulins)**, which are proteins that interact with antigens. These proteins are present as receptors on the B cells' surfaces. The antibody receptors on each B cell are specific for a particular antigen. These antigen-specific B cells also produce soluble copies of the cell surface antibodies; the soluble antibodies interact specifically with antigens in the body to neutralize them or target them for destruction. **Antibody-mediated immunity**, immunity resulting from the action of antibodies, is particularly effective against extracellular pathogens such as certain bacteria and soluble pathogen products such as toxins in the blood or lymph.

As we discussed, innate immunity is directed against features common to pathogen such as the peptidoglycan of all gram-positive bacteria or the lipopolysaccharide of all gram-negative bacteria. By contrast, adaptive immunity is directed to interactions with particular pathogen-specific macromolecules such as the M-protein antigen on a single strain of *Streptococcus pyogenes* (⮑ Section 33.2). Adaptive immunity is characterized by the properties of *specificity, memory,* and *tolerance.* None of these properties is found in the innate response.

Specificity

The **specificity** of the antigen–antibody or antigen–TCR interaction is dependent on the capacity of the lymphocyte cell receptor to interact with particular antigens. The innate host response challenges virtually any invading microorganism, even those pathogens the host has never before encountered. In the adaptive immune response, effective immunity cannot be detected for several days after the first contact with the pathogen. However, once the adaptive immune response is triggered by antigen contact, it is exclusively and specifically directed to the eliciting pathogen through recognition of its unique molecular features (**Figure 29.5a**).

Memory

The immune system must encounter antigen to stimulate production of detectable and effective antigen-activated antibodies or TCRs. A subsequent exposure to the same antigen stimulates rapid production of large quantities of the same T cells or antibodies. This capacity to respond more quickly and vigorously to subsequent exposures to the eliciting antigen is known as **immune memory** (Figure 29.5b). Immune memory provides the host with the ability to immediately resist previously encountered pathogens. We take advantage of immune memory by immunizing (inoculating, vaccinating) susceptible individuals with dead or weakened pathogens (or their products) to artificially stimulate and enhance immunity for a number of dangerous pathogens (⮑ Section 28.7).

Tolerance

Tolerance is the acquired *inability* to make an adaptive immune response directed to *self* antigens. Because all macromolecules in the host are also potential antigens, the host immune system must avoid recognizing host macromolecules; they would be damaged if recognized by antibodies or T cells. Thus, the adaptive immune response must develop the capacity to discriminate between *foreign* (nonself and dangerous) antigens and *host* (self and not dangerous) antigens (Figure 29.5c).

MiniQuiz

- Identify the antigen-specific cells in the cell-mediated and antibody-mediated immune responses.
- Distinguish between the terms immune memory and immune tolerance.
- In a host response to a pathogen, what would be the outcome of a breakdown of immune specificity, memory, or tolerance?

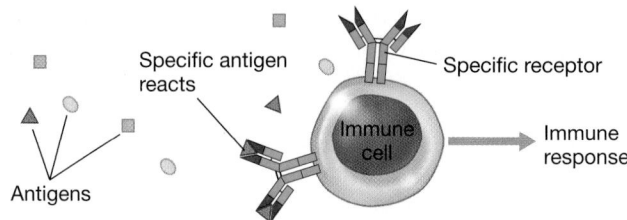

Specificity: Immune cells recognize and react with individual molecules (antigens) via direct molecular interactions.

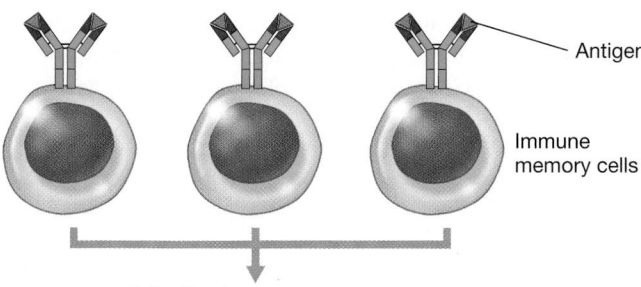

Collective immune response

Memory: The immune response to a specific antigen is *faster* and *stronger* upon subsequent exposure because the initial antigen exposure induced growth and division of antigen-reactive cells, resulting in multiple copies of antigen-reactive cells.

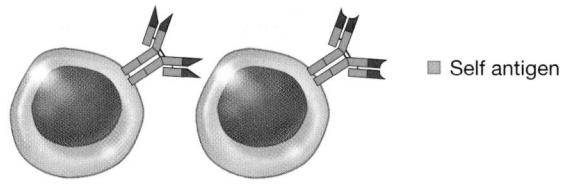

■ Self antigen

Immune cells specific for nonself antigens

Tolerance: Immune cells are not able to react with self antigen. Self-reactive cells are destroyed during development of the immune response.

Figure 29.5 **The adaptive immune response.** Key features of antibody-mediated and cell-mediated immunity are (*a*) specificity, (*b*) memory, and (*c*) tolerance.

Antigens and Antigen Presentation

The adaptive immune response recognizes a broad range of pathogen-derived macromolecules. The macromolecules are degraded and processed in host cells to produce antigens that are in turn presented to T cells. We first discuss antigens and then focus on the mechanisms of antigen processing and presentation to T cells.

29.3 Immunogens and Antigens

Antigens are substances that react with antibodies or TCRs. Most, but not all, antigens are **immunogens**, substances that induce an immune response. Here we examine the features of effective immunogens and then define the features of antigens that promote interactions with antibodies and TCRs.

Intrinsic Properties of Immunogens

Immunogens share several intrinsic properties that enable them to induce an adaptive immune response. First, *molecular size* is an important property of immunogenicity. For example, low-molecular-weight compounds called **haptens** cannot induce an immune response but can bind to antibodies. Because haptens are bound by antibodies, they are antigens even though they are not immunogenic. Haptens include sugars, amino acids, and other low-molecular-weight organic compounds. When coupled to a larger protein carrier, haptens become effective immunogens. Effective immunogens generally have a molecular weight of 10,000 or greater. Thus, sufficient molecular size is an indication of potential immunogenicity; this property and the other key properties discussed next are summarized in **Table 29.1**.

Complex, nonrepeating polymers such as proteins are effective immunogens. Complex carbohydrates can also be very good immunogens. In contrast, nucleic acids, simple polysaccharides with repeating subunits, and lipids, because they are composed of chains of identical or nearly identical monomers, tend to be poor immunogens. Thus, *sufficient molecular complexity* is another property of immunogenicity.

Large, complex macromolecules in insoluble or aggregated form (for example, proteins precipitated by heating) are usually excellent immunogens. The insoluble material is readily taken up by a phagocyte, leading to an adaptive immune response. By contrast, the soluble form of the same molecule is often a very poor immunogen; the soluble molecule is not ingested efficiently by phagocytes. Thus, *appropriate physical form* is another property of immunogenicity.

Extrinsic Properties of Immunogens

Although many substances are intrinsically immunogenic, *extrinsic* factors also influence immunogenicity. Three extrinsic factors important for immunogens include the *dose*, the *route* of administration, and the *foreign nature* of the immunogen to the host.

The dose of an immunogen administered to a host can be important for an effective immune response, but a broad range of doses ordinarily provides satisfactory immunity. In general, doses of 10 μg to 1 g are effective in most mammals. Doses of immunogen higher than 1 g or lower than 10 μg may not stimulate an immune response; extremely high or low doses may actually suppress a specific immune response by stimulating development of tolerance.

UNIT 9

Table 29.1 *Properties of immunogens*

Properties *intrinsic* to the immunogen	
Size	>10,000 molecular weight
Complexity	Polymers > monomers
Form	Aggregated > soluble

Properties *extrinsic* to the immunogen	
Dose	10 μg to 1g
Route	Parenteral > oral or topical
Foreignness	Nonself >> self

The route of administration of an immunogen is also important. Immunizations given by parenteral (outside of the gastrointestinal tract) routes, usually by injection, are normally more effective than those given topically or orally. When given by oral or topical routes, antigens may be significantly degraded before contacting a phagocyte.

Finally, an effective immunogen must be foreign with respect to the host. The adaptive immune system recognizes and eliminates only foreign antigens. Self antigens are not recognized; individuals are tolerant of their own self molecules.

Antigen Binding by Antibodies and T Cell Receptors

The antibody or TCR does not interact with the antigenic macromolecule as a whole, but only with a distinct portion of the molecule called an antigenic **epitope** (**Figure 29.6**). Epitopes include sugars, short peptides, and other organic molecules.

Antibodies interact with accessible epitopes. A sequence of four to six amino acids is the optimal size for an epitope. Thus, proteins, many of which consist of hundreds or even thousands of amino acids, are arrays of overlapping epitopes. In many cases, antibodies recognize epitopes composed of amino acids from two portions of the molecule that are distant in terms of their primary structure, but are brought together by folding into the secondary, tertiary, or quaternary structures of a macromolecule. These conformational epitopes add to the antigenic complexity of macromolecules. The surface of a bacterial cell or virus consists of a mosaic of proteins, polysaccharides, and other macromolecules, all with individual epitopes.

Although antibodies generally recognize epitopes expressed in native conformations on macromolecular surfaces, TCRs recognize epitopes only after the immunogens have been partially degraded, or *processed*. Antigen processing destroys the conformational structure of a macromolecule, generally breaking proteins into peptides of less than 20 amino acids long. As a result, TCRs recognize linear epitopes in the primary protein structure rather than the conformational epitopes recognized by antibodies. Processed antigens are then presented to T cells on the surface of specialized APCs or target cells, as we will discuss in Section 29.4.

Antibodies and TCRs can distinguish between closely related epitopes. For example, antibodies can distinguish between glucose and galactose sugars, which differ only in the orientation of a single hydroxyl group. However, specificity is not absolute, and an individual antibody or TCR may react to some extent with several different but structurally similar epitopes. The antigen that induced the antibody or TCR is called the *homologous* antigen, and the noninducing antigens that react with the antibody are called *heterologous* antigens. An interaction between an antibody or TCR and a heterologous antigen is called a *cross-reaction*.

29.4 Antigen Presentation to T Cells

T cells are thymus-derived lymphocytes that interact with antigens and activate the adaptive immune response through TCRs (Section 29.2). Here we examine how the TCR interacts with peptide antigen bound by a major histocompatibility complex (MHC) protein on an antigen-presenting phagocyte cell or on an infected target cell.

The T Cell Receptor

The TCR is a membrane-spanning protein that extends from the T cell surface into the extracellular environment. Each T cell has thousands of copies of the same TCR on its surface. A functional TCR consists of two polypeptides, an α chain and a β chain. Each polypeptide consists of several **domains**, regions of the protein that have defined structural and functional properties. Each chain has a variable (V) domain and a constant (C) domain (**Figure 29.7**). The V_α and V_β domains interact cooperatively to form an antigen–binding site. As we will see in Section 30.7, the adaptive immune response can generate TCRs that will bind nearly every known peptide antigen. Other antigens, such as complex polysaccharides, are not recognized by TCRs, but these may be bound by the immunoglobulin receptors on B cells. TCRs recognize and bind a peptide antigen only when it is bound to a *self* protein, the major histocompatibility complex protein.

Major Histocompatibility Complex Proteins

A linked set of genes found in all vertebrates encode the proteins of the **major histocompatibility complex (MHC)**. The MHC proteins in humans, called human leukocyte antigens or HLAs, were first identified as the major antigens responsible for immune-mediated organ transplant rejection. We now know, however, that MHC proteins function primarily as antigen-presenting molecules, binding pathogen-derived antigens and displaying these antigens for interaction with TCRs.

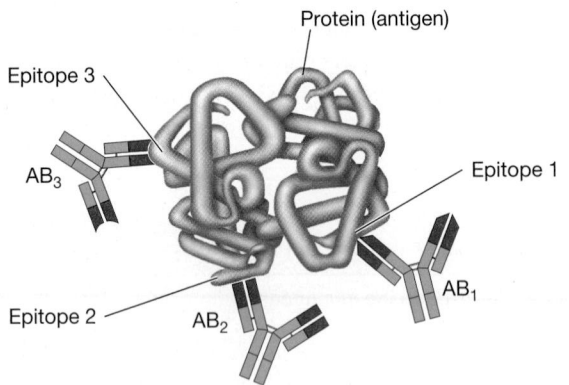

Figure 29.6 Antigens and epitopes for antibodies. Antigens may contain several different epitopes, each capable of reacting with a different antibody (AB). The epitope 1 recognized by AB_1 is a conformational epitope. Epitope 1 consists of two nonlinear parts of the folded polypeptide; the folding brings two distant portions of the polypeptide together to make a single epitope.

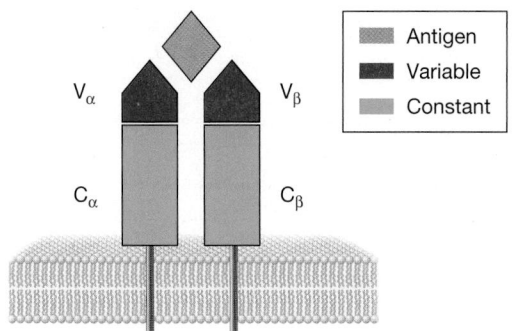

Figure 29.7 The T cell receptor. The V domains of the α chain and β chain combine to form the peptide antigen–binding site.

There are class I and class II MHC proteins. **Class I MHC proteins** are found on the surfaces of all nucleated cells. **Class II MHC proteins** are found only on the surface of B lymphocytes, macrophages, and dendritic cells, all of which are APCs. This differential cellular distribution relates to the functions of the class I and class II proteins.

A class I MHC protein consist of two polypeptides, a membrane-embedded alpha chain encoded in the MHC gene region and a smaller protein called *beta-2 microglobulin* ($\beta_2 m$) encoded by a non-MHC gene (**Figure 29.8a**). A class II MHC protein consists of two noncovalently linked polypeptides called α and β. Like class I α chains, these polypeptides are embedded in the cytoplasmic membrane and project outward from the cell surface (Figure 29.8b).

In different members of the same species, MHC proteins are not structurally identical. Different individuals usually have subtle differences in the amino acid sequence of homologous MHC proteins. These genetically encoded MHC variants, of which there are over 200 in humans, are called *polymorphisms*. Polymorphisms in MHC proteins are the major antigenic barriers for tissue transplantation from one individual to another; tissue transplants not matched for MHC identity are recognized as nonself and are rejected. We present the detailed molecular structure and genetic organization of the MHC genes and proteins in Chapter 30.

Antigen Presentation

The MHC proteins cannot be expressed on the cell surface unless they are complexed with peptide. These MHC–peptide complexes reflect the composition of the proteins inside the cell. For example, a cell that contains no pathogens or foreign antigens displays MHC proteins complexed with self peptides derived from the normal catabolism of proteins during cell growth. On the other hand, cells that have ingested foreign proteins or pathogens and cells infected with viruses produce peptides that also interact with MHC proteins. In this case, the MHC proteins expressed on the cell surface are complexed with foreign peptides. These MHC proteins with embedded peptides permit T cells to identify foreign antigens. T cells continually sample the molecular landscapes on the surface of other cells to identify cells carrying nonself antigens. The TCR on a given T cell binds only to MHC–peptide complexes that consist of foreign antigen embedded in the MHC structure; a T cell cannot interact with the foreign antigen unless

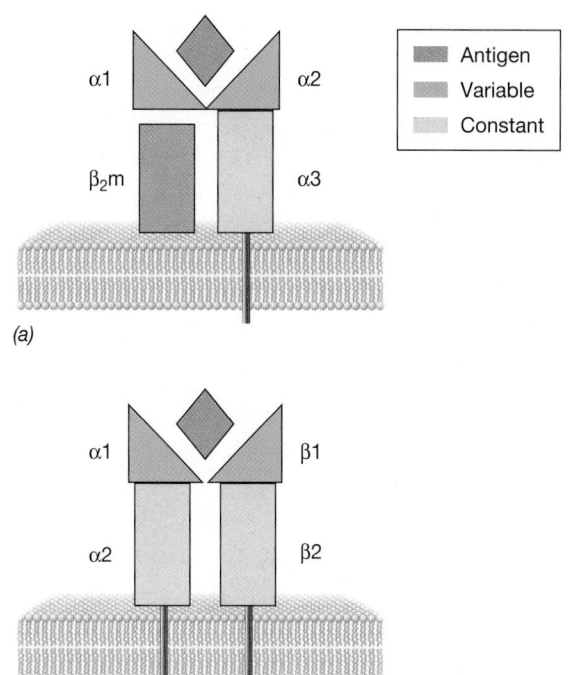

(a)

(b)

Figure 29.8 The MHC proteins. (a) Class I MHC protein. The α1 and α2 domains interact to form the peptide antigen–binding site. (b) Class II MHC protein. The α1 and β1 domains combine to form the peptide antigen–binding site.

the antigen is presented by an MHC protein. No T cells can react with the MHC–peptide complexes on uninfected cells because self-reactive T cells have been eliminated during the development of tolerance in the immune system.

How does this happen? Host cells can acquire nonself antigens through infection or phagocytosis. The host cells then degrade (process) the antigens to form small peptides. The processed antigen peptides are loaded into the MHC protein and the MHC–peptide complex is then inserted into the cytoplasmic membrane, to be recognized by T cells. Two distinct antigen-processing schemes are at work, one for MHC I antigen presentation and one for MHC II antigen presentation (**Figure 29.9**).

MHC I proteins present peptide antigens derived from pathogen proteins in the cytoplasm of nonphagocytic cells that have been infected by viruses and other intracellular pathogens; such infected nonphagocytic cells are called target cells (Figure 29.9a). Proteins derived from infecting viruses, for example, are taken up and digested in the cytoplasm in a structure called the *proteasome*. Peptides about ten amino acids long are transported into the endoplasmic reticulum (ER) through a pore formed by two proteins, called the *transporters associated with antigen processing (TAP)*. Once the peptides have entered the ER, they are bound by the MHC I protein, which has been assembled in the ER and held in place near the TAP site by a group of proteins called *chaperones* until a peptide is bound. The MHC I–peptide complex is then released from the chaperones and moves to the cell surface, where it integrates into the membrane and can be recognized by T cells. Thus, the MHC proteins act as a platform

Target cell

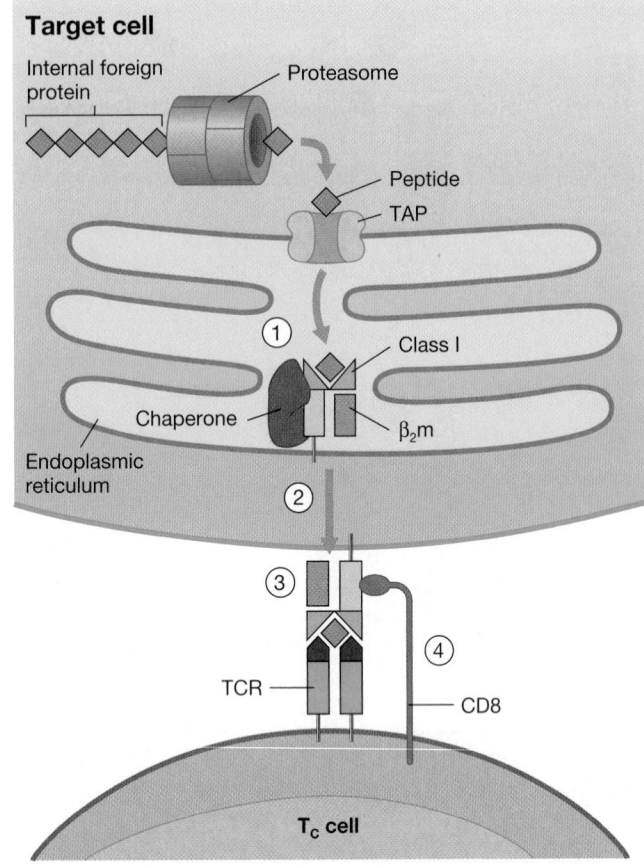

(a) **MHC I antigen presentation pathway**

Antigen-presenting cell

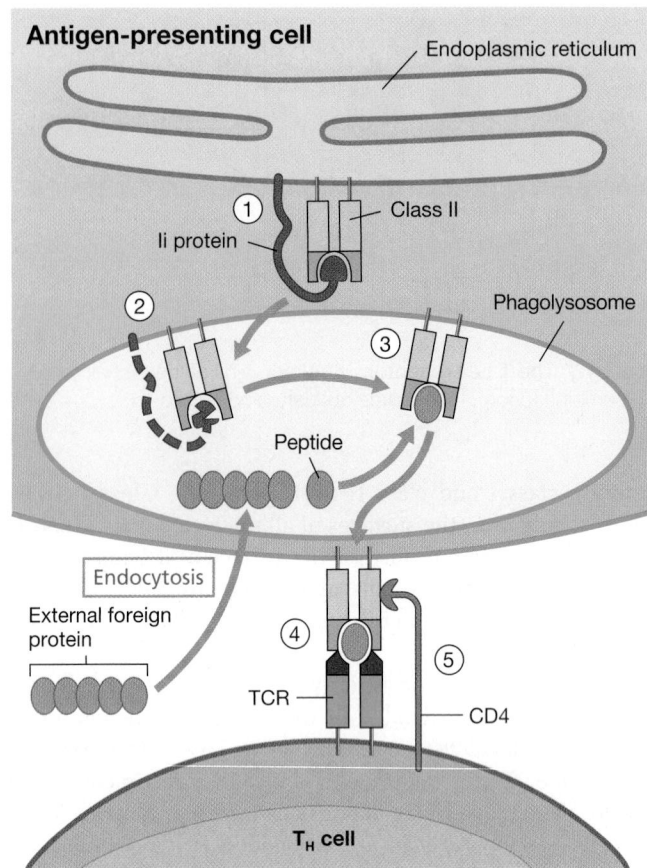

(b) **MHC II antigen presentation pathway**

Figure 29.9 Antigen presentation by MHC I and MHC II proteins. (a) MHC I proteins are stabilized in the endoplasmic reticulmn (ER) by chaperones until antigen is bound. ① Protein antigens manufactured within the cell are degraded by the proteasome in the cytoplasm and the peptide fragments are transported into the ER through a pore formed by the TAP proteins. The peptides bind to MHC I, ② are transported to the cell surface, and ③ interact with T cell receptors (TCRs) on the surface of T_C cells. ④ The CD8 coreceptor on the T_C cell engages MHC I, resulting in a stronger complex. The T_C cells then release cytokines and cytolytic toxins, killing the target cell.

(b) MHC II proteins in the ER are ① assembled with Ii, preventing MHC II from complexing with peptides in the ER. ② Lysosomes containing MHC II then fuse with phagosomes, forming phagolysosomes where the Ii and foreign proteins, imported from outside the cell by endocytosis, are digested. ③ The MHC II protein then binds to the digested foreign peptides, and the complex is transported to the cell surface, ④ where it interacts with TCRs and ⑤ the CD4 coreceptor on T_H cells. The T_H cells then release cytokines that act on other cells to promote an immune response.

to which the foreign antigen is bound. Next, the TCR on the surface of a T cell interacts with both antigen (nonself) and MHC protein (self) on the surface of the target cell. This T cell–target cell interaction induces specialized T-cytotoxic (T_C) cells to produce cytotoxic proteins called perforins that kill the virus-infected target cell (Section 29.5). Any nucleated cell can act as a target cell for T cells recognizing peptide–MHC I complexes.

The MHC class II proteins are the antigen-presenting proteins in a second pathway (Figure 29.9b). MHC II proteins are expressed exclusively in the phagocytic APCs, where they function to present peptides from engulfed extracellular pathogens such as bacteria. MHC class II proteins are initially assembled in the ER, much as MHC I proteins are assembled. However, differing from the assembly pathway of MHC I proteins, a chaperone protein called Ii, or invariant chain, binds to the MHC II protein, blocking peptide loading inside the ER. These MHC II–Ii complexes are transported from the ER to lysosomes. After phagocytosis of a pathogen, the phagosome containing the foreign

antigen fuses with the lysosome to form a phagolysosome. Here the foreign antigens as well as the Ii peptide are digested by lysosomal enzymes. The foreign peptides, generally about 11–15 amino acids long (slightly larger than MHC I–binding peptides), are bound in the newly opened MHC II antigen-binding site. The complex is transported to the cytoplasmic membrane, where it is displayed on the cell surface to specialized **T-helper (T_H) cells**. The T_H cells, through the TCR, recognize the MHC II–peptide complex. This interaction activates the T_H cells to secrete cytokines, stimulating antibody production by B cells or causing inflammation.

CD4 and CD8 Coreceptors

In addition to the TCR, each T cell expresses a unique cell surface protein that functions as a coreceptor. T_H cells express a CD4 protein coreceptor, and T_C cells express a CD8 protein coreceptor (Figure 29.9). When the TCR binds to the peptide–MHC complex, the coreceptor on the T cell also binds to the MHC protein on the antigen-bearing cell, strengthening the molecular interactions between the cells and enhancing activation of the T cell. CD4 binds only to the class II protein, strengthening T_H cell interaction with APCs that express MHC II protein. Likewise, CD8 binds only to the MHC I protein, enhancing the binding of T_C cells to MHC I–bearing target cells. The CD4 and CD8 proteins are also used for *in vitro* tests as T cell markers to differentiate T_H ($CD4^+$) cells from T_C ($CD8^+$) cells.

MiniQuiz

- Identify the cells that display MHC I and MHC II proteins on their surface.
- Define the sequence of events for processing and presenting antigens from both intracellular and extracellular pathogens.

 T Lymphocytes and Immunity

Antigen presentation activates precursor T lymphocytes to differentiate into T cells responsible for antigen-specific cell-mediated immunity. These functions include cell-mediated killing, inflammatory responses, and "help" for antibody-producing B cells. In the absence of antigen-activated T cells, there is little antigen-specific immunity and no immune memory.

29.5 T-Cytotoxic Cells and Natural Killer Cells

In the previous section we introduced two subsets of T cells, the T-cytotoxic cells and the T-helper cells. Here we examine the antigen-specific cell-killing function of the T-cytotoxic cells in detail. We also introduce the natural killer (NK) cell, a lymphocyte-like cell that uses another mechanism to recognize and kill cells infected with intracellular pathogens.

T-Cytotoxic Cells

T-cytotoxic (T_C) cells, also known as cytotoxic T lymphocytes (CTLs), are $CD8^+$ T cells that directly kill cells that display foreign surface antigens. As we discussed in Section 29.4, T_C cells recognize foreign antigens embedded in MHC I proteins. Cells displaying the foreign antigen are killed by the T_C cells. For example, a viral peptide embedded in MHC I, displayed on a virus-infected cell, marks the cell for interaction and killing by a T_C cell.

Contact between a T_C cell and the target cell is required for cell death. The point of initial contact is between the TCR and the antigen–MHC I complex. The CD8 protein on the T_C cell then binds the MHC I protein, strengthening the interaction. On contact with the target cell, granules in the T_C cell migrate to the contact site, where the contents of the granules are released (degranulation). The granules contain perforin, which enters the membrane of the target cell and forms a pore. T_C granules also contain granzymes. When granzymes enter the target cell through the pore created by perforins, the target cell undergoes *apoptosis*, or programmed cell death, characterized by death and degradation of the cell from within (**Figure 29.10**). The T_C cells, however, remain unaffected;

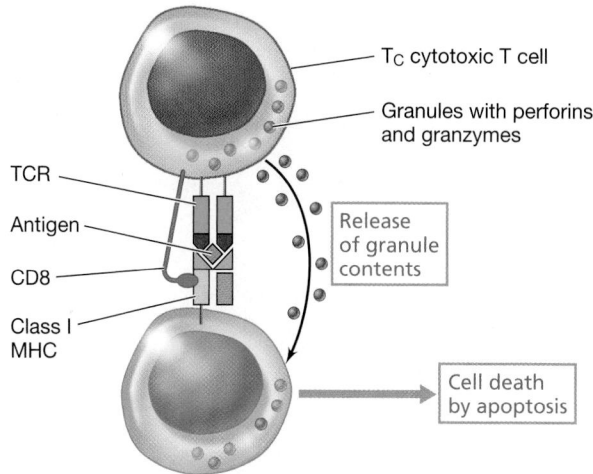

Figure 29.10 Cytotoxic T cells. T_C cells are activated by antigens presented on any cell together with MHC I protein. The T_C cells respond by releasing granules that contain perforin and granzymes, cytotoxins that perforate the target cell and cause apoptosis, respectively.

their membranes are not damaged by perforin. T_C cells kill only those cells displaying the foreign antigen because the granules are released only at the contact surface between the T_C and the antigen-bearing target cell. Cells lacking the antigen recognized by the T_C cells do not make contact and are not killed.

Natural Killer Cells

Natural killer cells (NK cells) are cytotoxic lymphocytes that are distinct from T cells and B cells. Nevertheless, NK cells resemble T_C cells in their ability to destroy cancer cells and cells infected with intracellular pathogens. NK cells also use perforin and granzymes to kill their targets, destroying cancer cells and virus-infected cells without prior exposure or contact with the foreign cells. NK cells recognize and destroy infected or aberrant cells using a two-receptor system. In the process, the number of NK cells does not increase, nor do they exhibit memory after interaction with target cells.

The molecular targets of NK cells are proteins on the surface of other cells. As NK cells circulate and interact with the cells in the body, they use special MHC I receptors to recognize MHC I proteins on normal, healthy cells. Binding of the NK receptors to MHC I deactivates the NK cell, turning off the perforin and granzyme killing mechanisms, a process called *licensing* (**Figure 29.11a**). Tumor cells or pathogen-infected cells, however, may express stress proteins on their surface; NK cells have receptors for many of these stress proteins. In addition, many tumor cells and virus-infected cells reduce or eliminate normal MHC I protein expression patterns to evade the antigen-specific immune response. Especially in the absence of the MHC licensing interaction, the stress receptors on NK cells engage stress proteins on target cells. The NK cell responds by releasing cytotoxic perforins and granzymes, thus destroying virus-infected or tumor cells that express disease-indicating stress proteins and no longer express the MHC proteins of healthy cells (Figure 29.11b).

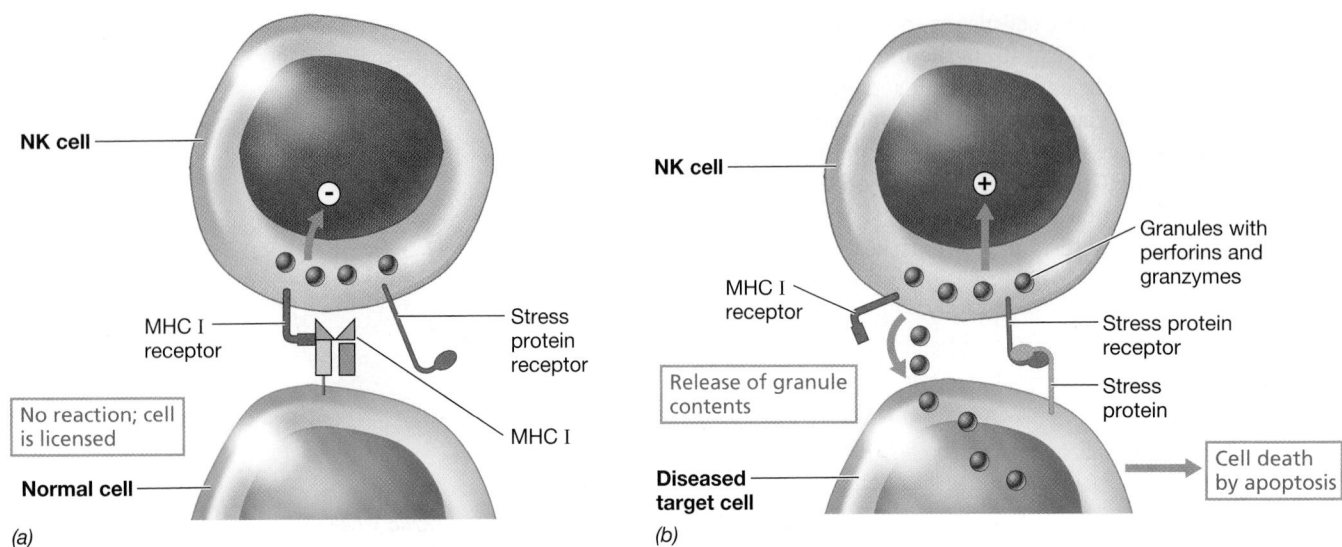

NK cell

MHC I receptor

No reaction; cell is licensed

Stress protein receptor

MHC I

Normal cell

(a)

NK cell

MHC I receptor

Release of granule contents

Granules with perforins and granzymes

Stress protein receptor

Stress protein

Diseased target cell

Cell death by apoptosis

(b)

Figure 29.11 Natural killer cells. Natural killer (NK) cells have two receptors, one that interacts with MHC I on healthy cells and another that interacts with cell stress proteins on tumor cells or pathogen-infected cells. *(a)* An MHC I interaction licenses the healthy cells, preventing the NK cell from releasing its contents. *(b)* Pathogen-infected cells or tumor cells often down-regulate MHC I expression and express stress proteins. In the absence of MHC I, the NK cell interacts with the stress protein and releases perforins and granzymes, killing the diseased cell.

MiniQuiz

- Identify and compare the targets and the recognition mechanisms used by T_C and NK cells.
- Describe the common effector system (the cell-killing system) used by T_C and NK cells.

29.6 T-Helper Cells

Here we focus on T_H cells. Interaction with an MHC–antigen complex activates $CD4^+$ T_H cells to produce cytokines. Cytokines, in turn, control the differentiation and activity of effector cells such as phagocytic macrophages, neutrophils, and antibody-producing B cells. Undifferentiated T_H cells differentiate into T_H1, T_H2, and T_H17 subsets. T_H1 and T_H2 cells play a role in adaptive immunity, promoting inflammation and antibody production respectively. T_H17 cells amplify the innate immune response.

TH1 Cells and Macrophage Activation

Macrophages play a central role as APCs in both antibody-mediated and cell-mediated immunity. As illustrated in **Figure 29.12**, macrophages engulf, process, and present antigen to T_H cells. Stimulated by T_H1 cells and the cytokines they produce, activated macrophages take up and kill foreign cells more efficiently than resting macrophages. T_H1-activated macrophages kill intracellular bacteria that normally multiply in nonactivated macrophages or other cell types. Although most bacteria taken into macrophages are killed and digested, some bacteria survive and multiply within macrophages. Bacteria that multiply in macrophages include *Mycobacterium tuberculosis*, *Mycobacterium leprae*, and *Listeria monocytogenes*, the bacteria that cause tuberculosis, leprosy, and listeriosis, respectively. Animals given a moderate dose of *M. tuberculosis* are able to over-come the infection and develop resistance because of the T cell–mediated immune response. The T cells active in the response are the T-inflammatory cells, the T_H1 subset. They activate macrophages and other nonspecific phagocytes by secreting cytokines, including IFN-γ (gamma interferon), GM-CSF (granulocyte–monocyte colony-stimulating factor), and

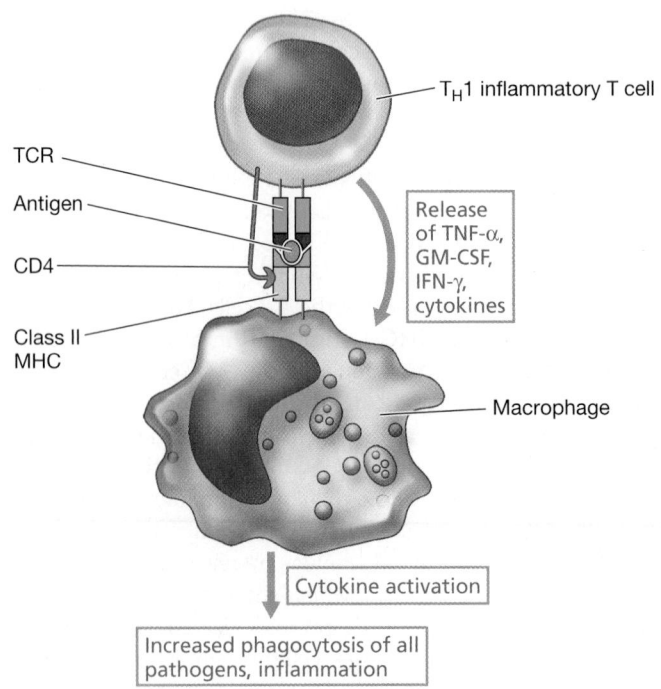

T_H1 inflammatory T cell

TCR

Antigen

CD4

Class II MHC

Release of TNF-α, GM-CSF, IFN-γ, cytokines

Macrophage

Cytokine activation

Increased phagocytosis of all pathogens, inflammation

Figure 29.12 T_H1 cells. T_H1 cells (T-inflammatory cells) are activated by antigens presented on macrophages in the context of MHC II protein. Activated T_H1 cells produce cytokines that stimulate the macrophages to increase phagocyte activity and promote inflammation.

TNF-α (Figure 29.12). Surprisingly, such immunized animals also phagocytose and kill unrelated organisms such as *Listeria*. Macrophages in the immunized animal have thus been activated to kill any secondary invader as effectively as they resist and kill the original pathogen.

T_H1-activated macrophages not only kill pathogen-infected cells, but also help destroy tumor cells. For example, tumor cells often produce tumor-specific antigens not found on normal cells. Tumor cells can be destroyed by macrophages activated by the T_H1 cells that react with the tumor-specific antigen. Transplantation rejection, a major problem encountered after organs or tissues are transplanted from one person to another, is also mediated by T_H1-activated macrophages. In this case, T_H1 cells recognize the nonself MHC proteins of the transplant, triggering macrophage activation and transplant destruction.

T_H2 Cells

T_H2 cells play a pivotal role in B cell activation and antibody production. As discussed in Section 29.2, B cells make antibodies. Differentiated B cells are coated with antibodies that act as antigen receptors. Antigen binds to the B cell antigen receptors, but the B cell does not immediately produce soluble antibodies. The antibody-bound antigen is first taken into the cell by endocytosis and degraded in the B cell. Peptides from the degraded antigen are then presented on the B cell's MHC II protein (**Figure 29.13**). In this way the B cell serves a dual role, first as an APC, and second as an antibody producer. As an APC, the B cell takes up and processes antigen into peptides and loads them into MHC II. The B cell then presents the MHC II–peptide to a T_H2 cell. The T_H2 cell responds by producing IL-4 (*inter*leukin-4) and IL-5, cytokines that activate the B cell. The activated B cell differentiates into a plasma cell that produces and secretes antibodies, as we will discuss in Section 29.8.

T_H17 Cells

T_H17 cells develop from undifferentiated or *naive* T_H cells and can be driven to differentiate by the cytokines IL-6 and TNF-β, produced by dendritic cells that encounter pathogens. The naive T_H cells that encounter IL-6 and TNF-β develop into T_H17 cells, which in turn produce IL-17 and other cytokines that attract neutrophils to the site of infection. Thus, the major function of T_H17 cells is to produce the IL-17 that draws neutrophils to infection sites. Because this event happens independent of antigen contact, T_H17 cells are amplifiers of innate immunity.

MiniQuiz
- Describe the role of T_H1 cells in activation of macrophages.
- Describe the role of T_H2 cells in activation of B cells.
- Describe the role of T_H17 cells in activation of neutrophils.

 ## IV Antibodies and Immunity

Here we concentrate on the role of B cells and antibodies in immunity. B cells, which are lymphocytes with immunoglobulin surface receptors, may present antigens to T cells and then

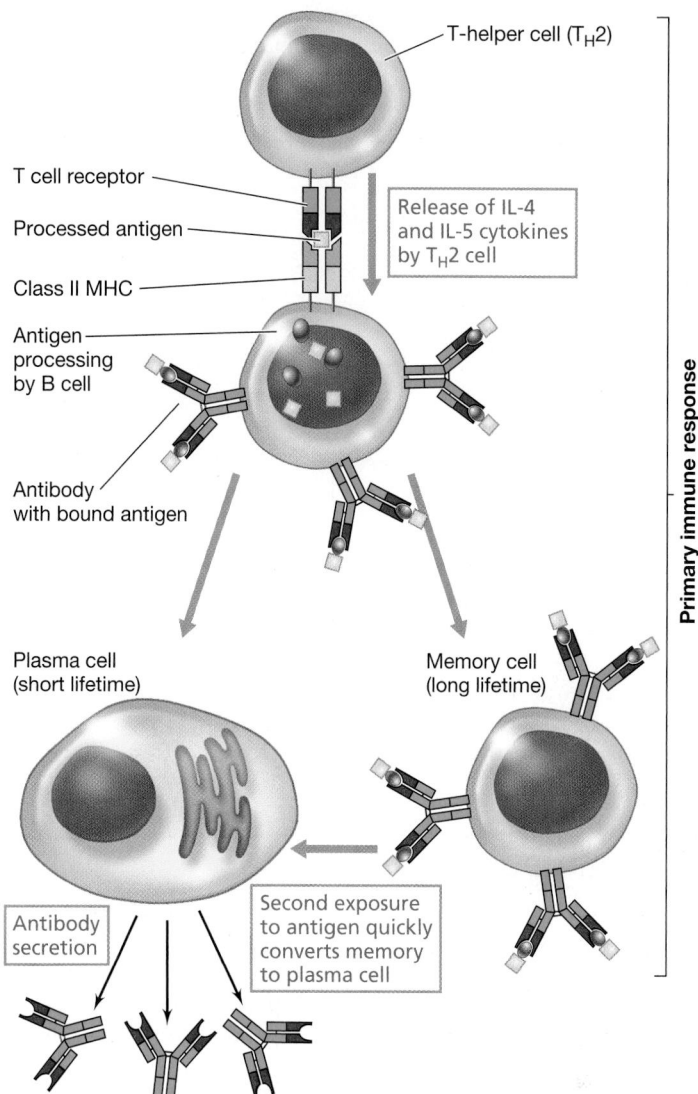

Figure 29.13 T cell–B cell interaction and antibody production. B cells initially function as antigen-presenting cells. They interact with antigen via the antigen-specific Ig receptor, promoting endocytosis of the antigen–antibody complex, and leading to antigen degradation and processing. After processing, antigen is presented to the T_H2 cell by the B cell's class II MHC molecule. The T_H2 cell is then activated to transcribe and translate genes for cytokines. The T cell cytokines then spur the same B cell to divide and form plasma cells (antibody producers) or memory cells. Plasma cells produce antibody. Memory cells quickly convert to plasma cells after a later antigen exposure.

differentiate to antibody-producing plasma cells. Antibody-mediated immunity results from the action of antibodies, proteins found in body fluids. Antibodies provide antigen-specific immunity that protects against extracellular pathogens and dangerous soluble proteins such as toxins. After considering the molecular structure of antibodies, we look at how B cells produce the great diversity of antibodies. We conclude by investigating the ability of antibodies to neutralize or destroy antigens within the animal body.

Table 29.2 *Properties of human immunoglobulins*

Class/ H chain[a]	Molecular weight/formula[b]	Serum (mg/ml)	Antigen- binding sites	Properties	Distribution
IgG γ	150,000 2(H + L)	13.5	2	Major circulating antibody; four subclasses: IgG1, IgG2, IgG3, IgG4; IgG1 and IgG3 activate complement	Extracellular fluid; blood and lymph; crosses placenta
IgM μ	970,000 (pentamer) 5[2(H + L)] + J	1.5	10	First antibody to appear after immunization; strong complement activator	Blood and lymph; monomer is B cell surface receptor
	175,000 (monomer) 2(H + L)	0	2		
IgA α	150,000 2(H + L)	3.5	2	Important circulating antibody	Secretions (saliva, colostrum, cellular and blood fluids); monomer in blood and dimer in secretions
	385,000 (secreted dimer) 2[2(H + L)] + J + SC	0.05	4	Major secretory antibody	
IgD δ	180,000 2(H + L)	0.03	2	Minor circulating antibody	Blood and lymph; B lymphocyte surfaces
IgE ε	190,000 2(H + L)	0.00005	2	Involved in allergic reactions and parasite immunity	Blood and lymph; C_H4 binds to mast cells and eosinophils

[a]All immunoglobulins may have either λ or κ light chain types, but not both.
[b]Based on the number and arrangement of heavy (H) and light (L) chains in each functional molecule. J is joining protein present in serum IgM and secretory IgA. SC is the secretory component found in secreted IgA.

29.7 Antibodies

Antibodies, or immunoglobulins (Ig), are protein molecules that interact specifically with antigenic epitopes. They are found in the serum and other body fluids such as mucosal secretions and even milk. Serum containing antigen-specific antibodies is called *antiserum*. Immunoglobulins are separated into five major classes based on their physical, chemical, and immunological properties: *IgG, IgA, IgM, IgD*, and *IgE* (**Table 29.2**).

Immunoglobulin G Structure

IgG is the most common circulating antibody, comprising about 80% of the serum immunoglobulins. IgG is composed of four polypeptide chains (**Figure 29.14**). Disulfide bridges (S—S bonds) connect the individual chains. In each IgG protein, two identical light chains of 25,000 molecular weight are paired with two identical heavy chains of 50,000 molecular weight, for a total molecular weight of 150,000. Each light chain has about 220 amino acids, and each heavy chain has about 440 amino acids. Each heavy chain interacts with a light chain to form a functional antigen-binding site. An IgG antibody, therefore, is *bivalent* because it contains two binding sites and can bind two identical epitopes.

Heavy Chains and Light Chains

Each IgG heavy chain is composed of several distinct protein domains (Figure 29.14). A heavy-chain variable domain is connected to three constant domains about 110 amino acids long. The amino acid sequence in the variable domain differs in each different antibody. The variable domain binds antigen. The three constant domains of each heavy chain are identical in all IgG molecules.

Each IgG light chain consists of two equally sized parts, a variable and a constant domain. The variable domain of a light chain interacts with the variable domain of a heavy chain to bind antigen. The amino acid sequence in the constant domain is the same in light chains of the same type.

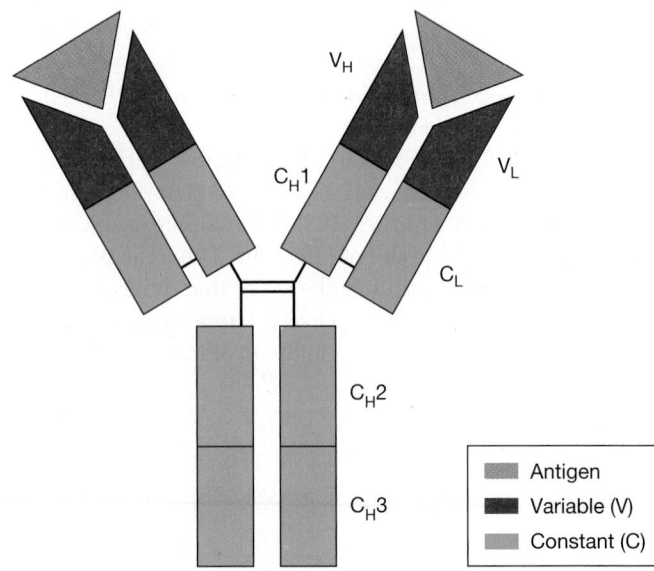

Figure 29.14 Immunoglobulin G structure. IgG consists of two heavy chains (50,000 molecular weight) and two light chains (25,000 molecular weight), with a total molecular weight of 150,000. One heavy and one light chain interact to form an antigen-binding unit. The variable domains of the heavy and light chains (V_H and V_L) bind antigen. The constant domains (C_H1, C_H2, C_H3, C_L) are identical in all IgG proteins. The chains are covalently joined with disulfide bonds.

The Antigen-Binding Site

The antigen-binding site of IgG and all other antibodies forms by cooperative interaction between the variable domains of heavy and light chains (**Figure 29.15**). The variable domains of the two chains interact, forming a receptor that binds antigen strongly but noncovalently. The measurable strength of binding of antibody to antigen is called *binding affinity*. A high-affinity antibody binds tightly to antigen.

Each individual's immune system has the capacity to recognize, or bind, countless antigens, and each antigen is bound by a unique antigen-binding site. To accommodate all possible antigens, each individual can produce billions of different antigen-binding sites in antibodies. How is this diversity in the antigen-binding site generated? As we discuss in the next section, new antibodies are con-

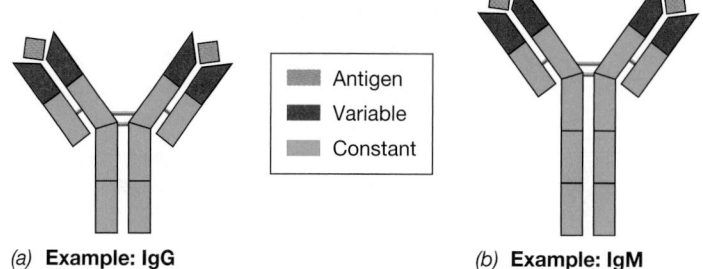

(a) **Example: IgG**   (b) **Example: IgM**

Figure 29.16 Immunoglobulin classes. All classes of Igs have V_H and V_L that bind antigen. (a) IgG, IgA, and IgD have three constant domains. (b) The heavy chains of IgM and IgE have a fourth constant domain.

stantly created through recombination and mutation events in approximately 300 genes that encode the variable domains. The heavy-chain and light-chain genes together encode each unique antibody expressed on each B cell before antigen contact. Antigen interaction with the B cell antibody stimulates the B cell to produce and secrete soluble copies of the preformed antibody.

Other Antibody Classes

Antibodies of the other classes differ from IgG. The class of a given antibody molecule is defined by the amino acid sequence of its heavy-chain constant domains. The heavy chain called gamma (γ) defines the IgG class; alpha (α) defines IgA; delta (δ) defines IgD; mu (μ) defines IgM; and epsilon (ε) defines IgE (Table 29.2). The constant domain sequences constitute three-fourths of the heavy chains of IgG, IgA, and IgD and four-fifths of the heavy chains of IgM and IgE (**Figure 29.16**).

The structure of IgM is shown in **Figure 29.17**. IgM is usually found as an aggregate of five immunoglobulin molecules attached by at least one J (joining) chain. IgM is the first class of Ig made in a typical immune response to a bacterial infection, but IgMs generally have low affinity (binding strength) for antigen. Overall antigen-binding strength is enhanced to some degree, however, by the high *valence* of the pentameric IgM molecule; ten binding sites are available for interaction with antigen (Table 29.2 and Figure 29.17). The combined strength of binding by the multiple antigen-binding sites on IgM is called *avidity*. Thus, IgM has *low* affinity but *high* avidity for antigen. Up to 10% of serum antibodies are IgM. IgM monomers are also found on the surface of B cells, where they bind antigen.

Dimers of IgA are present in body fluids such as saliva, tears, breast milk colostrum, and mucosal secretions from the gastrointestinal, respiratory, and genitourinary tracts. These mucosal surfaces are associated with mucosa-associated lymphoid tissue (MALT) that produces IgA. In an average adult, the mucosal surfaces total about 400 m^2 (compare to skin, about 6 m^2), and large amounts of secretory IgA are produced—about 10 g per day. By contrast, the serum IgG produced in an individual is about 5 g per day. Thus the total amount of secretory IgA produced by the body is higher than the amount of serum IgG. Secretory IgA has two IgA molecules covalently linked by a J chain peptide and a protein called the *secretory component* that aids in transport of IgA across membranes (**Figure 29.18**). IgA is also present in serum as a monomer (Table 29.2).

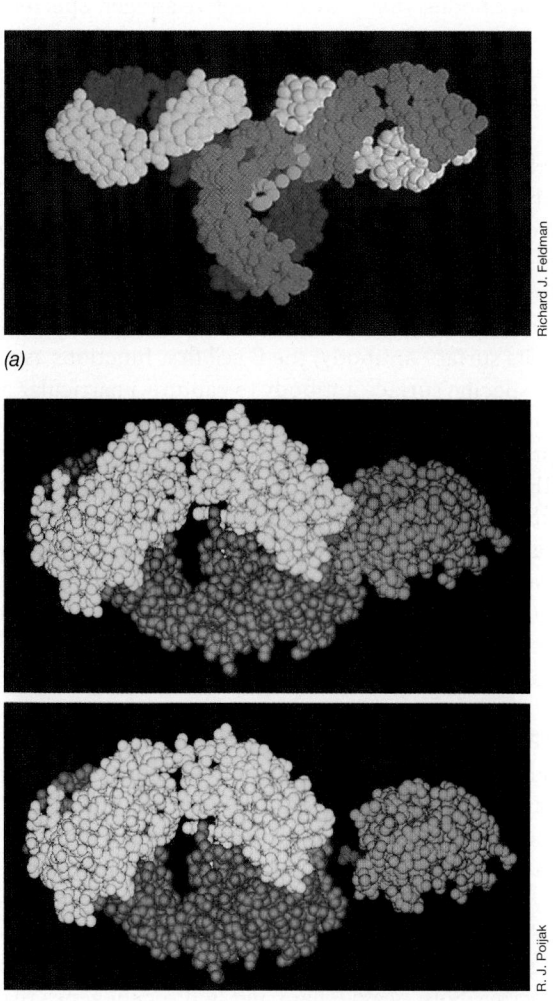

(a)

(b)

Richard J. Feldman

R. J. Poljak

Figure 29.15 Immunoglobulin structure and the antigen-binding site. (a) Space-filling model of an IgG molecule. The heavy chains are red and dark blue. The light chains are green and light blue. (b) Space-filling model of the binding interactions between an antigen and an immunoglobulin. The antigen (lysozyme) is green. The variable domain of the Ig heavy chain is blue; the light-chain variable domain is yellow. The amino acid in red is a glutamine in lysozyme. The glutamine fits into a pocket on the Ig molecule, but overall antigen–antibody interaction involves contacts between many other amino acids on the surfaces of both the Ig and the antigen. Reprinted with permission from *Science* 233:747 (1986) ©AAAS.

UNIT 9

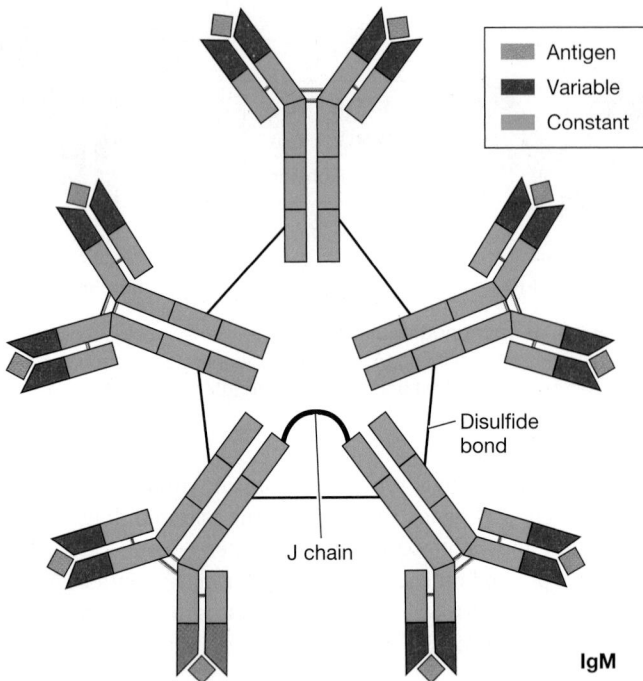

Figure 29.17 Immunoglobulin M. IgM is found in serum as a pentameric protein consisting of five IgM proteins covalently linked to one another via disulfide bonds and a J chain protein. Because it is a pentamer, IgM can bind up to 10 antigens, as shown.

IgE is found in extremely small amounts in serum (about 1 of every 50,000 serum Ig molecules is IgE). Most IgE is bound to cells. For example, IgE antibody, through its constant region, binds eosinophils, arming these granulocytes to target eukaryotic parasites like schistosomes and other worms. IgE also binds to tissue mast cells. Binding of antigen to the variable antigen-binding portions of IgE on mast cells causes release of the mast cell contents in a process called degranulation. Degranulation of mast cells triggers immediate-type hypersensitivities (allergies). The molecular weight of an IgE molecule is significantly higher than most other Igs (Table 29.2) because, like IgM, IgE has a fourth constant domain (Figure 29.16). On IgE, the additional constant region binds to eosinophils and mast cell surfaces, a critical step for activating the protective and allergic reactions associated with these cell types (᷂᷂ Figure 28.15).

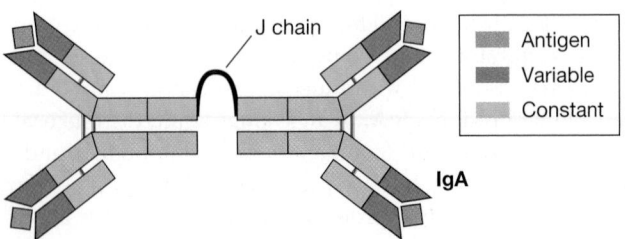

Figure 29.18 Immunoglobulin A. Secretory IgA (sIgA) is often found in body secretions as a dimer consisting of two IgA proteins covalently linked to one another via a joining (J) chain protein. A secretory component, not shown, aids in transport of IgA across mucosal membranes.

IgD, present in serum in low concentrations, has no known function. However, IgD, like IgM, is abundant on the surfaces of B cells, especially memory B cells.

29.8 Antibody Production

In this section, we examine a typical antibody response. At the cellular level, complex interactions between T cells and B cells produce antibodies that provide effective antigen-specific immunity. At the genetic level, B cells use unique mechanisms to generate unique antigen-binding receptors. A predictable sequence of events leads to antibody production after antigen exposure.

T Cell–B Cell Interactions

Antibody production is a direct response to antigen exposure. In the response, T cells and B cells interact through their respective antigen-specific cell-surface molecules, the TCR on the T cell and the surface antibody on the B cell (Figure 29.13). A differentiated B cell exposed to antigen for the first time binds the antigen through its surface antibody; the B cell first functions as an APC, using its specific surface antibody to capture a particular antigen. The antigen–antibody complex is then internalized, and the antigen is processed into peptides for loading onto MHC II proteins. The MHC–antigen complex then moves to the cell surface, where the complex is displayed for interaction with a T_H2 cell having an antigen-specific TCR on its surface. Formation of an MHC–antigen–TCR complex activates genes in the T_H2 cell, leading to cytokine production. The T_H2 cytokines stimulate the B cell, activating it to grow and differentiate into plasma cells that secrete antibodies targeted against the antigen.

Generation of Antigen Receptor Diversity

Each individual is capable of producing billions of different antibodies and TCRs, each aimed to interact with one of the countless antigens in our environment. How does the immune system produce all of these antigen-specific proteins? Immune receptor diversity is generated by a mechanism found only in B and T cells. Antibody production starts with stepwise rearrangements of the Ig-encoding genes. During development of B cells in the bone marrow, both heavy-chain and light-chain genes rearrange. The genes are recombined—individual gene pieces are mixed and matched in various combinations—by gene splicing and rearrangements in the differentiating B cells, a process called *somatic recombination*.

Figure 29.19 shows a typical rearrangement and expression pattern for one human light chain. The heavy-chain genes rearrange in an analogous, but more complex fashion; the heavy-chain gene complex has even more gene segments, allowing more recombinations and potential heavy chains, but only one rearrangement is produced in each B cell. The final result is a sin-

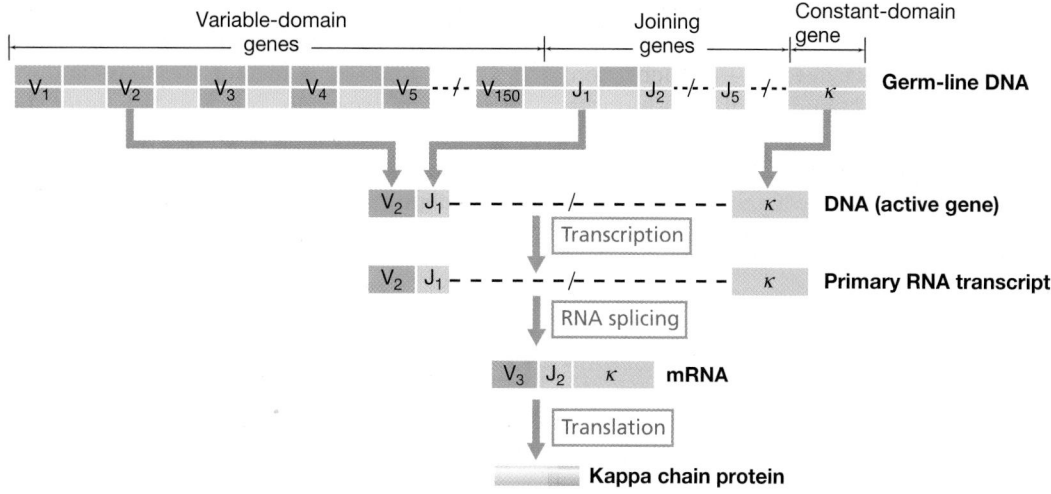

Figure 29.19 Immunoglobulin kappa chain gene rearrangement in human B cells. The gene segments are arranged in tandem in the kappa (κ) light-chain genes on chromosome 2. DNA rearrangements are completed in the maturing B cell. Any one of the 150 V (variable) sequences may combine with any one of the 5 J sequences. Thus, 750 (150 × 5) recombinations are possible, encoding 750 distinct kappa chains, but only one productive rearrangement occurs in each cell.

gle functional heavy-chain gene and a single functional light-chain gene. Each of these rearranged genes is transcribed, translated, and expressed to make, on the surface of the B cell, an antibody consisting of two heavy-chain proteins and two light-chain proteins.

Antigen exposure is necessary to stimulate the B cell to produce soluble antibodies and differentiate to plasma cells that will produce more soluble antibody copies. In addition, antigen exposure induces genetic hypermutation in B cells with productive antibody genes, which further modifies and diversifies the antibodies produced.

The large number of possible gene rearrangements coupled with the somatic hypermutation events after antigen exposure ensure almost unlimited antibody diversity. Similar rearrangements also occur during T cell development, resulting in the generation of considerable diversity in TCRs. T cells, however, do not use hypermutation to expand diversity.

Antibody Production and Immune Memory
Starting with a B cell, antibody production begins with antigen exposure and culminates with the production and secretion of an antigen-specific antibody according to the following sequence:

1. Antigens are spread via the lymphatic and blood circulatory systems to secondary lymphoid organs such as lymph nodes, spleen, and MALT (Section 28.1 and Figure 28.2). The route of antigen exposure influences the class of the antibodies produced. Intravenously injected antigen travels via the blood to the spleen, where IgM, IgG, and serum IgA antibodies are formed. Antigen introduced subcutaneously, intradermally, topically, or intraperitoneally is carried by the lymphatic system to the nearest lymph nodes, again stimulating production of IgM, IgG, and serum IgA. Antigen introduced to mucosal surfaces is delivered to the nearest MALT. For example, antigen delivered by mouth is delivered to the MALT in the

intestinal tract, preferentially stimulating production of secretory IgA in the gut.

2. Following the initial antigen exposure, each antigen-stimulated B cell multiplies and differentiates to form antibody-secreting plasma cells and memory cells (Figure 29.13). **Plasma cells** are relatively short-lived (less than 1 week), but produce and secrete large amounts of mostly IgM antibody in the **primary antibody response** (**Figure 29.20**). There is a latent period before specific antibody appears in the blood, followed by a gradual increase in antibody titer (antibody quantity), and then a slow decrease in the primary antibody response.

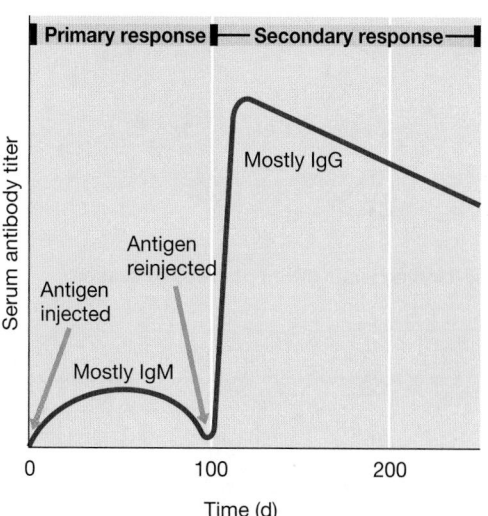

Figure 29.20 Primary and secondary antibody responses in serum. The antigen injected at day 0 and day 100 must be identical to induce a secondary response. The secondary response, also called a booster response, may be more than 10-fold greater than the primary response. Note the class switch from IgM production in the primary response to IgG production in the secondary response.

3. The **memory B cells** generated by the initial exposure to antigen may live for years. If there is a later reexposure to the immunizing antigen, memory B cells need no T cell activation; they quickly begin producing antibody and transform to plasma cells. The second and each subsequent exposure to antigen causes the antibody titer to rise rapidly to a level often 10–100 times greater than the titer following the first exposure. This rise in antibody titer is the **secondary antibody response**. The secondary response illustrates immune memory: a more rapid, more abundant antibody response than the primary response. The secondary response also switches from mostly IgM to another antibody class. In serum, the most common antibody switch is from IgM to IgG. This phenomenon is called *class switching* (Figure 29.20).

4. The titer slowly decreases over time, but subsequent exposures to the same antigen can cause another memory response. The rapid and strong memory response is the basis for the immunization procedure known as a "booster shot" (for example, the yearly rabies shot given to domestic animals). Periodic reimmunization maintains high levels of memory B cells and circulating antibody specific for a certain antigen, providing long-term active protection against individual infectious diseases.

MiniQuiz

- How do B cells act as APCs?
- How do T_H2 cells activate antigen-specific B cells?
- Explain the rationale for periodic reimmunizations in children and adults.

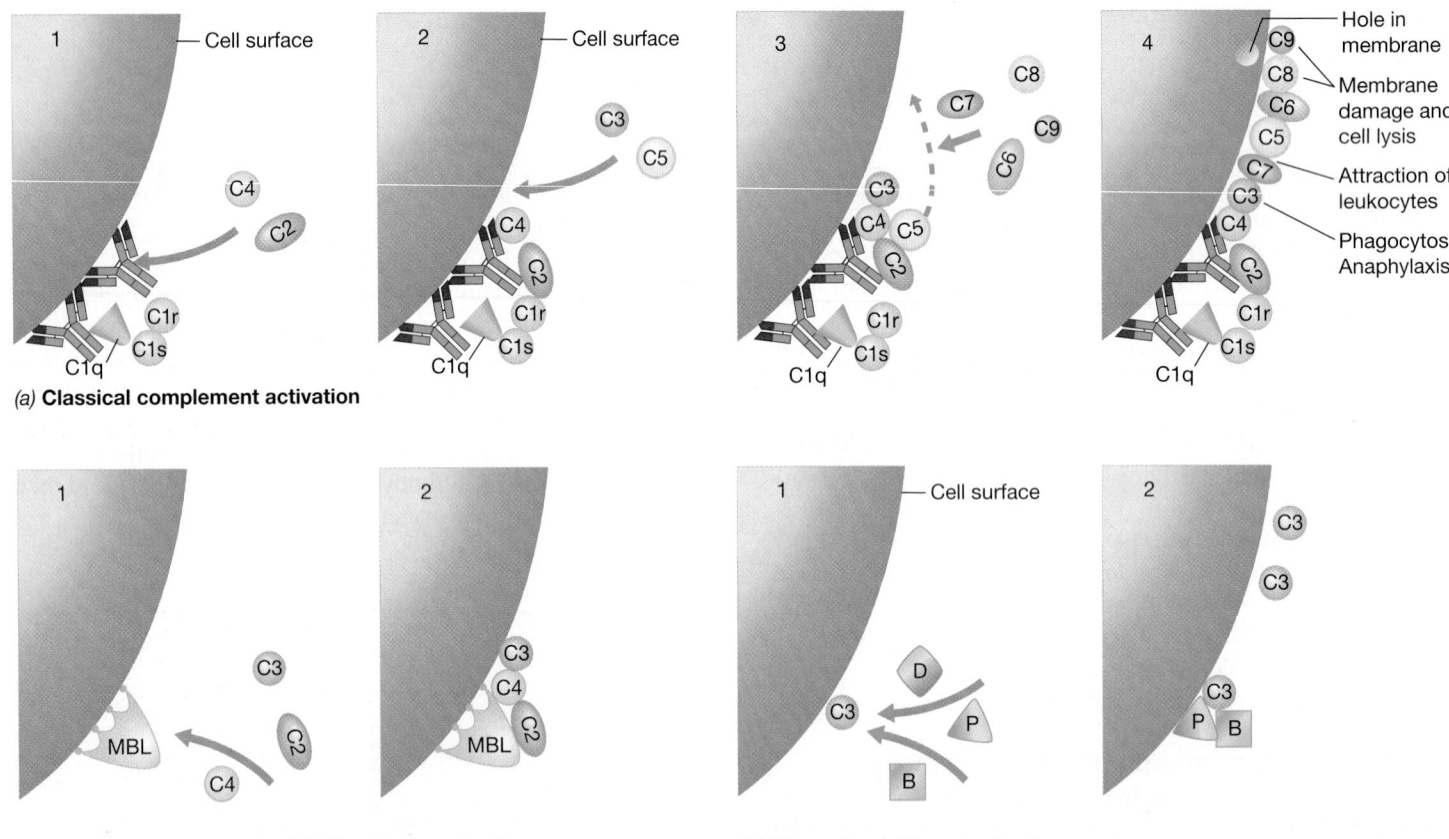

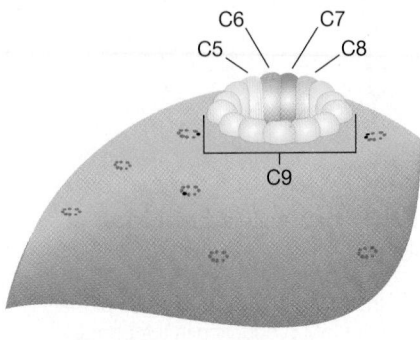

Figure 29.21 Complement activation. *(a)* The sequence, orientation, and activity of the components of the classical complement pathway as they interact to lyse a cell. 1. Binding of the antibody and the C1 protein complex (C1q, C1r, and C1s). 2. The C42 complex interacts with C3. 3. The C423 complex activates C5, which then binds an adjacent membrane site. 4. Sequential binding of C6, C7, C8, and C9 to C5 produces a pore in the membrane. C5–C9 is the membrane attack complex (MAC). *(b)* The mannose-binding lectin (MBL) pathway. MBL binds to the mannose on the bacterial membrane, and anchors formation of C423. The membrane-bound C3 activates C5, as in 3 above, and initiates formation of the MAC (4 above). *(c)* The alternative pathway. C3 bound to the cell attracts protein B and D, which activates C3. The C3B complex is further stabilized on the membrane by factor P (properdin). C3B then acts on C3 in the blood, causing more C3 to bind to the membrane. Bound C3 then activates C5, as in step 3 of the classical activation pathway above, and initiates formation of the MAC (4 above). *(d)* A schematic view of the pore formed by complement components C5 through C9.

29.9 Antibodies, Complement, and Pathogen Destruction

Complement is a group of sequentially interacting proteins that play an important effector role in both innate and adaptive immunity. Complement activity can be initiated by interactions with antigen–antibody complexes but can also be initiated by innate immune mechanisms. Complement proteins, reacting with one another and with target cell components, cause lysis of pathogen cells or mark cells for recognition by phagocytes, accelerating their destruction.

Classical Complement Activation and Cell Damage

Complement is a group of proteins, many with enzymatic activity. These proteins react in a prescribed sequential order after exposure to antigen–antibody complexes on a target cell. Complement activation may result in membrane damage and lysis of the target cell or enhanced phagocytosis of the target cell, a process called **opsonization**. Serum contains complement, and most antigen-bound IgG or IgM antibodies can bind complement (Table 29.2).

The individual proteins of complement are designated C1, C2, C3, and so on. Classical activation of complement occurs when IgG or IgM antibodies bind antigens, especially on cell surfaces. The antibodies are said to *fix* (bind) the ever-present complement proteins. The complement proteins react in a defined sequence, with activation of one complement component leading to activation of the next, and so on. The key steps, shown in **Figure 29.21**, start with binding of antibody to antigen (initiation) and binding of C1 components (C1q, C1r, and C1s) to the antibody–antigen complex, leading to C4-C2 deposition at an adjacent membrane site. This complex is a C3 convertase, an enzyme that cleaves C3 to C3a and C3b. The C3b cleavage product then binds to the convertase, forming a complex that initiates a C5-C6-C7 interaction at a second membrane site. C8 and C9 are then deposited with the C5-C6-C7 complex, resulting in membrane damage and cell lysis (**Figure 29.22**). The membrane-bound C5–9 components, called the *membrane attack complex* (*MAC*), insert at this membrane site to form a pore.

By-products of complement activation include chemoattractants called anaphylatoxins. They cause inflammatory reactions at the site of complement deposition. For example, when C3 is cleaved to C3a and C3b, C3b fixes to the target cell, as outlined above. Release of soluble C3a attracts and activates phagocytes, increasing phagocytosis. Reactions involving the C5a cleavage product lead to T cell attraction and cytokine release.

When activated by specific antibody, complement lyses many gram-negative bacteria. Gram-positive bacteria, on the other hand, are not killed by complement and specific antibodies. Gram-positive bacteria can, however, be destroyed through opsonization.

Opsonization

Opsonization is the enhancement of phagocytosis due to the deposition of antibody or complement on the surface of a pathogen or other antigen. For example, a bacterial cell is more likely to be phagocytosed when antibody binds antigen on its surface. If complement binds to an antibody–antigen complex on the cell surface, the cell is even more likely to be ingested. This is because most phagocytes, including neutrophils, macrophages, and B cells, have antibody receptors (FcR) as well as C3 receptors (C3R). These receptors bind the antibody constant domain and C3 complement protein, respectively. Normal phagocytic processes are enhanced about 10-fold by antibody binding and amplified another 10-fold by C3 binding.

Antibodies bound to surface antigens on gram-positive *Bacteria* activate the classical complement pathway and promote opsonization, leading to enhanced phagocytosis and pathogen destruction.

Complement Activation by the Mannose-Binding Lectin and Alternative Pathways

In addition to complement activation by the classical pathway, C3 can be deposited on membranes and the MAC can be activated by other methods. **Figure 29.23** outlines three major pathways that can activate the complement system. These pathways are the classical pathway, which we have already discussed, the mannose-binding lectin (MBL) pathway, and the alternative pathway.

The mannose-binding lectin (MBL) pathway depends on the activity of a serum MBL protein. MBL is a soluble PAMP (Section 29.1) that binds to mannose-containing polysaccharides found only on bacterial cell surfaces (Figure 29.21*b*). The MBL–polysaccharide complex resembles the C1 complexes of the classical complement system and fixes C4 and C2, again producing C3 convertase and binding C3b to C42. As before, this complex catalyzes formation of the C5–9 MAC and leads to lysis or opsonization of the bacterial cell.

The alternative pathway is a nonspecific complement activation mechanism using many of the classical complement pathway components and several unique serum proteins not associated with the classical complement pathway. Together they induce opsonization and activate the C5–9 MAC. The first step in alternative pathway activation is the binding of C3b produced by the

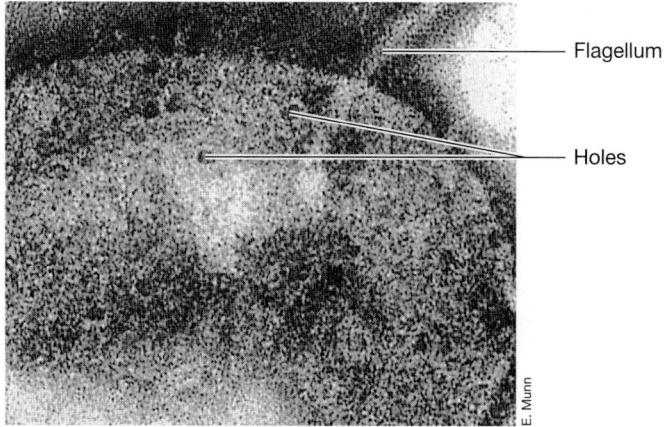

Figure 29.22 Complement activity on bacterial cells. This electron micrograph of *Salmonella enterica* serovar Paratyphi shows holes that formed in the bacterial cell envelope as a result of a reaction involving cell envelope antigens, specific antibody, and complement.

Flagellum

Holes

E. Munn

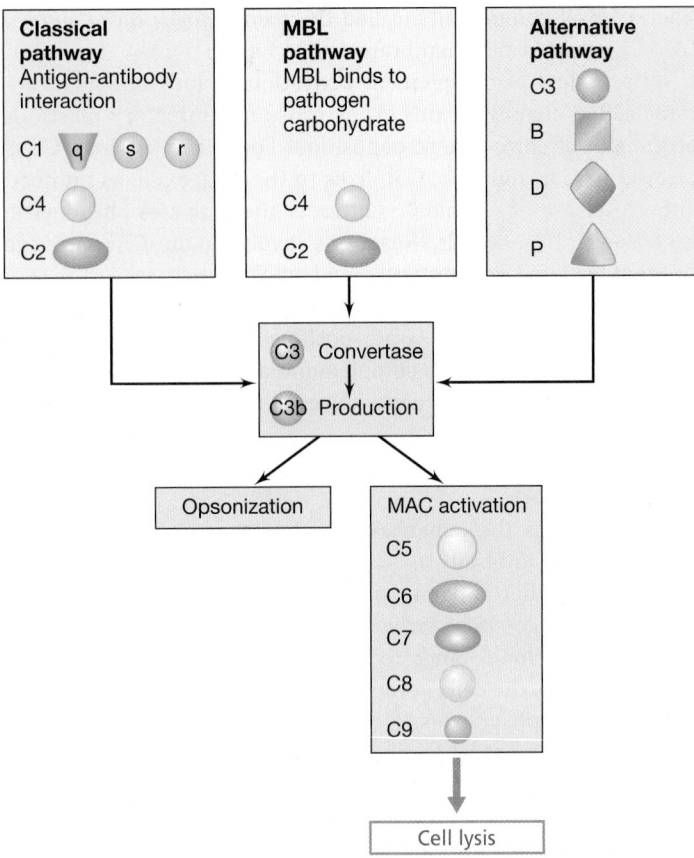

Figure 29.23 Activation of the complement system. The proteins involved in activation of the complement system in the classical pathway, mannose-binding lectin (MBL) pathway, and the alternative pathway are shown. The proteins interact in an ordered sequence from top to bottom, with each pathway independently leading to production of a C3 convertase that cleaves C3 to produce C3b, a protein necessary for initiation of the terminal complement-mediated outcomes of opsonization or cell lysis.

classical or MBL pathway to the bacterial cell surface. C3b on the membrane can then bind the alternative pathway serum protein factor B, which is cleaved by factor D to give soluble Ba and Bb. C3bBb complex is another C3 convertase. Factor P, or properdin, also binds to bacterial cell walls. P may join C3bBb to form C3bBbP. This is a very stable C3 convertase because it is fixed on the cell, as are the C3 convertases produced by the classical and MBL pathways (Figure 29.21c). C3bBbP then attracts more C3, which is deposited on the membrane, initiating the same reaction as the membrane-bound C423 complex of the classical complement pathway. The result is formation of the C5–9 MAC and cell destruction or enhanced opsonization via phagocyte C3 receptors.

Both the alternative pathway and the MBL pathway nonspecifically target bacterial invaders and lead to activation of the membrane attack complex and enhanced opsonization via formation of stable C3 convertases. MBL, factors B, D, and P, and classical complement proteins are part of the innate immune response. Neither the alternative pathway nor the MBL pathway requires prior antigen exposure or the presence of antibodies for activation. Through the alternative and MBL pathways, C3 convertase triggers formation of the C5–9 MAC or enhances opsonization via C3 receptors on phagocytes.

MiniQuiz

• Which antibody classes bind complement?

• What is meant by the term opsonization and how does the process help prevent bacterial disease?

• How does mannose-binding lectin interact with complement?

• How is the alternative pathway of complement interaction activated?

Big Ideas

29.1

Phagocytes use their membrane-bound pattern recognition receptors (PRRs) to recognize pathogen-associated molecular patterns (PAMPs). PRR–PAMP interactions activate phagocyte production of toxic oxygen-containing compounds that kill the pathogen as well as proteins that cause inflammation. Many pathogens have developed mechanisms to inhibit phagocytes.

29.2

Nonspecific phagocytes present antigen to specific T cells, triggering the production of antibodies and T_H1, T_H2, and T_C cells. T cells and antibodies react directly or indirectly to neutralize or destroy the antigen. The adaptive immune response is characterized by specificity for the antigen, the ability to respond more vigorously when reexposed to the same antigen (memory), and the acquired inability to interact with self antigens (tolerance).

29.3

Immunogens are foreign macromolecules that induce an immune response. Molecular size, complexity, and physical form are intrinsic properties of immunogens. When foreign immunogens are introduced into a host in an appropriate dose and route, they initiate an immune response. Antigens are molecules recognized by antibodies or TCRs. Antibodies recognize linear and conformational epitopes; TCRs recognize linear peptide epitopes.

29.4

T cells interact with antigen-bearing cells including dedicated APCs and pathogen-infected cells. At the molecular level, TCRs bind peptide antigens presented by MHC proteins on infected cells or APCs. These molecular interactions activate T cells to kill antigen-bearing cells or to induce inflammation or antibody production.

29.5

T-cytotoxic (T_C) cells recognize antigens on virus-infected host cells and tumor cells through antigen-specific TCRs. Antigen-specific recognition triggers killing via perforins and granzymes. Natural killer (NK) cells use the same effectors to kill virus-infected cells and tumors. NK cells, however, respond to the presence of stress proteins and the absence of normal MHC proteins on virus-infected cells and tumor cells. NK cells do not require antigen activation, nor do they exhibit memory.

29.6

T_H1 and T_H2 cells are essential activators of cell-mediated and antibody-mediated immune responses. Through the action of cytokines, T_H1 inflammatory cells activate macrophage effector cells and T_H2 helper cells activate B cells. T_H17 cells activate neutrophils.

29.7

Each immunoglobulin (antibody) protein consists of two heavy and two light chains. The antigen-binding site is formed by the interaction of the variable regions of one heavy and one light

chain. Each antibody class has different structural characteristics, expression patterns, and functional roles.

29.8

Antibody production is initiated when an antigen contacts an antigen-specific B cell. The antigen-reactive B cell processes the antigen and presents it to an antigen-specific T_H2 cell. The T_H2 cell becomes activated, producing cytokines that signal the antigen-specific B cell to clonally expand and differentiate to produce antibodies. Activated B cells live for years as memory cells and can rapidly expand and differentiate to produce high titers of antibodies after reexposure to antigen.

29.9

The complement system catalyzes bacterial opsonization and cell destruction. Complement is triggered by antibody interactions or by interactions with nonspecific activators such as mannose-binding lectin. Complement is a critical component in both innate and adaptive host defense.

Review of Key Terms

Adaptive immunity the acquired ability to recognize and destroy a particular pathogen or its products that is dependent on exposure to that pathogen

Antibody a soluble protein, produced by antigen-activated B cells and plasma cells, that interacts with antigen; also called immunoglobulin

Antibody-mediated immunity immunity resulting from the action of antibodies

Antigen a molecule capable of interacting with specific components of the immune system

Antigen-presenting cell (APC) a macrophage, dendritic cell, or B cell that processes antigens and presents them to a T-helper cell

B cell a lymphocyte that has immunoglobulin surface receptors, may present antigens to T cells, and may differentiate into a plasma cell, which produces immunoglobulin

Cell-mediated immunity immunity resulting from the action of antigen-specific T cells

Class I MHC protein an antigen-presenting molecule found on all nucleated vertebrate cells

Class II MHC protein an antigen-presenting molecule found on macrophages, B cells, and dendritic cells

Complement a series of proteins that react in a sequential manner with antibody–antigen complexes, mannose-binding lectin, or alternative activation pathway proteins to amplify or potentiate target cell destruction

Domain a region of a protein having a defined structure and function

Epitope the portion of an antigen that reacts with a specific antibody or T cell receptor

Hapten a low-molecular-weight molecule that combines with specific antibodies but is incapable of eliciting an immune response by itself

Immune memory the capacity to respond more quickly and vigorously to second and subsequent exposures to an eliciting antigen

Immunogen a molecule capable of eliciting an immune response

Immunoglobulin (Ig) a soluble protein produced by B cells and plasma cells that interacts with antigens; also called antibody

Innate immunity the noninducible ability to recognize and destroy a pathogen or its products that is not dependent upon previous exposure to a pathogen or its products

Major histocompatibility complex (MHC) a genetic complex responsible for encoding several cell surface proteins important in antigen presentation

Memory B cell a long-lived cell responsive to a specific antigen

Natural killer (NK) cell a specialized lymphocyte that recognizes and destroys foreign cells or infected host cells in a nonspecific manner

Neutrophil a leukocyte exhibiting phagocytic properties, a granular cytoplasm (granulocyte), and a multilobed nucleus; also called a polymorphonuclear leukocyte or PMN

Opsonization the enhancement of phagocytosis due to the deposition of antibody

or complement on the surface of a pathogen or other antigen

Pathogen-associated molecular pattern (PAMP) a repeating structural component of a microorganism or virus recognized by a pattern recognition receptor

Pattern recognition receptor (PRR) a membrane-bound protein that recognizes a pathogen-associated molecular pattern

Phagocyte a cell that recognizes, ingests, and degrades pathogens and pathogen products

Plasma cell a differentiated B cell that produces large amounts of antibodies

Primary antibody response the production of antibody after initial exposure to antigen; mostly of the IgM class

Secondary antibody response the production of antibody after a second and subsequent exposure to antigen; mostly of the IgG class

Specificity the ability of the immune response to interact with individual antigens

T cell a lymphocyte responsible for antigen-specific cellular interactions in the adaptive immune response

T cell receptor (TCR) an antigen-specific receptor protein on the surface of T cells

T-helper (T_H) cell a lymphocyte that interacts with MHC–antigen complexes through its T cell receptors

Tolerance the acquired inability to produce an immune response to specific antigens

Toll-like receptor (TLR) a pattern recognition receptor on phagocytes that interacts with a pathogen-associated molecular pattern

Review Questions

1. Identify some pathogen-associated molecular patterns (PAMPs) that are recognized by pattern recognition receptors (PRRs). What is the significance of the interactions between these molecules (Section 29.1)?

2. Describe the process of phagocytosis (Section 29.1).

3. Identify the three defining characteristics of the adaptive immune response (Section 29.2).

4. What is the difference between an immunogen and an antigen (Section 29.3)?

5. Describe the basic structure of class I and class II major histocompatibility complex (MHC) proteins. In what functional ways do they differ (Section 29.4)?

6. What are NK cells? How are they activated (Section 29.5)?

7. How do T_H cells differ from T_C cells? Differentiate between the functional roles of T_H1, T_H2, and T_H17 cells (Section 29.6).

8. Describe the structural and functional differences among the five major antibody classes (Section 29.7).

9. Describe the processes of primary and secondary antibody responses in serum (Section 29.8).

10. Describe the complement system. Is the order of protein interactions important? Why or why not? Identify the components of the mannose-binding lectin pathway for complement activation. Identify the components of the alternative pathway for complement activation (Section 29.9).

Application Questions

1. Describe the potential problems that would arise if a person had an acquired inability to phagocytose pathogens. Could the person survive in a normal environment such as a college campus? What defects in the phagocyte might cause lack of phagocytosis? Explain.

2. Specificity and tolerance are necessary qualities for an adaptive immune response. However, memory seems to be less critical, at least at first glance. Define the role of immune memory and explain how the production and maintenance of memory cells might benefit the host in the long term. Is memory a desirable trait for innate immunity? Explain.

3. What problems would arise if a person had a hereditary deficiency that resulted in an inability to present antigens to T_C cells? What would the problems be if the person had a deficiency in presenting

antigen to T_H1 cells? To T_H2 cells? To all T cells? What molecules might be deficient in each situation? Could a person having any one of these deficiencies survive in a normal environment? Explain for each.

4. Antibodies of the IgA class are probably more prevalent than those of the IgG class. Explain this and define the benefits this may have for the host.

5. Do you agree with the following statement? Complement is a critical component of antibody-mediated defense. Explain your answer. What might happen to persons who lack complement component C3? C5? Factor B (alternative pathway)? Mannose-binding lectin (MBL)?

 Need more practice? Test your understanding with Quantitative Questions; access additional study tools including tutorials, animations, and videos; and then test your knowledge with chapter quizzes and practice tests at **www.microbiologyplace.com**.

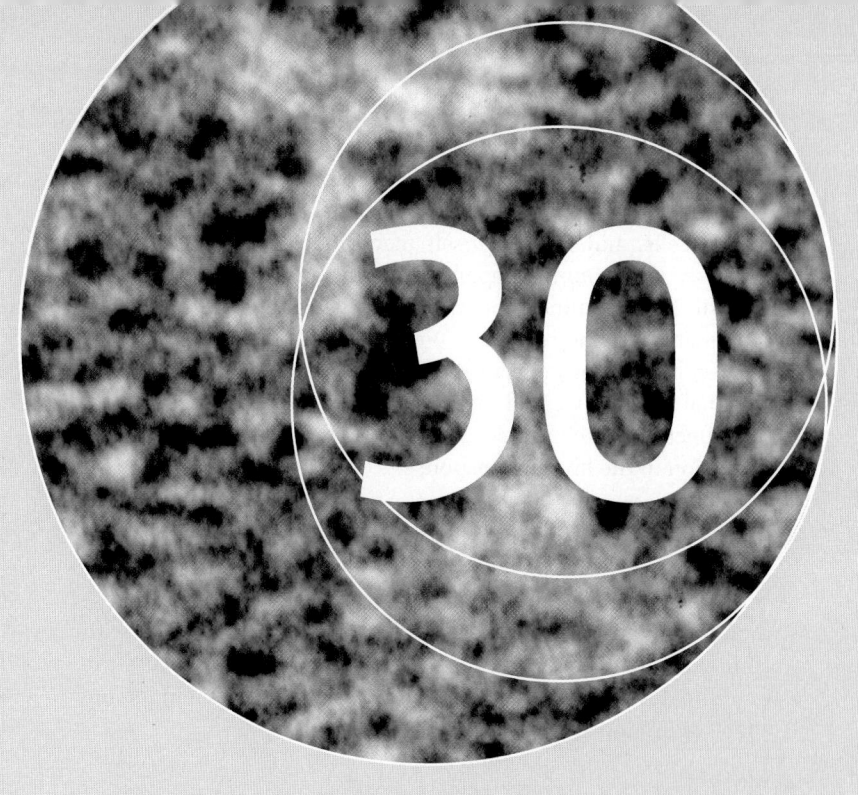

Molecular Immunology

Antibodies are key proteins in the adaptive immune response. The two molecules shown here display the typical Y-shape morphology of immunoglobulins. Two identical "arms" of the molecules bind antigens. The specificity of a given antibody stems from the unique sequence of amino acids in these regions.

The immune response employs receptor and effector proteins that target pathogens, their products, foreign cells such as cancer cells, and other nonself macromolecules. This chapter discusses the immune response proteins and their interactions. We will first examine proteins that target pathogens as part of the innate immune response. We then shift our attention to the antigen-binding proteins of the immunoglobulin superfamily. In the final section, we will investigate cytokines and chemokines, the effector proteins that control cell differentiation and cell activation in the immune response.

 Receptors and Immunity

Here we first examine pattern recognition receptors. These are innate immune response proteins that interact with common molecular targets on pathogens. We then discuss the immunoglobulin supergene family and its involvement in the adaptive immune response.

30.1 Innate Immunity and Pattern Recognition

Multicellular organisms must recognize and control pathogen infection. A basic system for *innate* recognition of pathogens is widely distributed in living organisms. Multicellular organisms from primitive plants to vertebrate animals have molecular recognition mechanisms that rapidly and effectively activate host defenses. Many invertebrates have genes homologous to the pattern recognition receptors found in higher animals.

Pathogen-Associated Molecular Patterns and Pattern Recognition Receptors

Pathogen-associated molecular patterns (PAMPs) are structural components common to a particular group of infectious agents. PAMPs are often macromolecules and include polysaccharides,

proteins, nucleic acids, or even lipids. The lipopolysaccharide (LPS) of the gram-negative bacterial cell wall (↩ Section 3.7) is an excellent example of a PAMP.

Pattern recognition receptors (PRRs) are a group of soluble and membrane-bound host proteins that interact with PAMPs. Soluble PRRs include the mannose-binding lectin (↩ Section 29.9; **R_`jc1.,/**). The PAMP recognized by mannose-binding lectin (MBL) is the sugar mannose, found as a repeating subunit in bacterial and fungal polysaccharides (mannose on mammalian cells is inaccessible to mannose-binding lectin). C-reactive protein, another soluble PRR, is an *acute phase protein* produced by the liver in response to inflammation. C-reactive protein interacts with the phosphorylcholine macromolecules of gram-positive bacterial cell walls. Both of these PRRs target pathogen surface PAMPs, and both bind complement proteins, leading to lysis or opsonization of the targeted cell.

Evolutionarily conserved membrane-bound PRRs promote phagocytosis. They are found on the surface of cells such as macrophages, monocytes, dendritic cells, and neutrophils, innate phagocytes that have the ability to engulf and destroy pathogens. PRRs were first recognized in the invertebrate *Drosophila* (the fruit fly), where they were called Toll receptors. Structural, functional, and evolutionary homologs of the Toll receptors, called **Toll-like receptors (TLRs)**, are widely expressed on mammalian innate immune cells. At least nine TLRs in humans interact with a variety of cell surface and soluble PAMPs from viruses, bacteria, and fungi. Table 30.1 identifies some of the known TLRs, several other PRRs, and their associated PAMPs.

Several TLRs interact with more than one PAMP. For example, TLR-4 is part of the innate immune response to bacterial LPS and also responds to molecules produced by damaged host cells called *heat shock proteins*. Both LPS and heat shock protein interact with TLR-4 via receptor proteins that, in turn, interact

Table 30.1 *Receptors and targets in the innate immune response*

Pattern recognition receptors (PRRs)	Pathogen-associated molecular patterns (PAMPs) and target organisms	Result of interaction
Mannose-binding lectin[a] (soluble)	Mannose-containing cell surface microbial components, as in gram-negative bacteria	Complement activation
C-reactive protein (soluble)	Components of gram-positive cell walls	
TLR-1[b] (Toll-like receptor 1)	Lipoproteins in mycobacteria	Signal transduction, phagocyte activation, and inflammation[c]
TLR-2	Peptidoglycan on gram-positive bacteria; zymosan in fungi	
TLR-3	dsRNA in viruses	
TLR-4	LPS (lipopolysaccharide) in gram-negative bacteria	
TLR-5	Flagellin in bacteria	
TLR-6	Lipoproteins in mycobacteria; zymosan in fungi	
TLR-7	ssRNA in viruses	
TLR-8	ssRNA in viruses	
TLR-9	Unmethylated CpG oligonucleotides in bacteria	

[a]The soluble PRRs are produced by liver cells in response to inflammatory cytokines.
[b]The Toll-like receptors are membrane-integrated PRRs expressed in phagocytes. TLR-1, -2, -4, -5, and -6 are in the cytoplasmic membrane. TLR-3, -7, -8, and -9 are found in intracellular organelle membranes such as in lysosomes.
[c] Toll-like receptors are all involved in phagocyte activation via signal transduction.

Leucine-Rich Repeats and the Immune Response

Proteins containing leucine-rich repeats are ancient innate immune receptors. Plants and invertebrates lack adaptive immunity and depend solely on the innate immune response to clear infections. Innate immunity uses germ line–encoded pattern recognition receptors (PRRs) to sense a wide variety of pathogen-associated molecular patterns (PAMPs) such as lipopolysaccharide, peptidoglycan, and microbial nucleic acids. Among these PRRs, leucine-rich repeat (LRR)-containing proteins are central to host defense in both plants and animals. Many LRR proteins have been linked to the initiation of innate immune reactions, including the Toll-like receptors (TLRs) of insects and mammals, Nod-like receptors (NLRs) of mammals, leucine-rich immune proteins (LRIMs) of mosquitoes, and R proteins in plants.

Drosophila Toll was the first PRR to be discovered, and deciphering the Toll pathway provided insight into innate immune mechanisms in other organisms. Drosophila Toll, however, does not interact directly with PAMPs but rather with a cleaved form of a cytokine-like molecule called spaetzle. Toll–spaetzle interaction triggers a signal transduction cascade, activating a transcription factor that induces the transcription of several genes encoding antimicrobial peptides. These peptides are produced in the fat body (the invertebrate equivalent of the liver) and released into the fly's blood, where they bind to and disrupt microbial surfaces, causing cell lysis. In contrast to Toll, mammalian TLRs and NLRs are true PRRs responsible for direct recognition of pathogens. A striking feature of TLR ligands is their molecular diversity, which includes proteins, nucleic acids, lipids, and polysaccharides. NLRs also detect a variety of PAMPs; however, NLRs recognize microbial molecules in the cell cytosol, whereas the membrane-bound TLRs sense PAMPs on the cell surface or in endosomes. Recognition of microorganisms by TLRs and NLRs activates downstream signaling pathways leading to innate and adaptive immune responses. LRIMs of the malaria mosquito Anopheles gambiae (Figure 1) are putative PRRs implicated in the killing of malaria parasites invading the mosquito midgut epithelium. A heterodimeric complex of two mosquito LRIMs controls the function of TEP1, a complement-like protein in the mosquito blood associated with parasite lysis. Structural analysis revealed that the interaction between LRIMs and TEP1 represents a new type of innate immune complex.

The established role of plant R genes in sensing microbial infections suggests that during evolution, LRR motifs were selected for PAMP recognition in the animal and plant lineages. The recognition of a diversity of PAMPs by LRR-containing proteins reflects the versatile nature of these motifs throughout the living world.

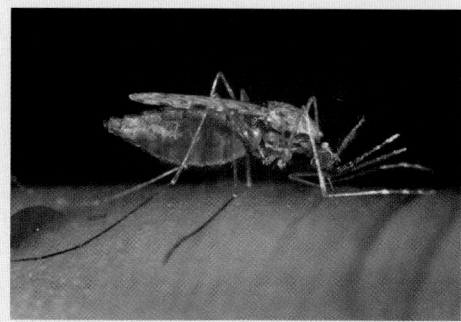

Figure 1 A female mosquito (*Anopheles gambiae*) feeding on a human host. Blood is visible in the mosquito's proboscis and abdomen.

with TLR-4. In other cases, the TLR binds directly to the PAMP without the interactions of receptor proteins, as is the case for TLR-5 and its target, flagellin.

Signal Transduction in Phagocytes

Interaction of a PAMP with the TLR triggers transmembrane *signal transduction*. Signal transduction pathways initiate gene transcription and translation of host-response proteins similarly to the membrane signal transduction mechanisms in prokaryotes (↩ Section 8.7). Activation of the innate response cells by signal transduction can result in enhanced phagocytosis and killing of pathogens or contribute to inflammation and tissue healing (↩ Section 28.5).

For example, a signal transduction pathway may be activated by the binding of LPS (a PAMP) to TLR-4 (a PRR) (**Dgespc1.,/**). TLR-4 then binds proteins in the cytosol, starting a cascade of reactions that activates transcription factors such as NFκB (nuclear factor kappa B), a protein that binds to specific regulatory sites on DNA, initiating transcription of downstream genes. Many of the NFκB-regulated genes encode host response proteins such as the cytokines that activate cells and initiate inflammation.

TLR-4 consists of three distinct protein domains, each with a separate function. The external domain of TLR-4 contains a binding site for LPS that is complexed with cell surface CD14 (Figure 30.1). A transmembrane domain in TLR-4 connects the external

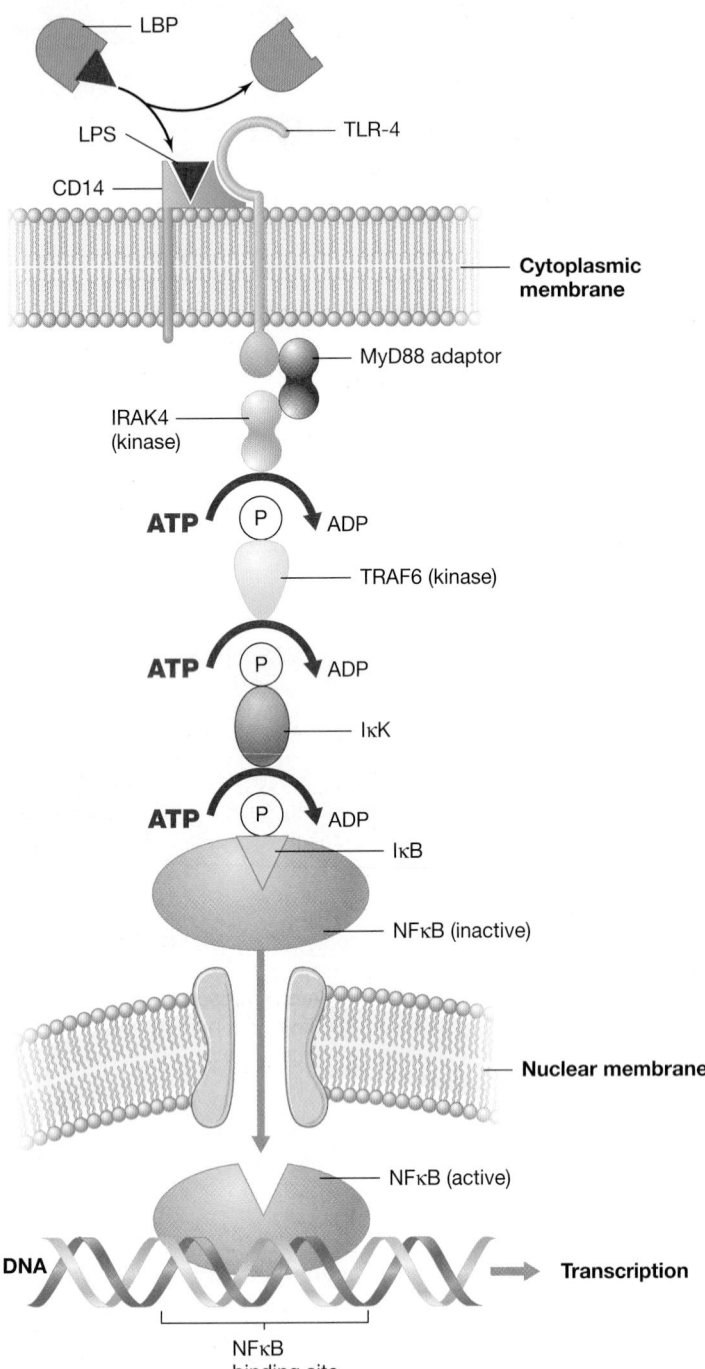

Figure 30.1 Signal transduction in innate immunity. Signal transduction is initiated when LPS, a PAMP, is bound by LBP (lipopolysaccharide-binding protein), which then transfers LPS to CD14 on the surface of a phagocyte. The LPS–CD14 complex then binds to the transmembrane TLR-4 receptor. The binding of TLR-4 initiates a series of reactions involving adaptor proteins and kinases, resulting in activation of the transcription factor NFκB. NFκB then diffuses across the nuclear membrane, binds to DNA, and initiates transcription of proteins essential for innate immunity.

domain to a cytoplasmic domain. Binding of the CD14–LPS complex by the external domain of TLR-4 causes a change in the conformation of a third TLR-4 domain extending into the cytoplasm, exposing a site that interacts with an adaptor protein,

MyD88. MyD88, in turn, is altered and binds a protein tyrosine kinase (PTK), IRAK4. PTKs work by catalyzing the transfer of energy-rich phosphates from ATP to the newly exposed tyrosines on the target protein. The phosphorylated IRAK4 initiates a *kinase* cascade, activating more proteins through ATP-mediated phosphorylation of TRAF6, another kinase. TRAF6 then phosphorylates IκK (inhibitor of kappa kinase). IκK then phosphorylates the IκB (inhibitor of kappa B) protein, causing it to dissociate from NFκB, which then diffuses across the nuclear membrane, binds to NFκB-binding motifs on DNA, and initiates transcription of downstream genes. As this example shows, signal transduction pathways initiate activation of transcription through ligand–receptor binding on the cell surface. The ligand–receptor interaction outside the cell induces the binding, recruitment, and concentration of the adaptor proteins and kinase enzymes inside the cell. A single kinase enzyme can then catalyze phosphorylation of many other signal cascade proteins, amplifying the effect of a single ligand–receptor interaction.

Signal transduction leading to activation of shared transcription factors and protein production is also the activation mechanism for lymphocytes in adaptive immunity, as we will discuss below.

MiniQuiz

• Identify the target of TLR-4 and the outcome for the host cell and the targeted pathogen.

• Define the general features of a signal transduction pathway starting with binding of a PAMP by a membrane-associated PRR.

30.2 Adaptive Immunity and the Immunoglobulin Superfamily

The **immunoglobulin gene superfamily** includes genes and their protein products that share structural, evolutionary, and functional features with immunoglobulin genes and proteins. The antigen-binding proteins in the adaptive immune response are part of this extended gene family.

As we discussed in Chapters 28 and 29, three different cell surface proteins interact directly with antigens during the adaptive immune response. These are the **immunoglobulins** (Igs or **antibodies**) produced by B cells that interact with antigens; the antigen-binding **T cell receptors** (TCRs) on the surface of T cells; and the proteins of the **major histocompatibility complex (MHC)** that process and present antigen. Each of these three antigen-binding proteins has a different location, structure, and function. MHC proteins, found on the surface of cells, present antigens to TCRs found exclusively on T cells (Section 29.4). Igs, found on the surface of B lymphocytes and in serum and mucosal secretions, interact directly with extracellular antigens (Section 29.7).

Structure and Evolution of Antigen-Binding Proteins

Ig, TCR, and MHC proteins share structural features and have evolved by duplication and selection of primordial antigen receptors. Some important Ig superfamily proteins are shown in

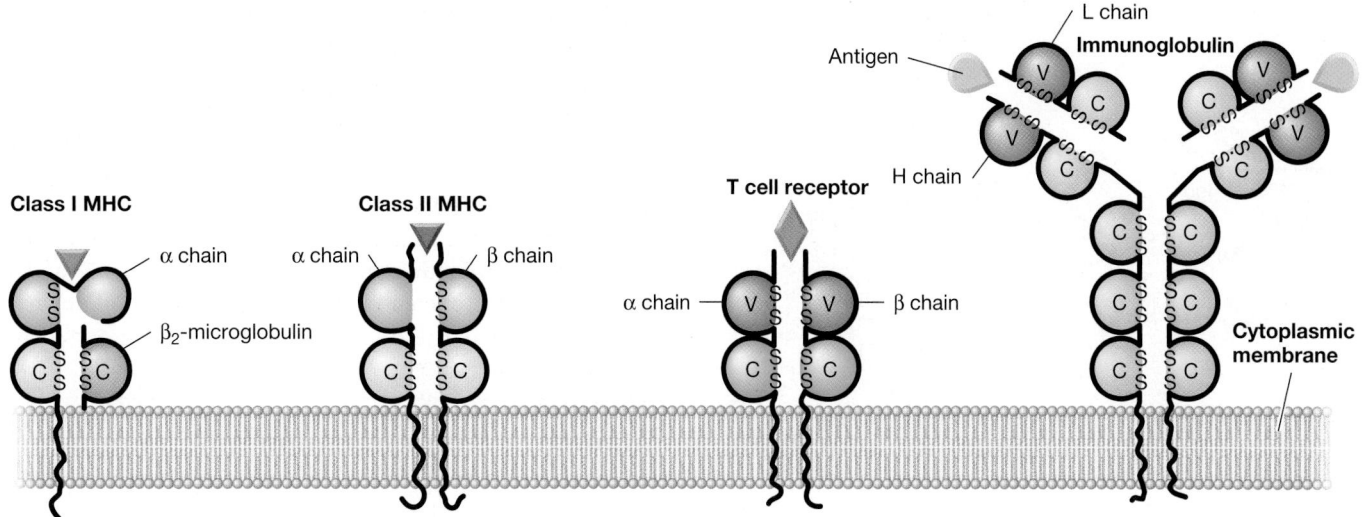

Figure 30.2 Immunoglobulin gene superfamily proteins. Constant domains have homologous amino acid sequences and higher-order structures. The Ig-like C domains in each protein chain indicate evolutionary relationships that identify the proteins as members of the Ig gene superfamily. The V domains of Igs and TCRs are also Ig domains, but the peptide-binding domains of MHC class I and class II proteins are not identifiable Ig domains because their structures vary considerably from the basic features of the Ig domain.

Figure 30.2. The proteins consist of a number of discrete domains. Each protein has at least one domain with a highly conserved amino acid sequence called a *constant (C) domain*. The C domain typically has about 100 amino acids with an intra-chain disulfide bond spanning 50–70 amino acids.

The *variable (V) domains* of TCR and Ig are about the same size as the constant domains, but V-domain structures can be considerably different from one another and from the C domains. C domains provide structural integrity for the antigen-binding molecules, attach the V domains to the cytoplasmic membrane, and give each protein its characteristic shape. C domains can also provide recognition sites for accessory molecules. For example, C domains of most IgG and all IgM proteins are bound by the C1q component of complement, a critical first step in initiating the complement activation sequence (๔๖ Section 29.9). Ig and TCR V domains have evolved to interact with a wide variety of antigens.

Likewise, MHC class I C domains bind to the accessory CD8 protein on T-cytotoxic (T_C) lymphocytes, and homologous MHC class II C domains bind CD4 on T-helper (T_H) cells. MHC I–CD8 and MHC II–CD4 interactions are critical steps for T cell activation and immune response development (๔๖ Section 29.4). By contrast, the V domains of MHC proteins have evolved independently of Ig and TCR V domains; they interact with non-self peptides, resulting in the MHC–peptide complex recognized as a foreign antigen by a TCR.

TCR, Ig, and MHC proteins each consist of two nonidentical polypeptides. The TCR consists of an alpha (α) and a beta (β) chain. MHC proteins also consist of two different polypeptide chains, again designated α and β (๔๖ Section 29.4). Igs have a separate heavy and light chain (๔๖ Section 29.7). These heterodimers are expressed on a cell surface and bind antigens. However, the specific function of each of these molecules is quite different. Igs can be anchored on B cell surfaces where they bind to pathogens and their products such as toxins. Igs are also produced in large quantities as soluble serum and mucosal proteins. TCRs, found exclusively on T lymphocytes, interact with antigenic peptides derived from processed pathogen proteins. These peptides are presented by the MHC proteins on target cells or specialized antigen-presenting cells, or APCs (as we discussed in Section 29.2, APCs include macrophages, dendritic cells, and B lymphocytes). Antigen-reactive T_C cells then kill the antigen-bearing cell; antigen-reactive T_H cells produce cytokines that activate the immune response (๔๖ Sections 29.5 and 29.6).

Signal Transduction in Antigen-Reactive Lymphocytes

As we have discussed, B cells and T cells interact with antigen through their Ig and TCR antigen receptors, respectively. As with the membrane-integrated PRRs in the innate immune response, the antigen-specific Igs and TCRs must transmit the signal from receptor binding across the cytoplasmic membrane to enhance transcription and activate the cell. B and T lymphocytes use the antigen-binding Ig and TCR proteins to transfer signals across the membrane by connecting to the common signal transduction pathways inside the cell. The antigen receptors, however, cannot directly connect to the signal transduction pathways because Igs and TCRs have very small cytoplasmic domains. These domains do not interact directly with the adaptor proteins common to signal transduction pathways. In addition, the cytoplasmic domains of neither Igs nor TCRs have cytoplasmic tyrosines that can be phosphorylated (**Figure 30.3**).

To get around this problem, both receptors associate with adaptor molecules that have *immune-based tyrosine-activation motifs* (*ITAMs*), possessing tyrosines that can be phosphorylated. These adaptor molecules are Igα and Igβ for immunoglobulins

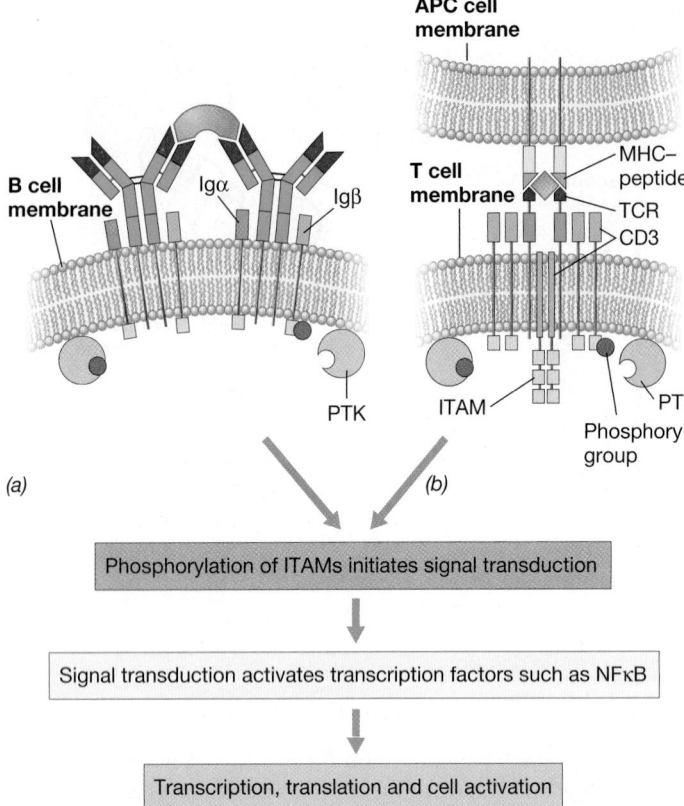

Phosphorylation of ITAMs initiates signal transduction

Signal transduction activates transcription factors such as NFκB

Transcription, translation and cell activation

Figure 30.3 Signal transduction in adaptive immunity. *(a)* The surface Ig on B cells associates with two adaptor proteins, Igα and Igβ, in the cytoplasmic membrane. Although Ig lacks ITAMs (immune-based tyrosine-activation motifs), the adaptor proteins contain multiple ITAMs that are exposed when antigen binds and cross-links Ig. Phosphorylation of the exposed ITAMs by PTK initiates signal cascades similar to those in innate immunity (Figure 30.1) and induces activation of transcription factors including NFκB. The transcription factors initiate transcription, leading to translation of proteins. *(b)* The TCR also associates with adaptor proteins, collectively called CD3. TCR interaction with the MHC–peptide complex exposes the ITAMs on the CD3 components, leading to phosphorylation, signal transduction, activation of transcription factors, and translation of T cell–specific proteins.

(Figure 30.3*a*) and the CD3 complex for TCRs (Figure 30.3*b*). The adaptors associate noncovalently with their respective antigen receptors in the membrane. Antigen binding by the Ig or TCR provokes conformational changes in the adaptor proteins. These changes expose the cytoplasmic ITAMs in the adaptor proteins, which are then phosphorylated by a family of protein–tyrosine kinases (PTKs), as shown in Figure 30.3*b*. As in signal transduction in the innate response, the kinase reaction initiates a cascade, culminating in the activation of NFκB and other transcription factors, and initiating transcription of downstream genes.

MiniQuiz

- Describe the structural features of the Ig constant domain.
- How do V-domain structures differ from C-domain structures?
- How are ITAMs in adaptor molecules influenced by Igs and TCRs?

Ⅱ The Major Histocompatibility Complex (MHC)

The major histocompatibility complex (MHC) is a group of genes found in all vertebrates. MHC proteins play a critical role in the presentation of processed antigens to other components of the immune system. The MHC spans about 4 Mbp on human chromosome 6 and is called the **human leukocyte antigen (HLA)** complex (**Figure 30.4**).

30.3 MHC Protein Structure

MHC proteins are the major barriers for tissue compatibility in transplantation, hence their name. Most individuals have different MHC alleles, producing variant MHC proteins; tissue transplanted from one individual to another is usually rejected by an immune response triggered by the MHC protein differences.

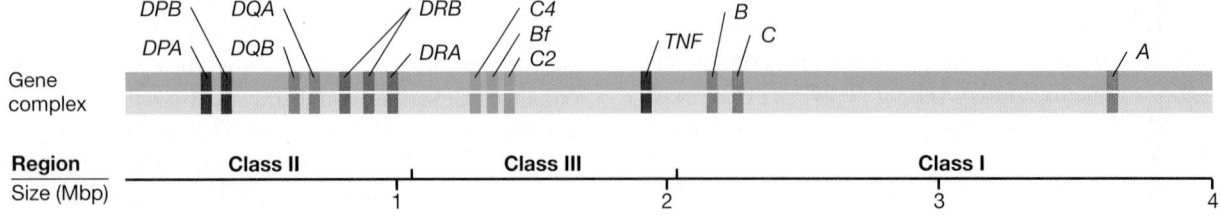

Figure 30.4 The human leukocyte antigen (HLA) gene map. The HLA complex, located on chromosome 6, is more than 4 million bases in length. The relative positions of some of the expressed genes are shown. Class II genes *DPA* and *DPB* encode class II proteins DPα and DPβ; *DQA* and *DQB* encode DQα and DQβ; *DRA* and two *DRB* loci encode DRα and DRβ; proteins. Other MHC genes designated as class III genes encode several proteins associated with immune-related functions. Not all class III genes are shown. *C4* and *C2* genes encode complement proteins C4 and C2 (🔗 Section 29.9). The *TNF* gene encodes a cytokine, tumor necrosis factor. The class I MHC proteins HLA-B, HLA-C, and HLA-A are encoded by genes *B*, *C*, and *A*. The class II loci DPA, DPB; DQA, DQB; DRA and two DRB, as well as the class I loci B, C, and A, are highly polymorphic and produce antigen-binding proteins.

This property of the MHC proteins hints at their natural function. MHC proteins are always expressed on cell surfaces as a complex with an embedded peptide. In normal cells, the embedded peptide is derived from breakdown products of cell metabolism. Thus, the MHC proteins hold embedded self peptides. In cells infected with a virus, however, some of the embedded peptides are derived from the virus. These viral peptides, when complexed with the MHC protein, look much like the slightly altered MHC proteins on a transplant. As a result, the MHC–virus peptide complexes are recognized as nonself and are targeted for destruction by T_C cells. The MHC proteins are designed to present peptides to T cells for screening and potential targeting (ᴄᴄ Section 29.4).

MHC proteins consist of two structural classes. *Class I MHC proteins* are found on the surfaces of all nucleated cells. As a rule, the class I proteins present peptide antigens to T_C cells. If class I–embedded peptides are recognized by T_C cells, the antigen-containing cell is targeted and directly destroyed (ᴄᴄ Section 29.5). *Class II MHC proteins* are found only on the surface of B lymphocytes, macrophages, and dendritic cells, the professional APCs (ᴄᴄ Section 29.4). Through the class II proteins, the APCs present antigens to the T_H cells, stimulating cytokine production that leads to antibody-mediated immune responses (ᴄᴄ Section 29.6).

Class I MHC Proteins

A **class I MHC protein** consists of two polypeptides (**Figure 30.5a**). The gene for the membrane-integrated alpha α chain is in the MHC gene region on chromosome 6. The other class I polypeptide is the noncovalently associated beta-2 microglobulin (β_2m). The three-dimensional structure of class I MHC protein reveals a distinctive shape that suggests how this protein interacts with the antigen peptide and the TCR simultaneously. The class I α chain folds to form a groove that is closed on both ends. The groove lies between two α-helices that straddle a β-sheet. In the endoplasmic reticulum, the MHC I groove is loaded with peptides of about 8 to 11 amino acids in length, derived from degraded endogenous proteins (Figure 30.5b). For example, viral proteins produced inside the cell are degraded into peptides and loaded into class I MHC proteins. The MHC–peptide complex then moves to the cell surface to be recognized by TCRs on T_C cells (ᴄᴄ Section 29.4).

Class II MHC Proteins

A **class II MHC protein** consists of two noncovalently linked, membrane-integrated polypeptides, α and β, found only on APCs. One α and one β polypeptide, expressed together, form a functional heterodimer (Figure 30.5c). Class II proteins may be arranged in pairs or trimers that enhance their stability. The α1 and β1 domains of the class II protein interact to form a peptide-binding site similar to the class I peptide-binding site. However, the ends of the groove are open, permitting the class II protein to bind and display peptides that may be significantly longer than 8–11 amino acids. Class II–binding peptides, generally 10 to 20 amino acids in length, are proteolytic fragments derived from exogenous pathogens internalized and processed by the APCs (ᴄᴄ Section 29.4). The APCs use the class II–peptide complex to interact with TCRs on T_H cells, leading to T_H activation (ᴄᴄ Section 29.6).

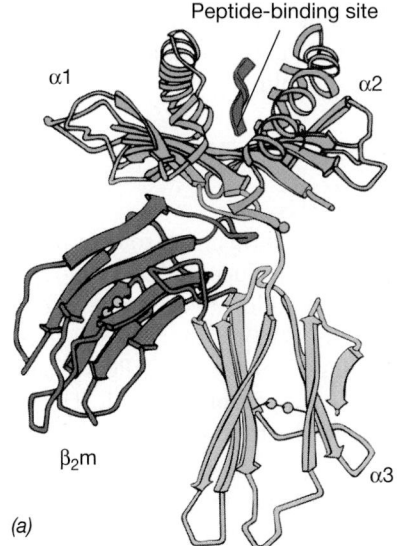

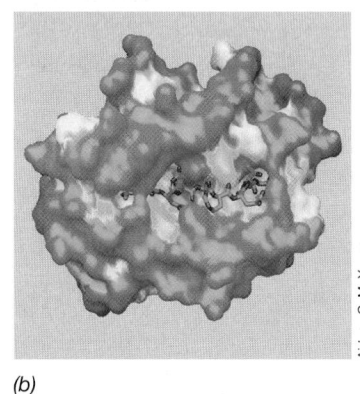

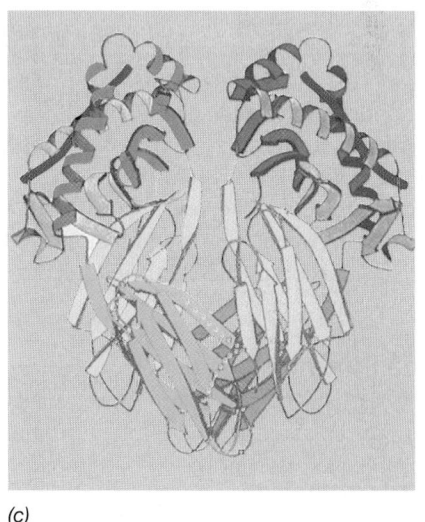

Figure 30.5 MHC protein structure. *(a)* The MHC class I protein. Beta-2 microglobulin (β_2m) binds noncovalently to the α chain. *(b)* An MHC I protein with a bound peptide, as seen from above. A nine-amino acid peptide is shown as a carbon backbone structure, embedded in a space-filling model of a mouse MHC I protein. *(c)* A class II protein dimer. The peptides and their position in the binding sites of the associated MHC II proteins are in brown. A helical shape indicates α-helix protein structure and a flat shape indicates a β-sheet (ᴄᴄ Figure 3.16).

30.4 MHC Polymorphism and Antigen Binding

The human MHC class I and class II genes encode peptide-binding proteins that bind antigen peptides for presentation to T cells. The many genetic variations of these proteins collectively bind all known peptides.

Polymorphism

Polymorphism is the occurrence in a population of multiple alleles (alternate forms of a gene) at a specific locus (the location of the gene on the chromosome) in frequencies that cannot be explained by recent random mutations. There are currently 767 HLA-A, 1178 HLA-B, and 439 HLA-C known alleles in the human population. Each person, however, has only two of these alleles at each locus; one allele is of paternal origin and one is of maternal origin. The two allelic proteins are expressed codominantly (equally). Thus, an individual usually displays six (three of maternal origin and three of paternal origin) of the many different genetically and structurally distinct alleles that encode class I proteins.

Likewise, highly polymorphic alleles encode class II proteins at the *HLA-DR*, *HLA-DP*, and *HLA-DQ* alpha- and beta-chain loci. Again, the class II gene products are expressed codominantly, resulting in expression of twelve alleles that encode distinct class II alpha and beta proteins.

These polymorphic variations in MHC proteins are major barriers to successful tissue transplants because the MHC proteins on the donor tissue (graft) are recognized as foreign antigens by the recipient's immune system. An immune response directed against the graft MHC proteins causes graft death and rejection. Tissue graft rejection, however, can be minimized by matching MHC alleles between donors and recipients. Control of rejection can also be accomplished through drugs that suppress the immune system.

Peptide Binding

Allelic variations in MHC proteins translate into amino acid changes concentrated in the antigen-binding groove, and each polymorphic variation of the MHC protein binds a different set of peptide antigens. The peptides bound by a single MHC protein share a common structural pattern, or peptide **motif**, and each different MHC protein binds a different motif. For example, for a certain class I protein, all of the peptides containing eight amino acids that it binds may have a phenylalanine at position 5 and a leucine at position 8. Thus, all peptides sharing the sequence X-X-X-X-phenylalanine-X-X-leucine (where X is any amino acid) would bind that MHC protein. Another MHC class I protein encoded by a different MHC allele binds a different motif, with nine amino acids and invariant amino acids tyrosine at position 2 and isoleucine at position 9 (X-tyrosine-X-X-X-X-X-X-isoleucine).

The invariant amino acids in each motif are *anchor residues*: They bind directly and specifically within an individual MHC–peptide binding groove. Thus, an individual MHC protein can bind and present many different peptides if the peptides contain the same anchor residues. Since each MHC protein binds a different motif with different anchor residues, the six possible MHC I proteins in an individual bind six different motifs. In this way, each individual can present a large number of different peptide antigens using the limited number of MHC I molecules available. MHC II proteins bind peptides in an analogous manner. As a result, within the human species, at least a few peptide antigens from each pathogen will display a motif that will be bound and presented by the MHC proteins. This system is very different from the mechanisms employed by Igs and TCRs that also bind antigens. Each Ig or TCR interacts very specifically with only a *single* antigen. As we shall see, these proteins employ a unique genetic mechanism to generate virtually unlimited diversity (Section 30.6).

 Antibodies

Antibodies are soluble proteins present in serum and other body fluids, where they function to neutralize and opsonize foreign antigens, or are cell surface antigen receptors on B lymphocytes. In this section, we look at the structure, antigen-binding function, genetic organization, and generation of diversity in the infinitely variable immunoglobulins.

30.5 Antibody Proteins and Antigen Binding

Antibodies, or immunoglobulins (Igs), are soluble proteins that interact with antigens and are produced by B lymphocytes. Antibodies consist of four polypeptides, two heavy chains (H) and two light chains (L) (Figure 30.2), arranged as a pair of heterodimers. Each heterodimer consists of a light-chain–heavy-chain pair and is a complete antigen-binding unit. The heavy and light chains are further divided into C (constant) and V (variable) domains. The C domains are responsible for common functions such as complement binding. The V domains of one H and one L chain interact to form an antigen-binding site (**Figure 30.6**). Here we examine the structural features of the V domains and the antigen-binding site.

Variable Domains

Amino acid sequences are considerably different in the V domains of different Igs (Figure 30.6). Amino acid variability is especially apparent in several **complementarity-determining regions**

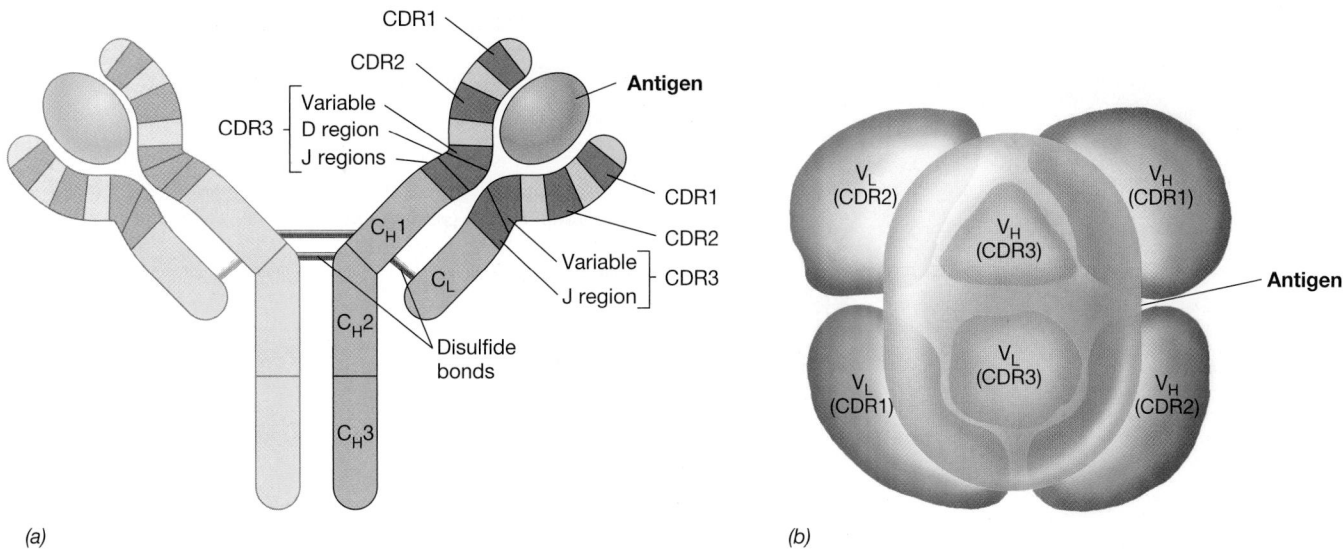

Figure 30.6 Antigen binding by immunoglobulin light and heavy chains. (a) One-half of an Ig is shown schematically, with a bound antigen. The V domains on the H and L chains are shown in red, with the antigen-binding CDR1, CDR2, and CDR3. C_H1, C_H2, and C_H3 are constant domains in the H chain, and C_L is the constant domain in the L chain. (b) The complementarity-determining regions (CDRs) from both H and L chains in part a are conformed to make a single antigen-binding site, or pocket, on the Ig. The site is shown from above. Red binding areas are from the H chain, and blue binding areas are from the L chain. The highly variable CDR3s from both H and L chains cooperate at the center of the site. An antigen is shown in gray, overlaying the site and contacting all CDRs. The actual shape of the site may be a shallow groove or a deep pocket, depending on the antibody–antigen pair involved.

(CDR). The three CDRs in each of the V domains provide most of the molecular contacts with antigen. CDR1 and CDR2 differ somewhat between different immunoglobulins, but the CDR3s differ dramatically from one another. The CDR3 of the heavy chain has a particularly complex structure, encoded within three distinct gene segments (see below). The CDR3 consists of the carboxy-terminal portion of the V domain, followed by a short "diversity" (D) segment of about three amino acids, and a longer "joining" (J) segment about 13-15 amino acids long. The light-chain CDR3 is similar, but lacks the D segment. The heavy- and light-chain CDRs cooperate in antigen binding.

Antigen Binding

The Ig three-dimensional structure was shown in Figure 29.15. Each antigen–antibody reaction requires the specific combination of the antigen with the variable domains of the associated heavy and light chains. The antigen-binding site of an antibody molecule measures about 2×3 nm, large enough to accommodate a small portion of the antigen, called an **epitope**, about 10 to 15 amino acids long. Antigen binding is ultimately a function of the Ig folding pattern of the heavy and light polypeptide chains. The Ig folds of the V region bring all six CDRs (CDR1, 2, and 3 from both heavy and light chains) together at the end of the Ig protein. The result is a unique and specific antigen-binding site (⮑ Figure 29.15 and Figure 30.6). In the next section, we examine the genetic mechanisms that generate the tremendous diversity found in the Ig proteins. Each antibody binds antigen with a characteristic affinity (binding strength). The affinity of an antibody is typically highest for the antigen for which it was selected, and antibodies usually do not bind other antigens. However, some antibodies will interact, usually weakly, with antigens other than the selecting antigen. This phenomenon is called a *cross reaction*.

30.6 Antibody Genes and Diversity

For most proteins, one gene encodes one protein. However, this is not the case with the heavy and light chains of immunoglobulins. Because the collection of antibodies in each individual must recognize and bind a wide variety of molecular structures, the immune system must generate almost unlimited antibody variation. Several mechanisms including somatic recombination, random heavy- and light-chain reassortment, and hypermutation all contribute to the almost limitless diversity generated from a relatively small, fixed number of Ig genes.

Immunoglobulin Genes

The gene encoding each immunoglobulin H or L chain is constructed from several gene segments. In each B cell, these gene segments undergo a series of somatic, random rearrangements (recombination followed by deletion of intervening sequences), to produce a single functional antibody gene derived from the pool of antibody genes. Molecular studies have verified this "genes in pieces" hypothesis by demonstrating that the V, D, and J gene segments encoding heavy-chain V domains, as well as the genes

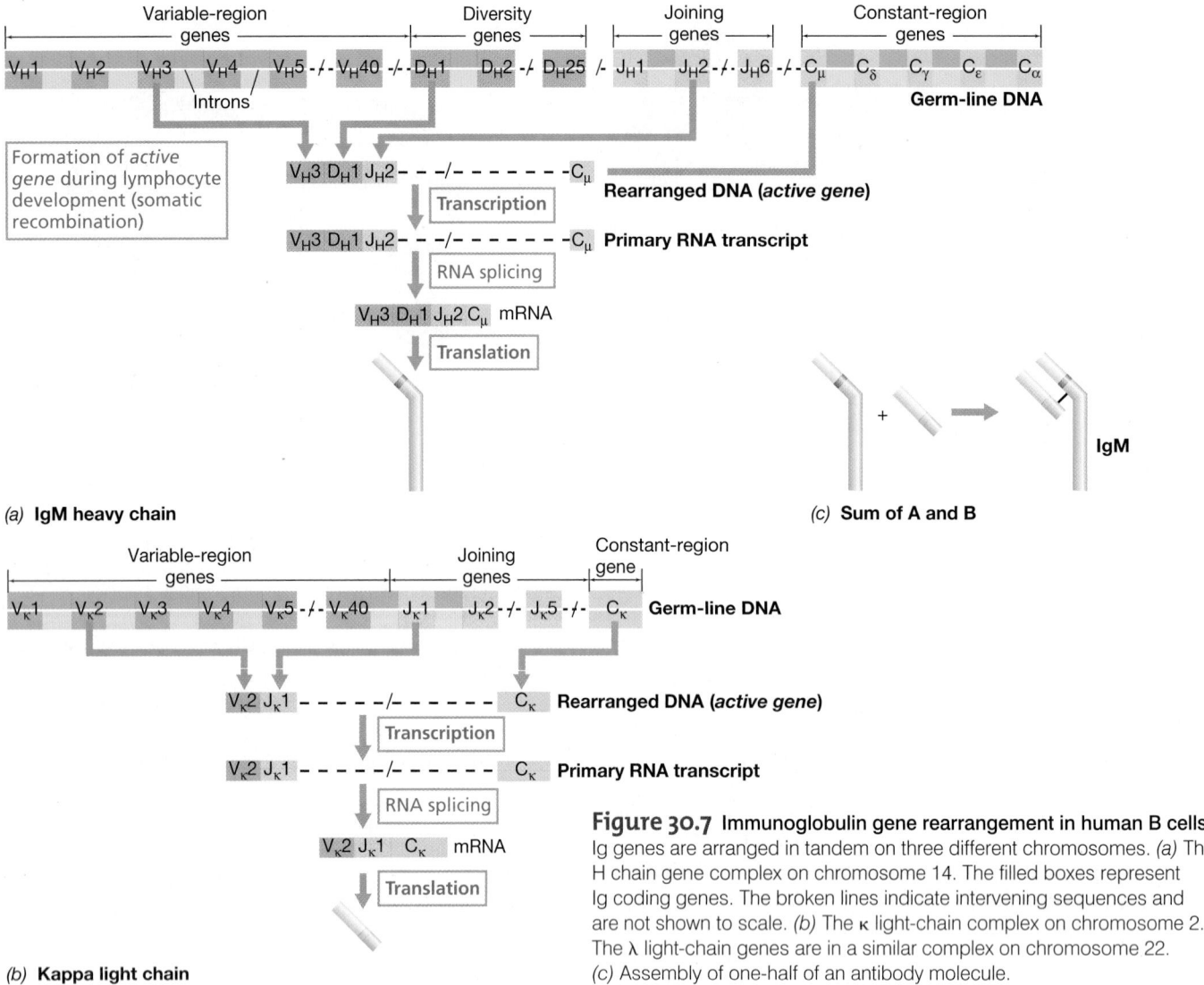

Figure 30.7 Immunoglobulin gene rearrangement in human B cells. Ig genes are arranged in tandem on three different chromosomes. *(a)* The H chain gene complex on chromosome 14. The filled boxes represent Ig coding genes. The broken lines indicate intervening sequences and are not shown to scale. *(b)* The κ light-chain complex on chromosome 2. The λ light-chain genes are in a similar complex on chromosome 22. *(c)* Assembly of one-half of an antibody molecule.

encoding C domains, are separated from one another in the genome. In each mature B cell the gene segments are brought together (somatically recombined) to form a single Ig heavy-chain gene (**Figure 30.7**). A single V gene encodes CDR1 and CDR2, whereas CDR3 is encoded by a mosaic of the 3′ end of the V gene, followed by the D and J genes.

In each B cell, only one protein-producing rearrangement occurs in the heavy- and light-chain genes. Called *allelic exclusion*, this mechanism ensures that each B cell produces only one Ig. Finally, the class-defining constant domains of Igs are encoded by separate C genes. Thus, four different gene segments, V, D, J, and C, recombine to form one functional heavy-chain gene. Similarly, light chains are encoded by recombination products of light-chain V, J, and C genes.

The gene segments required for all Igs exist in all cells but undergo recombination only in developing B lymphocytes. As shown in Figure 30.7, each B cell contains multiple kappa (κ) and a corresponding set of lambda (λ) light-chain V and J genes arranged in tandem. Each B cell also contains tandem V genes, D genes, and J genes for the heavy chains. In addition, the heavy-

chain constant-domain (C_H) genes and the light-chain constant-domain genes (C_L) are present. The V, D, J, and C genes are separated by noncoding sequences (introns) typical of gene arrangements in eukaryotes. Genetic recombination occurs in each B cell during its development. One each of the V, D, and J segments is randomly recombined to form a functional heavy-chain gene. On another chromosome, V and J segments are also randomly recombined to form a complete light-chain gene. The active gene, still containing an intervening sequence between the VDJ or VJ gene segments and the C gene segments, is transcribed, and the resulting primary RNA transcript is spliced to yield the final messenger RNA (mRNA). The mRNA is then translated to make the heavy and light chains of the Ig molecule.

Reassortment and VDJ Joining

Up to this point, all Ig diversity is generated from recombination of existing genes. In humans, for example, based on the numbers of genes at the kappa (κ) light-chain loci, there are 40 V × 5 J possible rearrangements, or 200 possible κ light chains. For the alternative lambda (λ) light chain, there are 30 V × 4 J = 120 possible chain combinations. About 6000 possible heavy chains can be formed by the rearrangement of 40 V × 25 D × 6 J genes. The final light chain and heavy chain produced by a given B cell result from reassortment of the heavy- and light-chain genes (Figure 30.7). Assuming that each heavy chain and light chain has an equal chance to be expressed in each cell, there are 6000 × 200 = 1,200,000 possible immunoglobulins with κ light chains and 6000 × 120 = 720,000 possible immunoglobulins with λ chains. In all, at least 1,920,000 possible antibodies can be expressed!

Additional diversity is generated by the DNA-joining mechanism. Joining of the V-D or D-J segments in the heavy chain or the V-J gene segments in the light chain is imprecise and frequently varies the sequence at these coding joints by a few nucleotides. Even more diversity is generated by additions of nucleotides at V-D and D-J coding joints on the heavy-chain genes, and at V-J coding joints in light-chain genes. Either random (N) or template-specific (P) nucleotides may be added. This N and P diversity at V-domain coding joints changes or adds amino acids in the CDR3 of both heavy and light chains.

Hypermutation

Finally, antibody diversity is expanded even more in B cells by **somatic hypermutation**, the mutation of Ig genes at much higher rates than the mutation rates observed in other genes. Somatic hypermutation of Ig genes is typically evident after a second exposure to an immunizing antigen. As we saw, a second exposure to antigen results in a change in the predominant antibody class produced, with a switch from IgM to IgG production (⮌ Section 29.8). Somatic hypermutation occurs only in the V regions of rearranged heavy- and light-chain genes. This process creates B cells bearing mutated receptors. These mutated B cells then compete for available antigen. This process selects B cells with receptors having higher antigen-binding strength (affinity) than the original B cell receptor. This *affinity maturation* process is one of the factors responsible for a dramatically stronger secondary immune response (⮌ Figure 29.20). The affinity maturation mechanism adds virtually unlimited possibilities to the generation of Ig diversity, making the potential antibody repertoire almost limitless.

MiniQuiz

- Describe the recombination events that produce a mature heavy-chain gene.
- Describe other somatic events that further enhance antibody diversity.

 IV T Cell Receptors

T cell receptors (TCRs) are cell surface antigen receptors on T cells that recognize peptide antigens embedded in MHC proteins. In this section, we look at the structure, antigen-binding function, and genetic organization of the TCRs.

30.7 T Cell Receptors: Proteins, Genes and Diversity

TCR proteins are integrated into the cytoplasmic membrane of T cells. TCRs consist of two polypeptides, the alpha (α) chain and the beta (β) chain. The α:β TCR specifically binds foreign peptides that are embedded in MHC molecules on the surface of APCs or target cells (⮌ Section 29.4). The TCR has the primary function of binding a foreign peptide, but it must do so in the context of the MHC protein. The TCRs accomplish this dual binding function through a binding site composed of the V domains of the α chain and β chain. The α-chain and β-chain V domains of TCRs contain CDR1, CDR2, and CDR3 segments that bind directly to the MHC-peptide antigen complex.

TCR Proteins

The three-dimensional structure of the TCR bound to MHC–peptide is shown in **Figure 30.8**. Both TCR and MHC proteins bind directly to peptide antigen. The MHC protein binds one face of the peptide, the MHC motif, whereas the TCR binds the other peptide face, the T cell epitope. The CDR regions of the TCR bind directly to the MHC–peptide complex, and each CDR has a specific binding function. The CDR3 regions of the TCR α chain and β chain bind with the antigen epitope; the CDR1 and CDR2 regions of the TCR α and β chains bind mainly to the MHC proteins.

TCR Genes and Diversity

Analogous to the H and L chains of immunoglobulins, the TCR α and β chains are encoded by distinct constant- and variable-domain gene segments. TCR V-region genes are arranged as a series of tandem segments. The α chain has about 80 V and 61 J genes, whereas the β chain has 50 V genes, 2 D genes, and 13 J genes (**Figure 30.9**). The β-chain V, D, and J genes and the α-chain V and J genes undergo recombination to form functional V-region genes. As in Igs, somatic mutations result from N and P diversity at V-D and D-J coding joints in the β chain and at the V-J coding joint in the α chain. Finally, the D region of the β chain can be transcribed in all three reading frames, leading to production of three separate transcripts from each D-region gene and creating greater diversity than would be expected from the D gene segments alone. As we discussed for reassortment of Ig H and L chains, individual α and β chains are produced by each T cell at random and joined to form a complete α:β heterodimer. The somatic hypermutation mechanisms responsible for another order of receptor diversity in Ig genes do not operate in T cells and do not generate additional TCR diversity. Potential TCR diversity, however, is still extensive, and on the order of 10^{15} different TCRs can be generated.

MiniQuiz

- Distinguish among the functions of the TCR CDR1, CDR2, and CDR3 segments.
- Which diversity-generating mechanisms are unique to TCRs? Which mechanisms are unique to Igs?

UNIT 9

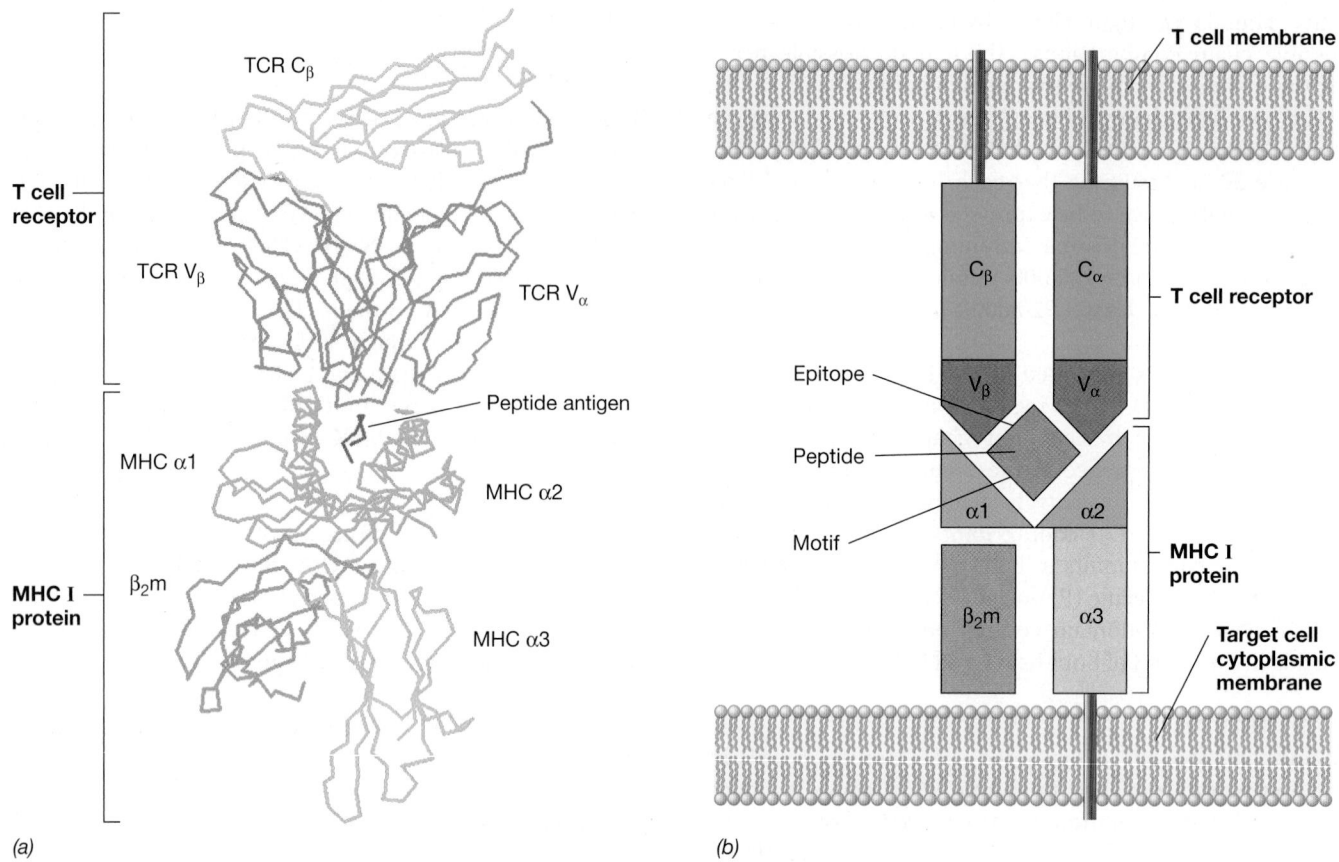

Figure 30.8 The TCR:MHC I–peptide complex. *(a)* A three-dimensional structure showing the orientation of TCR, peptide (brown), and MHC. This structure was derived from data deposited in the Protein Data Bank. *(b)* A diagram of the TCR:MHC–peptide structure. Note that the peptide is bound by both MHC and TCR proteins and has a distinct surface structure that interacts with each.

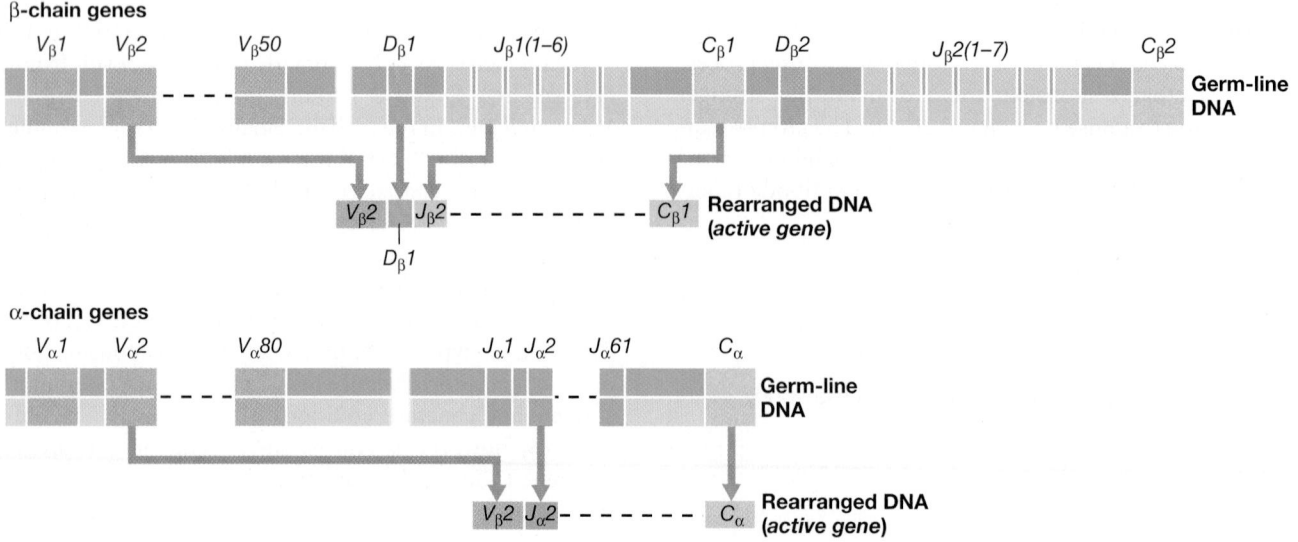

Figure 30.9 Organization of the human TCR α- and β-chain genes. The α-chain genes are located on chromosome 14 and the β-chain genes are on chromosome 6.

Ⓥ Molecular Switches in Immunity

Here we introduce the mechanisms that turn the immune response on or off. We first explore clonal selection, the mechanism by which antigen-reactive cells respond to foreign antigens while ignoring self antigens. Next we examine a sequence of molecular signals that are required for activating T cells or B cells. Finally, we introduce cytokines and chemokines, soluble proteins produced by activated cells to recruit and activate other cells in the immune response.

30.8 Clonal Selection and Tolerance

T cells must be able to discriminate between the dangerous non-self antigens and the harmless self antigens that compose our body tissues. Thus, T cells must acquire **tolerance**, or specific unresponsiveness to self antigens. To acquire tolerance, immune lymphocytes are maintained that interact only with the nonself antigens.

Clonal Selection

The **clonal selection** theory states that each antigen-reactive B cell or T cell has a cell surface receptor for a single antigen epitope. When stimulated by interaction with that antigen, each cell can replicate, and antigen-stimulated B and T cells grow and differentiate, producing a pool of cells that express the same antigen-specific receptors. A *clone* comprises the identical progeny of the initial antigen-reactive cell (**Figure 30.10**). Cells that have not interacted with antigen do not proliferate.

To respond to the seemingly infinite variety of antigens, a nearly infinite number of antigen-reactive cells are needed in the body. As we have discussed, the immune system can generate a nearly limitless number of antigen-specific B and T cell receptors. Inevitably, some of these receptors will have the potential to react with self antigens in the host. As a result, the immune system must eliminate or suppress these self-reactive cells, while at the same time selecting cells that may be useful against nonself antigens.

T Cell Selection and Tolerance

T cells undergo immune selection *for* potential antigen-reactive cells and selection *against* those cells that react strongly with self antigens. Selection against self-reactive cells results in the development of tolerance. The failure to develop tolerance may result in dangerous reactions to self antigens, a condition called *autoimmunity* (↩ Section 28.9).

Precursors of lymphocyte precursors that are destined to become T cells leave the bone marrow and enter the thymus, a primary lymphoid organ, via the bloodstream (**Figure 30.11**). During the process of T cell maturation in the thymus, immature T cells undergo a two-step selection process to (1) select potential antigen-reactive cells (positive selection) and (2) eliminate cells that react with self antigens (negative selection). **Positive selection** requires the interaction of new T cells in the thymus with the thymic self antigens; the peptide antigens in the thymus are of self origin. Using their TCRs, some T cells bind to MHC–peptide complexes on the thymic tissue. The T cells that do not bind MHC–peptide complexes undergo *apoptosis*, or programmed cell death, and are permanently eliminated. By contrast, those

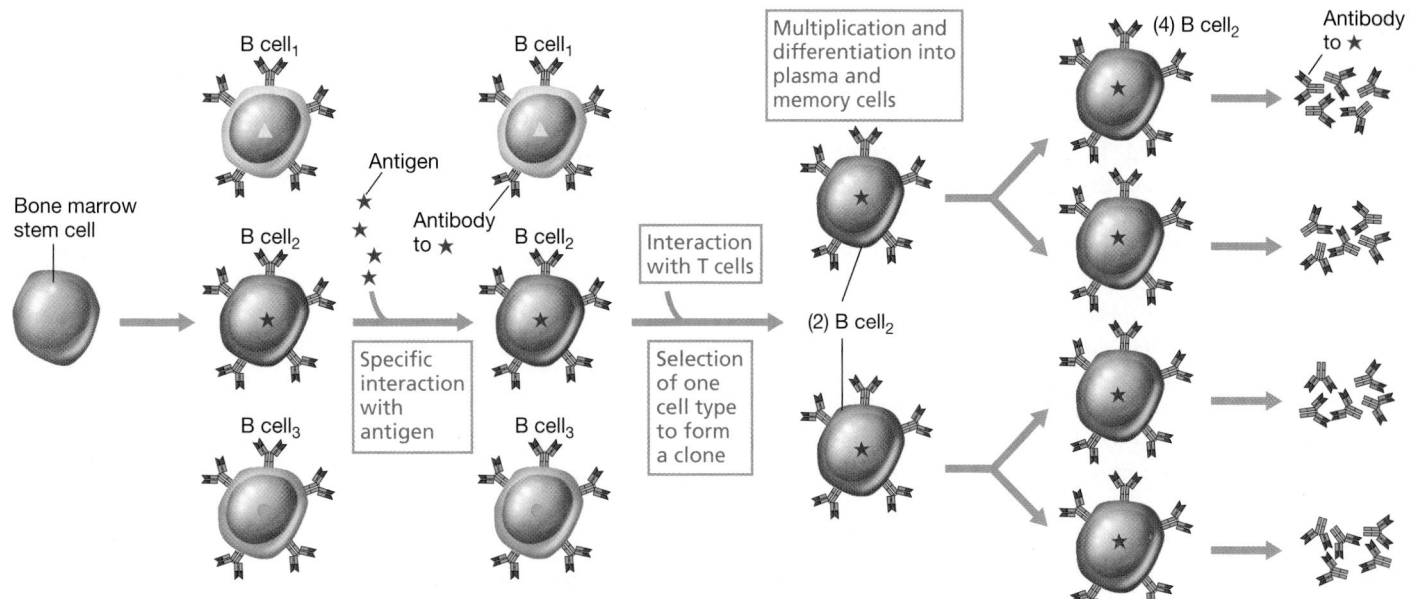

Figure 30.10 Clonal selection. Individual B cells, specific for a single antigen, proliferate and expand to form a clone after interaction with the specific antigen. The antigen drives selection and then proliferation of the individual antigen-specific B cell. Clonal copies of the original antigen-reactive cell have the same antigen-specific surface antibody. Continued exposure to antigen results in continued expansion of the clone.

UNIT 9

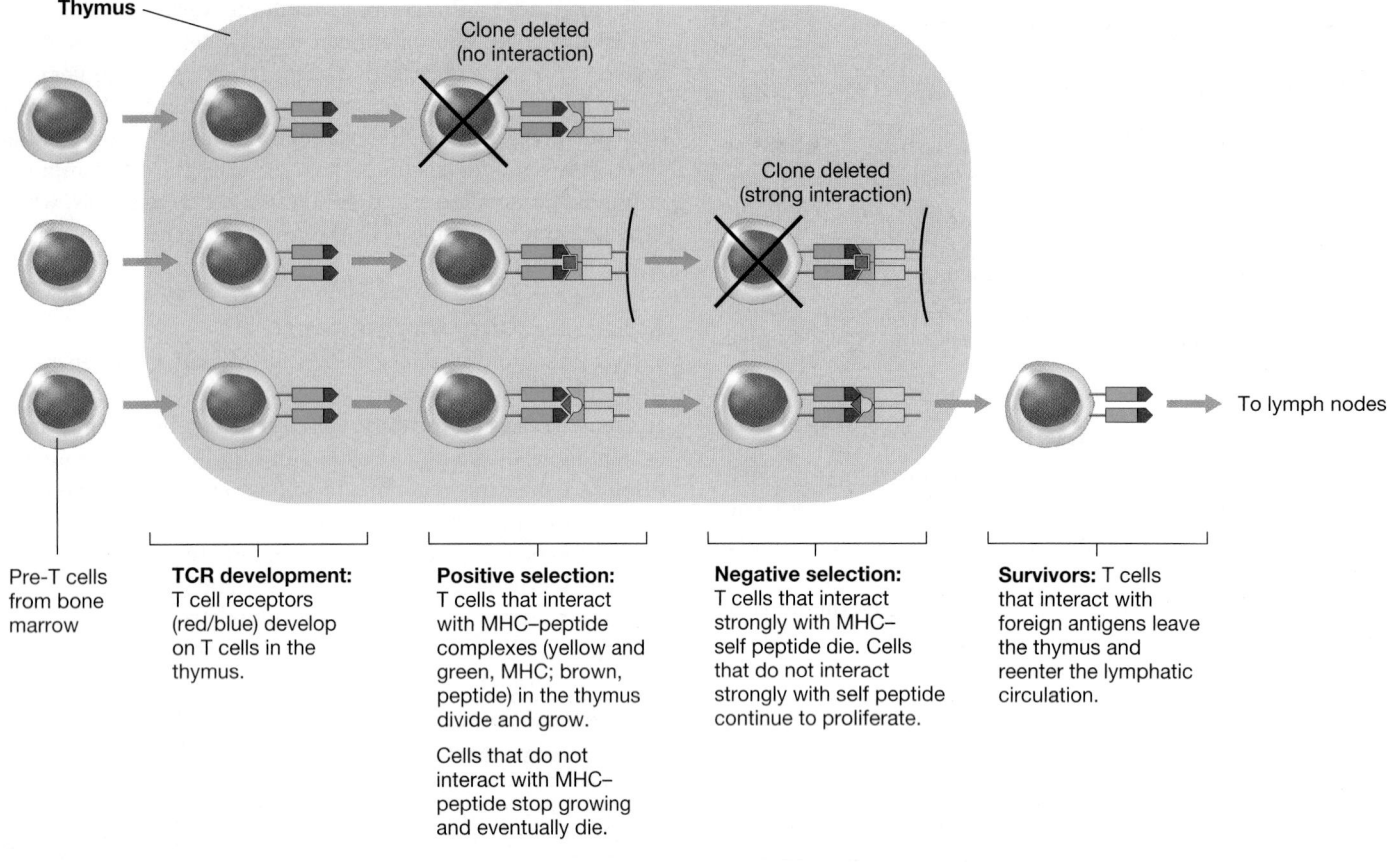

Figure 30.11 T cell selection and clonal deletion. T cells undergo selection for recognition of dangerous nonself antigens in the thymus.

T cells that bind thymic MHC proteins receive survival signals and continue to divide and grow. Positive selection retains T cells that recognize MHC–peptide and deletes T cells that do not recognize MHC–peptide and would therefore be unable to recognize MHC–peptide outside the thymus.

The second stage of T cell maturation is **negative selection**. Here the positively selected T cells continue to interact with thymic MHC–peptide. T cells that react with thymic self antigens are potentially dangerous if they react strongly with these antigens (autoimmunity). The strongly self-reactive T cells bind tightly to thymus cells, where they cannot divide and eventually die. TCRs that react less strongly with self MHC–peptide survive this selection and live. This two-stage thymic selection process for selecting self-tolerant, antigen-reactive T cells results in **clonal deletion**. Precursors of T cell clones that are either useless (do not bind) or harmful (bind too tightly) die in the thymus; more than 95% of all T cell precursors that enter the thymus do not survive the selection process.

The remaining selected T cells are destined to interact very strongly with nonself antigens. They are not destroyed in the thymus because their weak binding interactions with thymic self antigens signal them to proliferate. The selected and growing T cells leave the thymus and migrate to the spleen, mucosa-associated lymphoid tissue, and lymph nodes, where they can contact foreign antigens presented by B lymphocytes and other APCs.

B Cell Tolerance

The acquisition of immune tolerance in B cells is also necessary because antibodies produced by self-reactive B cells (autoantibodies) may cause autoimmunity and damage to host tissue. B cells also undergo a process of clonal deletion. Many self-reactive B cells are eliminated during development in the bone marrow, the primary lymphoid organ responsible for B cell development in mammals.

In addition to clonal deletion, **clonal anergy** (clonal unresponsiveness) also plays a role in final selection of the B cell repertoire. Some immature B cells are reactive to self antigens, but do not become activated even when exposed to high concentrations of self antigens. This is because B cell activation requires a second signal from T_H cells, as we shall now see. If no second signal is generated because the available T_H cells have been rendered tolerant to the antigen in the thymus, the B cell remains unresponsive.

MiniQuiz

- Distinguish between positive and negative T cell selection. How does positive and negative selection control the development of tolerance in T cells?
- Identify the role of the thymic cells in T cell selection.
- Distinguish between clonal deletion and clonal anergy in B cells.

30.9 T Cell and B Cell Activation

In addition to the critical interactions with antigen through Igs or TCRs, T and B cells require additional molecular signals for activation. Lack of these signals results in unresponsive cells, even if they are exposed to antigen. This mechanism helps to prevent autoimmunity.

T Cell Activation

As we have discussed, T cells that react with self antigens are deleted in the thymus. However, many self antigens are not expressed in the thymus. As a result, many T cell clones responsive to nonthymus antigens avoid clonal deletion in the thymus. These self-reactive T cells become anergic, but may persist as unresponsive T cells. The key to maintaining clonal anergy in these potentially dangerous self-reactive T cells is the signal mechanism used to activate T cells after they leave the thymus.

When positively and negatively selected T cells leave the thymus, they migrate to the secondary lymphoid organs (lymph nodes, spleen, and mucosa-associated lymphoid tissue; ↩ Section 28.1). These antigen-reactive T cells have not yet encountered specific antigen and are therefore naive or uncommitted T cells. Uncommitted T cells must be activated by an APC to become competent effector cells (↩ Section 29.4).

The first step in activation of uncommitted T cells is binding of the MHC–foreign peptide complex on the APC by the TCR (**Figure 30.12**). This first signal is absolutely required for activation. Without TCR interacting with MHC–peptide, a T_C cell cannot be activated. The next step requires the interaction of two more proteins, one found on the APC, called B7, and one found only on T cells, called CD28. The binding of B7 to CD28, a second signal, activates the T_C cell, making it an effector cell. In the absence of a B7–CD28 interaction, the T cell is not activated (Figure 30.12). A T_C cell that is activated will kill any target cell that displays antigen, even those cells that do not display CD28. After a T cell is activated, only the first signal (TCR binding to MHC–peptide) is necessary to induce killer activity. An analogous situation occurs with T_H cells.

T Cell Anergy

The requirement for a second activation signal has major implications for establishing and maintaining clonal anergy. Self antigens that are not found in the thymus are present on many other cells in the body. An uncommitted T_C cell that interacts with one of these self antigens found on a cell that is not an APC will receive only an MHC–peptide signal because non-APCs do not display the B7 protein necessary to complete the second signal. In the absence of the B7–CD28 interaction, a T_C cell that engages MHC–peptide is permanently anergized and can never be activated (Figure 30.12); the B7–CD28 second signal is absolutely required for activation. Uncommitted T_H lymphocytes are activated in the same way, also using the B7–CD28 coreceptor second signal.

B Cell Activation

The B cell also has independent signals other than antigen interaction for activation and antibody production. As compared to the activation signals for T cells, however, different signals

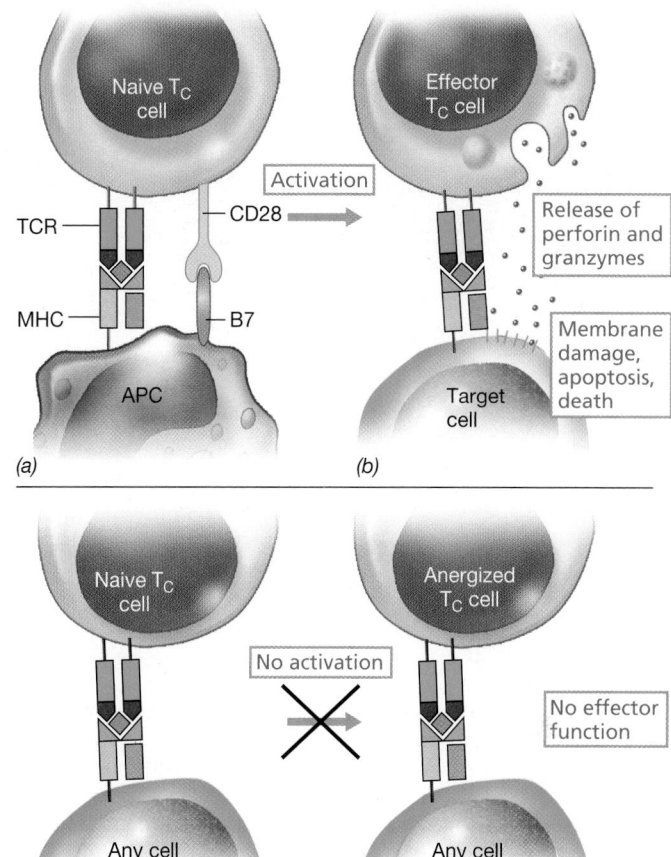

Figure 30.12 T cell activation. (a) A naive T_C cell interacts via TCR with the MHC–peptide complex on an APC. Antigen interaction via the TCR is the first required activation signal. The T_C cell also has a CD28 protein that interacts with a B7 protein on the APC. This interaction and binding is also required. The simultaneous interactions of the T_C cell and APC via both required signals activate the naive T cell. (b) The activated T_C cell is then capable of killing any target cell as long as TCR:MHC–peptide interactions take place. (c) A naive T_C cell interacts via the TCR with the MHC–peptide complex on any cell. Although the conditions for the first signal (interactions via TCR with the MHC–peptide complex) are met, the second signal cannot be generated because only APCs display the B7 protein. (d) In the absence of the second signal, the T_C cell becomes permanently unresponsive, or anergized.

activate B cells. As we have seen, B cells are responsible for antigen uptake, processing, and presentation as well as the production of specific antibodies (**Figure 30.13** and ↩ Section 29.8). The first signal for the B cell is antigen binding and cross-linking of surface immunoglobulin.

The second signal for B cell activation involves several molecules. The antigen–Ig interaction first signal generates a transmembrane signal that stimulates the B cell to express CD40 on its surface. Meanwhile, the B cell ingests the antigen bound on the Igs, processes the ingested antigen to peptides, and presents peptide antigen embedded in MHC II to neighboring T_H cells (both T_H1 and T_H2 cells can be involved in this process). A T_H cell with TCR reactivity with the presented antigen can then interact with the antigen-presenting B cell. Interaction via the

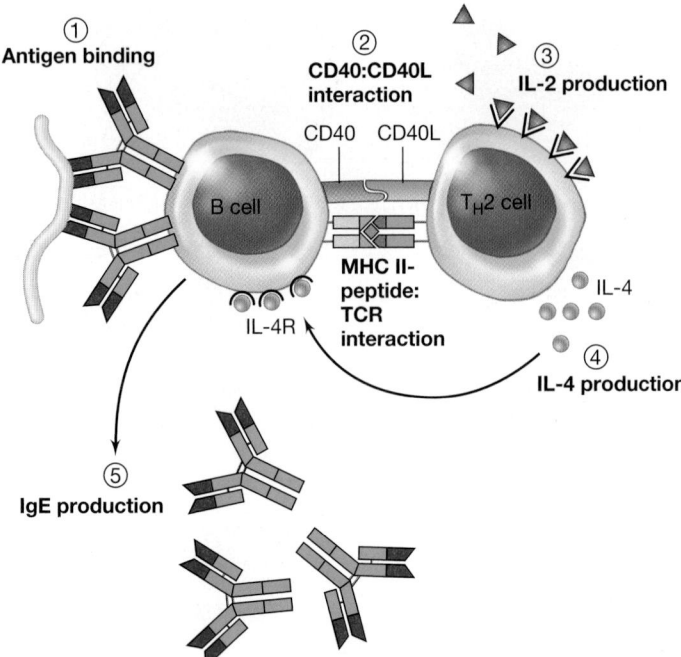

Figure 30.13 B cell activation: Signals and cytokines. (1) Antigen binds and cross-links the Ig receptors on a naive B cell. This signal stimulates the B cell to produce CD40 and express it on the cell surface. The B cell then processes the antigen and presents it to a T_H2 cell via MHC II. (2) The T_H2 cell interacts with the MHC II–peptide complex with its TCR. CD40L on the T_H2 cell then interacts with the B cell CD40. (3) These interactions stimulate the T_H2 cell to produce IL-2, which stimulates the same T_H2 cell (autocrine function). (4) The stimulated T_H2 cell can make a battery of cytokines, one of which is IL-4. IL-4 is a final activation signal for the B cell. (5) In this case, the cytokine-stimulated B cells then produce IgE. T_H2 cytokines stimulate activation, differentiation, and expansion of both T and B lymphocytes.

TCR:MHC II–peptide complex results in the expression of CD40L (CD40 ligand) by the T_H cells, which in turn binds to the B cell CD40. The CD40L–CD40 interaction initiates signal transduction in the T_H cell, leading to transcription of a number of T cell proteins, including IL-4 and other soluble cytokines. The cytokines, secreted by the T cell, interact with cytokine receptors on the B cell, completing the second signal for the B cells and stimulating antibody production. Thus, the complete second signal for B cell activation requires two interactions: the interaction of CD40 (T_H cell) with CD40L (B cell), as well as the interaction of cytokines (from the T_H cell) with the cytokine receptor (B cell).

After a B cell is activated, it no longer needs T cell interactions or cytokines to make antibody; antigen interaction alone can then stimulate antibody production. Some of these activated B cells will be transformed into plasma cells that will secrete large amounts of antibody in the primary immune response. Others remain as *memory B cells* and play a major role in the secondary immune response during subsequent exposure to an antigen (↩ Section 29.8).

MiniQuiz
- Define the activation signals for an uncommitted T cell.
- Define the activation signals for an uncommitted B cell.

30.10 Cytokines and Chemokines

Intercellular communication in the immune system is accomplished in many cases through a heterogeneous family of soluble effector proteins known as **cytokines** that are produced by leukocytes and other cells. Cytokines regulate cellular functions in immune cells and activate various cell types. The cytokines produced by lymphocytes are often called lymphokines or interleukins (ILs).

Cytokines secreted from one cell bind specific receptors on other cells. Some cytokines bind to receptors on the cell that produced them. Thus, these cytokines have autocrine (self-stimulatory) abilities. Other cytokines bind to receptors on other cells. Cytokine–receptor binding generally activates a signal transduction pathway, passing information across the cytoplasmic membrane to control activities such as transcription and protein synthesis. These signals ultimately result in cell differentiation and clonal proliferation.

Chemokines are a group of small proteins that function as chemoattractants for phagocytes and lymphocytes. Chemokines are produced by macrophages, lymphocytes, and other cells in response to bacterial products, viruses, and other cell-damaging agents. Chemokines attract phagocytes and T cells to the site of injury, stimulating an inflammatory response as well as potentiating a specific immune response.

Table 30.2 lists some important immune cytokines and chemokines, the cells that produce them, their most common target cells, and their most important biological effects. Over 50 cytokines are known, most of which are produced by either T cells or monocytes and macrophages. About 40 chemokines are known. First we examine the activity of cytokines involved in the induction of an antigen-specific, antibody-mediated immune response. We then look at cytokines produced by T_H1 cells that activate macrophages. Finally, we look at the action of the cytokines and chemokines involved in macrophage-mediated inflammation.

Cytokines and Antibody Production

B cells are responsible for antigen uptake, processing, and presentation as well as the production of specific antibodies. As we discussed in the previous section, B cells require two independent signals for activation and antibody production. B cells are activated by antigen binding to surface immunoglobulin (signal 1) followed by interaction between the B cell CD40 and CD40L on the T cell (Figure 30.13). The activated T_H cell responds by producing IL-2, which is secreted and bound by the IL-2R on the surface of the T_H cells. Thus, IL-2 can activate the same cell that secreted it. Under the influence of IL-2, the cell divides, making clonal copies. In the process, the T_H cell also makes other cytokines such as IL-4 and IL-5.

IL-4 then binds to the IL-4R on the original antigen-presenting B cell. The IL-4:IL-4R interaction stimulates the B cell to differentiate into a plasma cell, which ultimately produces antibodies (↩ Section 29.8). The IL-4 generated by the responding T cell is the second signal (signal 2) necessary for initiation of antibody production. In addition, the IL-4 interaction signals an immunoglobulin class switch. For example, IL-4:IL-4R interaction can switch antibody production by an affected B cell from IgM to IgE or IgG1.

Table 30.2 *Major immune cytokines and chemokines*

Cytokine (chemokine)	Major producer cells	Major target cells	Major effect
IL-4[a]	T$_H$2	B cells	Activation, proliferation, differentiation, IgG1 and IgE synthesis
IL-5	T$_H$2	B cells	Activation, proliferation, differentiation, IgA synthesis
IL-2	Naive T cells, T$_H$1, and T$_C$	T cells	Proliferation (often autocrine)
IFN-γ[b]	T$_H$1	Macrophages	Activation
GM-CSF[c]	T$_H$1	Macrophages	Growth and differentiation
TNF-α[d]	T$_H$1 Macrophages	Macrophages Vascular epithelium	Activation, production of pro-inflammatory cytokines Activation, inflammation
IL-1β	Macrophages	Vascular epithelium, lymphocytes	Activation, inflammation
IL-6	Macrophages, dendritic cells	Lymphocytes	Activation
IL-12	Macrophage	NK cells, naive T cells	Activation, enhances differentiation to T$_H$1
IL-17	T$_H$17	Neutrophils	Activation
CXCL8 (chemokine)	Macrophages	Neutrophils, basophils, T cells	Chemotactic factor
CCL2 (MCP-1[e]) (chemokine)	Macrophages	Macrophages, T cells	Chemotactic factor, activator

[a]IL, interleukin; [b]IFN, interferon; [c]GM-CSF, granulocyte, monocyte-colony stimulating factor; [d]TNF, tumor necrosis factor, [e]MCP, macrophage chemoattractant protein.

Alternatively, IL-5 produced by T$_H$ cells can bind an IL-5R on the antigen-presenting B cell. Parallel to the situation with IL-4:IL-4R interaction, the IL5:IL-5R also stimulates the B cell to differentiate into a plasma cell, which ultimately produces antibodies, and induces a class switch, but this time to IgA.

As these two examples show, IL-2, IL-4, and IL-5 cytokines are soluble mediators and activators for T lymphocytes and B cells. They interact to induce the antibody-mediated immune response; T$_H$ cells in different locations produce different B cell–activating cytokines to focus the antibody response to that environment. For example, T$_H$ cells near the skin produce more IL-4, inducing production of IgE, while T$_H$ cells in the gut produce IL-5, inducing production of secretory IgA. Thus, IL-4 and IL-5 not only control activation of the B cell, but also control the quality of the antibody response, directing the class switch from IgM to IgE, or IgA, respectively, thereby focusing the antibody production for a particular environment.

T$_H$1 and Macrophage Activation

Table 30.2 shows the activity of several cytokines produced by some T$_H$1 cells. These cytokines are important in the activation of macrophages. The cytokines IFN-γ (gamma interferon), GM-CSF (granulocyte–monocyte colony stimulating factor), and TNF-α (tumor necrosis factor alpha) are produced by antigen-activated T$_H$1 cells. These cytokines stimulate macrophage differentiation and activation.

Macrophages, Proinflammatory Cytokines, and Chemokines

Stimulated macrophages produce a number of cytokines and chemokines, many of which play a role in initiating inflammation. Some of the most important macrophage-produced

proinflammatory cytokines are IL-1β, TNF-α, IL-6, and IL-12. IL-1β and TNF-α induce activation of vascular endothelium. IL-6 activates lymphocytes, and all except IL-12 induce fever at the systemic level. IL-12 acts to stimulate natural killer (NK) cells and to induce naive T cells to differentiate to T$_H$1 cells (Table 30.2).

Chemokines produced by activated macrophages include CXCL8 and CCL2, also called MCP-1. CXCL8, also called IL-8, is produced by monocytes, macrophages, and other cells. CXCL8 is secreted by the affected cells and binds to receptors on T cells and neutrophils, where it acts as a chemoattractant. This results in a neutrophil-mediated inflammatory response followed by a specific immune response by the attracted T cells. As is the case for the cytokine receptors, engaged chemokine receptors on the target cells act through signal transduction pathways to induce activation of effector cells such as neutrophils or T cells.

CCL2 is produced by macrophages and other cells. CCL2 attracts basophils, eosinophils, monocytes, dendritic cells, natural killer cells, and T cells, stimulating production of more inflammatory mediators and potentially organizing an antigen-specific immune response.

MiniQuiz

- Identify the major cytokines and chemokines produced by T$_H$1 cells, T$_H$2 cells, and macrophages.
- What events stimulate cytokine and chemokine production?
- Identify the proinflammatory cytokines, the cells that produce them, and their effects on other cells.

Big Ideas

30.1
Interactions between PAMPs and host PRRs are integral components of the innate immune response. PRRs interact with PAMPS shared by various pathogens, activating complement and phagocytes to target and destroy pathogens. These interactions initiate signal transduction cascades that activate effector cells.

30.2
The Ig gene superfamily encodes proteins that are evolutionarily, structurally, and functionally related to immunoglobulins. The antigen-binding Igs, TCRs, and MHC proteins are members of this family. Antigen binding to Ig or TCR facilitates signal transduction through adaptor molecules containing ITAMs.

30.3
Class I MHC proteins are expressed on all nucleated cells and function to present endogenous antigenic peptides to TCRs on T_C cells. Class II MHC proteins are expressed only on APCs. They function to present exogenously derived peptide antigens to TCRs on T_H cells.

30.4
MHC genes encode proteins used to present peptide antigens to T cells. Class I and class II MHC genes are highly polymorphic. MHC class I and class II alleles encode proteins that bind and present peptides with conserved structural motifs.

30.5
The antigen-binding site of an Ig is composed of the V (variable) domains of one heavy chain and one light chain. Each heavy and each light chain contains three complementarity-determining regions, or CDRs, that are folded together to form the antigen-binding site.

30.6
Immunoglobulin diversity is generated by several mechanisms. Somatic recombination of gene segments allows shuffling of the various Ig gene segments. Random reassortment of the heavy- and light-chain genes, imprecise joining of VDJ and VJ gene segments, and hypermutation mechanisms contribute to nearly unlimited immunoglobulin diversity.

30.7
T cell receptors bind to peptide antigens presented by MHC proteins. The CDR3 regions of both the α chain and the β chain bind to the antigen epitope, whereas the CDR1 and CDR2 regions bind to the MHC protein. The V domain of the β chain of the TCR is encoded by V, D, and J gene segments. The V domain of the α chain of the TCR is encoded by V and J gene segments. TCR diversity is generated by a variety of genetic mechanisms, yielding practically unlimited TCR antigen-binding diversity.

30.8
The thymus is a primary lymphoid organ that provides an environment for the maturation of antigen-reactive T cells. Immature T cells that do not interact with MHC–peptide (positive selection) or react strongly with self antigens (negative selection) are eliminated by clonal deletion in the thymus. T cells that survive positive and negative selection leave the thymus and can participate in an effective immune response. B cell reactivity to self antigens is controlled through clonal deletion and anergy.

30.9
Many self-reactive T cells are deleted during development and maturation in the thymus. Uncommitted T cells are activated in the secondary lymphoid organs by first binding MHC–peptide with their TCRs (signal 1), followed by binding of the B7 APC protein to the CD28 T cell protein (signal 2). B cell activation is initiated by antigen interaction with surface immunoglobulin (signal 1), followed by interaction between the B cell CD40 protein and CD40L on the T cell to generate cytokine production (signal 2).

30.10
Cytokines produced by leukocytes and other cells are soluble mediators that regulate interactions between cells. Several cytokines, such as IL-2 and IL-4, affect lymphocytes and are critical components in the generation of specific immune responses. Other cytokines, such as IFN-γ and TNF-α, affect a wide variety of cell types. Chemokines produced by various cells are released in response to injury and are strong attractants for nonspecific inflammatory cells and T cells.

Review of Key Terms

Antibody a soluble protein, produced by B cells, that interacts with antigen; also called immunoglobulin

Chemokine a small, soluble protein that modulates inflammatory reactions and immunity

Class I MHC protein an antigen-presenting molecule found on all nucleated vertebrate cells

Class II MHC protein an antigen-presenting molecule found on macrophages, B cells, and dendritic cells (antigen-presenting cells)

Clonal anergy the inability to produce an immune response to specific antigens due to neutralization of effector cells

Clonal deletion for T cell selection in the thymus, the killing of useless or self-reactive clones

Clonal selection the production by a B or T cell of copies of itself after antigen interaction

Complementarity-determining region (CDR) a varying amino acid sequence within the variable domains of immunoglobulins or T cell receptors where contacts with antigen are made

Cytokine a small, soluble protein produced by a leukocyte that modulates inflammatory reactions and immunity

Epitope the portion of an antigen that is recognized by an immunoglobulin or a T cell receptor

Human leukocyte antigen (HLA) antigen-presenting protein encoded by a major histocompatibility complex gene in humans

Immunoglobulin (Ig) a soluble protein, produced by B cells, that interacts with antigen; also called antibody

Immunoglobulin gene superfamily a family of genes that are evolutionarily, structurally, and functionally related to immunoglobulins

Major histocompatibility complex (MHC) a genetic region that encodes several proteins important for antigen presentation and other host defense functions

Motif in antigen presentation, a conserved amino acid sequence found in all peptides that bind to a given MHC protein

Negative selection in T cell selection, the deletion of T cells that interact with self antigens in the thymus (see clonal deletion)

Pathogen-associated molecular pattern (PAMP) a structural component of a pathogen or pathogen product that is recognized by a pattern recognition receptor (PRR)

Pattern recognition receptor (PRR) a protein that recognizes a pathogen-associated molecular pattern (PAMP), such as a component of a microbial cell surface structure

Polymorphism in a population, the occurrence of multiple alleles for a gene locus at a higher frequency than can be explained by recent random mutations

Positive selection in T cell selection, the growth and development of T cells that

interact with self MHC–peptide in the thymus

Somatic hypermutation the mutation of immunoglobulin genes at rates higher than those observed in other genes

T cell receptor (TCR) the antigen-specific receptor protein on the surface of T cells

Tolerance the inability to produce an immune response to a specific antigen

Toll-like receptor (TLR) a pattern recognition receptor of phagocytes, structurally and functionally related to Toll receptors in *Drosophila*

Review Questions

1. Identify at least one soluble pattern recognition receptor (PRR), its interacting pathogen-associated molecular pattern (PAMP), and the resulting host response (Section 30.1).

2. Outline the structural features of the antigen-binding proteins (Section 30.2).

3. Compare and contrast the MHC class I and II proteins (Section 30.3).

4. Polymorphism implies that each different MHC protein binds a different peptide motif. For the MHC class I polymorphisms, how many different MHC proteins are expressed in an individual? By the entire human population (Section 30.4)?

5. Which Ig chains are used to construct a complete antigen-binding site? Which domains? Which CDRs (Section 30.5)?

6. Describe how reassortment and VDJ joining increase the diversity of antibodies (Section 30.6).

7. Describe the interaction of the TCR with peptide antigen and MHC protein. Be sure to identify the roles of the CDRs in the TCR (Section 30.7).

8. In TCRs, diversity can be generated by recombination and reassortment events as in Igs. As is the case in Igs, additional diversity is generated with somatic events such as N-region nucleotide additions and reading of the D segment in all three reading frames. Explain these diversity-generating mechanisms (Section 30.7).

9. Explain positive and negative selection of T cells (Section 30.8).

10. What molecular interactions are necessary for activation of uncommitted T cells? For activation of uncommitted B cells (Section 30.9)?

11. What are cytokines? Discuss the importance of cytokines in the immune system (Section 30.10).

Application Questions

1. Identify the consequences of a genetic mutation that eliminates a PRR by predicting the outcome for the host. Do this for at least one soluble PRR and one membrane-bound PRR.

2. Construct a table that lists the common features of proteins encoded by members of the Ig gene superfamily. For Igs, TCRs, and MHC proteins, identify the structural components that fit these common features.

3. Polymorphism implies that each different MHC protein binds a different peptide motif. However, for the MHC class I proteins, only 6 peptide motifs can be recognized in an individual, whereas over 350 motifs can be recognized by the entire human population. What advantage does this have for the population? For the individual?

4. Although genetic recombination events are important for generating significant diversity in the antigen-binding site of Igs, post-recombination somatic events may be even more important in achieving overall Ig diversity. Do you agree or disagree with this statement? Explain.

5. What would happen to the T cell repertoire in the absence of positive selection? In the absence of negative selection?

6. What would be the result of activation of all T cells that contact antigen? How does the multiple signal scheme prevent this from happening?

Need more practice? Test your understanding with quantitative questions; access additional study tools including tutorials, animations, and videos; and then test your knowledge with chapter quizzes and practice tests at **www.microbiologyplace.com**.

31

Diagnostic Microbiology and Immunology

The bacterium *Streptococcus pyogenes*, shown here, is the causative agent of a number of disease syndromes, including scarlet and rheumatic fevers and "strep throat." The ability to rapidly diagnose *S. pyogenes* infections is therefore an important job of the clinical microbiologist.

The clinical microbiologist detects, identifies, and characterizes the microorganisms that cause infectious disease from a variety of samples collected from sick hosts. Direct observation of pathogens acquired from clinical specimens is a very important tool for many infectious diseases. If direct observation cannot conclusively identify infectious agents, diagnostic approaches in clinical microbiology include evaluation of samples by growth-dependent techniques, molecular techniques, and immunoassays (**Figure 31.1**). Clinical laboratories grow, isolate, and identify most pathogenic bacteria within 48 hours of sampling. In some cases, indirect methods involving immunological and molecular procedures are used to identify pathogens. Indirect methods are particularly important for the rapid identification of bacteria that are difficult to isolate or grow and for identification of viruses and protozoa.

The usefulness of any diagnostic test depends on the test's specificity and sensitivity. **Specificity** is the ability of the test to recognize a single pathogen. Optimal specificity implies that the test is specific for a single pathogen, and will not identify any other pathogen. High specificity prevents false-positive results. For example, for the detection of *Neisseria gonorrhoeae*, the organism that causes gonorrhea, the specificity of Gram-stained smears of urethral exudate from men is about 99% and about 95% for endocervical exudates from women. Thus, the test is very specific for both men and women, although significantly more so for men; false-positive results are rare.

Sensitivity defines the lowest numbers of a pathogen or the lowest amount of a pathogen product that can be detected. The highest level of sensitivity requires that the test be capable of identifying a single organism or molecule. High sensitivity prevents false-negative reactions. For example, for the detection of *N. gonorrhoeae*, the sensitivity of Gram-stained smears of urethral exudate from men is about 90%, and about 50% for endocervical exudates from women. Thus, the test is a sensitive indicator for gonorrhea in men, but is much less sensitive for women; false-negative tests would be relatively common for women and, therefore, suspected cases of gonorrhea in women must be further evaluated by more sensitive methods, including culture techniques.

I Growth-Dependent Diagnostic Methods

The isolation and growth of pathogens from host specimens is an important step in defining the cause of many infectious diseases. Positive identification of a pathogen is coupled with antimicrobial drug susceptibility testing to devise a specific treatment plan.

31.1 Isolation of Pathogens from Clinical Specimens

Samples of tissues or fluids are collected for microbiological, immunological, and molecular biological analyses if a healthcare provider suspects a disease is caused by an infectious agent (Figure 31.1). Typical samples include blood, urine, feces, sputum, cerebrospinal fluid, or pus from a wound. Swabs may be used to obtain samples from suspected infected areas such as skin, nares,

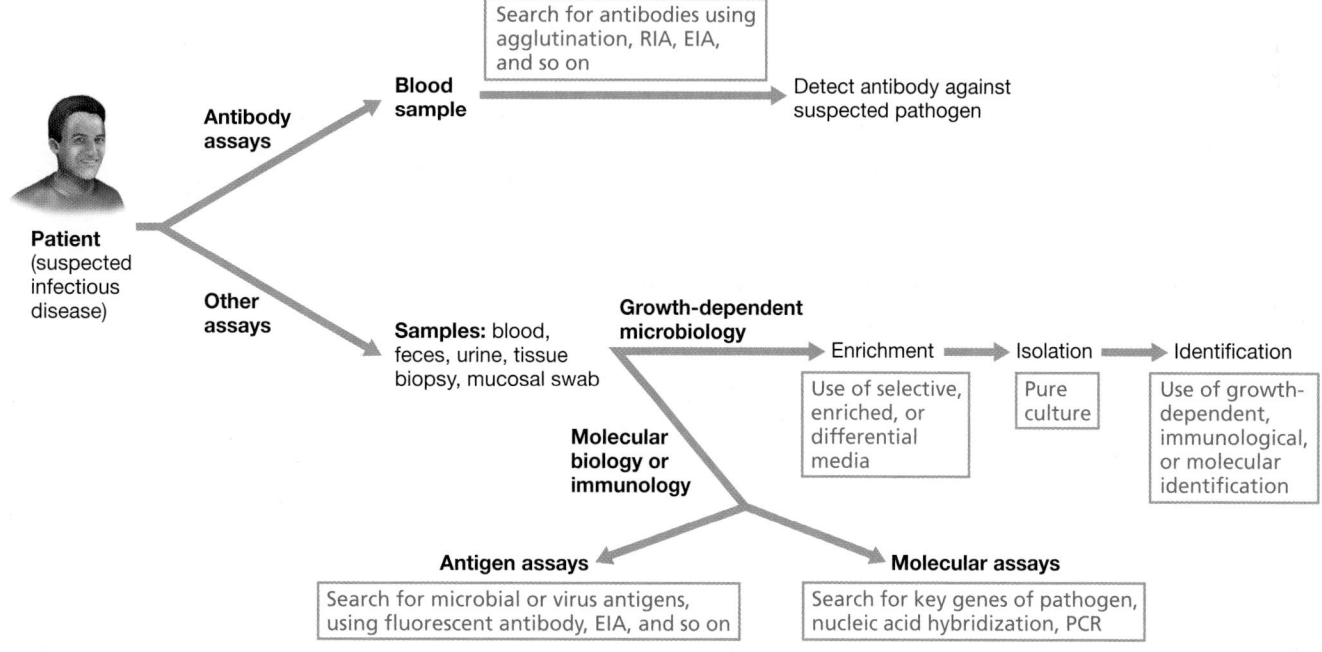

Figure 31.1 Laboratory identification of microbial pathogens. Diagnostic methods used for identification of infectious pathogens include growth-dependent microbiology assays, immunoassays, and molecular biology assays. Immunoassays can be used to measure patient immune responses, indicating pathogen exposure, or can be used to directly identify the pathogen in host tissue or culture.

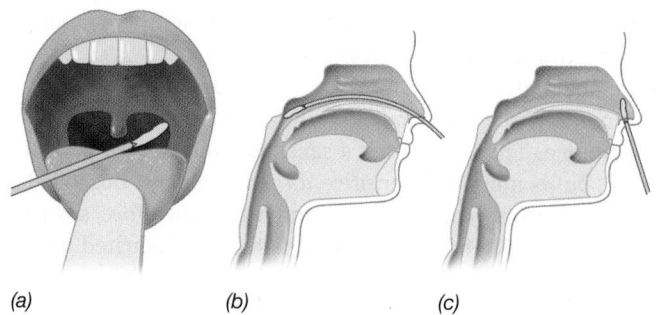

(a)   (b)   (c)

Figure 31.2 Methods for obtaining specimens from the upper respiratory tract. *(a)* Throat swab. *(b)* Nasopharyngeal swab passed through the nose. *(c)* Swabbing the inside of the nose.

or throat (**Figure 31.2**). The swab is then used to inoculate the surface of an agar plate or a tube of liquid culture medium. In some cases, a small piece of tissue (biopsy) may be obtained for culture. **Table 31.1** summarizes recommendations for initial culture of organisms isolated from clinical specimens.

If clinically relevant organisms are to be isolated and identified, the specimen must be obtained and handled properly to ensure that the pathogen survives. First, the specimen must be obtained from the actual site of the infection. The sample must also be taken aseptically to avoid contamination with irrelevant microorganisms. Next, the sample size must be large enough to ensure an inoculum sufficient for growth. Finally, the metabolic requirements for organism survival must be maintained during sampling, storage, and transport. For example, samples obtained from anoxic sites must be obtained, stored, and transported under anoxic conditions to ensure the survival of potential anaerobic pathogens. Sample storage and transportation time should be minimized and samples should be processed as soon as possible.

Direct Observation

Direct observation of pathogens acquired from clinical specimens growing in or on the host is a very important tool for diagnosis of many infectious diseases. Examples of the diagnostic power of direct observation of pathogens include the observation of acid-fast stained *Mycobacterium tuberculosis* in sputum from patients, proving infection with *M. tuberculosis* (Section 33.4). Likewise, infection with *Neisseria gonorrhoeae* can be diagnosed by direct observation of Gram-stained smears of a patient sample such as urethral exudates from men. The presence of gram-negative diplococci in clumps and in inclusions in neutrophils is diagnostic for the disease. In women, however, cervical smears often do not reveal the presence of the infecting organism and cultures must be done to establish or confirm a diagnosis of gonorrhea (Section 33.12 and see Figure 31.5). In these cases, the presence of the pathogen in the patient sample is considered unequivocal evidence for infection. Throughout this chapter and in Chapters 33–36, we will identify other situations when direct observation is a relevant diagnostic tool.

Growth Media and Culture

Enrichment culture, the use of selected culture media and incubation conditions to isolate microorganisms from samples (Section 22.1), is an important tool in the clinical laboratory. Most microorganisms of clinical importance can be grown, isolated, and identified using specialized growth media. Clinical samples are first grown on **general-purpose media**, media such as blood agar that support the growth of most aerobic and facultatively aerobic organisms (Figure 27.19). Organisms isolated from such media are often subcultured on more specialized media (Table 31.1). **Enriched media** containing specific growth factors enhance the growth of certain fastidious pathogens, such as *N. gonorrhoeae*. **Selective media** allow for some organisms to

---

Table 31.1 *Recommended enriched and selective media for primary isolation of pathogens*

Specimen	Media[a]				
	Blood agar	Enteric agar	CA	MTM	ANA
Fluids from chest, abdomen, pericardium, joint	+	+	+	−	+
Feces: rectal or enteric transport swabs[b]	+	+	+	−	−
Surgical tissue biopsies	+	+	−	−	+
Throat, sputum, tonsil, nasopharynx, lung, lymph nodes	+	+	+	−	−
Urethra, vagina, cervix	+	+	+	+	−
Urine	+	+	−	−	−
Blood[c]	+	+	+	−	+
Wounds, abscesses, exudates	+	+	+	−	+

[a]Blood agar, 5% whole sheep blood added to trypticase soy agar; enteric agar, either eosin–methylene blue (EMB) agar or MacConkey agar; CA, chocolate (heated blood) agar; MTM, modified Thayer–Martin agar; ANA, anaerobic agar, thioglycolate-containing blood agar or supplemented thioglycolate agar incubated anaerobically.
[b]Special enteric pathogen media, SMAC (MacConkey agar with sorbitol), is also used to culture fecal and enteric samples. SMAC is a selective and differential medium used for the isolation and identification of sorbitol-negative enteric pathogens such as enteropathogenic *Escherichia coli*.
[c]Blood is cultured initially in broth. Depending on the Gram stain characteristics of isolates, subculturing is done on MacConkey agar (gram-negative) or chocolate agar (gram-positive).
Source: Adapted from Murray, P.R., E.J. Baron, J.H. Jorgenson, M.L. Landry, and A. Pfaller. 2007. *Manual of Clinical Microbiology,* 9th edition. American Society for Microbiology, Washington, DC.

grow while inhibiting the growth of others due to the presence of inhibitory agents. Finally, **differential media** are specialized media that allow identification of organisms based on their growth, color, and appearance on the medium (see Figure 31.7).

Blood Culture

Bacteremia is the presence of microorganisms in the blood. Bacteremia is extremely uncommon in healthy individuals, normally occurring only transiently in response to invasive procedures such as tooth brushing, dental surgery, or trauma. The prolonged presence of bacteria in the blood is generally indicative of systemic infection.

Septicemia, or **sepsis**, is a blood infection by a virulent organism that enters the blood from a focus of infection, multiplies, and travels to various body tissues to initiate new infections. Septicemia can cause severe systemic symptoms, including fever and chills, followed by prostration. Severe cases of septicemia may result in *septic shock*, a life-threatening systemic condition characterized by severe reduction in blood pressure and multiple organ failures, including heart, kidneys, and lungs (⮌ Section 28.5). Blood cultures provide an immediate way of isolating and identifying the causal agent. Bacteria from blood cultures are commonly detected by indicators of microbial growth using automated culture systems, often followed by microscopic examination and subculture. The most common pathogens found in blood include gram-positive *Staphylococcus* spp. and *Enterococcus* spp., as well as gram-negative *Pseudomonas aeruginosa* and enteric bacteria, especially *Enterobacter* spp., *Escherichia coli*, and *Klebsiella pneumoniae*, and a variety of pathogenic fungi.

The standard blood culture procedure is to draw 20 ml of blood aseptically from a vein and inject it into two blood culture bottles containing an anticoagulant and a general-purpose culture medium. Some blood culture systems employ a chemical that lyses red and white blood cells, releasing intracellular pathogens. One bottle is incubated in air and one is incubated under anoxic conditions, and both are kept at 35°C for up to 5 days. Automated blood culture systems detect growth by monitoring carbon dioxide production and turbidity as often as every 10 minutes. Most clinically significant bacteria are recovered within 2 days, but detectable growth of fastidious organisms may take 3 to 5 days.

Up to 2–3% of blood cultures are contaminated by microorganisms introduced from the skin during blood sampling. Typically, these organisms include *Staphylococcus epidermidis*, coryneform bacteria, or propionibacteria. However, these organisms can also infect the heart (subacute bacterial endocarditis) or colonize intravascular devices such as artificial heart valves. Thus, the results of the blood culture must be reconciled with the clinical problem for an accurate diagnosis.

Urinary Tract Culture

Urinary tract infections are common, especially in women. Interpretation of microbiological findings from urine cultures can be confusing because the disease-causing agents are often members of the normal flora (for example, *E. coli*). In most cases, the urinary tract becomes infected by organisms that ascend into the bladder from the urethra. Urinary tract infections are also the most common form of healthcare-associated infections, often introduced through catheters (⮌ Section 32.7).

A significant urinary tract infection typically results in bacterial counts of 10^5 or more organisms per milliliter of a clean-voided midstream urine specimen. In the absence of infection, contamination of the urine from the external genitalia (almost unavoidable to some extent) results in less than 10^3 organisms per milliliter. The most common urinary tract pathogens are enteric bacteria, with *E. coli* accounting for about 90% of the cases. Other urinary tract pathogens include *Klebsiella, Enterobacter, Proteus, Pseudomonas, Staphylococcus saprophyticus*, and *Enterococcus. N. gonorrhoeae*, and *Chlamydia trachomatis*, the causal agent of nongonococcal urethritis, do not grow in the urine itself, but grow on the urethral epithelium. These common sexually transmitted infections are diagnosed by methods discussed later. This spectrum of urinary tract pathogens, however, is not limited to these organisms; the immune status of the host and the possibility of exposure to healthcare-associated pathogens may influence the species and strain of organisms responsible for urinary tract infections.

One method to screen for urinary tract infections is direct microscopic examination of urine to indicate *bacteriuria*, the presence of abnormal numbers of bacteria in the urine, but nearly all urine contains some level of bacterial growth. If colony counts are not done, bacteriuria can be monitored with the use of commercially available dipstick tests. For example, one dipstick test monitors the reduction of nitrate by detecting the reduction product, nitrite; this form of anaerobic respiration is common among enteric bacteria (⮌ Section 13.14). A positive test indicates high bacterial cell numbers and is indicated by a color change on the dipstick (**Figure 31.3**). Because significant nitrite production is produced in urine only when large numbers of organisms ($>10^5$ per milliliter) are present, the method is a rapid check for urinary tract infections. Other dipstick tests for urinary tract infections often used in conjunction with nitrate reduction detect esterase (produced by leukocytes responding to the infection) and peroxidase (produced by a variety of bacteria). Positive dipstick tests indicate infection and are followed by urine culture.

Figure 31.3 Urinalysis dipstick test. A control strip is shown underneath the test strip. From left to right, the strip measures abnormal levels of glucose, bilirubin, ketones, specific gravity, blood, pH, protein, urobilinogen, nitrite, and leukocytes (esterase) in a urine sample. Abnormal readings for esterase (trace positive, far right) and nitrite (strong positive, second from right) indicate bacteriuria. Subsequent culture of this sample indicated the presence of *Escherichia coli*.

John Martinko and Cheryl Broadie

UNIT 10

A Gram stain may also be done directly on urine samples exhibiting bacteriuria to identify the morphology of potential urinary tract pathogens. This method can be used to putatively identify gram-negative rods including the enteric bacteria, gram-negative cocci such as *Neisseria*, and gram-positive cocci such as *Enterococcus*. The Gram stain and other direct staining methods are also useful for direct detection of bacteria in other body fluids such as sputum and wound exudates.

To culture potential urinary tract pathogens, two media types are normally used. Blood agar, a general-purpose enriched medium, can be used for initial isolation. After isolation on general media, selective and differential enteric media such as MacConkey or eosin–methylene blue (EMB) agar permit the initial differentiation of lactose fermenters from non–lactose fermenters and inhibit the growth of possible contaminants, including gram-positive *Staphylococcus* species (**Figure 31.4**). Clinical microbiologists can often make a tentative identification of an isolate by observing the color and morphology of colonies of the suspected pathogen grown on the various media described in **Table 31.2**. This presumptive identification is followed by more detailed tests to make a definitive identification. Urine cultures can be done quantitatively by counting colonies on blood agar or a selective agar medium, using a calibrated loop delivering a specified amount of urine, usually 1 μl, as the inoculum for a plate.

If no bacterial growth is obtained despite persistent urinary tract infection symptoms, a clinician may request direct cultures for fastidious organisms such as *N. gonorrhoeae*, *C. trachomatis*, *Moraxella* spp., *Haemophilis ducreyi*, and mycoplasma.

Fecal Samples

Proper collection and preservation of feces is important for the isolation of intestinal pathogens. During storage, fecal material becomes more acidic, so extended delay between sampling and

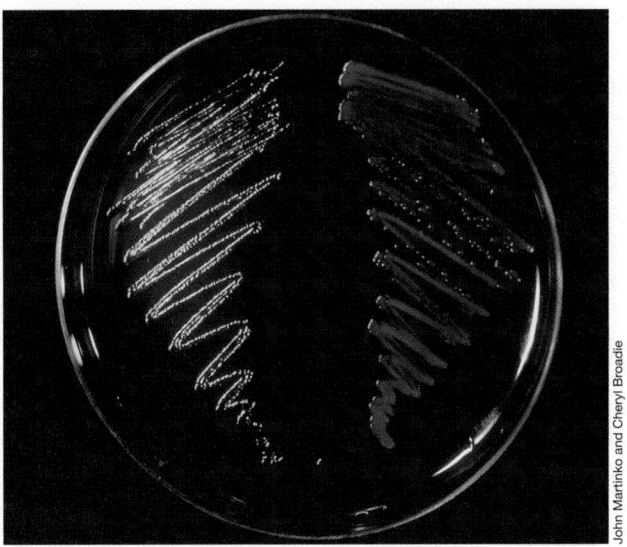

Figure 31.4 An eosin–methylene blue (EMB) agar plate. The plate shows a lactose fermenter, *Escherichia coli* (left), and a non–lactose fermenter, *Pseudomonas aeruginosa* (right). The green metallic sheen of the *E. coli* colonies is definitive for the lactose fermenters.

processing must be avoided. This is especially critical for the isolation of acid-sensitive *Shigella* and *Salmonella*.

Freshly collected fecal samples are placed in a sterile sealed container for transport to the laboratory. Bloody or pus-containing stools as well as stools from patients with suspected foodborne or waterborne infections are inoculated into a variety of selective media for isolation of individual bacteria. Intestinal eukaryotic pathogens are identified by direct microscopic observation of cysts in the stool sample or through antigen-detection assays

---

Table 31.2 *Colony characteristics of frequently isolated gram-negative rods cultured on various clinically useful media*

Organism	Agar media[a]				
	EMB	MC	SS	BS	HE
Escherichia coli	Dark center with greenish metallic sheen	Red or pink	Red to pink	Mostly inhibited	Yellow-pink
Enterobacter	Similar to *E. coli*, but colonies are larger	Red or pink	White or beige	Mucoid colonies with silver sheen	Yellow-pink
Klebsiella	Large, mucoid, brownish	Pink	Red to pink	Mostly inhibited	Yellow-pink
Proteus	Translucent, colorless	Transparent, colorless	Black center, clear periphery	Green	Clear
Pseudomonas	Translucent, colorless to gold	Transparent, colorless	Mostly inhibited	No growth	Clear
Salmonella	Translucent, colorless to gold	Translucent, colorless	Opaque	Black to dark green	Green or transparent with black centers
Shigella	Translucent, colorless to gold	Transparent, colorless	Opaque	Brown or inhibited	Green or transparent

[a]EMB, eosin–methylene blue agar; MC, MacConkey agar; SS, *Salmonella–Shigella* agar; BS, bismuth sulfite agar; HE, Hektoen enteric agar.
Source: Adapted from Murray, P.R., E.J. Baron, J.H. Jorgenson, M.L. Landry, and M.A. Pfaller. 2007. *Manual of Clinical Microbiology*, 8th edition. American Society for Microbiology, Washington, DC.

rather than by culture methods. Many laboratories also use a variety of selective and differential media and incubation conditions to identify *E. coli* O157:H7 and *Campylobacter*, two important intestinal pathogens typically acquired from contaminated food or water (⌒ Sections 36.9 and 36.10).

Wounds and Abscesses

Infections associated with traumatic injuries such as animal or human bites, burns, cuts, or the penetration of foreign objects must be carefully sampled to recover the relevant pathogen, and the results must be interpreted carefully to differentiate between infection and contamination. Wound infections and abscesses are frequently contaminated with normal flora, and swab samples from such lesions are frequently misleading. For abscesses and other purulent lesions, the best sampling method is to aspirate pus with a sterile syringe and needle following disinfection of the skin surface. Internal purulent lesions are sampled by biopsy or from tissues removed in surgery.

Several pathogens can be associated with wound infections. Because some of these are anaerobes, proper evaluation requires that samples be obtained, transported, and cultured under anoxic as well as oxic conditions. For example, potential pathogens commonly associated with purulent discharges from wound infections are *Staphylococcus aureus*, enteric bacteria, *Pseudomonas aeruginosa*, and anaerobes such as *Bacteroides* and *Clostridium* species. The major isolation media are blood agar, several selective media for enteric bacteria (Tables 31.1 and 31.2), and blood agar containing additional supplements and reducing agents for obligate anaerobes. Gram stains from such specimens are examined directly by microscopy.

Genital Specimens and Culture for Gonorrhea

In males, a purulent urethral discharge is usually indicative of a sexually transmitted infection (STI). These STIs are classified as nongonococcal or gonococcal urethritis. Nongonococcal urethritis is usually caused by *C. trachomatis*, *Ureaplasma urealyticum*, or *Trichomonas vaginalis* (⌒ Section 33.13). Gonococcal urethritis is caused by *N. gonorrhoeae* (⌒ Section 33.12). If no discharge is present, a sample can be obtained using a sterile narrow-diameter cotton swab that is inserted into the anterior urethra, left in place a few seconds to absorb any exudate, and then removed for observation, culture, and identification of *N. gonorrhoea* or another causative agent. Alternatively, a sample of the first morning urine from an infected individual usually contains viable cells of *N. gonorrhoeae*. In females, samples are usually obtained by swab from the cervix and the urethra.

N. gonorrhoeae is usually found as gram-negative diplococci, but can be pleomorphic. No similar microorganisms are observed among the normal flora of the urogenital tract. Thus, a Gram stain of a urethral, vaginal, or cervical smear showing gram-negative diplococci is diagnostic for gonorrhea. In acute gonorrhea, microscopic examination of purulent discharges usually reveals gram-negative diplococci in neutrophils (**Figure 31.5a**).

Clinical microbiology procedures are central to the diagnosis of gonorrhea. *N. gonorrhoeae* (referred to clinically as gonococcus) colonizes mucosal surfaces of the urethra, uterine cervix,

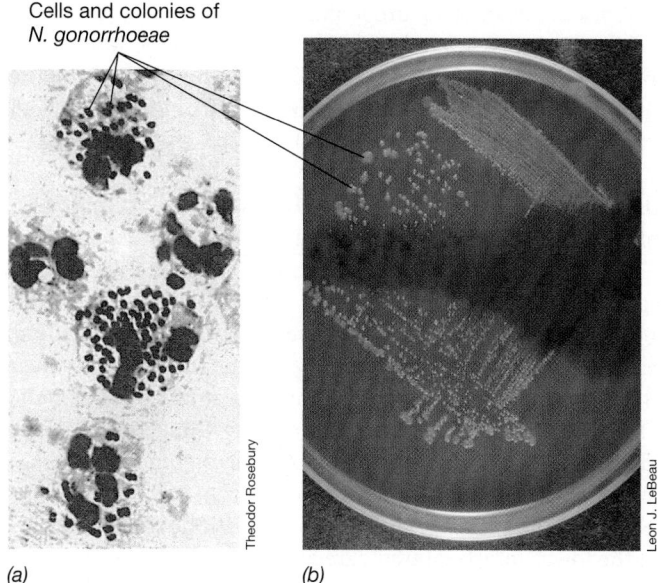

Cells and colonies of *N. gonorrhoeae*

(a)   (b)

Theodor Rosebury

Leon J. LeBeau

Figure 31.5 Identification of *Neisseria gonorrhoeae*. *(a)* Photomicrograph of *Neisseria gonorrhoeae* cells within human polymorphonuclear leukocytes from a urethral exudate. Note the paired diplococci (leader). *(b) N. gonorrhoeae* growing on Thayer–Martin agar. The plate has been stained in the middle with a reagent that turns colonies blue if cells contain cytochrome *c* (the oxidase test). *N. gonorrhoeae* colonies in contact with the reagent are blue, indicating that they are oxidase-positive.

anal canal, throat, and conjunctiva. The organism is sensitive to drying and therefore is transmitted almost exclusively by direct person-to-person contact, usually by sexual activity. Public health measures to control gonorrhea include identification of asymptomatic carriers, and this requires microbiological analysis.

A nonselective enriched medium for the isolation of *N. gonorrhoeae* contains heat-lysed blood and is called *chocolate agar* because of the deep brown appearance. The heated blood interacts with the media components, absorbing compounds that are normally toxic for *N. gonorrhoeae*. One of several selective media used for primary isolation is modified Thayer–Martin (MTM) agar (Figure 31.5b). This medium incorporates the antibiotics vancomycin, nystatin, trimethoprim, and colistin to suppress the growth of normal flora. These antibiotics have no effect on *N. gonorrhoeae* or *Neisseria meningitidis*, the cause of bacterial meningitis (⌒ Section 33.5).

Inoculated plates are incubated in a humid environment in an atmosphere containing 3–7% CO_2, required for growth of gonococci. The plates are examined after 31 and 48 hours and tested for their oxidase reaction because *Neisseria* species are oxidase-positive (Figure 31.5b). Oxidase-positive, gram-negative diplococci growing on chocolate agar or selective media are presumed to be gonococci if the inoculum was derived from genitourinary sources. Definitive identification of *N. gonorrhoeae* requires determination of carbohydrate utilization patterns and immunological or nucleic acid probe tests.

UNIT 10

Laboratory testing of urogenital samples for *N. gonorrhoeae* (and the often-associated *C. trachomatis*, ⮌ Section 33.13) is usually done using nucleic acid probe methods, DNA amplification via polymerase chain reaction (PCR), or other molecular methods (Sections 31.12 and 31.13).

Culture of Anaerobic Microorganisms

Obligate anaerobic bacteria are common causes of infection, and their identification requires special isolation and culture methods. In general, media for anaerobes do not differ greatly from those used for aerobes, except that they are (1) usually richer in organic constituents, (2) contain reducing agents (usually cysteine or thioglycolate) to remove oxygen, and (3) contain a redox indicator to indicate that conditions are anoxic. Collection, handling, and processing of specimens must exclude oxygen contamination because oxygen is toxic to obligately anaerobic organisms.

Several habitats in the body, such as portions of the oral cavity and the lower intestinal tract, are anoxic and support the growth of an anaerobic normal flora. Other parts of the body, however, can also become anoxic as a result of tissue injury or trauma that reduces blood supply and oxygen perfusion to the injured site. These anoxic sites can then be colonized by obligate anaerobes. In general, pathogenic anaerobic bacteria are part of the normal flora and are opportunistic pathogens. Two important exceptions are the pathogenic anaerobes *Clostridium tetani* (the cause of tetanus) and *Clostridium perfringens* (the cause of gas gangrene and one type of food poisoning), both endospore-forming bacteria that are predominantly soil organisms (⮌ Sections 34.9 and 36.7).

Isolation, growth, and identification of anaerobic pathogens are complicated by specimen contamination as well as the constant challenge of maintaining an anoxic environment during collection, transport, and culture. Samples collected by syringe aspiration or biopsy must be immediately placed in a tube containing oxygen-free gas, usually with a dilute salt solution containing a reducing agent such as thioglycolate and a redox indicator such as resazurin. Resazurin is colorless when reduced and becomes pink when oxidized, indicating oxygen contamination of the specimen. If an anaerobic transport tube is not available, the syringe itself can be used to transport the specimen; the needle is discarded and the syringe is plugged with a rubber stopper.

For anoxic incubation, agar plates are placed in a sealed jar, which is made anoxic either by replacing the atmosphere in the jar with an oxygen-free gas mixture (usually a mixture of nitrogen and carbon dioxide) or by removing oxygen from the enclosed vessel by some chemical means. For example, as shown in **Figure 31.6**, hydrogen is generated chemically in the jar. In the presence of a palladium catalyst, the hydrogen combines with the free oxygen in the vessel, forming water and removing the contaminating oxygen. Alternate means for providing anoxic conditions include the use of culture media containing reducing agents or the use of anoxic "glove boxes" filled with an oxygen-free gas such as nitrogen or hydrogen (⮌ Figure 5.28*b*).

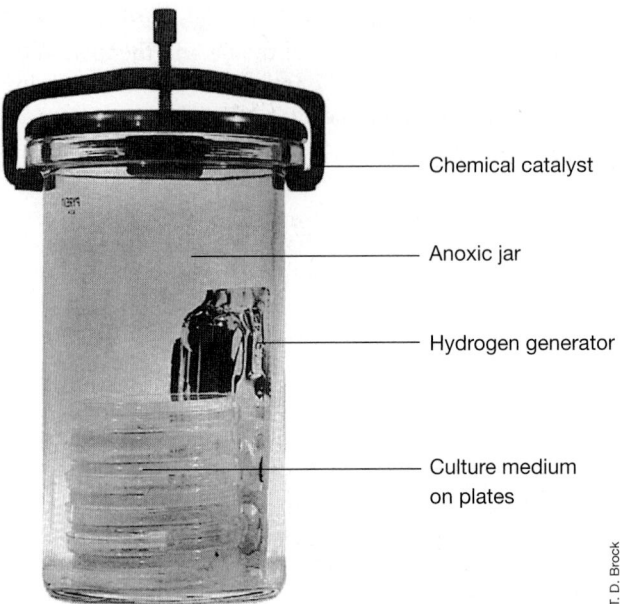

- Chemical catalyst
- Anoxic jar
- Hydrogen generator
- Culture medium on plates

T. D. Brock

Figure 31.6 Sealed jar for incubating cultures under anoxic conditions. The catalyst and hydrogen generator packet produce and maintain a reducing (anoxic) environment.

MiniQuiz

- Why do urine cultures almost always test positive for bacterial growth?
- Describe the methods used to maintain optimum conditions for the isolation of anaerobic pathogens.

31.2 Growth-Dependent Identification Methods

The clinical microbiologist must be able to identify the organism or organisms present if the inoculation of a general-purpose medium, one that supports the growth of most aerobic and facultatively aerobic organisms, results in bacterial growth. Many microorganisms recovered from clinical samples can be identified using growth-dependent assays. We consider these methods here.

Growth on Selective and Differential Media

Based on its growth characteristics on primary isolation media, a presumptive pathogen is typically subcultured onto specialized media designed to measure one of many different biochemical reactions. Some of these important biochemical tests are listed in **Table 31.3**. Specialized biochemical identification systems containing several different media, all in separate wells, can be inoculated at one time (**Figure 31.7**).

The media employed are selective, differential, or both. Eosin–methylene blue (EMB) agar, for example, is a widely used selective and differential medium for the isolation and differentiation of enteric bacteria. Methylene blue is a selective dye because it

Table 31.3 Important clinical diagnostic tests for bacteria

Test	Principle	Procedure	Most common use
Carbohydrate fermentation	Acid and/or gas is produced during fermentative growth with sugars or sugar alcohols.	Broth medium with carbohydrate and phenol red as pH indicator; inverted tube for gas	Enteric bacteria differentiation
Catalase	Enzyme decomposes hydrogen peroxide, H_2O_2.	Add a drop of H_2O_2 to dense culture and look for bubbles (O_2)	*Bacillus* (+) from *Clostridium* (−); *Streptococcus* (−) from *Micrococcus–Staphylococcus* (+)
Citrate utilization	Utilization of citrate as sole carbon source results in alkalinization of medium.	Citrate medium with bromthymol blue as pH indicator. Look for intense blue color (alkaline pH)	*Klebsiella–Enterobacter* (+) from *Escherichia* (−); *Edwardsiella* (−) from *Salmonella* (+)
Coagulase	Enzyme causes clotting of blood plasma.	Mix dense liquid suspension of bacteria with plasma, incubate, and look for fibrin clot	*Staphylococcus aureus* (+) from *Staphylococcus epidermidis* (−)
Decarboxylases (lysine, ornithine, arginine)	Decarboxylation of amino acid releases CO_2 and amine.	Medium enriched with amino acids. Bromcresol purple pH indicator becomes purple (alkaline pH) if there is enzyme action	Aid in determining bacterial group among the enteric bacteria
β-Galactosidase (ONPG) test	Orthonitrophenyl-β-galactoside (ONPG) is an artificial substrate for the enzyme. Hydrolysis of ONPG forms nitrophenol (yellow).	Incubate heavy suspension of lysed culture with ONPG. Look for yellow color	*Citrobacter* (+) from *Salmonella* (−). Identifying some *Shigella* and *Pseudomonas* species
Gelatin liquefaction	Many proteases hydrolyze gelatin and destroy the gel.	Incubate in broth with 12% gelatin. Cool to check for gel formation. If gelatin is hydrolyzed, tube remains liquid on cooling	Aid in identification of *Serratia, Pseudomonas, Flavobacterium, Clostridium*
Hydrogen sulfide (H_2S) production	H_2S is produced by breakdown of sulfur amino acids or reduction of thiosulfate.	H_2S detected in iron-rich medium from formation of black ferrous sulfide (many variants: Kligler's iron agar and triple sugar iron agar also detect carbohydrate fermentation)	Among enteric bacteria, to aid in identifying, *Salmonella, Edwardsiella*, and *Proteus*
Indole test	Tryptophan from proteins is converted to indole.	Detect indole in culture medium with dimethylaminobenzaldehyde (red color) or in colony smeared on paper containing dimethylamino-cinnamaldehyde (spot test; blue color)	Distinguish *Escherichia* (+) from most *Klebsiella* (−) and *Enterobacter* (−); *Edwardsiella* (+) from *Salmonella* (−); *Proteus vulgaris* (+) from *Proteus mirabilis* (−)
Methyl red test	Mixed-acid fermenters produce sufficient acid to lower pH below 4.3.	Glucose-broth medium. Add methyl red indicator to a sample after incubation	Differentiate *Escherichia* (+, culture red) from *Enterobacter* and *Klebsiella* (usually −, culture yellow)
Nitrate reduction	Nitrate (NO_3^-) as alternate electron acceptors is reduced to NO_2^- or N_2.	Broth with nitrate. After incubation, detect nitrate with α-naphthylamine-sulfanilic acid (red color). If negative, confirm that NO_3^- is still present by adding zinc dust to reduce NO_3^- to NO_2^-. If no color after zinc, then $NO_3^- \rightarrow N_2$	Aid in identification of enteric bacteria (usually +)
Oxidase test	Cytochrome *c* oxidizes artificial electron acceptor: tetramethyl *p*-phenylenediamine (Kovac's reagent), or dimethyl *p*-phenylenediamine (Gordon and McLeod's reagent).	Colonies are smeared on paper impregnated with reagent. Oxidase-positive colonies produce dark purple-black color in 10–15 sec with Kovac's reagent and blue color in 10–30 min with Gordon and McLeod's reagent	Differentiate *Neisseria* and *Moraxella* (+) from *Acinetobacter* (−); pseudomonads (+) and *Vibrionaceae* (+) from *Enterobacteriaceae* (−). Aid in identification of *Aeromonas* (+)
Oxidation–fermentation (OF) test	Some organisms produce acid only when growing aerobically.	Acid production in top part of sugar-containing culture tube; soft agar used to restrict mixing during incubation	Differentiate *Micrococcus* (acid produced aerobically only) from *Staphylococcus* (acid produced anaerobically). To characterize *Pseudomonas* (aerobic acid production) from enteric bacteria (acid produced anaerobically)

Table 31.3 *Important clinical diagnostic tests for bacteria (continued)*

Test	Principle	Procedure	Most common use
Phenylalanine deaminase test	Deamination produces phenylpyruvic acid, which is detected in a colorimetric test.	Medium enriched in phenylalanine. After growth, add ferric chloride reagent and look for green color	Characterize the genera *Proteus* and *Providencia*
Starch hydrolysis	Iodine-iodide mixture gives blue color with starch.	Grow organism on plate containing starch. Flood plate with Gram's iodine and look for clear zones around colonies	Identify typical starch hydrolyzers such as *Bacillus* spp.
Urease test	Urea, H_2N—CO—NH_2, is split to $2\ NH_3 + CO_2$.	Medium with 2% urea and phenol red indicator. Ammonia release raises pH, intense pink-red color	Distinguish *Klebsiella* (+) from *Escherichia* (−), and *Proteus* (+) from *Providencia* (−). To identify *Helicobacter pylori* (+)
Voges–Proskauer test	Acetoin is produced from sugar fermentation.	Chemical test for acetoin using α-naphthol	Separate *Klebsiella* and *Enterobacter* (+) from *Escherichia* (−). To characterize members of the genus *Bacillus*

inhibits the growth of gram-positive bacteria, and thus only gram-negative organisms can grow. EMB agar has an initial pH of 7.2 and contains lactose and sucrose, but not glucose, as energy sources. Acidification changes eosin, the differential media component, from colorless to red or black. Strong lactose-fermenting bacteria such as *Escherichia coli* acidify the medium and the colonies appear black with a greenish sheen. Butanediol-producing enteric bacteria such as *Klebsiella* or *Enterobacter* produce less acid, and colonies on EMB are pink to red. Colonies of non–lactose fermenters, such as *Salmonella*, *Shigella*, and *Pseudomonas*, are translucent or pink (Figure 31.4). Thus, EMB is preferentially selective for the growth of gram-negative bacteria and also differentiates among common enteric bacteria.

Differential media incorporate biochemical tests to measure the presence or absence of enzymes involved in catabolism of a specific substrate or substrates. For example, fermentation of sugars is measured by incorporating pH indicator dyes that change color on acidification (Figure 31.7*a*). Production of hydrogen or carbon dioxide during sugar fermentation is assayed by observing gas production either in gas collection vials or in agar (Figure 31.7*a*, *b*). Hydrogen sulfide (H_2S) production is assayed by growth in a medium containing ferric iron. If sulfide is produced, ferric iron reacts with H_2S to form ferrous sulfide (FeS), visible as a black precipitate (Figure 31.7*b*). In a medium containing citric acid (a tricarboxylic acid) as a carbon source, utilization by a cultured microorganism causes the pH to rise, and a dye changes color as conditions become alkaline (Figure 31.7*c*). Another testing method uses chromogenic substrates that alter the color of colonies of targeted organisms. For instance, MRSA ID agar, a proprietary selective and differential media, inhibits most methicillin-sensitive *Staphylococcus aureus* (MSSA), and most other bacteria and yeasts. Methicillin-resistant *Staphylococcus aureus* (MRSA), however, produces distinctive green colonies when grown on this medium. Fluorogenic media contain compounds that fluoresce when metabolized by target organisms. For example, fluorogenic media are used to identify *Escherichia coli* in water samples (Figures 35.2 and 35.3). Hundreds of differential tests are known, but only about 20 are used routinely (Figure 31.7*d*).

The biochemical reaction patterns for pathogens are stored in a computer databank. As the results of differential tests on an unknown pathogen are entered, the computer matches the characteristics of the unknown organism to metabolic patterns of known pathogens, allowing identification. As few as three or four key tests are sufficient to make an unambiguous identification of many pathogens. However, in some cases, more sophisticated identification procedures are required.

In addition to biochemical tests, analysis of cultured microorganisms may include several physical methods such as high-pressure liquid chromatography (HPLC) and gas–liquid chromatography (GLC) used to detect metabolites from anoxic microorganisms and cell wall fatty acids of *Mycobacterium* spp.

Identification and Diagnosis

Growth-dependent rapid identification systems are often used to identify enteric bacteria because these organisms are common causes of urinary tract and intestinal infections (Figure 31.7*d, e*). These systems consist of media that are selective and differential for groups of important pathogens or even for single bacterial species. For example, kits containing multiple media have been developed for identification of *Staphylococcus aureus*, *Streptococcus pyogenes*, *Neisseria gonorrhoeae*, *Haemophilus influenzae*, and *Mycobacterium tuberculosis*. Other kits are available for identification of the pathogenic fungi (eukaryotes) *Candida albicans* and *Cryptococcus neoformans* (Section 34.8).

The clinical microbiologist decides which diagnostic tests to use based on the origin of the clinical specimen, the basic characteristics of a pure culture of the specimen grown on general-purpose media (for example, morphology and Gram stain), and previous experience with similar cases.

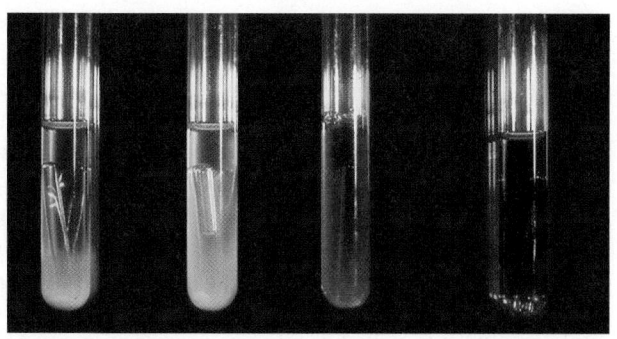

(a)

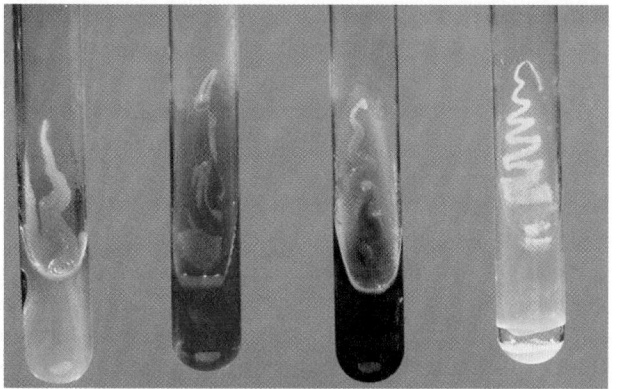

(b)

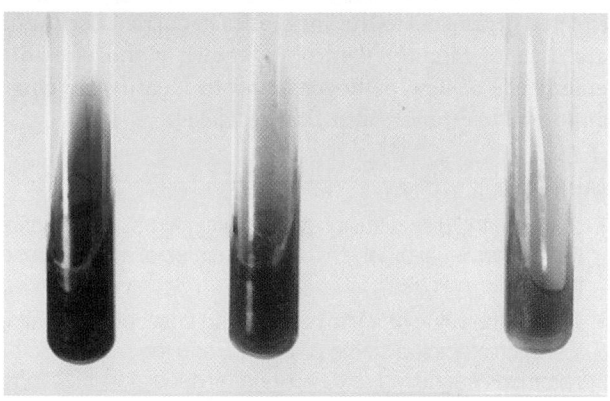

(c)

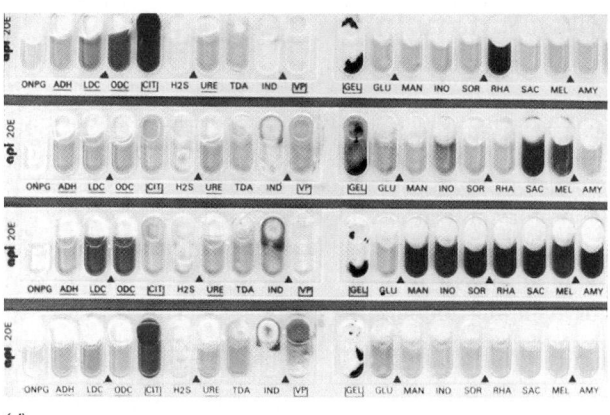

(d)

Figure 31.7 Growth-dependent diagnostic methods used for the identification of clinical isolates by color changes in various diagnostic media. *(a)* Use of a differential medium to assess sugar fermentation. Acid production is indicated by color change of the pH-indicating dye added to the liquid medium. If gas production occurs, a bubble appears in the inverted vial in each tube. From left to right: acid, acid and gas, negative, uninoculated. *(b)* A conventional diagnostic test for enteric bacteria in triple sugar iron (TSI) agar. The medium is inoculated both on the surface of the slant and by stabbing into the solid agar butt. The medium contains a small amount of glucose and a large amount of lactose and sucrose. Organisms able to ferment only the glucose cause acid formation only in the butt, whereas lactose- or sucrose-fermenting organisms cause acid formation throughout the slant. Gas formation is indicated by the breaking up of the agar in the butt. Hydrogen sulfide formation (either from protein degradation or from reduction of thiosulfate in the medium) is indicated by a blackening due to reaction of hydrogen sulfide with ferrous iron in the medium. From left to right: fermentation of glucose only; no reaction; hydrogen sulfide formation; fermentation of glucose and another sugar. *(c)* Measurement of citrate utilization by *Salmonella* on Simmons citrate agar. The change in pH causes a change in the color of the indicator dye. From left to right: positive, negative, uninoculated. *(d)* Media kits used for the rapid identification of clinical isolates. The principle is the same as in part a, but the whole arrangement has been miniaturized so that a number of tests can be run at the same time. Four separate strips, each with a separate culture, are shown. *(e)* Another arrangement of a miniaturized test kit. This one defines sugar utilization in nonfermentative organisms.

UNIT 10

31.3 Antimicrobial Drug Susceptibility Testing

Pathogens isolated from clinical specimens are identified to confirm medical diagnoses and to guide antimicrobial therapy. For many pathogens, appropriate and effective antimicrobial treatment is based on current experience and practices. For a select group of pathogens, however, decisions about appropriate antimicrobial therapy must be made on a case-by-case basis. Such pathogens include those for which antimicrobial drug resistance is common (for example, gram-negative enteric bacteria), those that cause life-threatening disease (for example, meningitis caused by *Neisseria meningitidis*), and those that require bacteriocidal rather than bacteriostatic drugs to prevent disease progression and tissue damage. Bacteriocidal agents are indicated, for example, for organisms that cause bacterial endocarditis, where total and rapid killing of the pathogen is critical for patient survival.

We discussed the basic principles for the measurement of antimicrobial activity in Chapter 26. The antimicrobial susceptibility of a culture can most easily be determined by an agar diffusion method or by using a tube dilution technique to determine the *minimum inhibitory concentration* (*MIC*) of an agent that is necessary to inhibit growth (Section 26.4). United States Food and Drug Administration regulations control the automated instruments used for susceptibility testing in the United States. Procedures and standards, including experimental end points for each organism and antibiotic, are constantly updated by the Clinical and Laboratory Standards Institute, a nonprofit organization that develops and establishes voluntary consensus standards for antibiotic testing, as well as other healthcare technologies (http://www.clsi.org).

The standard procedure for assessing antimicrobial activity is the *disc diffusion test* (**Figure 31.8a–e**). Agar media are inoculated by evenly spreading a defined density of a suspension of a pure culture on the agar surface. Filter paper discs containing a defined quantity (micrograms per disc) of an antimicrobial agent are then placed on the inoculated agar. After a specified period of incubation, the diameter of the inhibition zone around each disc is measured. **Table 31.4** presents zone sizes for several antibiotics. Inhibition zone diameters are then interpreted into susceptibility categories based on zone size. Standards for the efficacy of different antimicrobial agents against different bacterial pathogens are provided by the Food and Drug Administration or the Clinical and Laboratory Standards Institute.

The MIC procedure for antibiotic susceptibility testing employs an antibiotic dilution assay in agar (the standard test for anoxic microorganisms), in culture tubes (Figure 26.10), or in the wells of a microtiter plate (Figure 31.8*f*). Wells containing serial dilutions of antibiotics are inoculated with a standard inoculum of a test organism. Growth in the presence of each antibiotic is then observed by measuring turbidity. Antibiotic susceptibility is usually expressed as the highest dilution (lowest concentration) of antibiotic that completely inhibits growth. This defines the value of the MIC.

Etest (AB BIODISK, Solna, Sweden) is a non-diffusion-based technique that employs a preformed and predefined gradient of an antimicrobial agent immobilized on a plastic strip. The concentration gradient covers a MIC range across 15 twofold dilutions. When applied to the surface of an inoculated agar plate, the gradient transfers from the strip to the agar and remains stable for a period that covers the wide range of critical times associated with the growth characteristics of different microorganisms. After overnight incubation or longer, an elliptical zone of inhibition centered along the axis of the strip develops. The MIC value (in micrograms per milliliter) can be read at the point where the ellipse edge intersects the precalibrated Etest strip, providing a precise MIC (Figure 31.8*g*). This value can then be interpreted using current standards.

Many of the pathogens for which susceptibility testing is necessary are *healthcare-associated pathogens* or *nosocomial pathogens*, those acquired in hospitals and other healthcare settings. Hospital infection-control microbiologists generate and examine susceptibility data to generate periodic reports called **antibiograms**. These reports define the susceptibility of clinically isolated organisms to the antibiotics in current use. Antibiograms are used to monitor control of known pathogens, to track the emergence of new pathogens, and to identify the emergence of antibiotic resistance, all at the local level.

31.4 Safety in the Microbiology Laboratory

Clinical microbiology laboratories present significant biological hazards for workers. Standard laboratory practices for handling clinical samples have been established to prevent accidental laboratory infections. In the United States, every clinical and research institution that deals with human or primate tissue is required by law to have an occupational exposure control plan for handling bloodborne pathogens. This law was specifically designed to protect workers from infection by hepatitis B virus (HBV, the cause of infectious hepatitis, Section 33.11) and human immunodeficiency virus (HIV, the cause of acquired immunodeficiency syndrome [AIDS], Section 33.14). Implementation of these infection controls limits infection by all pathogens.

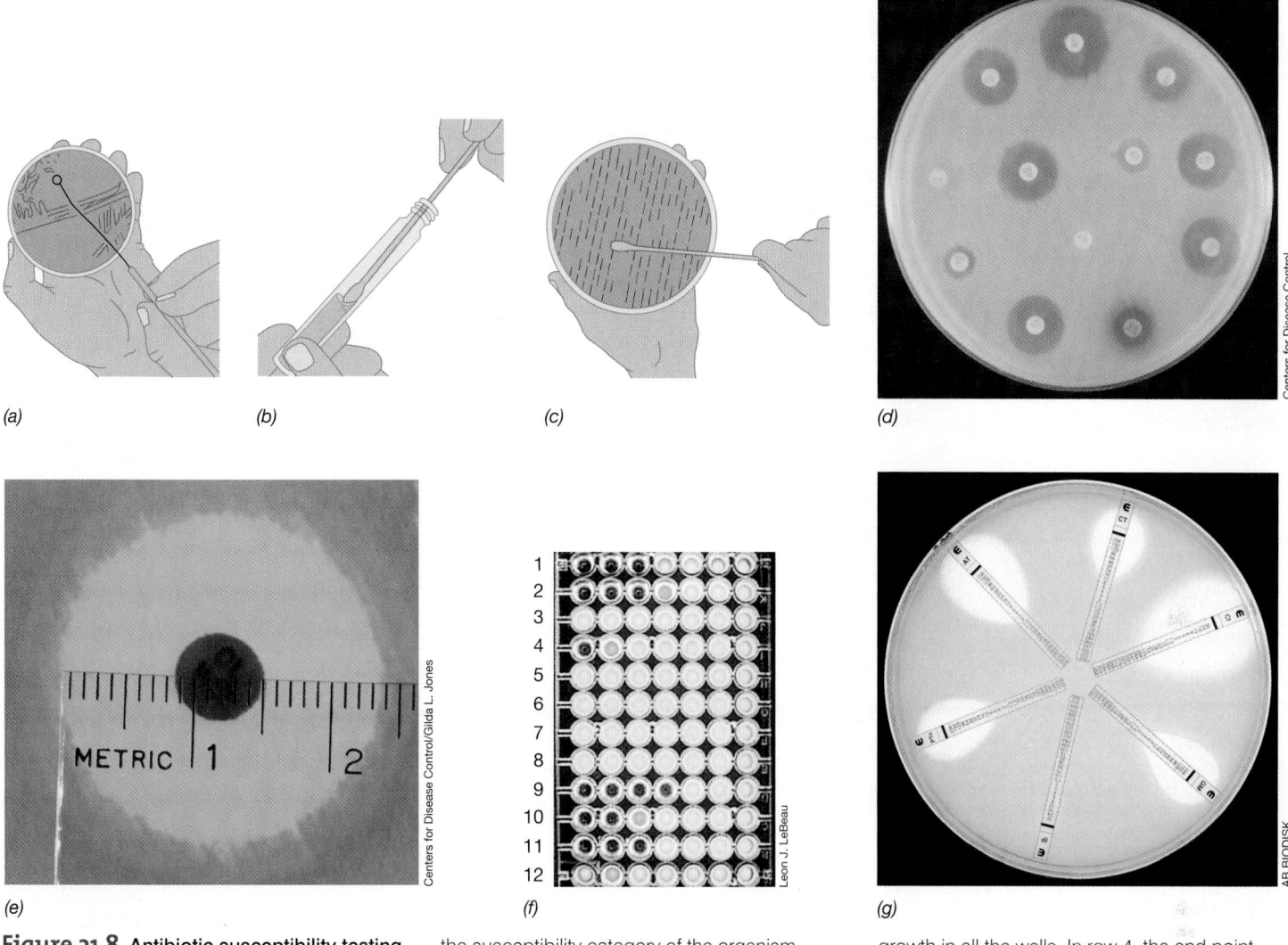

Figure 31.8 Antibiotic susceptibility testing.
Methods for determining the susceptibility of an organism to antibiotics. For the disc diffusion test, *(a)* isolated pure colonies are homogenized in a tube with an appropriate liquid medium to achieve a specified density compared to a turbidity standard. *(b)* A sterile cotton swab is dipped into the bacterial suspension and excess fluid removed by pressing the swab against the side of the tube. *(c)* The swab is streaked evenly over the surface of an appropriate agar medium. *(d)* Discs containing known amounts of different antibiotics are placed on the bacteria-inoculated agar surface. *(e)* After incubation, inhibition zones are observed and measured. From these data, the susceptibility category of the organism is determined by reference to an interpretive chart of zone sizes (Table 31.4). *(f)* Antibiotic susceptibility as determined by the broth dilution method. The organism is *Pseudomonas aeruginosa*. Each row has a different antibiotic. The microtiter plate enables automation of these tests. The end point is the well with the lowest concentration of antibiotic that shows no visible bacterial growth. The highest concentration of antibiotic is in the well at the left; serial twofold dilutions are made in the wells to the right. For example, in rows 1 and 2, the end point is the third well. In row 3, the antibiotic is ineffective at the concentrations tested, since there is bacterial growth in all the wells. In row 4, the end point is in the first well. *(g)* Antibiotic susceptibility determined by the Etest (AB BIODISK, Solna, Sweden) for different antibiotics (from 8 o'clock, PTc, piperacillin/tazobactam; AT, aztreonam; CT, cefotaxime; CI, ciprofloxacin; GM, gentamicin; IP, imipenem). Each strip is calibrated in terms of the minimum inhibitory concentration (MIC) in μg/ml starting with the lowest concentration from the center of the plate. The lowest concentration of antibiotic that inhibits bacterial growth is the MIC value for that particular agent (Section 26.4). For example, the MIC for cefotaxime (CT) is 16 μg/ml. This organism is resistant to imipenem (IP); MIC > 31 μg/ml.

Laboratory Safety

The two most common causes of laboratory accidents are ignorance and carelessness. Training and enforcement of established safety procedures, however, can prevent most accidents. Unfortunately, most laboratory-acquired infections do not result from identifiable exposures or accidents, but rather from routine handling of patient specimens. Infectious aerosols generated during processing of specimens are the most common causes of laboratory infections. Clinical laboratories follow the safety rules outlined here (required by law in the United States) to minimize the exposure of healthcare workers to infectious agents and thereby reduce the numbers of non-accident-associated laboratory infections. Strict adherence to safety rules ensures a safe and efficient laboratory environment that is in compliance with governmental regulations.

Table 31.4 *Standards for antimicrobial disc diffusion susceptibility tests*[a]

Antibiotic	Amount on disc	Inhibition zone diameter (mm)		
		Resistant	Intermediate	Susceptible
Ampicillin[b]	10 µg	13 or less	14–16	17 or more
Ampicillin[c]	10 µg	28 or less	—	29 or more
Ceftriaxone	30 µg	13 or less	14–20	21 or more
Chloramphenicol	30 µg	12 or less	13–17	18 or more
Clindamycin	2 µg	14 or less	15–20	21 or more
Erythromycin	15 µg	13 or less	14–22	23 or more
Gentamicin	10 µg	12 or less	13–14	15 or more
Methicillin[b]	5 µg	9 or less	10–13	14 or more
Nitrofurantoin	300 µg	14 or less	15–16	17 or more
Penicillin G[d]	10 units	28 or less	—	29 or more
Penicillin G[e]	10 units	14 or less	—	15 or more
Streptomycin	10 µg	6 or less	7–9	10 or more
Sulfonamide	10 µg	12 or less	13–16	17 or more
Tetracycline	30 µg	14 or less	15–18	19 or more
Trimethoprim–sulfamethoxazole	1.25/23.75 µg	10 or less	11–15	16 or more
Tobramycin	10 µg	12 or less	13–14	15 or more
Vancomycin[f]	10 µg	14 or less	15–16	17 or more
Vancomycin[g]	10 µg	14 or less (test for MIC)	—	15 or more

[a]Standards are defined and updated by Clinical and Laboratory Standards Institute (CLSI), an international nonprofit organization that develops voluntary consensus standards for antibiotic testing and other healthcare technologies (http://www.clsi.org)
[b]For *Enterobacteriaceae.*
[c]For staphylococci and highly penicillin-sensitive organisms.
[d]For staphylococci.
[e]For organisms such as enterococci that may cause some systemic infections treatable with high doses of penicillin G.
[f]For enterococci.
[g]For staphylococci.

1. Laboratories handling hazardous materials must restrict access. Only laboratory workers and essential support personnel should be allowed to enter the laboratory. These individuals must have knowledge of the biological risks in the laboratory and act accordingly.

2. Effective decontamination procedures must be in place. Infectious materials or wastes, including specimens, syringes and needles, inoculated media, bacterial cultures, tissue cultures, experimental animals, glassware, instruments, and surfaces must be fully decontaminated, without compromise. A 5.25% (full-strength) chlorine bleach solution or other approved disinfectant should be used to decontaminate spilled infectious material. All potentially infectious waste must be burned in a certified incinerator or handled by a licensed waste handler.

3. Personnel working with hazardous infectious agents or vaccines (for example, rabies, polio, or diphtheria-pertussis-tetanus vaccines) must be properly vaccinated against the agent. Persons working with human or primate tissue must be vaccinated against HBV.

4. All clinical specimens should be considered infectious and handled appropriately. This is especially important for preventing laboratory-acquired hepatitis because of the relative frequency of hepatitis viruses in blood specimens from the general population.

5. All pipetting must be done with mechanical pipetting devices (not by mouth).

6. Animals should be handled only by trained laboratory personnel. Anesthetics and tranquilizers should be used to avoid injury to both personnel and animals.

7. Laboratory personnel must wear laboratory coats or gowns, sealed shoes, rubber gloves, masks, eye protection, respiratory devices when needed, and other barrier devices appropriate for the level of exposure and the severity of the potential infection. Such barrier devices must also be properly decontaminated and stored after use. Laboratory personnel must also practice good personal hygiene with respect to handwashing. Eating and drinking, smoking, applying cosmetics or lip balm, or wearing contact lenses is not permitted in the clinical microbiology laboratory.

8. Because of the special risks associated with AIDS, all clinical (human) specimens should be treated as if they contain HIV. Protective gloves should be worn when handling specimens of any kind. Masks or full-face shields must be worn any time there is a possibility of generating an aerosol during specimen preparation. Needles must not be resheathed, bent, or broken; they should be placed in a labeled container designated expressly for this purpose that can be sealed and decontaminated before disposal.

These safety rules should be in effect in all laboratories that handle potential infectious agents. Specialized clinical laboratories may have additional rules and procedures in addition to these to ensure a safe work environment, as discussed below. In the final analysis, safety in the workplace is the responsibility of laboratory personnel. Any clinical laboratory has potential biohazards, but this environment is even more dangerous for untrained personnel or those who do not take the necessary precautions.

Biological Containment and Laboratory Biosafety Levels

The level of containment used to prevent accidental infections or accidental environmental contamination (escape) in clinical, research, and teaching laboratories must be adjusted to counter the biohazard potential of the organisms handled in the laboratory. Laboratories are classified according to their containment potential, or *biosafety level* (*BSL*), and are designated as *BSL-1, BSL-2, BSL-3,* or *BSL-4*. Personnel in laboratories working at all biosafety levels must follow good laboratory practices that ensure basic cleanliness and limit contamination, and laboratory surfaces must be decontaminated after each work shift or whenever spills occur. As in clinical laboratories, personnel cannot consume food or drink in the laboratory and must wash their hands when leaving the laboratory, and access to the laboratory must be restricted to laboratory personnel.

BSL-1 laboratories are the lowest level of containment. Work can be done on the open bench with organisms that present a low risk of infection; these organisms are not pathogens in normal individuals and include organisms such as *Bacillus subtilis*. Workers *should* be protected by barrier protection such as lab coats and gloves. An example of a BSL-1 facility is a teaching laboratory that does not use pathogens.

BSL-2 laboratories are designed to contain organisms that present a moderate risk of infection due to accidental ingestion, percutaneous injection, or exposure to mucous membranes via aerosols. Work with pathogens such as *Escherichia coli* or *Streptococcus pyogenes* is done in a BSL-2 laboratory or sometimes at higher containment levels. Normal procedures may be performed on bench tops and must adhere to all BSL-1 precautions, and additional barrier protection devices such as face and eye protection, gloves, and lab coats or gowns *must* be used. Procedures that generate large volumes of organisms or that may generate aerosols must be done in a biosafety cabinet. Most microbiology research, clinical, and teaching laboratories maintain BSL-2 containment standards.

BSL-3 laboratories are designed to contain emerging pathogens (♻ Section 32.10) and known pathogens that have a very high potential for causing serious infections, especially from aerosols. For example, if laboratory personnel handle extremely infectious airborne pathogens such as *Mycobacterium tuberculosis*, the causative agent of tuberculosis, the laboratory should be fitted with special features such as negatively pressurized rooms and air filters to prevent accidental release of the pathogen from the laboratory, in addition to BSL-1 and BSL-2 requirements. Biological safety cabinets are required for manipulations in a BSL-3 laboratory; work must not be done on the open bench. In some special cases, organisms that can normally be handled at BSL-2 must be handled at BSL-3. For example, *Staphylococcus aureus* can be handled on culture plates at BSL-2. However, when large quantities are grown, and especially when such quantities are centrifuged, work must be done in a BSL-3 facility to contain potential infectious aerosols. Specialized clinical, research, and teaching facilities must maintain BSL-3 safety levels.

BSL-4 laboratories are designed for maximum containment of life-threatening pathogens that have a high probability of transmission by aerosols and for which there is no effective immunization, treatment, or cure. In addition to BSL-1, BSL-2, and BSL-3 requirements, BSL-4 facilities require mechanisms for total isolation and physical containment of pathogens. Such mechanisms might include manipulation of cultures and clinical specimens through gloves in a sealed biological safety cabinet or by personnel wearing full-body, positive-pressure suits with air supplies (**Figure 31.9**). Examples of pathogens that must be manipulated in a BSL-4 facility include hemorrhagic fever viruses

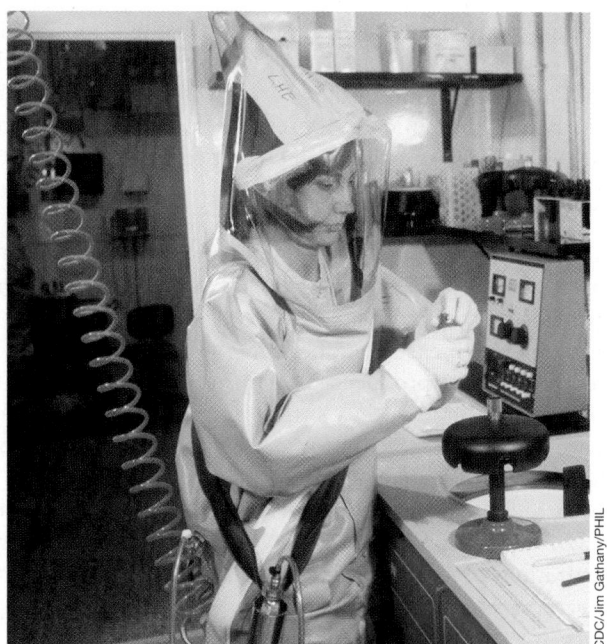

Figure 31.9 A worker in a BSL-4 (biological safety level 4) laboratory. BSL-4 is the highest level of biological control, affording maximum worker protection and pathogen containment. The worker has a whole-body sealed suit with an outside air supply and ventilation system. Air locks control all access to the laboratory. All material leaving the laboratory is autoclaved or chemically decontaminated.

UNIT 10

(Lassa, Marburg, Ebola, ⮌ Section 32.10) and drug-resistant *Mycobacterium tuberculosis* (⮌ Section 33.4). BSL-4 laboratories are usually associated with government facilities such as the Centers for Disease Control and Prevention (Atlanta, Georgia, USA) or university laboratories that specialize in infectious disease.

MiniQuiz

- What are the major precautions necessary to prevent spread of a bloodborne pathogen to laboratory personnel?
- What are the major causes of laboratory infections?
- Identify the biological hazard containment features of BSL-1, BSL-2, BSL-3, and BSL-4 laboratories.

II Immunology and Diagnostic Methods

Immunoassays are used in clinical, reference, and research laboratories to detect specific pathogens or pathogen products. When culture methods for pathogens are not routinely available or are prohibitively difficult to perform, as is the case with most viral infections or some bacterial pathogens, immunoassays often provide effective and relatively simple means to identify individual pathogens or exposure to pathogens.

31.5 Immunoassays for Infectious Disease

The immune response was discussed in Chapters 28–30. Many immunoassays utilize antibodies specific for pathogens or their products for *in vitro* tests designed to detect individual infectious agents. Patient immune responses can also be monitored to obtain evidence of exposure to and infection by a pathogen.

Antibody Titers

Isolation of a pathogen is not always possible or practical to confirm diagnosis of an infectious disease. An alternative approach that provides strong indirect evidence for infection by a particular pathogen is to measure antibody *titer* (quantity) directed to an antigen or antigens produced by the suspected pathogen. If an individual is infected with a suspected pathogen, the immune response—in this case, the antibody titer—to that pathogen should become elevated. Serial dilutions of patient serum are prepared and assayed by methods we will discuss in Sections 31.7–31.11. The **titer** is defined as the highest dilution (lowest concentration) of serum at which an antigen–antibody reaction is observed (**Figure 31.10**). These methods are called *serological tests* because they assay patient serum for antibody content.

A positive antibody titer indicates previous infection or exposure to a pathogen. For pathogens rarely found in a population, a single positive test for a pathogen-specific antibody without a follow-up test may indicate ongoing, active infection. This is the case, for example, for hantavirus (⮌ Section 34.2). In most cases, however, the mere presence of antibody does not indicate

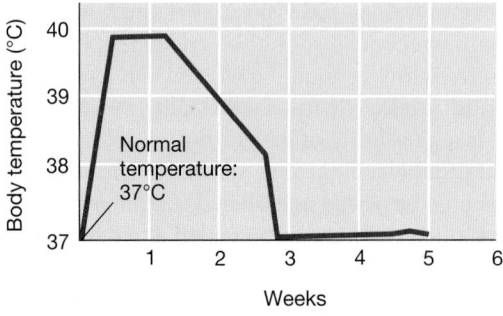

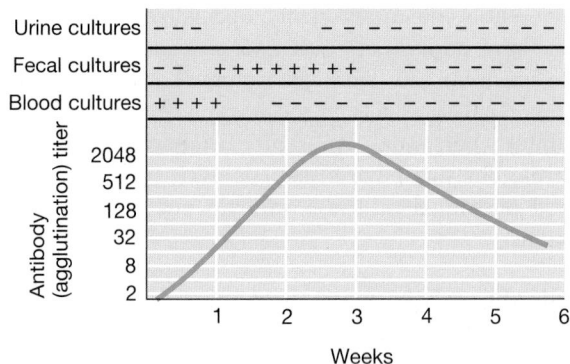

Figure 31.10 The course of infection in a typical untreated typhoid fever patient. Measurement of body temperature provides a measure of the course of clinical symptoms. The antibody titer was measured by determining the highest serum dilution (twofold series) causing agglutination of a test strain of *Salmonella enterica* serovar Typhi. Titer is shown as the *reciprocal* of the highest dilution showing an agglutination reaction. Presence of viable bacteria in blood, feces, and urine was determined from periodic cultures. Note that the pathogen clears from the blood as the antibody titer rises, while clearance from feces and urine requires a longer time. Body temperature gradually drops to normal as the antibody titer rises. The data given do not represent a single patient but are a composite of the pattern seen in large numbers of patients.

active infection. Antibody titers typically remain detectable for long periods after a previous infection has been resolved. To link an acute illness to a particular pathogen, it is essential to show a *rise* in antibody titer in serum samples taken from a patient during the acute disease and later during the convalescent phase of the disease. Frequently, the antibody titer is low during the acute stage of the infection and rises during convalescence (Figure 31.10). A rise in antibody titer is a strong indication that the illness is due to the suspected pathogen. In some cases, the presence of antibody in the serum may be due to a recent immunization. In fact, measurement of the rise in antibody titer following immunization is one of the best ways of determining that the immunization was effective.

Skin Tests

A number of pathogens induce a delayed-type hypersensitivity (DTH) response mediated by T$_H$1 cells (⮌ Section 28.9). For these pathogens, skin testing may be useful for determining exposure. As an example, a commonly used skin test is the *tuberculin test*, which consists of an intradermal injection of a soluble extract from cells of *Mycobacterium tuberculosis*. A positive

Table 31.5 *Immunological procedures for identification of infectious agents*

Pathogen/disease	Antigen	Procedure[a]
HIV/AIDS	Human immunodeficiency virus (HIV)	EIA Immunoblot
Borrelia burgdorferi (Lyme disease)	Flagellin Surface proteins	EIA Immunoblot
Brucella (brucellosis)	Cell wall antigen	Agglutination
Candida albicans (yeast infections)	Soluble extract of fungal proteins	Skin test
Corynebacterium diphtheriae (diphtheria)	Toxin	Skin test (Schick test)
Influenza virus (influenza)	Influenza virus suspensions such as nasopharyngeal exudate Nasopharynx cells containing influenza virus	EIA Immunofluorescence
Mycobacterium leprae (leprosy, or Hansen's disease)	Lepromin (soluble extract of bacterial proteins)	Skin test
Mycobacterium tuberculosis (tuberculosis)	Tuberculin (purified protein derivative, PPD)	Skin test
Neisseria meningitidis (meningitis)	Capsular polysaccharide	Passive hemagglutination (*N. meningitidis* polysaccharide adsorbed to red blood cells)
Pneumocystis jiroveci (lung infection)	*P. carinii* cells	Immunofluorescence
Rickettsial diseases (Q fever, typhus, Rocky Mountain spotted fever)	Killed rickettsial cells	Complement-based assay Cell agglutination tests EIA
Salmonella (gastroenteritis)	O and H antigen	Agglutination (Widal test) EIA
Streptococcus (group A) (strep throat, scarlet fever)	Streptolysin O (extoxin) DNase (extracellular protein)	Neutralization of hemolysis Neutralization of enzyme
Treponema pallidum (syphilis)	Cardiolipin-lecithin-cholesterol	Flocculation (Venereal Disease Research Laboratory [VDRL]) test
Vibrio cholerae (cholera)	O antigen	Agglutination Bacteriocidal test (in presence of complement); EIA

[a]Immunofluorescence tests use preformed antibody to detect the presence of the indicated pathogen in a patient specimen. Skin tests for *C. albicans*, *M. tuberculosis*, and *M. leprae* indicate T_H1-mediated delayed-type hypersensitivity. The *C. diphtheriae* Schick test detects serum antibodies with a toxin-neutralization skin test. All other tests measure serum antibody levels.

inflammatory reaction at the site of injection within 48 hours indicates current infection or previous exposure to *M. tuberculosis*. This test identifies responses caused by pathogen-specific inflammatory T_H1 cells (Figure 28.6). Skin tests are routinely used for aiding in diagnosis of tuberculosis, Hansen's disease (leprosy), some fungal diseases, and other infectious disease in which the antibody response is weak or nonexistent. Common immunodiagnostic tests for pathogens are shown in **Table 31.5**.

If a pathogen is extremely localized, there may be little induction of a systemic immune response and no rise in antibody titer or skin test reactivity, even if the pathogen is proliferating profusely at the site of infection. A good example is gonorrhea, caused by infection of mucosal surfaces with *Neisseria gonor-rhoeae*. As we will discuss in Section 33.12, gonorrhea does not elicit a systemic or protective immune response, there is no serum antibody titer or skin test reactivity, and reinfection of individuals is common.

MiniQuiz

• Define the term *titer*.

• Explain the reasons for changes in antibody titer for a single infectious agent, from the acute phase through the convalescent phase of the infection.

• Describe the method, time frame, and rationale for the tuberculin skin test. What component of the immune response does this test detect?

UNIT 10

31.6 Polyclonal and Monoclonal Antibodies

The immune response to a pathogen typically results in the production of immunoglobulins (Igs) directed at numerous antigenic determinants present on the pathogen (Section 29.7). Only a few of the many Igs are directed toward each antigenic determinant. The resulting antiserum is a complex mixture of different antibodies, or **polyclonal antibodies**. This antibody population is derived from many individual B cells. The serum is called *polyclonal antiserum*. Polyclonal antisera provide adequate immune protection to the host, but they are not precisely reproducible because they are the entire collection of the antibodies produced by an individual in response to a complex antigen.

Hybridomas and Monoclonal Antibodies

Each Ig is produced by a single B lymphocyte (Section 29.8). As a result, an *in vitro* B cell clone can produce limitless supplies of a single monospecific antibody; this is called a **monoclonal antibody**. Antibody-producing B cells, however, normally die after several weeks in cell culture (*in vitro*). To produce long-lived B cell clones, antibody-producing B cells are fused with B cell tumors called *myelomas*. Myelomas are capable of dividing indefinitely and are therefore immortal cell lines. The immortal cell lines that result from the B cell–myeloma fusion are hybrid cell lines called *hybridomas*. The hybridoma cell lines share the properties of both fusion partners. They grow indefinitely *in vitro* and produce antibodies (**Figure 31.11**).

To produce a monoclonal antibody, a mouse is immunized with the antigen of interest. During the next several weeks, antigen-specific B cells proliferate and begin producing antibodies in the mouse. Spleen or lymph node tissue, rich in B cells, is then removed from the mouse, and the B cells are fused with myeloma cells (Figure 31.11). Many cells fuse in culture and begin to grow, but only a small number are antibody-producing hybridomas. Hybridomas are selected from other cells by addition of *h*ypoxanthine, *a*minopterin, and *t*hymidine (HAT) to the *in vitro* cell culture medium. The HAT medium stops the growth of unfused myeloma cells because the myeloma cells, though able to grow indefinitely in cell culture, are unable to use the metabolites hypoxanthine and thymidine to bypass a metabolic block caused by aminopterin, a cell poison. By contrast, fused hybridoma cells can use hypoxanthine and thymidine to bypass the aminopterin block and grow normally in HAT medium; they receive the genes for use of hypoxanthine and thymidine from the B cell fusion partner. Unfused B cells die within a few days because they cannot divide in culture. Following fusion, the antibody-producing hybridoma clones must be identified.

An enzyme immunoassay (EIA) (Section 31.10) can be used to identify hybridomas that produce monoclonal antibodies. From a typical fusion, several distinct clones are isolated, each making a monoclonal antibody. Once the clones of interest are identified, they can be grown in the mouse as an antibody-producing tumor, or they can be grown in cell culture. Antibody can be harvested from the tumor or from the culture supernatant of the cell culture. Hybridomas can grow indefinitely or can be stored as frozen cells. The frozen cells can be thawed and grown in culture media

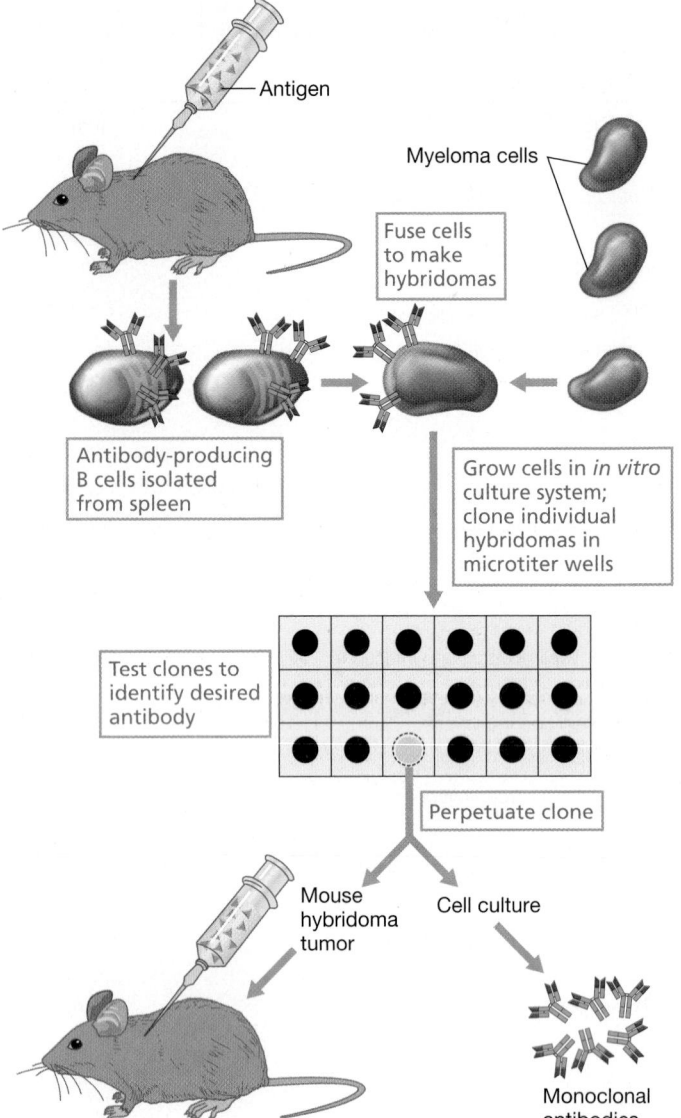

Figure 31.11 The hybridoma technique and production of monoclonal antibodies. The hybridoma can be indefinitely cultured or passed through animals as a tumor. The hybridoma cells are stored as frozen tumor cells that can be thawed and grown in tissue culture or in a suitable animal host.

at a later time to provide the desired monoclonal antibodies. Monoclonal antibodies have replaced polyclonal antibodies for many immunodiagnostic applications because they are highly specific bioreagents. **Table 31.6** compares the properties of polyclonal antibodies and monoclonal antibodies.

Diagnostic Uses

Both polyclonal and monoclonal antibodies are used for clinical diagnostic tests, immunological typing of bacteria, and identification of cells containing foreign surface antigens (for example, a virus-infected cell). Monoclonal antibodies have also been used in genetic engineering for identifying and measuring levels of gene products not detectable by other methods and also for increasing the specificity of existing clinical tests, including blood and tissue typing.

Table 31.6 *Characteristics of monoclonal and polyclonal antibody production*

Polyclonal	Monoclonal
Contains many antibodies recognizing many determinants on an antigen	Contains a single antibody recognizing only a single determinant
Various classes of antibodies are present (IgG, IgM, and so on)	Single class of antibody produced
Can make a specific antibody using only a highly purified antigen	Can make a specific antibody using an impure antigen
Reproducibility and standardization difficult	Highly reproducible

Table 31.7 *Types of antigen–antibody reactions*

Location of antigen	Accessory factors required	Reaction observed
Soluble	None	Precipitation
On cell or inert particle	None	Agglutination
Flagellum	None	Immobilization or agglutination
On bacterial cell	Complement	Lysis
On bacterial cell	Complement	Killing
On erythrocyte	Complement	Hemolysis
Toxin	None	Neutralization
Virus	None	Neutralization
On bacterial cell	Phagocyte, complement	Phagocytosis and opsonization

Because of their specificity, monoclonal antibodies are also used to detect and treat human cancers. Malignant cells contain surface antigens not expressed by normal cells. These tumor antigens are unique, tumor-specific cell proteins. Monoclonal antibodies prepared against the tumor antigens specifically target the malignant cells and have been used as vehicles to deliver toxins directly to them. Tumor-specific monoclonal antibodies covalently linked to toxins are now undergoing clinical trials. The specificity of monoclonal antibody treatments may greatly improve cancer therapy by offering an alternative to chemical and radiation treatments that damage normal host cells as well as cancer cells. www.microbiologyplace.com **Online Tutorial 31.1: Producing Monoclonal Antibodies**

MiniQuiz

- How can a polyclonal antibody preparation recognize a variety of antigenic determinants?
- What advantages do monoclonal antibodies have as compared with polyclonal antibodies? What are the advantages of polyclonal antibodies?

31.7 *In Vitro* Antigen–Antibody Reactions: Serology

The study of antigen–antibody reactions *in vitro* is called **serology**. When extended to diagnostic microbiology, serology means detection of pathogen-induced antibodies. Serological reactions are the basis for a number of diagnostic tests. Antigen–antibody reactions rely on the specific interaction of antigenic determinants with the *variable* region of the antibody molecule (Section 29.7). Various serological tests are used to detect either antigens or antibodies, depending on the properties of the antigen and on the conditions chosen for reaction (**Table 31.7**).

Specificity and Sensitivity

For serological tests, *specificity* means that the antibody–antigen reaction that is observed identifies exposure to a single pathogen. Thus, the antigen used to detect antibodies in patient serum must be unique to the pathogen in question, avoiding false-positive reactions. The *sensitivity* of some common serological

tests in terms of the amount of antibody necessary to detect antigen is shown in **Table 31.8**. The amount of antigen detected by each test system is proportional to the amount of antibody used. For example, immune precipitation reactions have very low sensitivity and require a large amount of antibody. The minimum antigen amount detected for this assay is 0.1–1.0 mg. Thus, precipitation tests are the least sensitive serological tests. By contrast, enzyme immunoassay (EIA) tests (Section 31.10) require 100,000 times less antibody and can detect 1 million times less antigen (0.1–1.0 ng quantities) than precipitation tests. EIA tests are among the most sensitive serological tests.

Neutralization

Neutralization is the interaction of antibody with antigen to block or distort the antigen sufficiently to reduce or eliminate its biological activity. Neutralization reactions can occur *in vitro* or *in vivo*.

Table 31.8 *Sensitivity of immunodiagnostic assays*

Assay	Sensitivity (μg of antibody per/ml)[a]
Precipitin reaction	
In fluids	24–160
In gels (double immunodiffusion)	24–160
Agglutination reactions	
Direct	0.4
Passive	0.08
Radioimmunoassay (RIA)	0.0008–0.008
Enzyme immunoassay (EIA)	0.0008–0.008
Immunofluorescence	8.0

[a]The smallest amount of antibody necessary to give a positive reaction in the presence of antigen.

For example, neutralization of a microbial toxin by specific antibody occurs when the toxin and specific antibody combine in such a way that the active portion of the toxin is blocked (**Figure 31.12**). Neutralization reactions can block the effects of many bacterial exotoxins, including many of those listed in Table 27.4. An antiserum containing an antibody that neutralizes a toxin is called an *antitoxin*. Antitoxin therapy is used to treat botulism, tetanus, and diphtheria, all diseases that result from the action of bacterial exotoxins.

Virus neutralization tests determine if an antibody present in patient serum can neutralize the infectivity of a virus. To perform the test, patient serum is mixed with a virus preparation. The mixture is used to infect blood cells or tissue culture cells; the infected cells are then monitored for cell death. If the inoculated cells survive, the antibody has neutralized the virus. For example, antibodies directed against the hemagglutinin and neuraminidase proteins of influenza viruses prevent the adsorption of the viruses to specific receptors on host cells, protecting them from cytopathic effects (↩ Section 33.8). A positive neutralization test indicates that the patient has antibodies and has been exposed to the virus. Neutralization tests have been developed for arboviruses and rabies virus as well as several others.

Precipitation

Precipitation results from the interaction of a soluble antibody with a soluble antigen to form an insoluble complex. Tests can be done in liquid, as in test tubes or capillary tubes, or can be done in agarose gel, as shown in **Figure 31.13**. Antibody molecules generally have two antigen-binding sites (that is, they are bivalent). Therefore, each antibody can bind two separate antigen molecules. If the antigen also has more than one available antibody-binding determinant, a precipitate may develop from aggregates

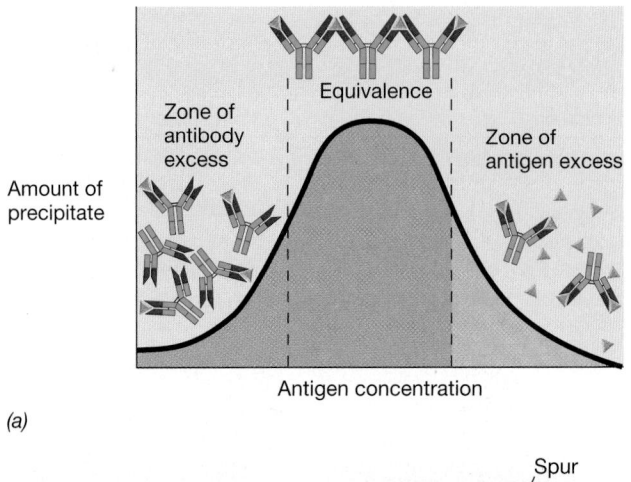

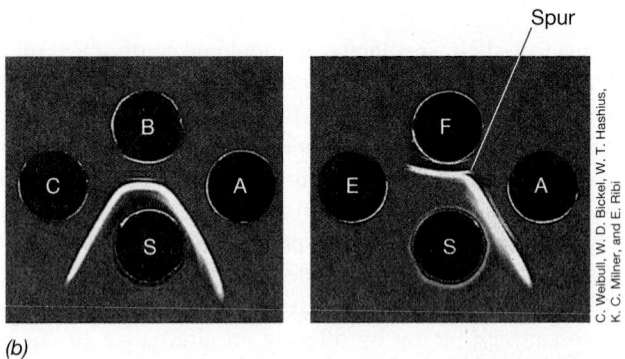

(b)

Figure 31.13 Precipitation reactions between soluble antigen and antibody. The graph *(a)* shows the extent of precipitation as a function of antigen and antibody concentration. *(b)* Precipitation in agar gel, a process called immunodiffusion. Wells labeled S contain antibodies to cells of *Proteus mirabilis*. Wells labeled A, B, and C contain soluble extracts of *P. mirabilis*. A line of identity is observed in the wells on the left. On the right, antigen E does not react, and antigen A shows partial identity with antigen F (see leader to spur).

of antibody and antigen molecules (Figure 31.13*a*). Because they are easily observed *in vitro*, precipitation reactions are very informative serological tests, especially for the quantitative measurement of antibody concentrations. Precipitation occurs maximally, however, only when there are optimal proportions of the two reacting substances. The presence of either excess antigen or excess antibody results in the formation of soluble immune complexes.

Precipitation reactions carried out in agarose gels, called *immunodiffusion tests*, are used to study the specificity of antigen–antibody reactions. In a few cases, they are used in the clinical laboratory as a diagnostic tool, especially in diagnoses of fungal infections such as coccidioidomycosis, histoplasmosis, blastomycosis, and paracoccidioidomycosis. In these tests and others like it, prepared antigen and patient antisera containing antibody are loaded into separate wells cut in an agarose gel. From the wells, the reagents diffuse outward, forming precipitation bands in the region where antibody interacts with antigen in optimal proportions (Figure 31.13*b*). The precipitation bands formed are characteristic for the reacting substances; two antigens reacting with an antiserum can be tested for molecular relationships by observing

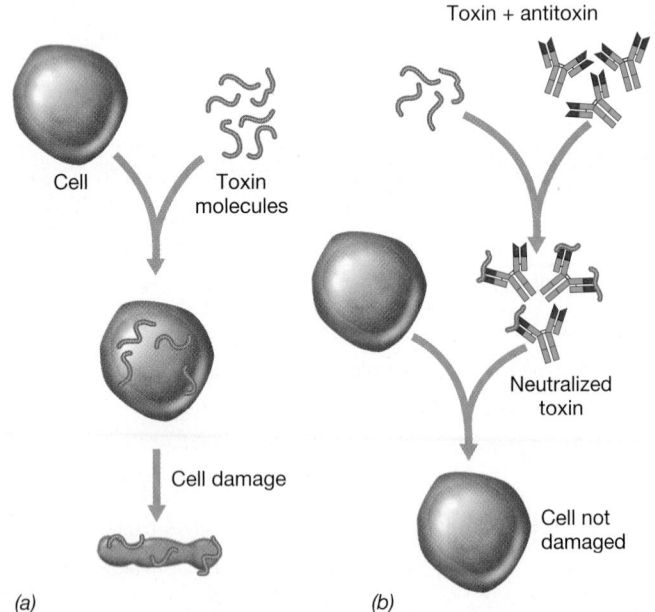

Figure 31.12 Neutralization of an exotoxin by an antitoxin antibody. *(a)* Untreated toxin results in cell destruction. *(b)* Antitoxin antibody neutralizes toxin and prevents cell destruction.

the bands formed when the two antigens are placed in adjacent wells equidistant from the antiserum well. For example, if two antigens in adjacent wells are identical, they will form a single, fused, precipitin band. This is called a *line of identity*. If, on the other hand, adjacent wells contain one antigen in common, but one well contains a second reacting antigen, a line of *partial identity* will form (Figure 31.13*b*). The extension of the precipitin line (representing a reaction between the antiserum and the second antigen) is called a *spur*. Immunodiffusion can thus be used to assess the relatedness of proteins obtained from different sources.

Unfortunately, the readily visible precipitation reactions are not very sensitive. Microgram quantities of specific antibody are necessary to visualize a precipitate (Table 31.8), whereas more sensitive diagnostic tests require only nanogram quantities. Consequently, with the exception of the clinical diagnostic immunodiffusion tests for fungal infections, immune precipitation assays are normally used only in research and reference laboratories.

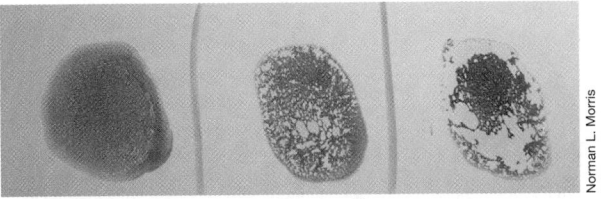

(a)

Blood type	Percentage of U.S. population	Serum	
		Anti A	Anti B
Type O	46	No aggl.	No aggl.
Type A	39	Aggl.	No aggl.
Type B	11	No aggl.	Aggl.
Type AB	4	Aggl.	Aggl.

(b)

Figure 31.14 Direct agglutination of human red blood cells for ABO blood typing. *(a)* The reaction on the left shows no agglutination. The reaction in the center shows the diffuse agglutination pattern that indicates a positive reaction for the B blood group. The reaction on the right shows the strong agglutination pattern with large, clumped agglutinates typical for the A blood group. *(b)* Table of expected blood grouping results for the U.S. population.

> ## MiniQuiz
> - In serological reactions, high specificity prevents false-positive reactions. High sensitivity prevents false-negative reactions. Explain.
> - Explain the principles of a neutralization reaction.
> - What are the minimum antigen and antibody requirements for a precipitation reaction?

31.8 Agglutination

Agglutination is the visible clumping of a particulate antigen when mixed with antibodies specific for the particulate antigen. Agglutination tests can be done in test tubes or in small-volume microtiter plates, or they can be done by mixing reagents on glass slides. Agglutination tests are about 100 times more sensitive than precipitation tests (Table 31.8) and are widely used in clinical and diagnostic laboratories; they are simple to perform, highly specific, inexpensive, rapid, and reasonably sensitive. Standardized agglutination tests are used for the identification of blood group (red blood cell) antigens as well as many pathogens and pathogen products.

Direct Agglutination

Direct agglutination results when soluble antibody causes clumping due to interaction with an antigen that is an integral part of the surface of a cell or other insoluble particle. Direct agglutination procedures are used for the identification of antigens found on the surface of red blood cells (erythrocytes). Agglutination of red blood cells is called *hemagglutination* and is the basis for human blood typing.

Red blood cells exhibit a variety of cell surface antigens, and individuals vary considerably with respect to the antigens present on their red blood cells. The major antigens on the surface of human red blood cells are called *A*, *B*, and *D*. D is also called *Rh* (*rhesus*). A and B antigens and antibodies are the basis for the ABO blood-typing assay. Red blood cells carrying the antigen visibly clump when mixed with specific antisera (**Figure 31.14**).

The antisera are obtained from human donors who have been immunized to A or B antigens by natural or artificial means.

For the A, B, and O blood types, individuals express codominant A and B alleles as one of the following antigen phenotypes: A, B, AB (one allele expressing the A antigen and one expressing the B antigen), or O (the absence of both A or B alleles). In addition, individuals make antibodies to most nonself blood group antigens. Type A individuals make antibodies to group B antigens, while type B individuals make antibodies to group A antigens. Type AB individuals have neither A nor B antibodies, but type O individuals have antibodies to both A and B antigens (Figure 31.14). These antibodies against A and B antigens are natural antibodies; they are produced by most individuals in response to ubiquitous related antigen sources such as enteric bacteria and food, and are not related to exposure to red blood cells from other individuals.

Blood typing using the A, B, and D antisera is done before blood transfusion to prevent red blood cell destruction that would occur if antibodies in the recipient's blood reacted with the red blood cells in the transfused blood, or vice versa. Antibody-coated red blood cells would likely undergo hemolysis (lysis of red blood cells) through the activity of complement (↩ Section 29.9), resulting in severe anemia.

Passive Agglutination

Passive agglutination is the agglutination of soluble antigens or antibodies that have been adsorbed or chemically coupled to cells or insoluble particles such as latex beads, charcoal particles, and red blood cells. The insolubilized antigen or antibody can then be detected by agglutination reactions. The cell or particle serves as an inert carrier. Passive agglutination reactions can be

UNIT 10

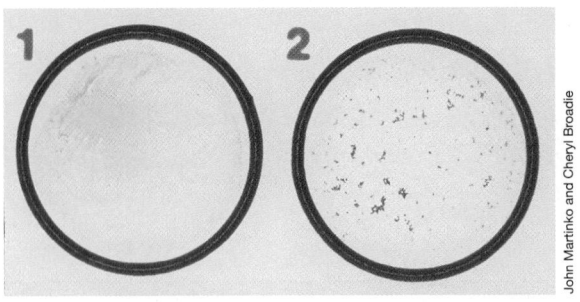

Figure 31.15 Latex bead agglutination test for *Staphylococcus aureus*. Panel 1 shows a negative control. Note the uniform pink color of the suspended latex beads coated with antibodies to protein A and clumping factor, two antigens found exclusively on the surface of *S. aureus* cells. Panel 2 shows the same suspension after a loopful of material from a bacterial colony was mixed into the suspension. The bright red clumps indicate that a positive agglutination reaction took place, revealing that the colony is *S. aureus*.

up to five times more sensitive than direct agglutination tests (Table 31.8), significantly increasing sensitivity.

The agglutination of antigen-coated or antibody-coated latex beads by complementary antibody or antigen from a patient is a typical rapid assay method. Small (0.8 μm) latex beads coated with a specific antigen are mixed with patient serum on a microscope slide and incubated for a short period. If patient antibody binds the antigen on the bead surface, the milky white latex suspension will become visibly clumped, indicating a positive agglutination reaction. Latex agglutination is also used to detect bacterial surface antigens by mixing a small amount of a bacterial colony with antibody-coated latex beads. For example, a commercially available suspension of latex beads coated with antibodies to protein A and clumping factor, two proteins found exclusively on the surface of *Staphylococcus aureus* cells, is specific for identification of clinical isolates of *S. aureus*. Unlike traditional growth-dependent tests for *S. aureus*, the latex bead assay takes only 30 seconds (**Figure 31.15**) and can be used directly on a clinical sample, such as the material from a purulent infection possibly caused by *S. aureus*. Latex bead agglutination assays have also been developed to identify other common pathogens, such as *Streptococcus pyogenes*, *Neisseria gonorrhoeae*, *Neisseria meningitidis*, *Haemophilus influenzae*, *Escherichia coli* O157:H7, and the fungi *Cryptococcus neoformans* and *Candida albicans*.

Latex agglutination tests are also used to detect serum antibodies directed against the body's own Ig, DNA, and other macromolecules. These self-reacting antibodies are associated with several autoimmune diseases; coupled with clinical information, their detection is an important finding in the diagnosis of autoimmune diseases (↩ Section 28.9).

Passive agglutination assays require no expensive equipment or particular expertise, and can be highly specific and very sensitive. In addition, the cost-effective nature of the assays makes them suitable for large-scale screening programs. These tests are therefore widely used in clinical and research applications.

MiniQuiz
- Distinguish between direct and passive agglutination. Which tests are more sensitive?
- What advantages do agglutination tests have over other immunoassays? What disadvantages?

31.9 Immunofluorescence

Antibodies chemically modified with fluorescent dyes can be used to detect antigens on intact cells. **Fluorescent antibodies** are widely used for diagnostic and research applications.

Fluorescent Methods

Antibodies can be covalently modified by fluorescent dyes such as rhodamine B, which fluoresces red, or fluorescein isothiocyanate, which fluoresces yellow-green. The attached dyes do not alter the specificity of the antibody but make it possible to detect the complex by use of a fluorescence microscope once it has bound to cell or tissue surface antigens (**Figure 31.16**). Cell-bound fluorescent antibodies emit a bright fluorescent color when excited with light of particular wavelengths. The emitted light is red-orange or yellow-green, depending on the dye used. Fluorescent antibodies are used in diagnostic microbiology because they permit the identification of a microorganism directly in a patient specimen (*in situ*), bypassing the need for the isolation and culture of the organism. The fluorescent antibody technique is also very useful in microbial ecology as a method for directly viewing and identifying microbial cells without prior isolation and culture (↩ Section 22.3).

Fluorescent antibody-staining methods can be either direct or indirect. In the *direct method*, the antibody that interacts with the surface antigen is itself covalently linked to the fluorescent dye. In the *indirect method*, the presence of a nonfluorescent antibody on the surface of a cell is detected by the use of a fluorescent antibody directed against the nonfluorescent antibody (**Figure 31.17**).

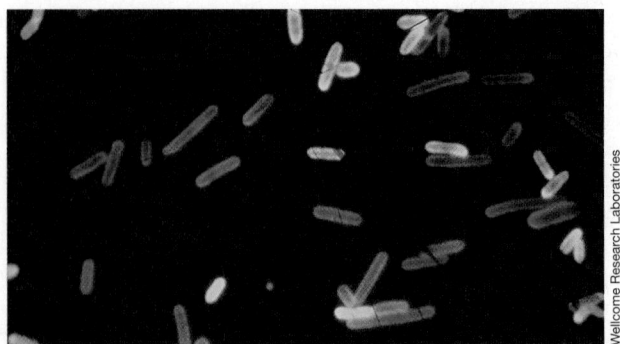

Figure 31.16 Fluorescent antibody reactions. Cells of *Clostridium septicum* were stained with antibody conjugated with fluorescein isothiocyanate, which fluoresces yellow-green. Cells of *Clostridium chauvoei* were stained with antibody conjugated with rhodamine B, which fluoresces red-orange.

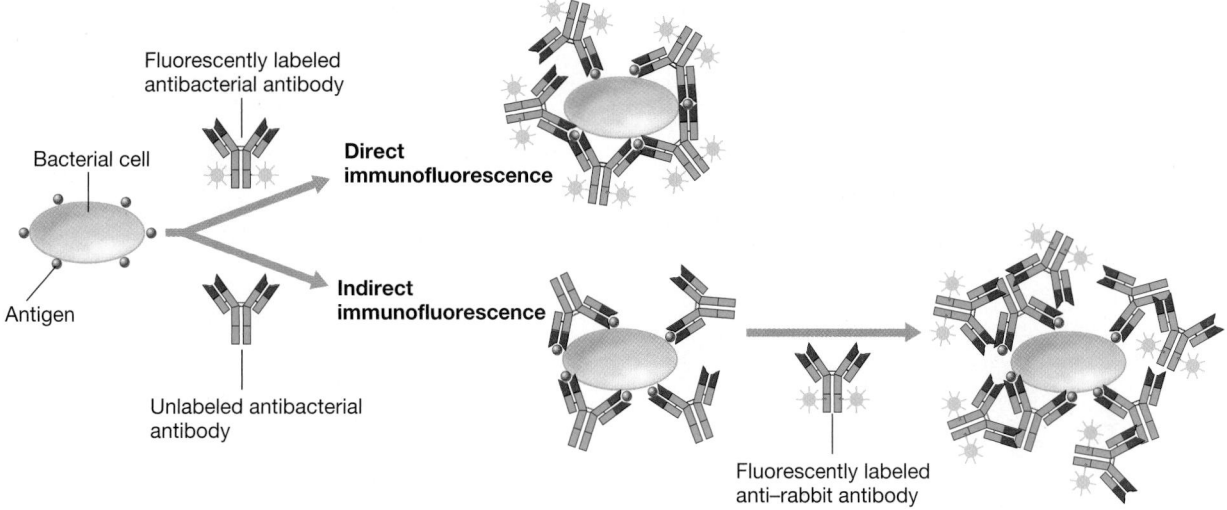

Figure 31.17 Fluorescent antibody methods for detection of microbial surface antigens.

Applications

In a typical test using fluorescent antibodies, a specimen containing a suspected pathogen is allowed to react with a specific fluorescent antibody and observed with a fluorescent microscope. If the pathogen contains surface antigens reactive with the antibody, the pathogen cells fluoresce (**Figure 31.18**).

Fluorescent antibodies can be applied directly to infected host tissues, permitting diagnosis long before primary isolation techniques yield a suspected pathogen. For example, for diagnosing legionellosis (or Legionnaires' disease), a form of infectious pneumonia, a positive identification can be made by staining biopsied lung tissue directly with fluorescent antibodies specific for cell wall antigens of *Legionella pneumophila* (Figure 31.18*a*), the causative agent of the disease. Likewise, a direct fluorescent antibody test detecting the capsule of *Bacillus anthracis* can be used

to confirm a diagnosis for anthrax (Figure 27.16*b*). Direct fluorescent antibody tests are also used to help diagnose viral infections (Figure 31.18*b*). The common respiratory pathogens influenza A and B, parainfluenza, respiratory syncytial virus (RSV), and adenovirus can be identified from respiratory tract specimens by direct fluorescent antibody methods. Fluorescent antibody methods can also be used to identify viruses grown in tissue or organ culture.

Fluorescent antibodies can also be used to separate mixtures of cells into relatively pure populations or to define the numbers of individual cell types in complex mixtures such as blood. For example, fluorescently labeled monoclonal antibodies directed against the CD4 and CD8 surface antigens of T lymphocytes are routinely used to identify and enumerate these cells in the blood (**Figure 31.19**). This assay is extremely important for patients

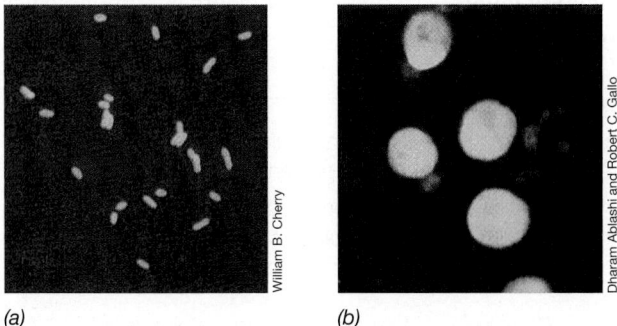

(a)   (b)

Figure 31.18 Fluorescent antibodies in clinical microbiology.
(a) Immunofluorescent stained cells of *Legionella pneumophila*, the cause of legionellosis. The specimen was taken from biopsied lung tissue. The individual organisms are 2–5 μm in length. *(b)* Detection of virus-infected cells by immunofluorescence. Human B lymphotrophic virus (HBLV)-infected spleen cells were incubated with serum containing antibodies to HBLV. Cells were then treated with fluorescein isothiocyanate–conjugated anti–human IgG antibodies. HBLV-infected cells fluoresce bright yellow. Unstained cells in the background did not react with the serum. Individual cells are about 10 μm in diameter.

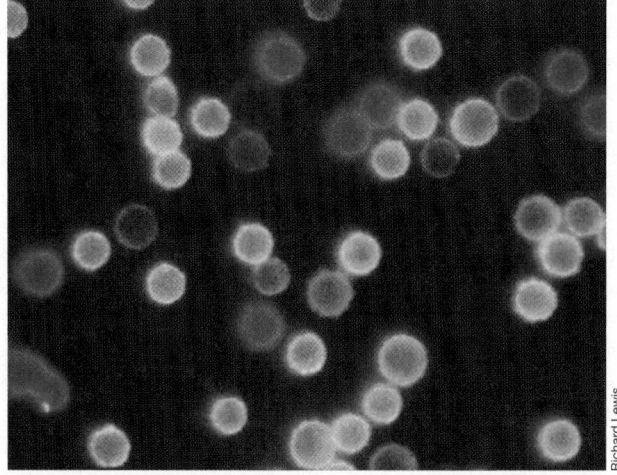

Figure 31.19 T lymphocytes stained with fluorescent-tagged monoclonal antibodies to specific surface markers. Yellow-green cells are T_C (CD8) cells; red-orange cells are T_H (CD4) cells. Individual cells are about 10–12 μm in diameter. Reprinted with permission from *Science* 239: Cover (February 12, 1988), © AAAS.

UNIT 10

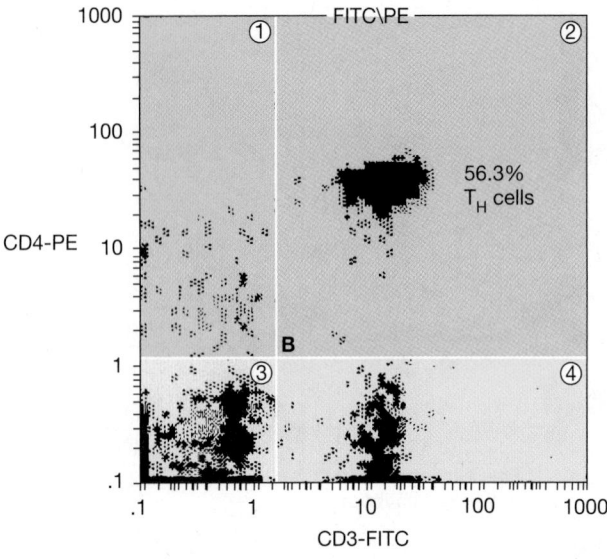

(a) **Cells from a healthy patient**

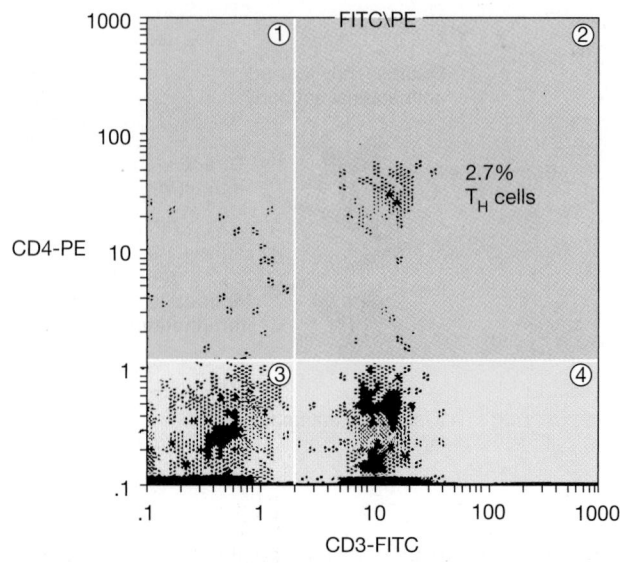

(b) **Cells from an AIDS patient**

Figure 31.20 CD3 and CD4 cell enumeration. Peripheral blood cells from a healthy human (a) and from a human with acquired immuno-deficiency syndrome (AIDS) (b) were assayed using a flow cytometer. Each dot represents a single cell. The cells were simultaneously labeled with monoclonal antibody to CD4 conjugated to phycoerythrin (PE) and with monoclonal antibody to CD3 conjugated to fluorescein isothiocyanate (FITC). CD3 is found on all T cells. CD4 is found only on T helper (T_H) cells. Quadrant 3 shows cells that were not stained with either antibody. Quadrant 1 shows cells stained with only anti-CD4. Quadrant 4 shows cells stained with only anti-CD3. Quadrant 2 shows cells stained with both anti-CD3 and anti-CD4. For the healthy patient in part a, 56.3% of the T cells were T_H cells, as shown by the dense staining pattern in quadrant 2. For the patient with AIDS in part b, only 2.7% of the total T cells were T_H cells, as indicated by the very light staining pattern in quadrant 2. Original data from Peter McConnachie, used with permission.

with human immunodeficiency virus/acquired immunodeficiency syndrome (HIV/AIDS). The CD4 T cell number and CD4/CD8 ratio change during the progression of AIDS. CD4 cell numbers and the CD4/CD8 ratio are indicators of disease progression. By determining these numbers, the clinician can follow the progress of the disease from HIV infection through clinical symptoms that define development of AIDS. The technique is also useful for monitoring the efficacy of drug therapy (ᴄᴐ Section 33.14).

Cells labeled with fluorescent antibodies can be visualized, counted, and separated with an instrument called a fluorescence cytometer or fluorescence-activated cell sorter (FACS). The FACS uses a laser beam to activate fluorescent antibody bound to cells, placing a charge on the labeled cells. In addition, the photometer records the mean fluorescence of the labeled cells. An electric field is then applied to the cell mixture. Fluorescing and nonfluorescing cells are deflected to opposite poles of the electric field, where each cell population is counted and deposited in a tube. The use of several antibodies, each labeled with a different fluorescent dye, can be used to simultaneously identify several cell markers. A typical application compares CD3 and CD4 surface proteins on T cells in healthy people and those with AIDS (**Figure 31.20**).

FACS analysis is also useful for research applications. For example, immunologists routinely use FACS methods to separate complex mixtures of immune cells. They can then study the properties of the highly enriched cell populations.

Under appropriate conditions, fluorescent antibodies yield rapid, highly specific information about a variety of clinical conditions. However, antibodies to surface antigens may cross-react between and among some bacterial species, some of which may be members of the normal flora. This is a major problem among enteric bacteria, for example, where cell wall lipopolysaccharide antigens are very similar. The clinical microbiologist must therefore perform controls using nonspecific sera and confirm positive immunofluorescent findings with other immunological or microbiological tests.

MiniQuiz

- Explain and compare direct and indirect fluorescent antibody assays, including the advantages and disadvantages of each.
- How are fluorescent antibodies used to identify specific cells in complex mixtures such as blood?

31.10 Enzyme Immunoassay and Radioimmunoassay

Enzyme immunoassay (EIA), or *enzyme-linked immunosorbent assay (ELISA)*, and **radioimmunoassay (RIA)** methods are very sensitive immunological assays and are therefore widely used in clinical and research applications. EIA and RIA employ covalently bonded enzymes and radioisotopes, respectively, to label

Procedure

1. Antibodies (Y) to virus bound to wells of microtiter plate

2. Add patient sample (secretions, serum, and so on) suspected of containing virus particles or virus antigens (⬡) and wash wells with buffer

3. Add antivirus antibody containing conjugated enzyme
   (E┬E)

4. Wash with buffer

5. Add substrate for enzyme and measure amount of colored product (●)

Results

Colored product

Quantitation

Amount of colored product produced is proportional to amount of antigen

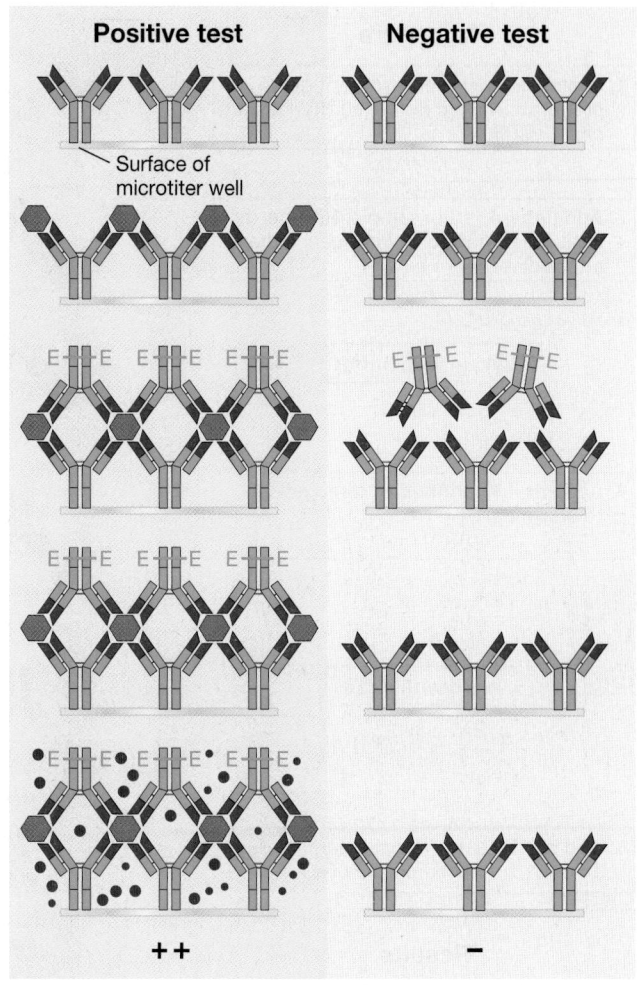

Positive test   **Negative test**

Surface of microtiter well

++   −

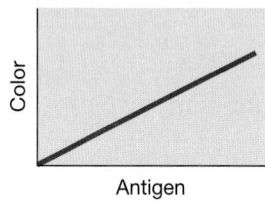

Color

Antigen

Figure 31.21 The direct EIA test. A direct EIA test can be used to detect antigenic pathogen components or antigenic metabolites in blood, urine, and other specimens.

antibody or antigen molecules. These molecules allow detection of very small quantities of antigen–antibody complexes (Table 31.8).

EIA

In EIA, an enzyme is covalently attached to an antibody molecule, creating an immunological tool with high specificity and high sensitivity. The enzyme's catalytic properties and the antibody's specificity are unaltered. Enzymes typically bound to antibodies include peroxidase, alkaline phosphatase, and β-galactosidase, all of which interact with substrates to form reaction products that can be detected in very low amounts.

Three EIA methods are commonly used for evaluation of specimens for infectious disease, one for detecting antigen (direct

EIA), another for detecting antibodies (indirect EIA), and another that detects antigen using a competition assay (competitive EIA). Direct EIA detects antigens such as virus particles from a blood or fecal sample (**Figure 31.21**). The specimen is added to the wells of a microtiter plate previously coated with antibodies specific for the antigen to be detected. If present in the sample, the virus particle will be bound by the antibodies. After unbound material is washed away, a second antibody containing a conjugated enzyme is added. The second antibody is also specific for the antigen, and it binds to other exposed antigenic determinants. Following a wash, the enzyme activity of the bound material in each microtiter well is determined by adding the substrate for the enzyme. The enzyme catalyzes the conversion of

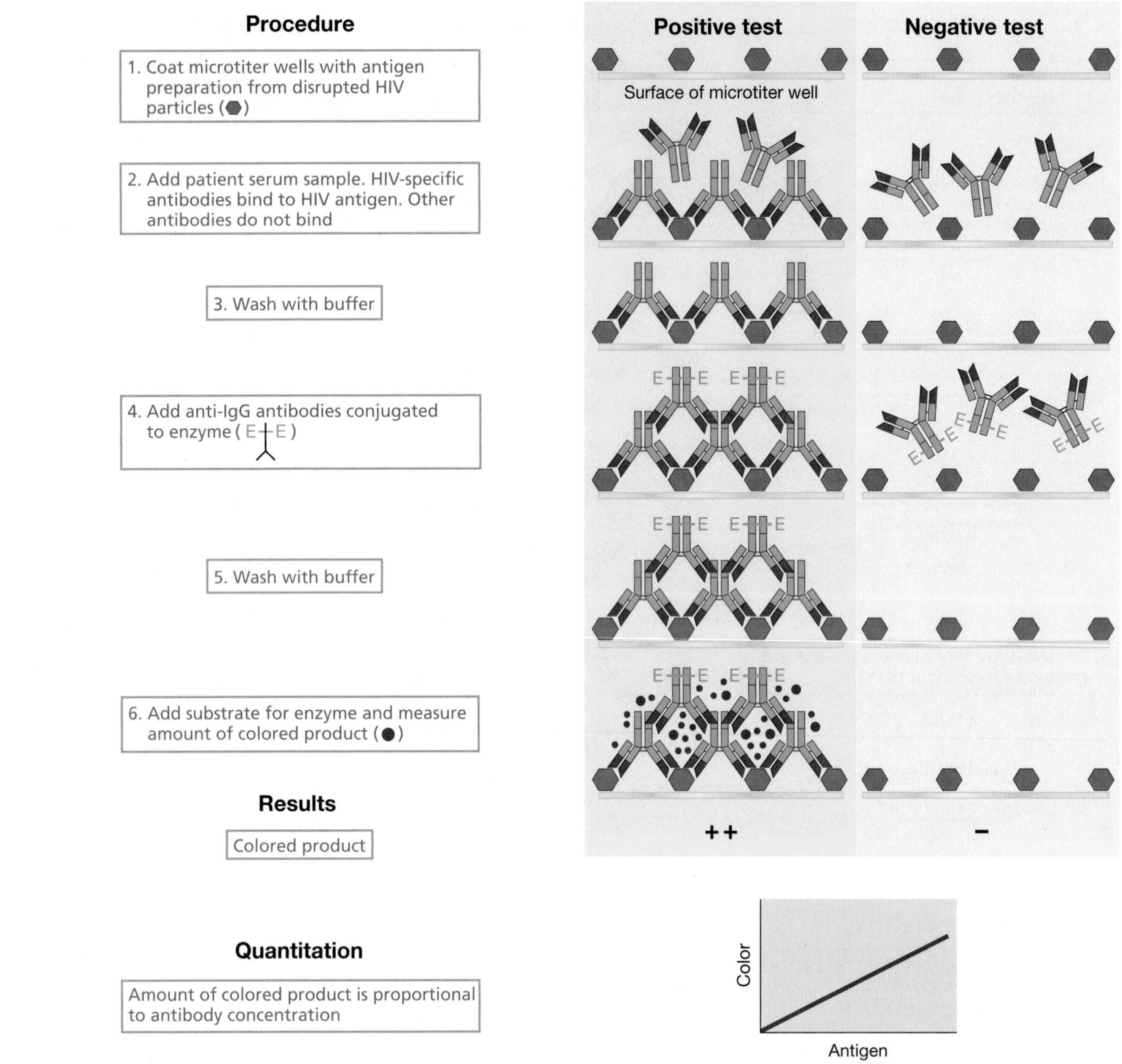

Procedure

1. Coat microtiter wells with antigen preparation from disrupted HIV particles (⬢)

2. Add patient serum sample. HIV-specific antibodies bind to HIV antigen. Other antibodies do not bind

3. Wash with buffer

4. Add anti-IgG antibodies conjugated to enzyme (E—E)

5. Wash with buffer

6. Add substrate for enzyme and measure amount of colored product (●)

Results

Colored product

Quantitation

Amount of colored product is proportional to antibody concentration

Figure 31.22 Indirect EIA test. An indirect EIA test is used in many applications including the detection of antibodies to HIV.

the substrate to a colored product, which is detected with a spectrophotometer. The color produced is proportional to the amount of antigen present.

To detect antibodies in human serum, an indirect EIA is employed. An indirect EIA is widely used to detect antibodies to human immunodeficiency virus (HIV) in human body fluids. This test illustrates the principal features of indirect EIA tests and is discussed below (**Figure 31.22**).

A third EIA method is used to identify antigens in clinical specimens. In addition to antigens from infectious diseases, the competitive EIA can be used to assay levels of drugs, hormones, and other compounds of interest in patient specimens. In the competitive EIA, a known amount of an antigen-specific antibody

is mixed and incubated with a patient specimen containing antigen (**Figure 31.23**). The complex is then added to antigen-coated wells on a microtiter plate. The plate is then washed, removing unbound antibody, including those antibodies that bound antigens in the patient specimen. A secondary antibody, coupled to enzyme, is then added, followed by substrate addition. The amount of color that develops in the sample is inversely proportional to the amount of antigen in the patient sample; the patient antigen in the specimen competes for the antibodies that were bound to the antigen-coated microtiter wells, thus the competition assay designation. In general, competition assays are more sensitive than either direct or indirect EIA assays.

Procedure

1. Mix patient specimen containing antigen (⬢) with known amount of antigen-specific antibody

2. Add specimen–antibody complex to antigen-coated microtiter well

3. Wash to remove unbound antibody

4. Add anti-IgG antibodies conjugated to enzyme (E–E)

3. Wash to remove unbound anti-Ig

6. Add substrate for enzyme and measure amount of colored product (●)

Results

Quantitation

Colored product is inversely proportional to antigen concentration in patient serum

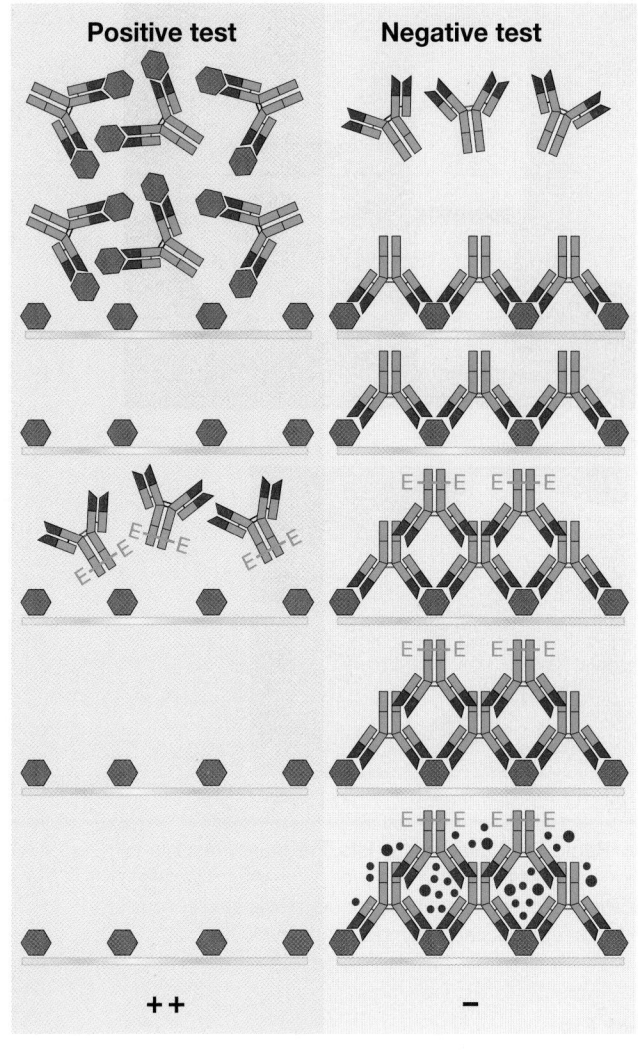

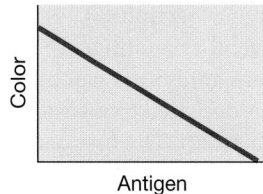

Figure 31.23 The competitive EIA test. A competitive EIA test can be used to detect antigenic pathogen components or antigenic metabolites in blood, urine, and other specimens. Competitive EIA is generally more sensitive than direct EIA for determining the concentration of antigens in specimens.

Modified rapid EIA procedures use reagents adsorbed to a fixed support material such as paper strips, nitrocellulose or plastic membranes, or plastic "dipsticks." These tests cause a color change on the strip or stick in a very short time. These rapid "point of care" tests are diagnostic aids for infectious diseases such as HIV/AIDS (⬅➔ Section 33.14) or "strep throat" (pharyngeal infection with *Streptococcus pyogenes*, ⬅➔ Section 33.2).

Additional applications for rapid tests include pregnancy tests and drug tests (**Figure 31.24**). In most of these tests, a body fluid, generally urine or blood, is applied to the reagent-support matrix. After reacting with the body fluids, the support is washed and developed with a second reagent that identifies antigen (direct test) or antibody (indirect test such as that for HIV) bound to the matrix. These tests are especially valuable where a small number of samples are to be analyzed at the site where the urine or blood is collected (for instance, away from a clinical laboratory) and can be performed by individuals lacking certified clinical laboratory skills. Results can be reported at the site, avoiding the need for delays in patient care or for follow-up visits to obtain test results. The drawback to these tests, however, is that they tend to be less specific than more elaborate tests. As a result, these point-of-care tests often need to be confirmed by standard laboratory tests.

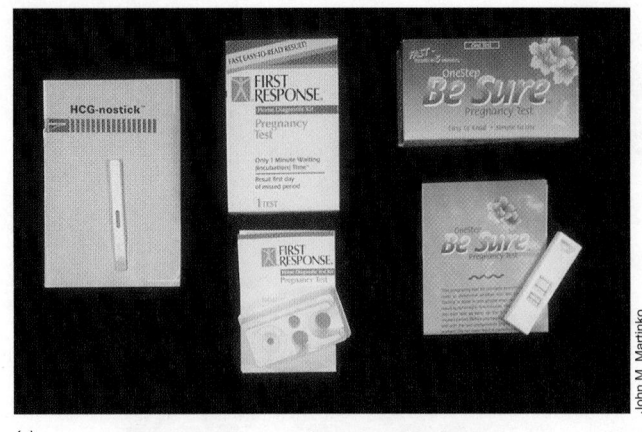

(a)

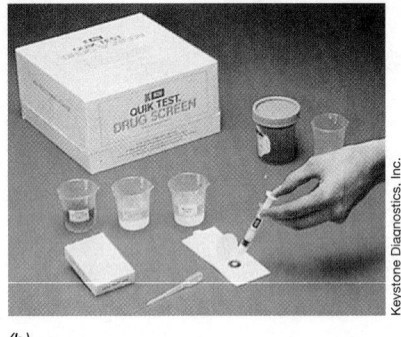

(b)

Figure 31.24 **Rapid EIA-based assay kits.** Pregnancy test kits *(a)* and the drug-testing kit *(b)* are available for point-of-care testing for these applications. Other kits are used for diagnosis of infectious diseases including streptococcal pharyngitis and HIV/AIDS.

The Indirect EIA

Initial infection with the virus that causes HIV/AIDS, the human immunodeficiency virus (HIV; ↩ Section 33.14) leads to the production of antibodies to several HIV antigens, in particular, those of the HIV envelope. These antibodies can be detected by an HIV EIA test, an indirect EIA designed to measure antibodies to HIV present in serum (Figure 31.22). The first-generation HIV test shown here illustrates the principles of the indirect EIA. For a discussion of later-generation HIV tests, see Section 33.14.

To carry out the HIV EIA test, microtiter plates are first coated with a preparation of disrupted HIV particles; about 200 ng of disrupted HIV is placed in each well. A diluted patient serum sample is then added, and the mixture is incubated to allow HIV-specific antibodies to bind to HIV antigens. To detect the presence of antigen–antibody complexes, a second antibody is then added. This second antibody is an enzyme-conjugated anti–human IgG preparation. The enzyme-conjugated anti–human IgG antibodies bind to patient HIV-specific IgG antibodies bound to the HIV antigen preparation. Next, the substrate for the conjugated enzyme is added and the enzyme activity is assayed. The color obtained in the enzyme assay is proportional to the amount of anti–human IgG antibody bound. The binding of the second antibody is an indication that antibodies from the patient's serum recognized the HIV antigens and that the patient has antibodies to HIV. This indicates

that the patient is infected with HIV. Control sera (known to be HIV-positive or HIV-negative) are assayed in parallel with patient samples to establish specificity (positive control) and measure the extent of background absorbance in the assay (negative control).

The HIV EIA test is a rapid, highly sensitive, extremely specific method for detecting exposure to HIV. Since EIAs in general are highly adaptable to mass screening and automation, the HIV EIA test and its later variants are used as standard screening methods for HIV exposure. Positive HIV EIA tests must be confirmed by an independent test, usually the HIV Western blot (immunoblot) test (Section 31.11). A positive HIV Western blot test after a positive HIV EIA test is considered proof of HIV infection.

A drawback to the HIV EIA test is the possibility of obtaining false-negative results. After HIV exposure, the immune system may take 6 weeks to a year to produce a detectable antibody titer. Therefore, individuals who have recently been infected with HIV may not yet be producing detectable amounts of antibody when they are tested. Another reason for a false-negative result in the HIV EIA test is the total destruction of the immune system seen in advanced cases of AIDS; if no immune cells are left in the body, no antibodies can be made and the EIA test is not useful. However, at this stage of disease, an AIDS diagnosis can be based on clinical information.

EIA Tests of Clinical Importance

Hundreds of other clinically useful EIAs have been developed. Some of these are direct EIAs for detecting antigens, including bacterial toxins such as cholera toxin, enteropathogenic *Escherichia coli* toxin, and *Staphylococcus aureus* enterotoxin. Viruses currently detected using direct EIA techniques include influenza H1N1, rotavirus, hepatitis viruses, rubella virus, bunyavirus, measles virus, mumps virus, and parainfluenza virus.

Indirect EIAs have been developed for detecting antibodies to a variety of clinically important bacteria. EIAs are available for detecting serum antibodies to *Salmonella* (gastrointestinal diseases), *Yersinia* (plague), *Brucella* (brucellosis), a variety of rickettsias (Rocky Mountain spotted fever, typhus, Q fever), *Vibrio cholerae* (cholera), *Mycobacterium tuberculosis* (tuberculosis), *Mycobacterium leprae* (leprosy), *Legionella pneumophila* (legionellosis), *Borrelia burgdorferi* (Lyme disease), and *Treponema pallidum* (syphilis), among others. EIAs have also been developed for detecting antibodies to *Candida* (yeast) and a variety of eukaryotic pathogens, including those causing amebiasis, Chagas' disease, schistosomiasis, toxoplasmosis, and malaria.

The speed, low cost, lack of hazardous waste, long shelf life, high specificity, and high sensitivity of EIA tests make them particularly useful immunodiagnostic tools. **www.microbiologyplace.com**
Online Tutorial 31.2: The ELISA Test

Radioimmunoassay

Radioimmunoassay (RIA) methods employ radioisotopes as antibody or antigen conjugates instead of the enzymes used in EIA. The isotope iodine-125 (^{125}I) is commonly used as the conjugate because antibodies or antigens can be readily iodinated (modified covalently with ^{125}I) without disrupting their immune specificity. RIA is used clinically to measure rare serum proteins such as human growth hormone, glucagon, vasopressin, testosterone, and insulin present in humans in extremely small amounts. RIAs are also used in some tests for illegal drugs.

Direct, indirect, or competitive RIAs have been developed using the same principles as the direct, indirect, and competitive EIAs; the major difference is that RIAs use radioactive isotopes instead of an enzyme–substrate detection system. RIA has the same sensitivity range as EIA (Table 31.8) and can also be performed as rapidly as EIA. However, the instruments used to detect radioactivity are specialized and costly. RIA also generates radioactive waste and the potential for worker and patient exposure hazards. Finally, the radioactive decay time (half-life) of the radioisotopes used for detection may limit the useful life of a test kit. As a result, RIA is often used only when an EIA is not sufficiently specific or sensitive. For example, RIA is often more useful than EIA for detecting levels of certain proteins and drug metabolites in serum because serum components inhibit some EIA enzyme–substrate or antigen–antibody interactions, such as the quantitation of IgE in patient serum, a test useful for diagnosing allergies. Thus, for certain applications, each test system has clear advantages.

MiniQuiz

- Why are EIA and RIA techniques more sensitive than immunoassays such as precipitation and agglutination?
- Compare direct EIA, indirect EIA, and competitive EIA with respect to their intended use for diagnostic purposes. Indicate the relative sensitivity of each method.

31.11 Immunoblots

Antibodies can also be used in the **immunoblot (Western blot)** method to identify individual specific proteins associated with specific pathogens, even in complex mixtures such as cell lysates or blood. The immunoblot method employs three techniques: (1) the separation of proteins on polyacrylamide gels, (2) the transfer (blotting) of proteins from gels to a nitrocellulose or nylon membrane, and (3) identification of the proteins by specific antibodies.

Immunoblot Procedures

In the first step of an immunoblot, described in **Figure 31.25**, a protein mixture is subjected to electrophoresis on a polyacrylamide

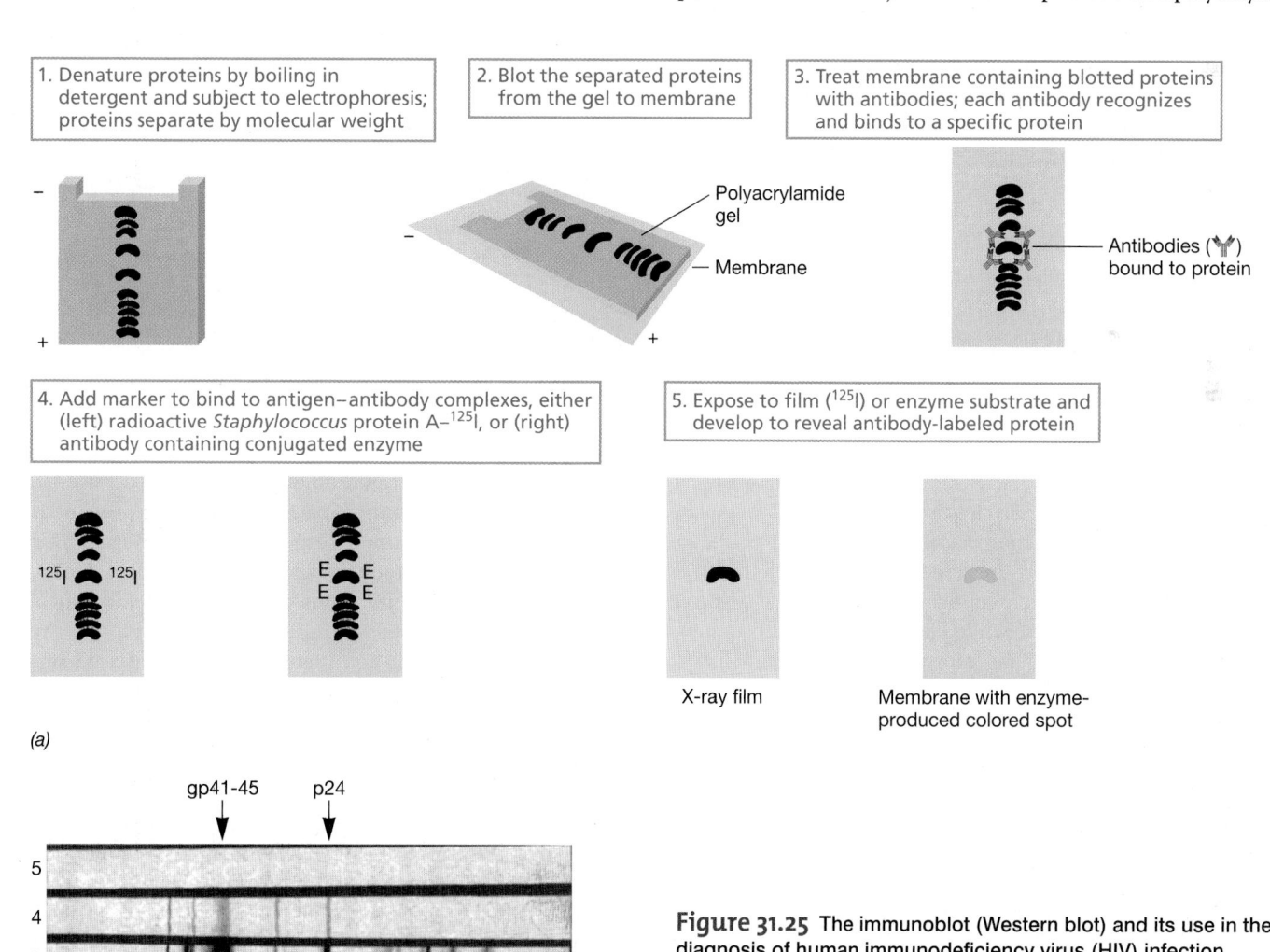

Figure 31.25 The immunoblot (Western blot) and its use in the diagnosis of human immunodeficiency virus (HIV) infection. (a) Protocol for an immunoblot. (b) Developed HIV immunoblot. The molecules p24 (capsid protein) and gp41 (envelope glycoprotein) are diagnostic for HIV. Lane 1, positive control serum (from known AIDS patients); lane 2, negative control serum (from healthy volunteer); lane 3, strong positive from patient sample; lane 4, weak positive from patient sample; lane 5, reagent blank to check for background binding.

gel. This separates the proteins into several distinct bands, each of which represents a single protein of specific molecular weight. An electrophoretic process is then used to elute the proteins from the gel and transfer them to a membrane. Antibodies specific for the pathogen components (antigen) of interest are then added to the membrane blot. Following an incubation period to allow the antibodies to bind, a radioactive marker that binds antigen–antibody complexes is added. A common radioactive marker is *Staphylococcus* protein A labeled with iodine-125 (^{125}I); protein A binds with high affinity to antibody. The bound radioactive marker location on the blot can be detected by exposing the membrane to X-ray film; the gamma rays emitted by the ^{125}I expose the film only at the bands that have formed the labeled antigen–antibody complexes (Figure 31.25). By comparing the location of the radioactive bands on the blot with the position of protein bands from control samples, a protein associated with a given pathogen can be positively identified.

Immunoblots are sometimes done using EIA technology rather than radioisotopes for detection of bound antigen–antibody complexes. Following treatment of the blotted proteins with specific antibody, the membrane is washed and then treated with a second antibody, which binds to the bound antibody. Covalently attached to this second antibody is an enzyme. The original antigen–antibody complexes are visualized when the enzyme is exposed to substrate: The product of the enzyme reaction leaves a colored product on the membrane at any spot where the enzyme-labeled secondary antibodies are bound to the antigen-reactive primary antibodies. As with the radioistope-labeled immunoblots, comparison of the location of the colored bands on the blot with the position of protein bands from control samples positively identifies the antigen of interest.

The immunoblot procedure can be used to detect either antigen (direct evidence for pathogen presence by detecting pathogen antigens in patient samples) or antibody (indirect evidence for pathogen exposure by detecting antibodies to pathogen).

The HIV Immunoblot

HIV immunoblots are generally less sensitive, more laborious, more time consuming, and more costly than the HIV EIA, so they are not used as HIV exposure screening tools. Immunoblots are, however, widely used for confirmation of HIV exposure. This is because the HIV EIA, while very sensitive, occasionally yields false-positive results. The more specific immunoblot is used to confirm positive EIA results.

Like the HIV EIA, the HIV immunoblot is an indirect test designed to detect the presence of antibodies to HIV in a serum sample. To perform the immunoblot, a purified preparation of HIV is treated with the detergent sodium dodecyl sulfate to solubilize HIV proteins and inactivate the virus. The HIV proteins are then resolved by polyacrylamide gel electrophoresis and blotted from the gel onto membranes (Figure 31.25). This technique separates at least seven major HIV proteins, and two of them, designated p24 and gp41, are specific proteins whose presence is diagnostic for HIV exposure. Protein p24 is an HIV capsid protein, and glycoprotein gp41 is an HIV envelope protein.

Membrane strips with the blotted proteins are available commercially for use by clinical laboratories.

The preblotted membrane strips are incubated with the patient serum sample. If the sample is HIV-positive, patient antibodies against HIV proteins will be present and will bind to the HIV proteins on the membrane. To detect whether antibodies from the serum sample have bound to HIV antigens, a detecting antibody, anti–human IgG conjugated to peroxidase enzyme, is added to the strips. If the detecting antibody binds, the activity of the conjugated enzyme, after addition of substrate, will form a brown band on the strip at the site of antibody binding. The patient is HIV-positive if the position of the bands resolved by exposure to the patient serum and a positive control serum are identical; a control negative serum is also analyzed in parallel and must show no bands (Figure 31.25*b*).

Although the intensity of the bands obtained in the HIV immunoblot varies somewhat from sample to sample (Figure 31.25*b*), the interpretation of an immunoblot is generally unequivocal, and thus the test is valuable for confirming positive EIA results for HIV and for eliminating false positives. An immunoblot assay is also used to confirm the specificity of antibody tests for infection by *Borrelia burgdorferi*, the organism that causes Lyme disease.

MiniQuiz

- What advantage does the immunoblot have over immunoassays such as EIA and RIA?
- Why is the immunoblot not used for general screening for HIV exposure?

Nucleic Acid–Based Diagnostic Methods

Extremely sensitive methods based on nucleic acid analyses are widely used in clinical microbiology to detect pathogens. These methods do not depend on pathogen isolation or growth or on the detection of an immune response to the pathogen. The methods depend on detection of species-specific nucleic acid sequences in 16S rRNA genes or species-specific genes. Most current systems are based on DNA amplification techniques. These include nucleic acid detection systems based on target amplification systems, probe amplification systems, and signal amplification systems.

31.12 Nucleic Acid Hybridization

Molecular methods use genotypic rather than phenotypic characteristics to identify specific pathogens. The success of genetic or DNA-based diagnostic procedures is based on several principles: (1) Nucleic acids can be readily isolated from infected tissues; (2) the nucleic acid sequence of a given pathogen's genome is unique, and so nucleic acid analysis can provide unequivocal identification; (3) nucleic acid sequences can be amplified to increase the amount of material available for analysis; (4) nucleic

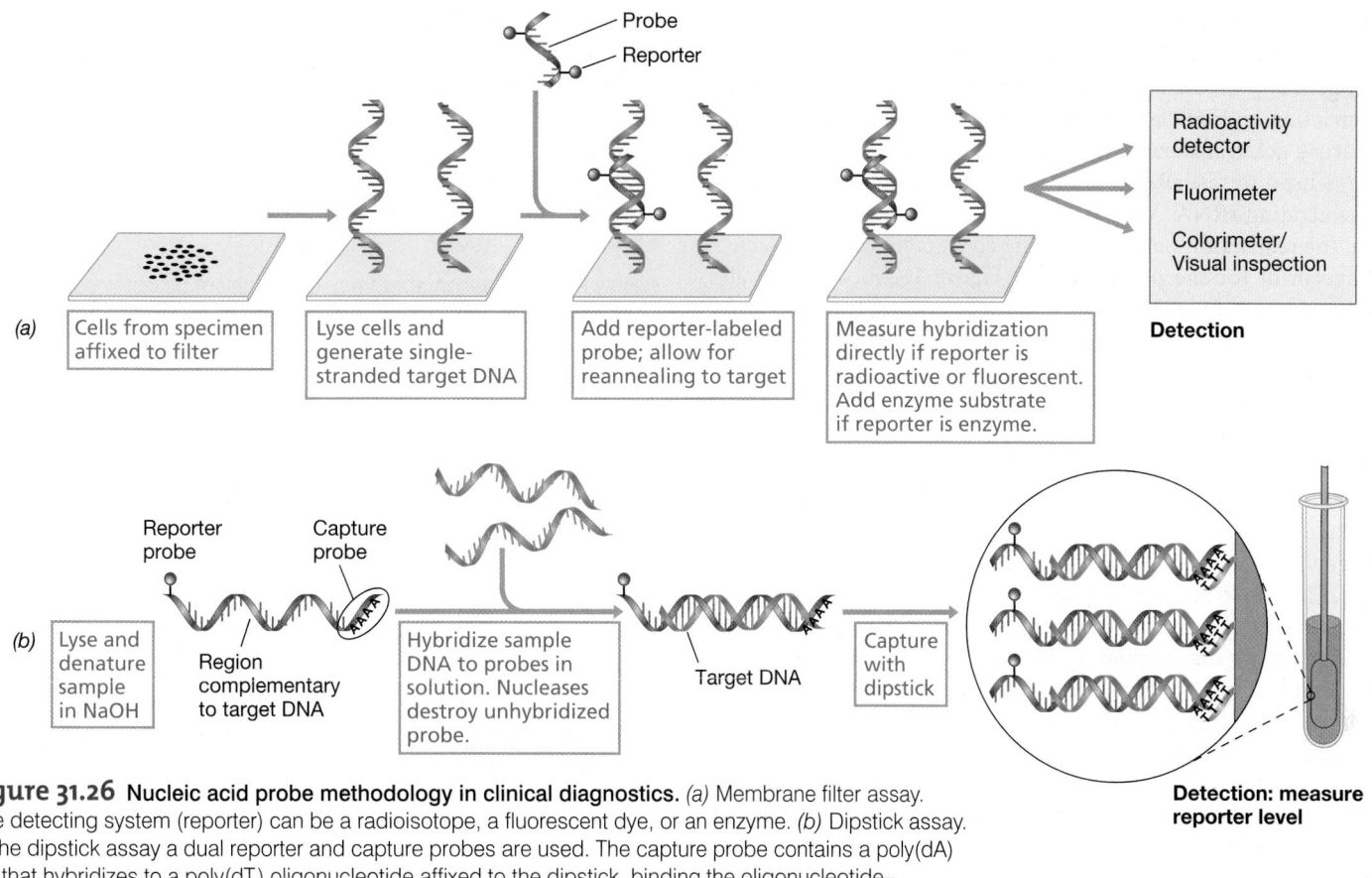

Figure 31.26 Nucleic acid probe methodology in clinical diagnostics. *(a)* Membrane filter assay. The detecting system (reporter) can be a radioisotope, a fluorescent dye, or an enzyme. *(b)* Dipstick assay. In the dipstick assay a dual reporter and capture probes are used. The capture probe contains a poly(dA) tail that hybridizes to a poly(dT) oligonucleotide affixed to the dipstick, binding the oligonucleotide–target-reporter complex. The complex can be detected as in part a.

acids, when present in sufficient amounts, can be readily visualized and measured.

Nucleic Acid Probes and Primers

Nucleic acid hybridization (↺ Section 11.2) is a critical technique for nucleic acid–based molecular methods. Instead of detecting a whole organism or its products, hybridization detects the presence of specific DNA sequences associated with a specific organism. To identify a microorganism through DNA analysis, the clinical microbiologist must have a unique **nucleic acid probe** for that microorganism. Nucleic acid probes typically consist of a single strand of DNA with a sequence unique to the gene of interest. A DNA probe oligonucleotide may be less than 100 bases or up to several kilobases in length. If a microorganism from a clinical specimen contains DNA or RNA sequences complementary to the probe, the probe will hybridize (following appropriate sample preparation to yield single-stranded DNA from the microorganism), forming a double-stranded molecule (**Figure 31.26**). To detect a reaction, the probe is labeled with a reporter molecule—a radioisotope, an enzyme, or a fluorescent compound that can be detected following hybridization. Depending on the reporter (radioisotopes and enzyme tags are the most sensitive), as little as 0.25 µg of DNA per sample can be detected.

DNA primers are even shorter pieces of sequence-specific DNA, designed specifically to hybridize to known species-specific genes. The short primer sequences are not used as probes, but are used as primers for DNA polymerase in the polymerase chain reaction (PCR) to amplify pathogen-specific DNA sequences.

Nucleic acid probes and primers offer several advantages over immunological assays. Nucleic acids are much more stable than proteins at high temperatures and at high pH, and are more resistant to organic solvents and other chemicals. Because of the relative chemical stability of the target nucleic acids, nucleic acid probe technology can even be used to positively identify organisms that are no longer viable. Additionally, some nucleic acid probes may be more specific than antibodies and can detect single-nucleotide differences between DNA sequences.

Small Subunit rRNA Phylogenetic Probe Assays

As we explained previously, small subunit (SSU) rRNA methods such as fluorescent *in situ* hybridization (FISH) methods can be used to identify various phylogenetic groups of microorganisms (↺ Section 16.9). A DNA probe can be designed to hybridize to ribosomal RNA (rRNA) of a particular genus or even species.

Advances in bacterial phylogenetics based on 16S rRNA sequences have allowed for the construction of species-specific and even strain-specific nucleic acid probes. Typing based on differences in rRNA sequences is called *ribotyping* (⟳ Section 16.9). Ribotyping reveals the unique DNA restriction patterns of rRNA genes when DNA from a particular organism is digested by restriction endonucleases. The digested DNA is separated on an agarose gel, transferred to a membrane, and a labeled rRNA probe is used to visualize the unique restriction patterns of the genes encoding rRNA. Within a species, and especially within a strain, the restriction pattern is highly conserved; it is a molecular fingerprint for the organism (⟳ Figure 16.18). Because all organisms have 16S rRNA, restriction pattern comparisons can be used to identify and track organisms responsible for disease outbreaks.

DNA Probe Assays

In clinical probe assays, colonies from plates or samples of infected tissue are treated with strong alkali, usually sodium hydroxide (NaOH), to lyse the cells and partially denature the pathogen DNA, forming single-stranded DNA molecules (Figure 31.26*a*). This mixture is then fixed to a matrix (filter or dipstick) or left in solution, and a labeled probe is added. Hybridization is then carried out by incubating at a temperature necessary to form a stable duplex between target DNA and probe DNA. The temperature used in each assay is governed by the length and nucleic acid composition of the probe and target DNA.

Following a wash to remove unhybridized probe DNA, the extent of hybridization is measured using the reporter molecule attached to the probe. This would require the measurement of radioactivity, enzyme activity, or fluorescence, depending on how the probe was labeled. Nucleic acid probes have been marketed for the identification of several major microbial pathogens and are used for the detection of *Neisseria gonorrhoeae* and *Chlamydia trachomatis* (Table 31.9; ⟳ Sections 33.12 and 33.13).

In addition to their use in clinical diagnostics, nucleic acid probes are widely used in food industries and food regulatory agencies. Probe detection systems can be used to monitor foods for contamination by pathogens such as *Salmonella* and *Staphylococcus*. In probe assays of food, an enrichment period is usually employed to allow low numbers of cells in the food to multiply to a detectable number. Probes designed for use in the food industry employ dipsticks precoated with pathogen-specific probe DNA to hybridize with pathogen DNA from the sample. Two-component probes that function as both a *reporter* probe and a *capture* probe are often used in these applications (Figure 31.26*b*). Following hybridization of the reporter to DNA from the target organism, the dipstick, which contains a sequence complementary to the capture probe (usually poly(dT) to capture poly(dA) on the probe), is inserted into the hybridization solution, where it binds the hybridized DNA. The detection system is then activated to visualize and quantify the hybridized DNA on the dipstick (Figure 31.26*a*).

Nucleic acid probes can be sensitive enough to detect less than 1 µg of nucleic acid per sample and can identify DNA extracted from about 10^6 bacterial cells or about 10^8 virus particles. Although molecular probes are not as sensitive as direct culture (where as few as 1–10 cells per sample can be detected), probe

methods are useful in situations where culture of organisms is difficult or even impossible. However, some applications of DNA technology rival the sensitivity of the culture method.

MiniQuiz

- What advantage does nucleic acid hybridization have over standard culture methods for identification of microorganisms? What disadvantages?
- How can information about a microorganism be obtained with a nucleic acid probe in the absence of standard growth-dependent assays?

31.13 Nucleic Acid Amplification

In Section 6.11 we discussed how the polymerase chain reaction (PCR) amplifies nucleic acids, forming multiple copies of target sequences. PCR techniques can use primers for a pathogen-specific gene to examine DNA derived from suspected infected tissue, even in the absence of an observable, culturable pathogen. As a result, PCR-based tests are widely used for identification of a number of individual pathogens and are particularly useful for identifying viral and intracellular infections, where culturing the responsible agents may be very difficult or even impossible.

PCR Testing and Analysis

PCR-based tests must include three basic components. First, DNA or RNA must be extracted from the sample to be tested. Second, the nucleic acid must be amplified using appropriate gene-specific nucleic acid primers. Short oligonucleotides (typically 15–31 nucleotides in length) are used as primers for PCR amplification of a specific gene or genes characteristic for a specific pathogen. Third, the amplified nucleic acid product (the amplicon) must be visualized, a procedure that can involve gel electrophoresis or other methods that visualize amplified DNA. A number of methods have been developed to increase amplicon-detection sensitivity as compared to gel electrophoresis. In one visualization method, DNA is dual-labeled during amplification. For the primer pair that targets a specific gene, one primer incorporates digoxigenin (DIG), a dTTP analog; the other primer is end-labeled with dinitrophenol (DNP). After amplification, the amplicon is mixed with blue latex beads coated with anti-DNP antibody; the amplicon will bind the blue beads. The beads are then exposed to a membrane coated with anti-DIG antibody; the blue beads containing the amplicon that has incorporated the DIG-labeled primer will localize to the antibody on the membrane, providing a blue band for a positive test. Several other visualization systems using biotin–streptavidin labeling or making use of enzyme-labeled monoclonal antibodies are also in use. These enhanced visualization systems are 10–100 times more sensitive than the gel electrophoresis visualization systems.

The presence of the appropriate amplified gene segment confirms the presence of the pathogen (**Figure 31.27**). Some pathogens for which either hybridization or PCR diagnostic methods are used for their identification are listed in **Table 31.9**.

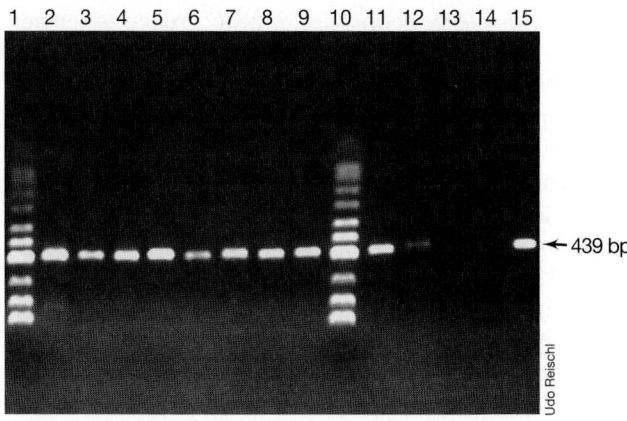

Figure 31.27 Polymerase chain reaction (PCR) analysis of patient sputum for *Mycobacterium tuberculosis* in the diagnosis of tuberculosis. Sputum samples from patients were used as a source of DNA. Amplification was initiated with a primer pair, which produced the indicated 439-base pair product when a pure culture of *M. tuberculosis* was used as the DNA source (lane 15). Lanes 2–9, 11, and 12 are from sputums positive for *M. tuberculosis* (lane 12 is a weak positive). Lanes 13 and 14 are from *M. tuberculosis*-negative sputum samples. Lanes 1 and 10 are molecular weight reference markers.

Reverse Transcriptase PCR and Real-Time PCR

The power of PCR has been extended by the development of related techniques such as *reverse transcriptase PCR (RT-PCR)* and *quantitative real-time PCR (qPCR)*. These techniques are routinely applied to the analysis of environmental samples in addition to their use for identification of pathogens in clinical samples.

RT-PCR uses pathogen-specific RNA to produce complementary DNA (cDNA) directly from patient samples, and can be used for detection of RNA retroviruses such as HIV and other RNA viruses. The first step in reverse transcriptase PCR (RT-PCR) is to use the enzyme reverse transcriptase (⮌ Section 21.11) to make a cDNA copy of an RNA sample. Next, PCR is used to amplify the cDNA. Expression of a particular gene from a pathogen may be monitored by isolating RNA and employing RT-PCR to make DNA copies of the corresponding gene(s). The amplified DNA can then be sequenced or probed for identification.

Many PCR tests employ qPCR; qPCR employs fluorescently labeled PCR products that yield an almost immediate result and avoids the need for postamplification nucleic acid purification and visualization. The accumulation of target DNA is monitored during the qPCR process. This is achieved by adding fluorescent probes to the PCR reaction mixture. Probe fluorescence increases upon binding to DNA. As the target DNA is amplified, the level of fluorescence increases proportionally. The fluorescent probes may be nonspecific or may be specific for the target DNA. For example, the dye SYBR Green binds *nonspecifically* to double-stranded DNA, but does not bind to single-stranded DNA or RNA; SYBR Green added to the PCR mixture becomes fluorescent only when bound, indicating that double-stranded DNA is present, in this case due to the amplification process

Table 31.9 *Pathogens identified with nucleic acid and PCR methods*

Pathogen	Diseases
Bacteria	
Campylobacter spp.	Food infections
Chlamydia trachomatis	Venereal syndromes; trachoma
Enterococcus spp.	Healthcare-associated infections
Escherichia coli (enteropathogenic strains)	Gastrointestinal disease
Haemophilus influenzae	Infectious meningitis
Legionella pneumophila	Pneumonia
Listeria monocytogenes	Listeriosis
Mycobacterium avium	Tuberculosis
Mycobacterium tuberculosis	Tuberculosis
Mycoplasma hominis	Urinary tract infection; pelvic inflammatory disease
Mycoplasma pneumoniae	Pneumonia
Neisseria gonorrhoeae	Gonorrhea
Neisseria meningitidis	Meningitis
Rickettsia spp.	Typhus, hemorrhagic fever, etc.
Salmonella spp.	Gastrointestinal disease
Shigella spp.	Gastrointestinal disease
Staphylococcus aureus	Purulent discharges (boils, blisters, pus-forming skin infections)
Streptococcus pyogenes	Scarlet fever; rheumatic fever; strep throat
Streptococcus pneumoniae	Pneumonia
Treponema pallidum	Syphilis
Fungi	
Blastomyces dermatitidis	Blastomycosis
Candida spp.	Candidiasis, thrush
Coccidioides immitis	Coccidioidomycosis
Histoplasma capsulatum	Histoplasmosis
Viruses	
Cytomegalovirus	Congenital viral infections
Epstein–Barr virus	Burkitt's lymphoma; mononucleosis
Hepatitis viruses A, B, C, D, E	Hepatitis
Herpes simplex virus (1 and 2)	Cold sores; genital herpes
Human immunodeficiency virus (HIV)	Acquired immunodeficiency syndrome (AIDS)
Human papillomavirus	Genital warts; cervical cancer
Influenza	Respiratory disease
Polyomavirus	Neurological disease
Rotavirus	Gastrointestinal disease
Protists	
Leishmania donovani	Leishmaniasis
Plasmodium spp.	Malaria
Pneumocystis jiroveci	Pneumonia
Trichomonas vaginalis	Trichomoniasis
Trypanosoma spp.	Trypanosomiasis

UNIT 10

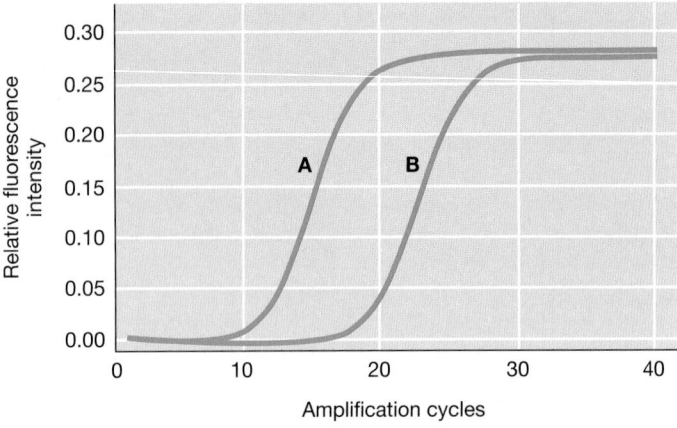

Figure 31.28 Quantitative real-time polymerase chain reaction (qPCR) of 16S RNA genes from *Desulfovibrio vulgaris*. DNA extracted from a laboratory culture was monitored for expression of 16S RNA (curve A) and *npt* (curve B), a kanamycin resistance marker, using gene-specific primers. SYBR Green, a fluorescent dye that fluoresces only when bound by double-stranded DNA, was mixed with the PCR mixture and used to visualize amplified DNA as it formed. The curve on the left (A) had 0.15 fluorescence units after 15 cycles, while the curve on the right (B) had 0.15 fluorescence units after 22 cycles, indicating that the16S RNA had a higher template abundance in this strain as compared to the abundance of the template for *npt*.

(Figure 31.28). *Gene-specific* fluorescent probes are made by attaching a fluorescent dye to a short DNA probe that matches the target sequence being amplified: the dye fluoresces only when double-stranded DNA of the correct sequence accumulates.

Because qPCR amplification can be monitored continuously, visualization by gel electrophoresis or other detection methods is not necessary to confirm amplification; detection of a gene diagnostic of a particular pathogen in a clinical sample may be performed in an hour or two instead of the usual overnight processing required by standard PCR. Moreover, by monitoring the *rate* of fluorescence increase in the PCR reaction, it is possible to accurately determine the *amount* of target DNA present in the original sample; qPCR can be used to assess the abundance of an organism in a sample by quantifying a gene characteristic for that particular organism.

MiniQuiz

- What advantage does nucleic acid amplification have over standard culture methods for identification of microorganisms? What disadvantages?

- How can information about a microorganism be obtained with a PCR assay in the absence of standard growth-dependent assays?

Big Ideas

31.1
Appropriate sampling and culture techniques are necessary to isolate and identify potential pathogens. The selection of sampling and culture conditions requires knowledge of the ecology, physiology, and nutrition of suspected pathogens.

31.2
Most pathogens exhibit unique metabolic patterns when grown on specialized selective and differential media. Growth-dependent patterns provide information necessary for accurate pathogen identification.

31.3
Antimicrobial drugs used for the treatment of infectious diseases must be tested for efficacy with clinical samples. Antimicrobial susceptibility is done to ensure appropriate therapy. Testing is based on the minimum inhibitory concentration of an agent necessary to completely inhibit growth of a pathogen.

31.4
Microbiology laboratory safety requires training, planning, and care to prevent contamination and possible infection of laboratory workers. Specific precautions and procedures must be in place to handle materials such as live cultures, inoculated culture media, used hypodermic needles, and patient specimens. Microorganisms are manipulated at four increasingly rigorous biosafety levels according to their degree of danger to laboratory workers.

31.5
An immune response is often a natural outcome of infection. Specific immune responses involving a rise in antibody titers and positive T cell-mediated skin tests can be used to provide evidence for past infections, current infections, and convalescence.

31.6
Polyclonal and monoclonal antibodies are used for research and diagnostic applications. Hybridoma technology reproducibly provides specific antibodies for a wide range of clinical, diagnostic, and research purposes.

31.7
Antigen–antibody binding is the basis for a number of serological tests. Specificity and sensitivity define the accuracy of individual assays. Neutralization and precipitation reactions produce visible results involving antigen–antibody interactions.

31.8
Direct agglutination tests are used for determination of blood types. Passive agglutination tests are available for identification of a variety of pathogens and pathogen-related products. Agglutination tests are rapid, relatively sensitive, highly specific, simple to perform, and inexpensive.

31.9

Fluorescent antibodies are used for quick, accurate identification of pathogens and other antigenic substances in tissue samples, blood, and other complex mixtures. Fluorescent antibody-based methods can be used for identification, enumeration, and sorting of a variety of prokaryotic and eukaryotic cell types.

31.10

EIA and RIA are the most sensitive immunological assays. Available for a variety of clinical and research applications, both techniques link a detection system, either an enzyme or a radioactive molecule, to an antibody or antigen, significantly enhancing sensitivity. EIA and RIA tests can be designed to detect either antibody (indirect tests) or antigen (direct tests).

31.11

Immunoblot, or Western blot, procedures detect antibodies to specific antigens or the antigens themselves. The antigens are separated by electrophoresis, transferred (blotted) to a membrane, and exposed to antibody. Immune complexes are visualized with enzyme-labeled or radioactive secondary antibodies. Immunoblots are extremely specific, but procedures are technically demanding, expensive, and time consuming.

31.12

Nucleic acid hybridization is used for identification of microorganisms. A nucleic acid sequence specific for the microorganism of interest must be available to design a probe. Various DNA-based methodologies are currently used in clinical, food, and research laboratories.

31.13

Gene amplification (PCR) methods are used for a wide variety of diagnostic tests and for analysis of environmental samples.

Review of Key Terms

Agglutination a reaction between antibody and particle-bound antigen resulting in visible clumping of the particles

Antibiogram a report indicating the susceptibility of clinically isolated microorganisms to the antibiotics in current use

Bacteremia the presence of bacteria in the blood

Differential media growth media that allow identification of microorganisms based on phenotypic properties

Enriched media media that allow metabolically fastidious microorganisms to grow because of the addition of specific growth factors

Enrichment culture the use of select culture media and incubation conditions to isolate microorganisms from natural samples

Enzyme immunoassay (EIA) a test that uses antibodies linked to enzymes to detect antigens or antibodies in body fluids

Fluorescent antibody an antibody molecule covalently modified with a fluorescent dye

that makes the antibody visible under fluorescent light

General-purpose media growth media that support the growth of most aerobic and facultatively aerobic organisms

Immunoblot (Western blot) the detection of specific proteins by separating them via electrophoresis, transferring them to a membrane, and adding specific antibodies

Monoclonal antibody a single type of antibody made by a single B cell hybridoma clone

Neutralization the interaction of antibody with antigen that reduces or blocks the biological activity of the antigen

Nucleic acid probe an oligonucleotide of unique sequence used as a hybridization probe for identifying specific genes

Polyclonal antibodies a variety of antibodies made by many different B cell clones

Precipitation a reaction between antibody and a soluble antigen resulting in a visible, insoluble complex

Radioimmunoassay (RIA) a test that employs radioactive antibody or antigen to detect antigen or antibody binding

Selective media media that enhance the growth of certain organisms while retarding the growth of others due to an added media component

Sensitivity the lowest amount of antigen that can be detected by an immunological assay

Sepsis a blood infection

Septicemia a blood infection

Serology the study of antigen–antibody reactions *in vitro*

Specificity the ability of an antibody or a lymphocyte to recognize a single antigen, or of a diagnostic test to identify a specific pathogen

Titer the quantity of antibody present in a solution

Review Questions

1. Describe the standard procedure for obtaining and culturing a throat culture and a blood sample. What special precautions must be taken while obtaining the blood culture (Section 31.1)?

2. What is the difference between bacteremia and septicemia (Section 31.1)?

3. Why is it important to process clinical specimens as rapidly as possible? What special procedures and precautions are necessary for the isolation and culture of anaerobes (Section 31.1)?

4. Differentiate between selective and differential media. Is eosin–methylene blue agar a selective medium or a differential medium? How and why is it used in a clinical laboratory (Section 31.2)?

5. Outline the principles of five important diagnostic tests for bacteria (Section 31.2).

6. Describe the differences between biosafety level 1, 2, 3, and 4 laboratories (Section 31.4).

7. Why does the antibody titer rise after infection? Is a high antibody titer indicative of an ongoing infection? Explain. Why is it necessary to obtain an acute and a convalescent blood sample to monitor infections (Section 31.5)?

8. What is the difference between monoclonal and polyclonal antibodies? Describe three uses of these antibodies (Section 31.6).

9. Describe a neutralization reaction with reference to microbial toxins and antisera (Section 31.7).

10. Agglutination tests are widely used for clinical diagnostic purposes. Why is this the case (Section 31.8)?

11. How are fluorescent antibodies used for the diagnosis of viral diseases? What advantages do fluorescent antibodies have over unlabeled antibodies (Section 31.9)?

12. Enzyme immunoassay (EIA) and radioimmunoassay (RIA) tests are extremely sensitive, as compared with agglutination. Why is this the case (Section 31.10)?

13. Why is the immunoblot (Western blot) procedure used to confirm screening tests that are positive for human immunodeficiency virus (HIV) (Section 31.11)?

14. What information is essential for the design of a pathogen-specific nucleotide probe? Where can one obtain such information? Is this information available for all pathogens (Section 31.12)?

15. Identify the advantages of using a DNA amplification test for the detection of a pathogen (Section 31.13).

Application Questions

1. A blood culture is positive for *Staphylococcus epidermidis*. Explain the finding. Is it likely that the patient has *S. epidermidis* bacteremia? Prepare a list of possibilities and questions for a discussion with the physician in charge. What additional information will be needed to confirm or rule out a bacteremia?

2. With respect to both short-term and long-term consequences, why is it a common medical practice to treat an infectious disease with antibiotics before isolating the suspected pathogen? After a pathogen has been isolated and identified, what further steps should be taken to confirm appropriate antibiotic susceptibility? Why are these measures rarely employed away from a hospital environment?

3. Explain the rationale for collecting serum specimens from patients during an acute infectious disease and about 2 weeks later (Section 31.5). What information would you expect to obtain from the serum of a recovering patient?

4. Compared with growth-dependent clinical diagnostic procedures, what are the advantages of rapid identification systems such as agglutination tests and immune-based detection systems such as EIA tests? What are the potential disadvantages of the rapid nonculture tests?

5. What are the major advantages of using DNA probes in diagnostic microbiology? What information is needed to design sequence-specific polymerase chain reaction (PCR) assay probes for a microorganism? Where can you find this information?

6. Define the procedures you would use to isolate and identify a new pathogen. Be sure to include growth-dependent assays, immunoassays, and molecular assays. Where would you report your findings? Which of your assays could be adapted to be used as a routine, high-throughput test for rapid clinical diagnosis?

Need more practice? Test your understanding with quantitative questions; access additional study tools including tutorials, animations, and videos; and then test your knowledge with chapter quizzes and practice tests at **www.microbiologyplace.com**.

Epidemiology

The human immunodeficiency virus (HIV), shown here in gold attached to human tissues, is the causative agent of HIV/AIDS. Epidemiological studies of AIDS cases that first appeared in large numbers in the 1980s quickly identified the major modes of transmission of the virus.

Many infectious diseases are adequately controlled in developed countries. In Chapter 1 we compared the current common causes of death in the United States with those at the beginning of the twentieth century (Figure 1.8). In most developed countries, infectious diseases now cause far fewer deaths than noninfectious diseases. Worldwide, however, infectious diseases remain serious public health problems, accounting for 23.5% of 58.7 million annual deaths. Even in developed countries, new infectious diseases such as H1N1 influenza ("swine flu") emerge unexpectedly, and previously controlled diseases such as tuberculosis reemerge. Others, such as HIV/AIDS and malaria, continue to be problems worldwide. Even in developed countries like the United States, deaths due to infectious diseases are increasing (**Figure 32.1**). Effective control of infectious diseases remains a worldwide challenge that requires scientific, medical, economic, sociological, political, and educational solutions. The occurrence and spread of infectious diseases are the focus of the epidemiologist.

I   Principles of Epidemiology

Epidemiology is the study of the occurrence, distribution, and determinants of health and disease in a population; it deals with **public health**, the health of the population as a whole. Infections may be transmitted to people by living or nonliving carriers. Infected persons may spread the pathogens to other members of the population, thus acting as living carriers themselves. Here we consider how pathogens spread through populations; we examine the principles of epidemiology and the application of these principles to the control of infectious diseases. In the next four chapters we will survey the diseases themselves.

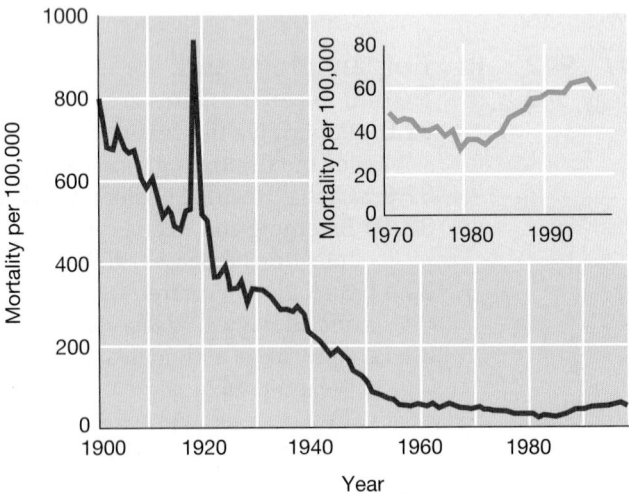

Figure 32.1 Deaths due to infectious disease in the United States. Although infectious disease death rates steadily declined throughout most of the twentieth century (except for the large numbers of deaths in 1918–1919 caused by the influenza pandemic), the death rate has increased significantly since 1980. Adapted from Hughes, J.M. 2001. Emerging Infectious Diseases: A CDC Perspective. *Emerg. Infect. Dis. 17:* 494–496.

32.1 The Science of Epidemiology

To cause disease, a pathogen must grow and reproduce in the host. For this reason, epidemiologists track the natural history of pathogens. In many cases an individual pathogen cannot grow outside the host; if the host dies, the pathogen also dies. Pathogens that kill the host before they move to a new host will become extinct. Most host-dependent pathogens therefore adapt to coexist with the host.

A well-adapted pathogen lives in balance with its host, taking what it needs for existence and causing only a minimum of harm. Such pathogens may cause **chronic infections** (long-term infections) in the host. When there is a balance between host and pathogen, both host and pathogen survive. On the other hand, a host whose resistance is compromised because of factors such as poor diet, age, and other stressors can be harmed (Section 27.12). In addition, new pathogens occasionally emerge to which the individual host, and sometimes the entire species, has not developed resistance. Such emerging pathogens often cause **acute infections**, characterized by rapid and dramatic onset. In these cases, pathogens can be selective forces in the evolution of the host, just as hosts, as they develop resistance, can be selective forces in the evolution of pathogens.

In some cases, the pathogen is not dependent on the host for survival. Such a pathogen can cause devastatingly acute disease, with no consequences for the pathogen. For example, organisms in the genus *Clostridium* occasionally infect humans, causing life-threatening diseases such as tetanus, botulism, gangrene, and certain gastrointestinal diseases. Host damage, and even death, causes no harm to populations of these pathogens because they are normal inhabitants of the soil and only accidentally and occasionally infect humans.

The epidemiologist traces the spread of a disease to identify its origin and mode of transmission. Epidemiological data are obtained by collecting disease information in a population. Data are gathered from disease-reporting surveillance networks, clinical records, and patient interviews, all with the goal of defining common factors for an illness. This is in contrast to individual patient treatment and diagnosis in the clinic or laboratory. Knowledge of both the population dynamics and clinical problems associated with a disease is needed to formulate effective public health measures for disease control.

MiniQuiz
- How does an epidemiologist differ from a microbiologist?
- Why do epidemiologists acquire population-based data about infectious diseases?

32.2 The Vocabulary of Epidemiology

Certain terms have specialized meanings in epidemiology. A disease is an **epidemic** when it simultaneously infects an unusually high number of individuals in a population; a **pandemic** is a widespread, usually worldwide, epidemic. By contrast, an **endemic disease** is one that is constantly present, usually at low incidence, in a population (**Figure 32.2**). A disease classified as endemic

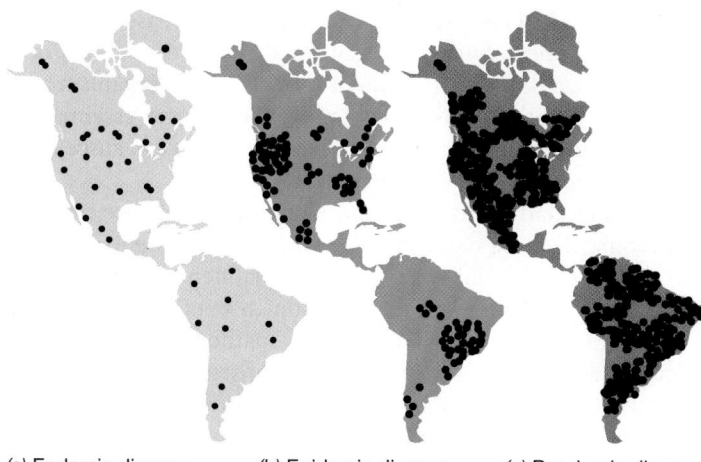

(a) Endemic disease   (b) Epidemic disease   (c) Pandemic disease

Figure 32.2 Endemic, epidemic, and pandemic disease. Each dot represents several cases of a particular disease. *(a)* Endemic diseases are present in the population in particular geographical areas. *(b)* Epidemic diseases show high incidence in a wider area, usually developing from an endemic focus. *(c)* Pandemic diseases are distributed worldwide. Diseases such as influenza are endemic in certain areas and develop into annual epidemics under appropriate circumstances, such as crowding. Epidemics may develop into pandemics.

implies that the pathogen may not be highly virulent or that the majority of host individuals in the population may be immune, resulting in the low disease incidence. Individuals infected with an endemic disease are **reservoirs** of infection, a source of infectious agents from which other individuals may be infected.

The **incidence** of a particular disease is the *number of new cases* in a population in a given time period. For example, in 2007 there were 37,503 new cases of acquired immunodeficiency syndrome (AIDS) in the United States, for an incidence of 12.53 new cases per 100,000 people per year. The **prevalence** of a given disease is the *total number of new and existing disease cases* in a population in a given time period. For example, within the United States there were about 492,000 persons living with AIDS at the end of 2006. Expressed another way, the prevalence of AIDS in this population was about 167 cases per 100,000 persons in 2006. Thus, *incidence* provides a record of new cases of a disease, whereas *prevalence* indicates the total disease burden in a population.

Sporadic cases of a disease occur one at a time in geographically separated areas, suggesting that the incidents are not related. A disease **outbreak**, on the other hand, is the appearance of numerous cases of the disease in a short period in an area previously experiencing only sporadic cases. Diseased individuals who show no symptoms or only mild symptoms have *subclinical infections.* Subclinically infected individuals are frequently **carriers** of the particular disease, with the pathogen reproducing within them and being shed into the environment.

Mortality and Morbidity

The incidence and prevalence of disease, as determined from statistical analyses of illness and death records, are indicators of the public health of a particular group such as the total global population or the population of a localized region, such as a city, state, or country. Public health conditions and concerns vary with location and time, and the assessment of public health at a given moment provides only a snapshot of a dynamic situation. Public health policies are designed to reduce incidence and prevalence of disease. Public health policies and laws can be evaluated by examining public health statistics over long time periods.

Mortality is the incidence of *death* in a population. Infectious diseases were the major causes of death in 1900 in all countries and geographic regions, but they are now less prevalent in developed countries. Noninfectious "lifestyle" diseases such as heart disease and cancer are now much more prevalent and cause higher mortality than do infectious diseases (ᗕᗒ Figure 1.8). However, this could change rapidly if public health measures were to break down. In developing countries, infectious diseases are still major causes of mortality (**Table 32.1** and Section 32.9).

Table 32.1 *Worldwide deaths due to infectious diseases, 2004[a]*

Disease	Deaths	Causative agent(s)
Respiratory infections[b]	4,259,000	Bacteria, viruses, fungi
Acquired immunodeficiency syndrome (AIDS)	2,040,000	Virus
Diarrheal diseases	2,163,000	Bacteria, viruses
Tuberculosis[c]	1,464,000	Bacterium
Malaria	889,000	Protist
Measles[c]	424,000	Virus
Meningitis, bacterial[c]	340,000	Bacterium
Pertussis (whooping cough)[c]	254,000	Bacterium
Tetanus[c]	163,000	Bacterium
Hepatitis (all types)[d]	159,000	Viruses
Syphilis	99,000	Bacterium
Trypanosomiasis (sleeping sickness)	52,000	Protist
Leishmaniasis	47,000	Protist
Schistosomiasis	41,000	Helminth
Dengue	18,000	Virus
Chagas' disease	11,000	Helminth
Japanese encephalitis	11,000	Virus
Chlamydia	9,000	Bacterium
Intestinal nematode infections	6,000	Helminth
Other communicable diseases	1,351,000	Various agents

[a]Globally, there were about 58.7 million deaths from all causes in 2004. About 13.8 million deaths, or 23.5%, were from communicable infectious diseases, nearly all in developing countries. Data show the 20 leading causes of death due to infectious diseases. The world population in 2004 was estimated at 6.4 billion. Data are from the World Health Organization (WHO), Geneva, Switzerland.
[b]For some acute respiratory agents such as influenza and *Streptococcus pneumoniae* there are effective vaccines; for others, such as colds, there are no vaccines.
[c]Diseases for which effective vaccines are available.
[d]Vaccines are available for hepatitis A virus and hepatitis B virus. There are no vaccines for other hepatitis agents.

Morbidity is the incidence of *disease* in populations and includes both fatal and nonfatal diseases. Morbidity statistics define the public health of a population more precisely than mortality statistics because many diseases have relatively low mortality. The major causes of illness are quite different from the major causes of death. For example, high-morbidity infectious diseases include acute respiratory diseases such as the common cold and acute digestive disorders. Both seldom directly cause death in developed countries. Thus, both of these diseases have high morbidity, but low mortality. On the other hand, Ebola virus infects only several hundred people worldwide every year, but the mortality in some outbreaks approaches 70%. Thus, Ebola has low morbidity, but high mortality.

Disease Progression

The progression of clinical symptoms for a typical acute infectious disease can be divided into stages:

1. *Infection:* The organism invades, colonizes, and grows in the host.

2. *Incubation period:* A period elapses between infection and the appearance of disease symptoms. Some diseases, like influenza, have very short incubation periods, measured in days; others, like AIDS, have longer ones, sometimes extending for years. The incubation period for a given disease is determined by inoculum size, virulence, the life cycle of the pathogen, resistance of the host, and distance of the site of entrance from the focus of infection. At the end of incubation, the first symptoms, such as headache and a feeling of illness, appear.

3. *Acute period:* The disease is at its height, with overt symptoms such as fever and chills.

4. *Decline period:* Disease symptoms subside. Any fever subsides, usually following a period of intense sweating, and a feeling of well-being develops. The decline period may be rapid (within 1 day), in which case it is said to occur by *crisis*, or it may be slower, extending over several days, in which case it is said to be by *lysis*.

5. *Convalescent period:* The patient regains strength and returns to normal.

During the later stages of the infection cycle, the immune mechanisms of the host become increasingly important, and in many cases complete recovery from the disease requires immunity.

MiniQuiz

- Distinguish between an endemic disease, an epidemic disease, and a pandemic disease.
- Distinguish between morbidity and mortality. Is host mortality advantageous for the pathogen?

32.3 Disease Reservoirs and Epidemics

Reservoirs are sites in which infectious agents remain viable and from which individuals may become infected. Reservoirs may be either animate or inanimate. Some pathogens whose reservoirs are inanimate are primarily saprophytic (living on dead matter)

and only incidentally infect humans and cause disease. For example, *Clostridium tetani*, the organism that causes tetanus, normally inhabits the soil. Infection of animals by this organism is an accidental event. That is, infection of a host is not essential for the bacterium's continued existence; in the absence of susceptible hosts, *C. tetani* still survives in nature.

For many pathogens, however, living organisms are the only reservoirs. In these cases, the reservoir host is essential for the life cycle of the infectious agent; maintenance of human pathogens of this kind requires person-to-person transmission. Many viral and bacterial respiratory pathogens and sexually transmitted pathogens require human hosts; the staphylococci and streptococci are examples of human-restricted pathogens, as are the agents that cause diphtheria, gonorrhea, and mumps. As we shall see, many pathogens that live their entire life cycle dependent on a single host species, especially humans, can be eradicated or at least controlled. **Table 32.2** lists some human infectious diseases with epidemic potential and their reservoirs.

Zoonosis

Some infectious diseases are caused by pathogens that reproduce in both humans and animals. A disease that primarily infects animals but is occasionally transmitted to humans is called a **zoonosis**. Animal-to-animal transmission of veterinary diseases can be high because public health measures are much less developed for animal populations than for humans. Occasionally, transmission of a zoonotic disease is from animal to human; person-to-person transfer of these pathogens is rare, but does occur. As we shall see later in this chapter, these occasional infections sometimes lead to virulent outbreaks of new infectious diseases. Factors leading to the emergence of zoonotic disease include the existence and propagation of the infectious agent in an animal host, the proper environment for propagation and transfer of the agent, and the presence of the new susceptible host species. When there is animal-to-human transmission, a new infectious disease may suddenly emerge in the exposed human population. For examples, see the Microbial Sidebars "Swine Flu—Pandemic (H1N1) 2009 Influenza" for a discussion of the recent influenza pandemic and "SARS as a Model of Epidemiological Success" for a discussion of severe acute respiratory syndrome (SARS) and other zoonotic epidemics. See also the discussion of hantaviruses in Section 34.2.

Control of a zoonotic disease in the human population does not usually eliminate the disease as a potential public health problem. Eradication of the human form of a zoonotic disease can generally be achieved only through elimination of the disease in the animal reservoir. This is because the essential maintenance of the pathogen depends on animal-to-animal transfer, and humans are incidental, nonessential hosts. For example, plague is primarily a disease of rodents. Effective control of plague is achieved by control of the infected rodent population and the insect (flea) that carries the pathogen to humans. These methods are more effective in preventing plague transmission than interventions such as vaccines in the incidental human host (⮎ Section 34.7). Zoonotic bovine tuberculosis is indistinguishable from human tuberculosis. Often spread from infected cattle to

Table 32.2 *Epidemic diseases: Agents, sources, reservoirs, and controls*

Disease	Causative agent[a]	Infection sources	Reservoirs	Control measures
Common-source epidemics[b]				
Anthrax	*Bacillus anthracis* (B)	Milk or meat from infected animals	Cattle, swine, goats, sheep, horses	Destruction of infected animals
Bacillary dysentery	*Shigella dysenteriae* (B)	Fecal contamination of food and water	Humans	Detection and control of carriers; oversight of food handlers; decontamination of water supplies
Botulism	*Clostridium botulinum* (B)	Soil-contaminated food	Soil	Proper preservation of food
Brucellosis	*Brucella melitensis* (B)	Milk or meat from infected animals	Cattle, swine, goats, sheep, horses	Pasteurization of milk; control of infection in animals
Cholera	*Vibrio cholerae* (B)	Fecal contamination of food and water	Humans	Decontamination of public water sources; immunization
E. coli O157:H7 food infection	*Escherichia coli* O157:H7 (B)	Fecal contamination of food and water	Humans, cattle	Decontamination of public water sources; oversight of food handlers; pasteurization of beverages
Giardiasis	*Giardia* spp. (P)	Fecal contamination of water	Wild mammals	Decontamination of public water sources
Hepatitis	Hepatitis A, B, C, D, E (V)	Infected humans	Humans	Decontamination of contaminated fluids and fomites; immunization if available (A and B)
Legionnaires' disease	*Legionella pneumophila* (B)	Contaminated water	High-moisture environments	Decontamination of air conditioning cooling towers, etc.
Paratyphoid	*Salmonella enterica* serovar Paratyphi (B)	Fecal contamination of food and water	Humans	Decontamination of public water sources; oversight of food handlers; immunization
Typhoid fever	*Salmonella enterica* serovar Typhi (B)	Fecal contamination of food and water	Humans	Decontamination of public water sources; oversight of food handlers; pasteurization of milk; immunization
Host-to-host epidemics				
Respiratory diseases				
Diphtheria	*Corynebacterium diphtheriae* (B)	Human cases and carriers; infected food and fomites	Humans	Immunization; quarantine of infected individuals
Hantavirus pulmonary syndrome	Hantavirus (V)	Inhalation of contaminated fecal material; contact	Rodents	Control of rodent population and exposure
Hemorrhagic fever	Ebola virus (V)	Infected body fluids	Unknown	Quarantine of active cases
Meningococcal meningitis	*Neisseria meningitidis* (B)	Human cases and carriers	Humans	Exposure treated with sulfadiazine for susceptible strains; immunization
Pneumococcal pneumonia	*Streptococcus pneumoniae* (B)	Human carriers	Humans	Antibiotic treatment; isolation of cases for period of communicability
Tuberculosis	*Mycobacterium tuberculosis* (B)	Sputum from human cases; infected milk	Humans, cattle	Treatment with antimycobacterial drugs; pasteurization of milk
Whooping cough	*Bordetella pertussis* (B)	Human cases	Humans	Immunization; case isolation
German measles	Rubella virus (V)	Human cases	Humans	Immunization; avoid contact between infected individuals and pregnant women
Influenza	Influenza virus (V)	Human cases	Humans, animals	Immunization
Measles	Measles virus (V)	Human cases	Humans	Immunization

Table 32.2 *Epidemic diseases: Agents, sources, reservoirs, and controls (continued)*

Disease	Causative agent[a]	Infection sources	Reservoirs	Control measures
Host-to host epidemics				
Sexually transmitted diseases[c]				
Acquired immunodeficiency syndrome (AIDS)	Human immunodeficiency virus (HIV)	Infected body fluids, especially blood and semen	Humans	Treatment with metabolic inhibitors (not curative)
Chlamydia	*Chlamydia trachomatis* (B)	Urethral, vaginal, and anal secretions	Humans	Testing for organism during routine pelvic examinations; chemotherapy of carriers and potential contacts; case tracing and treatment
Genital warts, cervical cancer	Human papilloma-virus (HPV)	Urethral and vaginal secretions	Humans	Immunization
Gonorrhea	*Neisseria gonorrhoeae* (B)	Urethral and vaginal secretions	Humans	Chemotherapy of carriers and potential contacts
Syphilis	*Treponema pallidum* (B)	Infected exudate or blood	Humans	Identification by serological tests; antibiotic treatment of seropositive individuals
Trichomoniasis	*Trichomonas vaginalis* (P)	Urethral, vaginal, and prostate secretions	Humans	Treatment of infected individuals and contacts with antimicrobial drugs
Vectorborne diseases				
Epidemic typhus	*Rickettsia prowazekii* (B)	Bite from infected louse	Humans, lice	Control louse population
Lyme disease	*Borrelia burgdorferi* (B)	Bite from infected tick	Rodents, deer, ticks	Avoid tick exposure; treat infected individuals with antibiotics
Malaria	*Plasmodium* spp. (P)	Bite from *Anopheles* mosquito	Humans, mosquito	Control mosquito population; treat and prevent human infections with antimalarial drugs
Plague	*Yersinia pestis* (B)	Bite from flea	Wild rodents	Control rodent populations; immunization
Rocky Mountain spotted fever	*Rickettsia rickettsii* (B)	Bite from infected tick	Ticks, rabbits, mice	Avoid tick exposure; treat infected individuals with antibiotics
Direct-contact zoonotic diseases				
Psittacosis	*Chlamydophila psittaci* (B)	Contact with birds or bird excrement	Wild and domestic birds	Avoid contact with birds; treat infected individuals with antibiotics
Rabies	Rabies virus (V)	Bite by carnivores, contact with infected neural tissue	Wild and domestic carnivores	Avoid animal bites; immunization of animal handlers and exposed individuals
Tularemia	*Francisella tularensis* (B)	Contact with rabbits	Rabbits	Avoid contact with rabbits; treat infected individuals with antibiotics

[a]B, Bacteria; V, virus; P, protist.
[b]Some common-source diseases can also be spread from host to host.
[c]Sexually transmitted diseases can also be controlled by effective use of condoms and by sexual abstinence.

humans, the disease was brought under control primarily by identifying and destroying infected animals. Pasteurization of milk was also of considerable importance because milk was the main mechanism of transmission of bovine tuberculosis to humans (↪ Section 33.4).

Certain infectious diseases are caused by organisms such as protists that have more complex life cycles including an obligate transfer from a nonhuman host to a human host, followed by an obligate transfer back to the nonhuman host (for example, malaria, ↪ Section 34.5). In such cases, the disease may potentially be controlled in either humans or the alternate animal host.

Carriers

A living *carrier* is a pathogen-infected individual who has a subclinical infection and shows no symptoms or only mild symptoms of clinical disease. Carriers are potential sources of infection for others. Carriers may be in the incubation period of the disease, in which case the carrier state precedes the development of actual symptoms. Respiratory infections such as colds and influenza, for example, are often spread via carriers who are unaware of their infection and so are not taking any precautions against infecting others. The carrier state lasts only a short time for carriers who develop acute disease. However, chronic carriers may spread disease for extended periods of time. Chronic carriers usually appear healthy. They may have recovered from a clinical disease but still harbor viable pathogens, or their infections may not be apparent.

Carriers can be identified using diagnostic techniques; culture or immunoassay surveys are conducted in populations to identify carriers. For example, skin testing with *Mycobacterium tuberculosis* antigens tests for delayed hypersensitivity. This reaction, easily detected in the skin test, reveals exposure and previous or current infection with *M. tuberculosis* and is widely used to identify previous infection and carriers of tuberculosis (⟳ Section 33.4). Other diseases in which carriers contribute to the spread of infection include hepatitis, typhoid fever, and AIDS. Culture or immunoassay surveys of food handlers and healthcare workers are sometimes used to identify individuals who are carriers and pose a risk as sources of infection.

A famous example of a chronic carrier was the woman known as Typhoid Mary, a cook in New York City in the early part of the twentieth century. Typhoid Mary (her real name was Mary Mallon) was employed as a cook during a typhoid fever epidemic in 1906. Investigations revealed that Mary was associated with a number of the typhoid outbreaks. She was the likely source of infection because her feces contained large numbers of the typhoid bacterium, *Salmonella enterica* serovar Typhi. She remained a carrier throughout her life, probably because her gallbladder was infected and continuously secreted organisms into her intestine. She refused to have her gallbladder removed and was imprisoned. Released on the pledge that she would not cook or handle food for others, Mary disappeared, changed her name, and continued to cook in restaurants and public institutions, leaving behind epidemic outbreaks of typhoid fever. After several years, she was again arrested and imprisoned and remained in custody until her death in 1938.

MiniQuiz

- What is a zoonotic disease?
- What is a disease reservoir?
- Distinguish between acute and chronic carriers. Provide an example of each.

32.4 Infectious Disease Transmission

Epidemiologists follow the transmission of a disease by correlating geographic, climatic, social, and demographic data with disease incidence. These correlations are used to identify possible modes of transmission. A disease limited to a restricted geographic location, for example, may suggest a particular carrier; malaria, a disease of tropical regions, is transmitted by mosquito species restricted to tropical regions. A marked seasonality or periodicity of a disease often indicates certain modes of transmission. Such is the case for influenza, whose incidence increases dramatically at the time of year when children are in school and come in close contact, increasing opportunities for person-to-person viral transmission.

Finally, pathogen survival depends on efficient host-to-host transmission. The mode of pathogen transmission is usually related to the preferred habitat of the pathogen in the body.

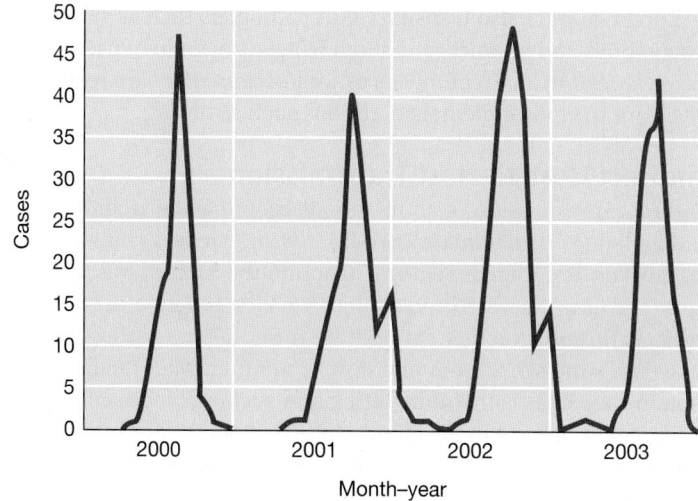

Figure 32.3 The incidence of California encephalitis in the United States by month and year. Note the sharp rise in late summer, followed by a complete decline in winter. The disease cycle follows the yearly cycle of the mosquito that is the vector for the infection. In 2003 there were 108 cases in 12 states. Data are from the Centers for Disease Control and Prevention, Atlanta, Georgia, USA.

Respiratory pathogens are typically airborne, for example, whereas intestinal pathogens are spread through fecal contamination of food or water. In some cases, environmental factors such as weather patterns may influence the survival of the pathogen. For example, California encephalitis, caused by single-stranded RNA bunyaviruses, occurs primarily during the summer and fall months and disappears every winter in a predictable cyclical pattern (**Figure 32.3**). The virus is transmitted from mosquito hosts that die during the winter months, causing the disease to disappear until the insect host reappears and again transmits the virus in the summer months. Virtually all mosquito-transmitted encephalitis viruses cause disease with the same seasonal pattern.

Pathogens can be classified by their mechanism of transmission, but all mechanisms have three stages in common: (1) escape from the host, (2) travel, and (3) entry into a new host. Pathogen transmission can be by direct or indirect mechanisms.

Direct Host-to-Host Transmission

Host-to-host transmission occurs when an infected host transmits a disease directly to a susceptible host without the assistance of an intermediate host or inanimate object. Upper respiratory infections such as the common cold and influenza are most often transmitted host to host by droplets resulting from sneezing or coughing. Many of these droplets, however, do not remain airborne for long. Transmission, therefore, requires close, although not necessarily intimate, person-to-person contact.

Some pathogens are extremely sensitive to environmental factors such as drying and heat and are unable to survive for significant periods of time away from the host. These pathogens, transmitted only by intimate person-to-person contact such as exchange of body fluids in sexual intercourse, include those responsible for sexually transmitted diseases including syphilis (*Treponema pallidum*) and gonorrhea (*Neisseria gonorrhoeae*).

UNIT 10

Direct contact also transmits skin pathogens such as staphylococci (boils and pimples) and fungi (ringworm). These pathogens often spread by indirect means as well because they are relatively resistant to environmental conditions such as drying.

Indirect Host-to-Host Transmission

Indirect transmission of an infectious agent can be facilitated by either living or inanimate carriers. Living carriers transmitting pathogens are called **vectors**. Commonly, arthropods (mites, ticks, or fleas) or vertebrates (dogs, cats, or rodents) act as vectors. Arthropod vectors may not be hosts for the pathogen, but may carry the agent from one host to another. Many arthropods obtain their nourishment by biting and sucking blood, and if the pathogen is present in the blood, the arthropod vector may ingest the pathogen and transmit it when biting another individual. In some cases viral pathogens replicate in the arthropod vector, which is then considered an alternate host. Such is the case for West Nile virus (⟳ Section 34.6). Such replication leads to an increase in pathogen numbers, increasing the probability that a subsequent bite will lead to infection.

Inanimate agents such as bedding, toys, books, and surgical instruments can also transmit disease. These inanimate objects are collectively called **fomites**. Food and water are potential disease **vehicles**. Fomites can also be disease vehicles, but major epidemics originating from a single-vehicle source are typically traced to common sources such as food or water because food and water are shared commodities consumed in large amounts by everyone.

Epidemics

Major epidemics are usually classified as either *common-source epidemics* or *host-to-host epidemics*. These two types of epidemics are contrasted in **Figure 32.4**. Table 32.2 summarizes the key epidemiological features of major epidemic diseases.

A **common-source epidemic** arises as the result of infection (or intoxication) of a large number of people from a contaminated common source such as food or water. Such epidemics are often caused by a breakdown in the sanitation of a central food or water distribution system. Foodborne and waterborne common-source epidemics are primarily intestinal diseases; the pathogen leaves the body in fecal material, contaminates food or water supplies due to improper sanitary procedures, and then enters the intestinal tract of the recipient during ingestion of the food or water. Waterborne and foodborne diseases are generally controlled by public health measures, which we discuss further in Chapters 35 and 36. A classic common-source epidemic is cholera. In 1855 the British physician John Snow showed that cholera spreads through drinking water. His studies correlating cholera incidence with water distribution systems in London demonstrated that cholera is spread by fecal contamination of a water supply. The infectious agent, the bacterium *Vibrio cholerae*, was transmitted through consumption of the contaminated common-source vehicle, water (⟳ Section 35.5).

The disease incidence for a common-source outbreak is characterized by a rapid rise to a peak incidence because a large number of individuals become ill within a relatively brief period of

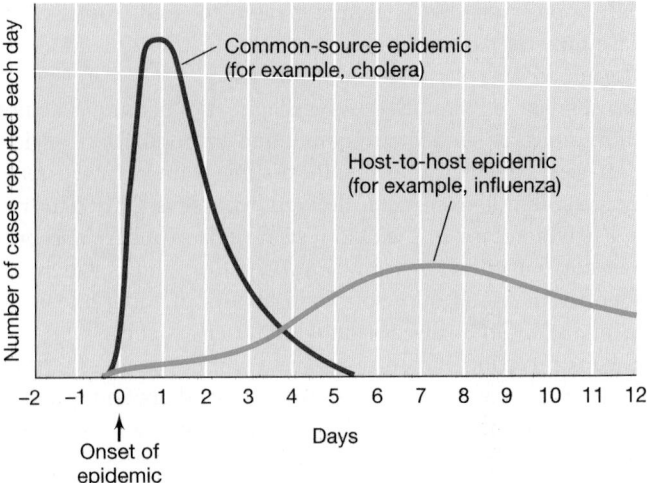

Figure 32.4 Types of epidemics. The shape of the curve that plots incidence of an epidemic disease against time identifies the likely type of the epidemic. If the vehicle of the disease is, for example, a common source such as contaminated food or water shared by the people who become infected, the curve rises sharply to a peak and then declines rapidly, but not as rapidly as the rise. Cases continue to be reported for a period approximately equal to the duration of one incubation period of the disease. If the vectors of the disease are infected hosts who transmit the disease to new hosts, the host-to-host propagation of the disease shows an incidence that rises relatively slowly as cases continue to be reported over a period equivalent to several incubation periods of the disease.

time (Figure 32.4). Assuming that the pathogen-contaminated common source is discovered and sanitized, the incidence of a common-source illness also declines rapidly, although the decline is less rapid than the rise. Cases continue to be reported for a period of time approximately equal to the duration of one incubation period of the disease.

In a **host-to-host epidemic**, the disease incidence shows a relatively slow, progressive rise (Figure 32.4) and a gradual decline. Cases continue to be reported over a period of time equivalent to several incubation periods of the disease. A host-to-host epidemic can be initiated by the introduction of a single infected individual into a susceptible population, with this individual infecting one or more people. The pathogen then replicates in susceptible individuals, reaches a communicable stage, and is transferred to other susceptible individuals, where it again replicates and becomes communicable. Influenza and chicken pox are examples of diseases that are typically spread in host-to-host epidemics. Chapter 33 discusses these and a number of other diseases propagated by host-to-host transmission.

MiniQuiz

- Distinguish between direct and indirect transmission of disease. Cite at least one example of each.
- For indirect disease transmission, distinguish among vectors, fomites, and vehicles.
- Distinguish between a common-source epidemic and a host-to-host epidemic. Cite at least one example of each.

32.5 The Host Community

The colonization of a susceptible host population by a pathogen may lead to explosive infections, transmission to uninfected hosts, and an epidemic. As the host population develops resistance, however, the spread of the pathogen is checked, and eventually a balance is reached in which host and pathogen populations are in equilibrium. In an extreme case, failure to reach equilibrium could result in death and eventual extinction of the host species. If the pathogen has no other host, then the extinction of the host also results in extinction of the pathogen. Thus, the evolutionary success of a pathogen may depend on its ability to establish a balanced equilibrium with the host population rather than its ability to destroy the host population. In most cases, the evolution of the host and the pathogen affect one another; that is, the host and pathogen *coevolve*.

Coevolution of a Host and a Pathogen

A classic example of host and pathogen coevolution began when a virus was intentionally introduced for purposes of controlling feral rabbits in Australia. Rabbits introduced into Australia from Europe in 1859 had spread until they were overrunning large parts of the continent, causing massive crop and vegetation damage.

Myxoma virus was introduced into Australia in 1950 to control the rabbit population. The virus is extremely virulent and usually causes a fatal infection. Mosquitoes and other biting insects spread it rapidly. Within several months, the virus infection had spread over a large area, rising to a peak incidence in the summer when the mosquito vectors were present, and then declining in the winter as mosquitoes disappeared. More than 95% of the infected rabbits died during the first year of the epidemic. However, when virus isolated from infected rabbits was characterized for virulence in newborn feral and laboratory rabbits, the viral isolates from the field were found to have reduced virulence; also the resistance of the feral rabbits was found to have increased dramatically. Within 6 years, rabbit mortality dropped to about 84% (**Figure 32.5**). In time, all the feral rabbits acquired the resistance factors. By the 1980s the rabbit population in Australia was nearing levels not seen since before the introduction of the myxoma virus, accompanied by widespread environmental destruction and pressure on native plants and animals.

In 1995, Australian authorities began controlled releases of another highly virulent rabbit pathogen, the rabbit hemorrhagic disease virus (RHDV), a single-stranded, positive-sense RNA virus (👉 Section 21.8). Because RHDV is spread by direct host-to-host contact and kills animals within days of initial infection, authorities believed the infections would kill all rabbits in a local population. Thus, they reasoned, the rabbit population would not be able to quickly develop resistance to RHDV, as it had to the arthropod-borne myxomatosis virus. Initial reports indicated that RHDV was very effective at reducing local rabbit populations. However, natural infection of some rabbits by an indigenous hemorrhagic fever virus conferred immune cross-resistance to the introduced RHDV. This unpredictable immune response limited the effectiveness of the control program in certain areas

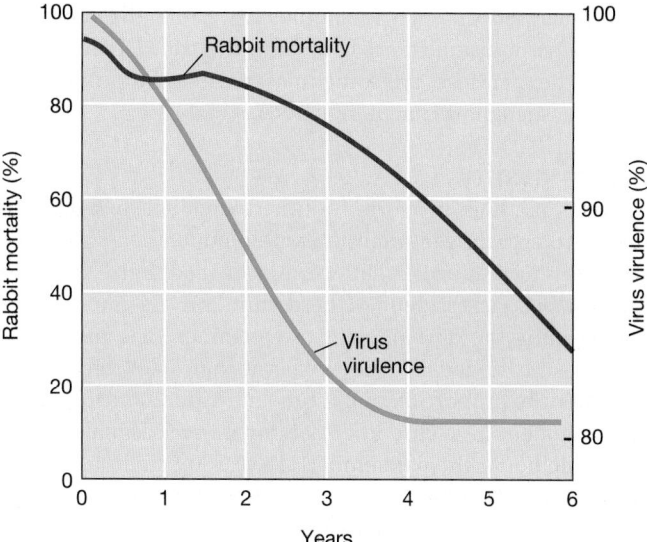

Figure 32.5 Myxoma virus, virulence, and Australian rabbit susceptibility. Data were collected after myxoma virus was introduced into Australia in 1950. Virus virulence is given as the average mortality in laboratory rabbits for virus recovered from the field each year. Rabbit mortality was determined by removing young feral rabbits from dens and infecting them with a viral strain that killed 90–95% of laboratory rabbits.

of Australia. Again, the host developed resistance to the control agent, moving the host–pathogen balance toward equilibrium.

Although coevolution of host and pathogen may be common in diseases that rely on host-to-host transmission, for pathogens that do not rely on host-to-host transmission, as we mentioned for *Clostridium*, there is no selection for decreased virulence to support mutual coexistence. Vectorborne pathogens usually transmitted by the bite of arthropods or ticks are also under no evolutionary pressure to spare the human host. As long as the vector can obtain its blood meal before the host dies, the pathogen can maintain a high level of virulence, decimating the human host in the process of infection. For example, the malaria parasites (*Plasmodium* spp.) show antigenic variations in their coat proteins that aid in avoiding the immune response of the host. This genetic ability to avoid the host responses increases pathogen virulence without regard to the susceptibility of the host. However, as we shall see for malaria, the host may develop disease-specific resistance under the constant evolutionary pressure exerted by a highly virulent pathogen (👉 Section 34.5).

Other evidence for the phenomenon of continually increasing pathogen virulence comes from studies of supervirulent diarrheal diseases in newborns. In hospital nurseries, *Escherichia coli* can cause severe diarrheal illness and even death, and virulence seems to increase with each passage of the pathogen through a hospital patient. The *E. coli* organisms replicate in one host and are then transferred to another patient through carriers such as healthcare providers or on fomites such as soiled bedding and furniture. Even if the host dies or does not transfer the disease by contact to others, the virulent *E. coli* strain infects other hosts through transmission by means other than direct person-to-person contact.

Extraordinary efforts such as completely washing the nursery and furniture with disinfectant, coupled with transferring nursery staff to other services, are sometimes necessary to interrupt the cycle of these highly virulent infections.

Herd Immunity

In general, if a high proportion of the individuals in a group are immune to a pathogen, then the whole population will be protected; this resistance to infection is called **herd immunity** (**Figure 32.6**). Assessment of herd immunity is important for understanding the development of epidemics. The more highly infectious a pathogen is (or the longer its period of infectivity is), the greater the proportion of immune individuals must be to prevent an epidemic. Such is the case for most seasonal influenza viruses, including the pandemic (H1N1) 2009 strain also known as swine flu. A lower proportion of immune individuals can prevent an epidemic by a less infectious agent or one with a brief period of infectivity. Mumps virus, which is less infectious than influenza virus, exhibits this pattern. In the absence of immunity, even poorly infectious agents can be transmitted person-to-person if susceptible hosts have repeated or constant contact with an infected individual. This appears to be the case for the transmission of H5N1 avian influenza among humans (⟳ Section 33.8).

The proportion of the population that must be immune to prevent infection in the rest of the population can be estimated from data derived from immunization programs. For example, studies of polio incidence in large populations in the United States indicate that if a population is 70% immunized, polio will be essentially absent from the population. The immunized individuals protect the rest of the population because they cannot pass on the pathogen; they break the cycle of infection because pathogens are unsuccessful in infecting them. For highly infectious diseases such as influenza and measles, up to 90–95% of the population must be immune to confer herd immunity.

A value of about 70% of the population immunized has also been estimated to confer herd immunity for diphtheria (Figure 32.6), but studies of several small diphtheria outbreaks indicate that in densely populated areas a much higher proportion of susceptible individuals must be immunized to prevent an epidemic. With diphtheria, an additional complication arises because immunized persons can harbor the pathogen and can thus be chronic carriers. This is because immunization protects against the effects of the diphtheria toxin, but not necessarily against infection by *Corynebacterium diphtheriae*, the bacterium that causes diphtheria (⟳ Section 33.3).

Cycles of Disease

Certain diseases occur in cycles. For example, influenza occurs in an annual cyclic pattern, causing epidemics propagated among school children and in other populations in which susceptible individuals are in close proximity. Influenza infectivity is high in crowded environments such as schools because the virus is transmitted by the respiratory route. Major epidemic strains of influenza virus change virtually every year, and as a result most children are highly susceptible to infection. On the introduction of virus into a school, an explosive, propagated epidemic results. Virtually every individual becomes infected and then becomes immune. As the

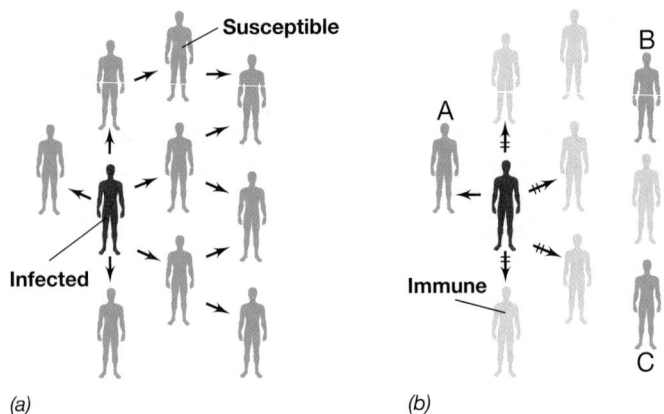

Figure 32.6 Herd immunity and transmission of infection. Immunity in some individuals protects individuals without immunity from infection. *(a)* In a population with no immunity, transfer of a pathogen from one infected individual can ultimately infect (arrows) all the individuals as newly infected individuals in turn transfer the pathogen to other individuals. *(b)* In a population that is only moderately dense and that has some immunity against a moderately transmissible pathogen such as *Corynebacterium diphtheriae* (causal agent of diphtheria), an infected individual cannot transfer the pathogen to all susceptible individuals because resistant individuals, immune from previous exposure or immunization, break the cycle of pathogen transmission: Susceptible individual A becomes infected, but susceptible individuals B and C are protected.

immune population increases, the epidemic subsides, but introduction of a new virus can propagate another epidemic.

MiniQuiz
- Explain coevolution of host and pathogen. Cite a specific example.
- How does herd immunity prevent a nonimmune individual from acquiring a disease? Give an example.

 Current Epidemics

Here we examine data collected by national and worldwide disease-surveillance programs that provide a picture of emerging disease patterns for acquired immunodeficiency syndrome (AIDS) and healthcare-associated infections. We discuss the swine flu influenza pandemic, officially known as pandemic (H1N1) 2009 influenza, in the Microbial Sidebar "Swine Flu— Pandemic (H1N1) 2009 Influenza."

32.6 The HIV/AIDS Pandemic

HIV/AIDS is a continuum of disease, starting with the infection of an individual with the human immunodeficiency virus (HIV), leading to the clinical disease, AIDS, a disease that attacks the immune system (⟳ Section 33.14).

HIV/AIDS Numbers

The first reported cases of AIDS were diagnosed in the United States in 1981. In the United States alone, more than 1 million cases have been reported, with over 500,000 deaths through

Swine Flu—Pandemic (H1N1) 2009 Influenza

A pandemic began in March 2009 with an influenza outbreak in Mexico. Influenza pandemics occur every 10 to 40 years. They result from antigenic variation in existing influenza A virus strains. Typical year-to-year antigenic variation, called antigenic drift, is caused by point mutations in the RNA of the influenza genome (Section 21.9). These mutations seldom cause pandemics, but do cause annual influenza outbreaks. Pandemic influenza strains arise from a much larger change in the viral genome termed antigenic shift. The influenza RNA genome consists of eight segments. Antigenic shift reassorts these segments. In the swine flu pandemic—officially, pandemic (H1N1) 2009—pigs, probably in Mexico, were simultaneously infected with swine influenza, bird influenza, and human influenza. During viral maturation, the viral RNA segments reproduced inside an influenza-infected cell are mixed together and packaged. A new, viable, infective virus must contain at least one copy of each RNA segment for the virus to infect another cell, replicate, and so on. To infect humans, the virus must also contain the proteins necessary for viral attachment to and invasion of human cells. Fortunately, even in such a favorable mixing pot as the pig, the packaging together of a new infective mixture such as this does not occur very often, and such reassorted viruses cause new pandemics only sporadically. But we can be sure that they will break out as new strains are mixed in susceptible animals and spread to susceptible human populations.

A virus that results from antigenic shift has the potential to contain antigens to which no human has had prior exposure; the only way we can obtain immunity to a new strain is to become infected and produce an immune response. This means that immunity to a new virus is nonexistent; the only way we can obtain immunity to a new strain is to become infected and produce an immune response. The bad news is that for the virus of pandemic (H1N1) 2009, almost no one less than 50 years of age has any immunity because they have never been exposed to similar viral strains. As a result, many of the deaths in this pandemic are of people younger than 50 who were healthy until they were infected by the virus. The good news is that pandemic (H1N1) 2009 is related to the 1957 pandemic called Asian flu, and, farther back, to the 1918 worldwide influenza pandemic that killed over 2,000,000 people worldwide. So peoples who are 50 or older probably have been infected with a strain related to pandemic (H1N1) 2009 and have immune cells and antibodies (immune memory, Section 29.8) that can respond to control its current pandemic virus.

Within six months of its emergence, pandemic (H1N1) 2009 had spread to almost every country in the world, causing significant mortality in most of those. The pattern of spread is similar to that of seasonal influenza, but has one significant difference. The virus began its spread from an initial focus of infection in Mexico and the southwestern United States in March, at the very end of the traditional winter flu season. Instead of dying out, as most seasonal flu outbreaks do at the end of winter, pandemic (H1N1) 2009 continued to spread through the summer months in the United States, especially in susceptible populations such as children at summer camps. This virus, like all influenza viruses, spreads very easily from person to person; it infects nearly all children and young adults exposed to it because they are all susceptible (Figure 1). The 2009–2010 flu season in the United States saw a predictable pattern of a highly communicable disease in a highly susceptible population, but the peak incidence occurred much sooner than normal. The highest incidence of influenza-like illness occurred in October and November and tapered off during the usual peak flu season in January through March (Figure 1). This early onset is undoubtedly because of the emergence of pandemic (H1N1) 2009 influenza at the very end of the last flu season. When children started back to school in the autumn, the concentration of susceptible individuals allowed rapid spread and explosive increases in case numbers. In the Southern Hemisphere, where the flu season is from roughly April to September, the pandemic strain spread with all of the characteristics of seasonal flu. Numbers of infected persons peaked and then tapered off as the season progressed.

Why has this strain spread so rapidly? The main reasons are probably the high infectivity of influenza and the mildness of the disease in most persons. The "barely sick" can spread the infection as they interact with others at work and school. As we will see later (Section 33.8), each of us is infected with an influenza virus about once every two years. This is the hallmark of a highly infectious pathogen that spreads very easily in a susceptible population.

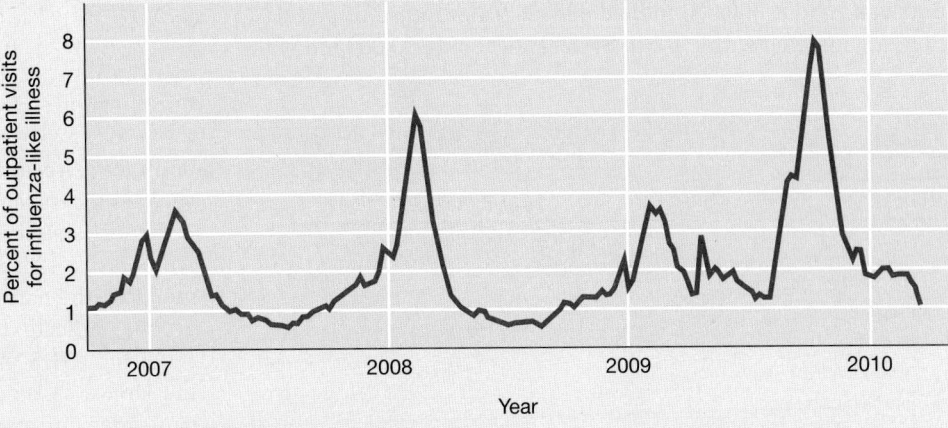

Figure 1 **Influenza-like illness, 2006–2010.** Pandemic (H1N1) 2009 influenza virus caused the high incidence of disease from the middle of 2009 through 2010. In the 2009–2010 flu season the peak incidence of disease was higher than in the three pervious seasons and occurred 3–4 months earlier than usual. Data are adapted from the Centers for Disease Control and Prevention, Atlanta, Georgia.

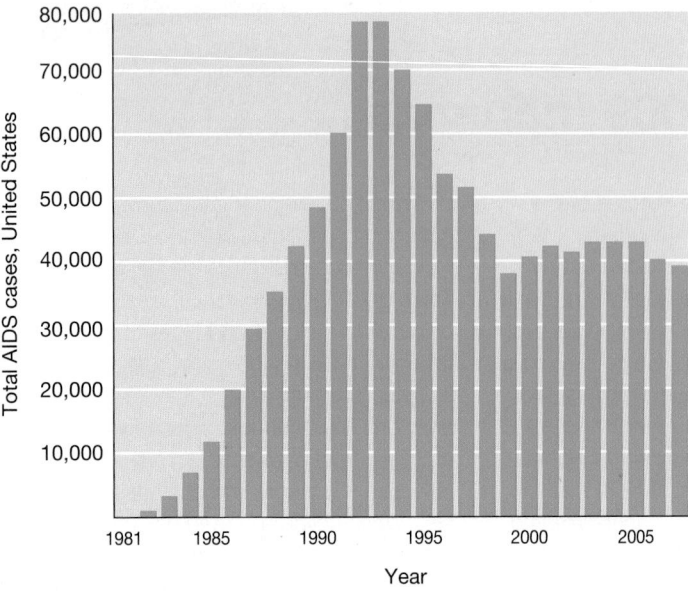

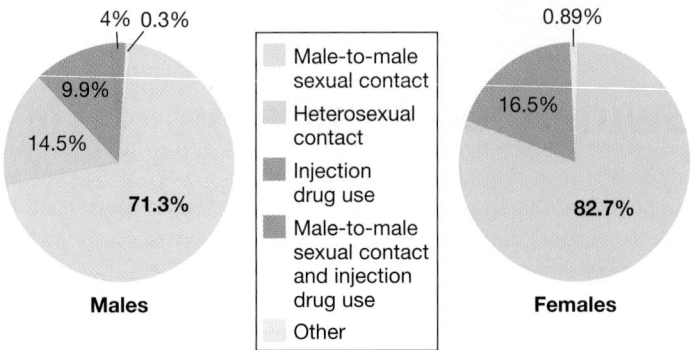

Figure 32.8 Distribution of AIDS cases by risk group and sex in adolescents and adults in the United States, 2007. Data were collected from 31,518 males and 10,977 females diagnosed with HIV/AIDS in 2007. Data are from the Centers for Disease Control and Prevention, Atlanta, Georgia.

Figure 32.7 Annual newly diagnosed cases of acquired immunodeficiency syndrome (AIDS) since 1981 in the United States. Cumulatively, there have been 998,255 cases of AIDS through 2005. Data are from the HIV/AIDS Surveillance Report, Centers for Disease Control and Prevention, Division of HIV/AIDS Prevention-Surveillance and Epidemiology, Atlanta, Georgia.

2007. A total of 56,300 new HIV infections and 37,503 new AIDS cases were diagnosed in 2007. Over 35,000 new AIDS cases have been diagnosed and reported every year since 1989 (**Figure 32.7**).

Worldwide, from 1981 through 2007, an estimated 80 million people have been infected with HIV. Probably more than 45 million people have already died from AIDS, and about 33 million are currently infected with HIV (**Table 32.3**). Globally, another 2.7 million individuals are infected each year. The Americas have up to 2.8 million HIV-infected individuals, and North America may have as many as 1.3 million. An estimated 455,636 persons in the United States are living with AIDS. Europe has about 2 million people living with HIV. Sub-Saharan Africa has 20.3 million infected people. In the sub-Saharan African country of Botswana about 300,000 individuals (23.9% of the adult population) are infected with HIV. In Swaziland, about 190,000 individ-

uals, or 26.1% of the adult population, are infected with HIV. Worldwide, AIDS caused about 2 million deaths in 2007, with 1.5 million of those deaths in sub-Saharan Africa.

The Epidemiology of HIV/AIDS

Case studies in the United States in the 1980s initially suggested a high AIDS prevalence among homosexual men and intravenous drug abusers. This indicated a transmissible agent, presumably transferred during sexual activity or by blood-contaminated needles. Individuals receiving blood or blood products were also at high risk: Hemophiliacs who required infusions of blood products, usually pooled from multiple donors, acquired AIDS, as did a small number of individuals who received blood transfusions or tissue transplants before 1982 (when blood-screening procedures were implemented). Today, almost none of the new AIDS cases can be attributed to HIV-contaminated blood products.

Soon after the discovery of HIV, laboratory immunosorbent assays and immunoblot tests (↩ Sections 31.10 and 31.11) were developed to detect antibodies to the virus in serum. Extensive surveys of HIV incidence and prevalence defined the spread of HIV and ensured that new cases would not be transmitted by blood transfusions. The pattern illustrated in **Figure 32.8** is typical of an agent transmissible by blood or other body fluids. The identification of defined high-risk groups implied that HIV was not transmitted from person to person by casual contact, such as the respiratory route, or by contaminated food or water. Instead, body fluids, primarily blood and semen, were identified as the vehicles for transmission of HIV.

Figure 32.8 shows that in the United States the number of AIDS cases is disproportionately high in men who have sex with men, but the patterns in women and in certain racial and ethnic groups indicate that homosexuality is not a prerequisite for acquiring AIDS. Among women, for example, heterosexuals are the largest risk group, whereas in African American and Hispanic men, intravenous drug use is linked to HIV infection nearly as often as homosexual activity. In fact, if we consider all risk groups, heterosexual activity is the fastest growing risk factor for HIV transmission among adults.

Table 32.3 *HIV/AIDS infections worldwide, 2007*[a]

Location	HIV/AIDS infections
The Americas	2.8 million
North America	1.3 million
Europe	2 million
Africa	21.7 million
Sub-Saharan Africa	20.3 million
East Asia and Pacific	1 million
South and Southeast Asia	6.4 million
Australia and New Zealand	15,000

[a]The total number of individuals infected with HIV/AIDS is estimated to be about 33 million. Data are from the World Health Organization.

The study of individuals who are at high risk for acquiring AIDS indicates that virtually all who acquire HIV today share two specific behavior patterns. First, they engage in activities (sex or drug use) in which body fluids, usually semen or blood, are transferred. Second, they exchange body fluids with multiple partners, either through sexual activity or through needle-sharing drug activity (or both). With each encounter they have a probability of receiving body fluids from an HIV-infected individual and therefore a chance of acquiring HIV infection.

The incidence of AIDS in hemophiliacs and blood transfusion recipients has been virtually eliminated. Rigorous screening of the blood supply has greatly reduced the medical transfer of HIV-contaminated blood. Blood clotting factors needed by hemophiliacs are also now risk free, either because they are genetically engineered products or because they have received a heat treatment sufficient to inactivate HIV.

In 2007, there were 87 new cases of pediatric AIDS in the United States. HIV can be transmitted to the fetus by infected mothers and probably also in mothers' milk. Infants born to HIV-infected mothers have maternally derived antibodies to HIV in their blood. However, a positive diagnosis of HIV infection in infants must wait a year or more after birth because about 70% of infants showing maternal HIV antibodies at birth are later found not to be infected with HIV.

Heterosexual transmission of HIV is the norm in Africa. In some regions, fewer men than women are infected with HIV. The identification of high-risk groups such as prostitutes has led to the development of health education campaigns. These campaigns inform the public of HIV transmission methods and define high-risk behaviors. Because no cure or effective immunization for AIDS is available, public health education remains the most effective approach to the control of HIV/AIDS. We discuss the pathology and therapy of HIV/AIDS in Section 33.14.

MiniQuiz

- Describe the major risk factors for acquiring HIV infection. Tailor your answer to your country of origin.
- Estimate the total number of individuals in the United States who now have AIDS and predict how many will be living with AIDS in the next 2 years.

32.7 Healthcare-Associated Infections

A **healthcare-associated infection (HAI)** is a local or systemic infection acquired by a patient in a healthcare facility, particularly during a stay in the facility. HAIs cause significant morbidity and mortality. About 5% of patients admitted to healthcare facilities acquire HAIs, also called *nosocomial infections* (*nosocomium* is the Latin word for "hospital"). About 1.7 million nosocomial infections occur annually in the United States, leading directly or indirectly to almost 100,000 deaths.

Some HAIs are acquired from patients with communicable diseases, but others are caused by pathogens that are selected and maintained within the hospital environment. Cross-infection from patient to patient or from healthcare personnel to patient presents a constant hazard. Healthcare-associated pathogens are often found as normal flora in either patients or healthcare staff.

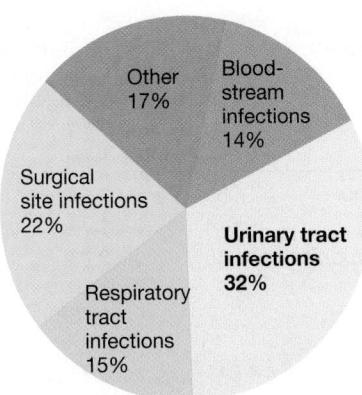

Figure 32.9 **Healthcare-associated infections.** About 1.7 million healthcare-associated infections occur annually in the United States. Data are from Klevens et al., Estimated Health Care-Associated Infections and Deaths in U.S. Hospitals, 2002. *Public Health Reports 122:* 160–166, 2007.

The Hospital Environment

Infectious diseases are spread easily and rapidly in hospitals for several reasons. (1) Many patients have low resistance to infectious disease; because of their illness they are compromised hosts (Section 27.12). For example, intensive care units provide care for the most acute and severe illnesses and account for 24.5% of total HAIs. (2) Healthcare facilities treat infectious disease patients; these patients may be pathogen reservoirs. (3) Multiple patients in rooms and wards increase the chance of cross-infection. (4) Healthcare personnel move from patient to patient, increasing the probability of transfer of pathogens. (5) Healthcare procedures such as hypodermic injection, spinal puncture, and removal of tissue samples (biopsy) or fluids (blood) breach the skin barrier and may introduce pathogens into the patient. (6) In maternity wards of hospitals, newborn infants are unusually susceptible to certain infections because they lack well-developed defense mechanisms. (7) Surgical procedures expose internal organs to sources of contamination; the stress of surgery may lower the resistance of the patient to infection. (8) Certain therapeutic drugs, such as steroids used for controlling inflammation, increase the susceptibility to infection. (9) Use of antibiotics to control infections selects for antibiotic-resistant organisms (Section 26.12).

Infection Sites

The most common sites of HAIs are shown in **Figure 32.9**. Of the 99,000 estimated deaths caused by HAIs in 2002, 36,000 were from pneumonia, 31,000 from bloodstream infections, 13,000 from urinary tract infections, 8,000 from surgical site infections, and 11,000 for all other sites. This distribution of infections and the numbers of deaths attributable to HAIs are representative for annual totals.

Healthcare-Associated Pathogens

Healthcare-associated pathogens preferentially infect several sites in the body, notably the urinary tract, bloodstream, and respiratory tract. A relatively small number of pathogens cause the

UNIT 10

Table 32.4 *Number of intensive care unit nosocomial infections in the United States, by site and organism*

Pathogen	Bloodstream	Respiratory tract	Urinary tract
Enterobacter spp.	1,083	4,444	1,560
Escherichia coli	514	1,725	5,393
Klebsiella pneumoniae	735	2,865	1,891
Haemophilus influenzae		1,738	
Pseudomonas aeruginosa	841	6,752	3,365
Staphylococcus aureus	2,758	7,205	497
Staphylococcus spp.	8,181		838
Enterococcus spp.	2,967	682	4,226
Candida albicans	1,090	1,862	4,856
Other pathogens	3,774	12,537	8,075
Total number[a]	21,943	39,810	30,701
Total %	23.7	43.1	32.2

[a]The total number of nosocomial infections in intensive care units during a recent 8-year period was 92,454. Data are from a National Nosocomial Infections Surveillance System Report, Centers for Disease Control and Prevention, Atlanta, Georgia.

majority of HAIs at these sites, but a number of other infectious agents can cause HAIs (**Table 32.4**).

One of the most important and widespread hospital pathogens is *Staphylococcus aureus*. It is the most common cause of pneumonia and the third most common cause of blood infections. *S. aureus* is also particularly problematic in nurseries. Many hospital strains of *S. aureus* are unusually virulent and are resistant to common antibiotics, making treatment very difficult. *S. aureus* and the other staphylococci constitute the largest cause of healthcare-associated blood infections and are also very prevalent in wound infections. Staphylococci are commonly found in the upper respiratory tract or on the skin, where they are a part of the normal flora in many individuals, including hospital patients and personnel.

Escherichia coli is the most common cause of urinary tract infections in hospitals, but *Enterococcus* spp., *Pseudomonas aeruginosa*, *Candida albicans*, and *Klebsiella pneumoniae* infections are also very common. *Enterococcus*, *E. coli*, and *K. pneumoniae* are normally found only in the human body. But *Candida* and *Pseudomonas* are opportunistic pathogens; they are commonly found in the environment, but cause disease only in individuals with compromised defenses. Isolates of *P. aeruginosa* from HAIs are often resistant to many different antibiotics, complicating treatment. *E. coli*, *Staphylococcus*, and *Enterococcus* also have potential for multiple drug resistance.

MiniQuiz

- Why are patients in healthcare facilities more susceptible than normal individuals to pathogens?
- What are the sources of HAI pathogens?

Epidemiology and Public Health

Here we identify some of the methods used to identify, track, contain, and eradicate infectious diseases within populations. We also identify some important current and future threats from infectious diseases.

32.8 Public Health Measures for the Control of Disease

Public health refers to the health of the general population and to the activities of public health authorities in the control of disease. The incidence and prevalence of many infectious diseases dropped dramatically during the past century, especially in developed countries, because of universal improvements in public health from advances in basic living conditions. Access to safe water and food, improved public sewage treatment, less crowded living conditions, and lighter workloads have contributed immeasurably to disease control, primarily by reducing exposure to infectious agents. Several historically important diseases, including smallpox, typhoid fever, diphtheria, brucellosis, and poliomyelitis, have been controlled and in some cases virtually eliminated by active, disease-specific public health measures such as quarantine and vaccination.

Controls Directed against Common Vehicles

The transmission of pathogens in food or water can be eliminated by preventing contamination of *common vehicles* of infection such as food or water. Water purification methods have dramatically reduced the incidence of typhoid fever. Laws controlling food purity and preparation have greatly decreased the probability of transmission of foodborne pathogens to humans. For example, the destruction of infected cattle and pasteurization of milk have virtually eliminated the spread of bovine tuberculosis in humans.

Transmission of respiratory pathogens, which are carried in the air, is difficult to prevent. Attempts at chemical disinfection of air have been unsuccessful. Air filtration is a viable method but is limited to small, enclosed areas. In Japan, many people wear face masks when they have upper respiratory infections to prevent transmission to others, but such methods, although effective, are voluntary and are difficult to institute as public health measures.

Controls Directed against the Reservoir

When the disease reservoir is primarily domestic animals, the infection of humans can be prevented if the disease is eliminated from the infected animal population. Immunization or destruction of infected animals may eliminate the disease in animals and, consequently, in humans. These procedures have nearly eliminated brucellosis and bovine tuberculosis in humans and have controlled bovine spongiform encephalitis (mad cow disease) in cattle in the United Kingdom, Canada, and the United States. In the process, the health of the domestic animal population is also improved.

When the disease reservoir is a wild animal, eradication is much more difficult. Rabies, for example, is a disease of both wild and domestic animals that is transmitted to domestic animals primarily from wild animals. Thus, control of rabies in domestic animals and in humans can be achieved by immunization of domestic animals. However, because the majority of rabies cases in the United States are in wild animals (↩ Section 34.1), eradication of rabies would require the immunization or destruction of all wild animal reservoirs, including such diverse species as raccoons, bats, skunks, and foxes. Although oral rabies immunization is practical and recommended for rabies control in restricted wild animal populations, its efficacy is untested in large, diverse animal populations such as the wild animal reservoir in the United States.

If insects such as the mosquito vectors that transmit malaria and West Nile fever are the disease reservoir, effective control of the disease can be accomplished by eliminating the reservoir with insecticides or other agents. The use of toxic or carcinogenic chemicals, however, must be balanced with environmental concerns. In some cases the elimination of one public health problem only creates another. For example, the insecticide dichlorodiphenyltrichloroethane (DDT) is very effective against mosquitoes and is credited with eradicating yellow fever and malaria in North America. DDT use, however, is currently banned in the United States because of environmental concerns. DDT is still used in many developing countries to control mosquito-borne diseases, but its use is declining worldwide.

When humans are the disease reservoir (as, for example, in HIV/AIDS), control and eradication can be difficult, especially if there are asymptomatic carriers. On the other hand, certain diseases that are limited to humans have no asymptomatic phase. If these can be prevented through immunization or treatment with antimicrobial drugs, the disease can be eradicated if those who have contracted the disease and all possible contacts are strictly quarantined, immunized, and treated. Such a strategy was successfully employed by the World Health Organization to eradicate smallpox and is currently being used to eradicate polio, as we discuss later.

Immunization

Smallpox, diphtheria, tetanus, pertussis (whooping cough), measles, mumps, rubella, and poliomyelitis have been controlled primarily by immunization. Diphtheria, for example, is no longer considered an endemic disease in the United States. Vaccines are available for a number of other infectious diseases (↩ Table 28.4). As we discussed in Section 32.5, 100% immunization is not necessary for disease control in a population, although the percentage needed to ensure disease control varies with the infectivity and virulence of the pathogen and with the living conditions of the population (for example, crowding).

Measles epidemics offer an example of the effects of herd immunity. The occasional resurgence of the highly contagious measles virus emphasizes the importance of maintaining appropriate immunization levels for a given pathogen. Until 1963, the year an effective measles vaccine was licensed, nearly every child in the United States acquired measles through natural infections, resulting in over 400,000 annual cases. After introduction of the vaccine, the number of annual measles infections decreased rapidly (↩ Figure 33.15). Case numbers reached a low of 1497 by 1983. However, by 1990, the percentage of children immunized against measles fell to 70%, and the number of new cases rose to 27,786. Within 3 years, a concerted effort to increase measles immunization levels to above 90% virtually eliminated indigenous measles transmission in the United States, and a total of only 312 measles cases were reported in 1993. Currently, about 100 cases of measles are reported each year in the United States, most due to infections imported by visitors from other countries.

In the United States, most children are now adequately immunized, but up to 80% of adults lack effective immunity to important infectious diseases because immunity from childhood vaccinations declines with time. When childhood diseases infect adults, they can have devastating effects. For example, if a woman contracts rubella (a vaccine-preventable viral disease) during pregnancy, the fetus may develop serious developmental and neurological disorders. Measles, mumps, and chicken pox are also more serious diseases in adults than in children.

All adults are advised to review their immunization status and check their medical records to ascertain dates of immunizations. This is particularly true for individuals who are traveling abroad. Tetanus immunizations, for example, must be renewed at least every 10 years to provide effective immunity. Surveys of adult populations have shown that more than 10% of adults under the age of 40 and over 50% of those over 60 are not adequately immunized. General recommendations for immunization were discussed in Section 28.7 and those for specific infections will be discussed in Chapters 33 through 36.

Quarantine and Isolation

Quarantine restricts the movement of a person with active infection to prevent spread of the pathogen to other people. The length of quarantine for a given disease is the longest period of communicability for that disease. To be effective, quarantine measures must prevent the infected individual from contacting unexposed individuals. Quarantine is not as severe a measure as strict isolation, which is used in hospitals for unusually infectious and dangerous diseases.

By international agreement, six diseases require quarantine: smallpox, cholera, plague, yellow fever, typhoid fever, and relapsing fever. Each is a very serious, particularly communicable disease. Spread of certain other highly contagious diseases such as Ebola hemorrhagic fever and meningitis may also be controlled by quarantine or isolation as outbreaks occur.

Surveillance

Surveillance is the observation, recognition, and reporting of diseases as they occur. **Table 32.5** lists the diseases currently under surveillance in the United States. Several of the epidemic diseases listed in Table 32.2 and Table 32.8 are not on the surveillance list. However, many of these diseases and other common diseases such as seasonal influenza are surveyed through regional laboratories that identify index cases—those cases of disease that exhibit unusually high incidence, new syndromes or characteristics, or are linked to new or evolving pathogens that have high potential for causing new epidemics.

Table 32.5 *Reportable infectious agents and diseases in the United States, 2010*

Diseases caused by *Bacteria*	Diseases caused by viruses
Anthrax	Arbovirus (neuroinvasive encephalitis and non-neuroinvasive disease)
Botulism	
Brucellosis	California serogroup
Chancroid	Eastern equine
Chlamydia trachomatis infection	Powassan
Cholera	St. Louis
Diphtheria	West Nile
Ehrlichiosis/Anaplasmosis	Western equine
Gonorrhea	Dengue
Haemophilus influenzae, invasive disease	Hantavirus pulmonary syndrome
Hansen's disease (leprosy)	Hepatitis A, B, C
Hemolytic uremic syndrome	HIV infection/AIDS
Legionellosis	Influenza
Listeriosis	Novel influenza A infection
Lyme disease	Pediatric mortality
Meningococcal disease	Measles
Pertussis	Mumps
Plague	Polio
Psittacosis	Rabies
Q fever	Rubella
Salmonellosis	Severe acute respiratory syndrome (SARS-CoV)
Shiga toxin–producing *Escherichia coli* (STEC)	Smallpox
Shigellosis	Varicella
Spotted fever rickettsiosis (Rocky Mountain spotted fever)	Viral hemorrhagic fevers
	Arena virus
Streptococcal toxic shock syndrome	Crimean–Congo hemorrhagic fever virus
Streptococcus pneumoniae, invasive disease	Ebola virus
Syphilis	Lassa virus
Tetanus	Marburg virus
Toxic shock syndrome (other than streptococcal)	Yellow fever
	Diseases caused by protists
Tuberculosis	Cryptosporidiosis
Tularemia	Cyclosporiosis
Typhoid fever	Malaria
Vancomycin-intermediate *Staphylococcus aureus* (VISA)	Giardiasis
	Disease caused by a helminth
Vancomycin-resistant *Staphylococcus aureus* (VRSA)	Trichinosis (tricinellosis)
Vibriosis	

The **Centers for Disease Control and Prevention (CDC)** in the United States, through the National Center for Infectious Diseases (NCID), operates a number of surveillance programs, as shown in **Table 32.6**. Many diseases are reportable to more than one surveillance program. Although redundant reporting may at first seem unnecessary, a disease may fall into several categories that affect healthcare plans and policies. For example, reporting of vancomycin-resistant staphylococci to the National Nosocomial

Infections Surveillance System (NNIS) and to CDC as a notifiable disease (Table 32.5) provides a national database. Using this information, a hospital infection team can formulate and implement plans for isolation, diagnosis, and drug-susceptibility testing of staphylococcal infections to identify antibiotic-resistant strains, stop their spread, and begin appropriate treatment.

Pathogen Eradication

A concerted disease eradication program was responsible for the eradication of naturally occurring smallpox. Smallpox was a disease with a reservoir consisting solely of the individuals with acute smallpox infections, and transmission was exclusively person to person. Infected individuals transmitted the disease through direct contact with previously unexposed individuals. Although smallpox, a viral disease, cannot be treated once acquired, immunization practices were very effective; vaccination with the related vaccinia virus conferred complete immunity. The World Health Organization (WHO) implemented a smallpox eradication plan in 1967. Because of the success of vaccination programs worldwide, endemic smallpox had already been confined to Africa, the Middle East, and the Indian subcontinent. WHO workers then vaccinated everyone in remaining endemic areas. Each subsequent outbreak or suspected outbreak was targeted by WHO teams that traveled to the outbreak site, quarantined individuals with active disease, and vaccinated all contacts. To break the chain of possible infection, they then immunized everyone who had contact with the contacts. This aggressive policy eliminated the active natural disease within a decade, and in 1980, WHO proclaimed the eradication of smallpox.

Polio, another viral disease that is largely preventable with an effective vaccine, is also targeted for eradication (endemic polio has been eradicated from the Western Hemisphere). Using much the same strategy to target polio as was used for smallpox, WHO undertook a widespread immunization program in 1988, concentrating efforts in remaining endemic areas. In all, over 2 billion individuals, mostly children, have been immunized, preventing an estimated 5 million cases of paralytic polio. By 2009, known endemic polio was restricted to Nigeria, India, Pakistan, and Afghanistan. Of 1500 polio cases reported in 2006, 1393 were in these countries. The remaining cases were spread among 11 countries in Africa, the Middle East, the Indian subcontinent, and Indonesia. Individual outbreaks are treated by immunization of all susceptible persons in the region of the outbreak.

Hansen's disease (leprosy), another disease restricted to humans, is also targeted for eradication. Active cases of Hansen's disease can now be effectively treated with a multidrug therapy that cures the patient and also prevents spread of *Mycobacterium leprae*, the causal agent (♻ Section 33.4).

Other communicable diseases are candidates for eradication. These include Chagas' disease (by treating active cases and destroying the insect vector of the *Trypanosoma cruzi* parasite in the American tropics) and dracunculiasis (by treating drinking water in Africa, Saudi Arabia, Pakistan, and other places in Asia to prevent transmission of *Dracunculus medinensis*, the Guinea helminth parasite). Eradication of syphilis may be possible because the disease is found only in humans and is treatable.

Table 32.6 *National Center for Infectious Diseases (NCID) surveillance systems for infectious disease notification and tracking in the United States*

Surveillance system (acronym)	Disease surveillance responsibility
121 Cities Mortality Reporting System	Influenza, pneumonia, all deaths
Active Bacterial Core Surveillance	Invasive bacterial diseases
BaCon Study	Bacterial contamination associated with blood transfusion
Border Infectious Disease Surveillance Project (BIDS)	Infectious disease along the U.S.–Mexican border
Dialysis Survey Network (DSN)	Vascular access infections and bacterial resistance in hemodialysis patients
Electronic Foodborne Outbreak Investigation and Reporting System (EFORS)	Foodborne outbreaks
EMERGEncy ID NET	Emerging infectious diseases
Foodborne Diseases Active Surveillance Network (FoodNet)	Foodborne disease
Global Emerging Infections Sentinel Network (GeoSentinel)	Global emerging diseases
Gonococcal Isolate Surveillance Project (GISP)	Antimicrobial resistance in *Neisseria gonorrhoeae*
Health Alert Network (HAN)	Health threat notification network, especially for bioterrorism
Integrated Disease Surveillance and Response (IDSR)	World Health Organization (WHO/AFRO) initiative for infectious diseases in Africa
Intensive Care Antimicrobial Resistance Epidemiology (ICARE)	Antimicrobial resistance and antimicrobial use in healthcare settings
International Network for the Study and Prevention of Emerging Antimicrobial Resistance (INSPEAR)	Global emergence of drug-resistant organisms
Laboratory Response Network (LRN)	Bioterrorism, chemical terrorism, and public health emergencies
Measles Laboratory Network	Measles in the Americas and the Caribbean
National Antimicrobial Resistance Monitoring System: Enteric Bacteria (NARMS)	Antimicrobial resistance in human nontyphoid *Salmonella*, *Escherichia coli* O157:H7, and *Campylobacter* isolates from agricultural and food sources
National Malaria Surveillance	Malaria in the United States
National Molecular Subtyping Network for Foodborne Disease Surveillance (PulseNet)	Molecular fingerprinting of foodborne bacteria
National Nosocomial Infections Surveillance System (NNIS)	Healthcare-associated infections
National Notifiable Diseases Surveillance System (NNDSS)	Reportable infectious diseases (see Table 32.5)

Table 32.6 *(continued)*

Surveillance system (acronym)	Disease surveillance responsibility
National Respiratory and Enteric Virus Surveillance System (NREVSS)	Respiratory syncytial virus (RSV), human parainfluenza viruses, respiratory and enteric adenoviruses, and rotavirus
National Surveillance System for Health Care Workers (NaSH)	Healthcare worker occupational infections
National Tuberculosis Genotyping and Surveillance Network	Tuberculosis genotyping repository
National West Nile Virus Surveillance System	West Nile virus
Public Health Laboratory Information System (PHLIS)	Notifiable diseases
Select Agent Program (SAP)	Potential bioterrorism agents
Surveillance for Emerging Antimicrobial Resistance Connected to Healthcare (SEARCH)	Emerging antimicrobial resistance in healthcare settings
Unexplained Deaths and Critical Illnesses Surveillance System	Emerging infectious diseases worldwide
United States Influenza Sentinel Physicians Surveillance Network	260 clinical sites that report incidence and prevalence of influenza infections
Viral Hepatitis Surveillance Program (VHSP)	Viral hepatitis
Waterborne-Disease Outbreak Surveillance System	Waterborne diseases

Diphtheria, caused by *Corynebacterium diphtheriae*, is no longer endemic in North America. The disease could be globally eradicated by application of the strict immunization protocols that have virtually eliminated it from North America. Rabies might be eradicated with oral baits that provide immunization of the wild carnivores that constitute the reservoir.

MiniQuiz
- Compare public measures for controlling infectious disease caused by insect reservoirs and by human carriers.
- Identify public health methods used to halt the spread of an epidemic disease.
- Outline the steps taken to eradicate smallpox and polio.

32.9 Global Health Considerations

The World Health Organization (WHO) has divided the world into six geographic regions for the purpose of collecting and reporting health information such as causes of morbidity and mortality. These geographic regions are Africa, the Americas (North America, the Caribbean, Central America, and South America), the eastern Mediterranean, Europe, Southeast Asia, and the Western Pacific. Here we compare mortality data from a

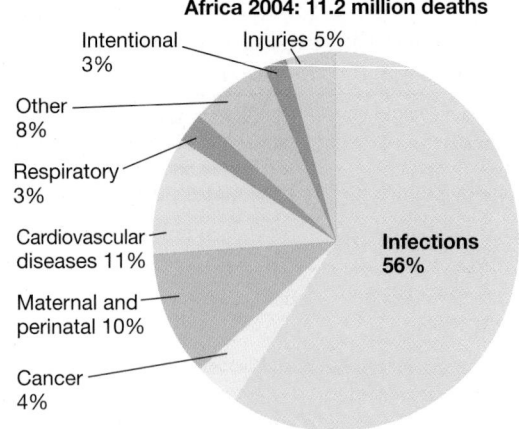

Africa 2004: 11.2 million deaths

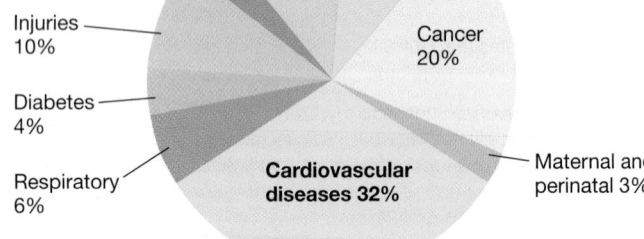

The Americas 2004: 6.2 million deaths

Figure 32.10 Causes of death in Africa and the Americas, 2004. There were 11.2 million deaths in Africa, with a population of 738 million. Infectious diseases caused 6.3 million deaths. There were 6.2 million deaths in the Americas, with a population of 874 million. Infectious diseases caused 611,000 deaths. Intentional deaths include murder, suicide, and war.

relatively developed region, the Americas, to that from a developing region, Africa.

Infectious Disease in the Americas and Africa: A Comparison

Here we compare health statistics for the Americas and Africa in 2004 when the worldwide population was about 6.4 billion. Worldwide, about 58.7 million individuals died, giving a mortality rate of 9.1 deaths per 1000 inhabitants per year. About 13.8 million, or 23.5%, of these deaths were attributable to infectious diseases. There were about 874 million people in the Americas, where there were about 6.2 million deaths, or about 7.1 deaths per 1000 inhabitants per year. In Africa, there were about 738 million people and about 11.2 million annual deaths, or about 15.2 deaths per 1000 inhabitants per year. These statistics illustrate the differences in overall mortality in high-income and low-income areas of the world, but examination of the causes of mortality in these regions is even more instructive.

Figure 32.10 indicates that most deaths in Africa were due to infectious diseases, whereas in the Americas, cancer and cardiovascular diseases were the leading causes of mortality. In Africa,

there were about 6.3 million deaths due to infectious diseases, but only 611,000 in the Americas. The African death toll due to infectious diseases was over 10% of the total deaths in the world. In developed countries, the dramatic reduction in death rates from infection over the last century (↺ Figure 1.8) is undoubtedly due to the advances in public health we have already discussed. Lack of resources in developing countries limits access to adequate sanitation, safe food and water, immunizations, healthcare, and medicines.

Travel to Endemic Areas

The high incidence of disease in many parts of the world is a concern for people traveling to such areas. However, travelers can be immunized against many of the diseases that are endemic in foreign countries. Recommendations for immunization for those traveling abroad are shown in **Table 32.7**. By international agreement, immunization certificates for yellow fever are required for travel to or from areas with endemic yellow fever. These areas include much of equatorial South America and Africa. Most other nonstandard immunizations are recommended only for people who are expected to be at high risk. In many parts of the world, travelers may be exposed to diseases for which there are no effective immunizations (for example, AIDS, Ebola hemorrhagic fever, dengue fever, amebiasis, encephalitis, malaria, and typhus). Travelers should take precautions such as avoiding unprotected sex, avoiding insect and animal bites, drinking only water that has been properly treated to kill all microorganisms, eating properly stored and prepared food, and undergoing antibiotic and chemotherapeutic programs for prophylaxis or for suspected exposure. For specific information regarding travel from the United States to any international destination, consult *CDC Information for Travelers* at http://wwwnc.cdc.gov/travel/.

Table 32.7 *Immunizations required or recommended for international travel*

Disease	Destination	Recommendation[a]
Yellow fever	Tropical and subtropical countries, especially in sub-Saharan Africa and South America	*Immunization required* for entry and exit from endemic regions
Rabies	Rural, mountainous, and upland areas	*Immunization recommended* if direct contact with wild carnivores is anticipated
Typhoid fever	Many African, Asian, and Central and South American countries	*Immunization recommended* in areas endemic for typhoid fever

[a]Vaccinations are generally recommended for diphtheria, pertussis, hepatitis A, hepatitis B, tetanus, polio, measles, mumps, rubella, and influenza as appropriate for the age of the traveler as well as the destination. Many U.S. citizens are immunized against these diseases through normal immunization practices. Requirements for specific vaccinations for each country are found at the website. Recommendations are also made for other appropriate infectious disease prevention measures, such as prophylactic drug therapy for malaria and plague prevention when visiting endemic areas. Yellow fever immunizations are required for travel to or from endemic areas.

Source: National Center for Infectious Diseases Travelers' Health, U.S. Department of Health and Human Services, http://wwwnc.cdc.gov/travel/

- Contrast mortality due to infectious diseases in Africa and the Americas.
- List a series of infectious diseases for which you have not been immunized and with which you could come into contact next year.

32.10 Emerging and Reemerging Infectious Diseases

Infectious diseases are global, dynamic health problems. Here we examine some recent patterns of infectious disease, some reasons for the changing patterns, and the methods used by epidemiologists to identify and deal with new threats to public health.

Emerging and Reemerging Diseases

The worldwide distribution of diseases can change dramatically and rapidly. Alterations in the pathogen, the environment, or the host population contribute to the spread of new diseases, with potential for high morbidity and mortality. Diseases that suddenly become prevalent are **emerging diseases**. Emerging diseases are not limited to "new" diseases, but also include **reemerging diseases** that were previously under control; reemerging diseases are especially a problem when antibiotics become less effective and public health systems fail. Recent dramatic examples of global emerging and reemerging disease are shown in **Figure 32.11**. Diseases with potential for emergence or reemergence are described in **Table 32.8**. The epidemic diseases listed in Table 32.2 also have the potential to emerge or reemerge as widespread epidemics and pandemics.

Emerging epidemic diseases are not a new phenomenon. Among the diseases that rapidly and sometimes catastrophically emerged in the past are syphilis (caused by *Treponema pallidum*) and plague (caused by *Yersinia pestis*). In the Middle Ages, up to one-third of all humans were killed by the plague epidemics that swept Europe, Asia, and Africa. Influenza caused a devastating worldwide pandemic in 1918–1919, claiming up to 100 million lives. In the 1980s, legionellosis (caused by *Legionella pneumophila*), acquired immunodeficiency syndrome (AIDS), and Lyme disease emerged as major new diseases. Important emerging pathogens in the last decade include West Nile virus and pandemic (H1N1) 2009 influenza, the strain that emerged in spring 2009 in Mexico and rapidly spread throughout the world (see the Microbial Sidebar "Swine Flu—Pandemic (H1N1) 2009 Influenza"). Health officials worldwide are also concerned about the potential for rapid emergence of another pandemic influenza developing from avian influenza (↩ Section 33.8).

Emergence Factors

Factors responsible for the emergence of new pathogens may be related to (1) human demographics and behavior; (2) technology and industry; (3) economic development and land use; (4) international travel and commerce; (5) microbial adaptation and change; (6) breakdown of public health standards; and (7) catastrophic events that upset the usual host–pathogen balance.

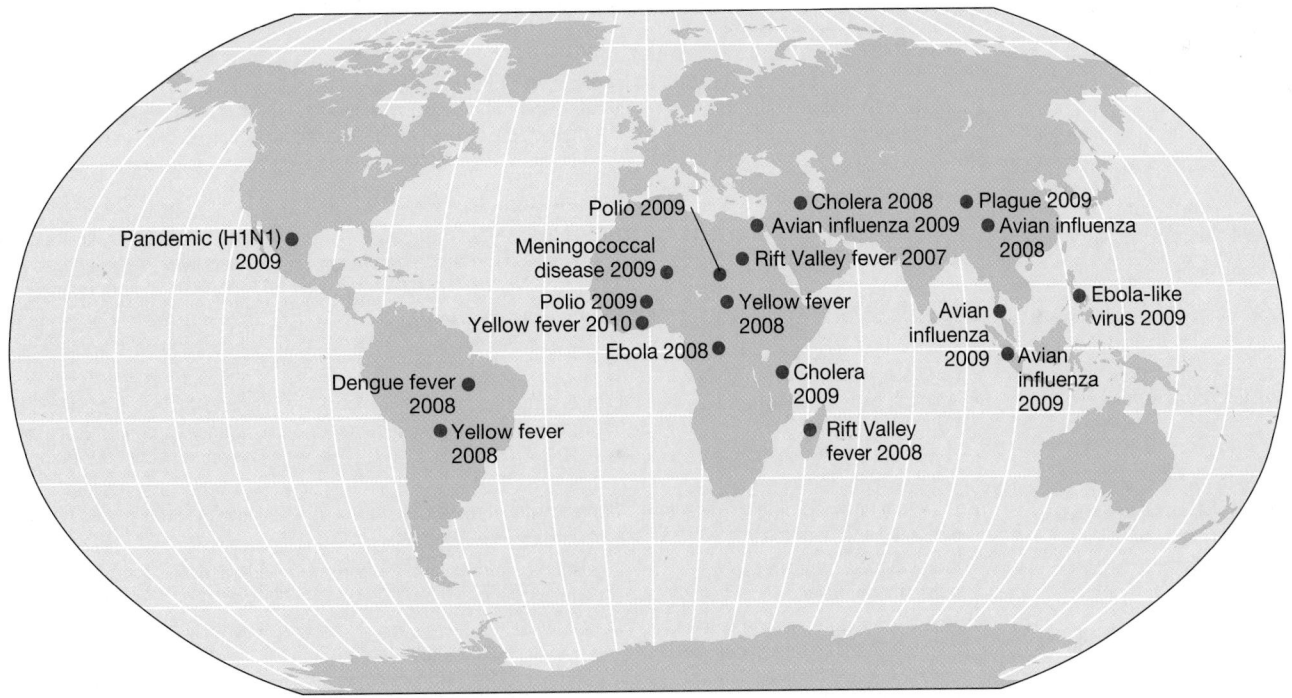

Figure 32.11 Recent outbreaks of emerging and reemerging infectious diseases. The diseases shown are local epidemics capable of producing widespread epidemics and pandemics. Not shown are established pandemic diseases such as HIV/AIDS and predictable annual epidemic diseases such as seasonal epidemic human influenza. Pandemic (H1N1) 2009 influenza, which presumably originated in Mexico, spread worldwide within 6 months of its recognition in March 2009.

Table 32.8 *Emerging and reemerging epidemic infectious diseases*

Agent	Disease and symptoms	Mode of transmission	Cause of emergence
Bacteria, rickettsias, and chlamydias			
Bacillus anthracis	Anthrax: respiratory distress, hemorrhage	Inhalation or contact with endospores	Bioterrorism
Borrelia burgdorferi	Lyme disease: rash, fever, neurological and cardiac abnormalities, arthritis	Bite of infective *Ixodes* tick	Increase in deer and human populations in wooded areas
Campylobacter jejuni	Campylobacter enteritis: abdominal pain, diarrhea, fever	Ingestion of contaminated food, water, or milk; fecal–oral spread from infected person or animal	Increased recognition; consumption of undercooked poultry
Chlamydia trachomatis	Trachoma, genital infections, conjunctivitis, infant pneumonia	Sexual intercourse	Increased sexual activity; poor sanitation
Escherichia coli O157:H7	Hemorrhagic colitis, thrombocytopenia, hemolytic uremic syndrome	Ingestion of contaminated food, especially undercooked beef and raw milk	Development of a new pathogen
Haemophilus influenzae biogroup *aegyptus*	Brazilian purpuric fever: purulent conjunctivitis, fever, vomiting	Discharges of infected persons; flies are suspected vectors	Possible increase in virulence due to mutation
Helicobacter pylori	Gastritis, peptic ulcers, possibly stomach cancer	Contaminated food or water, especially unpasteurized milk; contact with infected pets	Increased recognition
Legionella pneumophila	Legionnaires' disease: malaise, myalgia, fever, headache, respiratory illness	Air-cooling systems, water supplies	Recognition in an epidemic situation
Mycobacterium tuberculosis	Tuberculosis: cough, weight loss, lung lesions; infection can spread to other organ systems	Sputum droplets (exhaled through a cough or sneeze) from a person with active disease	Immunosuppression, immunodeficiency; antimicrobial drug resistance
Neisseria meningitidis	Bacterial meningitis	Person-to-person contact	Urbanization, breakdown or lack of local public health surveillance
Staphylococcus aureus	Abscesses, pneumonia, endocarditis, toxic shock	Contact with the organism in a purulent lesion or on the hands	Recognition in an epidemic situation; possibly mutation; antimicrobial drug resistance
Streptococcus pyogenes	Scarlet fever, rheumatic fever, toxic shock	Direct contact with infected persons or carriers; ingestion of contaminated foods	Change in virulence of the bacteria; possibly mutation
Vibrio cholerae	Cholera: severe diarrhea, rapid dehydration	Water contaminated with the feces of infected persons; food exposed to contaminated water	Poor sanitation and hygiene; possibly introduced via bilge water from cargo ships
Viruses			
Chikungunya virus (CHIKV)	Debilitating fever, nausea, muscle pain, chronic fatigue	Bite of an infected mosquito (*Aedes* spp. in Africa and Asia)	Poor mosquito control; outdoor exposure; rapid spread to nonimmune populations
Dengue	Hemorrhagic fever	Bite of an infected mosquito (primarily *Aedes aegypti*)	Poor mosquito control; increased urbanization in tropics; increased air travel
Filoviruses (Marburg, Ebola)	Fulminant, high mortality, hemorrhagic fever	Direct contact with infected blood, organs, secretions, and semen	Unknown; in Europe and the United States, virus-infected monkeys shipped from developing countries via air
Hendravirus	Respiratory and neurological disease in horses and humans	Contact with infected bats, horses	Human intrusion into natural environment
Hantaviruses	Abdominal pain, vomiting, hemorrhagic fever	Inhalation of aerosolized rodent urine and feces	Human intrusion into virus or rodent ecological niche
Hepatitis B	Nausea, vomiting, jaundice; chronic infection leads to hepatocellular carcinoma and cirrhosis	Contact with saliva, semen, blood, or vaginal fluids of an infected person; mode of transmission to children not known	Probably increased sexual activity and intravenous drug abuse; transfusion (before 1978)
Hepatitis C	Nausea, vomiting, jaundice; chronic infection leads to hepatocellular carcinoma and cirrhosis	Exposure (percutaneous) to contaminated blood or plasma; sexual transmission	Recognition through molecular virology applications; blood transfusion practices, especially in Japan

Table 32.8 (continued)

Agent	Disease and symptoms	Mode of transmission	Cause of emergence
Viruses			
Hepatitis E	Fever, abdominal pain, jaundice	Contaminated water	Newly recognized
Human immunodeficiency viruses: HIV-1 and HIV-2	HIV disease, including AIDS: severe immune system dysfunction, opportunistic infections	Sexual contact with or exposure to blood or tissues of an infected person; vertical transmission	Urbanization; changes in lifestyle or mores; increased intravenous drug use; international travel
Human papillomavirus	Skin and mucous membrane lesions (genital warts); strongly linked to cancer of the cervix and penis	Direct contact (sexual contact or contact with contaminated surfaces)	Increased surveillance and reporting
Human T cell lymphotrophic viruses (HTLV-I and HTLV-II)	Leukemias and lymphomas	Vertical transmission through blood or breast milk; exposure to contaminated blood products; sexual transmission	Increased intravenous drug abuse
Influenza	Fever, headache, cough, pneumonia	Airborne; especially in crowded, enclosed spaces	Animal–human virus reassortment; antigenic shift
Lassa	Fever, headache, sore throat, nausea	Contact with urine or feces of infected rodents	Urbanization and conditions favoring infestation by rodents
Measles	Fever, conjunctivitis, cough, red blotchy rash	Airborne; direct contact with respiratory secretions of infected persons	Deterioration of public health infrastructure supporting immunization
Monkeypox	Rash, lymphadenopathy, pulmonary distress	Direct contact with infected primates and other hosts	Travel to endemic areas, consumption and handling of infected primates and other hosts
Nipah virus	Hemorrhagic fever	Close contact with bats and pigs in Malaysia	Exposure to infected animals
Norwalk and Norwalk-like agents	Gastroenteritis, epidemic diarrhea	Most likely fecal–oral; vehicles may include drinking and swimming water, and uncooked foods	Increased recognition
Rabies	Acute viral encephalomyelitis	Bite of a rabid animal; contact with infected neural tissue	Introduction of infected host reservoir to new areas
Rift Valley	Febrile illness	Bite of an infective mosquito	Importation of infected mosquitoes or animals; development (dams, irrigation)
Rotavirus	Enteritis: diarrhea, vomiting, dehydration, and low-grade fever	Primarily fecal–oral; fecal–respiratory transmission can also occur	Increased recognition
Venezuelan equine encephalitis	Encephalitis	Bite of an infective mosquito	Movement of mosquitoes and hosts (horses)
West Nile virus	Meningitis, encephalitis	*Culex pipiens* mosquito and avian hosts	Agricultural development, increase in mosquito breeding areas, rapid spread to nonimmune populations
Yellow fever	Fever, headache, muscle pain, nausea, vomiting	Bite of an infective mosquito (*Aedes aegypti*)	Lack of mosquito control and vaccination; urbanization in tropics; increased air travel
Protists and fungi			
Candida	Candidiasis: fungal infections of the gastrointestinal tract, vagina, and oral cavity	Endogenous flora; contact with secretions or excretions from infected persons	Immunosuppression; medical devices (catheters); antibiotic use
Cryptococcus	Meningitis; sometimes infections of the lungs, kidneys, prostate, liver	Inhalation	Immunosuppression
Cryptosporidium	Cryptosporidiosis: infection of epithelial cells in the gastrointestinal and respiratory tracts	Fecal–oral, person-to-person, waterborne	Development near watershed areas; immunosuppression
Giardia intestinalis	Giardiasis; infection of the upper small intestine, diarrhea, bloating	Ingestion of fecally contaminated food or water	Inadequate control in water supply systems; immunosuppression; international travel
Microsporidia	Gastrointestinal illness, diarrhea; wasting in immunosuppressed persons	Unknown; probably ingestion of fecally contaminated food or water	Immunosuppression; recognition

▶

Table 32.8 *Emerging and reemerging epidemic infectious diseases (continued)*

Agent	Disease and symptoms	Mode of transmission	Cause of emergence
Protists and fungi			
Plasmodium	Malaria	Bite of an infective *Anopheles* mosquito	Urbanization; changing protist biology; environmental changes; drug resistance; air travel
Pneumocystis jiroveci	Acute pneumonia	Unknown; possibly reactivation of latent infection	Immunosuppression
Toxoplasma gondii	Toxoplasmosis; fever, lymphadenopathy, lymphocytosis	Exposure to feces of cats carrying the protists; sometimes foodborne	Immunosuppression; increase in cats as pets
Other agents			
Bovine prions	Bovine spongiform encephalitis (BSE, animal) and variant Creutzfeldt–Jakob disease (vCJD, human)	Foodborne	Consumption of contaminated beef

The demographics of human populations have changed dramatically in the last two centuries. In 1800, less than 2% of the world's population lived in urban areas. By contrast, today nearly one-half of the world's population lives in cities. The numbers, sizes, and population density in modern urban centers make disease transmission much easier. For example, dengue fever (Table 32.8) is now recognized as a serious hemorrhagic disease in tropical cities, largely due to the spread of dengue virus in the mosquito *Aedes aegypti*. The disease now spreads as an epidemic in tropical urban areas. Prior to 1950, dengue fever was rare, presumably because the virus was not easily spread among a more dispersed, smaller population.

Human behavior, especially in large population centers, also contributes to disease spread. For example, sexually promiscuous practices in population centers have been a major contributing factor to the spread of hepatitis and HIV/AIDS.

Technological advances and industrial development have a generally positive impact on living standards worldwide, but in some cases these advances have contributed to the spread of diseases. For example, although tremendous technological advances have been made in healthcare during the twentieth century, there has been a dramatic increase in healthcare-associated infections (Section 32.7). Antibiotic resistance in microorganisms is another negative outcome of modern healthcare practices. For example, vancomycin-resistant enterococci and staphylococci and drug-resistant *Streptococcus pneumoniae* and *Mycobacterium tuberculosis* are important emerging pathogens in developed countries.

Transportation, bulk processing, and central distribution methods have become increasingly important for quality assurance and economy in the food industry. However, these same factors can increase the potential for common-source epidemics when sanitation measures fail. For example, a single meat-processing plant spread *Escherichia coli* O157:H7 (Table 32.8) to people in eight states in 2009 in the United States. The contaminated food source, ground beef, was recalled and the epidemic was eventually stopped, but not before several people died. There was a similar incident with spinach contaminated by *E. coli* O157:H7 in runoff from a dairy farm in 2006. The *E. coli*–contaminated spinach was distributed nationally by a single packing plant and caused illness,

kidney failure, and a few deaths; this prompted a U.S. Food and Drug Administration recommendation that fresh spinach not be consumed for a time (Sections 36.5 and 36.9).

Economic development and changes in land use can also promote disease spread. For example, Rift Valley fever, a mosquito-borne viral infection, has been on the increase since the completion of the Aswan High Dam in Egypt in 1970. The dam flooded 2 million acres, and the enlarged shoreline increased breeding grounds for mosquitoes at the edge of the new reservoir. The first major epidemic of Rift Valley fever developed in Egypt in 1977, when an estimated 200,000 people became ill and 598 died. There have been several epidemic outbreaks in the area since then, and the disease has become endemic near the reservoir.

Lyme disease, the most common vectorborne disease in the United States, is on the rise largely due to changes in land use patterns (Section 34.4). Reforestation and the resulting increase in populations of deer and mice (the natural reservoirs for the disease-producing *Borrelia burgdorferi*) have resulted in greater numbers of infected ticks, the arthropod vector. In addition, larger numbers of homes and recreational areas in and near forests increase contact between the infected ticks and humans, consequently increasing disease incidence.

International travel and commerce also affect the spread of pathogens. For example, filoviruses (*Filoviridae*), a group of RNA viruses, cause fevers culminating in hemorrhagic disease in infected hosts. These untreatable viral diseases typically have a mortality rate above 20%. Most outbreaks have been restricted to equatorial central Africa, where the natural primate hosts and other vectors live. Travel of potential hosts to or from endemic areas is usually implicated in disease transmission. For example, one of the filoviruses was imported into Marburg, Germany, in 1967, with a shipment of African green monkeys used for laboratory work. The virus quickly spread from the primate host to some of the human handlers. Twenty-five people were initially infected, and six more developed disease as a result of contact with the human cases. Seven people died in this outbreak of what became known as the *Marburg virus*. This virus has reemerged in separate outbreaks since that time.

Table 32.9 *Virulence factors encoded by bacteriophages, plasmids, and transposons*

Genetic element	Organism	Virulence factors
Bacteriophage	*Streptococcus pyogenes*	Erythrogenic toxin
	Escherichia coli	Shiga-like toxin
	Staphylococcus aureus	Enterotoxins A, D, E, staphylokinase, toxic shock syndrome toxin-1 (TSST-1)
	Clostridium botulinum	Neurotoxins C, D, E
	Corynebacterium diphtheriae	Diphtheria toxin
Plasmid	*Escherichia coli*	Enterotoxins, pili colonization factor, hemolysin, urease, serum resistance factor, adherence factors, cell invasion factors
	Bacillus anthracis	Edema factor, lethal factor, protective antigen, poly-D-glutamic acid capsule
	Yersinia pestis	Coagulase, fibrinolysin, murine toxin
Transposon	*Escherichia coli*	Heat-stable enterotoxins, aerobactin siderophores, hemolysin and pili operons
	Shigella dysenteriae	Shiga toxin
	Vibrio cholerae	Cholera toxin

In 1989, another shipment of laboratory monkeys brought a different filovirus to Reston, Virginia, in the United States. Fortunately, the virus was not pathogenic for humans, but, having an effective respiratory transmission mode, the Reston virus infected and killed most of the monkeys at the Reston facility within days. These two filoviruses are closely related to the Ebola virus (Table 32.8).

Sporadic Ebola outbreaks in central Africa, often characterized by mortality rates greater than 50%, are caused by viral hemorrhagic fever pathogens. No immunization or therapy is available for prevention or treatment of disease resulting from infection by these viruses. These pathogens could potentially be spread via air travel throughout the world in a matter of days. A highly contagious respiratory agent such as the Reston virus that also has the high mortality potential of the Ebola virus could devastate population centers worldwide in a matter of weeks.

Pathogen adaptation and change can contribute to disease emergence. For example, nearly all RNA viruses, including influenza, HIV, and the hemorrhagic fever viruses, mutate rapidly. Because RNA viruses lack correction mechanisms for errors made during RNA replication, they incorporate genome mutations at an extremely high rate compared with most DNA viruses. The RNA viruses can present major epidemiological problems because of their changeable genomes.

Bacterial genetic mechanisms are capable of enhancing virulence and promoting emergence of new epidemics. Virulence-enhancing factors are often carried on mobile genetic elements such as bacteriophages, plasmids, and transposons. **Table 32.9** lists some virulence factors carried on these mobile genetic elements that contribute to pathogen emergence.

Drug resistance is another factor in the reemergence of some bacterial and viral pathogens. Although several drugs are effective against certain viral diseases, resistance to these drugs is very common, especially among the RNA viruses. For example, many strains of HIV develop resistance to azidothymidine (AZT) unless it is used in combination with other drugs (Section 33.14).

A breakdown of public health measures is sometimes responsible for the emergence or reemergence of diseases. For instance, cholera (caused by *Vibrio cholerae*) can be adequately controlled, even in endemic areas, by providing proper sewage disposal and water treatment. In 1991 an outbreak of cholera due to contaminated municipal water supplies in Peru was one of the first indications that the current cholera pandemic had reached the Americas (Section 35.5). In 1993, the municipal water supply of Milwaukee, Wisconsin, was contaminated with the chlorine-resistant protist *Cryptosporidium*, resulting in over 400,000 cases of intestinal disease, 4000 of which required hospitalization. Enhanced filtration systems rid the water supply of the pathogen (Section 35.6).

Inadequate public vaccination programs can lead to the resurgence of previously controlled diseases. For example, recent outbreaks of diphtheria in the former Soviet Union resulted from inadequate immunization of susceptible children due to the breakdown in public health infrastructures. Pertussis, another vaccine-preventable childhood respiratory disease, has increased recently in Eastern Europe and in the United States due to inadequate immunization among adults and children.

Finally, abnormal natural occurrences sometimes upset the usual host–pathogen balance. For example, hantavirus is a well-known human pathogen that occurs naturally in rodent populations, including some laboratory animals (Section 34.2). An abnormally high number of cases of human hantavirus infection leading to several deaths were reported in 1993 in the American Southwest and were linked to exposure to wild animal droppings. The likelihood of exposure to mice and droppings was increased due to a larger than normal wild mouse population resulting from near-record rainfall, a long growing season, and a mild winter. The favorable environmental conditions led to a dramatic increase in pathogen density. These factors enhanced probability of exposure for susceptible human hosts.

Addressing Emerging Diseases

Many of the emerging diseases we consider here are absent from the official notifiable disease list for the United States (Table 32.5). How then do public health officials define emerging diseases and prevent major epidemics? The keys for addressing emerging diseases are recognition of the disease and intervention to prevent pathogen transmission.

The first step in disease recognition is surveillance. Epidemic diseases that exhibit particular clinical syndromes warrant intensive public health surveillance. These syndromes are (1) acute respiratory diseases, (2) encephalitis and aseptic meningitis, (3) hemorrhagic fever, (4) acute diarrhea, (5) clusterings of high fever cases, (6) unusual clusterings of any disease or deaths, and (7) resistance to common drugs or treatment. New diseases are primarily recognized because of their epidemic incidence, clusterings, and syndromes. As the prevalence and pathology of an emerging disease are recognized, the disease is added to the notifiable disease list. For example, AIDS was recognized as a disease in 1981 and became a reportable disease in 1984. Likewise, the incidence of diseases due to Shiga toxin-producing *Escherichia coli* (STEC), such as *Escherichia coli* O157:H7, including hemolytic uremic syndrome, are increasing and became reportable in 1995 (Table 32.5).

Intervention to prevent spread of emerging infections must be a public health response employing various methods. Disease-specific intervention is the key to controlling individual outbreaks. Methods such as quarantine, immunization, and drug treatment must be applied to contain and isolate outbreaks of specific diseases. Finally, for vectorborne and zoonotic diseases, the nonhuman host or vector must be identified to allow intervention in the life cycle of the pathogen and interrupt transfer to humans. International public health surveillance and intervention programs were instrumental in controlling the emergence of severe acute respiratory syndrome (SARS), a disease that emerged rapidly, explosively, and unpredictably from a zoonotic source. On the other hand, even a rapid and focused surveillance control response was unsuccessful in containing the spread of pandemic (H1N1) 2009 influenza, and, within months after it was recognized, a worldwide pandemic was in progress (see the Microbial Sidebars "SARS as a Model of Epidemiological Success" and "Swine Flu—Pandemic (H1N1) 2009 Influenza").

MiniQuiz

- What factors are important in the emergence or reemergence of potential pathogens?
- Indicate general and specific methods that would be useful for dealing with emerging infectious diseases.

32.11 Biological Warfare and Biological Weapons

Biological warfare is the use of biological agents to incapacitate or kill a military or civilian population in an act of war or terrorism. Biological weapons have been used against targets in the United States, and biological weapon-making facilities are suspected to be in the hands of several governments as well as extremist groups.

Characteristics of Biological Weapons

Biological weapons are organisms or toxins that are (1) easy to produce and deliver, (2) safe for use by the offensive forces, and (3) able to incapacitate or kill individuals under attack in a reproducible and consistent manner. Many organisms or biological toxins fit these rather general criteria, and we discuss several of these below.

Although biological weapons are potentially useful in the hands of conventional military forces, the greatest likelihood of biological weapons use is probably by terrorist groups. This is in part due to the availibility and low cost of producing and propagating many of the organisms useful for biological warfare. Biological weapons are accessible to nearly every government and well-financed private organization.

Candidate Biological Weapons

Virtually all pathogenic bacteria or viruses are potentially useful for biological warfare, and several of the most likely candidate organisms are relatively simple to grow and disseminate. Commonly considered biological weapons agents are listed in **Table 32.10**. The most frequently mentioned candidate as a biological weapon is *Bacillus anthracis*, the causal agent of anthrax. We discuss anthrax in the next section.

Agents that have potential as biological weapons are classified into two categories by the Centers for Disease Control and Prevention. The highest level of threat comes from Category A agents. These can be easily disseminated by, for example, aerosols, or can be transmitted from person to person. These agents characteristically cause high mortality and consequently have high impact on public health. Preparations for attacks by such agents require a specific plan for each agent. Category A agents include *Bacillus anthracis; Clostridium botulinum* toxin, the agent that causes botulism (large amounts of preformed botulinum toxin delivered through a common vehicle such as drinking water could have devastating consequences because the lethal dose of botulinum toxin for a human is 2 μg or less); *Francisella tularensis*, the agent that cause tularemia ("rabbit fever"); *Yersinia pestis*, the organism responsible for plague; *Variola major*, the virus that cause smallpox, and the hemorrhagic fever viruses, including filoviruses such as Ebola and Marburg, and arenaviruses such as Lassa and Machupo.

Category B agents are moderately easy to spread, result in moderate morbidity and low mortality, and require specialized diagnostic and surveillance capabilities. Category A and B agents are identified in Table 32.10.

Smallpox

Smallpox virus, *Variola major*, has intimidating potential as a biological warfare agent because it can be easily spread by contact or aerosol spray and it has a mortality rate of 30% or more. Its potential for use as a biological weapon is considered low because the only known stocks of smallpox virus are in guarded repositories in the United States and Russia. A possibility remains, however, for terrorist groups or military forces to gain (or have) access to the smallpox virus. Because of this, the United States government has made provisions to immunize frontline healthcare and public safety personnel for smallpox. Although

Table 32.10 *Bioterrorism agents and diseases*

Bacteria and rickettsias

Bacillus anthracis (anthrax)	Category A[a]
Brucella sp. (brucellosis)	
Burkholderia mallei (glanders)	
Burkholderia pseudomallei (melioidosis)	
Chlamydophila psittaci (psittacosis)	
Vibrio cholerae (cholera)	
Clostridium botulinum toxin (botulism[b])	Category A[a]
Clostridium perfringens (Epsilon toxin[b])	
Coxiella burnetii (Q fever)	
Escherichia coli O157:H7 (gastrointestinal disease)	
Francisella tularensis (tularemia)	Category A[a]
Yersinia pestis (plague)	
Staphylococcus aureus enterotoxin B[b]	
Salmonella enterica serovar Typhi (typhoid fever)	
Salmonella sp. (salmonellosis)	
Shigella (shigellosis)	
Rickettsia prowazekii (typhus)	

Viral agents

Variola major (smallpox)	Category A[a]
Alphaviruses (viral encephalitis)	
Venezuelan equine encephalitis virus	
Eastern equine encephalitis virus	
Western equine encephalitis virus	
Nipah virus	
Viral hemorrhagic fevers viruses	Category A[a]
Filoviruses: Ebola, Marburg	
Arenaviruses: Lassa, Machupo	
Hantaviruses	

Protists

Cryptosporidium parvum (waterborne gastroenteritis)	

Plants

Ricinus communis (ricin toxin from castor bean[b])	

[a]Category A biological agents have the highest potential to be effective biological warfare agents. The other agents present a lesser threat and are designated as Category B biological agents.
[b]Preformed toxin; all other agents require infection.
Source: Information is from the Centers for Disease Control and Prevention, Atlanta, Georgia.

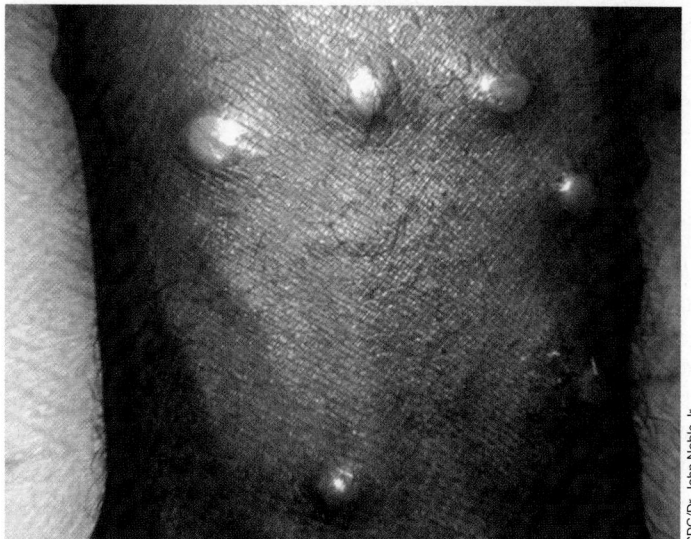

CDC/Dr. John Noble Jr.

Figure 32.12 Smallpox. The maculopapular lesions characteristic of smallpox infection are shown on a patient's forearm. Naturally occurring smallpox was eradicated by 1977.

with reduced immune competence, such as persons using anti-inflammatory steroid medications and persons with HIV/AIDS.

About 1 in 1000 of those vaccinated develops serious complications from the vaccine. These consequences include myocarditis and erythema multiforme, a toxic or allergic response to the vaccine. Generalized vaccinia (systemic vaccinia infection) occasionally occurs in individuals with skin conditions such as eczema. Life-threatening progressive vaccinia sometimes occurs in vaccinated individuals who are immunosuppressed due to therapy or disease. On average, one to two people per 1,000,000 who receive the vaccine will likely die from a vaccinia virus complication.

Because the smallpox virus is no longer found in nature and because the risks of the vaccine now outweigh the risk of contracting smallpox, the vaccine is no longer recommended for everyone, and over 90% of the worldwide population is inadequately vaccinated and susceptible to the disease. Preparations for a potential smallpox attack in the United States have included recommendations for immunization of certain individuals: persons having close contact with smallpox patients; workers evaluating, caring for, or transporting smallpox patients; laboratory personnel handling clinical specimens from smallpox patients; and other persons such as housekeeping personnel who might contact infectious materials from smallpox patients.

Delivery of Biological Weapons

Most organisms suitable for biological weapons use can be spread as an aerosol, providing simple, rapid, widespread dissemination leading to infection. Examples of several aerosol exposures are instructive.

In 1962, one of the last outbreaks of smallpox in a developed country occurred in Germany. A German worker developed smallpox after returning from Pakistan, where smallpox was endemic. The patient, who had a cough, was immediately hospitalized and

there is an extremely effective smallpox vaccine that uses the closely related vaccinia virus as the immunogen, this vaccine has not been in general use for almost 30 years because wild smallpox was eradicated worldwide by 1977.

Although vaccinia immunization is very effective, it carries significant risk. Normal vaccine reactions include formation of a pustule that resembles the lesions seen in smallpox (**Figure 32.12**); like smallpox lesions, the pustule forms a scab that falls off in 2 to 3 weeks, leaving a small scar. Many people have mild adverse reactions such as fevers and rashes. Vaccination is not recommended for persons with eczema or other chronic or acute skin conditions or heart disease, pregnant women, and those

UNIT 10

SARS as a Model of Epidemiological Success

Handling of the *severe acute respiratory syndrome* (SARS) epidemic early in this decade is an excellent example of epidemiological success. Like many other rapidly emerging diseases, SARS was viral and zoonotic in origin. Such characteristics have the potential to trigger explosive disease in humans when the infectious agents cross host species barriers. In many cases, the original viruses have been traced back to an animal host, but in others, the original host is unknown or is so ubiquitous that adequate vector control is nearly impossible, and thus disease persists. For example, West Nile virus is transmitted through mosquitoes that feed on infected birds. Although public health officials knew from the outset that West Nile disease would be seasonal and related to mosquitoes and infected birds, they could not prevent its spread. Thus, human West Nile cases spread quickly across the United States over a 5-year period, starting in Florida in 2001, and still occur seasonally across the country.

In contrast to West Nile disease, a different scenario describes the SARS epidemic. The SARS epidemic originated in late 2002 in Guangdong Province, China. By the following February, the virus had spread to 32 countries. Global travel provided the major vehicle for SARS dissemination. The etiology of SARS was quickly traced to a coronavirus derived from an animal source. The coronavirus entered the human food chain through exotic food animals such as civet cats. The SARS coronavirus (SARS-CoV), shown in Figure 1, originated in bats. Civet cats acquired the virus by consuming fruit contaminated by bats. SARS-CoV likely evolved over an extended period of time in bats and developed, quite by accident, the ability to infect civet cats and then humans.

Much like common cold viruses, SARS-CoV is a relatively hardy, easily spread RNA virus that is difficult to contain. Once in humans, SARS-CoV is very contagious because it can be spread in several ways,

including person-to-person by sneezing and coughing and by contact with contaminated fomites or feces. Ordinarily, a new coldlike virus would be of little concern, but SARS-CoV causes infections with significant morbidity and mortality. There have been about 8500 known SARS-CoV infections and over 800 deaths, for an overall mortality rate of nearly 10%. In persons over 65 years of age, the mortality rate approached 50%, attesting to SARS-CoV virulence as a human pathogen. About 20% of all SARS cases were in healthcare workers, demonstrating the high infectivity of the virus. Standard containment and infection control methods practiced by healthcare personnel were not effective in controlling spread of the disease. When this was realized, SARS patients were confined for the course of the disease in strict isolation in negative-pressure rooms. To prevent infection, healthcare workers wore respirators when working with SARS patients or when handling fomites (bed linens, eating utensils, and so on) contaminated with SARS-CoV.

The recognition and containment of the clinical disease was the start of an international response involving clinicians, scientists, and public officials. Almost immediately, travel to and from the endemic area was restricted, limiting further outbreaks. SARS-CoV isolation was achieved rapidly, and this information was used to develop the PCR tests used to track the disease. As laboratory work progressed, epidemiologists traced the virus back to the civet food source in China and stopped further transmission to humans by restricting the sale of civets and other foods from wild sources. These actions collectively stopped the outbreak.

SARS is an example of a serious infection that emerged very rapidly from a unique source. However, rapid isolation and characterization of the SARS pathogen, nearly instant development of worldwide notification procedures and diagnostic tests, and a

concerted effort to understand the biology and genetics of this novel pathogen quickly controlled the disease; there has not been another case of SARS since early 2004. The rapid emergence of SARS, and the equally rapid and successful international effort to identify and control the outbreak, provide a model for the control of emerging epidemics.

As international travel and trade expand, the chances for propagation and rapid dissemination of new exotic diseases will continue to increase. We should therefore anticipate the emergence of other serious infectious zoonotic diseases, including pandemic influenza. We hope that the lessons learned from the SARS epidemic will pay dividends when other emerging diseases appear.

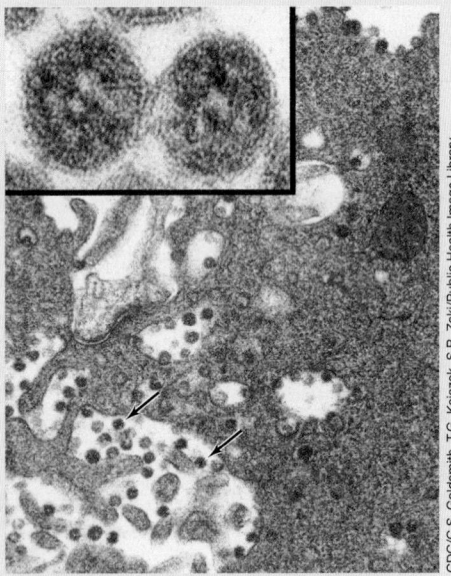

CDC/C.S. Goldsmith, T.G. Ksiazek, S.R. Zaki/Public Health Image Library

Figure 1 **Severe acute respiratory virus syndrome coronavirus (SARS-CoV).** The upper left panel shows isolated SARS-CoV virions. An individual virion is 132 nm in diameter. The large panel shows coronaviruses within the cytoplasmic membrane–bound vacuoles and in the rough endoplasmic reticulum of host cells. The virus replicates in the cytoplasm and exits the cell through the cytoplasmic vacuoles.

quarantined; the cough aerosolized the virus and caused illness and one death in 19 vaccinated individuals.

There have been planned bioterrorist attacks in the United States and other countries even before the anthrax attacks of 2001 (Section 32.12). In 1984 in The Dalles, Oregon (United States), cultists sprayed salad bars in ten restaurants with a culture of *Salmonella enterica* serovar Typhimurium, causing 751 cases of foodborne salmonellosis in a region that usually has fewer than 10 cases per year. In 1995, a radical political group released sarin nerve gas, a chemical weapon, into a Tokyo subway, killing several people and injuring many more. This incident is relevant to a discussion of biological weapons because this group also possessed anthrax cultures, bacteriological media, drone airplanes, and spray tanks.

Delivery of preformed bacterial toxins such as botulinum toxin or staphylococcal enterotoxin to large populations may be impractical because most exotoxins are proteins that lose effectiveness as they are diluted or denatured, and are destroyed in common sources such as drinking water. However, delivery of toxins could be aimed at selected individuals and small groups, or delivered randomly to instigate panic.

Prevention and Response to Biological Weapons

Proactive measures against the deployment of biological weapons have already begun with periodic updating of the international agreements of the 1972 Biological and Toxic Weapons Convention. The fifth and most recent update was in 2002. At the practical level, governments are now supporting the large-scale production and distribution of vaccines along with the development of strategic and tactical plans to prevent and contain biological weapons.

The United States government, through the Centers for Disease Control and Prevention (CDC), has devised and enhanced the Select Agent Program surveillance system to monitor possession and use of potential bioterrorism agents. The CDC Laboratory Response Network and the Health Alert Network have been upgraded to enhance their diagnostic capabilities and increase the reporting abilities of local and regional healthcare centers to rapidly identify bioterrorism events as well as emerging diseases.

MiniQuiz

- What characteristics make a pathogen or its products particularly useful as a biological weapon?
- Identify two infectious agents that could be effective biological weapons. How could the agents be disseminated?

32.12 Anthrax as a Biological Weapon

Bacillus anthracis is a Category A agent for biowarfare and bioterrorism. Here we discuss its unique properties, the diseases it causes, and methods for prevention, diagnosis, and treatment.

Biology and Growth

B. anthracis is a ubiquitous saprophytic soil inhabitant. It grows as an aerobic gram-positive rod, 1 μm in diameter and 3–4 μm in length. Like other species of the genus, *B. anthracis* produces

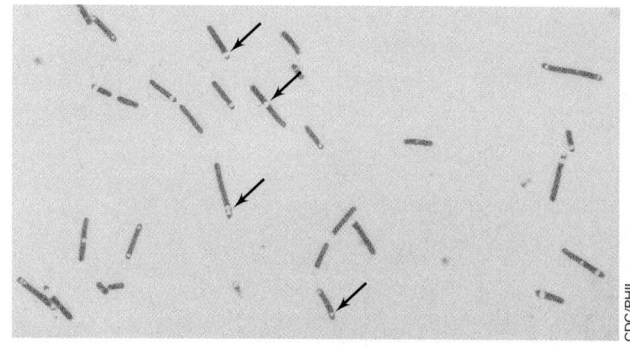

(a)

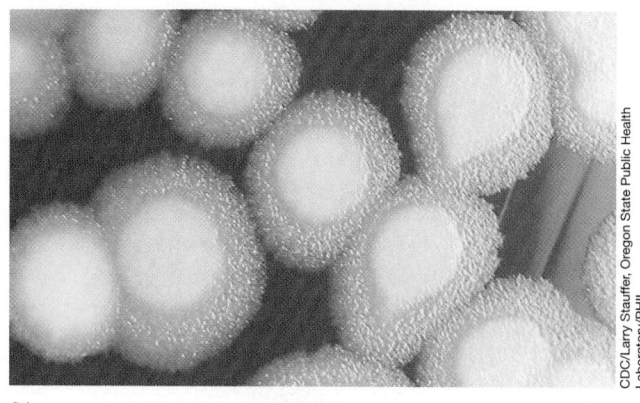

(b)

Figure 32.13 *Bacillus anthracis.* *(a) B. anthracis* is a gram-positive endospore-forming rod approximately 1 μm in diameter and 3–4 μm in length. Note the developing endospores (arrows). *(b) B. anthracis* colonies on blood agar. The nonhemolytic colonies take on a characteristic "ground glass" appearance.

endospores resistant to heat and drying (**Figure 32.13a**). Endospore formation enhances the ability to disseminate *B. anthracis* in aerosols. Viable endospores are sometimes recovered from contaminated animal products such as hides and fur. Growth on blood agar produces large colonies with a characteristic "ground glass" appearance (Figure 32.13b). Strains having a poly-D-glutamic acid capsule are resistant to phagocytosis.

Infection and Pathogenesis

B. anthracis endospores are the normal means of acquiring anthrax. The disease usually affects domestic animals, especially ungulates—cows, sheep, and goats. The number of infections in animals, although considerable, is not known. The animals acquire the disease from plants or soil in pastures. In humans and animals, there are three forms of the disease. *Cutaneous anthrax* is contracted when abraded skin is contaminated by *B. anthracis* endospores (**Figure 32.14a**). Cutaneous anthrax cases are rare in the United States. *Gastrointestinal anthrax* is contracted from consumption of endospore-contaminated plants or meat from animals infected with anthrax. Human gastrointestinal anthrax is rarely seen. *Pulmonary anthrax* is contracted when the endospores are inhaled. Inhalation of the endospores or the live bacteria results in pulmonary infections characterized by pulmonary and cerebral hemorrhage (Figure 32.14b). Untreated pulmonary

CDC/PHIL

CDC/Larry Stauffer, Oregon State Public Health Laboratory/PHIL

UNIT 10

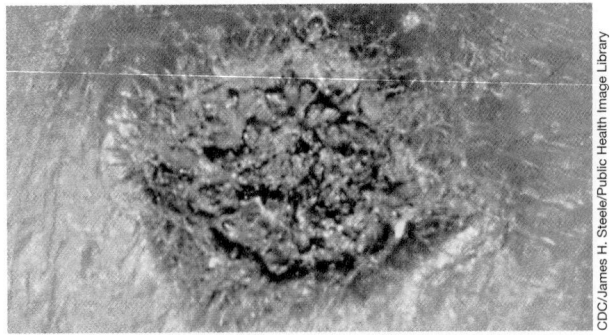

(a)

(b)

Figure 32.14 Anthrax. *(a)* Cutaneous anthrax. The blackened lesion on the forearm of a patient, about 2 cm in diameter, results from tissue necrosis. Cutaneous anthrax, even when untreated, usually is a localized, nonlethal infection. *(b)* Inhalation anthrax. The fixed and sectioned human brain shows hemorrhagic meningitis (dark coloration) due to a fatal case of inhalation anthrax.

anthrax has a mortality rate of nearly 100%. Fortunately, pulmonary anthrax cases, even in agricultural workers, are extremely rare. The most recent naturally acquired pulmonary anthrax infection in the United States occurred in 1976. However, several cases of pulmonary anthrax identified in 2001 were caused by bioterrorism attacks.

Pathogenesis results from inhalation of 8000–50,000 endospores of an encapsulated toxigenic strain. Pathogenic *B. anthracis* produces three proteins—*protective antigen* (PA), *lethal factor* (LF), and *edema factor* (EF). PA and LF form *lethal toxin*. PA and EF form *edema toxin*. PA is the cell-binding B component of these AB-type toxins (↩ Table 27.4). EF causes edema and LF causes cell death. Growth of *B. anthracis* in the lymph nodes and lymphatic tissues draining the lungs leads to edema and cell death, culminating in tissue destruction, shock, and death.

Clinical symptoms start with sore throat, fever, and muscle aches. After several days, symptoms escalate to include difficulty in breathing, followed by systemic shock. Fatality rates can approach 90% even when exposure is recognized and treatment

is started, and can be nearly 100% in cases for which treatment is not started until after the onset of symptoms.

Weaponized Anthrax

The term *weaponized* is applied to strains and preparations of *B. anthracis*, usually in endospore form, having properties that enhance dissemination and use as biological weapons. Such strains and preparations were developed in several countries in the post–World War II era, but overt development of new biological weapons was halted by international treaty in 1972. The physical characteristics of the weaponized anthrax preparations typically include a small particle size, usually interspersed with a very fine particulate agent such as talc. This small-particle, powdery form ensures that the endospores will spread easily by air currents. Thus, opening an envelope containing endospores or releasing the powder–endospore mixture into a ventilation system or other air current has the potential to contaminate surrounding areas and personnel.

A weaponized form of anthrax was used in a series of bioterrorism attacks in the United States in 2001. The anthrax attacks were carried out by mailing envelopes or packages containing weaponized anthrax endospores. The attacks were apparently directed at the news media (Florida) and the government (Washington, DC area). A third focus of attack, the Pennsylvania–New Jersey–New York area, had no defined single target, but disrupted mail service in the Northeast; some anthrax-contaminated mail facilities were still not in use 2 years later. In all, the attacks were responsible for 22 cases of anthrax, 11 of which were cutaneous anthrax, and 11 were pulmonary anthrax. Five deaths resulted from the attacks. The bioterrorist was a bioweapons laboratory worker.

The incidents in the United States were not the first or the most serious anthrax biological weapons infections. In 1979, *B. anthracis* spores were inadvertently released into the atmosphere from a biological weapons facility in Sverdlovsk, Russia. Less than 1 g of endospores was released, and everyone in the area surrounding the facility was immunized and given prophylactic antibiotic therapy as soon as the first anthrax case was diagnosed. Even with these quick reactive measures, 77 persons outside the facility contracted pulmonary anthrax and 66 died.

Vaccination, Prophylaxis, Treatment, and Diagnosis

Vaccination for anthrax has thus far been restricted to individuals who are considered at risk. This includes agricultural animal workers and military personnel. The current vaccine, called *anthrax vaccine adsorbed* (AVA), is prepared from a cell-free *B. anthracis* culture filtrate.

B. anthracis infection has a minimum incubation time of about eight days. Antibiotics can be used for treatment. Ciprofloxacin, a broad-spectrum quinolone antibiotic, is used against strains that are penicillin resistant, including many laboratory and biological weapons strains. Ciprofloxacin is also used prophylactically to treat potentially exposed individuals.

Rapid diagnostic tests are available to detect microbial endospores. However, positive identification of *B. anthracis* relies on culture techniques and direct observation of either infected tissues or cultured organisms. The characteristic ground-glass appearance on blood agar, coupled with the isolation of gram-positive endospore-forming rods growing in extended chains, is presumptive evidence for *B. anthracis* (Figure 32.13).

MiniQuiz

- What factors make *B. anthracis* an effective biological weapon?
- Indicate the steps you would use to identify the use of *B. anthracis* in a bioterror attack. Indicate treatment steps for potential victims.

Big Ideas

32.1

Epidemiology is the study of the occurrence, distribution, and determinants of health and disease in a population. To understand infectious disease, effects on both populations and individuals must be studied. The interactions of pathogens with hosts can be dynamic, affecting the long-term evolution and survival of all species involved.

32.2

An endemic disease is continually present at low incidence in a population. An epidemic disease is one that has increased to unusually high incidence in a population. Incidence is a record of new cases of a disease, whereas prevalence is a record of total cases of a disease in a population. Infectious diseases cause morbidity and may cause mortality. An infectious disease follows a predictable clinical pattern in the host.

32.3

Many pathogens exist only in humans and are maintained only by transmission from person to person. Some human pathogens, however, live mostly in soil, water, or animals. An understanding of disease reservoirs, carriers, and pathogen life cycles is critical for controlling disease.

32.4

Infectious diseases can be transmitted directly from one host to another host, indirectly from living vectors or inanimate objects (fomites), or from common-source vehicles such as food and water. Epidemics may be of host-to-host origin or originate from a common source.

32.5

For most infectious diseases, hosts and pathogens coevolve to reach a steady state that favors the continued survival of both host and pathogen. When a large proportion of a host population is immune to a given disease, disease spread is inhibited.

32.6

HIV/AIDS is a major worldwide public health problem. There is no effective cure or immunization to prevent AIDS. HIV/AIDS transmission can be controlled using public health surveillance and education.

32.7

Patients in healthcare facilities are unusually susceptible to infectious disease and are exposed to various infectious agents. Treatment of HAIs is complicated by reduced host resistance.

32.8

Food and water purity regulations, vector control, immunization, quarantine, isolation, and disease surveillance are public health measures that reduce the incidence of communicable diseases.

32.9

Infectious diseases account for 23.5% of all mortality worldwide. Most cases of infectious diseases are in developing countries. Control of infectious diseases can be accomplished by public health measures.

32.10

Changes in host, vector, or pathogen conditions, whether natural or artificial, can encourage the explosive emergence or reemergence of infectious diseases. Global surveillance and intervention programs must be in place to prevent new epidemics and pandemics.

32.11

Bioterrorism is a threat in a world of rapid international travel and easily accessible technical information. Biological agents can be used as weapons by military forces or by terrorist groups. Aerosols or common sources such as food and water are the most likely modes of delivery. Prevention and containment measures rely on a well-prepared public health infrastructure.

32.12

Bacillus anthracis has emerged as an important pathogen because of its use as a biological weapon. Highly infective weaponized endospore preparations have been used as bioterrorism agents. Pulmonary anthrax has a fatality rate of almost 100% in untreated individuals. Effective treatment relies on timely observation and diagnosis of symptoms. Treatment of pulmonary anthrax does not guarantee survival.

Review of Key Terms

Acute infection a short-term infection, usually characterized by dramatic onset

Biological warfare the use of biological agents to incapacitate or kill humans

Carrier a subclinically infected individual who may spread a disease

Centers for Disease Control and Prevention (CDC) the agency of the U.S. Public Health Service that tracks disease trends, provides disease information to the public and to healthcare professionals, and forms public policy regarding disease prevention and intervention

Chronic infection a long-term infection

Common-source epidemic an epidemic resulting from infection of a large number of people from a single contaminated source

Emerging disease an infectious disease whose incidence recently increased or whose incidence threatens to increase in the near future

Endemic disease a disease that is constantly present, usually in low numbers

Epidemic the occurrence of a disease in unusually high numbers in a localized population

Epidemiology the study of the occurrence, distribution, and determinants of health and disease in a population

Fomite an inanimate object that, when contaminated with a viable pathogen, can transfer the pathogen to a host

Healthcare-associated infection (HAI) a local or systemic infection acquired by a patient in a healthcare facility, particularly during a stay in the facility; also called nosocomial infection

Herd immunity the resistance of a population to a pathogen as a result of the immunity of a large portion of the population

Host-to-host epidemic epidemic resulting from person-to-person contact, characterized by a gradual rise and fall in number of cases

Incidence the number of new disease cases reported in a population in a given time period

Morbidity the incidence of illness in a population

Mortality the incidence of death in a population

Outbreak the occurrence of a large number of cases of a disease in a short period of time

Pandemic a worldwide epidemic

Prevalence the total number of new and existing disease cases reported in a population in a given time period

Public health the health of the population as a whole

Quarantine the restriction of the movement of individuals with highly contagious serious infections to prevent spread of the disease

Reemerging disease an infectious disease, thought to be under control, that produces a new epidemic

Reservoir a source of viable infectious agents from which individuals may be infected

Surveillance the observation, recognition, and reporting of diseases as they occur

Vector a living agent that transfers a pathogen (differs from genetic vector, discussed in Chapters 11 and 25)

Vehicle a nonliving source of pathogens that transmits the pathogens to large numbers of individuals; common vehicles are food and water

Zoonosis a disease that occurs primarily in animals, but can be transmitted to humans

Review Questions

1. Distinguish between *acute* and *chronic* infections (Section 32.1).

2. List the five most common causes of mortality due to infectious diseases throughout the world. Are any of these diseases preventable by immunization (Section 32.2)?

3. Define (a) reservoir of infections, (b) incidence of infections, (c) prevalence, (d) mortality, and (e) morbidity (Section 32.2).

4. Explain the difference between a chronic carrier and an acute carrier of an infectious disease (Section 32.3).

5. With examples, describe direct and indirect host-to-host transmission of diseases (Section 32.4).

6. What is herd immunity? How is it important in protecting a population against infectious diseases (Section 32.5)?

7. Identify the major risk factors for acquiring human immunodeficiency virus (HIV) infection in the United States. Does this pattern hold for other geographic regions (Section 32.6)?

8. What is HAI also known as? Name at least three sources of HAIs in healthcare environments (Section 32.7).

9. Describe the major medical and public health measures developed in the twentieth century that were instrumental in controlling the spread of infectious diseases in developed countries (Section 32.8).

10. Compare the contribution of infectious diseases to mortality in developed and developing countries (Section 32.9).

11. Review the major reasons for the emergence of new infectious diseases. What methods are available for identifying and controlling the emergence of new infectious diseases (Section 32.10)?

12. Describe the general properties of an effective biological warfare agent. How does smallpox meet these criteria? Identify other organisms that meet the basic requirements for a bioweapon (Section 32.11).

13. Describe the use of *Bacillus anthracis* as a biological weapon. Devise a plan to protect yourself against a *B. anthracis* attack (Section 32.12).

Application Questions

1. Smallpox, a disease that was limited to humans, was eradicated. Plague, a disease with a zoonotic reservoir in rodents (Table 32.2), can never be eradicated. Explain this statement and why you agree or disagree with the possibility of eradicating plague on a global scale. Devise a plan to eradicate plague in a limited environment such as a town or city. Be sure to use methods that involve the reservoir, the pathogen, and the host.

2. Acquired immunodeficiency syndrome (AIDS) is a disease that can be eliminated because it is propagated by person-to-person contact and there are no known animal reservoirs. Do you agree or disagree with this statement? Explain your answer. Design a program for eliminating AIDS in a developed country and in a developing country. How would these programs differ? What factors would work against the success of your program, both in terms of human behavior and in terms of the AIDS disease itself? Why are the numbers of HIV-infected and AIDS patients continuing to grow, especially in developing countries? HIV/AIDS incidence (new cases) in developed countries has been virtually unchanged in this century (Figure 32.7). The numbers of individuals living with AIDS, however, is increasing. Explain this contradiction.

3. Travel to developing countries involves some exposure to infectious diseases. What general precautions should you take before, during, and after visits to developing countries? Where can you obtain information concerning infectious diseases in a specific foreign country? When you return from a foreign country, are you a disease risk to your family or your associates? Explain.

4. Identify a specific pathogen that would be a suitable agent for effective biological warfare. Describe the properties of the pathogen in the context of its use as a biological weapon. Describe equipment and other resources necessary for growing large amounts of the pathogen. Identify a suitable delivery method. As you will propagate and deliver the pathogen, describe the precautions you will take to protect yourself. Now reverse your role. As a public health official at your university, describe how you would recognize and diagnose the disease caused by the agent. Indicate the measures you would take to treat the illnesses caused by the agent. How could you best limit the damage? Would quarantine and isolation methods be useful? What about immunization and antibiotics?

Need more practice? Test your understanding with quantitative questions; access additional study tools including tutorials, animations, and videos; and then test your knowledge with chapter quizzes and practice tests at **www.microbiologyplace.com**.

33

Person-to-Person Microbial Diseases

Cells of the bacterium *Neisseria gonorrhoeae*, sometimes called the gonococcus, are highlighted in blue in this scanning electron micrograph. *N.gonorrhoeae* causes the sexually transmitted disease gonorrhea and is transmitted only by intimate person-to-person contact.

Perhaps more than a million microbial species exist in nature, but only a few hundred species cause disease. Many microorganisms are closely associated with plants or animals, including humans, in beneficial relationships. However, pathogenic species have profoundly negative effects on host organisms. In this and the next three chapters, we examine representative human pathogens. We will investigate their biology as well as the diseases and their diagnosis, treatment, and prevention.

Our coverage is organized based on the pathogen's mode of transmission, which presents infectious disease in the context of the ecology of the pathogen. In this chapter we consider diseases transmitted from person to person. For example, influenza virus and streptococci cause diseases with overlapping symptoms, although the causal agents, one viral and one bacterial, are very different. Here these pathogens are discussed together because they are spread from person to person via a respiratory route. Using this approach, we will establish the connections between biologically diverse, but ecologically and pathogenically related, disease agents.

In Chapters 34 through 36, diseases whose modes of transmission require animal or arthropod vectors, or common sources such as soil, water, and food, will be examined.

Airborne Transmission of Diseases

Aerosols, such as those generated by a human sneeze (**Figure 33.1**), are important vehicles for person-to-person transmission of many infectious diseases. Respiratory diseases are spread in this fashion. For example, *Mycobacterium tuberculosis*, the bacterium that causes the disease tuberculosis, has spread in this way to infect

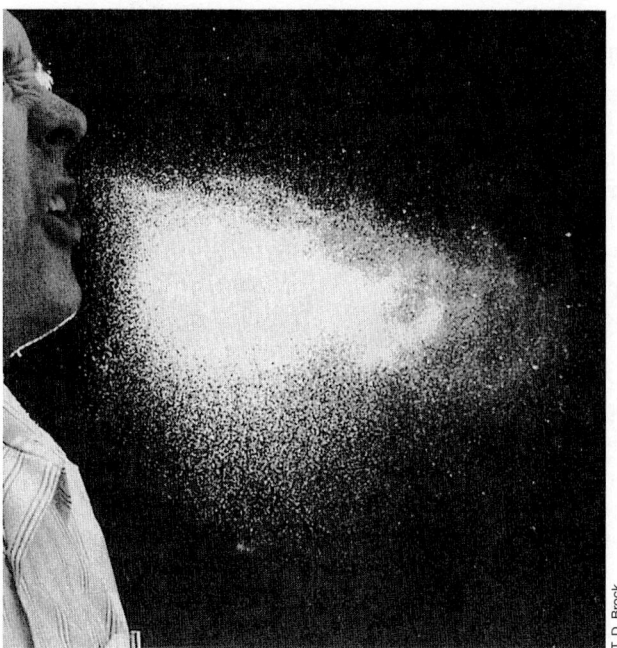

Figure 33.1 High-speed photograph of an unstifled sneeze.

at least one-third of the world's population. In addition, respiratory spread of influenza and cold viruses is so efficient that virtually everyone has been infected, sometimes several times a year, as in the case of colds.

33.1 Airborne Pathogens

Microorganisms found in air are derived from soil, water, plants, animals, people, and other sources. In outdoor air, soil organisms predominate. Indoors, the concentration of microorganisms is considerably higher than outdoors, especially for organisms that originate in the human respiratory tract.

Most microorganisms survive poorly in air. As a result, pathogens are effectively transmitted among humans only over short distances. Certain pathogens, however, survive under dry conditions and can remain alive in dust for long periods of time. Because of their thick, rigid cell walls, gram-positive bacteria (*Staphylococcus*, *Streptococcus*) are generally more resistant to drying than gram-negative bacteria. Likewise, the waxy layer of *Mycobacterium* cell walls resists drying and promotes survival. The endospores of endospore-forming bacteria are extremely resistant to drying but are not generally passed from human to human in the endospore form.

Large numbers of moisture droplets are expelled during sneezing (**Figure 33.1**), and a sizable number are expelled during coughing or simply talking. Each infectious droplet is about 10 μm in diameter and may contain one or two microbial cells or virions. The initial speed of the droplet movement is about 100 m/sec (more than 325 km/h) in a sneeze and ranges from 16 to 48 m/sec during coughing or shouting. The number of bacteria in a single sneeze varies from 10,000 to 100,000. Because of their small size, the moisture droplets evaporate quickly in the air, leaving behind a nucleus of organic matter and mucus to which bacterial cells are attached.

Respiratory Infections

Humans breathe about 500 million liters of air in a lifetime, much of it containing microorganism-laden dust. The speed at which air moves through the respiratory tract varies, and in the lower respiratory tract the rate is quite slow. As air slows down, particles in it stop moving and settle. Large particles settle first and the smaller ones later; only particles smaller than 3 μm travel as far as the bronchioles in the lower respiratory tract (**Figure 33.2**). Of course, most pathogens are much smaller than this, and different organisms characteristically colonize the respiratory tract at different levels. The upper and lower respiratory tracts offer decidedly different environments, favoring different microorganisms.

Bacterial and Viral Pathogens

Most human respiratory pathogens are transmitted from person to person because humans are the only reservoir for the pathogens; pathogen survival thus depends on person-to-person transmission. Here we discuss some of the pathogens that are transmitted primarily via the respiratory route. However, many of these such as *Streptococcus* spp., cold viruses, and influenza

UNIT 11

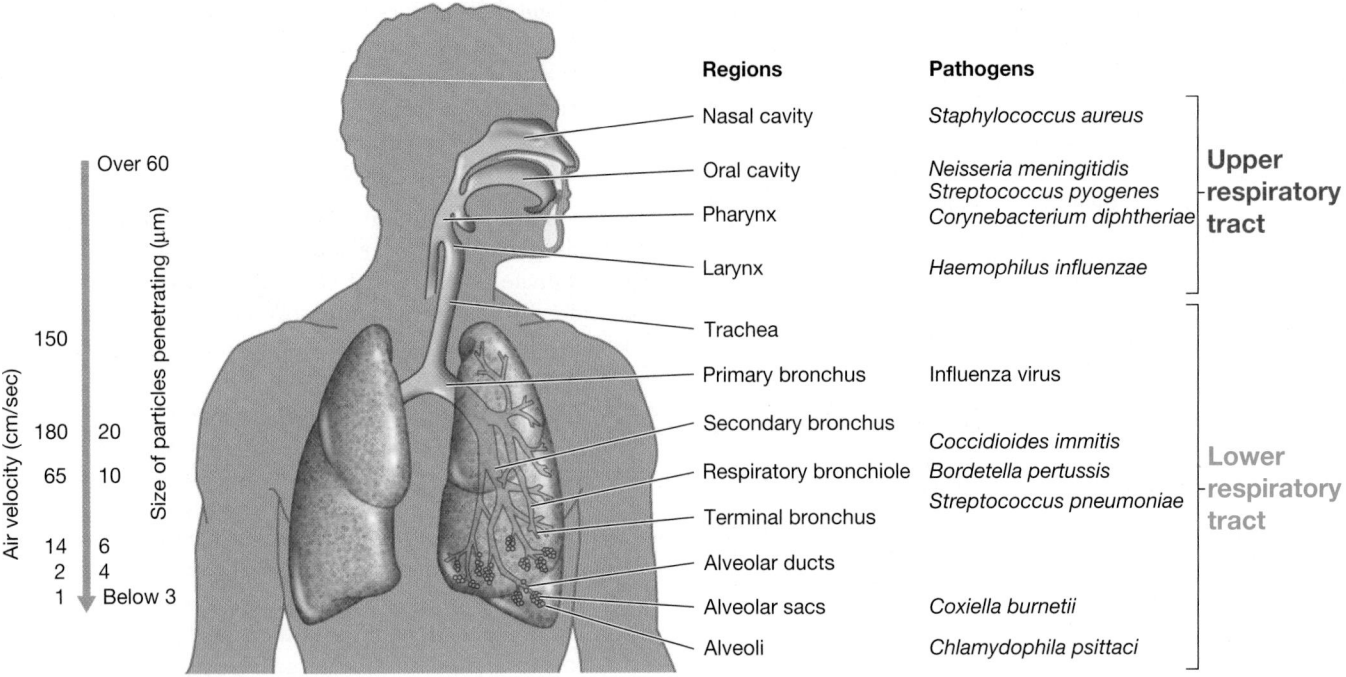

Figure 33.2 The respiratory system of humans. The microorganisms listed generally initiate infections at the indicated locations.

can also be transmitted via direct contact or on fomites. A few respiratory pathogens such as *Legionella pneumophila* (legionellosis, or Legionnaires' disease) are transmitted primarily from water or soil and thus do not require person-to-person propagation; we discuss these in Chapter 35. Bacterial and viral respiratory infections, serious in themselves, often initiate secondary problems that can be life-threatening. Thus, accurate and rapid diagnosis and treatment of respiratory infections can limit host damage. Many bacterial and viral pathogens can be controlled by immunization. Most respiratory bacterial pathogens respond readily to antibiotic therapy, but antiviral drug treatment options are generally limited.

MiniQuiz

- Identify the physical features of gram-positive bacteria that allow them to survive for long periods in air and dust.
- Identify pathogens more commonly found in the upper respiratory tract. Identify pathogens more commonly found in the lower respiratory tract.

33.2 Streptococcal Diseases

The bacteria *Streptococcus pyogenes* and *Streptococcus pneumoniae* (**Figure 33.3**) are important human respiratory pathogens; both organisms are transmitted by the respiratory route. *S. pneumoniae* is found in the respiratory flora of up to 40% of healthy individuals. Although endogenous strains do not cause disease in most normal individuals, they can cause severe respiratory disease in compromised individuals.

Streptococci are nonsporulating, homofermentative, aerotolerant, anaerobic gram-positive cocci (Section 18.1). Cells of *S. pyogenes* (Figure 33.3*a*) typically grow in elongated chains, as do many other members of the genus (Figure 18.3*b*). Pathogenic strains of *S. pneumoniae* typically grow in pairs or short chains, and virulent strains produce an extensive polysaccharide capsule (Figure 33.3*b*).

Streptococcus pyogenes: Epidemiology and Pathogenesis

Streptococcus pyogenes, also called *group A Streptococcus* (GAS)(Figure 33.3*a*), is frequently isolated from the upper respiratory tract of healthy adults. Although numbers of endogenous *S. pyogenes* are usually low, if host defenses are weakened or a new, highly virulent strain is introduced, acute suppurative (pus-forming) infections are possible. *S. pyogenes* is the cause of streptococcal pharyngitis, or "strep throat." Most isolates from clinical cases of streptococcal pharyngitis produce a toxin that lyses red blood cells in culture media, a condition called *β-hemolysis* (Figure 27.19*a*). Streptococcal pharyngitis is characterized by a severe sore throat, enlarged tonsils with exudate, tender cervical lymph nodes, a mild fever, and general malaise. *S. pyogenes* can also cause related infections of the middle ear (otitis media), the mammary glands (mastitis), infections of the superficial layers of the skin (pyoderma or impetigo) (impetigo can also be caused by *Staphylococcus aureus*) (**Figure 33.4**), erysipelas, an acute streptococcal skin infection (**Figure 33.5**), necrotizing fasciitis, an infection of subcutaneous tissue, and several conditions linked to the after effects of streptococcal infections.

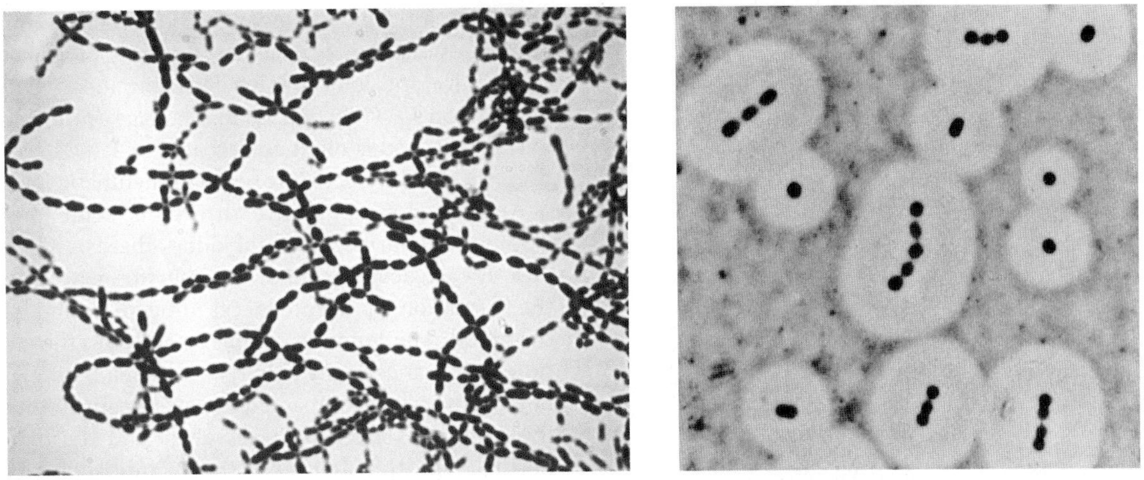

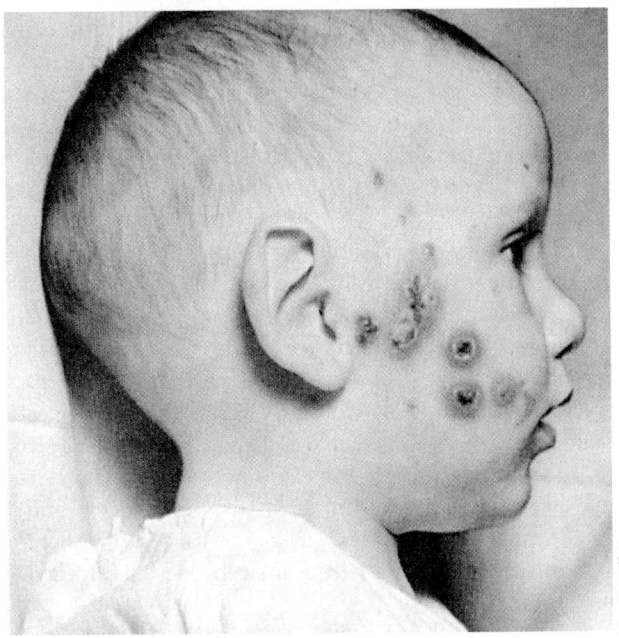

Figure 33.3 *Streptococcus* **pathogens.** *(a) Streptococcus pyogenes* grows in chains. The cells range in size from 1 to 2 μm in diameter. *(b)* India ink negative stain of *Streptococcus pneumoniae*. An extensive capsule surrounds the cells, which are about 0.5–1.2 μm in diameter.

About half of the clinical cases of severe sore throat are due to *Streptococcus pyogenes*, with most others due to viral infections. An accurate, rapid determination of the cause of the sore throat is important. If the sore throat is due to *S. pyogenes*, rapid, complete treatment of streptococcal sore throat is important because untreated streptococcal infections can lead to serious diseases such as scarlet fever, rheumatic fever, acute glomerulonephritis, and streptococcal toxic shock syndrome. On the other hand, if the sore throat is due to a virus, treatment with antibacterial

drugs (antibiotics) will be useless, and may promote antimicrobial drug resistance (↩ Section 26.12).

Certain GAS strains carry a lysogenic bacteriophage that encodes streptococcal pyrogenic exotoxin A (SpeA), SpeB, SpeC, and SpeF. These exotoxins are responsible for most of the symptoms of streptococcal toxic shock syndrome (STSS) and **scarlet fever** (**Figure 33.6**). Streptococcal pyrogenic exotoxins are superantigens that recruit large numbers of T cells to the infected tissues (↩ Section 28.10). Toxic shock results when the activated

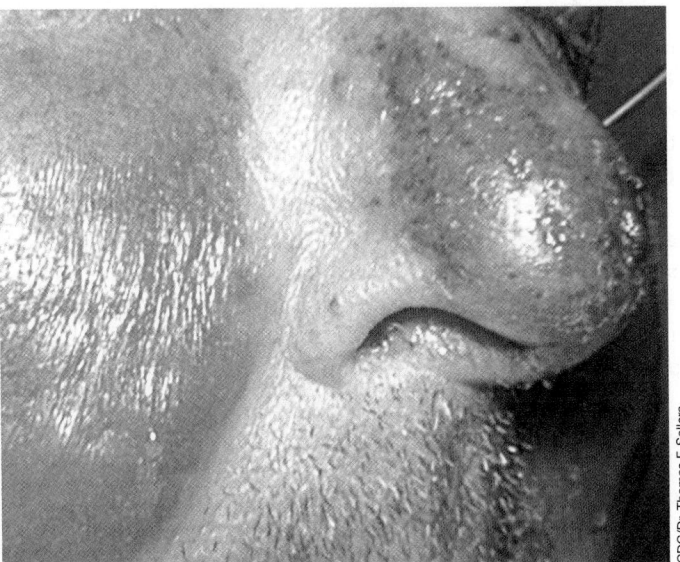

Figure 33.4 **Typical lesions of impetigo.** Impetigo is commonly caused by *Streptococcus pyogenes* or *Staphylococcus aureus*.

Figure 33.5 **Erysipelas.** Erysipelas is a *Streptococcus pyogenes* infection of the skin, shown here on the nose and cheeks, characterized by redness and distinct margins of infection. Other commonly-infected body sites include the ears and the legs.

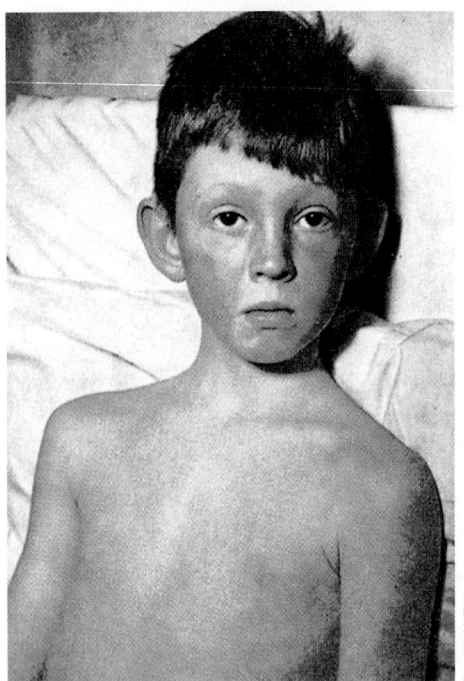

Franklin H. Top

Figure 33.6 Scarlet fever. The typical rash of scarlet fever results from the action of the pyrogenic exotoxins produced by *Streptococcus pyogenes*.

T cells secrete cytokines, which in turn activate large numbers of macrophages and neutrophils, causing local and systemic inflammation and tissue destruction.

Occasionally GAS causes fulminant (sudden and severe) invasive systemic infection such as cellulitis, a skin infection in subcutaneous layers, and necrotizing fasciitis, a rapid and progressive disease resulting in extensive destruction of subcutaneous tissue, muscle, and fat. Necrotizing fasciitis is responsible for the dramatic reports of "flesh-eating bacteria." In these cases, SpeA, SpeB, SpeC, and SpeF, as well as the bacterial cell surface M protein, function as superantigens. These diseases cause inflammation resulting in extensive tissue destruction.

Invasive streptococcal disease including cellulitis, necrotizing fasciitis, scarlet fever, and STSS occur in an estimated 11,000 patients per year. Death occurs in up to 15% of these patients (about 50% in STSS). In all of these cases, timely and adequate treatment of the GAS infection stops production of the superantigen and its effects.

Other Streptococcal Diseases

Untreated or insufficiently treated *S. pyogenes* infections may lead to other serious diseases, even in the absence of active infection. These severe nonsuppurative (non-pus-forming) poststreptococcal diseases usually occur about 1 to 4 weeks after the onset of a streptococcal infection. The immune response to the invading pathogen produces antibodies that cross-react with host tissue antigens on the heart, joints, and kidneys, resulting in damage to these tissues. The most serious of these diseases is **rheumatic fever** caused by rheumatogenic strains of *S. pyogenes*. These strains contain cell surface antigens that are similar to heart valve and joint antigens. Rheumatic fever is an autoimmune disease; antibodies directed against streptococcal antigens also react with heart valve and joint antigens, causing inflammation and tissue destruction (↩ Section 28.9). Damage to host tissues may be permanent, and is often exacerbated by later streptococcal infections that lead to recurring bouts of rheumatic fever.

Another nonsuppurative disease is acute poststreptococcal glomerulonephritis, a painful kidney disease. This immune complex disease develops following infection with *S. pyogenes* due to the formation of streptococcal antigen–antibody complexes in the blood. The immune complexes lodge in the glomeruli (filtration membranes of the kidney), causing inflammation of the kidney (nephritis) accompanied by severe pain. Within several days, the complexes are usually dissolved and the patient returns to normal. Unfortunately, even timely antibacterial treatment may not prevent glomerulonephritis. Only a few strains of *S. pyogenes*, so-called nephritogenic strains, produce this painful disease, but up to 15% of infections with nephritogenic strains cause glomerulonephritis (↩ Section 28.9).

Because infection induces strain-specific immunity, reinfection by a particular *S. pyogenes* strain is rare. However, there may be up to 150 different strains defined by distinct cell surface M proteins. Thus, an individual can be infected multiple times by different *S. pyogenes* strains. There are no available vaccines to prevent *S. pyogenes* infections.

Diagnosis of *Streptococcus pyogenes*

Several rapid antigen detection (RAD) systems have been developed for identification of *S. pyogenes*. Surface antigens are first extracted by enzymatic or chemical means directly from a swab of the patient's throat. The antigens are then detected using antibodies specific for surface proteins of *S. pyogenes* with immunological methods such as latex bead agglutination, fluorescent antibody staining, and enzyme immunoassay (EIA), methods described in Chapter 31. Using these methods, clinical specimens can be quickly processed, sometimes in just a few minutes. Rapid diagnostic tests allow the physician to initiate appropriate antibiotic therapy to treat GAS infections and prevent more serious disease.

A more accurate confirmation of GAS infection is a positive culture from the throat or lesion grown on sheep blood agar (↩ Figure 27.19*a*). Although the RAD tests are nearly as specific as throat cultures, they can be up to 40% less sensitive, leading to false-negative reports. Throat cultures take up to two days to process, hence the popularity of the RAD tests. Serology tests are the most sensitive tests available for identifying recent streptococcal infections. Patients are examined for the presence or increase of antibodies (rise in titer) to streptococcal antigens. The detection of new antibodies or an increase in the quantity of existing antibodies confirms a recent streptococcal infection (↩ Section 31.5).

Streptococcus pneumoniae

The other major pathogenic streptococcal species, *Streptococcus pneumoniae* (Figure 33.3*b*), causes invasive lung infections that often develop as secondary infections to other respiratory disorders. Strains of *S. pneumoniae* that are encapsulated are particularly pathogenic because they are potentially very invasive. Cells invade alveolar tissues (lower respiratory tract) in the lung, where the capsule enables the cells to resist phagocytosis and elicit a strong host inflammatory response. Reduced lung function, called *pneumonia*, can result from accumulation of recruited phagocytic cells and fluid. The *S. pneumoniae* cells can then spread from the focus of infection as a bacteremia, sometimes resulting in bone infections, middle ear infections, and endocarditis. Untreated invasive pneumococcal disease has a mortality rate of about 30%. Even with aggressive antimicrobial treatment, individuals hospitalized with pneumococcal pneumonia have up to 10% mortality.

Laboratory diagnosis of *S. pneumoniae* is based on the culture of gram-positive diplococci from either patient sputum or blood. There are over 90 different serotypes (antigenic capsule variants), and, as for *S. pyogenes*, infection induces immunity to only the infecting serotype of *S. pneumoniae*.

Prevention and Treatment

Effective vaccines are available for prevention of infection by the most common strains of *S. pneumoniae*. A vaccine for adults consists of a mixture of 23 capsular polysaccharides from the most prevalent pathogenic strains. The vaccine is recommended for the elderly, healthcare providers, individuals with compromised immunity, and others at high risk for respiratory infections. A vaccine containing seven capsular polysaccharides conjugated to diphtheria protein is recommended for children, age 2–23 months, to prevent ear infections (↩ Section 28.7). No vaccine is available for GAS.

Both GAS and *S. pneumoniae* can be treated with antibiotics. Penicillin G and its many derivatives are the agents of choice for treating GAS infections. Erythromycin and other antibacterial drugs are used in individuals who have penicillin allergies. *S. pneumoniae* infections respond quickly to penicillin G therapy, but up to 30% of pathogenic isolates now exhibit resistance to penicillin. Erythromycin and cefotaxime resistance is also found in some strains, and a few strains exhibit multiple drug resistance. Thus, each pathogenic isolate must be tested for antibiotic sensitivity. All strains are sensitive to vancomycin. Invasive disease such as pneumonia caused by drug-resistant *S. pneumoniae* is now a reportable disease in the United States; more than 3000 cases are reported annually.

MiniQuiz

- How does *Streptococcus pyogenes* infection cause rheumatic fever?
- What is the primary virulence factor for *Streptococcus pneumoniae*?

33.3 Diphtheria and Pertussis

Corynebacterium diphtheriae causes *diphtheria*, a severe respiratory disease that typically infects children. Diphtheria is preventable and treatable. *C. diphtheriae* is a gram-positive, nonmotile,

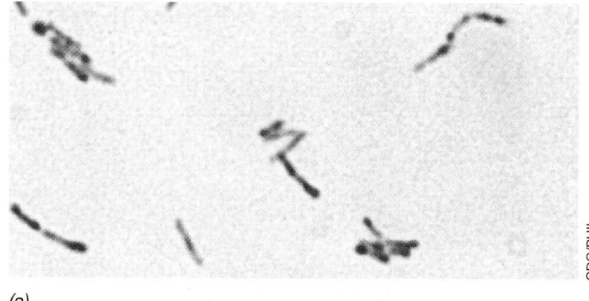

(a)

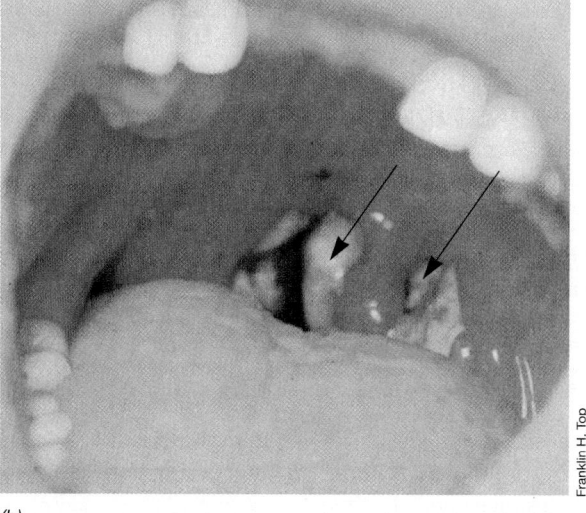

(b)

Figure 33.7 *Corynebacterium* **and diphtheria.** *(a)* Cells of *Corynebacterium diphtheriae* showing typical club-shaped appearance. The gram-positive cells are 0.5–1.0 μm in diameter and may be several micrometers in length. *(b)* Pseudomembrane (arrows) in an active case of diphtheria caused by the bacterium *C. diphtheriae*.

aerobic bacterium that forms irregular rods that may appear as club-shaped cells during growth (**Figure 33.7a**; ↩ Section 18.4). **Pertussis** or **whooping cough** is a serious respiratory disease caused by infection with *Bordetella pertussis*, a small, gram-negative, aerobic coccobacillus that is a member of the *Betaproteobacteria* (**Figure 33.8**). Pertussis affects mostly children but can cause serious respiratory disease for anyone. The disease is preventable and curable.

Diphtheria Epidemiology, Pathology, Prevention, and Treatment

Diphtheria was once a major childhood disease, but it is now rarely encountered because an effective vaccine is available. In the United States and other developed countries, the disease is virtually unknown. Worldwide, over 5000 fatal cases of diphtheria occur per year, largely because of a lack of effective immunization programs in less developed countries.

Corynebacterium diphtheriae enters the body, infecting the tissues of the throat and tonsils. The organism spreads from healthy carriers or infected individuals to susceptible individuals by airborne droplets. Previous infection or immunization provides

UNIT 11

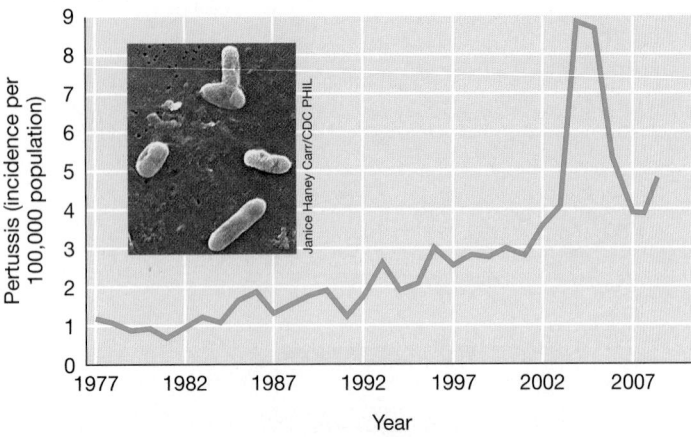

Figure 33.8 *Bordetella* **and pertussis.** The scanning electron micrograph (inset) shows the coccobacillus *Bordetella* sp. The variably shaped organisms range from 0.2 to 0.5 μm in diameter and are up to 1.0 μm in length. The graph shows the incidence of pertussis per 100,000 population caused by respiratory infection with *Bordetella pertussis*. There were 25,616 cases of pertussis in 2005, mostly in infants and school-age children, triple the number of 2001. After 2005, the incidence declined significantly, but was rising again by 2009. Data are from the Centers for Disease Control and Prevention, Atlanta, Georgia, USA.

resistance to the effects of the potent diphtheria exotoxin. Throat tissues respond to *C. diphtheriae* infection by forming a characteristic lesion called a *pseudomembrane* (Figure 33.7b), which consists of damaged host cells and cells of *C. diphtheriae*. Pathogenic strains of *C. diphtheriae* lysogenized by bacteriophage β produce a powerful exotoxin called diphtheria toxin that inhibits eukaryotic protein synthesis, leading to cell death (⮂ Figure 27.21). Death from diphtheria is usually due to a combination of the effects of partial suffocation and tissue destruction by exotoxin. In untreated infections, the toxin can cause systemic damage to the heart (about 25% of diphtheria patients develop myocarditis), kidneys, liver, and adrenal glands. *C. diphtheriae* isolated from the throat is diagnostic for diphtheria. Nasal or throat swabs are used to inoculate blood agar, tellurite medium, or the selective Loeffler's medium that inhibits the growth of most other respiratory pathogens.

Prevention of diphtheria is accomplished with a highly effective toxoid vaccine, part of the DTaP (diphtheria toxoid, tetanus toxoid, and acellular pertussis) vaccine (⮂ Section 28.7). Penicillin, erythromycin, and gentamicin are generally effective for stopping *C. diphtheriae* growth and further toxin production, but do not alter the effects of preformed toxin. Diphtheria antitoxin (an antiserum produced in horses) contains neutralizing antibodies, but is available only for serious acute cases of diphtheria. Early administration of both antibiotics and antitoxin is necessary for effective treatment of the acute disease.

Pertussis

Pertussis, also known as whooping cough, is an acute, highly infectious respiratory disease now observed frequently in children under 19 years of age. Infants less than 6 months of age, who are too young to be effectively vaccinated, have the highest incidence of disease and also have the most severe disease. *B. pertussis*

attaches to cells of the upper respiratory tract by producing a specific adherence factor called *filamentous hemagglutinin antigen*, which recognizes a complementary molecule on the surface of host cells. Once attached, *B. pertussis* grows and produces *pertussis exotoxin*. This potent toxin induces synthesis of cyclic adenosine monophosphate (cyclic AMP), which is at least partially responsible for the events that lead to host tissue damage.

B. pertussis also produces an endotoxin, which may induce some of the symptoms of whooping cough. Clinically, whooping cough is characterized by a recurrent, violent cough that can last up to 6 weeks. The spasmodic coughing gives the disease its name; a whooping sound results from the patient inhaling deep breaths to obtain sufficient air. Worldwide, there are up to 50 million cases and over 250,000 pertussis deaths each year, most in developing countries. *B. pertussis* is endemic worldwide and pertussis remains a problem, even in developed countries, usually due to inadequate immunization.

Pertussis Epidemiology

In the United States there has been an upward trend of *B. pertussis* infections and disease since the 1980s, reversing a trend that started with the introduction of an effective pertussis vaccine. In 1976, the year of lowest prevalence and incidence, there were only 1010 reported cases of pertussis. By contrast, in 2005, there were 25,616 cases. Although the numbers of infections have declined in recent years compared to the peak incidence in 2004–2005, the incidence is still significantly higher than in the 1990s (Figure 33.8). In the United States pertussis causes about 14 deaths per year. About 60% of recent cases were in adolescents and adults of all ages who lacked appropriate immunity. About 13% of cases were in children less than 6 months of age who had not yet received all of the recommended doses of pertussis vaccine. Up to 32% of coughs lasting 1 to 2 weeks or longer may be caused by *B. pertussis*. Pertussis is an endemic disease; incidence rises cyclically as populations become susceptible and are exposed to the pathogen. Lack of appropriate immunization at all ages may be adding to the overall higher incidence of pertussis as compared to recent decades.

Pertussis Diagnosis, Prevention, and Treatment

Diagnosis of whooping cough can be made by fluorescent antibody staining of a nasopharyngeal swab specimen or by actual culture of the organism. For best recovery of *B. pertussis*, a nasopharyngeal aspirate is inoculated directly onto a blood–glycerol–potato extract agar plate (although not selective, this rich medium supports good recovery of *B. pertussis*). The β-hemolytic colonies containing small gram-negative coccobacilli are tested for *B. pertussis* by a latex bead agglutination test or are stained with a fluorescent antibody specific for *B. pertussis* for positive identification. A polymerase chain reaction (PCR) test is considered the most sensitive and preferred diagnostic test. Improved diagnostic and reporting techniques may be one reason for the recent observed increase in pertussis cases in the United States, but the disease may still be underreported, especially in adolescents and adults.

A vaccine consisting of proteins derived from *B. pertussis* is part of the routinely administered DTaP vaccine. This vaccine is normally given to children at appropriate intervals beginning soon after birth (Section 28.7). The acellular pertussis vaccine has fewer side effects than the older pertussis vaccines and has caused no deaths. It is also recommended for adolescents and certain populations of adults (healthcare and childcare workers) as well as young children.

Worldwide, immunization programs should be targeted to children, but immunization of adolescents and adults should also be a priority because vaccinated individuals lose effective immunity within 10 years and can transmit *B. pertussis* to young children. Vaccination of a large percentage of the population is necessary to build herd immunity (Section 32.5).

Cultures of *B. pertussis* are killed by ampicillin, tetracycline, and erythromycin, although antibiotics alone do not seem to be sufficient to kill the pathogen *in vivo*: A patient with whooping cough remains infectious for up to 2 weeks following commencement of antibiotic therapy, indicating that the immune response may be more important than antibiotics for eliminating *B. pertussis* from the body.

MiniQuiz
- Is the pathogenesis of diphtheria due to infection? Is the pathogenesis of whooping cough due to infection?
- What measures can be taken to decrease the current incidence of pertussis in a population?

33.4 *Mycobacterium*, Tuberculosis, and Hansen's Disease

Tuberculosis (TB) is caused by the gram-positive, acid-fast bacillus *Mycobacterium tuberculosis* (Section 18.5). The German microbiologist Robert Koch isolated and described the causative agent in 1882 (Section 1.8). A related *Mycobacterium* species, *Mycobacterium leprae*, causes Hansen's disease (leprosy). All mycobacteria share acid-fast properties due to the waxy mycolic acid constituent of their cell wall. Mycolic acid allows these organisms to retain carbol-fuchsin, a red dye, after washing in 3% hydrochloric acid in alcohol (**Figure 33.9**; Section 18.5).

Tuberculosis Epidemiology

Mycobacterium tuberculosis is easily transmitted by the respiratory route; even normal conversation can spread the organism from person to person. At one time, TB was the most important infectious disease of humans and accounted for one-seventh of all deaths worldwide. Presently, over 13,000 new cases of TB and over 600 deaths occur each year in the United States. Worldwide, TB still accounts for over 1.4 million deaths per year, almost 11% of all deaths due to infectious disease (Table 32.1). About one-third of the world's population has been infected with *M. tuberculosis*. Many new TB cases in the United States occur in acquired immunodeficiency syndrome (AIDS) patients.

Tuberculosis Pathology

The interaction of the human host and the bacterium *M. tuberculosis* is determined both by the virulence of the strain and the resistance of the host. Cell-mediated immunity plays a critical role in

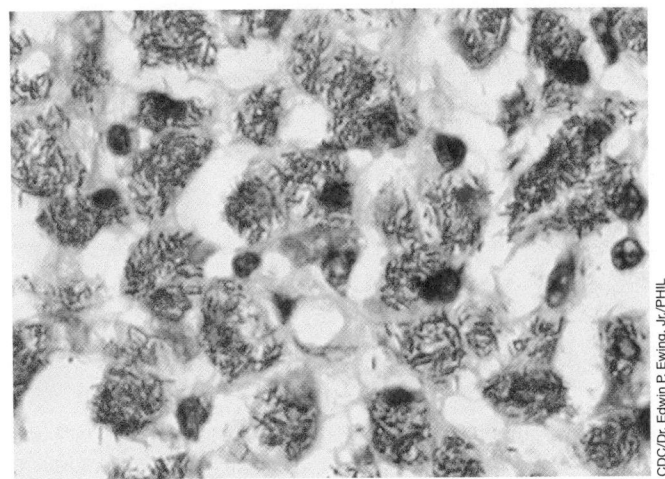

Figure 33.9 Mycobacteria. Here an acid-fast stained lymph node biopsy from a patient with HIV/AIDS displays *Mycobacteroum avium*. Multiple bacilli, stained red with carbol-fuchsin, are evident inside each cell. The individual rods are about 0.4 μm in diameter and up to 4 μm in length.

the prevention of active disease after infection. TB can be a *primary* infection (initial infection) or *postprimary* infection (reinfection). Primary infection typically results from inhalation of droplets containing viable *M. tuberculosis* bacteria from an individual with an active pulmonary infection. The inhaled bacteria settle in the lungs and grow. The host mounts an immune response to *M. tuberculosis*, resulting in a delayed-type hypersensitivity reaction (Section 28.9) and the formation of aggregates of activated macrophages, called *tubercles* (Figure 1.20). Mycobacteria often survive and grow within the macrophages, even with an ongoing immune response. In individuals with low resistance, the bacteria are not controlled and the pulmonary infection becomes acute, leading to extensive destruction of lung tissue, the spread of the bacteria to other parts of the body, and death. In these cases, *M. tuberculosis* survives both the low pH and the effects of the oxidative antibacterial products found in the lysosomes of phagocytes such as macrophages.

In most cases of TB, however, an obvious acute infection does not occur. The infection remains localized, is usually inapparent, and appears to end. But this initial infection hypersensitizes the individual to the bacteria or their products and consequently alters the response of the individual to subsequent or postprimary infections by *M. tuberculosis*. A diagnostic skin test, called the **tuberculin test**, can be used to measure this hypersensitivity. In a hypersensitive individual, tuberculin, a protein extract from *M. tuberculosis*, elicits a local immune inflammatory reaction within 1–3 days at the site of an intradermal injection. The reaction is characterized by induration (hardening) and edema (swelling) (Figure 28.6). An individual exhibiting this reaction is said to be *tuberculin-positive*, and many healthy adults show positive reactions as a result of previous inapparent infections. A positive tuberculin test does not indicate active disease, but only that the individual has been exposed to the organism in the past and has generated a cell-mediated immune response against *M. tuberculosis*.

UNIT 11

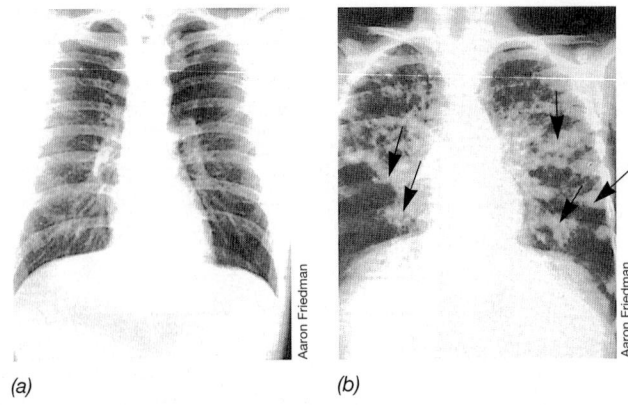

(a)   *(b)*

Figure 33.10 Tuberculosis X-ray. *(a)* Normal chest X-ray. The faint white lines are arteries and other blood vessels. *(b)* Chest X-ray of an advanced case of pulmonary tuberculosis; white patches (arrows) indicate areas of disease. These patches, or tubercles as they are called, may contain viable cells of *Mycobacterium tuberculosis*. Lung tissue and function is permanently destroyed by these lesions.

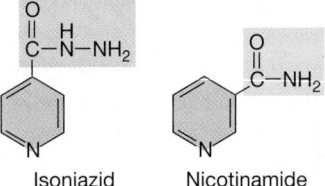

Isoniazid   Nicotinamide

Figure 33.11 Structure of isoniazid (isonicotinic acid hydrazide). Isoniazid is an effective chemotherapeutic agent for tuberculosis. Note the structural similarity to nicotinamide.

For most individuals, this cell-mediated immunity is protective and lifelong. However, some tuberculin-positive patients develop postprimary tuberculosis through reinfection from outside sources or as a result of reactivation of bacteria that have remained dormant in lung macrophages, often for years. For example, advanced age, malnutrition, overcrowding, stress, and hormonal changes may reduce effective immunity in untreated individuals and allow reactivation of dormant infections. Because latent *M. tuberculosis* can become activated many years after the initial exposure and immune response, individuals who have a positive tuberculin test are treated with antimicrobial agents for long periods of time. Postprimary mycobacterial infections often progress to chronic infections that result in destruction of lung tissue, followed by partial healing and calcification at the infection site. Chronic postprimary TB often results in a gradual spread of tubercular lesions in the lungs. Bacteria are found in the sputum in individuals with active disease, and areas of destroyed tissue can be seen in X-rays (**Figure 33.10**).

Tuberculosis Prevention and Treatment

Individuals who have active cases of TB may spread the disease simply by coughing or speaking. Because TB is highly contagious, the U.S. Occupational Safety and Health Administration has stringent requirements for the protection of healthcare workers who are responsible for TB patient care. For example, patients with infectious tuberculosis must be hospitalized in negative-pressure rooms. In addition, healthcare workers who have patient contact must be provided with personally fitted face masks having high-efficiency particulate air (HEPA) filters to prevent the passage of *M. tuberculosis* cells in sputum or on dust particles.

Antimicrobial therapy of TB has been a major factor in control of the disease. Streptomycin was the first effective antibiotic, but the real revolution in treatment came with the discovery of isonicotinic acid hydrazide, called *isoniazid* (INH) (**Figure 33.11**). This drug, specific for mycobacteria, is effective, inexpensive, relatively

nontoxic, and readily absorbed when given orally. Although the mode of action of isoniazid is not completely understood, it affects the synthesis of mycolic acid by *Mycobacterium*. Mycolic acid is a lipid that complexes with peptidoglycan in the mycobacterial cell wall.

Isoniazid probably functions as a growth factor analog of the structurally related molecule, nicotinamide. As such, isoniazid would be incorporated in place of nicotinamide and inactivate enzymes required for mycolic acid synthesis. Treatment of mycobacteria with very small amounts of isoniazid (as little as 5 picomoles [pmol] per 10^9 cells) results in complete inhibition of mycolic acid synthesis, and continued incubation results in loss of outer areas of the cell wall, a loss of cellular integrity, and death. Following treatment with isoniazid, mycobacteria lose their acid-fast properties, in keeping with the role of mycolic acid in this staining property.

Treatment is typically achieved with daily doses of isoniazid and rifampin for 2 months, followed by biweekly doses for a total of 9 months. This treatment eradicates the pathogen and prevents emergence of antibiotic-resistant organisms. Failure to complete the entire prescribed treatment may allow the infection to be reactivated, and reactivated organisms are often resistant to the original treatment drugs. Incomplete treatment encourages antibiotic resistance because a high rate of spontaneous mutations in surviving *M. tuberculosis* promotes rapid acquisition of resistance to single antibiotics. To ensure treatment and thus discourage development of antibiotic-resistant organisms, direct observation of treatment may be necessary for noncompliant individuals. In populations such as hospitals and nursing homes, where resistant mycobacterial strains are most likely to be present, patients are routinely treated with up to four drugs for 2 months, followed by rifampin–isoniazid treatment for a total of 6 months. Multiple drug therapy reduces the possibility that strains having resistance to more than one drug will emerge.

Resistance of *M. tuberculosis* to isoniazid and other drugs, however, is increasing, especially in AIDS patients. A number of strains that are resistant to both isoniazid and rifampin have already emerged. Treatment of these strains, called *multi-drug-resistant tuberculosis strains* (MDR TB), requires the use of second-line tuberculosis drugs that are generally more toxic, less effective, and more costly than rifampin and isoniazid. A World Health Organization (WHO) survey indicated that up to 20% of MDR TB strains are extensively drug-resistant (XDR TB) strains. XDR TB strains have resistance to virtually all TB drugs, including the second-line drugs. Preventing emergence of these strains requires better diagnostic and drug susceptibility tests in addition to new anti-TB treatment drugs and regimens.

In many countries, immunization with an attenuated strain of *Mycobacterium bovis*, the *bacillus Calmette-Guerin* (*BCG*) strain, is routine for prevention of TB. However, in the United States and other countries where the prevalence of *Mycobacterium tuberculosis* infection and disease is relatively low, immunization with BCG is discouraged. The live BCG vaccine induces a delayed-type hypersensitivity response, and all individuals who receive it develop a positive tuberculin test. This compromises the tuberculin test as a diagnostic and epidemiologic indicator for the spread of *M. tuberculosis* infection.

Mycobacterium leprae and Hansen's Disease (Leprosy)

Mycobacterium leprae, discovered by the Norwegian scientist G.A. Hansen in 1873, causes Hansen's disease, also known as *leprosy*. *M. leprae* is the only *Mycobacterium* species that has not been grown on artificial media. The armadillo is the only experimental animal that has been successfully used to grow *M. leprae* and achieve symptoms similar to those in the human disease.

The most serious form of Hansen's disease is *lepromatous* leprosy, characterized by folded, bulblike lesions on the body, especially on the face and extremities (**Figure 33.12**). The lesions are due to the growth of *M. leprae* cells in the skin and may contain up to 10^9 bacterial cells per gram of tissue. Like other mycobacteria, *M. leprae* from the lesions stain deep red with carbol-fuchsin in the acid-fast staining procedure, providing a rapid, definitive demonstration of active infection. Lepromatous leprosy has a very poor prognosis. In severe cases the disfiguring lesions lead to destruction of peripheral nerves and loss of motor function.

Many Hansen's disease patients exhibit less-pronounced lesions from which no bacterial cells can be recovered. These individuals have the *tuberculoid* form of the disease. Tuberculoid leprosy is characterized by a vigorous delayed-type hypersensitivity response (↩ Section 28.9) and a good prognosis for spontaneous

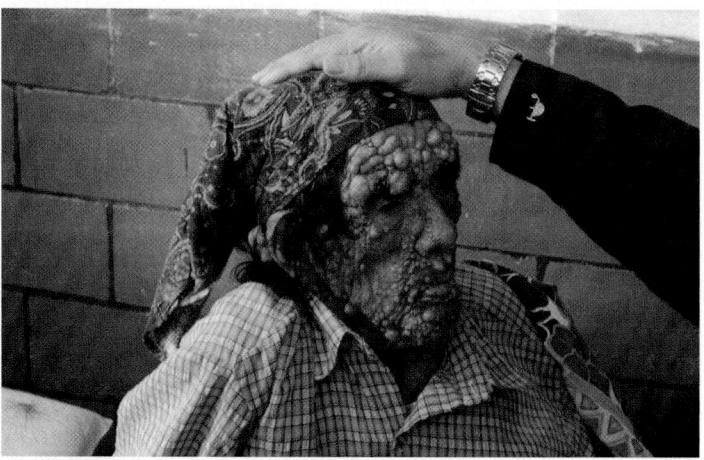

Figure 33.12 Lepromatous leprosy lesions on the skin. Lepromatous leprosy is caused by infection with *Mycobacterium leprae*. The lesions can contain up to 10^9 bacterial cells per gram of tissue, indicating an active uncontrolled infection with a poor prognosis.

recovery. Hansen's disease of either form, and the continuum of intermediate forms, is treated using a multiple drug therapy (MDT) protocol, which includes some combination of dapsone (4,4′-sulfonylbisbenzeneamine), rifampin, and clofazimine. As in TB, drug-resistant strains have appeared, especially after inadequate treatment or treatment with single drugs. Extended drug therapy of up to 1 year with a MDT protocol is required for eradication of the organism.

The pathogenicity of *M. leprae* is due to a combination of delayed hypersensitivity and the invasiveness of the organism. Transmission is by direct contact as well as respiratory routes, but Hansen's disease is not as highly contagious as TB. The time from exposure to onset of disease varies from several weeks to years, or even decades. During this time, *M. leprae* cells grow within macrophages, causing an intracellular infection that can result in large numbers of bacteria within the skin, leading to the characteristic lesions.

In many areas of the world, the incidence of Hansen's disease is very low. Worldwide, however, over 750,000 new cases of the disease are reported each year. About 100 cases are reported annually in the United States, mostly in southern states, or among immigrants from the Caribbean islands or Central America. Ninety percent of worldwide cases are in Madagascar, Mozambique, Tanzania, and Nepal. Up to 2 million people are permanently disabled as a result of Hansen's disease, but because of the chronic nature and long latent period of the disease, it may be unrecognized and unreported in as many as 12 million people.

Other Pathogenic *Mycobacterium* Species

A common pathogen of dairy cattle, *Mycobacterium bovis* is pathogenic for humans as well as other animals. *M. bovis* enters humans via the intestinal tract, typically from the ingestion of unpasteurized milk. After a localized intestinal infection, the organism eventually spreads to the respiratory tract and initiates the classic symptoms of TB. *M. bovis* is a different organism from *M. tuberculosis*, although the genomes of the two organisms are very similar. There is no observed difference in their infectivity and pathogenesis in humans, although the genome of *M. bovis* has several gene deletions compared with that of *M. tuberculosis*. Pasteurization of milk and elimination of diseased cattle have eradicated bovine-to-human transmission of TB in developed countries.

A number of other *Mycobacterium* species are also occasional human pathogens. For example, *M. kansasii*, *M. scrofulaceum*, *M. chelonae*, and a few other mycobacterial species can cause disease. Respiratory disease due to the *Mycobacterium avium* complex of organisms (including *M. avium* and *M. intracellulare*) is particularly dangerous in AIDS patients or other immune-compromised individuals; these opportunistic pathogens rarely infect healthy individuals (Figure 33.9).

MiniQuiz

- Why is *Mycobacterium tuberculosis* a widespread respiratory pathogen?

- Describe factors that contribute to drug resistance in mycobacterial infections.

33.5 *Neisseria meningitidis*, Meningitis, and Meningococcemia

Meningitis is an inflammation of the meninges, the membranes that line the central nervous system, especially the spinal cord and brain. Meningitis can be caused by viral, bacterial, fungal, or protist infections. Here we will deal with infectious bacterial meningitis caused by *Neisseria meningitidis* and a related infection, **meningococcemia**.

Neisseria meningitidis, often called *meningococcus*, is a gram-negative, nonsporulating, obligately aerobic, oxidase-positive, encapsulated diplococcus (**Figure 33.13**; ↩ Section 17.10), about 0.6–1.0 μm in diameter. At least 13 pathogenic strains of *N. meningitidis* are recognized. Antigenic differences in capsular polysaccharides distinguish each strain.

Epidemiology and Pathology

Meningococcal meningitis often occurs in epidemics, usually in closed populations such as military installations and college campuses. It typically strikes older school-age children and young adults. Up to 30% of individuals carry *N. meningitidis* in the nasopharynx with no apparent harmful effects. In epidemic situations, the prevalence of carriers may rise to 80%. The trigger for conversion from the asymptomatic carrier state to pathogenic acute infection is unknown.

In an acute meningococcus infection, the bacterium is transmitted to the host, usually via the airborne route, and attaches to the cells of the nasopharynx. Once there, the organism gains access to the bloodstream, causing bacteremia and upper respiratory tract symptoms. The bacteremia sometimes leads to fulminant meningococcemia, characterized by septicemia, intravascular coagulation, shock, and death in over 10% of cases. Meningitis is another possible serious outcome of infection. Meningitis is characterized by sudden onset of headache, vomiting, and stiff neck, and can progress to coma and death in a matter of hours. Up to 3% of acute meningococcal meningitis victims die.

In the United States, there were 1057 cases of serious meningococcal disease in 2008, the lowest number since 1977. The long-term decreased incidence indicates the success of widespread vaccination in susceptible populations. However, the mortality rate in recent years was over 10%.

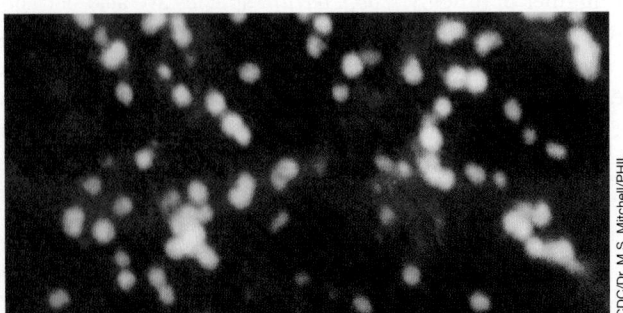

Figure 33.13 Fluorescent antibody stain of *Neisseria meningitidis*. The organism causes meningitis and meningococcemia. This specimen is from the cerebrospinal fluid of an infected patient. The individual cocci are about 0.6–1.0 μm in diameter.

Diagnosis, Prevention, and Treatment

Specimens isolated from nasopharyngeal swabs, blood, or cerebrospinal fluid are inoculated onto modified Thayer–Martin medium (↩ Figure 31.5*b*), a selective medium that suppresses the growth of most normal flora, but allows the growth of the pathogenic members of the genus, *N. meningitidis* and *Neisseria gonorrhoeae*. Colonies showing gram-negative diplococcus morphology and a positive oxidase test are presumptively identified as *Neisseria* (↩ Table 31.3). Due to the rapid onset of life-threatening symptoms, preliminary diagnosis is often based on clinical symptoms and treatment is started before culture tests confirm infection with *N. meningitidis*.

Penicillin G is the drug of choice for the treatment of *N. meningitidis* infections. However, resistant strains have been reported. Chloramphenicol is the accepted alternative agent for treatment of infections in penicillin-sensitive individuals. A number of broad-spectrum cephalosporins are also effective.

Naturally occurring strain-specific antibodies acquired by subclinical infections are effective for preventing infections in most adults. Vaccines consisting of purified polysaccharides or polysaccharides from the most prevalent pathogenic strains conjugated to diphtheria toxin are available and are used to immunize susceptible individuals. The vaccines are used to prevent infection in certain susceptible populations such as military recruits and students living in dormitories. In addition, rifampin is often used as a chemoprophylactic antimicrobial drug to eradicate the carrier state and prevent disease in close contacts of infected individuals.

Other Causes of Meningitis

A number of other organisms can also cause meningitis. Acute meningitis is usually caused by one of the pyogenic bacteria such as *Staphylococcus*, *Streptococcus*, or *Haemophilus influenzae*. *H. influenzae* primarily infects young children. An effective vaccine for preventing *H. influenzae* meningitis is available and is required in the United States for school-age children (↩ Section 28.7).

Several viruses also cause meningitis. Among these are herpes simplex virus, lymphocytic choriomeningitis virus, mumps virus, and a variety of enteroviruses. In general, viral meningitis is less severe than bacterial meningitis.

MiniQuiz

• Identify the symptoms and causes of meningitis.

• Describe the infection by *Neisseria meningitidis* and the resulting development of meningococcemia.

33.6 Viruses and Respiratory Infections

The most prevalent human infectious diseases are caused by viruses. Most viral diseases are acute, self-limiting infections, but some can be problematic in healthy adults. We begin here by describing measles, mumps, rubella, and chicken pox, all common, endemic viral diseases transmitted in infectious droplets by an airborne route.

Measles

Measles (*rubeola* or *7-day measles*) affects susceptible children as an acute, highly infectious, often epidemic disease. The measles virus is a paramyxovirus, a negative-strand RNA virus (Section 21.9) that enters the nose and throat by airborne transmission, quickly leading to systemic viremia. Symptoms start with nasal discharge and redness of the eyes. As the disease progresses, fever and cough appear and rapidly intensify, followed by a characteristic rash (**Figure 33.14**); symptoms generally persist for 7–10 days. Circulating antibodies to measles virus are measurable about 5 days after initiation of infection; the serum antibodies and T-cytotoxic lymphocytes combine to eliminate the virus from the system. Possible postinfection complications include middle ear infection, pneumonia, and, in rare cases, measles encephalomyelitis. Encephalomyelitis has a mortality rate of nearly 20% and can cause neurological disorders including a form of epilepsy. Of the 131 measles cases that occurred in 2008, 15 of the infected individuals were hospitalized.

Although once a common childhood illness, measles is generally limited now to rather isolated outbreaks in the United States because of widespread immunization programs begun in the

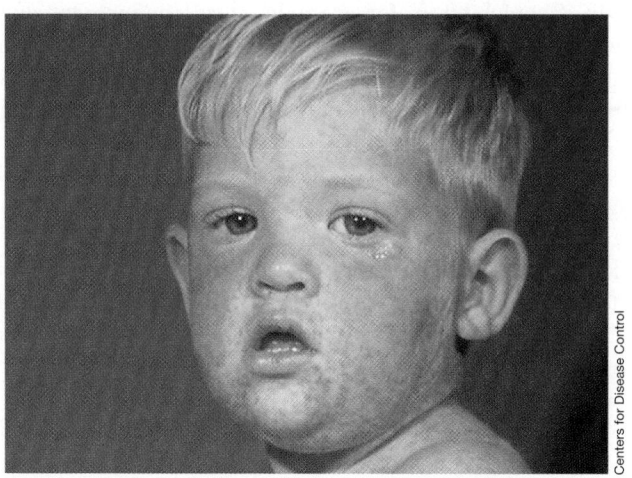

(a)

(b)

Figure 33.14 Measles in children. *(a)* The light pink rash starts on the head and neck, and *(b)* spreads to the chest, trunk, and limbs. Discrete papules coalesce into blotches as the rash progresses for several days.

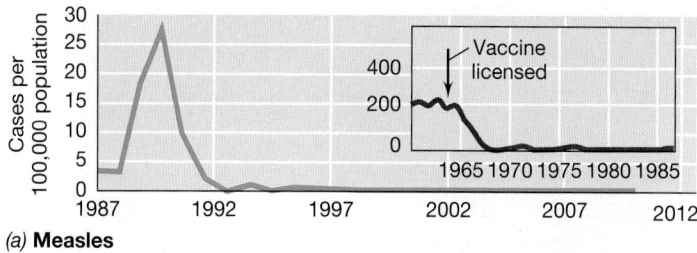

(a) **Measles**

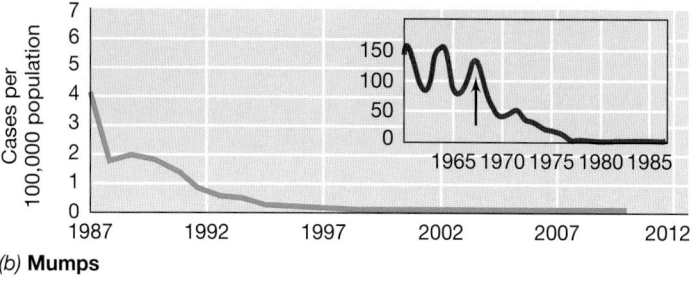

(b) **Mumps**

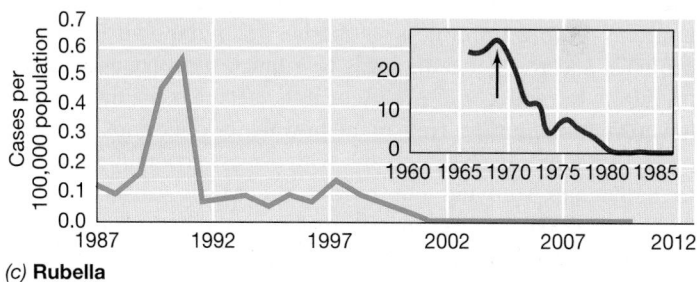

(c) **Rubella**

Figure 33.15 Viral diseases and vaccines. Major childhood viral diseases are now controlled by the MMR (measles, mumps, rubella) vaccine in the United States. Data are from the Centers for Disease Control and Prevention, Atlanta, Georgia.

mid-1960s (**Figure 33.15a**). Outbreaks generally occur only in populations that were not immunized or were inadequately immunized. Over 90% of the cases were either acquired outside the United States or were associated with contact with travelers to foreign countries. Worldwide, measles remains endemic and still causes over 400,000 annual deaths, mostly in children. Because the disease is highly infectious, all public school systems in the United States require proof of immunization before a child can enroll. Active immunization is done with an attenuated virus preparation as part of the MMR (measles, mumps, rubella) vaccine (Figure 28.14). A childhood case of measles generally confers lifelong immunity to reinfection.

Mumps

Mumps, like measles, is caused by a paramyxovirus and is also highly infectious. Mumps is spread by airborne droplets, and the disease is characterized by inflammation of the salivary glands, leading to swelling of the jaws and neck (**Figure 33.16**). The virus spreads through the bloodstream and may infect other organs,

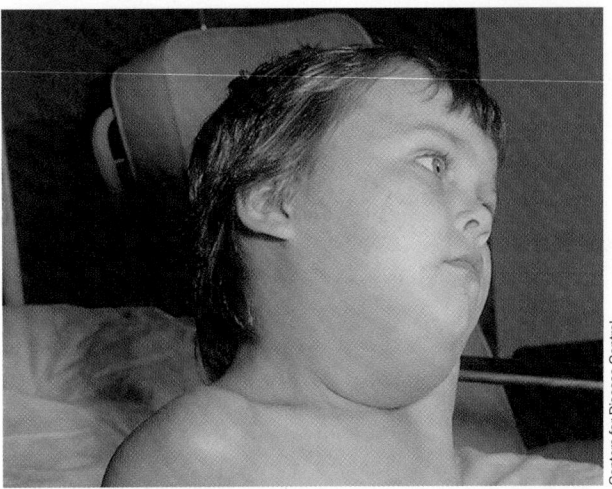

Figure 33.16 **Mumps.** Glandular swelling characterizes infection with the mumps virus.

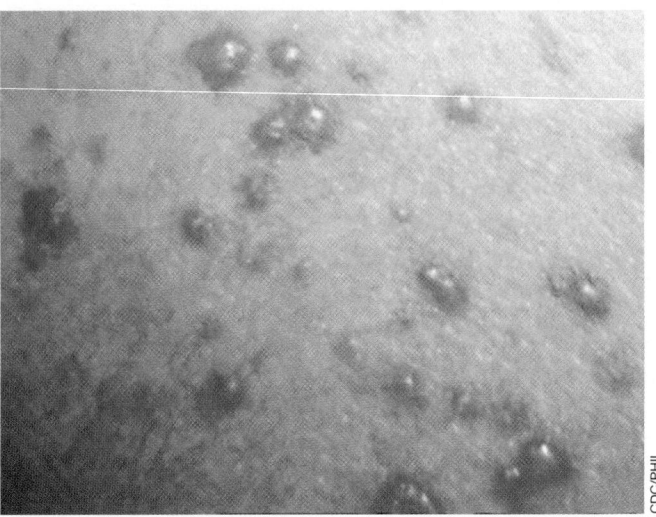

Figure 33.17 **Chicken pox.** Mild papular rash associated with the infection by varicella-zoster virus (VZV), the herpesvirus that causes chicken pox.

including the brain, testes, and pancreas. Severe complications may include encephalitis and, very rarely, sterility. The host immune response produces antibodies to mumps virus surface proteins, and this generally leads to a quick recovery and lasting immunity to reinfection. An attenuated vaccine is highly effective for preventing mumps. Hence, the prevalence of mumps in developed countries is usually very low, with disease generally restricted to individuals who did not receive the MMR vaccine (Figure 33.15b). In 2006, however, an outbreak centered in the midwestern United States involved more than 5000 cases, significantly up from a normal number of less than 300 cases per year since 2001. The outbreak affected mainly young adults (18–34). As a result, recommendations for immunizations were revised to target school-age children, healthcare workers, and adults at high risk.

Rubella

Rubella (*German measles* or *3-day measles*) is caused by a single-stranded, positive-sense RNA virus of the togavirus group (Section 9.11). Disease symptoms resemble measles but are generally milder. Rubella is less contagious than measles, and thus a significant proportion of the population has never been infected. During the first three months of pregnancy, however, rubella virus can infect the fetus by placental transmission and cause serious fetal abnormalities including stillbirth, deafness, heart and eye defects, and brain damage. Thus, women should not be immunized with the rubella vaccine or contract rubella during pregnancy. For this reason, routine childhood immunization against rubella should be practiced. An attenuated virus is administered as part of the MMR vaccine (Figure 28.14). The low incidence of cases since 2001, coupled with the high degree of protection from the vaccine and the relatively low infectivity of the virus, suggest that rubella is no longer endemic in the United States (Figure 33.15c).

Chicken Pox and Shingles

Chicken pox (*varicella*) is a common childhood disease caused by the varicella-zoster virus (VZV), a DNA herpesvirus (Section 21.14). VZV is highly contagious and is transmitted by infectious

droplets, especially when susceptible individuals are in close contact. In schoolchildren, for example, close confinement during the winter months leads to the spread of VZV through airborne droplets from infected classmates and through contact with contaminated fomites. The virus enters the respiratory tract, multiplies, and is quickly disseminated via the bloodstream, resulting in a systemic papular rash that quickly heals, rarely leaving disfiguring marks (**Figure 33.17**). An attenuated virus vaccine is now used in the United States. The reported annual incidence of chicken pox, now about 40,000 cases per year, is about one-fourth of the number of cases reported prior to 1995, the year the vaccine was licensed for use. Since 2003, VZV infections have been nationally notifiable, resulting in an increased number of reported cases.

VZV establishes a lifelong latent infection in nerve cells. The virus occasionally migrates from this reservoir to the skin surface, causing a painful skin eruption referred to as *shingles* (*zoster*). Shingles most commonly strikes immunosuppressed individuals or the elderly. The prophylactic use of human hyperimmune globulin prepared against the virus is useful for preventing the onset of symptoms of shingles. Such therapy is advised only for patients for whom secondary infections such as pneumonia or encephalitis, occasionally associated with shingles, may be life-threatening. To prevent shingles, a vaccine is recommended for individuals over 60 years of age. The vaccine stimulates antibody and T-cytotoxic cell immunity to VZV, keeping VZV from migrating out of nerve ganglia to skin cells.

MiniQuiz

- How do the genomes of the measles virus and the German measles virus differ?
- Describe the potential serious outcomes of infection by measles, mumps, rubella, and VZV viruses.
- Identify the effects of immunization on the incidence of measles, mumps, rubella, and chicken pox.

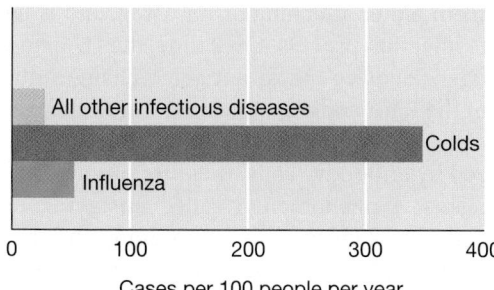

Figure 33.18 Colds and influenza. These viral diseases are the leading causes of acute infectious disease in the United States. This pattern is typical for recent years. Colds and influenza cause much higher morbidity as compared to all other infectious diseases.

33.7 Colds

Colds are the most common of infectious diseases. As shown in **Figure 33.18**, people acquire about ten colds for every other infectious disease, except influenza. Colds are viral infections that are transmitted via droplets spread from person to person in coughs, sneezes, and respiratory secretions. Colds are usually of short duration, lasting 1 week or less, and the symptoms are milder than other respiratory diseases such as influenza. **Table 33.1** compares the symptoms of colds and influenza.

Each person averages more than three colds per year throughout his or her lifetime (Figure 33.18). Cold symptoms include rhinitis (inflammation of the nasal region, especially the mucous membranes), nasal obstruction, watery nasal discharges, and a general feeling of malaise, usually without fever. *Rhinoviruses*, positive-sense, single-stranded RNA viruses of the picornavirus group (**Figure 33.19a** and ♀♂ Section 21.8), are the most common causes of colds. At least 115 different rhinoviruses have been identified. About 25% of colds are due to infections with other viruses. *Coronaviruses* (Figure 33.19b) cause 15% of all colds in adults. Adenoviruses, coxsackie viruses, respiratory syncytial viruses (RSV), and orthomyxoviruses are collectively responsible for about 10% of colds. Each of these viruses may also cause more serious disease. For example, one adenovirus strain produces a severe and sometimes lethal respiratory infection.

Colds generally induce a specific, local, neutralizing IgA antibody response. However, the number of potential infectious

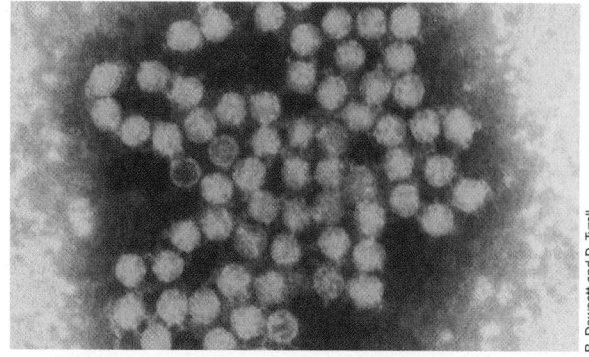

(a)

(b)

Figure 33.19 Common cold viruses. Transmission electron micrographs. *(a)* Human rhinovirus. Each rhinovirus virion is about 30 nm in diameter. *(b)* Human coronavirus. Each coronavirus virion is about 60 nm in diameter.

agents makes immunity due to previous exposure very unlikely. The sheer numbers of viruses that might cause a cold also preclude the development of useful vaccines.

Aerosol transmission of the virus is probably the major means of spreading colds, although experiments with volunteers suggest that direct contact and fomite contact are also methods of transmission. Most antiviral drugs are ineffective against the common cold, but a pyrazidine derivative (**Figure 33.20a**) has shown promise for preventing colds after virus exposure. In addition, new experimental antiviral drugs are being designed based on information derived from three-dimensional structures. For

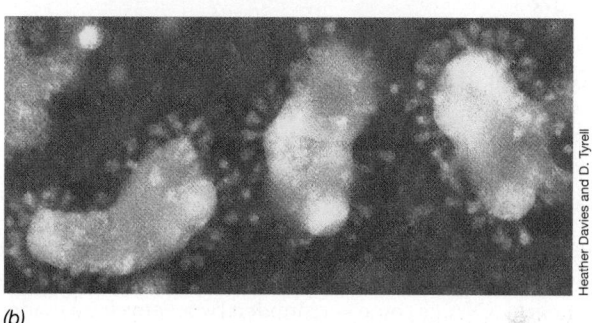

(a)

(b)

Figure 33.20 Experimental antirhinovirus drugs. *(a)* The structure of 3-methoxy-6-[4-(3-methylphenyl)]-1-piperazinyl. *(b)* The structure of WIN 52084, a receptor-blocking drug.

Table 33.1 *Colds and influenza*		
Symptoms	*Cold*	*Influenza*
Fever	Rare	Common (39–40°C); sudden onset
Headache	Rare	Common
General malaise	Slight	Common; often quite severe; can last several weeks
Nasal discharge	Common and abundant	Less common; usually not abundant
Sore throat	Common	Less common
Vomiting and/ or diarrhea	Rare	Common in children

example, the antirhinovirus drug WIN 52084 (Figure 33.20*b*) binds to the virus, changing its three-dimensional surface configuration and disrupting rhinovirus binding to the host cell receptor ICAM-1 (intercellular adhesion molecule-1), thus preventing infection. Alpha interferon, a cytokine, is also effective in preventing the onset of colds. Thus, there are several experimental possibilities for cold prevention and treatment, although none are widely accepted as effective and safe. Because colds are generally brief and self-limiting, treatment is aimed at controlling symptoms, especially nasal discharges, with antihistamine and decongestant drugs.

MiniQuiz

- Define the cause and symptoms of common colds.
- Discuss the possibilities for effective treatment and prevention of colds.

33.8 Influenza

Influenza is caused by an RNA virus of the orthomyxovirus group (♻ Section 21.9). Influenza virus is a single-stranded, negative-sense, helical RNA genome surrounded by an envelope made up of protein, a lipid bilayer, and external glycoproteins (**Figure 33.21**). There are three different types of influenza viruses: influenza A, influenza B, and influenza C. Here we consider only influenza A because it is the most important human pathogen.

Influenza Antigens and Genes

Each strain of influenza A virus can be identified by a unique set of surface glycoproteins. These glycoproteins are hemagglutinin (HA or H antigen) and neuraminidase (NA or N antigen). Each virus will have one type of HA and one type of NA on its surface. HA is important in the attachment of virus to the host cells. NA is instrumental for release of virus from host cells (Figure 33.21). Infection or immunization with an influenza strain results in production of IgA antibodies that are reactive with the HA and NA glycoproteins. When antibody binds to HA or NA, the virus

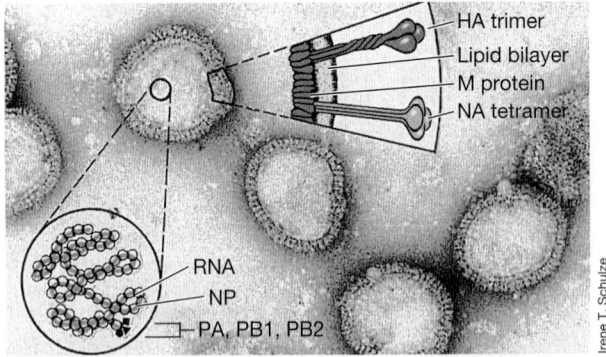

Figure 33.21 Electron micrograph of influenza virus. The photo shows the location of the major viral coat proteins and the nucleic acid. Each virion is about 100 nm in diameter. HA, hemagglutinin (three copies make up the HA coat spike); NA, neuraminidase (four copies make up the NA coat spike); M, coat protein; NP, nucleoprotein; PA, PB1, PB2, other internal proteins, some of which may have enzymatic functions.

is blocked from either attaching or releasing, and is neutralized, stopping the infection process. Over time, the HA and NA glycoprotein antigens acquire minor antigenic changes due to point mutations in the RNA coding sequences. These changes alter one or more amino acids in the glycoprotein, altering their ability to be recognized by antibody. Thus, these mutations create slightly altered antigens, a phenomenon called **antigenic drift**. As a result, immunity to a given virus strain diminishes as the strain mutates, and reinfection with the mutated strain can occur.

As we mentioned previously, the influenza A virus genome is single-stranded RNA. The RNA genome is arranged in a highly unusual manner; the genome is *segmented*, with single-stranded RNA genes found on each of eight distinct segments (♻ Section 21.9 and Figure 21.17*b*). During virus maturation in the host cell, the viral RNA segments are packaged randomly. To be infective, a virus must be packaged so it contains one copy of each of the eight gene segments.

Occasionally more than one strain of influenza infects a single animal at one time. In such a case, the two strains could infect a single cell, and gene segments from both viruses would be reproduced. When packaging occurs, the segments from the two strains may be mixed; an individual virus is likely to be a mosaic of the two infecting viruses, containing some, but not all, of the genes from each virus. In effect, the mixed-genome virus instantly becomes a new virus strain. This mixing of gene fragments between different strains of influenza virus is called *reassortment*.

Unique *reassortant viruses* result in **antigenic shift**, a major change in an antigen resulting from the total replacement of an RNA segment. Antigenic shift can immediately and completely change one or both of the major HA and NA viral glycoproteins and any of the other viral genes.

Influenza Epidemiology

Human influenza virus is transmitted from person to person through the air, primarily in droplets expelled during coughing and sneezing. The virus infects the mucous membranes of the upper respiratory tract and occasionally invades the lungs. Symptoms include a low-grade fever lasting 3–7 days, chills, fatigue, headache, and general aching (Table 33.1). Recovery is usually spontaneous and rapid. Most of the serious consequences of influenza infection occur from bacterial secondary infections in persons whose resistance has been lowered by the influenza infection. Especially in infants and elderly people, influenza is often followed by bacterial pneumonia; death, if it occurs, is usually due to the bacterial infection. Annually, influenza causes 3–5 million cases of severe illness and is implicated in 250,000–500,000 deaths worldwide.

Most infected individuals develop protective immunity to the infecting virus, making it impossible for a strain of the same antigenic type to cause widespread infection—an epidemic—until the virus encounters another susceptible population. Immunity is dependent on the production of secretory IgA antibodies and T-cytotoxic lymphocytes directed at HA and NA glycoproteins.

Influenza exists in human populations as an endemic viral disease, and severe localized influenza outbreaks occur every year

Table 33.2 *Influenza pandemics*

Year	Name	Strain
1889	Russian	H2N2
1900	Old Hong Kong	H3N8
1918	Spanish	H1N1
1957	Asian	H2N2
1968	Hong Kong	H3N2
2009	Swine	H1N1

from late autumn through the winter. Each year, antigenic drift results in some reduction of immunity in the population and is responsible for the recurrence of epidemics, severe widespread outbreaks, which occur in a 2- to 3-year cycle.

Influenza Pandemics

Pandemics, worldwide epidemics, are much less frequent than outbreaks and epidemics, occurring from 10 to 40 years apart (**Table 33.2**). They result from antigenic shift involving reassortment of viruses from two or more species. Virtually all of the major pandemics resulted from reassortment of avian influenza viruses and human influenza viruses in swine (**Figure 33.22**). Swine cells have receptors for both avian and human orthomyxoviruses and can bind and propagate both avian and human influenza strains. If swine are infected with both human and avian strains at the same time, the two unrelated viruses can reassort, resulting in antigenically unique viruses (antigenic shift) that can infect many humans because of a lack of host immunity. Reassortment with animal strains and infection into humans

occurs periodically but unpredictably, continually raising the possibility of a rapidly emerging, highly virulent influenza strain for which there is no preexisting immunity in the human population.

Worldwide deaths due to the influenza A "Spanish flu" pandemic of 1918 was about 50 million, with some estimates as high as 100 million people worldwide; up to 2 million deaths occurred in the United States (♻ Figure 32.1). Although there have been several pandemics during the last 130 years (Table 33.2), none has been as catastrophic as the 1918 flu. The virulence of the 1918 influenza is not fully understood, but appears to be due to the host response to the novel pathogen. This pathogen apparently stimulated production and release of large amounts of inflammatory cytokines, resulting in systemic inflammation and disease in susceptible individuals.

The 1957 outbreak of the so-called Asian flu also developed into a pandemic (**Figure 33.23**). The pandemic strain was a virulent mutant virus, differing antigenically from all previous strains. Immunity to this strain was not present, and the virus spread rapidly throughout the world. It first appeared in the interior of China in February 1957 and by April had spread to Hong Kong. From Hong Kong, the virus infected sailors on naval ships and emerged in San Diego, California. In May, an outbreak occurred in Newport, Rhode Island, on a naval vessel. From that time, outbreaks occurred continuously in various parts of the United States. The peak incidence occurred in October, when 22 million new cases developed.

Pandemic influenza A (H1N1) 2009 spread much more rapidly than Asian flu, starting from an original focus of infection in Mexico and spreading quickly to the United States, Europe, and Central and South America. The pandemic influenza A (H1N1)

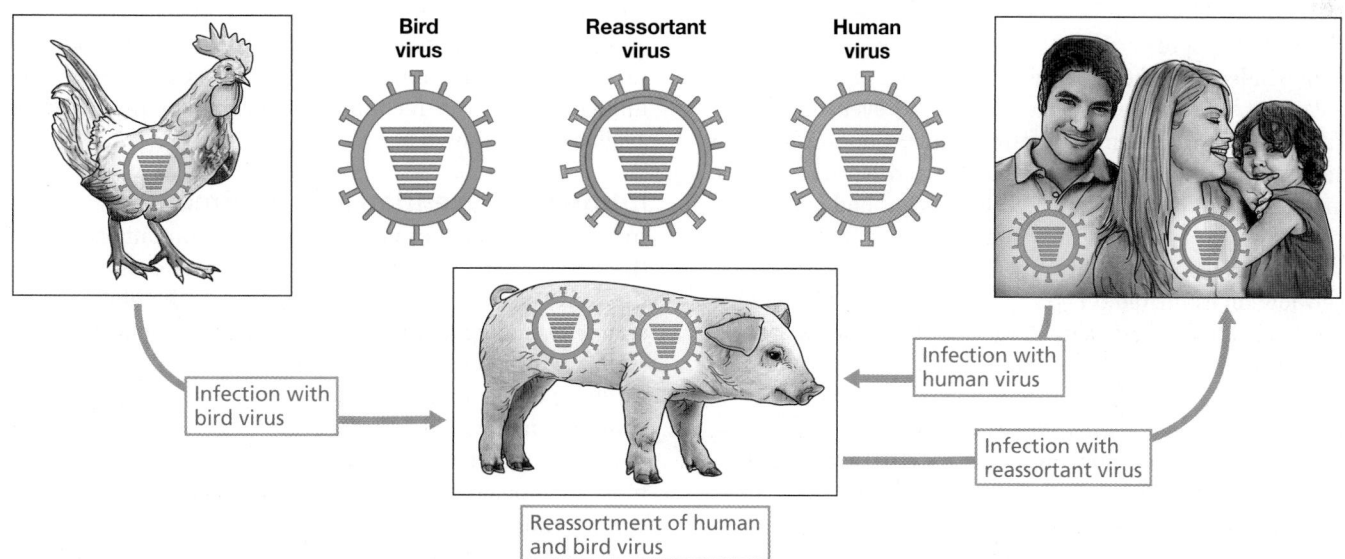

Figure 33.22 Influenza virus reassortment. Reassortments take place in swine. Influenza strains that originate in birds and humans can infect pigs. If a pig is infected at the same time with a bird virus and a human virus, the viruses can reassort. The reassortant virus may then infect humans. If the reassortant contains antigens that are unique, infections may cause pandemics.

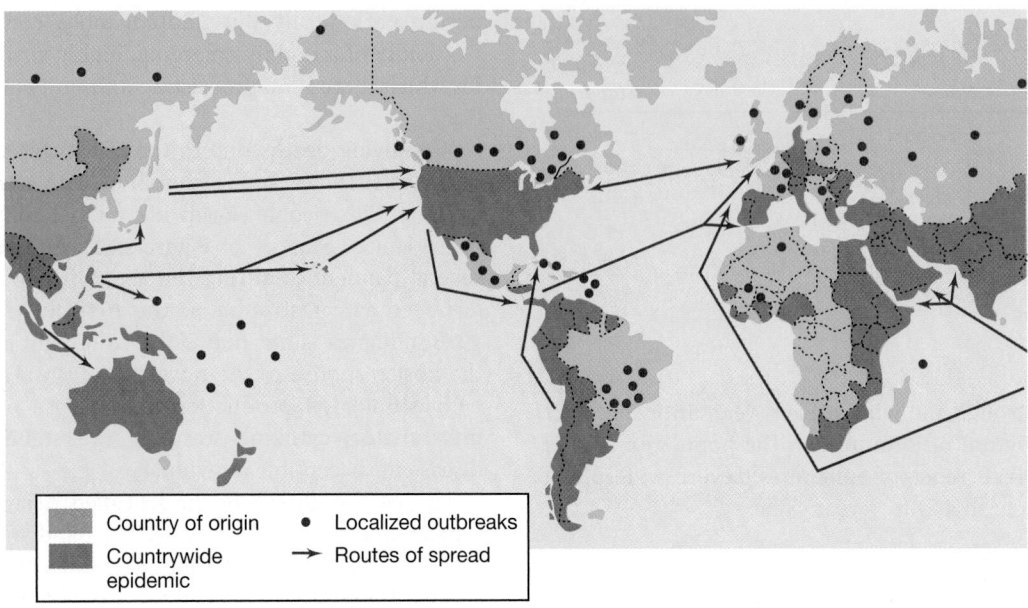

Figure 33.23 An influenza pandemic. The spread of the Asian influenza pandemic of 1957 is shown. The original epidemic focus of this pandemic was probably in China. Agricultural practices involving poultry and swine coupled with human interactions with these animals allowed the reassortment of influenza viral genomes from the three host species, producing a new strain for which there was no immune memory in humans.

2009 virus is sometimes called "swine flu" because the reassorted virus apparently developed in pigs (Figure 33.22). It is a reassortant virus consisting of RNA segments derived from human and bird influenza, and reassorted in swine. From the swine reservoir, it emerged to infect humans. First recognized in March 2009, the virus was declared a pandemic on June 11, 2009. By September 2009, the virus had spread worldwide. During the flu season in the Southern Hemisphere (May–October 2009), pandemic influenza A (H1N1) 2009 spread rapidly, causing widespread disease. Although pandemic influenza virus did not seem to be extraordinarily virulent, the pandemic was widespread even during the non-influenza-season summer months of 2009 (June–August) in Northern Hemisphere countries in Europe and North America, demonstrating that it is fully adapted to humans and can spread very easily. Even though the infection was prevalent in 2009–2010, the overall mortality rate for this pandemic strain was relatively low, an estimated 0.1–0.2%, perhaps only slightly higher than seasonal influenza mortality. A vaccine was made available in October 2009 to slow the advance of the pandemic.

A potentially devastating avian influenza, the influenza A H5N1 strain, also called *avian influenza,* appeared in Hong Kong in 1997, apparently jumping directly from the avian host to humans without the pig intermediate. H5N1 has now been reported in birds throughout Asia, Europe, the Middle East, and North Africa; it has not yet spread to birds in the Americas, Australia, or Antarctica. The H5N1 virus has reemerged several times over the last decade; the most recent outbreaks occurred in Egypt and Indonesia. Since 2003, 495 cases of human H5N1 infections have been confirmed worldwide, resulting in 292 deaths, an overall mortality rate of almost 60%. H5N1 is spread directly from avian species, usually domestic chickens or ducks,

to humans through prolonged contact or the eating of infected birds. At this time, avian influenza can be spread human to human only after prolonged close contact, but some reports indicate that H5N1 has infected swine. This event could set the stage for reassortment with human influenza strains that also infect swine. Such a reassortment could create a new and highly infective virus for which there is no immunity in humans, starting another influenza pandemic. Plans are in place both nationally and internationally to provide appropriate vaccines and support for potential pandemics initiated by this and other emergent influenza strains. A recombinant vaccine for the H5N1 virus is available on a limited basis.

Influenza Prevention and Treatment

Influenza epidemics can be controlled by immunization. However, the selection of appropriate strains for vaccines is complicated by the large number of existing strains and the ability of existing strains to undergo antigenic drift and antigenic shift. When new strains evolve, vaccines are not immediately available, but through careful worldwide surveillance, samples of the major emerging strains of influenza virus are usually obtained before there are epidemics. In the United States, immunization preparations are reformulated annually to target current prevalent strains. The targeted strains, chosen at the end of each influenza season, are grown in embryonated eggs and inactivated. The inactivated viral strains (two influenza A and one influenza B) are mixed to prepare a vaccine used for immunization prior to the next influenza season.

In general, influenza immunization is recommended for those individuals most likely to acquire the disease and develop serious secondary illnesses. Influenza immunization is currently recommended for everyone over 50 years of age, for those suffering from chronic debilitating diseases (for example, AIDS patients, chronic respiratory disease patients, and so on), and for health-care workers. Effective artificial immunity from the inactivated influenza vaccine lasts only a few years and is strain-specific. An attenuated live-virus vaccine is recommended for young adults, and may confer longer-lasting immunity.

Influenza A may also be controlled by use of antiviral drugs. The adamantanes, amantadine and rimantadine, are synthetic amines that inhibit viral replication. The neuraminidase inhibitors oseltamivir (Tamiflu) and zanamivir (Relenza) (🔗 Table 26.6) block release of newly replicated virions of influenza A and B and H5N1 avian virus. These drugs are used to treat ongoing influenza and shorten the course and severity of infection. They are most effective when given very early in the course of the infection. The adamantanes and oseltamivir also prevent the onset and spread of influenza.

Drug resistance has already occurred in some of the most dangerous influenza strains. Neither pandemic influenza A (H1N1) 2009 nor the H5N1 avian influenza is susceptible to the adamantanes. Although most influenza viruses are susceptible to the neuraminidase inhibitors, a few isolates of pandemic influenza A (H1N1) 2009 are resistant to oseltamivir.

Treatment of influenza symptoms with aspirin, especially in children, is not recommended. Aspirin treatment of influenza has been linked to development of *Reye's syndrome*, a rare but occasionally fatal complication involving the central nervous system.

MiniQuiz
- Distinguish between antigenic drift and antigenic shift in influenza.
- Discuss the possibilities for effective immunization programs for influenza and compare them to the possibilities for immunization for colds.

 # Direct-Contact Transmission of Diseases

Some pathogens are spread primarily by direct contact with an infected person or by contact with blood or excreta from an infected person. Many of the respiratory diseases we have discussed can also be spread by direct contact, but here we discuss other diseases spread primarily person to person through direct contact with infected individuals. These include staphylococcal infections, ulcers, and hepatitis.

33.9 *Staphylococcus*

The genus *Staphylococcus* contains pathogens of humans and other animals. Staphylococci commonly infect skin and wounds and may also cause pneumonia. Most staphylococcal infections result from the transfer of staphylococci in normal flora from an infected, asymptomatic individual to a susceptible individual.

Staphylococci are nonsporulating, gram-positive, facultatively aerobic cocci about 0.5–1.5 μm in diameter. They divide in multiple planes to form irregular clumps of cells (**Figure 33.24a**).

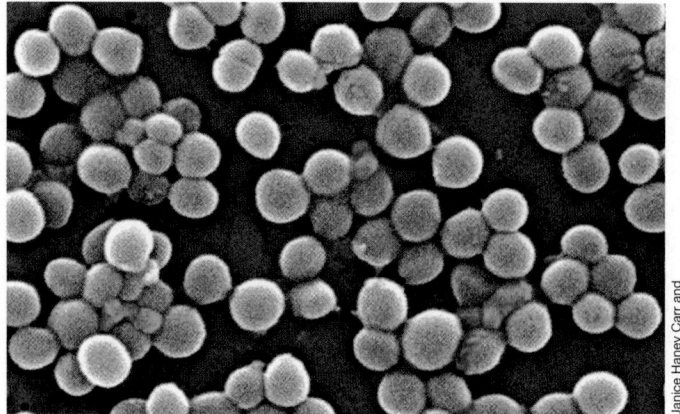

(a)

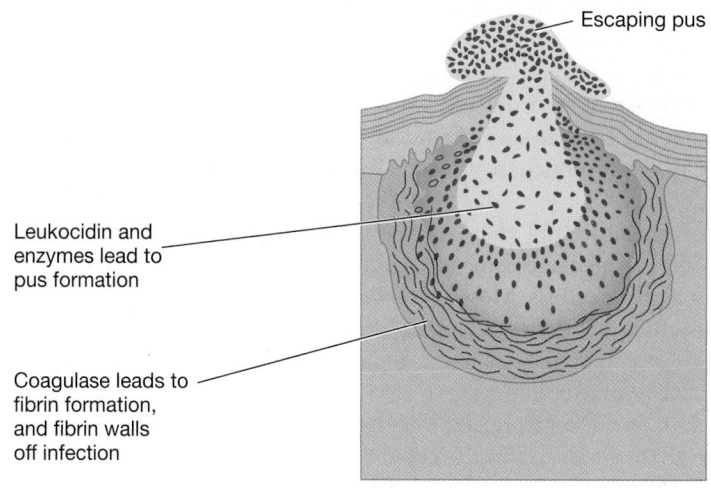

Escaping pus

Leukocidin and enzymes lead to pus formation

Coagulase leads to fibrin formation, and fibrin walls off infection

(b)

Figure 33.24 *Staphylococcus*. (a) *Staphylococcus aureus*. Each cell ranges in size from 0.5 to 1.5 μm in diameter. The cells divide in all planes, giving the appearance of a cluster of grapes. The genus name is taken from *staphylos*, Greek for grape. (b) The structure of a boil. Staphylococci initiate a localized skin infection and become walled off by coagulated blood and fibrin through the action of coagulase, a virulence factor. The ruptured boil releases pus, consisting of dead host cells and bacteria.

They are resistant to drying and tolerate high concentrations of salt (10% NaCl) when grown on artificial media. Staphylococci are readily dispersed in dust particles through the air and on surfaces. In humans, two species are important: *Staphylococcus epidermidis*, a nonpigmented species usually found on the skin or mucous membranes, and *Staphylococcus aureus*, a yellow-pigmented species. Both species are potential pathogens, but *S. aureus* is more commonly associated with human disease. Both species are frequently present in the normal microbial flora of the upper respiratory tract and the skin (Figure 33.2).

Epidemiology and Pathogenesis

Staphylococci cause diseases including acne, boils (Figure 33.24b), pimples, impetigo, pneumonia, osteomyelitis, carditis, meningitis, and arthritis. Many of these diseases are *pyogenic* (pus-forming). Healthy individuals are often carriers, and resident staphylococci in the upper respiratory tract or skin seldom cause disease. During the first week of life infants often become colonized from the mother or from another close human contact. Serious staphylococcal infections often occur when the resistance of the host is low because of hormonal changes, debilitating illness, wounds, or treatment with steroids or other drugs that compromise immunity.

All staphylococci produce catalase, an enzyme that converts hydrogen peroxide (H_2O_2) to water (H_2O) and oxygen (O_2). Catalase is not considered a virulence factor, but the catalase test distinguishes staphylococci from streptococci, which do not produce catalase (⮂ Section 5.18).

Those strains of *S. aureus* that cause human disease produce a variety of virulence factors (⮂ Table 27.4). At least four different *hemolysins* have been recognized, and a single strain often produces several. Hemolysins may cause cell lysis *in vivo*, and cause the red blood cell lysis seen around colonies on blood agar plates *in vitro*. Another virulence factor produced by *S. aureus* is *coagulase*, an enzyme that converts fibrin to fibrinogen, forming a localized clot. The production of coagulase is generally associated with pathogenicity. Clotting induced by coagulase results in the accumulation of fibrin around the bacterial cells, making it difficult for host defense agents to come into contact with the bacteria and preventing phagocytosis. Most *S. aureus* strains also produce *leukocidin*, a protein that destroys leukocytes. Production of leukocidin in skin lesions such as boils and pimples results in host cell destruction and is one of the factors responsible for pus. Some strains of *S. aureus* also produce proteolytic enzymes, hyaluronidase, fibrinolysin, lipase, ribonuclease, and deoxyribonuclease.

Certain strains of *S. aureus* are responsible for **toxic shock syndrome (TSS)**, a serious outcome of staphylococcal infection, characterized by high fever, rash, vomiting, diarrhea, and death. TSS was first recognized in women and was associated with use of highly absorbent tampons. In menstruating females, blood and mucus in the vagina can become colonized by *S. aureus* from the skin, and the presence of a tampon concentrates this material, creating ideal microbial growth conditions. Largely through education and alterations in materials used in tampons, TSS due to tampon use is now relatively rare. Over 70% of TSS cases, however, result in death. TSS is now seen in both men and women and is usually initiated by staphylococcal infections following surgery. The symptoms of TSS result from an exotoxin called *toxic shock syndrome toxin-1* (TSST-1). TSST-1 is a superantigen (⮂ Section 28.10) released by growing staphylococci. The toxin recruits large numbers of T cells, culminating in an inflammatory response characteristic of superantigen reactions. TSS may also be caused by superantigens from other pathogenic bacteria, including *Streptococcus pyogenes* (Section 33.2).

Staphylococcal *enterotoxin A*, another superantigen, causes a form of food poisoning. After ingestion of toxin-contaminated food, the toxin stimulates T cells localized along the intestine, resulting in a massive T cell response and release of inflammatory mediators. The final outcome is the severe but short-lived diarrhea and vomiting associated with staphylococcal food poisoning (⮂ Section 36.6).

Diagnosis, Prevention, and Treatment

Isolates from suspected staphylococcal infections can be cultured on enriched agar media such as blood agar. To reduce the growth of other gram-positive organisms, a selective medium with 7.5% NaCl is used. An enriched differential medium, mannitol salt agar, contains 7.5% NaCl, the sugar mannitol, and phenol red, a pH indicator. The medium is used to differentiate staphylococci. *S. aureus*, generally considered to be more pathogenic than other members of the genus, ferments mannitol, turning the medium yellow, while other species such as *S. epidermidis* do not ferment mannitol. For identification of methicillin-resistant *S. aureus* (MRSA), specialized and proprietary chromogenic media that are selective and differential are used. MRSA colonies appear blue, while others either do not grow or remain white.

Extensive use of antibiotics has resulted in the selection of resistant strains of *S. aureus* and *S. epidermidis*. Healthcare-associated infections from antibiotic-resistant staphylococci often occur in patients whose resistance is lowered due to other diseases, surgical procedures, or drug therapy (⮂ Section 32.7). Patients may acquire staphylococci from healthcare personnel who are asymptomatic carriers of drug-resistant strains. As a result, appropriate antimicrobial drug therapy for *S. aureus* infections is a major problem in healthcare environments. For example, over 100,000 cases of infection with MRSA are *reported* each year, mostly in healthcare facilities, but some studies estimate that the total number of MRSA infections may exceed 1 million each year. Up to 100,000 of these are more serious invasive MRSA infections. Up to 15% of MRSA infections are acquired in the community by individuals who have no association with healthcare as either patients or contacts. Although some community-acquired staphylococcal infections are still treatable with penicillin, MRSA and other antibiotic-resistant *S. aureus* strains are becoming more common. We now know that antibiotic resistance genes and virulence genes in staphylococci are often acquired by a single horizontal gene transfer involving the staphylococcal cassette chromosome (SCC), a mobile DNA element that carries *mecA*, the gene that confers methicillin resistance, and also carries several virulence factors. Because virulence and antibiotic resistance are closely linked, disease-producing isolates of *S. aureus*, regardless of their origin, must be checked for antibiotic susceptibility.

Prevention of staphylococcal infections is problematic because most individuals are asymptomatic carriers, and diseases such as acne and impetigo can be transmitted by simple contact with contaminated fingers. Prevention strategies include exclusion of infected individuals, including carriers, and elimination of the pathogen. In hospital environments such as surgical wards and nurseries, carriers of known pathogenic strains must be either excluded or treated with topical or systemic antimicrobial drugs to eliminate the pathogens. In one study, the noses of patients were swabbed and cultured before surgical procedures. Patients who were positive for *S. aureus* were treated intranasally with a topical antibiotic and a body wash with chlorhexidine, a topical antiseptic. Treated patients had a greater than 50% reduction in postoperative *S. aureus* infections compared to a control group.

33.10 *Helicobacter pylori* and Gastric Ulcers

Helicobacter pylori is a gram-negative, highly motile, spiral-shaped bacterium (**Figure 33.25**) related to *Campylobacter* (↻ Section 36.10). The organism is 2.5–3.5 μm long and 0.5–1.0 μm in diameter and has one to six polar flagella at one end. *H. pylori*, first identified in human intestinal biopsies in 1983, is a pathogen associated with gastritis, ulcers, and gastric cancers. This organism colonizes the non-acid-secreting mucosa of the stomach and the upper intestinal tract, including the duodenum (↻ Figure 27.9). Genetic studies of different strains of *H. pylori* indicate that this organism has been associated with humans at least since humans first migrated from Africa.

Epidemiology

Up to 80% of gastric ulcer patients have concomitant *H. pylori* infections, and up to 50% of asymptomatic adults in developing countries are chronically infected. Person-to-person contact and ingestion of contaminated food or water are the probable transmission methods for *H. pylori*. Although there is no known non-human reservoir of *H. pylori*, the organism has occasionally been recovered from cats kept as household pets, indicating that it can be spread to or from animals in close contact with humans. Infection occurs in high incidence within families, and the overall prevalence in the population increases with age. These factors suggest host-to-host transmission. However, infections with *H. pylori* sometimes also occur in epidemic clusters, suggesting transmission from common sources such as food or water.

Pathology, Diagnosis, and Treatment

H. pylori is a major preventable and treatable cause of many gastric ulcers. The bacterium is slightly invasive and colonizes the surfaces of the gastric mucosa, where it is protected from the effects of stomach acids by the gastric mucus layer. After mucosal colonization, a combination of pathogen products and host responses cause inflammation, tissue destruction, and ulceration. Pathogen products such as vacA (a cytotoxin), urease, and lipopolysaccharide may contribute to localized tissue destruction and ulceration. Antibodies to *H. pylori* are usually present in infected individuals, but are not protective and do not prevent colonization. Individuals who acquire *H. pylori* tend to have chronic infections unless they are treated with antibiotics. Chronic gastritis due to untreated *H. pylori* infection may lead to the development of gastric cancers.

Clinical signs of *H. pylori* infection include belching and stomach (epigastric) pain. Definitive diagnosis requires the recovery and culture or observation of *H. pylori* from a gastric ulcer biopsy. Serum antibodies indicate *H. pylori* infection, but because infections seem to be chronic and antibodies may persist for months after a given infection, *H. pylori* antibodies are not reliable indicators of acute, active disease. A simple *in vivo* diagnostic test, the urease test, is available for *H. pylori*. In this test, the patient ingests ^{13}C- or ^{14}C-labeled urea ($H_2N-CO-NH_2$). If urease is present, the urea will be hydrolyzed into carbon dioxide (CO_2) and amines. The presence of labeled CO_2 in the patient's exhaled breath indicates the presence of urease, produced almost exclusively by *H. pylori*. Recovery of *H. pylori* organisms or antigens from the stool is also indicative of infection.

Evidence for a causal association between *H. pylori* and gastric ulcers comes from antibiotic treatments for the disease. Most patients relapse within 1 year after long-term treatment of ulcers with antacid preparations. However, by treating ulcers as an infectious disease, permanent cures are often obtained. *H. pylori* infection is usually treated with a combination of drugs, including the antibacterial compound metronidazole, an antibiotic such as tetracycline or amoxicillin, and a bismuth-containing antacid preparation. The combination treatment, administered for 14 days, abolishes the *H. pylori* infection and provides a long-term cure. For their contributions to unraveling the connection between *H. pylori* and peptic and duodenal ulcers, the Australian scientists Robin Warren and Barry Marshall were awarded the 2005 Nobel Prize in Physiology or Medicine.

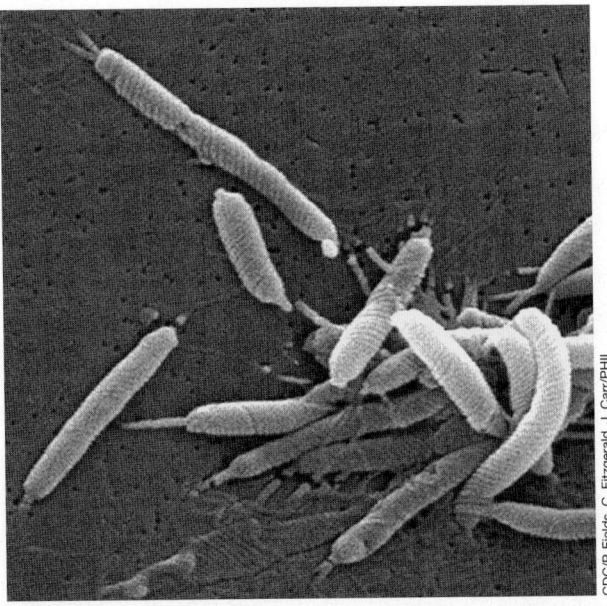

CDC/P. Fields, C. Fitzgerald, J. Carr/PHIL

Figure 33.25 *Helicobacter.* Scanning electron micrograph. Cells range in size from 1.5 to 10 μm in length and 0.3 to 1 μm in diameter. The organism in the lower left is 3.2 μm in length. Note the sheathed flagella.

UNIT 11

Table 33.3 *Hepatitis viruses*

Disease	Virus and genome	Vaccine	Clinical illness	Transmission route
Hepatitis A	*Hepatovirus* (HAV) ssRNA	Yes	Acute	Enteric
Hepatitis B	*Orthohepadnavirus* (HBV) dsDNA	Yes	Acute, chronic, oncogenic	Parenteral, sexual
Hepatitis C	*Hepacivirus* (HCV) ssRNA	No	Chronic, oncogenic	Parenteral
Hepatitis D	*Deltavirus* (HDV) ssRNA	No	Fulminant, only with HBV	Parenteral
Hepatitis E	*Caliciviridae* family (HEV) ssRNA	No	Fulminant disease in pregnant women	Enteric
Hepatitis G	*Flaviviridae* family (HGV) ssRNA	No	Asymptomatic	Parenteral

33.11 Hepatitis Viruses

Hepatitis is a liver inflammation commonly caused by an infectious agent. Hepatitis sometimes results in acute illness followed by destruction of functional liver anatomy and cells, a condition known as **cirrhosis**. Hepatitis due to infection can cause chronic or acute disease, and some forms lead to liver cancer. Although many viruses and a few bacteria can cause hepatitis, a restricted group of viruses is often associated with liver disease. Hepatitis viruses are phylogenetically diverse; none are genetically related, but all infect cells in the liver. **Table 33.3** characterizes the known hepatitis viruses.

Epidemiology

Hepatitis A virus (HAV) is transmitted from person to person or by ingestion of fecally contaminated food or water. Often called *infectious hepatitis*, HAV usually causes mild, even subclinical infections, but rare cases of severe liver disease occur. The most significant food vehicles for HAV are shellfish, usually oysters or clams harvested from water polluted by human fecal material. In recent years, HAV has also been transmitted in fresh produce. In 2003 a significant outbreak in the eastern United States was traced to eating raw or undercooked green onions. The general trend for numbers of HAV infections has moved steadily downward and is now at all-time record low levels, partly due to the availability of an effective vaccine (**Figure 33.26**). HAV causes

more cases of viral hepatitis than any other virus, and over 30% of individuals in the United States have antibodies to HAV, indicating a past infection.

Infection due to *hepatitis B virus* (HBV) is often called *serum hepatitis*. HBV is a hepadnavirus, a partially double-stranded DNA virus (↩ Section 21.11). The mature virus particle containing the viral genome is called a *Dane particle* (**Figure 33.27**). HBV causes acute, often severe disease that can lead to liver failure and death. Chronic HBV infection can lead to cirrhosis and liver cancer. HBV is usually transmitted by a parenteral (outside the gut) route, such as blood transfusion or through shared hypodermic needles contaminated with infected blood. HBV may also be transmitted through exchanges of body fluids, as in sexual intercourse. The number of new HBV infections is decreasing, again due to an effective vaccine. However, over 100,000 people worldwide and nearly 5000 people in the United States die each year due to complications such as cancers generated by chronic HBV infection.

Hepatitis D virus (HDV) is a defective virus that lacks genes for its own protein coat (↩ Section 21.11). HDV is also transmitted by parenteral routes, but because it is a defective virus, it cannot replicate and express a complete virus unless the cell is also infected with HBV. The HDV genome replicates independently

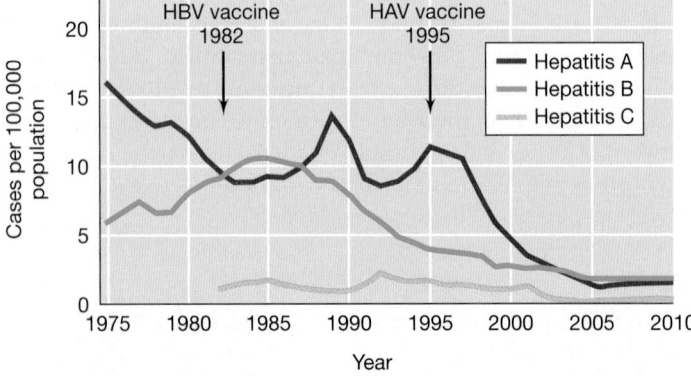

Figure 33.26 Hepatitis in the United States. The incidence of hepatitis is shown by viral agent. In 2007 there were 2979 reported cases of hepatitis A, 4519 reported cases of hepatitis B, and 845 reported cases of hepatitis C. Data obtained from the Centers for Disease Control and Prevention, Atlanta, Georgia.

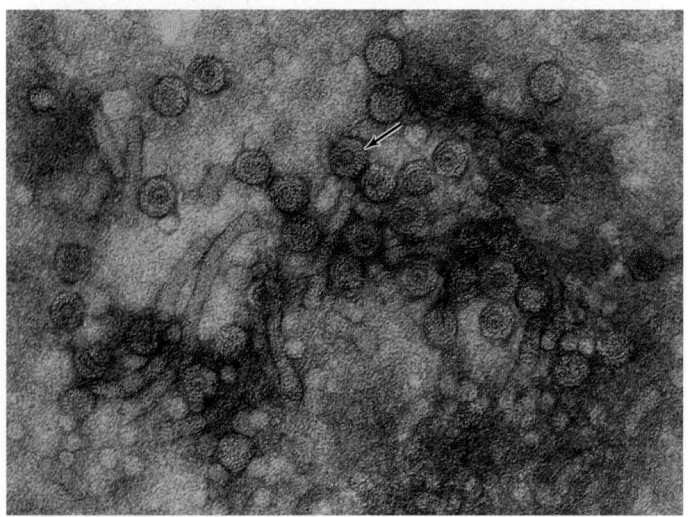

Figure 33.27 Hepatitis B virus (HBV). The arrow indicates a complete HBV particle, which is about 42 nm in diameter and is called a Dane particle.

but uses the protein coat of HBV for expression. Thus, HDV infections are always coinfections with HBV.

Hepatitis C virus (HCV) is also transmitted parenterally. HCV generally produces a mild or even asymptomatic disease at first, but up to 85% of individuals develop chronic hepatitis, with up to 20% leading to chronic liver disease and cirrhosis. Chronic infection leads to hepatocarcinoma (liver cancer) in 3–5% of infected individuals. The latency period for development of cancer can be several decades after the primary infection. Only a fraction of the estimated 25,000 annual new infections are recognized and reported in the United States (Figure 33.26). Large numbers of HCV-related deaths occur annually due to chronic HCV infections that develop into liver cancer. HCV-induced liver disease is the most common liver disease currently seen in clinical settings in the United States and accounts for up to 10,000 of the 25,000 annual deaths due to liver cancer, other chronic liver diseases, and cirrhosis.

Hepatitis E virus (HEV) transmits hepatitis via an enteric route. HEV causes an acute, self-limiting hepatitis that varies in severity from case to case but is often the cause of fulminant disease in pregnant women. HEV is endemic in Mexico as well as in tropical and subtropical regions of Africa and Asia.

Hepatitis G virus (HGV) is commonly found in the blood of patients with other forms of acute hepatitis, but HGV alone seems to cause very mild disease or is completely asymptomatic. Screening for HGV shows that up to 8.1% of blood donors may be positive for HGV, but because HGV is not associated with demonstrable clinical disease, the significance of these findings is not clear.

Pathology and Diagnosis

Hepatitis is an acute disease of the liver. Symptoms include fever; jaundice (production and release of excess bilirubin by the liver due to destruction of liver cells, resulting in yellowing of the skin and the whites of the eyes); hepatomegaly (liver enlargement); and cirrhosis (breakdown of the normal liver tissue architecture). Mild hepatitis is characterized by relatively minor elevation of liver enzymes such as alanine aminotransferase (ALT). Fulminant disease is characterized by a rapid onset of severe symptoms such as jaundice and cirrhosis and is often a life-threatening condition. All hepatitis viruses cause similar acute clinical diseases and cannot be readily distinguished based on the clinical findings alone. Chronic hepatitis infections, usually caused by HBV or HCV, are often asymptomatic or produce very mild symptoms, but can cause serious liver disease, even in the absence of hepatocarcinoma.

Diagnosis of hepatitis is based primarily on clinical findings and laboratory tests that determine liver function problems. Cirrhosis is diagnosed by visual examination of biopsied liver tissue. Virus-specific assays are also used to confirm diagnosis, identify the infectious agent, and determine a course of treatment. Direct culture of hepatitis viruses is usually not used for identification purposes, and HCV and HGV have not been successfully cultured.

Many of the molecular diagnostic tools discussed in Chapter 31 are used to diagnose hepatitis. Some of the most common methods for determining hepatitis identity are enzyme immunoassay (EIA) tests. Most hepatitis EIAs identify viral proteins in

blood specimens. However, several indirect EIA tests can detect IgM or IgG antibodies to HBV. IgM is associated with the primary immune response to HBV, while IgG is associated with the secondary response. Therefore, identification of the antibody class can determine whether the HBV infection is a new infection (IgM) or a chronic or latent infection (IgG). Other immune-based tests used for the detection of hepatitis viruses include immunoblots, immunoelectron microscopy (the use of an immune visualization technique such as EIA to enhance electron microscopy), and immunofluorescence. Polymerase chain reaction tests and DNA hybridization tests are also used for the detection of the viral genome in blood or in liver tissue obtained by biopsy.

Prevention and Treatment

Infection with HAV or HBV can be prevented with effective vaccines. HBV vaccination is recommended and in most cases is required for school-age children in the United States. No effective vaccines are available for the other hepatitis viruses.

Universal precautions are standards developed for personnel handling infectious waste and body fluids. Mandated by law for patient care and clinical laboratory facilities, they are designed to prevent infection by all parenterally transmitted hepatitis viruses (HBV, HCV, HDV, and HGV) as well as HIV (human immunodeficiency virus). The precautions prescribe a high level of vigilance and aseptic handling and containment procedures to deal with patients, body fluids, and infected waste materials (↪ Section 31.4). In addition to person-to-person transmission, HAV can be spread through contamination of common sources such as food and water. Hepatitis A outbreaks can be prevented by maintaining pathogen-free food and water supplies.

Pooled human immune gamma globulin can be used to prevent HAV infection if given soon after exposure. For postexposure prevention of HBV infection, specific hepatitis B immune globulin (HBIG), coupled with administration of the HBV vaccine, has been effective.

Most treatment of hepatitis is supportive, providing rest and time to allow liver damage to resolve and be repaired. In some cases, antiviral drugs are effective for treatment. Alpha interferon is effective against HCV when combined with the drug ribavirin in some patients. HBV can be treated with the antiviral drugs foscarnet, ribavirin, lamivudine, and ganciclovir.

MiniQuiz
- Describe the mode of transmission for hepatitis A virus, hepatitis B virus, and hepatitis C virus.
- Describe potential prevention and treatment methods for hepatitis A virus and hepatitis B virus.

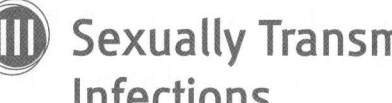

Sexually Transmitted Infections

Sexually transmitted infections, or **STIs**, also called *sexually transmitted diseases* (*STDs*) or *venereal diseases*, are caused by a wide variety of bacteria, viruses, protists, and even fungi

Table 33.4 *Sexually transmitted infections and treatment guidelines*

Disease	Causative organism(s)[a]	Recommended treatment[b]
Gonorrhea	*Neisseria gonorrhoeae* (B)	Cefixime or ceftriaxone, *and* azithromycin or doxycycline
Syphilis	*Treponema pallidum* (B)	Benzathine penicillin G
Chlamydia trachomatis infections	*Chlamydia trachomatis* (B)	Doxycycline or azithromycin
Nongonococcal urethritis	*C. trachomatis* (B) or *Ureaplasma urealyticum* (B) or *Mycoplasma genitalium* (B) or *Trichomonas vaginalis* (P)	Azithromycin
Lymphogranuloma venereum	*C. trachomatis* (B)	Doxycycline
Chancroid	*Haemophilus ducreyi* (B)	Azithromycin
Genital herpes	Herpes simplex 2 (V)	No known cure; symptoms can be controlled with acyclovir, valacyclovir, and other antiviral drugs
Genital warts	Human papillomavirus (HPV) (certain strains)	No known cure; symptomatic warts can be removed surgically, chemically, or by cryotherapy
Trichomoniasis	*Trichomonas vaginalis* (P)	Metronidazole
Acquired immunodeficiency syndrome (AIDS)	Human immunodeficiency virus (HIV)	No known cure; nucleotide base analogs, protease inhibitors, fusion inhibitors, and nonnucleoside reverse transcriptase inhibitors may slow disease progression
Pelvic inflammatory disease	*N. gonorrhoeae* (B) or *C. trachomatis* (B)	Cefotetan
Vulvovaginal candidiasis	*Candida albicans* (F)	Butoconazole

[a]B, bacterium; V, virus; P, protist; F, fungus.
[b]Recommendations of the U.S. Department of Health and Human Service, Public Health Service. For many treatment plans, there are a number of alternatives.

(**Table 33.4**). Unlike respiratory pathogens that are shed constantly in large numbers by an infected individual, sexually transmitted pathogens are generally found only in body fluids from the genitourinary tract that are exchanged during sexual activity. This is because sexually transmitted pathogens are typically very sensitive to environmental stressors such as drying, heat, and light. Their habitat, the human genitourinary tract, is a protected, moist environment. Thus, these pathogens preferentially and sometimes exclusively colonize the genitourinary tract.

Diagnosis and treatment of STIs is challenging for both social and biological reasons. First, it is difficult to identify the infection source and stop its spread; up to one-third of all STIs are in teenagers with multiple sex partners. Second, many STIs have minor symptoms; infected individuals often do not seek treatment. Third, social stigmas attached to STIs prevent many individuals from seeking prompt treatment. Prompt effective treatment of STIs, however, is important for a number of reasons. First, most STIs are curable, and all are controllable with appropriate intervention. Second, delay or lack of treatment can lead to long-term problems such as infertility, cancer, heart disease, degenerative nerve disease, birth defects, stillbirth, or destruction of the immune system.

Because transmission of STIs is limited to intimate physical contact, generally during sexual intercourse, their spread can be controlled by sexual abstinence (no exchange of body fluids) or by the use of barriers such as condoms that stop the exchange of body fluids during sexual activity.

STIs are very common and continue to pose social as well as medical problems. Here we discuss some prevalent STIs.

33.12 Gonorrhea and Syphilis

Gonorrhea and *syphilis* are preventable, treatable bacterial STIs. Because of differences in their symptoms, the overall pattern of disease differs between the two. Gonorrhea is very prevalent, and often asymptomatic, especially in women. The disease is often unrecognized and remains untreated. Syphilis, on the other hand, now has a low incidence (**Figure 33.28**). This is partly because syphilis exhibits very obvious symptoms in its primary stage and infected individuals usually seek immediate treatment.

Gonorrhea

Neisseria gonorrhoeae, often called the *gonococcus*, causes gonorrhea. *N. gonorrhoeae* is a gram-negative, nonsporulating, obligately aerobic, oxidase-positive diplococcus related biochemically and phylogenetically to *Neisseria meningitidis* (↩ Section 17.10 and Section 33.5). *N. gonorrhoeae* is killed rapidly by drying, sunlight, and ultraviolet light and normally does not survive away from

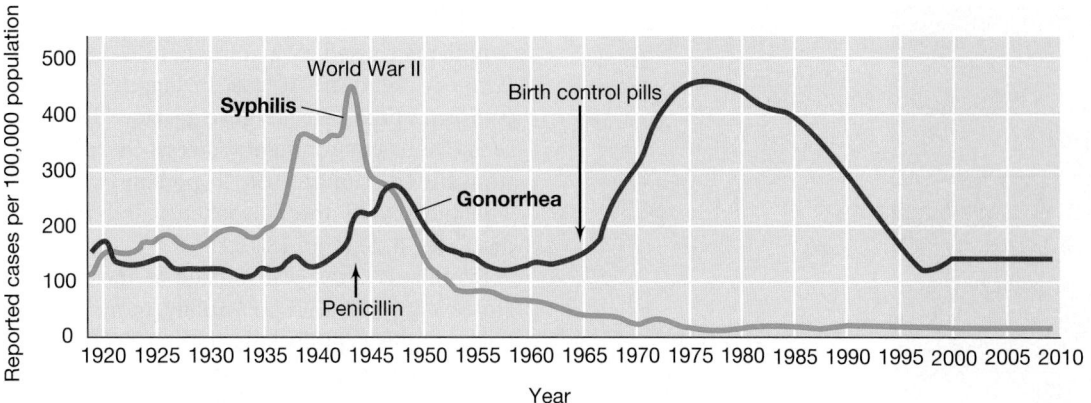

Figure 33.28 Reported cases of gonorrhea and syphilis in the United States. Note the downward trend in disease incidence after the introduction of antibiotics and the upward trend in the incidence of gonorrhea after the introduction of birth control pills. In 2007 there were 355,991 new cases of gonorrhea and 11,466 new cases of primary and secondary syphilis in the United States.

the mucous membranes of the genitourinary tract (**Figure 33.29**). Because of its extreme sensitivity to environmental conditions, *N. gonorrhoeae* can be transmitted only by intimate person-to-person contact. The pathogen enters the body by way of the mucous membranes of the genitourinary tract.

The symptoms of gonorrhea are quite different in the male and female. In females, gonorrhea is characterized by a mild vaginitis that is difficult to distinguish from vaginal infections caused by other organisms, and thus, the infection may easily go unnoticed. Complications from untreated gonorrhea in females, however, can lead to a condition known as pelvic inflammatory disease (PID). PID is a chronic inflammatory disease that can lead to long-term complications such as sterility. In the male, the organism causes a painful infection of the urethral canal. Complications from untreated gonorrhea affecting both males and females include

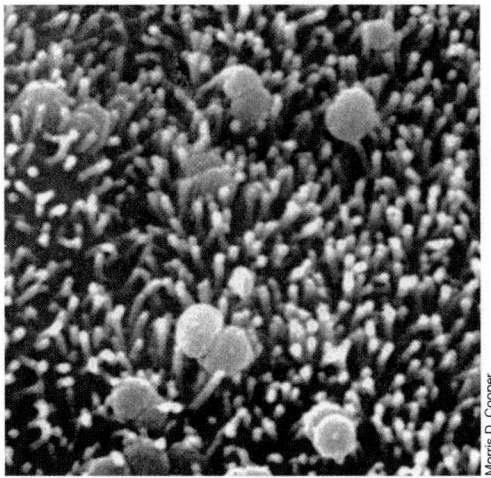

Figure 33.29 The causative agent of gonorrhea, *Neisseria gonorrhoeae*. The scanning electron micrograph of the microvilli of human fallopian tube mucosa shows how cells of *N. gonorrhoeae* attach to the surface of epithelial cells. Note the distinct diplococcus morphology. Cells of *N. gonorrhoeae* are about 0.8 μm in diameter.

damage to heart valves and joint tissues due to immune complex deposition.

In addition to gonorrhea, *N. gonorrhoeae* also causes eye infections in newborns. Infants born of infected mothers may acquire eye infections during birth. Therefore, prophylactic treatment of the eyes of all newborns with an ointment containing erythromycin is generally mandatory to prevent gonococcal infection in infants. We discussed the clinical microbiology and diagnosis of gonorrhea in Section 31.1.

Treatment of gonorrhea with penicillin was the method of choice until the 1980s when strains of *N. gonorrhoeae* resistant to penicillin arose. The quinolones ciprofloxacin, oflaxacin, or levofloxacin were also used, but by 2006, about 14% of *N. gonorrhoeae* strains isolated nationwide had developed resistance. Strains resistant to penicillin and quinolones respond to alternative antibiotic therapy with a single dose of the β-lactam antibiotics cefixime or ceftriaxone. An antichlamydial agent, normally azithromycin or doxycycline, is often given when treating gonorrhea because nearly 50% of gonorrhea patients are also infected with a harder to diagnose STI pathogen, *Chlamydia trachomatis* (Table 33.4; Section 33.13).

The incidence of gonococcus infection remains relatively high for the following reasons: (1) Although antibodies are produced, they are either not protective, or they are strain-specific and provide no cross-immunization. Therefore, effective acquired immunity does not exist and repeated reinfection is possible. In addition, within a single *N. gonorrhoeae* strain, antigenic switches can change immune response targets. *N. gonorrhoeae* can switch to alternate forms of opacity protein antigens (Opa) and surface pilin antigens, creating new serotypes and preventing effective immunity. (2) The use of oral contraceptives alters the local mucosal environment in favor of the pathogen. Oral contraceptives induce the body to mimic pregnancy, which results, among other things, in a lack of glycogen production in the vagina and a rise in the vaginal pH. Lactic acid bacteria normally found in the adult vagina (↩ Figure 27.12) fail to develop under such circumstances, facilitating colonization by *N. gonorrhoeae* transmitted

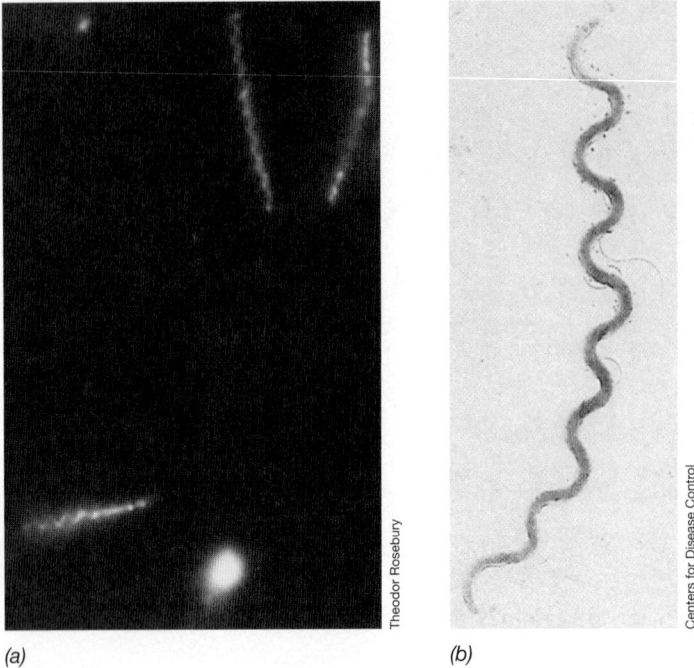

(a)   (b)

Figure 33.30 The syphilis spirochete, *Treponema pallidum*. *(a)* Dark-field microscopy of an exudate. *Treponema pallidum* cells measure 0.15 μm wide and 10–15 μm long. *(b)* Shadow-cast electron micrograph of a cell of *T. pallidum*. The endoflagella are typical of spirochetes.

from an infected partner. (3) Symptoms in the female are so mild that the disease may be unrecognized, and an infected female with multiple partners can infect many males. The disease can be controlled if the sexual contacts of infected persons are quickly identified and treated. But it is often difficult to obtain contact information and even more difficult to arrange treatment; the social stigmas associated with STIs are often obstacles to obtaining medical care.

Syphilis

Syphilis is caused by a spirochete, *Treponema pallidum. T. pallidum* is about 10–15 μm in length and extremely thin, about 0.15 μm in diameter (**Figure 33.30**). The spirochete is extremely sensitive to environmental stress; therefore, syphilis is normally transmitted from person to person by intimate sexual contact. The biology of the spirochetes and the genus *Treponema* is discussed in Section 18.16.

Syphilis is often transmitted at the same time as gonorrhea. However, syphilis is potentially more serious than gonorrhea. For example, syphilis kills about 100,000 people per year worldwide, whereas gonorrhea directly kills only about 1000 people per year. Largely because of differences in the symptoms and pathobiology of the two diseases, the incidence of syphilis in the United States is much lower than the incidence of gonorrhea. The incidence of syphilis, however, has increased in recent years, with over 10,000 new infections occurring each year, from a low of about 6000 in 1997.

The syphilis spirochete does not pass through unbroken skin, and initial infection most probably takes place through tiny breaks in the epidermal layer. In the male, initial infection is usually on the penis; in the female it is most often in the vagina, cervix, or perineal region. In about 10% of cases, infection is extragenital, usually in the oral region. During pregnancy, the organism can be transmitted from an infected woman to the fetus; the disease acquired by the infant is called **congenital syphilis**.

Syphilis is an extremely complex disease and, in an individual patient, may progress into any of three stages, but the disease always begins with a localized infection called *primary syphilis*. In primary syphilis, *T. pallidum* multiplies at the initial site of entry, and a characteristic primary lesion called a *chancre* forms within 2 weeks to 2 months (**Figure 33.31**). Dark-field microscopy of the syphilitic chancre exudate reveals the actively motile spirochetes (Figure 33.30*a*). In most cases the chancre heals spontaneously and *T. pallidum* disappears from the site. Some cells,

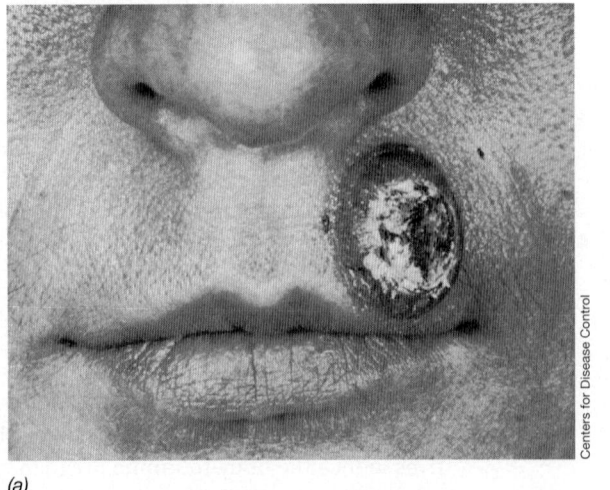
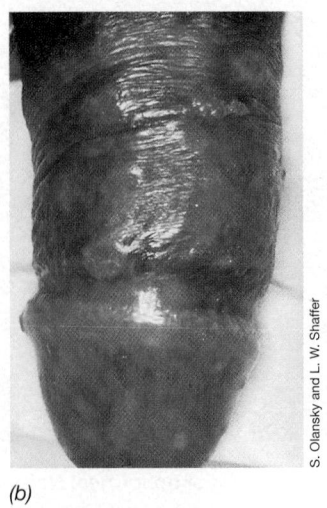

(a)   (b)

Figure 33.31 Primary syphilis lesions. *(a)* Chancre on lip. *(b)* Several chancres on penis. The chancre is the characteristic lesion of primary syphilis at the site of infection by *Treponema pallidum*. Patients who acquire such lesions generally seek medical intervention, and the obvious chancre hastens diagnosis and treatment.

however, spread from the initial site to various parts of the body, such as the mucous membranes, eyes, joints, bones, or central nervous system, where extensive multiplication occurs. A hypersensitivity reaction to the treponemes often takes place, revealed by the development of a generalized skin rash; this rash is the key symptom of *secondary syphilis*. At first, the secondary rash papules may contain *T. pallidum*, making them highly infectious. Eventually the spirochetes are cleared from the secondary lesions and infectivity is reduced.

The subsequent course of the disease in the absence of treatment is highly variable. About one-fourth of infected individuals undergo a spontaneous cure as demonstrated by a decrease in antibody titer. Another one-fourth exhibit no further symptoms, although static or elevated antibody titers indicate a persistent, chronic, active infection. About half of untreated patients develop *tertiary syphilis*, with symptoms ranging from relatively mild infections of the skin and bone to serious and even fatal infections of the cardiovascular system or central nervous system. Involvement of the nervous system can cause generalized paralysis or other severe neurological damage. Relatively low numbers of *T. pallidum* are present in individuals with tertiary syphilis; most of the symptoms probably result from inflammation due to delayed hypersensitivity reactions to the spirochetes.

Several tests used in laboratory diagnoses of syphilis were discussed in Chapter 31. The single most important physical sign of a primary syphilis infection, the chancre, is diagnostic for the disease. Infected individuals generally seek treatment for syphilis because of the highly visible chancre.

Penicillin is highly effective in syphilis therapy, and the primary and secondary stages of the disease can usually be controlled by a single injection of benzathine penicillin G. In tertiary syphilis, penicillin treatment must be extended for longer periods of time.

The incidence of primary and secondary syphilis in the United States has decreased over the last two decades and is near the lowest levels since record keeping began.

MiniQuiz
- Explain at least one potential reason for the high incidence of gonorrhea as compared with syphilis.
- Describe the progression of untreated gonorrhea and untreated syphilis. Do treatments produce a cure for each disease?

33.13 Chlamydia, Herpes, Trichomoniasis, and Human Papillomavirus

Chlamydia, herpes, trichomoniasis, and human papillomavirus infections are important STIs. These diseases are very prevalent in the population and are much more difficult to diagnose and treat than are syphilis and gonorrhea.

Chlamydia

A number of sexually transmitted diseases can be ascribed to infection by the obligate intracellular bacterium *Chlamydia trachomatis* (**Figure 33.32** and ⮑ Section 18.9). The total incidence of sexually transmitted *C. trachomatis* infections probably greatly outnumbers the incidence of gonorrhea. Over 1 million cases are now reported annually in the United States, but there may be more than 4 million new sexually transmitted infections with *C. trachomatis* every year. Chlamydia infection is the most prevalent STI and reportable communicable disease in the United States. *C. trachomatis* also causes a serious eye infection called *trachoma*, but the strains of *C. trachomatis* responsible for STIs

(a)   *(b)*

Figure 33.32 Cells of *Chlamydia trachomatis* (arrows) attached to human fallopian tube tissues. *(a)* Cells attached to the microvilli of a fallopian tube. *(b)* A damaged fallopian tube containing a cell of *C. trachomatis* (arrow) in the lesion.

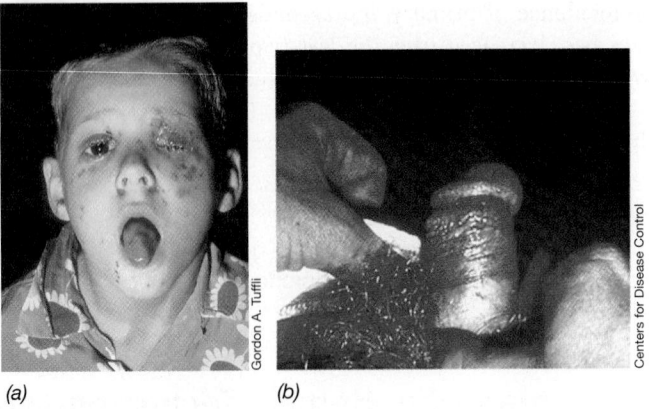

(a) *(b)*

Figure 33.33 Herpesvirus. *(a)* A severe case of herpes blisters on the face due to infection with herpes simplex 1 virus. *(b)* Herpes simplex 2 virus infection on the penis.

are distinct from those causing trachoma. Chlamydial infections may also be transmitted congenitally to the newborn in the birth canal, causing newborn conjunctivitis and pneumonia.

Nongonococcal urethritis (NGU) due to *C. trachomatis* is one of the most frequently observed sexually transmitted diseases in males and females, but the infections are often inapparent. In a small percentage of cases, chlamydial NGU leads to serious acute complications, including testicular swelling and prostate inflammation in men, and cervicitis and pelvic inflammatory disease in women. It can also cause fallopian tube damage in women; cells of *C. trachomatis* attach to microvilli of fallopian tube cells, enter, multiply, and eventually lyse the cells (Figure 33.32). Untreated NGU can cause infertility.

Chlamydial NGU is relatively difficult to diagnose by traditional isolation and identification methods, but the organism can be cultured. Samples obtained from a vaginal or pelvic swab or from discharges can be used in nucleic acid probe tests, nucleic acid amplification tests, fluorescent antibody tests, and enzyme immunoassay (EIA) tests for detecting *C. trachomatis* genes and antigens. If chlamydial infection is suspected, even in the absence of a positive diagnostic test, treatment is initiated with azithromycin or doxycycline (Table 33.4).

Chlamydial NGU is frequently observed as a secondary infection following gonorrhea. Both *Neisseria gonorrhoeae* and *C. trachomatis* are often transmitted to a new host in a single event; treatment of gonorrhea with cefixime or ceftriaxone is usually successful, but does not eliminate the chlamydia. Although cured of gonorrhea, these patients are still infected with chlamydia and eventually experience an apparent recurrence of gonorrhea that is instead a case of chlamydial NGU. Thus, patients treated for gonorrhea are also given azithromycin or doxycycline to treat the potential coinfection with the usually undiagnosed *C. trachomatis*.

Lymphogranuloma venereum is a sexually transmitted disease caused by distinct strains of *C. trachomatis* (LGV 1, 2, and 3). The disease occurs most frequently in males and is characterized by infection and swelling of the lymph nodes in and about the groin. From the infected lymph nodes, chlamydial cells may travel to the rectum and cause a painful inflammation of rectal tissues called proctitis. Lymphogranuloma venereum has the

potential to cause regional lymph node damage and the complications of proctitis. It is the only chlamydial infection that invades beyond the epithelial cell layer.

Herpes

Herpesviruses are a large group of complex double-stranded DNA viruses (Section 21.14), many of which are human pathogens. The herpes simplex viruses are responsible for cold sores and genital infections.

Herpes simplex 1 virus (HSV-1) infects the epithelial cells around the mouth and lips, causing cold sores (fever blisters) (**Figure 33.33a**). HSV-1 may, however, occasionally infect other body sites, including the anogenital regions. HSV-1 is spread via direct contact or through saliva. The incubation period of HSV-1 infections is short (3–5 days), and the lesions heal without treatment in 2–3 weeks. The virus is most likely spread primarily by contact with infectious lesions. Latent herpes infections are common, with the virus persisting in low numbers in nerve tissue. Recurrent acute herpes infections are due to a periodic triggering of virus activity by unknown or indeterminant causes such as coinfections and stress. Oral herpes caused by HSV-1 is quite common and apparently has no harmful effects on the host beyond the discomfort of the oral blisters.

Herpes simplex 2 virus (HSV-2) infections are associated primarily with the anogenital region, where the virus causes painful blisters on the penis of males (Figure 33.33b) or on the cervix, vulva, or vagina of females. HSV-2 infections are generally transmitted by direct sexual contact, and the disease is most easily transmitted when active blisters are present but may also be transmitted during asymptomatic periods, even when the infection is presumably latent. HSV-2 occasionally infects other sites such as the mucous membranes of the mouth and can also be transmitted to a newborn by contact with herpetic lesions in the birth canal at birth. The disease in the newborn varies from latent infections with no apparent damage to systemic disease resulting in brain damage or death. To avoid herpes infections in newborns, delivery by cesarean section is advised for pregnant women with genital herpes infections. The long-term effects of genital herpes infections are not yet fully understood. However, studies have indicated a significant correlation between genital herpes infections and cervical cancer in females.

Genital herpes infections are presently incurable, although a limited number of drugs have been successful in controlling the infectious blister stages. The guanine analog acyclovir (**Figure 33.34**), given orally and also applied topically, is particularly effective in limiting the shed of active virus from blisters and

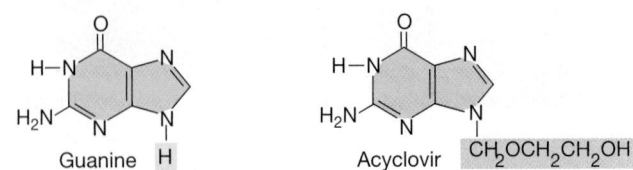

Guanine H Acyclovir CH₂OCH₂CH₂OH

Figure 33.34 Guanine and the guanine analog acyclovir. Acyclovir has been used therapeutically to control genital herpes (HSV-2) blisters.

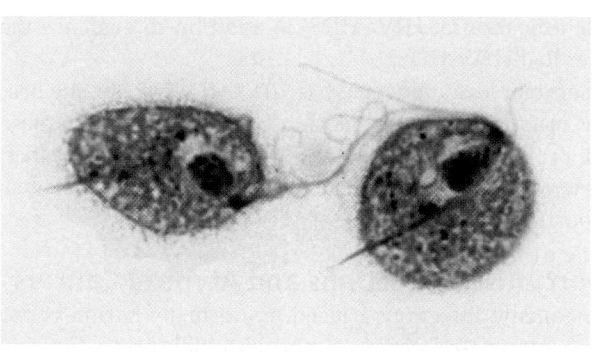

Figure 33.35 *Trichomonas vaginalis*. This flagellated protist causes trichomoniasis, a common sexually transmitted infection.

promoting the healing of blistering lesions. Acyclovir, valacyclovir, and vidarabine are nucleoside analogs that interfere with herpesvirus DNA polymerase, inhibiting viral DNA replication.

Trichomoniasis

Nongonococcal urethritis may also be caused by infections with the protist *Trichomonas vaginalis* (**Figure 33.35**). *T. vaginalis* does not produce the resting cells or cysts important to the life cycle of many protists. As a result, transmission is usually from person to person, generally by sexual intercourse. However, cells of *T. vaginalis* can survive for 1–2 hours on moist surfaces, 30–40 minutes in water, and up to 24 hours in urine or semen. Thus, *T. vaginalis* is sometimes transmitted by contaminated toilet seats, sauna benches, and towels. *T. vaginalis* infects the vagina in women, the prostate and seminal vesicles of men, and the urethra of both males and females.

Trichomoniasis can be asymptomatic in males. In women trichomoniasis is characterized by a vaginal discharge, vaginitis, and painful urination. The infection is more common in females; surveys indicate that 25–50% of sexually active women are infected; only about 5% of men are infected because of the killing action of prostatic fluids. The male partner of an infected female should be examined for *T. vaginalis* and treated if necessary because sexually active asymptomatic males can transmit the infection. Trichomoniasis is diagnosed by observation of the motile protists in a wet mount of fluid discharged from the patient. The antiprotozoal drug *metronidazole* is effective for treating trichomoniasis (Table 33.4).

Human Papillomavirus

Human papillomaviruses (HPV) comprise a family of double-stranded DNA viruses. Of more than 100 different strains, about 30 are transmitted sexually, and several of these cause genital warts and cervical cancer (↩ Chapter 28 Microbial Sidebar, "The Promise of New Vaccines"). About 20 million people in the United States are infected, and up to 80% of women over age 50 have had at least one HPV infection. Over 6 million people acquire new HPV infections annually. Almost 10,000 women develop cervical cancer, and about 3700 die each year.

Most HPV infections are asymptomatic, with some progressing to cause genital warts. Others cause cervical neoplasia (abnormalities in cells of the cervix), and a few progress to cervical cancers. Most HPV infections resolve spontaneously but, as with many viral infections, there is no adequate treatment or cure for active infections. Because of their potential as oncogenic viruses, an HPV vaccine has been developed. The vaccine is designed to provide immunity to the most oncogenic (cancer-causing) viral strains. The HPV vaccine is currently recommended for females 11–26 years of age. The vaccine has also been recommended for males because they develop HPV infection that can lead to anal and penile cancers. Perhaps more importantly, immunized males can no longer be carriers of HPV. The vaccine was designed to stop HPV infection, and, ultimately, to prevent cervical cancer, but the short time the vaccine has been in use precludes any definitive information concerning anticancer benefits.

MiniQuiz

- Describe pertinent clinical features and treatment protocols for chlamydia, herpes, trichomoniasis, and human papillomavirus.
- Why are these diseases more difficult to diagnose than gonorrhea or syphilis?

33.14 Acquired Immunodeficiency Syndrome: AIDS and HIV

Acquired immunodeficiency syndrome (AIDS) was recognized as a distinct disease in 1981. More than 1 million cases of AIDS have been reported since then in the United States alone, and more than 500,000 people have died from AIDS; over 1 million people are living with infection by human immunodeficiency virus (HIV), the cause of AIDS. Worldwide, more than 80 million people have been infected with HIV and at least 2.7 million are infected each year. Over 2 million people die each year, and over 45 million people have already died from AIDS (↩ Section 32.6).

HIV

HIV is divided into two types, *HIV-1* and *HIV-2*, which are genetically similar but distinct. HIV-2, discovered in west Africa in 1985, is less virulent than HIV-1 and causes a milder AIDS-like disease. Currently, more than 99% of global AIDS cases are due to HIV-1, and thus we focus on HIV-1 here.

HIV-1 is a retrovirus (↩ Section 21.11) that preferentially targets macrophages and T cells in the human immune system. Infection eventually leads to depletion of immune T cells, crippling host immune defenses. The genome contains 9749 nucleotides in each of its two identical single-stranded RNA molecules. Using the viral RNA as a template, reverse transcriptase in the intact virion catalyzes formation of a complementary single-stranded DNA molecule. The enzyme then converts the complementary DNA (cDNA) into double-stranded DNA, which integrates into the host cell genome (↩ Section 9.12).

In the United States, the numbers of newly diagnosed HIV infections are increasing, and declines in HIV morbidity and mortality attributable to combination antiretroviral therapy have ended, reversing the trends of decreasing AIDS prevalence and severity seen in the 1990s. The number of HIV-infected individuals will continue to rise unless curative treatment measures or prevention methods are discovered. We have already considered the epidemiology of AIDS (⟳ Section 32.6) and some diagnostic methods for identifying and tracking HIV infection (⟳ Sections 31.10–31.11). Here we focus on the pathogenesis of AIDS, the usual course of HIV infection, the effects of HIV on the immune system, and treatment regimens.

A Definition of HIV/AIDS

AIDS was first suspected of being a disease affecting the immune system because **opportunistic infections**, infections usually observed only in persons with dysfunctional immunity, were prevalent in certain populations. A definition of AIDS was adopted in 1993 by the Centers for Disease Control and Prevention and is used to define AIDS in the United States. In 1999, this definition was revised to include HIV infections, in part because over the last twenty years, advances in drug therapy have slowed the rate of disease progression to full-blown AIDS and significantly extended survival in HIV-infected individuals. The current case surveillance definition includes the AIDS definitions as below, but also adds criteria for defining HIV infection, now identifying the continuum from HIV infection to clinical AIDS as HIV/AIDS.

The case definition for AIDS includes those who (a) test positive for HIV and (b) meet one of the two following criteria:

1. A CD4 T cell number of less than 200/μl of whole blood (the normal count is 600–1000/μl) or a CD4 T cell/total lymphocytes percentage of less than 14%.

2. A CD4 T cell number of more than 200/μl; *and* any of the following fungal diseases including candidiasis, coccidioidomycosis, cryptococcosis, histoplasmosis, isosporiasis, *Pneumocystis jiroveci* pneumonia, cryptosporidiosis, or toxoplasmosis of the brain; bacterial diseases including pulmonary tuberculosis or other *Mycobacterium* spp. infections, or recurrent *Salmonella* septicemia; viral diseases including cytomegalovirus infection, HIV-related encephalopathy, HIV wasting syndrome, chronic ulcers, or bronchitis due to herpes simplex; or progressive multifocal leukoencephalopathy, malignant diseases such as invasive cervical cancer, Kaposi's sarcoma, Burkitt's lymphoma, primary lymphoma of the brain, or immunoblastic lymphoma; or recurrent pneumonia due to any agent.

The 1999 revision includes the following criteria for defining HIV infection in adults, adolescents, and children over 18 months of age.

1. A positive HIV antibody screening test (EIA or a rapid screening test), followed by a positive result on a more sensitive or more specific confirmatory test (Western blot or immunofluorescence test), or a positive virologic test, such as positive tests for HIV nucleic acids, plasma HIV RNA, HIV PCR, HIV p24 antigen test, or HIV isolation (viral culture).

2. If laboratory criteria are not met, proof of HIV infection includes prior diagnosis of HIV infection in the medical record, or laboratory or clinical conditions that fulfill the criteria for AIDS, as above.

Pediatric criteria for children less than 18 months of age are similar, with parallel laboratory or clinical conditions required for identification of HIV/AIDS, in addition to evidence that the mother had HIV/AIDS.

In summary, an individual has HIV/AIDS if he or she has laboratory tests that are positive for HIV or HIV antibodies, HIV genes, proteins, or virus; has a drastically reduced T-helper lymphocyte count; or has clinical evidence of at least one of a number of opportunistic infections or atypical cancers.

Opportunistic Infections and Atypical Cancers

Opportunistic infections caused by normally harmless protists, fungi, bacteria, and viruses occur with high prevalence in HIV/AIDS patients (**Figure 33.36**). The most common opportunistic disease in HIV/AIDS patients is pneumonia caused by the fungus *Pneumocystis jiroveci* (Figure 33.36d). *P. jiroveci* is found in the lungs of most people, but only causes disease in immunosuppressed patients. HIV/AIDS patients can have high antibody titers to *P. jiroveci*, but their lack of ability to mount an effective cellular response mediated by T-helper cells allows the fungus to grow in the lungs, causing pneumonia. Almost all of the opportunistic agents listed above or in Figure 33.36 are very difficult to treat. For example, many of the drugs used to treat infections with the eukaryotic fungi, protists, and opportunistic viruses have significant side effects for the host. Opportunistic bacterial infections due to mycobacteria are often from drug-resistant strains (Section 33.4). Patient compliance and follow-up are also problematic because mycobacterial treatment regimens may last up to 1 year.

A disease frequently seen in HIV/AIDS patients is Kaposi's sarcoma, an atypical cancer of the cells lining the blood vessels and characterized by purple patches on the surface of the skin, especially in the extremities (**Figure 33.37**). Kaposi's sarcoma is caused by coinfection of HIV and human herpesvirus 8 (HHV-8) and is 20,000-fold more prevalent in AIDS patients than in the general population.

HIV Pathogenesis

HIV infects cells that have CD4 cell surface protein. The two cell types most commonly infected are macrophages and a class of lymphocytes called T-helper (T_H) cells, both of which are important components of the immune system. Infected macrophages and T cells produce and release large numbers of HIV particles (**Figure 33.38**), which in turn infect other cells that display CD4.

In addition to CD4, HIV must interact with coreceptors on target cells. HIV infection normally occurs first in macrophages, a type of antigen-presenting cell (APC) that has a very low level of CD4 on its surface (**Figure 33.39**). At the cell surface, the macrophage CD4 molecule binds to the gp120 protein of HIV. The viral gp120 protein then interacts with the macrophage protein CCR5, a chemokine receptor. CCR5 is a coreceptor for HIV and, together with CD4, forms the docking site where the HIV envelope fuses with the host cytoplasmic membrane; this allows the insertion of the viral nucleocapsid into the cell. The CCR5 coreceptor is required for HIV binding to macrophages. Individuals who express a variant CCR5 protein do not bind HIV and do not acquire HIV infections. After HIV has infected the macrophage, a different form of gp120 is made, which in turn binds to a different coreceptor, the CXCR4 chemokine receptor on T cells.

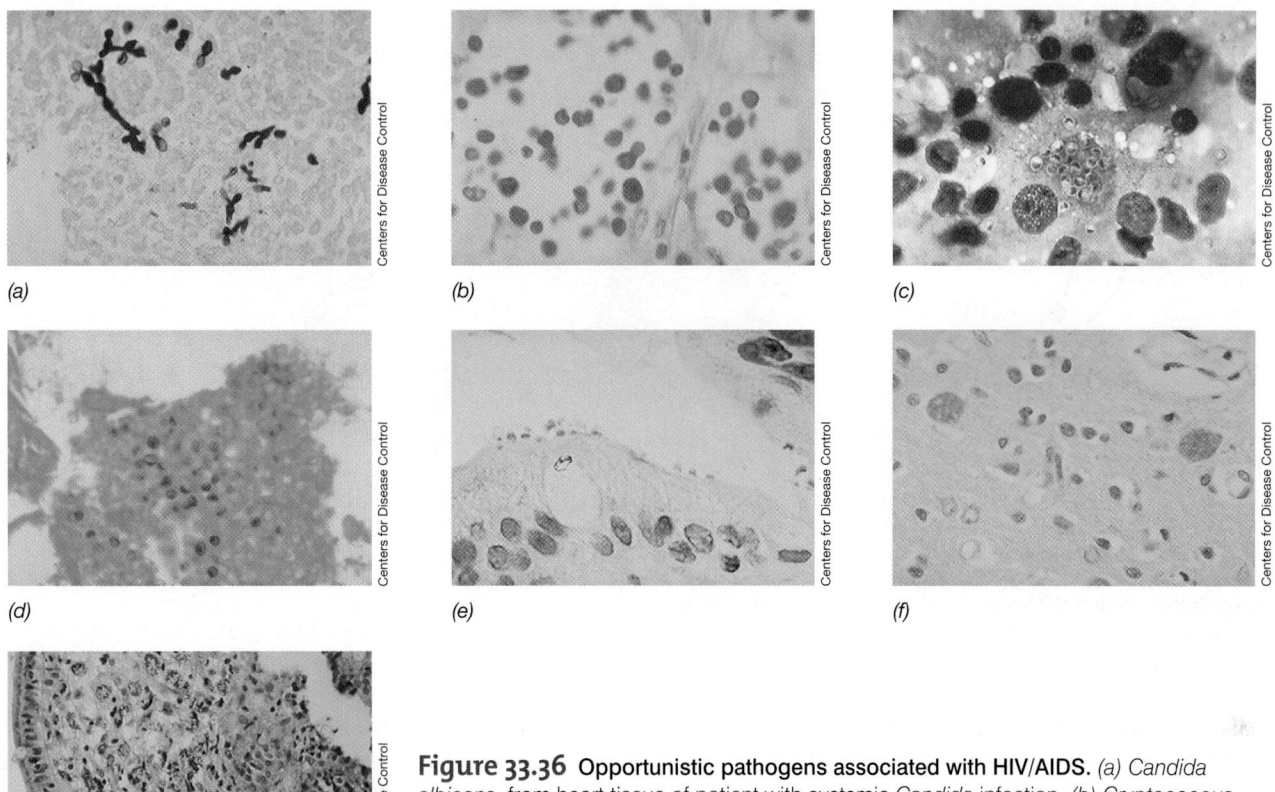

Figure 33.36 Opportunistic pathogens associated with HIV/AIDS. *(a) Candida albicans*, from heart tissue of patient with systemic *Candida* infection. *(b) Cryptococcus neoformans*, from liver tissue of a patient with cryptococcosis. *(c) Histoplasma capsulatum*, from liver tissue of patient with histoplasmosis. *(d) Pneumocystis jiroveci*, from patient with pulmonary pneumocytosis. *(e) Cryptosporidium* sp. from small intestine of a patient with cryptosporidiosis. *(f) Toxoplasma gondii*, from brain tissue of patient with toxoplasmosis. *(g) Mycobacterium* spp. infection of the small bowel (acid-fast stain).

HIV then enters and destroys the CD4 T-helper lymphocytes, the T$_H$1 and T$_H$2 cells that provide cell-mediated inflammatory responses and B cell help (⮌ Section 29.6). Thus, HIV infection starts in macrophages and progresses to a T cell infection. The result of HIV infection is the systematic destruction of macrophages and T cells, leading to a catastrophic breakdown of immunity; opportunistic infections can then develop unchecked.

In people with HIV/AIDS, CD4 lymphocytes become greatly reduced in number. However, HIV infection does not immediately kill the host cell. HIV can exist as a provirus (⮌ Figure 9.24) and not an infectious virion; under these conditions, the reverse-transcribed HIV genome, now DNA, is integrated into host

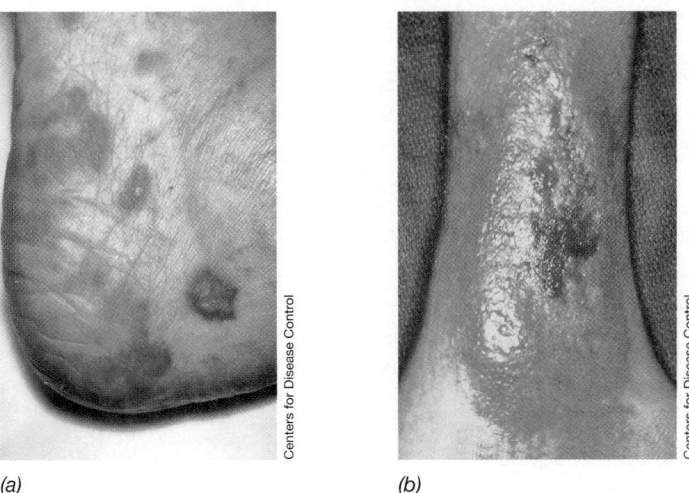

Figure 33.37 Kaposi's sarcoma. Lesions are shown as they appear on *(a)* the heel and lateral foot, and *(b)* the distal leg and ankle.

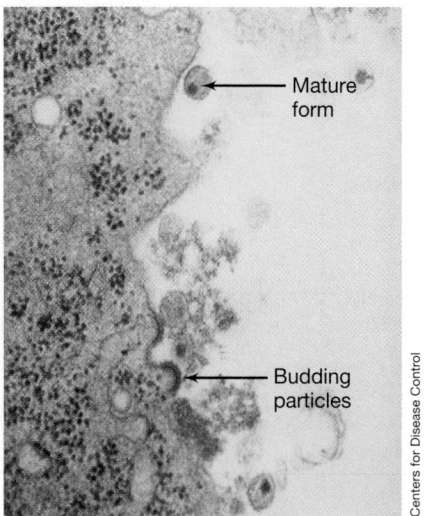

Figure 33.38 • Human lymphocyte releasing HIV. Transmission electron micrograph of a thin section. Cells were from a hemophiliac patient who developed AIDS. HIV particles are 90–120 nm in diameter.

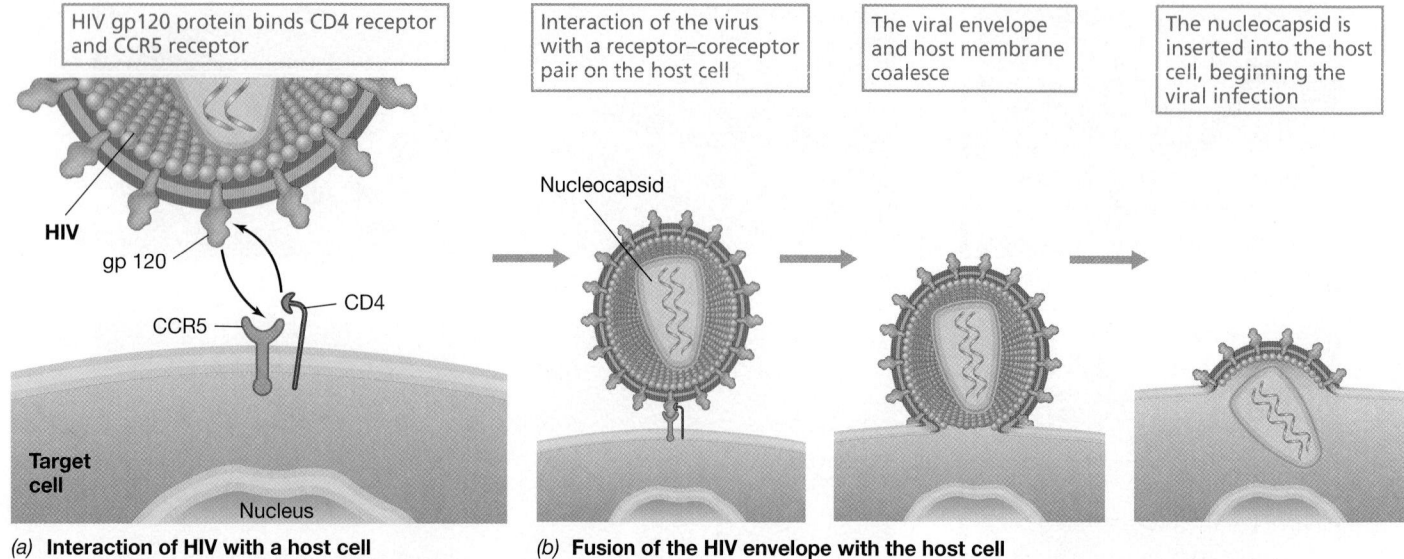

Figure 33.39 Infection of a CD4 target cell with the AIDS virus, HIV. (a) Recognition and binding of the virus by CCR5 and CD4 receptors. (b) The viral nucleocapsid eventually enters the cell.

chromosomal DNA. The cell may show no outward sign of infection, and HIV DNA can remain latent for long periods, replicating as the host DNA replicates. Eventually, virus synthesis occurs and new HIV particles are produced and released from the cell. T cells producing HIV no longer divide and eventually die.

Destruction of CD4 cells is accelerated following the processing of HIV antigens by infected T cells. Such cells insert molecules of gp120 from HIV particles into their cell surfaces. The embedded gp120 protein on the infected cells then binds CD4 on uninfected T cells. The infected and uninfected cells can then fuse to produce multinucleate giant cells called *syncytia*. One HIV-infected T cell may eventually bind and fuse with up to 50 uninfected T cells. Shortly after syncytia form, the fused cells lose immune function and die.

Ongoing HIV infection results in a progressive decline in CD4 cell numbers. In a healthy human, CD4 cells constitute about 70% of the total T cell pool; in HIV/AIDS patients, the number of CD4 cells steadily decreases, and by the time opportunistic infections begin to appear, CD4 cells may be almost absent (🔁 Figure 31.20 and **Figure 33.40**).

Outcomes of HIV Infection

The reduction of CD4 T cells has serious consequences for HIV/AIDS patients. As the number of CD4 cells declines, cytokine production falls, leading to the functional reduction of the immune response; the antibody-mediated and cell-mediated immune responses in HIV/AIDS patients are gradually destroyed. Systemic infections by opportunistic fungi and mycobacteria

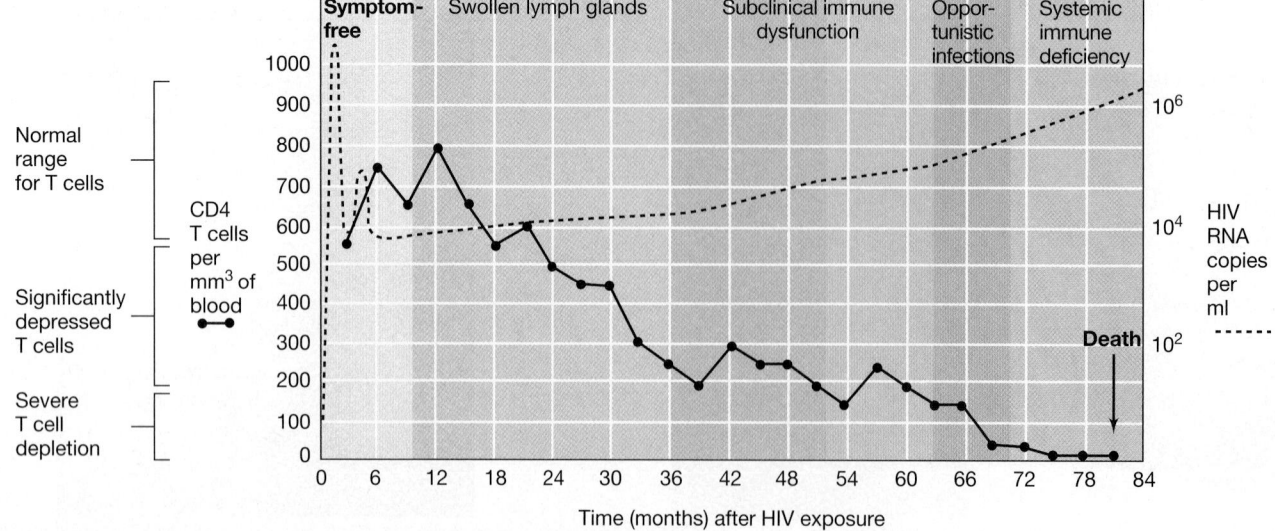

Figure 33.40 Decline of CD4 T lymphocytes and progress of HIV infection. During the typical progression of untreated AIDS, there is a gradual loss in the number and functional ability of the CD4 T cells, while the viral load, measured as HIV-specific RNA copies per milliliter of blood, gradually increases after an initial decline.

point to a loss of T cell–mediated immunity. Opportunistic viral and bacterial infections associated with HIV/AIDS indicate a decline in antibody production due to the loss of the T-helper cells necessary to stimulate antibody production by B cells.

The progression of untreated HIV infection to AIDS follows a predictable pattern. During the clinical latency period, a very active infection is in progress. First, there is an intense immune response to HIV: About 1 billion virions are destroyed each day and HIV numbers drop. However, this means that HIV is replicating at a very high rate; HIV replication causes the destruction of about 100 million CD4 T cells each day. Eventually, the immune response is overwhelmed, HIV levels increase, and the CD4 T cells are completely destroyed, crippling the immune response and allowing opportunistic pathogens to initiate infections. The example in Figure 33.40 documents T cell destruction and the increase in HIV over a typical time course for an untreated HIV infection culminating in AIDS.

Diagnosing HIV Infection

HIV infection can be diagnosed by identifying antibodies to the pathogen. The HIV EIA (♻ Figure 31.22) is used for large-scale screening of donated blood to prevent transfusion-associated HIV transmission. About 0.25% (2–3 per 1000) of all blood donated by volunteer donors in the United States tests HIV-positive in the EIA assay. A positive HIV EIA must be confirmed by a second procedure called an immunoblot (Western blot), a technique that combines the analytical tools of protein purification and immunology (♻ Section 31.11), or by immunofluorescence tests (♻ Section 31.9).

Point-of-care rapid tests are also used for preliminary screening of blood or other body fluids for the presence of antibodies to HIV. One test uses a single drop of patient blood and a single reagent. The reagent is a bifunctional recombinant fusion protein containing an antibody that binds a red blood cell antigen; this is coupled to gp41 HIV surface antigen. In a positive test, patient antibody to gp41 cross links the antibody-bound red blood cells, producing a visible agglutination. In another test, saliva is used as a source of secretory antibody to HIV. The saliva is expelled onto a cartridge containing immobilized HIV antigens. A second antibody, reactive with the bound antibody and conjugated to an enzyme, is then added. After addition of the enzyme substrate, a positive reaction shows a colored product. These rapid tests are designed to provide maximum convenience, speed (minutes instead of the hours or days required for EIA or immunoblot), extended shelf life, portability, and ease of use and interpretation. In general, however, the rapid tests are not as sensitive or specific as the standard HIV EIA and, as with the EIA test, positive rapid tests must be confirmed by HIV immunoblot or immunofluorescence tests.

These tests, no matter how sensitive or specific, fail to detect HIV-positive individuals who have recently acquired the virus and have not yet made a detectable antibody response. This period may be more than 6 weeks after exposure to HIV. Despite this drawback, these tests ensure the general safety of the blood supply, and the risk of contracting HIV through contaminated blood or blood products is now very low. Sexual contact with multiple partners and group intravenous drug use are the major risk factors for acquiring HIV infection (♻ Figure 32.8).

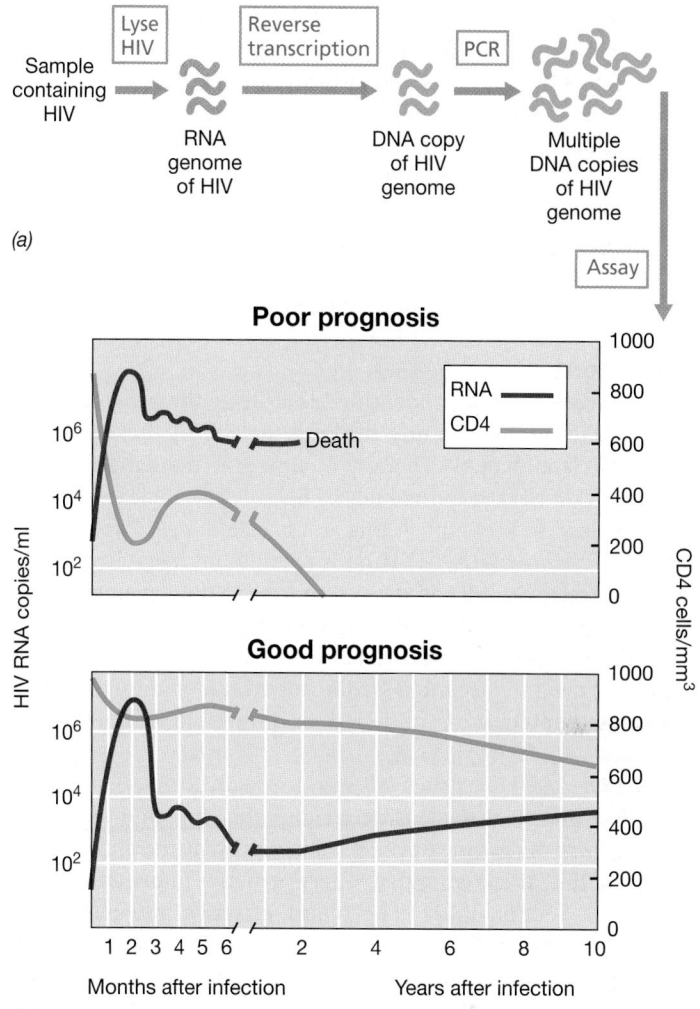

Figure 33.41 Monitoring of HIV load. *(a)* Detection of HIV by RT-PCR (reverse transcription–polymerase chain reaction). The HIV copies obtained are compared quantitatively with DNA copies from a control template that is amplified in the same RT-PCR amplification. HIV load is expressed as the number of HIV copies per milliliter of patient plasma. *(b)* Time course for HIV infection as monitored by HIV RT-PCR. Progression of infection is estimated based on viral load at successive times after infection. CD4 T cell counts are measured in cells per cubic millimeter. In the upper panel, a viral load greater than 10^4 copies per milliliter correlates with below normal CD4 cell numbers (normal = 600–1500/mm^3), indicating a poor prognosis and early death of the patient. In the lower panel, a viral load less than 10^4 copies per milliliter correlates with normal CD4 cell numbers, indicating a good prognosis and extended survival of the patient. Data are adapted from the Centers for Disease Control and Prevention, Atlanta, Georgia.

Several laboratory tests detect HIV RNA directly and quantitatively from blood samples. These tests use a virus-specific reverse transcriptase–polymerase chain reaction (RT-PCR). The RT-PCR estimates the number of viruses present in the blood, or the **viral load**. The RT-PCR test indicates the magnitude of HIV replication and correlates with the rate of CD4 T cell destruction, a direct indicator of the magnitude of destruction of the immune system. The RT-PCR test is not routinely used to screen for HIV because it is costly and technically demanding. After initial diagnosis of infection, however, the RT-PCR test is used to monitor progression of HIV/ AIDS and the effectiveness of chemotherapy (**Figure 33.41**).

The prognosis for an untreated HIV-infected individual is poor. Opportunistic pathogens or malignancies eventually kill most HIV/AIDS patients. Long-term studies indicate that the average person infected with HIV progresses through several stages of decreasing immune function, with CD4 cells dropping from a normal range of 600–1000/mm^3 of blood to near zero over a period of 5–7 years (Figure 33.40). Although the rate of decline in immune function varies from one HIV-infected individual to another, it is rare for an HIV-positive individual to live for more than 10 years without antiretroviral drug therapy.

Treatment

Several drugs have been identified that delay the progression of HIV/AIDS and significantly prolong the life of those infected with HIV (**Table 33.5**). Therapy is aimed at reducing the viral load of HIV-infected individuals to below detectable levels. The strategy used to accomplish this aim is called *highly active antiretroviral therapy* (HAART) and is carried out by giving at least three antiretroviral drugs at once to inhibit the development of drug-resistant HIV. Multiple drug therapy, however, is not a cure for HIV infection. In individuals who have no detectable viral load after drug treatment, a significant viral load returns if therapy is interrupted or discontinued, or if multiple drug resistance develops.

Effective anti-HIV drugs fall into four categories. The first of these is a group of nucleoside analogs that function as *reverse transcriptase inhibitors*. Reverse transcriptase is the enzyme that converts the single-stranded RNA genetic information into cDNA(↩ Section 9.12). The oldest effective anti-HIV drug, *azidothymidine* (AZT), is an inhibitor of HIV replication that closely resembles the nucleoside thymidine, but lacks the correct attachment site for the next base in a replicating nucleotide chain, resulting in termination of the growing DNA chain. Thus, AZT and the other nucleoside analogs are **nucleoside reverse transcriptase inhibitors (NRTIs)**, and they stop HIV replication

(a) **Azidothymidine**

(b) **Nevirapine**

(c) **Saquinavir**

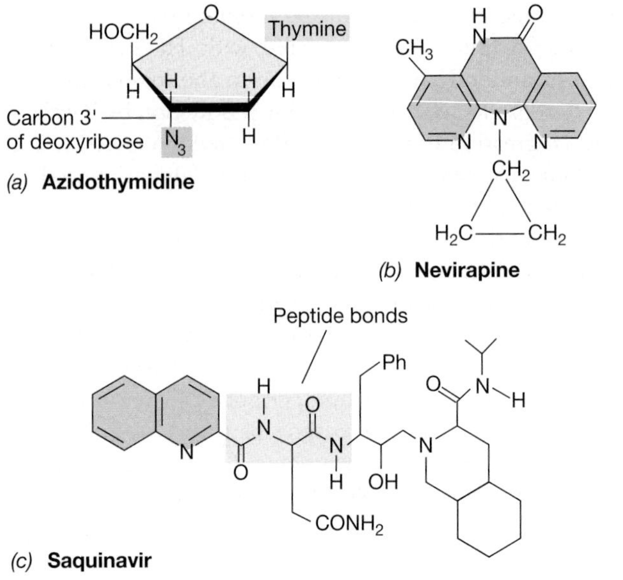

Figure 33.42 HIV/AIDS chemotherapeutic drugs. *(a)* Azidothymidine (AZT), a nucleoside reverse transcriptase inhibitor. This nucleoside analog is missing the −OH group on the 3′ carbon, causing nucleotide chain elongation to terminate when the analog is incorporated, inhibiting virus replication. *(b)* Nevirapine, a nonnucleoside reverse transcriptase inhibitor, binds directly to the catalytic site of HIV reverse transcriptase, also inhibiting elongation of the nucleotide chain. *(c)* Saquinavir, a protease inhibitor, was designed by computer modeling to fit the active site of the HIV protease. Saquinavir is a peptide analog: The tan highlighted area shows the region analogous to peptide bonds. Blocking the activity of HIV protease prevents the processing of HIV proteins and maturation of the virus.

(**Figure 33.42a**, Table 33.5). These drugs rapidly decrease the viral load when given to HIV-infected individuals. Drug-resistant strains of HIV arise within several weeks if only a single NRTI is administered, the result of viral mutation and selection.

Table 33.5 *Chemotherapeutic agents for HIV/AIDS treatment*	
Drug name	**Drug class and mechanism of activity**
Azidothymidine (AZT, ZDV, or zidovudine) Dideoxycytidine (ddC or zalcitabine) Dideoxyinosine (ddI or didanosine) Stavudine (d4T) Lamivudine (3TC)	*Nucleoside reverse transcriptase inhibitors (NRTIs)*; nucleoside analogs that inhibit reverse transcriptase; nucleotide chain synthesis terminator; increases survival time and reduces incidence of opportunistic infection in AIDS patients; toxic to bone marrow cells; may be used in combination with other drugs in highly active antiretroviral treatment (HAART) protocols
Efavirenz Nevirapine Delavirdine	*Nonnucleoside reverse transcriptase inhibitors (NNRTIs)*; bind directly to reverse transcriptase and disrupt the catalytic site; do not compete with nucleosides; may be used in combination with other drugs in HAART protocols
Indinavir Nelfinavir Saquinavir Ritonavir	*Protease inhibitors*; computer-designed peptide analogs designed to bind to the active site of the HIV protease, inhibiting processing of viral polypeptides and virus maturation; may be used in combination with other drugs in HAART protocols
Enfuvirtide	*Fusion inhibitor*; synthetic polypeptide that binds to the gp41 protein and inhibits the fusion of HIV membranes with host cytoplasmic membranes
Elvitegravir Raltegravir	*Integrase inhibitors*; drugs that inhibit integration of HIV DNA into host DNA

Although this process seems very rapid, HIV replicates very quickly (see above), and as little as four single-nucleotide mutations can result in resistance to a given nucleoside analog.

The second category of anti-HIV drugs is the **nonnucleoside reverse transcriptase inhibitors (NNRTIs)** (Figure 33.42*b*; Table 33.5). NNRTIs directly inhibit the activity of reverse transcriptase by interacting with the protein and altering the conformation of the catalytic site. Unfortunately, a single mutation in the reverse transcriptase gene is often sufficient to reduce the effectiveness of these drugs.

Another category of anti-HIV drugs are the **protease inhibitors (PI)** (Figure 33.42*c*; Table 33.5). The protease inhibitors are computer-designed peptide analogs that inhibit processing of viral polypeptides by binding to the active site of the processing enzyme, HIV protease (Figure 26.31); this inhibits virus maturation. As with the enzyme-targeted chemotherapy strategies aimed at reverse transcriptase, a single mutation in the HIV protease gene can cause drug resistance.

Another category of approved anti-HIV drugs is the drug *enfuvirtide*, a **fusion inhibitor** composed of a 36–amino acid synthetic peptide that functions by binding to the gp41 membrane protein of HIV (Table 33.5). Binding of the protein stops the conformational changes necessary for the fusion of the viral envelope and CD4 cytoplasmic membranes. Enfuvirtide is prescribed to HIV-positive individuals who have developed antiretroviral drug resistance or increased viral load after conventional therapy.

The **integrase inhibitors** *elvitegravir* and *raltegravir* are members of a new category of anti-HIV drugs. These drugs target integrase, the HIV protein that catalyzes the integration of viral dsDNA into host cell DNA. By interfering with integration of viral DNA into the host cell genome, the HIV replication cycle is interrupted. Raltegravir is approved for use and elvitegravir is undergoing clinical trials. The drugs will be used as part of HAART therapy, in combination with other drugs, and may also be particularly useful to treat patients who have developed resistance to all other classes of HIV drugs (Table 33.5).

A typical recommended HAART protocol for treatment of an individual with established HIV infection includes at least one protease inhibitor or one NNRTI, plus a combination of two NRTIs (Table 33.5). A resistant virus would, therefore, have to develop resistance to three drugs simultaneously; the probability of this occurring is very small. Thus, multiple drug therapy reduces the probability that a drug-resistant virus could emerge. This combination therapy is then monitored by RT-PCR to track changes in viral load. An effective HAART protocol reduces viral load to nondetectable levels (less than 500 copies of HIV per milliliter of blood) within several days (Figure 33.41). The therapy is continued and monitored for viral load indefinitely. If the viral load again reaches detectable limits, the drug cocktail is changed because an increase in viral load indicates the emergence of drug-resistant HIV.

In addition to drug resistance, some of the antiviral drugs are toxic to the host. In many cases, nucleoside analogs are not well tolerated by patients, presumably because they interfere with host functions such as cell division (Table 33.5). In general, the NNRTIs and the protease inhibitors are better tolerated because they target virus-specific functions. However, drug resistance and host toxicity are major problems in all forms of HIV therapy. Thus, new chemotherapeutic agents and drug protocols are constantly being developed and tailored to the needs of individual patients.

AIDS Immunization

A safe, effective HIV vaccine is a critical component of HIV/AIDS pandemic control. Such a vaccine could be used to prevent HIV infection and for therapy to cure infection. However, the extreme genetic variability of HIV has thus far hampered the development of such a universal vaccine.

A current strategy used for experimental preventive vaccines uses the prime–boost method. The "prime" component is a vaccine designed to initially stimulate a primary immune response; it is usually administered several weeks to months before another vaccine, the "boost," is administered. The boost is designed to stimulate a memory response, triggering a long-lasting and effective host immunity to HIV. In a study that showed the efficacy of this strategy for producing an HIV/AIDS vaccine, antibodies were induced to the HIV envelope protein gp120. These antibodies could then block CD4–gp120 interactions and inhibit infection. The prime vaccine was a recombinant canarypox vaccine expressing HIV subtype E gp120 protein as well as other HIV proteins. The boost vaccine was a recombinant gp120 subunit vaccine directed at subtypes B and E gp120 (Sections 15.13, 28.7, and 28.8). The B and E HIV subtypes chosen are common in Thailand, the site of the study. After Phase I and Phase II clinical trials (for safety and efficacy, respectively) in small numbers of individuals, a Phase III clinical trial with large numbers of individuals at risk for heterosexually transmitted HIV showed a moderate protective effect for preventing HIV infection. Unfortunately, this approach may not be successful for a universal vaccine because there are a number of gp120 subtypes, and the genes that encode them mutate frequently, forming antigenic variants that would not be recognized by antibodies made to the vaccine.

Clinical immunization trials are also proceeding with other subunit vaccines consisting of HIV envelope proteins engineered into vaccinia virus or adenovirus particles. Using these harmless viruses as expression vectors and vehicles for delivery of HIV antigens, several vaccines elicit a strong antibody and cellular immune response to HIV.

Other potential immunization candidates include killed intact HIV. These inactivated vaccines could only be used as therapeutic vaccines to treat individuals already infected with HIV because inactivation procedures may not deactivate 100% of the HIV; it would be unethical to expose uninfected individuals to even a small risk of HIV infection. Some laboratories are also exploring the possibilities of producing live attenuated virus for use as a therapeutic vaccine. This strategy is supported by the finding that individuals infected with HIV-2, a related virus that causes a milder form of AIDS with a very long latent period, prevents infection with HIV-1, the strain responsible for HIV/AIDS. However, there are potential risks with this strategy. For example, integrated virus could cause cancer, or mutations might reactivate virulence in the attenuated HIV vaccine.

UNIT 11

Successful HIV immunization protocols would probably not be useful for treating most patients that already have HIV/AIDS, because they may lack enough immune function to respond to a vaccine. Thus, despite considerable knowledge of the mechanisms of HIV infection and a clear understanding of the AIDS disease process, there is currently no proven medical intervention strategy for the prevention or cure of HIV/AIDS.

HIV/AIDS Prevention

Public education and avoidance of high-risk behavior remain the major tools used to prevent HIV/AIDS. HIV spread is linked to promiscuous sexual activities and other activities that involve exchange of body fluids, which include not only male homosexuality, but also female prostitution and intravenous drug use. In the United States the fastest growing method of transmission is between heterosexual partners. Prevention requires avoidance of the high-risk behaviors associated with exchange of body fluids. The U.S. Surgeon General has issued a report that makes specific recommendations that individuals can follow if they wish to reduce the likelihood of HIV infection. Among the recommendations are the following:

1. Avoid mouth contact with penis, vagina, or rectum.
2. Avoid all sexual activities that could cause cuts or tears in the linings of the rectum, vagina, or penis.

3. Avoid sexual activities with individuals from high-risk groups. These include prostitutes (both male and female); those who have multiple partners, particularly homosexual men and bisexual individuals; and intravenous drug users.
4. If a person has had sex with a member of one of the high-risk groups, a blood test should be done to determine if infection with HIV has occurred. The blood test should be repeated at intervals for a year or longer because of the lag time in the immune response. If the test is positive, the sexual partners of the HIV-positive individual must be protected by use of a condom during sexual intercourse. The use of condoms is also recommended for all extramarital sexual activity.

For more information about prevention of STIs and HIV/AIDS, contact the American Social Health Association STI Resource Center Hotline at 919-361-8488. The ASHA website is http://www.ashastd.org.

www.microbiologyplace.com **Online Tutorial 33.1 HIV Replication**

MiniQuiz
- Review the definition of HIV/AIDS. What symptoms of HIV/AIDS are shared by all HIV/AIDS patients?
- What are the current prevention and treatment guidelines for HIV/AIDS infection? Are they effective?

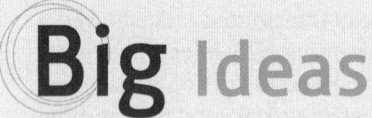

Big Ideas

33.1
Bacterial and viral respiratory pathogens are transmitted in air. Most respiratory pathogens are transferred from person to person via respiratory aerosols generated by coughing, sneezing, talking, or breathing.

33.2
Diseases caused by streptococci include streptococcal pharyngitis and pneumococcal pneumonia. *Streptococcus pyogenes* infections may develop from pharyngitis into serious conditions such as scarlet fever and rheumatic fever. Pneumonia caused by *Streptococcus pneumoniae* is a serious disease with high mortality. Definitive diagnosis for both pathogens is by culture. Infections with both pathogens are treatable with antimicrobial drugs, but drug-resistant strains of *S. pneumoniae* are common.

33.3
Diphtheria is an acute respiratory disease caused by the gram-positive bacterium *Corynebacterium diphtheriae*. Early childhood immunization is effective for preventing this very serious respiratory disease. Whooping cough is an endemic disease caused by *Bordetella pertussis*. Immunization of children, adolescents, and adults can control its propagation and spread.

33.4
Tuberculosis is one of the most prevalent and dangerous infectious diseases in the world. Its incidence is increasing in developed countries in part because of the emergence of drug-resistant strains of *Mycobacterium tuberculosis*. The pathology of tuberculosis and other mycobacterial diseases such as Hansen's disease (leprosy) is influenced by the cellular immune response.

33.5
Neisseria meningitidis is a common cause of meningococcemia and meningitis in young adults and occasionally occurs in epidemics in closed populations such as schools and military installations. Bacterial meningitis and meningococcemia are serious diseases with high mortality rates. Treatment and prevention strategies are in place to deal with epidemic outbreaks. Effective vaccines are available for the most prevalent pathogenic strains.

33.6
Viral respiratory diseases are highly infectious and may cause serious health problems. However, the common childhood viral diseases measles, mumps, rubella, and chicken pox are all controllable with appropriate immunizations.

33·7

Colds are the most common infectious viral diseases. Usually caused by a rhinovirus, colds are generally mild and self-limiting diseases. Each infection induces specific, protective immunity, but the large number of cold viruses precludes complete protective immunity or vaccines.

33·8

Influenza outbreaks occur annually due to the plasticity of the influenza genome. Influenza epidemics and pandemics occur periodically. Surveillance and immunization are used to prevent influenza.

33·9

Staphylococci are usually harmless inhabitants of the upper respiratory tract and skin, but several serious diseases can result from pyogenic infection or from the actions of staphylococcal superantigen exotoxins. Antibiotic resistance is common, even in community-acquired infections.

33·10

Helicobacter pylori infection appears to be a common cause of gastric ulcers. Gastric ulcers are now treated as an infectious disease, promoting a permanent cure.

33·11

Viral hepatitis can result in acute liver disease, which may be followed by cirrhosis, chronic liver disease. HBV and HCV can cause chronic infections leading to liver cancer. Vaccines are available for HAV and HBV. The incidence and prevalence of hepatitis has decreased significantly in the last 20 years in the United States, but viral hepatitis is still a major public health problem because of the high infectivity of the viruses and the lack of effective treatment options.

33·12

Gonorrhea and syphilis, caused by *Neisseria gonorrhoeae* and *Treponema pallidum*, respectively, are STIs with potential serious consequences if infections are not treated. In the United States, the incidence of gonorrhea has decreased in the last several years, but the incidence of syphilis has increased.

33·13

Chlamydia, the most prevalent STI, is caused by infection with the bacterium *Chlamydia trachomatis*. Untreated chlamydial nongonococcal urethritis causes serious complications in males and females. Herpes infections can be transmitted sexually and are caused by herpes simplex 1 and 2 viruses. There is no cure for herpes infections. *Trichomonas vaginalis* is a protist responsible for trichomoniasis, another STI. Human papillomaviruses cause widespread STIs that may lead to cancer. There is an effective HPV vaccine. In general, these STIs are widespread and are more difficult to diagnose and treat than gonorrhea or syphilis.

33·14

HIV/AIDS is one of the most prevalent infectious diseases in the human population. Human immunodeficiency virus destroys the immune system, and opportunistic pathogens then kill the host. There is no effective cure for HIV infection or AIDS. Antiviral drugs, however, may slow or stop the progress of AIDS. There is no effective vaccine for HIV. Prevention for the spread of HIV infection requires education and avoidance of high-risk behaviors involving exchange of body fluids.

Review of Key Terms

Antigenic drift a minor change in influenza virus antigens due to gene mutation

Antigenic shift a major change in influenza virus antigen due to gene reassortment

Cirrhosis breakdown of normal liver architecture, resulting in fibrosis

Congenital syphilis syphilis contracted by an infant from its mother during pregnancy

Fusion inhibitor a synthetic polypeptide that binds to viral glycoproteins, inhibiting fusion of viral and host cell membranes

Hepatitis liver inflammation, commonly caused by an infectious agent

Human papillomavirus (HPV) a sexually transmitted virus that causes genital warts, cervical neoplasia, and cancer

Integrase inhibitor drug that interrupts the HIV replication cycle by interfering with integrase, the HIV protein that catalyzes the integration of viral dsDNA into host cell DNA

Meningitis inflammation of the meninges (brain tissue), sometimes caused by *Neisseria*

meningitidis and characterized by sudden onset of headache, vomiting, and stiff neck, often progressing to coma within hours

Meningococcemia a rapidly progressing severe disease caused by *Neisseria meningitidis* and characterized by septicemia, intravascular coagulation, and shock

Nonnucleoside reverse transcriptase inhibitor (NNRTI) a nonnucleoside compound that inhibits the action of viral reverse transcriptase by binding directly to the catalytic site

Nucleoside reverse transcriptase inhibitor (NRTI) a nucleoside analog compound that inhibits the action of viral reverse transcriptase by competing with nucleosides

Opportunistic infection an infection usually observed only in an individual with a dysfunctional immune system

Pertussis (whooping cough) a disease caused by an upper respiratory tract infection with *Bordetella pertussis*, characterized by a deep persistent cough

Protease inhibitor (PI) a compound that inhibits the action of viral protease by binding directly to the catalytic site, preventing viral protein processing

Rheumatic fever an inflammatory autoimmune disease triggered by an immune response to infection by *Streptococcus pyogenes*

Scarlet fever characteristic reddish rash resulting from an exotoxin produced by *Streptococcus pyogenes*

Sexually transmitted infection (STI) an infection that is usually transmitted by sexual contact

Toxic shock syndrome (TSS) the acute systemic shock resulting from a host response to an exotoxin produced by *Staphylococcus aureus*

Tuberculin test a skin test for previous infection with *Mycobacterium tuberculosis*

Viral load a quantitative assessment of the amount of virus in a host organism, usually in the blood

Review Questions

1. Why do gram-positive bacteria cause respiratory diseases more frequently than gram-negative bacteria (Section 33.1)?

2. Discuss some of the diseases caused by streptococcal infections and the treatments available (Section 33.2).

3. Describe the causal agents and the symptoms of diphtheria and pertussis. Why has diphtheria incidence declined in the United States, while pertussis incidence is higher than a decade ago (Section 33.3)?

4. Describe the clinical features of Hansen's disease (Section 33.4).

5. Describe the symptoms of meningococcemia and meningitis. How are these diseases treated? What is the prognosis for each (Section 33.5)?

6. Compare and contrast the pathogenesis of measles, mumps, rubella, and chicken pox (Section 33.6).

7. What is the difference between a cold and the flu (Sections 33.7 and 33.8)?

8. Why is influenza such a common respiratory disease? How are influenza vaccines chosen (Section 33.8)?

9. Distinguish between pathogenic staphylococci and those that are part of the normal flora (Section 33.9).

10. Describe the evidence linking *Helicobacter pylori* to gastric ulcers. How would you treat an ulcer patient (Section 33.10)?

11. Describe the major pathogenic hepatitis viruses. How are they related to one another? How is each spread (Section 33.11)?

12. Why did the incidence of gonorrhea rise dramatically in the mid-1960s, while the incidence of syphilis actually decreased at the same time (Section 33.12)?

13. For the sexually transmitted diseases chlamydia, herpes, trichomoniasis, and human papillomavirus, describe the incidence of each. In each case, is treatment possible, and if so, is it an effective cure? Why or why not (Section 33.13)?

14. Describe how human immunodeficiency virus (HIV) effectively shuts down both humoral immunity and cell-mediated immunity. Why are vaccines to HIV so difficult to develop (Section 33.14)?

Application Questions

1. How can an epidemic of whooping cough be controlled? How can it be prevented? Since the incidence of this disease is no longer decreasing, apply your prevention methods to the current "mini-epidemic" (Section 33.3). Compare the incidence of pertussis to that of diphtheria. Since childhood immunizations generally include both vaccines, why is the incidence of pertussis rising, whereas that of diphtheria has remained very low? *Hint*: There may be several reasons, including available vaccines and vaccine practices, herd immunity, and increased susceptibility in various populations.

2. Why does active tuberculosis often lead to a permanent reduction in lung capacity, whereas most other respiratory diseases cause only temporary respiratory problems? Worldwide, the prevalence of tuberculosis infection is very high, but active disease is much lower. Please explain.

3. Your college roommate goes home for the weekend, becomes extremely ill, and is diagnosed with bacterial meningitis at a local hospital. Because he was away, university officials are not aware of his illness. What should you do to protect yourself against meningitis? Should you notify university health officials?

4. Measles, mumps, and rubella were once very common childhood diseases. However, outbreaks of these diseases are now regarded as serious incidents requiring immediate attention from public health

officials. Explain this shift in attitudes in the context of disease prevalence, availability of vaccines, and the potential health consequences of each disease in a college population.

5. Discuss the molecular biology of antigenic shift in influenza viruses and comment on the immunological consequences for the host. Why does antigenic shift prevent the production of a single universally effective vaccine for influenza control? Next, compare antigenic shift to antigenic drift. Which mechanism is more important for the evolution of the influenza virus? Which causes the greatest antigenic change? Which creates the biggest problems for vaccine developers? Which can lead to pandemic influenza? Why?

6. Arrange the hepatitis viruses in order of disease severity, both in the short term and in the long term.

7. As the director of your dormitory's public health advisory group, you are charged to present information on chlamydia, herpes, trichomoniasis, and human papillomavirus, all STIs. Besides this textbook, where can you get reliable information about STIs? Present information on prevention, symptoms, and treatment for each STI. Will your program for each disease overlap? For each of the diseases, discuss the practical, social, legal, and public health issues that must be considered to identify and notify the sexual partners of infected individuals.

Need more practice? Test your understanding with quantitative questions; access additional study tools including tutorials, animations, and videos; and then test your knowledge with chapter quizzes and practice tests at **www.microbiologyplace.com**.

Vectorborne and Soilborne Microbial Pathogens

The bacterium *Borrelia burgdorferi* causes Lyme disease. This organism, highlighted in green here, consists of tightly coiled, slender, flexuous cells. *B. burgdorferi* cells are transmitted to humans by ticks and thus the organism is a typical vectorborne pathogen.

Our focus in this chapter is on pathogenic microorganisms that cause diseases transmitted through animals, arthropods, and soil. Animal-transmitted pathogens have their origins in nonhuman vertebrates; infected animal populations can transmit infections to humans. Arthropods can act as vectors, spreading pathogens to new hosts via a bite. Humans are often accidental hosts in the life cycle of vectorborne pathogens, but they may also be a reservoir, as is the case for the *Plasmodium* spp. that cause malaria. Soilborne pathogens may be transmitted through contact with soil containing the pathogens, which may then enter the host through wounds or abrasions.

Animals can transmit agents of serious, sometimes fatal diseases such as rabies and hantavirus syndrome. Insects are vectors for human diseases such as plague, where humans are accidental hosts, and malaria, where humans are important disease reservoirs. Arthropod-transmitted diseases have killed millions of people, altered the course of human history, and even influenced human evolution. Fungal diseases and tetanus present unique problems because the pathogens that cause these diseases live in soil and cannot be effectively eliminated or contained.

I Animal-Transmitted Pathogens

A zoonosis is an animal disease transmissible to humans, generally by direct contact, aerosols, or bites. Immunization and veterinary care control many infectious diseases in domesticated animals, reducing the transfer of zoonotic pathogens to humans. However, feral (wild) animals neither receive veterinary care nor are they immunized, making them a source of potential infections. Diseases in animals may be **enzootic**, present endemically in certain populations, or **epizootic**, with incidence reaching epidemic proportions. Epizootic diseases often occur on a periodic, sometimes cyclic basis. Because of the unusually high prevalence of diseased animals in epizootic situations, the potential for transferring pathogens from infected animals to humans increases. We focus our discussion here on two important zoonotic pathogens, rabies and hantavirus, transmitted to humans by contact with infected vertebrates.

34.1 Rabies Virus

Rabies occurs in wild animals as an enzootic disease that can spread as a zoonotic disease to humans. The major enzootic reservoirs of the rabies virus in the United States are raccoons, skunks, coyotes, foxes, and bats. A small number of rabies cases also occur annually in domestic animals (**Figure 34.1**).

Epidemiology and Pathology

Rabies is a vaccine-preventable infectious disease in humans. Nevertheless, about 55,000 people, mostly children, die every year, primarily in developing countries in Asia and Africa where it is enzootic in domestic animals such as dogs. Annually, about 14 million people worldwide receive rabies postexposure prophylactic treatment after animal bites; in the United States, over 20,000 individuals receive postexposure prophylaxis.

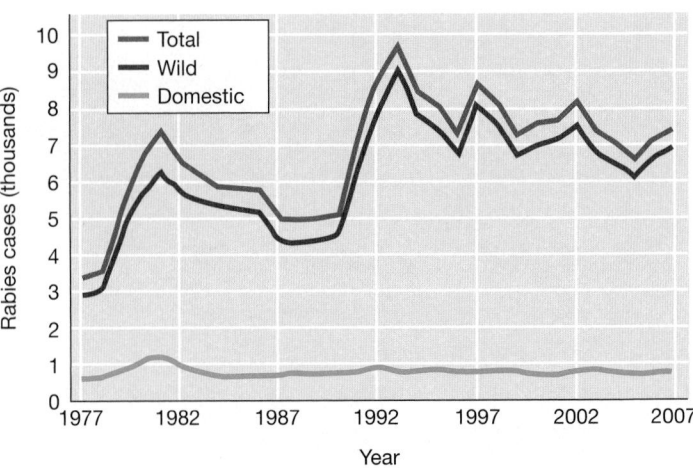

Figure 34.1 Rabies cases in wild and domestic animals in the United States. Rabies is an enzootic disease in wild animal populations, especially in raccoons in the eastern United States. The peaks in the numbers of infected wild animals are epizootic events. An epizootic event kills a large portion of the host population, and this smaller population supports less rabies transmission. The disease numbers fall, and the disease again becomes enzootic. As the population recovers from the epizootic event, rabies case numbers rise, endemic rabies becomes an epizootic disease, a large number of hosts die, and the cycle repeats. Over 500 cases of rabies are reported annually in domestic animals, nearly all acquired from contact with wild animals. Data are from the Centers for Disease Control and Prevention, Atlanta, Georgia, USA.

Rabies is caused by a rhabdovirus, a negative-strand RNA virus (↺ Section 21.9) that infects cells in the central nervous system of most warm-blooded animals, almost invariably leading to death once clinical symptoms have developed. The virus (**Figure 34.2a**) enters the body via virus-contaminated saliva through a wound from a bite or through contamination of mucous membranes. Rabies virus multiplies at the site of inoculation and travels to the central nervous system. The incubation period before the onset of symptoms is highly variable, depending on the host, the size, location, and depth of the inoculating wound, and the number of viral particles transmitted in the bite. In dogs, the incubation period averages 10–14 days. In humans, 9 months or more may pass before rabies symptoms become apparent.

The virus proliferates in the brain, especially in the thalamus and hypothalamus. Infection leads to fever, excitation, dilation of the pupils, excessive salivation, and anxiety. A fear of swallowing (hydrophobia) develops from uncontrollable spasms of the throat muscles. Death eventually results from respiratory paralysis. In humans, an *untreated* rabies infection that becomes symptomatic is almost always fatal.

Diagnosis and Treatment of Rabies

Rabies is diagnosed in the laboratory by examining tissue samples. Fluorescent antibodies or monoclonal antibody tests that recognize rabies virus in brain or corneal tissue are used to confirm a clinical diagnosis of rabies, either in a potentially rabid animal or in postmortem examination of a human or other animal (Figure 34.2a). Characteristic virus inclusions called *Negri bodies* in the

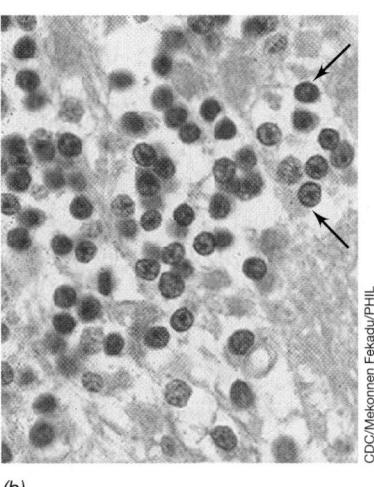

(a)   *(b)*

Figure 34.2 Rabies virus. *(a)* The bullet-shaped rabies viruses shown in this transmission electron micrograph of a tissue section from an infected animal are about 75 × 180 nm. *(b)* Pathology of rabies. A tissue section from the brain of a human rabies victim was stained with hematoxylin and eosin. Rabies virus causes characteristic cytoplasmic inclusions called Negri bodies, which contain rabies virus antigens. They are seen here as dark-stained, sharply differentiated, roughly spherical masses about 2–10 μm in diameter (arrows).

cytoplasm of nerve cells confirm rabies virus infection (Figure 34.2*b*). Reverse transcriptase–polymerase chain reaction (RT-PCR) testing and sequencing can also be used to identify rabies virus strains in clinical specimens.

Prevention of rabies in humans must start immediately after contact with potentially rabid animals because untreated rabies is almost always lethal. Guidelines for treating possible human exposure to rabies are shown in **Table 34.1**. A wild or stray animal suspected of being rabid should be captured, sacrificed, and immediately examined for evidence of the rabies virus. If a domestic animal, generally a dog, cat, or ferret, bites a human, especially if the bite is unprovoked, the animal is typically held in quarantine

for 10 days to check for clinical signs of rabies. If the animal exhibits rabies signs, or a definitive diagnosis of its illness cannot be made after 10 days, the human patient is passively immunized with rabies immune globulin (purified antibodies to rabies virus obtained from a hyperimmune individual) injected at both the site of the bite and intramuscularly. The patient is also immunized with a rabies virus vaccine. Because of the very slow progression of rabies in humans, this combination of passive and active immune therapy is nearly 100% effective, stopping the onset of the active disease. The French microbiologist Louis Pasteur developed the first rabies vaccine over 130 years ago (↩ Section 1.7).

Rabies is prevented largely through immunization. Inactivated rabies vaccines are used in the United States for both human and domestic animal immunizations. Inactivated and attenuated virus preparations are also used worldwide. Prophylactic rabies immunization is recommended for individuals at high risk, such as veterinarians, animal control personnel, animal researchers, and individuals who work in rabies research or rabies vaccine production laboratories.

Rabies Prevention

The rabies treatment strategy outlined here is extremely successful, and fewer than three cases of human rabies are reported in the United States each year, nearly always the result of bites by wild animals. Because domestic animals often have exposure to wild animals, all dogs and cats should be vaccinated against rabies beginning at 3 months of age. Booster inoculations should be given at least every 3 years. Other domestic animals, including large farm animals, are often immunized with rabies vaccines.

The key to effective rabies prevention and possible eradication, at least in the United States, lies in control of the infection in the rabies virus reservoir, primarily wild animals (Figure 34.1). If all or even most members of the disease reservoir are immune, the disease can be stopped and possibly eradicated. Oral subunit vaccines (↩ Section 15.13) consisting of vaccinia virus or canarypox virus with engineered genes that encode and express rabies

Table 34.1 *Guidelines for treating possible human exposure to rabies virus*	
Unprovoked bite by a domestic animal	
Animal suspected of rabies	*Animal not suspected of rabies*
1. Sacrifice animal and test for rabies.	1. Hold for 10 days. If no symptoms, do not treat human.
2. Begin treatment of human immediately.[a]	2. If symptoms develop, treat human immediately.[a]
Bite by wild carnivore (for example, skunk, bat, fox, raccoon, coyote)	
Regard animal as rabid	
1. Sacrifice animal and test for rabies.	
2. Begin treatment of human immediately.[a]	
Bite by wild rodent, squirrel, livestock, rabbit	
Consult local or state public health officials about possible recent cases of rabies transmitted by these animals (these animals rarely transmit rabies). If no reports, do not treat human.	

[a]All bites should be thoroughly cleansed with viricidal soap and water. Treatment for previously unvaccinated individuals is generally a combination of rabies immunoglobulin and rabies vaccine. Previously vaccinated individuals are not given rabies immunoglobulin, but are given another course of rabies vaccine.

coat proteins can be used to immunize populations of susceptible wild animals. The vaccines, administered in food "baits," have reduced the incidence and spread of rabies in limited geographic areas. This strategy targets and immunizes the wild animal rabies reservoir, an impossible task if the vaccines must be administered by traditional injection methods. By vaccinating the wild animal reservoir, it may be possible to eradicate the disease.

34.2 Hantavirus

Hantaviruses cause several severe diseases including **hantavirus pulmonary syndrome (HPS)**, an acute respiratory and cardiac disease, and **hemorrhagic fever with renal syndrome (HFRS)**, an acute disease characterized by shock and kidney failure. Both diseases are caused by hantavirus transmission from infected rodents. Hantavirus is named for Hantaan, Korea, the site of a hemorrhagic fever outbreak where the virus was first recognized as a human pathogen.

Biology, Epidemiology, and Pathology

The genus *Hantavirus* is a member of the *Bunyaviridae*, a family of enveloped, segmented, negative-strand RNA viruses (**Figure 34.3**; ↩ Section 21.9). The family includes viruses that cause either HPS or HFRS. Hantaviruses are related to hemorrhagic fever viruses such as Lassa fever virus and Ebola virus (↩ Section 32.10). Hantaviruses persistently infect rodents including mice, rats, lemmings, and voles, and are occasionally transmitted to humans from animal reservoirs.

Hantaviruses are handled with biosafety level 4 precautions (BSL-4; ↩ Section 31.4) because of the potential for life-threatening infections and the lack of effective cures or immunizations. Human cases of hantavirus and other BSL-4 viral pathogens are investigated by the Special Pathogens Branch of the Centers for Disease Control and Prevention, Atlanta, Georgia, USA (Microbial Sidebar, "Special Pathogens and Viral Hemorrhagic Fevers").

A significant hantavirus outbreak in the United States occurred near the Four Corners region of Arizona, Colorado, New Mexico, and Utah in 1993. The outbreak resulted from a rapid population expansion of the deer mouse (*Peromyscus maniculatus*) in the spring of 1993. The outbreak caused 32 deaths in 53 infected people, illustrating the potential danger of outbreaks due to pathogens that can be directly transmitted from animal reservoirs, sometimes under new or unusual circumstances. In total, there have been 485 cases of HPS with 160 deaths (33%) from 1993 through 2007 in the United States, mostly in western states. A 2004 outbreak in Brazil caused 85 cases and 34 deaths, for a case fatality rate of 40%. The HPS strains are more prevalent in the Americas. The HFRS strains are commonly implicated in outbreaks in Eurasia and generally have a lower mortality rate than HPS strains; some strains cause as little as 1% mortality. The incidence of infection with HFRS strains, however, is much higher; up to 200,000 infections are recognized annually, chiefly in China, Korea, and Russia. Continued investigation of hantaviruses will likely identify a number of other pathogenic strains.

Hantaviruses are most commonly transmitted by inhalation of virus-contaminated rodent excreta. Humans are accidental hosts and are infected only when they come into contact with rodents or their waste. A variety of environmental factors led to a perfect scenario for the spread of the virus to humans in the Four Corners outbreak of 1993. The outbreak started after a mild winter that was followed by abundant spring rains. These conditions produced an unusually high amount of plant growth. The vegetation provided abundant food and triggered a rodent population explosion in 1993. As a result, humans were more likely to be exposed to the mice and excreta that contained hantavirus. The virus is most commonly spread via aerosols in the form of dust

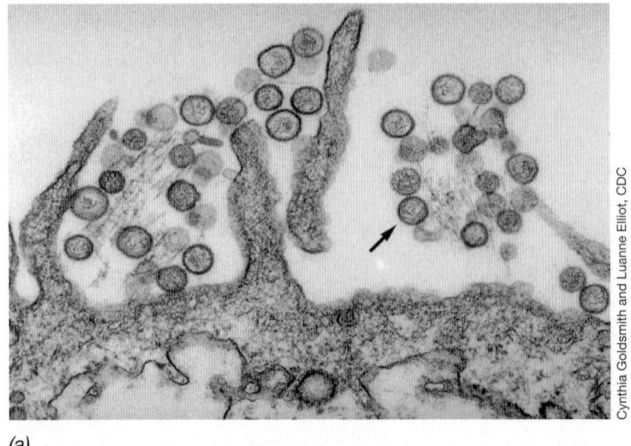
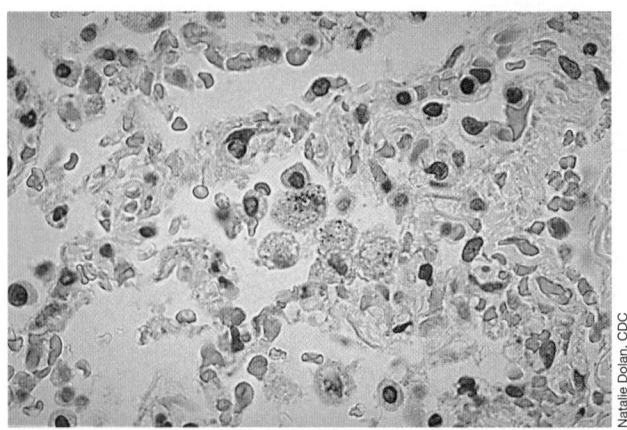

(a)   *(b)*

Cynthia Goldsmith and Luanne Elliot, CDC

Natalie Dolan, CDC

Figure 34.3 Hantavirus. *(a)* An electron micrograph of the Sin Nombre hantavirus. The arrow indicates one of several virions. The virus is approximately 0.1 μm in diameter. *(b)* Immunostaining of Andes hantavirus antigens in alveolar macrophages. Each granular dark blue–stained area indicates cellular infection of an individual macrophage (approximately 15 μm in diameter).

Special Pathogens and Viral Hemorrhagic Fevers

The Special Pathogens Branch of the Centers for Disease Control and Prevention (CDC) in Atlanta, Georgia, USA, specializes in the handling of a subgroup of dangerous pathogens, the hemorrhagic fever viruses. These agents cause viral hemorrhagic fevers (VHF) and include the hantaviruses (discussed in this chapter) and the filoviruses, such as Ebola (Figure 1), discussed in Section 32.10 in the context of emerging infectious diseases and in Section 32.11 in the context of potential biological warfare agents. These viruses warrant an entire program because they are some of the most lethal infectious agents known.

Hemorrhagic fever viruses are handled under biosafety level 4 (BSL-4) standards. BSL-4 is the highest level of biological containment available and is used only for work with agents that pose a high risk of life-threatening disease and for which no treatment exists. BSL-4 facilities require mechanisms for total isolation and physical containment of pathogens, such as sealed biological safety cabinets and positive-pressure suits for personnel handling cultures and clinical specimens (𝒸𝒫 Figure 31.9). The VHF agents are RNA viruses transmitted by the aerosol route from animal or arthropod hosts. They are not normally human pathogens and do not depend on humans for their survival. Outbreaks of VHFs occur irregularly, and infection from human to human by the aerosol route is inefficient. Most human-to-human transmission is the result of prolonged contact with an infected individual or with his or her blood or waste.

VHFs are characterized by severe symptoms that affect multiple organ systems. The virus damages the overall vascular system, and body functions such as oxygen and waste transport and temperature regulation go out of control. Hemorrhaging (bleeding), for which the viruses are named, is rarely the cause of death in VHFs. Instead, extensive damage to organ systems is typically the cause of death.

In the United States, the only endemic VHFs are caused by hantaviruses. With hantaviruses we understand the vectors and hosts and how to prevent human infections; rodent infestations must be contained and human contact with rodent excreta must be limited because hantavirus can survive for long periods in dried feces, saliva, or urine. However, with many hemorrhagic fever viruses, the vectors and mechanisms of transmission are unknown. For example, Ebola virus (Figure 1) is endemic in central Africa and spreads among humans, but at least four or five subtypes of the virus originate in still-unidentified animal hosts. Bats and various species of primates have been implicated as natural hosts and reservoirs, but definitive proof for the origin and reservoirs of Ebola viruses is still lacking.

Hantavirus diseases occur in two forms. Hantavirus hemolytic uremic syndrome causes about 30 deaths per year (15% mortality) in the United States. A total of 485 cases of hantavirus pulmonary syndrome and 160 deaths (33%) have been reported from 1993 to 2007. By contrast, there have been seven major Ebola outbreaks in Africa since 1976, the latest in 2007. In humans, Ebola hemorrhagic fever is devastating. In outbreaks involving more than 15 cases, mortality ranged from 29% to 88%. In total, 1709 people have acquired Ebola and 1146 have died (67% mortality). Obviously, if Ebola were to infect individuals in a densely populated area, the results could be devastating.

The overwhelmingly high mortality rates for VHFs makes them some of the most feared diseases, but vaccines and treatments are being developed. One effective vaccine consists of Ebola glycoprotein expressed by a genetically engineered innocuous virus, vesicular stomatitis virus. In trials with monkeys, this vaccine was completely effective, protecting against lethal challenges by both Ebola and Marburg VHF viruses.

Another successful approach to prevent infection by Ebola virus, aimed at postexposure control, uses RNAi, or *RNA interference* (𝒸𝒫 Section 7.10). The technique uses short pieces of synthetic double-stranded Ebola RNA to target infectious Ebola virus mRNA and induce its destruction, even after exposure to lethal doses of the virus. This technique was protective in monkey models and could be applied to treatment of other VHFs as well.

The Special Pathogens Branch of the CDC works with hemorrhagic fever viruses and outbreaks in this country and worldwide. It is charged with managing infected patients, developing diagnostic tools to identify the viruses, and gathering scientific and clinical information about the viruses, their diseases, and distribution. Its goal is to predict outbreaks, quickly identify them when they occur, predict viral behavior during the outbreak, and implement adequate measures to stop the outbreak. It is the first and only line of defense in the United States for identifying new and established VHF pathogens when outbreaks occur. For more information about the Special Pathogens Branch of the CDC and the pathogens and diseases it studies, consult the website at http://www.cdc.gov/ncidod/dvrd/spb/index.htm.

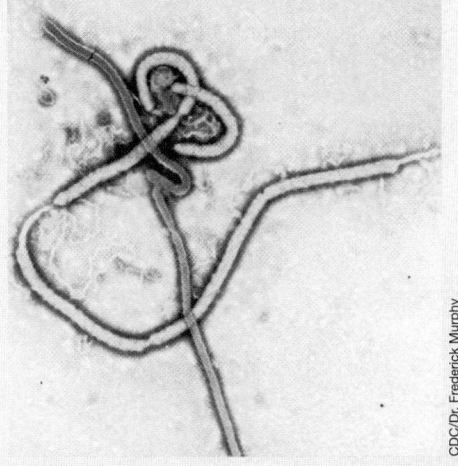

CDC/Dr. Frederick Murphy

Figure 1 Ebola virus. Transmission electron micrograph of a negatively stained preparation of Ebola virus.

generated from mouse droppings or dried urine. However, there have been rare reports of person-to-person transmission, as well as a few incidents of infection spread by a rodent bite.

HPS is characterized by a sudden onset of fever, myalgia (muscle pain), thrombocytopenia (reduction in the number of blood platelets), leukocytosis (an increase in the number of circulating leukocytes), and pulmonary capillary leakage. Death occurs within several days, usually due to shock and cardiac complications precipitated by pulmonary edema (leakage of fluid into the lungs, causing suffocation and heart failure). These symptoms are typical of the Sin Nombre hantavirus, which caused the Four Corners outbreak, but other symptoms may be evident, depending on the strain of virus causing the disease. For example, the Bayou strain common in rodents in the southeastern United States also causes kidney failure.

Diagnosis, Treatment, and Prevention

If hantavirus from candidate infections can be grown in tissue culture, the strain can be identified by serological techniques including a virus plaque–reduction neutralization assay. In this assay, patient serum is tested for antibodies that inhibit the formation of viral plaques in tissue culture. More commonly, EIAs (enzyme immunoassays; ↩ Section 31.10) are performed on patient blood to identify antibodies, indicating exposure and an immune response. The presence of the viral genome, indicating infection, can be detected with RT-PCR (↩ Section 31.13) using patient tissue or blood specimens.

There is no virus-specific treatment or vaccine for hantaviruses. Hantavirus infection can be prevented by avoiding rodent contact and rodent habitat. Destruction of mouse habitat, restricting food supplies (for example, keeping food in sealed containers), and aggressive rodent extermination measures are the accepted means of control. The long-term prognosis for disease eradication is poor because a high percentage of rodents in a given geographical area are infected with the local hantavirus strain. For example, retrospective serological testing of deer mice in the Four Corners area in 1993 indicated that 30% of the local mouse population carried the Sin Nombre hantavirus.

MiniQuiz

- Why are hantaviruses considered a major public health problem in the United States?
- Describe the spread of hantaviruses to humans. What are some effective measures for preventing infection by hantaviruses?

Ⅱ Arthropod-Transmitted Pathogens

Pathogens can be spread to hosts from the bite of a pathogen-infected arthropod vector. In many cases, such as in the rickettsial illnesses, Lyme disease, and plague, humans are accidental hosts for the pathogen. In other cases, however, infected humans are required hosts in the pathogen life cycle, as is the case for malaria.

34.3 Rickettsial Pathogens

The **rickettsias** are small bacteria that have an obligate intracellular existence in vertebrates, usually mammals, and are also associated with bloodsucking arthropods such as fleas, lice, or ticks. We discussed the biology of rickettsias in Section 17.13. Rickettsias cause diseases in humans and animals, the most important of which are typhus fever, spotted fever rickettsiosis (also called Rocky Mountain spotted fever), and ehrlichiosis. Rickettsias take their name from Howard Ricketts, a scientist at the University of Chicago who first discovered them. Ricketts died from infection with the rickettsia that causes typhus fever, *Rickettsia prowazekii*. Rickettsias have not been cultured in artificial media but can be cultured in laboratory animals, lice, mammalian tissue culture cells, and the yolk sac of chick embryos. In animals, growth takes place primarily in phagocytes.

Genomic analysis of the 1.1-Mbp genome of *R. prowazekii* indicates that these intracellular parasites are closely related to human mitochondria. Like the mitochondria, the rickettsial genome contains a minimal set of genes, most of which are directed at maintaining intracellular existence. For example, the rickettsias lack most of the genes necessary for independent energy metabolism and structural biosynthesis. The rickettsial genome also contains virulence genes related to the *virB* operon of the plant pathogen *Agrobacterium tumefaciens* (↩ Section 25.4). This operon encodes virulence factors for DNA transfer and protein export; in rickettsias, these factors allow the pathogen to use host cell systems for these functions.

Rickettsias are divided into three groups, based loosely on the clinical diseases they cause. The groups are (1) the *typhus group*, typified by *R. prowazekii*; (2) the *spotted fever group*, typified by *Rickettsia rickettsii*; and (3) the *ehrlichiosis group*, characterized by *Ehrlichia chaffeensis*.

The Typhus Group: *Rickettsia prowazekii*

Typhus is caused by *R. prowazekii*. Epidemic typhus is transmitted from person to person by the common body or head louse (**Figure 34.4**). Humans are the only known mammalian host for typhus. During World War I, an epidemic of typhus spread throughout Eastern Europe and caused almost 3 million deaths.

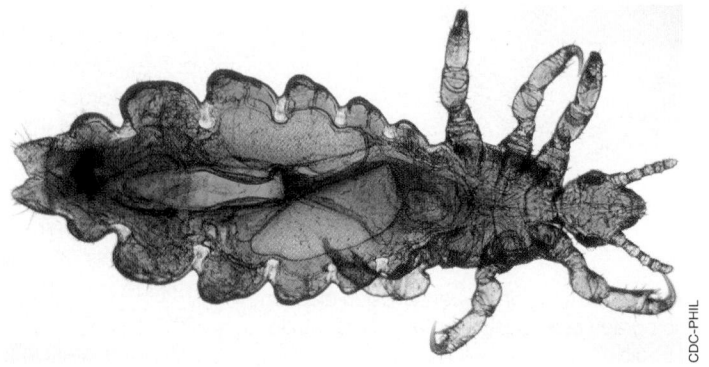

Figure 34.4 The human louse, *Pediculus humanus*. The female louse, about 3 mm long, can carry *Rickettsia prowazekii*, the agent that causes typhus. In addition, the body louse can carry *Borrelia recurrentis*, the agent of relapsing fever, and *Bartonella quintana*, the agent of trench fever.

Typhus has historically been a problem among military troops during wartime. Because of the unsanitary, cramped conditions characteristic of wartime military infantry operations, lice are spread easily among soldiers, and typhus is spread in epidemic proportions. Up until World War II, typhus caused more military deaths than combat.

Cells of *R. prowazekii* are introduced through the skin when a puncture caused by a louse bite becomes contaminated with louse feces, the major source of rickettsial cells. During an incubation period of 1–3 weeks, the organism multiplies inside cells lining the small blood vessels. Symptoms of typhus (fever, headache, and general body weakness) then begin to appear. Five to nine days later, a characteristic rash is observed in the armpits and generally spreads over the body, except for the face, palms of the hands, and soles of the feet. Complications from untreated typhus include damage to the central nervous system, lungs, kidneys, and heart. Epidemic typhus has a mortality rate of 6–30%. Tetracycline and chloramphenicol are most commonly used to control infections caused by *R. prowazekii*. *Rickettsia typhi*, the organism that causes murine typhus, is another important pathogen in the typhus group.

The Spotted Fever Group: *Rickettsia rickettsii*

Spotted fever rickettsiosis, commonly called Rocky Mountain spotted fever (RMSF), was first recognized in the western United States in about 1900 but is more prevalent today in the mid-South region (**Figure 34.5**). RMSF is caused by *R. rickettsii* and is transmitted to humans by various ticks, most commonly the dog and wood ticks. Over 2000 people now acquire the disease every year in the United States, a 103% increase since 2002. The increase may be the result of human encroachment in tick-infested areas due to recreational activities or housing developments. Increases may also have resulted from better diagnostic methods and general awareness of this enzootic disease. Humans acquire the pathogen when bitten by an infected tick.

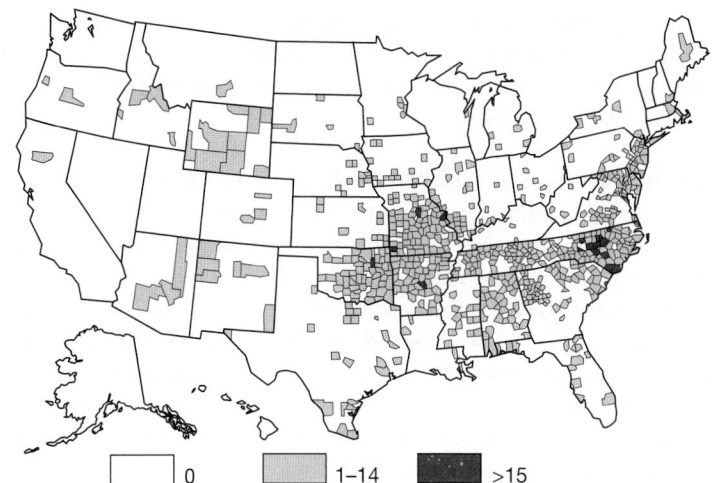

Figure 34.5 Spotted fever rickettsiosis (Rocky Mountain spotted fever) in the United States. The 2221 cases reported in 2007 are shown by county of origin. Cases were concentrated in the eastern and mid-South states west to Oklahoma.

The rickettsial agent is found in the salivary glands of the tick and in the female tick's ovaries. The agent is maintained in nature by transovarial transmission to larvae from the infected female.

Cells of *R. rickettsii*, unlike other rickettsias, grow within the nucleus of the host cell as well as in host cell cytoplasm (**Figure 34.6a, b**). Following an incubation period of 3–12 days, characteristic symptoms, including fever and a severe headache, occur. Within 3–5 days, a rash breaks out on the whole body (Figure 34.6c), generally accompanied by gastrointestinal problems such as diarrhea and vomiting. The clinical symptoms of RMSF persist for over 2 weeks if the disease is untreated. Tetracycline or

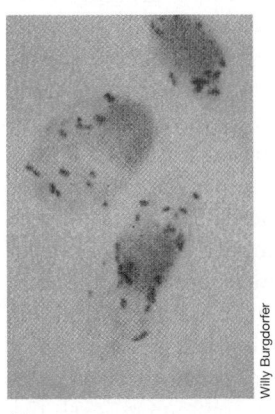
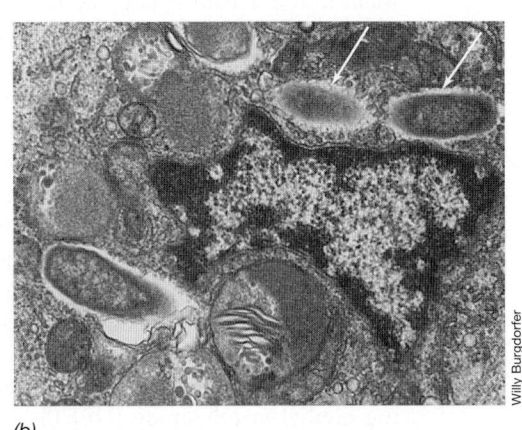
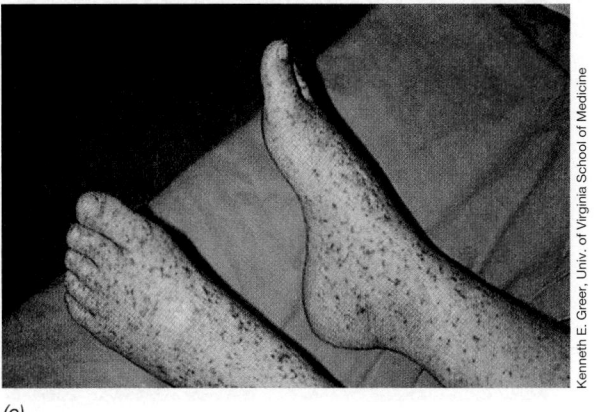

(a)   *(b)*   *(c)*

Figure 34.6 *Rickettsia rickettsii* **and spotted fever rickettsiosis.** *(a)* Cells of *R. rickettsii*, growing in the cytoplasm and nucleus of tick hemocytes. Individual cells are about 0.4 μm in diameter. *(b)* Transmission electron micrograph of *R. rickettsii* (arrows) in a granular hemocyte of an infected wood tick (*Dermacentor andersoni*). *(c)* Rash of spotted fever rickettsiosis on the feet. The whole-body rash is indicative of spotted fever rickettsiosis and helps distinguish this disease from typhus, in which the rash does not cover the whole body. Cases of spotted fever rickettsiosis are increasing in the U.S., especially in eastern and mid-South states (Figure 34.5).

UNIT 11

chloramphenicol generally promotes a prompt recovery from RMSF if administered early in the course of the infection, and treated patients have less than 1% mortality. Mortality is about 30% in untreated cases.

Ehrlichiosis and Tickborne Anaplasmosis

The *Ehrlichia* and related genera (⮌ Section 17.13) are responsible for two emerging tickborne diseases in the United States, *human monocytic ehrlichiosis* (*HME*) and *human granulocytic anaplasmosis* (*HGA*) (formerly called human granulocytic ehrlichiosis, or HGE). The rickettsias that cause HME are *Ehrlichia chaffeensis* and *Rickettsia sennetsu*. The rickettsias that cause HGA are *Ehrlichia ewingii* and *Anaplasma phagocytophilum*.

The onset of these clinically indistinguishable rickettsial diseases is characterized by flulike symptoms that can include fever, headache, malaise, and leukopenia (decreased number of leukocytes) or thrombocytopenia. Laboratory findings frequently document changes in liver function, characterized by an increase in the enzyme hepatic transaminase. Peripheral blood leukocytes have visible inclusions of cells, a diagnostic indicator for the diseases (**Figure 34.7**). The symptoms, except for the inclusions, are similar to other rickettsial infections, and the diseases can range from subclinical to fatal in outcome. Long-term complications for progressive untreated cases may include respiratory and renal insufficiency and serious neurological involvement.

Laboratory diagnosis of these rickettsial diseases is based on an indirect fluorescent antibody assay of patient serum and also on PCR tests of whole blood or serum to detect the presence of rickettsial DNA. Rickettsias can be observed in inclusions in granulocytes from the blood of patients with ehrlichiosis.

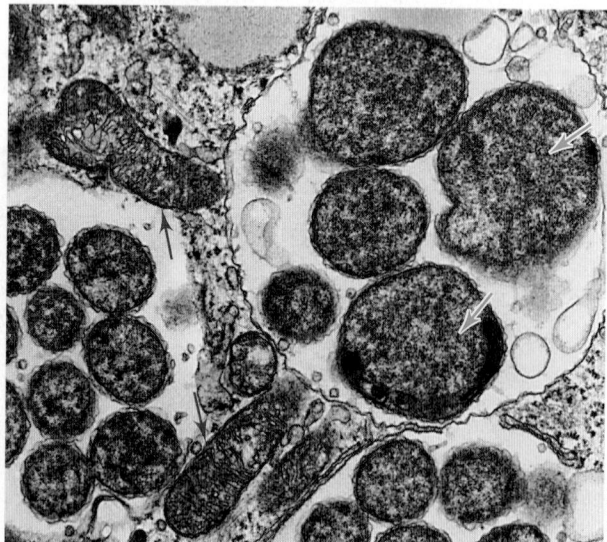

Figure 34.7 *Ehrlichia chaffeensis*, the causative agent of human monocytic ehrlichiosis (HME). The electron micrograph shows inclusions in a human monocyte that contains large numbers of *E. chaffeensis* cells. The blue arrows indicate two of the many bacteria in each inclusion. The *E. chaffeensis* cells are about 300–900 nm in diameter. Mitochondria are shown with red arrows.

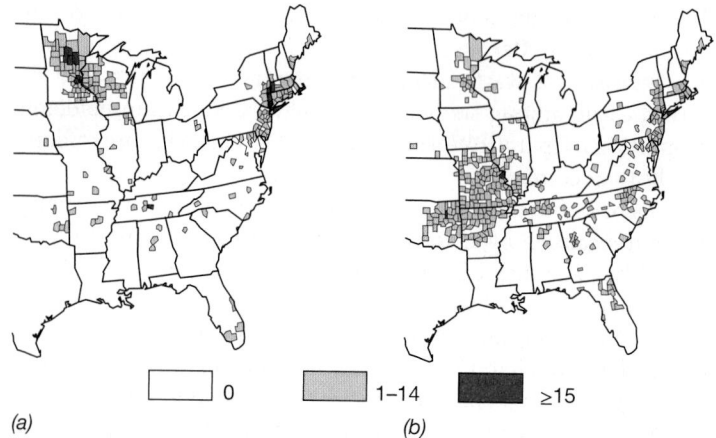

| 0 | 1–14 | ≥15 |

(a)   *(b)*

Figure 34.8 Anaplasmosis and ehrlichiosis in the United States, 2007. *(a)* Human granulocytic anaplasmosis, caused by *Anaplasma phagocytophilum*. There were 834 cases, concentrated in the Northeast and upper Midwest. *(b)* Human monocytic ehrlichiosis, caused by *Ehrlichia chaffeensis*. There were 828 cases, concentrated in the Northeast and lower Midwest. For each disease, only a few cases appeared in the West.

HGA and HME are spread by the bites of infected ticks. The mammalian reservoirs include deer and possibly rodents, in addition to the human hosts. Retrospective serological analyses in areas with relatively high incidence of tickborne disease indicate that HGA may be a more prevalent disease than spotted fever rickettsiosis. Many HGA infections are not properly identified because of the variable nature of the symptoms. However, since 1999 HGA and HME have been reportable diseases in the United States. Each is responsible for about half of all reported cases of ehrlichiosis. HGA occurs primarily in the upper Midwest and coastal New England, while HME is concentrated in the lower Midwest and the East Coast (**Figure 34.8**). Together, almost 2000 cases are reported each year, and this number is rising annually. As with spotted fever rickettsiosis, human encroachment into tick habitat coupled with updated case definitions and increased awareness of the disease may be leading to the increases in reported cases. Undoubtedly, however, the numbers reported are lower than the actual number of cases that occur. HGA and HME will be reported more frequently as physicians become more familiar with these emerging tickborne diseases.

Similar to other tickborne illnesses, humans are exposed to ehrlichiosis pathogens during outdoor activities in tick-infested areas. Golfers, hikers, and others who are recreationally or occupationally exposed to tick habitat are most prone to infection. Prevention of ehrlichiosis involves reducing exposure to ticks and tick bites by avoiding tick habitat, wearing tick-proof clothing, and applying appropriate insect repellents such as those containing diethyl-*m*-toluamide (DEET). At the community level, tick densities can be successfully reduced through areawide application of acaricides (chemicals specifically toxic for ticks and related arthropods) and removal of tick habitat, such as leaves and brush. Doxycycline, a semisynthetic tetracycline, is the antibiotic of choice for the treatment of HGA and HME.

Other Rickettsial Diseases

Q fever is a pneumonia-like infection caused by the obligate intracellular parasite *Coxiella burnetii*, a bacterium related to the rickettsias (⮂ Section 17.13). Although not transmitted to humans directly by an insect bite, the agent of Q fever is transmitted to animals such as sheep, cattle, and goats by insect bites. Various arthropod species are reservoirs and vectors for infection. Domestic animals generally have inapparent infections, but may shed large quantities of *C. burnetii* cells in their urine, feces, milk, and other body fluids. Infected animals or contaminated animal products such as wool, meat, and milk are potential sources for human infection. The resulting influenza-like illness is probably underreported and may progress to include prolonged fever, headache, chills, chest pains, pneumonia, and endocarditis. In the United States Q fever is most prevalent in states with large numbers of agricultural animals, especially in the Midwest and West. Over 150 cases are reported annually.

Laboratory diagnosis of infection with *C. burnetti* can be made by immunological tests designed to measure host antibodies to the pathogen. A complement fixation test and an immunofluorescence antibody test are widely used. *C. burnetii* infections respond to tetracycline, and therapy should be started quickly in any suspected human case of Q fever to prevent endocarditis and related heart damage. Finally, Q fever is a potential biological warfare agent (⮂ Section 32.11).

Scrub typhus, or *tsutsugamushi disease*, is restricted to Asia, the Indian subcontinent, and Australia and is caused by *Orientia tsutsugamushi*. Although the disease is similar to typhus, *O. tsutsugamushi* is transmitted by mites to rodent hosts. Humans are occasional accidental hosts.

Diagnosis and Control

In the past, rickettsial infections have been difficult to diagnose because the characteristic rash associated with many rickettsial diseases may be mistaken for measles, scarlet fever, or adverse drug reactions. Laboratory confirmation of rickettsial diseases can be done using pathogen-specific immunological and molecular biology reagents. These include antibody-based tests that detect rickettsial surface antigens by latex bead agglutination assays, immunofluorescent antibody assays, EIA, and by PCR-based nucleic acid assays.

Control of most rickettsial diseases requires control of the vectors: lice, fleas, and ticks. For humans traveling in wooded or grassy areas, the use of insect repellents containing DEET usually prevents tick attachment. Firmly attached ticks should be removed gently with forceps, care being taken to remove all the mouthparts of the insect. Although a vaccine is available for the prevention of typhus, the few cases reported do not warrant its general administration in the United States. No vaccines are currently available for the prevention of the more prevalent tickborne infections, spotted fever rickettsiosis, HGA, or HME.

MiniQuiz

• What are the arthropod vectors and animal hosts for typhus, spotted fever rickettsiosis, ehrlichiosis, and anaplasmosis?

• What precautions can be taken to prevent rickettsial infections?

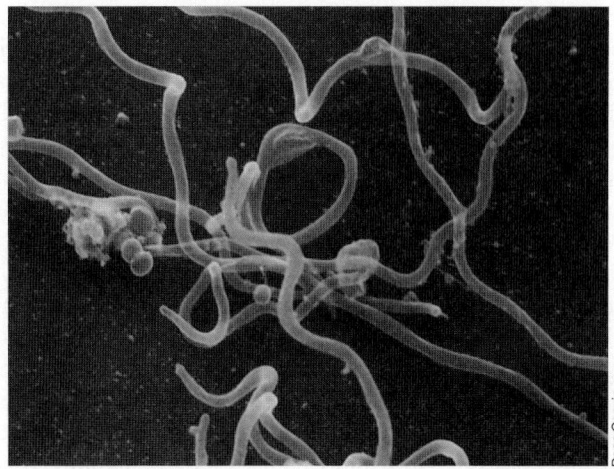

Figure 34.9 Scanning electron micrograph of the Lyme spirochete, *Borrelia burgdorferi*. A single cell is approximately 0.4 μm in diameter.

34.4 Lyme Disease and *Borrelia*

Lyme disease is an emerging tickborne disease that affects humans and other animals. Lyme disease was named for Old Lyme, Connecticut, where cases were first recognized, and is currently the most prevalent arthropod-borne disease in the United States. Lyme disease is caused by infection with a spirochete, *Borrelia burgdorferi* (**Figure 34.9**; ⮂ Section 18.16), transmitted by the bite of *Ixodes* spp. ticks. The ticks that carry *B. burgdorferi* feed on the blood of birds, domesticated animals, various wild animals, and humans.

Epidemiology

Deer and white-footed field mice are prime mammalian reservoirs of *B. burgdorferi* in the northeastern United States. These animals are parasitized by the deer tick, *Ixodes scapularis* (**Figure 34.10**). In other parts of the country, different species of rodents and ticks transmit the Lyme spirochete. For example, in the western United States, *Ixodes pacificus*, the Pacific black-legged tick, is the vector; the deer mouse and wood rat are common hosts.

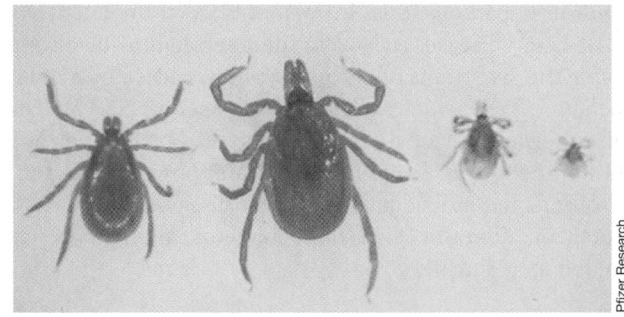

Figure 34.10 Deer ticks (*Ixodes scapularis*), the major vector of Lyme disease. Left to right, male and female adult ticks, nymph, and larva forms. The length of an adult female is about 3 mm. Although all forms feed on humans, the female nymphal and adult ticks are principally responsible for transmitting *Borrelia burgdorferi*.

UNIT 11

Lyme disease has also been identified in Europe and Asia. In Europe, the tick vector is *Ixodes ricinus*, which may also harbor *Borrelia garinii*, another organism that causes a Lyme disease–like illness. In Asian countries, *Borrelia afzelii* is transmitted by *Ixodes persulcatus*. In all cases, different local rodent host reservoirs have been identified. Thus, Lyme disease seems to have a broad geographic distribution; it is caused by closely related *Borrelia* pathogens transmitted to humans by tick vectors that have a number of mammalian hosts and reservoirs.

Ixodes spp. are smaller than many other ticks, making them easy to overlook (Figure 34.10). Unlike the vectors of other tick-borne diseases, a very high percentage of the deer ticks (up to 25% of nymphs and 50% of adults in certain regions of the Northeast) carry *B. burgdorferi*. Extended contact with the infected tick vectors increases the probability of disease transmission.

In the United States most cases of Lyme disease have been reported from the Northeast and upper Midwest, but cases have been observed in nearly every state. The number of Lyme disease cases is rising and over 30,000 are reported each year (**Figure 34.11**).

Pathology

Cells of *B. burgdorferi* are transmitted to humans while the tick is obtaining a blood meal (**Figure 34.12a**). A systemic infection develops, leading to the acute symptoms of Lyme disease, which include headache, backache, chills, and fatigue. In about 75% of cases, a large rash known as *erythema migrans* is observed at the site of the tick bite (Figure 34.12b, c). During this acute stage, Lyme disease is treatable with tetracycline or penicillin. Untreated Lyme disease may progress to a chronic stage weeks to months after the initial tick bite. Chronic untreated Lyme disease causes arthritis in 40–60% of patients. By contrast, only about 10% of patients develop arthritis if treated with antibiotics. Neurological involvement such as palsy, weakness in the limbs, and facial tics occurs in 15–20% and heart damage in about 8% of patients. In untreated cases, cells of *B. burgdorferi* infecting the central nervous system may lie dormant for long periods before causing additional chronic symptoms, including visual disturbances, facial paralysis, and seizures.

No toxins or other virulence factors have yet been identified in Lyme disease pathogenesis. In many respects, the latent symptoms of Lyme disease, especially the neurological involvement, resemble the symptoms of chronic syphilis, caused by a different spirochete, *Treponema pallidum* (Section 33.12). Unlike syphilis, however, Lyme disease is not spread by human contact. Small numbers of *B. burgdorferi* cells are, however, shed in the urine of infected individuals, and Lyme disease can occasionally spread from domestic animal populations, particularly cattle, through infected urine.

Diagnosis

Antibodies appear in response to *B. burgdorferi* 4–6 weeks after infection and can be detected by an indirect EIA or a fluorescent antibody assay. However, the most definitive serological test for Lyme disease is the Lyme immunoblot (Section 31.11). Because antibodies to the Lyme spirochete antigens persist for years after infection, the presence of antibodies does not neces-

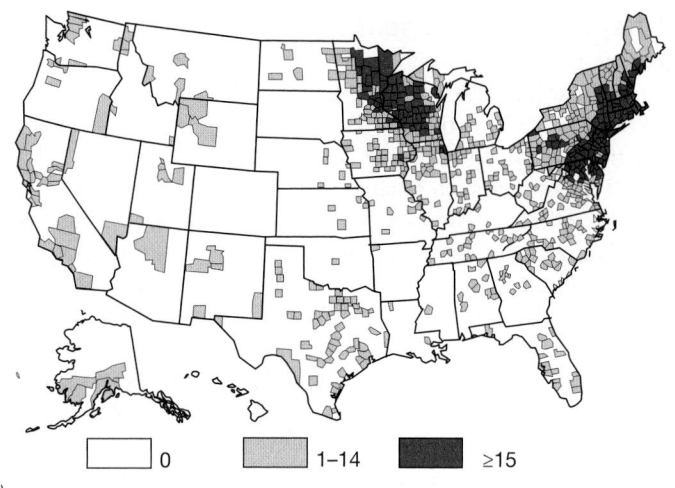

(a)

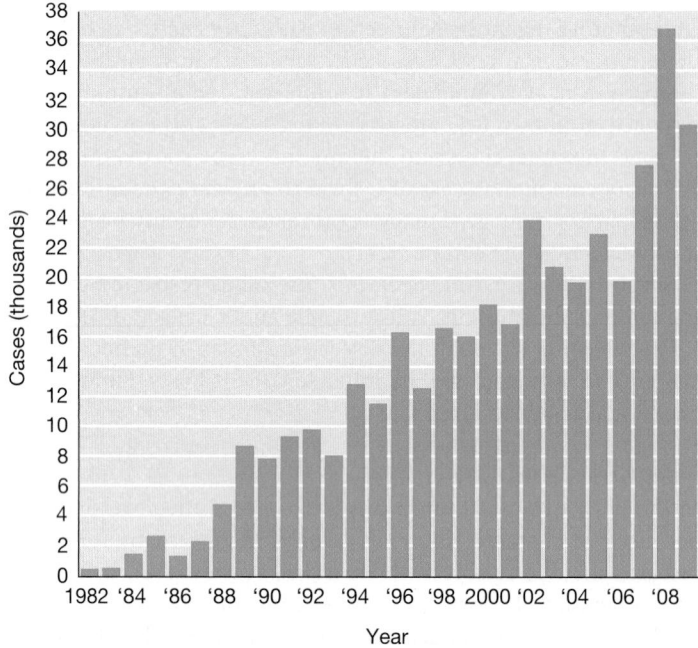

(b)

Figure 34.11 Lyme disease in the United States. *(a)* Lyme disease in the United States in 2007. There were 27,444 cases, concentrated in the Northeast and upper Midwest. *(b)* Number of reported cases of Lyme disease by year in the United States. Lyme disease is reported through the National Notifiable Diseases Surveillance System of the Centers for Disease Control and Prevention, Atlanta, Georgia.

sarily indicate recent infection. In addition, existing antibodies may not confer immunity to further infection.

A PCR assay has also been developed for the detection of *B. burgdorferi* in body fluids and tissues. Although rapid and sensitive, the PCR assay cannot differentiate between live *B. burgdorferi* in active disease and dead *B. burgdorferi* found in treated or inactive disease. *B. burgdorferi* can also be cultured from nearly 80% of the original erythema migrans lesions (Figure 34.11b, c), but culture is usually not done because *B. burgdorferi* grows very slowly *in vitro*, even on highly specialized media.

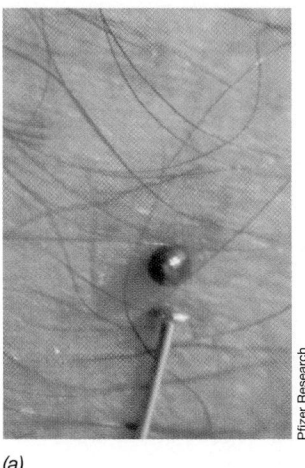
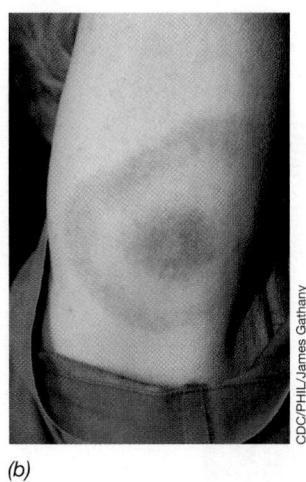
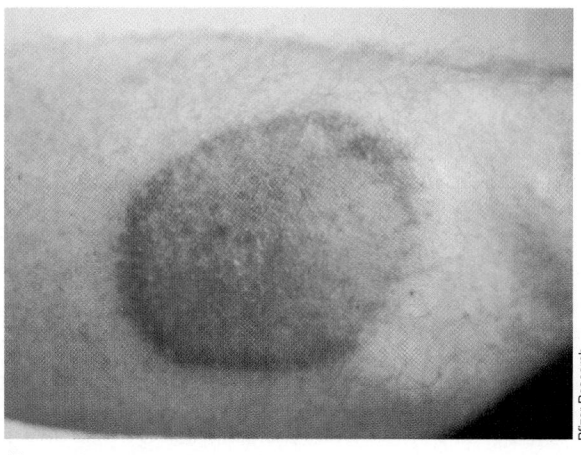

(a)   *(b)*   *(c)*

Figure 34.12 Lyme disease infection. *(a)* Deer tick obtaining a blood meal from a human. Characteristic rashes on the arm of a patient who acquired Lyme disease. The rash, known as erythema migrans (EM), starts at the site of a tick bite and grows in a circular, bull's-eye fashion *(b)*, or in a concentric circular fashion *(c)* over a period of several days. A typical EM example is about 5 cm in diameter.

In the end, Lyme disease is usually diagnosed clinically. If a patient has Lyme disease symptoms and other findings such as facial tics or arthritis, has had recent tick exposure, and exhibits erythema migrans, a presumptive diagnosis of Lyme disease is made and antibiotic treatment is initiated.

Prevention and Treatment

Prevention of Lyme disease requires proper precautions to prevent tick attachment. Insect repellents containing DEET are very effective. In tick-infested areas such as woods, tall grass, and brush, protective clothing such as long pants tucked into tight-fitting socks and boots, and a long-sleeved shirt with a snug collar and cuffs may prevent tick attachment. After spending time in a tick-infested environment, individuals should check themselves carefully for ticks and gently remove any attached ticks (including the head). An effective human Lyme disease vaccine is not available. Lyme disease vaccines are available, however, for immunization of susceptible domestic animals.

Treatment of early acute Lyme disease can be with doxycycline, amoxicillin (a β-lactam antibiotic) or an alternative antimicrobial compound for 20 to 30 days. For patients having neurological or cardiac symptoms due to *B. burgdorferi* infection, parenteral ceftriaxone is indicated. This β-lactam antibiotic crosses the blood–brain barrier, affecting spirochetes in the central nervous system. Lyme arthritis may respond to large doses of penicillin; even long-standing Lyme arthritis may be cured with doxycycline or amoxicillin plus probenicid, given for 30 days or longer.

MiniQuiz
- What are the primary symptoms of Lyme disease?
- Describe the incidence of Lyme disease over the last ten years.
- Outline methods for prevention of *Borrelia burgdorferi* infection.

34.5 Malaria and *Plasmodium*

Malaria is a disease caused by *Plasmodium* spp., a group of protists that are members of the alveolate group (⮌ Section 20.9). *Plasmodium* spp. cause malaria-like diseases in warm-blooded hosts; the complex protist life cycle includes an arthropod mosquito vector (**Figure 34.13**). The malaria protists are important human pathogens. Malaria has played an important role in the development and spread of human culture and has even affected human evolution. Malaria is still a significant human disease even though several effective treatments are available. Malaria infections annually occur in 350 to 500 million people worldwide, and each year nearly 1 million of these will die, making malaria one of the most common causes of death due to infectious disease worldwide (⮌ Table 32.1).

Four species, *Plasmodium vivax, P. falciparum, P. ovale*, and *P. malariae*, cause most malaria infections in humans. The most widespread disease is caused by *P. vivax*; the most serious disease is caused by *P. falciparum*. Humans are the only reservoirs for these four species. The protists carry out part of their life cycle in the human reservoir and part in the female *Anopheles* mosquito, the only vector that transmits *Plasmodium* spp. The vector spreads the protist from person to person.

Epidemiology

Anopheles mosquitoes live predominantly in the tropics and subtropics, and transmit *Plasmodium* spp. (**Figure 34.14**). In general, malaria is not a disease of temperate or colder regions. For example, malaria did not exist in the northern regions of North America prior to settlement by Europeans, but was a major problem in the southern United States, where appropriate mosquito habitat existed. The disease is associated with wet, low-lying areas where mosquitoes breed. The term *malaria* is derived from the Italian words meaning "bad air."

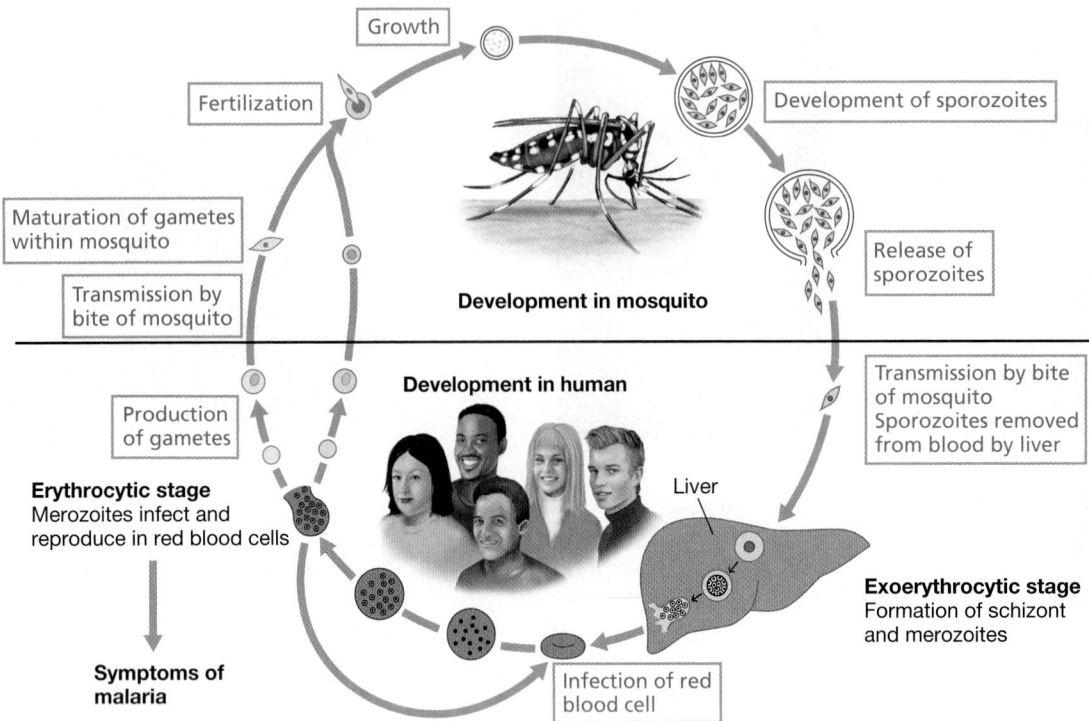

Figure 34.13 The life cycle of *Plasmodium vivax*. The protist genus *Plasmodium* comprises the malarial pathogens, all of which have a life cycle dependent on growth in both a warm-blooded host and the mosquito vector. Transmission of the protist to and from the warm-blooded host is done by the bite of a mosquito.

The life cycle of the malaria protist is complex (Figure 34.13). First, the human host is infected by plasmodial *sporozoites*, small, elongated cells produced in the mosquito that localize in the salivary gland of the insect. The mosquito injects saliva (containing an anticoagulant) along with the sporozoites into the human host when obtaining a blood meal. The sporozoites travel through the bloodstream to the liver, where they infect liver cells and remain quiescent or replicate and become enlarged in a stage called the *schizont*. The schizonts then segment into a number of small cells called *merozoites*, which leave the liver, again entering the blood

circulation. Some of the merozoites then infect red blood cells (erythrocytes).

The plasmodial life cycle in erythrocytes proceeds with division, growth, and release of merozoites; this results in destruction of the host red blood cells. *P. vivax* growth in red cells usually repeats at synchronized intervals of 48 hours. During this 48-hour period, the host experiences the defining clinical symptoms of malaria, characterized by chills followed by fever of up to 40°C (104°F). The chill–fever pattern coincides with the release of

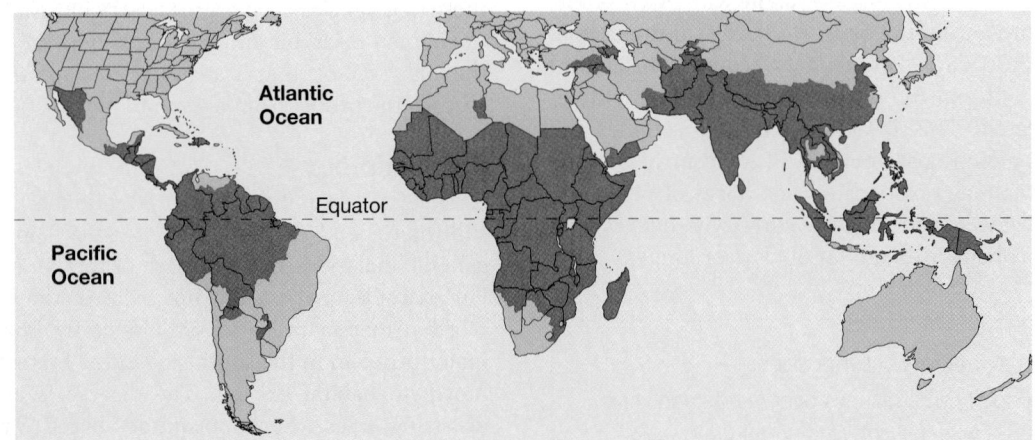

Figure 34.14 Worldwide endemic malaria regions. Malaria is an endemic disease worldwide in subtropical and tropical regions, highlighted in red. *Source:* Centers for Disease Control and Prevention, Atlanta, Georgia.

P. vivax merozoites from the erythrocytes during the synchronized asexual reproduction cycle. Vomiting and severe headache may accompany the chill–fever cycles, and over the longer term, characteristic symptomatic malaria generally alternates with asymptomatic periods. Because of the destruction of red blood cells, malaria generally causes anemia and some enlargement of the spleen (splenomegaly).

In the vertebrate host, merozoites develop into *gametocytes*, cells that infect only mosquitoes. The gametocytes are ingested when another *Anopheles* mosquito takes a blood meal from an infected person; they mature within the mosquito into *gametes*. Two gametes fuse, and a zygote forms. The zygote migrates by amoeboid motility to the outer wall of the insect's intestine where it enlarges and forms a number of sporozoites. These are released and reach the salivary gland of the mosquito from where they can be inoculated into another human, and the cycle begins again.

Diagnosis and Treatment

Conclusive diagnosis of malaria in humans requires the identification of *Plasmodium*-infected erythrocytes in blood smears (**Figure 34.15**). Fluorescent nucleic acid stains, nucleic acid probes, PCR assays, and antigen-detection methods (rapid diagnostic tests, RDTs) may all be used to verify *Plasmodium* infections or to differentiate between infections with various *Plasmodium* species.

Chemoprophylaxis for travel to endemic areas and treatment of malaria is usually accomplished with chloroquine. Chloroquine is the drug of choice for treating merozoites within red cells, but does not kill sporozoites. The closely related drug primaquine, however, eliminates sporozoites of *P. vivax* and *P. ovale* that may remain in liver cells. Treatment with both chloroquine and primaquine produces a radical cure. Even in individuals who have undergone radical drug treatment, however, malaria may recur years after the primary infection. Apparently not all of the sporozoites in the liver are eliminated; they reinitiate malaria months

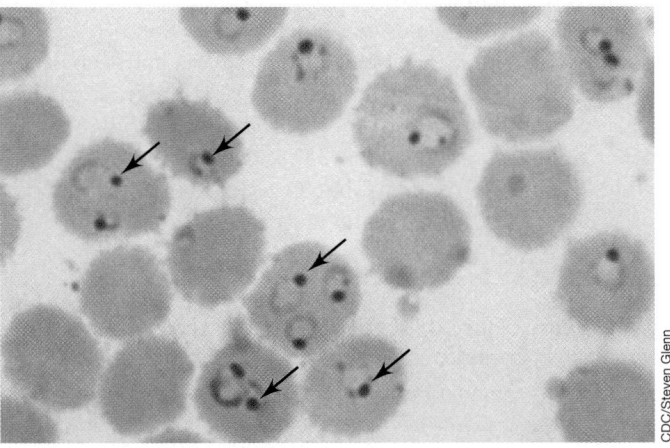

Figure 34.15 *Plasmodium falciparum.* This parasitic protist is one of several in the *Plasmodium* genus that causes malaria. Organisms (arrows) are shown growing inside human red blood cells. Uninfected red blood cells are about 6 μm in diameter. Infected red blood cells are slightly enlarged.

or years later by undergoing asexual reproduction (schizogony) and releasing a new generation of merozoites.

In many parts of the world *Plasmodium* strains have developed resistance to chloroquine or primaquine or both, and some strains have developed resistance to other drugs as well. In areas with known drug-resistant strains, mefloquin or doxycycline is prescribed for prophylaxis; a combination of atovaquone and proguanil (Malarone) is recommended for both treatment and prophylaxis. A new category of antimalarial drugs is comprised of synthetic derivatives of artemisinin, a natural compound containing reactive peroxide groups that form free radicals. These compounds are active *in vivo*. Even in the case of this relatively new drug, however, there are reports of artemisinin-resistant *Plasmodium* strains. A new experimental drug, NITD609, is unique in that it targets a parasite membrane transport protein. A single dose of this experimental compound is curative for *Plasmodium* infections in mice. Clinical trials in humans are in progress.

Prevention and Control

Antimalarial drug treatment is an inexpensive but short-term solution to malaria prevention and control, and drug-resistant strains of *Plasmodium* spp. further complicate matters. The most effective control measure is to interrupt the life cycle of the protist by eliminating one of the obligate hosts, the *Anopheles* mosquito.

Several approaches to mosquito control are possible. The first method requires *elimination of habitat* by drainage of swamps and similar breeding areas. During the 1930s, about 33,000 miles of ditches were constructed in 16 southern states in the United States, removing 544,000 acres of mosquito breeding area. Millions of gallons of oil were also spread on swamps to reduce the oxygen supply to mosquito larvae.

The second method of mosquito control requires *elimination of the mosquito* by insecticides, followed by treatment of patients with antimalarial drugs, thereby breaking the *Plasmodium* life cycle. The insecticide dichlorodiphenyltrichloroethane (DDT) was used to control larvae and adult mosquitoes. During World War II, the Public Health Service organized an Office of Mosquito Control in War Areas, and because many U.S. military bases were in the southern states, this organization carried out an extensive eradication program in the United States as well as overseas. In 1946 there were 48,610 cases of malaria in the United States when Congress established a 5-year malaria eradication program using drug prophylaxis and treatment for individuals, and DDT treatment of mosquito infestations. By 1953 there were only 1310 malaria cases. In 1934 there were about 4000 deaths from malaria; in 1952 there were only 25 deaths.

Although the overall public health threat from malaria in the United States is now minimal, very low numbers of endemic malaria cases have resurfaced in recent years as far north as New York City. Malaria incidence also increases due to cases imported by soldiers or immigrants from malaria-endemic areas. On average, there are about 1400 cases of malaria and less than ten deaths in the United States each year. Most cases are imported.

In other parts of the world, eradication has been much slower, but the same control measures are used. Reduction of mosquito

UNIT 11

habitat, control of mosquitoes by insecticides, chemoprophylaxis for potentially exposed individuals, and treatment of infected individuals with antimalarial drugs are still the major strategies for controlling malaria. Several malaria vaccines are in development, including synthetic peptide vaccines, recombinant particle vaccines, and DNA vaccines.

Malaria and Human Evolution

Malaria has been endemic in the tropical and subtropical regions of the world for thousands of years. In West Africans, resistance to malaria caused by *P. falciparum* is associated with the **sickle cell trait**, a genetic mutation that alters a red blood cell protein to form hemoglobin S; the amino acid sequence of hemoglobin S differs from that of normal hemoglobin at only a single amino acid in the protein. The neutral amino acid valine is substituted for the glutamic acid in the beta chain of normal hemoglobin A. As a result, mutated hemoglobin S binds oxygen less efficiently than normal hemoglobin A. Under conditions of low oxygen concentration, hemoglobin S forms long, thin aggregates that cause the red cell to change from a biconcave round cell to an elongated C-shaped cell, called a *sickle cell* (**Figure 34.16**). The elongated red blood cell is fragile, leading to rapid turnover of the defective red blood cells. Individuals who are homozygous for the sickle cell trait are particularly susceptible to changes in oxygen concentrations because they have lower numbers of oxygen-carrying red blood cells; they suffer from sickle cell anemia.

Individuals who are heterozygous for hemoglobin S have what is called *sickle cell trait*, but also have resistance to malaria as compared to normal individuals. In heterozygotes hemoglobin S can still produce sickled cells, but not as readily as in homozygotes. However, the growth of *P. falciparum* inside the red blood cell causes the heterozygous cells to sickle more easily than in uninfected heterozygous cells. The aggregated hemoglobin S in

sickled cells apparently disrupts the red blood cell cytoplasmic membrane, allowing potassium to diffuse from the cell. *P. falciparum* cannot grow in the low-potassium environment of the disrupted cell. Thus, persons with the sickle cell trait can live a more or less normal life and are resistant to malaria.

In many Mediterranean populations, a diverse group of genetic abnormalities affects hemoglobin production and efficiency. These are known collectively as the **thalassemias**. The thalassemias are also statistically and geographically associated with increased resistance to malaria associated with an increase in oxidants in red blood cells.

Hemoglobin S and thalassemias result from genetic mutations. These mutations cause red blood cell and oxygen-processing deficiencies that are deleterious in normal human populations. In individuals and populations exposed to *Plasmodium* infections and malaria, however, these mutations are positively selected; although the mutations cause red blood cell abnormalities and oxygen-processing deficiencies, they confer resistance to malaria and enhance the survival of those individuals carrying the mutation.

Another case in which *Plasmodium* spp. influences evolution involves the major histocompatibility complex (MHC) and the immune system. As discussed in Chapter 29, the MHC class I and class II proteins present antigens to T cells for initiation of an immune response. In malaria-prone equatorial West Africa, individuals are very likely to have one particular MHC class I gene and one particular set of class II genes. These selected MHC genes, common in the West African population, are virtually unknown in other human populations. Individuals who express these genes have as much resistance to severe malaria as those with the hemoglobin S trait. The MHC proteins encoded by these selected genes are exceptionally good antigen-presenting molecules for certain malarial antigens and initiate a strong protective immune response to *Plasmodium* spp. infection. As is the case with the hemoglobin variants, *Plasmodium* is a selective agent for certain MHC genes that enhance host survival. Individuals with MHC genes that confer malaria resistance have a measurable survival advantage and are more likely to live and pass the resistance-conferring genes on to their descendants.

Thus, *Plasmodium* infection causing malaria has been a selective agent in human evolution. Other pathogens, such as *Mycobacterium tuberculosis* (tuberculosis, ↩ Section 33.4) and *Yersinia pestis* (plague, Section 34.7), may also have promoted selective changes in humans, but in no case is the evidence as clear as it is for *Plasmodium* and malaria.

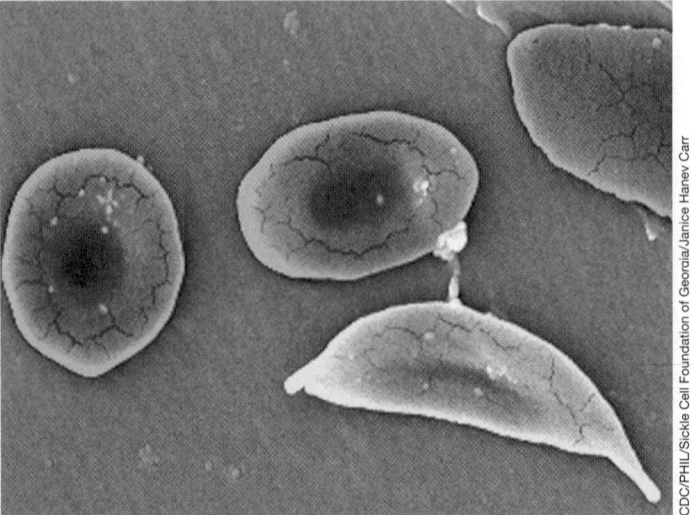

CDC/PHIL/Sickle Cell Foundation of Georgia/Janice Haney Carr

Figure 34.16 Sickled red blood cell. The red blood cells on the left and top center appear as normal round, biconcave disks. The cell in the lower center has the typical sickle shape seen in red blood cells under oxidative stress in patients who have sickle cell anemia.

MiniQuiz

- Which stages of the *Plasmodium* life cycle occur in humans, and which in the mosquito?
- What are the natural reservoirs and vectors for *Plasmodium* species? How can malaria be prevented or eradicated?
- Review genetic mechanisms responsible for malaria resistance. Why are genes known to confer malaria resistance not found in all humans?

34.6 West Nile Virus

West Nile virus (WNV) causes **West Nile fever**, a human viral disease transmitted through the bite of a mosquito (**Figure 34.17**). The virus can invade the nervous system of its warm-blooded host. WNV is a member of the flavivirus group and has a symmetrical, enveloped icosahedral capsid (Figure 34.17*b*) containing a positive-sense, single-stranded RNA genome of about 11,000 nucleotides (Section 21.8).

Epidemiology

WNV infection in humans was first identified in Uganda (Africa) in 1937. By the 1950s, the virus had spread to Egypt and Israel. In the 1990s there were WNV outbreaks in horses, birds, and humans in African and European countries. In 1999, the first cases were reported in the United States in the Northeast, around New York. A total of 63 cases occurred that year (**Figure 34.18**). In the United States from 1999 through 2001, there were 149 confirmed cases of human WNV disease, including 18 deaths,

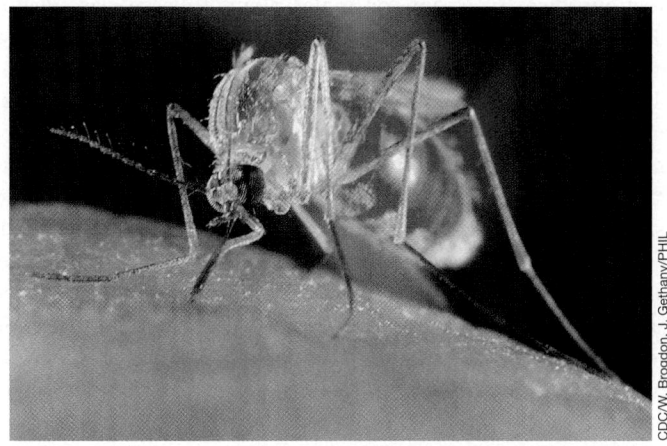

(a)

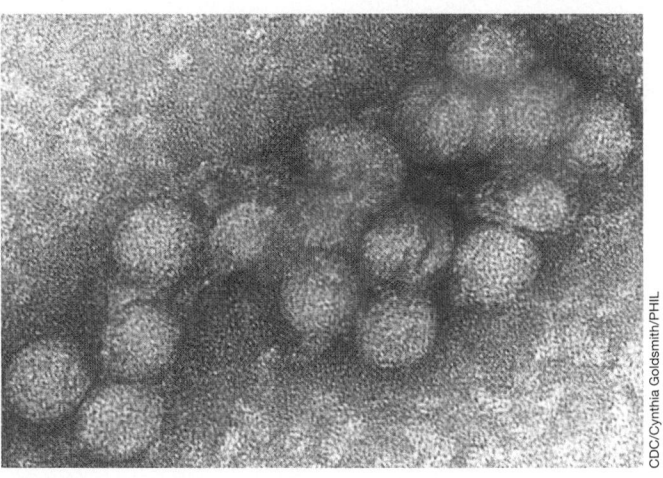

(b)

Figure 34.17 **West Nile virus.** *(a)* The mosquito *Culex quinquefasciatus*, shown here engorged with human blood, is a West Nile virus vector. *(b)* An electron micrograph of the West Nile virus. The icosahedral virion is about 40–60 nm in diameter.

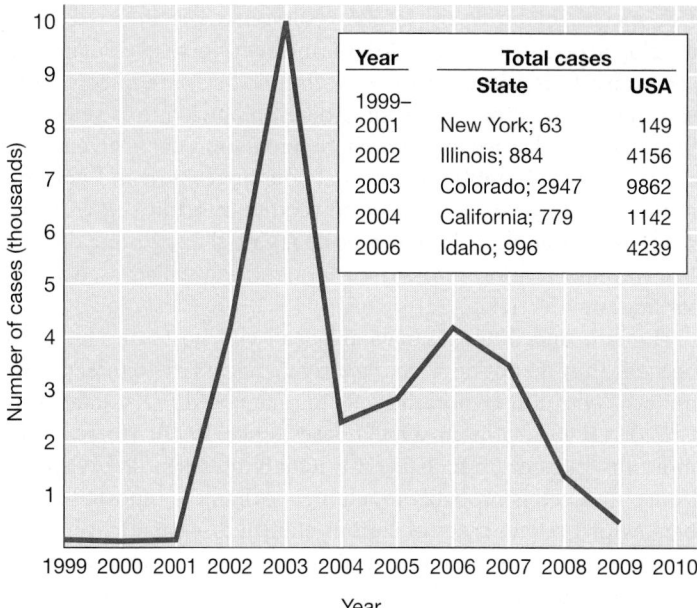

Year	Total cases	
	State	**USA**
1999– 2001	New York; 63	149
2002	Illinois; 884	4156
2003	Colorado; 2947	9862
2004	California; 779	1142
2006	Idaho; 996	4239

Figure 34.18 **West Nile virus in the United States.** The virus has caused about 30,000 cases of human disease and over 1100 deaths since 1999. The mosquito-borne virus infection was transmitted from east to west in annual avian epidemics until about 2007. West Nile virus is now endemic in mosquitoes and bird populations in the United States. Data are from the Centers for Disease Control and Prevention, Atlanta, Georgia.

almost all in the Northeast and along the Atlantic coast. Moving with the seasonal appearance and disappearance of the mosquito vectors, by 2002 this emerging disease had shifted from the East Coast to the Midwest, with a peak reported number of cases of 884 in Illinois and nationwide case totals of 4156. By contrast, Colorado had only 14 cases. The disease continued to move across North America, peaking at 9862 cases, centered in Colorado (2947 cases), and the upper Midwest in 2003. California had 3 cases. Overall case numbers declined after 2003, but the focus shifted to the far West and then to the northern mountain states; California had 779 cases in 2004; Idaho had 3 cases in 2004 and 13 in 2005. In 2006, the highest incidence of disease was in Idaho, with 996 cases. After 2006, there was no apparent focus of the disease. Since 2003, annual case numbers have declined, and in 2009 less than 500 cases were reported nationwide.

The human case numbers reflect the natural history of viral infection in mosquito vectors and birds; humans are incidentally infected as the virus spreads to new susceptible bird populations. As the epizootic disease in birds subsided, the human case numbers have also declined; WNV is now an enzootic disease in the surviving bird population in the United States. Paralleling the decrease of WNV in birds, human cases of West Nile fever have declined from a high of 9862 in 2003 to 361 in 2009.

WNV Transmission and Pathology

WNV normally causes active disease in some birds and is transferred to susceptible hosts by the bite of an infected mosquito. A number of mosquito species are known vectors, and at least

130 species of birds are WNV reservoirs. The infected birds develop a viremia lasting 1–4 days, and survivors develop lifelong immunity. Mosquitoes feeding on viremic birds are infected and can then infect other susceptible birds, renewing the cycle. The incidence of disease in the avian population in a given area decreases as susceptible avian hosts die or recover and develop immunity. However, the mosquito vectors transmit the WNV to new susceptible hosts in new areas, moving the epizootic disease in a wavelike fashion across the continent, as we discussed above for human infections.

Humans and other animals are dead-end hosts for the virus because they do not develop the viremia necessary to infect mosquitoes. The human mortality rate for diagnosed WNV infections is 3.9% (1143 deaths in 29,383 cases since 1999). Horses have mortality rates of up to 40%. The actual number of WNV infections, however, is probably much higher than the reported numbers, which reflect cases of serious clinical disease. Most human infections are asymptomatic or very mild and are not reported. After an incubation period of 3–14 days, about 20% of infected individuals develop West Nile fever, a mild illness lasting 3–6 days. Fever may be accompanied by headache, nausea, myalgia, rash, lymphadenopathy (swelling of lymph nodes), and malaise. Less than 1% of infected individuals develop serious neurological diseases such as West Nile encephalitis or meningitis. Diagnosed cases, however, have a much higher rate of serious disease, with adults over age 50 being more susceptible to serious effects such as neurological complications. Diagnosis of WNV disease includes assessment of clinical symptoms followed by confirmation with a positive EIA test for WNV antibodies in serum.

Prevention and Control of WNV

Like St. Louis encephalitis virus and other mosquito-borne viruses that cause encephalitis, transmission of WNV is seasonal in the United States and is dependent on exposure to the mosquito population. The primary means of control for WNV spread is by limiting exposure to the disease vector. Individuals should avoid mosquito habitat, wear appropriate mosquito-resistant clothing, and apply insect repellents containing DEET, as for Lyme disease. At the community level, WNV spread can be controlled by destruction of mosquito habitat and application of appropriate insecticides. There is no human vaccine, although several candidates are in development. Veterinary vaccines are available and widely used in horses. Treatment, as for most viral illnesses, is rest, fluids, and symptomatic relief of fever and pain. No antiviral drugs are known to be effective *in vivo* against WNV.

MiniQuiz
- Identify the vector and reservoir for West Nile virus.
- Trace the progress of West Nile virus in the United States since 1999.

34.7 Plague and *Yersinia*

Pandemic **plague** has caused more human deaths than any other infectious disease except for malaria and tuberculosis. Plague killed as much as one-third of Europe's population in individual pandemics in the Middle Ages.

Plague is caused by *Yersinia pestis*, a gram-negative, facultatively aerobic, rod-shaped bacterium (**Figure 34.19**) which is a member of the enteric bacteria group (↺ Section 17.11). Plague is a disease of domestic and wild rodents; rats are the primary disseminating host in urban communities. Humans are accidental hosts and are not critical for the maintenance of the pathogen. Fleas are intermediate hosts and vectors, spreading plague between the mammalian hosts (**Figure 34.20**). Most infected rats die soon after symptoms appear, but the small proportion of survivors develop a chronic infection, providing a persistent reservoir of virulent *Y. pestis*.

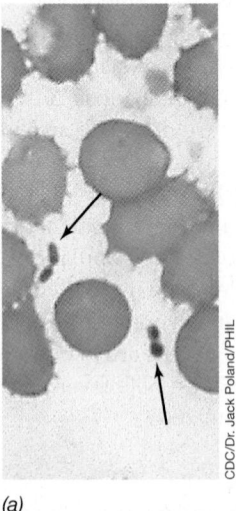
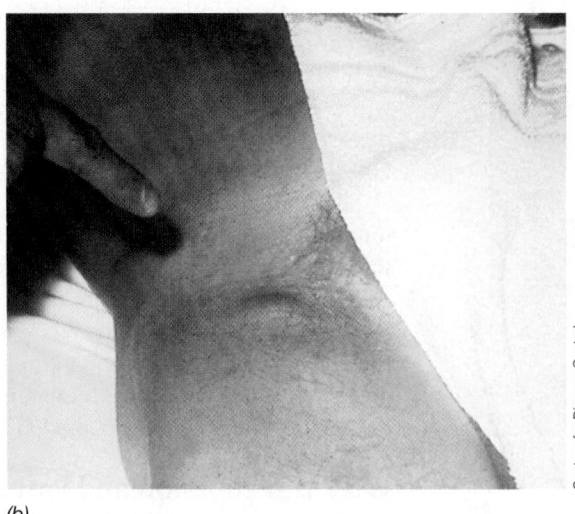
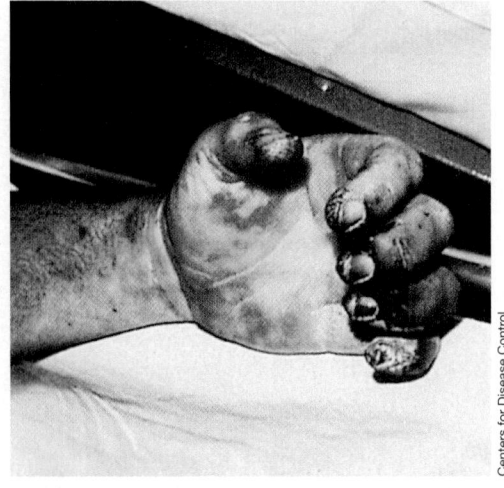

(a)   (b)   (c)

Figure 34.19 Plague in humans. *(a) Yersinia pestis*, the causative agent of plague, is a gram-negative rod, about 2 μm in length and up to 1 μm in diameter. The organisms in this blood smear (arrows) show the characteristic bipolar staining pattern. *(b)* A bubo formed in the groin. *(c)* Gangrene and sloughing of skin in the hand of a plague victim.

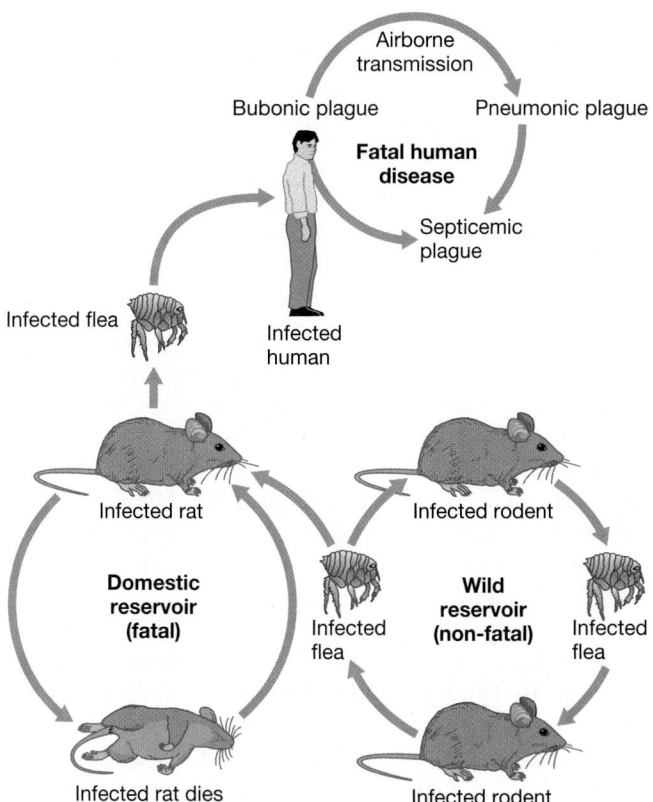

Figure 34.20 The epidemiology of plague due to *Yersinia pestis*. Plague in some wild rodents is a mild, self-limiting infection. In rodents that act as disseminating hosts (marmots, golden-mantled ground squirrels, prairie dogs, rats) and in humans, plague is often a fatal disease. Infected fleas desert the dead rodent host and look for another host, such as a human, an accidental host.

Epidemiology

Plague is endemic in countries in Africa, Asia, the Americas, and in south-central Eurasia. In 2003, greater than 98% of cases occurred in Africa. In the United States a handful of cases are diagnosed annually, mostly in the southwestern states, where the disease, called *sylvatic plague*, is enzootic among wild rodents. Plague is transmitted by several species of fleas, one of which is *Xenopsylla cheopis*, the rat flea. Fleas ingest *Y. pestis* by sucking blood from an infected animal. Cells multiply in the flea's intestine and are transmitted to a healthy animal in subsequent bites.

As the disease spreads, most of the host rats die and the infected fleas seek new hosts, including humans. Once in humans, cells of *Y. pestis* typically travel to the lymph nodes, where they cause swelling. The regional and pronounced swollen lymph nodes are called *buboes* and for this reason, the disease is often called *bubonic plague* (Figure 34.19*b*). The buboes become filled with *Y. pestis*, and encapsulated *Y. pestis* prevents phagocytosis and destruction by cells of the immune system. Secondary buboes form in peripheral lymph nodes, and cells eventually enter the bloodstream, causing septicemia. Multiple local hemorrhages produce dark splotches on the skin, giving plague its historical name, the "Black Death" (Figure 34.19*c*). If not treated prior to the septicemic stage, the symptoms of plague (lymph node swelling and pain, prostration, shock, and delirium) usually progress and cause death within 3–5 days.

Pathology of Plague

The pathogenesis of plague is not clearly understood, but cells of *Y. pestis* produce virulence factors that contribute to the disease process. The V and W antigens of *Y. pestis* cell walls are protein–lipoprotein complexes that inhibit phagocytosis. *Murine toxin*, an exotoxin that is lethal for mice, is produced by virulent strains of *Y. pestis*. Murine toxin is a respiratory inhibitor that blocks mitochondrial electron transport at coenzyme Q. It produces systemic shock, liver damage, and respiratory distress in mice. Although murine toxin is highly active in only certain animal species, it may be involved in human plague because these symptoms are also seen in affected humans. *Y. pestis* also produces a highly immunogenic endotoxin that may play a role in the disease process.

Pneumonic plague occurs when *Y. pestis* is either inhaled directly or reaches the lungs via the blood or lymphatic circulation. Symptoms are usually absent until the last day or two of the disease when large amounts of bloody sputum are produced. Untreated individuals rarely survive more than 2 days. Pneumonic plague is highly contagious and can spread rapidly via the person-to-person respiratory route if infected individuals are not immediately quarantined. *Septicemic plague* is the rapid spread of *Y. pestis* throughout the body via the bloodstream without the formation of buboes and usually causes death before a diagnosis can be made.

Treatment and Control

Bubonic plague can be successfully treated if rapidly diagnosed. *Y. pestis* infection is treated with streptomycin or gentamicin, given parenterally. Alternatively, doxycycline, ciprofloxacin, or chloramphenicol may be given intravenously. If treatment is started promptly, mortality from bubonic plague can be reduced to 1–5% of those infected. Pneumonic and septicemic plague can also be treated, but these forms progress so rapidly that antibiotic therapy, even if begun when symptoms first appear, is usually too late: Up to 90% of untreated pneumonic plague victims die. There were 47 cases of plague in the United States in 2000–2007. Mortality was less than 10%. Worldwide, there are usually about 2000 confirmed cases per year. *Y. pestis* is an organism that could be used for a bioterrorism attack (♺ Section 32.11), and oral doxycycline and ciprofloxacin are recommended as prophylactic antibiotics in this setting.

Plague control is accomplished through surveillance and control of animal reservoirs, vectors (fleas), and human contacts. Undoubtedly, improved public health practices and the control of rodent populations have limited human exposure to plague, especially in developed countries.

MiniQuiz
- Distinguish among sylvatic, bubonic, septicemic, and pneumonic plague.
- What are the insect vector, the natural host reservoir, and the treatment for plague?

Soilborne Pathogens

Several pathogenic microorganisms live in soil. Fungi are ubiquitous soil microorganisms, and a few are human and animal pathogens. Some bacteria are also important soilborne pathogens. In contrast to many person-to-person or vectorborne pathogens, soilborne pathogens are accidental agents of infection, with no life cycle dependency on the accidental host. Soil is an unlimited reservoir of these pathogens, and thus these pathogens cannot be eliminated.

34.8 Fungal Pathogens

Fungi in some form grow in nearly every ecological niche, but are most commonly found in nature as free-living saprophytes. Some fungi cause accidental, often opportunistic, and sometimes serious infections and disease. Individuals who have impaired immunity due to drug treatment or diseases such as human immunodeficiency virus/acquired immunodeficiency syndrome (HIV/AIDS) are more susceptible to opportunistic fungal pathogens. Increased numbers of serious fungal infections in recent years are probably due to the growing number of individuals who are immunosuppressed because of drug therapy or infection.

The fungi include the eukaryotic organisms commonly known as yeasts, which normally grow as single cells (**Figure 34.21***a*), and molds (mycelial forms), which grow in branching filaments (hyphae) with or without septa (cross walls) (Figure 34.21*b*). The taxonomy and biological diversity of these organisms were discussed in Sections 20.13–20.18. Fortunately, most fungi are harmless to humans. Only about 50 species cause human disease. In healthy individuals the overall incidence of serious fungal infections is rather low, although certain superficial fungal infections are quite common.

Epidemiology and Pathogenicity

Fungi cause disease through three major mechanisms: inappropriate immune responses; toxin production; and mycoses, or

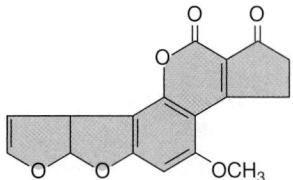

Figure 34.22 Structure of aflatoxin B1. This toxin is one of a group of related compounds produced by *Aspergillus flavus*.

growth of fungus in or on the body. First, some fungi trigger immune responses that result in allergic (hypersensitivity) reactions following exposure to specific fungal antigens. Reexposure to the same fungi, whether growing on the host or in the environment, may cause allergic symptoms. For example, *Aspergillus* spp. (⇄ Figure 20.28), a common saprophyte often found in nature as a leaf mold, produces potent allergens, often causing asthma and other hypersensitivity reactions. *Aspergillus* also has other mechanisms for producing disease.

A second fungal disease-producing mechanism involves the production and activity of *mycotoxins*, a large, diverse group of fungal exotoxins. The best-known examples of mycotoxins are the *aflatoxins* (**Figure 34.22**) produced by *Aspergillus flavus*, a species that commonly grows on improperly stored food, such as grain. Aflatoxins are highly toxic and carcinogenic, inducing tumors at high frequency in some animals, especially in birds that feed on contaminated grain. The direct role of aflatoxins in human disease is not well defined.

Pathogenicity genes in fungi can be transferred between related organisms by horizontal transfer of whole chromosomes comprising up to one-quarter of the genome. Demonstrated first in *Fusarium*, a plant pathogen that is an opportunistic pathogen in humans, horizontal transfer of pathogenicity genes can convert a nonpathogenic strain into a pathogen. This feature is probably shared by other fungal genera and may explain the ability of fungi to infect a wide variety of hosts; a pathogen lacking genes for invasion of a host can acquire invasion genes from a phylogenetically related, nonpathogenic saprophyte living on the host.

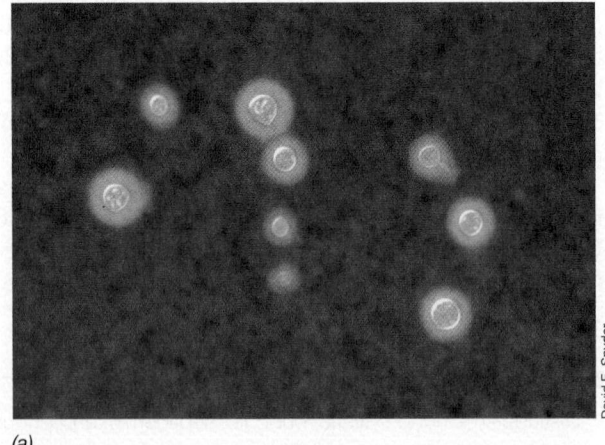

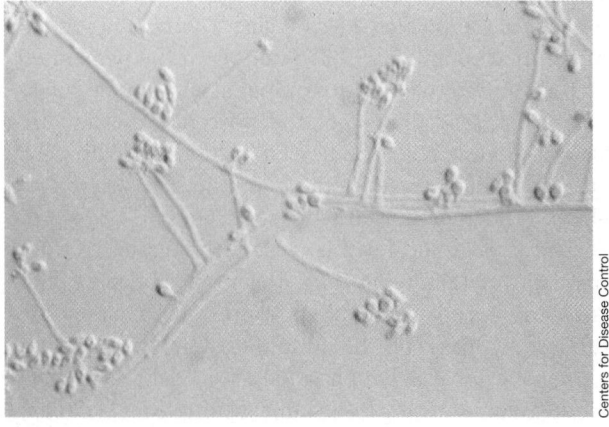

(a)   *(b)*

Figure 34.21 Typical forms of pathogenic fungi. *(a)* Yeast form of *Cryptococcus neoformans*, stained with India ink to show the capsule. The cells are from 4 to 20 μm in diameter. *(b)* *Sporothrix schenckii*, showing the branching, or hyphae, characteristic of the mold form of fungi. The round conidia are about 2 μm in diameter.

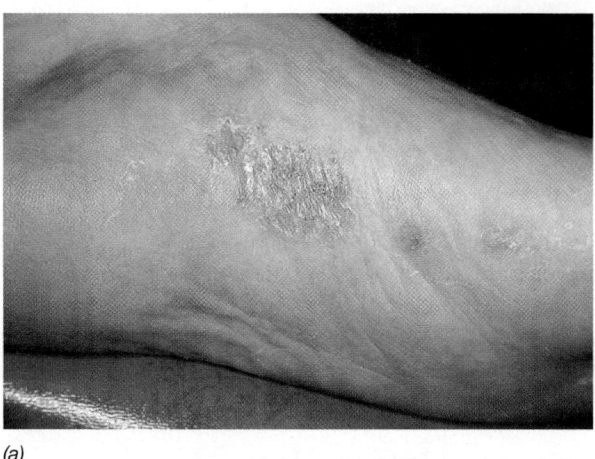
(a)

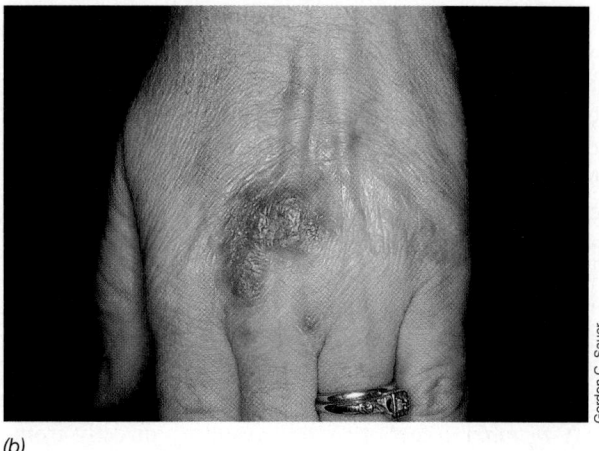
(b)

Figure 34.23 Fungal infections. *(a)* Superficial mycosis of the foot (athlete's foot) due to infection with *Trichophyton rubrum. (b)* Sporotrichosis, a subcutaneous infection due to *Sporothrix schenckii.*

Mycoses

The third fungal disease-producing mechanism is through infections called mycoses. The growth of a fungus on or in the body is called a **mycosis** (plural, mycoses). Mycoses are fungal infections that range in severity from relatively innocuous, superficial lesions to serious, life-threatening diseases.

Mycoses fall into three categories. The first of these are the *superficial mycoses*. In these diseases, fungi colonize the skin, hair, or nails, and infect only the surface layers (**Figure 34.23a**). **Table 34.2** lists some of the fungi that cause superficial mycoses.

In general, these diseases are benign and self-limiting. Some, such as *Trichophyton* infections of the feet (athlete's foot), are quite common. Spread is by personal contact with an infected person, by contact with contaminated surfaces such as bathtubs, shower stalls, or floors, or by contact with contaminated shared articles such as towels or bed linens. Treatment for severe cases is with topical application of miconazole nitrate or griseofulvin. Griseofulvin can also be administered orally. After entering the bloodstream, it passes to the skin where it can inhibit fungal growth.

Table 34.2 *Pathogenic fungi and diseases*

Disease	Causal organism	Site
Superficial mycoses (dermatomycoses)		
Ringworm	*Microsporum*	Scalp of children
Favus	*Trichophyton*	Scalp
Athlete's foot	*Epidermophyton, Trichophyton*	Between toes, skin
Jock itch	*Trichophyton, Epidermophyton*	Genital region
Keratitis	*Fusarium*	Eye (cornea)
Subcutaneous mycoses		
Sporotrichosis	*Sporothrix schenckii*	Arms, hands
Chromoblastomycosis	Several fungal genera	Legs, feet
Systemic mycoses		
Aspergillosis	*Aspergillus* spp.[a]	Lungs
Blastomycosis	*Blastomyces dermatitidis*	Lungs, skin
Candidiasis	*Candida albicans*[b]	Oral cavity, intestinal tract
Coccidioidomycosis	*Coccidioides immitis*[b]	Lungs
Cryptococcosis	*Cryptococcus neoformans*[b]	Lungs, meninges
Histoplasmosis	*Histoplasma capsulatum*[b]	Lungs
Pneumocystis pneumonia	*Pneumocystis jiroveci*[b]	Lungs

[a]*Aspergillus* can also cause allergies, toxemia, and limited infections.
[b]An opportunistic pathogen frequently implicated in the pathogenesis of HIV/AIDS.

The *subcutaneous mycoses* are a second category of fungal infections. They involve deeper layers of skin (Figure 34.23*b*) and are caused by a different group of organisms (Table 34.2). One disease in this category is sporotrichosis, an occupational hazard of agricultural workers, miners, and others who come into contact with the soil. The causal organism, *Sporothrix schenckii*, is a ubiquitous saprophyte on wood and in soil. Lesions are usually initiated by infection at a small wound or abrasion site, and *S. schenckii* can readily be isolated from the lesion and cultured *in vitro*. Treatment is with oral potassium iodide or oral ketoconazole.

The *systemic mycoses* are the third and most serious category of fungal infections. They involve fungal growth in internal organs of the body and are subclassified as primary or secondary infections. A *primary* infection is one resulting directly from the fungal pathogen in an otherwise normal, healthy individual. A *secondary* infection is one in a host that harbors a predisposing condition, such as antibiotic therapy or immunosuppression.

In the United States the most widespread primary fungal infections are histoplasmosis, caused by *Histoplasma capsulatum*, and coccidioidomycosis (San Joaquin Valley fever), caused by *Coccidioides immitis*. Both of these organisms normally live in soil and both cause respiratory disease. The host becomes infected by inhaling airborne spores that germinate and grow in the lungs. Histoplasmosis is primarily a disease of rural areas in the midwestern United States, especially in the Ohio and Mississippi River valleys. Most cases are mild and are often mistaken for more common respiratory infections. San Joaquin Valley fever is generally restricted to the desert regions of the southwestern United States. The fungus lives in desert soils, and the spores are disseminated on dry, windblown particles that are inhaled. In some areas in the southwestern United States, as many as 80% of the inhabitants may be infected, although most individuals suffer no apparent ill effects.

A number of systemic fungal infections, including histoplasmosis and coccidioidomycosis, are especially serious and common in individuals whose immune systems have been impaired, for example, by HIV/AIDS or by immunosuppressive drugs. These fungi are opportunistic pathogens; they cause serious infections only in individuals who have impaired defense mechanisms. These are secondary fungal diseases because normal individuals either do not get the disease or generally have a less severe form. Examples of other fungi involved as secondary opportunistic pathogens are given in Table 34.2.

Treatment and Control

Effective chemotherapy against systemic fungal infections is difficult because most antibiotics that inhibit fungi (which are eukaryotes) also affect their hosts (↩ Section 26.11). For example, one of the most effective antifungal agents, amphotericin B, is widely used to treat systemic fungal infections of humans but may cause serious side effects such as kidney toxicity.

Control of infections by elimination of fungal pathogens from the environment is impractical. As with many common-source pathogens, control of fungal growth cannot be achieved because there is a limitless reservoir. Exposure to fungi cannot be eliminated, but risks can be reduced by decontamination and indoor air filtration systems.

MiniQuiz
• Describe superficial, subcutaneous, and systemic mycoses.
• Distinguish between a primary and a secondary fungal disease.

34.9 Tetanus and *Clostridium tetani*

Tetanus is a serious, life-threatening disease. Although tetanus is preventable through immunization, 247 individuals acquired tetanus in the United States from 2000 through 2007. Among those, about 13% died. Worldwide, tetanus causes over 150,000 deaths per year, mostly in Africa and Southeast Asia, even though it is a vaccine-preventable infectious disease.

Biology and Epidemiology

Tetanus is caused by an exotoxin produced by *Clostridium tetani*, an obligately anaerobic, endospore-forming rod (↩ Section 18.2). The natural reservoir of *C. tetani* is soil, where it is a ubiquitous resident, although it is occasionally found in the gut of mammals, as are other *Clostridium* species.

Cells of *C. tetani* normally gain access to the body through a soil-contaminated wound, typically a deep puncture. In the wound, anoxic conditions allow germination of endospores, growth of the organism, and production of a potent exotoxin, the *tetanus toxin*. The organism is noninvasive; its sole method of causing disease is through the action of tetanus toxin on host cells. The incubation time is variable and may take from four days to several weeks, depending on the number of endospores inoculated at the time of injury. Tetanus is not transmitted from person to person.

Pathogenesis

We have already examined the activity of tetanus toxin at the cellular and molecular level (↩ Section 27.10). The toxin directly affects the release of inhibitory signaling molecules in the nervous system. These inhibitory signals control the "relaxation" phase of muscle contraction. The absence of inhibitory signaling molecules results in rigid paralysis of the voluntary muscles, often called *lockjaw* because it is observed first in the muscles of the jaw and face (**Figure 34.24**). Death is usually due to respiratory failure, and mortality is relatively high (usually over 10% even in developed countries).

Diagnosis, Control, Prevention, and Treatment

Diagnosis of tetanus is based on exposure, clinical symptoms, and, rarely, identification of the toxin in the blood or tissues of the patient. The organism may also be cultured from the wound, but success is highly variable.

The natural reservoir of *C. tetani* is the soil. Because *C. tetani* is an accidental pathogen in humans and is not dependent on humans or other animals for its propagation, there is no possibility for eradication. Therefore, control measures must focus on prevention.

Tetanus is a preventable disease. The existing toxoid vaccine is completely effective for disease prevention. Virtually all tetanus cases occur in individuals who were inadequately immunized. Individuals from 25 to 59 years of age are the fastest growing age

Royal College of Surgeons of Edinburgh

Figure 34.24 A soldier dying from tetanus. Note the rigid paralysis. This painting by Charles Bell is in the Royal College of Surgeons, Edinburgh, Scotland.

group for contracting tetanus, presumably because public health immunization programs target infants, school-age individuals, and seniors 60 years of age and older.

Appropriate treatment of serious cuts, lacerations, and punctures in individuals who have already been immunized with tetanus toxoid includes thorough cleaning of the wound, debridement (removal) of damaged tissue, and administration of a "booster" tetanus toxoid immunization. If the wound is severe and is contaminated by soil, treatment should also include administration of an antitoxin preparation, especially if the patient's immunization status is unknown or is out of date. The tetanus antitoxin is typically a pooled human anti-tetanus immunoglobulin (approved for human use worldwide) or a preparation of antibodies to tetanus made in horses (approved for use in many developing countries). Both of these preparations work by binding and neutralizing the tetanus exotoxin. The antitoxin is generally given intramuscularly, but intrathecal injection (injection into the sheath surrounding the spinal cord) is superior because the antitoxin can then get to the affected nerve root much more efficiently (᠄᠄ Section 27.10). These measures prevent active tetanus from occurring.

Acute symptomatic tetanus is treated with antibiotics, usually penicillin, to stop growth and toxin production by *C. tetani*, and antitoxin to prevent binding of newly released toxin to cells. Supportive therapy such as sedation, administration of muscle relaxants, and mechanical respiration may be necessary to control the effects of paralysis. Treatment cannot provide a reversal of symptoms, because toxin that is already bound to tissues cannot be neutralized. Even with antitoxin, antibiotics, and supportive therapy, tetanus patients have significant morbidity and mortality.

Other Endospore-Forming Pathogens in Soil

Several other species of *Clostridium* and *Bacillus*, all endospore-forming organisms, are pathogens and all are normally found in soil, making their eradication impossible. All cause disease because of their production of potent exotoxins. *C. tetani* is found almost exclusively in soil, but *C. botulinum*, *C. difficile*, and *C. perfringens* are occasionally found in the gut of humans and other animals as part of the normal microbial flora. *C. perfringens* and *C. botulinum* are important potential pathogens, but unlike *C. tetani*, they cause diseases transmitted by the foodborne route (᠄᠄ Section 36.7) rather than directly from soil. *C. difficile* is a commensal organism found in the human colon; it occasionally causes diarrhea. *Bacillus anthracis*, an important veterinary pathogen that has also been used in biowarfare, causes anthrax, and is also typically found in soils (᠄᠄ Section 32.12).

MiniQuiz
- Describe infection by *C. tetani* and the elaboration of tetanus toxin.
- Describe the steps necessary to prevent tetanus in an individual who has sustained a puncture wound.
- Describe treatment options for individuals with tetanus.

Big Ideas

34.1
Rabies occurs primarily in wild animals and is an important enzootic and epizootic disease that can cause serious zoonotic infections in humans, most frequently in developing countries. In the United States rabies is transmitted from the wild animal reservoir to domestic animals or, very rarely, to humans. Vaccination of domestic and wild animals is important for the control of rabies.

34.2
Hantaviruses are present worldwide in rodent populations and cause zoonotic diseases such as HPS and HFRS in humans. In the Americas, hantavirus infections have case fatality rates of over 30%.

34.3
Rickettsias are obligate intracellular parasitic bacteria transmitted to hosts by arthropod vectors. The incidence of spotted fever rickettsiosis, HGA, and HME is increasing due to several factors. Most rickettsial infections can be controlled by antibiotic therapy, but prompt recognition and diagnosis of these diseases remains difficult.

34.4
Lyme disease is the most prevalent arthropod-borne disease in the United States today. It is transmitted from several mammalian host vectors to humans by ticks. Prevention and treatment of Lyme disease are straightforward, but accurate and timely diagnosis of infection is essential.

34.5

Infections with *Plasmodium* spp. cause malaria, a widespread, mosquito-transmitted disease that causes significant morbidity and mortality in tropical and subtropical regions of the world. Malaria is also a selection factor for resistance genes in humans. Malaria is preventable with a combination of public health and chemotherapy measures, but no vaccines are available.

34.6

West Nile fever is a mosquito-borne viral disease. In the natural cycle of the pathogen, birds are infected with West Nile virus by the bite of infected mosquitoes. Humans and other vertebrates are occasional dead-end hosts. Most human infections are asymptomatic and undiagnosed, but complications in diagnosed infections cause about 3.9% mortality.

34.7

Plague can be transmitted to individuals who have contact with rodent populations and their parasitic fleas, the enzootic reser-

voirs for *Yersinia pestis*. A disseminated systemic infection or a pneumonic infection leads to rapid death, but the bubonic form is treatable with antibiotics.

34.8

Certain soilborne fungi produce disease in humans. Superficial, subcutaneous, and systemic mycoses are difficult to control because of a lack of antifungal drugs and the ubiquitous nature of the pathogens. Fungal infections may cause serious systemic disease, usually in individuals with impaired immunity, such as in HIV/AIDS patients.

34.9

Clostridium tetani is a ubiquitous soilborne microorganism that can cause tetanus, a disease characterized by toxin production and rigid paralysis. Tetanus has significant morbidity and mortality. Tetanus is preventable with appropriate immunization. Treatment for acute tetanus includes antibiotics, active and passive immunization, and supportive therapy.

Review of Key Terms

Enzootic an endemic disease present in an animal population

Epizootic an epidemic disease present in an animal population

Hantavirus pulmonary syndrome (HPS) an acute disease characterized by pneumonia, caused by rodent hantavirus

Hemorrhagic fever with renal syndrome (HFRS) an emerging acute disease characterized by shock and kidney failure, caused by rodent hantavirus

Lyme disease a tick-transmitted disease caused by the spirochete *Borrelia burgdorferi*

Malaria a disease characterized by recurrent episodes of fever and anemia, caused by the protist *Plasmodium* spp., usually transmitted between mammals through the bite of the *Anopheles* mosquito

Mycosis (plural, **mycoses**) an infection caused by a fungus

Plague an enzootic disease in rodents caused by *Yersinia pestis* that can be transferred to humans through the bite of a flea

Rabies a usually fatal neurological disease caused by the rabies virus usually transmitted by the bite or saliva of an infected animal

Rickettsias obligate intracellular bacteria of the genus *Rickettsia* responsible for diseases including typhus, spotted fever rickettsiosis, and ehrlichiosis

Sickle cell trait a genetic trait that confers resistance to malaria, but causes a reduction in the oxygen-carrying capacity of the blood by reducing the life expectancy of the affected red blood cells

spotted fever rickettsiosis a tick-transmitted disease caused by *Rickettsia rickettsii*,

characterized by fever, headache, rash, and gastrointestinal symptoms; formerly called Rocky Mountain spotted fever

Tetanus a disease characterized by rigid paralysis of the voluntary muscles, caused by an exotoxin produced by *Clostridium tetani*

Thalassemia a genetic trait that confers resistance to malaria, but causes a reduction in the efficiency of red blood cells by altering a red blood cell enzyme

Typhus a louse-transmitted disease caused by *Rickettsia prowazekii*, characterized by fever, headache, weakness, rash, and damage to the central nervous system and internal organs

West Nile fever a neurological disease caused by West Nile virus, a virus transmitted by mosquitoes from birds to humans

Zoonosis an animal disease transmitted to humans

Review Questions

1. Identify the animals most likely to carry rabies in the United States. Which immunization programs are in place for the treatment of rabies? Which immunization programs are in place for the prevention of rabies (Section 34.1)?

2. Describe the clinical features of hantavirus pulmonary syndrome (HPS) (Section 34.2).

3. What are the three categories of organisms that cause rickettsial diseases? Describe briefly their clinical features (Section 34.3).

4. Identify the most common reservoir and vector for Lyme disease in the United States. How can the spread of Lyme disease be controlled? How can Lyme disease be treated (Section 34.4)?

5. Describe the relationship between malaria and human evolution (Section 34.5).

6. Describe the spread of West Nile virus infections in the United States from 1999 to 2009. What animals are the primary hosts? Are humans productive alternate hosts? Explain (Section 34.6).

7. Describe the epidemiology and pathology of the plague (Section 34.7).

8. Identify the natural source of most fungal pathogens. How can fungal exposure be controlled? What particular problems, especially in terms of therapy, do fungi pose for the clinician (Section 34.8)?

9. Describe the invasiveness and toxicity of *Clostridium tetani*. Discuss the major mechanism of pathogenesis for tetanus and define measures for prevention and treatment (Section 34.9).

Application Questions

1. Describe the sequence of events you would take if a child received a bite (provoked or unprovoked) from a stray dog with no record of rabies immunization. Present one scenario in which you are able to capture and detain the dog and another for a dog that escapes. How would these procedures differ from a situation in which the child was bitten by a dog that had documented, up-to-date rabies immunizations?

2. Oral histories from Native Americans indicate the presence of hantavirus pulmonary syndrome prior to the "discovery" and definition of HPS in 1993. Explain these findings in terms of emerging zoonotic diseases and human land use practices.

3. Discuss at least three common properties of the disease agents and review the disease process for spotted fever rickettsiosis, typhus, and ehrlichiosis. Why is ehrlichiosis emerging as an important rickettsial disease? Compare its emergence to that of Lyme disease.

4. Malaria eradication has been a goal of public health programs for at least 100 years. What factors preclude our ability to eradicate malaria? If an effective vaccine was developed, could malaria be eradicated? Compare this possibility to the possibility of eradicating plague.

5. Devise a plan to prevent the spread of West Nile virus to humans in your community. Identify the costs involved in such a plan, both at the individual level and at the community level. Find out if a mosquito abatement program is active in your community. What methods, if any, are used in your area for the reduction of mosquito populations?

Need more practice? Test your understanding with quantitative questions; access additional study tools including tutorials, animations, and videos; and then test your knowledge with chapter quizzes and practice tests at **www.microbiologyplace.com**.

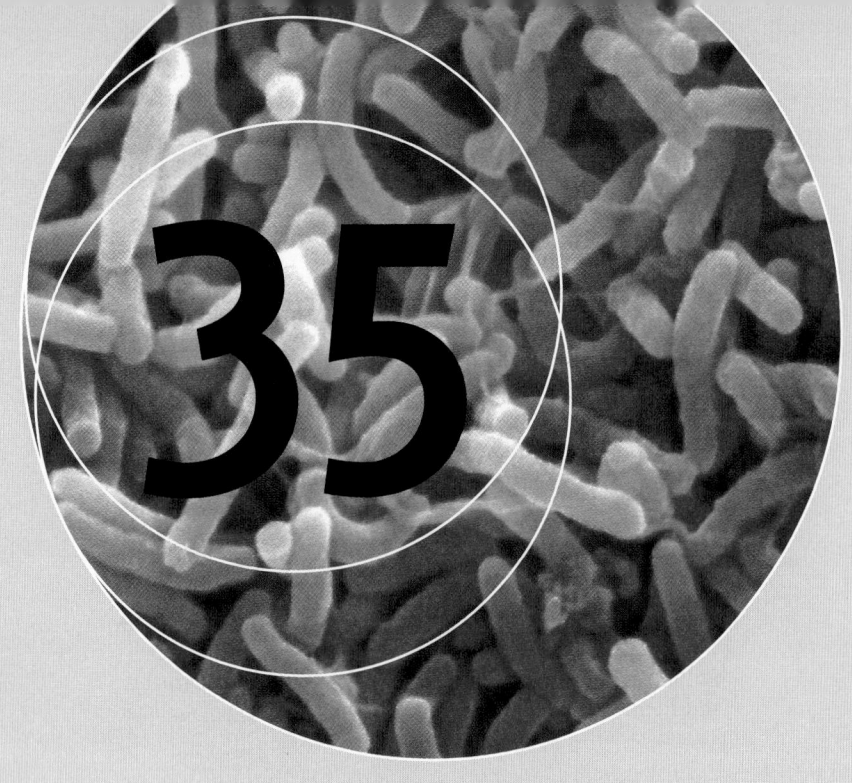

35

Wastewater Treatment, Water Purification, and Waterborne Microbial Diseases

Vibrio cholerae is a gram-negative enteric bacterium that causes the severe gastrointestional disease cholera. The organism is indigenous to coastal waters in many subtropical and tropical locations. Cholera is associated with contaminated water, and outbreaks are often traced to breakdowns in water and sewage treatment facilities.

lean water, free of biological and chemical contaminants, is essential for public health. Standard procedures to disinfect drinking water and remediate wastewater are in place in developed countries to achieve water quality. Water quality, however, is sometimes compromised, even in large-scale public wastewater and drinking water systems. Lapses in water quality can promote dramatic and even life-threatening spread of infectious disease. This chapter examines standard methods of water monitoring, treatment, and remediation. We also investigate the causes of some common waterborne diseases.

I Wastewater Microbiology and Water Purification

Water is the most important potential common source of infectious diseases and a potential source for chemically induced intoxications. This is because a single water source often serves large numbers of people, as, for example, in large cities. Everyone in these cities must use the available water, and contaminated water has the potential to spread disease to all exposed individuals. Water quality is therefore the most important single factor for ensuring public health. The methods commonly used to assess water quality depend on standard microbiological and chemical techniques. Waste purification and treatment protocols use physical, chemical, and biological means to identify, remove, and degrade pollutants.

35.1 Public Health and Water Quality

Even water that looks perfectly transparent and clean may be contaminated with pathogenic microorganisms that may pose a serious health hazard. It is impractical to screen water for every pathogenic organism that may be present, and a few nonpathogenic microorganisms are generally tolerable, and even unavoidable, in a water supply. However, water supplies can be sampled for the presence of specific *indicator microorganisms*, the presence of which signals potential contamination with pathogens.

Coliforms and Water Quality

A widely used indicator for microbial water contamination is the **coliform** group of microorganisms. Coliforms are useful indicators of water contamination because many of them inhabit the intestinal tract of humans and other animals in large numbers. Thus, the presence of coliforms in water may indicate fecal contamination. Coliforms are defined as facultatively aerobic, gram-negative, non-spore-forming, rod-shaped bacteria that ferment lactose with gas formation within 48 hours at 35°C. This operational definition of the coliform group includes taxonomically unrelated microorganisms. Many coliforms, however, are members of the enteric bacteria group (↩ Section 17.11). The coliform group includes a subgroup of thermotolerant bacteria known as fecal coliforms and includes the usually harmless *Enterobacter*; *Escherichia coli*, a common intestinal organism and occasional pathogen; and *Klebsiella pneumoniae*, a less common pathogenic intestinal inhabitant.

In general, the presence of fecal coliforms, especially *E. coli*, in a water sample indicates fecal contamination and indicates that the water is unsafe for human consumption. The presence or absence and the enumeration of fecal coliforms in water samples are standard parameters used for assessing water quality; fecal coliform detection methods are standardized and relatively easy to perform (see below). The absence of fecal coliforms, however, does not ensure good water quality. When excreted into water, the fecal coliforms eventually die, but some pathogens may not die as quickly. In addition, fecal coliform tests provide no information concerning the presence or absence of viruses or protists. Thus, fecal coliform tests are useful for identifying the presence of enteric bacterial pathogens, but can fail as an overall indicator of water quality.

Testing for Fecal Coliforms and *Escherichia coli*

Several procedures are used to test for fecal coliforms and *E. coli* in water samples. All tests assay the growth of organisms recovered from water samples. Common methods of enumerating the samples include the *most-probable-number* (*MPN*) procedure and the *membrane filter* (*MF*) procedure. The MPN procedure employs liquid culture medium in test tubes to which samples of drinking water are added. Growth in the culture vessels indicates microbial contamination of the water supply. For the MF procedure, at least 100 ml of the water sample is passed through a sterile membrane filter, trapping any bacteria on the filter surface. The filter is placed on a plate of eosin–methylene blue (EMB) culture medium, which is selective for gram-negative, lactose-fermenting microorganisms, including the coliforms (**Figure 35.1**; ↩ Figure 31.4). Following incubation, coliform colonies are counted, and from this value the number of coliforms in the original water sample can be calculated.

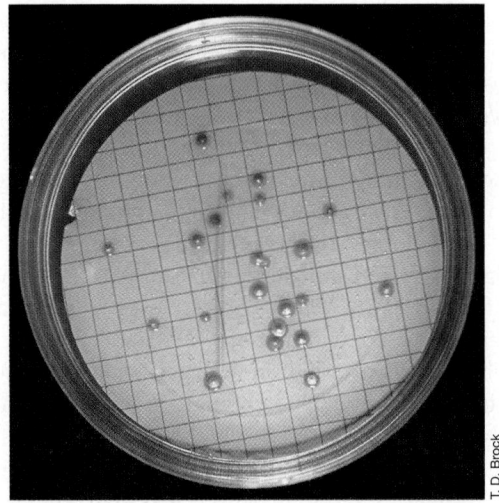

Figure 35.1 Coliform colonies growing on a membrane filter. A drinking water sample was passed through the filter. The filter was then placed on eosin–methylene blue (EMB) medium that is both selective and differential for lactose-fermenting bacteria (coliforms). The dark, shiny appearance of the colonies is characteristic of coliforms. Each colony developed from one viable coliform cell present in the original sample.

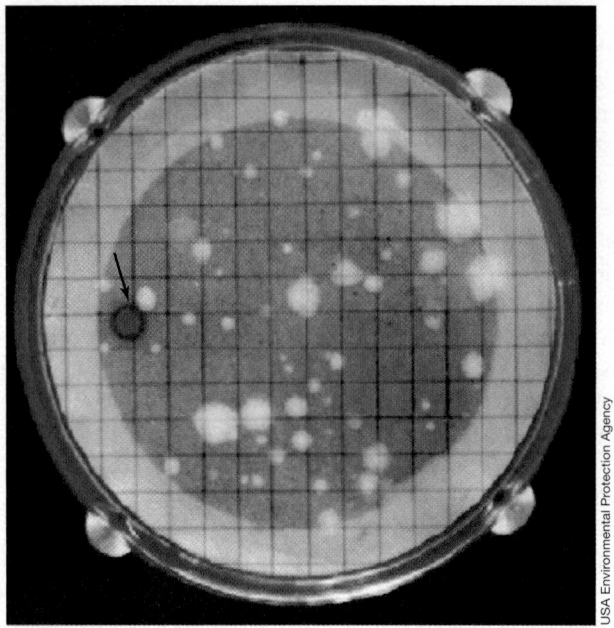

Figure 35.2 Total coliforms and *Escherichia coli*. A filter exposed to a drinking water sample was incubated at 35°C for 24 hours on MI media and examined under UV light. The single *E. coli* colony appears dark blue (arrow). The other colonies are coliforms that fluoresce and appear white to light blue.

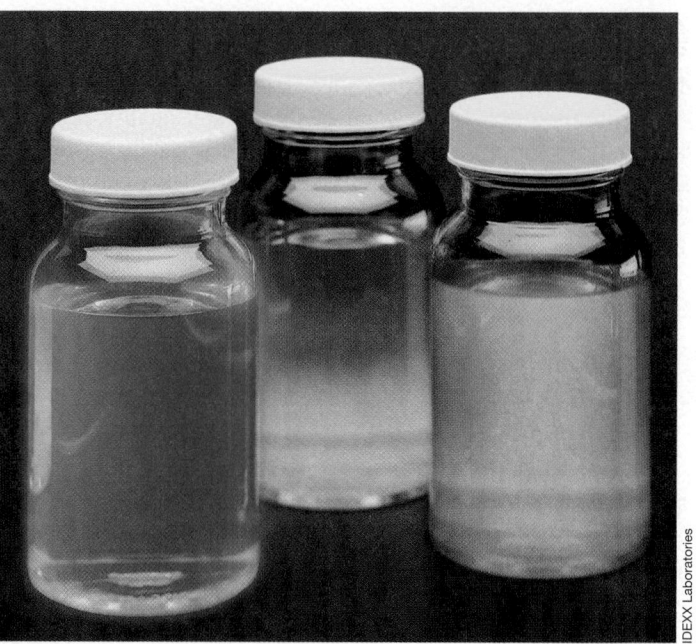

Figure 35.3 The IDEXX Colilert water quality test system. Colilert reagents are added to 100-ml water samples. After incubation for 24 h at 35–37°C, the samples develop yellow color if they contain coliform bacteria (right). Samples containing *Escherichia coli* develop yellow color and also develop blue fluorescence (left). Samples negative for coliform bacteria remain clear (center).

Selective media are used not only to detect total coliforms, but also to specifically identify *E. coli* at the same time. Designated as defined substrate tests, they are generally faster and usually more accurate than EMB agar tests. Defined substrate tests are based on the ability of coliforms and *E. coli* to metabolize certain substrates. For example, all coliforms, including *E. coli*, metabolize 4-methylumbelliferyl-β-D-galactopyranoside (MUG) using the enzyme β-galactosidase. If coliforms are present in a sample, MUG is metabolized to produce a fluorescent product visible under ultraviolet (UV) light (**Figure 35.2**). To distinguish total coliforms from *E. coli*, another enzyme–substrate reaction is used at the same time. *E. coli*, but not other coliforms, produces the enzyme β-glucuronidase, which metabolizes indoxyl β-D-glucuronide (IBDG) to a blue compound. The blue compound colors only growing *E. coli* colonies. As a result, *E. coli* colonies fluoresce and are also dark blue, and this differentiates them from colonies of other coliforms (Figure 35.2). The test uses a membrane filter method and media containing both MUG and IBDG (called *MI media*). The filter is overlaid on MI agar, incubated, and examined within 24 hours for blue colonies (*E. coli*) and fluorescent colonies (total coliforms).

A commonly used method for performing coliform counts is the IDEXX Colilert test system. This test system relies on the relative ability of β-galactosidase and β-glucuronidase in coliforms and *E. coli*, respectively, to utilize a proprietary substrate mix. Using the same principles as the selective media described for Figure 35.2, this method shows colored and fluorescent products for total coliforms and *E. coli* (**Figure 35.3**).

In properly regulated drinking water supply systems, total coliform and *E. coli* fecal coliform tests should be negative. A positive test indicates that a breakdown has occurred in the purification or distribution system. Drinking water standards in the United States are specified under law by the Safe Drinking Water Act and administered by the United States Environmental Protection Agency (EPA). This law provides minimum legal parameters for the development of safe drinking water standards. To be considered safe, no more than 5.0% of total samples (samples are 100 ml) can be coliform-positive in a 1-month test period. For water systems that collect fewer than 40 samples per month, no more than one sample can be coliform-positive each month. Samples that are coliform-positive must be analyzed for either fecal coliforms or *E. coli*. If the system has two consecutive coliform-positive samples, and one is also positive for *E. coli*, the system has a Maximum Contaminant Level (MCL) violation. MCL is the highest level of a contaminant that is allowed in drinking water.

Water utilities report coliform test results to the EPA, and if they do not meet the prescribed standards, the utilities must notify the public and take steps to correct the problem. Water purification utilities for smaller communities and even large cities sometimes fail to meet these standards.

Public Health and Drinking Water Purification

Intestinal infections due to waterborne pathogens are still common, even in developed countries, and some estimates indicate

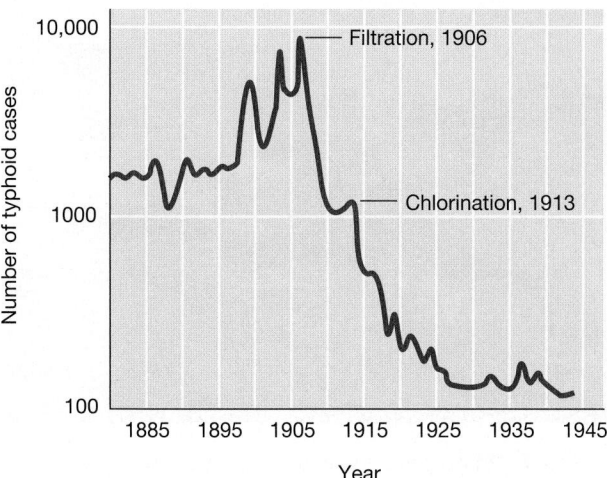

Figure 35.4 **The effect of water purification on the incidence of waterborne disease.** The graph shows the incidence of typhoid fever in Philadelphia, Pennsylvania. Note the dramatic reduction in the incidence of typhoid fever after the introduction of both filtration and chlorination.

that waterborne diseases impact the health of several million people each year in the United States alone. Water treatment practices, however, have significantly improved access to safe water, starting with public works projects coupled with the application and development of water microbiology in the early twentieth century.

Coliform-counting culture methods were developed and adapted around 1906. At the time, water purification was limited to *filtration* to reduce turbidity. Although filtration significantly decreased the microbial load of water, many microorganisms still passed through the filters. Around 1913, **chlorine** came into use as a disinfectant for large water supplies. Chlorine gas was an effective and inexpensive general disinfectant for drinking water, and its use quickly reduced the incidence of waterborne disease. **Figure 35.4** illustrates the dramatic drop in incidence of typhoid fever (caused by *Salmonella enterica* serovar Typhi) in a major American city after purification procedures using filtration and chlorination were introduced. Similar results were obtained in other cities.

Major improvements in public health in the United States, starting near the beginning of the twentieth century, were largely due to the adoption of water filtration and disinfection treatment procedures in large-scale, publicly operated wastewater and drinking water treatment plants. The effectiveness of filtration and chlorination was monitored by the coliform test. Public works engineering and microbiology were the most important contributors to the dramatic advances in public health in developed countries in the twentieth century.

MiniQuiz

- Why do the bacterial colonies recovered from drinking water that grow on MI media indicate fecal contamination of the water supply?

- What general procedures are used to reduce microbial numbers (microbial load) in water supplies?

35.2 Wastewater and Sewage Treatment

Wastewater is domestic sewage or liquid industrial waste that cannot be discarded in untreated form into lakes or streams due to public health, economic, environmental, and aesthetic considerations. Wastewater treatment employs physical and chemical methods as well as industrial-scale use of microorganisms. Wastewater enters a treatment plant and, following treatment, the **effluent water**—treated wastewater discharged from the wastewater treatment facility—is suitable for release into surface waters such as lakes and streams or to drinking water purification facilities.

Wastewater and Sewage

Wastewater from domestic sewage or industrial sources cannot be discarded in untreated form into lakes or streams. **Sewage** is liquid effluent contaminated with human or animal fecal materials. Wastewater may also contain potentially harmful inorganic and organic compounds as well as pathogenic microorganisms. Wastewater treatment can use physical, chemical, and biological (microbiological) processes to remove or neutralize contaminants.

On average, each person in the United States uses 100–200 gallons of water every day for washing, cooking, drinking, and sanitation. Wastewater collected from these activities must be treated to remove contaminants before it can be released into surface waters. About 16,000 publicly owned treatment works (POTW) operate in the United States. Most POTWs are fairly small, treating 1 million gallons (3.8 million liters) or less of wastewater per day. Collectively, however, these plants treat about 32 billion gallons of wastewater daily. Wastewater plants are usually constructed to handle both domestic and industrial wastes. Domestic wastewater is made up of sewage, "gray water" (the water resulting from washing, bathing, and cooking), and wastewater from small-scale food processing in homes and restaurants.

Industrial wastewater includes liquid discharged from the petrochemical, pesticide, food and dairy, plastics, pharmaceutical, and metallurgical industries. Industrial wastewater may contain toxic substances; some manufacturing and processing plants are required by the EPA to pretreat toxic or heavily contaminated discharges before they enter POTWs. Pretreatment may involve mechanical processes in which large debris is removed. Some wastewaters are pretreated biologically or chemically to remove highly toxic substances such as cyanide; heavy metals such as arsenic, lead, and mercury; or organic materials such as acrylamide, atrazine (a herbicide), and benzene. These substances are converted to less toxic forms by treatment with chemicals or microorganisms capable of neutralizing, oxidizing, precipitating, or volatilizing these wastes. The pretreated wastewater can then be released to the POTW.

Wastewater Treatment and Biochemical Oxygen Demand

The goal of a wastewater treatment facility is to reduce organic and inorganic materials in wastewater to a level that no longer supports microbial growth and to eliminate other potentially toxic materials. The efficiency of treatment is expressed in terms

WASTEWATER

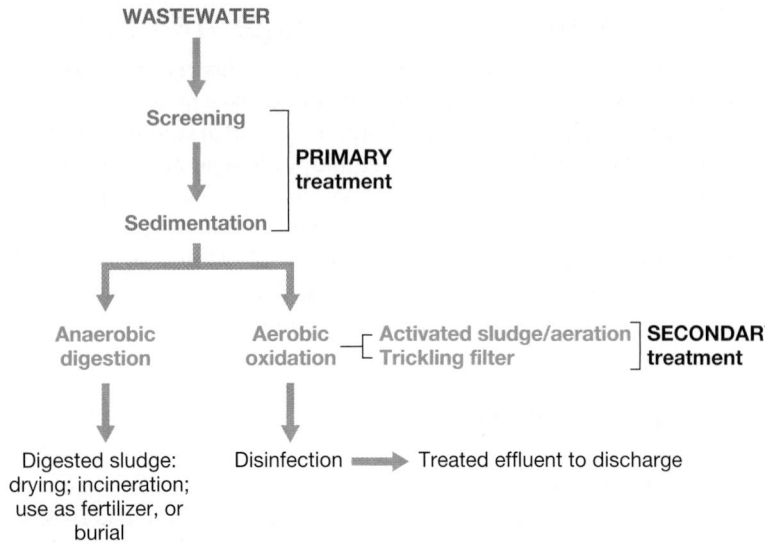

Anaerobic digestion → Digested sludge: drying; incineration; use as fertilizer, or burial

Aerobic oxidation — Activated sludge/aeration / Trickling filter — SECONDARY treatment

Disinfection ⟹ Treated effluent to discharge

Figure 35.5 **Wastewater treatment processes.** Effective water treatment plants use the primary and secondary treatment methods shown here. Tertiary treatment may also be used to reduce BOD levels in effluent water to undetectable levels.

Figure 35.6 **Primary treatment of wastewater.** Wastewater is pumped into the reservoir (left) where solids settle. As the water level rises, the water spills through the grates to successively lower levels. Water at the lowest level, now virtually free of solids, enters the spillway (arrow) and is pumped to a secondary treatment facility.

of a reduction in the **biochemical oxygen demand (BOD)**, the relative amount of dissolved oxygen consumed by microorganisms to completely oxidize all organic and inorganic matter in a water sample (♻ Section 23.8). High levels of organic and inorganic materials in the wastewater result in a high BOD.

Typical values for domestic wastewater, including sewage, are approximately 200 BOD units. For industrial wastewater from sources such as dairy plants, the values can be as high as 1500 BOD units. An efficient wastewater treatment facility reduces BOD levels to less than 5 BOD units in the final treated water. Wastewater facilities are designed to treat both low-BOD sewage and high-BOD industrial wastes.

Treatment is a multistep operation employing a number of independent physical and biological processes (**Figure 35.5**). *Primary, secondary*, and sometimes *tertiary* treatments are employed to reduce biological and chemical contamination in the wastewater, and each level of treatment employs more complex technologies.

Primary Wastewater Treatment

Primary wastewater treatment uses only physical separation methods to separate solid and particulate organic and inorganic materials from wastewater. Wastewater entering the treatment plant is passed through a series of grates and screens that remove large objects. The effluent is allowed to settle for a few hours. Solids settle to the bottom of the separation reservoir and the effluent is drawn off to be discharged or for further treatment (**Figure 35.6**).

Municipalities that provide only primary treatment discharge extremely polluted water with high BOD into adjacent waterways; high levels of soluble and suspended organic matter and other nutrients remain in water following primary treatment. These nutrients can trigger undesirable microbial growth, further reducing water quality. Most treatment plants employ secondary and even tertiary treatments to reduce the organic content of

the wastewater before release to natural waterways. Secondary treatment processes use both aerobic and anaerobic microbial digestion to further reduce organic nutrients in wastewater.

Secondary Anaerobic Wastewater Treatment

Secondary anaerobic wastewater treatment involves a series of degradative and fermentative reactions carried out by various prokaryotes under anoxic conditions. Anaerobic treatment is typically used to treat wastewater containing large quantities of insoluble organic matter (and therefore having a very high BOD) such as fiber and cellulose waste from food and dairy plants. The anaerobic degradation process is carried out in large, enclosed tanks called *sludge digesters* or *bioreactors* (**Figure 35.7**). The process requires the collective activities of many different types of prokaryotes. The major reactions are summarized in Figure 35.7c.

First, anaerobes use polysaccharidases, proteases, and lipases to digest suspended solids and large macromolecules into soluble components. These soluble components are then fermented to yield a mixture of fatty acids, hydrogen (H_2), and carbon dioxide (CO_2); the fatty acids are further fermented by the cooperative actions of syntrophic bacteria (♻ Section 14.5) to produce acetate, CO_2, and H_2. These products are then used as substrates by methanogenic *Archaea* (♻ Section 19.3), fermenting acetate to produce methane (CH_4) and CO_2, the major products of anoxic sewage treatment (Figure 35.7c). The CH_4 is burned off or used as fuel to heat and power the wastewater treatment plant.

Secondary Aerobic Wastewater Treatment

Secondary aerobic wastewater treatment uses oxidative degradation reactions carried out by microorganisms under aerobic conditions to treat wastewater containing low levels of organic materials. In general, wastewaters that originate from residential sources can be treated efficiently using only aerobic treatment. Several aerobic degradative processes can be used for wastewater treatment; *activated sludge* methods are the most common (**Figure 35.8a, b**). Here, wastewater is continuously mixed and

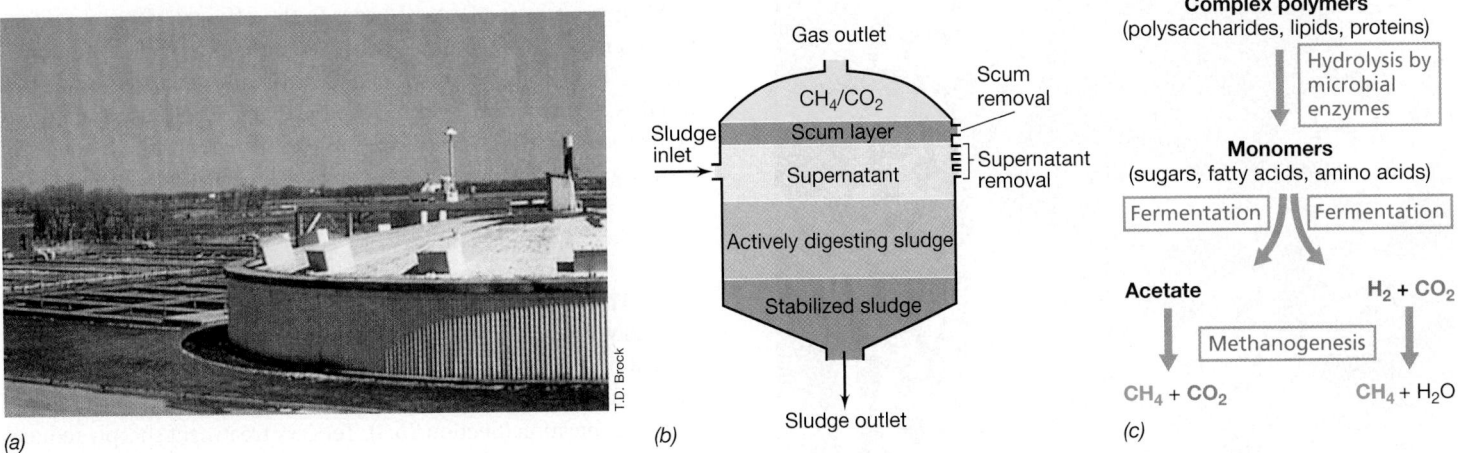

Figure 35.7 Secondary anaerobic wastewater treatment. *(a)* Anaerobic sludge digester. Only the top of the tank is shown; the remainder is underground. *(b)* Inner workings of a sludge digester. *(c)* Major microbial processes in anaerobic sludge digestion. Methane (CH_4) and carbon dioxide (CO_2) are the major products of anaerobic biodegradation.

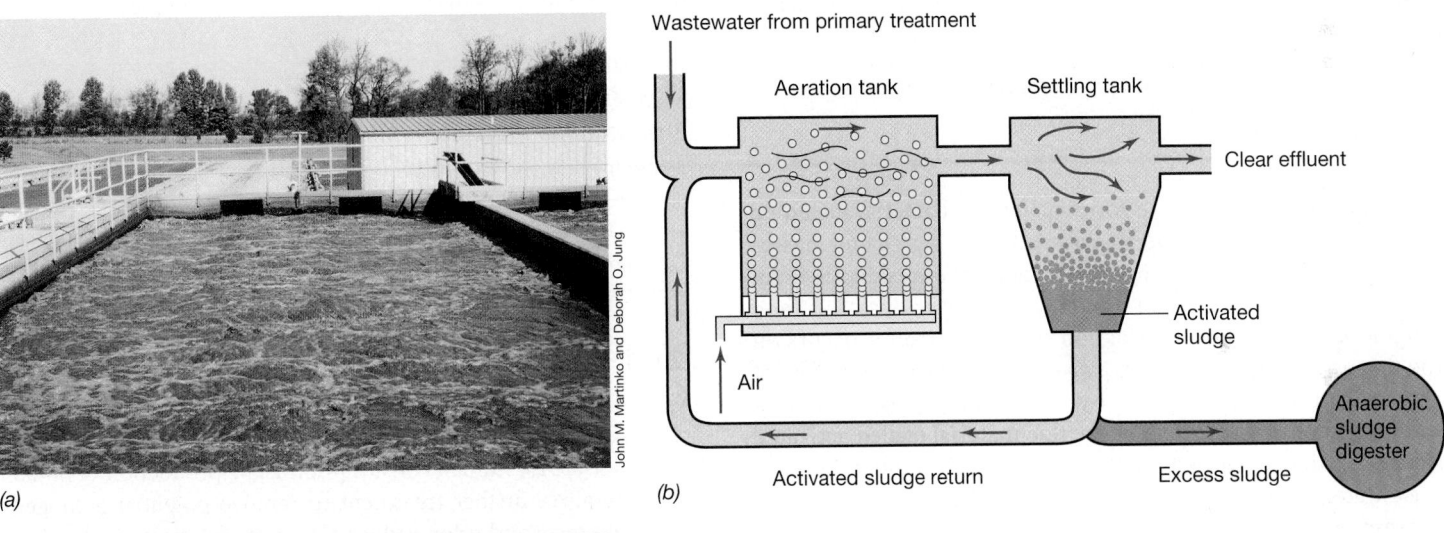

Figure 35.8 Secondary aerobic wastewater treatment processes. A treatment facility for a small city, Carbondale, Illinois, USA. Parts a and b show the activated sludge method. *(a)* Aeration tank of an activated sludge installation in a metropolitan wastewater treatment plant. The tank is 30 m long, 10 m wide, and 5 m deep. *(b)* Wastewater flow through an activated sludge installation. Recirculation of activated sludge to the aeration tank introduces microorganisms responsible for oxidative degradation of the organic components of the wastewater. *(c)* Trickling filter method. The booms rotate, distributing wastewater slowly and evenly on the rock bed. The rocks are 10–15 cm in diameter and the bed is 2 m deep.

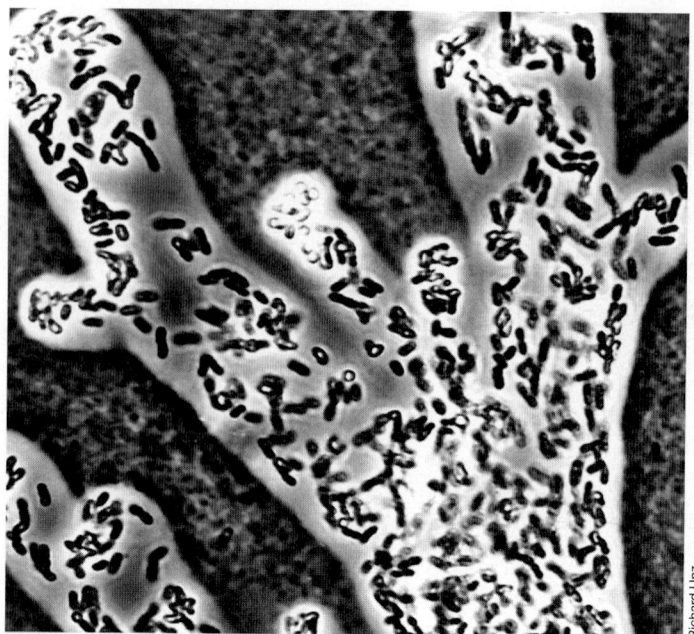

Figure 35.9 A wastewater floc formed by the bacterium *Zoogloea ramigera*. Floc formed in the activated sludge process consists of a large number of small, rod-shaped cells of *Z. ramigera* surrounded by a polysaccharide slime layer, arranged in characteristic fingerlike projections in this negative stain with India ink.

aerated in large tanks. Slime-forming aerobic bacteria, including *Zoogloea ramigera* and others, grow and form aggregated masses called flocs (**Figure 35.9**). The biology of *Zoogloea* is discussed in Section 17.7. Protists, small animals, filamentous bacteria, and fungi attach to the flocs. Oxidation of organic matter occurs on the floc as it is agitated and exposed to air. The aerated effluent containing the flocs is pumped into a holding tank or clarifier where the flocs settle. Some of the floc material (called activated sludge) is then returned to the aerator as inoculum for new wastewater, and the rest is pumped to the anaerobic sludge digester (Figure 35.7) or is removed, dried, and burned, or is used for fertilizer.

Wastewater normally stays in an activated sludge tank for 5–10 hours, a time too short for complete oxidation of all organic matter. However, during this time much of the soluble organic matter is adsorbed to the floc and incorporated by the microbial cells. The BOD of the liquid effluent is considerably reduced (up to 95%) when compared to the incoming wastewater; most of the material with high BOD is now in the settled flocs. The flocs can then be transferred to the anoxic sludge digester for conversion to CO_2 and CH_4.

The *trickling filter* method is also commonly used for secondary aerobic treatment (Figure 35.8c). A trickling filter is a bed of crushed rocks, about 2 m thick. Wastewater is sprayed on top of the rocks and slowly passes through the bed. The organic material in the wastewater adsorbs to the rocks, and microorganisms grow on the large, exposed rock surfaces. The complete mineralization of organic matter to CO_2, ammonia, nitrate, sulfate, and phosphate takes place in the extensive microbial biofilm that develops on the rocks.

Most treatment plants chlorinate the effluent after secondary treatment to further reduce the possibility of biological contamination. The treated effluent can then be discharged into streams or lakes. In the eastern United States, many wastewater treatment facilities use UV radiation to disinfect effluent water. Ozone (O_3), a strong oxidizing agent that is an effective bacteriocide and viricide, is also used for wastewater disinfection in some treatment plants in the United States.

Tertiary Wastewater Treatment

Tertiary wastewater treatment is any physicochemical or biological process employing bioreactors, precipitation, filtration, or chlorination procedures similar to those employed for drinking water purification (Section 35.3). Tertiary treatment sharply reduces levels of inorganic nutrients, especially phosphate, nitrite, and nitrate, from the final effluent.

Wastewater receiving tertiary treatment essentially contains no nutrients and cannot support extensive microbial growth. Tertiary treatment is the most complete method of treating sewage but has not been widely adopted due to the costs associated with such complete nutrient removal.

MiniQuiz

- What is biochemical oxygen demand (BOD)? Why is BOD reduction necessary in wastewater treatment?
- Identify primary, secondary (anoxic and oxic), and tertiary wastewater treatment methods.
- Other than treated water, what are the final products of wastewater treatment? How might these end products be used?

35.3 Drinking Water Purification

Wastewater treated by secondary methods can usually be discharged into rivers and streams. However, such water is not **potable** (safe for human consumption). The production of potable water requires further treatment to remove potential pathogens, eliminate taste and odor, reduce nuisance chemicals such as iron and manganese, and decrease **turbidity**, which is a measure of suspended solids. **Suspended solids** are small particles of solid pollutants that resist separation by ordinary physical means.

Physical and Chemical Purification

A typical city drinking water treatment installation is shown in **Figure 35.10a**. Figure 35.10b shows the process that purifies **raw water** (also called **untreated water**) that flows through the treatment plant. Raw water is first pumped from the source, in this case a river, to a sedimentation basin where anionic **polymers**, alum (aluminum sulfate), and chlorine are added. **Sediment**, including soil, sand, mineral particles, and other large particles, settles out. The sediment-free water is then pumped to a **clarifier** or coagulation basin, which is a large holding tank where **coagulation** takes place. The alum and anionic polymers form large particles from the much smaller suspended solids. After mixing, the particles continue to interact, forming large, aggregated masses, a process called **flocculation**. The large, aggregated

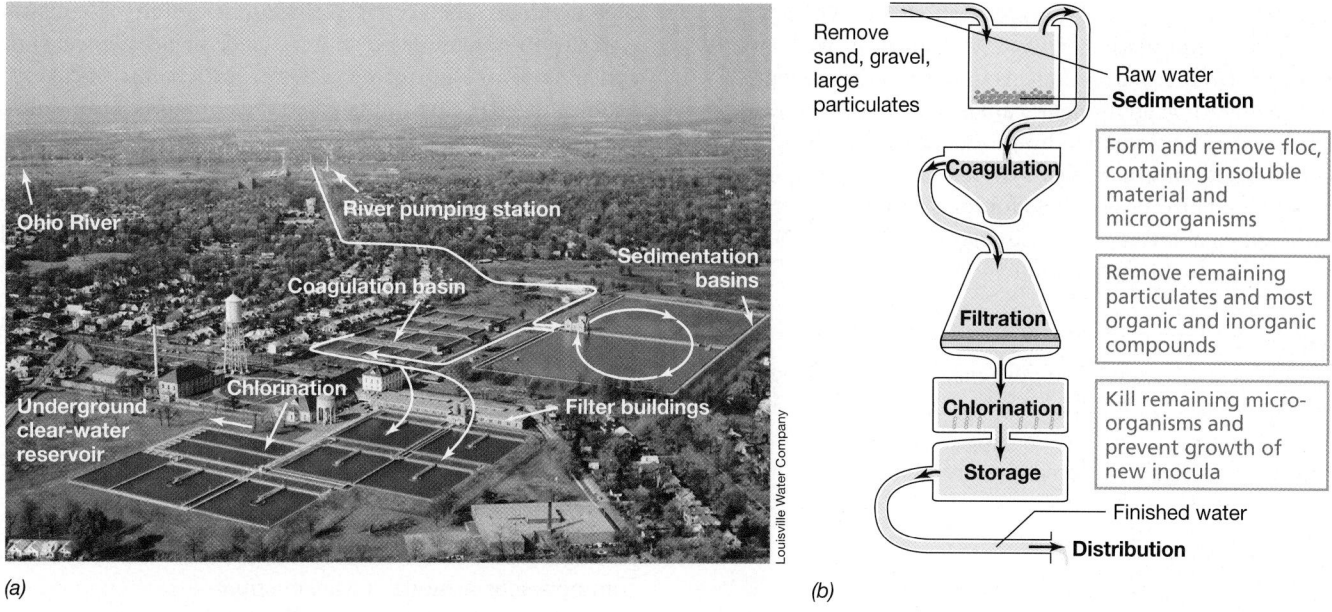

Figure 35.10 Water purification plant. *(a)* Aerial view of a water treatment plant in Louisville, Kentucky, USA. The arrows indicate direction of flow of water through the plant. *(b)* Schematic overview of a typical community water purification system.

particles (floc) settle out by gravity, trapping microorganisms and adsorbing suspended organic matter and sediment.

After coagulation, flocculation, and sedimentation, the clarified water undergoes **filtration** through a series of filters designed to remove organic and inorganic solutes, as well as remaining suspended particles and microorganisms. The filters typically consist of thick layers of sand, activated charcoal, and ion exchangers. When combined with previous purification steps, the filtered water is free of particulate matter, most organic and inorganic chemicals, and nearly all microorganisms.

Disinfection

Clarified, filtered water must be disinfected before it is released to the supply system as pure, potable **finished water**. **Primary disinfection** is the introduction of sufficient disinfectant into clarified, filtered water to kill existing microorganisms and inhibit further microbial growth. Chlorination is the most common method of primary disinfection. In sufficient doses, chlorine kills most microorganisms within 30 minutes. A few pathogenic protists such as *Cryptosporidium*, however, are not easily killed by chlorine treatment (Section 35.6). In addition to killing microorganisms, chlorine oxidizes and effectively neutralizes many organic compounds. Since most taste- and odor-producing chemicals are organic compounds, chlorination improves water taste and smell. Chlorine is added to water either from a concentrated solution of sodium hypochlorite or calcium hypochlorite, or as chlorine gas from pressurized tanks. Chlorine gas is commonly used in large water treatment plants because it is most amenable to automatic control. When dissolved in water, chlorine gas is extremely volatile and dissipates within hours from treated water. To maintain adequate levels of chlorine for primary disinfection, many municipal water treatment plants introduce ammonia gas with

the chlorine to form the stable, nonvolatile chlorine-containing compound **chloramine**, $HOCl + NH_3 \rightarrow NH_2Cl + H_2O$.

Chlorine is consumed when it reacts with organic materials. Therefore, sufficient quantities of chlorine must be added to finished water containing organic materials so that a small amount, called the *chlorine residual*, remains. The chlorine residual reacts to kill any remaining microorganisms. The water plant operator performs chlorine analyses on the treated water to determine the level of chlorine to be added for **secondary disinfection**, the maintenance of sufficient chlorine residual or other disinfectant residual in the water distribution system to inhibit microbial growth. A chlorine residual level of 0.2–0.6 mg/liter is suitable for most water supplies. After chlorine treatment, the now potable water is pumped to storage tanks from which it flows by gravity or pumps through a **distribution system** of storage tanks and supply lines to the consumer. Residual chlorine levels inhibit growth of bacteria in the finished water prior to reaching the consumer. It does not protect against catastrophic system failures such as a broken pipe in the distribution system. To maintain residual chlorine levels throughout the distribution system, most municipal water treatment plants also introduce ammonia gas with the chlorine to form chloramine.

UV radiation is also used as an effective means of disinfection. As we discussed in Section 26.2, UV radiation is used to treat secondarily treated effluent from water treatment plants. In Europe, UV irradiation is commonly used for drinking water applications, and it is increasingly used in the United States. For disinfection, UV light is generated from mercury vapor lamps. Their major energy output is at 253.7 nm, a wavelength that is bacteriocidal and may also kill cysts and oocysts of protists such as *Giardia* and *Cryptosporidium*, important eukaryotic pathogens in water (Section 35.6). Viruses, however, are more resistant.

UNIT 12

UV radiation has several advantages over chemical disinfection procedures like chlorination. First, UV irradiation is a physical process that introduces no chemicals into the water. Second, UV radiation–generating equipment can be used in existing flow systems. Third, no disinfection by-products are formed with UV disinfection. Especially in smaller systems where finished water is not pumped long distances or held for long periods (reducing the need for residual chlorine), UV disinfection may be preferable to reduce dependence on chlorination.

MiniQuiz

- Trace the treatment of water through a drinking water treatment plant, from the inlet to the final distribution point (faucet).
- What specific purposes do sedimentation, coagulation, filtration, and disinfection accomplish in the drinking water treatment process?

Waterborne Microbial Diseases

Common-source infectious diseases are caused by microbial contamination of materials shared by a large number of individuals. The most important common source of infectious disease is contaminated water; the failure of a single step in the drinking water purification process may result in the exposure of thousands or even millions of individuals to an infectious agent.

Common-source waterborne diseases are significant sources of morbidity and mortality, especially in developing countries. Even in developed countries, breakdowns in water treatment plants or the lack of access to clean water in times of emergency can contribute to the development of a waterborne disease outbreak.

Bacteria, viruses, and protists cause waterborne infectious diseases. Waterborne diseases begin as infections. Contaminated water may cause infection even if only a small number of microorganisms are present. Whether or not exposure to a pathogen causes disease is a function of the virulence of the pathogen and the general ability of the host to resist infection.

35.4 Sources of Waterborne Infection

Human pathogens can be transmitted through untreated or improperly treated water used for drinking and cooking. Another common source of disease transmission is through pathogen-contaminated water used for swimming and bathing.

Potable Water

Because everyone consumes water through drinking and cooking, water is a common source of pathogen dissemination and has a very high potential for the catastrophic spread of epidemic disease. As we have already discussed, water supplies in developed countries usually meet rigid quality standards, limiting the spread of waterborne diseases. Waterborne disease outbreaks, however, occasionally occur in developing countries due to lapses in water quality. Isolated outbreaks affecting low numbers

of individuals also occur from consumption of contaminated water from nonregulated sources (such as private wells) or from consumption of untreated water from streams or lakes. These sources may be contaminated by fecal material from humans or animals.

Microorganisms transmitted in drinking water generally grow in the intestines and leave the body in feces, which may in turn pollute water. If a new host consumes the water, the pathogen may colonize the host's intestine and cause disease. From 1974 to 2006 in the United States, 729 drinking water–associated disease outbreaks occurred—an average of about 23 outbreaks per year (**Figure 35.11a**). Bacterial, viral, and protist pathogens are occasionally transmitted in drinking water (**Table 35.1**). We discuss *Giardia* and *Cryptosporidium* in Section 35.6 and *Legionella* in Section 35.7, but several bacterial pathogens including pathogenic strains of *Escherichia coli* as well as gastrointestinal virus infections are discussed in Chapter 36, where we introduce these pathogens as common agents of foodborne infections, another common-source mode of transmission.

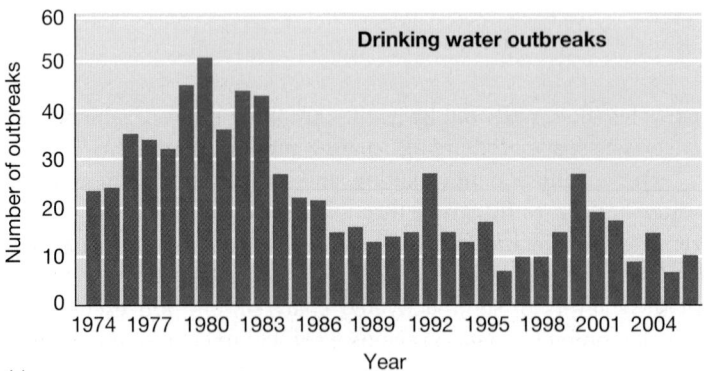

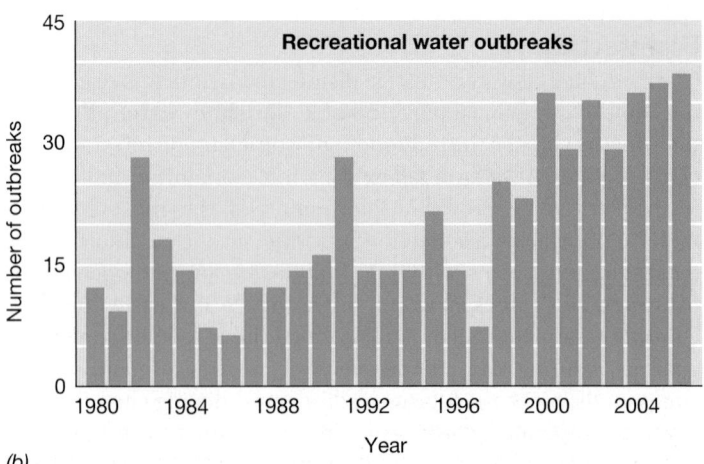

Figure 35.11 Waterborne disease outbreaks. Data were provided by the Centers for Disease Control and Prevention, Atlanta, Georgia, USA. *(a)* Reported drinking water disease outbreaks from 1974 to 2006. Of 729 outbreaks, about 90% were due to biological agents (bacteria, viruses, and protists). *(b)* Reported recreational water outbreaks from 1980 to 2006. Of 544 total outbreaks, almost all were due to biological agents.

Table 35.1 *Reported infectious disease outbreaks associated with drinking water in the United States, 2005–2006*[a]

Disease	Agent	Outbreaks	Cases
Cryptosporidiosis	*Cryptosporidium*	1	10
Giardiasis	*Giardia*	1	41
Legionellosis	*Legionella*	10	43
Acute gastrointestinal illness	*Escherichia coli* and *Campylobacter jejuni*	1	60
	Campylobacter jejuni	1	32
	Norovirus and *Campylobacter jejuni*	1	139
	Norovirus	2	196
	Hepatitis A	1	16
	Unknown[b]	2	75

[a]Data provided by the Centers for Disease Control and Prevention, Atlanta, Georgia. There were 20 outbreaks and 612 cases of infectious disease due to drinking water contamination by infectious agents. Four deaths occurred, all due to legionellosis. Regulated community-owned water systems were implicated in nine outbreaks. Seven outbreaks were due to noncommunity water systems such as those in some schools, churches, and lodges. Individual water supply systems such as private wells accounted for two outbreaks. Two outbreak sources could not be determined. Most outbreaks involving *Legionella* could not be attributed to water system or source deficiencies, but were most likely due to building and point-of-use factors, such as those that produced aerosols.
[b]Unknown infections were consistent with norovirus infection, but were not confirmed.

Recreational Water

Recreational waters include freshwater recreational areas such as ponds, streams, and lakes, as well as public swimming and wading pools. Recreational waters can also be sources of waterborne disease, and historically cause disease outbreaks at levels roughly comparable to those caused by drinking water (Figure 35.11*b*).

The operation of public swimming and wading pools is regulated by state and local health departments. The United States EPA establishes limits for bacteria in recreational freshwaters (monthly geometric mean for all samples of <33/100 ml for enterococci or <126/100 ml for *E. coli*) and marine waters (<35/100 ml for enterococci). Local and state governments have the authority to set standards above or below these guidelines, and many states use a single-sample maximum as well as the geometric mean for setting standards and defining levels of contamination that constitute violations. For example, the state of Indiana standard is 125 *E. coli* cells per 100 ml as a geometric mean, with a single-sample maximum of 235/100 ml. Thus, waters that exceed 235 *E. coli*, even if their geometric mean count was not greater than 125, would be in violation of Indiana's water standards. Private swimming pools, spas, and hot tubs are unregulated and are occasional sources of outbreaks of waterborne diseases.

Over a 27-year period (1980–2006), 544 waterborne disease outbreaks were from recreational waters in the United States, or about 20 outbreaks per year (Figure 35.11*b*). **Table 35.2** categorizes recreational water outbreaks according to the infectious diseases occurring in recent years.

Waterborne Infections in Developing Countries

Worldwide, waterborne infections are a much larger problem than in the United States and other developed countries. Developing countries often have inadequate water and sewage treatment facilities, and access to safe, potable water is limited. As a result, diseases such as cholera (Section 35.5), typhoid fever, and amebiasis (Section 35.8) are important public health problems in the developing world.

MiniQuiz

- Identify the bacteria, protists, and viruses commonly responsible for disease outbreaks due to drinking water contamination.
- Identify the bacteria, protists, and viruses commonly responsible for disease outbreaks due to recreational water contamination.

35.5 Cholera

Cholera is a severe diarrheal disease that is now largely restricted to the developing world. Cholera is an example of a major waterborne disease that can be controlled by application of appropriate public health measures for water treatment.

Biology and Epidemiology

Cholera is typically caused by ingestion of contaminated water containing *Vibrio cholerae*, a gram-negative, curved rod–shaped

Table 35.2 *Reported infectious disease outbreaks associated with recreational water in the United States, 2005–2006*[a]

Agent[b]	Outbreaks	Cases
Bacteria		
Campylobacter jejuni	1	6
Escherichia coli	3	10
Legionella	8	124
Leptospira	2	46
Pseudomonas aeruginosa	9	101
Shigella sonnei	4	41
Parasites		
Cryptosporidium	31	3751
Giardia intestinalis	1	11
Cryptosporidium and *Giardia*	1	55
Naegleria fowleri	1	2
Schistosoma	2	4
Virus		
Norovirus	5	99

[a]Data provided by the Centers for Disease Control and Prevention, Atlanta, Georgia. In all, 68 outbreaks occurred over 2 years.
[b]*Campylobacter jejuni*, *Escherichia coli*, *Shigella sonnei*, *Cryptosporidium*, *Giardia*, and norovirus outbreaks cause gastroenteritis. *Legionella* causes acute respiratory disease. *Leptospira* causes systemic infections and aseptic meningitis. *Pseudomonas aeruginosa* causes dermatitis. The amoeba *Naegleria fowleri* causes meningoencephalitis; all cases were fatal. *Schistosoma*, a helminth parasite, causes schistosomiasis, a disease characterized chiefly by parasitic infestations of venous vessels in the intestines and liver.

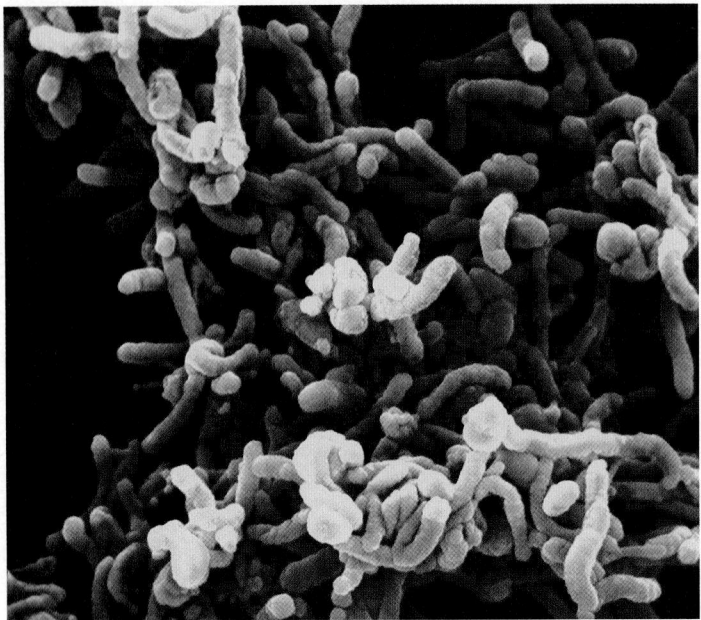

Figure 35.12 Cells of *Vibrio cholerae*. This colorized scanning electron micrograph shows a rod to curved rod morphology. The organism is about 0.3 μm in diameter and up to 2 μm in length.

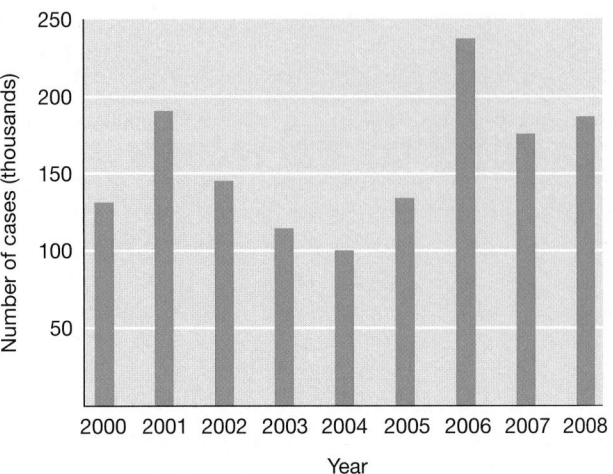

Figure 35.13 **Cholera cases.** The reported cholera cases from 2000 to 2008 show a generally increasing trend. Up to 90–95% of cholera cases are unreported. Over 95% of all reported cases occur in Africa. Data were provided by the World Health Organization.

Proteobacterium (**Figure 35.12**; ↵ Section 17.12). As with many waterborne diseases, cholera can also be associated with food consumption. For example, in the Americas, consumption of raw shellfish and raw vegetables has been associated with cholera. Presumably, vegetables washed in contaminated water and shellfish beds contaminated by untreated sewage transmitted the disease.

Since 1817, cholera has swept the world in seven major pandemics. Two distinct pandemic strains of *V. cholerae* are recognized, known as the *classic* and the *El Tor* biotypes. The *V. cholerae* O1 El Tor biotype started the seventh pandemic in Indonesia in 1961, and its spread continues to the present. This pandemic has caused over 5 million cases of cholera and more than 250,000 deaths and continues to be a major cause of morbidity and mortality, especially in developing countries; as is typical for infectious diseases, the highest prevalence of cholera is in developing countries, especially in Africa. In 1992, a genetic variant known as *V. cholerae* O139 Bengal arose in Bangladesh and caused an extensive epidemic. *V. cholerae* O139 Bengal has continued to spread since 1992, causing several major epidemics, and may be the agent of an eighth pandemic.

Cholera is endemic in Africa, Southeast Asia, the Indian subcontinent, and Central and South America. Epidemic cholera occurs frequently in areas where sewage treatment is either inadequate or absent. Worldwide, there were 190,130 reported cases and 5143 deaths reported in 2008, with over 98% of all reported cases occurring in Africa. About 100,000 cases or more have been reported annually since 2000, with a low of 95,560 cases in 2004, and a high of 236,896 cases in 2006 (**Figure 35.13**). The World Health Organization estimates that only 5–10% of cholera cases are reported, so the total incidence of cholera exceeds 1 million cases per year. Even in developed countries, the disease

is a threat. A handful of cases are reported each year in the United States, rarely caused by drinking water. Many recent cases are imported, often in food. A few cases are possibly from endemic sources; raw shellfish seems to be the most common vehicle, presumably because *V. cholerae* may be free-living in coastal waters in endemic areas, where the pathogen adheres to the marine microflora ingested by the shellfish (**Figure 35.14**).

Pathogenesis

The ingestion of 10^8–10^9 cholera vibrios is generally required to cause disease. The ingested *V. cholerae* cells attach to epithelial cells in the small intestine where they grow and release cholera toxin, a potent enterotoxin (↵ Figure 27.24). Studies in human volunteers have shown that stomach acidity is responsible for the large inoculum needed to initiate cholera; human volunteers

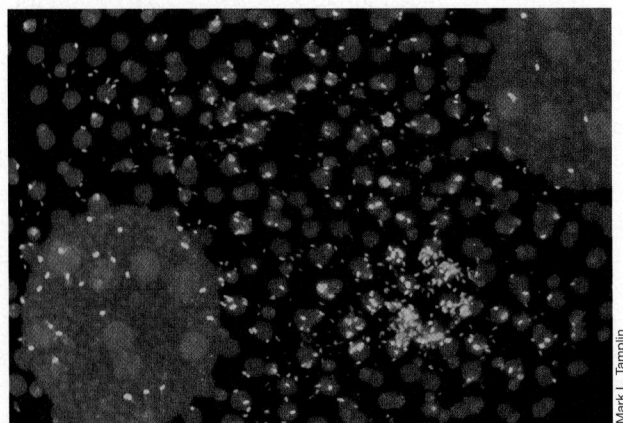

Figure 35.14 Cells of *Vibrio cholerae* attached to the surface of *Volvox*, a freshwater alga. The sample was from a cholera-endemic area in Bangladesh. The *V. cholerae* cells are stained green by a monoclonal antibody to bacterial cell surface proteins. The red color is due to the fluorescence of chlorophyll *a* in the algae.

given bicarbonate to neutralize gastric acidity developed cholera when given as few as 10^4 cells. Even lower cell numbers can initiate infection if *V. cholerae* is ingested with food, presumably because the food protects the vibrios from destruction by stomach acidity.

Cholera enterotoxin causes severe diarrhea that can result in dehydration and death unless the patient is given fluid and electrolyte therapy. The enterotoxin causes fluid losses of up to 20 liters (20 kg or 44 lb) per day. The mortality rate from *untreated* cholera is typically 25–50% and can be much higher under conditions of severe crowding and malnutrition.

Diagnosis and Prevention of Cholera

Cholera is diagnosed by the presence of the gram-negative, comma-shaped *V. cholerae* bacilli in the "rice water" stools (nearly liquid feces) of patients with severe diarrhea (**Figure 35.15**).

Immunization is not normally recommended for cholera prevention, but a whole-cell oral vaccine directed against the El Tor biotype is currently available for use in high-risk situations, such as after natural disasters that compromise water treatment and purification systems. The vaccine, as well as natural infection, provides effective but short-lived immunity. No vaccine protects against the new *V. cholerae* O139 Bengal serotype. Public health measures such as adequate sewage treatment and a reliable source of safe drinking water are the most important measures for preventing cholera. *V. cholerae* is eliminated from wastewater during proper sewage treatment and drinking water purification procedures. For individuals traveling in cholera-endemic areas, attention to personal hygiene and avoidance of untreated water or ice, raw food, and raw or undercooked fish or shellfish offer protection against contracting cholera.

Treatment of Cholera

Cholera treatment is simple, effective, and inexpensive. Intravenous or oral liquid and electrolyte replacement therapy [20 g of glucose, 4.2 g of sodium chloride (NaCl), 4.0 g of sodium bicarbonate (NaHCO$_3$), and 1.8 g of potassium chloride (KCl) dissolved in 1 liter of water] is the most effective means of cholera treatment. Oral treatment is preferred because no special equipment or sterile precautions are necessary. Effective fluid and electrolyte replacement reduces mortality to about 1%. Streptomycin or tetracycline may shorten the course of infection and the shedding of viable cells, but antibiotics are of little benefit without simultaneous fluid and electrolyte replacement.

MiniQuiz
- Identify the most likely means for acquiring cholera.
- Identify specific, effective methods for preventing and treating cholera.

35.6 Giardiasis and Cryptosporidiosis

Giardiasis and cryptosporidiosis are diseases caused by the protists *Giardia intestinalis* and *Cryptosporidium parvum*, respectively. These organisms continue to be problematic even in well-regulated water supplies because they are found in nearly all surface waters and are resistant to chlorine disinfection.

Giardiasis

Giardia intestinalis, also called *Giardia lamblia*, is a flagellated protist (𝒞𝒫 Section 20.7) that is usually transmitted to humans in fecally contaminated water, although foodborne and sexual transmission of giardiasis have also been documented. Giardiasis is an acute gastroenteritis caused by this organism. The protist cells, called *trophozoites* (**Figure 35.16a**), produce a resting stage called a **cyst** (Figure 35.16b). The cyst has a thick protective wall that allows the pathogen to resist drying and chemical disinfection. After a person ingests the cysts in contaminated water, the cysts germinate, attach to the intestinal wall, and cause the symptoms of giardiasis: an explosive, foul-smelling, watery diarrhea, intestinal cramps, flatulence, nausea, weight loss, and malaise. Symptoms may be acute or chronic. The foul-smelling diarrhea and the absence of blood or mucus in the stool distinguish giardiasis from bacterial or viral diarrheas. Many infected individuals exhibit no symptoms but act as carriers; *G. intestinalis* can establish itself in a stable, symptom-free relationship with its host.

G. intestinalis was the infectious agent in 1 of the 20 recent drinking water infectious disease outbreaks in the United States (Table 35.1). The thick-walled cysts are resistant to chlorine, and

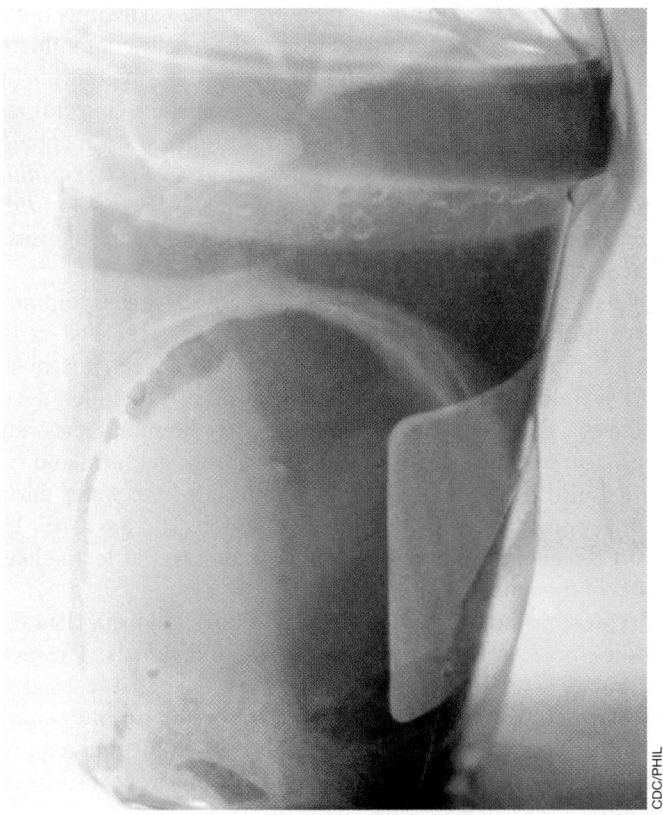

CDC/PHIL

Figure 35.15 A fecal sample from a cholera patient. The "rice-water" stool is nearly liquid. The solid material that has settled in a bottom layer is mucus. The stool from cholera patients is essentially isotonic with blood, containing high amounts of sodium (Na$^+$), potassium (K$^+$), and bicarbonate (HCO$_3^-$) ions, as well as large numbers of *Vibrio cholerae* cells (𝒞𝒫 Figure 27.24).

UNIT 12

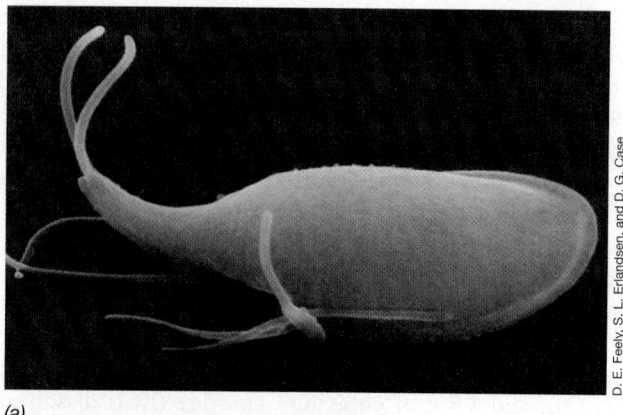

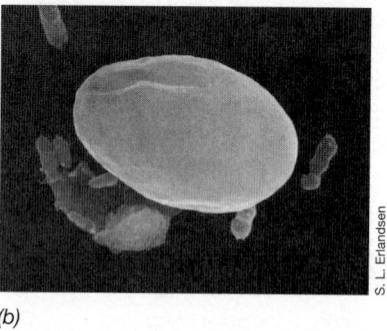

Figure 35.16 The parasite *Giardia*. Scanning electron micrographs. *(a)* A motile trophozoite. The trophozoite is about 15 μm in length. *(b)* A giardial cyst. The cyst is about 11 μm in length.

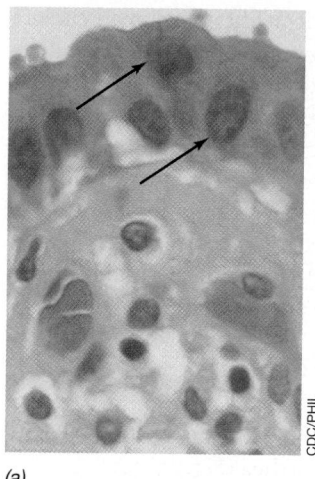

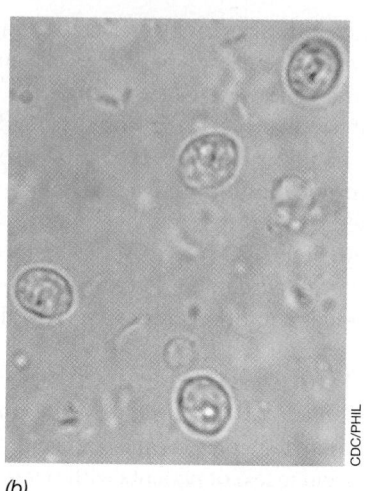

Figure 35.17 *Cryptosporidium*. *(a)* The arrows point to two of the many intracellular trophozoites embedded in human gastrointestinal epithelium. The trophozoites are 2–5 μm in diameter. *(b)* The thick-walled oocysts are 3–5 μm in diameter in this fecal sample.

most outbreaks have been associated with water systems that used only chlorination as a means of water purification. Water subjected to proper clarification and filtration followed by chlorination or other disinfection (Section 35.3) is generally free of *Giardia* cysts.

Giardiasis can also be contracted from ingestion of water from infected swimming pools or lakes (Table 35.2). *Giardia* cysts have been found in 97% of surface water sources (lakes, ponds, and streams) in the United States. Isolated cases of giardiasis have been associated with untreated drinking water in wilderness areas. Beavers and muskrats are frequent carriers of *Giardia* and may transmit cells or cysts to water supplies, making the water a possible source of human infection. As a safety precaution, water consumed from rivers and streams, for example, during a camping or hiking trip, should be filtered and treated with iodine or chlorine, or filtered and boiled. Boiling is the preferred method to ensure that water is free of pathogens.

Laboratory diagnostic methods include the demonstration of *Giardia* cysts in the stool or the demonstration of *Giardia* antigens in the stool using a direct EIA (enzyme immunosorbent assay). The drugs quinacrine, furazolidone, and metronidazole are useful for treating acute giardiasis.

Cryptosporidiosis

The protist *Cryptosporidium parvum* lives as a parasite in warm-blooded animals. The protists are small (2–5 μm), round cells that invade and grow intracellularly in mucosal epithelial cells of the stomach and intestine (**Figure 35.17a**). The protist produces thick-walled, chlorine-resistant, infective cells called *oocysts*, which are shed into water in high numbers in the feces of infected warm-blooded animals (Figure 35.17b). The infection is passed on when other animals consume the fecally contaminated water. *Cryptosporidium* cysts are highly resistant to chlorine (up to 14 times more resistant than chlorine-resistant *Giardia*) and UV radiation disinfection. Because of this property, sedimentation and filtration methods must be used to remove *Cryptosporidium* from water supplies. From 2005 through 2006, *Cryptosporidium* was responsible for 31 of the 68 recreational waterborne disease outbreaks (Table 35.2).

C. parvum was responsible for the largest single common-source outbreak of a waterborne disease ever recorded in the United States. In the spring of 1993 in Milwaukee, Wisconsin, USA, over 403,000 people in the population of 1.6 million developed a diarrheal illness that was traced to the municipal water supply. Spring rains and runoff from surrounding farmland had drained into Lake Michigan and overburdened the water purification system, leading to contamination by *C. parvum*. The protist is a significant intestinal parasite in dairy cattle, the likely source of the outbreak.

Cryptosporidiosis is usually a self-limiting mild diarrhea that subsides in 2 weeks or less in normal individuals. However, individuals with impaired immunity, such as that caused by HIV/AIDS, or the very young or old can develop serious complications. In the Milwaukee outbreak, about 4400 people required hospital care, and 50–100 died of complications from the disease, including severe dehydration.

The Milwaukee outbreak highlights the vulnerability of water purification systems, the need for constant water monitoring and surveillance, and the consequences of the failure of a large water supply system. In addition to the toll of human morbidity and mortality, the epidemic cost an estimated $96 million in medical costs and lost productivity.

Laboratory diagnostic methods for cryptosporidiosis include the demonstration of *Cryptosporidium* oocysts in the stool. Treatment is unnecessary for those with uncompromised immunity. For individuals undergoing immunosuppressive therapy (for example, prednisone), discontinuation of immunosuppressive drugs is recommended. Immunocompromised individuals should be given supportive therapy such as intravenous fluids and electrolytes.

MiniQuiz

• Explain the importance of cysts in the survival, infectivity, and chlorine resistance of both *Giardia* and *Cryptosporidium*.

• Why are protists often associated with waterborne diseases, even in developed countries? Outline steps to reduce their impact.

35.7 Legionellosis (Legionnaires' Disease)

Legionella pneumophila, the bacterium that causes legionellosis, is an important waterborne pathogen normally transmitted in aerosols rather than through drinking or recreational waters.

Biology and Epidemiology

Legionella pneumophila was first discovered as the pathogen that caused an outbreak of pneumonia during an American Legion convention in Philadelphia, Pennsylvania, USA, in the summer of 1976. *L. pneumophila* is a thin, gram-negative, obligately aerobic rod (**Figure 35.18**) with complex nutritional requirements, including an unusually high iron requirement. The organism can be isolated from terrestrial and aquatic habitats as well as from legionellosis patients.

L. pneumophila is present in small numbers in lakes, streams, and soil. It is relatively resistant to heating and chlorination, so it can spread through water distribution systems. It is commonly found in large numbers in cooling towers and evaporative condensers of large air conditioning systems. The pathogen grows in the water and is disseminated in humidified aerosols. Human

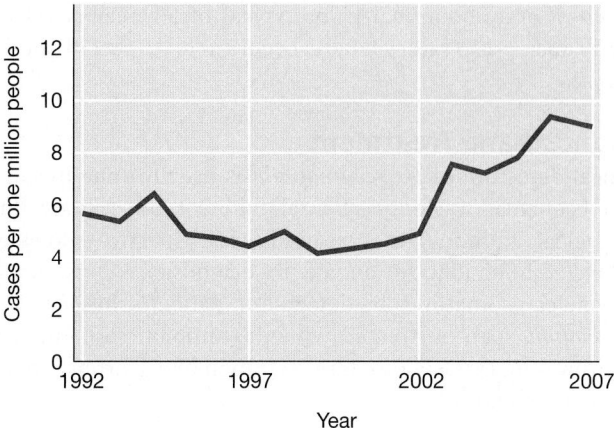

Figure 35.19 Incidence of Legionnaire's disease in the United States. In 2007, there were 2716 reported cases. Data are from the Centers for Disease Control and Prevention, Atlanta, Georgia.

infection is by way of airborne droplets, but the infection is not spread from person to person. Further evidence for this is the fact that annual outbreaks of legionellosis tend to peak in mid-to late summer months when air conditioners are extensively used.

L. pneumophila has also been found in hot water tanks and whirlpool spas, where it can grow to high numbers in warm (35–45°C), stagnant water. Epidemiological studies indicate that *L. pneumophila* infections occur at all times of the year, primarily as a result of aerosols generated by heating/cooling systems and common practices such as showering or bathing. Overall, the incidence of reported cases of legionellosis had been about 4–6 cases per million in the United States, but in the last several years the incidence has risen to nearly 8 cases per million. In 2007, there were 2716 reported cases (**Figure 35.19**). The increase in reported cases may be a result of an actual increase in infections or an increase in recognition and reporting; formerly, up to 90% of actual cases were probably not diagnosed or properly reported. Prevention of legionellosis can be accomplished by improving the maintenance and design of water-dependent cooling and heating systems and water delivery systems. The pathogen can be eliminated from water supplies by hyperchlorination or by heating water to greater than 63°C.

Pathogenesis

In the body, *L. pneumophila* invades and grows in alveolar macrophages and monocytes as an intracellular parasite. Infections are often asymptomatic or produce a mild cough, sore throat, mild headache, and fever. These mild, self-limiting cases, called *Pontiac fever*, are generally not treated and resolve in 2–5 days. Elderly individuals whose resistance has been previously compromised, however, often acquire more serious infections resulting in pneumonia. Certain serotypes of *L. pneumophila* (more than 10 are known) are strongly associated with the pneumonic form of the infection. Prior to the onset of pneumonia, intestinal disorders are common, followed by high fever, chills, and muscle aches. These symptoms precede the dry cough

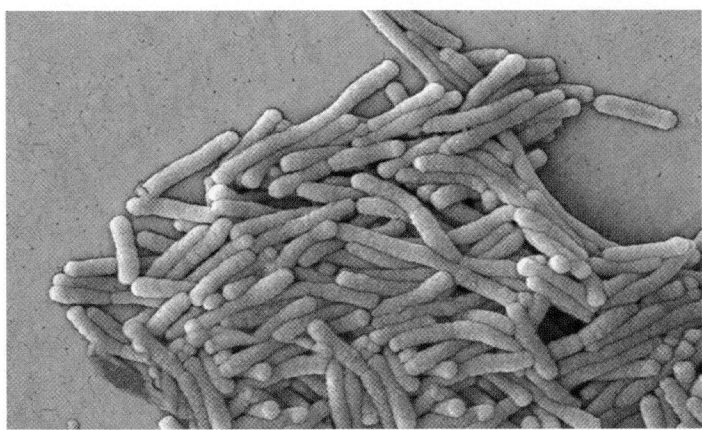

Figure 35.18 *Legionella pneumophila*. Colorized scanning electron micrograph of *L. pneumophila* cells. Cells are 0.3–0.6 μm in diameter and up to 2 μm in length.

and chest and abdominal pains typical of legionellosis. Death, usually due to respiratory failure, occurs in up to 10% of pneumonia cases.

Diagnosis and Treatment

Clinical detection of *L. pneumophila* is usually done by culture from bronchial washings, pleural fluid, or other body fluids. Serological (antibody) tests are used as retrospective evidence for *Legionella* infection. As an aid in diagnosis, *L. pneumophila* antigens can sometimes be detected in patient urine. *Legionella pneumophila* can be treated with the antibiotics rifampin and erythromycin. Intravenous administration of erythromycin is the treatment of choice.

MiniQuiz
- Indicate the source of *Legionella pneumophila*.
- Identify specific measures for control of *Legionella pneumophila*.

35.8 Typhoid Fever and Other Waterborne Diseases

Various bacteria, viruses, and protists can transmit common-source waterborne diseases. These diseases are a significant source of morbidity, especially in developing countries.

Typhoid Fever

On a global scale, probably the most important pathogenic bacteria transmitted by the water route are *Salmonella enterica* serovar Typhi, the organism causing typhoid fever, and *Vibrio cholerae*, the organism causing cholera, which we discussed previously (Section 35.5).

Although *S. enterica* ser. Typhi may also be transmitted by contaminated food (Section 36.8) and by direct contact from infected individuals, the most common and serious means of transmission worldwide is through water. Typhoid fever has been virtually eliminated in developed countries, primarily due to effective water treatment procedures. In the United States, there are fewer than 400 cases in most years, but, as described in Section 35.1, typhoid fever was a major public health threat before drinking water was routinely filtered and chlorinated (Figure 35.4). However, breakdown of water treatment methods, contamination of water during floods, earthquakes, and other disasters, or cross-contamination of water supply pipes from leaking sewer lines can propagate epidemics of typhoid fever, even in developed countries.

Viruses

Viruses can also be transmitted in water and cause human disease. Quite commonly, enteroviruses such as poliovirus, norovirus, and hepatitis A virus are shed into the water in fecal material. The most serious of these is poliovirus, but wild poliovirus has been eliminated from the Western Hemisphere and is endemic only in Nigeria, Afghanistan, Pakistan, and India. Although viruses can survive in water for relatively long periods, they are inactivated by disinfection with agents such as chlorine.

Amebiasis

Certain amoebae inhabit the tissues of humans and other vertebrates, usually in the oral cavity or intestinal tract, and some of these are pathogenic. We discussed the general properties of amoeboid protists in Section 20.12.

Worldwide, *Entamoeba histolytica* is a common pathogenic protist transmitted to humans, primarily by contaminated water and occasionally through contaminated food. *E. histolytica* is an anaerobic amoeba; the trophozoites lack mitochondria (**Figure 35.20**). Like *Giardia*, the trophozoites of *E. histolytica* produce cysts. Cysts ingested by humans germinate in the intestine, where amoebic cells grow both on and in intestinal mucosal cells. Many infections are asymptomatic, but continued growth may lead to invasion and ulceration of the intestinal mucosa, causing diarrhea and severe intestinal cramps. With further growth the amoebae can invade the intestinal wall, a condition called *dysentery*, characterized by intestinal inflammation, fever, and the passage of intestinal exudates, including blood and mucus.

If not treated, invasive trophozoites of *E. histolytica* can invade the liver and occasionally the lung and brain. Growth in these tissues can cause severe abscesses and death. Worldwide, up to 100,000 individuals die each year from invasive amebic dysentery. The disease is extremely common in tropical and subtropical countries worldwide, with at least 50 million people developing symptomatic diarrhea annually and up to 10-fold more having asymptomatic disease. In the United States, several hundred cases occur each year, mostly near international borders in the Southwest.

E. histolytica amebiasis can be treated with the drugs dehydroemetine for invasive disease and diloxanide furoate for certain asymptomatic cases, as in immunocompromised individuals, but amoebicidal drugs are not universally effective. Spontaneous cures do occur, suggesting that the host immune system plays a role in ending the infection. However, protective immunity is not an outcome of primary infection, and reinfection is common. The disease is kept at very low incidence in regions that practice adequate sewage treatment. Amoebic infestation due to exposure

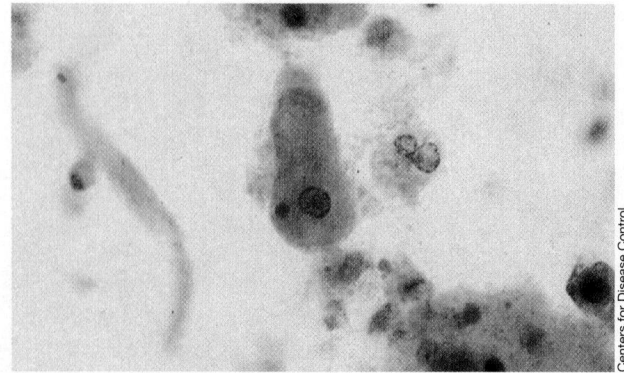

Figure 35.20 The trophozoite of *Entamoeba histolytica*, the amoeba that causes amebiasis. Note the discrete, darkly stained nucleus. The small red structures are red blood cells. The trophozoites range from 12 to 60 μm in length.

to improperly treated sewage and the use of untreated surface waters for drinking purposes are the usual causes of amebiasis.

Demonstration of *E. histolytica* cysts in the stool, trophozoites in tissue, or the positive results for antibodies to *E. histolytica* in the blood from an EIA (enzyme immunoassay) are used for the laboratory diagnosis of amebiasis.

Naegleria fowleri can also cause amebiasis, but in a very different form. *N. fowleri* is a free-living amoeba found in soil and in water runoff. *N. fowleri* infections usually result from swimming or bathing in warm, soil-contaminated water sources such as hot springs or lakes and streams in the summer. This free-living amoeba enters the body through the nose and burrows directly into the brain. Here, the organism propagates, causing extensive hemorrhage and brain damage (**Figure 35.21**). This condition is called **meningoencephalitis**. Death usually results within a week. From 1999 to 2003, there were 12 outbreaks, each a single individual who was infected by swimming or wading in a lake, pond, or stream in summer. All cases resulted in death.

Prevention can be accomplished by avoiding swimming in shallow, warm freshwater, such as farm ponds and shallow lakes and rivers in summer. Swimmers are advised to avoid stirring up bottom sediments, the natural habitat of the pathogen. Diagnosis of *N. fowleri* infection requires observation of the amoebae in the cerebrospinal fluid. If a definitive diagnosis can be done quickly, the drug amphotericin B is used to treat infections.

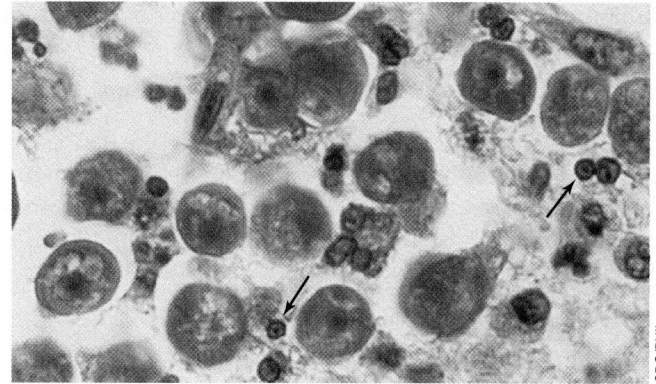

Figure 35.21 Trophozoites of *Naegleria fowleri* in brain tissue. This amoeba causes meningoencephalitis. Oval to round and amoeboid (irregularly shaped) trophozoites (arrows) are present as dark-stained structures with densely stained nuclei. There is extensive destruction of the surrounding brain tissue. Individual trophozoites are 10–35 μm long.

MiniQuiz

• Explain the impact of effective water hygiene on the spread of human diseases such as typhoid fever and polio spread by fecal contamination of water supplies.

• Describe public health measures that could be used to eliminate or reduce the number of cases of amebiasis due to *Entamoeba histolytica* or meningoencephalitis due to *Naegleria fowleri*.

Big Ideas

35.1
Drinking water quality is determined by counting coliform and fecal coliform bacteria using standardized techniques, a reliable indicator of fecal contamination in water supplies. Filtration and chlorination of water significantly decreases microbial numbers. Water purification methods are a major factor in improving public health in developed countries in the last century.

35.2
Sewage and industrial wastewater treatment reduces the BOD (biochemical oxygen demand) of wastewater. Primary, secondary, and tertiary wastewater treatment uses physical, biological, and physicochemical processes. After secondary or tertiary treatment, effluent water has significantly reduced BOD and is suitable for release into the environment.

35.3
Drinking water purification plants employ industrial-scale physical and chemical systems that remove or neutralize biological, inorganic, and organic contaminants from natural, community, and industrial sources. Water purification plants employ clarification, filtration, and disinfection processes to produce potable water.

35.4
Contaminated drinking water and recreational water are sources of waterborne pathogens. In the United States, the number of disease outbreaks due to these sources is relatively small in relation to the large number of exposures to water. Worldwide, lack of adequate water treatment facilities and access to clean water contribute significantly to the spread of infectious diseases.

35.5
Vibrio cholerae is the agent of cholera, an acute diarrheal disease that causes severe dehydration. Cholera occurs in pandemics. The current focus of the more than 1 million annual cases of cholera is in Africa. In endemic areas, avoidance of contaminated water and food are reasonable preventive measures. Oral rehydration and electrolyte replacement effectively treat the disease, reducing overall mortality to about 1%.

35.6
Giardiasis and cryptosporidiosis are spread by the chlorine-resistant cysts of *Giardia* and *Cryptosporidium* in drinking water and recreational water contaminated by the feces of infected humans or animals. Infection with either protist can cause acute

gastrointestinal illness and may lead to more serious disease in compromised individuals.

35.7

Legionella pneumophila is a respiratory pathogen that causes Pontiac fever and legionellosis, a more serious infection that may result in pneumonia. *L. pneumophila* grows to high numbers in warm water and is spread via aerosols. The prevalence of legionellosis is increasing and infections are underreported.

35.8

Typhoid fever, viral infections, and amebiasis are important waterborne diseases. Waterborne typhoid fever and viral illnesses, common diseases in developing countries, can be controlled by effective water treatment. Amebic dysentery caused by *Entamoeba histolytica* affects millions of people worldwide. Meningoencephalitis is a rare but usually fatal condition caused by *Naegleria fowleri* amebiasis.

Review of Key Terms

Biochemical oxygen demand (BOD) the relative amount of dissolved oxygen consumed by microorganisms for complete oxidation of organic and inorganic material in a water sample

Chloramine a disinfectant chemical manufactured on-site by combining chlorine and ammonia at precise ratios

Chlorine a chemical used in its gaseous state to disinfect water; a residual level is maintained throughout the distribution system

Clarifier a reservoir in which suspended solids in raw water are coagulated and removed through precipitation

Coagulation the formation of large insoluble particles from much smaller, colloidal particles by the addition of aluminum sulfate and anionic polymers

Coliforms facultatively aerobic, gram-negative, non-spore-forming, lactose-fermenting bacteria

Cyst an infectious form of a protist that is encased in a thick-walled, chemically and physically resistant coating

Distribution system water pipes, storage reservoirs, tanks, and other equipment used to deliver drinking water to consumers or store it before delivery

Effluent water treated wastewater discharged from a wastewater treatment facility

Filtration the removal of suspended particles from water by passing it through one or more permeable membranes or media (e.g., sand, anthracite, or diatomaceous earth)

Finished water water delivered to the distribution system after treatment

Flocculation the water treatment process after coagulation that uses gentle stirring to cause suspended particles to form larger, aggregated masses (flocs)

Meningoencephalitis invasion, inflammation, and destruction of brain tissue by the amoeba *Naegleria fowleri* or another pathogen

Polymer in water purification, a chemical in liquid form used as a coagulant in the clarification process to flocculate a suspension

Potable drinkable; safe for human consumption

Primary disinfection the introduction of sufficient chlorine or other disinfectant into clarified, filtered water to kill existing microorganisms and inhibit further microbial growth

Primary wastewater treatment physical separation of wastewater contaminants, usually by separation and settling

Raw water surface water or groundwater that has not been treated in any way (also called untreated water)

Secondary aerobic wastewater treatment oxidative reactions carried out by microorganisms under aerobic conditions to treat

wastewater containing low levels of organic materials

Secondary anaerobic wastewater treatment degradative and fermentative reactions carried out by microorganisms under anoxic conditions to treat wastewater containing high levels of insoluble organic materials

Secondary disinfection the maintenance of sufficient chlorine or other disinfectant residual in the water distribution system to inhibit microbial growth

Sediment soil, sand, minerals, and other large particles found in raw water

Sewage liquid effluents contaminated with human or animal fecal material

Suspended solid a small particle of solid pollutant that resists separation by ordinary physical means

Tertiary wastewater treatment the physicochemical or biological processing of wastewater to reduce levels of inorganic nutrients

Turbidity a measurement of suspended solids in water

Untreated water surface water or groundwater that has not been treated in any way (also called raw water)

Wastewater liquid derived from domestic sewage or industrial sources which cannot be discarded in untreated form into lakes or streams

Review Questions

1. Describe the principles of different tests for fecal coliforms (Section 35.1).

2. Trace wastewater treatment in a typical plant from incoming water to release. What is the overall reduction in the BOD for typical household wastewater? What is the overall reduction in the BOD for typical industrial wastewater (Section 35.2)?

3. Identify (stepwise) the process of purifying drinking water. What important contaminants are targeted by each step in the process (Section 35.3)?

4. Why are common sources of infection, such as contaminated water sources, a significant threat to public health (Section 35.4)?

5. How can one identify cholera without doing laboratory tests (Section 35.5)?

6. Why do giardiasis and cryptosporidiosis present a constant threat to public health (Section 35.6)?

7. Describe the clinical features of *Legionella* (Section 35.7).

8. Indicate the methods that are used to control typhoid fever, norovirus infection, and amebiasis in water systems in developed countries (Section 35.8).

Application Questions

1. Why is reduction of BOD in wastewater a primary goal of waste-water treatment? What are the consequences of releasing wastewater with a high BOD into local water sources such as lakes or streams?

2. In the United States, the federal government has defined, by law, a strict set of drinking water standards. Federal recreational water standards, however, are recommendations, and local governments can set more or less stringent standards. Explain why recreational water standards are flexible and devise recreational water standards for the area in which you live.

3. Worldwide, we are in the midst of the seventh cholera pandemic, and the eighth pandemic may be starting. Using sources such as the World Health Organization and the Centers for Disease Control and Prevention, define the status of the current pandemic with regard to its geographic distribution, endemic areas, and most recent outbreaks. Comment on methods that could be used to decrease the spread of cholera and control the annual epidemic outbreaks that occur in endemic areas. Can cholera be eradicated?

4. As a visitor to a country in which cholera is an endemic disease, what specific steps would you take to reduce your risk of cholera exposure? Will these precautions also prevent you from contracting other waterborne diseases? Which ones? Identify waterborne diseases for which your precautions may not prevent infection.

5. Why are surface waters contaminated with the cysts of various protists? What steps might public health officials take to remedy this problem?

6. Discuss the wastewater and drinking water treatment schemes that must be in place to control such diseases as typhoid fever. Is it possible to eliminate *Salmonella enterica* serovar Typhi, as has been effectively done for poliovirus? Would it be possible to eliminate *Naegleria fowleri* or *Entamoeba histolytica*? Explain.

7. Noroviruses are frequently causes of acute gastrointestinal disease outbreaks. How can control of these viruses be accomplished?

Need more practice? Test your understanding with Quantitative Questions; access additional study tools including tutorials, animations, and videos; and then test your knowledge with chapter quizzes and practice tests at **www.microbiologyplace.com**.

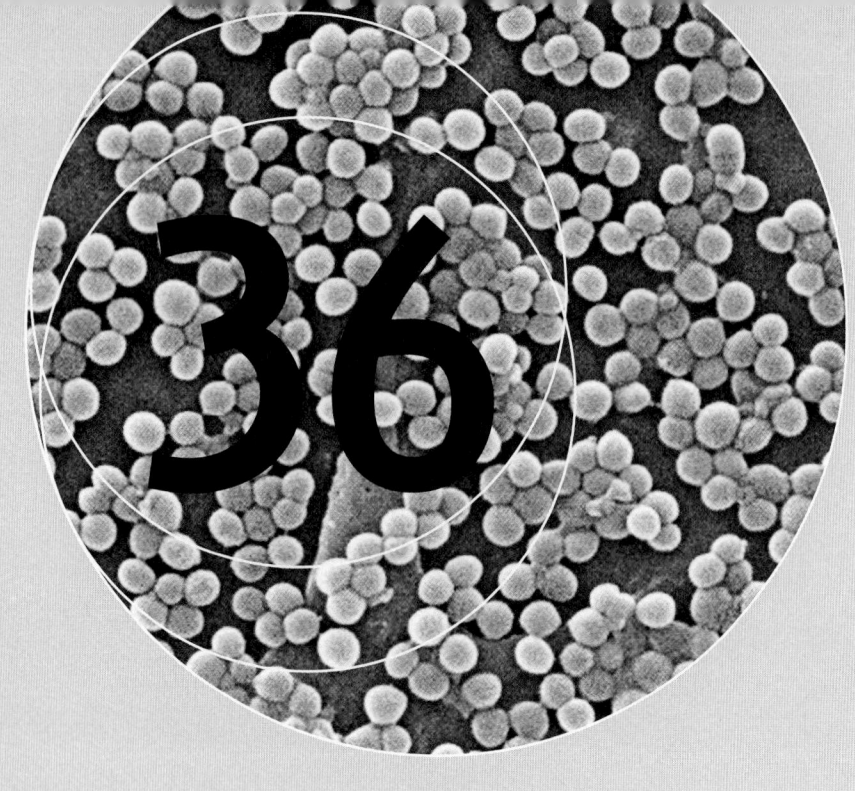

36

Food Preservation and Foodborne Microbial Diseases

Cells of the bacterium *Staphylococcus aureus* produce a toxin that causes severe intestinal distress. "Staph" food poisoning is a classic and common foodborne illness and is typically linked to contaminated foods left under conditions that allow for rapid growth of the organism.

Humans are constantly exposed to bacteria, fungi, and viruses in food as well as in air and water. The foods we eat, whether they are fresh, prepared, or even preserved, are seldom sterile and may be contaminated with spoilage microorganisms or occasionally with pathogens. On the other hand, microbial activity is important for the production of some foods. For example, cheese, buttermilk, sour cream, and yogurt are all produced by microbial fermentation. Sauerkraut is a fermented vegetable food. Certain sausages, pâtés, and liver spreads are produced by microbial fermentation. Cider vinegar is produced by the activities of the acetic acid bacteria, and alcoholic beverages are produced by fermentation. Some foods contain living microorganisms thought to confer health benefits. We discussed these foods, called probiotic foods, in the context of replacing or augmenting the normal microbial flora in the human gut.

Here we examine food preservation methods that limit unwanted microbial growth and food spoilage. We also look at microbial processes that aid in food preservation and, not incidentally, create a variety of fermented foods. Finally, we discuss microbial products and microorganisms that cause food poisoning and food infection.

I Food Preservation and Microbial Growth

Microorganisms are important spoilage agents in foods, causing food shortages and economic loss. Various methods, some utilizing desirable microbial growth, are used for controlling spoilage organisms.

36.1 Microbial Growth and Food Spoilage

Microorganisms, including a few human pathogens, colonize and grow on common foods. Foods provide a suitable medium for the growth of various microorganisms, and microbial growth often reduces food quality and availability.

Food Spoilage

Food spoilage is any change in the appearance, smell, or taste of a food product that makes it unacceptable to the consumer. Spoiled food may still be safe to eat, but is generally regarded as unpalatable and will not be purchased or readily consumed. Food spoilage causes losses to producers, distributors, and consumers in the form of reduced quality and quantity, and inevitably leads to higher prices.

Foods consist of organic materials that can be nutrients for the growth of chemoorganotrophic bacteria. The physical and chemical characteristics of the food determine its degree of susceptibility to microbial activity. With respect to spoilage, foods are classified into three major categories: (1) **perishable food**, including many fresh food items; (2) **semiperishable food**, such as potatoes and nuts; and (3) **stable** or **nonperishable food**, such as flour and sugar (**Table 36.1**).

The three food categories differ greatly with regard to their *moisture content*, which is related to **water activity, a_w** (Section

Table 36.1 *Food classification by storage potential*	
Food classification	**Examples**
Perishable	Meats, fish, poultry, eggs, milk, most fruits and vegetables
Semiperishable	Potatoes, some apples, and nuts
Nonperishable	Sugar, flour, rice, and dry beans

5.16). Water activity is a measure of the availability of water for use in metabolic processes. Nonperishable foods have low water activity and can generally be stored for considerable lengths of time without spoilage. Perishable and semiperishable foods, by contrast, typically have higher water activities. Thus, these foods must be stored under conditions that inhibit microbial growth.

Fresh foods are spoiled by many different bacteria and fungi. The chemical properties of foods vary widely, and each food is characterized by the nutrients it contains as well as other factors such as acidity or alkalinity. As a result, each fresh food is typically colonized and spoiled by a relatively restricted group of microorganisms; the spoilage organisms are those that can gain access to the food and use the available nutrients (**Table 36.2**).

For example, enteric bacteria such as *Salmonella, Shigella*, and *Escherichia*, all potential pathogens sometimes found in the gut of animals, often contaminate meat. At slaughter, intestinal contents containing live bacteria may be accidentally spilled during removal of the intestines and result in contamination of the carcass. These organisms can also contaminate produce through fecal contamination of water supplies. Likewise, lactic acid bacteria, the most common microorganisms in dairy products, are the major spoilers of milk and milk products. *Pseudomonas* species are found in both soil and animals and cause the spoilage of fresh foods of all types.

Growth of Microorganisms in Foods

Microbial growth in foods follows the normal pattern for bacterial growth (Section 5.7). The length of the lag phase depends on the properties of the contaminating microorganism and the food substrate. The time required for the population density to reach a significant level in a given food product depends on both the size of the initial inoculum and the rate of growth during the exponential phase. The rate of growth during the exponential phase depends on the temperature, the nutrient value of the food, and the suitability of other growth conditions.

Throughout much of the exponential growth phase, microbial numbers in a food product may be so low that no measurable effect can be observed, with only the last few population doublings leading to observable spoilage. Thus, for much of the period of microbial growth in a food there is no visible or easily detectable change in food quality; spoilage is usually observed only when the microbial population density is high.

MiniQuiz
- List the major food groups as categorized by water availability.
- Identify factors that lead to growth of microorganisms in food.

Table 36.2 *Microbial spoilage of fresh food*[a]

Food product	Type of microorganism	Common spoilage organisms, by genus
Fruits and vegetables	Bacteria	*Erwinia, Pseudomonas, Corynebacterium* (mainly vegetable pathogens; rarely spoil fruit)
	Fungi	*Aspergillus, Botrytis, Geotrichum, Rhizopus, Penicillium, Cladosporium, Alternaria, Phytophthora,* various yeasts
Fresh meat, poultry, eggs, and seafood	Bacteria	*Acinetobacter, Aeromonas, Pseudomonas, Micrococcus, Achromobacter, Flavobacterium, Proteus, Salmonella, Escherichia, Campylobacter, Listeria*
	Fungi	*Cladosporium, Mucor, Rhizopus, Penicillium, Geotrichum, Sporotrichum, Candida, Torula, Rhodotorula*
Milk	Bacteria	*Streptococcus, Leuconostoc, Lactococcus, Lactobacillus, Pseudomonas, Proteus*
High-sugar foods	Bacteria	*Clostridium, Bacillus, Flavobacterium*
	Fungi	*Saccharomyces, Torula, Penicillium*

[a]The organisms listed are the most commonly observed spoilage agents of fresh, perishable foods. Many of these genera include species that are human pathogens.

36.2 Food Preservation

Food storage and preservation methods slow the growth of microorganisms that spoil food and cause foodborne disease.

Cold

A crucial factor affecting microbial growth is temperature (↩ Section 5.12). In general, a lower storage temperature results in less microbial growth and slower spoilage. However, a number of psychrotolerant (cold-tolerant) microorganisms can grow, albeit slowly, at refrigerator temperatures (3–5°C). Therefore, storage of perishable food products for long periods of time (more than several days) is possible only at temperatures below freezing. Freezing and subsequent thawing, however, alter the physical structure, taste, and appearance of many foods such as leafy green vegetables like spinach and lettuce, making them unacceptable to the consumer. Freezing is widely used, however, for the preservation of solid foods such as meats and many fruits and vegetables. Freezers providing a temperature of –20°C are most commonly used. Storage for weeks or months is possible at –20°C, but microorganisms can still grow in pockets of liquid water trapped within the frozen food. For long-term storage, temperatures of –80°C [the temperature of solid carbon dioxide (CO_2), "dry ice"] are necessary. Because of the high equipment and energy costs necessary to maintain such low temperatures, ultracold freezing is not used for routine food storage.

Pickling and Acidity

Another factor affecting microbial growth in food is pH. Foods vary somewhat in pH, but most are neutral or acidic. Microorganisms differ in their ability to grow under acidic conditions, but conditions of pH 5 or less inhibit the growth of most spoilage organisms. Therefore, weak acids are often used for food preservation, a process called **pickling**. Vinegar, a dilute acetic acid fermentation product of the acetic acid bacteria, is usually added in the pickling process. Pickling methods usually mix the vinegar with large amounts of salt or sugar to decrease water availability (as discussed below) and further inhibit microbial growth. Common pickled foods include cucumbers (sweet, sour, and dill pickles), peppers, meats, fish, and fruits.

Drying and Dehydration

As we mentioned, water activity, or a_w, is a measure of the availability of water for use by microorganisms in metabolic processes. The a_w of pure water is 1.00; the molecules in pure water are loosely ordered and rearrange freely. When solute is added, the a_w decreases. As water molecules reorder around the solute, the free rearrangement of the solute-bound water molecules becomes energetically unfavorable. The microbial cells must then compete with solute for the reduced amount of free water. In general, bacteria are poor competitors for the remaining free water, but fungi are good competitors. In practice this means that high concentrations of solutes such as sugars or salts, which greatly reduce a_w, typically inhibit bacterial growth. For example, most bacteria are inhibited by a concentration of 7.5% sodium chloride (NaCl) (a_w of 0.957), with the exception of some gram-positive cocci, such as *Staphylococcus*. On the other hand, molds compete well for free water under conditions of low a_w and often grow well in high-sugar foods such as syrups.

Some commercially important foods are preserved by the addition of salt or sugar. Sugar is used mainly in fruits (jams, jellies, and preserves). Salted products are primarily meats and fish. Sausage and ham are preserved using various curing salts, including NaCl. Some meats also undergo a smoking process. Preserved meat products vary widely in a_w, depending on how much salt is added and how much the meat has been dried. Some cured meat products such as country ham or jerky can be kept at room temperature for extended periods of time. Others with higher a_w require refrigeration for long-term storage.

Microbial growth in foods can also be controlled by drying, which lowers water content and availability. Drying is used to preserve highly perishable foods such as meat, fish, milk, vegetables, fruit, and eggs. The least damaging physical method used to

Figure 36.1 Spray dryer. Industrial spray dryers are used to dry or concentrate large volumes of high-value liquid foods.

dry foods is the process of **lyophilization (freeze-drying)** in which foods are frozen and water is then removed under vacuum. This method is very expensive, however, and is used mainly for specialized applications such as preparation of military rations that may need to be stored for long periods even under wet or warm conditions.

Spray drying is the process of spraying, or atomizing, liquids such as milk in a heated atmosphere. The atomization produces small droplets, increasing the surface-to-volume ratio of the liquid, promoting rapid drying without destroying the food. This technology is widely used in the production of powdered milk, certain concentrated liquid dairy products such as evaporated milk, and concentrated food ingredients such as liquid flavorings (**Figure 36.1**).

Heating

Heat is used to reduce the bacterial load or even sterilize a food product; it is especially useful for the preservation of liquids and wet foods. *Pasteurization*, a process in which liquids are heated to a specified temperature for a precise time, was described in Section 26.1. Pasteurization does not sterilize liquids, but reduces the bacterial load of spoilage organisms and pathogens, significantly extending the shelf life of the liquid. Pasteurization can be done at 63°C (145°F) for 30 seconds or at 71°C (160°F) for 15 seconds. Typically, perishable liquids such as milk, fruit juices, and beer are pasteurized. Ultrahigh-temperature (UHT) processing, sometimes called ultrapasteurization, can be used to preserve the same liquids. UHT processing heats the liquid to 138°C (238°F) for 2 to 4 seconds. This treatment kills all microorganisms, extending the shelf life of liquids like milk to 6 months or

longer without the need for refrigeration. UHT-processed milk is common in Europe, but is not easily found in the United States.

Canning is a process in which food is sealed in a container such as a can or glass jar and then heated. In theory, canning should sterilize the food product, but this requires processing at the correct temperature for the correct length of time. However, when properly sealed and heated, most canned food should remain stable and unspoiled indefinitely at any temperature.

The temperature–time relationships for canning depend on the type of food, its pH, the size of the container, and the consistency or density of the food. Because heat must completely penetrate the food within the container, effective heating times must be longer for large containers or very dense foods. Acidic foods can often be canned effectively by heating just to boiling, 100°C, whereas nonacidic foods must be heated to autoclave temperatures (121°C). For some foods in large containers, times of 20–50 minutes must be used. Heating times long enough to guarantee absolute sterility of every container would make most foods unpalatable and could also reduce nutritional value. Even properly canned foods, therefore, may not be sterile. The process used for commercial canning, called *retort canning*, employs equipment similar to an autoclave to apply steam under pressure (↺ Section 26.1).

If live microorganisms remain in a can, growth of organisms can produce extensive amounts of gas and build pressure, resulting in bulges or, in extreme cases, explosion (**Figure 36.2**). The environment inside a can is anoxic, and some of the anaerobic bacteria that grow in canned foods are toxin producers of the genus *Clostridium* (Section 36.7). Food from a bulging can, therefore, should never be eaten. On the other hand, the lack of obvious gas production is not an absolute guarantee that canned food is safe to consume.

Aseptic Food Processing

Several foods in the United States and many more in Asia and Europe are now prepared and packaged under aseptic conditions. Foods processed and packaged aseptically can be stored at room temperature for months or longer without spoilage. Aseptic processing uses *flash heating*, a process using a rapid, short heating cycle, or sterilization by cooking. The processed foods are then packaged in aseptic containers, usually cardboard cartons lined with foil and plastic. The process may require "clean room" conditions similar to those in a hospital operating room. For example, incoming room air is filtered to limit contamination from spores and bacteria in the atmosphere. Special equipment is required to flash-heat and deliver the product aseptically into sterile packaging materials.

In the United States, fruit juices (juice boxes) and milk substitutes are often processed in this way. In many European countries, milk products are flash-heated to 133°C and packaged aseptically. Perishable food products prepared aseptically can be stored at room temperature for at least 6 months. This developing technology significantly increases product shelf life and eliminates the need for refrigeration for many products. However, the equipment and processing plants necessary for aseptic food processing are expensive.

(a)   (b)   (c)   (d)

Figure 36.2 Changes in sealed tin cans as a result of microbial spoilage. (a) A normal can. The top of the can is pulled in slightly due to the negative pressure (vacuum) inside. (b) Swelling resulting from minimal gas production. The top of the can bulges slightly. (c) Severe swelling due to extensive gas production. (d) The can shown in part c was dropped, and the gas pressure resulted in a violent explosion, tearing the lid apart.

High-Pressure Processing

High-pressure processing (HPP) is a technology that uses very high hydrostatic pressure (up to 100,000 lb/in^2) to kill most pathogens and spoilage organisms in packaged foods. Applications include several food types. Fruits and vegetables such as avocado products, salsas, chopped onions, applesauce, ready-to-eat meats, and juices can all be processed in bulk or in consumer packaging. The packaged foods are loaded into a vessel that is flooded with water and placed under pressure (**Figure 36.3**). Pressure treatment kills most foodborne pathogens, but does not kill endospores; the products are not absolutely sterile, but shelf life is increased from days to months.

Chemical Preservation

Over 3000 different compounds are used as food additives. These chemical additives are classified by the United States Food and Drug Administration as "generally recognized as safe" (GRAS) and find wide application in the food industry for enhancing or preserving texture, color, freshness, or flavor. A small number of these compounds are used to control microbial growth in food (**Table 36.3**). Many of these microbial growth inhibitors, such as sodium propionate and sodium benzoate, have been used for many years with no evidence of human toxicity. Others, such as nitrites (a carcinogen precursor) and ethylene oxide and propylene oxide (mutagens), are more controversial food additives because these compounds may adversely affect human health. The use of spoilage-retarding additives, however, significantly extends the useful shelf life of finished foods. Chemical food additives contribute significantly to an increase in quantity and in the perceived quality of available food items.

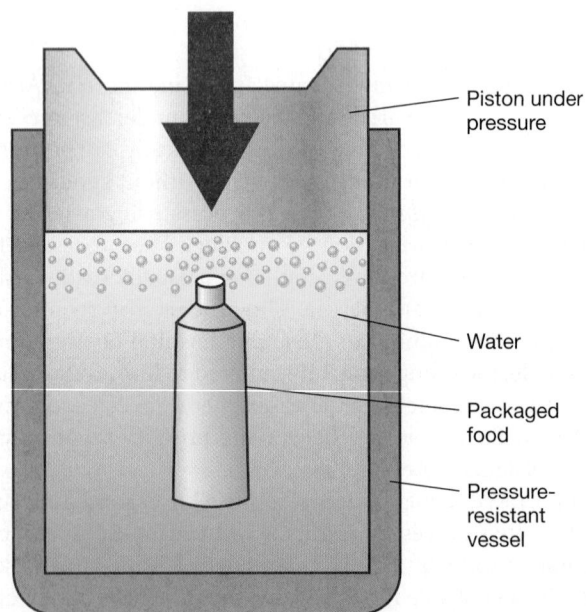

Figure 36.3 High-pressure processing (HPP) of food. Packaged foods are loaded into a vessel that is flooded with water and placed under very high hydrostatic pressure (up to 100,000 lb/in^2) to kill most pathogens and spoilage organisms in packaged foods.

Piston under pressure

Water

Packaged food

Pressure-resistant vessel

Table 36.3 *Chemical food preservatives*	
Chemical	**Food**
Sodium or calcium propionate	Bread
Sodium benzoate	Carbonated beverages, fruit, fruit juices, pickles, margarine, preserves
Sorbic acid	Citrus products, cheese, pickles, salads
Sulfur dioxide, sulfites, bisulfites	Dried fruits and vegetables, wine
Formaldehyde (from food-smoking process)	Meat, fish
Ethylene and propylene oxides	Spices, dried fruits, nuts
Sodium nitrite	Smoked ham, bacon

Because of the time and cost required for testing of any chemical proposed as a food preservative or additive, it is unlikely that many new chemicals will be added to the list of safe and approved chemical food preservatives listed in Table 36.3.

Irradiation

Irradiation of food with ionizing radiation is an effective method for reducing contamination by bacteria, fungi, and even insects (ᴄᴇ Section 26.2). **Table 36.4** lists foods for which radiation treatment has been approved in the United States. Foods including herbs, spices, and grains are routinely irradiated. Fresh meats and fish can be irradiated to limit contamination by *Escherichia coli* O157:H7 (ground beef), *Campylobacter jejuni* (poultry), and *Vibrio* spp. (seafood). In an attempt to limit foodborne disease outbreaks in fresh produce, irradiation was approved in 2008 to control foodborne pathogens in iceberg lettuce and spinach. In many countries throughout the world, spices, seafood, vegetables, grains, potatoes, sterilized meals, and meats are irradiated.

For food irradiation, gamma rays generated from radioactive cobalt (^{60}Co) or cesium (^{136}Cs), or from high-energy electrons produced by linear accelerators, are used as radiation sources. Alternatively, beta rays can be generated from an electron gun, analogous to but significantly more powerful than the electron beam generated by the cathode ray gun formerly used in television sets. In addition, X-rays can be generated with electron beams focused on metal foil. X-rays have much greater penetrating power than beta rays and are therefore useful for treating large-volume food preparations. Beta ray and X-ray sources can be switched on and off at will and do not require a radioactive source.

Irradiated food products receive a controlled radiation dose. This dose varies considerably for each food category and pur-

Figure 36.4 The radura, the international symbol for radiation. Packaging of foods treated with radiation must be labeled with the radura, the international symbol for radiation, as well as the statement "treated by irradiation" or "treated with radiation."

pose. For example, a dose of 44 kilograys (kGy) is used to sterilize meat products used on United States NASA space flights and is nearly ten times higher than the dose of 4.5 kGy used for control of pathogens in ground beef (Table 36.4). In the United States, a consumer product information label and the radura, the international symbol for radiation, must be affixed to foods that are irradiated in whole (**Figure 36.4**). Irradiated ingredients that are a major portion of a food product must be identified in the ingredients list, but the radura symbol does not need to be shown. Irradiated ingredients such as spices that are minor components of a finished food product do not have to be identified as irradiated.

MiniQuiz

- Identify food spoilage microorganisms that are also pathogens.
- Identify physical and chemical methods used for food preservation. How does each method limit growth of microorganisms?

36.3 Fermented Foods and Mushrooms

Many common foods and beverages are preserved, produced, or enhanced through the direct actions of microorganisms. Desirable microbial processes can produce significant alterations in raw foods; the product is called a *fermented food*. **Fermentation** is the anaerobic catabolism of organic compounds, generally carbohydrates, in the absence of an external electron acceptor (ᴄᴇ Section 4.8). Bacteria important in the fermented foods industry are the lactic acid bacteria, the acetic acid bacteria, and the propionic acid bacteria (**Table 36.5**). These bacteria do not grow below about pH 3.5, so food fermentation is a self-limiting process.

One of the most common fermented foods is yeast bread, in which the fermentation of simple sugars and grain carbohydrates by the yeast *Saccharomyces cerevisiae* (Table 36.5) produces

Table 36.4 *Irradiated foods by category, dose, and purpose*

Food category	Dose[a] (kGy)	Purpose
Fresh meat: ground beef	4.50	Reduce bacterial pathogens
Herbs, spices, enzymes, and flavorings	30.00	Sterilize
Pork	1.00	Reduce *Trichinella spiralis* protist
Meats used in NASA[b] space flight program	44.0	Sterilize
Poultry	3.00	Reduce bacterial pathogens
Wheat flour	0.50	Inhibit mold
White flour	0.15	Inhibit mold

[a]The highest value for recommended doses is given. One kGy (kilogray) is 1000 grays. One gray, an SI unit, is 1 joule of radiation absorbed by 1 kilogram of matter and is also equivalent to 100 rad. For the radiation requirements necessary to kill specific microorganisms, refer to Table 26.1.
[b]National Aeronautics and Space Administration, USA.

Table 36.5 *Fermented foods and fermentation microorganisms*[a]

Food category	Primary fermenting microorganism
Dairy foods	
Cheeses	*Lactococcus*
	Lactobacillus
	Streptococcus thermophilus
Fermented milk products	
Buttermilk	*Lactococcus*
Sour cream	*Lactococcus*
Yogurt	*Lactobacillus*
	Streptococcus thermophilus
Alcoholic beverages	*Zymomonas*
	Saccharomyces[b]
Yeast breads	*Saccharomyces cerevisiae*[c]
Meat products	
Dry sausages (pepperoni, salami) and semidry sausages (summer sausage, bologna)	*Pediococcus*
	Lactobacillus
	Micrococcus
	Staphylococcus
Vegetables	
Cabbage (sauerkraut)	*Leuconostoc*
	Lactobacillus
Cucumbers (pickles)	Lactic acid bacteria
Soy sauce	*Aspergillus*[d]
	Tetragenococcus halophilus
	Yeasts

[a]Unless otherwise noted, these are all species of *Firmicutes* except for *Micrococcus*, which is a genus of *Actinobacteria*. *Zymomonas* is a genus of *Alphaproteobacteria*.
[b]Yeast. Various *Saccharomyces* species are used in alcohol fermentations.
[c]Baker's yeast.
[d]A mold.

Figure 36.5 Fermented foods. Bread, sausage meats, cheeses, many dairy products, and fermented and pickled vegetables are food products that are produced or enhanced by fermentation reactions catalyzed by microorganisms.

carbon dioxide (CO_2), raising the bread and producing the holes in the finished loaf (**Figure 36.5**). Other *Saccharomyces*-fermented food products include wine, beer, and whiskey. Production of these beverages is often carried out at industrial scales (Sections 15.7 and 15.8).

Dairy Products

Fermented dairy products were originally developed to preserve milk, an economically important fresh food that normally undergoes rapid spoilage. Dairy products include cheese and other fermented milk products such as yogurt, buttermilk, and sour cream (Figure 36.5). Milk contains the disaccharide lactose. Lactose can be hydrolyzed by the enzyme lactase into glucose and galactose. These monosaccharides are fermented to the final product of lactic acid by the lactic acid bacteria (Table 36.5). This fermentation reaction produces a significant decrease in pH from the neutral or slightly basic pH of raw milk to a pH of less than 5.3 in cheeses and less than 4.6 in other fermented milk products.

Starter cultures of lactic acid bacteria are introduced into raw milk, and fermentation proceeds for a time depending on the desired product. For some cheeses, a second inoculum may be introduced to produce a second fermentation. For example, following lactic acid fermentation, Swiss-style (Emmentaler) cheeses are reinoculated with *Propionibacterium*. The secondary fermentation catabolizes lactic acid to propionic acid, acetic acid, and CO_2. The CO_2 produces the large holes that characterize Swiss-style cheeses. Secondary fermentations with *Lactobacillus* and the mold *Penicillium roqueforti* produce the blue veins and distinctive taste and aroma of blue cheese. Each cheese type is produced under carefully controlled conditions. Time of fermentation, temperature, the extent of aging, and the types of fermenting microorganisms are rigidly controlled to ensure a distinctive and reproducible product.

Meat Products

Fermented meat products fall into several categories. *Sausages* are generally made from pork, beef, or poultry. The most common are the dry sausages, such as salami and pepperoni, and the semidry sausages such as bolognas and summer sausages (Figure 36.5).

Sausages are made using a uniformly blended mixture of meat, salt, and seasonings. A starter culture of lactic acid bacteria is added, and fermentation reduces the pH of the mixture to below 5. After fermentation, sausages are often smoked and dried to a moisture content of about 30%. Dry sausages can be held at room temperature for extended periods of time. Semidry sausages such as summer sausage have a final moisture content of about 50% and are less resistant to spoilage, so they are generally refrigerated.

Fish, often mixed with rice, shrimp, and spices, are also fermented to make fish pastes and fish-flavored products.

Vegetables and Vegetable Products

The variety of specialty fermented vegetable foods is practically endless. The most economically important fermented vegetable foods are sauerkraut (fermented cabbage) and some types of pickles (fermented cucumbers). Peppers, olives, onions, tomatoes, and many fruits are also fermented.

Vegetables are often fermented in salt brine to enhance preservation and flavor. The salt also helps prevent the growth of unwanted organisms, the desired fermentative organisms being salt-tolerant. Fermentation may also improve digestibility by breaking down plant tissues. For example, fermented legume products (peas, beans, and lentils) have a marked reduction in the flatulence-producing oligosaccharides that characterize fresh legumes.

Soy Sauce

Soy sauce is a complex fermentation product made by fermentation of soybeans and wheat. A culture of the fungus *Aspergillus* (Table 36.5) is spread on a cooked wheat–soybean mixture, where it grows for 2–3 days. This preparation, known as koji, is then mixed with brine [17–19% sodium chloride (NaCl)] and fermentation proceeds for 2–4 months or more in large vats (**Figure 36.6**). The *Aspergillus* and various microorganisms, including *Lactobacillus* and *Pediococcus* and several other fungi, produce fermentation products from the brined koji that contribute to the desirable characteristics of the final product. After fermentation, the liquid sauce is filtered, pasteurized, and bottled as soy sauce.

Vinegar

Vinegar is produced by the conversion of ethyl alcohol to acetic acid by the acetic acid bacteria. Key genera of acetic acid bacteria include *Acetobacter* and *Gluconobacter* (Section 17.8). Vinegar is produced from dilute ethanol solutions; the usual starting material is wine, fermented rice, or alcoholic apple juice (hard cider). Vinegar can also be produced from a mixture of pure alcohol in water, in which case it is called distilled vinegar, the term "distilled" referring to the alcohol from which the product is made rather than the vinegar itself. Vinegar is used as a flavoring agent in salads and other foods, and because of its acidity, it is also used in pickling. Foods pickled with high concentrations of vinegar can be stored unrefrigerated for years.

Figure 36.6 Soy sauce fermentation. The vats, each about 1 m deep and 4 m in diameter, contain koji, a mixture of wheat and soybeans inoculated with *Aspergillus* mixed in salt brine. Fermentation proceeds for up to 1 year in the vats. The liquid is then filtered, pasteurized, and bottled as soy sauce.

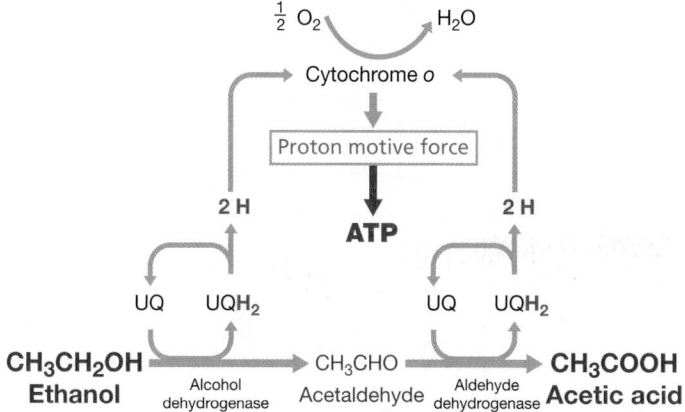

Figure 36.7 Vinegar production. The key process in vinegar production is the oxidation of ethanol to acetic acid. UQ, ubiquinone.

Acetic acid bacteria are strictly aerobic, but unlike most other aerobes, some species such as *Gluconobacter* do not oxidize organic electron donors completely to CO_2 and water (H_2O) (**Figure 36.7**). If ethyl alcohol is the electron donor, they oxidize it to acetic acid, which then accumulates in the medium. Acetic acid bacteria are acid-tolerant and are not killed by the low pH products that they generate. Because this process is aerobic, the demand for oxygen during growth is very high; the production of vinegar requires sufficient aeration of the medium.

Three processes are in use for vinegar production. The *open-vat method*, which is the original process, is still used in France where it was first developed. Wine is placed in shallow vats to facilitate exposure to the air. At the surface of the liquid, the acetic acid bacteria develop as a slimy layer. This process is not very efficient because the bacteria contact both the air and substrate only at the surface.

The second process is the *trickle method*. Alcoholic liquid is trickled over loosely packed beechwood twigs or shavings, arranged in a vat or column. The bacteria grow on the surface of the wood shavings, using the trickling liquid as a substrate. A stream of air enters at the bottom and passes upward, facilitating maximum contact between the bacteria, air, and substrate. The vat is called a *vinegar generator* (**Figure 36.8**), and the process is operated in a continuous fashion. The wood shavings in a vinegar generator are the support on which the bacteria grow to produce a biofilm, so they are not consumed and can last from 5 to 30 years, depending on the kind of alcoholic liquid used in the process.

Finally, the *bubble method* of vinegar production incorporates industrial incubation techniques to introduce and mix air into a fermenter containing the alcoholic substrate and inoculated with acetic acid bacteria. The bubble method is highly efficient; 90–98% of the alcohol is converted to acetic acid.

Acetic acid can easily be made chemically from alcohol, but the microbial product, *vinegar*, is a distinctive material. The flavor of vinegar is affected by other substances present in the starting material and produced during fermentation. For this reason, the microbial methods for vinegar production, especially the vinegar generator method, have not been supplanted by chemical processes.

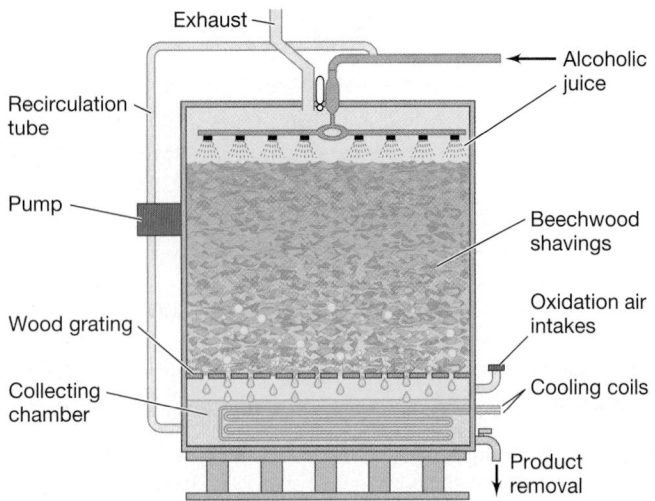

Figure 36.8 A vinegar generator. Alcoholic juice is trickled through wood shavings as air is passed upward. Acetic acid bacteria colonize the wood shavings, forming a biofilm that oxidizes alcohol to acetic acid. The dilute acetic acid pools in the collecting chamber, from where it is recycled through the generator. The pooled product is drained off when the acetic acid content reaches 4%, the minimum concentration to be labeled as *vinegar*.

Mushrooms

Several kinds of fungi are sources of human food, of which the most important are the mushrooms. Mushrooms are a group of filamentous fungi that form large, edible fruiting bodies (**Figure 36.9**). The fruiting body is called the mushroom and is formed through the association of a large number of individual fungal hyphae to form a mycelium.

The mushroom commercially grown in most parts of the world is the basidiomycete *Agaricus bisporus*, and it is generally cultivated in "mushroom farms." The organism is grown in special beds, usually in buildings where temperature and humidity are carefully controlled and exposure to light is severely limited. Beds are prepared by mixing soil with a material rich in organic

matter, such as horse manure, and the beds are then inoculated with a pure culture of the mushroom fungus that has been grown in large bottles on an organic-rich medium.

In the mushroom bed, the mycelium grows and spreads through the substrate, and after several weeks it is ready for the next step, the induction of mushroom formation. This is accomplished by adding a layer of soil to the surface of the bed. The appearance of mushrooms on the surface of the bed is called a *flush* (Figure 36.9*a*), and for freshness the mushrooms must be collected immediately upon flushing. After collection they are packaged and kept cool until brought to market.

Another cultured mushroom is the shiitake, *Lentinus edulus*. Shiitake mushrooms are cellulose-digesting fungi that grow on hardwood trees. They are cultivated on small logs (Figure 36.9*b*). The logs are hydrated by soaking in water. Plugs of mushroom culture are inoculated into small holes drilled in the logs. The fungus grows through the log, and after about a year forms a flush of fruiting bodies, the edible mushroom. Shiitake mushrooms are considered to have a superior taste to *Agaricus bisporus*, and therefore are more expensive.

MiniQuiz

- Identify important dairy, meat, and vegetable food products that are produced or enhanced by microbial fermentation or growth.
- Identify the microbial group or groups that are most important for food fermentations.
- Identify edible fungi.

Ⅱ Foodborne Disease, Microbial Sampling, and Epidemiology

If food is not decontaminated or preserved, pathogens may grow in it and cause foodborne diseases with significant morbidity and mortality. Like waterborne diseases, foodborne illnesses are common-source diseases. A single contaminated food source from

(a)

(b)

Figure 36.9 Mushroom production. *(a)* A flush of the mushroom *Agaricus bisporus*. *(b)* The shiitake mushroom *Lentinus edulus*.

a food-processing plant or a restaurant may affect a large number of people. In 2010, chicken feed contaminated with *Salmonella* used at two egg production farms in Iowa infected eggs distributed nationally, and caused over 1500 infections. Each year in the United States, there are an estimated 25,000 foodborne disease outbreaks. As many as 76 million Americans are affected, an estimated 13 million acquire significant illnesses, 325,000 are hospitalized, and 5000 people die from foodborne diseases each year. Most outbreaks are due to improper food handling and preparation by consumers and affect small numbers of individuals, usually in the home. Occasional outbreaks affect large numbers of individuals because they are caused by breakdowns in safe food handling and preparation at food-processing and distribution plants. Most foodborne illnesses are unreported because the connection between food and illness is not made.

Foodborne illness is largely preventable; appropriate monitoring of food sources and disease outbreaks provides the basis for protecting consumers. The food industry and the government set standards and monitor food sources to control and prevent foodborne disease.

36.4 Foodborne Disease and Microbial Sampling

The most prevalent foodborne diseases in the United States are classified as *food poisonings* (*FP*) or *food infections* (*FI*); some diseases fall into both categories. **Table 36.6** lists the microorganisms that cause these diseases. Special microbial sampling techniques are necessary to isolate and identify the pathogens and toxins responsible for foodborne diseases, and a variety of growth-dependent, immunological, and molecular techniques are used. Foodborne illnesses and outbreaks are reported to the Centers for Disease Control and Prevention through PulseNet and FoodNet reporting systems.

Foodborne Diseases

Food poisoning, also called **food intoxication**, is disease that results from ingestion of foods containing preformed microbial toxins. The microorganisms that produced the toxins do not have to grow in the host and are often not alive at the time the contaminated food is consumed; the ingestion and action of a bioactive toxin causes the illness. We previously discussed some of these toxins, notably the exotoxin of *Clostridium botulinum*

Table 36.6 *Annual foodborne disease estimates for the United States[a]*

Organism	Disease[b]	Number per year	Foods
Bacteria			
Bacillus cereus	FP and FI	27,000	Rice and starchy foods, high-sugar foods, meats, gravies, pudding, dry milk
Campylobacter jejuni	FI	1,963,000	Poultry, dairy
Clostridium perfringens	FP and FI	248,000	Meat and vegetables held at improper storage temperature
Escherichia coli O157:H7	FI	63,000	Meat, especially ground meat, raw vegetables
Other enteropathogenic *Escherichia coli*	FI	110,000	Meat, especially ground meat, raw vegetables
Listeria monocytogenes	FI	2,500	Refrigerated "ready to eat" foods
Salmonella spp.	FI	1,340,000	Poultry, meat, dairy, eggs
Staphylococcus aureus	FP	185,000	Meat, desserts
Streptococcus spp.	FI	50,000	Dairy, meat
Yersinia enterocolitica	FI	87,000	Pork, milk
All other bacteria	FP and FI	102,000	
Total bacteria		**4,177,500**	
Protists			
Cryptosporidium parvum	FI	30,000	Raw and undercooked meat
Cyclospora cayetanensis	FI	16,000	Fresh produce
Giardia intestinalis	FI	200,000	Contaminated or infected meat
Toxoplasma gondii	FI	113,000	Raw and undercooked meat
Total protists		**359,000**	
Viruses			
Noroviruses	FI	9,200,000	Shellfish, many other foods
All other viruses	FI	82,000	
Total viruses		**9,282,000**	
Total annual foodborne diseases		**13,818,500**	

[a]Estimates are based on data provided by the Centers for Disease Control and Prevention, Atlanta, Georgia, USA, and are typical of recent years.
[b]FP, food poisoning: FI, food infection.

(⮌ Figure 27.22) and the superantigen toxins of *Staphylococcus* and *Streptococcus* (⮌ Section 28.10). **Food infection** is ingestion of food containing sufficient numbers of viable pathogens to cause infection and disease in the host. We discuss major foodborne infections in Sections 36.8–36.12.

Microbial Sampling for Foodborne Disease

Along with nonpathogenic microorganisms that cause spoilage, pathogenic microorganisms may be present in fresh foods. Rapid diagnostic methods that do not require pathogen growth or culture have been developed to detect important food pathogens such as *Escherichia coli* O157:H7, *Salmonella*, *Staphylococcus*, and *Clostridium botulinum*. Molecular and immunology-based tests are used to identify both toxin and pathogen contamination of foods and other products such as drugs and cosmetics. We discussed the use of nucleic acid probes and the polymerase chain reaction for the detection of specific pathogens, including foodborne pathogens, in Sections 31.12 and 31.13. The presence of a foodborne pathogen or toxin is not sufficient to link a particular food to a specific foodborne disease outbreak; the suspect pathogen or toxin must be isolated and identified to establish its role in a foodborne illness.

Isolation and growth of pathogens from nonliquid foods usually require preliminary treatment to suspend microorganisms embedded or entrapped within the food. A standard method uses a specialized blender called a *stomacher* (**Figure 36.10**). The stomacher processes a wide variety of solid and semisolid samples such as fresh and processed meat, dry fruits, cereals, grains, seeds, cheese, cosmetics, and for biomedical applications, pharmaceutical products and tissue samples. The sample is sealed in a sterile bag. Paddles in the stomacher crush, blend, and homogenize the

samples under conditions that prevent contamination by other organisms. Although a traditional blender could also be used to process samples, the sealed bag stomacher system prevents contamination from outside sources, eliminates cleanup between each sample run, and eliminates generation of aerosols. The homogenized samples can then be analyzed in various ways.

Foods sampled for microorganisms or toxins should be examined as soon after processing as possible; if examination cannot begin within 1 hour of sampling, the food should be refrigerated. Frozen food should be thawed in the original container in a refrigerator and examined or cultured as soon as thawing is complete. In addition to identifying pathogens in food, disease investigators must obtain foodborne pathogens from the disease outbreak patients to establish a cause-and-effect relationship between the pathogen and the illness. In many cases, fecal samples can be cultured to recover suspected foodborne pathogens.

Food or patient samples can be inoculated onto enriched media, followed by transfer to differential or selective media for isolation and identification, as described for the isolation of human pathogens (⮌ Section 31.2). Final identification of foodborne pathogens is based on growth characteristics and biochemical reaction patterns. The use of molecular and genetic methods such as the polymerase chain reaction, enzyme immunoassays, nucleic acid probes, nucleic acid sequencing, pulsed-field gel electrophoresis (PFGE), and ribotyping may be used to identify specific organisms.

MiniQuiz
- Distinguish between food infection and food poisoning.
- Describe microbial sampling procedures for solid foods such as meat.

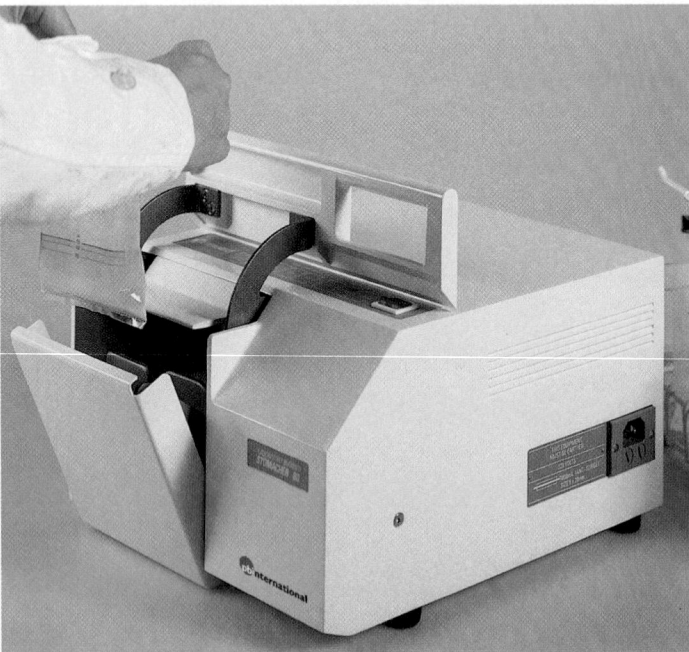

Figure 36.10 A stomacher. Paddles in this specialized blender homogenize the solid food sample in a sealed, sterile bag. The sample is suspended in a sterile solution.

36.5 Foodborne Disease Epidemiology

There are often clusters of cases of a foodborne disease in a particular place because microorganisms from a single common contaminated food, such as salads or hamburgers served from a home, school cafeteria, college dining hall, restaurant, or mess hall, are ingested by many individuals. In addition, central processing plants and central food distribution centers provide opportunities for contaminated foods to cause multiple disease outbreaks in far-flung locations, as when contaminated spinach grown in California caused outbreaks across the United States. We shall see how the food epidemiologist tracks outbreaks and determines their source, often down to the field, processing plant, or point-of-preparation facility in which the food was contaminated.

Spinach and *Escherichia coli* O157:H7

In 2006 an outbreak of illness associated with *Escherichia coli* O157:H7 occurred in the United States and was linked to the consumption of ready-to-eat packaged fresh spinach. The outbreak was quickly traced to a food-processing facility in California. First linked to the spinach product in September, the outbreak caused at least 199 infections. Of these, 102 individuals were hospitalized and 31 developed hemolytic uremic syndrome. At least three deaths were attributed to the outbreak.

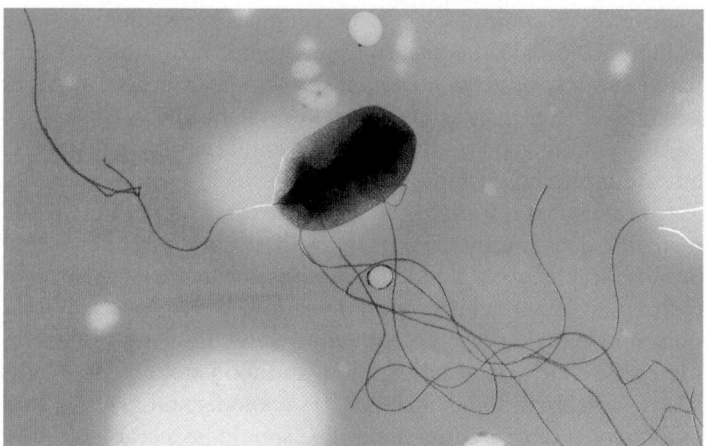

Figure 36.11 *Escherichia coli* O157:H7. The cell, about 1 μm in diameter, as it appears in a colorized transmission electron micrograph showing peritrichous flagella.

The remarkably short duration and rapid end to this epidemic—the first case was confirmed in late August and the last reported in early October—is a testament to efficiency and cooperation among public health facilities across the country. We discussed surveillance networks for infectious disease information in Chapter 32. In this case, two of these networks—FoodNet and PulseNet—were used to define the source and stop the outbreak. The contaminated spinach was distributed nationwide from the California processing plant, but most disease cases were not in the West. The two states affected most were Wisconsin, with 49 cases, and Ohio, with 25 cases; there were only 2 cases in California.

Because *E. coli* O157:H7 (**Figure 36.11**) has been well studied, public health officials were able to identify the strain found in the bagged spinach and determine its origin. They conclusively linked the outbreak to the bagged spinach, traced it back to the processing plant, and eventually traced it to an agricultural field in the vicinity of the processing plant. DNA from the organisms isolated from regional outbreaks was typed using pulsed-field gel electrophoresis (PFGE), a form of gel electrophoresis that better distinguishes between large molecules and is used in pathogen identification. The patterns obtained were then compared; the results showed that the same strain was responsible for the disease in various parts of the country. The common thread in the geographically isolated outbreaks was consumption of the suspected lots of bagged spinach originating from a single California facility.

The precise source of the outbreak, although it has been traced to a field near the processing plant, remains unknown. Feral pigs and domestic cattle are present in the vicinity of the identified field, and contaminated wells or surface waters used for irrigation may have introduced the pathogen into the fields and eventually into the spinach. The original source was almost certainly animal in origin, as *E. coli* is an enteric organism found naturally only in the intestine of animals.

The spinach epidemic, although serious and even deadly for some, was discovered, contained, and stopped very quickly. However, this incident also shows how centralized food-processing facilities can quickly spread disease to large and distant popula-

tions. Food hygiene standards and surveillance must be maintained at the highest possible level in central food-processing and distribution facilities.

Food Disease Reporting

In the United States foodborne outbreaks are reportable to the Centers for Disease Control and Prevention through *FoodNet*. Identification of particular organisms responsible for foodborne disease outbreaks is particularly important. A reporting system called *PulseNet International* is an international molecular subtyping network for foodborne disease surveillance. It consists of national and regional PulseNet organizations from the United States, Canada, Europe, Asia, Latin America, the Caribbean, and the Middle East. The organization collects and shares molecular subtyping data from PFGE DNA fingerprints of organisms implicated in foodborne disease outbreaks. Tracking the characteristics of foodborne illnesses and identifying the causal agents often allows epidemiologists to pinpoint the original source of contaminated food, as we discussed above.

MiniQuiz
- Identify the potential for a foodborne disease outbreak from a single contamination event at a centralized food-processing facility.
- Describe tracking of a foodborne disease outbreak.

Food Poisoning

Food poisoning can be caused by various bacteria and fungi. Here we consider *Staphylococcus* and *Clostridium*, the two genera responsible for the highest numbers of microbial food poisoning cases.

36.6 Staphylococcal Food Poisoning

Food poisoning is often caused by staphylococcal enterotoxin (SE) produced by the bacterium *Staphylococcus aureus*. Staphylococci are small, gram-positive cocci (**Figure 36.12**; ↩ Section 18.1) and, as we discussed in Section 33.9, they are normal members of the flora of the skin and upper respiratory tract of at least 20–30% of all humans, and are often opportunistic pathogens. *S. aureus* is frequently associated with food poisoning because it can grow in many common foods, and some strains produce several heat-stable enterotoxins. SE consumed in food produces gastroenteritis characterized by nausea, vomiting, and diarrhea, usually within 1–6 hours.

Epidemiology

Each year there are an estimated 185,000 cases of staphylococcal food poisoning in the United States (Table 36.6). The foods most commonly responsible are custard- and cream-filled baked goods, poultry, eggs, raw and processed meat, puddings, and creamy salad dressings. Salads prepared with mayonnaise-based dressings such as those containing shellfish, chicken, pasta, tuna, potato, egg, or meat are also commonly implicated. If these foods are refrigerated immediately after preparation, they usually remain

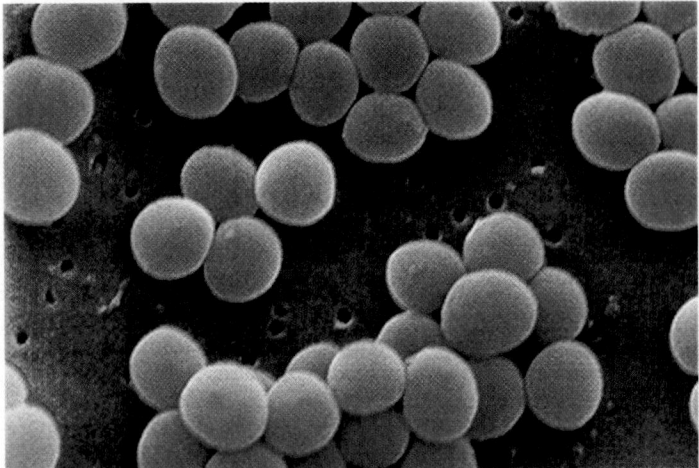

Figure 36.12 *Staphylococcus aureus.* In this colorized scanning electron micrograph, the individual gram-positive cocci are about 0.8 μm in diameter. Staphylococci divide in multiple planes, producing the appearance that gives the genus its name (from the Greek *staphyle*, bunch of grapes).

safe because *S. aureus* grows poorly at low temperatures. However, foods kept at room temperature in kitchens or outdoors at picnics can support rapid bacterial growth and enterotoxin production if contaminated with *S. aureus*. Even if the toxin-containing foods are heated before eating, the heat-stable toxin may remain active. Some SEs are stable for over 16 hours at 60°C, a temperature that would kill *S. aureus*. Live *S. aureus* need not be present in foods causing illness: The illness is solely due to the preformed SE.

Staphylococcal Enterotoxins
S. aureus strains produce up to 20 different but related SEs. Most strains of *S. aureus* produce only one or two of these toxins, and some strains are nonproducers. However, any one of the toxins can cause staphylococcal food poisoning. These enterotoxins are further classified as *superantigens*. Superantigens stimulate large numbers of T cells, which in turn release intercellular mediators called *cytokines*. In the intestine, superantigens activate a general inflammatory response that causes gastroenteritis and significant fluid loss due to diarrhea and vomiting (⮌ Section 28.10).

The *S. aureus* enterotoxins are called SEA, SEB, SEC, and SED and are encoded by the genes *SEA*, *SEB*, *SEC*, and *SED*. *SEB* and *SEC* are on the bacterial chromosome, *SEA* is on a lysogenic bacteriophage, and *SED* is on a plasmid. The *S. aureus* SE genes are genetically related. The phage- and plasmid-encoded genes are movable genetic elements that can transfer toxin production to nontoxigenic strains of *Staphyloccus* by horizontal gene transfer (⮌ Section 12.11).

Diagnosis, Treatment, and Prevention
Certain assays detect SEs in food, and other assays detect *S. aureus* exonuclease, an enzyme that degrades DNA, as a metabolite in food. These qualitative tests confirm that *S. aureus* is or has been present. To obtain quantitative data and determine the extent of bacterial contamination, bacterial plate counts are required. For

staphylococcal counts, a high-salt medium (either sodium chloride or lithium chloride at a final concentration of 7.5%) is used. Compared to most bacteria present in foods, staphylococci thrive in habitats with a high salt content and low water activity.

The symptoms of *S. aureus* food poisoning can be quite severe, but are typically self-limiting, usually resolving within 48 hours as the toxin passes from the body. Severe cases may require treatment for dehydration. Treatment with antibiotics is not useful because staphylococcal food poisoning is caused by a preformed toxin, not an active bacterial infection. Staphylococcal food poisoning can be prevented by proper sanitation and hygiene in food production, food preparation, and food storage. As a rule, foods susceptible to colonization by *S. aureus* and kept for several hours at temperatures above 4°C should be discarded rather than eaten.

MiniQuiz
- Identify the symptoms and mechanism of staphylococcal food poisoning.
- Why does antibiotic treatment not affect the outcome or the severity of disease in patients with staphylococcal food poisoning?

36.7 Clostridial Food Poisoning

Clostridium perfringens and *Clostridium botulinum* cause serious food poisoning. Members of the genus *Clostridium* are anaerobic endospore-forming rods (⮌ Section 18.2). Canning and cooking procedures kill living organisms but do not necessarily kill all endospores. Under appropriate anaerobic conditions, the endospores in food can germinate and produce toxin.

Clostridium perfringens Food Poisoning
C. perfringens is an anaerobic, gram-positive, endospore-forming rod commonly found in soil (**Figure 36.13**). *C. perfringens* is also found in sewage, primarily because it lives in small numbers in the intestinal tract of many humans and animals. *C. perfringens*

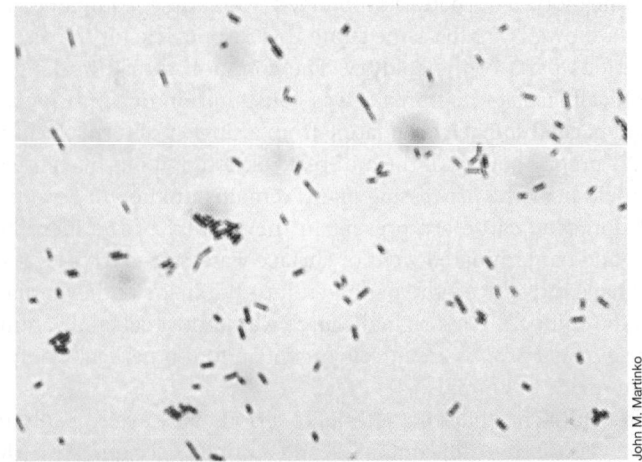

John M. Martinko

Figure 36.13 *Clostridium perfringens.* The Gram stain shows individual gram-positive rods about 1 μm in diameter.

is the most often reported cause of food poisoning in the United States, with an estimated 248,000 annual cases (Table 36.6).

Perfringens food poisoning requires the ingestion of a large dose of *C. perfringens* ($>10^8$ cells) in contaminated cooked or uncooked foods, usually high-protein foods such as meat, poultry, and fish. Large numbers of *C. perfringens* can grow in meat dishes cooked in bulk where heat penetration is often insufficient. Surviving *C. perfringens* endospores germinate under anoxic conditions, as in sealed containers such as jars or cans. The *C. perfringens* grows quickly in the food, especially if left to cool at 20–40°C for short time periods. However, the toxin is not yet present at this stage.

Ingested with contaminated food, the living *C. perfringens* begin to sporulate and produce toxin in the consumer's intestine (Table 27.4). The perfringens enterotoxin alters the permeability of the intestinal epithelium, leading to nausea, diarrhea, and intestinal cramps, usually with no fever. The onset of perfringens food poisoning begins about 7–15 hours after consumption of the contaminated food, but usually resolves within 24 hours. Fatalities are rare.

Diagnosis, Treatment, and Prevention

Diagnosis of perfringens food poisoning is made by isolation of *C. perfringens* from the feces or, more reliably, by a direct enzyme immunoassay to detect *C. perfringens* enterotoxin in feces. Because *C. perfringens* food poisoning is self-limiting, antibiotic treatment is not indicated. Supportive therapy—fluids and electrolyte replacement—may be used in serious cases. Prevention of perfringens food poisoning requires preventing contamination of raw and cooked foods and proper heating of all foods during cooking and canning. Cooked foods should be refrigerated as soon as possible to rapidly lower temperatures and inhibit *C. perfringens* growth.

Botulism

Botulism is a severe, often fatal, food poisoning caused by the consumption of food containing the exotoxin produced by *C. botulinum*. This bacterium normally inhabits soil or water, but its endospores may contaminate raw foods. If the foods are properly processed so that the *C. botulinum* endospores are removed or killed, no problem arises; however, if viable endospores remain in the food, they may germinate and produce botulinum toxin. Ingesting even a small amount of this neurotoxin can be dangerous.

We discussed the nature and activity of botulinum toxin in Section 27.10 (Figure 27.22). Botulinum toxin is a neurotoxin that causes flaccid paralysis, usually affecting the autonomic nerves that control body functions such as respiration and heartbeat. At least seven distinct botulinum toxins are known. Because the toxins are destroyed by heat (80°C for 10 minutes), thoroughly cooked food, even though contaminated with toxin, is totally harmless.

Most cases of foodborne botulism are caused by eating processed foods contaminated with *C. botulinum* endospores. Typically, such foods are consumed without cooking after processing. For example, nonacid, home-canned vegetables such as corn and beans are often used without cooking when making cold salads. Smoked and fresh fish, vacuum-packed in plastic, are also often

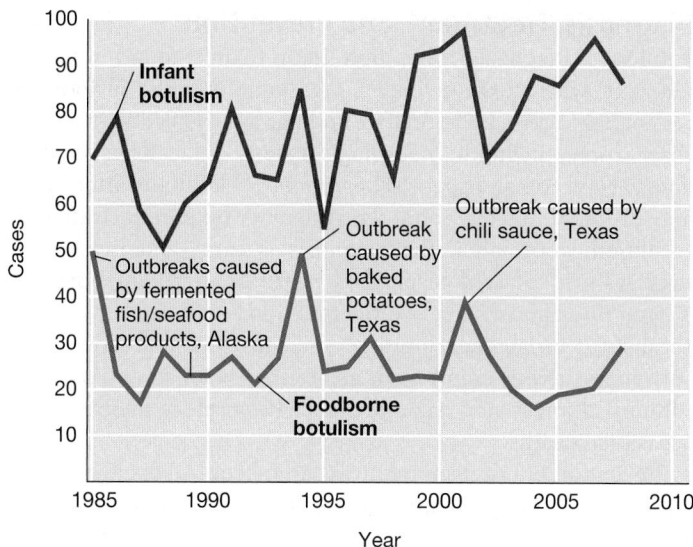

Figure 36.14 Botulism in the United States. Both foodborne and infant botulism are shown. In years with high numbers of cases, major outbreaks that account for the increase are indicated. Data are from the Centers for Disease Control and Prevention, Atlanta, Georgia, USA.

eaten without cooking. Under such conditions, viable *C. botulinum* endospores may germinate, and the vegetative cells may produce sufficient toxin to cause severe food poisoning. An average of 25 cases of foodborne botulism, about 18% of all botulism, occurred annually in the United States between 2000 and 2007 (**Figure 36.14**).

The majority of botulism cases occur following *infection* with *C. botulinum*. For example, infant botulism occurs after newborns ingest endospores of *C. botulinum* (Figure 36.14). In most cases, the source cannot be identified because *C. botulinum* endospores are widespread. If the infant's normal flora is not developed or if the infant is undergoing antibiotic therapy, ingested endospores can germinate in the infant's intestine, triggering *C. botulinum* growth and toxin production. Most cases of infant botulism occur between the first week of life and 2 months of age, rarely occurring in children older than 6 months, presumably because the normal intestinal flora is more developed. Over 60% of all botulism cases in the United States are in infants. An average of 86 cases of infant botulism occurred annually in the United States from 2000 to 2007 (Figure 36.14). Wound botulism can also occur from infection, presumably from endospores in contaminating material introduced via a parenteral route. Wound botulism is most commonly associated with illicit injectable drug use; in the United States an average of about 29 cases occurred annually from 2000 to 2007.

All forms of botulism are quite rare, with at most six cases occurring per 10 million individuals per year in the United States. Botulism, however, is a very serious disease because of the high mortality associated with the disease; about 16% of all foodborne cases are fatal. Death occurs from respiratory paralysis or cardiac arrest due to the paralyzing action of the botulinum neurotoxin.

Diagnosis, Treatment, and Prevention

Botulism is diagnosed when botulinum toxin is found in patient serum or when toxin or live *C. botulinum* is found in food the patient has ingested. Laboratory findings are coupled with clinical observations, including neurological signs of localized paralysis (impaired vision and speech) beginning 18–24 hours after ingestion of contaminated food. Treatment is by administration of botulinum antitoxin if the diagnosis is early, and mechanical ventilation for flaccid respiratory paralysis. In infant botulism, *C. botulinum* and toxin are often found in bowel contents. Infant botulism is usually self-limiting, and most infants recover with only supportive therapy, such as assisted ventilation. Antitoxin administration is not recommended. Respiratory failure causes occasional deaths.

Prevention of botulism requires careful control of canning and preservation methods. Susceptible foods should be heated to destroy endospores; boiling for 20 minutes destroys the toxin. Home-prepared foods are the most common source of foodborne botulism outbreaks.

MiniQuiz

- Describe the events that lead to *Clostridium perfringens* food poisoning. What is the likely outcome of the poisoning?
- Describe the development of botulism in adults and infants. What is the likely outcome of botulism?

Food Infection

Food infection results from ingestion of food containing sufficient numbers of viable pathogens to cause infection and disease in the host. Food infection is very common (Table 36.6), and we begin with a common bacterial cause, *Salmonella*. Many food infection agents can also cause waterborne diseases.

36.8 Salmonellosis

Salmonellosis is a gastrointestinal disease typically caused by foodborne *Salmonella* infection. Symptoms begin after the pathogen colonizes the intestinal epithelium. *Salmonella* are gram-negative, facultatively aerobic, motile rods related to *Escherichia coli* and other enteric bacteria (⮌ Section 17.11). *Salmonella* normally inhabits the animal intestine and is thus found in sewage.

The nomenclature of the *Salmonella* spp. is based on taxonomic schemes that differentiate strains by virtue of biochemical, serological, and molecular (nucleic acid–based) characteristics. The accepted species name for the pathogenic members of the genus is *Salmonella enterica*. Based on nucleic acid analyses, there are seven evolutionary groups or subspecies of *S. enterica*. Most human pathogens fall into group I, designated as a single subspecies, *S. enterica* subspecies *enterica*. Finally, each subspecies may be divided into *serovars* (serological variations, also called *serotypes*). Thus, the organism formally named *Salmonella enterica* subspecies *enterica* serovar Typhi is usually called *Salmonella enterica* serovar Typhi and is often abbreviated to *Salmonella Typhi*. *S. enterica* ser. Typhi causes the serious human disease typhoid fever but is rare in the United States. Most of the 500 or

so annual foodborne cases caused by *S. enterica* ser. Typhi are acquired outside the United States.

A number of other *S. enterica* serovars also cause foodborne gastroenteritis. In all, over 1400 *Salmonella* serovars cause disease in humans. *S. enterica* serovars Typhimurium and Enteritidis are the most common agents of foodborne salmonellosis in humans.

Epidemiology and Pathogenesis

The incidence of salmonellosis has been steady over the last decade, with about 40,000–45,000 documented cases each year (**Figure 36.15**). However, less than 4% of salmonellosis cases are probably reported, so the incidence of salmonellosis is probably over 1 million cases every year (Table 36.6).

The ultimate sources of the foodborne salmonellas are the intestinal tracts of humans and other warm-blooded animals, and there are several routes by which these bacteria may enter the food supply. The bacteria may reach food through fecal contamination from food handlers. Food production animals such as chickens, pigs, and cattle may harbor *Salmonella* serovars that are pathogenic to humans, and the bacteria may be carried through to finished fresh foods such as eggs, meat, and dairy products. *Salmonella* food infections are often traced to products such as custards, cream cakes, meringues, pies, and eggnog made with uncooked eggs. Other foods commonly implicated in salmonellosis outbreaks are meats and meat products such as meat pies, cured but uncooked sausages and meats, poultry, milk, and milk products.

The most common salmonellosis is enterocolitis. Ingestion of food containing viable *Salmonella* results in colonization of the small and large intestine. Onset of the disease occurs 8–48 hours after ingestion. Symptoms include the sudden onset of headache,

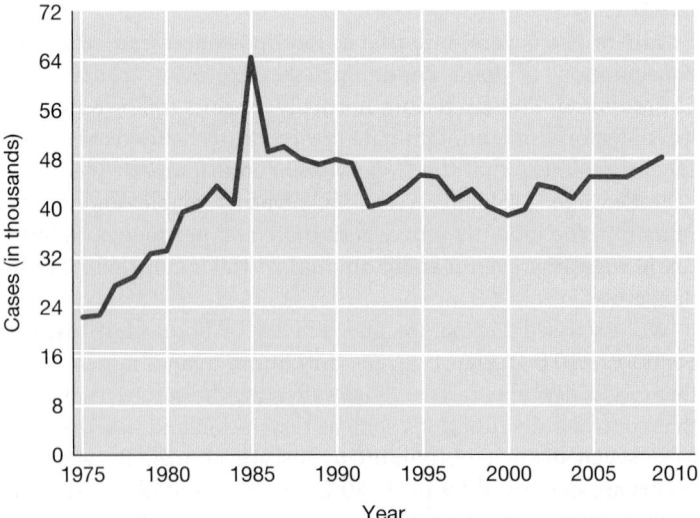

Figure 36.15 Reported cases of salmonellosis in the United States. Most cases of salmonellosis are foodborne. The total number of reported cases in 2007 was 47,995. The high incidence in 1985 was caused by contamination of pasteurized milk that was mixed with raw (unprocessed) milk in a dairy processing plant in Illinois. Data are from the Centers for Disease Control and Prevention, Atlanta, Georgia, USA.

chills, vomiting, and diarrhea, followed by a fever that lasts a few days. The disease normally resolves without intervention in 2–5 days. After recovery, however, patients may shed *Salmonella* in feces for several weeks. Some patients recover and remain asymptomatic, but shed organisms for months or even years; they are chronic carriers (♻ Section 32.3). A few serovars of *Salmonella* may also cause septicemia (a blood infection) and enteric or typhoid fever, a disease characterized by systemic infection and high fever lasting several weeks. Mortality can approach 15% in untreated typhoid fever.

The pathogenesis of *Salmonella* infections starts with uptake of the organisms from the gut. *Salmonella* ingested in food or water invades phagocytes and grows as an intracellular pathogen, spreading to adjacent cells as host cells die. After invasion, pathogenic *Salmonella* uses a combination of endotoxins, enterotoxins, and cytotoxins to damage and kill host cells (see the Chapter 27 Microbial Sidebar "Virulence in *Salmonella*"), leading to the classic symptoms of salmonellosis.

Diagnosis, Treatment, and Prevention

Foodborne salmonellosis is diagnosed from observation of clinical symptoms, history of recent food consumption, and culturing the organism from feces. Selective and differential media are used to identify *Salmonella* and discriminate it from other gram-negative rods (♻ Table 31.2). Tests for the presence of *Salmonella* are commonly used on animal food products, such as raw meat, poultry, eggs, and powdered milk. *Salmonella* has also been found, however, in nonmeat and nondairy food, including produce (cantaloupes and tomatoes) and peanut butter. Tests for *Salmonella* in food include several rapid tests, but even rapid tests rely on culture-based enrichment procedures to increase *Salmonella* numbers to testable levels. The established standard used by PulseNet for epidemiological investigations is pulsed-field gel electrophoresis (PFGE; Section 36.5). This molecular typing technique can discriminate between various *Salmonella* serovars.

For enterocolitis, treatment is usually unnecessary, and antibiotic treatment does not shorten the course of the disease or eliminate the carrier state. Antibiotic treatment, however, significantly reduces the length and severity of septicemia and typhoid fever. Mortality due to typhoid fever can be reduced to less than 1% with appropriate antibiotic therapy. Multi-drug-resistant *Salmonella* are a significant clinical problem.

Properly cooked foods heated to at least 70°C are generally safe if consumed immediately, held at 50°C, or stored immediately at 4°C. Any foods that become contaminated by an infected food handler can support the growth of *Salmonella* if the foods are held for long periods of time, especially without heating or refrigeration. *Salmonella* infections are more common in summer than in winter, probably because warm environmental conditions generally favor the growth of microorganisms in foods.

Although local laws and enforcement vary, because of the lengthy carrier state, infected individuals are often banned from work as food handlers until their feces are negative for *Salmonella* in three successive cultures.

MiniQuiz
- Describe salmonellosis food infection. How does it differ from food poisoning?
- How might *Salmonella* contamination of food production animals be contained?

36.9 Pathogenic *Escherichia coli*

Most strains of *Escherichia coli* are common members of the enteric microflora in the human colon and are not pathogenic. A few strains, however, are potential foodborne pathogens. There are about 200 known pathogenic *E. coli* strains, all of which act on the intestine. Several are characterized by their production of potent enterotoxins and may cause life-threatening diarrheal disease and urinary tract infections. The pathogenic strains are divided based on the type of toxin they produce and the specific diseases they cause.

Shiga Toxin–Producing *Escherichia coli* (STEC)

Shiga toxin–producing *Escherichia coli* (STEC) produce *verotoxin*, an enterotoxin similar to the Shiga toxin produced by *Shigella dysenteriae* (♻ Table 27.4). Formerly known as enterohemorrhagic *E. coli* (EHEC), the most widely distributed STEC is *E. coli* O157:H7 (Figure 36.11). Up to 90% of all STEC infections are caused by *E. coli* O157:H7. After a person ingests food or water containing STEC, the bacteria grow in the small intestine and produce verotoxin. Verotoxin causes both hemorrhagic (bloody) diarrhea and kidney failure. *E. coli* O157:H7 causes an estimated 60,000 infections and 50 deaths from foodborne disease in the United States each year (Table 36.6). STEC strains are the leading cause of hemolytic uremic syndrome and kidney failure, with 292 cases reported in 2007, about half in children under 5 years of age.

About 40% of STEC infections are caused by the consumption of contaminated uncooked or undercooked meat, particularly mass-processed ground beef. *E. coli* O157:H7 is a member of the normal microbiome in healthy cattle; it can enter the human food chain if meat is contaminated with intestinal contents during slaughter and processing. In several major outbreaks in the United States caused by *E. coli* O157:H7, infected ground beef from regional distribution centers was the source of contamination. Infected meat products caused disease in several states. Another outbreak was caused by processed and cured, but uncooked beef in ready-to-eat sausages. The source of contamination was the beef, and the *E. coli* O157:H7 probably originated from slaughtered beef carcasses.

In 2003, the Food Safety and Inspection Service of the United States Department of Agriculture reported 20 positive results of 6584 samples (0.03%) of ground beef analyzed for *E. coli* O157:H7. *E. coli* O157:H7 has also been implicated in food infection outbreaks from dairy products, fresh fruit, and raw vegetables. Contamination of the fresh foods by fecal material, typically from cattle carrying the *E. coli* O157:H7 strain, has been implicated in several of these cases, as we discussed in Section 36.5.

Because *E. coli* O157:H7 grows in the intestines and is found in fecal material, it is also a potential source of waterborne gastrointestinal disease. Several outbreaks have also occurred in day-care facilities, where the presumed route of exposure is oral–fecal contamination.

Other Pathogenic *Escherichia coli*

Children in developing countries often contract diarrheal disease caused by *E. coli*. *E. coli* can also be the cause of "traveler's diarrhea," a common enteric infection causing watery diarrhea in travelers to developing countries. The primary causal agents are the enterotoxigenic *E. coli* (ETEC). The ETEC strains usually produce one of two heat-labile, diarrhea-producing enterotoxins. In studies of United States citizens traveling in Mexico, the infection rate with ETEC is often greater than 50%. The prime vehicles are foods such as fresh vegetables (for example, lettuce in salads) and water. The very high infection rate in travelers is due to contamination of local public water supplies. The local population is usually resistant to the infecting strains, presumably because they have acquired resistance to the endemic ETEC strains. Secretory IgA antibodies in the bowel prevent colonization of the pathogen in local residents, but the organism readily infects the nonimmune travelers and causes disease.

Enteropathogenic *E. coli* (EPEC) strains cause diarrheal diseases in infants and small children but do not cause invasive disease or produce toxins. Enteroinvasive *E. coli* (EIEC) strains cause invasive disease in the colon, producing watery, sometimes bloody diarrhea. The EIEC strains are taken up by phagocytes, but escape lysis in the phagolysosomes, grow in the cytoplasm, and move into other cells in much the same way as pathogenic *Salmonella* strains. This invasive disease causes diarrhea and is common in developing countries.

Diagnosis and Treatment

Illness from *E. coli* O157:H7 and other STEC strains is a reportable infectious disease in the United States. The general pattern established for diagnosis, treatment, and prevention of infection by *E. coli* O157:H7 reflects current procedures used for all of the pathogenic *E. coli* strains. Laboratory diagnosis requires culture from the feces and identification of the O (lipopolysaccharide) and H (flagellar) antigens and toxins by serology. Identification of strains is also done using DNA analyses such as restriction fragment length polymorphism and PFGE. *E. coli* O157:H7 outbreaks are reported through FoodNet and PulseNet to the Centers for Disease Control and Prevention.

Treatment of *E. coli* O157:H7 and other STEC infections includes supportive care and monitoring of renal function, blood hemoglobin, and platelets. Antibiotics may be harmful because they may cause the release of large amounts of verotoxin from dying *E. coli* cells. For other pathogenic *E. coli* infections, treatment usually includes supportive therapy and, for severe cases and invasive disease, antimicrobial drugs to shorten and eliminate infection.

Prevention

The most effective way to prevent infection with foodborne STEC is to make sure that meat is cooked thoroughly, which means that it should appear gray or brown and juices should be clear. As we discussed above (Section 36.2), the United States has approved the irradiation of ground meat as an acceptable means of eliminating or reducing food infection bacteria, largely because *E. coli* O157:H7 has been implicated in several foodborne epidemics. To process foods such as ground beef, large-scale production plants may mix and grind meat from hundreds or even thousands of animals together; the grinding process could distribute the pathogens from a single infected animal throughout the meat. Short of cooking, penetrating radiation is considered the only effective means to ensure decontamination.

In general, proper food handling, water purification, and appropriate hygiene prevent the spread of pathogenic *E. coli*. Raw foods should be washed thoroughly. Traveler's diarrhea can be prevented by avoiding consumption of local water and uncooked foods.

MiniQuiz

- Describe the pathology of *Escherichia coli* food infections due to STEC, ETEC, EPEC, and EIEC strains.
- Why is *E. coli* O157:H7 considered a dangerous and reportable pathogen?

36.10 *Campylobacter*

Species of *Campylobacter* are the most common reported cause of bacterial foodborne infections in the United States. Cells of *Campylobacter* species are gram-negative, motile, curved rods to spiral-shaped bacteria that grow at reduced oxygen tension as microaerophiles (⌇⌇ Section 17.19). Several pathogenic species, *Campylobacter jejuni* (**Figure 36.16**), *C. coli*, and *C. fetus*, are recognized. *C. jejuni* and *C. coli* account for almost 2 million annual cases of bacterial diarrhea (Table 36.6). *C. fetus* is a major cause of sterility and spontaneous abortion in cattle and sheep.

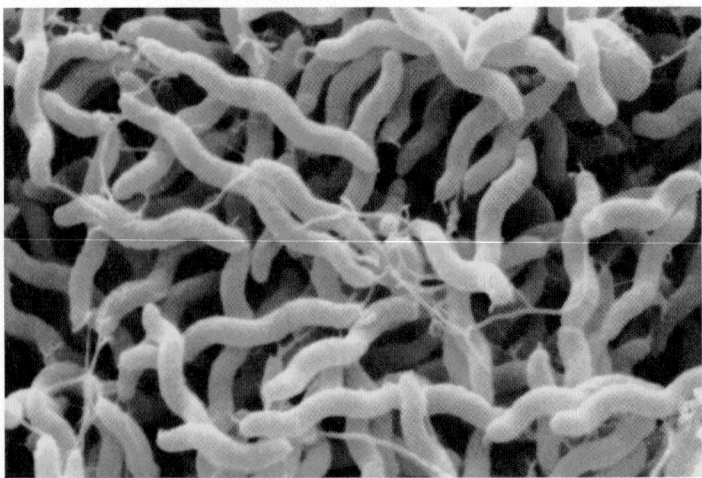

Figure 36.16 *Campylobacter jejuni*. The gram-negative curved rods shown in this colorized scanning electron micrograph are about 1 μm in diameter.

Epidemiology and Pathology

Campylobacter is transmitted to humans via contaminated food, most frequently in poultry, pork, raw shellfish, or in surface waters. *C. jejuni* is a normal resident in the intestinal tract of poultry; virtually all chickens and turkeys are normally colonized with this organism. According to the United States Department of Agriculture, up to 90% of turkey and chicken carcasses and over 30% of hog carcasses may be contaminated with *Campylobacter*. Beef, on the other hand, is rarely a vehicle for this pathogen. *Campylobacter* species also infect domestic animals such as dogs, causing a milder form of diarrhea than that observed in humans. *Campylobacter* infections in infants are frequently traced to infected domestic animals, especially dogs.

After a person ingests cells of *Campylobacter*, the organism multiplies in the small intestine, invades the epithelium, and causes inflammation. Because *C. jejuni* is sensitive to gastric acid, cell numbers as high as 10^4 may be required to initiate infection. However, this number may be reduced to less than 500 if the bacteria are ingested in food, or are ingested by a person taking medication to reduce stomach acid production. *Campylobacter* infection causes a high fever (usually greater than 104°F or 40°C), headache, malaise, nausea, abdominal cramps, and profuse diarrhea with watery, frequently bloody, stools. The disease subsides in about 7–10 days. Spontaneous recovery from *Campylobacter* infections is often complete, but relapses occur in up to 25% of cases.

Diagnosis, Treatment, and Prevention

Diagnosis of *Campylobacter* food infection requires isolation of the organism from stool samples and identification by growth-dependent tests, immunological assays, or molecular tests. Serious *C. jejuni* infections are often seen in infants. In these cases, diagnosis is important; selective media and specific immunological methods have been developed for positive identification of this organism. Erythromycin and quinolone treatment may be useful early in severe diarrheal disease. Adequate personal hygiene, proper washing of uncooked poultry (and any kitchenware coming in contact with uncooked poultry), and thorough cooking of meat eliminate *Campylobacter* contamination.

As with other foodborne infections, epidemiologic investigations are based on PFGE analysis of recovered organisms. Data shared on PulseNet are used to track the spread of *Campylobacter* and determine its origin.

MiniQuiz
- Describe the pathology of *Campylobacter* food infection. What is the likely outcome?
- How might *Campylobacter* contamination of food production animals be controlled?

36.11 Listeriosis

Listeria monocytogenes causes **listeriosis**, a gastrointestinal food infection that may lead to bacteremia and meningitis. *L. monocytogenes* is a short, gram-positive, nonsporulating coccobacillus

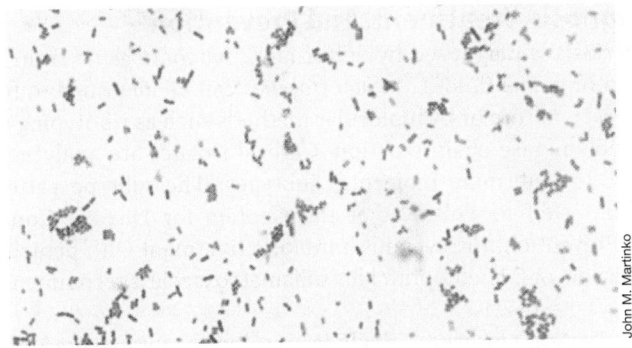

Figure 36.17 *Listeria monocytogenes.* This Gram stain shows gram-positive coccobacilli, about 0.5 μm in diameter.

that is acid-, salt- and cold-tolerant and facultatively aerobic (**Figure 36.17**; ↩ Section 18.1).

Epidemiology and Pathology

L. monocytogenes is found widely in soil and water; virtually no food source is safe from possible *L. monocytogenes* contamination. Food can become contaminated at any stage during food production or processing. Food preservation by refrigeration, which ordinarily slows microbial growth, is ineffective in limiting growth of this psychrotolerant organism. Ready-to-eat meats, fresh soft cheeses, unpasteurized dairy products, and inadequately pasteurized milk are the major food vehicles for this pathogen, even when foods are properly stored at refrigerator temperature (4°C).

L. monocytogenes is an intracellular pathogen. It enters the body through the gastrointestinal tract in contaminated food. Phagocytes take up the pathogen in a phagolysosome. This triggers production of listeriolysin O, which lyses the phagolysosome and releases *L. monocytogenes* into the cytoplasm. Here it multiplies and produces ActA, a surface protein that induces host cell actin polymerization, which moves the pathogen to the cytoplasmic membrane. At the cytoplasmic membrane, the complex pushes out, forming protrusions called filopods. The filopods are then ingested by surrounding cells and the cycle starts again. This mechanism allows *L. monocytogenes* to move from cell to cell without exposure to antibodies, complement, or neutrophils. Specific immunity to *L. monocytogenes* is through cell-mediated T_H1 inflammatory cells (↩ Section 29.6). Particularly susceptible populations include the elderly, pregnant women, newborns, and immunosuppressed individuals [for example, transplant patients undergoing steroid therapy and acquired immunodeficiency syndrome (AIDS) patients].

Although exposure to *L. monocytogenes* is undoubtedly very common, there are only about 2500 estimated cases of clinical listeriosis each year, and fewer than 1000 are reported. Nearly all diagnosed cases require hospitalization. Acute listeriosis is rare and is characterized by septicemia, often leading to meningitis, with a mortality rate of 20% or higher. About 30–40 listeriosis deaths are reported annually in the United States.

UNIT 12

Diagnosis, Treatment, and Prevention

Listeriosis is diagnosed by culturing *L. monocytogenes* from the blood or spinal fluid. *L. monocytogenes* can be identified in food by direct culture or by molecular methods such as ribotyping and the polymerase chain reaction. Clinical isolates are analyzed by PFGE to determine molecular subtypes. The subtype patterns are reported to PulseNet at the Centers for Disease Control and Prevention. Intravenous antibiotic treatment with penicillin, ampicillin, or trimethoprim plus sulfamethoxazole is recommended for invasive disease.

Prevention measures include recalling contaminated food and taking steps to limit *L. monocytogenes* contamination at the food-processing site. Because *L. monocytogenes* is susceptible to heat and radiation, raw food and food-handling equipment can be readily decontaminated. However, without pasteurizing or cooking the finished food product, the risk of contamination cannot be eliminated because of the widespread distribution of the pathogen.

Individuals who are immunocompromised should avoid unpasteurized dairy products and ready-to-eat processed meats. Pregnant women should also avoid foods that may transmit *L. monocytogenes* because spontaneous abortion is a frequent outcome of listeriosis.

MiniQuiz

- What is the likely outcome of *Listeria monocytogenes* exposure in normal individuals?

- What populations are most susceptible to serious disease from *L. monocytogenes* infection? Why?

36.12 Other Foodborne Infectious Diseases

Over 200 other microorganisms, viruses, and other infectious agents such as prions contribute to foodborne diseases, and we consider a few of them here.

Bacteria

Table 36.6 lists several bacteria that cause human foodborne disease that we have not covered in this chapter. *Yersinia enterocolitica* is commonly found in the intestines of domestic animals and causes foodborne infections due to contaminated meat and dairy products. The most serious consequence of *Y. enterocolitica* infection is enteric fever, a severe life-threatening infection. *Bacillus cereus* produces two enterotoxins that cause diarrhea and vomiting. The organism grows in foods such as rice, pasta, meats, or sauces that are cooked and left at room temperature to cool slowly. Endospores of this gram-positive rod germinate and toxin is produced. Reheating may kill the *B. cereus*, but the toxin may remain active. *B. cereus* may also cause a food infection similar to that caused by *Clostridium perfringens*. *Shigella* species can cause severe invasive gastroenteritis called *shigellosis*. About 20,000 cases of shigellosis are reported each year in the United States, with up to 150 million cases worldwide. Most *Shigella* infections are the result of fecal–oral contamination, but food and water are occasional vehicles. Several members of the *Vibrio* genus

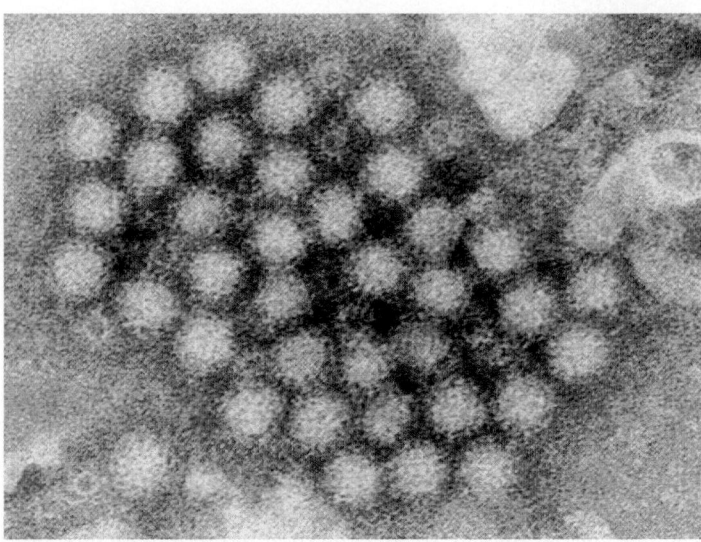

Figure 36.18 Human norovirus. The virus was isolated from a patient with diarrhea. Individual norovirus particles have an indistinct rough outer edge and are about 27 nm in diameter.

cause food poisoning in persons who consume contaminated shellfish.

Viruses

The largest number of annual foodborne infections is thought to be caused by viruses. In general, viral foodborne illness consists of gastroenteritis characterized by diarrhea, often accompanied by nausea and vomiting. Recovery is spontaneous and rapid, usually within 24–48 hours ("24-hour bug"). Noroviruses (**Figure 36.18**) are responsible for most of these mild foodborne infections in the United States (Table 36.6), accounting for over 9 million of the estimated 13 million annual cases of foodborne disease. Rotavirus, astrovirus, and hepatitis A collectively cause 100,000 cases of foodborne disease each year. These viruses inhabit the gut and are often transmitted to food or water with fecal matter. As with many foodborne infections, proper food handling, handwashing, and a source of clean water to prepare fresh foods are essential to prevent infection.

Protists

Important foodborne protist diseases are listed in Table 36.6. Protists including *Giardia intestinalis*, *Cryptosporidium parvum* (🔗 Figures 35.16 and 35.17), and *Cyclospora cayetanensis* (**Figure 36.19a**) can be spread in foods contaminated by fecal matter in untreated water used to wash, irrigate, or spray crops. Fresh foods such as fruits are often implicated as the source of these protists. We discussed giardiasis and cryptosporidiosis as waterborne diseases (🔗 Section 35.6). Cyclosporiasis is an acute gastroenteritis and is an important emerging disease. In the United States, most cases are acquired by eating fresh produce imported from other countries.

Toxoplasma gondii is a protist spread through cat feces, but is also found in raw or undercooked meat. In most individuals, toxoplasmosis is a mild, self-limiting gastroenteritis. However,

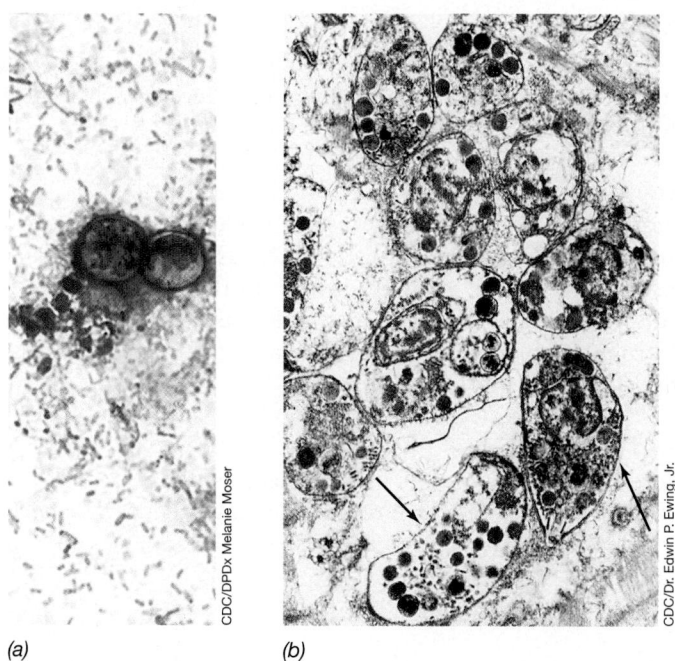

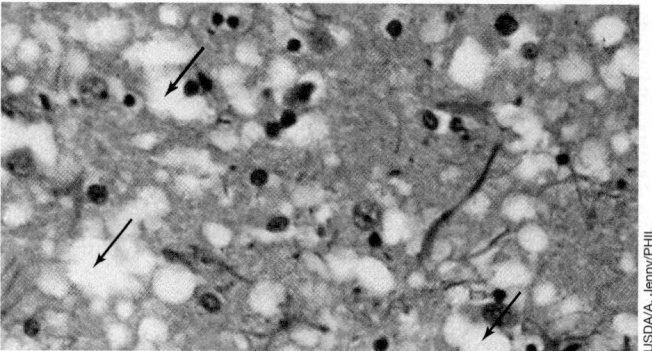

Figure 36.20 A brain section from a cow with bovine spongiform encephalopathy (BSE). The vacuoles, appearing as holes (arrows), give the brains of infected animals a distinct spongelike appearance.

(a) (b)

Figure 36.19 Protists transmitted in food. (a) *Cyclospora cayetanensis* oocysts in a stool sample from an affected patient. The oocysts, stained red with safranin, are about 8–10 μm in diameter. (b) Tachyzoites of *Toxoplasma gondii*, an intracellular parasite. In this transmission electron micrograph, the tachyzoites (arrows) are in a cystlike structure in a cardiac myocyte. Tachyzoites are generally elongated to crescent in form, about 4–7 μm long by 2–4 μm wide.

prenatal infection of the fetus can lead to serious acute toxoplasmosis resulting in tissue involvement, cyst formation, and complications such as myocarditis, blindness, and stillbirth. Immunocompromised individuals such as people with acquired immunodeficiency syndrome (AIDS) may develop acute toxoplasmosis. *T. gondii* grows intracellularly and forms structures called *tachyzoites* (Figure 36.19*b*) that eventually lyse the cell and infect nearby cells, resulting in tissue destruction. Tachyzoites can cross the placenta and infect the fetus. Toxoplasma infections in compromised hosts can be treated with the antiprotist drug pyrimethamine.

Prions, BSE, and nvCJD

Prions are proteins, presumably of host origin, that adopt novel conformations, inhibiting normal protein function and causing degeneration of neural tissue (Section 9.15). Human prion diseases are characterized by neurological symptoms including progressive depression, loss of motor coordination, and dementia.

A foodborne prion disease in humans known as new variant Creutzfeldt–Jakob Disease (nvCJD) has been linked to consumption of meat products from cattle afflicted with *bovine spongiform encephalopathy* (*BSE*), a prion disease commonly called "mad

cow disease." A slow-acting degenerative nervous system disorder, nvCJD has a latent period that may extend for years after exposure to the BSE prion. Nearly 200 people in Great Britain and other European countries have acquired nvCJD. However, nvCJD linked to domestic meat consumption has not been observed in the United States. BSE prions consumed in meat products from affected cattle trigger human protein analogs to assume an altered conformation, resulting in protein dysfunction and disease (Figure 9.28). The terminal stages of both BSE and nvCJD are characterized by large vacuoles in brain tissue, giving the brain a "spongy" appearance, from which BSE derives its name (**Figure 36.20**).

In the United Kingdom and Europe, about 180,000 cattle were diagnosed with BSE and destroyed in the 1990s. Brains of slaughtered animals are routinely tested for BSE in the United States, and several cattle with BSE have been found in Canadian and U.S. herds. In Europe and North America, all cattle known or suspected to have BSE have been destroyed. Bans on cattle feeds containing cattle meat and bone meal appear to have stopped the development of new cases of BSE in Europe and have kept the incidence of this disease very low in North America. The infecting prions were probably transferred to food production animals through meat and bone meal feed derived from infected cattle or other animals not approved for human consumption.

Diagnosis of BSE is done by testing using a prion-susceptible mouse strain or by immunohistochemical or micrographic analysis of biopsied neural tissue (Figure 36.20).

MiniQuiz

- Identify the viruses, bacteria, and protists most likely to cause foodborne illnesses.
- How might prion contamination of food production animals be prevented in the United States?

Big Ideas

36.1
The growth of contaminating microorganisms causes most food spoilage. The potential for microbial food spoilage depends on the nutrient value and water content of the food. Microbial spoilage limits the shelf life of perishable and semiperishable foods. Some food spoilage microorganisms are also pathogens.

36.2
Microbial growth in foods must be limited to reduce spoilage and prevent disease. The growth of microorganisms in perishable foods can be controlled by refrigeration, freezing, canning, pickling, dehydration, aseptic processing, chemical preservation, and irradiation.

36.3
Microbial fermentation is used for preparing, preserving, and enhancing foods including breads, dairy products, meats, fruits, and vegetables.

36.4
Foodborne diseases include food poisoning and food infection. Food poisoning results from the action of microbial toxins, and food infections are due to the growth of microorganisms in the body. Specialized techniques are used to sample and identify microorganisms that cause foodborne disease outbreaks.

36.5
Tracking of foodborne disease outbreaks uses microbiological information disseminated via database-sharing networks. Identification of common characteristics of foodborne pathogens from seemingly isolated foodborne outbreaks can pinpoint the origin of foodborne contamination and track the spread of the disease.

36.6
Staphylococcal food poisoning results from the ingestion of a preformed staphylococcal enterotoxin, a superantigen produced by *Staphylococcus aureus* as it grows in food. In some cases, *S. aureus* cannot be cultured from toxin-containing food. Proper food preparation, handling, and storage can prevent staphylococcal food poisoning.

36.7
Clostridium food poisoning results from ingestion of toxins produced by microbial growth in foods or from microbial growth followed by toxin production in the body. Perfringens food poisoning is quite common and is usually a self-limiting gastrointestinal disease. Botulism is a rare but serious disease, with significant mortality.

36.8
More than 1 million cases of salmonellosis occur every year in the United States. Infection results from ingestion of *Salmonella* introduced into food from food production animals or food handlers.

36.9
Pathogenic *Escherichia coli* cause many food infections. Contamination of foods from fecal material spreads strains pathogenic to humans. Good hygiene practices and specific antibacterial measures such as irradiation of ground beef can control these pathogens.

36.10
Campylobacter infection is the most prevalent foodborne bacterial infection in the United States. Though usually self-limiting, this disease affects nearly 2 million people per year.

36.11
Listeria monocytogenes is an environmentally ubiquitous microorganism. In healthy individuals, *Listeria* seldom causes infection. However, in immunocompromised individuals, *Listeria* can cause serious disease and even death.

36.12
Over 200 different infectious agents cause foodborne disease. Viruses cause the most foodborne illnesses. Bacteria, protists, and prions also cause significant foodborne illness.

Review of Key Terms

Botulism food poisoning due to ingestion of food containing botulinum toxin produced by *Clostridium botulinum*

Canning sealing food in a container and heating to destroy living organisms and endospores

Fermentation the anaerobic catabolism of organic compounds, generally carbohydrates, in the absence of an external electron acceptor

Food infection a microbial infection resulting from the ingestion of pathogen-contaminated food followed by growth of the pathogen in the host

Food poisoning (food intoxication) a disease caused by the ingestion of food that contains preformed microbial toxins

Food spoilage a change in the appearance, smell, or taste of a food that makes it unacceptable to the consumer

Irradiation the exposure of food to ionizing radiation for the purpose of inhibiting growth of microorganisms and insect pests or to retard ripening

Listeriosis a gastrointestinal food infection caused by *Listeria monocytogenes* that may lead to bacteremia and meningitis

Lyophilization (freeze-drying) the removal of all water from frozen food under vacuum

Nonperishable (stable) food food of low water activity that has an extended shelf life and is resistant to spoilage by microorganisms

Perishable food fresh food generally of high water activity that has a very short shelf life due to potential for spoilage by growth of microorganisms

Pickling acidifying food to prevent microbial growth and spoilage

Salmonellosis enterocolitis or other gastrointestinal disease caused by any of over 1400 variants of *Salmonella* spp.

Semiperishable food food of intermediate water activity that has a limited shelf life due to potential for spoilage by growth of microorganisms

Water activity (a_w) a measure of the availability of water for use in metabolic processes

Review Questions

1. Identify and define the three major categories of food perishability (Section 36.1).

2. Identify the major methods used to preserve food. Provide an example of a food preserved by each method (Section 36.2).

3. Describe the various ways of sampling food to detect microorganisms (Section 36.4).

4. Distinguish between food infection and food poisoning (Section 36.4).

5. Identify the organizations that track foodborne disease in the United States and internationally (Section 36.5).

6. Describe the pathogenesis and treatment of staphylococcal food poisoning (Section 36.6).

7. Identify the two major types of clostridial food poisoning. Which is most prevalent? Which is most dangerous? Why (Section 36.7)?

8. What are the possible sources of *Salmonella* spp. that cause food infections (Section 36.8)?

9. Describe the various strains of pathogenic *Escherichia coli* and their related diseases (Section 36.9).

10. *Campylobacter* causes more foodborne infections than any other bacterium. Identify at least one reason why this is true (Section 36.10).

11. Identify the food sources of *Listeria monocytogenes* infections. Identify the individuals who are at high risk for listeriosis (Section 36.11).

12. Describe the significant clinical feature of prion diseases (Section 36.12).

Application Questions

1. Identify optimum storage conditions for perishable, semiperishable, and nonperishable food products. Consider economic factors such as the cost of preservatives, storage space, and the intrinsic value of the food item.

2. For a food of your choice, devise a way to preserve the food by lowering the water activity without drying.

3. Perfringens food poisoning involves ingestion of *Clostridium perfringens* followed by growth and sporulation in the intestine of the host. Sporulation triggers toxin production. Is this disease truly a food poisoning, or might it be classified as a food infection? Explain.

4. Improperly handled potato salads are often the source of staphylococcal food poisoning or salmonellosis. Explain the means by which a potato salad could become contaminated with either *Staphylococcus aureus* or *Salmonella* spp.

5. *Clostridium botulinum* requires an anoxic environment for production of botulinum toxin. Identify methods of food preservation that create the anoxic environment necessary for growth of *C. botulinum*. Conversely, identify methods of food preservation that create an aerobic environment and prevent the growth of *C. botulinum*. What other factors influence the growth of *C. botulinum*?

6. Indicate the precautions necessary to prevent infection with pathogenic *Escherichia coli*. Concentrate on *E. coli* O157:H7 and safe food handling, cooking, and consumption practices.

7. Devise a plan to eliminate *Campylobacter* from a poultry flock or from the finished poultry product. Explain the benefits of *Campylobacter*-free poultry and explain the problems that your plan might encounter.

8. Listeriosis normally occurs only when there is a breakdown in T_H1 cell–mediated immunity. Indicate why this is so. Devise a vaccine to protect against listeriosis. Would your vaccine be an inactivated bacterial strain or product, or would it be an attenuated organism? Explain. Would your vaccine be of use in the listeriosis-prone population?

9. Indicate reasons for the high incidence of viral foodborne disease, especially with noroviruses. Devise a plan to eliminate noroviruses from the food supply.

10. Indicate the problems inherent in tracking a latent infectious agent such as the BSE prion. Can prion diseases be eliminated? If so, how?

Need more practice? Test your understanding with Quantitative Questions; access additional study tools including tutorials, animations, and videos; and then test your knowledge with chapter quizzes and practice tests at **www.microbiologyplace.com**.

Appendix 1

Energy Calculations in Microbial Bioenergetics

The information in Appendix 1 is intended to help calculate changes in free energy accompanying chemical reactions carried out by microorganisms. It begins with definitions of the terms required to make such calculations and proceeds to show how knowledge of redox state, atomic and charge balance, and other factors are necessary to calculate free-energy problems successfully.

I. Definitions

1. ΔG^0 = standard free-energy change of the reaction at 1 atm pressure and 1 M concentrations; ΔG = free-energy change under the conditions specified; $\Delta G^{0\prime}$ = free-energy change under standard conditions at pH 7.

2. Calculation of ΔG^0 for a chemical reaction from the free energy of formation, G_f^0, of products and reactants:

$$\Delta G^0 = \sum \Delta G_f^0(\text{products}) - \sum \Delta G_f^0(\text{reactants})$$

   That is, sum the ΔG_f^0 of products, sum the ΔG_f^0 of reactants, and subtract the latter from the former.

3. For energy-yielding reactions involving H^+, converting from standard conditions (pH 0) to cellular conditions (pH 7):

$$\Delta G^{0\prime} = \Delta G^0 + m\Delta G_f^0(H^+)$$

   where m is the net number of protons in the reaction (m is negative when more protons are consumed than formed) and $\Delta G_f^0(H^+)$ is the free energy of formation of a proton at pH 7 (−39.83 kJ) at 25°C.

4. Effect of concentrations on ΔG: With soluble substrates, the concentration ratios of products formed to exogenous substrates used are generally equal to or greater than 10^{-2} at the beginning of growth and equal to or less than 10^{-2} at the end of growth. From the relation between ΔG and the equilibrium constant (see item 8), it can be calculated that ΔG for the free-energy yield in these situations differs from the free-energy yield under standard conditions by no more than 11.7 kJ, and so to a first approximation, free-energy yields under standard conditions can be used in most situations. However, if H_2 is a product, H_2-consuming bacteria may consume the H_2 and keep its concentration so low that the free-energy yield of the reaction is significantly affected. Thus, in the fermentation of ethanol to acetate and H_2 by syntrophic bacteria ($C_2H_5OH + H_2O \rightarrow C_2H_3O_2^- + 2\,H_2 + H^+$), the $\Delta G^{0\prime}$ (at 1 atm H_2 is +9.68 kJ, but at 10^{-4} atm H_2 it is −36.03 kJ. With H_2-consuming bacteria present, therefore, ethanol fermentation by syntrophic bacteria converts from an endergonic to an exergonic reaction. (See also item 9.)

5. Reduction potentials: By convention, electrode equations are written as *reductions*; that is, the direction is oxidant + $ne^- \rightarrow$ reductant, where n is the number of electrons transferred. The standard reduction potential (E_0) of the hydrogen electrode, $2\,H^+ + 2\,e^- \rightarrow H_2$, is set by definition at 0.0 V at 1 atm pressure of H_2 gas and 1.0 M H^+ at 25°C. E_0' is the standard reduction potential at pH 7. See also Table A1.2.

6. Relation of free energy to reduction potential:

$$\Delta G^{0\prime} = -nF'\Delta E_0'$$

   where n is the number of electrons transferred, F is the Faraday constant (96.48 kJ/V), and $\Delta E_0'$ is the E_0' of the electron-*accepting* couple minus the E_0' of the electron-*donating* couple.

7. Equilibrium constant, K. For the generalized reaction $aA + bB \leftrightarrow cC + dD$,

$$K = \frac{[C]^c[D]^d}{[A]^a[B]^b}$$

   where A, B, C, and D represent reactants and products; a, b, c, and d represent number of molecules of each; and brackets indicate concentrations. This is true only when the chemical system is in equilibrium.

8. Relation of equilibrium constant, K, to free-energy change. At constant temperature, pressure, and pH,

$$\Delta G = \Delta G^{0\prime} + RT \ln K$$

   where R is a constant (8.29 J/mol/kelvin) and T is the absolute temperature (on the Kelvin scale).

9. Two substances can react in a redox reaction even if the standard potentials are unfavorable, provided that the concentrations are appropriate.

   Assume that normally the reduced form of A would donate electrons to the oxidized form of B. However, if the concentration of the reduced form of A was low and the concentration of the reduced form of B was high, it would be possible for the reduced form of B to donate electrons to the oxidized form of A. Thus, the reaction would proceed in the direction opposite that predicted from standard potentials. A practical example of this is the utilization of H^+ as an electron acceptor to produce H_2. Normally, H_2 production in fermentative bacteria is not extensive because H^+ is a poor electron acceptor; the E_0' of the $2\,H^+/H_2$ pair is −0.41 V. However, if the concentration of H_2 is kept low by its continual removal (for example, by methanogenic *Archaea*, which use $H_2 + CO_2$ to produce methane, CH_4, or by many other anaerobes capable of consuming H_2 anaerobically), the potential will be more positive and then H^+ will be a suitable electron acceptor.

II. Oxidation State or Number

1. The oxidation state of an element in an elementary substance (for example, H_2, O_2) is zero.

2. The oxidation state of the ion of an element is equal to its charge (for example, $Na^+ = +1$, $Fe^{3+} = +3$, $O^{2-} = -2$).

3. The sum of oxidation numbers of all atoms in a neutral molecule is zero. Thus, H_2O is neutral because it has two H at $+1$ each and one O at -2.

4. In an ion, the sum of oxidation numbers of all atoms is equal to the charge on that ion. Thus, in the OH^- ion, $O(-2) + H(+1) = -1$.

5. In compounds, the oxidation state of O is almost always -2 and that of H is $+1$.

6. In simple carbon compounds, the oxidation state of C can be calculated by adding up the H and O atoms present and using the oxidation states of these elements as given in item 5, because in a neutral compound the sum of all oxidation numbers must be zero. Thus, the oxidation state of carbon in methane, CH_4, is -4 (4 H at $+1$ each $= +4$); in carbon dioxide, CO_2, the oxidation state of carbon is $+4$ (2 O at -2 each $= -4$).

7. In organic compounds with more than one C atom, it may not be possible to assign a specific oxidation number to each C atom, but it is still useful to calculate the oxidation state of the compound as a whole. The same conventions are used. Thus, the oxidation state of carbon in glucose, $C_6H_{12}O_6$, is zero (12 H at $+1 = 12$; 6 O at $-2 = -12$) and the oxidation state of carbon in ethanol, C_2H_6O, is -2 each (6 H at $+1 = +6$; one O at -2).

8. In all oxidation–reduction reactions there is a balance between the oxidized and reduced products. To calculate an oxidation–reduction balance, the number of molecules of each product is multiplied by its oxidation state. For instance, in calculating the oxidation–reduction balance for the alcoholic fermentation (glucose $\rightarrow 2\,C_2H_6O + 2\,CO_2$), there are two molecules of ethanol at -4 (for a total of -8) and two molecules of CO_2 at $+4$ (for a total of $+8$), so the net balance is zero. When constructing model reactions, it is useful to first calculate redox balances to be certain that the reaction is possible.

III. Calculating Free-Energy Yields for Hypothetical Reactions

Energy yields can be calculated either from free energies of formation of the reactants and products or from differences in reduction potentials of electron-donating and electron-accepting partial reactions.

Calculations from Free Energy

Free energies of formation are given in **Table A1.1**. The procedure to use for calculating energy yields of reactions follows.

1. **Balancing reactions.** In all cases, it is essential to ascertain that the coupled oxidation–reduction reaction is *balanced*. Balancing involves three things: (a) the *total number of each kind of atom* must be identical on both sides of the equation; (b) there must be an *ionic balance* so that when positive and negative ions are added up on the right side of the equation, the total ionic charge (whether positive, negative, or neutral) exactly balances the ionic charge on the left side of the equation; and (c) there must be an *oxidation–reduction balance* so that all the electrons removed from one substance are transferred to another substance. In general, when constructing balanced reactions, one proceeds in the reverse of the three steps just listed. Usually, if steps (c) and (b) have been properly handled, step (a) becomes correct automatically.

2. **Examples:** (a) What is the balanced reaction for the oxidation of H_2S to SO_4^{2-} with O_2? First, decide how many electrons are involved in the oxidation of H_2S to SO_4^{2-}. This can be most easily calculated from the oxidation states of the compounds, using the rules given previously. Because H has an oxidation state of $+1$, the oxidation state of S in H_2S is -2. Because O has an oxidation state of -2, the oxidation state of S in SO_4^{2-} is $+6$ (because it is an ion, and using the rules given in items 4 and 5 of the previous section). Thus, the oxidation of H_2S to SO_4^{2-} involves an *eight-electron transfer* (from -2 to $+6$). Because each O atom can accept two electrons (the oxidation state of O in O_2 is zero, but in H_2O is -2), this means that two molecules of molecular oxygen, O_2, are required to provide sufficient electron-accepting capacity. Thus, at this point, we know that the reaction requires 1 H_2S and 2 O_2 on the left side of the equation, and 1 SO_4^{2-} on the right side. To achieve an ionic balance, we must have two positive charges on the right side of the equation to balance the two negative charges of SO_4^{2-}. Thus, 2 H^+ must be added to the right side of the equation, making the overall reaction

$$H_2S + 2\,O_2 \longrightarrow SO_4^{2-} + 2\,H^+$$

By inspection, it can be seen that this equation is also balanced in terms of the total number of atoms of each kind on each side of the equation.

(b) What is the balanced reaction for the oxidation of H_2S to SO_4^{2-} with Fe^{3+} as electron acceptor? We have just ascertained that the oxidation of H_2S to SO_4^{2-} is an eight-electron transfer. Because the reduction of Fe^{3+} to Fe^{2+} is only a one-electron transfer, 8 Fe^{3+} will be required. At this point, the reaction looks like

$$H_2S + 8\,Fe^{3+} \longrightarrow 8\,Fe^{2+} + SO_4^{2-} \quad \text{(not balanced)}$$

We note that the ionic balance is incorrect. We have 24 positive charges on the left and 14 positive charges on the right (16+ from Fe, 2− from sulfate). To equalize the charges, we add 10 H^+ on the right. Now our equation looks like

$$H_2S + 8\,Fe^{3+} \longrightarrow 8\,Fe^{2+} + 10\,H^+ + SO_4^{2-} \quad \text{(not balanced)}$$

To provide the necessary hydrogen for the H^+ and oxygen for the sulfate, we add 4 H_2O to the left and find that the equation is now balanced:

$$H_2S + 4\,H_2O + 8\,Fe^{3+} \longrightarrow$$
$$8\,Fe^{2+} + 10\,H^+ + SO_4^{2-} \quad \text{(balanced)}$$

Table A1.1 *Free energies of formation (G_f^0) for some substances (kJ/mol)*[a]

Carbon compound	Carbon compound	Metal	Nonmetal	Nitrogen compound
CO, −137.34	Glutamine, −529.7	Cu^+, +50.28	H_2, 0	N_2, 0
CO_2, −394.4	Glyceraldehyde, −437.65	Cu^{2+}, +64.94	H^+, 0 at pH 0; −39.83 at pH 7 (−5.69 per pH unit)	NO, +86.57
CH_4, −50.75	Glycerate, −658.1	CuS, −49.02	O_2, 0	NO_2, +51.95
H_2CO_3, −623.16	Glycerol, −488.52	Fe^{2+}, −78.87	OH^-, −157.3 at pH 14; −198.76 at pH 7; −237.57 at pH 0	NO_2^-, −37.2
HCO_3^-, −586.85	Glycine, −314.96	Fe^{3+}, −4.6	H_2O, −237.17	NO_3^-, −111.34
CO_3^{2-}, −527.90	Glycolate, −530.95	$FeCO_3$, −673.23	H_2O_2, −134.1	NH_3, −26.57
Acetaldehyde, −139.9	Glyoxalate, −468.6	FeS_2, −150.84	PO_4^{3-}, −1026.55	NH_4^+, −79.37
Acetate, −369.41	Guanine, +46.99	$FeSO_4$, −829.62	Se^0, 0	N_2O, +104.18
Acetone, −161.17	α-Ketoglutarate, −797.55	PbS, −92.59	H_2Se, −77.09	N_2H_4, +128
Alanine, −371.54	Lactate, −517.81	Mn^{2+}, −227.93	SeO_4^{2-}, −439.95	
Arginine, −240.2	Lactose, −1515.24	Mn^{3+}, −82.12	S^0, 0	
Aspartate, −700.4	Malate, −845.08	MnO_4^-, −506.57	SO_3^{2-}, −486.6	
Benzene, +124.5	Mannitol, −942.61	MnO_2, −456.71	SeO_4^{2-}, −744.6	
Benzoic acid, −245.6	Methanol, −175.39	$MnSO_4$, −955.32	$S_2O_3^{2-}$, −513.4	
n-Butanol, −171.84	Methionine, −502.92	HgS, −49.02	H_2S, −27.87	
Butyrate, −352.63	Methylamine, −40.0	MoS_2, −225.42	HS^-, +12.05	
Caproate, −335.96	Oxalate, −674.04	ZnS, −198.60	S^{2-}, +85.8	
Citrate, −1168.34	Palmitic acid, −305			
o-Cresol, −37.1	Phenol, −47.6			
Crotonate, −277.4	*n*-Propanol, −175.81			
Cysteine, −339.8	Propionate, −361.08			
Dimethylamine, −3.3	Pyruvate, −474.63			
Ethanol, −181.75	Ribose, −757.3			
Formaldehyde, −130.54	Succinate, −690.23			
Formate, −351.04	Sucrose, −370.90			
Fructose, −951.38	Toluene, +114.22			
Fumarate, −604.21	Trimethylamine, −37.2			
Gluconate, −1128.3	Tryptophan, −112.6			
Glucose, −917.22	Urea, −203.76			
Glutamate, −699.6	Valerate, −344.34			

[a]Values for free energy of formation of various compounds can be found in Dean, J. A. 1973. *Lange's Handbook of Chemistry*, 11th edition. McGraw-Hill, New York; Garrels, R. M., and C. L. Christ. 1965. *Solutions, Minerals, and Equilibria*. Harper & Row, New York; Burton, K. 1957. In Krebs, H. A., and H. L. Komberg. Energy transformation in living matter, *Ergebnisse der Physiologie* (appendix): Springer-Verlag, Berlin; and Thauer, R. K., K. Jungermann, and H. Decker. 1977. Energy conservation in anaerobic chemotrophic bacteria. *Bacteriol. Rev. 41:* 100–180.

In general, ionic balance can be achieved by adding H^+ or OH^- to the left or right side of the equation, and because all reactions take place in an aqueous medium, H_2O molecules can be added where needed. Whether H^+ or OH^- is added generally depends on whether the reaction is taking place under acidic or alkaline conditions.

3. **Calculation of energy yield for balanced equations from free energies of formation.** Once an equation has been balanced, the free-energy yield can be calculated by inserting the values for the free energy of formation of each reactant and product from Table A1.1 and using the formula in item 2 of the first section of this appendix. For instance, for the equation

$$H_2S + 2\,O_2 \longrightarrow SO_4^{2-} + 2\,H^+$$

G_f^0 values: $(-27.87) + (0) \longrightarrow$
$$(-744.6) + 2(-39.83) \quad \text{(assuming pH 7)}$$

$$\Delta G^{0\prime} = [(-744.6) + 2(-39.83)] - [(-27.87) + (0)]$$
$$= -796.39 \text{ kJ}$$

The $G_f{}^0$ values for the products (right side of reaction) are summed and subtracted from the $G_f{}^0$ values for the reactants (left side of reaction), taking care to ensure that the arithmetic signs are correct. From the data in Table A1.1, a wide variety of free-energy yields for reactions of microbiological interest can be calculated.

Calculation of Free-Energy Yield from Reduction Potential

Reduction potentials of some important redox pairs are given in **Table A1.2**. The amount of energy that can be released from two half reactions can be calculated from the *differences* in reduction potentials of the two reactions and from the number of electrons transferred. The farther apart the two half reactions are, and the greater the number of electrons transferred, the more energy released.

The conversion of potential difference to free energy is given by the formula $\Delta G^{0\prime} = -nF\Delta E_0{}'$, where n is the number of electrons, F is the Faraday constant (96.48 kJ/V), and $\Delta E_0{}'$ is the difference in reduction potentials. Thus, the 2 H^+/H_2 couple has a

potential of -0.41 V and the $\frac{1}{2}$ O_2/H_2O pair has a potential of $+0.82$ V, and so the potential difference is 1.23 V, which (because two electrons are involved) is equivalent to a free-energy yield (ΔG^0) of -237.34 kJ. On the other hand, the potential difference between the 2 H^+/H_2 and the $NO_3{}^-/NO_2{}^-$ reactions is less, 0.84 V, which is equivalent to a free-energy yield of -162.08 kJ.

Because many biochemical reactions are two-electron transfers, it is often useful to give energy yields for two-electron reactions, even if more electrons are involved. Thus, the $SO_4{}^{2-}/H_2$ redox pair involves eight electrons, and complete reduction of $SO_4{}^{2-}$ with H_2 requires 4 H_2 (equivalent to eight electrons). From the reduction potential difference between 2 H^+/H_2 and $SO_4{}^{2-}/H_2S$ (0.19 V), a free-energy yield of -146.64 kJ is calculated, or -36.66 kJ per two electrons. By convention, reduction potentials are given for conditions in which equal concentrations of oxidized and reduced forms are present. In actual practice, the concentrations of these two forms may be quite different. As discussed earlier in this appendix (Section I, item 9), it is possible to couple half reactions even if the potential difference is unfavorable, provided the concentrations of the reacting species are appropriate.

Table A1.2 *Microbiologically important reduction potentials*[a]

Redox pair	$E_0{}'$ (V)	Redox pair	$E_0{}'$ (V)
$SO_4{}^{2-}/HSO_3{}^-$	-0.52	Acrylyl-CoA/propionyl-CoA	-0.015
CO_2/formate	-0.43	Glycine/acetate$^-$ + $NH_4{}^+$	-0.010
2 H^+/H_2	-0.41	$S_4O_6{}^{2-}/S_2O_3{}^{2-}$	$+0.024$
$S_2O_3{}^{2-}/HS^- + HSO_3{}^-$	-0.40	Fumarate^{2-}/succinate^{2-}	$+0.033$
Ferredoxin ox/red	-0.39	Cytochrome b ox/red	$+0.035$
Flavodoxin ox/red[b]	-0.37	Ubiquinone ox/red	$+0.113$
NAD^+/NADH	-0.32	$AsO_4{}^{3-}/AsO_3{}^{3-}$	$+0.139$
Cytochrome c_3 ox/red	-0.29	Dimethyl sulfoxide (DMSO)/dimethyl sulfide (DMS)	$+0.16$
CO_2/acetate$^-$	-0.29	$Fe(OH)_3 + HCO_3{}^-/FeCO_3$ (Fe^{3+}/Fe^{2+}, pH 7)	$+0.20$
S^0/HS^-	-0.27	$S_3O_6{}^{2-}/S_2O_3{}^{2-} + HSO_3{}^-$	$+0.225$
CO_2/CH_4	-0.24	Cytochrome c_1 ox/red	$+0.23$
FAD/FADH	-0.22	$NO_2{}^-/NO$	$+0.36$
$SO_4{}^{2-}/HS^-$	-0.217	Cytochrome a_3 ox/red	$+0.385$
Acetaldehyde/ethanol	-0.197	Chlorobenzoate$^-$/benzoate$^-$ + HCl	$+0.297$
Pyruvate$^-$/lactate$^-$	-0.19	$NO_3{}^-/NO_2{}^-$	$+0.43$
FMN/FMNH	-0.19	$SeO_4{}^{2-}/SeO_3{}^{2-}$	$+0.475$
Dihydroxyacetone phosphate/glycerolphosphate	-0.19	Fe^{3+}/Fe^{2+} (pH 2)	$+0.77$
$HSO_3{}^-/S_3O_6{}^{2-}$	-0.17	Mn^{4+}/Mn^{2+}	$+0.798$
Flavodoxin ox/red[b]	-0.12	O_2/H_2O	$+0.82$
$HSO_3{}^-/HS^-$	-0.116	$ClO_3{}^-/Cl^-$	$+1.03$
Menaquinone ox/red	-0.075	NO/N_2O	$+1.18$
Adenosine phosphosulfate/AMP + $HSO_3{}^-$	-0.060	N_2O/N_2	$+1.36$
Rubredoxin ox/red	-0.057		

[a]Data from Thauer, R. K., K. Jungermann, and K. Decker, 1977. Energy conservation in anaerobic chemotrophic bacteria. *Bacteriol. Rev. 41*: 100–180.
[b]Separate potentials are given for each electron transfer in this potentially two-electron transfer.

Appendix 2

Bergey's Manual of Systematic Bacteriology, Second Edition

List of Genera and Higher-Order Taxa[a]

Domain *Archaea*

Phylum I. *Crenarchaeota*
  Class I. *Thermoprotei*
    Order I. *Thermoproteales*
      Family I. *Thermoproteaceae*
        Genus I. *Thermoproteus*
        Genus II. *Caldivirga*
        Genus III. *Pyrobaculum*
        Genus IV. *Thermocladium*
        Genus V. *Vulcanisaeta*
      Family II. *Thermofilaceae*
        Genus I. *Thermofilum*
    Order II. *Caldisphaerales*
      Family I. *Caldisphaeraceae*
        Genus I. *Caldisphaera*
    Order III. *Desulfurococcales*
      Family I. *Desulfurococcaceae*
        Genus I. *Desulfurococcus*
        Genus II. *Acidilobus*
        Genus III. *Aeropyrum*
        Genus IV. *Ignicoccus*
        Genus V. *Staphylothermus*
        Genus VI. *Stetteria*
        Genus VII. *Sulfophobococcus*
        Genus VIII. *Thermodiscus*
        Genus IX. *Thermosphaera*
      Family II. *Pyrodictiaceae*
        Genus I. *Pyrodictium*
        Genus II. *Hyperthermus*
        Genus III. *Pyrolobus*
    Order IV. *Sulfolobales*
      Family I. *Sulfolobaceae*
        Genus I. *Sulfolobus*
        Genus II. *Acidianus*
        Genus III. *Metallosphaera*
        Genus IV. *Stygiolobus*
        Genus V. *Sulfurisphaera*
        Genus VI. *Sulfurococcus*
Phylum II. *Euryarchaeota*
  Class I. *Methanobacteria*
    Order I. *Methanobacteriales*
      Family I. *Methanobacteriaceae*
        Genus I. *Methanobacterium*
        Genus II. *Methanobrevibacter*
        Genus III. *Methanosphaera*

        Genus IV. *Methanothermobacter*
      Family II. *Methanothermaceae*
        Genus I. *Methanothermus*
  Class II. *Methanococci*
    Order I. *Methanococcales*
      Family I. *Methanococcaceae*
        Genus I. *Methanococcus*
        Genus II. *Methanothermococcus*
      Family II. *Methanocaldococcaceae*
        Genus I. *Methanocaldococcus*
        Genus II. *Methanotorris*
  Class III. *Methanomicrobia*
    Order I. *Methanomicrobiales*
      Family I. *Methanomicrobiaceae*
        Genus I. *Methanomicrobium*
        Genus II. *Methanoculleus*
        Genus III. *Methanofollis*
        Genus IV. *Methanogenium*
        Genus V. *Methanolacinia*
        Genus VI. *Methanoplanus*
      Family II. *Methanocorpusculaceae*
        Genus I. *Methanocorpusculum*
      Family III. *Methanospirillaceae*
        Genus I. *Methanospirillum*
      Genera *Incertae sedis*[b]
        Genus I. *Methanocalculus*
    Order II. *Methanosarcinales*
      Family I. *Methanosarcinaceae*
        Genus I. *Methanosarcina*
        Genus II. *Methanococcoides*
        Genus III. *Methanohalobium*
        Genus IV. *Methanohalophilus*
        Genus V. *Methanolobus*
        Genus VI. *Methanomethylovorans*
        Genus VII. *Methanimicrococcus*
        Genus VIII. *Methanosalsum*
      Family II. *Methanosaetaceae*
        Genus I. *Methanosaeta*
  Class IV. *Halobacteria*
    Order I. *Halobacteriales*
      Family I. *Halobacteriaceae*
        Genus I. *Halobacterium*
        Genus II. *Haloarcula*
        Genus III. *Halobaculum*
        Genus IV. *Halobiforma*
        Genus V. *Halococcus*
        Genus VI. *Haloferax*
        Genus VII. *Halogeometricum*
        Genus VIII. *Halomicrobium*
        Genus IX. *Halorhabdus*
        Genus X. *Halorubrum*

        Genus XI. *Halosimplex*
        Genus XII. *Haloterrigena*
        Genus XIII. *Natrialba*
        Genus XIV. *Natrinema*
        Genus XV. *Natronobacterium*
        Genus XVI. *Natronococcus*
        Genus XVII. *Natronomonas*
        Genus XVIII. *Natronorubrum*
  Class V. *Thermoplasmata*
    Order I. *Thermoplasmatales*
      Family I. *Thermoplasmataceae*
        Genus I. *Thermoplasma*
      Family II. *Picrophilaceae*
        Genus I. *Picrophilus*
      Family III. *Ferroplasmaceae*
        Genus I. *Ferroplasma*
  Class VI. *Thermococci*
    Order I. *Thermococcales*
      Family I. *Thermococcaceae*
        Genus I. *Thermococcus*
        Genus II. *Palaeococcus*
        Genus III. *Pyrococcus*
  Class VII. *Archaeoglobi*
    Order I. *Archaeoglobales*
      Family I. *Archaeoglobaceae*
        Genus I. *Archaeoglobus*
        Genus II. *Ferroglobus*
        Genus III. *Geoglobus*
  Class VIII. *Methanopyri*
    Order I. *Methanopyrales*
      Family I. *Methanopyraceae*
        Genus I. *Methanopyrus*

Domain *Bacteria*

Phylum I. *Aquificae*
  Class I. *Aquificae*
    Order I. *Aquificales*
      Family I. *Aquificaceae*
        Genus I. *Aquifex*
        Genus II. *Calderobacterium*
        Genus III. *Hydrogenobaculum*
        Genus IV. *Hydrogenobacter*
        Genus V. *Hydrogenothermus*
        Genus VI. *Persephonella*
        Genus VII. *Sulfurihydrogenibium*
        Genus VIII. *Thermocrinis*
      Genera *Incertae sedis*[b]
        Genus I. *Balnearium*
        Genus II. *Desulfurobacterium*
        Genus III. *Thermovibrio*

---

[a]*Bergey's Manual of Systematic Bacteriology*, second edition, consists of 5 volumes. Since not all volumes were published at the same time, the list of genera and higher-order taxa shown here are the organisms recognized as of 2010. Genera or higher-order taxa in quotation marks are recognized taxa whose names have not yet been validated. Because bacterial taxonomy is a work in progress, updates to the list shown here occur as new genera and species are described and as new data support new taxonomic arrangements. For further discussion on bacterial taxonomy, see Sections 16.10–16.13. For a current list of validly published genus and species names of prokaryotes, refer to the URL http://www.bacterio.cict.fr/. This website, the most complete, accurate, and up-to-date site on prokaryotic taxonomy, is maintained by Jean Euzéby, Société de Bactériologie Systématique et Vétérinaire, Toulouse, France. In addition to thanking Dr. Euzéby, the authors of *BBOM* 13/e also give hearty thanks to Dr. Barny Whitman, Editorial Director of *Bergey's Manual*, University of Georgia, for help with this taxonomic outline. Any errors in the outline presented here are solely the responsibility of *BBOM* 13/e authors.

[b]Taxa of uncertain affiliation. *Incertae sedis,* Latin for "of uncertain placement."

Phylum II. *Thermotogae*
  Class I. *Thermotogae*
    Order I. *Thermotogales*
      Family I. *Thermotogaceae*
        Genus I. *Thermotoga*
        Genus II. *Fervidobacterium*
        Genus III. *Geotoga*
        Genus IV. *Marinitoga*
        Genus V. *Petrotoga*
        Genus VI. *Thermosipho*
Phylum III. *Thermodesulfobacteria*
  Class I. *Thermodesulfobacteria*
    Order I. *Thermodesulfobacteriales*
      Family I. *Thermodesulfobacteriaceae*
        Genus I. *Thermodesulfobacterium*
        Genus II. *Thermodesulfatator*
Phylum IV. *Deinococcus-Thermus*
  Class I. *Deinococci*
    Order I. *Deinococcales*
      Family I. *Deinococcaceae*
        Genus I. *Deinococcus*
    Order II. *Thermales*
      Family I. *Thermaceae*
        Genus I. *Thermus*
        Genus II. *Marinithermus*
        Genus III. *Meiothermus*
        Genus IV. *Oceanithermus*
        Genus V. *Vulcanithermus*
Phylum V. *Chrysiogenetes*
  Class I. *Chrysiogenetes*
    Order I. *Chrysiogenales*
      Family I. *Chrysiogenaceae*
        Genus I. *Chrysiogenes*
Phylum VI. *Chloroflexi*
  Class I. *Chloroflexi*
    Order I. *Chloroflexales*
      Family I. *Chloroflexaceae*
        Genus I. *Chloroflexus*
        Genus II. *Chloronema*
        Genus III. *Heliothrix*
        Genus IV. *Roseiflexus*
      Family II. *Oscillochloridaceae*
        Genus I. *Oscillochloris*
    Order II. *Herpetosiphonales*
      Family I. *Herpetosiphonaceae*
        Genus I. *Herpetosiphon*
  Class II. *Anaerolineae*
    Order I. *Anaerolineales*
      Family I. *Anaerolineaceae*
        Genus I. *Anaerolinea*
        Genus II. *Caldilinea*
Phylum VII. *Thermomicrobia*
  Class I. *Thermomicrobia*
    Order I. *Thermomicrobiales*
      Family I. *Thermomicrobiaceae*
        Genus I. *Thermomicrobium*
Phylum VIII. *Nitrospirae*
  Class I. *Nitrospira*
    Order I. *Nitrospirales*
      Family I. *Nitrospiraceae*
        Genus I. *Nitrospira*
        Genus II. *Leptospirillum*
        Genus III. *Magnetobacterium*
        Genus IV. *Thermodesulfovibrio*
Phylum IX. *Deferribacteres*
  Class I. *Deferribacteres*
    Order I. *Deferribacterales*
      Family I. *Deferribacteraceae*
        Genus I. *Deferribacter*

Genus II. *Denitrovibrio*
Genus III. *Flexistipes*
Genus IV. *Geovibrio*
Genera *Incertae sedis*[b]
Genus I. *Synergistes*
Genus II. *Caldithrix*
Phylum X. *Cyanobacteria*
  Class I. *Cyanobacteria*
    Subsection I. *Subsection 1*
      Family I. Family 1.1
        Form genus I. *Chamaesiphon*[c]
        Form genus II. *Chroococcus*
        Form genus III. *Cyanobacterium*
        Form genus IV. *Cyanobium*
        Form genus V. *Cyanothece*
        Form genus VI. *Dactylococcopsis*
        Form genus VII. *Gloeobacter*
        Form genus VIII. *Gloeocapsa*
        Form genus IX. *Gloeothece*
        Form genus X. *Microcystis*
        Form genus XI. *Prochlorococcus*
        Form genus XII. *Prochloron*
        Form genus XIII. *Synechococcus*
        Form genus XIV. *Synechocystis*
    Subsection II. *Subsection 2*
      Family I. Family 2.1
        Form genus I. *Cyanocystis*
        Form genus II. *Dermocarpella*
        Form genus III. *Stanieria*
        Form genus IV. *Xenococcus*
      Family II. Family 2.2
        Form genus I. *Chroococcidiopsis*
        Form genus II. *Myxosarcina*
        Form genus III. *Pleurocapsa*
    Subsection III. *Subsection 3*
      Family I. Family 3.1
        Form genus I. *Arthrospira*
        Form genus II. *Borzia*
        Form genus III. *Crinalium*
        Form genus IV. *Geitlerinema*
        Genus V. *Halospirulina*
        Form genus VI. *Leptolyngbya*
        Form genus VII. *Limnothrix*
        Form genus VIII. *Lyngbya*
        Form genus IX. *Microcoleus*
        Form genus X. *Oscillatoria*
        Form genus XI. *Planktothrix*
        Form genus XII. *Prochlorothrix*
        Form genus XIII. *Pseudanabaena*
        Form genus XIV. *Spirulina*
        Form genus XV. *Starria*
        Form Genus XVI. *Symploca*
        Genus XVII. *Trichodesmium*
        Form genus XVIII. *Tychonema*
    Subsection IV. *Subsection 4*
      Family I. Family 4.1
        Form genus I. *Anabaena*
        Form genus II. *Anabaenopsis*
        Form genus III. *Aphanizomenon*
        Form genus IV. *Cyanospira*
        Form genus V. *Cylindrospermopsis*
        Form genus VI. *Cylindrospermum*
        Form genus VII. *Nodularia*
        Form genus VIII. *Nostoc*
        Form genus IX. *Scytonema*
      Family II. Family 4.2
        Form genus I. *Calothrix*
        Form genus II. *Rivularia*
        Form genus III. *Tolypothrix*

    Subsection V. *Subsection 5*
      Family I. Family 5.1
        Form genus I. *Chlorogloeopsis*
        Form genus II. *Fischerella*
        Form genus III. *Geitleria*
        Form genus IV. *Iyengariella*
        Form genus V. *Nostochopsis*
        Form genus VI. *Stigonema*
Phylum XI. *Chlorobi*
  Class I. *Chlorobia*
    Order I. *Chlorobiales*
      Family I. *Chlorobiaceae*
        Genus I. *Chlorobium*
        Genus II. *Ancalochloris*
        Genus III. *Chlorobaculum*
        Genus IV. *Chloroherpeton*
        Genus V. *Pelodictyon*
        Genus VI. *Prosthecochloris*
Phylum XII. *Proteobacteria*
  Class I. *Alphaproteobacteria*
    Order I. *Rhodospirillales*
      Family I. *Rhodospirillaceae*
        Genus I. *Rhodospirillum*
        Genus II. *Azospirillum*
        Genus III. *Inquilinus*
        Genus IV. *Magnetospirillum*
        Genus V. *Phaeospirillum*
        Genus VI. *Rhodocista*
        Genus VII. *Rhodospira*
        Genus VIII. *Rhodovibrio*
        Genus IX. *Roseospira*
        Genus X. *Skermanella*
        Genus XI. *Thalassospira*
        Genus XII. *Tistrella*
      Family II. *Acetobacteraceae*
        Genus I. *Acetobacter*
        Genus II. *Acidiphilium*
        Genus III. *Acidisphaera*
        Genus IV. *Acidocella*
        Genus V. *Acidomonas*
        Genus VI. *Asaia*
        Genus VII. *Craurococcus*
        Genus VIII. *Gluconacetobacter*
        Genus IX. *Gluconobacter*
        Genus X. *Kozakia*
        Genus XI. *Muricoccus*
        Genus XII. *Paracraurococcus*
        Genus XIII. *Rhodopila*
        Genus XIV. *Roseococcus*
        Genus XV. *Rubritepida*
        Genus XVI. *Stella*
        Genus XVII. *Teichococcus*
        Genus XVIII. *Zavarzinia*
    Order II. *Rickettsiales*
      Family I. *Rickettsiaceae*
        Genus I. *Rickettsia*
        Genus II. *Orientia*
      Family II. *Anaplasmataceae*
        Genus I. *Anaplasma*
        Genus II. *Aegyptianella*
        Genus III. *Cowdria*
        Genus IV. *Ehrlichia*
        Genus V. *Neorickettsia*
        Genus VI. *Wolbachia*
        Genus VII. *Xenohaliotis*
      Family III. *Holosporaceae*
        Genus I. *Holospora*
      Genera *Incertae sedis*[b]
        Genus I. *Caedibacter*

[c]The taxonomic position of the cyanobacteria in *Bergey's Manual* is left open. The term "form genus" refers to a group of cyanobacteria with very characteristic morphology found worldwide. However, not all isolates of such a type may actually fit into the same genus. In some cases, pure cultures of form genera are not available.

Genus II. *Lyticum*
Genus III. *Odyssella*
Genus IV. *Pseudocaedibacter*
Genus V. *Symbiotes*
Genus VI. *Tectibacter*
Order III. *Rhodobacterales*
  Family I. *Rhodobacteraceae*
    Genus I. *Rhodobacter*
    Genus II. *Ahrensia*
    Genus III. *Albidovulum*
    Genus IV. *Amaricoccus*
    Genus V. *Antarctobacter*
    Genus VI. *Gemmobacter*
    Genus VII. *Hirschia*
    Genus VIII. *Hyphomonas*
    Genus IX. *Jannaschia*
    Genus X. *Ketogulonicigenium*
    Genus XI. *Leisingera*
    Genus XII. *Maricaulis*
    Genus XIII. *Methylarcula*
    Genus XIV. *Oceanicaulis*
    Genus XV. *Octadecabacter*
    Genus XVI. *Pannonibacter*
    Genus XVII. *Paracoccus*
    Genus XVIII. *Pseudorhodobacter*
    Genus XIX. *Rhodobaca*
    Genus XX. *Rhodothalassium*
    Genus XXI. *Rhodovulum*
    Genus XXII. *Roseibium*
    Genus XXIII. *Roseinatronobacter*
    Genus XXIV. *Roseivivax*
    Genus XXV. *Roseobacter*
    Genus XXVI. *Roseovarius*
    Genus XXVII. *Rubrimonas*
    Genus XXVIII. *Ruegeria*
    Genus XXIX. *Sagittula*
    Genus XXX. *Silicibacter*
    Genus XXXI. *Staleya*
    Genus XXXII. *Stappia*
    Genus XXXIII. *Sulfitobacter*
Order IV. *Sphingomonadales*
  Family I. *Sphingomonadaceae*
    Genus I. *Sphingomonas*
    Genus II. *Blastomonas*
    Genus III. *Erythrobacter*
    Genus IV. *Erythromicrobium*
    Genus V. *Erythromonas*
    Genus VI. *Novosphingobium*
    Genus VII. *Porphyrobacter*
    Genus VIII. *Rhizomonas*
    Genus IX. *Sandaracinobacter*
    Genus X. *Sphingobium*
    Genus XI. *Sphingopyxis*
    Genus XII. *Zymomonas*
Order V. *Caulobacterales*
  Family I. *Caulobacteraceae*
    Genus I. *Caulobacter*
    Genus II. *Asticcacaulis*
    Genus III. *Brevundimonas*
    Genus IV. *Phenylobacterium*
Order VI. *Rhizobiales*
  Family I. *Rhizobiaceae*
    Genus I. *Rhizobium*
    Genus II. *Agrobacterium*
    Genus III. *Allorhizobium*
    Genus IV. *Carbophilus*
    Genus V. *Chelatobacter*
    Genus VI. *Ensifer*

    Genus VII. *Sinorhizobium*
  Family II. *Aurantimonadaceae*
    Genus I. *Aurantimonas*
    Genus II. *Fulvimarina*
  Family III. *Bartonellaceae*
    Genus I. *Bartonella*
  Family IV. *Brucellaceae*
    Genus I. *Brucella*
    Genus II. *Mycoplana*
    Genus III. *Ochrobactrum*
  Family V. *Phyllobacteriaceae*
    Genus I. *Phyllobacterium*
    Genus II. *Aminobacter*
    Genus III. *Aquamicrobium*
    Genus IV. *Fluvibacter*
    Genus V. *Candidatus* "Liberibacter"[d]
    Genus VI. *Mesorhizobium*
    Genus VII. *Nitratireductor*
    Genus VIII. *Pseudaminobacter*
  Family VI. *Methylocystaceae*
    Genus I. *Methylocystis*
    Genus II. *Albibacter*
    Genus III. *Methylopila*
    Genus IV. *Methylosinus*
    Genus V. *Terasakiella*
  Family VII. *Beijerinckiaceae*
    Genus I. *Beijerinckia*
    Genus II. *Chelatococcus*
    Genus III. *Methylocapsa*
    Genus IV. *Methylocella*
  Family VIII. *Bradyrhizobiaceae*
    Genus I. *Bradyrhizobium*
    Genus II. *Afipia*
    Genus III. *Agromonas*
    Genus IV. *Blastobacter*
    Genus V. *Bosea*
    Genus VI. *Nitrobacter*
    Genus VII. *Oligotropha*
    Genus VIII. *Rhodoblastus*
    Genus IX. *Rhodopseudomonas*
  Family IX. *Hyphomicrobiaceae*
    Genus I. *Hyphomicrobium*
    Genus II. *Ancalomicrobium*
    Genus III. *Ancylobacter*
    Genus IV. *Angulomicrobium*
    Genus V. *Aquabacter*
    Genus VI. *Azorhizobium*
    Genus VII. *Blastochloris*
    Genus VIII. *Devosia*
    Genus IX. *Dichotomicrobium*
    Genus X. *Filomicrobium*
    Genus XI. *Gemmiger*
    Genus XII. *Labrys*
    Genus XIII. *Methylorhabdus*
    Genus XIV. *Pedomicrobium*
    Genus XV. *Prosthecomicrobium*
    Genus XVI. *Rhodomicrobium*
    Genus XVII. *Rhodoplanes*
    Genus XVIII. *Seliberia*
    Genus XIX. *Starkeya*
    Genus XX. *Xanthobacter*
  Family X. *Methylobacteriaceae*
    Genus I. *Methylobacterium*
    Genus II. *Microvirga*
    Genus III. *Protomonas*
    Genus IV. *Roseomonas*
  Family XI. *Rhodobiaceae*
    Genus I. *Rhodobium*

    Genus II. *Roseospirillum*
Order VII. *Parvularculales*
  Family I. *Parvularculaceae*
    Genus I. *Parvularcula*
Class II. *Betaproteobacteria*
Order I. *Burkholderiales*
  Family I. *Burkholderiaceae*
    Genus I. *Burkholderia*
    Genus II. *Cupriavidus*
    Genus III. *Lautropia*
    Genus IV. *Limnobacter*
    Genus V. *Pandoraea*
    Genus VI. *Paucimonas*
    Genus VII. *Polynucleobacter*
    Genus VIII. *Ralstonia*
    Genus IX. *Thermothrix*
    Genus X. *Wautersia*
  Family II. *Oxalobacteraceae*
    Genus I. *Oxalobacter*
    Genus II. *Duganella*
    Genus III. *Herbaspirillum*
    Genus IV. *Janthinobacterium*
    Genus V. *Massilia*
    Genus VI. *Oxalicibacterium*
    Genus VII. *Telluria*
  Family III. *Alcaligenaceae*
    Genus I. *Alcaligenes*
    Genus II. *Achromobacter*
    Genus III. *Bordetella*
    Genus IV. *Brackiella*
    Genus V. *Derxia*
    Genus VI. *Kerstersia*
    Genus VII. *Oligella*
    Genus VIII. *Pelistega*
    Genus IX. *Pigmentiphaga*
    Genus X. *Sutterella*
    Genus XI. *Taylorella*
  Family IV. *Comamonadaceae*
    Genus I. *Comamonas*
    Genus II. *Acidovorax*
    Genus III. *Alicycliphilus*
    Genus IV. *Brachymonas*
    Genus V. *Caldimonas*
    Genus VI. *Delftia*
    Genus VII. *Diaphorobacter*
    Genus VIII. *Hydrogenophaga*
    Genus IX. *Hylemonella*
    Genus X. *Lampropedia*
    Genus XI. *Macromonas*
    Genus XII. *Ottowia*
    Genus XIII. *Polaromonas*
    Genus XIV. *Ramlibacter*
    Genus XV. *Rhodoferax*
    Genus XVI. *Variovorax*
    Genus XVII. *Xenophilus*
  Genera *Incertae sedis*[b]
    Genus I. *Aquabacterium*
    Genus II. *Ideonella*
    Genus III. *Leptothrix*
    Genus IV. *Roseateles*
    Genus V. *Rubrivivax*
    Genus VI. *Schlegelella*
    Genus VII. *Sphaerotilus*
    Genus VIII. *Tepidimonas*
    Genus IX. *Thiomonas*
    Genus X. *Xylophilus*
Order II. *Hydrogenophilales*
  Family I. *"Hydrogenophilaceae"*

[d]In bacterial taxonomy, candidatus status is given to organisms known to exist by 16S rRNA gene sequencing and other key properties, but which are not yet in pure culture. Some candidatus organisms are in laboratory culture but not pure culture.

Genus I. *Teredinibacter*
Order XI. *Vibrionales*
  Family I. *Vibrionaceae*
    Genus I. *Vibrio*
    Genus II. *Allomonas*
    Genus III. *Catenococcus*
    Genus IV. *Enterovibrio*
    Genus V. *Grimontia*
    Genus VI. *Listonella*
    Genus VII. *Photobacterium*
    Genus VIII. *Salinivibrio*
Order XII. *Aeromonadales*
  Family I. *Aeromonadaceae*
    Genus I. *Aeromonas*
    Genus II. *Oceanimonas*
    Genus III. *Oceanisphaera*
    Genus IV. *Tolumonas*
  Family II. *Incertae sedis:*[b] *Succinivibrionaceae*
    Genus I. *Succinivibrio*
    Genus II. *Anaerobiospirillum*
    Genus III. *Ruminobacter*
    Genus IV. *Succinimonas*
Order XIII. *Enterobacteriales*
  Family I. *Enterobacteriaceae*
    Genus I. *Escherichia*
    Genus II. *Alterococcus*
    Genus III. *Arsenophonus*
    Genus IV. *Brenneria*
    Genus V. *Buchnera*
    Genus VI. *Budvicia*
    Genus VII. *Buttiauxella*
    Genus VIII. *Calymmatobacterium*
    Genus IX. *Cedecea*
    Genus X. *Citrobacter*
    Genus XI. *Edwardsiella*
    Genus XII. *Enterobacter*
    Genus XIII. *Erwinia*
    Genus XIV. *Ewingella*
    Genus XV. *Hafnia*
    Genus XVI. *Klebsiella*
    Genus XVII. *Kluyvera*
    Genus XVIII. *Leclercia*
    Genus XIX. *Leminorella*
    Genus XX. *Moellerella*
    Genus XXI. *Morganella*
    Genus XXII. *Obesumbacterium*
    Genus XXIII. *Pantoea*
    Genus XXIV. *Pectobacterium*
    Genus XXV. *Phlomobacter*
    Genus XXVI. *Photorhabdus*
    Genus XXVII. *Plesiomonas*
    Genus XXVIII. *Pragia*
    Genus XXIX. *Proteus*
    Genus XXX. *Providencia*
    Genus XXXI. *Rahnella*
    Genus XXXII. *Raoultella*
    Genus XXXIII. *Saccharobacter*
    Genus XXXIV. *Salmonella*
    Genus XXXV. *Samsonia*
    Genus XXXVI. *Serratia*
    Genus XXXVII. *Shigella*
    Genus XXXVIII. *Sodalis*
    Genus XXXIX. *Tatumella*
    Genus XL. *Trabulsiella*
    Genus XLI. *Wigglesworthia*
    Genus XLII. *Xenorhabdus*
    Genus XLIII. *Yersinia*
    Genus XLIV. *Yokenella*
Order XIV. *Pasteurellales*
  Family I. *Pasteurellaceae*
    Genus I. *Pasteurella*
    Genus II. *Actinobacillus*

Genus III. *Gallibacterium*
Genus IV. *Haemophilus*
Genus V. *Lonepinella*
Genus VI. *Mannheimia*
Genus VII. *Phocoenobacter*

Class IV. *Deltaproteobacteria*
Order I. *Desulfurellales*
  Family I. *Desulfurellaceae*
    Genus I. *Desulfurella*
    Genus II. *Hippea*
Order II. *Desulfovibrionales*
  Family I. *Desulfovibrionaceae*
    Genus I. *Desulfovibrio*
    Genus II. *Bilophila*
    Genus III. *Lawsonia*
  Family II. *Desulfomicrobiaceae*
    Genus I. *Desulfomicrobium*
  Family III. *Desulfohalobiaceae*
    Genus I. *Desulfohalobium*
    Genus II. *Desulfomonas*
    Genus III. *Desulfonatronovibrio*
    Genus IV. *Desulfothermus*
  Family IV. *Desulfonatronumaceae*
    Genus I. *Desulfonatronum*
Order III. *Desulfobacterales*
  Family I. *Desulfobacteraceae*
    Genus I. *Desulfobacter*
    Genus II. *Desulfatibacillum*
    Genus III. *Desulfobacterium*
    Genus IV. *Desulfobacula*
    Genus V. *Desulfobotulus*
    Genus VI. *Desulfocella*
    Genus VII. *Desulfococcus*
    Genus VIII. *Desulfofaba*
    Genus IX. *Desulfofrigus*
    Genus X. *Desulfomusa*
    Genus XI. *Desulfonema*
    Genus XII. *Desulforegula*
    Genus XIII. *Desulfosarcina*
    Genus XIV. *Desulfospira*
    Genus XV. *Desulfotignum*
  Family II. *Desulfobulbaceae*
    Genus I. *Desulfobulbus*
    Genus II. *Desulfocapsa*
    Genus III. *Desulfofustis*
    Genus IV. *Desulforhopalus*
    Genus V. *Desulfotalea*
  Family III. *Nitrospinaceae*
    Genus I. *Nitrospina*
Order IV. *Desulfarcales*
  Family I. *Desulfarculaceae*
    Genus I. *Desulfarculus*
Order V. *Desulfuromonales*
  Family I. *Desulfuromonaceae*
    Genus I. *Desulfuromonas*
    Genus II. *Desulfuromusa*
    Genus III. *Malonomonas*
    Genus IV. *Pelobacter*
  Family II. *Geobacteraceae*
    Genus I. *Geobacter*
    Genus II. *Trichlorobacter*
Order VI. *Syntrophobacterales*
  Family I. *Syntrophobacteraceae*
    Genus I. *Syntrophobacter*
    Genus II. *Desulfacinum*
    Genus III. *Desulforhabdus*
    Genus IV. *Desulfovirga*
    Genus V. *Thermodesulforhabdus*
  Family II. *Syntrophaceae*
    Genus I. *Syntrophus*
    Genus II. *Desulfobacca*
    Genus III. *Desulfomonile*

Genus IV. *Smithella*
Order VII. *Bdellovibrionales*
  Family I. *Bdellovibrionaceae*
    Genus I. *Bdellovibrio*
    Genus II. *Bacteriovorax*
    Genus III. *Micavibrio*
    Genus IV. *Vampirovibrio*
Order VIII. *Myxococcales*
  Suborder I. *Cystobacterineae*
    Family I. *Cystobacteraceae*
      Genus I. *Cystobacter*
      Genus II. *Anaeromyxobacter*
      Genus III. *Archangium*
      Genus IV. *Hyalangium*
      Genus V. *Melittangium*
      Genus VI. *Stigmatella*
    Family II. *Myxococcaceae*
      Genus I. *Myxococcus*
      Genus II. *Corallococcus*
      Genus III. *Pyxicoccus*
  Suborder II. *Sorangiineae*
    Family I. *Polyangiaceae*
      Genus I. *Polyangium*
      Genus II. *Byssophaga*
      Genus III. *Chondromyces*
      Genus IV. *Haploangium*
      Genus V. *Jahnia*
      Genus VI. *Sorangium*
  Suborder III. *Nannocystineae*
    Family I. *Nannocystaceae*
      Genus I. *Nannocystis*
      Genus II. *Plesiocystis*
    Family II. *Haliangiaceae*
      Genus I. *Haliangium*
    Family III. *Kocueriaceae*
      Genus I. *Kocueria*

Class V. *Epsilonproteobacteria*
Order I. *Campylobacterales*
  Family I. *Campylobacteraceae*
    Genus I. *Campylobacter*
    Genus II. *Arcobacter*
    Genus III. *Dehalospirillum*
    Genus IV. *Sulfurospirillum*
  Family II. *Helicobacteraceae*
    Genus I. *Helicobacter*
    Genus II. *Sulfurimonas*
    Genus III. *Thiovulum*
    Genus IV. *Wolinella*
  Family III. *Nautiliaceae*
    Genus I. *Nautilia*
    Genus II. *Caminibacter*
  Family IV. *Hydrogenimonaceae*
    Genus I. *Hydrogenimonas*

Phylum XIII. *Firmicutes*
Class I. *Bacilli*
Order I. *Bacillales*
  Family I. *Bacillaceae*
    Genus I. *Bacillus*
    Genus II. *Alkalibacillus*
    Genus III. *Amphibacillus*
    Genus IV. *Anoxybacillus*
    Genus V. *Cerasibacillus*
    Genus VI. *Filobacillus*
    Genus VII. *Geobacillus*
    Genus VIII. *Gracilibacillus*
    Genus IX. *Halobacillus*
    Genus X. *Halolactibacillus*
    Genus XI. *Lentibacillus*
    Genus XII. *Marinococcus*
    Genus XIII. *Oceanobacillus*
    Genus XIV. *Paraliobacillus*
    Genus XV. *Pontibacillus*

Family XIII. *Incertae sedis*[b]
  Genus I. *Anaerovorax*
  Genus II. *Mogibacterium*
Family XIV. *Incertae sedis*[b]
  Genus I. *Anaerobranca*
Family XV. *Incertae sedis*[b]
  Genus I. *Aminobacterium*
  Genus II. *Aminomonas*
  Genus III. *Anaerobaculum*
  Genus IV. *Dethiosulfovibrio*
  Genus V. *Thermanaerovibrio*
Family XVI. *Incertae sedis*[b]
  Genus I. *Carboxydocella*
Family XVII. *Incertae sedis*[b]
  Genus I. *Sulfobacillus*
  Genus II. *Thermaerobacter*
Family XVIII. *Incertae sedis*[b]
  Genus I. *Symbiobacterium*
Family XIX. *Incertae sedis*[b]
  Genus I. *Acetoanaerobium*
Order II. *Halanaerobiales*
  Family I. *Halanaerobiaceae*
    Genus I. *Halanaerobium*
    Genus II. *Halocella*
    Genus III. *Halothermothrix*
  Family II. *Halobacteroidaceae*
    Genus I. *Halobacteroides*
    Genus II. *Acetohalobium*
    Genus III. *Halanaerobacter*
    Genus IV. *Halonatronum*
    Genus V. *Natroniella*
    Genus VI. *Orenia*
    Genus VII. *Selenihalanaerobacter*
    Genus VIII. *Sporohalobacter*
Order III. *Thermoanaerobacterales*
  Family I. *Thermoanaerobacteraceae*
    Genus I. *Thermoanaerobacter*
    Genus II. *Ammonifex*
    Genus III. *Caldanaerobacter*
    Genus IV. *Carboxydothermus*
    Genus V. *Gelria*
    Genus VI. *Moorella*
    Genus VII. *Thermacetogenium*
    Genus VIII. *Thermanaeromonas*
  Family II. *Thermodesulfobiaceae*
    Genus I. *Thermodesulfobium*
    Genus II. *Coprothermobacter*
  Family III. *Incertae sedis*[b]
    Genus I. *Caldicellulosiruptor*
    Genus II. *Thermoanaerobacterium*
    Genus III. *Thermosediminibacter*
    Genus IV. *Thermovenabulum*
  Family IV. *Incertae sedis*[b]
    Genus I. *Mahella*
Class III. *Erysipelotrichi*
  Order I. *Erysipelotrichales*
    Family I. *Erysipelotrichaceae*
      Genus I. *Erysipelothrix*
      Genus II. *Allobaculum*
      Genus III. *Bulleidia*
      Genus IV. *Catenibacterium*
      Genus V. *Coprobacillus*
      Genus VI. *Holdemania*
      Genus VII. *Solobacterium*
      Genus VIII. *Turicibacter*
Phylum XIV. *Actinobacteria*
  Class I. *Actinobacteria*
    Order I. *Actinomycetales*
      Family I. *Actinomycetaceae*
        Genus I. *Actinomyces*
        Genus II. *Actinobaculum*
        Genus III. *Arcanobacterium*

        Genus IV. *Mobiluncus*
        Genus V. *Varibaculum*
    Order II. *Bifidobacteriales*
      Family I. *Bifidobacteriaceae*
        Genus I. *Bifidobacterium*
        Genus II. *Aeriscardovia*
        Genus III. *Falcivibrio*
        Genus IV. *Gardnerella*
        Genus V. *Parascardovia*
        Genus VI. *Scardovia*
    Order III. *Catenulisporales*
      Family I. *Catenulisporaceae*
        Genus I. *Catellatospora*
      Family II. *Actinospicaceae*
        Genus I. *Actinospica*
    Order IV. *Corynebacteriales*
      Family I. *Corynebacteriaceae*
        Genus I. *Corynebacterium*
        Genus II. *Turicella*
      Family II. *Dietziaceae*
        Genus I. *Dietzia*
      Family III. *Mycobacteriaceae*
        Genus I. *Mycobacterium*
      Family IV. *Nocardiaceae*
        Genus I. *Nocardia*
        Genus II. *Gordonia*
        Genus III. *Millisia*
        Genus IV. *Rhodococcus*
        Genus V. *Skermania*
        Genus VI. *Smaragdicoccus*
        Genus VII. *Williamsia*
      Family V. *Segniliparaceae*
        Genus I. *Segniliparus*
      Family VI. *Tsukamurellaceae*
        Genus I. *Tsukamurella*
    Order V. *Frankiales*
      Family I. *Frankiaceae*
        Genus I. *Frankia*
      Family II. *Acidothermaceae*
        Genus I. *Acidothermus*
      Family III. *Cryptosporangiaceae*
        Genus I. *Cryptosporangium*
        Genus II. *Fodinicola*
      Family IV. *Geodermatophilaceae*
        Genus I. *Geodermatophilus*
        Genus II. *Blastococcus*
        Genus III. *Modestobacter*
      Family V. *Nakamurellaceae*
        Genus I. *Nakamurella*
        Genus II. *Humicoccus*
      Family VI. *Sporichthyaceae*
        Genus I. *Sporichthya*
    Order VI. *Glycomycetales*
      Family I. *Glycomycetacae*
        Genus I. *Glycomyces*
        Genus II. *Stackebrandtia*
    Order VII. *Jiangellales*
      Family I. *Jiangellaceae*
        Genus I. *Jiangella*
    Order VIII. *Kineosporiales*
      Family I. *Kineosporiaceae*
        Genus I. *Kineosporia*
        Genus II. *Kineococcus*
        Genus III. *Quadrisphaera*
    Order IX. *Micrococcales*
      Family I. *Micrococcaceae*
        Genus I. *Micrococcus*
        Genus II. *Acaricomes*
        Genus III. *Arthrobacter*
        Genus IV. *Citricoccus*
        Genus V. *Kocuria*
        Genus VI. *Nesterenkonia*

        Genus VII. *Renibacterium*
        Genus VIII. *Rothia*
        Genus IX. *Yaniella*
        Genus X. *Zhihengliuella*
      Family II. *Beutenbergiaceae*
        Genus I. *Beutenbergia*
        Genus II. *Salana*
        Genus III. *Serinibacter*
      Family III. *Bogoriellaceae*
        Genus I. *Bogoriella*
        Genus II. *Georgenia*
      Family IV. *Brevibacteriaceae*
        Genus I. *Brevibacterium*
      Family V. *Cellulomonadaceae*
        Genus I. *Cellulomonas*
        Genus II. *Actinotalea*
        Genus III. *Demequina*
        Genus IV. *Oerskovia*
        Genus V. *Tropheryma*
      Family VI. *Dermabacteraceae*
        Genus I. *Dermabacter*
        Genus II. *Brachybacterium*
        Genus III. *Demetria*
        Genus IV. *Dermacoccus*
        Genus V. *Kytococcus*
      Family VII. *Dermatophilaceae*
        Genus I. *Dermatophilus*
        Genus II. *Kineosphaera*
      Family VIII. *Intrasporangiaceae*
        Genus I. *Intrasporangium*
        Genus II. *Arsenicicoccus*
        Genus III. *Humihabitans*
        Genus IV. *Janibacter*
        Genus V. *Knoellia*
        Genus VI. *Kribbia*
        Genus VII. *Lapillicoccus*
        Genus VIII. *Ornithinicoccus*
        Genus IX. *Ornithinimicrobium*
        Genus X. *Oryzihumus*
        Genus XI. *Phycicoccus*
        Genus XII. *Serinicoccus*
        Genus XIII. *Terrabacter*
        Genus XIV. *Terracoccus*
        Genus XV. *Tetrasphaera*
      Family IX. *Jonesiaceae*
        Genus I. *Jonesia*
        Genus II. *Kribbella*
      Family X. *Microbacteriaceae*
        Genus I. *Microbacterium*
        Genus II. *Agreia*
        Genus III. *Agrococcus*
        Genus IV. *Agromyces*
        Genus V. *Clavibacter*
        Genus VI. *Cryobacterium*
        Genus VII. *Curtobacterium*
        Genus VIII. *Frigoribacterium*
        Genus IX. *Frondihabitans*
        Genus X. *Gulosibacter*
        Genus XI. *Humibacter*
        Genus XII. *Labedella*
        Genus XIII. *Leifsonia*
        Genus XIV. *Leucobacter*
        Genus XV. *Microcella*
        Genus XVI. *Microterricola*
        Genus XVII. *Mycetocola*
        Genus XVIII. *Okibacterium*
        Genus XIX. *Phycicola*
        Genus XX. *Plantibacter*
        Genus XXI. *Pseudoclavibacter*
        Genus XXII. *Rathayibacter*
        Genus XXIII. *Rhodoglobus*
        Genus XXIV. *Salinibacterium*

Order IV. *Incertae sedis*[b]
 Genus I. *Thermonema*
Order V. *Incertae sedis*[b]
 Genus I. *Toxothrix*
Phylum XXII. "*Verrucomicrobia*"
 Class I. *Verrucomicrobiae*
  Order I. *Verrucomicrobiales*
   Family I. *Verrucomicrobiaceae*
    Genus I. *Verrucomicrobium*
    Genus II. *Prosthecobacter*

Family II. "*Akkermansiaceae*"
 Genus I. *Akkermansia*
Family III. "*Rubritaleaceae*"
 Genus I. *Rubritalea*
 Genus II. *Persicirhabdus*
 Genus III. *Roseibacillus*
Class II. *Opitutae*
 Order I. *Opitutales*
  Family I. *Opitutaceae*
   Genus I. *Opitutus*
   Genus II. *Alterococcus*

Order II. *Puniceicoccales*
 Family I. *Puniceicoccaceae*
  Genus I. *Puniceicoccus*
  Genus II. *Cerasicoccus*
  Genus III. *Coraliomargarita*
  Genus IV. *Pelagicoccus*
Class III. "*Spartobacteria*"
 Order I. "*Chthoniobacterales*"
  Family I. "*Chthoniobacteraceae*"
   Genus I. "*Chthoniobacter*"
   Genus II. '*Candidatus* Xiphinematobacter'[d]

Glossary

Only the major terms and concepts are included. If a term is not here, consult the index.

ABC (ATP-binding cassette) transporter A membrane transport system consisting of three proteins, one of which hydrolyzes ATP, one of which binds the substrate, and one of which functions as the transport channel through the membrane.

Abscess A localized infection characterized by production of pus.

Acetogen A bacterium that carries out acetogenesis.

Acetogenesis Energy metabolism in which acetate is produced from either H_2 plus CO_2 or from organic compounds.

Acetotrophic Acetate consuming.

Acetyl-CoA pathway A pathway of autotrophic CO_2 fixation and acetate oxidation widespread in obligate anaerobes including methanogens, acetogens, and sulfate-reducing bacteria.

Acetylene reduction assay A method of measuring activity of nitrogenase by substituting acetylene for the natural substrate of the enzyme, N_2. Acetylene is reduced to ethylene or ethane, depending on the nitrogenase system involved.

Acid-fastness A property of *Mycobacterium* species; cells stained with basic fuchsin dye resist decolorization with acidic alcohol.

Acid mine drainage Acidic water containing H_2SO_4 derived from the microbial oxidation of iron sulfide minerals.

Acidophile An organism that grows best at acidic pH values.

Acridine orange A nonspecific fluorescent dye used to stain microbial cells in a natural sample.

Activation energy The energy needed to make substrate molecules more reactive; enzymes function by lowering activation energy.

Activator protein A regulatory protein that binds to specific sites on DNA and stimulates transcription; involved in positive control.

Active immunity An immune state achieved by self-production of antibodies. Compare with *passive immunity*.

Active site The portion of an enzyme that is directly involved in binding substrate(s).

Active transport The energy-dependent process of transporting substances into or out of the cell without chemically changing the transported substances.

Acute infection A short-term infection, usually characterized by dramatic onset.

Adaptive immunity (antigen-specific immunity) The acquired ability to recognize and destroy a particular pathogen or its products; dependent on previous exposure to the pathogen or its products.

Adenosine triphosphate (ATP) A nucleotide that is the primary form in which chemical energy is conserved and utilized in cells.

Adherence This property allows cells to stick to host surfaces.

Aerobe An organism that grows in the presence of O_2; may be facultative, obligate, or microaerophilic.

Aerobic secondary wastewater treatment Digestive reactions carried out by microorganisms under aerobic conditions to treat wastewater containing low levels of organic materials.

Aerosol Suspension of particles in airborne water droplets.

Aerotolerant anaerobe An anaerobic microorganism whose growth is not inhibited by O_2.

Agglutination A reaction between antibody and particle-bound antigen resulting in visible clumping of the particles.

Algae Phototrophic eukaryotic micro- and macroorganisms.

Alkaliphile An organism that grows best at high pH.

Allele A sequence variant of a given gene.

Allergy A harmful immune reaction, usually caused by a foreign antigen in food, pollen, or chemicals, which results in immediate-type or delayed-type hypersensitivity.

Allosteric enzyme An enzyme that contains two combining sites, the active site (where the substrate binds) and the allosteric site (where an effector molecule binds).

Amoeboid movement A type of motility in which cytoplasmic streaming moves the organism forward.

Amino acid One of the 22 monomers that make up proteins; chemically, a two-carbon carboxylic acid containing an amino group and a characteristic substituent on the alpha carbon.

Aminoacyl-tRNA synthetase An enzyme that catalyzes the attachment of the correct amino acid to the correct tRNA.

Aminoglycoside An antibiotic such as streptomycin, containing amino sugars linked by glycosidic bonds.

Anabolic reactions (anabolism) The biochemical processes involved in the synthesis of cell constituents from simpler molecules, usually requiring energy.

Anaerobe An organism that grows in the absence of O_2; some may even be killed by O_2 (obligate or strict anaerobes).

Anaerobic respiration Use of an electron acceptor other than O_2 in an electron transport–based oxidation leading to a proton motive force.

Anammox Anoxic ammonia oxidation.

Anaphylaxis (anaphylactic shock) A violent allergic reaction caused by an antigen–antibody reaction.

Anergy The inability to produce an immune response to specific antigens due to neutralization of effector cells.

Anoxic Oxygen-free. Usually used in reference to a microbial habitat.

Anoxic secondary wastewater treatment Digestive and fermentative reactions carried out by microorganisms under anoxic conditions to treat wastewater containing high levels of insoluble organic materials.

Anoxygenic photosynthesis The use of light energy to synthesize ATP by cyclic photophosphorylation without O_2 production.

Antenna pigments Light-harvesting chlorophylls or bacteriochlorophylls in photocomplexes that funnel energy to the reaction center.

Antibiogram A report indicating the sensitivity of clinically isolated microorganisms to the antibiotics in current use.

Antibiotic A chemical substance produced by a microorganism that kills or inhibits the growth of another microorganism.

Antibiotic resistance The acquired ability of a microorganism to grow in the presence of an antibiotic to which the microorganism is usually sensitive.

Antibody A soluble protein produced by B lymphocytes and plasma cells that interacts specifically with antigen; also called *immunoglobulin*.

Antibody-mediated immunity Immunity resulting from direct interaction with antibodies; also called *humoral immunity*.

Anticodon A sequence of three bases in transfer RNA that base-pairs with a codon during protein synthesis.

Antigen A molecule capable of interacting with specific components of the immune system.

Antigen-presenting cell (APC) A macrophage, dendritic cell, or B cell that presents processed antigen peptides to a T cell.

Antigenic determinant (epitope) The portion of an antigen that interacts with an immunoglobulin or T cell receptor.

Antigenic drift In influenza virus, minor changes in viral proteins (antigens) due to gene mutation.

Antigenic shift In influenza virus, major changes in viral proteins (antigens) due to gene reassortment.

Antimicrobial Harmful to microorganisms by either killing or inhibiting growth.

Antimicrobial agent A chemical that kills or inhibits the growth of microorganisms.

Antimicrobial drug resistance The acquired ability of a microorganism to grow in the presence of an antimicrobial drug to which the microorganism is usually susceptible.

Antiparallel In reference to double-stranded nucleic acids, the two strands run in opposite directions; one strand runs $5' \rightarrow 3'$, the complementary strand $3' \rightarrow 5'$.

Antiseptic (germicide) A chemical agent that kills or inhibits growth of microorganisms and is sufficiently nontoxic to be applied to living tissues.

Antiserum A serum containing antibodies.

Antitoxin An antibody that specifically interacts with and neutralizes a toxin.

Apoptosis Programmed cell death.

Archaea Phylogenetically related prokaryotes distinct from *Bacteria*.

Artificial chromosome A single copy vector that can carry extremely long inserts of DNA and is widely used for cloning segments of large genomes.

Aseptic technique The manipulation of sterile instruments or culture media in such a way as to maintain sterility.

Aspartame A nonnutritive sweetener composed of the amino acids aspartate and phenylalanine, the latter as a methyl ester.

ATP Adenosine triphosphate, the principal energy carrier of the cell.

ATPase (ATP synthase) A multiprotein enzyme complex embedded in the cytoplasmic membrane that catalyzes the synthesis of ATP coupled to dissipation of the proton motive force.

Attenuation In a pathogen, a decrease or loss of virulence. Also, a mechanism for controlling gene expression. Typically, transcription is terminated after initiation but before a full-length mRNA is produced.

Autoantibody An antibody that reacts to self antigens.

Autoclave A sealed sterilizing device that destroys microorganisms with temperature and steam under pressure.

Autoimmunity The immune reactions of a host against its own self antigens.

Autoinducer A small signal molecule that takes part in quorum sensing.

Autoinduction A gene regulatory mechanism involving small, diffusible signal molecules that are produced in larger amounts as population size increases.

Autolysis The lysis of a cell brought about by the activity of the cell itself.

Autoradiography Detection of radioactivity in a sample, for example, a cell or gel, by placing it in contact with a photographic film.

Autotroph An organism able to grow on CO_2 as sole source of carbon.

Auxotroph An organism that has developed a nutritional requirement through mutation. Contrast with *prototroph*.

B cell A lymphocyte that has immunoglobulin surface receptors, may present antigens to T cells, and may form plasma cells, which produce immunoglobulin.

Bacteremia The presence of microorganisms in the blood.

Bacteria Phylogenetically related prokaryotes distinct from *Archaea*.

Bacterial artificial chromosome (BAC) Circular artificial chromosome with bacterial origin of replication.

Bacteriochlorophyll A pigment of phototrophic organisms consisting of light-sensitive magnesium tetrapyrroles.

Bacteriocidal agent An agent that kills bacteria.

Bacteriocins Agents produced by certain bacteria that inhibit or kill closely related species.

Bacteriocyte A specialized insect cell in which bacterial symbionts reside.

Bacteriome A specialized region in several insect groups that contains insect bacteriocyte cells packed with bacterial symbionts.

Bacteriophage A virus that infects prokaryotic cells.

Bacteriorhodopsin A protein containing retinal that is found in the membranes of certain extremely halophilic *Archaea* and that is involved in light-mediated ATP synthesis.

Bacteriostatic agent An agent that inhibits bacterial growth.

Bacteroid A swollen, deformed *Rhizobium* cell found in the root nodule; capable of nitrogen fixation.

Banded iron formation Iron oxide–rich ancient sedimentary rocks containing zones of oxidized iron (Fe^{3+}) formed by oxidation of Fe^{2+} by O_2 produced by cyanobacteria.

Basal body The "motor" portion of the bacterial flagellum, embedded in the cytoplasmic membrane and wall.

Base composition In reference to nucleic acids, the proportion of the total bases consisting of guanine plus cytosine or thymine plus adenine base pairs. Usually expressed as a guanine plus cytosine (GC) value, for example, 60% GC.

Batch culture A closed-system microbial culture of fixed volume.

β-Lactam antibiotic An antibiotic such as penicillin that contains the four-membered heterocyclic β-lactam ring.

Binary fission Cell division whereby a cell grows by intercalary growth to twice its minimum size and then divides to form two cells.

Binomial system The system devised by Linnaeus for naming organisms by giving them a genus name and a species epithet.

Biocatalysis The use of microorganisms to synthesize a product or carry out a specific chemical transformation.

Biochemical oxygen demand (BOD) The amount of dissolved oxygen consumed by microorganisms for complete oxidation of organic and inorganic material in a water sample.

Biofilm Microbial colonies encased in an adhesive, usually polysaccharide material and attached to a surface.

Biofuel A fuel made by microorganisms from the fermentation of carbon-rich feedstocks.

Biogeochemistry The study of microbially mediated chemical transformations of geochemical interest, for example, nitrogen or sulfur cycling.

Bioinformatics The use of computer programs to analyze, store, and access DNA and protein sequences.

Biological warfare The use of biological agents to kill or incapacitate a population.

Bioluminescence The enzymatic production of visible light by living organisms.

Bioremediation The use of microorganisms to remove or detoxify toxic or unwanted chemicals in an environment.

Biosynthesis The production of needed cellular constituents from other (usually simpler) molecules.

Biosynthetic penicillin The production of a particular form of penicillin by supplying the producing organism with specific side-chain precursors.

Biotechnology The use of organisms, typically genetically altered, in industrial, medical, or agricultural applications.

Biotransformation In industrial microbiology, the use of microorganisms to convert a substance to a chemically modified form.

Black smoker A deep-sea hydrothermal vent emitting superheated 250–400°C water and minerals.

Bone marrow A primary lymphoid organ containing the pluripotent precursor cells for all blood and immune cells, including B cells.

Botulism Food poisoning due to ingestion of food containing botulinum toxin produced by *Clostridium botulinum*.

Brewing The manufacture of alcoholic beverages such as beer and ales from the fermentation of malted grains.

Broad-spectrum antibiotic An antibiotic that acts on both gram-positive and gram-negative *Bacteria*.

Calvin cycle The series of biosynthetic reactions by which most photosynthetic organisms convert CO_2 to organic compounds.

Canning The process of sealing food in a closed container and heating to destroy living organisms and endospores.

Capsid The protein shell that surrounds the genome of a virus.

Capsomere The subunit of the virus capsid.

Capsule A dense, well-defined polysaccharide or protein layer closely surrounding a cell.

Carboxysome Polyhedral cellular inclusions of crystalline ribulose bisphosphate carboxylase (RubisCO), the key enzyme of the Calvin cycle.

Carcinogen A substance that causes the initiation of tumor formation. Frequently a mutagen.

Cardinal temperatures The minimum, maximum, and optimum growth temperatures for a given organism.

Carotenoid A hydrophobic accessory pigment present along with chlorophyll in photosynthetic membranes.

Carrier Subclinically infected individual who may spread a disease.

Cassette mutagenesis Creating mutations by the insertion of a DNA cassette.

Catabolic reactions (catabolism) The biochemical processes involved in the breakdown of organic or inorganic compounds, usually leading to the production of energy.

Catabolite repression The suppression of alternative catabolic pathways by a preferred source of carbon and energy.

Catalysis An increase in the rate of a chemical reaction.

Catalyst A substance that promotes a chemical reaction without itself being changed in the end.

CD4 cells T-helper cells. They are targets for HIV infection.

Cell The fundamental unit of living matter.

Cell-mediated immunity An immune response generated by interactions with antigen-specific T cells. Compare with *antibody-mediated immunity*.

Cell wall A rigid layer present outside the cytoplasmic membrane that confers structural strength on the cell and protection from osmotic lysis.

Centers for Disease Control and Prevention (CDC) An agency of the United States Public Health Service that tracks disease trends, provides disease information to the public and to healthcare professionals, and forms public policy regarding disease prevention and intervention.

Chaperonin (molecular chaperone) A protein that helps other proteins fold or refold from a partly denatured state.

Chemiosmosis The use of ion gradients, especially proton gradients, across membranes to generate ATP.

Chemokine A small, soluble protein produced by a variety of cells that modulates inflammatory reactions and immunity.

Chemolithotroph An organism that obtains its energy from the oxidation of inorganic compounds.

Chemoorganotroph An organism that obtains its energy from the oxidation of organic compounds.

Chemostat A continuous culture device controlled by the concentration of limiting nutrient and dilution rate.

Chemotaxis Movement toward or away from a chemical.

Chemotherapeutic agent An antimicrobial agent that can be used internally.

Chemotherapy Treatment of infectious disease with chemicals or antibiotics.

Chitin A polymer of *N*-acetylglucosamine commonly found in the cell walls of fungi.

Chloramine A water purification chemical made by combining chlorine and ammonia at precise ratios.

Chlorination A highly effective disinfectant procedure for drinking water using chlorine gas or other chlorine-containing compounds as disinfectant.

Chlorine A chemical used in its gaseous state to disinfect water. A residual level is maintained throughout the distribution system.

Chlorophyll A pigment of phototrophic organisms consisting of light-sensitive magnesium tetrapyrroles.

Chloroplast The chlorophyll-containing organelle of phototrophic eukaryotes.

Chlorosome A cigar-shaped structure enclosed by a nonunit membrane and containing the light-harvesting bacteriochlorophyll (*c*, *d*, or *e*) in green sulfur bacteria and in *Chloroflexus*.

Chromogenic Producing color; for example, a chromogenic colony is a pigmented colony.

Chromosomal island A bacterial chromosome region of foreign origin that contains clustered genes for some extra property such as virulence or symbiosis.

Chromosome A genetic element carrying genes essential to cellular function. Prokaryotes typically have a single chromosome consisting of a circular DNA molecule. Eukaryotes typically have several chromosomes, each containing a linear DNA molecule.

Chronic infection A long-term infection.

-cidal Suffix indicating killing; for example, a bacteriocidal agent kills bacteria. Compare with *-static*.

Ciliate A protist characterized in part by rapid motility driven by numerous short appendages called cilia.

Cilium Short, filamentous structure that beats with many others to make a cell move.

Cirrhosis Breakdown of the normal liver architecture resulting in fibrosis.

Cistron A gene as defined by the *cis-trans* test; a segment of DNA (or RNA) that encodes a single polypeptide chain.

Citric acid cycle A cyclical series of reactions resulting in the conversion of acetate to CO_2 and NADH. Also called the *tricarboxylic acid cycle* or *Krebs cycle*.

Clarifier (coagulation basin) A reservoir in which the suspended solids of raw water are coagulated and removed.

Class I MHC protein An antigen-presenting molecule found on all nucleated vertebrate cells.

Class II MHC protein An antigen-presenting molecule found on macrophages, B lymphocytes, and dendritic cells in vertebrates.

Clonal anergy The inability to produce an immune response to specific antigens due to neutralization of effector cells.

Clonal deletion For T cell selection in the thymus, the killing of useless or self-reactive clone precursors.

Clonal selection A theory that each B or T lymphocyte, when stimulated by antigen, divides to form a clone of itself.

Clone In immunology, a copy of an antigen-reactive lymphocyte, usually in large numbers. Also, a number of copies of a DNA fragment obtained by allowing an inserted DNA fragment to be replicated by a phage or plasmid.

Cloning vectors Genetic elements into which genes can be recombined and replicated.

Coagulation The formation of large insoluble particles from much smaller, colloidal particles by the addition of aluminum sulfate and anionic polymers.

Coccoid Sphere-shaped.

Coccus A spherical bacterium.

Codon A sequence of three bases in messenger RNA that encodes a specific amino acid.

Codon bias The nonrandom usage of multiple codons encoding the same amino acid.

Codon usage The relative proportions of different codons encoding the same amino acid; it varies in different organisms.

Coenocytic The presence of multiple nuclei in fungal hyphae without septa.

Coenzyme A low-molecular-weight molecule that participates in an enzymatic reaction by accepting and donating electrons or functional groups. Examples: NAD^+, FAD.

Coevolution Evolution that proceeds jointly in a pair of intimately associated species owing to the effects each has on the other.

Coliform Gram-negative, nonsporulating, facultatively aerobic rod that ferments lactose with gas formation within 48 hours at 35°C.

Colonial The growth form of certain protists and green algae in which several cells live together and cooperate for feeding, motility, or reproduction; an early form of multicellularity.

Colonization The multiplication of a microorganism after it has attached to host tissues or other surfaces.

Colony A macroscopically visible population of cells growing on solid medium, arising from a single cell.

Cometabolism The metabolic transformation of a substance while a second substance serves as primary energy or carbon source.

Commensalism A type of symbiosis in which only one of two organisms in a relationship benefits.

Commodity chemicals Chemicals such as ethanol that have low monetary value and thus are sold primarily in bulk.

Common-source epidemic An epidemic resulting from infection of a large number of people from a single contaminated source.

Community Two or more cell populations coexisting in a certain area at a given time.

Compatible solutes Organic compounds (or potassium ions) that serve as cytoplasmic solutes to balance water relations for cells growing in environments of high salt or sugar.

Competence The ability to take up DNA and become genetically transformed.

Complement A series of proteins that react sequentially with antibody–antigen complexes, mannose-binding lectin, or alternate activation pathway proteins to amplify or potentiate target cell destruction.

Complement fixation The consumption of complement by an antibody–antigen reaction.

Complementarity-determining region (CDR) A varying amino acid sequence within the variable domains of immunoglobulins or T cell receptors where most molecular contacts with antigen are made.

Complementary Nucleic acid sequences that can base-pair with each other.

Complex medium Any culture medium whose precise chemical composition is unknown. Also called undefined media.

Concatemer A DNA molecule consisting of two or more separate molecules linked end to end to form a long, linear structure.

Congenital syphilis Syphilis contracted by an infant from its mother during pregnancy.

Conidia Asexual spores of fungi.

Conjugation The transfer of genes from one prokaryotic cell to another by a mechanism involving cell-to-cell contact.

Consensus sequence A nucleic acid sequence in which the base present in a given position is that base most commonly found when many experimentally determined sequences are compared.

Consortium A two-membered (or more) bacterial culture (or natural assemblage) in which each organism benefits from the others.

Contagious Transmissible.

Cortex The region inside the spore coat of an endospore, around the core.

Covalent bond A nonionic chemical bond formed by a sharing of electrons between two atoms.

Crenarchaeota A phylum of *Archaea* that contains both hyperthermophilic and cold-dwelling organisms.

Crista (plural, cristae) An inner membrane in a mitochondrion; a site of respiration.

Culture A particular strain or kind of organism growing in a laboratory medium.

Culture medium An aqueous solution of various nutrients suitable for the growth of microorganisms.

Cutaneous Relating to the skin.

Cyanobacteria Prokaryotic oxygenic phototrophs containing chlorophyll *a* and phycobilins.

Cyclic AMP A regulatory nucleotide that participates in catabolite repression.

Cyst A resting stage formed by some bacteria and protists in which the whole cell is surrounded by a thick-walled chemically and physically resistant coating; not the same as a spore or endospore.

Cytochrome An iron-containing porphyrin complexed with proteins, which functions as an electron carrier in the electron transport system.

Cytokine A small, soluble protein produced by a leukocyte that modulates inflammatory reactions and immunity.

Cytoplasm The fluid portion of a cell, bounded by the cell membrane.

Cytoplasmic membrane A semipermeable barrier that separates the cell interior (cytoplasm) from the environment.

Cytoskeleton Cellular scaffolding typical of eukaryotic cells in which microtubules, microfilaments, and intermediate filaments define the cell's shape.

DAPI A nonspecific fluorescent dye used to stain microbial cells in a natural sample to obtain total cell numbers.

Decontamination Treatment that renders an object or inanimate surface safe to handle.

Deep sea Marine waters below a depth of 1000 m.

Defective virus A virus that relies on another virus, the helper virus, to provide some of its components.

Defined medium Any culture medium whose exact chemical composition is known. Compare with *complex medium*.

Degeneracy In relation to the genetic code, the fact that more than one codon can code for the same amino acid.

Delayed hypersensitivity An inflammatory allergic response mediated by T lymphocytes.

Deletion The removal of a portion of a gene.

Denaturation The irreversible destruction of a macromolecule, as for example, the destruction of a protein by heat.

Denaturing gradient gel electrophoresis (DGGE) An electrophoretic technique capable of separating nucleic acid fragments of the same size that differ in sequence.

Dendritic cell A type of leukocyte having phagocytic and antigen-presenting properties, found in various body tissues; transports antigen to lymph nodes and spleen.

Denitrification Anaerobic respiration in which nitrate is reduced to nitrogen gases under anoxic conditions.

Dental caries Tooth decay resulting from bacterial infection.

Dental plaque Bacterial cells encased in a matrix of extracellular polymers and salivary products, found on the teeth.

Deoxyribonucleic acid (DNA) A polymer of nucleotides connected via a phosphate–deoxyribose sugar backbone; the genetic material of cells and some viruses.

Desiccation Drying.

Dideoxynucleotide A nucleotide lacking the 3′-hydroxyl group on the deoxyribose sugar. Used in the Sanger method of DNA sequencing.

Differential media A growth medium that allows identification of microorganisms based on phenotypic properties.

Differentiation The modification of a cell in terms of structure and/or function occurring during the course of development.

Dipicolinic acid A substance unique to endospores that confers heat resistance on these structures.

Diploid In eukaryotes, an organism or cell with two chromosome complements, one derived from each haploid gamete.

Disease An injury to a host organism, caused by a pathogen or other factor, that affects the host organism's function.

Disinfectant An antimicrobial agent used only on inanimate objects.

Disinfection The elimination of pathogens from inanimate objects or surfaces.

Disproportionation The splitting of a chemical compound into two new compounds, one more oxidized and one more reduced than the original compound.

Distilled alcoholic beverage A beverage containing alcohol concentrated by distillation.

Distribution system Water pipes, storage reservoirs, tanks, and other means used to deliver drinking water to consumers or store it before delivery.

Divisome A complex of proteins that directs cell division processes in prokaryotes.

DNA Deoxyribonucleic acid, the genetic material of cells and some viruses.

DNA cassette An artificially designed segment of DNA that usually carries a gene for resistance to an antibiotic or some other convenient marker and is flanked by convenient restriction sites.

DNA–DNA hybridization The experimental determination of genomic similarity by measuring the extent of hybridization of DNA from the genome of one organism with that of another.

DNA gyrase An enzyme found in most prokaryotes that introduces negative supercoils in DNA.

DNA library See *gene library*.

DNA ligase An enzyme that seals nicks in the backbone of DNA.

DNA polymerase An enzyme that synthesizes a new strand of DNA in the 5′ → 3′ direction using an antiparallel DNA strand as a template.

DNA vaccine A vaccine that uses the DNA of a pathogen to elicit an immune response.

Domain (1) The highest level of biological classification. The three domains of biological organisms are the *Bacteria*, the *Archaea*, and the *Eukarya*. (2) A region of a protein having a defined structure and function.

Doubling time The time needed for a population to double. See also *generation time*.

Downstream position Refers to nucleic acid sequences on the 3′ side of a given site on the DNA or RNA molecule. Compare with *upstream position*.

Early protein A protein synthesized soon after virus infection and before replication of the virus genome.

Ecology Study of the interrelationships between organisms and their environments.

Ecosystem A dynamic complex of organisms and their physical environment interacting as a functional unit.

Ecotype A population of genetically identical cells sharing a particular resource within an ecological niche.

Effluent water Treated wastewater discharged from a wastewater treatment facility.

Ehrlichiosis One of a group of emerging tick-transmitted diseases caused by rickettsias of the *Ehrlichia* genus.

Electron acceptor A substance that accepts electrons during an oxidation–reduction reaction.

Electron donor A compound that donates electrons in an oxidation–reduction reaction.

Electron transport phosphorylation Synthesis of ATP involving a membrane-associated electron transport chain and the creation of a proton motive force. Also called *oxidative phosphorylation*.

Electrophoresis Separation of charged molecules in an electric field.

Electroporation The use of an electric pulse to enable cells to take up DNA.

ELISA See *enzyme immunoassay*.

Emerging disease Infectious disease whose incidence recently increased or whose incidence threatens to increase in the near future.

Enantiomer One form of a molecule that is the mirror image of another form of the same molecule.

Endemic disease A disease that is constantly present in low numbers in a population, usually in low numbers. Compare with *epidemic*.

Endergonic reaction A chemical reaction requiring an input of energy to proceed.

Endocytosis A process in which a particle such as a virus is taken intact into an animal cell. Phagocytosis and pinocytosis are two kinds of endocytosis.

Endoplasmic reticulum An extensive array of internal membranes in eukaryotes.

Endospore A differentiated cell formed within the cells of certain gram-positive bacteria that is extremely resistant to heat as well as to other harmful agents.

Endosymbiosis The engulfment of one cell type by another cell type and the subsequent and stable association of the two cells.

Endosymbiotic hypothesis The idea that a chemoorganotrophic bacterium and a cyanobacterium were stably incorporated into another cell type to give rise, respectively, to the mitochondria and chloroplasts of modern-day eukaryotes.

Endotoxin The lipopolysaccharide portion of the cell envelope of certain gram-negative *Bacteria*, which is a toxin when solubilized. Compare with *exotoxin*.

Enriched media Media that allow metabolically fastidious organisms to grow because of the addition of specific growth factors.

Enrichment bias A problem with enrichment cultures in which "weed" species tend to dominate in the enrichment, often to the exclusion of the most abundant or ecologically significant organisms in the inoculum.

Enrichment culture Use of selective culture media and incubation conditions to isolate specific microorganisms from natural samples.

Enteric bacteria A large group of gram-negative, rod-shaped *Bacteria* characterized by a facultatively aerobic metabolism and commonly found in the intestines of animals.

Enterotoxin A protein that is released extracellularly by a microorganism as it grows and that produces immediate damage to the small intestine of the host.

Entropy A measure of the degree of disorder in a system; entropy always increases in a closed system.

Enveloped In reference to a virus, having a lipoprotein membrane surrounding the virion.

Environmental genomics (metagenomics) Genomic analysis of pooled DNA from an environmental sample without first isolating or identifying the individual organisms.

Enzootic An endemic disease present in an animal population.

Enzyme A catalyst, usually composed of protein, that promotes specific reactions or groups of reactions.

Enzyme immunoassay (EIA) A test that uses antibodies to detect antigens or antibodies in body fluids. Also called enzyme-linked immunosorbent assay (ELISA).

Epidemic A disease occurring in an unusually high number of individuals in a population at the same time. Compare with *endemic*.

Epidemiology The study of the occurrence, distribution, and determinants of health and disease in a population.

Epilimnion The warmer and less dense surface waters of a stratified lake.

Epitope The portion of an antigen that is recognized by an immunoglobulin or a T cell receptor.

Epizootic An epidemic disease present in an animal population.

***Escherichia coli* O157:H7** An enterotoxigenic strain of *E. coli* spread by fecal contamination of animal or human origin to food and water.

Eukarya The phylogenetic domain containing all eukaryotic organisms.

Eukaryote A cell or organism having a unit membrane–enclosed nucleus and usually other organelles; a member of the *Eukarya*.

Euryarchaeota A phylum of *Archaea* that contains primarily methanogens, extreme halophiles, *Thermoplasma*, and some marine hyperthermophiles.

Evolution Descent with modification; DNA sequence variation and the inheritance of that variation.

Evolutionary distance In phylogenetic trees, the sum of the physical distance on a tree separating organisms; this distance is inversely proportional to evolutionary relatedness.

Exergonic reaction A chemical reaction that proceeds with the liberation of energy.

Exoenzyme An enzyme produced by a microorganism and then excreted into the environment.

Exon The coding sequence in a split gene. Contrast with *introns*, the intervening noncoding regions.

Exotoxin A protein that is released extracellularly by a microorganism as it grows and that produces immediate host cell damage. Compare with *endotoxin*.

Exponential growth Growth of a microbial population in which the cell number doubles within a fixed time period.

Exponential phase A period during the growth cycle of a population in which growth increases at an exponential rate.

Expression The ability of a gene to function within a cell in such a way that the gene product is formed.

Expression vector A cloning vector that contains the necessary regulatory sequences allowing transcription and translation of a cloned gene or genes.

Extein The portion of a protein that remains and has biological activity after the splicing out of any inteins.

Extracellular matrix (ECM) Proteins and polysaccharides that surround an animal cell and in which the cell is embedded.

Extreme halophile An organism whose growth is dependent on large concentrations (generally >9%) of NaCl.

Extreme piezophile A piezophilic organism unable to grow at a pressure of 1 atm and typically requiring several hundred atmospheres of pressure for growth.

Extremophile An organism that grows optimally under one or more chemical or physical extremes, such as high or low temperature or pH.

Extremozyme An enzyme able to function in one or more chemical or physical extremes, for example, high temperature or low pH.

Facultative Indicates that an organism is able to grow in either the presence or absence of an environmental factor (for example, "facultative aerobe").

FAME Fatty acid methyl ester; a technique for identifying microorganisms by their fatty acids.

Fatty acid An organic acid containing a carboxylic acid group and a hydrocarbon chain of various lengths; major components of lipids.

Feedback inhibition A decrease in the activity of the first enzyme of a biochemical pathway caused by buildup of the final product of the pathway.

Fermentation Anaerobic catabolism of an organic compound in which the compound serves as both an electron donor and an electron acceptor and in which ATP is usually produced by substrate-level phosphorylation.

Fermentation (industrial) A large-scale microbial process.

Fermenter An organism that carries out the process of fermentation.

Fermentor A growth vessel, usually quite large, used to culture microorganisms for the production of some commercially valuable product.

Ferredoxin An electron carrier of very negative reduction potential; small protein containing iron–sulfur clusters.

Fever An abnormal increase in body temperature.

Filamentous In the form of very long rods, many times longer than wide.

Filtration The removal of suspended particles from water by passing it through one or more permeable membranes or media (e.g., sand, anthracite, or diatomaceous earth).

Fimbria (plural, fimbriae) Short, filamentous structure on a bacterial cell; although flagella-like in structure, it is generally present in many copies and not involved in motility. Plays a role in adherence to surfaces and in the formation of pellicles. See also *pilus*.

Finished water Water delivered to the distribution system after treatment.

FISH Fluorescent *in situ* hybridization; a method employing a fluorescent dye covalently bonded to a specific nucleic acid probe for identifying or tracking organisms in the environment.

Fitness The capacity of an organism to survive and reproduce as compared to competing organisms.

Flagellum (plural, flagella) A thin, filamentous organ of motility in prokaryotes that functions by rotating. In motile eukaryotes, the flagellum, if present, moves by a whiplike motion.

Flavoprotein A protein containing a derivative of riboflavin, which functions as an electron carrier in the electron transport system.

Flocculation The water treatment process after coagulation that uses gentle stirring to cause suspended particles to form larger, aggregated masses (flocs).

Flow cytometry A technique for counting and examining microscopic particles by suspending them in a stream of fluid and passing them by an electronic detection device.

Fluorescent Having the ability to emit light of a certain wavelength when activated by light of another wavelength.

Fluorescent antibody An antibody molecule covalently modified with a fluorescent dye that makes the antibody visible under fluorescent light.

Fluorescent *in situ* hybridization (FISH) A method employing a fluorescent dye covalently bonded to a specific nucleic acid probe for identifying or tracking organisms in the environment.

Fomite Inanimate object that, when contaminated with a viable pathogen, can transfer the pathogen to a host.

Food infection A microbial infection resulting from the ingestion of pathogen-contaminated food followed by growth of the pathogen in the host.

Food poisoning (food intoxication) Disease caused by the ingestion of food that contains preformed microbial toxins.

Food spoilage Any change in a food product that makes it unacceptable to the consumer.

Frameshift A type of mutation. Because the genetic code is read three bases at a time, if reading begins at either the second or third base of a codon, a faulty product usually results.

Free energy (G) Energy available to do work; $G^{0\prime}$ is free energy under standard conditions.

Fruiting body A macroscopic reproductive structure produced by some fungi (for example, mushrooms) and some *Bacteria* (for example, myxobacteria), each distinct in size, shape, and coloration.

FtsZ A protein that forms a ring along the mid-cell division plane to initiate cell division.

Fungi Nonphototrophic eukaryotic microorganisms that contain rigid cell walls.

Fungicidal agent An agent that kills fungi.

Fungistatic agent An agent that inhibits fungal growth.

Fusion inhibitor A synthetic polypeptide that binds to viral glycoproteins, inhibiting fusion of viral and host cell membranes.

Fusion protein A protein that is the result of fusing two different proteins together by merging their coding sequences into a single gene.

Gametes In eukaryotes, the haploid germ cells that result from meiosis.

Gas vesicle A gas-filled structure made of protein; confers buoyancy on a cell when present in the cytoplasm in large numbers.

GC ratio In DNA (or RNA) from any organism, the percentage of the total nucleic acid that consists of guanine plus cytosine bases (expressed as mol% GC).

Gel An inert polymer, usually made of agarose or polyacrylamide, used for separating macromolecules such as nucleic acids and proteins by electrophoresis.

Gel electrophoresis A technique for separation of nucleic acid molecules by passing an electric current through a gel made of agarose or polyacrylamide.

Gene A unit of heredity; a segment of DNA (or RNA in some viruses) specifying a particular protein or polypeptide chain, or a tRNA or an rRNA.

Gene chip Small, solid-state supports to which genes or portions of genes are affixed and arrayed spatially in a known pattern (also called *microarrays*).

Gene cloning See *molecular cloning*.

Gene disruption (also called gene knockout) The inactivation of a gene by insertion of a DNA fragment that interrupts the coding sequence.

Gene expression Transcription of a gene followed by translation of the resulting mRNA into protein(s).

Gene family Genes that are related in sequence to each other as the result of a common evolutionary origin.

Gene fusion A structure created by joining together segments of two separate genes, in particular when the regulatory region of one gene is joined to the coding region of a reporter gene.

Gene library A collection of cloned DNA fragments that contains all the genetic information for a particular organism.

General-purpose medium A growth medium that supports the growth of most aerobic and facultatively aerobic organisms.

Generation time The time required for a cell population to double. See also *doubling time*.

Gene therapy Treatment of a disease caused by a dysfunctional gene by introduction of a normally functioning copy of the gene.

Genetically modified organism (GMO) An organism whose genome has been altered using genetic engineering. The abbreviation is also used in terms such as GM crops and GM foods.

Genetic code The correspondence between nucleic acid sequence and amino acid sequence of proteins.

Genetic element A structure that carries genetic information, such as a chromosome, a plasmid, or a virus genome.

Genetic engineering The use of *in vitro* techniques in the isolation, manipulation, alteration, and expression of DNA (or RNA) and in the development of genetically modified organisms.

Genetic map The arrangement of genes on a chromosome.

Genetics Heredity and variation of organisms.

Genome The total complement of genetic information of a cell or a virus.

Genomics The discipline that maps, sequences, analyzes, and compares genomes.

Genotype The complete genetic makeup of an organism; the complete description of a cell's genetic information. Compare with *phenotype*.

Genus A taxonomic group of related species.

Germicide (antiseptic) A chemical agent that kills or inhibits growth of microorganisms and is sufficiently nontoxic to be applied to living tissues.

Glycocalyx Polysaccharide components outside of the bacterial cell wall; usually a loose network of polymer fibers extending outward from the cell.

Glycolysis Reactions of the Embden–Meyerhof–Parnas pathway in which glucose is converted to pyruvate.

Glycosidic bond A type of covalent bond that links sugar units together in a polysaccharide.

Glyoxylate cycle A series of reactions including some citric acid cycle reactions that are used for aerobic growth on C_2 or C_3 organic acids.

Gonococcus *Neisseria gonorrhoeae*, the gram-negative diplococcus that causes the disease gonorrhea.

Gram-negative cells A major phylogenetic lineage of prokaryotic cells with a cell wall containing relatively little peptidoglycan, and an outer membrane composed of lipopolysaccharide, lipoprotein, and other complex macromolecules; stain pink in the Gram stain.

Gram-positive cells A major phylogenetic lineage of prokaryotic cells containing mainly peptidoglycan in their cell wall; stain purple in the Gram stain.

Gram stain A differential staining technique in which cells stain either pink (gram-negative) or purple (gram-positive), depending upon their structural makeup.

Green fluorescent protein (GFP) A protein that fluoresces green and is widely used in genetic analysis.

Green sulfur bacteria Anoxygenic phototrophs containing chlorosomes and bacteriochlorophyll *c*, *d*, or *e* as light-harvesting chlorophyll.

Group translocation An energy-dependent transport system in which the substance transported is chemically modified during the process of being transported by a series of proteins.

Growth In microbiology, an increase in cell number.

Growth factor analog A chemical agent that is related to and blocks the uptake or utilization of a growth factor.

Growth rate The rate at which growth occurs, usually expressed as the generation time.

Guild A group of metabolically related organisms.

HAART (highly active antiretroviral therapy) The treatment of HIV infection with two or more antiretroviral drugs at once to inhibit the development of drug resistance.

Habitat An environment within an ecosystem where a microbial community could reside.

Halophile An organism requiring salt (NaCl) for growth.

Halorhodopsin A light-driven chloride pump that accumulates Cl^- within the cytoplasm.

Halotolerant Capable of growing in the presence of NaCl, but not requiring it.

Hantavirus pulmonary syndrome (HPS) An emerging acute viral disease characterized by pneumonia, caused by rodent hantavirus.

Haploid An organism or cell containing only one set of chromosomes.

Hapten A low-molecular-weight substance not inducing antibody formation itself but still able to combine with a specific antibody.

Healthcare-associated infection (nosocomial infection) An infection contracted in a healthcare-associated setting.

Heat shock proteins Proteins induced by high temperature (or certain other stresses) that protect against high temperature, especially by refolding partially denatured proteins or by degrading them.

Heat shock response Response to high temperature that includes the synthesis of heat shock proteins together with other changes in gene expression.

Heliobacteria Anoxygenic phototrophs containing bacteriochlorophyll *g*.

Helix A spiral structure in a macromolecule that contains a repeating pattern.

Helper virus A virus that provides some necessary components for a defective virus.

Hemagglutination Agglutination of red blood cells.

Hemolysins Bacterial toxins capable of lysing red blood cells.

Hemolysis Lysis of red blood cells.

Hemorrhagic fever with renal syndrome (HFRS) An emerging acute disease characterized by shock and kidney failure, caused by rodent hantavirus.

HEPA filter A high-efficiency particulate air filter used in laboratories and industry to remove particles, including microorganisms, from intake or exhaust air flow.

Hepadnavirus A virus whose DNA genome replicates by way of an RNA intermediate.

Hepatitis Liver inflammation, commonly caused by an infectious agent.

Herd immunity The resistance of a group to a pathogen as a result of the immunity of a large proportion of the group to that pathogen.

Herpes simplex The virus that causes both genital herpes and cold sores.

Heterocyst A differentiated cyanobacterial cell that carries out nitrogen fixation.

Heteroduplex A DNA double helix composed of single strands from two different DNA molecules.

Heterofermentative Describes lactic acid bacteria capable of making more than one fermentation product.

Heterotroph An organism that requires organic carbon as its carbon source; also a chemoorganotroph.

Hfr cell A cell with the F plasmid integrated into the chromosome.

Histones Basic proteins that protect and compact the DNA in eukaryotes and some *Archaea*.

Homoacetogens *Bacteria* that produce acetate as the sole product of sugar fermentation or from $H_2 + CO_2$. Also called *acetogens*.

Homofermentative In reference to lactic acid bacteria, producing only lactic acid as a fermentation product.

Homologous Describes genes related in sequence to an extent that implies common genetic ancestry; includes both orthologs and paralogs.

Homologous antigen An antigen that reacts with the antibody it has induced.

Horizontal gene transfer The transfer of genetic information between organisms as opposed to its vertical inheritance from parental organism(s).

Host (or host cell) An organism or cell type capable of supporting the growth of a virus or other parasite.

Host-to-host epidemic An epidemic resulting from host-to-host contact, characterized by a gradual rise and fall in disease incidence.

Human artificial chromosome (HAC) An artificial chromosome with a human centromere sequence array.

Human granulocytic anaplasmosis (HGA) Rickettsiosis caused by *Ehrlichia ewingii* or *Anaplasma phagocytophilum*.

Human leukocyte antigen (HLA) An antigen-presenting protein encoded by a major histocompatibility complex gene in humans.

Human monocytic ehrlichiosis (HME) A rickettsiosis caused by *Ehrlichia chaffeensis* or *Rickettsia sennetsu*.

Human papillomavirus (HPV) A sexually transmitted virus that causes genital warts, cervical neoplasia, and cancer.

Humoral immunity An immune response involving antibodies.

Humus Dead organic matter.

Hybridization Base pairing of single strands of DNA or RNA from two different (but related) sources to give a hybrid double helix.

Hybridoma The fusion of an immortal (tumor) cell with a lymphocyte to produce an immortal lymphocyte.

Hydrogenase An enzyme, widely distributed in anaerobic microorganisms, capable of taking up or evolving H_2.

Hydrogen bond A weak chemical bond between a hydrogen atom and a second, more electronegative element, usually an oxygen or nitrogen atom.

Hydrogenosome An organelle of endosymbiotic origin in the cytoplasm of certain anaerobic eukaryotes that functions to oxidize pyruvate to $H_2 + CO_2 +$ acetate.

Hydrolysis Breakdown of a polymer into smaller units, usually monomers, by addition of water; digestion.

Hydrophobic interactions Attractive forces between molecules due to the close positioning of nonhydrophilic portions of the two molecules.

Hydrothermal vents Warm or hot water–emitting springs associated with crustal spreading centers on the seafloor.

Hydroxypropionate pathway An autotrophic pathway found in *Chloroflexus* and a few *Archaea*.

Hypersensitivity An immune reaction causing damage to the host, caused either by antigen–antibody reactions or cellular immune processes. See *allergy*.

Hyperthermophile A prokaryote having a growth temperature optimum of 80°C or higher.

Hypervariable region A varying amino acid sequence within the variable domains of immunoglobulins or T cell receptors where most molecular contacts with antigen are made (also known as a *complementarity-determining region*).

Hypolimnion The colder, more dense, and often anoxic bottom waters of a stratified lake.

Icosahedron A geometrical shape occurring in many virus particles, with 20 triangular faces and 12 corners.

Immediate hypersensitivity An allergic response mediated by vasoactive products released from tissue mast cells.

Immobilized enzyme An enzyme attached to a solid support over which substrate is passed and converted to product.

Immune Able to resist infectious disease.

Immune memory The capacity to respond more quickly and vigorously to second and subsequent exposures to an eliciting antigen.

Immunity The ability of an organism to resist infection.

Immunization (vaccination) Inoculation of a host with inactive or weakened pathogens or pathogen products to stimulate protective immunity.

Immunoblot (Western blot) The detection of specific proteins by separating them via electrophoresis, transferring them to a membrane, and adding specific antibodies.

Immunodeficiency Having a dysfunctional or completely nonfunctional immune system.

Immunogen A molecule capable of eliciting an immune response.

Immunoglobulin (Ig) A soluble protein produced by B cells and plasma cells that interacts with antigen; also called *antibody*.

Immunoglobulin gene superfamily A family of genes that are evolutionarily, structurally, and functionally related to immunoglobulins.

Incidence The number of new disease cases reported in a population in a given time period.

Induced enzyme An enzyme subject to induction.

Induced mutation A mutation caused by external agents such as mutagenic chemicals or radiation.

Induction Production of an enzyme in response to a signal (often the presence of the substrate for the enzyme).

Industrial microbiology The large-scale use of microorganisms to make products of commercial value.

Infection Growth of an organism within a host.

Infection thread In the formation of root nodules, a cellulosic tube through which *Rhizobium* cells travel to reach and infect root cells.

Inflammation A nonspecific reaction to noxious stimuli such as toxins and pathogens, characterized by redness (erythema), swelling (edema), pain, and heat, usually localized at the site of infection.

Informational macromolecule Any large polymeric molecule that carries genetic information, including DNA, RNA, and protein.

Inhibition In reference to growth, the reduction of microbial growth because of a decrease in the number of organisms present or alterations in the microbial environment.

Innate immunity (nonspecific immunity) The noninducible ability to recognize and destroy an individual pathogen or its products that does not rely on previous exposure to a pathogen or its products.

Inoculum Cell material used to initiate a microbial culture.

Insertion A genetic phenomenon in which a piece of DNA is inserted into the middle of a gene.

Insertion sequence (IS) The simplest type of transposable element, which carries only genes involved in transposition.

In silico The use of computers to perform sophisticated analyses.

Integrase The enzyme that inserts cassettes into an integron.

Integrase inhibitor A drug that interrupts the HIV replication cycle by interfering with integrase, the HIV protein that catalyzes the integration of viral dsDNA into host cell DNA.

Integrating vector A cloning vector that can be inserted into a host chromosome.

Integration The process by which a DNA molecule becomes incorporated into another genome.

Integron A genetic element that collects and expresses genes carried on mobile cassettes.

Intein An intervening sequence in a protein; a segment of a protein that can splice itself out.

Interactome The total set of interactions between proteins (or other macromolecules) in an organism.

Intercalary growth In cell division, enlargement of a cell at several growing points.

Interferon Cytokine proteins produced by virus-infected cells that induce signal transduction in nearby cells, resulting in transcription of antiviral genes and expression of antiviral proteins.

Interleukin (IL) Soluble cytokine or chemokine mediator secreted by leukocytes.

Intermediate filament A filamentous polymer of fibrous keratin proteins, supercoiled into thicker fibers, that functions in maintaining cell shape and the positioning of certain organelles in the eukaryotic cell.

Interspecies hydrogen transfer The process by which organic matter is degraded by the interaction of several groups of microorganisms in which H_2 production and H_2 consumption are closely coupled.

Introns The intervening noncoding sequences in a split gene. Contrast with *exons*, the coding sequences.

Invasion The ability of a pathogen to enter into host cells or tissues, spread, and cause disease.

In vitro In glass, away from the living organism.

In vivo In the body, in a living organism.

Ionophore A compound that can cause the leakage of ions across membranes.

Irradiation In food microbiology, the exposure of food to ionizing radiation to inhibit microorganisms and insect pests or to retard growth or ripening.

Isomers Two molecules that have the same molecular formula but that differ structurally.

Isotopes Different forms of the same element containing the same number of protons and electrons but differing in the number of neutrons.

Isotopic fractionation Discrimination by enzymes against the heavier isotope of the various isotopes of carbon or sulfur, leading to enrichment of the lighter isotopes.

Jaundice The production and release of excess bilirubin in the liver due to destruction of liver cells, resulting in yellowing of the skin and whites of the eye.

Joule (J) A unit of energy equal to 10^7 ergs; 1000 joules equal 1 kilojoule (kJ).

Kilobase (kb) A 1000-base fragment of nucleic acid. A *kilobase pair* (kbp) is a fragment containing 1000 base pairs.

Kinase An enzyme that adds a phosphoryl group, usually from ATP, to a compound.

Koch's postulates A set of criteria for proving that a given microorganism causes a given disease.

Lag phase The period after inoculation of a culture before growth begins.

Lagging strand The new strand of DNA that is synthesized in short pieces during DNA replication and then joined together later.

Laser tweezers A device used to obtain pure cultures in which a single cell is optically trapped with a laser and moved away from contaminating organisms into sterile growth medium.

Late protein A protein synthesized later in virus infection after replication of the virus genome.

Latent virus A virus present in a cell, yet not causing any detectable effect.

Lateral gene transfer The transfer of genes from a cell to another cell that is not its offspring. Also called *horizontal gene transfer*.

Leaching Removal of valuable metals from ores by microbial action.

Leading strand The new strand of DNA that is synthesized continuously during DNA replication.

Leghemoglobin An O_2-binding protein found in root nodules.

Leukocidin A substance able to destroy phagocytes.

Leukocyte A nucleated cell found in the blood; a white blood cell.

Lichen A fungus and an alga (or a cyanobacterium) living in symbiotic association.

Lipids Water-insoluble organic molecules important in structure of the cytoplasmic membrane and (in some organisms) the cell wall. See also *phospholipid*.

Lipopolysaccharide (LPS) Complex lipid structure containing unusual sugars and fatty acids found in most gram-negative *Bacteria* and constituting the chemical structure of the outer membrane.

Listeriosis A gastrointestinal food infection caused by *Listeria monocytogenes* that may lead to bacteremia and meningitis.

Lophotrichous Having a tuft of polar flagella.

Lower respiratory tract Trachea, bronchi, and lungs.

Luminescence The production of light.

Lyme disease An emerging tick-transmitted disease caused by the spirochete *Borrelia burgdorferi*.

Lymph A fluid similar to blood that lacks red blood cells and travels through a separate circulatory system (the lymphatic system) containing lymph nodes.

Lymphocyte A subset of leukocytes found in the blood that are involved in the adaptive immune response.

Lyophilization (freeze-drying) The process of removing all water from frozen food under vacuum.

Lysin An antibody that induces lysis.

Lysis Loss of cellular integrity with release of cytoplasmic contents.

Lysogen A prokaryote containing a prophage. See also *temperate virus*.

Lysogenic pathway After virus infection, a series of steps that leads to a state (lysogeny) in which the viral genome is replicated as a provirus along with that of the host.

Lysogeny A state following virus infection in which the viral genome is replicated as a provirus along with the genome of the host.

Lysosome An organelle containing digestive enzymes for hydrolyses of proteins, fats, and polysaccharides.

Lytic pathway A series of steps after virus infection that lead to virus replication and the destruction (lysis) of the host cell.

Macromolecule A large molecule (polymer) formed by the connection of a number of small molecules (monomers); proteins, nucleic acids, lipids, and polysaccharides in a cell.

Macrophage A large leukocyte found in tissues that has phagocytic and antigen-presenting capabilities.

Magnetosome A small particle of Fe_3O_4 present in cells that exhibit magnetotaxis (magnetic bacteria).

Magnetotaxis The directed movement of bacterial cells by a magnetic field.

Major histocompatibility complex (MHC) A genetic region with genes that encode several proteins important for antigen presentation and other host defense functions.

Malaria An insect-transmitted disease characterized by recurrent episodes of fever and anemia; caused by the protist *Plasmodium* spp., usually transmitted between mammals through the bite of the *Anopheles* mosquito.

Malignant In reference to a tumor, an infiltrating metastasizing growth no longer under normal growth control.

Mast cells Tissue cells adjoining blood vessels throughout the body that contain granules with inflammatory mediators.

Medium (plural, media) In microbiology, the nutrient solution(s) used to grow microorganisms.

Megabase (Mb) One million nucleotide bases (or base pairs, abbreviated Mbp).

Meiosis A specialized form of nuclear division that halves the diploid number of chromosomes to the haploid number, for gametes of eukaryotic cells.

Membrane Any thin sheet or layer. See especially *cytoplasmic membrane*.

Memory (immune memory) The ability to rapidly produce large quantities of specific immune cells or antibodies after subsequent exposure to a previously encountered antigen.

Memory B cell Long-lived B cell responsive to an individual antigen.

Meningitis Inflammation of the meninges (brain tissue), sometimes caused by *Neisseria meningitidis* and characterized by sudden onset of headache, vomiting, and stiff neck, often progressing to coma within hours.

Meningococcemia A fulminant disease caused by *Neisseria meningitidis* and characterized by septicemia, intravascular coagulation, and shock.

Meningoencephalitis The invasion, inflammation, and destruction of brain tissue by the amoeba *Naegleria fowleri* or a variety of other pathogens.

Mesophile An organism living in the temperature range near that of warm-blooded animals and usually showing a growth temperature optimum between 25 and 40°C.

Messenger RNA (mRNA) An RNA molecule that contains the genetic information to encode one or more polypeptides.

Metabolism All biochemical reactions in a cell, both anabolic and catabolic.

Metabolome The total complement of small molecules and metabolic intermediates of a cell or organism.

Metagenome The total genetic complement of all the cells present in a particular environment.

Metagenomics See *environmental genomics*.

Metazoa Multicellular animals.

Methanogen A methane-producing member of the *Archaea*.

Methanogenesis The biological production of methane (CH_4).

Methanotroph An organism capable of oxidizing methane.

Methylotroph An organism capable of oxidizing organic compounds that do not contain carbon–carbon bonds; if able to oxidize CH_4, also a methanotroph.

Microaerophile An organism requiring O_2 but at a level lower than that in air.

Microarray Small, solid-state supports to which genes or portions of genes are affixed and arrayed spatially in a known pattern (also called a *gene chip*).

Microautoradiography (MAR) Measurement of the uptake of radioactive substrates by visually observing the cells in an exposed photograph emulsion.

Microbial ecology The study of microorganisms in their natural environments.

Microbial leaching The extraction of valuable metals such as copper from sulfide ores by microbial activities.

Microbial plastics Polymers consisting of microbially produced (and thus biodegradable) substances, such as polyhydroxyalkanoates.

Microelectrode A small glass electrode for measuring pH or specific compounds such as O_2 or H_2S that can be immersed into a microbial habitat at microscale intervals.

Microenvironment The immediate physical and chemical surroundings of a microorganism.

Microfilament A filamentous polymer of the protein actin that helps maintain the shape of a eukaryotic cell.

Micrometer One-millionth of a meter, or 10^{-6} m (abbreviated μm), the unit used for measuring microorganisms.

Microorganism A microscopic organism consisting of a single cell or cell cluster, also including the viruses, which are not cellular.

Microtubule A filamentous polymer of the proteins α-tubulin and β-tubulin that functions in eukaryotic cell shape and motility.

Minimum inhibitory concentration (MIC) The minimum concentration of a substance necessary to prevent microbial growth.

Minus (negative)-strand nucleic acid An RNA or DNA strand that has the opposite sense of (would be complementary to) the mRNA of a virus.

Missense mutation A mutation in which a single codon is altered so that one amino acid in a protein is replaced with a different amino acid.

Mitochondrion A eukaryotic organelle responsible for the processes of respiration and electron transport phosphorylation.

Mitosis The normal form of nuclear division in eukaryotic cells in which chromosomes are replicated and partitioned into two daughter nuclei.

Mixotroph An organism that uses organic compounds as carbon sources but uses inorganic compounds as electron donors for energy metabolism.

Modification enzyme An enzyme that chemically alters bases within a restriction enzyme recognition site and thus prevents the site from being cut.

Molds Filamentous fungi.

Molecular chaperone A protein that helps other proteins fold or refold properly.

Molecular clock A gene, such as for ribosomal RNA, whose DNA sequence can be used as a comparative temporal measure of evolutionary divergence.

Molecular cloning The isolation and incorporation of a fragment of DNA into a vector where it can be replicated.

Molecule Two or more atoms chemically bonded to one another.

Monoclonal antibody A single type of antibody produced from a single clone of B cells. This antibody has uniform structure and specificity.

Monocytes Circulating white blood cells that contain many lysosomes and can differentiate into macrophages.

Monomer A building block of a polymer.

Monophyletic In phylogeny, a group descended from one ancestor.

Monotrichous Having a single polar flagellum.

Morbidity The incidence of illness in a population.

Morphology The shape of an organism.

Mortality The incidence of death in a population.

Most-probable-number (MPN) technique The serial dilution of a natural sample to determine the highest dilution yielding growth.

Motif A conserved amino acid sequence found in all peptide antigens that bind to a given MHC protein.

Motility The property of movement of a cell under its own power.

Mucous membrane Layers of epithelial cells that interact with the external environment.

Mucus Soluble glycoproteins secreted by epithelial cells that coat the mucous membrane.

Multilocus sequence typing (MLST) A taxonomic tool for classifying organisms on the basis of gene sequence variations in several housekeeping genes.

Mushroom The aboveground fruiting body, or basidiocarp, of basidiomycete fungi.

Mutagen An agent that induces mutation, such as radiation or certain chemicals.

Mutant An organism whose genome carries a mutation.

Mutation An inheritable change in the base sequence of the genome of an organism.

Mutator strain A mutant strain in which the rate of mutation is increased.

Mutualism A type of symbiosis in which both organisms in the relationship benefit.

Mycorrhiza A symbiotic association between a fungus and the roots of a plant.

Mycosis Any infection caused by a fungus.

Myeloma A malignant tumor of a plasma cell (antibody-producing cell).

Natural killer (NK) cell A specialized lymphocyte that recognizes and destroys foreign cells or infected host cells in a nonspecific manner.

Natural penicillin The parent penicillin structure, produced by cultures of *Penicillium* not supplemented with side-chain precursors.

Negative control A mechanism for regulating gene expression in which a repressor protein prevents transcription of genes.

Negative selection In T cell selection, the deletion of T cells that interact with self antigens in the thymus. See *clonal deletion*.

Negative strand A nucleic acid strand that has the opposite sense to (is complementary to) the mRNA.

Negative-strand virus A virus with a single-stranded genome that has the opposite sense to (is complementary to) the viral mRNA.

Neutralization An interaction of antibody with antigen that reduces or blocks the biological activity of the antigen.

Neutrophil (polymorphonuclear leukocyte, PMN) A type of leukocyte exhibiting phagocytic properties, a granular cytoplasm (granulocyte), and a multilobed nucleus.

Neutrophile An organism that grows best around pH 7.

Niche In ecological theory, an organism's residence in a community, including both biotic and abiotic factors.

Nitrification The microbial oxidation of ammonia to nitrate (NH_3 to NO_3^-).

Nitrifying bacteria (nitrifiers) Chemolithotrophic *Bacteria* and *Archaea* that catalyze nitrification.

Nitrogen fixation Reduction of nitrogen gas to ammonia ($N_2 + 8\,H \rightarrow 2\,NH_3 + H_2$) by the enzyme nitrogenase.

Nod factors Oligosaccharides produced by root nodule bacteria that help initiate the plant–bacterial symbiosis.

Nodule A tumorlike structure produced by the roots of symbiotic nitrogen-fixing plants. Contains the nitrogen-fixing microbial component of the symbiosis.

Noncoding RNA An RNA molecule that is not translated into protein.

Nonnucleoside reverse transcriptase inhibitor (NNRTI) A nonnucleoside compound that inhibits the action of retroviral reverse transcriptase by binding directly to the catalytic site.

Nonperishable (stable) foods Foods of low water activity that have an extended shelf life and are resistant to spoilage by microorganisms.

Nonpolar Possessing hydrophobic (water-repelling) characteristics and not easily dissolved in water.

Nonsense codon Another name for a stop codon.

Nonsense mutation A mutation in which the codon for an amino acid is changed to a stop codon.

Normal microflora Microorganisms that are usually found associated with healthy body tissue.

Northern blot A hybridization procedure where RNA is in the gel and DNA or RNA is the probe. Compare with *Southern blot* and *immunoblot*.

Nosocomial infection (healthcare-associated infection) An infection contracted in a healthcare setting.

Nucleic acid A polymer of nucleotides. See *deoxyribonucleic acid* and *ribonucleic acid*.

Nucleic acid probe A strand of nucleic acid that can be labeled and used to hybridize to a complementary molecule from a mixture of other nucleic acids. In clinical microbiology or microbial ecology, a short oligonucleotide of unique sequence used as a hybridization probe for identifying specific genes.

Nucleocapsid The complete complex of nucleic acid and protein packaged in a virus particle.

Nucleoid The aggregated mass of DNA that makes up the chromosome of prokaryotic cells.

Nucleoside A nucleotide minus phosphate.

Nucleoside reverse transcriptase inhibitor (NRTI) A nucleoside analog compound that inhibits the action of viral reverse transcriptase by competing with nucleosides.

Nucleosome A spherical complex of eukaryotic DNA plus histones.

Nucleotide A monomeric unit of nucleic acid, consisting of a sugar, a phosphate, and a nitrogenous base.

Nucleus A membrane-enclosed structure in eukaryotes containing the genetic material (DNA) organized in chromosomes.

Nutrient A substance taken by a cell from its environment and used in catabolic or anabolic reactions.

Obligate Indicates an environmental condition always required for growth (for example, "obligate anaerobe").

Oligonucleotide A short nucleic acid molecule, either obtained from an organism or synthesized chemically.

Oligotrophic Describes a habitat in which nutrients are in low supply.

Oncogene A gene whose expression causes formation of a tumor.

Open reading frame (ORF) A sequence of DNA or RNA that could be translated to give a polypeptide.

Operator A specific region of the DNA at the initial end of a gene, where the repressor protein binds and blocks mRNA synthesis.

Operon One or more genes transcribed into a single RNA and under the control of a single regulatory site.

Operon fusion A gene fusion in which a coding sequence that retains its own translational signals is fused to the transcriptional signals of another gene.

Opportunistic infection An infection usually observed only in an individual with a dysfunctional immune system.

Opportunistic pathogen An organism that causes disease in the absence of normal host resistance.

Opsonization The enhancement of phagocytosis due to the deposition of antibody or complement on the surface of a pathogen or other antigen.

Organelle A bilayer membrane–enclosed structure such as the mitochondrion found in eukaryotic cells.

Ortholog A gene found in one organism that is similar to that in another organism but differs because of speciation. See also *paralog*.

Osmophile An organism that grows best in the presence of high levels of solute, typically a sugar.

Osmosis The diffusion of water through a membrane from a region of low solute concentration to one of higher concentration.

Outbreak The occurrence of a large number of cases of a disease in a short period of time.

Outer membrane A phospholipid- and polysaccharide-containing unit membrane that lies external to the peptidoglycan layer in cells of gram-negative *Bacteria*.

Overlapping genes Two or more genes in which part or all of one gene is embedded in the other.

Oxic Containing oxygen; aerobic. Usually used in reference to a microbial habitat.

Oxidation A process by which a compound gives up electrons (or H atoms) and becomes oxidized.

Oxidation–reduction (redox) reaction A pair of reactions in which one compound becomes oxidized while another becomes reduced and takes up the electrons released in the oxidation reaction.

Oxidative (electron transport) phosphorylation The nonphototrophic production of ATP at the expense of a proton motive force formed by electron transport.

Oxygenase An enzyme that catalyzes the incorporation of oxygen from O_2 into organic or inorganic compounds.

Oxygenic photosynthesis The use of light energy to synthesize ATP and NADPH by noncyclic photophosphorylation with the production of O_2 from water.

Palindrome A nucleotide sequence on a DNA molecule in which the same sequence is found on each strand but in the opposite direction.

Pandemic A worldwide epidemic.

Paralog A gene within an organism whose similarity to one or more other genes in the same organism is the result of gene duplication (compare with *ortholog*).

Parasite An organism able to live in or on a host and cause disease.

Parasitism A symbiotic relationship between two organisms in which the host organism is harmed in the process.

Passive immunity Immunity resulting from transfer of antibodies or immune cells from an immune to a nonimmune individual.

Pasteurization The use of controlled heat to reduce the microbial load, including disease-producing microorganisms and spoilage microorganisms, in heat-sensitive liquids.

Pathogen A disease-causing microorganism.

Pathogen-associated molecular pattern (PAMP) A repeating structural component of a microbial cell or virus recognized by a pattern recognition receptor.

Pathogenicity The ability of a pathogen to cause disease.

Pathogenicity island A bacterial chromosome region of foreign origin that contains clustered genes for virulence.

Pattern recognition receptor (PRR) A protein in a phagocyte membrane that recognizes a pathogen-associated molecular pattern, such as a component of a microbial cell surface structure.

Pathway engineering The assembly of a new or improved biochemical pathway, using genes from one or more organisms.

Penicillin A class of antibiotics that inhibit bacterial cell wall synthesis; characterized by a β-lactam ring.

Pentose phosphate pathway A major metabolic pathway for the production and catabolism of pentoses (C_5 sugars).

Peptide bond A type of covalent bond joining amino acids in a polypeptide.

Peptidoglycan The rigid layer of the cell walls of *Bacteria*, a thin sheet composed of *N*-acetylglucosamine, *N*-acetylmuramic acid, and a few amino acids.

Periplasm The area between the cytoplasmic membrane and the outer membrane in gram-negative *Bacteria*.

Perishable food Fresh food generally of high water activity that has a very short shelf life due to potential for spoilage by growth of microorganisms.

Peritrichous flagellation In flagellar arrangements, having flagella attached to many places on the cell surface.

Peroxisome An organelle that functions to rid the cell of toxic substances such as peroxides, alcohols, and fatty acids.

Pertussis (whooping cough) A disease caused by an upper respiratory tract infection with *Bordetella pertussis*, characterized by a deep persistent cough.

pH The negative logarithm of the hydrogen ion (H^+) concentration of a solution.

Phage See *bacteriophage*.

Phagemid A cloning vector that can replicate either as a plasmid or as a bacteriophage.

Phagocyte One of a group of cells that recognizes, ingests, and degrades pathogens and pathogen products.

Phagocytosis A mechanism for ingesting particulate food in which a portion of the cytoplasmic membrane surrounds the particle and brings it into the cell.

Phenotype The observable characteristics of an organism, such as color, motility, or morphology. Compare with *genotype*.

Phosphodiester bond A type of covalent bond linking nucleotides together in a polynucleotide.

Phospholipid A lipid containing a substituted phosphate group and two fatty acid chains on a glycerol backbone.

Photoautotroph An organism able to use light as its sole source of energy and CO_2 as its sole carbon source.

Photoheterotroph An organism using light as a source of energy and organic compounds as a carbon source.

Photophosphorylation The synthesis of energy-rich phosphate bonds in ATP using light energy.

Photosynthesis The series of reactions in which ATP is synthesized by light-driven reactions and CO_2 is fixed into cell material. See also *anoxygenic photosynthesis* and *oxygenic photosynthesis*.

Phototaxis Movement of a cell toward light.

Phototroph An organism that obtains energy from light.

Phycobilin The light-capturing open chain tetrapyrrole component of phycobiliproteins.

Phycobiliprotein The accessory pigment complex in cyanobacteria that contains phycocyanin or phycoerythrin coupled to proteins.

Phycobilisome Aggregates of phycobiliproteins.

Phylogenetic probe An oligonucleotide, sometimes made fluorescent by attachment of a dye, complementary in sequence to some ribosomal RNA signature sequence.

Phylogeny The evolutionary (natural) history of organisms.

Phylotype One or more organisms with the same or related sequences of a phylogenetic marker gene.

Phylum A major lineage of cells in one of the three domains of life.

Phytanyl A branched-chain hydrocarbon containing 20 carbon atoms, commonly found in the lipids of *Archaea*.

Phytopathogen A microorganism that causes plant disease.

Pickling The process of acidifying food, typically with acetic acid, to prevent microbial growth and spoilage.

Piezophile An organism that lives optimally at high hydrostatic pressure.

Piezotolerant An organism able to tolerate high hydrostatic pressure but growing best at 1 atm.

Pilus (plural, pili) A fimbria-like structure that is present on fertile cells, both Hfr and F^+, and is involved in DNA transfer during conjugation. Sometimes called a sex pilus. See also *fimbria*.

Pinocytosis In eukaryotes, phagocytosis of soluble molecules.

Plague An endemic disease in rodents caused by *Yersinia pestis* that is occasionally transferred to humans through the bite of a flea.

Plaque A zone of lysis or cell inhibition caused by virus infection on a lawn of cells.

Plasma The liquid portion of the blood containing proteins and other solutes.

Plasma cell A large, differentiated, short-lived B lymphocyte specializing in abundant (but short-term) antibody production.

Plasmid An extrachromosomal genetic element that is not essential for growth and has no extracellular form.

Plate count A viable counting method in which the number of colonies on a plate is used as a measure of cell number.

Platelet A noncellular disc-shaped structure containing protoplasm, found in large numbers in blood and functioning in the blood-clotting process.

Plus-strand nucleic acid An RNA or DNA strand that has the same sense as the mRNA of a virus.

Point mutation A mutation that involves a single base pair.

Polar Possessing hydrophilic characteristics and generally water-soluble.

Polar flagellation In flagellar arrangements, having flagella attached at one end or both ends of the cell.

Poly-β-hydroxybutyrate (PHB) A common storage material of prokaryotic cells consisting of a polymer of β-hydroxybutyrate (PHB) or other β-alkanoic acids (PHA).

Polyclonal antibodies A mixture of antibodies made by many different B cell clones.

Polyclonal antiserum A mixture of antibodies to a variety of antigens or to a variety of determinants on a single antigen.

Polymer A large molecule formed by polymerization of monomeric units. In water purification, a chemical in liquid form used as a coagulant to produce flocculation in the clarification process.

Polymerase chain reaction (PCR) Artificial amplification of a DNA sequence by repeated cycles of strand separation and replication.

Polymorphism In a population, the occurrence of multiple alleles for a gene locus at a higher frequency than can be explained by recent random mutations.

Polymorphonuclear leukocyte (PMN) Motile white blood cells containing many lysosomes and specializing in phagocytosis. Characterized by a distinct segmented nucleus. Also, a neutrophil.

Polynucleotide A polymer of nucleotides bonded to one another by phosphodiester bonds.

Polypeptide Several amino acids linked together by peptide bonds.

Polyprotein A large protein expressed from a single gene and subsequently cleaved to form several individual proteins.

Polysaccharide A long chain of monosaccharides (sugars) linked by glycosidic bonds.

Polyvalent vaccine A vaccine that immunizes against more than one disease.

Population A group of organisms of the same species in the same place at the same time.

Porins Protein channels in the outer membrane of gram-negative *Bacteria* through which small to medium-sized molecules can flow.

Positive control A mechanism for regulating gene expression in which an activator protein functions to promote transcription of genes.

Positive selection In T cell selection, the stimulation of growth and development of T cells that interact with self MHC protein in the thymus.

Positive strand A nucleic acid strand that has the same sense as the mRNA.

Positive-strand virus A virus with a single-stranded genome that has the same complementarity as the viral mRNA.

Potable In water purification, drinkable; safe for human consumption.

Precipitation A reaction between antibody and soluble antigen resulting in visible antibody–antigen complexes.

Prevalence The total number of new and existing disease cases reported in a population in a given time period.

Pribnow box The consensus sequence TATAAT located approximately 10 base pairs upstream from the transcriptional start site. A binding site for RNA polymerase.

Primary adaptive immune response The production of antibodies or immune T cells on first exposure to antigen; the antibodies are mostly of the IgM class.

Primary antibody response Antibodies made on first exposure to antigen; mostly of the class IgM.

Primary disinfection The introduction of sufficient chlorine or other disinfectant into clarified, filtered water to kill existing microorganisms and inhibit further microbial growth.

Primary endosymbiosis Acquisition of the alpha-proteobacterial ancestor of the mitochondrion or of the cyanobacterial ancestor of the chloroplast by another kind of cell.

Primary lymphoid organ An organ in which precursor lymphoid cells develop into mature lymphocytes.

Primary metabolite A metabolite excreted during the exponential growth phase.

Primary producer An organism that synthesizes new organic material from CO_2. Also an *autotroph*.

Primary structure In an informational macromolecule, such as a polypeptide or a nucleic acid, the precise sequence of monomeric units.

Primary transcript An unprocessed RNA molecule that is the direct product of transcription.

Primary wastewater treatment The physical separation of wastewater contaminants, usually by separation and settling.

Primase The enzyme that synthesizes the RNA primer used in DNA replication.

Primer A short length of DNA or RNA used to initiate synthesis of a new DNA strand.

Prion An infectious protein whose extracellular form contains no nucleic acid.

Probe See *nucleic acid probe*.

Probiotic A live microorganism that, when administered to a host, may confer a health benefit.

Prochlorophyte A prokaryotic oxygenic phototroph that contains chlorophylls *a* and *b* but lacks phycobilins.

Prokaryote A cell or organism lacking a nucleus and other membrane-enclosed organelles and usually having its DNA in a single circular molecule. Members of the *Bacteria* and the *Archaea*.

Promoter The site on DNA where the RNA polymerase binds and begins transcription.

Prophage The state of the genome of a temperate virus when it is replicating in synchrony with that of the host, typically integrated into the host genome. See *provirus*.

Prophylactic Treatment, usually immunological or chemotherapeutic, designed to protect an individual from a future attack by a pathogen.

Prostheca A cytoplasmic extrusion bounded by the cell wall, such as a bud, hypha, or stalk.

Prosthetic group The tightly bound, nonprotein portion of an enzyme; not the same as a *coenzyme*.

Protease inhibitor A compound that inhibits the action of viral protease by binding directly to the catalytic site, preventing viral protein processing.

Protein A polymeric molecule consisting of one or more polypeptides.

Protein fusion A gene fusion in which two coding sequences are fused so that they share the same transcriptional and translational start sites.

Protein splicing Removal of intervening sequences from a protein.

Proteobacteria A large phylum of *Bacteria* that includes many of the common gram-negative bacteria, including *Escherichia coli*.

Proteome The total set of proteins encoded by a genome or the total protein complement of an organism.

Proteomics The large-scale or genome-wide study of the structure, function, and regulation of the proteins of an organism.

Proteorhodopsin A light-sensitive retinal-containing protein found in some marine *Bacteria* that catalyzes ATP formation.

Protist A unicellular eukaryotic microorganism; may be flagellated or aflagellate, phototrophic or nonphototrophic, and most lack cell walls; includes algae and protozoa.

Proton motive force A source of energy resulting from the separation of protons from hydroxyl ions across the cytoplasmic membrane, generating a membrane potential.

Protoplasm The complete cellular contents, cytoplasmic membrane, cytoplasm, and nucleus/nucleoid of a cell.

Protoplast A cell from which the wall has been removed.

Prototroph The parent from which an auxotrophic mutant has been derived. Contrast with *auxotroph*.

Protozoa Unicellular eukaryotic microorganisms that lack cell walls.

Provirus (prophage) The genome of a temperate virus when it is replicating in step with, and often integrated into, the host chromosome.

Psychrophile An organism able to grow at low temperatures and showing a growth temperature optimum of <15°C.

Psychrotolerant Able to grow at low temperature but having a growth temperature optimum of >20°C.

Public health The health of the population as a whole.

Pure culture A culture containing a single kind of microorganism.

Purine One of the nitrogen bases of nucleic acids that contain two fused rings; adenine and guanine.

Purple nonsulfur bacteria A group of phototrophic bacteria that contain bacteriochlorophyll *a* or *b*, grow best as photoheterotrophs, and have a relatively low tolerance for H_2S.

Purple sulfur bacteria A group of phototrophic bacteria containing bacteriochlorophylls *a* or *b* and characterized by the ability to oxidize H_2S and store elemental sulfur inside the cells (or, in the genera *Ectothiorhodospira* and *Halorhodospira,* outside the cell).

Pyogenic Pus-forming; causing abscesses.

Pyrimidine One of the nitrogen bases of nucleic acids that contain a single ring; cytosine, thymine, and uracil.

Pyrite A common iron ore, FeS_2.

Pyrogenic Fever-inducing.

Quarantine The practice of restricting the movement of individuals with highly contagious serious infections to prevent spread of the disease.

Quaternary structure In proteins, the number and arrangement of individual polypeptides in the final protein molecule.

Quinolones Synthetic antibacterial compounds that interact with DNA gyrase and prevent supercoiling of bacterial DNA.

Quorum sensing A regulatory system that monitors the population size and controls gene expression based on cell density.

Rabies A usually fatal neurological disease caused by the rabies virus that is usually transmitted by the bite or saliva of an infected carnivore.

Radioimmunoassay (RIA) A test assay employing radioactive antibody or antigen for the detection of antigen or antibody binding.

Radioisotope An isotope of an element that undergoes spontaneous decay with the release of radioactive particles.

Raw water Surface water or groundwater that has not been treated in any way (also called untreated water).

Reaction center A photosynthetic complex containing chlorophyll (or bacteriochlorophyll) and other components, within which occurs the initial electron transfer reactions of photophosphorylation.

Reading-frame shift See *frameshift*.

Recalcitrant Resistant to microbial attack.

Recombinant DNA A DNA molecule containing DNA originating from two or more sources.

Recombination The process by which DNA molecules from two separate sources exchange sections or are brought together into a single DNA molecule.

Redox See *oxidation–reduction reaction*.

Reduction A process by which a compound accepts electrons to become reduced.

Reduction potential (E_0') The inherent tendency, measured in volts, of the oxidized compound of a redox pair to become reduced.

Reductive dechlorination The removal of Cl as Cl^- from an organic compound by reducing the carbon atom from C–Cl to C–H.

Reemerging disease An infectious disease, thought to be under control, that produces a new epidemic.

Regulation Processes that control the rates of synthesis of proteins, such as induction and repression.

Regulatory nucleotide A nucleotide that functions as a signal rather than being incorporated into RNA or DNA.

Regulon A set of operons that are all controlled by the same regulatory protein (repressor or activator).

Replacement vector A cloning vector, such as a bacteriophage, in which some of the DNA of the vector can be replaced with foreign DNA.

Replication The synthesis of DNA using DNA as a template.

Replication fork The site on the chromosome where DNA replication occurs and where the enzymes replicating the DNA are bound to untwisted, single-stranded DNA.

Replicative form A double-stranded DNA molecule that is an intermediate in the replication of single-stranded DNA viruses.

Reporter gene A gene incorporated into a vector because the product it encodes is easy to detect.

Repression Prevention of the synthesis of an enzyme in response to a signal.

Repressor protein A regulatory protein that binds to specific sites on DNA and blocks transcription; involved in negative control.

Reservoir A source of viable infectious agents from which individuals may be infected.

Resolution In microbiology, the ability to distinguish two objects as distinct and separate under the microscope.

Respiration Catabolic reactions producing ATP in which either organic or inorganic compounds are primary electron donors and organic or inorganic compounds are ultimate electron acceptors.

Response regulator protein One of the members of a two-component system; a regulatory protein that is phosphorylated by a sensor protein (see *sensor kinase protein*).

Restriction enzymes (restriction endonucleases) Enzymes that recognize and cleave specific DNA sequences, generating either blunt or single-stranded (sticky) ends.

Restriction map A map showing the location of restriction enzyme cut sites on a segment of DNA.

Retrovirus A virus whose RNA genome has a DNA intermediate as part of its replication cycle.

Reverse citric acid cycle A mechanism for autotrophy in green sulfur bacteria and several nonphototrophic prokaryotes.

Reverse DNA gyrase A topoisomerase present in all hyperthermophilic prokaryotes that introduces positive supercoils in DNA.

Reverse electron transport The energy-dependent movement of electrons against the thermodynamic gradient to form a strong electron donor from a weaker electron donor.

Reverse transcriptase The enzyme that makes a DNA copy using RNA as template.

Reverse transcription The process of copying information found in RNA into DNA.

Reversion An alteration in DNA that reverses the effects of a prior mutation.

Rheumatic fever An inflammatory autoimmune disease triggered by an immune response to infection by *Streptococcus pyogenes*.

Rhizosphere The region immediately adjacent to plant roots.

Ribonucleic acid (RNA) A polymer of nucleotides connected via a phosphate–ribose backbone; involved in protein synthesis or as genetic material of some viruses.

Ribosomal Database Project (RDP) A large database of small subunit ribosomal RNA gene sequences that can be retrieved electronically and used in comparative ribosomal RNA gene sequence studies.

Ribosomal RNA (rRNA) The type of RNA found in the ribosome; some rRNAs participate actively in the process of protein synthesis.

Ribosome A structure composed of RNAs and proteins upon which new proteins are made.

Riboswitch An RNA domain, usually in an mRNA molecule, that can bind a specific small molecule and alter its secondary structure; this in turn controls translation of the mRNA.

Ribotyping A means of identifying microorganisms from analysis of DNA fragments generated from restriction enzyme digestion of genes encoding their 16S rRNA.

Ribozyme An RNA molecule that can catalyze a chemical reaction.

Ribulose monophosphate pathway A reaction series in certain methylotrophs in which formaldehyde is assimilated into cell material using ribulose monophosphate as the C_1 acceptor molecule.

Rickettsias Obligate intracellular bacteria that cause disease, including typhus, spotted fever rickettsiosis, and ehrlichiosis.

RNA Ribonucleic acid; functions in protein synthesis as messenger RNA, transfer RNA, and ribosomal RNA.

RNA editing Changing the coding sequence of an RNA molecule by altering, adding, or removing bases.

RNA interference (RNAi) A response that is triggered by the presence of double-stranded RNA and results in the degradation of ssRNA homologous to the inducing dsRNA.

RNA life A hypothetical ancient life form lacking DNA and protein, in which RNA had both a genetic coding and a catalytic function.

RNA polymerase An enzyme that synthesizes RNA in the $5' \rightarrow 3'$ direction using an antiparallel $3' \rightarrow 5'$ DNA strand as a template.

RNA processing The conversion of a precursor RNA to its mature form.

RNA replicase An enzyme that can produce RNA from an RNA template.

Rocky Mountain spotted fever See *spotted fever rickettsiosis*.

Rolling circle replication A mechanism, used by some plasmids and viruses, of replicating circular DNA, starting by nicking and unrolling one strand and using the other, still circular strand as a template for DNA synthesis.

Root nodule A tumorlike growth on certain plant roots that contains symbiotic nitrogen-fixing bacteria.

RubisCO The acronym for ribulose bisphosphate carboxylase, a key enzyme of the Calvin cycle.

Rumen The forestomach of ruminant animals in which cellulose digestion occurs.

S-layer A paracrystalline outer wall layer composed of protein or glycoprotein and found in many prokaryotes.

Salmonellosis Enterocolitis or other gastrointestinal disease caused by any of more than 1400 variants of *Salmonella* species.

Sanitizers Agents that reduce, but may not eliminate, microbial numbers to a safe level.

Scale-up The conversion of an industrial process from a small laboratory setup to a large commercial fermentation.

Scarlet fever A disease characterized by high fever and a reddish skin rash resulting from an exotoxin produced by cells of *Streptococcus pyogenes*.

Screening A procedure that permits the identification of organisms by phenotype or genotype, but does not inhibit or enhance the growth of particular phenotypes or genotypes.

Secondary adaptive immune response The enhanced production of antibodies or immune T cells on second and subsequent exposures to antigen; the antibodies are mostly of the IgG class.

Secondary aerobic wastewater treatment Oxidative reactions carried out by microorganisms under aerobic conditions to treat wastewater containing low levels of organic materials.

Secondary anaerobic wastewater treatment Degradative and fermentative reactions carried out by microorganisms under anoxic conditions to treat wastewater containing high levels of insoluble organic materials.

Secondary antibody response Antibodies made on second (subsequent) exposure to antigen; mostly of the class IgG.

Secondary disinfection The maintenance of sufficient chlorine or other disinfectant residual in the water distribution system to inhibit microbial growth.

Secondary endosymbiosis Acquisition by a mitochondrion-containing eukaryotic cell of the chloroplasts of a red or green algal cell.

Secondary fermentation A fermentation in which the substrates are the fermentation products of some other organism.

Secondary metabolite A product excreted by a microorganism in the late exponential growth phase and the stationary phase.

Secondary structure The initial pattern of folding of a polypeptide or a polynucleotide, usually the result of hydrogen bonding.

Sediment (1) In water purification, the soil, sand, minerals, and other large particles found in raw water. (2) In large bodies of water (lakes, the oceans), the materials (mud, rock, and the like) that form the bottom surface of the water body.

Selection Placing organisms under conditions that favor or inhibit the growth of those with a particular phenotype or genotype.

Selective medium A growth medium that enhances the growth of certain organisms while inhibiting the growth of others due to an added media component.

Selective toxicity The ability of a compound to inhibit or kill pathogenic microorganisms without adversely affecting the host.

Self-splicing intron An intron that possesses ribozyme activity and splices itself out.

Semiconservative replication DNA synthesis yielding new double helices, each consisting of one parental and one progeny strand.

Semiperishable food Food of intermediate water activity that has a limited shelf life due to potential for spoilage by growth of microorganisms.

Semisynthetic penicillin   A natural penicillin that has been chemically altered.

Sensitivity   In immunodiagnostics, the lowest amount of antigen that can be detected in an immunological assay.

Sensor kinase protein   One of the members of a two-component system; a kinase found in the cell membrane that phosphorylates itself in response to an external signal and then passes the phosphoryl group to a response regulator protein (see *response regulator protein*).

Septicemia (sepsis)   A bloodborne systemic infection.

Sequencing   In reference to nucleic acids, deducing the order of nucleotides in a DNA or RNA molecule.

Serine pathway   A reaction series in certain methylotrophs in which formaldehyde is assimilated into cell material by way of the amino acid serine.

Serology   The study of antigen–antibody reactions *in vitro*.

Serum   The fluid portion of blood remaining after the blood cells and materials responsible for clotting are removed.

Sewage   Liquid effluents contaminated with human or animal fecal material.

Sexually transmitted infection (STI)   An infection that is usually transmitted by sexual contact.

Shine–Dalgarno sequence   A short stretch of nucleotides on a prokaryotic mRNA molecule upstream of the translational start site that binds to ribosomal RNA and thereby brings the ribosome to the initiation codon on the mRNA.

Short interfering RNA (siRNA)   Short double-stranded RNA molecules that trigger RNA interference.

Shotgun cloning   Making a gene library by random cloning of DNA fragments.

Shotgun sequencing   Sequencing of DNA from previously cloned small fragments of a genome in a random fashion followed by computational methods to reconstruct the entire genome sequence.

Shuttle vector   A cloning vector that can replicate in two different organisms; used for moving DNA between unrelated organisms.

Sickle cell trait   A genetic trait that confers resistance to malaria, but causes a reduction in the oxygen-carrying capacity of the blood by reducing the life expectancy of the affected red blood cells.

Siderophore   An iron chelator that can bind iron present at very low concentrations.

Signal sequence   A special N-terminal sequence of approximately 20 amino acids that signals that a protein should be exported across the cytoplasmic membrane.

Signal transduction   Indirect transmission of an external signal to a target in the cell. See *two-component regulatory system*.

Silent mutation   A change in DNA sequence that has no effect on the phenotype.

Simple transport system   A transporter that consists of only a membrane-spanning protein and is typically driven by energy from the proton motive force.

Single-cell protein   Protein derived from microbial cells for use as food or a food supplement.

Site-directed mutagenesis   Construction *in vitro* of a gene with a specific mutation.

16S rRNA   A large polynucleotide (~1500 bases) that functions as a part of the small subunit of the ribosome of prokaryotes (*Bacteria* and *Archaea*) and from whose sequence evolutionary relationships can be obtained; eukaryotic counterpart is 18S rRNA.

Slime layer   A diffuse layer of polymer fibers, typically polysaccharides, that forms an outer surface layer on the cell.

Slime molds   Nonphototrophic eukaryotic microorganisms lacking cell walls, which aggregate to form fruiting structures (cellular slime molds) or simply masses of protoplasm (acellular slime molds).

Small subunit (SSU) RNA   Ribosomal RNA from the 30S ribosomal subunit of *Bacteria* and *Archaea* or the 40S ribosomal subunit of eukaryotes, that is, 16S or 18S ribosomal RNA, respectively.

Solfatara   A hot, sulfur-rich, generally acidic environment commonly inhabited by hyperthermophilic *Archaea*.

Somatic hypermutation   The mutation of immunoglobulin genes at rates higher than those observed in other genes.

Southern blot   A hybridization procedure where DNA is in the gel and RNA or DNA is the probe. Compare with *Northern blot* and *immunoblot*.

Species   Defined in microbiology as a collection of strains that all share the same major properties and differ in one or more significant properties from other collections of strains; defined phylogenetically as a monophyletic, exclusive group based on DNA sequence.

Species abundance   The proportion of each species in a community.

Species richness   The total number of different species present in a community.

Specificity   (1) The ability of the immune response to interact with individual antigens. (2) The ability of a diagnostic or research test to identify a specific pathogen.

Spheroplast   A spherical, osmotically sensitive cell derived from a bacterium by loss of some but not all of the rigid wall layer. If all of the rigid wall layer has been completely lost, the structure is called a *protoplast*.

Spirilla (singular, spirillum)   Spiral-shaped cells.

Spirochete   A slender, tightly coiled gram-negative bacterium characterized by possession of endoflagella used for motility.

Spliceosome   A complex of ribonucleoproteins that catalyze the removal of introns from RNA primary transcripts.

Splicing   The RNA-processing step by which introns are removed and exons joined.

Spontaneous generation   The hypothesis that living organisms can originate from nonliving matter.

Spontaneous mutation   A mutation that occurs "naturally" without the help of mutagenic chemicals or radiation.

Spore   A general term for resistant resting structures formed by many prokaryotes and fungi.

Sporozoa   Nonmotile parasitic protozoa.

Spotted fever rickettsiosis   A tick-transmitted disease caused by *Rickettsia rickettsii*, characterized by fever, headache, rash, and gastrointestinal symptoms; formerly called Rocky Mountain spotted fever.

Stalk   An elongate structure, either cellular or excreted, that anchors a cell to a surface.

Stable isotope probing (SIP)   A method for characterizing an organism that incorporates a particular substrate by feeding the substrate in ^{13}C form and then isolating ^{13}C-enriched DNA and analyzing the genes.

Start codon   A special codon, usually AUG, that signals the start of a protein.

-static   Suffix indicating inhibition of growth. For example, a bacteriostatic agent inhibits bacterial growth. Compare with *-cidal*.

Stationary phase   The period during the growth cycle of a microbial population in which growth ceases.

Stem cell   A cell that can develop into a number of final cell types.

Stereoisomers   Mirror-image forms of two molecules having the same molecular and structural formulas.

Sterilant (sterilizer, sporicide)   A chemical agent that destroys all foms of microbial life.

Sterile   Free of all living organisms and viruses.

Sterilization   The killing or removal of all living organisms and viruses from a growth medium.

Sterols   Hydrophobic multiringed structures that strengthen the cytoplasmic membrane of eukaryotic cells and a few prokaryotes.

Stickland reaction   The fermentation of an amino acid pair in which one amino acid serves as an electron donor and a second serves as an electron acceptor.

Stop codon   A codon that signals the end of a protein.

Strain   A population of cells of a single species all descended from a single cell; a clone.

Stringent response   A global regulatory control that is activated by amino acid starvation or energy deficiency.

Stroma   The inner membrane surrounding the lumen of the chloroplast.

Stromatolite A laminated microbial mat, typically built from layers of filamentous and other microorganisms; may fossilize.

Substrate The molecule that undergoes a specific reaction with an enzyme.

Substrate-level phosphorylation The synthesis of energy-rich phosphate bonds through reaction of inorganic phosphate with an activated organic substrate.

Sulfate-reducing and sulfur-reducing bacteria Two groups of *Bacteria* that respire anaerobically with SO_4^{2-} and S^0, respectively, as electron acceptors, producing H_2S.

Superantigen A pathogen product capable of eliciting an inappropriately strong immune response by stimulating greater than normal numbers of T cells.

Supercoil Highly twisted form of circular DNA.

Superoxide anion (O_2^-) A derivative of O_2 capable of oxidative destruction of cell components.

Suppressor A mutation that restores a wild-type phenotype without altering the original mutation, usually arising by mutation in another gene.

Surveillance Observation, recognition, and reporting of diseases as they occur.

Suspended solid A small particle of solid pollutant that resists separation by ordinary physical means.

Symbiosis An intimate relationship between two organisms, often developed through prolonged association and coevolution.

Synthetic DNA A DNA molecule that has been made by a chemical process in a laboratory.

Syntrophy The cooperation of two or more organisms to anaerobically degrade a substance neither can degrade alone.

Systematics The study of the diversity of organisms and their relationships; includes taxonomy and phylogeny.

Systemic Not localized in the body; an infection disseminated widely through the body.

Taxis A movement toward or away from a stimulus.

Taxonomy The science of identification, classification, and nomenclature.

T cell A lymphocyte responsible for antigen-specific cellular interactions. T cells are divided into functional subsets including T_C (cytotoxic) cells and T_H (helper) cells. T_H cells are further subdivided into T_H1 (inflammatory) cells and T_H2 (helper) cells, which aid B cells in antibody formation.

T cell receptor The antigen-specific receptor protein on the surface of T lymphocytes.

T-DNA The segment of the *Agrobacterium* Ti plasmid that is transferred to plant cells.

Teichoic acid A phosphorylated polyalcohol found in the cell wall of some gram-positive *Bacteria*.

Telomerase An enzyme complex that replicates DNA at the end of eukaryotic chromosomes.

Temperate virus A virus whose genome is able to replicate along with that of its host without causing cell death in a state called lysogeny.

Termination Stopping the elongation of an RNA molecule at a specific site.

Tertiary structure The final folded structure of a polypeptide that has previously attained secondary structure.

Tertiary wastewater treatment The physicochemical or biological processing of wastewater to reduce levels of inorganic nutrients.

Tetanus A disease involving rigid paralysis of the voluntary muscles caused by an exotoxin produced by *Clostridium tetani*.

Tetracycline A member of a class of antibiotics characterized by a four-membered naphthacene ring.

Thalassemia A genetic trait that confers resistance to malaria, but causes a reduction in the efficiency of red blood cells by altering a red blood cell enzyme.

T-helper (T_H) cells Lymphocytes that interact with MHC–peptide complexes through their T cell receptor (TCR).

Thermocline The zone of water in a stratified lake in which temperature and oxygen concentration drop precipitously with depth.

Thermophile An organism with a growth temperature optimum between 45 and 80°C.

Thermosome A heat-shock (chaperonin) protein complex that functions to refold partially heat-denatured proteins in hyperthermophiles.

Thylakoid A membrane layer containing the photosynthetic pigments in chloroplasts and in cyanobacteria.

Thymus The primary lymphoid organ responsible for development of T cells.

Ti plasmid A conjugative plasmid present in the bacterium *Agrobacterium tumefaciens* that can transfer genes into plants.

Titer In immunology, the quantity of antibody present in a solution.

Tolerance The acquired inability to produce an immune response to a specific antigen.

Toll-like receptor (TLR) One of a family of pattern recognition receptors (PRRs) found on phagocytes, structurally and functionally related to Toll receptors in *Drosophila*, that recognize a pathogen-associated molecular pattern (PAMP).

Toxicity The ability of an organism to cause disease by means of a preformed toxin that inhibits host cell function or kills host cells.

Toxic shock syndrome (TSS) Acute systemic shock resulting from host response to an exotoxin produced by *Staphylococcus aureus*.

Toxigenicity The degree to which an organism is able to elicit toxic symptoms.

Toxin A microbial substance able to induce host damage.

Toxoid A toxin modified so that it is no longer toxic but is still able to induce antibody formation.

Transcription The synthesis of an RNA molecule complementary to one of the two strands of a double-stranded DNA molecule.

Transcriptome The complement of all RNAs produced in an organism under a specific set of conditions.

Transduction The transfer of host genes from one cell to another by a virus.

Transfection The transformation of a prokaryotic cell by DNA or RNA from a virus. Used also to describe the process of genetic transformation in eukaryotic cells.

Transfer RNA (tRNA) A small RNA molecule used in translation that possesses an anticodon at one end and has the corresponding amino acid attached to its other end.

Transformation (1) The transfer of genetic information via free DNA. (2) A process, sometimes initiated by infection with certain viruses, whereby a normal animal cell becomes a cancer cell.

Transgenic organism A plant or animal with foreign DNA inserted into its genome.

Transition A mutation in which a pyrimidine base is replaced by another pyrimidine or a purine is replaced by another purine.

Translation The synthesis of protein using the genetic information in a messenger RNA as a template.

Transmissible spongiform encephalopathy (TSE) A degenerative disease of the brain caused by prion infection.

Transpeptidation The formation of peptide bonds between the short peptides present in peptidoglycan, the cell wall polymer of *Bacteria*.

Transporters Membrane proteins that function to transport substances into and out of the cell.

Transposable element A genetic element with the ability to move (transpose) from one site to another on host DNA molecules.

Transposase An enzyme that catalyzes the insertion of DNA segments into other DNA molecules.

Transposon A type of transposable element that carries other genes in addition to those involved in transposition; often these genes confer selectable phenotypes such as antibiotic resistance.

Transposon mutagenesis The insertion of a transposon into a gene; this inactivates the host gene, leading to a mutant phenotype, and also confers the phenotype associated with the transposon gene.

Transversion A mutation in which a pyrimidine base is replaced by a purine or vice versa.

Tuberculin test A skin test for previous infection with *Mycobacterium tuberculosis*.

Turbidity A measurement of suspended solids in water.

Two-component regulatory system A regulatory system containing a sensor protein and a response regulator protein (see *sensor kinase protein* and *response regulator protein*).

Typhus A louse-transmitted disease caused by *Rickettsia prowazekii*, causing fever, headache, weakness, rash, and damage to the central nervous system and internal organs.

Universal phylogenetic tree A tree that shows the evolutionary position of representatives of all domains of living organisms.

Untreated water Surface water or groundwater that has not been treated in any way (also called raw water).

Upper respiratory tract The nasopharynx, oral cavity, and throat.

Upstream position Refers to nucleic acid sequences on the 5′ side of a given site on a DNA or RNA molecule. Compare with *downstream position*.

Vaccination (immunization) The inoculation of a host with inactive or weakened pathogens or pathogen products to stimulate protective immunity.

Vaccine An inactivated or weakened pathogen, or an innocuous pathogen product used to stimulate protective immunity.

Vacuole A small space in a cell that contains fluid and is surrounded by a membrane. In contrast to a vesicle, a vacuole is not rigid.

Vector (1) A self-replicating DNA molecule that carries DNA segments between organisms and can be used as a cloning vector to carry cloned genes or other DNA segments for genetic engineering. (2) A living agent, usually an insect or other animal, able to carry pathogens from one host to another.

Vector vaccine A vaccine made by inserting genes from a pathogenic virus into a relatively harmless carrier virus.

Vehicle A nonliving source of pathogens that infect large numbers of individuals; common vehicles are food and water.

Viable Alive; able to reproduce.

Viable count A measurement of the concentration of live cells in a microbial population.

Viral load The number of viral genome copies in the tissue of an infected host, providing a quantitative assessment of the amount of virus in the host.

Viricidal agent An agent that stops viral replication and activity.

Virion A virus particle; the virus nucleic acid surrounded by a protein coat and in some cases other material.

Viristatic agent An agent that inhibits viral replication.

Viroid Small, circular, single-stranded RNA that causes certain plant diseases.

Virulence The relative ability of a pathogen to cause disease.

Virulent virus A virus that lyses or kills the host cell after infection; a nontemperate virus.

Virus A genetic element that contains either RNA or DNA and that replicates in cells; has an extracellular form.

Volatile fatty acids (VFAs) The major fatty acids (acetate, propionate, and butyrate) produced during fermentation in the rumen.

Wastewater The liquid derived from domestic sewage or industrial sources, which cannot be discarded in untreated form into lakes or streams.

Water activity (a_w) An expression of the relative availability of water in a substance. Pure water has an a_w of 1.000.

Western blot See *immunoblot*.

West Nile fever A neurological disease caused by West Nile virus, a virus transmitted by mosquitoes from birds to humans.

Wild type A strain of microorganism isolated from nature. The usual or native form of a gene or organism.

Winogradsky column A glass column packed with mud and overlaid with water to mimic an aquatic environment in which various bacteria develop over a period of months.

Wobble In reference to protein synthesis, a less rigid form of base pairing allowed only in codon–anticodon pairing.

Xenobiotic A completely synthetic chemical compound not naturally occurring on Earth.

Xerophile An organism adapted to growth at very low water potentials.

Yeast The single-celled growth form of various fungi.

Yeast artificial chromosome (YAC) A genetically engineered chromosome with yeast origin of replication and CEN sequence.

Zoonosis A disease, primarily of animals, that is occasionally transmitted to humans.

Zygote In eukaryotes, the single diploid cell resulting from the union of two haploid gametes.

Photo Credits

Chapter 1 Opener: Steve Gschmeissner/Photo Researchers; 1.1: Paul V. Dunlap; 1.2a: L.K. Kimble and M.T. Madigan; 1.2b,c: M.T. Madigan; 1.5a: Douglas E. Caldwell, University of Saskatchewan; 1.5b: From R. Amann, J. Snaidr, M. Wagner, W. Ludwig, and K.H. Schleifer, 1996. In situ visualization of high genetic diversity in a natural bacterial community. Journal of Bacteriology 178:3496-3500, Fig. 2b. © 1996 American Society for Microbiology. Photo: Jiri Snaidr; 1.5c: Ricardo Guerrero; 1.6a: Image produced by M. Jentoft-Nilsen, F. Hasler, D. Chesters (NASA/Goddard) and T. Nielsen (Univ. of Hawaii)/NASA Headquarters; 1.7a: Norbert Pfennig, University of Konstanz, Germany; 1.7b: Thomas D. Brock; 1.9a: Joe Burton; 1.10: Steve Gschmeissner/Photo Researchers; 1.11a: John A. Breznak, Michigan State University; 1.11b: U.S. Department of Energy; 1.12: Library of Congress; 1.13a: Thomas D. Brock; 1.13b: Library of Congress; 1.13c: Brian J. Ford; 1.14: Drawing by Ferdinand Cohn, originally published in Hedwigia 5:161-166 (1866); 1.17a: Pearson Science; 1.17b: M.T. Madigan; 1.18: Images from the History of Medicine, The National Library of Medicine; 1.20: Robert Koch, 1884. "Die Aetiologie der Tuberkulose." Mittheilungen aus dem Kaiserlichen Gesundheitsamte 2:1-88; 1.21a: Photograph by Lesley A. Robertson for the Kluyver Laboratory Museum, Delft University of Technology, Delft, The Netherlands; 1.21b: Paintings by Henriette Wilhelmina Beijerinck, photographed by Lesley A. Robertson for the Kluyver Laboratory Museum, Delft University of Technology, Delft, The Netherlands; 1.22a: From Sergei Winogradsky, Microbiologie du Sol, portion of Plate IV. Paris, France: Masson et Cie Editeurs, 1949. Reproduced by permission of Dunod Editeur, Paris, France; 1.22b: Sergei Winogradsky, Microbiologie du Sol. Paris, France: Masson, 1949; 1.MS.1: Walter Hesse, 1884. "Uber quantitative Bestimmung der in der luft enthaltenen Mikroorganismen," in H. Struck (ed.), Mittheilungen aus dem Kaiserlichen Gesundheitsamte. Verlag August Hirschwald; 1.MS.2: Paul V. Dunlap.

Chapter 2 Opener: Norbert Pfennig, University of Konstanz, Germany; 2.1a: LEO Electron Microscopy; 2.2a: Thomas D. Brock; 2.2b: Norbert Pfennig, University of Konstanz, Germany; 2.4b: Leon J. Le Beau, University of Illinois at Chicago; 2.4c: Molecular Probes; 2.5: M.T. Madigan; 2.6a,b: Richard W. Castenholz, University of Oregon; 2.6c: Nancy J. Trun, National Cancer Institute; 2.7a: Linda Barnett and James Barnett, University of East Anglia, U.K.; 2.7b: Suzanne V. Kelly, Scottsdale Community College; 2.8a: Subramanian Karthikeyan, University of Saskatchewan; 2.8b: Gernot Arp, University of Gottingen, Gottingen, Germany, and Christian Boker, Carl Zeiss Jena, Germany; 2.9: ZELMI, TU-Berlin, Germany; 2.10a: Stanley C. Holt, University of Texas Health Science Center; 2.10b: Robin Harris; 2.10c: F. Rudolf Turner, Indiana University; 2.12a: John Bozzola and Michael T. Madigan; 2.12b: Reinhard Rachel and Karl O. Stetter, Archives of Microbiology 128:288-293 (1981). © 1981 by Springer-Verlag GmbH & Co. KG; 2.12c: Samuel F. Conti and Thomas D. Brock; 2.13a: Erskine Palmer, CDC; 2.13b: A. Dale Kaiser, Stanford University; 2.14a: Edward Kellenberger, Werner Villiger, and Jan Hobot; 2.14b: Birgit Arnold-Schulz-Gahmen, University of Basel, Switzerland; 2.15: Michael W. Davidson/The Florida State University Research Foundation; 2.20a: Douglas E. Caldwell, University of Saskatchewan; 2.20b: Hans-Dietrich Babenzien, Institute of Freshwater Ecology and Inland Fisheries, Neuglobsow, Germany; 2.21a: Hans Hippe, Deutsche Sammlung von Mikroorganismen und Zellkulturen GmbH, Braunschweig, Germany; 2.21b: Thomas D. Brock; 2.22: Richard W. Castenholz, University of Oregon; 2.23: James T. Staley, University of Washington; 2.24: Reproduced by permission from J.A. Breznak, Biology of nonpathogenic, host-associated spirochetes. CRC Critical Reviews of Microbiology 2:457-489 (1973). Original micrographs from R. Joseph and E. Canale-Parola, Axial fibrils of anaerobic spirochetes: ultrastructure and chemical characteristics. Archives of Microbiology 81:146-168 (1972); 2.25a: Norbert Pfennig, University of Konstanz, Germany; 2.25b: M.T. Madigan; 2.26: Michael J. Daly, Uniformed Services University of the Health Sciences; 2.27: Reinhard Rachel and Karl O. Stetter, University of Regensburg, Regensburg, Germany; 2.29: Karl O. Stetter and Reinhard Rachel, University of Regensburg, Germany; 2.30: William D. Grant, University of Leicester, U.K.; 2.31: Thomas D. Brock; 2.33a: Birke/mauritius images/age fotostock; 2.33b: MYCOsearch; 2.33c, 2.34: M.T. Madigan.

Chapter 3 Opener: Nicholas Blackburn, Marine Biological Laboratory, University of Copenhagen, Denmark; 3.1.1-3.1.5: Norbert Pfennig, University of Konstanz, Germany; 3.1.6: Thomas D. Brock; 3.2a: Esther R. Angert, Cornell University; 3.2b: Heide Schulz/Univ of CA Davis; 3.4b: Gerhard Wanner, University of Munich, Germany; 3.15b: Leon J. Le Beau, University of Illinois at Chicago; 3.15c: J.L. Pate; 3.15d: Thomas D. Brock and Samuel F. Conti; 3.15e,f: Akiko Umeda and K. Amako; 3.17a: Leon J. Le Beau, University of Illinois at Chicago; 3.20b: Terry J. Beveridge, University of Guelph, Guelph, Ontario; 3.20c: Georg E. Schulz; 3.22: Susan F. Koval, University of Western Ontario; 3.23a: Elliot Juni, University of Michigan; 3.23b: M.T. Madigan; 3.23c: Frank B. Dazzo and Richard Heinzen; 3.24: J.P. Duguid and J.F. Wilkinson; 3.25: Charles C. Brinton, Jr., University of Pittsburgh; 3.26b (top): Michael T. Madigan; 3.26b (bottom): Mercedes Berlanga and International Microbiology; 3.27a: M.T. Madigan; 3.27b: Norbert Pfennig, University of Konstanz, Germany; 3.28a: Stefan Spring, Technical University of Munich, Germany; 3.28b: Richard Blakemore and W. O'Brien; 3.28c: Dennis A. Bazylinski, Iowa State University; 3.29: Thomas D. Brock; 3.30a: A.E. Walsby, University of Bristol, Bristol, England; 3.30b: S. Pellegrini and Maria Grilli Caiola; 3.31a: Reproduced from A.E. Konopka et al., Isolation and characterization of gas vesicles from Microcyclus aquaticus. Archives of Microbiology 112:133-140 (March 1, 1977). © 1977 by Springer-Verlag GmbH & Co. KG; 3.32a-c, 3.33: Hans Hippe, Deutsche Sammlung von Mikroorganismen und Zellkulturen GmbH, Braunschweig, Germany; 3.34a-d: Judith F.M. Hoeniger and C.L. Headley; 3.35a: H.S. Pankratz, T.C. Beaman, and Philipp Gerhardt; 3.35b: Kirsten Price, Harvard University; 3.38: Elnar Leifson; 3.39: Carl E. Bauer, Indiana University; 3.40a: R. Jarosch; 3.40b: Norbert Pfennig, University of Konstanz, Germany; 3.41a: David De Rosier; 3.42: Ken F. Jarrell; 3.45a,b: Richard W. Castenholz, University of Oregon; 3.45c, d: Mark J. McBride, University of Wisconsin, Milwaukee; 3.48f: Nicholas Blackburn, Marine Biological Laboratory, University of Copenhagen, Denmark; 3.49a: Norbert Pfennig, University of Konstanz, Germany; 3.49b: Carl E. Bauer, Indiana University; 3.MS.1: Luke Alderwick, University of Birmingham, UK.

Chapter 4 Opener: James A. Shapiro, University of Chicago; 4.3a-d, 4.5c: James A. Shapiro, University of Chicago; 4.7, 4.16b: Richard J. Feldmann, National Institutes of Health; 4.20b: Siegfried Engelbrecht-Vandré; 4.MS.1: Pearson Science; 4.T02a,b: Cheryl L. Broadie and John Vercillo, Southern Illinois University at Carbondale.

Chapter 5 Opener: Christine-Jacobs Wagner; 5.2b: T. den Blaauwen and Nanne Nanninga, University of Amsterdam, The Netherlands; 5.4b: Alex Formstone; 5.4c: Christine-Jacobs Wagner; 5.5b: Akiko Umeda and K. Amako; 5.15: Deborah O. Jung and M.T. Madigan; 5.20a-c: John Gosink and James T. Staley, University of Washington; 5.20d: M.T. Madigan; 5.21a: Katherine M. Brock; 5.21b, 5.22a,b: Thomas D. Brock; 5.23: Nancy L. Spear 5.27a: Deborah O. Jung and M.T. Madigan; 5.27b: Coy Laboratory Products; 5.30: Thomas D. Brock; 5.MS.1: Yue Jiang, Hong Kong Baptist University, Hong Kong.

Chapter 6 Opener: Huntington Potter and David Dressler; 6.5: Stephen P. Edmondson and Elizabeth Parker; 6.11: Huntington Potter and David Dressler; 6.25b: Sarah French; 6.34b: Reprinted with permission from M. Ruff et al., Class II aminoacyl transfer RNA synthetases: crystal structure of yeast aspartyl-tRNA synthetase complexed with tRNA(Asp). Science 252:1682-1689 (1991). © 1991 American Association for the Advancement of Science. Photo by Dino Moras.

Chapter 7 Opener: Jan Karlseder, The Salk Institute for Biological Studies; 7.2: Katsu Murakami, The Pennsylvania State University; 7.8: Elisabeth Pierson, FNWI-Radboud University Nijmegen, Pearson Science; 7.10c: Jan Karlseder, The Salk Institute for Biological Studies.

Chapter 8 Opener and 8.10: Reprinted with permission from S. Schultz et al., Crystal structure of a CAP-DNA complex: The DNA is bent by 90 degrees. Science 253:1001-1007 (1991). © 1991 by the American Association for the Advancement of Science. Photo by Thomas A. Steitz and Steve C. Schultz; 8.3b: Stephen P. Edmondson, Southern Illinois University at Carbondale; 8.19: Timothy C. Johnston, Murray State University.

Chapter 9 Opener: Omikron/Photo Researchers; 9.2a: John T. Finch, Medical Research Council/Laboratory of Molecular Biology, Cambridge, U.K.; 9.4c: W.F. Noyes; 9.4d: Timothy S. Baker and Norman H. Olson, Purdue University; 9.5a: P.W. Choppin and W. Stoeckenius; 9.5b: M. Wurtz; 9.6b: Jack Parker; 9.7 (left): Paul Kaplan; 9.7 (right): Thomas D. Brock; 9.17: A. Dale Kaiser, Stanford University; 9.25: Biao Ding & Yijun Qi.

Chapter 10 Opener: Thomas D. Brock; 10.1a: Thomas D. Brock; 10.1b: Peter T. Borgia, Southern Illinois University School of Medicine; 10.1c: Shiladitya DasSarma, Priya Arora, Lone Simonsen; 10.2, 10.8: Thomas D. Brock; 10.17: Charles C. Brinton, Jr., University of Pittsburgh; 10.26: Masaki Shioda and S. Takayanago.

Chapter 11 Opener: M.T. Madigan; 11.2a: Elizabeth Parker; 11.2b: Jack Parker; 11.3: S. Alex Lim, David C. Schwartz, and Eileen T. Dimalanta, University of Wisconsin at Madison; 11.4: Laurie Ann Achenbach, Southern Illinois University at Carbondale; 11.9: Jason A. Kahana and Pamela A. Silver, Harvard Medical School; 11.12: Daniel L. Nickrent; 11.13 (left): Norbert Pfennig, University of Konstanz, Germany; 11.13 (middle): Hans Hippe, Deutsche Sammlung von Mikroorganismen und Zellkulturen GmbH, Braunschweig, Germany; 11.13 (right): M.T. Madigan; 11.20b: Jack Parker; 11.MS.1: Alex Valm and Gary Borisy, Marine Biological Lab, Woods Hole, MA.

Chapter 12 Opener: SPL/Photolibrary; 12.2b: M.T. Madigan; 12.10a: GeneChip® Human Genome U133 Plus 2.0 Array, Affymetrix; 12.10b: Affymetrix; 12.11: Jack Parker; 12.MS.1: Microworks/Photolibrary.com.

Chapter 13 Opener: Niels Ulrik Frigaard; 13.5a: Yuuji Tsukii, Protist Information Server (protist.i.hosei.ac.jp); 13.6b: Simon Scheuring; 13.7: Niels Ulrik Frigaard; 13.10c: Kaori Ohki, Tokai University, Shimizu, Japan; 13.12: M.T. Madigan; 13.13a: George Feher, University of California at San Diego; 13.13b: Marianne Schiffer and

Oceanographic Institution; 19.26: Gertraud Rieger, R. Hermann, Reinhard Rachel, and Karl O. Stetter, University of Regensburg, Germany; 19.27: Suzette L. Pereira, Ohio State University.

Chapter 20 Opener: Jörg Piper; 20.2: E. Guth, T. Hashimoto, and S.F. Conti; 20.3b,c: Don W. Fawcett, M.D., Harvard Medical School; 20.4a: Helen Shio and Miklos Muller, The Rockefeller University; 20.5a: Thomas D. Brock; 20.5b: A. Wellma/ NaturimBild//Blickwinkel/age fotostock; 20.6: T. Slankis and S. Gibbs, McGill University; 20.7: Jian-ming Li and Nancy Martin, University of Louisville School of Medicine; 20.8: SPL/Photo Researchers; 20.9: Ohad Medalia and Wolfgang Baumeister; 20.10: Rupal Thazhath and Jacek Gaertig, University of Georgia; 20.11b: Melvin S. Fuller; 20.13a: Michael Abbey/ Photo Researchers; 20.13b: Steve J. Upton, Kansas State University; 20.14: Blaine Mathison, CDC; 20.15: Oxford Scientific/ Photolibrary; 20.16a: M.T. Madigan; 20.16b: Sydney Tamm; 20.17: Steve J. Upton, Kansas State University; 20.18: Irena Kaczmarska-Ehrman, Mount Allison University; 20.19a: Rita R. Colwell, National Science Foundation; 20.19b,c: North Carolina State University Center for Applied Ecology; 20.20a: Mae Melvin, CDC; 20.20b: Silvia Botero Kleiven, The Swedish Institute for Infectious Disease Control; 20.21a: Jörg Piper; 20.21b-d: Irena Kaczmarska-Ehrman, Mount Allison University; 20.22: Andrew Syred/Photo Researchers; 20.23: M. Haberey; 20.24: Stephen Sharnoff (sharnoffphotos.com); 20.25: Kenneth B. Raper; 20.27a: MYCOsearch; 20.28a: Cheryl L. Broadie, Southern Illinois University at Carbondale; 20.28b: CDC; 20.29: M.T. Madigan; 20.30: J. Forsdyke/ SPL/Photo Researchers; 20.32: Forest Brem; 20.33a: Alena Kubátová (http://botany.natur.cuni.cz/cs/sbirka-kultur-hub-ccf); 20.33b: Hossler/Custom Medical Stock Photo; 20.34: Thomas D. Brock; 20.37: Samuel F. Conti and Thomas D. Brock; 20.38a: Shutterstock; 20.38b: U.S. Department of Agriculture; 20.39: Jean Lecomte/Biosphoto/Peter Arnold; 20.40: Richard W. Castenholz, University of Oregon; 20.41a: Arthur M. Nonomura; 20.41b: Thomas D. Brock; 20.41c: Ralf Wagner (dr-ralf-wagner.de); 20.41d: NaturimBild/blickwinkel/Alamy; 20.41e: Aurora M. Nedelcu; 20.42a: Guillaume Dargaud (www.gdargaud.net); 20.42b: Yuuji Tsukii, Protist Information Server (protist.i.hosei.ac.jp), Hosei University, Japan.

Chapter 21 Opener: CDC; 21.1: R.C. Valentine; 21.8: F. Grundy and Martha Howe; 21.10a,b: Mark Young; 21.10c: Claire Geslin; 21.10d: David Prangishvili, Institut Pasteur; 21.11: Jed Fuhrman, University of Southern California; 21.13a: Arthur J. Olson, Molecular Graphics Laboratory, Scripps Research Institute; 21.14a, 21.15: CDC; 21.17a: P.W. Choppin and W. Stoeckenius; 21.18a, b: Timothy S. Baker and Norman H. Olson, Purdue University; 21.21a: CDC; 21.22: Timothy S. Baker and Xiaodong Yan, Purdue University; 21.23: Alexander

Eb and Jerome Vinograd; 21.26: R.W. Horne; 21.27: D. Dales and F. Fenner; 21.MS.1: D. Raoult, CNRS, Marseille, France.

Chapter 22 Opener: From R. Amann, J. Snaidr, M. Wagner, W. Ludwig, and K.-H. Schleifer, 1996. In situ visualization of high genetic diversity in a natural bacterial community. Journal of Bacteriology 178:3496-3500, Fig. 2b. © 1996 American Society for Microbiology. Photo: Jiri Snaidr; 22.2b: Norbert Pfennig, University of Konstanz, Germany; 22.3a: James A. Shapiro, University of Chicago; 22.3b: Marie Asao, Deborah O. Jung, and Michael T. Madigan; 22.6a,b: Marc Mussman and Michael Wagner; 22.6c: Willm Martens-Habbena; 22.7: Molecular Probes; 22.8: Daniel Gage; 22.9: Reproduced by permission of the American Society for Microbiology from A.T. Nielsen et al., Identification of a novel group of bacteria in sludge from a deteriorated biological phosphorus removal reactor. Applied Environmental Microbiology 65:1251-1258 (1999), fig. 5B (left, right). Image: Alex T. Nielsen, Technical University of Denmark, Lyngby, Denmark; 22.10a: David A. Stahl, Northwestern University; 22.10b: From R. Amann, J. Snaidr, M. Wagner, W. Ludwig, and K.-H. Schleifer, 1996. In situ visualization of high genetic diversity in a natural bacterial community. Journal of Bacteriology 178:3496-3500, Fig. 2b. © 1996 American Society for Microbiology. Photo: Jiri Snaidr; 22.11: Marc Mussmann and Michael Wagner; 22.13: Jennifer A. Fagg and Michael J. Ferris, Montana State University; 22.15: Alexander Loy and Michael Wagner; 22.21: Niels Peter Revsbech; 22.27: Michael Wagner; 22.28: Colin J. Murrell.

Chapter 23 Opener: Christian Jeanthon, Centre National de la Recherche Scientifique, France; 23.1: Hans Paerl; 23.4a: Frank B. Dazzo, Michigan State University; 23.4b: Thomas D. Brock; 23.5a: C.T. Huang, Karen Xu, Gordon McFeters, and Philip S. Stewart; 23.5b: Cindy E. Morris, INRA, Centre de Recherche d'Avignon, France. Previously published in Applied and Environmental Microbiology 63:1570-1576; 23.5c: J.M. Sanchez, J. Lidel Lope and Ricardo Amils; 23.6b: Rodney M. Donlan and Emerging Infectious Diseases; 23.8: Matthew Parsek and Brad Borlee; 23.9a: Jesse Dillon and David A. Stahl; 23.9b: David M. Ward, Montana State University. Reproduced with permission of the American Society for Microbiology; 23.10b: M.T. Madigan; 23.12: T.R.G. Gray, University of Essex, Colchester, U.K.; 23.14a: Esta van Heerden; 23.14b: Terry C. Hazen; 23.16b: Thomas D. Brock; 23.17: NASA photo processed by Otis Brown and Robert Evans, obtained through Dawn Cardascia, Earth Science Support Office; 23.18a: Alexandra Z. Worden and Mya E. Breitbart, Scripps Institution of Oceanography, University of California at San Diego; 23.18b: Sallie Chrisholm; 23.19a: Hans W. Paerl, University of North Carolina at Chapel Hill; 23.19b: Alexandra Z. Worden and Brian P. Palenik, Scripps Institution of Oceanography, University of California at San Diego; 23.20: Vladimir Yurkov; 23.22:

Daniela Nicastro; 23.23: Jed Fuhrman; 23.26: Hideto Takami, Japan Marine Science and Technology Center, Kanagawa, Japan; 23.28a: Douglas Bartlett; 23.31: Woods Hole Oceanographic Institution; 23.33: Christian Jeanthon, Centre National de la Recherche Scientifique, France; 23.32: Deborah Kelley, University of Washington.

Chapter 24 Opener: Jörg Bollmann; 24.3: Evan Solomon; 24.6a: John A. Breznak, Michigan State University; 24.6b,c: Monica Lee and Stephen H. Zinder; 24.10: J. M. Sanchez, J. Lidel Lope and Ricardo Amils; 24.11a: Ravin Donald, Northern Arizona University; 24.11b, 24.12: Thomas D. Brock; 24.13a: Jörg Bollmann; 24.13b: M.L. Cros Miguel and J.M. Fortuño Alós; 24.14a: Jörg Piper; 24.15: Thomas D. Brock; 24.17: Ashanti Goldfields, Ghana; 24.20a,b: U.S. Environmental Protection Agency Headquarters; 24.20c: Bassam Lahoud, Lebanese American University; 24.21: Thomas D. Brock; 24.22: iStockphoto; 24.25c: Helmut Brandl, University of Zurich, Switzerland; 24.MS.1: Eye of Science/Photo Researchers.

Chapter 25 Opener: Jörg Graf; 25.1a: Thomas D. Brock; 25.1b: M.T. Madigan; 25.2: Thomas D. Brock; 25.4: J. Overmann and H.van Gemerden; 25.5: Gerhard Wanner and Jörg Overlmann, Ultrastructural Characterization of the Prokaryotic Symbiosis in "Chlorochromatium aggregatum." Journal of Bacteriology, May 2008, p. 3721-3730, Vol. 190, No. 10. © 2008, American Society for Microbiology. Reproduced by permission. 25.7: Joe Burton; 25.8: Ben B. Bohlool, University of Hawaii; 25.9: Joe Burton; 25.11a: Ben B. Bohlool, University of Hawaii; 25.11b-d: Reproduced with permission from G. Truchet et al., Sulphated lipo-oligosaccharide signals of *Rhizobium meliloti* elicit root nodule organogenesis in alfalfa. Nature 351:670-673 (1991). © 1991 Macmillan Magazines Limited. Photo by Jacques Vasse, Jean Denarie, and Georges Truchet; 25.15: B. Dreyfus, Institut de Recherche pour le Developpement (ORSTOM), Dakar, Senegal; 25.16, 25.17: J.-H. Becking, Wageningen Agricultural University, Wageningen, Netherlands; 25.18: Jo Handelsman, University of Wisconsin at Madison; 25.21a: Photo by Jacob R. Schramm; 25.21b: D.J. Read, University of Sheffield, England; 25.23: S.A. Wilde; 25.25.1: iStockphoto; 25.25.2: Bernard Swain; 25.25.3: Nancy L. Spear; 25.25.4: iStockphoto; 25.26b: Sharisa D. Beck, Southern Illinois University at Carbondale; 25.33: Amparo Latorre; 25.34c: iStockphoto; 25.36a: Chris Frazee and Margaret J. Mcfall-Ngai, University of Wisconsin; 25.36b: Margaret J. Mcfall-Ngai, University of Wisconsin; 25.37a: Dudley Foster, Woods Hole Oceanographic Institution; 25.37b: Carl Wirsen, Woods Hole Oceanographic Institution; 25.38a: Reproduced from C.M. Cavanaugh et al., Prokaryotic cells in the hydrothermal vent tube worm *Riftia pachyptila* Jones: possible chemoautotrophic symbionts. Science 213:340-342 (July 17, 1981), Fig. 1b. © 1981 American Association for the Advancement of

Science. Photo by Colleen M. Cavanaugh, Harvard University; 25.38b: Reprinted with permission from Nature 302:58-61, Fig. 3a. © 1983 Macmillan Magazines Limited. Photo: Colleen M. Cavanaugh, Harvard University; 25.39a: Michele Maltz and Jörg Graf; 25.40: Jörg Graf; 25.41: Kazuhiko Koike and Kiroshi Yamashita; 25.42: Ernesto Weil; 25.MS: Michael Poulsen and Cameron Currie.

Chapter 26 Opener: Carlos Pedros-Alio and Thomas D. Brock; 26.3c, 26.4: John M. Martinko; 26.6: Thomas D. Brock; 26.7: John M. Martinko; 26.8a: Carlos Pedros-Alio and Thomas D. Brock; 26.8b: Janice Carr and Rob Weyant, HIP, NCID, CDC; 26.10: Thomas D. Brock; 26.MS.1: World Health Organization. 2010. Multidrug and extensively drug-resistant TB (M/XDR-TB): 2010 global report on surveillance and response. WHO/HTM/ TB/2010.3.

Chapter 27 Opener: Thomas D. Brock; 27.5: Thomas J. Lie, University of Washington; 27.7: C. Lai, Max A. Listgarten, and B. Rosan; 27.8: Isaac L. Schechmeister and John J. Bozzola, Southern Illinois University at Carbondale; 27.10: Dwayne C. Savage and R.V.H. Blumershine; 27.12b: John Durham/Photo Researchers; 27.15a: Edward T. Nelson, J.D. Clements, and R.A. Finkelstein; 27.15b: J. William Costerton, Montana State University; 27.16: Larry Stauffer, Oregon State Public Health Laboratory, CDC; 27.17: Richard Facklam, CDC; 27.18: James A. Roberts; 27.19a: Thomas D. Brock; 27.19b: Leon J. Le Beau, University of Illinois at Chicago; 27.25: Arthur O. Tzianabos and R.D. Millham; 27.MS.1: BSIP/Photolibrary.com; 27. MS.2: Janice Haney Carr, CDC.

Chapter 28 Opener: A.B. Dowsett/SPL/ Photo Researchers; 28.6, 28.10, 28.16, 28.MS.1: CDC.

Chapter 29 Opener: Reproduced with permission from Science 239, Cover (February 12, 1988). © 1988 American Association for the Advancement of Science. Micrograph by Richard S. Lewis; 29.1a: John M. Martinko and M.T. Madigan; 29.1b: Division of Parasitic Diseases, NCID, CDC; 29.1c: Joe Millar, CDC; 29.1d: John M. Martinko and M.T. Madigan; 29.2: J.G. Hirsch; 29.15a: Richard J. Feldmann, National Institutes of Health; 29.15b: Reproduced with permission from A.G. Amit et al., Three-dimensional structure of an antigen-antibody complex at 2.8 A resolution. Science 233:747-753 (August 15, 1986), Fig. 3. © 1986 American Association for the Advancement of Science. Images: Roberto J. Poljak; 29.22: E. Munn.

Chapter 30 Opener: Klaus Boller/Photo Researchers; 30.5a: Don C. Wiley, Howard Hughes Medical Institute; 30.5b: Aideen C.M. Young, Albert Einstein College of Medicine, Bronx, New York; 30.5c: Reproduced by permission from J.H. Brown et al., Three-dimensional structure of the human class II histocompatibility antigen HLA-DR1. Nature 364:33-39 (1993). © 1993 Macmillan Magazines Limited. Image by Don C. Wiley, Harvard

Index